Compiled by

ALAN SPOONER

CLARENDON PRESS · OXFORD

Oxford University Press, Walton Street, Oxford OX2 6DP

Oxford New York Toronto
Delhi Bombay Calcutta Madras Karachi
Kuala Lumpur Singapore Hong Kong Tokyo
Nairobi Dar es Salaam Cape Town
Melbourne Auckland Madrid

and associated companies in
Berlin Ibadan

Oxford is a trade mark of Oxford University Press

Published in the United States by
Oxford University Press Inc., New York

First published 1992 as The Oxford Minireference Thesaurus

This edition first published 1993
Reprinted 1993, 1994

British Library Cataloguing in Publication Data
Data available

Library of Congress Cataloging in Publication Data
Data available

ISBN 0-19-869221-8

Typeset by Wyvern Typesetting Ltd.
Printed in Great Britain by
Clays Ltd.
Bungay, Suffolk

Preface

A THESAURUS helps you find the words you need to express yourself more effectively and more interestingly. This thesaurus is designed to combine within its small format maximum ease of use with maximum helpfulness. It includes a larger range of synonyms and other information than might be expected in a book of this size. Headwords are arranged in a simple alphabetical sequence, and the organization of each entry is straightforward and largely self-explanatory. Normally, you will find what you want under the headword you look up, though cross-references may be necessary if you need opposites, or if you want a larger range of words to choose from.

A thesaurus should be used with caution. Firstly, no synonym list should be regarded as 'complete'. Many could be extended, some almost indefinitely. Consider, for example, the variety of words we might substitute for (say) *good* or *pleasant*. Secondly, seldom in English are two words completely interchangeable. So-called synonyms may convey distinct nuances of meaning, or belong to different contexts, or carry different signals about the writer or intended reader – and so on. I hope, therefore, that this volume will be a useful resource, but not just as a lifeless repository of 'words to use': my main hope is that a thesaurus – even a small one like this – will prompt us to *think* about language, and enable us to exploit more fully our own knowledge and understanding of its complex processes.

ALAN SPOONER

The Nottingham Trent University
May 1993

The Little Oxford Thesaurus

COMPILER AND EDITOR-IN-CHIEF
Alan Spooner

MANAGING EDITOR
Sara Tulloch

ASSISTANT EDITORS
Anne Knight   Christine Cowley

Using the thesaurus

In this thesaurus you will find

Headwords

The words you want to look up are printed in bold and arranged in a single alphabetical sequence. In addition, there may be sub-heads in bold at the end of main entries for derived forms and phrases.

Synonyms

Synonyms are listed alphabetically, except that distinct senses of a headword are numbered and treated separately.

Under some headwords, in addition to the lists of synonyms given there, a cross-reference printed in SMALL CAPITALS takes you to another entry to provide an extended range of synonyms. These cross-references are marked by the arrowhead symbol ▷.

Related words

Lists of words which are not synonyms but which have a common relationship to the headword (eg, kinds of vehicle listed under *vehicle*) are printed in italic, flagged by the symbol □.

Antonyms

Cross-references printed in SMALL CAPITALS introduce you to lists of opposites. These cross-references are preceded by the abbreviation *Opp.*

Part-of-speech labels

Part-of-speech labels are given throughout. (See list of abbreviations.) Under each headword, uses as *adjective, adverb, noun,* and *verb* are separated by the symbol •.

Illustrative phrases

Meanings of less obvious senses are indicated by illustrative phrases printed in *italic.*

Usage warnings

Usage markers in *italic* precede words which are normally informal, derogatory, etc. (See list of abbreviations.)

Abbreviations used in this thesaurus

Parts of speech

adj	adjective
adv	adverb
int	interjection
n	noun
prep	preposition
vb	verb

Other abbreviations

derog	normally used in a derogatory, negative, or uncomplimentary sense
fem	feminine
inf	normally used informally
joc	normally jocular or joking
old use	old-fashioned or obsolete
opp	opposites, antonyms
plur	plural
poet	poetic
sl	slang
Amer	word or phrase usually regarded as American usage.
Fr	word or phrase common in English contexts, but still identifiably French.
Ger	ditto German
Gr	ditto Greek
It	ditto Italian
Lat	ditto Latin
Scot	word or phrase usually regarded as Scottish usage.
▷	This symbol shows that you will find relevant information if you go to the word indicated.

A

abandon *vb* **1** evacuate, leave, quit, vacate, withdraw from. **2** break with, desert, *inf* dump, forsake, *inf* give someone the brush-off, jilt, leave behind, *inf* leave in the lurch, maroon, renounce, repudiate, strand, *inf* throw over, *inf* wash your hands of. **3** *abandon a claim.* abdicate, cancel, cede, *sl* chuck in, discard, discontinue, disown, *inf* ditch, drop, finish, forfeit, forgo, give up, postpone, relinquish, resign, scrap, surrender, waive, yield.

abbey *n* cathedral, church, convent, friary, monastery, nunnery, priory.

abbreviate *vb* abridge, compress, condense, curtail, cut, digest, edit, précis, prune, reduce, shorten, summarize, trim, truncate. *Opp* LENGTHEN.

abdicate *vb* renounce the throne, *inf* step down. ▷ ABANDON, RESIGN.

abduct *vb* carry off, kidnap, *inf* make away with, seize.

abhor *vb* detest, execrate, loathe, recoil from, shudder at. ▷ HATE.

abhorrent *adj* abominable, detestable, disgusting, execrable, horrible, loathsome, nauseating, obnoxious, odious, offensive, repellent, repugnant, repulsive, revolting. ▷ HATEFUL. *Opp* ATTRACTIVE.

abide *vb* **1** accept, bear, brook, endure, put up with, stand, *inf* stomach, submit to, suffer, tolerate. **2** ▷ STAY. **abide by** ▷ OBEY.

ability *n* aptitude, bent, brains, capability, capacity, cleverness, competence, expertise, facility, faculty, flair, genius, gift, intelligence, knack, *inf* know-how, knowledge, means, potential, potentiality, power, proficiency, prowess, qualification, resources, scope, skill, strength, talent, training, wit.

ablaze *adj* afire, aflame, aglow, alight, blazing, burning, fiery, flaming, incandescent, lit up, on fire, raging. ▷ BRILLIANT.

able *adj* **1** accomplished, adept, capable, clever, competent, effective, efficient, experienced, expert, gifted, *inf* handy, intelligent, masterly, practised, proficient, qualified, skilful, skilled, strong, talented, trained. *Opp* INCOMPETENT. **2** allowed, at liberty, authorized, available, eligible, equipped, fit, free, permitted, prepared, ready, willing. *Opp* UNABLE.

abnormal *adj* aberrant, anomalous, atypical, *inf* bent, bizarre, curious, deformed, deviant, distorted, eccentric, erratic, exceptional, extraordinary, freak, funny, heretical, idiosyncratic, irregular, *inf* kinky, malformed, odd, peculiar, perverse, perverted, queer, singular, strange, uncharacteristic, uncommon, unexpected, unnatural, unorthodox, unrepresentative, untypical, unusual, wayward, weird. *Opp* NORMAL.

abolish *vb* abrogate, annul, cancel, delete, destroy, dispense with, do away with, eliminate, end, eradicate, finish, *inf* get rid of, invalidate, liquidate, nullify, overturn, put an end to, quash, repeal, remove, rescind, revoke, suppress, terminate, withdraw. *Opp* CREATE.

abominable *adj* abhorrent, appalling, atrocious, awful, base, beastly, brutal, contemptible, cruel, despicable, detestable, disgusting, distasteful, dreadful, execrable, foul, hateful, heinous, horrible, immoral, inhuman, inhumane, loathsome, nasty, nauseating, obnoxious, odious, offensive, repellent, repugnant, repulsive, revolting, terrible, vile. ▷ UNPLEASANT. *Opp* PLEASANT.

abort *vb* **1** be born prematurely, die, miscarry. **2** *abort take-off.* call off, end, halt, nullify, stop, terminate.

abortion *n* **1** miscarriage, premature birth, termination of pregnancy. **2** ▷ MONSTER.

abortive *adj* fruitless, futile, ineffective, ineffectual, pointless, stillborn, unavailing, unfruitful, unproductive, unsuccessful, useless, vain. *Opp* SUCCESSFUL.

abound *vb* be plentiful, flourish, prevail, swarm, teem, thrive.

abrasive *adj* biting, caustic, galling, grating, harsh, hurtful, irritating, rough, sharp. ▷ UNKIND. *Opp* KIND.

abridge *vb* abbreviate, compress, condense, curtail, cut, digest, edit, précis, prune, reduce, shorten, summarize, telescope, truncate. *Opp* EXPAND.

abridged *adj* abbreviated, bowdlerized, censored, compact, concise, condensed, cut, edited, *inf* potted, shortened, telescoped, truncated.

abrupt *adj* **1** disconnected, hasty, headlong, hurried, precipitate, quick, rapid, sudden, swift, unexpected, unforeseen, unpredicted. **2** *an abrupt drop.* precipitous, sharp, sheer, steep. **3** *an abrupt manner.* blunt, brisk, brusque, curt, discourteous, gruff, impolite, rude, snappy, terse, unceremonious, uncivil, ungracious. *Opp* GENTLE, GRADUAL.

absent *adj* **1** away, *sl* bunking off, gone, missing, off, out, playing truant, *inf* skiving. *Opp* PRESENT. **2** ▷ ABSENT-MINDED.

absent-minded *adj* absent, absorbed, abstracted, careless, day-dreaming, distracted, dreamy, far-away, forgetful, heedless, impractical, inattentive, oblivious, preoccupied, scatterbrained, thoughtless, unaware, unheeding, unthinking, vague, withdrawn, wool-gathering. *Opp* ALERT.

absolute *adj* **1** categorical, certain, complete, conclusive, decided, definite, downright, entire, full, genuine, implicit, inalienable, indubitable, infallible, *inf* out-and-out, perfect, positive, pure, sheer, supreme, sure, thorough, total, unadulterated, unalloyed, unambiguous, unconditional, unequivocal, unmitigated, unmixed, unqualified, unquestionable, unreserved, unrestricted, utter. **2** *absolute ruler.* almighty, autocratic, despotic, dictatorial, omnipotent, sovereign, totalitarian, tyrannical, undemocratic. **3** *absolute opposites. inf* dead, diametrical, exact, precise.

absorb *vb* **1** assimilate, consume, devour, digest, drink in, fill up with, hold, imbibe, incorporate, ingest, mop up, receive, retain, soak up, suck up, take in, utilize. *Opp* EMIT. **2** *absorb a blow.* cushion, deaden, lessen, reduce, soften. **3** *absorb a person.* captivate, engage, engross, enthrall, fascinate, involve, occupy, preoccupy. ▷ INTEREST. **absorbed** ▷ INTERESTED.

absorbent *adj* absorptive, permeable, pervious, porous, spongy. *Opp* IMPERVIOUS.

absorbing *adj* engrossing, fascinating, gripping, riveting, spellbinding. ▷ INTERESTING.

abstain *vb*

abstain from avoid, cease, decline, deny yourself, desist from, eschew, forgo, give up, go without, refrain from, refuse, reject, renounce, resist, shun, withhold from.

abstemious *adj* ascetic, austere, frugal, moderate, restrained, self-denying, self-disciplined, sober, sparing, teetotal, temperate. *Opp* SELF-INDULGENT.

abstract *adj* **1** abstruse, academic, hypothetical, indefinite, intangible, intellectual, metaphysical, notional, philosophical, theoretical, unpractical, unreal, unrealistic. *Opp* CONCRETE. **2** *abstract art*. non-pictorial, non-representational, symbolic. • *n* digest, outline, précis, résumé, summary, synopsis.

abstruse *adj* complex, cryptic, deep, devious, difficult, enigmatic, esoteric, hard, incomprehensible, mysterious, mystical, obscure, perplexing, problematical, profound, puzzling, recherché, recondite, unfathomable. *Opp* OBVIOUS.

absurd *adj* anomalous, crazy, daft, eccentric, farcical, foolish, grotesque, idiotic, illogical, incongruous, irrational, laughable, ludicrous, nonsensical, outlandish, paradoxical, preposterous, ridiculous, risible, senseless, silly, stupid, surreal, unreasonable, untenable, zany. ▷ FUNNY, MAD. *Opp* RATIONAL.

abundant *adj* ample, bounteous, bountiful, copious, excessive, flourishing, full, generous, lavish, liberal, luxuriant, overflowing, plenteous, plentiful, prodigal, profuse, rampant, rank, rich, well-supplied. *Opp* SCARCE.

abuse *n* **1** assault, cruel treatment, ill-treatment, maltreatment, misappropriation, misuse, perversion. **2** *verbal abuse*. curse, execration, imprecation, insult, invective, obloquy, obscenity, slander, vilification, vituperation. • *vb* **1** batter, damage, exploit, harm, hurt, ill-treat, injure, maltreat, manhandle, misemploy, misuse, molest, rape, spoil, treat roughly. **2** *abuse verbally*. affront, berate, be rude to, *inf* call names, castigate, criticize, curse, defame, denigrate, disparage, insult, inveigh against, libel, malign, revile, slander, *inf* slate, *inf* smear, sneer at, swear at, traduce, upbraid, vilify, vituperate, wrong. ▷ COMPLIMENT.

abusive *adj* acrimonious, angry, censorious, contemptuous, critical, cruel, defamatory, denigrating, derisive, derogatory, disparaging, harsh, hurtful, impolite, injurious, insulting, libellous, obscene, offensive, opprobrious, pejorative, profane, rude, scathing, scornful, scurrilous, slanderous, vituperative. *Opp* KIND, POLITE.

abysmal *adj* **1** bottomless, boundless, deep, extreme, immeasurable, incalculable, infinite, profound, unfathomable, vast. **2** ▷ BAD.

abyss *n inf* bottomless pit, chasm, crater, fissure, gap, gulf, hole, opening, pit, rift, void.

academic *adj* **1** educational, pedagogical, scholastic. **2** bookish, brainy, clever, erudite, highbrow, intelligent, learned, scholarly, studious, well-read. **3** *academic study*. abstract, conjectural, hypothetical, impractical, intellectual, notional, pure, speculative, theoretical, unpractical. • *n inf* egghead, highbrow, intellectual, scholar, thinker.

accelerate *vb* **1** *inf* do a spurt, *inf* get a move on, go faster, hasten, increase speed, pick up speed, quicken, speed up. **2** bring on, expedite, promote, spur on, step up, stimulate.

accent *n* **1** brogue, cadence, dialect, enunciation, inflection, intonation, pronunciation, sound, speech pattern, tone. **2** accentuation, beat, emphasis, force, prominence, pulse, rhythm, stress.

accept *vb* **1** acquire, get, *inf* jump at, receive, take, welcome. **2** acknowledge, admit, assume, bear, put up with, reconcile yourself to, resign yourself to, submit to, suffer, tolerate, undertake. **3** *accept an argument.* abide by, accede to, acquiesce in, adopt, agree to, approve, believe in, be reconciled to, consent to, defer to, grant, recognize, *inf* stomach, *inf* swallow, take in, *inf* wear, yield to. *Opp* REJECT.

acceptable *adj* **1** agreeable, appreciated, gratifying, pleasant, pleasing, welcome, worthwhile. **2** adequate, admissible, appropriate, moderate, passable, satisfactory, suitable, tolerable, unexceptionable. *Opp* UNACCEPTABLE.

acceptance *n* acquiescence, agreement, approval, consent, willingness. *Opp* REFUSAL.

accepted *adj* acknowledged, agreed, axiomatic, canonical, common, indisputable, recognized, standard, undeniable, undisputed, universal, unquestioned. *Opp* CONTROVERSIAL.

accessible *adj* approachable, at hand, attainable, available, close, convenient, *inf* get-at-able, *inf* handy, reachable, ready, within reach. *Opp* INACCESSIBLE.

accessory *n* **1** addition, adjunct, appendage, attachment, component, extension, extra, fitting. **2** ▷ ACCOMPLICE.

accident *n* **1** blunder, chance, coincidence, contingency, fate, fluke, fortune, hazard, luck, misadventure, mischance, misfortune, mishap, mistake, *inf* pot luck, serendipity. **2** calamity, catastrophe, collision, *inf* contretemps, crash, derailment, disaster, *inf* pile-up, *sl* shunt, wreck.

accidental *adj* adventitious, arbitrary, casual, chance, coincidental, *inf* fluky, fortuitous, fortunate, haphazard, inadvertent, lucky, random, unconscious, unexpected, unforeseen, unfortunate, unintended, unintentional, unlooked-for, unlucky, unplanned, unpremeditated. *Opp* INTENTIONAL.

acclaim *vb* applaud, celebrate, cheer, clap, commend, exalt, extol, hail, honour, laud, praise, salute, welcome.

accommodate *vb* **1** aid, assist, equip, fit, furnish, help, oblige, provide, serve, suit, supply. **2** *accommodate guests.* billet, board, cater for, entertain, harbour, hold, house, lodge, provide for, *inf* put up, quarter, shelter, take in. **3** *accommodate yourself to new surroundings.* accustom, adapt, reconcile. **accommodating** ▷ CONSIDERATE.

accommodation *n* board, home, housing, lodgings, pied-à-terre, premises, shelter. □ *apartment, barracks,* inf *bedsit, bedsitter, billet, boarding house,* inf *digs, flat, guest house, hall of residence, hos-*

tel, hotel, inn, lodge, married quarters, motel, pension, rooms, self-catering, timeshare, youth hostel. ▷ HOUSE.

accompany *vb* **1** attend, chaperon, conduct, convoy, escort, follow, go with, guard, guide, look after, partner, squire, *inf* tag along with, travel with, usher. **2** be associated with, be linked with, belong with, coexist with, coincide with, complement, occur with, supplement.

accompanying *adj* associated, attached, attendant, complementary, concomitant, connected, related.

accomplice *n* abettor, accessory, ally, assistant, associate, collaborator, colleague, confederate, conspirator, helper, *inf* henchman, partner.

accomplish *vb* achieve, attain, *inf* bring off, carry off, carry out, carry through, complete, conclude, consummate, discharge, do successfully, effect, execute, finish, fulfil, perform, realize, succeed in.

accomplished *adj* adept, expert, gifted, polished, proficient, skilful, talented.

accomplishment *n* ability, attainment, expertise, gift, skill, talent.

accord *n* agreement, concord, congruence, harmony, rapport, unanimity, understanding.

account *n* **1** bill, calculation, check, computation, invoice, receipt, reckoning, *inf* score, statement, tally. **2** chronicle, commentary, description, diary, explanation, history, log, memoir, narration, narrative, portrayal, record, report, statement, story, tale, version, *inf* write-up. **3** *of no account.* advantage, benefit, concern, consequence, consideration, importance, interest, merit, profit, significance, standing, use, value, worth. **account for** ▷ EXPLAIN.

accumulate *vb* accrue, agglomerate, aggregate, amass, assemble, bring together, build up, collect, come together, gather, grow, heap up, hoard, increase, mass, multiply, pile up, stack up, *inf* stash away, stockpile, store up. *Opp* DISPERSE.

accumulation *n inf* build-up, collection, conglomeration, gathering, growth, heap, hoard, mass, pile, stock, stockpile, store, supply.

accurate *adj* authentic, careful, certain, correct, exact, factual, faithful, faultless, meticulous, minute, nice, perfect, precise, reliable, right, scrupulous, sound, *inf* spot-on, strict, sure, true, truthful, unerring, veracious. *Opp* INACCURATE.

accusation *n* allegation, alleged offence, charge, citation, complaint, denunciation, impeachment, indictment, summons.

accuse *vb* arraign, attack, blame, bring charges against, censure, charge, condemn, denounce, hold responsible, impeach, impugn, incriminate, indict, inform against, make allegations against, *inf* point the finger at, prosecute, summons, tax. *Opp* DEFEND.

accustomed *adj* common, conventional, customary, established, expected, familiar, habitual, normal, ordinary, prevailing, regular, routine, set, traditional, usual, wonted. **get accustomed** ▷ ADAPT.

ache *n* anguish, discomfort, distress, hurt, pain, pang, smart, soreness, suffering, throbbing, twinge. • *vb* **1** be painful, be sore, hurt, pound, smart, sting, suffer, throb. **2** ▷ DESIRE.

achieve *vb* **1** accomplish, attain, bring off, carry out, complete, conclude, consummate, discharge, do successfully, effect, engineer, execute, finish, fulfil, manage, perform, succeed in. **2** *achieve fame.* acquire, earn, gain, get, obtain, procure, reach, score, win.

acid *adj* sharp, sour, stinging, tangy, tart, vinegary. *Opp* BLAND, SWEET.

acknowledge *vb* **1** accede, accept, acquiesce, admit, affirm, agree, allow, concede, confess, confirm, declare, endorse, grant, own up to, profess, yield. *Opp* DENY. **2** *acknowledge a greeting.* answer, notice, react to, reply to, respond to, return. **3** *acknowledge a friend.* greet, hail, recognize, salute, *inf* say hello to. *Opp* IGNORE.

acme *n* apex, crown, culmination, height, highest point, maximum, peak, pinnacle, summit, top, zenith. *Opp* NADIR.

acquaint *vb* advise, announce, apprise, brief, disclose, divulge, enlighten, inform, make aware, make familiar, notify, reveal, tell.

acquaintance *n* **1** awareness, familiarity, knowledge, understanding. **2** ▷ FRIEND.

acquire *vb* buy, come by, earn, get, obtain, procure, purchase, receive, secure.

acquisition *n* accession, addition, *inf* buy, gain, possession, prize, property, purchase.

acquit *vb* absolve, clear, declare innocent, discharge, dismiss, exculpate, excuse, exonerate, find innocent, free, *inf* let off, liberate, release, reprieve, set free, vindicate. *Opp* CONDEMN. **acquit yourself** ▷ BEHAVE.

acrid *adj* bitter, caustic, harsh, pungent, sharp, unpleasant.

acrimonious *adj* abusive, acerbic, angry, bad-tempered, bitter, caustic, censorious, churlish, cutting, hostile, hot-tempered, ill-natured, ill-tempered, irascible, mordant, peevish, petulant, quarrelsome, rancorous, sarcastic, sharp, spiteful, tart, testy, venomous, virulent, waspish. *Opp* PEACEABLE.

act *n* **1** achievement, action, deed, effort, enterprise, exploit, feat, move, operation, proceeding, step, undertaking. **2** *act of parliament.* bill [= *draft act*], decree, edict, law, order, regulation, statute. **3** *a stage act.* item, performance, routine, sketch, turn. • *vb* **1** behave, carry on, conduct yourself, deport yourself. **2** function, have an effect, operate, serve, take effect, work. **3** *Act now!* do something, get involved, make a move, react, take steps. ▷ BEGIN. **4** *act a role.* appear (as), assume the character of, *derog* camp it up, characterize, dramatize, enact, *derog* ham it up, imitate, impersonate, mime, mimic, *derog* overact, perform, personify, play, portray, pose as, represent, seem to be, simulate. ▷ PRETEND.

acting *adj* deputy, interim, provisional, stand-by, stopgap, substitute, surrogate, temporary, vice-.

action *n* **1** act, deed, effort, endeavour, enterprise, exploit, feat, measure, performance, proceeding, process, step, undertaking, work. **2** activity, drama, energy, enterprise, excitement, exercise, exertion, initiative, liveliness, motion, movement, vigour, vitality. **3** *action of a play.* events, happenings, incidents, story. **4** *action of a watch.* functioning, mechanism, operation, working, works. **5** *military action.* ▷ BATTLE.

activate *vb* actuate, animate, arouse, energize, excite, fire, galvanize, *inf* get going, initiate, mobilize, motivate, prompt, rouse, set in motion, set off, start, stimulate, stir, trigger.

active *adj* **1** agile, animated, brisk, bustling, busy, dynamic, energetic, enterprising, enthusiastic, functioning, hyperactive, live, lively, militant, moving, nimble, *inf* on the go, restless, spirited, sprightly, strenuous, vigorous, vital, vivacious, working. **2** *active support.* assiduous, committed, dedicated, devoted, diligent, employed, engaged, enthusiastic, hard-working, industrious, involved, occupied, sedulous, staunch, zealous. *Opp* INACTIVE.

activity *n* **1** action, animation, bustle, commotion, energy, excitement, hurly-burly, hustle, industry, life, liveliness, motion, movement, stir. **2** hobby, interest, job, occupation, pastime, project, pursuit, scheme, task, undertaking, venture. ▷ WORK.

actor, actress *n* artist, artiste, lead, leading lady, performer, player, star, supporting actor, trouper, walk-on part. ▷ ENTERTAINER. **actors** cast, company, troupe.

actual *adj* authentic, bona fide, certain, confirmed, corporeal, current, definite, existing, factual, genuine, indisputable, in existence, legitimate, living, material, real, realistic, tangible, true, truthful, unquestionable, verifiable. *Opp* IMAGINARY.

acute *adj* **1** narrow, pointed, sharp. **2** *acute pain.* cutting, excruciating, exquisite, extreme, fierce, intense, keen, piercing, racking, severe, sharp, shooting, sudden, violent. **3** *an acute mind.* alert, analytical, astute, canny, *inf* cute, discerning, incisive, intelligent, keen, observant, penetrating, perceptive, percipient, perspicacious, quick, sharp, shrewd, *inf* smart, subtle. ▷ CLEVER. **4** *an acute problem.* compelling, crucial, decisive, immediate, important, overwhelming, pressing, serious, urgent, vital. **5** *an acute illness.* critical, sudden. *Opp* CHRONIC, DULL, STUPID.

adapt *vb* **1** acclimatize, accommodate, accustom, adjust, attune, become conditioned, become hardened, become inured, fit, get accustomed (to), get used (to), habituate, harmonize, orientate, reconcile, suit, tailor, turn. **2** *adapt to a new use.* alter, amend, change, convert, metamorphose, modify, process, rearrange, rebuild, reconstruct, refashion, remake, remodel, reorganize, reshape, transform, vary. ▷ EDIT.

add *vb* annex, append, attach, combine, integrate, join, put together, *inf* tack on, unite. *Opp* DEDUCT. **add to** ▷ INCREASE. **add up (to)** ▷ TOTAL.

addict *n* **1** alcoholic, *sl* dopefiend, *sl* junkie, *inf* user. **2** *a TV addict.* ▷ ENTHUSIAST.

addiction *n* compulsion, craving, dependence, fixation, habit, obsession.

addition *n* **1** adding up, calculation, computation, reckoning, totalling, *inf* totting up. **2** accession, accessory, accretion, addendum, additive, adjunct, admixture, afterthought, amplification, annexe, appendage, appendix, appurtenance, attachment, continuation, development, enlargement, expansion, extension, extra, increase, increment, postscript, supplement.

additional *adj* added, extra, further, increased, more, new, other, spare, supplementary.

address *n* **1** directions, location, whereabouts. **2** *deliver an address.* discourse, disquisition, harangue, homily, lecture, oration, sermon, speech, talk. • *vb* **1** accost, apostrophize, approach, *inf* buttonhole, engage in conversation, greet, hail, salute, speak to, talk to. **2** *address an audience.* give a speech to, harangue, lecture. **address yourself to** ▷ TACKLE.

adept *adj* accomplished, clever, competent, expert, gifted, practised, proficient. ▷ SKILFUL. *Opp* UNSKILFUL.

adequate *adj* acceptable, all right, average, competent, fair, fitting, good enough, middling, *inf* OK, passable, presentable, respectable, satisfactory, *inf* so-so, sufficient, suitable, tolerable. *Opp* INADEQUATE.

adhere *vb* bind, bond, cement, cling, fuse, glue, gum, paste, stick. ▷ FASTEN.

adherent *n* aficionado, devotee, disciple, fan, follower, *inf* hanger-on, supporter.

adhesive *adj* glued, gluey, gummed, self-adhesive, sticky.

adjoining *adj* abutting, adjacent, bordering, closest, contiguous, juxtaposed, nearest, neighbouring, next, touching. *Opp* DISTANT.

adjourn *vb* break off, defer, discontinue, dissolve, interrupt, postpone, prorogue, put off, suspend.

adjournment *n* break, delay, interruption, pause, postponement, prorogation, recess, stay, stoppage, suspension.

adjust *vb* **1** adapt, alter, amend, arrange, balance, change, convert, correct, modify, position, put right, rectify, refashion, regulate, remake, remodel, reorganize, reshape, set, set to rights, tailor, temper, tune, vary. **2** acclimatize, accommodate, accustom, conform, fit, habituate, harmonize, reconcile yourself.

administer *vb* **1** administrate, command, conduct affairs, control, direct, govern, head, lead, manage, organize, oversee, preside over, regulate, rule, run, superintend, supervise. **2** *administer justice.* apply, carry out, execute, implement, prosecute. **3** *administer medicine.* deal out, dispense, distribute, dole out, give, hand out, measure out, mete out, provide, supply.

administrator *n* boss, bureaucrat, civil servant, controller, director, executive, head, manager, managing director, *derog* mandarin, organizer, superintendent. ▷ CHIEF.

admirable *adj* awe-inspiring, commendable, creditable, deserving, enjoyable, estimable, excellent, exemplary, fine, great, honourable, laudable, likeable, lovable, marvellous, meritorious, pleasing, praiseworthy, valued, wonderful, worthy. *Opp* CONTEMPTIBLE.

admiration *n* appreciation, approval, awe, commendation, esteem, hero-worship, high regard, honour, praise, respect. *Opp* CONTEMPT.

admire *vb* applaud, appreciate, approve of, be delighted by, commend, enjoy, esteem, have a high opinion of, hero-worship, honour, idolize, laud, like, look up to, love, marvel at, praise, respect, revere, think highly of, value, venerate, wonder at. ▷ LOVE. *Opp* HATE.

admiring ▷ COMPLIMENTARY, RESPECTFUL.

admission *n* **1** access, admittance, entrance, entrée, entry. **2** acceptance, acknowledgement, affirmation, agreement, avowal, concession, confession, declaration, disclosure, profession, revelation. *Opp* DENIAL.

admit *vb* **1** accept, allow in, grant access, let in, provide a place (in), receive, take in. **2** *admit guilt*. accept, acknowledge, agree, allow, concede, confess, declare, disclose, divulge, grant, own up, profess, recognize, reveal, say reluctantly. *Opp* DENY.

adolescence *n* boyhood, girlhood, growing up, puberty, *inf* your teens, youth.

adolescent *adj* boyish, girlish, immature, juvenile, pubescent, puerile, teenage, youthful. • *n* boy, girl, juvenile, minor, *inf* teenager, youngster, youth.

adopt *vb* **1** accept, appropriate, approve, back, champion, choose, embrace, endorse, espouse, follow, *inf* go for, patronize, support, take on, take up. **2** befriend, foster, stand by, take in, *inf* take under your wing.

adore *vb* adulate, dote on, glorify, honour, idolize, love, revere, reverence, venerate, worship. ▷ ADMIRE. *Opp* HATE.

adorn *vb* beautify, decorate, embellish, garnish, grace, ornament, trim.

adrift *adj* **1** afloat, anchorless, drifting, floating. **2** aimless, astray, directionless, lost, purposeless, rootless.

adult *adj* developed, full-grown, full-size, grown-up, marriageable, mature, nubile, of age. *Opp* IMMATURE.

adulterate *vb* alloy, contaminate, corrupt, debase, defile, dilute, *inf* doctor, pollute, taint, thin, water down, weaken.

advance *n* betterment, development, evolution, forward movement, growth, headway, improvement, progress. • *vb* **1** approach, bear down, come near, forge ahead, gain ground, go forward, make headway, make progress, *inf* make strides, move forward, press ahead, press on, proceed, progress, *inf* push on. *Opp* RETREAT. **2** *science advances*. develop, evolve, grow, improve, increase, prosper, thrive. **3** *advance your career*. accelerate, assist, benefit, boost, expedite, facilitate, further, help the progress of, promote. *Opp* HINDER. **4** *advance a theory*. adduce, cite, furnish, give, present, propose, submit, suggest. **5** *advance money*. lend, loan, offer, pay, proffer, provide, supply. *Opp* WITHHOLD.

advanced *adj* **1** latest, modern, sophisticated, ultra-modern, up-to-date. **2** *advanced ideas*. avant-garde, contemporary, experimental, forward-looking, futuristic, imaginative, innovative, inventive, new, novel, original, pioneering, progressive, revolutionary, trend-setting, unconventional, unheard-of, *inf* way-out. **3** *advanced maths*. complex, complicated, difficult, hard, higher. **4** *advanced for her age*. grown-up, mature, precocious, sophisticated, well-developed. *Opp* BACKWARD, BASIC, OLD.

advantage *n* **1** aid, asset, assistance, benefit, boon, convenience, gain, help, profit, service, use, usefulness. **2** *have an advantage.* dominance, edge, *inf* head start, superiority. **take advantage of** ▷ EXPLOIT.

advantageous *adj* beneficial, constructive, favourable, gainful, helpful, invaluable, positive, profitable, salutary, useful, valuable, worthwhile. ▷ GOOD. *Opp* USELESS.

adventure *n* **1** chance, enterprise, escapade, exploit, feat, gamble, incident, occurrence, operation, risk, undertaking, venture. **2** danger, excitement, hazard.

adventurous *adj* **1** audacious, bold, brave, courageous, daredevil, daring, enterprising, *derog* foolhardy, heroic, intrepid, *derog* rash, *derog* reckless, valiant, venturesome. **2** *an adventurous trip.* challenging, dangerous, difficult, eventful, exciting, hazardous, perilous, risky. *Opp* UNADVENTUROUS.

adversary *n* antagonist, attacker, enemy, foe, opponent, rival. *Opp* FRIEND.

adverse *adj* **1** antagonistic, attacking, censorious, critical, derogatory, disapproving, hostile, hurtful, inimical, negative, uncomplimentary, unfavourable, unfriendly, unkind, unsympathetic. **2** *adverse conditions.* contrary, deleterious, detrimental, disadvantageous, harmful, inappropriate, inauspicious, opposing, prejudicial, uncongenial, unfortunate, unpropitious. *Opp* FAVOURABLE.

advertise *vb* announce, broadcast, display, flaunt, make known, market, merchandise, notify, *inf* plug, proclaim, promote, promulgate, publicize, *inf* push, show off, *inf* spotlight, tout.

advertisement *n inf* advert, announcement, bill, *inf* blurb, *TV* break, circular, commercial, hand out, leaflet, notice, placard, *inf* plug, poster, promotion, publicity, *old use* puff, sign, *inf* small ad.

advice *n* admonition, caution, counsel, guidance, help, opinion, recommendation, suggestion, tip, view, warning. ▷ NEWS.

advisable *adj* expedient, judicious, politic, prudent, recommended, sensible. ▷ WISE. *Opp* SILLY.

advise *vb* **1** admonish, advocate, caution, counsel, encourage, enjoin, exhort, guide, instruct, prescribe, recommend, suggest, urge, warn. **2** ▷ INFORM.

adviser *n* cicerone, confidant(e), consultant, counsellor, guide, mentor.

advocate *n* **1** apologist, backer, champion, proponent, supporter. **2** ▷ LAWYER. • *vb* argue for, back, champion, endorse, favour, recommend, speak for, uphold.

aerodrome *n* airfield, airport, airstrip, landing-strip.

aesthetic *adj* artistic, beautiful, cultivated, in good taste, sensitive tasteful. *Opp* UGLY.

affair *n* **1** activity, business, concern, issue, interest, matter, operation, project, question, subject, topic, transaction, undertaking. **2** circumstance, episode, event, happening, incident, occasion, occurrence, proceeding, thing. **3** *love affair.* affaire, amour, attachment, intrigue, involvement, liaison, relationship, romance.

affect *vb* **1** act on, agitate, alter, attack, change, concern, disturb, grieve, have an effect on, have an impact on, *inf* hit, impinge on, impress, influence, modify, move, pertain to, perturb, relate to, stir, touch, transform, trouble, upset. **2** *affect an accent.* adopt, assume, feign, *inf* put on. ▷ PRETEND.

affectation *n* artificiality, insincerity, mannerism, posturing, pretension. ▷ PRETENCE.

affected *adj* **1** artificial, contrived, insincere, mannered, *inf* put on, studied, unnatural. ▷ PRETENTIOUS. **2** *affected by disease.* afflicted, attacked, damaged, distressed, hurt, infected, injured, poisoned, stricken, troubled.

affection *n* amity, attachment, feeling, fondness, friendliness, friendship, liking, partiality, regard, *inf* soft spot, tenderness, warmth. ▷ LOVE. *Opp* HATRED.

affectionate *adj* caring, doting, fond, kind, tender, warm. ▷ LOVING. *Opp* ALOOF.

affinity *n* closeness, compatibility, fondness, kinship, like-mindedness, likeness, liking, rapport, relationship, resemblance, similarity, sympathy.

affirm *vb* assert, attest, aver, avow, confirm, declare, maintain, state, swear, testify.

affirmation *n* assertion, avowal, confirmation, declaration, oath, promise, pronouncement, statement, testimony.

affirmative *adj* agreeing, assenting, concurring, confirming, consenting, positive. *Opp* NEGATIVE.

afflict *vb* affect, annoy, bedevil, beset, bother, burden, cause suffering to, distress, grieve, harass, harm, hurt, oppress, pain, pester, plague, rack, torment, torture, trouble, try, vex, worry, wound.

affluent *adj* **1** flourishing, *inf* flush, *sl* loaded, moneyed, *usu derog* plutocratic, prosperous, rich, wealthy, *inf* well-heeled, well-off, well-to-do. **2** *affluent life-style.* expensive, gracious, lavish, luxurious, opulent, pampered, self-indulgent, sumptuous. *Opp* POOR.

afford *vb* **1** be rich enough, find enough, have the means, manage to give, sacrifice, spare, *inf* stand. **2** ▷ PROVIDE.

afloat *adj* aboard, adrift, at sea, floating, on board ship, under sail.

afraid *adj* **1** aghast, agitated, alarmed, anxious, apprehensive, *inf* chicken, cowardly, cowed, craven, daunted, diffident, faint-hearted, fearful, frightened, hesitant, horrified, horror-struck, intimidated, jittery, nervous, panicky, panic-stricken, pusillanimous, reluctant, scared, terrified, terror-stricken, timid, timorous, trembling, uneasy, unheroic, *inf* windy, *inf* yellow. *Opp* CONFIDENT, FEARLESS. **2** [*inf*] *I'm afraid I'm late.* apologetic, regretful, sorry, unhappy. **be afraid** ▷ FEAR.

afterthought *n* addendum, addition, appendix, extra, postscript.

age *n* **1** advancing years, decrepitude, dotage, old age, senescence, senility. **2** *a bygone age.* days, epoch, era, generation, period, time. **3** [*inf*] *ages ago.* aeon, lifetime, long time. • *vb* decline, degenerate, develop, grow older, look older, mature, mellow, ripen. **aged** ▷ OLD.

agenda *n* list, plan, programme, schedule, timetable.

agent *n* broker, delegate, emissary, envoy, executor, *old use* functionary, go-between, intermediary, mediator, middleman, negotiator, proxy, representative, spokesman, spokeswoman, surrogate, trustee.

aggravate *vb* **1** add to, augment, compound, exacerbate, exaggerate, heighten, increase, inflame, intensify, magnify, make more serious, make worse, worsen. *Opp* ALLEVIATE. **2** [Some think this use wrong.] ▷ ANNOY.

aggressive *adj* antagonistic, assertive, attacking, bellicose, belligerent, bullying, *sl* butch, contentious, destructive, hostile, jingoistic, *sl* macho, militant, offensive, provocative, pugnacious, pushful, *inf* pushy, quarrelsome, violent, warlike, zealous. *Opp* DEFENSIVE, PEACEABLE.

aggressor *n* assailant, attacker, belligerent, instigator, invader.

agile *adj* acrobatic, active, adroit, deft, fleet, graceful, limber, lissom, lithe, lively, mobile, nimble, quick-moving, sprightly, spry, supple, swift. *Opp* CLUMSY, SLOW.

agitate *vb* **1** beat, churn, convulse, ferment, froth up, ruffle, shake, stimulate, stir, toss, work up. **2** alarm, arouse, confuse, discomfit, disconcert, disturb, excite, fluster, incite, perturb, rouse, shake up, stir up, trouble, unnerve, unsettle, upset, worry. *Opp* CALM. **agitated** ▷ EXCITED, NERVOUS.

agitator *n Fr* agent provocateur, demagogue, firebrand, rabble-rouser, revolutionary, trouble-maker.

agonize *vb* be in agony, hurt, labour, struggle, suffer, worry, wrestle.

agony *n* anguish, distress, suffering, torment, torture. ▷ PAIN.

agree *vb* **1** accede, accept, acknowledge, acquiesce, admit, allow, assent, be willing, concede, consent, covenant, grant, make a contract, pledge yourself, promise, undertake. **2** accord, be unanimous, be united, coincide, concur, conform, correspond, fit, get on, harmonize, match, *inf* see eye to eye, suit. *Opp* DISAGREE. **agree on** ▷ CHOOSE. **agree with** ▷ ENDORSE.

agreeable *adj* acceptable, delightful, enjoyable, nice. ▷ PLEASANT. *Opp* DISAGREEABLE.

agreement *n* **1** accord, affinity, compatibility, compliance, concord, conformity, congruence, consensus, consent, consistency, correspondence, harmony, similarity, sympathy, unanimity, unity. **2** acceptance, alliance, armistice, arrangement, bargain, bond, compact, concordat, contract, convention, covenant, deal, *Fr* entente, pact, pledge, protocol, settlement, treaty, truce, understanding. *Opp* DISAGREEMENT.

agricultural *adj* **1** agrarian, bucolic, pastoral, rural. **2** *agricultural land.* cultivated, farmed, planted, productive, tilled.

agriculture *n* agronomy, crofting, cultivation, farming, growing, husbandry, tilling.

aground *adj* beached, grounded, helpless, high-and-dry, marooned, shipwrecked, stranded, stuck.

aid *n* advice, assistance, avail, backing, benefit, collaboration, contribution, cooperation, donation, encouragement, funding, grant, guidance, help, loan, patronage, prop, relief, sponsorship, subsidy, succour, support. • *vb* abet, assist, back, befriend, benefit, collaborate with, contribute to, co-operate with, encourage, facilitate, forward, help, *inf* lend a hand, profit, promote, prop up, *inf* rally round, relieve, subsidize, succour, support, sustain.

ailing *adj* diseased, feeble, infirm, poorly, sick, suffering, unwell, weak. ▷ ILL.

ailment *n* affliction, disease, disorder, infirmity, malady, sickness. ▷ ILLNESS.

aim *n* ambition, aspiration, cause, design, desire, destination, direction, dream, end, focus, goal, hope, intent, intention, mark, object, objective, plan, purpose, target, wish. • *vb* **1** address, beam, direct, fire at, focus, level, line up, point, send, sight, take aim, train, turn, zero in on. **2** *aim to win.* aspire, attempt, design, endeavour, essay, intend, mean, plan, propose, resolve, seek, strive, try, want, wish.

aimless *adj* chance, directionless, pointless, purposeless, rambling, random, undisciplined, unfocused, wayward. *Opp* PURPOSEFUL.

air *n* **1** airspace, atmosphere, ether, heavens, sky, *poet* welkin. **2** *fresh air.* breath, breeze, draught, oxygen, waft, wind, *poet* zephyr. **3** *air of authority.* ambience, appearance, aspect, aura, bearing, character, demeanour, effect, feeling, impression, look, manner, mien, mood, quality, style. • *vb* **1** aerate, dry off, freshen, refresh, ventilate. **2** *air opinions.* articulate, disclose, display, exhibit, express, give vent to, make known, make public, put into words, show off, vent, voice.

aircraft *n old use* flying-machine. □ *aeroplane, airliner, airship, balloon, biplane, bomber, delta-wing, dirigible, fighter, flying boat, glider, gunship, hang-glider, helicopter, jet, jumbo, jump-jet, microlight, monoplane, plane, seaplane, turboprop, VTOL (vertical take-off and landing).*

airman *n* aviator, flier, pilot.

airport *n* aerodrome, airfield, air strip, heliport, landing-strip, runway.

airy *adj* blowy, breezy, draughty, fresh, open, spacious, ventilated. *Opp* STUFFY.

aisle *n* corridor, gangway, passage, passageway.

akin *adj* allied, related, similar.

alarm *n* **1** alert, signal, warning. □ *alarm-clock, bell, fire-alarm, gong, siren, tocsin, whistle.* **2** anxiety, apprehension, consternation, dismay, distress, fright, nervousness, panic, trepidation, uneasiness. ▷ FEAR. • *vb* agitate, daunt, dismay, distress, disturb, panic, *inf* put the wind up, scare, shock, startle, surprise, unnerve, upset, worry. ▷ FRIGHTEN. *Opp* REASSURE.

alcohol *n sl* bevvy, *inf* booze, drink, hard stuff, intoxicant, liquor, spirits, wine.

alcoholic *adj* brewed, distilled, fermented, *sl* hard, inebriating, intoxicating, spirituous, *inf* strong. • *n* addict, dipsomaniac, drunkard, inebriate, toper. *Opp* TEETOTALLER.

alert *adj* active, agile, alive (to), attentive, awake, careful, circumspect, eagle-eyed, heedful, lively, observant, on the alert, on the lookout, *inf* on the qui vive, on the watch, on your guard, on your toes, perceptive, quick, ready, responsive, sensitive, sharp-eyed, vigilant, wary, watchful, wide-awake. *Opp* ABSENT-MINDED, INATTENTIVE. • *vb* advise, alarm, caution, forewarn, give the alarm, inform, make aware, notify, signal, tip off, warn.

alibi *n* excuse, explanation.

alien *adj* exotic, extra-terrestrial, foreign, outlandish, remote, strange, unfamiliar. • *n* foreigner, newcomer, outsider, stranger.

alight *adj* ablaze, afire, aflame, blazing, bright, burning, fiery, ignited, illuminated, lit up, live, on fire, shining. • *vb* come down, come to rest, descend, disembark, dismount, get down, get off, land, perch, settle, touch down.

align *vb* **1** arrange in line, line up, place in line, straighten up. **2** *align with the opposition.* affiliate, agree, ally, associate, co-operate, join, side, sympathize.

alike *adj* akin, analogous, close, cognate, comparable, corresponding, equivalent, identical, indistinguishable, like, matching, parallel, related, resembling, similar, the same, twin, uniform. *Opp* DISSIMILAR.

alive *adj* **1** active, animate, breathing, existent, existing, extant, flourishing, in existence, live, living, *old use* quick, surviving. **2** *alive to new ideas.* ▷ ALERT. *Opp* DEAD.

allay *vb* alleviate, assuage, calm, check, compose, diminish, ease, lessen, lull, mitigate, moderate, mollify, pacify, quell, quench, quiet, quieten, reduce, relieve, slake (*thirst*), soften, soothe, subdue. *Opp* STIMULATE.

allegation *n* accusation, assertion, charge, claim, complaint, declaration, deposition, statement, testimony.

allege *vb* adduce, affirm, assert, asseverate, attest, aver, avow, claim, contend, declare, depose, insist, maintain, make a charge, plead, state.

allegiance *n* devotion, duty, faithfulness, *old use* fealty, fidelity, loyalty, obedience.

allergic *adj* antagonistic, antipathetic, averse, disinclined, hostile, incompatible (with), opposed.

alleviate *vb* abate, allay, ameliorate, assuage, check, diminish, ease, lessen, lighten, make lighter, mitigate, moderate, pacify, palliate, quell, quench, reduce, relieve, slake (*thirst*), soften, soothe, subdue, temper. *Opp* AGGRAVATE.

alliance *n* affiliation, agreement, association, bloc, bond, cartel, coalition, combination, compact, concordat, confederation, connection, consortium, covenant, entente, federation, guild, league, marriage, pact, partnership, relationship, syndicate, treaty, understanding, union.

allot *vb* allocate, allow, apportion, assign, award, deal out, *inf* dish out, *inf* dole out, dispense, distribute, divide out, give out, grant, mete out, provide, ration, set aside, share out.

allow *vb* **1** approve, authorize, bear, consent to, enable, endure, grant permission for, let, license, permit, *inf* put up with, sanction, *inf* stand, suffer, support, tolerate. *Opp* FORBID. **2** acknowledge, admit, concede, grant, own. **3** ▷ ALLOT. *Opp* DENY.

allowance *n* **1** allocation, allotment, amount, measure, portion, quota, ration, share. **2** alimony, annuity, grant, maintenance, payment, pension, pocket money, subsistence. **3** *allowance on the full price.* deduction, discount, rebate, reduction, remittance, subsidy. **make allowances for** ▷ TOLERATE.

alloy *n* admixture, aggregate, amalgam, blend, combination, composite, compound, fusion, mixture.

allude *vb* **allude to** hint at, make an allusion to, mention, refer to, speak of, suggest, touch on.

allure *vb* attract, beguile, bewitch, cajole, charm, coax, decoy, draw, entice, fascinate, inveigle, lead on, lure, magnetize, persuade, seduce, tempt.

allusion *n* hint, mention, reference, suggestion.

ally *n* abettor, accessory, accomplice, associate, backer, collaborator, colleague, companion, comrade, confederate, friend, helper, helpmate, *inf* mate, partner, supporter. *Opp* ENEMY. • *vb* affiliate, amalgamate, associate, band together, collaborate, combine, confederate, cooperate, form an alliance, fraternize, join, join forces, league, *inf* link up, marry, merge, side, *inf* team up, unite.

almighty *adj* **1** all-powerful, omnipotent, supreme. **2** ▷ BIG.

almost *adv* about, all but, approximately, around, as good as, just about, nearly, not quite, practically, virtually.

alone *adj* apart, by yourself, deserted, desolate, forlorn, friendless, isolated, lonely, lonesome, on your own, separate, single, solitary, solo, unaccompanied, unassisted.

aloof *adj* chilly, cold, cool, detached, disinterested, dispassionate, distant, formal, frigid, haughty, impressive, inaccessible, indifferent, remote, reserved, reticent, self-contained, self-possessed, *inf* standoffish, supercilious, unapproachable, unconcerned, undemonstrative, unemotional, unforthcoming, unfriendly, uninvolved, unresponsive, unsociable, unsympathetic. *Opp* FRIENDLY, SOCIABLE.

aloud *adv* audibly, clearly, distinctly, out loud.

also *adv* additionally, besides, furthermore, in addition, moreover, *joc* to boot, too.

alter *vb* adapt, adjust, amend, change, convert, edit, emend, enlarge, modify, reconstruct, reduce, reform, remake, remodel, reorganize, reshape, revise, transform, vary.

alteration *n* adaptation, adjustment, amendment, change, conversion, difference, modification, reorganization, revision, transformation.

alternate *vb* come alternately, follow each other, interchange, oscillate, replace each other, rotate, *inf* see-saw, substitute for each other, take turns.

alternative *n* **1** choice, option, selection. **2** back-up, replacement, substitute.

altitude *n* elevation, height.

altogether *adv* absolutely, completely, entirely, fully, perfectly, quite, thoroughly, totally, utterly, wholly.

always *adv* consistently, constantly, continually, continuously, endlessly, eternally, everlastingly, evermore, forever, invariably, perpetually, persistently, regularly, repeatedly, unceasingly, unfailingly, unremittingly.

amalgamate *vb* affiliate, ally, associate, band together, blend, coalesce, combine, come together, compound, confederate, form an alliance, fuse, integrate, join, join forces, league, *inf* link up, marry, merge, mix, put together, synthesize, *inf* team up, unite. *Opp* SPLIT.

amateur *adj* inexperienced, lay, unpaid, unqualified. ▷ AMATEURISH. • *n* dabbler, dilettante, enthusiast, layman, non-professional. *Opp* PROFESSIONAL.

amateurish *adj* clumsy, crude, *inf* do-it-yourself, incompetent, inept, inexpert, *inf* rough-and-ready, second-rate, shoddy, unpolished, unprofessional, unskilful, unskilled, untrained. *Opp* SKILLED.

amaze *vb* astonish, astound, awe, bewilder, confound, confuse, daze, disconcert, dumbfound, *inf* flabbergast, perplex, *inf* rock, shock, stagger, startle, stun, stupefy, surprise. **amazed** ▷ SURPRISED.

amazing *adj* astonishing, astounding, awe-inspiring, breathtaking, exceptional, exciting, extraordinary, *inf* fantastic, incredible, miraculous, notable, phenomenal, prodigious, remarkable, *inf* sensational, shocking, special, staggering, startling, stunning, stupendous, unusual, *inf* wonderful. *Opp* ORDINARY.

ambassador *n* agent, attaché, *Fr* chargé d'affaires, consul, diplomat, emissary, envoy, legate, nuncio, plenipotentiary, representative.

ambiguous *adj* ambivalent, confusing, enigmatic, equivocal, indefinite, indeterminate, puzzling, uncertain, unclear, vague, woolly. ▷ UNCERTAIN. *Opp* DEFINITE.

ambition *n* **1** commitment, drive, energy, enterprise, enthusiasm, *inf* go, initiative, *inf* push, pushfulness, self-assertion, thrust, zeal. **2** aim, aspiration, desire, dream, goal, hope, ideal, intention, object, objective, target, wish.

ambitious *adj* **1** assertive, committed, eager, energetic, enterprising, enthusiastic, go-ahead, *inf* go-getting, hard-working, industrious, keen, *inf* pushy, zealous. **2** *ambitious ideas. inf* big, far-reaching, grand, grandiose, large-scale, unrealistic. *Opp* APATHETIC.

ambivalent *adj* ambiguous, back-handed (*compliment*), confusing, doubtful, equivocal, inconclusive, inconsistent, indefinite, self-contradictory, *inf* two-faced, unclear, uncommitted, unresolved, unsettled. ▷ UNCERTAIN.

ambush *n* ambuscade, attack, *snare, surprise attack, trap.* • *vb* attack, ensnare, entrap, intercept, lie in wait for, pounce on, surprise, swoop on, trap, waylay.

amenable *adj* accommodating, acquiescent, adaptable, agreeable, biddable, complaisant, compliant, co-operative, deferential, docile, open-minded, persuadable, responsive, submissive, tractable, willing. *Opp* OBSTINATE.

amend *vb* adapt, adjust, alter, ameliorate, change, convert, correct, edit, emend, improve, make better, mend, modify, put right, rectify, reform, remedy, reorganize, reshape, revise, transform, vary.

amiable *adj* affable, agreeable, amicable, friendly, genial, good-natured, kind-hearted, kindly, likeable, well-disposed. *Opp* UNFRIENDLY.

ammunition *n* buckshot, bullet, cartridge, grenade, missile, projectile, round, shell, shrapnel.

amoral *adj* lax, loose, unethical, unprincipled, without standards. ▷ IMMORAL. *Opp* MORAL.

amorous *adj* affectionate, ardent, carnal, doting, enamoured, erotic, fond, impassioned, loving, lustful, passionate, *sl* randy, sexual, *inf* sexy. *Opp* COLD.

amount *n* aggregate, bulk, entirety, extent, lot, mass, measure, quantity, quantum, reckoning, size, sum, supply, total, value, volume, whole. • *vb* **amount to** add up to, aggregate, be equivalent to, come to, equal, make, mean, total.

ample *adj* abundant, bountiful, broad, capacious, commodious, considerable, copious, extensive, fruitful, generous, great, large, lavish, liberal, munificent, plentiful, profuse, roomy, spacious, substantial, unstinting, voluminous. ▷ BIG, PLENTY. *Opp* INSUFFICIENT.

amplify *vb* **1** add to, augment, broaden, develop, dilate upon, elaborate, enlarge, expand, expatiate on, extend, fill out, lengthen, make fuller, make longer, supplement. **2** *amplify sound.* boost, heighten, increase, intensify, magnify, make louder, raise the volume. *Opp* DECREASE.

amputate *vb* chop off, cut off, dock, lop off, poll, pollard, remove, sever, truncate. ▷ CUT.

amuse *vb* absorb, beguile, cheer (up), delight, divert, engross, enliven, entertain, gladden, interest, involve, make laugh, occupy, please, raise a smile, *inf* tickle. *Opp* BORE. **amusing** ▷ ENJOYABLE, FUNNY.

amusement *n* **1** delight, enjoyment, fun, hilarity, laughter, mirth. ▷ MERRIMENT. **2** distraction, diversion, entertainment, game, hobby, interest, joke, leisure activity, pastime, play, pleasure, recreation, sport.

anaemic *adj* bloodless, colourless, feeble, frail, pale, pallid, pasty, sallow, sickly, unhealthy, wan, weak.

analogy *n* comparison, likeness, metaphor, parallel, resemblance, similarity, simile.

analyse *vb* anatomize, assay, break down, criticize, dissect, evaluate, examine, interpret, investigate, scrutinize, separate out, take apart, test.

analysis *n* breakdown, critique, dissection, enquiry, evaluation, examination, interpretation, investigation, *inf* post-mortem, scrutiny, study, test. *Opp* SYNTHESIS.

analytical *adj* analytic, critical, *inf* in-depth, inquiring, investigative, logical, methodical, penetrating, questioning, rational, searching, systematic. *Opp* SUPERFICIAL.

anarchy *n* bedlam, chaos, confusion, disorder, disorganization, insurrection, lawlessness, misgovernment, misrule, mutiny, pandemonium, riot. *Opp* ORDER.

ancestor *n* antecedent, forebear, forefather, forerunner, precursor, predecessor, progenitor.

ancestry *n* blood, derivation, descent, extraction, family, genealogy, heredity, line, lineage, origin, parentage, pedigree, roots, stock, strain.

anchor *vb* berth, make fast, moor, secure, tie up. ▷ FASTEN.

anchorage *n* harbour, haven, marina, moorings, port, refuge, sanctuary, shelter.

ancient *adj* **1** aged, antediluvian, antiquated, antique, archaic, elderly, fossilized, obsolete, old, old-fashioned, outmoded, out-of-date, passé, superannuated, time-worn, venerable. **2** *ancient times.* bygone, earlier, early, former, *poet* immemorial, *inf* olden, past, prehistoric, primeval, primitive, primordial, remote, *old use* of yore. *Opp* MODERN.

angel *n* archangel, cherub, divine messenger, seraph.

angelic *adj* **1** beatific, blessed, celestial, cherubic, divine, ethereal, heavenly, holy, seraphic, spiritual. **2** *angelic behaviour.* exemplary, innocent, pious, pure, saintly, unworldly, virtuous. ▷ GOOD. *Opp* DEVILISH.

anger *n* angry feelings, annoyance, antagonism, bitterness, choler, displeasure, exasperation, fury, hostility, indignation, ire, irritability, outrage, passion, pique, rage, rancour, resentment, spleen, tantrum, temper, vexation, wrath. • *vb inf* aggravate, antagonize, *sl* bug, displease, *inf* drive mad, enrage, exasperate, incense, incite, inflame, infuriate, irritate, madden, make angry, *inf* make someone's blood boil, *inf* needle, outrage, pique, provoke, *inf* rile, vex. ▷ ANNOY. *Opp* PACIFY.

angle *n* **1** bend, corner, crook, nook, point. **2** *a new angle.* approach, outlook, perspective, point of view, position, slant, standpoint, viewpoint. • *vb* bend, bevel, chamfer, slant, turn, twist.

angry *adj inf* aerated, apoplectic, bad-tempered, bitter, *inf* bristling, *inf* choked, choleric, cross, disgruntled, enraged, exasperated, excited, fiery, fuming, furious, heated, hostile, *inf* hot under the collar, ill-tempered, incensed, indignant, infuriated, *inf* in high dudgeon, irascible, irate, livid, mad, outraged, provoked, raging, *inf* ratty, raving, resentful, riled, seething, smouldering, *inf* sore, splenetic, *sl* steamed up, stormy, tempestuous, vexed, *inf* ugly, *inf* up in arms, wild, wrathful. ▷ ANNOYED. *Opp* CALM. be angry, become angry *inf* be in a paddy, *inf* blow up, boil, bridle, bristle, flare up, *inf* fly off the handle, fulminate, fume, *inf* get steamed up, lose your temper, rage, rant, rave, *inf* see red, seethe, snap, storm. **make angry** ▷ ANGER.

anguish *n* agony, anxiety, distress, grief, heartache, misery, pain, sorrow, suffering, torment, torture, tribulation, woe.

angular *adj* bent, crooked, indented, jagged, sharp-cornered, zigzag. *Opp* STRAIGHT.

animal *adj* beastly, bestial, brutish, carnal, fleshly, inhuman, instinctive, physical, savage, sensual, subhuman, wild. • *n* beast, being, brute, creature, *plur* fauna, organism, *plur* wildlife.
□ *amphibian, arachnid, biped, carnivore, herbivore, insect, invertebrate, mammal, marsupial, mollusc, monster, omnivore, pet, quadruped, reptile, rodent, scavenger, vertebrate.* ▷ BIRD, FISH, INSECT.
□ *aardvark, antelope, ape, armadillo, baboon, badger, bear, beaver, bison, buffalo, camel, caribou, cat, chamois, cheetah, chimpanzee, chinchilla, chipmunk, coypu, deer, dog, dolphin, donkey, dormouse, dromedary, elephant, elk, ermine, ferret, fox, frog, gazelle, gerbil, gibbon, giraffe, gnu, goat, gorilla, grizzly bear, guinea-pig, hamster, hare, hedgehog, hippopotamus, horse, hyena, ibex, impala, jackal, jaguar, jerboa, kangaroo, koala, lemming, lemur, leopard, lion, llama, lynx, marmoset, marmot, marten, mink, mongoose, monkey, moose, mouse, musquash, ocelot, octopus, opossum, orang-utan, otter, panda, panther, pig, platypus, polar bear, pole-cat, porcupine, porpoise, rabbit, rat, reindeer, rhinoceros, roe, salamander, scorpion, seal,*

sea-lion, sheep, shrew, skunk, snake, spider, squirrel, stoat, tapir, tiger, toad, vole, wallaby, walrus, weasel, whale, wildebeest, wolf, wolverine, wombat, yak, zebra.

animate *adj* alive, breathing, conscious, feeling, live, living, sentient. ▷ ANIMATED. *Opp* INANIMATE. • *vb* activate, arouse, brighten up, *inf* buck up, cheer up, encourage, energize, enliven, excite, exhilarate, fire, galvanize, incite, inspire, invigorate, kindle, liven up, make lively, move, *inf* pep up, *inf* perk up, quicken, rejuvenate, revitalize, revive, rouse, spark, spur, stimulate, stir, urge, vitalize.

animated *adj* active, alive, bright, brisk, bubbling, busy, cheerful, eager, ebullient, energetic, enthusiastic, excited, exuberant, gay, impassioned, lively, passionate, quick, spirited, sprightly, vibrant, vigorous, vivacious, zestful. *Opp* LETHARGIC.

animation *n* activity, briskness, eagerness, ebullience, energy, enthusiasm, excitement, exhilaration, gaiety, high spirits, life, liveliness, *inf* pep, sparkle, spirit, sprightliness, verve, vigour, vitality, vivacity, zest. *Opp* LETHARGY.

animosity *n* acerbity, acrimony, animus, antagonism, antipathy, asperity, aversion, bad blood, bitterness, dislike, enmity, grudge, hate, hatred, hostility, ill will, loathing, malevolence, malice, malignancy, malignity, odium, rancour, resentment, sarcasm, sharpness, sourness, spite, unfriendliness, venom, vindictiveness, virulence. *Opp* FRIENDLINESS.

annex *vb* acquire, appropriate, conquer, occupy, purloin, seize, take over, usurp.

annihilate *vb* abolish, destroy, eliminate, eradicate, erase, exterminate, extinguish, extirpate, *inf* finish off, *inf* kill off, *inf* liquidate, nullify, obliterate, raze, slaughter, wipe out.

annotation *n* comment, commentary, elucidation, explanation, footnote, gloss, interpretation, note.

announce *vb* **1** advertise, broadcast, declare, disclose, divulge, give notice of, intimate, make public, notify, proclaim, promulgate, propound, publicize, publish, put out, report, reveal, state. **2** *announce a speaker.* introduce, lead into, preface, present.

announcement *n* advertisement, bulletin, communiqué, declaration, disclosure, intimation, notification, proclamation, promulgation, publication, report, revelation, statement.

announcer *n* anchorman, anchorwoman, broadcaster, commentator, compère, disc jockey, DJ, *poet* harbinger, herald, master of ceremonies, *inf* MC, messenger, newscaster, newsreader, reporter, town crier.

annoy *vb inf* aggravate, antagonize, *inf* badger, be an annoyance to, bother, *sl* bug, chagrin, displease, distress, drive mad, exasperate, fret, gall, *inf* get at, *inf* get on your nerves, grate, harass, harry, infuriate, irk, irritate, jar, madden, make cross, molest, *inf* needle, *inf* nettle, offend, peeve, pester, pique, *inf* plague, provoke, put out, rankle, rile, *inf* rub up the wrong way, ruffle, *inf* spite, tease, trouble, try (someone's patience), upset, vex, worry. ▷ ANGER. *Opp* PLEASE.

annoyance *n* **1** chagrin, crossness, displeasure, exasperation, irritation, pique, vexation.

▷ ANGER. 2 *Noise is an annoyance.* *inf* aggravation, bother, harassment, irritant, nuisance, offence, *inf* pain in the neck, pest, provocation, worry.

annoyed *adj* chagrined, cross, displeased, exasperated, *inf* huffy, irritated, jaundiced, *inf* miffed, *inf* needled, *inf* nettled, offended, *inf* peeved, piqued, *inf* put out, *inf* riled, *inf* shirty, *inf* sore, upset, vexed. ▷ ANGRY. *Opp* PLEASED. be annoyed *inf* go off in a huff, take offence, *inf* take umbrage.

annoying *adj* *inf* aggravating, bothersome, displeasing, exasperating, galling, grating, inconvenient, infuriating, irksome, irritating, jarring, maddening, offensive, provocative, provoking, tiresome, troublesome, trying, upsetting, vexatious, vexing, wearisome, worrying.

anoint *vb* 1 embrocate, grease, lubricate, oil, rub, smear. 2 bless, consecrate, dedicate, hallow, sanctify.

anonymous *adj* 1 incognito, nameless, unacknowledged, unidentified, unknown, unnamed, unspecified, unsung. 2 *anonymous letters.* unattributed, unsigned. 3 *anonymous style.* characterless, impersonal, nondescript, unidentifiable, unrecognizable, unremarkable.

answer *n* 1 acknowledgement, *inf* comeback, reaction, rejoinder, reply, response, retort, riposte. 2 explanation, outcome, solution. 3 *answer to a charge.* countercharge, defence, plea, rebuttal, refutation, vindication. • *vb* 1 acknowledge, give an answer, react, rejoin, reply, respond, retort, return. 2 explain, resolve, solve. 3 *answer a charge.* counter, defend yourself against, disprove, rebut, refute. 4 *answer a need.* correspond to, echo, fit, match up to, meet, satisfy, serve, suffice, suit. **answer back** ▷ ARGUE.

antagonism *n* antipathy, dissension, enmity, friction, opposition, rancour, rivalry, strife. ▷ HOSTILITY.

antagonize *vb* alienate, anger, annoy, embitter, estrange, irritate, make an enemy of, offend, provoke, *inf* put off, upset.

anthem *n* canticle, chant, chorale, hymn, introit, paean, psalm.

anthology *n* collection, compendium, compilation, digest, miscellany, selection, treasury.

anticipate *vb* 1 forestall, obviate, preclude, pre-empt, prevent. 2 [Many think this use incorrect.] ▷ FORESEE.

anticlimax *n* bathos, *inf* comedown, *inf* damp squib, disappointment, *inf* let-down.

antics *n* buffoonery, capers, clowning, escapades, foolery, fooling, *inf* larking-about, pranks, *inf* skylarking, tomfoolery, tricks.

antidote *n* antitoxin, corrective, countermeasure, cure, drug, neutralizing agent, remedy.

antiquarian *n* antiquary, antiques expert, collector, dealer.

antiquated *adj* aged, anachronistic, ancient, antediluvian, archaic, dated, medieval, obsolete, old, old-fashioned, *inf* out, outdated, outmoded, out-of-date, passé, *inf* past it, *inf* prehistoric, *inf* primeval, primitive, quaint, superannuated, unfashionable. ▷ ANTIQUE. *Opp* NEW.

antique *adj* antiquarian, collectible, historic, old-fashioned, traditional, veteran, vintage. ▷ ANTIQUATED. • *n* bygone, collectible, collector's item, curio, curiosity, *Fr* objet d'art, rarity.

antiquity *n* classical times, days gone by, former times, *inf* olden days, the past.

antiseptic *adj* aseptic, clean, disinfectant, disinfected, germ free, germicidal, hygienic, medicated, sanitized, sterile, sterilized, sterilizing, unpolluted.

antisocial *adj* alienated, anarchic, disagreeable, disorderly, disruptive, misanthropic, nasty, obnoxious, offensive, rebellious, rude, troublesome, uncooperative, undisciplined, unruly, unsociable. ▷ UNFRIENDLY. *Opp* SOCIABLE.

anxiety *n* **1** angst, apprehension, concern, disquiet, distress, doubt, dread, fear, foreboding, fretfulness, misgiving, nervousness, qualm, scruple, strain, stress, tension, uncertainty, unease, worry. **2** *anxiety to succeed.* desire, eagerness, enthusiasm, impatience, keenness, longing, solicitude, willingness.

anxious 1 afraid, agitated, alarmed, apprehensive, concerned, distracted, distraught, distressed, disturbed, edgy, fearful, *inf* fraught, fretful, *inf* jittery, nervous, *inf* nervy, *inf* on edge, overwrought, perturbed, restless, tense, troubled, uneasy, upset, watchful, worried. **2** *anxious to succeed.* avid, careful, desirous, *inf* desperate, *inf* dying, eager, impatient, intent, *inf* itching, keen, longing, solicitous, willing, yearning. **be anxious** ▷ WORRY.

apathetic *adj* casual, cool, dispassionate, dull, emotionless, half-hearted, impassive, inactive, indifferent, indolent, languid, lethargic, listless, passive, phlegmatic, slow, sluggish, tepid, torpid, unambitious, uncommitted, unconcerned, unenterprising, unenthusiastic, unfeeling, uninterested, uninvolved, unmotivated, unresponsive. *Opp* ENTHUSIASTIC.

apathy *n* coolness, inactivity, indifference, lassitude, lethargy, listlessness, passivity, torpor. *Opp* ENTHUSIASM.

apex *n* **1** crest, crown, head, peak, pinnacle, point, summit, tip, top, vertex. **2** *apex of your career.* acme, apogee, climax, consummation, crowning moment, culmination, height, high point, zenith. *Opp* NADIR.

aphrodisiac *adj* arousing, erotic, *inf* sexy, stimulating.

apologetic *adj* ashamed, blushing, conscience-stricken, contrite, penitent, red-faced, regretful, remorseful, repentant, rueful, sorry. *Opp* UNREPENTANT.

apologize *vb* ask pardon, be apologetic, express regret, make an apology, repent, say sorry.

apology *n* acknowledgement, confession, defence, excuse, explanation, justification, plea.

apostle *n* crusader, disciple, evangelist, follower, messenger, missionary, preacher, propagandist, proselytizer, teacher.

appal *vb* alarm, disgust, dismay, distress, harrow, horrify, nauseate, outrage, revolt, shock, sicken, terrify, unnerve. ▷ FRIGHTEN. **appalling** ▷ ATROCIOUS, BAD, FRIGHTENING.

apparatus *n* appliance, *inf* contraption, device, equipment, gadget, *inf* gear, implement, instrument, machine, machinery, mechanism, *inf* set-up, system, *inf* tackle, tool, utensil.

apparent *adj* blatant, clear, conspicuous, detectable, discernible, evident, manifest, noticeable, observable, obvious, ostensible, overt, patent, perceptible, recognizable, self-explanatory, unconcealed, unmistakable, visible. *Opp* HIDDEN.

apparition *n* chimera, ghost, hallucination, illusion, manifestation, phantasm, phantom, presence, shade, spectre, spirit, *inf* spook, vision, wraith.

appeal *n* **1** application, call, *Fr* cri de coeur, cry, entreaty, petition, plea, prayer, request, solicitation, supplication. **2** allure, attractiveness, charisma, charm, *inf* pull, seductiveness. • *vb* ask earnestly, beg, beseech, call, canvass, cry out, entreat, implore, invoke, petition, plead, pray, request, solicit, supplicate. **appeal to** ▷ ATTRACT.

appear *vb* **1** arise, arrive, attend, begin, be published, be revealed, be seen, *inf* bob up, come, come into view, come out, *inf* crop up, enter, develop, emerge, *inf* heave into sight, loom, materialize, occur, originate, show, *inf* show up, spring up, surface, turn up. **2** *I appear to be wrong.* look, seem, transpire, turn out. **3** *appear in a play.* ▷ PERFORM.

appearance *n* **1** arrival, advent, emergence, presence, rise. **2** *a smart appearance.* air, aspect, bearing, demeanour, exterior, impression, likeness, look, mien, semblance.

appease *vb* assuage, calm, conciliate, humour, mollify, pacify, placate, propitiate, quiet, reconcile, satisfy, soothe, *inf* sweeten, tranquillize, win over. *Opp* ANGER.

appendix *n* addendum, addition, annexe, codicil, epilogue, postscript, rider, supplement.

appetite *n* craving, demand, desire, eagerness, fondness, greed, hankering, hunger, keenness, longing, lust, passion, predilection, proclivity, relish, *inf* stomach, taste, thirst, urge, willingness, wish, yearning, *inf* yen, zeal, zest.

appetizing *adj* delicious, *inf* moreish, mouthwatering, tasty, tempting.

applaud *vb* acclaim, approve, *inf* bring the house down, cheer, clap, commend, compliment, congratulate, eulogize, extol, *inf* give someone a hand, give someone an ovation, hail, laud, praise, salute. *Opp* CRITICIZE.

applause *n* acclaim, acclamation, approval, cheering, clapping, éclat, ovation, plaudits. ▷ PRAISE.

appliance *n* apparatus, contraption, device, gadget, implement, instrument, machine, mechanism, tool, utensil.

applicant *n* aspirant, candidate, competitor, entrant, interviewee, participant, postulant.

apply *vb* **1** administer, affix, bring into contact, lay on, put on, rub on, spread, stick. ▷ FASTEN. **2** *rules apply to all.* appertain, be relevant, have a bearing (on), pertain, refer, relate. **3** *apply common sense.* bring into use, employ, exercise, implement, practise, use, utilize, wield. **apply for** ▷ REQUEST. **apply yourself** ▷ CONCENTRATE.

appoint *vb* **1** arrange, authorize, decide on, determine, establish, fix, ordain, prescribe, settle. **2** *appoint you to do a job.* assign, choose, co-opt, delegate, depute, designate, detail, elect, make an appointment, name, nominate, *inf* plump for, select, settle on, vote for.

appointment *n* **1** arrangement, assignation, consultation, date, engagement, fixture, interview, meeting, rendezvous, session, *old* *use* tryst. **2** choice, choosing, commissioning, election, naming, nomination, selection. **3** job, office, place, position, post, situation.

appreciate *vb* **1** admire, applaud, approve of, be grateful for, be sensitive to, cherish, commend, enjoy, esteem, favour, find worthwhile, like, praise, prize, rate highly, regard highly, respect, sympathize with, treasure, value, welcome. **2** *appreciate the facts.* acknowledge, apprehend, comprehend, know, realize, recognize, see, understand. **3** *value appreciates.* build up, escalate, gain, go up, grow, improve, increase, inflate, mount, rise, soar, strengthen. *Opp* DEPRECIATE, DESPISE, DISREGARD.

apprehensive *adj* afraid, concerned, disturbed, edgy, fearful, *inf* jittery, nervous, *inf* nervy, *inf* on edge, troubled, uneasy, worried. ▷ ANXIOUS. *Opp* FEARLESS.

apprentice *n* beginner, learner, novice, probationer, pupil, starter, tiro, trainee.

approach *n* **1** advance, advent, arrival, coming, movement, nearing. **2** access, doorway, entrance, entry, passage, road, way in. **3** *your approach to work.* attitude, course, manner, means, method, mode, *Lat* modus operandi, procedure, style, system, technique, way. **4** *an approach for help.* appeal, application, invitation, offer, overture, proposal, proposition. • *vb* **1** advance, bear down, catch up, come near, draw near, gain (on), loom, move towards, near, progress. *Opp* RETREAT. **2** *approach a task.* ▷ BEGIN. **3** *approach someone for help.* ▷ CONTACT.

approachable *adj* accessible, affable, informal, kind, open, relaxed, sympathetic, *inf* unstuffy, well-disposed. ▷ FRIENDLY. *Opp* ALOOF.

appropriate *adj* applicable, apposite, apropos, apt, becoming, befitting, compatible, correct, decorous, deserved, due, felicitous, fit, fitting, germane, happy, just, *old use* meet, opportune, pertinent, proper, relevant, right, seasonable, seemly, suitable, tactful, tasteful, timely, well-judged, well-suited, well-timed. *Opp* INAPPROPRIATE. • *vb* annex, arrogate, commandeer, confiscate, expropriate, gain control of, *inf* hijack, requisition, seize, take, take over, usurp. ▷ STEAL.

approval *n* **1** acclaim, acclamation, admiration, applause, appreciation, approbation, commendation, esteem, favour, liking, plaudits, praise, regard, respect, support. *Opp* DISAPPROVAL. **2** acceptance, acquiescence, agreement, assent, authorization, *inf* blessing, confirmation, consent, endorsement, *inf* go-ahead, *inf* green light, licence, mandate, *inf* OK, permission, ratification, sanction, seal, stamp, support, *inf* thumbs up, validation. *Opp* REFUSAL.

approve *vb* accede to, accept, affirm, agree to, allow, assent to, authorize, *inf* back, *inf* bless, confirm, consent to, countenance, endorse, *inf* give your blessing to, *inf* go along with, pass, permit, ratify, *inf* rubber-stamp, sanction, sign, subscribe to, support, tolerate, uphold, validate. *Opp* REFUSE, VETO. approve of ▷ ADMIRE.

approximate *adj* close, estimated, imprecise, inexact, loose, near, rough. *Opp* EXACT. • *vb* **approximate to** approach, be close to, be similar to, border on, come near to, equal roughly, look like, resemble, simulate, verge on.

approximately *adv* about, approaching, around, *Lat* circa, close to, just about, loosely, more or less, nearly, *inf* nigh on, *inf* pushing, roughly, round about.

aptitude *n* ability, bent, capability, facility, fitness, flair, gift, suitability, talent. ▷ SKILL.

arbitrary *adj* **1** capricious, casual, chance, erratic, fanciful, illogical, indiscriminate, irrational, random, subjective, unplanned, unpredictable, unreasonable, whimsical, wilful. *Opp* METHODICAL. **2** *arbitrary rule.* absolute, autocratic, despotic, dictatorial, high-handed, imperious, summary, tyrannical, tyrannous, uncompromising.

arbitrate *vb* adjudicate, decide the outcome, intercede, judge, make peace, mediate, negotiate, pass judgement, referee, settle, umpire.

arbitration *n* adjudication, *inf* good offices, intercession, judgement, mediation, negotiation, settlement.

arbitrator *n* adjudicator, arbiter, go-between, intermediary, judge, mediator, middleman, negotiator, ombudsman, peacemaker, referee, *inf* trouble-shooter, umpire.

arch *n* arc, archway, bridge, vault. • *vb* arc, bend, bow. ▷ CURVE.

archetype *n* classic, example, ideal, model, original, paradigm, pattern, precursor, prototype, standard.

archives *n* annals, chronicles, documents, history, libraries, memorials, museums, papers, records, registers.

ardent *adj* eager, enthusiastic, fervent, hot, impassioned, intense, keen, passionate, warm, zealous. *Opp* APATHETIC.

arduous *adj* backbreaking, demanding, exhausting, gruelling, heavy, herculean, laborious, onerous, punishing, rigorous, severe, strenuous, taxing, tiring, tough, uphill. ▷ DIFFICULT. *Opp* EASY.

area *n* **1** acreage, breadth, expanse, extent, patch, sheet, size, space, square-footage, stretch, surface, tract, width. **2** district, environment, environs, locality, neighbourhood, part, precinct, province, quarter, region, sector, terrain, territory, vicinity, zone. **3** *an area of study.* field, sphere, subject.

argue *vb* **1** answer back, *inf* bandy words, bargain, bicker, debate, deliberate, demur, differ, disagree, discuss, dispute, dissent, expostulate, fall out, feud, fight, haggle, have an argument, *inf* have words, object, protest, quarrel, remonstrate, *inf* row, spar, squabble, take exception, wrangle. **2** *argue a case.* assert, claim, contend, demonstrate, hold, maintain, make a case, plead, prove, reason, show, suggest.

argument *n* **1** altercation, bickering, clash, conflict, controversy, difference (of opinion), disagreement, dispute, expostulation, feud, fight, protest, quarrel, remonstration, row, *inf* set-to, squabble, *inf* tiff, wrangle. **2** consultation, debate, defence, deliberation, dialectic, discussion, exposition, polemic. **3** *argument of a lecture.* abstract, case, contention, gist, hypothesis, idea, outline, plot, reasoning, summary, synopsis, theme, thesis, view.

arid *adj* **1** barren, desert, dry, fruitless, infertile, lifeless, parched, sterile, torrid, unproductive, waste, waterless. *Opp* FRUITFUL. **2** *arid work.* boring, dreary, dull, pointless, tedious, uninspired, uninteresting, vapid.

arise *vb* come up, crop up, get up, rise. ▷ APPEAR.

aristocrat *n* grandee, lady, lord, noble, nobleman, noblewoman, patrician. ▷ PEER.

aristocratic *adj inf* blueblooded, courtly, élite, gentle, highborn, lordly, noble, patrician, princely, royal, thoroughbred, titled, upper class.

arm *n* appendage, bough, branch, extension, limb, offshoot, projection. • *vb* equip, fortify, furnish, provide, supply. **arms** ▷ WEAPON(S).

armed services *plur n* force, forces, troops. □ *air force, army, militia, navy.* □ *cavalry, infantry.* □ *battalion, brigade, cohort, company, corps, foreign legion, garrison, legion, patrol, platoon, rearguard, regiment, reinforcements, squad, squadron, task-force, vanguard.* ▷ FIGHTER, RANK, SOLDIER.

armistice *n* agreement, cease-fire, peace, treaty, truce.

armoury *n* ammunition-dump, arsenal, depot, magazine, ordnance depot, stockpile.

aroma *n* bouquet, fragrance, odour, perfume, redolence, savour, scent, smell, whiff.

arouse *vb* awaken, call forth, encourage, foment, foster, kindle, provoke, quicken, stimulate, stir up, *inf* whip up. ▷ CAUSE. *Opp* ALLAY.

arrange *vb* **1** adjust, align, array, categorize, classify, collate, display, dispose, distribute, grade, group, lay out, line up, marshal, order, organize, *inf* pigeon-hole, position, put in order, range, rank, set out, sift, sort (out), space out, systematize, tabulate, tidy up. **2** *arrange a party.* bring about, contrive, coordinate, devise, manage, organize, plan, prepare, see to, settle, set up. **3** *arrange music.* adapt, harmonize, orchestrate, score, set.

arrangement *n* **1** adjustment, alignment, design, disposition, distribution, grouping, layout, marshalling, organization, planning, positioning, setting out, spacing, tabulation. ▷ ARRAY. **2** agreement, bargain, compact, contract, deal, pact, scheme, settlement, terms, understanding. **3** *musical arrangement.* adaptation, harmonization, orchestration, setting, version.

array *n* arrangement, assemblage, collection, demonstration, display, exhibition, formation, *inf* line-up, muster, panoply, parade, presentation, show, spectacle. • *vb* **1** adorn, apparel, attire, clothe, deck, decorate, dress, equip, fit out, garb, rig out, robe, wrap. **2** ▷ ARRANGE.

arrest *n* apprehension, capture, detention, seizure. • *vb* **1** bar, block, check, delay, end, halt, hinder, impede, inhibit, interrupt, obstruct, prevent, restrain, retard, slow, stem, stop. **2** *arrest a suspect.* apprehend, *inf* book, capture, catch, *inf* collar, detain, have up, hold, *inf* nab, *inf* nick, *inf* pinch, *inf* run in, seize, take into custody, take prisoner.

arrival *n* **1** advent, appearance, approach, coming, entrance, homecoming, landing, return, touchdown. **2** *new arrivals.* caller, newcomer, visitor.

arrive *vb* **1** appear, come, disembark, drive up, drop in, enter, get in, land, make an entrance, *inf* roll in, *inf* roll up, show up, touch down, turn up. **2** ▷ SUCCEED. **arrive at** ▷ REACH.

arrogant *adj* boastful, brash, brazen, bumptious, cavalier, *inf* cocky, conceited, condescending, disdainful, egotistical, haughty, *inf* high and mighty, high-handed, imperious, impudent, insolent, lofty, lordly, overbearing, patronizing, pompous, presumptuous, proud, scornful, self-admiring, self-important, smug, snobbish, *inf* snooty, *inf* stuck-up, supercilious, superior, vain. *Opp* MODEST.

arsonist *n* fire-raiser, incendiary, pyromaniac.

art *n* **1** aptitude, artistry, cleverness, craft, craftsmanship, dexterity, expertise, facility, knack, proficiency, skilfulness, skill, talent, technique, touch, trick. **2** artwork, craft, fine art. □ *architecture, batik, carpentry, cloisonné, collage, crochet, drawing, embroidery, enamelling, engraving, etching, fashion design, graphics, handicraft, illustration, jewellery, knitting, linocut, lithography, marquetry, metalwork, modelling, monoprint, needlework, origami, painting, patchwork, photography, pottery, printmaking, sculpture, sewing, sketching, spinning, weaving, woodcut, woodwork.*

artful *adj* astute, canny, clever, crafty, cunning, deceitful, designing, devious, *inf* fly, *inf* foxy, ingenious, knowing, scheming, shrewd, skilful, sly, smart, sophisticated, subtle, tricky, wily. *Opp* NAÏVE.

article **1** item, object, thing. **2** *magazine article.* ▷ WRITING.

articulate *adj* clear, coherent, comprehensible, distinct, eloquent, expressive, fluent, *derog* glib, intelligible, lucid, understandable, vocal. *Opp* INARTICULATE. ● *vb* ▷ SPEAK.

articulated *adj* bending, flexible, hinged, jointed.

artificial *adj* **1** fabricated, made-up, man-made, manufactured, synthetic, unnatural. **2** *artificial style.* affected, assumed, bogus, concocted, contrived, counterfeit, factitious, fake, false, feigned, forced, imitation, insincere, laboured, mock, *inf* phoney, pretended, pseudo, *inf* put on, sham, simulated, spurious, unreal. *Opp* NATURAL.

artist *n* craftsman, craftswoman. □ *architect, carpenter, cartoonist, commercial artist, designer, draughtsman, draughtswoman, engraver, goldsmith, graphic designer, illustrator, mason, painter, photographer, potter, printer, sculptor, silversmith, smith, weaver.* ▷ ENTERTAINER, MUSICIAN, PERFORMER.

artistic *adj* aesthetic, attractive, beautiful, creative, cultured, decorative, *inf* designer, imaginative, ornamental, tasteful. *Opp* UGLY.

ascend *vb* climb, come up, defy gravity, fly, go up, levitate, lift off, make an ascent, mount, move up, rise, scale, slope up, soar, take off. *Opp* DESCEND.

ascent *n* ascension, climb, gradient, hill, incline, ramp, rise, slope. *Opp* DESCENT.

ascertain *vb* confirm, determine, discover, establish, find out, identify, learn, make certain, make sure, settle, verify.

ascetic *adj* abstemious, austere, celibate, chaste, frugal, harsh, hermit-like, plain, puritanical, restrained, rigorous, selfcontrolled, self-denying, selfdisciplined, severe, spartan, strict, temperate. *Opp* SELF-INDULGENT.

ash *n* burnt remains, cinders, clinker, embers.

ashamed *adj* **1** abashed, apologetic, chagrined, chastened, conscience-stricken, contrite, discomfited, distressed, guilty, humbled, humiliated, mortified, penitent, red-faced, remorseful, repentant, rueful, shamefaced, sorry, upset. **2** *ashamed of your nakedness.* bashful, blushing, demure, diffident, embarrassed, modest, prudish, self-conscious, sheepish, shy. *Opp* SHAMELESS.

ask *vb* appeal, apply, badger, beg, beseech, catechize, crave, demand, enquire, entreat, implore, importune, inquire, interrogate, invite, petition, plead, pose a question, pray, press, query, question, quiz, request, require, seek, solicit, sue, supplicate. **ask for** ▷ ATTRACT.

asleep *adj* comatose, *inf* dead to the world, dormant, dozing, *inf* fast off, hibernating, inactive, inattentive, *inf* in the land of nod, *sl* kipping, napping, *inf* off, *inf* out like a light, resting, sedated, sleeping, slumbering, snoozing, *inf* sound off, unconscious, under sedation. ▷ NUMB. *Opp* AWAKE.

aspect *n* **1** angle, attribute, characteristic, circumstance, detail, element, facet, feature, quality, side, standpoint, viewpoint. **2** air, appearance, attitude, bearing, countenance, demeanour, expression, face, look, manner, mien, visage. **3** *a southern aspect.* direction, orientation, outlook, position, prospect, situation, view.

asperity *n* abrasiveness, acerbity, acidity, acrimony, astringency, bitterness, churlishness, crossness, harshness, hostility, irascibility, irritability, peevishness, rancour, roughness, severity, sharpness, sourness, venom, virulence. *Opp* MILDNESS.

aspiration *n* aim, ambition, craving, desire, dream, goal, hope, longing, objective, purpose, wish, yearning.

aspire *vb* **aspire to** aim for, crave, desire, dream of, hope for, long for, pursue, seek, set your sights on, strive after, want, wish for, yearn for. **aspiring** ▷ POTENTIAL.

assail *vb* assault, bombard, pelt, set on. ▷ ATTACK.

assault *n* battery, *inf* GBH, mugging, rape. ▷ ATTACK. • *vb* abuse, assail, *inf* beat up, *inf* do over, fall on, fight, fly at, jump on, lash out at, *inf* lay into, mob, molest, mug, *inf* pitch into, pounce on, rape, rush at, set about, set on, strike at, violate, *inf* wade into. ▷ ATTACK.

assemble *vb* **1** come together, congregate, convene, converge, crowd, flock, gather, group, herd, join up, meet, rally round, swarm, throng round. **2** accumulate, amass, bring together, collect, gather, get together, marshal, mobilize, muster, pile up, rally, round up. **3** build, construct, erect, fabricate, fit together, make, manufacture, piece together, produce, put together. *Opp* DISMANTLE, DISPERSE.

assembly *n* assemblage, conclave, conference, congregation, congress, convention, convocation, council, gathering, meeting, parliament, rally, synod. ▷ CROWD.

assent *n* acceptance, accord, acquiescence, agreement, approbation, approval, compliance, consent,

inf go-ahead, permission, sanction, willingness. *Opp* REFUSAL.
• *vb* accede, accept, acquiesce, agree, approve, be willing, comply, concede, concur, consent, express agreement, give assent, say 'yes', submit, yield.
Opp REFUSE.

assert *vb* affirm, allege, argue, asseverate, attest, claim, contend, declare, emphasize, insist, maintain, proclaim, profess, protest, state, stress, swear, testify. **assert yourself** ▷ INSIST.

assertive *adj* aggressive, assured, authoritative, bold, *inf* bossy, certain, confident, decided, decisive, definite, dogmatic, domineering, emphatic, firm, forceful, insistent, peremptory, *derog* opinionated, positive, *derog* pushy, self-assured, strong, strong-willed, *derog* stubborn, uncompromising.
Opp SUBMISSIVE.

assess *vb* appraise, assay (*metal*), calculate, compute, consider, determine, estimate, evaluate, fix, gauge, judge, price, reckon, review, *inf* size up, value, weigh up, work out.

asset *n* advantage, aid, benefit, blessing, boon, *inf* godsend, good, help, profit, resource, strength, support. **assets** capital, effects, estate, funds, goods, holdings, means, money, possessions, property, resources, savings, securities, valuables, wealth, *inf* worldly goods.

assign *vb* **1** allocate, allot, apportion, consign, dispense, distribute, give, hand over, share out.
2 *assign to a job.* appoint, authorize, delegate, designate, nominate, ordain, prescribe, put down, select, specify, stipulate. **3** *assign my success to luck.* accredit, ascribe, attribute, credit.

assignment *n* chore, duty, errand, job, mission, obligation, post, project, responsibility, task. ▷ WORK.

assist *vb* abet, advance, aid, back, benefit, boost, collaborate, cooperate, facilitate, further, help, *inf* lend a hand, promote, *inf* rally round, reinforce, relieve, second, serve, succour, support, sustain, work with. *Opp* HINDER.

assistance *n* aid, backing, benefit, collaboration, contribution, cooperation, encouragement, help, patronage, reinforcement, relief, sponsorship, subsidy, succour, support. *Opp* HINDRANCE.

assistant *n* abettor, accessory, accomplice, acolyte, aide, ally, associate, auxiliary, backer, collaborator, colleague, companion, comrade, confederate, deputy, helper, helpmate, *inf* henchman, mainstay, *derog* minion, partner, *inf* right-hand man, *inf* right-hand woman, second, second-in-command, stand-by, subordinate, supporter.

associate *n* ▷ ASSISTANT, FRIEND.
• *vb* **1** ally yourself, be friends, combine, consort, fraternize, *inf* gang up, *inf* go around (with), *sl* hang out (with), *inf* hob nob (with), keep company, join up, link up, make friends, mingle, mix, side, socialize.
Opp DISSOCIATE. **2** *associate snow with winter.* bracket together, connect, put together, relate, *inf* tie up.

association *n* affiliation, alliance, amalgamation, body, brotherhood, cartel, clique, club, coalition, combination, company, confederation, consortium, cooperative, corporation, federation, fellowship, group, league, marriage, merger, organization, partnership, party, society, syndicate, trust, union. ▷ FRIENDSHIP.

assorted *adj* different, differing, *old use* divers, diverse, heterogeneous, manifold, miscellaneous, mixed, motley, multifarious, sundry, varied, various.

assortment *n* agglomeration, array, choice, collection, diversity, farrago, jumble, medley, mélange, miscellany, *inf* mishmash, *inf* mixed bag, mixture, pot-pourri, range, selection, variety.

assume *vb* **1** believe, deduce, expect, guess, *inf* have a hunch, have no doubt, imagine, infer, presume, presuppose, suppose, surmise, suspect, take for granted, think, understand. **2** *assume duties.* accept, embrace, take on, undertake. **3** *assume an air of.* acquire, adopt, affect, don, dress up in, fake, feign, pretend, put on, simulate, try on, wear.

assumption *n* belief, conjecture, expectation, guess, hypothesis, premise, premiss, supposition, surmise, theory.

assurance *n* commitment, guarantee, oath, pledge, promise, vow, undertaking, word (of honour).

assure *vb* convince, give a promise, guarantee, make sure, persuade, pledge, promise, reassure, swear, vow. **assured** ▷ CONFIDENT.

astonish *vb* amaze, astound, baffle, bewilder, confound, daze, *inf* dazzle, dumbfound, electrify, *inf* flabbergast, leave speechless, nonplus, shock, stagger, startle, stun, stupefy, surprise, take aback, take by surprise, *inf* take your breath away, *sl* wow. **astonishing** ▷ AMAZING.

astound *vb* ▷ ASTONISH.

astray *adv* adrift, amiss, awry, lost, off course, *inf* off the rails, wide of the mark, wrong.

astute *adj* acute, adroit, artful, canny, clever, crafty, cunning, discerning, *inf* fly, *inf* foxy, guileful, ingenious, intelligent, knowing, observant, perceptive, perspicacious, sagacious, sharp, shrewd, sly, subtle, wily. *Opp* STUPID.

asylum *n* cover, haven, refuge, retreat, safety, sanctuary, shelter.

asymmetrical *adj* awry, crooked, distorted, irregular, lop-sided, unbalanced, uneven, *inf* wonky. *Opp* SYMMETRICAL.

atheist *n* heathen, pagan, sceptic, unbeliever.

athletic *adj* acrobatic, active, energetic, fit, muscular, powerful, robust, sinewy, *inf* sporty, *inf* strapping, strong, sturdy, vigorous, well-built, wiry. *Opp* WEAK.

athletics *n* field events, track events. □ *cross-country, decathlon, discus, high jump, hurdles, javelin, long jump, marathon, pentathlon, pole-vault, relay, running, shot, sprint, triple jump.*

atmosphere *n* **1** aerospace, air, ether, heavens, ionosphere, sky, stratosphere, troposphere. **2** ambience, aura, character, climate, environment, feeling, mood, spirit, tone, *inf* vibes, vibrations.

atom *n inf* bit, crumb, grain, iota, jot, molecule, morsel, particle, scrap, speck, spot, trace.

atone *vb* answer, be punished, compensate, do penance, expiate, make amends, make reparation, make up (for), pay the penalty, pay the price, recompense, redeem yourself, redress.

atrocious *adj* abominable, appalling, barbaric, bloodthirsty, brutal, brutish, callous, cruel, diabolical, dreadful, evil, execrable, fiendish, frightful, grim, gruesome, hateful, heartless, heinous, hideous, horrendous, horrible, horrific, horrifying, inhuman, merciless, monstrous, nauseating, revolting, sadistic, savage, shocking, sickening, terrible, vicious, vile, villainous, wicked.

atrocity *n* crime, cruelty, enormity, offence, outrage. ▷ EVIL.

attach *vb* **1** add, affix, anchor, append, bind, combine, connect, couple, fix, join, link, secure, stick, tie, unite, weld. ▷ FASTEN. *Opp* DETACH. **2** ascribe, assign, associate, attribute, impute, place, relate to. **attached** ▷ LOVING.

attack *n* **1** aggression, ambush, assault, battery, blitz, bombardment, broadside, cannonade, charge, counter-attack, foray, incursion, invasion, offensive, onset, onslaught, pre-emptive strike, raid, rush, sortie, strike. **2** *verbal attack.* abuse, censure, criticism, diatribe, impugnment, invective, outburst, tirade. **3** *attack of coughing.* bout, convulsion, fit, outbreak, paroxysm, seizure, spasm, stroke, *inf* turn. • *vb* **1** ambush, assail, assault, *inf* beat up, *inf* blast, bombard, charge, counterattack, descend on, *inf* do over, engage, fall on, fight, fly at, invade, jump on, lash out at, *inf* lay into, mob, mug, *inf* pitch into, pounce on, raid, rush, set about, set on, storm, strike at, *inf* wade into. **2** *attack verbally.* abuse, censure, criticize, denounce, impugn, inveigh against, libel, malign, round on, slander, snipe at, traduce, vilify. *Opp* DEFEND. **3** *attack a task.* ▷ BEGIN.

attacker *n* aggressor, assailant, critic, detractor, enemy, intruder, invader, mugger, opponent, persecutor, raider, slanderer. ▷ FIGHTER.

attain *vb* accomplish, achieve, acquire, arrive at, complete, earn, fulfil, gain, get, grasp, *inf* make, obtain, *inf* pull (something) off, procure, reach, realize, secure, touch, win.

attempt *n* assault, bid, effort, endeavour, *inf* go, start, try, undertaking. • *vb* aim, aspire, do your best, endeavour, essay, exert yourself, *inf* have a crack, *inf* have a go, make a bid, make an assault, make an effort, put yourself out, seek, *inf* spare no effort, strive, *inf* sweat blood, tackle, try, undertake, venture.

attend *vb* **1** appear, be present, go (to), frequent, present yourself, *inf* put in an appearance, visit. **2** accompany, chaperon, conduct, escort, follow, guard, usher. **3** *attend carefully.* concentrate, follow, hear, heed, listen, mark, mind, note, notice, observe, pay attention, think, watch. **attend to** assist, care for, help, look after, mind, minister to, nurse, see to, take care of, tend, wait on.

attendant *n* assistant, escort, helper, usher. ▷ SERVANT.

attention *n* **1** alertness, awareness, care, concentration, concern, diligence, heed, notice, recognition, thought, vigilance. **2** *kind attention.* attentiveness, civility, consideration, courtesy, gallantry, good manners, kindness, politeness, regard, respect, thoughtfulness.

attentive *adj* **1** alert, awake, concentrating, heedful, intent, observant, watchful. *Opp* INATTENTIVE. **2** ▷ POLITE. *Opp* RUDE.

attire *n* accoutrements, apparel, array, clothes, clothing, costume, dress, finery, garb, garments, *inf* gear, *old use* habit, outfit, raiment, wear, *old use* weeds. • *vb* ▷ DRESS.

attitude *n* 1 air, approach, aspect, bearing, behaviour, carriage, demeanour, disposition, frame of mind, manner, mien, mood, posture, stance. 2 *political attitudes.* approach, belief, feeling, opinion, orientation, outlook, position, standpoint, thought, view, viewpoint.

attract *vb* 1 allure, appeal to, beguile, bewitch, bring in, captivate, charm, decoy, enchant, entice, fascinate, *sl* get someone going, interest, inveigle, lure, magnetize, seduce, tempt, *sl* turn someone on. 2 *a magnet attracts iron.* drag, draw, pull, tug at. 3 *attract attention.* ask for, cause, court, encourage, generate, incite, induce, invite, provoke, seek out, *inf* stir up. *Opp* REPEL.

attractive *adj* adorable, alluring, appealing, appetizing, becoming, bewitching, captivating, *inf* catchy (*tune*), charming, *inf* cute, delightful, desirable, disarming, enchanting, endearing, engaging, enticing, enviable, fascinating, fetching, flattering, glamorous, good-looking, gorgeous, handsome, hypnotic, interesting, inviting, irresistible, lovable, lovely, magnetic, personable, pleasing, prepossessing, pretty, quaint, seductive, sought-after, stunning, *inf* taking, tasteful, tempting, winning, winsome. ▷ BEAUTIFUL. *Opp* REPULSIVE.

attribute *n* characteristic, feature, property, quality, trait. • *vb* accredit, ascribe, assign, blame, charge, credit, impute, put down, refer, trace back.

audacious *adj* adventurous, courageous, daring, fearless, *derog* foolhardy, intrepid, *derog* rash, *derog* reckless, venturesome. ▷ BOLD. *Opp* TIMID.

audacity *n* boldness, *inf* cheek, effrontery, forwardness, impertinence, impudence, presumptuousness, rashness, *inf* sauce, temerity. ▷ COURAGE.

audible *adj* clear, detectable, distinct, high, loud, noisy, recognizable. *Opp* INAUDIBLE.

audience *n* assembly, congregation, crowd, gathering, house, listeners, meeting, onlookers, *TV* ratings, spectators, *inf* turn-out, viewers.

auditorium *n* assembly room, concert-hall, hall, theatre.

augment *vb* add to, amplify, boost, eke out, enlarge, expand, extend, fill out, grow, increase, intensify, magnify, make larger, multiply, raise, reinforce, strengthen, supplement, swell. *Opp* DECREASE.

augur *vb* bode, forebode, foreshadow, forewarn, give an omen, herald, portend, predict, promise, prophesy, signal.

augury *n* forecast, forewarning, omen, portent, prophecy, sign, warning.

auspicious *adj* favourable, *inf* hopeful, lucky, positive, promising, propitious. *Opp* OMINOUS.

austere *adj* 1 abstemious, ascetic, chaste, cold, economical, exacting, forbidding, formal, frugal, grave, hard, harsh, hermit-like, parsimonious, puritanical, restrained, rigorous, self-denying, self-disciplined, serious, severe, sober, spartan, stern, *inf* strait-laced, strict, thrifty, unpampered. **2** *austere dress.* modest, plain, simple, unadorned, unfussy. *Opp* LUXURIOUS, ORNATE.

authentic *adj* accurate, actual, bona fide, certain, dependable, factual, genuine, honest, legitimate, original, real, reliable, true, trustworthy, truthful, undisputed, valid, veracious. ▷ AUTHORITATIVE. *Opp* FALSE.

authenticate *vb* certify, confirm, corroborate, endorse, substantiate, validate, verify.

author *n* **1** composer, dramatist, novelist, playwright, poet, scriptwriter. ▷ WRITER. **2** architect, begetter, creator, designer, father, founder, initiator, inventor, maker, mover, organizer, originator, parent, planner, prime mover, producer.

authoritarian *adj* autocratic, *inf* bossy, despotic, dictatorial, dogmatic, domineering, strict, tyrannical.

authoritative *adj* approved, certified, definitive, dependable, official, recognized, sanctioned, scholarly. ▷ AUTHENTIC.

authority *n* **1** approval, authorization, consent, licence, mandate, permission, permit, sanction, warrant. **2** charge, command, control, domination, force, influence, jurisdiction, might, power, prerogative, right, sovereignty, supremacy, sway, weight. **3** *authority on wine.* *inf* boffin, *inf* buff, connoisseur, expert, scholar, specialist. **the authorities** administration, government, management, officialdom, *inf* powers that be.

authorize *vb* accede to, agree to, allow, approve, *inf* back, commission, consent to, empower, endorse, entitle, legalize, license, make official, mandate, *inf* OK, pass, permit, ratify, *inf* rubber-stamp, sanction, sign the order, sign the warrant, validate. **authorized** ▷ OFFICIAL.

automatic *adj* **1** conditioned, habitual, impulsive, instinctive, involuntary, natural, reflex, spontaneous, unconscious, unintentional, unthinking. **2** automated, computerized, electronic, mechanical, programmable, programmed, robotic, self-regulating, unmanned.

autonomous *adj* free, independent, self-determining, self-governing, sovereign.

auxiliary *adj* additional, ancillary, assisting, *inf* back-up, emergency, extra, helping, reserve, secondary, spare, subordinate, subsidiary, substitute, supplementary, supporting, supportive.

available *adj* accessible, at hand, convenient, disposable, free, handy, obtainable, procurable, ready, to hand, uncommitted, unengaged, unused, usable. *Opp* INACCESSIBLE.

avaricious *adj* acquisitive, covetous, grasping, greedy, mercenary, miserly.

avenge *vb* exact punishment, *inf* get your own back, repay, requite, take revenge.

average *adj* common, commonplace, everyday, mediocre, medium, middling, moderate, normal, regular, *inf* run of the mill, typical, unexceptional, usual. ▷ ORDINARY. *Opp* EXCEPTIONAL. • *n* mean, mid-point, norm, standard. • *vb* equalize, even out, normalize, standardize.

averse *adj* antipathetic, disinclined, hostile, opposed, reluctant resistant, unwilling.

aversion *n* antagonism, antipathy, dislike, distaste, hostility, reluctance, repugnance, unwillingness. ▷ HATRED.

avert *vb* change the course of, deflect, draw off, fend off, parry, prevent, stave off, turn aside, turn away, ward off.

avoid *vb* abstain from, be absent from, *inf* beg the question, *inf* bypass, circumvent, dodge, *inf* duck, elude, escape, eschew, evade, fend off, find a way round, get out of the way of, *inf* get round, *inf* give a wide berth to, help (*can't help it*), ignore, keep away from, keep clear of, refrain from, run away from, shirk, shun, side-step, skirt round, *inf* skive off, steer clear of. *Opp* SEEK.

await *vb* be ready for, expect, hope for, lie in wait for, look out for, wait for.

awake *adj* **1** aware, conscious, insomniac, open-eyed, restless, sleepless, *inf* tossing and turning, wakeful, wide awake. **2** ▷ ALERT. *Opp* ASLEEP.

awaken *vb* alert, animate, arouse, awake, call, excite, kindle, revive, rouse, stimulate, stir up, wake, waken.

award *n* badge, cap, cup, decoration, endowment, grant, medal, prize, reward, scholarship, trophy. • *vb* accord, allot, assign, bestow, confer, decorate with, endow, give, grant, hand over, present.

aware *adj* acquainted, alive (to), appreciative, attentive, cognizant, conscious, conversant, familiar, heedful, informed, knowledgeable, mindful, observant, responsive, sensible, sensitive, versed. *Opp* IGNORANT, INSENSITIVE.

awe *n* admiration, amazement, apprehension, dread, fear, respect, reverence, terror, veneration, wonder.

awe-inspiring *adj* awesome, *old use* awful, breathtaking, dramatic, grand, imposing, impressive, magnificent, marvellous, overwhelming, solemn, *inf* stunning, stupendous, sublime, wondrous. ▷ FRIGHTENING, WONDERFUL. *Opp* INSIGNIFICANT.

awful *adj* **1** ▷ AWE-INSPIRING. **2** *awful weather*. ▷ BAD.

awkward *adj* **1** blundering, bungling, clumsy, gauche, gawky, *inf* ham-fisted, inelegant, inept, inexpert, maladroit, uncoordinated, ungainly, ungraceful, unskilful, wooden. **2** *an awkward load*. bulky, cumbersome, inconvenient, unmanageable, unwieldy. **3** *an awkward problem*. annoying, difficult, perplexing, thorny, *inf* ticklish, troublesome, trying, vexatious, vexing. **4** *an awkward silence*. embarrassing, touchy, tricky, uncomfortable, uneasy. **5** *awkward children*. *inf* bloody-minded, *sl* bolshie, defiant, disobedient, disobliging, exasperating, intractable, misbehaving, naughty, obstinate, perverse, *inf* prickly, rebellious, refractory, rude, stubborn, touchy, uncooperative, undisciplined, unruly, wayward. *Opp* COOPERATIVE, EASY, NEAT.

awning *n* canopy, flysheet, screen, shade, shelter, tarpaulin.

axe *n* battleaxe, chopper, cleaver, hatchet, tomahawk. • *vb* cancel, cut, discharge, discontinue, dismiss, eliminate, get rid of, *inf* give the chop to, make redundant, rationalize, remove, sack, terminate, withdraw.

axle *n* rod, shaft, spindle.

developed, undeveloped. *Opp* FORWARD.

B

baby *n* babe, child, infant, newborn, toddler.

babyish *adj* childish, immature, infantile, juvenile, puerile, simple. *Opp* MATURE.

back *adj* dorsal, end, hind, hinder, hindmost, last, rear, rearmost. • *n* **1** end, hindquarters, posterior, rear, stern, tail, tail-end. **2** reverse, verso. *Opp* FRONT. • *vb* **1** back away, back off, back-pedal, backtrack, *inf* beat a retreat, give way, go backwards, move back, recede, recoil, retire, retreat, reverse. *Opp* ADVANCE. **2** ▷ SUPPORT. **back down** ▷ RETREAT. **back out** ▷ WITHDRAW.

backer *n* advocate, *inf* angel, benefactor, patron, promoter, sponsor, supporter.

background *n* **1** circumstances, context, history, *inf* lead-up, setting, surroundings. **2** breeding, culture, education, experience, grounding, milieu, tradition, training, upbringing.

backing *n* **1** aid, approval, assistance, encouragement, endorsement, funding, grant, help, investment, loan, patronage, sponsorship, subsidy, support. **2** *musical backing*. accompaniment, orchestration, scoring.

backward *adj* **1** regressive, retreating, retrograde, retrogressive, reverse. **2** afraid, bashful, coy, diffident, hesitant, inhibited, modest, reluctant, reserved, reticent, self-effacing, shy, timid, unassertive, unforthcoming. **3** *a backward pupil*. disadvantaged, handicapped, immature, late-starting, retarded, slow, subnormal, under-

bad *adj* [*Bad* describes anything we don't like. Possible synonyms are almost limitless.] **1** *bad men, deeds*. abhorrent, base, beastly, blameworthy, corrupt, criminal, cruel, dangerous, delinquent, deplorable, depraved, detestable, evil, guilty, immoral, infamous, malevolent, malicious, malignant mean, mischievous, nasty, naughty, offensive, regrettable, reprehensible, rotten, shameful, sinful, unworthy, vicious, vile, villainous, wicked, wrong. **2** *a bad accident*. appalling, awful, calamitous, dire, disastrous, distressing, dreadful, frightful, ghastly, grave hair-raising, hideous, horrible, painful, serious, severe, shocking terrible, unfortunate, unpleasant, violent. **3** *bad driving, work*. abominable, abysmal, appalling, atrocious, awful, cheap, *inf* chronic, defective, deficient, diabolical, disgraceful, dreadful, egregious, execrable, faulty, feeble, *inf* grotty, hopeless, imperfect, inadequate, incompetent, incorrect, ineffective, inefficient, inferior, *inf* lousy pitiful, poor, *inf* ropy, shoddy, *inf* sorry, substandard, unsound, unsatisfactory, useless, weak, worthless. **4** *bad conditions*. adverse, deleterious, detrimental, discouraging, *inf* frightful, harmful, harsh, hostile, inappropriate, inauspicious, prejudicial, uncongenial, unfortunate, unhelpful, unpropitious. **5** *bad smell*. decayed, decomposing, diseased, foul, loathsome, mildewed, mouldy, nauseating, noxious, objectionable, obnoxious, odious, offensive, polluted, putrid, rancid repellent, repulsive, revolting, sickening, rotten, smelly, sour, spoiled, tainted, vile. **6** *I feel bad.* ▷ ILL. *Opp* GOOD.

badge *n* chevron, crest, device, emblem, insignia, logo, mark, medal, sign, symbol, token.

bad-tempered *adj* acrimonious, angry, bilious, cantankerous, churlish, crabbed, cross, *inf* crotchety, disgruntled, disobliging, dyspeptic, fretful, gruff, grumbling, grumpy, hostile, hot-tempered, ill-humoured, ill-tempered, irascible, irritable, malevolent, malign, moody, morose, peevish, petulant, quarrelsome, querulous, rude, scowling, short-tempered, shrewish, snappy, *inf* stroppy, sulky, sullen, testy, truculent, unfriendly, unsympathetic. *Opp* GOOD-TEMPERED.

baffle *vb* **1** *inf* bamboozle, bemuse, bewilder, confound, confuse, defeat, *inf* floor, *inf* flummox, foil, frustrate, mystify, perplex, puzzle, *inf* stump, thwart. **baffling** ▷ INEXPLICABLE.

bag *n* basket, carrier, carrier-bag, case, handbag, haversack, holdall, reticule, rucksack, sack, satchel, shopping-bag, shoulder-bag. ▷ BAGGAGE. • *vb* capture, catch, ensnare, snare.

baggage *n* accoutrements, bags, belongings, *inf* gear, impedimenta, paraphernalia. ▷ LUGGAGE.

bait *n* allurement, attraction, bribe, carrot, decoy, enticement, inducement, lure, temptation. • *vb* annoy, goad, harass, hound, jeer at, *inf* needle, persecute, pester, provoke, tease, torment.

balance *n* **1** scales, weighing-machine. **2** equilibrium, equipoise, poise, stability, steadiness. **3** correspondence, equality, equivalence, evenness, parity, symmetry. **4** *spend a bit & save the balance.* difference, excess, remainder, residue, rest, surplus. • *vb* **1** cancel out, compensate for, counteract, counterbalance, counterpoise, equalize, even up, level, make steady, match, neutralize, offset, parallel, stabilize, steady. **2** keep balanced, keep in equilibrium, poise, steady, support. **balanced** ▷ EVEN, IMPARTIAL, STABLE.

bald *adj* **1** baldheaded, bare, hairless, smooth, thin on top. **2** *bald truth.* direct, forthright, plain, simple, stark, straightforward, unadorned, uncompromising.

bale *n* bunch, bundle, pack, package, truss. • *vb* **bale out** eject, escape, jump out, parachute down.

ball *n* **1** drop, globe, globule, orb, shot, sphere, spheroid. **2** dance, disco, party, social.

balloon *n* airship, dirigible, hot-air balloon. • *vb* ▷ BILLOW.

ballot *n* election, plebiscite, poll, referendum, vote.

ban *n* boycott, embargo, interdiction, moratorium, prohibition, proscription, taboo, veto. • *vb* banish, bar, debar, disallow, exclude, forbid, interdict, make illegal, ostracize, outlaw, prevent, prohibit, proscribe, put a ban on, restrict, stop, suppress, veto. *Opp* PERMIT.

banal *adj* boring, clichéd, cliché-ridden, commonplace, *inf* corny, dull, hackneyed, humdrum, obvious, *inf* old hat, ordinary, overused, pedestrian, platitudinous, predictable, stereotyped, trite, unimaginative, uninteresting, unoriginal, vapid. *Opp* INTERESTING.

band *n* **1** belt, border, fillet, hoop, line, loop, ribbon, ring, strip, stripe, swathe. **2** association, body, clique, club, company, crew, flock, gang, herd, horde, party, society, troop. ▷ GROUP. **3** [*music*] ensemble, group, orchestra.

bandage *n* dressing, gauze, lint, plaster.

bandit *n* brigand, buccaneer, desperado, footpad, gangster, gunman, highwayman, hijacker, marauder, outlaw, pirate, robber, thief.

bandy *adj* bandy-legged, bowed, bow-legged. • *vb bandy words.* exchange, interchange, pass, swap, throw, toss. ▷ ARGUE.

bang *n* **1** blow, *inf* box, bump, collision, cuff, knock, punch, slam, smack, stroke, thump, wallop, whack. ▷ HIT. **2** blast, boom, clap, crash, explosion, pop, report, thud, thump. ▷ SOUND.

banish *vb* deport, drive out, eject, evict, excommunicate, exile, expatriate, expel, ostracize, oust, outlaw, rusticate, send away, ship away, transport. **2** ban, bar, *inf* black, debar, eliminate, exclude, forbid, get rid of, make illegal, prohibit, proscribe, put an embargo on, remove, restrict, stop, suppress, veto.

bank *n* **1** camber, declivity, dike, earthwork, embankment, gradient, incline, mound, ramp, rampart, ridge, rise, slope, tilt. **2** *river bank.* brink, edge, margin, shore, side. **3** *bank of controls.* array, collection, display, file, group, line, panel, rank, row, series. • *vb* **1** cant, heel, incline, lean, list, pitch, slant, slope, tilt, tip. **2** *bank money.* deposit, save.

bankrupt *adj inf* broke, failed, *sl* gone bust, gone into liquidation, insolvent, ruined, spent up, wound up. ▷ POOR. *Opp* SOLVENT.

banner *n* banderole, colours, ensign, flag, pennant, pennon, standard, streamer.

banquet *n inf* binge, *sl* blow-out, dinner, feast, repast, *inf* spread. ▷ MEAL.

banter *n* badinage, chaffing, joking, persiflage, pleasantry, raillery, repartee, ribbing, ridicule, teasing, word-play.

bar *n* **1** beam, girder, pole, rail, railing, rod, shaft, stake, stick, strut. **2** barricade, barrier, check, deterrent, hindrance, impediment, obstacle, obstruction. **3** band, belt, line, streak, strip, stripe. **4** *bar of soap.* block, cake, chunk, hunk, ingot, lump, nugget, piece, slab, wedge. **5** *drink in a bar.* café, canteen, counter, inn, lounge, pub, public house, saloon, taproom, tavern, wine bar. • *vb* **1** ban, banish, debar, exclude, forbid to enter, keep out, ostracize, outlaw, prevent from entering, prohibit, proscribe. **2** *bar the way.* arrest, block, check, deter, halt, hinder, impede, obstruct, prevent, stop, thwart.

barbarian *adj* ▷ BARBARIC. • *n* boor, churl, heathen, hun, ignoramus, lout, pagan, philistine, savage, vandal, *sl* yob.

barbaric *adj* barbarous, brutal, brutish, crude, inhuman, primitive, rough, savage, uncivil, uncivilized, uncultivated, wild. ▷ CRUEL. *Opp* CIVILIZED.

bare *adj* **1** bald, denuded, exposed, naked, nude, stark-naked, stripped, unclad, unclothed, uncovered, undressed. **2** *bare moor.* barren, bleak, desolate, featureless, open, treeless, unwooded, windswept. **3** *bare trees.* defoliated, leafless, shorn. **4** *a bare room.* austere, empty, plain, simple, unadorned, undecorated, unfurnished, vacant. **5** *a bare wall.* blank, clean, unmarked. **6** *bare facts.* direct, explicit, hard, honest literal, open, plain, straightforward, unconcealed, undisguised, unembellished. **7** *the bare minimum.* basic, essential, just adequate, just sufficient, minimal, minimum. • *vb* betray, bring to light, disclose, expose, lay bare, make known, publish, reveal, show, uncover, undress, unmask, unveil.

bargain *n* **1** agreement, arrangement, compact, contract, covenant, deal, negotiation, pact, pledge, promise, settlement, transaction, treaty, understanding. **2** *bargain in the sales.* *inf* giveaway, good buy, good deal, loss-leader, reduced item, *inf* snip, special offer. • *vb* argue, barter, discuss terms, do a deal, haggle, negotiate. **bargain for** ▷ EXPECT.

bark *vb* **1** growl, yap. **2** *bark your shin.* abrade, chafe, graze, rub, score, scrape, scratch.

barmaid, barman *ns* attendant, server, steward, stewardess, waiter, waitress.

barracks *n* accommodation, billet, camp, garrison, lodging, quarters.

barrage *n* **1** ▷ BARRIER. **2** *barrage of gunfire.* assault, attack, battery, bombardment, cannonade, fusillade, gunfire, onslaught, salvo, storm, volley.

barrel *n* butt, cask, churn, cistern, drum, hogshead, keg, tank, tub, tun, water-butt.

barren *adj* **1** arid, bare, desert, desolate, dried-up, dry, empty, infertile, lifeless, non-productive, treeless, uncultivated, unproductive, unprofitable, untilled, useless, waste. **2** childless, fruitless, infertile, sterile, sterilized, unfruitful. *Opp* FERTILE.

barricade *n* ▷ BARRIER. • *vb* bar, block off, defend, obstruct.

barrier *n* **1** bar, barrage, barricade, blockade, boom, bulwark, dam, earthwork, fortification, embankment, fence, frontier, hurdle, obstacle, obstruction, palisade, railing, rampart, stockade, wall. **2** *barrier to progress.* check, drawback, handicap, hindrance, impediment, limitation, restriction, stumbling-block.

barter *vb* bargain, deal, exchange, negotiate, swap, trade, traffic.

base *adj* contemptible, cowardly, degrading, depraved, despicable, detestable, dishonourable, evil, ignoble, immoral, inferior, low, mean, scandalous, selfish, shabby, shameful, sordid, undignified, unworthy, vulgar, vile. ▷ WICKED. • *n* **1** basis, bed, bedrock, bottom, core, essentials, foot, footing, foundation, fundamentals, groundwork, infrastructure, pedestal, plinth, rest, root, stand, substructure, support, underpinning. **2** camp, centre, depot, headquarters, post, starting-point, station. • *vb* build, construct, establish, found, ground, locate, position, post, secure, set up, station.

basement *n* cellar, crypt, vault.

bashful *adj* abashed, backward, blushing, coy, demure, diffident, embarrassed, faint-hearted, inhibited, meek, modest, nervous, reserved, reticent, retiring, self-conscious, self-effacing, shamefaced, sheepish, shy, timid, timorous, uneasy, unforthcoming. *Opp* ASSERTIVE.

basic *adj* central, chief, crucial, elementary, essential, foremost, fundamental, important, intrinsic, key, main, necessary, primary, principal, radical, underlying, vital. *Opp* UNIMPORTANT.

basin *n* bath, bowl, container, dish, pool, sink, stoup.

basis *n* base, core, footing, foundation, ground, infrastructure, premise, principle, starting-point, support, underpinning.

bask *vb* enjoy, feel pleasure, glory, lie, lounge, luxuriate, relax, sunbathe, wallow.

basket *n* bag, hamper, pannier, punnet, skip, trug.

bastard *n* illegitimate child, *old use* love-child, natural child.

bat *n* club, racket, racquet.

bath *n* douche, jacuzzi, pool, sauna, shower, *inf* soak, *inf* tub, wash.

bathe *vb* **1** clean, cleanse, immerse, moisten, rinse, soak, steep, swill, wash. **2** *bathe in the sea.* go swimming, paddle, plunge, splash about, swim, *inf* take a dip.

bathos *n* anticlimax, *inf* come-down, disappointment, *inf* let-down. *Opp* CLIMAX.

baton *n* cane, club, cudgel, rod, staff, stick, truncheon.

batter *vb* beat, bludgeon, cudgel, keep hitting, pound. ▷ HIT.

battery *n* **1** artillery-unit, emplacement. **2** *electric battery.* accumulator, cell. **3** *assault and battery.* assault, attack, *inf* beating-up, blows, mugging, onslaught, thrashing, violence.

battle *n* action, air-raid, Armageddon, attack, blitz, brush, campaign, clash, combat, conflict, contest, confrontation, crusade, *inf* dogfight, encounter, engagement, fight, fray, hostilities, offensive, pitched battle, pre-emptive strike, quarrel, *inf* shoot-out, siege, skirmish, strife, struggle, war, warfare. • *vb* ▷ FIGHT.

battlefield *n* arena, battleground, theatre of war.

bawdy *adj* broad, earthy, erotic, lusty, *inf* naughty, racy, *inf* raunchy, ribald, *inf* sexy, *inf* spicy. [*derog synonyms*] *inf* blue, coarse, dirty, immoral, improper, indecent, indecorous, indelicate, lascivious, lecherous, lewd, licentious, obscene, pornographic, prurient, risqué, rude, salacious, smutty, suggestive, titillating, vulgar. *Opp* PROPER.

bawl *vb* cry, roar, shout, thunder, wail, yell, yelp.

bay *n* **1** bight, cove, creek, estuary, fjord, gulf, harbour, indentation, inlet, ria, sound. **2** alcove, booth, compartment, niche, nook, opening, recess.

bazaar *n* auction, boot-sale, bring-and-buy, fair, fête, jumble sale, market, sale.

be *vb* **1** be alive, breathe, endure, exist, live. **2** *be here all day.* continue, dwell, inhabit, keep going, last, occupy a position, persist, remain, stay, survive. **3** *the next event will be tomorrow.* arise, befall, come about, happen, occur, take place. **4** *want to be a writer.* become, develop into.

beach *n* bank, coast, coastline, foreshore, littoral, sand, sands, seashore, seaside, shore, *poet* strand.

beacon *n* bonfire, fire, flare, light, lighthouse, pharos, signal.

bead *n* blob, drip, drop, droplet, globule, jewel, pearl.

beaker *n* cup, glass, goblet, jar, mug, tankard, tumbler.

beam *n* **1** bar, board, boom, brace, girder, joist, plank, post, rafter, spar, stanchion, stud, support, timber. **2** *beam of light.* gleam, pencil, ray, shaft, stream. • *vb* **1** aim, broadcast, direct, emit, radiate, send out, shine, transmit. **2** *beam* *happily*. grin, laugh, look radiant, radiate happiness, smile.

bear *vb* **1** carry, hold, prop up, shoulder, support, sustain, take. **2** *bear an inscription*. display, exhibit, have, possess, show. **3** *bear gifts*. bring, carry, convey, deliver, fetch, move, take, transfer, transport. **4** *bear pain*. abide, accept, brook, cope with, endure, live with, permit, *inf* put up with, reconcile yourself to, *inf* stand, *inf* stomach, suffer, sustain, tolerate, undergo. **5** *bear children, fruit*. breed, *old use* bring forth, develop, engender, generate, give birth to, produce, spawn, yield. **bear out** ▷ CONFIRM. **bear up** ▷ SURVIVE. **bear witness** ▷ TESTIFY.

bearable *adj* acceptable, endurable, supportable, survivable, sustainable, tolerable.

bearing *n* **1** air, appearance, aspect, attitude, behaviour, carriage, demeanour, deportment, look, manner, mien, poise, posture, presence, stance, style. **2** *evidence had no bearing*. applicability, application, connection, import, pertinence, reference, relation, relationship, relevance, significance. **bearings** aim, course, direction, line, location, orientation, path, position, road, sense of direction, tack, track, way, whereabouts.

beast *n* brute, creature, monster, savage. ▷ ANIMAL.

beastly *adj* abominable, barbaric, bestial, brutal, cruel, savage. ▷ VILE.

beat *n* **1** accent, pulse, rhythm, stress, tempo, throb. **2** *policeman's beat*. course, itinerary, journey, path, rounds, route, way.
• *vb* **1** batter, bludgeon, buffet, cane, clout, cudgel, flail, flog, hammer, knock about, lash, *inf* lay into, manhandle, pound, punch, scourge, strike, *inf* tan, thrash, thump, trounce, *inf* wallop, whack, whip. ▷ HIT. **2** *beat eggs*. agitate, blend, froth up, knead, mix, pound, stir, whip, whisk. **3** *heart was beating*. flutter, palpitate, pound, pulsate, race, throb, thump. **5** *beat an opponent*. best, conquer, crush, defeat, excel, get the better of, *inf* lick, master, outclass, outdistance, outdo, outpace, outrun, outwit, overcome, overpower, overthrow, overwhelm, rout, subdue, surpass, *inf* thrash, trounce, vanquish, win against, worst. **beat up** ▷ ATTACK.

beautiful *adj* admirable, aesthetic, alluring, appealing, artistic, attractive, becoming, bewitching, brilliant, captivating, charming, *old use* comely, dainty, decorative, delightful, elegant, enjoyable, exquisite, *old use* fair, fascinating, fetching, fine, good-looking, glamorous, glorious, gorgeous, graceful, handsome, irresistible, lovely, magnificent, neat, picturesque, pleasing, pretty, pulchritudinous, quaint, radiant, ravishing, scenic, seductive, sensuous, sexy, spectacular, splendid, stunning, superb, tasteful, tempting. *Opp* UGLY.

beautify *vb* adorn, bedeck, deck, decorate, embellish, garnish, make beautiful, ornament, prettify, *derog* tart up, *inf* titivate. *Opp* DISFIGURE.

beauty *n* allure, appeal, attractiveness, charm, elegance, fascination, glamour, glory, grace, handsomeness, loveliness, magnificence, picturesqueness, prettiness, pulchritude, radiance, splendour.

becalmed *adj* helpless, idle, motionless, still, unmoving.

beckon *vb* gesture, motion, signal, summon, wave.

become *vb* **1** be transformed into, change into, develop into, grow into, mature into, metamorphose into, turn into. **2** *Red becomes you.* be appropriate to, be becoming to, befit, enhance, fit, flatter, harmonize with, set off, suit. **becoming** ▷ ATTRACTIVE, SUITABLE.

bed *n* **1** resting-place. □ *air-bed, berth, bunk, cot, couch, couchette, cradle, crib, divan, four-poster, hammock, pallet, palliasse, truckle bed, waterbed.* **2** *bed of concrete.* base, foundation, groundwork, layer, substratum. **3** *river bed.* bottom, channel, course, watercourse. **4** *flower bed.* border, garden, patch, plot.

bedclothes *n* bedding, bed linen. □ *bedspread, blanket, bolster, continental quilt, counterpane, coverlet, duvet, eiderdown, electric-blanket, mattress, pillow, pillowcase, pillowslip, quilt, sheet, sleeping-bag.*

bedraggled *adj* dirty, dishevelled, drenched, messy, muddy, scruffy, sodden, soiled, stained, unkempt, untidy, wet, wringing. *Opp* SMART.

beer *n* ale, bitter, lager, mild, porter, stout.

befall *vb* be the outcome, *old use* betide, chance, come about, *inf* crop up, eventuate, happen, occur, take place, *inf* transpire.

before *adv* already, earlier, in advance, previously, sooner.

befriend *vb inf* chat up, *inf* gang up with, get to know, make friends with, make the acquaintance of, *inf* pal up with.

beg *vb* **1** *inf* cadge, scrounge, solicit, sponge. **2** *beg a favour.* ask, beseech, cajole, crave, entreat, implore, importune, petition, plead, pray, request, supplicate, wheedle.

beget *vb* breed, bring about, cause, create, engender, father, generate, give rise to, procreate, produce, propagate, result in, sire, spawn.

beggar *n* cadger, destitute person, down-and-out, homeless person, mendicant, pauper, ragamuffin, scrounger, sponger, tramp, vagrant. ▷ POOR.

begin *vb* **1** activate, approach, attack, be first, broach, commence, conceive, create, embark on, enter into, found, *inf* get cracking, *inf* get going, inaugurate, initiate, inspire, instigate, introduce, kindle, launch, lay the foundations, lead off, move into, move off, open, originate, pioneer, precipitate, provoke, set about, set in motion, set off, set out, set up, *inf* spark off, start, *inf* take steps, take the initiative, take up, touch off, trigger off, undertake. **2** *Spring begins gradually.* appear, arise, break out, come into existence, crop up, emerge, get going, happen, materialize, originate, spring up. *Opp* END.

beginner *n* **1** creator, founder, initiator, inspiration, instigator, originator, pioneer. **2** *only a beginner.* apprentice, fresher, greenhorn, inexperienced person, initiate, learner, novice, recruit, starter, tiro, trainee.

beginning *n* **1** birth, commencement, conception, creation, dawn, embryo, emergence, establishment, foundation, genesis, germ, inauguration, inception, initiation, instigation, introduction, launch, onset, opening, origin, outset, point of departure, rise, source, start, starting-point, threshold. **2** *beginning of a book.*

preface, prelude, prologue. *Opp* END.

begrudge *vb* be bitter about, covet, envy, grudge, mind, object to, resent.

behave *vb* 1 acquit yourself, act, *inf* carry on, comport yourself, conduct yourself, function, operate, perform, react, respond, run, work. 2 *told to behave.* act properly, be good, be on best behaviour.

behaviour *n* actions, attitude, bearing, comportment, conduct, courtesy, dealings, demeanour, deportment, manners, performance, reaction, response, ways.

behead *vb* decapitate, guillotine.

behold *vb* descry, discern, espy, look at, note, notice, see, set eyes on, view.

being *n* 1 actuality, essence, existence, life, living, reality, solidity, substance. 2 animal, creature, individual, person, spirit, soul.

belated *adj* behindhand, delayed, last-minute, late, overdue, posthumous, tardy, unpunctual.

belch *vb* 1 break wind, *inf* burp, emit wind. 2 *belch smoke.* discharge, emit, erupt, fume, gush, send out, smoke, spew out, vomit.

belief *n* 1 acceptance, assent, assurance, certainty, confidence, credence, reliance, security, sureness, trust. 2 *religious belief.* attitude, conviction, creed, doctrine, dogma, ethos, faith, feeling, ideology, morality, notion, opinion, persuasion, principles, religion, standards, tenets, theories, views. *Opp* SCEPTICISM.

believe *vb* 1 accept, be certain of, count on, credit, depend on, endorse, have faith in, reckon on, rely on, subscribe to, *inf* swallow, swear by, trust. *Opp* DISBELIEVE. 2 assume, consider, *inf* dare say, feel, gather, guess, hold, imagine, judge, know, maintain, postulate, presume, speculate, suppose, take it for granted, think. **make believe** ▷ IMAGINE.

believer *n* adherent, devotee, disciple, fanatic, follower, proselyte, supporter, upholder, zealot. *Opp* ATHEIST.

belittle *vb* be unimpressed by, criticize, decry, denigrate, deprecate, depreciate, detract from, disparage, minimize, *inf* play down, slight, speak slightingly of, underrate, undervalue. *Opp* EXAGGERATE, FLATTER, PRAISE.

bell *n* alarm, carillon, chime, knell, peal, signal. ▷ RING.

belligerent *adj* aggressive, antagonistic, argumentative, bellicose, bullying, combative, contentious, defiant, disputatious, fierce, hawkish, hostile, jingoistic, martial, militant, militaristic, provocative, pugnacious, quarrelsome, violent, warlike, warmongering, warring. ▷ UNFRIENDLY. *Opp* PEACEABLE.

belong *vb* 1 be owned (by), go (with), pertain (to), relate (to). 2 be at home, feel welcome, have a place. 3 *belong to a club.* be affiliated with, be a member of, be connected with, *inf* be in with.

belongings *n* chattels, effects, *inf* gear, goods, impedimenta, possessions, property, things.

belt *n* 1 band, circle, loop. 2 *belt round the waist.* cincture, cummerbund, girdle, girth, sash, strap, waistband, *old use* zone. 3 *green belt.* area, district, line, stretch, strip, swathe, tract, zone.

bemuse *vb* befuddle, bewilder, confuse, mix up, muddle, perplex, puzzle, stupefy.

bench *n* 1 form, pew, seat, settle. 2 counter, table, work-bench, work-table. 3 *magistrate's bench*. court, courtroom, judge, magistrate, tribunal.

bend *n* angle, arc, bow, corner, crank, crook, curvature, curve, flexure, loop, turn, turning, twist, zigzag. • *vb* 1 arch, be flexible, bow, buckle, coil, contort, crook, curl, curve, deflect, distort, divert, flex, fold, *inf* give, loop, mould, refract, shape, turn, twist, warp, wind, yield. 2 bow, crouch, curtsy, duck, genuflect, kneel, lean, stoop.

benefactor *n inf* angel, backer, *derog* do-gooder, donor, *inf* fairy godmother, patron, philanthropist, promoter, sponsor, supporter, well-wisher.

beneficial *adj* advantageous, benign, constructive, favourable, fruitful, good, health-giving, healthy, helpful, improving, nourishing, nutritious, positive, productive, profitable, rewarding, salubrious, salutary, supportive, useful, valuable, wholesome. *Opp* HARMFUL.

beneficiary *n* heir, heiress, inheritor, legatee, recipient, successor (*to title*).

benefit *n* 1 advantage, asset, blessing, boon, convenience, gain, good thing, help, privilege, prize, profit, service, use. *Opp* DISADVANTAGE. 2 *unemployment benefit*. aid, allowance, assistance, *inf* dole, *inf* hand-out, grant, payment, social security, welfare. • *vb* advance, advantage, aid, assist, better, boost, do good to, enhance, further, help, improve, profit, promote, serve.

benevolent *adj* altruistic, beneficent, benign, caring, charitable, compassionate, considerate, friendly, generous, helpful, humane, humanitarian, kind-hearted, kindly, liberal, magnanimous, merciful, philanthropic, supportive, sympathetic, unselfish, warm-hearted. ▷ KIND. *Opp* UNKIND.

benign *adj* gentle, harmless, kind. ▷ BENEFICIAL, BENEVOLENT.

bent *adj* 1 angled, arched, bowed, buckled, coiled, contorted, crooked, curved, distorted, folded, hunched, looped, twisted, warped. 2 [*inf*] *a bent dealer*. corrupt, criminal, dishonest, illegal, immoral, untrustworthy, wicked. *Opp* HONEST, STRAIGHT. • *n* ▷ APTITUDE, BIAS.

bequeath *vb* endow, hand down, leave, make over, pass on, settle, will.

bequest *n* endowment, gift, inheritance, legacy, settlement.

bereavement *n* death, loss.

bereft *adj* deprived, destitute, devoid, lacking, robbed, wanting.

berserk *adj inf* beside yourself, crazed, crazy, demented, deranged, frantic, frenetic, frenzied, furious, infuriated, insane, mad, maniacal, rabid, violent, wild. *Opp* CALM. **go berserk** ▷ RAGE, RAMPAGE.

berth *n* 1 bed, bunk, hammock. 2 *berth for ships*. anchorage, dock, harbour, haven, landing-stage, moorings, pier, port, quay, slipway, wharf. • *vb* anchor, dock, drop anchor, land, moor, tie up. **give a wide berth to** ▷ AVOID.

beseech *vb* ask, beg, entreat, implore, importune, plead, supplicate.

besiege *vb* beleaguer, beset, blockade, cut off, encircle, encompass, hem in, isolate, pester, plague, siege, surround.

best *adj* choicest, excellent, finest, first-class, foremost, incomparable, leading, matchless, optimum, outstanding, pre-eminient, superlative, supreme, top, unequalled, unrivalled, unsurpassed.

bestial *adj* animal, beast-like, beastly, brutal, brutish, inhuman, subhuman. ▷ SAVAGE.

bestow *vb* award, confer, donate, give, grant, present.

bet *n inf* flutter, gamble, *inf* punt, speculation, stake, wager. • *vb* bid, chance, do the pools, enter a lottery, gamble, *inf* have a flutter, hazard, lay bets, *inf* punt, risk, speculate, stake, venture, wager.

betray *vb* **1** be a Judas to, be a traitor to, be false to, cheat, conspire against, deceive, denounce, desert, double-cross, give away, *inf* grass on, incriminate, inform against, inform on, jilt, let down, *inf* rat on, report, *inf* sell down the river, sell out, *inf* shop, *inf* tell tales about, *inf* turn Queen's evidence on. **2** *betray secrets.* disclose, divulge, expose, give away, indicate, let out, let slip, manifest, reveal, show, tell.

better *adj* **1** preferable, recommended, superior. **2** convalescent, cured, fitter, healed, healthier, improved, *inf* on the mend, progressing, recovered, recovering, restored. • *vb* ▷ IMPROVE, SURPASS.

beware *vb* avoid, be alert, be careful, be cautious, be on your guard, guard (against), heed, keep clear (of), look out, mind, shun, steer away (from), take care, take heed, take precautions, watch out, *inf* watch your step.

bewilder *vb* baffle, *inf* bam boozle, bemuse, confound, confuse, daze, disconcert, disorientate, distract, floor, *inf* flummox, mislead, muddle, mystify, perplex, puzzle, stump.

bewitch *vb* captivate, cast a spell on, charm, enchant, enrapture, fascinate, spellbind.

bias *n* **1** aptitude, bent, inclination, leaning, liking, partiality, penchant, predilection, predisposition, preference, proclivity, proneness, propensity, tendency. **2** [*derog*] bigotry, chauvinism, favouritism, imbalance, injustice, nepotism, one-sidedness, partiality, partisanship, prejudice, racism, sexism, unfairness. • *vb* ▷ INFLUENCE.

biased *adj* bigoted, blinkered, chauvinistic, distorted, emotive, influenced, interested, jaundiced, loaded, one-sided, partial, partisan, prejudiced, racist, sexist, slanted, tendentious, unfair, unjust, warped. *Opp* UNBIASED.

bicycle *n inf* bike, cycle, penny-farthing, *inf* push-bike, racer, tandem, *inf* two-wheeler.

bid *n* **1** offer, price, proposal, proposition, tender. **2** *a bid to win.* attempt, *inf* crack, effort, endeavour, *inf* go, try, venture. • *vb* **1** make an offer, offer, proffer, propose, tender. **2** ▷ COMMAND.

big *adj* **1** above average, *inf* almighty, ample, astronomical, bold, broad, Brobdingnagian, bulky, burly, capacious, colossal, commodious, considerable, elephantine, enormous, extensive, fat, formidable, gargantuan, generous, giant, gigantic, grand, great, gross, heavy, hefty, high, huge, *inf* hulking, husky, immeasurable, immense, impressive, *inf* jumbo, *inf* king-sized, large, largish, lofty, long, mammoth, massive, mighty, monstrous, monumental, mountainous, overgrown, oversized, prodigious, roomy, sizeable, spacious, stupendous, substantial, swingeing (*increase*), tall, *inf* terrific, thick, *inf* thumping, tidy (*sum*), titanic, towering, *inf* tremendous, vast, voluminous, weighty, *inf* whacking, *inf* whopping, wide. **2** *a big decision.* grave, important, influential, leading, main, major, momentous, notable, powerful, prime, principal, prominent, serious, significant. **3** *a big number.* ▷ INFINITE. **4** *a big name.* ▷ FAMOUS. **5** *a big noise.* ▷ LOUD. *Opp* SMALL.

bigot *n* chauvinist, fanatic, prejudiced person, racist, sexist, zealot.

bigoted *adj* intolerant, one-sided, partial, prejudiced. ▷ BIASED.

bill *n* **1** account, invoice, receipt, statement, tally. **2** advertisement, broadsheet, bulletin, circular, handbill, handout, leaflet, notice, placard, poster, sheet. **3** *a Parliamentary bill.* draft law, proposed law. **4** *a bird's bill.* beak, mandible.

billow *vb* balloon, belly, bulge, fill out, heave, puff out, rise, roll, surge, swell, undulate.

bind *vb* **1** attach, clamp, combine, connect, fuse, hitch, hold together, join, lash, link, rope, secure, strap, tie, truss, unify, unite, weld. ▷ FASTEN. **2** *bind a wound.* bandage, cover, dress, encase, swathe, wrap. **3** *bound to obey.* compel, constrain, force, necessitate, oblige, require. **binding** ▷ COMPULSORY, FORMAL.

biography *n* autobiography, life, life-story, memoirs, recollections. ▷ WRITING.

bird *n inf* birdie, chick, cock, *joc* feathered friend, fledgling, fowl, hen, nestling. □ *gamebird, plur poultry, seabird, wader, waterfowl, wildfowl.* □ *albatross, auk, bittern, blackbird, budgerigar, bullfinch, bunting, bustard, buzzard, canary, cassowary, chaffinch, chiff-chaff, chough, cockatoo, coot, cormorant, corncrake, crane, crow, cuckoo, curlew, dabchick, dipper, dove, duck, dunnock, eagle, egret, emu, falcon, finch, flamingo, flycatcher, fulmar, goldcrest, goldfinch, goose, grebe, greenfinch, grouse, gull, hawk, heron, hoopoe, hornbill, humming bird, ibis, jackdaw, jay, kingfisher, kestrel, kite, kiwi, kookaburra, lapwing, lark, linnet, macaw, magpie, martin, mina bird, moorhen, nightingale, nightjar, nuthatch, oriole, osprey, ostrich, ousel, owl, parakeet, parrot, partridge, peacock, peewit, pelican, penguin, peregrine, petrel, pheasant, pigeon, pipit, plover, ptarmigan, puffin, quail, raven, redbreast, redstart, robin, rook, sandpiper, seagull, shearwater, shelduck, shrike, skua, skylark, snipe, sparrow, sparrowhawk, spoonbill, starling, stonechat, stork, swallow, swan, swift, teal, tern, thrush, tit, toucan, turkey, turtle-dove, vulture, wagtail, warbler, waxwing, wheatear, woodcock, woodpecker, wren, yellowhammer.*

birth *n* **1** childbirth, confinement, delivery, labour, nativity, parturition. **2** ancestry, background, blood, breeding, derivation, descent, extraction, family, genealogy, line, lineage, parentage, pedigree, race, stock, strain. **3** ▷ BEGINNING. **give birth** bear, calve, farrow, foal. ▷ BEGIN.

bisect *vb* cross, cut in half, divide, halve, intersect.

bit *n* **1** atom, bite, block, chip, chunk, crumb, division, dollop, fraction, fragment, gobbet, grain, helping, hunk, iota, lump, modicum, morsel, mouthful, part, particle, piece, portion, sample, scrap, section, segment, share, slab, slice, snippet, soupçon, speck, spot, taste, titbit, trace. **2** *Wait a bit.* flash, instant, *inf* jiffy, minute, moment, second, *inf* tick, time, while.

bite *n* **1** nip, pinch, sting. **2** *a bite to eat.* morsel, mouthful, nibble, snack, taste. ▷ BIT. ● *vb* **1** champ, chew, crunch, cut into, gnaw, masticate, munch, nibble, nip, rend, snap, tear at, wound. **2** *An insect bit me.* pierce, sting. **3** *The screw won't bite.* grip, hold.

bitter *adj* **1** acid, acrid, harsh, sharp, sour, unpleasant. **2** *a bitter experience.* calamitous, dire, distasteful, distressing, galling, hateful, heartbreaking, painful, poignant, sorrowful, unhappy, unwelcome, upsetting. **3** *bitter remarks.* acrimonious, acerbic, angry, cruel, cynical, embittered, envious, hostile, jaundiced, jealous, malicious, rancorous, resentful, savage, sharp, spiteful, stinging, vicious, violent, waspish. **4** *a bitter wind.* biting, cold, fierce, freezing, perishing, piercing, raw. *Opp* KIND, MILD, PLEASANT.

bizarre *adj* curious, eccentric, fantastic, freakish, grotesque, odd, outlandish, outré, surreal, weird. ▷ STRANGE. *Opp* ORDINARY.

black *adj* blackish, coal-black, dark, dusky, ebony, funereal, gloomy, inky, jet, jet-black, moonless, murky, pitch-black, pitch-dark, raven, sable, sooty, starless, unlit. ● *vb* **1** blacken, polish. **2** ▷ BLACKLIST.

blackleg *n sl* scab, strikebreaker, traitor.

blacklist *vb* ban, bar, blackball, boycott, debar, disallow, exclude, ostracize, preclude, proscribe, put an embargo on, refuse to handle, repudiate, snub, veto.

blade *n* dagger, edge, knife, razor, scalpel, vane. ▷ SWORD.

blame *n* accountability, accusation, castigation, censure, charge, complaint, condemnation, criticism, culpability, fault, guilt, imputation, incrimination, liability, onus, *inf* rap, recrimination, reprimand, reproach, reproof, responsibility, *inf* stick, stricture. ● *vb* accuse, admonish, censure, charge, chide, condemn, criticize, denounce, *inf* get at, hold responsible, incriminate, rebuke, reprehend, reprimand, reproach, reprove, round on, scold, tax, upbraid. *Opp* EXCUSE.

blameless *adj* faultless, guiltless, innocent, irreproachable, moral, unimpeachable, upright. *Opp* GUILTY.

bland *adj* affable, amiable, banal, boring, calm, characterless, dull, flat, gentle, insipid, mild, nondescript, smooth, soft, soothing, suave, tasteless, trite, unappetizing, unexciting, uninspiring, uninteresting, vapid, watery, weak, *inf* wishy-washy. *Opp* INTERESTING.

blank *adj* **1** bare, clean, clear, empty, plain, spotless, unadorned, unmarked, unused, void. **2** *a blank look.* apathetic, baffled, baffling, dead, *inf* deadpan, emotionless, expressionless, featureless, glazed, immobile, impassive, inane, inscrutable, lifeless, poker-faced, uncomprehending, unresponsive, vacant, vacuous. ● *n* **1** emptiness, nothingness, vacuity, vacuum, void. **2** *blanks on a form.* box, break, gap, line, space.

blaspheme *vb* curse, execrate, imprecate, profane, swear.

blasphemous *adj* disrespectful, godless, impious, irreligious, irreverent, profane, sacrilegious, sinful, ungodly, wicked. *Opp* REVERENT.

blast *n* 1 gale, gust, wind. 2 blare, din, noise, racket, roar, sound. 3 ▷ EXPLOSION. • *vb* ▷ ATTACK, EXPLODE. **blast off** ▷ LAUNCH.

blatant *adj* apparent, bare-faced, bold, brazen, conspicuous, evident, flagrant, glaring, obtrusive, obvious, open, overt, shameless, stark, unconcealed, undisguised, unmistakable, visible. *Opp* HIDDEN.

blaze *n* conflagration, fire, flame, flare-up, holocaust, inferno, outburst. • *vb* burn, erupt, flame, flare.

bleach *vb* blanch, discolour, etiolate, fade, lighten, pale, peroxide (*hair*), whiten.

bleak *adj* bare, barren, blasted, cheerless, chilly, cold, comfortless, depressing, desolate, dismal, dreary, exposed, grim, hopeless, joyless, sombre, uncomfortable, unpromising, windswept, wintry. *Opp* COMFORTABLE, WARM.

bleary *adj* blurred, *inf* blurry, cloudy, dim, filmy, fogged, foggy, fuzzy, hazy, indistinct, misty, murky, obscured, smeary, unclear, watery. *Opp* CLEAR.

blemish *n* blotch, blot, chip, crack, defect, deformity, disfigurement, eyesore, fault, flaw, imperfection, mark, mess, smudge, speck, stain, ugliness. □ *birthmark, blackhead, blister, callus, corn, freckle, mole, naevus, pimple, pustule, scar, spot, verruca, wart, whitlow,* sl *zit.* • *vb* deface, disfigure, flaw, mar, mark, scar, spoil, stain, tarnish.

blend *n* alloy, amalgam, amalgamation, combination, composite, compound, concoction, fusion, mélange, mix, mixture, synthesis, union. • *vb* 1 amalgamate, coalesce, combine, commingle, compound, fuse, harmonize, integrate, intermingle, intermix, meld, merge, mingle, synthesize, unite. 2 *blend in a bowl.* beat, mix, stir together, whip, whisk.

bless *vb* 1 anoint, consecrate, dedicate, grace, hallow, make sacred, ordain, sanctify. 2 *bless God's name.* adore, exalt, extol, glorify, magnify, praise. *Opp* CURSE.

blessed *adj* 1 adored, divine, hallowed, holy, revered, sacred, sanctified. 2 ▷ HAPPY.

blessing 1 benediction, consecration, grace, prayer. 2 approbation, approval, backing, concurrence, consent, leave, permission, sanction, support. 3 *Warmth is a blessing.* advantage, asset, benefit, boon, comfort, convenience, *inf* godsend, help. *Opp* CURSE, MISFORTUNE.

blight *n* affliction, ailment, *old use* bane, cancer, canker, curse, decay, disease, evil, illness, infestation, misfortune, *old use* pestilence, plague, pollution, rot, scourge, sickness, trouble. • *vb* ▷ SPOIL.

blind *adj* 1 blinded, eyeless, sightless, unseeing. □ *astigmatic, colour-blind, long-sighted, myopic, near-sighted, short-sighted, suffering from cataract, suffering from glaucoma, visually handicapped.* 2 *blind devotion.* blinkered, heedless, ignorant, inattentive, indifferent, indiscriminate, insensible, insensitive, irrational, mindless, oblivious, prejudiced, unaware, unobservant, unreasoning. • *n* awning, cover, curtain, screen, shade, shutters. • *vb* 1 dazzle, make blind. 2 ▷ DECEIVE.

blink *vb* coruscate, flash, flicker, flutter, gleam, glimmer, shimmer, sparkle, twinkle, wink.

bliss *n* blessedness, delight, ecstasy, euphoria, felicity, gladness, glee, happiness, heaven, joy, paradise, rapture. ▷ PLEASURE. *Opp* MISERY.

bloated *adj* dilated, distended, enlarged, inflated, puffy, swollen.

block *n* **1** bar, brick, cake, chock, chunk, hunk, ingot, lump, mass, piece, slab. **2** ▷ BLOCKAGE. • *vb* **1** bar, barricade, *inf* bung up, choke, clog, close, congest, constrict, dam, fill, impede, jam, obstruct, plug, stop up. **2** *block a plan*. deter, halt, hamper, hinder, hold back, prevent, prohibit, resist, *inf* scotch, *inf* stonewall, stop, thwart.

blockage *n* barrier, block, bottleneck, congestion, constriction, delay, hang-up, hindrance, impediment, jam, obstacle, obstruction, resistance, stoppage.

blond, blonde *adj* bleached, fair, flaxen, golden, light, platinum, silvery, yellow.

bloodshed *n* bloodletting, butchery, carnage, killing, massacre, murder, slaughter, slaying, violence.

bloodthirsty *adj* barbaric, brutal, feral, ferocious, fierce, homicidal, inhuman, murderous, pitiless, ruthless, sadistic, sanguinary, savage, vicious, violent, warlike. ▷ CRUEL. *Opp* HUMANE.

bloody *adj* **1** bleeding, bloodstained, raw. **2** *a bloody battle*. cruel, gory, sanguinary. ▷ BLOODTHIRSTY.

bloom *n* **1** blossom, bud, floret, flower. **2** *bloom of youth*. beauty, blush, flush, glow, prime. • *vb* be healthy, blossom, *poet* blow, bud, burgeon, *inf* come out, develop, flourish, flower, grow, open, prosper, sprout, thrive. *Opp* FADE.

blot *n* **1** blob, blotch, mark, smear, smirch, smudge, *inf* splodge, spot, stain. **2** *blot on the landscape*. blemish, defect, eyesore, fault, flaw, ugliness. • *vb* bespatter, blemish, blotch, blur, disfigure, mar, mark, smudge, spoil, spot, stain. **blot out** ▷ OBLITERATE. **blot your copybook** ▷ MISBEHAVE.

blotchy *adj* blemished, brindled, discoloured, inflamed, marked, patchy, smudged, spotty, streaked, uneven.

blow *n* **1** bang, bash, *inf* belt, *inf* biff, box (*ears*), buffet, bump, clip, clout, clump, concussion, hit, jolt, knock, punch, rap, slap, *inf* slosh, smack, *inf* sock, stroke, swat, swipe, thump, thwack, wallop, welt, whack, whop. **2** *a sad blow*. affliction, *inf* bombshell, calamity, disappointment, disaster, misfortune, shock, surprise, upset. • *vb* blast, breathe, exhale, fan, puff, waft, whine, whirl, whistle. **blow up 1** dilate, enlarge, expand, fill, inflate, pump up. **2** exaggerate, magnify, make worse, overstate. **3** blast, bomb, burst, detonate, dynamite, erupt, explode, go off, set off, shatter. **4** [*inf*] erupt, get angry, lose your temper, rage.

blue *adj* **1** aquamarine, azure, cerulean, cobalt, indigo, navy, sapphire, sky-blue, turquoise, ultramarine. **2** ▷ BAWDY. **3** ▷ SAD. • *vb* ▷ SQUANDER.

blueprint *n* basis, design, draft, model, outline, pattern, pilot, plan, project, proposal, prototype, scheme.

bluff *vb* cozen, deceive, delude, dupe, fool, hoodwink, mislead.

blunder *n inf* boob, *inf* botch, *inf* clanger, *sl* cock-up, error, fault, *Fr* faux pas, gaffe, howler, indiscretion, miscalculation, misjudgement, mistake, slip, slip-up, solecism. • *vb* be clumsy, *inf* botch up, bumble, bungle, *inf* drop a clanger, err, flounder, *inf* foul up, *sl* goof, go wrong, *inf* make a hash of something, make a mistake, mess up, miscalculate, misjudge, *inf* put your foot in it, slip up, stumble.

blunt *adj* 1 dull, rounded, thick, unpointed, unsharpened, worn. *Opp* SHARP. 2 *blunt criticism.* abrupt, bluff, brusque, candid, curt, direct, downright, forthright, frank, honest, insensitive, outspoken, plain-spoken, rude, straightforward, tactless, unceremonious, undiplomatic. *Opp* TACTFUL. • *vb* abate, allay, anaesthetize, dampen, deaden, desensitize, dull, lessen, numb, soften, take the edge off, weaken. *Opp* SHARPEN.

blur *vb* bedim, befog, blear, cloud, conceal, confuse, darken, dim, fog, mask, muddle, obscure, smear, unfocus.

blurred *adj* bleary, blurry, clouded, cloudy, confused, dim, faint, foggy, fuzzy, hazy, ill-defined, indefinite, indistinct, misty, nebulous, out of focus, smoky, unclear, unfocused, vague. *Opp* CLEAR.

blurt *vb* **blurt out** be indiscreet, *inf* blab, burst out with, come out with, cry out, disclose, divulge, exclaim, *inf* give the game away, let out, let slip, reveal, *inf* spill the beans, tell, utter.

blush *vb* be ashamed, colour, flush, glow, go red, redden.

blustering *adj* angry, boasting, boisterous, bragging, bullying, crowing, defiant, domineering, hectoring, noisy, ranting, self-assertive, showing-off, storming, swaggering, threatening, vaunting, violent. *Opp* MODEST.

blustery *adj* gusty, squally, unsettled, windy.

board *n* 1 blockboard, chipboard, clapboard, panel, plank, plywood, scantling, sheet, slab, slat, timber, weatherboard. 2 *board of directors.* cabinet, committee, council, directorate, jury, panel. • *vb* 1 accommodate, billet, feed, house, lodge, put up, quarter, stay. 2 *board a bus.* catch, embark (on), enter, get on, go on board.

boast *vb sl* be all mouth, *inf* blow your own trumpet, bluster, brag, crow, exaggerate, gloat, praise yourself, *sl* shoot a line, show off, *inf* sing your own praises, swagger, *inf* swank, *inf* talk big, vaunt.

boaster *n inf* big-head, *inf* big mouth, braggadocio, braggart, *inf* loudmouth, *inf* poser, show-off, swaggerer, swank.

boastful *adj inf* big-headed, bragging, *inf* cocky, conceited, egotistical, ostentatious, proud, puffed up, swaggering, swanky, swollen-headed, vain, vainglorious. *Opp* MODEST.

boat *n* craft, ship. ▷ VESSEL.

boatman *n* bargee, coxswain, ferryman, gondolier, lighterman, oarsman, rower, waterman, yachtsman. ▷ SAILOR.

bob *vb* be agitated, bounce, dance, hop, jerk, jig about, jolt, jump, leap, move about, nod, oscillate, shake, toss about, twitch. **bob up** ▷ APPEAR.

body *n* 1 anatomy, being, build, figure, form, frame, individual, physique, shape, substance, torso, trunk. 2 cadaver, carcass, corpse, mortal remains, mummy, relics, remains, *sl* stiff. 3 association, band, committee, company, corporation, society. ▷ GROUP. 4 *body of material*. accumulation, agglomeration, collection, corpus, mass.

bodyguard *n* defender, guard, minder, protector.

bog *n* fen, marsh, marshland, mire, morass, mudflats, peat bog, quagmire, quicksands, salt-marsh, *old use* slough, swamp, wetlands. **get bogged down** be hindered, get into difficulties, get stuck, grind to a halt, sink.

bogus *adj* counterfeit, fake, false, fictitious, fraudulent, imitation, *inf* phoney, sham, spurious. *Opp* GENUINE.

Bohemian *adj inf* arty, *old use* beatnik, bizarre, eccentric, hippie, informal, nonconformist, off-beat, unconventional, unorthodox, *inf* way-out, weird.

boil *n* abscess, blister, carbuncle, chilblain, eruption, gathering, gumboil, inflammation, pimple, pock, pustule, sore, spot, tumour, ulcer, *sl* zit. • *vb* **1** cook, heat, simmer, stew. **2** bubble, effervesce, foam, seethe, steam. **3** ▷ RAGE.

boisterous *adj* animated, disorderly, exuberant, irrepressible, lively, loud, noisy, obstreperous, riotous, rollicking, rough, rowdy, stormy, tempestuous, tumultuous, undisciplined, unrestrained, unruly, uproarious, wild. *Opp* CALM.

bold *adj* **1** adventurous, audacious, brave, confident, courageous, daredevil, daring, dauntless, enterprising, fearless, *derog* foolhardy, forceful, gallant, hardy, heroic, intrepid, *inf* plucky, *derog* rash, *derog* reckless, resolute, self-confident, unafraid, valiant, valorous, venturesome. **2** [*derog*] *a bold request*. brash, brazen, *inf* cheeky, forward, fresh, impertinent, impudent, insolent, pert, presumptuous, rude, saucy, shameless, unashamed. **3** *bold colours, writing*. big, bright, clear, conspicuous, eye-catching, large, obvious, prominent, pronounced, showy, striking, strong, vivid. *Opp* FAINT, TIMID.

bolster *n* cushion, pillow. • *vb* ▷ SUPPORT.

bolt *n* **1** arrow, dart, missile, projectile. **2** peg, pin, rivet, rod, screw. **3** *bolt on a door*. bar, catch, fastening, latch, lock. • *vb* **1** bar, close, fasten, latch, lock, secure. **2** *The animals bolted*. abscond, dart away, dash away, escape, flee, fly, run off, rush off. **3** *bolt food*. ▷ EAT. **bolt from the blue** ▷ SURPRISE.

bomb *n* bombshell, explosive. ▷ WEAPON. • *vb* ▷ BOMBARD.

bombard *vb* **1** assail, assault, attack, batter, blast, blitz, bomb, fire at, pelt, pound, shell, shoot at, strafe. **2** badger, beset, harass, importune, pester, plague.

bombardment *n* attack, barrage, blast, blitz, broadside, burst, cannonade, discharge, fusillade, hail, salvo, volley.

bombastic *adj* extravagant, grandiloquent, grandiose, high-flown, inflated, magniloquent, pompous, turgid.

bond *n* **1** chain, cord, fastening, fetters, handcuffs, manacles, restraints, rope, shackles. **2** *bond of friendship*. affiliation, affinity, attachment, connection, link, relationship, tie, unity. **3** *a legal bond*. agreement, compact, contract, covenant, guarantee, legal document, pledge, promise, word. • *vb* ▷ STICK.

bondage *n* enslavement, serfdom, servitude, slavery, subjection, thraldom, vassalage.

bonus *n* **1** bounty, commission, dividend, gift, gratuity, hand-out, honorarium, largesse, payment, *inf* perk, reward, supplement, tip. **2** addition, advantage, benefit, extra, *inf* plus.

bony *adj* angular, emaciated, gangling, gawky, lanky, lean, scraggy, scrawny, skinny, thin, ungainly. *Opp* GRACEFUL, PLUMP.

book *n* booklet, copy, edition, hardback, paperback, publication, tome, volume, work. □ *album, annual, anthology, atlas, bestiary,* old use *chap-book, compendium, concordance, diary, dictionary, digest, directory, encyclopaedia, fiction, gazetteer, guidebook, handbook, hymnal, hymn-book, jotter, ledger, lexicon, libretto, manual, manuscript, missal, notebook, omnibus, picture-book, prayer-book, primer, psalter, reading book, reference book,* music *score, scrap-book, scroll, sketch-book, text-book, thesaurus, vade mecum.* ▷ WRITING. • *vb* **1** *book for speeding.* arrest, take your name, write down details. **2** *book in advance.* arrange, buy, engage, order, organize, reserve, sign up.

booklet *n* brochure, leaflet, pamphlet, paperback.

boom *n* **1** bang, blast, crash, explosion, reverberation, roar, rumble. ▷ SOUND. **2** *boom in trade.* bonanza, boost, expansion, growth, improvement, increase, spurt, upsurge, upturn. ▷ PROSPERITY. **3** *boom across a river.* ▷ BARRIER. • *vb* **1** ▷ SOUND. **2** ▷ PROSPER.

boorish *adj* barbarian, ignorant, ill-bred, ill-mannered, loutish, oafish, philistine, uncultured, vulgar. *Opp* CULTURED.

boost *n* aid, encouragement, fillip, impetus, help, lift, push, stimulus. • *vb* advance, aid, assist, augment, bolster, build up, buoy up, encourage, enhance, enlarge, expand, foster, further, give an impetus to, heighten, help, improve, increase, inspire, lift, promote, push up, raise, support, sustain. *Opp* DEPRESS.

booth *n* box, carrel, compartment, cubicle, hut, kiosk, stall, stand.

booty *n* contraband, gains, haul, loot, pickings, pillage, plunder, spoils, *inf* swag, takings, trophies, winnings.

border *n* **1** brim, brink, edge, edging, frame, frieze, frill, fringe, hem, margin, perimeter, periphery, rim, surround, verge. **2** borderline, boundary, frontier, limit. **3** *flower border.* bed, herbaceous border. • *vb* abut on, adjoin, be adjacent to, be alongside, join, share a border with, touch.

bore *vb* **1** burrow, drill, mine, penetrate, sink, tunnel. ▷ PIERCE. **2** *bore listeners.* alienate, depress, jade, *inf* leave cold, tire, *inf* turn off, weary. *Opp* INTEREST.

boring *adj* arid, commonplace, dead, dreary, dry, dull, flat, humdrum, long-winded, monotonous, prolix, repetitious, repetitive, soporific, stale, tedious, tiresome, trite, uneventful, unexciting, uninspiring, uninteresting, vapid, wearisome, wordy. *Opp* INTERESTING.

born *adj* congenital, genuine, instinctive, natural, untaught.

borrow *vb* adopt, appropriate, be lent, *inf* cadge, copy, crib, make use of, obtain, pirate, plagiarize, *inf* scrounge, *inf* sponge, take, use, usurp. *Opp* LEND.

boss *n* employer, head. ▷ CHIEF.

bossy *adj* aggressive, assertive, authoritarian, autocratic, bullying, despotic, dictatorial, domineering, exacting, hectoring, high-handed, imperious, lordly, magisterial, masterful, officious, oppressive, overbearing, peremptory, *inf* pushy, self-assertive, tyrannical. *Opp* SERVILE.

bother *n* **1** ado, difficulty, disorder, disturbance, fuss, *inf* hassle, problem, *inf* to-do. **2** annoyance, inconvenience, irritation, nuisance, pest, trouble, worry. • *vb* **1** annoy, bewilder, concern, confuse, disconcert, dismay, disturb, exasperate, harass, *inf* hassle, inconvenience, irk, irritate, molest, nag, perturb, pester, plague, trouble, upset, vex, worry. **2** be concerned, be worried, care, mind, take trouble. ▷ TROUBLE.

bottle *n* flask. □ *carafe, carboy, decanter, flagon, jar, jeroboam, magnum, phial, pitcher, vial, wine-bottle.* **bottle up** ▷ SUPPRESS.

bottom *adj* deepest, least, lowest, minimum. • *n* **1** base, bed, depth, floor, foot, foundation, lowest point, nadir, pedestal, substructure, underneath, underside. *Opp* TOP. **2** basis, essence, grounds, heart, origin, root, source. **3** *your bottom. vulg* arse, backside, behind, *inf* bum, buttocks, *joc* posterior, rear, rump, seat, *inf* sit-upon.

bottomless *adj* deep, immeasurable, unfathomable, unplumbable.

bounce *vb* bob, bound, bump, jump, leap, move about, rebound, recoil, ricochet, spring.

bound *adj* **1** certain, committed, compelled, constrained, destined, doomed, duty-bound, fated, forced, obligated, obliged, pledged, required, sure. **2** *bound with rope.* ▷ BIND. • *vb* bob, bounce, caper, frisk, frolic, gambol, hop, hurdle, jump, leap, pounce, romp, skip, spring, vault. **bound for** aimed at, directed towards, going to, heading for, making for, off to, travelling towards.

boundary *n* border, borderline, bounds, brink, circumference, confines, demarcation, edge, end, extremity, fringe, frontier, interface, limit, margin, perimeter, threshold, verge.

boundless *adj* endless, everlasting, immeasurable, incalculable, inexhaustible, infinite, limitless, unbounded, unconfined, unflagging, unlimited, unrestricted, untold. ▷ VAST. *Opp* FINITE.

bounty *n* alms, altruism, beneficence, benevolence, charity, generosity, giving, goodness, kindness, largesse, liberality, munificence, philanthropy, unselfishness.

bouquet *n* **1** arrangement, bunch, buttonhole, corsage, garland, nosegay, posy, spray, wreath. **2** *bouquet of wine.* ▷ SMELL.

bout *n* **1** attack, fit, period, run, spell, stint, stretch, time, turn. **2** battle, combat, competition, contest, encounter, engagement, fight, match, round, *inf* set-to, struggle.

bow *vb* **1** bend, bob, curtsy, genuflect, incline, kowtow, nod, prostrate yourself, salaam, stoop. **2** ▷ SUBMIT.

bowels *n* **1** entrails, guts, *inf* innards, insides, intestines, viscera, vitals. **2** core, depths, heart, inside.

bower *n* alcove, arbour, bay, gazebo, grotto, hideaway, pavilion, pergola, recess, retreat, sanctuary, shelter, summer-house.

bowl *n* basin, bath, casserole, container, dish, pan, pie-dish, tureen. • *vb* fling, hurl, lob, pitch, throw, toss.

box *n* carton, case, chest, container, crate. □ *bin, caddy, canister, cartridge, casket, coffer, coffin, pack, package, punnet, tea-chest, tin, trunk.* • *vb inf* engage in fisticuffs, fight, punch, scrap, spar. ▷ HIT.

boxer *n* prize-fighter, sparring partner. □ *bantamweight, cruiserweight, featherweight, flyweight, heavyweight, lightweight, middleweight, welterweight.*

boy *n derog* brat, *inf* kid, lad, schoolboy, son, *derog* stripling, *derog* urchin, youngster, youth.

boycott *n* ban, blacklist, embargo, prohibition. • *vb* avoid, black, blackball, blacklist, exclude, *inf* give the cold-shoulder to, ignore, make unwelcome, ostracize, outlaw, prohibit, spurn, stay away from.

bracing *adj* crisp, exhilarating, health-giving, invigorating, refreshing, restorative, stimulating, tonic.

brag *vb* crow, gloat, show off. ▷ BOAST.

brain *n* cerebrum, *inf* grey matter, intellect, intelligence, mind, *inf* nous, reason, sense, understanding, wisdom, wit.

brainwash *vb* condition, indoctrinate, re-educate.

branch *n* **1** arm, bough, limb, prong, shoot, sprig, stem, twig. **2** department, division, office, offshoot, part, ramification, section, subdivision, wing. • *vb* diverge, divide, fork, ramify, split, subdivide. **branch out** ▷ DIVERSIFY.

brand *n* kind, label, line, make, sort, trademark, type, variety. • *vb* **1** burn, identify, label, mark, scar, stamp, tag. **2** censure, characterize, denounce, discredit, stigmatize, vilify.

brash *adj* brazen, bumptious, insolent, rash, reckless, rude, self-assertive. ▷ ARROGANT.

bravado *n* arrogance, bluster, braggadocio, machismo, swagger.

brave *adj* adventurous, audacious, bold, chivalrous, cool, courageous, daring, dauntless, determined, fearless, gallant, game, *sl* gutsy, heroic, indomitable, intrepid, lion-hearted, *derog* macho, noble, *inf* plucky, resolute, spirited, stalwart, stoical, stout-hearted, tough, unafraid, uncomplaining, undaunted, unshrinking, valiant, valorous, venturesome. *Opp* COWARDLY.

bravery *n* audacity, boldness, *sl* bottle, courage, daring, dauntlessness, determination, fearlessness, fibre, firmness, fortitude, gallantry, *inf* grit, *inf* guts, heroism, intrepidity, mettle, *inf* nerve, *inf* pluck, prowess, resolution, spirit, *sl* spunk, stoicism, tenacity, valour, will-power. *Opp* COWARDICE.

brawl *n* affray, altercation, *inf* bust-up, clash, *inf* dust-up, fracas, fray, *inf* free-for-all, mêlée, *inf* punch-up, quarrel, row, scrap, scuffle, *inf* set-to, tussle. • *vb* ▷ FIGHT.

brazen *adj* barefaced, blatant, cheeky, defiant, flagrant, impertinent, impudent, insolent, rude, shameless, unabashed, unashamed. *Opp* SHAMEFACED.

breach *n* **1** aperture, break, chasm, crack, fissure, gap, hole, opening, rent, space, split. **2** alienation, difference, disagreement, divorce, drifting apart, estrangement, quarrel, rift, rupture, schism, separation, split.

3 *breach of law.* contravention, failure, infringement, offence, transgression, violation.

bread *n* □ *brioche, cob, croissant, French bread, loaf, roll, stick of bread, toast.* ▷ FOOD.

break *n* 1 breach, breakage, burst, chink, cleft, crack, crevice, cut, fissure, fracture, gap, gash, hole, leak, opening, rent, rift, rupture, slit, split, tear. 2 *break from work. inf* breather, breathing-space, hiatus, interlude, intermission, interval, *inf* let-up, lull, pause, respite, rest, tea-break. 3 *break in service.* disruption, halt, interruption, lapse, suspension. • *vb* 1 breach, burst, *inf* bust, chip, crack, crumple, crush, damage, demolish, fracture, fragment, knock down, ruin, shatter, shiver, smash, *inf* smash to smithereens, snap, splinter, split, squash, wreck. ▷ DESTROY. 2 *break the law.* contravene, defy, disobey, disregard, fail to observe, flout, go back on, infringe, transgress, violate. 3 *break a record.* beat, better, do more than, exceed, excel, go beyond, outdo, outstrip, pass, surpass. **break down** ▷ ANALYSE, DEMOLISH. **break in** ▷ INTERRUPT, INTRUDE. **break off** ▷ FINISH. **break out** ▷ ESCAPE. **break through** ▷ PENETRATE. **break up** ▷ DISINTEGRATE.

breakdown *n* 1 collapse, destruction, disintegration,downfall, failure, fault, hitch, malfunction, ruin, stoppage. 2 analysis, classification, detailing, dissection, itemization, *inf* rundown.

breakthrough *n* advance, development, discovery, find, improvement, innovation, invention, leap forward, progress, revolution, success.

breakwater *n* groyne, jetty, mole, pier, sea-defence.

breath *n* breeze, gust, murmur, pant, puff, sigh, stir, waft, whiff, whisper.

breathe *vb* 1 exhale, inhale, pant, puff, respire, suspire. 2 hint, let out, tell, whisper.

breathless *adj* exhausted, gasping, out of breath, panting, *inf* puffed, *inf* puffing and blowing, tired out, wheezy, winded.

breed *n* ancestry, clan, family, kind, line, lineage, nation, pedigree, progeny, race, sort, species, stock, strain, type, variety. • *vb* 1 bear young ones, beget young ones, increase, multiply, procreate, produce young, propagate (*plants*), raise young ones, reproduce. 2 *breed contempt.* arouse, cause, create, cultivate, develop, engender, foster, generate, induce, nourish, nurture, occasion.

breeze *n* air-current, breath, draught, waft, wind, *poet* zephyr.

breezy *adj* airy, *inf* blowy, draughty, fresh, gusty, windy.

brevity *n* briefness, compactness, compression, conciseness, concision, curtness, economy, incisiveness, pithiness, shortness, succinctness, terseness.

brew *n* blend, compound, concoction, drink, hash, infusion, liquor, mixture, potion, preparation, punch, stew. • *vb* 1 boil, cook, ferment, infuse, make, simmer, steep, stew. 2 *brew mischief.* concoct, contrive, *inf* cook up, develop, devise, foment, hatch, plan, plot, prepare, scheme, stir up.

bribe *n sl* backhander, bribery, *inf* carrot, enticement, *sl* graft, gratuity, incentive, inducement, *inf* payola, protection money, *inf* sweetener, tip. • *vb* buy off, corrupt, entice, *inf* grease your palm, influence, offer a bribe, pervert, reward, suborn, tempt, tip.

brick *n* block, breeze-block, cube, set, sett, stone.

bridge *n* arch, connection, crossing, link, span, way over. □ *aqueduct, Bailey bridge, drawbridge, flyover, footbridge, overpass, pontoon bridge, suspension bridge, swing bridge, viaduct.* • *vb* connect, cross, fill, join, link, pass over, span, straddle, tie together, traverse, unite.

bridle *vb* check, control, curb, restrain.

brief *adj* **1** cursory, ephemeral, evanescent, fast, fleeting, hasty, limited, little, momentary, passing, quick, sharp, short, short-lived, temporary, transient, transitory. **2** *brief comment.* abbreviated, abridged, compact, compendious, compressed, concise, condensed, crisp, curt, curtailed, incisive, laconic, pithy, shortened, succinct, terse, thumbnail, to the point. *Opp* LONG. • *n* **1** advice, briefing, data, description, directions, information, instructions, orders, outline, plan. **2** *a barrister's brief.* argument, case, defence, dossier, summary. • *vb* advise, coach, direct, enlighten, *inf* fill someone in, give someone the facts, guide, inform, instruct, prepare, prime, *inf* put someone in the picture.

briefs *n* camiknickers, knickers, panties, pants, shorts, trunks, underpants.

brigand *n* bandit, buccaneer, desperado, footpad, gangster, highwayman, marauder, outlaw, pirate, robber, ruffian, thief.

bright *adj* **1** ablaze, aglow, alight, beaming, blazing, burnished, colourful, dazzling, flashing, *derog* flashy, fresh, *derog* gaudy, glaring, gleaming, glistening, glittering, glossy, glowing, incandescent, lambent, light, luminous, lustrous, pellucid, polished, radiant, refulgent, resplendent, scintillating, shimmering, shining, shiny, showy, sparkling, twinkling, vivid. **2** *bright sky.* clear, cloudless, fair, sunny. **3** *bright prospects.* auspicious, favourable, good, hopeful, optimistic, rosy. **4** *a bright smile.* ▷ CHEERFUL. **5** *bright ideas.* ▷ CLEVER. *Opp* DULL.

brighten *vb* **1** cheer (up), enliven, gladden, illuminate, light up, liven up, *inf* perk up, revitalize, smarten up. **2** *The sky brightened.* become sunny, clear up, lighten.

brilliant *adj* **1** coruscating, dazzling, glaring, glittering, glorious, intense, resplendent, scintillating, shining, showy, sparkling, splendid, vivid. ▷ BRIGHT. *Opp* DULL. **2** [*inf*] *a brilliant game.* ▷ EXCELLENT.

brim *n* brink, circumference, edge, limit, lip, margin, perimeter, periphery, rim, top, verge.

bring *vb* **1** bear, carry, convey, deliver, fetch, take, transfer, transport. **2** *bring a friend.* accompany, conduct, escort, guide, lead, usher. **3** *The play brought great applause.* attract, cause, create, draw, earn, engender, generate, get, give rise to, induce, lead to, occasion, produce, prompt, provoke, result in. **bring about** ▷ CREATE. **bring in** ▷ EARN, INTRODUCE. **bring off** ▷ ACHIEVE. **bring on** ▷ ACCELERATE, CAUSE. **bring out** ▷ EMPHASIZE, PRODUCE. **bring up** ▷ EDUCATE, RAISE.

brink *n* bank, border, boundary, brim, circumference, edge, fringe, limit, lip, margin, perimeter, periphery, rim, skirt, threshold, verge.

brisk *adj* 1 active, alert, animated, bright, businesslike, bustling, busy, crisp, decisive, energetic, fast, keen, lively, nimble, quick, rapid, *inf* snappy, *inf* spanking (*pace*), speedy, spirited, sprightly, spry, vigorous. *Opp* LEISURELY. 2 *a brisk wind.* bracing, enlivening, fresh, invigorating, refreshing, stimulating.

bristle *n* barb, hair, prickle, quill, spine, stubble, thorn, whisker, wire. • *vb* become angry, become defensive, become indignant, bridle, flare up.

brittle *adj* breakable, crackly, crisp, crumbling, delicate, easily broken, fragile, frail, frangible, weak. *Opp* FLEXIBLE, RESILIENT.

broad *adj* 1 ample, capacious, expansive, extensive, great, large, open, roomy, spacious, sweeping, vast, wide. 2 *broad daylight.* clear, full, open, plain, undisguised. 3 *broad outline.* general, imprecise, indefinite, inexact, non-specific, sweeping, undetailed, vague. 4 *broad tastes.* all-embracing, catholic, comprehensive, eclectic, encyclopaedic, universal, wide-ranging. 5 *broad humour.* bawdy, *sl* blue, coarse, earthy, improper, impure, indecent, indelicate, racy, ribald, suggestive, vulgar. ▷ BROAD-MINDED. *Opp* FINITE, NARROW.

broadcast *n* programme, relay, show, telecast, transmission. • *vb* 1 advertise, announce, circulate, disseminate, make known, make public, proclaim, promulgate, publish, relay, report, send out, spread about, televise, transmit. 2 *broadcast seed.* scatter, sow at random.

broadcaster *n* anchor-man, announcer, commentator, compère, disc jockey, DJ, linkman, newsreader, presenter. ▷ ENTERTAINER.

broaden *vb* branch out, build up, develop, diversify, enlarge, expand, extend, increase, open up, spread, widen. *Opp* LIMIT.

broad-minded *adj* all-embracing, balanced, broad, catholic, comprehensive, cosmopolitan, eclectic, enlightened, liberal, open-minded, permissive, tolerant, unbiased, unbigoted, unprejudiced, unshockable. *Opp* NARROW-MINDED.

brochure *n* booklet, broadsheet, catalogue, circular, folder, handbill, leaflet, pamphlet, prospectus, tract.

brooch *n* badge, clasp, clip, fastening.

brood *n* children, clutch (*of eggs*), family, issue, litter, offspring, progeny, young. • *vb* 1 hatch, incubate, sit on. 2 *brood over mistakes.* agonize, dwell (on), *inf* eat your heart out, fret, mope, sulk, worry. ▷ THINK.

brook *n* beck, burn, channel, *poet* rill, rivulet, runnel, stream, watercourse. • *vb* ▷ TOLERATE.

browbeat *vb* badger, bully, coerce, cow, hector, intimidate, tyrannize. ▷ FRIGHTEN.

brown *adj* beige, bronze, buff, chestnut, chocolate, dun, fawn, khaki, ochre, russet, sepia, tan, tawny, terracotta, umber. • *vb* bronze, burn, colour, grill, tan, toast.

browse *vb* 1 crop grass, eat, feed, graze, pasture. 2 *browse in a book.* dip in, flick through, leaf through, look through, peruse, read here and there, scan, skim, thumb through.

bruise *n* black eye, bump, contusion, discoloration, *inf* shiner, welt. • *vb* blacken, crush, damage, discolour, injure, knock, mark. ▷ WOUND.

brush *n* 1 besom, broom. 2 *brush with police.* ▷ CONFLICT. • *vb* 1 comb, groom, scrub, sweep, tidy, whisk. 2 *just brushed the gatepost.* graze, touch. **brush aside** ▷ DISMISS, DISREGARD. **brush-off** ▷ REBUFF. **brush up** ▷ REVISE.

brutal *adj* atrocious, barbaric, barbarous, beastly, bestial, bloodthirsty, bloody, brutish, callous, cold-blooded, cruel, dehumanized, ferocious, hard-hearted, heartless, inhuman, inhumane, merciless, murderous, pitiless, remorseless, ruthless, sadistic, savage, uncivilized, unfeeling, vicious, violent, wild. ▷ UNKIND. *Opp* HUMANE.

brutalize *vb* dehumanize, harden, inure, make brutal.

brute *adj* crude, irrational, mindless, physical, rough, stupid, unfeeling, unthinking. ▷ BRUTISH. • *n* 1 beast, creature, dumb animal. ▷ ANIMAL. 2 [*inf*] *a cruel brute.* barbarian, bully, devil, lout, monster, ruffian, sadist, savage, swine.

brutish *adj* animal, barbaric, barbarous, beastly, bestial, boorish, brutal, coarse, cold-blooded, crude, cruel, *inf* gross, inhuman, insensitive, loutish, mindless, savage, senseless, stupid, subhuman, uncouth, unintelligent, unthinking. *Opp* HUMANE.

bubble *n* air-pocket, blister, hollow, vesicle. • *vb* boil, effervesce, fizz, fizzle, foam, froth, gurgle, seethe, sparkle. **bubbles** effervescence, fizz, foam, froth, head, lather, suds.

bubbly *adj* carbonated, effervescent, fizzy, foaming, seething, sparkling. ▷ LIVELY.

buccaneer *n* adventurer, bandit, brigand, corsair, marauder, pirate, privateer, robber.

bucket *n* can, pail, scuttle, tub.

buckle *n* catch, clasp, clip, fastener, fastening, hasp. • *vb* 1 clasp, clip, do up, fasten, hitch up, hook up, secure. 2 bend, bulge, cave in, collapse, contort, crumple, curve, dent, distort, fold, twist, warp.

bud *n* shoot, sprout. • *vb* begin to grow, burgeon, develop, shoot, sprout. **budding** ▷ POTENTIAL, PROMISING.

budge *vb* 1 change position, give way, move, shift, stir, yield. 2 *can't budge him.* alter, change, dislodge, influence, move, persuade, propel, push, remove, shift, sway.

budget *n* accounts, allocation of funds, allowance, estimate, financial planning, funds, means, resources. • *vb* allocate money, allot resources, allow (for), estimate expenditure, plan your spending, provide (for), ration your spending.

buff *n* ▷ ENTHUSIAST. • *vb* burnish, clean, polish, rub, shine, smooth.

buffer *n* bulwark, bumper, cushion, fender, pad, safeguard, screen, shield, shock-absorber.

buffet *n* 1 bar, café, cafeteria, counter, snack-bar. 2 *a stand-up buffet.* ▷ MEAL. • *vb* ▷ HIT.

bug *n* 1 ▷ INSECT, MICROBE. 2 *bug in a computer program.* breakdown, defect, error, failing, fault, flaw, *inf* gremlin, imperfection, malfunction, mistake, *inf* snarl-up, virus. • *vb* 1 intercept, interfere with, listen in to, spy on, tap. 2 [*s*... *Untidiness bugs me.* ▷ ANNOY.

build *vb* assemble, construct, develop, erect, fabricate, form, found, *inf* knock together, make, put together, put up, raise, rear, set up. **build up** ▷ INTENSIFY.

builder *n* bricklayer, construction worker, labourer.

building *n* construction, edifice, erection, piece of architecture, *inf* pile, premises, structure. □ *arcade, barn, barracks, basilica, boat-house, bungalow, cabin, castle, cathedral, chapel, chateau, church, cinema, college, complex, cottage, dovecote, factory, farm-house, flats, fort, fortress, garage, gazebo, gymnasium, hall, hangar, hotel, house, inn, library, lighthouse, mansion, mausoleum, mill, monastery, monument, mosque, museum, observatory, outbuilding, outhouse, pagoda, palace, pavilion, pier, power-station, prison, pub, public house, restaurant, school, shed, shop, silo, skyscraper, stable, storehouse, studio, summer-house, synagogue, temple, theatre, tower, villa, warehouse, windmill.*

bulb *n* **1** corm, tuber. □ *amaryllis, bluebell, crocus, daffodil, freesia, hyacinth, lily, narcissus, snowdrop, tulip.* **2** *electric bulb.* lamp, light.

bulbous *adj* bloated, bulging, convex, distended, ovoid, pear-shaped, pot-bellied, rotund, rounded, spherical, swollen, tuberous.

bulge *n* bump, distension, hump, knob, lump, projection, protrusion, protuberance, rise, swelling. • *vb* belly, billow, dilate, distend, enlarge, expand, project, protrude, stick out, swell.

bulk *n* **1** amplitude, bigness, body, dimensions, extent, immensity, largeness, magnitude, mass, size, substance, volume, weight. **2** *the bulk of the work. inf* best part, greater part, majority, preponderance.

bulky *adj* awkward, chunky, cumbersome, large, unwieldy. ▷ BIG.

bulletin *n* account, announcement, communication, communiqué, dispatch, message, newsflash, notice, proclamation, report, statement.

bullion *n* bar, ingot, nugget, solid gold, solid silver.

bull's-eye *n* bull, centre, mark, middle, target.

bully *vb* bludgeon, browbeat, coerce, cow, domineer, frighten, harass, hector, intimidate, oppress, persecute, *inf* pick on, *inf* push around, terrorize, threaten, torment, tyrannize.

bulwark *n* defence, earthwork, fortification, parapet, protection, rampart, redoubt, wall. ▷ BARRIER.

bump *n* **1** bang, blow, buffet, collision, crash, knock, smash, thud, thump. **2** bulge, distension, hump, knob, lump, projection, protrusion, protuberance, rise, swelling, tumescence, welt. • *vb* **1** bang, collide with, crash into, jar, knock, ram, slam, smash into, strike, thump, wallop. ▷ HIT. **2** bounce, jerk, jolt, shake. **bump into** ▷ MEET. **bump off** ▷ KILL.

bumptious *adj* arrogant, *inf* big-headed, boastful, brash, *inf* cocky, conceited, egotistic, forward, immodest, officious, overbearing, over-confident, pompous, presumptuous, pretentious, *inf* pushy, self-assertive, self-important, smug, *inf* stuck-up, *inf* snooty, swaggering, vain, vainglorious, vaunting. *Opp* MODEST.

bumpy *adj* **1** bouncy, jarring, jerky, jolting. **2** *a bumpy road.* broken, irregular, jagged, knobbly, lumpy, pitted, rocky, rough, rutted, stony, uneven. *Opp* SMOOTH.

bunch *n* **1** batch, bundle, clump, cluster, collection, heap, lot, number, pack, quantity, set, sheaf, tuft. **2** *bunch of flowers.* bouquet, posy, spray. **3** [*inf*] *bunch of friends.* band, crowd, gang, gathering, mob, party, team, troop. ▷ GROUP. • *vb* assemble, cluster, collect, congregate, crowd, flock, gather, group, herd, huddle, mass, pack. *Opp* DISPERSE.

bundle *n* bag, bale, bunch, carton, collection, pack, package, packet, parcel, sheaf, truss. • *vb* bale, bind, enclose, fasten, pack, package, roll, tie, truss, wrap. **bundle out** ▷ EJECT.

bung *n* cork, plug, stopper. • *vb* ▷ THROW.

bungle *vb* blunder, botch, *sl* cock up, *inf* foul up, fluff, *inf* make a hash of, *inf* make a mess of, *inf* mess up, mismanage, *inf* muck up, *inf* muff, ruin, *inf* screw up, spoil.

buoy *n* beacon, float, marker, mooring buoy, signal. • *vb* **buoy up** ▷ RAISE.

buoyant *adj* **1** floating, light. **2** *a buoyant mood.* ▷ CHEERFUL.

burden *n* **1** cargo, encumbrance, load, weight. **2** *burden of guilt.* affliction, albatross, anxiety, care, cross, duty, handicap, millstone, obligation, onus, problem, responsibility, sorrow, trial, trouble, worry. • *vb* afflict, bother, encumber, hamper, handicap, impose on, load (with), *inf* lumber (with), oppress, overload (with), *inf* saddle (with), strain, tax, trouble, weigh down, worry.

burdensome *adj* bothersome, difficult, exacting, hard, heavy, onerous, oppressive, taxing, tiring, troublesome, trying, wearisome, wearying, weighty, worrying. *Opp* EASY.

bureau **1** desk, writing-desk. **2** *travel bureau.* agency, counter, department, office, service.

bureaucracy *n* administration, government, officialdom, paperwork, *inf* red tape, regulations.

burglar *n* cat-burglar, housebreaker, intruder, robber. ▷ THIEF.

burglary *n* break-in, forcible entry, house-breaking, larceny, pilfering, robbery, stealing, theft, thieving.

burgle *vb* break in, pilfer, rob. ▷ STEAL.

burial *n* entombment, funeral, interment, obsequies.

burlesque *n* caricature, imitation, mockery, parody, pastiche, satire, *inf* send-up, spoof, *inf* take-off, travesty.

burly *adj* athletic, beefy, brawny, heavy, hefty, hulking, husky, muscular, powerful, stocky, stout, *inf* strapping, strong, sturdy, thickset, tough, well-built. *Opp* THIN.

burn *n* blister, charring. • *vb* **1** be alight, blaze, flame, flare, flash, flicker, glow, smoke, smoulder, spark, sparkle. **2** carbonize, consume, cremate, destroy by fire, ignite, incinerate, kindle, light, reduce to ashes, set fire to, set on fire. **3** *burn your skin.* blister, brand, char, scald, scorch, sear, shrivel, singe, sting, toast. ▷ FIRE, HEAT.

burning *adj* **1** ablaze, afire, aflame, alight, blazing, flaming, glowing, incandescent, lit up, on fire, raging, smouldering. **2** *burning pain.* biting, blistering, boiling, fiery, hot, inflamed, scalding, scorching, searing, smarting, stinging. **3** *burning chemicals.* acid, caustic, corrosive. **4** *a burning smell.* acrid, pungent, reeking scorching, smoky. **5** *a burning desire*. acute, ardent, consuming, eager, fervent, flaming, frenzied, heated, impassioned, intense, passionate, red-hot, vehement. **6** *a burning issue*. crucial, important, pertinent, pressing, relevant, urgent, vital.

burrow *n* earth, excavation, hole, retreat, set, shelter, tunnel, warren. • *vb* delve, dig, excavate, mine, tunnel.

burst *vb* **1** break, crack, disintegrate, erupt, explode, force open, give way, open suddenly, part suddenly, puncture, rupture, shatter, split, tear. **2** ▷ RUSH.

bury *vb* cover, embed, enclose, engulf, entomb, immerse, implant, insert, inter, lay to rest, plant, put away, secrete, sink, submerge. ▷ HIDE.

bus *n old use* charabanc, coach, double-decker, minibus, *old use* omnibus.

bushy *adj* bristling, bristly, dense, fluffy, fuzzy, hairy, luxuriant, rough, shaggy, spreading, sticking out, tangled, thick, thick-growing, unruly, untidy.

business *n* **1** affair, concern, duty, function, issue, matter, obligation, problem, question, responsibility, subject, task, topic. **2** calling, career, craft, employment, industry, job, line of work, occupation, profession, pursuit, trade, vocation, work. **3** buying and selling, commerce, dealings, industry, marketing, merchandising, selling, trade, trading, transactions. **4** company, concern, corporation, enterprise, establishment, firm, organization, *inf* outfit, partnership, practice, *inf* set-up, venture.

businesslike *adj* careful, efficient, hard-headed, logical, methodical, neat, orderly, practical, professional, prompt, systematic, well-organized. *Opp* DISORGANIZED.

businessman, businesswoman *ns* dealer, entrepreneur, executive, financier, industrialist, magnate, manager, merchant, trader, tycoon.

bustle *n* activity, agitation, commotion, excitement, flurry, fuss, haste, hurly-burly, hurry, hustle, movement, restlessness, scurry, stir, *inf* to-do, *inf* toing and froing. • *vb* dart, dash, fuss, hasten, hurry, hustle, make haste, move busily, rush, scamper, scramble, scurry, scuttle, *inf* tear, whirl.

busy *adj* **1** active, assiduous, bustling about, committed, dedicated, diligent, employed, energetic, engaged, engrossed, *inf* hard at it, immersed, industrious, involved, keen, occupied, *inf* on the go, pottering, preoccupied, slaving, *inf* tied up, tireless, *inf* up to your eyes, working. *Opp* IDLE. **2** *busy shops*. bustling, frantic, full, hectic, lively.

busybody *n* gossip, meddler, *inf* Nosey Parker, scandalmonger, snooper, spy. **be a busybody** ▷ INTERFERE.

butt *n* **1** haft, handle, shaft, stock. **2** *water butt*. barrel, cask, water-butt. **3** *cigar butt*. end, remains, remnant, stub. **4** *butt of ridicule*. end, mark, object, subject, target, victim. • *vb* buffet, bump, jab, knock, poke, prod, punch, push, ram, shove, strike, thump. ▷ HIT. **butt in** ▷ INTERRUPT.

buttocks *n vulg* arse, backside, behind, bottom, *vulg* bum, *Amer* butt, *Fr* derrière, fundament, haunches, hindquarters, *joc* posterior, rear, rump, seat.

buttress *n* pier, prop, support. • *vb* brace, prop up, reinforce, shore up, strengthen, support.

buxom *adj* ample, bosomy, *vulg* chesty, full-figured, healthy-looking, plump, robust, rounded, voluptuous. *Opp* THIN.

buy *vb* acquire, come by, gain, get, get on hire purchase, *inf* invest in, obtain, pay for, procure, purchase. *Opp* SELL.

buyer *n* client, consumer, customer, puchaser, shopper.

bypass *vb* avoid, circumvent, dodge, evade, find a way round, get out of, go round, ignore, neglect, omit, sidestep, skirt.

by-product *n* adjunct, complement, consequence, corollary, repercussion, result, side-effect.

bystander *n* eyewitness, looker-on, observer, onlooker, passer-by, spectator, watcher, witness.

C

cabin *n* **1** bothy, chalet, cottage, hut, lodge, shack, shanty, shed, shelter. **2** *cabin on a ship.* berth, compartment, deck-house, quarters.

cable *n* **1** chain, cord, flex, guy, hawser, lead, line, mooring, rope, wire. **2** *news by cable.* message, telegram, wire.

cacophonous *adj* atonal, discordant, dissonant, harsh, noisy, unmusical. *Opp* HARMONIOUS.

cacophony *n* atonality, caterwauling, din, discord, disharmony, dissonance, harshness, jangle, noise, racket, row, rumpus, tumult. *Opp* HARMONY.

cadence *n* accent, beat, inflection, intonation, lilt, metre, pattern, rhythm, rise and fall, sound, stress, tune.

cadet *n* beginner, learner, recruit, tiro, trainee.

cadge *vb* ask, beg, scrounge, sponge.

café *n* bar, bistro, brasserie, buffet, cafeteria, canteen, coffee bar, coffee house, coffee shop, diner, restaurant, snack-bar, take-away, tea-room, tea-shop.

cage *n* aviary, coop, enclosure, hutch, pen, pound. • *vb* ▷ CONFINE.

cajole *vb inf* butter up, coax, flatter, inveigle, persuade, seduce.

cake *n* **1** bun, gateau. **2** *cake of soap.* bar, block, chunk, cube, loaf, lump, mass, piece, slab. • *vb* **1** coat, clog, cover, encrust, make dirty, make muddy. **2** coagulate, congeal, consolidate, dry, harden, solidify, thicken.

calamitous *adj* awful, cataclysmic, catastrophic, deadly, devastating, dire, disastrous, distressful, dreadful, fatal, ghastly, ruinous, serious, terrible, tragic, unfortunate, unlucky, woeful.

calamity *n* accident, affliction, cataclysm, catastrophe, disaster, misadventure, mischance, misfortune, mishap, tragedy, tribulation.

calculate *vb* add up, ascertain, assess, compute, count, determine, do sums, enumerate, estimate, evaluate, figure out, find out, gauge, judge, reckon, total, value, weigh, work out. **calculated** ▷ DELIBERATE. **calculating** ▷ CRAFTY.

calibre *n* **1** bore, diameter, gauge, measure, size. **2** ability, capability, capacity, character, competence, distinction, excellence, genius, gifts, importance, merit, proficiency, quality, skill, stature, talent, worth.

call *n* 1 bellow, cry, exclamation, roar, scream, shout, yell. 2 bidding, invitation, signal, summons. 3 *social call.* stay, stop, visit. 4 *no call for it.* cause, demand, excuse, justification, need, occasion, request, requirement. • *vb* 1 bellow, clamour, cry out, exclaim, hail, roar, shout, yell. 2 *call on friends.* drop in, socialize, visit. 3 *called her 'Jane'.* baptize, christen, dub, name. 4 *play called 'Lear'.* entitle, title. 5 *call me at 7.* arouse, awaken, get someone up, rouse, wake, waken. 6 *call a meeting.* convene, gather, invite, order, summon. 7 *call by phone.* contact, dial, phone, ring, telephone. **call for** ▷ FETCH, REQUEST. **call off** ▷ CANCEL. **call someone names** ▷ INSULT.

calligraphy *n* copperplate, handwriting, illumination, lettering, penmanship, script.

calling *n* business, career, employment, job, line of work, métier, occupation, profession, pursuit, trade, vocation, work.

callous *adj* apathetic, cold, cold-hearted, cool, dispassionate, hard-bitten, *inf* hard-boiled, hardened, hard-hearted, *inf* hard-nosed, heartless, inhuman, insensitive, merciless, pitiless, ruthless, *inf* thick-skinned, uncaring, unconcerned, unemotional, unfeeling, unsympathetic. ▷ CRUEL. *Opp* SENSITIVE.

callow *adj* adolescent, *inf* born yesterday, *inf* green, immature, inexperienced, innocent, juvenile, naïve, raw, unsophisticated, *inf* wet behind the ears, young. *Opp* MATURE.

calm *adj* 1 airless, even, flat, glassy, *poet* halcyon (*days*), like a millpond, motionless, placid, quiet, slow-moving, smooth, still, unclouded, unwrinkled, windless. 2 collected, *derog* complacent, composed, controlled, cool, dispassionate, equable, impassive, imperturbable, *inf* laid-back, level-headed, moderate, pacific, passionless, patient, peaceful, poised, quiet, relaxed, restful, restrained, sedate, self-possessed, sensible, serene, tranquil, undemonstrative, unemotional, unexcitable, *inf* unflappable, unhurried, unperturbed, unruffled, untroubled. *Opp* EXCITABLE, STORMY. • *n* flat sea, peace, quietness, stillness, tranquillity. ▷ CALMNESS. • *vb* appease, compose, control, cool, lull, mollify, pacify, placate, quieten, sedate, settle down, smooth, sober down, soothe, tranquillize. *Opp* DISTURB.

calmness *n derog* complacency, composure, equability, equanimity, imperturbability, level-headedness, peace of mind, sangfroid, self-possession, serenity, *inf* unflappability. *Opp* ANXIETY, EXCITEMENT.

camouflage *n* blind, cloak, concealment, cover, disguise, façade, front, guise, mask, pretence, protective colouring, screen, veil. • *vb* cloak, conceal, cover up, disguise, hide, mask, obscure, screen, veil.

camp *n* bivouac, camping-ground, campsite, encampment, settlement.

campaign *n* action, battle, crusade, drive, effort, fight, manoeuvre, movement, offensive, operation, push, struggle, war.

campus *n* grounds, setting, site.

canal *n* channel, waterway.

cancel *vb* abandon, abolish, abort, abrogate, annul, call off, countermand, cross out, delete, drop, eliminate, erase, expunge, frank (*stamps*), give up, invalidate, override, overrule, postpone, quash, repeal, repudiate, rescind, revoke, scrap, *inf* scrub, wipe out, write off. **cancel out** ▷ NEUTRALIZE.

cancer *n* canker, carcinoma, growth, malignancy, melanoma, tumour.

candid *adj* blunt, direct, fair, forthright, frank, honest, ingenuous, just, *inf* no-nonsense, objective, open, out-spoken, plain, sincere, straight, straightforward, transparent, true, truthful, unbiased, undisguised, unequivocal, unflattering, unprejudiced. *Opp* INSINCERE.

candidate *n* applicant, aspirant, competitor, contender, contestant, entrant, nominee, *inf* possibility, pretender (*to throne*), runner, suitor.

cane *n* bamboo, rod, stick. • *vb* ▷ THRASH.

canoe *n* dug-out, kayak.

canopy *n* awning, cover, covering, shade, shelter, umbrella.

canvass *n* campaign, census, enquiry, examination, investigation, market research, opinion poll, poll, probe, scrutiny, survey. • *vb* ask for, campaign, *inf* drum up support, electioneer, seek, solicit.

canyon *n* defile, gap, gorge, gulch, pass, ravine, valley.

cap *n* covering, lid, top. ▷ HAT. • *vb* ▷ COVER.

capable *adj* able, accomplished, adept, clever, competent, effective, effectual, efficient, experienced, expert, gifted, *inf* handy, intelligent, masterly, practised, proficient, qualified, skilful, skilled, talented, trained. *Opp* INCAPABLE. **capable of** apt to, disposed to, equal to, liable to.

capacity *n* **1** content, dimensions, magnitude, room, size, volume. **2** ability, acumen, capability, cleverness, competence, intelligence, potential, power, skill, talent, wit. **3** *in an official capacity*. appointment, duty, function, job, office, place, position, post, province, responsibility, role.

cape *n* **1** cloak, coat, cope, mantle, robe, shawl, wrap. **2** head, headland, peninsula, point, promontory.

caper *vb* bound, cavort, dance, frisk, frolic, gambol, hop, jig about, jump, leap, play, prance, romp, skip, spring.

capital *adj* **1** chief, controlling, first, foremost, important, leading, main, paramount, pre-eminent, primary, principal. **2** *capital letters*. big, block, initial, large, upper-case. **3** ▷ EXCELLENT. • *n* **1** chief city, centre of government. **2** assets, cash, finance, funds, investments, money, principal, property, *sl* (the) ready, resources, riches, savings, stock, wealth, *inf* the wherewithal.

capitulate *vb* acquiesce, be defeated, concede, desist, fall, give in, relent, submit, succumb, surrender, *inf* throw in the towel, yield.

capricious *adj* changeable, erratic, fanciful, fickle, fitful, flighty, impulsive, inconstant, mercurial, moody, quirky, uncertain, unpredictable, unreliable, unstable, variable, wayward, whimsical. *Opp* STEADY.

capsize *vb* flip over, invert, keel over, overturn, tip over, turn over *inf* turn turtle, turn upside down.

capsule *n* lozenge, medicine, pill, tablet.

captain *n* **1** boss, chief, head, leader. **2** commander, master, officer in charge, pilot, skipper.

caption *n* description, explanation, heading, headline, superscription, title.

captivate *vb* attract, beguile, bewitch, charm, delight, enamour, enchant, enrapture, enslave, ensnare, enthral, entrance, fascinate, hypnotize, infatuate, mesmerize, seduce, *inf* steal your heart, *inf* turn your head, win. *Opp* DISGUST.

captive *adj* caged, captured, chained, confined, detained, enslaved, ensnared, fettered, gaoled, imprisoned, incarcerated, jailed, restricted, secure, taken prisoner, *inf* under lock and key. *Opp* FREE. • *n* convict, detainee, hostage, internee, prisoner, slave.

captivity *n* bondage, confinement, custody, detention, duress, imprisonment, incarceration, internment, protective custody, remand, restraint, servitude, slavery. ▷ PRISON. *Opp* FREEDOM.

capture *n* apprehension, arrest, seizure. • *vb* apprehend, arrest, *inf* bag, bind, catch, *inf* collar, corner, ensnare, entrap, *inf* get, *inf* nab, net, *inf* nick, overpower, secure, seize, snare, take prisoner, trap. ▷ CONQUER. *Opp* LIBERATE.

car *n* automobile, *inf* banger, *joc* bus, *joc* jalopy, motor, motor car, *sl* wheels. □ *cab, convertible, coupé, Dormobile, estate, fastback, hatchback, jeep, Land Rover, limousine, Mini, panda car, patrol car, police car, saloon, shooting brake, sports car, taxi, tourer.* ▷ VEHICLE.

carcass *n* **1** body, cadaver, corpse, meat, remains. **2** *carcass of a car.* framework, hulk, remains, shell, skeleton, structure.

card *n* cardboard, pasteboard. □ *bank card, birthday card, business card, calling card, credit card, get-well card, greetings card, identity card, invitation, membership card, notelet, picture postcard, playing-card, postcard, union card, Valentine, visiting card.*

care *n* **1** attention, carefulness, caution, circumspection, concentration, concern, diligence, exactness, forethought, heed, interest, meticulousness, pains, prudence, solicitude, thoroughness, thought, vigilance, watchfulness. **2** anxiety, burden, concern, difficulty, hardship, problem, responsibility, sorrow, stress, tribulation, vexation, woe, worry. ▷ TROUBLE. **3** *left in my care.* charge, control, custody, guardianship, keeping, management, protection, safekeeping, ward. • *vb* be troubled, bother, concern yourself, mind, worry. **care for** ▷ LOVE, TEND.

career *n* business, calling, craft, employment, job, livelihood, living, métier, occupation, profession, trade, vocation, work. • *vb* ▷ RUSH.

carefree *adj* **1** blasé, casual, cheery, contented, debonair, easy, easy-going, happy-go-lucky, indifferent, insouciant, *inf* laid-back, light-hearted, nonchalant, relaxed, unconcerned, unworried. ▷ HAPPY. **2** *carefree holiday.* leisured, peaceful, quiet, relaxing, restful, trouble-free, untroubled. *Opp* ANXIOUS.

careful *adj* **1** alert, attentive, cautious, chary, circumspect, heedful, mindful, observant, prudent, solicitous, thoughtful, vigilant, wary, watchful. **2** *careful work*. accurate, conscientious, deliberate, diligent, exhaustive, fastidious, *derog* fussy, judicious, methodical, meticulous, neat, orderly, organized, painstaking, particular, precise, punctilious, responsible, rigorous, scrupulous, systematic, thorough, well-organized. *Opp* CARELESS. **be careful** ▷ BEWARE.

careless *adj* **1** absent-minded, heedless, ill-considered, imprudent, inattentive, incautious, inconsiderate, irresponsible, negligent, rash, reckless, thoughtless, uncaring, unguarded, unthinking, unwary. **2** *careless work*. casual, confused, cursory, disorganized, hasty, imprecise, inaccurate, jumbled, messy, perfunctory, scatter-brained, shoddy, slapdash, slipshod, *inf* sloppy, slovenly, thoughtless, untidy. *Opp* CAREFUL.

carelessness *n* haste, inattention, irresponsibility, negligence, recklessness, *inf* sloppiness, slovenliness, thoughtlessness, untidiness. *Opp* CARE.

caress *vb* cuddle, embrace, fondle, hug, kiss, make love to, *sl* neck with, nuzzle, pat, pet, rub against, smooth, stroke, touch.

caretaker *n* custodian, janitor, keeper, porter, superintendent, warden, watchman.

careworn *adj* gaunt, grim, haggard. ▷ WEARY.

cargo *n* consignment, freight, goods, lading (*bill of lading*), load, merchandise, payload, shipment.

caricature *n* burlesque, cartoon, parody, satire, *inf* send-up, spoof, *inf* take-off, travesty. • *vb* burlesque, distort, exaggerate, imitate, lampoon, make fun of, mimic, mock, overact, overdo, parody, ridicule, satirize, *inf* send up, *inf* take off.

caring *n* concern, kindness, nursing, solicitude.

carnage *n* blood-bath, bloodshed, butchery, havoc, holocaust, killing, massacre, pogrom, shambles, slaughter.

carnal *adj* animal, bodily, erotic, fleshly, natural, physical, sensual, sexual. ▷ LUSTFUL. *Opp* SPIRITUAL.

carnival *n* celebration, fair, festival, festivity, fête, fiesta, fun and games, gala, jamboree, merrymaking, pageant, parade, procession, revelry, show, spectacle.

carp *vb* cavil, find fault, *inf* go on, *inf* gripe, grumble, object, pick holes, quibble, *inf* split hairs, whinge. ▷ COMPLAIN.

carpentry *n* joinery, woodwork.

carriage *n* **1** coach. ▷ VEHICLE. **2** bearing, comportment, demeanour, gait, manner, mien, posture, presence, stance.

carrier *n* **1** bearer, conveyor, courier, delivery-man, delivery-woman, dispatch rider, errand-boy, errand-girl, haulier, messenger, porter, postman, runner. **2** *carrier of a disease*. contact, host, transmitter.

carry *vb* **1** bring, *inf* cart, communicate, ferry, fetch, haul, lead, lift, *inf* lug, manhandle, move, relay, remove, ship, shoulder, take, transfer, transmit, transport. ▷ CONVEY. **2** *carry weight*. bear, hold up, maintain, support. **3** *carry a penalty*. demand, entail, involve, lead to, occasion, require, result in. **carry on** ▷ CONTINUE. **carry out** ▷ DO.

cart *n* barrow, dray, truck, wagon, wheelbarrow. • *vb* ▷ CARRY.

carton *n* box, cartridge, case, container, pack, package, packet.

cartoon *n* animation, caricature, comic strip, drawing, sketch.

cartridge *n* 1 canister, capsule, case, cassette, container, cylinder, tube. 2 *cartridge for a gun.* magazine, round, shell.

carve *vb* 1 slice. ▷ CUT. 2 *carve stone. inf* chip away at, chisel, engrave, fashion, hew, incise, sculpture, shape.

cascade *n* cataract, deluge, flood, gush, torrent, waterfall. • *vb* ▷ POUR.

case *n* 1 box, cabinet, carton, casket, chest, container, crate, pack, packaging, suitcase, trunk. ▷ LUGGAGE. 2 *case of mistaken identity.* example, illustration, instance, occurrence, specimen, state of affairs. 3 *rules don't apply in his case.* circumstances, condition, context, plight, predicament, situation, state. 4 *a legal case.* action, argument, cause, dispute, inquiry, investigation, lawsuit, suit.

cash *n* banknotes, bills, change, coins, currency, *inf* dough, funds, hard money, legal tender, money, notes, *inf* (the) ready, *inf* the wherewithal. • *vb* exchange for cash, realize, sell. **cash in on** ▷ PROFIT.

cashier *n* accountant, banker, check-out person, clerk, teller, treasurer. • *vb* ▷ DISMISS.

cask *n* barrel, butt, hogshead, tub, tun, vat.

cast *n* 1 ▷ SCULPTURE. 2 *cast of a play.* characters, company, dramatis personae, performers, players, troupe. • *vb* 1 bowl, chuck, drop, fling, hurl, impel, launch, lob, pelt, pitch, project, scatter, shy, sling, throw, toss. 2 *cast a sculpture.* form, found, mould, shape. ▷ SCULPTURE. **cast off** ▷ SHED, UNTIE.

castaway *adj* abandoned, deserted, exiled, marooned, rejected, shipwrecked, stranded.

caste *n* class, degree, estate, grade, level, position, rank, standing, station, status, stratum.

castigate *vb* censure, chasten, chastise, *old use* chide, correct, discipline, lash, punish, rebuke, reprimand, scold, *inf* tell off. ▷ CRITICIZE.

castle *n* château, citadel, fort, fortress, mansion, palace, stately home, stronghold, tower.

castrate *vb* emasculate, geld, neuter, spay, sterilize, unsex.

casual *adj* 1 accidental, chance, erratic, fortuitous, incidental, irregular, promiscuous, random, serendipitous, sporadic, unexpected, unforeseen, unintentional, unplanned, unpremeditated, unstructured, unsystematic. *Opp* DELIBERATE. 2 *casual attitude.* apathetic, blasé, careless, *inf* couldn't-care-less, easy-going, *inf* free-and-easy, lackadaisical, *inf* laid-back, lax, negligent, nonchalant, offhand, relaxed, *inf* slap-happy, *inf* throwaway, unconcerned, unenthusiastic, unimportant, unprofessional. *Opp* ENTHUSIASTIC. 3 *casual clothes.* comfortable, informal. *Opp* FORMAL.

casualty *n* dead person, death, fatality, injured person, injury, loss, victim, wounded person.

cat *n* kitten, *inf* moggy, *inf* pussy, tabby, tom, tomcat.

catacombs *n* crypt, sepulchre, tomb, underground passage, vault.

catalogue *n* brochure, directory, index, inventory, list, record, register, roll, schedule, table. • *vb* classify, codify, file, index, list, make an inventory of, record, register, tabulate.

catapult *vb* fire, fling, hurl, launch. ▷ THROW.

cataract *n* cascade, falls, rapids, torrent, waterfall.

catastrophe *n* blow, calamity, cataclysm, crushing blow, *débâcle, devastation, disaster, fiasco, holocaust, mischance, mishap, ruin, ruination, tragedy, upheaval.* ▷ MISFORTUNE.

catch *n* **1** bag, booty, capture, haul, net, prey, prize, take. **2** *suspected a catch.* difficulty, disadvantage, drawback, obstacle, problem, snag, trap, trick. **3** *catch on a door.* bolt, clasp, clip, fastener, fastening, hasp, hook, latch, lock. • *vb* **1** clutch, ensnare, entrap, grab, grasp, grip, hang on to, hold, hook, net, seize, snare, snatch, take, tangle, trap. **2** *catch a thief.* apprehend, arrest, capture, *inf* cop, corner, detect, discover, expose, intercept, *inf* nab, *inf* nobble, stop, surprise, take by surprise, unmask. **3** *caught me unawares.* come upon, discover, find, surprise. **4** *catch a bus.* be in time for, get on. **5** *catch a cold.* become infected by, contract, get. **catch on** ▷ SUCCEED, UNDERSTAND. **catch-phrase** ▷ SAYING. **catch-22** ▷ DILEMMA. **catch up** ▷ OVERTAKE.

catching *adj* communicable, contagious, infectious, spreading, transmissible, transmittable.

catchy *adj* attractive, haunting, memorable, popular, singable, tuneful.

categorical *adj* absolute, authoritative, certain, complete, decided, decisive, definite, direct, dogmatic, downright, emphatic, explicit, express, firm, forceful, *inf* out-and-out, positive, strong, total, unambiguous, unconditional, unequivocal, unmitigated, unqualified, unreserved, utter, vigorous. *Opp* TENTATIVE.

category *n* class, classification, division, grade, group, head, heading, kind, order, rank, ranking, section, sector, set, sort, type, variety.

cater *vb* cook, make arrangements, minister, provide, provision, serve, supply.

catholic *adj* all-embracing, all-inclusive, broad, broad-minded, comprehensive, cosmopolitan, eclectic, general, liberal, universal, varied, wide, wide-ranging.

cattle *plur n* beef, bullocks, bulls, calves, cows, heifers, livestock, oxen, steers, stock.

catty *adj inf* bitchy, ill-natured, malevolent, malicious, mean, nasty, rancorous, sly, spiteful, venomous, vicious. ▷ UNKIND. *Opp* KIND.

cause *n* **1** basis, beginning, genesis, grounds, motivation, motive, occasion, origin, reason, root, source, spring, stimulus. **2** agent, author, *old use* begetter, creator, initiator, inspiration, inventor, originator, producer. **3** *cause of his lateness.* excuse, explanation, pretext, reason. **4** *a good cause.* aim, belief, concern, end, ideal, object, purpose, undertaking. • *vb* **1** arouse, awaken, begin, bring about, bring on, create, effect, effectuate, engender, foment, generate, give rise to, incite, kindle, lead to, occasion, precipitate, produce, provoke, result in, set off, spark off, stimulate, trigger off, *inf* whip up. **2** compel, force, induce, motivate.

caustic *adj* **1** acid, astringent, burning, corrosive, destructive. **2** *caustic criticism.* acidulous, acrimonious, biting, bitter, critical, cutting, mordant, pungent, sarcastic, scathing, severe, sharp, stinging, trenchant, virulent, waspish. *Opp* MILD.

caution *n* **1** alertness, attentiveness, care, carefulness, circumspection, discretion, forethought, heed, heedfulness, prudence, vigilance, wariness, watchfulness. **2** *let off with a caution.* admonition, caveat, *inf* dressing-down, injunction, reprimand, *inf* talking-to, *inf* ticking-off, warning. • *vb* **1** advise, alert, counsel, forewarn, inform, *inf* tip off, warn. **2** *cautioned by the police.* admonish, censure, give a warning, reprehend, reprimand, *inf* tell off, *inf* tick off.

cautious *adj* **1** alert, attentive, careful, heedful, prudent, scrupulous, vigilant, watchful. **2** *cautious comments. inf* cagey, calculating, chary, circumspect, deliberate, discreet, gingerly, grudging, guarded, hesitant, judicious, non-committal, restrained, suspicious, tactful, tentative, unadventurous, wary, watchful. *Opp* RECKLESS.

cavalcade *n* march-past, parade, procession, spectacle, troop.

cave *n* cavern, cavity, den, grotto, hole, pothole, underground chamber. • *vb* **cave in** ▷ COLLAPSE, SURRENDER.

cavity *n* cave, crater, dent, hole, hollow, pit.

cease *vb* break off, call a halt, conclude, cut off, desist, discontinue, end, finish, halt, *inf* kick (*a habit*), *inf* knock off, *inf* lay off, leave off, *inf* pack in, *inf* pack up, refrain, stop, terminate. *Opp* BEGIN.

ceaseless *adj* chronic, constant, continual, continuous, endless, everlasting, incessant, interminable, never-ending, non-stop, permanent, perpetual, persistent, relentless, unending, unremitting, untiring. *Opp* INTERMITTENT, TEMPORARY.

celebrate *vb* **1** be happy, have a celebration, let yourself go, *inf* live it up, make merry, *inf* paint the town red, rejoice, revel, *old use* wassail. **2** *celebrate an anniversary.* commemorate, hold, honour, keep, observe, remember. **3** *celebrate a wedding.* officiate at, solemnize. **celebrated** ▷ FAMOUS.

celebration *n* banquet, binge, carnival, commemoration, feast, festivity, *inf* jamboree, *joc* jollification, merry-making, observance, *inf* orgy, party, *inf* rave-up, revelry, *church* service, *inf* shindig, solemnization. □ *anniversary, birthday, festival, fête, gala, jubilee, remembrance, reunion, wedding.*

celebrity *n* **1** ▷ FAME. **2** big name, *inf* bigwig, dignitary, famous person, idol, notability, personality, public figure, star, superstar, VIP, worthy.

celestial *adj* **1** astronomical, cosmic, galactic, interplanetary, interstellar, starry, stellar, universal. **2** *celestial beings.* angelic, blissful, divine, ethereal, godlike, heavenly, seraphic, spiritual, sublime, supernatural, transcendental, visionary.

celibacy *n* bachelorhood, chastity, continence, purity, self-restraint, spinsterhood, virginity.

celibate *adj* abstinent, chaste, continent, immaculate, single, unmarried, unwedded, virgin. • *n* bachelor, spinster, virgin.

cell *n* cavity, chamber, compartment, cubicle, den, enclosure, living space, prison, room, space, unit.

cellar *n* basement, crypt, vault, wine-cellar.

cemetery *n* burial-ground, churchyard, graveyard, necropolis.

censor *vb* amend, ban, bowdlerize, *inf* clean up, cut, edit, exclude, expurgate, forbid, prohibit, remove.

censorious *adj* fault-finding, *inf* holier-than-thou, judgemental, moralistic, Pharisaical, self-righteous. ▷ CRITICAL.

censure *n* accusation, admonition, blame, castigation, condemnation, criticism, denunciation, diatribe, disapproval, *inf* dressing-down, harangue, rebuke, reprimand, reproach, reprobation, reproof, *inf* slating, stricture, *inf* talking-to, *inf* telling-off, tirade, verbal attack, vituperation. • *vb* admonish, berate, blame, *inf* carpet, castigate, caution, chide, condemn, criticize, denounce, lecture, rebuke, reproach, reprove, scold, take to task, *sl* tear (someone) off a strip, *inf* tell off, *inf* tick off, upbraid.

census *n* count, survey, tally.

central *adj* **1** focal, inner, innermost, interior, medial, middle. **2** *central facts.* chief, crucial, essential, fundamental, important, key, main, major, overriding, pivotal, primary, principal, vital. *Opp* PERIPHERAL.

centralize *vb* amalgamate, bring together, concentrate, rationalize, streamline, unify. *Opp* DISPERSE.

centre *n* bull's-eye, core, focal point, focus, heart, hub, inside, interior, kernel, middle, midpoint, nucleus, pivot. *Opp* PERIMETER. • *vb* concentrate, converge, focus.

centrifugal *adj* dispersing, diverging, moving outwards, scattering, spreading. *Opp* CENTRIPETAL.

centripetal *adj* converging. *Opp* CENTRIFUGAL.

cereal *n* corn, grain. □ *barley, corn on the cob, maize, millet, oats, rice, rye, sweet corn, wheat.*

ceremonial *adj* celebratory, dignified, liturgical, majestic, official, ritual, ritualistic, solemn, stately. ▷ FORMAL. *Opp* INFORMAL.

ceremonious *adj* civil, courteous, courtly, dignified, formal, grand, *derog* pompous, proper, punctilious, *derog* starchy. ▷ POLITE. *Opp* CASUAL.

ceremony *n* **1** celebration, commemoration, *inf* do, event, formal occasion, function, occasion, parade, reception, rite, ritual, service, solemnity. **2** ceremonial, decorum, etiquette, formality, grandeur, pageantry, pomp, pomp and circumstance, protocol, ritual, spectacle.

certain *adj* **1** adamant, assured, confident, constant, convinced, decided, determined, firm, invariable, positive, resolved, satisfied, settled, stable, steady, sure, undoubting, unshakable, unwavering. **2** *certain proof.* absolute, authenticated, categorical, certified, clear, clear-cut, conclusive, convincing, definite, dependable, established, genuine, guaranteed, incontestable, incontrovertible, indubitable, infallible, irrefutable, known, official, plain, reliable, settled, sure, true, trustworthy, unarguable, undeniable, undisputed, undoubted, unmistakable, unquestionable, valid, verifiable. **3** *certain disaster.* destined, fated, guaranteed, imminent, inescapable, inevitable, inexorable, predestined, predictable, unavoidable.

4 *certain to pay up*. bound, compelled, obliged, required, sure. 5 *certain people*. individual, particular, some, specific, unnamed, unspecified. *Opp* UNCERTAIN. **be certain** ▷ KNOW. **for certain** ▷ DEFINITELY. **make certain** ▷ ENSURE.

certainty *n* 1 actuality, certain fact, *inf* foregone conclusion, foreseeable outcome, inevitability, necessity, *inf* sure thing. 2 assertiveness, assurance, authority, certitude, confidence, conviction, knowledge, positiveness, proof, sureness, truth, validity. *Opp* DOUBT.

certificate *n* authorization, award, credentials, degree, diploma, document, guarantee, licence, pass, permit, qualification, warrant.

certify *vb* 1 affirm, asseverate, attest, authenticate, aver, avow, bear witness, confirm, declare, endorse, guarantee, notify, sign, swear, testify, verify, vouch, vouchsafe, warrant, witness. 2 *certify as competent*. authorize, charter, commission, franchise, license, recognize, validate.

chain *n* 1 bonds, coupling, fetters, handcuffs, irons, links, manacles, shackles. 2 *chain of events*. column, combination, concatenation, cordon, line, progression, row, sequence, series, set, string, succession, train. • *vb* bind, clap in irons, fetter, handcuff, link, manacle, shackle, tether, tie. ▷ FASTEN.

chair *n* armchair, carver, deckchair, dining-chair, easy chair, recliner, rocking-chair, throne. ▷ SEAT. • *vb* ▷ PRESIDE.

chairperson *n* chair, chairman, chairwoman, convenor, director, leader, moderator, organizer, president, speaker.

challenge *vb* 1 accost, confront, *inf* have a go at, take on, tax. 2 *challenge to duel*. dare, defy, *old use* demand satisfaction, provoke, summon. 3 *challenge a decision*. argue against, call in doubt, contest, dispute, dissent from, impugn, object to, oppose, protest against, query, question, take exception to.

challenging *adj* inspiring, stimulating, testing, thought-provoking, worthwhile. ▷ DIFFICULT. *Opp* EASY.

chamber *n* cavity, cell, compartment, niche, nook, space. ▷ ROOM.

champion *adj* great, leading, record-breaking, supreme, top, unrivalled, victorious, winning, world-beating. • *n* 1 conqueror, hero, medallist, prize-winner, record-breaker, superman, superwoman, titleholder, victor, winner. 2 *champion of the poor*. backer, defender, guardian, patron, protector, supporter, upholder, vindicator. 3 [*old use*] *champion in lists*. challenger, contender, contestant, fighter, knight, warrior. • *vb* ▷ SUPPORT.

championship *n* competition, contest, series, tournament.

chance *adj* accidental, adventitious, casual, coincidental, *inf* fluky, fortuitous, fortunate, haphazard, inadvertent, incidental, lucky, random, unexpected, unforeseen, unfortunate, unlooked-for, unplanned, unpremeditated. *Opp* DELIBERATE. • *n* 1 accident, coincidence, destiny, fate, fluke, fortune, gamble, hazard, luck, misfortune, serendipity. 2 *chance of rain*. danger, liability, likelihood, possibility, probability, prospect, risk. 3 occasion, opportunity, time, turn. • *vb* 1 ▷ RISK. 2 ▷ HAPPEN.

chancy *adj* dangerous, *inf* dicey, *inf* dodgy, hazardous, *inf* iffy, insecure, precarious, risky, speculative, ticklish, tricky, uncertain, unpredictable, unsafe. *Opp* SAFE.

change *n* 1 adaptation, adjustment, alteration, break, conversion, deterioration, development, difference, diversion, improvement, innovation, metamorphosis, modification, modulation, mutation, new look, rearrangement, refinement, reformation, reorganization, revolution, shift, substitution, swing, transfiguration, transformation, transition, translation, transmogrification, transmutation, transposition, *inf* turn-about, U-turn, variation, variety, vicissitude. 2 *small change*. ▷ CASH. • *vb* 1 acclimatize, accommodate, accustom, adapt, adjust, affect, alter, amend, convert, diversify, influence, modify, process, rearrange, reconstruct, refashion, reform, remodel, reorganize, reshape, restyle, tailor, transfigure, transform, translate, transmogrify, transmute, vary. 2 *opinions change*. alter, be transformed, *inf* chop and change, develop, fluctuate, metamorphose, move on, mutate, shift, vary. 3 *change one thing for another*. alternate, displace, exchange, replace, substitute, switch, swop, transpose. 4 *change money*. barter, convert, trade in. **change into** ▷ BECOME. **change someone's mind** ▷ CONVERT. **change your mind** ▷ RECONSIDER.

changeable *adj* capricious, chequered (*career*), erratic, fickle, fitful, fluctuating, fluid, inconsistent, inconstant, irregular, mercurial, mutable, protean, shifting, temperamental, uncertain, unpredictable, unreliable, unsettled, unstable, unsteady, *inf* up and down, vacillating, variable, varying, volatile, wavering. *Opp* CONSTANT.

channel *n* 1 aqueduct, canal, conduit, course, dike, ditch, duct, groove, gully, gutter, moat, overflow, pipe, sluice, sound, strait, trench, trough, watercourse, waterway. ▷ STREAM. 2 avenue, means, medium, path, route, way. 3 *TV channel*. *inf* side, station, waveband, wavelength. • *vb* conduct, convey, direct, guide, lead, pass on, route, send, transmit.

chant *n* hymn, plainsong, psalm. ▷ SONG. • *vb* intone. ▷ SING.

chaos *n* anarchy, bedlam, confusion, disorder, disorganization, lawlessness, mayhem, muddle, pandemonium, shambles, tumult, turmoil. *Opp* ORDER.

chaotic *adj* anarchic, confused, deranged, disordered, disorderly, disorganized, haphazard, *inf* haywire, *inf* higgledy-piggledy, jumbled, lawless, muddled, rebellious, riotous, *inf* shambolic, *inf* topsy-turvy, tumultuous, uncontrolled, ungovernable, unruly, untidy, *inf* upside-down. *Opp* ORDERLY.

char *vb* blacken, brown, burn, carbonize, scorch, sear, singe.

character *n* 1 distinctiveness, flavour, idiosyncracy, individuality, integrity, peculiarity, quality, stamp, taste, uniqueness. ▷ CHARACTERISTIC. 2 *a forceful character*. attitude, constitution, disposition, individuality, make-up, manner, nature, personality, reputation, temper, temperament. 3 *a famous character*. figure, human being, individual, person, personality, *inf* type. 4 *She's a* *character! inf* case, comedian, comic, eccentric, *inf* nut-case, oddity, *derog* weirdo. **5** *character in a play*. part, persona, portrayal, role. **6** *written characters*. cipher, figure, hieroglyphic, ideogram, letter, mark, rune, sign, symbol, type.

characteristic *adj* **1** [*of an individual*] distinctive, distinguishing, essential, idiosyncratic, individual, particular, peculiar, recognizable, singular, special, specific, symptomatic, unique. **2** [*of a kind*] representative, typical. • *n* attribute, distinguishing feature, feature, hallmark, idiosyncracy, mark, peculiarity, property, quality, symptom, trait.

characterize *vb* brand, delineate, depict, describe, differentiate, distinguish, draw, identify, individualize, mark, portray, present, recognize, typify.

charade *n* absurdity, deceit, deception, fabrication, farce, make-believe, masquerade, mockery, *inf* play-acting, pose, pretence, *inf* put-up job, sham.

charge *n* **1** cost, expenditure, expense, fare, fee, payment, postage, price, rate, terms, toll, value. **2** *in my charge*. care, command, control, custody, guardianship, jurisdiction, keeping, protection, responsibility, safe-keeping, supervision, trust. **3** *criminal charges*. accusation, allegation, imputation, indictment. **4** *cavalry charge*. action, assault, attack, drive, incursion, invasion, offensive, onslaught, raid, rush, sally, sortie, strike. • *vb* **1** ask for, debit, exact, levy, make you pay, require. **2** accuse, blame, impeach, indict, prosecute, tax. **3** *charge with a duty*. burden, commit, empower, entrust, give, impose on. **4** *charged us to do our best*. ask, command, direct, enjoin, exhort, instruct. **5** *charge an enemy*. assail, assault, attack, *inf* fall on, rush, set on, storm, *inf* wade into.

charitable *adj* bountiful, generous, humanitarian, liberal, munificent, open-handed, philanthropic, unsparing. ▷ KIND. *Opp* MEAN.

charity *n* **1** affection, altruism, benevolence, bounty, caring, compassion, consideration, generosity, goodness, helpfulness, humanity, kindness, love, mercy, philanthropy, self-sacrifice, sympathy, tender-heartedness, unselfishness, warm-heartedness. **2** *old use* alms, alms-giving, bounty, donation, financial support, gift, *inf* handout, largesse, offering, patronage, poor relief. **3** good cause, the needy, the poor.

charm *n* **1** allure, appeal, attractiveness, charisma, fascination, hypnotic power, lovable nature, lure, magic, magnetism, power, pull, seductiveness. ▷ BEAUTY. **2** *magic charm*. curse, enchantment, incantation, magic, mumbo-jumbo, sorcery, spell, witchcraft, wizardry. **3** *charm on a bracelet*. amulet, lucky charm, mascot, ornament, talisman, trinket. • *vb* allure, attract, beguile, bewitch, cajole, captivate, cast a spell on, decoy, delight, disarm, enchant, enrapture, enthral, entrance, fascinate, hold spellbound, hypnotize, intrigue, lure, mesmerize, please, seduce, soothe, win over. **charming** ▷ ATTRACTIVE.

chart *n* diagram, graph, map, plan, sketch-map, table.

charter *vb* **1** employ, engage, hire, lease, rent. **2** ▷ CERTIFY.

chase *vb* drive, follow, go after, hound, hunt, pursue, run after, track, trail.

chasm *n* abyss, canyon, cleft, crater, crevasse, drop, fissure, gap, gulf, hole, hollow, opening, pit, ravine, rift, split, void.

chaste *adj* **1** abstinent, celibate, *inf* clean, continent, good, immaculate, inexperienced, innocent, moral, pure, sinless, uncorrupted, undefiled, unmarried, virgin, virginal, virtuous. *Opp* IMMORAL **2** *chaste dress.* austere, becoming, decent, decorous, maidenly, modest, plain, restrained, severe, simple, tasteful, unadorned. *Opp* INDECENT.

chasten *vb* **1** restrain, subdue. ▷ HUMILIATE. **2** ▷ CHASTISE.

chastise *vb* castigate, chasten, correct, discipline, penalize, rebuke, scold. ▷ PUNISH, REPRIMAND.

chastity *n* abstinence, celibacy, continence, innocence, integrity, maidenhood, morality, purity, restraint, sinlessness, virginity, virtue. *Opp* LUST.

chat *n* chatter, *inf* chin-wag, *inf* chit-chat, conversation, gossip, *inf* heart-to-heart. • *vb* chatter, converse, gossip, *inf* natter, prattle. ▷ TALK. **chat up** ▷ WOO.

chauvinist *n* bigot, *inf* MCP (= *male chauvinist pig*), patriot, sexist, xenophobe.

cheap *adj* **1** bargain, budget, cut-price, *inf* dirt-cheap, discount, economical, economy, fair, inexpensive, *inf* knock-down, low-priced, reasonable, reduced, *inf* rock-bottom, sale, under-priced. **2** *cheap quality.* base, inferior, poor, second-rate, shoddy, *inf* tatty, tawdry, *inf* tinny, *inf* trashy, worthless. **3** *a cheap insult.* contemptible, crude, despicable, facile, glib, ill-bred, ill-mannered, mean, silly, tasteless, unworthy, vulgar. *Opp* EXPENSIVE, WORTHY.

cheapen *vb* belittle, debase, degrade, demean, devalue, discredit, downgrade, lower the tone (of), popularize, prostitute, vulgarize.

cheat *n* **1** charlatan, cheater, *inf* con-man, counterfeiter, deceiver, double-crosser, extortioner, forger, fraud, hoaxer, impersonator, impostor, mountebank, *inf* phoney, *inf* quack, racketeer, rogue, *inf* shark, swindler, trickster, *inf* twister. **2** artifice, bluff, chicanery, *inf* con, confidence trick, deceit, deception, *sl* fiddle, fraud, hoax, imposture, lie, misrepresentation, pretence, *inf* put-up job, *inf* racket, *inf* rip-off, ruse, sham, swindle, *inf* swizz, treachery, trick. • *vb* **1** bamboozle, beguile, bilk, *inf* con, deceive, defraud, *sl* diddle, *inf* do, double-cross, dupe, *sl* fiddle, *inf* fleece, fool, hoax, hoodwink, outwit, *inf* rip off, rob, *inf* short-change, swindle, take in, trick. **2** *cheat in an exam.* copy, crib, plagiarize.

check *adj* ▷ CHEQUERED. • *n* **1** break, delay, halt, hesitation, hiatus, interruption, pause, stop, stoppage, suspension. **2** *medical check.* check-up, examination, *inf* going-over, inspection, investigation, *inf* once-over, scrutiny, test. • *vb* **1** arrest, bar, block, bridle, control, curb, delay, foil, govern, halt, hamper, hinder, hold back, impede, inhibit, keep in check, obstruct, regulate, rein, repress, restrain, retard, slow down, stem, stop, stunt (*growth*), thwart. **2** *check answers.* *Amer* check out, compare, cross-check, examine, inspect, investigate, monitor, research, scrutinize, test, verify.

cheek *n* audacity, boldness, brazenness, effrontery, impertinence, impudence, insolence, presumptuousness, rudeness, shamelessness, temerity.

cheeky *adj* arrogant, audacious, bold, brazen, cool, discourteous, disrespectful, flippant, forward, impertinent, impolite, impudent, insolent, insulting, irreverent, mocking, pert, presumptuous, rude, *inf* saucy, shameless, *inf* tongue-in-cheek. *Opp* RESPECTFUL.

cheer *n* **1** acclamation, applause, cry of approval, encouragement, hurrah, ovation, shout of approval. **2** ▷ HAPPINESS. • *vb* **1** acclaim, applaud, clap, encourage, shout, yell. *Opp* JEER. **2** comfort, console, delight, encourage, exhilarate, gladden, make cheerful, please, solace, uplift. *Opp* SADDEN. **cheer someone up** ▷ COMFORT, ENTERTAIN. **cheer up** ▷ BRIGHTEN. **Cheer up!** *inf* buck up, look happy, *inf* perk up, smile, *sl* snap out of it, take heart.

cheerful *adj* animated, bouncy, bright, buoyant, cheery, *inf* chirpy, contented, convivial, delighted, elated, festive, gay, genial, glad, gleeful, good-humoured, hearty, hopeful, jaunty, jocund, jolly, jovial, joyful, joyous, jubilant, laughing, light, light-hearted, lively, merry, optimistic, *inf* perky, pleased, positive, rapturous, sparkling, spirited, sprightly, sunny, warm-hearted. ▷ HAPPY. *Opp* BAD-TEMPERED, CHEERLESS.

cheerless *adj* bleak, comfortless, dark, depressing, desolate, dingy, disconsolate, dismal, drab, dreary, dull, forbidding, forlorn, frowning, funereal, gloomy, grim, joyless, lacklustre, melancholy, miserable, mournful, sober, sombre, sullen, sunless, uncongenial, unhappy, uninviting, unpleasant, unpromising, woeful, wretched. ▷ SAD. *Opp* CHEERFUL.

chemical *n* compound, element, substance.

chemist *n old use* apothecary, *Amer* drug-store, pharmacist, pharmacy.

chequered *adj* **1** check, criss-cross, in squares, like a chessboard, patchwork, tartan, tessellated. **2** *chequered career.* ▷ CHANGEABLE.

cherish *vb* be fond of, care for, cosset, foster, hold dear, keep safe, look after, love, nourish, nurse, nurture, prize, protect, treasure, value.

chest *n* **1** box, caddy, case, casket, coffer, crate, strongbox, trunk. **2** breast, rib-cage, thorax.

chew *vb* bite, champ, crunch, gnaw, grind, masticate, munch, nibble. ▷ EAT. **chew over** ▷ CONSIDER.

chick *n* fledgling, nestling.

chicken *n* bantam, broiler, cockerel, fowl, hen, pullet, rooster.

chief *adj* **1** arch, best, first, greatest, head, highest, in charge, leading, major, most experienced, most honoured, most important, oldest, outstanding, premier, principal, senior, supreme, top, unequalled, unrivalled. **2** *chief facts.* basic, cardinal, central, dominant, especial, essential, foremost, fundamental, high-priority, indispensable, key, main, necessary, overriding, paramount, predominant, primary, prime, salient, significant, substantial, uppermost, vital, weighty. *Opp* UNIMPORTANT.

• *n* administrator, authority-figure, *inf* bigwig, *inf* boss, captain, chairperson, chieftain, commander, commanding officer, commissioner, controller, director, employer, executive, foreman, forewoman, *inf* gaffer, *Amer inf* godfather, governor, head, king, leader, manager, managing director, master, mistress, *inf* number one, officer, organizer, overseer, owner, president, principal, proprietor, ring-leader, ruler, superintendent, supervisor, *inf* supremo.

chiefly *adv* especially, essentially, generally, in particular, mainly, mostly, particularly, predominantly, primarily, principally, usually.

child *n* **1** adolescent, *inf* babe, baby, *Scot* bairn, *inf* bambino, boy, *derog* brat, girl, *derog* guttersnipe, infant, juvenile, *inf* kid, lad, lass, minor, newborn, *inf* nipper, offspring, *inf* stripling, toddler, *inf* tot, *derog* urchin, youngster, youth. **2** daughter, descendant, heir, issue, offspring, progeny, son.

childhood *n* adolescence, babyhood, boyhood, girlhood, infancy, minority, schooldays, *inf* teens, youth.

childish *adj* babyish, credulous, foolish, immature, infantile, juvenile, puerile. ▷ SILLY. *Opp* MATURE.

childlike *adj* artless, frank, *inf* green, guileless, ingenuous, innocent, naïve, natural, simple, trustful, unaffected, unsophisticated. *Opp* ARTFUL.

chill *n* ▷ COLD. • *vb* cool, freeze, keep cold, make cold, refrigerate. *Opp* WARM.

chilly *adj* **1** cold, cool, crisp, fresh, frosty, icy, *inf* nippy, *inf* parky, raw, sharp, wintry. **2** *a chilly greeting.* aloof, cool, dispassionate, frigid, hostile, ill-disposed, remote, reserved, *inf* standoffish, unforthcoming, unfriendly, unresponsive, unsympathetic, unwelcoming. *Opp* WARM.

chime *n* carillon, peal, striking, tintinnabulation, tolling. • *vb* ▷ RING.

chimney *n* flue, funnel, smoke-stack.

china *n* porcelain. ▷ CROCKERY.

chink *n* **1** cleft, crack, cranny, crevice, cut, fissure, gap, opening, rift, slit, slot, space, split. **2** ▷ SOUND.

chip *n* **1** bit, flake, fleck, fragment, piece, scrap, shard, shaving, shiver, slice, sliver, splinter, wedge. **2** *a chip in a cup.* crack, damage, flaw, gash, nick, notch, scratch, snick. • *vb* break, crack, damage, gash, nick, notch, scratch, splinter. **chip away** ▷ CHISEL. **chip in** ▷ CONTRIBUTE, INTERRUPT.

chisel *vb* carve, *inf* chip away, cut, engrave, fashion, model, sculpture, shape.

chivalrous *adj* bold, brave, chivalric, courageous, courteous, courtly, gallant, generous, gentlemanly, heroic, honourable, knightly, noble, polite, respectable, true, trustworthy, valiant, valorous, worthy. *Opp* COWARDLY, RUDE.

choice *adj* ▷ EXCELLENT. • *n* **1** alternative, dilemma, need to choose, option. **2** *make your choice.* choosing, decision, election, liking, nomination, pick, preference, say, vote. **3** *a choice of food.* array, assortment, diversity, miscellany, mixture, range, selection, variety.

choke *vb* **1** asphyxiate, garrotte, smother, stifle, strangle, suffocate, throttle. **2** *choke in smoke.* cough, gag, gasp, retch. **3** *choked with traffic.* block, *inf* bung up, clog, close, congest, constrict, dam, fill, jam, obstruct, smother, stop up. **choke back** ▷ SUPPRESS.

choose *vb* adopt, agree on, appoint, decide on, determine on, distinguish, draw lots for, elect, establish, fix on, identify, isolate, name, nominate, opt for, pick out, *inf* plump for, prefer, select, settle on, show a preference for, single out, vote for.

choosy *adj* dainty, discerning, discriminating, exacting, fastidious, finical, finicky, fussy, *inf* hard to please, nice, particular, pernickety, *inf* picky, selective. *Opp* INDIFFERENT.

chop *vb* cleave, cut, hack, hew, lop, slash, split. ▷ CUT. **chop and change** ▷ CHANGE.

chopper *n* axe, cleaver.

choppy *adj* roughish, ruffled, turbulent, uneven, wavy. *Opp* SMOOTH.

chore *n* burden, drudgery, duty, errand, job, task, work.

chorus *n* **1** choir, choral society, vocal ensemble. **2** *join in the chorus.* refrain, response.

christen *vb* anoint, baptize, call, dub, name.

chronic *adj* **1** ceaseless, constant, continuing, deep-rooted, habitual, incessant, incurable, ineradicable, ingrained, lasting, lifelong, lingering, long-lasting, long-lived, long-standing, never-ending, non-stop, permanent, persistent, unending. *Opp* ACUTE, TEMPORARY. **2** [*inf*] *chronic driving.* ▷ BAD.

chronicle *n* account, annals, archive, chronology, description, diary, history, journal, narrative, record, register, saga, story.

chronological *adj* consecutive, in order, sequential.

chronology *n* **1** almanac, calendar, diary, journal, log, schedule, timetable. **2** *establish the chronology.* dating, order, sequence, timing.

chubby *adj* buxom, dumpy, plump, podgy, portly, rotund, round, stout, tubby. ▷ FAT. *Opp* THIN.

chunk *n* bar, block, brick, chuck, *inf* dollop, hunk, lump, mass, piece, portion, slab, wad, wedge, *inf* wodge.

church *n* abbey, basilica, cathedral, chapel, convent, monastery, nunnery, parish church, priory.

churchyard *n* burial-ground, cemetery, graveyard.

chute *n* channel, incline, ramp, rapid, slide, slope.

cinema *n* films, *old use* flicks, *Amer* motion pictures, *inf* movies, *inf* pictures.

circle *n* **1** annulus, band, circlet, disc, hoop, ring. □ *belt, circuit, circulation, circumference, circumnavigation, coil, cordon, curl, curve, cycle, ellipse, girdle, globe, gyration, lap, loop, orb, orbit, oval, revolution, rotation, round, sphere, spiral, tour, turn, wheel, whirl, whorl.* **2** circle of friends. association, band, body, clique, club, company, fellowship, fraternity, gang, party, set, society. ▷ GROUP. • *vb* **1** circulate, circumnavigate, circumscribe, coil, compass, corkscrew, curl, curve, go round, gyrate, loop, orbit, pirouette, pivot, reel, revolve, rotate, spin, spiral, swirl, swivel, tour, turn, wheel, whirl, wind. **2** *trees circle the lawn.* encircle, enclose, encompass, girdle, hem in, ring, skirt, surround.

circuit *n* journey round, lap, orbit, revolution, tour.

circuitous *adj* curving, devious, indirect, labyrinthine, meandering, oblique, rambling, roundabout, serpentine, tortuous, twisting, winding, zigzag. *Opp* DIRECT.

circular *adj* **1** annular, discoid, ringlike, round. **2** *circular conversation.* circumlocutory, cyclic, periphrastic, repeating, repetitive, roundabout, tautologous.
• *n* advertisement, leaflet, letter, notice, pamphlet.

circulate *vb* **1** go round, move about, move round, orbit. ▷ CIRCLE. **2** *circulate gossip.* advertise, disseminate, distribute, issue, make known, noise abroad, promulgate, publicize, publish, *inf* put about, send round, spread about.

circulation *n* **1** flow, movement, pumping, recycling. **2** broadcasting, diffusion, dissemination, distribution, promulgation, publication, spreading, transmission. **3** *newspaper circulation.* distribution, sales-figures.

circumference *n* border, boundary, circuit, edge, exterior, fringe, limit, margin, outline, outside, perimeter, periphery, rim, verge.

circumstance *n* affair, event, happening, incident, occasion, occurrence. **circumstances** **1** background, causes, conditions, considerations, context, contingencies, details, factors, facts, influences, particulars, position, situation, state of affairs, surroundings. **2** finances, income, resources.

circumstantial *adj* conjectural, deduced, inferred, unprovable. *Opp* PROVABLE.

cistern *n* bath, container, reservoir, tank.

citadel *n* acropolis, bastion, castle, fort, fortification, fortress, garrison, stronghold, tower.

cite *vb* adduce, advance, *inf* bring up, enumerate, mention, name, quote, *inf* reel off, refer to, specify.

citizen *n* burgess, commoner, denizen, dweller, freeman, householder, inhabitant, national, native, passport-holder, ratepayer, resident, subject, taxpayer, voter.

city *n* capital, conurbation, metropolis, town, urban district.

civil *adj* **1** affable, civilized, considerate, courteous, obliging, respectful, urbane, well-bred, well-mannered. ▷ POLITE. *Opp* IMPOLITE. **2** *civil administration.* civilian, domestic, internal, national. ▷ MILITARY. **3** *civil liberties.* communal, public, social, state. **civil rights** freedom, human rights, legal rights, liberty, political rights. **civil servant** administrator, bureaucrat, *derog* mandarin.

civilization *n* achievements, attainments, culture, customs, mores, organization, refinement, sophistication, urbanity, urbanization.

civilize *vb* cultivate, domesticate, educate, enlighten, humanize, improve, make better, organize, refine, socialize, urbanize.

civilized *adj* advanced, cultivated, cultured, democratic, developed, domesticated, educated, enlightened, humane, orderly, polite, refined, sociable, social, sophisticated, urbane, urbanized, well-behaved, well-run. *Opp* UNCIVILIZED.

claim *vb* 1 ask for, collect, command, demand, exact, insist on, request, require, take. 2 affirm, allege, argue, assert, attest, contend, declare, insist, maintain, pretend, profess, state.

clairvoyant *adj* extra-sensory, oracular, prophetic, psychic, telepathic. • *n* fortune-teller, oracle, prophet, seer, sibyl, soothsayer.

clamber *vb* climb, crawl, move awkwardly, scramble.

clammy *adj* close, damp, dank, humid, moist, muggy, slimy, sticky, sweaty, wet.

clamour *n* babel, commotion, din, hubbub, hullabaloo, noise, outcry, racket, row, screeching, shouting, storm, uproar. • *vb* call out, cry out, exclaim, shout, yell.

clan *n* family, house, tribe.

clannish *adj derog* cliquish, close, close-knit, insular, isolated, narrow, united.

clap *n* bang, crack, crash, report, smack. ▷ SOUND. • *vb* 1 applaud, *sl* put your hands together, show approval. 2 *clap on the back.* ▷ HIT.

clarify *vb* 1 clear up, define, elucidate, explain, explicate, gloss, illuminate, make clear, simplify, *inf* spell out, throw light on. *Opp* CONFUSE. 2 *clarify wine.* cleanse, clear, filter, purify, refine. *Opp* CLOUD.

clash *vb* 1 bang, clang, clank, crash, resonate, ring. ▷ SOUND. 2 ▷ CONFLICT. 3 *The events clashed.* ▷ COINCIDE.

clasp *n* 1 brooch, buckle, catch, clip, fastener, fastening, hasp, hook, pin. 2 cuddle, embrace, grasp, grip, hold, hug. • *vb* 1 ▷ FASTEN. 2 cling to, clutch, embrace, enfold, grasp, grip, hold, hug, squeeze. 3 *clasp your hands.* hold together, wring.

class *n* 1 category, classification, division, domain, genre, genus, grade, group, kind, league, order, quality, rank, set, sort, species, sphere, type. 2 *social class.* caste, degree, descent, extraction, grouping, lineage, pedigree, standing, station, status. □ *aristocracy, bourgeoisie, commoners, (the) commons, gentry, lower class, middle class, nobility, proletariat, ruling class, serfs, upper class, upper-middle class, (the) workers, working class.* 3 class in school. band, form, *Amer* grade, group, set, stream, year. • *vb* ▷ CLASSIFY.

classic *adj* 1 abiding, ageless, deathless, enduring, established, exemplary, flawless, ideal, immortal, lasting, legendary, masterly, memorable, notable, outstanding, perfect, time-honoured, undying, unforgettable, *inf* vintage. ▷ EXCELLENT. *Opp* COMMONPLACE, EPHEMERAL. 2 *a classic case.* archetypal, characteristic, copybook, definitive, model, paradigmatic, regular, standard, typical, usual. *Opp* UNUSUAL. • *n* masterpiece, masterwork, model.

classical *adj* 1 ancient, Attic, Greek, Hellenic, Latin, Roman. 2 *classical style.* austere, dignified, elegant, pure, restrained, simple, symmetrical, well-proportioned. 3 *classical music.* established, harmonious, highbrow, serious.

classification *n* categorization, codification, ordering, organization, systematization, tabulation, taxonomy. ▷ CLASS.

classify *vb* arrange, bracket together, catalogue, categorize, class, grade, group, order, organize, *inf* pigeon-hole, put into sets, sort, systematize, tabulate. **classified** ▷ SECRET.

clause *n* article, condition, item, paragraph, part, passage, provision, proviso, section, subsection.

claw *n* nail, talon. • *vb* graze, injure, lacerate, maul, rip, scrape, scratch, slash, tear.

clean *adj* **1** decontaminated, dirt-free, disinfected, hygienic, immaculate, laundered, perfect, polished, sanitary, scrubbed, spotless, sterile, sterilized, tidy, unadulterated, unsoiled, unstained, unsullied, washed, wholesome. **2** *clean water*. clarified, clear, distilled, fresh, pure, purified, unpolluted. **3** *clean paper*. blank, new, plain, uncreased, unmarked, untouched, unused. **4** *a clean edge*. neat, regular, smooth, straight, tidy. **5** *a clean fight*. chivalrous, fair, honest, honourable, sporting, sportsmanlike. **6** *clean fun*. chaste, decent, good, innocent, moral, respectable, upright, virtuous. *Opp* DIRTY. • *vb* cleanse, clear up, tidy up, wash. □ *bath, bathe, brush, buff, decontaminate, deodorize, disinfect, dry-clean, dust, filter, flush, groom, hoover, launder, mop, polish, purge, purify, rinse, sand-blast, sanitize, scour, scrape, scrub, shampoo, shower, soap, sponge, spring-clean, spruce up, sterilize, swab, sweep, swill, vacuum, wipe, wring out.* *Opp* CONTAMINATE. **make a clean breast of** ▷ CONFESS.

clean-shaven *adj* beardless, shaved, shaven, shorn, smooth.

clear *adj* **1** clean, colourless, crystalline, glassy, limpid, pellucid, pure, transparent. **2** *clear weather*. cloudless, fair, fine, sunny, starlit, unclouded. *Opp* CLOUDY. **3** *clear colours*. bright, lustrous, shining, sparkling, strong, vivid. **4** *clear conscience*. blameless, easy, guiltless, innocent, quiet, satisfied, sinless, undisturbed, untarnished, untroubled, unworried. **5** *clear handwriting*. bold, clean, definite, distinct, explicit, focused, legible, positive, recognizable, sharp, simple, visible, well-defined. **6** *clear sound*. audible, clarion (*call*), distinct, penetrating, sharp. **7** *clear instructions*. clear-cut, coherent, comprehensible, explicit, intelligible, lucid, perspicuous, precise, specific, straightforward, unambiguous, understandable, unequivocal, well-presented. **8** *clear case of cheating*. apparent, blatant, clear-cut, conspicuous, evident, glaring, indisputable, manifest, noticeable, obvious, palpable, perceptible, plain, pronounced, straightforward, unconcealed, undisguised, unmistakable. *Opp* UNCERTAIN. **9** *clear space*. empty, free, open, passable, uncluttered, uncrowded, unhampered, unhindered, unimpeded, unobstructed. • *vb* **1** disappear, evaporate, fade, melt away, vanish. **2** become clear, brighten, clarify, lighten, uncloud. **3** clean, make clean, make transparent, polish, wipe. **4** *clear weeds*. disentangle, eliminate, get rid of, remove, strip. **5** *clear a drain*. clean out, free, loosen, open up, unblock, unclog. **6** *clear of blame*. absolve, acquit, exculpate, excuse, exonerate, free, *inf* let off, liberate, release, vindicate. **7** *clear a building*. empty, evacuate. **8** *clear a fence*. bound over, jump, leap over, pass over, spring over, vault. **clear away** ▷ REMOVE. **clear off** ▷ DEPART. **clear up** ▷ CLEAN, EXPLAIN.

clearing *n* gap, glade, opening, space.

cleave *vb* divide, halve, rive, slit, split. ▷ CUT.

clench *vb* **1** clamp up, close tightly, double up, grit (*your teeth*), squeeze tightly. **2** clasp, grasp, grip, hold.

clergyman *n* archbishop, ayatollah, bishop, canon, cardinal, chaplain, churchman, cleric, curate, deacon, *fem* deaconess, dean, divine, ecclesiastic, evangelist, friar, guru, imam, *inf* man of the cloth, minister, missionary, monk, padre, parson, pastor, preacher, prebend, prelate, priest, rabbi, rector, vicar. *Opp* LAYMAN.

clerical *adj* **1** *clerical and administrative work.* office, secretarial, *inf* white-collar. **2** *a clerical collar* canonical, ecclesiastical, episcopal, ministerial, monastic, pastoral, priestly, rabbinical, sacerdotal, spiritual.

clerk *n* assistant, bookkeeper, computer operator, copyist, filing clerk, office boy, office girl, office worker, *inf* pen-pusher, receptionist, recorder, scribe, secretary, shorthand-typist, stenographer, typist, word-processor operator.

clever *adj* able, academic, accomplished, acute, adept, adroit, apt, artful, artistic, astute, *inf* brainy, bright, brilliant, canny, capable, *derog* crafty, creative, *derog* cunning, *inf* cute, *inf* deep, deft, dextrous, discerning, expert, *derog* foxy, gifted, guileful, *inf* handy, imaginative, ingenious, intellectual, intelligent, inventive, judicious, keen, knowing, knowledgeable, observant, penetrating, perceptive, percipient, perspicacious, precocious, quick, quick-witted, rational, resourceful, sagacious, sensible, sharp, shrewd, skilful, skilled, slick, *derog* sly, smart, subtle, talented, *derog* wily, wise, witty. *Opp* STUPID, UNSKILFUL. **clever person** *inf* egghead, expert, genius, *derog* know-all, mastermind, prodigy, sage, *derog* smart alec, *derog* smart-arse, virtuoso, wizard.

cleverness *n* ability, acuteness, astuteness, brilliance, *derog* cunning, expertise, ingenuity, intellect, intelligence, mastery, quickness, sagacity, sharpness, shrewdness, skill, subtlety, talent, wisdom, wit. *Opp* STUPIDITY.

cliché *n* banality, *inf* chestnut, commonplace, hackneyed phrase, platitude, stereotype, truism, well-worn phrase.

client *n* plur clientele, consumer, customer, patient, patron, shopper, user.

cliff *n* bluff, crag, escarpment, precipice, rock-face, scar, sheer drop.

climate *n* **1** ▷ WEATHER. **2** *climate of opinion.* ambience, atmosphere, aura, disposition, environment, feeling, mood, spirit, temper, trend.

climax *n* **1** acme, apex, apogee, crisis, culmination, head, highlight, high point, peak, summit, zenith. *Opp* BATHOS. **2** *sexual climax.* orgasm.

climb *n* ascent, grade, gradient, hill, incline, pitch, rise, slope. • *vb* **1** ascend, clamber up, defy gravity, go up, levitate, lift off, mount, move up, scale, shin up, soar, swarm up, take off. **2** incline, rise, slope up. **3** *climb a mountain.* conquer, reach the top of. **climb down** ▷ DESCEND.

clinch *vb* agree, close, complete, conclude, confirm, decide, determine, finalize, make certain of, ratify, secure, settle, shake hands on, sign, verify.

cling *vb* adhere, attach, fasten, fix, hold fast, stick. **cling to** ▷ EMBRACE.

clinic *n* health centre, infirmary, medical centre, sick-bay, surgery.

clip *n* **1** ▷ FASTENER. **2** *clip from a film*. bit, cutting, excerpt, extract, fragment, part, passage, portion, quotation, section, snippet, trailer. • *vb* **1** pin, staple. ▷ FASTEN. **2** crop, dock, prune, shear, snip, trim. ▷ CUT.

cloak *n* **1** cape, cope, mantle, poncho, robe, wrap. ▷ COAT. **2** ▷ COVER. • *vb* cover, disguise, mantle, mask, screen, shroud, veil, wrap. ▷ HIDE.

clock *n* time-piece. □ *alarm-clock, chronometer, dial, digital clock, grandfather clock, hourglass, pendulum clock, sundial, watch.*

clog *vb* block, *inf* bung up, choke, close, congest, dam, fill, impede, jam, obstruct, plug, stop up.

close *adj* **1** accessible, adjacent, adjoining, at hand, convenient, handy, near, neighbouring, point-blank. **2** *close friends*. affectionate, attached, dear, devoted, familiar, fond, friendly, intimate, loving, *inf* thick. **3** *close comparison*. alike, analogous, comparable, compatible, corresponding, related, resembling, similar. **4** *a close crowd*. compact, compressed, congested, cramped, crowded, dense, *inf* jam-packed, packed, thick. **5** *close scrutiny*. attentive, careful, concentrated, detailed, minute, painstaking, precise, rigorous, searching, thorough. **6** *close with information*. confidential, private, reserved, reticent, secretive, taciturn. **7** *close with money*. illiberal, mean, *inf* mingy, miserly, niggardly, parsimonious, penurious, stingy, tight, tight-fisted, ungenerous. **8** *close atmosphere*. airless, confined, fuggy, humid, muggy, oppressive, stale, stifling, stuffy, suffocating, sweltering, unventilated, warm. *Opp* DISTANT, OPEN. • *n* **1** cessation, completion, conclusion, culmination, end, finish, stop, termination. **2** cadence, coda, finale. **3** *close of a play*. denouement, last act. • *vb* **1** bolt, fasten, lock, make inaccessible, padlock, put out of bounds, seal, secure, shut. **2** *close a road*. bar, barricade, block, make impassable, obstruct, seal off, stop up. **3** *close proceedings*. complete, conclude, culminate, discontinue, end, finish, stop, terminate, *inf* wind up. **4** *close a gap*. fill, join up, make smaller, reduce, shorten. *Opp* OPEN.

closed *adj* **1** fastened, locked, sealed, shut. **2** completed, concluded, done with, ended, finished, over, resolved, settled, tied up.

clot *n* embolism, lump, mass, thrombosis. • *vb* coagulate, coalesce, congeal, curdle, make lumps, set, solidify, stiffen, thicken.

cloth *n* fabric, material, stuff, textile. □ *astrakhan, bouclé, brocade, broderie anglaise, buckram, calico, cambric, candlewick, canvas, cashmere, cheesecloth, chenille, chiffon, chintz, corduroy, cotton, crepe, cretonne, damask, denim, dimity, drill, drugget, elastic, felt, flannel, flannelette, gabardine, gauze, georgette, gingham, hessian, holland, lace, lamé, lawn, linen, lint, mohair, moiré, moquette, muslin, nankeen, nylon, oilcloth, oilskin, organdie, organza, patchwork, piqué, plaid, plissé, plush, polycotton, polyester, poplin, rayon, sackcloth, sacking, sailcloth, sarsenet, sateen, satin, satinette, seersucker, serge, silk, stockinet, taffeta, tapes-*

try, tartan, terry, ticking, tulle, tussore, tweed, velour, velvet, velveteen, viscose, voile, winceyette, wool, worsted.

clothe *vb* accoutre, apparel, array, attire, cover, deck, drape, dress, fit out, garb, *inf* kit out, outfit, robe, swathe, wrap up. *Opp* STRIP. **clothe yourself in** ▷ WEAR.

clothes *plur n* apparel, attire, *inf* clobber, clothing, costume, dress, ensemble, finery, garb, garments, *inf* gear, *inf* get-up, outfit, *old use* raiment, *inf* rig-out, *sl* togs, trousseau, underclothes, uniform, vestments, wardrobe, wear, weeds. □ *anorak, apron, blazer, blouse, bodice, breeches, caftan, cagoule, cape, cardigan, cassock, chemise, chuddar, cloak, coat, crinoline, culottes, décolletage, doublet, dress, dressing-gown, duffel coat, dungarees, frock, gaiters, gauntlet, glove, gown, greatcoat, gym-slip, habit, housecoat, jacket, jeans, jerkin, jersey, jodhpurs, jumper, kilt, knickers, leg-warmers, leotard, livery, loincloth, lounge suit, mackintosh, mantle, miniskirt, muffler, neck-tie, négligé, nightclothes, nightdress, oilskins, overalls, overcoat, pants, parka, pinafore, poncho, pullover, pyjamas, raincoat, robe, rompers, sari, sarong, scarf, shawl, shirt, shorts, singlet, skirt, slacks, smock, sock, sou'wester, spats, stocking, stole, suit, surplice, sweater, sweatshirt, tail-coat, tie, tights, trousers, trunks, t-shirt, tunic, tutu, uniform, waistcoat, wet-suit, wind-cheater, wrap, yashmak.* ▷ HAT, SHOE, UNDERCLOTHES.

cloud *n* billow, haze, mist, rain cloud, storm cloud. • *vb* blur, conceal, cover, darken, dull, eclipse, enshroud, hide, mantle, mist up, obfuscate, obscure, screen, shroud, veil.

cloudless *adj* bright, clear, starlit, sunny, unclouded. *Opp* CLOUDY.

cloudy *adj* **1** dark, dismal, dull, gloomy, grey, leaden, lowering, overcast, sullen, sunless. *Opp* CLOUDLESS. **2** *cloudy windows.* blurred, blurry, dim, misty, opaque, steamy, unclear. **3** *cloudy liquid.* hazy, milky, muddy, murky. *Opp* CLEAR.

clown *n* buffoon, comedian, comic, fool, funnyman, jester, joker. ▷ IDIOT.

club *n* **1** bat, baton, bludgeon, cosh, cudgel, mace, staff, stick, truncheon. **2** association, brotherhood, circle, company, federation, fellowship, fraternity, group, guild, league, order, organization, party, set, sisterhood, society, sorority, union. • *vb* ▷ HIT. **club together** ▷ COMBINE.

clue *n* hint, idea, indication, indicator, inkling, key, lead, pointer, sign, suggestion, suspicion, tip, tip-off, trace.

clump *n* bunch, bundle, cluster, collection, mass, shock (*of hair*), thicket, tuft. ▷ GROUP.

clumsy *adj* **1** awkward, blundering, bumbling, bungling, fumbling, gangling, gawky, graceless, *inf* ham-fisted, heavy-handed, hulking, inelegant, lumbering, maladroit, shambling, uncoordinated, ungainly, ungraceful, unskilful. *Opp* SKILFUL. **2** amateurish, badly-made, bulky, cumbersome, heavy, inconvenient, inelegant, large, ponderous, rough, shapeless, unmanageable, unwieldy. *Opp* NEAT. **3** *a clumsy remark.* boorish, gauche, ill-judged, inappropriate, indelicate, indiscreet, inept, insensitive, tactless, uncouth, undiplomatic, unsubtle, unsuitable.

cluster *n* assembly, batch, bunch, clump, collection, crowd, gathering, knot. ▷ GROUP. • *vb* ▷ GATHER.

clutch *n* clasp, control, evil embrace, grasp, grip, hold, possession, power. • *vb* catch, clasp, cling to, grab, grasp, grip, hang on to, hold on to, seize, snatch, take hold of.

clutter *n* chaos, confusion, disorder, jumble, junk, litter, lumber, mess, mix-up, muddle, odds and ends, rubbish, tangle, untidiness. • *vb* be scattered about, fill, lie about, litter, make untidy, *inf* mess up, muddle, strew.

coach *n* **1** bus, carriage, *old use* charabanc. **2** *games coach*. instructor, teacher, trainer, tutor. • *vb* direct, drill, exercise, guide, instruct, prepare, teach, train, tutor.

coagulate *vb* clot, congeal, curdle, *inf* jell, set, solidify, stiffen, thicken.

coarse *adj* **1** bristly, gritty, hairy, harsh, lumpy, prickly, rough, scratchy, sharp, stony, uneven, unfinished. *Opp* FINE, SOFT. **2** *coarse language*. bawdy, blasphemous, boorish, common, crude, earthy, foul, immodest, impolite, improper, impure, indecent, indelicate, offensive, ribald, rude, smutty, uncouth, unrefined, vulgar. *Opp* REFINED.

coast *n* beach, coastline, littoral, seaboard, seashore, seaside, shore. • *vb* cruise, drift, free-wheel, glide, sail, skim, slide, slip.

coastal *adj* maritime, nautical, naval, seaside.

coat *n* **1** □ *anorak, blazer, cagoule, cardigan, dinner-jacket, doublet, duffel coat, greatcoat, jacket, jerkin, mackintosh, overcoat, raincoat, tail-coat, tunic, tuxedo, waistcoat, wind-cheater*. **2** *an animal's coat*. fleece, fur, hair, hide, pelt, skin. **3** *coat of paint*. coating, cover, film, finish, glaze, layer, membrane, overlay, patina, sheet, veneer, wash. ▷ COVERING. • *vb* ▷ COVER. **coat of arms** ▷ CREST.

coax *vb* allure, beguile, cajole, charm, decoy, entice, induce, inveigle, manipulate, persuade, tempt, urge, wheedle.

cobble *vb* **cobble together** botch, knock up, make, mend, patch up, put together.

code *n* **1** etiquette, laws, manners, regulations, rule-book, rules, system. **2** *message in code*. cipher, secret language, signals, sign-system.

coerce *vb* bludgeon, browbeat, bully, compel, constrain, dragoon, force, frighten, intimidate, press-gang, pressurize, terrorize.

coercion *n* browbeating, brute force, bullying, compulsion, conscription, constraint, duress, force, intimidation, physical force pressure, *inf* strong-arm tactics, threats.

coffer *n* box, cabinet, case, casket, chest, crate, trunk.

cog *n* ratchet, sprocket, tooth.

cogent *adj* compelling, conclusive convincing, effective, forceful, forcible, indisputable, irresistible, logical, persuasive, potent, powerful, rational, sound, strong, unanswerable, weighty, well-argued. ▷ COHERENT. *Opp* IRRATIONAL.

cohere *vb* bind, cake, cling together, coalesce, combine, consolidate, fuse, hang together, hold together, join, stick together, unite.

coherent *adj* articulate, cohering, cohesive, connected, consistent, integrated, logical, lucid, orderly, organized, rational, reasonable, reasoned, sound, structured, systematic, unified, united, well-ordered, well-structured. ▷ COGENT. *Opp* INCOHERENT.

coil *n* circle, convolution, corkscrew, curl, helix, kink, loop, ring, roll, screw, spiral, twirl, twist, vortex, whirl, whorl. • *vb* bend, curl, entwine, loop, roll, snake, spiral, turn, twine, twirl, twist, wind, writhe.

coin *n* 1 bit, piece. 2 [*plur*] cash, change, coppers, loose change, silver, small change. ▷ MONEY. • *vb* 1 forge, make, mint, mould, stamp. 2 *coin a name.* conceive, concoct, create, devise, dream up, fabricate, hatch, introduce, invent, make up, originate, produce, think up.

coincide *vb* accord, agree, be congruent, be identical, be in unison, be the same, clash, coexist, come together, concur, correspond, fall together, happen together, harmonize, line up, match, square, synchronize, tally.

coincidence *n* 1 accord, agreement, coexistence, concurrence, conformity, congruence, congruity, correspondence, harmony, similarity. 2 *meet by coincidence.* accident, chance, fluke, luck.

cold *adj* 1 arctic, biting, bitter, bleak, chill, chilly, cool, crisp, cutting, draughty, freezing, fresh, frosty, glacial, heatless, ice-cold, icy, inclement, keen, *inf* nippy, numbing, *inf* parky, penetrating, perishing, piercing, polar, raw, shivery, Siberian, snowy, unheated, wintry. 2 *cold hands.* blue with cold, chilled, dead, frostbitten, frozen, numbed, shivering, shivery. 3 *a cold heart.* aloof, apathetic, callous, cold-blooded, cool, cruel, distant, frigid, hard, hard-hearted, heartless, indifferent, inhospitable, inhuman, insensitive, passionless, phlegmatic, reserved, standoffish, stony, uncaring, unconcerned, undemonstrative, unemotional, unenthusiastic, unfeeling, unkind, unresponsive, unsympathetic. ▷ UNFRIENDLY. *Opp* HOT, KIND. • *n* 1 chill, coldness, coolness, freshness, iciness, low temperature, wintriness. *Opp* HEAT. 2 *cold in the head.* catarrh, *inf* flu, influenza, *inf* the sniffles. **feel the cold** freeze, quiver, shake, shiver, shudder, suffer from hypothermia, tremble.

cold-blooded *adj* barbaric, brutal, callous, hard-hearted, inhuman, inhumane, merciless, pitiless, ruthless, savage. ▷ CRUEL. *Opp* HUMANE.

cold-hearted *adj* apathetic, cool, dispassionate, frigid, heartless, impassive, impersonal, indifferent, insensitive, thick-skinned, uncaring, unemotional, unfeeling, unkind, unresponsive, unsympathetic. ▷ UNFRIENDLY. *Opp* FRIENDLY.

collaborate *vb* 1 band together, cooperate, join forces, *inf* pull together, team up, work together. 2 [*derog*] collude, connive, conspire, join the opposition, *inf* rat, turn traitor.

collaboration *n* 1 association, concerted effort, cooperation, partnership, tandem, teamwork. 2 [*derog*] collusion, connivance, conspiracy, treachery.

collaborator *n* 1 accomplice, ally, assistant, associate, co-author, colleague, confederate, fellow-worker, helper, helpmate, partner, *joc* partner-in-crime, teammate. **2** *collaborator with an enemy.* blackleg, *inf* Judas, quisling, *inf* scab, traitor, turncoat.

collapse *n* break-down, break-up, cave-in, destruction, downfall, end, fall, ruin, ruination, subsidence, wreck. • *vb* **1** break down, break up, buckle, cave in, crumble, crumple, deflate, disintegrate, double up, fall apart, fall down, fall in, fold up, give in, *inf* go west, sink, subside, tumble down. **2** *collapse in the heat.* become ill, *inf* bite the dust, black out, *inf* crack up, faint, founder, *inf* go under, *inf* keel over, pass out, *old use* swoon. **3** *sales collapsed.* become less, crash, deteriorate, diminish, drop, fail, slump, worsen.

collapsible *adj* adjustable, folding, retractable, telescopic.

colleague *n* associate, business partner, fellow-worker. ▷ COLLABORATOR.

collect *vb* **1** accumulate, agglomerate, aggregate, amass, assemble, bring together, cluster, come together, concentrate, congregate, convene, converge, crowd, forgather, garner, gather, group, harvest, heap, hoard, lay up, muster, pile up, put by, rally, reserve, save, scrape together, stack up, stockpile, store. *Opp* DISPERSE. **2** *collect money for charity.* be given, raise, secure, take. **3** *collect goods from a shop.* acquire, bring, fetch, get, load up, obtain, pick up. **collected** ▷ CALM.

collection *n* **1** accumulation, array, assemblage, assortment, cluster, conglomeration, heap, hoard, mass, pile, set, stack, store. ▷ GROUP. **2** *old use* alms-giving, flag-day, free-will offering, offertory, voluntary contributions, *inf* whip-round.

collective *adj* combined, common, composite, co-operative, corporate, democratic, group, joint, shared, unified, united. *Opp* INDIVIDUAL.

college *n* academy, conservatory, institute, polytechnic, school, university.

collide *vb* **collide with** bump into, cannon into, crash into, knock, meet, run into, slam into, smash into, strike, touch. ▷ HIT.

collision *n* accident, bump, clash, crash, head-on collision, impact, knock, pile-up, scrape, smash, wreck.

colloquial *adj* chatty, conversational, everyday, informal, slangy, vernacular. *Opp* FORMAL.

colonist *n* colonizer, explorer, pioneer, settler. *Opp* NATIVE.

colonize *vb* occupy, people, populate, settle in, subjugate.

colony *n* **1** dependency, dominion, possession, protectorate, province, settlement, territory. **2** ▷ GROUP.

colossal *adj* Brobdingnagian, elephantine, enormous, gargantuan, giant, gigantic, herculean, huge, immense, *inf* jumbo, mammoth, massive, mighty, monstrous, monumental, prodigious, titanic, towering, vast. ▷ BIG. *Opp* SMALL.

colour *n* **1** coloration, colouring, hue, pigment, pigmentation, shade, tincture, tinge, tint, tone. □ *amber, azure, beige, black, blue, bronze, brown, buff, carroty, cherry, chestnut, chocolate, cobalt, cream, crimson, dun, fawn, gilt, gold, golden, green, grey, indigo, ivory, jet-black, khaki, lavender, maroon, mauve, navy blue, ochre, olive, orange, pink, puce, purple, red, rosy, russet, sandy, scarlet, silver, tan, tawny, turquoise, vermil-*

ion, violet, white, yellow. 2 *colour in your cheeks.* bloom, blush, flush, glow, rosiness, ruddiness. • *vb* 1 colour-wash, crayon, dye, paint, pigment, shade, stain, tinge, tint. 2 blush, bronze, brown, burn, flush, redden, tan. *Opp* FADE. 3 *coloured by prejudice.* affect, bias, distort, impinge on, influence, pervert, prejudice, slant, sway. **colours** ▷ FLAG.

colourful *adj* 1 bright, brilliant, chromatic, gaudy, irridescent, multicoloured, psychedelic, showy, vibrant. 2 *colourful personality.* dashing, distinctive, dynamic, eccentric, energetic, exciting, flamboyant, flashy, florid, glamorous, unusual, vigorous. 3 *colourful description.* graphic, interesting, lively, picturesque, rich, stimulating, striking, telling, vivid. *Opp* COLOURLESS.

colouring *n* colourant, dye, pigment, pigmentation, stain, tincture. ▷ COLOUR.

colourless *adj* 1 albino, ashen, blanched, faded, grey, monochrome, neutral, pale, pallid, sickly, wan, *inf* washed out, waxen. ▷ WHITE. 2 bland, boring, characterless, dingy, dismal, dowdy, drab, dreary, dull, insipid, lacklustre, lifeless, ordinary, tame, uninspiring, uninteresting, vacuous, vapid. *Opp* COLOURFUL.

column *n* 1 pilaster, pile, pillar, pole, post, prop, shaft, support, upright. 2 *newspaper column.* article, feature, leader, leading article, piece. 3 *column of soldiers.* cavalcade, file, line, procession, queue, rank, row, string, train.

comb *vb* 1 arrange, groom, neaten, smarten up, spruce up, tidy, untangle. 2 *comb the house.* hunt through, ransack, rummage through, scour, search thoroughly.

combat *n* action, battle, bout, clash, conflict, contest, duel, encounter, engagement, fight, skirmish, struggle, war, warfare. • *vb* battle against, contend against, contest, counter, defy, face up to, grapple with, oppose, resist, stand up to, strive against, struggle against, tackle, withstand. ▷ FIGHT.

combination *n* aggregate, alloy, amalgam, blend, compound, concoction, concurrence, conjunction, fusion, marriage, mix, mixture, synthesis, unification. 2 alliance, amalgamation, association, coalition, confederacy, confederation, consortium, conspiracy, federation, grouping, link-up, merger, partnership, syndicate, union.

combine *vb* 1 add together, amalgamate, bind, blend, bring together, compound, fuse, incorporate, integrate, intertwine, interweave, join, link, *inf* lump together, marry, merge, mingle, mix, pool, put together, synthesize, unify, unite. *Opp* DIVIDE. 2 *combine as a team.* ally, associate, band together, club together, coalesce, connect, cooperate, form an alliance, gang together, gang up, join forces, team up. *Opp* DISPERSE.

combustible *adj* flammable, inflammable. *Opp* INCOMBUSTIBLE.

come *vb* 1 advance, appear, approach, arrive, draw near, enter, get to, move (towards), near, reach, visit. 2 *take what comes.* happen, materialize, occur, put in an appearance, show up. **come about** ▷ HAPPEN. **come across** ▷ FIND. **come apart** ▷ DISINTEGRATE. **come clean** ▷ CONFESS. **come out with** ▷ SAY. **come round** ▷ RECOVER. **come up** ▷ ARISE. **come upon** ▷ FIND.

comedian *n* buffoon, clown, comic, fool, humorist, jester, joker, wag. ▷ ENTERTAINER.

comedy *n* buffoonery, clowning, facetiousness, farce, hilarity, humour, jesting, joking, satire, slapstick, wit.

comfort *n* 1 aid, cheer, consolation, encouragement, help, moral support, reassurance, relief, solace, succour, sympathy. 2 *living in comfort.* abundance, affluence, contentment, cosiness, ease, luxury, opulence, plenty, relaxation, well-being. *Opp* DISCOMFORT, POVERTY. • *vb* assuage, calm, cheer up, console, ease, encourage, gladden, hearten, help, reassure, relieve, solace, soothe, succour, sympathize with.

comfortable *adj* 1 *inf* comfy, convenient, cosy, easy, padded, reassuring, relaxing, roomy, snug, soft, upholstered, warm. 2 *comfortable clothes.* informal, loose-fitting, well-fitting, well-made. 3 *a comfortable life.* affluent, agreeable, contented, happy, homely, luxurious, pleasant, prosperous, relaxed, restful, serene, tranquil, untroubled, well-off. *Opp* UNCOMFORTABLE.

comic *adj* absurd, amusing, comical, diverting, droll, facetious, farcical, funny, hilarious, humorous, hysterical, jocular, joking, laughable, ludicrous, *inf* priceless, *inf* rich, ridiculous, sarcastic, sardonic, satirical, side-splitting, silly, uproarious, waggish, witty. *Opp* SERIOUS. • *n* 1 ▷ COMEDIAN. 2 ▷ MAGAZINE.

command *n* 1 behest, bidding, commandment, decree, directive, edict, injunction, instruction, mandate, order, requirement, ultimatum, writ. 2 authority, charge, control, direction, government, jurisdiction, management, oversight, power, rule, sovereignty, supervision, sway. 3 *command of a language.* grasp, knowledge, mastery. • *vb* 1 adjure, *old use* bid, charge, compel, decree, demand, direct, enjoin, instruct, ordain, order, prescribe, request, require. 2 *command a ship.* administer, be in charge of, control, direct, govern, have authority over, head, lead, manage, reign over, rule, supervise.

commandeer *vb* appropriate, confiscate, hijack, impound, requisition, seize, sequester, take over.

commander *n* captain, commandant, commanding-officer, general, head, leader, officer-in-charge. ▷ CHIEF.

commemorate *vb* be a memorial to, be a reminder of, celebrate, honour, immortalize, keep alive the memory of, memorialize, pay your respects to, pay homage to, pay tribute to, remember, salute, solemnize.

commence *vb* embark on, enter on, inaugurate, initiate, launch, open, set off, set out, set up, start. ▷ BEGIN. *Opp* FINISH.

commend *vb* acclaim, applaud, approve of, compliment, congratulate, eulogize, extol, praise, recommend. *Opp* CRITICIZE.

commendable *adj* admirable, creditable, deserving, laudable, meritorious, praiseworthy, worthwhile. ▷ GOOD. *Opp* DEPLORABLE.

comment *vb* animadversion, annotation, clarification, commentary, criticism, elucidation, explanation, footnote, gloss, interjection, interpolation, mention, note, observation, opinion, reaction, reference, remark, statement. • *vb* animadvert, criticize, elucidate, explain, interject, interpolate interpose, mention, note, observe, opine, remark, say, state.

commentary *n* **1** account, broadcast, description, report. **2** *commentary on a poem.* analysis, criticism, critique, discourse, elucidation, explanation, interpretation, notes, review.

commentator *n* announcer, broadcaster, journalist, reporter.

commerce *n* business, buying and selling, dealings, financial transactions, marketing, merchandising, trade, trading, traffic, trafficking.

commercial *adj* business, economic, financial, mercantile, monetary, money-making, pecuniary, profitable, profit-making, trading. • *n inf* advert, advertisement, *inf* break, *inf* plug.

commiserate *vb* be sorry (for), be sympathetic, comfort, condole, console, feel (for), grieve, mourn, show sympathy (for), sympathize. *Opp* CONGRATULATE.

commission *n* **1** appointment, promotion, warrant. **2** *commission to do a job.* booking, order, request. **3** *commission on a sale.* allowance, *inf* cut, fee, percentage, *inf* rake-off, reward. **4** ▷ COMMITTEE.

commit *vb* **1** be guilty of, carry out, do, enact, execute, perform, perpetrate. **2** *commit to safekeeping.* consign, deliver, deposit, entrust, give, hand over, put away, transfer. **commit yourself** ▷ PROMISE.

commitment *n* **1** assurance, duty, guarantee, liability, pledge, promise, undertaking, vow, word. **2** *commitment to a cause.* adherence, dedication, determination, devotion, involvement, loyalty, zeal. **3** *social commitments.* appointment, arrangement, engagement.

committed *adj* active, ardent, *inf* card-carrying, dedicated, devoted, earnest, enthusiastic, fervent, firm, keen, passionate, resolute, single-minded, staunch, unwavering, wholehearted, zealous. *Opp* APATHETIC.

committee *n* body, council, panel. □ *assembly, board, cabinet, caucus, commission, convention, junta, jury, parliament, quango, synod, think-tank, working party.* ▷ GROUP, MEETING.

common *adj* **1** average, *inf* common or garden, conventional, customary, daily, everyday, familiar, frequent, habitual, normal, ordinary, plain, popular, prevalent, regular, routine, *inf* run-of-the-mill, standard, stock, traditional, typical, undistinguished, unexceptional, unsurprising, usual, well-known, widespread, workaday. ▷ COMMONPLACE. **2** *common knowledge.* accepted, collective, communal, general, joint, mutual, open, popular, public, shared, universal. **3** *the common people.* lower class, lowly, plebeian, proletarian. **4** [*inf*] *Don't be common!* boorish, churlish, coarse, crude, disreputable, ill-bred, inferior, loutish, low, rude, uncouth, unrefined, vulgar, *inf* yobbish. *Opp* ARISTOCRATIC, DISTINCTIVE, UNUSUAL. • *n* heath, park, parkland.

commonplace *adj* banal, boring, forgettable, hackneyed, humdrum, mediocre, obvious, ordinary, pedestrian, plain, platitudinous, predictable, prosaic, routine, standard, trite, unexciting, unremarkable. ▷ COMMON. *Opp* MEMORABLE. • *n* ▷ PLATITUDE.

commotion *n inf* ado, agitation, *inf* bedlam, bother, brawl, *inf* brouhaha, *inf* bust-up, chaos, clamour, confusion, contretemps, din, disorder, disturbance, excitement, ferment, flurry, fracas, fray, furore, fuss, hubbub, hullabaloo, incident, *inf* kerfuffle, noise, *inf* palaver, pandemonium, *inf* punch-up, quarrel, racket, riot, row, rumpus, sensation, *inf* shemozzle, *inf* stir, *inf* to-do, tumult, turbulence, turmoil, unrest, upheaval, uproar, upset.

communal *adj* collective, common, general, joint, mutual, open, public, shared. *Opp* PRIVATE.

communicate *vb* **1** commune, confer, converse, correspond, discuss, get in touch, interrelate, make contact, speak, talk, write (to). **2** *communicate information.* advise, announce, broadcast, convey, declare, disclose, disseminate, divulge, express, get across, impart, indicate, inform, intimate, make known, mention, network, notify, pass on, proclaim, promulgate, publish, put across, put over, relay, report, reveal, say, show, speak, spread, state, transfer, transmit, write. **3** *communicate a disease.* give, infect someone with, pass on, spread, transfer, transmit. **4** *The passage communicates with the kitchen.* be connected, lead (to).

communication *n* **1** communicating, communion, contact, interaction, *old use* intercourse. □ *announcement, bulletin, cable, card, communiqué, conversation, correspondence, dialogue, directive, dispatch, document, fax, gossip,* inf *grapevine, information, intelligence, intimation, letter,* inf *memo, memorandum, message, news, note, notice, proclamation, report, rumour, signal, statement, talk, telegram, transmission, wire, word of mouth, writing.* □ *CB, computer, intercom, radar, telegraph, telephone, teleprinter, walkie-talkie.* **2** *mass communication.* mass media, the media. □ *advertising, broadcasting, cable television, magazines, newspapers, the press, radio, satellite, telecommunication, television.*

communicative *adj* articulate, *inf* chatty, frank, informative, open, out-going, responsive, sociable. ▷ TALKATIVE. *Opp* SECRETIVE

community *n* colony, commonwealth, commune, country, kibbutz, nation, society, state. ▷ GROUP.

commute *vb* **1** adjust, alter, curtail, decrease, lessen, lighten, mitigate, reduce, shorten. **2** ▷ TRAVEL.

compact *adj* **1** close-packed, compacted, compressed, consolidated, dense, firm, heavy, packed, solid, tight-packed. *Opp* LOOSE. **2** handy, neat, portable, small. **3** abbreviated, abridged, brief, compendious, compressed, concentrated, condensed, short, small, succinct, terse. ▷ CONCISE. *Opp* LARGE. • *n* ▷ AGREEMENT.

companion *n* accomplice, assistant, associate, chaperone, colleague, comrade, confederate, confidant(e), consort, *inf* crony, escort, fellow, follower, *inf* henchman, mate, partner, stalwart. ▷ FRIEND, HELPER.

company *n* **1** companionship, fellowship, friendship, society. **2** [*inf* *company for tea.* callers, guests, visitors. **3** *mixed company.* assemblage, association, band, body, circle, club, community, coterie, crew, crowd, ensemble, entourage gang, gathering, society, throng, troop, troupe (*of actors*). 4 *trading company*. business, cartel, concern, conglomerate, consortium, corporation, establishment, firm, house, line, organization, partnership, *inf* set-up, syndicate, union. ▷ GROUP.

comparable *adj* analogous, cognate, commensurate, compatible, corresponding, equal, equivalent, matching, parallel, proportionate, related, similar, twin. *Opp* DISSIMILAR.

compare *vb* check, contrast, correlate, draw parallels (between), equate, juxtapose, liken, make comparisons, make connections (between), measure (against), relate (to), set side by side, weigh (against). **compare with** ▷ EQUAL.

comparison *n* analogy, comparability, contrast, correlation, difference, distinction, juxtaposition, likeness, parallel, relationship, resemblance, similarity.

compartment *n* alcove, area, bay, berth, booth, cell, chamber, *inf* cubbyhole, cubicle, division, hole, kiosk, locker, niche, nook, partition, pigeonhole, section, slot, space, subdivision.

compatible *adj* **1** harmonious, like-minded, similar, well-matched. ▷ FRIENDLY. **2** *compatible claims*. accordant, congruent, consistent, consonant, matching, reconcilable. *Opp* INCOMPATIBLE.

compel *vb* bind, bully, coerce, constrain, dragoon, drive, exact, force, impel, make, necessitate, oblige, order, press, press-gang, pressurize, require, *inf* shanghai, urge.

compendium *n* abridgement, abstract, anthology, collection, condensation, digest, handbook, summary.

compensate *vb* **1** atone, *inf* cough up, expiate, indemnify, make amends, make good, make reparation, make restitution, make up for, pay back, pay compensation, recompense, redress, reimburse, remunerate, repay, requite. **2** counterbalance, counterpoise, even up, neutralize, offset.

compensation *n* amends, damages, indemnity, recompense, refund, reimbursement, reparation, repayment, restitution.

compère *n* anchor-man, announcer, disc jockey, host, hostess, linkman, Master of Ceremonies, MC, presenter.

compete *vb* **1** be a contestant, enter, participate, perform, take part, take up the challenge. **2** be in competition, conflict, contend, emulate, oppose, rival, strive, struggle, undercut, vie. ▷ FIGHT. *Opp* CO-OPERATE. **compete with** ▷ RIVAL.

competent *adj* able, acceptable, accomplished, adept, adequate, capable, clever, effective, effectual, efficient, experienced, expert, fit, *inf* handy, practical, proficient, qualified, satisfactory, skilful, skilled, trained, workmanlike, worthwhile. *Opp* INCOMPETENT.

competition *n* **1** competitiveness, conflict, contention, emulation, rivalry, struggle. **2** challenge, championship, contest, event, game, heat, match, quiz, race, rally, series, tournament, trial.

competitive *adj* **1** aggressive, antagonistic, combative, contentious, cut-throat, hard-fought, keen, lively, sporting, well-fought. **2** *competitive prices*. average, comparable with others, fair, moderate, reasonable, similar to others.

competitor *n* adversary, antagonist, candidate, challenger, contender, contestant, entrant, finalist, opponent, participant, rival.

compile *vb* accumulate, amass, arrange, assemble, collate, collect, compose, edit, gather, marshal, organize, put together.

complain *vb inf* beef, *inf* bellyache, *sl* bind, carp, cavil, find fault, fuss, *inf* gripe, groan, *inf* grouch, grouse, grumble, lament, moan, object, protest, wail, whine, *inf* whinge. *Opp* PRAISE. **complain about** ▷ CRITICIZE.

complaint *n* **1** accusation, *inf* beef, charge, condemnation, criticism, grievance, *inf* gripe, grouse, grumble, moan, objection, protest, stricture, whine, whinge. **2** *a medical complaint.* affliction, ailment, disease, disorder, infection, malady, malaise, sickness, upset. ▷ ILLNESS.

complaisant *adj* accommodating, acquiescent, amenable, biddable, compliant, cooperative, deferential, docile, obedient, obliging, pliant, polite, submissive, tractable, willing. *Opp* OBSTINATE.

complement *n* **1** completion, *inf* finishing touch, perfection. **2** *a full complement.* aggregate, capacity, quota, sum, total. • *vb* add to, complete, make whole, perfect, round off, top up.

complementary *adj* interdependent, matching, reciprocal, toning, twin.

complete *adj* **1** comprehensive, entire, exhaustive, full, intact, total, unabbreviated, unabridged, uncut, unedited, unexpurgated, whole. **2** accomplished, achieved, completed, concluded, done, ended, finished, over. ▷ PERFECT. **3** *a complete disaster.* absolute, arrant, downright, extreme, *inf* out-and-out, outright, pure, rank, sheer, thorough, thoroughgoing, total, unmitigated, unmixed, unqualified, utter, *inf* wholesale. *Opp* INCOMPLETE. • *vb* **1** accomplish, achieve, carry out, clinch, close, conclude, crown, do, end, finalize, finish, fulfil, perfect, perform, round off, terminate, *inf* top off, *inf* wind up. **2** *complete forms.* answer, fill in.

complex *adj* complicated, composite, compound, convoluted, elaborate, *inf* fiddly, heterogeneous, intricate, involved, *inf* knotty (*problem*), labyrinthine, manifold, mixed, multifarious, multiple, multiplex, ornate, perplexing, problematical, sophisticated, tortuous, *inf* tricky. *Opp* SIMPLE.

complexion *n* appearance, colour, colouring, look, pigmentation, skin, texture.

complicate *vb* compound, confound, confuse, elaborate, entangle, make complicated, mix up, muddle, *inf* screw up, *inf* snarl up, tangle, twist. *Opp* SIMPLIFY.

complicated ▷ COMPLEX.

complication *n* complexity, confusion, convolution, difficulty, dilemma, intricacy, *inf* mix-up, obstacle, problem, ramification, set-back, snag, tangle.

compliment *n* accolade, admiration, appreciation, approval, commendation, congratulations, encomium, eulogy, felicitations, flattery, honour, panegyric, plaudits, praise, testimonial, tribute. • *vb* applaud, commend, congratulate, *inf* crack up, eulogize, extol, felicitate, flatter, give credit, laud, pay homage to, praise, salute, speak highly of. *Opp* INSULT.

complimentary *adj* admiring, appreciative, approving, commendatory, congratulatory, encomiastic, eulogistic, favourable, flattering, *derog* fulsome, generous, laudatory, panegyrical, rapturous, supportive. *Opp* ABUSIVE, CONTEMPTUOUS, CRITICAL.

comply *vb* abide (by), accede, accord, acquiesce, adhere (to), agree, assent, be in accordance, coincide, concur, consent, correspond (to), defer, fall in (with), fit in, follow, fulfil, harmonize, keep (to), match, meet, obey, observe, perform, respect, satisfy, square (with), submit, suit, yield. ▷ CONFORM. *Opp* DEFY.

component *n* bit, constituent, element, essential part, ingredient, item, part, piece, *inf* spare, spare part, unit.

compose *vb* **1** build, compile, constitute, construct, fashion, form, frame, make, put together. **2** *compose music.* arrange, create, devise, imagine, make up, produce, write. **3** *compose yourself.* calm, control, pacify, quieten, soothe, tranquillize. be composed of ▷ COMPRISE. **composed** ▷ CALM.

composition *n* **1** assembly, constitution, creation, establishment, formation, formulation, *inf* make-up, setting up. **2** balance, configuration, layout, organization, structure. **3** *a literary composition.* article, essay, story. ▷ WRITING. **4** *a musical composition.* opus, piece, work. ▷ MUSIC.

compound *adj* complex, complicated, composite, intricate, involved, multiple. *Opp* SIMPLE. • *n* **1** alloy, amalgam, blend, combination, composite, composition, fusion, mixture, synthesis. **2** *compound for cattle. Amer* corral, enclosure, pen, run. • *vb* ▷ COMBINE, COMPLICATE.

comprehend *vb* appreciate, apprehend, conceive, discern, fathom, follow, grasp, know, perceive, realize, see, take in, *inf* twig, understand.

comprehensible *adj* clear, easy, intelligible, lucid, meaningful, plain, self-explanatory, simple, straightforward, understandable. *Opp* INCOMPREHENSIBLE.

comprehensive *adj* all-embracing, broad, catholic, compendious, complete, detailed, encyclopaedic, exhaustive, extensive, far-reaching, full, inclusive, indiscriminate, sweeping, thorough, total, universal, wholesale, wide-ranging. *Opp* SELECTIVE.

compress *vb* abbreviate, abridge, compact, concentrate, condense, constrict, contract, cram, crush, flatten, *inf* jam, précis, press, shorten, squash, squeeze, stuff, summarize, telescope, truncate. *Opp* EXPAND. **compressed** ▷ COMPACT, CONCISE.

comprise *vb* be composed of, comprehend, consist of, contain, cover, embody, embrace, include, incorporate, involve.

compromise *n* bargain, concession, *inf* give-and-take, *inf* halfway house, middle course, middle way, settlement. • *vb* **1** concede a point, go to arbitration, make concessions, meet halfway, negotiate a settlement, reach a formula, settle, *inf* split the difference, strike a balance. **2** *compromise your reputation.* damage, discredit, disgrace, dishonour, imperil, jeopardize, prejudice, risk, undermine, weaken. **compromising** ▷ SHAMEFUL.

compulsion *n* **1** coercion, duress, force, necessity, restriction, restraint. **2** *compulsion to smoke.* addiction, drive, habit, impulse, pressure, urge.

compulsive *adj* **1** besetting, compelling, driving, instinctive, involuntary, irresistible, overpowering, overwhelming, powerful, uncontrollable, urgent. **2** *compulsive drinker.* addicted, habitual, incorrigible, incurable, obsessive, persistent.

compulsory *adj* binding, contractual, *Fr* de rigueur, enforceable, essential, imperative, imposed, incumbent, indispensable, inescapable, mandatory, necessary, obligatory, official, prescribed, required, requisite, set, statutory, stipulated, unavoidable. *Opp* OPTIONAL.

compunction *n* contrition, hesitation, pang of conscience, qualm, regret, remorse, scruple, self-reproach.

compute *vb* add up, ascertain, assess, calculate, count, determine, estimate, evaluate, measure, reckon, total, work out.

computer *n* mainframe, micro, microcomputer, mini-computer, PC, personal computer, robot, word-processor.

comrade *n* associate, colleague, companion. ▷ FRIEND.

conceal *vb* blot out, bury, camouflage, cloak, cover up, disguise, envelop, gloss over, hide, hush up, keep dark, keep quiet, keep secret, mask, obscure, screen, secrete, suppress, veil. *Opp* REVEAL. **concealed** ▷ HIDDEN.

concede *vb* accept, acknowledge, admit, agree, allow, confess, grant, make a concession, own, profess, recognize. **concede defeat** capitulate, *inf* cave in, cede, give in, resign, submit, surrender, yield.

conceit *n* arrogance, boastfulness, egotism, self-admiration, self-esteem, self-love, vanity. ▷ PRIDE.

conceited *adj* arrogant, *inf* bigheaded, boastful, bumptious, *inf* cocksure, *inf* cocky, egocentric, egotistic(al), grand, haughty, *inf* high and mighty, immodest, narcissistic, overweening, pleased with yourself, proud, self-centred, self-important, self-satisfied, smug, snobbish, *inf* snooty, *inf* stuck-up, supercilious, *inf* swollen-headed, *inf* toffee-nosed, vain, vainglorious. *Opp* MODEST.

conceive *vb* **1** become pregnant. **2** *conceive a plan.* conjure up, contrive, create, design, devise, *inf* dream up, envisage, evolve, form, formulate, frame, germinate, hatch, imagine, initiate, invent, make up, originate, plan, plot, produce, realize, suggest, think up, visualize, work out. ▷ THINK.

concentrate *n* distillation, essence, extract. • *vb* **1** apply yourself, attend, be absorbed, be attentive, engross yourself, think, work hard. **2** accumulate, centralize, centre, cluster, collect, congregate, converge, crowd, focus, gather, mass. *Opp* DISPERSE. **3** *concentrate a liquid.* condense, reduce, thicken. *Opp* DILUTE. **concentrated 1** ▷ INTENSIVE. **2** condensed, evaporated, reduced, strong, thick, undiluted.

conception *n* **1** begetting, conceiving, fathering, fertilization, genesis, impregnation, initiation, origin. ▷ BEGINNING. **2** ▷ IDEA.

concern *n* **1** attention, care, charge, consideration, heed, interest, regard. **2** *no concern of yours.* affair, business, involvement, matter, problem, responsibility, task.

3 *matter for concern.* anxiety, burden, disquiet, distress, fear, malaise, solicitude, worry. 4 *business concern.* business, company, corporation, establishment, enterprise, firm, organization.
• *vb* affect, be important to, be relevant to, interest, involve, matter to, pertain to, refer to, relate to.

concerned *adj* 1 *concerned parents.* bothered, caring, distressed, disturbed, fearful, perturbed, solicitous, touched, troubled, uneasy, unhappy, upset, worried. ▷ ANXIOUS. 2 *the people concerned.* connected, implicated, interested, involved, referred to, relevant. ▷ RESPONSIBLE.

concerning *prep* about, apropos of, germane to, involving, re, regarding, relating to, relevant to, with reference to, with regard to.

concert *n* performance, programme, show. ▷ ENTERTAINMENT, MUSIC.

concerted *adj* collaborative, collective, combined, cooperative, joint, mutual, shared, united.

concession *n* adjustment, allowance, reduction.

concise *adj* brief, compact, compendious, compressed, concentrated, condensed, epigrammatic, laconic, pithy, short, small, succinct, terse. ▷ ABRIDGED. *Opp* DIFFUSE.

conclude *vb* 1 cease, close, complete, culminate, end, finish, round off, stop, terminate. 2 assume, decide, deduce, gather, infer, judge, reckon, suppose, surmise. ▷ THINK.

conclusion *n* 1 close, completion, culmination, end, epilogue, finale, finish, peroration, rounding-off, termination. 2 answer, belief, decision, deduction, inference, interpretation, judgement, opinion, outcome, resolution, result, solution, upshot, verdict.

conclusive *adj* certain, convincing, decisive, definite, persuasive, unambiguous, unanswerable, unequivocal, unquestionable. *Opp* INCONCLUSIVE.

concoct *vb* contrive, cook up, counterfeit, devise, fabricate, feign, formulate, hatch, invent, make up, plan, prepare, put together, think up.

concord *n* agreement, euphony, harmony, peace.

concrete *adj* actual, definite, existing, factual, firm, material, objective, palpable, physical, real, solid, substantial, tactile, tangible, touchable, visible. *Opp* ABSTRACT.

concur *vb* accede, accord, agree, assent. ▷ COMPLY.

concurrent *adj* coexisting, coinciding, concomitant, contemporaneous, contemporary, overlapping, parallel, simultaneous, synchronous.

condemn *vb* 1 blame, castigate, censure, criticize, damn, decry, denounce, deplore, deprecate, disapprove of, disparage, execrate, rebuke, reprehend, reprove, revile, *inf* slam, *inf* slate, upbraid. *Opp* COMMEND. 2 convict, find guilty, judge, pass judgement, prove guilty, punish, sentence. *Opp* ACQUIT.

condense *vb* 1 abbreviate, abridge, compress, contract, curtail, précis, reduce, shorten, summarize, synopsize. *Opp* EXPAND. 2 *condense a liquid.* concentrate, distil, reduce, solidify, thicken. *Opp* DILUTE.

condensation *n* haze, mist, precipitation, *inf* steam, water-drops.

condescend *vb* deign, demean yourself, humble yourself, lower yourself, stoop. **condescending** ▷ HAUGHTY.

condition *n* **1** case, circumstance, *inf* fettle, fitness, form, health, *inf* nick, order, shape, situation, state, *inf* trim, working order. **2** limitation, obligation, prerequisite, proviso, qualification, requirement, requisite, restriction, stipulation, terms. **3** *medical condition.* ▷ ILLNESS. • *vb* acclimatize, accustom, brainwash, educate, mould, prepare, re-educate, *inf* soften up, teach, train.

conditional *adj* dependent, limited, provisional, qualified, restricted, safeguarded, *inf* with strings attached. *Opp* UNCONDITIONAL.

condone *vb* allow, connive at, disregard, endorse, excuse, forgive, ignore, let someone off, overlook, pardon, tolerate.

conducive *adj* advantageous, beneficial, encouraging, favourable, helpful, supportive. **be conducive to** ▷ ENCOURAGE.

conduct *n* **1** actions, attitude, bearing, behaviour, comportment, demeanour, deportment, manners, ways. **2** *conduct of affairs.* administration, control, direction, discharge, government, guidance, handling, leading, management, operation, organization, regulation, running, supervision. • *vb* **1** administer, be in charge of, chair, command, control, direct, escort, govern, handle, head, look after, manage, organize, oversee, preside over, regulate, rule, run, steer, superintend, supervise, usher. **2** *conduct me home.* accompany, escort, guide, lead, pilot, take, usher. **3** *conduct electricity.* carry, channel, convey, transmit. **conduct yourself** ▷ BEHAVE.

confer *vb* **1** accord, award, bestow, give, grant, honour with, impart, invest, present. **2** compare notes, consult, converse, debate, deliberate, discourse, discuss, exchange ideas, *inf* put your heads together, seek advice. ▷ TALK.

conference *n* colloquium, congress, consultation, convention, council, deliberation, discussion, forum, seminar, symposium. ▷ MEETING.

confess *vb* acknowledge, admit, be truthful, *inf* come clean, concede, disclose, divulge, *inf* make a clean breast (of), own up, unbosom yourself, unburden yourself.

confession *n* acknowledgement, admission, declaration, disclosure, expression, profession, revelation.

confide *vb* consult, speak confidentially, *inf* spill the beans, open your heart, *inf* tell all, tell secrets, unbosom yourself, trust.

confidence *n* **1** belief, certainty, credence, faith, hope, optimism, positiveness, reliance, trust. **2** aplomb, assurance, boldness, composure, conviction, firmness, nerve, panache, self-assurance, self-confidence, self-possession, self-reliance, spirit, verve. *Opp* DOUBT, HESITATION. **have confidence in** ▷ TRUST.

confident *adj* **1** certain, convinced, hopeful, optimistic, positive, sanguine, sure, trusting. **2** *a confident person.* assertive, assured, bold, *derog* cocksure, composed, cool, definite, fearless, secure, self-assured, self-confident, self-possessed, self-reliant, unafraid. *Opp* DOUBTFUL.

confidential *adj* **1** classified, *inf* hush-hush, intimate, *inf* off the record, personal, private, restricted, secret, suppressed, top secret. **2** *confidential secretary*. personal, private, trusted.

confine *vb* bind, box in, cage, circumscribe, constrain, *inf* coop up, cordon off, cramp, curb, detain, enclose, gaol, hedge in, hem in, *inf* hold down, immure, incarcerate, isolate, keep in, limit, localize, restrain, restrict, rope off, shut in, shut up, surround, wall up. ▷ IMPRISON. *Opp* FREE.

confirm *vb* **1** authenticate, back up, bear out, corroborate, demonstrate, endorse, establish, fortify, give credence to, justify, lend force to, prove, reinforce, settle, show, strengthen, substantiate, support, underline, witness to, vindicate. **2** *confirm a deal*. authorize, *inf* clinch, formalize, guarantee, make legal, make official, ratify, sanction, validate, verify.

confiscate *vb* appropriate, commandeer, expropriate, impound, remove, seize, sequester, sequestrate, take away, take possession of.

conflict *n* **1** antagonism, antipathy, contention, contradiction, difference, disagreement, discord, dissension, friction, hostility, incompatibility, inconsistency, opposition, strife. **2** altercation, battle, *inf* brush, clash, combat, confrontation, contest, dispute, encounter, engagement, feud, fight, quarrel, row, *inf* set-to, skirmish, struggle, war, warfare, wrangle. • *vb* **1** *inf* be at odds, be at variance, be incompatible, clash, compete, contend, contradict, contrast, *inf* cross swords, differ, disagree, oppose each other. ▷ FIGHT, QUARREL.

conform *vb* acquiesce, agree, be good, behave conventionally, blend in, *inf* do what you are told, fit in, *inf* keep in step, obey, *inf* see eye to eye, *inf* toe the line. ▷ COMPLY.

conformist *n* conventional person, traditionalist, yes-man. *Opp* REBEL.

conformity *n* complaisance, compliance, conventionality, obedience, orthodoxy, submission, uniformity.

confront *vb* accost, argue with, attack, brave, challenge, defy, encounter, face up to, meet, oppose, resist, stand up to, take on, withstand. *Opp* AVOID.

confuse *vb* **1** disarrange, disorder, distort, entangle, garble, jumble, *inf* mess up, mingle, mix up, muddle, tangle, *inf* throw into disarray, upset. **2** *rules confuse me*. agitate, baffle, befuddle, bemuse, bewilder, confound, disconcert, disorientate, distract, *inf* flummox, fluster, mislead, mystify, perplex, puzzle, *inf* rattle, *inf* throw. **3** *confuse twins*. fail to distinguish. **confusing** ▷ PUZZLING.

confused *adj* **1** chaotic, disordered, disorderly, disorganized, *inf* higgledy-piggledy, jumbled, messy, mixed up, muddled, *inf* screwed-up, *sl* shambolic, *inf* topsy-turvy, twisted. **2** *confused ideas*. aimless, contradictory, disconnected, disjointed, garbled, incoherent, inconsistent, irrational, misleading, obscure, rambling, unclear, unsound, unstructured, woolly. **3** *confused mind*. addled, addle-headed, baffled, bewildered, dazed, disorientated, distracted, flustered, fuddled, *inf* in a tizzy, inebriated, muddle-headed, *inf* muzzy, mystified, nonplussed, perplexed, puzzled. ▷ MAD. *Opp* ORDERLY.

confusion *n* **1** *inf* ado, anarchy, bedlam, bother, chaos, clutter, commotion, confusion, din, disorder, disorganization, disturbance, fuss, hubbub, hullabaloo, jumble, maelstrom, *inf* mayhem, mêlée, mess, *inf* mix-up, muddle, pandemonium, racket, riot, rumpus, shambles, tumult, turbulence, turmoil, upheaval, uproar, welter, whirl. **2** *mental confusion.* bemusement, bewilderment, disorientation, distraction, mystification, perplexity, puzzlement. *Opp* ORDER.

congeal *vb* clot, coagulate, coalesce, condense, curdle, freeze, harden, *inf* jell, set, solidify, stiffen, thicken.

congenial *adj* acceptable, agreeable, amicable, companionable, compatible, genial, kindly, suitable, sympathetic, understanding, well-suited. ▷ FRIENDLY. *Opp* UNCONGENIAL.

congenital *adj* hereditary, inborn, inbred, inherent, inherited, innate, natural.

congested *adj* blocked, choked, clogged, crammed, crowded, full, jammed, obstructed, overcrowded, stuffed. *Opp* CLEAR.

congratulate *vb* applaud, compliment, felicitate, praise.

congregate *vb* assemble, cluster, collect, come together, convene, converge, crowd, forgather, gather, get together, mass, meet, muster, rally, rendezvous, swarm, throng. ▷ GROUP.

conjure *vb* bewitch, charm, compel, enchant, invoke, raise, rouse, summon. **conjure up** ▷ PRODUCE.

conjuring *n* illusions, legerdemain, magic, sleight of hand, tricks, wizardry.

connect *vb* **1** attach, combine, couple, engage, fix, interlock, join, link, put on, switch on, tie, turn on, unite. ▷ FASTEN. **2** associate, bracket together, compare, make a connection between, put together, relate, tie up. *Opp* SEPARATE.

connection *n* affinity, association, bond, coherence, contact, correlation, correspondence, interrelationship, link, relationship, relevance, tie, *inf* tie-up, unity. *Opp* SEPARATION.

conquer *vb* **1** annex, beat, best, capture, checkmate, crush, defeat, get the better of, humble, *inf* lick, master, occupy, outdo, overcome, overpower, overrun, overthrow, overwhelm, possess, prevail over, quell, rout, seize, silence, subdue, subject, subjugate, succeed against, surmount, take, *inf* thrash, triumph over, vanquish, worst. ▷ WIN. **2** *conquer a mountain.* climb, reach the top of.

conquest *n* annexation, appropriation, capture, defeat, domination, invasion, occupation, overthrow, subjection, subjugation, *inf* takeover. ▷ VICTORY.

conscience *n* compunction, ethics, honour, fairness, misgivings, morality, morals, principles, qualms, reservations, scruples, standards.

conscientious *adj* accurate, attentive, careful, diligent, dutiful, exact, hard-working, high-minded, honest, meticulous, painstaking, particular, punctilious, responsible, rigorous, scrupulous, serious, thorough. *Opp* CARELESS.

conscious *adj* **1** alert, awake, aware, compos mentis, sensible. **2** *a conscious act.* calculated, deliberate, intended, intentional, knowing, planned, premeditated, self-conscious, studied, voluntary, waking, wilful. *Opp* UNCONSCIOUS.

consecrate *vb* bless, dedicate, devote, hallow, make sacred, sanctify. *Opp* DESECRATE.

consecutive *adj* continuous, following, one after the other, running (*3 days running*), sequential, succeeding, successive.

consent *n* acquiescence, agreement, approval, assent, concurrence, imprimatur, permission, seal of approval. • *vb* accede, acquiesce, agree, approve, comply, concede, concur, conform, submit, undertake, yield. *Opp* REFUSE. **consent to** ▷ ALLOW.

consequence *n* **1** aftermath, by-product, corollary, effect, end, *inf* follow-up, issue, outcome, repercussion, result, sequel, side-effect, upshot. **2** *of no consequence.* account, concern, importance, moment, note, significance, value, weight.

consequent *adj* consequential, ensuing, following, resultant, resulting, subsequent.

conservation *n* careful management, economy, good husbandry, maintenance, preservation, protection, safeguarding, saving, upkeep. *Opp* DESTRUCTION.

conservationist *n* ecologist, environmentalist, *inf* green, preservationist.

conservative *adj* **1** conventional, die-hard, hidebound, moderate, narrow-minded, old-fashioned, reactionary, sober, traditional, unadventurous. **2** *conservative estimate.* cautious, moderate, reasonable, understated, unexaggerated. **3** *conservative politics.* right-of-centre, right-wing, Tory. *Opp* PROGRESSIVE. • *n* conformist, die-hard, reactionary, right-winger, Tory, traditionalist.

conserve *vb* be economical with, hold in reserve, keep, look after, maintain, preserve, protect, safeguard, save, store up, use sparingly. *Opp* DESTROY, WASTE.

consider *vb* **1** *inf* chew over, cogitate, contemplate, deliberate, discuss, examine, meditate, mull over, muse, ponder, puzzle over, reflect, ruminate, study, *inf* turn over, weigh up. ▷ THINK. **2** believe, deem, judge, reckon.

considerable *adj* appreciable, big, biggish, comfortable, fairly important, fairly large, noteworthy, noticeable, perceptible, reasonable, respectable, significant, sizeable, substantial, *inf* tidy (amount), tolerable, worthwhile. *Opp* NEGLIGIBLE.

considerate *adj* accommodating, altruistic, attentive, caring, charitable, cooperative, friendly, generous, gracious, helpful, kind, kind-hearted, kindly, neighbourly, obliging, polite, sensitive, solicitous, sympathetic, tactful, thoughtful, unselfish. *Opp* SELFISH.

consign *vb* commit, convey, deliver, devote, entrust, give, hand over, pass on, relegate, send, ship, transfer.

consignment *n* batch, cargo, delivery, load, lorry-load, shipment, van-load.

consist *vb* **consist of** add up to, amount to, be composed of, be made of, comprise, contain, embody, include, incorporate, involve.

consistent *adj* **1** constant, dependable, faithful, predictable, regular, reliable, stable, steadfast, steady, unchanging, undeviating, unfailing, uniform, unvarying.

2 *The stories are consistent.* accordant, compatible, congruous, consonant, in accordance, in agreement, in harmony, of a piece. *Opp* INCONSISTENT.

console *vb* calm, cheer, comfort, ease, encourage, hearten, relieve, solace, soothe, succour, sympathize with.

consolidate *vb* make secure, make strong, reinforce, stabilize, strengthen. *Opp* WEAKEN.

consort *vb* **consort with** accompany, associate with, befriend, be friends with, be seen with, fraternize with, *inf* gang up with, keep company with, mix with.

conspicuous *adj* apparent, blatant, clear, discernible, distinguished, dominant, eminent, evident, flagrant, glaring, impressive, manifest, marked, notable, noticeable, obtrusive, obvious, ostentatious, outstanding, patent, perceptible, plain, prominent, pronounced, self-evident, shining (*example*), showy, striking, unconcealed, unmistakable, visible. *Opp* INCONSPICUOUS.

conspiracy *n* cabal, collusion, connivance, *inf* frame-up, insider dealing, intrigue, machinations, plot, *inf* racket, scheme, stratagem, treason.

conspirator *n* plotter, schemer, traitor, *inf* wheeler-dealer.

conspire *vb* be in league, collude, combine, connive, cooperate, hatch a plot, have designs, intrigue, plot, scheme.

constant *adj* 1 ceaseless, chronic, consistent, continual, continuous, endless, eternal, everlasting, fixed, immutable, incessant, invariable, neverending, non-stop, permanent, perpetual, persistent, predictable, regular, relentless, repeated, stable, steady, sustained, unbroken, unchanging, unending, unflagging, uniform, uninterrupted, unremitting, unvarying. 2 *a constant friend.* dedicated, dependable, determined, devoted, faithful, firm, indefatigable, loyal, reliable, resolute, staunch, steadfast, tireless, true, trustworthy, trusty, unswerving, unwavering. *Opp* CHANGEABLE.

constitute *vb* appoint, bring together, compose, comprise, create, establish, form, found, inaugurate, make (up), set up.

construct *vb* assemble, build, create, engineer, erect, fabricate, fashion, fit together, form, *inf* knock together, make, manufacture, pitch (*tent*), produce, put together, put up, set up. *Opp* DEMOLISH.

construction *n* 1 assembly, building, creation, erecting, erection, manufacture, production, putting-up, setting-up. 2 building, edifice, erection, structure.

constructive *adj* advantageous, beneficial, cooperative, creative, helpful, positive, practical, productive, useful, valuable, worthwhile. *Opp* DESTRUCTIVE.

consult *vb* confer, debate, discuss, exchange views, *inf* put your heads together, refer (to), seek advice, speak (to), *inf* talk things over. ▷ QUESTION.

consume *vb* 1 devour, digest, drink, eat, *inf* gobble up, *inf* guzzle, *inf* put away, swallow. 2 *consume energy.* absorb, deplete, drain, eat into, employ, exhaust, expend, swallow up, use up, utilize.

contact *n* connection, join, junction, touch, union. ▷ COMMUNICATION. • *vb* apply to, approach, call on, communicate with, correspond with, *inf* drop a line to, get hold of, get in touch with, make overtures to, notify, phone, ring, sound out, speak to, talk to, telephone.

contagious *adj* catching, communicable, infectious, spreading, transmittable.

contain *vb* **1** accommodate, enclose, hold. **2** be composed of, comprise, consist of, embody, embrace, include, incorporate, involve. **3** *contain your anger.* check, control, curb, hold back, keep back, limit, repress, restrain, stifle.

container *n* holder, receptacle, repository, vessel. ▷ BAG, BARREL, BOTTLE, BOWL, BOX, CUP, DISH, GLASS, LUGGAGE, POT.

contaminate *vb* adulterate, befoul, corrupt, debase, defile, dirty, foul, infect, poison, pollute, soil, spoil, stain, sully, taint. *Opp* PURIFY.

contemplate *vb* **1** eye, gaze at, look at, observe, regard, stare at, survey, view, watch. ▷ SEE. **2** cogitate, consider, day-dream, deliberate, examine, meditate, mull over, muse, plan, ponder, reflect, ruminate, study, work out. ▷ THINK. **3** envisage, expect, intend, propose.

contemporary *adj* **1** *contemporary with me at school.* coeval, coexistent, coinciding, concurrent, contemporaneous, simultaneous, synchronous. **2** *contemporary music.* current, fashionable, the latest, modern, newest, novel, present-day, *inf* trendy, topical, up-to-date, *inf* with-it.

contempt *n* abhorrence, contumely, derision, detestation, disdain, disgust, dislike, disrespect, loathing, ridicule, scorn. ▷ HATRED. *Opp* ADMIRATION. **feel contempt for** ▷ DESPISE.

contemptible *adj* base, beneath contempt, despicable, detestable, discreditable, disgraceful, dishonourable, disreputable, ignominious, inferior, loathsome, low-down, mean, odious, pitiful, *inf* shabby, shameful, worthless, wretched. ▷ HATEFUL. *Opp* ADMIRABLE.

contemptuous *adj* arrogant, belittling, condescending, derisive, disdainful, dismissive, disrespectful, haughty, *inf* holier-than-thou, imperious, insolent, insulting, jeering, lofty, patronizing, sarcastic, scathing, scornful, sneering, *sl* snide, snobbish, *inf* snooty, *sl* snotty, supercilious, superior, withering. *Opp* RESPECTFUL. **be contemptuous of** ▷ DESPISE.

contend *vb* **1** compete, contest, cope, dispute, grapple, oppose, rival, strive, struggle, vie. ▷ FIGHT, QUARREL. **2** *contend that you're innocent.* affirm, allege, argue, assert, claim, declare, maintain, plead.

content *adj* ▷ CONTENTED. • *n* **1** constituent, element, ingredient, part. **2** ▷ CONTENTMENT. • *vb* ▷ SATISFY.

contented *adj* cheerful, comfortable, *derog* complacent, content, fulfilled, gratified, peaceful, pleased, relaxed, satisfied, serene, smiling, smug, uncomplaining, untroubled, well-fed. ▷ HAPPY. *Opp* DISSATISFIED.

contentment *n* comfort, content, contentedness, ease, fulfilment, relaxation, satisfaction, serenity, smugness, tranquillity, well-being. ▷ HAPPINESS. *Opp* DISSATISFACTION.

contest *n* ▷ COMPETITION, FIGHT. • *vb* **1** compete for, contend for, fight for, *inf* make a bid for, strive for, struggle for, take up the challenge of, vie for. **2** *contest a decision.* argue against, challenge, debate, dispute, doubt, oppose, query, question, refute, resist.

contestant *n* candidate, competitor, contender, entrant, opponent, participant, player, rival.

context *n* background, environment, frame of reference, framework, milieu, position, setting, situation, surroundings.

continual *adj* eternal, everlasting, frequent, limitless, ongoing, perennial, perpetual, recurrent, regular, repeated. ▷ CONTINUOUS. *Opp* OCCASIONAL.

continuation *n* **1** continuance, extension, maintenance, prolongation, protraction, resumption. **2** addition, appendix, postscript, sequel, supplement.

continue *vb* **1** carry on, endure, go on, keep on, last, linger, persevere, persist, proceed, pursue, remain, stay, *inf* stick at, survive, sustain. **2** *continue after lunch.* *inf* pick up the threads, recommence, restart, resume. **3** *continue a series.* extend, keep going, keep up, lengthen, maintain, prolong.

continuous *adj* ceaseless, constant, continuing, endless, incessant, interminable, lasting, never-ending, non-stop, permanent, persistent, relentless, *inf* round-the-clock, *inf* solid, sustained, unbroken, unceasing, unending, uninterrupted, unremitting. ▷ CHRONIC, CONTINUAL. *Opp* INTERMITTENT.

contour *n* curve, form, outline, relief, shape.

contract *n* agreement, bargain, bond, commitment, compact, concordat, covenant, deal, indenture, lease, pact, settlement, treaty, understanding, undertaking. • *vb* **1** become denser, become smaller, close up, condense, decrease, diminish, draw together, dwindle, fall away, lessen, narrow, reduce, shrink, shrivel, slim down, thin out, wither. *Opp* EXPAND. **2** agree, arrange, close a deal, covenant, negotiate a deal, promise, sign an agreement, undertake. **3** *contract a disease.* become infected by, catch, develop, get.

contraction *n* **1** diminution, narrowing, shortening, shrinkage, shrivelling. **2** abbreviation, diminutive, shortened form.

contradict *vb* argue with, challenge, confute, controvert, deny, disagree with, dispute, gainsay, impugn, oppose, speak against.

contradictory *adj* antithetical, conflicting, contrary, different, discrepant, incompatible, inconsistent, irreconcilable, opposed, opposite. *Opp* COMPATIBLE.

contraption *n* apparatus, contrivance, device, gadget, invention, machine, mechanism.

contrary *adj* **1** conflicting, contradictory, converse, different, opposed, opposite, other, reverse. **2** *contrary winds.* adverse, hostile, inimical, opposing, unfavourable. **3** *a contrary child.* awkward, cantankerous, defiant, difficult, disobedient, disobliging, disruptive, intractable, obstinate, perverse, rebellious, *inf* stroppy, stubborn, subversive, uncooperative, unhelpful, wayward, wilful. *Opp* HELPFUL.

contrast *n* antithesis, comparison, difference, differentiation, disparity, dissimilarity, distinction, divergence, foil, opposition. *Opp* SIMILARITY. • *vb* **1** compare, differentiate, discriminate, distinguish, emphasize differences, make a distinction, set one against the other. **2** be set off (by), clash, conflict, deviate (from), differ (from). **contrasting** ▷ DISSIMILAR.

contribute *vb* add, bestow, *inf* chip in, donate, *inf* fork out, furnish, give, present, provide, put up, subscribe, supply. **contribute to** ▷ SUPPORT.

contribution *n* **1** donation, fee, gift, grant, *inf* hand-out, offering, payment, sponsorship, subscription. **2** addition, encouragement, input, support. ▷ HELP.

contributor *n* **1** backer, benefactor, donor, giver, helper, patron, sponsor, subscriber, supporter. **2** ▷ WRITER.

control *n* **1** administration, authority, charge, command, curb, direction, discipline, government, grip, guidance, influence, jurisdiction, leadership, management, mastery, orderliness, organization, oversight, power, regulation, restraint, rule, strictness, supervision, supremacy, sway. **2** button, dial, handle, key, lever, switch. • *vb* **1** administer, *inf* be at the helm, be in charge, *inf* boss, command, conduct, cope with, deal with, direct, dominate, engineer, govern, guide, handle, have control of, lead, look after, manage, manipulate, order about, oversee, regiment, regulate, rule, run, superintend, supervise. **2** *control animals.* check, confine, contain, curb, hold back, keep in check, master, repress, restrain, subdue, supress.

controversial *adj* **1** arguable, controvertible, debatable, disputable, doubtful, problematical, questionable. *Opp* ACCEPTED. **2** argumentative, contentious, dialectic, litigious, polemical, provocative.

controversy *n* altercation, argument, confrontation, contention, debate, disagreement, dispute, dissension, issue, polemic, quarrel, war of words, wrangle.

convalesce *vb* get better, improve, make progress, mend, recover, recuperate, regain strength.

convalescent *adj* getting better, healing, improving, making progress, *inf* on the mend, recovering, recuperating.

convene *vb* bring together, call, convoke, summon. ▷ GATHER.

convenient *adj* accessible, appropriate, at hand, available, commodious, expedient, handy, helpful, labour-saving, nearby, neat, opportune, serviceable, suitable, timely, usable, useful. *Opp* INCONVENIENT.

convention *n* **1** custom, etiquette, formality, matter of form, practice, rule, tradition. **2** ▷ ASSEMBLY.

conventional *adj* **1** accepted, accustomed, commonplace, correct, customary, decorous, expected, formal, habitual, mainstream, orthodox, prevalent, received, *inf* run-of-the-mill, standard, straight, traditional, unadventurous, unimaginative, unoriginal, unsurprising. ▷ ORDINARY. **2** [*derog*] bourgeois, conservative, hidebound, pedestrian, reactionary, rigid, stereotyped, *inf* stuffy. *Opp* UNCONVENTIONAL.

converge *vb* coincide, combine, come together, join, link up, meet, merge, unite. *Opp* DIVERGE.

conversation *n inf* chat, *inf* chin-wag, colloquy, communication, conference, dialogue, discourse, discussion, exchange of views, gossip, *inf* heart-to-heart, intercourse, *inf* natter, palaver, phone-call, *inf* powwow, tête-à-tête. ▷ TALK.

convert *vb* change someone's mind, convince, persuade, re-educate, reform, regenerate, rehabilitate, save, win over. ▷ CHANGE.

convey *vb* **1** bear, bring, carry, conduct, deliver, export, ferry, fetch, forward, import, move, send, shift, ship, shuttle, take, taxi, transfer, transport. **2** *convey a message.* communicate, disclose, impart, imply, indicate, mean, relay, reveal, signify, tell, transmit.

convict *n* condemned person, criminal, culprit, felon, malefactor, prisoner, wrongdoer. • *vb* condemn, declare guilty, prove guilty, sentence. *Opp* ACQUIT.

conviction *n* **1** assurance, certainty, confidence, firmness. **2** *religious conviction.* belief, creed, faith, opinion, persuasion, position, principle, tenet, view.

convince *vb* assure, *inf* bring round, convert, persuade, prove to, reassure, satisfy, sway, win over.

convincing ▷ PERSUASIVE.

convulsion *n* **1** disturbance, eruption, outburst, tremor, turbulence, upheaval. **2** [*medical*] attack, fit, paroxysm, seizure, spasm.

convulsive *adj* jerky, shaking, spasmodic, *inf* twitchy, uncontrolled, uncoordinated, violent, wrenching.

cook *vb* concoct, heat up, make, prepare, warm up. □ *bake, barbecue, boil, braise, brew, broil, casserole, coddle, fry, grill, pickle, poach, roast, sauté, scramble, simmer, steam, stew, toast.* **cook up** ▷ PLOT.

cooking *n* baking, catering, cookery, cuisine.

cool *adj* **1** chilled, chilly, coldish, iced, refreshing, unheated. ▷ COLD. *Opp* HOT. **2** calm, collected, composed, dignified, elegant, *inf* laid-back, level-headed, phlegmatic, quiet, relaxed, self-possessed, sensible, serene, unexcited, unflustered, unruffled, urbane. ▷ BRAVE. **3** [*derog*] aloof, apathetic, cold-blooded, dispassionate, distant, frigid, half-hearted, indifferent, lukewarm, negative, offhand, reserved, *inf* stand-offish, unconcerned, unemotional, unenthusiastic, unfriendly, uninvolved, unresponsive, unsociable, unwelcoming. *Opp* PASSIONATE. **4** [*inf*] *cool customer.* ▷ INSOLENT. • *vb* **1** chill, freeze, ice, refrigerate. *Opp* HEAT. **2** *cool your enthusiasm.* abate, allay, assuage, calm, dampen, diminish, lessen, moderate, *inf* pour cold water on, quiet, temper. *Opp* INFLAME.

cooperate *vb* act in concert, collaborate, combine, conspire, help each other, *inf* join forces, *inf* pitch in, *inf* play along, *inf* play ball, *inf* pull together, support each other, unite, work as a team, work together. *Opp* COMPETE.

cooperation *n* assistance, collaboration, cooperative effort, coordination, help, joint action, mutual support, teamwork. *Opp* COMPETITION.

cooperative *adj* **1** accommodating, comradely, constructive, hard-working, helpful to each other, keen, obliging, supportive, united, willing, working as a team. **2** *cooperative effort.* collective, combined, communal, concerted, coordinated, corporate, joint, shared.

cope *vb* get by, make do, manage, survive, win through. **cope with** ▷ ENDURE, MANAGE.

copious *adj* abundant, ample, bountiful, extravagant, generous, great, huge, inexhaustible, large, lavish, liberal, luxuriant, overflowing, plentiful, profuse, unsparing, unstinting. *Opp* SCARCE.

copy *n* **1** carbon copy, clone, counterfeit, double, duplicate, facsimile, fake, forgery, imitation, likeness, model, pattern, photocopy, print, replica, representation, reproduction, tracing, transcript, twin, Xerox. **2** *copy of a book.* edition, volume. • *vb* **1** borrow, counterfeit, crib, duplicate, emulate, follow, forge, imitate, photocopy, plagiarize, print, repeat, reproduce, simulate, transcribe. **2** ape, imitate, impersonate, mimic, parrot.

cord *n* cable, catgut, lace, line, rope, strand, string, twine, wire.

cordon *n* barrier, chain, fence, line, ring, row. **cordon off** ▷ ISOLATE.

core *n* **1** centre, heart, inside, middle, nucleus. **2** *core of a problem.* central issue, crux, essence, gist, heart, kernel, *sl* nitty-gritty, nub.

cork *n* bung, plug, stopper.

corner *n* **1** angle, crook, joint. **2** bend, crossroads, intersection, junction, turn, turning. **3** *a quiet corner.* hideaway, hiding-place, hole, niche, nook, recess, retreat. • *vb* capture, catch, trap.

corporation *n* **1** company, concern, enterprise, firm, organization. **2** council, local government.

corpse *n* body, cadaver, carcass, mortal remains, remains, skeleton, *sl* stiff.

correct *adj* **1** accurate, authentic, confirmed, exact, factual, faithful, faultless, flawless, genuine, literal, precise, reliable, right, strict, true, truthful, verified. **2** *correct manners.* acceptable, appropriate, fitting, just, normal, proper, regular, standard, suitable, tactful, unexceptionable, well-mannered. *Opp* WRONG. • *vb* **1** adjust, alter, cure, *inf* debug, put right, rectify, redress, remedy, repair. **2** *correct pupils' work.* assess, mark. **3** ▷ REPRIMAND.

correspond *vb* accord, agree, be congruous, be consistent, coincide, concur, conform, correlate, fit, harmonize, match, parallel, square, tally. **corresponding** ▷ EQUIVALENT. **correspond with** communicate with, send letters to, write to.

correspondence *n* letters, memoranda, *inf* memos, messages, notes, writings.

correspondent *n* contributor, journalist, reporter. ▷ WRITER.

corridor *n* hall, hallway, passage, passageway.

corrode *vb* **1** consume, eat into, erode, oxidize, rot, rust, tarnish. **2** crumble, deteriorate, disintegrate, tarnish.

corrugated *adj* creased, *inf* crinkly, fluted, furrowed, lined, puckered, ribbed, ridged, wrinkled.

corrupt *adj inf* bent, bribable, criminal, crooked, debauched, decadent, degenerate, depraved, *inf* dirty, dishonest, dishonourable, dissolute, evil, false, fraudulent, illegal, immoral, iniquitous, low, perverted, profligate, rotten, sinful, unethical, unprincipled, unscrupulous, unsound, untrustworthy, venal, vicious, wicked. *Opp* HONEST. • *vb* **1** bribe, divert, *inf* fix, influence, pervert, suborn, subvert. **2** *corrupt the innocent.* debauch, deprave, lead astray, make corrupt, tempt, seduce.

cosmetics *n* make-up, toiletries.

cosmic *adj* boundless, endless, infinite, limitless, universal.

cosmopolitan *adj* international, multicultural, sophisticated, urbane. *Opp* PROVINCIAL.

cost *n* amount, charge, expenditure, expense, fare, figure, outlay, payment, price, rate, tariff, value. • *vb* be valued at, be worth, fetch, go for, realize, sell for, *inf* set you back.

costume *n* apparel, attire, clothing, dress, fancy-dress, garb, garments, *inf* get-up, livery, outfit, period dress, raiment, robes, uniform, vestments. ▷ CLOTHES.

cosy *adj* comfortable, *inf* comfy, easy, homely, intimate, reassuring, relaxing, restful, secure, snug, soft, warm. *Opp* UNCOMFORTABLE.

council *n* committee, conclave, convention, convocation, corporation, gathering, meeting. ▷ ASSEMBLY.

counsel *n* ▷ LAWYER. • *vb* advise, discuss (with), give help, guide, listen to your views.

count *vb* **1** add up, calculate, check, compute, enumerate, estimate, figure out, keep account of, *inf* notch up, number, reckon, score, take stock of, tell, total, *inf* tot up, work out. **2** be important, have significance, matter, signify. **count on** ▷ EXPECT.

countenance *n* air, appearance, aspect, demeanour, expression, face, features, look, visage. • *vb* ▷ APPROVE.

counter *n* **1** bar, sales-point, service-point, table. **2** chip, disc, marker, piece, token. • *vb* answer, *inf* come back at, contradict, defend yourself against, hit back at, parry, react to, rebut, refute, reply to, resist, ward off.

counteract *vb* act against, annul, be an antidote to, cancel out, counterbalance, fight against, foil, invalidate, militate against, negate, neutralize, offset, oppose, resist, thwart, withstand, work against.

counterbalance *vb* balance, compensate for, counteract, counterpoise, counterweight, equalize.

counterfeit *adj* artificial, bogus, copied, ersatz, fake, false, feigned, forged, fraudulent, imitation, make-believe, meretricious, pastiche, *inf* phoney, *inf* pretend, *inf* pseudo, sham, simulated, spurious, synthetic. *Opp* GENUINE. • *vb* copy, fake, falsify, feign, forge, imitate, pretend, *inf* put on, sham, simulate.

countless *adj* endless, immeasurable, incalculable, infinite, innumerable, limitless, many, measureless, myriad, numberless, numerous, unnumbered, untold. *Opp* ▷ FINITE.

country *n* **1** canton, commonwealth, domain, empire, kingdom, land, nation, people, power, principality, realm, state, territory. **2** *open country.* countryside, green belt, landscape, scenery.

couple *n* brace, duo, pair, twosome. • *vb* **1** combine, connect, fasten, hitch, join, link, match, pair, unite, yoke. ▷ MATE.

coupon *n* tear-off slip, ticket, token, voucher.

courage *n* audacity, boldness, *sl* bottle, bravery, daring, dauntlessness, determination, fearlessness, fibre, firmness, fortitude, gallantry, *inf* grit, *inf* guts, heroism, indomitability, intrepidity, mettle, *inf* nerve, patience, *inf* pluck, prowess, resolution, spirit, *sl* spunk, stoicism, tenacity, valour, will-power. *Opp* COWARDICE.

courageous *adj* audacious, bold, brave, cool, daring, dauntless, determined, fearless, gallant, game, *sl* gutsy, heroic, indomitable, intrepid, lion-hearted, noble, *inf* plucky, resolute, spirited, stalwart, stoical, stout-hearted, tough, unafraid, uncomplaining, undaunted, unshrinking, valiant, valorous. *Opp* COWARDLY.

course *n* 1 bearings, circuit, direction, line, orbit, path, route, track, way. 2 *course of events.* advance, continuation, development, movement, passage, passing, progress, progression, succession. 3 *course of lectures.* curriculum, programme, schedule, sequence, series, syllabus.

court *n* 1 assizes, bench, court martial, high court, lawcourt, magistrates' court. 2 entourage, followers, palace, retinue. 3 ▷ COURTYARD. • *vb* 1 *inf* ask for, attract, invite, provoke, seek, solicit. 2 date, *inf* go out with, make advances to, make love to, try to win, woo.

courteous *adj* civil, considerate, gentlemanly, ladylike, urbane, well-bred, well-mannered. ▷ POLITE.

courtier *n* attendant, follower, lady, lord, noble, page, steward.

courtyard *n* court, enclosure, forecourt, patio, *inf* quad, quadrangle, yard.

cover *n* 1 ▷ COVERING. 2 binding, case, dust-jacket, envelope, file, folder, portfolio, wrapper. 3 camouflage, cloak, concealment, cover-up, deception, disguise, façade, front, hiding-place, mask, pretence, refuge, sanctuary, shelter, smokescreen. 3 *air cover.* defence, guard, protection, support. • *vb* 1 blot out, bury, camouflage, cap, carpet, cloak, clothe, cloud, coat, conceal, curtain, disguise, drape, dress, encase, enclose, enshroud, envelop, face, hide, hood, mantle, mask, obscure, overlay, overspread, plaster, protect, screen, shade, sheathe, shield, shroud, spread over, surface, tile, veil, veneer, wrap up. 2 *cover expenses.* be enough for, match, meet, pay for, suffice for. 3 *talk covered many subjects.* comprise, contain, deal with, embrace, encompass, include, incorporate, involve, treat.

covering *n* blanket, canopy, cap, carpet, casing, cladding, cloak, coat, coating, cocoon, cover, crust, facing, film, incrustation, layer, lid, mantle, outside, pall, rind, roof, screen, sheath, sheet, shell, shield, shroud, skin, surface, tarpaulin, top, veil, veneer, wrapping. ▷ BEDCLOTHES.

coward *n* *inf* chicken, craven, deserter, runaway, *inf* wimp.

cowardice *n* cowardliness, desertion, evasion, faint-heartedness, *inf* funk, shirking, spinelessness, timidity. ▷ FEAR. *Opp* COURAGE.

cowardly *adj* abject, afraid, base, chicken-hearted, cowering, craven, dastardly, faint-hearted, fearful, *inf* gutless, *inf* lily-livered, pusillanimous, spineless, submissive, timid, timorous, unchivalrous, ungallant, unheroic, *inf* wimpish, *sl* yellow. ▷ FRIGHTENED. *Opp* COURAGEOUS.

cower *vb* cringe, crouch, flinch, grovel, hide, quail, shiver, shrink, skulk, tremble.

coy *adj* arch, bashful, coquettish, demure, diffident, embarrassed, evasive, hesitant, modest, reserved, reticent, retiring, self-conscious, sheepish, shy, timid, unforthcoming. *Opp* BOLD.

crack *n* **1** breach, break, chink, chip, cleavage, cleft, cranny, craze, crevice, fissure, flaw, fracture, gap, opening, rift, rupture, slit, split. **2** ▷ JOKE. • *vb* **1** break, chip, fracture, snap, splinter, split. **2** ▷ HIT, SOUND. **crack up** ▷ DISINTEGRATE.

craft *n* **1** handicraft, job, skilled work, technique, trade. ▷ CRAFTSMANSHIP, CUNNING. **2** *a sea-going craft.* ▷ VESSEL. • *vb* ▷ MAKE.

craftsmanship *n* art, artistry, cleverness, craft, dexterity, expertise, handiwork, knack, *inf* know-how, workmanship. ▷ SKILL.

crafty *adj* artful, astute, calculating, canny, cheating, clever, conniving, cunning, deceitful, designing, devious, *inf* dodgy, *inf* foxy, furtive, guileful, ingenious, knowing, machiavellian, manipulative, scheming, shifty, shrewd, sly, *inf* sneaky, tricky, wily. *Opp* HONEST, NAÏVE.

craggy *adj* jagged, rocky, rough, rugged, steep, uneven.

cram *vb* **1** compress, crowd, crush, fill, force, jam, overcrowd, overfill, pack, press, squeeze, stuff. **2** ▷ STUDY.

cramped *adj* close, crowded, narrow, restricted, tight, uncomfortable. *Opp* ROOMY.

crane *n* davit, derrick, hoist.

crash *n* **1** bang, boom, clash, explosion. ▷ SOUND. **2** accident, bump, collision, derailment, disaster, impact, knock, pile-up, smash, wreck. **3** *crash on the stock market* collapse, depression, failure, fall. • *vb* **1** bump, collide, knock, lurch pitch, smash. ▷ HIT. **2** collapse, crash-dive, dive, fall, plummet, plunge, topple.

crate *n* box, carton, case, packing-case, tea-chest.

crater *n* abyss, cavity, chasm, hole, hollow, opening, pit.

crawl *vb* **1** clamber, creep, edge, inch, slither, squirm, worm, wriggle. **2** [*inf*] be obsequious, cringe, fawn, flatter, grovel, *sl* suck up, toady.

craze *n* diversion, enthusiasm, fad, fashion, infatuation, mania, novelty, obsession, passion, pastime, rage, *inf* thing, trend, vogue.

crazy *adj* **1** berserk, crazed, delirious, demented, deranged, frantic, frenzied, hysterical, insane, lunatic, *inf* potty, *inf* scatty, unbalanced, unhinged, wild. ▷ MAD. **2** *crazy comedy.* daft, eccentric, farcical, idiotic, *inf* knockabout, ludicrous, ridiculous, *sl* wacky, zany. ▷ ABSURD. **3** *crazy ideas.* confused, foolish, ill-considered, illogical, impractical, irrational, senseless, silly, unrealistic, unreasonable, unwise. ▷ STUPID. **4** ▷ ENTHUSIASTIC. *Opp* SENSIBLE.

creamy *adj* milky, oily, rich, smooth, thick, velvety.

crease *n* corrugation, crinkle, fold, furrow, groove, line, pleat, pucker, ridge, ruck, tuck, wrinkle • *vb* crimp, crinkle, crumple, crush, fold, furrow, pleat, pucker, ridge, ruck, rumple, wrinkle.

create *vb old use* beget, begin, be the creator of, breed, bring about, bring into existence, build, cause, compose, conceive, concoct, constitute, construct, design, devise, *inf* dream up, engender, engineer, establish, father, forge, form, found, generate, give rise to, hatch, imagine, institute, invent, make up, manufacture, occasion, originate, produce, set up, shape, sire, think up. ▷ MAKE. *Opp* DESTROY.

creation *n* 1 beginning, birth, building, conception, constitution, construction, establishing, formation, foundation, generation, genesis, inception, institution, making, origin, procreation, production, shaping. 2 achievement, brainchild, concept, effort, handiwork, invention, product, work of art. *Opp* DESTRUCTION.

creative *adj* artistic, clever, fecund, fertile, imaginative, ingenious, inspired, inventive, original, positive, productive, resourceful, talented. *Opp* DESTRUCTIVE.

creator *n* architect, artist, author, begetter, builder, composer, craftsman, designer, deviser, discoverer, initiator, inventor, maker, manufacturer, originator, painter, parent, photographer, potter, producer, sculptor, smith, weaver, writer.

creature *n* beast, being, brute, mortal being, organism. ▷ ANIMAL.

credentials *n* authorization, documents, identity card, licence, passport, permit, proof of identity, warrant.

credible *adj* believable, conceivable, convincing, imaginable, likely, persuasive, plausible, possible, reasonable, tenable, thinkable, trustworthy. *Opp* INCREDIBLE.

credit *n* approval, commendation, distinction, esteem, fame, glory, honour, *inf* kudos, merit, praise, prestige, recognition, reputation, status, tribute. • *vb* 1 accept, believe, *inf* buy, count on, depend on, endorse, have faith in, reckon on, rely on, subscribe to, *inf* swallow, swear by, trust. *Opp* DOUBT. 2 *credit you with sense.* ascribe to, assign to, attach to, attribute to. 3 *credit £10 to my account.* add, enter. *Opp* DEBIT.

creditable *adj* admirable, commendable, estimable, good, honourable, laudable, meritorious, praiseworthy, respectable, well thought of, worthy. *Opp* UNWORTHY.

credulous *adj* easily taken in, *inf* green, gullible, *inf* soft, trusting, unsuspecting. ▷ NAÏVE. *Opp* SCEPTICAL.

creed *n* belief, conviction, doctrine, dogma, faith, principle, teaching, tenet.

creek *n* bay, cove, estuary, harbour, inlet.

creep *vb* crawl, edge, inch, move quietly, move slowly, pussyfoot, slink, slip, slither, sneak, steal, tiptoe, worm, wriggle, writhe.

creepy *adj* disturbing, eerie, frightening, ghostly, hair-raising, macabre, ominous, scary, sinister, spine-chilling, *inf* spooky, supernatural, threatening, uncanny, unearthly, weird.

crest *n* 1 comb, plume, tuft. 2 *crest of a hill.* apex, brow, crown, head, peak, pinnacle, ridge, summit, top. 3 badge, coat of arms, design, device, emblem, heraldic device, insignia, seal, shield, sign, symbol.

crevice *n* break, chink, cleft, crack, cranny, fissure, furrow, groove, rift, slit, split.

crew *n* band, company, gang, party, team. ▷ GROUP.

crime *n* delinquency, dishonesty, *old use* felony, illegality, lawbreaking, lawlessness, misconduct, misdeed, misdemeanour, offence, *inf* racket, sin, transgression of the law, violation, wrongdoing. □ *abduction, arson, assassination, blackmail, burglary, extortion, hijacking, hooliganism, kidnapping, manslaughter, misappropriation, mugging, murder, pilfering, piracy, poaching, rape, robbery, shop-lifting, smuggling, stealing, terrorism, theft, treason, vandalism.*

criminal *adj inf* bent, corrupt, *inf* crooked, culpable, dishonest, felonious, illegal, illicit, indictable, lawless, nefarious, *inf* shady, unlawful. ▷ WICKED, WRONG. *Opp* LAWFUL. ● *n inf* baddy, convict, *inf* crook, culprit, delinquent, desperado, felon, gangster, knave, lawbreaker, malefactor, miscreant, offender, outlaw, recidivist, ruffian, scoundrel, *old use* transgressor, villain, wrongdoer. □ *bandit, brigand, buccaneer, defaulter, gunman, highwayman,* sl *hoodlum, hooligan, pickpocket, racketeer, receiver, swindler, thug,* ▷ CRIME.

cringe *vb* blench, cower, crouch, dodge, duck, flinch, grovel, quail, quiver, recoil, shrink back, shy away, tremble, wince.

cripple *vb* **1** disable, dislocate, fracture, hamper, hamstring, incapacitate, lame, maim, mutilate, paralyse, weaken. **2** damage, make useless, put out of action, sabotage, spoil. **crippled** ▷ HANDICAPPED.

crisis *n* calamity, catastrophe, climax, critical moment, danger, difficulty, disaster, emergency, predicament, problem, turning point.

crisp *adj* **1** breakable, brittle, crackly, crispy, crunchy, fragile, friable, hard and dry. **2** ▷ BRACING, BRISK.

criterion *n* measure, principle, standard, touchstone, yardstick.

critic *n* **1** analyst, authority, commentator, judge, pundit, reviewer. **2** attacker, detractor.

critical *adj* **1** captious, carping, censorious, criticizing, deprecatory, depreciatory, derogatory, disapproving, disparaging, fault-finding, hypercritical, judgemental, *inf* nit-picking, *derog* Pharisaical, scathing, slighting, uncomplimentary, unfavourable. *Opp* COMPLIMENTARY. **2** analytical, discerning, discriminating, intelligent, judicious, perceptive, probing, sharp. **3** *critical moment.* basic, crucial, dangerous, decisive, important, key, momentous, pivotal, vital. *Opp* UNIMPORTANT.

criticism *n* **1** censure, condemnation, diatribe, disapproval, disparagement, reprimand, reproach, stricture, tirade, verbal attack. **2** *literary criticism.* analysis, appraisal, appreciation, assessment, commentary, critique, elucidation, evaluation, judgement, valuation.

criticize *vb* **1** belittle, berate, blame, carp, *inf* cast aspersions on, castigate, censure, *old use* chide, condemn, complain about, decry, disapprove of, disparage, fault, find fault with, *inf* flay, *inf* get at, impugn, *inf* knock, *inf* lash, *inf* pan, *inf* pick holes in, *inf* pitch into, *inf* rap, rate, rebuke, reprimand, satirize, scold, *inf* slam, *inf* slate, snipe at. *Opp* PRAISE. **2** analyse, appraise, assess, evaluate, discuss, judge, review.

crockery *n* ceramics, china, crocks, dishes, earthenware, porcelain, pottery, tableware. □ *basin, bowl, coffee-cup, coffee-pot, cup, din-*

ner plate, dish, jug, milk-jug, mug, plate, Amer *platter, pot, sauceboat, saucer, serving dish, side plate, soup bowl, sugar-bowl, teacup, teapot,* old use *trencher, tureen.*

crook *n* **1** angle, bend, corner, hook. **2** ▷ CRIMINAL.

crooked *adj* **1** angled, askew, awry, bendy, bent, bowed, contorted, curved, curving, deformed, gnarled, lopsided, misshapen, off-centre, tortuous, twisted, twisty, warped, winding, zigzag. ▷ INDIRECT. **2** ▷ CRIMINAL.

crop *n* gathering, harvest, produce, sowing, vintage, yield. • *vb* bite off, browse, clip, graze, nibble, shear, snip, trim. ▷ CUT. **crop up** ▷ ARISE.

cross *adj* bad-tempered, cantankerous, crotchety, *inf* grumpy, ill-tempered, irascible, irate, irritable, peevish, short-tempered, testy, tetchy, upset, vexed. ▷ ANGRY, ANNOYED. *Opp* GOOD-TEMPERED. • *n* **1** intersection, X. **2** *a cross to bear.* affliction, burden, difficulty, grief, misfortune, problem, sorrow, trial, tribulation, trouble, worry. **3** *cross of breeds.* amalgam, blend, combination, cross-breed, half-way house, hybrid, mixture, mongrel. • *vb* **1** criss-cross, intersect, meet, zigzag. **2** *cross a river.* bridge, ford, go across, pass over, span, traverse. **3** *cross someone.* annoy, block, frustrate, hinder, impede, interfere with, oppose, stand in the way of, thwart. **cross out** ▷ CANCEL. **cross swords** ▷ CONFLICT.

crossing *vb* **1** bridge, causeway, flyover, ford, level-crossing, overpass, pedestrian crossing, pelican crossing, subway, stepping-stones, underpass, zebra crossing. **2** *sea crossing.* ▷ JOURNEY.

crossroads *n* interchange, intersection, junction.

crouch *vb* bend, bow, cower, cringe, duck, kneel, squat, stoop.

crowd *n* **1** army, assemblage, assembly, bunch, circle, cluster, collection, company, crush, flock, gathering, horde, host, mass, mob, multitude, pack, press, rabble, swarm, throng. ▷ GROUP. **2** *a football crowd.* audience, gate, spectators. • *vb* assemble, bundle, cluster, collect, compress, congregate, cram, crush, flock, gather, get together, herd, huddle, jam, jostle, mass, muster, overcrowd, pack, *inf* pile, press, push, squeeze, swarm, throng.

crowded *adj* congested, cramped, full, jammed, *inf* jam-packed, jostling, overcrowded, overflowing, packed, swarming, teeming, thronging. *Opp* EMPTY.

crown *n* **1** circlet, coronet, diadem, tiara. **2** *crown of a hill.* apex, brow, head, peak, ridge, summit, top. • *vb* **1** anoint, appoint, enthrone, install. **2** cap, complete, conclude, consummate, culminate, finish off, perfect, round off, top.

crucial *adj* central, critical, decisive, essential, important, major, momentous, pivotal, serious. *Opp* UNIMPORTANT.

crude *adj* **1** natural, raw, unprocessed, unrefined. **2** *crude work.* amateurish, awkward, bungling, clumsy, inartistic, incompetent, inelegant, inept, makeshift, primitive, rough, rudimentary, unpolished, unskilful, unworkmanlike. *Opp* REFINED. **3** ▷ VULGAR.

cruel *adj* atrocious, barbaric, barbarous, beastly, bestial, bloodthirsty, bloody, brutal, callous, cold-blooded, cold-hearted, diabolical, ferocious, fiendish, fierce, flinty, grim, hard, hard-hearted, harsh, heartless, hellish, implacable, inexorable, inhuman, inhumane, malevolent, merciless, murderous, pitiless, relentless, remorseless, ruthless, sadistic, savage, severe, sharp, spiteful, stern, stony-hearted, tyrannical, unfeeling, unjust, unkind, unmerciful, unrelenting, vengeful, venomous, vicious, violent. *Opp* KIND.

cruelty *n* barbarity, bestiality, bloodthirstiness, brutality, callousness, cold-bloodedness, ferocity, hard-heartedness, heartlessness, inhumanity, malevolence, ruthlessness, sadism, savagery, unkindness, viciousness, violence.

cruise *n* sail, voyage. ▷ TRAVEL.

crumb *n* bit, bite, fragment, grain, morsel, particle, scrap, shred, sliver, speck.

crumble *vb* break into pieces, break up, crush, decay, decompose, deteriorate, disintegrate, fall apart, fragment, grind, perish, pound, powder, pulverize.

crumbly *adj* friable, granular, powdery. *Opp* SOLID.

crumple *vb* crease, crinkle, crush, dent, fold, mangle, pucker, rumple, wrinkle.

crunch *vb* break, champ, chew, crush, grind, masticate, munch, scrunch, smash, squash.

crusade *n* campaign, drive, holy war, jehad, movement, struggle, war.

crush *n* congestion, jam. ▷ CROWD. • *vb* **1** break, bruise, compress, crumple, crunch, grind, mangle, mash, pound, press, pulp, pulverize, shiver, smash, splinter, squash, squeeze. **2** *crush opponents.* humiliate, mortify, overwhelm, quash, rout, thrash, vanquish. ▷ CONQUER.

crust *n* incrustation, outer layer, outside, rind, scab, shell, skin, surface. ▷ COVERING.

crux *n* centre, core, crucial issue, essence, heart, nub.

cry *n* battle-cry, bellow, call, caterwaul, ejaculation, exclamation, hoot, howl, outcry, roar, scream, screech, shout, shriek, whoop, yell, yelp, yowl. • *vb* bawl, blubber, grizzle, howl, keen, shed tears, snivel, sob, wail, weep, whimper, whinge. **cry off** ▷ WITHDRAW. **cry out** ▷ SHOUT.

crypt *n* basement, catacomb, cellar, grave, sepulchre, tomb, undercroft, vault.

cryptic *adj* arcane, cabalistic, coded, concealed, enigmatic, esoteric, hidden, mysterious, mystical, obscure, occult, perplexing, puzzling, recondite, secret, unclear, unintelligible, veiled. *Opp* INTELLIGIBLE.

cuddle *vb* caress, clasp lovingly, dandle, embrace, fondle, hold closely, huddle against, hug, kiss, make love, nestle against, nurse, pet, snuggle up to.

cudgel *n* baton, bludgeon, cane, club, cosh, stick, truncheon. • *vb* batter, beat, bludgeon, cane, *inf* clobber, cosh, pound, pummel, thrash, thump, *inf* thwack. ▷ HIT.

cue *n* hint, prompt, reminder, sign, signal.

culminate *vb* build up to, climax, conclude, reach a finale, rise to a peak. ▷ END.

culpable *adj* blameworthy, criminal, guilty, knowing, liable, punishable, reprehensible, wrong. ▷ DELIBERATE. *Opp* INNOCENT.

culprit *n* delinquent, malefactor, miscreant, offender, troublemaker, wrongdoer. ▷ CRIMINAL.

cult *n* **1** craze, fan-club, fashion, following, devotees, party, school, trend, vogue. **2** *religious cult.* ▷ DENOMINATION.

cultivate *vb* **1** dig, farm, fertilize, hoe, manure, mulch, plough, prepare, rake, till, turn, work. **2** grow, plant, produce, raise, sow, take cuttings, tend. **3** *cultivate a friendship.* court, develop, encourage, foster, further, improve, promote, pursue, try to achieve.

cultivated *adj* **1** agricultural, farmed, planted, prepared, tilled. **2** ▷ CULTURED.

cultivation *n* agriculture, agronomy, breeding, culture, growing, farming, gardening, horticulture, husbandry, nurturing.

cultural *adj* aesthetic, artistic, civilized, civilizing, educational, elevating, enlightening, highbrow, improving, intellectual.

culture *n* **1** art, background, civilization, customs, education, learning, mores, traditions, way of life. **2** ▷ CULTIVATION.

cultured *adj* artistic, civilized, cultivated, discriminating, educated, elegant, erudite, highbrow, knowledgeable, polished, refined, scholarly, sophisticated, well-bred, well-educated, well-read. *Opp* IGNORANT.

cunning *adj* **1** artful, devious, dodgy, guileful, insidious, knowing, machiavellian, sly, subtle, tricky, wily. ▷ CRAFTY. **2** adroit, astute, ingenious, skilful. ▷ CLEVER. • *n* **1** artfulness, chicanery, craft, craftiness, deceit, deception, deviousness, duplicity, guile, slyness, trickery. **2** cleverness, expertise, ingenuity, skill.

cup *n* **1** beaker, bowl, chalice, glass, goblet, mug, tankard, teacup, tumbler, wine-glass. **2** award, prize, trophy.

cupboard *n* cabinet, chiffonier, closet, dresser, filing-cabinet, food-cupboard, larder, locker, sideboard, wardrobe.

curable *adj* operable, remediable, treatable. *Opp* INCURABLE.

curb *vb* bridle, check, contain, control, deter, hamper, hinder, hold back, impede, inhibit, limit, moderate, repress, restrain, restrict, subdue, suppress. *Opp* ENCOURAGE.

curdle *vb* clot, coagulate, congeal, go lumpy, go sour, thicken.

cure *n* **1** antidote, corrective, medication, nostrum, palliative, panacea, prescription, remedy, restorative, solution, therapy, treatment. ▷ MEDICINE. **2** deliverance, healing, recovery, recuperation, restoration, revival. • *vb* alleviate, correct, counteract, ease, *inf* fix, heal, help, mend, palliate, put right, rectify, relieve, remedy, repair, restore, solve, treat. *Opp* AGGRAVATE.

curiosity *n* inquisitiveness, interest, interference, meddling, nosiness, prying, snooping.

curious *adj* **1** inquiring, inquisitive, interested, probing, puzzled, questioning, searching. **2** interfering, intrusive, meddlesome, *inf* nosy, prying. **3** ▷ STRANGE. **be curious** ▷ PRY.

curl *n* bend, coil, curve, kink, loop, ringlet, scroll, spiral, swirl, turn, twist, wave, whorl. • *vb* **1** bend, coil, corkscrew, curve, entwine, loop, spiral, turn, twine, twist, wind, wreathe, writhe. **2** *curl your hair.* crimp, frizz, perm.

curly *adj* crimped, curled, curling, frizzy, fuzzy, kinky, permed, wavy. *Opp* STRAIGHT.

current *adj* **1** alive, contemporary, continuing, existing, extant, fashionable, living, modern, ongoing, present, present-day, prevailing, prevalent, reigning, remaining, surviving, *inf* trendy, up-to-date. **2** *current passport.* usable, valid. *Opp* OLD. • *n* course, draught, drift, flow, jet, river, stream, tide, trend, undercurrent, undertow.

curriculum *n* course, programme of study, syllabus.

curse *n* blasphemy, exclamation, expletive, imprecation, malediction, oath, obscenity, profanity, swearword. ▷ EVIL. *Opp* BLESSING. • *vb* blaspheme, damn, fulminate, swear, utter curses. *Opp* BLESS. **cursed** ▷ HATEFUL.

cursory *adj* brief, careless, casual, desultory, fleeting, hasty, hurried, perfunctory, quick, slapdash, superficial. *Opp* THOROUGH.

curt *adj* abrupt, blunt, brief, brusque, concise, crusty, gruff, laconic, monosyllabic, offhand, rude, sharp, short, snappy, succinct, tart, terse, unceremonious, uncommunicative, ungracious. ▷ RUDE. *Opp* EXPANSIVE.

curtail *vb* abbreviate, abridge, break off, contract, cut short, decrease, diminish, *inf* dock, guillotine, halt, lessen, lop, prune, reduce, restrict, shorten, stop, terminate, trim, truncate. *Opp* EXTEND.

curtain *n* blind, drape, drapery, hanging, screen. • *vb* drape, mask, screen, shroud, veil. ▷ HIDE.

curtsy *vb* bend the knee, bow, genuflect, salaam.

curve *n* arc, arch, bend, bow, bulge, camber, circle, convolution, corkscrew, crescent, curl, curvature, cycloid, loop, meander, spiral, swirl, trajectory, turn, twist, undulation, whorl. • *vb* arc, arch, bend, bow, bulge, camber, coil, corkscrew, curl, loop, meander, snake, spiral, swerve, swirl, turn, twist, wind. ▷ CIRCLE.

curved *adj* concave, convex, convoluted, crescent, crooked, curvilinear, curving, curvy, rounded, serpentine, shaped, sinuous, sweeping, swelling, tortuous, turned, undulating, whorled.

cushion *n* bean-bag, bolster, hassock, headrest, pad, pillow. • *vb* absorb, bolster, deaden, lessen, mitigate, muffle, protect from, reduce the effect of, soften, support.

custodian *n* caretaker, curator, guardian, keeper, overseer, superintendent, warden, warder, *inf* watch-dog, watchman.

custody *n* **1** care, charge, guardianship, keeping, observation, possession, preservation, protection, safe-keeping. **2** *in police custody.* captivity, confinement, detention, imprisonment, incarceration, remand.

custom *n* **1** convention, etiquette, fashion, form, formality, habit, institution, manner, observance, policy, practice, procedure, routine, tradition, usage, way, wont. **2** *A shop needs custom.* business, buyers, customers, patronage, support, trade.

customary *adj* accepted, accustomed, common, commonplace, conventional, established, everyday, expected, fashionable, general, habitual, normal, ordinary, popular, prevailing, regular, routine, traditional, typical, usual, wonted. *Opp* UNUSUAL.

customer *n* buyer, client, consumer, patron, purchaser, shopper. *Opp* SELLER.

cut *n* **1** gash, graze, groove, incision, laceration, nick, notch, opening, rent, rip, slash, slice, slit, snick, snip, split, stab, tear. ▷ INJURY. **2** *cut in prices.* cut-back, decrease, fall, lowering, reduction, saving. • *vb* **1** amputate, axe, carve, chip, chisel, chop, cleave, clip, crop, dice, dissect, divide, dock, engrave, fell, gash, gouge, grate, graze, guillotine, hack, halve, hew, incise, knife, lacerate, lance, lop, mince, mow, nick, notch, open, pare, pierce, poll, pollard, prune, reap, rive, saw, scalp, score, sever, share, shave, shear, shred, slash, slice, slit, snick, snip, split, stab, subdivide, trim, whittle, wound. **2** abbreviate, abridge, bowdlerize, censor, condense, curtail, digest, edit, précis, shorten, summarize, truncate. ▷ REDUCE. **cut and dried** ▷ DEFINITE. **cut in** ▷ INTERRUPT. **cut off** ▷ REMOVE, STOP. **cut short** ▷ CURTAIL.

cutlery *n inf* eating irons. □ *breadknife, butter-knife, carving knife, cheese knife, dessert-spoon, fish knife, fish fork, fork, knife, ladle, salad servers, spoon, steak knife, tablespoon, teaspoon.*

cutter *n* □ *axe, billhook, chisel, chopper, clippers, harvester, guillotine, lawnmower, mower, saw, scalpel, scissors, scythe, secateurs, shears, sickle.* ▷ KNIFE.

cutting *adj* acute, biting, caustic, incisive, keen, mordant, sarcastic, satirical, sharp, trenchant. ▷ HURTFUL.

cycle *n* **1** circle, repetition, revolution, rotation, round, sequence, series. **2** bicycle, *inf* bike, moped, *inf* motor bike, motor cycle, penny-farthing, scooter, tandem, tricycle. • *vb* ▷ TRAVEL.

cyclic *adj* circular, recurring, repeating, repetitive, rotating.

cynical *adj* doubting, *inf* hard, incredulous, misanthropic, mocking, negative, pessimistic, questioning, sceptical, sneering. *Opp* OPTIMISTIC.

D

dabble *vb* **1** dip, paddle, splash, wet. **2** *dabble in a hobby.* potter about, tinker, work casually.

dabbler *n* amateur, dilettante, potterer.

dagger *n* bayonet, blade, *old use* dirk, knife, kris, poniard, stiletto.

daily *adj* diurnal, everyday, quotidian, regular.

dainty *adj* **1** charming, delicate, exquisite, fine, graceful, meticulous, neat, nice, pretty, skilful. **2** choosy, discriminating, fastidious, finicky, fussy, genteel, mincing, sensitive, squeamish, well-mannered. **3** *a dainty morsel.* appealing, appetizing, choice, delectable, delicious. *Opp* CLUMSY, GROSS.

dally *vb* dawdle, delay, *inf* dilly-dally, hang about, idle, linger, loaf, loiter, play about, procrastinate, saunter, *old use* tarry, waste time.

dam *n* bank, barrage, barrier, dike, embankment, wall, weir. • *vb* block, check, hold back, obstruct, restrict, stanch, stem, stop.

damage *n* destruction, devastation, harm, havoc, hurt, injury, loss, mutilation, sabotage. • *vb* **1** blemish, break, buckle, burst, *inf* bust, chip, crack, cripple, deface, destroy, disable, disfigure, *inf* do mischief to, flaw, fracture, harm, hurt, immobilize, impair, incapacitate, injure, make inoperative, make useless, mar, mark, mutilate, *inf* play havoc with, ruin, rupture, sabotage, scar, scratch, spoil, strain, vandalize, warp, weaken, wound, wreck. **damaged** ▷ FAULTY. **damages** ▷ COMPENSATION. **damaging** ▷ HARMFUL.

damn *vb* attack, berate, castigate, censure, condemn, criticize, curse, denounce, doom, execrate, sentence, swear at.

damnation *n* doom, everlasting fire, hell, perdition, ruin. *Opp* SALVATION.

damp *adj* clammy, dank, dewy, dripping, drizzly, foggy, humid, misty, moist, muggy, perspiring, rainy, soggy, steamy, sticky, sweaty, unaired, unventilated, wet, wettish. *Opp* DRY. • *vb* **1** dampen, humidify, moisten, sprinkle. **2** ▷ DISCOURAGE.

dance *n* choreography, dancing. □ *ball, barn-dance, ceilidh, disco, discothηque,* inf *hop,* inf *knees-up, party,* inf *shindy, social, square dance.* □ *ballet, ballroom dancing, break-dancing, country dancing, disco dancing, flamenco dancing, folk dancing, Latin-American dancing, limbo dancing, morris dancing, old-time dancing, tap-dancing.* □ *bolero, cancan, conga, fandango, fling, foxtrot, gavotte, hornpipe, jig, mazurka, minuet, polka, polonaise, quadrille, quickstep, reel, rumba, square dance, tango, waltz.* • *vb* caper, cavort, frisk, frolic, gambol, hop about, jig, jive, jump, leap, prance, rock, skip, *joc* trip the light fantastic, whirl.

danger *n* **1** crisis, distress, hazard, insecurity, jeopardy, menace, peril, pitfall, trouble, uncertainty. **2** *danger of frost.* chance, liability, possibility, risk, threat.

dangerous *adj* **1** alarming, breakneck, *inf* chancy, critical, destructive, explosive, grave, *sl* hairy, harmful, hazardous, insecure, menacing, *inf* nasty, noxious, perilous, precarious, reckless, risky, threatening, toxic, uncertain, unsafe. **2** *dangerous men.* desperate, ruthless, treacherous, unmanageable, unpredictable, violent, volatile, wild. *Opp* HARMLESS.

dangle *vb* be suspended, depend, droop, flap, hang, sway, swing, trail, wave about.

dank *adj* chilly, clammy, damp, moist, unaired.

dappled *adj* blotchy, brindled, dotted, flecked, freckled, marbled, motley, mottled, particoloured, patchy, pied, speckled, spotted, stippled, streaked, varicoloured, variegated.

dare *vb* **1** gamble, have the courage, risk, take a chance, venture. **2** challenge, defy, provoke, taunt. **daring** ▷ BOLD.

dark *adj* **1** black, blackish, cheerless, clouded, cloudy, coal-black, dim, dingy, dismal, drab, dreary, dull, dusky, funereal, gloomy, glowering, glum, grim, inky, moonless, murky, overcast, pitch-black, pitch-dark, *poet* sable, shadowy, shady, sombre, starless, stygian, sullen, sunless, tenebrous, unilluminated, unlighted, unlit. **2** *dark colours.* dense, heavy, strong. **3** *dark complexion.* black, brown, dark-skinned, dusky, swarthy, tanned. 4 ▷ HIDDEN, MYSTERIOUS. *Opp* LIGHT, PALE.

darken *vb* **1** become overcast, cloud over. **2** blacken, dim, eclipse, obscure, overshadow, shade. *Opp* LIGHTEN.

darling *n inf* apple of your eye, beloved, *inf* blue-eyed boy, dear, dearest, favourite, honey, love, loved one, pet, sweet, sweetheart, true love.

dart *n* arrow, bolt, missile, shaft. • *vb* bound, fling, flit, fly, hurtle, leap, move suddenly, shoot, spring, *inf* whiz, *inf* zip. ▷ DASH.

dash *n* **1** chase, race, run, rush, sprint, spurt. • *vb* **1** bolt, chase, dart, fly, hasten, hurry, move quickly, race, run, rush, speed, sprint, tear, *inf* zoom. **2** ▷ HIT.

dashing *adj* animated, dapper, dynamic, elegant, lively, smart, spirited, stylish, vigorous.

data *plur n* details, evidence, facts, figures, information, statistics.

date *n* **1** day. ▷ TIME. **2** *date with a friend.* appointment, assignation, engagement, fixture, meeting, rendezvous. **out-of-date** ▷ OBSOLETE. **up-to-date** ▷ MODERN.

daunt *vb* alarm, depress, deter, discourage, dishearten, dismay, intimidate, overawe, put off, unnerve. ▷ FRIGHTEN. *Opp* ENCOURAGE.

dawdle *vb* be slow, dally, delay, *inf* dilly-dally, hang about, idle, lag behind, linger, loaf about, loiter, move slowly, straggle, *inf* take it easy, *inf* take your time, trail behind. *Opp* HURRY.

dawn *n* day-break, first light, *inf* peep of day, sunrise. ▷ BEGINNING.

day *n* **1** daylight, daytime, light. **2** age, epoch, era, period, time.

day-dream *n* dream, fantasy, hope, illusion, meditation, pipe-dream, reverie, vision, wool-gathering. • *vb* dream, fantasize, imagine, meditate.

daze *vb* benumb, paralyse, shock, stun, stupefy. ▷ AMAZE.

dazzle *vb* blind, confuse, disorientate. **dazzling** ▷ BRILLIANT.

dead *adj* **1** cold, dead and buried, deceased, departed, *inf* done for, inanimate, inert, killed, late, lifeless, perished, rigid, stiff. *Opp* ALIVE. **2** *dead language.* died out, extinct, obsolete. **3** *dead with cold.* deadened, insensitive, numb, paralysed, without feeling. **4** *dead battery, engine.* burnt out, defunct, flat, inoperative, not going, not working, no use, out of order, unresponsive, used up, useless, worn out. **5** *a dead party.* boring, dull, moribund, slow, uninteresting. *Opp* LIVELY. **6** *dead centre.* ▷ EXACT. **dead person** ▷ CORPSE. **dead to the world** ▷ ASLEEP.

deaden *vb* **1** anaesthetize, desensitize, dull, numb, paralyse. **2** blunt, check, cushion, damp, diminish, hush, lessen, mitigate, muffle, mute, quieten, reduce, smother, soften, stifle, suppress, weaken.

deadlock *n* halt, impasse, stalemate, standstill, stop, stoppage, tie.

deadly *adj* dangerous, destructive, fatal, lethal, mortal, noxious, terminal. ▷ HARMFUL. *Opp* HARMLESS.

deafen *vb* make deaf, overwhelm.

deafening ▷ LOUD.

deal *n* **1** agreement, arrangement, bargain, contract, pact, settlement, transaction, understanding. **2** amount, quantity, volume. • *vb* **1** allot, apportion, assign, dispense, distribute, divide, *inf* dole out, give out, share out. **2** *deal someone a blow*. administer, apply, deliver, give, inflict, mete out. **3** *deal in stocks and shares*. buy and sell, do business, trade, traffic. deal with ▷ MANAGE, TREAT.

dealer *n* agent, broker, distributor, merchant, retailer, shopkeeper, stockist, supplier, trader, tradesman, vendor, wholesaler.

dear *adj* **1** adored, beloved, close, darling, intimate, loved, precious, treasured, valued, venerated. ▷ LOVABLE. *Opp* HATEFUL. **2** costly, exorbitant, expensive, high-priced, over-priced, *inf* pricey. *Opp* CHEAP. • *n* ▷ DARLING.

death *n* **1** decease, demise, dying, loss, passing. ▷ END. **2** casualty, fatality. **put to death** ▷ EXECUTE.

debase *vb* belittle, commercialize, degrade, demean, depreciate, devalue, diminish, lower the tone of, pollute, reduce the value of, ruin, soil, spoil, sully, vulgarize.

debatable *adj* arguable, contentious, controversial, controvertible, disputable, doubtful, dubious, moot (*point*), open to doubt, open to question, problematical, questionable, uncertain, unsettled, unsure. *Opp* CERTAIN.

debate *n* argument, conference, consultation, controversy, deliberation, dialectic, discussion, disputation, dispute, polemic. • *vb* argue, *inf* chew over, consider, deliberate, discuss, dispute, *inf* mull over, question, reflect on, weigh up, wrangle.

debit *vb* cancel, remove, subtract, take away. *Opp* CREDIT.

debris *n* bits, detritus, flotsam, fragments, litter, pieces, remains, rubbish, rubble, ruins, waste, wreckage.

debt *n* account, arrears, bill, debi dues, indebtedness, liability, obligation, score, what you owe. **in debt** bankrupt, defaulting, insolvent. ▷ POOR.

decadent *adj* corrupt, debased, debauched, declining, degenerate dissolute, immoral, self-indulgen *Opp* MORAL.

decay *vb* atrophy, break down, co rode, crumble, decompose, degen erate, deteriorate, disintegrate, di solve, fall apart, fester, go bad, go off, mortify, moulder, oxidize, pe ish, putrefy, rot, shrivel, spoil, waste away, weaken, wither.

deceit *n* artifice, cheating, chicanery, craftiness, cunning, dece fulness, dishonesty, dissimulatio double-dealing, duplicity, guile, hypocrisy, insincerity, lying, mis representation, pretence, sham, slyness, treachery, trickery, unde handedness, untruthfulness. ▷ DECEPTION. *Opp* HONESTY.

deceitful *adj* cheating, crafty, cu ning, deceiving, deceptive, desig ing, dishonest, double-dealing, duplicitous, false, fraudulent, fur ive, hypocritical, insincere, lying secretive, shifty, sneaky, treache ous, *inf* tricky, *inf* two-faced, underhand, unfaithful, untrustworthy, wily. *Opp* HONEST.

deceive *vb inf* bamboozle, be an impostor, beguile, betray, blind, bluff, cheat, *inf* con, defraud, delude, *inf* diddle, double-cross, dupe, fool, *inf* fox, *inf* have on, hoax, hoodwink, *inf* kid, *inf* lead on, lie, mislead, mystify, *inf* outsmart, outwit, pretend, swindle, *inf* take for a ride, *inf* take in, trick.

decelerate *vb* brake, decrease speed, go slower, lose speed, slov down. *Opp* ACCELERATE.

decent *adj* **1** acceptable, appropriate, becoming, befitting, chaste, courteous, decorous, delicate, fitting, honourable, modest, polite, presentable, proper, pure, respectable, seemly, sensitive, suitable, tasteful. *Opp* INDECENT. **2** [*inf*] *a decent meal.* agreeable, nice, pleasant, satisfactory. ▷ GOOD. *Opp* BAD.

deception *n* bluff, cheat, *inf* con, confidence trick, cover-up, deceit, fake, feint, *inf* fiddle, fraud, hoax, imposture, lie, pretence, ruse, sham, stratagem, subterfuge, swindle, trick, wile. ▷ DECEIT.

deceptive *adj* ambiguous, deceiving, delusive, dishonest, distorted, equivocal, evasive, fallacious, false, fraudulent, illusory, insincere, lying, mendacious, misleading, specious, spurious, treacherous, unreliable, wrong. *Opp* GENUINE.

decide *vb* adjudicate, arbitrate, choose, conclude, determine, elect, fix on, judge, make up your mind, opt for, pick, reach a decision, resolve, select, settle. **decided** ▷ DEFINITE.

decipher *vb* disentangle, *inf* figure out, read, work out. ▷ DECODE.

decision *n* conclusion, decree, finding, judgement, outcome, result, ruling, verdict.

decisive *adj* **1** conclusive, convincing, crucial, final, influential, positive, significant. **2** *decisive action.* certain, confident, decided, definite, determined, firm, forceful, forthright, incisive, resolute, strong-minded, sure, unhesitating. *Opp* TENTATIVE.

declaration *n* affirmation, announcement, assertion, avowal, confirmation, deposition, disclosure, edict, manifesto, notice, proclamation, profession, promulgation, pronouncement, protestation, revelation, statement, testimony.

declare *vb* affirm, announce, assert, attest, avow, broadcast, certify, claim, confirm, contend, disclose, emphasize, insist, maintain, make known, proclaim, profess, pronounce, protest, report, reveal, show, state, swear, testify, *inf* trumpet forth, witness. ▷ SAY.

decline *n* decrease, degeneration, deterioration, diminuendo, downturn, drop, fall, falling off, loss, recession, reduction, slump, worsening. • *vb* **1** decrease, degenerate, deteriorate, die away, diminish, drop away, dwindle, ebb, fail, fall off, flag, lessen, peter out, reduce, shrink, sink, slacken, subside, tail off, taper off, wane, weaken, wilt, worsen. *Opp* IMPROVE. **2** *decline an invitation.* abstain from, forgo, refuse, reject, *inf* turn down, veto. *Opp* ACCEPT.

decode *vb inf* crack, decipher, explain, figure out, interpret, make out, read, solve, understand, unravel, unscramble.

decompose *vb* break down, decay, disintegrate, go off, moulder, putrefy, rot.

decorate *vb* **1** adorn, array, beautify, *old use* bedeck, colour, deck, *inf* do up, embellish, embroider, festoon, garnish, make beautiful, ornament, paint, paper, *derog* prettify, refurbish, renovate, smarten up, spruce up, *derog* tart up, trim, wallpaper. **2** give a medal to, honour, reward.

decoration **1** accessories, adornment, arabesque, elaboration, embellishment, finery, flourishes, ornament, ornamentation, trappings, trimmings. **2** award, badge, colours, medal, order, ribbon, star.

decorative *adj* elaborate, fancy, non-functional, ornamental, ornate. *Opp* FUNCTIONAL.

decorous *adj* appropriate, becoming, befitting, correct, dignified, fitting, genteel, polite, presentable, proper, refined, respectable, sedate, seemly, staid, suitable, well-behaved. ▷ DECENT. *Opp* INDECOROUS.

decorum *n* correctness, decency, dignity, etiquette, good form, good manners, gravity, modesty, politeness, propriety, protocol, respectability, seemliness.

decoy *n* bait, distraction, diversion, enticement, inducement, lure, red herring, stool-pigeon, trap. • *vb* allure, attract, bait, draw, entice, inveigle, lead, lure, seduce, tempt, trick.

decrease *n* abatement, contraction, curtailment, cut, cut-back, decline, de-escalation, diminuendo, diminution, downturn, drop, dwindling, easing-off, ebb, fall, falling off, lessening, lowering, reduction, shrinkage, wane. *Opp* INCREASE. • *vb* **1** abate, curtail, cut, ease off, lower, reduce, slim down, turn down. **2** condense, contract, decline, die away, diminish, dwindle, fall off, lessen, peter out, shrink, slacken, subside, *inf* tail off, taper off, wane. *Opp* INCREASE.

decree *n* act, command, declaration, dictate, dictum, directive, edict, enactment, fiat, injunction, judgement, law, mandate, order, ordinance, proclamation, promulgation, regulation, ruling, statute. • *vb* command, decide, declare, determine, dictate, direct, ordain, order, prescribe, proclaim, promulgate, pronounce, rule.

decrepit *adj* battered, broken down, derelict, dilapidated, feeble, frail, infirm, ramshackle, tumble-down, weak, worn out. ▷ OLD.

dedicate *vb* **1** commit, consecrate, devote, give, hallow, pledge, sanctify, set apart. **2** *dedicate a book*. address, inscribe. **dedicated** ▷ KEEN, LOYAL.

dedication *n* **1** adherence, allegiance, commitment, devotion, enthusiasm, faithfulness, fidelity, loyalty, single-mindedness, zeal. **2** inscription.

deduce *vb* conclude, divine, draw the conclusion, extrapolate, gather, glean, infer, *inf* put two and two together, reason, surmise, *sl* suss out, understand, work out.

deduct *vb inf* knock off, subtract, take away. *Opp* ADD.

deduction *n* **1** allowance, decrease, diminution, discount, reduction, removal, subtraction, withdrawal. **2** conclusion, finding, inference, reasoning, result.

deed *n* **1** accomplishment, achievement, act, action, adventure, effort, endeavour, enterprise, exploit, feat, performance, stunt, undertaking. **2** ▷ DOCUMENT.

deep *adj* **1** abyssal, bottomless, chasmic, fathomless, profound, unfathomable, unplumbed, yawning. **2** *deep feelings*. earnest, extreme, genuine, heartfelt, intense, serious, sincere. **3** *deep in thought*. absorbed, concentrating, engrossed, immersed, lost, preoccupied, rapt, thoughtful. **4** *deep matters*. abstruse, arcane, esoteric, intellectual, learned, obscure, recondite. ▷ DIFFICULT. **5** *deep sleep*. heavy, sound. **6** *deep colour*. dark, rich, strong, vivid. **7** *deep sound*. bass, booming, growling, low, low-pitched, resonant, reverberating, sonorous. *Opp* SHALLOW, SUPERFICIAL, THIN.

deface *vb* blemish, damage, disfigure, harm, impair, injure, mar, mutilate, ruin, spoil, vandalize.

defeat *n* beating, conquest, downfall, *inf* drubbing, failure, humiliation, *inf* licking, overthrow, *inf* put-down, rebuff, repulse, reverse, rout, setback, subjugation, thrashing, trouncing. *Opp* VICTORY. • *vb* baulk, beat, best, be victorious over, check, checkmate, *inf* clobber, confound, conquer, crush, destroy, *inf* flatten, foil, frustrate, get the better of, *sl* hammer, *inf* lay low, *inf* lick, master, outdo, outvote, outwit, overcome, overpower, overthrow, overwhelm, prevail over, put down, quell, repulse, rout, ruin, *inf* smash, stop, subdue, subjugate, suppress, *inf* thrash, thwart, triumph over, trounce, vanquish, whip, win a victory over. *Opp* LOSE. **be defeated** ▷ LOSE. **defeated** ▷ UNSUCCESSFUL.

defect *n* blemish, *computing* bug, deficiency, error, failing, fault, flaw, imperfection, inadequacy, irregularity, lack, mark, mistake, shortcoming, shortfall, spot, stain, want, weakness, weak point. • *vb* change sides, desert, go over.

defective *adj* broken, deficient, faulty, flawed, *inf* gone wrong, imperfect, incomplete, *inf* on the blink, unsatisfactory, wanting, weak. *Opp* PERFECT.

defence *n* **1** cover, deterrence, guard, protection, safeguard, security, shelter, shield. ▷ BARRIER. **2** alibi, apologia, apology, case, excuse, explanation, justification, plea, testimony, vindication.

defenceless *adj* exposed, helpless, impotent, insecure, powerless, unguarded, unprotected, vulnerable, weak.

defend *vb* **1** cover, fight for, fortify, guard, keep safe, preserve, protect, safeguard, screen, secure, shelter, shield, *inf* stick up for, watch over. **2** argue for, champion, justify, plead for, speak up for, stand by, stand up for, support, uphold, vindicate. *Opp* ATTACK.

defendant *n* accused, appellant, offender, prisoner.

defensive *adj* **1** cautious, defending, protective, wary, watchful. **2** apologetic, faint-hearted, self-justifying. *Opp* AGGRESSIVE.

defer *vb* **1** adjourn, delay, hold over, lay aside, postpone, prorogue (*parliament*), put off, *inf* shelve, suspend. **2** ▷ YIELD.

deference *n* acquiescence, compliance, obedience, submission. ▷ RESPECT.

defiant *adj* aggressive, antagonistic, belligerent, bold, brazen, challenging, daring, disobedient, headstrong, insolent, insubordinate, mutinous, obstinate, rebellious, recalcitrant, refractory, self-willed, stubborn, truculent, uncooperative, unruly, unyielding. *Opp* COOPERATIVE.

deficient *adj* defective, inadequate, insufficient, lacking, meagre, scanty, scarce, short, sketchy, unsatisfactory, wanting, weak. *Opp* ADEQUATE, EXCESSIVE.

defile *vb* contaminate, corrupt, degrade, desecrate, dirty, dishonour, foul, infect, make dirty, poison, pollute, soil, stain, sully, taint, tarnish.

define *vb* **1** be the boundary of, bound, circumscribe, delineate, demarcate, describe, determine, fix, limit, mark off, mark out, outline, specify. **2** *define a word.* clarify, explain, formulate, give the meaning of, interpret, spell out.

definite *adj* apparent, assured, categorical, certain, clear, clear-cut, confident, confirmed, cut-and-dried, decided, determined, discernible, distinct, emphatic, exact, explicit, express, fixed, incisive, marked, noticeable, obvious, particular, perceptible, plain, positive, precise, pronounced, settled, specific, sure, unambiguous, unequivocal, unmistakable, well-defined. *Opp* VAGUE.

definitely *adv* beyond doubt, certainly, doubtless, for certain, indubitably, positively, surely, unquestionably, without doubt, without fail.

definition *n* **1** clarification, elucidation, explanation, interpretation. **2** clarity, clearness, focus, precision, sharpness.

definitive *adj* agreed, authoritative, complete, conclusive, correct, decisive, final, last (*word*), official, permanent, reliable, settled, standard, ultimate, unconditional. *Opp* PROVISIONAL.

deflect *vb* avert, deviate, divert, fend off, head off, intercept, parry, prevent, sidetrack, swerve, switch, turn aside, veer, ward off.

deformed *adj* bent, buckled, contorted, crippled, crooked, defaced, disfigured, distorted, gnarled, grotesque, malformed, mangled, misshapen, mutilated, twisted, ugly, warped.

defraud *vb inf* con, *inf* diddle, embezzle, *inf* fleece, rob, swindle. ▷ CHEAT.

deft *adj* adept, adroit, agile, clever, dextrous, expert, handy, neat, *inf* nifty, nimble, proficient, quick, skilful. *Opp* CLUMSY.

defy *vb* **1** challenge, confront, dare, disobey, face up to, flout, *inf* kick against, rebel against, refuse to obey, resist, stand up to, withstand. **2** baffle, beat, defeat, elude, foil, frustrate, repel, repulse, resist, thwart, withstand.

degenerate *adj* ▷ CORRUPT. • *vb* become worse, decline, deteriorate, *inf* go to the dogs, regress, retrogress, sink, slip, weaken, worsen. *Opp* IMPROVE.

degrade *vb* **1** cashier, demote, depose, downgrade. **2** abase, brutalize, cheapen, corrupt, debase, dehumanize, deprave, desensitize, dishonour, harden, humiliate, mortify. **degrading** ▷ SHAMEFUL.

degree *n* **1** calibre, class, grade, order, position, rank, standard, standing, station, status. **2** extent, intensity, level, measure.

deify *vb* idolize, treat as a god, venerate, worship.

deign *vb* concede, condescend, demean yourself, lower yourself, stoop, vouchsafe.

deity *n* creator, divinity, god, goddess, godhead, idol, immortal, power, spirit, supreme being.

dejected *adj* depressed, disconsolate, dispirited, down, downcast, downhearted, heavy-hearted, in low spirits. ▷ SAD.

delay *n* check, deferment, deferral, filibuster, hiatus, hitch, hold-up, interruption, moratorium, pause, postponement, setback, stay (*of execution*), stoppage, wait. • *vb* **1** check, defer, detain, halt, hinder, hold over, hold up, impede, keep back, keep waiting, make late, obstruct, postpone, put back, put off, retard, set back, slow down, stay, stop, suspend.

2 be late, be slow, *inf* bide your time, dally, dawdle, *inf* dilly-dally, *inf* drag your feet, *inf* get bogged down, hang about, hang back, hang fire, hesitate, lag, linger, loiter, mark time, pause, *inf* play for time, procrastinate, stall, *old use* tarry, temporize, vacillate, wait. *Opp* HURRY.

delegate *n* agent, ambassador, emissary, envoy, go-between, legate, messenger, nuncio, plenipotentiary, representative, spokesperson. • *vb* appoint, assign, authorize, charge, commission, depute, designate, empower, entrust, mandate, nominate.

delegation *n* commission, deputation, mission.

delete *vb* blot out, cancel, cross out, cut out, edit out, efface, eliminate, eradicate, erase, expunge, obliterate, remove, rub out, strike out, wipe out.

deliberate *adj* **1** arranged, calculated, cold-blooded, conscious, contrived, culpable, designed, intended, intentional, knowing, malicious, organized, planned, prearranged, preconceived, premeditated, prepared, purposeful, studied, thought out, wilful, worked out. **2** careful, cautious, circumspect, considered, diligent, measured, methodical, orderly, painstaking, regular, slow, thoughtful, unhurried, watchful. *Opp* HASTY, INSTINCTIVE. • *vb* ▷ THINK.

delicacy *n* **1** accuracy, care, cleverness, daintiness, discrimination, exquisiteness, fineness, finesse, fragility, intricacy, precision, sensitivity, subtlety, tact. **2** *delicacies to eat.* rarity, speciality, treat.

delicate *adj* **1** dainty, diaphanous, easily broken, easily damaged, elegant, exquisite, fine, flimsy, fragile, frail, gauzy, gentle, feathery, intricate, light, sensitive, slender, soft, tender. *Opp* TOUGH. **2** *delicate work.* accurate, careful, clever, deft, precise, skilled. *Opp* CLUMSY. **3** *delicate flavour, colour.* faint, mild, muted, pale, slight, subtle. **4** *delicate health.* feeble, puny, sickly, squeamish, unhealthy, weak. **5** *delicate problem.* awkward, confidential, embarrassing, private, problematical, prudish, *inf* sticky, ticklish, touchy. **6** *delicate handling.* considerate, diplomatic, discreet, judicious, prudent, sensitive, tactful. *Opp* CRUDE.

delicious *adj* appetizing, choice, delectable, enjoyable, luscious, *inf* mouth-watering, *inf* nice, palatable, savoury, *inf* scrumptious, succulent, tasty, tempting, toothsome, *sl* yummy.

delight *n* bliss, delectation, ecstasy, enchantment, enjoyment, felicity, gratification, happiness, joy, paradise, pleasure, rapture, satisfaction. • *vb* amuse, bewitch, captivate, charm, cheer, divert, enchant, enrapture, entertain, enthral, entrance, fascinate, gladden, gratify, please, ravish, thrill, transport. *Opp* DISMAY. **delighted** ▷ HAPPY, PLEASED.

delightful *adj* agreeable, attractive, captivating, charming, congenial, delectable, diverting, enjoyable, *inf* nice, pleasant, pleasing, pleasurable, rewarding, satisfying, spell-binding. ▷ BEAUTIFUL.

delinquent *n* culprit, defaulter, hooligan, lawbreaker, malefactor, miscreant, offender, roughneck, ruffian, *inf* tear-away, vandal, wrongdoer, young offender. ▷ CRIMINAL.

delirious *adj inf* beside yourself, crazy, demented, deranged, distracted, ecstatic, excited, feverish, frantic, frenzied, hysterical, incoherent, irrational, light-headed, rambling, wild. ▷ DRUNK, MAD. *Opp* SANE, SOBER.

deliver *vb* **1** bear, bring, carry, cart, convey, distribute, give out, hand over, make over, present, purvey, supply, surrender, take round, transfer, transport, turn over. **2** *deliver a lecture.* announce, broadcast, express, give, make, read. ▷ SPEAK. **3** *deliver a blow.* administer, aim, deal, direct, fire, inflict, launch, strike, throw. ▷ HIT. **4** ▷ RESCUE.

delivery *n* **1** conveyance, dispatch, distribution, shipment, transmission, transportation. **2** *a delivery of goods.* batch, consignment. **3** *delivery of a speech.* enunciation, execution, implementation, performance, presentation. **4** childbirth, confinement, parturition.

deluge *n* downpour, flood, inundation, rainfall, rainstorm, rush, spate. • *vb* drown, engulf, flood, inundate, overwhelm, submerge, swamp.

delusion *n* dream, fantasy, hallucination, illusion, mirage, misconception, mistake, self-deception.

delve *vb* burrow, dig, explore, investigate, probe, research, search.

demand *n old use* behest, claim, command, desire, expectation, importunity, insistence, need, order, request, requirement, requisition, want. • *vb* call for, claim, cry out for, exact, expect, insist on, necessitate, order, request, require, requisition, want. ▷ ASK. **demanding** ▷ DIFFICULT, IMPORTUNATE. **in demand** ▷ POPULAR.

demean *vb* abase, cheapen, debase, degrade, disgrace, humble, humiliate, lower, make (yourself) cheap, *inf* put (yourself) down, sacrifice (your) pride, undervalue. **demeaning** ▷ SHAMEFUL.

democratic *adj* **1** classless, egalitarian. **2** chosen, elected, elective, popular, representative. *Opp* TOTALITARIAN.

demolish *vb* break down, bulldoze, dismantle, flatten, knock down, level, pull down, raze, tear down, topple, undo, wreck. ▷ DESTROY. *Opp* BUILD.

demon *n* devil, evil spirit, fiend, goblin, imp, spirit.

demonstrable *adj* conclusive, confirmable, evident, incontrovertible, indisputable, irrefutable, palpable, positive, provable, undeniable, unquestionable, verifiable.

demonstrate *vb* **1** confirm, describe, display, embody, establish, evince, exemplify, exhibit, explain, expound, express, illustrate, indicate, manifest, prove, represent, show, substantiate, teach, typify, verify. **2** lobby, march, parade, picket, protest, rally.

demonstration *n* **1** confirmation, description, display, evidence, exhibition, experiment, expression, illustration, indication, manifestation, presentation, proof, representation, show, substantiation, test, trial, verification. **2** *inf* demo, march, parade, picket, protest, rally, sit-in, vigil.

demonstrative *adj* affectionate, effusive, emotional, fulsome, loving, open, uninhibited, unreserved, unrestrained. *Opp* RETICENT.

demote *vb* downgrade, put down, reduce, relegate. *Opp* PROMOTE.

demure *adj* bashful, coy, diffident, modest, prim, quiet, reserved, reticent, retiring, sedate, shy, sober, staid. *Opp* CONCEITED.

den *n* hideaway, hide-out, hiding-place, hole, lair, private place, retreat, sanctuary, secret place, shelter.

denial *n* abnegation, contradiction, disavowal, disclaimer, negation, refusal, refutation, rejection, renunciation, repudiation, veto. *Opp* ADMISSION.

denigrate *vb* belittle, blacken the reputation of, criticize, decry, disparage, impugn, malign, *inf* put down, *inf* run down, sneer at, speak slightingly of, traduce, *inf* turn your nose up, vilify. ▷ DESPISE. *Opp* PRAISE.

denomination *n* **1** category, class, classification, designation, kind, size, sort, species, type, value. **2** church, communion, creed, cult, order, persuasion, schism, school, sect.

denote *vb* be the sign for, designate, express, indicate, mean, represent, signal, signify, stand for, symbolize.

denouement *n* climax, *inf* pay-off, resolution, solution, *inf* sorting out, *inf* tidying up, unravelling. ▷ END.

denounce *vb* accuse, attack verbally, betray, blame, brand, censure, complain about, condemn, criticize, declaim against, decry, fulminate against, *inf* hold forth against, impugn, incriminate, inform against, inveigh against, pillory, report, reveal, stigmatize, *inf* tell off, vilify, vituperate. *Opp* PRAISE.

dense *adj* **1** close, compact, concentrated, heavy, impassable, impenetrable, *inf* jam-packed, lush, massed, packed, solid, thick, tight, viscous. *Opp* THIN. **2** ▷ STUPID.

dent *n* concavity, depression, dimple, dint, dip, hollow, indentation, pit. • *vb* bend, buckle, crumple, knock in.

denude *vb* bare, defoliate, deforest, expose, remove, strip, unclothe, uncover. *Opp* CLOTHE.

deny *vb* **1** contradict, controvert, disagree with, disclaim, disown, dispute, gainsay, negate, oppose, rebuff, refute, reject, repudiate. *Opp* AGREE. **2** begrudge, deprive of, disallow, refuse, withhold. *Opp* GRANT. **deny yourself** ▷ ABSTAIN.

depart *vb* **1** abscond, begin a journey, *inf* check out, *inf* clear off, decamp, disappear, embark, emigrate, escape, exit, go away, *sl* hit the road, leave, make off, *inf* make tracks, *inf* make yourself scarce, migrate, move away, move off, *inf* push off, quit, retire, retreat, run away, run off, *sl* scarper, *sl* scram, set forth, set off, set out, start, take your leave, vanish, withdraw. **2** ▷ DEVIATE. **departed** ▷ DEAD.

department *n* **1** branch, division, office, part, section, sector, subdivision, unit. **2** [*inf*] *not my department.* area, concern, domain, field, function, job, line, province, responsibility, specialism, sphere.

departure *n* disappearance, embarkation, escape, exit, exodus, going, retirement, retreat, withdrawal. *Opp* ARRIVAL.

depend *vb* **depend on** *inf* bank on, be dependent on, count on, hinge on, need, pivot on, put your faith in, *inf* reckon on, rely on, rest on, trust.

dependable *adj* conscientious, consistent, faithful, honest, regular, reliable, safe, sound, steady, true, trustworthy, unfailing. *Opp* UNRELIABLE.

dependence *n* **1** confidence, need, reliance, trust. **2** ▷ ADDICTION.

dependent *adj* **dependent on** **1** conditional on, connected with, controlled by, determined by, liable to, relative to, subject to, vulnerable to. *Opp* INDEPENDENT. **2** *dependent on drugs.* addicted to, enslaved by, *inf* hooked on, reliant on.

depict *vb* delineate, describe, draw, illustrate, narrate, outline, paint, picture, portray, represent, reproduce, show, sketch.

deplete *vb* consume, cut, decrease, drain, lessen, reduce, use up. *Opp* INCREASE.

deplorable *adj* awful, blameworthy, discreditable, disgraceful, disreputable, dreadful, execrable, lamentable, regrettable, reprehensible, scandalous, shameful, shocking, unfortunate, unworthy. ▷ BAD. *Opp* COMMENDABLE.

deplore *vb* **1** grieve for, lament, mourn, regret. **2** ▷ CONDEMN.

deploy *vb* arrange, bring into action, distribute, manage, position, use systematically, utilize.

deport *vb* banish, exile, expatriate, expel, remove, send abroad, transport.

depose *vb* demote, dethrone, dismiss, displace, get rid of, oust, remove, *inf* topple.

deposit *n* **1** advance payment, down-payment, initial payment, part-payment, retainer, security, stake. **2** accumulation, alluvium, dregs, layer, lees, precipitate, sediment, silt, sludge. • *vb* **1** drop, *inf* dump, lay down, leave, *inf* park, place, precipitate, put down, set down. **2** *deposit money.* bank, pay in, save.

depot *n* **1** arsenal, base, cache, depository, dump, hoard, store, storehouse. **2** *bus depot.* garage, headquarters, station, terminus.

deprave *vb* brutalize, corrupt, debase, degrade, influence, pervert. **depraved** ▷ CORRUPT.

depreciate *vb* **1** become less, decrease, deflate, drop, fall, go down, lessen, lower, reduce, slump, weaken. *Opp* APPRECIATE. **2** ▷ DISPARAGE.

depress *vb* **1** burden, cast down, discourage, dishearten, dismay, dispirit, enervate, grieve, lower the spirits of, make sad, oppress, sadden, tire, upset, weary. *Opp* CHEER. **2** *depress the market.* bring down, deflate, make less active, push down, undermine, weaken. *Opp* BOOST. **depressed, depressing** ▷ SAD.

depression *n* **1** *inf* blues, dejection, desolation, despair, despondency, gloom, glumness, heaviness, hopelessness, low spirits, melancholy, misery, pessimism, sadness, weariness. *Opp* HAPPINESS. **2** cavity, concavity, dent, dimple, dip, excavation, hole, hollow, impression, indentation, pit, recess, rut, sunken area. *Opp* BUMP. **3** *economic depression.* decline, hard times, recession, slump. *Opp* BOOM, HIGH.

deprive *vb* **deprive of** deny, dispossess of, prevent from using, refuse, rob of, starve of, strip of, take away, withdraw, withhold. **deprived** ▷ POOR.

deputize *vb* **deputize for** act as deputy, act as stand-in for, cover for, do the job of, replace, represent, stand in for, substitute for, take over from, understudy.

deputy *n* agent, ambassador, assistant, delegate, emissary, *inf* fill-in, locum, proxy, relief, replacement, representative, reserve, second-in-command, spokesperson, *inf* stand-in, substitute, supply, surrogate, understudy, vice-captain, vice-president.

derelict *adj* abandoned, broken down, decrepit, deserted, desolate, dilapidated, forgotten, forlorn, forsaken, neglected, overgrown, ruined, run-down, tumbledown, uncared-for, untended.

derivation *n* ancestry, descent, etymology, extraction, origin, root. ▷ BEGINNING.

derive *vb* acquire, borrow, collect, crib, draw, extract, gain, gather, get, glean, *inf* lift, obtain, pick up, procure, receive, secure, take. **be derived** ▷ ORIGINATE.

descend *vb* **1** climb down, come down, drop, fall, go down, move down, plummet, plunge, sink, swoop down. **2** decline, dip, incline, slant, slope. **3** alight, disembark, dismount, get down, get off. *Opp* ASCEND. **be descended** ▷ ORIGINATE. **descend on** ▷ ATTACK.

descendant *n* child, heir, scion, successor. *Opp* ANCESTOR. **descendants** family, issue, line, lineage, offspring, posterity, progeny, *old use* seed.

descent *n* **1** declivity, dip, drop, fall, incline, slant, slope, way down. *Opp* ASCENT. **2** *aristocratic descent.* ancestry, background, blood, derivation, extraction, family, genealogy, heredity, lineage, origin, parentage, pedigree, stock, strain.

describe *vb* **1** characterize, define, delineate, depict, detail, explain, express, give an account of, narrate, outline, portray, present, recount, relate, report, represent, sketch, speak of, tell about. **2** *describe a circle.* draw, mark out, trace.

description *n* account, characterization, commentary, definition, delineation, depiction, explanation, narration, outline, portrait, portrayal, report, representation, sketch, story, word-picture.

descriptive *adj* colourful, detailed, explanatory, expressive, graphic, illustrative, pictorial, vivid.

desecrate *vb* abuse, contaminate, corrupt, debase, defile, degrade, dishonour, pervert, pollute, profane, treat blasphemously, treat disrespectfully, treat irreverently, vandalize, violate, vitiate. *Opp* REVERE.

desert *adj* arid, barren, desolate, dry, infertile, isolated, lonely, sterile, uncultivated, unfrequented, uninhabited, waterless, wild. *Opp* FERTILE. • *n* dust bowl, wasteland, wilderness. • *vb* **1** abandon, betray, forsake, give up, jilt, leave, *inf* leave in the lurch, maroon, quit, *inf* rat on, renounce, strand, vacate, *inf* walk out on, *inf* wash your hands of. **2** abscond, decamp, defect, go absent, run away. **deserted** ▷ EMPTY, LONELY.

deserter *n* absconder, absentee, apostate, backslider, betrayer, defector, escapee, fugitive, outlaw, renegade, runaway, traitor, truant, turncoat.

deserve *vb* be good enough for, be worthy of, earn, justify, merit, rate, warrant. **deserving** ▷ WORTHY.

design *n* **1** blueprint, conception, draft, drawing, model, pattern, plan, proposal, prototype, sketch. **2** mark, style, type, version. **3** arrangement, composition, configuration, form, pattern, shape. **4** *wander without design.* aim, end, goal, intention, object, objective, purpose, scheme. • *vb* conceive, construct, contrive, create, delineate, devise, draft, draw, draw up, fashion, form, intend, invent, lay out, make, map out, originate, outline, plan, plot, project, propose, scheme, shape, sketch, think up. **designing** ▷ CRAFTY. **have designs** ▷ PLOT.

designer *n* architect, artist, author, contriver, creator, deviser, inventor, originator.

desire *n* **1** ache, ambition, appetite, craving, fancy, hankering, hunger, *inf* itch, longing, requirement, thirst, urge, want, wish, yearning, *inf* yen. **2** avarice, covetousness, cupidity, greed, miserliness, rapacity. **3** *sexual desire.* ardour, libido, love, lust, passion. • *vb* ache for, ask for, aspire to, covet, crave, dream of, fancy, hanker after, *inf* have a yen for, hope for, hunger for, *inf* itch for, like, long for, lust after, need, pine for, prefer, pursue, *inf* set your heart on, set your sights on, strive after, thirst for, want, wish for, yearn for.

desolate *adj* **1** abandoned, bare, barren, benighted, bleak, cheerless, depressing, deserted, dismal, dreary, empty, forsaken, gloomy, *inf* god-forsaken, inhospitable, isolated, lonely, remote, unfrequented, uninhabited, wild, windswept. **2** bereft, companionless, dejected, depressed, despairing, disconsolate, distressed, forlorn, forsaken, inconsolable, lonely, melancholy, miserable, neglected, solitary, suicidal, wretched. ▷ SAD. *Opp* CHEERFUL.

despair *n* anguish, dejection, depression, desperation, despondency, hopelessness, pessimism, resignation, wretchedness. ▷ MISERY. • *vb* give in, give up, lose heart, lose hope, quit, surrender. *Opp* HOPE.

desperate *adj* **1** *inf* at your wits' end, beyond hope, despairing, inconsolable, wretched. **2** *desperate situation.* acute, bad, critical, dangerous, drastic, grave, hopeless, irretrievable, pressing, serious, severe, urgent. **3** *desperate criminals.* dangerous, foolhardy, impetuous, rash, reckless, violent, wild. **4** ▷ ANXIOUS.

despise *vb* be contemptuous of, condemn, deride, disapprove of, disdain, feel contempt for, hate, have a low opinion of, look down on, *inf* put down, scorn, sneer at, spurn, undervalue. ▷ DENIGRATE. *Opp* ADMIRE.

despondent *adj* dejected, depressed, discouraged, disheartened, down, downcast, *inf* down in the mouth, melancholy, morose, pessimistic, sad, sorrowful. ▷ MISERABLE.

despotic *adj* absolute, arbitrary, authoritarian, autocratic, dictatorial, domineering, oppressive, totalitarian, tyrannical. *Opp* DEMOCRATIC.

destination *n* goal, objective, purpose, stopping-place, target, terminus.

destined *adj* **1** foreordained, ineluctable, inescapable, inevitable, intended, ordained, predestined, predetermined, preordained, unavoidable. **2** *destined to fail.* bound, certain, doomed, fated, meant.

destiny *n* chance, doom, fate, fortune, karma, kismet, lot, luck, providence.

destitute *adj* bankrupt, deprived, down-and-out, homeless, impecunious, impoverished, indigent, insolvent, needy, penniless, poverty-stricken, *inf* skint. ▷ POOR. *Opp* WEALTHY.

destroy *vb* abolish, annihilate, blast, break down, burst, *inf* bust, crush, *inf* decimate, demolish, devastate, devour, dismantle, dispose of, do away with, eliminate, eradicate, erase, exterminate, extinguish, extirpate, finish off, flatten, fragment, get rid of, knock down, lay waste, level, liquidate, make useless, nullify, pull down, pulverize, put out of existence, raze, root out, ruin, sabotage, sack, scuttle, shatter, smash, stamp out, undo, uproot, vaporize, wipe out, wreck, write off. ▷ DEFEAT, END, KILL. *Opp* CONSERVE, CREATE.

destruction *n* annihilation, damage, *inf* decimation, demolition, depredation, devastation, elimination, end, eradication, erasure, extermination, extinction, extirpation, havoc, holocaust, liquidation, overthrow, pulling down, ruin, ruination, shattering, smashing, undoing, uprooting, wiping out, wrecking. ▷ KILLING. *Opp* CONSERVATION, CREATION.

destructive *adj* adverse, antagonistic, baleful, baneful, calamitous, catastrophic, damaging, dangerous, deadly, deleterious, detrimental, devastating, disastrous, fatal, harmful, injurious, internecine, lethal, malignant, negative, pernicious, pestilential, ruinous, violent. *Opp* CONSTRUCTIVE.

detach *vb* cut loose, cut off, disconnect, disengage, disentangle, divide, free, isolate, part, pull off, release, remove, segregate, separate, sever, take off, tear off, uncouple, undo, unfasten, unfix, unhitch. *Opp* ATTACH. **detached** ▷ ALOOF, IMPARTIAL, SEPARATE.

detail *n* aspect, circumstance, complexity, complication, component, element, fact, factor, feature, ingredient, intricacy, item, *plur* minutiae, nicety, particular, point, refinement, respect, specific, technicality.

detailed *adj inf* blow-by-blow, complete, complex, comprehensive, descriptive, exact, exhaustive, full, *derog* fussy, giving all details, *derog* hair-splitting, intricate, itemized, minute, particularized, specific. *Opp* GENERAL.

detain *vb* **1** arrest, capture, confine, gaol, hold, imprison, intern. **2** buttonhole, delay, hinder, hold up, impede, keep, keep waiting, restrain, retard, slow, stop, waylay.

detect *vb* ascertain, become aware of, diagnose, discern, discover, expose, feel, *inf* ferret out, find, hear, identify, locate, note, notice, observe, perceive, *inf* put your finger on, recognize, reveal, scent, see, sense, sight, smell, sniff out, spot, spy, taste, track down, uncover, unearth, unmask.

detective *n* investigator, policeman, policewoman, *inf* private eye, sleuth, *inf* snooper.

detention *n* captivity, confinement, custody, imprisonment, incarceration, internment.

deter *vb* check, daunt, discourage, dismay, dissuade, frighten off, hinder, impede, intimidate, obstruct, prevent, put off, repel, send away, stop, *inf* turn off, warn off. *Opp* ENCOURAGE.

deteriorate *vb* crumble, decay, decline, degenerate, depreciate, disintegrate, fall off, get worse, *inf* go downhill, lapse, relapse, slip, weaken, worsen. *Opp* IMPROVE.

determination *n inf* backbone, commitment, courage, dedication, doggedness, drive, firmness, fortitude, *inf* grit, *inf* guts, perseverance, persistence, pertinacity, resoluteness, resolution, resolve, single-mindedness, spirit, steadfastness, *derog* stubbornness, tenacity, will-power.

determine *vb* **1** arbitrate, clinch, conclude, decide, establish, find out, identify, judge, settle. **2** choose, decide on, fix on, resolve, select. **3** *What determined your choice?* affect, condition, dictate, govern, influence, regulate.

determined *adj* adamant, assertive, bent (*on success*), certain, convinced, decided, decisive, definite, dogged, firm, insistent, intent, *derog* obstinate, persistent, pertinacious, purposeful, resolute, resolved, single-minded, steadfast, strong-minded, strong-willed, *derog* stubborn, sure, tenacious, tough, unwavering. *Opp* IRRESOLUTE.

deterrent *n* barrier, caution, check, curb, difficulty, discouragement, disincentive, dissuasion, hindrance, impediment, obstacle, restraint, threat, *inf* turn-off, warning. *Opp* ENCOURAGEMENT.

detest *vb* abhor, abominate, despise, execrate, loathe. ▷ HATE.

detour *n* deviation, diversion, indirect route, roundabout route. **make a detour** ▷ DEVIATE.

detract *vb* **detract from** diminish, lessen, lower, reduce, take away from.

detrimental *adj* damaging, deleterious, disadvantageous, harmful, hurtful, inimical, injurious, prejudicial, unfavourable. *Opp* ADVANTAGEOUS.

devastate *vb* **1** damage severely, demolish, destroy, flatten, lay waste, level, obliterate, overwhelm, ravage, raze, ruin, sack, waste, wreck. **2** ▷ DISMAY.

develop *vb* **1** advance, age, arise, *inf* blow up, come into existence, evolve, get better, grow, flourish, improve, mature, move on, progress, ripen. *Opp* REGRESS. **2** *develop habits.* acquire, contract, cultivate, evolve, foster, get, pick up. **3** *develop ideas.* amplify, augment, elaborate, enlarge on, expatiate on, unfold, work up. **4** *business developed.* branch out, build up, diversify, enlarge, expand, extend, increase, swell.

development *n* **1** advance, betterment, change, enlargement, evolution, expansion, extension, *inf* forward march, furtherance, gain, growth, improvement, increase, progress, promotion, regeneration, reinforcement, spread. **2** happening, incident, occurrence, outcome, result, upshot. **3** *industrial development.* building, conversion, exploitation, use.

deviate *vb* branch off, depart, digress, diverge, divert, drift, err, go astray, go round, make a detour, stray, swerve, turn aside, turn off, vary, veer, wander.

device *n* **1** apparatus, appliance, contraption, contrivance, gadget, implement, instrument, invention, machine, tool, utensil. **2** dodge, expedient, gambit, gimmick, manoeuvre, plan, ploy, ruse, scheme, stratagem, stunt, tactic, trick, wile. **3** *heraldic device.* badge, crest, design, figure, logo, motif, shield, sign, symbol, token.

levil *n* demon, fiend, imp, spirit. **The Devil** the Adversary, Beelzebub, the Evil One, Lucifer, Mephistopheles, *inf* Old Nick, the Prince of Darkness, Satan.

levilish *adj* demoniac(al), demonic, diabolic(al), fiendish, hellish, impish, infernal, inhuman, Mephisophelian, satanic. ▷ EVIL. *Opp* ANGELIC.

levious *adj* **1** circuitous, crooked, deviating, indirect, periphrastic, rambling, round-about, sinuous, tortuous, wandering, winding. **2** [*derog*] calculating, cunning, deceitful, evasive, insincere, misleading, scheming, *inf* slippery, sly, sneaky, treacherous, underhand, wily. ▷ DISHONEST. *Opp* DIRECT.

evise *vb* arrange, conceive, concoct, contrive, *inf* cook up, create, design, engineer, form, formulate, frame, imagine, invent, make up, plan, plot, prepare, project, scheme, think out, think up, work out.

evoted *adj* committed, dedicated, enthusiastic, faithful, loving, staunch, true, unswerving, wholehearted, zealous. ▷ LOYAL. *Opp* DISLOYAL, HALF-HEARTED.

evotee *n inf* addict, aficionado, *inf* buff, enthusiast, fan, follower, *l* freak, supporter.

evotion *n* allegiance, attachment, commitment, dedication, devotedness, enthusiasm, fanaticism, fervour, loyalty, zeal. ▷ LOVE, PIETY.

devour *vb* consume, demolish, eat up, engulf, swallow up, take in. ▷ DESTROY, EAT.

devout *adj* god-fearing, godly, holy, religious, sincere, spiritual. ▷ PIOUS. *Opp* IRRELIGIOUS.

dexterous *adj* adroit, agile, deft, nimble, quick, sharp, skilful. ▷ CLEVER. *Opp* CLUMSY.

diabolical *adj* evil, fiendish, inhuman, satanic, wicked. ▷ DEVILISH. *Opp* SAINTLY.

diagnose *vb* detect, determine, distinguish, find, identify, isolate, name, pinpoint, recognize.

diagnosis *n* analysis, conclusion, explanation, identification, interpretation, opinion, pronouncement, verdict.

diagram *n* chart, drawing, figure, flow-chart, graph, illustration, outline, picture, plan, representation, sketch, table.

dial *n* clock, digital display, face, instrument, pointer, speedometer.

dialect *n* accent, argot, brogue, cant, creole, idiom, jargon, language, patois, phraseology, pronunciation, register, slang, speech, tongue, vernacular.

dialogue *n inf* chat, *inf* chin-wag, colloquy, communication, conference, conversation, debate, discourse, discussion, duologue, exchange, interchange, *old use* intercourse, meeting, oral communication, talk, *inf* tête-à-tête.

diary *n* annals, appointment book, calendar, chronicle, engagement book, journal, log, record.

dictate *vb* **1** read aloud, speak slowly. **2** command, decree, direct, enforce, give orders, impose, *inf* lay down the law, make the rules, ordain, order, prescribe, state categorically.

dictator *n* autocrat, *inf* Big Brother, despot, tyrant. ▷ RULER.

dictatorial *adj* absolute, arbitrary, authoritarian, autocratic, *inf* bossy, despotic, dogmatic, dominant, domineering, illiberal, imperious, intolerant, omnipotent, oppressive, overbearing, repressive, totalitarian, tyrannical, undemocratic. *Opp* DEMOCRATIC.

dictionary *n* concordance, glossary, lexicon, thesaurus, vocabulary, wordbook.

didactic *adj* instructive, lecturing, pedagogic, pedantic.

die *vb* **1** *inf* bite the dust, *inf* breathe your last, cease to exist, come to the end, decease, depart, expire, fall, *inf* give up the ghost, *sl* kick the bucket, lay down your life, lose your life, pass away, *sl* peg out, perish, *sl* pop off, *sl* snuff it, starve. **2** decline, decrease, die away, disappear, droop, dwindle, ebb, end, fade, fail, fizzle out, go out, languish, lessen, peter out, stop, subside, vanish, wane, weaken, wilt, wither.

diet *n* fare, food, intake, nourishment, nutriment, nutrition, sustenance. • *vb* abstain, *inf* cut down, deny yourself, fast, lose weight, ration yourself, reduce, slim.

differ *vb* **1** be different, be distinct, contrast, deviate, diverge, show differences, vary. **2** argue, be at odds, be at variance, clash, conflict, contradict, disagree, dispute, dissent, fall out, *inf* have a difference, oppose each other, quarrel, take issue with each other. *Opp* AGREE.

difference *n* **1** alteration, change, comparison, contrast, development, deviation, differential, differentiation, discrepancy, disparity, dissimilarity, distinction, diversity, incompatibility, incongruity, inconsistency, modification, nuance, unlikeness, variation, variety. *Opp* SIMILARITY. **2** argument, clash, conflict, controversy, debate, disagreement, disharmony, dispute, dissent, quarrel, strife, tiff, wrangle. *Opp* AGREEMENT.

different *adj* **1** assorted, clashing, conflicting, contradictory, contrasting, deviating, discordant, discrepant, disparate, dissimilar, distinguishable, divergent, diverse, heterogeneous, ill-matched, incompatible, inconsistent, miscellaneous, mixed, multifarious, opposed, opposite, *inf* poles apart, several, sundry, unlike, varied, various. *Opp* SIMILAR. **2** abnormal, altered, anomalous, atypical, bizarre, changed, distinct, distinctive, eccentric, extraordinary, fresh, individual, irregular, new, original, particular, peculiar, personal, revolutionary, separate, singular, special, specific, strange, uncommon, unconventional, unique, unorthodox, unusual. *Opp* CONVENTIONAL.

differentiate *vb* contrast, discriminate, distinguish, tell apart.

difficult *adj* **1** abstruse, advanced, baffling, complex, complicated, deep, *inf* dodgy, enigmatic, hard, intractable, intricate, involved, *inf* knotty, *inf* nasty, obscure, perplexing, problematical, *inf* thorny, ticklish, tricky. **2** arduous, awkward, backbreaking, burdensome, challenging, daunting, demanding, exacting, exhausting, formidable, gruelling, heavy, herculean, *inf* killing, laborious, onerous, punishing, rigorous, severe, strenuous, taxing, tough, uphill. **3** *difficult children.*

annoying, disruptive, fussy, headstrong, intractable, obstinate, obstreperous, refractory, stubborn, tiresome, troublesome, trying, uncooperative, unfriendly, unhelpful, unresponsive, unruly. *Opp* COOPERATIVE, EASY.

difficulty *n* adversity, challenge, complication, dilemma, embarrassment, enigma, *inf* fix, *inf* hang-up, hardship, *inf* hiccup, hindrance, hurdle, impediment, *inf* jam, *inf* mess, obstacle, perplexity, *inf* pickle, pitfall, plight, predicament, problem, puzzle, quandary, snag, *inf* spot, straits, *inf* stumbling-block, tribulation, trouble, *inf* vexed question.

diffident *adj* backward, bashful, coy, distrustful, doubtful, fearful, hesitant, hesitating, inhibited, insecure, introvert, meek, modest, nervous, private, reluctant, reserved, retiring, self-effacing, sheepish, shrinking, shy, tentative, timid, timorous, unadventurous, unassuming, underconfident, unsure, withdrawn. *Opp* CONFIDENT.

diffuse *adj* digressive, discursive, long-winded, loose, meandering, rambling, spread out, unstructured, vague, *inf* waffly, wandering. ▷ WORDY. *Opp* CONCISE. • *vb* ▷ SPREAD.

dig *vb* **1** burrow, delve, excavate, gouge, hollow, mine, quarry, scoop, tunnel. **2** cultivate, fork over, *inf* grub up, till, trench, turn over. **3** jab, nudge, poke, prod, punch, shove, thrust. **dig out** ▷ FIND. **dig up** disinter, exhume.

digest *n* ▷ SUMMARY. • *vb* **1** absorb, assimilate, dissolve, ingest, process, utilize. ▷ EAT. **2** consider, ponder, study, take in, understand.

digit *n* **1** figure, integer, number, numeral. **2** finger, toe.

dignified *adj* august, becoming, calm, courtly, decorous, distinguished, elegant, exalted, formal, grand, grave, imposing, impressive, lofty, lordly, majestic, noble, proper, refined, regal, sedate, serious, sober, solemn, stately, tasteful, upright. ▷ PROUD. *Opp* UNBECOMING.

dignitary *n inf* high-up, important person, luminary, notable, official, *inf* VIP, worthy.

dignity *n* calmness, courtliness, decorum, elegance, eminence, formality, glory, grandeur, *Lat* gravitas, gravity, greatness, honour, importance, majesty, nobility, propriety, regality, respectability, seriousness, sobriety, solemnity, stateliness. ▷ PRIDE.

digress *vb* depart, deviate, diverge, drift, get off the subject, *inf* go off at a tangent, *inf* lose the thread, ramble, stray, veer, wander.

dilapidated *adj* badly maintained, broken down, crumbling, decayed, decrepit, derelict, falling apart, falling down, in disrepair, in ruins, neglected, ramshackle, rickety, ruined, *inf* run-down, shaky, tottering, tumbledown, uncared-for.

dilemma *n inf* catch-22, deadlock, difficulty, doubt, embarrassment, *inf* fix, impasse, *inf* jam, *inf* mess, *inf* pickle, plight, predicament, problem, quandary, *inf* spot, stalemate.

diligent *adj* assiduous, busy, careful, conscientious, constant, devoted, earnest, energetic, hardworking, indefatigable, industrious, meticulous, painstaking, persevering, persistent, pertinacious, punctilious, scrupulous, sedulous, studious, thorough, tireless. *Opp* LAZY.

dilute *vb* adulterate, reduce the strength of, thin, water down, weaken. *Opp* CONCENTRATE.

dim *adj* **1** bleary, blurred, clouded, cloudy, dark, dingy, dull, faint, fogged, foggy, fuzzy, gloomy, grey, hazy, ill-defined, imperceptible, indistinct, indistinguishable, misty, murky, nebulous, obscure, obscured, pale, shadowy, sombre, unclear, vague, weak. **2** ▷ STUPID. *Opp* BRIGHT. • *vb* **1** blacken, cloud, darken, dull, make dim, mask, obscure, shade, shroud. **2** become dim, fade, go out, lose brightness, lower. *Opp* BRIGHTEN. **take a dim view** ▷ DISAPPROVE.

dimensions *plur n* capacity, extent, magnitude, measurements, proportions, scale, scope, size. ▷ MEASUREMENT.

diminish *vb* **1** abate, become less, contract, curtail, decline, decrease, depreciate, die down, dwindle, ease off, ebb, fade, lessen, *inf* let up, lower, peter out, recede, reduce, shorten, shrink, shrivel, slow down, subside, wane, *inf* wind down. ▷ CUT. *Opp* INCREASE. **2** belittle, cheapen, demean, deprecate, devalue, disparage, minimize, undervalue. *Opp* EXAGGERATE.

diminutive *adj* microscopic, midget, miniature, minuscule, minute, tiny, undersized. ▷ SMALL.

din *n* blaring, clamour, clangour, clatter, commotion, crash, hubbub, hullabaloo, noise, outcry, pandemonium, racket, roar, row, rumpus, shouting, tumult, uproar. ▷ SOUND.

dingy *adj* colourless, dark, depressing, dim, dirty, discoloured, dismal, drab, dreary, dull, faded, gloomy, grimy, murky, old, seedy, shabby, smoky, soiled, sooty, worn. *Opp* BRIGHT.

dining-room *n* cafeteria, carvery, refectory, restaurant.

dinner *n* banquet, feast. ▷ MEAL.

dip *n* **1** concavity, declivity, dent, depression, fall, hole, hollow, incline, slope. **2** *dip in the sea.* bathe, dive, immersion, plunge, soaking, swim. • *vb* **1** decline, descend, dive, fall, go down, sag, sink, slope down, slump, subside. **2** douse, drop, duck, dunk, immerse, lower, plunge, submerge. **take a dip** ▷ BATHE.

diplomacy *n* adroitness, delicacy, discretion, finesse, negotiation, skill, tact, tactfulness.

diplomat *n* ambassador, consul, government representative, negotiator, official, peacemaker, politician, representative, tactician.

diplomatic *adj* careful, considerate, delicate, discreet, judicious, polite, politic, prudent, sensitive, subtle, tactful, thoughtful, understanding. *Opp* TACTLESS.

direct *adj* **1** non-stop, shortest, straight, unbroken, undeviating, uninterrupted, unswerving. **2** blunt, candid, categorical, clear, decided, explicit, express, forthright, frank, honest, open, outspoken, plain, point-blank, sincere, straightforward, *derog* tactless, to the point, unambiguous, uncomplicated, *derog* undiplomatic, unequivocal, uninhibited, unqualified, unreserved. **3** *direct experience.* empirical, firsthand, *inf* from the horse's mouth, personal. **4** *direct opposites.* absolute, complete, diametrical, exact, head-on, *inf* out-and-out, utter. *Opp* INDIRECT. • *vb* **1** address, escort, guide, indicate the way, point, route, send, show the way, tell the way, usher. **2** aim, focus, level, target, train, turn. **3** administer, be in charge of, command, conduct, control, govern, handle, lead, manage, mastermind, oversee, regulate, rule, run, stage-manage, superintend, supervise, take charge of. **4** *direct someone to do something.* advise, bid, charge, command, counsel, enjoin, instruct, order, require, tell.

direction *n* aim, approach, (compass) bearing, course, orientation, path, point of the compass, road, route, tack, track, way. **directions** guidance, guidelines, instructions, orders, plans.

director *n* administrator, *inf* boss, executive, governor, manager, managing director, organizer, president, principal. ▷ CHIEF.

directory *n* catalogue, index, list, register.

dirt *n* **1** dust, excrement, filth, garbage, grime, impurity, mess, mire, muck, ooze, ordure, pollution, slime, sludge, smut, soot, stain. ▷ OBSCENITY, RUBBISH. **2** clay, earth, loam, mud, soil.

dirty *adj* **1** befouled, begrimed, besmirched, bespattered, black, dingy, dusty, filthy, foul, grimy, grubby, marked, messy, mucky, muddy, nasty, scruffy, shabby, slatternly, smeary, smudged, soiled, sooty, sordid, spotted, squalid, stained, sullied, tarnished, travel-stained, uncared for, unclean, untidy, unwashed. **2** *dirty water.* cloudy, contaminated, impure, muddy, murky, poisoned, polluted, tainted, untreated. **3** *dirty tactics.* dishonest, dishonourable, illegal, *inf* low-down, mean, rough, treacherous, unfair, ungentlemanly, unscrupulous, unsporting, unsportsmanlike. ▷ CORRUPT. **4** *dirty talk.* coarse, crude, improper, indecent, offensive, rude, smutty, vulgar. ▷ OBSCENE. *Opp* CLEAN. • *vb* befoul, foul, make dirty, mark, *inf* mess up, smear, smudge, soil, spatter, spot, stain, streak, tarnish. ▷ DEFILE. *Opp* CLEAN.

disability *n* affliction, complaint, defect, disablement, handicap, impairment, incapacity, infirmity, weakness.

disable *vb* cripple, damage, debilitate, enfeeble, *inf* hamstring, handicap, immobilize, impair, incapacitate, injure, lame, maim, make useless, mutilate, paralyse, put out of action, ruin, weaken. **disabled** ▷ HANDICAPPED.

disadvantage *n* drawback, handicap, hardship, hindrance, impediment, inconvenience, liability, *inf* minus, nuisance, privation, snag, trouble, weakness.

disagree *vb* argue, bicker, clash, conflict, contend, differ, dispute, dissent, diverge, fall out, fight, quarrel, squabble, wrangle. **disagree with** ▷ OPPOSE.

disagreeable *adj* disgusting, distasteful, nasty, objectionable, obnoxious, offensive, *inf* off-putting, repellent, sickening, unsavoury. ▷ UNPLEASANT. *Opp* PLEASANT.

disagreement *n* altercation, argument, clash, conflict, contention, controversy, debate, difference, discrepancy, disharmony, disparity, dispute, dissension, dissent, divergence, incompatibility, inconsistency, misunderstanding, opposition, quarrel, squabble, strife, *inf* tiff, variance, wrangle. *Opp* AGREEMENT.

disappear *vb* **1** become invisible, cease to exist, clear, die out, disperse, dissolve, dwindle, ebb, evanesce, evaporate, fade, melt away, recede, vanish, vaporize, wane. ▷ DIE. **2** depart, escape, flee, fly, go, pass out of sight, run away, walk away, withdraw. *Opp* APPEAR.

disappoint *vb* be worse than expected, chagrin, *inf* dash your hopes, disenchant, disillusion, dismay, displease, dissatisfy, fail to satisfy, *inf* let down, upset, vex. ▷ FRUSTRATE. *Opp* SATISFY. **disappointed** disillusioned, frustrated, *inf* let down, unsatisfied. ▷ SAD.

disapproval *n* anger, censure, condemnation, criticism, disapprobation, disfavour, dislike, displeasure, dissatisfaction, hostility, reprimand, reproach. *Opp* APPROVAL.

disapprove *vb* **disapprove of** be displeased by, belittle, blame, censure, condemn, criticize, denounce, deplore, deprecate, dislike, disparage, frown on, jeer at, look askance at, make unwelcome, object to, regret, reject, *inf* take a dim view of, take exception to. *Opp* APPROVE. **disapproving** ▷ CRITICAL.

disarm *vb* **1** demilitarize, demobilize, disband troops, make powerless, take weapons from. **2** charm, mollify, pacify, placate.

disaster *n* accident, act of God, blow, calamity, cataclysm, catastrophe, crash, débacle, failure, fiasco, *inf* flop, *inf* mess-up, misadventure, mischance, misfortune, mishap, reverse, tragedy, *inf* wash-out. *Opp* SUCCESS.

disastrous *adj* appalling, awful, calamitous, cataclysmic, catastrophic, crippling, destructive, devastating, dire, dreadful, fatal, ruinous, terrible, tragic. *Opp* SUCCESSFUL.

disbelieve *vb* be sceptical of, discount, discredit, doubt, have no faith in, mistrust, reject, suspect. *Opp* BELIEVE. **disbelieving** ▷ INCREDULOUS.

disc *n* **1** circle, counter, plate, token. **2** album, CD, LP, record, single. **3** [*computing*] CD-ROM, disk, diskette, floppy disk, hard disk.

discard *vb* abandon, cast off, *inf* chuck away, dispense with, dispose of, *inf* ditch, dump, eliminate, get rid of, jettison, junk, reject, scrap, shed, throw away, toss out.

discern *vb* be aware of, be sensitive to, detect, discover, discriminate, distinguish, make out, mark, notice, observe, perceive, recognize, spy. ▷ SEE. **discerning** ▷ PERCEPTIVE.

discernible *adj* detectable, distinguishable, measurable, perceptible. ▷ NOTICEABLE.

discharge *n* **1** release, dismissal. **2** emission, excretion, ooze, pus, secretion, suppuration. • *vb* **1** belch, eject, emit, expel, exude, give off, give out, pour out, produce, release, secrete, send out, spew, spit out. **2** *discharge guns.* detonate, explode, fire, let off, shoot. **3** *discharge employees.* dismiss, fire, make redundant, remove, sack, throw out. **4** *discharge a prisoner.* absolve, acquit, allow to leave, clear, dismiss, excuse, exonerate, free, let off, liberate, pardon, release. **5** *discharge duties.* accomplish, carry out, execute, fulfil, perform.

disciple *n* acolyte, adherent, admirer, apostle, apprentice, devotee, follower, learner, proselyte, pupil, scholar, student, supporter.

disciplinarian *n* authoritarian, autocrat, despot, dictator, *inf* hard-liner, *inf* hard taskmaster, martinet, *inf* slave-driver, *inf* stickler, tyrant.

discipline *n* **1** control, drilling, indoctrination, instruction, management, strictness, system, training. **2** good behaviour, obedience, order, orderliness, routine, self-control, self-restraint. • *vb* **1** break in, coach, control, drill, educate, govern, indoctrinate, instruct, keep in check, manage, restrain, school, train. **2** castigate, chasten, chastise, correct, penalize, punish, rebuke, reprimand, reprove, scold. **disciplined** ▷ OBEDIENT.

disclaim *vb* deny, disown, forswear, reject, renounce, repudiate. *Opp* ACKNOWLEDGE.

disclose *vb* divulge, expose, let out, make known. ▷ REVEAL.

discolour *vb* bleach, dirty, fade, mark, spoil the colour of, stain, tarnish, tinge.

discomfort *n* ache, care, difficulty, distress, hardship, inconvenience, irritation, soreness, uncomfortableness, uneasiness. ▷ PAIN. *Opp* COMFORT.

disconcert *vb* agitate, bewilder, confuse, discomfit, distract, disturb, fluster, nonplus, perplex, *inf* put off, puzzle, *inf* rattle, ruffle, throw off balance, trouble, unsettle, upset, worry. *Opp* REASSURE.

disconnect *vb* break off, cut off, detach, disengage, divide, part, sever, switch off, take away, turn off, uncouple, undo, unhitch, unhook, unplug. **disconnected** ▷ INCOHERENT.

discontented *adj* annoyed, disgruntled, displeased, dissatisfied, *inf* fed up, restless, sulky, unhappy, unsettled.

discord *n* **1** argument, conflict, contention, difference of opinion, disagreement, disharmony, dispute, friction, incompatibility, strife. ▷ QUARREL. **2** [*music*] cacophony, clash, jangle. ▷ NOISE. *Opp* HARMONY.

discordant *adj* **1** conflicting, contrary, differing, disagreeing, dissimilar, divergent, incompatible, incongruous, inconsistent, opposed, opposite. ▷ QUARRELSOME. **2** atonal, cacophanous, clashing, dissonant, grating, grinding, harsh, jangling, jarring, shrill, strident, tuneless, unmusical. *Opp* HARMONIOUS.

discount *n* abatement, allowance, concession, cut, deduction, *inf* mark-down, rebate, reduction. • *vb* disbelieve, dismiss, disregard, gloss over, ignore, overlook, reject.

discourage *vb* **1** cow, damp, dampen, daunt, demoralize, depress, disenchant, dishearten, dismay, dispirit, frighten, inhibit, intimidate, overawe, *inf* put down, *inf* put off, scare, *inf* throw cold water on, unman, unnerve. **2** *discourage vandalism.* check, deflect, deter, dissuade, hinder, prevent, put an end to, repress, restrain, slow down, stop, suppress. *Opp* ENCOURAGE.

discouragement *n* constraint, *inf* damper, deterrent, disincentive, hindrance, impediment, obstacle, restraint, setback. *Opp* ENCOURAGEMENT.

discourse *n* **1** ▷ CONVERSATION. **2** dissertation, essay, monograph, paper, speech, thesis, treatise. ▷ WRITING. • *vb* ▷ SPEAK.

discover *vb* ascertain, bring to light, come across, detect, *inf* dig up, disclose, *inf* dredge up, explore, expose, *inf* ferret out, find, hit on, identify, learn, light upon, locate, notice, observe, perceive, recognize, reveal, search out, spot, *sl* sus out, track down, turn up, uncover, unearth. ▷ INVENT. *Opp* HIDE.

discoverer *n* creator, explorer, finder, initiator, inventor, originator, pioneer, traveller.

discovery *n* breakthrough, conception, detection, disclosure, exploration, *inf* find, innovation, invention, recognition, revelation.

discredit *vb* attack, calumniate, challenge, defame, disbelieve, disgrace, dishonour, disprove, *inf* explode, prove false, raise doubts about, refuse to believe, ruin the reputation of, show up, slander, slur, smear, vilify.

discreet *adj* careful, cautious, chary, circumspect, considerate, delicate, diplomatic, guarded, judicious, low-key, mild, muted, polite, politic, prudent, restrained, sensitive, soft, subdued, tactful, thoughtful, understated, wary. *Opp* INDISCREET.

discrepancy *n* conflict, difference, disparity, dissimilarity, divergence, incompatibility, incongruity, inconsistency, variance. *Opp* SIMILARITY.

discretion *n* circumspection, diplomacy, good sense, judgement, maturity, prudence, responsibility, sensitivity, tact, wisdom. *Opp* TACTLESSNESS.

discriminate *vb* **1** differentiate, distinguish, draw a distinction, separate, tell apart. **2** be biased, be intolerant, be prejudiced, show discrimination. **discriminating** ▷ PERCEPTIVE.

discrimination *n* **1** discernment, good taste, insight, judgement, perceptiveness, refinement, selectivity, subtlety, taste. **2** [*derog*] bias, bigotry, chauvinism, favouritism, intolerance, male chauvinism, prejudice, racialism, racism, sexism, unfairness. *Opp* IMPARTIALITY.

discuss *vb* argue about, confer about, consider, consult about, debate, deliberate, examine, *inf* put heads together about, talk about, *inf* weigh up the pros and cons of, write about. ▷ TALK.

discussion *n* argument, colloquy, confabulation, conference, consideration, consultation, conversation, debate, deliberation, dialogue, discourse, examination, exchange of views, *inf* powwow, symposium. ▷ TALK.

disdainful *adj* contemptuous, jeering, mocking, scornful, sneering, supercilious, superior. ▷ PROUD.

disease *n* affliction, ailment, blight, *inf* bug, complaint, *inf* condition, contagion, disorder, infection, infirmity, malady, plague, sickness. ▷ ILLNESS.

diseased *adj* ailing, infirm, sick, unwell. ▷ ILL.

disembark *vb* alight, debark, detrain, get off, go ashore, land. *Opp* EMBARK.

disfigure *vb* blemish, damage, deface, deform, distort, impair, injure, make ugly, mar, mutilate, ruin, scar, spoil. *Opp* BEAUTIFY.

disgrace *n* **1** blot, contumely, degradation, discredit, dishonour, disrepute, embarrassment, humiliation, ignominy, obloquy, odium, opprobrium, scandal, shame, slur, stain, stigma. **2** ▷ OUTRAGE.

disgraceful *adj* contemptible, degrading, dishonourable, embarrassing, humiliating, ignominious, shameful, shaming, wicked. ▷ BAD.

disgruntled *adj* annoyed, cross, disaffected, disappointed, discontented, dissatisfied, *inf* fed up, grumpy, moody, sulky, sullen. ▷ BAD-TEMPERED.

disguise *n* camouflage, cloak, costume, cover, fancy dress, front, *inf* get-up, impersonation, make-up, mask, pretence, smoke-screen. • *vb* blend into the background, camouflage, conceal, cover up, dress up, falsify, gloss over, hide, make inconspicuous, mask, misrepresent, screen, shroud, veil. **disguise yourself as** ▷ IMPERSONATE.

disgust *n* abhorrence, antipathy, aversion, contempt, detestation, dislike, distaste, hatred, loathing, nausea, outrage, repugnance, repulsion, revulsion, sickness. • *vb* appal, be distasteful to, displease, horrify, nauseate, offend, outrage, put off, repel, revolt, sicken, shock, *inf* turn your stomach. *Opp* PLEASE. **disgusting** ▷ HATEFUL.

dish *n* **1** basin, bowl, casserole, container, plate, *old use* platter, tureen. **2** concoction, food, item on the menu, recipe. **dish out** ▷ DISTRIBUTE. **dish up** ▷ SERVE.

dishearten *vb* depress, deter, discourage, dismay, put off, sadden. *Opp* ENCOURAGE. **disheartened** ▷ SAD.

dishevelled *adj* bedraggled, disarranged, disordered, knotted, matted, messy, ruffled, rumpled, *inf* scruffy, slovenly, tangled, tousled, uncombed, unkempt, untidy. *Opp* NEAT.

dishonest *adj inf* bent, cheating, corrupt, criminal, crooked, deceitful, deceiving, deceptive, devious, dishonourable, disreputable, false, fraudulent, hypocritical, immoral, insincere, lying, mendacious, misleading, perfidious, *inf* shady, *inf* slippery, specious, swindling, thieving, treacherous, *inf* two-faced, *inf* underhand, unethical, unprincipled, unscrupulous, untrustworthy, untruthful. *Opp* HONEST.

dishonour *n inf* black mark, blot, degradation, discredit, disgrace, humiliation, ignominy, indignity, loss of face, obloquy, opprobrium, reproach, scandal, shame, slander, slur, stain, stigma. *Opp* HONOUR. • *vb* **1** abuse, affront, debase, defile, degrade, disgrace, offend, profane, shame, slight. **2** ▷ RAPE.

dishonourable *adj* base, blameworthy, compromising, despicable, discreditable, disgraceful, disgusting, dishonest, disloyal, disreputable, ignoble, ignominious, improper, infamous, mean, outrageous, perfidious, reprehensible, scandalous, shabby, shameful, shameless, treacherous, unchivalrous, unethical, unprincipled, unscrupulous, untrustworthy, unworthy, wicked. ▷ CORRUPT. *Opp* HONOURABLE.

disillusion *vb* disabuse, disappoint, disenchant, enlighten, reveal the truth to, undeceive.

disinfect *vb* cauterize, chlorinate, clean, cleanse, decontaminate, fumigate, purge, purify, sanitize, sterilize.

disinfectant *n* antiseptic, decontaminant, fumigant, germicide.

disinherit *vb* cut off, cut out of a will, deprive someone of his/her birthright, deprive someone of his/her inheritance.

disintegrate *vb* break into pieces, break up, come apart, crack up, crumble, decay, decompose, degenerate, deteriorate, fall apart, lose coherence, moulder, rot, shatter, smash, splinter.

disinterested *adj* detached, dispassionate, impartial, impersonal, neutral, objective, unbiased, uninvolved, unprejudiced. *Opp* BIASED.

disjointed *adj* aimless, broken up, confused, desultory, disconnected, dislocated, disordered, disunited, divided, incoherent, jumbled, loose, mixed up, muddled, rambling, separate, split up, unconnected, uncoordinated, wandering. *Opp* COHERENT.

dislike *n* animus, antagonism, antipathy, aversion, contempt, detestation, disapproval, disfavour, disgust, distaste, hatred, hostility, ill will, loathing, repugnance, revulsion. • *vb* avoid, despise, detest, disapprove of, feel dislike for, scorn, *inf* take against. ▷ HATE. *Opp* LOVE.

dislocate *vb* disengage, disjoint, displace, misplace, *inf* put out, put out of joint.

disloyal *adj* apostate, faithless, false, insincere, perfidious, recreant, renegade, seditious, subversive, treacherous, treasonable, *inf* two-faced, unfaithful, unreliable, untrue, untrustworthy. *Opp* LOYAL.

disloyalty *n* betrayal, double-dealing, duplicity, faithlessness, falseness, inconstancy, infidelity, perfidy, treachery, treason, unfaithfulness. *Opp* LOYALTY.

dismal *adj* bleak, cheerless, depressing, dreary, dull, funereal, gloomy, grey, grim, joyless, miserable, sombre, wretched. ▷ SAD.

dismantle *vb* demolish, knock down, strike, strip down, take apart, take down. *Opp* ASSEMBLE.

dismay *n* agitation, alarm, anxiety, apprehension, astonishment, consternation, depression, disappointment, discouragement, distress, dread, gloom, horror, pessimism, surprise. ▷ FEAR. • *vb* alarm, appal, daunt, depress, devastate, disappoint, discompose, discourage, disgust, dishearten, dispirit, distress, horrify, scare, shock, take aback, unnerve. ▷ FRIGHTEN. *Opp* PLEASE.

dismiss *vb* **1** disband, discard, free, let go, *inf* pack off, release, send away, *inf* send packing. **2** belittle, brush aside, discount, disregard, drop, give up, *inf* pooh-pooh, reject, repudiate, set aside, shelve, shrug off, wave aside. **3** *dismiss a worker*. *inf* axe, banish, cashier, disband, discharge, *inf* fire, get rid of, give notice to, *inf* give someone his/her cards, give the push to, lay off, make redundant, sack.

disobedient *adj* anarchic, contrary, defiant, delinquent, disorderly, disruptive, fractious, headstrong, insubordinate, intractable, mutinous, obdurate, obstinate, obstreperous, perverse, rebellious, recalcitrant, refractory, riotous, selfwilled, stubborn, uncontrollable, undisciplined, ungovernable, unmanageable, unruly, wayward, wild, wilful. ▷ NAUGHTY. *Opp* OBEDIENT.

disobey *vb* **1** be disobedient, mutiny, protest, rebel, revolt, rise up, strike. **2** break, contravene, defy, disregard, flout, ignore, infringe, oppose, rebel against, resist, transgress, violate. *Opp* OBEY.

disorder *n* **1** anarchy, chaos, clamour, confusion, disarray, disorderliness, disorganization, disturbance, fighting, fracas, fuss, jumble, lawlessness, mess, muddle, rumpus, *inf* shambles, tangle, tumult, untidiness, uproar. ▷ COMMOTION. *Opp* ORDER. 2 ▷ ILLNESS.

disorderly *adj* ▷ DISOBEDIENT, DISORGANIZED.

disorganized *adj* aimless, careless, chaotic, confused, disorderly, haphazard, illogical, jumbled, messy, muddled, rambling, scatter-brained, *inf* slapdash, *inf* slipshod, *inf* sloppy, slovenly, straggling, unmethodical, unplanned, unstructured, unsystematic, untidy. *Opp* SYSTEMATIC.

disown *vb* cast off, disclaim knowledge of, renounce, repudiate.

disparage *vb* belittle, demean, depreciate, discredit, insult, *inf* put down, slight, undervalue. ▷ CRITICIZE. **disparaging** ▷ UNCOMPLIMENTARY.

dispassionate *adj* calm, composed, cool, equable, even-tempered, level-headed, sober. ▷ IMPARTIAL, UNEMOTIONAL. *Opp* EMOTIONAL.

dispatch *n* bulletin, communiqué, document, letter, message, report. • *vb* 1 consign, convey, forward, mail, post, send, ship, transmit. 2 ▷ KILL.

dispense *vb* 1 allocate, allot, apportion, assign, deal out, disburse, distribute, dole out, give out, issue, measure out, mete out, parcel out, provide, ration out, share. 2 *dispense medicine*. make up, prepare, supply. **dispense with** ▷ OMIT, REMOVE.

disperse *vb* 1 break up, decentralize, devolve, disband, dismiss, dispel, dissipate, distribute, divide up, drive away, send away, send in different directions, separate, spread, stray. *Opp* GATHER. 2 disappear, dissolve, melt away, scatter, spread out, vanish.

displace *vb* 1 disarrange, dislocate, dislodge, disturb, misplace, move, put out of place, shift. 2 crowd out, depose, dispossess, evict, expel, oust, replace, succeed, supersede, supplant, take the place of, unseat, usurp.

display *n* 1 array, demonstration, exhibition, manifestation, pageant, parade, presentation, show, spectacle. 2 ceremony, ostentation, pageantry, pomp, showing off. • *vb* advertise, air, betray, demonstrate, disclose, exhibit, expose, flaunt, flourish, give evidence of, parade, present, produce, put on show, reveal, set out, show, show off, unfold, unfurl, unveil, vaunt. *Opp* HIDE.

displease *vb* anger, offend, *inf* put out, upset. ▷ ANNOY.

disposable *adj* 1 at your disposal, available, spendable, usable. 2 biodegradable, expendable, non-returnable, replaceable, *inf* throw-away.

dispose *vb* adjust, arrange, array, distribute, group, order, organize, place, position, put, set out, situate. **disposed** ▷ LIABLE. **dispose of** ▷ DESTROY, DISCARD.

disproportionate *adj* excessive, incommensurate, incongruous, inequitable, inordinate, out of proportion, unbalanced, uneven, unreasonable. *Opp* PROPORTIONAL.

disprove *vb* confute, contradict, controvert, demolish, discredit, *inf* explode, invalidate, negate, rebut, refute, show to be wrong. *Opp* PROVE.

dispute *n* ▷ QUARREL. • *vb* argue against, challenge, contest, contradict, controvert, deny, disagree with, doubt, fault, gainsay, impugn, object to, oppose, *inf* pick holes in, quarrel with, query, question, raise doubts about, take exception to. ▷ DEBATE. *Opp* ACCEPT.

disqualify *vb* bar, debar, declare ineligible, exclude, preclude, prohibit, reject, turn down.

disregard *vb* brush aside, despise, discount, dismiss, disobey, exclude, *inf* fly in the face of, forget, ignore, leave out, *inf* make light of, miss out, neglect, omit, overlook, pass over, pay no attention to, *inf* pooh-pooh, reject, shrug off, skip, slight, snub, turn a blind eye to. *Opp* HEED.

disreputable *adj* dishonest, dishonourable, *inf* dodgy, dubious, infamous, questionable, raffish, *inf* shady, suspect, suspicious, unconventional, unreliable, unsound, untrustworthy. *Opp* REPUTABLE.

disrespectful *adj* bad-mannered, blasphemous, derisive, discourteous, disparaging, impolite, impudent, inconsiderate, insolent, insulting, irreverent, mocking, scornful, uncivil, uncomplimentary, unmannerly. ▷ RUDE. *Opp* RESPECTFUL.

disrupt *vb* agitate, break up, confuse, disconcert, dislocate, disorder, disturb, interfere with, interrupt, intrude on, spoil, throw into disorder, unsettle, upset.

dissatisfaction *n* annoyance, chagrin, disappointment, discontentment, dismay, displeasure, disquiet, exasperation, frustration, irritation, malaise, mortification, regret, unhappiness. *Opp* SATISFACTION.

dissatisfied *adj* disaffected, disappointed, discontented, disgruntled, displeased, fed up, frustrated, unfulfilled, unsatisfied. ▷ UNHAPPY. *Opp* CONTENTED.

dissident *n derog* agitator, apostate, dissenter, independent thinker, non-conformer, protester, rebel, recusant, *inf* refusenik, revolutionary. *Opp* CONFORMIST.

dissimilar *adj* antithetical, clashing, conflicting, contrasting, different, disparate, distinct, distinguishable, divergent, diverse, heterogeneous, incompatible, irreconcilable, opposite, unlike, unrelated, various. *Opp* SIMILAR.

dissipate *vb* **1** break up, diffuse, disappear, disperse, scatter. **2** distribute, fritter away, spread about, squander, throw away, use up, waste. **dissipated** ▷ IMMORAL.

dissociate *vb* back away, cut off, detach, disengage, distance, divorce, isolate, segregate. ▷ SEPARATE. *Opp* ASSOCIATE.

dissolve *vb* **1** become liquid, decompose, deliquesce, dematerialize, diffuse, disappear, disintegrate, disperse, liquefy, melt away, vanish. **2** *dissolve a meeting.* adjourn, break up, cancel, disband, dismiss, divorce, end, sever, split up, suspend, terminate, *inf* wind up.

dissuade *vb* dissuade from advise against, argue out of, deter from, discourage from, persuade not to, put off, remonstrate against, warn against. *Opp* PERSUADE.

distance *n* **1** breadth, extent, gap, *inf* haul, interval, journey, length, measurement, mileage, range, reach, separation, space, span, stretch, width. **2** aloofness, coolness, haughtiness, isolation, remoteness, separation, *inf* standoffishness, unfriendliness.
• *vb* **distance yourself** be unfriendly, detach yourself, dissociate yourself, keep away, keep

your distance, remove yourself, separate yourself, set yourself apart, stay away. *Opp* INVOLVE.

distant *adj* 1 far, far-away, far-flung, *inf* god-forsaken, inaccessible, outlying, out-of-the-way, remote, removed. *Opp* CLOSE. 2 aloof, cool, formal, frigid, haughty, reserved, reticent, stiff, unapproachable, unenthusiastic, unfriendly, withdrawn. *Opp* FRIENDLY.

distasteful *adj* disgusting, displeasing, nasty, nauseating, objectionable, offensive, *inf* off-putting, repugnant, revolting, unpalatable. ▷ UNPLEASANT. *Opp* PLEASANT.

distinct *adj* 1 apparent, clear, clear-cut, definite, evident, noticeable, obvious, palpable, patent, perceptible, plain, precise, recognizable, sharp, unambiguous, unequivocal, unmistakable, visible, well-defined. *Opp* INDISTINCT. 2 contrasting, detached, different, discrete, dissimilar, distinguishable, individual, separate, special, *Lat* sui generis, unconnected, unique.

distinction *n* 1 contrast, difference, differentiation, discrimination, dissimilarity, distinctiveness, dividing line, division, individuality, particularity, peculiarity, separation. *Opp* SIMILARITY. 2 *distinction of being first.* celebrity, credit, eminence, excellence, fame, glory, greatness, honour, importance, merit, prestige, renown, reputation, superiority.

distinctive *adj* characteristic, different, distinguishing, idiosyncratic, individual, inimitable, original, peculiar, personal, singular, special, striking, typical, uncommon, unique. *Opp* COMMON.

distinguish *vb* 1 choose, decide, differentiate, discriminate, judge, make a distinction, separate, tell apart. 2 ascertain, determine, discern, know, make out, perceive, pick out, recognize, see, single out, tell. **distinguished** ▷ FAMOUS.

distort *vb* 1 bend, buckle, contort, deform, misshape, twist, warp, wrench. 2 alter, exaggerate, falsify, garble, misrepresent, pervert, slant, tamper with, twist, violate. **distorted** ▷ GNARLED, FALSE.

distract *vb* bewilder, bother, confound, confuse, deflect, disconcert, distress, divert, harass, mystify, perplex, puzzle, rattle, sidetrack, trouble, worry. **distracted** ▷ DISTRAUGHT, MAD.

distraction *n* 1 disturbance, diversion, interference, interruption, temptation, *inf* upset. 2 agitation, befuddlement, bewilderment, confusion, delirium, frenzy, insanity, madness. 3 ▷ DIVERSION.

distraught *adj* agitated, *inf* beside yourself, distracted, distressed, disturbed, emotional, excited, frantic, hysterical, overcome, overwrought, troubled, upset, worked up. ▷ ANXIOUS. *Opp* CALM.

distress *n* adversity, affliction, angst, anguish, anxiety, danger, desolation, difficulty, discomfort, dismay, fright, grief, heartache, misery, pain, poverty, privation, sadness, sorrow, stress, suffering, torment, tribulation, trouble, unhappiness, woe, worry, wretchedness. ▷ PAIN. • *vb* afflict, alarm, bother, *inf* cut up, dismay, disturb, frighten, grieve, harass, harrow, hurt, make miserable, oppress, pain, perplex, perturb, plague, sadden, scare, shake, shock, terrify, torment, torture, trouble, upset, vex, worry, wound. *Opp* COMFORT.

distribute *vb* allocate, allot, apportion, arrange, assign, circulate, deal out, deliver, *inf* dish out, dispense, disperse, dispose of, disseminate, divide out, *inf* dole out, give out, hand round, issue, mete out, partition, pass round, scatter, share out, spread, strew, take round. *Opp* COLLECT.

district *n* area, community, department, division, locality, neighbourhood, parish, part, partition, precinct, province, quarter, region, sector, territory, vicinity, ward, zone.

distrust *vb* be distrustful of, disbelieve, doubt, have misgivings about, have qualms about, mistrust, question, suspect. *Opp* TRUST.

distrustful *adj* cautious, chary, cynical, disbelieving, distrusting, doubtful, dubious, sceptical, suspicious, uncertain, uneasy, unsure, wary. *Opp* TRUSTFUL.

disturb *vb* **1** agitate, alarm, annoy, bother, discompose, disrupt, distract, distress, excite, fluster, frighten, hassle, interrupt, intrude on, perturb, pester, ruffle, scare, shake, startle, stir up, trouble, unsettle, upset, worry. **2** confuse, disorder, interfere with, jumble up, *inf* mess about with, move, muddle, rearrange, reorganize. **disturbed** ▷ DISTRAUGHT.

disturbance *n* disruption, interference, upheaval, upset. ▷ COMMOTION.

disunited *adj* divided, opposed, polarized, split. *Opp* UNITED.

disunity *n* difference, disagreement, discord, disharmony, disintegration, division, fragmentation, incoherence, opposition, polarization. *Opp* UNITY.

disused *adj* abandoned, archaic, closed, dead, discarded, discontinued, idle, neglected, obsolete, superannuated, unused, withdrawn. ▷ OLD. *Opp* CURRENT.

ditch *n* aqueduct, channel, dike, drain, gully, gutter, moat, trench, watercourse. • *vb* ▷ ABANDON.

dive *vb* crash-dive, descend, dip, drop, duck, fall, go snorkelling, go under, jump, leap, nosedive, pitch, plummet, plunge, sink, submerge, subside, swoop.

diverge *vb* branch, deviate, divide, fork, go off at a tangent, part, radiate, ramify, separate, split, spread, subdivide. ▷ DIFFER. *Opp* CONVERGE.

diverse *adj* assorted, different, dissimilar, distinct, divergent, diversified, heterogeneous, miscellaneous, mixed, multifarious, varied, various.

diversify *vb* branch out, broaden out, develop, divide, enlarge, expand, extend, spread out, vary.

diversion *n* **1** detour, deviation. **2** amusement, distraction, entertainment, fun, game, hobby, interest, pastime, play, recreation, relaxation, sport.

divert *vb* **1** alter, avert, change direction, deflect, deviate, rechannel, redirect, reroute, shunt, sidetrack, switch, turn aside. **2** amuse, beguile, cheer up, delight, distract, engage, entertain, keep happy, occupy, recreate, regale. **diverting** ▷ FUNNY.

divide *vb* **1** branch, detach, diverge, fork, move apart, part, separate, sunder. **2** allocate, allot, apportion, break up, cut up, deal out, dispense, distribute, dole out, give out, halve, measure out, mete out, parcel out, pass round, share out. **3** *divide a party.* cause disagreement in, disunite, polarize, split. **4** *divide into sets.* arrange, categorize, classify, grade, group, sort out, subdivide. *Opp* GATHER, UNITE.

divine *adj* angelic, celestial, godlike, hallowed, heavenly, holy, immortal, mystical, religious, sacred, saintly, seraphic, spiritual, superhuman, supernatural, transcendental. *Opp* MORTAL. • *n* ▷ CLERGYMAN. • *vb* ▷ PROPHESY.

divinity *n* **1** ▷ GOD. **2** religion, religious studies, theology.

division *n* **1** allocation, allotment, apportionment, cutting up, dividing, partition, segmentation, separation, splitting. **2** disagreement, discord, disunity, feud, quarrel, rupture, schism, split. **3** alcove, compartment, part, recess, section, segment. **4** *division between rooms, lands.* border, borderline, boundary line, demarcation, divider, dividing wall, fence, frontier, margin, partition, screen. **5** *division of a business.* branch, department, section, subdivision, unit.

divorce *n* annulment, *inf* break-up, decree nisi, dissolution, separation, *inf* split-up. • *vb* annul marriage, dissolve marriage, part, separate, *inf* split up.

dizziness *n* faintness, giddiness, light-headedness, vertigo.

dizzy *adj* bewildered, confused, dazed, faint, giddy, light-headed, muddled, reeling, shaky, swimming, unsteady, *inf* woozy.

do *vb* **1** accomplish, achieve, bring about, carry out, cause, commit, complete, effect, execute, finish, fulfil, implement, initiate, instigate, organize, perform, produce, undertake. **2** *do the garden.* arrange, attend to, cope with, deal with, handle, look after, manage, work at. **3** *do sums.* answer, give your mind to, puzzle out, solve, think out, work out. **4** *Will this do?* be acceptable, be enough, be satisfactory, be sufficient, be suitable, satisfy, serve, suffice. **5** *Do as you like.* act, behave, conduct yourself, perform. **do away with** ▷ ABOLISH. **do up** ▷ DECORATE, FASTEN.

docile *adj* cooperative, domesticated, obedient, submissive, tractable. ▷ TAME.

dock *n* berth, boatyard, dockyard, dry dock, harbour, haven, jetty, landing-stage, marina, pier, port, quay, slipway, wharf. • *vb* **1** anchor, berth, drop anchor, land, moor, put in, tie up. **2** ▷ CUT.

doctor *n* general practitioner, *inf* GP, *inf* medic, medical officer, medical practitioner, *inf* MO, physician, *derog* quack, surgeon.

doctrine *n* axiom, belief, conviction, *Lat* credo, creed, dogma, maxim, orthodoxy, postulate, precept, principle, teaching, tenet, theory, thesis.

document *n* certificate, charter, chronicle, deed, diploma, form, instrument, legal document, licence, manuscript, *inf* MS, paper, parchment, passport, policy, printout, record, typescript, visa, warrant, will. • *vb* ▷ RECORD.

documentary *adj* **1** authenticated, chronicled, recorded, substantiated, written. **2** factual, historical, non-fiction, real life.

dodge *n* contrivance, device, knack, manoeuvre, ploy, *sl* racket, ruse, scheme, stratagem, subterfuge, trick, *inf* wheeze. • *vb* **1** avoid, duck, elude, escape, evade, fend off, move out of the way, sidestep, swerve, turn away, veer, weave. **2** *dodge work.* shirk, *inf* skive, *inf* wriggle out of. **3** *dodge a question.* equivocate, fudge, hedge, quibble, *inf* waffle.

dog *n* bitch, *inf* bow-wow, *derog* cur, dingo, hound, mongrel, pedigree, pup, puppy, whelp. • *vb* ▷ FOLLOW.

dogma *n* article of faith, belief, conviction, creed, doctrine, orthodoxy, precept, principle, teaching, tenet, truth.

dogmatic *adj* assertive, arbitrary, authoritarian, authoritative, categorical, certain, dictatorial, doctrinaire, *inf* hard-line, hidebound, imperious, inflexible, intolerant, legalistic, narrow-minded, obdurate, opinionated, pontifical, positive. ▷ STUBBORN. *Opp* AMENABLE.

dole *n* [*inf*] benefit, income support, social security, unemployment benefit. **dole out** ▷ DISTRIBUTE. **on the dole** ▷ UNEMPLOYED.

doll *n inf* dolly, figure, marionette, puppet, rag doll.

domestic *adj* **1** family, household, in the home, private. **2** *domestic air service.* indigenous, inland, internal, national.

domesticated *adj* house-broken, house-trained, tame, tamed, trained. *Opp* WILD.

dominant *adj* **1** biggest, chief, commanding, conspicuous, eye-catching, highest, imposing, largest, main, major, obvious, outstanding, pre-eminent, prevailing, primary, principal, tallest, uppermost, widespread. **2** ascendant, controlling, dominating, domineering, governing, influential, leading, powerful, predominant, presiding, reigning, ruling, supreme.

dominate *vb* **1** be dominant, be in the majority, control, direct, govern, influence, lead, manage, master, monopolize, outnumber, preponderate, prevail, rule, subjugate, take control, tyrannize. **2** dwarf, look down on, overshadow, tower over.

domineering *adj* authoritarian, autocratic, *inf* bossy, despotic, dictatorial, high-handed, oppressive, overbearing, *inf* pushy, strict, tyrannical. *Opp* SUBMISSIVE.

donate *vb* contribute, give, grant, hand over, make a donation, present, subscribe, supply.

donation *n* alms, contribution, freewill offering, gift, offering, present, subscription.

donor *n* backer, benefactor, contributor, giver, philanthropist, provider, sponsor, supplier, supporter.

doom *n* destiny, end, fate, fortune, karma, kismet, lot.

doomed *adj* **1** condemned, destined, fated, intended, ordained, predestined. **2** *a doomed enterprise.* accursed, bedevilled, cursed, damned, hopeless, ill-fated, ill-starred, luckless, star-crossed, unlucky.

door *n* barrier, doorway, entrance, exit, French window, gate, gateway, opening, portal, postern, revolving door, swing door, way out.

dormant *adj* **1** asleep, comatose, hibernating, inactive, inert, passive, quiescent, quiet, resting, sleeping. **2** *dormant talent.* hidden, latent, potential, unrevealed, untapped, unused. *Opp* ACTIVE.

dose *n* amount, dosage, measure, portion, prescribed amount, quantity. • *vb* administer, dispense, prescribe.

dossier *n* file, folder, records, set of documents.

dot *n* decimal point, fleck, full stop, iota, jot, mark, point, speck, spot. • *vb* fleck, mark with dots, punctuate, speckle, spot, stipple.

dote *vb* **dote on** adore, idolize, worship. ▷ LOVE.

double *adj* coupled, doubled, dual, duple, duplicated, paired, twin, twofold, two-ply. • *n* clone, copy, counterpart, *Ger* doppelgänger, duplicate, *inf* look-alike, opposite, *inf* spitting image, twin. • *vb* duplicate, increase, multiply by two, reduplicate, repeat. **double back** ▷ RETURN. **double up** ▷ COLLAPSE.

double-cross *vb* cheat, deceive, let down, trick. ▷ BETRAY.

doubt *n* **1** agnosticism, anxiety, apprehension, confusion, cynicism, diffidence, disbelief, disquiet, distrust, fear, hesitation, incredulity, indecision, misgiving, mistrust, perplexity, qualm, reservation, scepticism, suspicion, worry. **2** *doubt about meaning.* ambiguity, difficulty, dilemma, problem, query, question, uncertainty. *Opp* CERTAINTY. • *vb* be dubious, be sceptical about, disbelieve, distrust, fear, feel uncertain about, have doubts about, have misgivings about, have reservations about, hesitate, lack confidence, mistrust, query, question, suspect. *Opp* TRUST.

doubtful *adj* **1** agnostic, cynical, diffident, disbelieving, distrustful, dubious, hesitant, incredulous, sceptical, suspicious, tentative, uncertain, unclear, unconvinced, undecided, unsure. **2** *a doubtful decision.* ambiguous, debatable, dubious, equivocal, *inf* iffy, inconclusive, problematical, questionable, suspect, vague, worrying. **3** *a doubtful ally.* irresolute, uncommitted, unreliable, untrustworthy, vacillating, wavering. *Opp* CERTAIN, DEPENDABLE.

dowdy *adj* colourless, dingy, drab, dull, *inf* frumpish, old-fashioned, shabby, *inf* sloppy, slovenly, *inf* tatty, unattractive, unstylish. *Opp* SMART.

downfall *n* collapse, defeat, overthrow, ruin, undoing.

downhearted *adj* dejected, depressed, discouraged, *inf* down, downcast, miserable, unhappy. ▷ SAD.

downward *adj* declining, descending, downhill, easy, falling, going down. *Opp* UPWARD.

downy *adj* feathery, fleecy, fluffy, furry, fuzzy, soft, velvety, woolly.

drab *adj* cheerless, colourless, dingy, dismal, dowdy, dreary, dull, flat, gloomy, grey, grimy, lacklustre, shabby, sombre, unattractive, uninteresting. *Opp* BRIGHT.

draft *n* **1** first version, notes, outline, plan, rough version, sketch. **2** *bank draft.* cheque, order, postal order. • *vb* block out, compose, delineate, draw up, outline, plan, prepare, put together, sketch out, work out, write a draft of.

drag *vb* **1** draw, haul, lug, pull, tow, trail, tug. **2** *time drags.* be boring, crawl, creep, go slowly, linger, loiter, lose momentum, move slowly, pass slowly.

drain *n* channel, conduit, culvert, dike, ditch, drainage, drainpipe, duct, gutter, outlet, pipe, sewer, trench, water-course. • *vb* **1** bleed, clear, draw off, dry out, empty, evacuate, extract, pump out, remove, tap, take off. **2** drip, ebb, leak out, ooze, seep, strain, trickle. **3** *drain resources.* consume, deplete, exhaust, sap, spend, use up.

drama *n* **1** acting, dramatics, dramaturgy, histrionics, improvisation, stagecraft, theatre, theatricals, thespian arts. **2** comedy, dramatization, farce, melodrama, musical, opera, operetta, pantomime, performance, play, production, screenplay, script, show, stage version, TV version, tragedy. **3** *real-life drama.* action, crisis, excitement, suspense, turmoil.

dramatic *adj* **1** histrionic, stage, theatrical, thespian. **2** *dramatic gestures.* exaggerated, flamboyant, large, overdone, showy. **3** ▷ EXCITING.

dramatist *n* dramaturge, playwright, scriptwriter.

dramatize *vb* **1** adapt, make into a play. **2** exaggerate, make too much of, overdo, overplay, overstate.

drape *n old use* arras, curtain, drapery, hanging, screen, tapestry, valance. • *vb* cover, decorate, festoon, hang, swathe.

drastic *adj* desperate, dire, draconian, extreme, far-reaching, forceful, harsh, radical, rigorous, severe, strong, vigorous.

draught *n* **1** breeze, current, movement, puff, wind. **2** *draught of ale.* dose, drink, gulp, measure, pull, swallow, *inf* swig.

draw *n* **1** attraction, enticement, lure, *inf* pull. **2** dead-heat, deadlock, stalemate, tie. **3** competition, lottery, raffle. • *vb* **1** drag, haul, lug, pull, tow, tug. **2** *draw a crowd.* allure, attract, bring in, coax, entice, invite, lure, persuade, pull in, win over. **3** *draw a sword.* extract, remove, take out, unsheathe, withdraw. **4** *draw lots.* choose, pick, select. **5** *draw a conclusion.* arrive at, come to, deduce, formulate, infer, work out. **6** *draw water.* drain, let (*blood*), pour, pump, syphon, tap. **7** *draw 1-1.* be equal, finish equal, tie. **8** *draw pictures.* depict, map out, mark out, outline, paint, pen, pencil, portray, represent, sketch, trace. **draw out** ▷ EXTEND. **draw up** ▷ DRAFT, HALT.

drawback *n* defect, difficulty, disadvantage, hindrance, hurdle, impediment, obstacle, obstruction, problem, snag, stumbling block.

drawing *n* cartoon, design, graphics, illustration, outline, sketch. ▷ PICTURE.

dread *n* anxiety, apprehension, awe, *inf* cold feet, dismay, fear, *inf* the jitters, nervousness, perturbation, qualm, trepidation, uneasiness, worry. • *vb* be afraid of, shrink from, view with horror. ▷ FEAR.

dreadful *adj* alarming, appalling, awful, dire, distressing, evil, fearful, frightful, ghastly, grisly, gruesome, harrowing, hideous, horrible, horrifying, indescribable, monstrous, shocking, terrible, tragic, unspeakable, upsetting, wicked. ▷ BAD, FRIGHTENING.

dream *n* **1** daydream, delusion, fantasy, hallucination, illusion, mirage, nightmare, reverie, trance, vision. **2** ambition, aspiration, ideal, pipe-dream, wish. • *vb* conjure up, daydream, fancy, fantasize, hallucinate, have a vision, imagine, think. **dream up** ▷ INVENT.

dreary *adj* bleak, boring, cheerless, depressing, dismal, dull, gloomy, joyless, sombre, uninteresting. ▷ MISERABLE.

dregs *n* deposit, grounds (*of coffee*), lees, precipitate, remains, residue, sediment.

drench *vb* douse, drown, flood, inundate, saturate, soak, souse, steep, wet thoroughly.

dress *n* **1** apparel, attire, clothing, costume, garb, garments, *inf* gear, *inf* get-up, outfit, *old use* raiment. ▷ CLOTHES. **2** frock, gown, robe, shift. • *vb* **1** array, attire, clothe, cover, fit out, provide clothes for, put clothes on, robe. **2** *dress a wound.* attend to, bandage, bind up, care for, put a dressing on, tend, treat. *Opp* UNCOVER.

dressing *n* bandage, compress, plaster, poultice.

dribble *vb* **1** drool, slaver, slobber. **2** drip, flow, leak, ooze, run, seep, trickle.

drift *n* **1** accumulation, bank, dune, heap, mound, pile, ridge. **2** *drift of a speech.* ▷ GIST. • *vb* **1** be carried, coast, float, meander, move casually, move slowly, ramble, roam, rove, stray, waft, walk aimlessly, wander. **2** *snow drifts.* accumulate, gather, make drifts, pile up.

drill *n* **1** discipline, exercises, instruction, practice, *sl* square-bashing, training. • *vb* **1** coach, discipline, exercise, indoctrinate, instruct, practise, rehearse, school, teach, train. **2** bore, penetrate, perforate, pierce.

drink *n* **1** beverage, *inf* bevvy, *inf* cuppa, *inf* dram, draught, glass, *inf* gulp, *inf* night-cap, *inf* nip, pint, *joc* potation, sip, swallow, swig, *inf* tipple, tot. **2** alcohol, *inf* booze, *joc* grog, *joc* liquid refreshment, liquor. □ *ale, beer, bourbon, brandy, champagne, chartreuse, cider, cocktail, Cognac, crème de menthe, gin, Kirsch, lager, mead, perry,* inf *plonk, port, punch, rum, schnapps, shandy, sherry, vermouth, vodka, whisky, wine.* • *vb* **1** gulp, guzzle, imbibe, *inf* knock back, lap, partake of, *old use* quaff, sip, suck, swallow, swig, *inf* swill. **2** *inf* booze, carouse, get drunk, *inf* indulge, tipple, tope.

drip *n* bead, dribble, drop, leak, splash, spot, tear, trickle. • *vb* dribble, drizzle, drop, fall in drips, leak, plop, splash, sprinkle, trickle, weep.

drive *n* **1** excursion, jaunt, journey, outing, ride, run, *inf* spin, trip. **2** aggressiveness, ambition, determination, energy, enterprise, enthusiasm, *inf* get-up-and-go, impetus, industry, initiative, keenness, motivation, persistence, *inf* push, vigour, vim, zeal. **3** campaign, crusade, effort. • *vb* **1** bang, dig, hammer, hit, impel, knock, plunge, prod, push, ram, sink, stab, strike, thrust. **2** coerce, compel, constrain, force, oblige, press, urge. **3** *drive a car.* control, direct, guide, handle, herd, manage, pilot, propel, send, steer. ▷ TRAVEL. **drive out** ▷ EXPEL.

droop *vb* be limp, bend, dangle, fall, flop, hang, sag, slump, wilt, wither.

drop *n* **1** bead, blob, bubble, dab, drip, droplet, globule, pearl, spot, tear. **2** dash, *inf* nip, small quantity, *inf* tot. **3** *a steep drop.* declivity, descent, dive, escarpment, fall, incline, plunge, precipice, scarp. **4** *a drop in price.* cut, decrease, reduction, slump. *Opp* RISE. • *vb* **1** collapse, descend, dip, dive, fall, go down, jump down, lower, nosedive, plummet, *inf* plump, plunge, sink, slump, subside, swoop, tumble. **2** *drop from a team.* eliminate, exclude, leave out, omit. **3** *drop a friend.* abandon, desert, discard, *inf* dump, forsake, give up, jilt, leave, reject, scrap, shed.

drop behind ▷ LAG. **drop in on** ▷ VISIT. **drop off** ▷ SLEEP.

drown *vb* **1** engulf, flood, immerse, submerge, swamp. ▷ KILL. **2** *noise drowned my voice.* be louder than, overpower, overwhelm, silence.

drowsy *adj* dozing, dozy, heavy-eyed, listless, *inf* nodding off, sleepy, sluggish, somnolent, soporific, tired, weary. *Opp* LIVELY.

drudgery *n* chore, donkey-work, *inf* grind, labour, slavery, *inf* slog, toil, travail. ▷ WORK.

drug *n* **1** cure, medicament, medication, medicine, *old use* physic, remedy, treatment. **2** *inf* dope, narcotic, opiate. □ *analgesic, antidepressant, barbiturate, hallucinogen, pain-killer, sedative, stimulant, tonic, tranquillizer.* □ *caffeine, cannabis, cocaine, digitalis, hashish, heroin, insulin, laudanum, marijuana, morphia, nicotine, opium, phenobarbitone, quinine.* • *vb* anaesthetize, *inf* dope, dose, give a drug to, *inf* knock out, medicate, poison, sedate, stupefy, tranquillize, treat.

drum *n* **1** ▷ BARREL. **2** □ *bass-drum, bongo-drum, kettledrum, side-drum, snare-drum, tambour, tenor-drum,* plur *timpani, tom-tom.*

drunk *adj* delirious, fuddled, incapable, inebriate, inebriated, intoxicated, maudlin. [*slang*] blotto, bombed, boozed-up, canned, high, legless, merry, paralytic, pickled, pie-eyed, pissed, plastered, sloshed, soused, sozzled, stoned, tanked, tiddly, tight, tipsy. *Opp* SOBER.

drunkard *n* alcoholic, *inf* boozer, *sl* dipso, dipsomaniac, drunk, *inf* sot, tippler, toper, *sl* wino. *Opp* TEETOTALLER.

dry *adj* **1** arid, baked, barren, dead, dehydrated, desiccated, moistureless, parched, scorched, shrivelled, sterile, thirsty, waterless. *Opp* WET. **2** *a dry book.* boring, dreary, dull, flat, prosaic, stale, tedious, tiresome, uninspired, uninteresting. *Opp* LIVELY. **3** *dry humour. inf* dead-pan, droll, expressionless, laconic, lugubrious, unsmiling. • *vb* become dry, dehumidify, dehydrate, desiccate, go hard, make dry, parch, shrivel, wilt, wither.

dual *adj* binary, coupled, double, duplicate, linked, paired, twin.

dubious *adj* **1** ▷ DOUBTFUL. **2** *a dubious character. inf* fishy, *inf* shady, suspect, suspicious, unreliable, untrustworthy.

duck *vb* **1** avoid, bend, bob down, crouch, dip down, dodge, evade, sidestep, stoop, swerve, take evasive action. **2** immerse, plunge, push under, submerge.

due *adj* **1** in arrears, outstanding, owed, owing, payable, unpaid. **2** *due consideration.* adequate, appropriate, decent, deserved, expected, fitting, just, mature, merited, proper, requisite, right, rightful, scheduled, sufficient, suitable, wellearned. • *n* deserts, entitlement, merits, reward, rights. **dues** ▷ DUTY.

dull *adj* **1** dim, dingy, dowdy, drab, dreary, faded, flat, gloomy, lacklustre, lifeless, matt, plain, shabby, sombre, subdued. **2** *a dull sky.* cloudy, dismal, grey, heavy, leaden, murky, overcast, sullen, sunless. **3** *a dull sound.* deadened, indistinct, muffled, muted. **4** *a dull student.* dense, dim, dim-witted, obtuse, slow, *inf* thick, unimaginative, unintelligent, unresponsive. ▷ STUPID. **5** *a dull edge.* blunt, blunted, unsharpened. **6** *dull talk.*

boring, commonplace, dry, monotonous, prosaic, stodgy, tame, tedious, unexciting, uninteresting. *Opp* BRIGHT, SHARP.

dumb *adj* inarticulate, *inf* mum, mute, silent, speechless, tongue-tied, unable to speak.

dummy *n* **1** copy, counterfeit, duplicate, imitation, mock-up, model, reproduction, sample, sham, simulation, substitute, toy. **2** doll, figure, manikin, puppet.

dump *n* **1** junkyard, rubbish-heap, tip. **2** *arms dump.* arsenal, cache, depot, hoard, store. • *vb* deposit, discard, dispose of, *inf* ditch, drop, empty out, get rid of, jettison, offload, *inf* park, place, put down, reject, scrap, throw away, throw down, tip, unload.

dune *n* drift, hillock, hummock, mound, sand-dune.

dungeon *n* old use donjon, gaol, keep, lock-up, oubliette, pit, prison, vault.

duplicate *adj* alternative, copied, corresponding, identical, matching, second, twin. • *n* carbon copy, clone, copy, double, facsimile, imitation, likeness, *inf* look-alike, match, photocopy, photostat, replica, reproduction, twin, Xerox. • *vb* copy, do again, double up on, photocopy, print, repeat, reproduce, Xerox.

durable *adj* enduring, hard-wearing, heavy-duty, indestructible, long-lasting, permanent, resilient, stout, strong, substantial, thick, tough. *Opp* IMPERMANENT, WEAK.

dusk *n* evening, gloaming, gloom, sundown, sunset, twilight.

dust *n* dirt, grime, grit, particles, powder.

dusty *adj* **1** chalky, crumbly, dry, fine, friable, gritty, powdery, sandy, sooty. **2** *a dusty room.* dirty, filthy, grimy, grubby, mucky, uncleaned, unswept.

dutiful *adj* attentive, careful, compliant, conscientious, devoted, diligent, faithful, hard-working, loyal, obedient, obliging, punctilious, reliable, responsible, scrupulous, thorough, trustworthy, willing. *Opp* IRRESPONSIBLE.

duty *n* **1** allegiance, faithfulness, loyalty, obedience, obligation, onus, responsibility, service. **2** assignment, business, charge, chore, function, job, office, role, stint, task, work. **3** charge, customs, dues, fee, impost, levy, tariff, tax, toll.

dwarf *adj* ▷ SMALL. • *n* midget, pigmy. • *vb* dominate, look bigger than, overshadow, tower over.

dwell *vb* abide, be accommodated, live, lodge, reside, stay. **dwell in** ▷ INHABIT.

dwelling *n* abode, domicile, habitation, home, lodging, quarters, residence. ▷ HOUSE.

dying *adj* declining, expiring, fading, failing, moribund, obsolescent. *Opp* ALIVE.

dynamic *adj* active, committed, driving, eager, energetic, enterprising, enthusiastic, forceful, *inf* go-ahead, *derog* go-getting, high-powered, lively, motivated, powerful, pushful, *derog* pushy, spirited, vigorous, zealous. *Opp* APATHETIC.

E

eager *adj* agog, animated, anxious (*to please*), ardent, avid, bursting, committed, craving, desirous, earnest, enthusiastic, excited, fervent, fervid, hungry, impatient, intent, interested, *inf* itching, keen, *inf* keyed up, longing, motivated, passionate, *inf* raring (*to* go), voracious, yearning, zealous. *Opp* APATHETIC.

eagerness *n* alacrity, anxiety, appetite, ardour, avidity, commitment, desire, earnestness, enthusiasm, excitement, fervour, hunger, impatience, intentness, interest, keenness, longing, motivation, passion, thirst, zeal. *Opp* APATHY.

early *adj* **1** advance, ahead of time, before time, first, forward, premature. *Opp* LATE. **2** ancient, antiquated, initial, original, primeval, primitive. ▷ OLD. *Opp* RECENT.

earn *vb* **1** be paid, *inf* bring in, *inf* clear, draw, fetch in, gain, get, *inf* gross, make, make a profit of, net, obtain, pocket, realize, receive, *inf* take home, work for, yield. **2** attain, be worthy of, deserve, merit, qualify for, warrant, win.

earnest *adj* **1** assiduous, committed, conscientious, dedicated, determined, devoted, diligent, eager, hard-working, industrious, involved, purposeful, resolved, zealous. *Opp* CASUAL. **2** grave, heartfelt, impassioned, serious, sincere, sober, solemn, thoughtful, well-meant.

earnings *n* income, salary, stipend, wages. ▷ PAY.

earth *n* clay, dirt, ground, humus, land, loam, soil, topsoil.

earthenware *n* ceramics, china, crockery, *inf* crocks, porcelain, pots, pottery.

earthly *adj* corporeal, human, material, materialistic, mortal, mundane, physical, secular, temporal, terrestrial, worldly. *Opp* SPIRITUAL.

earthquake *n* quake, shock, tremor, upheaval.

earthy *adj* bawdy, coarse, crude, down to earth, frank, lusty, ribald, uninhibited. ▷ OBSCENE.

ease *n* **1** aplomb, calmness, comfort, composure, contentment, enjoyment, happiness, leisure, luxury, peace, quiet, relaxation, repose, rest, serenity, tranquillity. **2** dexterity, easiness, effortlessness, facility, nonchalance, simplicity, skill, speed, straightforwardness. *Opp* DIFFICULTY. • *vb* **1** allay, alleviate, assuage, calm, comfort, decrease, lessen, lighten, mitigate, moderate, pacify, quell, quieten, reduce, relax, relieve, slacken, soothe, tranquillize. **2** edge, guide, inch, manoeuvre, move gradually, slide, slip, steer.

easy *adj* **1** carefree, comfortable, contented, cosy, *inf* cushy, effortless, leisurely, light, painless, peaceful, pleasant, relaxed, relaxing, restful, serene, soft, tranquil, undemanding, unexacting, unhurried, untroubled. **2** clear, elementary, facile, foolproof, *inf* idiot-proof, manageable, plain, simple, straightforward, uncomplicated, understandable, user-friendly. **3** ▷ EASYGOING. *Opp* DIFFICULT.

easygoing *adj* accommodating, affable, amenable, calm, carefree, casual, cheerful, docile, even-tempered, flexible, forbearing, *inf* free and easy, friendly, genial, *inf* happy-go-lucky, indulgent, informal, *inf* laid-back, *derog* lax, lenient, liberal, mellow, natural, nonchalant, open, patient, permissive, placid, relaxed, tolerant, unexcitable, unruffled, *derog* weak. *Opp* STRICT.

eat *vb* consume, devour, digest, feed on, ingest, live on, *old use* partake of, swallow. □ *bite, bolt, champ, chew, crunch, gnaw, gobble, gorge, gormandize, graze, grind, gulp, guzzle,* inf *make a pig of yourself, masticate, munch, nibble, overeat, peck,* inf *scoff,* inf *slurp,* inf *stuff (yourself), taste,* inf *tuck in,* inf *wolf.* □ *banquet, breakfast, dine, feast, lunch, snack,* old use *sup.* **eat away, eat into** ▷ ERODE.

eatable *adj* digestible, edible, fit to eat, good, palatable, safe to eat, wholesome. ▷ TASTY. *Opp* INEDIBLE.

ebb *vb* fall, flow back, go down, recede, retreat, subside. ▷ DECLINE.

eccentric *adj* **1** aberrant, abnormal, anomalous, atypical, bizarre, cranky, curious, freakish, grotesque, idiosyncratic, *sl* kinky, odd, outlandish, out of the ordinary, peculiar, preposterous, quaint, queer, quirky, singular, strange, unconventional, unusual, *sl* wacky, *inf* way-out, *inf* weird, *inf* zany. ▷ ABSURD, MAD. **2** *eccentric circles.* irregular, off-centre. • *n inf* character, *inf* crackpot, crank, *inf* freak, individualist, nonconformist, *inf* oddball, oddity, *inf* weirdie, *inf* weirdo.

echo *vb* **1** resound, reverberate, ring, sound again. **2** ape, copy, duplicate, emulate, imitate, mimic, mirror, reiterate, repeat, reproduce, say again.

eclipse *vb* **1** block out, blot out, cloud, darken, dim, extinguish, obscure, veil. ▷ COVER. **2** excel, outdo, outshine, overshadow, *inf* put in the shade, surpass, top.

economic *adj* budgetary, business, financial, fiscal, monetary, money-making, trading.

economical *adj* **1** careful, cheese-paring, frugal, parsimonious, provident, prudent, sparing, thrifty. ▷ MISERLY. *Opp* WASTEFUL. **2** *an economical meal.* cheap, cost-effective, inexpensive, low-priced, money-saving, reasonable, *inf* value-for-money. *Opp* EXPENSIVE.

economize *vb* be economical, cut back, retrench, save, *inf* scrimp, skimp, spend less, *inf* tighten your belt. *Opp* SQUANDER.

economy *n* **1** frugality, *derog* meanness, *derog* miserliness, parsimony, providence, prudence, saving, thrift. *Opp* WASTE. **2** *the national economy.* budget, economic affairs, wealth. **3** ▷ BREVITY.

ecstasy *n* bliss, delight, delirium, elation, enthusiasm, euphoria, exaltation, fervour, frenzy, gratification, happiness, joy, rapture, thrill, trance, *old use* transport.

ecstatic *adj* blissful, delighted, delirious, elated, enraptured, enthusiastic, euphoric, exhilarated, exultant, fervent, frenzied, gleeful, joyful, orgasmic, overjoyed, *inf* over the moon, rapturous, transported. ▷ HAPPY.

eddy *n* circular movement, maelstrom, swirl, vortex, whirl, whirlpool, whirlwind. • *vb* move in circles, spin, swirl, turn, whirl.

edge *n* **1** border, boundary, brim, brink, circumference, frame, kerb, limit, lip, margin, outline, perimeter, periphery, rim, side, verge. **2** *edge of town.* outlying parts, outskirts, suburbs. **3** *edge on a knife.* acuteness, keenness, sharpness. **4** *edge of a curtain.* edging, fringe, hem, selvage. • *vb* **1** bind, border, fringe, hem, make an edge for, trim. **2** *edge away*. crawl, creep, inch, move stealthily, sidle, slink, steal, work your way, worm.

edible *adj* digestible, eatable, fit to eat, palatable, safe to eat, wholesome. ▷ TASTY. *Opp* INEDIBLE.

edit *vb* adapt, alter, amend, arrange, assemble, compile, get ready, modify, organize, prepare, put together, select, supervise the production of. □ *abridge, annotate, bowdlerize, censor, clean up, condense, copy-edit, correct, cut, dub, emend, expurgate, format, polish, proof-read, rearrange, rephrase, revise, rewrite, select, shorten, splice* (film).

edition *n* **1** copy, issue, number. **2** impression, printing, print-run, publication, version.

educate *vb* bring up, civilize, coach, counsel, cultivate, discipline, drill, edify, enlighten, guide, improve, inculcate, indoctrinate, inform, instruct, lecture, nurture, rear, school, teach, train, tutor.

educated *adj* cultured, enlightened, erudite, knowledgeable, learned, literate, numerate, sophisticated, trained, well-bred, well-read.

education *n* coaching, curriculum, enlightenment, guidance, indoctrination, instruction, schooling, syllabus, teaching, training, tuition. □ *academy, college, conservatory, polytechnic, sixth-form college, tertiary college, university.* ▷ SCHOOL, TEACHING.

eerie *adj inf* creepy, frightening, ghostly, mysterious, *inf* scary, spectral, *inf* spooky, strange, uncanny, unearthly, unnatural, weird.

effect *n* **1** aftermath, conclusion, consequence, impact, influence, issue, outcome, repercussion, result, sequel, upshot. **2** feeling, illusion, impression, sensation, sense. • *vb* accomplish, achieve, bring about, bring in, carry out, cause, create, effectuate, enforce, execute, implement, initiate, make, produce, put into effect, secure.

effective *adj* **1** able, capable, competent, effectual, efficacious, functional, impressive, potent, powerful, productive, proficient, real, serviceable, strong, successful, useful, worthwhile. ▷ EFFICIENT. **2** *an effective argument*. cogent, compelling, convincing, meaningful, persuasive, striking, telling. *Opp* INEFFECTIVE.

effeminate *adj* camp, effete, girlish, *inf* pansy, *inf* sissy, unmanly, weak, womanish. *Opp* MANLY.

effervesce *vb* bubble, ferment, fizz, foam, froth, sparkle.

effervescent *adj* bubbling, bubbly, carbonated, fizzy, foaming, frothy, gassy, sparkling.

efficient *adj* businesslike, cost-effective, economic, productive, streamlined, thrifty. ▷ EFFECTIVE. *Opp* INEFFICIENT.

effort *n* **1** application, diligence, *inf* elbow grease, endeavour, exertion, industry, labour, pains, strain, stress, striving, struggle, toil, *old use* travail, trouble, work. **2** *a brave effort*. attempt, endeavour, go, try, venture. **3** *a successful effort*. accomplishment, achievement, exploit, feat, job, outcome, product, production, result.

effusive *adj* demonstrative, ebullient, enthusiastic, exuberant, fulsome, gushing, lavish, *inf* over the top, profuse, voluble. *Opp* RETICENT.

egoism *n* egocentricity, egotism, narcissism, pride, self-centredness, self-importance, self-interest, selfishness, self-love, self-regard, vanity.

egotistical *adj* egocentric, self-admiring, self-centred, selfish. ▷ CONCEITED.

eject *vb* **1** banish, *inf* boot out, *inf* bundle out, deport, discharge, dismiss, drive out, evict, exile, expel, get rid of, *inf* kick out, oust, push out, put out, remove, sack, send out, shoot out, *inf* shove out, throw out, turn out. **2** ▷ EMIT.

elaborate *adj* **1** complex, complicated, detailed, exhaustive, intricate, involved, meticulous, minute, painstaking, thorough, well worked out. **2** *elaborate décor.* baroque, busy, Byzantine, decorative, fancy, fantastic, fussy, grotesque, intricate, ornamental, ornamented, ornate, rococo, showy. *Opp* SIMPLE. • *vb* add to, adorn, amplify, complicate, decorate, develop, embellish, enlarge on, enrich, expand, expatiate on, fill out, flesh out, give details of, improve on, ornament. *Opp* SIMPLIFY.

elapse *vb* go by, lapse, pass, slip by.

elastic *adj* bendy, bouncy, ductile, expandable, flexible, plastic, pliable, pliant, resilient, rubbery, *inf* springy, stretchable, *inf* stretchy, yielding. *Opp* RIGID.

elderly *adj* ageing, *inf* getting on, oldish. ▷ OLD.

elect *adj* [*goes after noun*] *president elect.* [*synonyms used after noun*] designate, to be; [*synonyms before noun*] chosen, elected, prospective, selected. • *vb* adopt, appoint, choose, name, nominate, opt for, pick, select, vote for.

election *n* ballot, choice, plebiscite, poll, referendum, selection, vote, voting.

electioneer *vb* campaign, canvass.

electorate *n* constituents, electors, voters.

electric *adj* **1** battery-operated, electrical, mains-operated. **2** *electric atmosphere.* electrifying. ▷ EXCITING.

electricity *n* current, energy, power, power supply.

elegant *adj* artistic, beautiful, chic, courtly, cultivated, dapper, debonair, dignified, exquisite, fashionable, fine, genteel, graceful, gracious, handsome, luxurious, modish, noble, pleasing, *inf* plush, *inf* posh, refined, smart, soigné(e), sophisticated, splendid, stately, stylish, suave, tasteful, urbane, well-bred. *Opp* INELEGANT.

elegy *n* dirge, lament, requiem.

element *n* **1** component, constituent, detail, essential, factor, feature, fragment, hint, ingredient, part, piece, small amount, trace, unit. **2** *in your element.* domain, environment, habitat, medium, sphere, territory. **elements** ▷ RUDIMENTS, WEATHER.

elementary *adj* basic, early, first, fundamental, initial, introductory, primary, principal, rudimentary, simple, straightforward, uncomplicated, understandable. ▷ EASY. *Opp* ADVANCED.

elevate *vb* exalt, hold up, lift, make higher, promote, rear. ▷ RAISE. **elevated** ▷ HIGH, NOBLE.

elicit *vb* bring out, call forth, derive, draw out, evoke, extort, extract, get, obtain, wrest, wring.

eligible *adj* acceptable, allowed, appropriate, authorized, available, competent, equipped, fit, fitting, proper, qualified, suitable, worthy. *Opp* INELIGIBLE.

eliminate *vb* **1** abolish, annihilate, delete, destroy, dispense with, do away with, eject, end, eradicate, exterminate, extinguish, finish off, get rid of, put an end to, remove, stamp out. ▷ KILL. **2** cut out, drop, exclude, knock out, leave out, omit, reject.

élite *n* aristocracy, best, *inf* cream, first-class people, flower, meritocracy, nobility, top people, *inf* upper crust.

eloquent *adj* articulate, expressive, fluent, forceful, *derog* glib, moving, persuasive, plausible, powerful, unfaltering. *Opp* INARTICULATE.

elude *vb* avoid, circumvent, dodge, *inf* duck, escape, evade, foil, get away from, *inf* give (someone) the slip, shake off, slip away from.

elusive *adj* **1** *inf* always on the move, evasive, fugitive, hard to find, slippery. **2** *elusive meaning.* ambiguous, baffling, deceptive, hard to pin down, indefinable, intangible, puzzling, shifting.

emaciated *adj* anorectic, atrophied, bony, cadaverous, gaunt, haggard, shrivelled, skeletal, skinny, starved, underfed, undernourished, wasted away, wizened. ▷ THIN.

emancipate *vb* deliver, discharge, enfranchise, free, give rights to, let go, liberate, loose, manumit, release, set free, unchain. *Opp* ENSLAVE.

embankment *n* bank, causeway, dam, earthwork, mound, rampart.

embark *vb* board, depart, go aboard, leave, set out. *Opp* DISEMBARK. **embark on** ▷ BEGIN.

embarrass *vb* abash, chagrin, confuse, discomfit, discompose, disconcert, discountenance, disgrace, distress, fluster, humiliate, *inf* make you blush, mortify, *inf* put you on the spot, shame, *inf* show up, upset. **embarrassed** ▷ ASHAMED. **embarrassing** ▷ AWKWARD, SHAMEFUL.

embellish *vb* adorn, beautify, deck, decorate, embroider, garnish, ornament, *sl* tart up, *inf* titivate. ▷ ELABORATE.

embezzle *vb* appropriate, misapply, misappropriate, peculate, *inf* put your hand in the till, take fraudulently. ▷ STEAL.

embezzlement *n* fraud, misappropriation, misuse of funds, peculation, stealing, theft.

embittered *adj* acid, bitter, disillusioned, envious, rancorous, resentful, sour. ▷ ANGRY.

emblem *n* badge, crest, device, image, insignia, mark, regalia, seal, sign, symbol, token.

embody *vb* **1** exemplify, express, incarnate, manifest, personify, reify, represent, stand for, symbolize. **2** bring together, combine, comprise, embrace, enclose, gather together, include, incorporate, integrate, involve, take in, unite.

embrace *vb* **1** clasp, cling to, cuddle, enfold, fondle, grasp, hold, hug, kiss, snuggle up to. **2** *embrace new ideas.* accept, espouse, receive, take on, welcome. **3** ▷ EMBODY.

embryonic *adj* early, immature, just beginning, rudimentary, underdeveloped, undeveloped, unformed. *Opp* MATURE.

emerge *vb* appear, arise, be revealed, come out, come to light, come to notice, emanate, *old* *use* issue forth, leak out, *inf* pop up, proceed, surface, transpire, *inf* turn out.

emergency *n* crisis, danger, difficulty, exigency, predicament, serious situation.

emigrate *vb* depart, go abroad, leave, quit, relocate, resettle, set out.

eminent *adj* august, celebrated, conspicuous, distinguished, elevated, esteemed, exalted, familiar, famous, great, high-ranking, honoured, illustrious, important, notable, noted, noteworthy, outstanding, pre-eminent, prominent, renowned, well-known. *Opp* LOWLY.

emit *vb* belch, discharge, disgorge, ejaculate, eject, exhale, expel, exude, give off, give out, issue, radiate, send out, spew out, spout, transmit, vent, vomit.

emotion *n* agitation, excitement, feeling, fervour, passion, sentiment, warmth. ▷ ANGER, LOVE, etc.

emotional *adj* **1** ardent, demonstrative, enthusiastic, excited, fervent, fiery, heated, hot-headed, impassioned, intense, irrational, moved, passionate, romantic, stirred, touched, warm-hearted, *inf* worked up. ▷ ANGRY, LOVING, etc. **2** *emotional language.* affecting, biased, emotive, heartfelt, heart-rending, inflammatory, loaded, moving, pathetic, poignant, prejudiced, provocative, sentimental, stirring, subjective, tear-jerking, tender, touching. *Opp* UNEMOTIONAL.

emphasis *n* accent, attention, force, gravity, importance, intensity, priority, prominence, strength, stress, urgency, weight.

emphasize *vb* accent, accentuate, bring out, dwell on, focus on, foreground, give emphasis to, highlight, impress, insist on, make obvious, *inf* play up, point up, *inf* press home, *inf* rub it in, show clearly, spotlight, stress, underline, underscore.

emphatic *adj* affirmative, assertive, categorical, confident, dogmatic, definite, firm, forceful, insistent, positive, pronounced, resolute, strong, uncompromising, unequivocal. *Opp* TENTATIVE.

empirical *adj* experiential, experimental, observed, practical, pragmatic. *Opp* THEORETICAL.

employ *vb* **1** commission, engage, enlist, have on the payroll, hire, pay, sign up, take on, use the services of. **2** apply, use, utilize.

employed *adj* active, busy, earning, engaged, hired, involved, in work, occupied, practising, working. ▷ BUSY. *Opp* UNEMPLOYED.

employee *n old use* hand, *inf* underling, worker. **employees** staff, workforce.

employer *n* boss, chief, *inf* gaffer, *inf* governor, head, manager, owner, proprietor, taskmaster.

employment *n* business, calling, craft, job, line, livelihood, living, métier, occupation, profession, pursuit, trade, vocation, work.

empty *adj* **1** bare, blank, clean, clear, deserted, desolate, forsaken, vacant, hollow, unfilled, unfurnished, uninhabited, unladen, unoccupied, unused, vacant, void. *Opp* FULL. **2** *empty threats.* futile, idle, impotent, ineffective, insincere, meaningless, pointless, purposeless, senseless, silly, unreal, worthless. • *vb* clear, discharge, drain, eject, evacuate, exhaust, pour out, remove, take out, unload, vacate, void. *Opp* FILL.

enable *vb* aid, allow, approve, assist, authorize, charter, empower, entitle, equip, facilitate, franchise, help, license, make it possible, permit, provide the means, qualify, sanction. *Opp* PREVENT.

enchant *vb* allure, beguile, bewitch, captivate, cast a spell on, charm, delight, enrapture, enthral, entrance, fascinate, hypnotize, mesmerize, spellbind. **enchanting** ▷ ATTRACTIVE.

enchantment *n* charm, conjuration, magic, sorcery, spell, witchcraft, wizardry. ▷ DELIGHT.

enclose *vb* bound, box, cage, case, cocoon, conceal, confine, contain, cover, encase, encircle, encompass, enfold, envelop, fence in, hedge in, hem in, immure, insert, limit, package, parcel up, pen, restrict, ring, secure, sheathe, shut in, shut up, surround, wall in, wall up, wrap. ▷ IMPRISON.

enclosure *n* **1** arena, cage, compound, coop, corral, court, courtyard, farmyard, field, fold, paddock, pen, pound, ring, run, sheepfold, stockade, sty, yard. **2** *enclosure in an envelope.* contents, inclusion, insertion.

encounter *n* **1** confrontation, meeting. **2** [*military*] battle, brush, clash, dispute, skirmish, struggle. ▷ FIGHT. • *vb* chance upon, clash with, come upon, confront, contend with, *inf* cross swords with, face, grapple with, happen upon, have an encounter with, meet, *inf* run into.

encourage *vb* **1** abet, advocate, animate, applaud, cheer, *inf* egg on, embolden, give hope to, hearten, incite, inspire, invite, persuade, prompt, rally, reassure, rouse, spur on, support, urge. **2** *encourage sales.* aid, be an incentive to, be conducive to, boost, engender, foster, further, generate, help, increase, induce, promote, stimulate. *Opp* DISCOURAGE.

encouragement *n* applause, approval, boost, cheer, exhortation, incentive, incitement, inspiration, reassurance, *inf* shot in the arm, stimulation, stimulus, support. *Opp* DISCOURAGEMENT.

encouraging *adj* comforting, heartening, hopeful, inspiring, optimistic, positive, promising, reassuring. ▷ FAVOURABLE.

encroach *vb* enter, impinge, infringe, intrude, invade, make inroads, trespass, violate.

end *n* **1** boundary, edge, extreme, extremity, limit, pole, tip. **2** cessation, close, coda, completion, conclusion, culmination, curtain (*of play*), denouement (*of plot*), ending, expiration, expiry, finale, finish, *inf* pay-off, resolution. **3** *journey's end.* destination, home, termination, terminus. **4** *end of a queue.* back, rear, tail. **5** *end of your life.* destruction, destiny, doom, extinction, fate, passing, ruin. ▷ DEATH. **6** *an end in view.* aim, aspiration, consequence, design, effect, intention, objective, outcome, plan, purpose, result, upshot. *Opp* BEGINNING. • *vb* **1** abolish, break off, bring to an end, complete, conclude, cut off, destroy, discontinue, *inf* drop, eliminate, exterminate, finalize, *inf* get rid of, halt, phase out, *inf* put an end to, *inf* round off, ruin, scotch, terminate, *inf* wind up. **2** break up, cease, close, come to an end, culminate, die, disappear, expire, fade away, finish, *inf* pack up, reach a climax, stop. *Opp* BEGIN.

endanger *vb* expose to risk, imperil, jeopardize, put at risk, threaten. *Opp* PROTECT.

endearing *adj* appealing, attractive, captivating, charming, disarming, enchanting, engaging, likable, lovable, sweet, winning, winsome. *Opp* REPULSIVE.

endeavour *vb* aim, aspire, attempt, do your best, exert yourself, strive, try.

endless *adj* **1** boundless, immeasurable, inexhaustible, infinite, limitless, measureless, unbounded, unfailing, unlimited. **2** abiding, ceaseless, constant, continual, continuous, enduring, eternal, everlasting, immortal, incessant, interminable, never-ending, nonstop, perpetual, persistent, unbroken, undying, unending, uninterrupted.

endorse *vb* **1** advocate, agree with, approve, authorize, *inf* back, condone, confirm, *inf* OK, sanction, set your seal of approval to, subscribe to, support. **2** *endorse a cheque.* countersign, sign.

endurance *n* determination, fortitude, patience, perseverance, persistence, pertinacity, resolution, stamina, staying-power, strength, tenacity.

endure *vb* **1** carry on, continue, exist, last, live on, persevere, persist, prevail, remain, stay, survive. **2** bear, cope with, experience, go through, *inf* put up with, stand, *inf* stick, *inf* stomach, submit to, suffer, *sl* sweat it out, tolerate, undergo, weather, withstand.
enduring ▷ ENDLESS.

enemy *n* adversary, antagonist, assailant, attacker, competitor, foe, opponent, opposition, the other side, rival, *inf* them. *Opp* FRIEND.

energetic *adj* active, animated, brisk, dynamic, enthusiastic, fast, forceful, hard-working, high-powered, indefatigable, lively, powerful, quick-moving, spirited, strenuous, tireless, unflagging, vigorous, zestful. *Opp* LETHARGIC.

energy *n* **1** animation, ardour, *inf* dash, drive, dynamism, élan, enthusiasm, exertion, fire, force, forcefulness, *inf* get-up-and-go, *inf* go, life, liveliness, might, *inf* pep, spirit, stamina, strength, verve, vigour, *inf* vim, vitality, vivacity, zeal, zest. *Opp* LETHARGY. **2** fuel, power.

enforce *vb* administer, apply, carry out, compel, execute, implement, impose, inflict, insist on, prosecute, put into effect, require, stress. *Opp* WAIVE.

engage *vb* **1** contract with, employ, enlist, hire, recruit, sign up, take on. **2** *cogs engage.* bite, fit together, interlock. **3** *engage to do something.* ▷ PROMISE. **4** *engaged me in gossip.* ▷ OCCUPY. **5** *engage in sport.* ▷ PARTICIPATE.

engaged *adj* **1** affianced, betrothed, *old use* plighted, *old use* promised, *old use* spoken for. **2** ▷ BUSY.

engagement *n* **1** betrothal, promise to marry, *old use* troth. **2** *social engagements.* appointment, arrangement, commitment, date, fixture, meeting, obligation, rendezvous. **3** ▷ BATTLE.

engine *n* **1** machine, motor. □ *diesel, electric, internal-combustion, jet, outboard, petrol, steam, turbine, turbo-jet, turbo-prop.* **2** locomotive.

engineer *n* mechanic, technician. • *vb* ▷ CONSTRUCT, DEVISE.

engrave *vb* carve, chisel, etch, inscribe. ▷ CUT.

enigma *n* conundrum, mystery, *inf* poser, problem, puzzle, riddle.

enjoy *vb* **1** admire, appreciate, bask in, be happy in, delight in, *inf* go in for, indulge in, *inf* lap up, luxuriate in, rejoice in, relish, revel in, savour, take pleasure from, take pleasure in. ▷ LIKE. **2** benefit from, experience, have, take advantage of, use. **enjoy yourself** celebrate, *inf* gad about, *inf* have a fling, have a good time, make merry.

enjoyable *adj* agreeable, amusing, delicious, delightful, diverting, entertaining, gratifying, likeable, *inf* nice, pleasurable, rewarding, satisfying. ▷ PLEASANT. *Opp* UNPLEASANT.

enlarge *vb* amplify, augment, blow up, broaden, build up, develop, dilate, distend, diversify, elongate, expand, extend, fill out, grow, increase, inflate, lengthen, magnify, multiply, spread, stretch, swell, supplement, wax, widen. *Opp* DECREASE. **enlarge on** ▷ ELABORATE.

enlighten *vb* edify, illuminate, inform, make aware. ▷ TEACH.

enlist *vb* **1** conscript, engage, enrol, impress, muster, recruit, sign up. **2** *enlist in the army.* enrol, enter, join up, register, sign on, volunteer. **3** *enlist help.* ▷ OBTAIN.

enliven *vb* animate, arouse, brighten, cheer up, energize, inspire, *inf* pep up, quicken, rouse, stimulate, vitalize, wake up.

enormous *adj* Brobdingnagian, colossal, elephantine, gargantuan, giant, gigantic, gross, huge, hulking, immense, *inf* jumbo, mammoth, massive, mighty, monstrous, mountainous, prodigious, stupendous, titanic, towering, tremendous, vast. ▷ BIG. *Opp* SMALL.

enough *adj* adequate, ample, as much as necessary, sufficient.

enquire *vb* ask, beg, demand, entreat, implore, inquire, query, question, quiz, request. **enquire about** ▷ INVESTIGATE.

enrage *vb* incense, inflame, infuriate, madden, provoke. ▷ ANGER.

enslave *vb* disenfranchise, dominate, make slaves of, subject, subjugate, take away the rights of. *Opp* EMANCIPATE.

ensure *vb* confirm, guarantee, make certain, make sure, secure.

entail *vb* call for, demand, give rise to, involve, lead to, necessitate, require.

enter *vb* **1** arrive, come in, get in, go in, infiltrate, invade, move in, step in. *Opp* DEPART. **2** dig into, penetrate, pierce, puncture, push into. **3** *enter a contest.* engage in, enlist in, enrol in, *inf* go in for, join, participate in, sign up for, take part in, take up, volunteer for. **4** *enter names on a list.* add, inscribe, insert, note down, put down, record, register, set down, sign, write. *Opp* REMOVE. **enter into** ▷ BEGIN.

enterprise *n* **1** adventure, effort, endeavour, operation, programme, project, undertaking, venture. **2** adventurousness, ambition, boldness, courage, daring, determination, drive, energy, *inf* get-up-and-go, initiative, *inf* push. **3** business, company, concern, firm, organization.

enterprising *adj* adventurous, ambitious, bold, courageous, daring, determined, eager, energetic, enthusiastic, *inf* go-ahead, *derog* go-getting, hard-working, imaginative, indefatigable, industrious, intrepid, keen, purposeful, *inf* pushful, *derog* pushy, resource-

ful, spirited, venturesome, vigorous, zealous.
Opp UNADVENTUROUS.

entertain *vb* **1** amuse, cheer up, delight, divert, keep amused, make laugh, occupy, please, regale, *inf* tickle. *Opp* BORE. **2** *entertain friends.* accommodate, be host to, be hostess to, cater for, give hospitality to, *inf* put up, receive, treat, welcome. **3** *entertain an idea.* accept, agree to, approve, consent to, consider, contemplate, harbour, support, take seriously. *Opp* IGNORE. **entertaining** ▷ INTERESTING.

entertainer *n* artist, artiste, performer. □ *acrobat, actor, actress, ballerina, broadcaster, busker, clown, comedian, comic, compère, conjurer, dancer, disc jockey, DJ, impersonator, jester, juggler, lion-tamer, magician, matador, mime artist, minstrel, singer, stunt man, toreador, trapeze artist, trouper, ventriloquist.* ▷ MUSICIAN.

entertainment *n* **1** amusement, distraction, diversion, enjoyment, fun, night-life, pastime, play, pleasure, recreation, sport. **2** divertissement, exhibition, extravaganza, performance, presentation, production, show, spectacle. □ *ballet, bullfight, cabaret, casino, ceilidh, cinema, circus, concert, dance, disco, discothèque, fair, firework display, flower show, gymkhana, motor show, nightclub, pageant, pantomime, play, radio, recital, recitation, revue, rodeo, son et lumière, tattoo, television, variety show, waxworks, zoo.* ▷ DANCE, DRAMA, MUSIC, SPORT.

enthusiasm *n* **1** ambition, ardour, avidity, commitment, drive, eagerness, excitement, exuberance, *derog* fanaticism, fervour, gusto, keenness, panache, passion, relish, spirit, verve, zeal, zest. *Opp* APATHY. **2** craze, diversion, *inf* fad, hobby, interest, passion, pastime.

enthusiast *n* addict, adherent, admirer, aficionado, *inf* buff, champion, devotee, fan, fanatic, *sl* fiend, *sl* freak, lover, supporter, zealot.

enthusiastic *adj* ambitious, ardent, avid, committed, *inf* crazy, delighted, devoted, eager, earnest, ebullient, energetic, excited, exuberant, fervent, fervid, hearty, impassioned, interested, involved, irrepressible, keen, lively, *inf* mad (about), *inf* mad keen, motivated, optimistic, passionate, positive, rapturous, raring (*to go*), spirited, unqualified, unstinting, vigorous, wholehearted, zealous.
Opp APATHETIC. **be enthusiastic** enthuse, get excited, *inf* go into raptures, *inf* go overboard, rave.

entice *vb* allure, attract, cajole, coax, decoy, inveigle, lead on, lure, persuade, seduce, tempt, trap, wheedle.

entire *adj* complete, full, intact, sound, total, unbroken, undivided, uninterrupted, whole.

entitle *vb* **1** call, christen, designate, dub, name, style, term, title. **2** *A licence entitles you to drive.* allow, authorize, empower, enable, justify, license, permit, qualify, warrant.

entitlement *n* claim, ownership, prerogative, right, title.

entity *n* article, being, object, organism, thing, whole.

entrails *n* bowels, guts, *inf* innards, inner organs, *inf* insides, intestines, viscera.

entrance *n* **1** access, admission, admittance. **2** appearance, arrival, coming, entry. **3** door, doorway, gate, gateway, ingress, opening,

portal, turnstile, way in.
4 ante-room, entrance hall, foyer, lobby, passage, passageway, porch, vestibule. *Opp* EXIT.

entrant *n* applicant, candidate, competitor, contender, contestant, entry, participant, player, rival.

entreat *vb* ask, beg, beseech, implore, importune, petition, sue, supplicate. ▷ REQUEST.

entry *n* **1** insertion, item, jotting, listing, note, record.
2 ▷ ENTRANCE. **3** ▷ ENTRANT.

envelop *vb* cloak, cover, enclose, enfold, enshroud, enwrap, shroud, swathe, veil, wrap. ▷ HIDE.

envelope *n* cover, sheath, wrapper, wrapping.

enviable *adj* attractive, covetable, desirable, favourable, sought-after.

envious *adj* begrudging, bitter, covetous, dissatisfied, *inf* green-eyed, *inf* green with envy, grudging, jaundiced, jealous, resentful.

environment *n* circumstances, conditions, context, ecosystem, environs, habitat, location, milieu, setting, situation, surroundings, territory.

envisage *vb* anticipate, contemplate, dream of, envision, fancy, forecast, foresee, imagine, picture, predict, visualize.

envy *n* bitterness, covetousness, cupidity, desire, discontent, dissatisfaction, ill-will, jealousy, longing, resentment. • *vb* begrudge, grudge, resent.

ephemeral *adj* brief, evanescent, fleeting, fugitive, impermanent, momentary, passing, short-lived, temporary, transient, transitory. *Opp* PERMANENT.

epidemic *adj* general, pandemic, prevalent, spreading, universal, widespread. • *n* outbreak, pestilence, plague, rash, upsurge.

episode *n* **1** affair, event, happening, incident, matter, occurrence.
2 chapter, instalment, part, passage, scene, section.

epitome *n* **1** archetype, embodiment, essence, exemplar, incarnation, personification, quintessence, representation, type.
2 ▷ SUMMARY.

equal *adj* balanced, coextensive, commensurate, congruent, correspondent, egalitarian, even, fair, identical, indistinguishable, interchangeable, level, like, matched, matching, proportionate, regular, the same, symmetrical, uniform. ▷ EQUIVALENT. *Opp* UNEQUAL. • *n* clone, compeer, counterpart, equivalent, fellow, peer, twin. • *vb* **1** balance, correspond to, draw with, tie with. **2** *No one equals Caruso.* be in the same class as, compare with, match, parallel, resemble, rival, vie with.

equality *n* **1** balance, congruence, correspondence, equivalence, identity, similarity, uniformity. *Opp* BIAS. **2** *social equality.* egalitarianism, even-handedness, fairness, justice, parity. *Opp* INEQUALITY.

equalize *vb* balance, catch up, compensate, even up, level, make equal, match, regularize, *inf* square, standardize.

equate *vb* assume to be equal, compare, juxtapose, liken, match, parallel, set side by side.

equilibrium *n* balance, equanimity, equipoise, evenness, poise, stability, steadiness, symmetry.

equip *vb* accoutre, arm, array, attire, caparison, clothe, dress, fit out, fit up, furnish, *inf* kit out, outfit, provide, stock, supply.

equipment *n* accoutrements, apparatus, appurtenances, *sl* clobber, furnishings, *inf* gear, *inf* hardware, implements, instruments, kit, machinery, materials, outfit, paraphernalia, plant, *inf* rig, *inf* stuff, supplies, tackle, *inf* things, tools, trappings. ▷ CLOTHES.

equivalent *adj* alike, analogous, comparable, corresponding, fair, interchangeable, parallel, proportionate, *Lat* pro rata, similar, synonymous. ▷ EQUAL.

equivocal *adj* ambiguous, circumlocutory, equivocating, evasive, noncommittal, oblique, periphrastic, questionable, roundabout, suspect.

equivocate *vb inf* beat about the bush, be equivocal, dodge the issue, fence, *inf* have it both ways, hedge, prevaricate, quibble, waffle.

era *n* age, date, day, epoch, period, time.

eradicate *vb* eliminate, erase, get rid of, root out, uproot. ▷ DESTROY.

erase *vb* cancel, cross out, delete, eradicate, expunge, efface, obliterate, rub out, wipe away, wipe off. ▷ REMOVE.

erect *adj* perpendicular, rigid, standing, straight, upright, vertical. • *vb* build, construct, elevate, establish, lift up, make upright, pitch (*a tent*), put up, raise, set up.

erode *vb* abrade, corrode, eat away, eat into, gnaw away, grind down, wash away, wear away.

erotic *adj* amatory, amorous, aphrodisiac, arousing, lubricious, lustful, *sl* randy, *sl* raunchy, seductive, sensual, venereal, voluptuous. ▷ SEXY.

err *vb* be mistaken, be naughty, *sl* boob, *inf* get it wrong, go astray, go wrong, misbehave, miscalculate, sin, *inf* slip up, transgress.

errand *n* assignment, commission, duty, job, journey, mission, task, trip.

erratic *adj* aberrant, capricious, changeable, fickle, fitful, fluctuating, inconsistent, irregular, shifting, spasmodic, sporadic, uneven, unpredictable, unreliable, unstable, unsteady, variable, wayward. *Opp* REGULAR. **2** aimless, directionless, haphazard, meandering, wandering.

error *n inf* bloomer, blunder, *sl* boob, *Lat* corrigendum, *Lat* erratum, fallacy, falsehood, fault, flaw, gaffe, *inf* howler, inaccuracy, inconsistency, inexactitude, lapse, misapprehension, miscalculation, misconception, misprint, mistake, misunderstanding, omission, oversight, sin, *inf* slip-up, solecism, transgression, *old use* trespass, wrongdoing. ▷ WRONG.

erupt *vb* be discharged, be emitted, belch, break out, burst out, explode, gush, issue, pour out, shoot out, spew, spout, spurt, vomit.

eruption *n* burst, discharge, emission, explosion, outbreak, outburst, rash.

escapade *n* adventure, exploit, *inf* lark, mischief, practical joke, prank, scrape, stunt.

escape *n* **1** bolt, breakout, departure, flight, flit, getaway, jailbreak, retreat, running away. **2** discharge, emission, leak, leakage, seepage. **3** *escape from reality.* avoidance, distraction, diversion, escapism, evasion, relaxation, relief. • *vb* **1** abscond, *inf* beat it, bolt, break free, break out, *inf* cut and run, decamp, disappear, *sl* do a bunk, elope, flee, fly, get away, *inf* give someone the slip, *sl* scarper, run away, slip away, *inf* slip the net, *inf* take to your heels, *inf* turn tail. **2** discharge, drain, leak, ooze, pour out, run out, seep. **3** *escape the nasty jobs.* avoid, dodge, duck, elude, evade, get away from, shirk, *sl* skive off.

escapism *n* day-dreaming, fantasy, pretence, unreality, wishful thinking.

escort *n* **1** bodyguard, convoy, guard, guide, pilot, protection, protector, safe-conduct. **2** *royal escort.* attendant, entourage, retinue, train. **3** *escort at a dance.* chaperon, companion, *inf* date, partner. • *vb* accompany, attend, chaperon, conduct, guard, *inf* keep an eye on, *inf* keep tabs on, look after, protect, shepherd, stay with, usher, watch.

essence *n* **1** centre, character, core, cornerstone, crux, essential quality, heart, kernel, life, meaning, nature, pith, quiddity, quintessence, soul, spirit, substance. **2** concentrate, decoction, elixir, extract, flavouring, fragrance, perfume, scent, tincture.

essential *adj* basic, characteristic, chief, crucial, elementary, fundamental, important, indispensable, inherent, innate, intrinsic, irreplaceable, key, leading, main, necessary, primary, principal, quintessential, requisite, vital. *Opp* INESSENTIAL.

establish *vb* **1** base, begin, constitute, construct, create, decree, found, form, inaugurate, initiate, institute, introduce, organize, originate, set up, start. **2** *establish yourself in a job.* confirm, ensconce, entrench, install, lodge, secure, settle, station. **3** *establish facts.* accept, agree, authenticate, certify, confirm, corroborate, decide, demonstrate, fix, prove, ratify, recognize, show to be true, substantiate, verify.

established *adj* deep-rooted, deep-seated, indelible, ineradicable, ingrained, long-lasting, long-standing, permanent, proven, reliable, respected, rooted, secure, traditional, well-known, well-tried. *Opp* NEW.

establishment *n* **1** composition, constitution, creation, formation, foundation, inauguration, inception, institution, introduction, setting up. **2** *a well-run establishment.* business, company, concern, enterprise, factory, household, institution, office, organization, shop.

estate *n* **1** area, development, domain, land. **2** assets, belongings, capital, chattels, effects, fortune, goods, inheritance, lands, possessions, property, wealth.

esteem *n* admiration, credit, estimation, favour, honour, regard, respect, reverence, veneration. • *vb* ▷ RESPECT.

estimate *n* appraisal, approximation, assessment, calculation, conjecture, estimation, evaluation, guess, *inf* guesstimate, judgement, opinion, price, quotation, reckoning, specification, valuation. • *vb* appraise, assess, calculate, compute, conjecture, consider, count up, evaluate, gauge, guess, judge, project, reckon, surmise, think out, weigh up, work out.

estimation *n* appraisal, appreciation, assessment, calculation, computation, consideration, estimate, evaluation, judgement, opinion, rating, view.

estuary *n* creek, *Scot* firth, fjord, inlet, *Scot* loch, river mouth.

eternal *adj* ceaseless, deathless, endless, everlasting, heavenly, immeasurable, immortal, infinite, lasting, limitless, measureless, never-ending, permanent, perpetual, timeless, unchanging, undying, unending, unlimited. ▷ CONTINUAL. *Opp* OCCASIONAL, TRANSIENT.

eternity *n* afterlife, eternal life, immortality, infinity, perpetuity.

ethical *adj* decent, fair, good, honest, just, moral, noble, principled, righteous, upright, virtuous. *Opp* IMMORAL.

ethnic *adj* cultural, folk, national, racial, traditional, tribal.

etiquette *n* ceremony, civility, code of behaviour, conventions, courtesy, decency, decorum, form, formalities, manners, politeness, propriety, protocol, rules of behaviour, standards of behaviour.

evacuate *vb* **1** clear, deplete, drain, move out, remove, send away, void. **2** abandon, decamp from, desert, empty, forsake, leave, pull out of, quit, relinquish, vacate, withdraw from.

evade *vb* **1** avoid, *inf* chicken out of, circumvent, dodge, duck, elude, escape from, fend off, flinch from, get away from, shirk, shrink from, shun, sidestep, *inf* skive, steer clear of, turn your back on. **2** *evade a question.* fudge, hedge, parry. ▷ EQUIVOCATE. *Opp* CONFRONT.

evaluate *vb* apprise, assess, calculate value of, estimate, judge, value, weigh up.

evaporate *vb* dehydrate, desiccate, disappear, disperse, dissipate, dissolve, dry up, evanesce, melt away, vanish, vaporize.

evasive *adj* ambiguous, *inf* cagey, circumlocutory, deceptive, devious, disingenuous, equivocal, equivocating, inconclusive, indecisive, indirect, *inf* jesuitical, misleading, noncommittal, oblique, prevaricating, roundabout, *inf* shifty, sophistical, uninformative. *Opp* DIRECT.

even *adj* **1** flat, flush, horizontal, level, plane, smooth, straight, true. **2** *even pulse.* consistent, constant, equalized, measured, metrical, monotonous, proportional, regular, rhythmical, symmetrical, unbroken, uniform, unvarying. **3** *even scores.* balanced, equal, identical, level, matching, the same. **4** ▷ EVEN-TEMPERED. *Opp* IRREGULAR. **even out** ▷ FLATTEN. **even up** ▷ EQUALIZE. **get even** ▷ RETALIATE.

evening *n* dusk, *poet* eventide, *poet* gloaming, nightfall, sundown, sunset, twilight.

event *n* **1** affair, business, chance, circumstance, contingency, episode, eventuality, experience, happening, incident, occurrence. **2** conclusion, consequence, effect, issue, outcome, result, upshot. **3** activity, ceremony, entertainment, function, occasion. **4** *sporting event.* bout, championship, competition, contest, engagement, fixture, game, match, meeting, tournament.

even-tempered *adj* balanced, calm, composed, cool, equable, even, impassive, imperturbable, pacific, peaceable, peaceful, placid, poised, reliable, self-possessed, serene, stable, steady, tranquil, unemotional, unexcitable, unruffled. *Opp* EXCITABLE.

eventual *adj* concluding, consequent, destined, due, ensuing, expected, final, last, overall, probable, resultant, resulting, ultimate.

everlasting *adj* ceaseless, deathless, endless, eternal, immortal, incorruptible, infinite, lasting, limitless, measureless, never-ending, permanent, perpetual, persistent, timeless, unchanging, undying, unending. *Opp* TRANSIENT.

evermore *adv* always, eternally, for ever, unceasingly.

evict *vb* dislodge, dispossess, eject, expel, *sl* give (someone) the boot, *inf* kick out, oust, put out, remove, throw out, *inf* turf out, turn out.

evidence *n* attestation, certification, confirmation, corroboration, data, demonstration, deposition, documentation, facts, grounds, information, proof, sign, statement, statistics, substantiation, testimony. **give evidence** ▷ TESTIFY.

evident *adj* apparent, certain, clear, discernible, manifest, noticeable, obvious, palpable, patent, perceptible, plain, self-explanatory, unambiguous, undeniable, unmistakable, visible. *Opp* UNCERTAIN.

evil *adj* **1** amoral, atrocious, base, black-hearted, blasphemous, corrupt, criminal, cruel, depraved, devilish, diabolical, dishonest, fiendish, foul, harmful, hateful, heinous, hellish, immoral, impious, infamous, iniquitous, irreligious, machiavellian, malevolent, malicious, malignant, nefarious, pernicious, perverted, reprobate, satanic, sinful, sinister, treacherous, ungodly, unprincipled, unrighteous, vicious, vile, villainous, wicked, wrong. ▷ BAD. *Opp* GOOD. **2** *evil smell.* foul, nasty, pestilential, poisonous, troublesome, unspeakable, vile. ▷ UNPLEASANT. *Opp* PLEASANT. • *n* **1** amorality, blasphemy, corruption, criminality, cruelty, depravity, dishonesty, fiendishness, heinousness, immorality, impiety, iniquity, *old use* knavery, malevolence, malice, mischief, pain, sin, sinfulness, suffering, treachery, turpitude, ungodliness, unrighteousness, vice, viciousness, villainy, wickedness, wrongdoing. ▷ CRIME. **2** *Poverty is an evil.* affliction, bane, calamity, catastrophe, curse, disaster, enormity, hardship, harm, ill, misfortune, wrong.

evocative *adj* atmospheric, convincing, descriptive, emotive, graphic, imaginative, provoking, realistic, stimulating, suggestive, vivid.

evoke *vb* arouse, awaken, call up, conjure up, elicit, excite, inspire, invoke, kindle, produce, provoke, raise, rouse, stimulate, stir up, suggest, summon up.

evolution *n* advance, development, emergence, formation, growth, improvement, maturation maturing, progress, unfolding.

evolve *vb* derive, descend, develop, emerge, grow, improve, mature, modify gradually, progress, unfold.

exact *adj* **1** accurate, correct, dead (*centre*), detailed, faithful, faultless, flawless, meticulous, painstaking, precise, punctilious, right, rigorous, scrupulous, specific, *inf* spot-on, strict, true, truthful, veracious. *Opp* IMPRECISE. **2** *exact copy.* identical, indistinguishable, literal, perfect. • *vb* claim, compel, demand, enforce, extort, extract, get, impose, insist on, obtain, require. **exacting** ▷ DIFFICULT.

exaggerate *vb* **1** amplify, embellish, embroider, enlarge, inflate, *inf* lay it on thick, magnify, make too much of, maximize, overdo, overemphasize, overestimate, overstate, *inf* pile it on, *inf* play up. *Opp* MINIMIZE. **2** ▷ CARICATURE. **exaggerated** ▷ EXCESSIVE.

exalt *vb* boost, elevate, lift, promote, raise, uplift. ▷ PRAISE. **exalted** ▷ HIGH.

examination *n* **1** analysis, appraisal, assessment, audit, catechism, *inf* exam, inspection, investigation, *inf* oral, paper, post-mortem, review, scrutiny, study, survey, test, *inf* viva, *Lat* viva voce. **2** [*medical*] *inf* check-up, scan. **3** *police examination.* cross-examination, enquiry, inquiry, inquisition, interrogation, probe, questioning, trial.

examine *vb* **1** analyse, appraise, audit (*accounts*), check, *inf* check out, explore, inquire into, inspect, investigate, peruse, probe, research, scan, scrutinize, sift, sort out, study, *sl* sus out, test, vet, weigh up. **2** *examine a witness.* catechize, cross-examine, cross-question, *inf* grill, interrogate, *inf* pump, question, sound out, try.

example *n* **1** case, illustration, instance, occurrence, sample, specimen. **2** *example to follow.* ideal, lesson, model, paragon, pattern, prototype. **make an example of** ▷ PUNISH.

exasperate *vb inf* aggravate, drive mad, gall, infuriate, irk, irritate, *inf* needle, pique, provoke, rile, vex. ▷ ANNOY.

excavate *vb* burrow, dig, gouge out, hollow out, mine, scoop out, unearth.

exceed *vb* beat, be more than, do more than, go beyond, go over, outnumber, outshine, outstrip, overstep, overtake, pass, transcend. ▷ EXCEL.

exceedingly *adv* amazingly, especially, exceptionally, excessively, extraordinarily, extremely, outstandingly, specially, unusually, very.

excel *vb* beat, be excellent, better, do best, eclipse, outclass, outdo, outshine, shine, stand out, surpass, top. ▷ EXCEED.

excellent *adj inf* ace, admirable, *inf* brilliant, *old use* capital, champion, choice, consummate, *sl* cracking, distinguished, esteemed, estimable, exceptional, exemplary, extraordinary, *inf* fabulous, *inf* fantastic, fine, first-class, first-rate, flawless, gorgeous, great, high-class, ideal, impressive, magnificent, marvellous, model, notable, outstanding, perfect, *inf* phenomenal, remarkable, *inf* smashing, splendid, sterling, *inf* stunning, *inf* super, superb, superlative, supreme, surpassing, *inf* terrific, *inf* tip-top, *old use* top-hole, *inf* top-notch, top-ranking, *inf* tremendous, unequalled, wonderful. *Opp* BAD.

except *vb* exclude, leave out, omit.

exception *n* **1** exclusion, omission, rejection. **2** abnormality, anomaly, departure, deviation, eccentricity, freak, irregularity, oddity, peculiarity, quirk, rarity. **take exception** ▷ OBJECT.

exceptional *adj* **1** aberrant, abnormal, anomalous, atypical, curious, deviant, eccentric, extraordinary, extreme, isolated, memorable, notable, odd, out-of-the-ordinary, peculiar, phenomenal, quirky, rare, remarkable, singular, solitary, special, strange, surprising, uncommon, unconventional, unexpected, unheard-of, unique, unparalleled, unprecedented, unpredictable, untypical, unusual. 2 ▷ EXCELLENT. *Opp* ORDINARY.

excerpt *n* citation, clip, extract, fragment, highlight, part, passage, quotation, section, selection.

excess *n* **1** abundance, glut, overabundance, overflow, *inf* overkill, profit, redundancy, superabundance, superfluity, surfeit, surplus. *Opp* SCARCITY. **2** debauchery, dissipation, extravagance, intemperance, over-indulgence, profligacy, wastefulness. *Opp* MODERATION.

excessive *adj* **1** disproportionate, exaggerated, extravagant, extreme, fanatical, immoderate, inordinate, intemperate, needless, overdone, prodigal, profligate, profuse, superfluous, undue, unnecessary, unneeded, wasteful. ▷ HUGE. *Opp* INADEQUATE. **2** *excessive prices.* exorbitant, extortionate, unjustifiable, unrealistic, unreasonable. *Opp* MODERATE.

exchange *n* deal, interchange, reciprocity, replacement, substitution, *inf* swap, switch.
• *vb* bargain, barter, change, convert (*currency*), interchange, reciprocate, replace, substitute, *inf* swap, switch, *inf* swop, trade, trade in, traffic. **exchange words** ▷ TALK.

excitable *adj inf* bubbly, chattery, edgy, emotional, explosive, fidgety, fiery, highly-strung, hot-tempered, irrepressible, jumpy, lively, mercurial, nervous, passionate, quick-tempered, restive, temperamental, unstable, volatile. *Opp* CALM.

excite *vb* **1** agitate, amaze, animate, arouse, awaken, discompose, disturb, elate, electrify, enthral, exhilarate, fluster, *inf* get going, incite, inflame, interest, intoxicate, make excited, move, perturb, provoke, rouse, stimulate, stir up, thrill, titillate, *inf* turn on, upset, urge, *inf* wind up, *inf* work up. **2** *excite interest.* activate, cause, elicit, encourage, engender, evoke, fire, generate, kindle, motivate, produce, set off, whet. *Opp* CALM.

excited *adj* agitated, boisterous, delirious, eager, enthusiastic, excitable, exuberant, feverish, frantic, frenzied, heated, *inf* het up, hysterical, impassioned, intoxicated, lively, moved, nervous, overwrought, restless, spirited, vivacious, wild. *Opp* APATHETIC.

exciting *adj* cliff-hanging, dramatic, electric, electrifying, eventful, fast-moving, galvanizing gripping, heady, hair-raising, inspiring, intoxicating, *inf* nail-biting, provocative, riveting, rousing, sensational, spectacular, spine-tingling, stimulating, stirring, suspenseful, tense, thrilling. ▷ AMAZING. *Opp* BORING.

excitement *n* action, activity, adventure, agitation, animation, commotion, delirium, drama, eagerness, enthusiasm, furore, fuss, heat, intensity, *inf* kicks, passion, stimulation, suspense, tension, thrill, unrest.

exclaim *vb* bawl, bellow, blurt out, call, cry out, *old use* ejaculate shout, utter, vociferate, yell. ▷ SAY.

exclamation *n* bellow, call, cry, *old use* ejaculation, expletive, interjection, oath, shout, swear-word, utterance, vociferation, yell

exclude *vb* ban, banish, bar, blacklist, debar, disallow, disown, eject except, excommunicate, expel, forbid, interdict, keep out, leave out, lock out, omit, ostracize, oust, outlaw, prohibit, proscribe, put an embargo on, refuse, reject, repudiate, rule out, shut out, veto. ▷ REMOVE. *Opp* INCLUDE.

exclusive *adj* 1 limiting, restricted, sole, unique, unshared. 2 *an exclusive club.* clannish, classy, closed, fashionable, *sl* posh, private, restrictive, select, selective, snobbish, *inf* up-market.

excreta *plur n* droppings, dung, excrement, faeces, manure, sewage, waste matter.

excrete *vb* defecate, evacuate the bowels, go to the lavatory, relieve yourself.

excursion *n* cruise, expedition, jaunt, journey, outing, ramble, tour, trip, voyage. ▷ TRAVEL.

excuse *n* alibi, apology, defence, explanation, extenuation, justification, mitigation, palliation, plea, pretext, rationalization, reason, vindication. • *vb* 1 apologize for, condone, disregard, explain away, forgive, ignore, justify, mitigate, overlook, pardon, pass over, sanction, tolerate, vindicate, warrant. 2 absolve, acquit, clear, discharge, exculpate, exempt, exonerate, free, let off, *inf* let off the hook, liberate, release. *Opp* BLAME.

execute *vb* 1 accomplish, achieve, bring off, carry out, complete, discharge, do, effect, enact, finish, implement, perform, *inf* pull off. 2 kill, put to death. □ *behead, burn, crucify, decapitate, electrocute, garrotte, gas, guillotine, hang, lynch, shoot, stone.*

executive *n* administrator, *inf* boss, director, manager, officer. ▷ CHIEF.

exemplary *adj* admirable, commendable, faultless, flawless, ideal, model, perfect, praiseworthy, unexceptionable.

exemplify *vb* demonstrate, depict, embody, illustrate, personify, represent, show, symbolize, typify.

exempt *vb* except, exclude, excuse, free, let off, *inf* let off the hook, liberate, release, spare.

exercise *n* 1 action, activity, aerobics, callisthenics, effort, exertion, games, gymnastics, PE, sport, *inf* warm-up, *inf* work-out. 2 *military exercises.* discipline, drill, manoeuvres, operation, practice, training. • *vb* 1 apply, bring to bear, display, effect, employ, execute, exert, expend, implement, put to use, show, use, utilize, wield. 2 *exercise your body.* discipline, drill, exert, jog, keep fit, practise, train, *inf* work out. 3 ▷ WORRY.

exertion *n* action, effort, endeavour, strain, striving, struggle. ▷ WORK.

exhaust *n* discharge, effluent, emission, fumes, gases, smoke. • *vb* 1 consume, deplete, dissipate, drain, dry up, empty, expend, finish off, *inf* run through, sap, spend, use up, void. 2 debilitate, enervate, *inf* fag, fatigue, prostrate, tax, tire, wear out, weary. **exhausted** ▷ BREATHLESS, WEARY.

exhausting *adj* arduous, backbreaking, crippling, debilitating, demanding, difficult, enervating, fatiguing, gruelling, hard, laborious, punishing, severe, strenuous, taxing, tiring, wearying.

exhaustion *n* debility, fatigue, lassitude, tiredness, weakness, weariness.

exhaustive *adj inf* all-out, careful, comprehensive, full-scale, intensive, meticulous, thorough. *Opp* INCOMPLETE.

exhibit *vb* 1 arrange, display, offer, present, put up, set up, show. 2 *exhibit knowledge.* air, betray, brandish, demonstrate, disclose, evidence, express, *derog* flaunt, indicate, manifest, *derog* parade, reveal, *derog* show off. *Opp* HIDE.

exhibition *n* demonstration, display, *inf* expo, exposition, presentation, show.

exhilarating *adj* bracing, cheering, enlivening, exciting, invigorating, refreshing, rejuvenating, stimulating, tonic, uplifting. ▷ HAPPY.

exhort *vb* advise, encourage, harangue, *inf* give a pep talk to, lecture, sermonize, urge.

exile *n* **1** banishment, deportation, expatriation, expulsion, transportation. **2** deportee, displaced person, émigré, expatriate, outcast, refugee, wanderer. • *vb* ban, banish, bar, deport, drive out, eject, evict, expatriate, expel, oust, send away, transport.

exist *vb* **1** be, be found, be in existence, be real, happen, occur. **2** abide, continue, endure, hold out, keep going, last, live, remain alive, subsist, survive. **existing** ▷ ACTUAL, CURRENT, LIVING.

existence *n* actuality, being, continuance, life, living, persistence, reality, survival.

exit *n* **1** barrier, door, doorway, egress, gate, gateway, opening, portal, way out. **2** *a hurried exit.* departure, escape, evacuation, exodus, flight, leave-taking, retreat, withdrawal. • *vb* ▷ DEPART.

exorbitant *adj* disproportionate, excessive, extortionate, extravagant, high, inordinate, outrageous, profiteering, prohibitive, *inf* sky-high, *inf* steep, *inf* stiff, *inf* swingeing, top, unjustifiable, unrealistic, unreasonable, unwarranted. ▷ EXPENSIVE. *Opp* REASONABLE.

exotic *adj* **1** alien, faraway, foreign, remote, romantic, unfamiliar, wonderful. **2** bizarre, colourful, different, exciting, extraordinary, foreign-looking, novel, odd, outlandish, peculiar, rare, singular, strange, striking, unfamiliar, unusual, weird. *Opp* ORDINARY.

expand *vb* **1** amplify, augment, broaden, build up, develop, diversify, elaborate, enlarge, extend, fill out, heighten, increase, make bigger, make longer, prolong. **2** become bigger, dilate, distend, grow, increase, lengthen, open out, stretch, swell, thicken, widen. *Opp* CONTRACT.

expanse *n* area, breadth, extent, range, sheet, space, spread, stretch, sweep, surface, tract.

expansive *adj* **1** affable, amiable, communicative, effusive, extrovert, friendly, genial, open, outgoing, sociable, well-disposed. ▷ TALKATIVE. *Opp* TACITURN. **2** ▷ BROAD. *Opp* NARROW.

expect *vb* **1** anticipate, await, bank on, bargain for, be prepared for, contemplate, count on, envisage, forecast, foresee, have faith in, hope for, imagine, look forward to, plan for, predict, prophesy, reckon on, wait for. **2** *expect obedience.* consider necessary, demand, insist on, look for, rely on, require, want. **3** *I expect he'll come.* assume, believe, conjecture, guess, imagine, judge, presume, presuppose, suppose, surmise, think. **expected** ▷ PREDICTABLE.

expectant *adj* **1** eager, hopeful, *inf* keyed up, *inf* on tenterhooks, optimistic, ready. **2** *inf* expecting, pregnant.

expedient *adj* advantageous, advisable, appropriate, apropos, beneficial, convenient, desirable, helpful, judicious, opportune, politic, practical, pragmatic, profitable, propitious, prudent, right, sensible, suitable, to your advantage, useful, worthwhile. • *n* contrivance, device, *inf* dodge, manoeuvre, means, measure, method, *inf* ploy, recourse, resort, ruse, scheme, stratagem, tactics.

expedition *n* crusade, excursion, exploration, journey, mission, pilgrimage, quest, raid, safari, tour, trek, trip, undertaking, voyage.

expel *vb* **1** ban, banish, cast out, *inf* chuck out, dismiss, drive out, eject, evict, exile, exorcise, *inf* fire, *inf* kick out, oust, remove, *inf* sack, send away, throw out, turn out, *inf* turf out. **2** *expel fumes.* belch, discharge, emit, exhale, give out, push out, send out, spew out.

expend *vb* consume, disburse, *sl* dish out, employ, pay out, spend, use.

expendable *adj* disposable, inessential, insignificant, replaceable, *inf* throw-away, unimportant.

expense *n* charge, cost, disbursement, expenditure, fee, outgoings, outlay, overheads, payment, price, rate, spending.

expensive *adj* costly, dear, generous, high-priced, over-priced, precious, *inf* pricey, *inf* steep, *inf* up-market, valuable. ▷ EXORBITANT. *Opp* CHEAP.

experience *n* **1** familiarity, involvement, observation, participation, practice, taking part. **2** background, expertise, *inf* know-how, knowledge, *Fr* savoir faire, skill, understanding, wisdom. **3** *a nasty experience.* adventure, circumstance, episode, event, happening, incident, occurrence, ordeal, trial. • *vb* encounter, endure, face, go through, have a taste of, know, meet, practise, sample, suffer, test out, try, undergo. **experienced** ▷ EXPERT.

experiment *n* demonstration, investigation, *inf* practical, proof, research, test, trial, try-out. • *vb* do experiments, examine, investigate, make tests, probe, research, test, try out.

experimental *adj* **1** exploratory, on trial, pilot, provisional, tentative, trial. **2** *experimental evidence.* empirical, experiential, proved, tested.

expert *adj* able, *inf* ace, *inf* brilliant, capable, competent, *inf* crack, experienced, knowing, knowledgeable, master, masterly, practised, professional, proficient, qualified, skilful, skilled, sophisticated, specialized, trained, well-versed, worldly-wise. ▷ CLEVER. *Opp* UNSKILFUL. • *n inf* ace, authority, connoisseur, *inf* dab hand, genius, *derog* know-all, master, *inf* old hand, professional, pundit, specialist, veteran, virtuoso, *derog* wiseacre, *inf* wizard. *Opp* AMATEUR.

expertise *n* adroitness, dexterity, expertness, judgement, *inf* know-how, knowledge, *Fr* savoir faire, skill.

expire *vb* become invalid, cease, come to an end, discontinue, finish, *inf* run out, terminate. ▷ DIE.

explain *vb* **1** clarify, clear up, decipher, decode, define, demonstrate, describe, disentangle, elucidate, expound, *inf* get across, *inf* get over, gloss, illustrate, interpret, make clear, make plain, provide an explanation, resolve, shed light on, simplify, solve, *inf* sort out, spell out, teach, translate, unravel. **2** *explain a mistake.* account for, excuse, give reasons for, justify, legitimatize, legitimize, make excuses for, rationalize, vindicate.

explanation *n* **1** account, analysis, clarification, definition, demonstration, description, elucidation, exegesis, explication, exposition, gloss, illustration, interpretation, key, meaning, rubric, significance, solution, translation. **2** cause, excuse, justification, motivation, motive, rationalization, reason, vindication.

explanatory *adj* descriptive, expository, helpful, illuminating, illustrative, interpretive, revelatory.

explicit *adj* categorical, clear, definite, detailed, direct, exact, express, frank, graphic, manifest, open, outspoken, patent, plain, positive, precise, put into words, said, specific, *inf* spelt out, spoken, stated, straightforward, unambiguous, unconcealed, unequivocal, unhidden, unreserved, well-defined. *Opp* IMPLICIT.

explode *vb* **1** backfire, blast, blow up, burst, detonate, erupt, go off, make an explosion, set off, shatter. **2** *explode a theory.* debunk, destroy, discredit, disprove, put an end to, rebut, refute, reject.

exploit *n* achievement, adventure, attainment, deed, enterprise, feat. • *vb* **1** build on, capitalize on, *inf* cash in on, develop, make capital out of, make use of, profit by, profit from, trade on, work on, use, utilize. **2** *exploit people.* *inf* bleed, enslave, ill-treat, impose on, keep down, manipulate, *inf* milk, misuse, oppress, *inf* rip off, *inf* squeeze dry, take advantage of, treat unfairly, withhold rights from.

explore *vb* **1** break new ground, probe, prospect, reconnoitre, scout, search, survey, tour, travel through. **2** *explore a problem.* analyse, examine, inspect, investigate, look into, probe, research, scrutinize, study.

explosion *n* **1** bang, blast, boom, burst, clap, crack, detonation, discharge, eruption, firing, report. **2** *explosion of anger.* fit, outbreak, outburst, *inf* paddy, paroxysm, spasm.

explosive *adj* dangerous, highly-charged, liable to explode, sensitive, unstable, volatile. *Opp* STABLE. • *n* cordite, dynamite, gelignite, gunpowder, TNT.

exponent *n* **1** executant, interpreter, performer, player. **2** advocate, champion, defender, expounder, presenter, propagandist, proponent, supporter, upholder.

expose *vb* bare, betray, dig up, disclose, display, exhibit, lay bare, reveal, show (up), uncover, unearth, unmask. ▷ REVEAL. *Opp* HIDE.

express *vb* air, articulate, disclose, give vent to, make known, phrase, put into words, release, vent, ventilate, voice, word. ▷ COMMUNICATE.

expression *n* **1** cliché, formula, phrase, phraseology, remark, statement, term, turn of phrase, usage, utterance, wording. ▷ SAYING. **2** articulation, confession, declaration, disclosure, revelation, statement. **3** *expression in your voice.* accent, depth, emotion, expressiveness, feeling, intensity, intonation, nuance, pathos, sensibility, sensitivity, sympathy, tone, understanding. **4** *expression on your face.* air, appearance, aspect, countenance, face, look, mien. □ *beam, frown, glare, glower, grimace, grin, laugh, leer, long face, lour,* *lower, poker-face, pout, scowl, smile, smirk, sneer, wince, yawn.*

expressionless *adj* blank, *inf* dead-pan, emotionless, empty, glassy, impassive, inscrutable, poker-faced, straight-faced, uncommunicative, wooden. **2** boring, dull, flat, monotonous, uninspiring, unmodulated, unvarying. *Opp* EXPRESSIVE.

expressive *adj* **1** indicative, meaningful, mobile, revealing, sensitive, significant, striking, suggestive, telling. **2** articulate, eloquent, lively, modulated, varied. *Opp* EXPRESSIONLESS.

exquisite *adj* delicate, elegant, fine, intricate, refined, skilful, well-crafted. ▷ BEAUTIFUL. *Opp* CRUDE.

extend *vb* **1** add to, broaden, build up, develop, draw out, enlarge, expand, increase, keep going, lengthen, make longer, open up, pad out, perpetuate, prolong, protract, *inf* spin out, spread, stretch, widen. **2** *extend a deadline.* defer, delay, postpone, put back, put off. **3** *extend your hand.* give, hold out, offer, outstretch, present, proffer, put out, raise, reach out, stick out, stretch out. **4** *The garden extends to the fence.* continue, range, reach.

extensive *adj* broad, comprehensive, expansive, far-ranging, far-reaching, sweeping, vast, wide, widespread. ▷ LARGE.

extent *n* amount, area, bounds, breadth, compass, degree, dimensions, distance, expanse, length, limit, magnitude, measure, measurement, proportions, quantity, range, reach, scale, scope, size, space, spread, sweep, width.

exterior *adj* external, outer, outside, outward, superficial. • *n* coating, covering, façade, front, outside, shell, skin, surface. *Opp* INTERIOR.

exterminate *vb* annihilate, destroy, eliminate, eradicate, extirpate, get rid of, obliterate, put an end to, root out, terminate. ▷ KILL.

external *adj* exterior, outer, outside, outward, superficial. *Opp* INTERNAL.

extinct *adj* burnt out, dead, defunct, died out, exterminated, extinguished, gone, inactive, vanished. ▷ OLD. *Opp* LIVING.

extinguish *vb* blow out, damp down, douse, put out, quench, slake, smother, snuff out, switch off. ▷ DESTROY. *Opp* KINDLE.

extort *vb* blackmail, bully, coerce, exact, extract, force, obtain by force.

extra *adj* accessory, added, additional, ancillary, auxiliary, excess, further, left-over, more, other, reserve, spare, superfluous, supernumerary, supplementary, surplus, temporary, unneeded, unused, unwanted.

extract *n* **1** concentrate, concentration, decoction, distillation, essence, quintessence. **2** abstract, citation, *inf* clip, clipping, cutting, excerpt, passage, quotation, selection. • *vb* **1** draw out, extricate, pull out, remove, take out, withdraw. **2** *extract a confession.* extort, force out, *inf* winkle out, *inf* worm out, wrench, wrest, wring. **3** *extract what you need.* choose, cull, derive, distil, gather, glean, quote, select. ▷ OBTAIN.

extraordinary *adj* abnormal, amazing, astonishing, astounding, awe-inspiring, bizarre, breathtaking, curious, exceptional, extreme, fantastic, *inf* funny, incredible, marvellous, miraculous, mysterious, mystical, notable, noteworthy, odd, outstanding, peculiar, *inf* phenomenal, prodigious, queer, rare, remarkable, *inf* sensational, signal, singular, special, staggering, strange, striking, stunning, stupendous, surprising, *inf* unbelievable, uncommon, unheard-of, unimaginable, unique, unprecedented, unusual, *inf* weird, wonderful. *Opp* ORDINARY.

extravagance *n* excess, immoderation, improvidence, lavishness, overindulgence, overspending, prodigality, profligacy, self-indulgence, wastefulness. *Opp* ECONOMY.

extravagant *adj* exaggerated, excessive, flamboyant, grandiose, immoderate, improvident, lavish, outrageous, overblown, overdone, pretentious, prodigal, profligate, profuse, reckless, self-indulgent, *inf* showy, spendthrift, uneconomical, unreasonable, unthrifty, wasteful. ▷ EXPENSIVE. *Opp* ECONOMICAL.

extreme *adj* **1** acute, drastic, excessive, greatest, intensest, maximum, severest, *inf* terrific, utmost. ▷ EXTRAORDINARY. **2** distant, endmost, farthest, furthest, furthermost, last, outermost, remotest, ultimate, uttermost. **3** *extreme opinions.* absolute, avant-garde, exaggerated, extravagant, extremist, fanatical, *inf* hard-line, immoderate, intemperate, intransigent, left-wing, militant, obsessive, outrageous, radical, right-wing, uncompromising, *inf* way-out, zealous. • *n* bottom, bounds, edge, end, extremity, left wing, limit, maximum, minimum, opposite, pole, right wing, top, ultimate.

extroverted *adj* active, confident, exhibitionist, outgoing, positive. ▷ SOCIABLE. *Opp* INTROVERTED.

exuberant *adj* **1** animated, boisterous, *inf* bubbly, buoyant, eager, ebullient, effervescent, energetic, enthusiastic, excited, exhilarated, exultant, high-spirited, irrepressible, lively, spirited, sprightly, vivacious. ▷ CHEERFUL. **2** *exuberant decoration.* baroque, exaggerated, highly-decorated, ornate, overdone, rich, rococo. **3** *exuberant growth.* abundant, copious, lush, luxuriant, overflowing, profuse, rank, teeming. *Opp* AUSTERE.

exultant *adj* delighted, ecstatic, elated, joyful, jubilant, *inf* on top of the world, overjoyed, rejoicing. ▷ EXUBERANT.

eye *n* **1** eyeball, *inf* peeper. **2** discernment, perception, sight, vision. • *vb* contemplate, examine, inspect, look at, observe, regard, scrutinize, study, watch. ▷ SEE.

eye-witness *n* bystander, looker-on, observer, onlooker, passer-by, spectator, watcher, witness.

F

fabric *n* **1** material, stuff, textile. ▷ CLOTH. **2** *fabric of a building.* constitution, construction, framework, make-up, structure, substance.

fabulous *adj* **1** fabled, fairy-tale, fanciful, fictitious, imaginary, legendary, mythical, story-book. **2** ▷ EXCELLENT.

face *n* **1** appearance, countenance, features, lineaments, look, *sl* mug, *old use* physiognomy, visage.

▷ EXPRESSION. **2** *face of building*. aspect, covering, exterior, façade, facet, front, outside, side, surface. • *vb* **1** be opposite, front, look towards, overlook. **2** *face danger*. appear before, brave, come to terms with, confront, cope with, defy, encounter, experience, face up to, meet, oppose, square up to, stand up to, tackle. **3** *face a wall with plaster*. clad, coat, cover, dress, finish, overlay, sheathe, veneer.

facetious *adj* cheeky, flippant, impudent, irreverent. ▷ FUNNY.

facile *adj* **1** cheap, easy, effortless, hasty, obvious, quick, simple, superficial, unconsidered. **2** *facile talker*. fluent, glib, insincere, plausible, ready, shallow, slick, *inf* smooth.

facility *n* **1** adroitness, alacrity, ease, expertise, fluency, *derog* glibness, skill, smoothness. **2** *a useful facility*. amenity, convenience, help, provision, resource, service.

fact *n* actuality, certainty, *Fr* fait accompli, reality, truth. *Opp* FICTION. **the facts** circumstances, data, details, evidence, information, *sl* the lowdown, particulars, statistics.

factor *n* aspect, cause, circumstance, component, consideration, constituent, contingency, detail, determinant, element, fact, influence, ingredient, item, parameter, part, particular.

factory *n* assembly line, forge, foundry, manufacturing plant, mill, plant, refinery, shop-floor, works, workshop.

factual *adj* **1** accurate, *Lat* bona fide, circumstantial, correct, demonstrable, empirical, faithful, genuine, matter-of-fact, objective, plain, prosaic, provable, realistic, straightforward, true, unadorned, unbiased, undistorted, unemotional, unimaginative, unvarnished, valid, verifiable, well-documented. *Opp* FALSE. **2** *a factual film*. biographical, documentary, historical, real-life. *Opp* FICTIONAL.

faculty *n* ability, aptitude, capability, capacity, flair, genius, gift, knack, power, talent.

fade *vb* **1** blanch, bleach, darken, dim, discolour, dull, etiolate, grow pale, whiten. *Opp* BRIGHTEN. **2** become less, decline, decrease, diminish, disappear, dwindle, evanesce, fail, melt away, vanish, wane, weaken. **3** *flowers fade*. droop, flag, perish, shrivel, wilt, wither.

fail *vb* **1** abort, be a failure, be unsuccessful, break down, close down, come to an end, *inf* come to grief, come to nothing, *sl* conk out, *inf* crash, cut out, fall through, *inf* fizzle out, *inf* flop, *inf* fold, fold up, founder, give up, go bankrupt, *inf* go bust, go out of business, meet with disaster, miscarry, misfire, *inf* miss out, peter out, stop working. **2** *fail in health*. decay, decline, deteriorate, diminish, disappear, dwindle, ebb, fade, get worse, give out, melt away, vanish, wane, weaken. **3** *fail to do something*. forget, neglect, omit. **4** *fail someone*. abandon, disappoint, *inf* let down. *Opp* IMPROVE, SUCCEED.

failing *n* blemish, defect, fault, flaw, foible, imperfection, shortcoming, weakness, weak spot.

failure *n* **1** abandonment, defeat, disappointment, disaster, downfall, fiasco, *inf* flop, loss, miscarriage, *inf* wash-out, wreck. **2** breakdown, collapse, crash, stoppage. **3** *failure to do your duty*.

dereliction, neglect, omission, remissness. *Opp* SUCCESS.

faint *adj* 1 blurred, blurry, dim, faded, feeble, hazy, ill-defined, indistinct, misty, muzzy, pale, pastel (*colours*), shadowy, unclear, vague. 2 *faint smell.* delicate, slight. 3 *faint sounds.* distant, hushed, low, muffled, muted, soft, stifled, subdued, thin, weak. 4 *faint in the head.* dizzy, exhausted, feeble, giddy, light-headed, unsteady, vertigious, weak, *inf* woozy. *Opp* CLEAR, STRONG. • *vb* become unconscious, black out, collapse, *inf* flake out, *inf* keel over, pass out, swoon.

fair *adj* 1 blond, blonde, flaxen, golden, light, yellow. 2 *fair weather.* bright, clear, clement, cloudless, dry, favourable, fine, pleasant, sunny. *Opp* DARK. 3 *a fair decision.* disinterested, even-handed, fair-minded, honest, honourable, impartial, just, lawful, legitimate, nonpartisan, open-minded, proper, right, unbiased, unprejudiced, upright. *Opp* UNJUST. 4 *a fair standard.* acceptable, adequate, average, indifferent, mediocre, middling, moderate, ordinary, passable, reasonable, respectable, satisfactory, *inf* so-so, tolerable. *Opp* UNACCEPTABLE. 5 ▷ BEAUTIFUL. • *n* 1 amusement-park, fairground, fun-fair. 2 bazaar, carnival, exhibition, festival, fête, gala, market, sale, show.

fairly *adv* moderately, pretty, quite, rather, reasonably, somewhat, tolerably, up to a point.

faith *n* 1 assurance, belief, certitude, confidence, credence, reliance, sureness, trust. *Opp* DOUBT. 2 conviction, creed, devotion, doctrine, dogma, persuasion, religion.

faithful *adj* 1 constant, dependable, devoted, dutiful, honest, loyal, reliable, staunch, steadfast, trusted, trusty, trustworthy, unswerving. 2 *a faithful account.* accurate, close, consistent, exact, factual, literal, precise. ▷ TRUE. *Opp* FALSE.

fake *adj* artificial, bogus, concocted, counterfeit, ersatz, factitious, false, fictitious, forged, fraudulent, imitation, invented, made-up, mock, *sl* phoney, pretended, sham, simulated, spurious, synthetic, trumped-up, unfounded, unreal. *Opp* GENUINE. • *n* 1 copy, counterfeit, duplicate, forgery, hoax, imitation, replica, reproduction, sham, simulation. 2 charlatan, cheat, fraud, hoaxer, humbug, impostor, mountebank, *sl* phoney, quack. • *vb* affect, copy, counterfeit, dissemble, falsify, feign, forge, fudge, imitate, make believe, mock up, pretend, put on, reproduce, sham, simulate.

fall *n* 1 collapse, crash, decline, decrease, depreciation, descent, dip, dive, downswing, downturn, drop, lowering, nosedive, plunge, reduction, slant, slump, tumble. 2 *fall of a fortress.* capitulation, capture, defeat, overthrow, seizure, submission, surrender. • *vb* 1 collapse, *inf* come a cropper, crash down, dive, drop down, founder, go down, keel over, overbalance, pitch, plummet, plunge, sink, slump, spiral, stumble, topple, trip over, tumble. 2 become less, become lower, decline, decrease, diminish, dwindle, ebb, lessen, subside. 3 descend, drop, fall away, slope down. 4 *curtains fell in folds.* be suspended, cascade, dangle, dip down, hang. 5 *silence fell.* come,come about, happen, occur, settle. 6 ▷ DIE.

7 ▷ SURRENDER. **fall apart** ▷ DISINTEGRATE. **fall back** ▷ RETREAT. **fall behind** ▷ LAG. **fall down, fall in** ▷ COLLAPSE. **fall off** ▷ DECLINE. **fall out** ▷ QUARREL. **fall through** ▷ FAIL.

fallacy *n* delusion, error, flaw, miscalculation, misconception, mistake, solecism.

fallible *adj* erring, frail, human, imperfect, liable to make mistakes, uncertain, unpredictable, unreliable, weak. *Opp* INFALLIBLE.

fallow *adj* dormant, resting, uncultivated, unplanted, unsown, unused.

false *adj* **1** deceptive, distorted, erroneous, fabricated, fallacious, faulty, fictitious, flawed, imprecise, inaccurate, incorrect, inexact, invalid, misleading, mistaken, spurious, unfactual, unsound, untrue, wrong. ▷ FAKE. **2** *false friends.* deceitful, dishonest, disloyal, double-dealing, double-faced, faithless, lying, treacherous, unfaithful, unreliable, untrustworthy. *Opp* TRUE. **false name** ▷ PSEUDONYM.

falsehood *n* fabrication, *inf* fib, fiction, lie, prevarication, *inf* story, untruth, *sl* whopper.

falsify *vb* alter, *inf* cook (*the books*), counterfeit, distort, exaggerate, fake, forge, *inf* fudge, imitate, misrepresent, mock up, oversimplify, pervert, simulate, slant, tamper with, tell lies about, twist.

falter *vb* **1** become weaker, flag, flinch, hesitate, hold back, lose confidence, pause, quail, stagger, stumble, totter, vacillate, waver. *Opp* PERSIST. **2** stammer, stutter. **faltering** ▷ HESITANT.

fame *n* acclaim, celebrity, distinction, eminence, glory, honour, illustriousness, importance, *inf* kudos, name, *derog* notoriety, pre-eminence, prestige, prominence, public esteem, renown, reputation, repute, *inf* stardom.

familiar *adj* **1** accustomed, common, conventional, current, customary, everyday, frequent, habitual, mundane, normal, ordinary, predictable, regular, routine, stock, traditional, usual, well-known. *Opp* STRANGE. **2** *familiar language. inf* chatty, close, confidential, *derog* forward, *inf* free-and-easy, *derog* impudent, informal, intimate, near, *derog* presumptuous, relaxed, sociable, unceremonious. ▷ FRIENDLY. *Opp* FORMAL. **familiar with** acquainted with, *inf* at home with, aware of, conscious of, expert in, informed about, knowledgeable about, trained in, versed in.

family *n* **1** brood, children, *inf* flesh and blood, generation, issue, kindred, *old use* kith and kin, litter, *inf* nearest and dearest, offspring, progeny, relations, relatives, *inf* tribe. **2** ancestry, blood, clan, dynasty, extraction, forebears, genealogy, house, line, lineage, pedigree, race, strain, tribe. □ *ancestor, descendant.* □ *aunt, brother, child, cousin, daughter, father, fiancé(e), forefather, foster-child, foster-parent, godchild, godparent, grandchild, grandparent, guardian, husband,* Amer *junior, kinsman, kinswoman, mother, nephew, next-of-kin, niece, parent, sibling, sister, son, step-child, step-parent, uncle, ward, widow, widower, wife.*

famine *n* dearth, hunger, lack, malnutrition, scarcity, shortage, starvation, want. *Opp* PLENTY.

famished *adj* craving, famishing, hungry, *inf* peckish, ravenous, starved, starving.

famous *adj* acclaimed, big, celebrated, distinguished, eminent, exalted, famed, glorious, great, historic, honoured, illustrious, important, legendary, lionized, notable, noted, *derog* notorious, outstanding, popular, prominent, proverbial, renowned, revered, time-honoured, venerable, well-known, world-famous. *Opp* UNKNOWN.

fan *n* **1** blower, extractor, propeller, ventilator. **2** *a soccer fan.* addict, admirer, aficionado, *inf* buff, devotee, enthusiast, fanatic, *inf* fiend, follower, *inf* freak, lover, supporter. ▷ FANATIC.

fanatic *n* activist, adherent, bigot, extremist, fiend, freak, maniac, militant, zealot.

fanatical *adj* bigoted, excessive, extreme, fervid, fervent, immoderate, irrational, maniacal, militant, obsessive, overenthusiastic, passionate, rabid, single-minded, zealous. *Opp* MODERATE.

fanciful *adj* capricious, chimerical, fancy, fantastic, illusory, imaginary, imagined, make-believe, unrealistic, whimsical.

fancy *adj* decorative, elaborate, embellished, embroidered, intricate, ornamented, ornate. ▷ FANCIFUL. • *n* ▷ IMAGINATION, WHIM. • *vb* **1** conjure up, dream of, envisage, imagine, picture, visualize. ▷ THINK. **2** be attracted to, crave, *inf* have a yen for, like, long for, prefer, want, wish for. ▷ DESIRE.

fantastic *adj* **1** absurd, amazing, elaborate, exaggerated, extraordinary, extravagant, fabulous, fanciful, far-fetched, grotesque, imaginative, implausible, incredible, odd, quaint, remarkable, rococo, strange, surreal, unbelievable, unlikely, unrealistic, weird. **2** ▷ EXCELLENT. *Opp* ORDINARY.

fantasy *n* chimera, day-dream, delusion, dream, fancy, hallucination, illusion, imagination, invention, make-believe, mirage, pipe-dream, reverie, vision. *Opp* REALITY.

far *adj* distant, far-away, far-off, outlying, remote. *Opp* NEAR.

farcical *adj* absurd, foolish, ludicrous, preposterous, ridiculous, silly. ▷ FUNNY.

fare *n* **1** charge, cost, fee, payment, price, ticket. **2** *festive fare.* ▷ FOOD.

farewell *adj* goodbye, last, leaving, parting, valedictory. • *n* departure, leave-taking, *inf* send-off, valediction. ▷ GOODBYE.

farm *n* farmhouse, farmstead, *old use* grange. □ *arable farm, croft, dairy farm, fish farm, fruit farm, livestock farm, organic farm, plantation, poultry farm, ranch, smallholding.*

farming *n* agriculture, agronomy, crofting, cultivation, food-production, husbandry.

fascinate *vb* allure, attract, beguile, bewitch, captivate, charm, delight, enchant, engross, enthral, entice, entrance, hypnotize, interest, mesmerize, rivet, spellbind. **fascinating** ▷ ATTRACTIVE.

fashion *n* **1** convention, manner, method, mode, way. **2** craze, cut, *inf* fad, line, look, pattern, rage, style, taste, trend, vogue.

fashionable *adj Fr* [à] la mode, chic, contemporary, current, elegant, *inf* in, in vogue, the latest, modern, modish, popular, smart, *inf* snazzy, sophisticated, stylish, tasteful, *inf* trendy, up-to-date, *inf* with it. *Opp* UNFASHIONABLE.

fast *adv* at full tilt, briskly, in no time, post-haste, quickly, rapidly, swiftly. • *adj* 1 breakneck, brisk, expeditious, express, hasty, headlong, high-speed, hurried, lively, *inf* nippy, precipitate, quick, rapid, smart, *inf* spanking, speedy, supersonic, swift, unhestitating. *Opp* SLOW. **2** *fast on the rocks.* attached, bound, fastened, firm, fixed, immobile, immovable, secure, tight. **3** *fast colours.* indelible, lasting, permanent, stable. **4** *fast living.* ▷ IMMORAL. • *vb* abstain, deny yourself, diet, go hungry, go without food, starve. *Opp* INDULGE.

fasten *vb* affix, anchor, attach, batten, bind, bolt, buckle, button, chain, clamp, clasp, cling, close, connect, couple, do up, fix, grip, hitch, hook, knot, join, lace, lash, latch on, link, lock, make fast, moor, nail, padlock, paste, peg, pin, rivet, rope, screw down, seal, secure, solder, staple, strap, tack, tape, tether, tie, unite, weld. ▷ STICK. *Opp* UNDO.

fastener *n* bond, connection, connector, coupling, fastening, link, linkage. □ *anchor, bolt, buckle, button, catch, chain, clamp, clasp, clip, dowel, dowel-pin, drawing-pin, glue, gum, hasp, hook, knot, lace, latch, lock, mooring, nail, padlock, painter, paste, peg, pin, rivet, rope, safety-pin, screw, seal, Sellotape, solder, staple, strap, string, tack, tape, tether, tie, toggle, Velcro, wedge, zip.*

fastidious *adj* choosy, dainty, delicate, discriminating, finical, finicky, fussy, hard to please, nice, particular, *inf* pernickety, *inf* picky, selective, squeamish.

fat *adj* 1 bloated, *inf* broad in the beam, bulky, chubby, corpulent, dumpy, flabby, fleshy, gross, heavy, massive, obese, overweight, paunchy, plump, podgy, portly, pot-bellied, pudgy, rotund, round, solid, squat, stocky, stout, thick, tubby, weighty, well-fed. ▷ BIG. **2** *fat meat.* fatty, greasy, oily. *Opp* LEAN. • *n* □ *adipose tissue, blubber, butter, dripping, grease, lard, margarine, oil, suet.*

fatal *adj* 1 deadly, final, incurable, lethal, malignant, mortal, terminal. **2** ▷ DISASTROUS.

fatality *n* casualty, death, loss.

fate *n* 1 chance, destiny, doom, fortune, karma, kismet, lot, luck, nemesis, *inf* powers above, predestination, providence, the stars. **2** death, demise, destruction, disaster, downfall, end, ruin.

fated *adj* certain, cursed, damned, decreed, destined, doomed, foreordained, inescapable, inevitable, intended, predestined, predetermined, preordained, sure.

father *n* begetter, *inf* dad, *inf* daddy, *inf* pa, *inf* papa, parent, *old use* pater, *inf* pop, sire.

fatigue *n* debility, exhaustion, feebleness, languor, lassitude, lethargy, tiredness, weakness, weariness. • *vb* debilitate, drain, enervate, exhaust, tire, weaken, weary.

fatigued ▷ WEARY.

fault *n* 1 blemish, defect, deficiency, demerit, failing, failure, fallacy, flaw, foible, frailty, imperfection, inaccuracy, malfunction, snag, weakness. **2** blunder, *inf* boob, error, failing, *Fr* faux pas, gaffe, *inf* howler, indiscretion, lapse, miscalculation, misconduct, misdeed, mistake, negligence, offence, omission, oversight, peccadillo, shortcoming, sin, slip, transgression, *old use* trespass, vice, wrongdoing. **3** *It was my fault.* accountability, blame, culpability, guilt, liability, responsibility. • *vb* ▷ CRITICIZE.

faultless *adj* accurate, correct, exemplary, flawless, ideal, in mint condition, irreproachable, sinless, unimpeachable. ▷ PERFECT. *Opp* FAULTY.

faulty *adj* broken, damaged, defective, deficient, flawed, illogical, imperfect, inaccurate, incomplete, incorrect, inoperative, invalid, not working, out of order, shop-soiled, unusable, useless. *Opp* FAULTLESS.

favour *n* **1** acceptance, approbation, approval, bias, favouritism, friendliness, goodwill, grace, liking, partiality, preference, support. **2** *Do me a favour.* benefit, courtesy, gift, good deed, good turn, indulgence, kindness, service. • *vb* **1** approve of, be in sympathy with, champion, choose, commend, esteem, *inf* fancy, *inf* go for, like, opt for, prefer, show favour to, think well of, value. *Opp* DISLIKE. **2** abet, advance, back, be advantageous to, befriend, forward, promote, support. ▷ HELP. *Opp* HINDER.

favourable *adj* **1** advantageous, appropriate, auspicious, beneficial, benign, convenient, following (*wind*), friendly, generous, helpful, kind, opportune, positive, promising, propitious, reassuring, suitable, supportive, sympathetic, understanding, well-disposed. **2** *a favourable review.* approving, commendatory, complimentary, congratulatory, encouraging, enthusiastic, laudatory. **3** *a favourable reputation.* agreeable, desirable, enviable, good, pleasing, satisfactory. *Opp* UNFAVOURABLE.

favourite *adj* beloved, best, choice, chosen, dearest, esteemed, ideal, liked, loved, popular, preferred, selected, well-liked. • *n* **1** choice, pick, preference. **2** *inf* apple of your eye, darling, idol, pet.

fear *n* alarm, anxiety, apprehension, apprehensiveness, awe, concern, consternation, cowardice, cravenness, diffidence, dismay, doubt, dread, faint-heartedness, foreboding, fright, *inf* funk, horror, misgiving, nervousness, panic, phobia, qualm, suspicion, terror, timidity, trepidation, uneasiness, worry. ▷ PHOBIA. *Opp* COURAGE. • *vb* be afraid of, dread, quail at, shrink from, suspect, tremble at, worry about.

fearful *adj* **1** alarmed, apprehensive, frightened, nervous, panic-stricken, scared, terrified, timid. ▷ AFRAID. *Opp* FEARLESS. **2** ▷ FEARSOME.

fearless *adj* bold, brave, dauntless, intrepid, resolute, stoical, unafraid, unconcerned, undaunted, valiant, valorous. ▷ COURAGEOUS. *Opp* FEARFUL.

fearsome *adj* appalling, awe-inspiring, awesome, daunting, dreadful, fearful, frightful, intimidating, terrible, terrifying. ▷ FRIGHTENING.

feasible *adj* **1** achievable, attainable, easy, possible, practicable, practical, realizable, viable, workable. *Opp* IMPRACTICAL. **2** *a feasible scenario.* credible, likely, plausible, reasonable. *Opp* IMPLAUSIBLE.

feast *n* banquet, *sl* blow-out, dinner, *inf* spread. ▷ MEAL. • *vb* dine, gorge, gormandize, *inf* wine and dine. ▷ EAT.

feat *n* accomplishment, achievement, act, action, attainment, deed, exploit, performance.

feather *n* plume, quill. **feathers** down, plumage.

feathery *adj* downy, fluffy, light, wispy.

feature *n* **1** aspect, attribute, characteristic, circumstance, detail, facet, hall mark, idiosyncrasy, mark, peculiarity, point, property, quality, trait. **2** *newspaper feature.* article, column, item, piece, report, story. • *vb* **1** emphasize, focus on, give prominence to, highlight, *inf* play up, present, promote, show up, *inf* spotlight, *inf* star, stress. **2** *feature in a film.* act, appear, figure, participate, perform, play a role, star, take a part. **features** ▷ FACE.

fee *n* bill, charge, cost, dues, emolument, fare, payment, price, remuneration, subscription, sum, tariff, terms, toll, wage.

feeble *adj* **1** ailing, debilitated, decrepit, delicate, enfeebled, exhausted, faint, fragile, frail, helpless, ill, impotent, inadequate, ineffective, infirm, languid, listless, poorly, powerless, puny, sickly, slight, useless, weak. *Opp* STRONG. **2** effete, feckless, hesitant, incompetent, indecisive, ineffectual, irresolute, *inf* namby-pamby, spineless, vacillating, weedy, wimpish, *inf* wishy-washy. **3** *feeble excuses.* flimsy, insubstantial, lame, paltry, poor, tame, thin, unconvincing.

feed *vb* **1** cater for, give food to, nourish, nurture, provender, provide for, provision, strengthen, suckle, support, sustain, *inf* wine and dine. **2** dine, eat, fare, graze, pasture. **feed on** ▷ EAT.

feel *vb* **1** caress, finger, fondle, handle, hold, manipulate, maul, *inf* paw, pet, stroke, touch. **2** *feel your way.* explore, fumble, grope. **3** *feel the cold.* be aware of, be conscious of, detect, discern, experience, know, notice, perceive, sense, suffer, undergo. **4** *It feels cold.* appear, give a feeling of, seem. **5** *feel something's true.* believe, consider, deem, guess, *inf* have a feeling, *inf* have a hunch, intuit, judge, think.

feeling *n* **1** sensation, sense of touch, sensitivity. **2** ardour, emotion, fervour, passion, sentiment, warmth. **3** *religious feelings.* attitude, belief, consciousness, guess, hunch, idea, impression, instinct, intuition, notion, opinion, perception, thought, view. **4** *a feeling for music.* fondness, responsiveness, sensibility, sympathy, understanding. **5** [*inf*] *a party feeling.* aura, atmosphere, mood, tone, *inf* vibrations.

fell *vb* bring down, chop down, cut down, flatten, *inf* floor, knock down, mow down, prostrate. ▷ KILL.

female *adj* ▷ FEMININE. *Opp* MALE. • *n* □ *aunt,* old use *damsel, daughter,* old use *débutante, fiancé, girl, girlfriend, grandmother, lady,* inf *lass, lesbian,* old use *maid,* old use *maiden,* old use *mistress, mother, niece, sister, spinster,* old use or sexist *wench, wife, woman.* □ *bitch, cow, doe, ewe, hen, lioness, mare, nanny-goat, sow, tigress, vixen.*

feminine *adj derog of men* effeminate, female, *derog* girlish, ladylike, womanly. *Opp* MASCULINE.

fen *n* bog, lowland, marsh, morass, quagmire, slough, swamp.

fence *n* barricade, barrier, fencing, hedge, hurdle, obstacle, paling, palisade, railing, rampart, stockade, wall, wire. • *vb* **1** bound, circumscribe, confine, coop up, encircle, enclose, hedge in, immure, pen, restrict, surround, wall in. **2** ▷ FIGHT.

fend *vb* **fend for yourself** care for yourself, do for yourself, *inf* get along, *inf* get by, look after yourself, *inf* scrape along, support yourself, survive. **fend off** ▷ REPEL.

ferment *n* ▷ COMMOTION. • *vb* **1** boil, bubble, effervesce, *inf* fizz, foam, froth, rise, seethe, work. **2** agitate, excite, foment, incite, instigate, provoke, rouse, stir up.

ferocious *adj* bestial, bloodthirsty, brutal, cruel, feral, fiendish, fierce, harsh, inhuman, merciless, murderous, pitiless, sadistic, savage, vicious, wild. *Opp* GENTLE.

ferry *n* ▷ VESSEL. • *vb* carry, export, fetch, import, shift, ship, shuttle, take across, taxi, transport. ▷ CONVEY.

fertile *adj* abundant, fecund, fertilized, flourishing, fruitful, lush, luxuriant, productive, prolific, rich, teeming, well-manured. *Opp* STERILE.

fertilize *vb* **1** impregnate, inseminate, pollinate. **2** cultivate, dress, enrich, feed, make fertile, manure, mulch, nourish, top-dress.

fertilizer *n* compost, dressing, dung, manure, mulch, nutrient.

fervent *adj* animated, ardent, avid, burning, committed, devout, eager, earnest, emotional, enthusiastic, excited, fanatical, fervid, fiery, frenzied, heated, impassioned, intense, keen, passionate, rapturous, spirited, vehement, vigorous, warm, wholehearted, zealous. *Opp* COOL.

fervour *n* ardour, eagerness, energy, enthusiasm, excitement, fervency, fire, heat, intensity, keenness, passion, sparkle, spirit, vehemence, vigour, warmth, zeal.

fester *vb* become infected, become inflamed, become poisoned, decay, discharge, gather, go bad, go septic, mortify, ooze, putrefy, rot, run, suppurate, ulcerate.

festival *n* anniversary, carnival, celebration, commemoration, fair, feast, fête, fiesta, gala, holiday, jamboree, jubilee. ▷ FESTIVITY.

festive *adj* celebratory, cheerful, cheery, convivial, gay, gleeful, jolly, jovial, joyful, joyous, lighthearted, merry, uproarious. ▷ HAPPY.

festivity *n* celebration, conviviality, entertainment, feasting, festive occasion, *inf* jollification, jollity, jubilation, merrymaking, merriment, mirth, rejoicing, revelry, revels. ▷ PARTY.

fetch *vb* **1** bear, bring, call for, carry, collect, convey, get, import, obtain, pick up, retrieve, transfer, transport. **2** *fetch a good price.* be bought for, bring in, earn, go for, make, produce, raise, realize, sell for. **fetching** ▷ ATTRACTIVE.

feud *n* animosity, antagonism, *inf* bad blood, conflict, dispute, enmity, grudge, hostility, rivalry, strife, vendetta. ▷ QUARREL.

fever *n* delirium, feverishness, high temperature.

feverish *adj* **1** burning, febrile, fevered, flushed, hot, inflamed, trembling. *Opp* COOL. **2** *feverish activity.* agitated, excited, frantic, frenetic, frenzied, hectic, hurried, impatient, passionate, restless.

few *adj inf* few and far between, hardly any, inadequate, infrequent, rare, scarce, sparse, sporadic, *inf* thin on the ground, uncommon. *Opp* MANY.

fibre *n* **1** filament, hair, strand, thread. **2** *moral fibre.* backbone, character, determination, spirit, tenacity, toughness. ▷ COURAGE.

fickle *adj* capricious, changeable, changing, disloyal, erratic, faithless, flighty, inconsistent, inconstant, mercurial, mutable, treacherous, undependable, unfaithful, unpredictable, unreliable, unstable, unsteady, *inf* up and down, vacillating, variable, volatile. *Opp* CONSTANT.

fiction *n* concoction, deception, fabrication, fantasy, figment of the imagination, flight of fancy, invention, lies, story-telling, *inf* tall story. ▷ WRITING. *Opp* FACT.

fictional *adj* fabulous, fanciful, imaginary, invented, legendary, made-up, make-believe, mythical, story-book. *Opp* FACTUAL.

fictitious *adj* apocryphal, assumed, fabricated, deceitful, fraudulent, imagined, invented, made-up, spurious, unreal, untrue. ▷ FALSE. *Opp* GENUINE.

fiddle *vb* interfere, meddle, play about, tamper. ▷ FIDGET. **fiddling** ▷ TRIVIAL.

fidget *vb* be restless, *inf* fiddle about, fret, frisk about, fuss, jerk about, *inf* jiggle, *inf* mess about, move restlessly, *inf* play about, shuffle, squirm, twitch, worry, wriggle about.

fidgety *adj* agitated, frisky, impatient, jittery, jumpy, nervous, on edge, restive, restless, *inf* twitchy, uneasy. *Opp* CALM.

field *n* **1** arable land, clearing, enclosure, grassland, *poet* glebe, green, *old use* mead, meadow, paddock, pasture. **2** *a games field*. arena, ground, pitch, playing-field, recreation ground, stadium. **3** *field of activity*. area, *inf* department, domain, province, sphere, subject, territory.

fiend *n* **1** demon, devil, evil spirit, goblin, hobgoblin, imp, Satan, spirit. **2** ▷ FANATIC.

fierce *adj* **1** angry, barbaric, barbarous, bloodthirsty, bloody, brutal, cold-blooded, cruel, dangerous, fearsome, ferocious, fiendish, fiery, homicidal, inhuman, merciless, murderous, pitiless, ruthless, sadistic, savage, untamed, vicious, violent, wild. **2** *fierce opposition*. active, aggressive, competitive, eager, heated, furious, intense, keen, passionate, relentless, strong, unrelenting. *Opp* GENTLE.

fiery *adj* **1** ablaze, afire, aflame, aglow, blazing, burning, fierce, flaming, glowing, heated, hot, incandescent, raging, red, red-hot. **2** *a fiery temper*. angry, ardent, choleric, excitable, fervent, furious, hot-headed, intense, irascible, irritable, livid, mad, passionate, touchy, violent. *Opp* COOL.

fight *n* action, affray, attack, battle, bout, brawl, *inf* brush, *inf* bust-up, clash, combat, competition, conflict, confrontation, contest, counter-attack, dispute, dogfight, duel, *inf* dustup, encounter, engagement, feud, *old use* fisticuffs, fracas, fray, *inf* free-for-all, hostilities, joust, match, mêlée, *inf* punch-up, raid, riot, rivalry, row, scramble, scrap, scrimmage, scuffle, *inf* set-to, skirmish, squabble, strife, struggle, tussle, war, wrangle. ▷ QUARREL. • *vb* **1** attack, battle, box, brawl, *inf* brush, clash, compete, conflict, contend, do battle, duel, engage, exchange blows, fence, feud, grapple, have a fight, joust, quarrel, row, scrap, scuffle, skirmish, spar, squabble, stand up (to), strive, struggle, *old use* tilt, tussle, wage war, wrestle. **2** *fight a decision*. campaign against, contest, defy, oppose, protest against, resist, take a stand against.

fighter *n* aggressor, antagonist, attacker, belligerent, campaigner, combatant, contender, contestant, defender. □ *archer, boxer,* inf *brawler, champion, duellist, freedom fighter, gladiator, guerrilla, gunman, knight, marine, marksman, mercenary, partisan, prize-fighter, pugilist, sniper, swordsman, terrorist, warrior, wrestler.* ▷ SOLDIER.

figure *n* **1** amount, cipher, digit, integer, number, numeral, sum, symbol, value. **2** diagram, drawing, graph, illustration, outline, picture, plate, representation. **3** *plump figure.* body, build, form, outline, physique, shape, silhouette. **4** *bronze figure.* ▷ SCULPTURE. **5** *well-known figure.* ▷ PERSON. • *vb* ▷ FEATURE. **figure out** ▷ CALCULATE, UNDERSTAND. **figures** ▷ STATISTICS.

file *n* **1** binder, box-file, case, cover, documentation, document-case, dossier, folder, portfolio, ring-binder. **2** *single file.* column, line, procession, queue, rank, row, stream, string, train. • *vb* **1** arrange, categorize, classify, enter, pigeon-hole, organize, put away, record, register, store, systematize. **2** *file through a door.* march, parade, proceed in a line, stream, troop.

fill *vb* **1** be full of, block, *inf* bung up, caulk, clog, close up, cram, crowd, flood, inflate, jam, load, obstruct, pack, plug, refill, replenish, seal, stock up, stop up, *inf* stuff, *inf* top up. *Opp* EMPTY. **2** *fill a need.* answer, fulfil, furnish, meet, provide, satisfy, supply. **3** *fill a post.* execute, hold, occupy, take over, take up. **fill out** ▷ SWELL.

filling *n* contents, *inf* innards, insides, padding, stuffing, wadding.

film *n* **1** coat, coating, cover, covering, haze, layer, membrane, mist, overlay, screen, sheet, skin, slick, tissue, veil. **2** cartoon, *old use* flick, motion picture, movie, picture, video, videotape.

filter *n* colander, gauze, membrane, mesh, riddle, screen, sieve, strainer. • *vb* clarify, filtrate, percolate, purify, refine, screen, sieve, sift, strain.

filth *n* decay, dirt, effluent, garbage, grime, *inf* gunge, impurity, muck, mud, ordure, pollution, putrescence, refuse, rubbish, scum, sewage, slime, sludge, trash. ▷ EXCRETA.

filthy *adj* **1** begrimed, caked, defiled, dirty, disgusting, dusty, foul, grimy, grubby, impure, messy, mucky, muddy, nasty, polluted, scummy, slimy, smelly, soiled, sooty, sordid, squalid, stinking, tainted, uncleaned, unkempt, unwashed, vile. **2** ▷ OBSCENE. *Opp* CLEAN.

final *adj* clinching, closing, concluding, conclusive, decisive, dying, end, eventual, finishing, last, settled, terminal, terminating, ultimate. *Opp* INITIAL.

finalize *vb* clinch, complete, conclude, settle, *inf* sew up, *inf* wrap up.

finance *n* accounting, banking, business, commerce, economics, investment, stocks and shares. • *vb* back, fund, guarantee, invest in, pay for, provide money for, subsidize, support, underwrite. **finances** assets, bank account, budget, capital, cash, funds, holdings, income, money, resources, wealth, *inf* the wherewithal.

financial *adj* economic, fiscal, monetary, pecuniary.

find *vb* **1** acquire, arrive at, become aware of, *inf* bump into, chance upon, come across, come upon, detect, diagnose, discover, dig out, dig up, encounter, espy, expose, *inf* ferret out, happen on, hit on, identify, learn, light on, locate, meet, note, notice, observe, *inf* put your finger on, reach, recognize, reveal, spot, stumble on, uncover, unearth. **2** *find lost property.* get back, recover, rediscover, regain, repossess, retrieve, trace, track down. **3** *found me a job.* give, pass on, procure, provide, supply. *Opp* LOSE.

finding *n* conclusion, decision, decree, judgement, pronouncement, verdict.

fine *adj* **1** admirable, beautiful, choice, classic, commendable, excellent, first-class, handsome, noble, select, superior, worthy. ▷ GOOD. **2** *fine workmanship.* consummate, craftsmanlike, meticulous, skilful, skilled. **3** *fine sand.* minute, powdery, soft. **4** *fine fabric.* dainty, delicate, exquisite, flimsy, fragile, silky. **5** *a fine point.* acute, keen, narrow, sharp, slender, slim, thin. **6** *a fine distinction.* fine-drawn, discriminating, hair-splitting, nice, precise, subtle. **7** *fine weather.* bright, clear, cloudless, dry, fair, nice, pleasant, sunny. • *n* charge, forfeit, penalty.

finish *n* **1** cessation, close, completion, conclusion, culmination, end, ending, finale, resolution, result, termination. **2** *finish on furniture.* appearance, completeness, gloss, lustre, patina, perfection, polish, shine, smoothness, surface, texture. • *vb* **1** accomplish, achieve, break off, bring to an end, cease, clinch, complete, conclude, discontinue, end, finalize, fulfil, halt, pack up, perfect, phase out, reach the end, round off, say goodbye, sign off, stop, take your leave, terminate, *inf* wind up, *inf* wrap up. **2** consume, drink up, eat up, empty, exhaust, expend, get through, *inf* polish off, *inf* say goodbye to, use up. **finish off** ▷ KILL.

finite *adj* bounded, calculable, controlled, countable, definable, defined, determinate, fixed, known, limited, measurable, numbered, rationed, restricted. *Opp* INFINITE.

fire *n* **1** blaze, burning, combustion, conflagration, flames, holocaust, inferno, pyre. **2** fireplace, grate, hearth. □ *boiler, bonfire, brazier, convector, electric fire, forge, furnace, gas fire, immersion-heater, incinerator, kiln, oven, radiator, stove.* **3** *fire in your veins.* ▷ PASSION. • *vb* **1** bake, burn, heat, ignite, kindle, light, put a light to, set alight, set fire to, spark off. **2** animate, awaken, enkindle, enliven, excite, incite, inflame, inspire, motivate, rouse, stimulate, stir. **3** *fire a gun or missile.* catapult, detonate, discharge, explode, launch, let off, propel, set off, shoot, trigger off. **4** *fire a worker.* dismiss, make redundant, sack, throw out. **fire at** ▷ BOMBARD. **hang fire** ▷ DELAY.

fireproof *adj* flameproof, incombustible, non-flammable. *Opp* INFLAMMABLE.

fire-raiser *n* arsonist, pyromaniac.

firm *adj* **1** compact, compressed, congealed, dense, hard, inelastic, inflexible, rigid, set, solid, stable, stiff, unyielding. **2** *firm on the* *rocks*. anchored, embedded, fast, fastened, fixed, immovable, secure, steady, tight. **3** *firm convictions*. adamant, decided, determined, dogged, obstinate, persistent, resolute, unshakeable, unwavering. **4** *a firm price*. agreed, settled, unchangeable. **5** *firm friends*. constant, dependable, devoted, faithful, loyal, reliable. • *n* business, company, concern, corporation, establishment, organization, partnership.

first *adj* **1** cardinal, chief, dominant, foremost, head, highest, key, leading, main, outstanding, paramount, predominant, primary, prime, prinicpal, top, uppermost. **2** *first steps*. basic, elementary, fundamental, initial, introductory, preliminary, rudimentary. **3** *first inhabitants*. aboriginal, archetypal, earliest, eldest, embryonic, oldest, original, primeval. **first-class, first-rate** ▷ EXCELLENT.

fish *n* □ *brill, brisling, carp, catfish, chub, cod, coelacanth, conger, cuttlefish, dab, dace, eel, flounder, goldfish, grayling, gudgeon, haddock, hake, halibut, herring, jellyfish, lamprey, ling, mackerel, minnow, mullet, perch, pike, pilchard, piranha, plaice, roach, salmon, sardine, sawfish, shark, skate, sole, sprat, squid, starfish, stickleback, sturgeon, swordfish,* inf *tiddler, trout, tuna, turbot, whitebait, whiting*. • *vb* angle, go fishing, trawl.

fisher *n* angler, fisherman, trawlerman.

fit *adj* **1** adapted, adequate, applicable, apposite, appropriate, apropos, apt, becoming, befitting, correct, decent, equipped, fitting, good enough, proper, right, satisfactory, seemly, sound, suitable, suited, timely. **2** able, capable, competent, in good form, on form, prepared, ready, strong, well enough. ▷ HEALTHY. *Opp* UNFIT. • *n* attack, bout, convulsion, eruption, explosion, outbreak, outburst, paroxysm, seizure, spasm, spell. • *vb* **1** accord with, become, be fitting for, conform with, correspond to, correspond with, go with, harmonize with, suit. **2** *fit things into place*. arrange, assemble, build, construct, dovetail, install, interlock, join, match, position, put in place, put together. **fit out, fit up** ▷ EQUIP.

fix *n inf* catch-22, corner, difficulty, dilemma, *inf* hole, *inf* jam, mess, *inf* pickle, plight, predicament, problem, quandary. • *vb* **1** attach, bind, connect, embed, implant, install, join, link, make firm, plant, position, secure, stabilize, stick. ▷ FASTEN. **2** *fix a price*. agree, appoint, arrange, arrive at, conclude, confirm, decide, define, establish, finalize, name, ordain, set, settle, sort out, specify. **3** *fix a broken window*. correct, make good, mend, put right, rectify, remedy, repair.

fixture *n* date, engagement, event, game, match, meeting.

fizz *vb* bubble, effervesce, fizzle, foam, froth, hiss, sizzle, sparkle, sputter.

fizzy *adj* bubbly, effervescent, foaming, sparkling.

flag *n* banner, bunting, colours, ensign, jack, pennant, pennon, standard, streamer. • *vb* **1** ▷ SIGNAL. **2** *enthusiasm flagged*. ▷ DECLINE.

flake *n* bit, chip, leaf, scale, scurf, shaving, slice, sliver, splinter, wafer.

flame *n* blaze, light, tongue. ▷ FIRE. • *vb* ▷ FLARE.

flap *vb* beat, flutter, oscillate, slap, sway, swing, thrash about, thresh about, wag, waggle, wave about.

flare *vb* **1** blaze, brighten, burst out, erupt, flame, shine. ▷ BURN. **2** ▷ WIDEN.

flash *vb* coruscate, dazzle, flicker, glare, glint, glitter, light up, reflect, scintillate, shine, spark, sparkle, twinkle. ▷ BURN.

flat *adj* **1** calm, even, horizontal, level, smooth, unbroken, unruffled. **2** outstretched, prone, prostrate, recumbent, spread-eagled, spread out, supine. **3** *a flat voice.* bland, boring, dead, dry, dull, featureless, insipid, lacklustre, lifeless, monotonous, spiritless, stale, tedious, tired, unexciting, uninteresting, unmodulated, unvarying. **4** *a flat tyre.* blown out, burst, deflated, punctured.
• *n* apartment, bedsitter, flatlet, maisonette, penthouse, rooms, suite.

flatten *vb* **1** compress, even out, iron out, level out, press, roll, smooth, straighten. **2** crush, demolish, devastate, level, raze, run over, squash, trample. ▷ DESTROY. **3** *flatten an opponent.* fell, floor, knock down, prostrate. ▷ DEFEAT.

flatter *vb* be flattering to, *inf* butter up, compliment, court, curry favour with, fawn on, humour, *inf* play up to, praise, *sl* suck up to, *inf* toady to. *Opp* INSULT. **flattering** ▷ COMPLIMENTARY, OBSEQUIOUS.

flatterer *n inf* crawler, *inf* creep, groveller, lackey, sycophant, time-server, *inf* toady, *inf* yes-man.

flattery *n* adulation, blandishments, *inf* blarney, *inf* boot-licking, cajolery, fawning, *inf* flannel, insincerity, obsequiousness, servility, *inf* soft soap, sycophancy, unctuousness.

flavour *n* **1** savour, taste. ▷ FLAVOURING. **2** air, ambience, atmosphere, aura, character, characteristic, feel, feeling, property, quality, spirit, stamp, style.
• *vb* add flavour to, add taste to, season, spice.

flavouring *n* additive, essence, extract, seasoning.

flaw *n* break, defect, error, fallacy, fault, imperfection, inaccuracy, loophole, mistake, shortcoming, slip, split, weakness. ▷ BLEMISH. **flawed** ▷ IMPERFECT.

flawless *adj* accurate, clean, faultless, immaculate, mint, pristine, sound, spotless, undamaged, unmarked. ▷ PERFECT. *Opp* IMPERFECT.

flee *vb* abscond, *inf* beat a retreat, *sl* beat it, bolt, clear off, cut and run, decamp, disappear, escape, fly, get away, hurry off, *inf* make a run for it, make off, retreat, run away, *sl* scarper, take flight, *inf* take to your heels, vanish, withdraw.

fleet *n* armada, convoy, flotilla, navy, squadron, task force.

fleeting *adj* brief, ephemeral, evanescent, fugitive, impermanent, momentary, mutable, passing, short, short-lived, temporary, transient, transitory. *Opp* PERMANENT.

flesh *n* carrion, fat, meat, muscle, tissue.

flex *n* cable, cord, extension, lead, wire. • *vb* ▷ BEND.

flexible *adj* **1** bendable, *inf* bendy, elastic, flexile, floppy, giving, limp, lithe, plastic, pliable, pliant, rubbery, soft, springy, stretchy, supple, whippy, willowy, yielding. **2** adjustable, alterable, fluid, mutable, open, provisional, variable. **3** *a flexible person.* accommodating, adaptable, amenable, compliant, conformable, co-operative, docile, easygoing, malleable, open-minded, responsive, tractable, willing. *Opp* RIGID.

flicker *vb* blink, flap, flutter, glimmer, gutter, quiver, shake, shimmer, sparkle, tremble, twinkle, vibrate, waver.

flight *n* **1** journey, trajectory. **2** ▷ ESCAPE.

flimsy *adj* **1** breakable, brittle, delicate, fine, fragile, frail, insubstantial, light, loose, slight, thin, weak. **2** *a flimsy building.* decrepit, dilapidated, gimcrack, jerry-built, makeshift, rickety, shaky, tottering, wobbly. **3** *a flimsy argument.* feeble, implausible, inadequate, superficial, trivial, unbelievable, unconvincing, unsatisfactory. *Opp* STRONG.

flinch *vb* blench, cower, cringe, dodge, draw back, duck, falter, jerk away, jump, quail, quake, recoil, shrink back, shy away, start, swerve, wince. **flinch from** ▷ EVADE.

fling *vb* bowl, *inf* bung, cast, *inf* chuck, heave, hurl, launch, lob, pelt, pitch, propel, send, *inf* shy, *inf* sling, throw, toss.

flippant *adj* cheeky, facetious, facile, *inf* flip, frivolous, light-hearted, shallow, superficial, thoughtless, unserious. *Opp* SERIOUS.

flirt *n female* coquette, *male* philanderer, *inf* tease. • *vb sl* chat someone up, lead someone on, make love, philander, toy with someone's affections.

flirtatious *adj* amorous, coquettish, flirty, *derog* philandering, playful, *derog* promiscuous, teasing.

float *vb* **1** be poised, be suspended, bob, drift, glide, hang, hover, sail, swim, waft. **2** *float a ship.* launch. *Opp* SINK.

flock *n* assembly, congregation, crowd, drove, gathering, herd, horde, multitude, swarm. ▷ GROUP. • *vb* ▷ GATHER.

flog *vb* beat, birch, cane, chastise, flagellate, flay, lash, scourge, thrash, whip. ▷ HIT.

flood *n* **1** cataract, deluge, downpour, flash-flood, inundation, overflow, rush, spate, stream, tidal wave, tide, torrent. **2** abundance, excess, glut, plethora, quantity, superfluity, surfeit, surge. • *vb* cover, deluge, drown, engulf, fill up, immerse, inundate, overflow, overwhelm, saturate, sink, submerge, swamp.

floor *n* **1** floorboards, flooring. **2** deck, level, storey, tier.

flop *vb* **1** collapse, dangle, droop, drop, fall, flag, flap about, hang down, sag, slump, topple, tumble, wilt. **2** ▷ FAIL.

floppy *adj* dangling, droopy, flabby, hanging, loose, limp, pliable, soft. ▷ FLEXIBLE. *Opp* RIGID.

flounder *vb* **1** blunder, flail, fumble, grope, move clumsily, plunge about, stagger, struggle, stumble, tumble, wallow. **2** falter, get confused, make mistakes, talk aimlessly.

flourish *n* ▷ GESTURE. • *vb* **1** be fruitful, be successful, bloom, blossom, boom, burgeon, develop, do well, flower, grow, increase, *inf* perk up, progress, prosper, strengthen, succeed, thrive. **2** *flourish an umbrella.* brandish, flaunt, gesture with, shake, swing, twirl, wag, wave, wield.

flow *n* cascade, course, current, drift, ebb, effusion, flood, gush, outpouring, spate, spurt, stream, tide, trickle. • *vb* bleed, cascade, course, dribble, drift, drip, ebb, flood, flush, glide, gush, issue, leak, move in a stream, ooze, overflow, pour, purl, ripple, roll, run, seep, spill, spring, spurt, squirt, stream, swirl, trickle, well, well up.

flower *n* 1 bloom, blossom, bud, floret, petal. □ *begonia, bluebell, buttercup, campanula, campion, candytuft, carnation, catkin, celandine, chrysanthemum, coltsfoot, columbine, cornflower, cowslip, crocus, crowfoot, cyclamen, daffodil, dahlia, daisy, dandelion, forget-me-not, foxglove, freesia, geranium, gladiolus, gypsophila, harebell, hollyhock, hyacinth, iris, jonquil, kingcup, lilac, lily, lupin, marguerite, marigold, montbretia, nasturtium, orchid, pansy, pelargonium, peony, periwinkle, petunia, phlox, pink, polyanthus, poppy, primrose, rhododendron, rose, saxifrage, scabious, scarlet pimpernel, snowdrop, speedwell, sunflower, tulip, violet, wallflower, water-lily.* • *vb* bloom, blossom, *poet* blow, bud, burgeon, come out, have flowers, open out, unfold. ▷ FLOURISH. **bunch of flowers** arrangement, bouquet, corsage, garland, posy, spray, wreath.

fluctuate *vb* alternate, be unsteady, change, go up and down, oscillate, seesaw, shift, swing, vacillate, vary, waver.

fluent *adj* articulate, effortless, eloquent, expressive, *derog* facile, felicitous, flowing, *derog* glib, natural, polished, ready, smooth, voluble, unhesitating. *Opp* HESITANT.

fluff *n* down, dust, feathers, floss, fuzz, thistledown.

fluffy *adj* downy, feathery, fibrous, fleecy, furry, fuzzy, hairy, light, silky, soft, velvety, wispy, woolly.

fluid *adj* 1 aqueous, flowing, gaseous, liquefied, liquid, melted, molten, running, *inf* runny, sloppy, watery. *Opp* SOLID. 2 *a fluid situation.* adjustable, alterable, changing, flexible, mutable, open, variable, undefined. • *n* gas, liquid, liquor, plasma, vapour.

fluke *n* accident, chance, serendipity, stroke of good luck, twist of fate.

flush *vb* 1 blush, colour, glow, go red, redden. 2 *flush a lavatory.* clean out, cleanse, flood, *inf* pull the plug, rinse out, wash out. 3 *flush from a hiding-place.* chase out, drive out, expel, send up.

fluster *vb* agitate, bewilder, bother, distract, flurry, perplex, put off, put out, *inf* rattle, *inf* throw, upset. ▷ CONFUSE.

flutter *vb* bat (*eyelid*), flap, flicker, flit, fluctuate, move agitatedly, oscillate, palpitate, quiver, shake, tremble, twitch, vacillate, vibrate, wave.

fly *vb* 1 ascend, flit, glide, hover, rise, sail, soar, swoop, take flight, take wing. 2 *fly a plane.* aviate, pilot, take off in. 3 *fly a flag.* display, flap, flutter, hang up, hoist, raise, show, wave. 4 *fly from danger.* flee, hurry, move quickly, run. ▷ ESCAPE. **fly at** ▷ ATTACK. **fly in the face of** ▷ DISREGARD.

flying *n* aeronautics, air-travel, aviation, flight, *inf* jetting.

foam *n* 1 bubbles, effervescence, froth, head (*on beer*), lather, scum, spume, suds. 2 sponge. • *vb* boil, bubble, effervesce, fizz, froth, lather, make foam.

focus *n* 1 clarity, correct adjustment, sharpness. 2 centre, core, focal point, heart, hub, pivot, target. • *vb* aim, centre, concentrate, direct attention, fix attention, home in, spotlight.

fog *n* bad visibility, cloud, haze, miasma, mist, smog, vapour.

foggy *adj* blurred, blurry, clouded, cloudy, dim, hazy, indistinct, misty, murky, obscure. *Opp* CLEAR.

foil *vb* baffle, block, check, circumvent, frustrate, halt, hamper, hinder, obstruct, outwit, prevent, stop, thwart. ▷ DEFEAT.

foist *vb inf* fob off, get rid of, impose, offload, palm off.

fold *n* **1** bend, corrugation, crease, crinkle, furrow, gather, hollow, knife-edge, line, pleat, pucker, wrinkle. **2** *fold for sheep.* ▷ ENCLOSURE. • *vb* **1** bend, crease, crimp, crinkle, double over, jack-knife, overlap, pleat, ply, pucker, tuck in, turn over. **2** close, collapse, let down, put down, shut. **3** *fold in your arms.* clasp, clip, embrace, enclose, enfold, entwine, envelop, hold, hug, wrap. **4** *business folded.* ▷ FAIL.

folk *n* clan, nation, people, the population, the public, race, society, tribe.

follow *vb* **1** accompany, chase, come after, dog, escort, go after, hound, hunt, keep pace with, pursue, replace, shadow, stalk, succeed, supersede, supplant, *inf* tag along with, tail, take the place of, track, trail. **2** *follow a path.* keep to, trace. **3** *follow rules.* abide by, adhere to, attend to, comply with, conform to, heed, honour, obey, observe, pay attention to, stick to, submit to, take notice of. **4** *follow my example.* adopt, be guided by, conform to, copy, imitate, mimic, mirror. **5** *follow an argument.* appreciate, comprehend, grasp, keep up with, take in, understand. **6** *follow football.* admire, be a fan of, keep abreast of, know about, take an interest in, support. **7** *It doesn't follow.* be inevitable, be logical, come about, ensue, happen, have the consequence, mean, result. **following** ▷ SUBSEQUENT.

folly *n* foolishness, insanity, lunacy, madness. ▷ STUPIDITY.

foment *vb* arouse, incite, instigate, kindle, provoke, rouse, stir up. ▷ STIMULATE.

fond *adj* **1** adoring, affectionate, caring, loving, tender, warm. **2** *a fond hope.* ▷ FOOLISH. **be fond of** ▷ LOVE.

fondle *vb* caress, cuddle, handle, pat, pet, snuggle, squeeze, touch.

food *n* aliment, *old use* bread, *old use* comestibles, cooking, cuisine, delicacies, diet, *inf* eatables, *inf* eats, fare, feed, fodder, foodstuff, forage, *inf* grub, *inf* junk food, *old use* meat, *sl* nosh, nourishment, nutriments, provender, provisions, rations, recipe, refreshments, sustenance, swill, *old use* tuck, *old use* viands, *old use* victuals. ▷ MEAL.

fool *n* **1** [*most synonyms inf*] ass, blockhead, booby, buffoon, dimwit, dope, dunce, dunderhead, dupe, fat-head, half-wit, ignoramus, mug, muggins, mutt, ninny, nit, nitwit, simpleton, sucker, twerp, wally. ▷ IDIOT. **2** clown, comedian, comic, coxcomb, entertainer, jester. • *vb inf* bamboozle, bluff, cheat, *inf* con, cozen, deceive, defraud, delude, dupe, fleece, gull, *inf* have on, hoax, hoodwink, *inf* kid, mislead, *inf* string along, swindle, take in, tease, trick. **fool about** ▷ MISBEHAVE.

foolish *adj* absurd, asinine, brainless, childish, crazy, daft, *inf* dopey, *inf* dotty, fatuous, feather-brained, feeble-minded, *old* *use* fond, frivolous, *inf* half-baked, hare-brained, idiotic, illogical, immature, inane, infantile, irrational, *inf* jokey, laughable, lighthearted, ludicrous, mad, meaningless, mindless, misguided, naïve, nonsensical, playful, pointless, preposterous, ridiculous, scatterbrained, *inf* scatty, senseless, shallow, silly, simple, simple-minded, simplistic, *inf* soppy, stupid, thoughtless, unintelligent, unreasonable, unsound, unwise, witless. *Opp* WISE.

foot *n* **1** claw, hoof, paw, trotter. **2** ▷ BASE.

footprint *n* footmark, spoor, track.

forbid *vb* ban, bar, debar, deny, deter, disallow, exclude, interdict, make illegal, outlaw, preclude, prevent, prohibit, proscribe, refuse, rule out, say no to, stop, veto. *Opp* ALLOW.

forbidden *adj* **1** against the law, taboo, unlawful, wrong. **2** *a forbidden area.* closed, out of bounds, restricted, secret.

forbidding *adj* gloomy, grim, menacing, ominous, stern, threatening, uninviting, unwelcoming. ▷ UNFRIENDLY. *Opp* FRIENDLY.

force *n* **1** aggression, *inf* arm-twisting, coercion, compulsion, constraint, drive, duress, effort, might, power, pressure, strength, vehemence, vigour, violence. **2** effect, energy, impact, intensity, momentum, shock. **3** *a military force.* army, body, group, troops. **4** *force of an argument.* cogency, effectiveness, persuasiveness, rightness, thrust, validity, weight. • *vb* **1** *inf* bulldoze, coerce, compel, constrain, drive, impel, impose on, make, oblige, order, press-gang, pressurize. **2** *force a door.* break open, burst open, prise open, smash, use force on, wrench. **3** *force something on someone.* impose, inflict.

foreboding *n* anxiety, apprehension, dread, fear, feeling, foreshadowing, forewarning, intimation, intuition, misgiving, omen, portent, premonition, presentiment, suspicion, warning, worry.

forecast *n* augury, expectation, outlook, prediction, prognosis, prognostication, projection, prophecy. • *vb* ▷ FORESEE.

forefront *n* avant-garde, front, lead, vanguard.

foreign *adj* **1** distant, exotic, faraway, outlandish, remote, strange, unfamiliar, unknown. **2** alien, external, immigrant, imported, incoming, international, outside, overseas, visiting. **3** *foreign ideas.* extraneous, odd, uncharacteristic, unnatural, untypical, unusual, unwanted. *Opp* NATIVE.

foreigner *n* alien, immigrant, newcomer, outsider, overseas visitor, stranger. *Opp* NATIVE.

foremost *adj* first, leading, main, primary, supreme. ▷ CHIEF.

forerunner *n* advance messenger, harbinger, herald, precursor, predecessor. ▷ ANCESTOR.

foresee *vb* anticipate, envisage, expect, forecast, picture. ▷ FORETELL.

foresight *n* anticipation, caution, far-sightedness, forethought, looking ahead, perspicacity, planning, preparation, prudence, readiness, vision.

forest *n* coppice, copse, jungle, plantation, trees, woodland, woods.

foretaste *n* advance warning, augury, example, foreknowledge, forewarning, indication, omen, premonition, preview, sample, specimen, *inf* tip-off, trailer, *inf* try-out.

foretell *vb* augur, *old use* bode, forebode, foreshadow, forewarn, give a foretaste of, herald, portend, predict, presage, prognosticate, prophesy, signify. ▷ FORESEE.

forethought *n* anticipation, caution, far-sightedness, foresight, looking ahead, perspicacity, planning, preparation, prudence, readiness, vision.

forewarning *n* advance warning, augury, omen, premonition, *inf* tip-off. ▷ FORETASTE.

forfeit *n* confiscation, damages, fee, fine, penalty, sequestration. • *vb* abandon, give up, let go, lose, pay up, relinquish, renounce, surrender.

forge *n* furnace, smithy, workshop. • *vb* 1 beat into shape, cast, construct, hammer out, manufacture, mould, shape, work. 2 coin, copy, counterfeit, fake, falsify, imitate, make illegally, reproduce. **forge ahead** ▷ ADVANCE.

forgery *n* copy, counterfeit, *inf* dud, fake, fraud, imitation, *inf* phoney, replica, reproduction.

forget *vb* 1 be forgetful, dismiss from your mind, disregard, fail to remember, ignore, leave out, lose track (of), miss out, neglect, omit, overlook, skip, suffer from amnesia, unlearn. 2 be without, leave behind, lose. *Opp* REMEMBER.

forgetful *adj* absent-minded, amnesiac, careless, distracted, inattentive, neglectful, negligent, oblivious, preoccupied, unconscious, unmindful, unreliable, vague, *inf* woolly-minded.

forgivable *adj* allowable, excusable, justifiable, negligible, pardonable, petty, understandable, venial. *Opp* UNFORGIVABLE.

forgive *vb* 1 absolve, acquit, clear, exculpate, excuse, exonerate, indulge, *inf* let off, pardon, spare. 2 *forgive a crime.* condone, ignore, make allowances for, overlook, pass over.

forgiveness *n* absolution, amnesty, clemency, compassion, exculpation, exoneration, grace, indulgence, leniency, mercy, pardon, reprieve, tolerance. *Opp* RETRIBUTION.

forgiving *adj* clement, compassionate, forbearing, generous, magnanimous, merciful, tolerant, understanding. ▷ KIND. *Opp* VENGEFUL.

forgo *vb* abandon, abstain from, do without, forswear, give up, go without, omit, pass up, relinquish, renounce, sacrifice, turn down, waive.

forked *adj* branched, cleft, divergent, divided, fork-like, pronged, split, V-shaped.

forlorn *adj* abandoned, alone, bereft, deserted, forsaken, friendless, lonely, outcast, solitary, unloved. ▷ SAD.

form *n* 1 appearance, arrangement, cast, character, configuration, design, format, framework, genre, guise, kind, manifestation, manner, model, mould, nature, pattern, plan, semblance, sort, species, structure, style, system, type, variety. 2 *human form.* anatomy, body, build, figure, frame, outline, physique, shape, silhouette. 3 *your form in school.* class, grade, group, level, set, stream, tutor-group. 4 *good form.* behaviour, convention, custom, etiquette, fashion, manners, practice. 5 *an application form.* document, paper. 6 *in good form.* condition, *inf* fettle, fitness, health, performance, spirits. 7 ▷ SEAT. • *vb* 1 bring into existence, cast, constitute, construct, create, design, establish, forge, found, give form to, make, model, mould, organize, produce, shape. 2 appear, arise, come into existence, develop, grow, materialize, take shape. 3 *form a team*. act as, compose, comprise, make up, serve as. 4 *form a habit*. acquire, cultivate, develop, get.

formal *adj* 1 aloof, ceremonial, ceremonious, conventional, cool, correct, customary, dignified, *inf* dressed-up, orthodox, *inf* posh, *derog* pretentious, proper, punctilious, ritualistic, solemn, sophisticated, stately, *inf* starchy, stiff, stiff-necked, unbending, unfriendly. 2 *formal language*. academic, impersonal, official, precise, reserved, specialist, stilted, technical, unemotional. 3 *a formal agreement*. binding, contractual, enforceable, legal, *inf* signed and sealed. 4 *a formal design*. calculated, geometrical, orderly, organized, regular, rigid, symmetrical. *Opp* INFORMAL.

format *n* appearance, design, layout, plan, shape, size, style.

former *adj* bygone, departed, ex-, last, late, old, one-time, past, previous, prior, recent. **the former** earlier, first, first-mentioned. *Opp* LATTER.

formidable *adj* awe-inspiring, awesome, challenging, daunting, difficult, dreadful, fearful, frightening, intimidating, large-scale, *inf* mind-boggling, onerous, overwhelming, prodigious, taxing. *Opp* EASY.

formula *n* 1 form of words, ritual, rubric, spell, wording. 2 *formula for success*. blueprint, method, prescription, procedure, recipe, rule, technique, way.

formulate *vb* 1 articulate, codify, define, express clearly, set out in detail, specify, systematize. 2 concoct, create, devise, evolve, form, invent, map out, originate, plan, work out.

forsake *vb* abandon, break off from, desert, forgo, forswear, give up, jettison, jilt, leave, quit, renounce, repudiate, surrender, throw over, *inf* turn your back on, vacate.

fort *n* camp, castle, citadel, fortification, fortress, garrison, stronghold, tower.

forthright *adj* blunt, candid, decisive, direct, outspoken, plain-speaking, straightforward, unequivocal, unhesitating, uninhibited. ▷ FRANK. *Opp* EVASIVE.

fortify *vb* 1 buttress, defend, garrison, protect, reinforce, secure against attack, shore up. 2 bolster, boost, brace, buoy up, cheer, embolden, encourage, hearten, invigorate, lift the morale of, reassure, stiffen the resolve of, strengthen, support, sustain. *Opp* WEAKEN.

fortitude *n* backbone, bravery, courage, determination, endurance, firmness, heroism, patience, resolution, stoicism, tenacity, valour, will-power. ▷ COURAGE. *Opp* COWARDICE.

fortunate *adj* auspicious, blessed, favourable, lucky, opportune, propitious, prosperous, providential, timely. ▷ HAPPY.

fortune *n* 1 accident, chance, destiny, fate, fortuity, karma, kismet, luck, providence. 2 affluence, assets, estate, holdings, inheritance, means, *inf* millions, money, opulence, *inf* pile, possessions, property, prosperity, riches, treasure, wealth.

fortune-teller *n* clairvoyant, crystal-gazer, futurologist, oracle, palmist, prophet, seer, soothsayer, star-gazer, sybil.

forward *adj* **1** advancing, front, frontal, head-first, leading, onward, progressive. **2** *forward planning*. advance, early, forward-looking, future, well-advanced. **3** *a forward child*. advanced, assertive, bold, brazen, cheeky, confident, familiar, *inf* fresh, impertinent, impudent, insolent, over-confident, precocious, presumptuous, pushful, *inf* pushy, shameless, uninhibited. *Opp* BACKWARD. • *vb* **1** dispatch, expedite, freight, post on, re-address, send, send on, ship, transmit, transport. **2** *forward your career*. accelerate, advance, encourage, facilitate, foster, further, hasten, help along, *inf* lend a helping hand to, promote, speed up, support. ▷ HELP. *Opp* HINDER.

foster *vb* **1** advance, cultivate, encourage, further, nurture, promote, stimulate. ▷ HELP. **2** *foster a child*. adopt, bring up, care for, look after, maintain, nourish, nurse, raise, rear, take care of.

foul *adj* **1** bad, contaminated, disagreeable, disgusting, fetid, filthy, hateful, impure, infected, loathsome, nasty, nauseating, nauseous, noisome, obnoxious, offensive, polluted, putrid, repellent, repugnant, repulsive, revolting, rotten, sickening, smelly, squalid, stinking, vile. ▷ DIRTY, SMELLING. **2** *foul crimes*. abhorrent, abominable, atrocious, beastly, cruel, evil, ingnominious, monstrous, scandalous, shameful, vicious, villainous, violent, wicked. **3** *foul language*. abusive, bawdy, blasphemous, coarse, common, crude, impolite, improper, indecent, insulting, licentious, offensive, rude, uncouth, vulgar. ▷ OBSCENE. **4** *foul weather*. foggy, rainy, rough, stormy, violent, windy. ▷ UNPLEASANT. **5** *foul play*. against the rules, dishonest, forbidden, illegal, invalid, prohibited, unfair, unsportsmanlike. *Opp* CLEAN, FAIR. • *n* infringement, violation. • *vb* ▷ DIRTY. **foul up** ▷ MUDDLE.

found *vb* **1** begin, bring about, create, endow, establish, fund, *inf* get going, inaugurate, initiate, institute, organize, originate, provide money for, raise, set up, start. **2** base, build, construct, erect, ground, rest, set.

foundation *n* **1** beginning, endowment, establishment, founding, inauguration, initiation, institution, organizing, setting up, starting. **2** base, basement, basis, bottom, cornerstone, foot, footing, substructure, underpinning. **3** *foundations of science*. basic principle, element, essential, fundamental, origin, *plur* rudiments.

founder *vb* abort, be wrecked, *inf* come to grief, fail, fall through, go down, miscarry, sink.

fountain *n* font, fount, fountainhead, jet, source, spout, spray, spring, well, well-spring.

foyer *n* ante-room, entrance, entrance hall, hall, lobby, reception.

fraction *n* division, part, portion, section, subdivision.

fracture *n* break, breakage, chip, cleavage, cleft, crack, fissure, gap, opening, rent, rift, rupture, split. • *vb* breach, break, cause a fracture in, chip, cleave, crack, rupture, separate, split, suffer a fracture in.

fragile *adj* ▷ FRAIL.

fragment *n* atom, bit, chip, crumb, *plur* debris, morsel, part, particle, piece, portion, remnant, scrap, shard, shiver, shred, sliver, *plur* smithereens, snippet, speck. • *vb* ▷ BREAK.

fragmentary *adj inf* bitty, broken, disconnected, disintegrated, disjointed, fragmented, imperfect, in bits, incoherent, incomplete, in fragments, partial, scattered, scrappy, sketchy, uncoordinated. *Opp* COMPLETE.

fragrance *n* aroma, bouquet, nose (*of wine*), odour, perfume, redolence, scent, smell.

fragrant *adj* aromatic, odorous, perfumed, redolent, scented, sweet-smelling.

frail *adj* breakable, brittle, dainty, delicate, easily damaged, feeble, flimsy, fragile, insubstantial, light, *derog* puny, rickety, slight, thin, unsound, unsteady, vulnerable, weak, *derog* weedy. ▷ ILL. *Opp* STRONG.

frame *n* **1** bodywork, chassis, construction, scaffolding, structure. ▷ FRAMEWORK. **2** *photo frame.* border, case, casing, edge, edging, mount, mounting. • *vb* **1** box in, enclose, mount, set off, surround. **2** ▷ COMPOSE. **frame of mind** ▷ ATTITUDE.

framework *n* bare bones, frame, outline, plan, shell, skeleton, support, trellis.

frank *adj* blunt, candid, direct, downright, explicit, forthright, genuine, *inf* heart-to-heart, honest, ingenuous, *inf* no-nonsense, open, outright, outspoken, plain, plain-spoken, revealing, serious, sincere, straightforward, straight from the heart, to the point, trustworthy, truthful, unconcealed, undisguised, unreserved. *Opp* INSINCERE.

frantic *adj* agitated, anxious, berserk, *inf* beside yourself, crazy, delirious, demented, deranged, desperate, distraught, excitable, feverish, *inf* fraught, frenetic, frenzied, furious, hectic, hurried, hysterical, mad, overwrought, panicky, rabid, uncontrollable, violent, wild, worked up. *Opp* CALM.

fraud *n* **1** cheating, chicanery, *inf* con-trick, counterfeit, deceit, deception, dishonesty, double-dealing, duplicity, fake, forgery, hoax, imposture, pretence, *inf* put-up job, ruse, sham, *inf* sharp practice, swindle, trick, trickery. **2** charlatan, cheat, *inf* con-man, hoaxer, humbug, impostor, mountebank, *sl* phoney, *inf* quack, rogue, scoundrel, swindler.

fraudulent *adj inf* bent, bogus, cheating, corrupt, counterfeit, criminal, *inf* crooked, deceitful, devious, *inf* dirty, dishonest, double-dealing, duplicitous, fake, false, forged, illegal, lying, *sl* phoney, sham, specious, swindling, underhand, unscrupulous. *Opp* HONEST.

fray *n* brawl, commotion, conflict, disturbance, fracas, mêlée, quarrel, rumpus. ▷ FIGHT.

frayed *adj* chafed, rough-edged, tattered, threadbare, unravelled, worn. ▷ RAGGED.

freak *adj* aberrant, abnormal, anomalous, atypical, bizarre, exceptional, extraordinary, freakish, odd, peculiar, queer, rare, unaccountable, unforeseeable, unpredictable, unusual, weird. *Opp* NORMAL. • *n* **1** aberration, abnormality, abortion, anomaly, curiosity, deformity, irregularity, monster, monstrosity, mutant, oddity, *inf* one-off, quirk, rarity, sport, variant. **2** ▷ FANATIC.

free *adj* **1** able, allowed, at leisure, at liberty, idle, independent, loose, not working, uncommitted, unconfined, unconstrained, unencumbered, unfixed, unrestrained, untrammelled. **2** *free from slavery.* emancipated, freeborn, let go, liberated, released, unchained, unfettered, unshackled. **3** *a free country.* autonomous, democratic, independent, self-governing, sovereign. **4** *free access.* accessible, clear, open, permitted, unhindered, unimpeded, unrestricted. **5** *free gifts.* complimentary, gratis, *sl* on the house, unasked-for, unsolicited, without charge. **6** *free space.* available, empty, uninhabited, unoccupied, vacant. **7** *free with money.* bounteous, casual, charitable, generous, lavish, liberal, munificent, ready, unstinting, willing. • *vb* **1** absolve, acquit, clear, deliver, discharge, disenthral, emancipate, enfranchise, exculpate, exonerate, let go, let off, let out, liberate, loose, make free, manumit, pardon, parole, ransom, release, reprieve, rescue, save, set free, spare, turn loose, unchain, unfetter, unleash, unlock, unloose. *Opp* CONFINE. **2** *free tangled ropes.* clear, disengage, disentangle, extricate, loose, unbind, undo, unknot, untie. *Opp* TANGLE. **free and easy** ▷ INFORMAL.

freedom *n* **1** autonomy, independence, liberty, self-determination, self-government, sovereignty. *Opp* CAPTIVITY. **2** deliverance, emancipation, exemption, immunity, liberation, release. **3** *freedom to choose.* ability, *Fr* carte blanche, discretion, free hand, latitude, leeway, leisure, licence, opportunity, permission, power, privilege, right, scope.

freeze *vb* **1** become ice, become solid, congeal, harden, ice over, ice up, solidify, stiffen. **2** chill, cool, make cold, numb. **3** *freeze food.* chill, deep-freeze, dry-freeze, ice, refrigerate. **4** *freeze the frame.* fix, hold, immobilize, keep still, paralyse, peg, petrify, stand still, stick, stop. **freezing** ▷ COLD.

freight *n* cargo, consignment, goods, haul, load, merchandise, payload, shipment.

frenzy *n* agitation, delirium, derangement, excitement, fever, fit, fury, hysteria, insanity, lunacy, madness, mania, outburst, paroxysm, passion, turmoil.

frequent *adj* common, constant, continual, countless, customary, everyday, familiar, habitual, incessant, innumerable, many, normal, numerous, ordinary, persistent, recurrent, recurring, regular, reiterative, repeated, usual. *Opp* INFREQUENT. • *vb* ▷ HAUNT.

fresh *adj* **1** additional, alternative, different, extra, just arrived, new, recent, supplementary, unfamiliar, up-to-date. **2** alert, energetic, healthy, invigorated, lively, *inf* perky, rested, revived, sprightly, spry, tingling, vigorous, vital. **3** *a fresh recruit.* callow, *inf* green, inexperienced, naïve, raw, unsophisticated, untried, *inf* wet behind the ears. **4** *fresh water.* clear, drinkable, potable, pure, refreshing, sweet, uncontaminated. **5** *fresh air.* airy, circulating, cool, unpolluted, ventilated. **6** *a fresh wind.* bracing, breezy, invigorating, moderate, sharp, stiff, strongish. **7** *fresh food.* healthy, natural, newly gathered, unprocessed, untreated, wholesome. **8** *fresh sheets.* clean, crisp, laundered, untouched, unused, washed-and-ironed. **9** *fresh colours.* bright, clean, glowing, just painted, renewed, restored, sparkling, unfaded, vivid. *Opp* OLD, STALE.

fret *vb* **1** agonize, be anxious, brood, lose sleep, worry. **2** ▷ ANNOY.

fretful *adj* anxious, distressed, disturbed, edgy, irritable, irritated, jittery, peevish, petulant, restless, testy, touchy, worried. ▷ BAD-TEMPERED. *Opp* CALM.

friction *n* **1** abrading, abrasion, attrition, chafing, fretting, grating, resistance, rubbing, scraping. **2** ▷ CONFLICT.

friend *n* acquaintance, associate, *inf* buddy, *inf* chum, companion, comrade, confidant(e), *inf* crony, intimate, *inf* mate, *inf* pal, partner, pen-friend, playfellow, playmate, supporter, well-wisher. ▷ ALLY, LOVER. *Opp* ENEMY. **be friends** ▷ ASSOCIATE. **make friends with** ▷ BEFRIEND.

friendless *adj* abandoned, alienated, alone, deserted, estranged, forlorn, forsaken, isolated, lonely, ostracized, shunned, shut out, solitary, unattached, unloved.

friendliness *n* benevolence, camaraderie, conviviality, devotion, esteem, familiarity, goodwill, helpfulness, hospitality, kindness, neighbourliness, regard, sociability, warmth. *Opp* HOSTILITY.

friendly *adj* accessible, affable, affectionate, agreeable, amiable, amicable, approachable, attached, benevolent, benign, *inf* chummy, civil, close, clubbable, companionable, compatible, comradely, conciliatory, congenial, convivial, cordial, demonstrative, expansive, favourable, genial, good-natured, gracious, helpful, hospitable, intimate, kind, kind-hearted, kindly, likeable, *inf* matey, neighbourly, outgoing, *inf* pally, sympathetic, tender, *inf* thick, warm, welcoming, well-disposed. ▷ FAMILIAR, LOVING, SOCIABLE. *Opp* UNFRIENDLY.

friendship *n* affection, alliance, amity, association, attachment, closeness, comradeship, fellowship, fondness, harmony, intimacy, rapport, relationship. ▷ FRIENDLINESS, LOVE. *Opp* HOSTILITY.

fright *n* **1** jolt, scare, shock, surprise. **2** alarm, apprehension, consternation, dismay, dread, fear, horror, panic, terror, trepidation.

frighten *vb* agitate, alarm, appal, browbeat, bully, cow, *inf* curdle your blood, daunt, dismay, distress, harrow, horrify, intimidate, make afraid, *inf* make your blood run cold, make your hair stand on end, menace, panic, persecute, *inf* petrify, *inf* put the wind up, scare, *inf* scare stiff, shake, shock, startle, terrify, terrorize, threaten, traumatize, tyrannize, unnerve, upset. ▷ DISCOURAGE. *Opp* REASSURE.

frightened *adj* afraid, aghast, alarmed, anxious, appalled, apprehensive, *inf* chicken, cowardly, craven, daunted, fearful, harrowed, horrified, horror-struck, panicky, panic-stricken, petrified, scared, shocked, terrified, terror-stricken, trembling, unnerved, upset, *inf* windy.

frightening *adj* alarming, appalling, blood-curdling, *inf* creepy, daunting, dire, dreadful, eerie, fearful, fearsome, formidable, ghostly, grim, hair-raising, horrifying, intimidating, petrifying, scary, sinister, spine-chilling, *inf* spooky, terrifying, traumatic, uncanny, unnerving, upsetting, weird, worrying. ▷ FRIGHTFUL.

frightful *adj* 1 awful, ghastly, grisly, gruesome, harrowing, hideous, horrible, horrid, horrific, macabre, shocking, terrible. ▷ FRIGHTENING. 2 ▷ BAD.

fringe *n* 1 borders, boundary, edge, limits, marches, margin, outskirts, perimeter, periphery. 2 border, edging, flounce, frill, gathering, ruffle, trimming, valance.

frisky *adj* active, animated, coltish, frolicsome, high-spirited, jaunty, lively, perky, playful, skittish, spirited, sprightly.

frivolity *n* childishness, facetiousness, flippancy, levity, light-heartedness, nonsense, playing about, silliness, triviality. ▷ FUN.

frivolous *adj* casual, childish, facetious, flighty, *inf* flip, flippant, foolish, inconsequential, insignificant, irresponsible, jocular, joking, minor, nugatory, paltry, petty, pointless, puerile, ridiculous, shallow, silly, stupid, superficial, trifling, trivial, trumpery, unimportant, unserious, vacuous, worthless. *Opp* SERIOUS.

frock *n* dress, gown, robe.

frolic *vb* caper, cavort, curvet, dance, frisk about, gambol, have fun, hop about, *inf* horse about, jump about, lark around, leap about, *inf* make whoopee, play about, prance, revel, rollick, romp, skip, skylark, sport.

front *adj* facing, first, foremost, leading, most advanced. • *n* 1 anterior, bow (*of ship*), façade, face, facing, forefront, foreground, frontage, head, nose, obverse, van, vanguard. 2 battle area, danger zone, front line. 3 *a brave front.* appearance, aspect, bearing, blind, *inf* cover-up, demeanour, disguise, expression, look, mask, pretence, show. *Opp* BACK.

frontal *adj* direct, facing, headon, oncoming, straight.

frontier *n* border, borderline, boundary, bounds, limit, marches, pale.

froth *n* bubbles, effervescence, foam, head (*on beer*), lather, scum, spume, suds.

frown *vb inf* give a dirty look, glare, glower, grimace, knit your brows, look sullen, lour, lower, scowl. **frown on** ▷ DISAPPROVE.

fruit *n* □ *apple, apricot, avocado, banana, berry, bilberry, blackberry, cherry, coconut, crabapple, cranberry, currant, damson, date, fig, gooseberry, grape, grapefruit, greengage, guava, hip, kiwi fruit, lemon, lichee, lime, litchi, loganberry, lychee, mango, medlar, melon, mulberry, nectarine, olive, orange, papaw, pawpaw, peach, pear, pineapple, plum, pomegranate, prune, quince, raisin, raspberry, satsuma, sloe, strawberry, sultana, tangerine, tomato, ugli.*

fruitful *adj* 1 abundant, bounteous, bountiful, copious, fecund, fertile, flourishing, lush, luxurious, plenteous, productive, profuse, prolific, rich. 2 advantageous, beneficial, effective, gainful, profitable, rewarding, successful, useful, well-spent, worthwhile. *Opp* FRUITLESS.

fruitless *adj* 1 barren, sterile, unfruitful, unproductive. 2 abortive, bootless, disappointing, futile, ineffective, ineffectual, pointless, profitless, unavailing, unprofitable, unrewarding, unsuccessful, useless, vain. *Opp* FRUITFUL.

frustrate *vb* baffle, balk, baulk, block, check, disappoint, discourage, foil, halt, hamstring, hinder, impede, inhibit, nullify, prevent, *inf* scotch, stop, stymie, thwart. ▷ DEFEAT. *Opp* ENCOURAGE.

frustrated *adj* disappointed, embittered, loveless, lovesick, resentful, thwarted, unfulfilled, unsatisfied.

fuel *n* □ *anthracite, butane, charcoal, coal, coke, derv, diesel, electricity, gas, gasoline, kindling, logs, methylated spirit, nuclear fuel, oil, paraffin, peat, petrol, propane, tinder, wood.* • *vb* encourage, feed, inflame, keep going, nourish, put fuel on, stoke up, supply with fuel.

fugitive *adj* ▷ TRANSIENT. • *n* deserter, escapee, escaper, refugee, renegade, runaway.

fulfil *vb* **1** accomplish, achieve, bring about, bring off, carry off, carry out, complete, consummate, discharge, do, effect, effectuate, execute, implement, make come true, perform, realize. **2** *fulfil a need.* answer, comply with, conform to, meet, obey, respond to, satisfy.

full *adj* **1** brimming, bursting, *inf* chock-a-block, *inf* chock-full, congested, crammed, crowded, filled, jammed, *inf* jam-packed, loaded, overflowing, packed, replete, solid, stuffed, topped-up, well-filled, well-stocked, well-supplied. **2** *a full stomach.* gorged, sated, satiated, satisfied, well-fed. **3** *the full story.* complete, comprehensive, detailed, entire, exhaustive, plenary, thorough, total, unabridged, uncensored, uncut, unedited, unexpurgated, whole. **4** *full speed.* extreme, greatest, highest, maximum, top, utmost. **5** *a full figure.* ample, broad, buxom, fat, large, plump, rounded, voluptuous, well-built. **6** *a full skirt.* baggy, generous, voluminous, wide. *Opp* EMPTY, INCOMPLETE, SMALL.

full-grown *adj* adult, grown-up, mature, ready, ripe.

fumble *vb* grope at, feel, handle awkwardly, mishandle, stumble, touch clumsily.

fume *vb* emit fumes, smoke, smoulder. **fuming** ▷ ANGRY.

fumes *plur n* exhaust, fog, gases, pollution, smog, smoke, vapour.

fun *n* amusement, clowning, diversion, enjoyment, entertainment, festivity, *inf* fooling around, frolic, *inf* fun-and-games, gaiety, games, *inf* high jinks, high spirits, horseplay, jocularity, jokes, joking, *joc* jollification, jollity, laughter, merriment, merrymaking, mirth, pastimes, play, playfulness, pleasure, pranks, recreation, romp, *inf* skylarking, sport, teasing, tomfoolery. ▷ FRIVOLITY. **make fun of** ▷ MOCK.

function *n* **1** aim, purpose, *Fr* raison d'être, use. ▷ JOB. **2** *an official function.* affair, ceremony, *inf* do, event, occasion, party, reception. • *vb* act, behave, go, operate, perform, run, work.

functional *adj* functioning, practical, serviceable, useful, utilitarian, working. *Opp* DECORATIVE.

fund *n* cache, hoard, *inf* kitty, mine, pool, reserve, reservoir, stock, store, supply, treasure-house. **funds** capital, endowments, investments, reserves, resources, riches, savings, wealth. ▷ MONEY.

fundamental *adj* axiomatic, basic, cardinal, central, crucial, elementary, essential, important, key, main, necessary, primary, prime, principal, quintessential, rudimentary, underlying. *Opp* INESSENTIAL.

funeral *n* burial, cremation, entombment, exequies, interment, obsequies, Requiem Mass, wake.

funereal *adj* dark, depressing, dismal, gloomy, grave, mournful, sepulchral, solemn, sombre. ▷ SAD. *Opp* CHEERFUL.

funnel *n* chimney, smoke-stack. • *vb* channel, direct, filter, pour.

funny *adj* **1** absurd, amusing, comic, comical, crazy, *inf* daft, diverting, droll, eccentric, entertaining, facetious, farcical, foolish, grotesque, hilarious, humorous, *inf* hysterical, ironic, jocose, jocular, *inf* killing, laughable, ludicrous, mad, merry, nonsensical, preposterous, *inf* priceless, *inf* rich, ridiculous, risible, sarcastic, sardonic, satirical, *inf* side-splitting, silly, slapstick, uproarious, waggish, witty, zany. *Opp* SERIOUS. **2** ▷ PECULIAR.

fur *n* bristles, coat, down, fleece, hair, hide, pelt, skin, wool.

furious *adj* **1** boiling, enraged, fuming, incensed, infuriated, irate, livid, mad, raging, savage, wrathful. ▷ ANGRY. **2** *furious activity.* agitated, fierce, frantic, frenzied, intense, tempestuous, tumultuous, turbulent, violent, wild. *Opp* CALM.

furnish *vb* **1** decorate, equip, fit out, fit up, *inf* kit out. **2** *furnish information.* afford, give, grant, provide, supply.

furniture *n* antiques, chattels, effects, equipment, fitments, fittings, fixtures, furnishings, household goods, *inf* movables, possessions. □ *armchair, bed, bench, bookcase, bunk, bureau, cabinet, chair, chesterfield, chest of drawers, chiffonier, commode, cot, couch, cradle, cupboard, cushion, desk, divan, drawer, dresser, dressing-table, easel, fender, filing-cabinet, fireplace, mantelpiece, ottoman, overmantel, pelmet, pew, pouffe, rocking-chair, seat, settee, sideboard, sofa, stool, suite, table, trestle-table, wardrobe, workbench.*

furrow *n* channel, corrugation, crease, cut, ditch, drill, fissure, fluting, gash, groove, hollow, line, rut, score, scratch, track, trench, wrinkle.

furrowed *adj* **1** creased, crinkled, corrugated, fluted, grooved, ploughed, ribbed, ridged, rutted, scored. **2** *furrowed brow.* frowning, lined, worried, wrinkled. *Opp* SMOOTH.

furry *adj* bristly, downy, feathery, fleecy, fuzzy, hairy, woolly.

further *adj* accessory, additional, another, auxiliary, extra, fresh, more, new, other, spare, supplementary.

furthermore *adv* additionally, also, besides, moreover, too.

furtive *adj* clandestine, concealed, conspiratorial, covert, deceitful, disguised, hidden, mysterious, private, secret, secretive, shifty, sly, *inf* sneaky, stealthy, surreptitious, underhand, untrustworthy. ▷ CRAFTY. *Opp* BLATANT.

fury *n* ferocity, fierceness, force, intensity, madness, power, rage, savagery, tempestuousness, turbulence, vehemence, violence, wrath. ▷ ANGER.

fuse *vb* amalgamate, blend, coalesce, combine, commingle, compound, consolidate, join, meld, melt, merge, mix, solder, unite, weld.

fusillade *n* barrage, burst, firing, outburst, salvo, volley.

fuss *n* ▷ COMMOTION. • *vb* agitate, bother, complain, *inf* create, fidget, *inf* flap, *inf* get worked up, grumble, make a commotion, worry.

fussy *adj* **1** carping, choosy, difficult, discriminating, *inf* faddy, fastidious, *inf* finicky, hard to please, niggling, *inf* nit-picking, particular, *inf* pernickety, scrupulous, squeamish. **2** *fussy decorations.* Byzantine, complicated, detailed, elaborate, fancy, ornate, overdone, rococo.

futile *adj* abortive, absurd, barren, bootless, empty, foolish, forlorn, fruitless, hollow, impotent, ineffective, ineffectual, pointless, profitless, silly, sterile, unavailing, unproductive, unprofitable, unsuccessful, useless, vain, wasted, worthless. *Opp* FRUITFUL.

future *adj* approaching, awaited, coming, destined, expected, forthcoming, impending, intended, planned, prospective, subsequent, unborn. • *n* expectations, outlook, prospects, time to come, tomorrow. *Opp* PAST.

fuzz *n* down, floss, fluff, hair.

fuzzy *adj* **1** downy, feathery, fleecy, fluffy, frizzy, furry, linty, woolly. **2** bleary, blurred, cloudy, dim, faint, hazy, ill-defined, indistinct, misty, obscure, out of focus, shadowy, unclear, unfocused, vague. *Opp* CLEAR.

G

gadget *n* apparatus, appliance, contraption, contrivance, device, implement, instrument, invention, machine, tool, utensil.

gag *n* ▷ JOKE. • *vb* check, curb, keep quiet, muffle, muzzle, prevent from speaking, quiet, silence, stifle, still, suppress.

gaiety *n* brightness, cheerfulness, colourfulness, delight, exhilaration, felicity, glee, happiness, high spirits, hilarity, jollity, joyfulness, joyousness, light-heartedness, liveliness, merriment, merrymaking, mirth.

gain *n* achievement, acquisition, advantage, asset, attainment, benefit, dividend, earnings, income, increase, proceeds, profit, return, revenue, winnings, yield. *Opp* LOSS. • *vb* **1** acquire, bring in, capture, collect, earn, garner, gather in, get, harvest, make, net, obtain, pick up, procure, profit, realize, reap, receive, win. *Opp* LOSE. **2** *gain your objective.* achieve, arrive at, attain, get to, reach, secure. *Opp* MISS. **gain on** approach, catch up with, close the gap, close with, go faster than, leave behind, overhaul, overtake.

gainful *adj* advantageous, beneficial, fruitful, lucrative, paid, productive, profitable, remunerative, rewarding, useful, worthwhile.

gala *n* carnival, celebration, fair, festival, festivity, fête, *inf* jamboree, party.

gale *n* blast, cyclone, hurricane, outburst, storm, tempest, tornado, typhoon, wind.

gallant *adj* attentive, chivalrous, courageous, courteous, courtly, dashing, fearless, gentlemanly, gracious, heroic, honourable, intrepid, magnanimous, noble, polite, valiant, well-bred. ▷ BRAVE. *Opp* VILLAINOUS.

gallows *n* gibbet, scaffold.

gamble *vb* back, bet, chance, draw lots, game, *inf* have a flutter, hazard, lay bets, risk money, speculate, stake money, *inf* take a chance, take risks, *inf* try your luck, venture, wager.

game *adj* ▷ BRAVE, WILLING. • *n* **1** amusement, diversion, entertainment, frolic, fun, jest, joke, *inf* lark, *inf* messing about, pastime, play, playing, recreation, romp, sport. **2** competition, contest, match, round, tournament. ▷ SPORT. **3** animals, game-birds, prey, quarry. **give the game away** ▷ REVEAL.

gang *n* band, crew, crowd, mob, pack, ring, team. ▷ GROUP. **gang together, gang up** ▷ COMBINE.

gangster *n* bandit, brigand, criminal, *inf* crook, desperado, gunman, hoodlum, hooligan, mafioso, mugger, racketeer, robber, ruffian, thug, tough.

gaol *n* borstal, cell, custody, dungeon, guardhouse, jail, *Amer* penitentiary, prison. • *vb* confine, detain, imprison, incarcerate, intern, *inf* send down, send to prison, *inf* shut away, shut up.

gaoler *n* guard, jailer, prison officer, *sl* screw, warder.

gap *n* **1** aperture, breach, break, cavity, chink, cleft, crack, cranny, crevice, gulf, hole, opening, rent, rift, rip, space, void. **2** breathing-space, discontinuity, hiatus, interlude, intermission, interruption, interval, lacuna, lapse, lull, pause, recess, respite, rest, suspension, wait. **3** *gap between political parties.* difference, disagreement, discrepancy, disparity, distance, divergence, division, incompatibility, inconsistency.

gape *vb* **1** open, part, split, yawn. **2** *inf* gawp, gaze, *inf* goggle, stare.

garbage *n* debris, detritus, junk, litter, muck, refuse, scrap, trash, waste. ▷ RUBBISH.

garble *vb* corrupt, distort, falsify, misconstrue, misquote, misrepresent, mutilate, pervert, slant, twist, warp. ▷ CONFUSE.

garden *n* allotment, patch, plot, yard. □ *arbour, bed, border, herbaceous border, lawn, orchard, patio, pergola, rock garden, rose garden, shrubbery, terrace, vegetable garden, walled garden, water garden, window-box.* **gardens** grounds, park.

gardening *n* cultivation, horticulture.

garish *adj* bright, Brummagem, cheap, crude, flamboyant, flashy, gaudy, harsh, loud, lurid, meretricious, ostentatious, raffish, showy, startling, tasteless, tawdry, vivid, vulgar. *Opp* DRAB, TASTEFUL.

garment *n* apparel, attire, clothing, costume, dress, garb, habit, outfit. ▷ CLOTHES.

garrison *n* **1** contingent, detachment, force, unit. **2** barracks, camp, citadel, fort, fortification, fortress, station, stronghold.

gas *n* exhalation, exhaust, fumes, miasma, vapour.

gash *vb* chop, cleave, cut, incise, lacerate, score, slash, slit, split, wound.

gasp *vb* blow, breathe with difficulty, choke, fight for breath, gulp, *inf* huff and puff, pant, puff, snort, wheeze. **gasping** ▷ BREATHLESS, THIRSTY.

gate *n* access, barrier, door, entrance, entry, exit, gateway, kissing-gate, opening, passage, *poet* portal, portcullis, turnstile, way in, way out, wicket, wicket-gate.

gather *vb* **1** accumulate, amass, assemble, bring together, build up, cluster, collect, come together, concentrate, congregate, convene, crowd, flock, forgather, get together, group, grow, heap up, herd, hoard, huddle together, marshal, mass, meet, mobilize, muster, rally, round up, pick up, pile up, stockpile, store up, swarm, throng. *Opp* DISPERSE. **2** *gather flowers.* cull, garner, glean, harvest, pick, pluck, reap. **3** *I gather he's ill.* assume, be led to believe, conclude, deduce, guess, infer, learn, surmise, understand.

gathering *n* assembly, conclave, congress, convention, convocation, function, *inf* get-together, meeting, party, rally, social. ▷ GROUP.

gaudy *adj* bright, Brummagem, cheap, crude, flamboyant, flashy, garish, harsh, loud, lurid, meretricious, ostentatious, raffish, showy, startling, tasteless, tawdry, vivid, vulgar. *Opp* DRAB, TASTEFUL.

gauge *n* **1** bench-mark, criterion, guide-line, measurement, norm, standard, test, yardstick. **2** capacity, dimensions, extent, measure, size, span, thickness, width. • *vb* ▷ ESTIMATE, MEASURE.

gaunt *adj* **1** bony, cadaverous, emaciated, haggard, hollow-eyed, lanky, lean, pinched, raw-boned, scraggy, scrawny, skeletal, starving, underweight, wasted away. ▷ THIN. *Opp* PLUMP. **2** *a gaunt ruin.* bare, bleak, desolate, dreary, forbidding, grim, stark, stern, unfriendly. *Opp* ATTRACTIVE.

gawky *adj* awkward, blundering, clumsy, gangling, gauche, gawky, inept, lumbering, maladroit, uncoordinated, ungainly, ungraceful, unskilful. *Opp* GRACEFUL.

gay *adj* **1** animated, bright, carefree, cheerful, colourful, festive, fun-loving, jolly, jovial, joyful, light-hearted, lively, merry, sparkling, sunny, vivacious. ▷ HAPPY. **2** ▷ HOMOSEXUAL.

gaze *vb* contemplate, gape, look, regard, stare, view, wonder (at).

gear *n* accessories, accoutrements, apparatus, appliances, baggage, belongings, equipment, *inf* get-up, harness, implements, instruments, kit, luggage, materials, paraphernalia, rig, stuff, tackle, things, tools, trappings. ▷ CLOTHES.

gem *n* gemstone, jewel, precious stone, *sl* sparkler.

general *adj* **1** accepted, accustomed, collective, common, communal, conventional, customary, everyday, familiar, habitual, normal, ordinary, popular, prevailing, prevalent, public, regular, *inf* run-of-the-mill, shared, typical, usual. **2** *general discussion.* across-the-board, all-embracing, blanket, broad-based, catholic, comprehensive, diversified, encyclopaedic, extensive, far-ranging, far-reaching, global, heterogeneous, hybrid, inclusive, sweeping, universal, wholesale, wide-ranging, widespread, worldwide. **3** *a general idea.* approximate, broad, ill-defined, imprecise, indefinite, inexact, in outline, loose, simplified, superficial, unclear, undefined, unspecific, vague. *Opp* SPECIFIC.

generally *adv* as a rule, broadly, chiefly, commonly, in the main, mainly, mostly, normally, on the whole, predominantly, principally, usually.

generate *vb* beget, breed, bring about, cause, create, engender, father, give rise to, make, originate, procreate, produce, propagate, sire, spawn, *inf* whip up.

generosity *n* bounty, largesse, liberality, munificence, philanthropy.

generous *adj* **1** benevolent, big-hearted, bounteous, bountiful, charitable, disinterested, forgiving, *inf* free, impartial, kind, liberal, magnanimous, munificent, noble, open, open-handed, philanthropic, public-spirited, unmercenary, unprejudiced, unselfish, unsparing, unstinting. **2** *generous gifts.* handsome, princely, undeserved, unearned, valuable. ▷ EXPENSIVE. **3** *generous portions.* abundant, ample, copious, lavish, plentiful, sizeable, substantial. ▷ BIG. *Opp* MEAN, SELFISH.

genial *adj* affable, agreeable, amiable, cheerful, convivial, cordial, easygoing, good-natured, happy, jolly, jovial, kindly, pleasant, relaxed, sociable, sunny, warm, warmhearted. ▷ FRIENDLY. *Opp* UNFRIENDLY.

genitals *n* genitalia, *inf* private parts, pudenda, sex organs.

genius *n* **1** ability, aptitude, bent, brains, brilliance, capability, flair, gift, intellect, intelligence, knack, talent, wit. **2** academic, *inf* egghead, expert, intellectual, *derog* know-all, mastermind, thinker, virtuoso.

genteel *adj derog* affected, chivalrous, courtly, gentlemanly, ladylike, mannered, overpolite, patrician, *inf* posh, refined, stylish, *inf* upper-crust. ▷ POLITE.

gentle *adj* **1** amiable, biddable, compassionate, docile, easygoing, good-tempered, harmless, humane, kind, kindly, lenient, loving, meek, merciful, mild, moderate, obedient, pacific, passive, peace-loving, pleasant, quiet, soft-hearted, sweet-tempered, sympathetic, tame, tender. **2** *gentle music.* low, muted, peaceful, reassuring, relaxing, soft, soothing. **3** *gentle wind.* balmy, delicate, faint, light, soft, warm. **4** *a gentle hint.* indirect, polite, subtle, tactful. **5** *a gentle hill.* easy, gradual, imperceptible, moderate, slight, steady. *Opp* HARSH, SEVERE.

genuine *adj* **1** actual, authentic, authenticated, *Lat* bona fide, legitimate, original, proper, *sl* pukka, real, sterling, veritable. **2** *genuine feelings.* candid, devout, earnest, frank, heartfelt, honest, sincere, true, unaffected, unfeigned. *Opp* FALSE.

germ *n* **1** basis, beginning, embryo, genesis, cause, nucleus, origin, root, seed, source, start. **2** bacterium, *inf* bug, microbe, micro-organism, virus.

germinate *vb* begin to grow, bud, develop, grow, root, shoot, spring up, sprout, start growing, take root.

gesture *n* action, flourish, gesticulation, indication, motion, movement, sign, signal. • *vb* gesticulate, indicate, motion, sign, signal. □ *beckon, bow, nod, point, salute, shake your head, shrug, smile, wave, wink.*

get *vb* **1** acquire, be given, bring, buy, come by, come in possession of, earn, fetch, gain, get hold of, inherit, *inf* land, *inf* lay hands on, obtain, pick up, procure, purchase, receive, retrieve, secure, take, win. **2** *get her by phone.* contact, get in touch with, reach, speak to. **3** *get a cold.* catch, come down with, contract, develop, fall ill with, suffer from. **4** *get a criminal.* apprehend, arrest, capture, catch, *inf* collar, *inf* nab, *sl* pinch, seize. **5** *get him to help.* cajole, cause, induce, influence, persuade prevail on, *inf* twist someone's arm, wheedle. **6** *get tea*. cook, make ready, prepare. **7** *get what he means*. absorb, appreciate, apprehend, comprehend, fathom, follow, glean, grasp, know, take in, understand, work out. **8** *get what he says*. catch, distinguish, hear, make out. **9** *get somewhere*. arrive, come, go, journey, reach, travel. **10** *get cold*. become, grow, turn. **get across** ▷ COMMUNICATE. **get ahead** ▷ PROSPER. **get at** ▷ CRITICIZE. **get away** ▷ ESCAPE. **get down** ▷ DESCEND. **get in** ▷ ENTER. **get off** ▷ DESCEND. **get on** ▷ PROSPER. **get out** ▷ LEAVE. **get together** ▷ GATHER.

getaway *n* escape, flight, retreat.

ghastly *adj* appalling, awful, deathlike, dreadful, frightening, frightful, grim, grisly, gruesome, hideous, horrible, macabre, nasty, shocking, terrible, upsetting. ▷ UNPLEASANT.

ghost *n* apparition, banshee, *inf* bogey, *Ger* doppelgänger, ghoul, hallucination, illusion, phantasm, phantom, poltergeist, shade, shadow, spectre, spirit, *inf* spook, vision, visitant, wraith. **give up the ghost** ▷ DIE.

ghostly *adj* creepy, disembodied, eerie, frightening, illusory, phantasmal, scary, sinister, spectral, *inf* spooky, supernatural, uncanny, unearthly, weird, wraith-like.

giant *adj* ▷ GIGANTIC. • *n* colossus, Goliath, leviathan, monster, ogre, superhuman, titan, *inf* whopper.

giddiness *n* dizziness, faintness, unsteadiness, vertigo.

giddy *adj* dizzy, faint, light-headed, reeling, silly, spinning, unbalanced, unsteady, vertiginous.

gift *n* **1** benefaction, bonus, bounty, charity, contribution, donation, favour, *inf* give-away, grant, gratuity, *inf* hand-out, honorarium, largesse, offering, present, tip. **2** ability, aptitude, bent, capability, capacity, facility, flair, genius, knack, power, strength, talent.

gifted *adj* able, capable, expert, skilful, skilled, talented. ▷ CLEVER.

gigantic *adj* Brobdingnagian, colossal, elephantine, enormous, gargantuan, giant, herculean, huge, immense, *inf* jumbo, *inf* king-size, mammoth, massive, mighty, monstrous, prodigious, titanic, towering, vast. ▷ BIG. *Opp* SMALL.

giggle *vb* snicker, snigger, titter. ▷ LAUGH.

gimcrack *adj* cheap, *inf* cheap and nasty, flimsy, rubbishy, shoddy, tawdry, trashy, trumpery, useless, worthless.

gimmick *n* device, ploy, ruse, stratagem, stunt, subterfuge, trick.

girder *n* bar, beam, joist, rafter.

girdle *n* band, belt, corset, waistband. • *vb* ▷ SURROUND.

girl *n sl* bird, *old use* damsel, daughter, débutante, girlfriend, fiancé, hoyden, lass, *old use* maid, *old use* maiden, *inf* miss, schoolgirl, tomboy, virgin, *old use or sexist* wench. ▷ WOMAN.

girth *n* circumference, measurement round, perimeter.

gist *n* core, direction, drift, essence, general sense, main idea, meaning, nub, pith, point, quintessence, significance.

give *vb* **1** accord, allocate, allot, allow, apportion, assign, award, bestow, confer, contribute, deal out, *inf* dish out, distribute, *inf* dole out, donate, endow, entrust, *inf* fork out, furnish, give away, give out, grant, hand over, lend, let (someone) have, offer, pass over, pay, present, provide, ration out, render, share out, supply. **2** *give information.* deliver, display, express, impart, issue, notify, publish, put across, put into words, reveal, set out, show, tell, transmit. **3** *give a shout.* emit, let out, utter, voice. **4** *give medicine.* administer, dispense, dose with, impose, inflict, mete out, prescribe. **5** *give a party.* arrange, organize, provide, put on, run, set up. **6** *give trouble.* cause, create, engender, occasion. **7** *give under pressure.* be flexible, bend, buckle, collapse, distort, fail, fall apart, give way, warp, yield. *Opp* RECEIVE, TAKE. **give away** ▷ BETRAY. **give in** ▷ SURRENDER. **give off**, **give out** ▷ EMIT. **give up** ▷ ABANDON, SURRENDER.

glad *adj* **1** content, delighted, gratified, joyful, overjoyed, pleased. ▷ HAPPY. *Opp* GLOOMY. **2** *glad to help.* disposed, eager, inclined, keen, ready, willing. *Opp* RELUCTANT.

glamorize *vb* idealize, romanticize.

glamorous *adj* alluring, appealing, colourful, dazzling, enviable, exciting, exotic, fascinating, glittering, prestigious, romantic, smart, spectacular, wealthy. ▷ BEAUTIFUL.

glamour *n* allure, appeal, attraction, brilliance, charm, excitement, fascination, glitter, high-life, lustre, magic, romance. ▷ BEAUTY.

glance *vb* glimpse, have a quick look, peek, peep, scan, skim, *sl* take a dekko. ▷ LOOK.

glare *vb* **1** frown, *inf* give a nasty look, glower, *inf* look daggers, lour, lower, scowl, stare angrily. **2** blaze, dazzle, flare, reflect, shine. ▷ LIGHT. **glaring** ▷ BRIGHT.

glass *n* **1** crystal, glassware. **2** glazing, pane, plate-glass, window. **3** looking-glass, mirror, reflector. **4** beaker, drinking-glass, goblet, tumbler, wine-glass. **5** optical instrument. □ *binoculars, field-glasses, goggles, magnifying glass, microscope, opera-glasses, telescope, spyglass.* **glasses** *inf* specs, spectacles. □ *bifocals, contact-lenses, eyeglass, lorgnette, monocle, pince-nez, reading glasses, sun-glasses, trifocals.*

glasshouse *n* conservatory, greenhouse, hothouse, orangery, vinery.

glassy *adj* **1** glazed, gleaming, glossy, icy, polished, shining, shiny, smooth, vitreous. **2** *glassy stare.* ▷ EXPRESSIONLESS.

glaze *vb* burnish, enamel, gloss, lacquer, polish, shellac, shine, varnish.

gleam *vb* flash, glimmer, glint, glisten, glow, reflect, shine. ▷ LIGHT. **gleaming** ▷ BRIGHT.

gleeful *adj* cheerful, delighted, ecstatic, exuberant, exultant, gay, jovial, joyful, jubilant, overjoyed, pleased, rapturous, triumphant. ▷ HAPPY. *Opp* SAD.

glib *adj* articulate, facile, fast-talking, fluent, insincere, plausible, quick, ready, shallow, slick, smooth, smooth-tongued, suave, superficial, unctuous. ▷ TALKATIVE. *Opp* INARTICULATE, SINCERE.

glide *vb* coast, drift, float, fly, freewheel, glissade, hang, hover, move smoothly, sail, skate, ski, skid, skim, slide, slip, soar, stream.

glimpse *n* glance, look, peep, sight, *inf* squint, view.
• *vb* discern, distinguish, espy, get a glimpse of, make out, notice,

observe, see briefly, sight, spot, spy.

glisten *vb* flash, gleam, glimmer, glint, glitter, reflect, shine. ▷ LIGHT.

glitter *vb* coruscate, flash, scintillate, spark, sparkle, twinkle. ▷ LIGHT. **glittering** ▷ BRIGHT.

gloat *vb* boast, brag, *inf* crow, exult, glory, rejoice, *inf* rub it in, show off, triumph.

global *adj* broad, far-reaching, international, pandemic, total, universal, wide-ranging, worldwide. *Opp* LOCAL.

globe *n* 1 ball, globule, orb, sphere. 2 earth, planet, world.

gloom *n* blackness, cloudiness, darkness, dimness, dullness, dusk, murk, murkiness, obscurity, semi-darkness, shade, shadow, twilight. ▷ DEPRESSION.

gloomy *adj* 1 cheerless, cloudy, dark, depressing, dim, dingy, dismal, dreary, dull, glum, grim, heavy, joyless, murky, obscure, overcast, shadowy, shady, sombre. 2 *a gloomy mood.* depressed, downhearted, lugubrious, mournful, pessimistic, saturnine. ▷ SAD. *Opp* CHEERFUL.

glorious *adj* 1 celebrated, distinguished, eminent, famed, famous, heroic, illustrious, noble, noted, renowned, triumphant. 2 *glorious weather.* beautiful, bright, brilliant, dazzling, delightful, excellent, fine, gorgeous, grand, impressive, lovely, magnificent, majestic, marvellous, outstanding, pleasurable, resplendent, spectacular, splendid, *inf* super, superb, wonderful. *Opp* ORDINARY.

glory *n* 1 credit, distinction, eminence, fame, honour, *inf* kudos, praise, prestige, renown, repute, reputation, success, triumph. 2 *glory to God.* adoration, exaltation, glorification, gratitude, homage, praise, thanksgiving, veneration, worship. 3 *glory of sunrise.* brightness, brilliance, grandeur, magnificence, majesty, radiance, splendour, wonder. ▷ BEAUTY.

gloss *n* 1 brightness, brilliance, burnish, finish, glaze, gleam, lustre, polish, sheen, shine, varnish. 2 annotation, comment, definition, elucidation, exegesis, explanation, footnote, marginal note, note, paraphrase. • *vb* annotate, comment on, define, elucidate, explain, interpret, paraphrase. **gloss over** ▷ CONCEAL.

glossary *n* dictionary, phrase-book, vocabulary, word-list.

glossy *adj* bright, burnished, glassy, glazed, gleaming, glistening, lustrous, polished, reflective, shiny, silky, sleek, smooth, waxed. *Opp* DULL.

glove *n* gauntlet, mitt, mitten.

glow *n* 1 burning, fieriness, heat, incandescence, luminosity, lustre, phosphorescence, radiation, red-heat, redness. 2 ardour, blush, enthusiasm, fervour, flush, passion, rosiness, warmth. • *vb* blush, flush, gleam, incandesce, light up, phosphoresce, radiate heat, redden, smoulder, warm up. ▷ LIGHT.

glower *vb* frown, glare, lour, lower, scowl, stare angrily.

glowing *adj* 1 aglow, bright, hot, incandescent, lambent, luminous, phosphorescent, radiant, red, red-hot, white-hot. 2 *glowing praise.* complimentary, enthusiastic, fervent, passionate, warm.

glue *n* adhesive, cement, fixative, gum, paste, sealant, size, wallpaper-paste. • *vb* affix, bond, cement, fasten, fix, gum, paste, seal, stick.

glum *adj* cheerless, displeased, gloomy, grim, heavy, joyless, lugubrious, moody, mournful, *inf* out of sorts, saturnine, sullen. ▷ SAD. *Opp* CHEERFUL.

glut *n* abundance, excess, overabundance, overflow, overprovision, plenty, superfluity, surfeit, surplus. *Opp* SCARCITY.

glutton *n joc* good trencherman, gormandizer, gourmand, *inf* greedy-guts, guzzler, *inf* pig.

gluttonous *adj* gormandizing, greedy, *inf* hoggish, *inf* piggish, insatiable, ravenous, voracious.

gnarled *adj* bent, bumpy, contorted, crooked, distorted, knobbly, knotted, knotty, lumpy, rough, rugged, twisted, warped.

gnaw *vb* bite, chew, erode, wear away. ▷ EAT.

go *n* attempt, chance, *inf* crack, opportunity, *inf* shot, *inf* stab, try, turn. • *vb* **1** advance, begin, be off, commence, decamp, depart, disappear, embark, escape, get away, get going, get moving, get out, get under way, leave, make off, move, *inf* nip along, pass along, pass on, proceed, retire, retreat, run, set off, set out, *inf* shove off, start, take off, take your leave, vanish, *old use* wend your way, withdraw. ▷ RUN, TRAVEL, WALK. **2** die, fade, fail, give way. **3** extend, lead, reach, stretch. **4** *car won't go.* act, function, operate, perform, run, work. **5** *bomb went bang.* give off, make, produce, sound. **6** *Time goes slowly.* elapse, lapse, pass. **7** *go sour.* become, grow, turn. **8** *Milk goes in the fridge.* belong, feel at home, have a proper place, live. go away ▷ DEPART. **go down** ▷ DESCEND, SINK. **go in for** ▷ LIKE. **go into** ▷ INVESTIGATE. **go off** ▷ EXPLODE. **go on** ▷ CONTINUE. **go through** ▷ SUFFER. **go to** ▷ VISIT. **go together** ▷ MATCH. **go with** ▷ ACCOMPANY. **go without** ▷ ABSTAIN.

goad *vb* badger, *inf* chivvy, egg on, *inf* hassle, needle, prick, prod, prompt, spur, urge. ▷ STIMULATE.

go-ahead *adj* ambitious, enterprising, forward-looking, progressive, resourceful. • *n* approval, *inf* green light, permission, sanction, *inf* say-so, *inf* thumbs-up.

goal *n* aim, ambition, aspiration, design, end, ideal, intention, object, objective, purpose, target.

gobble *vb* bolt, devour, gulp, guzzle. ▷ EAT.

go-between *n* agent, broker, envoy, intermediary, liaison, mediator, messenger, middleman, negotiator. **act as go-between** ▷ MEDIATE.

god, goddess *ns* deity, divinity, godhead, spirit. **God** the Almighty, the Creator, the supreme being. **the gods** the immortals, the pantheon, the powers above.

godsend *n inf* bit of good luck, blessing, boon, gift, miracle, *inf* stroke of good fortune, windfall.

golden *adj* **1** aureate, gilded, gilt. **2** *golden hair.* blond, blonde, flaxen, yellow.

good *adj* **1** acceptable, admirable, agreeable, appropriate, approved of, commendable, delightful, enjoyable, esteemed, *inf* fabulous, fair, *inf* fantastic, fine, gratifying, happy, *inf* incredible, lovely, marvellous, nice, perfect, *inf* phenomenal, pleasant, pleasing, praiseworthy, proper, remarkable, right, satisfactory, *inf* sensational, sound, splendid, suitable, *inf* super, superb, useful, valid, valuable, wonderful, worthy. ▷ EXCELLENT. **2** *a good person.* angelic, benevolent, caring, charitable, chaste, considerate, decent, dependable, dutiful, ethical, friendly, helpful, holy, honest, honourable, humane, incorruptible, innocent, just, law-abiding, loyal, merciful, moral, noble, obedient, personable, pure, reliable, religious, righteous, saintly, sound, *inf* straight, thoughtful, true, trustworthy, upright, virtuous, well-behaved, well-mannered, worthy. ▷ KIND. **3** *a good worker.* able, accomplished, capable, conscientious, efficient, gifted, proficient, skilful, skilled, talented. ▷ CLEVER. **4** *good work.* careful, competent, correct, creditable, efficient, meritorious, neat, orderly, presentable, professional, thorough, well-done. **5** *good food.* beneficial, delicious, eatable, healthy, nourishing, nutritious, tasty, well-cooked, wholesome. **6** *a good book.* classic, exciting, great, interesting, readable, well-written. *Opp* BAD. **good-humoured** ▷ GOOD-TEMPERED. **good-looking** ▷ HANDSOME. **good-natured** ▷ GOOD-TEMPERED. **good person** *inf* angel, *inf* jewel, philanthropist, *inf* saint, Samaritan, worthy. **goods 1** belongings, chattels, effects, possessions, property. **2** commodities, freight, load, merchandise, produce, stock, wares.

goodbye *n* farewell, departure, leave-taking, parting words, send-off, valediction. □ *adieu, adios, arrivederci, au revoir, auf Wiedersehen, bon voyage, ciao, cheerio, so long.*

good-tempered *adj* accommodating, amenable, amiable, benevolent, benign, cheerful, cheery, considerate, cooperative, cordial, friendly, genial, good-humoured, good-natured, helpful, in a good mood, obliging, patient, pleasant, relaxed, smiling, sympathetic, thoughtful, willing. ▷ KIND. *Opp* BAD-TEMPERED.

gorge *vb* be greedy, fill up, gormandize, guzzle, indulge yourself, *inf* make a pig of yourself, overeat, *inf* stuff yourself. ▷ EAT.

gorgeous *adj* colourful, dazzling, glorious, magnificent, resplendent, showy, splendid, sumptuous. ▷ BEAUTIFUL.

gory *adj* blood-stained, bloody, grisly, gruesome, sanguinary, savage.

gospel *n* creed, doctrine, good news, good tidings, message, religion, revelation, teaching, testament.

gossip *n* **1** casual talk, chatter, *inf* the grapevine, hearsay, prattle, rumour, scandal, small talk, *inf* tattle, *inf* tittle-tattle. **2** *inf* blab, busybody, chatterbox, *inf* Nosey Parker, rumourmonger, scandalmonger, tell-tale. • *vb inf* blab, chat, chatter, *inf* natter, prattle, spread scandal, *inf* tattle, tell tales, *inf* tittle-tattle. ▷ TALK.

gouge *vb* chisel, dig, gash, hollow, incise, scoop. ▷ CUT.

gourmet *n Fr* bon viveur, connoisseur, epicure, gastronome, *derog* gourmand.

govern *vb* **1** administer, be in charge of, command, conduct affairs, control, direct, guide, head, lead, look after, manage, oversee, preside over, reign, rule, run, steer, superintend, supervise. **2** *govern your anger.* bridle, check, control, curb, discipline, keep in check, keep under control, master, regulate, restrain, tame.

government *n* administration, authority, bureaucracy, conduct of state affairs, constitution, control, direction, domination, management, oversight, regime, regulation, rule, sovereignty, supervision, surveillance, sway. □ *commonwealth, democracy, dictatorship, empire, federation, kingdom, monarchy, oligarchy, republic.*

gown *n* dress, frock. ▷ CLOTHES.

grab *vb* appropriate, arrogate, *inf* bag, capture, catch, clutch, *inf* collar, commandeer, expropriate, get hold of, grasp, hold, *inf* nab, pluck, seize, snap up, snatch, usurp.

grace *n* **1** attractiveness, beauty, charm, ease, elegance, fluidity, gracefulness, loveliness, poise, refinement, softness, tastefulness. **2** *God's grace.* beneficence, benevolence, compassion, favour, forgiveness, goodness, graciousness, kindness, love, mercy. **3** *grace before meals.* blessing, prayer, thanksgiving.

graceful *adj* **1** agile, balletic, deft, dignified, easy, elegant, flowing, fluid, natural, nimble, pliant, slender, slim, smooth, supple, willowy. ▷ BEAUTIFUL. **2** *graceful compliments.* courteous, courtly, delicate, kind, polite, refined, suave, tactful, urbane. *Opp* GRACELESS.

graceless *adj* **1** awkward, clumsy, gangling, gawky, inelegant, maladroit, uncoordinated, ungainly. ▷ CLUMSY. *Opp* GRACEFUL. **2** *graceless manners.* boorish, gauche, inept, tactless, uncouth. ▷ RUDE.

gracious *adj* **1** affable, agreeable, civilized, cordial, courteous, dignified, elegant, friendly, good-natured, pleasant, polite, with grace. ▷ KIND. **2** clement, compassionate, forgiving, generous, indulgent, lenient, magnanimous, pitying, sympathetic. ▷ MERCIFUL. **3** *gracious living.* affluent, expensive, lavish, luxurious, opulent, self-indulgent, sumptuous.

grade *n* category, class, condition, degree, echelon, estate, level, mark, notch, point, position, quality, rank, rung, situation, standard, standing, status, step. • *vb* **1** arrange, categorize, classify, differentiate, group, organize, range, size, sort. **2** *grade students' work.* assess, evaluate, mark, rank, rate.

gradient *n* ascent, bank, declivity, hill, incline, rise, slope.

gradual *adj* continuous, easy, even, gentle, leisurely, moderate, regular, slow, steady, unhurried, unspectacular. *Opp* SUDDEN.

graduate *vb* **1** become a graduate, be successful, get a degree, pass, qualify. **2** *graduate a measuring-rod.* calibrate, divide into graded sections, gradate, mark off, mark with a scale.

graft *vb* implant, insert, join, splice.

grain *n* **1** atom, bit, crumb, fleck, fragment, granule, iota, jot, mite, molecule, morsel, mote, particle, scrap, seed, speck, trace. **2** ▷ CEREAL.

grand *adj* **1** aristocratic, august, dignified, eminent, glorious, great, important, imposing, impressive, lordly, magnificent, majestic, noble, opulent, palatial, regal, royal, splendid, stately, sumptuous, superb. ▷ BIG. **2** [*derog*] haughty, *inf* high-and-mighty, lofty, patronizing, pompous, posh, *inf* upper crust. ▷ GRANDIOSE. *Opp* MODEST.

grandiloquent *adj* bombastic, elaborate, florid, flowery, fustian, high-flown, inflated, melodramatic, ornate, poetic, pompous, rhetorical, turgid. ▷ GRANDIOSE. *Opp* SIMPLE.

grandiose *adj* affected, ambitious, exaggerated, extravagant, flamboyant, *inf* flashy, grand, highfalutin, ostentatious, overdone, *inf* over the top, pretentious, showy. ▷ GRANDILOQUENT. *Opp* MODEST.

grant *n* allocation, allowance, annuity, award, benefaction, bursary, concession, contribution, donation, endowment, expenses, gift, honorarium, investment, loan, pension, scholarship, sponsorship, subsidy, subvention. • *vb* **1** allocate, allot, allow, assign, award, bestow, confer, donate, give, pay, provide, supply. **2** *grant that I'm right.* accede, accept, acknowledge, admit, agree, concede, consent, vouchsafe.

graph *n* chart, column-graph, diagram, grid, pie chart, table.

graphic *adj* clear, descriptive, detailed, lifelike, lucid, photographic, plain, realistic, representational, vivid, well-drawn.

grapple *vb* clutch (at), grab, seize, tackle, wrestle. ▷ GRASP, FIGHT. **grapple with** *grapple with a problem.* attend to, come to grips with, contend with, cope with, deal with, engage with, get involved with, handle, *inf* have a go at, manage, try to solve.

grasp *vb* **1** catch, clasp, clutch, get hold of, *inf* get your hands on, grab, grapple with, grip, hang on to, hold, *inf* nab, seize, snatch, take hold of. **2** *grasp an idea.* appreciate, apprehend, comprehend, *inf* cotton on to, follow, *inf* get the drift of, get the hang of, get the point of, learn, master, realize, take in, understand. **grasping** ▷ GREEDY.

grass *n* downland, field, grassland, green, lawn, meadow, pasture, playing-field, prairie, savannah, steppe, *poet* sward, turf, veld. • *vb* ▷ INFORM.

grate *n* fireplace, hearth. • *vb* cut, grind, rasp, shred, triturate. **grate on** ▷ ANNOY. **grating** ▷ ANNOYING, HARSH.

grateful *adj* appreciative, beholden, gratified, indebted, obliged, thankful. *Opp* UNGRATEFUL.

gratify *vb* delight, fulfil, indulge, pander to, please, satisfy.

gratis *adj* complimentary, free, free of charge, gratuitous, without charge.

gratitude *n* appreciation, gratefulness, thankfulness, thanks.

gratuitous *adj* **1** ▷ GRATIS. **2** *gratuitous insults.* baseless, groundless, inappropriate, needless, unasked-for, uncalled-for, undeserved, unjustifiable, unmerited, unnecessary, unprovoked, unsolicited, unwarranted. *Opp* JUSTIFIABLE.

gratuity *n* bonus, *inf* perk, *Fr* pourboire, present, recompense, reward, tip.

grave *adj* **1** acute, critical, crucial, dangerous, important, *inf* life and death, major, momentous, perilous, pressing, serious, severe, significant, terminal (*illness*), threatening, urgent, vital, weighty, worrying. **2** *a grave offence.* criminal, indictable, punishable. **3** *a grave look.* dignified, earnest, grim, long-faced, pensive, sedate, serious, severe, sober, solemn, sombre, subdued, thoughtful, unsmiling. ▷ SAD. *Opp* CHEERFUL, TRIVIAL. • *n* barrow, burial-place, crypt, *inf* last resting-place, mausoleum, sepulchre, tomb, tumulus, vault. ▷ GRAVESTONE.

gravel *n* grit, pebbles, shingle, stones.

gravestone *n* headstone, memorial, monument, tombstone.

graveyard *n* burial-ground, cemetery, churchyard, necropolis.

gravity *n* **1** acuteness, danger, importance, magnitude, momentousness, seriousness, severity, significance, weightiness. **2** *behave with gravity*. ceremony, dignity, earnestness, *Lat* gravitas, pomp, reserve, sedateness, sobriety, solemnity. **3** *force of gravity*. attraction, gravitation, heaviness, ponderousness, pull, weight.

graze *n* abrasion, laceration, raw spot, scrape, scratch. ▷ WOUND.

grease *n* fat, lubrication, oil.

greasy *adj* **1** buttery, fatty, oily, slippery, slithery, smeary, waxy. **2** *greasy manner*. fawning, flattering, fulsome, grovelling, ingratiating, slick, *inf* smarmy, sycophantic, toadying, unctuous.

great *adj* **1** colossal, enormous, extensive, giant, gigantic, grand, huge, immense, large, massive, prodigious, *inf* tremendous, vast. ▷ BIG. **2** *great pain*. acute, considerable, excessive, extreme, intense, marked, pronounced. ▷ SEVERE. **3** *great events*. grand, imposing, large-scale, momentous, serious, significant, spectacular, weighty. ▷ IMPORTANT. **4** *great music*. brilliant, classic, *inf* fabulous, famous, *inf* fantastic, fine, first-rate, outstanding, wonderful. ▷ EXCELLENT. **5** *a great athlete*. able, celebrated, distinguished, eminent, gifted, notable, noted, prominent, renowned, talented, well-known. ▷ FAMOUS. **6** *a great friend*. chief, close, dedicated, devoted, faithful, fast, loyal, main, true, valued. **7** *a great reader*. active, ardent, assiduous, eager, enthusiastic, frequent, habitual, keen, passionate, zealous. **8** ▷ GOOD. *Opp* SMALL, UNIMPORTANT.

greed *n* **1** appetite, craving, gluttony, gormandizing, hunger, insatiability, intemperance, overeating, ravenousness, self-indulgence, voraciousness, voracity. **2** *greed for wealth*. acquisitiveness, avarice, covetousness, cupidity, desire, rapacity, self-interest. ▷ SELFISHNESS.

greedy *adj* **1** famished, gluttonous, gormandizing, *inf* hoggish, hungry, insatiable, intemperate, omnivorous, *inf* piggish, ravenous, self-indulgent, starving, voracious. *Opp* ABSTEMIOUS. **2** *greedy for wealth*. acquisitive, avaricious, avid, covetous, desirous, eager, grasping, materialistic, mean, mercenary, miserly, *inf* money-grubbing, rapacious, selfish. *Opp* UNSELFISH. **be greedy** ▷ GORGE. **greedy person** ▷ GLUTTON.

green *adj* **1** grassy, greenish, leafy, verdant. □ *emerald, grass-green, jade, khaki, lime, olive, pea-green, turquoise.* **2** ▷ IMMATURE.

greenery *n* foliage, leaves, plants, vegetation.

greet *vb* accost, acknowledge, address, give a greeting to, hail, receive, salute, *inf* say hello to, usher in, welcome.

greeting *n* salutation, reception, welcome. **greetings** compliments, congratulations, felicitations, good wishes, regards.

grey *adj* ashen, blackish, colourless, greying, grizzled, grizzly, hoary, leaden, livid, pearly, silver, silvery, slate-grey, smoky, sooty, whitish. ▷ GLOOMY.

grid *n* framework, grating, grille, lattice, network.

grief *n* affliction, anguish, dejection, depression, desolation, despondency, distress, heartache, heartbreak, melancholy, misery, mourning, pain, regret, remorse, sadness, sorrow, suffering, tragedy, unhappiness, woe, wretchedness. ▷ PAIN. *Opp* HAPPINESS. **come to grief** ▷ FAIL.

grievance *n* **1** calamity, damage, hardship, harm, indignity, injury, injustice. **2** allegation, *inf* bone to pick, charge, complaint, *inf* gripe, objection.

grieve *vb* **1** afflict, cause grief, depress, dismay, distress, hurt, pain, sadden, upset, wound. *Opp* PLEASE. **2** be in mourning, feel grief, *inf* eat your heart out, fret, lament, mope, mourn, suffer, wail, weep. *Opp* REJOICE.

grim *adj* alarming, appalling, awful, cruel, dire, dour, dreadful, fearsome, fierce, forbidding, formidable, frightening, frightful, frowning, ghastly, grisly, gruesome, harsh, hideous, horrible, *inf* horrid, inexorable, inflexible, joyless, louring, menacing, merciless, ominous, pitiless, relentless, ruthless, savage, severe, sinister, stark, stern, sullen, surly, terrible, threatening, unattractive, uncompromising, unfriendly, unpleasant, unrelenting, unsmiling, unyielding. ▷ GLOOMY. *Opp* CHEERFUL.

grime *n* dirt, dust, filth, grit, muck, scum, soot.

grind *vb* **1** abrade, comminute, crumble, crush, erode, granulate, grate, mill, pound, powder, pulverize, rasp, triturate. **2** file, polish, sand, sandpaper, scrape, sharpen, smooth, wear away, whet. **3** *grind your teeth*. gnash, grate, grit, rub together. **grind away** ▷ WORK. **grind down** ▷ OPPRESS.

grip *n* clasp, clutch, grasp, handclasp, hold, purchase, stranglehold. ▷ CONTROL. • *vb* **1** clasp, clutch, get a grip of, grab, grasp, hold, seize, take hold of. **2** *grip the imagination*. absorb, compel, engage, engross, enthral, entrance, fascinate, hypnotize, mesmerize, rivet, spellbind. **come to grips with** ▷ TACKLE.

grisly *adj* appalling, awful, bloody, disgusting, dreadful, fearful, frightful, ghastly, ghoulish, gory, grim, gruesome, hair-raising, hideous, horrible, *inf* horrid, horrifying, macabre, nauseating, repellent, repulsive, revolting, sickening, terrible.

gristly *adj* leathery, rubbery, tough, uneatable.

gritty *adj* abrasive, dusty, grainy, granular, gravelly, harsh, rasping, rough, sandy.

groan *vb* **1** cry out, lament, moan, sigh, wail, whimper, whine. **2** ▷ COMPLAIN.

groom *n* **1** ostler, stable-lad, stableman. **2** bridegroom, husband. • *vb* **1** brush, clean, make neat, neaten, preen, smarten up, spruce up, tidy, *inf* titivate. **2** *groom someone for a job*. coach, drill, educate, get ready, prepare, prime, train up, tutor.

groove *n* channel, cut, fluting, furrow, gouge, gutter, hollow, indentation, rut, score, scratch, slot, striation, track.

grope *vb* cast about, feel about, fish, flounder, fumble, search blindly.

gross *adj* **1** bloated, massive, obese, overweight, repellent, repulsive, revolting. ▷ FAT. **2** churlish, coarse, crude, rude, unrefined, unsophisticated, vulgar. **3** *gross injustice.* blatant, flagrant, glaring, manifest, monstrous, obvious, outrageous, shameful. **4** *gross income.* before tax, inclusive, overall, total, whole.

grotesque *adj* absurd, bizarre, curious, deformed, distorted, fantastic, freakish, gnarled, incongruous, ludicrous, macabre, malformed, misshapen, monstrous, outlandish, preposterous, queer, ridiculous, strange, surreal, twisted, ugly, unnatural, weird.

ground *n* **1** clay, dirt, earth, loam, mud, soil. **2** area, land, property, surroundings, terrain. **3** campus, estate, garden, park. **4** *sports ground.* arena, court, field, pitch, playground, playing-field, recreation ground, stadium. **5** *grounds for complaint.* argument, base, basis, case, cause, excuse, evidence, foundation, justification, motive, proof, rationale, reason. • *vb* **1** base, establish, found, set, settle. **2** coach, educate, instruct, prepare, teach, train, tutor. **3** beach, run ashore, shipwreck, strand, wreck.

groundless *adj* baseless, chimerical, false, gratuitous, hypothetical, illusory, imaginary, irrational, motiveless, needless, speculative, suppositional, uncalled for, unfounded, unjustifiable, unjustified, unproven, unreasonable, unsound, unsubstantiated, unsupported, unwarranted.

group *n* **1** [*people*] alliance, assemblage, assembly, association, band, bevy, body, brotherhood, *inf* bunch, cadre, cartel, caste, caucus, circle, clan, class, *derog* clique, club, cohort, colony, committee, community, company, conclave, congregation, consortium, contingent, corps, coterie, coven, crew, crowd, delegation, faction, family, federation, force, fraternity, gang, gathering, group, guild, horde, host, knot, league, meeting, *derog* mob, multitude, number, organization, party, phalanx, picket, platoon, posse, *derog* rabble, ring, sect, *derog* shower, sisterhood, society, squad, squadron, swarm, team, throng, troop, troupe, union, unit. **2** [*things, animals*] accumulation, agglomeration, assemblage, assortment, batch, battery (*guns*), brood (*chicks*), bunch, bundle, category, class, clump, cluster, clutch (*eggs*), collection, combination, conglomeration, constellation, convoy, covey (*birds*), fleet, flock, gaggle (*geese*), galaxy, grouping, heap, herd, hoard, host, litter, mass, pack, pile, pride (*lions*), school, set, shoal (*fish*), species. **3** ▷ MUSICIAN. • *vb* **1** arrange, assemble, assort, bracket together, bring together, categorize, classify, collect, deploy, gather, herd, marshal, order, organize, put together, set out, sort. **2** associate, band, cluster, come together, congregate, crowd, flock, gather, get together, herd, make groups, swarm, team up, throng.

grovel *vb* abase yourself, be humble, cower, *inf* crawl, *inf* creep, cringe, demean yourself, fawn, flatter, ingratiate yourself, *inf* kowtow, *inf* lick someone's boots, prostrate yourself, snivel, *inf* suck up, *inf* toady. **grovelling** ▷ OBSEQUIOUS.

grow *vb* **1** augment, become bigger, broaden, build up, burgeon, come to life, develop, emerge, enlarge, evolve, expand, extend, fill out, flourish, flower, germinate, improve, increase, lengthen, live, make progress, mature, multiply, mushroom, progress, proliferate, prosper, put on growth, ripen, rise, shoot up, spread, spring up, sprout, survive, swell, thicken, thrive. **2** *grow roses*. cultivate, farm, help along, nurture, produce, propagate, raise. **3** *grow older*. become, get, turn.

grown-up *adj* adult, fully-grown, mature, well-developed.

growth *n* **1** accretion, advance, augmentation, broadening, burgeoning, development, enlargement, evolution, expansion, extension, flowering, getting bigger, growing, improvement, increase, maturation, maturing, progress, proliferation, prosperity, spread, success. **2** crop, harvest, plants, produce, vegetation, yield. **3** cancer, cyst, excrescence, lump, swelling, tumour.

grub *n* **1** caterpillar, larva, maggot. **2** ▷ FOOD. • *vb* ▷ DIG.

grudge *n* ▷ RESENTMENT. • *vb* begrudge, covet, envy, resent.

grudging *adj* cautious, envious, guarded, half-hearted, hesitant, jealous, reluctant, resentful, secret, unenthusiastic, ungracious, unkind, unwilling. *Opp* ENTHUSIASTIC.

gruelling *adj* arduous, backbreaking, crippling, demanding, exhausting, fatiguing, laborious, punishing, severe, stiff, strenuous, taxing, tiring, tough, uphill, wearying. ▷ DIFFICULT. *Opp* EASY.

gruesome *adj* appalling, awful, bloody, disgusting, dreadful, fearful, fearsome, frightful, ghastly, ghoulish, gory, grim, grisly, hair-raising, hideous, horrible, *inf* horrid, horrific, horrifying, macabre, repellent, repugnant, revolting, shocking, sickening, terrible.

gruff *adj* **1** guttural, harsh, hoarse, husky, rasping, rough, throaty. **2** ▷ BAD-TEMPERED.

grumble *vb inf* beef, fuss, *inf* gripe, *inf* grouch, grouse, make a fuss, *inf* moan, object, protest, *inf* whinge. ▷ COMPLAIN.

guarantee *n* assurance, bond, oath, obligation, pledge, promise, surety, undertaking, warranty, word of honour. • *vb* **1** assure, certify, give a guarantee, pledge, promise, swear, undertake, vouch, vow. **2** ensure, make sure of, reserve, secure, stake a claim to.

guard *n* bodyguard, *inf* bouncer, custodian, escort, guardian, *sl* heavy, lookout, *sl* minder, patrol, picket, *sl* screw, security-guard, sentinel, sentry, warder, watchman. • *vb* be on guard over, care for, defend, keep safe, keep watch on, look after, mind, oversee, patrol, police, preserve, prevent from escaping, protect, safeguard, secure, shelter, shield, stand guard over, supervise, tend, watch, watch over. **on your guard** ▷ ALERT.

guardian *n* **1** adoptive parent, foster-parent. **2** champion, custodian, defender, keeper, preserver, protector, trustee, warden. ▷ GUARD.

guess *n* assumption, conjecture, estimate, feeling, *sl* guesstimate, guesswork, hunch, hypothesis, intuition, opinion, prediction, *inf* shot in the dark, speculation, supposition, surmise, suspicion, theory. • *vb* assume, conclude, conjecture, divine, estimate, expect, fancy, feel, have a hunch, have a theory, *inf* hazard a guess, hypothesize, imagine, intuit, judge, make a guess, postulate, predict, *inf* reckon, speculate, suppose, surmise, suspect, think likely, work out.

guest *n* **1** caller, company, visitor. **2** *hotel guests.* boarder, customer, lodger, patron, resident, tenant.

guidance *n* advice, briefing, counselling, direction, guidelines, guiding, help, instruction, leadership, management, *inf* spoon-feeding, *inf* taking by the hand, teaching, tips.

guide *n* **1** courier, escort, leader, navigator, pilot. **2** adviser, counsellor, director, guru, mentor. **3** atlas, directory, gazetteer, guidebook, handbook, *Lat* vade mecum. • *vb* **1** conduct, direct, escort, lead, manoeuvre, navigate, pilot, shepherd, show the way, steer, supervise, usher. **2** advise, brief, control, counsel, educate, give guidance to, govern, help along, influence, instruct, regulate, *inf* take by the hand, teach, train, tutor. *Opp* MISLEAD.

guilt *n* **1** blame, blameworthiness, criminality, culpability, fault, guiltiness, liability, responsibility, sinfulness, wickedness, wrongdoing. **2** *a look of guilt.* bad conscience, contriteness, contrition, dishonour, guilty feelings, penitence, regret, remorse, self-accusation, self-reproach, shame, sorrow. *Opp* INNOCENCE.

guiltless *adj* above suspicion, blameless, clear, faultless, free, honourable, immaculate, innocent, in the right, irreproachable, pure, sinless, untarnished, untroubled, virtuous. *Opp* GUILTY.

guilty *adj* **1** at fault, blameable, blameworthy, culpable, in the wrong, liable, reprehensible, responsible. **2** *a guilty look.* apologetic, ashamed, conscience-stricken, contrite, penitent, *inf* red-faced, regretful, remorseful, repentant, rueful, shamefaced, sheepish, sorry. *Opp* GUILTLESS, SHAMELESS.

gullible *adj* credulous, easily taken in, *inf* green, impressionable, inexperienced, innocent, naïve, suggestible, trusting, unsophisticated, unsuspecting, unwary. *Opp* WARY.

gulp *n* mouthful, swallow, *inf* swig. • *vb* **1** bolt down, gobble, swallow, *inf* wolf. ▷ EAT. **2** *inf* knock back, quaff, *inf* swig. ▷ DRINK. **3** *gulp back tears.* check, choke back, stifle, suppress.

gumption *n* cleverness, *inf* common sense, enterprise, initiative, judgement, *inf* nous, resourcefulness, sense, wisdom.

gun *n plur* artillery, firearm. □ *airgun, automatic, blunderbuss, cannon, machine-gun, mortar, musket, pistol, revolver, rifle, shot-gun,* plur *small arms, sub-machine-gun, tommy-gun.* **gun down** ▷ SHOOT.

gunfire *n* cannonade, cross-fire, firing, gunshots, salvo.

gunman *n* assassin, bandit, criminal, desperado, fighter, gangster, killer, murderer, sniper, terrorist.

gurgle *vb* babble, bubble, burble, ripple, purl, splash.

gush *n* burst, cascade, eruption, flood, flow, jet, outpouring, overflow, rush, spout, spurt, squirt, stream, tide, torrent. • *vb* **1** come in a gush, cascade, flood, flow freely, overflow, pour, run, rush, spout, spurt, squirt, stream, well up. **2** be enthusiastic, be sentimental, bubble over, fuss, *inf* go on, prattle on, talk on. **gushing** ▷ EFFUSIVE, SENTIMENTAL.

gusto *n* appetite, delight, enjoyment, enthusiasm, excitement, liveliness, pleasure, relish, satisfaction, spirit, verve, vigour, zest.

gut *vb* 1 clean, disembowel, draw, eviscerate, remove the guts of. 2 *gut a building.* clear, despoil, empty, loot, pillage, plunder, ransack, ravage, remove the contents of, sack, strip.

guts *plur n* 1 alimentary canal, belly, bowels, entrails, *inf* innards, insides, intestines, stomach, viscera. 2 ▷ COURAGE.

gutter *n* channel, conduit, ditch, drain, duct, guttering, sewer, sluice, trench, trough.

gypsy *n* nomad, Romany, traveller, wanderer.

gyrate *vb* circle, pirouette, revolve, rotate, spin, spiral, swivel, turn, twirl, wheel, whirl.

H

habit *n* 1 convention, custom, pattern, policy, practice, routine, rule, usage, *old use* wont. 2 attitude, bent, disposition, inclination, manner, mannerism, penchant, predisposition, proclivity, propensity, quirk, tendency, way. 3 *bad habit.* addiction, compulsion, craving, dependence, fixation, obsession, vice.

habitable *adj* in good repair, inhabitable, liveable, usable. *Opp* UNINHABITABLE.

habitual *adj* 1 accustomed, common, conventional, customary, established, expected, familiar, fixed, frequent, natural, normal, ordinary, predictable, regular, ritual, routine, set, settled, standard, traditional, typical, usual, *old use* wonted. 2 addictive, besetting, chronic, established, ineradicable, ingrained, obsessive, persistent, recurrent. 3 *habitual smokers.* addicted, conditioned, confirmed, dependent, hardened, *inf* hooked, inveterate, persistent.

hack *vb* carve, chop, gash, hew, mangle, mutilate, slash. ▷ CUT.

hackneyed *adj* banal, clichéd, cliché-ridden, commonplace, conventional, *inf* corny, familiar, feeble, obvious, overused, pedestrian, platitudinous, predictable, stale, stereotyped, stock, threadbare, tired, trite, uninspired, unoriginal. *Opp* NEW.

haggard *adj inf* all skin and bone, careworn, drawn, emaciated, exhausted, gaunt, hollow-cheeked, hollow-eyed, pinched, run-down, scraggy, scrawny, shrunken, thin, tired out, ugly, unhealthy, wasted, weary, withered, worn out, *inf* worried to death. *Opp* HEALTHY.

haggle *vb* argue, bargain, barter, discuss terms, negotiate, quibble, wrangle. ▷ QUARREL.

hail *vb* 1 accost, address, call to, greet, signal to. 2 ▷ ACCLAIM.

hair *n* 1 beard, bristles, curls, fleece, fur, hank, locks, mane, *inf* mop, moustache, shock, tresses, whiskers. 2 coiffure, cut, haircut, *inf* hair-do, hairstyle, style. □ *bob, braid, bun, crew-cut, dreadlocks, fringe, Mohican,* inf *perm, permanent wave, pigtail, plait, pony-tail, quiff, ringlets, short back and sides, sideboards, sideburns, tonsure, topknot.* 3 *false hair.* hair-piece, toupee, wig.

hairdresser *n* barber, coiffeur, coiffeuse, hair-stylist.

hairless *adj* bald, bare, clean-shaven, naked, shaved, shaven, smooth. *Opp* HAIRY.

hairy *adj* bearded, bristly, downy, feathery, fleecy, furry, fuzzy, hirsute, long-haired, shaggy, stubbly, woolly. *Opp* HAIRLESS.

half-hearted *adj* apathetic, cool, easily distracted, feeble, indifferent, ineffective, lackadaisical, listless, lukewarm, nonchalant, passive, perfunctory, phlegmatic, uncaring, uncommitted, unconcerned, unenthusiastic, unreliable, wavering, weak, *inf* wishy-washy. *Opp* ENTHUSIASTIC.

hall *n* **1** auditorium, concert-hall, lecture room, theatre. **2** corridor, entrance-hall, foyer, hallway, lobby, passage, passageway, vestibule.

hallowed *adj* blessed, consecrated, dedicated, holy, honoured, revered, reverenced, sacred, sacrosanct, worshipped.

hallucinate *vb* day-dream, dream, fantasize, *inf* have a trip, have hallucinations, *inf* see things, see visions.

hallucination *n* apparition, chimera, day-dream, delusion, dream, fantasy, figment of the imagination, illusion, mirage, vision. ▷ GHOST.

halt *n* break, cessation, close, end, interruption, pause, standstill, stop, stoppage, termination. • *vb* **1** arrest, block, break off, cease, check, curb, end, impede, obstruct, stop, terminate. **2** come to a halt, come to rest, desist, discontinue, draw up, pull up, quit, stop, wait. *Opp* START. **halting** ▷ HESITANT, IRREGULAR.

halve *vb* bisect, cut by half, cut in half, decrease, divide into halves, lessen, reduce by half, share equally, split in two.

hammer *n* mallet, sledge-hammer. • *vb inf* bash, batter, beat, drive, knock, pound, smash, strike. ▷ DEFEAT, HIT.

hamper *vb* baulk, block, curb, curtail, delay, encumber, entangle, fetter, foil, frustrate, handicap, hinder, hold back, hold up, impede, inhibit, interfere with, obstruct, prevent, restrain, restrict, retard, shackle, slow down, thwart, trammel. *Opp* HELP.

hand *n* **1** fist, *sl* mitt, palm, *inf* paw. **2** *hand on a dial.* index, indicator, pointer. **3** [*old use*] *factory hands.* ▷ WORKER. • *vb* convey, deliver, give, offer, pass, present, submit. **at hand** ▷ HANDY. **give a hand** ▷ HELP. **hand down** ▷ BEQUEATH. **hand over** ▷ SURRENDER. **hand round** ▷ DISTRIBUTE. **lend a hand** ▷ HELP. **to hand** ▷ HANDY.

handicap *n* **1** barrier, burden, disadvantage, difficulty, drawback, encumbrance, hindrance, impediment, inconvenience, limitation, *inf* minus, nuisance, obstacle, problem, restraint, restriction, shortcoming, stumbling-block. *Opp* ADVANTAGE. **2** defect, disability, impairment. • *vb* be a handicap to, burden, check, curb, disable, disadvantage, encumber, hamper, hinder, hold back, impede, limit, restrain, restrict, retard, trammel. *Opp* HELP.

handicapped *adj* [*Some synonyms may cause offence*] autistic, bedridden, blind, crippled, deaf, disabled, disadvantaged, dumb, dyslexic, incapacitated, invalid, lame, limbless, maimed, mute, paralysed, paraplegic, retarded, slow, spastic, unsighted. ▷ ILL.

handiwork *n* achievement, creation, doing, invention, production, responsibility, work.

handle *n* grip, haft, handgrip, helve, hilt, knob, stock (*of rifle*). • *vb* **1** caress, feel, finger, fondle, grasp, hold, *inf* maul, pat, *inf* paw, stroke, touch, treat. **2** *handle situations, people*. conduct, contend with, control, cope with, deal with, direct, guide, look after, manage, manipulate, tackle, treat. ▷ ORGANIZE. **3** *car handles well*. manoeuvre, operate, respond, steer, work. **4** *handle goods*. deal in, do trade in, market, sell, stock, touch, traffic in.

handsome *adj* **1** admirable, attractive, beautiful, comely, elegant, fair, fine-looking, good-looking, personable, tasteful. *Opp* UGLY. **2** *handsome gift*. big, bountiful, generous, goodly, gracious, large, liberal, magnanimous, munificent, sizeable, unselfish, valuable. *Opp* MEAN.

handy *adj* **1** convenient, easy to use, helpful, manageable, practical, serviceable, useful, well-designed, worth having. **2** *handy with tools*. adept, capable, clever, competent, practical, proficient, skilful. **3** *keep tools handy*. accessible, at hand, available, close at hand, easy to reach, get-at-able, nearby, reachable, ready, to hand. *Opp* AWKWARD, INACCESSIBLE.

hang *vb* **1** be suspended, dangle, depend, droop, flap, flop, sway, swing, trail down. **2** *hang washing*. attach, drape, fasten, fix, peg up, pin up, stick up, suspend. **3** *hang in the air*. drift, float, hover. **hang about** ▷ DAWDLE. **hang back** ▷ HESITATE. **hanging** ▷ PENDENT. **hangings** ▷ DRAPE. **hang on** ▷ WAIT. **hang on to** ▷ KEEP.

hank *n* coil, length, loop, piece, skein.

hanker *vb* ache, covet, crave, desire, fancy, *inf* have a yen, hunger, itch, long, pine, thirst, want, wish, yearn.

haphazard *adj* accidental, adventitious, arbitrary, casual, chance, chaotic, confusing, disorderly, disorganized, fortuitous, *inf* higgledy-piggledy, *inf* hit-or-miss, illogical, irrational, random, serendipitous, unforeseen, unplanned, unstructured, unsystematic. *Opp* ORDERLY.

happen *vb* arise, befall, *old use* betide, chance, come about, crop up, emerge, follow, materialize, occur, result, take place, *inf* transpire, *inf* turn out. **happen on** ▷ FIND.

happening *n* accident, affair, chance, circumstance, episode, event, incident, occasion, occurrence, phenomenon.

happiness *n* bliss, cheer, cheerfulness, contentment, delight, ecstasy, elation, enjoyment, euphoria, exhilaration, exuberance, felicity, gaiety, gladness, glee, *inf* heaven, high spirits, joy, joyfulness, joyousness, jubilation, light-heartedness, merriment, pleasure, pride, rapture, well-being. *Opp* SADNESS.

happy *adj* **1** beatific, blessed, blissful, *poet* blithe, buoyant, cheerful, cheery, contented, delighted, ecstatic, elated, enraptured, euphoric, exhilarated, exuberant, exultant, felicitous, festive, gay, glad, gleeful, good-humoured, gratified, grinning, halcyon (*days*), *inf* heavenly, high-spirited, idyllic, jocose, jocular, jocund, joking, jolly, jovial, joyful, joyous, jubilant, laughing, light-hearted, lively, merry, *inf* on top of the world, overjoyed, *inf* over the moon, pleased, proud, radiant, rapturous, rejoicing, relaxed, satisfied, smiling, *inf* starry-eyed, sunny, thrilled, triumphant. *Opp* SAD. **2** *a happy accident*. advantageous, appropriate, apt, auspicious, beneficial, convenient, favourable, felicitous, fortuitous, fortunate, lucky, opportune, propitious, timely, welcome, well-timed.

harangue *n* diatribe, exhortation, lecture, *inf* pep talk, tirade. ▷ SPEECH. • *vb* chivvy, encourage, exhort, lecture, pontificate, preach, sermonize. ▷ SPEAK, TALK.

harass *vb* annoy, attack, badger, bait, bother, chivvy, disturb, harry, *inf* hassle, hound, irritate, molest, nag, persecute, pester, *inf* pick on, *inf* plague, torment, trouble, vex, worry.

harassed *adj inf* at the end of your tether, careworn, distraught, distressed, exhausted, frayed, pressured, strained, stressed, tired, weary, worn out.

harbour *n* anchorage, dock, haven, jetty, landing-stage, marina, mooring, pier, port, quay, safe haven, shelter, wharf. • *vb* 1 conceal, give asylum to, give refuge to, give sanctuary to, hide, protect, shelter, shield. 2 *harbour a grudge*. cherish, cling on to, hold on to, keep in mind, maintain, nurse, nurture, retain.

hard *adj* 1 adamantine, compact, compressed, dense, firm, flinty, frozen, hardened, impenetrable, impervious, inflexible, rigid, rocky, solid, solidified, steely, stiff, stony, unbreakable, unyielding. 2 *hard labour*. arduous, backbreaking, exhausting, fatiguing, formidable, gruelling, harsh, heavy, laborious, onerous, rigorous, severe, stiff, strenuous, taxing, tiring, tough, uphill, wearying. 3 *a hard problem*. baffling, complex, complicated, confusing, difficult, enigmatic, insoluble, intricate, involved, knotty, perplexing, puzzling, tangled, *inf* thorny. 4 *a hard heart*. callous, cold, cruel, *inf* hard-boiled, hardhearted, harsh, heartless, hostile, inflexible, intolerant, merciless, obdurate, pitiless, ruthless, severe, stern, strict, unbending, unfeeling, unfriendly, unkind. 5 *a hard blow*. forceful, heavy, powerful, strong, violent. 6 *hard times*. austere, bad, calamitous, disagreeable, distressing, grim, intolerable, painful, unhappy, unpleasant. 7 *a hard worker*. assiduous, conscientious, devoted, indefatigable, industrious, keen, persistent, unflagging, untiring, zealous. *Opp* EASY, SOFT. **hard-headed** ▷ BUSINESSLIKE. **hard-hearted** ▷ CRUEL. **hard up** ▷ POOR. **hard-wearing** ▷ DURABLE.

harden *vb* bake, cake, clot, coagulate, congeal, freeze, gel, jell, ossify, petrify, reinforce, set, solidify, stiffen, strengthen, toughen. *Opp* SOFTEN.

hardly *adv* barely, faintly, only just, rarely, scarcely, seldom, with difficulty.

hardship *n* adversity, affliction, austerity, bad luck, deprivation, destitution, difficulty, distress, misery, misfortune, need, privation, suffering, *inf* trials and tribulations, trouble, unhappiness, want.

hardware *n* equipment, implements, instruments, ironmongery, machinery, tools.

hardy *adj* 1 durable, fit, healthy, hearty, resilient, robust, rugged, strong, sturdy, tough, vigorous. *Opp* TENDER. 2 ▷ BOLD.

harm *n* abuse, damage, detriment, disadvantage, disservice, havoc, hurt, inconvenience, injury, loss, mischief, misfortune, pain, unhappiness, *inf* upset, wrong. ▷ EVIL. • *vb* abuse, be harmful to, damage, hurt, ill-treat, impair, injure, maltreat, misuse, ruin, spoil, wound. *Opp* BENEFIT.

harmful *adj* addictive, bad, baleful, damaging, dangerous, deadly, deleterious, destructive, detrimental, disadvantageous, evil, fatal, hurtful, injurious, lethal, malign, negative, noxious, pernicious, poisonous, prejudicial, ruinous, unfavourable, unhealthy, unpleasant, unwholesome. *Opp* BENEFICIAL, HARMLESS.

harmless *adj* acceptable, benign, gentle, innocent, innocuous, inoffensive, mild, non-addictive, non-toxic, safe, tame, unobjectionable. *Opp* HARMFUL.

harmonious *adj* **1** concordant, consonant, *inf* easy on the ear, euphonious, harmonizing, melodious, musical, sweet-sounding, tonal, tuneful. *Opp* DISCORDANT. **2** *a harmonious meeting*. agreeable, amicable, compatible, congenial, congruous, cooperative, friendly, integrated, like-minded, sympathetic.

harmonize *vb* agree, balance, be in harmony, blend, cooperate, coordinate, correspond, go together, match, suit each other, tally, tone in.

harmony *n* **1** assonance, concord, consonance, euphony, tunefulness. **2** accord, agreement, amity, balance, compatibility, conformity, congruence, cooperation, friendliness, goodwill, like-mindedness, peace, rapport, sympathy, togetherness, understanding. *Opp* DISCORD.

harness *n* equipment, *inf* gear, straps, tackle. • *vb* control, domesticate, exploit, keep under control, make use of, mobilize, tame, use, utilize.

harsh *adj* **1** abrasive, bristly, coarse, hairy, rough, scratchy. **2** *harsh sounds*. cacophonous, croaking, croaky, disagreeable, discordant, dissonant, grating, gravelly, grinding, gruff, guttural, hoarse, husky, irritating, jarring, rasping, raucous, rough, screeching, shrill, squawking, stertorous, strident, unpleasant. **3** *harsh colours,* light. bright, brilliant, dazzling, gaudy, glaring, lurid. **4** *harsh smell*. acrid, bitter, sour, unpleasant. **5** *harsh conditions*. arduous, austere, comfortless, difficult, hard, severe, stressful, tough. **6** *harsh criticism, treatment*. abusive, acerbic, bitter, blunt, brutal, cruel, Draconian, frank, hard-hearted, hurtful, impolite, merciless, outspoken, pitiless, severe, sharp, stern, strict, uncivil, unforgiving, unkind, unrelenting, unsympathetic, untempered. *Opp* GENTLE.

harvest *n* crop, gathering-in, produce, reaping, return, yield. • *vb* bring in, collect, garner, gather, glean, mow, pick, reap, take in.

hash *n* **1** goulash, stew. **2** *inf* botch, confusion, farrago, *inf* hotchpotch, jumble, mess, *inf* mishmash, mixture. **make a hash of** ▷ BUNGLE.

hassle *n* altercation, argument, bother, confusion, difficulty, disagreement, disturbance, fighting, fuss, harassment, inconvenience, making difficulties, nuisance, persecution, problem, struggle, trouble, upset. • *vb* ▷ HARASS, QUARREL.

haste *n* dispatch, hurry, impetuosity, precipitateness, quickness, rashness, recklessness, rush, urgency. ▷ SPEED.

hasty *adj* **1** abrupt, fast, foolhardy, headlong, hot-headed, hurried, ill-considered, immediate, impetuous, impulsive, incautious, instantaneous, *inf* pell-mell, precipitate, quick, rapid, rash, reckless, speedy, sudden, summary (*justice*), swift. **2** *hasty work.* brief, careless, cursory, hurried, ill-considered, perfunctory, rushed, short, slapdash, superficial, thoughtless, unthinking. *Opp* CAREFUL, SLOW.

hat *n* head-dress. □ *Balaclava, bearskin, beret, biretta, boater, bonnet, bowler, busby, cap, coronet, crash-helmet, crown, deerstalker, diadem, fez, fillet, headband, helmet, hood, mitre, skullcap, sombrero, sou'-wester, stetson, sun-hat, tiara, top hat, toque, trilby, turban, wig, wimple, yarmulke.*

hatch *vb* **1** brood, incubate. **2** conceive, concoct, contrive, *inf* cook up, design, devise, *inf* dream up, formulate, invent, plan, plot, scheme, think up.

hate *n* **1** ▷ HATRED. **2** *a pet hate.* abomination, aversion, *Fr* bête noir, dislike, loathing. • *vb* abhor, abominate, be averse to, be hostile to, be revolted by, *inf* can't bear, *inf* can't stand, deplore, despise, detest, dislike, execrate, fear, find intolerable, loathe, object to, recoil from, resent, scorn, shudder at. *Opp* LIKE, LOVE.

hateful *adj* abhorred, abhorrent, abominable, accursed, awful, contemptible, cursed, *inf* damnable, despicable, detestable, disgusting, distasteful, execrable, foul, hated, heinous, horrible, *inf* horrid, loathsome, nasty, nauseating, obnoxious, odious, offensive, repellent, repugnant, repulsive, revolting, vile. ▷ EVIL. *Opp* LOVABLE.

hatred *n* abhorrence, animosity, antagonism, antipathy, aversion, contempt, detestation, dislike, enmity, execration, hate, hostility, ill-will, intolerance, loathing, misanthropy, odium, repugnance, revulsion. *Opp* LOVE.

haughty *adj* arrogant, boastful, bumptious, cavalier, *inf* cocky, conceited, condescending, disdainful, egotistical, *inf* high-and-mighty, *inf* hoity-toity, imperious, lofty, lordly, offhand, patronizing, pompous, presumptuous, pretentious, proud, self-admiring, self-important, smug, snobbish, *inf* snooty, *inf* stuck-up, supercilious, superior, *inf* uppish, vain. *Opp* MODEST.

haul *vb* carry, cart, convey, drag, draw, heave, *inf* lug, move, pull, tow, trail, transport, tug.

haunt *vb* **1** frequent, *inf* hang around, keep returning to, loiter about, patronize, spend time at, visit regularly. **2** *haunt the mind.* beset, linger in, obsess, plague, prey on, torment.

have *vb* **1** be in possession of, keep, maintain, own, possess, use, utilize. **2** *house has six rooms.* comprise, consist of, contain, embody, hold, include, incorporate, involve. **3** *have fun, illness.* be subject to, endure, enjoy, experience, feel, go through, know, live through, put up with, suffer, tolerate, undergo. **4** *have presents.* accept, acquire, be given, gain, get, obtain, procure, receive. **5** *thieves had the lot. inf* get away with, remove, retain, secure, steal, take. **6** *have a snack.* consume, eat, drink, partake of, swallow. **7** *have a party.* arrange, hold, organize, prepare, set up. **8** *have guests.* be host to, cater for, entertain, put up. **have on** ▷ HOAX. **have to** be compelled to, be forced to, have an obligation to, must, need to, ought to, should. **have up** ▷ ARREST.

haven *n* asylum, refuge, retreat, safety, sanctuary, shelter. ▷ HARBOUR.

havoc *n* carnage, chaos, confusion, damage, desolation, destruction, devastation, disorder, disruption, *inf* mayhem, *inf* rack and ruin, ruin, *inf* shambles, upset, waste, wreckage.

hazard *n* chance, danger, jeopardy, peril, risk, threat. • *vb* dare, gamble, jeopardize, risk, stake, take a chance with, venture.

hazardous *adj* chancy, dangerous, *inf* dicey, fraught with danger, parlous, perilous, precarious, risky, *inf* ticklish, *inf* tricky, uncertain, unpredictable, unsafe. *Opp* SAFE.

haze *n* cloud, film, fog, mist, steam, vapour.

hazy *adj* **1** blurred, blurry, clouded, cloudy, dim, faint, foggy, fuzzy, indefinite, milky, misty, obscure, unclear. **2** ▷ VAGUE. *Opp* CLEAR.

head *adj* ▷ CHIEF. • *n* **1** brain, cranium, skull. **2** *head for figures.* ability, brains, capacity, imagination, intelligence, intellect, mind, understanding. **3** *head of a mountain.* apex, crown, highest point, peak, summit, top, vertex. **4** boss, director, employer, leader, manager, ruler. ▷ CHIEF. **5** *head of a school.* headmaster, headmistress, head teacher, principal. **6** *head of a river.* ▷ SOURCE. • *vb* **1** be in charge of, command, control, direct, govern, guide, lead, manage, rule, run, superintend, supervise. **2** *head for home.* aim, go, make, *inf* make a beeline, point, set out, start, steer, turn. head off ▷ DEFLECT. **lose your head** ▷ PANIC. **off your head** ▷ MAD.

heading *n* caption, headline, rubric, title.

headquarters *n* administration, base, depot, head office, *inf* HQ, main office, *inf* nerve-centre.

heal *vb* **1** become healthy, get better, improve, knit, mend, recover, recuperate, unite. **2** cure, make better, minister to, nurse, rejuvenate, remedy, renew, restore, revitalize, tend, treat. **3** *heal differences.* patch up, put right, reconcile, repair, settle.

health *n* **1** condition, constitution, fettle, form, shape, trim. **2** *the picture of health.* fitness, robustness, soundness, strength, vigour, well-being.

healthy *adj* **1** active, blooming, fine, fit, flourishing, good, *inf* hale-and-hearty, hearty, *inf* in fine fettle, in good shape, lively, perky, robust, sound, strong, sturdy, vigorous, well. **2** bracing, health-giving, hygienic, invigorating, salubrious, sanitary, wholesome. *Opp* ILL, UNHEALTHY.

heap *n* accumulation, assemblage, bank, collection, hill, hoard, mass, mound, mountain, pile, stack. • *vb* accumulate, amass, bank up, collect, gather, hoard, mass, pile, stack, stockpile, store. **heaps** ▷ PLENTY.

hear *vb* **1** attend to, catch, *old use* hearken to, heed, listen to, overhear, pay attention to, pick up. **2** *hear evidence.* examine, investigate, judge, try. **3** *hear news.* be told, discover, find out, gather, get, *inf* get wind of, learn, receive.

hearing *n* case, inquest, inquiry, trial.

heart *n* **1** *sl* ticker. **2** centre, core, crux, essence, focus, hub, inside, kernel, marrow, middle, *sl* nitty-gritty, nub, nucleus, pith. **3** affection, compassion, concern, courage, feeling, goodness, humanity, kindness, love, pity, sensitivity, sympathy, tenderness, understanding, warmth.

heartbreaking *adj* bitter, distressing, grievous, heart-rending, pitiful, tragic.

heartbroken *adj* broken-hearted, dejected, desolate, despairing, dispirited, grieved, inconsolable, miserable, *inf* shattered. ▷ SAD.

hearten *vb* boost, cheer up, encourage, strengthen, uplift.

heartless *adj* callous, cold, icy, inhuman, pitiless, ruthless, steely, stony, unconcerned, unemotional, unkind, unsympathetic. ▷ CRUEL.

hearty *adj* **1** enthusiastic, exuberant, friendly, genuine, healthy, heartfelt, lively, positive, robust, sincere, spirited, strong, vigorous, warm. *Opp* HALF-HEARTED. **2** *a hearty dinner*. ▷ BIG.

heat *n* **1** calorific value, fever, fieriness, glow, hotness, incandescence, warmth. **2** closeness, heatwave, high temperature, hot weather, humidity, sultriness, torridity, warmth. **3** *heat of the moment*. anger, ardour, eagerness, enthusiasm, excitement, fervour, feverishness, fury, impetuosity, violence. ▷ PASSION. *Opp* COLD. • *vb* bake, blister, boil, burn, cook, *inf* frizzle, fry, grill, inflame, make hot, melt, reheat, roast, scald, scorch, simmer, sizzle, smoulder, steam, stew, swelter, toast, warm. *Opp* COOL. **heated** ▷ FERVENT, HOT.

heath *n* common land, moor, moorland, open country, waste land, wilderness.

heathen *adj* atheistic, barbaric, godless, idolatrous, infidel, irreligious, pagan, philistine, savage, unenlightened. • *n* atheist, barbarian, heretic, idolater, infidel, pagan, philistine, savage, sceptic, unbeliever.

heave *vb* **1** drag, draw, haul, hoist, lift, lug, move, pull, raise, tow, tug. **2** ▷ THROW. **heave into sight** ▷ APPEAR. **heave up** ▷ VOMIT.

heaven *n* **1** after-life, Elysium, eternal rest, the hereafter, the next world, nirvana, paradise. **2** bliss, contentment, delight, ecstasy, felicity, happiness, joy, perfection, pleasure, rapture, Utopia. *Opp* HELL.

heavenly *adj* angelic, beatific, beautiful, blissful, celestial, delightful, divine, exquisite, glorious, lovely, other-worldly, *inf* out of this world, saintly, spiritual, sublime, unearthly, wonderful.

heavy *adj* **1** bulky, burdensome, compact, concentrated, dense, hefty, immovable, large, leaden, massive, ponderous, unwieldy, weighty. ▷ BIG, FAT. **2** *heavy work*. arduous, demanding, difficult, hard, exhausting, laborious, onerous, strenuous, tough. **3** *heavy rain*. penetrating, pervasive, severe, torrential. **4** *a heavy crop*. abundant, copious, laden, loaded, profuse, thick. **5** *a heavy heart*. burdened, depressed, gloomy, miserable, sorrowful. ▷ SAD. **6** [*inf*] *a heavy lecture*. deep, dull, intellectual, intense, serious, tedious, wearisome. *Opp* LIGHT. **heavy-handed** ▷ CLUMSY. **heavy-hearted** ▷ SAD.

hectic *adj* animated, boisterous, brisk, bustling, busy, chaotic, excited, feverish, frantic, frenetic, frenzied, hurried, hyperactive, lively, mad, overactive, restless, riotous, rumbustious, *inf* rushed off your feet, turbulent, wild. *Opp* LEISURELY.

hedge *n* barrier, fence, hedgerow, screen. • *vb inf* beat about the bush, be evasive, equivocate, *inf* hum and haw, quibble, stall, temporize, waffle. **hedge in** ▷ ENCLOSE.

hedonistic *adj* epicurean, extravagant, intemperate, luxurious, pleasure-loving, self-indulgent, sensual, sybaritic, voluptuous. *Opp* PURITANICAL.

heed *vb* attend to, bear in mind, concern yourself about, consider, follow, keep to, listen to, mark, mind, note, notice, obey, observe, pay attention to, regard, take notice of. *Opp* DISREGARD.

heedful *adj* attentive, careful, concerned, considerate, mindful, observant, sympathetic, taking notice, vigilant, watchful. *Opp* HEEDLESS.

heedless *adj* blind, careless, deaf, inattentive, inconsiderate, neglectful, oblivious, reckless, regardless, thoughtless, uncaring, unconcerned, unmindful, unobservant, unsympathetic. *Opp* HEEDFUL.

heel *vb* careen, incline, lean, list, tilt, tip.

hefty *adj* beefy, brawny, bulky, burly, heavy, heavyweight, hulking, husky, large, massive, mighty, muscular, powerful, robust, rugged, solid, substantial, *inf* strapping, strong, tough. ▷ BIG. *Opp* SLIGHT.

height *n* **1** altitude, elevation, level, tallness, vertical measurement. **2** crag, fell, hill, mound, mountain, peak, prominence, ridge, summit, top. **3** *height of your career.* acme, apogee, climax, crest, culmination, extreme, high point, maximum, peak, pinnacle, zenith.

heighten *vb* add to, amplify, augment, boost, build up, elevate, enhance, improve, increase, intensify, lift up, magnify, make higher, maximize, raise, reinforce, sharpen, strengthen, supplement. *Opp* LOWER, REDUCE.

hell *n* **1** eternal punishment, Hades, infernal regions, lower regions, nether world, *sl* the other place, underworld. **2** ▷ MISERY. *Opp* HEAVEN.

help *n* advice, aid, assistance, avail, backing, benefit, boost, collaboration, contribution, cooperation, encouragement, friendship, guidance, moral support, patronage, relief, remedy, succour, support. *Opp* HINDRANCE. • *vb* **1** abet, advise, aid, aid and abet, assist, back, befriend, be helpful, boost, collaborate, contribute, cooperate, encourage, facilitate, forward, further the interests of, *inf* give a hand, *inf* lend a hand, profit, promote, prop up, *inf* rally round, serve, side with, *derog* spoonfeed, stand by, subsidize, succour, support, take pity on. *Opp* HINDER. **2** *linctus helps a cough.* alleviate, benefit, cure, ease, improve, lessen, make easier, relieve, remedy. **3** *can't help it.* ▷ AVOID, PREVENT.

helper *n* abettor, accessory, accomplice, ally, assistant, associate, collaborator, colleague, confederate, deputy, helpmate, *inf* henchman, partner, *inf* right-hand man, second, supporter, *inf* willing hands.

helpful *adj* **1** accommodating, benevolent, caring, considerate, constructive, cooperative, favourable, friendly, helping, kind, neighbourly, obliging, practical, supportive, sympathetic, thoughtful, willing. **2** *a helpful comment.*

advantageous, beneficial, informative, instructive, profitable, valuable, useful, worthwhile. **3** *a helpful tool.* convenient, easy to use, handy, manageable, practical, serviceable, useful, well designed, worth having. *Opp* UNHELPFUL, USELESS.

helping *adj* ▷ HELPFUL.
• *n* amount, *inf* dollop, plateful, portion, ration, serving, share.

helpless *adj* abandoned, crippled, defenceless, dependent, deserted, destitute, disabled, exposed, feeble, handicapped, impotent, incapable, in difficulties, infirm, lame, marooned, powerless, stranded, unprotected, vulnerable. *Opp* INDEPENDENT.

herald *n* **1** announcer, courier, messenger, town crier. **2** *herald of spring.* forerunner, harbinger, omen, precursor, sign.
• *vb* advertise, announce, indicate, make known, proclaim, promise, publicize. ▷ FORETELL.

herd *n* bunch, flock, mob, pack, swarm, throng. ▷ GROUP.
• *vb* assemble, collect, congregate, drive, gather, group together, round up, shepherd.

hereditary *adj* **1** ancestral, bequeathed, family, handed down, inherited, passed down, passed on, willed. **2** congenital, constitutional, genetic, inborn, inbred, inherent, inheritable, innate, native, natural, transmissible, transmittable.

heresy *n* blasphemy, dissent, idolatry, nonconformity, rebellion, *inf* stepping out of line, unorthodox ideas.

heretic *n* apostate, blasphemer, dissenter, free-thinker, iconoclast, nonconformist, rebel, renegade, unorthodox thinker.
Opp BELIEVER.

heretical *adj* apostate, atheistic, blasphemous, dissenting, free-thinking, heathen, iconoclastic, idolatrous, impious, irreligious, nonconformist, pagan, rebellious, unorthodox. *Opp* ORTHODOX.

heritage *n* birthright, culture, history, inheritance, legacy, past, tradition.

hermit *n* anchoress, anchorite, eremite, monk, recluse, solitary.

hero, heroine *ns* champion, conqueror, daredevil, exemplar, ideal, idol, luminary, protagonist, star, superman, *inf* superstar, superwoman, victor, winner.

heroic *adj* adventurous, audacious, bold, brave, chivalrous, courageous, daring, dauntless, doughty, epic, fearless, gallant, herculean, intrepid, lion-hearted, noble, selfless, staunch, steadfast, stout-hearted, superhuman, unafraid, valiant, valorous.
Opp COWARDLY.

hesitant *adj* cautious, diffident, dithering, faltering, halfhearted, halting, hesitating, indecisive, irresolute, nervous, *inf* shilly-shallying, shy, stammering, stumbling, stuttering, tentative, timid, uncertain, uncommitted, undecided, underconfident, unsure, vacillating, wary, wavering.
Opp DECISIVE, FLUENT.

hesitate *vb* **1** be hesitant, be indecisive, *inf* be in two minds, delay, demur, *inf* dilly-dally, dither, equivocate, falter, halt, hang back, haver, *inf* hum and haw, pause, put it off, *inf* shilly-shally, shrink back, teeter, temporize, think twice, vacillate, wait, waver.
2 stammer, stumble, stutter.

hesitation *n* caution, delay, diffidence, dithering, doubt, indecision, irresolution, nervousness, reluctance, *inf* shilly-shallying, uncertainty, vacillation, wavering.

hidden *adj* **1** camouflaged, concealed, covered, disguised, enclosed, invisible, obscured, out of sight, private, shrouded, *inf* under wraps, undetectable, unnoticeable, unseen, veiled. *Opp* VISIBLE. **2** *hidden meaning.* abstruse, arcane, coded, covert, cryptic, dark, esoteric, implicit, mysterious, mystical, obscure, occult, recondite, secret, unclear. *Opp* OBVIOUS.

hide *n* fur, leather, pelt, skin. • *vb* **1** blot out, bury, camouflage, cloak, conceal, cover, curtain, disguise, eclipse, enclose, mantle, mask, obscure, put away, put out of sight, screen, secrete, shelter, shroud, veil, wrap up. **2** *go into hiding. inf* go to ground, *inf* hole up, keep hidden, *inf* lie low, lurk, shut yourself away, take cover. **3** *hide facts.* censor, *inf* hush up, repress, silence, suppress, withhold.

hideous *adj* appalling, beastly, disgusting, dreadful, frightful, ghastly, grim, grisly, grotesque, gruesome, macabre, nauseous, odious, repellent, repulsive, revolting, shocking, sickening, terrible. ▷ UGLY. *Opp* BEAUTIFUL.

hiding-place *n* den, haven, hide, hideaway, *inf* hide-out, *inf* hidey-hole, lair, refuge, retreat, sanctuary.

hierarchy *n* grading, ladder, *inf* pecking-order, ranking, scale, sequence, series, social order, system.

high *adj* **1** elevated, extending upwards, high-rise, lofty, raised, soaring, tall, towering. **2** aristocratic, chief, distinguished, eminent, exalted, important, leading, powerful, prominent, royal, top, upper. **3** *high prices.* dear, excessive, exorbitant, expensive, extravagant, outrageous, *inf* steep, unreasonable. **4** *high winds.* exceptional, extreme, great, intense, *inf* stiff, stormy, strong. **5** *a high reputation.* favourable, good, noble, respected, virtuous. **6** *high sounds.* acute, high-pitched, penetrating, piercing, sharp, shrill, soprano, squeaky, treble.*Opp* LOW. **high-and-mighty** ▷ ARROGANT. **high-class** ▷ EXCELLENT. **high-handed** ▷ ARROGANT. **high-minded** ▷ MORAL. **high-powered** ▷ POWERFUL. **high-speed** ▷ FAST. **high-spirited** ▷ LIVELY.

highbrow *adj* **1** academic, bookish, brainy, cultured, intellectual, *derog* pretentious, sophisticated. **2** *highbrow books.* classical, cultural, deep, difficult, educational, improving, serious. *Opp* LOWBROW.

highlight *n* best moment, climax, high spot, peak, top point.

hilarious *adj* boisterous, cheerful, cheering, entertaining, jolly, jovial, lively, merry, mirthful, rollicking, side-splitting, uproarious. ▷ FUNNY.

hill *n* **1** elevation, eminence, foothill, height, hillock, hillside, hummock, knoll, mound, mount, mountain, peak, prominence, ridge, summit. □ *brae, down, fell, pike, stack, tor, wold.* **2** acclivity, ascent, declivity, drop, gradient, incline, ramp, rise, slope.

hinder *vb* arrest, bar, be a hindrance to, check, curb, delay, deter, endanger, frustrate, get in the way of, hamper, handicap, hit, hold back, hold up, impede, keep back, limit, obstruct, oppose, prevent, restrain, restrict, retard, sabotage, slow down, slow up, stand in the way of, stop, thwart. *Opp* HELP.

hindrance *n* bar, barrier, burden, check, curb, deterrent, difficulty, disadvantage, *inf* drag, drawback, encumbrance, handicap, hitch, impediment, inconvenience, limitation, obstacle, obstruction, restraint, restriction, snag, stumbling-block. *Opp* HELP.

hinge *n* articulation, joint, pivot. • *vb* depend, hang, rest, revolve, turn.

hint *n* **1** allusion, clue, idea, implication, indication, inkling, innuendo, insinuation, pointer, shadow, sign, suggestion, tip, *inf* tip-off. **2** *a hint of herbs.* dash, taste, tinge, touch, trace, undertone, whiff. • *vb* allude, give a hint, imply, indicate, insinuate, intimate, mention, suggest, tip off.

hire *vb* book, charter, employ, engage, lease, pay for the use of, rent, sign on, take on. **hire out** lease out, let, rent out, take payment for.

hiss *vb* buzz, fizz, purr, rustle, sizzle, whir, whizz.

historic *adj* celebrated, eminent, epoch-making, famed, famous, important, momentous, notable, outstanding, remarkable, renowned, significant, well-known. *Opp* INSIGNIFICANT.

historical *adj* actual, authentic, documented, factual, real, real-life, recorded, true, verifiable. *Opp* FICTITIOUS.

history *n* **1** antiquity, bygone days, heritage, historical events, the old days, the past. **2** annals, biography, chronicles, diaries, narratives, records.

histrionic *adj* actorish, dramatic, theatrical.

hit *n* **1** blow, bull's eye, collision, impact, shot, stroke. **2** success, triumph, *inf* winner. • *vb* **1** bang, bash, baste, batter, beat, belt, biff, birch, box, bludgeon, buffet, bump, butt, cane, cannon into, clap, clip, clobber, clock, clonk, clout, club, collide with, cosh, crack, crash into, cudgel, cuff, dash, deliver a blow, drive, elbow, flagellate, flail, flick, flip, flog, hammer, head, head-butt, impact, jab, jar, jog, kick, knee, knock, lam, lambaste, lash, nudge, pat, poke, pound, prod, pummel, punch, punt, putt, ram, rap, run into, scourge, slam, slap, slog, slosh, slug, smack, smash, smite, sock, spank, stab, strike, stub, swat, swipe, tan, tap, thrash, thump, thwack, wallop, whack, wham, whip. **2** *The slump hit sales.* affect, attack, bring disaster to, check, damage, do harm to, harm, have an effect on, hinder, hurt, make suffer, ruin. **hit back** ▷ RETALIATE. **hit on** ▷ DISCOVER.

hoard *n* accumulation, cache, collection, fund, heap, pile, reserve, stockpile, store, supply, treasure-trove. • *vb* accumulate, amass, assemble, collect, gather, keep, lay in, lay up, mass, pile up, put away put by, save, stockpile, store, treasure. *Opp* SQUANDER, USE.

hoarse *adj* croaking, grating, gravelly, growling, gruff, harsh, husky rasping, raucous, rough, throaty.

hoax *n* cheat, *inf* con, confidence trick, deception, fake, fraud, humbug, imposture, joke, *inf* leg-pull, practical joke, spoof, swindle, trick. • *vb* bluff, cheat, *inf* con, cozen, deceive, defraud, delude, dupe, fool, gull, *inf* have on, hoodwink, lead on, mislead, *inf* pull someone's leg, swindle, *inf* take for a ride, take in, trick. ▷ TEASE.

hoaxer *n inf* con-man, impostor, joker, practical joker, trickster. ▷ CHEAT.

hobble *vb* dodder, falter, limp, shuffle, stagger, stumble, totter. ▷ WALK.

hobby *n* amateur interest, avocation, diversion, interest, pastime, pursuit, recreation, relaxation, sideline.

hoist *n* block-and-tackle, crane, davit, jack, lift, pulley, tackle, winch, windlass. • *vb* elevate, heave, lift, pull up, raise, winch up.

hold *n* **1** clasp, clutch, foothold, grasp, grip, purchase, toehold. **2** *a hold over someone.* ascendancy, authority, control, dominance, influence, leverage, mastery, power, sway. • *vb* **1** bear, carry, catch, clasp, clench, cling to, clutch, cradle, embrace, enfold, grasp, grip, hang on to, have, hug, keep, possess, retain, seize, support, take. **2** *hold a suspect.* arrest, confine, coop up, detain, imprison, keep in custody, restrain. **3** *hold an opinion.* believe in, stick to, subscribe to, swear to. **4** *hold a pose.* continue, keep up, maintain, occupy, preserve, retain, sustain. **5** *hold a party.* celebrate, conduct, convene, have, organize. **6** *jug holds a litre.* contain, enclose, have a capacity of, include. **7** *My offer holds.* be unaltered, carry on, continue, endure, hold out, keep on, last, persist, remain unchanged, stay. **hold back** ▷ RESTRAIN. **hold forth** ▷ SPEAK, TALK. **hold out** ▷ OFFER, PERSIST. **hold over, hold up** ▷ DELAY. **hold-up** ▷ ROBBERY.

hole *n* **1** abyss, burrow, cave, cavern, cavity, chamber, chasm, crater, dent, depression, excavation, fault, fissure, hollow, indentation, niche, pit, pocket, pot-hole, recess, shaft, tunnel. **2** aperture, breach, break, chink, crack, cut, eyelet, fissure, gap, gash, leak, opening, orifice, perforation, puncture, rip, slit, slot, split, tear, vent.

holiday *n* bank holiday, break, day off, furlough, half-term, leave, recess, respite, rest, sabbatical, time off, vacation.

holiness *n* devotion, divinity, faith, godliness, piety, *derog* religiosity, sacredness, saintliness, *derog* sanctimoniousness, sanctity, venerability.

hollow *adj* **1** empty, unfilled, vacant, void. **2** cavernous, concave, deep, depressed, dimpled, indented, recessed, sunken. **3** *a hollow laugh, victory.* cynical, false, futile, insincere, insubstantial, meaningless, pointless, valueless, worthless. • *n* bowl, cave, cavern, cavity, concavity, crater, dent, depression, dimple, dint, dip, dish, excavation, furrow, hole, indentation, pit, trough. ▷ VALLEY. **hollow out** ▷ EXCAVATE.

holocaust *n* **1** conflagration, firestorm, inferno. **2** annihilation, bloodbath, destruction, devastation, extermination, genocide, massacre, pogrom.

holy *adj* **1** blessed, consecrated, dedicated, devoted, divine, hallowed, heavenly, revered, sacred, sacrosanct, venerable. **2** *holy pilgrims.* devout, faithful, God-fearing, godly, immaculate, *derog* pietistic, pious, prayerful, pure, religious, reverent, reverential, righteous, saintly, *derog* sanctimonious, sinless, unsullied. *Opp* IRRELIGIOUS.

home *n* **1** abode, accommodation, base, domicile, dwelling, dwelling-place, habitation, household, lodging, quarters, residence. ▷ HOUSE. **2** birthplace, native land. **3** *derog* institution. □ old use *almshouse, convalescent home,* *hospice, nursing-home,* old use *poorhouse, rest home, retirement home, retreat, shelter*.

homeless *adj* abandoned, destitute, dispossessed, down-and-out, evicted, exiled, forsaken, itinerant, nomadic, outcast, rootless, unhoused, vagrant, wandering. • *plur n* beggars, refugees, tramps, vagabonds, vagrants.

homely *adj* comfortable, congenial, cosy, easygoing, friendly, informal, intimate, modest, natural, relaxed, simple, unaffected, unassuming, unpretentious, unsophisticated. ▷ FAMILIAR. *Opp* FORMAL, SOPHISTICATED.

homogeneous *adj* akin, alike, comparable, compatible, consistent, identical, indistinguishable, matching, similar, uniform, unvarying. *Opp* DIFFERENT.

homosexual *adj inf* camp, gay, lesbian, *derog* queer.

honest *adj* above-board, blunt, candid, conscientious, direct, equitable, fair, forthright, frank, genuine, good, honourable, impartial, incorruptible, just, law-abiding, legal, legitimate, moral, *inf* on the level, open, outspoken, plain, principled, pure, reliable, respectable, scrupulous, sincere, square (*deal*), straight, straightforward, trustworthy, trusty, truthful, unbiased, unequivocal, unprejudiced, upright, veracious, virtuous. *Opp* DISHONEST.

honesty *n* **1** fairness, goodness, honour, integrity, morality, probity, rectitude, reliability, scrupulousness, sense of justice, trustworthiness, truthfulness, uprightness, veracity, virtue. *Opp* DECEIT. **2** bluntness, candour, directness, frankness, outspokenness, plainness, sincerity, straightforwardness.

honorary *adj* nominal, titular, unofficial, unpaid.

honour *n* **1** acclaim, accolade, compliment, credit, esteem, fame, good name, *inf* kudos, regard, renown, reputation, repute, respect, reverence, veneration. **2** distinction, duty, importance, pleasure, privilege. **3** *a sense of honour*. decency, dignity, honesty, integrity, loyalty, morality, nobility, principle, rectitude, righteousness, sincerity, uprightness, virtue. • *vb* acclaim, admire, applaud, celebrate, commemorate, commend, dignify, esteem, give credit to, glorify, pay homage to, pay respects to, pay tribute to, praise, remember, respect, revere, reverence, show respect to, sing the praises of, value, venerate, worship.

honourable *adj* admirable, chivalrous, creditable, decent, estimable, ethical, fair, good, high-minded, irreproachable, just, law-abiding, loyal, moral, noble, principled, proper, reputable, respectable, respected, righteous, sincere, *inf* straight, trustworthy, trusty, upright, venerable, virtuous, worthy. ▷ HONEST. *Opp* DISHONOURABLE.

hoodwink *vb* bluff, cheat, *inf* con, cozen, deceive, defraud, delude, dupe, fool, gull, *inf* have on, hoax, lead on, mislead, *inf* pull the wool over someone's eyes, swindle, *inf* take for a ride, take in, trick.

hook *n* barb, crook, peg. ▷ FASTENER. • *vb* **1** ▷ FASTEN. **2** *hook a fish*. capture, catch, take.

hooligan *n* bully, delinquent, hoodlum, lout, mugger, rough, ruffian, *inf* tearaway, thug, tough, trouble-maker, vandal, *inf* yob. ▷ CRIMINAL.

hoop *n* band, circle, girdle, loop, ring.

hop *vb* bound, caper, dance, jump, leap, limp, prance, skip, spring, vault.

hope *n* 1 ambition, aspiration, craving, day-dream, desire, dream, longing, wish, yearning. 2 *hope of better weather.* assumption, conviction, expectation, faith, likelihood, optimism, promise, prospect. • *vb inf* anticipate, aspire, be hopeful, believe, contemplate, count on, desire, expect, foresee, have faith, have hope, look forward (to), trust, wish. *Opp* DESPAIR.

hopeful *adj* 1 assured, confident, expectant, optimistic, positive, sanguine. 2 *hopeful signs.* auspicious, cheering, encouraging, favourable, heartening, promising, propitious, reassuring. *Opp* HOPELESS.

hopefully *adv* 1 confidently, expectantly, optimistically, with hope. 2 [*inf*] *Hopefully I'll be better tomorrow.* all being well, most likely, probably. [Many think this use of *hopefully* is wrong]

hopeless *adj* 1 defeatist, demoralized, despairing, desperate, disconsolate, fatalist, negative, pessimistic, resigned, wretched. 2 *a hopeless situation.* daunting, depressing, impossible, incurable, irremediable, irreparable, irreversible. 3 [*inf*] *He's hopeless!* feeble, inadequate, incompetent, inefficient, poor, useless, weak, worthless. *Opp* HOPEFUL.

horde *n* band, crowd, gang, mob, swarm, throng, tribe. ▷ GROUP.

horizontal *adj* even, flat, level, lying down, prone, prostrate, supine. *Opp* VERTICAL.

horrible *adj* awful, beastly, disagreeable, dreadful, ghastly, hateful, horrid, loathsome, macabre, nasty, objectionable, odious, offensive, revolting, terrible, unkind. ▷ HORRIFIC, UNPLEASANT. *Opp* PLEASANT.

horrific *adj* appalling, atrocious, blood-curdling, disgusting, dreadful, frightening, frightful, grisly, gruesome, hair-raising, harrowing, horrendous, horrifying, nauseating, shocking, sickening, spine-chilling, unacceptable, unnerving, unthinkable.

horrify *vb* alarm, appal, disgust, frighten, harrow, nauseate, outrage, scare, shock, sicken, stun, terrify, unnerve. **horrifying** ▷ HORRIFIC.

horror *n* 1 abhorrence, antipathy, aversion, detestation, disgust, dislike, dismay, distaste, dread, fear, hatred, loathing, panic, repugnance, revulsion, terror. 2 awfulness, frightfulness, ghastliness, gruesomeness, hideousness.

horse *n* bronco, carthorse, *old use* charger, cob, colt, filly, foal, *childish* gee-gee, gelding, hack, hunter, *old use* jade, mare, mount, mule, mustang, *inf* nag, *old use* palfrey, piebald, pony, racehorse, roan, skewbald, stallion, steed, warhorse.

horseman, horsewoman *ns* cavalryman, equestrian, jockey, rider.

hospitable *adj* cordial, courteous, generous, gracious, receptive, sociable, welcoming. ▷ FRIENDLY. *Opp* INHOSPITABLE.

hospital *n* clinic, convalescent home, dispensary, health centre, hospice, infirmary, medical centre, nursing home, sanatorium, sick bay.

hospitality *n* 1 accommodation, catering, entertainment. 2 cordiality, courtesy, friendliness, generosity, sociability, warmth, welcome.

host *n* **1** army, crowd, mob, multitude, swarm, throng, troop. ▷ GROUP. **2** ▷ COMPÈRE.

hostage *n* captive, pawn, prisoner, surety.

hostile *adj* **1** aggressive, antagonistic, antipathetic, attacking, averse, bellicose, belligerent, combative, confrontational, ill-disposed, inhospitable, inimical, malevolent, militant, opposed, oppressive, pugnacious, resentful, rival, unfriendly, unsympathetic, unwelcoming, warlike, warring. ▷ ANGRY. *Opp* FRIENDLY. **2** *hostile conditions*. adverse, contrary, opposing, unfavourable, unhelpful, unpropitious. ▷ BAD. *Opp* FAVOURABLE.

hostility *n* aggression, animosity, animus, antagonism, bad feeling, belligerence, confrontation, dissension, enmity, estrangement, friction, incompatibility, malevolence, malice, opposition, pugnacity, rancour, resentment, strife, unfriendliness. ▷ HATRED. *Opp* FRIENDSHIP. **hostilities** ▷ WAR.

hot *adj* **1** baking, blistering, boiling, burning, close, fiery, flaming, humid, oppressive, *inf* piping, red-hot, roasting, scalding, scorching, searing, sizzling, steamy, stifling, sultry, summery, sweltering, thermal, torrid, tropical, warm, white-hot. **2** *hot temper*. ardent, eager, emotional, excited, fervent, fervid, feverish, fierce, heated, hotheaded, impatient, impetuous, inflamed, intense, passionate, violent. **3** *hot taste*. acrid, biting, gingery, peppery, piquant, pungent, spicy, strong. *Opp* COLD, COOL. **hot-tempered** ▷ BAD-TEMPERED. **hot under the collar** ▷ ANGRY.

hotel *n* guest house, hostel, *joc* hostelry, inn, lodge, motel, pension. ▷ ACCOMMODATION.

hound *n* ▷ DOG. • *vb* annoy, badger, chase, harass, harry, hunt, nag, persecute, pester, pursue.

house *n old use* abode, domicile, dwelling, dwelling-place, habitation, home, homestead, household, place, residence.
□ *apartment,* inf *back-to-back, bungalow, chalet, cottage, council house, croft, detached house, farmhouse, flat, grange, hovel, homestead, hut, igloo, lodge, maisonette, manor, manse, mansion, penthouse,* inf *prefab, public house, rectory,* inf *semi, semi-detached house, shack, shanty, terraced house, thatched house,* inf *two-up two-down, vicarage, villa.*
• *vb* accommodate, billet, board, domicile, harbour, keep, lodge, place, *inf* put up, quarter, shelter, take in.

household *n* establishment, family, home, ménage, *inf* set-up.

hovel *n* cottage, *inf* dump, hole, hut, shack, shanty, shed.

hover *vb* **1** be suspended, drift, float, flutter, fly, hang, poise. **2** be indecisive, dally, dither, *inf* hang about, hang around, hesitate, linger, loiter, pause, vacillate, wait about, waver.

howl *vb* bay, bellow, cry, roar, shout, ululate, wail, yowl.

hub *n* axis, centre, core, focal point, focus, heart, middle, nucleus, pivot.

huddle *n* ▷ GROUP. • *vb* **1** cluster, converge, crowd, flock, gather, group, heap, herd, jam, jumble, pile, press, squeeze, swarm, throng. **2** cuddle, curl up, hug, nestle, snuggle.

hue *n* cast, complexion, dye, nuance, shade, tincture, tinge, tint, tone. ▷ COLOUR. **hue and cry** ▷ OUTCRY.

hug *vb* clasp, cling to, crush, cuddle, embrace, enfold, fold in your arms, hold close, huddle together, nestle together, nurse, snuggle against, squeeze.

huge *adj* **1** Brobdingnagian, colossal, elephantine, enormous, gargantuan, giant, gigantic, *inf* hulking, immense, imposing, impressive, *inf* jumbo, majestic, mammoth, massive, mighty, *inf* monster, monstrous, monumental, mountainous, prodigious, stupendous, titanic, towering, *inf* tremendous, vast, weighty, *inf* whopping. ▷ BIG. **2** *huge number*. ▷ INFINITE. *Opp* SMALL.

hulk *n* **1** body, carcass, frame, hull, shell, wreck. **2** *a clumsy hulk*. lout, lump, oaf.

hulking *adj* awkward, bulky, cumbersome, heavy, ungainly, unwieldy. ▷ BIG, CLUMSY.

hull *n* body, framework, structure.

hum *vb* buzz, drone, murmur, purr, sing, thrum, vibrate, whirr. **hum and haw** ▷ HESITATE.

human *adj* **1** anthropoid, hominid, hominoid, mortal. **2** *human feeling*. kind, rational, reasonable, sensible, sensitive, sympathetic, thoughtful. ▷ HUMANE. *Opp* INHUMAN. **human beings** folk, humanity, mankind, men and women, mortals, people.

humane *adj* altruistic, benevolent, charitable, civilized, compassionate, feeling, forgiving, good, human, humanitarian, kind-hearted, loving, magnanimous, merciful, philanthropic, pitying, refined, sympathetic, tender, understanding, unselfish, warm-hearted. ▷ KIND. *Opp* INHUMANE.

humble *adj* **1** deferential, docile, meek, modest, *derog* obsequious, polite, reserved, respectful, self-effacing, *derog* servile, submissive, subservient, *derog* sycophantic, unassertive, unassuming, unostentatious, unpresuming, unpretentious. *Opp* PROUD. **2** *humble birth*. base, commonplace, ignoble, inferior, insignificant, low, lowly, mean, obscure, ordinary, plebeian, poor, simple, undistinguished, unimportant, unprepossessing, unremarkable. • *vb* ▷ HUMILIATE.

humid *adj* clammy, damp, dank, moist, muggy, steamy, sticky, sultry, sweaty.

humiliate *vb* abase, abash, break, break someone's spirit, bring someone down, chagrin, chasten, crush, deflate, degrade, demean, discredit, disgrace, embarrass, humble, make someone ashamed, *inf* make someone eat humble pie, *inf* make someone feel small, mortify, *inf* put someone down, *inf* put someone in his/her place, shame, *inf* show someone up, *inf* take someone down a peg. **humiliating** ▷ SHAMEFUL.

humiliation *n* abasement, chagrin, degradation, discredit, disgrace, dishonour, embarrassment, ignominy, indignity, loss of face, mortification, obloquy, shame.

humility *n* deference, humbleness, lowliness, meekness, modesty, self-abasement, self-effacement, *derog* servility, shyness, unpretentiousness. *Opp* PRIDE.

humorous *adj* absurd, amusing, comic, comical, diverting, droll, entertaining, facetious, farcical, funny, hilarious, *inf* hysterical, ironic, jocose, jocular, *inf* killing, laughable, merry, *inf* priceless, risible, sarcastic, sardonic, satirical, *inf* side-splitting, slapstick, uproarious, waggish, whimsical, witty, zany. *Opp* SERIOUS.

humour *n* **1** absurdity, badinage, banter, comedy, drollness, facetiousness, fun, incongruity, irony, jesting, jocularity, jokes, joking, merriment, quips, raillery, repartee, satire, *inf* sense of fun, waggishness, wit, witticism, wittiness. **2** *in a good humour.* disposition, frame of mind, mood, spirits, state of mind, temper.

hump *n* bulge, bump, curve, growth, hunch, knob, lump, node, projection, protrusion, protuberance, swelling, tumescence. **2** *hump in the ground.* barrow, hillock, hummock, mound, rise, tumulus. • *vb* **1** arch, bend, crook, curl, curve, hunch, raise. **2** *hump a load.* drag, heave, hoist, lift, lug, raise, shoulder.

hunch *n* **1** ▷ HUMP. **2** feeling, guess, idea, impression, inkling, intuition, premonition, presentiment, suspicion. • *vb* arch, bend, crook, curl, curve, huddle, hump, raise, shrug.

hunger *n* **1** appetite, craving, greed, ravenousness, voracity. **2** deprivation, famine, lack of food, malnutrition, starvation, want. • *vb* ▷ DESIRE.

hungry *adj* aching, avid, covetous, craving, eager, emaciated, famished, famishing, greedy, longing, *inf* peckish, ravenous, starved, starving, underfed, undernourished, voracious.

hunt *n* chase, pursuit, quest, search. ▷ HUNTING. • *vb* **1** chase, course, dog, ferret, hound, poach, pursue, stalk, track, trail. **2** *hunt for lost property. inf* check out, enquire after, ferret out, look for, rummage, search for, seek, trace, track down.

hunter *n* huntsman, huntswoman, predator, stalker, trapper.

hunting *n* blood-sports, coursing, poaching, stalking, trapping.

hurdle *n* **1** barricade, barrier, fence, hedge, jump, obstacle, wall. **2** bar, check, complication, difficulty, handicap, hindrance, impediment, obstruction, problem, restraint, snag, stumbling block.

hurl *vb* cast, catapult, chuck, dash, fire, fling, heave, launch, *inf* let fly, pelt, pitch, project, propel, send, shy, sling, throw, toss.

hurricane *n* cyclone, storm, tempest, tornado, typhoon, whirlwind.

hurry *n* ▷ HASTE. • *vb* **1** *inf* belt, *inf* buck up, chase, dash, dispatch, *inf* fly, *inf* get a move on, hasten, hurtle, hustle, make haste, move quickly, rush, *inf* shift, speed, *inf* step on it, work faster. **2** *hurry a process.* accelerate, expedite, press on with, quicken, speed up. *Opp* DELAY. **hurried** ▷ HASTY.

hurt *vb* **1** ache, be painful, burn, pinch, smart, sting, suffer pain, throb, tingle. **2** *hurt physically.* abuse, afflict, agonize, bruise, cause pain to, cripple, cut, disable, injure, maim, misuse, mutilate, torture, wound. **3** *hurt mentally.* affect, aggrieve, be hurtful to, *inf* cut to the quick, depress, distress, grieve, humiliate, insult, offend, pain, sadden, torment, upset. **4** *hurt things.* damage, harm, impair, mar, ruin, sabotage, spoil.

hurtful *adj* biting, cruel, cutting, damaging, derogatory, detrimental, distressing, hard to bear, harmful, injurious, malicious, nasty, painful, sarcastic, scathing, spiteful, uncharitable, unkind, upsetting, vicious, wounding. *Opp* KIND.

hurtle *vb* charge, chase, dash, fly, plunge, race, rush, shoot, speed, tear.

hush *int* be quiet! be silent! *inf* hold your tongue! *sl* pipe down! *inf* shut up! • *vb* ▷ SILENCE. **hush up** ▷ SUPPRESS.

hustle *vb* **1** bustle, hasten, hurry, jostle, rush, scamper, scurry. **2** *hustled me away.* coerce, compel, force, push, shove, thrust.

hut *n* cabin, den, hovel, lean-to, shack, shanty, shed, shelter.

hybrid *n* amalgam, combination, composite, compound, cross, cross-breed, half-breed, mixture, mongrel.

hygiene *n* cleanliness, health, sanitariness, sanitation, wholesomeness.

hygienic *adj* aseptic, clean, disinfected, germ-free, healthy, pure, salubrious, sanitary, sterile, sterilized, unpolluted, wholesome. *Opp* UNHEALTHY.

hypnotic *adj* fascinating, irresistible, magnetic, mesmeric, mesmerizing, sleep-inducing, soothing, soporific, spellbinding.

hypnotize *vb* bewitch, captivate, cast a spell over, dominate, enchant, entrance, fascinate, gain power over, magnetize, mesmerize, *inf* put to sleep, spellbind, *inf* stupefy.

hypocrisy *n* cant, deceit, deception, double-dealing, double standards, double-talk, double-think, duplicity, falsity, *inf* humbug, inconsistency, insincerity.

hypocritical *adj* deceptive, double-dealing, double-faced, duplicitous, false, inconsistent, insincere, Pharisaical, *inf* phoney, self-deceiving, self-righteous, *inf* two-faced.

hypothesis *n* conjecture, guess, postulate, premise, proposition, speculation, supposition, theory, thesis.

hypothetical *adj* academic, alleged, assumed, conjectural, groundless, imaginary, presumed, putative, speculative, supposed, suppositional, theoretical, unreal.

hysteria *n* frenzy, hysterics, madness, mania, panic.

hysterical *adj* berserk, beside yourself, crazed, delirious, demented, distraught, frantic, frenzied, irrational, mad, overemotional, rabid, raving, uncontrollable, wild.

I

ice *n* black ice, floe, frost, glacier, iceberg, icicle, rime.

icy *adj* **1** arctic, chilling, freezing, frosty, frozen, glacial, polar, Siberian. ▷ COLD. **2** *icy roads.* glassy, greasy, slippery, *inf* slippy.

idea *n* **1** abstraction, attitude, belief, concept, conception, conjecture, construct, conviction, doctrine, hypothesis, notion, opinion, philosophy, principle, sentiment, teaching, tenet, theory, thought, view. **2** *a bright idea.* brainwave, design, fancy, guess, inspiration, plan, proposal, scheme, suggestion. **3** *idea of a poem.* intention, meaning, point. **4** *idea of what to expect.* clue, guidelines, impression, inkling, intimation, model, pattern, perception, suspicion, vision.

ideal *adj* 1 best, classic, complete, excellent, faultless, model, optimum, perfect, supreme, unsurpassable. 2 *an ideal world*. chimerical, dream, hypothetical, illusory, imaginary, unattainable, unreal, Utopian, visionary. • *n* 1 acme, criterion, epitome, exemplar, model, paragon, pattern, standard. 2 ▷ PRINCIPLE.

idealistic *adj* high-minded, impractical, over-optimistic, quixotic, romantic, starry-eyed, unrealistic. *Opp* REALISTIC.

idealize *vb* apotheosize, deify, exalt, glamorize, glorify, *inf* put on a pedestal, romanticize. ▷ IDOLIZE.

identical *adj* alike, comparable, congruent, corresponding, duplicate, equal, equivalent, indistinguishable, interchangeable, like, matching, the same, similar, twin. *Opp* DIFFERENT.

identifiable *adj* detectable, discernible, distinctive, distinguishable, familiar, known, named, noticeable, perceptible, recognizable, unmistakable. *Opp* UNIDENTIFIABLE.

identify *vb* 1 distinguish, label, mark, name, pick out, pinpoint, *inf* put a name to, recognize, single out, specify, spot. 2 *identify an illness*. detect, diagnose, discover. **identify with** empathize with, feel for, *inf* put yourself in the shoes of, relate to, sympathize with.

identity *n* 1 *inf* ID, name. 2 character, distinctiveness, individuality, nature, particularity, personality, selfhood, singularity, uniqueness.

ideology *n* assumptions, beliefs, creed, convictions, ideas, philosophy, principles, tenets, theories, underlying attitudes.

idiom *n* argot, cant, choice of words, dialect, expression, jargon, language, manner of speaking, parlance, phrase, phraseology, phrasing, turn of phrase, usage.

idiomatic *adj* colloquial, natural, vernacular, well-phrased.

idiosyncrasy *n* characteristic, eccentricity, feature, habit, individuality, mannerism, oddity, peculiarity, quirk, trait.

idiosyncratic *adj* characteristic, distinctive, eccentric, individual, odd, peculiar, personal, quirky, singular, unique. *Opp* COMMON.

idiot *n* [*most synonyms inf*] ass, blockhead, bonehead, booby, chump, clot, cretin, dim-wit, dolt, dope, duffer, dumb-bell, dummy, dunce, dunderhead, fat-head, fool, half-wit, ignoramus, imbecile, moron, nincompoop, ninny, nit-wit, simpleton, twerp, twit.

idiotic *adj* absurd, asinine, crazy, foolish, half-witted, imbecile, insane, irrational, mad, moronic, nonsensical, ridiculous, senseless. ▷ STUPID. *Opp* SENSIBLE.

idle *adj* 1 dormant, inactive, inoperative, in retirement, not working, redundant, retired, unemployed, unoccupied, unproductive, unused. 2 apathetic, good-for-nothing, indolent, lackadaisical, lazy, shiftless, slothful, slow, sluggish, torpid, uncommitted, work-shy. 3 *idle speculation*. casual, frivolous, futile, pointless, worthless. *Opp* BUSY. • *vb* be lazy, dawdle, do nothing, *inf* hang about, *inf* kill time, laze, loaf, loll, lounge about, *inf* mess about, potter, slack, stagnate, take it easy, vegetate. *Opp* WORK.

idler *n inf* good-for-nothing, *inf* layabout, *inf* lazybones, loafer, malingerer, shirker, *inf* skiver, slacker, sluggard, wastrel.

idol *n* **1** deity, effigy, fetish, god, graven image, icon, statue. **2** *pop idol.* celebrity, *inf* darling, favourite, hero, *inf* pin-up, star, *inf* superstar.

idolize *vb* adore, adulate, hero-worship, lionize, look up to, revere, reverence, venerate, worship. ▷ IDEALIZE.

idyllic *adj* Arcadian, bucolic, charming, delightful, happy, idealized, lovely, pastoral, peaceful, perfect, picturesque, rustic, unspoiled.

ignite *vb* burn, catch fire, fire, kindle, light, set alight, set on fire, spark off, touch off.

ignoble *adj* base, churlish, cowardly, despicable, disgraceful, dishonourable, infamous, low, mean, selfish, shabby, uncharitable, unchivalrous, unworthy. *Opp* NOBLE.

ignorance *n* inexperience, innocence, unawareness, unconsciousness, unfamiliarity. ▷ STUPIDITY. *Opp* KNOWLEDGE.

ignorant *adj* **1** ill-informed, innocent, lacking knowledge, oblivious, unacquainted, unaware, unconscious, unfamiliar (with), uninformed, unwitting. **2** benighted, *inf* clueless, illiterate, uncouth, uncultivated, uneducated, unenlightened, unlettered, unscholarly, unsophisticated. ▷ IMPOLITE, STUPID. *Opp* CLEVER, KNOWLEDGEABLE.

ignore *vb* disobey, disregard, leave out, miss out, neglect, omit, overlook, pass over, reject, *inf* shut your eyes to, skip, slight, snub, take no notice of, *inf* turn a blind eye to.

ill *adj* **1** ailing, bad, bedridden, bilious, *inf* dicky, diseased, feeble, frail, *inf* funny, *inf* groggy, indisposed, infected, infirm, invalid, nauseous, nauseated, *inf* off-colour, *inf* out of sorts, pasty, poorly, queasy, queer, *inf* seedy, sick, sickly, suffering, *inf* under the weather, unhealthy, unwell, valetudinarian, weak. *Opp* HEALTHY. **2** *ill effects.* bad, damaging, detrimental, evil, harmful, injurious, unfavourable, unfortunate, unlucky. *Opp* GOOD. • *plur n* the infirm, invalids, patients, the sick, sufferers, victims. **be ill** ail, languish, sicken. **ill-advised** ▷ MISGUIDED. **ill-bred** ▷ RUDE. **ill-fated** ▷ UNLUCKY. **ill-humoured** ▷ BAD-TEMPERED. **ill-mannered** ▷ RUDE. **ill-natured** ▷ UNKIND. **ill-omened** ▷ UNLUCKY. **ill-tempered** ▷ BAD-TEMPERED. **ill-treat** ▷ MISTREAT.

illegal *adj* actionable, against the law, banned, black-market, criminal, felonious, forbidden, illicit, invalid, irregular, outlawed, prohibited, proscribed, unauthorized, unconstitutional, unlawful, unlicensed, wrong, wrongful. ▷ ILLEGITIMATE. *Opp* LEGAL.

illegible *adj* indecipherable, indistinct, obscure, unclear, unreadable. *Opp* LEGIBLE.

illegitimate *adj* **1** against the rules, improper, inadmissible, incorrect, invalid, irregular, spurious, unauthorized, unjustifiable, unreasonable, unwarranted. ▷ ILLEGAL. **2** bastard, born out of wedlock, natural. *Opp* LEGITIMATE.

illiterate *adj* unable to read, uneducated, unlettered. ▷ IGNORANT. *Opp* LITERATE.

illness *n* abnormality, affliction, ailment, allergy, attack, blight, *inf* bug, complaint, condition, contagion, disability, disease, disorder, epidemic, fever, fit, health problem, indisposition, infection, infirmity, malady, malaise, pestilence, plague, sickness, *inf* trouble, *inf* turn, *inf* upset, weakness. ▷ WOUND.

illogical *adj* absurd, fallacious, inconsequential, inconsistent, invalid, irrational, senseless, unreasonable, unsound. ▷ SILLY. *Opp* LOGICAL.

illuminate *vb* **1** brighten, decorate with lights, light up, make brighter, reveal. **2** clarify, clear up, elucidate, enlighten, explain, explicate, throw light on.

illusion *n* **1** apparition, conjuring trick, day-dream, deception, delusion, dream, fancy, fantasy, figment of the imagination, hallucination, mirage. **2** *under an illusion.* error, false impression, misapprehension, misconception, mistake.

illusory *adj* chimerical, deceptive, deluding, delusive, fallacious, false, illusive, imagined, misleading, mistaken, sham, unreal, untrue. ▷ IMAGINARY. *Opp* REAL.

illustrate *vb* **1** demonstrate, elucidate, exemplify, explain, instance, show. **2** adorn, decorate, embellish, illuminate, ornament. **3** depict, draw pictures of, picture, portray.

illustration *n* **1** case in point, demonstration, example, exemplar, instance, sample, specimen. **2** decoration, depiction, diagram, drawing, figure, photograph, picture, sketch. ▷ IMAGE.

image *n* **1** imitation, likeness, projection, reflection, representation. ▷ PICTURE. **2** carving, effigy, figure, icon, idol, statue. **3** *the image of her mother*. counterpart, double, likeness, spitting-image, twin.

imaginary *adj* fabulous, fanciful, fictional, fictitious, hypothetical, imagined, insubstantial, invented, legendary, made-up, mythical, mythological, non-existent, supposed, unreal, visionary. ▷ ILLUSORY. *Opp* REAL.

imagination *n* artistry, creativity, fancy, ingenuity, insight, inspiration, inventiveness, *inf* mind's eye, originality, resourcefulness, sensitivity, thought, vision.

imaginative *adj* artistic, attractive, beautiful, clever, creative, fanciful, ingenious, innovative, inspired, inspiring, inventive, original, poetic, resourceful, sensitive, thoughtful, unusual, visionary, vivid. *Opp* UNIMAGINATIVE.

imagine *vb* **1** conceive, conjure up, *inf* cook up, create, dream up, envisage, fancy, fantasize, invent, make believe, make up, picture, pretend, see, think of, think up, visualize. **2** assume, believe, conjecture, guess, infer, judge, presume, suppose, surmise, suspect, think.

imitate *vb* **1** ape, burlesque, caricature, counterfeit, duplicate, echo, guy, mimic, parody, parrot, portray, reproduce, satirize, send up, simulate, *inf* take off, travesty. ▷ IMPERSONATE. **2** copy, emulate, follow, match, model yourself on.

imitation *adj* artificial, copied, counterfeit, dummy, ersatz, man-made, mock, model, *inf* phoney, reproduction, sham, simulated, synthetic. *Opp* REAL. • *n* **1** copying, duplication, emulation, mimicry, repetition. **2** *inf* clone, copy, counterfeit, dummy, duplicate, fake, forgery, impersonation, impression, likeness, *inf* mock-up, model, parody, reflection, replica, reproduction, sham, simulation, *inf* take-off, toy, travesty.

immature *adj* adolescent, babyish, backward, callow, childish, *inf* green, inexperienced, infantile, juvenile, new, puerile, undeveloped, unripe, young, youthful. *Opp* MATURE.

immediate *adj* **1** instant, instantaneous, prompt, quick, speedy, sudden, swift, unhesitating, unthinking. **2** *immediate need.* current, present, pressing, top-priority, urgent. **3** *immediate neighbours.* adjacent, close, closest, direct, near, nearest, neighbouring, next.

immediately *adv* at once, directly, forthwith, instantly, now, promptly, *inf* right away, straight away, unhesitatingly.

immense *adj* Brobdingnagian, colossal, elephantine, enormous, gargantuan, giant, gigantic, great, huge, *inf* hulking, immeasurable, imposing, impressive, incalculable, *inf* jumbo, large, mammoth, massive, mighty, *inf* monster, monstrous, monumental, mountainous, prodigious, stupendous, titanic, towering, *inf* tremendous, vast, *inf* whopping. ▷ BIG. *Opp* SMALL.

immerse *vb* bathe, dip, drench, drown, duck, dunk, inundate, lower, plunge, sink, submerge. **immersed** ▷ BUSY, INTERESTED.

immersion *n* baptism, dipping, ducking, plunge, submersion.

immigrant *n* alien, arrival, incomer, newcomer, outsider, settler.

imminent *adj* about to happen, approaching, close, coming, foreseeable, forthcoming, impending, looming, menacing, near, threatening.

immobile *adj* **1** ▷ IMMOVABLE. **2** frozen, inexpressive, inflexible, rigid. *Opp* MOBILE.

immobilize *vb* cripple, damage, disable, make immobile, paralyse, put out of action, sabotage, stop.

immoral *adj* abandoned, base, conscienceless, corrupt, debauched, degenerate, depraved, dishonest, dissipated, dissolute, evil, *inf* fast, impure, indecent, irresponsible, licentious, loose, low, profligate, promiscuous, *inf* rotten, sinful, unchaste, unethical, unprincipled, unscrupulous, vicious, villainous, wanton, wrong. ▷ WICKED. *Opp* MORAL. **immoral person** blackguard, cheat, degenerate, liar, libertine, profligate, rake, reprobate, scoundrel, sinner, villain, wrongdoer.

immortal *adj* **1** ageless, ceaseless, deathless, endless, eternal, everlasting, incorruptible, indestructible, never-ending, perpetual, sempiternal, timeless, unchanging, undying, unending, unfading. **2** *immortal beings.* divine, godlike, legendary, mythical. *Opp* MORTAL.

immortalize *vb* apotheosize, beatify, canonize, commemorate, deify, enshrine, keep alive, make immortal, make permanent, memorialize, perpetuate.

immovable *adj* **1** anchored, fast, firm, fixed, immobile, immobilized, motionless, paralysed, riveted, rooted, secure, set, settled, solid, static, stationary, still, stuck, unmoving. **2** ▷ IMMUTABLE.

immune *adj* exempt, free, immunized, inoculated, invulnerable, protected, resistant, safe, unaffected, vaccinated. *Opp* VULNERABLE.

immunize *vb* inoculate, vaccinate.

immutable *adj* constant, dependable, enduring, eternal, fixed, invariable, lasting, obdurate, permanent, perpetual, reliable, settled, stable, steadfast, unalterable, unchangeable, unswerving, unvarying. ▷ RESOLUTE. *Opp* CHANGEABLE.

impact *n* **1** bang, blow, bump, collision, concussion, contact, crash, knock, smash. **2** bearing, consequence, effect, force, impression, influence, repercussions, reverberations, shock, thrust.
• *vb* ▷ HIT.

impair *vb* cripple, damage, harm, injure, mar, ruin, spoil, weaken.

impale *vb* pierce, run through, skewer, spear, spike, spit, stab, stick, transfix.

impartial *adj* balanced, detached, disinterested, dispassionate, equitable, even-handed, fair, fair-minded, just, neutral, non-partisan, objective, open-minded, unbiased, uninvolved, unprejudiced. *Opp* BIASED.

impartiality *n* balance, detachment, disinterest, fairness, justice, neutrality, objectivity, open-mindedness. *Opp* BIAS.

impassable *adj* blocked, closed, obstructed, unusable.

impatient *adj* **1** anxious, eager, keen, impetuous, precipitate, *inf* raring. **2** agitated, chafing, edgy, fidgety, fretful, irritable, nervous, restive, restless, uneasy. **3** *an impatient manner*. abrupt, brusque, curt, hasty, intolerant, irascible, irritable, quick-tempered, short-tempered, snappish, snappy, testy. *Opp* APATHETIC, PATIENT.

impede *vb* arrest, bar, be an impediment to, check, curb, delay, deter, frustrate, get in the way of, hamper, handicap, hinder, *inf* hit, hold back, hold up, keep back, limit, obstruct, oppose, prevent, restrain, restrict, retard, sabotage, slow down, slow up, stand in the way of, stop, thwart. *Opp* HELP.

impediment *n* **1** bar, barrier, burden, check, curb, deterrent, difficulty, disadvantage, *inf* drag, drawback, encumbrance, hindrance, inconvenience, limitation, obstacle, obstruction, restraint, restriction, snag, stumbling-block. **2** ▷ HANDICAP.

impending *adj* about to happen, approaching, close, coming, foreseeable, forthcoming, imminent, looming, menacing, near, *inf* on the horizon, threatening.

impenetrable *adj* **1** dense, hard, resilient, solid, strong. ▷ IMPERVIOUS. **2** impregnable, invincible, inviolable, invulnerable, safe, secure, unassailable, unconquerable. *Opp* VULNERABLE. **3** *impenetrable language*. inaccessible, incomprehensible, inscrutable, unfathomable, *inf* unget-at-able. *Opp* ACCESSIBLE.

imperceptible *adj* faint, gradual, inappreciable, inaudible, indistinguishable, infinitesimal, insignificant, invisible, microscopic, minute, negligible, slight, small, subtle, tiny, unclear, undetectable, unnoticeable, vague. ▷ SMALL. *Opp* PERCEPTIBLE.

imperceptive *adj* impercipient, inattentive, slow, uncritical, undiscriminating, unobservant, unresponsive. ▷ STUPID. *Opp* PERCEPTIVE.

imperfect *adj* blemished, broken, chipped, cracked, damaged, defective, deficient, faulty, flawed, incomplete, incorrect, marred, partial, patchy, shop-soiled, spoilt, unfinished, wanting. *Opp* PERFECT.

imperfection *n* blemish, damage, defect, deficiency, error, failing, fault, flaw, foible, frailty, inadequacy, infirmity, peccadillo, shortcoming, weakness. *Opp* PERFECTION.

impermanent *adj* changing, destructible, ephemeral, evanescent, fleeting, momentary, passing, shifting, short-lived, temporary, transient, transitory, unstable. ▷ CHANGEABLE. *Opp* PERMANENT.

impersonal *adj* aloof, businesslike, cold, cool, correct, detached, disinterested, dispassionate, distant, formal, hard, inhuman, mechanical, objective, official, remote, stiff, unapproachable, unemotional, unfriendly, unprejudiced, unsympathetic, without emotion, wooden. *Opp* FRIENDLY.

impersonate *vb* disguise yourself as, do impressions of, dress up as, masquerade as, mimic, pass yourself off as, portray, pose as, pretend to be, *inf* take off. ▷ IMITATE.

impertinent *adj* bold, brazen, cheeky, *inf* cocky, *inf* cool, discourteous, disrespectful, forward, fresh, impolite, impudent, insolent, insubordinate, insulting, irreverent, pert, saucy. ▷ RUDE. *Opp* RESPECTFUL.

impervious *adj* **1** hermetic, impenetrable, impermeable, nonporous, solid, waterproof, water-repellent, watertight. *Opp* POROUS. **2** ▷ RESISTANT.

impetuous *adj* abrupt, careless, eager, hasty, headlong, hotheaded, impulsive, incautious, offhand, precipitate, quick, rash, reckless, speedy, spontaneous, *inf* spur-of-the-moment, *inf* tearing, thoughtless, unplanned, unpremeditated, unthinking, violent. *Opp* CAUTIOUS.

impetus *n* boost, drive, encouragement, energy, fillip, force, impulse, incentive, inspiration, momentum, motivation, power, push, spur, stimulation, stimulus, thrust.

impiety *n* blasphemy, godlessness, irreverence, profanity, sacrilege, sinfulness, ungodliness, unrighteousness, wickedness. *Opp* PIETY.

impious *adj* blasphemous, godless, irreligious, irreverent, profane, sacrilegious, sinful, unholy. ▷ WICKED. *Opp* PIOUS.

implausible *adj* doubtful, dubious, far-fetched, feeble, improbable, questionable, suspect, unconvincing, unlikely, unreasonable, weak. *Opp* PLAUSIBLE.

implement *n* apparatus, appliance, contrivance, device, gadget, instrument, mechanism, tool, utensil. • *vb* accomplish, achieve, bring about, carry out, effect, enforce, execute, fulfil, perform, put into effect, put into practice, realize, try out.

implicate *vb* associate, concern, connect, embroil, enmesh, ensnare, entangle, entrap, include, incriminate, inculpate, involve, show involvement in.

implication *n* **1** hidden meaning, hint, innuendo, insinua-tion, overtone, purport, significance. **2** *implication in crime.* association, connection, embroilment, entanglement, inclusion, involvement.

implicit *adj* **1** hinted at, implied, indirect, inherent, insinuated, tacit, understood, undeclared, unexpressed, unsaid, unspoken, unstated, unvoiced. *Opp* EXPLICIT. **2** *implicit faith.* ▷ ABSOLUTE.

imply *vb* **1** hint, indicate, insinuate, intimate, mean, point to, suggest. **2** ▷ SIGNIFY.

impolite *adj* discourteous, disrespectful, ill-bred, ill-mannered, uncivil, vulgar. ▷ RUDE. *Opp* POLITE.

import *vb* bring in, buy in, introduce, ship in. ▷ CONVEY.

important *adj* **1** basic, big, cardinal, central, chief, consequential, critical, epoch-making, essential, foremost, fundamental, grave, historic, key, main, major, momentous, newsworthy, noteworthy, once in a lifetime, outstanding, pressing, primary, principal, rare, salient, serious, signal, significant, strategic, substantial, urgent, valuable, vital, weighty. **2** celebrated, distinguished, eminent, famous, great, high-ranking, influential, known, leading, notable, noted, powerful, pre-eminent, prominent, renowned, top-level, well-known. *Opp* UNIMPORTANT. **be important** ▷ MATTER.

importunate *adj* demanding, impatient, insistent, persistent, pressing, relentless, urgent, unremitting.

importune *vb* badger, harass, hound, pester, plague, plead with, press, solicit, urge. ▷ ASK.

impose *vb* charge with, decree, dictate, enforce, exact, fix, foist, force, inflict, insist on, introduce, lay, levy, prescribe, set.

impose on ▷ BURDEN, EXPLOIT.

imposing ▷ IMPRESSIVE.

impossible *adj* hopeless, impracticable, impractical, inconceivable, insoluble, insuperable, insurmountable, *inf* not on, out of the question, unachievable, unattainable, unimaginable, unobtainable, unthinkable, unviable, unworkable. *Opp* POSSIBLE.

impotent *adj* debilitated, decrepit, emasculated, enervated, helpless, inadequate, incapable, incompetent, ineffective, ineffectual, inept, infirm, powerless, unable. ▷ WEAK. *Opp* POTENT.

impracticable *adj* not feasible, unachievable, unworkable, useless. ▷ IMPOSSIBLE. *Opp* PRACTICABLE.

impractical *adj* academic, idealistic, quixotic, romantic, theoretical, unrealistic, visionary. *Opp* PRACTICAL.

imprecise *adj* ambiguous, approximate, careless, estimated, fuzzy, guessed, hazy, ill-defined, inaccurate, inexact, inexplicit, loose, *inf* sloppy, undefined, unscientific, vague, *inf* waffly, *inf* woolly. *Opp* PRECISE.

impregnable *adj* impenetrable, invincible, inviolable, invulnerable, safe, secure, strong, unassailable, unconquerable. *Opp* VULNERABLE.

impress *vb* **1** affect, be memorable to, excite, influence, inspire, leave its mark on, move, persuade, *inf* stick in the mind of, stir, touch. **2** *impress a mark.* emboss, engrave, imprint, mark, print, stamp.

impression *n* **1** effect, impact, influence, mark. **2** belief, consciousness, fancy, feeling, hunch, idea, memory, notion, opinion, recollection, sense, suspicion, view. **3** dent, hollow, imprint, indentation, mark, print, stamp. **4** imitation, impersonation, mimicry, parody, *inf* take-off. **5** *impression of a book.* edition, printing, reprint.

impressionable *adj* easily influenced, gullible, inexperienced, naïve, persuadable, receptive, responsive, suggestible, susceptible.

impressive *adj* affecting, august, awe-inspiring, awesome, commanding, distinguished, evocative, exciting, formidable, grand, *derog* grandiose, great, imposing, magnificent, majestic, memorable, moving, powerful, redoubtable, remarkable, splendid, stately, stirring, striking, touching. ▷ BIG. *Opp* INSIGNIFICANT.

imprison *vb* cage, commit to prison, confine, detain, gaol, immure, incarcerate, intern, jail, keep in custody, keep under house arrest, *inf* keep under lock and key, lock away, lock up, *inf* put away, remand, *inf* send down, shut in, shut up. *Opp* FREE.

imprisonment *n* confinement, custody, detention, duress, gaol, house arrest, incarceration, internment, jail, remand, restraint.

improbable *adj* absurd, doubtful, dubious, far-fetched, *inf* hard to believe, implausible, incredible, preposterous, questionable, unbelievable, unconvincing, unexpected, unlikely. *Opp* PROBABLE.

impromptu *adj inf* ad-lib, extempore, extemporized, improvised, impulsive, made-up, offhand, *inf* off the cuff, *inf* off the top of your head, *inf* on the spur of the moment, spontaneous, unplanned, unpremeditated, unprepared, unrehearsed, unscripted. ▷ IMPULSIVE. *Opp* REHEARSED.

improper *adj* **1** ill-judged, ill-timed, inappropriate, incorrect, infelicitous, inopportune, irregular, mistaken, out of place, uncalled-for, unfit, unseemly, unsuitable, unwarranted. ▷ WRONG. **2** ▷ INDECENT. *Opp* PROPER.

impropriety *n* inappropriateness, incorrectness, indecency, indelicacy, infelicity, insensitivity, irregularity, rudeness, unseemliness. ▷ OBSCENITY. *Opp* PROPRIETY.

improve *vb* **1** advance, develop, get better, grow, increase, *inf* look up, move on, progress, *inf* take a turn for the better. **2** *improve after illness.* convalesce, *inf* pick up, rally, recover, recuperate, revive, strengthen, *inf* turn the corner. **3** *improve your ways.* ameliorate, amend, better, correct, enhance, enrich, make better, mend, polish (up), rectify, refine, reform, revise. **4** *improve a home.* decorate, extend, modernize, rebuild, recondition, refurbish, renovate, repair, touch up, update, upgrade. *Opp* WORSEN.

improvement *n* **1** advance, amelioration, betterment, correction, development, enhancement, gain, increase, progress, rally, recovery, reformation, upswing, upturn. **2** *home improvements.* alteration, extension, *inf* face-lift, modernization, modification, renovation.

improvise *vb* **1** *inf* ad-lib, concoct, contrive, devise, invent, make do, make up, *inf* throw together. **2** extemporize, perform impromptu, play by ear, vamp.

impudent *adj* audacious, bold, *inf* cheeky, disrespectful, forward, *inf* fresh, impertinent, insolent, pert, presumptuous, saucy. ▷ RUDE. *Opp* RESPECTFUL.

impulse *n* **1** drive, force, impetus, motive, pressure, push, stimulus, thrust. **2** caprice, desire, instinct, urge, whim.

impulsive *adj* automatic, emotional, hare-brained, hasty, headlong, hot-headed, impetuous, instinctive, intuitive, involuntary, madcap, precipitate, rash, reckless, *inf* snap, spontaneous, *inf* spur-of-the-moment, sudden, thoughtless, unconscious, unplanned, unpremeditated, unthinking, wild. ▷ IMPROMPTU. *Opp* DELIBERATE.

impure *adj* **1** adulterated, contaminated, defiled, foul, infected, polluted, tainted, unclean, unwholesome. ▷ DIRTY. **2** ▷ INDECENT.

impurity *n* contamination, defilement, infection, pollution, taint. ▷ DIRT.

inaccessible *adj* cut off, deserted, desolate, godforsaken, impassable, impenetrable, inconvenient, isolated, lonely, *inf* off the beaten track, outlying, out of reach, out-of-the-way, private, remote, solitary, unavailable, unfrequented, *inf* unget-at-able, unobtainable, unreachable, unusable. *Opp* ACCESSIBLE.

inaccurate *adj* erroneous, fallacious, false, faulty, flawed, imperfect, imprecise, incorrect, inexact, misleading, mistaken, unfaithful, unreliable, unsound, untrue, vague, wrong. *Opp* ACCURATE.

inactive *adj* asleep, dormant, hibernating, idle, immobile, inanimate, indolent, inert, languid, lazy, lethargic, out of action, passive, quiescent, quiet, sedentary, sleepy, slothful, slow, sluggish, somnolent, torpid, unemployed, unoccupied, vegetating. *Opp* ACTIVE.

inadequate *adj* deficient, disappointing, faulty, imperfect, incompetent, incomplete, ineffective, insufficient, limited, meagre, mean, niggardly, *inf* pathetic, scanty, scarce, *inf* skimpy, sparse, unacceptable, unsatisfactory, unsuitable. *Opp* ADEQUATE.

inadvisable *adj* foolish, ill-advised, imprudent, misguided, unwise. ▷ SILLY. *Opp* WISE.

inanimate *adj* cold, dead, dormant, immobile, inactive, insentient, lifeless, motionless, spiritless, unconscious. *Opp* ANIMATE.

inappropriate *adj* ill-judged, ill-suited, ill-timed, improper, inapplicable, inapposite, incompatible, incongruous, incorrect, inept, inopportune, irrelevant, out of place, tactless, tasteless, unbecoming, unbefitting, unfit, unseasonable, unseemly, unsuitable, unsuited, untimely, wrong. *Opp* APPROPRIATE.

inarticulate *adj* dumb, faltering, halting, hesitant, mumbling, mute, shy, silent, speechless, stammering, stuttering, tongue-tied, voiceless. ▷ INCOHERENT. *Opp* ARTICULATE.

inattentive *adj* absent-minded, abstracted, careless, daydreaming, distracted, dreaming, drifting, heedless, *inf* in a world of your own, lacking concentration, negligent, preoccupied, rambling, remiss, slack, unobservant, vague, wandering, wool-gathering. *Opp* ATTENTIVE.

inaudible *adj* imperceptible, mumbled, quiet, silenced, silent, stifled, undetectable, undistinguishable, unheard. ▷ FAINT. *Opp* AUDIBLE.

incapable *adj* **1** clumsy, helpless, impotent, inadequate, incompetent, ineffective, ineffectual, inept, powerless, stupid, unable, unfit, unqualified, useless, weak. *Opp* CAPABLE. **2** ▷ DRUNK.

incentive *n* bait, *inf* carrot, encouragement, enticement, impetus, incitement, inducement, lure, motivation, reward, stimulus, *inf* sweetener.

incessant *adj* ceaseless, chronic, constant, continual, continuous, endless, eternal, everlasting, interminable, never-ending, non-stop, perennial, permanent, perpetual, persistent, relentless, unbroken, unceasing, unending, unremitting. *Opp* INTERMITTENT, TEMPORARY.

incident *n* **1** affair, circumstance, episode, event, fact, happening, occasion, occurrence, proceeding. **2** *a nasty incident.* accident, confrontation, disturbance, fight, scene, upset. ▷ COMMOTION.

incidental *adj* accidental, adventitious, attendant, casual, chance, fortuitous, inessential, minor, odd, random, secondary, serendipitous, subordinate, subsidiary, unplanned. *Opp* ESSENTIAL.

incipient *adj* beginning, developing, early, embryonic, growing, new, rudimentary, starting.

incisive *adj* acute, clear, concise, cutting, decisive, direct, penetrating, percipient, precise, sharp, telling, trenchant. *Opp* VAGUE.

incite *vb* awaken, encourage, excite, fire, foment, inflame, inspire, prompt, provoke, rouse, spur on, stimulate, stir, urge, whip up, work up.

inclination *n* affection, bent, bias, disposition, fondness, habit, instinct, leaning, liking, partiality, penchant, predilection, predisposition, preference, proclivity, propensity, readiness, tendency, trend, willingness. ▷ DESIRE.

incline *n* acclivity, ascent, declivity, descent, drop, grade, gradient, hill, pitch, ramp, rise, slope. • *vb* angle, ascend, bank, bend, bow, descend, drop, gravitate, lean, rise, slant, slope, tend, tilt, tip, veer. **inclined (to)** ▷ LIABLE.

include *vb* **1** add in, blend in, combine, comprehend, comprise, consist of, contain, embody, embrace, encompass, incorporate, involve, make room for, mix, subsume, take in. **2** *The price includes tea.* allow for, cover, take into account. *Opp* EXCLUDE.

incoherent *adj* confused, disconnected, disjointed, disordered, disorganized, garbled, illogical, incomprehensible, inconsistent, irrational, jumbled, mixed up, muddled, rambling, scrambled, unclear, unconnected, unstructured, unsystematic. ▷ INARTICULATE. *Opp* COHERENT.

incombustible *adj* fireproof, fire-resistant, flameproof, non-flammable. *Opp* COMBUSTIBLE.

income *n* earnings, gain, interest, pay, pension, proceeds, profits, receipts, return, revenue, salary, takings, wages. *Opp* EXPENSE.

incoming *adj* **1** approaching, arriving, entering, coming, landing, new, next, returning. **2** *incoming tide.* flowing, rising. *Opp* OUTGOING.

incompatible *adj* antipathetic, at variance, clashing, conflicting, contradictory, contrasting, different, discordant, discrepant, incongruous, inconsistent, irreconcilable, mismatched, opposed, unsuited. *Opp* COMPATIBLE.

incompetent *adj* **1** bungling, clumsy, feckless, gauche, helpless, *inf* hopeless, incapable, ineffective, ineffectual, inefficient, inexperienced, maladroit, unfit, unqualified, unskilled, untrained. **2** bungled, inadequate, inexpert, unacceptable, unsatisfactory, unskilful, useless. *Opp* COMPETENT.

incomplete *adj* abbreviated, abridged, *inf* bitty, deficient, edited, expurgated, faulty, fragmentary, imperfect, insufficient, partial, selective, shortened, sketchy, unfinished, unpolished, wanting. *Opp* COMPLETE.

incomprehensible *adj* abstruse, arcane, baffling, beyond comprehension, cryptic, deep, enigmatic, esoteric, illegible, impenetrable, indecipherable, meaningless, mysterious, mystifying, obscure, opaque, *inf* over my head, perplexing, puzzling, recondite, strange, too difficult, unclear, unfathomable, unintelligible. *Opp* COMPREHENSIBLE.

inconceivable *adj* implausible, impossible to understand, incredible, *inf* mind-boggling, staggering, unbelievable, undreamed-of, unimaginable, unthinkable. *Opp* CREDIBLE.

inconclusive *adj* ambiguous, equivocal, indecisive, indefinite, interrogative, open, open-ended, questionable, uncertain, unconvincing, unresolved, *inf* up in the air. *Opp* CONCLUSIVE.

incongruous *adj* clashing, conflicting, contrasting, discordant, ill-matched, ill-suited, inappropriate, incompatible, inconsistent, irreconcilable, odd, out of keeping, out of place, surprising, uncoordinated, unsuited. ▷ ABSURD. *Opp* COMPATIBLE.

inconsiderate *adj* careless, cruel, heedless, insensitive, intolerant, negligent, rude, self-centred, selfish, tactless, thoughtless, uncaring, unconcerned, unfriendly, ungracious, unhelpful, unkind, unsympathetic, unthinking. *Opp* CONSIDERATE.

inconsistent *adj* capricious, changeable, erratic, fickle, inconstant, patchy, unpredictable, unreliable, unstable, *inf* up-and-down, variable. ▷ INCOMPATIBLE. *Opp* CONSISTENT.

inconspicuous *adj* camouflaged, concealed, discreet, hidden, insignificant, in the background, invisible, modest, ordinary, out of sight, plain, restrained, retiring, self-effacing, small, unassuming, unobtrusive, unostentatious. *Opp* CONSPICUOUS.

inconvenience *n* annoyance, bother, discomfort, disruption, drawback, encumbrance, hindrance, impediment, irritation, nuisance, trouble. • *vb* annoy, bother, discommode, disturb, incommode, irk, irritate, *inf* put out, trouble.

inconvenient *adj* annoying, awkward, bothersome, cumbersome, difficult, embarrassing, ill-timed, inopportune, irksome, irritating, tiresome, troublesome, unsuitable, untimely, untoward, unwieldy. *Opp* CONVENIENT.

incorporate *vb* admit, combine, comprehend, comprise, consist of, contain, embody, embrace, encompass, include, involve, mix in, subsume, take in, take into account, unite. *Opp* EXCLUDE.

incorrect *adj* erroneous, fallacious, false, faulty, imprecise, improper, inaccurate, inexact, mendacious, misinformed, misleading, mistaken, specious, untrue. *Opp* CORRECT.

incorrigible *adj* confirmed, *inf* dyed-in-the-wool, habitual, hardened, *inf* hopeless, impenitent, incurable, inveterate, irredeemable, obdurate, shameless, unalterable, unreformable, unrepentant. ▷ WICKED.

incorruptible *adj* **1** honest, honourable, just, moral, sound, *inf* straight, true, trustworthy, unbribable, upright. *Opp* CORRUPT. **2** ▷ EVERLASTING.

increase *n* addition, amplification, augmentation, boost, build-up, crescendo, development, enlargement, escalation, expansion, extension, gain, growth, increment, inflation, intensification, proliferation, rise, spread, upsurge, upturn. • *vb* **1** add to, advance, amplify, augment, boost, broaden, build up, develop, enlarge, expand, extend, improve, lengthen, magnify, make bigger, maximize, multiply, prolong, put up, raise, *inf* step up, strengthen, stretch, swell, widen. **2** escalate, gain, get bigger, grow, intensify, proliferate, *inf* snowball, spread, wax. *Opp* DECREASE.

incredible *adj* beyond belief, far-fetched, implausible, impossible, improbable, inconceivable, miraculous, surprising, unbelievable, unconvincing, unimaginable, unlikely, untenable, unthinkable. ▷ EXTRAORDINARY. *Opp* CREDIBLE.

incredulous *adj* disbelieving, distrustful, doubtful, dubious, mistrustful, questioning, sceptical, suspicious, unbelieving, uncertain, unconvinced. *Opp* CREDULOUS.

incriminate *vb* accuse, blame, charge, embroil, implicate, inculpate, indict, involve, *inf* point the finger at. *Opp* EXCUSE.

incur *vb* earn, expose yourself to, get, lay yourself open to, provoke, run up, suffer.

incurable *adj* **1** fatal, hopeless, inoperable, irremediable, irreparable, terminal, untreatable. *Opp* CURABLE. **2** ▷ INCORRIGIBLE.

indebted *adj old use* beholden, bound, grateful, obliged, thankful, under an obligation.

indecent *adj inf* blue, coarse, crude, dirty, immodest, impolite, improper, impure, indelicate, insensitive, naughty, obscene, offensive, risqué, rude, *inf* sexy, *inf* smutty, suggestive, titillating, unprintable, unrepeatable, unsuitable, vulgar. ▷ INDECOROUS. *Opp* DECENT.

indecisive *adj* doubtful, equivocal, evasive, *inf* in two minds, irresolute, undecided. ▷ HESITANT, INDEFINITE. *Opp* DECISIVE. **be indecisive** ▷ HESITATE.

indecorous *adj* churlish, ill-bred, inappropriate, *inf* in bad taste, tasteless, unbecoming, uncouth, undignified, unseemly, vulgar. ▷ INDECENT. *Opp* DECOROUS.

indefensible *adj* incredible, insupportable, unjustifiable, unpardonable, unreasonable, unsound, untenable, vulnerable, weak. ▷ WRONG.

indefinite *adj* ambiguous, blurred, confused, dim, general, ill-defined, imprecise, indeterminate, inexact, inexplicit, *inf* leaving it open, neutral, obscure, uncertain, unclear, unsettled, unspecific, unspecified, unsure, vague. ▷ INDECISIVE. *Opp* DEFINITE.

indelible *adj* fast, fixed, indestructible, ineffaceable, ineradicable, ingrained, lasting, unfading, unforgettable. ▷ PERMANENT.

indentation *n* cut, dent, depression, dimple, dip, furrow, groove, hollow, indent, mark, nick, notch, pit, recess, score, serration, toothmark, zigzag.

independence *n* **1** autonomy, freedom, individualism, liberty, nonconformity, self-confidence, self-reliance, self-sufficiency. **2** autarchy, home rule, self-determination, self-government, self-rule, sovereignty.

independent *adj* **1** carefree, *inf* footloose, free, freethinking, individualistic, nonconformist, non-partisan, open-minded, private, self-confident, self-reliant, separate, spontaneous, unbeholden, unbiased, uncommitted, unconventional, unprejudiced, untrammelled, without ties. **2** autonomous, liberated, neutral, non-aligned, self-determining, self-governing, sovereign.

indescribable *adj* beyond words, indefinable, inexpressible, stunning, unspeakable, unutterable.

indestructible *adj* durable, enduring, eternal, everlasting, immortal, imperishable, ineradicable, lasting, permanent, shatterproof, solid, strong, tough, toughened, unbreakable.

index *n* **1** catalogue, directory, guide, key, register, table (*of contents*). **2** ▷ INDICATOR.

indicate *vb* announce, betoken, convey, communicate, denote, describe, designate, display, evidence, express, give an indication (*of*), give notice of, imply, intimate, make known, manifest, mean, notify, point out, register, reveal, say, show, signal, signify, specify, spell, stand for, suggest, symbolize, warn.

indication *n* augury, clue, evidence, forewarning, hint, inkling, intimation, omen, portent, sign, signal, suggestion, symptom, token, warning.

indicator *n* clock, dial, display, gauge, index, instrument, marker, meter, needle, pointer, screen, sign, signal.

indifferent *adj* **1** aloof, apathetic, blasé, bored, casual, cold, cool, detached, disinterested, dispassionate, distant, half-hearted, impassive, incurious, insouciant, neutral, nonchalant, not bothered, uncaring, unconcerned, unemotional, unenthusiastic, unexcited, unimpressed, uninterested, uninvolved, unmoved. ▷ IMPARTIAL. *Opp* ENTHUSIASTIC. **2** commonplace, fair, mediocre, middling, moderate, *inf* nothing to write home about, *inf* poorish, undistinguished, unexciting. ▷ ORDINARY. *Opp* EXCELLENT.

indigestion *n* dyspepsia, flatulence, heartburn.

indignant *adj inf* aerated, annoyed, cross, disgruntled, exasperated, furious, heated, infuriated, *inf* in high dudgeon, irate, irked, irritated, livid, mad, *inf* miffed, *inf* peeved, piqued, provoked, *inf* put out, riled, sore, upset, vexed. ▷ ANGRY.

indirect *adj* **1** *inf* all round the houses, ambagious, bendy, circuitous, devious, erratic, long, meandering, oblique, rambling, roundabout, roving, tortuous, twisting, winding, zigzag. **2** *an indirect insult.* ambiguous, backhanded, circumlocutory, disguised, equivocal, euphemistic, evasive, implicit, implied, oblique. *Opp* DIRECT.

indiscreet *adj* careless, foolish, ill-advised, ill-considered, ill-judged, impolite, impolitic, incautious, injudicious, insensitive, tactless, undiplomatic, unguarded, unthinking, unwise. *Opp* DISCREET.

indiscriminate *adj* aimless, careless, casual, confused, desultory, general, haphazard, *inf* hit or miss, imperceptive, miscellaneous, mixed, promiscuous, random, uncritical, undifferentiated, undiscerning, undiscriminating, uninformed, unplanned, unselective, unsystematic, wholesale. *Opp* SELECTIVE.

indispensable *adj* basic, central, compulsory, crucial, essential, imperative, important, key, mandatory, necessary, needed, obligatory, required, requisite, vital. *Opp* UNNECESSARY.

indisputable *adj* absolute, accepted, acknowledged, axiomatic, beyond doubt, certain, clear, definite, evident, incontestable, incontrovertible, indubitable, irrefutable, positive, proved, proven, self-evident, sure, unanswerable, unarguable, undeniable, undisputed, undoubted, unimpeachable, unquestionable. *Opp* DEBATABLE.

indistinct *adj* **1** bleary, blurred, confused, dim, dull, faint, fuzzy, hazy, ill-defined, indefinite, misty, obscure, shadowy, unclear, vague. **2** deadened, muffled, mumbled, muted, slurred, unintelligible, woolly. *Opp* DISTINCT.

indistinguishable *adj* alike, identical, interchangeable, the same, twin, undifferentiated. *Opp* DIFFERENT.

individual *adj* characteristic, different, distinct, distinctive, exclusive, idiosyncratic, individualistic, particular, peculiar, personal, private, separate, singular, special, specific, unique. *Opp* COLLECTIVE, GENERAL. • *n* ▷ PERSON.

indoctrinate *vb* brainwash, implant, instruct, re-educate, train. ▷ TEACH.

induce *vb* **1** coax, encourage, incite, influence, inspire, motivate, persuade, press, prevail on, stimulate, sway, *inf* talk into, tempt, urge. *Opp* DISCOURAGE. **2** *induce a fever.* bring on, cause, effect, engender, generate, give rise to, lead to, occasion, produce, provoke.

inducement *n* attraction, bait, bribe, encouragement, enticement, incentive, spur, stimulus, *inf* sweetener.

indulge *vb* be indulgent to, cosset, favour, give in to, gratify, humour, mollycoddle, pamper, pander to, spoil, *inf* spoonfeed, treat. *Opp* DEPRIVE. **indulge in** ▷ ENJOY. **indulge yourself** be self-indulgent, drink too much, eat too much, give in to temptation, overdo it, overeat, spoil yourself, succumb, yield.

indulgent *adj* compliant, easy-going, fond, forbearing, forgiving, genial, kind, lenient, liberal, over-generous, patient, permissive, tolerant. *Opp* STRICT.

industrious *adj* assiduous, busy, conscientious, diligent, dynamic, earnest, energetic, enterprising, hard-working, involved, keen, laborious, persistent, pertinacious, productive, sedulous, tireless, unflagging, untiring, zealous. *Opp* LAZY.

industry *n* **1** business, commerce, manufacturing, production, trade. **2** activity, application, commitment, determination, diligence, dynamism, effort, energy, enterprise, industriousness, keenness, labour, perseverance, persistence, sedulousness, tirelessness, toil, zeal. ▷ WORK. *Opp* LAZINESS.

inedible *adj* bad for you, harmful, indigestible, nauseating, *inf* off, poisonous, rotten, tough, uneatable, unpalatable, unwholesome. *Opp* EDIBLE.

ineffective *adj* **1** fruitless, futile, *inf* hopeless, inept, unconvincing, unproductive, unsuccessful, useless, vain, worthless. **2** disorganized, feckless, feeble, idle, impotent, inadequate, incapable, incompetent, ineffectual, inefficient, powerless, shiftless, unenterprising, weak. *Opp* EFFECTIVE.

inefficient *adj* **1** extravagant, prodigal, uneconomic, wasteful. **2** ▷ INEFFECTIVE. *Opp* EFFICIENT.

inelegant *adj* awkward, clumsy, crude, gauche, graceless, inartistic, rough, uncouth, ungainly, unpolished, unskilful, unsophisticated, unstylish. ▷ UGLY. *Opp* ELEGANT.

ineligible *adj* disqualified, inappropriate, *inf* out of the running, *inf* ruled out, unacceptable, unauthorized, unfit, unqualified, unsuitable, unworthy. *Opp* ELIGIBLE.

inept *adj* **1** awkward, bumbling, bungling, clumsy, gauche, incompetent, inexpert, maladroit, unskilful, unskilled. **2** ▷ INAPPROPRIATE.

inequality *n* contrast, difference, discrepancy, disparity, dissimilarity, imbalance, incongruity, prejudice. *Opp* EQUALITY.

inert *adj* apathetic, dormant, idle, immobile, inactive, inanimate, lifeless, passive, quiescent, quiet, slow, sluggish, static, stationary, still, supine, torpid. *Opp* LIVELY.

inertia *n* apathy, deadness, idleness, immobility, inactivity, indolence, lassitude, laziness, lethargy, listlessness, numbness, passivity, sluggishness, torpor. *Opp* LIVELINESS.

inessential *adj* dispensable, expendable, minor, needless, non-essential, optional, ornamental, secondary, spare, superfluous, unimportant, unnecessary. *Opp* ESSENTIAL.

inevitable *adj* assured, *inf* bound to happen, certain, destined, fated, ineluctable, inescapable, inexorable, ordained, predictable, sure, unavoidable. ▷ RELENTLESS.

inexcusable *adj* ▷ UNFORGIVABLE.

inexpensive *adj* ▷ CHEAP.

inexperienced *adj* *inf* born yesterday, callow, *inf* green, immature, inexpert, innocent, naïve, new, probationary, raw, unaccustomed, unfledged, uninitiated, unskilled, unsophisticated, untried, *inf* wet behind the ears, young. *Opp* EXPERT.

inexplicable *adj* baffling, bewildering, confusing, enigmatic, incomprehensible, inscrutable, insoluble, mysterious, mystifying, perplexing, puzzling, strange, unaccountable, unexplainable, unfathomable, unsolvable. *Opp* STRAIGHTFORWARD.

infallible *adj* certain, dependable, faultless, foolproof, impeccable, perfect, reliable, sound, sure, trustworthy, unbeatable, unerring, unfailing. *Opp* FALLIBLE.

infamous *adj* disgraceful, disreputable, ill-famed, notorious, outrageous, well-known. ▷ WICKED.

infant *n* baby, *inf* toddler, *inf* tot. ▷ CHILD.

infantile *adj* [*derog*] adolescent, babyish, childish, immature, juvenile, puerile. ▷ SILLY. *Opp* MATURE.

infatuated *adj* besotted, charmed, enchanted, *inf* head over heels, in love, obsessed, *inf* smitten.

infatuation *n* *inf* crush, obsession, passion. ▷ LOVE.

infect *vb* **1** blight, contaminate, defile, poison, pollute, spoil, taint. **2** affect, influence, inspire, touch. **infected** ▷ SEPTIC.

infection *n* blight, contagion, contamination, epidemic, pestilence, pollution, virus. ▷ ILLNESS.

infectious *adj* catching, communicable, contagious, spreading, transmissible, transmittable.

infer *vb* assume, conclude, deduce, derive, draw a conclusion, extrapolate, gather, guess, reach the conclusion, surmise, understand, work out.

inferior *adj* **1** humble, junior, lesser, lower, lowly, mean, menial, secondary, second-class, servile, subordinate, subsidiary, unimportant. **2** cheap, indifferent, mediocre, poor, shoddy, tawdry, *inf* tinny. *Opp* SUPERIOR. • *n* ▷ SUBORDINATE.

infertile *adj* barren, sterile, unfruitful, unproductive.

infest *vb* infiltrate, overrun, pervade, plague. **infested** alive, crawling, swarming, teeming, verminous.

infidelity *n* **1** adultery, unfaithfulness. **2** ▷ DISLOYALTY.

infiltrate *vb* enter secretly, insinuate, intrude, penetrate, spy on.

infinite *adj* astronomical, big, boundless, countless, endless, eternal, everlasting, immeasurable, immense, incalculable, indeterminate, inestimable, inexhaustible, innumerable, interminable, limitless, multitudinous, never-ending, numberless, perpetual, uncountable, undefined, unending, unfathomable, unlimited, unnumbered, untold. ▷ HUGE. *Opp* FINITE.

infinity *n* endlessness, eternity, infinite distance, infinite quantity, infinitude, perpetuity, space.

infirm *adj* bedridden, crippled, elderly, feeble, frail, lame, old, poorly, senile, sickly, unwell. ▷ ILL, WEAK. *Opp* HEALTHY.

inflame *vb* arouse, encourage, excite, fire, foment, goad, ignite, incense, incite, kindle, madden, provoke, rouse, stimulate, stir up, work up. ▷ ANGER. *Opp* COOL. **inflamed** ▷ PASSIONATE, SEPTIC.

inflammable *adj* burnable, combustible, flammable, volatile. *Opp* INCOMBUSTIBLE.

inflammation *n* abscess, boil, infection, irritation, redness, sore, soreness, swelling.

inflate *vb* **1** blow up, dilate, distend, enlarge, puff up, pump up, swell. **2** ▷ EXAGGERATE.

inflexible *adj* **1** adamantine, firm, hard, hardened, immovable, rigid, solid, stiff, unbending, unyielding. **2** adamant, entrenched, fixed, immutable, inexorable, intractable, intransigent, obdurate, obstinate, *inf* pig-headed, refractory, resolute, rigorous, strict, stubborn, unalterable, unchangeable, uncompromising, unhelpful. *Opp* FLEXIBLE.

inflict *vb* administer, apply, deal out, enforce, force, impose, mete out, perpetrate, wreak.

influence *n* ascendancy, authority, control, direction, dominance, effect, guidance, hold, impact, leverage, power, pressure, pull, sway, weight. • *vb* **1** affect, bias, change, control, direct, dominate, exert influence on, guide, impinge on, impress, manipulate, modify, motivate, move, persuade, prejudice, prompt, put pressure on, stir, sway. **2** *influence a judge.* bribe, corrupt, lead astray, suborn, tempt.

influential *adj* authoritative, compelling, controlling, convincing, dominant, effective, far-reaching, forceful, guiding, important, inspiring, leading, moving, persuasive, potent, powerful, prestigious, significant, strong, telling, weighty. *Opp* UNIMPORTANT.

influx *n* flood, flow, inflow, inundation, invasion, rush, stream.

inform *vb* **1** advise, apprise, brief, communicate to, enlighten, *inf* fill in, give information to, instruct, leak, notify, *inf* put in the picture, teach, tell, *inf* tip off. **2** *inf* blab, give information, *sl* grass, *sl* peach, *sl* rat, *inf* sneak, *inf* split on, *inf* tell, *inf* tell tales. **inform against** ▷ BETRAY. **informed** ▷ KNOWLEDGEABLE.

informal *adj* **1** approachable, casual, comfortable, cosy, easy, easy-going, everyday, familiar, free and easy, friendly, homely, natural, ordinary, relaxed, simple, unceremonious, unofficial, unpretentious, unsophisticated. **2** *informal language.* chatty, colloquial, personal, slangy, vernacular. **3** *an informal design.* asymmetrical, flexible, fluid, intuitive, irregular, spontaneous. *Opp* FORMAL.

information *n* **1** announcement, briefing, bulletin, communication, enlightenment, instruction, message, news, report, statement, *old use* tidings, *inf* tip-off, word. **2** data, database, dossier, evidence, facts, intelligence, knowledge, statistics.

informative *adj* communicative, edifying, educational, enlightening, factual, giving information, helpful, illuminating, instructive, meaningful, revealing, useful. *Opp* MEANINGLESS.

informer *n sl* grass, informant, spy, *sl* stool-pigeon, *inf* tell-tale, traitor.

infrequent *adj* exceptional, intermittent, irregular, occasional, *inf* once in a blue moon, rare, spasmodic, uncommon, unusual. *Opp* FREQUENT.

infringe *vb* breach, break, contravene, defy, disobey, disregard, flout, ignore, overstep, sin against, transgress, violate.

ingenious *adj* adroit, artful, astute, brilliant, clever, complex, crafty, creative, cunning, deft, imaginative, inspired, intelligent, intricate, inventive, neat, original, resourceful, shrewd, skilful, *inf* smart, subtle, talented. *Opp* UNIMAGINATIVE.

ingenuous *adj* artless, childlike, frank, guileless, honest, innocent, naïve, open, plain, simple, sincere, trusting, unaffected, uncomplicated, unsophisticated. *Opp* SOPHISTICATED.

ingredient *n* component, constituent, element, factor, *plur* makings, part.

inhabit *vb old use* abide in, colonize, dwell in, live in, make your home in, occupy, people, populate, possess, reside in, settle in, set up home in.

inhabitable *adj* habitable, in good repair, liveable, usable. *Opp* UNINHABITABLE.

inhabitant *n* citizen, *old use* denizen, dweller, inmate, native, occupant, occupier, *plur* population, resident, settler, tenant, *plur* townsfolk, *plur* townspeople.

inherent *adj* built-in, congenital, essential, fundamental, hereditary, immanent, inborn, inbred, indwelling, ingrained, intrinsic, native, natural.

inherit *vb* be the inheritor of, be left, *inf* come into, receive as an inheritance, succeed to. **inherited** ▷ HEREDITARY.

inheritance *n* bequest, birthright, estate, fortune, heritage, legacy, patrimony.

inhibit *vb* bridle, check, control, curb, discourage, frustrate, hinder, hold back, prevent, quell, repress, restrain. **inhibited** ▷ REPRESSED, SHY.

inhibition *n* **1** bar, barrier, check, constraint, curb, impediment, interference, restraint, stricture. **2** blockage, diffidence, *inf* hang-up, repression, reserve, self-consciousness, shyness.

inhospitable *adj* antisocial, reclusive, reserved, solitary, standoffish, unkind, unsociable, unwelcoming. ▷ UNFRIENDLY. **2** bleak, cold, comfortless, desolate, grim, hostile, lonely. *Opp* HOSPITABLE.

inhuman *adj* animal, barbaric, barbarous, bestial, bloodthirsty, brutish, diabolical, fiendish, merciless, pitiless, ruthless, savage, unnatural, vicious. ▷ INHUMANE. *Opp* HUMAN.

inhumane *adj* cold-hearted, cruel, hard, hard-hearted, heartless, inconsiderate, insensitive, uncaring, uncharitable, uncivilized, unfeeling, unkind, unsympathetic. ▷ INHUMAN. *Opp* HUMANE.

initial *adj* beginning, commencing, earliest, first, inaugural, incipient, introductory, opening, original, primary, starting. *Opp* FINAL.

initiate *vb* activate, actuate, begin, commence, enter upon, get going, get under way, inaugurate, instigate, institute, introduce, launch, originate, set going, set in motion, set up, start, take the initiative, trigger.

initiative *n* ambition, drive, dynamism, enterprise, *inf* get-up-and-go, inventiveness, lead, leadership, originality, resourcefulness. **take the initiative** ▷ INITIATE.

injection *n inf* fix, inoculation, *inf* jab, vaccination.

injure *vb* break, crush, cut, damage, deface, disfigure, harm, hurt, ill-treat, impair, mar, ruin, spoil, vandalize. ▷ WOUND.

injurious *adj* **1** damaging, deleterious, destructive, detrimental, harmful, insalubrious, painful, ruinous. **2** ▷ ABUSIVE.

injury *n* damage, harm, hurt, mischief. ▷ WOUND.

injustice *n* bias, bigotry, discrimination, dishonesty, favouritism, illegality, inequality, inequity, one-sidedness, oppression, partiality, partisanship, prejudice, unfairness, unlawfulness, wrong, wrongness. *Opp* JUSTICE.

inn *n old use* hostelry, hotel, *inf* local, pub, tavern.

inner *adj* central, concealed, hidden, innermost, inside, interior, internal, intimate, inward, mental, middle, private, secret. *Opp* OUTER.

innocence *n* **1** goodness, honesty, incorruptibility, purity, righteousness, sinlessness, virtue. **2** [*derog*] gullibility, inexperience, naïvety, simple-mindedness.

innocent *adj* **1** above suspicion, angelic, blameless, chaste, childlike, faultless, free from blame, guiltless, harmless, honest, immaculate, incorrupt, inoffensive, pure, righteous, sinless, spotless, untainted, virginal, virtuous. *Opp* CORRUPT, GUILTY. **2** artless, childlike, credulous, *inf* green, guileless, gullible, inexperienced, ingenuous, naïve, simple, simple-minded, trusting, unsophisticated.

innovation *n* change, departure, invention, new feature, novelty, reform, revolution.

innovator *n* discoverer, experimenter, inventor, pioneer, reformer, revolutionary.

innumerable *adj* countless, many, numberless, uncountable, untold. ▷ INFINITE.

inquest *n* hearing. ▷ INQUIRY.

inquire *vb* ask, explore, investigate, seek information, *inf* probe, search, survey. ▷ ENQUIRE.

inquiry *n* cross-examination, examination, inquest, inquisition, interrogation, investigation, poll, *inf* post-mortem, *inf* probe, referendum, review, study, survey.

inquisitive *adj* curious, impertinent, indiscreet, inquiring, interfering, intrusive, investigative, meddlesome, meddling, *inf* nosy, probing, prying, questioning, sceptical, searching, *inf* snooping, spying. **be inquisitive** ▷ PRY.

insane *adj inf* crazy, deranged, lunatic, *inf* mental, psychotic, unbalanced, unhinged. ▷ MAD. *Opp* SANE.

inscription *n* dedication, engraving, epigraph, superscription, writing.

insect *n inf* bug, *inf* creepy-crawly. □ *ant, aphid, bee, beetle, blackfly, butterfly, cicada, cockchafer, cockroach, crane-fly, cricket, daddy-long-legs, damselfly, dragonfly, earwig, firefly, fly, glow-worm, gnat, grasshopper, hornet, ladybird, locust, mantis, mayfly, midge, mosquito, moth, sawfly, termite, tsetse (fly), wasp, weevil.*

insecure *adj* **1** dangerous, flimsy, loose, precarious, rickety, rocky, shaky, unsafe, unsound, unstable, unsteady, unsupported, weak, wobbly. **2** *an insecure feeling.* anxious, apprehensive, defenceless, exposed, open, uncertain, underconfident, unprotected, vulnerable, worried. *Opp* SECURE.

insensible *adj* anaesthetized, benumbed, *inf* dead to the world, inert, insensate, insentient, knocked out, numb, *inf* out, senseless, unaware, unconscious. *Opp* CONSCIOUS.

insensitive *adj* **1** anaesthetized, dead, numb, unresponsive. **2** boorish, callous, crass, cruel, imperceptive, obtuse, tactless, *inf* thick-skinned, thoughtless, uncaring, unfeeling, unsympathetic. *Opp* SENSITIVE.

inseparable *adj* always together, attached, indissoluble, indivisible, integral.

insert *vb* drive in, embed, implant, intercalate, interject, interleave, interpolate, interpose, introduce, place in, *inf* pop in, push in, put in, stick in, tuck in.

inside *adj* central, indoor, inner, innermost, interior, internal. • *n* bowels, centre, contents, core, heart, indoors, interior, lining, middle. *Opp* OUTSIDE. **insides** ▷ ENTRAILS.

insidious *adj* creeping, deceptive, furtive, pervasive, secretive, stealthy, subtle, surreptitious, treacherous, underhand. ▷ CRAFTY.

insignificant *adj* forgettable, inconsiderable, irrelevant, insubstantial, lightweight, meaningless, minor, negligible, paltry, small, trifling, trivial, undistinguished, unimpressive, valueless, worthless, unimportant. *Opp* SIGNIFICANT.

insincere *adj* artful, crafty, deceitful, deceptive, devious, dishonest, disingenuous, dissembling, false, feigned, flattering, *inf* foxy, hollow, hypocritical, lying, *inf* mealy-mouthed, mendacious, perfidious, *inf* phoney, pretended, *inf* put on, *inf* smarmy, sycophantic, treacherous, *inf* two-faced, untrue, untruthful, wily. *Opp* SINCERE.

insist *vb* **1** assert, asseverate, aver, avow, declare, emphasize, hold, maintain, state, stress, swear, take an oath, vow. **2** assert yourself, be assertive, command, persist, *inf* put your foot down, stand firm, *inf* stick to your guns. **insist on** ▷ DEMAND.

insistent *adj* assertive, demanding, dogged, emphatic, firm, forceful, importunate, inexorable, obstinate, peremptory, persistent, relentless, repeated, resolute, stubborn, unrelenting, unremitting, urgent.

insolence *n* arrogance, boldness, *inf* cheek, defiance, disrespect, effrontery, impertinence, impudence, incivility, insubordination, *inf* lip, presumptuousness, rudeness, *inf* sauce.

insolent *adj* arrogant, audacious, bold, brazen, *inf* cheeky, contemptuous, defiant, disdainful, disrespectful, forward, *inf* fresh, impertinent, impolite, impudent, insubordinate, insulting, offensive, pert, presumptuous, saucy, shameless, sneering, uncivil. ▷ RUDE. *Opp* POLITE.

insoluble *adj* baffling, enigmatic, incomprehensible, inexplicable, mystifying, puzzling, strange, unaccountable, unanswerable, unfathomable, unsolvable. *Opp* SOLUBLE.

insolvent *adj* bankrupt, *inf* bust, failed, ruined. ▷ POOR.

inspect *vb* check, examine, *sl* give it the once over, investigate, peruse, pore over, scan, scrutinize, study, survey, vet.

inspection *n* check, check-up, examination, *inf* going-over, investigation, review, scrutiny, survey.

inspector *n* controller, examiner, investigator, official, scrutineer, superintendent, supervisor, tester.

inspiration *n* **1** creativity, genius, imagination, muse. **2** enthusiasm, impulse, incitement, influence, motivation, prompting, spur, stimulation, stimulus. **3** *a sudden inspiration.* brainwave, idea, insight, revelation, thought.

inspire *vb* activate, animate, arouse, awaken, *inf* egg on, encourage, energize, enthuse, fire, galvanize, influence, inspirit, instigate, kindle, motivate, prompt, provoke, quicken, reassure, set off, spark off, spur, stimulate, stir, support.

instability *n* capriciousness, change, changeableness, fickleness, fluctuation, flux, impermanence, inconstancy, insecurity, mutability, precariousness, shakiness, transience, uncertainty, unpredictability, unreliability, unsteadiness, *inf* ups-and-downs, vacillation, variability, variations, weakness. *Opp* STABILITY.

install *vb* ensconce, establish, fit, fix, instate, introduce, place, plant, position, put in, settle, set up, situate, station. *Opp* REMOVE.

instalment *n* **1** payment, rent, rental. **2** chapter, episode, part.

instance *n* case, example, exemplar, illustration, occurrence, precedent, sample.

instant *adj* direct, fast, immediate, instantaneous, on-the-spot, prompt, quick, rapid, speedy, split-second, swift, unhesitating, urgent. • *n* flash, *inf* jiffy, moment, point of time, second, split second, *inf* tick, *inf* trice, *inf* twinkling.

instigate *vb* activate, begin, be the instigator of, bring about, cause, encourage, foment, generate, incite, initiate, inspire, kindle, prompt, provoke, set up, start, stimulate, stir up, urge, *inf* whip up.

instigator *n* agitator, fomenter, inciter, initiator, inspirer, leader, mischief-maker, provoker, ringleader, troublemaker.

instil *vb inf* din into, imbue, implant, inculcate, indoctrinate, infuse, ingrain, inject, insinuate, introduce.

instinct *n* bent, faculty, feel, feeling, guesswork, hunch, impulse, inclination, instinctive urge, intuition, presentiment, propensity, sixth-sense, the subconscious, tendency, urge.

instinctive *adj* automatic, congenital, constitutional, *inf* gut, impulsive, inborn, inbred, inherent, innate, instinctual, intuitive, involuntary, irrational, mechanical, native, natural, reflex, spontaneous, subconscious, unconscious, unreasoning, unthinking, visceral. *Opp* DELIBERATE.

institute *n* ▷ INSTITUTION. • *vb* begin, create, establish, fix up, found, inaugurate, initiate, introduce, launch, open, organize, originate, pioneer, set up, start.

institution *n* **1** creation, establishing, formation, founding, inauguration, inception, initiation, introduction, launching, opening, setting-up. **2** academy, asylum, college, establishment, foundation, home, hospital, institute, organization, school, *inf* set-up. **3** convention, custom, habit, practice, ritual, routine, rule, tradition.

instruct *vb* **1** coach, drill, educate, indoctrinate, inform, lecture, prepare, school, teach, train, tutor. **2** authorize, brief, charge, command, direct, enjoin, give the order, order, require, tell.

instruction *n* **1** briefing, coaching, demonstration, drill, education, guidance, indoctrination, lecture, lesson, schooling, teaching, training, tuition, tutorial, tutoring. **2** authorization, brief, charge, command, direction, directive, order, requisition.

instructive *adj* didactic, edifying, educational, enlightening, helpful, illuminating, improving, informational, informative, instructional, revealing.

instructor *n* adviser, coach, trainer, tutor. ▷ TEACHER.

instrument *n* apparatus, appliance, contraption, device, equipment, gadget, implement, machine, mechanism, tool, utensil. **musical instrument** □ *accordion, bagpipes, banjo, bassoon, bugle, castanets, celesta, cello, clarinet, clavichord, clavier, concertina, cor anglais, cornet, cymbals, double-bass, drum, dulcimer, euphonium, fiddle, fife, flugelhorn, flute, fortepiano, French horn, glockenspiel, gong, guitar, harmonica, harmonium, harp, harpsichord, horn, hurdy-gurdy, kettledrum, keyboard, lute, lyre, mouth-organ, oboe, organ, piano, piccolo, pipes, recorder, saxophone, sitar, spinet, synthesizer, tambourine, timpani, triangle, trombone, trumpet, tuba, tubular bells, ukulele, viol, viola, violin, virginals, xylophone, zither.*

instrumental *adj* active, advantageous, beneficial, contributory, helpful, influential, supportive, useful, valuable.

insubordinate *adj* defiant, disobedient, insurgent, mutinous, rebellious, riotous, seditious, undisciplined, unruly. ▷ IMPERTINENT. *Opp* OBEDIENT.

insufficient *adj* deficient, disappointing, inadequate, incomplete, little, meagre, mean, niggardly, *inf* pathetic, poor, scanty, scarce, short, skimpy, sparse, unsatisfactory. *Opp* EXCESSIVE, SUFFICIENT.

insular *adj* closed, cut-off, isolated, limited, narrow, narrow-minded, parochial, provincial, remote, separated. *Opp* BROADMINDED, COSMOPOLITAN.

insulate *vb* **1** cocoon, cover, cushion, enclose, isolate, lag, protect, shield, surround, wrap up. **2** cut off, detach, isolate, keep apart, quarantine, segregate, separate.

insult *n* abuse, affront, aspersion, *inf* cheek, contumely, defamation, impudence, indignity, insulting behaviour, libel, *inf* put-down, rudeness, slander, slight, slur, snub. • *vb* abuse, affront, be rude to, *inf* call names, *inf* cock a snook at, defame, dishonour, disparage, libel, mock, offend, outrage, patronize, revile, slander, slang, slight, sneer at, snub, *inf* thumb your nose at, vilify. *Opp* COMPLIMENT. **insulting** ▷ RUDE.

insuperable *adj* insurmountable, overwhelming, unconquerable. ▷ IMPOSSIBLE.

insurance *n* assurance, cover, indemnification, indemnity, policy, protection, security.

insure *vb* cover yourself, indemnify, protect, take out insurance.

intact *adj* complete, entire, integral, solid, sound, unbroken, undamaged, whole. ▷ PERFECT.

intangible *adj* abstract, airy, disembodied, elusive, ethereal, evanescent, fleeting, impalpable, imperceptible, imponderable, incorporeal, indefinite, insubstantial, invisible, shadowy, unreal, vague. *Opp* TANGIBLE.

integral *adj* **1** basic, constituent, essential, fundamental, indispensable, intrinsic, irreplaceable, necessary, requisite. *Opp* INESSENTIAL. **2** *an integral unit.* attached, complete, full, indivisible, whole. *Opp* SEPARATE.

integrate *vb* amalgamate, assemble, blend, bring together, coalesce, combine, consolidate, desegregate, fuse, harmonize, join, knit, merge, mix, put together, unify, unite, weld. *Opp* SEPARATE.

integrity *n* **1** decency, fidelity, goodness, honesty, honour, incorruptibility, loyalty, morality, principle, probity, rectitude, reliability, righteousness, sincerity, trustworthiness, uprightness, veracity, virtue. **2** ▷ UNITY.

intellect *n* *inf* brains, cleverness, genius, mind, rationality, reason, sense, understanding, wisdom, *old use* wit. ▷ INTELLIGENCE.

intellectual *adj* **1** academic, *inf* bookish, cerebral, cultured, educated, scholarly, studious, thinking, thoughtful. ▷ INTELLIGENT. **2** cultural, deep, difficult, educational, highbrow, improving, thought-provoking. • *n* academic, *inf* egghead, genius, highbrow, intellectual person, *inf* mastermind, *inf* one of the intelligentsia, savant, thinker.

intelligence *n* **1** ability, acumen, alertness, astuteness, brainpower, *inf* brains, brightness, brilliance, capacity, cleverness, discernment, genius, *inf* grey matter, insight, intellect, judgement, keenness, mind, *inf* nous, perceptiveness, perspicaciousness, perspicacity, quickness, reason, sagacity, sense, sharpness, shrewdness, understanding, wisdom, wit, wits. **2** data, facts, information, knowledge, *inf* low-down, news, notification, report, *inf* tip-off, warning. **3** espionage, secret service, spying.

intelligent *adj* able, acute, alert, astute, brainy, bright, brilliant, *inf* canny, clever, discerning, educated, intellectual, knowing, penetrating, perceptive, percipient, perspicacious, profound, quick, ratiocinative, rational, reasonable, sagacious, sensible, sharp, shrewd, *inf* smart, thinking, thoughtful, trenchant, wise, *inf* with it, witty. *Opp* STUPID.

intelligible *adj* clear, comprehensible, decipherable, fathomable, legible, logical, lucid, meaningful, plain, straightforward, unambiguous, understandable. *Opp* INCOMPREHENSIBLE.

intend *vb* aim, aspire, contemplate, design, determine, have in mind, mean, plan, plot, propose, purpose, resolve, scheme.

intense *adj* **1** ardent, burning, consuming, deep, eager, earnest, emotional, fanatical, fervent, fervid, impassioned, passionate, powerful, profound, serious, strong, towering, vehement, violent, zealous. *Opp* COOL, HALF-HEARTED. **2** *intense pain.* acute, agonizing, excruciating, extreme, fierce, great, harsh, keen, severe, sharp. *Opp* SLIGHT.

intensify *vb* add to, aggravate, augment, become greater, boost, build up, deepen, emphasize, escalate, fire, focus, fuel, heighten, *inf* hot up, increase, magnify, make greater, quicken, raise, redouble, reinforce, sharpen, *inf* step up, strengthen, whet. *Opp* REDUCE.

intensive *adj inf* all-out, comprehensive, concentrated, detailed, exhaustive, high-powered, thorough, unremitting.

intent *adj* absorbed, attentive, bent, committed, concentrated, concentrating, determined, eager, engrossed, enthusiastic, firm, focused, keen, occupied, preoccupied, resolute, set, steadfast, watchful, zealous. *Opp* CASUAL.
• *n* ▷ INTENTION.

intention *n* aim, ambition, design, end, goal, intent, object, objective, plan, point, purpose, target.

intentional *adj* calculated, conscious, contrived, deliberate, designed, intended, knowing, planned, pre-arranged, preconceived, premeditated, prepared, studied, wilful. *Opp* ACCIDENTAL.

intercept *vb* ambush, arrest, block, catch, check, cut off, deflect, head off, impede, interrupt, obstruct, stop, thwart, trap.

intercourse *n* **1** communication, conversation, dealings, interaction, traffic. **2** *sexual intercourse.* carnal knowledge, coition, coitus, congress, copulation, intimacy, love-making, mating, rape, sex, union.

interest *n* **1** attention, attentiveness, care, commitment, concern, curiosity, involvement, notice, regard, scrutiny. **2** *of no interest.* consequence, importance, moment, note, significance, value. **3** *leisure interests.* activity, diversion, hobby, pastime, preoccupation, pursuit, relaxation.
• *vb* absorb, appeal to, arouse the curiosity of, attract, capture the imagination of, captivate, concern, divert, enchant, engage, engross, entertain, enthral, fascinate, intrigue, involve, occupy, preoccupy, stimulate, *inf* turn on. ▷ EXCITE. *Opp* BORE.

interested *adj* **1** absorbed, attentive, curious, engrossed, enthusiastic, immersed, intent, involved, keen, occupied, preoccupied, rapt, responsive, riveted. *Opp* UNINTERESTED. **2** concerned, involved, partial. ▷ BIASED. *Opp* DISINTERESTED.

interesting *adj* absorbing, appealing, attractive, challenging, compelling, curious, engaging, engrossing, entertaining, enthralling, fascinating, gripping, imaginative, important, intriguing, inviting, original, piquant, *often ironic* riveting, spellbinding, unpredictable, unusual, varied. *Opp* BORING.

interfere *vb* be a busybody, butt in, interrupt, intervene, intrude, meddle, obtrude, *inf* poke your nose in, pry, snoop, *inf* stick your oar in, tamper. **interfere with** ▷ OBSTRUCT. **interfering** ▷ NOSY.

interim *adj* half-time, halfway, provisional, stopgap, temporary.

interior *adj* ▷ INTERNAL. • *n* centre, core, depths, heart, inside, middle, nucleus.

interlude *n* entràcte, intermezzo, intermission. ▷ INTERVAL.

intermediary *n* agent, ambassador, arbiter, arbitrator, broker, go-between, mediator, middleman, negotiator, referee, spokesperson, umpire.

intermediate *adj* average, *inf* betwixt and between, halfway, intermediary, intervening, mean, medial, median, middle, midway, *inf* neither one thing nor the other, neutral, *inf* sitting on the fence, transitional.

intermittent *adj* broken, discontinuous, erratic, fitful, irregular, occasional, *inf* on and off, periodic, random, recurrent, spasmodic, sporadic. *Opp* CONTINUOUS.

internal *adj* **1** inner, inside, interior. *Opp* EXTERNAL. **2** confidential, hidden, intimate, inward, personal, private, secret, undisclosed.

international *adj* cosmopolitan, global, intercontinental, universal, worldwide.

interpret *vb* clarify, clear up, construe, decipher, decode, define, elucidate, explain, explicate, expound, gloss, make clear, make sense of, paraphrase, render, rephrase, reword, simplify, sort out, translate, understand, unravel, work out.

interpretation *n* clarification, definition, elucidation, explanation, gloss, paraphrase, reading, rendering, translation, understanding, version.

interrogation *n* cross-examination, debriefing, examination, grilling, inquisition, questioning, *inf* third degree.

interrogative *adj* asking, inquiring, inquisitive, interrogatory, investigatory, questioning.

interrupt *vb* **1** *inf* barge in, break in, butt in, *inf* chime in, *inf* chip in, cut in, disrupt, disturb, heckle, hold up, interfere, intervene, intrude, punctuate, obstruct, spoil. **2** break off, call a halt to, cut off, cut short, discontinue, halt, stop, suspend, terminate.

interruption *n* break, check, disruption, division, gap, halt, hiatus, interference, intrusion, stop, suspension. ▷ INTERVAL.

intersect *vb* bisect each other, converge, criss-cross, cross, divide, meet, pass across each other.

interval *n* **1** adjournment, break, *inf* breather, breathing-space, delay, distance, gap, hiatus, interruption, lapse, lull, opening, pause, recess, respite, rest, space, void, wait. **2** entràcte, interlude, intermezzo, intermission.

intervene *vb* **1** come between, elapse, happen, occur, pass. **2** arbitrate, butt in, intercede, interfere, interpose, interrupt, intrude, mediate, *inf* step in.

interview *n* appraisal, audience, duologue, formal discussion, meeting, questioning, selection procedure, vetting. • *vb* appraise, ask questions, evaluate, examine, interrogate, question, sound out, vet.

interweave *vb* criss-cross, entwine, interlace, intertwine, knit, tangle, weave together.

intestines *plur n* bowels, entrails, innards, insides, offal.

intimate *adj* **1** affectionate, close, familiar, informal, loving, sexual. ▷ FRIENDLY. **2** *intimate details*. confidential, detailed, exhaustive, personal, private, secret. • *n* ▷ FRIEND. • *vb* ▷ INDICATE.

intimidate *vb* alarm, browbeat, bully, coerce, cow, daunt, dismay, frighten, hector, make afraid, menace, overawe, persecute, petrify, scare, terrify, terrorize, threaten, tyrannize.

intolerable *adj* excruciating, impossible, insufferable, insupportable, unacceptable, unbearable, unendurable. *Opp* TOLERABLE.

intolerant *adj* biased, bigoted, chauvinistic, classist, discriminatory, dogmatic, illiberal, narrow-minded, one-sided, opinionated, prejudiced, racist, sexist, uncharitable, unsympathetic, xenophobic. *Opp* TOLERANT.

intonation *n* accent, delivery, inflection, modulation, pronunciation, sound, speech pattern, tone.

intoxicate *vb* addle, inebriate, make drunk, stupefy. **intoxicated** ▷ DRUNK, EXCITED. **intoxicating** ▷ ALCOHOLIC, EXCITING.

intricate *adj* complex, complicated, convoluted, delicate, detailed, elaborate, entangled, fancy, *inf* fiddly, involved, *inf* knotty, labyrinthine, ornate, sophisticated, tangled, tortuous. *Opp* SIMPLE.

intrigue *n* ▷ PLOT. • *vb* **1** appeal to, arouse the curiosity of, attract, beguile, captivate, capture the interest of, engage, engross, excite the curiosity of, fascinate, interest, stimulate, *inf* turn on. *Opp* BORE.

intrinsic *adj* basic, essential, fundamental, immanent, inborn, inbred, in-built, inherent, native, natural, proper, real.

introduce *vb* **1** acquaint, make known, present. **2** announce, give an introduction to, lead into, preface. **3** add, advance, bring in, bring out, broach, create, establish, inaugurate, initiate, inject, insert, interpose, launch, make available, offer, phase in, pioneer, put forward, set up, start, suggest, usher in. ▷ BEGIN.

introduction *n* foreword, *inf* intro, *inf* lead-in, opening, overture, preamble, preface, prelude, prologue. ▷ BEGINNING.

introductory *adj* basic, early, first, fundamental, inaugural, initial, opening, prefatory, preliminary, preparatory, starting. *Opp* FINAL.

introverted *adj* contemplative, introspective, inward-looking, meditative, pensive, quiet, reserved, retiring, self-contained, shy, thoughtful, unsociable, withdrawn. *Opp* EXTROVERTED.

intrude *vb* break in, butt in, eavesdrop, encroach, gatecrash, interfere, interpose, interrupt, intervene, join uninvited, obtrude, *inf* snoop.

intruder *n* **1** eavesdropper, gatecrasher, infiltrator, interloper, *inf* uninvited guest. **2** burglar, housebreaker, invader, prowler, raider, robber, *inf* snooper, thief, trespasser.

intuition *n* insight, perceptiveness, percipience. ▷ INSTINCT.

invade *vb* descend on, encroach on, enter, impinge on, infest, infringe, march into, occupy, overrun, penetrate, raid, subdue, violate. ▷ ATTACK.

invalid *adj* **1** null and void, out-of-date, unacceptable, unusable, void, worthless. **2** fallacious, false, illogical, incorrect, irrational, spurious, unconvincing, unfounded, unreasonable, unscientific, unsound, untenable, untrue, wrong. *Opp* VALID. **3** ▷ ILL.
• *n* cripple, incurable, patient, sufferer, valetudinarian.

invaluable *adj* incalculable, inestimable, irreplaceable, precious, priceless, useful. ▷ VALUABLE. *Opp* WORTHLESS.

invariable *adj* certain, changeless, constant, eternal, even, immutable, inflexible, permanent, predictable, regular, reliable, rigid, solid, stable, steady, unalterable, unchangeable, unchanging, unfailing, uniform, unvarying, unwavering. *Opp* VARIABLE.

invasion *n* **1** encroachment, incursion, infiltration, inroad, intrusion, onslaught, raid, violation. ▷ ATTACK. **2** colony, flood, horde, infestation, spate, stream, swarm, throng.

invasive *adj* burgeoning, increasing, mushrooming, profuse, proliferating, relentless, unstoppable.

invent *vb* coin, conceive, concoct, construct, contrive, *inf* cook up, create, design, devise, discover, *inf* dream up, fabricate, formulate, *inf* hit upon, imagine, improvise, make up, originate, plan, put together, think up, trump up.

invention *n* **1** brainchild, coinage, contrivance, creation, design, discovery. **2** contraption, device, gadget. **3** deceit, fabrication, falsehood, fantasy, fiction, figment, lie. **4** ▷ INVENTIVENESS.

inventive *adj* clever, creative, enterprising, fertile, imaginative, ingenious, innovative, inspired, original, resourceful. *Opp* BANAL.

inventiveness *n* creativity, genius, imagination, ingenuity, inspiration, invention, originality, resourcefulness.

inventor *n* architect, author, *inf* boffin, creator, designer, discoverer, maker, originator.

inverse *adj* opposite, reversed, transposed.

invert *vb* capsize, overturn, reverse, transpose, turn upside down, upset.

invest *vb* **1** buy stocks and shares, play the market, speculate. **2** lay out, put to work, *inf* sink, use profitably, venture. **invest in** ▷ BUY.

investigate *vb* analyse, consider, enquire about, examine, explore, follow up, gather evidence about, *inf* go into, inquire into, look into, probe, research, scrutinize, sift (*evidence*), study, *inf* suss out, weigh up.

investigation *n* enquiry, examination, inquiry, inquisition, inspection, *inf* post-mortem, *inf* probe, quest, research, review, scrutiny, search, study, survey.

invidious *adj* discriminatory, objectionable, offensive, undesirable, unfair, unjust, unwarranted.

invigorating *adj* bracing, enlivening, exhilarating, fresh, healthful, health-giving, healthy, refreshing, rejuvenating, revitalizing, salubrious, stimulating, tonic, vitalizing. *Opp* EXHAUSTING.

invincible *adj* impregnable, indestructible, indomitable, insuperable, invulnerable, strong, unassailable, unbeatable, unconquerable, unstoppable.

invisible *adj* camouflaged, concealed, covered, disguised, hidden, imperceptible, inconspicuous, obscured, out of sight, secret, undetectable, unnoticeable, unnoticed, unseen. *Opp* VISIBLE.

invite *vb* **1** ask, encourage, request, summon, urge. **2** attract, entice, solicit, tempt. **inviting** ▷ ATTRACTIVE.

invoice *n* account, bill, list, statement.

invoke *vb* appeal to, call for, cry out for, entreat, implore, pray for, solicit, supplicate.

involuntary *adj* automatic, conditioned, impulsive, instinctive, mechanical, reflex, spontaneous, unconscious, uncontrollable, unintentional, unthinking, unwitting. *Opp* DELIBERATE.

involve *vb* **1** comprise, contain, embrace, entail, hold, include, incorporate, take in. **2** affect, concern, interest, touch. **3** *involve in crime.* embroil, implicate, include, incriminate, inculpate, *inf* mix up. **involved** ▷ BUSY, COMPLEX.

involvement *n* **1** activity, interest, participation. **2** association, complicity, entanglement, partnership.

ironic *adj* derisive, double-edged, ironical, mocking, sarcastic, satirical, wry.

irony *n* double meaning, mockery, paradox, sarcasm, satire.

irrational *adj* absurd, arbitrary, biased, crazy, emotional, emotive, illogical, insane, mad, nonsensical, prejudiced, senseless, subjective, surreal, unconvincing, unintelligent, unreasonable, unreasoning, unsound, unthinking, wild. ▷ SILLY. *Opp* RATIONAL.

irregular *adj* **1** erratic, fitful, fluctuating, halting, haphazard, intermittent, occasional, random, spasmodic, sporadic, unequal, unpredictable, unpunctual, variable, varying, wavering. **2** abnormal, anomalous, eccentric, exceptional, extraordinary, illegal, improper, odd, peculiar, quirky, unconventional, unofficial, unplanned, unscheduled, unusual. **3** *irregular surface.* broken, bumpy, jagged, lumpy, patchy, pitted, ragged, rough, uneven, up and down. *Opp* REGULAR.

irrelevant *adj inf* beside the point, extraneous, immaterial, impertinent, inapplicable, inapposite, inappropriate, inessential, mal apropos, *inf* neither here nor there, pointless, unconnected, unnecessary, unrelated. *Opp* RELEVANT.

irreligious *adj* agnostic, atheistic, godless, heathen, humanist, impious, irreverent, pagan, sinful, uncommitted, unconverted, ungodly, unrighteous, wicked. *Opp* RELIGIOUS.

irreparable *adj* hopeless, incurable, irrecoverable, irremediable, irretrievable, irreversible, lasting, permanent, unalterable.

irreplaceable *adj* inimitable, priceless, unique. ▷ RARE.

irrepressible *adj* boisterous, bouncy, *inf* bubbling, buoyant, ebullient, resilient, uncontrollable, ungovernable, uninhibited, unmanageable, unrestrainable, unstoppable, vigorous. ▷ LIVELY. *Opp* LETHARGIC.

irresistible *adj* compelling, inescapable, inexorable, irrepressible, not to be denied, overpowering, overriding, overwhelming, persuasive, powerful, relentless, seductive, strong, unavoidable, uncontrollable. *Opp* WEAK.

irresolute *adj* doubtful, fickle, flexible, *inf* hedging your bets, indecisive, open to compromise, tentative, uncertain, undecided, vacillating, wavering, weak, weak-willed. ▷ HESITANT. *Opp* RESOLUTE.

irresponsible *adj* antisocial, careless, conscienceless, devil-may-care, feckless, immature, immoral, inconsiderate, negligent, rash, reckless, selfish, shiftless, thoughtless, unethical, unreliable, unthinking, untrustworthy, wild. *Opp* RESPONSIBLE.

irreverent *adj* blasphemous, disrespectful, impious, irreligious, profane, sacrilegious, ungodly, unholy. ▷ RUDE. *Opp* REVERENT.

irrevocable *adj* binding, final, fixed, hard and fast, immutable, irreparable, irretrievable, irreversible, permanent, settled, unalterable, unchangeable.

irrigate *vb* flood, inundate, supply water to, water.

irritable *adj* bad-tempered, cantankerous, choleric, crabby, cross, crotchety, crusty, curmudgeonly, dyspeptic, easily annoyed, edgy, fractious, grumpy, ill-humoured, ill-tempered, impatient, irascible, oversensitive, peevish, pettish, petulant, *inf* prickly, querulous, *inf* ratty, short-tempered, snappy, testy, tetchy, touchy, waspish. ▷ ANGRY. *Opp* EVEN-TEMPERED.

irritate *vb* **1** cause irritation, itch, rub, tickle, tingle. **2** ▷ ANNOY.

island *n plur* archipelago, atoll, coral reef, isle, islet.

isolate *vb* cloister, cordon off, cut off, detach, exclude, insulate, keep apart, place apart, quarantine, seclude, segregate, separate, sequester, set apart, shut off, shut out, single out. **isolated** ▷ SOLITARY.

issue *n* **1** affair, argument, controversy, dispute, matter, point, problem, question, subject, topic. **2** conclusion, consequence, effect, end, impact, outcome, *inf* payoff, repercussions, result, upshot. **3** *issue of a magazine.* copy, edition, instalment, number, printing, publication, version. • *vb* **1** appear, come out, emerge, erupt, flow out, gush, leak, rise, spring. **2** announce, bring out, broadcast, circulate, declare, disseminate, distribute, give out, make public, print, produce, promulgate, publicize, publish, put out, release, send out, supply.

itch *n* **1** irritation, prickling, tickle, tingling. **2** ache, craving, desire, hankering, hunger, impatience, impulse, longing, need, restlessness, thirst, urge, wish, yearning, *inf* yen. • *vb* **1** be irritated, prickle, tickle, tingle. **2** ▷ DESIRE.

item *n* **1** article, bit, component, contribution, entry, ingredient, lot, matter, object, particular, thing. **2** *item in a newspaper.* account, article, feature, notice, piece, report.

J

jab *vb* dig, elbow, nudge, poke, prod, stab, thrust. ▷ HIT.

jacket *n* casing, cover, covering, envelope, folder, sheath, skin, wrapper, wrapping. ▷ COAT.

jaded *adj* **1** ▷ WEARY. **2** bored, *inf* fed up, gorged, listless, sated, satiated, *inf* sick and tired, surfeited. *Opp* LIVELY.

jagged *adj* angular, barbed, broken, chipped, denticulate, indented, irregular, notched, ragged, rough, serrated, sharp, snagged, spiky, toothed, uneven, zigzag. *Opp* SMOOTH.

jail, jailer *ns* ▷ GAOL, GAOLER.

jam *n* **1** blockage, bottleneck, congestion, crush, obstruction, press, squeeze, stoppage, throng. ▷ CROWD. **2** difficulty, dilemma, embarrassment, *inf* fix, *inf* hole, *inf* hot water, *inf* pickle, plight, predicament, quandary, tight corner, trouble. **3** conserve, jelly, marmalade, preserve. • *vb* **1** block, *inf* bung up, clog, congest, cram, crowd, crush, fill, force, pack, obstruct, overcrowd, ram, squash, squeeze, stop up, stuff. **2** prop, stick, wedge.

jar *n* amphora, carafe, container, crock, ewer, glass, jug, mug, pitcher, pot, receptacle, urn, vessel. ▷ BOTTLE. *vb* **1** jerk, jog, jolt, *inf* rattle, shake, shock, upset. **2** *That noise jars on me.* grate, grind, *inf* jangle. ▷ ANNOY. **jarring** ▷ HARSH.

jargon *n* argot, cant, creole, dialect, idiom, language, patois, slang, vernacular.

jaunt *n* excursion, expedition, outing, tour, trip. ▷ JOURNEY.

jaunty *adj* alert, breezy, brisk, buoyant, bright, carefree, *inf* cheeky, debonair, frisky, lively, perky, spirited, sprightly. ▷ HAPPY.

jazzy *adj* **1** animated, rhythmic, spirited, swinging, syncopated, vivacious. ▷ LIVELY. *Opp* SEDATE. **2** *jazzy colours.* bold, clashing, contrasting, flashy, gaudy, loud.

jealous *adj* **1** bitter, covetous, envious, *inf* green-eyed, *inf* green with envy, grudging, jaundiced, resentful. **2** *jealous of your reputation.* careful, possessive, protective, vigilant, watchful.

jeer *vb* barrack, boo, chaff, deride, disapprove, gibe, heckle, hiss, *inf* knock, laugh, make fun (of), mock, scoff, sneer, taunt, *inf* twit. ▷ RIDICULE. *Opp* CHEER.

jeopardize *vb* endanger, gamble, imperil, menace, put at risk, risk, threaten, venture.

jerk *vb* jar, jiggle, jog, jolt, lurch, move jerkily, move suddenly, pluck, pull, *inf* rattle, shake, tug, tweak, twist, twitch, wrench, *inf* yank.

jerky *adj* bouncy, bumpy, convulsive, erratic, fitful, jolting, jumpy, rough, shaky, spasmodic, *inf* stopping and starting, twitchy, uncontrolled, uneven. *Opp* STEADY

jest *n, vb* ▷ JOKE.

jester *n* buffoon, clown, comedian, comic, fool, joker.

jet *adj* ▷ BLACK. • *n* **1** flow, fountain, gush, rush, spout, spray, spurt, squirt, stream. **2** nozzle, sprinkler.

jetty *n* breakwater, groyne, landing-stage, mole, pier, quay, wharf.

jewel *n* brilliant, gem, gemstone, ornament, precious stone, *inf* rock, *inf* sparkler. ▷ JEWELLERY. □ *amber, cairngorm, carnelian, coral, diamond, emerald, garnet, ivory, jade, jasper, jet, lapis lazuli, moonstone, onyx, opal, pearl, rhinestone, ruby, sapphire, topaz, turquoise.*

jeweller *n* goldsmith, silversmith.

jewellery *n* gems, jewels, ornaments, treasure, *inf* sparklers. □*bangle, beads, bracelet, brooch, chain, charm, clasp, cuff-links, earring, locket, necklace, pendant, pin, ring, signet ring, tie-pin, watch, watch-chain.*

jilt *vb* abandon, break with, desert, *inf* ditch, drop, *inf* dump, forsake, *inf* give someone the brush-off, leave behind, *inf* leave in the lurch, renounce, repudiate, *inf* throw over, *inf* wash your hands of.

jingle *n* **1** doggerel, rhyme, song, tune, verse. **2** chinking, clinking, jangling, ringing, tinkling, tintinnabulation. • *vb* chime, chink, clink, jangle, ring, tinkle. ▷ SOUND.

job *n* **1** activity, assignment, charge, chore, duty, errand, function, housework, mission, operation, project, pursuit, responsibility, role, stint, task, undertaking, work. **2** appointment, business, calling, career, craft, employment, livelihood, living, métier, occupation, position, post, profession, sinecure, trade, vocation.

jobless *adj* out of work, redundant, unemployed, unwaged.

jocular *adj* cheerful, gay, glad, gleeful, happy, jocund, jokey, joking, jolly, jovial, joyous, jubilant, merry, overjoyed, rejoicing. *Opp* SAD, SERIOUS.

jog *vb* **1** bounce, jar, jerk, joggle, jolt, knock, nudge, shake. ▷ HIT. **2** *jog the memory*. activate, arouse, prompt, refresh, remind, set off, stimulate, stir. **3** *jog round the park*. exercise, lope, run, trot.

join *n* connection, joint, knot, link, mend, seam. • *vb* **1** add, amalgamate, attach, combine, connect, couple, dock, dovetail, fit, fix, juxtapose, knit, link, marry, merge, put together, splice, tack on, unite, yoke. ▷ FASTEN. *Opp* SEPARATE. **2** abut, adjoin, border on, come together, converge, meet, touch, verge on. **3** *join a crowd*. accompany, associate with, follow, go with, *inf* latch on to, tag along with, team up with. **4** *join a club*. affiliate with, become a member of, enlist in, enrol in, participate in, register for, sign up for, subscribe to, volunteer for. *Opp* LEAVE.

joint *adj* collaborative, collective, combined, common, communal, concerted, cooperative, corporate, general, mutual, shared, united. *Opp* SEPARATE. • *n* articulation, connection, hinge, junction, union.

joist *n* beam, girder, rafter.

joke *n* *inf* crack, funny story, *inf* gag, *old use* jape, jest, laugh, *inf* one-liner, pleasantry, pun, quip, wisecrack, witticism. • *vb* banter, be facetious, clown, fool about, have a laugh, jest, make jokes, quip, tease.

jolly *adj* cheerful, delighted, gay, glad, gleeful, grinning, high-spirited, jocose, jocular, jocund, joking, jovial, joyful, joyous, jubilant, laughing, merry, playful, rejoicing, rosy-faced, smiling, sportive. ▷ HAPPY. *Opp* SAD.

jolt *vb* **1** bounce, bump, jar, jerk, jog, shake, twitch. ▷ HIT. **2** *jolted me into action*. astonish, disturb, nonplus, shake up, shock, startle, stun, surprise.

jostle *vb* crowd in on, hustle, press, push, shove.

jot *vb* **jot down** note, scribble, take down. ▷ WRITE.

journal *n* **1** gazette, magazine, monthly, newsletter, newspaper, paper, periodical, publication, review, weekly. **2** account, annals, chronicle, diary, dossier, history, log, memoir, record, scrapbook.

journalist *n* broadcaster, columnist, contributor, correspondent, *derog* hack, *inf* newshound, newspaperman, newspaperwoman, pressman, reporter, writer.

journey *n* excursion, expedition, itinerary, jaunt, mission, odyssey, outing, peregrination, progress, route, tour, transition, travelling, trip, wandering. □ *cruise, drive, flight, hike, joy-ride, pilgrimage, ramble, ride, run, safari, sail, sea crossing, sea passage, trek, voyage, walk.* • *vb* ▷ TRAVEL.

joy *n* bliss, cheer, cheerfulness, delight, ecstasy, elation, euphoria, exaltation, exhilaration, exultation, felicity, gaiety, gladness, glee, gratification, happiness, high spirits, hilarity, jocularity, joviality, joyfulness, joyousness, jubilation, light-heartedness, merriment, mirth, pleasure, rapture, rejoicing, triumph. *Opp* SORROW.

joyful *adj* buoyant, cheerful, delighted, ecstatic, elated, euphoric, exhilarated, exultant, gay, glad, gleeful, jocund, jolly, jovial, joyous, jubilant, light-hearted, merry, overjoyed, pleased, rapturous, rejoicing, triumphant. ▷ HAPPY. *Opp* SAD.

jubilee *n* anniversary, celebration, commemoration, festival.

judge *n* **1** *sl* beak, justice, magistrate. **2** adjudicator, arbiter, arbitrator, moderator, referee, umpire. **3** *judge of wine.* authority, connoisseur, critic, expert, reviewer. • *vb* **1** condemn, convict, examine, pass judgement on, pronounce judgement on, punish, sentence, try. **2** adjudicate, mediate, moderate, referee, umpire. **3** believe, conclude, consider, decide, decree, deem, determine, estimate, gauge, guess, reckon, rule, suppose. **4** *judge others.* appraise, assess, criticize, evaluate, give your opinion of, rate, rebuke, scold, sit in judgement on, size up, weigh.

judgement *n* **1** arbitration, award, conclusion, conviction, decision, decree, *old use* doom, finding, outcome, penalty, punishment, result, ruling, verdict. **2** *use your judgement.* acumen, common sense, discernment, discretion, discrimination, expertise, good sense, reason, wisdom. ▷ INTELLIGENCE. **3** *in my judgement.* assessment, belief, estimation, evaluation, idea, impression, mind, notion, opinion, point of view, valuation.

judicial *adj* **1** forensic, legal, official. **2** ▷ JUDICIOUS.

judicious *adj* appropriate, astute, careful, circumspect, considered, diplomatic, discerning, discreet, discriminating, enlightened, expedient, judicial, politic, prudent, sage, sensible, shrewd, sober, thoughtful, well-advised, well-judged. ▷ WISE.

jug *n* bottle, carafe, container, decanter, ewer, flagon, flask, jar, pitcher, vessel.

juggle *vb* alter, *inf* cook, *inf* doctor, falsify, *inf* fix, manipulate, misrepresent, move about, rearrange, rig, tamper.

juice *n* drink, extract, fluid, liquid, sap.

juicy *adj* lush, moist, soft, *inf* squelchy, succulent, wet. *Opp* DRY.

jumble *n* chaos, clutter, confusion, disarray, disorder, farrago, hotchpotch, mess, muddle, tangle. • *vb* confuse, disarrange, disorganize, *inf* mess up, mingle, mix up, muddle, shuffle, tangle. *Opp* ARRANGE.

jump *n* **1** bounce, bound, hop, leap, pounce, skip, spring, vault. **2** ditch, fence, gap, gate, hurdle, obstacle. **3** *a jump in prices.* ▷ RISE. • *vb* **1** bounce, bound, caper, dance, frisk, frolic, gambol, hop, leap, pounce, prance, skip, spring. **2** *jump a fence.* clear, hurdle, vault. **3** *jump in surprise.* flinch, recoil, start, wince. **jump on** ▷ ATTACK. **make someone jump** ▷ STARTLE.

junction *n* confluence, connection, corner, crossroads, interchange, intersection, joining, juncture, *inf* link-up, meeting, points, T-junction, union.

jungle *n* forest, rain-forest, tangle, undergrowth, woods.

junior *adj* inferior, lesser, lower, minor, secondary, subordinate, subsidiary, younger. *Opp* SENIOR.

junk *n* clutter, debris, flotsam-and-jetsam, garbage, litter, lumber, oddments, odds and ends, refuse, rubbish, rummage, scrap, trash, waste. • *vb* ▷ DISCARD.

just *adj* apt, deserved, equitable, ethical, even-handed, fair, fair-minded, honest, impartial, justified, lawful, legal, legitimate, merited, neutral, proper, reasonable, rightful, right-minded, unbiased, unprejudiced, upright. ▷ MORAL. *Opp* UNJUST.

justice *n* **1** equity, even-handedness, fair-mindedness, fair play, impartiality, integrity, legality, neutrality, objectivity, right. ▷ MORALITY. **2** the law, legal proceedings, the police, punishment, retribution, vengeance.

justifiable *adj* acceptable, allowable, defensible, excusable, forgivable, justified, lawful, legitimate, pardonable, permissible, reasonable, understandable, warranted. *Opp* UNJUSTIFIABLE.

justify *vb* condone, defend, exculpate, excuse, exonerate, explain, explain away, forgive, legitimate, legitimize, pardon, rationalize, substantiate, support, sustain, uphold, validate, vindicate, warrant.

jut *vb* beetle, extend, overhang, poke out, project, protrude, stick out. *Opp* RECEDE.

juvenile *adj* **1** babyish, childish, immature, infantile, puerile, unsophisticated. **2** adolescent, *inf* teenage, underage, young, youthful. *Opp* MATURE.

K

keen *adj* **1** active, ambitious, anxious, ardent, assiduous, avid, bright, clever, committed, dedicated, devoted, diligent, eager, enthusiastic, fervent, fervid, industrious, intelligent, intense, intent, interested, motivated, passionate, quick, zealous. **2** *a keen knife.* knife-edged, piercing, razor-sharp, sharp, sharpened. **3** *a keen wit.* acerbic, acid, acute, biting, clever, cutting, discerning, incisive, lively, mordant, observant, pungent, rapier-like, sarcastic, satirical, scathing, shrewd, sophisticated, stinging. **4** *keen eyesight*. acute, clear, fine, perceptive, sensitive. **5** *a keen wind*. bitter, cold, extreme, icy, intense, penetrating, severe. **6** *keen prices*. competitive, low, rock-bottom. *Opp* APATHETIC, DULL.

keep *vb* **1** accumulate, amass, conserve, guard, hang on to, hoard, hold, maintain, preserve, protect, put aside, put away, retain, safeguard, save, store, stow away, withhold. **2** *keep going*. carry on, continue, do again and again, do for a long time, keep on, persevere in, persist in. **3** *keep left*. remain, stay. **4** *keep a family*. be responsible for, care for, cherish, feed, finance, foster, guard, have charge of, look after, maintain, manage, mind, own, pay for, protect, provide for, subsidize, support, take charge of, tend, watch over. **5** *keep a birthday*. celebrate, commemorate, mark, observe, solemnize. **6** *food keeps in the fridge*. be preserved, be usable, last, survive, stay fresh, stay good. **7** *won't keep you*. block, check, confine, curb, delay, detain, deter, get in the way of, hamper, hinder, hold up, impede, imprison, obstruct, prevent, restrain, retard. **keep still** ▷ STAY. **keep to** ▷ FOLLOW, OBEY. **keep up** ▷ PROLONG, SUSTAIN.

keeper *n* caretaker, curator, custodian, gaoler, guard, guardian, *inf* minder, warden, warder.

kernel *n* centre, core, essence, heart, middle, nub, pith.

key *n* **1** answer, clarification, clue, explanation, indicator, pointer, secret, solution. **2** *key to a map*. glossary, guide, index, legend.

keyboard *n* □ *accordion, celesta, clavichord, clavier, fortepiano, harmonium, harpsichord, organ, piano,* old use *pianoforte, spinet, synthesizer, virginals.*

kick *vb* boot, heel, punt. ▷ HIT.

kidnap *vb* abduct, capture, carry off, run away with, seize, snatch.

kill *vb* annihilate, assassinate, be the killer of, *sl* bump off, butcher, cull, decimate, destroy, *inf* dispatch, *inf* do away with, *sl* do in, execute, exterminate, *inf* finish off, *inf* knock off, liquidate, martyr, massacre, murder, put down, put to death, put to sleep, slaughter, slay, *inf* snuff out, take life, *Amer sl* waste. □ *behead, brain, choke, crucify, decapitate, disembowel, drown, electrocute, eviscerate, garrotte, gas, guillotine, hang, knife, lynch, poison, pole-axe, shoot, smother, stab, starve, stifle, stone, strangle, suffocate, throttle.*

killer *n* assassin, butcher, cutthroat, destroyer, executioner, exterminator, gunman, *sl* hit man, murderer, slayer.

killing *n* annihilation, assassination, bloodbath, bloodshed, butchery, carnage, decimation, destruction, elimination, eradication, euthanasia, execution, extermination, extinction, fratricide, genocide, homicide, infanticide, liquidation, manslaughter, martyrdom, massacre, matricide, murder, parricide, patricide, pogrom, regicide, slaughter, sororicide, suicide, unlawful killing, uxoricide.

kin *n* clan, family, *inf* folks, kindred, kith and kin, relations, relatives.

kind *adj* accommodating, affable, affectionate, agreeable, altruistic, amenable, amiable, amicable, approachable, attentive, avuncular, beneficent, benevolent, benign, bountiful, brotherly, caring, charitable, comforting, compassionate, congenial, considerate, cordial, courteous, encouraging, fatherly, favourable, friendly, generous, genial, gentle, good-natured, good-tempered, gracious, helpful, hospitable, humane, humanitarian, indulgent, kind-hearted, kindly, lenient, loving, merciful, mild, motherly, neighbourly, nice, obliging, patient, philanthropic, pleasant, polite, public-spirited, sensitive, sisterly, soft-hearted, sweet, sympathetic, tactful, tender, tender-hearted, thoughtful, tolerant, understanding, unselfish, warm, warm-hearted, well-intentioned, well-meaning, well-meant. *Opp* UNKIND. • *n* brand, breed, category, class, description, family, form, genre, genus, make, manner, nature, persuasion, race, set, sort, species, style, type, variety.

kindle *vb* **1** burn, fire, ignite, light, set alight, set fire to, spark off. **2** ▷ AROUSE.

king *n* **1** crowned head, His Majesty, monarch, ruler, sovereign. **2** ▷ CHIEF.

kingdom *n* country, empire, land, monarchy, realm.

kink *n* **1** bend, coil, crimp, crinkle, curl, knot, loop, tangle, twist, wave. **2** ▷ QUIRK.

kiosk *n* booth, stall. □ *bookstall, news-stand, telephone-box.*

kiss *vb* caress, embrace, *sl* neck, osculate, *old use* spoon. ▷ TOUCH.

kit *n* accoutrements, apparatus, appurtenances, baggage, effects, equipment, gear, *joc* impedimenta, implements, luggage, outfit, paraphernalia, rig, supplies, tackle, tools, tools of the trade, utensils.

kitchen *n* cookhouse, galley, kitchenette, scullery.

knack *n* ability, adroitness, aptitude, art, bent, dexterity, expertise, facility, flair, genius, gift, habit, intuition, *inf* know-how, skill, talent, trick, *inf* way.

knapsack *n* backpack, haversack, rucksack.

knead *vb* manipulate, massage, pound, press, pummel, squeeze, work.

kneel *vb* bend, bow, crouch, fall to your knees, genuflect, stoop.

knickers *n old use* bloomers, boxer-shorts, briefs, drawers, panties, pants, shorts, trunks, underpants.

knife *n* blade. □ *butter-knife, carving-knife, clasp-knife, cleaver, dagger, flick-knife, machete, penknife, pocket-knife, scalpel, sheath knife.* • *vb* cut, pierce, slash, stab. ▷ KILL, WOUND.

knit *vb* **1** crochet, weave. **2** bind, combine, connect, fasten, heal, interlace, interweave, join, knot, link, marry, mend, tie, unite. **knit your brow** ▷ FROWN.

knob *n* boss, bulge, bump, handle, lump, projection, protrusion, protuberance, stud, swelling.

knock *vb* **1** bang, *inf* bash, buffet, bump, pound, rap, smack, *old use* smite, strike, tap, thump. ▷ HIT. **2** ▷ CRITICIZE. **knock down** ▷ DEMOLISH. **knock off** ▷ CEASE. **knock out** ▷ STUN.

knot *n* **1** bond, bow, ligature, tangle, tie. **2** ▷ GROUP. • *vb* bind, do up, entangle, entwine, join, knit, lash, link, tether, tie, unite. ▷ FASTEN. *Opp* UNTIE.

know *vb* **1** be certain, have confidence, have no doubt. **2** *know facts.* be cognizant of, be familiar with, be knowledgeable about, comprehend, discern, have experience of, have in mind, realize, remember, understand. 3 *know a person*. be acquainted with, be a friend of, be friends with. 4 differentiate, distinguish, identify, make out, perceive, recognize, see.

know-all *n* expert, pundit, *inf* show-off, wiseacre.

knowing *adj* astute, clever, conspiratorial, cunning, discerning, experienced, expressive, meaningful, perceptive, shrewd. ▷ CUNNING, KNOWLEDGEABLE. *Opp* INNOCENT.

knowledge *n* 1 data, facts, information, intelligence, *sl* low-down. 2 acquaintance, awareness, background, cognition, competence, consciousness, education, erudition, experience, expertise, familiarity, grasp, insight, *inf* know-how, learning, lore, memory, science, scholarship, skill, sophistication, technique, training. *Opp* IGNORANCE.

knowledgeable *adj Fr* au fait, aware, cognizant, conversant, educated, enlightened, erudite, experienced, expert, familiar (with), *sl* genned up, informed, *inf* in the know, learned, scholarly, versed (in), well-informed. *Opp* IGNORANT.

L

label *n* brand, docket, hallmark, identification, imprint, logo, marker, sticker, tag, ticket, trademark. • *vb* brand, call, categorize, class, classify, define, describe, docket, identify, mark, name, pigeon-hole, stamp, tag.

laborious *adj* 1 arduous, back-breaking, difficult, exhausting, fatiguing, gruelling, hard, heavy, herculean, onerous, stiff, strenuous, taxing, tiresome, tough, uphill, wearisome, wearying. *Opp* EASY. 2 *a laborious style*. artificial, contrived, forced, heavy, laboured, overdone, overworked, pedestrian, ponderous, strained, unnatural. *Opp* FLUENT.

labour *n* 1 *inf* donkey-work, drudgery, effort, exertion, industry, navvying, *inf* pains, slavery, strain, *inf* sweat, toil, work. 2 employees, *old use* hands, wage-earners, workers, workforce. 3 childbirth, contractions, delivery, labour pains, parturition, *old use* travail. • *vb* drudge, exert yourself, navvy, *inf* slave, strain, strive, struggle, *inf* sweat, toil, travail. ▷ WORK. laboured ▷ LABORIOUS.

labourer *n* blue-collar worker, employee, *old use* hand, manual worker, *inf* navvy, wage-earner, worker.

labour-saving *adj* convenient, handy, helpful, time-saving.

labyrinth *n* complex, jungle, maze, network, tangle.

lace *n* 1 filigree, mesh, net, netting, openwork, tatting, web. 2 bootlace, cord, shoelace, string, thong. • *vb* ▷ FASTEN.

lacerate *vb* claw, cut, gash, graze, mangle, rip, scrape, scratch, slash, tear. ▷ WOUND.

lack *n* absence, dearth, deficiency, deprivation, famine, insufficiency, need, paucity, privation, scarcity, shortage, want. *Opp* PLENTY. • *vb* be lacking in, be short of, be without, miss, need, require, want.

lacking *adj* defective, deficient, inadequate, insufficient, short, unsatisfactory, weak. ▷ STUPID.

laden *adj* burdened, *inf* chock-full, fraught, full, hampered, loaded, oppressed, piled high, weighed down.

lady *n* 1 wife, woman. ▷ FEMALE. 2 aristocrat, peeress. ▷ TITLE.

ladylike *adj* aristocratic, courtly, cultured, dainty, decorous, elegant, genteel, modest, noble, polished, posh, prim and proper, *derog* prissy, refined, respectable, well-born, well-bred. ▷ POLITE.

lag *vb* 1 be slow, *inf* bring up the rear, come last, dally, dawdle, delay, drop behind, fall behind, go too slow, hang about, hang back, idle, linger, loiter, saunter, straggle, trail. 2 *lag pipes*. insulate, wrap up.

lair *n* den, hide-out, hiding-place, refuge, resting-place, retreat, shelter.

lake *n* boating-lake, lagoon, lido, *Scot* loch, mere, pool, pond, reservoir, sea, tarn, water.

lame *adj* 1 crippled, disabled, halting, handicapped, hobbled, hobbling, incapacitated, limping, maimed, spavined. 2 *a lame leg*. dragging, game, *inf* gammy, injured, stiff. 3 *a lame excuse*. feeble, flimsy, inadequate, poor, tame, thin, unconvincing, weak. • *vb* cripple, disable, hobble, incapacitate, maim. **be lame** ▷ LIMP.

lament *n* dirge, elegy, lamentation, moaning, monody, mourning, requiem, threnody. • *vb* bemoan, bewail, complain, cry, deplore, express your sorrow, grieve, keen, mourn, regret, shed tears, sorrow, wail, weep.

lamentable *adj* deplorable, regrettable, unfortunate, unhappy. ▷ SAD.

lamentation *n* complaints, crying, grief, grieving, lamenting, moaning, mourning, regrets, sobbing, tears, wailing, weeping.

lamp *n* bulb, fluorescent lamp, headlamp, lantern, standard lamp, street light, torch. ▷ LIGHT.

land *n* 1 coast, ground, landfall, shore, *joc* terra firma. 2 *lie of the land*. geography, landscape, terrain, topography. 3 country, fatherland, homeland, motherland, nation, region, state, territory. 4 farmland, earth, soil. 5 *land you own*. estate, grounds, property. • *vb* 1 alight, arrive, berth, come ashore, come to rest, disembark, dismount, dock, end a journey, get down, go ashore, light, reach landfall, settle, touch down. 2 *land yourself a job*. ▷ GET.

landing *n* 1 docking, re-entry, return, splashdown, touchdown. 2 alighting, arrival, deplaning, disembarkation. 3 ▷ LANDING-STAGE.

landing-stage *n* berth, dock, harbour, jetty, landing, pier, quay, wharf.

landlady, landlord *ns* 1 host, hostess, hotelier, *old use* innkeeper, licensee, publican, restaurateur. 2 landowner, lessor, letter, manager, manageress, owner, proprietor.

landmark *n* 1 feature, guidepost, high point, identification, visible feature. 2 milestone, new era, turning-point, watershed.

landscape *n* aspect, countryside, outlook, panorama, prospect, rural scene, scene, scenery, view, vista.

language *n* 1 parlance, speech, tongue. □ *argot, cant, colloquialism, dialect, formal language, idiolect, idiom, informal language, jargon, journalese, lingua franca,* inf *lingo, patois, register, slang, vernacular*. 2 linguistics. □ *etymology, lexicography,* *orthography, philology, phonetics, psycholinguistics, semantics, semiotics, sociolinguistics.* **3** *computer language.* code, system of signs.

languid *adj* apathetic, *inf* droopy, feeble, inactive, inert, lackadaisical, lazy, lethargic, slow, sluggish, torpid, unenthusiastic, weak. *Opp* ENERGETIC.

languish *vb* decline, flag, lose momentum, mope, pine, slow down, stagnate, suffer, sulk, waste away, weaken, wither. *Opp* FLOURISH.

lank *adj* **1** drooping, lifeless, limp, long, straight, thin. **2** ▷ LANKY.

lanky *adj* angular, awkward, bony, gangling, gaunt, lank, lean, long, scraggy, scrawny, skinny, tall, thin, ungraceful, weedy. *Opp* GRACEFUL, STURDY.

lap *n* **1** knees, thighs. **2** circle, circuit, course, orbit, revolution. • *vb* ▷ DRINK.

lapse *n* **1** backsliding, blunder, decline, error, failing, fault, flaw, mistake, omission, relapse, shortcoming, slip, *inf* slip-up, temporary failure, weakness. **2** *lapse of time.* break, gap, hiatus, *inf* hold-up, intermission, interruption, interval, lacuna, lull, pause. • *vb* **1** decline, deteriorate, diminish, drop, fall, sink, slide, slip, slump, subside. **2** *My membership lapsed.* become invalid, expire, finish, run out, stop, terminate.

large *adj* above average, abundant, ample, big, bold, broad, bulky, burly, capacious, colossal, commodious, considerable, copious, elephantine, enormous, extensive, fat, formidable, gargantuan, generous, giant, gigantic, grand, great, heavy, hefty, high, huge, *inf* hulking, immense, immeasurable, impressive, incalculable, infinite, *inf* jumbo, *inf* king-sized, largish, lofty, long, mammoth, massive, mighty, monstrous, monumental, mountainous, outsize, overgrown, oversized, prodigious, *inf* roomy, sizeable, spacious, substantial, swingeing (*increase*), tall, thick, *inf* thumping, *inf* tidy (*sum*), titanic, towering, *inf* tremendous, vast, voluminous, weighty, *inf* whacking, *inf* whopping, wide. *Opp* SMALL.

larva *n* caterpillar, grub, maggot.

lash *n* ▷ WHIP. • *vb* **1** beat, birch, cane, flail, flog, scourge, strike, thrash, whip. ▷ HIT. **2** ▷ CRITICIZE.

last *adj* closing, concluding, final, furthest, hindmost, latest, most recent, rearmost, terminal, terminating, ultimate. *Opp* FIRST. • *vb* carry on, continue, endure, hold, hold out, keep on, linger, live, persist, remain, stay, survive, *inf* wear well. *Opp* DIE, FINISH. **lasting** ▷ PERMANENT.

late *adj* **1** behindhand, belated, delayed, dilatory, overdue, slow, tardy, unpunctual. **2** *a late edition.* current, last, new, recent, up-to-date. **3** *the late king.* dead, deceased, departed, ex-, former, past, previous.

latent *adj* dormant, hidden, invisible, potential, undeveloped, undiscovered.

latitude *n inf* elbow-room, freedom, leeway, liberty, room, scope, space.

latter *adj* closing, concluding, last, last-mentioned, later, recent, second. *Opp* FORMER.

lattice *n* criss-cross, framework, grid, mesh, trellis.

laugh *vb* beam, be amused, burst into laughter, chortle, chuckle, *sl* fall about, giggle, grin, guffaw, roar with laughter, simper, smile, smirk, sneer, snicker, snigger, *inf* split your sides, titter. **laugh at** ▷ RIDICULE.

laughable *adj* absurd, derisory, ludicrous, preposterous, ridiculous. ▷ FUNNY.

laughing-stock *n* butt, figure of fun, victim.

laughter *n* chuckling, giggling, guffawing, hilarity, *inf* hysterics, laughing, laughs, merriment, mirth, snickering, sniggering, tittering. ▷ RIDICULE.

launch *vb* **1** begin, embark on, establish, float, found, inaugurate, initiate, open, organize, set in motion, set off, set up, start. **2** blast off, catapult, dispatch, fire, propel, send off, set off, shoot.

lavatory *n* bathroom, cloakroom, convenience, *inf* Gents, *inf* Ladies, latrine, *inf* loo, *inf* men's room, *childish* potty, *old use* privy, public convenience, toilet, urinal, water-closet, WC, *inf* women's room.

lavish *adj* **1** abundant, bountiful, copious, exuberant, free, generous, liberal, luxuriant, luxurious, munificent, opulent, plentiful, profuse, sumptuous, unselfish, unsparing, unstinting. **2** excessive, extravagant, improvident, prodigal, self-indulgent, wasteful. *Opp* ECONOMICAL.

law *n* **1** act, bill [= *draft law*], bylaw, commandment, decree, directive, edict, injunction, mandate, measure, order, ordinance, pronouncement, regulation, rule, statute. **2** *laws of science.* axiom, formula, postulate, principle, proposition, theory. **3** *laws of decency.* code, convention, practice. **4** *court of law.* justice, litigation.

law-abiding *adj* compliant, decent, disciplined, good, honest, obedient, orderly, peaceable, peaceful, respectable, well-behaved. *Opp* LAWLESS.

lawful *adj* allowable, allowed, authorized, constitutional, documented, just, justifiable, legal, legitimate, permissible, permitted, prescribed, proper, recognized, regular, right, rightful, valid. *Opp* ILLEGAL.

lawless *adj* anarchic, anarchical, badly-behaved, chaotic, disobedient, disorderly, illdisciplined, insubordinate, mutinous, rebellious, riotous, rowdy, seditious, turbulent, uncontrolled, undisciplined, ungoverned, unregulated, unrestrained, unruly, wild. ▷ WICKED. *Opp* LAW-ABIDING.

lawlessness *n* anarchy, chaos, disobedience, disorder, mob-rule, rebellion, rioting. *Opp* ORDER.

lawyer *n* advocate, barrister, counsel, legal representative, member of the bar, solicitor.

lax *adj* careless, casual, easygoing, flexible, indulgent, lenient, loose, neglectful, negligent, permissive, relaxed, remiss, slack, slipshod, unreliable, vague. *Opp* STRICT.

laxative *n* aperient, enema, purgative.

lay *vb* **1** apply, arrange, deposit, leave, place, position, put down, rest, set down, set out, spread. **2** *lay foundations.* build, construct, establish. **3** *lay the blame on someone.* ascribe, assign, attribute, burden, charge, impose, plant, *inf* saddle. **4** *lay plans.* concoct, create, design, organize, plan, set up. **lay bare** ▷ REVEAL. **lay bets** ▷ GAMBLE. **lay by** ▷ STORE. **lay down the law** ▷ DICTATE. **lay in** ▷ STORE. **lay into** ▷ ATTACK. **lay low** ▷ DEFEAT. **lay off something**

▷ CEASE. **lay someone off** ▷ DISMISS. **lay to rest** ▷ BURY. **lay up** ▷ STORE. **lay waste** ▷ DESTROY.

layer *n* **1** coat, coating, covering, film, sheet, skin, surface, thickness. **2** *layer of rock.* seam, stratum, substratum. *in layers* laminated, layered, sandwiched, stratified.

layman *n* **1** amateur, non-specialist, untrained person. *Opp* PROFESSIONAL. **2** [*church*] layperson, member of the congregation, parishioner, unordained person. *Opp* CLERGYMAN.

laze *vb* be lazy, do nothing, idle, lie about, loaf, lounge, relax, sit about, unwind.

laziness *n* dilatoriness, idleness, inactivity, indolence, lethargy, loafing, lounging about, shiftlessness, slackness, sloth, slowness, sluggishness, torpor. *Opp* INDUSTRY.

lazy *adj* **1** dilatory, easily pleased, easygoing, idle, inactive, indolent, languid, lethargic, listless, shiftless, *inf* skiving, slack, slothful, slow, sluggish, torpid, unenterprising, work-shy. **2** peaceful, quiet, relaxing. *Opp* ENERGETIC, INDUSTRIOUS. **be lazy** ▷ LAZE. **lazy person** ▷ SLACKER.

lead *n* **1** direction, example, guidance, leadership, model, pattern, precedent. **2** *lead on a crime.* clue, hint, line, tip, tip-off. **3** *lead in a race.* first place, front, spearhead, van, vanguard. **4** *lead in a play.* chief part, hero, heroine, principal, protagonist, starring role, title role. **5** cable, flex, wire. **6** *dog's lead.* chain, leash, strap. • *vb* **1** conduct, draw, escort, guide, influence, pilot, prompt, show the way, steer, usher. **2** be in charge of, captain, command, direct, govern, head, manage, preside over, rule, *inf* skipper, superintend, supervise. **3** be first, be in front, be in the lead, go first, head the field. **4** *lead the field.* beat, defeat, excel, outdo, outstrip, precede, surpass, vanquish. *Opp* FOLLOW. **lead astray** ▷ MISLEAD. **leading** ▷ CHIEF, INFLUENTIAL. **lead off** ▷ BEGIN.

leader *n* **1** ayatollah, boss, captain, chieftain, commander, conductor, courier, demagogue, director, figure-head, godfather, guide, head, patriarch, premier, prime minister, principal, ringleader, superior, *inf* supremo. ▷ CHIEF, RULER. **2** *leader in a newspaper.* editorial, leading article.

leaf *n* **1** blade, foliage, frond, greenery. **2** folio, page, sheet.

leaflet *n* advertisement, bill, booklet, brochure, circular, flyer, folder, handbill, handout, notice, pamphlet.

league *n* alliance, association, coalition, confederation, *derog* conspiracy, federation, fraternity, guild, society, union. ▷ GROUP. **be in league with** ▷ CONSPIRE.

leak *n* **1** discharge, drip, emission, escape, exudation, leakage, oozing, seepage, trickle. **2** aperture, break, chink, crack, crevice, cut, fissure, flaw, hole, opening, perforation, puncture, rent, split, tear. **3** *security leak.* disclosure, revelation. • *vb* **1** discharge, drip, escape, exude, ooze, percolate, seep, spill, trickle. **2** *leak secrets.* disclose, divulge, give away, let out, let slip, *inf* let the cat out of the bag about, make known, pass on, reveal, *inf* spill the beans about.

leaky *adj* cracked, dripping, holed, perforated, punctured.

lean *adj* angular, bony, emaciated, gangling, gaunt, hungry-looking, lanky, long, rangy, scraggy, scrawny, skinny, slender, slim, spare, thin, weedy, wiry. *Opp* FAT. • *vb* **1** bank, careen, heel over, incline, keel over, list, slant, slope, tilt, tip. **2** loll, prop yourself up, recline, rest, support yourself.

leaning *n* bent, bias, favouritism, inclination, instinct, liking, partiality, penchant, predilection, preference, propensity, readiness, taste, tendency, trend.

leap *vb* **1** bound, clear (*a fence*), hop over, hurdle, jump, leapfrog, skip over, spring, vault. **2** caper, cavort, dance, frisk, frolic, gambol, hop, prance, romp. **3** *leap on someone.* ambush, attack, pounce.

learn *vb* acquire, ascertain, assimilate, become aware of, become proficient in, be taught, *inf* catch on, commit to memory, discover, find out, gain, gain understanding of, gather, grasp, master, memorize, *inf* mug up, pick up, remember, study, *inf* swot up. **learned** ▷ ACADEMIC, EDUCATED.

learner *n* apprentice, beginner, cadet, initiate, L-driver, novice, pupil, scholar, starter, student, trainee, tiro.

learning *n* culture, education, erudition, information, knowledge, lore, scholarship, wisdom.

lease *n* agreement, contract. • *vb* charter, hire out, let, rent out, sublet.

least *adj* fewest, lowest, minimum, negligible, poorest, slightest, smallest, tiniest.

leather *n* chamois, hide, skin, suede.

leave *n* **1** authorization, consent, dispensation, liberty, licence, permission, sanction. **2** *leave from work.* absence, free time, furlough, holiday, recess, sabbatical, time off, vacation. • *vb* **1** *inf* be off, *inf* check out, decamp, depart, disappear, *sl* do a bunk, escape, exeunt [*they go out*], exit [*he/she goes out*], get away, get out, go away, go out, *sl* hop it, *inf* pull out, *sl* push off, retire, retreat, run away, say goodbye, set off, *inf* take off, take your leave, withdraw. **2** abandon, desert, evacuate, forsake, vacate. **3** *leave your job.* *inf* chuck in, *inf* drop out of, give up, quit, relinquish, renounce, resign from, retire from, *inf* walk out of, *inf* wash your hands of. **4** *leave it there.* allow to stay, *inf* let alone, let be. **5** *left it here.* deposit, place, position, put down, set down. **6** *left it somewhere.* forget, lose, mislay. **7** *leave it to you.* assign, cede, consign, entrust, refer, relinquish. **8** *leave in a will.* bequeath, hand down, will. **leave off** ▷ STOP. **leave out** ▷ OMIT.

lecture *n* **1** address, discourse, disquisition, instruction, lesson, paper, speech, talk, treatise. **2** *lecture on bad manners.* diatribe, harangue, sermon. ▷ REPRIMAND. • *vb* **1** be a lecturer, teach. **2** discourse, give a lecture, harangue, *inf* hold forth, pontificate, preach, sermonize, speak, talk formally. **3** ▷ REPRIMAND.

lecturer *n* don, fellow, instructor, professor, speaker, teacher, tutor.

ledge *n* mantel, overhang, projection, ridge, shelf, sill, step, window-sill.

left *adj, n* **1** left-hand, port [= *left facing bow of ship*], sinistral. **2** *left wing in politics.* communist, Labour, leftist, liberal, Marxist, progressive, radical, *derog* red, revolutionary, socialist. *Opp* RIGHT.

leg *n* 1 limb, member, *inf* peg, *inf* pin, shank. □ *ankle, calf, foot, hock, knee, shin, thigh.* 2 brace, column, pillar, prop, support, upright. 3 *leg of a journey.* lap, length, part, section, stage, stretch. **pull someone's leg** ▷ HOAX.

legacy *n* bequest, endowment, estate, inheritance.

legal *adj* 1 above-board, acceptable, admissible, allowable, allowed, authorized, constitutional, just, lawful, legalized, legitimate, licensed, licit, permitted, permissible, proper, regular, right, rightful, valid. *Opp* ILEGAL. 2 *legal proceedings.* forensic, judicial, judiciary.

legalize *vb* allow, authorize, legitimate, legitimize, license, make legal, normalize, permit, regularize, validate. *Opp* BAN.

legend *n* epic, folk-tale, myth, saga, tradition. ▷ STORY.

legendary *adj* 1 apocryphal, epic, fabled, fabulous, fictional, fictitious, imaginary, invented, made-up, mythical, non-existent, story-book, traditional. 2 *a legendary name.* ▷ FAMOUS.

legible *adj* clear, decipherable, distinct, intelligible, neat, plain, readable, understandable. *Opp* ILLEGIBLE.

legitimate *adj* 1 authentic, genuine, proper, real, regular, true. ▷ LEGAL. 2 *a legitimate deception.* ethical, just, justifiable, moral, proper, reasonable, right. *Opp* ILLEGITIMATE. • *vb* ▷ LEGALIZE.

leisure *n* breathing-space, ease, holiday, liberty, opportunity, quiet, recreation, relaxation, relief, repose, respite, rest, spare time, time off.

leisurely *adj* easy, gentle, lingering, peaceful, relaxed, relaxing, restful, unhurried. ▷ SLOW. *Opp* BRISK.

lend *vb* advance, loan. *Opp* BORROW.

length *n* 1 distance, extent, footage, measure, measurement, mileage, reach, size, span, stretch. 2 duration, period, stretch, term.

lengthen *vb* continue, drag out, draw out, enlarge, elongate, expand, extend, get longer, increase, make longer, *inf* pad out, prolong, protract, pull out, stretch. *Opp* SHORTEN.

lenient *adj* charitable, easygoing, forbearing, forgiving, gentle, humane, indulgent, merciful, mild, permissive, soft, soft-hearted, sparing, tolerant. ▷ KIND. *Opp* STRICT.

less *adj* fewer, reduced, shorter, smaller. *Opp* MORE.

lessen *vb* 1 assuage, cut, deaden, decrease, ease, lighten, lower, make less, minimize, mitigate, reduce, relieve, tone down. 2 abate, become less, decline, decrease, die away, diminish, dwindle, ease off, let up, moderate slacken, subside, tail off, weaken. *Opp* INCREASE.

lesson *n* 1 class, drill, instruction laboratory, lecture, practical, seminar, session, task, teaching, tutorial, workshop. 2 *a moral lesson.* admonition, example, moral, warning.

let *vb* 1 agree to, allow to, authorize to, consent to, enable to, give permission to, license to, permit to, sanction to. 2 *let a house.* charter, contract out, hire, lease, rent. **let alone, let be** ▷ LEAVE. **let go, let loose** ▷ LIBERATE. **let off** ▷ FIRE. **let out** ▷ LIBERATE. **let**

someone off ▷ ACQUIT. **let up** ▷ LESSEN.

letdown *n* anti-climax, disappointment, disillusionment, *inf* wash-out.

lethal *adj* deadly, fatal, mortal, poisonous.

lethargic *adj* apathetic, comatose, dull, heavy, inactive, indifferent, indolent, languid, lazy, listless, phlegmatic, sleepy, slow, slothful, sluggish, torpid. ▷ WEARY. *Opp* ENERGETIC.

lethargy *n* apathy, idleness, inactivity, indolence, inertia, laziness, listlessness, slothfulness, slowness, sluggishness, torpor, weariness. *Opp* ENERGY.

letter *n* **1** character, consonant, sign, symbol, vowel. **2** *old use* billet-doux, card, communication, dispatch, epistle, message, missive, note, postcard. **letters** correspondence, junk mail, mail, post.

level *adj* **1** even, flat, flush, horizontal, plane, regular, smooth, straight, true, uniform. **2** horizontal. **3** *level scores.* balanced, even, equal, matching, *inf* neck-and-neck, the same. *Opp* UNEVEN. • *n* **1** altitude, depth, elevation, height, value. **2** degree, echelon, grade, plane, position, rank, *inf* rung on the ladder, stage, standard, standing, status. **3** *level in a building.* floor, storey. • *vb* **1** even out, flatten, rake, smooth. **2** bulldoze, demolish, destroy, devastate, knock down, lay low, raze, tear down, wreck. **level-headed** ▷ SENSIBLE.

lever *vb* force, prise, wrench.

liable *adj* **1** accountable, answerable, blameworthy, responsible. **2** *liable to fall over.* apt, disposed, inclined, in the habit of, likely, minded, predisposed, prone, ready, susceptible, tempted, vulnerable, willing.

liaison *n* **1** communication, contact, cooperation, liaising, linkage, links, mediation, relationship, tie. **2** ▷ AFFAIR.

liar *n* deceiver, false witness, *inf* fibber, perjurer, *inf* story-teller.

libel *n* calumny, defamation, denigration, insult, lie, misrepresentation, obloquy, scandal, slander, slur, smear, vilification. • *vb* blacken the name of, calumniate, defame, denigrate, disparage, malign, misrepresent, slander, slur, smear, write lies about, traduce, vilify.

libellous *adj* calumnious, cruel, damaging, defamatory, disparaging, false, hurtful, insulting, lying, malicious, mendacious, scurrilous, slanderous, untrue, vicious.

liberal *adj* **1** abundant, ample, bounteous, bountiful, copious, free, generous, lavish, munificent, open-handed, plentiful, unstinting. **2** *liberal attitudes.* big-hearted, broad-minded, charitable, easy-going, enlightened, fair-minded, humanitarian, indulgent, impartial, latitudinarian, lenient, magnanimous, open-minded, permissive, philanthropic, tolerant, unbiased, unbigoted, unopinionated, unprejudiced, unselfish. *Opp* NARROW-MINDED. **3** *liberal politics.* progressive, radical, reformist. *Opp* CONSERVATIVE.

liberalize *vb* broaden, ease, enlarge, make more liberal, moderate, open up, relax, soften, widen.

liberate *vb* deliver, discharge, disenthral, emancipate, enfranchise, free, let go, let loose, let out, loose, manumit, ransom, release, rescue, save, set free, untie. *Opp* CAPTURE, SUBJUGATE.

liberty *n* autonomy, emancipation, independence, liberation, release, self-determination, self-rule. ▷ FREEDOM. **at liberty** ▷ FREE.

licence *n* **1** certificate, credentials, document, papers, permit, warrant. **2** ▷ FREEDOM.

license *vb* **1** allow, approve, authorize, certify, commission, empower, entitle, give a licence to, permit, sanction, validate. **2** buy a licence for, make legal.

lid *n* cap, cover, covering, top.

lie *n* deceit, dishonesty, disinformation, fabrication, falsehood, falsification, *inf* fib, fiction, invention, misrepresentation, prevarication, untruth, *inf* whopper. *Opp* TRUTH. • *vb* **1** *inf* be economical with the truth, bluff, commit perjury, deceive, falsify the facts, *inf* fib, perjure yourself, prevaricate, tell lies. **2** be horizontal, be prone, be prostrate, be recumbent, be supine, lean back, lounge, recline, repose, rest, sprawl, stretch out. **3** *The house lies in a valley.* be, be found, be located, be situated, exist. **lie low** ▷ HIDE.

life *n* **1** being, existence, living. **2** activity, animation, dash, élan, energy, enthusiasm, exuberance, *inf* go, liveliness, soul, sparkle, spirit, sprightliness, verve, vigour, vitality, vivacity, zest. **3** autobiography, biography, memoir, story.

lifeless *adj* **1** comatose, dead, deceased, inanimate, inert, insensate, insensible, killed, motionless, unconscious. **2** *lifeless desert.* arid, bare, barren, desolate, empty, sterile, uninhabited, waste. **3** *a lifeless performance.* apathetic, boring, dull, flat, heavy, lacklustre, lethargic, slow, torpid, unexciting, wooden. *Opp* LIVELY, LIVING.

lifelike *adj* authentic, convincing, faithful, graphic, natural, photographic, realistic, true-to-life, vivid. *Opp* UNREALISTIC.

lift *n* elevator, hoist. • *vb* **1** buoy up, carry, elevate, heave up, hoist, jack up, pick up, pull up, raise, rear. **2** ascend, fly, lift off, rise, soar. **3** boost, cheer, enhance, improve, promote. **4** ▷ STEAL.

light *adj* **1** lightweight, portable, underweight, weightless. *Opp* HEAVY. **2** bright, illuminated, lit-up, well-lit. *Opp* DARK. **3** *light work.* ▷ EASY. **4** *a light wind.* ▷ GENTLE. **5** *a light touch.* ▷ DELICATE. **6** *light colours.* ▷ PALE. **7** *a light heart.* ▷ CHEERFUL. **8** *light traffic.* ▷ SPARSE. • *n* **1** beam, blaze, brightness, brilliance, effulgence, flare, flash, fluorescence, glare, gleam, glint, glitter, glow, halo, illumination, incandescence, luminosity, lustre, phosphorescence, radiance, ray, reflection, scintillation, shine, sparkle, twinkle. □ *candlelight, daylight, firelight, gaslight, moonlight, starlight, sunlight, torchlight, twilight.* □ *arc light, beacon, bulb, candelabra, candle, chandelier, electric light, flare, floodlight, fluorescent lamp, headlamp, headlight, lamp, lantern, laser, lighthouse, lightship, neon light, pilot light, searchlight, spotlight, standard lamp, street light, strobe, stroboscope, taper, torch, traffic lights.* • *vb* **1** fire, ignite, kindle, put a match to, set alight, set fire to, switch on. *Opp* EXTINGUISH. **2** ▷ LIGHTEN. **bring to light** ▷ DISCOVER. **give light, reflect light** be bright, be luminous, be phosphorescent, blaze, blink, burn, coruscate, dazzle, flash, flicker, glare, gleam, glimmer, glint, glisten, glitter, glow, radiate, reflect, scintillate, shimmer, shine, spark, sparkle, twinkle. **light-headed** ▷ DIZZY. **light-hearted** ▷ CHEERFUL. **light up** ▷ LIGHTEN. **shed light on** ▷ EXPLAIN.

lighten *vb* **1** cast light on, floodlight, illuminate, irradiate, light up, shed light on, shine on. **2** *The sky lightened.* become lighter, brighten, cheer up, clear. **3** ▷ LESSEN.

lighthouse *n* beacon, light, lightship, warning-light.

like *adj* akin to, analogous to, close to, cognate with, comparable to, congruent with, corresponding to, equal to, equivalent to, identical to, parallel to, similar to. • *n* liking, partiality, predilection, preference. • *vb* admire, approve of, appreciate, be attracted to, be fond of, be interested in, be keen on, be partial to, be pleased by, delight in, enjoy, find pleasant, *sl* go for, *inf* go in for, have a high regard for, *inf* have a weakness for, prefer, relish, revel in, take pleasure in, *inf* take to, welcome. ▷ LOVE. *Opp* HATE.

likeable *adj* admirable, attractive, charming, congenial, endearing, interesting, lovable, nice, personable, pleasant, pleasing. ▷ FRIENDLY. *Opp* HATEFUL.

likelihood *n* chance, hope, possibility, probablility, prospect.

likely *adj* **1** anticipated, expected, feasible, foreseeable, plausible, possible, predictable, probable, reasonable, unsurprising. **2** *a likely candidate.* able, acceptable, appropriate, convincing, credible, favourite, fitting, hopeful, promising, qualified, suitable, *inf* tipped to win. **3** *likely to come.* apt, disposed, inclined, liable, prone, ready, tempted, willing. *Opp* UNLIKELY.

liken *vb* compare, equate, juxtapose, match.

likeness *n* **1** affinity, analogy, compatibility, congruity, correspondence, resemblance, similarity. *Opp* DIFFERENCE. **2** copy, depiction, drawing, duplicate, facsimile, image, model, picture, portrait, replica, representation, reproduction, study.

liking *n* affection, affinity, appetite, eye, fondness, inclination, partiality, penchant, predilection, predisposition, preference, propensity, *inf* soft spot, taste, weakness. ▷ LOVE. *Opp* HATRED.

limb *n* appendage, member, offshoot, projection. □ *arm, bough, branch, flipper, foreleg, forelimb, leg, wing.*

limber *vb* **limber up** exercise, get ready, loosen up, prepare, warm up.

limbo *n* **in limbo** abandoned, forgotten, in abeyance, left out, neglected, neither one thing nor the other, *inf* on hold, *inf* on the back burner, unattached.

limit *n* **1** border, boundary, bounds, brink, confines, demarcation line, edge, end, extent, extreme point, frontier, perimeter. **2** ceiling, check, curb, cut-off point, deadline, inhibition, limitation, maximum, restraint, restriction, stop, threshold. • *vb* bridle, check, circumscribe, confine, control, curb, define, fix, hold in check, put a limit on, ration, restrain, restrict. **limited** ▷ FINITE, INADEQUATE.

limitation *n* **1** ▷ LIMIT. **2** defect, deficiency, fault, inadequacy, shortcoming, weakness.

limitless *adj* boundless, countless, endless, everlasting, immeasurable, incalculable, inexhaustible, infinite, innumersable, never-ending, numberless, perpetual, renewable, unbounded, unconfined, unending, unimaginable, unlimited, unrestricted. ▷ VAST. *Opp* FINITE.

limp *adj inf* bendy, drooping, flabby, flaccid, flexible, *inf* floppy, lax, loose, pliable, sagging, slack, soft, weak, wilting, yielding. ▷ WEARY. *Opp* RIGID. • *vb* be lame, falter, hobble, hop, stagger, totter.

line *n* **1** band, borderline, boundary, contour, contour line, dash, mark, streak, striation, strip, stripe, stroke, trail. **2** corrugation, crease, *inf* crow's feet, fold, furrow, groove, score, wrinkle. **3** cable, cord, flex, hawser, lead, rope, string, thread, wire. **4** chain, column, cordon, crocodile, file, procession, queue, rank, row, series. **5** *railway line.* branch, main line, route, service, track. • *vb* **1** mark with lines, rule, score, streak, striate, underline. **2** *line the street.* border, edge, fringe. **3** *line a garment.* cover the inside, insert a lining, pad, reinforce. **line up** ▷ ALIGN, QUEUE.

linger *vb* continue, dally, dawdle, delay, dither, endure, hang about, hover, idle, lag, last, loiter, pause, persist, procrastinate, remain, *inf* shilly-shally, stay, stay behind, *inf* stick around, survive, temporize, wait about. *Opp* HURRY.

lining *n* inner coat, inner layer, interfacing, liner, padding.

link *n* **1** bond, connection, connector, coupling, join, joint, linkage, tie, yoke. ▷ FASTENER. **2** affiliation, alliance, association, communication, interdependence, liaison, partnership, relationship, *inf* tie-up, twinning, union. • *vb* **1** amalgamate, associate, attach, compare, concatenate, connect, couple, interlink, join, juxtapose, make a link, merge, network, relate, see a link, twin, unite, yoke. ▷ FASTEN.

lip *n* brim, brink, edge, rim.

liquefy *vb* become liquid, dissolve, liquidize, melt, run, thaw. *Opp* SOLIDIFY.

liquid *adj* aqueous, flowing, fluid, liquefied, melted, molten, running, *inf* runny, sloppy, *inf* sloshy, thin, watery, wet. *Opp* SOLID. • *n* fluid, juice, liquor, solution, stock.

liquidate *vb* annihilate, destroy, *inf* do away with, *inf* get rid of, remove, silence, wipe out. ▷ KILL.

liquor *n* **1** alcohol, *sl* booze, *sl* hard stuff, intoxicants, *sl* shorts, spirits, strong drink. **2** ▷ LIQUID.

list *n* catalogue, column, directory, file, index, inventory, listing, register, roll, roster, rota, schedule, shopping-list, table. • *vb* **1** catalogue, enumerate, file, index, itemize, make a list of, note, record, register, tabulate, write down. **2** bank, careen, heel, incline, keel over, lean, slant, slope, tilt, tip.

listen *vb* attend, concentrate, eavesdrop, *old use* hark, hear, heed, *inf* keep your ears open, lend an ear, overhear, pay attention, take notice.

listless *adj* apathetic, enervated, feeble, heavy, languid, lazy, lethargic, lifeless, phlegmatic, sluggish, tired, torpid, unenthusiastic, uninterested, weak. ▷ WEARY. *Opp* LIVELY.

literal *adj* close, exact, faithful, matter of fact, plain, prosaic, strict, unimaginative, verbatim, word for word.

literary *adj* 1 cultured, educated, erudite, imaginative, learned, refined, scholarly, well-read, widely-read. 2 *literary style*. ornate, *derog* pedantic, poetic, polished, rhetorical, *derog* self-conscious, sophisticated, stylish.

literate *adj* 1 educated, wellread. 2 accurate, correct, readable, well-written.

literature *n* books, brochures, circulars, creative writing, handbills, handouts, leaflets, pamphlets, papers, writings. □ *autobiography, biography, comedy, crime fiction, criticism, drama, epic, essay, fantasy, fiction, folktale, journalism, myth and legend, novels, parody, poetry, propaganda, prose, romance, satire, science fiction, tragedy*. ▷ WRITING.

lithe *adj* agile, flexible, limber, lissom, loose-jointed, pliable, pliant, supple. *Opp* STIFF.

litter *n* bits and pieces, clutter, debris, fragments, garbage, jumble, junk, mess, odds and ends, refuse, rubbish, trash, waste. • *vb* clutter, fill with litter, make untidy, *inf* mess up, scatter, strew.

little *adj* 1 *inf* baby, bantam, compact, concise, diminutive, *inf* dinky, dwarf, exiguous, fine, fractional, infinitesimal, lean, lilliputian, microscopic, midget, *inf* mini, miniature, minuscule, minute, narrow, petite (*woman*), *inf* pint-sized, *inf* pocket-sized, *inf* poky, portable, pygmy, short, slender, slight, small, *inf* teeny, thin, tiny, toy, undergrown, undersized, *inf* wee, *inf* weeny. *Opp* BIG. 2 *little food*. inadequate, insufficient, meagre, mean, *inf* measly, miserly, modest, niggardly, parsimonious, *inf* piddling, scanty, skimpy, stingy, ungenerous, unsatisfactory. 3 *of little importance*. inconsequential, insignificant, minor, negligible, nugatory, slim (*chance*), slight, trifling, trivial, unimportant.

live *adj* 1 ▷ LIVING. 2 *a live fire*. ▷ ALIGHT. 3 *a live issue*. contemporary, current, important, pressing, relevant, topical, vital. *Opp* DEAD. • *vb* 1 breathe, continue, endure, exist, flourish, function, last, persist, remain, stay alive, survive. *Opp* DIE. 2 be accommodated, dwell, lodge, reside, room, stay. 3 *live on £20 a week*. fare, *inf* get along, keep going, make a living, pay the bills, subsist. **live in** ▷ INHABIT. **live on** ▷ EAT.

liveliness *n* activity, animation, boisterousness, bustle, dynamism, energy, enthusiasm, exuberance, *inf* go, gusto, high spirits, spirit, sprightliness, verve, vigour, vitality, vivacity, zeal. *Opp* APATHY.

lively *adj* active, agile, alert, animated, boisterous, bubbly, bustling, busy, cheerful, colourful, dashing, eager, energetic, enthusiastic, exciting, expressive, exuberant, frisky, gay, high-spirited, irrepressible, jaunty, jazzy, jolly, merry, nimble, *inf* perky, playful, quick, spirited, sprightly, stimulating, strong, vigorous, vital, vivacious, vivid, *inf* zippy. ▷ HAPPY. *Opp* APATHETIC.

livestock *n* cattle, farm animals.

living *adj* active, actual, alive, animate, breathing, existing, extant, flourishing, functioning, live, living, *old use* quick, sentient, surviving, vigorous, vital. ▷ LIVELY. *Opp* DEAD, EXTINCT. • *n* income, livelihood, occupation, subsistence, way of life.

load *n* 1 burden, cargo, consignment, freight, lading, lorry-load, shipment, van-load. 2 *load of responsibility*. *inf* albatross, anxiety, care, *inf* cross, encumbrance, *inf* millstone, onus, trouble, weight, worry. • *vb* 1 burden, encumber, fill, heap, overwhelm, pack, pile, ply, saddle, stack, stow, weigh down. 2 *load a gun.* charge, prime. **loaded** ▷ BIASED, LADEN, WEALTHY.

loafer *n* idler, *inf* good-for-nothing, layabout, *inf* lazybones, lounger, shirker, *sl* skiver, vagrant, wastrel.

loan *n* advance, credit, mortgage. • *vb* advance, allow, credit, lend.

loathe *vb* abhor, abominate, be averse to, be revolted by, despise, detest, dislike, execrate, find intolerable, hate, object to, recoil from, resent, scorn, shudder at. *Opp* LOVE.

lobby *n* 1 ante-room, corridor, entrance hall, entry, foyer, hall, hallway, porch, reception, vestibule. 2 *environmental lobby.* campaign, campaigners, pressure-group, supporters. • *vb* persuade, petition, pressurize, try to influence, urge.

local *adj* 1 adjacent, adjoining, nearby, neighbouring, serving the locality. 2 *local politics.* community, limited, narrow, neighbourhood, parochial, particular, provincial, regional. *Opp* GENERAL, NATIONAL. • *n* 1 inhabitant, resident, townsman, townswoman. 2 ▷ PUB.

locality *n* area, catchment area, community, district, location, neighbourhood, parish, region, residential area, town, vicinity, zone.

localize *vb* concentrate, confine, contain, enclose, keep within bounds, limit, narrow down, pin down, restrict. *Opp* SPREAD.

locate *vb* 1 come across, detect, discover, find, identify, *inf* lay your hands on, *inf* run to earth, search out, track down, unearth. 2 build, establish, find a place for, found, place, position, put, set up, site, situate, station.

location *n* 1 locale, locality, place, point, position, setting, site, situation, spot, venue, whereabouts. 2 *film locations.* background, scene, setting.

lock *n* bar, bolt, catch, clasp, fastening, hasp, latch, padlock. • *vb* bolt, close, fasten, padlock, seal, secure, shut. **lock away** ▷ IMPRISON. **lock out** ▷ EXCLUDE. **lock up** ▷ IMPRISON.

lodge *n* ▷ HOUSE. • *vb* 1 accommodate, billet, board, house, *inf* put up. 2 abide, dwell, live, *inf* put up, reside, stay, stop. 3 *lodge a complaint.* enter, file, make formally, put on record, record, register, submit.

lodger *n* boarder, guest, inmate, paying guest, resident, tenant.

lodgings *n* accommodation, apartment, billet, boarding-house, *inf* digs, lodging-house, *sl* pad, quarters, rooms, shelter, *sl* squat, temporary home.

lofty *adj* 1 elevated, high, imposing, majestic, noble, soaring, tall, towering. 2 ▷ ARROGANT.

log *n* 1 timber, wood. 2 account, diary, journal, record.

logic *n* clarity, deduction, intelligence, logical thinking, ratiocination, rationality, reasonableness, reasoning, sense, validity.

logical *adj* clear, cogent, coherent, consistent, deductive, intelligent, methodical, rational, reasonable, sensible, sound, *inf* step-by-step, structured, systematic, valid, well-reasoned, well-thought-out, wise. *Opp* ILLOGICAL.

loiter *vb* be slow, dally, dawdle, hang back, linger, *inf* loaf about, *inf* mess about, skulk, *inf* stand about, straggle.

lone *adj* isolated, separate, single, solitary, solo, unaccompanied. ▷ LONELY.

lonely *adj* **1** abandoned, alone, forlorn, forsaken, friendless, lonesome, loveless, neglected, outcast, reclusive, retiring, solitary, unsociable, withdrawn. ▷ SAD. **2** *inf* cut off, deserted, desolate, distant, faraway, isolated, *inf* off the beaten track, out of the way, remote, secluded, unfrequented, uninhabited.

long *adj* big, drawn out, elongated, endless, extended, extensive, great, interminable, large, lasting, lengthy, longish, prolonged, protracted, slow, stretched, sustained, time-consuming, unending. • *vb* crave, hanker, have a longing (for), hunger, *inf* itch, pine, thirst, wish, yearn. **long for** ▷ DESIRE. **long-lasting, long-lived** ▷ PERMANENT. **long-standing** ▷ OLD. **long-suffering** ▷ PATIENT. **long-winded** ▷ TEDIOUS.

longing *n* appetite, craving, desire, hankering, hunger, *inf* itch, need, thirst, urge, wish, yearning, *inf* yen.

look *n* **1** gaze, glance, glimpse, observation, peek, peep, sight, *inf* squint, view. **2** air, appearance, aspect, attractiveness, bearing, beauty, complexion, countenance, demeanour, expression, face, looks, manner, mien. • *vb* **1** behold, *inf* cast your eye, consider, contemplate, examine, eye, gape, *inf* gawp, gaze, glance, glimpse, goggle, have a look, inspect, observe, ogle, pay attention (to), peek, peep, peer, read, regard, scan, scrutinize, see, skim through, squint, stare, study, survey, *sl* take a dekko, take note (of), view, watch. **2** *The house looks south.* face, overlook. **3** *look pleased.* appear, seem. **look after** ▷ TEND. **look down on** ▷ DESPISE. **look for** ▷ SEEK. **look into** ▷ INVESTIGATE. **look out** ▷ BEWARE. **look up to** ▷ ADMIRE.

look-out *n* guard, sentinel, sentry, watchman.

loom *vb* arise, appear, dominate, emerge, hover, impend, materialize, menace, rise, stand out, stick up, take shape, threaten, tower.

loop *n* bend, bow, circle, coil, curl, eye, hoop, kink, noose, ring, turn, twist, whorl. • *vb* bend, coil, curl, entwine, make a loop, turn, twist, wind.

loophole *n* escape, *inf* get-out, *inf* let-out, outlet, way out.

loose *adj* **1** detachable, detached, disconnected, independent, insecure, loosened, movable, moving, scattered, shaky, unattached, unconnected, unfastened, unsteady, wobbly. **2** *loose animals.* at large, at liberty, escaped, free, free-range, released, roaming, uncaged, unconfined, unfettered, unrestricted, untied. **3** *loose hair.* dangling, hanging, spread out, straggling, trailing. **4** *loose clothing.* baggy, *inf* floppy, loose-fitting, slack, unbuttoned. **5** *loose thinking.* broad, careless, casual, diffuse, general, ill-defined, illogical, imprecise, inexact, informal, lax, rambling, rough, *inf* sloppy, unscientific, unstructured, vague. *Opp* PRECISE, SECURE, TIGHT. **6** ▷ IMMORAL. • *vb* ▷ FREE, LOOSEN.

loosen *vb* **1** ease off, free, let go, loose, make loose, relax, release, separate, slacken, unfasten, unloose, untie. ▷ UNDO. **2** become loose, come adrift, open up. *Opp* TIGHTEN.

loot *n* booty, contraband, haul, *inf* ill-gotten gains, plunder, prize, spoils, *inf* swag, takings. • *vb* despoil, pillage, plunder, raid, ransack, ravage, rifle, rob, sack, steal from.

lopsided *adj* askew, asymmetrical, awry, *inf* cockeyed, crooked, one-sided, tilting, unbalanced, unequal, uneven.

lord *n* aristocrat, noble, peer. □ *baron, count, duke, earl,* old use *thane, viscount.* ▷ RULER.

lose *vb* **1** be deprived of, cease to have, drop, find yourself without, forfeit, forget, leave (somewhere), mislay, misplace, miss, part with, stray from. *Opp* FIND. **2** admit defeat, be defeated, capitulate, *inf* come to grief, fail, get beaten, get thrashed, succumb, suffer defeat. *Opp* WIN. **3** *lose your chance.* fritter, let slip, squander, waste. **4** *lose pursuers.* escape from, evade, get rid of, give the slip, leave behind, outrun, shake off, throw off. **losing** ▷ UNSUCCESSFUL.

loser *n* the defeated, runner-up, the vanquished. *Opp* WINNER.

loss *n* bereavement, damage, defeat, deficit, depletion, deprivation, destruction, diminution, disappearance, erosion, failure, forfeiture, impairment, privation, reduction, sacrifice. *Opp* GAIN. **losses** casualties, deaths, death toll, fatalities.

lost *adj* **1** abandoned, departed, destroyed, disappeared, extinct, forgotten, gone, irrecoverable, irretrievable, left behind, mislaid, misplaced, missing, strayed, untraceable, vanished. **2** absorbed, day-dreaming, dreamy, distracted, engrossed, preoccupied, rapt. **3** corrupt, damned, fallen. ▷ WICKED.

lot *n a lot in a sale.* ▷ ITEM. **a lot of, lots of** ▷ PLENTY. **draw lots** ▷ GAMBLE. **the lot** all (of), everything, the whole thing, *inf* the works.

lotion *n* balm, cream, embrocation, liniment, ointment, pomade, salve, unguent.

lottery *n* **1** raffle, sweepstake. **2** gamble, speculation, venture.

loud *adj* **1** audible, blaring, booming, clamorous, clarion (*call*), deafening, ear-splitting, echoing, fortissimo, high, noisy, penetrating, piercing, raucous, resounding, reverberant, reverberating, roaring, rowdy, shrieking, shrill, sonorous, stentorian, strident, thundering, thunderous, vociferous. **2** *loud colours.* ▷ GAUDY. *Opp* QUIET.

lounge *n* drawing-room, front room, living-room, parlour, salon, sitting-room. • *vb* be idle, be lazy, dawdle, hang about, idle, *inf* kill time, laze, loaf, lie around, loiter, *inf* loll about, *inf* mess about, *inf* mooch about, relax, *inf* skive, slouch, slump, sprawl, stand about, take it easy, vegetate, waste time.

lout *n* boor, churl, rude person, oaf, *inf* yob.

lovable *adj* adorable, appealing, attractive, charming, cuddly, *inf* cute, *inf* darling, dear, enchanting, endearing, engaging, fetching, likeable, lovely, pleasing, taking, winning, winsome. *Opp* HATEFUL.

love *n* **1** admiration, adoration, adulation, affection, ardour, attachment, attraction, desire, devotion, fancy, fervour, fondness infatuation, liking, passion, regard, tenderness, warmth. ▷ FRIENDSHIP. **2** beloved, darling, dear, dearest, loved one. ▷ LOVER. • *vb* **1** admire, adore, be charmed by, be fond of, be infatuated by, be in love with, care for, cherish, desire, dote on, fancy, *inf* have a crush on, have a passion for, idolize, lose your heart to, lust after, treasure, value, want, worship. ▷ LIKE. *Opp* HATE. **in love** besotted, devoted, enamoured, fond, *inf* head over heels, infatuated. **love affair** affair, amour, courtship, intrigue, liaison, relationship, romance. **make love** be intimate, *inf* canoodle, caress, copulate, court, cuddle, embrace, flirt, fornicate, have intercourse, have sex, kiss, mate, *sl* neck, *inf* pet, philander, *old use* spoon, woo. ▷ SEX.

loved *adj* beloved, cherished, darling, dear, dearest, esteemed, favourite, precious, treasured, valued, wanted.

loveless *adj* cold, frigid, heartless, passionless, undemonstrative, unfeeling, unloving, unresponsive. *Opp* LOVING. ▷ UNLOVED.

lovely *adj* appealing, attractive, charming, delightful, enjoyable, fine, good, nice, pleasant, pretty, sweet. ▷ BEAUTIFUL. *Opp* NASTY.

lover *n* admirer, boyfriend, companion, concubine, fiancé(e), *old use* follower, friend, gigolo, girlfriend, *inf* intended, mate, mistress, *old use* paramour, suitor, sweetheart, *sl* toy boy, valentine.

lovesick *adj* frustrated, languishing, lovelorn, pining.

loving *adj* admiring, adoring, affectionate, amorous, ardent, attached, brotherly, caring, close, concerned, dear, demonstrative, devoted, doting, fatherly, fond, friendly, inseparable, kind, maternal, motherly, passionate, paternal, protective, sisterly, tender, warm. ▷ FRIENDLY, SEXY. *Opp* LOVELESS.

low *adj* **1** flat, low-lying, sunken. **2** *low trees*. little, short, squat, stumpy, stunted. **3** *low status*. abject, base, degraded, humble, inferior, junior, lesser, lower, lowly, menial, miserable, modest, servile. **4** *low behaviour*. churlish, coarse, common, cowardly, crude, *old use* dastardly, disreputable, ignoble, mean, nasty, vulgar, wicked. ▷ IMMORAL. **5** *low sounds*. gentle, indistinct, muffled, murmurous, muted, pianissimo, quiet, soft, subdued, whispered. **6** *low notes*. bass, deep, reverberant. *Opp* HIGH. **in low spirits** ▷ SAD. **low point** ▷ NADIR.

lowbrow *adj* accessible, easy, ordinary, pop, popular, *derog* rubbishy, simple, straightforward, *derog* trashy, *derog* uncultured, undemanding, unpretentious, unsophisticated. *Opp* HIGHBROW.

lower *vb* **1** dip, drop, haul down, let down, take down. **2** *lower prices*. bring down, cut, decrease, discount, lessen, mark down, reduce, *inf* slash. **3** *lower the volume*. abate, diminish, quieten, tone down, turn down. **4** *lower yourself*. abase, belittle, debase, degrade, demean, discredit, disgrace, humble, humiliate, stoop. *Opp* RAISE.

lowly *adj* base, humble, insignificant, little-known, low, low-born, meek, modest, obscure, unimportant. ▷ ORDINARY. *Opp* EMINENT.

loyal *adj* committed, constant, dedicated, dependable, devoted, dutiful, faithful, honest, patriotic, reliable, sincere, stable, staunch, steadfast, steady, true, trustworthy, trusty, unswerving, unwavering. *Opp* DISLOYAL.

loyalty *n* allegiance, constancy, dedication, dependability, devotion, duty, faithfulness, fealty, fidelity, honesty, patriotism, reliability, staunchness, steadfastness, trustworthiness. *Opp* DISLOYALTY.

lubricate *vb* grease, oil.

luck *n* **1** accident, chance, coincidence, destiny, fate, *inf* fluke, fortune, serendipity. **2** *wished her luck*. *inf* break, good fortune, happiness, prosperity, success.

lucky *adj* **1** accidental, appropriate, chance, *inf* fluky, fortuitous, opportune, providential, timely, unintentional, unplanned, welcome. **2** blessed, favoured, fortunate, successful. ▷ HAPPY. **3** *lucky number*. advantageous, auspicious. *Opp* UNLUCKY.

luggage *n* baggage, belongings, *inf* gear, impedimenta, paraphernalia, *inf* things. □ *bag, basket, box, brief-case, case, chest, hamper, handbag, hand luggage, haversack, holdall, knapsack, pannier,* old use *portmanteau, purse, rucksack, satchel, suitcase, trunk, wallet.*

lukewarm *adj* **1** room temperature, tepid, warm. **2** apathetic, cool, half-hearted, indifferent, unenthusiastic.

lull *n* break, calm, delay, gap, halt, hiatus, interlude, interruption, interval, lapse, *inf* let-up, pause, respite, rest, silence. • *vb* calm, hush, pacify, quell, quieten, soothe, subdue, tranquillize.

lumber *n* **1** beams, boards, planks, timber, wood. **2** bits and pieces, clutter, jumble, junk, litter, odds and ends, rubbish, trash, *inf* white elephants. • *vb* **1** blunder, move clumsily, shamble, trudge. **2** ▷ BURDEN.

luminous *adj* bright, glowing, luminescent, lustrous, phosphorescent, radiant, refulgent, shining. ▷ LIGHT.

lump *n* **1** ball, bar, bit, block, cake, chunk, clod, clot, cube, *inf* dollop, gob, gobbet, hunk, ingot, mass, nugget, piece, slab, wad, wedge, *inf* wodge. **2** boil, bulge, bump, carbuncle, cyst, excrescence, growth, hump, knob, node, nodule, protrusion, protuberance, spot, swelling, tumescence, tumour. • *vb* **lump together** ▷ COMBINE.

lunacy *n* delirium, dementia, derangement, frenzy, hysteria, illogicality, insanity, madness, mania, psychosis, unreason. ▷ STUPIDITY.

lunatic *adj* ▷ MAD. • *n* *inf* crackpot, *inf* crank, *inf* loony, madman, madwoman, maniac, *inf* mental case, *inf* nutcase, *inf* nutter, psychopath, psychotic.

lunge *vb* **1** jab, stab, strike, thrust. **2** charge, dash, dive, lurch, plunge, pounce, rush, spring, throw yourself.

lurch *vb* heave, lean, list, lunge, pitch, plunge, reel, roll, stagger, stumble, sway, totter, wallow. **leave in the lurch** ▷ ABANDON.

lure *vb* allure, attract, bait, charm, coax, decoy, draw, entice, induce, inveigle, invite, lead on, persuade, seduce, tempt.

lurid *adj* **1** bright, gaudy, glaring, glowing, striking, vivid. **2** ▷ SENSATIONAL.

lurk *vb* crouch, hide, lie in wait, lie low, prowl, skulk, steal.

luscious *adj* appetizing, delectable, delicious, juicy, mouthwatering, rich, succulent, sweet, tasty.

lust *n* **1** carnality, concupiscence, desire, lasciviousness, lechery, libido, licentiousness, passion, sensuality, sexuality. **2** appetite, craving, greed, hunger, itch, longing.

lustful *adj* carnal, concupiscent, erotic, lascivious, lecherous, lewd, libidinous, licentious, on heat, passionate, *sl* randy, salacious, sensual, *sl* turned on. ▷ SEXY.

lustrous *adj* burnished, glazed, gleaming, glossy, metallic, polished, reflective, sheeny, shiny.

luxuriant *adj* **1** abundant, ample, copious, dense, exuberant, fertile, flourishing, green, lush, opulent, plenteous, plentiful, profuse, prolific, rank, rich, teeming, thick, thriving, verdant. **2** ▷ ORNATE. *Opp* SPARSE.

luxurious *adj* comfortable, costly, expensive, extravagant, grand, hedonistic, lavish, lush, magnificent, opulent, pampered, *inf* plush, *inf* posh, rich, *inf* ritzy, self-indulgent, splendid, sumptuous, sybaritic, voluptuous. *Opp* SPARTAN.

luxury *n* affluence, comfort, ease, enjoyment, extravagance, grandeur, hedonism, high living, indulgence, magnificence, opulence, pleasure, relaxation, self-indulgence, splendour, sumptuousness, voluptuousness.

lying *adj* crooked, deceitful, deceptive, dishonest, double-dealing, duplicitous, false, hypocritical, inaccurate, insincere, mendacious, misleading, perfidious, unreliable, untrustworthy, untruthful. *Opp* TRUTHFUL. • *n* deceit, deception, dishonesty, duplicity, falsehood, *inf* fibbing, hypocrisy, mendacity, perfidy, perjury, prevarication.

lyrical *adj* emotional, expressive, impassioned, inspired, melodious, musical, poetic, rapturous, rhapsodic, song-like, sweet, tuneful. *Opp* PROSAIC.

M

macabre *adj* eerie, fearsome, frightful, ghoulish, grim, grisly, grotesque, gruesome, morbid, *inf* sick, unhealthy, weird.

machine *n* appliance, contraption, contrivance, device, engine, gadget, implement, instrument, mechanism, motor, robot, tool. ▷ MACHINERY.

machinery *n* **1** apparatus, equipment, gear, machines, plant. **2** constitution, method, organization, procedure, structure, system.

mackintosh *n* anorak, cape, mac, sou'wester, waterproof. ▷ COAT.

mad *adj* **1** *inf* batty, berserk, *inf* bonkers, *inf* certified, *inf* crackers, crazed, crazy, *inf* daft, delirious, demented, deranged, disordered, distracted, *inf* dotty, eccentric, fanatical, frantic, frenzied, hysterical, insane, irrational, *inf* loony, lunatic, maniacal, manic, *inf* mental, moonstruck, *Lat* non compos mentis, *inf* nutty, *inf* off your head, *inf* off your rocker, *inf* out of your mind, possessed, *inf* potty, psychotic, *inf* queer in the head, *inf* round the bend, *inf* round the twist, *inf* screwy, *inf* touched, unbalanced, unhinged, unstable, *inf* up the pole, wild. *Opp* SANE. **2** *a mad comedy.* ▷ ABSURD. **3** ▷ ANGRY. **4** ▷ ENTHUSIASTIC.

madden *vb* anger, craze, derange, *inf* drive crazy, enrage, exasperate, excite, incense, inflame, infuriate, irritate, make you mad, *inf* make your blood boil, *inf* make you see red, provoke, *inf* send you round the bend, unhinge, vex.

madman, madwoman *ns* *inf* crackpot, *inf* crank, eccentric, *inf* loony, lunatic, maniac, *inf* mental case, *inf* nutcase, *inf* nutter, psychopath, psychotic.

madness *n* delirium, dementia, derangement, eccentricity, folly, frenzy, hysteria, illogicality, insanity, lunacy, mania, mental illness, psychosis, unreason. ▷ STUPIDITY.

magazine *n* **1** comic, journal, monthly, newspaper, pamphlet, paper, periodical, publication, quarterly, weekly. **2** *magazine of weapons.* ammunition dump, armoury, arsenal, storehouse.

magic *adj* **1** conjuring, miraculous, mystic, necromantic, supernatural. **2** bewitching, charming, enchanting, entrancing, magical, spellbinding. • *n* **1** black magic, charm, enchantment, hocus-pocus, incantation, *inf* mumbo-jumbo, necromancy, occultism, sorcery, spell, voodoo, witchcraft, witchery, wizardry. **2** conjuring, illusion, legerdemain, sleight of hand, trickery, tricks.

magician *n* conjuror, enchanter, enchantress, illusionist, magus, necromancer, sorcerer, *old use* warlock, witch, witch-doctor, wizard.

magnetic *adj* alluring, attractive, bewitching, captivating, charismatic, charming, compelling, engaging, enthralling, entrancing, fascinating, hypnotic, inviting, irresistible, seductive, spellbinding. *Opp* REPULSIVE.

magnetism *n* allure, appeal, attractiveness, charisma, charm, drawing power, fascination, irresistibility, lure, power, pull, seductiveness.

magnificent *adj* awe-inspiring, beautiful, distinguished, excellent, fine, glorious, gorgeous, grand, grandiose, great, imposing, impressive, majestic, marvellous, noble, opulent, *inf* posh, regal, resplendent, rich, spectacular, splendid, stately, sumptuous, superb, wonderful. *Opp* ORDINARY.

magnify *vb* **1** amplify, augment, *inf* blow up, enlarge, expand, increase, intensify, make larger. *Opp* SHRINK. **2** *magnify difficulties.* *inf* blow up out of proportion, dramatize, exaggerate, heighten, inflate, make too much of, maximize, overdo, overestimate, overstate. *Opp* MINIMIZE.

magnitude *n* bigness, enormousness, extent, greatness, immensity, importance, size.

mail *n* correspondence, letters, parcels, post. • *vb* dispatch, forward, post, send.

maim *vb* cripple, disable, hamstring, handicap, incapacitate, lame, mutilate. ▷ WOUND.

main *adj* basic, biggest, cardinal, central, chief, critical, crucial, dominant, dominating, essential, first, foremost, fundamental, greatest, largest, leading, major, most important, outstanding, paramount, predominant, pre-eminent, prevailing, primary, prime, principal, special, strongest, supreme, top, vital. *Opp* MINOR.

mainly *adv* above all, as a rule, chiefly, especially, essentially, first and foremost, generally, in the main, largely, mostly, normally, on the whole, predominantly, primarily, principally, usually.

maintain *vb* **1** carry on, continue, hold to, keep going, keep up, perpetuate, persevere in, persist in, preserve, retain, stick to, sustain. **2** *maintain a car.* care for, keep in good condition, look after, service, take care of. **3** *maintain a family.* feed, keep, pay for, provide for, stand by, support. **4** *maintain your innocence.* affirm, allege, argue, assert, aver, claim, contend, declare, defend, insist, proclaim, profess, state, uphold.

maintenance *n* **1** care, conservation, looking after, preservation, repairs, servicing, upkeep. **2** alimony, allowance, contribution, subsistence, support.

majestic *adj* august, awe-inspiring, awesome, dignified, distinguished, elevated, exalted, glorious, grand, grandiose, imperial, imposing, impressive, kingly, lofty, lordly, magisterial, magnificent, monumental, noble, pompous, princely, queenly, regal, royal, splendid, stately, striking, sublime.

majesty *n* awesomeness, dignity, glory, grandeur, kingliness, loftiness, magnificence, nobility, pomp, royalty, splendour, stateliness, sublimity.

major *adj* bigger, chief, considerable, extensive, greater, important, key, larger, leading, outstanding, principal, serious, significant. ▷ MAIN. *Opp* MINOR.

majority *n* **1** *inf* best part, *inf* better part, bulk, greater number, mass, preponderance. **2** adulthood, coming of age, manhood, maturity, womanhood. **be in the majority** ▷ DOMINATE.

make *n* brand, kind, model, sort, type, variety. • *vb* **1** assemble, beget, bring about, build, compose, constitute, construct, contrive, craft, create, devise, do, engender, erect, execute, fabricate, fashion, forge, form, frame, generate, invent, make up, manufacture, mass-produce, originate, produce, put together, think up. **2** *make dinner.* concoct, cook, *inf* fix, prepare. **3** *make clothes.* knit, *inf* run up, sew, weave. **4** *make an effigy.* carve, cast, model, mould, sculpt, shape. **5** *make a speech.* deliver, pronounce, utter. ▷ SPEAK. **6** *made her chairperson.* appoint, elect, nominate, ordain. **7** *make P into B.* alter, change, convert, modify, transform, turn. **8** *make a fortune.* earn, gain, get, obtain, receive. **9** *make a good employee.* become, change into, grow into, turn into. **10** *make your objective.* accomplish, achieve, arrive at, attain, catch, get to, reach, win. **11** *2 + 2 makes 4.* add up to, amount to, come to, total. **12** *make rules.* agree, arrange, codify, establish, decide on, draw up, fix, write. **13** *make her happy.* cause to become, render. **14** *make trouble.* bring about, carry out, cause, give rise to, provoke, result in. **15** *make them obey.* coerce, compel, constrain, force, induce, oblige, order, pressurize, prevail on, require. **make amends** ▷ COMPENSATE. **make believe** ▷ IMAGINE. **make fun of** ▷ RIDICULE. **make good** ▷ PROSPER. **make love** ▷ LOVE. **make off** ▷ DEPART. **make off with** ▷ STEAL. **make out** ▷ UNDERSTAND. **make up** ▷ INVENT. **make up for** ▷ COMPENSATE. **make up your mind** ▷ DECIDE.

make-believe *adj* fanciful, feigned, imaginary, made-up, mock, *inf* pretend, pretended, sham, simulated, unreal.

• *n* dream, fantasy, play-acting, pretence, self-deception, unreality.

maker *n* architect, author, builder, creator, director, manufacturer, originator, producer.

makeshift *adj* emergency, improvised, provisional, stopgap, temporary.

maladjusted *adj* disturbed, muddled, neurotic, unbalanced.

male *adj* manly, masculine, virile. • *n* □ *bachelor,* inf *bloke, boy, boyfriend,* inf *bridegroom, brother, chap,* inf *codger, father, fellow, gentleman, groom,* inf *guy, husband, lad, lover, man, son,* inf *squire, uncle, widower.* □ *buck, bull, cock, dog, ram, stallion, tom(cat).* *Opp* FEMALE.

malefactor *n* delinquent, lawbreaker, offender, villain, wrongdoer. ▷ CRIMINAL.

malice *n* animosity, *inf* bitchiness, bitterness, *inf* cattiness, enmity, hatred, hostility, ill-will, malevolence, maliciousness, malignity, nastiness, rancour, spite, spitefulness, vengefulness, venom, viciousness, vindictiveness.

malicious *adj inf* bitchy, bitter, *inf* catty, evil, evil-minded, hateful, ill-natured, malevolent, malignant, mischievous, nasty, rancorous, revengeful, sly, spiteful, vengeful, venomous, vicious, villainous, vindictive, wicked. *Opp* KIND.

malignant *adj* dangerous, deadly, destructive, fatal, harmful, injurious, life-threatening, poisonous, *inf* terminal, uncontrollable, virulent. ▷ MALICIOUS.

malleable *adj* ductile, plastic, pliable, soft, tractable, workable. *Opp* BRITTLE.

malnutrition *n* famine, hunger, starvation, undernourishment.

man *n* **1** [*either sex*] ▷ MANKIND. **2** ▷ MALE. • *vb* cover, crew, provide staff for, staff.

manage *vb* **1** administer, be in charge of, be manager of, command, conduct, control, direct, dominate, govern, head, lead, look after, mastermind, operate, organize, oversee, preside over, regulate, rule, run, superintend, supervise, take care of, take control of, take over. **2** *manage your jobs.* accomplish, achieve, bring about, carry out, contend with, cope with, deal with, do, finish, get through, handle, manipulate, muddle through, perform, sort out, succeed in, undertake. **3** *Can you manage?* cope, fend for yourself, *inf* make it, scrape by, shift for yourself, succeed, survive. **4** *can manage £10.* afford, spare.

manageable *adj* **1** acceptable, convenient, easy to manage, handy, neat, reasonable. *Opp* AWKWARD. **2** amenable, compliant, controllable, disciplined, docile, governable, submissive, tame, tractable. ▷ OBEDIENT. *Opp* DISOBEDIENT.

manager, manageress *ns* administrator, *inf* boss, chief, controller, director, executive, foreman, forewoman, governor, head, organizer, overseer, proprietor, ruler, superintendent, supervisor. ▷ CHIEF.

mandatory *adj* compulsory, essential, necessary, needed, obligatory, required, requisite.

mangle *vb* butcher, cripple, crush, cut, damage, deform, disfigure, hack, injure, lacerate, maim, maul, mutilate, ruin, spoil, squash, tear, wound.

mangy *adj* dirty, filthy, moth-eaten, nasty, scabby, scruffy, shabby, slovenly, squalid, *inf* tatty, unkempt, wretched.

manhandle *vb* **1** carry, haul, heave, hump, lift, manoeuvre, move, pull, push. **2** abuse, batter, *inf* beat up, ill-treat, knock about, maltreat, mistreat, misuse, *inf* rough up, treat roughly.

mania *n* craving, craze, enthusiasm, fad, fetish, frenzy, infatuation, obsession, passion, preoccupation, rage. ▷ MADNESS.

maniac *n* lunatic, psychopath, psychotic. ▷ MADMAN.

manifest *adj* apparent, blatant, clear, conspicuous, discernible, evident, explicit, glaring, noticeable, obvious, patent, plain, recognizable, undisguised, visible. • *vb* ▷ SHOW.

manifesto *n* declaration, policy statement.

manipulate *vb* **1** feel, massage, rub, stimulate. **2** *manipulate people.* control, direct, engineer, exploit, guide, handle, influence, manage, manoeuvre, orchestrate, steer.

mankind *n Lat* homo sapiens, human beings, humanity, humankind, the human race, man, men and women, mortals, people.

manly *adj* chivalrous, gallant, heroic, *inf* macho, male, mannish, masculine, strong, swashbuckling, vigorous, virile. ▷ BRAVE. *Opp* EFFEMINATE.

man-made *adj* artificial, imitation, manufactured, mass-produced, processed, simulated, synthetic, unnatural. *Opp* NATURAL.

manner *n* **1** approach, fashion, means, method, mode, procedure, process, style, technique, way. **2** air, aspect, attitude, bearing, behaviour, character, conduct, demeanour, deportment, disposition, look, mien. **3** *all manner of things.* genre, kind, sort, type, variety.

manners behaviour, breeding, civility, conduct, courtesy, decorum, etiquette, gentility, politeness, protocol, refinement, social graces.

mannerism *n* characteristic, habit, idiosyncrasy, peculiarity, quirk, trait.

manoeuvre *n* device, dodge, gambit, intrigue, move, operation, plan, plot, ploy, ruse, scheme, stratagem, strategy, tactics, trick. • *vb* contrive, engineer, guide, jockey, manipulate, move, navigate, negotiate, pilot, steer.

manoeuvres army exercise, operation, training.

mansion *n* castle, château, manor, manor-house, palace, stately home, villa. ▷ HOUSE.

mantle *n* cape, cloak, covering, hood, shawl, shroud, wrap. • *vb* ▷ COVER.

manufacture *vb* assemble, build, construct, create, fabricate, make, mass-produce, prefabricate, process, put together, *inf* turn out. **manufactured** ▷ MAN-MADE.

manufacturer *n* factory-owner, industrialist, maker, producer.

manure *n* compost, dung, fertilizer, *inf* muck.

manuscript *n* document, papers, script. ▷ BOOK.

many *adj* abundant, assorted, copious, countless, diverse, frequent, innumerable, multifarious, myriad, numberless, numerous, profuse, *inf* umpteen, uncountable, untold, varied, various. *Opp* FEW.

map *n* chart, diagram, plan.

mar *vb* blight, blot, damage, deface, disfigure, harm, hurt, impair, ruin, spoil, stain, taint, tarnish, wreck.

marauder *n* bandit, buccaneer, invader, pirate, plunderer, raider.

march *n* cortège, demonstration, march-past, parade, procession, progress. • *vb* file, pace, parade, step, stride, troop. ▷ WALK.

margin *n* **1** border, boundary, brink, edge, frieze, perimeter, periphery, rim, side, verge. **2** allowance, latitude, leeway, room, scope, space.

marginal *adj* borderline, doubtful, insignificant, minimal, negligible, peripheral, slight, small, unimportant.

marital *adj* conjugal, matrimonial, nuptial.

mark *n* **1** blemish, blot, blotch, dent, dot, fingermark, *plur* graffiti, impression, line, marking, pockmark, print, scar, scratch, scribble, smear, smudge, smut, *inf* splotch, spot, stain, *plur* stigmata, streak, trace, vestige. **2** *mark of breeding*. characteristic, feature, indication, indicator, marker, token. **3** *identifying mark*. badge, brand, device, emblem, fingerprint, hallmark, label, seal, sign, stamp, standard, symbol, trademark. • *vb* **1** blemish, blot, brand, bruise, cut, damage, deface, dent, dirty, disfigure, draw on, make a mark on, mar, scar, scratch, scrawl over, scribble on, smudge, spot, stain, stamp, streak, tattoo, write on. **2** *mark pupils' work*. appraise, assess, correct, evaluate, grade, tick. **3** *mark my words*. attend to, heed, listen to, mind, note, notice, observe, pay attention to, take note of, take seriously, *inf* take to heart, watch.

market *n* auction, bazaar, exchange, fair, marketplace, sale. ▷ SHOP. • *vb* advertise, deal in, make available, merchandise, peddle, promote, put on the market, retail, sell, *inf* tout, trade, trade in, try to sell, vend.

marksman *n* crack shot, gunman, sharpshooter, sniper.

maroon *vb* abandon, cast away, desert, forsake, isolate, leave, put ashore, strand.

marriage *n* **1** matrimony, partnership, union, wedlock. **2** nuptials, union, wedding. □ *bigamy, monogamy, polygamy*.

marriageable *adj* adult, mature, nubile.

marry *vb* espouse, *inf* get hitched, *inf* get spliced, join in matrimony, *inf* tie the knot, unite, wed.

marsh *n* bog, fen, marshland, morass, mud, mudflats, quagmire, quicksands, saltings, saltmarsh, *old use* slough, swamp, wetland.

marshal *vb* arrange, assemble, collect, deploy, draw up, gather, group, line up, muster, organize, set out.

martial *adj* aggressive, bellicose, belligerent, militant, military, pugnacious, soldierly, warlike. *Opp* PEACEABLE.

marvel *n* miracle, phenomenon, wonder. • *vb* **marvel at** admire, applaud, be amazed by, be astonished by, be surprised by, gape at, praise, wonder at.

marvellous *adj* admirable, amazing, astonishing, astounding, breathtaking, excellent, exceptional, extraordinary, *inf* fabulous, *inf* fantastic, glorious, impressive, incredible, magnificent, miraculous, out of the ordinary, *inf* out of this world, phenomenal, praiseworthy, prodigious, remarkable, *inf* sensational, spectacular, splendid, staggering, stunning, stupendous, *inf* super, superb, surprising, *inf* terrific, unbelievable, wonderful, wondrous. *Opp* ORDINARY.

masculine *adj* boyish, *inf* butch, gentlemanly, heroic, *inf* macho, male, manly, mannish, muscular, powerful, strong, vigorous, virile. *Opp* FEMININE.

mash *vb* beat, crush, grind, mangle, pound, pulp, pulverize, smash, squash.

mask *n* camouflage, cloak, cover, cover-up, disguise, façade, front, guise, screen, shield, veil, visor. • *vb* blot out, camouflage, cloak, conceal, cover, disguise, hide, obscure, screen, shield, shroud, veil.

masonry *n* bricks, brickwork, stone, stonework.

mass *adj* comprehensive, general, large-scale, popular, universal, wholesale, widespread. • *n* **1** accumulation, agglomeration, aggregation, body, bulk, *inf* chunk, collection, concretion, conglomeration, *inf* dollop, heap, hoard, *inf* hunk, *inf* load, lot, lump, mound, mountain, pile, profusion, quantity, stack, volume. **2** ▷ GROUP. • *vb* accumulate, aggregate, amass, assemble, collect, congregate, convene, flock together, gather, marshal, meet, mobilize, muster, pile up, rally.

massacre *vb* annihilate, slaughter. ▷ KILL.

massage *vb* knead, manipulate, rub.

mast *n* aerial, flagpole, maypole, pylon, transmitter.

master *n* **1** *inf* boss, employer, governor, keeper, lord, overseer, owner, person in charge, proprietor, ruler, taskmaster. ▷ CHIEF. **2** captain, skipper. **3** *master of an art.* *inf* ace, authority, expert, genius, mastermind, maestro, virtuoso. **4** ▷ TEACHER. • *vb* **1** become expert in, *inf* get off by heart, *inf* get the hang of, grasp, know, learn, understand. **2** break in, bridle, check, conquer, control, curb, defeat, dominate, *inf* get the better of, govern, manage, overcome, overpower, quell, regulate, repress, rule, subdue, subjugate, suppress, tame, triumph over, vanquish.

masterly *adj* accomplished, adroit, consummate, dexterous, excellent, expert, masterful, matchless, practiced, proficient, skilful, skilled, unsurpassable.

mastermind *n* architect, brains, conceiver, contriver, creator, engineer, expert, genius, intellectual, inventor, manager, originator, planner, prime mover. • *vb* carry through, conceive, contrive, devise, direct, engineer, execute, manage, organize, originate, plan, plot. ▷ MANAGE.

masterpiece *n* best work, *Fr* chef-d'oeuvre, classic, *inf* hit, *Lat* magnum opus, masterwork, *Fr* pièce de résistance.

match *n* **1** bout, competition, contest, duel, game, test match, tie, tournament, tourney. **2** *met my match.* complement, counterpart, double, equal, equivalent, twin. **3** *a good match.* combination, fit, pair, similarity. **4** *a love match.* alliance, friendship, marriage, partnership, relationship, union. • *vb* **1** agree, accord, be compatible, be equivalent, be the same, be similar, blend, coincide, combine, compare, coordinate, correspond, fit, *inf* go together, harmonize, suit, tally, tie in, tone in. *Opp* CONTRAST. **2** ally, combine, fit, join, link up, marry, mate, pair off, pair up, put together, team up. *Opp* SPLIT. **matching** ▷ SIMILAR.

mate *n* **1** *inf* better half, companion, consort, helpmeet, husband, partner, spouse, wife. ▷ FRIEND. **2** assistant, associate, collaborator, colleague, helper. • *vb* become partners, copulate, couple, have intercourse, *inf* have sex, join, marry, *inf* pair up, unite, wed.

material *adj* concrete, corporeal, palpable, physical, solid, substantial, tangible. • *n* **1** fabric, stuff, textile. ▷ CLOTH. **2** components, constituents, content, data, facts, ideas, information, matter, notes, resources, statistics, stuff, subject matter, substance, supplies, things.

materialize *vb* appear, become visible, emerge, occur, take shape, *inf* turn up.

mathematics *n* mathematical science, *inf* maths, number work. □ *addition, algebra, arithmetic, calculus, division, geometry, multiplication, operational research, statistics, subtraction, trigonometry.*

matted *adj* knotted, tangled, uncombed, unkempt. ▷ DISHEVELLED.

matter *n* **1** body, material, stuff, substance. **2** discharge, pus, suppuration. **3** *a matter of life and death.* affair, business, concern, episode, event, fact, incident, issue, occurrence, question, situation, subject, thing, topic. **4** *What's the matter?* difficulty, problem, trouble, upset, worry. • *vb* be important, be of consequence, be significant, count, make a difference, mean something, signify. **matter-of-fact** ▷ PROSAIC.

mature *adj* **1** adult, advanced, experienced, full-grown, grown-up, nubile, of age, perfect, sophisticated, well-developed. **2** mellow, ready, ripe, seasoned. *Opp* IMMATURE. • *vb* age, come to fruition, develop, grow up, mellow, reach maturity, ripen.

maturity *n* adulthood, completion, majority, mellowness, perfection, readiness, ripeness.

maul *vb* claw, injure, *inf* knock about, lacerate, mangle, manhandle, mutilate, paw, savage, treat roughly, wound.

maximize *vb* **1** add to, augment, build up, increase, make the most of. **2** inflate, magnify, overdo, overstate. ▷ EXAGGERATE. *Opp* MINIMIZE.

maximum *adj* biggest, extreme, full, fullest, greatest, highest, largest, maximal, most, peak, supreme, top, topmost, utmost, uttermost. • *n* apex, ceiling, climax, extreme, highest point, peak, pinnacle, top, upper limit, zenith. *Opp* MINIMUM.

maybe *adv* conceivably, perhaps, possibly.

maze *n* complex, confusion, convolution, labyrinth, network, tangle, web.

meadow *n* field, *old use* mead, paddock, pasture.

meagre *adj* deficient, inadequate, insufficient, lean, mean, paltry, poor, puny, scanty, skimpy, slight, sparse, thin, unsatisfying. ▷ SMALL. *Opp* GENEROUS.

meal *n inf* blow-out, *old use* collation, repast, *inf* spread. □ *banquet, barbecue, breakfast, buffet, dinner,* inf *elevenses, feast, high tea, lunch, luncheon, picnic, snack,* *supper, take-away, tea, tea-break,* old use *tiffin.*

mean *adj* **1** beggarly, *inf* cheese-paring, close, close-fisted, illiberal, *inf* mingy, miserly, niggardly, parsimonious, *inf* penny-pinching, selfish, sparing, stingy, *inf* tight, tight-fisted, ungenerous. **2** *a mean disposition.* callous, churlish, contemptible, cruel, despicable, hard-hearted, ignoble, ill-tempered, malicious, nasty, shabby, shameful, small-minded, *inf* sneaky, spiteful, uncharitable, unkind, vicious. **3** *a mean dwelling.* base, common, humble, inferior, insignificant, low, lowly, miserable, poor, shabby, squalid, wretched. *Opp* GENEROUS, VALUABLE. • *vb* **1** augur, betoken, communicate, connote, convey, denote, *inf* drive at, express, foretell, *inf* get over, herald, hint at, imply, indicate, intimate, portend, presage, refer to, represent, say, show, signal, signify, specify, spell out, stand for, suggest, symbolize. **2** *I mean to succeed.* aim, desire, have in mind, hope, intend, plan, propose, purpose, want, wish. **3** *The job means long hours.* entail, involve, necessitate.

meander *vb* ramble, rove, snake, twist and turn, wander, wind, zigzag. **meandering** ▷ TWISTY.

meaning *n* connotation, content, definition, denotation, drift, explanation, force, gist, idea, implication, import, importance, interpretation, message, point, purport, purpose, relevance, sense, significance, signification, substance, thrust, value.

meaningful *adj* deep, eloquent, expressive, meaning, pointed, positive, pregnant, relevant, serious, significant, suggestive, telling, telltale, weighty, worthwhile. *Opp* MEANINGLESS.

meaningless *adj* **1** absurd, coded, incomprehensible, incoherent, inconsequential, irrelevant, nonsensical, pointless, senseless. **2** *meaningless compliments.* empty, flattering, hollow, insincere, shallow, silly, sycophantic, vacuous, worthless. *Opp* MEANINGFUL.

means *n* **1** ability, capacity, channel, course, fashion, machinery, manner, medium, method, mode, process, way. **2** *private means.* ▷ WEALTH.

measurable *adj* appreciable, considerable, perceptible, quantifiable, reasonable, significant. *Opp* NEGLIGIBLE.

measure *n* **1** allocation, allowance, amount, amplitude, extent, magnitude, portion, quantity, quota, range, ration, scope, size, unit. ▷ MEASUREMENT. **2** criterion, *inf* litmus test, standard, test, touchstone, yardstick. **3** *measures to curb crime.* act, action, bill, control, course of action, expedient, law, means, procedure, step. • *vb* assess, calculate, calibrate, compute, count, determine, estimate, gauge, judge, mark out, meter, plumb (*depth*), quantify, rank, rate, reckon, survey, take measurements of, weigh. **measure out** ▷ DISPENSE.

measurement *n* amount, calculation, dimensions, extent, figure, mensuration, size. ▷ MEASURE. □ *area, breadth, bulk, capacity, depth, distance, height, length, mass, speed, time, volume, weight, width.* □ *acreage, footage, mileage, tonnage.*

meat *n* flesh. ▷ FOOD. □ *bacon, beef, chicken, game, gammon, ham, lamb, mutton, pork, poultry, turkey, veal, venison.* □ *brawn, breast,* *burger, brisket, chine, chops, chuck, cutlet, fillet, flank, hamburger, leg, loin, mince, offal, paté, potted meat, rib, rissole, rump, sausage, scrag, shoulder, silverside, sirloin, spare-rib, steak, topside, tripe.*

mechanic *n* engineer, technician.

mechanical *adj* **1** automated, automatic, machine-driven, technological. **2** cold, habitual, impersonal, inhuman, instinctive, lifeless, matter-of-fact, perfunctory, reflex, routine, soulless, unconscious, unemotional, unfeeling, unimaginative, uninspired, unthinking. *Opp* HUMAN.

mechanize *vb* automate, bring up to date, computerize, equip with machines, modernize.

medal *n* award, decoration, honour, medallion, prize, reward, trophy.

medallist *n* champion, victor, winner.

meddle *vb inf* be a busybody, butt in, interfere, *inf* poke your nose in, pry, snoop, tamper.

mediate *vb* act as go-between, act as mediator, arbitrate, intercede, liaise, negotiate.

mediator *n* arbiter, arbitrator, broker, conciliator, go-between, intercessor, intermediary, judge, liaison officer, middleman, moderator, negotiator, peacemaker, referee, umpire.

medicinal *adj* curative, healing, medical, remedial, restorative, therapeutic.

medicine *n* **1** healing, surgery, therapeutics, therapy, treatment. **2** cure, dose, drug, medicament, medication, nostrum, panacea, *old use* physic, prescription, remedy, treatment. □ *anaesthetic, antibiotic, antidote, antiseptic, aspirin, capsule, gargle, herbal remedy, iodine, inhaler, linctus, lotion, lozenge, narcotic, ointment, painkiller, pastille, penicillin, pill, sedative, suppository, tablet, tonic, tranquillizer.*

mediocre *adj* amateurish, average, *inf* common-or-garden, commonplace, everyday, fair, indifferent, inferior, medium, middling, moderate, ordinary, passable, pedestrian, poorish, *inf* run-of-the-mill, second-rate, *inf* so-so, undistinguished, unexceptional, unexciting, uninspired, unremarkable, weakish. *Opp* OUTSTANDING.

meditate *vb* be lost in thought, brood, cerebrate, chew over, cogitate, consider, contemplate, deliberate, mull things over, muse, ponder, pray, reflect, ruminate, think, turn over.

meditation *n* cerebration, contemplation, deliberation, musing, prayer, reflection, rumination, thought, yoga.

meditative *adj* brooding, contemplative, pensive, prayerful, rapt, reflective, ruminative, thoughtful.

medium *adj* average, intermediate, mean, medial, median, mid, middle, middling, mid-sized, midway, moderate, normal, ordinary, standard, usual. • *n* **1** average, centre, compromise, mean, middle, midpoint, norm. **2** agency, approach, channel, form, means, method, mode, vehicle, way. **3** clairvoyant, seer, spiritualist. **the media, mass media** ▷ COMMUNICATION.

meek *adj* acquiescent, compliant, deferential, docile, forbearing, gentle, humble, long-suffering, lowly, mild, modest, non-militant, obedient, patient, peaceable, quiet, resigned, retiring, self-effacing,

shy, soft, spineless, submissive, tame, timid, tractable, unambitious, unassuming, unprotesting, weak, *inf* wimpish.
Opp AGGRESSIVE.

meet *vb* **1** *inf* bump into, chance upon, collide with, come across, confront, contact, encounter, face, happen on, have a meeting with, run across, run into, see. **2** be introduced to, make the acquaintance of. **3** come and fetch, greet, *inf* pick up, rendezvous with, welcome. **4** assemble, collect, come together, congregate, convene, forgather, gather, have a meeting, muster, rally, rendezvous. **5** *The ends don't meet.* abut, adjoin, come together, connect, converge, cross, intersect, join, link up, merge, touch, unite. **6** *meet a request.* acquiesce in, agree to, answer, comply with, deal with, fulfil, *inf* measure up to, observe, pay, satisfy, settle, take care of. **7** *meet difficulties.* encounter, endure, experience, go through, suffer, undergo.

meeting *n* **1** assembly, gathering, *inf* get-together, *inf* powwow. □ *audience, board, briefing, cabinet, caucus, committee, conclave, conference, congregation, congress, convention, council, discussion group, forum, prayer meeting, rally, seminar, service, synod.* **2** appointment, assignation, date, engagement, *inf* get-together, rendezvous, *old use* tryst. **3** *chance meeting.* confrontation, contact, encounter. **4** *meeting of lines, roads.* confluence (*of rivers*), convergence, crossing, crossroads, intersection, joining, junction, T-junction, union.

melancholy *adj* cheerless, dejected, depressed, depressing, despondent, disconsolate, dismal, dispirited, dispiriting, *inf* down, down-hearted, forlorn, gloomy, glum, joyless, lifeless, low, lugubrious, melancholic, miserable, moody, morose, mournful, sombre, sorrowful, unhappy, woebegone, woeful. ▷ SAD.
Opp CHEERFUL. • *n* ▷ SADNESS.

mellow *adj* **1** mature, rich, ripe, smooth, sweet. **2** *mellow mood.* agreeable, amiable, comforting, cordial, genial, gentle, happy, kindly, peaceful, pleasant, reassuring, soft, subdued, warm.
Opp HARSH. • *vb* age, develop, improve with age, mature, ripen, soften, sweeten.

melodious *adj* dulcet, *inf* easy on the ear, euphonious, harmonious, lyrical, mellifluous, melodic, sweet, tuneful.

melodramatic *adj* emotional, exaggerated, *inf* hammy, histrionic, overdone, overdrawn, *inf* over the top, sensationalized, sentimental, theatrical.

melody *n* air, song, strain, subject, theme, tune.

melt *vb* deliquesce, dissolve, liquefy, soften, thaw, unfreeze.
melt away ▷ DISAPPEAR.

member *n* associate, colleague, fellow, life-member, paid-up member.

memorable *adj* catchy (*tune*), distinguished, extraordinary, haunting, historic, impressive, indelible, ineradicable, never-to-be-forgotten, notable, outstanding, remarkable, striking, unforgettable.

memorial *n* cairn, cenotaph, gravestone, headstone, monument, plaque, reminder, statue, tablet, tomb.

memorize *n* commit to memory, *inf* get off by heart, learn, learn by rote, learn parrot-fashion, remember, retain.

memory *n* 1 ability to remember, recall, retention. 2 impression, recollection, reminder, reminiscence, souvenir. 3 *memory of the dead.* fame, honour, name, remembrance, reputation, respect.

menace *n* danger, peril, threat, warning. • *vb* alarm, bully, cow, intimidate, terrify, terrorize, threaten. ▷ FRIGHTEN.

mend *vb* 1 fix, patch up, put right, rectify, remedy, renew, renovate, repair, restore. □ *beat out, darn, patch, replace parts, sew up, solder, stitch up, touch up, weld.* 2 *mend your ways.* ameliorate, amend, correct, cure, improve, make better, reform, revise. 3 *mend after illness.* convalesce, get better, heal, improve, recover, recuperate.

menial *adj* base, boring, common, degrading, demeaning, humble, inferior, insignificant, low, lowly, servile, slavish, subservient, unskilled, unworthy.
• *n inf* dogsbody, lackey, minion, slave, underling. ▷ SERVANT.

mental *adj* 1 abstract, cerebral, cognitive, conceptual, intellectual, rational, theoretical. 2 *mental illness.* emotional, psychological, subjective, temperamental. ▷ MAD.

mentality *n* attitude, bent, character, disposition, frame of mind, inclination, *inf* make-up, outlook, personality, predisposition, propensity, psychology, set, temperament, way of thinking.

mention *vb* acknowledge, allude to, animadvert on, bring up, broach, cite, comment on, disclose, draw attention to, enumerate, hint at, *inf* let drop, let out, make known, make mention, name, note, observe, pay tribute to, point out, quote, recognize, refer to, remark, report, reveal, say, speak about, touch on, write about.

mercenary *adj* acquisitive, avaricious, covetous, grasping, greedy, *inf* money-mad, venal.
• *n* ▷ FIGHTER.

merchandise *n* commodities, goods, items for sale, produce, products, stock. • *vb* ▷ ADVERTISE.

merchant *n* broker, dealer, distributor, retailer, salesman, seller, shopkeeper, stockist, supplier, trader, tradesman, tradeswoman, vendor, wholesaler.

merciful *adj* beneficent, benevolent, charitable, clement, compassionate, forbearing, forgiving, generous, gracious, humane, humanitarian, indulgent, kind, kind-hearted, kindly, lenient, liberal, magnanimous, mild, pitying, *inf* soft, soft-hearted, sympathetic, tender-hearted, tolerant.
Opp MERCILESS.

merciless *adj* barbaric, barbarous, brutal, callous, cold, cruel, cut-throat, hard, hard-hearted, harsh, heartless, indifferent, inexorable, inflexible, inhuman, inhumane, intolerant, malevolent, pitiless, relentless, remorseless, rigorous, ruthless, savage, severe, stern, stony-hearted, strict, tyrannical, unbending, unfeeling, unforgiving, unkind, unmerciful, unrelenting, unremitting, vicious.
Opp MERCIFUL.

mercy *n* beneficence, benignity, charity, clemency, compassion, feeling, forbearance, forgiveness, generosity, grace, humaneness, humanity, indulgence, kind-heartedness, kindness, leniency, love, pity, quarter, sympathy, understanding.

merge *vb* 1 amalgamate, blend, coalesce, combine, come together, confederate, consolidate, fuse, integrate, join together, link up, mingle, mix, pool, put together, unite. **2** *motorways merge.* converge, join, meet. *Opp* SEPARATE.

merit *n* credit, distinction, excellence, good, goodness, importance, quality, strength, talent, value, virtue, worth, worthiness. • *vb* be entitled to, be worthy of, deserve, earn, have a right to, incur, justify, rate, warrant.

meritorious *adj* admirable, commendable, estimable, exemplary, honourable, laudable, praiseworthy, worthy.

merriment *n* amusement, cheerfulness, conviviality, exuberance, gaiety, glee, good cheer, high spirits, hilarity, jocularity, joking, jollity, joviality, *inf* larking about, laughter, levity, light-heartedness, liveliness, mirth, vivacity. ▷ MERRYMAKING.

merry *adj* bright, *inf* bubbly, carefree, cheerful, cheery, *inf* chirpy, convivial, festive, fun-loving, gay, glad, hilarious, jocular, jolly, jovial, joyful, joyous, light-hearted, lively, mirthful, rollicking, spirited, vivacious. ▷ HAPPY. *Opp* SERIOUS.

merrymaking *n* carousing, celebration, conviviality, festivity, frolic, fun, *inf* fun and games, *inf* jollification, *inf* junketing, merriment, revelry, roistering, sociability, *old use* wassailing. ▷ PARTY.

mesh *n* grid, lace, lacework, lattice, lattice-work, net, netting, network, reticulation, screen, sieve, tangle, tracery, trellis, web, webbing.

mess *n* **1** chaos, clutter, confusion, dirt, disarray, disorder, hotchpotch, jumble, litter, *inf* mishmash, muddle, *inf* shambles, tangle, untidiness. ▷ CONFUSION, DIRT. **2** *made a mess of it. inf* botch, failure, *inf* hash, *inf* mix-up. **3** *got into a mess.* difficulty, dilemma, *inf* fix, *inf* jam, *inf* pickle, plight, predicament, problem, trouble. • *vb* **mess about** amuse yourself, loaf, loiter, lounge about, *inf* monkey about, *inf* muck about, *inf* play about. **make a mess of** ▷ BUNGLE, MUDDLE. **mess up** ▷ MUDDLE. **mess up a job** ▷ BUNGLE.

message *n* announcement, bulletin, cable, communication, communiqué, dispatch, information, intelligence, letter, memo, memorandum, missive, news, note, notice, report, statement, *old use* tidings.

messenger *n* bearer, carrier, courier, dispatch-rider, emissary, envoy, errand-boy, errand-girl, go-between, harbinger, herald, intermediary, legate, Mercury, messenger-boy, messenger-girl, nuncio, postman, runner.

messy *adj* blowzy, careless, chaotic, cluttered, dirty, dishevelled, disorderly, filthy, grubby, mucky, muddled, *inf* shambolic, slapdash, sloppy, slovenly, unkempt, untidy. *Opp* NEAT.

metallic *adj* **1** gleaming, lustrous, shiny. **2** clanking, clinking, ringing.

metaphorical *adj* allegorical, figurative, non-literal, symbolic. *Opp* LITERAL.

method *n* **1** approach, fashion, *inf* knack, manner, means, methodology, mode, *Lat* modus operandi, plan, procedure, process, programme, recipe, scheme, style, technique, trick, way. **2** arrangement, design, discipline, neatness, order, orderliness, organization, pattern, routine, structure, system.

methodical *adj* businesslike, careful, deliberate, disciplined, logical, meticulous, neat, ordered, orderly, organized, painstaking, precise, rational, regular, routine, structured, systematic, tidy. *Opp* DISORGANIZED.

meticulous *adj* accurate, exact, exacting, fastidious, *inf* finicky, painstaking, particular, perfectionist, precise, punctilious, scrupulous, thorough. *Opp* CARELESS.

microbe *n* bacillus, bacterium, *inf* bug, germ, micro-organism, virus.

middle *adj* central, centre, half-way, inner, inside, intermediate, intervening, mean, medial, median, mid, middle-of-the-road, midway, neutral. • *n* bull's eye, centre, core, crown (*of road*), focus, half-way point, heart, hub, inside, middle position, midpoint, midst, nucleus.

middling *adj* average, fair, *inf* fair to middling, indifferent, mediocre, moderate, modest, *inf* nothing to write home about, ordinary, passable, run-of-the-mill, *inf* so-so, unremarkable. *Opp* OUTSTANDING.

might *n* capability, capacity, energy, force, muscle, potency, power, strength, superiority, vigour.

mighty *adj* brawny, dominant, doughty, energetic, enormous, forceful, great, hefty, muscular, potent, powerful, robust, *inf* strapping, strong, sturdy, vigorous, weighty. ▷ BIG. *Opp* WEAK.

migrate *vb* emigrate, go, immigrate, move, relocate, resettle, settle, travel.

mild *adj* **1** affable, amiable, conciliatory, docile, easygoing, equable, forbearing, forgiving, gentle, good-tempered, harmless, indulgent, inoffensive, kind, kindly, lenient, meek, merciful, modest, non-violent, pacific, peaceable, placid, quiet, *inf* soft, soft-hearted, submissive, sympathetic, tractable, unassuming, understanding, yielding. **2** *mild weather*. balmy, calm, clement, fair, peaceful, pleasant, serene, temperate, warm. **3** *a mild illness*. insignificant, minor, modest, slight, trivial, unimportant. **4** *mild flavour*. bland, delicate, faint, mellow, soothing, subtle. *Opp* SEVERE, STRONG.

mildness *n* affability, amiability, clemency, docility, forbearance, gentleness, kindness, leniency, moderation, placidity, softness, sympathy, tenderness. *Opp* ASPERITY.

militant *adj* active, aggressive, assertive, attacking, combative, fierce, hostile, positive, pugnacious. *Opp* PASSIVE. • *n* activist, extremist, *inf* hawk, partisan.

militaristic *adj* bellicose, belligerent, combative, fond of fighting, hawkish, hostile, pugnacious, warlike. *Opp* PEACEABLE.

military *adj* armed, belligerent, combatant, enlisted, fighting, martial, uniformed, warlike. *Opp* CIVIL.

militate *vb* militate against cancel out, counter, counteract, countervail, discourage, hinder, oppose, prevent, resist.

milk *vb* bleed, drain, exploit, extract, tap, wring.

milky *adj* chalky, cloudy, misty, opaque, whitish. *Opp* CLEAR.

mill *n* **1** factory, foundry, plant, shop, works, workshop. **2** crusher, grinder, quern, watermill, windmill. • *vb* crush, granulate, grate, grind, pound, powder, pulverize.

mill about move aimlessly, seethe, swarm, throng, wander.

mimic *n* caricaturist, imitator, impersonator, impressionist. • *vb* ape, caricature, copy, do impressions of, echo, imitate, impersonate, lampoon, look like, *inf* make fun of, mirror, mock, parody, parrot, pretend to be, reproduce, ridicule, satirize, simulate, sound like, *inf* take off.

mind *n* **1** astuteness, brain, brainpower, brains, cleverness, *inf* grey matter, head, insight, intellect, intelligence, judgement, memory, mental power, perception, psyche, rationality, reason, reasoning, remembrance, sagacity, sapience, sense, shrewdness, thinking, understanding, wisdom, wit, wits. **2** attitude, belief, bias, disposition, humour, inclination, intention, opinion, outlook, persuasion, plan, point of view, position, view, viewpoint, way of thinking, wishes. • *vb* **1** attend to, care for, guard, keep an eye on, look after, take care, take charge of, watch. **2** *mind the warning.* be careful about, beware of, heed, listen to, look out for, mark, note, obey, pay attention to, remember, take notice of, watch out for. **3** *won't mind if he's late.* be annoyed, be bothered, be offended, be resentful, bother, care, complain, disapprove, grumble, object, take offence, worry. **be in two minds** ▷ HESITATE. **make up your mind** ▷ DECIDE. **out of your mind** ▷ MAD.

mindful *adj* alert, attentive, aware, conscious, heedful, *inf* on the lookout, vigilant, watchful. ▷ CAREFUL. *Opp* CARELESS.

mindless *adj* brainless, fatuous, idiotic, obtuse, senseless, thick, thoughtless, unintelligent, unthinking, witless. ▷ STUPID. *Opp* INTELLIGENT.

mine *n* **1** coalfield, colliery, excavation, opencast mine, pit, quarry, shaft, tunnel, working. **2** *mine of information.* fund, repository, source, store, storehouse, supply, treasury, vein, wealth. • *vb* dig, excavate, extract, quarry, remove, scoop out, unearth.

mineral *n* metal, ore, rock.

mingle *vb* amalgamate, associate, blend, circulate, combine, commingle, get together, fraternize, *inf* hobnob, intermingle, intermix, join, merge, mix, move about, *inf* rub shoulders, socialize, unite.

miniature *adj* baby, diminutive, dwarf, pocket, pygmy, reduced, scaled-down, small-scale, tiny, toy. ▷ SMALL.

minimal *adj* least, minimum, negligible, nominal, slightest, smallest, token.

minimize *vb* **1** cut down, decrease, diminish, lessen, pare, prune, reduce. **2** *minimize problems.* belittle, decry, devalue, depreciate, gloss over, make light of, play down, underestimate, undervalue. *Opp* MAXIMIZE.

minimum *adj* bottom, least, littlest, lowest, minimal, minutest, nominal, *inf* rock bottom, slightest, smallest. • *n* least, lowest, minimum amount, minimum quantity, nadir. *Opp* MAXIMUM.

minister *n* ▷ CLERGYMAN, OFFICIAL. • *vb* **minister to** aid, assist, attend to, care for, help, look after, nurse, see to, support, wait on.

minor *adj* inconsequential, inferior, insignificant, lesser, little, negligible, petty, secondary, smaller, subordinate, subsidiary, trivial, unimportant. ▷ SMALL.

Opp MAJOR. • *n* ▷ ADOLESCENT, CHILD.

minstrel *n* balladeer, bard, entertainer, jongleur, musician, singer, troubadour.

mint *adj* brand-new, first-class, fresh, immaculate, new, perfect, unblemished, unmarked, unused. • *n* fortune, heap, *inf* packet, pile, stack, unlimited supply, vast amount. • *vb* cast, coin, forge, make, manufacture, produce, stamp out, strike.

minute *adj* diminutive, dwarf, infinitesimal, insignificant, lilliputian, microscopic, *inf* mini, miniature, minuscule, *inf* pint-sized, pocket, pygmy, tiny. ▷ SMALL.

minutes *plur n* log, notes, proceedings, record, résumé, summary, transactions.

miracle *n* marvel, miraculous event, mystery, wonder.

miraculous *adj* abnormal, extraordinary, incredible, inexplicable, magic, magical, mysterious, paranormal, phenomenal, preternatural, remarkable, supernatural, unaccountable, unbelievable, unexplainable. ▷ MARVELLOUS.

mirage *n* delusion, hallucination, illusion, vision.

mire *n* bog, fen, marsh, morass, mud, ooze, quagmire, quicksand, slime, *old use* slough, swamp. ▷ DIRT.

mirror *n* glass, looking-glass, reflector, speculum. • *vb* echo, reflect, repeat, send back.

misadventure *n* accident, calamity, catastrophe, disaster, ill fortune, mischance, misfortune, mishap.

misanthropic *adj* anti-social, cynical, mean, nasty, surly, unfriendly, unpleasant, unsociable. *Opp* PHILANTHROPIC.

misappropriate *vb* defalcate, embezzle, expropriate, peculate. ▷ STEAL.

misbehave *vb* be a nuisance, be bad, behave badly, be mischievous, *inf* blot your copybook, *inf* carry on, commit an offence, default, disobey, do wrong, err, fool about, make mischief, *inf* mess about, *inf* muck about, offend, *inf* play about, play up, *inf* raise Cain, sin, transgress.

misbehaviour *n* badness, delinquency, disobedience, disorderliness, horseplay, indiscipline, insubordination, mischief, mischief-making, misconduct, misdemeanour, naughtiness, rowdyism, rudeness, sin, vandalism, wrongdoing.

miscalculate *vb inf* boob, err, *inf* get it wrong, go wrong, make a mistake, miscount, misjudge, misread, overestimate, overrate, overvalue, *inf* slip up, underestimate, underrate.

miscarriage *n* **1** abortion, premature birth, termination of pregnancy. **2** *miscarriage of justice.* breakdown, collapse, defeat, error, failure, perversion.

miscarry *vb* **1** abort, *inf* lose a baby, suffer a miscarriage. **2** break down, *inf* come to grief, come to nothing, fail, fall through, founder, go wrong, misfire. *Opp* SUCCEED.

miscellaneous *adj* assorted, different, *old use* divers, diverse, heterogeneous, manifold, mixed, motley, multifarious, sundry, varied, various.

miscellany *n* assortment, diversity, gallimaufry, hotchpotch, jumble, medley, mélange, *inf* mixed bag, mixture, pot-pourri *inf* ragbag, variety.

mischief *n* **1** devilment, devilry, escapade, impishness, misbehaviour, misconduct, *inf* monkey business, naughtiness, playfulness, prank, rascality, roguishness, scrape, *inf* shenanigans, trouble. **2** damage, difficulty, evil, harm, hurt, injury, misfortune, trouble.

mischievous *adj* annoying, badly behaved, boisterous, disobedient, elvish, fractious, frolicsome, full of mischief, impish, lively, naughty, playful, *inf* puckish, rascally, roguish, sportive, uncontrollable, *inf* up to no good. ▷ WICKED. *Opp* WELL-BEHAVED.

miser *n* hoarder, miserly person, niggard, *inf* Scrooge, *inf* skinflint. *Opp* SPENDTHRIFT.

miserable *adj* **1** broken-hearted, crestfallen, *inf* cut up, dejected, depressed, desolate, despairing, despondent, disappointed, disconsolate, dismayed, dispirited, distressed, doleful, *inf* down, downcast, downhearted, forlorn, friendless, gloomy, glum, grief-stricken, heartbroken, hopeless, in low spirits, *inf* in the doldrums, *inf* in the dumps, joyless, lachrymose, languishing, lonely, low, melancholy, mournful, moping, sad, sorrowful, suicidal, tearful, uneasy, unfortunate, unhappy, unlucky, woebegone, woeful, wretched. **2** churlish, cross, disagreeable, discontented, *inf* grumpy, ill-natured, mean, miserly, morose, pessimistic, sour, sulky, sullen, surly, taciturn, unfriendly, unhelpful, unsociable. **3** *miserable living conditions*. abject, awful, bad, deplorable, destitute, disgraceful, distressing, heart-breaking, hopeless, impoverished, inadequate, inhuman, lamentable, pathetic, pitiable, pitiful, poor, shameful, sordid, soul-destroying, squalid, vile, uncivilized, uncomfortable, vile, worthless, wretched. **4** *miserable weather*. cheerless, damp, depressing, dismal, dreary, grey, inclement, sunless, unpleasant, wet. *Opp* HAPPY, PLEASANT.

miserly *adj* avaricious, *inf* cheese-paring, *inf* close, *inf* close-fisted, covetous, economical, grasping, greedy, mean, mercenary, mingy, niggardly, parsimonious, penny-pinching, penurious, sparing, stingy, *inf* tight, *inf* tight-fisted. *Opp* GENEROUS.

misery *n* **1** angst, anguish, anxiety, bitterness, dejection, depression, despair, desperation, despondency, discomfort, distress, dolour, gloom, grief, heartache, heartbreak, *inf* hell, hopelessness, melancholy, sadness, sorrow, suffering, unhappiness, woe, wretchedness. *Opp* HAPPINESS. **2** adversity, affliction, deprivation, destitution, hardship, indigence, misfortune, need, oppression, penury, poverty, privation, squalor, suffering, *inf* trials and tribulations, tribulation, trouble, want, wretchedness.

misfire *vb* abort, fail, fall through, *inf* flop, founder, go wrong, miscarry. *Opp* SUCCEED.

misfortune *n* accident, adversity, affliction, bad luck, blow, calamity, catastrophe, contretemps, curse, disappointment, disaster, evil, hard luck, hardship, ill-luck, misadventure, mischance, mishap, reverse, setback, tragedy, trouble, vicissitude.

misguided *adj* erroneous, foolish, ill-advised, ill-judged, inappropriate, incorrect, inexact, misinformed, misjudged, misled, mistaken, unfounded, unjust, unsound, unwise. ▷ WRONG.

misjudge *vb* get wrong, guess wrongly, *inf* jump to the wrong conclusion, make a mistake, misinterpret, overestimate, overvalue, underestimate, undervalue. ▷ MISCALCULATE.

mislay *vb* lose, mislocate, misplace, put in the wrong place.

mislead *vb* bluff, confuse, delude, fool, give misleading information to, give a wrong impression to, lead astray, *inf* lead up the garden path, lie to, misdirect, misguide, misinform, outwit, *inf* take for a ride, take in, *inf* throw off the scent, trick. ▷ DECEIVE. **misleading** ▷ DECEPTIVE, PUZZLING.

miss *vb* **1** absent yourself from, avoid, be absent from, be too late for, dodge, escape, evade, fail to keep, forget, forgo, let go, lose, play truant from, *inf* skip, *inf* skive off. **2** *miss a target*. be wide of, fail to hit, fall short of. **3** *miss absent friends*. feel nostalgia for, grieve for, lament, long for, need, pine for, want, yearn for. **miss out** ▷ OMIT.

misshapen *adj* awry, bent, contorted, corkscrew, crippled, crooked, crumpled, deformed, disfigured, distorted, gnarled, grotesque, knotted, malformed, monstrous, screwed up, tangled, twisted, twisty, ugly, warped. *Opp* PERFECT.

missile *n* projectile. □ *arrow, ballistic missile, bomb, boomerang, brickbat, bullet, dart, grenade, guided missile, rocket, shell, shot, torpedo*. ▷ WEAPON.

missing *adj* absent, disappeared, lost, mislaid, *inf* skiving, straying, truant, unaccounted-for. *Opp* PRESENT.

mission *n* **1** delegation, deputation, expedition, exploration, journey, sortie, task-force, voyage. **2** *mission in life*. aim, assignment, calling, commitment, duty, function, goal, job, life's work, métier, objective, occupation, profession, purpose, quest, undertaking, vocation. **3** *evangelical mission*. campaign, crusade, holy war.

missionary *n* campaigner, crusader, evangelist, minister, preacher, proselytizer.

mist *n* **1** cloud, drizzle, fog, haze, smog, vapour. **2** condensation, film, steam.

mistake *n inf* bloomer, blunder, *inf* boob, *inf* botch, *inf* clanger, erratum, error, false step, fault, *Fr* faux pas, gaffe, *inf* howler, inaccuracy, indiscretion, lapse, misapprehension, miscalculation, misconception, misjudgement, misprint, misspelling, misunderstanding, omission, oversight, slip, slip-up, solecism, wrong move. • *vb* confuse, *inf* get the wrong end of the stick, get wrong, misconstrue, misinterpret, misjudge, misread, misunderstand, mix up, *inf* take the wrong way.

mistaken *adj* erroneous, distorted, false, faulty, ill-judged, inaccurate, inappropriate, incorrect, inexact, misguided, misinformed, unfounded, unjust, unsound. ▷ WRONG. *Opp* CORRECT.

mistimed *adj* badly timed, early, inconvenient, inopportune, late, unseasonable, untimely. *Opp* OPPORTUNE.

mistreat *vb* abuse, batter, damage, harm, hurt, ill-treat, ill-use, injure, *inf* knock about, maltreat, manhandle, misuse, molest, treat roughly.

mistress *n* [*mostly old use*] **1** chief, head, keeper, owner, person in charge, proprietor. **2** ▷ TEACHER. **3** ▷ LOVER.

mistrust *n* apprehension, *inf* chariness, distrust, doubt, misgiving, reservation, scepticism, suspicion, uncertainty, unsureness, wariness. • *vb* be sceptical about, be suspicious of, be wary of, disbelieve, distrust, doubt, fear, have doubts about, have misgivings about, have reservations about, question, suspect. *Opp* TRUST.

misty *adj* bleary, blurred, blurry, clouded, cloudy, dim, faint, foggy, fuzzy, hazy, indistinct, murky, obscure, opaque, shadowy, smoky, steamy, unclear, vague. *Opp* CLEAR.

misunderstand *vb inf* get the wrong end of the stick, get wrong, misapprehend, miscalculate, misconceive, misconstrue, mishear, misinterpret, misjudge, misread, miss the point of, mistake, mistranslate. *Opp* UNDERSTAND.

misunderstanding *n* **1** error, failure of understanding, false impression, misapprehension, miscalculation, misconception, misconstruction, misinterpretation, misjudgement, misreading, mistake, *inf* mix up, wrong idea. **2** argument, *inf* contretemps, controversy, difference of opinion, disagreement, discord, dispute. ▷ QUARREL.

misuse *n* abuse, careless use, corruption, ill-treatment, ill-use, maltreatment, misapplication, misappropriation, mishandling, mistreatment, perversion. • *vb* **1** damage, harm, mishandle, treat carelessly. **2** *misuse an animal.* abuse, batter, damage, harm, hurt, ill-treat, ill-use, injure, *inf* knock about, maltreat, manhandle, mistreat, molest, treat roughly. **3** *misuse funds.* fritter away, misappropriate, squander, use wrongly, waste.

mitigate *vb* abate, allay, alleviate, decrease, ease, extenuate, lessen, lighten, make milder, moderate, palliate, qualify, reduce, relieve, soften, *inf* take the edge off, temper, tone down. *Opp* AGGRAVATE.

mix *n* amalgam, assortment, blend, combination, compound, range, variety. • *vb* **1** alloy, amalgamate, blend, coalesce, combine, commingle, compound, confuse, diffuse, emulsify, fuse, homogenize, integrate, intermingle, join, jumble up, make a mixture, meld, merge, mingle, mix up, muddle, put together, shuffle, stir together, unite. *Opp* SEPARATE. **2** *mix with people.* ▷ SOCIALIZE.

mixed *adj* **1** assorted, different, diverse, heterogeneous, miscellaneous, varied, various. **2** *mixed with other things.* adulterated, alloyed, diluted, impure. **3** *mixed ingredients.* amalgamated, combined, composite, hybrid, integrated, joint, mongrel, united. **4** *mixed feelings.* ambiguous, ambivalent, confused, equivocal, muddled, uncertain.

mixture *n* **1** alloy, amalgam, amalgamation, association, assortment, blend, collection, combination, composite, compound, concoction, conglomeration, emulsion, farrago, fusion, gallimaufry, *inf* hotchpotch, intermingling, jumble, medley, mélange, merger, mess, mingling, miscellany, *inf* mishmash, mix, *inf* motley collection, pastiche, potpourri, selection, suspension, synthesis, variety. **2** cross-breed, half-caste, hybrid, mongrel.

moan *n* complaint, grievance, lament, lamentation. • *vb* **1** complain, grieve, *inf* grouse, grumble, lament. **2** cry, groan, keen, sigh, ululate, wail, weep, whimper, whine. ▷ SOUND.

mob *n inf* bunch, crowd, gang, herd, horde, host, multitude, pack, press, rabble, riot, *inf* shower, swarm, throng. ▷ GROUP. • *vb* besiege, crowd round, hem in, jostle, surround, swarm round, throng round.

mobile *adj* 1 itinerant, motorized, movable, portable, transportable, travelling, unfixed. 2 able to move, active, agile, independent, moving, nimble, *inf* on the go, *inf* up and about. 3 *mobile features.* animated, changeable, changing, expressive, flexible, fluid, plastic, shifting. *Opp* IMMOVABLE.

mobilize *vb* activate, assemble, call up, conscript, enlist, enrol, gather, get together, levy, marshal, muster, organize, rally, stir up, summon.

mock *adj* artificial, counterfeit, ersatz, fake, false, imitation, make-believe, man-made, *inf* pretend, simulated, sham, substitute. • *vb* decry, deride, disparage, flout, gibe at, insult, jeer at, lampoon, laugh at, make fun of, make sport of, parody, poke fun at, ridicule, satirize, scoff at, scorn, *inf* send up, sneer at, tantalize, taunt, tease, travesty. ▷ MIMIC.

mockery *n* derision, insults, jeering, laughter, ridicule, scorn. □ *burlesque, caricature, lampoon, parody, sarcasm, satire,* inf *send-up,* inf *spoof,* inf *take-off, travesty.*

mocking *adj* contemptuous, derisive, disparaging, disrespectful, insulting, irreverent, jeering, rude, sarcastic, satirical, scornful, taunting, teasing, uncomplimentary, unkind. *Opp* RESPECTFUL.

mode *n* 1 approach, configuration, manner, medium, method, *Lat* modus operandi, procedure, set-up, system, technique, way. 2 ▷ FASHION.

model *adj* 1 imitation, miniature, scaled-down, toy. 2 *model pupil.* exemplary, ideal, perfect, unequalled. • *n* 1 archetype, copy, dummy, effigy, facsimile, image, imitation, likeness, miniature, *inf* mock-up, paradigm, prototype, replica, representation, scale model, toy. 2 *model of excellence.* byword, epitome, example, exemplar, ideal, nonpareil, paragon, pattern, standard, yardstick. 3 *artist's model.* poser, sitter, subject. 4 *latest model.* brand, design, kind, mark, type, version. 5 *fashion model.* mannequin. • *vb* carve, fashion, form, make, mould, sculpt, shape. **model yourself on** ▷ IMITATE.

moderate *adj* 1 average, balanced, calm, cautious, commonsensical, cool, deliberate, fair, judicious, medium, middle, *inf* middle-of-the-road, middling, modest, normal, ordinary, rational, reasonable, respectable, sensible, sober, steady, temperate, unexceptional, usual. *Opp* EXTREME. 2 *moderate winds.* gentle, light, mild. • *vb* 1 abate, become less extreme, decline, decrease, die down, ease off, subside. 2 blunt, calm, check, curb, dull, ease, keep down, lessen, make less extreme, mitigate, modify, modulate, mollify, reduce, regulate, restrain, slacken, subdue, temper, tone down.

moderately *adv* comparatively, fairly, passably, *inf* pretty, quite, rather, reasonably, somewhat, to some extent.

moderation *n* balance, caution, common sense, fairness, reasonableness, restraint, reticence, sobriety, temperance.

modern *adj* advanced, avant-garde, contemporary, current, fashionable, forward-looking, fresh, futuristic, in vogue, latest, modish, new, newfangled, novel, present, present-day, progressive, recent, stylish, *inf* trendy, up-to-date, up-to-the-minute, *inf* with it. *Opp* OLD.

modernize *vb* bring up-to-date, *inf* do up, improve, make modern, rebuild, redesign, redo, refurbish, regenerate, rejuvenate, renovate, revamp, update.

modest *adj* **1** diffident, humble, inconspicuous, lowly, meek, plain, quiet, reserved, restrained, reticent, retiring, self-effacing, simple, unassuming, unobtrusive, unostentatious, unpretentious. *Opp* CONCEITED. **2** bashful, chaste, coy, decent, demure, discreet, proper, seemly, self-conscious, shamefaced, shy, simple. **3** *a modest income.* limited, medium, middling, moderate, normal, ordinary, reasonable, unexceptional. *Opp* EXCESSIVE.

modesty *n* **1** humbleness, humility, lowliness, meekness, reserve, restraint, reticence, self-effacement, simplicity. *Opp* OSTENTATION. **2** *modesty about undressing.* bashfulness, coyness, decency, demureness, discretion, propriety, seemliness, self-consciousness, shame, shyness.

modify *vb* adapt, adjust, alter, amend, change, convert, improve, reconstruct, redesign, remake, remodel, reorganize, revise, reword, rework, transform, vary. ▷ MODERATE.

modulate *vb* adjust, balance, change key, change the tone, lower the tone, moderate, regulate, soften, tone down.

moist *adj* affected by moisture, clammy, damp, dank, dewy, humid, misty, *inf* muggy, rainy, *inf* runny, steamy, watery, wettish. ▷ WET. *Opp* DRY.

moisten *vb* damp, dampen, humidify, make moist, moisturize, soak, spray, wet. *Opp* DRY.

moisture *n* condensation, damp, dampness, dankness, dew, humidity, liquid, precipitation, spray, steam, vapour, water, wet, wetness.

molest *vb* abuse, accost, annoy, assault, attack, badger, bother, disturb, harass, harry, hassle, hector, ill-treat, interfere with, irk, irritate, manhandle, mistreat, *inf* needle, persecute, pester, plague, set on, tease, torment, vex, worry.

molten *adj* fluid, liquid, liquefied, melted, soft.

moment *n* **1** flash, instant, *inf* jiffy, minute, second, split second, *inf* tick, *inf* trice, *inf* twinkling of an eye, *inf* two shakes. **2** *a historic moment.* hour, juncture, occasion, opportunity, point in time, stage, time.

momentary *adj* brief, ephemeral, evanescent, fleeting, fugitive, hasty, passing, quick, short, short-lived, temporary, transient, transitory. *Opp* PERMANENT.

momentous *adj* consequential, critical, crucial, decisive, epoch-making, fateful, grave, historic, important, portentous, serious, significant, weighty. *Opp* UNIMPORTANT.

monarch *n* crowned head, emperor, empress, king, potentate, queen, ruler, tsar.

monarchy *n* empire, domain, kingdom, realm.

money *n* affluence, arrears, assets, bank-notes, *inf* bread, capital, cash, change, cheque, coin, copper, credit card, credit transfer, currency, damages, debt, dividend, *inf* dough, dowry, earnings, endowment, estate, expenditure, finance, fortune, fund, grant, income, interest, investment, legal tender, loan, *inf* lolly, *old use* lucre, mortgage, *inf* nest-egg, notes, outgoings, patrimony, pay, penny, pension, pocket-money, proceeds, profit, *inf* the ready, remittance, resources, revenue, riches, salary, savings, silver, sterling, takings, tax, traveller's cheque, wage, wealth, *inf* the wherewithal, winnings.

mongrel *n* cross-breed, cur, half-breed, hybrid, mixed breed.

monitor *n* **1** detector, guardian, prefect, supervisor, watchdog. **2** *TV monitor.* screen, set, television, TV, VDU, visual display unit. • *vb* audit, check, examine, *inf* keep an eye on, oversee, record, supervise, trace, track, watch.

monk *n* brother, friar, hermit.

monkey *n* ape, primate, simian. □ *baboon, chimpanzee, gibbon, gorilla, marmoset, orang-utan.*

monopolize *vb* control, *inf* corner the market in, dominate, have a monopoly of, *inf* hog, keep for yourself, own, shut others out of, take over. *Opp* SHARE.

monotonous *adj* boring, colourless, dreary, dull, featureless, flat, level, repetitious, repetitive, soporific, tedious, tiresome, tiring, toneless, unchanging, uneventful, unexciting, uniform, uninteresting, unvarying, wearisome. *Opp* INTERESTING.

monster *n* abortion, beast, bogeyman, brute, demon, devil, fiend, freak, giant, horror, monstrosity, monstrous creature, mutant, ogre, troll.

monstrous *adj* **1** colossal, elephantine, enormous, gargantuan, giant, gigantic, great, huge, hulking, immense, *inf* jumbo, mammoth, mighty, prodigious, titanic, towering, tremendous, vast. ▷ BIG. **2** *a monstrous crime.* abhorrent, atrocious, awful, beastly, brutal, cruel, devilish, dreadful, disgusting, evil, ghoulish, grisly, gross, gruesome, heinous, hideous, *inf* horrendous, horrible, horrific, horrifying, inhuman, nightmarish, obscene, outrageous, repulsive, shocking, terrible, ugly, villainous, wicked. ▷ EVIL.

monument *n* cairn, cenotaph, cross, gravestone, headstone, mausoleum, memorial, obelisk, pillar, prehistoric remains, relic, reminder, shrine, tomb, tombstone.

monumental *adj* **1** awe-inspiring, awesome, classic, enduring, epoch-making, grand, historic, impressive, lasting, large-scale, major, memorable, unforgettable. ▷ BIG. **2** *a monumental plaque.* commemorative, memorial.

mood *n* **1** attitude, disposition, frame of mind, humour, inclination, nature, spirit, state of mind, temper, vein. **2** atmosphere, feeling, tone. ▷ ANGRY, HAPPY, SAD, etc. **in the mood** ▷ READY.

moody *adj* abrupt, bad-tempered, cantankerous, capricious, changeable, crabby, cross, crotchety, depressed, depressive, disgruntled, erratic, fickle, gloomy, grumpy, *inf* huffy, ill-humoured, inconstant, irritable, melancholy, mercurial, miserable, morose, peevish, petulant, *inf* short, short-tempered, snappy, sulky, sullen, temperamental, testy, *inf* touchy, unpredictable, unreliable, unstable, volatile. ▷ SAD.

moor *n* fell, heath, moorland, wasteland. • *vb* anchor, berth, dock, make fast, secure, tie up. ▷ FASTEN.

mope *vb* be sad, brood, despair, grieve, languish, *inf* moon, pine, sulk.

moral *adj* **1** blameless, chaste, decent, ethical, good, highminded, honest, honourable, incorruptible, innocent, irreproachable, just, law-abiding, noble, principled, proper, pure, respectable, responsible, right, righteous, sinless, trustworthy, truthful, upright, upstanding, virtuous. **2** *a moral tale.* allegorical, cautionary, didactic, moralistic, moralizing. *Opp* IMMORAL. • *n* lesson, maxim, meaning, message, point, precept, principle, teaching. **morals** ▷ MORALITY. **moral tale** allegory, cautionary tale, fable, parable.

morale *n* attitude, cheerfulness, confidence, *Fr* esprit de corps, *inf* heart, mood, self-confidence, self-esteem, spirit, state of mind.

morality *n* behaviour, conduct, decency, ethics, ethos, fairness, goodness, honesty, ideals, integrity, justice, morals, principles, propriety, rectitude, righteousness, rightness, scruples, standards, uprightness, virtue.

moralize *vb* lecture, philosophize, pontificate, preach, sermonize.

morbid *adj* black (*humour*), brooding, dejected, depressed, ghoulish, gloomy, grim, grotesque, gruesome, lugubrious, macabre, melancholy, monstrous, morose, pathological, pessimistic, *inf* sick, sombre, unhappy, unhealthy, unpleasant, unwholesome. *Opp* CHEERFUL.

more *adj* added, additional, extra, further, increased, longer, new, other, renewed, supplementary. *Opp* LESS.

moreover *adv* also, as well, besides, further, furthermore, in addition, *old use* to boot, too.

morose *adj* bad-tempered, churlish, depressed, gloomy, glum, grim, humourless, ill-natured, melancholy, moody, mournful, pessimistic, saturnine, sour, sulky, sullen, surly, taciturn, unhappy, unsociable. ▷ SAD. *Opp* CHEERFUL.

morsel *n* bite, crumb, fragment, gobbet, mouthful, nibble, piece, sample, scrap, small amount, soupçon, spoonful, taste, titbit. ▷ BIT.

mortal *adj* **1** ephemeral, human, passing, temporal, transient. *Opp* IMMORTAL. **2** *mortal sickness.* deadly, fatal, lethal, terminal. **3** *mortal enemies.* deadly, implacable, irreconcilable, remorseless, sworn, unrelenting. • *n* creature, human being, man, person, soul, woman.

mortality *n* **1** corruptibility, humanity, impermanence, transience. **2** *infant mortality.* death-rate, dying, fatalities, loss of life.

mortify *vb* abash, chagrin, chasten, *inf* crush, deflate, embarrass, humble, humiliate, *inf* put down, shame.

mostly *adv* chiefly, commonly, generally, largely, mainly, normally, predominantly, primarily, principally, typically, usually.

moth-eaten *adj* antiquated, decrepit, holey, mangy, ragged, shabby, *inf* tatty. ▷ OLD.

mother *n old use* dam, *inf* ma, *inf* mamma, *old use* mater, *inf* mum, *inf* mummy, parent. • *vb* care for, cherish, coddle, comfort, cuddle, fuss over, indulge, look after, love, nourish, nurse, nurture, pamper, protect, spoil, take care of.

motherly *adj* caring, kind, maternal, protective. ▷ LOVING.

motif *n* decoration, design, device, figure, idea, leitmotif, ornament, pattern, symbol, theme.

motion *n* action, activity, agitation, change, commotion, development, evolution, move, movement, progress, rise and fall, shift, stir, stirring, to and fro, travel, travelling, trend. • *vb* ▷ GESTURE.

motionless *adj* at rest, calm, frozen, immobile, inanimate, inert, lifeless, paralysed, peaceful, resting, stagnant, static, stationary, still, stock-still, unmoving. *Opp* MOVING.

motivate *vb* activate, actuate, arouse, cause, drive, egg on, encourage, excite, galvanize, goad, incite, induce, influence, inspire, instigate, move, occasion, persuade, prompt, provoke, push, rouse, spur, stimulate, stir, urge.

motive *n* aim, ambition, cause, drive, encouragement, end, enticement, grounds, impulse, incentive, incitement, inducement, inspiration, instigation, intention, lure, motivation, object, provocation, purpose, push, rationale, reason, spur, stimulation, stimulus, thinking.

motor *n* ▷ ENGINE, VEHICLE. • *vb* drive, go by car. ▷ TRAVEL.

mottled *adj* blotchy, brindled, dappled, flecked, freckled, marbled, patchy, spattered, speckled, spotted, spotty, streaked, streaky, variegated.

motto *n* adage, aphorism, catchphrase, maxim, precept, proverb, rule, saw, saying, slogan.

mould *n* blight, fungus, growth, mildew. • *vb* cast, fashion, forge, form, model, *inf* sculpt, shape, stamp, work.

mouldy *adj* carious, damp, decaying, decomposing, fusty, mildewed, mouldering, musty, putrefying, rotten, stale.

mound *n* bank, dune, elevation, heap, hill, hillock, hummock, hump, knoll, pile, stack, tumulus.

mount *n* ▷ MOUNTAIN. • *vb* **1** ascend, clamber up, climb, fly up, go up, rise, rocket upwards, scale, shoot up, soar. *Opp* DESCEND. **2** *mount a horse*. get astride, get on, jump onto. **3** *savings mount*. accumulate, build up, escalate, expand, get bigger, grow, increase, intensify, multiply, pile up, swell. *Opp* DECREASE. **4** *mount a picture*. display, exhibit, frame, install, prepare, put in place, put on, set up.

mountain *n* alp, arête, *Scot* ben, elevation, eminence, height, hill, mound, mount, peak, prominence, range, ridge, sierra, summit, tor, volcano.

mountainous *adj* alpine, craggy, daunting, formidable, high, hilly, precipitous, rocky, rugged, steep, towering. ▷ BIG.

mourn *vb* bemoan, bewail, fret, go into mourning, grieve, keen, lament, mope, pine, regret, wail, weep. *Opp* REJOICE.

mournful *adj* dismal, distressed, distressing, doleful, funereal, gloomy, grief-stricken, grieving, heartbreaking, heartbroken, lamenting, lugubrious, melancholy, plaintive, plangent, sad, sorrowful, tearful, tragic, unhappy, woeful. *Opp* CHEERFUL.

mouth *n* 1 *inf* chops, *sl* gob, jaws, *sl* kisser, lips, maw, muzzle, palate. 2 *mouth of cave.* aperture, door, doorway, entrance, exit, gate, gateway, inlet, opening, orifice, outlet, vent, way in. 3 *mouth of a river.* delta, estuary, outflow. • *vb* articulate, enunciate, form, pronounce. ▷ SAY.

mouthful *n* bite, gobbet, gulp, morsel, sip, spoonful, swallow, taste.

movable *adj* adjustable, changeable, detachable, floating, mobile, portable, transferable, transportable, unfixed, variable. *Opp* IMMOVABLE.

move *n* 1 act, action, deed, device, dodge, gambit, manoeuvre, measure, movement, ploy, ruse, step, stratagem, *inf* tack, tactic. 2 *a career move.* change, changeover, relocation, shift, transfer. 3 *your move.* chance, go, opportunity, turn. • *vb* 1 *move about.* be agitated, be astir, budge, change places, change position, fidget, flap, roll, shake, shift, stir, swing, toss, tremble, turn, twist, twitch, wag, *inf* waggle, wave, *inf* wiggle. 2 *move along.* cruise, fly, jog, journey, make headway, make progress, march, pass, proceed, travel, walk. 3 *move quickly.* bolt, *inf* bowl along, canter, career, dash, dart, flit, flounce, fly, gallop, hasten, hurry, hurtle, hustle, *inf* nip, race, run, rush, shoot, speed, stampede, streak, sweep along, sweep past, *inf* tear, *inf* zip, *inf* zoom. 4 *move slowly.* amble, crawl, dawdle, drift, stroll. 5 *move gracefully.* dance, flow, glide, skate, skim, slide, slip, sweep. 6 *move awkwardly.* dodder, falter, flounder, lumber, lurch, pitch, shuffle, stagger, stumble, sway, totter, trip, trundle. 7 *move stealthily.* crawl, creep, edge, slink, slither. 8 *move things.* carry, export, import, shift, ship, relocate, transfer, transplant, transport, transpose. 9 *moved him to act.* encourage, impel, influence, inspire, persuade, prompt, stimulate, urge. 10 *move the crowd's feelings.* affect, arouse, enrage, fire, impassion, rouse, stir, touch. 11 *moved to improve the situation.* act, do something, make a move, take action. **move away** ▷ DEPART. **move back** ▷ RETREAT. **move down** ▷ DESCEND. **move in** ▷ ENTER. **move round** ▷ CIRCULATE, ROTATE. **move towards** ▷ APPROACH. **move up** ▷ ASCEND.

movement *n* 1 action, activity, migration, motion, shifting, stirring. ▷ GESTURE, MOVE. 2 *movement towards green issues.* change, development, drift, evolution, progress, shift, swing, tendency, trend. 3 *a political movement.* campaign, crusade, drive, faction, group, organization, party. 4 *military movements.* exercise, operation.

movie *n* film, *inf* flick, motion picture.

moving *adj* 1 active, alive, astir, dynamic, flowing, going, in motion, mobile, movable, on the move, travelling, under way. *Opp* MOTIONLESS. 2 *a moving tale.* affecting, emotional, emotive, exciting, heart-rending, heartwarming, inspirational, inspiring, pathetic, poignant, spine-tingling, stirring, *inf* tear-jerking, thrilling, touching.

mow *vb* clip, cut, scythe, shear, trim.

muck *n* dirt, droppings, dung, excrement, faeces, filth, grime, *inf* gunge, manure, mess, mire, mud, ooze, ordure, rubbish, scum, sewage, slime, sludge.

mucky *adj* dirty, filthy, foul, grimy, grubby, messy, muddy, scummy, slimy, soiled, sordid, squalid. *Opp* CLEAN.

mud *n* clay, dirt, mire, muck, ooze, silt, slime, sludge, slurry, soil.

muddle *n* chaos, clutter, confusion, disorder, *inf* hotchpotch, jumble, mess, *inf* mishmash, *inf* mix up, *inf* shambles, tangle, untidiness. • *vb* **1** bemuse, bewilder, confound, confuse, disorient, disorientate, mislead, perplex, puzzle. *Opp* CLARIFY. **2** disarrange, disorder, disorganize, entangle, *inf* foul up, jumble, make a mess of, *inf* mess up, mix up, scramble, shuffle, tangle. *Opp* TIDY.

muddy *adj* **1** caked, dirty, filthy, messy, mucky, soiled. **2** *muddy water*. cloudy, impure, misty, opaque. **3** *muddy ground*. boggy, marshy, sloppy, sodden, soft, spongy, waterlogged, wet. *Opp* CLEAN, FIRM.

muffle *vb* **1** cloak, conceal, cover, enclose, enfold, envelop, shroud, swathe, wrap up. **2** *muffle noise*. damp, dampen, deaden, disguise, dull, hush, mask, mute, quieten, silence, soften, stifle, still, suppress, tone down.

muffled *adj* damped, deadened, dull, fuzzy, indistinct, muted, silenced, stifled, suppressed, unclear, woolly. *Opp* CLEAR.

mug *n* beaker, cup, *old use* flagon, pot, tankard. • *vb* assault, beat up, jump on, molest, rob, set on, steal from. ▷ ATTACK. **mug up** ▷ LEARN.

mugger *n* attacker, hooligan, robber, ruffian, thief, thug. ▷ CRIMINAL.

mugging *n* attack, robbery, street crime. ▷ CRIME.

muggy *adj* clammy, close, damp, humid, moist, oppressive, steamy, sticky, stuffy, sultry, warm.

multiple *adj* complex, compound, double, many, numerous, plural, quadruple, quintuple, triple.

multiplicity *n* abundance, array, complex, diversity, number, plurality, profusion, variety.

multiply *vb* **1** double, quadruple, quintuple, *inf* times, triple. **2** become numerous, breed, increase, proliferate, propagate, reproduce, spread.

multitude *n* crowd, host, large number, legion, lots, mass, myriad, swarm, throng. ▷ GROUP.

mumble *vb* be inarticulate, murmur, mutter, speak indistinctly, swallow your words.

munch *vb* bite, chew, champ, chomp, crunch, eat, gnaw, masticate.

mundane *adj* banal, common, commonplace, down-to-earth, dull, everyday, familiar, human, material, physical, practical, quotidian, routine, temporal, worldly. ▷ ORDINARY. *Opp* EXTRAORDINARY, SPIRITUAL.

municipal *adj* borough, city, civic, community, district, local, public, town, urban.

murder *n* assassination, fratricide, genocide, homicide, infanticide, killing, manslaughter, matricide, parricide, patricide, regicide, sororicide, unlawful killing, uxoricide. • *vb* ▷ KILL.

murderer *n* assassin, *inf* butcher, cutthroat, gunman, homicide, killer, slayer.

murderous *adj* barbarous, bloodthirsty, bloody, brutal, cruel, dangerous, deadly, fell, ferocious, fierce, homicidal, inhuman, pitiless, ruthless, savage, vicious, violent.

murky *adj* clouded, cloudy, dark, dim, dismal, dreary, dull, foggy, funereal, gloomy, grey, misty, muddy, obscure, overcast, shadowy, sombre. *Opp* CLEAR.

murmur *n* background noise, buzz, drone, grumble, hum, mutter, rumble, susurration, undertone, whisper. • *vb* drone, hum, moan, mumble, mutter, rumble, speak in an undertone, whisper. ▷ GRUMBLE, TALK.

muscular *adj* athletic, *inf* beefy, brawny, broad-shouldered, burly, hefty, *inf* hulking, husky, powerful, powerfully built, robust, sinewy, *inf* strapping, strong, sturdy, tough, well-built, well-developed, wiry. *Opp* WEAK.

muse *vb* cogitate, consider, contemplate, deliberate, meditate, mull over, ponder, reflect, ruminate, study, think.

mushy *adj* pulpy, spongy, squashy. ▷ SOFT.

music *n* harmony. □ *blues, chamber music, choral music, classical music, dance music, disco music, folk, instrumental music, jazz, orchestral music, plain-song, pop, ragtime, reggae, rock, soul, swing.* □ *anthem, ballad, cadenza, calypso, canon, cantata, canticle, carol, chant, concerto, dance, dirge, duet, étude, fanfare, fugue, hymn, improvisation, intermezzo, lullaby, march, musical, nocturne, nonet, octet, opera, operetta, oratorio, overture, prelude, quartet, quintet, rhapsody, rondo, scherzo, sea shanty, septet, sextet, sonata, song, spiritual, symphony, toccata, trio.*

musical *adj* euphonious, harmonious, lyrical, melodious, pleasant, sweet-sounding, tuneful. **musical instrument** ▷ INSTRUMENT.

musician *n* composer, music-maker, performer, player, singer. □ *accompanist, bass, bugler, cellist, clarinettist, conductor, contralto, drummer, fiddler, flautist, guitarist, harpist, instrumentalist, maestro, minstrel, oboist, organist, percussionist, pianist, piper, soloist, soprano, tenor, timpanist, treble, trombonist, trumpeter, violinist, virtuoso, vocalist.* **musicians** □ *band, choir, chorus, consort, duet, duo, ensemble, group, nonet, octet, orchestra, quartet, quintet, septet, sextet, trio.*

muster *vb* assemble, call together, collect, come together, convene, convoke, gather, get together, group, marshal, mobilize, rally, round up, summon.

musty *adj* airless, damp, dank, fusty, mildewed, mildewy, mouldy, smelly, stale, stuffy, unventilated.

mutant *n* abortion, anomaly, deviant, freak, monster, monstrosity, sport, variant.

mutation *n* alteration, deviance, evolution, metamorphosis, modification, transfiguration, transformation, transmutation, variation. ▷ CHANGE.

mute *adj* dumb, quiet, silent, speechless, tacit, taciturn, tight-lipped, tongue-tied, voiceless. • *vb* damp, dampen, deaden, dull, hush, make quieter, mask, muffle, quieten, silence, soften, stifle, still, suppress, tone down.

mutilate *vb* cripple, damage, deface, disable, disfigure, dismember, injure, lame, maim, mangle, mar, spoil, vandalize, wound.

mutinous *adj* contumacious, defiant, disobedient, insubordinate, insurgent, insurrectionary, rebellious, refractory, revolutionary, seditious, subversive, ungovernable, unmanageable, unruly. *Opp* OBEDIENT.

mutiny *n* defiance, disobedience, insubordination, insurgency, insurrection, rebellion, revolt, revolution, sedition, subversion, unruliness, uprising. • *vb* agitate, be mutinous, disobey, rebel, revolt, rise up, strike.

mutter *vb* drone, grumble, mumble, murmur, speak in an undertone, whisper. ▷ GRUMBLE, TALK.

mutual *adj* common, interactive, joint, reciprocal, reciprocated, requited, shared.

muzzle *n* jaws, mouth, nose, snout. • *vb* censor, gag, restrain, silence, stifle, suppress.

mysterious *adj* arcane, baffling, bewildering, bizarre, confusing, cryptic, curious, dark, enigmatic, incomprehensible, inexplicable, inscrutable, insoluble, magical, miraculous, mystical, mystifying, obscure, perplexing, puzzling, recondite, secret, strange, uncanny, unexplained, unfathomable, unknown, weird. *Opp* STRAIGHTFORWARD.

mystery *n* conundrum, enigma, miracle, problem, puzzle, question, riddle, secret.

mystical *adj* abnormal, arcane, cabalistic, ineffable, metaphysical, mysterious, occult, other-worldly, preternatural, religious, spiritual, supernatural. *Opp* MUNDANE.

mystify *vb* baffle, *inf* bamboozle, *inf* beat, bewilder, confound, confuse, *inf* flummox, fool, hoax, perplex, puzzle, *inf* stump.

myth *n* **1** allegory, fable, legend, mythology, symbolism. **2** fabrication, falsehood, fiction, invention, make-believe, pretence, untruth.

mythical *adj* **1** allegorical, fabled, fabulous, legendary, mythic, mythological, poetic, symbolic. **2** false, fanciful, fictional, imaginary, invented, make-believe, nonexistent, pretended, unreal. *Opp* REAL.

N

nadir *n* bottom, depths, low point, zero. *Opp* ZENITH.

nag *n* ▷ HORSE. • *vb* annoy, badger, chivvy, find fault with, goad, *inf* go on at, harass, hector, *inf* henpeck, keep complaining, pester, *inf* plague, scold, worry.

nail *n* pin, spike, stud, tack. ▷ FASTEN.

naïve *adj* artless, *inf* born yesterday, candid, childlike, credulous, *inf* green, guileless, gullible, inexperienced, ingenuous, innocent, open, simple, simple-minded, stupid, trustful, trusting, unsophisticated, unsuspecting, unwary. *Opp* ARTFUL.

naked *adj* bare, denuded, disrobed, exposed, in the nude, nude, stark-naked, stripped, unclothed, unconcealed, uncovered, undraped, undressed.

name *n* **1** alias, appellation, Christian name, first name, forename, given name, *inf* handle, identity, nickname, nom de plume, pen name, personal name, pseudonym, sobriquet, surname, title. **2** denomination, designation, epithet, term. • *vb* **1** baptize, call, christen, dub, style. **2** *name a book.* entitle, label. **3** *named him man of the match.* appoint, choose, commission, delegate, designate, elect, nominate, select, single out, specify. **named** ▷ SPECIFIC.

nameless *adj* **1** anonymous, incognito, unheard-of, unidentified, unnamed, unsung. **2** *nameless horrors.* dreadful, horrible, indescribable, inexpressible, shocking, unmentionable, unspeakable, unutterable.

nap *n* catnap, doze, *inf* forty winks, rest, *inf* shut-eye, siesta, sleep, snooze.

narrate *vb* chronicle, describe, detail, recount, rehearse, relate, repeat, report, retail, tell, unfold.

narration *n* commentary, reading, recital, recitation, relation, storytelling, telling, voiceover.

narrative *n* account, chronicle, description, history, report, story, tale, *inf* yarn.

narrator *n* author, chronicler, raconteur, reporter, storyteller.

narrow *adj* attenuated, close, confined, constricted, constricting, cramped, enclosed, fine, limited, restricted, slender, slim, thin, tight. *Opp* WIDE.

narrow-minded *adj* biased, bigoted, conservative, conventional, hidebound, illiberal, inflexible, insular, intolerant, narrow, old-fashioned, parochial, petty, prejudiced, prim, prudish, puritanical, reactionary, rigid, small-minded, straitlaced, *inf* stuffy. *Opp* BROAD-MINDED.

nasty *adj* [*Nasty* refers to anything you do not like. The range of synonyms is almost limitless: we give only a selection here.] bad, beastly, dangerous, difficult, dirty, disagreeable, disgusting, distasteful, foul, hateful, horrible, loathsome, *sl* lousy, objectionable, obnoxious, obscene, *inf* off-putting, repulsive, revolting, severe, sickening, unkind, unpleasant. *Opp* NICE.

nation *n* civilization, community, country, domain, land, people, population, power, race, realm, society, state, superpower.

national *adj* **1** ethnic, popular, racial. **2** *a national emergency.* countrywide, general, nationwide, state, widespread. • *n* citizen, inhabitant, native, resident, subject.

nationalism *n* chauvinism, jingoism, loyalty, patriotism, xenophobia.

native *adj* **1** aboriginal, indigenous, local, original. **2** *native wit.* congenital, hereditary, inborn, inbred, inherent, inherited, innate, mother (*wit*), natural. • *n* aborigine, life-long resident.

natural *adj* **1** common, everyday, habitual, normal, ordinary, predictable, regular, routine, standard, typical, usual. **2** *natural feelings.* healthy, hereditary, human, inborn, inherited, innate, instinctive, intuitive, kind, maternal, native, paternal, proper, right. **3** *a natural smile.* artless, authentic, candid, genuine, guileless, sincere, spontaneous, unaffected, unpretentious, unselfconscious, unstudied. **4** *natural resources.* crude (*oil*), raw, unadulterated, unprocessed, unrefined. **5** *a natural leader.* born, congenital, untaught. *Opp* UNNATURAL.

nature *n* 1 countryside, creation, ecology, environment, natural history, scenery, wildlife.
2 attributes, character, complexion, constitution, disposition, essence, humour, make-up, manner, personality, properties, quality, temperament, traits.
3 category, description, kind, sort, species, type, variety.

naughty *adj* 1 bad, badly-behaved, bad-mannered, boisterous, contrary, defiant, delinquent, disobedient, disorderly, disruptive, fractious, headstrong, impish, impolite, incorrigible, insubordinate, intractable, misbehaved, mischievous, obstinate, obstreperous, perverse, playful, puckish, rascally, rebellious, refractory, roguish, rude, self-willed, stubborn, troublesome, uncontrollable, undisciplined, ungovernable, unmanageable, unruly, wayward, wicked, wild, wilful. 2 [*inf*] cheeky, improper, ribald, risqué, shocking, *inf* smutty, vulgar. ▷ OBSCENE. *Opp* POLITE, WELL BEHAVED.

nauseate *vb* disgust, offend, repel, revolt, sicken.

nauseous *adj* disgusting, foul, loathsome, nauseating, offensive, repulsive, revolting, sickening, stomach-turning.

nautical *adj* marine, maritime, naval, seafaring, seagoing, yachting.

navigate *vb* captain, direct, drive, guide, handle, manoeuvre, map-read, pilot, sail, skipper, steer.

navy *n* armada, convoy, fleet, flotilla.

near *adj* 1 abutting, adjacent, adjoining, bordering, close, connected, contiguous, immediate, nearby, neighbouring, next-door. 2 *Christmas is near.* approaching, coming, forthcoming, imminent, impending, looming, *inf* round the corner. 3 *near friends.* close, dear, familiar, intimate, related. *Opp* DISTANT.

nearly *adv* about, all but, almost, approaching, approximately, around, as good as, close to, just about, not quite, practically, roughly, virtually.

neat *adj* 1 adroit, clean, dainty, deft, dexterous, elegant, *inf* natty, orderly, organized, pretty, *inf* shipshape, smart, *inf* spick and span, spruce, straight, systematic, tidy, trim, uncluttered, well-kept. 2 accurate, expert, methodical, meticulous, precise, skilful. 3 *neat alcohol.* pure, *inf* straight, unadulterated, undiluted. *Opp* CLUMSY, UNTIDY.

necessary *adj* compulsory, destined, essential, fated, imperative, important, indispensable, ineluctable, inescapable, inevitable, inexorable, mandatory, needed, needful, obligatory, predestined, required, requisite, unavoidable, vital. *Opp* UNNECESSARY.

necessity *n* 1 compulsion, essential, inevitability, *inf* must, need, obligation, prerequisite, requirement, requisite, *Lat* sine qua non. 2 beggary, destitution, hardship, indigence, need, penury, poverty, privation, shortage, suffering, want.

need *n* call, demand, lack, requirement, want. ▷ NECESSITY. • *vb* be short of, call for, crave, demand, depend on, lack, miss, rely on, require, want.

needless *adj* excessive, gratuitous, pointless, redundant, superfluous, unnecessary.

needy *adj* badly off, destitute, *inf* hard up, impecunious, impoverished, indigent, necessitous, penurious, poverty-stricken, underpaid. ▷ POOR.

negate *vb* annul, cancel out, deny, gainsay, invalidate, nullify, oppose.

negative *adj* adversarial, antagonistic, *inf* anti, contradictory, destructive, disagreeing, dissenting, grudging, nullifying, obstructive, opposing, pessimistic, uncooperative, unenthusiastic, unresponsive, unwilling. • *n* denial, no, refusal, rejection, veto. *Opp* POSITIVE.

neglect *n* carelessness, dereliction of duty, disregard, inadvertence, inattention, indifference, negligence, oversight, slackness. • *vb* abandon, be remiss about, disregard, forget, ignore, leave alone, let slide, lose sight of, miss, omit, overlook, pay no attention to, shirk, skip. **neglected** ▷ DERELICT.

negligent *adj* careless, forgetful, heedless, inattentive, inconsiderate, indifferent, irresponsible, lax, offhand, reckless, remiss, slack, sloppy, slovenly, thoughtless, uncaring, unthinking. *Opp* CAREFUL.

negligible *adj* imperceptible, inconsequential, inconsiderable, insignificant, minor, nugatory, paltry, petty, slight, small, tiny, trifling, trivial, unimportant. *Opp* CONSIDERABLE.

negotiate *vb* arbitrate, bargain, come to terms, confer, deal, discuss terms, haggle, intercede, make arrangements, mediate, parley, transact.

negotiation *n* arbitration, bargaining, conciliation, debate, diplomacy, discussion, mediation, parleying, transaction.

negotiator *n* agent, ambassador, arbitrator, broker, conciliator, diplomat, go-between, intercessor, intermediary, mediator, middleman.

neighbourhood *n* area, community, district, environs, locality, place, purlieus, quarter, region, surroundings, vicinity, zone.

neighbouring *adj* adjacent, adjoining, attached, bordering, close, closest, connecting, contiguous, near, nearby, nearest, next-door, surrounding.

neighbourly *adj* civil, considerate, friendly, helpful, kind, sociable, thoughtful, well-disposed.

nerve *n* coolness, determination, firmness, fortitude, resolution, resolve, will-power. ▷ COURAGE.

nervous *adj* afraid, agitated, anxious, apprehensive, disturbed, edgy, excitable, fearful, fidgety, flustered, fretful, highly-strung, ill-at-ease, *inf* in a tizzy, insecure, *inf* jittery, *inf* jumpy, *inf* nervy, neurotic, on edge, *inf* on tenterhooks, *inf* rattled, restive, restless, ruffled, shaky, shy, strained, tense, timid, *inf* touchy, *inf* twitchy, uneasy, unnerved, unsettled, *inf* uptight, worried. ▷ FRIGHTENED. *Opp* CALM.

nestle *vb* cuddle, curl up, huddle, nuzzle, snuggle.

net *n* lace, lattice-work, mesh, netting, network, web. • *vb* **1** catch, capture, enmesh, ensnare, trammel, trap. **2** *net £200 a week.* accumulate, bring in, clear, earn, get, make, realize, receive, *inf* take home.

network *n* **1** *inf* criss-cross, grid, labyrinth, lattice, maze, mesh, net, netting, tangle, tracery, web. **2** complex, organization, system.

neurosis *n* abnormality, anxiety, depression, mental condition, obsession, phobia.

neurotic *adj* anxious, distraught, disturbed, irrational, maladjusted, nervous, obsessive, overwrought, unbalanced, unstable.

neuter *adj* ambiguous, ambivalent, asexual, indeterminate, uncertain. • *vb* castrate, *inf* doctor, emasculate, geld, spay, sterilize.

neutral *adj* **1** detached, disinterested, dispassionate, fair, impartial, indifferent, non-aligned, non-belligerent, non-partisan, objective, unaffiliated, unaligned, unbiased, uncommitted, uninvolved, unprejudiced. *Opp* BIASED. **2** *neutral colours.* characterless, colourless, dull, drab, indefinite, indeterminate, intermediate, neither one thing nor the other, pale, vague. *Opp* DISTINCTIVE.

neutralize *vb* annul, cancel out, compensate for, counteract, counterbalance, invalidate, make ineffective, make up for, negate, nullify, offset, wipe out.

new *adj* **1** brand-new, clean, different, fresh, mint, strange, unfamiliar, unheard of, untried, unused. **2** *new ideas.* advanced, contemporary, current, different, fashionable, latest, modern, modernistic, newfangled, novel, original, recent, revolutionary, *inf* trendy, up-to-date. **3** *new data.* added, additional, changed, extra, further, just arrived, supplementary, unexpected, unknown. *Opp* OLD.

newcomer *n* alien, arrival, immigrant, new boy, new girl, outsider, settler, stranger.

news *n* account, advice, announcement, bulletin, communication, communiqué, dispatch, headlines, information, intelligence, *inf* the latest, message, newscast, newsflash, newsletter, notice, press-release, proclamation, report, rumour, statement, *old use* tidings, word.

newspaper *n inf* daily, gazette, journal, paper, periodical, *inf* rag, tabloid.

next *adj* **1** adjacent, adjoining, closest, nearest, neighbouring, next-door. **2** *the next moment.* following, soonest, subsequent, succeeding.

nice *adj* **1** accurate, careful, delicate, discriminating, exact, fine, hair-splitting, meticulous, precise, punctilious, scrupulous, subtle. **2** *nice manners.* dainty, elegant, fastidious, fussy, particular, *inf* pernickety, polished, refined, well-mannered. **3** [*inf:* In this sense, nice refers to anything which you like. The range of synonyms is almost limitless: we give only a selection here.] acceptable, agreeable, amiable, attractive, beautiful, delicious, delightful, friendly, good, gratifying, kind, likeable, pleasant, satisfactory, welcome. *Opp* NASTY.

niche *n* alcove, corner, hollow, nook, recess.

nickname *n* alias, sobriquet.

niggardly *adj* mean, miserly, parsimonious, stingy.

nimble *adj* acrobatic, active, adroit, agile, brisk, deft, dextrous, limber, lithe, lively, *inf* nippy, quick-moving, sprightly, spry, swift. *Opp* CLUMSY.

nip *vb* bite, clip, pinch, snag, snap at, squeeze.

nobility *n* **1** dignity, glory, grandeur, greatness, high-mindedness, integrity, magnanimity, morality, nobleness, uprightness, virtue, worthiness. **2** *the nobility.* aristocracy, élite, gentry, nobles, peerage, the ruling classes, *inf* the upper crust.

noble *adj* **1** aristocratic, *inf* blue-blooded, courtly, distinguished, élite, gentle, high-born, high-ranking, patrician, princely, royal, thoroughbred, titled, upper-class. **2** *noble deeds.* brave, chivalrous, courageous, gallant, glorious, heroic. **3** *noble thoughts.* elevated, high-flown, honourable, lofty, magnanimous, moral, upright, virtuous, worthy. **4** *noble music.* dignified, elegant, grand, great, imposing, impressive, magnificent, majestic, splendid, stately. *Opp* BASE, COMMON.
• *n* aristocrat, gentleman, gentlewoman, grandee, lady, lord, nobleman, noblewoman, patrician, peer, peeress.

nod *vb* bend, bob, bow. **nod off** ▷ SLEEP.

noise *n inf* babel, bawling, bedlam, blare, cacophony, *inf* caterwauling, clamour, clangour, clatter, commotion, din, discord, *inf* fracas, hubbub, *inf* hullabaloo, outcry, pandemonium, *inf* racket, row, *inf* rumpus, screaming, screeching, shouting, shrieking, tumult, uproar, yelling. ▷ SOUND. *Opp* SILENCE.

noiseless *adj* inaudible, mute, muted, quiet, silent, soft, soundless, still. *Opp* NOISY.

noisy *adj* blaring, boisterous, booming, cacophonous, chattering, clamorous, deafening, discordant, dissonant, ear-splitting, fortissimo, harsh, loud, raucous, resounding, reverberating, rowdy, screaming, screeching, shrieking, shrill, strident, talkative, thunderous, tumultuous, unmusical, uproarious, vociferous. *Opp* NOISELESS.

nomadic *adj* itinerant, peripatetic, roving, travelling, vagrant, wandering, wayfaring.

nominal *adj* **1** formal, in name only, ostensible, self-styled, *inf* so-called, supposed, theoretical, titular. **2** *a nominal sum.* insignificant, minimal, minor, small, token.

nominate *vb* appoint, choose, designate, elect, name, propose, put forward, *inf* put up, recommend, select, specify.

non-existent *adj* chimerical, fictional, fictitious, hypothetical, imaginary, imagined, legendary, made-up, mythical, unreal. *Opp* REAL.

nonplus *vb* amaze, baffle, confound, disconcert, dumbfound, flummox, perplex, puzzle, render speechless.

nonsense *n* **1** [Most synonyms *inf*] balderdash, bilge, boloney, bosh, bunk, bunkum, claptrap, codswallop, double-Dutch, drivel, eyewash, fiddlesticks, foolishness, gibberish, gobbledegook, mumbo jumbo, piffle, poppycock, rot, rubbish, silliness, stuff and nonsense, stupidity, tommy-rot, trash, tripe, twaddle. **2** *The plan was a nonsense.* absurdity, inanity, mistake, nonsensical idea.

nonsensical *adj* absurd, asinine, crazy, *inf* daft, fatuous, foolish, idiotic, illogical, impractical, inane, incomprehensible, irrational, laughable, ludicrous, mad, meaningless, preposterous, ridiculous, senseless, stupid, unreasonable. ▷ SILLY. *Opp* SENSIBLE.

non-stop *adj* ceaseless, constant, continual, continuous, endless, eternal, incessant, interminable, perpetual, persistent, *inf* round-the-clock, steady, unbroken, unending, uninterrupted, unremitting.

norm *n* criterion, measure, model, pattern, rule, standard, type, yardstick.

normal *adj* **1** accepted, accustomed, average, common, commonplace, conventional, customary, established, everyday, familiar, general, habitual, natural, ordinary, orthodox, predictable, prosaic, quotidian, regular, routine, *inf* run-of-the-mill, standard, typical, universal, unsurprising, usual. **2** *a normal person.* balanced, healthy, rational, reasonable, sane, stable, *inf* straight, well-adjusted. *Opp* ABNORMAL.

normalize *vb* legalize, regularize, regulate, return to normal.

nose *n* **1** nostrils, proboscis, snout. **2** *nose of a boat.* bow, front, prow. • *vb* enter cautiously, insinuate yourself, intrude, nudge your way, penetrate, probe, push, shove. **nose about** ▷ PRY.

nostalgia *n* longing, memory, pining, regret, reminiscence, sentiment, sentimentality, yearning.

nostalgic *adj* emotional, maudlin, regretful, romantic, sentimental, wistful, yearning.

nosy *adj* curious, eavesdropping, inquisitive, interfering, meddlesome, prying.

notable *adj* celebrated, conspicuous, distinctive, distinguished, eminent, evident, extraordinary, famous, illustrious, important, impressive, memorable, noted, noteworthy, noticeable, obvious, outstanding, pre-eminent, prominent, rare, remarkable, renowned, singular, striking, uncommon, unforgettable, unusual, well-known. *Opp* ORDINARY.

note *n* **1** billet-doux, chit, communication, correspondence, epistle, jotting, letter, *inf* memo, memorandum, message, postcard. **2** annotation, comment, cross-reference, explanation, footnote, gloss, jotting, marginal note. **3** *a note in your voice.* feeling, quality, sound, tone. **4** *a £5 note.* banknote, bill, currency, draft. • *vb* **1** enter, jot down, record, scribble, write down. **2** *note mentally.* detect, discern, discover, feel, find, heed, mark, mind, notice, observe, pay attention to, remark, register, see, spy, take note of. **noted** ▷ FAMOUS.

noteworthy *adj* exceptional, extraordinary, rare, remarkable, uncommon, unique, unusual. *Opp* ORDINARY.

nothing *n cricket* duck, *tennis* love, *old use* naught, *football* nil, nought, zero, *sl* zilch.

notice *n* **1** advertisement, announcement, handbill, handout, intimation, leaflet, message, note, notification, placard, poster, sign, warning. **2** attention, awareness, cognizance, consciousness, heed, note, regard. • *vb* be aware, detect, discern, discover, feel, find, heed, make out, mark, mind, note, observe, pay attention to, perceive, register, remark, see, spy, take note. **give notice** ▷ NOTIFY, WARN.

noticeable *adj* appreciable, audible, clear, clear-cut, considerable, conspicuous, detectable, discernible, distinct, distinguishable, manifest, marked, measurable, notable, observable, obtrusive, obvious, overt, palpable, perceivable, perceptible, plain, prominent, pronounced, salient, significant, striking, unconcealed, unmistakable, visible. *Opp* IMPERCEPTIBLE.

notify *vb* acquaint, advise, alert, announce, apprise, give notice, inform, make known, proclaim, publish, report, tell, warn.

notion *n* apprehension, belief, concept, conception, fancy, hypothesis, idea, impression, *inf* inkling, opinion, sentiment, theory, thought, understanding, view.

notorious *adj* disgraceful, disreputable, flagrant, ill-famed, infamous, outrageous, overt, patent, scandalous, shocking, talked about, undisguised, undisputed, well-known. ▷ FAMOUS, WICKED.

nourish *vb* feed, maintain, nurse, nurture, provide for, strengthen, support, sustain. **nourishing** ▷ NUTRITIOUS.

nourishment *n* diet, food, goodness, nutrient, nutriment, nutrition, sustenance, *old use* victuals.

novel *adj* different, fresh, imaginative, innovative, new, odd, original, rare, singular, startling, strange, surprising, uncommon, unconventional, unfamiliar, untested, unusual. *Opp* FAMILIAR. • *n* best-seller, *inf* blockbuster, fiction, novelette, novella, romance, story. ▷ WRITING.

novelty *n* **1** freshness, newness, oddity, originality, strangeness, surprise, unfamiliarity, uniqueness. **2** bauble, curiosity, gimmick, knick-knack, ornament, souvenir, trifle, trinket.

novice *n* amateur, apprentice, beginner, *inf* greenhorn, inexperienced person, initiate, learner, probationer, tiro, trainee.

now *adv* at once, at present, here and now, immediately, instantly, just now, nowadays, promptly, *inf* right now, straight away, today.

noxious *adj* corrosive, foul, harmful, nasty, noisome, objectionable, poisonous, polluting, sulphureous, sulphurous, unwholesome.

nub *n* centre, cord, crux, essence, gist, heart, kernel, nucleus, pith, point.

nucleus *n* centre, core, heart, kernel, middle.

nude *adj* bare, disrobed, exposed, in the nude, naked, stark-naked, stripped, unclothed, uncovered, undressed.

nudge *vb* bump, dig, elbow, hit, jab, jog, jolt, poke, prod, push, shove, touch.

nuisance *n* annoyance, bother, burden, inconvenience, irritant, irritation, *inf* pain, pest, plague, trouble, vexation, worry.

nullify *vb* abolish, annul, cancel, do away with, invalidate, negate, neutralize, quash, repeal, rescind, revoke, stultify.

numb *adj* anaesthetized, *inf* asleep, benumbed, cold, dead, deadened, frozen, immobile, insensible, insensitive, paralysed, senseless, suffering from pins and needles. *Opp* SENSITIVE. • *vb* anaesthetize, benumb, deaden, desensitize, drug, dull, freeze, immobilize, make numb, paralyse, stun, stupefy.

number *n* **1** digit, figure, integer, numeral, unit. **2** aggregate, amount, *inf* bunch, collection, crowd, multitude, quantity, sum, total. ▷ GROUP. **3** *musical number.* item, piece, song. **4** *a number of a magazine.* copy, edition, impression, issue, printing, publication. • *vb* add up to, total, work out at. ▷ COUNT.

numerous *adj* abundant, copious, countless, endless, incalculable, infinite, innumerable, many, multitudinous, myriad, numberless, plentiful, several, uncountable, untold. *Opp* FEW.

nun *n* abbess, mother-superior, novice, prioress, sister.

nurse *n* **1** district-nurse, *old use* matron, sister. **2** nanny, nursemaid. • *vb* **1** care for, look after, minister to, nurture, tend, treat. **2** breast-feed, feed, suckle, wet-nurse. **3** cherish, coddle, cradle, cuddle, dandle, hold, hug, mother, pamper.

nursery *n* **1** crèche, kindergarten, nursery school. **2** garden centre, market garden.

nurture *vb* bring up, cultivate, educate, feed, look after, nourish, nurse, rear, tend, train.

nut *n* kernel. □ *almond, brazil, cashew, chestnut, cob-nut, coconut, filbert, hazel, peanut, pecan, pistachio, walnut.*

nutrient *n* fertilizer, goodness, nourishment.

nutriment *n* food, goodness, nourishment, nutrition, sustenance.

nutritious *adj* alimentary, beneficial, good for you, health-giving, healthy, nourishing, sustaining, wholesome.

O

oasis *n* **1** spring, watering-hole, well. **2** asylum, haven, refuge, resort, retreat, safe harbour, sanctuary.

oath *n* **1** assurance, avowal, guarantee, pledge, promise, undertaking, vow, word of honour. **2** blasphemy, curse, exclamation, expletive, *inf* four-letter word, imprecation, malediction, obscenity, profanity, swearword.

obedient *adj* acquiescent, amenable, biddable, compliant, conformable, deferential, disciplined, docile, duteous, dutiful, law-abiding, manageable, submissive, subservient, tamed, tractable, well-behaved, well-trained. *Opp* DISOBEDIENT.

obese *adj* corpulent, gross, overweight. ▷ FAT.

obey *vb* abide by, accept, acquiesce in, act in accordance with, adhere to, agree to, be obedient to, be ruled by, bow to, carry out, comply with, conform to, defer to, do what you are told, execute, follow, fulfil, give in to, heed, honour, implement, keep to, mind, observe, perform, *inf* stick to, submit to, take orders from. *Opp* DISOBEY.

object *n* **1** article, body, entity, item, thing. **2** aim, end, goal, intent, intention, objective, point, purpose, reason. **3** *object of ridicule.* butt, destination, target. • *vb* argue, be opposed, carp, cavil, complain, demur, disapprove, dispute, dissent, expostulate, *sl* grouse, grumble, make an objection, *inf* mind, *inf* moan, oppose, protest, quibble, raise objections, raise questions, remonstrate, take a stand, take exception. *Opp* ACCEPT, AGREE.

objection *n* argument, cavil, challenge, complaint, demur, demurral, disapproval, exception, opposition, outcry, protest, query, question, quibble, refusal, remonstration.

objectionable *adj* abhorrent, detestable, disagreeable, disgusting, dislikeable, displeasing, distasteful, foul, hateful, insufferable, intolerable, loathsome, nasty, nauseating, noisome, obnoxious, odious, offensive, *inf* off-putting, repellent, repugnant, repulsive, revolting, sickening, unacceptable, undesirable, unwanted. ▷ UNPLEASANT. *Opp* ACCEPTABLE.

objective *adj* 1 detached, disinterested, dispassionate, factual, impartial, impersonal, neutral, open-minded, outward-looking, rational, scientific, unbiased, uncoloured, unemotional, unprejudiced. 2 *objective evidence.* empirical, existing, observable, real. *Opp* SUBJECTIVE. • *n* aim, ambition, aspiration, design, destination, end, goal, hope, intent, intention, object, point, purpose, target.

obligation *n* commitment, compulsion, constraint, contract, duty, liability, need, requirement, responsibility. ▷ PROMISE. *Opp* OPTION.

obligatory *adj* binding, compulsory, essential, mandatory, necessary, required, requisite, unavoidable. *Opp* OPTIONAL.

oblige *vb* 1 coerce, compel, constrain, force, make, require. 2 *Please oblige me.* accommodate, gratify, indulge, please. **obliged** ▷ BOUND, GRATEFUL. **obliging** ▷ HELPFUL, POLITE.

oblique *adj* 1 angled, askew, aslant, canted, declining, diagonal, inclined, leaning, listing, raked, rising, skewed, slanted, slanting, slantwise, sloping, tilted. 2 *an oblique insult.* backhanded, circuitous, circumlocutory, devious, implicit, implied, indirect, roundabout. ▷ EVASIVE. *Opp* DIRECT.

obliterate *vb* blot out, cancel, cover over, delete, destroy, efface, eliminate, eradicate, erase, expunge, extirpate, leave no trace of, rub out, wipe out.

oblivion *n* 1 anonymity, darkness, disregard, extinction, limbo, neglect, obscurity. 2 amnesia, coma, forgetfulness, ignorance, insensibility, obliviousness, unawareness, unconsciousness.

oblivious *adj* forgetful, heedless, ignorant, insensible, insensitive, unacquainted, unaware, unconscious, unfeeling, uninformed, unmindful, unresponsive. *Opp* AWARE.

obscene *adj* abominable, bawdy, *inf* blue, coarse, corrupting, crude, debauched, degenerate, depraved, dirty, disgusting, distasteful, filthy, foul, foul-mouthed, gross, immodest, immoral, improper, impure, indecent, indecorous, indelicate, *inf* kinky, lecherous, lewd, loathsome, nasty, *inf* off-colour, offensive, outrageous, perverted, pornographic, prurient, repulsive, ribald, risqué, rude, salacious, scatalogical, scurrilous, shameful, shameless, shocking, *inf* sick, smutty, suggestive, unchaste, vile, vulgar. ▷ OBJECTIONABLE, SEXY. *Opp* DECENT.

obscenity *n* abomination, blasphemy, coarseness, dirtiness, evil, filth, foulness, grossness, immorality, impropriety, indecency, lewdness, licentiousness, offensiveness, outrage, perversion, pornography, profanity, scurrility, vileness. ▷ SWEARWORD.

obscure *adj* 1 blurred, clouded, concealed, covered, dark, dim, faint, foggy, hazy, hidden, inconspicuous, indefinite, indistinct, masked, misty, murky, nebulous, secret, shadowy, shady, shrouded, unclear, unlit, unrecognizable, vague, veiled. *Opp* CLEAR. 2 *an obscure joke.* arcane, baffling, complex, cryptic, delphic, enigmatic, esoteric, incomprehensible, mystifying, perplexing, puzzling, recherché, recondite, strange. *Opp* OBVIOUS. **3** *an obscure poet.* forgotten, minor, undistinguished, unfamiliar, unheard of, unimportant, unknown, unnoticed. *Opp* FAMOUS. • *vb* block out, blur, cloak, cloud, conceal, cover, darken, disguise, eclipse, envelop, hide, make obscure, mask, obfuscate, overshadow, screen, shade, shroud, veil. *Opp* CLARIFY.

obsequious *adj* abject, *inf* bootlicking, crawling, cringing, deferential, effusive, fawning, flattering, fulsome, *inf* greasy, grovelling, ingratiating, insincere, mealy-mouthed, menial, *inf* oily, servile, *inf* smarmy, submissive, subservient, sycophantic, unctuous. **be obsequious** ▷ GROVEL.

observant *adj* alert, astute, attentive, aware, careful, eagle-eyed, heedful, mindful, on the lookout, *inf* on the qui vive, perceptive, percipient, quick, sharp-eyed, shrewd, vigilant, watchful, with eyes peeled. *Opp* INATTENTIVE.

observation *n* **1** attention (to), examination, inspection, monitoring, scrutiny, study, surveillance, viewing, watching. **2** comment, note, opinion, reaction, reflection, remark, response, sentiment, statement, thought, utterance.

observe *vb* **1** consider, contemplate, detect, discern, examine, *inf* keep an eye on, look at, monitor, note, notice, perceive, regard, scrutinize, see, spot, spy, stare at, study, view, watch, witness. **2** *observe rules.* abide by, adhere to, comply with, conform to, follow, heed, honour, keep, obey, pay attention to, respect. **3** *observe Easter.* celebrate, commemorate, keep, mark, recognize, remember, solemnize. **4** *observed that it was badly acted.* animadvert (on), comment, declare, explain, make an observation, mention, reflect, remark, say, state.

observer *n* beholder, bystander, commentator, eyewitness, looker-on, onlooker, spectator, viewer, watcher, witness.

obsess *vb* become an obsession with, bedevil, consume, control, dominate, grip, haunt, monopolize, plague, possess, preoccupy, rule, take hold of.

obsession *n* addiction, *inf* bee in your bonnet, conviction, fetish, fixation, *inf* hang-up, *inf* hobby-horse, *Fr* idée fixe, infatuation, mania, passion, phobia, preoccupation, *sl* thing.

obsessive *adj* addictive, compulsive, consuming, controlling, dominating, haunting, passionate.

obsolescent *adj* ageing, aging, declining, dying out, fading, going out of use, losing popularity, moribund, *inf* on the way out, waning.

obsolete *adj* anachronistic, antiquated, antique, archaic, dated, dead, discarded, disused, extinct, old-fashioned, *inf* old hat, outdated, out-of-date, outmoded, passé, primitive, superannuated, superseded, unfashionable. ▷ OLD. *Opp* CURRENT.

obstacle *n* bar, barricade, barrier, block, blockage, catch, check, difficulty, hindrance, hurdle, impediment, obstruction, problem, restriction, snag, *inf* stumbling-block.

obstinate *adj* adamant, *sl* bloody-minded, defiant, determined, dogged, firm, headstrong, immovable, inflexible, intractable, intransigent, *inf* mulish, obdurate, persistent, pertinacious, perverse, *inf* pig-headed, refractory, resolute, rigid, self-willed, single-minded, *inf* stiff-necked, stubborn, tenacious, uncooperative, unreasonable, unyielding, wilful, wrong-headed. *Opp* AMENABLE.

obstreperous *adj* awkward, boisterous, disorderly, irrepressible, naughty, rough, rowdy, *inf* stroppy, turbulent, uncontrollable, undisciplined, unmanageable, unruly, vociferous, wild. ▷ NOISY. *Opp* WELL-BEHAVED.

obstruct *vb* arrest, bar, block, bring to a standstill, check, curb, delay, deter, frustrate, halt, hamper, hinder, hold up, impede, inhibit, interfere with, interrupt, occlude, prevent, restrict, retard, slow down, stand in the way of, *inf* stonewall, stop, *inf* stymie, thwart. *Opp* HELP.

obtain *vb* **1** acquire, attain, be given, bring, buy, capture, come by, come into possession of, earn, elicit, enlist (*help*), extort, extract, find, gain, get, get hold of, *inf* lay your hands on, *inf* pick up, procure, purchase, receive, secure, seize, take possession of, win. **2** *rules still obtain.* apply, be in force, be in use, be relevant, be valid, exist, prevail, stand.

obtrusive *adj* blatant, conspicuous, forward, importunate, inescapable, interfering, intrusive, meddling, meddlesome, noticeable, out of place, prominent, unwanted, unwelcome. ▷ OBVIOUS. *Opp* INCONSPICUOUS.

obtuse *adj* dense, dull, imperceptive, slow, slow-witted. ▷ STUPID. *Opp* CLEVER.

obviate *vb* avert, forestall, make unnecessary, preclude, prevent, remove, take away.

obvious *adj* apparent, bald, blatant, clear, clear-cut, conspicuous, distinct, evident, eye-catching, flagrant, glaring, gross, inescapable, intrusive, manifest, notable, noticeable, obtrusive, open, overt, palpable, patent, perceptible, plain, prominent, pronounced, recognizable, self-evident, self-explanatory, straightforward, unconcealed, undisguised, undisputed, unmistakable, visible. *Opp* HIDDEN, OBSCURE.

occasion *n* **1** chance, circumstance, moment, occurrence, opportunity, time. **2** *no occasion for rudeness.* call, cause, excuse, grounds, justification, need, reason. **3** *a happy occasion.* affair, celebration, ceremony, event, function, *inf* get-together, happening, incident, occurrence, party.

occasional *adj* casual, desultory, fitful, infrequent, intermittent, irregular, odd, *inf* once in a while, periodic, random, rare, scattered, spasmodic, sporadic, uncommon, unpredictable. *Opp* FREQUENT, REGULAR.

occult *adj* ▷ SUPERNATURAL. • *n* black arts, black magic, cabbalism, diabolism, occultism, sorcery, the supernatural, witchcraft.

occupant *n* denizen, householder, incumbent, inhabitant, lessee, lodger, occupier, owner, resident, tenant.

occupation *n* **1** incumbency, lease, occupancy, possession, residency, tenancy, tenure, use. **2** appropriation, colonization, conquest, invasion, oppression, seizure, subjection, subjugation, suzerainty, *inf* takeover, usurpation. **3** appointment, business, calling, career, employment, job, *inf* line, métier, position, post, profession, situation, trade, vocation, work. **4** *leisure occupation.* activity, diversion, entertainment, hobby, interest, pastime, pursuit, recreation.

occupy *vb* 1 dwell in, inhabit, live in, move into, reside in, take up residence in, tenant. 2 *occupy space.* fill, take up, use, utilize. 3 capture, colonize, conquer, garrison, invade, overrun, possess, subjugate, take over, take possession of. 4 *occupy your time.* absorb, busy, divert, engage, engross, involve, preoccupy. **occupied** ▷ BUSY.

occur *vb* appear, arise, befall, be found, chance, come about, come into being, *old use* come to pass, *inf* crop up, develop, exist, happen, manifest itself, materialize, *inf* show up, take place, *inf* transpire, *inf* turn out, *inf* turn up.

occurrence *n* affair, case, circumstance, development, event, happening, incident, manifestation, matter, occasion, phenomenon, proceeding.

odd *adj* 1 *odd numbers.* uneven. *Opp* EVEN. 2 *an odd sock.* extra, left over, *inf* one-off, remaining, single, spare, superfluous, surplus, unmatched, unused. 3 *odd jobs.* casual, irregular, miscellaneous, occasional, part-time, random, sundry, varied, various. 4 *odd behaviour.* abnormal, anomalous, atypical, bizarre, *inf* cranky, curious, deviant, different, eccentric, exceptional, extraordinary, freak, funny, idiosyncratic, incongruous, inexplicable, *inf* kinky, outlandish, out of the ordinary, peculiar, puzzling, queer, rare, singular, strange, uncharacteristic, uncommon, unconventional, unexpected, unusual, weird. *Opp* NORMAL.

oddments *plur n* bits, bits and pieces, fragments, *inf* junk, leftovers, litter, odds and ends, offcuts, remnants, scraps, shreds, unwanted pieces.

odious *adj* detestable, execrable, loathsome, offensive, repugnant, repulsive. ▷ HATEFUL.

odorous *adj* fragrant, odoriferous, perfumed, scented. ▷ SMELLING.

odour *n* aroma, bouquet, fragrance, nose, redolence, scent, smell, stench, *inf* stink.

odourless *adj* deodorized, unscented. *Opp* ODOROUS.

offence *n* 1 breach, crime, fault, felony, infringement, lapse, malefaction, misdeed, misdemeanour, outrage, peccadillo, sin, transgression, trespass, violation, wrong, wrongdoing. 2 anger, annoyance, disgust, displeasure, hard feelings, indignation, irritation, pique, resentment, *inf* upset. **give offence** ▷ OFFEND.

offend *vb* 1 affront, anger, annoy, cause offence, chagrin, disgust, displease, embarrass, give offence, hurt your feelings, insult, irritate, make angry, *inf* miff, outrage, pain, provoke, *inf* put your back up, revolt, rile, sicken, slight, snub, upset, vex. 2 *offend against the law.* do wrong, transgress, violate. **be offended** be annoyed, *inf* take umbrage.

offender *n* criminal, culprit, delinquent, evil-doer, guilty party, lawbreaker, malefactor, miscreant, outlaw, sinner, transgressor, wrongdoer.

offensive *adj* 1 abusive, annoying, antisocial, coarse, detestable, disagreeable, disgusting, displeasing, disrespectful, embarrassing, foul, impolite, improper, indecent, insulting, loathsome, nasty, nauseating, nauseous, noxious, objectionable, obnoxious, *inf* off-putting, repugnant, revolting, rude, sickening, unpleasant, unsavoury, vile, vulgar. ▷ OBSCENE. *Opp* PLEASANT.

2 *offensive action*. aggressive, antagonistic, attacking, belligerent, hostile, threatening, warlike. *Opp* PEACEABLE. • *n* ▷ ATTACK.

offer *n* bid, proposal, proposition, suggestion, tender. • *vb* 1 bid, extend, give the opportunity of, hold out, make an offer of, make available, proffer, put forward, put up, suggest. 2 *offer to help*. come forward, propose, *inf* show willing, volunteer.

offering *n* contribution, donation, gift, oblation, offertory, present, sacrifice.

offhand *adj* 1 abrupt, aloof, careless, cavalier, cool, curt, offhanded, perfunctory, unceremonious, uncooperative, uninterested. ▷ CASUAL. 2 ▷ IMPROMPTU.

office *n* 1 bureau, room, workplace, workroom. 2 appointment, assignment, commission, duty, function, job, occupation, place, position,post, responsibility, role, situation, work.

officer *n* 1 adjutant, aide-de-camp, CO, commandant, commanding officer. 2 *police officer*. constable, PC, policeman, policewoman, WPC. 3 ▷ OFFICIAL.

official *adj* accredited, approved, authentic, authoritative, authorized, bona fide, certified, formal, lawful, legal, legitimate, licensed, organized, proper, recognized, true, trustworthy, valid. ▷ FORMAL. • *n* administrator, agent, appointee, authorized person, bureaucrat, dignitary, executive, functionary, mandarin, officer, organizer, representative, responsible person. □ *bailiff, captain, chief, clerk of court, commander, commissioner, consul, customs officer, director, elder (*of church*), equerry, governor, manager, marshal, mayor, mayoress, minister, monitor, ombudsman, overseer, prefect, president, principal, proctor, proprietor, registrar, sheriff, steward, superintendent, supervisor, usher.*

officiate *vb* adjudicate, be in charge, be responsible, chair (*a meeting*), conduct, have authority, manage, preside, referee, *inf* run (*a meeting*), umpire.

officious *adj inf* bossy, bumptious, *inf* cocky, dictatorial, forward, impertinent, interfering, meddlesome, meddling, overzealous, *inf* pushy, self-appointed, self-important.

offset *vb* cancel out, compensate for, counteract, counterbalance, make amends for, make good, make up for, redress. ▷ BALANCE.

offshoot *n* branch, by-product, derivative, development, *inf* spin-off, subsidiary product.

offspring *n* [*sing*] baby, child, descendant, heir, successor. [*plur*] brood, family, fry, issue, litter, progeny, *old use* seed, spawn, young.

often *adv* again and again, *inf* all the time, commonly, constantly, continually, frequently, generally, habitually, many times, regularly, repeatedly, time after time, time and again, usually.

oil *vb* grease, lubricate.

oily *adj* 1 buttery, fat, fatty, greasy, oleaginous. 2 *an oily manner*. ▷ OBSEQUIOUS.

ointment *n* balm, cream, embrocation, emollient, liniment, lotion, paste, salve, unguent.

old *adj* 1 ancient, antediluvian, antiquated, antique, crumbling, decayed, decaying, decrepit, dilapidated, early, historic, medieval, obsolete, primitive, quaint, ruined, superannuated, timeworn, venerable, veteran, vintage.

▷ OLD-FASHIONED. **2** *old times.* bygone, classical, forgotten, former, (*time*) immemorial, *old use* olden, past, prehistoric, previous, primeval, primitive, primordial, remote. **3** *old people.* advanced in years, aged, *inf* doddery, elderly, geriatric, *inf* getting on, grey-haired, hoary, *inf* in your dotage, long-lived, oldish, *inf* past it, senile. **4** *old customs.* age-old, enduring, established, lasting, long-standing, time-honoured, traditional, well-established. **5** *old clothes.* moth-eaten, ragged, scruffy, shabby, threadbare, worn, worn-out. **6** *old bread.* dry, stale. **7** *old tickets.* cancelled, expired, invalid, used. **8** *an old hand.* experienced, expert, familiar, mature, practised, skilled, veteran. *Opp* NEW, YOUNG. **old age** *inf* declining years, decrepitude, *inf* dotage, senility. **old person** centenarian, *inf* fogey, *inf* fogy, nonagenarian, octogenarian, pensioner, septuagenarian.

old-fashioned *adj* anachronistic, antiquated, archaic, backward-looking, conventional, dated, fusty, hackneyed, narrow-minded, obsolete, old, *inf* old hat, outdated, out-of-date, out-of-touch, outmoded, passé, pedantic, prim, proper, prudish, reactionary, time-honoured, traditional, unfashionable. *Opp* MODERN. **old-fashioned person** *inf* fogey, *inf* fogy, *inf* fuddy-duddy, pedant, reactionary, *inf* square.

omen *n* augury, auspice, foreboding, forewarning, harbinger, indication, portent, premonition, presage, prognostication, sign, token, warning, *inf* writing on the wall.

ominous *adj* baleful, dire, fateful, forbidding, foreboding, grim, ill-omened, ill-starred, inauspicious, lowering, menacing, portentous, prophetic, sinister, threatening, unfavourable, unlucky, unpromising, unpropitious, warning. *Opp* AUSPICIOUS.

omission *n* **1** deletion, elimination, exception, excision, exclusion. **2** failure, gap, neglect, negligence, oversight, shortcoming.

omit *vb* **1** cross out, cut, dispense with, drop, edit out, eliminate, erase, except, exclude, ignore, jump, leave out, miss out, overlook, pass over, reject, skip, strike out. **2** fail, forget, neglect.

omnipotent *adj* all-powerful, almighty, invincible, supreme, unconquerable.

oncoming *adj* advancing, approaching, facing, looming, nearing.

onerous *adj* burdensome, demanding, heavy, laborious, taxing. ▷ DIFFICULT.

one-sided *adj* **1** biased, bigoted, partial, partisan, prejudiced. **2** *one-sided game.* ill-matched, unbalanced, unequal, uneven.

onlooker *n* bystander, eye-witness, looker-on, observer, spectator, watcher, witness.

only *adj* lone, one, single, sole, solitary, unique. • *adv* barely, exclusively, just, merely, simply, solely.

ooze *vb* bleed, discharge, emit, exude, leak, secrete, seep, weep.

opaque *adj* cloudy, dark, dim, dull, filmy, hazy, impenetrable, muddy, murky, obscure, turbid, unclear. *Opp* CLEAR.

open *adj* **1** agape, ajar, gaping, unbolted, unfastened, unlocked, unsealed, unwrapped, wide, wide-open, yawning. **2** accessible, available, exposed, free, public, revealed, unenclosed, unprotected, unrestricted. **3** *open space*. bare, broad, clear, empty, extensive, spacious, treeless, uncrowded, undefended, unfenced, unobstructed, vacant. **4** *open arms*. extended, outstretched, spread out, unfolded. **5** *open nature*. artless, candid, communicative, flexible, frank, generous, guileless, honest, innocent, magnanimous, open-minded, responsive, sincere, straightforward, transparent, uninhibited. **6** *open defiance*. apparent, barefaced, blatant, conspicuous, downright, evident, flagrant, obvious, outspoken, overt, plain, unconcealed, undisguised, visible. **7** *an open question*. arguable, debatable, moot, problematical, unanswered, undecided, unresolved, unsettled. *Opp* CLOSED, HIDDEN. • *vb* **1** unbar, unblock, unbolt, unclose, uncork, undo, unfasten, unfold, unfurl, unlatch, unlock, unroll, unseal, untie, unwrap. **2** become open, gape, yawn. **3** *open proceedings*. activate, begin, commence, establish, *inf* get going, inaugurate, initiate, *inf* kick off, launch, set in motion, set up, start. *Opp* CLOSE.

opening *adj* first, inaugural, initial, introductory. *Opp* FINAL. • *n* **1** aperture, breach, break, chink, cleft, crack, crevice, cut, door, doorway, fissure, gap, gash, gate, gateway, hatch, hole, leak, mouth, orifice, outlet, rent, rift, slit, slot, space, split, tear, vent. **2** beginning, birth, commencement, dawn, inauguration, inception, initiation, launch, outset, start. **3** *a business opening*. *inf* break, chance, opportunity, way in.

operate *vb* **1** act, function, go, perform, run, work. **2** *operate a machine*. control, deal with, drive, handle, manage, use, work. **3** *operate on a patient*. do an operation, perform surgery.

operation *n* **1** control, direction, function, functioning, management, operating, performance, running, working. **2** action, activity, business, campaign, effort, enterprise, exercise, manoeuvre, movement, procedure, proceeding, process, project, transaction, undertaking, venture. **3** [*medical*] biopsy, surgery, transplant.

operational *adj* functioning, going, in operation, in use, in working order, operating, operative, running, *inf* up and running, usable, working.

operative *adj* **1** ▷ OPERATIONAL. **2** *the operative word*. crucial, important, key, principal, relevant, significant. • *n* ▷ WORKER.

opinion *n* assessment, attitude, belief, comment, conclusion, conjecture, conviction, estimate, feeling, guess, idea, impression, judgement, notion, perception, point of view, sentiment, theory, thought, view, viewpoint, way of thinking.

opponent *n* adversary, antagonist, challenger, competitor, contender, contestant, enemy, foe, opposer, opposition, rival. *Opp* ALLY.

opportune *adj* advantageous, appropriate, auspicious, beneficial, convenient, favourable, felicitous, fortunate, good, happy, lucky, propitious, right, suitable, timely, well-timed. *Opp* INCONVENIENT.

opportunity *n inf* break, chance, moment, occasion, opening, possibility, time.

oppose *vb* argue with, attack, be at variance with, be opposed to, challenge, combat, compete against, confront, contend with, contest, contradict, controvert, counter, counterattack, defy, disagree with, disapprove of, dissent from, face, fight, object to, obstruct, *inf* pit your wits against, quarrel with, resist, rival, stand up to, *inf* take a stand against, take issue with, withstand. *Opp* SUPPORT. **opposed** ▷ HOSTILE, OPPOSITE.

opposite *adj* **1** antithetical, conflicting, contradictory, contrasting, converse, different, hostile, incompatible, inconsistent, opposed, opposing, rival. **2** contrary, reverse. **3** *your opposite* number. corresponding, equivalent, facing, matching, similar. • *n* antithesis, contrary, converse, reverse.

opposition *n* antagonism, antipathy, competition, defiance, disapproval, enmity, hostility, objection, resistance, scepticism, unfriendliness. ▷ OPPONENT. *Opp* SUPPORT.

oppress *vb* abuse, afflict, burden, crush, depress, encumber, enslave, exploit, grind down, harass, intimidate, keep under, maltreat, overburden, persecute, pressurize, *inf* ride roughshod over, subdue, subjugate, terrorize, *inf* trample on, tyrannize, weigh down.

oppressed *adj* browbeaten, downtrodden, enslaved, exploited, misused, persecuted, subjugated, tyrannized.

oppression *n* abuse, despotism, enslavement, exploitation, harassment, injustice, maltreatment, persecution, pressure, subjection, subjugation, suppression, tyranny.

oppressive *adj* **1** brutal, cruel, despotic, harsh, repressive, tyrannical, undemocratic, unjust. **2** airless, close, heavy, hot, humid, muggy, stifling, stuffy, suffocating, sultry.

optimism *n* buoyancy, cheerfulness, confidence, hope, idealism, positiveness. *Opp* PESSIMISM.

optimistic *adj* buoyant, cheerful, confident, expectant, hopeful, idealistic, *inf* looking on the bright side, positive, sanguine. *Opp* PESSIMISTIC.

optimum *adj* best, finest, first-class, first-rate, highest, ideal, maximum, most favourable, perfect, prime, superlative, top.

option *n* alternative, chance, choice, election, possibility, selection.

optional *adj* avoidable, discretionary, dispensable, elective, inessential, possible, unnecessary, unforced, voluntary. *Opp* COMPULSORY.

oral *adj* by mouth, said, spoken, unwritten, uttered, verbal, vocal, voiced.

oratory *n* declamation, eloquence, enunciation, fluency, *inf* gift of the gab, grandiloquence, magniloquence, rhetoric, speaking, speech making.

orbit *n* circuit, course, path, revolution, trajectory. • *vb* circle, encircle, go round, travel round.

orbital *adj* circular, encircling.

orchestrate *vb* **1** arrange, compose. **2** ▷ ORGANIZE.

ordeal *n* affliction, anguish, difficulty, distress, hardship, misery, *inf* nightmare, pain, suffering, test, torture, trial, tribulation, trouble.

order *n* **1** arrangement, array, classification, codification, disposition, lay-out, *inf* line-up, neatness, organization, pattern, progression, sequence, series, succession, system, tidiness. **2** calm, control, discipline, good behaviour, government, harmony, law and order, obedience, orderliness, peace, peacefulness, quiet, rule. **3** *social orders.* caste, category, class, degree, group, hierarchy, kind, level, rank, sort, status. **4** *in good order.* condition, repair, state. **5** *orders to be obeyed.* command, decree, direction, directive, edict, fiat, injunction, instruction, law, mandate, ordinance, regulation, requirement, rule. **6** *an order for goods.* application, booking, commission, demand, mandate, request, requisition, reservation. **7** *religious orders.* association, brotherhood, community, fraternity, group, guild, lodge, sect, sisterhood, society, sodality, sorority. *Opp* DISORDER. • *vb* **1** arrange, categorize, classify, codify, lay out, organize, put in order, sort out, tidy up. **2** *old use* bid, charge, command, compel, decree, demand, direct, enjoin, instruct, ordain, require, tell. **3** apply for, ask for, book, reserve, requisition, send away for.

orderly *adj* **1** careful, methodical, neat, organized, regular, symmetrical, systematic, tidy, well-arranged, well-organized, well-prepared. *Opp* CONFUSED, DISORGANIZED. **2** civilized, controlled, decorous, disciplined, law-abiding, peaceable, polite, restrained, well-behaved, well-mannered. *Opp* UNDISCIPLINED.

ordinary *adj* accustomed, average, common, *inf* common or garden, commonplace, conventional, customary, established, everyday, fair, familiar, habitual, humble, *inf* humdrum, indifferent, mediocre, medium, middling, moderate, modest, mundane, nondescript, normal, orthodox, passable, pedestrian, plain, prosaic, quotidian, reasonable, regular, routine, *inf* run-of-the-mill, satisfactory, simple, *inf* so-so, standard, stock, traditional, typical, undistinguished, unexceptional, unexciting, unimpressive, uninspired, uninteresting, unpretentious, unremarkable, unsurprising, usual, well-known, workaday. *Opp* EXTRAORDINARY.

organic *adj* **1** animate, biological, growing, live, living, natural. **2** *an organic whole.* coherent, coordinated, evolving, integral, integrated, methodical, organized, structured, systematic.

organism *n* animal, being, cell, creature, living thing, plant.

organization *n* **1** arrangement, categorization, classification, codification, composition, coordination, logistics, organizing, planning, regimentation, *inf* running, structuring. **2** *a business organization.* alliance, association, body, business, club, combine, company, concern, confederation, conglomerate, consortium, corporation, federation, firm, group, institute, institution, league, network, *inf* outfit, party, society, syndicate, union.

organize *vb* **1** arrange, catalogue, categorize, classify, codify, compose, coordinate, group, order, *inf* pigeon-hole, put in order, rearrange, regiment, *inf* run, sort, sort out, structure, systematize, tabulate, tidy up. *Opp* JUMBLE. **2** build, coordinate, create, deal with, establish, make arrangements for, manage, mobilize, orchestrate, plan, put together, run, *inf* see to, set up, take care of.

organized ▷ OFFICIAL, SYSTEMATIC.

orgy *n* Bacchanalia, *inf* binge, debauch, *inf* fling, party, *inf* rave-up, revel, revelry, Saturnalia, *inf* spree.

orient *vb* acclimatize, accommodate, accustom, adapt, adjust, condition, familiarize, orientate, position.

oriental *adj* Asiatic, eastern, far-eastern.

origin *n* **1** base, basis, beginning, birth, cause, commencement, cradle, creation, dawn, derivation, foundation, fount, fountainhead, genesis, inauguration, inception, launch, outset, provenance, root, source, start, well-spring. *Opp* END. **2** *humble origins.* ancestry, background, descent, extraction, family, genealogy, heritage, lineage, parentage, pedigree, start in life, stock.

original *adj* **1** aboriginal, archetypal, earliest, first, initial, native, primal, primitive, primordial. **2** *original antiques.* actual, authentic, genuine, real, true, unique. **3** *original ideas.* creative, first-hand, fresh, imaginative, ingenious, innovative, inspired, inventive, new, novel, resourceful, thoughtful, unconventional, unfamiliar, unique, unusual. *Opp* HACKNEYED.

originate *vb* **1** arise, be born, be derived, be descended, begin, come, commence, crop up, derive, emanate, emerge, issue, proceed, spring up, start, stem. **2** beget, be the inventor of, bring about, coin, conceive, create, design, discover, engender, found, give birth to, inaugurate, initiate, inspire, institute, introduce, invent, launch, mastermind, pioneer, produce, think up.

ornament *n* accessory, adornment, bauble, beautification, decoration, embellishment, embroidery, enhancement, filigree, frill, frippery, garnish, gewgaw, ornamentation, tracery, trimming, trinket. ▷ JEWELLERY. • *vb* adorn, beautify, deck, decorate, dress up, elaborate, embellish, emblazon, emboss, embroider, enhance, festoon, garnish, prettify, trim.

ornamental *adj* attractive, decorative, fancy, flashy, pretty, showy.

ornate *adj* arabesque, baroque, *inf* busy, decorated, elaborate, fancy, florid, flamboyant, flowery, fussy, luxuriant, ornamented, overdone, pretentious, rococo. *Opp* PLAIN.

orphan *n* foundling, stray, waif.

orthodox *adj* accepted, accustomed, approved, authorized, common, conformist, conservative, conventional, customary, established, mainstream, normal, official, ordinary, prevailing, recognized, regular, standard, traditional, usual, well-established. *Opp* UNCONVENTIONAL.

ostensible *adj* alleged, apparent, outward, pretended, professed, *inf* put-on, reputed, specious, supposed, visible. *Opp* REAL.

ostentation *n* affectation, display exhibitionism, flamboyance, *inf* flashiness, flaunting, parade, pretention, pretentiousness, self-advertisement, show, showing-off, *inf* swank. *Opp* MODESTY.

ostentatious *adj* flamboyant, *inf* flashy, pretentious, showy, *inf* swanky, vainglorious. ▷ BOASTFUL. *Opp* MODEST.

ostracize *vb* avoid, banish, *inf* black, blackball, blacklist, boycott, cast out, cold-shoulder, *inf* cut, *inf* cut dead, exclude, excommunicate, expel, isolate, reject, *inf* send to Coventry, shun, shut out, snub. *Opp* BEFRIEND.

oust *vb* banish, drive out, eject, expel, *inf* kick out, remove, replace, *inf* sack, supplant, take over from, unseat.

outbreak *n* epidemic, *inf* flare-up, plague, rash, upsurge.

outburst *n* attack, effusion, eruption, explosion, fit, flood, outbreak, outpouring, paroxysm, rush, spasm, surge, upsurge.

outcast *n* castaway, displaced person, exile, leper, outlaw, outsider, pariah, refugee, reject, untouchable.

outcome *n* conclusion, consequence, effect, end-product, result, sequel, upshot.

outcry *n* cry of disapproval, dissent, hue and cry, objection, opposition, protest, protestation, remonstrance.

outdo *vb* beat, defeat, exceed, excel, *inf* get the better of, outbid, outdistance, outrun, outshine, outstrip, outweigh, overcome, surpass, top, trump.

outdoor *adj* alfresco, open-air, out of doors, outside.

outer *adj* **1** exterior, external, outside, outward, superficial, surface. **2** distant, further, outlying, peripheral, remote. *Opp* INNER.

outfit *n* **1** accoutrements, attire, costume, ensemble, equipment, garb, *inf* gear, *inf* get-up, *inf* rig, suit, trappings, *inf* turn-out. **2** ▷ ORGANIZATION.

outgoing *adj* **1** *outgoing president.* departing, emeritus, ex-, former, last, leaving, past, retiring. **2** *outgoing tide.* ebbing, falling, retreating. **3** ▷ SOCIABLE. *Opp* INCOMING. **outgoings** ▷ EXPENSE.

outing *n* excursion, expedition, jaunt, picnic, ride, tour, trip.

outlast *vb* outlive, survive.

outlaw *n* bandit, brigand, criminal, deserter, desperado, fugitive, highwayman, marauder, outcast, renegade, robber. • *vb* ban, exclude, forbid, prohibit, proscribe. ▷ BANISH.

outlet *n* **1** channel, discharge, duct, egress, escape route, exit, mouth, opening, orifice, safety valve, vent, way out. **2** ▷ SHOP.

outline *n* **1** abstract, *inf* bare bones, diagram, digest, draft, framework, plan, précis, résumé, *inf* rough idea, *inf* rundown, scenario, skeleton, sketch, summary, synopsis, thumbnail sketch. **2** contour, figure, form, profile, shadow, shape, silhouette. • *vb* delineate, draft, give the outline, give the gist of, plan out, précis, rough out, sketch out, summarize.

outlook *n* **1** aspect, panorama, scene, sight, vantage point, view, vista. **2** *your mental outlook.* angle, attitude, frame of mind, opinion, perspective, point of view, position, slant, standpoint, viewpoint. **3** *the weather outlook.* expectations, forecast, *inf* look-out, prediction, prognosis, prospect.

outlying *adj* distant, far-away, far-flung, far-off, outer, outermost, remote. *Opp* CENTRAL.

output *n* achievement, crop, harvest, production, productivity, result, yield.

outrage *n* **1** atrocity, crime, *inf* disgrace, enormity, indignity, outrageous act, scandal, *inf* sensation, violation. **2** anger, bitterness, disgust, fury, horror, indignation, resentment, revulsion, shock, wrath. • *vb* ▷ ANGER.

outrageous *adj* 1 abominable, atrocious, barbaric, beastly, bestial, criminal, cruel, disgraceful, disgusting, execrable, infamous, iniquitous, monstrous, nefarious, notorious, offensive, preposterous, revolting, scandalous, shocking, unspeakable, unthinkable, vile, villainous, wicked. 2 *outrageous prices.* excessive, extortionate, extravagant, immoderate, unreasonable. *Opp* REASONABLE.

outside *adj* 1 exterior, external, facing, outer, outward, superficial, surface, visible. 2 *outside interference.* alien, extraneous, foreign. 3 *outside chance.* ▷ REMOTE. • *n* appearance, case, casing, exterior, façade, face, front, look, shell, skin, surface.

outsider *n* alien, foreigner, *inf* gatecrasher, guest, immigrant, interloper, intruder, invader, newcomer, non-member, non-resident, outcast, stranger, trespasser, visitor.

outskirts *plur n* borders, edge, environs, fringe, margin, outer areas, periphery, purlieus, suburbs. *Opp* CENTRE.

outspoken *adj* blunt, candid, direct, explicit, forthright, frank, plain-spoken, tactless, unambiguous, undiplomatic, unequivocal, unreserved. ▷ HONEST. *Opp* EVASIVE.

outstanding *adj* 1 above the rest, celebrated, conspicuous, distinguished, dominant, eminent, excellent, exceptional, extraordinary, first-class, first-rate, great, important, impressive, memorable, notable, noteworthy, noticeable, predominant, pre-eminent, prominent, remarkable, singular, special, striking, superior, top rank, unrivalled. ▷ FAMOUS. *Opp* ORDINARY. 2 ▷ OVERDUE.

outward *adj* apparent, evident, exterior, external, manifest, noticeable, observable, obvious, ostensible, outer, outside, superficial, surface, visible.

outwit *vb* deceive, dupe, fool, *inf* get the better of, gull, hoax, hoodwink, make a fool of, outfox, outmanoeuvre, *inf* outsmart, *inf* put one over on, *inf* take in, trick. ▷ CHEAT.

oval *adj* egg-shaped, ellipsoidal, elliptical, oviform, ovoid.

ovation *n* acclaim, acclamation, applause, cheering, plaudits, praise.

overcast *adj* black, clouded, cloudy, dark, dismal, dull, gloomy, grey, leaden, lowering, murky, sombre, starless, stormy, sunless, threatening. *Opp* CLOUDLESS.

overcoat *n* greatcoat, mackintosh, top-coat, trench-coat.

overcome *adj* at a loss, beaten, *inf* bowled over, *inf* done in, exhausted, overwhelmed, prostrate, speechless. • *vb* ▷ OVERTHROW.

overcrowded *adj* congested, crammed, crawling, filled to capacity, full, jammed, *inf* jampacked, overloaded, packed.

overdue *adj* 1 belated, delayed, late, slow, tardy, unpunctual. *Opp* EARLY. 2 *overdue bills.* due, outstanding, owing, unpaid, unresolved, unsettled.

overeat *vb* be greedy, eat too much, feast, gorge, gormandize, *inf* guzzle, indulge yourself, *inf* make a pig of yourself, overindulge, *inf* stuff yourself.

overflow *vb* brim over, flood, pour over, run over, spill, well up.

overgrown *adj* **1** outsize, oversized. ▷ BIG. **2** *overgrown garden.* overrun, rank, tangled, uncut, unkempt, untidy, untrimmed, unweeded, weedy, wild.

overhang *vb* beetle, bulge, jut, project, protrude, stick out.

overhaul *vb* **1** check over, examine, *inf* fix, inspect, mend, rebuild, recondition, refurbish, renovate, repair, restore, service. **2** ▷ OVERTAKE.

overhead *adj* aerial, elevated, high, overhanging, raised, upper.

overlook *vb* **1** fail to notice, forget, leave out, miss, neglect, omit. **2** condone, disregard, excuse, gloss over, ignore, let pass, make allowances for, pardon, pass over, pay no attention to, *inf* shut your eyes to, *inf* turn a blind eye to, *inf* write off. **3** *overlook a lake.* face, front, have a view of, look at, look down on, look on to.

overpower *vb* ▷ OVERTHROW.

overpowering *adj* compelling, consuming, inescapable, insupportable, irrepressible, irresistible, overriding, overwhelming, powerful, strong, unbearable, uncontrollable, unendurable.

oversee *vb* administer, be in charge of, control, direct, invigilate, *inf* keep an eye on, preside over, superintend, supervise, watch over.

oversight *n* **1** carelessness, dereliction of duty, error, failure, fault, mistake, omission. **2** *oversight of a job.* administration, control, direction, management, supervision, surveillance.

overstate *vb inf* blow up out of proportion, embroider, exaggerate, magnify, make too much of, maximize, overemphasize, overstress.

overt *adj* apparent, blatant, clear, evident, manifest, obvious, open, patent, plain, unconcealed, undisguised, visible. *Opp* SECRET.

overtake *vb* catch up with, gain on, leave behind, outdistance, outpace, outstrip, overhaul, pass.

overthrow *n* conquest, defeat, destruction, mastery, rout, subjugation, suppression, unseating. • *vb* beat, bring down, conquer, crush, deal with, defeat, depose, dethrone, get the better of, *inf* lick, master, oust, overcome, overpower, overturn, overwhelm, rout, *inf* send packing, subdue, *inf* topple, triumph over, unseat, vanquish, win against.

overtone *n* association, connotation, hint, implication, innuendo, reverberation, suggestion, undertone.

overturn *vb* **1** capsize, flip, invert, keel over, knock over, spill, tip over, topple, turn over, *inf* turn turtle, turn upside down, up-end, upset. **2** ▷ OVERTHROW.

overwhelm *vb* **1** engulf, flood, immerse, inundate, submerge, swamp. **2** ▷ OVERTHROW. **overwhelming** ▷ OVERPOWERING.

owe *vb* be in debt, have debts.

owing *adj* due, outstanding, overdue, owed, payable, unpaid, unsettled. **owing to** because of, caused by, on account of, resulting from, thanks to, through.

own *vb* be the owner of, have, hold, possess. **own up** ▷ CONFESS.

owner *n* freeholder, holder, landlady, landlord, possessor, proprietor.

P

pace *n* **1** step, stride. **2** *a fast pace.* gait, *inf* lick, movement, quickness, rate, speed, tempo, velocity. • *vb* ▷ WALK.

pacify *vb* appease, assuage, calm, conciliate, humour, mollify, placate, quell, quieten, soothe, subdue, tame, tranquillize. *Opp* ANGER.

pack *n* **1** bale, box, bundle, package, packet, parcel. **2** backpack, duffel bag, haversack, kitbag, knapsack, rucksack. **3** ▷ GROUP. • *vb* **1** bundle (up), fill, load, package, parcel up, put, put together, store, stow, wrap up. **2** compress, cram, crowd, jam, overcrowd, press, ram, squeeze, stuff, tamp down, wedge. **pack off** ▷ DISMISS. **pack up** ▷ FINISH.

pact *n* agreement, alliance, armistice, arrangement, bargain, compact, concord, concordat, contract, covenant, deal, *Fr* entente, league, peace, settlement, treaty, truce, understanding.

pad *n* **1** cushion, filler, hassock, kneeler, padding, pillow, stuffing, wad. **2** jotter, memo pad, notebook, stationery, writing pad. • *vb* cushion, fill, line, pack, protect, stuff, upholster, wad. **pad out** ▷ EXTEND.

padding *n* **1** filling, protection, upholstery, stuffing, wadding. **2** prolixity, verbiage, verbosity, *inf* waffle, wordiness.

paddle *n* oar, scull. • *vb* **1** propel, row, scull. **2** dabble, splash about, wade.

paddock *n* enclosure, field, meadow, pasture.

pagan *adj* atheistic, godless, heathen, idolatrous, infidel, irreligious, polytheistic, unchristian. • *n* atheist, heathen, infidel, savage, unbeliever.

page *n* **1** folio, leaf, recto, sheet, side, verso. **2** errand-boy, messenger, page-boy.

pageant *n* ceremony, display, extravaganza, parade, procession, show, spectacle, tableau.

pageantry *n* ceremony, display, formality, grandeur, magnificence, pomp, ritual, show, spectacle, splendour.

pain *n* ache, aching, affliction, agony, anguish, cramp, crick, discomfort, distress, headache, hurt, irritation, ordeal, pang, smart, smarting, soreness, spasm, stab, sting, suffering, tenderness, throb, throes, toothache, torment, torture, twinge. • *vb* ▷ HURT.

painful *adj* **1** aching, *inf* achy, agonizing, burning, cruel, excruciating, *old use* grievous, hard to bear, hurting, inflamed, piercing, raw, severe, sharp, smarting, sore, *inf* splitting (*head*), stabbing, stinging, tender, throbbing. **2** distressing, harrowing, hurtful, laborious, *inf* traumatic, trying, unpleasant, upsetting, vexing. **3** *a painful decision.* difficult, hard, troublesome, uncongenial. *Opp* PAINLESS. **be painful** ▷ HURT.

painkiller *n* anaesthetic, analgesic, anodyne, palliative, sedative.

painless *adj* comfortable, easy, effortless, pain-free, simple, trouble-free, undemanding. *Opp* PAINFUL.

paint *n* colour, colouring, dye, pigment, stain, tint. □ *distemper, emulsion, enamel, gloss paint, lacquer, matt paint, oil-colour, oil-paint, oils, pastel, primer, tempera, under-*

coat, varnish, water-colour, whitewash. • *vb* **1** apply paint to, coat, colour, cover, daub, decorate, dye, enamel, gild, lacquer, redecorate, stain, tint, touch up, varnish, whitewash. **2** delineate, depict, describe, picture, portray, represent.

painter *n* artist, decorator, illustrator, miniaturist.

painting *n* fresco, landscape, miniature, mural, oil-painting, portrait, still-life, water-colour.

pair *n* brace, couple, duet, duo, mates, partners, partnership, set of two, twins, twosome. • *vb* **pair off**, **pair up** couple, double up, find a partner, get together, join up, *inf* make a twosome, match up, *inf* pal up, team up.

palace *n* castle, château, mansion, official residence, stately home. ▷ HOUSE.

palatable *adj* acceptable, agreeable, appetizing, easy to take, eatable, edible, nice to eat, pleasant, tasty. *Opp* UNPALATABLE.

palatial *adj* aristocratic, grand, large-scale, luxurious, majestic, opulent, *inf* posh, splendid, stately, up-market.

pale *adj* **1** anaemic, ashen, blanched, bloodless, cadaverous, colourless, corpse-like, *inf* deathly, drained, etiolated, ghastly, ghostly, ill-looking, pallid, pasty, *inf* peaky, sallow, sickly, unhealthy, wan, *inf* washed-out, *inf* whey-faced, white, whitish. **2** *pale colours.* bleached, dim, faded, faint, light, pastel, subtle, weak. *Opp* BRIGHT. • *vb* become pale, blanch, blench, dim, etiolate, fade, lighten, lose colour, whiten.

pall *n* cloth, mantle, shroud, veil. ▷ COVERING. • *vb* become boring, become uninteresting, cloy, irritate, jade, sate, satiate, weary.

palliative *adj* alleviating, calming, reassuring, sedative, soothing. • *n* painkiller, sedative, tranquillizer.

palpable *adj* apparent, corporeal, evident, manifest, obvious, patent, physical, real, solid, substantial, tangible, touchable, visible. *Opp* INTANGIBLE.

palpitate *vb* beat, flutter, pound, pulsate, quiver, shiver, throb, tremble, vibrate.

paltry *adj* contemptible, inconsequential, insignificant, petty, *inf* piddling, pitiable, puny, trifling, unimportant, worthless. ▷ SMALL. *Opp* IMPORTANT.

pamper *vb* coddle, cosset, humour, indulge, mollycoddle, overindulge, pet, spoil, spoonfeed.

pamphlet *n* booklet, brochure, bulletin, catalogue, circular, flyer, folder, handbill, handout, leaflet, notice, tract.

pan *n* container, utensil. □ *billycan, casserole, frying-pan, pot, saucepan, skillet.* • *vb* ▷ CRITICIZE.

panache *n* animation, brio, confidence, dash, élan, energy, enthusiasm, flair, flamboyance, flourish, savoir-faire, self-assurance, spirit, style, swagger, verve, zest.

pandemonium *n* babel, bedlam, chaos, confusion, hubbub, noise, rumpus, turmoil, uproar. ▷ COMMOTION.

pander *n* go-between, *inf* pimp, procurer. • *vb* **pander to** bow to, cater for, fulfil, gratify, humour, indulge, please, provide, satisfy.

pane *n* glass, light, panel, sheet of glass, window.

panel *n* **1** insert, pane, panelling, rectangle, *plur* wainscot. **2** committee, group, jury, team.

panic *n* alarm, consternation, *inf* flap, horror, hysteria, stampede, terror. • *vb* become panic-stricken, *inf* fall apart, *inf* flap, *inf* go to pieces, *inf* lose your head, *inf* lose your nerve, overreact, stampede. ▷ FEAR.

panic-stricken *adj* alarmed, *inf* beside yourself, disorientated, frantic, frenzied, horrified, hysterical, *inf* in a cold sweat, *inf* in a tizzy, jumpy, overexcited, panicky, panic-struck, terror-stricken, undisciplined, unnerved, worked-up. ▷ FRIGHTENED. *Opp* CALM.

panorama *n* landscape, perspective, prospect, scene, view, vista.

panoramic *adj* commanding, extensive, scenic, sweeping, wide.

pant *vb* blow, breathe quickly, gasp, *inf* huff and puff, puff, wheeze. **panting** ▷ BREATHLESS.

pants *n* **1** *old use* bloomers, boxer shorts, briefs, camiknickers, drawers, knickers, panties, pantihose, shorts, *inf* smalls, trunks, underpants, *inf* undies, Y-fronts. **2** ▷ TROUSERS.

paper *n* **1** folio, leaf, sheet. □ *A4 (A1, A2, etc), card, cardboard, cartridge paper, foolscap, manila, notepaper, papyrus, parchment, postcard, quarto, stationery, tissue-paper, toilet paper, tracing-paper, vellum, wallpaper, wrapping-paper, writing-paper.* **2** certificate, credentials, deed, document, form, *inf* ID, identification, licence, record. **3** *the daily paper. inf* daily, journal, newspaper, *inf* rag, tabloid. **4** *an academic paper.* article, discourse, dissertation, essay, monograph, thesis, treatise.

parable *n* allegory, exemplum, fable, moral tale. ▷ WRITING.

parade *n* cavalcade, ceremony, column, cortge, display, file, march-past, motorcade, pageant, procession, review, show, spectacle. • *vb* **1** assemble, file past, form up, line up, make a procession, march past, present yourself, process. ▷ WALK. **2** ▷ DISPLAY.

paradise *n* Eden, Elysium, heaven, Shangri-La, Utopia.

paradox *n* absurdity, anomaly, contradiction, incongruity, inconsistency, self-contradiction.

paradoxical *adj* absurd, anomalous, conflicting, contradictory, illogical, improbable, incongruous, self-contradictory.

parallel *adj* **1** equidistant. **2** *parallel events.* analogous, cognate, contemporary, corresponding, equivalent, like, matching, similar. • *n* **1** analogue, counterpart, equal, likeness, match. **2** analogy, comparison, correspondence, equivalence, kinship, resemblance, similarity. • *vb* be parallel to, be parallel with, compare with, correspond to, duplicate, echo, equate with, keep pace with, match, remind you of, run alongside.

paralyse *vb* anaesthetize, cripple deactivate, deaden, desensitize, disable, freeze, halt, immobilize, incapacitate, lame, numb, petrify, stop, stun.

paralysed *adj* crippled, dead, desensitized, disabled, handicapped, immobile, immovable, incapacitated, lame, numb, palsied, paralytic, paraplegic, rigid, unusable, useless.

paralysis *n* deadness, immobility numbness, palsy, paraplegia.

paraphernalia *plur n* accessories, apparatus, baggage, belongings, chattels, *inf* clobber, effects, equipment, gear, impedimenta, materials, *inf* odds and ends, possessions, property, *inf* rig, stuff, tackle, things, trappings.

paraphrase *vb* explain, interpret, put into other words, rephrase, restate, reword, rewrite, translate.

parcel *n* bale, box, bundle, carton, case, pack, package, packet. **parcel out** ▷ DIVIDE. **parcel up** ▷ PACK.

parch *vb* bake, burn, dehydrate, desiccate, dry, scorch, shrivel, wither. **parched** ▷ DRY, THIRSTY.

pardon *n* absolution, amnesty, condonation, discharge, exculpation, exoneration, forgiveness, indulgence, mercy, release, reprieve. • *vb* absolve, condone, exculpate, excuse, exonerate, forgive, free, grant pardon, let off, overlook, release, remit, reprieve, set free, spare.

pardonable *adj* allowable, condonable, excusable, forgivable, justifiable, minor, negligible, petty, understandable, venial (*sin*). *Opp* UNFORGIVABLE.

parent *n* begetter, father, guardian, mother, procreator, progenitor.

parentage *n* ancestry, birth, descent, extraction, family, line, lineage, pedigree, stock.

park *n* common, gardens, green, recreation ground. □ *amusement park, arboretum, botanical gardens, car-park, estate, national park, nature reserve, parkland, reserve, theme park.* • *vb* deposit, leave, place, position, put, station, store. **park yourself** ▷ SETTLE.

parliament *n* assembly, conclave, congress, convocation, council, diet, government, legislature, lower house, senate, upper house.

parody *n* burlesque, caricature, distortion, imitation, lampoon, mimicry, satire, *inf* send-up, *inf* spoof, *inf* take-off, travesty. • *vb* ape, burlesque, caricature, guy, imitate, lampoon, mimic, satirize, *inf* send up, *inf* take off, travesty. ▷ RIDICULE.

parry *vb* avert, block, deflect, evade, fend off, push away, repel, repulse, stave off, ward off.

part *n* **1** bit, branch, component, constituent, department, division, element, fraction, fragment, ingredient, parcel, particle, percentage, piece, portion, ramification, scrap, section, sector, segment, shard, share, single item, subdivision, unit. **2** department, faction, party, section, subdivision, unit. **3** *part of a book.* chapter, episode. **4** *part of a town.* area, district, neighbourhood, quarter, region, sector, vicinity. **5** *part of the body.* limb, member, organ. **6** *part in a play.* cameo, character, role. • *vb* **1** cut off, detach, disconnect, divide, pull apart, separate, sever, split, sunder. *Opp* JOIN. **2** break away, depart, go away, leave, part company, quit, say goodbye, separate, split up, take leave, withdraw. *Opp* MEET. **part with** ▷ RELINQUISH. **take part** ▷ PARTICIPATE.

partial *adj* **1** imperfect, incomplete, limited, qualified, unfinished. *Opp* COMPLETE. **2** *partial judge.* biased, one-sided, partisan, prejudiced, unfair. *Opp* IMPARTIAL. **be partial to** ▷ LIKE.

participate *vb* assist, be active, be involved, contribute, cooperate, engage, enter, help, join in, partake, share, take part.

participation *n* activity, assistance, complicity, contribution, cooperation, engagement, involvement, partnership, sharing.

particle *n* 1 bit, crumb, dot, drop, fragment, grain, hint, iota, jot, mite, morsel, *old use* mote, piece, scintilla, scrap, shred, sliver, *inf* smidgen, speck, trace. 2 atom, electron, molecule, neutron.

particular *adj* 1 distint, idiosyncratic, individual, peculiar, personal, singular, specific, uncommon, unique, unmistakable. 2 *particular with detail.* exact, nice, painstaking, precise, rigorous, scrupulous, thorough. 3 *gave particular pleasure.* especial, exceptional, important, marked, notable, noteworthy, outstanding, significant, special, unusual. 4 *particular about food.* choosy, critical, discriminating, fastidious, finical, finicky, fussy, meticulous, nice, *inf* pernickety, selective. *Opp* GENERAL, EASYGOING. **particulars** circumstances, details, facts, information, *sl* low-down.

parting *n* departure, farewell, going away, leave-taking, leaving, saying goodbye, separation, splitting up, valediction.

partisan *adj* biased, bigoted, blinkered, devoted, factional, fanatical, narrow-minded, one-sided, partial, prejudiced, sectarian, unfair. *Opp* IMPARTIAL. • *n* adherent, devotee, fanatic, follower, freedom fighter, guerrilla, resistance fighter, supporter, underground fighter, zealot.

partition *n* 1 break-up, division, separation, splitting up. 2 barrier, panel, room-divider, screen, wall. • *vb* cut up, divide, parcel out, separate off, share out, split up, subdivide.

partner *n* 1 accessory, accomplice, ally, assistant, associate, *inf* bedfellow, collaborator, colleague, companion, comrade, confederate, helper, *inf* mate, *sl* sidekick. 2 consort, husband, mate, spouse, wife.

partnership *n* 1 affiliation, alliance, association, combination, company, confederation, cooperative, syndicate. 2 collaboration, complicity, cooperation. 3 marriage, relationship, union.

party *n* 1 celebration, *inf* do, festivity, function, gathering, *inf* get-together, *inf* jollification, *inf* knees-up, merrymaking, *inf* rave-up, *inf* shindig, social gathering. □ *ball, banquet, barbecue, ceilidh, dance,* inf *disco, discothque, feast,* inf *hen-party, house-warming, orgy, picnic, reception, reunion,* inf *stag-party, tea-party, wedding.* 2 *a political party.* alliance, association, bloc, cabal, *inf* camp, caucus, clique, coalition, faction, junta, league, sect, side. ▷ GROUP.

pass *n* 1 canyon, col, cut, defile, gap, gorge, gully, opening, passage, ravine, valley, way through. 2 *identity pass.* authority, authorization, clearance, *inf* ID, licence, passport, permission, permit, safe-conduct, ticket, warrant. • *vb* 1 go beyond, go by, move on, move past, outstrip, overhaul, overtake, proceed, progress, *inf* thread your way. 2 *time passes.* disappear, elapse, fade, go away, lapse, tick by, vanish. 3 *pass drinks.* circulate, deal out, deliver, give, hand over, offer, present, share, submit supply, transfer. 4 *pass a resolution.* agree, approve, authorize, confirm, decree, enact, establish, ordain, pronounce, ratify, validate. 5 *I pass! inf* give in, opt out, say nothing, waive your rights. **pass away** ▷ DIE. **pass on**

▷ TRANSFER. **pass out** ▷ FAINT. **pass over** ▷ IGNORE.

passable *adj* **1** acceptable, adequate, admissible, allowable, all right, fair, indifferent, mediocre, middling, moderate, not bad, ordinary, satisfactory, *inf* so-so, tolerable. *Opp* UNACCEPTABLE. **2** clear, navigable, open, traversable, unblocked, unobstructed, usable. *Opp* IMPASSABLE.

passage *n* **1** corridor, entrance, hall, hallway, lobby, passageway, vestibule. **2** *passage of time.* advance, flow, lapse, march, movement, moving on, passing, progress, progression, transition. **3** *sea passage.* crossing, cruise, voyage. ▷ JOURNEY. **4** *through passage.* pass, route, thoroughfare, tunnel, way through. ▷ OPENING. **5** *passage from a book.* citation, episode, excerpt, extract, paragraph, part, piece, portion, quotation, scene, section, selection.

passenger *n* commuter, rider, traveller, voyager.

passer-by *n* bystander, onlooker, witness.

passion *n* appetite, ardour, avidity, avidness, commitment, craving, craze, desire, drive, eagerness, emotion, enthusiasm, fanaticism, fervency, fervour, fire, flame, frenzy, greed, heat, hunger, infatuation, intensity, keenness, love, lust, mania, obsession, strong feeling, suffering, thirst, urge, urgency, vehemence, zeal, zest.

passionate *adj* ardent, aroused, avid, burning, committed, eager, emotional, enthusiastic, excited, fanatical, fervent, fiery, frenzied, greedy, heated, hot, hungry, impassioned, infatuated, inflamed, intense, lustful, manic, obsessive, roused, sexy, strong, urgent, vehement, violent, worked up, zealous. *Opp* APATHETIC.

passive *adj* apathetic, complaisant, compliant, deferential, docile, impassive, inert, inactive, long-suffering, malleable, non-violent, patient, phlegmatic, pliable, quiescent, receptive, resigned, sheepish, submissive, supine, tame, tractable, unassertive, unmoved, unresisting, yielding. ▷ CALM. *Opp* ACTIVE.

past *adj* bygone, dead, *inf* dead and buried, earlier, ended, finished, forgotten, former, gone, historical, late, olden (*days*), *inf* over and done with, previous, recent, sometime. • *n* antiquity, days gone by, former times, history, old days, olden days, past times. *Opp* FUTURE.

paste *n* **1** adhesive, fixative, glue, gum. **2** pâté, spread. • *vb* fix, glue, stick. ▷ FASTEN.

pastiche *n* blend, composite, compound, *inf* hotchpotch, mess, miscellany, mixture, *inf* motley collection, patchwork, selection.

pastime *n* activity, amusement, avocation, distraction, diversion, entertainment, fun, game, hobby, leisure activity, occupation, play, recreation, relaxation, sport.

pastoral *adj* **1** agrarian, agricultural, Arcadian, bucolic, country, farming, idyllic, outdoor, provincial, rural, rustic. ▷ PEACEFUL. *Opp* URBAN. **2** *pastoral duties.* clerical, ecclesiastical, parochial, ministerial, priestly.

pasture *n* field, grassland, grazing, mead, meadow, paddock, pasturage.

pat *vb* caress, dab, slap, stroke, tap. ▷ TOUCH.

patch *n* darn, mend, piece, reinforcement, repair. • *vb* cover, darn, fix, mend, reinforce, repair, sew up, stitch up.

patchy *adj inf* bitty, blotchy, changing, dappled, erratic, inconsistent, irregular, speckled, spotty, uneven, unpredictable, variable, varied, varying. *Opp* UNIFORM.

patent *adj* apparent, blatant, clear, conspicuous, evident, flagrant, manifest, obvious, open, plain, self-evident, transparent, undisguised, visible.

path *n* **1** alley, bridle-path, bridleway, esplanade, footpath, footway, pathway, pavement, *Amer* sidewalk, towpath, track, trail, walk, walkway, way. ▷ ROAD. **2** approach, course, direction, flight path, orbit, route, trajectory, way.

pathetic *adj* **1** affecting, distressing, emotional, emotive, heartbreaking, heart-rending, lamentable, moving, piteous, pitiable, pitiful, plaintive, poignant, stirring, touching, tragic. ▷ SAD. **2** ▷ INADEQUATE.

pathos *n* emotion, feeling, pity, poignancy, sadness, tragedy.

patience *n* **1** calmness, composure, endurance, equanimity, forbearance, fortitude, leniency, long-suffering, resignation, restraint, self-control, serenity, stoicism, toleration, *inf* unflappability. **2** *work with patience*. assiduity, determination, diligence, doggedness, endurance, firmness, perseverance, persistence, pertinacity, *inf* stickability, tenacity.

patient *adj* **1** accommodating, acquiescent, calm, compliant, composed, docile, easygoing, even-tempered, forbearing, forgiving, lenient, long-suffering, mild, philosophical, quiet, resigned, self-possessed, serene, stoical, submissive, tolerant, uncomplaining. **2** *a patient worker*. assiduous, determined, diligent, dogged, persevering, persistent, steady, tenacious, unhurried, untiring. *Opp* IMPATIENT. • *n* case, invalid, outpatient, sufferer.

patriot *n derog* chauvinist, loyalist, nationalist, *derog* xenophobe.

patriotic *adj derog* chauvinistic, *derog* jingoistic, loyal, nationalistic, *derog* xenophobic.

patriotism *n derog* chauvinism, *derog* jingoism, loyalty, nationalism, *derog* xenophobia.

patrol *n* **1** beat, guard, policing, sentry-duty, surveillance, vigilance, watch. **2** guard, lookout, patrolman, sentinel, sentry, watchman. • *vb* be on patrol, defend, guard, inspect, keep a lookout, make the rounds, police, protect, stand guard, tour, walk the beat, watch over.

patron *n* **1** advocate, *inf* angel, backer, benefactor, champion, defender, helper, philanthropist, promoter, sponsor, subscriber, supporter. **2** *patron of a shop*. client, customer, frequenter, *inf* regular, shopper.

patronage *n* backing, business, custom, help, sponsorship, support, trade.

patronize *vb* **1** back, be a patron of, bring trade to, buy from, deal with, encourage, frequent, give patronage to, shop at, support. **2** *be patronizing towards*. humiliate, *inf* look down on, *inf* look down your nose at, *inf* put down, talk down to. **patronizing** ▷ SUPERIOR.

pattern *n* **1** arrangement, decoration, design, device, figuration, figure, motif, ornamentation, sequence, shape, system, tessellation. **2** archetype, criterion, example, exemplar, guide, ideal, model, norm, original, paragon, precedent, prototype, sample, specimen, standard, yardstick.

pause *n* break, *inf* breather, breathing space, caesura, check, delay, gap, halt, hesitation, hiatus, hold-up, interlude, intermission, interruption, interval, lacuna, lapse, *inf* let-up, lull, moratorium, respite, rest, standstill, stop, stoppage, suspension, wait. • *vb* break off, delay, falter, halt, hang back, have a pause, hesitate, hold, mark time, rest, stop, *inf* take a break, *inf* take a breather, wait.

pave *vb* asphalt, concrete, cover with paving, flag, *old use* macadamize, *inf* make up, surface, tarmac, tile. **pave the way** ▷ PREPARE.

pavement *n* footpath, *Amer* sidewalk. ▷ PATH.

pay *n* cash in hand, compensation, dividend, earnings, emoluments, fee, gain, honorarium, income, money, payment, profit, recompense, reimbursement, remittance, return, salary, settlement, stipend, take-home pay, wages. • *vb* **1** *inf* cough up, *inf* fork out, give, grant, hand over, proffer, recompense, remunerate, requite, spend, *inf* stump up. **2** *pay debts.* bear the cost of, clear, compensate, *inf* foot, honour, indemnify, meet, pay back, pay off, pay up, refund, reimburse, repay, settle. **3** *crime doesn't pay.* avail, benefit, be profitable, pay off, produce results, prove worthwhile, yield a return. **4** *pay for mistakes.* be punished, suffer. ▷ ATONE. **pay back** ▷ RETALIATE.

payment *n* advance, alimony, allowance, charge, commission, compensation, contribution, cost, deposit, disbursement, donation, expenditure, fare, fee, figure, fine, instalment, loan, outgoings, outlay, pocket-money, premium, price, ransom, rate, remittance, reward, royalty, *inf* sub, subscription, subsistence, supplement, surcharge, tip, toll, wage. *Opp* INCOME.

peace *n* **1** accord, agreement, amity, conciliation, concord, friendliness, harmony, order. *Opp* CONFLICT. **2** alliance, armistice, cease-fire, pact, treaty, truce. *Opp* WAR. **3** *peace of mind.* calm, calmness, peace and quiet, peacefulness, placidity, quiet, repose, serenity, silence, stillness, tranquillity. *Opp* ANXIETY.

peaceable *adj* amicable, civil, conciliatory, cooperative, friendly, gentle, harmonious, inoffensive, mild, non-violent, pacific, peace-loving, placid, temperate, understanding. *Opp* QUARRELSOME.

peaceful *adj* balmy, calm, easy, gentle, pacific, placid, pleasant, quiet, relaxing, restful, serene, slow-moving, soothing, still, tranquil, undisturbed, unruffled, untroubled. *Opp* NOISY, STORMY.

peacemaker *n* adjudicator, appeaser, arbitrator, conciliator, diplomat, intercessor, intermediary, mediator, reconciler, referee, umpire.

peak *n* **1** apex, brow, cap, crest, crown, eminence, hill, mountain, pinnacle, point, ridge, summit, tip, top. **2** *peak of your career.* acme, apogee, climax, consummation, crisis, crown, culmination, height, highest point, zenith.

peal *n* carillon, chime, chiming, clangour, knell, reverberation, ringing, tintinnabulation, toll. • *vb* chime, clang, resonate, ring, ring the changes, sound, toll.

peasant *n* [*derog*] boor, bumpkin, churl, oaf, rustic, serf, swain, village idiot, yokel.

pebbles *plur n* cobbles, gravel, stones.

peculiar *adj* **1** aberrant, abnormal, anomalous, atypical, bizarre, curious, deviant, eccentric, exceptional, freakish, funny, odd, offbeat, outlandish, out of the ordinary, quaint, queer, quirky, surprising, strange, uncommon, unconventional, unusual, weird. **2** *your peculiar style.* characteristic, different, distinctive, identifiable, idiosyncratic, individual, natural, particular, personal, private, singular, special, unique, unmistakable. *Opp* COMMON, ORDINARY.

peculiarity *n* abnormality, characteristic, difference, distinctiveness, eccentricity, foible, idiosyncrasy, individuality, mannerism, oddity, outlandishness, quirk, singularity, speciality, trait, uniqueness.

pedantic *adj* **1** academic, bookish, donnish, dry, formal, humourless, learned, old-fashioned, pompous, scholarly, schoolmasterly, stiff, stilted, *inf* stuffy. **2** *inf* by the book, doctrinaire, exact, fastidious, fussy, inflexible, *inf* nit-picking, precise, punctilious, strict, unimaginative. *Opp* INFORMAL, LAX.

peddle *vb inf* flog, hawk, market, *inf* push, sell, traffic in, vend.

pedestrian *adj* **1** pedestrianized, traffic-free. **2** banal, boring, dreary, dull, commonplace, flat-footed, lifeless, mundane, prosaic, run-of-the-mill, tedious, unimaginative, uninteresting. ▷ ORDINARY. • *n inf* foot-slogger, foot-traveller, stroller, walker.

pedigree *adj* pure-bred, thoroughbred. • *n* ancestry, blood, descent, extraction, family, family history, genealogy, line, lineage, parentage, roots, stock, strain.

pedlar *n old use* chapman, *inf* cheapjack, *old use* colporteur, door-to-door salesman, hawker, *inf* pusher, seller, street-trader, trafficker, vendor.

peel *n* coating, rind, skin. • *vb* denude, flay, hull, pare, skin, strip. ▷ UNDRESS.

peep *vb* **1** glance, have a look, peek, squint. ▷ LOOK. **2** ▷ SHOW.

peer *n* aristocrat, grandee, noble, nobleman, noblewoman, patrician, titled person. □ *baron, baroness, countess, duchess, duke, earl, lady, lord, marchioness, marquis, viscount, viscountess.* • *vb* have a look, look earnestly, spy, squint. ▷ LOOK. **peers** **1** aristocracy, nobility, peerage. **2** colleagues, compeers, confrères, equals, fellows, peer-group.

peevish *adj* cantankerous, churlish, crabby, crusty, curmudgeonly, grumpy, illhumoured, irritable, petulant, querulous, testy, touchy, waspish. ▷ BAD-TEMPERED.

peg *n* bolt, dowel, pin, rod, stick, thole-pin. • *vb* ▷ FASTEN.

pelt *n* coat, fur, hide, skin. • *vb* assail, bombard, shower, strafe. ▷ THROW.

pen *n* **1** coop, *Amer* corral, enclosure, fold, hutch, pound. **2** ball-point, biro, felt-tip, fountain pen, *old use* quill.

penalize *vb* discipline, fine, impose a penalty on, punish.

penalty *n* fine, forfeit, price. ▷ PUNISHMENT. **pay the penalty** ▷ ATONE.

penance *n* amends, atonement, contrition, penitence, punishment, reparation. **do penance** ▷ ATONE.

pendent *adj* dangling, hanging, loose, pendulous, suspended, swaying, swinging, trailing.

pending *adj* about to happen, forthcoming, *inf* hanging fire, imminent, impending, *inf* in the offing, undecided, waiting.

penetrate *vb* **1** bore through, break through, drill into, enter, get into, get through, infiltrate, lance, make a hole, perforate, pierce, probe, puncture, stab, stick in. **2** *damp penetrates.* filter through, impregnate, percolate through, permeate, pervade, seep into, suffuse.

penitent *adj* apologetic, conscience-stricken, contrite, regretful, remorseful, repentant, rueful, shamefaced, sorry. *Opp* UNREPENTANT.

pennon *n* banner, flag, pennant, standard, streamer.

pension *n* annuity, benefit, old age pension, superannuation.

pensive *adj* brooding, cogitating, contemplative, day-dreaming, *inf* far-away, *inf* in a brown study, lost in thought, meditative, reflective, ruminative, thoughtful.

penury *n* beggary, destitution, impoverishment, indigence, lack, need, poverty, scarcity, want.

people *n* **1** folk, human beings, humanity, humans, individuals, ladies and gentlemen, mankind, men and women, mortals, persons. **2** citizenry, citizens, community, electorate, *inf* grass roots, *Gr* hoi polloi, nation, *inf* the plebs, populace, population, the public, society, subjects. **3** *your own people.* clan, family, kinsmen, kith and kin, nation, race, relations, relatives, tribe. • *vb* colonize, fill, inhabit, occupy, overrun, populate, settle.

perceive *vb* **1** become aware of, catch sight of, descry, detect, discern, discover, distinguish, espy, glimpse, hear, identify, make out, notice, note, observe, recognize, see, spot. **2** appreciate, apprehend, comprehend, deduce, feel, *inf* figure out, gather, grasp, infer, know, realize, sense, understand.

perceptible *adj* appreciable, audible, detectable, discernible, distinct, distinguishable, evident, identifiable, manifest, marked, notable, noticeable, observable, obvious, palpable, perceivable, plain, recognizable, unmistakable, visible. *Opp* IMPERCEPTIBLE.

perception *n* appreciation, apprehension, awareness, cognition, comprehension, consciousness, discernment, insight, instinct, intuition, knowledge, observation, perspective, realization, recognition, sensation, sense, understanding, view.

perceptive *adj* acute, alert, astute, attentive, aware, clever, discerning, discriminating, observant, penetrating, percipient, perspicacious, quick, responsive, sensitive, sharp, sharp-eyed, shrewd, sympathetic, understanding. ▷ INTELLIGENT.

perch *n* rest, resting-place, roost. • *vb* balance, rest, roost, settle, sit.

percussion *n* □ *bell, castanets, celesta, chime bar, cymbal, glockenspiel, gong, kettledrum, maracas, rattle, tambourine, timpani, triangle, tubular bells, vibraphone, whip, wood block, xylophone.* ▷ DRUM.

perdition *n* damnation, doom, downfall, hell, hellfire, ruin, ruination.

perfect *adj* 1 absolute, complete, completed, consummate, excellent, exemplary, faultless, finished, flawless, ideal, immaculate, incomparable, matchless, mint, superlative, unbeatable, undamaged, unexceptionable, unqualified, whole. 2 blameless, irreproachable, pure, sinless, spotless, unimpeachable. 3 accurate, authentic, correct, exact, faithful, immaculate, impeccable, precise, tailor-made, true. *Opp* IMPERFECT. • *vb* bring to fruition, bring to perfection, carry through, complete, consummate, effect, execute, finish, fulfil, make perfect, realize, *inf* see through.

perfection *n* 1 beauty, completeness, excellence, faultlessness, flawlessness, ideal, precision, purity, wholeness. *Opp* IMPERFECTION. 2 *the perfection of a plan.* accomplishment, achievement, completion, consummation, end, fruition, fulfilment, realization.

perforate *vb* bore through, drill, penetrate, pierce, prick, punch, puncture, riddle.

perform *vb* 1 accomplish, achieve, bring about, carry on, carry out, commit, complete, discharge, dispatch, do, effect, execute, finish, fulfil, *inf* pull off. 2 behave, function, go, operate, run, work. 3 *perform on stage.* act, appear, dance, feature, figure, take part. 4 *perform a play, song.* enact, mount, present, play, produce, put on, render, represent, serenade, sing, stage.

performance *n* 1 accomplishment, achievement, carrying out, completion, doing, execution, fulfilment. 2 act, behaviour, conduct, deception, exhibition, exploit, feat, play-acting, pretence. 3 *stage performance.* acting, début, impersonation, interpretation, play, playing, portrayal, presentation, production, rendition, representation. □ *concert, dress rehearsal, first night, last night, matinée, première, preview, rehearsal, show, sketch, turn.*

performer *n* actor, actress, artist, artiste, player, singer, star, *inf* superstar, thespian, trouper. ▷ ENTERTAINER.

perfume *n* 1 aroma, bouquet, fragrance, odour, scent. ▷ SMELL. 2 after-shave, eau de Cologne, scent, toilet water.

perfunctory *adj* apathetic, automatic, brief, cursory, dutiful, fleeting, half-hearted, hurried, inattentive, indifferent, mechanical, offhand, routine, superficial, uncaring, unenthusiastic, uninterested, uninvolved, unthinking. *Opp* ENTHUSIASTIC.

perhaps *adv* conceivably, maybe, *old use* peradventure, *old use* perchance, possibly.

peril *n* danger, hazard, insecurity, jeopardy, risk, susceptibility, threat, vulnerability.

perilous *adj* dangerous, hazardous, insecure, risky, uncertain, unsafe, vulnerable. *Opp* SAFE.

perimeter *n* border, borderline, boundary, bounds, circumference, confines, edge, fringe, frontier, limit, margin, periphery, verge.

period *n* 1 duration, interval, phase, season, session, span, spell, stage, stint, stretch, term, while. 2 aeon, age, epoch, era. ▷ TIME.

periodic *adj* cyclical, intermittent, occasional, recurrent, repeated, spasmodic, sporadic.

peripheral *adj* 1 distant, on the perimeter, outer, outermost, outlying. 2 borderline, incidental, inessential, irrelevant, marginal, minor, nonessential, secondary, tangential, unimportant, unnecessary. *Opp* CENTRAL.

perish *vb* **1** be destroyed, be killed, die, expire, fall, lose your life, meet your death, pass away. **2** crumble away, decay, decompose, disintegrate, go bad, rot.

perishable *adj* biodegradable, destructible, liable to perish, unstable. *Opp* PERMANENT.

perjury *n* bearing false witness, lying, mendacity.

permanent *adj* abiding, ceaseless, changeless, chronic, constant, continual, continuous, durable, endless, enduring, eternal, everlasting, fixed, immutable, incessant, incurable, indestructible, indissoluble, ineradicable, interminable, invariable, irreparable, irreversible, lasting, lifelong, long-lasting, never-ending, non-stop, ongoing, perennial, perpetual, persistent, stable, steady, unalterable, unceasing, unchanging, undying, unending. *Opp* TEMPORARY.

permeate *vb* diffuse, filter through, flow through, impregnate, infiltrate, penetrate, percolate, pervade, saturate, soak through, spread through.

permissible *adj* acceptable, admissible, allowable, allowed, excusable, lawful, legal, legitimate, licit, permitted, proper, right, sanctioned, tolerable, valid, venial (*sin*). *Opp* UNACCEPTABLE.

permission *n* agreement, approbation, approval, aquiescence, assent, authority, authorization, consent, dispensation, franchise, *inf* go-ahead, *inf* green light, leave, licence, *inf* rubber stamp, sanction, seal of approval, stamp of approval, support. ▷ PERMIT.

permissive *adj* aquiescent, consenting, easygoing, indulgent, latitudinarian, lenient, liberal, libertarian, tolerant.

permit *n* authority, authorization, certification, charter, licence, order, pass, passport, ticket, visa, warrant. • *vb* admit, agree to, allow, approve of, authorize, consent to, endorse, enfranchise, give an opportunity for, give permission for, give your blessing to, legalize, license, make possible, sanction, *old use* suffer, support, tolerate.

perpendicular *adj* at right angles, erect, plumb, straight up and down, upright, vertical.

perpetual *adj* abiding, ageless, ceaseless, chronic, constant, continual, continuous, endless, enduring, eternal, everlasting, frequent, immortal, immutable, incessant, incurable, indestructible, ineradicable, interminable, invariable, lasting, long-lasting, never-ending, non-stop, ongoing, perennial, permanent, persistent, protracted, recurrent, recurring, repeated, *old use* sempiternal, timeless, unceasing, unchanging, undying, unending, unfailing, unremitting. *Opp* TEMPORARY.

perpetuate *vb* continue, eternalize, eternize, extend, immortalize, keep going, maintain, make permanent, preserve.

perplex *vb* baffle, *inf* bamboozle, befuddle, bewilder, confound, confuse, disconcert, distract, dumbfound, muddle, mystify, nonplus, puzzle, *inf* stump, *inf* throw, worry.

perquisite *n* benefit, bonus, *inf* consideration, emolument, extra, fringe benefit, gratuity, *inf* perk, tip.

persecute *vb* abuse, afflict, annoy, badger, bother, bully, discriminate against, harass, hector, hound, ill-treat, intimidate, maltreat, martyr, molest, oppress, pester, *inf* put the screws on, suppress, terrorize, torment, torture, trouble, tyrannize, victimize, worry.

persist *vb* be diligent, be steadfast, carry on, continue, endure, go on, *inf* hang on, hold out, *inf* keep at it, keep going, *inf* keep it up, keep on, last, linger, persevere, *inf* plug away, remain, *inf* soldier on, stand firm, stay, *inf* stick at it. *Opp* CEASE.

persistent *adj* **1** ceaseless, chronic, constant, continual, continuous, endless, eternal, everlasting, incessant, interminable, lasting, long-lasting, never-ending, obstinate, permanent, perpetual, persisting, recurrent, recurring, remaining, repeated, unending, unrelenting, unrelieved, unremitting. *Opp* BRIEF, INTERMITTENT. **2** assiduous, determined, dogged, hard-working, indefatigable, patient, persevering, pertinacious, relentless, resolute, steadfast, steady, stubborn, tenacious, tireless, unflagging, untiring, unwavering, zealous. *Opp* LAZY.

person *n* adolescent, adult, baby, being, *inf* body, character, child, *inf* customer, figure, human, human being, individual, infant, mortal, personage, soul, *inf* type, woman. ▷ MAN, PEOPLE, WOMAN.

persona *n* character, exterior, façade, guise, identity, image, part, personality, role, self-image.

personal *adj* **1** distinct, distinctive, exclusive, idiosyncratic, individual, inimitable, particular, peculiar, private, special, unique, your own. *Opp* GENERAL. **2** *a personal appearance.* actual, in person, in the flesh, live, physical. **3** *personal letters.* confidential, friendly, individual, informal, intimate, private, secret. *Opp* PUBLIC. **4** *personal friends.* bosom, close, dear, familiar, intimate, known. **5** *personal remarks.* belittling, critical, derogatory, disparaging, insulting, offensive, pejorative, rude, slighting, unfriendly. **6** *personal knowledge.* direct, empirical, experiential, firsthand.

personality *n* **1** attractiveness, character, charisma, charm, disposition, identity, individuality, magnetism, *inf* make-up, nature, persona, psyche, temperament. **2** *inf* big name, celebrity, idol, luminary, name, public figure, star, superstar.

personification *n* allegorical representation, embodiment, epitome, human likeness, incarnation, living image, manifestation.

personify *vb* allegorize, embody, epitomize, exemplify, give human shape to, incarnate, manifest, personalize, represent, stand for, symbolize, typify.

personnel *n* employees, manpower, people, staff, workforce, workers.

perspective *n* angle, approach, attitude, outlook, point of view, position, prospect, slant, standpoint, view, viewpoint.

persuade *vb* bring round, cajole, coax, convert, convince, entice, exhort, importune, induce, influence, inveigle, press, prevail upon, prompt, talk into, tempt, urge, use persuasion, wheedle (into), win over. *Opp* DISSUADE.

persuasion *n* **1** argument, blandishment, brainwashing, cajolery, coaxing, conditioning, enticement, exhortation, inducement, persuading, propaganda, reasoning. **2** affiliation, belief, conviction, creed, denomination, faith, religion, sect.

persuasive *adj* cogent, compelling, conclusive, convincing, credible, effective, efficacious, eloquent, forceful, influential, logical, plausible, potent, reasonable, sound, strong, telling, unarguable, valid, watertight. *Opp* UNCONVINCING.

pertain *vb* appertain, apply, be relevant, have bearing, have reference, have relevance, refer. **pertain to** affect, concern.

pertinent *adj* apposite, appropriate, apropos, apt, fitting, germane, relevant, suitable. *Opp* IRRELEVANT.

perturb *vb* agitate, alarm, bother, confuse, discompose, discomfit, disconcert, disquiet, distress, disturb, fluster, frighten, make anxious, ruffle, scare, shake, trouble, unnerve, unsettle, upset, vex, worry. *Opp* REASSURE.

peruse *vb* examine, inspect, look over, read, run your eye over, scan, scrutinize, study.

pervade *vb* affect, diffuse, fill, filter through, flow through, impregnate, penetrate, percolate, permeate, saturate, spread through, suffuse.

pervasive *adj* general, inescapable, insidious, omnipresent, penetrating, permeating, pervading, prevalent, rife, ubiquitous, universal, widespread.

perverse *adj* adamant, contradictory, contrary, disobedient, fractious, headstrong, illogical, inappropriate, inflexible, intractable, intransigent, obdurate, obstinate, peevish, *inf* pig-headed, rebellious, refractory, self-willed, stubborn, tiresome, uncooperative, unhelpful, unreasonable, wayward, wilful, wrong-headed. *Opp* REASONABLE.

perversion *n* **1** corruption, distortion, falsification, misrepresentation, misuse, twisting. **2** aberration, abnormality, depravity, deviance, deviation, immorality, impropriety, *inf* kinkiness, perversity, unnaturalness, vice, wickedness.

pervert *n* debauchee, degenerate, deviant, perverted person, profligate. • *vb* **1** bend, deflect, distort, divert, falsify, misrepresent, perjure, subvert, twist, undermine. **2** *pervert a witness.* bribe, corrupt, lead astray.

perverted *adj* abnormal, amoral, bad, corrupt, debauched, degenerate, depraved, deviant, dissolute, eccentric, evil, immoral, improper, *inf* kinky, profligate, sick, twisted, unnatural, unprincipled, warped, wicked, wrong. ▷ OBSCENE. *Opp* NATURAL.

pessimism *n* cynicism, despair, despondency, fatalism, gloom, hopelessness, negativeness, resignation, unhappiness. *Opp* OPTIMISM.

pessimistic *adj* bleak, cynical, defeatist, despairing, despondent, fatalistic, gloomy, hopeless, melancholy, morbid, negative, resigned, unhappy. ▷ SAD. *Opp* OPTIMISTIC.

pest *n* **1** annoyance, bane, bother, curse, irritation, nuisance, *inf* pain in the neck, *inf* thorn in your flesh, trial, vexation. **2** *inf* bug, *inf* creepy-crawly, insect, parasite, *plur* vermin.

pester *n* annoy, badger, bait, besiege, bother, *inf* get under someone's skin, harass, harry, *inf* hassle, irritate, molest, nag, nettle, plague, provoke, torment, trouble, worry.

pestilence *n* blight, curse, epidemic, pandemic, plague, scourge. ▷ ILLNESS.

pet *n inf* apple of your eye, darling, favourite, idol. • *vb* caress, cuddle, fondle, kiss, nuzzle, pat, stroke. ▷ TOUCH.

petition *n* appeal, application, entreaty, list of signatures, plea, request, solicitation, suit, supplication. • *vb* appeal to, call upon, deliver a petition to, entreat, importune, solicit, sue, supplicate. ▷ ASK.

petty *adj* **1** inconsequential, insignificant, minor, niggling, small, trivial, trifling. ▷ UNIMPORTANT. *Opp* IMPORTANT. **2** *petty complaints.* grudging, mean, nit-picking, small-minded, ungenerous. *Opp* GENEROUS.

phase *n* development, period, season, spell, stage, state, step. ▷ TIME. **phase in** ▷ INTRODUCE. **phase out** ▷ FINISH.

phenomenal *adj* amazing, astonishing, astounding, exceptional, extraordinary, *inf* fantastic, incredible, marvellous, *inf* mind-boggling, miraculous, notable, outstanding, prodigious, rare, remarkable, *inf* sensational, singular, staggering, stunning, unbelievable, unorthodox, unusual, *inf* wonderful. *Opp* ORDINARY.

phenomenon *n* **1** circumstance, event, experience, fact, happening, incident, occasion, occurrence, sight. **2** *an unusual phenomenon.* curiosity, marvel, miracle, phenomenal person, phenomenal thing, prodigy, rarity, sensation, spectacle, wonder.

philanthropic *adj* altruistic, beneficent, benevolent, bountiful, caring, charitable, generous, humane, humanitarian, magnanimous, munificent, public-spirited, ungrudging. ▷ KIND. *Opp* MISANTHROPIC.

philanthropist *n* altruist, benefactor, donor, giver, *inf* Good Samaritan, humanitarian, patron, provider, sponsor.

philistine *adj* boorish, ignorant, lowbrow, materialistic, uncivilized, uncultivated, uncultured, unenlightened, unlettered, vulgar.

philosopher *n* sage, student of philosophy, thinker.

philosophical *adj* **1** abstract, academic, analytical, erudite, esoteric, ideological, impractical, intellectual, learned, logical, metaphysical, rational, reasoned, scholarly, theoretical, thoughtful, wise. **2** calm, collected, composed, detached, equable, imperturbable, judicious, patient, reasonable, resigned, serene, sober, stoical, unemotional, unruffled. *Opp* EMOTIONAL.

philosophize *vb* analyse, moralize, pontificate, preach, rationalize, reason, sermonize, theorize, think things out.

philosophy *n* **1** epistemology, ideology, logic, metaphysics, rationalism, thinking. **2** *philosophy of life.* attitude, convictions, outlook, set of beliefs, tenets, values, viewpoint, wisdom.

phlegmatic *adj* apathetic, cold, cool, frigid, impassive, imperturbable, indifferent, lethargic, passive, placid, slow, sluggish, stoical, stolid, torpid, undemonstrative, unemotional, unenthusiastic, unfeeling, uninvolved, unresponsive. *Opp* EXCITABLE.

phobia *n* anxiety, aversion, dislike, dread, *inf* hang-up, hatred, horror, loathing, neurosis, obsession, repugnance, revulsion. ▷ FEAR. □ *agoraphobia* (open space), *arachnophobia* (spiders), *claustrophobia* (enclosed space), *xenophobia* (foreigners).

phone *vb* call, dial, *inf* give a buzz, ring, telephone.

phoney *adj* affected, artificial, assumed, bogus, cheating, contrived, counterfeit, deceitful, ersatz, factitious, fake, faked, false, fictitious, fraudulent, hypocritical, imitation, insincere, mock, pretended, *inf* pseudo, *inf* put-on, *inf* put-up, sham, spurious, synthetic, trick, unreal. *Opp* REAL.

photocopy *vb* copy, duplicate, photostat, print off, reproduce, *inf* run off.

photograph *n* enlargement, exposure, negative, *inf* photo, picture, plate, positive, print, shot, slide, *inf* snap, snapshot, transparency. • *vb* film, shoot, snap, take a photograph of.

photographic *adj* **1** accurate, exact, faithful, graphic, lifelike, naturalistic, realistic, representational, true to life. **2** *photographic memory*. pictorial, retentive, visual.

phrase *n* clause, expression. ▷ SAYING. • *vb* ▷ SAY.

phraseology *n* diction, expression, idiom, language, parlance, phrasing, style, turn of phrase, wording.

physical *adj* actual, bodily, carnal, concrete, corporal, corporeal, earthly, fleshly, incarnate, material, mortal, palpable, physiological, real, solid, substantial, tangible. *Opp* INTANGIBLE, SPIRITUAL.

physician *n* consultant, doctor, general practitioner, *inf* GP, *inf* medic, medical practitioner, specialist.

physiological *adj* anatomical, bodily, physical. *Opp* PSYCHOLOGICAL.

physique *n* body, build, figure, form, frame, muscles, physical condition, shape.

pick *n* **1** choice, election, option, preference, selection. **2** best, cream, élite, favourite, flower, pride. • *vb* **1** cast (*actor*), choose, decide on, elect, fix on, make a choice of, name, nominate, opt for, prefer, select, settle on, single out, vote for. **2** *pick flowers*. collect, cull, cut, gather, harvest, pluck, pull off, take. **pick on** ▷ BULLY. **pick up** ▷ GET, IMPROVE.

pictorial *adj* diagrammatic, graphic, illustrated, realistic, representational, vivid.

picture *n* **1** delineation, depiction, image, likeness, outline, portrayal, profile, representation. □ *abstract, cameo, caricature, cartoon, collage, design, doodle, drawing, engraving, etching, fresco,* plur *graffiti,* plur *graphics, icon, identikit, illustration, landscape, montage, mosaic, mural, oil-painting, old master, painting, photofit, photograph, pin-up, plate, portrait, print, reproduction, self-portrait, silhouette, sketch, slide,* inf *snap, snapshot, still life, transfer, transparency, triptych,* Fr *trompe l'oeil, video, vignette.* **2** film, movie, moving picture, video. ▷ FILM. • *vb* **1** caricature, delineate, depict, display, doodle, draw, engrave, etch, evoke, film, illustrate, outline, paint, photograph, portray, print, represent, show, sketch, video. **2** *picture the future*. conceive, describe, dream up, envisage, envision, fancy, imagine, see in your mind's eye, think up, visualize.

picturesque *adj* **1** attractive, charming, colourful, idyllic, lovely, pleasant, pretty, quaint, scenic, *inf* story-book. ▷ BEAUTIFUL. *Opp* UNATTRACTIVE. **2** *picturesque language.* colourful, descriptive, expressive, graphic, imaginative, poetic, vivid. *Opp* PROSAIC.

pie *n* flan, pasty, patty, quiche, tart, tartlet, turnover, vol-au-vent.

piece *n* **1** bar, bit, bite, block, chip, chunk, crumb, division, *inf* dollop, fraction, fragment, grain, helping, hunk, length, lump, morsel, part, particle, portion, quantity, remnant, sample, scrap, section, segment, shard, share, shred, slab, slice, sliver, snippet, speck, stick, tablet, *inf* titbit, wedge. **2** component, constituent, element, spare part, unit. **3** *piece of music, work.* article, composition, example, instance, item, number, passage, specimen, work. ▷ MUSIC, WRITING. **piece together** ▷ ASSEMBLE.

pied *adj* dappled, flecked, mottled, particoloured, patchy, piebald, spotted, variegated.

pier *n* **1** breakwater, jetty, landing-stage, quay, wharf. ▷ DOCK. **2** buttress, column, pile, pillar, post, support, upright.

pierce *vb* bayonet, bore through, cut, drill, enter, go through, impale, jab, lance, make a hole in, penetrate, perforate, poke through, prick, punch, puncture, riddle, skewer, spear, spike, spit, stab, stick into, thrust into, transfix, tunnel through, wound. **piercing** ▷ SHARP.

piety *n* dedication, devotion, devotedness, devoutness, faith, godliness, holiness, piousness, religion, *derog* religiosity, saintliness, sanctity. *Opp* IMPIETY.

pig *n* boar, hog, *inf* piggy, piglet, runt, sow, swine.

pile *n* **1** abundance, accumulation, agglomeration, collection, concentration, conglomeration, deposit, heap, hoard, *inf* load, mass, mound, *inf* mountain, plethora, quantity, stack, stockpile, supply, *inf* tons. **2** column, pier, post, support, upright. • *vb* accumulate, amass, assemble, bring together, build up, collect, concentrate, deposit, gather, heap, hoard, load, mass, stack up, stockpile, store.

pilfer *vb inf* filch, *inf* pinch, rob, shoplift. ▷ STEAL.

pilgrim *n* crusader, *old use* palmer. ▷ TRAVELLER.

pill *n* bolus, capsule, lozenge, pastille, pellet, pilule, tablet. ▷ MEDICINE.

pillage *n* buccaneering, depredation, despoliation, devastation, looting, marauding, piracy, plunder, plundering, ransacking, rape, rapine, robbery, robbing, sacking, stealing, stripping. • *vb* despoil, devastate, loot, maraud, plunder, raid, ransack, ravage, raze, rob, sack, steal, strip, vandalize.

pillar *n* baluster, caryatid, column, pier, pilaster, pile, post, prop, shaft, stanchion, support, upright.

pilot *n* **1** airman, *old use* aviator, captain, flier. **2** coxswain, helmsman, navigator, steersman. • *vb* conduct, convey, direct, drive, fly, guide, lead, navigate, shepherd, steer.

pimple *n* blackhead, boil, eruption, pustule, spot, swelling, *sl* zit. **pimples** acne, rash.

pin *n old use* bodkin, bolt, brooch, clip, dowel, drawing-pin, hatpin, nail, peg, rivet, safety-pin, spike, staple, thole, tiepin. • *vb* clip, nail, pierce, staple, tack, transfix. ▷ FASTEN.

pinch *vb* 1 crush, grip, hurt, nip, press, squeeze, tweak. 2 ▷ STEAL.

pine *vb* mope, mourn, sicken, waste away. **pine for** ▷ WANT.

pinnacle *n* 1 acme, apex, cap, climax, consummation, crest, crown, crowning point, height, highest point, peak, summit, top, zenith. 2 *pinnacle on a roof.* spire, steeple, turret.

pioneer *n* 1 colonist, discoverer, explorer, frontiersman, frontierswoman, pathfinder, settler, trail-blazer. 2 innovator, inventor, originator, pace-maker, trend-setter. • *vb* begin, *inf* bring out, create, develop, discover, establish, experiment with, found, inaugurate, initiate, institute, introduce, invent, launch, open up, originate, set up, start.

pious *adj* 1 dedicated, devoted, devout, faithful, god-fearing, godly, good, holy, moral, religious, reverent, reverential, saintly, sincere, spiritual, virtuous. *Opp* IMPIOUS. 2 [*derog*] *inf* goody-goody, *inf* holier-than-thou, hypocritical, insincere, mealy-mouthed, pietistic, Pharisaical, sanctimonious, self-righteous, self-satisfied, *inf* smarmy, unctuous. *Opp* SINCERE.

pip *n* 1 pit, seed, stone. 2 mark, spot, star. 3 bleep, blip, sound, stroke.

pipe *n* conduit, channel, duct, hose, hydrant, line, main, pipeline, piping, tube. • *vb* 1 carry along a pipe, carry along a wire, channel, convey, deliver, supply, transmit. 2 *pipe a tune.* blow, play, sound, *inf* tootle, whistle. **pipe up** ▷ SPEAK. **piping** ▷ HOT, SHRILL.

piquant *adj* 1 appetizing, pungent, salty, sharp, spicy, tangy, tart, tasty. *Opp* BLAND. 2 *a piquant notion.* arresting, exciting, interesting, provocative, stimulating. *Opp* BANAL.

pirate *n* buccaneer, *old use* corsair, marauder, privateer, sea rover. ▷ THIEF. • *vb* ▷ PLAGIARIZE.

pit *n* 1 abyss, chasm, crater, depression, ditch, excavation, hole, hollow, pothole, rut, trench, well. 2 coal-mine, colliery, mine, mineshaft, quarry, shaft, working.

pitch *n* 1 bitumen, tar. 2 *pitch of a roof.* angle, gradient, incline, slope, steepness, tilt. 3 *musical pitch.* frequency, tuning. 4 *soccer pitch.* arena, ground, playing-field, stadium. • *vb* 1 erect, put up, raise, set up. 2 *pitch stones.* bowl, *inf* bung, cast, *inf* chuck, fire, fling, heave, hurl, launch, lob, sling, throw, toss. 3 *pitch into the water.* dive, drop, fall headlong, plunge, plummet, *inf* take a nosedive, topple. **pitch about** ▷ TOSS. **pitch in** ▷ COOPERATE. **pitch into** ▷ ATTACK.

piteous *adj* affecting, distressing, heartbreaking, heart-rending, lamentable, miserable, moving, pathetic, pitiable, pitiful, plaintive, poignant, touching, woeful, wretched. ▷ SAD.

pitfall *n* catch, danger, difficulty, hazard, peril, snag, trap.

pitiful *adj* 1 abject, contemptible, deplorable, hopeless, inadequate, incompetent, insignificant, laughable, mean, *inf* miserable, *inf* pathetic, pitiable, ridiculous, sorry, trifling, unimportant, useless, worthless. *Opp* ADMIRABLE. 2 ▷ PITEOUS.

pitiless *adj* bloodthirsty, brutal, callous, cruel, ferocious, hard, heartless, inexorable, inhuman, merciless, relentless, ruthless, sadistic, unfeeling, unrelenting, unrelieved, unremitting, unsympathetic. *Opp* MERCIFUL.

pitted *adj* dented, eaten away, eroded, *inf* holey, marked, pockmarked, rough, scarred, uneven. *Opp* SMOOTH.

pity *n* charity, clemency, commiseration, compassion, condolence, feeling, forbearance, forgiveness, grace, humanity, kindness, leniency, love, mercy, regret, *old use* ruth, softness, sympathy, tenderness, understanding, warmth. *Opp* CRUELTY. • *vb inf* bleed for, commiserate with, *inf* feel for, feel sorry for, show pity for, sympathize with, weep for.

pivot *n* axis, axle, centre, fulcrum, gudgeon, hinge, hub, pin, point of balance, spindle, swivel. • *vb* hinge, revolve, rotate, spin, swivel, turn, twirl, whirl.

placard *n* advert, advertisement, bill, notice, poster, sign.

placate *vb* appease, calm, conciliate, humour, mollify, pacify, soothe.

place *n* **1** area, country, district, locale, location, locality, locus, neighbourhood, part, point, position, quarter, region, scene, setting, site, situation, *inf* spot, town, venue, vicinity, *inf* whereabouts. **2** *a place in society.* condition, degree, estate, function, grade, job, mission, niche, office, position, rank, role, standing, station, status. **3** *a place to live.* ▷ HOUSE. **4** *a place to sit.* ▷ SEAT. • *vb* **1** deposit, dispose, *inf* dump, lay, leave, locate, pinpoint, plant, position, put down, rest, set down, set out, settle, situate, stand, station, *inf* stick. **2** arrange, categorize, class, classify, grade, order, position, put in order, rank, sort. **3** *can't place it.* identify, put a name to, put into context, recognize, remember.

placid *adj* **1** collected, composed, cool, equable, even-tempered, imperturbable, level-headed, mild, phlegmatic, restful, sensible, stable, steady, unexcitable. **2** calm, motionless, peaceful, quiet, tranquil, unruffled, untroubled. *Opp* EXCITABLE, STORMY.

plagiarize *vb* appropriate, borrow, copy, *inf* crib, imitate, infringe copyright, *inf* lift, pirate, purloin, reproduce. ▷ STEAL.

plague *n* **1** affliction, bane, blight, calamity, contagion, epidemic, infection, outbreak, pandemic, pestilence. ▷ ILLNESS. **2** infestation, invasion, nuisance, scourge, swarm, visitation. • *vb* afflict, annoy, be a nuisance to, bother, distress, disturb, harass, harry, hound, irritate, molest, *inf* nag, persecute, pester, torment, torture, trouble, vex, worry.

plain *adj* **1** apparent, audible, certain, clear, comprehensible, definite, distinct, evident, intelligible, legible, lucid, manifest, obvious, patent, simple, transparent, unambiguous, understandable, unmistakable, visible, well-defined. *Opp* OBSCURE. **2** *plain speech.* basic, blunt, candid, direct, downright, explicit, forthright, frank, honest, informative, outspoken, plain-spoken, prosaic, sincere, straightforward, unequivocal, unvarnished. **3** *plain living.* austere, drab, everyday, frugal, homely, modest, ordinary, simple, Spartan, stark, unadorned, unattractive, undecorated, unexciting, unprepossessing, unpretentious, unremarkable, workaday. *Opp* SOPHISTICATED. • *n* grassland, pasture, pampas, prairie, savannah, steppe, tundra, veld.

plaintive *adj* doleful, melancholy, mournful, plangent, sorrowful, wistful. ▷ SAD.

plan *n* **1** *inf* bird's-eye view, blueprint, chart, design, diagram, drawing, layout, map, representation, sketch-map. **2** *a plan of action.* aim, course of action, design, formula, idea, intention, method, plot, policy, procedure, programme, project, proposal, proposition, *inf* scenario, scheme, strategy, system. • *vb* **1** arrange, concoct, contrive, design, devise, draw up a plan, formulate, invent, map out, *inf* mastermind, organize, outline, plot, prepare, scheme, think out, work out. **2** *I plan to go away.* aim, conspire, contemplate, envisage, expect, intend, mean, propose, think of. **planned** ▷ DELIBERATE.

plane *adj* even, flat, flush, level, smooth, uniform. • *n* **1** flat surface, level, surface. **2** ▷ AIRCRAFT.

planet *n* globe, orb, satellite, sphere, world. □ *Earth, Jupiter, Mars, Mercury, Neptune, Pluto, Saturn, Uranus, Venus.*

plank *n* beam, board, planking, timber.

planning *n* arrangement, design, drafting, forethought, organization, preparation, setting up, thinking out.

plant *n* greenery, growth, undergrowth, vegetation. □ *annual, bulb, cactus, cereal, climber, fern, flower, fungus, grass, herb, lichen, moss, perennial, shrub, tree, vegetable, vine, water-plant, weed.* ▷ FLOWER, TREE, VEGETABLE. **2** *a manufacturing plant.* factory, foundry, mill, shop, works, workshop. **3** *industrial plant.* apparatus, equipment, machinery, machines. • *vb* **1** bed out, set out, sow, transplant. **2** locate, place, position, put, situate, station.

plaster *n* **1** mortar, stucco. **2** dressing, sticking-plaster. • *vb* apply, bedaub, coat, cover, daub, smear, spread.

plastic *adj* ductile, flexible, malleable, pliable, shapable, soft, supple, workable. □ *bakelite, celluloid, polystyrene, polythene, polyurethane, polyvinyl, PVC, vinyl.*

plate *n* **1** *old use* charger, dinner-plate, dish, platter, salver, side-plate, soup-plate, *old use* trencher. **2** lamina, lamination, layer, leaf, pane, panel, sheet, slab, stratum. **3** *plates in a book.* illustration, *inf* photo, photograph, picture, print. **4** *a dental plate.* dentures, false teeth. • *vb* anodize, coat, cover, electroplate, galvanize (*with zinc*), gild (*with gold*).

platform **1** dais, podium, rostrum, stage, stand. **2** *political platform.* ▷ POLICY.

platitude *n* banality, cliché, commonplace, truism.

plausible *adj* **1** acceptable, believable, conceivable, credible, imaginable, likely, logical, persuasive, possible, probable, rational, reasonable, sensible, tenable, thinkable. *Opp* IMPLAUSIBLE. **2** deceptive, glib, meretricious, misleading, specious, smooth, sophistical.

play *n* **1** amusement, diversion, entertainment, frivolity, fun, *inf* fun and games, *inf* horseplay, joking, make-believe, merrymaking, playing, pretending, recreation, revelry, *inf* skylarking, sport. **2** *play in moving parts.* flexibility, freedom, freedom of movement, *inf* give, latitude, leeway, looseness, movement, tolerance. 3 ▷ DRAMA. • *vb* 1 amuse yourself, caper, cavort, disport yourself, enjoy yourself, fool about, frisk, frolic, gambol, *inf* have a good time, have fun, *inf* mess about, romp, sport. 2 *play a game.* join in, participate, take part. 3 *play an opponent.* challenge, compete against, oppose, rival, take on, vie with. 4 *play a role.* act, depict, impersonate, perform, portray, pretend to be, represent, take the part of. 5 *play an instrument.* make music on, perform on, strum. 6 *play the radio.* have on, listen to, operate, put on, switch on. play about ▷ MISBEHAVE. **play along, play ball** ▷ COOPERATE. **play down** ▷ MINIMIZE. **play for time** ▷ DELAY. **play it by ear** ▷ IMPROVISE. **play up** ▷ MISBEHAVE. **play up to** ▷ FLATTER.

player *n* 1 athlete, competitor, contestant, participant, sportsman, sportswoman. 2 actor, actress, artiste, entertainer, instrumentalist, musician, performer, soloist, Thespian, trouper. ▷ ENTERTAINER, MUSICIAN.

playful *adj* active, cheerful, coltish, facetious, flirtatious, frisky, frolicsome, fun-loving, good-natured, high-spirited, humorous, impish, jesting, *inf* jokey, joking, kittenish, light-hearted, lively, mischievous, puckish, roguish, skittish, spirited, sportive, sprightly, teasing, *inf* tongue-in-cheek, vivacious, waggish. *Opp* SERIOUS.

plea *n* 1 appeal, entreaty, invocation, petition, prayer, request, solicitation, suit, supplication. 2 argument, excuse, explanation, justification, pretext, reason.

plead *vb* 1 appeal, ask, beg, beseech, cry out, demand, entreat, implore, importune, petition, request, seek, solicit, supplicate. 2 allege, argue, assert, aver, declare, maintain, reason, swear.

pleasant *adj* acceptable, affable, agreeable, amiable, approachable, attractive, balmy, beautiful, charming, cheerful, congenial, decent, delicious, delightful, enjoyable, entertaining, excellent, fine, friendly, genial, gentle, good, gratifying, *inf* heavenly, hospitable, kind, likeable, lovely, mellow, mild, nice, palatable, peaceful, pleasing, pleasurable, pretty, reassuring, relaxed, satisfying, soothing, sympathetic, warm, welcome, welcoming. *Opp* ANNOYING, UNPLEASANT.

please *vb* 1 amuse, cheer up, content, delight, divert, entertain, give pleasure to, gladden, gratify, humour, make happy, satisfy, suit. 2 *Do what you please.* ▷ WANT.

pleasing ▷ PLEASANT.

pleased *adj inf* chuffed, *derog* complacent, contented, delighted, elated, euphoric, glad, grateful, gratified, *sl* over the moon, satisfied, thankful, thrilled. ▷ HAPPY. *Opp* ANNOYED.

pleasure *n* 1 bliss, comfort, contentment, delight, ecstasy, enjoyment, euphoria, fulfilment, gladness, gratification, happiness, joy, rapture, satisfaction, solace. 2 amusement, diversion, entertainment, fun, luxury, recreation, self-indulgence.

pleat *n* crease, flute, fold, gather, tuck.

plebiscite *n* ballot, poll, referendum, vote.

pledge *n* 1 assurance, covenant, guarantee, oath, pact, promise, undertaking, vow, warranty, word. 2 *a pledge left at a pawnbroker's.* bail, bond, collateral, deposit, pawn, security, surety. • *vb* agree, commit yourself, contract, give your word, guarantee, promise, swear, undertake, vouch, vouchsafe, vow.

plenary *adj* full, general, open.

plentiful *adj* abounding, abundant, ample, bounteous, bountiful, bristling, bumper (*crop*), copious, generous, inexhaustible, lavish, liberal, overflowing, plenteous, profuse, prolific. *Opp* SCARCE. **be plentiful** ▷ ABOUND.

plenty *n* abundance, adequacy, affluence, cornucopia, excess, fertility, flood, fruitfulness, glut, *inf* heaps, *inf* lashings, *inf* loads, a lot, *inf* lots, *inf* masses, much, more than enough, *inf* oceans, *inf* oodles, *inf* piles, plenitude, plentifulness, plethora, prodigality, profusion, prosperity, quantities, *inf* stacks, sufficiency, superabundance, surfeit, surplus, *inf* tons, wealth. *Opp* SCARCITY.

pliable *adj* **1** bendable, *inf* bendy, ductile, flexible, plastic, pliant, springy, supple. **2** *a pliable character.* adaptable, compliant, docile, easily influenced, easily led, easily persuaded, impressionable, manageable, persuadable, receptive, responsive, succeptible, suggestible, tractable, yielding.

plod *vb* **1** slog, tramp, trudge. ▷ WALK. **2** drudge, grind on, labour, persevere, *inf* peg away, *inf* plug away, toil. ▷ WORK.

plot *n* **1** acreage, allotment, area, estate, garden, lot, parcel, patch, smallholding, tract. **2** *a plot of a novel.* chain of events, narrative, organization, outline, scenario, story, story-line, thread. **3** *a subversive plot.* cabal, conspiracy, intrigue, machination, plan, scheme. • *vb* **1** chart, compute, draw, map out, mark, outline, plan, project. **2** *plot to rob a bank.* collude, conspire, have designs, intrigue, machinate, scheme. **3** *plot a crime.* arrange, *inf* brew, conceive, concoct, *inf* cook up, design, devise, dream up, hatch.

pluck *n* ▷ COURAGE. *vb* **1** collect, gather, harvest, pick, pull off, remove. **2** grab, jerk, pull, seize, snatch, tear away, tweak, yank. **3** *pluck a chicken.* denude, remove feathers from, strip. **4** *pluck a violin.* play pizzicato, strum, twang.

plug *n* **1** bung, cork, stopper. **2** ▷ ADVERTISEMENT. • *vb* **1** block up, *inf* bung up, close, cork, fill, jam, seal, stop up, stuff up. **2** advertise, commend, mention frequently, promote, publicize, puff, recommend. **plug away** ▷ WORK.

plumb *adv* **1** accurately, *inf* dead, exactly, precisely, *inf* slap. **2** perpendicularly, vertically. • *vb* fathom, measure, penetrate, probe, sound.

plumbing *n* heating system, pipes, water-supply.

plume *n* feather, *plur* plumage, quill.

plump *adj* ample, buxom, chubby, dumpy, overweight, podgy, portly, pudgy, *inf* roly-poly, rotund, round, squat, stout, tubby, *inf* well-upholstered. ▷ FAT. *Opp* THIN. **plump for** ▷ CHOOSE.

plunder *n* booty, contraband, loot, pickings, pillage, prize, spoils, swag, takings. • *vb* capture, despoil, devastate, lay waste, loot, maraud, pillage, raid, ransack, ravage, rifle, rob, sack, seize, spoil, steal from, strip, vandalize.

plunge *vb* **1** descend, dip, dive, drop, engulf, fall, fall headlong, hurtle, immerse, jump, leap, lower, nosedive, pitch, plummet, sink, submerge, swoop, tumble. **2** force, push, stick, thrust.

poach *vb* **1** hunt, steal. **2** ▷ COOK.

pocket *n* bag, container, pouch, receptacle. • *vb* ▷ TAKE.

pod *n* case, hull, shell.

poem *n inf* ditty, *inf* jingle, piece of poetry, rhyme, verse. □ *ballad, ballade, doggerel, eclogue, elegy, epic, epithalamium, haiku, idyll, lay, limerick, lyric, nursery-rhyme, pastoral, ode, sonnet, vers libre.* ▷ VERSE.

poet *n* bard, lyricist, minstrel, poetaster, rhymer, rhymester, sonneteer, versifier. ▷ WRITER.

poetic *adj* emotive, *derog* flowery, imaginative, lyrical, metrical, musical, poetical. *Opp* PROSAIC.

poignant *adj* affecting, distressing, heartbreaking, heartfelt, heart-rending, moving, painful, pathetic, piquant, piteous, pitiful, stirring, tender, touching, upsetting. ▷ SAD.

point *n* **1** apex, peak, prong, sharp end, spike, spur, tine, tip. **2** *a point in space.* location, place, position, site, situation, spot. **3** *a point in time.* instant, juncture, moment, second, stage, time. **4** *decimal point.* dot, full stop, mark, speck, spot. **5** *the point of an argument.* aim, burden, crux, drift, end, essence, gist, goal, heart, import, intention, meaning, motive, nub, object, objective, pith, purpose, quiddity, relevance, significance, subject, substance, theme, thrust, use, usefulness. **6** *points to raise.* aspect, detail, idea, item, matter, particular, question, thought, topic. **7** *good points in her character.* attribute, characteristic, facet, feature, peculiarity, property, quality, trait. • *vb* **1** call attention to, direct attention to, draw attention to, indicate, point out, show, signal. **2** aim, direct, guide, lead, steer. **pointed** ▷ SHARP. **to the point** ▷ RELEVANT.

pointer *n* arrow, hand (*of clock*), indicator.

pointless *adj* aimless, fatuous, fruitless, futile, inane, ineffective, senseless, silly, unproductive, useless, vain, worthless. ▷ STUPID.

poise *n* aplomb, assurance, balance, calmness, composure, coolness, dignity, equanimity, equilibrium, equipoise, imperturbability, presence, sang-froid, self-confidence, self-control, self-possession, serenity, steadiness. • *vb* balance, be poised, hover, keep in balance, support, suspend.

poised *adj* **1** balanced, hovering, in equilibrium, standing, steady, teetering, wavering. **2** *poised to begin.* keyed up, prepared, ready, set, standing by, waiting. **3** *a poised performer.* assured, calm, composed, cool, cool-headed, dignified, self-confident, self-possessed, serene, suave, *inf* unflappable, unruffled, urbane.

poison *n* bane, toxin, venom. • *vb* **1** adulterate, contaminate, infect, pollute, taint. ▷ KILL. **2** *poison the mind.* corrupt, defile, deprave, envenom, pervert, prejudice, subvert, warp. **poisoned** ▷ DIRTY, POISONOUS.

poisonous *adj* deadly, fatal, infectious, lethal, mephitic, miasmic, mortal, noxious, poisoned, septic, toxic, venomous, virulent.

poke *vb* butt, dig, elbow, goad, jab, jog, nudge, prod, stab, stick, thrust. ▷ HIT. **poke about** ▷ SEARCH. **poke fun at**

▷ RIDICULE. **poke out** ▷ PROTRUDE.

poky *adj* confined, cramped, inconvenient, restrictive, uncomfortable. ▷ SMALL. *Opp* SPACIOUS.

polar *adj* antarctic, arctic, freezing, glacial, icy, Siberian. ▷ COLD.

polarize *vb* diverge, divide, move to opposite positions, separate, split.

pole *n* **1** bar, beanpole, column, flag-pole, mast, post, rod, shaft, spar, staff, stake, standard, stick, stilt, upright. **2** *opposite poles.* end, extreme, limit. **poles apart** ▷ DIFFERENT.

police *n* sl the Bill, constabulary, *sl* the fuzz, *inf* the law, police force, policemen. • *vb* control, guard, keep in order, keep the peace, monitor, oversee, patrol, protect, provide a police presence, supervise, watch over.

policeman, policewoman *ns* *inf* bobby, constable, *sl* cop, *sl* copper, detective, *Fr* gendarme, inspector, officer, PC, police constable, *sl* rozzer, woman police constable, WPC.

policy *n* **1** approach, code of conduct, custom, guidelines, *inf* line, method, practice, principles, procedure, protocol, regulations, rules, stance, strategy, tactics. **2** intentions, manifesto, plan of action, platform, programme, proposals.

polish *n* **1** brightness, brilliance, finish, glaze, gleam, gloss, lustre, sheen, shine, smoothness, sparkle. **2** beeswax, French polish, oil, shellac, varnish, wax. **3** *His manners show polish.* *inf* class, elegance, finesse, grace, refinement, sophistication, style, suavity, urbanity. • *vb* brighten, brush up, buff up, burnish, French-polish, gloss, rub down, rub up, shine, smooth, wax. **polish off** ▷ FINISH. **polish up** ▷ IMPROVE.

polished *adj* **1** bright, burnished, glassy, gleaming, glossy, lustrous, shining, shiny. **2** *polished manners.* civilized, *inf* classy, cultivated, cultured, debonair, elegant, expert, faultless, fine, finished, flawless, genteel, gracious, impeccable, perfect, perfected, polite, *inf* posh, refined, soigné(e), sophisticated, suave, urbane. *Opp* ROUGH.

polite *adj* agreeable, attentive, chivalrous, civil, considerate, correct, courteous, courtly, cultivated, deferential, diplomatic, discreet, euphemistic, formal, gallant, genteel, gentlemanly, gracious, ladylike, obliging, polished, proper, respectful, tactful, thoughtful, well-bred, well-mannered, well-spoken. *Opp* RUDE.

political *adj* **1** administrative, civil, diplomatic, governmental, legislative, parliamentary, state. **2** activist, factional, militant, partisan, party-political. □ *anarchist, capitalist, communist, conservative, democrat, fascist, Labour, leftist, left-wing, liberal, Marxist, moderate, monarchist, nationalist, Nazi, parliamentarian, radical, republican, revolutionary, rightist, right-wing, socialist, Tory,* old use *Whig.*

politics *n* diplomacy, government, political affairs, political science, public affairs, statecraft, statesmanship. □ *anarchy, capitalism, communism, democracy, dictatorship, martial law, monarchy, oligarchy, republic.*

poll *n* **1** ballot, election, vote. **2** canvass, census, plebiscite, referendum, survey. • *vb* ballot, canvass, question, sample. survey.

pollute *vb* adulterate, befoul, blight, contaminate, corrupt, defile, dirty, foul, infect, poison, soil, taint.

pomp *n* brilliance, ceremonial, ceremony, display, formality, glory, grandeur, magnificence, ostentation, pageantry, ritual, show, solemnity, spectacle, splendour.

pompous *adj* affected, arrogant, bombastic, conceited, grandiloquent, grandiose, haughty, *inf* highfalutin, imperious, long-winded, magisterial, ornate, ostentatious, overbearing, pedantic, pontifical, posh, pretentious, self-important, sententious, showy, smug, snobbish, *inf* snooty, *inf* stuck-up, *inf* stuffy, supercilious, turgid, vain, vainglorious. ▷ PROUD. *Opp* MODEST.

ponderous *adj* **1** awkward, bulky, burdensome, cumbersome, heavy, hefty, huge, massive, unwieldy, weighty. *Opp* LIGHT. **2** *a ponderous style.* dreary, dull, elephantine, heavy-handed, humourless, inflated, laboured, lifeless, long-winded, overdone, pedestrian, plodding, prolix, slow, stilted, stodgy, tedious, tiresome, verbose, *inf* windy. *Opp* LIVELY.

pool *n* lagoon, lake, mere, oasis, paddling-pool, pond, puddle, swimming-pool, tarn. • *vb* ▷ COMBINE.

poor *adj* **1** badly off, bankrupt, beggarly, *inf* broke, deprived, destitute, disadvantaged, *inf* down-and-out, *inf* hard up, homeless, impecunious, impoverished, in debt, indigent, insolvent, necessitous, needy, *inf* on your uppers, penniless, penurious, poverty-stricken, *sl* skint, straitened, underpaid, underprivileged. **2** *poor soil.* barren, exhausted, infertile, sterile, unfruitful, unproductive. **3** *a poor salary.* inadequate, insufficient, low, meagre, mean, scanty, small, sparse, unprofitable, unrewarding. **4** *poor in health. inf* below par, poorly. ▷ ILL. **5** *poor quality.* amateurish, bad, cheap, defective, deficient, disappointing, faulty, imperfect, inferior, low-grade, mediocre, paltry, second-rate, shoddy, substandard, unacceptable, unsatisfactory, useless, worthless. **6** *poor child!* forlorn, hapless, ill-fated, luckless, miserable, pathetic, pitiable, sad, unfortunate, unhappy, unlucky, wretched. *Opp* GOOD, LARGE, LUCKY, RICH. • *plur n* beggars, the destitute, down-and-outs, the homeless, paupers, tramps, the underprivileged, vagrants, wretches.

populace *n* commonalty, *derog Gk* hoi polloi, masses, people, public, *derog* rabble, *derog* riff-raff.

popular *adj* **1** accepted, acclaimed, *inf* all the rage, approved, celebrated, famous, fashionable, favoured, favourite, *inf* in, in demand, liked, lionized, loved, renowned, sought-after, *inf* trendy, well-known, well-liked, well-received. *Opp* UNPOPULAR. **2** *popular opinion.* average, common, conventional, current, democratic, general, of the people, ordinary, predominant, prevailing, representative, standard, universal.

popularize *vb* **1** make popular, promote, spread. **2** *popularize classics.* make easy, simplify, *derog* tart up.

populate *vb* colonize, dwell in, fill, inhabit, live in, occupy, overrun, people, reside in, settle.

population *n* citizenry, citizens, community, denizens, folk, inhabitants, natives, occupants, people, populace, public, residents.

populous *adj* crowded, full, heavily populated, jammed, overcrowded, overpopulated, packed, swarming, teeming.

porch *n* doorway, entrance, lobby, portico.

pore *vb* **pore over** examine, go over, peruse, read, scrutinize, study.

pornographic *adj* arousing, *inf* blue, erotic, explicit, exploitative, sexual, sexy, titillating. ▷ OBSCENE.

porous *adj* absorbent, cellular, holey, penetrable, permeable, pervious, spongy. *Opp* IMPERVIOUS.

port *n* anchorage, dock, dockyard, harbour, haven, marina, mooring, sea-port.

portable *adj* compact, convenient, easy to carry, handy, light, lightweight, manageable, mobile, movable, pocket, pocket-sized, small, transportable. *Opp* UNWIELDY.

porter *n* **1** caretaker, concierge, door-keeper, doorman, gatekeeper, janitor, security-guard, watchman. **2** baggage-handler, bearer, carrier.

portion *n* allocation, allowance, bit, chunk, division, fraction, fragment, helping, hunk, measure, part, percentage, piece, quantity, quota, ration, scrap, section, segment, serving, share, slice, sliver, subdivision, wedge. **portion out** ▷ SHARE.

portrait *n* depiction, image, likeness, picture, portrayal, profile, representation, self-portrait. ▷ PICTURE.

portray *vb* **1** delineate, depict, describe, evoke, illustrate, paint, picture, represent, show. **2** ▷ IMPERSONATE.

pose *n* **1** attitude, position, posture, stance. **2** act, affectation, attitudinizing, façade, masquerade, pretence. • *vb* **1** keep still, model, sit, strike a pose. **2** attitudinize, *inf* be a poser, be a poseur, posture, *inf* put on airs, show off. **3** *pose a question.* advance, ask, broach, posit, postulate, present, put forward, submit, suggest. **pose as** ▷ IMPERSONATE.

poser *n* **1** dilemma, enigma, problem, puzzle, question, riddle. **2** ▷ POSEUR.

poseur *n* attitudinizer, exhibitionist, fraud, impostor, masquerader, *inf* phoney, *inf* poser, pretender, *inf* show-off.

posh *adj inf* classy, elegant, fashionable, formal, grand, lavish, luxurious, ostentation, rich, showy, smart, snobbish, stylish, sumptuous, *inf* swanky, *inf* swish.

position *n* **1** locality, location, locus, place, placement, point, reference, site, situation, spot, whereabouts. **2** *an awkward position.* circumstances, condition, predicament, situation, state. **3** *position of the body.* angle, pose, posture, stance. **4** *intellectual position.* assertion, attitude, contention, hypothesis, opinion, outlook, perspective, principle, proposition, standpoint, thesis, view, viewpoint. **5** *position in a firm.* appointment, degree, employment, function, grade, job, level, niche, occupation, place, post, rank, role, standing, station, status, title. • *vb* arrange, deploy, dispose, fix, locate, place, put, settle, site, situate, stand, station.

positive *adj* **1** affirmative, assured, categorical, certain, clear, conclusive, confident, convinced, decided, definite, emphatic, explicit, firm, incontestable, incontrovertible, irrefutable, real, sure, undeniable, unequivocal. **2** *positive advice.* beneficial, constructive, helpful, optimistic, practical, useful, worthwhile. *Opp* NEGATIVE.

possess *vb* **1** be in possession of, enjoy, have, hold, own. **2** be gifted with, embody, embrace, include. **3** *possess territory.* acquire, control, dominate, govern, invade, occupy, rule, seize, take over. **4** *possess a person.* bewitch, captivate, cast a spell over, charm, enthral, haunt, hypnotize, obsess.

possessions *plur n* assets, belongings, chattels, effects, estate, fortune, goods, property, riches, things, wealth, worldly goods.

possessive *adj* clinging, dominating, domineering, jealous, overbearing, proprietorial, protective, selfish. ▷ GREEDY.

possibility *n* capability, chance, danger, feasibility, likelihood, odds, opportunity, plausibility, potential, potentiality, practicality, probability, risk.

possible *adj* achievable, admissible, attainable, conceivable, credible, *inf* doable, feasible, imaginable, likely, obtainable, *inf* on, plausible, potential, practicable, practical, probable, prospective, realizable, reasonable, tenable, thinkable, viable, workable. *Opp* IMPOSSIBLE.

possibly *adv* God willing, *inf* hopefully, if possible, maybe, *old use* peradventure, *old use* perchance, perhaps.

post *n* **1** baluster, bollard, brace, capstan, column, gate-post, leg, newel, pale, paling, picket, pier, pile, pillar, pole, prop, pylon, shaft, stake, stanchion, standard, starting-post, strut, support, upright, winning-post. **2** *a sentry's post.* location, place, point, position, station. **3** *post in a firm.* appointment, assignment, employment, function, job, occupation, office, place, position, situation, task, work. **4** airmail, cards, delivery, junk mail, letters, mail, packets, parcels, postcards. • *vb* **1** advertise, announce, display, pin up, proclaim, promulgate, publicize, publish, put up, stick up. **2** *post a letter.* dispatch, mail, send, transmit. **3** *post a sentry.* appoint, assign, locate, place, position, set, situate, station.

poster *n* advertisement, announcement, bill, broadsheet, circular, display, flyer, notice, placard, sign.

posterity *n* descendants, future generations, heirs, issue, offspring, progeny, successors.

postpone *vb* adjourn, defer, delay, extend, hold over, keep in abeyance, lay aside, put back, put off, *inf* put on ice, *inf* put on the back burner, *inf* shelve, stay, suspend, temporize.

postscript *n* addendum, addition, afterthought, codicil (*to will*), epilogue, *inf* PS.

postulate *vb* assume, hypothesize, posit, propose, suppose, theorize.

posture *n* **1** appearance, bearing, carriage, deportment, pose, position, stance. **2** ▷ ATTITUDE.

posy *n* bouquet, bunch of flowers, buttonhole, corsage, nosegay, spray.

pot *n* basin, bowl, casserole, cauldron, container, crock, crucible, dish, jar, pan, saucepan, stewpot, teapot, urn, vessel.

potent *adj* **1** effective, forceful, formidable, influential, intoxicating (*drink*), mighty, overpowering, overwhelming, powerful, puissant, strong, vigorous. ▷ STRONG. **2** *a potent argument.* ▷ PERSUASIVE. *Opp* WEAK.

potential *adj* **1** aspiring, budding, embryonic, future, *inf* hopeful, intending, latent, likely, possible, probable, promising, prospective, *inf* would-be. **2** *potential disaster*. imminent, impending, looming, threatening. • *n* aptitude, capability, capacity, possibility, resources.

potion *n* brew, concoction, decoction, dose, draught, drink, drug, elixir, liquid, medicine, mixture, philtre, potation, tonic.

potter *vb* dabble, do odd jobs, fiddle about, loiter, mess about, tinker, work.

pottery *n* ceramics, china, crockery, crocks, earthenware, porcelain, stoneware, terracotta.

pouch *n* bag, pocket, purse, reticule, sack, wallet.

poultry *n* □ *bantam, chicken, duck, fowl, goose, guinea-fowl, hen, pullet, turkey.*

pounce *vb* ambush, attack, drop on, jump on, leap on, seize, snatch, spring at, strike, swoop down on, take by surprise.

pound *n* compound, corral, enclosure, pen. • *vb* batter, beat, crush, grind, hammer, knead, mash, powder, pulp, pulverize, smash. ▷ HIT.

pour *vb* **1** cascade, course, discharge, disgorge, flood, flow, gush, run, spew, spill, spout, spurt, stream. **2** *pour wine*. decant, empty, serve, tip.

poverty *n* **1** beggary, bankruptcy, debt, destitution, hardship, impecuniousness, indigence, insolvency, necessity, need, penury, privation, want. **2** *a poverty of talent*. absence, dearth, insufficiency, lack, paucity, scarcity, shortage. *Opp* WEALTH.

powder *n* dust, particles, talc. • *vb* **1** atomize, comminute, crush, granulate, grind, pound, pulverize, reduce to powder. **2** besprinkle, coat, cover with powder, dredge, dust, sprinkle.

powdered *adj* **1** ▷ POWDERY. **2** dehydrated, dried, freeze-dried.

powdery *adj* chalky, crumbly, crushed, disintegrating, dry, dusty, fine, friable, granular, granulated, ground, loose, powdered, pulverized, sandy. *Opp* SOLID, WET.

power *n* **1** ability, capability, capacity, competence, drive, energy, faculty, force, might, muscle, potential, skill, talent, vigour. **2** *power to arrest*. authority, privilege, right. **3** *power of a tyrant*. ascendancy, *inf* clout, command, control, dominance, domination, dominion, influence, mastery, omnipotence, oppression, potency, rule, sovereignty, supremacy, sway. ▷ STRENGTH. *Opp* WEAKNESS.

powerful *adj* authoritative, cogent, commanding, compelling, consuming, convincing, dominant, dynamic, effective, effectual, energetic, forceful, high-powered, influential, invincible, irresistible, mighty, muscular, omnipotent, overpowering, overwhelming, persuasive, potent, sovereign, vigorous, weighty. ▷ STRONG. *Opp* POWERLESS.

powerless *adj* defenceless, disabled, feeble, helpless, impotent, incapable, incapacitated, ineffective, ineffectual, paralysed, unable, unfit. ▷ WEAK. *Opp* POWERFUL.

practicable *adj* achievable, attainable, *inf* doable, feasible, performable, possible, practical, realistic, sensible, viable, workable. *Opp* IMPRACTICABLE.

practical *adj* **1** applied, empirical, experimental. **2** businesslike,

capable, competent, down-to-earth, efficient, expert, hard-headed, matter-of-fact, *inf* no-nonsense, pragmatic, proficient, realistic, sensible, skilled. **3** *a practical tool.* convenient, functional, handy, usable, useful, utilitarian. **4** ▷ PRACTICABLE. *Opp* IMPRACTICAL, THEORETICAL.
practical joke ▷ TRICK.

practically *adv* almost, close to, just about, nearly, to all intents and purposes, virtually.

practice *n* **1** action, actuality, application, doing, effect, operation, reality, use. **2** *inf* dummy-run, exercise, practising, preparation, rehearsal, *inf* run-through, training, *inf* try-out, *inf* work-out. **3** *common practice.* convention, custom, habit, modus operandi, routine, tradition, way, wont. **4** *a doctor's practice.* business, office, work.

practise *vb* **1** do exercises, drill, exercise, prepare, rehearse, train, warm up, *inf* work out. **2** *practise what you preach.* apply, carry out, do, engage in, follow, make a practice of, perform, put into practice.

praise *n* **1** acclaim, acclamation, accolade, admiration, adulation, applause, approbation, approval, commendation, compliment, congratulation, encomium, eulogy, homage, honour, ovation, panegyric, plaudits, testimonial, thanks, tribute. **2** *praise to God.* adoration, devotion, glorification, worship. • *vb* **1** acclaim, admire, applaud, cheer, clap, commend, compliment, congratulate, *inf* crack up, eulogize, extol, give a good review of, marvel at, offer praise to, pay tribute to, *inf* rave about, recommend, *inf* say nice things about, show approval of. *Opp* CRITICIZE. **2** *praise God.* adore, exalt, glorify, honour, laud, magnify, worship. *Opp* CURSE.

praiseworthy *adj* admirable, commendable, creditable, deserving, laudable, meritorious, worthy. ▷ GOOD. *Opp* BAD.

pram *n* baby-carriage, *old use* perambulator, push-chair.

prance *vb* bound, caper, cavort, dance, frisk, frolic, gambol, hop, jig about, jump, leap, play, romp, skip, spring.

prattle *vb* babble, blather, chatter, gabble, maunder, *inf* rattle on, *inf* witter on.

pray *vb* beseech, call upon, invoke, say prayers, supplicate. ▷ ASK.

prayer *n* collect, devotion, entreaty, invocation, litany, meditation, petition, praise, supplication.

prayer-book *n* breviary, missal.

preach *vb* **1** deliver a sermon, evangelize, expound, proselytize, spread the Gospel. **2** expatiate, give moral advice, harangue, *inf* lay down the law, lecture, moralize, pontificate, sermonize.

preacher *n* cleric, crusader, divine, ecclesiastic, evangelist, minister, missionary, moralist, pastor, revivalist. ▷ CLERGYMAN.

prearranged *adj* arranged beforehand, fixed, planned, predetermined, prepared, rehearsed, thought out. *Opp* SPONTANEOUS.

precarious *adj* dangerous, *inf* dicey, *inf* dodgy, dubious, hazardous, insecure, perilous, risky, rocky, shaky, slippery, treacherous, uncertain, unreliable, unsafe, unstable, unsteady, vulnerable, wobbly. *Opp* SAFE.

precaution *n* anticipation, defence, insurance, preventive measure, protection, provision, safeguard, safety measure.

precede *vb* be in front of, come before, go ahead, go before, go in front, herald, introduce, lead, lead into, pave the way for, preface, prefix, start, usher in. *Opp* FOLLOW.

precious *adj* **1** costly, expensive, invaluable, irreplaceable, priceless, valuable. *Opp* WORTHLESS. **2** adored, beloved, darling, dear, loved, prized, treasured, valued, venerated.

precipice *n* bluff, cliff, crag, drop, escarpment, precipitous face, rock.

precipitate *adj* breakneck, hasty, headlong, meteoric, premature. ▷ QUICK. • *vb* accelerate, advance, bring on, cause, encourage, expedite, further, hasten, hurry, incite, induce, instigate, occasion, provoke, spark off, trigger off.

precipitation *n* □ *dew, downpour, drizzle, hail, rain, rainfall, shower, sleet, snow, snowfall.*

precipitous *adj* abrupt, perpendicular, sharp, sheer, steep, vertical.

precise *adj* **1** accurate, clear-cut, correct, defined, definite, distinct, exact, explicit, fixed, measured, right, specific, unambiguous, unequivocal, well-defined. *Opp* IMPRECISE. **2** *precise work.* careful, critical, exacting, fastidious, faultless, finicky, flawless, meticulous, nice, perfect, punctilious, rigorous, scrupulous. *Opp* CARELESS.

preclude *vb* avert, avoid, bar, debar, exclude, forestall, frustrate, impede, make impossible, obviate, pre-empt, prevent, prohibit, rule out, thwart.

precocious *adj* advanced, forward, gifted, mature, quick. ▷ CLEVER. *Opp* BACKWARD.

preconception *n* assumption, bias, expectation, preconceived idea, predisposition, prejudgement, prejudice, presupposition.

predatory *adj* acquisitive, avaricious, covetous, extortionate, greedy, hunting, marauding, pillaging, plundering, preying, rapacious, ravenous, voracious.

predecessor *n* ancestor, antecedent, forebear, forefather, forerunner, precursor.

predetermined *adj* **1** fated, destined, doomed, ordained, predestined. **2** agreed, prearranged, preplanned, recognized, *inf* set up.

predicament *n* crisis, difficulty, dilemma, embarrassment, emergency, *inf* fix, impasse, *inf* jam, *inf* mess, *inf* pickle, plight, problem, quandary, situation, state.

predict *vb* augur, forebode, forecast, foresee, foreshadow, foretell, foretoken, forewarn, hint, intimate, presage, prognosticate, prophesy, tell fortunes.

predictable *adj* anticipated, certain, expected, foreseeable, foreseen, likely, *inf* on the cards, probable, sure, unsurprising. *Opp* UNPREDICTABLE.

predominant *adj* ascendant, chief, dominating, leading, main, preponderant, prevailing, prevalent, primary, ruling, sovereign.

predominate *vb* be in the majority, control, dominate, *inf* have the upper hand, hold sway, lead, outnumber, outweigh, preponderate, prevail, reign, rule.

pre-eminent *adj* distinguished, eminent, excellent, incomparable, matchless, outstanding, peerless, supreme, unrivalled, unsurpassed.

pre-empt *vb* anticipate, appropriate, arrogate, expropriate, forestall, seize, take over.

preface *n* exordium, foreword, introduction, *inf* lead-in, overture, preamble, prelude, proem, prolegomenon, prologue. • *vb* begin, introduce, lead into, open, precede, prefix, start.

prefer *vb* advocate, *inf* back, be partial to, choose, fancy, favour, *inf* go for, incline towards, like, like better, opt for, pick out, *inf* plump for, *inf* put your money on, recommend, select, single out, think preferable, vote for, want.

preferable *adj* advantageous, better, better-liked, chosen, desirable, favoured, likely, nicer, preferred, recommended, wanted. *Opp* OBJECTIONABLE.

preference *n* **1** choice, fancy, favourite, liking, option, pick, selection, wish. **2** favouritism, inclination, partiality, predilection, prejudice, proclivity.

preferential *adj* advantageous, better, biased, favourable, favoured, privileged, showing favouritism, special, superior.

pregnant *adj* **1** carrying a child, expectant, *inf* expecting, gestating, gravid, parturient, *old use* with child. **2** *pregnant remark*. ▷ MEANINGFUL.

prejudice *n* bias, bigotry, chauvinism, discrimination, dogmatism, fanaticism, favouritism, intolerance, jingoism, leaning, narrow-mindedness, partiality, partisanship, predilection, predisposition, prejudgement, racialism, racism, sexism, unfairness, xenophobia. *Opp* TOLERANCE. • *vb* **1** bias, colour, incline, influence, interfere with, make prejudiced, predispose, sway. **2** *prejudice your chances*. damage, harm, injure, ruin, spoil, undermine.

prejudiced *adj* biased, bigoted, chauvinist, discriminatory, illiberal, intolerant, jaundiced, jingoistic, leading (*question*), loaded, narrow-minded, one-sided, parochial, partial, partisan, racist, sexist, tendentious, unfair, xenophobic. *Opp* IMPARTIAL. **prejudiced person** bigot, chauvinist, fanatic, racist, sexist, zealot.

prejudicial *adj* damaging, deleterious, detrimental, disadvantageous, harmful, inimical, injurious, unfavourable.

preliminary *adj* advance, earliest, early, experimental, exploratory, first, inaugural, initial, introductory, opening, prefatory, preparatory, qualifying, tentative, trial. • *n* ▷ PRELUDE.

prelude *n* beginning, *inf* curtain-raiser, exordium, introduction, opener, opening, overture, preamble, precursor, preface, preliminary, preparation, proem, prolegomenon, prologue, start, starter, *inf* warm-up. *Opp* CONCLUSION, POSTSCRIPT.

premature *adj* abortive, before time, early, hasty, ill-timed, precipitate, *inf* previous, too early, too soon, undeveloped, untimely. *Opp* LATE.

premeditated *adj* calculated, conscious, considered, contrived, deliberate, intended, intentional, planned, prearranged, preconceived, predetermined, preplanned, studied, wilful. *Opp* SPONTANEOUS.

premiss *n* assertion, assumption, basis, grounds, hypothesis, proposition, supposition, theorem.

premonition *n* anxiety, fear, foreboding, forewarning, *inf* funny feeling, *inf* hunch, indication, intuition, misgiving, omen, portent, presentiment, suspicion, warning, worry.

preoccupied *adj* **1** absorbed, engaged, engrossed, immersed, interested, involved, obsessed, sunk, taken up, wrapped up. **2** absent-minded, abstracted, day-dreaming, distracted, faraway, inattentive, lost in thought, musing, pensive, pondering, rapt, reflecting, thoughtful.

preparation *n* arrangements, briefing, *inf* gearing up, getting ready, groundwork, making provision, measures, organization, plans, practice, preparing, setting up, spadework, training.

prepare *vb* **1** arrange, cook, devise, *inf* do what's necessary, *inf* fix up, get ready, make arrangements, make ready, organize, pave the way, plan, process, set up. ▷ MAKE. **2** *prepare for exams.* *inf* cram, practise, revise, study, *inf* swot. **3** *prepare pupils for exams.* brief, coach, educate, equip, instruct, rehearse, teach, train, tutor. **prepared** ▷ PRE-ARRANGED, READY. **prepare yourself** be prepared, be ready, brace yourself, discipline yourself, fortify yourself, steel yourself.

preposterous *adj* bizarre, excessive, extreme, grotesque, monstrous, outrageous, surreal, unreasonable, unthinkable. ▷ ABSURD.

prerequisite *adj* compulsory, essential, indispensable, mandatory, necessary, obligatory, prescribed, required, requisite, specified, stipulated. *Opp* OPTIONAL. • *n* condition, essential, necessity, precondition, proviso, qualification, requirement, requisite, *Lat* sine qua non, stipulation.

prescribe *n* advise, assign, command, demand, dictate, direct, fix, impose, instruct, lay down, ordain, order, recommend, require, specify, stipulate, suggest.

presence *n* **1** attendance, closeness, companionship, company, nearness, propinquity, proximity, society. **2** air, appearance, aura, bearing, comportment, demeanour, impressiveness, mien, personality, poise, selfassurance, self-possession.

present *adj* **1** adjacent, at hand, close, here, in attendance, nearby. **2** contemporary, current, existing, extant, present-day, up-to-date. • *n* **1** *inf* here and now, today. **2** alms, bonus, bounty, charity, contribution, donation, endowment, gift, grant, gratuity, handout, offering, tip. • *vb* **1** award, bestow, confer, dispense, distribute, donate, give, hand over, offer. **2** *present evidence.* adduce, bring forward, demonstrate, display, exhibit, furnish, proffer, put forward, reveal, set out, show, submit. **3** *present a guest.* announce, introduce, make known. **4** *present a play.* act, bring out, perform, put on, stage. **present yourself** ▷ ATTEND, REPORT.

presentable *adj* acceptable, adequate, all right, clean, decent, decorous, fit to be seen, good enough, neat, passable, proper, respectable, satisfactory, suitable, tidy, tolerable, *inf* up to scratch, worthy.

presently *adv old use* anon, before long, by and by, *inf* in a jiffy, shortly, soon.

preserve *n* **1** conserve, jam, jelly, marmalade. **2** *wildlife preserve.* reservation, reserve, sanctuary. • *vb* care for, conserve, defend, guard, keep, lay up, look after, maintain, perpetuate, protect, retain, safeguard, save, secure, stockpile, store, support, sustain, uphold, watch over. *Opp* DESTROY. **2** *preserve food.* bottle, can, chill, cure, dehydrate, dry, freeze, freeze-dry, irradiate, jam, pickle, refrigerate, salt, tin. **3** *preserve a corpse.* embalm, mummify.

preside *vb* be in charge, chair, officiate, take charge, take the chair. **preside over** ▷ GOVERN.

press *n* newspapers, magazines, the media. • *vb* **1** apply pressure to, compress, condense, cram, crowd, crush, depress, force, gather, *inf* jam, push, shove, squash, squeeze, subject to pressure. **2** *press laundry.* flatten, iron, smooth. **3** *press someone to stay.* ask, beg, bully, coerce, constrain, dragoon, entreat, exhort, implore, importune, induce, *inf* lean on, persuade, pressure, pressurize, put pressure on, request, require, urge. **pressing** ▷ URGENT.

pressure *n* **1** burden, compression, force, heaviness, load, might, power, stress, weight. **2** *pressure of modern life.* adversity, affliction, constraints, demands, difficulties, exigencies, *inf* hassle, hurry, oppression, problems, strain, stress, urgency. • *vb* ask, beg, bully, coerce, constrain, dragoon, entreat, exhort, implore, importune, induce, *inf* lean on, persuade, press, pressurize, put pressure on, request, require, urge.

prestige *n* cachet, celebrity, credit, distinction, eminence, esteem, fame, glory, good name, honour, importance, influence, *inf* kudos, regard, renown, reputation, respect, standing, stature, status.

prestigious *adj* acclaimed, august, celebrated, creditable, distinguished, eminent, esteemed, estimable, famed, famous, highly-regarded, high-ranking, honourable, honoured, important, influential, pre-eminent, renowned, reputable, respected, significant, wellknown. *Opp* INSIGNIFICANT.

presume *vb* **1** assume, believe, conjecture, gather, guess, hypothesize, imagine, infer, postulate, suppose, surmise, suspect, *inf* take for granted, *inf* take it, think. **2** *He presumed to correct me.* be presumptuous enough, dare, have the effrontery, make bold, take the liberty, venture.

presumptuous *adj* arrogant, bold, brazen, *inf* cheeky, conceited, forward, impertinent, impudent, insolent, overconfident, *inf* pushy, shameless, unauthorized, unwarranted. ▷ PROUD.

pretence *n* act, acting, affectation, appearance, artifice, camouflage, charade, counterfeiting, deception, disguise, display, dissembling, dissimulation, excuse, façade, falsification, feigning, feint, fiction, front, guise, hoax, *inf* humbug, hypocrisy, insincerity, invention, lying, make-believe, masquerade, pose, posing, posturing, pretext, ruse, sham, show, simulation, subterfuge, trickery, wile. ▷ DECEIT.

pretend *vb* **1** act, affect, allege, behave insincerely, bluff, counterfeit, deceive, disguise, dissemble, dissimulate, fake, feign, fool, hoax, hoodwink, imitate, impersonate, *inf* kid, lie, *inf* make out, mislead, play-act, play a part, perform, pose, posture, profess, purport, put on an act, sham, simulate, take someone in, trick. **2** ▷ IMAGINE. **3** ▷ CLAIM.

pretender *n* aspirant, claimant, rival, suitor.

pretentious *adj* affected, *inf* arty, conceited, exaggerated, extravagant, grandiose, *inf* highfalutin, immodest, inflated, ostentatious, overblown, *inf* over the top, pompous, showy, *inf* snobbish, superficial. *Opp* UNPRETENTIOUS.

pretext *n* cloak, cover, disguise, excuse, pretence.

pretty *adj* appealing, attractive, *inf* bonny, charming, *inf* cute, dainty, delicate, *inf* easy on the eye, fetching, good-looking, lovely, nice, pleasing, *derog* pretty-pretty, winsome. ▷ BEAUTIFUL. *Opp* UGLY. • *adv* fairly, moderately, quite, rather, reasonably, somewhat, tolerably. *Opp* VERY.

prevail *vb* be prevalent, hold sway, predominate, preponderate, succeed, triumph, *inf* win the day. ▷ WIN. **prevailing** ▷ PREVALENT.

prevalent *adj* accepted, ascendant, chief, common, commonest, current, customary, dominant, dominating, effectual, established, extensive, familiar, fashionable, general, governing, influential, main, mainstream, normal, ordinary, orthodox, pervasive, popular, powerful, predominant, prevailing, principal, ruling, ubiquitous, universal, usual, widespread. *Opp* UNUSUAL.

prevaricate *vb inf* beat about the bush, be evasive, cavil, deceive, equivocate, *inf* fib, hedge, lie, mislead, quibble, temporize.

prevent *vb* anticipate, avert, avoid, baffle, bar, block, check, control, curb, deter, fend off, foil, forestall, frustrate, hamper, *inf* head off, *inf* help (*can't help it*), hinder, impede, inhibit, inoculate against, intercept, *inf* nip in the bud, obstruct, obviate, preclude, pre-empt, prohibit, *inf* put a stop to, restrain, save, stave off, stop, take precautions against, thwart, ward off. ▷ FORBID. *Opp* ENCOURAGE.

preventive *adj* anticipatory, counteractive, deterrent, obstructive, precautionary, pre-emptive, preventative.

previous *adj* **1** above-mentioned, aforementioned, aforesaid, antecedent, earlier, erstwhile, foregoing, former, past, preceding, prior. *Opp* SUBSEQUENT. **2** ▷ PREMATURE.

prey *n* kill, quarry, victim. • *vb* **prey on** eat, feed on, hunt, kill, live off. ▷ EXPLOIT.

price *n* **1** amount, charge, cost, *inf* damage, expenditure, expense, fare, fee, figure, outlay, payment, rate, sum, terms, toll, value, worth. **2** *Give me a price.* estimate, offer, quotation, valuation. ▷ PAYMENT. **pay the price for** ▷ ATONE.

priceless *adj* **1** costly, dear, expensive, incalculable, inestimable, invaluable, irreplaceable, precious, *inf* pricey, rare, valuable. *Opp* WORTHLESS. **2** ▷ FUNNY.

prick *vb* **1** bore into, jab, lance, perforate, pierce, punch, puncture, riddle, stab, sting. **2** ▷ STIMULATE.

prickle *n* **1** barb, bristle, bur, needle, spike, spine, thorn. **2** irritation, itch, pricking, prickling, tingle, tingling. • *vb* irritate, itch, make your skin crawl, scratch, sting, tingle.

prickly *adj* **1** barbed, bristly, rough, scratchy, sharp, spiky, spiny, stubbly, thorny, unshaven. *Opp* SMOOTH. **2** ▷ IRRITABLE.

pride *n* **1** *Fr* amour propre, gratification, happiness, honour, pleasure, satisfaction, self-respect, self-satisfaction. ▷ DIGNITY. **2** *her pride and joy.* jewel, treasure, treasured possession. 3 *pride before a fall.* arrogance, being proud, *inf* big-headedness, boastfulness, conceit, egotism, haughtiness, hubris, megalomania, narcissism, overconfidence, presumption, self-admiration, self-esteem, self-importance, self-love, smugness, snobbery, snobbishness, vainglory, vanity. *Opp* HUMILITY.

priest *n* confessor, Druid, lama, minister, preacher. ▷ CLERGYMAN.

priggish *adj* conservative, fussy, *inf* goody-goody, haughty, moralistic, prudish, self-righteous, sententious, stiff-necked, *inf* stuffy. ▷ PRIM.

prim *adj* demure, fastidious, formal, inhibited, narrow-minded, precise, *inf* prissy, proper, prudish, *inf* starchy, strait-laced. *Opp* BROADMINDED.

primal *adj* 1 early, earliest, first, original, primeval, primitive, primordial. 2 ▷ PRIMARY.

primarily *adv* basically, chiefly, especially, essentially, firstly, fundamentally, generally, mainly, mostly, particularly, predominantly, pre-eminently, principally.

primary *adj* basic, cardinal, chief, dominant, first, foremost, fundamental, greatest, important, initial, leading, main, major, outstanding, paramount, predominant, pre-eminent, primal, prime, principal, supreme, top.

prime *adj* 1 best, first-class, first-rate, foremost, select, superior, top, top-quality. ▷ EXCELLENT. 2 ▷ PRIMARY. • *vb* get ready, prepare.

primitive *adj* 1 aboriginal, ancient, barbarian, early, first, prehistoric, primeval, savage, uncivilized, uncultivated. 2 *primitive technology.* antediluvian, backward, basic, *inf* behind the times, crude, elementary, obsolete, rough, rudimentary, simple, simplistic, undeveloped. ▷ OLD. 3 *primitive art.* childlike, crude, naïve, unpolished, unrefined, unsophisticated. *Opp* ADVANCED, SOPHISTICATED.

principal *adj* basic, cardinal, chief, dominant, dominating, first, foremost, fundamental, greatest, highest, important, key, leading, main, major, outstanding, paramount, preeminent, predominant, prevailing, primary, prime, starring, supreme, top. • *n* 1 ▷ CHIEF. 2 *the principal in a play.* diva, hero, heroine, lead, leading role, prima ballerina, protagonist, star.

principle *n* 1 assumption, axiom, belief, creed, criterion, doctrine, dogma, ethic, idea, ideal, maxim, notion, precept, proposition, rule, standard, teaching, tenet, truism, truth, values. 2 *a person of principle.* conscience, high-mindedness, honesty, honour, ideals, integrity, morality, probity, scruples, standards, uprightness, virtue. **principles** basics, elements, essentials, fundamentals, laws, philosophy, theory.

print *n* 1 impression, imprint, indentation, mark, stamp. 2 characters, fount, lettering, letters, printing, text, type, typeface. 3 copy, duplicate, engraving, etching, facsimile, linocut, lithograph, monoprint, photograph, reproduction, silk screen, woodcut. ▷ PICTURE. • *vb* 1 copy, impress, imprint, issue, publish, run off, stamp. 2 ▷ WRITE.

prior *adj* earlier, erstwhile, former, late, old, onetime, previous.

priority *n* first place, greater importance, precedence, preference, prerogative, right-of-way, seniority, superiority, urgency.

prise *vb* force, lever, prize, wrench.

prison *n old use* approved school, *old use* Borstal, cell, *sl* clink, custody, detention centre, dungeon, gaol, guardhouse, house of correction, jail, *inf* lock-up, *Amer* penitentiary, oubliette, reformatory, *sl* stir, youth custody centre. ▷ CAPTIVITY.

prisoner *n* captive, convict, detainee, *inf* gaolbird, hostage, inmate, internee, lifer, *inf* old lag, *inf* trusty.

privacy *n* concealment, isolation, monasticism, quietness, retirement, retreat, seclusion, secrecy, solitude.

private *adj* **1** exclusive, individual, particular, personal, privately owned, reserved. **2** classified, confidential, *inf* hush-hush, *inf* off the record, restricted, secret, top secret, undisclosed. **3** *a private meeting.* clandestine, closed, covert, intimate, surreptitious. **4** *a private hideaway.* concealed, hidden, inaccessible, isolated, little-known, quiet, secluded, sequestered, solitary, unknown, withdrawn. *Opp* PUBLIC.

privilege *n* advantage, benefit, concession, entitlement, exemption, freedom, immunity, licence, prerogative, right.

privileged *adj* **1** advantaged, authorized, élite, entitled, favoured, honoured, immune, licensed, powerful, protected, sanctioned, special, superior. **2** ▷ WEALTHY.

prize *n* accolade, award, jackpot, *inf* purse, reward, trophy, winnings. • *vb* **1** appreciate, approve of, cherish, esteem, hold dear, like, rate highly, regard, revere, treasure, value. **2** ▷ PRISE.

probable *adj* believable, convincing, credible, expected, feasible, likely, *inf* odds-on, plausible, possible, predictable, presumed, undoubted, unquestioned. *Opp* IMPROBABLE.

probationer *n* apprentice, beginner, inexperienced worker, learner, novice, tiro.

probe *n* enquiry, examination, exploration, inquiry, investigation, research, scrutiny, study. • *vb* **1** delve, dig, penetrate, plumb, poke, prod. **2** *probe a problem.* examine, explore, go into, inquire into, investigate, look into, research into, scrutinize, study.

problem *n* **1** brain-teaser, conundrum, enigma, mystery, *inf* poser, puzzle, question, riddle. **2** *a worrying problem.* burden, *inf* can of worms, complication, difficulty, dilemma, dispute, *inf* facer, *inf* headache, *inf* hornet's nest, predicament, quandary, set-back, snag, trouble, worry.

problematic *adj* complicated, controversial, debatable, difficult, disputed, doubtful, enigmatic, hard to deal with, *inf* iffy, intractable, moot (*point*), problematical, puzzling, questionable, sensitive, taxing, *inf* tricky, uncertain, unsettling, worrying. *Opp* STRAIGHTFORWARD.

procedure *n* approach, conduct, course of action, *inf* drill, formula, method, methodology, *Lat* modus operandi, plan of action, policy, practice, process, routine, scheme, strategy, system, technique, way.

proceed *vb* **1** advance, carry on, continue, follow, forge ahead, *inf* get going, go ahead, go on, make headway, make progress, move along, move forward, *inf* press on, progress. **2** arise, be derived, begin, develop, emerge, grow, originate, spring up, start. ▷ RESULT.

proceedings *plur n* 1 events, *inf* goings-on, happenings, things. 2 *legal proceedings.* action, lawsuit, procedure, process. 3 *proceedings of a meeting.* annals, business, dealings, *inf* doings, matters, minutes, records, report, transactions.

proceeds *plur n* earnings, gain, gate, income, profit, receipts, returns, revenue, takings. ▷ MONEY.

process *n* 1 function, method, operation, procedure, proceeding, system, technique. 2 *process of ageing.* course, development, evolution, experience, progression. • *vb* 1 alter, change, convert, deal with, make usable, manage, modify, organize, prepare, refine, transform, treat. 2 ▷ PARADE.

procession *n* cavalcade, chain, column, cortège, file, line, march, march-past, motorcade, pageant, parade, sequence, string, succession, train.

proclaim *vb* 1 announce, advertise, assert, declare, give out, make known, profess, promulgate, pronounce, publish. 2 ▷ DECREE.

procrastinate *vb* be dilatory, be indecisive, dally, defer a decision, delay, *inf* dilly-dally, dither, *inf* drag your feet, equivocate, evade the issue, hesitate, *inf* hum and haw, pause, *inf* play for time, postpone, put things off, *inf* shilly-shally, stall, temporize, vacillate, waver.

procure *vb* acquire, buy, come by, find, get, *inf* get hold of, *inf* lay your hands on, obtain, *inf* pick up, purchase, requisition.

prod *vb* dig, elbow, goad, jab, nudge, poke, push, urge on. ▷ HIT, URGE.

prodigal *adj* excessive, extravagant, immoderate, improvident, irresponsible, lavish, profligate, reckless, self-indulgent, wasteful. *Opp* THRIFTY.

prodigy *n* curiosity, freak, genius, marvel, miracle, phenomenon, rarity, sensation, talent, virtuoso, *inf* whizz kid, wonder, *Ger* Wunderkind.

produce *n* crop, harvest, output, yield. ▷ PRODUCT. • *vb* 1 assemble, bring out, cause, compose, conjure up, construct, create, cultivate, develop, fabricate, form, generate, give rise to, grow, initiate, invent, make, manufacture, originate, provoke, result in, supply, think up, turn out, yield. 2 *produce evidence.* advance, bring out, disclose, display, exhibit, furnish, introduce, offer, present, provide, put forward, reveal, show, supply, throw up. 3 *produce children.* bear, beget, breed, give birth to, raise, rear. 4 *produce a play.* direct, mount, present, put on, stage.

product *n* 1 artefact, by-product, commodity, end-product, goods, merchandise, output, produce, production. 2 consequence, effect, fruit, issue, outcome, result, upshot, yield.

productive *adj* 1 beneficial, busy, constructive, creative, effective, efficient, gainful (*employment*), inventive, profitable, profitmaking, remunerative, rewarding, useful, valuable, worthwhile. 2 *a productive garden.* abundant, bounteous, bountiful, fecund, fertile, fruitful, lush, prolific, vigorous. *Opp* UNPRODUCTIVE.

profess *vb* **1** affirm, announce, assert, asseverate, aver, confess, confirm, declare, maintain, state, vow. **2** *profess to be an expert.* allege, claim, make out, pretend, purport.

profession *n* **1** business, calling, career, craft, employment, job, line of work, métier, occupation, trade, vocation, work. **2** *profession of love.* acknowledgement, affirmation, announcement, assertion, avowal, confession, declaration, statement, testimony.

professional *adj* **1** able, authorized, educated, experienced, expert, knowledgeable, licensed, official, proficient, qualified, skilled, trained. *Opp* AMATEUR. **2** full-time, paid. **3** *professional work.* businesslike, competent, conscientious, efficient, masterly, proper, skilful, thorough, well-done. *Opp* UNPROFESSIONAL. • *n* expert, professional player, professional worker.

proficient *adj* able, accomplished, adept, capable, competent, efficient, expert, gifted, professional, skilled, talented. *Opp* INCOMPETENT.

profile *n* **1** contour, outline, shape, side view, silhouette. **2** *personal profile.* account, biography, *Lat* curriculum vitae, sketch, study.

profit *n* advantage, benefit, excess, gain, interest, proceeds, return, revenue, surplus, yield. • *vb* **1** advance, avail, benefit, further the interests of, pay, serve. ▷ HELP. **2** *profit from a sale.* capitalize (on), *inf* cash in, earn money, gain, *inf* make a killing, make a profit, make money. **profit by**, **profit from** ▷ EXPLOIT.

profitable *adj* advantageous, beneficial, commercial, enriching, fruitful, gainful, lucrative, money-making, paying, productive, profit-making, remunerative, rewarding, useful, valuable, well-paid, worthwhile. *Opp* UNPROFITABLE.

profiteer *n* black-marketeer, exploiter, extortionist, racketeer. • *vb* exploit, extort, fleece, overcharge.

profligate *adj* **1** abandoned, debauched, degenerate, depraved, dissolute, immoral, libertine, licentious, loose, perverted, promiscuous, sinful, sybaritic, unprincipled, wanton. ▷ WICKED. **2** extravagant, prodigal, reckless, spendthrift, wasteful.

profound *adj* **1** deep, heartfelt, intense, sincere. **2** *a profound discussion.* abstruse, arcane, erudite, esoteric, imponderable, informed, intellectual, knowledgeable, learned, penetrating, philosophical, recondite, sagacious, scholarly, serious, thoughtful, wise. **3** *profound silence.* absolute, complete, extreme, fundamental, perfect, thorough, total, unqualified. *Opp* SUPERFICIAL.

profuse *adj* abundant, ample, bountiful, copious, extravagant, exuberant, generous, lavish, luxuriant, plentiful, productive, prolific, superabundant, thriving, unsparing, unstinting. *Opp* MEAN, SPARSE.

programme *n* **1** agenda, bill of fare, calendar, curriculum, *inf* line-up, listing, menu, plan, routine, prospectus, schedule, scheme, syllabus, timetable. **2** *a TV programme.* broadcast, performance, presentation, production, show, transmission.

progress *n* **1** advance, breakthrough, development, evolution, forward movement, furtherance, gain, growth, headway, improvement, march (*of time*), maturation, progression, *inf* step forward. 2 journey, route, travels, way. 3 *progress in a career*. advancement, betterment, elevation, promotion, rise, *inf* step up. • *vb* advance, *inf* come on, develop, *inf* forge ahead, go forward, go on, make headway, make progress, move forward, press forward, press on, proceed, prosper. ▷ IMPROVE. *Opp* REGRESS, STAGNATE.

progression *n* 1 ▷ PROGRESS. 2 chain, concatenation, course, flow, order, row, sequence, series, string, succession.

progressive *adj* 1 accelerating, advancing, continuing, continuous, developing, escalating, gradual, growing, increasing, ongoing, steady. 2 *progressive ideas*. advanced, avant-garde, contemporary, dynamic, enterprising, forward-looking, *inf* go-ahead, modernistic, radical, reformist, revisionist, revolutionary, up-to-date. *Opp* CONSERVATIVE.

prohibit *vb* ban, bar, block, censor, check, *inf* cut out, debar, disallow, exclude, foil, forbid, hinder, impede, inhibit, interdict, make illegal, outlaw, place an embargo on, preclude, prevent, proscribe, restrict, rule out, shut out, stop, taboo, veto. *Opp* ALLOW.

prohibitive *adj* discouraging, excessive, exorbitant, impossible, *inf* out of reach, out of the question, unreasonable, unthinkable.

project *n* activity, assignment, contract, design, enterprise, idea, job, piece of research, plan, programme, proposal, scheme, task, undertaking, venture. • *vb* 1 concoct, contrive, design, devise, invent, plan, propose, scheme, think up. 2 beetle, bulge, extend, jut out, overhang, protrude, stand out, stick out. 3 *project into space*. cast, *inf* chuck, fling, hurl, launch, lob, propel, shoot, throw. 4 *project light*. cast, flash, shine, throw out. 5 *project future profits*. estimate, forecast, predict.

proliferate *vb* burgeon, flourish, grow, increase, multiply, mushroom, reproduce, thrive.

prolific *adj* 1 abundant, bounteous, bountiful, copious, fruitful, numerous, plenteous, profuse, rich. 2 *a prolific writer*. creative, fertile, productive. *Opp* UNPRODUCTIVE.

prolong *vb* delay, *inf* drag out, draw out, elongate, extend, increase, keep up, lengthen, make longer, *inf* pad out, protract, *inf* spin out, stretch out. *Opp* SHORTEN.

prominent *adj* 1 conspicuous, discernible, distinguishable, evident, eye-catching, large, notable, noticeable, obtrusive, obvious, pronounced, recognizable, salient, significant, striking. *Opp* INCONSPICUOUS. 2 bulging, jutting out, projecting, protruding, protuberant, sticking out. 3 celebrated, distinguished, eminent, familiar, foremost, illustrious, important, leading, major, much-publicized, noted, outstanding, public, renowned. ▷ FAMOUS. *Opp* UNKNOWN.

promiscuous *adj* casual, haphazard, indiscriminate, irresponsible, non-selective, random, undiscriminating. ▷ IMMORAL. *Opp* MORAL.

promise *n* 1 assurance, commitment, compact, contract, covenant, guarantee, oath, pledge, undertaking, vow, word, word of honour. 2 *actor with promise*. capability, expectation(s), latent ability, potential, promising qualities, talent. • *vb* 1 agree, assure, commit yourself, consent, contract, engage, give a promise, give your word, guarantee, pledge, swear, take an oath, undertake, vow. 2 *The clouds promise rain.* augur, *old use* betoken, forebode, foretell, hint at, indicate, presage, prophesy, show signs of, suggest.

promising *adj* auspicious, budding, encouraging, favourable, hopeful, likely, optimistic, propitious, talented, *inf* up-and-coming.

promontory *n* cape, foreland, headland, peninsula, point, projection, ridge, spit, spur.

promote *vb* 1 advance, elevate, exalt, give promotion, move up, prefer, raise, upgrade. 2 *promote a product.* advertise, back, boost, champion, encourage, endorse, further, help, make known, market, patronize, *inf* plug, popularize, publicize, *inf* push, recommend, sell, speak for, sponsor, support. ▷ HELP.

promoter *n* backer, champion, patron, sponsor, supporter.

promotion *n* 1 advancement, elevation, preferment, rise, upgrading. 2 *promotion of a product.* advertising, backing, encouragement, furtherance, marketing, publicity, recommendation, selling, sponsorship.

prompt *adj* eager, efficient, expeditious, immediate, instantaneous, on time, punctual, timely, unhesitating, willing. ▷ QUICK. *Opp* UNPUNCTUAL. • *n* cue, line, reminder. • *vb* advise, coax, egg on, encourage, exhort, help, incite, influence, inspire, jog the memory, motivate, nudge, persuade, prod, provoke, remind, rouse, spur, stimulate, urge.

prone *adj* 1 face down, flat, horizontal, lying, on your front, prostrate, stretched out. *Opp* SUPINE. 2 *prone to colds.* apt, disposed, given, inclined, liable, likely, predisposed, subject, susceptible, tending, vulnerable. *Opp* IMMUNE.

prong *n* point, spike, spur, tine.

pronounce *vb* 1 articulate, aspirate, enunciate, express, put into words, say, sound, speak, utter, vocalize, voice. 2 *pronounce judgement.* announce, assert, asseverate, declare, decree, judge, make known, proclaim, state. ▷ SPEAK.

pronounced *adj* clear, conspicuous, decided, definite, distinct, evident, inescapable, marked, noticeable, obvious, prominent, recognizable, striking, unambiguous, undisguised, unmistakable, well-defined.

pronunciation *n* accent, articulation, delivery, diction, elocution, enunciation, inflection, intonation, modulation, speech.

proof *n* 1 authentication, certification, confirmation, corroboration, demonstration, evidence, facts, grounds, substantiation, testimony, validation, verification. 2 *the proof of the pudding.* criterion, judgement, measure, test, trial.

prop *n* brace, buttress, crutch, post, stay, strut, support, truss, upright. • *vb* 1 bolster, brace, buttress, hold up, reinforce, shore up, support, sustain. 2 lean, rest, stand.

propaganda *n* advertising, brainwashing, disinformation, indoctrination, persuasion, publicity.

propagate *vb* 1 breed, generate, increase, multiply, produce, proliferate, reproduce. 2 *propagate ideas.* circulate, disseminate, pass on, promote, promulgate, publish, spread, transmit. **3** *propagate plants.* grow from seed, layer, sow, take cuttings.

propel *vb* drive, force, impel, launch, move, *inf* pitchfork, push, send, set in motion, shoot, spur, thrust, urge.

propeller *n* rotor, screw, vane.

proper *adj* **1** becoming, conventional, decent, decorous, delicate, dignified, formal, genteel, gentlemanly, grave, in good taste, ladylike, modest, polite, *derog* prim, *derog* prudish, respectable, sedate, seemly, serious, solemn, tactful, tasteful. **2** acceptable, accepted, advisable, apposite, appropriate, apropos, apt, deserved, fair, fitting, just, lawful, legal, normal, orthodox, rational, sensible, suitable, unexceptionable, usual, valid. **3** *the proper time.* accurate, correct, exact, precise, right. **4** *the proper place.* allocated, distinctive, individual, own, particular, reserved, separate, special, unique. *Opp* IMPROPER.

property *n* **1** assets, belongings, capital, chattels, effects, fortune, *inf* gear, goods, holdings, patrimony, possessions, resources, riches, wealth. **2** acreage, buildings, estate, land, premises. **3** attribute, characteristic, feature, hallmark, idiosyncrasy, oddity, peculiarity, quality, quirk, trait.

prophecy *n* augury, crystal-gazing, divination, forecast, foretelling, fortune-telling, oracle, prediction, prognosis, prognostication, vaticination.

prophesy *vb* augur, bode, divine, forecast, foresee, foreshadow, foretell, portend, predict, presage, prognosticate, promise, vaticinate.

prophet *n* clairvoyant, forecaster, fortune-teller, oracle, seer, sibyl, soothsayer.

prophetic *adj* apocalyptic, far-seeing, oracular, predictive, prescient, prognostic, prophesying, sibylline.

propitious *adj* advantageous, auspicious, favourable, fortunate, happy, lucky, opportune, promising, providential, rosy, timely, well-timed.

proportion *n* **1** balance, comparison, correlation, correspondence, distribution, equivalence, ratio, statistical relationship. **2** allocation, fraction, part, percentage, piece, quota, ration, section, share. ▷ NUMBER, QUANTITY. **proportions** dimensions, extent, magnitude, measurements, size, volume.

proportional *adj* analogous, balanced, commensurate, comparable, corresponding, equitable, in proportion, just, proportionate, relative, symmetrical. *Opp* DISPROPORTIONATE.

proposal *n* bid, declaration, draft, motion, offer, plan, project, proposition, recommendation, scheme, statement, suggestion, tender.

propose *vb* **1** advance, ask for, *inf* come up with, present, propound, put forward, recommend, submit, suggest. **2** aim, have in mind, intend, mean, offer, plan, purpose. **3** *propose a candidate.* nominate, put forward, put up, sponsor.

propriety *n* appropriateness, aptness, correctness, courtesy, decency, decorum, delicacy, dignity, etiquette, fairness, fitness, formality, gentility, good form, good manners, gravity, justice, modesty, politeness, *derog* prudishness, refinement, respectability, sedateness, seemliness, sensitivity, suitability, tact, tastefulness. *Opp* IMPROPRIETY.

prosaic *adj* **1** clear, direct, down to earth, factual, matter-of-fact, plain, simple, straightforward, to the point, unadorned, understandable, unemotional, unsentimental, unvarnished. **2** [*derog*] characterless, clichéd, commonplace, dry, dull, flat, hackneyed, lifeless, monotonous, mundane, pedestrian, prosy, routine, stereotyped, trite, unfeeling, unimaginative, uninspired, uninspiring, unpoetic, unromantic. ▷ ORDINARY. *Opp* POETIC.

prosecute *vb* **1** accuse, arraign, bring an action against, bring to trial, charge, indict, institute legal proceedings against, prefer charges against, put on trial, sue, take legal proceedings against, take to court. **2** ▷ PURSUE.

prospect *n* **1** aspect, landscape, outlook, panorama, perspective, scene, seascape, sight, spectacle, view, vista. **2** *prospect of fine weather.* anticipation, chance, expectation, hope, likelihood, opportunity, possibility, probability, promise. • *vb* explore, quest, search, survey.

prospective *adj* anticipated, approaching, awaited, coming, expected, forthcoming, future, imminent, impending, intended, likely, looked-for, negotiable, pending, possible, potential, probable.

prospectus *n* announcement, brochure, catalogue, leaflet, manifesto, pamphlet, programme, scheme, syllabus.

prosper *vb* become prosperous, be successful, *inf* boom, burgeon, develop, do well, fare well, flourish, *inf* get ahead, *inf* get on, *inf* go from strength to strength, grow, *inf* make good, *inf* make your fortune, profit, progress, strengthen, succeed, thrive. *Opp* FAIL.

prosperity *n* affluence, *inf* bonanza, *inf* boom, good fortune, growth, opulence, plenty, profitability, riches, success, wealth.

prosperous *adj* affluent, *inf* blooming, *inf* booming, buoyant, expanding, flourishing, fruitful, healthy, moneyed, moneymaking, productive, profitable, prospering, rich, successful, thriving, vigorous, wealthy, *inf* well-heeled, well-off, well-to-do. *Opp* UNSUCCESSFUL.

prostitute *n old use* bawd, call girl, *old use* camp follower, *old use* courtesan, *old use* harlot, *inf* hooker, streetwalker, *old use* strumpet, *inf* tart, toy boy, trollop, whore. • *vb* cheapen, debase, degrade, demean, devalue, lower, misuse.

prostrate *adj* ▷ OVERCOME, PRONE. • *vb prostrate yourself* abase yourself, bow, kneel, kowtow, lie flat, submit. ▷ GROVEL.

protagonist *n* chief actor, contender, contestant, hero, heroine, lead, leading figure, principal, title role.

protect *vb* care for, cherish, conserve, defend, escort, guard, harbour, insulate, keep, keep safe, look after, mind, preserve, provide cover for, safeguard, screen, secure, shield, stand up for, support, take care of, tend, watch over. *Opp* ENDANGER, NEGLECT.

protection *n* **1** care, conservation, custody, defence, guardianship, patronage, preservation, safekeeping, safety, security, tutelage. **2** barrier, buffer, bulwark, cloak, cover, guard, insulation, screen, shelter, shield.

protective *adj* 1 fireproof, insulating, preservative, protecting, sheltering, shielding, waterproof. 2 *protective parents.* careful, defensive, heedful, jealous, paternalistic, possessive, solicitous, vigilant, watchful.

protector *n* benefactor, bodyguard, champion, defender, guard, guardian, *sl* minder, patron.

protest *n* 1 complaint, cry of disapproval, demur, demurral, dissent, exception, grievance, *inf* gripe, *inf* grouse, grumble, objection, opposition, outcry, protestation, remonstrance. 2 *inf* demo, demonstration, march, rally. • *vb* 1 appeal, argue, challenge a decision, complain, cry out, expostulate, express disapproval, fulminate, *inf* gripe, *inf* grouse, grumble, make a protest, *inf* moan, object, remonstrate, take exception. 2 demonstrate, *inf* hold a demo, march. 3 *protest your innocence.* affirm, assert, asseverate, aver, declare, insist on, profess, swear.

protracted *adj* endless, extended, interminable, long-drawn-out, long-winded, never-ending, prolonged, spun-out. ▷ LONG. *Opp* SHORT.

protrude *vb* balloon, bulge, extend, jut out, overhang, poke out, project, stand out, stick out, stick up, swell.

protruding *adj* bulbous, bulging, distended, gibbous, humped, jutting, overhanging, projecting, prominent, protuberant, swollen, tumescent.

proud *adj* 1 appreciative, delighted, glad, gratified, happy, honoured, pleased, satisfied. 2 *a proud bearing.* brave, dignified, independent, self-respecting. 3 *a proud history.* august, distinguished, glorious, great, honourable, illustrious, noble, reputable, respected, splendid, worthy. 4 [*derog*] arrogant, *inf* big-headed, boastful, bumptious, *inf* cocky, *inf* cocksure, conceited, disdainful, egocentric, egotistical, grand, haughty, *inf* high and mighty, immodest, lordly, narcissistic, self-centred, self-important, self-satisfied, smug, snobbish, *inf* snooty, *inf* stuck-up, supercilious, *inf* swollen-headed, *inf* toffee-nosed, vain, vainglorious. *Opp* MODEST.

provable *adj* demonstrable, verifiable. *Opp* UNPROVABLE.

prove *vb* ascertain, assay, attest, authenticate, *inf* bear out, certify, check, confirm, corroborate, demonstrate, establish, explain, justify, show to be true, substantiate, test, verify. *Opp* DISPROVE.

proven *adj* accepted, proved, reliable, tried and tested, trustworthy, undoubted, unquestionable, valid, verified. *Opp* DOUBTFUL, THEORETICAL.

proverb *n* adage, maxim, *old use* saw. ▷ SAYING.

proverbial *adj* aphoristic, axiomatic, clichéd, conventional, customary, famous, legendary, time-honoured, traditional, well-known.

provide *vb* afford, allot, allow, arrange for, cater, contribute, donate, endow, equip, *inf* fix up with, *inf* fork out, furnish, give, grant, lay on, lend, make provision, offer, present, produce, purvey, spare, stock, supply, yield.

providence *n* destiny, divine intervention, fate, fortune, karma, kismet.

provident *adj* careful, economical, far-sighted, forward-looking, frugal, judicious, prudent, thrifty.

providential *adj* felicitous, fortunate, happy, lucky, opportune, timely.

provincial *adj* **1** local, regional. *Opp* NATIONAL. **2** [*derog*] backward, boorish, bucolic, insular, narrow-minded, parochial, rural, rustic, small-minded, uncultivated, uncultured, unsophisticated. *Opp* COSMOPOLITAN.

provisional *adj* conditional, interim, stopgap, temporary, tentative, transitional. *Opp* DEFINITIVE, PERMANENT.

provisions *plur n* food, foodstuff, groceries, provender, rations, requirements, stocks, stores, subsistence, supplies, *old use* victuals.

proviso *n* condition, exception, limitation, provision, qualification, requirement, restriction, rider, stipulation.

provocation *n inf* aggravation, cause, challenge, grievance, grounds, incentive, incitement, inducement, justification, motivation, motive, reason, stimulus, taunts, teasing.

provocative *adj* **1** alluring, arousing, erotic, pornographic, *inf* raunchy, seductive, sensual, sensuous, *inf* sexy, tantalizing, tempting. **2** *inf* aggravating, annoying, infuriating, irksome, irritating, maddening, provoking, teasing, vexing.

provoke *vb* **1** activate, arouse, awaken, bring about, call forth, cause, elicit, encourage, excite, foment, generate, give rise to, induce, initiate, inspire, instigate, kindle, motivate, promote, prompt, spark off, start, stimulate, stir up, urge on, work up. **2** *inf* aggravate, anger, annoy, enrage, exasperate, gall, *inf* get on your nerves, goad, incense, incite, inflame, infuriate, insult, irk, irritate, madden, offend, outrage, pique, rile, rouse, tease, torment, upset, vex, *inf* wind up, worry. *Opp* PACIFY.

prowess *n* **1** ability, adeptness, adroitness, aptitude, cleverness, competence, dexterity, excellence, expertise, genius, mastery, proficiency, skill, talent. **2** *prowess in battle.* boldness, bravery, courage, daring, doughtiness, gallantry, heroism, mettle, spirit, valour.

prowl *vb* creep, lurk, roam, rove, skulk, slink, sneak, steal. ▷ WALK.

proximity *n* **1** closeness, nearness, propinquity. **2** locality, neighbourhood, vicinity.

prudent *adj* advisable, careful, cautious, circumspect, discreet, economical, far-sighted, frugal, judicious, politic, proper, provident, reasonable, sagacious, sage, sensible, shrewd, thoughtful, thrifty, vigilant, watchful, wise. *Opp* UNWISE.

prudish *adj* decorous, easily shocked, illiberal, intolerant, narrow-minded, old-fashioned, *inf* old-maidish, priggish, prim, *inf* prissy, proper, puritanical, rigid, shockable, strait-laced, strict. *Opp* BROAD-MINDED.

prune *vb* clip, cut back, lop, pare down, trim. ▷ CUT.

pry *vb* be curious, be inquisitive, be nosy, delve, *inf* ferret, inquire, interfere, intrude, investigate, meddle, *inf* nose about, peer, poke about, *inf* poke your nose in, search, *inf* snoop, *inf* stick your nose in. **prying** ▷ INQUISITIVE.

pseudonym *n* alias, assumed name, false name, incognito, nickname, *Fr* nom de plume, pen-name, sobriquet, stage name.

psychic *adj* clairvoyant, extrasensory, magical, mental, metaphysical, mystic, occult, preternatural, psychical, spiritual, supernatural, telepathic. • *n* astrologer, clairvoyant, crystal-gazer, fortune-teller, medium, mind-reader, spiritualist, telepathist.

psychological *adj* cerebral, emotional, mental, subconscious, subjective, subliminal, unconscious. *Opp* PHYSIOLOGICAL.

pub *n old use* alehouse, bar, *sl* boozer, cocktail lounge, *old use* hostelry, inn, *inf* local, public house, saloon, tavern, wine bar.

puberty *n* adolescence, growing-up, juvenescence, pubescence, sexual maturity, *inf* teens.

public *adj* **1** accessible, available, common, familiar, free, known, open, shared, unconcealed, unrestricted, visible, well-known. **2** *public support.* civic, civil, collective, communal, community, democratic, general, majority, national, popular, social, universal. **3** *a public figure.* ▷ PROMINENT. *Opp* PRIVATE. • *n the public.* citizens, the community, the country, the nation, people, the populace, society, voters.

publication *n* **1** appearance, issuing, printing, production. ▷ BOOK, MAGAZINE. **2** advertising, announcement, broadcasting, declaration, disclosure, dissemination, proclamation, promulgation, publicizing, reporting.

publicity *n* **1** attention, *inf* ballyhoo, fame, *inf* hype, limelight, notoriety. **2** advertising, marketing, promotion. ▷ ADVERTISEMENT.

publicize *vb* advertise, *sl* hype, *inf* plug, promote, *old use* puff. ▷ PUBLISH.

publish *vb* **1** bring out, circulate, issue, make available, print, produce, put on sale, release. **2** *publish secrets.* advertise, announce, break the news about, broadcast, communicate, declare, disclose, disseminate, divulge, issue a statement about, *inf* leak, make known, make public, proclaim, promulgate, publicize, *inf* put about, report, reveal, spread.

pucker *vb* compress, contract, crease, crinkle, draw together, purse, screw up, squeeze, tighten, wrinkle.

puerile *adj* babyish, boyish, childish, immature, infantile, juvenile. ▷ SILLY.

puff *n* **1** blast, blow, breath, draught, flurry, gust, whiff, wind. **2** *a puff of smoke.* cloud, wisp. • *vb* **1** blow, breathe heavily, gasp, huff, pant, wheeze. **2** *puff at a cigar. inf* drag, draw, inhale, pull, smoke, suck. **3** *sails puffed by the wind.* balloon, billow, distend, enlarge, inflate, rise, swell.

pugnacious *adj* aggressive, antagonistic, argumentative, bellicose, belligerent, combative, contentious, disputatious, excitable, fractious, hostile, hottempered, litigious, militant, unfriendly, warlike. ▷ QUARRELSOME. *Opp* PEACEABLE.

pull *vb* **1** drag, draw, haul, lug, tow, trail. *Opp* PUSH. **2** jerk, tug, pluck, rip, wrench, *inf* yank. **3** *pull a tooth.* extract, pull out, remove, take out. **pull off** ▷ DETACH. **pull out** ▷ WITHDRAW. **pull round** ▷ RECOVER. **pull someone's leg** ▷ TEASE. **pull through** ▷ RECOVER. **pull together** ▷ COOPERATE. **pull up** ▷ HALT.

pulp *n* mash, mush, paste, pap, puré. • *vb* crush, liquidize, mash, pound, pulverize, purée, smash, squash.

pulsate *vb* beat, drum, oscillate, palpitate, pound, pulse, quiver, reverberate, throb, tick, vibrate.

pulse *n* beat, drumming, oscillation, pounding, pulsation, rhythm, throb, ticking, vibration.

pump *vb* drain, draw off, empty, force, raise, siphon. **pump up** blow up, fill, inflate.

punch *vb* 1 beat, *sl* biff, box, *inf* clout, cuff, jab, poke, prod, pummel, slog, *sl* slug, *sl* sock, strike, thump. ▷ HIT. 2 ▷ PIERCE.

punctual *adj* in good time, *inf* on the dot, on time, prompt. *Opp* UNPUNCTUAL.

punctuate *vb* 1 insert punctuation, point. 2 *punctuated by applause.* break, interrupt, intersperse, *inf* pepper.

punctuation *n* marks, points, stops. □ *accent, apostrophe, asterisk, bracket, caret, cedilla, colon, comma, dash, exclamation mark, full stop, hyphen, question mark, quotation marks, speech marks, semicolon.*

puncture *n* blow-out, burst, burst tyre, *inf* flat, flat tyre, hole, leak, opening, perforation, pin-prick, rupture. • *vb* deflate, go through, let down, penetrate, perforate, pierce, prick, rupture.

pungent *adj* 1 aromatic, hot, peppery, piquant, seasoned, sharp, spicy, strong, tangy. 2 acid, acrid, astringent, caustic, *inf* chemically, harsh, sour, stinging. 3 *pungent criticism.* biting, bitter, incisive, mordant, sarcastic, scathing, trenchant.

punish *vb* castigate, chasten, chastise, correct, discipline, exact retribution from, impose punishment on, inflict punishment on, *inf* make an example of, pay back, penalize, *inf* rap over the knuckles, scold, *inf* teach someone a lesson.

punishment *n* chastisement, correction, discipline, forfeit, imposition, *inf* just deserts, penalty, punitive measure, retribution, revenge, sentence. □ *banishment, beating, the birch,* old use *Borstal, the cane, capital punishment, cashiering, confiscation of property, corporal punishment, detention, excommunication, execution, exile, fine, flogging, gaol,* inf *hiding, imprisonment, jail, keelhauling, lashing, pillory, prison, probation, scourging, spanking, the stocks, torture, whipping.*

punitive *adj* disciplinary, penal, retaliatory, retributive, revengeful, vindictive.

puny *adj* diminutive, dwarf, feeble, frail, sickly, stunted, underdeveloped, undernourished, undersized. ▷ SMALL. *Opp* LARGE, STRONG.

pupil *n* apprentice, beginner, disciple, follower, learner, novice, protégé(e), scholar, schoolboy, schoolchild, schoolgirl, student, tiro.

puppet *n* doll, dummy, finger-puppet, glove-puppet, hand-puppet, marionette, string-puppet.

purchase *n* 1 acquisition, *inf* buy (*a good buy*), investment. 2 grasp, grip, hold, leverage, support. • *vb* acquire, buy, get, invest in, obtain, pay for, procure, secure.

pure *adj* 1 authentic, genuine, neat, real, solid, sterling, straight, unadulterated, unalloyed, undiluted. 2 *pure food.* eatable, germ-free, hygienic, natural, pasteurized, uncontaminated, untainted, wholesome. **3** *pure water*. clean, clear, distilled, drinkable, fresh, potable, sterile, unpolluted. **4** *pure in morals*. blameless, chaste, decent, good, impeccable, innocent, irreproachable, maidenly, modest, moral, proper, sinless, stainless, virginal, virtuous. **5** *pure genius*. absolute, complete, downright, *inf* out-and-out, perfect, sheer, thorough, total, true, unmitigated, unqualified, utter. **6** *pure science*. abstract, academic, conjectural, conceptual, hypothetical, speculative, theoretical. *Opp* IMPURE, PRACTICAL.

purgative *n* aperient, cathartic, enema, laxative, purge.

purge *vb* **1** clean out, cleanse, clear, depurate, empty, purify, wash out. **2** *purge your opponents*. eject, eliminate, eradicate, expel, get rid of, liquidate, oust, remove, root out.

purify *vb* clarify, clean, cleanse, decontaminate, depurate, disinfect, distil, filter, fumigate, make pure, purge, refine, sanitize, sterilize.

puritan *n* fanatic, *derog* killjoy, moralist, *derog* prude, zealot.

puritanical *adj* ascetic, austere, moralistic, narrow-minded, pietistic, prim, proper, prudish, rigid, self-denying, self-disciplined, severe, stern, stiff-necked, strait-laced, strict, temperate, unbending, uncompromising. *Opp* HEDONISTIC.

purpose *n* **1** aim, ambition, aspiration, design, end, goal, hope, intent, intention, motivation, motive, object, objective, outcome, plan, point, rationale, result, target, wish. **2** determination, devotion, drive, firmness, persistence, resolution, resolve, steadfastness, tenacity, will, zeal. **3** *purpose of a tool*. advantage, application, benefit, good (*what's the good of it?*), point, practicality, use, usefulness, utility, value. • *vb* ▷ INTEND.

purposeful *adj* calculated, decided, decisive, deliberate, determined, devoted, firm, persistent, positive, resolute, steadfast, *derog* stubborn, tenacious, unwavering, wilful, zealous. ▷ INTENTIONAL. *Opp* HESITANT.

purposeless *adj* aimless, bootless, empty, gratuitous, meaningless, pointless, senseless, unnecessary, useless, vacuous, wanton. *Opp* MEANINGFUL, USEFUL.

purposely *adv* consciously, deliberately, intentionally, knowingly, on purpose, wilfully.

purse *n* bag, handbag, moneybag, pocketbook, pouch, wallet.

pursue *vb* **1** chase, follow, go after, go in pursuit of, harry, hound, hunt, keep up with, run after, shadow, stalk, *inf* tail, trace, track down, trail. **2** aim for, aspire to, be committed to, carry on, conduct, continue, dedicate yourself to, engage in, follow up, *inf* go for, persevere in, persist in, proceed with, prosecute, *inf* stick with, strive for, try for. **3** *pursue truth*. inquire into, investigate, quest after, search for, seek.

pursuit *n* **1** chase, chasing, following, harrying, *inf* hue and cry, hunt, hunting, pursuing, shadowing, stalking, tracking down, trail. **2** *leisure pursuits*. activity, employment, enthusiasm, hobby, interest, obsession, occupation, pastime, pleasure, speciality, specialization.

push *vb* **1** advance, drive, force, hustle, impel, jostle, move, nudge, poke, press, prod, propel, set in motion, shove, thrust. **2** *push a button*. depress, press. **3** *push into* *a space.* compress, cram, crowd, crush, insert, jam, pack, put, ram, squash, squeeze. **4** *push someone to act.* browbeat, bully, coerce, compel, constrain, dragoon, encourage, force, hurry, importune, incite, induce, influence, *inf* lean on, motivate, nag, persuade, pressurize, prompt, put pressure on, spur, stimulate, urge. **5** *push a new product.* advertise, boost, make known, market, *inf* plug, promote, publicize. *Opp* PULL. **push around** ▷ BULLY. **push off** ▷ DEPART. **push on** ▷ ADVANCE.

put *vb* **1** arrange, assign, commit, consign, deploy, deposit, dispose, fix, hang, lay, leave, locate, park, place, *inf* plonk, position, rest, set down, settle, situate, stand, station. **2** *put a question.* express, formulate, frame, phrase, say, state, utter, voice, word, write. **3** *put a proposal.* advance, bring forward, offer, outline, present, propose, submit, suggest, tender. **4** *put blame on someone.* attach, attribute, cast, fix, impose, inflict, lay, *inf* pin. **put across** ▷ COMMUNICATE. **put back** ▷ RETURN. **put by** ▷ SAVE. **put down** ▷ KILL, SUPPRESS. **put in** ▷ INSERT, INSTALL. **put off** ▷ POSTPONE. **put out** ▷ EJECT, EXTINGUISH. **put over** ▷ COMMUNICATE. **put right** ▷ REPAIR. **put someone up** ▷ ACCOMMODATE. **put up** ▷ RAISE. **put your foot down** ▷ INSIST. **put your foot in it** ▷ BLUNDER.

putative *adj* alleged, assumed, conjectural, presumed, reputed, rumoured, supposed, suppositious.

putrefy *vb* decay, decompose, go bad, go off, moulder, rot, spoil.

putrid *adj* bad, corrupt, decaying, decomposing, fetid, foul, mouldy, putrefying, rotten, rotting, spoilt.

puzzle *n inf* brain-teaser, conundrum, difficulty, dilemma, enigma, mystery, paradox, *inf* poser, problem, quandary, question, riddle. • *vb* baffle, bewilder, confuse, confound, *inf* floor, *inf* flummox, mystify, nonplus, perplex, set thinking, *inf* stump, *inf* stymie, worry. **puzzle out** ▷ SOLVE. **puzzle over** ▷ CONSIDER.

puzzling *adj* ambiguous, baffling, bewildering, confusing, cryptic, enigmatic, impenetrable, inexplicable, insoluble, *inf* mind-boggling, mysterious, mystifying, perplexing, strange, unaccountable, unanswerable, unfathomable, worrying. *Opp* STRAIGHTFORWARD.

pygmy *adj* dwarf, lilliputian, midget, tiny. ▷ SMALL.

Q

quadrangle *n* cloisters, courtyard, enclosure, *inf* quad, yard.

quagmire *n* bog, fen, marsh, mire, morass, mud, quicksand, *old use* slough, swamp.

quail *vb* back away, be apprehensive, blench, cower, cringe, falter, flinch, quake, recoil, show fear, shrink, tremble, wince.

quaint *adj* antiquated, antique, charming, curious, eccentric, fanciful, fantastic, odd, offbeat, old-fashioned, old-world, outlandish, peculiar, picturesque, strange, *inf* twee, unconventional, unexpected, unfamiliar, unusual, whimsical.

quake *vb* convulse, heave, move, quaver, quiver, rock, shake, shiver, shudder, stagger, sway, tremble, vibrate, wobble.

qualification *n* **1** ability, aptitude, capability, capacity, certification, competence, eligibility, experience, fitness, *inf* know-how, knowledge, proficiency, quality, skill, suitability, training. **2** certificate, degree, diploma, doctorate, first degree, Master's degree, matriculation. **3** *agree without qualification.* caveat, condition, exception, limitation, modification, proviso, reservation, restriction.

qualified *adj* **1** able, capable, certificated, competent, equipped, experienced, expert, fit, practised, professional, proficient, skilled, trained, well-informed. *Opp* UNSKILLED. **2** *qualified applicants.* appropriate, eligible, suitable. **3** *qualified praise.* cautious, conditional, equivocal, guarded, half-hearted, limited, modified, provisional, reserved, restricted. *Opp* UNCONDITIONAL.

qualify *vb* **1** authorize, empower, entitle, equip, fit, make eligible, permit, sanction. **2** become eligible, get through, *inf* make the grade, meet requirements, pass. **3** *qualify your praise.* abate, lessen, limit, mitigate, moderate, modulate, restrain, restrict, soften, temper, weaken.

quality *n* **1** calibre, class, condition, excellence, grade, rank, sort, standard, status, value, worth. **2** *personal qualities.* attribute, characteristic, distinction, feature, mark, peculiarity, property, trait.

quandary *n inf* catch-22, *inf* cleft stick, confusion, difficulty, dilemma, perplexity, plight, predicament, uncertainty.

quantity *n* aggregate, amount, bulk, consignment, dosage, dose, expanse, extent, length, load, lot, magnitude, mass, measurement, number, part, portion, proportion, quantum, sum, total, volume, weight. ▷ MEASURE.

quarrel *n* altercation, argument, bickering, clash, conflict, confrontation, contention, controversy, debate, difference, disagreement, discord, disharmony, dispute, dissension, division, feud, *inf* hassle, misunderstanding, row, *inf* ructions, rupture, *inf* scene, schism, *inf* slanging match, split, squabble, strife, *inf* tiff, vendetta, wrangle. • *vb* argue, *inf* be at loggerheads, *inf* be at odds, bicker, clash, conflict, contend, *inf* cross swords, differ, disagree, dispute, dissent, *inf* fall out, feud, haggle, misunderstand one another, *inf* row, squabble, wrangle. ▷ FIGHT. **quarrel with** ▷ DISPUTE.

quarrelsome *adj* aggressive, angry, argumentative, bad-tempered, cantankerous, choleric, contentious, contrary, cross, defiant, disagreeable, dyspeptic, explosive, fractious, impatient, irascible, irritable, petulant, peevish, querulous, quick-tempered, *inf* stroppy, testy, truculent, volatile, unfriendly. ▷ PUGNACIOUS. *Opp* PEACEABLE.

quarry *n* **1** game, kill, object, prey, victim. **2** excavation, mine, pit, working. • *vb* dig out, excavate, extract, mine.

quarter *n* area, district, division, locality, neighbourhood, part, region, section, sector, territory, vicinity, zone. • *vb* accommodate, billet, board, house, lodge, *inf* put up, shelter, station. **quarters** abode, accommodation, barracks, billet, domicile, dwelling-place, home, housing, living quarters, lodgings, residence, rooms, shelter.

quash *vb* **1** abolish, annul, cancel, invalidate, overrule, overthrow, reject, rescind, reverse, revoke. **2** ▷ QUELL.

quaver *vb* falter, fluctuate, oscillate, pulsate, quake, quiver, shake, shiver, shudder, tremble, vibrate, waver.

quay *n* berth, dock, harbour, jetty, landing-stage, pier, wharf.

queasy *adj* bilious, *inf* green, *inf* groggy, nauseated, nauseous, *inf* poorly, *inf* queer, sick, unwell. ▷ ILL.

queer *adj* **1** aberrant, abnormal, anomalous, atypical, bizarre, curious, different, eerie, exceptional, extraordinary, *inf* fishy, freakish, *inf* funny, incongruous, inexplicable, irrational, mysterious, odd, offbeat, outlandish, peculiar, puzzling, quaint, remarkable, *inf* rum, singular, strange, unaccountable, uncanny, uncommon, unconventional, unexpected, unnatural, unorthodox, unusual, weird. **2** *inf* cranky, deviant, eccentric, questionable, *inf* shady (*customer*), *inf* shifty, suspect, suspicious. ▷ MAD. *Opp* NORMAL. **3** ▷ ILL. **4** ▷ HOMOSEXUAL.

quell *vb* **1** crush, overcome, put down, quash, repress, subdue, suppress. **2** *quell fears.* allay, alleviate, calm, mitigate, moderate, mollify, pacify, soothe, tranquillize.

quench **1** allay, appease, cool, sate, satisfy, slake. **2** *quench a fire.* damp down, douse, extinguish, put out, smother, snuff out, stifle, suppress.

quest *n* crusade, expedition, exploration, hunt, mission, pilgrimage, pursuit, search, voyage of discovery. • *vb* **quest after** ▷ SEEK.

question *n* **1** *inf* brain-teaser, conundrum, demand, enquiry, inquiry, *inf* poser, query, request, riddle. **2** *an unresolved question.* argument, controversy, debate, difficulty, dispute, doubt, misgiving, mystery, objection, problem, puzzle, uncertainty. • *vb* **1** ask, catechize, cross-examine, cross-question, debrief, enquire of, examine, *inf* grill, inquire of, interrogate, interview, probe, *inf* pump, quiz. **2** *question a decision.* argue over, be sceptical about, call into question, cast doubt upon, challenge, dispute, doubt, enquire about, impugn, inquire about, object to, oppose, quarrel with, query.

questionable *adj* arguable, borderline, debatable, disputable, doubtful, dubious, *inf* iffy, moot, problematical, *inf* shady (*customer*), suspect, suspicious, uncertain, unclear, unprovable, unreliable.

questionnaire *n* catechism, opinion poll, question sheet, quiz, survey, test.

queue *n* chain, concatenation, column, *inf* crocodile, file, line, line-up, procession, row, string, succession, tail-back, train. • *vb* fall in, form a queue, line up, wait in a queue.

quibble *n* ▷ OBJECTION. • *vb* *inf* bandy words, be evasive, carp, cavil, equivocate, *inf* nit-pick, object, pettifog, *inf* split hairs, wrangle.

quick *adj* **1** breakneck, brisk, expeditious, express, fast, *old use* fleet, headlong, high-speed, *inf* nippy, precipitate, rapid, *inf* smart (*pace*), *inf* spanking, speedy, swift. **2** *a quick reaction.* adroit, agile, animated, brisk, deft, dexterous, energetic, lively, nimble, spirited, spry, vigorous. **3** *a quick response.* abrupt, early, hasty, hurried, immediate, instant, instantaneous, perfunctory, precipitate, prompt, punctual, ready, sudden, summary, unhesitating. **4** *a quick mind.* acute, alert, apt, astute, bright, clever, intelligent, perceptive, quick-witted, sharp, shrewd, smart. *Opp* SLOW. **5** *a quick rest.* brief, fleeting, momentary, passing, perfunctory, short, short-lived, temporary, transitory. **6** [*old use*] *the quick and the dead.* ▷ ALIVE. *Opp* SLOW.

quicken *vb* **1** accelerate, expedite, hasten, hurry, go faster, speed up. **2** ▷ AROUSE.

quiet *adj* **1** inaudible, noiseless, silent, soundless. **2** *quiet music.* hushed, low, pianissimo, soft, *It* sotto voce. **3** *a quiet person.* composed, contemplative, contented, gentle, introverted, meditative, meek, mild, modest, peaceable, reserved, retiring, shy, taciturn, thoughtful, uncommunicative, unforthcoming, unsociable, withdrawn. **4** *a quiet life.* cloistered, sheltered, tranquil, unadventurous, unexciting, untroubled. **5** *a quiet place.* isolated, lonely, peaceful, private, secluded, sequestered, undisturbed, unfrequented. **6** *quiet weather.* calm, motionless, placid, restful, serene, still. *Opp* BUSY, NOISY, RESTLESS.

quieten *vb* **1** calm, compose, hush, lull, pacify, sedate, soothe, subdue, tranquillize. **2** deaden, dull, muffle, mute, reduce the volume of, silence, soften, stifle, suppress, tone down.

quirk *n* aberration, caprice, crotchet, eccentricity, idiosyncrasy, kink, oddity, peculiarity, trick, whim.

quit *vb* **1** abandon, decamp from, depart from, desert, exit from, forsake, go away from, leave, walk out (on), withdraw. **2** abdicate, discontinue, drop, give up, leave, *inf* pack it in, relinquish, renounce, repudiate, resign from, retire from, withdraw from. **3** [*inf*] *Quit pushing!* cease, desist from, discontinue, leave off, stop.

quite *adv* [NB: the two senses are almost opposite.] **1** *Yes, I've quite finished.* absolutely, altogether, completely, entirely, perfectly, thoroughly, totally, unreservedly, utterly, wholly. **2** *quite good, but not perfect.* comparatively, fairly, moderately, *inf* pretty, rather, relatively, somewhat, to some extent.

quits *adj* equal, even, level, repaid, revenged, square.

quiver *vb* flicker, fluctuate, flutter, oscillate, palpitate, pulsate, quake, quaver, shake, shiver, shudder, tremble, vibrate, wobble.

quixotic *adj* fanciful, foolhardy, idealistic, impracticable, impractical, romantic, *inf* starry-eyed, unrealistic, unrealizable, unselfish, Utopian, visionary. *Opp* REALISTIC.

quiz *n* competition, exam, examination, questioning, questionnaire, quiz-game, test. • *vb* ▷ QUESTION.

quizzical *adj* amused, comical, curious, intrigued, perplexed, puzzled, queer, questioning.

quota *n* allocation, allowance, apportionment, assignment, *inf* cut, part, portion, proportion, ration, share.

quotation *n* **1** allusion, citation, *inf* clip, cutting, excerpt, extract, passage, piece, reference, selection. **2** estimate, price, tender, valuation.

quote *vb* **1** cite, instance, mention, produce a quotation from, refer to, repeat, reproduce. **2** *quote a price*. estimate, tender.

R

rabble *n* crowd, gang, herd, *Gk* hoi polloi, horde, mob, *inf* riffraff, swarm, throng. ▷ GROUP.

race *n* **1** breed, clan, ethnic group, family, folk, genus, kind, lineage, nation, people, species, stock, tribe, variety. **2** chase, competition, contention, contest, heat, rivalry. □ *cross-country, greyhound race, horse-race, hurdles, marathon, motor-race, regatta, relay, road-race, rowing, scramble, speedway, sprint, steeple-chase, stock-car race, swimming, track event.* • *vb* **1** *I'll race you!* compete with, contest with, have a race with, try to beat. **2** *race along*. career, dash, *inf* fly, gallop, hasten, hurry, move fast, run, rush, speed, sprint, *inf* tear, *inf* zip, *inf* zoom.

racetrack *n* cinder-track, circuit, dog-track, lap, racecourse.

racial *adj* ethnic, folk, genetic, national, tribal.

racism *n* bias, bigotry, chauvinism, discrimination, intolerance, prejudice, racialism, xenophobia. □ *anti-Semitism, apartheid.*

racist *adj* biased, bigoted, chauvinist, discriminatory, intolerant, prejudiced, racialist, xenophobic. □ *anti-Semitic.*

rack *n* frame, framework, holder, scaffold, scaffolding, shelf, stand, support. • *vb* ▷ TORTURE.

radiant *adj* **1** beaming, bright, brilliant, effulgent, gleaming, glorious, glowing, incandescent, luminous, phosphorescent, refulgent, shining. **2** *The bride was radiant.* ▷ HAPPY.

radiate *vb* beam, diffuse, emanate, emit, give off, gleam, glow, send out, shed, shine, spread, transmit.

radical *adj* **1** basic, cardinal, deep-seated, elementary, essential, fundamental, primary, principal, profound. **2** complete, comprehensive, drastic, entire, exhaustive, thorough, thoroughgoing. **3** *radical politics*. extreme, extremist, fanatical, far-reaching, revolutionary, *derog* subversive. *Opp* MODERATE, SUPERFICIAL.

radio *n* CB, *sl* ghettoblaster, portable, receiver, set, *inf* transistor, transmitter, walkie-talkie, *old use* wireless. • *vb* broadcast, send out, transmit.

rafter *n* beam, girder, joist.

rage *n* ▷ ANGER. • *vb* be angry, boil, fume, go berserk, lose control, rave, *inf* see red, seethe, storm.

ragged *adj* **1** chafed, frayed, in ribbons, old, patched, patchy, ravelled, rent, ripped, rough, rough-edged, shabby, shaggy, tattered, tatty, threadbare, torn, unkempt, unravelled, untidy, worn out. **2** *ragged line*. denticulated, disorganized, erratic, irregular, jagged, serrated, uneven, zigzag.

rags *plur n* bits and pieces, cloths, fragments, old clothes, remnants, ribbons, scraps, shreds, tatters.

raid *n* assault, attack, blitz, foray, incursion, inroad, invasion, onslaught, sally, sortie, strike, surprise attack, swoop. • *vb* 1 assault, attack, descend on, invade, pounce on, rush, storm, swoop down on. 2 loot, maraud, pillage, plunder, ransack, rifle, rob, sack, steal from, strip.

raider *n* attacker, brigand, invader, looter, marauder, outlaw, pillager, pirate, plunderer, ransacker, robber, rustler, thief.

railway *n* line, permanent way, *Amer* railroad, rails, track. □ *branch line, cable railway, funicular, light railway, main line, metro, mineral line, monorail, mountain railway, narrow gauge, rack-and-pinion, rapid transit system, siding, standard gauge, tramway, tube, underground.* ▷ TRAIN.

rain *n* cloudburst, deluge, downpour, drizzle, precipitation, raindrops, rainfall, rainstorm, shower, squall. • *vb inf* bucket down, drizzle, pelt, pour, *inf* rain cats and dogs, spit, teem.

rainy *adj* damp, drizzly, showery, wet.

raise *vb* 1 elevate, heave up, hoist, hold up, jack up, lift, loft, pick up, put up, rear. 2 *raise prices.* augment, boost, increase, inflate, put up, *inf* up. 3 *raise to a higher rank.* exalt, prefer, promote, upgrade. 4 *raise a monument.* build, construct, create, erect, set up. 5 *raise hopes.* activate, arouse, awaken, build up, buoy up, encourage, engender, enlarge, excite, foment, foster, heighten, incite, kindle, motivate, provoke, rouse, stimulate, uplift. 6 *raise animals, children, crops.* breed, bring up, care for, cultivate, educate, farm, grow, look after, nurture, produce, propagate, rear. 7 *raise money.* amass, collect, get, make, receive, solicit. 8 *raise questions.* advance, bring up, broach, express, instigate, introduce, mention, moot, originate, pose, present, put forward, suggest. *Opp* LOWER, REDUCE. **raise from the dead** ▷ RESURRECT. **raise the alarm** ▷ WARN.

rally *n* 1 assembly, *inf* demo, demonstration, gathering, march, mass meeting, protest. 2 ▷ COMPETITION. • *vb* 1 assemble, convene, get together, marshal, muster, organize, round up, summon. 2 come together, reassemble, reform, regroup. 3 *rally after illness.* ▷ RECOVER.

ram *vb* 1 bump, butt, collide with, crash into, slam into, smash into, strike. ▷ HIT. 2 compress, cram, crowd, crush, drive, force, jam, pack, press, push, squash, squeeze, tamp down, wedge.

ramble *n* hike, tramp, trek, walk. • *vb* 1 hike, range, roam, rove, tramp, trek, stroll, wander. ▷ WALK. 2 digress, drift, *inf* lose the thread, maunder, *inf* rabbit on, *inf* rattle on, talk aimlessly, wander, *inf* witter on.

rambling *adj* 1 circuitous, indirect, labyrinthine, meandering, roundabout, tortuous, twisting, wandering, winding, zigzag. *Opp* DIRECT. 2 aimless, circumlocutory, confused, diffuse, digressive, disconnected, discursive, disjointed, illogical, incoherent, jumbled, muddled, periphrastic, unstructured, verbose, wordy. *Opp* COHERENT. 3 *a rambling house.* asymmetrical, extensive, irregular, large, sprawling, straggling, straggly. *Opp* COMPACT.

ramification *n* branch, by-product, complication, consequence, division, effect, extension, implication, offshoot, result, subdivision, upshot.

ramp *n* acclivity, gradient, incline, rise, slope.

rampage *n* frenzy, riot, tumult, uproar, vandalism, violence. • *vb* behave violently, go berserk, go wild, lose control, race about, run amok, run riot, rush about, storm about. **on the rampage** ▷ WILD.

ramshackle *adj* broken-down, crumbling, decrepit, derelict, dilapidated, flimsy, jerry-built, rickety, ruined, run-down, shaky, tottering, tumbledown, unsafe, unstable, unsteady. *Opp* SOLID.

random *adj* accidental, adventitious, aimless, arbitrary, casual, chance, fortuitous, haphazard, *inf* hit-or-miss, indiscriminate, irregular, serendipitous, stray, unconsidered, unplanned, unpremeditated, unspecific, unsystematic. *Opp* DELIBERATE, SYSTEMATIC.

range *n* **1** area, compass, distance, extent, field, gamut, limit, orbit, radius, reach, scope, span, spectrum, sphere, spread, sweep. **2** *a wide range of goods.* diversity, selection, variety. **3** *range of mountains.* chain, file, line, rank, row, series, string, tier. • *vb* **1** differ, extend, fluctuate, go, reach, run the gamut, spread, stretch, vary. **2** ▷ RANK. **3** ▷ ROAM.

rank *adj* **1** *rank growth.* ▷ ABUNDANT. **2** *rank smell.* ▷ SMELLING. • *n* **1** column, file, formation, line, order, queue, row, series, tier. **2** birth, blood, caste, class, condition, degree, echelon, estate, grade, level, position, standing, station, status, stratum, title. • *vb* arrange, array, assort, categorize, class, classify, grade, graduate, line up, order, organize, range, rate, set out in order, sort.

ransack *vb* **1** comb, explore, go through, rake through, rummage through, scour, search, *inf* turn upside down. **2** *ransack a shop.* despoil, loot, pillage, plunder, raid, ravage, rob, sack, strip, wreck.

ransom *n* payment, *inf* payoff, price, redemption. • *vb* buy the release of, deliver, redeem.

rap *vb* **1** knock, strike, tap. ▷ HIT. **2** ▷ CRITICIZE.

rape *n* **1** assault, sexual attack. **2** ▷ PILLAGE. • *vb* assault, defile, deflower, dishonour, force yourself on, *inf* have your way with, *old use* ravish, violate.

rapid *adj* alacritous, breakneck, brisk, expeditious, express, fast, *old use* fleet, hasty, headlong, high-speed, hurried, immediate, impetuous, instant, instantaneous, *inf* lightning, *inf* nippy, precipitate, prompt, quick, smooth, speedy, swift, unchecked, uninterrupted. *Opp* SLOW.

rapids *plur n* cataract, current, waterfall, white water.

rapture *n* bliss, delight, ecstasy, elation, euphoria, exaltation, happiness, joy, pleasure, thrill, transport.

rare *adj* abnormal, atypical, curious, exceptional, extraordinary, *inf* few and far between, infrequent, irreplaceable, limited, occasional, odd, out of the ordinary, peculiar, scarce, singular, special, strange, surprising, uncommon, unfamiliar, unusual. *Opp* COMMON.

rascal *n* blackguard, *old use* bounder, devil, good-fornothing, imp, knave, mischiefmaker, miscreant, ne'er-do-well, rapscallion, rogue, *inf* scally-wag, scamp, scoundrel, troublemaker, villain, wastrel. ▷ CRIMINAL.

rash *adj* careless, foolhardy, harebrained, hasty, headlong, headstrong, heedless, hotheaded, hurried, ill-advised, ill-considered, impetuous, imprudent, impulsive, incautious, indiscreet, injudicious, madcap, precipitate, reckless, risky, thoughtless, unthinking, wild. *Opp* CAREFUL. • *n* **1** efflorescence, eruption, spots. **2** *a rash of thefts.* ▷ OUTBREAK.

rasp *vb* **1** abrade, file, grate, rub, scrape. **2** *rasp orders.* croak, screech, speak hoarsely. ▷ SPEAK. **rasping** ▷ HARSH.

rate *n* **1** gait, pace, speed, tempo, velocity. **2** amount, charge, cost, fare, fee, figure, payment, price, scale, tariff, wage. • *vb* **1** appraise, assess, class, classify, compute, consider, estimate, evaluate, gauge, grade, judge, measure, prize, put a price on, rank, reckon, regard, value, weigh. **2** *rate a prize.* be worthy of, deserve, merit. **3** ▷ REPRIMAND.

rather *adv* **1** fairly, moderately, *inf* pretty, quite, relatively, slightly, somewhat. **2** *would rather have tea than coffee.* more willingly, preferably, sooner.

ratify *vb* approve, authorize, confirm, endorse, sanction, sign, validate, verify.

rating *n* classification, evaluation, grade, grading, mark, order, placing, ranking.

ratio *n* balance, correlation, correspondence, fraction, percentage, proportion, relationship.

ration *n* allocation, allotment, allowance, amount, helping, measure, percentage, portion, quota, share. • *vb* allocate, allot, apportion, conserve, control, distribute fairly, dole out, give out, limit, parcel out, restrict, share equally. **rations** food, necessaries, necessities, provisions, stores, supplies.

rational *adj* balanced, clear-headed, commonsense, enlightened, intelligent, judicious, logical, lucid, normal, ratiocinative, reasonable, reasoned, reasoning, sane, sensible, sound, thoughtful, wise. *Opp* IRRATIONAL.

rationale *n* argument, case, cause, excuse, explanation, grounds, justification, logical basis, principle, reason, reasoning, theory, vindication.

rationalize *vb* **1** account for, be rational about, elucidate, excuse, explain, justify, make rational, provide a rationale for, ratiocinate, think through, vindicate. **2** ▷ REORGANIZE.

rattle *vb* **1** clatter, vibrate. **2** agitate, jar, joggle, *inf* jiggle about, jolt, shake about. **3** [*inf*] *rattled him by booing.* alarm, discomfit, discompose, disconcert, disturb, fluster, frighten, make nervous, put off, unnerve, upset, worry. **rattle off** ▷ RECITE. **rattle on** ▷ RAMBLE, TALK.

raucous *adj* ear-splitting, harsh, husky, grating, jarring, noisy, rasping, rough, screeching, shrill, squawking, strident.

ravage *vb* damage, despoil, destroy, devastate, lay waste, loot, pillage, plunder, raid, ransack, ruin, sack, spoil, wreak havoc on, wreck.

rave *vb* **1** be angry, fulminate, fume, rage, rant, roar, storm, thunder. **2** be enthusiastic, enthuse, *inf* go into raptures, *inf* gush, rhapsodize.

ravenous *adj* famished, hungry, insatiable, ravening, starved, starving, voracious. ▷ GREEDY.

ravish *vb* **1** bewitch, captivate, capture, charm, delight, enchant, entrance, spellbind, transport. **2** ▷ RAPE. **ravishing** ▷ BEAUTIFUL.

raw *adj* **1** fresh, rare (*steak*), uncooked, underdone, unprepared, wet (*fish*). **2** *raw materials.* crude, natural, unprocessed, unrefined, untreated. **3** *raw recruits. inf* green, ignorant, immature, inexperienced, innocent, new, unseasoned, untrained, untried. **4** *raw skin.* bloody, chafed, grazed, inflamed, painful, red, rough, scraped, scratched, sensitive, sore, tender, vulnerable. **5** *raw wind.* ▷ COLD.

ray *n* **1** bar, beam, laser, pencil, shaft, streak, stream. **2** *a ray of hope.* flicker, gleam, glimmer, hint, indication, scintilla, sign, trace.

raze *vb* bulldoze, demolish, destroy, flatten, level, tear down.

razor *n* □ *cut-throat razor, disposable razor, electric razor, safety razor.*

reach *n* compass, distance, orbit, range, scope, sphere. • *vb* **1** achieve, arrive at, attain, come to, get hold of, get to, go as far as, grasp, *inf* make, take, touch. **2** *reach for the salt.* put out your hand, stretch, try to get. **3** *reach me by phone.* communicate with, contact, get in touch with. **reach out** ▷ EXTEND.

react *vb* act, answer, behave, conduct yourself, reciprocate, reply, respond, retaliate, retort, take revenge. **react to** ▷ COUNTER.

reaction *n* answer, backlash, *inf* come-back, counter, countermove, effect, feedback, parry, reciprocation, reflex, rejoinder, reply, reprisal, response, retaliation, retort, revenge, riposte.

reactionary *adj* conservative, die-hard, old-fashioned, rightist, right-wing, *inf* stick-in-the-mud, traditionalist, unprogressive. *Opp* PROGRESSIVE.

read *vb* **1** devour, *inf* dip into, glance at, interpret, look over, peruse, pore over, review, scan, skim, study. **2** *can't read the handwriting.* decipher, decode, interpret, make out, understand.

readable *adj* **1** absorbing, compulsive, easy, engaging, enjoyable, entertaining, gripping, interesting, stimulating, well-written. *Opp* BORING. **2** clear, comprehensible, decipherable, distinct, intelligible, legible, neat, plain, understandable. *Opp* ILLEGIBLE.

readily *adv* cheerfully, eagerly, easily, effortlessly, freely, gladly, happily, promptly, quickly, ungrudgingly, unhesitatingly, voluntarily, willingly.

ready *adj* **1** accessible, *inf* all set, arranged, at hand, available, complete, convenient, done, finalized, finished, fit, obtainable, prepared, primed, ripe, set, set up, waiting. **2** *ready to help.* agreeable, consenting, content, disposed, eager, equipped, *inf* game, glad, inclined, in the mood, keen, *inf* keyed up, liable, likely, minded, of a mind, open, organized, pleased, poised, predisposed, primed, *inf* psyched up, raring (*to go*), trained, willing. **3** *ready wit.* acute, adroit, alert, apt, facile, immediate, prompt, quick, quick-witted, rapid, sharp, smart, speedy. *Opp* SLOW, UNPREPARED.

real *adj* **1** actual, authentic, certain, corporeal, everyday, existing, factual, genuine, material, natural, ordinary, palpable, physical, pure, realistic, tangible, visible. **2** authenticated, *Lat* bona fide, legal, legitimate, official, valid, verifiable. **3** *real friends*. dependable, positive, reliable, sound, true, trustworthy, worthy. **4** *real grief*. earnest, heartfelt, honest, sincere, truthful, unaffected, undoubted, unfeigned, unquestionable. *Opp* FALSE.

realism *n* **1** authenticity, fidelity, naturalism, verisimilitude. **2** *realism in business*. clear-sightedness, common sense, objectivity, practicality, pragmatism.

realistic *adj* **1** businesslike, clear-sighted, commonsense, down-to-earth, feasible, hardheaded, *inf* hard-nosed, levelheaded, logical, matter-of-fact, *inf* no-nonsense, objective, possible, practicable, practical, pragmatic, rational, sensible, tough, unemotional, unsentimental, viable, workable. **2** *realistic pictures*. authentic, convincing, faithful, graphic, lifelike, natural, recognizable, representational, true-to-life, truthful, vivid. **3** *realistic prices*. acceptable, adequate, fair, genuine, justifiable, moderate, reasonable. *Opp* UNREALISTIC.

reality *n* actuality, authenticity, certainty, empirical knowledge, experience, fact, life, *inf* nitty-gritty, real life, the real world, truth, verity. *Opp* FANTASY.

realize *vb* **1** accept, appreciate, apprehend, be aware of, become conscious of, *inf* catch on to, comprehend, conceive of, *inf* cotton on to, grasp, know, perceive, recognize, see, sense, *inf* twig, understand, *inf* wake up to. **2** *realize an ambition*. accomplish, achieve, bring about, complete, effect, effectuate, fulfil, implement, make a reality of, obtain, perform, put into effect. **3** *realize a price*. *inf* bring in, *inf* clear, earn, fetch, make, net, obtain, produce.

realm *n* country, domain, empire, kingdom, monarchy, principality.

reap *vb* **1** cut, garner, gather in, glean, harvest, mow. **2** *reap a reward*. acquire, bring in, collect, get, obtain, receive, win.

rear *adj* back, end, hind, hinder, hindmost, last, rearmost. *Opp* FRONT. • *n* **1** back, end, stern (*of ship*), tail-end. **2** ▷ BUTTOCKS. • *vb* **1** breed, bring up, care for, cultivate, educate, feed, look after, nurse, nurture, produce, raise, train. **2** *rear your head*. elevate, hold up, lift, raise, uplift. **3** ▷ BUILD.

rearrange *vb* change round, regroup, reorganize, switch round, swop round, transpose. ▷ CHANGE.

rearrangement *n* anagram, reorganization, transposition. ▷ CHANGE.

reason *n* **1** apology, argument, case, cause, defense, excuse, explanation, grounds, incentive, justification, motive, occasion, pretext, rationale, vindication. **2** brains, common sense, *inf* gumption, intelligence, judgement, logic, mind, *inf* nous, perspicacity, rationality, reasonableness, sanity, sense, understanding, wisdom, wit. ▷ REASONING. **3** *reason for living*. aim, goal, intention, motivation, motive, object, objective, point, purpose, spur, stimulus. • *vb* **1** act rationally, calculate, cerebrate, conclude, consider, deduce, estimate, figure out, hypothesize, infer, intellectualize, judge, *inf* put two and two together, ratiocinate, resolve, theorize, think, use your head, work out. **2** *I reasoned with her*. argue, debate, discuss, expostulate, remonstrate.

reasonable *adj* **1** calm, helpful, honest, intelligent, rational, realistic, sane, sensible, sincere, sober, thinking, thoughtful, unemotional, wise. **2** *reasonable argument*. arguable, believable, credible, defensible, justifiable, logical, plausible, practical, reasoned, sound, tenable, viable, well-thought-out. **3** *reasonable prices*. acceptable, appropriate, average, cheap, competitive, conservative, fair, inexpensive, moderate, ordinary, proper, right, suitable, tolerable, unexceptionable. *Opp* IRRATIONAL.

reasoning *n* analysis, argument, case, *derog* casuistry, cerebration, deduction, dialectic, hypothesis, line of thought, logic, proof, rationalization, *derog* sophistry, theorizing, thinking.

reassure *vb* assure, bolster, buoy up, calm, cheer, comfort, encourage, give confidence to, hearten, *inf* set someone's mind at rest, support, uplift. *Opp* ALARM, THREATEN. **reassuring** ▷ SOOTHING, SUPPORTIVE.

rebel *adj* ▷ REBELLIOUS. • *n* anarchist, apostate, dissenter, freedom fighter, heretic, iconoclast, insurgent, malcontent, maverick, mutineer, nonconformist, recusant, resistance fighter, revolutionary, schismatic. • *vb* be a rebel, disobey, dissent, fight, *inf* kick over the traces, mutiny, refuse to obey, revolt, rise up, *inf* run riot, *inf* take a stand. *Opp* CONFORM. **rebel against** ▷ DEFY.

rebellion *n* contumacy, defiance, disobedience, insubordination, insurgency, insurrection, mutiny, rebelliousness, resistance, revolt, revolution, rising, schism, sedition, uprising.

rebellious *adj inf* bolshie, breakaway, contumacious, defiant, difficult, disaffected, disloyal, disobedient, incorrigible, insubordinate, insurgent, intractable, malcontent, mutinous, obstinate, quarrelsome, rebel, recalcitrant, refractory, resistant, revolting, revolutionary, seditious, uncontrollable, ungovernable, unmanageable, unruly, wild. *Opp* OBEDIENT.

rebirth *n* reawakening, regeneration, renaissance, renewal, resurgence, resurrection, return, revival.

rebound *vb inf* backfire, *inf* boomerang, bounce, misfire, recoil, ricochet, spring back.

rebuff *n inf* brush-off, check, discouragement, refusal, rejection, slight, snub. • *vb* cold-shoulder, decline, discourage, refuse, reject, repulse, slight, snub, spurn, turn down.

rebuild *n* reassemble, reconstruct, recreate, redevelop, refashion, regenerate, remake. ▷ RECONDITION.

rebuke *vb* admonish, castigate, censure, chide, reprehend, reproach, reprove, scold, upbraid. ▷ REPRIMAND.

recall *vb* **1** bring back, call in, summon, withdraw. **2** ▷ REMEMBER.

recede *vb* abate, decline, dwindle, ebb, fall back, go back, lessen, regress, retire, retreat, return, shrink back, sink, slacken, subside, wane, withdraw.

receipt *n* **1** account, acknowledgement, bill, proof of purchase, sales slip, ticket. **2** *receipt of goods*. acceptance, delivery, reception, receipts gains, gate, income, proceeds, profits, return, takings.

receive *vb* **1** accept, acquire, be given, be paid, be sent, collect, come by, come into, derive, earn, gain, get, gross, inherit, make, net, obtain, take. **2** *receive an injury.* bear, be subjected to, endure, experience, meet with, suffer, sustain, undergo. **3** *receive visitors.* accommodate, admit, entertain, greet, let in, meet, show in, welcome. *Opp* GIVE.

recent *adj* brand-new, contemporary, current, fresh, just out, latest, modern, new, novel, present-day, up-to-date, young. *Opp* OLD.

reception *n* **1** greeting, response, welcome. **2** ▷ PARTY.

receptive *adj* amenable, favourable, flexible, interested, open, open-minded, responsive, susceptible, sympathetic, tractable, welcoming, well-disposed. *Opp* RESISTANT.

recess *n* **1** alcove, apse, bay, cavity, corner, cranny, hollow, indentation, niche, nook. **2** adjournment, break, *inf* breather, breathing-space, interlude, intermission, interval, respite, rest, time off.

recession *n* decline, depression, downturn, slump.

recipe *n* directions, formula, instructions, method, plan, prescription, procedure, technique.

reciprocal *adj* corresponding, exchanged, joint, mutual, requited, returned, shared.

reciprocate *vb* exchange, give the same in return, match, repay, requite, return.

recital *n* **1** concert, performance, programme. **2** *recital of events.* account, description, narration, narrative, recounting, rehearsal, relation, repetition, story, telling. ▷ RECITATION.

recitation *n* declaiming, declamation, delivery, monologue, narration, performance, presentation, reading, *old use* rendition, speaking, telling.

recite *vb* articulate, declaim, deliver, narrate, perform, present, quote, *inf* rattle off, recount, reel off, rehearse, relate, repeat, speak, tell.

reckless *adj* **1** brash, careless, *inf* crazy, daredevil, *inf* devil-may-care, foolhardy, harebrained, *inf* harum-scarum, hasty, heedless, impetuous, imprudent, impulsive, inattentive, incautious, indiscreet, injudicious, irresponsible, *inf* mad, madcap, negligent, rash, thoughtless, unconsidered, unwise, wild. *Opp* CAREFUL. **2** *reckless criminals.* dangerous, desperate, hardened, violent.

reckon *vb* **1** add up, appraise, assess, calculate, compute, count, enumerate, estimate, evaluate, figure out, gauge, number, tally, total, value, work out. **2** ▷ THINK.

reclaim *vb* **1** get back, *inf* put in for, recapture, recover, regain. **2** *reclaim derelict land.* make usable, redeem, regenerate, reinstate, rescue, restore, salvage, save.

recline *vb* lean back, lie, loll, lounge, repose, rest, sprawl, stretch out.

recluse *n* anchoress, anchorite, hermit, loner, monk, nun, solitary.

recognizable *adj* detectable, distinctive, distinguishable, identifiable, known, noticeable, perceptible, undisguised, unmistakable, visible.

recognize *vb* **1** detect, diagnose, discern, distinguish, identify, know, name, notice, perceive, pick out, place (*can't place him*), *inf* put a name to, recall, recollect, remember, see, spot. **2** *recognize your faults.* accept, acknowledge, admit to, appreciate, be aware of, concede, confess, grant, realize, understand. **3** *recognize someone's rights.* approve of, *inf* back, endorse, legitimize, ratify, sanction, support, validate.

recoil *vb* blench, draw back, falter, flinch, jerk back, jump, quail, shrink, shy away, start, wince. ▷ REBOUND.

recollect *vb* hark back to, recall, reminisce about, summon up, think back to. ▷ REMEMBER.

recommend *vb* **1** advise, advocate, counsel, exhort, prescribe, propose, put forward, suggest, urge. **2** applaud, approve of, *inf* back, commend, favour, *inf* plug, praise, *inf* push, *inf* put in a good word for, speak well of, support, vouch for. ▷ ADVERTISE.

recommendation *n* advice, advocacy, approbation, approval, *inf* backing, commendation, counsel, favourable mention, reference, seal of approval, support, testimonial.

reconcile *vb* bring together, *old use* conciliate, harmonize, make friendly again, placate, reunite, settle differences between. **be reconciled to** accept, adjust to, resign yourself to, submit to.

recondition *vb* make good, overhaul, rebuild, renew, renovate, repair, restore.

reconnaissance *n* examination, exploration, inspection, investigation, observation, *inf* recce, reconnoitring, scouting, spying, survey.

reconnoitre *vb inf* case, *inf* check out, examine, explore, gather intelligence (about), inspect, investigate, patrol, scout, scrutinize, spy, survey, *sl* suss out.

reconsider *vb* be converted, change your mind, come round, reappraise, reassess, re-examine, rethink, review your position, think better of.

reconstruct *vb* act out, mock up, recreate, rerun. ▷ REBUILD.

record *n* **1** account, annals, archives, catalogue, chronicle, diary, documentation, dossier, file, journal, log, memorandum, minutes, narrative, note, register, report, transactions. **2** best performance, best time. **3** ▷ RECORDING. • *vb* **1** chronicle, document, enter, inscribe, list, log, minute, note, put down, register, set down, take down, transcribe, write down. **2** *record on tape.* keep, preserve, tape, tape-record, video.

recording *n* performance, release. □ *album, audio-tape, cassette, CD, compact disc, digital recording, disc, long-playing record, LP, mono recording, record, single, stereo recording, tape, tape-recording, tele-recording, video, video-cassette, video disc, videotape.*

record-player *n* CD player, gramophone, midi system, *old use* phonograph, record deck, turntable.

recount *vb* communicate, describe, detail, impart, narrate, recite, relate, report, tell, unfold.

recover *vb* **1** find, get back, get compensation for, make good, make up for, recapture, reclaim, recoup, regain, repossess, restore, retrieve, salvage, trace, track down, win back. **2** *inf* be on the mend, come round, convalesce, *inf* get back on your feet, get better, heal, improve, mend, *inf* pull round, *inf* pull through, rally, recuperate, regain your strength, revive, survive, *inf* take a turn for the better.

recovery *n* **1** recapture, reclamation, repossession, restoration, retrieval, salvage, salvaging. **2** *recovery from illness.* advance, convalescence, cure, deliverance, healing, improvement, progress, rally, recuperation, revival, upturn.

recreation *n* amusement, distraction, diversion, enjoyment, entertainment, fun, games, hobby, leisure, pastime, play, pleasure, refreshment, relaxation, sport.

recrimination *n* accusation, *inf* come-back, counter-attack, counter-charge, reprisal, retaliation, retort.

recruit *n* apprentice, beginner, conscript, *inf* greenhorn, initiate, learner, neophyte, *inf* new boy, new girl, novice, tiro, trainee. *Opp* VETERAN. • *vb* advertise for, conscript, draft in, engage, enlist, enrol, *old use* impress, mobilize, muster, register, sign on, sign up, take on.

rectify *vb* amend, correct, cure, *inf* fix, make good, put right, repair, revise.

recumbent *adj* flat, flat on your back, horizontal, lying down, prone, reclining, stretched out, supine. *Opp* UPRIGHT.

recuperate *vb* convalesce, get better, heal, improve, mend, rally, regain strength, revive. ▷ RECOVER.

recur *vb* be repeated, come back again, happen again, persist, reappear, repeat, return.

recurrent *adj* chronic, cyclical, frequent, intermittent, iterative, periodic, persistent, recurring, regular, repeated, repetitive, returning. ▷ CONTINUAL.

recycle *vb* reclaim, recover, retrieve, reuse, salvage, use again.

red *adj* bloodshot, blushing, embarrassed, fiery, flaming, florid, flushed, glowing, inflamed, rosy, rubicund, ruddy. □ *auburn, blood-red, brick-red, cardinal, carmine, carroty, cerise, cherry, chestnut, crimson, damask, flame-coloured, foxy, magenta, maroon, orange, pink, rose, roseate, ruby, scarlet, titian, vermilion, wine-coloured.* **red herring** ▷ DECOY.

redden *vb* blush, colour, flush, glow.

redeem *vb* buy back, cash in, exchange for cash, reclaim, recover, re-purchase, trade in, win back. ▷ LIBERATE. **redeem yourself** ▷ ATONE.

redolent *adj* **1** aromatic, fragrant, perfumed, scented, smelling. **2** *redolent of the past.* reminiscent, suggestive.

reduce *vb* **1** abate, abbreviate, abridge, clip, compress, curtail, cut, cut back, cut down, decimate, decrease, detract from, devalue, dilute, diminish, *inf* dock (*wages*), *inf* ease up on, halve, impair, lessen, limit, lower, make less, minimize, moderate, narrow, prune, shorten, shrink, simplify, *inf* slash, slim down, tone down, trim, truncate, weaken, whittle. **2** become less, contract, dwindle, shrink. **3** *reduce a liquid.* concentrate, condense, thicken. **4** *reduce to rubble.* break up, destroy, grind, pulp, pulverize, triturate. **5** *reduce to poverty.* degrade, demote, downgrade, humble, impoverish, move down, put down, ruin. *Opp* INCREASE, RAISE.

reduction *n* **1** contraction, curtailment, *inf* cutback, deceleration (*of speed*), decimation, decline, decrease, diminution, drop, impairment, lessening, limitation, loss, moderation, narrowing, remission, shortening, shrinkage, weakening. **2** *reduction in price.* concession, cut, depreciation, devaluation, discount, rebate, refund. *Opp* INCREASE.

redundant *adj* excessive, inessential, non-essential, superfluous, supernumerary, surplus, too many, unnecessary, unneeded, unwanted. *Opp* NECESSARY.

reek *n* stench, stink. ▷ SMELL.

reel *n* bobbin, spool. • *vb* falter, lurch, pitch, rock, roll, spin, stagger, stumble, sway, totter, waver, whirl, wobble. **reel off** ▷ RECITE.

refer *vb* **refer to 1** allude to, bring up, cite, comment on, draw attention to, make reference to, mention, name, point to, quote, speak of, specify, touch on. **2** *refer one person to another.* direct to, guide to, hand over to, pass on to, recommend to, send to. **3** *refer to the dictionary.* consult, go to, look up, resort to, study, turn to.

referee *n* adjudicator, arbiter, arbitrator, judge, mediator, umpire.

reference *n* **1** allusion, citation, example, illustration, instance, intimation, mention, note, quotation, referral, remark. **2** endorsement, recommendation, testimonial.

refill *vb* fill up, refuel, renew, replenish, top up.

refine *vb* **1** clarify, cleanse, clear, decontaminate, distil, process, purify, treat. **2** *refine manners.* civilize, cultivate, improve, perfect, polish.

refined *adj* **1** aristocratic, civilized, courteous, courtly, cultivated, cultured, delicate, dignified, discerning, discriminating, educated, elegant, fastidious, genteel, gentlemanly, gracious, ladylike, nice, polished, polite, *inf* posh, precise, *derog* pretentious, *derog* prissy, sensitive, sophisticated, stylish, subtle, tasteful, *inf* upper-crust, urbane, well-bred, well brought-up. *Opp* RUDE. **2** *refined oil.* clarified, distilled, processed, purified, treated. *Opp* CRUDE.

refinement *n* **1** breeding, *inf* class, courtesy, cultivation, delicacy, discernment, discrimination, elegance, finesse, gentility, graciousness, polish, *derog* pretentiousness, sensitivity, sophistication, style, subtlety, taste, urbanity. **2** *refinements in design.* alteration, change, enhancement, improvement, modification, perfection.

reflect *vb* **1** echo, mirror, return, send back, shine back, throw back. **2** brood, cerebrate, *inf* chew things over, consider, contemplate, deliberate, meditate, ponder, remind yourself, reminisce, ruminate. ▷ THINK. **3** *Her success reflects her hard work.* bear witness to, correspond to, demonstrate, evidence, exhibit, illustrate, indicate, match, point to, reveal, show.

reflection *n* **1** echo, image, likeness. **2** *reflection of hard work.* demonstration, evidence, indication, manifestation, result. **3** *no reflection on you.* aspersion, censure, criticism, discredit, imputation, reproach, shame, slur. **4** *time for reflection.* cerebration, cogitation, contemplation, deliberation, meditation, pondering, rumination, self-examination, study, thinking, thought.

reflective *adj* 1 glittering, lustrous, reflecting, shiny, silvery. 2 ▷ THOUGHTFUL.

reform *vb* 1 ameliorate, amend, become better, better, change, convert, correct, improve, make better, mend, put right, reconstruct, rectify, remodel, reorganize, save. 2 *reform a system.* purge, reconstitute, regenerate, revolutionize.

refrain *vb* **refrain from** abstain from, avoid, cease, desist from, do without, eschew, forbear, leave off, *inf* quit, renounce, stop.

refresh *vb* 1 cool, energize, enliven, fortify, freshen, invigorate, *inf* perk up, quench the thirst of, reanimate, rejuvenate, renew, restore, resuscitate, revitalize, revive, slake (*thirst*). 2 *refresh the memory.* activate, awaken, jog, remind, prod, prompt, stimulate.

refreshing *adj* 1 bracing, cool, enlivening, exhilarating, inspiriting, invigorating, restorative, reviving, stimulating, thirst-quenching, tingling, tonic. *Opp* EXHAUSTING. 2 *a refreshing change.* different, fresh, interesting, new, novel, original, unexpected, unfamiliar, unforeseen, unpredictable, welcome. *Opp* BORING.

refreshments *n* drinks, eatables, *inf* eats, *inf* nibbles, snack. ▷ DRINK, FOOD.

refrigerate *vb* chill, cool, freeze, ice, keep cold.

refuge *n* asylum, *inf* bolt-hole, cover, harbour, haven, *inf* hideaway, hideout, *inf* hidey-hole, hiding-place, protection, retreat, safety, sanctuary, security, shelter, stronghold.

refugee *n* displaced person, émigré, exile, fugitive, outcast, runaway.

refund *n* rebate, repayment. • *vb* give back, pay back, recoup, reimburse, repay, return.

refusal *n inf* brush-off, denial, disagreement, disapproval, rebuff, rejection, veto. *Opp* ACCEPTANCE.

refuse *n* detritus, dirt, garbage, junk, litter, rubbish, trash, waste. • *vb* baulk at, decline, deny, disallow, *inf* jib at, *inf* pass up, rebuff, reject, repudiate, say no to, spurn, turn down, veto, withhold. *Opp* ACCEPT, GRANT.

refute *vb* counter, discredit, disprove, negate, prove wrong, rebut.

regain *vb* be reunited with, find, get back, recapture, reclaim, recoup, recover, repossess, retake, retrieve, return to, win back.

regal *adj derog* haughty, imperial, kingly, lordly, majestic, noble, palatial, *derog* pompous, princely, queenly, royal, stately. ▷ SPLENDID.

regard *n* 1 gaze, look, scrutiny, stare. 2 attention, care, concern, consideration, deference, heed, notice, reference, respect, sympathy, thought. 3 admiration, affection, appreciation, approbation, approval, deference, esteem, favour, honour, love, respect, reverence, veneration. • *vb* 1 behold, contemplate, eye, gaze at, keep an eye on, look at, note, observe, scrutinize, stare at, view, watch. 2 *regarded me as a liability.* account, consider, deem, esteem, judge, look upon, perceive, rate, reckon, respect, think of, value, view, weigh up.

regarding *prep* about, apropos, concerning, connected with, involving, on the subject of, pertaining to, *inf* re, respecting, with reference to, with regard to.

regardless *adj* **regardless of** careless about, despite, heedless of, indifferent to, neglectful of, notwithstanding, unconcerned about, unmindful of.

regime *n* administration, control, discipline, government, leadership, management, order, reign, rule, system.

regiment *vb* arrange, control, discipline, organize, regulate, systematize.

region *n* area, country, department, district, division, expanse, land, locality, neighbourhood, part, place, province, quarter, sector, territory, tract, vicinity, zone.

register *n* archives, catalogue, chronicle, diary, directory, file, index, inventory, journal, ledger, list, record, roll, tally. • *vb* **1** enlist, enrol, enter your name, join, sign on. **2** *register a complaint.* catalogue, enter, list, log, make official, minute, present, record, set down, submit, write down. **3** *register emotion.* betray, display, divulge, express, indicate, manifest, reflect, reveal, show. **4** *register in a hotel. inf* check in, sign in. **5** *register what someone says.* keep in mind, make a note of, mark, notice, take account of.

regress *vb* backslide, degenerate, deteriorate, fall back, go back, move backwards, retreat, retrogress, revert, slip back. *Opp* PROGRESS.

regret *n* **1** bad conscience, compunction, contrition, guilt, penitence, pricking of conscience, remorse, repentance, self-accusation, self-condemnation, self-reproach, shame. **2** disappointment, grief, sadness, sorrow, sympathy. • *vb* accuse yourself, bemoan, be regretful, be sad, bewail, deplore, deprecate, feel remorse, grieve (about), lament, mourn, repent (of), reproach yourself, rue, weep (over).

regretful *adj* apologetic, ashamed, conscience-stricken, contrite, disappointed, guilty, penitent, remorseful, repentant, rueful, sorry. ▷ SAD. *Opp* UNREPENTANT.

regrettable *adj* deplorable, disappointing, distressing, lamentable, reprehensible, sad, shameful, undesirable, unfortunate, unhappy, unlucky, unwanted, upsetting, woeful, wrong.

regular *adj* **1** consistent, constant, equal, even, fixed, measured, ordered, predictable, recurring, repeated, rhythmic, steady, symmetrical, systematic, uniform, unvarying. □ *daily, hourly, monthly, weekly, yearly.* **2** *a regular procedure.* accustomed, common, commonplace, conventional, customary, established, everyday, familiar, frequent, habitual, known, normal, official, ordinary, orthodox, prevailing, proper, routine, scheduled, standard, traditional, typical, usual. **3** *a regular supporter.* dependable, faithful, reliable. *Opp* IRREGULAR. • *n inf* faithful, frequenter, habitué, regular customer, patron.

regulate *vb* **1** administer, conduct, control, direct, govern, manage, monitor, order, organize, oversee, restrict, supervise. **2** *regulate temperature.* adjust, alter, balance, change, get right, moderate, modify, set, vary.

regulation *n* by-law, commandment, decree, dictate, directive, edict, law, order, ordinance, requirement, restriction, rule, ruling, statute.

rehearsal *n* dress rehearsal, *inf* dry run, exercise, practice, preparation, *inf* read-through, *inf* run-through, *inf* try-out.

rehearse *vb* drill, go over, practise, prepare, *inf* run over, *inf* run through, try out.

rehearsed *adj* calculated, practised, pre-arranged, premeditated, prepared, scripted, studied, thought out. *Opp* IMPROMPTU.

reign *n* administration, ascendancy, command, empire, government, jurisdiction, kingdom, monarchy, power, rule, sovereignty. • *vb* be king, be on the throne, be queen, command, govern, have power, hold sway, rule, *inf* wear the crown.

reincarnation *n* rebirth, return to life, transmigration.

reinforce *vb* **1** back up, bolster, buttress, fortify, give strength to, hold up, prop up, stay, stiffen, srrengthen, support, toughen. **2** *reinforce an army.* add to, assist, augment, help, increase the size of, provide reinforcements for, supplement.

reinforcements *plur n* additional troops, auxiliaries, back-up, help, reserves, support.

reinstate *vb* recall, rehabilitate, restore, take back, welcome back. *Opp* DISMISS.

reject *vb* **1** cast off, discard, discount, dismiss, eliminate, exclude, jettison, *inf* junk, put aside, scrap, send back, throw away, throw out. **2** *reject friends.* disown, *inf* drop, *inf* give someone the cold shoulder, jilt, rebuff, renounce, repel, repudiate, repulse, *inf* send packing, shun, spurn, turn your back on. **3** *reject an invitation.* brush aside, decline, refuse, say no to, turn down, veto. *Opp* ACCEPT, ADOPT.

rejoice *vb* be happy, celebrate, delight, exult, glory, revel, triumph. *Opp* GRIEVE.

relapse *n* degeneration, deterioration, recurrence (*of illness*), regression, reversion, *inf* set-back, worsening. • *vb* backslide, degenerate, deteriorate, fall back, have a relapse, lapse, regress, retreat, revert, sink back, slip back, weaken.

relate *vb* **1** communicate, describe, detail, divulge, impart, make known, narrate, present, recite, recount, rehearse, report, reveal, tell. **2** ally, associate, compare, connect, consider together, coordinate, correlate, couple, join, link. **relate to** **1** appertain to, apply to, bear upon, be relevant to, concern, *inf* go with, pertain to, refer to. **2** *relate to other people.* be friends with, empathize with, fraternize with, handle, have a relationship with, identify with, socialize with, understand.

related *adj* affiliated, akin, allied, associated, cognate, comparable, connected, consanguineous, interconnected, interdependent, interrelated, joined, joint, linked, mutual, parallel, reciprocal, relative, similar, twin. ▷ RELEVANT. *Opp* UNRELATED.

relation *n* **1** *old use* kinsman, *old use* kinswoman, *plur* kith and kin, member of the family, relative. ▷ FAMILY. **2** *relation of a story.* ▷ NARRATION.

relationship *n* **1** affiliation, affinity, association, attachment, bond, closeness, connection, consanguinity, correlation, correspondence, interconnection, interdependence, kinship, link, parallel, pertinence, rapport, ratio, tie, understanding. ▷ SIMILARITY. *Opp* CONTRAST. **2** affair, *inf* intrigue, *inf* liaison, love affair, romance, sexual relations. ▷ FRIENDSHIP.

relative *adj* ▷ RELATED, RELEVANT. **relative to** commensurate (with), comparative, proportional, proportionate. *Opp* UNRELATED. • *n* ▷ RELATION.

relax *vb* **1** be easy, be relaxed, calm down, cool down, feel at home, *inf* let go, *inf* put your feet up, rest, *inf* slow down, *inf* take it easy, unbend, unwind. *Opp* TENSION. **2** *relax your vigilance.* abate, curb, decrease, diminish, ease off, lessen, loosen, mitigate, moderate, reduce, release, relieve, slacken, soften, temper, *inf* tone down, unclench, unfasten, weaken. *Opp* INCREASE.

relaxation *n* **1** ease, informality, loosening up, relaxing, repose, rest, unwinding. ▷ RECREATION. *Opp* TENSION. **2** abatement, alleviation, diminution, lessening, *inf* let-up, mitigation, moderation, remission, slackening, weakening. *Opp* INCREASE.

relaxed *adj derog* blasé, calm, carefree, casual, comfortable, contented, cool, cosy, easygoing, *inf* free and easy, friendly, good-humoured, happy, *inf* happy-go-lucky, informal, insouciant, *inf* laid-back, *derog* lax, leisurely, light-hearted, nonchalant, peaceful, reassuring, restful, serene, *derog* slack, tranquil, unconcerned, unhurried, untroubled. *Opp* TENSE.

relay *n* **1** shift, turn. **2** *live relay.* broadcast, programme, transmission. • *vb* broadcast, communicate, pass on, send out, spread, televise, transmit.

release *vb* **1** acquit, allow out, deliver, discharge, dismiss, emancipate, excuse, exonerate, free, let go, let loose, let off, liberate, loose, pardon, rescue, save, set free, set loose, unchain, unfasten, unfetter, unleash, unshackle, untie. *Opp* DETAIN. **2** fire off, launch, let fly, let off, send off. **3** *release information.* circulate, disseminate, distribute, issue, make available, present, publish, put out, send out, unveil.

relegate *vb* **1** consign to a lower position, demote, downgrade, put down. **2** banish, dispatch, exile.

relent *vb* acquiesce, become more lenient, be merciful, capitulate, give in, give way, relax, show pity, soften, weaken, yield.

relentless *adj* **1** dogged, fierce, hard-hearted, implacable, incessant, inexorable, intransigent, merciless, obdurate, obstinate, pitiless, remorseless, ruthless, uncompromising, unfeeling, unforgiving, unmerciful, unyielding. ▷ CRUEL. **2** unceasing, unrelieved, unstoppable, unyielding. ▷ CONTINUAL.

relevant *adj* appertaining, applicable, apposite, appropriate, apropos, apt, connected, essential, fitting, germane, linked, material, pertinent, proper, related, relative, significant, suitable, suited, to the point. *Opp* IRRELEVANT.

reliable *adj* certain, conscientious, consistent, constant, dependable, devoted, efficient, faithful, honest, infallible, loyal, predictable, proven, punctilious, regular, reputable, responsible, safe, solid, sound, stable, staunch, steady, sure, trusted, trustworthy, trusty, unchanging, unfailing. *Opp* UNRELIABLE.

relic *n* heirloom, heritage, inheritance, keepsake, memento, remains, reminder, remnant, souvenir, survival, token, vestige.

relief *n* abatement, aid, alleviation, assistance, assuagement, comfort, cure, deliverance, diversion, ease, easement, help, *inf* let-up, mitigation, palliation, relaxation, release, remedy, remission, respite, rest.

relieve *vb* abate, alleviate, anaesthetize, assuage, bring relief to, calm, comfort, console, cure, diminish, disburden, disencumber, dull, ease, lessen, lift, lighten, make less, mitigate, moderate, palliate, reduce, relax, release, rescue, soften, soothe, unburden. ▷ HELP. *Opp* INTENSIFY.

religion *n* **1** belief, creed, divinity, doctrine, dogma, *derog* pietism, theology. **2** creed, cult, denomination, faith, persuasion, sect. □ *Buddhism, Christianity, Hinduism, Islam, Judaism, Sikhism, Taoism, Zen.*

religious *adj* **1** devotional, divine, holy, sacramental, sacred, scriptural, theological. *Opp* SECULAR. **2** church-going, committed, dedicated, devout, God-fearing, godly, *derog* pietistic, pious, *derog* religiose, reverent, righteous, saintly, *derog* sanctimonious, spiritual. *Opp* IRRELIGIOUS. **3** *religious wars.* bigoted, doctrinal, fanatical, sectarian, schismatic.

relinquish *vb* concede, give in, hand over, part with, submit, surrender, yield.

relish *n* **1** appetite, delight, enjoyment, enthusiasm, gusto, pleasure, zest. **2** flavour, piquancy, savour, tang, taste. • *vb* appreciate, delight in, enjoy, like, love, revel in, savour, take pleasure in.

reluctant *adj* averse, disinclined, grudging, hesitant, loath, unenthusiastic, unwilling. *Opp* EAGER.

rely *vb* **rely on** *inf* bank on, count on, depend on, have confidence in, lean on, put your faith in, *inf* swear by, trust.

remain *vb old use* abide, be left, carry on, continue, endure, keep on, linger, live on, persevere, persist, stay, *inf* stay put, survive, tarry, wait. **remaining** ▷ RESIDUAL.

remainder *n* balance, excess, extra, remnant, residue, residuum, rest, surplus. ▷ REMAINS.

remains *plur n* **1** crumbs, debris, detritus, dregs, fragments, *inf* leftovers, oddments, *inf* odds and ends, offcuts, remainder, remnants, residue, rubble, ruins, scraps, traces, vestiges, wreckage. **2** *historic remains.* heirloom, heritage, inheritance, keepsake, memento, monument, relic, reminder, souvenir, survival. **3** *human remains.* ashes, body, bones, carcass, corpse.

remake *vb* piece together, rebuild, reconstitute, reconstruct, redo. ▷ RENEW.

remark *n* comment, mention, observation, opinion, reflection, statement, thought, utterance, word. • *vb* **1** assert, comment, declare, mention, note, observe, pass comment, reflect, say, state. **2** heed, mark, notice, observe, perceive, see, take note of.

remarkable *adj* amazing, astonishing, astounding, conspicuous, curious, different, distinguished, exceptional, extraordinary, important, impressive, marvellous, memorable, notable, noteworthy, odd, out-of-the-ordinary, outstanding, peculiar, phenomenal, prominent, signal, significant, singular, special, strange, striking, surprising, *inf* terrific, *inf* tremendous, uncommon, unforgettable, unusual, wonderful. *Opp* ORDINARY.

remedy *n inf* answer, antidote, corrective, countermeasure, cure, cure-all, drug, elixir, medicament, medication, medicine, nostrum, palliative, panacea, prescription, redress, relief, restorative, solution, therapy, treatment.
• *vb* alleviate, *inf* ameliorate, answer, control, correct, counteract, *inf* fix, heal, help, mend, mitigate, palliate, put right, rectify, redress, relieve, repair, solve, treat. ▷ CURE.

remember *vb* **1** be mindful of, have a memory of, have in mind, keep in mind, recognize. **2** learn, memorize, retain. **3** *remember old times.* be nostalgic about, hark back to, recall, recollect, reminisce about, review, summon up, tell stories about, think back to. **4** *remember Christmas.* celebrate, commemorate, observe. *Opp* FORGET.

remind *vb* give a reminder to, jog the memory of, nudge, prompt.

reminder *n* **1** aide-mémoire, cue, hint, *inf* memo, memorandum, mnemonic, note, *inf* nudge, prompt, *inf* shopping list. **2** heirloom, inheritance, keepsake, memento, relic, souvenir, survival.

reminisce *vb* be nostalgic, hark back, look back, recall, remember, review, tell stories, think back.

reminiscence *n* account, anecdote, memoir, memory, recollection, remembrance.

reminiscent *adj* evocative, nostalgic, recalling, redolent, suggestive.

remiss *adj* careless, dilatory, forgetful, irresponsible, lax, negligent, slack, thoughtless. *Opp* CAREFUL.

remit *vb* **1** *remit a debt.* cancel, let off, settle. **2** abate, decrease, ease off, lessen, relax, slacken. **3** dispatch, forward, send, transmit. ▷ PAY.

remittance *n* allowance, fee, payment.

remnants *plur n* bits, fragments, *inf* leftovers, oddments, offcuts, residue, scraps, traces, vestiges. ▷ REMAINS.

remodel *vb* ▷ RENEW.

remorse *n* bad conscience, compunction, contrition, grief, guilt, mortification, pangs of conscience, penitence, pricking of conscience, regret, repentance, sadness, self-accusation, self-reproach, shame, sorrow.

remorseful *adj* ashamed, conscience-stricken, contrite, grief-stricken, guilt-ridden, guilty, penitent, regretful, repentant, rueful, sorry. *Opp* UNREPENTANT.

remorseless *adj* dogged, implacable, inexorable, intransigent, merciless, obdurate, pitiless, relentless, ruthless, uncompromising, unforgiving, unkind, unmerciful, unremitting. ▷ CRUEL.

remote *adj* **1** alien, cut off, desolate, distant, far-away, foreign, God-forsaken, hard to find, inaccessible, isolated, lonely, outlying, out of reach, out of reach, out of the way, secluded, solitary, unfamiliar, unfrequented, *inf* unget-at-able, unreachable. *Opp* CLOSE. **2** *a remote chance.* doubtful, implausible, improbable, negligible, outside, poor, slender, slight, small, unlikely. *Opp* SURE. **3** *a remote manner.* abstracted, aloof, cold, cool, detached, haughty, preoccupied, reserved, standoffish, uninvolved, withdrawn. *Opp* FRIENDLY.

removal *n* **1** relocation, removing, taking away, transfer, transportation. **2** elimination, eradication, extermination, liquidation, purge, purging. ▷ KILLING. **3** *removal from a job.* deposition, dethronement, dislodgement, dismissal, displacement, ejection, expulsion, *inf* firing, making redundant, ousting, redundancy, *inf* sacking, transference, unseating. **4** *removal of teeth.* drawing, extraction, pulling, taking out, withdrawal.

remove *vb* **1** abolish, abstract, amputate (*limb*), banish, clear away, cut off, cut out, delete, depose, detach, disconnect, dismiss, dispense with, displace, dispose of, do away with, eject, eliminate, eradicate, erase, evict, excise, exile, expel, expunge, *inf* fire, *inf* get rid of, *inf* kick out, kill, oust, purge, root out, rub out, *inf* sack, send away, separate, strike out, sweep away, take out, throw out, turn out, undo, unfasten, uproot, wash off, wipe (*tape-recording*), wipe out. **2** *remove furniture.* carry away, convey, move, take away, transfer, transport. **3** *remove a tooth.* draw out, extract, pull out, take out. **4** *remove clothes.* doff (*a hat*), peel off, strip off, take off.

rend *vb* cleave, lacerate, pull apart, rip, rupture, shred, split, tear.

render *vb* **1** cede, deliver, furnish, give, hand over, offer, present, proffer, provide, surrender, tender, yield. **2** *render a song.* execute, interpret, perform, play, produce, sing. **3** *rendered me speechless.* cause to be, make.

rendezvous *n* appointment, assignation, date, engagement, meeting, meeting-place, *old use* tryst.

renegade *n* apostate, backslider, defector, deserter, fugitive, heretic, mutineer, outlaw, rebel, runaway, traitor, turncoat.

renege *vb* **renege on** abjure, abrogate, *inf* back out of, break, default on, fail to keep, go back on, *sl* rat on, repudiate, *sl* welsh on.

renew *vb* **1** bring up to date, *inf* do up, *inf* give a face-lift to, improve, mend, modernize, overhaul, recondition, reconstitute, recreate, redecorate, redesign, redevelop, redo, refit, refresh, refurbish, regenerate, reintroduce, rejuvenate, remake, remodel, renovate, repaint, repair, replace, replenish, restore, resume, resurrect, revamp, revitalize, revive, touch up, transform, update. **2** *renew an activity.* come back to, pick up again, recommence, restart, resume, return to. **3** *renew vows.* confirm, reaffirm, reiterate, repeat, restate.

renounce *vb* **1** abandon, abjure, abstain from, declare your opposition to, deny, desert, discard, disown, eschew, forgo, forsake, forswear, give up, reject, repudiate, spurn. **2** *renounce the throne.* abdicate, *inf* quit, relinquish, resign, surrender.

renovate *vb* ▷ RENEW.

renovation *n* improvement, modernization, overhaul, reconditioning, redevelopment, refit, refurbishment, renewal, repair, restoration, transformation, updating.

renowned *adj* celebrated, distinguished, eminent, illustrious, noted, prominent, well-known. ▷ FAMOUS.

rent *n* **1** fee, hire, instalment, payment, rental. **2** *a rent in a garment.* ▷ SPLIT. • *vb* charter, farm out, hire, lease, let.

reorganize *vb* rationalize, rearrange, re-deploy, reshuffle, restructure.

repair *vb* **1** *inf* fix, mend, overhaul, patch up, put right, rectify, refit, restore, service. ▷ RENEW. **2** darn, patch, sew up.

repay *vb* **1** compensate, give back, pay back, recompense, refund, reimburse, remunerate, settle. **2** avenge, get even, *inf* get your own back, reciprocate, requite, retaliate, return, revenge.

repeal *vb* abolish, abrogate, annul, cancel, invalidate, nullify, rescind, reverse, revoke.

repeat *vb* **1** do again, duplicate, redo, rehearse, replay, replicate, reproduce, re-run, show again. **2** echo, quote, recapitulate, re-echo, regurgitate, reiterate, restate, retell, say again.

repel *vb* **1** check, drive away, fend off, fight off, hold off, *inf* keep at bay, parry, push away, rebuff, repulse, resist, ward off, withstand. **2** *repel water.* be impermeable to, exclude, keep out, reject. **3** *cruelty repels us.* alienate, be repellent to, disgust, nauseate, offend, *inf* put off, revolt, sicken, *inf* turn off. *Opp* ATTRACT.

repellent *adj* **1** impermeable, impervious, resistant, unsusceptible. **2** ▷ REPULSIVE.

repent *vb* bemoan, be repentant about, bewail, feel repentance for, lament, regret, reproach yourself for, rue.

repentance *n* contrition, guilt, penitence, regret, remorse, self-accusation, self-reproach, shame, sorrow.

repentant *adj* apologetic, ashamed, conscience-stricken, contrite, grief-stricken, guilt-ridden, guilty, penitent, regretful, remorseful, rueful, sorry. *Opp* UNREPENTANT.

repertory *n* collection, repertoire, repository, reserve, stock, store, supply.

repetitive *adj* boring, incessant, iterative, monotonous, recurrent, repeated, repeating, repetitious, tautologous, tedious, unchanging, unvaried. ▷ CONTINUAL.

replace *vb* **1** make good, put back, reinstate, restore, return. **2** be a replacement for, come after, follow, oust, succeed, supersede, supplant, take over from, take the place of. ▷ DEPUTIZE. **3** *replace worn parts.* change, renew, substitute.

replacement *n inf* fill-in, proxy, stand-in, substitute, successor, understudy.

replenish *vb* fill up, refill, renew, restock, top up.

replete *adj inf* bursting, crammed, gorged, *inf* jam-packed, overloaded, sated, stuffed. ▷ FULL.

replica *n inf* carbon copy, clone, copy, duplicate, facsimile, imitation, likeness, model, reconstruction, reproduction.

reply *n* acknowledgement, answer, *inf* come-back, reaction, rejoinder, response, retort, riposte. • *vb* answer, give a reply, react, rejoin, respond. **reply to** ▷ ACKNOWLEDGE, COUNTER.

report *n* **1** account, announcement, article, communication, communiqué, description, dispatch, narrative, news, record, statement, story, *inf* write-up. **2** backfire, bang, blast, boom, crack, detonation, discharge, explosion, noise. • *vb* **1** announce, broadcast, circulate, communicate, declare, describe, disclose, divulge, document, give an account of, notify, present a report on, proclaim, publish, put out, record, recount, reveal, state, tell. **2** *report for duty.* announce yourself, check in, clock in, introduce yourself, make yourself known, present yourself, sign in. **3** *report someone to the police.* complain about, denounce, inform against, *inf* tell on.

reporter *n* columnist, commentator, correspondent, journalist, newscaster, newsman, newswoman, newspaperman, newspaperwoman, news presenter, photojournalist.

repose *n* calm, calmness, comfort, ease, inactivity, peace, peacefulness, poise, quiescence, quiet, quietness, relaxation, respite, rest, serenity, stasis, stillness, tranquillity. ▷ SLEEP. *Opp* ACTIVITY.

reprehensible *adj* blameworthy, culpable, deplorable, disgraceful, immoral, objectionable, regrettable, remiss, shameful, unworthy, wicked. ▷ GUILTY. *Opp* INNOCENT.

represent *vb* **1** act out, assume the guise of, be an example of, embody, enact, epitomize, exemplify, exhibit, express, illustrate, impersonate, incarnate, masquerade as, personify, pose as, present, pretend to be, stand for, symbolize, typify. **2** characterize, define, delineate, depict, describe, draw, paint, picture, portray, reflect, show, sketch. **3** act for, speak for, stand up for.

representation *n* depiction, figure, icon, image, imitation, likeness, model, picture, portrait, portrayal, resemblance, semblance, statue.

representative *adj* **1** archetypal, average, characteristic, illustrative, normal, typical. *Opp* ABNORMAL. **2** *representative government.* chosen, democratic, elected, elective, popular. *Opp* TOTALITARIAN. • *n* **1** delegate, deputy, proxy, spokesman, spokeswoman, stand-in, substitute. **2** agent, *inf* rep, salesman, salesperson, saleswoman, *inf* traveller. **3** ambassador, consul, diplomat, emissary, envoy, legate. **4** *Amer* congressman, councillor, Member of Parliament, MP, ombudsman.

repress *vb* **1** control, crush, curb, keep down, limit, oppress, overcome, put down, quell, restrain, subdue, subjugate. **2** *repress emotion.* *inf* bottle up, frustrate, inhibit, stifle, suppress.

repressed *adj* **1** cold, frigid, frustrated, inhibited, neurotic, *inf* prim and proper, tense, unbalanced, undemonstrative, *inf* uptight. *Opp* UNINHIBITED. **2** *repressed emotion.* *inf* bottled up, hidden, latent, subconscious, suppressed, unconscious, unfulfilled.

repression *n* **1** authoritarianism, censorship, coercion, control, despotism, dictatorship, oppression, restraint, subjugation, totalitarianism, tyranny. **2** *repression of emotion.* *inf* bottling up, frustration, inhibition, suffocation, suppression.

repressive *adj* authoritarian, autocratic, brutal, coercive, cruel, despotic, dictatorial, fascist, harsh, illiberal, oppressive, restricting, severe, totalitarian, tyrannical, undemocratic, unenlightened. *Opp* LIBERAL.

reprieve *n* amnesty, pardon, postponement, respite, stay of execution. • *vb* commute a sentence, forgive, let off, pardon, postpone execution, set free, spare.

reprimand *n* admonition, castigation, censure, condemnation, criticism, *inf* dressing-down, *inf* going-over, *inf* lecture, lesson, *inf* rap on the knuckles, rebuke, remonstration, reproach, reproof, scolding, *inf* slap on the wrist, *inf* slating, *inf* talking-to, *inf* telling-off, *inf* ticking-off, upbraiding, *inf* wigging. • *vb* admonish, berate, blame, *inf* carpet, castigate, censure, chide, condemn, correct, criticize, disapprove of, *inf* dress down, find fault with, *inf* haul over the coals, *inf* lecture, *inf* rap, rate, *inf* read the riot act to, rebuke, reprehend, reproach, reprove, scold, *inf* slate, *inf* take to task, *inf* teach a lesson, *inf* tell off, *inf* tick off, upbraid. *Opp* PRAISE.

reprisal *n* counter-attack, getting even, redress, repayment, retaliation, retribution, revenge, vengeance.

reproach *n* blame, disapproval, disgrace, scorn. • *vb* censure, criticize, scold, show disapproval of, upbraid. ▷ REPRIMAND. *Opp* PRAISE.

reproachful *adj* admonitory, censorious, condemnatory, critical, disapproving, disparaging, reproving, scornful, withering.

reproduce *vb* **1** copy, counterfeit, duplicate, forge, imitate, mimic, photocopy, print, redo, reissue, reprint, simulate. ▷ REPEAT. **2** beget young, breed, increase, multiply, procreate, produce offspring, propagate, regenerate, spawn.

reproduction *n* **1** breeding, cloning, increase, multiplying, procreation, proliferation, propagation, spawning. **2** *inf* carbon copy, clone, copy, duplicate, facsimile, fake, forgery, imitation, likeness, print, replica.

repudiate *vb* **1** deny, disagree with, dispute, rebuff, refute, reject, scorn, turn down. *Opp* ACKNOWLEDGE. **2** *repudiate an agreement.* abrogate, discard, disown, go back on, recant, renounce, rescind, retract, reverse, revoke.

repugnant *adj* ▷ REPULSIVE.

repulsive *adj* abhorrent, abominable, beastly, disagreeable, disgusting, distasteful, distressing, foul, gross, hateful, hideous, loathsome, nasty, nauseating, nauseous, objectionable, obnoxious, odious, offensive, *inf* off-putting, repellent, repugnant, revolting, *inf* sick, sickening, unattractive, unpalatable, unpleasant, unsavoury, unsightly, vile. ▷ UGLY. *Opp* ATTRACTIVE.

reputable *adj* creditable, dependable, esteemed, famous, good, highly regarded, honourable, honoured, prestigious, reliable, respectable, respected, trustworthy, unimpeachable, *inf* up-market, well-thought-of, worthy. *Opp* DISREPUTABLE.

reputation *n* character, fame, name, prestige, recognition, renown, repute, standing, stature, status.

reputed *adj* alleged, assumed, believed, considered, deemed, famed, judged, purported, reckoned, regarded, rumoured, said, supposed, thought.

request *n* appeal, application, call, demand, entreaty, petition, plea, prayer, question, requisition, solicitation, suit, supplication. • *vb* adjure, appeal, apply (for), ask, beg, beseech, call for, claim, demand, desire, entreat, implore, importune, invite, petition, pray for, require, requisition, seek, solicit, supplicate.

require *vb* **1** be missing, be short of, depend on, lack, need, want. **2** *require a response.* call for, coerce, command, compel, direct, force, insist, instruct, make, oblige, order, put pressure on. ▷ REQUEST. **required** ▷ REQUISITE.

requirement *n* condition, demand, essential, necessity, need, precondition, prerequisite, provision, proviso, qualification, *Lat* sine qua non, stipulation.

requisite *adj* compulsory, essential, imperative, indispensable, mandatory, necessary, needed, obligatory, prescribed, required, set, stipulated. *Opp* OPTIONAL.

requisition *n* application, authorization, demand, mandate, order, request, voucher. • *vb* **1** demand, order, *inf* put in for, request. **2** appropriate, commandeer, confiscate, expropriate, occupy, seize, take over, take possession of.

rescue *n* deliverance, emancipation, freeing, liberation, recovery, release, relief, salvage. • *vb* **1** deliver, emancipate, extricate, free, let go, liberate, loose, ransom, release, save, set free. **2** get back, recover, retrieve, salvage.

research *n* analysis, enquiry, examination, experimentation, exploration, fact-finding, inquiry, investigation, *inf* probe, scrutiny, searching, study. • *vb inf* check out, *inf* delve into, experiment, investigate, *inf* probe, search, study.

resemblance *n* affinity, closeness, coincidence, comparability, comparison, conformity, congruity, correspondence, equivalence, likeness, similarity, similitude.

resemble *vb* approximate to, bear resemblance to, be similar to, compare with, look like, mirror, sound like, *inf* take after.

resent *vb* begrudge, be resentful about, dislike, envy, feel bitter about, grudge, grumble at, object to, *inf* take exception to, *inf* take umbrage at.

resentful *adj* aggrieved, annoyed, begrudging, bitter, disgruntled, displeased, embittered, envious, grudging, hurt, indignant, irked, jaundiced, jealous, malicious, offended, *inf* peeved, *inf* put out, spiteful, unfriendly, ungenerous, upset, vexed, vindictive. ▷ ANGRY.

resentment *n* animosity, bitterness, discontent, envy, grudge, hatred, hurt, ill-will, indignation, irritation, jealousy, malevolence, malice, pique, rancour, spite, unfriendliness, vexation, vindictiveness. ▷ ANGER.

reservation *n* **1** condition, doubt, hedging, hesitation, misgiving, proviso, qualification, qualm, reluctance, reticence, scepticism, scruple. **2** *hotel reservation.* appointment, booking. **3** *a wildlife reservation.* ▷ RESERVE.

reserve *n* **1** cache, fund, hoard, *inf* nest-egg, reservoir, savings, stock, stockpile, store, supply. **2** *inf* back-up, deputy, *plur* reinforcements, replacement, stand-by, *inf* stand-in, substitute, understudy. **3** *a wildlife reserve.* enclave, game park, preserve, protected area, reservation, safari-park, sanctuary. **4** aloofness, caution, modesty, quietness, reluctance, reticence, self-consciousness, self-effacement, shyness, *derog* stand-offishness, taciturnity, timidity. • *vb* **1** earmark, hoard, hold back, keep, keep back, preserve, put aside, retain, save, set aside, stockpile, store up. **2** *reserve seats. inf* bag, book, order, pay for. **reserved** ▷ RETICENT.

reside *vb* **reside in** dwell in, inhabit, live in, lodge in, occupy, settle in.

residence *n old use* abode, address, domicile, dwelling, dwelling-place, habitation, home, quarters, seat. ▷ HOUSE.

resident *adj* in residence, livingin, permanent, remaining, staying. • *n* citizen, denizen, dweller, householder, houseowner, inhabitant, *inf* local, native.

residual *adj* abiding, continuing, left over, outstanding, persisting, remaining, surviving, unconsumed, unused.

resign *vb* abandon, abdicate, *sl* chuck in, forsake, give up, leave, quit, relinquish, renounce, retire, stand down, step down, surrender, vacate. **resigned** ▷ PATIENT. **resign yourself to** ▷ ACCEPT.

resilient *adj* **1** bouncy, elastic, firm, plastic, pliable, rubbery, springy, supple. *Opp* BRITTLE. **2** *a resilient person.* adaptable, buoyant, irrepressible, strong, tough, unstoppable. *Opp* VULNERABLE.

resist *vb* avoid, be resistant to, check, confront, counteract, defy, face up to, hinder, *inf* hold out against, *inf* hold your ground against, impede, inhibit, keep at bay, oppose, prevent, rebuff, refuse, stand up to, withstand. ▷ FIGHT. *Opp* ASSIST, YIELD.

resistant *adj* defiant, hostile, intransigent, invulnerable, obstinate, opposed, refractory, stubborn, uncooperative, unresponsive, unyielding. **resistant to** against, impervious to, invulnerable to, opposed to, proof against, repellent of, unaffected by, unsusceptible to, unyielding to. *Opp* SUSCEPTIBLE.

resolute *adj* adamant, bold, committed, constant, courageous, decided, decisive, determined, dogged, firm, immovable, immutable, indefatigable, *derog* inflexible, *derog* obstinate, persevering, persistent, pertinacious, relentless, resolved, single-minded, staunch, steadfast, strong-minded, strong-willed, *derog* stubborn, tireless, unbending, undaunted, unflinching, unshakable, unswerving, untiring, unwavering. *Opp* IRRESOLUTE.

resolution *n* **1** boldness, commitment, constancy, determination, devotion, doggedness, firmness, *derog* obstinacy, perseverance, persistence, pertinacity, purposefulness, resolve, single-mindedness, staunchness, steadfastness, *derog* stubbornness, tenacity, will-power. ▷ COURAGE. **2** commitment, oath, pledge, promise, undertaking, vow. **3** *resolution at a meeting.* decision, motion, proposal, proposition, statement. **4** *resolution of a problem.* answer, denouement, disentangling, resolving, settlement, solution, sorting out.

resolve *n* ▷ RESOLUTION. • *vb* **1** agree, conclude, decide formally, determine, elect, fix, make a decision, make up your mind, opt, pass a resolution, settle, undertake, vote. **2** *resolve a problem*. answer, clear up, disentangle, figure out, settle, solve, sort out, work out.

resonant *adj* booming, echoing, full, pulsating, resounding, reverberant, reverberating, rich, ringing, sonorous, thunderous, vibrant, vibrating.

resort *n* **1** alternative, course of action, expedient, option, recourse, refuge, remedy, reserve. **2** *a seaside resort*. holiday town, retreat, spa, *old use* watering-place. • *vb* **resort to** adopt, *inf* fall back on, have recourse to, make use of, turn to, use. **2** frequent, go to, *inf* hang out in, haunt, invade, patronize, visit.

resound *vb* boom, echo, pulsate, resonate, reverberate, ring, rumble, thunder, vibrate. **resounding** ▷ RESONANT.

resourceful *adj* clever, creative, enterprising, imaginative, ingenious, innovative, inspired, inventive, original, skilful, *inf* smart, talented. *Opp* SHIFTLESS.

resources *plur n* **1** assets, capital, funds, possessions, property, reserves, riches, wealth. ▷ MONEY. **2** *natural resources*. materials, raw materials.

respect *n* **1** admiration, appreciation, awe, consideration, courtesy, deference, esteem, homage, honour, liking, love, politeness, regard, reverence, tribute, veneration. **2** *perfect in every respect*. aspect, attribute, characteristic, detail, element, facet, feature, particular, point, property, quality, trait, way. • *vb* admire, appreciate, be polite to, defer to, esteem, have high regard for, honour, look up to, pay homage to, revere, reverence, show respect to, think well of, value, venerate. *Opp* DESPISE.

respectable *adj* **1** decent, genteel, honest, honourable, law-abiding, refined, respected, unimpeachable, upright, worthy. **2** *respectable clothes*. chaste, clean, decorous, dignified, modest, presentable, proper, seemly. *Opp* DISREPUTABLE. **3** *a respectable sum*. ▷ CONSIDERABLE.

respectful *adj* admiring, civil, considerate, cordial, courteous, deferential, dutiful, gentlemanly, gracious, humble, ladylike, obliging, polite, proper, reverent, reverential, *derog* servile, subservient, thoughtful, well-mannered. *Opp* DISRESPECTFUL.

respective *adj* individual, own, particular, personal, relevant, separate, several, special, specific.

respite *n* break, *inf* breather, delay, hiatus, holiday, intermission, interruption, interval, *inf* let-up, lull, pause, recess, relaxation, relief, remission, rest, time off, time out, vacation.

resplendent *adj* brilliant, dazzling, glittering, shining, splendid ▷ BRIGHT.

respond *vb* **respond to** **1** acknowledge, answer, counter, give a response to, react to, reciprocate, reply to. **2** *respond to need* ▷ SYMPATHIZE.

response *n* acknowledgement, answer, *inf* comeback, counter, counterblast, feedback, reaction, rejoinder, reply, retort, riposte.

responsible *adj* **1** at fault, culpable, guilty, liable, to blame. **2** *a responsible person*. accountable, answerable, concerned, conscientious, creditable, dependable, diligent, dutiful, ethical, honest, in charge, law-abiding, loyal, mature, moral, reliable, sensible, sober, steady, thinking, thoughtful, trustworthy, unselfish. *Opp* IRRESPONSIBLE.

IBLE. **3** *a responsible job.* burdensome, decision-making, executive, *inf* front-line, important, managerial, *inf* top. *Opp* MENIAL.

responsive *adj* alert, alive, aware, impressionable, interested, open, perceptive, receptive, sensitive, sharp, sympathetic, warm-hearted, wideawake, willing. *Opp* UNINTERESTED.

rest *n* **1** break, *inf* breather, holiday, breathing-space, comfort, ease, hiatus, holiday, idleness, inactivity, indolence, interlude, intermission, interval, leisure, *inf* let-up, *inf* lie-down, *inf* loafing, lull, nap, pause, quiet, recess, relaxation, relief, remission, repose, respite, siesta, tea-break, time off, vacation. ▷ SLEEP. **2** base, brace, bracket, holder, prop, stand, support, trestle, tripod. **3** ▷ REMAINDER. • *vb* **1** be still, doze, have a rest, idle, laze, lie back, lie down, lounge, nod off, *inf* put your feet up, recline, relax, snooze, *inf* take a nap, *inf* take it easy, unwind. ▷ SLEEP. **2** lean, place, position, prop, set, stand, support. **3** *It all rests on the weather.* depend, hang, hinge, rely, turn. **come to rest** ▷ HALT.

restaurant *n* eating-place. □ *bistro, brasserie, buffet, café, cafeteria, canteen, carvery, diner, dining-room, grill, refectory, snack-bar, steak-house.*

restful *adj* calm, calming, comfortable, leisurely, peaceful, quiet, relaxed, relaxing, reposeful, soothing, still, tranquil, undisturbed, unhurried, untroubled. *Opp* EXHAUSTING.

restless *adj* **1** agitated, anxious, edgy, excitable, fidgety, highly-strung, impatient, *inf* jittery, jumpy, nervous, *inf* on tenterhooks, restive, skittish, uneasy, worked up, worried. ▷ ACTIVE. **2** *a restless night.* disturbed, interrupted, sleepless, *inf* tossing and turning, troubled, uncomfortable, unsettled. *Opp* RESTFUL.

restore *vb* **1** bring back, give back, make restitution, put back, reinstate, replace, return. **2** *restore antiques.* clean, *inf* do up, fix, *inf* make good, mend, rebuild, recondition, reconstruct, refurbish, renew, renovate, repair, touch up. **3** *restore good relations.* re-establish, rehabilitate, reinstate, reintroduce, rekindle, revive. **4** *restore to health.* cure, nurse, rejuvenate, resuscitate, revitalize.

restrain *vb* **1** check, control, curb, govern, hold back, inhibit, keep back, keep under control, limit, regulate, rein in, repress, restrict, stifle, stop, strait-jacket, subdue, suppress. **2** arrest, bridle, confine, detain, fetter, handcuff, harness, imprison, incarcerate, jail, *inf* keep under lock and key, lock up, manacle, muzzle, pinion, tie up. **restrained** ▷ CALM, DISCREET.

restrict *vb* circumscribe, confine, control, cramp, delimit, enclose, impede, imprison, inhibit, keep within bounds, limit, regulate, shut. ▷ RESTRAIN. *Opp* FREE.

restriction *n* ban, check, constraint, control, curb, curfew, inhibition, limit, limitation, proviso, qualification, regulation, restraint, rule, stipulation.

result *n* **1** conclusion, consequence, effect, end-product, fruit, issue, outcome, repercussion, sequel, upshot. **2** *result of a trial.* decision, judgement, verdict. **3** *result of a sum.* answer, product, score, total. • *vb* arise, be produced, come about, develop, emanate, emerge, ensue, eventuate, follow, happen, issue, occur, proceed, spring, stem, take place, turn out. **result in** ▷ CAUSE.

resume *vb* begin again, carry on, continue, *inf* pick up the threads, proceed, recommence, reconvene, re-open, restart.

resumption *n* continuation, recommencement, re-opening, *inf* restart.

resurrect *vb* breathe new life into, bring back, raise (from the dead), reawaken, restore, resuscitate, revitalize, revive. ▷ RENEW.

retain *vb* **1** *inf* hang on to, hold, hold back, keep, keep control of, maintain, preserve, reserve, save. *Opp* LOSE. **2** *retain moisture.* absorb, soak up. **3** *retain facts.* keep in mind, learn, memorize, remember. *Opp* FORGET.

retaliate *vb* avenge yourself, be revenged, counter-attack, exact retribution, *inf* get even, *inf* get your own back, *inf* give tit for tat, hit back, pay back, repay, revenge yourself, seek retribution, *inf* settle a score, strike back, *inf* take an eye for an eye, take revenge, wreak vengeance.

retaliation *n* counter-attack, reprisal, retribution, revenge, vengeance.

retard *vb* check, handicap, hinder, hold back, hold up, impede, obstruct, postpone, put back, set back, slow down. ▷ DELAY. **retarded** ▷ BACK-WARD.

reticent *adj* aloof, *derog* antisocial, bashful, cautious, *derog* cold, cool, demure, diffident, discreet, distant, modest, quiet, remote, reserved, restrained, retiring, secretive, self-conscious, self-effacing, shy, silent, *derog* standoffish, taciturn, timid, uncommunicative, unemotional, undemonstrative, unforthcoming, unresponsive, unsociable, withdrawn. *Opp* DEMONSTRATIVE.

retinue *n* attendants, company, entourage, followers, *inf* hangers-on, servants, suite, train.

retire *vb* **1** give up, leave, quit, resign. **2** *retire from society.* become reclusive, cloister yourself, go away, go into retreat, retreat from the world, sequester yourself, withdraw. **3** aestivate, go to bed, hibernate, *sl* hit the hay. ▷ SLEEP.

retort *n* answer, *inf* comeback, rebuttal, rejoinder, reply, response, retaliation, riposte. • *vb* answer, counter, react, rejoin, reply, respond, retaliate, return.

retract *vb* **1** draw in, pull back, pull in. **2** abandon, cancel, disclaim, disown, forswear, *inf* have second thoughts about, recant, renounce, repeal, repudiate, rescind, reverse, revoke, withdraw.

retreat *n* **1** departure, escape, evacuation, exit, flight, retirement, withdrawal. **2** *a secluded retreat.* asylum, den, haven, *inf* hideaway, hideout, hiding-place, refuge, resort, sanctuary, shelter. • *vb* **1** back away, back down, climb down, decamp, depart, evacuate, fall back, flee, give ground, go away, leave, move back, pull back, retire, *inf* run away, take flight, *inf* take to your heels, *inf* turn tail, withdraw. **2** *the floods retreated.* ebb, flow back, recede, shrink back. *Opp* ADVANCE.

retribution *n* compensation, *Lat* quid pro quo, recompense, redress, reprisal, retaliation, revenge, *old use* satisfaction, vengeance. *Opp* FORGIVENESS.

retrieve *vb* bring back, come back with, fetch back, find, get back, make up for, recapture, reclaim, recoup, recover, regain, repossess, rescue, restore, return, salvage, save, take back, trace, track down.

retrograde *adj* backward, negative, regressive, retreating, retrogressive, reverse.

retrospective *adj* backward-looking, looking back, looking behind, nostalgic, with hindsight.

return *n* **1** advent, arrival, homecoming, reappearance, re-entry. **2** *return to normality.* re-establishment (of), regression, reversion. **3** *return of a problem* recrudescence, recurrence, re-emergence, repetition. **4** *return of stolen goods.* replacing, restitution, restoration, retrieval. **5** *return on an investment.* benefit, earnings, gain, income, interest, proceeds, profit, yield. • *vb* **1** backtrack, come back, do a U-turn, double back, go back, reassemble, reconvene, re-enter, regress, retrace your steps, revert, turn back. **2** put back, readdress, repatriate, replace, restore, send back. **3** *return money.* exchange, give back, refund, reimburse, repay. **4** *return a verdict. inf* come up with, deliver, give, proffer, report. **5** *The problem returned. inf* crop up again, happen again, reappear, recur, resurface.

reveal *vb* announce, bare, betray, bring to light, communicate, confess, declare, denude, dig up, disclose, display, divulge, exhibit, expose, *inf* give the game away, lay bare, leak, *inf* let on, *inf* let out, *inf* let slip, make known, open, proclaim, produce, publish, show, show up, *inf* spill the beans about, *inf* take the wraps off, tell, uncover, undress, unearth, unfold, unmask, unveil. *Opp* HIDE.

revel *n* carnival, festival, fête, *inf* jamboree, *inf* rave-up, *inf* spree. ▷ REVELRY. • *vb* carouse, celebrate, *inf* have a spree, have fun, indulge in revelry, *inf* live it up, make merry, *inf* paint the town red. **revel in** ▷ ENJOY.

revelation *n* admission, announcement, communiqué, confession, declaration, disclosure, discovery, exposé, exposure, information, *inf* leak, news, proclamation, publication, revealing, unmasking, unveiling.

revelry *n* carousing, celebration, conviviality, debauchery, festivity, fun, gaiety, *inf* high jinks, jollification, jollity, *inf* junketing, *inf* living it up, merry-making, revelling, revels, roistering, *inf* spree. ▷ PARTY.

revenge *n* reprisal, retaliation, retribution, spitefulness, vengeance, vindictiveness. • *vb* avenge, repay. **be revenged** ▷ RETALIATE.

revenue *n* gain, income, interest, money, proceeds, profits, receipts, returns, takings, yield.

reverberate *vb* boom, echo, pulsate, resonate, resound, ring, rumble, throb, thunder, vibrate.

revere *vb* admire, adore, adulate, beatify, esteem, feel reverence for, glorify, honour, idolize, pay homage to, praise, respect, reverence, value, venerate, worship. *Opp* DESPISE.

reverence *n* admiration, adoration, adulation, awe, deference, devotion, esteem, glorification, homage, honour, idolization, praise, respect, veneration, worship.

reverent *adj* adoring, awed, awestruck, deferential, devoted, devout, pious, prayerful, religious, respectful, reverential, solemn, worshipful. *Opp* IRREVERENT.

reverie *n inf* brown study, daydream, dream, fantasy, meditation.

reverse *adj* back, back-to-front, backward, contrary, converse, inverse, inverted, opposite, rear. • *n* 1 antithesis, contrary, converse, opposite. 2 back, rear, underside, verso, wrong side. 3 defeat, difficulty, disaster, failure, mishap, misfortune, problem, reversal, setback, *inf* upset, vicissitude. • *vb* 1 change, invert, overturn, transpose, turn round, turn upsidedown. 2 *reverse a car.* back, drive backwards, go into reverse. 3 *reverse a decision.* abandon, annul, cancel, countermand, invalidate, negate, nullify, overturn, quash, recant, repeal, rescind, retract, revoke, undo.

review *n* 1 examination, *inf* look back, *inf* post-mortem, reappraisal, reassessment, recapitulation, reconsideration, re-examination, report, retrospective, study, survey. 2 *book review.* appreciation, assessment, commentary, criticism, critique, evaluation, judgement, notice, *inf* write-up. • *vb* 1 appraise, assess, consider, evaluate, examine, *inf* go over, inspect, reassess, recapitulate, reconsider, re-examine, scrutinize, study, survey, take stock of, *inf* weigh up. 2 *review a book.* criticize, write a review of.

revise *vb* 1 adapt, alter, change, correct, edit, emend, improve, modify, overhaul, *inf* polish up, reconsider, rectify, *inf* redo, *inf* rehash, rephrase, revamp, reword, rework, rewrite, update. 2 *revise for exams.* brush up, *inf* cram, learn, study, *inf* swot.

revival *n* advance, progress, quickening, reanimation, reawakening, rebirth, recovery, renaissance, renewal, restoration, resurgence, resurrection, resuscitation, return, revitalization, upsurge.

revive *vb* 1 awaken, come back to life, *inf* come round, *inf* come to, quicken, rally, reawaken, recover, resurrect, rouse, waken. *Opp* RELAPSE. 2 bring back to life, *inf* cheer up, freshen, invigorate, refresh, renew, restore, resuscitate, revitalize, strengthen. *Opp* WEARY.

revolt *n* civil war, coup, coup d'état, insurrection, mutiny, putsch, rebellion, reformation, revolution, rising, *inf* take-over, uprising. • *vb* 1 disobey, dissent, mutiny, rebel, riot, rise up. 2 appal, disgust, nauseate, offend, outrage, repel, sicken, upset.

revolting ▷ OFFENSIVE.

revolution *n* 1 ▷ REVOLT. 2 circuit, cycle, gyration, orbit, rotation, spin, turn. 3 change, reorganization, reorientation, shift, transformation, *inf* turn-about, upheaval, *inf* upset, *inf* U-turn.

revolutionary *adj* 1 insurgent, mutinous, rebel, rebellious, seditious, subversive. 2 *revolu-tionary ideas.* avant-garde, challenging, creative, different, experimental, extremist, innovative, new, novel, progressive, radical, *inf* unheard-of, upsetting. *Opp* CONSERVATIVE. • *n* anarchist, extremist, freedom fighter, insurgent, mutineer, rebel, terrorist.

revolve *vb* circle, go round, gyrate, orbit, pirouette, pivot, reel, rotate, spin, swivel, turn, twirl, wheel, whirl.

revulsion *n* abhorrence, aversion, disgust, hatred, loathing, nausea, outrage, repugnance.

reward *n* award, bonus, bounty, compensation, decoration, favour, honour, medal, payment, prize, recompense, remuneration, requital, return, tribute. *Opp* PUNISHMENT. • *vb* compensate, decorate, give a reward to, honour, recompense, remunerate, repay. *Opp* PENALIZE, PUNISH. **rewarding** ▷ PROFITABLE, WORTHWHILE.

rhapsodize *vb* be expansive, effuse, enthuse, *inf* go into raptures.

rhetoric *n derog* bombast, eloquence, expressiveness, *inf* gift of the gab, grandiloquence, magniloquence, oratory, rhetorical language, *derog* speechifying.

rhetorical *adj* [*most synonyms derog*] artifical, bombastic, florid, *inf* flowery, fustian, grandiloquent, grandiose, highflown, insincere, oratorical, ornate, pretentious, verbose, wordy.

rhyme *n* doggerel, jingle, poem. ▷ VERSE.

rhythm *n* accent, beat, measure, metre, movement, pattern, pulse, stress, tempo, throb, time.

rhythmic *adj* beating, measured, metrical, predictable, pulsing, regular, repeated, steady, throbbing. *Opp* IRREGULAR.

ribald *adj* bawdy, coarse, disrespectful, earthy, naughty, racy, rude, scurrilous, *inf* smutty, vulgar. ▷ OBSCENE.

ribbon *n* band, braid, line, strip, stripe, tape, trimming. **in ribbons** ▷ RAGGED.

rich *adj* **1** affluent, *inf* flush, *inf* loaded, moneyed, opulent, plutocratic, prosperous, wealthy, *inf* well-heeled, well-off, well-to-do. *Opp* POOR. **2** *rich furnishings*. costly, elaborate, expensive, lavish, luxurious, precious, priceless, splendid, sumptuous, valuable. **3** *rich land*. fecund, fertile, fruitful, lush, productive. **4** *a rich harvest*. abundant, ample, bountiful, copious, plenteous, plentiful, profuse, prolific, teeming. **5** *rich colours*. deep, full, intense, strong, vibrant, vivid, warm. **6** *rich food*. cloying, creamy, fat, fattening, fatty, full-flavoured, heavy, highly-flavoured, luscious, sumptuous, sweet. **rich person** billionaire, capitalist, millionaire, plutocrat, tycoon.

riches *plur n* affluence, fortune, means, money, opulence, plenty, possessions, prosperity, resources, wealth.

rickety *adj* dilapidated, flimsy, frail, insecure, ramshackle, shaky, tottering, tumbledown, unsteady, wobbly. ▷ WEAK.

rid *vb* clear, deliver (from), free, purge, rescue, save. **get rid of** ▷ DESTROY, REMOVE.

riddle *n* **1** *inf* brain-teaser, conundrum, enigma, mystery, *inf* poser, problem, puzzle, question. **2** filter, screen, sieve. • *vb* **1** filter, screen, sieve, sift, strain. **2** *riddle with holes*. honeycomb, *inf* pepper, perforate, pierce, puncture.

ride *n* ▷ JOURNEY. • *vb* **1** *ride a bike*. control, handle, manage, sit on, steer. **2** *ride on a bike*. be carried, free-wheel, pedal. **3** *ride on a horse*. amble, canter, gallop, trot. ▷ TRAVEL.

ridge *n* arête, bank, crest, edge, embankment, escarpment. ▷ HILL.

ridicule *n* badinage, banter, burlesque, caricature, contumely, derision, invective, jeering, jibing, lampoon, laughter, mockery, parody, raillery, *inf* ribbing, sarcasm, satire, scorn, sneers, taunts, teasing. • *vb* be sarcastic, be satirical about, burlesque, caricature, chaff, deride, gibe at, guy, hold up to ridicule, jeer at, jibe at, joke about, lampoon, laugh at, make fun of, make jokes about, mimic, mock, parody, pillory, *inf* poke fun at, *inf* rib, satirize, scoff at, *inf* send up, sneer at, subject to ridicule, *inf* take the mickey, taunt, tease, travesty.

ridiculous *adj* absurd, amusing, comic, comical, *inf* crazy, *inf* daft, eccentric, farcical, foolish, grotesque, hilarious, illogical, irrational, laughable, ludicrous, mad, nonsensical, preposterous, senseless, silly, unbelievable, unreasonable, weird, *inf* zany. ▷ FUNNY, STUPID. *Opp* SENSIBLE.

rife *adj* abundant, common, endemic, prevalent, widespread.

rift *n* 1 breach, break, chink, cleft, crack, fracture, gap, gulf, opening, split. 2 *a rift between friends.* alienation, conflict, difference, disagreement, disruption, division, opposition, schism, separation.

rig *n* 1 ▷ RIGGING. 2 *oil rig.* platform. 3 [*inf*] *sporting rig.* apparatus, clothes, equipment, gear, kit, outfit, stuff, tackle. • *vb* **rig out** equip, fit out, kit out, outfit, provision, set up, supply.

rigging *n* rig, tackle. □ *halyards, ropes and pulleys, sails.*

right *adj* 1 decent, ethical, fair, good, honest, honourable, just, law-abiding, lawful, moral, principled, responsible, righteous, right-minded, upright, virtuous. 2 *right answers.* accurate, apposite, appropriate, apt, correct, exact, factual, faultless, fitting, genuine, perfect, precise, proper, sound, suitable, true, truthful, valid, veracious. 3 *the right way.* advantageous, beneficial, best, convenient, good, normal, preferable, preferred, recommended, sensible, usual. 4 *your right side.* right-hand, starboard [= *right facing bow of ship*]. 5 *right wing in politics.* conservative, fascist, reactionary, Tory. *Opp* LEFT, WRONG. • *n* 1 decency, equity, ethics, fairness, goodness, honesty, integrity, justice, morality, propriety, reason, truth, virtue. 2 *right to free speech.* entitlement, facility, freedom, liberty, prerogative, privilege. 3 *right to give orders.* authority, commission, franchise, licence, position, power, title. • *vb* 1 amend, correct, make amends for, put right, rectify, redress, remedy, repair, set right. 2 pick up, set upright, stand upright, straighten up.

righteous *adj* blameless, ethical, God-fearing, good, guiltless, *derog* holier-than-thou, honest, just, law-abiding, moral, pure, *derog* sanctimonious, upright, upstanding, virtuous. *Opp* SINFUL.

rightful *adj* authorized, *Lat* bona fide, correct, just, lawful, legal, legitimate, licensed, licit, proper, real, true, valid. *Opp* ILLEGAL.

rigid *adj* 1 adamantine, firm, hard, inelastic, inflexible, set, solid, steely, stiff, strong, unbending, wooden. ▷ OBSTINATE. 2 *rigid discipline.* harsh, intransigent, punctilious, stern, strict, uncompromising, unkind, unrelenting, unyielding. ▷ RIGOROUS. *Opp* FLEXIBLE.

rigorous *adj* 1 conscientious, demanding, exact, exacting, meticulous, painstaking, precise, punctilious, rigid, scrupulous, strict, stringent, structured, thorough, tough, uncompromising, undeviating, unsparing, unswerving. *Opp* LAX. **2** *rigorous climate.* extreme, hard, harsh, inclement, inhospitable, severe, unfriendly, unpleasant. ▷ COLD. *Opp* MILD.

rim *n* brim, brink, circumference, edge, lip, perimeter, periphery.

rind *n* crust, husk, outer layer, peel, skin.

ring *n* **1** annulus, band, bracelet, circle, circlet, collar, corona, eyelet, girdle, halo, hoop, loop, O, ringlet. **2** *boxing ring.* arena, enclosure, rink. **3** *drugs ring.* association, band, gang, mob, organization, syndicate. ▷ GROUP. **4** *the ring of a bell.* boom, buzz, chime, clang, clink, *inf* ding-a-ling, jangle, jingle, knell, peal, ping, resonance, reverberation, tinkle, tintinnabulation, tolling. **5** *give me a ring sometime. inf* bell, *inf* buzz, call, *inf* tinkle. • *vb* **1** bind, circle, embrace, encircle, enclose, encompass, gird, surround. **2** boom, buzz, chime, clang, clink, jangle, jingle, peal, ping, resonate, resound, reverberate, sound (the knell), tinkle, toll. **3** call, *inf* give a buzz, phone, ring up, telephone.

rinse *vb* bathe, clean, cleanse, drench, flush, sluice, swill, wash.

riot *n* affray, anarchy, brawl, chaos, commotion, demonstration, disorder, disturbance, fracas, fray, hubbub, imbroglio, insurrection, lawlessness, mass protest, mêlé, mutiny, pandemonium, *inf* punch-up, revolt, rioting, riotous behaviour, rising, row, *inf* rumpus, *inf* shindy, strife, tumult, turmoil, unrest, uproar, violence. • *vb* brawl, create a riot, *inf* go on the rampage, *inf* go wild, mutiny, rampage, rebel, revolt, rise up, run riot, *inf* take to the streets. ▷ FIGHT.

riotous *adj* anarchic, boisterous, chaotic, disorderly, lawless, mutinous, noisy, obstreperous, rampageous, rebellious, rowdy, tumultuous, uncivilized, uncontrollable, undisciplined, ungovernable, unrestrained, unruly, uproarious, violent, wild. *Opp* ORDERLY.

rip *vb* gash, lacerate, pull apart, rend, rupture, shred, slit, split, tear.

ripe *adj* mature, mellow, ready to use.

ripen *vb* age, become riper, come to maturity, develop, mature, mellow.

ripple *n* ▷ WAVE. • *vb* agitate, disturb, make waves, purl, ruffle, stir.

rise *n* **1** acclivity, ascent, bank, camber, climb, elevation, hill, hump, incline, ramp, ridge, slope. **2** *a rise in prices.* escalation, gain, increase, increment, jump, leap, upsurge, upswing, upturn, upward movement. • *vb* **1** arise, ascend, climb, fly up, go up, jump, leap, levitate, lift, lift off, mount, soar, spring, take off. **2** get to your feet, get up, stand up. **3** *prices rise each year.* escalate, grow, increase, spiral. **4** *cliffs rise above us.* loom, stand out, stick up, tower. **rise up** ▷ REBEL.

risk *n* **1** chance, likelihood, possibility. **2** danger, gamble, hazard, peril, speculation, uncertainty, venture. • *vb* **1** chance, dare, endanger, hazard, imperil, jeopardize. **2** *risk money.* gamble, speculate, venture.

risky *adj inf* chancy, *inf* dicey, hazardous, *inf* iffy, perilous, precarious, unsafe. ▷ DANGEROUS. *Opp* SAFE.

ritual *n* ceremonial, ceremony, custom, formality, liturgy, observance, practice, rite, routine, sacrament, service, set procedure, solemnity, tradition.

rival *n* adversary, antagonist, challenger, competitor, contender, contestant, enemy, opponent, opposition. • *vb* **1** challenge, compete with, contend with, contest, emulate, oppose, struggle with, undercut, vie with. *Opp* COOPERATE. **2** be as good as, compare with, equal, match, measure up to.

rivalry *n* antagonism, competition, competitiveness, conflict, contention, feuding, opposition, strife. *Opp* COOPERATION.

river *n* brook, rivulet, stream, watercourse, waterway. □ *channel, confluence, delta, estuary, lower reaches, mouth, source, tributary, upper reaches.*

road *n* roadway, route, way. □ *alley, arterial road, avenue, boulevard, bridle-path, bridle-way, bypass, byway, byroad, cart-track, causeway, clearway, crescent, cul-de-sac, drive, driveway, dual carriageway,* Amer *freeway, highway, lane, motorway, one-way street, path, pathway, ring road, service road, side-road, side-street, slip-road, street, thoroughfare, tow-path, track, trail, trunk road,* old use *turnpike.*

roam *vb* amble, drift, meander, prowl, ramble, range, rove, saunter, stray, stroll, *inf* traipse, travel, walk, wander.

roar *vb* bellow, cry out, growl, howl, shout, snarl, thunder, yell, yowl. ▷ SOUND.

rob *vb* burgle, *inf* con, defraud, hold up, loot, *inf* mug, *old use* mulct, pick pockets, pilfer from, pillage, plunder, ransack, rifle, steal from. ▷ STEAL.

robber *n* bandit, brigand, burglar, cat burglar, *inf* con-man, defrauder, embezzler, *old use* highwayman, housebreaker, looter, *inf* mugger, pickpocket, pirate, shoplifter, swindler, thief.

robbery *n* breaking and entering, burglary, *inf* con, confidence trick, embezzlement, fraud, hijacking, *inf* hold-up, larceny, looting, *inf* mugging, pilfering, pillage, plunder, sacking, *inf* scrumping, shoplifting, stealing, *inf* stick-up, theft, thieving.

robe *n* cloak, dress, frock, gown. □ *bathrobe, caftan, cassock, dressing-gown, habit, housecoat, kimono, peignoir, surplice, vestment.* • *vb* ▷ DRESS.

robot *n* android, automated machine, automaton, bionic man, bionic woman, computerized machine, mechanical man.

robust *adj* **1** athletic, brawny, fit, *inf* hale and hearty, hardy, healthy, hearty, muscular, powerful, rugged, sound, strong, sturdy, tough, vigorous. **2** durable, serviceable, strongly-made, well-made. *Opp* WEAK.

rock *n* boulder, crag, ore, outcrop, scree, stone. □ *igneous, metamorphic, sedimentary.* □ *basalt, chalk, clay, flint, gneiss, granite, gravel, lava, limestone, marble, obsidian, pumice, quartz, sandstone, schist, shale, slate, tufa, tuff.* • *vb* **1** lurch, move to and fro, pitch, reel, roll, shake, sway, swing, toss, totter, wobble. **2** ▷ SHOCK.

rocky *adj* **1** barren, inhospitable, pebbly, rough, rugged, stony. **2** ▷ UNSTEADY.

rod *n* bar, baton, cane, dowel, pole, rail, shaft, spoke, staff, stick, strut, wand.

rogue *n* blackguard, charlatan, cheat, *inf* con-man, fraud, *old* *use* knave, mischief-maker, *inf* quack, rapscallion, rascal, ruffian, scoundrel, swindler, trickster, villain, wastrel, wretch. ▷ CRIMINAL.

role *n* **1** character, impersonation, lines, part, portrayal. **2** *role in a business.* contribution, duty, function, job, position, post, task.

roll *n* **1** cylinder, drum, reel, scroll, spool, tube. **2** catalogue, directory, index, inventory, list, listing, record, register. • *vb* **1** go round, gyrate, move round, revolve, rotate, run, somersault, spin, tumble, turn, twirl, whirl. **2** coil, curl, furl, make into a roll, twist, wind, wrap. **3** *roll the lawn.* flatten, level off, level out, smooth. **4** *ship rolled in the storm.* lumber, lurch, pitch, reel, rock, stagger, sway, toss, totter, wallow, welter. **rolling** ▷ WAVY. **roll in, roll up** ▷ ARRIVE.

romance *n* **1** idyll, love story, novel. ▷ WRITING. **2** adventure, colour, excitement, fascination, glamour, mystery. **3** affair, amour, attachment, intrigue, liaison, love affair, relationship.

romantic *adj* **1** colourful, dreamlike, exotic, fabulous, fairy-tale, glamorous, idyllic, nostalgic, picturesque. **2** *romantic feelings.* affectionate, amorous, emotional, erotic, loving, passionate, *inf* sexy, *inf* soppy, tender. **3** *romantic fiction.* emotional, escapist, heartwarming, nostalgic, reassuring, sentimental, *derog* sloppy, tender, unrealistic. **4** *romantic ideals.* chimerical, *inf* head in the clouds, idealistic, illusory, impractical, improbable, quixotic, starry-eyed, unworkable, Utopian, visionary. *Opp* REALISTIC.

room *n* **1** *inf* elbow-room, freedom, latitude, leeway, margin, scope, space, territory. **2** *a room in a house.* apartment, cell, *old use* chamber. □ *ante-room, attic, audience chamber, bathroom, bedroom, boudoir, cell, cellar, chapel, classroom, cloakroom, conservatory, corridor, dining-room, dormitory, drawing-room, dressing-room, gallery, guest-room, hall, kitchen, kitchenette, laboratory, landing, larder, laundry, lavatory, library, living-room, loft, lounge, music-room, nursery, office, pantry, parlour, passage, play-room, porch, salon, saloon, scullery, sick-room, sitting-room, spare room, state-room, store-room, studio, study, toilet, utility room, waiting-room, ward, washroom, WC, workroom, workshop.*

roomy *adj* capacious, commodious, large, sizeable, spacious, voluminous. ▷ BIG. *Opp* SMALL.

root *n* **1** radicle, rhizome, rootlet, tap root, tuber. **2** *the root of a problem.* base, basis, bottom, cause, foundation, fount, origin, seat, source, starting-point. • *vb* **root out** ▷ REMOVE.

rope *n* cable, cord, line, strand, string. □ *halyard, hawser, lanyard, lariat, lasso, tether.* • *vb* bind, hitch, lash, moor, tether, tie. ▷ FASTEN.

rot *n* **1** corrosion, corruption, decay, decomposition, deterioration, disintegration, dry rot, mould, mouldiness, putrefaction, wet rot. **2** *What rot!* ▷ NONSENSE. • *vb* become rotten, corrode, crumble, decay, decompose, degenerate, deteriorate, disintegrate, fester, go bad, *inf* go off, perish, putrefy, rust, spoil.

rota *n* list, roster, schedule, timetable.

rotary *adj* gyrating, revolving, rotating, rotatory, spinning, turning, twirling, twisting, whirling.

rotate *vb* **1** go round, gyrate, have a rotary movement, move round, pirouette, pivot, reel, revolve, roll, spin, swivel, turn, turn anticlockwise, turn clockwise, twiddle, twirl, twist, wheel, whirl. **2** *rotate duties.* alternate, pass round, share out, take in turn, take turns.

rotten *adj* **1** bad, corroded, crumbling, decayed, decaying, decomposed, disintegrating, foul, mouldering, mouldy, *inf* off, overripe, perished, putrid, rusty, smelly, tainted, unfit for consumption, unsound. *Opp* SOUND. **2** ▷ IMMORAL.

rough *adj* **1** broken, bumpy, coarse, craggy, irregular, knobbly, jagged, lumpy, pitted, rocky, rugged, rutted, stony, uneven. **2** *rough skin.* bristly, callused, chapped, coarse, hairy, harsh, leathery, ragged, scratchy, shaggy, unshaven, wrinkled. **3** *a rough sea.* agitated, choppy, stormy, tempestuous, turbulent, violent, wild. **4** *a rough voice.* cacophonous, discordant, grating, gruff, harsh, hoarse, husky, rasping, raucous, strident, unmusical, unpleasant. *Opp* SMOOTH. **5** *rough manners, a rough fellow.* badly-behaved, bluff, blunt, brusque, churlish, ill-bred, impolite, loutish, rowdy, rude, surly, *inf* ugly, uncivil, uncivilized, undisciplined, unfriendly. **6** *rough treatment.* brutal, cruel, painful, ruffianly,thuggish, violent. **7** *rough work.* amateurish, careless, clumsy, crude, hasty, imperfect, inept, *inf* rough and ready, unfinished, unpolished, unskilful. **8** *a rough estimate.* approximate, general, hasty, imprecise, inexact, sketchy, vague. *Opp* EXACT, GENTLE, SMOOTH.

roughly *adv* about, approximately, around, close to, nearly.

round *adj* **1** [*two-dimensional*] annular, circular, curved, disc-shaped, hoop-shaped, orbicular, ring-shaped. **2** [*three-dimensional*] ball-shaped, bulbous, cylindrical, globelike, globoid, globular, orb-shaped, spherical, spheroid. **3** *a round figure.* ample, full, plump, rotund, rounded, well-padded. ▷ FAT. ● *n* bout, contest, game, heat, stage. ● *vb* skirt, travel round, turn. **round off** ▷ COMPLETE. **round on** ▷ ATTACK. **round the bend** ▷ MAD. **round the clock** ▷ CONTINUOUS. **round up** ▷ ASSEMBLE.

roundabout *adj* circuitous, circular, devious, indirect, long, meandering, oblique, rambling, tortuous, twisting, winding. *Opp* DIRECT. ● *n* **1** carousel, merry-go-round, *old use* whirligig. **2** traffic island. **round-shouldered** *adj* hunch-backed, humpbacked, stooping.

rouse *vb* **1** arise, arouse, awaken, call, get up, wake up. **2** *rouse to a frenzy.* agitate, animate, electrify, excite, galvanize, goad, incite, inflame, provoke, spur on, stimulate, stir up, *inf* wind up, work up.

rout *vb* conquer, crush, overpower, overwhelm, put to flight, *inf* send packing. ▷ DEFEAT.

route *n* course, direction, itinerary, journey, path, road, way.

routine *adj* accustomed, commonplace, customary, everyday, familiar, habitual, normal, ordinary, perfunctory, planned, run-of-the-mill, scheduled, uneventful, well-rehearsed. ● *n* **1** course of action, custom, *inf* drill, habit, method, pattern, plan, practice, procedure, schedule, system, way. **2** *comedy routine.* act, number, performance, programme, set piece.

row *n* **1** [rhyme with *crow*] chain, column, cordon, file, line, queue, rank, sequence, series, string, tier. **2** [rhyme with *cow*] ado, commotion, fuss, hubbub, hullabaloo, *inf* racket, rumpus, tumult, uproar. ▷ NOISE. **3** altercation, argument, controversy, disagreement, dispute, fight, fracas, *inf* ructions, *inf* slanging match, squabble. ▷ QUARREL. • *vb* **1** [rhyme with *crow*] *row a boat.* move, propel, scull. **2** [rhyme with *cow*] ▷ QUARREL.

rowdy *adj* badly-behaved, boisterous, disorderly, ill-disciplined, irrepressible, lawless, obstreperous, riotous, rough, turbulent, undisciplined, unruly, violent, wild. ▷ NOISY. *Opp* QUIET.

royal *adj* imperial, kingly, majestic, princely, queenly, regal, stately. • *n* [*inf*] member of royal family. □ *consort, Her/His Majesty, Her/His Royal Highness, king, monarch, prince, princess, queen, queen mother, regent, sovereign.* ▷ NOBLE.

rub *vb* **1** caress, knead, massage, smooth, stroke. **2** abrade, chafe, graze, scrape, wear away. **3** *rub clean.* buff, burnish, polish, scour, scrub, shine, wipe. **rub it in** ▷ EMPHASIZE. **rub out** ▷ ERASE. **rub up the wrong way** ▷ ANNOY.

rubbish *n* **1** debris, detritus, dregs, dross, filth, flotsam and jetsam, garbage, junk, leavings, *inf* left-overs, litter, lumber, muck, *inf* odds and ends, offal, offcuts, refuse, rejects, rubble, scrap, slops, sweepings, trash, waste. **2** ▷ NONSENSE.

rubble *n* broken bricks, debris, fragments, remains, ruins, wreckage.

ruddy *adj* fresh, flushed, glowing, healthy, red, sunburnt.

rude *adj* **1** abrupt, abusive, bad-mannered, bad-tempered, blasphemous, blunt, boorish, brusque, cheeky, churlish, coarse, common, condescending, contemptuous, discourteous, disparaging, disrespectful, foul, graceless, gross, ignorant, ill-bred, ill-mannered, impertinent, impolite, improper, impudent, in bad taste, inconsiderate, indecent, insolent, insulting, loutish, mocking, naughty, oafish, offensive, off-hand, patronizing, peremptory, personal (*remarks*), saucy, scurrilous, shameless, tactless, unchivalrous, uncivil, uncomplimentary, uncouth, ungracious, *old use* unmannerly, unprintable, vulgar. ▷ OBSCENE. *Opp* POLITE. **2** *rude workmanship.* awkward, basic, bumbling, clumsy, crude, inartistic, primitive, rough, rough-hewn, simple, unpolished, unskilful, unsophisticated, unsubtle. *Opp* SOPHISTICATED. **be rude to** ▷ INSULT.

rudeness *n* abuse, *inf* backchat, bad manners, boorishness, *inf* cheek, churlishness, condescension, contempt, discourtesy, disrespect, ill-breeding, impertinence, impudence, incivility, insolence, insults, oafishness, tactlessness, uncouthness, vulgarity.

rudiments *plur n* basic principles, basics, elements, essentials, first principles, foundations, fundamentals.

rudimentary *adj* basic, crude, elementary, embryonic, immature, initial, introductory, preliminary, primitive,

provisional, undeveloped.
Opp ADVANCED.

ruffian *n inf* brute, bully, desperado, gangster, hoodlum, hooligan, lout, mugger, rogue, scoundrel, thug, *inf* tough, villain, *inf* yob.

ruffle *vb* 1 agitate, disturb, ripple, stir. 2 *ruffle your hair*. derange, disarrange, dishevel, disorder, *inf* mess up, rumple, tangle, tousle. 3 *ruffle your composure*. annoy, confuse, disconcert, disquiet, fluster, irritate, *inf* nettle, *inf* rattle, *inf* throw, unnerve, unsettle, upset, vex, worry.
Opp SMOOTH.

rug *n* blanket, coverlet, mat, matting.

rugged *adj* 1 bumpy, craggy, irregular, jagged, pitted, rocky, rough, stony, uneven. 2 *rugged conditions*. arduous, difficult, hard, harsh, onerous, rough, severe, tough. 3 *rugged good looks*. burly, hardy, husky, muscular, robust, rough, strong, sturdy, ungraceful, unpolished, weather-beaten.

ruin *n* bankruptcy, breakdown, collapse, *inf* crash, destruction, downfall, end, failure, fall, ruination, undoing, wreck. • *vb* damage, demolish, destroy, devastate, flatten, overthrow, shatter, spoil, wreck. **ruins** debris, havoc, remains, rubble, wreckage.

ruined *adj* crumbling, derelict, dilapidated, fallen down, in ruins, ramshackle, ruinous, tumbledown, uninhabitable, unsafe, wrecked.

ruinous *adj* 1 apocalyptic, calamitous, cataclysmic, catastrophic, crushing, destructive, devastating, dire, disastrous, fatal, harmful, injurious, pernicious, shattering. 2 ▷ RUINED.

rule *n* 1 axiom, code, decree, *plur* guidelines, law, ordinance, practice, precept, principle, regulation, ruling, statute.
2 administration, ascendancy, authority, command, control, domination, dominion, empire, government, influence, jurisdiction, management, mastery, oversight, power, regime, reign, sovereignty, supervision, supremacy, sway.
3 *as a general rule*. convention, custom, norm, routine, standard.
• *vb* 1 administer, be the ruler of, command, control, direct, dominate, govern, guide, hold sway, lead, manage, predominate, reign, run, superintend. 2 adjudicate, decide, decree, deem, determine, find, judge, pronounce, resolve.
rule out ▷ EXCLUDE.

ruler *n* administrator, *inf* Big Brother, law-maker, leader, manager. □ *autocrat, Caesar, demagogue, despot, dictator, doge, emir, emperor, empress, governor, kaiser, king, lord, monarch, potentate, president, prince, princess, queen, rajah, regent, satrap, sovereign, sultan, suzerain, triumvirate* [= three ruling jointly], *tyrant, tzar, viceroy*. ▷ CHIEF.

rumour *n* chat, *inf* chit-chat, gossip, hearsay, *inf* low-down, news, prattle, report, scandal, *inf* tittle-tattle, whisper.

run *n* 1 canter, dash, gallop, jog, marathon, race, sprint, trot. 2 *a run in the car*. drive, excursion, jaunt, journey, joyride, ride, *inf* spin, trip. 3 *run of bad luck*. chain, sequence, series, stretch. 4 *chicken run*. compound, coop, enclosure, pen. • *vb* 1 bolt, canter, career, dash, gallop, hare, hurry, jog, race, rush, scamper, scoot, scurry, scuttle, speed, sprint, tear, trot. 2 *buses run hourly*. go, operate, ply, provide a service, travel. 3 *car runs well*. behave, function, perform, work. 4 *water runs downhill.* cascade, dribble, flow, gush, leak, pour, spill, stream, trickle. 5 *Who runs the country?* administer, conduct, control, direct, govern, look after, maintain, manage, rule, supervise. **run across** ▷ MEET. **run after** ▷ PURSUE. **run away** ▷ ESCAPE. **run into** ▷ MEET.

runner *n* 1 athlete, competitor, entrant, hurdler, jogger, participant, sprinter. 2 courier, dispatch-rider, errand-boy, errand-girl, messenger. 3 *plant sends out runners.* offshoot, shoot, sprout, sucker, tendril.

runny *adj* fluid, free-flowing, liquid, running, thin, watery. *Opp* SOLID, VISCOUS.

rupture *n* 1 breach, break, burst, cleavage, fracture, puncture, rift, split. 2 *rupture between friends.* break-up, disunity, schism, separation. 3 [*medical*] hernia. • *vb* break, burst, fracture, part, separate, split.

rural *adj* agrarian, agricultural, Arcadian, bucolic, countrified, pastoral, rustic, sylvan. *Opp* URBAN.

rush *n* 1 bustle, dash, haste, hurry, panic, pressure, race, scramble, speed, turmoil, urgency. 2 *rush of water.* cataract, flood, gush, spate, surge. 3 *rush of people.* charge, onslaught, stampede. • *vb* bolt, burst, bustle, canter, career, charge, dash, fly, gallop, *inf* get a move on, hare, hasten, hurry, jog, make haste, move fast, race, run, scamper, *inf* scoot, scramble, scurry, scuttle, shoot, speed, sprint, stampede, *inf* step on it, *inf* tear, trot, *inf* zoom.

rust *vb* become rusty, corrode, crumble away, oxidize, rot.

rustic *adj* 1 ▷ RURAL. 2 *rustic simplicity.* artless, clumsy, crude, naïve, *derog* oafish, plain, rough, simple, uncomplicated, uncultured, unpolished, unsophisticated.

rusty *adj* 1 corroded, oxidized, rotten, tarnished. 2 [*inf*] *My French is rusty.* dated, forgotten, out of practice, unused.

rut *n* 1 channel, furrow, groove, indentation, pothole, track, trough, wheel-mark. 2 *in a rut.* dead end, habit, pattern, routine, treadmill.

ruthless bloodthirsty, brutal, callous, cruel, dangerous, ferocious, fierce, hard, heartless, inexorable, inhuman, merciless, pitiless, relentless, sadistic, unfeeling, unrelenting, unsympathetic, vicious, violent. *Opp* MERCIFUL.

S

sabotage *n* disruption, vandalism, wilful damage, wrecking. • *vb* cripple, damage, destroy, disable, disrupt, incapacitate, put out of action, *inf* throw a spanner in the works (of), vandalize, wreck.

sack *n* 1 bag, pouch. 2 *inf* the boot, *inf* the chop, dismissal, firing, redundancy, *inf* your cards. • *vb* 1 *inf* axe, discharge, dismiss, *inf* fire, give someone notice, *inf* give someone the boot, *inf* give someone the chop, *inf* give someone the sack, lay off, make redundant. 2 ▷ DESTROY, PLUNDER. **get the sack** be dismissed, be sacked, *inf* get your cards, get your marching orders, lose your job.

sacred *adj* blessed, blest, consecrated, dedicated, divine, godly, hallowed, holy, religious, revered, sacrosanct, sanctified, venerable, venerated. *Opp* SECULAR.

sacrifice *n* immolation, oblation, offering, propitiation, votive offering. • *vb* **1** immolate, kill, offer up, slaughter, yield up. **2** abandon, forfeit, forgo, give up, let go, lose, relinquish, renounce, surrender.

sacrilege *n* blasphemy, desecration, disrespect, heresy, impiety, irreverence, profanation.

sacrilegious *adj* atheistic, blasphemous, disrespectful, heretical, impious, irreligious, irreverent, profane, ungodly. ▷ WICKED. *Opp* REVERENT.

sacrosanct *adj* inviolable, inviolate, protected, respected, secure, untouchable. ▷ SACRED.

sad *adj* **1** abject, *inf* blue, brokenhearted, careworn, cheerless, crestfallen, dejected, depressed, desolate, despairing, desperate, despondent, disappointed, disconsolate, discontented, discouraged, disgruntled, disheartened, disillusioned, dismal, dispirited, dissatisfied, distressed, doleful, dolorous, *inf* down, downcast, downhearted, dreary, forlorn, friendless, funereal, gloomy, glum, grave, grief-stricken, grieving, grim, guilty, heartbroken, *inf* heavy, heavy-hearted, homesick, hopeless, in low spirits, *inf* in the doldrums, joyless, lachrymose, lonely, *inf* long-faced, *inf* low, lugubrious, melancholy, miserable, moody, moping, morose, mournful, pathetic, penitent, pessimistic, piteous, pitiable, pitiful, plaintive, poignant, regretful, rueful, saddened, serious, sober, sombre, sorrowful, sorry, tearful, troubled, unhappy, unsatisfied, upset, wistful, woebegone, woeful, wretched. **2** *sad news*. calamitous, deplorable, depressing, disastrous, discouraging, dispiriting, distressing, grievous, heartbreaking, heart-rending, lamentable, morbid, moving, painful, regrettable, *inf* tear-jerking, touching, tragic, unfortunate, unsatisfactory, unwelcome, upsetting. **3** *a sad state of disrepair*. ▷ UNSATISFACTORY. *Opp* HAPPY.

sadden *vb* aggrieve, *inf* break someone's heart, deject, depress, disappoint, discourage, dishearten, dismay, dispirit, distress, grieve, make sad, upset. *Opp* CHEER.

sadistic *adj* barbarous, beastly, brutal, inhuman, monstrous, perverted, pitiless, ruthless, vicious. ▷ CRUEL.

sadness *n* bleakness, care, dejection, depression, desolation, despair, despondency, disappointment, disillusionment, dissatisfaction, distress, dolour, gloom, glumness, grief, heartbreak, heaviness, homesickness, hopelessness, joylessness, loneliness, melancholy, misery, moping, moroseness, mournfulness, pessimism, poignancy, regret, ruefulness, seriousness, soberness, sombreness, sorrow, tearfulness, trouble, unhappiness, wistfulness, woe. *Opp* HAPPINESS.

safe *adj* **1** defended, foolproof, guarded, immune, impregnable, invulnerable, protected, secured, shielded. ▷ SECURE. *Opp* VULNERABLE. **2** *inf* alive and well, *inf* all right, *inf* in one piece, intact, sound, undamaged, unharmed, unhurt, uninjured, unscathed, well, whole. **3** *safe drivers*. cautious, circumspect, dependable, reliable, trustworthy.

4 *safe pets*. docile, friendly, harmless, innocuous, tame. **5** *safe to drink*. decontaminated, drinkable, eatable, fit for human consumption, fresh, good, non-poisonous, non-toxic, pasteurized, potable, pure, purified, uncontaminated, unpolluted, wholesome. **6** *safe vehicle*. airworthy, roadworthy, seaworthy, tried and tested. *Opp* DANGEROUS. **make safe** ▷ SECURE. **safe keeping** care, charge, custody, guardianship, keeping, protection.

safeguard *vb* care for, defend, keep safe, look after, protect, shelter, shield.

safety *n* **1** cover, immunity, invulnerability, protection, refuge, sanctuary, security, shelter. **2** *safety of air travel*. dependability, harmlessness, reliability.

sag *vb* be limp, bend, dip, droop, fall, flop, hang down, sink, slump. ▷ DROP.

sail *n* **1** canvas. □ *foresail, gaffsail, jib, lateen sail, lugsail, mainsail, mizzen, spinnaker, spritsail, topsail.* **2** cruise, sea-passage, voyage. ▷ JOURNEY. • *vb* **1** captain, navigate, paddle, pilot, punt, row, skipper, steer. **2** cruise, go sailing, put to sea, set sail, steam. ▷ TRAVEL.

sailor *n* mariner, seafarer, *old use* sea dog, seaman. □ *able seaman, bargee, boatman, boatswain, bosun, captain, cox, coxswain,* plur *crew, deck-hand, helmsman, mate, midshipman, navigator, pilot, rating, rower, skipper, yachtsman, yachtswoman.*

saintly *adj* angelic, blessed, blest, chaste, godly, holy, innocent, moral, pious, pure, religious, righteous, seraphic, sinless, virginal, virtuous. ▷ GOOD. *Opp* SATANIC.

sake *n* account, advantage, behalf, benefit, gain, good, interest, welfare.

salary *n* compensation, earnings, emolument, income, pay, payment, remuneration, stipend, wages.

sale *n* marketing, selling, trade, traffic, transaction, vending. □ *auction, bazaar, closing-down sale, fair, jumble sale, market, rummage sale, spring sale.*

salesperson *n* assistant, auctioneer, representative, salesman, saleswoman, shop-boy, shop-girl, shopkeeper.

saliva *n inf* dribble, *inf* spit, spittle, sputum.

sallow *adj* anaemic, bloodless, colourless, etiolated, pale, pallid, pasty, unhealthy, wan, yellowish.

salt *adj* brackish, briny, saline, salted, salty, savoury. *Opp* FRESH.

salubrious *adj* health-giving, healthy, hygienic, invigorating, nice, pleasant, refreshing, sanitary, wholesome. *Opp* UNHEALTHY.

salute *n* acknowledgement, gesture, greeting, salutation, wave. • *vb* accost, acknowledge, address, greet, hail, honour, pay respects to, recognize. ▷ GESTURE.

salvage *n* **1** reclamation, recovery, rescue, retrieval, salvation, saving. **2** recyclable material, waste. • *vb* conserve, preserve, reclaim, recover, recycle, redeem, rescue, retrieve, reuse, save, use again.

salvation *n* deliverance, escape, help, preservation, redemption, rescue, saving, way out. *Opp* DAMNATION.

salve *n* balm, cream, demulcent, embrocation, emolient, liniment, lotion, ointment, unguent. • *vb* alleviate, appease, assuage, comfort, ease, mitigate, mollify, soothe.

same *adj* 1 actual, identical, self-same. 2 *two women wearing the same jacket.* analogous, comparable, consistent, corresponding, duplicate, equal, equivalent, indistinguishable, interchangeable, matching, parallel, similar, synonymous [= *having same meaning*], twin, unaltered, unchanged, uniform, unvaried. *Opp* DIFFERENT.

sample *n* bit, demonstration, example, foretaste, free sample, illustration, indication, instance, model, pattern, representative piece, selection, snippet, specimen, taste, trailer (*of film*), trial offer. • *vb* experience, inspect, take a sample of, taste, test, try.

sanatorium *n* clinic, convalescent home, hospital, nursing home, rest-home.

sanctify *vb* beatify, bless, canonize, consecrate, hallow, justify, purify.

sanctimonious *adj* canting, holier-than-thou, hypocritical, insincere, moralizing, pharisaical, pietistic, *sl* pi, pious, self-righteous, sententious, *inf* smarmy, smug, superior, unctuous.

sanction *n* agreement, approval, authorization, *inf* blessing, confirmation, consent, encouragement, endorsement, legalization, licence, permission, ratification, support, validation. • *vb* agree to, allow, approve, authorize, confirm, consent to, endorse, *inf* give your blessing to, give permission for, legalize, legitimize, licence, permit, ratify, support, validate.

sanctity *adj* divinity, godliness, grace, holiness, piety, sacredness, saintliness.

sanctuary *n* 1 asylum, haven, protection, refuge, retreat, safety, shelter. 2 *wildlife sanctuary.* conservation area, park, preserve, reservation, reserve. 3 *a holy sanctuary.* chapel, church, holy of holies, holy place, sanctum, shrine, temple.

sands *plur n* beach, seaside, shore, *poet* strand.

sane *adj inf* all there, balanced, *Lat* compos mentis, *inf* in your right mind, level-headed, lucid, normal, of sound mind, rational, reasonable, sensible, sound, stable, well-balanced. *Opp* MAD.

sanguine *adj* buoyant, cheerful, confident, expectant, hopeful, *inf* looking on the bright side, optimistic, positive. *Opp* PESSIMISTIC.

sanitary *adj* aseptic, bacteria-free, clean, disinfected, germfree, healthy, hygienic, pure, salubrious, sterile, sterilized, uncontaminated, unpolluted, wholesome. *Opp* UNHEALTHY.

sanitation *n* drainage, drains, lavatories, sanitary arrangements, sewage disposal, sewers.

sap *n* fluid, life-blood, moisture, vigour, vitality, vital juices. • *vb* bleed, drain. ▷ EXHAUST.

sarcasm *n* acerbity, asperity, contumely, derision, irony, malice, mockery, ridicule, satire, scorn.

sarcastic *adj* acerbic, acidulous, biting, caustic, contemptuous, cutting, demeaning, derisive, disparaging, hurtful, ironic, ironical, mocking, satirical, scathing, sharp, sneering, spiteful, taunting, trenchant, venomous, vitriolic, withering, wounding. ▷ HUMOROUS.

sardonic *adj* bitter, black, cruel, cynical, grim, heartless, malicious, mordant, wry. ▷ HUMOROUS.

sash *n* band, belt, cummerbund, girdle, waistband.

satanic *adj* demonic, devilish, diabolical, fiendish, hellish, infernal, Mephistophelian. ▷ WICKED. *Opp* SAINTLY.

satchel *n* bag, pouch, school-bag, shoulder-bag.

satellite *n* **1** moon, planet. **2** *man-made satellite.* spacecraft, sputnik.

satire *n* burlesque, caricature, derision, invective, irony, lampoon, mockery, parody, ridicule, satirical comedy, scorn, *inf* send-up, *inf* spoof, *inf* take-off, travesty. ▷ WRITING.

satirical *adj* critical, derisive, disparaging, disrespectful, ironic, irreverent, mocking, scornful. ▷ HUMOROUS, SARCASTIC.

satirize *vb* be satirical about, burlesque, caricature, criticize, deride, hold up to ridicule, lampoon, laugh at, make fun of, mimic, mock, parody, pillory, *inf* send up, *inf* take off, travesty. ▷ RIDICULE.

satisfaction *n* comfort, content, contentment, delight, enjoyment, fulfilment, gratification, happiness, joy, pleasure, pride, self-satisfaction. *Opp* DISSATISFACTION.

satisfactory *adj* acceptable, adequate, *inf* all right, competent, fair, *inf* good enough, *inf* not bad, passable, pleasing, satisfying, sufficient, suitable, tolerable, *inf* up to scratch. *Opp* UNSATISFACTORY.

satisfy *vb* appease, assuage, comfort, comply with, content, fill, fulfil, gratify, make happy, meet, pacify, placate, please, put an end to, quench, sate, satiate, serve (*a need*), settle, slake (*thirst*), solve, supply. *Opp* FRUSTRATE. **satisfied** ▷ CONTENT.

saturate *vb* drench, fill, impregnate, permeate, soak, souse, steep, suffuse, waterlog, wet.

sauce *n* **1** condiment, gravy, ketchup, relish. **2** ▷ INSOLENCE.

saucepan *n* cauldron, pan, pot, skillet, stockpot.

savage *adj* **1** barbarian, barbaric, cannibal, heathen, pagan, primitive, uncivilized, uncultivated, uneducated. *Opp* CIVILIZED. **2** *savage beasts.* feral, fierce, undomesticated, untamed, wild. **3** *savage attack.* angry, atrocious, barbarous, beastly, bestial, blistering, bloodthirsty, bloody, brutal, callous, cold-blooded, cruel, demonic, diabolical, ferocious, fierce, heartless, inhuman, merciless, murderous, pitiless, ruthless, sadistic, unfeeling, vicious, violent. *Opp* TAME. • *n* barbarian, beast, brute, cannibal, fiend, savage person. • *vb* attack, bite, claw, lacerate, maul, mutilate.

save *vb* **1** be sparing with, collect, conserve, economize, hoard, hold back, hold on to, invest, keep, *inf* lay aside, *inf* put by, put in a safe place, reserve, retain, scrape together, set aside, *inf* stash away, store up, take care of, use wisely. *Opp* WASTE. **2** *save from captivity.* bail out, deliver, free, liberate, ransom, redeem, release, rescue, set free. **3** *save from destruction.* recover, retrieve, salvage. **4** *save from danger.* defend, deliver, guard, keep safe, preserve, protect, safeguard, screen, shelter, shield. **5** *saved me from looking a fool.* check, deter, preclude, prevent, spare, stop. *Opp* ABANDON.

saving *n* **1** economizing, frugality, parsimony, prudence, *inf* scrimping and scraping, thrift. **2** cut, discount, economy, reduction.

savings *n* capital, funds, investments, *inf* nest-egg, reserves, resources, riches, wealth.

saviour *n* **1** champion, defender, deliverer, *inf* friend in need, guardian, liberator, rescuer. **2** [*theological*] Christ, Our Lord, The Messiah, The Redeemer.

savour *n* flavour, piquancy, relish, smell, tang, taste, zest. • *vb* appreciate, delight in, enjoy, relish, smell, taste.

savoury *adj* appetizing, delicious, flavoursome, piquant, salty. ▷ TASTY. *Opp* SWEET.

saw *n* **1** □ *chain-saw, hack-saw, jigsaw, ripsaw.* **2** [*old use*] *just an old saw.* ▷ SAYING. • *vb* ▷ CUT.

say *vb* affirm, allege, announce, answer, articulate, assert, asseverate, aver, *old use* bruit abroad, *inf* come out with, comment, communicate, convey, declare, disclose, divulge, ejaculate, enunciate, exclaim, express, intimate, maintain, mention, mouth, phrase, pronounce, *inf* put it about, read out, recite, rejoin, remark, repeat, reply, report, respond, retort, reveal, signify, state, suggest, tell, utter. ▷ SPEAK, TALK, TELL.

saying *n* adage, aphorism, apophthegm, axiom, catch-phrase, catchword, clichA, dictum, epigram, expression, formula, maxim, motto, phrase, precept, proverb, quotation, remark, *old use* saw, slogan, statement, tag, truism, watchword.

scab *n* clot of blood, crust, sore.

scale *n* **1** dandruff, flake, plate, scurf. **2** *remove scale from teeth.* caking, coating, crust, deposit, encrustation, *inf* fur, plaque, tartar. **3** *the scale on a thermometer.* calibration, gradation, graduation. **4** *the social scale.* hierarchy, ladder, order, ranking, spectrum. **5** *small/large scale.* proportion, ratio. ▷ SIZE. **6** *musical scale.* sequence, series. □ *chromatic scale, diatonic scale, major scale, minor scale.* • *vb* ascend, clamber up, climb, go up, mount. **scales** balance, weighing-machine.

scamper *vb* dash, frisk, frolic, gambol, hasten, hurry, play, romp, run, rush, scuttle.

scan *vb* **1** check, examine, explore, eye, gaze at, investigate, look at, pore over, scrutinize, search, stare at, study, survey, view, watch. **2** *scan the papers.* flip through, glance at, read quickly, skim, thumb through.

scandal *n* **1** discredit, disgrace, dishonour, disrepute, embarrassment, ignominy, infamy, notoriety, obloquy, outrage, reproach, sensation, shame. **2** calumny, defamation, gossip, innuendo, libel, rumour, slander, slur, *inf* smear, *inf* tittle-tattle.

scandalize *vb* affront, appal, disgust, horrify, offend, outrage, shock, upset.

scandalous *adj* **1** disgraceful, disgusting, dishonourable, disreputable, ignominious, immodest, immoral, improper, indecent, indecorous, infamous, licentious, notorious, outrageous, shameful, shocking, sinful, sordid, unmentionable, unspeakable, wicked. **2** *a scandalous lie.* calumnious, defamatory, libellous, scurrilous, slanderous, untrue.

scansion *n* metre, prosody, rhythm. ▷ VERSE.

scanty *adj* **1** inadequate, insufficient, meagre, mean, *sl* measly, *inf* mingy, minimal, scant, scarce, *inf* skimpy, sparing, sparse, stingy. ▷ SMALL. *Opp* PLENTIFUL. **2** *scanty clothes.* indecent, revealing, *inf* see-through, thin.

scapegoat *n* dupe, *sl* fall guy, *inf* front, whipping-boy, victim.

scar *n* blemish, brand, burn, cicatrice, cicatrix, cut, disfigurement, injury, mark, scab, scratch. ▷ WOUND. • *vb* blemish, brand, burn, damage, deface, disfigure, injure, leave a scar on, mark, scratch, spoil.

scarce *adj inf* few and far between, *inf* hard to come by, *inf* hard to find, inadequate, infrequent, in short supply, insufficient, lacking, meagre, rare, scant, scanty, sparse, *inf* thin on the ground, uncommon, unusual. *Opp* PLENTIFUL.

scarcely *adv* barely, hardly, only just.

scarcity *n* dearth, famine, inadequacy, insufficiency, lack, need, paucity, poverty, rarity, shortage, want. *Opp* PLENTY.

scare *n* alarm, jolt, shock, start. ▷ FRIGHT. • *vb* **1** alarm, dismay, intimidate, make someone afraid, *inf* make someone jump, menace, panic, shake, shock, startle, terrorize, threaten, unnerve. ▷ FRIGHTEN. *Opp* REASSURE.

scarf *n* headscarf, muffler, shawl, stole.

scary *adj* [*inf*] creepy, eerie, hair-raising, horrible, scaring, unnerving. ▷ FRIGHTENING.

scathing *adj* biting, caustic, critical, humiliating, mordant, satirical, savage, scornful, tart, withering. *Opp* COMPLIMENTARY.

scatter *vb* **1** break up, disband, disintegrate, dispel, disperse, divide, send in all directions, separate. **2** *scatter seeds.* broadcast, disseminate, intersperse, shed, shower, sow, spread, sprinkle, strew, throw about. *Opp* GATHER.

scatterbrained *adj* absent-minded, careless, crazy, disorganized, forgetful, frivolous, hare-brained, inattentive, muddled, *inf* not with it, *inf* scatty, thoughtless, unreliable, unsystematic, vague. ▷ SILLY.

scavenge *vb* forage, rummage, scrounge, search.

scenario *n* design, framework, layout, outline, plan, scheme, storyline, structure, summary.

scene *n* **1** area, background, context, locale, locality, location, place, position, setting, site, situation, spot, whereabouts. **2** *a beautiful scene.* picture, sight, spectacle. ▷ SCENERY. **3** *scene from a film.* act, chapter, *inf* clip, episode, part, section, sequence. **4** *a nasty scene.* altercation, argument, *inf* carry-on, commotion, disturbance, furore, fuss, quarrel, row, tantrum, *inf* to-do, *inf* upset.

scenery *n* **1** landscape, outlook, panorama, prospect, scene, terrain, view, vista. **2** *stage scenery.* backdrop, flats, set, setting.

scenic *adj* attractive, beautiful, breathtaking, grand, impressive, lovely, panoramic, picturesque, pretty, spectacular.

scent *n* **1** aroma, bouquet, fragrance, nose, odour, perfume, redolence, smell. **2** after-shave, eau de cologne, lavender water, perfume. **3** *an animal's scent.* spoor, track, trail. • *vb* ▷ SMELL. **scented** ▷ SMELLING.

sceptic *n* agnostic, cynic, doubter, *inf* doubting Thomas, disbeliever, scoffer, unbeliever. *Opp* BELIEVER.

sceptical *adj* agnostic, cynical, disbelieving, distrustful, doubting, dubious, incredulous, mistrustful, questioning, scoffing, suspicious, uncertain, unconvinced, unsure. *Opp* CONFIDENT.

scepticism *n* agnosticism, cynicism, disbelief, distrust, doubt, dubiety, incredulity, lack of confidence, mistrust, suspicion. *Opp* FAITH.

schedule *n* agenda, calendar, diary, itinerary, list, plan, programme, register, scheme, timetable. • *vb* appoint, arrange, assign, book, earmark, fix a time, organize, outline, plan, programme, time, timetable.

scheme *n* **1** approach, blueprint, design, draft, idea, method, plan, procedure, programme, project, proposal, scenario, strategy, system. **2** *a dishonest scheme.* conspiracy, *inf* dodge, intrigue, machinations, manoeuvre, plot, *inf* ploy, *inf* racket, ruse, stratagem, subterfuge, tactic. **3** *colour scheme.* arrangement, design. • *vb* collude, connive, conspire, *inf* cook something up, *inf* hatch a plot, intrigue, machinate, manoeuvre, plan, plot.

scholar *n* academic, *inf* egghead, expert, highbrow, intellectual, professor, pundit, savant. ▷ PUPIL.

scholarly *adj* **1** academic, bookish, *inf* brainy, *inf* deep, erudite, highbrow, intellectual, knowledgeable, learned, lettered, widely-read. **2** *scholarly treatise.* documented, researched, rigorous, scientific, well-argued, well-informed.

scholarship *n* **1** academic achievement, education, erudition, intellectual attainment, knowledge, learning, research, schooling, scientific rigour, wisdom. **2** *a scholarship to Oxford.* award, bursary, endowment, exhibition, fellowship, grant.

school *n* **1** educational institution. □ *academy, boarding-school, coeducational school, college, comprehensive (school), first school, grammar school, high school, infant school, junior school, kindergarten, nursery school, playgroup, preparatory school, primary school, public school, secondary school, seminary.* **2** *a school of whales.* shoal. ▷ GROUP. • *vb* ▷ EDUCATE.

science *n* organized knowledge, systematic study. □ *acoustics, aeronautics, agricultural science, anatomy, anthropology, artifical intelligence, astronomy, astrophysics, behavioural science, biochemistry, biology, biophysics, botany, chemistry, climatology, computer science, cybernetics, dietetics, domestic science, dynamics, earth science, ecology, economics, electronics, engineering, entomology, environmental science, food science, genetics, geographical science, geology, geophysics, hydraulics, information technology, life science, linguistics, materials science, mathematics, mechanics, medical science, metallurgy, meteorology, microbiology, mineralogy, ornithology, pathology, pharmacology, physics, physiology, political science, psychology, robotics, sociology, space technology, telecommunications, thermodynamics, toxicology, veterinary science, zoology.*

scientific *adj* analytical, methodical, meticulous, orderly, organized, precise, rational, regulated, rigorous, systematic.

scientist *n inf* boffin, researcher, scientific expert, technologist.

scintillating *adj* brilliant, clever, coruscating, dazzling, effervescent, flashing, glittering, liveley, sparkling, vivacious, witty. *Opp* DULL.

scoff *vb* **1** belittle, be sarcastic, be scornful, deride, disparage, gibe, jeer, jibe, laugh, mock, *inf* poke fun, ridicule, sneer, taunt, tease. **2** ▷ EAT.

scold *vb* admonish, berate, blame, *inf* carpet, castigate, censure, chide, criticize, disapprove of, find fault with, *inf* jump down someone's throat, *inf* lecture, *inf* nag, rate, rebuke, reprehend, reprimand, reproach, reprove, *inf* slate, *inf* tell off, *inf* tick off, upbraid.

scoop *n* **1** bailer, ladle, shovel, spoon. **2** *news scoop.* exclusive, inside story, *inf* latest, revelation. *vb* dig, excavate, gouge, hollow, scrape, shovel, spoon.

scope *n* **1** ambit, area, breadth, compass, competence, extent, field, limit, range, reach, span, sphere, terms of reference. **2** *scope for expansion.* capacity, chance, *inf* elbow-room, freedom, latitude, leeway, liberty, opportunity, outlet, room, space, spread.

scorch *vb* blacken, brand, burn, char, heat, roast, sear, singe.

score *n* **1** account, amount, count, marks, points, reckoning, result, sum, tally, total. **2** *score on furniture.* cut, groove, incision, line, mark, nick, scrape, scratch, slash. • *vb* **1** account for, achieve, add up, *inf* chalk up, earn, gain, *inf* knock up, make, tally, win. **2** *score a groove.* cut, engrave, gouge, incise, mark, scrape, scratch, slash. **3** *score music.* orchestrate, write out. **settle a score** ▷ RETALIATE.

scorn *n* contempt, contumely, derision, detestation, disdain, disgust, dislike, dismissal, disparagement, disrespect, jeering, mockery, rejection, ridicule, scoffing, sneering, taunt-ing. *Opp* ADMIRATION. • *vb* be scornful about, contemn, deride, despise, disapprove of, disdain, dislike, dismiss, disparage, hate, insult, jeer at, laugh at, look down on, make fun of, mock, reject, ridicule, *inf* scoff at, sneer at, spurn, taunt, *inf* turn up your nose at. *Opp* ADMIRE.

scornful *adj* condescending, contemptuous, contumelious, deprecative, derisive, disdainful, dismissive, disparaging, disrespectful, haughty, insulting, jeering, mocking, patronizing, sarcastic, satirical, scathing, scoffing, sneering, *inf* snide, *inf* snooty, supercilious, superior, taunting, withering. *Opp* RESPECTFUL.

scoundrel *n* blackguard, blighter, bounder, cad, good-for-nothing, heel, knave, miscreant, rascal, rogue, ruffian, scallywag, scamp, villain, wretch.

scour *vb* **1** abrade, buff up, burnish, clean, cleanse, polish, rub, scrape, scrub, shine, wash. **2** *scour the house.* comb, forage through, hunt through, rake through, ransack, rummage through, search, *inf* turn upside down.

scourge *n* **1** affliction, bane, curse, evil, misery, misfortune, plague, torment, woe. **2** ▷ WHIP. • *vb* beat, belt, flagellate, flog, horsewhip, lash, whip.

scout *n* lookout, spy. • *vb* explore, get information, hunt around, investigate, look about, reconnoitre, search, *inf* snoop, spy.

scowl *vb* frown, glower, grimace, *inf* look daggers, lower.

scraggy *adj* bony, emaciated, gaunt, lanky, lean, scrawny, skinny, starved, thin, underfed. *Opp* PLUMP.

scramble *n* commotion, confusion, *inf* free-for-all, haste, hurry, mêlé, race, rush, scrimmage, struggle. • *vb* **1** clamber, climb, crawl, grope, move awkwardly, scrabble. **2** *scramble for gold.* compete, contend, dash, fight, hasten, hurry, jostle, push, run, rush, scuffle, strive, struggle, tussle, vie. **3** *scramble a message.* confuse, jumble, mix up.

scrap *n* **1** atom, bit, crumb, fraction, fragment, grain, hint, iota, jot, mite, molecule, morsel, particle, piece, rag, scintilla, shard, shred, sliver, snippet, speck, trace. **2** *inf* junk, leavings, litter, odds and ends, offcuts, refuse, rejects, remains, remnants, residue, rubbish, salvage, waste. **3** *a friendly scrap.* argument, quarrel, scuffle, *inf* set-to, squabble, tiff, tussle, wrangle. ▷ FIGHT. • *vb* **1** abandon, cancel, discard, *inf* ditch, drop, give up, jettison, throw away, write off. **2** *scrap over trifles.* argue, bicker, flare up, quarrel, spar, squabble, tussle, wrangle. ▷ FIGHT.

scrape *n* **1** abrasion, graze, injury, laceration, scratch, scuff, wound. **2** *an awkward scrape.* difficulty, escapade, *inf* kettle of fish, piece of mischief, plight, prank, predicament, trouble. • *vb* **1** abrade, bark, bruise, damage, graze, injure, lacerate, scratch, scuff, skin, wound. **2** *scrape clean.* clean, file, rasp, rub, scour, scrub. **scrape together** ▷ COLLECT.

scrappy *adj* bitty, careless, disjointed, fragmentary, hurriedly done, imperfect, incomplete, inconclusive, sketchy, slipshod, unfinished, unpolished, unsatisfactory. *Opp* PERFECT.

scratch *n* abrasion, damage, dent, gash, gouge, graze, groove, indentation, injury, laceration, line, mark, score, scoring, scrape, scuff, wound. • *vb* abrade, claw at, cut, damage the surface of, dent, gash, gouge, graze, groove, incise, injure, lacerate, mark, rub, scarify, score, scrape, scuff, wound. **up to scratch** ▷ SATISFACTORY.

scrawl *vb* doodle, scribble, write hurriedly. ▷ WRITE.

scream *n* & *vb* bawl, caterwaul, cry, howl, roar, screech, shout, shriek, squeal, wail, yell, yowl.

screen *n* **1** blind, curtain, divider, partition. **2** camouflage, concealment, cover, disguise, protection, shelter, shield, smokescreen. **3** *sift through a screen.* filter, mesh, riddle, sieve, strainer. • *vb* **1** divide, partition off, subdivide, wall off. **2** camouflage, cloak, conceal, cover, disguise, guard, hide, mask, protect, safeguard, shade, shelter, shield, shroud, veil. **3** *screen employees for security.* *inf* check out, examine, investigate, process, sift out, vet.

screw *n* **1** bolt, screw-bolt. **2** rotation, spiral, turn, twist. • *vb* rotate, turn, twist.

screw down ▷ FASTEN. **screw up** ▷ BUNGLE, TWIST.

scribble *vb* ▷ SCRAWL.

scribe *n* amanuensis, clerk, copyist, secretary, transcriber, writer.

script *n* **1** calligraphy, handwriting, penmanship. **2** *script of a play.* libretto, screenplay, text, words.

scripture *n* bible, holy writ, sacred writings, Word of God. □ *Bhagavad-Gita,* inf *the Good Book, the Gospel, Holy Bible, Koran, Upanishad.*

scrounge *vb* beg, cadge, importune.

scrub *vb* 1 brush, clean, rub, scour, wash. 2 ▷ CANCEL.

scruffy *adj* bedraggled, dirty, dishevelled, disordered, dowdy, frowsy, messy, ragged, scrappy, shabby, slatternly, slovenly, tatty, ungroomed, unkempt, untidy, worn out. *Opp* SMART.

scruple *n* compunction, conscience, doubt, hesitation, misgiving, qualm, reluctance, *inf* second thought. • *vb* [*usu neg*] be reluctant, have a con-science (about), have scruples (about), hesitate, hold back (from), *inf* think twice (about).

scrupulous *adj* 1 careful, cautious, conscientious, diligent, exacting, fastidious, *inf* finicky, meticulous, minute, neat, painstaking, precise, punctilious, rigid, rigorous, strict, systematic, thorough. 2 *scrupulous honesty*. ethical, fair-minded, honest, honourable, just, moral, principled, proper, upright, upstanding. *Opp* UNSCRUPULOUS.

scrutinize *vb* analyse, check, examine, *inf* go over with a toothcomb, inspect, investigate, look closely at, *inf* probe, sift, study.

scrutiny *n* analysis, examination, inspection, investigation, probing, search, study.

sculpture *n* three-dimensional art. □ *bas-relief, bronze, bust, carving, caryatid, cast, effigy, figure, figurine, maquette, marble, moulding, plaster cast, relief, statue, statuette.* • *vb* carve, cast, chisel, fashion, form, hew, model, mould, *inf* sculpt, shape.

scum *n* dirt, film, foam, froth, impurities, suds.

scurrilous *adj* abusive, calumnious, coarse, defamatory, derogatory, disparaging, foul, indecent, insulting, libellous, low, obscene, offensive, opprobrious, scabrous, shameful, slanderous, vile, vulgar.

sea *adj* aquatic, marine, maritime, nautical, naval, ocean-going, oceanic, salt-water, seafaring, seagoing. • *n inf* briny, *poet* deep, lake, *old use* main, ocean.

seal *n* 1 sea-lion, walrus. 2 *royal seal*. badge, coat of arms, crest, emblem, escutcheon, impression, imprint, mark, monogram, sign, stamp, symbol, token. • *vb* 1 close, fasten, lock, make airtight, make watertight, plug, secure, shut, stick down, stop up. 2 *seal an agreement*. affirm, authenticate, *inf* clinch, conclude, confirm, corroborate, decide, endorse, finalize, guarantee, ratify, settle, sign, validate, verify.

seam *n* 1 join, stitching. 2 *seam of coal*. bed, layer, lode, stratum, thickness, vein.

seamy *adj* disreputable, distasteful, nasty, repulsive, shameful, sordid, squalid, unattractive, unpleasant, unsavoury, unwholesome.

search *n* check, enquiry, examination, hunt, inspection, investigation, look, *inf* probe, pursuit, quest, scrutiny. • *vb* 1 cast about, explore, ferret about, hunt, investigate, *inf* leave no stone unturned, look, nose about, poke about, prospect, pry, seek. 2 *search suspects*. check, examine, *inf* frisk, inspect, scrutinize. 3 *search a house*. comb, go through, ransack, rifle, rummage through, scour. **searching** ▷ INQUISITIVE, THOROUGH.

seaside *n* beach, coast, coastal resort, sands, sea-coast, sea-shore, shore.

season *n* period, phase, time. • *vb* 1 add seasoning to, flavour, *inf* pep up, salt, spice. **2** *season wood.* age, harden, mature, ripen.

seasonable *adj* appropriate, apt, convenient, favourable, fitting, normal, opportune, propitious, suitable, timely, well-timed.

seasoning *n* additives, condiments, flavouring, relish, zest. □ *dressing, herbs, mustard, pepper, relish, salt, sauce, spice, vinegar.*

seat *n* **1** place, sitting-place. □ *armchair, bench, carver, chair, chaise longue, couch, deck-chair, dining-chair, easy chair,* Fr *fauteuil, form, pew, pillion, pouffe, reclining chair, rocking-chair, saddle, settee, settle, sofa, squab, stall, stool, throne, window seat.* **2** *a country seat.* ▷ RESIDENCE. **3** ▷ BUTTOCKS. **seat yourself** ▷ SIT.

secateurs *plur n* clippers, cutters, pruning shears.

secluded *adj* cloistered, concealed, cut off, hidden, inaccessible, isolated, lonely, monastic, *inf* off the beaten track, private, remote, retired, screened, sequestered, sheltered, shut away, solitary, unfrequented, unvisited. *Opp* PUBLIC.

seclusion *n* concealment, hiding, isolation, loneliness, privacy, remoteness, retirement, separation, shelter, solitariness.

second *adj* added, additional, alternative, another, complementary, duplicate, extra, following, further, later, matching, next, other, repeated, subsequent, twin. • *n* **1** flash, instant, *inf* jiffy, moment, *inf* tick, *inf* twinkling, *inf* wink. **2** *second in a fight.* assistant, deputy, helper, *inf* number two, *inf* right-hand man, right-hand woman, second-in-command, *inf* stand-in, subordinate, supporter, understudy, vice-. • *vb* **1** aid, assist, back, encourage, give approval to, help, promote, side with, sponsor, support. **2** *second to another job.* move, reassign, relocate, shift, transfer.

secondary *adj* **1** alternative. ancillary, auxiliary, *inf* backup, extra, inessential, inferior, lesser, lower, minor, nonessential, reinforcing, reserve, second, second-rate, spare, subordinate, subsidiary, supporting, supportive, supplementary, unimportant. **2** *secondary sources.* copied, derivative, second-hand, unoriginal.

second-hand *adj* **1** *inf* hand-me-down, old, used, worn. *Opp* NEW. **2** *second-hand experience.* indirect, secondary, vicarious. *Opp* DIRECT.

second-rate *adj* commonplace, indifferent, inferior, low-grade, mediocre, middling, ordinary, poor, second-best, second-class, undistinguished, unexciting, uninspiring.

secret *adj* **1** clandestine, concealed, covert, disguised, hidden, *inf* hushed up, *inf* hush-hush, invisible, private, secluded, shrouded, stealthy, undercover, underground, unknown. ▷ SECRETIVE. **2** *secret papers.* classified, confidential, inaccessible, intimate, personal, restricted, sensitive, top-secret, undisclosed, unpublished. **3** *secret meanings.* arcane, cryptic, encoded, esoteric, incomprehensible, mysterious, occult, recondite. **4** *secret about his private life.* ▷ SECRETIVE. *Opp* OPEN, PUBLIC.

secretary *n* amanuensis, clerk, filing-clerk, personal assistant, scribe, shorthand-typist, stenographer, typist, word-processor operator.

secrete *vb* **1** cloak, conceal, cover up, disguise, enshroud, hide, mask, put away, put into hiding. **2** *secrete fluid.* discharge, emit, excrete, exude, give off, leak, ooze, produce, release.

secretion *n* discharge, emission, escape, excretion, leakage, release.

secretive *adj* close-lipped, enigmatic, furtive, mysterious, quiet, reserved, reticent, secret, shifty, silent, taciturn, tight-lipped, uncommunicative, unforthcoming, withdrawn. *Opp* COMMUNICATIVE.

sect *n* cult, denomination, faction, order, party, persuasion. ▷ GROUP.

sectarian *adj* bigoted, clannish, cliquish, cultic, denominational, dogmatic, exclusive, factional, fanatical, inflexible, narrow, narrow-minded, partial, partisan, prejudiced, rigid, schismatic.

section *n* bit, branch, chapter, compartment, component, department, division, element, fraction, fragment, group, instalment, leg (*of journey*), part, passage, piece, portion, quarter, sample, sector, segment, slice, stage, subdivision, subsection.

sector *n* area, district, division, part, quarter, region, zone. ▷ SECTION.

secular *adj* civil, earthly, lay, material, mundane, non-religious, temporal, terrestrial, worldly. *Opp* RELIGIOUS.

secure *adj* **1** cosy, defended, guarded, immune, impregnable, invulnerable, protected, safe, sheltered, shielded, snug, unharmed, unhurt, unscathed. **2** *secure doors.* bolted, burglarproof, closed, fast, fastened, fixed, foolproof, immovable, locked, shut, solid, tight, unyielding. **3** *secure faith.* certain, confident, firm, stable, steady, strong, sure, unquestioning. • *vb* **1** defend, guard, make safe, preserve, protect, shelter, shield. **2** anchor, attach, bolt, close, fix, lock, make fast, screw down, tie down. ▷ FASTEN. **3** *secure a loan.* acquire, be promised, come by, gain, get, obtain, procure, win.

sedate *adj* calm, collected, composed, controlled, conventional, cool, decorous, deliberate, dignified, equable, even-tempered, formal, grave, imperturbable, level-headed, peaceful, *derog* prim, proper, quiet, sensible, serene, serious, slow, sober, solemn, staid, strait-laced, tranquil, unruffled. *Opp* LIVELY. • *vb* calm, put to sleep, tranquillize, treat with sedatives.

sedative *adj* anodyne, calming, lenitive, narcotic, relaxing, soothing, soporific, tranquillizing. • *n* anodyne, barbiturate, calmative, depressant, narcotic, opiate, sleeping-pill, soporific, tranquillizer.

sedentary *adj* desk-bound, immobile, inactive, seated, sitting down. *Opp* ACTIVE.

sediment *n* deposit, dregs, grounds, lees, precipitate, remains, residue, *inf* sludge.

sedition *n* agitation, incitement, insurrection, mutiny, rabble-rousing, revolt, treachery, treason. ▷ REBELLION.

seduce *vb* **1** allure, beguile, charm, corrupt, deceive, decoy, deprave, ensnare, entice, inveigle, lead astray, lure, mislead, tempt. **2** debauch, deflower, dishonour, rape, ravish, *old use* ruin, violate.

seduction *n* **1** allurement, attraction, charm, temptation. **2** rape, ravishing. ▷ SEX.

seductive *adj* alluring, appealing, attractive, bewitching, captivating, charming, coquettish, enchanting, enticing, flirtatious, inviting, irresistible, persuasive, provocative, tantalizing, tempting, *inf* sexy. *Opp* REPULSIVE.

see *vb* **1** behold, catch sight of, descry, discern, discover, distinguish, espy, glimpse, identify, look at, make out, mark, note, notice, observe, perceive, recognize, regard, sight, spot, spy, view, watch, witness. **2** *see what someone means.* appreciate, apprehend, comprehend, fathom, follow, *inf* get the hang of, grasp, know, perceive, realize, take in, understand. **3** *see problems ahead.* anticipate, conceive, envisage, foresee, foretell, imagine, picture, visualize. **4** *see what can be done.* consider, decide, discover, investigate, mull over, reflect on, think about, weigh up. **5** *see a play.* attend, be a spectator at, watch. **6** *seeing him tonight.* court, go out with, *inf* have a date with, meet, socialize with, visit, woo. **7** *see you home.* accompany, conduct, escort. **8** *saw fighting in the war.* endure, experience, go through, suffer, survive, undergo. **9** *Guess who I saw today!* encounter, face, meet, run into, talk to, visit. **see to** ▷ ORGANIZE.

seed *n* **1** egg, embryo, germ, ovule, ovum, semen, spawn, sperm, spore. **2** *seed in fruit.* pip, pit, stone. • *vb* ▷ SOW.

seek *vb* aim at, apply for, ask for, aspire to, beg for, demand, desire, go after, hope for, hunt for, inquire after, look for, pursue, quest after, request, search for, solicit, strive after, try for, want, wish for.

seem *vb* appear, feel, give an impression of being, have an appearance of being, look, pretend to be, sound.

seep *vb* dribble, drip, exude, flow, leak, ooze, percolate, run, soak, trickle.

seer *n* clairvoyant, fortune-teller, oracle, prophet, prophetess, psychic, sibyl, soothsayer, vaticinator.

seethe *vb* be agitated, be angry, boil, bubble, erupt, foam, froth up, rise, simmer, stew, surge.

segment *n* bit, compartment, department, division, element, fraction, fragment, part, piece, portion, quarter, section, sector, slice, subdivision, subsection, wedge.

segregate *vb* compartmentalize, cut off, exclude, isolate, keep apart, put apart, separate, sequester, set apart, shut out.

segregation *n* **1** apartheid, discrimination. **2** isolation, quarantine, seclusion, separation.

seize *vb* **1** abduct, apprehend, arrest, capture, catch, clutch, *inf* collar, detain, grab, grasp, grip, hold, *inf* nab, pluck, possess, snatch, take, take into custody, take prisoner. **2** *seize a country.* annex, invade. **3** *seize property.* appropriate, commandeer, confiscate, hijack, impound, steal, take away. *Opp* RELEASE. **seize up** ▷ STICK.

seizure *n* **1** abduction, annexation, appropriation, arrest, capture, confiscation, hijacking, invasion, sequestration, theft, usurpation. **2** [*medical*] apoplexy, attack, convulsion, epileptic fit, fit, paroxysm, spasm, stroke.

seldom *adv* infrequently, occasionally, rarely.

select *adj* best, choice, chosen, élite, excellent, exceptional, exclusive, favoured, finest, first-class, first-rate, *inf* hand-picked, preferred, prime, privileged, rare, selected, special, top-quality. *Opp* ORDINARY. • *vb* appoint, cast (*actor for role*), choose, decide on, elect, nominate, opt for, pick, prefer, settle on, single out, vote for.

selection *n* **1** choice, option, pick, preference. **2** *a selection of goods.* assortment, range, variety. **3** *selection from the classics.* excerpts, extracts, passages, quotations.

selective *adj* careful, *inf* choosy, discerning, discriminating, particular, specialized. *Opp* COMPREHENSIVE, IMPERCEPTIVE.

self-confident *adj* assertive, assured, collected, cool, fearless, independent, outgoing, poised, positive, self-assured, self-possessed, self-reliant, sure of yourself. ▷ BOLD. *Opp* SELF-CONSCIOUS.

self-conscious *adj* awkward, bashful, blushing, coy, diffident, embarrassed, ill at ease, insecure, nervous, reserved, self-effacing, sheepish, shy, uncomfortable, unnatural. ▷ TIMID. *Opp* SELF-CONFIDENT.

self-contained *adj* **1** complete, independent, separate. **2** aloof, cold, reserved, self-reliant, uncommunicative, undemonstrative, unemotional.

self-control *n* calmness, composure, coolness, patience, resolve, restraint, self-command, self-denial, self-discipline, self-possession, self-restraint, will-power.

self-denial *n* abstemiousness, fasting, moderation, self-abnegation, self-sacrifice, temperance, unselfishness. *Opp* SELF-INDULGENCE.

self-employed *adj* freelance, independent.

self-esteem *n* **1** ▷ SELF-RESPECT. **2** arrogance, *inf* big-headedness, conceit, egotism, overconfidence, self-admiration, self-importance, self-love, smugness, vanity.

self-explanatory *adj* apparent, axiomatic, blatant, clear, conspicuous, eye-catching, flagrant, glaring, inescapable, manifest, obvious, patent, plain, recognizable, self-evident, understandable, unmistakable, visible.

self-governing *adj* autonomous, free, independent, sovereign.

self-important *adj* arrogant, bombastic, conceited, grandiloquent, haughty, magisterial, ostentatious, pompous, pontifical, pretentious, self-centred, sententious, smug, *inf* snooty, *inf* stuck-up, supercilious, vainglorious.

self-indulgence *n* extravagance, gluttony, greed, hedonism, pleasure, profligacy, self-gratification. ▷ SELFISHNESS. *Opp* SELF-DENIAL.

self-indulgent *adj* dissipated, epicurean, extravagant, gluttonous, gourmandizing, greedy, hedonistic, immoderate, intemperate, pleasure-loving, profligate, sybaritic. ▷ SELFISH. *Opp* ABSTEMIOUS.

selfish *adj* acquisitive, avaricious, covetous, demanding, egocentric, egotistic, grasping, greedy, inconsiderate, mean, mercenary, miserly, self-absorbed, self-centred, self-indulgent, self-interested, self-seeking, self-serving, *inf* stingy, thoughtless, uncaring, ungenerous, unhelpful, unsympathetic, worldly. *Opp* UNSELFISH.

selfishness *n* acquisitiveness, avarice, covetousness, egotism, greed, meanness, miserliness, niggardliness, possessiveness, self-indulgence, self-interest, self-love, self-regard, *inf* stinginess, thoughtlessness.

self-reliant *adj* autonomous, independent, self-contained, self-sufficient, self-supporting.

self-respect *n Fr* amour propre, dignity, honour, integrity, morale, pride, self-confidence, self-esteem.

self-righteous *adj* complacent, *inf* goody-goody, *inf* holier-than-thou, mealy-mouthed, pharisaical, pietistic, pious, pompous, priggish, proud, sanctimonious, self-important, self-satisfied, sleek, smug, superior, vain.

self-sufficient *adj* autonomous, independent, self-reliant, self-supporting.

self-willed *adj* determined, dogged, forceful, headstrong, inflexible, intractable, intransigent, *inf* mulish, obstinate, *inf* pig-headed, single-minded, *inf* stiff-necked, stubborn, uncontrollable, uncooperative, wilful.

sell *vb* **1** auction, barter, deal in, exchange, give in part-exchange, handle, hawk, *inf* keep, *inf* knock down, offer for sale, peddle, *inf* put under the hammer, retail, sell off, stock, tout, trade, *inf* trade in (*traded in my car*), traffic in, vend. **2** *sell hard.* advertise, market, merchandise, package, promote, *inf* push.

seller *n* dealer, merchant, stockist, supplier, trader, vendor. □ *agent, barrow-boy, broker,* old use *colporteur, costermonger,* old use *hawker, market-trader, pedlar,* inf *rep, representative, retailer, salesman, salesperson, saleswoman, shop assistant, shopkeeper, storekeeper, street trader, tradesman, traveller, wholesaler.* ▷ SHOP.

seminal *adj* basic, constructive, creative, fertile, formative, imaginative, important, influential, innovative, new, original, primary, productive.

send *vb* **1** address, consign, convey, deliver, direct, dispatch, fax, forward, mail, post, remit, ship, transmit. **2** *send a rocket to the moon.* fire, launch, project, propel, release, shoot. **send away** ▷ DISMISS. **send down** ▷ IMPRISON. **send for** ▷ SUMMON. **send-off** *n* ▷ GOODBYE. **send out** ▷ EMIT. **send round** ▷ CIRCULATE. **send up** ▷ PARODY.

senile *adj* declining, doddery, *inf* in your dotage, old, *derog* past it.

senior *adj* chief, elder, higher, high-ranking, major, older, principal, revered, superior, well-established. *Opp* JUNIOR.

sensation *n* **1** awareness, feeling, perception, sense. **2** *affair caused a sensation.* commotion, excitement, furore, outrage, scandal, stir, thrill.

sensational *adj* **1** blood-curdling, hair-raising, lurid, melodramatic, overwritten, scandal-mongering, shocking, startling, stimulating, violent. **2** [*inf*] *a sensational result.* amazing, astonishing, astounding, breathtaking, electrifying, exciting, extraordinary, *inf* fabulous, *inf* fantastic, *inf* great, incredible, marvellous, remarkable, spectacular, spine-tingling, stirring, superb, surprising, thrilling, unbelievable, unexpected, wonderful.

sense *n* **1** awareness, consciousness, faculty, feeling, sensation. □ *hearing, sight, smell, taste,* *touch.* **2** brains, cleverness, gumption, intellect, intelligence, intuition, judgement, logic, *inf* nous, perception, reason, reasoning, understanding, wisdom, wit. **3** *the sense of a message.* coherence, connotations, denotation, *inf* drift, gist, import, intelligibility, interpretation, meaning, message, point, purport, significance, signification, substance. • *vb* be aware (of), detect, discern, divine, feel, guess, *inf* have a hunch, intuit, notice, perceive, *inf* pick up vibes, realize, respond to, suspect, understand. ▷ FEEL, HEAR, SEE, SMELL, TASTE. **make sense of** ▷ UNDERSTAND.

senseless *adj* **1** anaesthetized, asleep, comatose, insensate, insensible, knocked out, numb, *inf* out like a light, stunned, unconscious. **2** absurd, crazy, fatuous, meaningless, pointless, purposeless, silly. ▷ STUPID.

sensible *adj* **1** calm, commonsense, commonsensical, cool, discreet, discriminating, intelligent, judicious, level-headed, logical, prudent, rational, realistic, reasonable, reasoned, sage, sane, serious-minded, sound, straightforward, thoughtful, wise. *Opp* STUPID. **2** *sensible phenomena.* corporeal, existent, material, palpable, perceptible, physical, real, tangible, visible. **3** *sensible clothes.* comfortable, functional, *inf* no-nonsense, practical, useful. *Opp* FASHIONABLE, IMPRACTICAL. **sensible of** acquainted with, alert to, alive to, appreciative of, aware of, cognizant of, in touch with, mindful of, responsive to, *inf* wise to.

sensitive *adj* **1** considerate, perceptive, reactive, receptive, responsive, susceptible, sympathetic, tactful, thoughtful, understanding. **2** *a sensitive temperament.* emotional, hypersensitive, impressionable, temperamental, thin-skinned, touchy, volatile, vulnerable. **3** *sensitive skin.* delicate, fine, fragile, painful, soft, sore, tender. **4** *a sensitive topic.* confidential, controversial, delicate, *inf* tricky, secret. *Opp* INSENSITIVE. **sensitive to** affected by, attuned to, aware of, considerate of, perceptive about, receptive to, responsive to, understanding about.

sensual *adj* animal, bodily, carnal, fleshly, physical, pleasure-loving, self-indulgent, voluptuous, worldly. ▷ SEXY. *Opp* ASCETIC.

sensuous *adj* beautiful, emotional, gratifying, lush, luxurious, rich, richly embellished.

sentence *n* **1** exclamation, question, statement, thought, utterance. **2** decision, judgement, pronouncement, punishment, ruling. • *vb* condemn, pass judgement on, pronounce sentence on.

sentiment *n* **1** attitude, belief, idea, judgement, opinion, outlook, thought, view. **2** *sentiment of a poem.* emotion, feeling, sensibility.

sentimental *adj* **1** compassionate, emotional, nostalgic, romantic, soft-hearted, sympathetic, tearful, tender, warm-hearted, *inf* weepy. **2** [*derog*] gushing, *inf* gushy, indulgent, insincere, maudlin, mawkish, *inf* mushy, overdone, over-emotional, *inf* sloppy, *inf* soppy, *inf* sugary, tearjerking, *inf* treacly, unrealistic, *sl* yucky. *Opp* CYNICAL.

sentimentality *n* bathos, emotionalism, insincerity, *inf* kitsch, mawkishness, nostalgia, *inf* slush.

sentry *n* guard, lookout, patrol, picket, sentinel, watch, watchman.

separable *adj* detachable, distinguishable, fissile, removable.

separate *adj* apart, autonomous, cloistered, cut off, detached, different, discrete, disjoined, distinct, divided, divorced, fenced off, free-standing, independent, individual, isolated, particular, peculiar, secluded, segregated, separated, shut off, solitary, unattached, unconnected, unique, unrelated, unshared, withdrawn. • *vb* 1 break up, cut off, detach, disconnect, disengage, disentangle, disjoin, dismember, dissociate, divide, fence off, fragment, hive off, isolate, keep apart, part, pull apart, segregate, sever, split, sunder, take apart, uncouple, unfasten, unhook, unravel, unyoke. 2 *Thepaths separate here.* bifurcate, branch, diverge, fork. 3 *separated the grain from the chaff.* abstract, distinguish, filter out, remove, set apart, sift out, winnow. 4 *He separated from his partner.* become estranged, disband, divorce, part company, *inf* split up. *Opp* COMBINE, UNITE.

separation *n* 1 amputation, cutting off, detachment, disconnection, dismemberment, dissociation, division, fission, fragmentation, parting, rift, severance, splitting. *Opp* CONNECTION. 2 *separation of partners.* break, *inf* break-up, divorce, estrangement, rift, split. *Opp* UNION.

septic *adj* diseased, festering, infected, inflamed, poisoned, purulent, putrefying, putrid, suppurating.

sequel *n* consequence, continuation, development, *inf* follow-up, issue, outcome, result, upshot.

sequence *n* 1 arrangement, chain, concatenation, course, cycle, line, order, procession, programme, progression, range, row, run, sequence, series, set, string, succession, train. 2 *a sequence from a film. inf* clip, episode, excerpt, extract, scene, section.

serene *adj* 1 calm, idyllic, peaceful, placid, pleasing, quiet, restful, still, tranquil, unclouded, undisturbed, unperturbed, unruffled, untroubled. 2 *serene temperament.* collected, composed, contented, cool, easy-going, equable, even-tempered, imperturbable, pacific, peaceable, poised, self-possessed, *inf* unflappable. *Opp* BOISTEROUS, EXCITABLE.

series *n* 1 arrangement, chain, concatenation, course, cycle, line, order, procession, programme, progression, range, row, run, sequence, set, string, succession, train. 2 *TV series.* mini-series, serial, *inf* soap, soap-opera.

serious *adj* 1 dignified, grave, grim, humourless, long-faced, pensive, poker-faced, sedate, sober, solemn, sombre, staid, stern, straight-faced, thoughtful, unsmiling. *Opp* CHEERFUL. 2 *serious discussion.* deep, earnest, heavy, honest, important, intellectual, momentous, profound, significant, sincere, weighty. 3 *serious illness.* acute, appalling, awful, calamitous, critical, dangerous, dreadful, frightful, ghastly, grievous, hideous, horrible, *inf* life-and-death, nasty, severe, shocking, terrible, unfortunate, unpleasant, urgent, violent. *Opp* TRIVIAL. 4 *serious worker.* careful, committed, conscientious, diligent, hard-working.

sermon *n* address, discourse, homily, lecture, lesson, talk.

serpentine *adj* labyrinthine, meandering, roundabout, sinuous, snaking, tortuous, twisting, vermicular, winding. *Opp* STRAIGHT.

serrated *adj* cogged, crenellated, denticulate, indented, jagged, notched, saw-like, toothed, zigzag. *Opp* STRAIGHT.

servant *n* assistant, attendant, *derog* dogsbody, *inf* domestic, *derog* drudge, helper, *derog* hireling, *derog* menial, *old use* servitor, *inf* skivvy, slave, *old use* vassal. □ *au pair, barmaid, barman, batman,* inf *boots, butler, chamber-maid,* inf *char, charwoman, chauffeur, chef, cleaner,* old use *coachman, commissionaire, cook,* inf *daily, errand boy, factotum,* derog *flunkey, footman, governess, groom, home help, houseboy, housemaid, housekeeper, kitchenmaid,* derog *lackey, lady-in-waiting, maid, maidservant, major-domo, manservant, nanny, page, parlour-maid,* old use *postilion,* old use *retainer,* plur *retinue, scout, scullery maid,* old use *scullion,* old use *seneschal, slave, steward, stewardess, valet, waiter, waitress.*

serve *vb* **1** aid, accommodate, assist, attend, *inf* be at someone's beck and call, further, help, look after, minister to, wait upon, work for. **2** *serve in the forces.* be employed, do your duty, enlist, fight, sign on. **3** *serve goods.* deal out, distribute, dole out, give out, make available, provide, sell, supply. **4** *serve at table.* carve, *inf* dish up, officiate, wait. **5** *serve a sentence.* complete, endure, go through, pass, spend, survive.

service *n* **1** aid, assistance, benefit, favour, help, kindness, office. **2** *service of the community.* attendance (on), employment (by), ministering (to), work (for). **3** *a bus service.* business, organization, provision, system, timetable. **4** *give the car a service.* check-over, maintenance, overhaul, repair, servicing. **5** *church service.* ceremony, liturgy, meeting, rite, ritual, worship. □ *baptism, christening, communion, compline, evensong, funeral, marriage, Mass, matins, Requiem Mass, vespers.* • *vb service a vehicle.* check, maintain, mend, overhaul, repair, tune.

serviceable *adj* dependable, durable, functional, hard-wearing, lasting, practical, strong, tough, usable.

servile *adj* abject, acquiescent, base, *inf* boot-licking, craven, cringing, deferential, fawning, flattering, grovelling, humble, ingratiating, low, menial, obsequious, slavish, submissive, subservient, sycophantic, *inf* time-serving, toadying, unctuous. *Opp* BOSSY. **be servile** ▷ GROVEL.

serving *n* helping, plateful, portion, ration.

session *n* **1** assembly, conference, discussion, hearing, meeting, sitting. **2** *a session at the baths.* period, term, time.

set *adj* **1** *set price.* advertised, agreed, arranged, defined, definite, fixed, prearranged, predetermined, prepared, scheduled, standard. **2** *set in your ways.* established, invariable, predictable, regular, stable, unchanging, unvarying. • *n* **1** batch, bunch, category, class, clique, collection, combination, kind, series, sort. ▷ GROUP. **2** *a TV set.* apparatus, receiver. **3** *set for a play.* scene, scenery, setting, stage. • *vb* **1** arrange, assign, deploy, deposit, dispose, lay, leave, locate, lodge, park, place, plant, *inf* plonk, put, position, rest, set down, set out, settle, situate, stand, station. **2** *set a clock.* adjust, correct, put right, rectify, regulate. 3 *set a post in concrete.* embed, fasten, fix. 4 *set like concrete.* become firm, congeal, *inf* gel, harden, *inf* jell, stiffen, take shape. 5 *set a problem.* ask, express, formulate, frame, phrase, pose, present, put forward, suggest, write. 6 *set a target.* allocate, allot, appoint, decide, designate, determine, establish, identify, name, ordain, prescribe, settle. **set about** ▷ ATTACK, BEGIN. **set free** ▷ LIBERATE. **set off** ▷ DEPART, EXPLODE. **set on** ▷ ATTACK. **set on fire** ▷ IGNITE. **set out** ▷ DEPART. **set up** ▷ ESTABLISH.

set-back *n inf* blow, check, complication, defeat, delay, difficulty, disappointment, hindrance, *inf* hitch, hold-up, impediment, misfortune, obstacle, problem, relapse, reverse, snag, upset.

settee *n* chaise longue, couch, sofa.

setting *n* 1 background, context, environment, environs, frame, habitat, locale, location, place, position, site, surroundings. 2 *setting for a play.* backcloth, backdrop, scene, scenery, set.

settle *vb* 1 arrange, conclude, deal with, decide, organize, put in order, straighten out. 2 alight, come to rest, land, light, *inf* make yourself comfortable, *inf* park yourself, pause, rest, roost, sit down. 3 *settle things in place.* assign, deploy, deposit, dispose, lay, locate, lodge, park, place, plant, position, put, rest, set, set down, situate, stand, station. 4 *the dust settled.* calm down, clear, compact, go down, sink, subside. 5 *settle what to do.* agree, choose, decide, establish, fix. 6 *settle differences.* end, negotiate, put an end to, reconcile, resolve, sort out, square. 7 *settle debts.* clear, discharge, pay, pay off. 8 *settle new territory.* become established in, colonize, immigrate, make your home in, occupy, people, set up home in, stay in.

settlement *n* 1 camp, colony, community, encampment, kibbutz, outpost, post, town, village. 2 agreement, arbitration, arrangement, contract, payment.

settler *n* colonist, frontiersman, immigrant, newcomer, pioneer, squatter.

sever *vb* 1 amputate, break, cut off, detach, disconnect, disjoin, part, remove, separate, split, terminate. ▷ CUT. 2 *sever a relationship.* abandon, break off, discontinue, end, put an end to, suspend, terminate.

several *adj* assorted, certain, different, divers, a few, a handful of, many, miscellaneous, a number of, some, sundry, a variety of, various.

severe *adj* 1 aloof, brutal, cold, cold-hearted, cruel, disapproving, dour, exacting, forbidding, glowering, grave, grim, hard, harsh, inexorable, merciless, obdurate, pitiless, relentless, rigorous, stern, stony, strict, unbending, uncompromising, unkind, unsmiling, unsympathetic, unyielding. 2 *severe illness.* acute, critical, dangerous, drastic, fatal, great, intense, keen, life-threatening, mortal, nasty, serious, sharp, terminal, troublesome. 3 *severe penalties.* draconian, extreme, maximum, oppressive, punitive, stringent. 4 *severe weather.* adverse, bad, inclement, violent, *inf* wicked. ▷ COLD, STORMY. 5 *a severe challenge.* arduous, demanding, difficult, onerous, punishing, taxing, tough. **6** *severe style*. austere, bare, chaste, plain, simple, spartan, stark, unadorned. *Opp* FRIENDLY, MILD, ORNATE.

sew *vb* baste, darn, hem, mend, repair, stitch, tack.

sewage *n* effluent, waste.

sewer *n* drain, drainage, sanitation, septic tank, soak-away.

sewing *n* dressmaking, embroidery, mending, needlepoint, needlework, tapestry.

sex *n* **1** gender, sexuality. **2** carnal knowledge, coition, coitus, congress, consummation of marriage, copulation, coupling, fornication, *inf* going to bed, incest, intercourse, intimacy, love-making, masturbation, mating, orgasm, perversion, rape, seduction, sexual intercourse, sexual relations, union. **have sex (with)** be intimate (with), consummate marriage, copulate (with), couple (with), fornicate (with), have sexual intercourse (with), make love (to), mate (with), rape, ravish, *sl* screw, seduce, unite (with).

sexism *n inf* chauvinism, discrimination, prejudice.

sexual *adj* **1** genital, procreative, progenitive, reproductive. **2** ▷ SEXY.

sexuality *n* gender. □ *bisexuality, hermaphroditism, heterosexuality, homosexuality*.

sexy *adj* **1** amorous, carnal, concupiscent, erotic, lascivious, lecherous, *derog* lewd, libidinous, *derog* lubricious, lustful, passionate, provocative, *derog* prurient, *inf* randy, seductive, sensual, sexual, *inf* sultry, venereal, voluptuous. **2** attractive, *inf* beddable, desirable, *sl* dishy, flirtatious. **3** *sexy books*. aphrodisiac, arousing, pornographic, *sl* raunchy, salacious, *inf* steamy, suggestive, titillating, *inf* torrid. ▷ OBSCENE.

shabby *adj* **1** bedraggled, dilapidated, dingy, dirty, dowdy, drab, faded, frayed, *inf* grubby, mangy, *inf* moth-eaten, ragged, run-down, *inf* scruffy, seedy, tattered, *inf* tatty, threadbare, unattractive, worn, worn-out. *Opp* SMART. **2** *shabby behaviour*. base, contemptible, despicable, disagreeable, discreditable, dishonest, dishonourable, disreputable, ignoble, *inf* low-down, mean, nasty, shameful, shoddy, unfair, unfriendly, ungenerous, unkind, unworthy. *Opp* HONOURABLE.

shack *n* cabin, hovel, hut, lean-to, shanty, shed.

shade *n* **1** ▷ SHADOW. **2** awning, blind, canopy, covering, curtain, parasol, screen, shelter, shield, umbrella, Venetian blind. **3** *a shade of blue*. colour, hue, intensity, tinge, tint, tone. **4** *shades of meaning*. degree, difference, nicety, nuance, variation. • *vb* **1** camouflage, conceal, cover, hide, mask, obscure, protect, screen, shield, shroud, veil. **2** *shade with pencil*. black out, block in, cross-hatch, darken, fill in, make dark.

shadow *n* **1** darkness, dimness, dusk, gloom, obscurity, penumbra, semi-darkness, shade, umbra. **2** *The sun casts shadows*. outline, shape, silhouette. **3** *a shadow of doubt*. ▷ HINT. • *vb* dog, follow, hunt, *inf* keep tabs on, keep watch on, pursue, stalk, *inf* tag onto, tail, track, trail, watch.

shadowy *adj* **1** dark, dim, faint, hazy, ill-defined, indefinite, indistinct, nebulous, obscure, unclear, unrecognizable, vague. ▷ GHOSTLY. **2** ▷ SHADY.

shady *adj* **1** *poet* bosky, cool, dark, dim, dusky, gloomy, leafy,

shaded, shadowy, sheltered, sunless. *Opp* SUNNY. **2** *a shady character.* devious, dishonest, disreputable, dubious, *inf* fishy, questionable, shifty, suspicious, unreliable, untrustworthy.
Opp HONEST.

shaft *n* **1** arrow, column, handle, helve, pillar, pole, post, rod, shank, stanchion, stem, stick, upright. **2** duct, mine, pit, tunnel, well, working. **3** *shaft of light.* beam, gleam, laser, pencil, ray, streak.

shaggy *adj* bushy, dishevelled, fibrous, fleecy, hairy, hirsute, matted, rough, tousled, unkempt, unshorn, untidy, woolly.
Opp SMOOTH.

shake *vb* **1** convulse, heave, jump, quake, quiver, rattle, rock, shiver, shudder, sway, throb, totter, tremble, vibrate, waver, wobble. **2** *shake your umbrella.* agitate, brandish, flourish, gyrate, jar, jerk, *inf* jiggle, *inf* joggle, jolt, oscillate, sway, swing, twirl, twitch, vibrate, wag, *inf* waggle, wave, *inf* wiggle. **3** *The bad news shook us.* alarm, distress, disturb, frighten, perturb, *inf* rattle, shock, startle, *inf* throw, unnerve, unsettle, upset. ▷ SURPRISE.

shaky *adj* **1** decrepit, dilapidated, feeble, flimsy, frail, insecure, precarious, ramshackle, rickety, rocky, shaking, unreliable, unsound, unsteady, weak, wobbly. **2** *a shaky voice.* faltering, quavering, quivering, trembling, tremulous. **3** *a shaky start.* nervous, tentative, uncertain, underconfident, unimpressive, unpromising. *Opp* STEADY, STRONG.

shallow *adj* **1** *shallow water.* [There are no apt synonyms for this sense.] **2** *shallow argument.* empty, facile, foolish, frivolous, glib, insincere, puerile, silly, simple, *inf* skin-deep, slight, superficial, trivial, unconvincing, unscholarly, unthinking.
Opp DEEP.

sham *adj* artificial, bogus, counterfeit, ersatz, fake, false, fictitious, fraudulent, imitation, make-believe, mock, *inf* pretend, pretended, simulated, synthetic.
• *n* counterfeit, fake, fiction, fraud, hoax, imitation, make-believe, pretence, *inf* put-up job, simulation.
• *vb* counterfeit, fake, feign, imitate, make believe, pretend, simulate.

shambles *plur n* **1** battlefield, scene of carnage, slaughterhouse. **2** [*inf*] chaos, confusion, devastation, disorder, mess, muddle, *inf* pigsty, *inf* tip.

shame *n* **1** chagrin, degradation, discredit, disgrace, dishonour, distress, embarrassment, guilt, humiliation, ignominy, infamy, loss of face, mortification, obloquy, opprobrium, remorse, stain, stigma, vilification. **2** *a shame to mistreat him so.* outrage, pity, scandal, wickedness. • *vb* abash, chagrin, chasten, discomfit, disconcert, discountenance, disgrace, embarrass, humble, humiliate, make someone ashamed, mortify, *inf* put someone in his/her place, *inf* show someone up.

shamefaced *adj* **1** abashed, ashamed, chagrined, *inf* hang-dog, humiliated, mortified, penitent, *inf* red-faced, remorseful, repentant, sorry. **2** bashful, coy, embarrassed, modest, self-conscious, sheepish, shy, timid.
Opp SHAMELESS.

shameful *adj* **1** *a shameful crime.* base, contemptible, deplorable, disgraceful, infamous, ignoble, low, mean, outrageous, reprehensible, scandalous, unworthy. ▷ WICKED. 2 *shameful to be found out.* compromising, degrading, demeaning, discreditable, dishonourable, embarrassing, humiliating, ignominious, *sl* infra dig, inglorious, lowering, mortifying, undignified. *Opp* HONOURABLE.

shameless *adj* 1 barefaced, bold, brazen, cheeky, cool, defiant, flagrant, hardened, impenitent, impudent, incorrigible, insolent, unabashed, unashamed, unrepentant. 2 *shameless nudity.* frank, honest, immodest, indecorous, improper, open, rude, shocking, unblushing, unconcealed, undisguised, unselfconscious. *Opp* SHAMEFACED.

shape *n* 1 body, build, figure, physique, profile, silhouette. 2 *geometrical shape.* configuration, figure, form, format, model, mould, outline, pattern. □ [two-dimensional] *circle, diamond, ellipse, heptagon, hexagon, lozenge, oblong, octagon, oval, parallelogram, pentagon, polygon, quadrant, quadrilateral, rectangle, rhomboid, rhombus, ring, semicircle, square, trapezium, trapezoid, triangle.* [three-dimensional] *cone, cube, cylinder, decahedron, hemisphere, hexahedron, octahedron, polyhedron, prism, pyramid, sphere.* • *vb* adapt, adjust, carve, cast, cut, fashion, form, frame, give shape to, model, mould, *inf* sculpt, sculpture, whittle.

shapeless *adj* 1 amorphous, formless, indeterminate, irregular, nebulous, undefined, unformed, unstructured, vague. 2 *a shapeless figure.* deformed, distorted, *inf* dumpy, flat, misshapen, twisted, unattractive, unshapely. *Opp* SHAPELY.

shapely *adj* attractive, comely, *inf* curvaceous, elegant, good-looking, graceful, neat, trim, *inf* voluptuous, well-proportioned. *Opp* SHAPELESS.

share *n* allocation, allotment, allowance, bit, cut, division, due, fraction, helping, part, percentage, piece, portion, proportion, quota, ration, serving, *sl* whack. • *vb* 1 allocate, allot, apportion, deal out, distribute, divide, dole out, *inf* go halves or shares (with), halve, partake of, portion out, ration out, share out, split. 2 *share work.* be involved, cooperate, join, participate, take part. **shared** ▷ JOINT.

sharp *adj* 1 acute, arrow-shaped, cutting, fine, jagged, keen, knife-edged, needle-sharp, pointed, razor-sharp, sharpened, spiky, tapering. 2 *sharp bend, drop.* abrupt, acute, angular, hairpin, marked, precipitous, sheer, steep, sudden, surprising, unexpected, vertical. 3 *sharp focus.* clear, defined, distinct, focused, well-defined. 4 *a sharp storm.* extreme, heavy, intense, serious, severe, sudden, violent. 5 *sharp frost.* biting, bitter, keen, nippy. ▷ COLD. 6 *sharp pain.* acute, excruciating, painful, stabbing, stinging. 7 *sharp rejoinder.* acerbic, acid, acidulous, barbed, biting, caustic, critical, cutting, hurtful, incisive, malicious, mocking, mordant, sarcastic, sardonic, scathing, spiteful, tart, trenchant, unkind, venomous, vitriolic. 8 *sharp mind.* acute, agile, alert, artful, astute, bright, clever, crafty, *inf* cute, discerning, incisive, intelligent, observant, penetrating, perceptive, probing, quick-witted, searching, shrewd, *inf* smart. 9 *sharp eyes.* attentive, eagle-eyed, observant, *inf* peeled (*keep your eyes peeled*), quick, watchful, wide-open. **10** *sharp taste, smell*. acid, acrid, bitter, caustic, hot, piquant, pungent, sour, spicy, tangy, tart. **11** *sharp sound*. clear, detached, ear-splitting, high, high-pitched, penetrating, piercing, shrieking, shrill, staccato, strident. *Opp* BLUNT, DULL, SLIGHT.

sharpen *vb* file, grind, hone, make sharp, strop, whet. *Opp* BLUNT.

sharpener *n* file, grindstone, hone, pencil-sharpener, strop, whetstone.

shatter *vb* blast, break, break up, burst, crack, crush, dash to pieces, demolish, destroy, disintegrate, explode, pulverize, shiver, smash, *inf* smash to smithereens, splinter, split, wreck. **shattered** ▷ SURPRISED, WEARY.

sheaf *n* bundle, bunch, file, ream.

shear *vb* clip, strip, trim. ▷ CUT.

sheath *n* casing, covering, scabbard, sleeve.

sheathe *vb* cocoon, cover, encase, enclose, put away, put in a sheath, wrap.

shed *n* hut, hutch, lean-to, outhouse, penthouse, potting-shed, shack, shelter, storehouse. • *vb* abandon, cast off, discard, drop, let fall, moult, pour off, scatter, shower, spill, spread, throw off. **shed light** ▷ SHINE.

sheen *n* brightness, burnish, glaze, gleam, glint, gloss, lustre, patina, polish, radiance, reflection, shimmer, shine.

sheep *n* ewe, lamb, mutton, ram, wether.

sheepish *adj* abashed, ashamed, bashful, coy, docile, embarrassed, guilty, meek, mortified, reticent, self-conscious, self-effacing, shamefaced, shy, timid. *Opp* SHAMELESS.

sheer *adj* **1** absolute, arrant, complete, downright, out-and-out, plain, pure, simple, thoroughgoing, total, unadulterated, unalloyed, unmitigated, unmixed, unqualified, utter. **2** *a sheer cliff*. abrupt, perpendicular, precipitous, steep, vertical. **3** *sheer silk*. diaphanous, filmy, fine, flimsy, gauzy, gossamer, *inf* see-through, thin, translucent, transparent.

sheet *n* **1** bedsheet, duvet cover. **2** [*paper*] folio, leaf, page. **3** [*glass, etc*] pane, panel, plate. **4** [*ice, etc*] blanket, coating, covering, film, lamina, layer, membrane, skin, veneer. **5** [*water*] area, expanse, surface.

shelf *n* ledge, shelving.

shell *n* **1** carapace (*of tortoise*), case, casing, covering, crust, exterior, façade, hull, husk, outside, pod. **2** *fired shells at them*. cartridge, projectile. • *vb* attack with gunfire, barrage, bomb, bombard, fire at, shoot at, strafe.

shellfish *n* bivalve, crustacean, mollusc. □ *barnacle, clam, cockle, conch, crab, crayfish, cuttlefish, limpet, lobster, mussel, oyster, prawn, scallop, shrimp, whelk, winkle.*

shelter *n* **1** asylum, cover, haven, lee, protection, refuge, safety, sanctuary, security. **2** barrier, concealment, cover, fence, hut, roof, screen, shield. **3** *seek shelter for the night*. accommodation, lodging, home, housing, resting-place. ▷ HOUSE. **4** *air-raid shelter*. bunker. • *vb* **1** defend, enclose, guard, keep safe, protect, safeguard, screen, secure, shade, shield. **2** *shelter a runaway*. accommodate, give shelter to, harbour, hide, *inf* put up. **sheltered** ▷ QUIET.

shelve *vb* **1** defer, hold in abeyance, lay aside, postpone, put off, put on ice. **2** ▷ SLOPE.

shield *n* **1** barrier, bulwark, defence, guard, protection, safeguard, screen, shelter. **2** *a warrior's shield.* buckler, *heraldry* escutcheon. • *vb* cover, defend, guard, keep safe, protect, safeguard, screen, shade, shelter.

shift *n* **1** adjustment, alteration, change, move, switch, transfer, transposition. **2** *night shift.* crew, gang, group, period, *inf* stint, team, workforce. • *vb* adjust, alter, budge, change, displace, reposition, switch, transfer, transpose. ▷ MOVE. **shift for yourself** ▷ MANAGE.

shiftless *adj* idle, indolent, ineffective, inefficient, inept, irresponsible, lazy, unambitious, unenterprising. *Opp* RESOURCEFUL.

shifty *adj* artful, canny, crafty, cunning, deceitful, designing, devious, dishonest, evasive, *inf* foxy, furtive, scheming, secretive, *inf* shady, *inf* slippery, sly, treacherous, tricky, untrustworthy, wily. *Opp* STRAIGHTFORWARD.

shimmer *vb* flicker, glimmer, glisten, ripple. ▷ SHINE.

shine *n* brightness, burnish, coruscation, glaze, gleam, glint, gloss, glow, luminosity, lustre, patina, phosphorescence, polish, radiance, reflection, sheen, shimmer, sparkle, varnish. • *vb* **1** beam, be luminous, blaze, coruscate, dazzle, emit light, flare, flash, glare, gleam, glint, glisten, glitter, glow, phosphoresce, radiate, reflect, scintillate, shed light, shimmer, sparkle, twinkle. **2** *used to shine at maths.* be brilliant, be clever, do well, excel, *inf* make your mark, stand out. **3** *shine your shoes.* brush, buff up, burnish, clean, polish, rub up. **shining** ▷ BRIGHT, CONSPICUOUS.

shingle *n* **1** gravel, pebbles, stones. **2** *roofing shingle.* tile.

shiny *adj* bright, brilliant, burnished, gleaming, glistening, glossy, glowing, luminous, lustrous, phosphorescent, polished, reflective, rubbed, shimmering, shining, sleek, smooth. *Opp* DULL.

ship *n* boat. ▷ VESSEL. • *vb* carry, ls *inf* cart, convey, deliver, ferry, freight, move, send, transport.

shirk *vb* avoid, dodge, duck, evade, get out of, neglect, shun. **shirk work** be lazy, malinger, *inf* skive, slack.

shiver *n* flutter, frisson, quiver, rattle, shake, shudder, thrill, tremor, vibration. • *vb* chatter, flap, flutter, quake, quaver, quiver, rattle, shake, shudder, tremble, twitch, vibrate.

shock *n* **1** blow, collision, concussion, impact, jolt, thud. **2** *came as a shock. inf* bombshell, surprise, *inf* thunderbolt. **3** *state of shock.* dismay, distress, fright, trauma, upset. • *vb* **1** alarm, amaze, astonish, astound, confound, daze, dismay, distress, dumbfound, frighten, *inf* give someone a turn, jar, jolt, numb, paralyse, petrify, rock, scare, shake, stagger, startle, stun, stupefy, surprise, *inf* throw, traumatize, unnerve. **2** *Sadism shocks us.* appal, disgust, horrify, nauseate, offend, outrage, repel, revolt, scandalize, sicken.

shoddy *adj* **1** cheap, flimsy, *inf* gimcrack, inferior, jerry-built, meretricious, nasty, poor quality, *inf* rubbishy, second-rate, shabby, *sl* tacky, *inf* tatty, tawdry, *inf* trashy. **2** *shoddy work.* careless, messy, negligent, slipshod, *inf* sloppy, slovenly, untidy. *Opp* SUPERIOR.

shoe *n* plur footwear. □ *boot, bootee, brogue, clog, espadrille,* inf *flip-flop, galosh, gum-boot,* inf *lace-up, moccasin, plimsoll, pump, sabot, sandal,* inf *slip-on, slipper, trainer, wader, wellington.*

shoemaker *n* bootmaker, cobbler.

shoot *n* branch, bud, new growth, offshoot, sprout, sucker, twig. • *vb* **1** *shoot a gun*. aim, discharge, fire. **2** *shoot the enemy*. aim at, bombard, fire at, gun down, hit, hunt, kill, *inf* let fly at, open fire on, *inf* pick off, shell, snipe at, strafe, *inf* take pot-shots at. **3** *shoot from your chair*. bolt, dart, dash, fly, hurtle, leap, move quickly, race, run, rush, speed, spring, streak. **4** *plants shoot in the spring*. bud, burgeon, develop, flourish, grow, put out shoots, spring up, sprout.

shop *n* boutique, cash-and-carry, department store, *old use* emporium, establishment, market, outlet, retailer, seller, store, wholesaler. □ *baker, betting shop, bookshop, butcher, chandler, chemist, confectioner, couturier, creamery, dairy, delicatessen, draper, fishmonger, florist, garden-centre, greengrocer, grocer, haberdasher, herbalist, hypermarket, ironmonger, jeweller, launderette, minimarket, newsagent, off-licence, outfitter, pawnbroker, pharmacy, post office, poulterer, stationer, supermarket, tailor, take-away, tobacconist, toyshop, video shop, vintner, watchmaker.*

shopkeeper *n* dealer, merchant, retailer, salesgirl, salesman, saleswoman, stockist, storekeeper, supplier, trader, tradesman.

shopper *n* buyer, customer, patron.

shopping *n* **1** buying, *inf* spending-spree. **2** goods, purchases.

shopping-centre *n* arcade, complex, hypermarket, mall, precinct.

shore *n* bank, beach, coast, edge, foreshore, sands, seashore, seaside, shingle, strand. • *vb* **shore up** ▷ SUPPORT.

short *adj* **1** diminutive, *inf* dumpy, dwarfish, little, midget, *fem* petite, *derog* pint-sized, slight, small, squat, *inf* stubby, *inf* stumpy, stunted, tiny, *inf* wee, undergrown. **2** *a short visit*. brief, cursory, curtailed, ephemeral, fleeting, momentary, passing, quick, short-lived, temporary, transient, transitory. **3** *a short book*. abbreviated, abridged, compact, concise, cut, pocket, shortened, succinct. **4** *in short supply*. deficient, inadequate, insufficient, lacking, limited, low, meagre, scanty, scarce, sparse, wanting. **5** *a short manner*. abrupt, bad-tempered, blunt, brusque, cross, curt, gruff, grumpy, impolite, irritable, laconic, sharp, snappy, taciturn, terse, testy, uncivil, unfriendly, unkind, unsympathetic. *Opp* EXPANSIVE, LONG, PLENTIFUL, TALL. **cut short** ▷ SHORTEN.

shortage *n* absence, dearth, deficiency, deficit, insufficiency, lack, paucity, poverty, scarcity, shortfall, want. *Opp* PLENTY.

shortcoming *n* bad habit, defect, deficiency, drawback, failing, failure, fault, flaw, foible, imperfection, limitation, vice, weakness, weak point.

shorten *vb* abbreviate, abridge, compress, condense, curtail, cut, cut down, cut short, diminish, dock, lop, précis, prune, reduce, shrink, summarize, take up (*clothes*), telescope, trim, truncate. *Opp* LENGTHEN.

shortly *adv old use* anon, before long, by and by, directly, presently, soon.

short-sighted *adj* **1** myopic, near-sighted. **2** unadventurous, unimaginative, without vision.

short-tempered *adj* abrupt, acerbic, brusque, crabby, cross, crusty, curt, gruff, irascible, irritable, peevish, peremptory, shrewish, snappy, testy, touchy, waspish. *Opp* BAD-TEMPERED.

shot *n* **1** ball, bullet, discharge, missile, pellet, projectile, round, *inf* slug. **2** *heard a shot.* bang, blast, crack, explosion, report. **3** *a first-class shot.* marksman, markswoman, sharpshooter. **4** *give it a shot.* attempt, chance, *inf* crack, effort, endeavour, *inf* go, hit, kick, *inf* stab, stroke, try. **5** *photographic shot.* angle, photograph, picture, scene, sequence, snap, snapshot.

shout *vb* bawl, bellow, *inf* belt, call, cheer, clamour, cry out, exclaim, howl, rant, roar, scream, screech, shriek, talk loudly, vociferate, whoop, yell, yelp, yowl. *Opp* WHISPER.

shove *vb inf* barge, crowd, drive, elbow, hustle, impel, jostle, nudge, press, prod, push, shoulder, thrust.

shovel *vb* clear, dig, scoop, shift.

show *n* **1** drama, performance, play, presentation, production. ▷ ENTERTAINMENT. **2** *flower show.* competition, demonstration, display, exhibition, *inf* expo, exposition, fair, presentation. **3** *show of strength.* appearance, demonstration, façade, illusion, impression, pose, pretence, threat. **4** *just for show.* affectation, exhibitionism, flamboyance, ostentation, pretentiousness, showing off. • *vb* **1** bare, betray, demonstrate, display, divulge, exhibit, expose, make public, make visible, manifest, open up, present, produce, reveal, uncover. **2** *Let your feelings show.* appear, be seen, be visible, catch the eye, come out, emerge, make an appearance, materialize, *inf* peep through, stand out, stick out. **3** *show the way.* conduct, direct, escort, guide, indicate, lead, point out, steer, usher. **4** *show kindness.* accord, bestow, confer, grant, treat with. **5** *This photo shows us at work.* depict, give a picture of, illustrate, picture, portray, represent, symbolize. **6** *Show me how.* clarify, describe, elucidate, explain, instruct, make clear, teach, tell. **7** *Tests show I was right.* attest, bear out, confirm, corroborate, demonstrate, evince, exemplify, manifest, prove, substantiate, verify, witness. **show off** ▷ BOAST. **show up** ▷ ARRIVE, HUMILIATE.

showdown *n* confrontation, crisis, *inf* decider, decisive encounter, *inf* moment of truth.

shower *n* **1** drizzle, sprinkling. ▷ RAIN. **2** douche, shower-bath. • *vb* **1** deluge, drop, rain, spatter, splash, spray, sprinkle. **2** *shower with gifts.* heap, inundate, load, overwhelm.

show-off *n inf* big-head, boaster, braggart, conceited person, egotist, exhibitionist, *inf* poser, poseur, *inf* showman, swaggerer.

showy *adj* bright, conspicuous, elaborate, fancy, flamboyant, flashy, florid, fussy, garish, gaudy, lavish, *inf* loud, lurid, ornate, ostentatious, *inf* over the top, pretentious, striking, trumpery, vulgar. *Opp* DISCREET.

shred *n* atom, bit, fragment, grain, hint, iota, jot, piece, scintilla, scrap, sliver, snippet, speck, trace. • *vb* cut to shreds, destroy, grate, rip up, scrap, tear. **shreds** rags, ribbons, strips, tatters.

shrewd *adj* acute, artful, astute, calculating, *inf* canny, clever, crafty, cunning, discerning, discriminating, *inf* foxy, ingenious, intelligent, knowing, observant, perceptive, percipient, perspicacious, quick-witted, sage, sharp, sly, smart, wily, wise. *Opp* STUPID.

shriek *vb* cry, scream, screech, squawk, squeal.

shrill *adj* ear-splitting, high, high-pitched, jarring, penetrating, piercing, piping, screaming, screeching, screechy, sharp, shrieking, strident, treble, whistling. *Opp* GENTLE, SONOROUS.

shrine *n* altar, chapel, holy of holies, holy place, place of worship, reliquary, sanctum, tomb. ▷ CHURCH.

shrink *vb* **1** become smaller, contract, decrease, diminish, dwindle, lessen, make smaller, narrow, reduce, shorten. ▷ SHRIVEL. *Opp* EXPAND. **2** *shrink with fear.* back off, cower, cringe, flinch, hang back, quail, recoil, retire, shy away, wince, withdraw. *Opp* ADVANCE.

shrivel *vb* become parched, become wizened, dehydrate, desiccate, curl, droop, dry out, dry up, pucker up, wilt, wither, wrinkle. ▷ SHRINK.

shroud *n* blanket, cloak, cloud, cover, mantle, mask, pall, veil, winding-sheet. • *vb* camouflage, cloak, conceal, cover, disguise, enshroud, envelop, hide, mask, screen, swathe, veil, wrap up.

shrub *n* bush, tree. □ *berberis, blackthorn, broom, bryony, buckthorn, buddleia, camellia, daphne, forsythia, gorse, heather, hydrangea, japonica, jasmine, lavender, lilac, myrtle, privet, rhododendron, rosemary, rue, viburnum.*

shudder *vb* be horrified, convulse, jerk, quake, quiver, rattle, shake, shiver, squirm, tremble, vibrate.

shuffle *vb* **1** confuse, disorganize, intermix, intersperse, jumble, mix, mix up, rearrange, reorganize. **2** *shuffle along.* drag your feet, scrape, shamble, slide. ▷ WALK.

shun *vb* avoid, disdain, eschew, flee, *inf* give the cold shoulder to, keep clear of, rebuff, reject, shy away from, spurn, steer clear of, turn away from. *Opp* SEEK.

shut *vb* bolt, close, fasten, latch, lock, push to, replace, seal, secure, slam. **shut in** ▷ CONFINE, IMPRISON. **shut off** ▷ ISOLATE. **shut out** ▷ EXCLUDE. **shut up** ▷ CONFINE, IMPRISON, SILENCE.

shutter *n* blind, louvre, screen.

shy *adj* apprehensive, backward, bashful, cautious, chary, coy, diffident, hesitant, inhibited, introverted, modest, *inf* mousy, nervous, reserved, reticent, retiring, self-conscious, self-effacing, sheepish, timid, timorous, underconfident, wary, withdrawn. *Opp* ASSERTIVE, UNINHIBITED. • *vb* ▷ THROW.

sibling *n* brother, sister, twin. ▷ FAMILY.

sick *adj* **1** afflicted, ailing, bedridden, diseased, ill, indisposed, infirm, *inf* laid up, *inf* poorly, *inf* queer, sickly, *inf* under the weather, unhealthy, unwell. ▷ ILL. **2** airsick, bilious, carsick, likely to vomit, nauseated, nauseous, queasy, seasick, squeamish. **3** *sick* *of rudeness.* annoyed (by), bored (with), disgusted (by), distressed (by), *inf* fed up (with), glutted (with), nauseated (by), sated (with), sickened (by), tired, troubled (by), upset (by), weary. 4 [*inf*] *a sick joke.* ▷ MORBID. **be sick** ▷ VOMIT.

sicken *vb* 1 *inf* catch a bug, fail, fall ill, take sick, weaken. 2 appal, be sickening to, disgust, make someone sick, nauseate, offend, repel, revolt, *inf* turn someone off, *inf* turn someone's stomach. **sickening** ▷ REPULSIVE.

sickly *adj* 1 ailing, anaemic, delicate, drawn, feeble, frail, ill, pale, pallid, *inf* peaky, unhealthy, wan, weak. ▷ ILL. *Opp* HEALTHY. 2 *sickly sentiment.* cloying, maudlin, mawkish, *inf* mushy, nasty, nauseating, obnoxious, *inf* off-putting, syrupy, treacly, unpleasant. *Opp* REFRESHING.

sickness *n* biliousness, nausea, queasiness, vomiting. ▷ ILLNESS.

side *n* 1 *sides of a cube.* elevation, face, facet, flank, surface. 2 *side of the road.* border, boundary, brim, brink, edge, fringe, limit, margin, perimeter, rim, verge. 3 *sides in a debate.* angle, aspect, attitude, perspective, point of view, position, school of thought, slant, standpoint, view, viewpoint. 4 *sides in a quarrel.* army, camp, faction, interest, party, sect, team. • *vb* **side with** ally with, favour, form an alliance with, *inf* go along with, join up with, partner, prefer, support, team up with. ▷ HELP.

sidestep *vb* avoid, circumvent, dodge, *inf* duck, evade, skirt round.

sidetrack *vb* deflect, distract, divert.

sideways *adj* 1 crabwise, indirect, lateral, oblique. 2 *a sideways glance.* covert, sidelong, sly, *inf* sneaky, unobtrusive.

siege *n* blockade. • *vb* ▷ BESIEGE.

sieve *n* colander, riddle, screen, strainer. • *vb* ▷ SIFT.

sift *vb* 1 filter, riddle, screen, separate, sieve, strain. 2 *sift evidence.* analyse, examine, investigate, pick out, review, scrutinize, select, sort out, weed out, winnow.

sigh *n* breath, exhalation, murmur, suspiration. ▷ SOUND.

sight *n* 1 eyesight, seeing, vision, visual perception. 2 *within sight.* field of vision, gaze, range, view, visibility. 3 *a brief sight of it.* glimpse, look. 4 *an impressive sight.* display, exhibition, scene, show, show-piece, spectacle. • *vb* behold, descry, discern, distinguish, espy, glimpse, make out, notice, observe, perceive, recognize, see, spot. **catch sight of** ▷ SEE.

sightseer *n* globe-trotter, holiday-maker, tourist, tripper, visitor.

sign *n* 1 augury, forewarning, hint, indication, indicator, intimation, omen, pointer, portent, presage, warning. ▷ SIGNAL. 2 *sign that someone was here.* clue, *inf* giveaway, indication, manifestation, marker, proof, reminder, spoor (of animal), suggestion, symptom, token, trace, vestige. 3 *put up a sign.* advertisement, notice, placard, poster, publicity, signboard. 4 *identifying sign.* badge, brand, cipher, device, emblem, flag, hieroglyph, ideogram, ideograph, insignia, logo, mark, monogram, rebus, symbol, trademark. • *vb* 1 autograph, countersign, endorse, inscribe, write. 2 ▷ SIGNAL. **sign off** ▷ FINISH. **sign on** ▷ ENLIST. **sign over** ▷ TRANSFER.

signal *n* **1** communication, cue, gesticulation, gesture, *inf* goahead, indication, motion, sign, signal, *inf* tip-off, token, warning. □ *alarm-bell, beacon, bell, burglar-alarm, buzzer, flag, flare, gong, green light, indicator, password, red light, reveille, rocket, semaphore signal, siren, smoke-signal, tocsin,* old use *trafficator, traffic-lights, warning-light, whistle, winker.* **2** *radio signal.* broadcast, emission, output, transmission, waves. • *vb* beckon, communicate, flag, gesticulate, give or send a signal, indicate, motion, notify, sign, wave. ▷ GESTURE.

signature *n* autograph, endorsement, mark, name.

signet *n* seal, stamp.

significance *n* denotation, force, idea, implication, import, importance, message, point, purport, relevance, sense, signification, usefulness, value, weight. ▷ MEANING.

significant *adj* **1** eloquent, expressive, indicative, informative, knowing, meaningful, pregnant, revealing, suggestive, symbolic, *inf* tell-tale. **2** *significant event.* big, consequential, considerable, historic, important, influential, memorable, newsworthy, noteworthy, relevant, salient, serious, sizeable, valuable, vital, worthwhile. *Opp* INSIGNIFICANT.

signify *vb* **1** announce, be a sign of, betoken, communicate, connote, convey, denote, express, foretell, impart, imply, indicate, intimate, make known, reflect, reveal, signal, suggest, symbolize, tell, transmit. **2** *It doesn't signify.* be significant, count, matter, merit consideration.

signpost *n* finger-post, pointer, road-sign, sign.

silence *n* **1** calm, calmness, hush, noiselessness, peace, quiet, quietness, quietude, soundlessness, stillness, tranquillity. *Opp* NOISE. **2** *Her silence puzzled us.* dumbness, muteness, reticence, speechlessness, taciturnity, uncommunicativeness. • *vb* **1** gag, hush, keep quiet, make silent, muzzle, repress, shut up, suppress. **2** *silence engine noise.* damp, deaden, muffle, mute, quieten, smother, stifle. **Silence!** Be quiet! Be silent! *inf* Hold your tongue! Hush! Keep quiet! *inf* Pipe down! Shut up! Stop talking!

silent *adj* **1** hushed, inaudible, muffled, muted, noiseless, quiet, soundless. **2** dumb, laconic, *inf* mum, reserved, reticent, speechless, taciturn, tight-lipped, tongue-tied, uncommunicative, unforthcoming, voiceless. **3** *silent listeners.* attentive, rapt, restrained, still. **4** *silent agreement.* implied, implicit, mute, tacit, understood, unexpressed, unspoken, unuttered. *Opp* EXPLICIT, NOISY, TALKATIVE. **be silent** keep quiet, *inf* pipe down, say nothing, *inf* shut up.

silhouette *n* contour, form, outline, profile, shadow, shape.

silky *adj* delicate, fine, glossy, lustrous, satiny, sleek, smooth, soft, velvety.

silly *adj* **1** absurd, asinine, brainless, childish, crazy, daft, *inf* dopey, *inf* dotty, fatuous, feather-brained, feeble-minded, flighty, foolish, *old use* fond, frivolous, grotesque, *inf* half-baked, hare-brained, idiotic, ill-advised, illogical, immature, impractical, imprudent, inadvisable, inane, infantile, irrational, *inf* jokey, laughable, light-hearted, ludicrous, mad, meaningless, mindless, misguided, naïve, nonsensical, playful, pointless, preposterous, ridiculous, scatter-brained, *inf* scatty, senseless, shallow, simple, simple-minded, simplistic, *inf* soppy, stupid, thoughtless, unintelligent, unreasonable, unsound, unwise, wild, witless. *Opp* SERIOUS, WISE. **2** [*inf*] *knocked silly*. ▷ UNCONSCIOUS.

silt *n* alluvium, deposit, mud, ooze, sediment, slime, sludge.

silvan *adj* arboreal, leafy, tree-covered, wooded.

similar *adj* akin, alike, analogous, comparable, compatible, congruous, co-ordinating, corresponding, equal, equivalent, harmonious, homogeneous, identical, indistinguishable, like, matching, parallel, related, resembling, the same, toning, twin, uniform, well-matched. *Opp* DIFFERENT.

similarity *n* affinity, closeness, congruity, correspondence, equivalence, homogeneity, kinship, likeness, match, parallelism, relationship, resemblance, sameness, similitude, uniformity. *Opp* DIFFERENCE.

simmer *vb* boil, bubble, cook, seethe, stew.

simple *adj* **1** artless, basic, candid, childlike, elementary, frank, fundamental, guileless, homely, honest, humble, ingenuous, innocent, lowly, modest, *derog* naïve, natural, simple-minded, *derog* silly, sincere, unaffected, unassuming, unpretentious, unsophisticated. *Opp* SOPHISTICATED. **2** *simple instructions*. clear, comprehensible, direct, easy, fool-proof, intelligible, lucid, straightforward, uncomplicated, understandable. *Opp* COMPLEX. **3** *a simple dress*. austere, classical, plain, severe, stark, unadorned, unembellished. *Opp* ORNATE.

simplify *vb* clarify, explain, make simple, paraphrase, prune, *inf* put in words of one syllable, streamline, unravel, untangle. *Opp* COMPLICATE.

simplistic *adj* [*always derog*] facile, inadequate, naïve, over-simple, oversimplified, shallow, silly, superficial.

simulate *vb* act, counterfeit, dissimulate, enact, fake, feign, imitate, *inf* mock up, play-act, pretend, reproduce, sham.

simultaneous *adj* coinciding, concurrent, contemporaneous, parallel, synchronized, synchronous.

sin *n* blasphemy, corruption, depravity, desecration, devilry, error, evil, fault, guilt, immorality, impiety, iniquity, irreverence, misdeed, offence, peccadillo, profanation, sacrilege, sinfulness, transgression, *old use* trespass, ungodliness, unrighteousness, vice, wickedness, wrong, wrongdoing. • *vb* be guilty of sin, blaspheme, do wrong, err, fall from grace, go astray, lapse, misbehave, offend, stray, transgress.

sincere *adj* candid, direct, earnest, frank, genuine, guileless, heartfelt, honest, open, real, serious, simple, *inf* straight, straightforward, true, truthful, unaffected, unfeigned, upright, wholehearted. *Opp* INSINCERE.

sincerity *n* candour, directness, earnestness, frankness, genuineness, honesty, honour, integrity, openness, straightforwardness, trustworthiness, truthfulness, uprightness.

sinewy *adj* brawny, muscular, strapping, tough, wiry. ▷ STRONG.

sinful *adj* bad, blasphemous, corrupt, damnable, depraved, erring, evil, fallen, guilty, immoral, impious, iniquitous, irreligious, irreverent, profane, sacrilegious, ungodly, unholy, unrighteous, vile, wicked, wrong, wrongful. *Opp* RIGHTEOUS.

sing *vb* carol, chant, chirp, chorus, croon, descant, hum, intone, serenade, trill, vocalize, warble, whistle, yodel.

singe *vb* blacken, burn, char, scorch, sear.

singer *n* songster, vocalist. □ *alto, balladeer, baritone, bass, carol-singer, castrato, counter-tenor,* plur *choir, choirboy, choirgirl, chorister,* plur *chorus, coloratura, contralto, crooner,* It *diva, folk singer, minstrel, opera singer, pop star, precentor,* It *prima donna, soloist, soprano, tenor, treble, troubadour*.

single *adj* **1** exclusive, individual, isolated, lone, odd, one, only, personal, separate, singular, sole, solitary, unique, unparalleled. **2** *a single person*. celibate, *inf* free, unattached, unmarried. • *vb* **single out** ▷ CHOOSE.

single-handed *adj* alone, independent, solitary, unaided, unassisted, without help.

single-minded *adj* dedicated, determined, devoted, dogged, *derog* fanatical, *derog* obsessive, persevering, resolute, steadfast, tireless, unswerving, unwavering.

singular *adj* **1** ▷ SINGLE. **2** abnormal, curious, different, distinct, eccentric, exceptional, extraordinary, odd, outstanding, peculiar, rare, remarkable, strange, unusual. ▷ DISTINCTIVE. *Opp* COMMON.

sinister *adj* **1** dark, disquieting, disturbing, evil, forbidding, foreboding, frightening, gloomy, inauspicious, malevolent, malignant, menacing, minatory, ominous, threatening, upsetting. **2** *sinister motives*. bad, corrupt, criminal, dishonest, furtive, illegal, nefarious, questionable, *inf* shady, suspect, treacherous, unworthy, villainous.

sink *n* basin, stoup, washbowl. • *vb* **1** collapse, decline, descend, diminish, disappear, droop, drop, dwindle, ebb, fade, fail, fall, go down, go lower, plunge, set (*sun sets*), slip down, subside, vanish, weaken. **2** be engulfed, be submerged, founder, go down, go under. **3** *sink a ship*. scupper, scuttle. **4** *sink a borehole*. bore, dig, drill, excavate.

sinner *n* evil-doer, malefactor, miscreant, offender, reprobate, transgressor, wrongdoer.

sip *vb* drink, lap, sample, taste.

sit *vb* **1** be seated, perch, rest, seat (yourself), settle, squat, take a seat, *inf* take the weight off your feet. **2** *sit for a portrait*. pose. **3** *sit an exam*. be a candidate in, *inf* go in for, take, write. **4** *Parliament sat for 12 hours*. assemble, be in session, convene, gather, get together, meet.

site *n* area, campus, ground, location, place, plot, position, setting, situation, spot. • *vb* ▷ SITUATE.

sitting-room *n* drawing-room, living-room, lounge.

situate *vb* build, establish, found, install, locate, place, position, put, set up, site, station.

situation *n* **1** area, locale, locality, location, place, position, setting, site, spot. **2** *an awkward situation*. case, circumstances, condition, *inf* kettle of fish, plight, position, predicament, state of affairs. **3** *situations vacant.* employment, job, place, position, post.

size *n* amount, area, bigness, breadth, bulk, capacity, depth, dimensions, extent, gauge, height, immensity, largeness, length, magnitude, mass, measurement, proportions, scale, scope, volume, weight, width. ▷ MEASURE. • *vb* **size up** ▷ ASSESS.

sizeable *adj* considerable, decent, generous, largish, significant, worthwhile. ▷ BIG.

skate *vb* glide, skim, slide.

skeleton *n* bones, frame, framework, structure.

sketch *n* **1** description, design, diagram, draft, drawing, outline, picture, plan, *inf* rough, skeleton, vignette. **2** *comic sketch.* performance, playlet, scene, skit, turn. • *vb* depict, draw, indicate, outline, portray, represent. **sketch out** ▷ OUTLINE.

sketchy *adj* bitty, crude, cursory, hasty, hurried, imperfect, incomplete, inexact, perfunctory, rough, scrappy, undeveloped, unfinished, unpolished. *Opp* DETAILED, PERFECT.

skid *vb* aquaplane, glide, go out of control, slide, slip.

skilful *adj* able, accomplished, adept, adroit, apt, artful, capable, competent, consummate, crafty, cunning, deft, dexterous, experienced, expert, gifted, handy, ingenious, masterful, masterly, practised, professional, proficient, qualified, shrewd, smart, talented, trained, versatile, versed, workmanlike. ▷ CLEVER. *Opp* UNSKILFUL.

skill *n* ability, accomplishment, adroitness, aptitude, art, artistry, capability, cleverness, competence, craft, cunning, deftness, dexterity, experience, expertise, facility, flair, gift, handicraft, ingenuity, knack, mastery, professionalism, proficiency, prowess, shrewdness, talent, technique, training, versatility, workmanship.

skilled *adj* experienced, expert, qualified, trained, versed. ▷ SKILFUL.

skim *vb* **1** aquaplane, coast, fly, glide, move lightly, plane, sail, skate, ski, skid, slide, slip. **2** *skim a book.* dip into, leaf through, look through, read quickly, scan, skip, thumb through.

skin *n* casing, coat, coating, complexion, covering, epidermis, exterior, film, fur, hide, husk, integument, membrane, outside, peel, pelt, rind, shell, surface. • *vb* excoriate, flay, pare, peel, shell, strip.

skin-deep *adj* insubstantial, shallow, superficial, trivial, unimportant.

skinny *adj* bony, emaciated, gaunt, half-starved, lanky, scraggy, wasted. ▷ THIN.

skip *vb* **1** bound, caper, cavort, dance, frisk, gambol, hop, jump, leap, prance, romp, spring. **2** *skip the boring bits.* avoid, forget, ignore, leave out, miss out, neglect, omit, overlook, pass over, skim through. **3** *skip lessons.* be absent from, cut, miss, play truant from.

skirmish *n* brush, fight, fray, scrimmage, *inf* set-to, tussle. • *vb* ▷ FIGHT.

skirt *vb* avoid, border, bypass, circle, encircle, go round, pass round, *inf* steer clear of, surround.

skit *n* burlesque, parody, satire, sketch, spoof, *inf* take-off.

sky *n* air, atmosphere, *poet* blue, *poet* empyrean, *poet* firmament, *poet* heavens, space, stratosphere, *poet* welkin.

slab *n* block, chunk, hunk, lump, piece, slice, wedge, *inf* wodge.

slack *adj* **1** drooping, limp, loose, sagging, soft. *Opp* TIGHT. **2** *slack attitude.* careless, dilatory, disorganized, easy-going, flaccid, idle, inattentive, indolent, lax, lazy, listless, neglectful, negligent, permissive, relaxed, remiss, slothful, unbusinesslike, uncaring, undisciplined. *Opp* RIGOROUS. **3** *slack trade.* inactive, quiet, slow, slow-moving, sluggish. *Opp* BUSY. • *vb* be lazy, idle, malinger, neglect your duty, shirk, *inf* skive.

slacken *vb* **1** ease off, loosen, relax, release. **2** *slacken speed.* abate, decrease, ease, lessen, lower, moderate, reduce, slow down.

slacker *n inf* good-for-nothing, idler, lazy person, malingerer, *sl* skiver, sluggard.

slake *vb* allay, assuage, cool, ease, quench, relieve, satisfy.

slam *vb* **1** bang, shut. ps5 **2** [*inf*] ▷ CRITICIZE.

slander *n* backbiting, calumny, defamation, denigration, insult, libel, lie, misrepresentation, obloquy, scandal, slur, smear, vilification. • *vb* blacken the name of, calumniate, defame, denigrate, disparage, libel, malign, misrepresent, slur, smear, spread tales about, tell lies about, traduce, vilify.

slanderous *adj* abusive, calumnious, cruel, damaging, defamatory, disparaging, false, hurtful, insulting, libellous, lying, malicious, mendacious, scurrilous, untrue, vicious.

slang *n* argot, cant, jargon. • *vb* ▷ INSULT. **slanging match** ▷ QUARREL.

slant *n* **1** angle, bevel, camber, cant, diagonal, gradient, incline, list, pitch, rake, ramp, slope, tilt. **2** *slant on a problem.* approach, attitude, perspective, point of view, standpoint, view, viewpoint. **3** *slant to the news.* bias, distortion, emphasis, imbalance, one-sidedness, prejudice. • *vb* **1** be at an angle, be skewed, incline, lean, shelve, slope, tilt. **2** *slant the news.* bias, colour, distort, prejudice, twist, weight. **slanting** ▷ OBLIQUE.

slap *vb* smack, spank. ▷ HIT.

slash *vb* gash, slit. ▷ CUT.

slaughter *n* bloodshed, butchery, carnage, killing, massacre, murder. • *vb* annihilate, butcher, massacre, murder, slay. ▷ KILL.

slaughterhouse *n* abattoir, shambles.

slave *n old use* bondslave, drudge, serf, thrall, vassal. ▷ SERVANT. • *vb* drudge, exert yourself, grind away, labour, *inf* sweat, toil, *inf* work your fingers to the bone. ▷ WORK.

slave-driver *n* despot, hard taskmaster, tyrant.

slaver *vb* dribble, drool, foam at the mouth, salivate, slobber, spit.

slavery *n* bondage, captivity, enslavement, serfdom, servitude, subjugation, thraldom, vassalage. *Opp* FREEDOM.

slavish *adj* **1** abject, cringing, fawning, grovelling, humiliating, menial, obsequious, servile, submissive. **2** *slavish imitation.* close, flattering, strict, sycophantic, unimaginative, unoriginal. *Opp* INDEPENDENT.

slay *vb* assassinate, bump off, butcher, destroy, dispatch, execute, exterminate, *inf* finish off, martyr, massacre, murder, put down, put to death, slaughter. ▷ KILL.

sleazy *adj* cheap, contemptible, dirty, disreputable, low-class, mean, mucky, run-down, seedy, slovenly, sordid, squalid, unprepossessing.

sledge *n* bob-sleigh, sled, sleigh, toboggan.

sleek *adj* **1** brushed, glossy, graceful, lustrous, shining, shiny, silken, silky, smooth, soft, trim, velvety, well-groomed. *Opp* UNTIDY. **2** *a sleek look.* complacent, contented, fawning, self-satisfied, *inf* slimy, *inf* smarmy, smug, suave, thriving, unctuous, well-fed.

sleep *n inf* beauty sleep, catnap, coma, dormancy, doze, *inf* forty winks, hibernation, *sl* kip, *inf* nap, repose, rest, *inf* shut-eye, siesta, slumber, snooze, torpor, unconsciousness. • *vb* aestivate, be sleeping, be unconscious, catnap, *inf* doss down, doze, *inf* drop off, drowse, fall asleep, go to bed, *inf* have forty winks, hibernate, *sl* kip, *inf* nod off, rest, slumber, snooze, *inf* take a nap. **sleeping** ▷ ASLEEP.

sleepiness *n* drowsiness, lassitude, lethargy, somnolence, tiredness, torpor.

sleepless *adj* awake, conscious, disturbed, insomniac, restless, *inf* tossing and turning, wakeful, watchful, wide awake. *Opp* ASLEEP.

sleepwalker *n* noctambulist, somnambulist.

sleepy *adj* **1** comatose, *inf* dopey, drowsy, heavy, lethargic, ready to sleep, sluggish, somnolent, soporific, tired, torpid, weary. **2** *a sleepy village.* boring, dull, inactive, quiet, restful, slowmoving, unexciting. *Opp* LIVELY.

slender *adj* **1** fine, graceful, lean, narrow, slender, slight, svelte, sylphlike, trim. ▷ THIN. **2** *slender thread.* feeble, fragile, tenuous. **3** *slender means.* inadequate, meagre, scanty, small. *Opp* FAT, LARGE.

slice *n* carving, layer, piece, rasher, shaving, sliver, wedge. • *vb* carve, shave off. ▷ CUT.

slick *adj* **1** adroit, artful, clever, cunning, deft, dextrous, efficient, quick, skilful, smart. **2** *a slick talker.* glib, meretricious, plausible, *inf* smarmy, smooth, smug, specious, suave, superficial, *inf* tricky, unctuous, untrustworthy, urbane, wily. **3** *slick hair.* glossy, oiled, plastered down, shiny, sleek, smooth.

slide *n* **1** avalanche, landslide, landslip. **2** *photographic slide.* transparency. • *vb* aquaplane, coast, glide, glissade, plane, skate, ski, skid, skim, slip, slither, toboggan.

slight *adj* **1** imperceptible, inadequate, inconsequential, inconsiderable, insignificant, insufficient, little, minor, negligible, scanty, slim (*chance*), small, superficial, trifling, trivial, unimportant. **2** *slight build.* delicate, diminutive, flimsy, fragile, frail, petite, sickly, slender, slim, svelte, sylphlike, thin, tiny, weak. *Opp* BIG. • *n, vb* ▷ INSULT.

slightly *adv* hardly, moderately, only just, scarcely. *Opp* VERY.

slim *adj* **1** fine, graceful, lean, narrow, slender, svelte, sylphlike, trim. ▷ THIN. **2** *a slim chance.* little, negligible, remote, slight, unlikely. • *vb* become slimmer, diet, lose weight, reduce.

slime *n* muck, mucus, mud, ooze, sludge.

slimy *adj* clammy, greasy, mucous, muddy, oily, oozy, *inf* slippy, slippery, slithery, *inf* squidgy, *inf* squishy, wet.

sling *vb* cast, *inf* chuck, fling, heave, hurl, launch, *inf* let fly, lob, pelt, pitch, propel, shoot, shy, throw, toss.

slink *vb* creep, edge, move guiltily, prowl, skulk, slither, sneak, steal.

slinky *adj* [*inf*] *a slinky dress.* clinging, close-fitting, graceful, *inf* sexy, sinuous, sleek.

slip *n* **1** accident, *inf* bloomer, blunder, error, fault, *Fr* faux pas, impropriety, inaccuracy, indiscretion, lapse, miscalculation, mistake, oversight, *inf* slip of the pen, slip of the tongue, *inf* slip-up. **2** *slip of paper.* note, piece, sheet, strip. • *vb* **1** aquaplane, coast, glide, glissade, move out of control, skate, ski, skid, skim, slide, slip, slither, stumble, trip. **2** *slipped into the room.* creep, edge, move quietly, slink, sneak, steal. **give someone the slip** ▷ ESCAPE. **let slip** ▷ REVEAL. **slip away, slip the net** ▷ ESCAPE. **slip up** ▷ BLUNDER.

slippery *adj* **1** glassy, greasy, icy, lubricated, oily, slimy, *inf* slippy, slithery, smooth, wet. **2** *a slippery customer.* crafty, cunning, devious, evasive, *inf* hard to pin down, shifty, sly, *inf* smarmy, smooth, sneaky, specious, *inf* tricky, unreliable, untrustworthy, wily.

slipshod *adj* careless, disorganized, lax, messy, slapdash, *inf* sloppy, slovenly, untidy.

slit *n* aperture, breach, break, chink, cleft, crack, cut, fissure, gap, gash, hole, incision, opening, rift, slot, split, tear, vent. • *vb* cut, gash, slice, split, tear.

slither *vb* creep, glide, *inf* skitter, slide, slink, slip, snake, worm.

sliver *n* chip, flake, shard, shaving, snippet, splinter, strip. ▷ PIECE.

slobber *vb* dribble, drool, salivate, slaver.

slogan *n* battle-cry, catch-phrase, catchword, jingle, motto, war-cry, watchword. ▷ SAYING.

slope *n* angle, bank, bevel, camber, cant, gradient, hill, incline, pitch, rake, ramp, scarp, slant, tilt □ [upwards] *acclivity, ascent, rise.* □ [downwards] *decline, declivity, descent, dip, drop, fall.* • *vb* ascend bank, decline, descend, dip, fall, incline, lean, pitch, rise, shelve, slant, tilt, tip. **sloping** ▷ OBLIQUE.

sloppy *adj* **1** liquid, messy, runny, *inf* sloshy, slushy, *inf* splashing about, squelchy, watery, wet. **2** *sloppy work.* careless, dirty, disorganized, lax, messy, slapdash, slipshod, slovenly, unsystematic, untidy. **3** ▷ SENTIMENTAL.

slot *n* **1** aperture, breach, break, channel, chink, cleft, crack, cut, fissure, gap, gash, groove, hole, incision, opening, rift, slit, split, vent. **2** *slot on a schedule.* place, position, space, spot, time.

sloth *n* apathy, idleness, indolence, inertia, laziness, lethargy, sluggishness, torpor.

slouch *vb* droop, hunch, loaf, loll, lounge, sag, shamble, slump, stoop.

slovenly *adj* careless, *inf* couldn't-care-less, disorganized, lax, messy, shoddy, slapdash slatternly, *inf* sloppy, slovenly, thoughtless, unmethodical, untidy. *Opp* CAREFUL.

slow *adj* **1** careful, cautious, crawling, dawdling, delayed, deliberate, dilatory, gradual, lagging, late, lazy, leisurely, lingering, loitering, measured, moderate, painstaking, plodding, protracted, slow-moving, sluggardly, sluggish, steady, tardy, torpid, unhurried, unpunctual. **2** *slow learner*. backward, dense, dim, dull, obtuse, *inf* thick. ▷ STUPID. **3** *slow worker*. phlegmatic, reluctant, unenthusiastic, unwilling. ▷ SLUGGISH. *Opp* FAST. • *vb* **slow down** brake, decelerate, *inf* ease up, go slower, hold back, reduce speed. **be slow** ▷ DAWDLE, DELAY.

sludge *n* mire, muck, mud, ooze, precipitate, sediment, silt, slime, slurry, slush.

sluggish *adj* apathetic, dull, idle, inactive, indolent, inert, lazy, lethargic, lifeless, listless, phlegmatic, slothful, torpid, unresponsive. ▷ SLOW. *Opp* LIVELY.

sluice *vb* flush, rinse, swill, wash.

slumber *n, vb* ▷ SLEEP.

slump *n* collapse, crash, decline, depression, dip, downturn, drop, fall, falling-off, plunge, recession, trough. *Opp* BOOM. • *vb* **1** collapse, crash, decline, dive, drop, fall off, plummet, plunge, recede, sink, slip, *inf* take a nosedive, worsen. *Opp* PROSPER. **2** *slump in a chair*. be limp, collapse, droop, flop, hunch, loll, lounge, sag, slouch, subside.

slur *n* affront, aspersion, calumny, imputation, innuendo, insinuation, insult, libel, slander, smear, stigma. • *vb* garble, lisp, mumble.

slurry *n* mud, ooze, slime.

sly *adj* artful, *inf* canny, *inf* catty, conniving, crafty, cunning, deceitful, designing, devious, disingenuous, *inf* foxy, furtive, guileful, insidious, knowing, scheming, secretive, *inf* shifty, shrewd, *inf* sneaky, *inf* snide, stealthy, surreptitious, treacherous, tricky, underhand, wily. *Opp* CANDID, OPEN.

smack *vb* pat, slap, spank. ▷ HIT.

small *adj* **1** *inf* baby, bantam, compact, concise, cramped, diminutive, *inf* dinky, dwarf, exiguous, fine, fractional, infinitesimal, lean, lilliputian, little, microscopic, midget, *inf* mini, miniature, minuscule, minute, narrow, petite, *inf* pint-sized, *inf* pocket-sized, *inf* poky, portable, pygmy, short, slender, slight, *inf* teeny, thin, tiny, toy, undergrown, undersized, *inf* wee, *inf* weeny. **2** *small helpings*. inadequate, insufficient, meagre, mean, *inf* measly, miserly, modest, niggardly, parsimonious, *inf* piddling, scanty, skimpy, stingy, ungenerous, unsatisfactory. **3** *a small problem*. inconsequential, insignificant, minor, negligible, nugatory, slim (*chance*), slight, trifling, trivial, unimportant. *Opp* BIG. **small arms** ▷ WEAPON.

small-minded *adj* bigoted, grudging, hidebound, illiberal, intolerant, narrow, narrow-minded, old-fashioned, parochial, petty, prejudiced, rigid, selfish, trivial, unimaginative. ▷ MEAN. *Opp* BROAD-MINDED.

smart *adj* **1** acute, adept, artful, astute, bright, clever, crafty, *inf* cute, discerning, ingenious, intelligent, perceptive, perspicacious, quick, quickwitted, shrewd, *sl* streetwise. *Opp* DULL. **2** *smart appearance*. bright, chic, clean, dapper, *inf* dashing, elegant, fashionable, fresh, modish, *inf* natty, neat, *inf* posh, *inf* snazzy, *Fr* soigné, spruce, stylish, tidy, trim, well-dressed, well-groomed, well-looked-after. *Opp* SCRUFFY. **3** *a smart pace.* brisk, *inf* cracking, fast, forceful, quick, rapid, *inf* rattling, speedy, swift. **4** *a smart blow.* painful, sharp, stinging, vigorous. • *vb* ▷ HURT.

smash *vb* **1** *smash to pieces.* crumple, crush, demolish, destroy, shatter, squash, wreck. ▷ BREAK. **2** *smash into a wall.* bang, bash, batter, bump, collide, crash, hammer, knock, pound, ram, slam, strike, thump, wallop. ▷ HIT.

smear *n* **1** blot, daub, mark, smudge, stain, streak. **2** *a smear on your name.* aspersion, calumny, defamation, imputation, innuendo, insinuation, libel, slander, slur, stigma, vilification. • *vb* **1** dab, daub, plaster, rub, smudge, spread, wipe. **2** *smear a reputation.* attack, besmirch, blacken, calumniate, defame, discredit, libel, malign, slander, stigmatize, tarnish, vilify.

smell *n* odour, redolence. □ [pleasant] *aroma, bouquet, fragrance, incense, nose, scent.* □ [unpleasant] *fetor, mephitis, miasma, perfume,* inf *pong, pungency, reek, stench, stink, whiff.* • *vb* **1** *inf* get a whiff of, scent, sniff. **2** *onions smell. inf* hum, *inf* pong, reek, stink, whiff.

smelling *adj* [*Smelling* is usually used in combination: sweet-smelling, etc.] **1** *pleasant-smelling.* aromatic, fragrant, musky, odoriferous, odorous, perfumed, redolent, scented, spicy. **2** *unpleasant-smelling.* fetid, foul, gamy, *inf* high, malodorous, mephitic, miasmic, musty, noisome, *inf* off, *sl* pongy, pungent, putrid, rank, reeking, rotten, smelly, stinking, *inf* whiffy. *Opp* ODOURLESS.

smelly *adj* ▷ SMELLING.

smile *n, vb* beam, grin, leer, laugh, simper, smirk, sneer.

smoke *n* **1** air pollution, exhaust, fog, fumes, gas, smog, steam, vapour. **2** cigar, cigarette, cheroot, *inf* fag, pipe, tobacco. • *vb* **1** emit smoke, fume, reek, smoulder. **2** *smoke cigars.* inhale, puff at.

smoky *adj* clouded, dirty, foggy, grimy, hazy, sooty. *Opp* CLEAR.

smooth *adj* **1** even, flat, horizontal, level, plane, regular, unbroken, unruffled. **2** *smooth sea.* calm, peaceful, placid, quiet, restful. **3** *a smooth finish.* burnished, glassy, glossy, polished, satiny, shiny, silken, silky, sleek, soft, velvety. **4** *smooth progress.* comfortable, easy, effortless, fluent, steady, uncluttered, uneventful, uninterrupted, unobstructed. **5** *a smooth taste.* agreeable, bland, mellow, mild, pleasant, soft, soothing. **6** *a smooth mixture.* creamy, flowing, runny. **7** *a smooth talker.* convincing, facile, glib, insincere, plausible, polite, self-assured, self-satisfied, slick, smug, sophisticated, suave, untrustworthy, urbane. *Opp* ROUGH. • *vb* buff up, burnish, even out, file, flatten, iron, level, level off, plane, polish, press, roll out, sand down, sandpaper.

smother *vb* **1** asphyxiate, choke, cover, kill, snuff out, stifle, strangle, suffocate, throttle. **2** ▷ SUPPRESS.

smoulder *vb* burn, smoke. **smouldering** ▷ ANGRY.

smudge *vb* blot, blur, dirty, mark, smear, stain, streak.

smug *adj* complacent, conceited, *inf* holier-than-thou, pleased, priggish, self-important, selfrighteous, self-satisfied, sleek, superior. *Opp* HUMBLE.

snack *n* bite, *inf* elevenses, *inf* nibble, refreshments. ▷ MEAL.

snack-bar *n* buffet, café, cafeteria, fast-food restaurant, transport café.

snag *n* catch, complication, difficulty, drawback, hindrance, hitch, impediment, obstacle, obstruction, problem, set-back, *inf* stumbling-block. • *vb* catch, jag, rip, tear.

snake *n* ophidian, serpent. □ *adder, anaconda, boa constrictor, cobra, copperhead, flying-snake, grass snake, mamba, python, rattlesnake, sand snake, sea snake, sidewinder, tree snake, viper.* • *vb* crawl, creep, meander, twist and turn, wander, worm, zigzag. **snaking** ▷ TWISTY.

snap *adj* ▷ SUDDEN. • *vb* **1** break, crack, fracture, give way, part, split. **2** *snap your fingers.* click, crack, pop. **3** *dog snapped at me.* bite, gnash, nip, snatch. **4** *snap orders.* bark, growl, *inf* jump down someone's throat, snarl, speak angrily. **snare** *n* ambush, booby-trap, *old use* gin, noose, springe, trap. • *vb* capture, catch, decoy, ensnare, entrap, net, trap.

snarl *vb* **1** bare the teeth, growl. **2** *snarl up rope.* confuse, entangle, jam, knot, tangle, twist.

snatch *vb* **1** catch, clutch, grab, grasp, lay hold of, pluck, seize, take, wrench away, wrest away. **2** abduct, kidnap, remove, steal.

sneak *vb* **1** creep, move stealthily, prowl, skulk, slink, stalk, steal. **2** [*inf*] *sneak on someone. sl* grass, inform (against), report, *sl* snitch, *inf* tell tales (about).

sneaking *adj* furtive, half-formed, intuitive, lurking, nagging, *inf* niggling, persistent, private, secret, uncomfortable, unconfessed, undisclosed, unproved, worrying.

sneaky *adj* cheating, contemptible, crafty, deceitful, despicable, devious, dishonest, furtive, *inf* low-down, mean, nasty, shady, *inf* shifty, sly, treacherous, underhand, unorthodox, unscrupulous, untrustworthy. *Opp* STRAIGHTFORWARD.

sneer *vb* be contemptuous, be scornful, boo, curl your lip, hiss, hoot, jeer, laugh, mock, scoff, sniff. **sneer at** ▷ DENIGRATE, RIDICULE.

sniff *vb* **1** *inf* get a whiff of, scent, smell. **2** ▷ SNIVEL. **3** ▷ SNEER.

snigger *vb* chuckle, giggle, laugh, snicker, titter.

snip *vb* clip, dock, nick, nip. ▷ CUT.

snipe *vb* fire, shoot, *inf* take potshots. **snipe at** ▷ CRITICIZE.

snippet *n* fragment, morsel, particle, scrap, shred, snatch. ▷ PIECE.

snivel *vb* blubber, cry, grizzle, mule, sob, sniff, sniffle, snuffle, whimper, whine, *inf* whinge. ▷ WEEP.

snobbish *adj* affected, condescending, disdainful, élitist, haughty, highfalutin, *inf* hoity-toity, lofty, lordly, patronizing, pompous, *inf* posh, presumptuous, pretentious, *inf* putting on airs, self-important, smug, *inf* snooty, *inf* stuck-up, supercilious, superior, *inf* toffee-nosed. ▷ CONCEITED. *Opp* UNPRETENTIOUS.

snoop *vb* be inquisitive, butt in, do detective work, interfere, intrude, investigate, meddle, *inf* nose about, pry, sneak, spy, *inf* stick your nose in.

snooper *n* busybody, detective, investigator, meddler, sneak, spy.

snout *n* face, muzzle, nose, nozzle, proboscis, trunk.

snub *vb* be rude to, brush off, cold-shoulder, disdain, humiliate, insult, offend, *inf* put someone down, rebuff, reject, scorn, *inf* squash.

snuff *vb* extinguish, put out. **snuff it** ▷ DIE. **snuff out** ▷ KILL.

snug *adj* **1** comfortable, *inf* comfy, cosy, enclosed, friendly, intimate, protected, reassuring, relaxed, relaxing, restful, safe, secure, sheltered, soft, warm. **2** *a snug fit.* close-fitting, exact, well-tailored.

soak *vb* bathe, *inf* dunk, drench, immerse, marinate, penetrate, permeate, pickle, saturate, souse, steep, submerge, wet thoroughly. **soaked, soaking** ▷ WET. **soak up** ▷ ABSORB.

soar *vb* **1** ascend, climb, float, fly, glide, hang, hover, rise, tower. **2** *prices soared.* escalate, increase, rise, rocket, shoot up, spiral.

sob *vb* blubber, cry, gasp, snivel, *inf* sob your heart out, whimper. ▷ WEEP.

sober *adj* **1** calm, clear-headed, composed, dignified, grave, in control, level-headed, lucid, peaceful, quiet, rational, sedate, sensible, serene, serious, solemn, steady, subdued, tranquil, unexciting. *Opp* SILLY. **2** *sober habits.* abstemious, moderate, *inf* on the wagon, restrained, self-controlled, staid, teetotal, temperate. *Opp* DRUNK. **3** *sober dress.* colourless, drab, dull, plain, sombre.

sociable *adj* affable, approachable, *old use* clubbable, companionable, convivial, extroverted, friendly, gregarious, hospitable, neighbourly, outgoing, warm, welcoming. ▷ SOCIAL. *Opp* UNFRIENDLY, WITHDRAWN.

social *adj* **1** civilized, collaborative, gregarious, organized. **2** *social events.* collective, communal, community, general, group, popular, public. ▷ SOCIABLE. *Opp* SOLITARY. • *n* dance, disco, *inf* do, gathering, *inf* get-together, party, reception, reunion, soirée. ▷ PARTY.

socialize *vb* associate, be sociable, entertain, fraternize, get together, *inf* go out together, join in, keep company, mix, relate.

society *n* **1** civilization, the community, culture, the human family, mankind, nation, people, the public. **2** *the society of our friends.* camaraderie, companionship, company, fellowship, friendship, togetherness. **3** *a secret society, etc.* academy, alliance, association, brotherhood, circle, club, confraternity, fraternity, group, guild, league, organization, sisterhood, sodality, sorority, union.

sofa *n* chaise longue, couch, settee, sofa bed. ▷ SEAT.

soft *adj* **1** compressible, crumbly, cushiony, elastic, flabby, flexible, floppy, limp, malleable, mushy, plastic, pliable, pliant, pulpy, spongy, springy, squashable, squashy, squeezable, supple, tender, yielding. **2** *soft ground.* boggy, marshy, muddy, sodden, waterlogged. **3** *a soft bed.* comfortable, cosy. **4** *soft texture.* downy, feathery, fleecy, fluffy, furry, satiny, silky, sleek, smooth, velvety. **5** *soft music.* dim, faint, low, mellifluous, muted, peaceful, quiet, relaxing, restful, soothing, subdued. **6** *soft breeze.* balmy, delicate, gentle, light, mild, pleasant, warm. **7** [*inf*] *a soft option.* easy, undemanding.

8 *soft feelings.* ▷ SOFT-HEARTED. *Opp* HARD, HARSH, VIOLENT.

soften *vb* **1** abate, alleviate, buffer, cushion, deaden, decrease, deflect, diminish, lower, make softer, mellow, mitigate, moderate, muffle, pacify, palliate, quell, quieten, reduce the impact of, subdue, temper, tone down, turn down. **2** dissolve, fluff up, lighten, liquefy, make softer, melt. **3** *soften in attitude.* become softer, concur, ease up, give in, give way, *inf* let up, relax, succumb, weaken, yield. *Opp* HARDEN, INTENSIFY.

soft-hearted *adj* benign, compassionate, conciliatory, easygoing, generous, indulgent, kind-hearted, *derog* lax, lenient, merciful, permissive, sentimental, *inf* soft, sympathetic, tender, tender-hearted, tolerant, understanding. ▷ KIND. *Opp* CRUEL.

soggy *adj* drenched, dripping, heavy (*soil*), saturated, soaked, sodden, *inf* sopping, wet through. ▷ WET. *Opp* DRY.

soil *n* clay, dirt, earth, ground, humus, land, loam, marl, topsoil. • *vb* befoul, besmirch, blacken, contaminate, defile, dirty, make dirty, muddy, pollute, smear, stain, sully, tarnish.

solace *n* comfort, consolation, reassurance, relief. • *vb* ▷ CONSOLE.

soldier *n* fighter, fighting man, fighting woman, *old use* man at arms, serviceman, servicewoman, warrior. □ *cadet, cavalryman, centurion, commando, conscript, guardsman, gunner, infantryman, lancer, marine, mercenary, NCO, officer, paratrooper, private, recruit, regular, rifleman, sapper, sentry, trooper,* plur *troops, warrior.* ▷ FIGHTER, RANK. • *vb* **soldier on** ▷ PERSIST.

sole *adj* exclusive, individual, lone, one, only, single, singular, solitary, unique.

solemn *adj* **1** earnest, gloomy, glum, grave, grim, long-faced, reserved, sedate, serious, sober, sombre, staid, straight-faced, thoughtful, unsmiling. *Opp* CHEERFUL. **2** *a solemn occasion.* august, awe-inspiring, awesome, ceremonial, ceremonious, dignified, ecclesiastical, formal, grand, holy, important, imposing, impressive, liturgical, momentous, pompous, religious, ritualistic, stately. *Opp* FRIVOLOUS.

solicit *vb* appeal for, ask for, beg, entreat, importune, petition, seek.

solicitous *adj* **1** attentive, caring, concerned, considerate, sympathetic. **2** ▷ ANXIOUS.

solid *adj* **1** concrete, hard, impenetrable, impermeable, rigid, unmoving. **2** *a solid crowd.* compact, crowded, dense, jammed, packed. **3** *solid gold.* authentic, genuine, pure, real, unadulterated, unalloyed, unmixed. **4** *a solid hour.* continual, continuous, entire, unbroken, uninterrupted, unrelieved, whole. **5** *solid foundations.* firm, fixed, immovable, robust, sound, stable, steady, stout, strong, sturdy, substantial, unbending, unyielding, well-made. **6** *a solid shape.* cubic, rounded, spherical, thick, three-dimensional. **7** *solid evidence.* authoritative, cogent, coherent, convincing, genuine, incontrovertible, indisputable, irrefutable, provable, proven, real, sound, tangible, weighty. **8** *solid support.* complete, dependable, effective, like-minded, reliable, stalwart, strong, trustworthy, unanimous, undivided, united, unwavering, vigorous. *Opp* FLUID, FRAGMENTARY, WEAK.

solidarity *n* accord, agreement, coherence, cohesion, concord, harmony, like-mindedness, unanimity, unity. *Opp* DISUNITY.

solidify *vb* cake, clot, coagulate, congeal, crystallize, freeze, harden, jell, set, thicken. *Opp* LIQUEFY.

soliloquy *n* monologue, speech.

solitary *adj* **1** alone, anti-social, cloistered, companionless, friendless, isolated, lonely, lonesome, reclusive, unsociable, withdrawn. **2** *a solitary survivor*. individual, one, only, single, sole. **3** *a solitary place*. desolate, distant, hidden, inaccessible, isolated, out-of-the-way, private, remote, secluded, sequestered, unfrequented, unknown. *Opp* NUMEROUS, PUBLIC, SOCIAL. • *n* anchorite, hermit, *inf* loner, recluse.

solitude *n* aloneness, friendlessness, isolation, loneliness, privacy, remoteness, retirement, seclusion.

solo *adv* alone, individually, on your own, unaccompanied.

soloist *n* performer, player, singer. ▷ MUSIC.

soluble *adj* **1** explicable, manageable, solvable, tractable, understandable. **2** *soluble in water*. dispersing, dissolving, melting. *Opp* INSOLUBLE.

solution *n* **1** answer, clarification, conclusion, denouement, elucidation, explanation, explication, key, outcome, resolution, solving, unravelling, working out. **2** *a chemical solution*. blend, compound, emulsion, infusion, mixture, suspension.

solve *vb* answer, clear up, *inf* crack, decipher, elucidate, explain, explicate, figure out, find the solution to, interpret, puzzle out, resolve, unravel, work out.

solvent *adj* creditworthy, in credit, profitable, reliable, self-supporting, solid, sound, viable. *Opp* BANKRUPT.

sombre *adj* black, bleak, cheerless, dark, dim, dismal, doleful, drab, dreary, dull, funereal, gloomy, grave, grey, joyless, lowering, lugubrious, melancholy, morose, mournful, serious, sober. ▷ SAD. *Opp* CHEERFUL.

somewhat *adv* fairly, moderately, *inf* pretty, quite, rather, *inf* sort of.

song *n* air, *inf* ditty, *inf* hit, lyric, number, tune. □ *anthem, aria, ballad, blues, calypso, cantata, canticle, carol, chant, chorus, descant, folk-song, hymn, jingle,* Ger *lied* [plur *lieder*], *lullaby, madrigal, nursery rhyme, plainsong, pop song, psalm, reggae, rock, serenade, shanty, soul, spiritual, wassail.*

sonorous *adj* deep, full, loud, powerful, resonant, resounding, reverberant, rich, ringing. *Opp* SHRILL.

soon *adv old use* anon, *inf* any minute now, before long, *inf* in a minute, presently, quickly, shortly, straight away.

sooner *adv* **1** before, earlier. **2** *sooner have tea than coffee*. preferably, rather.

soot *n* dirt, grime.

soothe *vb* allay, appease, assuage, calm, comfort, compose, ease, mollify, pacify, quiet, relieve, salve, settle, still, tranquillize.

soothing *adj* **1** balmy, balsamic, comforting, demulcent, emollient, healing, lenitive, mild, palliative. **2** *soothing music*. calming, gentle, peaceful, pleasant, reassuring, relaxing, restful, serene.

sophisticated *adj* **1** adult, *sl* cool, cosmopolitan, cultivated, cultured, elegant, fashionable, *inf* grown-up, mature, polished, *inf* posh, *derog* pretentious, refined, stylish, urbane, worldly. *Opp* UNSOPHISTICATED. **2** *sophisticated ideas.* advanced, clever, complex, complicated, elaborate, hard to understand, ingenious, intricate, involved, subtle. *Opp* PRIMITIVE, SIMPLE.

soporific *adj* boring, deadening, hypnotic, sedative, sleep-inducing, sleepy, somnolent. *Opp* LIVELY.

sorcerer *n* conjuror, enchanter, enchantress, magician, magus, medicine man, necromancer, sorceress, *old use* warlock, witch, witch-doctor, wizard.

sorcery *n* black magic, charms, conjuring, diabolism, incantations, magic, *inf* mumbo jumbo, necromancy, the occult, spells, voodoo, witchcraft, wizardry.

sordid *adj* **1** dingy, dirty, disreputable, filthy, foul, miserable, *inf* mucky, nasty, offensive, polluted, putrid, ramshackle, seamy, seedy, *inf* sleazy, *inf* slummy, squalid, ugly, unclean, undignified, unpleasant, unsanitary, wretched. *Opp* CLEAN. **2** *sordid dealings.* avaricious, base, corrupt, covetous, degenerate, despicable, dishonourable, ignoble, ignominious, immoral, mean, mercenary, rapacious, selfish, *inf* shabby, shameful, unethical, unscrupulous. *Opp* HONOURABLE.

sore *adj* **1** aching, burning, chafing, delicate, hurting, inflamed, painful, raw, red, sensitive, smarting, stinging, tender. **2** aggrieved, hurt, irked, *inf* peeved, *inf* put out, resentful, upset, vexed. ▷ ANNOYED. • *n* abrasion, abscess, boil, bruise, burn, carbuncle, gall, gathering, graze, infection, inflammation, injury, laceration, pimple, rawness, redness, scrape, spot, swelling, ulcer. ▷ WOUND. **make sore** abrade, burn, bruise, chafe, chap, gall, graze, hurt, inflame, lacerate, redden, rub.

sorrow *n* **1** affliction, anguish, dejection, depression, desolation, despair, desperation, despondency, disappointment, discontent, disgruntlement, dissatisfaction, distress, dolour, gloom, glumness, grief, heartache, heartbreak, heaviness, homesickness, hopelessness, loneliness, melancholy, misery, misfortune, mourning, sad feelings, sadness, suffering, tearfulness, tribulation, trouble, unhappiness, wistfulness, woe, wretchedness. *Opp* HAPPINESS. **2** *sorrow for wrongdoing.* apologies, guilt, penitence, regret, remorse, repentance. • *vb* agonize, be sorrowful, be sympathetic, bewail, grieve, lament, mourn, weep. *Opp* REJOICE.

sorrowful *adj* broken-hearted, dejected, disconsolate, distressed, doleful, grief stricken, heartbroken, long faced, lugubrious, melancholy, miserable, mournful, regretful, rueful, saddened, sombre, tearful, unhappy, upset, woebegone, woeful, wretched. ▷ SAD, SORRY. *Opp* HAPPY.

sorry *adj* **1** apologetic, ashamed, conscience-stricken, contrite, guilt-ridden, penitent, regretful, remorseful, repentant, shamefaced. **2** *sorry for the homeless.* compassionate, concerned, merciful, pitying, sympathetic, understanding.

sort *n* 1 brand, category, class, classification, description, form, genre, group, kind, make, mark, nature, quality, set, type, variety. 2 breed, class, family, genus, race, species, strain, stock, variety. • *vb* arrange, assort, catalogue, categorize, classify, divide, file, grade, group, order, organize, put in order, rank, systematize, tidy. *Opp* MIX. **sort out** 1 choose, *inf* put on one side, segregate, select, separate, set aside. 2 *sort out a problem.* attend to, clear up, cope with, deal with, find an answer to, grapple with, handle, manage, organize, put right, resolve, solve, straighten out, tackle.

soul *n* 1 psyche, spirit. 2 [*inf*] *poor soul!* ▷ PERSON.

soulful *adj* deeply felt, eloquent, emotional, expressive, fervent, heartfelt, inspiring, moving, passionate, profound, sincere, spiritual, stirring, uplifting, warm. *Opp* SOULLESS.

soulless *adj* cold, inhuman, insincere, mechanical, perfunctory, routine, spiritless, superficial, trite, unemotional, unfeeling, uninspiring, unsympathetic. *Opp* SOULFUL.

sound *adj* 1 durable, fit, healthy, hearty, *inf* in good shape, robust, secure, solid, strong, sturdy, tough, undamaged, uninjured, unscathed, vigorous, well, whole. 2 *sound food.* eatable, edible, fit for human consumption, good, wholesome. 3 *sound ideas.* balanced, coherent, commonsense, correct, convincing, judicious, logical, orthodox, prudent, rational, reasonable, reasoned, sane, sensible, well-founded, wise. 4 *a sound business.* dependable, established, profitable, recognized, reliable, reputable, safe, secure, trustworthy, viable. *Opp* BAD, WEAK. • *n* din, noise, resonance, timbre, tone. □ [Most of these words can be used as either *nouns* or *verbs.*] *bang, bark, bawl, bay, bellow, blare, blast, bleat, bleep, boo, boom, bray, buzz, cackle, caw, chime, chink, chirp, chirrup, chug, clack, clamour, clang, clank, clap, clash, clatter, click, clink, cluck, coo, crack, crackle, crash, creak, croak, croon, crow, crunch, cry, drone, echo, explosion, fizz, grate, grizzle, groan, growl, grunt, gurgle, hiccup, hiss, honk, hoot, howl, hum, jabber, jangle, jeer, jingle, lisp, low, miaow, moan, moo, murmur, neigh, patter, peal, ping, pip, plop, pop, purr, quack, rattle, report, reverberation, ring, roar, rumble, rustle, scream, screech, shout, shriek, sigh, sizzle, skirl, slam, slurp, smack, snap, snarl, sniff, snore, snort, sob, splutter, squawk, squeak, squeal, squelch, swish, throb, thud, thump, thunder, tick, ting, tinkle, toot, trumpet, twang, tweet, twitter, wail, warble, whimper, whine, whinny, whir, whistle, whiz, whoop, woof, yap, yell, yelp, yodel, yowl.* ▷ NOISE. • *vb* 1 become audible, be heard, echo, make a noise, resonate, resound, reverberate. 2 *sound a signal.* activate, cause, create, make, make audible, produce, pronounce, set off, utter. **sound out** check, examine, inquire into, investigate, measure, plumb, probe, research, survey, test, try.

soup *n* broth, consommé, stock.

sour *n* 1 acid, acidic, acidulous, bitter, citrus, lemony, pungent, sharp, tangy, tart, unripe, vinegary. 2 *sour milk.* bad, curdled, *inf* off, rancid, stale, turned. 3 *sour remarks.* acerbic, bad-tempered, bitter, caustic, cross, crusty, curmudgeonly, cynical, disaffected, disagreeable, grudging, grumpy, ill-natured, irritable, jaundiced, peevish, petulant, snappy, testy, unpleasant.

source *n* **1** author, begetter, cause, creator, derivation, informant, initiator, originator, root, starting-point. **2** *source of river*. head, origin, spring, start, well-head, well-spring. ▷ BEGINNING.

souvenir *n* heirloom, keepsake, memento, relic, reminder.

sovereign *adj* **1** absolute, all-powerful, dominant, highest, royal, supreme, unlimited. **2** *sovereign state*. autonomous, independent, self-governing. • *n* emperor, empress, king, monarch, prince, princess, queen. ▷ RULER.

sow *vb* broadcast, disseminate, plant, scatter, seed, spread.

space *adj* extraterrestrial, interplanetary, interstellar, orbiting. • *n* **1** emptiness, endlessness, ionosphere, infinity, stratosphere, the universe. **2** *space to move about*. *inf* elbow-room, expanse, freedom, latitude, leeway, margin, room, scope, spaciousness. **3** *an empty space*. area, blank, break, chasm, concourse, distance, duration, gap, hiatus, hole, intermission, interval, lacuna, lapse, opening, place, spell, stretch, time, vacuum, wait. • *vb space things out*. ▷ ARRANGE.

spacious *adj* ample, broad, capacious, commodious, extensive, large, open, roomy, sizeable, vast, wide. ▷ BIG. *Opp* SMALL.

span *n* breadth, compass, distance, duration, extent, interval, length, period, reach, scope, stretch, term, width. • *vb* arch over, bridge, cross, extend across, go over, pass over, reach over, straddle, stretch over, traverse.

spank *vb* slap, slipper, smack. ▷ HIT, PUNISH.

spar *vb* box, exchange blows, scrap, shadow-box. ▷ FIGHT.

spare *adj* **1** additional, auxiliary, extra, free, inessential, in reserve, leftover, odd, remaining, superfluous, supernumerary, supplementary, surplus, unnecessary, unneeded, unused, unwanted. *Opp* NECESSARY. **2** *a spare figure*. ▷ THIN. • *vb* **1** be merciful to, deliver, forgive, free, have mercy on, let go, let off, liberate, pardon, redeem, release, reprieve, save. **2** *spare money, time, etc.* afford, allow, donate, give, give up, manage, part with, provide, sacrifice. **sparing** ▷ ECONOMICAL, MISERLY.

spark *n* flash, flicker, gleam, glint, scintilla, sparkle. • *vb* **spark off** ignite, kindle. ▷ PROVOKE.

sparkle *vb* burn, coruscate, flash, flicker, gleam, glint, glitter, reflect, scintillate, shine, spark, twinkle, wink. ▷ LIGHTEN.

sparkling *adj* **1** brilliant, flashing, glinting, glittering, scintillating, shining, shiny, twinkling. ▷ BRIGHT. *Opp* DULL. **2** *sparkling drinks*. aerated, bubbling, bubbly, carbonated, effervescent, fizzy, foaming.

sparse *adj inf* few and far between, inadequate, light, little, meagre, scanty, scarce, scattered, sparing, spread out, thin, *inf* thin on the ground. *Opp* PLENTIFUL.

spartan *adj* abstemious, ascetic, austere, bare, bleak, disciplined, frugal, hard, harsh, plain, rigid, rigorous, severe, simple, stern, strict. *Opp* LUXURIOUS.

spasm *n* attack, contraction, convulsion, eruption, fit, jerk, outburst, paroxysm, seizure, *plur* throes, twitch.

spasmodic *adj inf* by fits and starts, erratic, fitful, intermittent, interrupted, irregular, jerky, occasional, *inf* on and off, periodic, sporadic. *Opp* CONTINUOUS, REGULAR.

spate *n* cataract, flood, flow, gush, inundation, onrush, outpouring, rush, torrent.

spatter *vb* bespatter, besprinkle, daub, pepper, scatter, shower, slop, speckle, splash, splatter, spray, sprinkle.

speak *vb* answer, argue, articulate, ask, communicate, complain, converse, declaim, declare, deliver a speech, discourse, ejaculate, enunciate, exclaim, express yourself, fulminate, harangue, hold a conversation, hold forth, object, *inf* pipe up, plead, pronounce words, read aloud, recite, say something, soliloquize, *inf* speechify, talk, tell, use your voice, utter, verbalize, vocalize, voice. ▷ SAY, TALK. **speak about** ▷ MENTION. **speak to** ▷ ADDRESS. **speak your mind** be honest, say what you think, speak honestly, speak out, state your opinion, voice your thoughts.

speaker *n* lecturer, mouthpiece, orator, public speaker, spokesperson.

spear *n* assegai, harpoon, javelin, lance, pike.

special *adj* **1** different, distinguished, exceptional, extraordinary, important, infrequent, momentous, notable, noteworthy, odd, *inf* out-of-the-ordinary, rare, red-letter (*day*), remarkable, significant, strange, uncommon, unconventional, unorthodox, unusual. *Opp* ORDINARY. **2** *Petrol has a special smell.* characteristic, distinctive, idiosyncratic, memorable, peculiar, singular, unique, unmistakable. **3** *my special chair.* especial, individual, particular, personal. **4** *a special tool for the job.* bespoke, proper, specific, specialized.

specialist *n* **1** authority, connoisseur, expert, fancier (*pigeon fancier*), master, professional, *inf* pundit, researcher. **2** [*medical*] consultant. ▷ MEDICINE.

speciality *n inf* claim to fame, expertise, field, forte, genius, *inf* line, specialization, special knowledge, special skill, strength, strong point, talent.

specialize *vb* **specialize in** be a specialist in, be best at, concentrate on, devote yourself to, have a reputation for.

specialized *adj* esoteric, expert, specialist, unfamiliar.

species *n* breed, class, genus, kind, race, sort, type, variety.

specific *adj* clear-cut, defined, definite, detailed, exact, explicit, express, fixed, identi-fied, individual, itemized, known, named, particular, peculiar, precise, predetermined, special, specified, unequivocal. *Opp* GENERAL.

specify *vb* be specific about, define, denominate, detail, enumerate, establish, identify, itemize, list, name, particularize, *inf* set out, spell out, stipulate.

specimen *n* exemplar, example, illustration, instance, model, pattern, representative, sample.

specious *adj* casuistic, deceptive, misleading, plausible, seductive.

speck *n* bit, crumb, dot, fleck, grain, mark, mite, *old use* mote, particle, speckle, spot, trace.

speckled *adj* blotchy, brindled, dappled, dotted, flecked, freckled, mottled, patchy, spattered (with), spotted, spotty, sprinkled (with), stippled.

spectacle *n* ceremonial, ceremony, colourfulness, display, exhibition, extravaganza, grandeur, magnificence, ostentation, pageantry, parade, pomp, show, sight, spectacular effects, splendour. **spectacles** ▷ GLASS.

spectacular *adj* beautiful, breathtaking, colourful, dramatic, elaborate, eye-catching, impressive, magnificent, *derog* ostentatious, sensational, showy, splendid, stunning.

spectator *n* plur audience, beholder, bystander, *plur* crowd, eye-witness, looker-on, observer, onlooker, passer-by, viewer, watcher, witness.

spectre *n* apparition, ghost, phantom, presentiment, vision, wraith. ▷ SPIRIT.

spectrum *n* compass, extent, gamut, orbit, range, scope, series, span, spread, sweep, variety.

speculate *vb* **1** conjecture, consider, hypothesize, make guesses, meditate, ponder, reflect, ruminate, surmise, theorize, weigh up, wonder. ▷ THINK. **2** *speculate in shares.* gamble, invest speculatively, *inf* play the market, take a chance, wager.

speculative *adj* **1** abstract, based on guesswork, conjectural, doubtful, *inf* gossipy, hypothetical, notional, suppositional, suppositious, theoretical, unfounded, uninformed, unproven, untested. *Opp* PROVEN. **2** *speculative investments. inf* chancy, *inf* dicey, *inf* dodgy, hazardous, *inf* iffy, risky, uncertain, unpredictable, unreliable, unsafe. *Opp* SAFE.

speech *n* **1** articulation, communication, declamation, delivery, diction, elocution, enunciation, expression, pronunciation, speaking, talking, using words, utterance. **2** dialect, idiolect, idiom, jargon, language, parlance, register, tongue. **3** *a public speech.* address, discourse, disquisition, harangue, homily, lecture, oration, paper, presentation, sermon, *inf* spiel, talk, tirade. **4** *speech in a play.* dialogue, lines, monologue, soliloquy.

speechless *adj* dumb, dumbfounded, dumbstruck, inarticulate, *inf* mum, mute, nonplussed, silent, thunderstruck, tongue-tied, voiceless. *Opp* TALKATIVE.

speed *n* **1** pace, rate, tempo, velocity. **2** alacrity, briskness, celerity, dispatch, expeditiousness, fleetness, haste, hurry, quickness, rapidity, speediness, swiftness. • *vb* **1** *inf* belt, *inf* bolt, *inf* bowl along, canter, career, dash, dart, flash, flit, fly, gallop, *inf* go like the wind, hasten, hurry, hurtle, make haste, move quickly, *inf* nip, *inf* put your foot down, race, run, rush, shoot, sprint, stampede, streak, tear, *inf* zoom. **2** *speed on the road.* break the speed limit, go too fast. **speed up** ▷ ACCELERATE.

speedy *adj* **1** expeditious, fast, *old use* fleet, nimble, quick, rapid, swift. **2** *a speedy exit.* hasty, hurried, immediate, precipitate, prompt, unhesitating. *Opp* SLOW.

spell *n* **1** bewitchment, charm, conjuration, conjuring, enchantment, incantation, magic formula, sorcery, witchcraft, witchery. **2** *the spell of the theatre.* allure, captivation, charm, enthralment, fascination, glamour, magic. **3** *a spell of rain.* interval, period, phase, season. **4** *a spell at the wheel.* session, stint, stretch, term, time, tour of duty, turn, watch. • *vb* augur, bode, foretell, indicate, mean, portend, presage, signal, signify, suggest. **spell out** ▷ CLARIFY.

spellbound *adj* bewitched, captivated, charmed, enchanted, enthralled, entranced, fascinated, hypnotized, mesmerized, overcome, overpowered, transported.

spend *vb* **1** *inf* blue, consume, *inf* cough up, disburse, expend, exhaust, *inf* fork out, fritter, *inf* get through, invest, *inf* lash out, pay out, *inf* shell out, *inf* splash out, *inf* splurge, squander. **2** *spend time.* devote, fill, occupy, pass, use up, waste.

spendthrift *n inf* big spender, prodigal, profligate, wasteful person, wastrel. *Opp* MISER.

sphere *n* **1** ball, globe, globule, orb, spheroid. **2** *sphere of influence.* area, department, discipline, domain, field, province, range, scope, speciality, subject, territory. **3** *social sphere.* caste, class, domain, milieu, position, rank, society, station, stratum, walk of life.

spherical *adj* ball-shaped, globe-shaped, globular, rotund, round, spheric, spheroidal.

spice *n* **1** flavouring, piquancy, relish, seasoning. □ *allspice, bayleaf, capsicum, cardamom, cassia, cayenne, chilli, cinnamon, cloves, coriander, curry powder, ginger, grains of paradise, juniper, mace, nutmeg, paprika, pepper, pimento, poppy seed, saffron, sesame, turmeric.* **2** *add spice to life.* colour, excitement, gusto, interest, *inf* lift, *inf* pep, sharpness, stimulation, vigour, zest.

spicy *adj* aromatic, fragrant, gingery, highly flavoured, hot, peppery, piquant, pungent, seasoned, spiced, tangy, zestful. *Opp* BLAND.

spike *n* barb, nail, pin, point, projection, prong, skewer, spine, stake, tine. • *vb* impale, perforate, pierce, skewer, spear, spit, stab, stick.

spill *vb* **1** overturn, slop, splash about, tip over, upset. **2** brim, flow, overflow, run, pour. **3** *lorry spilled its load.* discharge, drop, scatter, shed, tip.

spin *vb* **1** gyrate, pirouette, revolve, rotate, swirl, turn, twirl, twist, wheel, whirl. **2** *head was spinning.* be giddy, reel, suffer vertigo, swim. **spin out** ▷ PROLONG.

spindle *n* axis, axle, pin, rod, shaft.

spine *n* **1** backbone, spinal column, vertebrae. **2** *hedgehog's spines.* barb, bristle, needle, point, prickle, prong, quill, spike, spur, thorn.

spineless *adj* cowardly, craven, faint-hearted, feeble, helpless, irresolute, *inf* lily-livered, pusillanimous, *inf* soft, timid, unheroic, weedy, *inf* wimpish. ▷ WEAK. *Opp* BRAVE.

spiral *adj* cochlear, coiled, corkscrew, turning, whorled. • *n* coil, curl, helix, screw, whorl. • *vb* **1** turn, twist. **2** *spiralling prices.* ▷ FALL, RISE.

spire *n* flηche, pinnacle, steeple.

spirit *n* **1** *Lat* anima, breath, mind, psyche, soul. **2** *supernatural spirits.* apparition, *inf* bogy, demon, devil, genie, ghost, ghoul, gremlin, hobgoblin, imp, incubus, nymph, phantasm, phantom, poltergeist, *poet* shade, shadow, spectre, *inf* spook, sprite, sylph, vision, visitant, wraith, zombie. **3** *spirit of a poem.* aim, atmosphere, essence, feeling, heart, intention, meaning, mood, purpose, sense. **4** *fighting spirit.*

animation, bravery, cheerfulness, confidence, courage, daring, determination, dynamism, energy, enthusiasm, fire, fortitude, *inf* get-up-and-go, *inf* go, *inf* guts, heroism, liveliness, mettle, morale, motivation, optimism, pluck, resolve, valour, verve, vivacity, will-power, zest. **5** ▷ ALCOHOL.

spirited *adj* active, animated, assertive, brave, brisk, buoyant, courageous, daring, determined, dynamic, energetic, enterprising, enthusiastic, frisky, gallant, *inf* gutsy, intrepid, lively, mettlesome, plucky, positive, resolute, sparkling, sprightly, vigorous, vivacious. *Opp* SPIRITLESS.

spiritless *adj* apathetic, cowardly, defeatist, despondent, dispirited, dull, irresolute, lacklustre, languid, lethargic, lifeless, listless, melancholy, negative, passive, slow, unenterprising, unenthusiastic. *Opp* SPIRITED.

spiritual *adj* devotional, divine, eternal, heavenly, holy, incorporeal, inspired, other-worldly, religious, sacred, unworldly, visionary. *Opp* TEMPORAL.

spit *n* **1** dribble, saliva, spittle, sputum. • *vb* dribble, expectorate, salivate, splutter. **spit out** ▷ DISCHARGE. **spitting image** ▷ TWIN.

spite *n* animosity, animus, antagonism, *inf* bitchiness, bitterness, *inf* cattiness, gall, grudge, hate, hatred, hostility, ill-feeling, ill will, malevolence, malice, maliciousness, malignity, rancour, resentment, spleen, venom, vindictiveness. • *vb* ▷ ANNOY.

spiteful *adj* acid, acrimonious, *inf* bitchy, bitter, *inf* catty, cruel, cutting, hateful, hostile, hurtful, ill-natured, invidious, malevolent, malicious, nasty, poisonous, punitive, rancorous, resentful, revengeful, sharp, *inf* snide, sour, unforgiving, venomous, vicious, vindictive. *Opp* KIND.

splash *vb* **1** bespatter, besprinkle, shower, slop, *inf* slosh, spatter, spill, splatter, spray, sprinkle, squirt, wash. **2** *splash about in water.* bathe, dabble, paddle, wade. **3** *splash news across the front page.* blazon, display, exhibit, flaunt, *inf* plaster, publicize, show, spread. **splash out** ▷ SPEND.

splay *vb* make a V-shape, slant, spread.

splendid *adj* admirable, awe-inspiring, beautiful, brilliant, costly, dazzling, dignified, elegant, fine, first-class, glittering, glorious, gorgeous, grand, great, handsome, imposing, impressive, lavish, luxurious, magnificent, majestic, marvellous, noble, ornate, *derog* ostentatious, palatial, *inf* posh, refulgent, regal, resplendent, rich, royal, *derog* showy, spectacular, *inf* splendiferous, stately, sublime, sumptuous, *inf* super, superb, supreme, wonderful. ▷ EXCELLENT.

splendour *n* beauty, brilliance, ceremony, costliness, display, elegance, *inf* glitter, glory, grandeur, luxury, magnificence, majesty, nobility, ostentation, pomp, pomp and circumstance, refulgence, richness, show, spectacle, stateliness, sumptuousness.

splice *vb* bind, conjoin, entwine, join, knit, marry, tie together, unite.

splinter *n* chip, flake, fragment, shard, shaving, shiver, sliver. • *vb* chip, crack, fracture, shatter, shiver, smash, split. ▷ BREAK.

split *n* **1** break, chink, cleavage, cleft, crack, cranny, crevice, fissure, furrow, gash, groove, leak, opening, rent, rift, rip, rupture, slash, slit, tear. **2** breach, dichotomy, difference, dissension, divergence of opinion, division, divorce, estrangement, schism, separation. ▷ QUARREL. • *vb* **1** break up, disintegrate, divide, divorce, go separate ways, move apart, separate. **2** *split logs.* burst, chop, cleave, crack, rend, rip apart, rip open, slash, slice, slit, splinter, tear. ▷ CUT. **3** *split profits.* allocate, allot, apportion, distribute, divide, halve, share. **4** *road splits.* bifurcate, branch, diverge, fork. **split on** ▷ INFORM.

spoil *vb* **1** blight, blot, blotch, bungle, damage, deface, destroy, disfigure, *inf* dish, harm, injure, *inf* make a mess of, mar, *inf* mess up, ruin, stain, undermine, undo, upset, vitiate, worsen, wreck. *Opp* IMPROVE. **2** *food spoiled in the heat.* become useless, curdle, decay, decompose, go bad, *inf* go off, moulder, perish, putrefy, rot, *inf* turn. **3** *spoil children.* coddle, cosset, dote on, indulge, make a fuss of, mollycoddle, over-indulge, pamper.

spoken *adj* oral, unwritten, verbal, *Lat* viva voce. *Opp* WRITTEN.

spokesperson *n* mouthpiece, representative, spokesman, spokeswoman.

sponge *vb* **1** clean, cleanse, mop, rinse, sluice, swill, wash, wipe. **2** [*inf*] *sponge on friends.* be dependent (on), cadge (from), scrounge (from).

spongy *adj* absorbent, compressible, elastic, giving, porous, soft, springy, yielding. *Opp* SOLID.

sponsor *n inf* angel, backer, benefactor, donor, patron, promoter, supporter. • *vb* back, be a sponsor of, finance, fund, help, patronize, promote, subsidize, support, underwrite.

sponsorship *n* aegis, auspices, backing, benefaction, funding, guarantee, patronage, promotion, support.

spontaneous *adj* **1** *inf* ad lib, extempore, impromptu, impulsive, *inf* off-the-cuff, unplanned, unpremeditated, unprepared, unrehearsed, voluntary. **2** *a spontaneous reaction.* automatic, instinctive, instinctual, involuntary, mechanical, natural, reflex, unconscious, unconstrained, unforced, unthinking. *Opp* PREMEDITATED.

spooky *adj* creepy, eerie, frightening, ghostly, haunted, mysterious, scary, uncanny, unearthly, weird.

spool *n* bobbin, reel.

spoon *n* dessert-spoon, ladle, tablespoon, teaspoon.

spoon-feed *vb* cosset, help, indulge, mollycoddle, pamper, spoil.

spoor *n* footprints, scent, traces, track.

sporadic *adj* erratic, fitful, intermittent, irregular, occasional, periodic, scattered, separate, unpredictable.

sport *n* **1** activity, amusement, diversion, enjoyment, entertainment, exercise, fun, games, pastime, play, pleasure, recreation. □ *aerobics, angling, archery, athletics, badminton, base-ball, basketball, billiards, blood sports, bobsleigh, bowls, boxing, canoeing, climbing, cricket, croquet, cross-country, curling, darts, decathlon, discus, fishing, football, gliding, golf, gymnastics, hockey, hunting, hurdling, ice-hockey, javelin, jog-*

ging, keep-fit, lacrosse, marathon, martial arts, mountaineering, netball, orienteering, pentathlon, inf *ping-pong, polo, pool, potholing, quoits, racing, rock-climbing, roller-skating, rounders, rowing, Rugby, running, sailing, shooting, shot, show-jumping, skating, skiing, skin-diving, sky-diving, snooker, soccer, squash, street-hockey, surfing, surf-riding, swimming, table-tennis, tennis, tobogganing, trampolining, volley-ball, water-polo, water-skiing, windsurfing, winter sports, wrestling, yachting.* ▷ ATHLETICS, RACE. **2** badinage, banter, humour, jesting, joking, merriment, raillery, teasing. • *vb* **1** caper, cavort, divert yourself, frisk about, frolic, gambol, lark about, rollick, romp, skip about. **2** *sport new clothes.* display, exhibit, flaunt, show off, wear.

sporting *adj* considerate, fair, generous, good-humoured, honourable, sportsmanlike.

sportive *adj* coltish, frisky, kittenish, light-hearted, playful, waggish. ▷ SPRIGHTLY.

sportsperson *n* contestant, participant, player, sportsman, sportswoman.

sporty *adj* **1** active, athletic, energetic, fit, vigorous. **2** *sporty clothes.* casual, informal, *inf* loud, rakish, showy, *inf* snazzy.

spot *n* **1** blemish, blot, blotch, discoloration, dot, fleck, mark, patch, smudge, speck, speckle, stain, stigma. **2** *spot on the skin.* birthmark, boil, freckle, *plur* impetigo, mole, naevus, pimple, pock, pockmark, *plur* rash, sty, whitlow, *sl* zit. **3** *spots of rain.* bead, blob, drop. **4** *spot for a picnic.* locale, locality, location, neighbourhood, place, point, position, scene, setting, site, situation. **5** *an awkward spot.* difficulty, dilemma, embarrassment, mess, predicament, quandary, situation. **6** *spot of bother.* bit, small amount, *inf* smidgen. • *vb* **1** blot, discolour, fleck, mark, mottle, smudge, spatter, speckle, splash, spray, stain. **2** ▷ SEE.

spotless *adj* **1** clean, fresh, immaculate, laundered, unmarked, unspotted, white. **2** *spotless reputation.* blameless, faultless, flawless, immaculate, innocent, irreproachable, pure, unblemished, unsullied, untarnished, *inf* whiter than white.

spotty *adj* blotchy, dappled, flecked, freckled, mottled, pimply, pock-marked, pocky, spattered, speckled, speckly, *inf* splodgy, spotted.

spouse *n* better half, *old use* helpmate, husband, partner, wife.

spout *n* duct, fountain, gargoyle, geyser, jet, lip, nozzle, outlet, rose (*of watering-can*), spray, waterspout. • *vb* **1** discharge, emit, erupt, flow, gush, jet, pour, shoot, spew, spit, spurt, squirt, stream. **2** ▷ TALK.

sprawl *vb* **1** flop, lean back, lie, loll, lounge, recline, relax, slouch, slump, spread out, stretch out. **2** be scattered, branch out, spread, straggle.

spray *n* **1** drizzle, droplets, fountain, mist, shower, splash, sprinkling. **2** *spray of flowers.* arrangement, bouquet, branch, bunch, corsage, posy, sprig. **3** *spray for paint.* aerosol, atomizer, spray-gun, sprinkler, vaporizer. • *vb* diffuse, disperse, scatter, shower, spatter, splash, spread in droplets, sprinkle.

spread *n* 1 broadcasting, broadening, development, diffusion, dispensing, dispersal, dissemination, distribution, expansion, extension, growth, increase, passing on, proliferation, promotion, promulgation. 2 *spread of a bird's wings*. breadth, compass, extent, size, span, stretch, sweep. 3 ▷ MEAL. • *vb* 1 arrange, display, lay out, open out, unfold, unfurl, unroll. 2 broaden, enlarge, expand, extend, fan out, get bigger, get longer, get wider, lengthen, *inf* mushroom, proliferate, straggle, widen. 3 *spread news*. advertise, broadcast, circulate, diffuse, dispense, disperse, disseminate, distribute, divulge, give out, make known, pass on, pass round, proclaim, promote, promulgate, publicize, publish, scatter, sow, transmit. 4 *spread butter*. apply, cover a surface with, smear.

spree *n inf* binge, debauch, escapade, *inf* fling, frolic, *inf* orgy, outing, revel. ▷ REVELRY.

sprightly *adj* active, agile, animated, brisk, *inf* chipper, energetic, jaunty, lively, nimble, *inf* perky, playful, quickmoving, spirited, sportive, spry, vivacious. *Opp* LETHARGIC.

spring *n* 1 bounce, buoyancy, elasticity, give, liveliness, resilience. 2 *clock spring*. coil, mainspring. 3 *spring of water*. fount, fountain, geyser, source (*of river*), spa, well, well-spring. • *vb* bounce, bound, hop, jump, leap, pounce, vault. **spring from** come from, derive from, proceed from, stem from. **spring up** appear, arise, burst forth, come up, develop, emerge, germinate, grow, shoot up, sprout.

springy *adj* bendy, elastic, flexible, pliable, resilient, spongy, stretchy, supple. *Opp* RIGID.

sprinkle *vb* drip, dust, pepper, scatter, shower, spatter, splash, spray, strew.

sprint *vb* dash, *inf* hare, race, speed, *inf* tear. ▷ RUN.

sprout *n* bud, shoot. • *vb* bud, come up, develop, emerge, germinate, grow, shoot up, spring up.

spruce *adj* clean, dapper, elegant, groomed, *inf* natty, neat, *inf* posh, smart, tidy, trim, well-dressed, well-groomed, *inf* well-turned-out. *Opp* SCRUFFY. • *vb* **spruce up** ▷ TIDY.

spur *n* 1 encouragement, goad, impetus, incentive, incitement, inducement, motivation, motive, prod, prompting, stimulus, urging. 2 *motorway spur*. branch, projection. • *vb* animate, egg on, encourage, impel, incite, motivate, pressure, pressurize, prick, prod, prompt, provide an incentive, stimulate, urge.

spurn *vb* disown, give (someone) the cold shoulder, jilt, rebuff, reject, renounce, repel, repudiate, repulse, shun, snub, turn your back on.

spy *n* contact, double agent, fifth columnist, *sl* grass, infiltrator, informant, informer, *inf* mole, private detective, secret agent, snooper, stool-pigeon, undercover agent. • *vb* 1 be a spy, be engaged in spying, eavesdrop, gather intelligence, inform, *inf* snoop. 2 ▷ SEE. **spy on** keep under surveillance, *inf* tail, trail, watch.

spying *n* counter-espionage, detective work, eavesdropping, espionage, intelligence, snooping, surveillance.

squabble *vb* argue, bicker, clash, *inf* row, wrangle. ▷ QUARREL.

squalid *adj* 1 dingy, dirty, disgusting, filthy, foul, mean, mucky, nasty, poverty-stricken, repulsive, run-down, *inf* sleazy, slummy, sordid, ugly, uncared-for, unpleasant, unsalubrious, wretched. *Opp* CLEAN. 2 *squalid behaviour.* corrupt, degrading, dishonest, dishonourable, disreputable, immoral, scandalous, *inf* shabby, shameful, unethical, unworthy. *Opp* HONOURABLE.

squander *vb inf* blow, *inf* blue, dissipate, *inf* fritter, misuse, spend unwisely, *inf* splurge, use up, waste. *Opp* SAVE.

square *adj* 1 perpendicular, rectangular, right-angled. 2 *a square deal. inf* above-board, decent, equitable, ethical, fair, honest, honourable, proper, *inf* right and proper, *inf* straight. • *n* 1 piazza, plaza. 2 [*inf*] *an old-fashioned square.* bourgeois, conformist, conservative, conventional person, die-hard, *inf* fuddy-duddy, *inf* old fogy, *inf* stick-in-the-mud, traditionalist. • *vb square an account.* ▷ SETTLE. **squared** chequered, criss-crossed, marked in squares.

squash *vb* 1 compress, crumple, crush, flatten, mangle, mash, pound, press, pulp, smash, stamp on, tamp down, tread on. 2 *squash into a room.* cram, crowd, pack, push, ram, shove, squeeze, stuff, thrust, wedge. 3 *squash an uprising.* control, put down, quash, quell, repress, suppress. 4 *squash with a look.* humiliate, *inf* put down, silence, snub.

squashy *adj* mashed up, mushy, pulpy, shapeless, soft, spongy, squelchy, yielding. *Opp* FIRM.

squat *adj* burly, dumpy, plump, podgy, short, stocky, thick, thickset. *Opp* TALL. • *vb* crouch, sit.

squeamish *adj inf* choosy, dainty, fastidious, finicky, over-scrupulous, particular, *inf* pernickety, prim, *inf* prissy, prudish, scrupulous.

squeeze *vb* 1 clasp, compress, crush, embrace, enfold, exert pressure on, flatten, grip, hug, mangle, pinch, press, squash, stamp on, tread on, wring. 2 cram, crowd, pack, push, ram, shove, squash, stuff, tamp, thrust, wedge. **squeeze out** expel, extrude, force out.

squirm *vb* twist, wriggle, writhe.

squirt *vb* ejaculate, eject, gush, jet, send out, shoot, spit, spout, splash, spray, spurt.

stab *n* 1 blow, cut, jab, prick, puncture, thrust, wound, wounding. 2 *stab of pain.* ▷ PAIN. • *vb* bayonet, cut, injure, jab, lance, perforate, pierce, puncture, skewer, spike, stick, thrust, transfix, wound. **have a stab at** ▷ TRY.

stability *n* balance, constancy, durability, equilibrium, firmness, immutability, permanence, reliability, solidity, soundness, steadiness, strength. *Opp* INSTABILITY.

stabilize *vb* balance, become stable, give stability to, keep upright, make stable, settle. *Opp* UPSET.

stable *adj* 1 balanced, firm, fixed, solid, sound, steady, strong, sturdy. 2 constant, continuing, durable, established, immutable, lasting, long-lasting, permanent, predictable, resolute, steadfast, unchanging, unwavering. 3 *a stable personality.* balanced, even-tempered, reasonable, sane, sensible. *Opp* UNSTABLE.

stack *n* 1 accumulation, heap, hill, hoard, mound, mountain, pile, quantity, stock, stockpile, store. 2 chimney, pillar, smokestack. 3 *stack of hay. old use* cock, haycock, haystack, rick, stook. • *vb* accumulate, amass, assemble, build up, collect, gather, heap, load, mass, pile, *inf* stash away, stockpile.

stadium *n* amphitheatre, arena, ground, sports-ground.

staff *n* 1 baton, cane, crook, crosier, flagstaff, pike, pole, rod, sceptre, shaft, stake, standard, stave, stick, token, wand. 2 *staff of a business.* assistants, crew, employees, *old use* hands, personnel, officers, team, workers, workforce. • *vb* man, provide with staff, run.

stage *n* 1 apron, dais, performing area, platform, podium, proscenium, rostrum. 2 *stage of a journey.* juncture, leg, phase, period, point, time. • *vb* arrange, *inf* get up, mount, organize, perform, present, produce, *inf* put on, set up, stage-manage.

stagger *vb* 1 falter, lurch, pitch, reel, rock, stumble, sway, teeter, totter, walk unsteadily, waver, wobble. 2 *price staggered us.* alarm, amaze, astonish, astound, confuse, dismay, dumbfound, flabbergast, shake, shock, startle, stun, stupefy, surprise, worry.

stagnant *adj* motionless, sluggish, stale, standing, static, still, without movement. *Opp* MOVING.

stagnate *vb* achieve nothing, become stale, be stagnant, degenerate, deteriorate, idle, languish, stand still, stay still, vegetate. *Opp* PROGRESS.

stain *n* 1 blemish, blot, blotch, discoloration, mark, smear, smutch, speck, spot. 2 *a wood stain.* colouring, dye, paint, pigment, tinge, tint, varnish. • *vb* 1 blacken, blemish, blot, contaminate, dirty, discolour, make dirty, mark, smudge, soil, tarnish. 2 *stain your reputation.* besmirch, damage, defile, disgrace, shame, spoil, sully, taint. 3 *stain wood.* colour, dye, paint, tinge, tint, varnish.

stair *n* riser, step, tread. **stairs** escalator, flight of stairs, staircase, stairway, steps.

stake *n* 1 paling, palisade, pike, pile, pillar, pole, post, rod, sceptre, shaft, stake, standard, stave, stick, upright. 2 *a gambler's stake.* bet, pledge, wager. • *vb* 1 fasten, hitch, secure, tether, tie up. 2 *stake a claim.* establish, put on record, state. 3 *stake my life on it.* bet, chance, gamble, hazard, risk, venture, wager. **stake out** define, delimit, demarcate, enclose, fence in, mark off, outline.

stale *adj* 1 dry, hard, limp, mouldy, musty, *inf* off, old, *inf* past its best, tasteless. 2 *stale ideas.* banal, clichéd, familiar, hackneyed, old-fashioned, out-of-date, overused, stock, threadbare, *inf* tired, trite, uninteresting, unoriginal, worn out. *Opp* FRESH.

stalemate *n* deadlock, impasse, standstill.

stalk *n* branch, shaft, shoot, stem, trunk, twig. • *vb* 1 chase, dog, follow, haunt, hound, hunt, pursue, shadow, tail, track, trail. 2 *stalk about.* prowl, rove, stride, strut. ▷ WALK.

stall *n* booth, compartment, kiosk, stand, table. • *vb* be obstructive, delay, hang back, haver, hesitate, pause, *inf* play for time, postpone, prevaricate, procrastinate, put off, stonewall, stop, temporize, waste time.

stalwart *adj* courageous, dependable, determined, faithful, indomitable, intrepid, redoubtable, reliable, resolute, robust, staunch, steadfast, sturdy, tough, trustworthy, valiant. ▷ BRAVE, STRONG. *Opp* WEAK.

stamina *n* endurance, energy, *inf* grit, indomitability, resilience, staunchness, staying power, *inf* stickability.

stammer *vb* falter, hesitate, hem and haw, splutter, stumble, stutter. ▷ TALK.

stamp *n* **1** brand, die, hallmark, impression, imprint, print, punch, seal. **2** *stamp of genius.* characteristic, mark, sign. **3** *stamp on a letter.* franking, postage stamp. • *vb* **1** bring down, strike, thump. **2** *stamp a mark.* brand, emboss, engrave, impress, imprint, label, mark, print, punch. **stamp on** ▷ SUPPRESS. **stamp out** ▷ ELIMINATE.

stampede *n* charge, dash, flight, panic, rout, rush, sprint. • *vb* **1** bolt, career, charge, dash, gallop, panic, run, rush, sprint, *inf* take to your heels, tear. **2** *stampede cattle.* frighten, panic, rout, scatter.

stand *n* **1** base, pedestal, rack, support, tripod, trivet. **2** booth, kiosk, stall. **3** grandstand, terraces. • *vb* **1** arise, get to your feet, get up, rise. **2** *Stand it on the floor.* arrange, deposit, erect, locate, place, position, put up, set up, situate, station, upend. **3** *Trees stand along the avenue.* be, be situated, exist. **4** *My offer stands.* be unchanged, continue, remain valid, stay. **5** *Can't stand onions.* abide, bear, endure, put up with, suffer, tolerate, *inf* wear. **stand by** ▷ SUPPORT. **stand for** ▷ SYMBOLIZE. **stand in for** ▷ DEPUTIZE. **stand out** ▷ SHOW. **stand up for** ▷ PROTECT, SUPPORT. **stand up to** ▷ RESIST.

standard *adj* accepted, accustomed, approved, average, basic, classic, common, conventional, customary, definitive, established, everyday, familiar, habitual, normal, official, ordinary, orthodox, popular, prevailing, prevalent, recognized, regular, routine, set, staple (*diet*), stock, traditional, typical, universal, usual. *Opp* UNUSUAL. • *n* **1** archetype, benchmark, criterion, example, exemplar, gauge, grade, guide, guideline, ideal, level of achievement, measure, measurement, model, paradigm, pattern, requirement, rule, sample, specification, touchstone, yardstick. **2** average, level, mean, norm. **3** *standard of a regiment.* banner, colours, ensign, flag, pennant. **4** *lamp standard.* column, pillar, pole, post, support, upright. **standards** ▷ MORALITY.

standardize *vb* average out, conform to a standard, equalize, homogenize, normalize, regiment, stereotype, systematize.

standoffish *adj* aloof, antisocial, cold, cool, distant, frosty, haughty, remote, reserved, reticent, retiring, secretive, self-conscious, *inf* snooty, taciturn, unapproachable, uncommunicative, unforthcoming, unfriendly, unsociable, withdrawn. *Opp* FRIENDLY.

standpoint *n* angle, attitude, belief, opinion, perspective, point of view, position, stance, vantage point, view, viewpoint.

standstill *n inf* dead end, deadlock, halt, *inf* hold-up, impasse, jam, stalemate, stop, stoppage.

staple *adj* basic, chief, important, main, principal. ▷ STANDARD.

star *n* **1** celestial body, sun. □ *asteroid, comet, evening star,* *falling star, lodestar, morning star, nova, shooting star, supernova.* 2 asterisk, pentagram. 3 *TV star.* attraction, big name, celebrity, *It* diva, *inf* draw, idol, leading lady, leading man, personage, *It* prima donna, starlet, superstar. ▷ PERFORMER.

starchy *adj* aloof, conventional, formal, prim, stiff. ▷ UNFRIENDLY.

stare *vb* gape, *inf* gawp, gaze, glare, goggle, look fixedly, peer. **stare at** contemplate, examine, eye, scrutinize, study, watch.

stark *adj* 1 austere, bare, bleak, depressing, desolate, dreary, gloomy, grim. 2 *stark contrast.* absolute, clear, complete, obvious, perfect, plain, sharp, sheer, thoroughgoing, total, unqualified, utter.

start *n* 1 beginning, birth, commencement, creation, dawn, establishment, founding, fount, inauguration, inception, initiation, institution, introduction, launch, onset, opening, origin, outset, point of departure, setting out, spring, springboard. *Opp* FINISH. 2 *an unfair start.* advantage, edge, head-start, opportunity. 3 *bank-loan gave me a start.* assistance, backing, financing, help, *inf* leg-up, *inf* send-off, sponsorship. 4 *a nasty start.* jump, shock, surprise. • *vb* 1 depart, embark, *inf* get going, *inf* get under way, *sl* hit the road, *inf* kick off, leave, move off, proceed, set off, set out. 2 *start something.* activate, beget, begin, commence, create, embark on, engender, establish, found, *inf* get cracking on, *inf* get off the ground, *inf* get the ball rolling, give birth to, inaugurate, initiate, instigate, institute, introduce, launch, open, originate, pioneer, set in motion, set up. *Opp* FINISH. 3 *start at sudden noise.* blench, draw back, flinch, jerk, jump, quail, recoil, shy, spring up, twitch, wince. **make someone start** ▷ STARTLE.

startle *vb* agitate, alarm, catch unawares, disturb, frighten, give you a start, jolt, make you jump, make you start, scare, shake, shock, surprise, take aback, take by surprise, upset. **startling** ▷ SURPRISING.

starvation *n* deprivation, famine, hunger, malnutrition, undernourishment, want.

starve *vb* die of starvation, go hungry, go without, perish. **starve yourself** diet, fast, go on hunger strike, refuse food. **starving** ▷ HUNGRY.

state *n* 1 *plur* circumstances, condition, fitness, health, mood, *inf* shape, situation. 2 agitation, excitement, *inf* flap, panic, plight, predicament, *inf* tizzy. 3 *a sovereign state.* land, nation. ▷ COUNTRY. • *vb* affirm, announce, assert, asseverate, aver, communicate, declare, express, formulate, proclaim, put into words, report, specify, submit, testify, voice. ▷ SAY, SPEAK, TALK.

stately *adj* august, dignified, distinguished, elegant, formal, grand, imperial, imposing, impressive, lofty, majestic, noble, pompous, regal, royal, solemn, splendid, striking. *Opp* INFORMAL. **stately home** ▷ MANSION.

statement *n* account, affirmation, announcement, annunciation, assertion, bulletin, comment, communication, communiqué, declaration, disclosure, explanation, message, notice, proclamation, proposition, report, testament, testimony, utterance.

statesman *n* diplomat, politician.

static *adj* constant, fixed, immobile, immovable, inert, invariable, motionless, passive, stable, stagnant, stationary, steady, still, unchanging, unmoving. *Opp* MOBILE, VARIABLE.

station *n* **1** calling, caste, class, degree, employment, level, location, occupation, place, position, post, rank, situation, standing, status. **2** *fire station.* base, depot, headquarters, office. **3** *radio station.* channel, company, transmitter, wavelength. **4** *railway station.* halt, platform, stopping-place, terminus, train station. • *vb* assign, garrison, locate, place, position, put, site, situate, spot, stand.

stationary *adj* at a standstill, at rest, halted, immobile, immovable, motionless, parked, pausing, standing, static, still, stock-still, unmoving. *Opp* MOVING.

stationery *n* paper, office supplies, writing materials.

statistics *n* data, figures, information, numbers.

statue *n* carving, figure, statuette. ▷ SCULPTURE.

statuesque *adj* dignified, elegant, imposing, impressive, poised, stately, upright.

stature *n* **1** build, height, size, tallness. **2** *artist of international stature.* esteem, greatness, recognition. ▷ STATUS.

status *n* class, degree, eminence, grade, importance, level, position, prestige, prominence, rank, reputation, significance, standing, station, stature, title.

staunch *adj* ▷ STEADFAST.

stay *n* **1** holiday, *old use* sojourn, stop, stop-over, visit. **2** *stay of execution.* ▷ DELAY. **3** ▷ SUPPORT. • *vb* **1** *old use* bide, carry on, continue, endure, *inf* hang about, hold out, keep on, last, linger, live on, loiter, persist, remain, survive, *old use* tarry, wait. **2** *stay in a hotel. old use* abide, be accommodated, be a guest, be housed, board, dwell, live, lodge, reside, settle, *old use* sojourn, stop, visit. **3** *stay judgement.* ▷ DELAY.

steadfast *adj* committed, constant, dedicated, dependable, determined, devoted, faithful, firm, loyal, patient, persevering, reliable, resolute, resolved, single-minded, sound, stalwart, staunch, steady, true, trustworthy, trusty, unchanging, unfaltering, unflinching, unswerving, unwavering. ▷ STRONG. *Opp* UNRELIABLE.

steady *adj* **1** balanced, confident, fast, firm, immovable, poised, safe, secure, settled, solid, stable, substantial. **2** *a steady flow.* ceaseless, changeless, consistent, constant, continuous, dependable, endless, even, incessant, invariable, never-ending, nonstop, perpetual, persistent, regular, reliable, repeated, rhythmic, *inf* round-the-clock, unbroken, unchanging, undeviating, unfaltering, unhurried, uniform, uninterrupted, unrelieved, unremitting, unvarying. *Opp* UNSTEADY. **3** ▷ STEADFAST. • *vb* **1** balance, brace, hold steady, keep still, make steady, secure, stabilize, support. **2** *steady your nerves.* calm, control, soothe, tranquillize.

steal *vb* **1** annex, appropriate, arrogate, burgle, commandeer, confiscate, embezzle, expropriate, *inf* filch, hijack, *inf* knock off, *inf* lift, loot, *inf* make off with, misappropriate, *inf* nick, peculate, pick pockets, pilfer, pillage, *inf* pinch, pirate, plagiarize, plunder, poach, purloin, *inf* rip you off, rob, seize, shop-lift, *inf* sneak, *inf* snitch, *inf* swipe, take, thieve, usurp, walk off with. **2** *steal quietly upstairs*. creep, move stealthily, slink, slip, sneak, tiptoe.

stealing *n* robbery, theft, thieving. □ *break-in, burglary, embezzlement, fraud, hijacking, housebreaking, larceny, looting, misappropriation, mugging, peculation, pilfering, pillage, piracy, plagiarism, plundering, poaching, purloining, scrumping, shop-lifting.*

stealthy *adj* clandestine, concealed, covert, disguised, furtive, imperceptible, inconspicuous, quiet, secret, secretive, *inf* shifty, sly, *inf* sneaky, surreptitious, underhand, unobtrusive. *Opp* BLATANT.

steam *n* condensation, haze, mist, smoke, vapour.

steamy *adj* **1** blurred, clouded, cloudy, fogged over, foggy, hazy, misted over, misty. **2** *a steamy atmosphere*. close, damp, humid, moist, muggy, *inf* sticky, sultry, sweaty, sweltering. **3** [*inf*] *steamy sex scenes*. ▷ SEXY.

steep *adj* **1** abrupt, bluff, headlong, perpendicular, precipitous, sharp, sheer, sudden, vertical. *Opp* GRADUAL. **2** *steep prices*. ▷ EXPENSIVE. • *vb* ▷ SOAK.

steeple *n* pinnacle, point, spire.

steer *vb* be at the wheel, control, direct, drive, guide, navigate, pilot. **steer clear of** ▷ AVOID.

stem *n* peduncle, shoot, stalk, stock, trunk, twig. • *vb* arise, come, derive, develop, emanate, flow, issue, originate, proceed, result, spring, sprout. **2** *to stem the flow*. ▷ CHECK.

stench *n* mephitis, *inf* pong, reek, stink. ▷ SMELL.

step *n* **1** footfall, footstep, pace, stride, tread. **2** doorstep, rung, stair, tread. **3** *a step forward*. advance, move, movement, progress, progression. **4** *step in a process*. action, initiative, manoeuvre, measure, phase, procedure, stage. • *vb* put your foot, stamp, stride, trample, tread. ▷ WALK. **steps** ladder, stairs, staircase, stairway, stepladder. **step down** ▷ RESIGN. **step in** ▷ ENTER, INTERVENE. **step on it** ▷ HURRY. **step up** ▷ INCREASE. **take steps** ▷ BEGIN.

stereoscopic *adj* solid-looking, three-dimensional, *inf* 3-D.

stereotype *n* formula, model, pattern, stereotyped idea.

stereotyped *adj* clichéd, conventional, formalized, hackneyed, predictable, standard, standardized, stock, typecast, unoriginal.

sterile *adj* **1** arid, barren, childless, dry, fruitless, infertile, lifeless, unfruitful, unproductive. **2** *sterile bandage*. antiseptic, aseptic, clean, disinfected, germ-free, hygienic, pure, sanitary, sterilized, uncontaminated, uninfected, unpolluted. **3** *a sterile attempt*. abortive, fruitless, hopeless, pointless, unprofitable, useless. *Opp* FERTILE, FRUITFUL, SEPTIC.

sterilize *vb* **1** clean, cleanse, decontaminate, depurate, disinfect, fumigate, make sterile, pasteurize, purify. **2** *sterilize animals*. castrate, caponize, emasculate, geld, neuter, perform a vasectomy on, spay, vasectomize.

stern *adj* adamant, austere, authoritarian, critical, dour, forbidding, frowning, grim, hard, harsh, inflexible, obdurate, resolute, rigid, rigorous, severe, strict, stringent, tough, unbending, uncompromising, unrelenting, unremitting. ▷ SERIOUS. *Opp* SOFT-HEARTED. • *n stern of ship*. aft, back, rear end.

stew *n* casserole, goulash, hash, hot-pot, ragout. • *vb* boil, braise, casserole, simmer. ▷ COOK.

steward, stewardess *ns* 1 attendant, waiter. ▷ SERVANT. 2 marshal, officer, official.

stick *n* branch, stalk, twig. □ *bar, baton, cane, club, hockey-stick, pike, pole, rod, staff, stake, walking-stick, wand*. • *vb* **1** bore, dig, impale, jab, penetrate, pierce, pin, poke, prick, prod, punch, puncture, run through, spear, spike, spit, stab, thrust, transfix. **2** *stick with glue*. adhere, affix, agglutinate, bind, bond, cement, cling, coagulate, fuse together, glue, gum, paste, solder, weld. ▷ FASTEN. **3** *stick in your mind*. be fixed, continue, endure, keep on, last, linger, persist, remain, stay. **4** *stick in mud*. become trapped, get bogged down, seize up, jam, wedge. **5** ▷ TOLERATE. **stick at** ▷ PERSIST. **stick in** ▷ PENETRATE. **stick out** ▷ PROTRUDE. **stick together** ▷ UNITE. **stick up** ▷ PROTRUDE. **stick up for** ▷ DEFEND. **stick with** ▷ SUPPORT.

sticky *adj* **1** adhesive, glued, gummed, self-adhesive. **2** *sticky paint*. gluey, glutinous, *inf* gooey, gummy, tacky, viscous. **3** *sticky weather*. clammy, close, damp, dank, humid, moist, muggy, steamy, sultry, sweaty. *Opp* DRY.

stiff *adj* **1** compact, dense, firm, hard, heavy, inelastic, inflexible, rigid, semi-solid, solid, solidified, thick, tough, unbending, unyielding, viscous. **2** *stiff joints*. arthritic, immovable, painful, paralysed, rheumatic, taut, tight. **3** *a stiff task*. arduous, challenging, difficult, exacting, exhausting, hard, laborious, tiring, tough, uphill. **4** *stiff opposition*. determined, dogged, obstinate, powerful, resolute, stubborn, unyielding, vigorous. **5** *a stiff manner*. artificial, awkward, clumsy, cold, forced, formal, graceless, haughty, inelegant, laboured, mannered, pedantic, self-conscious, standoffish, starchy, stilted, *inf* stuffy, tense, turgid, ungainly, unnatural, wooden. **6** *stiff penalties*. cruel, drastic, excessive, harsh, hurtful, merciless, pitiless, punishing, punitive, relentless, rigorous, severe, strict. **7** *stiff wind*. brisk, fresh, strong. **8** *stiff drink*. alcoholic, potent, strong. *Opp* EASY, RELAXED, SOFT.

stiffen *vb* become stiff, clot, coagulate, congeal, dry out, harden, jell, set, solidify, thicken, tighten, toughen.

stifle *vb* **1** asphyxiate, choke, smother, strangle, suffocate, throttle. **2** *stifle laughter*. check, control, curb, dampen, deaden, keep back, muffle, restrain, suppress, withhold. **3** *stifle free speech*. crush, destroy, extinguish, kill off, quash, repress, silence, stamp out, stop.

stigma *n* blot, brand, disgrace, dishonour, mark, reproach, shame, slur, stain, taint.

stigmatize *vb* brand, condemn, defame, denounce, disparage, label, mark, pillory, slander, vilify.

still *adj* at rest, calm, even, flat, hushed, immobile, inert, lifeless, motionless, noiseless, pacific, peaceful, placid, quiet, restful, serene, silent, smooth, soundless, stagnant, static, stationary, tranquil, unmoving, unruffled, untroubled, windless. *Opp* ACTIVE,

NOISY. • *vb* allay, appease, assuage, calm, lull, make still, pacify, quieten, settle, silence, soothe, subdue, suppress, tranquillize. *Opp* AGITATE.

stimulant *n* anti-depressant, drug, *inf* pick-me-up, restorative, *inf* reviver, *inf* shot in the arm, tonic. ▷ STIMULUS.

stimulate *vb* activate, arouse, awaken, cause, encourage, excite, fan, fire, foment, galvanize, goad, incite, inflame, inspire, instigate, invigorate, kindle, motivate, prick, prompt, provoke, quicken, rouse, set off, spur, stir up, titillate, urge, whet. *Opp* DISCOURAGE.

stimulating *adj* arousing, challenging, exciting, exhilarating, inspirational, inspiring, interesting, intoxicating, invigorating, provocative, provoking, rousing, stirring, thought-provoking, titillating. *Opp* UNINTERESTING.

stimulus *n* challenge, encouragement, fillip, goad, incentive, inducement, inspiration, prompting, provocation, spur, stimulant. *Opp* DISCOURAGEMENT.

sting *n* bite, prick, stab. ▷ PAIN. • *vb* **1** bite, nip, prick, wound. **2** smart, tingle. ▷ HURT.

stingy *adj* **1** avaricious, cheese-paring, close, close-fisted, covetous, mean, mingy, miserly, niggardly, parsimonious, penny-pinching, tight-fisted, ungenerous. **2** *stingy helpings*. inadequate, insufficient, meagre, *inf* measly, scanty. ▷ SMALL. *Opp* GENEROUS.

stink *n, vb* ▷ SMELL.

stipulate *vb* demand, insist on, make a stipulation, require, specify.

stipulation *n* condition, demand, prerequisite, proviso, requirement, specification.

stir *n* ▷ COMMOTION. • *vb* **1** agitate, beat, blend, churn, mingle, mix, move about, scramble, whisk. **2** *stir from sleep*. arise, bestir yourself, *inf* get a move on, *inf* get going, get up, move, rise, *inf* show signs of life, *inf* stir your stumps. **3** *stir emotions*. activate, affect, arouse, awaken, challenge, disturb, electrify, excite, exhilarate, fire, impress, inspire, kindle, move, resuscitate, revive, rouse, stimulate, touch, upset.

stirring *adj* affecting, arousing, challenging, dramatic, electrifying, emotional, emotion-charged, emotive, exciting, exhilarating, heady, impassioned, inspirational, inspiring, interesting, intoxicating, invigorating, moving, provocative, provoking, rousing, spirited, stimulating, stirring, thought-provoking, thrilling, titillating, touching. *Opp* UNEXCITING.

stitch *vb* darn, mend, repair, sew, tack.

stock *adj* accustomed, banal, clichéd, common, commonplace, conventional, customary, expected, hackneyed, ordinary, predictable, regular, routine, run-of-the-mill, set, standard, staple, stereotyped, *inf* tired, traditional, trite, unoriginal, usual. *Opp* UNEXPECTED. • *n* **1** cache, hoard, reserve, reservoir, stockpile, store, supply. **2** *stock of a shop*. commodities, goods, merchandise, range, wares. **3** *farm stock*. animals, beasts, cattle, flocks, herds, livestock. **4** *ancient stock*. ancestry, blood, breed, descent, dynasty, extraction, family, forebears, genealogy, line, lineage, parentage, pedigree. **5** *meat stock.* broth, soup. • *vb* carry, deal in, handle, have available, *inf* keep, keep in stock, market, offer, provide, sell, supply, trade in. **out of stock** sold out, unavailable. **take stock** ▷ REVIEW.

stockade *n* fence, paling, palisade, wall.

stockings *n* nylons, panti-hose, socks, tights.

stockist *n* merchant, retailer, seller, shopkeeper, supplier.

stocky *adj* burly, compact, dumpy, heavy-set, short, solid, squat, stubby, sturdy, thickset. *Opp* THIN.

stodgy *adj* **1** filling, heavy, indigestible, lumpy, soggy, solid, starchy. *Opp* SUCCULENT. **2** *a stodgy lecture.* boring, dull, ponderous, *inf* stuffy, tedious, tiresome, turgid, unexciting, unimaginative, uninteresting. *Opp* LIVELY.

stoical *adj* calm, cool, disciplined, impassive, imperturbable, long-suffering, patient, philosophical, phlegmatic, resigned, stolid, uncomplaining, unemotional, unexcitable, *inf* unflappable. *Opp* EXCITABLE.

stoke *vb* fuel, keep burning, mend, put fuel on, tend.

stole *n* cape, shawl, wrap.

stolid *adj* bovine, dull, heavy, immovable, impassive, lumpish, phlegmatic, unemotional, unexciting, unimaginative, wooden. ▷ STOICAL. *Opp* LIVELY.

stomach *n* abdomen, belly, *inf* guts, *inf* insides, *derog* paunch, *derog* pot, *inf* tummy. • *vb* ▷ TOLERATE.

stomach-ache *n* colic, *inf* collywobbles, *inf* gripes, *inf* tummy-ache.

stone *n* **1** boulder, cobble, *plur* gravel, pebble, rock, *plur* scree. ▷ ROCK. **2** block, flagstone, sett, slab. **3** *a memorial stone.* gravestone, headstone, memorial, monolith, obelisk, tablet. **4** *precious stone.* ▷ JEWEL. **5** *stone in fruit.* pip, pit, seed.

stony *adj* **1** pebbly, rocky, rough, shingly. **2** *stony silence.* adamant, chilly, cold, coldhearted, expressionless, frigid, hard, *inf* hard-boiled, heartless, hostile, icy, indifferent, insensitive, merciless, pitiless, steely, stony-hearted, uncaring, unemotional, unfeeling, unforgiving, unfriendly, unresponsive, unsympathetic.

stooge *n* butt, dupe, *inf* fall-guy, lackey, puppet.

stoop *vb* **1** bend, bow, crouch, duck, hunch your shoulders, kneel, lean, squat. **2** condescend, degrade yourself, deign, humble yourself, lower yourself, sink.

stop *n* **1** ban, cessation, close, conclusion, end, finish, halt, shutdown, standstill, stoppage, termination. **2** *a stop for refreshments.* break, destination, pause, resting-place, stage, station, stopover, terminus. **3** *a stop at a hotel.* holiday, *old use* sojourn, stay, vacation, visit. • *vb* **1** break off, call a halt to, cease, conclude, cut off, desist from, discontinue, end, finish, halt, *inf* knock off, leave off, *inf* pack in, pause, quit, refrain from, rest from, suspend, terminate. **2** *stop the flow.* bar, block, check, curb, cut off, delay, frustrate, halt, hamper, hinder, immobilize, impede, intercept, interrupt, *inf* nip in the bud, obstruct, put a stop to, stanch, staunch, stem, suppress, thwart. **3** *stop in a hotel.* be a guest, have a holiday, *old use* sojourn, spend time, stay, visit. 4 *stop a gap*. *inf* bung up, close, fill in, plug, seal. 5 *stop a thief*. arrest, capture, catch, detain, hold, seize. 6 *the rain stopped*. be over, cease, come to an end, finish, peter out. 7 *the bus stopped*. come to rest, draw up, halt, pull up.

stopper *n* bung, cork, plug.

store *n* 1 accumulation, cache, fund, hoard, quantity, reserve, reservoir, stock, stockpile, supply. ▷ STOREHOUSE. 2 *a grocery store*. outlet, retail business, retailers, supermarket. ▷ SHOP. • *vb* accumulate, aggregate, deposit, hoard, keep, lay by, lay in, lay up, preserve, put away, reserve, save, set aside, *inf* stash away, stockpile, stock up, stow away.

storehouse *n* depository, repository, storage, store, store-room. □ *armoury, arsenal, barn, cellar, cold-storage, depot, granary, larder, pantry, safe, silo, stock-room, strong-room, treasury, vault, warehouse*.

storey *n* deck, floor, level, stage, tier.

storm *n* 1 disturbance, onslaught, outbreak, outburst, stormy weather, tempest, tumult, turbulence. □ *blizzard, cloudburst, cyclone, deluge, dust-storm, electrical storm, gale, hailstorm, hurricane, mistral, monsoon, rainstorm, sand-storm, simoom, sirocco, snowstorm, squall, thunderstorm, tornado, typhoon, whirlwind*. 2 *storm of protest*. ▷ CLAMOUR. • *vb* ▷ ATTACK.

stormy *adj* angry, blustery, choppy, fierce, furious, gusty, raging, rough, squally, tempestuous, thundery, tumultuous, turbulent, vehement, violent, wild, windy. *Opp* CALM.

story *n* 1 account, anecdote, chronicle, fiction, history, narration, narrative, plot, recital, record, scenario, tale, yarn. □ *allegory, children's story, crime story, detective story, epic, fable, fairy-tale, fantasy, folk-tale, legend, mystery, myth, novel, parable, romance, saga, science fiction, SF, thriller*, inf *whodunit*. 2 *story in newspaper*. article, dispatch, exclusive, feature, news item, piece, report, scoop. 3 falsehood, *inf* fib, lie, tall story, untruth.

storyteller *n* author, biographer, narrator, raconteur, teller.

stout *adj* 1 *inf* beefy, big, bulky, burly, *inf* chubby, corpulent, fleshy, heavy, *inf* hulking, overweight, plump, portly, solid, stocky, *inf* strapping, thick-set, *inf* tubby, well-built. ▷ FAT. *Opp* THIN. 2 *stout rope*. durable, reliable, robust, sound, strong, sturdy, substantial, thick, tough. 3 *stout fighter*. bold, brave, courageous, fearless, gallant, heroic, intrepid, plucky, resolute, spirited, valiant. *Opp* WEAK.

stove *n* boiler, cooker, fire, furnace, heater, oven, range.

stow *vb* load, pack, put away, *inf* stash away, store.

straggle *vb* be dispersed, be scattered, dangle, dawdle, drift, fall behind, lag, loiter, meander, ramble, scatter, spread out, stray, string out, trail, wander. **straggling** ▷ DISORGANIZED, LOOSE.

straight *adj* 1 aligned, direct, flat, linear, regular, smooth, true, unbending, undeviating, unswerving. 2 neat, orderly, organized, right, *inf* shipshape, sorted out, spruce, tidy. 3 *a straight sequence*. consecutive, continuous, non-stop, perfect, sustained, unbroken, uninterrupted, unrelieved. 4 ▷ STRAIGHTFORWARD. *Opp* CROOKED, INDIRECT, UNTIDY.

straight away at once, directly, immediately, instantly, now, without delay.

straighten *vb* disentangle, make straight, put straight, rearrange, sort out, tidy, unbend, uncurl, unravel, untangle, untwist.

straightforward *adj* blunt, candid, direct, easy, forthright, frank, genuine, honest, intelligible, lucid, open, plain, simple, sincere, straight, truthful, uncomplicated. *Opp* DEVIOUS.

strain *n* **1** anxiety, difficulty, effort, exertion, hardship, pressure, stress, tension, worry. **2** *genetic strain.* ▷ ANCESTRY. • *vb* **1** haul, heave, make taut, pull, stretch, tighten, tug. **2** *strain to succeed.* attempt, endeavour, exert yourself, labour, make an effort, strive, struggle, toil, try. **3** *strain yourself.* exercise, exhaust, overtax, *inf* push to the limit, stretch, tax, tire out, weaken, wear out, weary. **4** *strain a muscle.* damage, hurt, injure, overwork, pull, rick, sprain, tear, twist, wrench, **5** *strain liquid.* clear, drain, draw off, filter, percolate, purify, riddle, screen, separate, sieve, sift.

strained *adj* **1** artificial, constrained, distrustful, embarrassed, false, forced, insincere, self-conscious, stiff, tense, uncomfortable, uneasy, unnatural. **2** *strained look.* drawn, tired, weary. **3** *strained interpretation.* far-fetched, incredible, laboured, unlikely, unreasonable. *Opp* NATURAL, RELAXED.

strainer *n* colander, filter, riddle, sieve.

strand *n* fibre, filament, string, thread, wire. • *vb* **1** abandon, desert, forsake, leave stranded, lose, maroon. **2** *strand a ship.* beach, ground, run aground, wreck. **stranded** ▷ AGROUND, HELPLESS.

strange *adj* **1** abnormal, astonishing, atypical, bizarre, curious, eerie, exceptional, extraordinary, fantastic, *inf* funny, grotesque, irregular, odd, out-of-the-ordinary, outré, peculiar, quaint, queer, rare, remarkable, singular, surprising, surreal, uncommon, unexpected, unheard-of, unique, unnatural, untypical, unusual. **2** *strange neighbours. inf* cranky, eccentric, sinister, unconventional, weird, *inf* zany. **3** *a strange problem.* baffling, bewildering, inexplicable, insoluble, mysterious, mystifying, perplexing, puzzling, unaccountable. **4** *strange places.* alien, exotic, foreign, little-known, off the beaten track, outlandish, outof-the-way, remote, unexplored, unmapped. **5** *strange experience.* different, fresh, new, novel, unaccustomed, unfamiliar. *Opp* FAMILIAR, ORDINARY.

strangeness *n* abnormality, bizarreness, eccentricity, eeriness, extraordinariness, irregularity, mysteriousness, novelty, oddity, oddness, outlandishness, peculiarity, quaintness, queerness, rarity, singularity, unconventionality, unfamiliarity.

stranger *n* alien, foreigner, guest, newcomer, outsider, visitor.

strangle *vb* **1** asphyxiate, choke, garotte, smother, stifle, suffocate, throttle. **2** *strangle a cry.* ▷ SUPPRESS.

strangulate *vb* bind, compress, constrict, squeeze.

strangulation *n* asphyxiation, garotting, suffocation.

strap *n* band, belt, strop, tawse, thong, webbing. • *vb* ▷ FASTEN.

stratagem *n* artifice, device, *inf* dodge, manoeuvre, plan, ploy, ruse, scheme, subterfuge, tactic, trick.

strategic *adj* advantageous, critical, crucial, deliberate, key, planned, politic, tactical, vital.

strategy *n* approach, design, manoeuvre, method, plan, plot, policy, procedure, programme, *inf* scenario, scheme, tactics.

stratum *n* layer, seam, table, thickness, vein.

straw *n* corn, stalks, stubble.

stray *adj* 1 abandoned, homeless, lost, roaming, roving, wandering. 2 *stray bullets*. accidental, casual, chance, haphazard, isolated, lone, occasional, odd, random, single. • *vb* 1 get lost, get separated, go astray, meander, move about aimlessly, ramble, range, roam, rove, straggle, wander. 2 *stray from the point*. deviate, digress, diverge, drift, get off the subject, *inf* go off at a tangent, veer.

streak *n* 1 band, bar, dash, line, mark, score, smear, stain, stria, striation, strip, stripe, stroke, vein. 2 *a selfish streak*. component, element, strain, touch, trace. 3 *streak of good luck*. period, run, series, spate, spell, stretch, time. • *vb* 1 mark with streaks, smear, smudge, stain, striate. 2 *streak past*. dart, dash, flash, fly, gallop, hurtle, move at speed, rush, *inf* scoot, speed, sprint, tear, *inf* whip, zoom.

streaky *adj* barred, lined, smeary, smudged, streaked, striated, stripy, veined.

stream *n* 1 beck, brook, brooklet, burn, channel, freshet, *poet* rill, river, rivulet, streamlet, watercourse. 2 cascade, cataract, current, deluge, effluence, flood, flow, fountain, gush, jet, outpouring, rush, spate, spurt, surge, tide, torrent. • *vb* cascade, course, deluge, flood, flow, gush, issue, pour, run, spill, spout, spurt, squirt, surge, well.

streamer *n* banner, flag, pennant, pennon, ribbon.

streamlined *adj* 1 aerodynamic, elegant, graceful, hydrodynamic, sleek, smooth. 2 ▷ EFFICIENT.

street *n* avenue, roadway, terrace. ▷ ROAD.

strength *n* 1 brawn, capacity, condition, energy, fitness, force, health, might, muscle, power, resilience, robustness, sinew, stamina, stoutness, sturdiness, toughness, vigour. 2 *strength of purpose*. *inf* backbone, commitment, courage, determination, firmness, *inf* grit, perseverance, persistence, resolution, resolve, spirit, tenacity. *Opp* WEAKNESS.

strengthen *vb* 1 bolster, boost, brace, build up, buttress, encourage, fortify, harden, hearten, increase, make stronger, prop up, reinforce, stiffen, support, tone up, toughen. 2 *strengthen an argument*. back up, consolidate, corroborate, enhance, justify, substantiate. *Opp* WEAKEN.

strenuous *adj* 1 arduous, backbreaking, burdensome, demanding, difficult, exhausting, gruelling, hard, laborious, punishing, stiff, taxing, tough, uphill. *Opp* EASY. 2 *strenuous efforts*. active, committed, determined, dogged, dynamic, eager, energetic, herculean, indefatigable, laborious, pertinacious, resolute, spirited, strong, tenacious, tireless, unremitting, vigorous, zealous. *Opp* CASUAL.

stress *n* 1 anxiety, difficulty, distress, hardship, pressure, strain, tenseness, tension, trauma, worry. **2** accent, accentuation, beat, emphasis, importance, significance, underlining, urgency, weight. • *vb* **1** accent, accentuate, assert, draw attention to, emphasize, feature, highlight, insist on, lay stress on, mark, put stress on, repeat, spotlight, underline, underscore. **2** *stressed by work*. burden, distress, overstretch, pressurize, pressure, push to the limit, tax, weigh down.

stressful *adj* anxious, difficult, taxing, tense, tiring, traumatic, worrying. *Opp* RELAXED.

stretch *n* **1** period, spell, stint, term, time, tour of duty. **2** *stretch of country*. area, distance, expanse, length, span, spread, sweep, tract. • *vb* **1** broaden, crane (*your neck*), dilate, distend, draw out, elongate, enlarge, expand, extend, flatten out, inflate, lengthen, open out, pull out, spread out, swell, tauten, tighten, widen. **2** *stretch into the distance*. be unbroken, continue, disappear, extend, go, reach out, spread. **3** *stretch resources*. overextend, overtax, *inf* push to the limit, strain, tax.

strew *vb* disperse, distribute, scatter, spread, sprinkle.

strict *adj* **1** austere, authoritarian, autocratic, firm, harsh, merciless, *inf* no-nonsense, rigorous, severe, stern, stringent, tyrannical, uncompromising. *Opp* EASYGOING. **2** *strict rules*. absolute, binding, defined, *inf* hard and fast, inflexible, invariable, precise, rigid, stringent, tight, unchangeable. *Opp* FLEXIBLE. **3** *strict truthfulness*. accurate, complete, correct, exact, meticulous, perfect, precise, right, scrupulous.

stride *n* pace, step. • *vb* ▷ WALK.

strident *adj* clamorous, discordant, grating, harsh, jarring, loud, noisy, raucous, screeching, shrill, unmusical. *Opp* SOFT.

strife *n* animosity, arguing, bickering, competition, conflict, discord, disharmony, dissention, enmity, friction, hostility, quarrelling, rivalry, unfriendliness. ▷ FIGHT. *Opp* CO-OPERATION.

strike *n* **1** go-slow, industrial action, stoppage, walk-out, withdrawal of labour. **2** assault, attack, bombardment. • *vb* **1** bang against, bang into, beat, collide with, hammer, impel, knock, rap, run into, smack, smash into, *inf* thump, *inf* whack. ▷ ATTACK, HIT. **2** *strike a match*. ignite, light. **3** *tragedy struck us forcibly*. affect, afflict, *inf* come home to, impress, influence. **4** *clock struck one*. chime, ring, sound. **5** *strike for more pay*. *inf* come out, *inf* down tools, stop work, take industrial action, withdraw labour, work to rule. **6** *strike a flag, tent*. dismantle, lower, pull down, remove, take down.

striking *adj* affecting, amazing, arresting, conspicuous, distinctive, extraordinary, glaring, imposing, impressive, memorable, noticeable, obvious, out-of-the-ordinary, outstanding, prominent, showy, stunning, telling, unmistakable, unusual. *Opp* INCONSPICUOUS.

string *n* **1** cable, cord, fibre, line, rope, twine. **2** chain, file, line, procession, progression, queue, row, sequence, series, stream, succession, train. • *vb string together* connect, join, line up, link, thread. **stringed instruments** strings. □ *banjo, cello,* *clavichord, double-bass,* inf *fiddle, guitar, harp, harpsichord, lute, lyre, piano, sitar, spinet, ukulele, viola, violin, zither.*

stringy *adj* chewy, fibrous, gristly, sinewy, tough. *Opp* TENDER.

strip *n* band, belt, fillet, lath, line, narrow piece, ribbon, shred, slat, sliver, stripe, swathe. • *vb* **1** bare, clear, decorticate, defoliate, denude, divest, *old use* doff, excoriate, flay, lay bare, peel, remove the covering, remove the paint, remove the skin, skin, uncover. *Opp* COVER. **2** *strip to the waist.* bare yourself, disrobe, expose yourself, get undressed, uncover yourself. *Opp* DRESS. **strip down** ▷ DISMANTLE. **strip off** ▷ UNDRESS.

stripe *n* band, bar, chevron, line, ribbon, streak, striation, strip, stroke, swathe.

striped *adj* banded, barred, lined, streaky, striated, stripy.

strive *vb* attempt, *inf* do your best, endeavour, make an effort, strain, struggle, try. ▷ FIGHT.

stroke *n* **1** action, blow, effort, knock, move, swipe. ▷ HIT. **2** *a stroke of the pen.* flourish, gesture, line, mark, movement, sweep. **3** [*medical*] apoplexy, attack, embolism, fit, seizure, spasm, thrombosis. • *vb* caress, fondle, massage, pass your hand over, pat, pet, rub, soothe, touch.

stroll *n, vb* amble, meander, saunter, wander. ▷ WALK.

strong *adj* **1** durable, hard, hardwearing, heavy-duty, impregnable, indestructible, permanent, reinforced, resilient, robust, sound, stout, substantial, thick, unbreakable, well-made. **2** *strong physique.* athletic, *inf* beefy, *inf* brawny, burly, fit, *inf* hale and hearty, hardy, *inf* hefty, *inf* husky, mighty, muscular, powerful, robust, sinewy, stalwart, *inf* strapping, sturdy, tough, well-built, wiry. **3** *strong personality.* assertive, committed, determined, domineering, dynamic, energetic, forceful, independent, reliable, resolute, stalwart, steadfast, *inf* stout, strong-minded, strong-willed, tenacious, vigorous. ▷ STUBBORN. **4** *strong commitment.* active, assiduous, deep-rooted, deep-seated, *derog* doctrinaire, *derog* dogmatic, eager, earnest, enthusiastic, fervent, fierce, firm, genuine, intense, keen, loyal, passionate, positive, rabid, sedulous, staunch, true, vehement, zealous. **5** *strong government.* decisive, dependable, *derog* dictatorial, fearless, *derog* tyrannical, unswerving, unwavering. **6** *strong measures.* aggressive, draconian, drastic, extreme, harsh, high-handed, ruthless, severe, tough, unflinching, violent. **7** *a strong army.* formidable, invincible, large, numerous, powerful, unconquerable, well-armed, well-equipped, well-trained. **8** *strong colour, light.* bright, brilliant, clear, dazzling, garish, glaring, vivid. **9** *strong taste, smell.* concentrated, highly-flavoured, hot, intense, noticeable, obvious, overpowering, prominent, pronounced, pungent, sharp, spicy, unmistakable. **10** *strong evidence.* clear-cut, cogent, compelling, convincing, evident, influential, persuasive, plain, solid, telling, undisputed. **11** *strong drink.* alcoholic, concentrated, intoxicating, potent, undiluted. *Opp* WEAK.

stronghold *n* bastion, bulwark, castle, citadel, *old use* fastness, fort, fortification, fortress, garrison.

structure *n* 1 arrangement, composition, configuration, constitution, design, form, formation, *inf* make-up, order, organization, plan, shape, system. 2 complex, construction, edifice, erection, fabric, framework, pile, superstructure. ▷ BUILDING. • *vb* arrange, build, construct, design, form, frame, give structure to, organize, shape, systematize.

struggle *n* 1 challenge, difficulty, effort, endeavour, exertion, labour, problem. 2 ▷ FIGHT. • *vb* 1 endeavour, exert yourself, labour, make an effort, move violently, strain, strive, toil, try, work hard, wrestle, wriggle about, writhe about. 2 *struggle through mud.* flail, flounder, stumble, wallow. 3 ▷ FIGHT.

stub *n* butt, end, remains, remnant, stump. • *vb* ▷ HIT.

stubble *n* 1 stalks, straw. 2 beard, bristles, *inf* five-o'clock shadow, hair, roughness.

stubbly *adj* bristly, prickly, rough, unshaven.

stubborn *adj* defiant, determined, difficult, disobedient, dogged, dogmatic, headstrong, inflexible, intractable, intransigent, mulish, obdurate, obstinate, opinionated, persistent, pertinacious, *inf* pig-headed, recalcitrant, refractory, rigid, self-willed, tenacious, uncompromising, uncooperative, uncontrollable, unmanageable, unreasonable, unyielding, wayward, wilful. *Opp* AMENABLE.

stuck *adj* 1 bogged down, cemented, fast, fastened, firm, fixed, glued, immovable. 2 *stuck on a problem.* baffled, beaten, held up, *inf* stumped, *inf* stymied.

stuck-up *adj* arrogant, *inf* bigheaded, bumptious, *inf* cocky, conceited, condescending, *inf* high-and-mighty, patronizing, proud, self-important, snobbish, *inf* snooty, supercilious, *inf* toffee-nosed. *Opp* MODEST.

student *n* apprentice, disciple, learner, postgraduate, pupil, scholar, schoolchild, trainee, undergraduate.

studied *adj* calculated, conscious, contrived, deliberate, intentional, planned, premeditated.

studious *adj* academic, assiduous, attentive, bookish, brainy, earnest, hard-working, intellectual, scholarly, serious-minded, thoughtful.

study *vb* 1 analyse, consider, contemplate, enquire into, examine, give attention to, investigate, learn about, look closely at, peruse, ponder, pore over, read carefully, research, scrutinize, survey, think about, weigh. 2 *study for exams. inf* cram, learn, *inf* mug up, read, *inf* swot, work.

stuff *n* 1 ingredients, matter, substance. 2 fabric, material, textile. ▷ CLOTH. 3 *all sorts of stuff.* accoutrements, articles, belongings, *inf* bits and pieces, *inf* clobber, effects, *inf* gear, impedimenta, junk, objects, *inf* paraphernalia, possessions, *inf* tackle, things. • *vb* 1 compress, cram, crowd, force, jam, pack, press, push, ram, shove, squeeze, stow, thrust, tuck. 2 *stuff a cushion.* fill, line, pad. **stuff yourself** ▷ EAT.

stuffing *n* 1 filling, lining, padding, quilting, wadding. 2 *stuffing in poultry.* forcemeat, seasoning.

stuffy *adj* 1 airless, close, fetid, fuggy, fusty, heavy, humid, muggy, musty, oppressive, stale, steamy, stifling, suffocating, sultry, unventilated, warm. *Opp* AIRY. **2** [*inf*] *a stuffy old bore.* boring, conventional, dreary, dull, formal, humourless, narrow-minded, old-fashioned, pompous, prim, staid, *inf* stodgy, strait-laced. *Opp* LIVELY.

stumble *vb* **1** blunder, flounder, lurch, miss your footing, reel, slip, stagger, totter, trip, tumble. ▷ WALK. **2** *stumble in speech.* become tongue-tied, falter, hesitate, pause, stammer, stutter.

stumbling-block *n* bar, difficulty, hindrance, hurdle, impediment, obstacle, snag.

stump *vb* baffle, bewilder, *inf* catch out, confound, confuse, defeat, *inf* flummox, mystify, outwit, perplex, puzzle, *inf* stymie. **stump up** ▷ PAY.

stun *vb* **1** daze, knock out, knock senseless, make unconscious. **2** amaze, astonish, astound, bewilder, confound, confuse, dumbfound, flabbergast, numb, shock, stagger, stupefy. **stunning** ▷ BEAUTIFUL, STUPENDOUS.

stunt *n inf* dare, exploit, feat, trick. • *vb stunt growth.* ▷ CHECK.

stupendous *adj* amazing, colossal, enormous, exceptional, extraordinary, huge, incredible, marvellous, miraculous, notable, phenomenal, prodigious, remarkable, *inf* sensational, singular, special, staggering, stunning, tremendous, unbelievable, wonderful. *Opp* ORDINARY.

stupid *adj* [*Most synonyms derog*] **1** addled, bird-brained, bone-headed, bovine, brainless, clueless, cretinous, dense, dim, doltish, dopey, drippy, dull, dumb, empty-headed, feather-brained, feeble-minded, foolish, gormless, half-witted, idiotic, ignorant, imbecilic, imperceptive, ineducable, lacking, lumpish, mindless, moronic, naïve, obtuse, puerile, senseless, silly, simple, simple-minded, slow, slow in the uptake, slow-witted, subnormal, thick, thick-headed, thick-skulled, thickwitted, unintelligent, unthinking, unwise, vacuous, weak in the head, witless. **2** *a stupid thing to do.* absurd, asinine, barmy, crack-brained, crass, crazy, fatuous, feeble, futile, half-baked, hare-brained, ill-advised, inane, irrational, irrelevant, irresponsible, laughable, ludicrous, lunatic, mad, nonsensical, pointless, rash, reckless, ridiculous, risible, scatter-brained, thoughtless, unjustifiable. **3** *stupid after a knock on the head.* dazed, in a stupor, semi-conscious, sluggish, stunned, stupefied. *Opp* INTELLIGENT. **stupid person** ▷ FOOL.

stupidity *n* absurdity, crassness, denseness, dullness, *inf* dumbness, fatuity, fatuousness, folly, foolishness, futility, idiocy, ignorance, imbecility, inanity, lack of intelligence, lunacy, madness, mindlessness, naïvety, pointlessness, recklessness, silliness, slowness, thoughtlessness. *Opp* INTELLIGENCE.

stupor *n* coma, daze, inertia, lassitude, lethargy, numbness, shock, state of insensibility, torpor, trance, unconsciousness.

sturdy *adj* **1** athletic, brawny, burly, hardy, healthy, hefty, husky, muscular, powerful, robust, stalwart, stocky, *inf* strapping, vigorous, well-built. **2** *sturdy shoes, etc.* durable, solid, sound, substantial, tough, well-made. **3** *sturdy opposition.* determined, firm, indomitable, resolute, staunch, steadfast, uncompromising, vigorous. ▷ STRONG. *Opp* WEAK.

stutter *vb* stammer, stumble. ▷ TALK.

style *n* **1** dash, elegance, flair, flamboyance, panache, polish, refinement, smartness, sophistication, stylishness, taste. **2** *not my style.* approach, character, custom, habit, idiosyncrasy, manner, method, way. **3** *style in writing.* diction, mode, phraseology, phrasing, register, sentence structure, tenor, tone, wording. **4** *style in clothes.* chic, cut, design, dress-sense, fashion, look, mode, pattern, shape, tailoring, type, vogue.

stylish *adj Fr* à la mode, chic, *inf* classy, contemporary, *inf* dapper, elegant, fashionable, modern, modish, *inf* natty, *inf* posh, smart, *inf* snazzy, sophisticated, *inf* trendy, up-to-date. *Opp* OLD-FASHIONED.

subconscious *adj* deep-rooted, hidden, inner, intuitive, latent, repressed, subliminal, suppressed, unacknowledged, unconscious. *Opp* CONSCIOUS.

subdue *vb* check, curb, hold back, keep under, moderate, quieten, repress, restrain, suppress, temper. ▷ SUBJUGATE.

subdued *adj* **1** chastened, crestfallen, depressed, downcast, grave, reflective, repressed, restrained, serious, silent, sober, solemn, thoughtful. ▷ SAD. *Opp* EXCITED. **2** *subdued music.* calm, hushed, low, mellow, muted, peaceful, placid, quiet, soft, soothing, toned down, tranquil, unobtrusive.

subject *adj* **1** captive, dependent, enslaved, oppressed, ruled, subjugated. **2** *subject to interference.* exposed, liable, prone, susceptible, vulnerable. *Opp* FREE. • *n* **1** citizen, dependant, national, passport-holder, taxpayer, voter. **2** *subject for discussion.* affair, business, issue, matter, point, proposition, question, theme, thesis, topic. **3** *subject of study.* area, branch of knowledge, course, discipline, field. □ *anatomy, archaeology, architecture, art, astronomy, biology, business, chemistry, computing, craft, design, divinity, domestic science, drama, economics, education, electronics, engineering, English, environmental science, ethnology, etymology, geography, geology, heraldry, history, languages, Latin, law, linguistics, literature, mathematics, mechanics, medicine, metallurgy, metaphysics, meteorology, music, natural history, oceanography, ornithology, penology, pharmacology, pharmacy, philology, philosophy, photography, physics, physiology, politics, psychology, religious studies, science, scripture, social work, sociology, sport, surveying, technology, theology, topology, zoology.* • *vb* **1** *subject a thing to scrutiny.* expose, lay open, submit. **2** ▷ SUBJUGATE.

subjective *adj* biased, emotional, *inf* gut (*reaction*), idiosyncratic, individual, instinctive, intuitive, personal, prejudiced, self-centred. *Opp* OBJECTIVE.

subjugate *vb* beat, conquer, control, crush, defeat, dominate, enslave, enthral, *inf* get the better of, master, oppress, overcome, overpower, overrun, put down, quash, quell, subdue, subject, tame, triumph over, vanquish.

sublimate *vb* channel, convert, divert, idealize, purify, redirect, refine.

sublime *adj* ecstatic, elated, elevated, exalted, great, heavenly, high, high-minded, lofty, noble, spiritual, transcendent. *Opp* BASE.

submerge *vb* **1** cover with water, dip, drench, drown, *inf* dunk, engulf, flood, immerse, inundate, overwhelm, soak, swamp. **2** dive, go down, go under, plummet, sink, subside.

submission *n* **1** acquiescence, capitulation, compliance, giving in, surrender, yielding. ▷ SUBMISSIVENESS. **2** contribution, entry, offering, presentation, tender. **3** *a legal submission.* argument, claim, contention, idea, proposal, suggestion, theory.

submissive *adj* accommodating, acquiescent, amenable, biddable, *derog* boot-licking, compliant, deferential, docile, humble, meek, obedient, obsequious, passive, pliant, resigned, servile, slavish, supine, sycophantic, tame, tractable, unassertive, uncomplaining, unresisting, weak, yielding. *Opp* ASSERTIVE.

submissiveness *n* acquiescence, assent, compliance, deference, docility, humility, meekness, obedience, obsequiousness, passivity, resignation, servility, submission, subservience, tameness.

submit *vb* **1** accede, bow, capitulate, concede, give in, *inf* knuckle under, succumb, surrender, yield. **2** *submit a proposal.* advance, enter, give in, hand in, offer, present, proffer, propose, propound, put forward, state, suggest. **submit to** ▷ ACCEPT, OBEY.

subordinate *adj* inferior, junior, lesser, lower, menial, minor, secondary, subservient, subsidiary. • *n* aide, assistant, dependant, employee, inferior, junior, menial, *inf* underling. ▷ SERVANT.

subscribe *vb* **subscribe to** **1** contribute to, covenant to, donate to, give to, patronize, sponsor, support. **2** *subscribe to a magazine.* be a subscriber to, buy regularly, pay a subscription to. **3** *subscribe to a theory.* advocate, agree with, approve of, *inf* back, believe in, condone, consent to, endorse, *inf* give your blessing to.

subscriber *n* patron, regular customer, sponsor, supporter.

subscription *n* fee, due, payment, regular contribution, remittance.

subsequent *adj* coming, consequent, ensuing, following, future, later, next, resultant, resulting, succeeding, successive. *Opp* PREVIOUS.

subside *vb* **1** abate, calm down, decline, decrease, die down, diminish, dwindle, ebb, fall, go down, lessen, melt away, moderate, quieten, recede, shrink, slacken, wear off. **2** *subside into a chair.* collapse, descend, lower yourself, settle, sink. *Opp* RISE.

subsidiary *adj* additional, ancillary, auxiliary, complementary, contributory, inferior, lesser, minor, secondary, subordinate, supporting.

subsidize *vb* aid, back, finance, fund, give subsidy to, maintain, promote, sponsor, support, underwrite.

subsidy *n* aid, backing, financial help, funding, grant, maintenance, sponsorship, subvention, support.

substance *n* **1** actuality, body, concreteness, corporeality, reality, solidity. **2** chemical, fabric, make-up, material, matter, stuff. **3** *substance of an argument.* core, essence, gist, import, meaning, significance, subject-matter, theme. **4** [*old use*] *a person of substance.* ▷ WEALTH.

substandard *adj inf* below par, disappointing, inadequate, inferior, poor, shoddy, unworthy.

substantial *adj* **1** durable, hefty, massive, solid, sound, stout, strong, sturdy, well-built, well-made. **2** big, consequential, considerable, generous, great, large, significant, sizeable, worthwhile. *Opp* FLIMSY, SMALL.

substitute *adj* **1** acting, deputy, relief, reserve, stand-by, surrogate, temporary. **2** alternative, ersatz, imitation. • *n* alternative, deputy, locum, proxy, relief, replacement, reserve, stand-in, stopgap, substitution, supply, surrogate, understudy. • *vb* **1** change, exchange, interchange, replace, *inf* swop, *inf* switch. **2** *substitute for an absentee.* act as a substitute, cover, deputize, double, stand in, supplant, take the place of, take over the role of, understudy.

subtle *adj* **1** delicate, elusive, faint, fine, gentle, mild, slight, unobtrusive. **2** *subtle argument.* arcane, clever, indirect, ingenious, mysterious, recondite, refined, shrewd, sophisticated, tactful, understated. ▷ CUNNING. *Opp* OBVIOUS.

subtract *vb* debit, deduct, remove, take away, take off. *Opp* ADD.

suburban *adj* residential, outer, outlying.

suburbs *n* fringes, outer areas, outskirts, residential areas, suburbia.

subversive *adj* challenging, disruptive, insurrectionary, questioning, radical, seditious, traitorous, treacherous, treasonous, undermining, unsettling. ▷ REVOLUTIONARY. *Opp* CONSERVATIVE, ORTHODOX.

subvert *vb* challenge, corrupt, destroy, disrupt, overthrow, overturn, pervert, ruin, undermine, upset, wreck.

subway *n* tunnel, underpass.

succeed *vb* **1** accomplish your objective, *inf* arrive, be a success, do well, flourish, *inf* get on, *inf* get to the top, *inf* make it, prosper, thrive. **2** be effective, *inf* catch on, produce results, work. **3** be successor to, come after, follow, inherit from, replace, take over from. *Opp* FAIL. **succeeding** ▷ SUBSEQUENT.

success *n* **1** fame, good fortune, prosperity, wealth. **2** *success of a plan.* accomplishment, achievement, attainment, completion, effectiveness, successful outcome. **3** *a great success. inf* hit, *inf* sensation, triumph, victory, *inf* winner. *Opp* FAILURE.

successful *adj* **1** booming, effective, effectual, flourishing, fruitful, lucrative, money-making, productive, profitable, profit-making, prosperous, rewarding, thriving, useful, well-off. **2** best-selling, celebrated, famed, famous, high-earning, leading, popular, top, unbeaten, victorious, well-known, winning. *Opp* UNSUCCESSFUL.

succession *n* chain, flow, line, procession, progression, run, sequence, series, string.

successive *adj* consecutive, continuous, in succession, succeeding, unbroken, uninterrupted.

successor *n* heir, inheritor, replacement.

succinct *adj* brief, compact, concise, condensed, epigrammatic, pithy, short, terse, to the point. *Opp* WORDY.

succulent *adj* fleshy, juicy, luscious, moist, mouthwatering, palatable, rich.

succumb *vb* accede, be overcome, capitulate, give in, give up, give way, submit, surrender, yield. *Opp* SURVIVE.

suck *vb* **suck up** absorb, draw up, pull up, soak up. **suck up to** ▷ FLATTER.

sudden *adj* **1** abrupt, brisk, hasty, hurried, impetuous, impulsive, precipitate, quick, rash, *inf* snap, swift, unconsidered, unplanned, unpremeditated. *Opp* SLOW. **2** *a sudden shock.* acute, sharp, startling, surprising, unannounced, unexpected, unforeseeable, unforeseen, unlooked-for. *Opp* PREDICTABLE.

suds *n* bubbles, foam, froth, lather, soapsuds.

sue *vb* **1** indict, institute legal proceedings against, proceed against, prosecute, summons, take legal action against. **2** *sue for peace.* ▷ ENTREAT.

suffer *vb* **1** bear, cope with, endure, experience, feel, go through, live through, *inf* put up with, stand, tolerate, undergo, withstand. **2** *suffer from a wound.* ache, agonize, feel pain, hurt, smart. **3** *suffer for a crime.* atone, be punished, make amends, pay.

suffice *vb* answer, be sufficient, *inf* do, satisfy, serve.

sufficient *adj* adequate, enough, satisfactory. *Opp* INSUFFICIENT.

suffocate *vb* asphyxiate, choke, smother, stifle, stop breathing, strangle, throttle.

sugar *n* □ *brown sugar, cane sugar, caster sugar, demerara, glucose, granulated sugar, icing sugar, lump sugar, molasses, sucrose, sweets, syrup, treacle.* • *vb* sweeten.

sugary *adj* **1** glazed, iced, sugared, sweetened. ▷ SWEET. **2** *sugary sentiments.* cloying, honeyed, sickly. ▷ SENTIMENTAL.

suggest *vb* **1** advise, advocate, counsel, moot, move, propose, propound, put forward, raise, recommend, urge. **2** call to mind, communicate, hint, imply, indicate, insinuate, intimate, make you think (of), mean, signal.

suggestion *n* **1** advice, counsel, offer, plan, prompting, proposal, recommendation, urging. **2** breath, hint, idea, indication, intimation, notion, suspicion, touch, trace.

suggestive *adj* **1** evocative, expressive, indicative, reminiscent, thought-provoking. **2** ▷ INDECENT.

suicidal *adj* **1** hopeless, *inf* kamikaze, self-destructive. **2** ▷ DESOLATE.

suit *n* outfit. ▷ CLOTHES. • *vb* **1** accommodate, be suitable for, conform to, fill your needs, fit in with, gratify, harmonize with, match, please, satisfy, tally with. *Opp* DISPLEASE. **2** *That colour suits you.* become, fit, look good on.

suitable *adj* acceptable, applicable, apposite, appropriate, apt, becoming, befitting, congenial, convenient, correct, decent, decorous, fit, fitting, handy, *old use* meet, opportune, pertinent, proper, relevant, right, satisfactory, seemly, tasteful, timely, well-chosen, well-judged, well-timed. *Opp* UNSUITABLE.

sulk *vb* be sullen, brood, mope, pout.

sullen *adj* **1** anti-social, bad-tempered, brooding, churlish, crabby, cross, disgruntled, dour, glum, grim, grudging, ill-humoured, lugubrious, moody, morose, *inf* out of sorts, petulant, pouting, resentful, silent, sour, stubborn, sulking, sulky, surly, uncommunicative, unforgiving, unfriendly, unhappy, unsociable. ▷ SAD. **2** *a sullen sky*. cheerless, dark, dismal, dull, gloomy, grey, leaden, sombre. *Opp* CHEERFUL.

sultry *adj* **1** close, hot, humid, *inf* muggy, oppressive, steamy, stifling, stuffy, warm. *Opp* COLD. **2** *sultry beauty*. erotic, mysterious, passionate, provocative, seductive, sensual, sexy, voluptuous.

sum *n* aggregate, amount, number, quantity, reckoning, result, score, tally, total, whole. • *vb* **sum up** ▷ SUMMARIZE.

summarize *vb* abridge, condense, digest, encapsulate, give the gist, make a summary, outline, précis, *inf* recap, recapitulate, reduce, review, shorten, simplify, sum up. *Opp* ELABORATE.

summary *n* abridgement, abstract, condensation, digest, epitome, gist, outline, précis, recapitulation, reduction, résumé, review, summation, summing-up, synopsis.

summery *adj* bright, sunny, tropical, warm. ▷ HOT. *Opp* WINTRY.

summit *n* **1** apex, crown, head, height, peak, pinnacle, point, top. *Opp* BASE. **2** *summit of success*. acme, apogee, climax, culmination, high point, zenith. *Opp* NADIR.

summon *vb* **1** command, demand, invite, order, send for, subpoena. **2** assemble, call, convene, convoke, gather together, muster, rally.

sunbathe *vb* bake, bask, *inf* get a tan, sun yourself, tan.

sunburnt *adj* blistered, bronzed, brown, peeling, tanned, weather-beaten.

sundry *adj* assorted, different, *old use* divers, miscellaneous, mixed, various.

sunken *adj* **1** submerged, underwater, wrecked. **2** *sunken cheeks*. concave, depressed, drawn, hollow, hollowed.

sunless *adj* cheerless, cloudy, dark, dismal, dreary, dull, gloomy, grey, overcast, sombre. *Opp* SUNNY.

sunlight *n* daylight, sun, sunbeams, sunshine.

sunny *adj* **1** bright, clear, cloudless, fair, fine, summery, sunlit, sunshiny, unclouded. *Opp* SUNLESS. **2** ▷ CHEERFUL.

sunrise *n* dawn, day-break.

sunset *n* dusk, evening, gloaming, nightfall, sundown, twilight.

sunshade *n* awning, canopy, parasol.

superannuated *adj* **1** discharged, *inf* pensioned off, *inf* put out to grass, old, retired. **2** discarded, disused, obsolete, thrown out, worn out. ▷ OLD.

superannuation *n* annuity, pension.

superb *adj* admirable, excellent, fine, first-class, first-rate, grand, impressive, marvellous, superior. ▷ SPLENDID. *Opp* INFERIOR.

superficial *adj* **1** cosmetic, external, exterior, on the surface, outward, shallow, skin-deep, slight, surface, unimportant. **2** careless, casual, cursory, desultory, facile, frivolous, hasty, hurried, inattentive, lightweight, *inf* nodding (*acquaintance*), oversimplified, passing, perfunctory, simple-minded, simplistic, sweeping (*generalization*), trivial, unconvincing, uncritical, undiscriminating, unquestioning, unscholarly, unsophisticated. *Opp* ANALYTICAL, DEEP.

superfluous *adj* excess, excessive, extra, needless, redundant, spare, superabundant, surplus, unnecessary, unneeded, unwanted. *Opp* NECESSARY.

superhuman *adj* **1** god-like, herculean, heroic, phenomenal, prodigious. **2** *superhuman powers.* divine, higher, metaphysical, supernatural.

superimpose *vb* overlay, place on top of.

superintend *vb* administer, be in charge of, be the supervisor of, conduct, control, direct, look after, manage, organize, oversee, preside over, run, supervise, watch over.

superior *adj* **1** better, *inf* classier, greater, higher, higher-born, loftier, more important, more impressive, nobler, senior, *inf* up-market. **2** *superior quality.* choice, exclusive, fine, first-class, first-rate, select, top, unrivalled. **3** *superior attitude.* arrogant, condescending, contemptuous, disdainful, élitist, haughty, *inf* high-and-mighty, lofty, paternalistic, patronizing, self-important, smug, snobbish, *inf* snooty, stuck-up, supercilious. *Opp* INFERIOR.

superlative *adj* best, choicest, consummate, excellent, finest, first-rate, incomparable, matchless, peerless, *inf* tip-top, *inf* top-notch, unrivalled, unsurpassed. ▷ SUPREME.

supernatural *adj* abnormal, ghostly, inexplicable, magical, metaphysical, miraculous, mysterious, mystic, occult, other-worldly, paranormal, preternatural, psychic, spiritual, uncanny, unearthly, unnatural, weird.

superstition *n* delusion, illusion, myth, *inf* old wives' tale, superstitious belief.

superstitious *adj* credulous, groundless, illusory, irrational, mythical, traditional, unfounded, unprovable.

supervise *vb* administer, be in charge of, be the supervisor of, conduct, control, direct, govern, invigilate (*an exam*), *inf* keep an eye on, lead, look after, manage, organize, oversee, preside over, run, superintend, watch over.

supervision *n* administration, conduct, control, direction, government, invigilation, management, organization, oversight, running, surveillance.

supervisor *n* administrator, chief, controller, director, executive, foreman, *inf* gaffer, head, inspector, invigilator, leader, manager, organizer, overseer, superintendent, timekeeper.

supine *adj* **1** face upwards, flat on your back, prostrate, recumbent. *Opp* PRONE. **2** ▷ PASSIVE.

supplant *vb* displace, dispossess, eject, expel, oust, replace, supersede, *inf* step into the shoes of, *inf* topple, unseat.

supple *adj* bending, *inf* bendy, elastic, flexible, flexile, graceful, limber, lithe, plastic, pliable, pliant, resilient, soft. *Opp* RIGID.

supplement *n* **1** additional payment, excess, surcharge. **2** *a newspaper supplement, etc.* addendum, addition, annexe, appendix, codicil, continuation, endpiece, extra, insert, postscript, sequel. • *vb* add to, augment, boost, complement, extend, reinforce, *inf* top up.

supplementary *adj* accompanying, added, additional, ancillary, auxiliary, complementary, excess, extra, new, supportive.

supplication *n* appeal, entreaty, petition, plea, prayer, request, solicitation.

supplier *n* dealer, provider, purveyor, retailer, seller, shopkeeper, vendor, wholesaler.

supply *n* cache, hoard, quantity, reserve, reservoir, stock, stockpile, store. **2** equipment, food, necessities, provisions, rations, shopping. **3** *a regular supply.* delivery, distribution, provision, provisioning. • *vb* cater to, contribute, deliver, distribute, donate, endow, equip, feed, furnish, give, hand over, pass on, produce, provide, purvey, sell, stock.

support *n* **1** aid, approval, assistance, backing, back-up, bolstering, contribution, cooperation, donation, encouragement, fortifying, friendship, help, interest, loyalty, patronage, protection, reassurance, reinforcement, sponsorship, succour. **2** brace, bracket, buttress, crutch, foundation, frame, pillar, post, prop, sling, stanchion, stay, strut, substructure, trestle, truss, underpinning. **3** *financial support.* expenses, funding, keep, maintenance, subsistence, upkeep. • *vb* **1** bear, bolster, buoy up, buttress, carry, give strength to, hold up, keep up, prop up, provide a support for, reinforce, shore up, strengthen, underlie, underpin. **2** *support someone in trouble.* aid, assist, back, be faithful to, champion, comfort, defend, encourage, favour, fight for, give support to, help, rally round, reassure, side with, speak up for, stand by, stand up for, stay with, *inf* stick up for, *inf* stick with, take someone's part. **3** *support a family.* bring up, feed, finance, fund, keep, look after, maintain, nourish, provide for, sustain. **4** *support a charity.* be a supporter of, be interested in, contribute to, espouse (*a cause*), follow, give to, patronize, pay money to, sponsor, subsidize, work for. **5** *support a point of view.* accept, adhere to, advocate, agree with, allow, approve, argue for, confirm, corroborate, defend, endorse, explain, justify, promote, ratify, substantiate, uphold, validate, verify. *Opp* SUBVERT, WEAKEN. **support yourself** lean, rest.

supporter *n* **1** adherent, admirer, advocate, aficionado, apologist, champion, defender, devotee, enthusiast, *inf* fan, fanatic, follower, seconder, upholder, voter. **2** ally, assistant, collaborator, helper, henchman, second.

supportive *adj* caring, concerned, encouraging, helpful, favourable, heartening, interested, kind, loyal, positive, reassuring, sustaining, sympathetic, understanding. *Opp* SUBVERSIVE.

suppose *vb* **1** accept, assume, believe, conclude, conjecture, expect, guess, infer, judge, postulate, presume, presuppose, speculate, surmise, suspect, take for granted, think. **2** daydream, fancy, fantasize, hypothesize, imagine, maintain, postulate, pretend, theorize. **supposed** ▷ HYPOTHETICAL, PUTATIVE. **supposed to** due to, expected to, having a duty to, meant to, required to.

supposition *n* assumption, belief, conjecture, fancy, guess, *inf* guesstimate, hypothesis, inference, notion, opinion, presumption, speculation, surmise, theory, thought.

suppress *vb* **1** conquer, *inf* crack down on, crush, end, finish off, halt, overcome, overthrow, put an end to, put down, quash, quell, stamp out, stop, subdue. 2 *suppress emotion*. bottle up, censor, choke back, conceal, cover up, hide, hush up, keep quiet about, keep secret, muffle, mute, obstruct, prohibit, repress, restrain, silence, smother, stamp on, stifle, strangle.

supremacy *n* ascendancy, dominance, domination, dominion, lead, mastery, predominance, pre-eminence, sovereignty, superiority.

supreme *adj* best, choicest, consummate, crowning, culminating, excellent, finest, first-rate, greatest, highest, incomparable, matchless, outstanding, paramount, peerless, predominant, pre-eminent, prime, principal, superlative, surpassing, *inf* tip-top, top, *inf* top-notch, ultimate, unbeatable, unbeaten, unparalleled, unrivalled, unsurpassable, unsurpassed.

sure *adj* 1 assured, certain, confident, convinced, decided, definite, persuaded, positive. 2 *sure to come*. bound, certain, compelled, obliged, required. 3 *a sure fact*. accurate, clear, convincing, guaranteed, indisputable, inescapable, inevitable, infallible, proven, reliable, true, unchallenged, undeniable, undisputed, undoubted, verifiable. 4 *a sure ally*. dependable, effective, established, faithful, firm, infallible, loyal, reliable, resolute, safe, secure, solid, steadfast, steady, trustworthy, trusty, undeviating, unerring, unfailing, unfaltering, unflinching, unswerving, unwavering. *Opp* UNCERTAIN.

surface *n* 1 coat, coating, covering, crust, exterior, façade, integument, interface, outside, shell, skin, veneer. 2 *cube has six surfaces*. face, facet, plane, side. 3 *a working surface*. bench, table, top, worktop. • *vb* 1 appear, arise, *inf* come to light, come up, *inf* crop up, emerge, materialize, rise, *inf* pop up. 2 coat, cover, laminate, veneer.

surfeit *n* excess, flood, glut, overabundance, overindulgence, oversupply, plethora, superfluity, surplus.

surge *n* burst, gush, increase, onrush, onset, outpouring, rush, upsurge. ▷ WAVE. • *vb* billow, eddy, flow, gush, heave, make waves, move irresistibly, push, roll, rush, stampede, stream, sweep, swirl, well up.

surgery *n* 1 biopsy, operation. 2 *a doctor's surgery*. clinic, consulting room, health centre, infirmary, medical centre, sick-bay.

surly *adj* bad-tempered, boorish, cantankerous, churlish, crabby, cross, crotchety, *inf* crusty, curmudgeonly, dyspeptic, gruff, *inf* grumpy, ill-natured, ill-tempered, irascible, miserable, morose, peevish, rough, rude, sulky, sullen, testy, touchy, uncivil, unfriendly, ungracious, unpleasant. *Opp* FRIENDLY.

surmise *vb* assume, believe, conjecture, expect, fancy, gather, guess, hypothesize, imagine, infer, judge, postulate, presume, presuppose, sense, speculate, suppose, suspect, take for granted, think.

surpass *vb* beat, better, do better than, eclipse, exceed, excel, go beyond, leave behind, *inf* leave standing, outclass, outdistance, outdo, outperform, outshine, outstrip, overshadow, top, transcend, worst.

surplus *n* balance, excess, extra, glut, oversupply, remainder, residue, superfluity, surfeit.

surprise *n* 1 alarm, amazement, astonishment, consternation, dismay, incredulity, stupefaction, wonder. 2 *a complete surprise.* blow, *inf* bolt from the blue, *inf* bombshell, *inf* eye-opener, jolt, shock. • *vb* 1 alarm, amaze, astonish, astound, disconcert, dismay, dumbfound, flabbergast, nonplus, rock, shock, stagger, startle, stun, stupefy, *inf* take aback, take by surprise, *inf* throw. 2 *surprise someone doing wrong.* capture, catch out, *inf* catch red-handed, come upon, detect, discover, take unawares.

surprised *adj* alarmed, amazed, astonished, astounded, disconcerted, dismayed, dumbfounded, flabbergasted, incredulous, *inf* knocked for six, nonplussed, *inf* shattered, shocked, speechless, staggered, startled, struck dumb, stunned, taken aback, taken by surprise, *inf* thrown, thunderstruck.

surprising *adj* alarming, amazing, astonishing, astounding, disconcerting, extraordinary, frightening, incredible, *inf* offputting, shocking, staggering, startling, stunning, sudden, unexpected, unforeseen, unlooked-for, unplanned, unpredictable, upsetting. *Opp* PREDICTABLE.

surrender *n* capitulation, giving in, resignation, submission. • *vb* 1 acquiesce, capitulate, *inf* cave in, collapse, concede, fall, *inf* give in, give up, give way, give yourself up, resign, submit, succumb, *inf* throw in the towel, *inf* throw up the sponge, yield. 2 *surrender your ticket.* deliver up, give up, hand over, part with, relinquish. 3 *surrender your rights.* abandon, cede, renounce, waive.

surreptitious *adj* clandestine, concealed, covert, crafty, disguised, furtive, hidden, private, secret, secretive, shifty, sly, *inf* sneaky, stealthy, underhand. *Opp* BLATANT.

surround *vb* besiege, beset, cocoon, cordon off, encircle, enclose, encompass, engulf, environ, girdle, hem in, hedge in, ring, skirt, trap, wrap.

surrounding *adj* adjacent, adjoining, bordering, local, nearby, neighbouring.

surroundings *n* ambience, area, background, context, environment, location, milieu, neighbourhood, setting, vicinity.

surveillance *n* check, observation, reconnaissance, scrutiny, supervision, vigilance, watch.

survey *n* appraisal, assessment, census, count, evaluation, examination, inquiry, inspection, investigation, review, scrutiny, study, triangulation. • *vb* 1 appraise, assess, estimate, evaluate, examine, inspect, investigate, look over, review, scrutinize, study, view, weigh up. 2 do a survey of, map out, measure, plan out, plot, reconnoitre, triangulate.

survival *n* continuance, continued existence, persistence.

survive *vb* 1 *inf* bear up, carry on, continue, endure, keep going, last, live, persist, remain. 2 *survive disaster.* come through, live through, outlast, outlive, pull through, weather, withstand. *Opp* SUCCUMB.

susceptible *adj* affected (by), disposed, given, inclined, liable, open, predisposed, prone, responsive, sensitive, vulnerable. *Opp* RESISTANT.

suspect *adj* doubtful, dubious, inadequate, questionable, *inf* shady, suspected, suspicious, unconvincing, unreliable, unsatisfactory, untrustworthy. • *vb* 1 call into question, disbelieve, distrust, doubt, have suspicions about, mistrust. 2 *suspect that she's lying*. believe, conjecture, consider, guess, imagine, infer, presume, speculate, suppose, surmise, think.

suspend *vb* 1 dangle, hang, swing. 2 *suspend work*. adjourn, break off, defer, delay, discontinue, freeze, hold in abeyance, hold up, interrupt, postpone, put off, *inf* put on ice, shelve. 3 *suspend from duty*. debar, dismiss, exclude, expel, lay off, lock out, send down.

suspense *n* anticipation, anxiety, apprehension, doubt, drama, excitement, expectancy, expectation, insecurity, irresolution, nervousness, not knowing, tension, uncertainty, waiting.

suspicion *n* 1 apprehension, apprehensiveness, caution, distrust, doubt, dubiety, dubiousness, *inf* funny feeling, guess, hesitation, *inf* hunch, impression, misgiving, mistrust, presentiment, qualm, scepticism, uncertainty, wariness. 2 *suspicion of a smile*. glimmer, hint, inkling, shadow, suggestion, tinge, touch, trace.

suspicious *adj* 1 apprehensive, *inf* chary, disbelieving, distrustful, doubtful, dubious, incredulous, in doubt, mistrustful, sceptical, uncertain, unconvinced, uneasy, wary. *Opp* TRUSTFUL. 2 *suspicious character*. disreputable, dubious, *inf* fishy, peculiar, questionable, *inf* shady, suspect, suspected, unreliable, untrustworthy. *Opp* TRUSTWORTHY.

sustain *vb* 1 continue, develop, elongate, extend, keep alive, keep going, keep up, maintain, prolong. 2 ▷ SUPPORT.

sustenance *n* eatables, edibles, food, foodstuffs, nourishment, nutriment, provender, provisions, rations, *old use* victuals.

swag *n* booty, loot, plunder, takings.

swagger *vb* parade, strut. ▷ WALK.

swallow *vb* consume, *inf* down, gulp down, guzzle, ingest, take down. ▷ DRINK, EAT. **swallow up** absorb, assimilate, enclose, enfold, make disappear. ▷ SWAMP.

swamp *n* bog, fen, marsh, marshland, morass, mud, mudflats, quagmire, quicksand, saltmarsh, *old use* slough, wetlands. • *vb* deluge, drench, engulf, envelop, flood, immerse, inundate, overcome, overwhelm, sink, submerge, swallow up.

swampy *adj* boggy, marshy, muddy, soft, soggy, unstable, waterlogged, wet. *Opp* DRY, FIRM.

swarm *n* cloud, crowd, hive, horde, host, multitude. ▷ GROUP. • *vb* cluster, congregate, crowd, flock, gather, mass, move in a swarm, throng. **swarm up** ▷ CLIMB. **swarm with** ▷ TEEM.

swarthy *adj* brown, dark, dark-complexioned, dark-skinned, dusky, tanned.

swashbuckling *adj* adventurous, aggressive, bold, daredevil, daring, dashing, *inf* macho, manly, swaggering. *Opp* TIMID.

sway *vb* 1 bend, fluctuate, lean from side to side, oscillate, rock, roll, swing, undulate, wave. 2 *sway opinions*. affect, bias, bring round, change (someone's mind), convert, convince, govern, influence, persuade, win over. 3 *sway from a chosen path*. divert, go off course, swerve, veer, waver.

swear *vb* 1 affirm, asseverate, attest, aver, avow, declare, give your word, insist, pledge, promise, state on oath, take an oath, testify, vouchsafe, vow. 2 blaspheme, curse, execrate, imprecate, use swearwords, utter profanities.

swearword *n inf* bad language, blasphemy, curse, execration, expletive, *inf* four-letter word, imprecation, oath, obscenity, profanity, swearing.

sweat *vb* 1 *inf* glow, perspire, swelter. 2 ▷ WORK.

sweaty *adj* clammy, damp, moist, perspiring, sticky, sweating.

sweep *vb* brush, clean, clear, dust, tidy up. **sweep along** ▷ MOVE. **sweep away** ▷ REMOVE. **sweeping** ▷ GENERAL, SUPERFICIAL.

sweet *adj* 1 aromatic, fragrant, honeyed, luscious, mellow, perfumed, sweetened, sweetscented, sweet-smelling. 2 [*derog*] *sickly sweet*. cloying, saccharine, sentimental, sickening, sickly, sugary, syrupy, treacly. 3 *sweet sounds*. dulcet, euphonious, harmonious, heavenly, mellifluous, melodious, musical, pleasant, silvery, soothing, tuneful. 4 *a sweet nature*. affectionate, amiable, attractive, charming, dear, endearing, engaging, friendly, genial, gentle, gracious, lovable, lovely, nice, pretty, unselfish, winning. *Opp* ACID, BITTER, NASTY, SAVOURY. • *n* 1 *inf* afters, dessert, pudding. 2 [*usu plur*] *old use* bon-bons, *Amer* candy, confectionery, *inf* sweeties, *old use* sweetmeats. □ *acid drop, barley sugar, boiled sweet, bull's-eye, butterscotch, candy, candyfloss, caramel, chewing-gum, chocolate, fondant, fruit pastille, fudge, humbug, liquorice, lollipop, marshmallow, marzipan, mint, nougat, peppermint, rock, toffee, Turkish delight.*

sweeten *vb* 1 make sweeter, sugar. 2 *sweeten your temper*. appease, assuage, calm, mellow, mollify, pacify, soothe.

sweetener *n* □ *artificial sweetener, honey, saccharine, sugar, sweetening, syrup.*

swell *vb* 1 balloon, become bigger, belly, billow, blow up, bulge, dilate, distend, enlarge, expand, fatten, fill out, grow, increase, inflate, mushroom, puff up, rise. 2 *swell numbers*. augment, boost, build up, extend, increase, make bigger, raise, step up. *Opp* SHRINK.

swelling *n* blister, boil, bulge, bump, distension, enlargement, excrescence, hump, inflammation, knob, lump, node, nodule, prominence, protrusion, protuberance, tumescence, tumour.

sweltering *adj* humid, muggy, oppressive, steamy, sticky, stifling, sultry, torrid, tropical. ▷ HOT.

swerve *vb* career, change direction, deviate, diverge, dodge about, sheer off, swing, take avoiding action, turn aside, veer, wheel.

swift *adj* agile, brisk, expeditious, fast, *old use* fleet, fleet-footed, hasty, hurried, nimble, *inf* nippy, prompt, quick, rapid, speedy, sudden. *Opp* SLOW.

swill *vb* 1 bathe, clean, rinse, sponge down, wash. 2 ▷ DRINK.

swim *vb* bathe, dive in, float, go swimming, *inf* take a dip.

swimming-bath *n* baths, leisure-pool, lido, swimming-pool.

swim-suit *n* bathing-costume, bathing-dress, bathing-suit, bikini, swimwear, trunks.

swindle *n* cheat, chicanery, *inf* con, confidence trick, deception, double-dealing, fraud, knavery, *inf* racket, *inf* rip-off, *inf* sharp practice, *inf* swizz, trickery. • *vb inf* bamboozle, cheat, *inf* con, cozen, deceive, defraud, *inf* diddle, *inf* do, double-cross, dupe, exploit, *inf* fiddle, *inf* fleece, fool, gull, hoax, hoodwink, mulct, *inf* pull a fast one on you, *inf* rook, *inf* take you for a ride, trick, *inf* welsh (*on a bet*).

swindler *n* charlatan, cheat, cheater, *inf* con man, counterfeiter, double-crosser, extortioner, forger, fraud, hoaxer, impostor, knave, mountebank, quack, racketeer, scoundrel, *inf* shark, trickster, *inf* twister.

swing *n* change, fluctuation, movement, oscillation, shift, variation. • *vb* **1** be suspended, dangle, flap, fluctuate, hang loose, move from side to side, move to and fro, oscillate, revolve, rock, roll, sway, swivel, turn, twirl, wave about. **2** *swing opinion.* affect, bias, bring round, change (someone's mind), convert, convince, govern, influence, persuade, win over. **3** *support swung to the opposition.* change, move across, shift, transfer, vary. **4** *swing from a path.* deviate, divert, go off course, swerve, veer, waver, zigzag.

swipe *vb* **1** lash out at, strike, swing at. ▷ HIT. **2** ▷ STEAL.

swirl *vb* boil, churn, circulate, curl, eddy, move in circles, seethe, spin, surge, twirl, twist, whirl.

switch *n* circuit-breaker, light-switch, power-point. • *vb* change, divert, exchange, redirect, replace, reverse, shift, substitute, *inf* swap, transfer, turn.

swivel *vb* gyrate, pirouette, pivot, revolve, rotate, spin, swing, turn, twirl, wheel.

swoop *vb* descend, dive, drop, fall, fly down, lunge, plunge, pounce. **swoop on** ▷ RAID.

sword *n* blade, broadsword, cutlass, dagger, foil, kris, rapier, sabre, scimitar.

sycophantic *adj* flattering, servile, *inf* smarmy, toadyish, unctuous. ▷ FLATTER.

syllabus *n* course, curriculum, outline, programme of study.

symbol *n* mark, sign, token. □ *badge, brand, character, cipher, coat of arms, crest, emblem, figure, hieroglyph, ideogram, ideograph, image, insignia, letter, logo, logotype, monogram, motif, number, numeral, pictogram, pictograph, trademark.*

symbolic *adj* allegorical, emblematic, figurative, meaningful, metaphorical, representative, significant, suggestive, symptomatic, token (*gesture*).

symbolize *vb* be a sign of, betoken, communicate, connote, denote, epitomize, imply, indicate, mean, represent, signify, stand for, suggest.

symmetrical *adj* balanced, even, proportional, regular. *Opp* ASYMMETRICAL.

sympathetic *adj* benevolent, caring, charitable, comforting, compassionate, commiserating, concerned, consoling, empathetic, friendly, humane, interested, kind-hearted, kindly, merciful, pitying, soft-hearted, solicitous, sorry, supportive, tender, tolerant, understanding, warm. *Opp* UNSYMPATHETIC.

sympathize *vb inf* be on the same wavelength, be sorry, be sympathetic, comfort, commiserate, condole, console, empathize, feel, grieve, have sympathy, identify (with), mourn, pity, respond, show sympathy, understand.

sympathy *n* affinity, commiseration, compassion, concern, condolence, consideration, empathy, feeling, fellow-feeling, kindness, mercy, pity, rapport, solicitousness, tenderness, understanding.

symptom *n* characteristic, evidence, feature, indication, manifestation, mark, marker, sign, warning, warning-sign.

symptomatic *adj* characteristic, indicative, representative, suggestive, typical.

synthesis *n* amalgamation, blend, coalescence, combination, composite, compound, fusion, integration, union. *Opp* ANALYSIS.

synthetic *adj* artificial, bogus, concocted, counterfeit, ersatz, fabricated, fake, *inf* made-up, man-made, manufactured, mock, *inf* phoney, simulated, spurious, unnatural. *Opp* GENUINE, NATURAL.

syringe *n* hypodermic, needle.

system *n* **1** network, organization, *inf* set-up, structure. **2** approach, arrangement, logic, method, methodology, *Lat* modus operandi, order, plan, practice, procedure, process, routine, rules, scheme, technique. **3** *system of government.* constitution, regime. **4** *system of knowledge.* categorization, classification, code, discipline, philosophy, science, set of principles, theory.

systematic *adj* according to plan, businesslike, categorized, classified, codified, constitutional, co-ordinated, logical, methodical, neat, ordered, orderly, organized, planned, rational, regimented, routine, scientific, structured, tidy, well-arranged, well-organized, well-rehearsed, well-run. *Opp* UNSYSTEMATIC.

systematize *vb* arrange, catalogue, categorize, classify, codify, make systematic, organize, rationalize, regiment, standardize, tabulate.

T

table *n* **1** bench, board, counter, desk, gate-leg table, kitchen table, worktop. **2** *table of information.* agenda, catalogue, chart, diagram, graph, index, inventory, list, register, schedule, tabulation, timetable. • *vb* bring forward, lay on the table, offer, proffer, propose, submit.

tablet *n* **1** capsule, drop, lozenge, medicine, pastille, pellet, pill. **2** *tablet of soap.* bar, block, chunk, piece, slab. **3** *tablet of stone.* gravestone, headstone, memorial, plaque, plate, tombstone.

taboo *adj* banned, censored, disapproved of, forbidden, interdicted, prohibited, proscribed, unacceptable, unlawful, unmentionable, unnamable. ▷ RUDE. • *n* anathema, ban, curse, interdiction, prohibition, proscription, taboo subject.

tabulate *vb* arrange as a table, catalogue, index, list, pigeon-hole, set out in columns, systematize.

tacit *adj* implicit, implied, silent, undeclared, understood, unexpressed, unsaid, unspoken, unvoiced.

taciturn *adj* mute, quiet, reserved, reticent, silent, tight-lipped, uncommunicative, unforthcoming. *Opp* TALKATIVE.

tack *n* **1** drawing-pin, nail, pin, tintack. **2** *the wrong tack.* approach, bearing, course, direction, heading, line, policy, procedure, technique. • *vb* **1** nail, pin. ▷ FASTEN. **2** sew, stitch. **3** *tack in a yacht.* beat against the wind, change course, go about, zigzag. **tack on** ▷ ADD.

tackle *n* **1** accoutrements, apparatus, *inf* clobber, equipment, fittings, gear, implements, kit, outfit, paraphernalia, rig, rigging, tools. **2** *a football tackle.* attack, block, challenge, interception, intervention. • *vb* **1** address (yourself to), apply yourself to, attempt, attend to, combat, *inf* come to grips with, concentrate on, confront, cope with, deal with, engage in, face up to, focus on, get involved in, grapple with, handle, *inf* have a go at, manage, set about, settle down to, sort out, take on, undertake. **2** *tackle an opponent.* attack, challenge, intercept, stop, take on.

tacky *adj* adhesive, gluey, *inf* gooey, gummy, sticky, viscous, wet. *Opp* DRY.

tact *n* adroitness, consideration, delicacy, diplomacy, discernment, discretion, finesse, judgement, perceptiveness, politeness, savoir-faire, sensitivity, tactfulness, thoughtfulness, understanding. *Opp* TACTLESSNESS.

tactful *adj* adroit, appropriate, considerate, courteous, delicate, diplomatic, discreet, judicious, perceptive, polite, politic, sensitive, thoughtful, understanding. *Opp* TACTLESS.

tactical *adj* artful, calculated, clever, deliberate, planned, politic, prudent, shrewd, skilful, strategic.

tactics *n* approach, campaign, course of action, design, device, manoeuvre, manoeuvring, plan, ploy, policy, procedure, ruse, scheme, stratagem, strategy.

tactless *adj* blundering, blunt, boorish, bungling, clumsy, discourteous, gauche, heavyhanded, hurtful, impolite, impolitic, inappropriate, inconsiderate, indelicate, indiscreet, inept, insensitive, maladroit, misjudged, thoughtless, uncivil, uncouth, undiplomatic, unkind. ▷ RUDE. *Opp* TACTFUL.

tactlessness *n* boorishness, clumsiness, gaucherie, indelicacy, indiscretion, ineptitude, insensitivity, lack of diplomacy, misjudgement, thoughtlessness, uncouthness. ▷ RUDENESS. *Opp* TACT.

tag *n* **1** docket, label, marker, name tag, price tag, slip, sticker, tab, ticket. **2** *a Latin tag.* ▷ SAYING. • *vb* identify, label, mark, ticket. **tag along with** ▷ FOLLOW.

tail *n* appendage, back, brush (*of fox*), buttocks, end, extremity, rear, rump, scut (*of rabbit*), tail-end. • *vb* dog, follow, hunt, pursue, shadow, stalk, track, trail. **tail off** ▷ DECLINE.

taint *vb* **1** adulterate, contaminate, defile, dirty, infect, poison, pollute, soil. **2** *taint a reputation.* besmirch, blacken, blemish, damage, dishonour, harm, ruin, slander, smear, spoil, stain, sully, tarnish.

take *vb* **1** acquire, bring, carry away, *inf* cart off, catch, clasp, clutch, fetch, gain, get, grab, grasp, grip, hold, pick up, pluck, remove, secure, seize, snatch, transfer. **2** *take prisoners*. abduct, arrest, capture, catch, corner, detain, ensnare, entrap, secure. **3** *take property*. appropriate, get away with, pocket. ▷ STEAL. **4** *take 2 from 4*. deduct, eliminate, subtract, take away. **5** *take passengers*. accommodate, carry, contain, have room for, hold. **6** *take a partner*. accompany, conduct, convey, escort, ferry, guide, lead, transport. **7** *take a taxi*. engage, hire, make use of, travel by, use. **8** *take a subject*. have lessons in, learn about, read, study. **9** *can't take pain*. abide, accept, bear, brook, endure, receive, *inf* stand, *inf* stomach, suffer, tolerate, undergo, withstand. **10** *take food, drink*. consume, drink, eat, have, swallow. **11** *It takes courage to own up*. necessitate, need, require, use up. **12** *take a new name*. adopt, assume, choose, select. **take aback** ▷ SURPRISE. **take after** ▷ RESEMBLE. **take against** ▷ DISLIKE. **take back** ▷ WITHDRAW. **take in** ▷ ACCOMMODATE, DECEIVE, UNDERSTAND. **take life** ▷ KILL. **take off** ▷ IMITATE. **take off, take out** ▷ REMOVE. **take on, take up** ▷ UNDERTAKE. **take over** ▷ USURP. **take part** ▷ PARTICIPATE. **take place** ▷ HAPPEN. **take to task** ▷ REPRIMAND. **take up** ▷ BEGIN, OCCUPY.

take-over *n* amalgamation, combination, incorporation, merger.

takings *n* earnings, gains, gate, income, proceeds, profits, receipts, revenue.

tale *n* account, anecdote, chronicle, narration, narrative, relation, report, *sl* spiel, story, yarn. ▷ WRITING.

talent *n* ability, accomplishment, aptitude, brilliance, capacity, expertise, facility, faculty, flair, genius, gift, ingenuity, knack, *inf* know-how, prowess, skill, strength, versatility.

talented *adj* able, accomplished, artistic, brilliant, distinguished, expert, gifted, inspired, proficient, skilful, skilled, versatile. ▷ CLEVER. *Opp* UNSKILFUL.

talisman *n* amulet, charm, fetish, mascot.

talk *n* **1** baby-talk, *inf* blarney, chat, *inf* chin-wag, *inf* chit-chat, confabulation, conference, conversation, dialogue, discourse, discussion, gossip, intercourse, language, palaver, *inf* powwow, *inf* tattle, *inf* tittle-tattle, words. **2** *a public talk*. address, diatribe, exhortation, harangue, lecture, oration, *inf* peptalk, presentation, sermon, speech, tirade. • *vb* **1** address one another, articulate ideas, commune, communicate, confer, converse, deliver a speech, discourse, discuss, enunciate, exchange views, have a conversation, *inf* hold forth, lecture, negotiate, *inf* pipe up, pontificate, preach, pronounce words, say something, sermonize, speak, tell, use language, use your voice, utter, verbalize, vocalize. □ *babble, bawl, bellow, blab, blether, blurt out, breathe, burble, call out, chat, chatter, clamour, croak, cry, drawl, drone, gabble, gas, gibber, gossip, grunt, harp, howl, intone, jabber, jaw, jeer, lisp, maunder, moan, mumble, murmur, mutter, natter, patter, prattle, pray, preach, rabbit on, rant, rasp, rattle on, rave, roar, scream, screech, shout, shriek, slur, snap, snarl, speak in an undertone, splutter, spout, squeal, stammer, stutter, utter, vociferate, wail, whimper, whine, whinge, whisper, witter, yell.* **2** *talk French*.

communicate in, express yourself in, pronounce, speak. **3** *get someone to talk.* confess, give information, *inf* grass, inform, *inf* let on, *inf* spill the beans, *inf* squeal, *inf* tell tales. ▷ SAY, SPEAK. **talk about** ▷ DISCUSS. **talk to** ▷ ADDRESS.

talkative *adj* articulate, *inf* chatty, communicative, effusive, eloquent, expansive, garrulous, glib, gossipy, long-winded, loquacious, open, prolix, unstoppable, verbose, vocal, voluble, wordy. *Opp* TACITURN. **talkative person** chatter-box, *sl* gas-bag, gossip, *sl* wind-bag.

tall *adj* colossal, giant, gigantic, high, lofty, soaring, towering. ▷ BIG. *Opp* SHORT.

tally *n* addition, count, reckoning, record, sum, total. • *vb* **1** accord, agree, coincide, concur, correspond, match up, square. **2** *tally up the bill.* add, calculate, compute, count, reckon, total, work out.

tame *adj* **1** amenable, biddable, broken in, compliant, disciplined, docile, domesticated, gentle, manageable, meek, mild, obedient, safe, subdued, submissive, tamed, tractable, trained. **2** *tame animals.* approachable, bold, fearless, friendly, sociable, unafraid. **3** *a tame story.* bland, boring, dull, feeble, flat, insipid, lifeless, tedious, unadventurous, unexciting, uninspiring, uninteresting, vapid, *inf* wishy-washy. *Opp* EXCITING, WILD. • *vb* break in, conquer, curb, discipline, domesticate, house-train, humble, keep under, make tame, master, mollify, mute, quell, repress, subdue, subjugate, suppress, temper, tone down, train.

tamper *vb* **tamper with** alter, *inf* fiddle about with, interfere with, make adjustments to, meddle with, tinker with.

tan *n* sunburn, suntan. • *vb* bronze, brown, burn, colour, darken, get tanned.

tang *n* acidity, *inf* bite, *inf* edge, *inf* nip, piquancy, pungency, savour, sharpness, spiciness, zest.

tangible *adj* actual, concrete, corporeal, definite, material, palpable, perceptible, physical, positive, provable, real, solid, substantial, tactile, touchable. *Opp* INTANGIBLE.

tangle *n* coil, complication, confusion, jumble, jungle, knot, labyrinth, mass, maze, mesh, mess, muddle, scramble, twist, web. • *vb* **1** complicate, confuse, entangle, entwine, *inf* foul up, intertwine, interweave, muddle, ravel, scramble, *inf* snarl up, twist. **2** *tangle fish in a net.* catch, enmesh, ensnare, entrap, trap. *Opp* DISENTANGLE, FREE. **3** *tangle with criminals.* become involved with, confront, cross. **tangled** ▷ DISHEVELLED, INTRICATE.

tangy *adj* acid, appetizing, bitter, fresh, piquant, pungent, refreshing, sharp, spicy, strong, tart. *Opp* BLAND.

tank *n* **1** aquarium, basin, cistern, reservoir. **2** *army tank.* armoured vehicle.

tanned *adj* brown, sunburnt, suntanned, weather-beaten.

tantalize *vb* bait, entice, frustrate, *inf* keep on tenterhooks, lead on, plague, provoke, taunt, tease, tempt, titillate, torment.

tap *n* **1** *Amer* faucet, spigot, stopcock, valve. **2** knock, rap. • *vb* knock, rap, strike. ▷ HIT.

tape *n* **1** band, belt, binding, braid, fillet, ribbon, strip, stripe.

2 audiotape, cassette, magnetic tape, tape-recording, videotape.

taper *n* candle, lighter, spill. • *vb* attenuate, become narrower, narrow, thin.

taper off ▷ DECLINE.

target *n* 1 aim, ambition, end, goal, hope, intention, objective, purpose. 2 *target of attack.* butt, object, quarry, victim.

tariff *n* 1 charges, menu, price-list, schedule. 2 *tariff on imports.* customs, duty, excise, impost, levy, tax, toll.

tarnish *vb* 1 blacken, corrode, dirty, discolour, soil, spoil, stain, taint. 2 *tarnish a reputation.* blemish, blot, calumniate, defame, denigrate, disgrace, dishonour, mar, ruin, spoil, stain, sully.

tarry *vb* dawdle, delay, *inf* hang about, hang back, linger, loiter, pause, procrastinate, temporize, wait.

tart *adj* 1 acid, acidic, acidulous, astringent, biting, citrus, harsh, lemony, piquant, pungent, sharp, sour, tangy. 2 *a tart rejoinder.* ▷ SHARP. *Opp* BLAND, SWEET. • *n* 1 flan, pastry, pasty, patty, pie, quiche, tartlet, turnover. 2 ▷ PROSTITUTE.

task *n* activity, assignment, burden, business, charge, chore, duty, employment, enterprise, errand, imposition, job, mission, requirement, test, undertaking, work. **take to task** ▷ REPRIMAND.

taste *n* 1 character, flavour, relish, savour. 2 bit, bite, morsel, mouthful, nibble, piece, sample, titbit. 3 *an acquired taste.* appetite, appreciation, choice, fancy, fondness, inclination, judgement, leaning, liking, partiality, preference. 4 *a person of taste.* breeding, cultivation, culture, discernment, discretion, discrimination, education, elegance, fashion sense, finesse, good judgement, perception, perceptiveness, polish, refinement, sensitivity, style, tastefulness. • *vb* nibble, relish, sample, savour, sip, test, try. **in bad taste** ▷ TASTELESS. **in good taste** ▷ TASTEFUL.

tasteful *adj* aesthetic, artistic, attractive, charming, *Fr* comme il faut, correct, cultivated, decorous, dignified, discerning, discreet, discriminating, elegant, fashionable, in good taste, judicious, *inf* nice, polite, proper, refined, restrained, sensitive, smart, stylish, tactful, well-judged. *Opp* TASTELESS.

tasteless *adj* 1 cheap, coarse, crude, *inf* flashy, garish, gaudy, graceless, improper, inartistic, in bad taste, indecorous, indelicate, inelegant, injudicious, in poor taste, *inf* kitsch, *inf* loud, meretricious, ugly, unattractive, uncouth, uncultivated, undiscriminating, unfashionable, unimaginative, unpleasant, unrefined, unseemly, unstylish, vulgar. *Opp* TASTEFUL. 2 *tasteless food.* bland, characterless, flavourless, insipid, mild, uninteresting, watered-down, watery, weak, *inf* wishy-washy. *Opp* TASTY.

tasty *adj* appetizing, delectable, delicious, flavoursome, luscious, *inf* mouth-watering, *inf* nice, palatable, *inf* scrumptious, toothsome, *sl* yummy. □ *acid, bitter, creamy, fruity, hot, meaty, peppery, piquant, salty, savoury, sharp, sour, spicy, sugary, sweet, tangy, tart.* *Opp* TASTELESS.

tattered *adj* frayed, ragged, rent, ripped, shredded, tatty, threadbare, torn, worn out. *Opp* SMART.

tatters *plur n* bits, pieces, rags, ribbons, shreds, torn pieces.

tatty *adj* 1 frayed, old, patched, ragged, ripped, scruffy, shabby, tattered, torn, threadbare, untidy, worn out. 2 ▷ TAWDRY. *Opp* SMART.

taunt *vb* annoy, goad, insult, jeer at, reproach, tease, torment. ▷ RIDICULE.

taut *adj* firm, rigid, stiff, strained, stretched, tense, tight. *Opp* SLACK.

tautological *adj* long-winded, otiose, pleonastic, prolix, redundant, repetitious, repetitive, superfluous, tautologous, verbose, wordy. *Opp* CONCISE.

tautology *n* duplication, long-windedness, pleonasm, prolixity, repetition, verbiage, verbosity, wordiness.

tavern *n old use* alehouse, bar, hostelry, inn, *inf* local, pub, public house.

tawdry *adj inf* Brummagem, cheap, common, eye-catching, fancy, *inf* flashy, garish, gaudy, inferior, meretricious, poor quality, showy, tasteless, tatty, tinny, vulgar, worthless. *Opp* TASTEFUL.

tax *n* charge, due, duty, imposition, impost, levy, tariff, *old use* tribute. □ *airport tax, community charge, corporation tax, customs, death duty, estate duty, excise, income tax, poll tax, property tax, rates,* old use *tithe, toll, value added tax. vb* 1 assess, exact, impose a tax on, levy a tax on. 2 *tax someone's patience.* burden, exhaust, make heavy demands on, overwork, pressure, pressurize, strain, try. ▷ TIRE. **tax with** accuse of, blame for, censure for, charge with, reproach for, reprove for.

taxi *n* cab, *old use* hackney carriage, minicab.

teach *vb* advise, brainwash, coach, counsel, demonstrate to, discipline, drill, edify, educate, enlighten, familiarize with, give lessons in, ground in, impart knowledge to, implant knowledge in, inculcate habits in, indoctrinate, inform, instruct, lecture, school, train, tutor.

teacher *n* adviser, educator, guide. □ *coach, counsellor, demonstrator, don, governess, guru, headteacher, housemaster, housemistress, instructor, lecturer, maharishi, master, mentor, mistress, pedagogue, preacher, preceptor, professor, pundit, schoolmaster, schoolmistress, schoolteacher, trainer, tutor.*

teaching *n* 1 education, guidance, instruction, training. □ *brainwashing, briefing, coaching, computer-aided learning, counselling, demonstration, familiarization, grounding, indoctrination, lecture, lesson, practical, preaching, rote learning, schooling, seminar, tuition, tutorial, work experience, workshop.* 2 *religious teachings.* doctrine, dogma, gospel, precept, principle, tenet.

team *n* club, crew, gang, *inf* line up, side. ▷ GROUP.

tear *n* 1 [rhymes with *fear*] droplet, tear-drop. [*plur*] *inf* blubbering, crying, sobs, weeping. 2 [rhymes with *bear*] cut, fissure, gap, gash, hole, laceration, opening, rent, rip, slit, split. • *vb* claw, gash, lacerate, mangle, pierce, pull apart, rend, rip, rive, rupture, scratch, sever, shred, slit, snag, split. **shed tears** ▷ WEEP.

tearful *adj inf* blubbering, crying, emotional, in tears, lachrymose, snivelling, sobbing, weeping, *inf* weepy, wet-cheeked, whimpering. ▷ SAD.

tease *vb inf* aggravate, annoy, badger, bait, chaff, goad, harass, irritate, laugh at, make fun of, mock, *inf* needle, *inf* nettle, pester, plague, provoke, *inf* pull someone's leg, *inf* rib, tantalize, taunt, torment, vex, worry. ▷ RIDICULE.

teasing *n* badinage, banter, chaffing, joking, mockery, provocation, raillery, *inf* ribbing, ridicule, taunts.

technical *adj* **1** complicated, detailed, esoteric, expert, professional, specialized. **2** *technical skill.* engineering, industrial, mechanical, technological, scientific.

technician *n* engineer, mechanic, skilled worker, *plur* technical staff.

technique *n* **1** approach, dodge, knack, manner, means, method, mode, procedure, routine, system, trick, way. **2** *an artist's technique.* art, artistry, cleverness, craft, craftsmanship, expertise, facility, *inf* know-how, proficiency, skill, talent, workmanship.

technological *adj* advanced, automated, computerized, electronic, scientific.

tedious *adj* banal, boring, dreary, *inf* dry-as-dust, dull, endless, *inf* humdrum, irksome, laborious, long-drawn-out, long-winded, monotonous, prolonged, repetitious, slow, soporific, tiresome, tiring, unexciting, uninteresting, vapid, wearing, wearisome, wearying. *Opp* INTERESTING.

tedium *n* boredom, dreariness, dullness, ennui, long-windedness, monotony, repetitiousness, slowness, tediousness.

teem *vb* **1** abound (in), be alive (with), be full (of), be infested, be overrun (by), *inf* bristle, *inf* crawl, proliferate, seethe, swarm with. **2** ▷ RAIN.

teenager *n* adolescent, boy, girl, juvenile, minor, youngster, youth.

teetotal *adj* abstemious, abstinent, *sl* on the wagon, restrained, self-denying, self-disciplined, temperate.

teetotaller *n* abstainer, non-drinker.

telegram *n* cable, cablegram, fax, telex, wire.

telepathic *adj* clairvoyant, psychic.

telephone *n inf* blower, carphone, handset, phone. • *vb inf* buzz, call, dial, *inf* give someone a buzz, *inf* give someone a call, *inf* give someone a tinkle, phone, ring, ring up.

telescope *vb* abbreviate, collapse, compress, elide, shorten.

telescopic *adj* adjustable, collapsible, expanding, extending, retractable.

televise *vb* broadcast, relay, send out, transmit.

television *n inf* the box, monitor, receiver, *inf* small screen, *inf* telly, video.

tell *vb* **1** acquaint with, advise, announce, assure, communicate, describe, disclose, divulge, explain, impart, inform, make known, narrate, notify, portray, promise, recite, recount, rehearse, relate, reveal, utter. ▷ SPEAK, TALK. **2** *tell the difference.* calculate, comprehend, decide, discover, discriminate, distinguish, identify, notice, recognize, see. **3** *told me what to do.* command, direct, instruct, order. **tell off** ▷ REPRIMAND.

teller *n* **1** author, narrator, raconteur, storyteller. **2** *teller in a bank.* bank clerk, cashier.

telling *adj* considerable, effective, influential, potent, powerful, significant, striking, weighty.

temper *n* **1** attitude, character, disposition, frame of mind, humour, *inf* make-up, mood, personality, state of mind, temperament. **2** *watch your temper.* anger, churlishness, fit of anger, fury, hot-headedness, ill-humour, irascibility, irritability, *inf* paddy, passion, peevishness, petulance, rage, surliness, tantrum, unpredictability, volatility, *inf* wax, wrath. **3** *keep your temper.* calmness, composure, *sl* cool, coolness, equanimity, sang-froid, self-control, self-possession. • *vb* **1** assuage, lessen, mitigate, moderate, modify, modulate, reduce, soften, soothe, tone down. **2** *temper steel.* harden, strengthen, toughen.

temperament *n* attitude, character, *old use* complexion, disposition, frame of mind, *old use* humour, *inf* make-up, mood, nature, personality, spirit, state of mind, temper.

temperamental *adj* **1** characteristic, congenital, constitutional, inherent, innate, natural. **2** *temperamental moods.* capricious, changeable, emotional, erratic, excitable, explosive, fickle, highly-strung, impatient, inconsistent, inconstant, irascible, irritable, mercurial, moody, neurotic, passionate, sensitive, touchy, undependable, unpredictable, unreliable, *inf* up and down, variable, volatile.

temperance *n* abstemiousness, continence, moderation, self-discipline, self-restraint, sobriety, teetotalism.

temperate *adj* calm, controlled, disciplined, moderate, reasonable, restrained, self-possessed, sensible, sober, stable, steady. *Opp* EXTREME.

tempest *n* cyclone, gale, hurricane, tornado, tumult, typhoon, whirlwind. ▷ STORM.

tempestuous *adj* fierce, furious, tumultuous, turbulent, vehement, violent, wild. ▷ STORMY. *Opp* CALM.

temple *n* church, house of god, mosque, pagoda, place of worship, shrine, synagogue.

tempo *n* beat, pace, rate, rhythm, pulse, speed.

temporal *adj* earthly, fleshly, impermanent, material, materialistic, mortal, mundane, non-religious, passing, secular, sublunary, terrestrial, transient, transitory, worldly. *Opp* SPIRITUAL.

temporary *adj* **1** brief, ephemeral, evanescent, fleeting, fugitive, impermanent, interim, makeshift, momentary, passing, provisional, short, short-lived, short-term, stop-gap, transient, transitory. **2** *temporary captain.* acting. *Opp* PERMANENT.

tempt *vb* allure, attract, bait, bribe, cajole, captivate, coax, decoy, entice, fascinate, inveigle, lure, offer incentives, persuade, seduce, tantalize, woo. **tempting** ▷ APPETIZING, ATTRACTIVE.

temptation *n* allure, allurement, appeal, attraction, cajolery, coaxing, draw, enticement, fascination, inducement, lure, persuasion, pull, seduction, snare, wooing.

tenable *adj* arguable, believable, conceivable, credible, creditable, defendable, defensible, feasible, justifiable, legitimate, logical, plausible, rational, reasonable, sensible, sound, supportable, understandable, viable. *Opp* INDEFENSIBLE.

tenacious *adj* determined, dogged, firm, intransigent, obdurate, obstinate, persistent, pertinacious, resolute, single-minded, steadfast, strong, stubborn, tight, uncompromising, unfaltering, unshakeable, unswerving, unwavering, unyielding. *Opp* WEAK.

tenant *n* inhabitant, leaseholder, lessee, lodger, occupant, occupier, resident.

tend *vb* **1** attend to, care for, cherish, cultivate, guard, keep, *inf* keep an eye on, look after, manage, mind, minister to, mother, protect, supervise, take care of, watch. **2** *tend the sick.* nurse, treat. **3** *tend to fall asleep.* be biased, be disposed, be inclined, be liable, be prone, have a tendency, incline.

tendency *n* bias, disposition, drift, inclination, instinct, leaning, liability, partiality, penchant, predilection, predisposition, proclivity, proneness, propensity, readiness, susceptibility, trend.

tender *adj* **1** dainty, delicate, fleshy, fragile, frail, green, immature, soft, succulent, vulnerable, weak, young. **2** *tender meat.* chewable, eatable, edible. **3** *a tender wound.* aching, inflamed, painful, sensitive, smarting, sore. **4** *a tender love-song.* emotional, heartfelt, moving, poignant, romantic, sentimental, stirring, touching. **5** *tender care.* affectionate, amorous, caring, compassionate, concerned, considerate, fond, gentle, humane, kind, loving, merciful, pitying, soft-hearted, sympathetic, tender-hearted, warm-hearted. *Opp* TOUGH, UNSYMPATHETIC.

tense *adj* **1** rigid, strained, stretched, taut, tight. **2** *a tense person.* anxious, apprehensive, edgy, excited, fidgety, highly-strung, intense, jittery, jumpy, *inf* keyed-up, nervous, on edge, *inf* on tenterhooks, overwrought, restless, strained, stressed, *inf* strung up, touchy, uneasy, *sl* uptight, worried. **3** *a tense situation.* exciting, fraught, *inf* nail-biting, nerve-racking, stressful, worrying. *Opp* RELAXED.

tension *n* **1** pull, strain, stretching, tautness, tightness. **2** *the tension of waiting.* anxiety, apprehension, edginess, excitement, nervousness, stress, suspense, unease, worry. *Opp* RELAXATION.

tent *n* □ *bell tent, big-top, frame tent, marquee, ridge tent, tepee, trailer tent, wigwam.*

tentative *adj* cautious, diffident, doubtful, experimental, exploratory, half-hearted, hesitant, inconclusive, indecisive, indefinite, nervous, preliminary, provisional, shy, speculative, timid, uncertain, uncommitted, unsure, *inf* wishy-washy. *Opp* DECISIVE.

tenuous *adj* attenuated, fine, flimsy, fragile, insubstantial, slender, slight, weak. ▷ THIN. *Opp* STRONG.

tepid *adj* **1** lukewarm, warm. **2** *a tepid response.* ▷ APATHETIC.

term *n* **1** duration, period, season, span, spell, stretch, time. **2** *a school term. Amer* semester, session. **3** *technical terms.* appellation, designation, epithet, expression, name, phrase, saying, title, word. **terms 1** conditions, particulars, provisions, provisos, specifications, stipulations. **2** *a hotel's terms.* charges, fees, prices, rates, schedule, tariff.

terminal *adj* deadly, fatal, final, incurable, killing, lethal, mortal. • *n* **1** keyboard, VDU, workstation. **2** *passenger terminal.*

airport, terminus. **3** *electric terminal.* connection, connector, coupling.

terminate *vb* bring to an end, cease, come to an end, discontinue, end, finish, *inf* pack in, phase out, stop, *inf* wind up. ▷ END. *Opp* BEGIN.

terminology *n* argot, cant, choice of words, jargon, language, nomenclature, phraseology, special terms, technical language, vocabulary.

terminus *n* destination, last stop, station, terminal, termination.

terrain *n* country, ground, land, landscape, territory, topography.

terrestrial *adj* earthly, mundane, ordinary, sublunary.

terrible *adj* **1** acute, appalling, awful, *inf* beastly, distressing, dreadful, fearful, fearsome, formidable, frightening, frightful, ghastly, grave, gruesome, harrowing, hideous, horrendous, horrible, horrific, horrifying, insupportable, intolerable, loathsome, nasty, nauseating, outrageous, revolting, shocking, terrific, terrifying, unbearable, vile. **2** ▷ BAD.

terrific *adj Terrific* may mean *causing terror* (▷ TERRIBLE). It is more often used *informally* of anything which is *extreme* in its own way: *a terrific problem* ▷ EXTREME; *terrific size* ▷ BIG; *a terrific party* ▷ EXCELLENT; *a terrific storm* ▷ VIOLENT.

terrify *vb* appal, dismay, horrify, *inf* make your blood run cold, petrify, shock, terrorize. ▷ FRIGHTEN. **terrified** ▷ FRIGHTENED. **terrifying** ▷ FRIGHTENING.

territory *n* area, colony, *old use* demesne, district, domain, dominion, enclave, jurisdiction, land, neighbourhood, precinct, preserve, province, purlieu, region, sector, sphere, state, terrain, tract, zone. ▷ COUNTRY.

terror *n* alarm, awe, consternation, dismay, dread, fright, *inf* funk, horror, panic, shock, trepidation. ▷ FEAR.

terrorist *n* assassin, bomber, desperado, gunman, hijacker.

terrorize *vb* browbeat, bully, coerce, cow, intimidate, menace, persecute, terrify, threaten, torment, tyrannize. ▷ FRIGHTEN.

terse *adj* abrupt, brief, brusque, compact, concentrated, concise, crisp, curt, epigrammatic, incisive, laconic, pithy, short, *inf* short and sweet, *inf* snappy, succinct, to the point. *Opp* VERBOSE.

test *n* analysis, appraisal, assay, assessment, audition, *inf* check-over, *inf* check-up, evaluation, examination, inspection, interrogation, investigation, probation, quiz, screen-test, trial, *inf* try-out. • *vb* analyse, appraise, assay, assess, audition, check, evaluate, examine, experiment with, inspect, interrogate, investigate, probe, *inf* put someone through their paces, put to the test, question, quiz, screen, try out.

testify *vb* affirm, attest, aver, bear witness, declare, give evidence, proclaim, state on oath, swear, vouch, witness.

testimonial *n* character reference, commendation, recommendation, reference.

testimony *n* affidavit, assertion, declaration, deposition, evidence, statement, submission.

tether *n* chain, cord, fetter, halter, lead, leash, painter, restraint, rope. • *vb* chain up, fetter, keep on a tether, leash, restrain, rope, secure, tie up. ▷ FASTEN.

text *n* **1** argument, content, contents, matter, subject matter, wording. **2** *a literary text.* book, textbook, work. ▷ WRITING. **3** *a text from scripture.* line, motif, passage, quotation, sentence, theme, topic, verse.

textile *n* fabric, material, stuff. ▷ CLOTH.

texture *n* appearance, composition, consistency, feel, grain, quality, surface, tactile quality, touch, weave.

thank *vb* acknowledge, express thanks, say thank you, show gratitude.

thankful *adj* appreciative, contented, glad, grateful, happy, indebted, pleased, relieved. *Opp* UNGRATEFUL.

thankless *adj* bootless, futile, profitless, unappreciated, unrecognized, unrewarded, unrewarding. *Opp* PROFITABLE.

thanks *plur n* acknowledgement, appreciation, gratefulness, gratitude, recognition, thanksgiving. **thanks to** as a result of, because of, owing to, through.

thaw *vb* become liquid, defrost, de-ice, heat up, melt, soften, uncongeal, unfreeze, unthaw, warm up. *Opp* FREEZE.

theatre *n* **1** auditorium, hall, opera-house, playhouse. **2** acting, dramaturgy, histrionic arts, show business, thespian arts. □ *ballet, masque, melodrama, mime, musical, music-hall, opera, pantomime, play.* ▷ DRAMA, ENTERTAINMENT, PERFORMANCE.

theatrical *adj* **1** dramatic, histrionic, thespian. **2** [*derog*] *a theatrical exit.* affected, artificial, calculated, demonstrative, exaggerated, forced, *inf* hammy, melodramatic, ostentatious, overacted, overdone, *inf* over the top, pompous, self-important, showy, stagy, stilted, unconvincing, unnatural. *Opp* NATURAL.

theft *n* burglary, larceny, pilfering, robbery, thievery. ▷ STEALING.

theme *n* **1** argument, core, essence, gist, idea, issue, keynote, matter, point, subject, text, thesis, thread, topic. **2** *a musical theme.* air, melody, motif, subject, tune.

theology *n* divinity, religion, religious studies.

theoretical *adj* abstract, academic, conjectural, doctrinaire, hypothetical, ideal, notional, pure (*science*), putative, speculative, suppositious, unproven, untested. *Opp* PRACTICAL, PROVEN.

theorize *vb* conjecture, form a theory, guess, hypothesize, speculate.

theory *n* **1** argument, assumption, belief, conjecture, explanation, guess, hypothesis, idea, notion, speculation, supposition, surmise, thesis, view. **2** *theory of a subject.* laws, principles, rules, science. *Opp* PRACTICE.

therapeutic *adj* beneficial, corrective, curative, healing, healthy, helpful, medicinal, remedial, restorative, salubrious. *Opp* HARMFUL.

therapist *n* counsellor, healer, physiotherapist, psychoanalyst, psychotherapist.

therapy *n* cure, healing, remedy, tonic, treatment. □ *chemotherapy, group therapy, hydrotherapy, hypnotherapy, occupational therapy, physiotherapy, psychotherapy, radiotherapy.* ▷ MEDICINE.

therefore *adv* accordingly, consequently, hence, so, thus.

thesis *n* **1** argument, assertion, contention, hypothesis, idea, opinion, postulate, premise, premiss, proposition, theory, view. **2** *a research thesis.* disquisition, dissertation, essay, monograph, paper, tract, treatise.

thick *adj* **1** broad, *inf* bulky, chunky, stout, sturdy, wide. ▷ FAT. **2** *a thick layer.* deep, heavy, substantial, woolly. **3** *a thick crowd.* compact, dense, impassable, impenetrable, numerous, packed, solid. **4** *thick liquid.* clotted, coagulated, concentrated, condensed, firm, glutinous, heavy, jellied, sticky, stiff, viscid, viscous. **5** *thick growth.* abundant, bushy, luxuriant, plentiful. **6** *thick with visitors.* alive, bristling, *inf* chock-full, choked, covered, crammed, crawling, crowded, filled, full, jammed, swarming, teeming. *Opp* THIN.

thicken *vb* coagulate, clot, concentrate, condense, congeal, firm up, gel, jell, reduce, solidify, stiffen.

thickness *n* **1** breadth, density, depth, fatness, viscosity, width. **2** *a thickness of paint, rock.* coating, layer, seam, stratum.

thief *n* bandit, brigand, burglar, cat-burglar, cutpurse, embezzler, footpad, highwayman, housebreaker, kleptomaniac, looter, mugger, peculator, pickpocket, pilferer, pirate, plagiarist, poacher, purloiner, robber, safe-cracker, shoplifter, stealer, swindler. ▷ CRIMINAL.

thieving *adj* dishonest, light-fingered, rapacious. • *n* ▷ STEALING.

thin *adj* **1** anorexic, attenuated, bony, cadaverous, emaciated, fine, flat-chested, gangling, gaunt, lanky, lean, narrow, pinched, rangy, scraggy, scrawny, skeletal, skinny, slender, slight, slim, small, spare, spindly, underfed, undernourished, underweight, wiry. *Opp* FAT. **2** *a thin layer.* delicate, diaphanous, filmy, fine, flimsy, gauzy, insubstantial, light, *inf* see-through, shallow, sheer (*silk*), superficial, translucent, wispy. **3** *a thin crowd.* meagre, scanty, scarce, scattered, sparse. **4** *thin liquid.* dilute, flowing, fluid, runny, sloppy, watery, weak. **5** *thin atmosphere.* rarefied. **6** *a thin excuse.* feeble, implausible, tenuous, transparent, unconvincing. *Opp* DENSE, STRONG, THICK. • *vb* dilute, water down, weaken. **thin out** **1** become less dense, decrease, diminish, disperse. **2** make less dense, prune, reduce, trim, weed out.

thing *n* **1** apparatus, artefact, article, body, contrivance, device, entity, gadget, implement, invention, item, object, utensil. **2** affair, circumstance, deed, event, eventuality, happening, incident, occurrence, phenomenon. **3** *a thing on your mind.* concept, detail, fact, factor, feeling, idea, point, statement, thought. **4** *a thing to be done.* act, action, chore, deed, job, responsibility, task. **5** [*inf*] *a thing about snakes.* aversion, fixation, *inf* hang-up, mania, neurosis, obsession, passion, phobia, preoccupation. **things** **1** baggage, belongings, clothing, equipment, *inf* gear, luggage, possessions, *inf* stuff. **2** *How are things?* circumstances, conditions, life.

think *vb* **1** attend, brood, chew things over, cogitate, concentrate, consider, contemplate, day-dream, deliberate, dream, dwell (on), fantasize, give thought (to), meditate, *inf* mull over, muse, ponder, *inf* rack your brains, reason, reflect, remind yourself of, reminisce, ruminate, use your intelligence, work things out, worry. **2** *Do you think it's true?* accept, admit, assume, be convinced, believe, be under the impression, conclude, deem, estimate, feel, guess, have faith, imagine, judge, presume, reckon, suppose, surmise. **think better of** ▷ RECONSIDER. **thinking** ▷ INTELLIGENT, THOUGHTFUL. **think up** ▷ DEVISE.

thinker *n inf* brain, innovator, intellect, inventor, *inf* mastermind, philosopher, sage, savant, scholar.

thirst *n* **1** drought, dryness, thirstiness. **2** *thirst for knowledge.* appetite, craving, desire, eagerness, hunger, itch, longing, love (of), lust, passion, urge, wish, yearning, *inf* yen. • *vb* be thirsty, crave, have a thirst, hunger, long, strive (after), wish, yearn. **thirst for** ▷ WANT.

thirsty *adj* **1** arid, dehydrated, dry, *inf* gasping, panting, parched. **2** *thirsty for news.* avid, craving, desirous, eager, greedy, hankering, itching, longing, voracious, yearning.

thorn *n* barb, bristle, needle, prickle, spike, spine.

thorny *adj* **1** barbed, bristly, prickly, scratchy, sharp, spiky, spiny. **2** ▷ DIFFICULT.

thorough *adj* **1** assiduous, attentive, careful, comprehensive, conscientious, deep, detailed, diligent, efficient, exhaustive, extensive, full, *inf* in-depth, methodical, meticulous, minute, observant, orderly, organized, painstaking, particular, penetrating, probing, scrupulous, searching, systematic, thoughtful, watchful. *Opp* SUPERFICIAL. **2** *a thorough rascal.* absolute, arrant, complete, downright, out-and-out, perfect, proper, sheer, thoroughgoing, total, unmitigated, unmixed, unqualified, utter.

thought *n* **1** *inf* brainwork, brooding, *inf* brown study, cerebration, cogitation, concentration, consideration, contemplation, daydreaming, deliberation, intelligence, introspection, meditation, mental activity, musing, pensiveness, ratiocination, rationality, reason, reasoning, reflection, reverie, rumination, study, thinking, worrying. **2** *a clever thought.* belief, concept, conception, conclusion, conjecture, conviction, idea, notion, observation, opinion. **3** *no thought of gain.* aim, design, dream, expectation, hope, intention, objective, plan, prospect, purpose, vision. **4** *a kind thought.* attention, concern, consideration, kindness, solicitude, thoughtfulness.

thoughtful *adj* **1** absorbed, abstracted, anxious, attentive, brooding, contemplative, dreamy, grave, introspective, meditative, pensive, philosophical, rapt, reflective, serious, solemn, studious, thinking, wary, watchful, worried. **2** *thoughtful work.* careful, conscientious, diligent, exhaustive, intelligent, methodical, meticulous, observant, orderly, organized, painstaking, rational, scrupulous, sensible, systematic, thorough. **3** *a thoughtful kindness.* attentive, caring, compassionate, concerned, considerate, friendly, good-natured, helpful, obliging, public-spirited, solicitous, unselfish. ▷ KIND. *Opp* THOUGHTLESS.

thoughtless *adj* **1** absent-minded, careless, forgetful, hasty, heedless, ill-considered, impetuous, inadvertent, inattentive, injudicious, irresponsible, mindless, negligent, rash, reckless, *inf* scatter-brained, unobservant, unthinking. ▷ STUPID. **2** *a thoughtless insult.* cruel, heartless, impolite, inconsiderate, insensitive, rude, selfish, tactless, uncaring, undiplomatic, unfeeling. ▷ UNKIND. *Opp* THOUGHTFUL.

thrash *vb* beat, birch, cane, flay, flog, lash, scourge, whip. ▷ DEFEAT, HIT.

thread *n* **1** fibre, filament, hair, strand. □ *cotton, line, silk, string, thong, twine, wool, yarn.* **2** *thread of a story.* argument, continuity, course, direction, drift, line of thought, plot, story line, tenor, theme. • *vb* put on a thread, string together. **thread your way** file, pass, pick your way, wind.

threadbare *adj* frayed, old, ragged, scruffy, shabby, tattered, tatty, worn, worn-out.

threat *n* **1** commination, intimidation, menace, warning. **2** *threat of rain.* danger, forewarning, intimation, omen, portent, presage, risk, warning.

threaten *vb* **1** browbeat, bully, cow, intimidate, make threats against, menace, pressurize, terrorize. ▷ FRIGHTEN. *Opp* REASSURE. **2** *clouds threaten rain.* forebode, foreshadow, forewarn of, give warning of, portend, presage, warn of. **3** *the recession threatens jobs.* endanger, imperil, jeopardize, put at risk.

threatening *adj* forbidding, grim, impending, looming, menacing, minatory, ominous, portentous, sinister, stern, *inf* ugly, unfriendly, worrying. *Opp* SUPPORTIVE.

three *n* triad, trio, triplet, triumvirate.

three-dimensional *adj* in the round, rounded, sculptural, solid, stereoscopic.

threshold *n* **1** doorstep, doorway, entrance, sill. **2** *threshold of a new era.* ▷ BEGINNING.

thrifty *adj* careful, *derog* close-fisted, economical, frugal, *derog* mean, *derog* niggardly, parsimonious, provident, prudent, skimping, sparing. *Opp* EXTRAVAGANT.

thrill *n* adventure, *inf* buzz, excitement, frisson, *inf* kick, pleasure, sensation, shiver, suspense, tingle, titillation, tremor. • *vb* arouse, delight, electrify, excite, galvanize, rouse, stimulate, stir, titillate. **thrilling** ▷ EXCITING.

thriller *n* crime story, detective story, mystery, *inf* whodunit. ▷ WRITING.

thrive *vb* be vigorous, bloom, boom, burgeon, *inf* come on, develop strongly, do well, expand, flourish, grow, increase, *inf* make strides, prosper, succeed. *Opp* DIE. **thriving** ▷ PROSPEROUS, VIGOROUS.

throat *n* gullet, neck, oesophagus, uvula, windpipe.

throaty *adj* deep, gravelly, gruff, guttural, hoarse, husky, rasping, rough, thick.

throb *vb* beat, palpitate, pound, pulsate, pulse, vibrate.

throe *n* convulsion, effort, fit, labour, *plur* labour-pains, pang, paroxysm, spasm. ▷ PAIN.

thrombosis *n* blood-clot, embolism.

throng *n* assembly, crowd, crush, gathering, horde, jam, mass, mob, multitude, swarm. ▷ GROUP.

throttle *vb* asphyxiate, choke, smother, stifle, strangle, suffocate. ▷ KILL.

throw *vb* **1** bowl, *inf* bung, cast, *inf* chuck, fling, heave, hurl, launch, lob, pelt, pitch, propel, put (*the shot*), send, *inf* shy, *inf* sling, toss. **2** *throw light.* cast, project, shed. **3** *throw a rider.* dislodge, floor, shake off, throw down, throw off, unseat, upset. **4** ▷ DISCONCERT. **throw away** ▷ DISCARD. **throw out** ▷ EXPEL. **throw up** ▷ PRODUCE, VOMIT.

throw-away *adj* **1** cheap, disposable. **2** *throw-away remark.* casual, offhand, passing, unimportant.

thrust *vb* butt, drive, elbow, force, impel, jab, lunge, plunge, poke, press, prod, propel, push, ram, send, shoulder, shove, stab, stick, urge.

thug *n* assassin, *inf* bully-boy, delinquent, desperado, gangster, *inf* hoodlum, hooligan, killer, mugger, *inf* rough, ruffian, *inf* tough, trouble-maker, vandal, *inf* yob. ▷ CRIMINAL.

thunder *n* clap, crack, peal, roll, rumble. ▷ SOUND.

thunderous *adj* booming, deafening, reverberant, reverberating, roaring, rumbling. ▷ LOUD.

thus *adv* accordingly, consequently, for this reason, hence, so, therefore.

thwart *vb* baffle, baulk, block, check, foil, frustrate, hinder, impede, obstruct, prevent, stand in the way of, stop, stump.

ticket *n* **1** coupon, pass, permit, token, voucher. **2** *price ticket.* docket, label, marker, tab, tag.

ticklish *adj* **1** *inf* giggly, responsive to tickling, sensitive. **2** *a ticklish problem.* awkward, delicate, difficult, risky, *inf* thorny, touchy, tricky.

tide *n* current, drift, ebb and flow, movement, rise and fall.

tidiness *n* meticulousness, neatness, order, orderliness, organization, smartness, system. *Opp* DISORDER.

tidy *adj* **1** neat, orderly, presentable, shipshape, smart, *inf* spick and span, spruce, straight, trim, uncluttered, well-groomed, well-kept. **2** *tidy habits.* businesslike, careful, house-proud, methodical, meticulous, organized, systematic, well-organized. *Opp* UNTIDY. • *vb* arrange, clean up, groom, make tidy, neaten, put in order, rearrange, reorganize, set straight, smarten, spruce up, straighten, titivate. *Opp* MUDDLE.

tie *vb* **1** bind, chain, do up, hitch, interlace, join, knot, lash, moor, rope, secure, splice, tether, truss up. ▷ FASTEN. *Opp* UNTIE. **2** *tie in a race.* be equal, be level, be neck and neck, draw.

tier *n* course (*of bricks*), layer, level, line, order, range, rank, row, stage, storey, stratum, terrace.

tight *adj* **1** close, fast, firm, fixed, immovable, secure, snug. **2** *a tight lid.* airtight, close-fitting, hermetic, impermeable, impervious, leak-proof, sealed, waterproof, watertight. **3** *tight supervision.* harsh, inflexible, precise, rigorous, severe, strict, stringent. **4** *tight ropes.* rigid, stiff, stretched, taut, tense. **5** *a tight space.* compact, constricted, crammed, cramped, crowded, dense, inadequate, limited, packed, small. **6** ▷ DRUNK. **7** ▷ MISERLY. *Opp* FREE, LOOSE.

tighten *vb* **1** become tighter, clamp down, close, close up, constrict, harden, make tighter, squeeze, stiffen, tense. ▷ FASTEN. **2** *tighten ropes.* pull tighter, stretch, tauten. **3** *tighten screws.*

give another turn to, screw up. *Opp* LOOSEN.

till *vb* cultivate, dig, farm, plough, work.

tilt *vb* **1** angle, bank, cant, careen, heel over, incline, keel over, lean, list, slant, slope, tip. **2** *tilt with lances.* joust, thrust. ▷ FIGHT.

timber *n* beam, board, boarding, deal, lath, log, lumber, plank, planking, post, softwood, tree, tree trunk. ▷ WOOD.

time *n* **1** date, hour, instant, juncture, moment, occasion, opportunity, point. **2** duration, interval, period, phase, season, semester, session, spell, stretch, term, while. □ *aeon, century, day, decade, eternity, fortnight, hour, lifetime, minute, month, second, week, weekend, year.* **3** *time of Nero.* age, days, epoch, era, period. **4** *time in music.* beat, measure, rhythm, tempo. • *vb* **1** choose a time for, estimate, fix a time for, judge, organize, plan, schedule, timetable. **2** *time a race.* clock, measure the time of.

timeless *adj* ageless, deathless, eternal, everlasting, immortal, immutable, indestructible, permanent, unchanging, undying, unending.

timely *adj* appropriate, apt, fitting, suitable.

timepiece *n* □ *chronometer, clock, digital clock, digital watch, hourglass, stop-watch, sundial, timer, watch, wrist-watch.*

timetable *n* agenda, calendar, curriculum, diary, list, programme, roster, rota, schedule.

timid *adj* afraid, apprehensive, bashful, chicken-hearted, cowardly, coy, diffident, fainthearted, fearful, modest, *inf* mousy, nervous, pusillanimous, reserved, retiring, scared, sheepish, shrinking, shy, spineless, tentative, timorous, unadventurous, unheroic, wimpish. ▷ FRIGHTENED. *Opp* BOLD.

tingle *n* **1** itch, itching, pins and needles, prickling, stinging, throb, throbbing, tickle, tickling. **2** *a tingle of excitement.* quiver, sensation, shiver, thrill. • *vb* itch, prickle, sting, tickle.

tinker *vb* dabble, fiddle, fool about, interfere, meddle, *inf* mess about, *inf* play about, tamper, try to mend, work amateurishly.

tinny *adj* cheap, flimsy, inferior, insubstantial, poor-quality, shoddy, tawdry.

tinsel *n* decoration, glitter, gloss, show, sparkle, tinfoil.

tint *n* colour, colouring, dye, hue, shade, stain, tincture, tinge, tone, wash.

tiny *adj* diminutive, dwarf, imperceptible, infinitesimal, insignificant, lilliputian, microscopic, midget, *inf* mini, miniature, minuscule, minute, negligible, pygmy, *inf* teeny, unimportant, *inf* wee, *inf* weeny. ▷ SMALL. *Opp* BIG.

tip *n* **1** apex, cap, crown, end, extremity, ferrule, finial, head, nib, peak, pinnacle, point, sharp end, summit, top, vertex. **2** *tip for a waiter. inf* baksheesh, gift, gratuity, inducement, money, *inf* perk, present, reward, service-charge, *inf* sweetener. **3** *useful tips.* advice, clue, forecast, hint, information, pointer, prediction, suggestion, tip-off, warning. **4** *rubbish tip.* dump, rubbish-heap. • *vb* **1** careen, incline, keel, lean, list, slant, slope, tilt. **2** drop off, dump, empty, pour out, spill, unload, upset. **3** *tip a waiter.* give a tip to, remunerate, reward. **tip over** ▷ OVERTURN.

tire *vb* 1 become bored, become tired, flag, grow weary, weaken. 2 debilitate, drain, enervate, exhaust, fatigue, *inf* finish, *sl* knacker, make tired, overtire, sap, *inf* shatter, *inf* take it out of, tax, wear out, weary. *Opp* REFRESH. **tired** ▷ WEARY. **tired of** bored with, *inf* fed up with, impatient with, sick of. **tiring** ▷ EXHAUSTING.

tiredness *n* drowsiness, exhaustion, fatigue, inertia, jet-lag, lassitude, lethargy, listlessness, sleepiness, weariness.

tireless *adj* determined, diligent, dogged, dynamic, energetic, hard-working, indefatigable, persistent, pertinacious, resolute, sedulous, unceasing, unfaltering, unflagging, untiring, unwavering, vigorous. *Opp* LAZY.

tiresome *adj* 1 boring, dull, monotonous, tedious, tiring, unexciting, uninteresting, wearisome, wearying. *Opp* EXCITING. 2 *tiresome delays.* annoying, bothersome, distracting, exasperating, inconvenient, infuriating, irksome, irritating, maddening, petty, troublesome, trying, unwelcome, upsetting, vexatious, vexing.

tiring *adj* debilitating, demanding, difficult, exhausting, fatiguing, hard, laborious, strenuous, taxing, wearying. *Opp* REFRESHING.

tissue *n* 1 fabric, material, structure, stuff, substance. 2 *tissue-paper.* □ *lavatory paper, napkin, paper handkerchief, serviette, toilet paper, tracing-paper.*

title *n* 1 caption, heading, headline, inscription, name, rubric. 2 appellation, designation, form of address, office, position, rank, status. □ *Baron, Baroness, Count, Countess, Dame, Doctor, Dr, Duchess, Duke, Earl, Lady, Lord, Marchioness, Marquis, Master, Miss, Mr, Mrs, Ms, Professor, Rev, Reverend, Sir, Viscount, Viscountess.* ▷ RANK, ROYAL. 3 *title to an inheritance.* claim, deed, entitlement, interest, ownership, possession, prerogative, right. • *vb* call, designate, entitle, give a title to, label, name, tag.

titled *adj* aristocratic, noble, upper class.

titter *vb* chortle, chuckle, giggle, snicker, snigger. ▷ LAUGH.

titular *adj* formal, nominal, official, putative, *inf* so-called, theoretical, token. *Opp* ACTUAL.

toast *vb* 1 brown, grill. ▷ COOK. 2 *toast a guest.* drink a toast to, drink the health of, drink to, honour, pay tribute to, raise your glass to.

tobacco *n* □ *cigar, cigarette, pipe tobacco, plug, snuff.*

together *adv* all at once, at the same time, collectively, concurrently, consecutively, continuously, cooperatively, hand in hand, in chorus, in unison, jointly, shoulder to shoulder, side by side, simultaneously.

toil *n inf* donkey work, drudgery, effort, exertion, industry, labour, work. • *vb* drudge, exert yourself, grind away, *inf* keep at it, labour, *inf* plug away, *inf* slave away, struggle, *inf* sweat. ▷ WORK.

toilet *n* 1 convenience, latrine, lavatory, *sl* loo, *old use* privy, urinal, water closet, WC. 2 [*old use*] *make your toilet.* dressing, grooming, making up, washing.

token *adj* cosmetic, dutiful, emblematic, insincere, nominal, notional, perfunctory, representative, superficial, symbolic. *Opp* GENUINE. • *n* 1 badge, emblem, evidence, expression, indication, mark, marker, proof, reminder, sign, symbol, testimony. **2** *a token of esteem.* keepsake, memento, reminder, souvenir. **3** *a bus token.* coin, counter, coupon, disc, voucher.

tolerable *adj* **1** acceptable, allowable, bearable, endurable, sufferable, supportable. **2** *tolerable food.* adequate, all right, average, fair, mediocre, middling, *inf* OK, ordinary, passable, satisfactory. *Opp* INTOLERABLE.

tolerance *n* **1** broad-mindedness, charity, fairness, forbearance, forgiveness, lenience, open-mindedness, openness, patience, permissiveness. **2** *tolerance of others.* acceptance, sufferance, sympathy (towards), toleration, understanding. **3** *tolerance in moving parts.* allowance, clearance, deviation, fluctuation, play, variation.

tolerant *adj* big-hearted, broad-minded, charitable, easygoing, fair, forbearing, forgiving, generous, indulgent, *derog* lax, lenient, liberal, magnanimous, open-minded, patient, permissive, *derog* soft, sympathetic, understanding, unprejudiced. *Opp* INTOLERANT.

tolerate *vb* abide, accept, admit, bear, brook, concede, condone, countenance, endure, *inf* lump (*I'll have to lump it!*), make allowances for, permit, *inf* put up with, sanction, *inf* stick, *inf* stand, *inf* stomach, suffer, *inf* take, undergo, *inf* wear, weather.

toll *n* charge, dues, duty, fee, levy, payment, tariff, tax. • *vb* chime, peal, ring, sound, strike.

tomb *n* burial-chamber, burial-place, catacomb, crypt, grave, gravestone, last resting-place, mausoleum, memorial, monument, sepulchre, tombstone, vault.

tonality *n* key, tonal centre.

tone *n* **1** accent, colouring, expression, feel, inflection, intonation, manner, modulation, note, phrasing, pitch, quality, sonority, sound, timbre. **2** *tone of a poem, place.* air, atmosphere, character, effect, feeling, mood, spirit, style, temper, vein. **3** *colour tone.* colour, hue, shade, tinge, tint, tonality. **tone down** ▷ SOFTEN. **tone in** ▷ HARMONIZE. **tone up** ▷ STRENGTHEN.

tongue *n* dialect, idiom, language, parlance, patois, speech, talk, vernacular.

tongue-tied *adj* dumb, dumbfounded, inarticulate, *inf* lost for words, mute, silent, speechless.

tonic *n* boost, cordial, dietary supplement, fillip, *inf* pick-me-up, refresher, restorative, stimulant.

tool *n* apparatus, appliance, contraption, contrivance, device, gadget, hardware, implement, instrument, invention, machine, mechanism, utensil, weapon. □ [carpentry] *auger, awl, brace and bit, bradawl, chisel, clamp, cramp, drill, file, fretsaw, gimlet, glass-paper, hacksaw, hammer, jigsaw, mallet, pincers, plane, pliers, power-drill, rasp, sander, sandpaper, saw, screwdriver, spokeshave, T-square, vice, wrench.* □ [gardening] *billhook, dibber, fork, grass-rake, hoe, lawnmower, mattock, pruning knife, pruning shears, rake, roller, scythe, secateurs, shears, sickle, spade, Strimmer, trowel.* □ [various] *axe, bellows, chainsaw, chopper, clippers, crowbar, cutter, hatchet, jack, ladder, lever, penknife, pick, pickaxe, pitchfork, pocket-knife, scis-*

sors, shovel, sledge-hammer, spanner, tape-measure, tongs, tweezers.

tooth *n* □ *canine, eye-tooth, fang, incisor, molar, tusk, wisdom tooth.* **false teeth** bridge, denture, dentures, plate.

toothed *adj* cogged, crenellated, denticulate, indented, jagged, serrated. *Opp* SMOOTH.

top *adj inf* ace, best, choicest, finest, first, foremost, greatest, highest, incomparable, leading, maximum, most, peerless, preeminent, prime, principal, supreme, topmost, unequalled, winning. *n* **1** acme, apex, apogee, crest, crown, culmination, head, height, high point, peak, pinnacle, summit, tip, vertex, zenith. **2** *top of a table.* surface. **3** *top of a jar.* cap, cover, covering, lid, stopper. *Opp* BOTTOM. • *vb* **1** complete, cover, decorate, finish off, garnish, surmount. **2** beat, be higher than, better, cap, exceed, excel, outdo, outstrip, surpass, transcend.

topic *n* issue, matter, point, question, subject, talking-point, text, theme, thesis.

topical *adj* contemporary, current, recent, timely, up-to-date.

topography *n* features, geography, *inf* lie of the land.

topple *vb* **1** bring down, fell, knock down, overturn, throw down, tip over, upset. **2** collapse, fall, overbalance, totter, tumble. **3** *topple a rival.* oust, overthrow, unseat. ▷ DEFEAT.

torch *n* bicycle lamp, brand, electric lamp, flashlight, lamp, *old use* link.

torment *n* affliction, agony, anguish, distress, harassment, misery, ordeal, persecution, plague, scourge, suffering, torture, vexation, woe, worry, wretchedness. ▷ PAIN. • *vb* afflict, annoy, bait, be a torment to, bedevil, bother, bully, distress, harass, inflict pain on, intimidate, *inf* nag, persecute, pester, plague, tease, torture, vex, victimize, worry. ▷ HURT.

torpid *adj* apathetic, dormant, dull, inactive, indolent, inert, lackadaisical, languid, lethargic, lifeless, listless, passive, phlegmatic, slothful, slow, slow-moving, sluggish, somnolent, spiritless. *Opp* LIVELY.

torrent *n* cascade, cataract, deluge, downpour, effusion, flood, flow, gush, inundation, outpouring, overflow, rush, spate, stream, tide.

torrential *adj* copious, heavy, relentless, soaking, teeming, violent.

tortuous *adj* bent, circuitous, complicated, contorted, convoluted, corkscrew, crooked, curling, curvy, devious, indirect, involved, labyrinthine, mazy, meandering, roundabout, serpentine, sinuous, turning, twisted, twisting, twisty, wandering, winding, zigzag. *Opp* DIRECT, STRAIGHT.

torture *n* **1** cruelty, degradation, humiliation, inquisition, persecution, punishment, torment. **2** affliction, agony, anguish, distress, misery, pain, plague, scourge, suffering. • *vb* **1** be cruel to, brainwash, bully, cause pain to, degrade, dehumanize, humiliate, hurt, inflict pain on, intimidate, persecute, rack, torment, victimize. **2** *tortured by doubts.* afflict, agonize, annoy, bedevil, bother, distress, harass, *inf* nag, pester, plague, tease, vex, worry.

toss *vb* **1** bowl, cast, *inf* chuck, fling, flip, heave, hurl, lob, pitch, shy, sling, throw. **2** *toss about in a storm.* bob, dip, flounder, lurch, move restlessly, pitch, plunge, reel, rock, roll, shake, twist and turn, wallow, welter, writhe, yaw.

total *adj* 1 complete, comprehensive, entire, full, gross, overall, whole. 2 *total disaster*. absolute, downright, out-and-out, outright, perfect, sheer, thorough, thoroughgoing, unalloyed, unmitigated, unqualified, utter. • *n* aggregate, amount, answer, lot, sum, totality, whole. • *vb* 1 add up to, amount to, come to, make. 2 add up, calculate, compute, count, find the sum of, find the total of, reckon up, totalize, *inf* tot up, work out.

totalitarian *adj* absolute, arbitrary, authoritarian, autocratic, despotic, dictatorial, fascist, illiberal, one-party, oppressive, tyrannous, undemocratic, unrepresentative. *Opp* DEMOCRATIC.

totter *vb* dodder, falter, reel, rock, stagger, stumble, teeter, topple, tremble, waver, wobble. ▷ WALK.

touch *n* 1 feeling, texture, touching. 2 brush, caress, contact, dab, pat, stroke, tap. 3 *an expert's touch*. ability, capability, experience, expertise, facility, feel, flair, gift, knack, manner, sensitivity, skill, style, technique, understanding, way. 4 *a touch of salt*. bit, dash, drop, hint, intimation, small amount, suggestion, suspicion, taste, tinge, trace. • *vb* 1 be in contact with, brush, caress, contact, cuddle, dab, embrace, feel, finger, fondle, graze, handle, hit, kiss, lean against, manipulate, massage, nuzzle, pat, paw, pet, push, rub, stroke, tap, tickle. 2 *touch the emotions*. affect, arouse, awaken, concern, disturb, impress, influence, inspire, move, stimulate, stir, upset. 3 *touch 100 m.p.h.* attain, reach, rise to. 4 *I can't touch her skill*. be in the same league as, *inf* come up to, compare with, equal, match, parallel, rival. **touched** ▷ EMOTIONAL, MAD. **touching** ▷ EMOTIONAL. **touch off** ▷ BEGIN, IGNITE. **touch on** ▷ MENTION. **touch up** ▷ IMPROVE.

touchy *adj* edgy, highly strung, hypersensitive, irascible, irritable, jittery, jumpy, nervous, oversensitive, peevish, querulous, quick-tempered, sensitive, short-tempered, snappy, temperamental, tense, testy, tetchy, thin-skinned, unpredictable, waspish.

tough *adj* 1 durable, hard-wearing, indestructible, lasting, rugged, sound, stout, strong, substantial, unbreakable, well-built, well-made. 2 *tough physique*. *inf* beefy, brawny, burly, hardy, muscular, robust, stalwart, strong, sturdy. 3 *tough opposition*. invulnerable, merciless, obdurate, obstinate, resilient, resistant, resolute, ruthless, stiff, stubborn, tenacious, unyielding. 4 *a tough taskmaster*. cold, cool, *inf* hard-boiled, hardened, *inf* hard-nosed, inhuman, severe, stern, stony, uncaring, unsentimental, unsympathetic. 5 *tough meat*. chewy, hard, gristly, leathery, rubbery, uneatable. 6 *tough work*. arduous, demanding, difficult, exacting, exhausting, gruelling, hard, laborious, stiff, strenuous, taxing, troublesome. 7 *a tough problem*. baffling, intractable, *inf* knotty, mystifying, perplexing, puzzling, *inf* thorny. *Opp* EASY, TENDER, WEAK.

toughen *vb* harden, make tougher, reinforce, strengthen.

tour *n* circular tour, drive, excursion, expedition, jaunt, journey, outing, peregrination, ride, trip. • *vb* do the rounds of, explore, go round, make a tour of, visit. ▷ TRAVEL.

tourist *n* day-tripper, holiday-maker, sightseer, traveller, tripper, visitor.

tournament *n* championship, competition, contest, event, match, meeting, series.

tow *vb* drag, draw, haul, lug, pull, trail, tug.

tower *n* □ *belfry, campanile, castle, fort, fortress, keep, minaret, pagoda, skyscraper, spire, steeple, turret.* • *vb* ascend, dominate, loom, rear, rise, soar, stand out, stick up.

towering *adj* **1** colossal, gigantic, high, huge, imposing, lofty, mighty, soaring. ▷ TALL. **2** *a towering rage.* extreme, fiery, immoderate, intemperate, intense, mighty, overpowering, passionate, unrestrained, vehement, violent.

town *n* borough, city, community, conurbation, municipality, settlement, township, urban district, village.

toxic *adj* dangerous, deadly, harmful, lethal, noxious, poisonous. *Opp* HARMLESS.

trace *n* **1** clue, evidence, footprint, *inf* give-away, hint, indication, intimation, mark, remains, sign, spoor, token, track, trail, vestige. **2** ▷ BIT. • *vb* **1** detect, discover, find, get back, recover, retrieve, seek out, track down. ▷ TRACK. **2** *trace an outline.* copy, draw, go over, make a copy of, mark out, sketch. **kick over the traces** ▷ REBEL.

track *n* **1** footmark, footprint, mark, scent, spoor, trace, trail, wake (*of ship*). **2** *a farm track.* bridle-path, bridleway, cart-track, footpath, path, route, trail, way. ▷ ROAD. **3** *a racing track.* circuit, course, dirt-track, race-track. **4** *railway track.* branch, branch line, line, mineral line, permanent way, rails, railway, route, tramway. • *vb* chase, dog, follow, hound, hunt, pursue, shadow, stalk, tail, trace, trail. **make tracks** ▷ DEPART. **track down** ▷ TRACE.

trade *n* **1** barter, business, buying and selling, commerce, dealing, exchange, industry, market, marketing, merchandising, trading, traffic, transactions. **2** *a skilled trade.* calling, career, craft, employment, job, *inf* line, occupation, profession, pursuit, work. • *vb* buy and sell, do business, have dealings, market goods, merchandise, retail, sell, traffic (in). **trade in** ▷ EXCHANGE. **trade on** ▷ EXPLOIT.

trader *n* broker, buyer, dealer, merchant, retailer, roundsman, salesman, seller, shopkeeper, stockist, supplier, tradesman, trafficker (*in illegal goods*), vendor.

tradition *n* **1** convention, custom, habit, institution, practice, rite, ritual, routine, usage. **2** *popular tradition.* belief, folklore.

traditional *adj* **1** accustomed, conventional, customary, established, familiar, habitual, historic, normal, orthodox, regular, time-honoured, typical, usual. *Opp* UNCONVENTIONAL. **2** *traditional stories.* folk, handed down, old, oral, popular, unwritten. *Opp* MODERN.

traffic *n* conveyance, movements, shipping, transport, transportation. ▷ VEHICLE. • *vb* ▷ TRADE.

tragedy *n* adversity, affliction, *inf* blow, calamity, catastrophe, disaster, misfortune. *Opp* COMEDY.

tragic *adj* **1** appalling, awful, calamitous, catastrophic, depressing, dire, disastrous, dreadful, fatal, fearful, hapless, ill-fated, ill-omened, ill-starred, inauspicious, lamentable, terrible, tragical, unfortunate, unlucky. **2** *a tragic expression.* bereft, distressed, funereal, grief-stricken, hurt, pathetic, piteous, pitiful, sorrowful, woeful, wretched. ▷ SAD. *Opp* COMIC.

trail *n* **1** evidence, footmarks, footprints, marks, scent, signs, spoor, traces, wake (*of ship*). **2** path, pathway, route, track. ▷ ROAD. • *vb* **1** dangle, drag, draw, haul, pull, tow. **2** chase, follow, hunt, pursue, shadow, stalk, tail, trace, track down. **3** ▷ DAWDLE.

train *n* **1** carriage, coach, diesel, *inf* DMU, electric train, express, intercity, local train, railcar, steam train, stopping train. **2** *train of servants.* cortège, entourage, escort, followers, guard, line, retainers, retinue, staff, suite. **3** *train of events.* ▷ SEQUENCE. • *vb* **1** coach, discipline, drill, educate, instruct, prepare, school, teach, tutor. **2** do exercises, exercise, *inf* get fit, practise, prepare yourself, rehearse, *inf* work out. **3** ▷ AIM.

trainee *n* apprentice, beginner, cadet, learner, *inf* L-driver, novice, pupil, starter, student, tiro, unqualified person.

trainer *n* coach, instructor, teacher, tutor.

trait *n* attribute, characteristic, feature, idiosyncrasy, peculiarity, property, quality, quirk.

traitor *n* apostate, betrayer, blackleg, collaborator, defector, deserter, double-crosser, fifth columnist, informer, *inf* Judas, quisling, renegade, turncoat.

tramp *n* **1** hike, march, trek, trudge, walk. **2** *a homeless tramp.* beggar, *inf* destitute person, *inf* dosser, *inf* down and out, drifter, homeless person, rover, traveller, vagabond, vagrant, wanderer. • *vb inf* footslog, hike, march, plod, stride, toil, traipse, trek, trudge, *sl* yomp. ▷ WALK.

trample *vb* crush, flatten, squash, *inf* squish, stamp on, step on, tread on, walk over.

trance *n inf* brown study, daydream, daze, dream, ecstasy, hypnotic state, rapture, reverie, semiconsciousness, spell, stupor, unconsciousness.

tranquil *adj* **1** calm, halcyon (*days*), peaceful, placid, quiet, restful, serene, still, undisturbed, unruffled. *Opp* STORMY. **2** *a tranquil mood.* collected, composed, dispassionate, *inf* laid-back, sedate, sober, unemotional, unexcited, untroubled. *Opp* EXCITED.

tranquillizer *n* barbiturate, bromide, narcotic, opiate, sedative.

transaction *n* agreement, bargain, business, contract, deal, negotiation, proceeding.

transcend *vb* beat, exceed, excel, outdo, outstrip, rise above, surpass, top.

transcribe *vb* copy out, render, reproduce, take down, translate, transliterate, write out.

transfer *vb* bring, carry, change, convey, deliver, displace, ferry, hand over, make over, move, pass on, pass over, relocate, remove, second, shift, sign over, take, transplant, transport, transpose.

transform *vb* adapt, alter, change, convert, improve, metamorphose, modify, mutate, permute, rebuild, reconstruct, remodel, revolutionize, transfigure, translate, transmogrify, transmute, turn.

transformation *n* adaptation, alteration, change, conversion, improvement, metamorphosis, modification, mutation, reconstruction, revolution, transfiguration, transition, translation, transmogrification, transmutation, *inf* turn-about.

transgression *n* crime, error, fault, lapse, misdeed, misdemeanour, offence, sin, wickedness, wrongdoing.

transient *adj* brief, ephemeral, evanescent, fleeting, fugitive, impermanent, momentary, passing, *inf* quick, short, short-lived, temporary, transitory. *Opp* PERMANENT.

transit *n* conveyance, journey, movement, moving, passage, progress, shipment, transfer, transportation, travel.

transition *n* alteration, change, change-over, conversion, development, evolution, modification, movement, progress, progression, shift, transformation, transit.

translate *vb* change, convert, decode, elucidate, explain, express, gloss, interpret, make a translation, paraphrase, render, reword, spell out, transcribe. ▷ TRANSFORM.

translation *n* decoding, gloss, interpretation, paraphrase, rendering, transcription, transliteration, version.

translator *n* interpreter, linguist.

transmission *n* **1** broadcasting, communication, diffusion, dissemination, relaying, sending out. **2** *transmission of goods.* carriage, carrying, conveyance, dispatch, sending, shipment, shipping, transfer, transference, transport, transportation.

transmit *vb* **1** convey, dispatch, disseminate, forward, pass on, post, send, transfer, transport. **2** *transmit a message.* broadcast, cable, communicate, emit, fax, phone, radio, relay, telephone, telex, wire. *Opp* RECEIVE.

transparent *adj* **1** clear, crystalline, diaphanous, filmy, gauzy, limpid, pellucid, *inf* see-through, sheer, translucent. **2** *transparent honesty.* ▷ CANDID.

transplant *vb* displace, move, relocate, reposition, resettle, shift, transfer, uproot.

transport *n* carrier, conveyance, haulage, removal, shipment, shipping, transportation. □ *aircraft, barge, boat, bus, cable-car, car, chair-lift, coach, cycle, ferry, horse, lorry, Metro, minibus,* old use *omnibus, ship, space-shuttle, taxi, train, tram, van.* □ *air, canal, railway, road, sea, waterways.* ▷ VEHICLE, VESSEL. • *vb* **1** bear, carry, convey, fetch, haul, move, remove, send, shift, ship, take, transfer. **2** ▷ DEPORT.

transpose *vb* change, exchange, interchange, metathesize, move round, rearrange, reverse, substitute, swap, switch, transfer.

transverse *adj* crosswise, diagonal, oblique.

trap *n* ambush, booby-trap, deception, gin, mantrap, net, noose, pitfall, ploy, snare, trick. • *vb* ambush, arrest, capture, catch, catch out, corner, deceive, dupe, ensnare, entrap, inveigle, net, snare, trick.

trappings *plur n* accessories, accompaniments, accoutrements, adornments, appointments, decorations, equipment, finery, fittings, furnishings, *inf* gear, ornaments, paraphernalia, *inf* things, trimmings.

trash *n* **1** debris, garbage, junk, litter, refuse, rubbish, sweepings, waste. **2** ▷ NONSENSE.

travel *n* globe-trotting, moving around, peregrination, touring, tourism, travelling, wandering. □ *cruise, drive, excursion, expedition, exploration, flight, hike, holiday, journey, march, migration, mission, outing, pilgrimage, ramble, ride, safari, sail, sea-passage, tour, trek, trip, visit, voyage, walk.* • *vb inf* gad about, *inf* gallivant, journey, make a trip, move, proceed, progress, roam, *poet* rove, voyage, wander. ▷ GO. □ *aviate, circumnavigate (the world), commute, cruise, cycle, drive, emigrate, fly, free-wheel, hike, hitchhike, march, migrate, motor, navigate, paddle, pedal, pilot, punt, ramble, ride, row, sail, shuttle, steam, tour, trek, walk.*

traveller *n* **1** astronaut, aviator, commuter, cosmonaut, cyclist, driver, flyer, migrant, motorcyclist, motorist, passenger, pedestrian, sailor, voyager, walker. **2** *a company traveller. inf* rep, representative, salesman, saleswoman. **3** *overseas travellers.* explorer, globe-trotter, hiker, hitchhiker, holidaymaker, pilgrim, rambler, stowaway, tourist, tripper, wanderer, wayfarer. **4** *live as travellers.* gypsy, itinerant, nomad, tinker, tramp, vagabond.

travelling *adj* homeless, itinerant, migrant, migratory, mobile, nomadic, peripatetic, restless, roaming, roving, touring, vagrant, wandering.

treacherous *adj* **1** deceitful, disloyal, double-crossing, double-dealing, duplicitous, faithless, false, perfidious, sneaky, unfaithful, untrustworthy. **2** *treacherous conditions.* dangerous, deceptive, hazardous, misleading, perilous, risky, shifting, unpredictable, unreliable, unsafe, unstable. *Opp* LOYAL, RELIABLE.

treachery *n* betrayal, dishonesty, disloyalty, double-dealing, duplicity, faithlessness, infidelity, perfidy, untrustworthiness. ▷ TREASON. *Opp* LOYALTY.

tread *vb* **tread on** crush, squash underfoot, stamp on, step on, trample, walk on. ▷ WALK

treason *n* betrayal, high treason, mutiny, rebellion, sedition. ▷ TREACHERY.

treasure *n* cache, cash, fortune, gold, hoard, jewels, riches, treasure trove, valuables, wealth. • *vb* adore, appreciate, cherish, esteem, guard, keep safe, love, prize, rate highly, value, venerate, worship.

treasury *n* bank, exchequer, hoard, repository, storeroom, treasure-house, vault.

treat *n* entertainment, gift, outing, pleasure, surprise. • *vb* **1** attend to, behave towards, care for, look after, use. **2** *treat a topic.* consider, deal with, discuss, tackle. **3** *treat a patient, wound.* cure, dress, give treatment to, heal, medicate, nurse, prescribe medicine for, tend. **4** *treat food.* process. **5** *treat a friend to dinner.* entertain, give a treat, pay for, provide for, regale.

treatise *n* disquisition, dissertation, essay, monograph, pamphlet, paper, thesis, tract. ▷ WRITING.

treatment *n* **1** care, conduct, dealing (with), handling, management, manipulation, organization, reception, usage, use. **2** *treatment of illness.* cure, first aid, healing, nursing, remedy, therapy. ▷ MEDICINE, THERAPY.

treaty *n* agreement, alliance, armistice, compact, concordat, contract, covenant, convention, *inf* deal, entente, pact, peace, protocol, settlement, truce, understanding.

tree *n* bush, sapling, standard. □ *bonsai, conifer, cordon, deciduous tree, espalier, evergreen, pollard, standard.* □ *ash, banyan, baobab, bay, beech, birch, cacao, cedar, chestnut, cypress, elder, elm, eucalyptus, fir, fruit-tree, gum-tree, hawthorn, hazel, holly, horse-chestnut, larch, lime, maple, oak, olive, palm, pine, plane, poplar, redwood, rowan, sequoia, spruce, sycamore, tamarisk, tulip tree, willow, yew.*

tremble *vb* quail, quake, quaver, quiver, rock, shake, shiver, shudder, vibrate, waver.

tremendous *adj* alarming, appalling, awful, fearful, fearsome, frightening, frightful, horrifying, shocking, startling, terrible, terrific. ▷ BIG, EXCELLENT, REMARKABLE.

tremor *n* **1** agitation, hesitation, quavering, quiver, shaking, trembling, vibration. **2** earthquake, seismic disturbance.

tremulous *adj* **1** agitated, anxious, excited, frightened, jittery, jumpy, nervous, timid, uncertain. *Opp* CALM. **2** quivering, shaking, shivering, trembling, *inf* trembly, vibrating. *Opp* STEADY.

trend *n* **1** bent, bias, direction, drift, inclination, leaning, movement, shift, tendency. **2** *latest trend.* craze, *inf* fad, fashion, mode, *inf* rage, style, *inf* thing, vogue, way.

trendy *adj inf* all the rage, contemporary, fashionable, *inf* in, latest, modern, stylish, up-to-date, voguish. *Opp* OLD-FASHIONED.

trespass *vb* encroach, enter illegally, intrude, invade.

trial *n* **1** case, court martial, enquiry, examination, hearing, inquisition, judicial proceeding, lawsuit, tribunal. **2** attempt, check, *inf* dry run, experiment, rehearsal, test, testing, trial run, *inf* try-out. **3** *a sore trial.* affliction, burden, difficulty, hardship, nuisance, ordeal, *sl* pain in the neck, *inf* pest, problem, tribulation, trouble, worry.

triangular *adj* three-cornered, three-sided.

tribe *n* clan, dynasty, family, group, horde, house, nation, pedigree, people, race, stock, strain.

tribute *n* accolade, appreciation, commendation, compliment, eulogy, glorification, homage, honour, panegyric, praise, recognition, respect, testimony. **pay tribute to** ▷ HONOUR.

trick *n* **1** illusion, legerdemain, magic, sleight of hand. **2** *deceitful trick.* cheat, *inf* con, deceit, deception, fraud, hoax, imposture, joke, *inf* leg-pull, manoeuvre, ploy, practical joke, prank, pretence, ruse, scheme, stratagem, stunt, subterfuge, swindle, trap, trickery, wile. **3** *clever trick.* art, craft, device, dodge, expertise, gimmick, knack, *inf* know-how, secret, skill, technique. **4** *a trick of speech.* characteristic, habit, idiosyncrasy, mannerism, peculiarity, way. • *vb inf* bamboozle, bluff, catch out, cheat, *inf* con, cozen, deceive, defraud, *inf* diddle, dupe, fool, hoax, hoodwink, *inf* kid, mislead, outwit, *inf* pull your leg, swindle, *inf* take in.

trickery *n* bluffing, cheating, chicanery, deceit, deception, dishonesty, double-dealing, duplicity, fraud, *inf* funny business, guile, *inf* hocus-pocus, *inf* jiggery-pokery, knavery, *inf* skulduggery, slyness, swindling, trick.

trickle *vb* dribble, drip, drizzle, drop, exude, flow slowly, leak, ooze, percolate, run, seep. *Opp* GUSH.

trifle *vb* behave frivolously, dabble, fiddle, fool about, play about. **trifling** ▷ TRIVIAL.

trill *vb* sing, twitter, warble, whistle.

trim *adj* compact, neat, orderly, *inf* shipshape, smart, spruce, tidy, well-groomed, well-kept, well-ordered. *Opp* UNTIDY. • *vb* **1** clip, crop, cut, dock, pare down, prune, shape, shear, shorten, snip, tidy. **2** ▷ DECORATE.

trip *n* day out, drive, excursion, expedition, holiday, jaunt, journey, outing, ride, tour, visit, voyage. • *vb* **1** blunder, catch your foot, fall, stagger, stumble, totter, tumble. **2** *trip along.* caper, dance, frisk, gambol, run, skip. **make a trip** ▷ TRAVEL.

trite *adj* banal, commonplace, ordinary, pedestrian, predictable, uninspired, uninteresting.

triumph *n* **1** accomplishment, achievement, conquest, coup, *inf* hit, knockout, master-stroke, *inf* smash hit, success, victory, *inf* walk-over, win. **2** *return in triumph.* celebration, elation, exultation, joy, jubilation, rapture. • *vb* be victorious, carry the day, prevail, succeed, take the honours, win. **triumph over** ▷ DEFEAT.

triumphant *adj* **1** conquering, dominant, successful, victorious, winning. *Opp* UNSUCCESSFUL. **2** boastful, *inf* cocky, elated, exultant, gleeful, gloating, immodest, joyful, jubilant, proud, triumphal.

trivial *adj inf* fiddling, *inf* footling, frivolous, inconsequential, inconsiderable, inessential, insignificant, little, meaningless, minor, negligible, paltry, pettifogging, petty, *inf* piddling, *inf* piffling, silly, slight, small, superficial, trifling, trite, unimportant, worthless. *Opp* IMPORTANT.

trophy *n* **1** booty, loot, mementoes, rewards, souvenirs, spoils. **2** *a sporting trophy.* award, cup, laurels, medal, palm, prize.

trouble *n* **1** adversity, affliction, anxiety, burden, difficulty, distress, grief, hardship, illness, inconvenience, misery, misfortune, pain, problem, sadness, sorrow, suffering, trial, tribulation, unhappiness, vexation, worry. **2** *crowd trouble.* bother, commotion, conflict, discontent, discord, disorder, dissatisfaction, disturbance, fighting, fuss, misbehaviour, misconduct, naughtiness, row, strife, turmoil, unpleasantness, unrest, violence. **3** *engine trouble.* breakdown, defect, failure, fault, malfunction. **4** *took the trouble to get it right.* care, concern, effort, exertion, labour, pains, struggle, thought. • *vb* afflict, agitate, alarm, anguish, annoy, bother, cause trouble to, concern, discommode, distress, disturb, exasperate, grieve, harass, *inf* hassle, hurt, impose on, inconvenience, interfere with, irk, irritate, molest, nag, pain, perturb, pester, plague, *inf* put out, ruffle, threaten, torment, upset, vex, worry. *Opp* REASSURE. **troubled** ▷ WORRIED.

troublemaker *n Fr* agent provocateur, agitator, criminal, culprit, delinquent, hooligan, malcontent, mischief-maker, offender, rabble-rouser, rascal, ringleader, ruffian, scandalmonger, *inf* stirrer, vandal, wrongdoer.

troublesome *adj* annoying, badly-behaved, bothersome, disobedient, disorderly, distressing, inconvenient, irksome, irritating, naughty, *inf* pestiferous, pestilential, rowdy, tiresome, trying, uncooperative, unruly, upsetting, vexatious, vexing, wearisome, worrisome, worrying. *Opp* HELPFUL.

trousers *n inf* bags, breeches, corduroys, culottes, denims, dungarees, jeans, jodhpurs, *old use* knickerbockers, *inf* Levis, overalls, *Amer* pants, plus-fours, shorts, ski-pants, slacks, *Scot* trews, trunks.

truancy *n* absenteeism, desertion, malingering, shirking, *inf* skiving.

truant *n* absentee, deserter, dodger, idler, malingerer, runaway, shirker, *inf* skiver. **play truant** be absent, desert, malinger, *inf* skive, stay away.

truce *n* agreement, armistice, cease-fire, moratorium, pact, peace, suspension of hostilities, treaty.

true *adj* **1** accurate, actual, authentic, confirmed, correct, exact, factual, faithful, faultless, flawless, genuine, literal, proper, real, realistic, right, veracious, verified, veritable. *Opp* FALSE. **2** *a true friend*. constant, dedicated, dependable, devoted, faithful, firm, honest, honourable, loyal, reliable, responsible, sincere, staunch, steadfast, steady, trustworthy, trusty, upright. **3** *the true owner*. authorized, legal, legitimate, rightful, valid. **4** *true aim*. accurate, exact, perfect, precise, *inf* spot-on, unerring, unswerving. *Opp* INACCURATE.

truncheon *n* baton, club, cudgel, staff, stick.

trunk *n* **1** bole, shaft, stalk, stem, stock. **2** *a person's trunk*. body, frame, torso. **3** *an elephant's trunk*. nose, proboscis. **4** *a clothes' trunk*. box, case, casket, chest, coffer, crate, locker, suitcase.

trust *n* **1** assurance, belief, certainty, certitude, confidence, conviction, credence, faith, reliance. **2** *a position of trust*. responsibility, trusteeship. • *vb* **1** *inf* bank on, believe in, be sure of, confide in, count on, depend on, have confidence in, have faith in, *inf* pin your hopes on, rely on. **2** assume, expect, hope, imagine, presume, suppose, surmise. *Opp* DOUBT.

trustful *adj* confiding, credulous, gullible, innocent, trusting, unquestioning, unsuspecting, unsuspicious, unwary. *Opp* DISTRUSTFUL.

trustworthy *adj* constant, dependable, ethical, faithful, honest, honourable, *inf* loyal, moral, on the level, principled, reliable, responsible, *inf* safe, sensible, sincere, steadfast, steady, straightforward, true, *old use* trusty, truthful, upright. *Opp* DECEITFUL.

truth *n* **1** facts, reality. *Opp* LIE. **2** accuracy, authenticity, correctness, exactness, factuality, genuineness, integrity, reliability, truthfulness, validity, veracity, verity. **3** *an accepted truth*. axiom, fact, maxim, truism.

truthful *adj* accurate, candid, correct, credible, earnest, factual, faithful, forthright, frank, honest, proper, realistic, reliable, right, sincere, *inf* straight, straightforward, true, trustworthy, valid, veracious, unvarnished. *Opp* DISHONEST.

try *n* attempt, *inf* bash, *inf* crack, effort, endeavour, experiment, *inf* go, *inf* shot, *inf* stab, test, trial. • *vb* 1 aim, attempt, endeavour, essay, exert yourself, make an effort, strain, strive, struggle, venture. 2 *try something new.* appraise, *inf* check out, evaluate, examine, experiment with, *inf* have a go at, *inf* have a stab at, investigate, test, try out, undertake. **trying** ▷ ANNOYING, TIRESOME. **try someone's patience** ▷ ANNOY.

tub *n* barrel, bath, butt, cask, drum, keg, pot, vat.

tube *n* capillary, conduit, cylinder, duct, hose, main, pipe, spout, tubing.

tuck *vb* cram, gather, insert, push, put away, shove, stuff. *tuck in* ▷ EAT.

tuft *n* bunch, clump, cluster, tuffet, tussock.

tug *vb* drag, draw, haul, heave, jerk, lug, pluck, pull, tow, twitch, wrench, *inf* yank.

tumble *vb* 1 collapse, drop, fall, flop, pitch, roll, stumble, topple, trip up. 2 *tumble things into a heap.* disarrange, dump, jumble, mix up, rumple, shove, spill, throw carelessly, toss.

tumbledown *adj* badly maintained, broken down, crumbling, decrepit, derelict, dilapidated, ramshackle, rickety, ruined, shaky, tottering.

tumult *n* ado, agitation, chaos, commotion, confusion, disturbance, excitement, ferment, fracas, frenzy, hubbub, hullabaloo, rumpus, storm, tempest, upheaval, uproar, welter.

tumultuous *adj* agitated, boisterous, confused, excited, frenzied, hectic, passionate, stormy, tempestuous, turbulent, unrestrained, unruly, violent, wild. *Opp* CALM.

tune *n* air, melody, motif, song, strain, theme. • *vb* adjust, calibrate, regulate, set, temper.

tuneful *adj inf* catchy, euphonious, mellifluous, melodic, melodious, musical, pleasant, singable, sweet-sounding. *Opp* TUNELESS.

tuneless *adj* atonal, boring, cacophonous, discordant, dissonant, harsh, monotonous, unmusical. *Opp* TUNEFUL.

tunnel *n* burrow, gallery, hole, mine, passage, passageway, shaft, subway, underpass. • *vb* burrow, dig, excavate, mine, penetrate.

turbulent *adj* 1 agitated, boisterous, confused, disordered, excited, hectic, passionate, restless, seething, turbid, unrestrained, violent, volatile, wild. 2 *a turbulent crowd.* badly-behaved, disorderly, lawless, obstreperous, riotous, rowdy, undisciplined, unruly. 3 *turbulent weather.* blustery, bumpy, choppy (*sea*), rough, stormy, tempestuous, violent, wild, windy. *Opp* CALM.

turf *n* grass, grassland, green, lawn, *poet* sward.

turgid *adj* affected, bombastic, flowery, fulsome, grandiose, high-flown, overblown, pompous, pretentious, stilted, wordy. *Opp* ARTICULATE.

turmoil *n inf* bedlam, chaos, commotion, confusion, disorder, disturbance, ferment, *inf* hubbub, *inf* hullabaloo, pandemonium, riot, row, rumpus, tumult, turbulence, unrest, upheaval, uproar, welter. *Opp* CALM.

turn *n* **1** circle, coil, curve, cycle, loop, pirouette, revolution, roll, rotation, spin, twirl, twist, whirl. **2** angle, bend, change of direction, corner, deviation, *inf* dogleg, hairpin bend, junction, loop, meander, reversal, shift, turning-point, *inf* U-turn, zigzag. **3** *your turn in a game.* chance, *inf* go, innings, opportunity, shot, stint. **4** *a comic turn.* ▷ PERFORMANCE. **5** *a nasty turn.* ▷ ILLNESS. • *vb* **1** circle, coil, curl, gyrate, hinge, loop, move in a circle, orbit, pivot, revolve, roll, rotate, spin, spiral, swivel, twirl, twist, whirl, wind, yaw. **2** bend, change direction, corner, deviate, divert, go round a corner, negotiate a corner, steer, swerve, veer, wheel. **3** *turn a pumpkin into a coach.* adapt, alter, change, convert, make, modify, remake, remodel, transfigure, transform. **4** *turn to and fro.* squirm, twist, wriggle, writhe. **turn aside** ▷ DEVIATE. **turn down** ▷ REJECT. **turn into** ▷ BECOME. **turn off** ▷ DEVIATE, DISCONNECT, REPEL. **turn on** ▷ ATTRACT, CONNECT. **turn out** ▷ EXPEL, HAPPEN, PRODUCE. **turn over** ▷ CONSIDER, OVERTURN. **turn tail** ▷ ESCAPE. **turn up** ▷ ARRIVE, DISCOVER.

turning-point *n* crisis, crossroads, new direction, revolution, watershed.

turnover *n* business, cash-flow, efficiency, output, production, productivity, profits, revenue, throughput, yield.

twiddle *vb* fiddle with, fidget with, fool with, mess with, twirl, twist.

twig *n* branch, offshoot, shoot, spray, sprig, sprout, stalk, stem, stick, sucker, tendril.

twilight *n* dusk, evening, eventide, gloaming, gloom, halflight, nightfall, sundown, sunset.

twin *adj* balancing, corresponding, double, duplicate, identical, indistinguishable, *inf* look-alike, matching, paired, similar, symmetrical. • *n* clone, counterpart, double, duplicate, *inf* look-alike, match, pair, *inf* spitting image.

twirl *vb* **1** gyrate, pirouette, revolve, rotate, spin, turn, twist, wheel, whirl, wind. **2** *twirl an umbrella.* brandish, twiddle, wave.

twist *n* **1** bend, coil, curl, kink, knot, loop, tangle, turn, zigzag. **2** *a twist to a story.* revelation, surprise ending. • *vb* **1** bend, coil, corkscrew, curl, curve, loop, revolve, rotate, screw, spin, spiral, turn, weave, wind, wreathe, wriggle, writhe, zigzag. **2** *twist ropes.* entangle, entwine, intertwine, interweave, tangle. **3** *twist a lid off.* jerk, wrench, wrest. **4** *twist out of shape.* buckle, contort, crinkle, crumple, distort, screw up, warp, wrinkle. **5** *twist meaning.* alter, change, falsify, misquote, misrepresent. **twisted** ▷ CONFUSED, PERVERTED, TWISTY.

twisty *adj* bending, bendy, circuitous, coiled, contorted, crooked, curving, curvy, *inf* in and out, indirect, looped, meandering, misshapen, rambling, roundabout, serpentine, sinuous, snaking, tortuous, twisted, twisting, *inf* twisting and turning, winding, zigzag. *Opp* STRAIGHT.

twitch *n* blink, convulsion, flutter, jerk, jump, spasm, tic, tremor. • *vb* fidget, flutter, jerk, jump, start, tremble.

two *n* couple, duet, duo, match, pair, twosome.

type *n* **1** category, class, classification, description, designation, form, genre, group, kind, mark, set, sort, species, variety. **2** *He was the very type of evil.* embodiment,

epitome, example, model, pattern, personification, standard.
3 *printed in large type*. characters, font, fount, lettering, letters, print, printing, typeface.

typical *adj* 1 characteristic, distinctive, particular, representative, special. 2 *a typical day*. average, conventional, normal, ordinary, orthodox, predictable, standard, stock, unsurprising, usual. *Opp* UNUSUAL.

tyrannical *adj* absolute, authoritarian, autocratic, *inf* bossy, cruel, despotic, dictatorial, domineering, harsh, high-handed, illiberal, imperious, oppressive, overbearing, ruthless, severe, totalitarian, tyrannous, undemocratic, unjust. *Opp* DEMOCRATIC, LIBERAL.

tyrant *n* autocrat, despot, dictator, *inf* hard taskmaster, oppressor, slave-driver. ▷ RULER.

U

ugly *adj* 1 deformed, disfigured, disgusting, dreadful, frightful, ghastly, grim, grisly, grotesque, gruesome, hideous, horrible, *inf* horrid, ill-favoured, loathsome, misshapen, monstrous, nasty, objectionable, odious, offensive, repellent, repulsive, revolting, shocking, sickening, terrible. 2 *ugly furniture*. displeasing, inartistic, inelegant, plain, tasteless, unattractive, unpleasant, unprepossessing, unsightly. 3 *an ugly mood*. angry, cross, dangerous, forbidding, hostile, menacing, ominous, sinister, surly, threatening, unfriendly. *Opp* BEAUTIFUL.

ulterior *adj* concealed, covert, hidden, personal, private, secondary, secret, undeclared, underlying, undisclosed, unexpressed. *Opp* OVERT.

ultimate *adj* 1 closing, concluding, eventual, extreme, final, furthest, last, terminal, terminating. 2 *ultimate truth*. basic, fundamental, primary, root, underlying.

umpire *n* adjudicator, arbiter, arbitrator, judge, linesman, moderator, official, *inf* ref, referee.

unable *adj* impotent, incompetent, powerless, unfit, unprepared, unqualified. *Opp* ABLE.

unacceptable *adj* bad, forbidden, illegal, improper, inadequate, inadmissible, inappropriate, inexcusable, insupportable, intolerable, invalid, taboo, unsatisfactory, unsuitable, wrong. *Opp* ACCEPTABLE.

unaccompanied *adj* alone, lone, single-handed, sole, solo, unaided, unescorted.

unaccountable *adj* ▷ INEXPLICABLE.

unaccustomed *adj* ▷ STRANGE.

unadventurous *adj* 1 cautious, cowardly, spiritless, timid, unimaginative. 2 *an unadventurous life*. cloistered, limited, protected, sheltered, unexciting. *Opp* ADVENTUROUS.

unalterable *adj* ▷ IMMUTABLE.

unambiguous *adj* ▷ DEFINITE.

unanimous *adj* ▷ UNITED.

unasked *adj* ▷ UNINVITED.

unassuming *adj* ▷ MODEST.

unattached *adj* autonomous, *inf* available, free, independent, separate, single, uncommitted, unmarried, *inf* unspoken for.

unattractive *adj* characterless, colourless, displeasing, dull, inartistic, inelegant, nasty, objectionable, *inf* off-putting, plain, repellent, repulsive, tasteless, uninviting, unpleasant, unprepossessing, unsightly. ▷ UGLY. *Opp* ATTRACTIVE.

unauthorized *adj* illegal, illegitimate, illicit, irregular, unapproved, unlawful, unofficial. *Opp* OFFICIAL.

unavoidable *adj* certain, compulsory, destined, fated, fixed, ineluctable, inescapable, inevitable, inexorable, mandatory, necessary, obligatory, predetermined, required, sure, unalterable.

unaware *adj* ▷ IGNORANT.

unbalanced *adj* **1** asymmetrical, irregular, lopsided, off-centre, shaky, uneven, unstable, wobbly. **2** biased, bigoted, one-sided, partial, partisan, prejudiced, unfair, unjust. **3** *unbalanced mind.* ▷ MAD.

unbearable *adj* insufferable, insupportable, intolerable, overpowering, overwhelming, unacceptable, unendurable. *Opp* TOLERABLE.

unbeatable *adj* ▷ INVINCIBLE.

unbecoming *adj* dishonourable, improper, inappropriate, indecorous, indelicate, offensive, tasteless, unattractive, unbefitting, undignified, ungentlemanly, unladylike, unseemly, unsuitable. *Opp* DECOROUS.

unbelievable *adj* ▷ INCREDIBLE.

unbelieving *adj* ▷ INCREDULOUS.

unbend *vb* **1** straighten, uncurl, untwist. **2** loosen up, relax, rest, unwind.

unbending *adj* ▷ INFLEXIBLE.

unbiased *adj* balanced, disinterested, enlightened, even-handed, fair, impartial, independent, just, neutral, non-partisan, objective, open-minded, reasonable, *inf* straight, unbigoted, undogmatic, unprejudiced. *Opp* BIASED.

unbreakable *adj* ▷ INDESTRUCTIBLE.

unbroken *adj* ▷ CONTINUOUS, WHOLE.

uncalled-for *adj* ▷ UNNECESSARY.

uncared-for *adj* ▷ DERELICT.

uncaring *adj* ▷ CALLOUS.

unceasing *adj* ▷ CONTINUOUS.

uncertain *adj* **1** ambiguous, arguable, *inf* chancy, confusing, conjectural, cryptic, enigmatic, equivocal, hazardous, hazy, *inf* iffy, imprecise, incalculable, inconclusive, indefinite, indeterminate, problematical, puzzling, questionable, risky, speculative, *inf* touch and go, unclear, unconvincing, undecided, undetermined, unforeseeable, unknown, unresolved, woolly. **2** *uncertain what to believe.* agnostic, ambivalent, doubtful, dubious, *inf* hazy, insecure, *inf* in two minds, self-questioning, unconvinced, undecided, unsure, vague, wavering. **3** *an uncertain climate.* changeable, erratic, fitful, inconstant, irregular, precarious, unpredictable, unreliable, unsettled, variable. *Opp* CERTAIN.

unchanging *adj* ▷ CONSTANT.

uncharitable *adj* ▷ UNKIND.

uncivilized *adj* anarchic, antisocial, backward, barbarian, barbaric, barbarous, brutish, crude, disorganized, illiterate, Philistine, primitive, savage, uncultured, uneducated, unenlightened, unsophisticated, wild. *Opp* CIVILIZED.

unclean *adj* ▷ DIRTY.

unclear *adj* ▷ UNCERTAIN.

unclothed *adj* ▷ NAKED.

uncomfortable *adj* **1** bleak, cold, comfortless, cramped, hard, inconvenient, lumpy, painful.

2 *uncomfortable clothes*. formal, restrictive, stiff, tight, tight-fitting. 3 *an uncomfortable silence*. awkward, distressing, embarrassing, nervous, restless, troubled, uneasy, worried. *Opp* COMFORTABLE.

uncommon *adj* ▷ UNUSUAL.

uncommunicative *adj* ▷ TACITURN.

uncomplimentary *adj* censorious, critical, deprecatory, depreciatory, derogatory, disapproving, disparaging, pejorative, scathing, slighting, unfavourable, unflattering. ▷ RUDE. *Opp* COMPLIMENTARY.

uncompromising *adj* ▷ INFLEXIBLE.

unconcealed *adj* ▷ OBVIOUS.

unconditional *adj* absolute, categorical, complete, full, outright, total, unequivocal, unlimited, unqualified, unreserved, unrestricted, wholehearted, *inf* with no strings attached. *Opp* CONDITIONAL.

uncongenial *adj* alien, antipathetic, disagreeable, incompatible, unattractive, unfriendly, unpleasant, unsympathetic. *Opp* CONGENIAL.

unconquerable *adj* ▷ INVINCIBLE.

unconscious *adj* 1 anaesthetized, *inf* blacked-out, comatose, concussed, *inf* dead to the world, insensible, *inf* knocked out, *inf* knocked silly, oblivious, *inf* out for the count, senseless, sleeping. 2 blind, deaf, ignorant, oblivious, unaware. 3 *unconscious humour*. accidental, inadvertent, unintended, unintentional, unwitting. 4 *an unconscious reaction*. automatic, *sl* gut, impulsive, instinctive, involuntary, reflex, spontaneous, unthinking. 5 *an unconscious desire*. repressed, subconscious, subliminal, suppressed. *Opp* CONSCIOUS.

unconsciousness *n inf* blackout, coma, faint, oblivion, sleep.

uncontrollable *adj* ▷ UNDISCIPLINED.

unconventional *adj* abnormal, atypical, *inf* cranky, eccentric, exotic, futuristic, idiosyncratic, independent, inventive, nonconforming, non-standard, odd, off-beat, original, peculiar, progressive, revolutionary, strange, surprising, unaccustomed, unorthodox, *inf* way-out, wayward, weird, zany. *Opp* CONVENTIONAL.

unconvincing *adj* implausible, improbable, incredible, invalid, spurious, unbelievable, unlikely. *Opp* PERSUASIVE.

uncooperative *adj* lazy, obstructive, recalcitrant, selfish, unhelpful, unwilling. *Opp* COOPERATIVE.

uncover *vb* bare, come across, detect, dig up, disclose, discover, disrobe, exhume, expose, locate, reveal, show, strip, take the wraps off, undress, unearth, unmask, unveil, unwrap. *Opp* COVER.

undamaged *adj* ▷ PERFECT.

undefended *adj* defenceless, exposed, helpless, insecure, unarmed, unfortified, unguarded, unprotected, vulnerable, weaponless. *Opp* SECURE.

undemanding *adj* ▷ EASY.

undemonstrative *adj* ▷ ALOOF.

underclothes *plur n* lingerie, *inf* smalls, underclothing, undergarments, underthings, underwear, *inf* undies. □ *bra, braces, brassière, briefs, camiknickers, corset, drawers, garter, girdle, knickers, panties, pantihose, pants, petticoat, slip, suspenders, tights,* *trunks, underpants, underskirt, vest.*

undercurrent *n* atmosphere, feeling, hint, sense, suggestion, trace, undertone.

underestimate *vb* belittle, depreciate, dismiss, disparage, minimize, miscalculate, misjudge, underrate, undervalue. *Opp* EXAGGERATE.

undergo *vb* bear, be subjected to, endure, experience, go through, live through, put up with, *inf* stand, submit yourself to, suffer, withstand.

underground *adj* **1** buried, hidden, subterranean, sunken. **2** clandestine, revolutionary, secret, subversive, unofficial, unrecognized.

undergrowth *n* brush, bushes, ground cover, plants, vegetation.

undermine *vb* burrow under, destroy, dig under, erode, excavate, mine under, ruin, sabotage, sap, subvert, tunnel under, undercut, weaken, wear away.

underprivileged *adj* deprived, destitute, disadvantaged, downtrodden, impoverished, needy, oppressed. ▷ POOR. *Opp* PRIVILEGED.

undersea *adj* subaquatic, submarine, underwater.

understand *vb* **1** appreciate, apprehend, be conversant with, *inf* catch on, comprehend, *inf* cotton on to, decipher, decode, fathom, figure out, follow, gather, *inf* get, *inf* get to the bottom of, grasp, interpret, know, learn, make out, make sense of, master, perceive, realize, recognize, see, take in, *inf* twig. **2** *understand animals.* be in sympathy with, empathize with, sympathize with.

understanding *n* **1** ability, acumen, brains, cleverness, discernment, insight, intellect, intelligence, judgement, penetration, perceptiveness, percipience, sense, wisdom. **2** *understanding of a problem.* appreciation, apprehension, awareness, cognition, comprehension, grasp, knowledge. **3** *understanding between people.* accord, agreement, compassion, consensus, consent, consideration, empathy, fellow feeling, harmony, kindness, mutuality, sympathy, tolerance. **4** *a formal understanding.* arrangement, bargain, compact, contract, deal, entente, pact, settlement, treaty.

understate *vb* belittle, *inf* make light of, minimize, *inf* play down, *inf* soft-pedal. *Opp* EXAGGERATE.

undertake *vb* **1** agree, attempt, consent, covenant, guarantee, pledge, promise, try. **2** *undertake a task.* accept responsibility for, address, approach, attend to, begin, commence, commit yourself to, cope with, deal with, embark on, grapple with, handle, manage, tackle, take on, take up.

undertaking *n* **1** affair, business, enterprise, project, task, venture. **2** agreement, assurance, contract, guarantee, pledge, promise, vow.

undervalue *vb* belittle, depreciate, dismiss, disparage, minimize, miscalculate, misjudge, underestimate, underrate.

underwater *adj* subaquatic, submarine, undersea.

undeserved *adj* unearned, unfair, unjustified, unmerited, unwarranted.

undesirable *adj* ▷ OBJECTIONABLE.

undisciplined *adj* anarchic, chaotic, disobedient, disorderly, disorganized, intractable, rebellious, uncontrollable, uncontrolled, ungovernable, unmanageable, unruly, unsystematic, untrained, wild, wilful. *Opp* OBEDIENT.

undiscriminating *adj* ▷ IMPERCEPTIVE.

undisguised *adj* ▷ OBVIOUS.

undistinguished *adj* ▷ ORDINARY.

undo *vb* **1** detach, disconnect, disengage, loose, loosen, open, part, separate, unbind, unbuckle, unbutton, unchain, unclasp, unclip, uncouple, unfasten, unfetter, unhook, unleash, unlock, unpick, unpin, unscrew, unseal, unshackle, unstick, untether, untie, unwrap, unzip. **2** *undo someone's good work.* annul, cancel out, destroy, mar, nullify, quash, reverse, ruin, spoil, undermine, vitiate, wipe out, wreck.

undoubted *adj* ▷ INDISPUTABLE.

undoubtedly *adv* certainly, definitely, doubtless, indubitably, of course, surely, undeniably, unquestionably.

undress *vb* disrobe, divest yourself, *inf* peel off, shed your clothes, strip off, take off your clothes, uncover yourself.

undressed *adj* ▷ NAKED.

undue *adj* ▷ EXCESSIVE.

undying *adj* ▷ ETERNAL.

uneasy *adj* anxious, apprehensive, awkward, concerned, distressed, distressing, disturbed, edgy, fearful, insecure, jittery, nervous, restive, restless, tense, troubled, uncomfortable, unsettled, upsetting, worried.

uneducated *adj* ▷ IGNORANT.

unemotional *adj* apathetic, clinical, cold, cool, dispassionate, frigid, hard-hearted, heartless, impassive, indifferent, objective, unfeeling, unmoved, unresponsive. *Opp* EMOTIONAL.

unemployed *adj* jobless, laid off, on the dole, out of work, redundant, *inf* resting, unwaged. ▷ IDLE.

unendurable *adj* ▷ UNBEARABLE.

unenthusiastic *adj* ▷ APATHETIC, UNINTERESTED.

unequal *adj* **1** different, differing, disparate, dissimilar, uneven, varying. **2** *unequal treatment.* biased, prejudiced, unjust. **3** *an unequal contest.* ill-matched, one-sided, unbalanced, uneven, unfair. *Opp* EQUAL, FAIR.

unequalled *adj* incomparable, inimitable, matchless, peerless, supreme, surpassing, un-matched, unparalleled, unrivalled, unsurpassed.

unethical *adj* ▷ IMMORAL.

uneven *adj* **1** bent, broken, bumpy, crooked, irregular, jagged, jerky, pitted, rough, rutted, undulating, wavy. **2** *an uneven rhythm.* erratic, fitful, fluctuating, inconsistent, spasmodic, unpredictable, variable, varying. **3** *an uneven load.* asymmetrical, lopsided, unsteady. **4** *uneven contest.* ill-matched, one-sided, unbalanced, unequal, unfair. *Opp* EVEN.

uneventful *adj* ▷ UNEXCITING.

unexciting *adj* boring, dreary, dry, dull, humdrum, monotonous, predictable, quiet, repetitive, routine, soporific, straightforward, tedious, trite, uneventful, uninspiring, uninteresting, vapid, wearisome. ▷ ORDINARY. *Opp* EXCITING.

unexpected *adj* accidental, chance, fortuitous, sudden, surprising, unforeseen, unhoped-for, unlooked-for, unplanned, unpredictable, unusual. *Opp* PREDICTABLE.

unfair *adj* ▷ UNJUST.

unfaithful *adj* deceitful, disloyal, double-dealing, duplicitous, faithless, false, fickle, inconstant, perfidious, traitorous, treacherous, treasonable, unreliable, untrue, untrustworthy. *Opp* FAITHFUL.

unfaithfulness *n* **1** duplicity, perfidy, treachery, treason. **2** adultery, infidelity.

unfamiliar *adj* ▷ STRANGE.

unfashionable *adj* dated, obsolete, old-fashioned, *inf* out, outmoded, passé, superseded, unstylish. *Opp* FASHIONABLE.

unfasten *vb* ▷ UNDO.

unfavourable *adj* **1** adverse, attacking, contrary, critical, disapproving, discouraging, hostile, ill-disposed, inauspicious, negative, opposing, uncomplimentary, unfriendly, unhelpful, unkind, unpromising, unpropitious, unsympathetic. **2** *an unfavourable reputation.* bad, undesirable, unenviable, unsatisfactory. *Opp* FAVOURABLE.

unfeeling *adj* ▷ CALLOUS.

unfinished *adj* imperfect, incomplete, rough, sketchy, uncompleted, unpolished. *Opp* PERFECT.

unfit *adj* **1** ill-equipped, inadequate, incapable, incompetent, unsatisfactory, useless. **2** *unfit for family viewing.* improper, inappropriate, unbecoming, unsuitable, unsuited. **3** *an unfit athlete.* feeble, flabby, out of condition, unhealthy. ▷ ILL. *Opp* FIT.

unflagging *adj* ▷ TIRELESS.

unflinching *adj* ▷ RESOLUTE.

unforeseen *adj* ▷ UNEXPECTED.

unforgettable *adj* ▷ MEMORABLE.

unforgivable *adj* inexcusable, mortal (*sin*), reprehensible, shameful, unjustifiable, unpardonable, unwarrantable. *Opp* FORGIVABLE.

unfortunate *adj* ▷ UNLUCKY.

unfriendly *adj* aggressive, aloof, antagonistic, antisocial, cold, cool, detached, disagreeable, distant, forbidding, frigid, haughty, hostile, ill-disposed, ill-natured, impersonal, indifferent, inhospitable, menacing, nasty, obnoxious, offensive, remote, reserved, rude, sour, standoffish, *inf* starchy, stern, supercilious, threatening, unapproachable, uncivil, uncongenial, unenthusiastic, unforthcoming, unkind, unneighbourly, unresponsive, unsociable, unsympathetic, unwelcoming. *Opp* FRIENDLY.

ungainly *adj* ▷ AWKWARD.

ungodly *adj* ▷ IRRELIGIOUS.

ungovernable *adj* ▷ UNDISCIPLINED.

ungrateful *adj* displeased, ill-mannered, rude, selfish, unappreciative, unthankful. *Opp* GRATEFUL.

unhappy *adj* **1** dejected, depressed, dispirited, down, downcast, gloomy, miserable, mournful, sorrowful. ▷ SAD. **2** *unhappy about losing.* bad-tempered, disaffected, discontented, disgruntled, disillusioned, displeased, dissatisfied, *inf* fed up, *inf* grumpy, morose, sulky, sullen, unsatisfied. **3** *an unhappy choice.* ▷ UNSATISFACTORY.

unhealthy *adj* **1** ailing, debilitated, delicate, diseased, feeble, frail, infected, infirm, *inf* poorly, sick, sickly, suffering, unwell, valetudinary, weak. ▷ ILL. **2** *unhealthy conditions.* deleterious, detrimental, dirty, harmful, insalubrious, insanitary, noxious, polluted, unhygienic, unwholesome. *Opp* HEALTHY.

unheard-of *adj* ▷ UNUSUAL.

unhelpful *adj* disobliging, inconsiderate, negative, slow, uncivil, uncooperative, unwilling. *Opp* HELPFUL.

unhygienic *adj* ▷ UNHEALTHY.

unidentifiable *adj* anonymous, camouflaged, disguised, hidden, undetectable, unidentified, unknown, unrecognizable. *Opp* IDENTIFIABLE.

unidentified *adj* anonymous, incognito, mysterious, nameless, unfamiliar, unknown, unmarked, unnamed, unrecognized, unspecified. *Opp* SPECIFIC.

uniform *adj* consistent, even, homogeneous, identical, indistinguishable, predictable, regular, same, similar, single, standard, unbroken, unvaried, unvarying. *Opp* DIFFERENT. • *n* costume, livery, outfit.

unify *vb* amalgamate, bring together, coalesce, combine, consolidate, fuse, harmonize, integrate, join, merge, unite, weld together. *Opp* SEPARATE.

unimaginative *adj* banal, boring, clichéd, derivative, dull, hackneyed, inartistic, insensitive, obvious, ordinary, prosaic, stale, trite, ugly, uninspired, uninteresting, unoriginal. *Opp* IMAGINATIVE.

unimportant *adj* ephemeral, forgettable, immaterial, inconsequential, inconsiderable, inessential, insignificant, irrelevant, lightweight, minor, negligible, peripheral, petty, secondary, slight, trifling, trivial, valueless, worthless. ▷ SMALL. *Opp* IMPORTANT.

uninhabitable *adj* condemned, in bad repair, unliveable, unusable. *Opp* HABITABLE.

uninhabited *adj* abandoned, deserted, desolate, empty, tenantless, uncolonized, unoccupied, unpeopled, unpopulated, untenanted, vacant.

uninhibited *adj* abandoned, candid, casual, easygoing, frank, informal, natural, open, outgoing, outspoken, relaxed, spontaneous, unbridled, unconstrained, unrepressed, unreserved, unrestrained, unselfconscious, wild. *Opp* REPRESSED.

unintelligent *adj* ▷ STUPID.

unintelligible *adj* ▷ INCOMPREHENSIBLE.

unintentional *adj* accidental, fortuitous, inadvertent, involuntary, unconscious, unintended, unplanned, unwitting. *Opp* INTENTIONAL.

uninterested *adj* apathetic, bored, incurious, indifferent, lethargic, passive, phlegmatic, unconcerned, unenthusiastic, uninvolved, unresponsive. *Opp* INTERESTED.

uninteresting *adj* boring, dreary, dry, dull, flat, monotonous, obvious, predictable, tedious, unexciting, uninspiring, vapid, wearisome. ▷ ORDINARY. *Opp* INTERESTING.

uninterrupted *adj* ▷ CONTINUOUS.

uninvited *adj* **1** unasked, unbidden, unwelcome. **2** *an uninvited comment.* gratuitous, unsolicited, voluntary.

uninviting *adj* ▷ UNATTRACTIVE.

union *n* **1** alliance, amalgamation, association, coalition, confederation, conjunction, federation, integration, joining together, merger, unanimity, unification, unity. **2** amalgam, blend, combination, combining, compound, fusion, grafting, marrying, mixture, synthesis, welding. **3** marriage, matrimony, partnership, wedlock.

unique *adj* distinctive, incomparable, lone, *inf* one-off, peculiar, peerless, *inf* second to none, single, singular, unequalled, unparalleled, unrepeatable, unrivalled.

unit *n* component, constituent, element, entity, item, module, part, piece, portion, section, segment, whole.

unite *vb* ally, amalgamate, associate, blend, bring together, coalesce, collaborate, combine, commingle, confederate, connect, consolidate, conspire, cooperate, couple, federate, fuse, go into partnership, harmonize, incorporate, integrate, interlock, join, join forces, link, link up, marry, merge, mingle, mix, stick together, tie up, unify, weld together. ▷ MARRY. *Opp* SEPARATE.

united *adj* agreed, allied, coherent, collective, common, concerted, coordinated, corporate, harmonious, integrated, joint, like-minded, mutual, *inf* of one mind, shared, *inf* solid, unanimous, undivided. *Opp* DISUNITED. **be united** ▷ AGREE.

unity *n* accord, agreement, coherence, concord, consensus, harmony, integrity, like-mindedness, oneness, rapport, solidarity, unanimity, wholeness. *Opp* DISUNITY.

universal *adj* all-embracing, all-round, boundless, common, comprehensive, cosmic, general, global, international, omnipresent, pandemic, prevailing, prevalent, total, ubiquitous, unbounded, unlimited, widespread, worldwide.

universe *n* cosmos, creation, the heavens, *old use* macrocosm.

unjust *adj* biased, bigoted, indefensible, inequitable, one-sided, partial, partisan, prejudiced, undeserved, unfair, unjustified, unlawful, unmerited, unreasonable, unwarranted, wrong, wrongful. *Opp* JUST.

unjustifiable *adj* excessive, immoderate, indefensible, inexcusable, unacceptable, unconscionable, unforgivable, unjust, unreasonable, unwarranted. *Opp* JUSTIFIABLE.

unkind *adj* abrasive, *inf* beastly, callous, caustic, cold-blooded, discourteous, disobliging, hard, hard-hearted, harsh, heartless, hurtful, ill-natured, impolite, inconsiderate, inhuman, inhumane, insensitive, malevolent, malicious, mean, merciless, nasty, pitiless, relentless, rigid, rough, ruthless, sadistic, savage, selfish, severe, sharp, spiteful, stern, tactless, thoughtless, uncaring, uncharitable, unchristian, unfeeling, unfriendly, unpleasant, unsympathetic, unthoughtful, vicious. ▷ ANGRY, CRITICAL, CRUEL. *Opp* KIND.

unknown *adj* **1** anonymous, disguised, incognito, mysterious, nameless, strange, unidentified, unnamed, unrecognized, unspecified. **2** *an unknown country.* alien, foreign, uncharted, undiscovered, unexplored, unfamiliar, unmapped. **3** *an unknown actor.* humble, insignificant, little-known, lowly, obscure, undistinguished, unheard-of, unimportant. *Opp* FAMOUS.

unlawful *adj* ▷ ILLEGAL.

unlikely *adj* **1** dubious, far-fetched, implausible, improbable, incredible, suspect, suspicious, *inf* tall (*story*), unbelievable, unconvincing, unthinkable. **2** *an* *unlikely possibility*. distant, doubtful, faint, *inf* outside, remote, slight. *Opp* LIKELY.

unlimited *adj* ▷ BOUNDLESS.

unload *vb* disburden, discharge, drop off, *inf* dump, empty, offload, take off, unpack. *Opp* LOAD.

unloved *adj* abandoned, discarded, forsaken, hated, loveless, lovelorn, neglected, rejected, spurned, uncared-for, unvalued, unwanted. *Opp* LOVED.

unlucky *adj* **1** accidental, calamitous, chance, disastrous, dreadful, tragic, unfortunate, untimely, unwelcome. **2** *an unlucky person*. *inf* accident-prone, hapless, luckless, unhappy, unsuccessful, wretched. **3** *an unlucky number*. cursed, ill-fated, ill-omened, ill-starred, inauspicious, jinxed, ominous, unfavourable. *Opp* LUCKY.

unmanageable *adj* ▷ UNDISCIPLINED.

unmarried *adj inf* available, celibate, *inf* free, single, unwed. **unmarried person** bachelor, celibate, spinster.

unmentionable *adj* ▷ TABOO.

unmistakable *adj* ▷ DEFINITE, OBVIOUS.

unnamed *adj* ▷ UNIDENTIFIED.

unnatural *adj* **1** abnormal, bizarre, eccentric, eerie, extraordinary, fantastic, freak, freakish, inexplicable, magic, magical, odd, outlandish, preternatural, queer, strange, supernatural, unaccountable, uncanny, unusual, weird. **2** *unnatural feelings*. callous, cold-blooded, cruel, hard-hearted, heartless, inhuman, inhumane, monstrous, perverse, perverted, sadistic, savage, stony-hearted, unfeeling, unkind. **3** *unnatural behaviour*. actorish, affected, bogus, contrived, fake, feigned, forced, insincere, laboured, mannered, *inf* out of character, overdone, *inf* phoney, pretended, *inf* pseudo, *inf* put on, self-conscious, stagey, stiff, stilted, theatrical, uncharacteristic, unspontaneous. **4** *unnatural materials*. artificial, fabricated, imitation, man-made, manufactured, simulated, synthetic. *Opp* NATURAL.

unnecessary *adj* dispensable, excessive, expendable, extra, inessential, needless, nonessential, redundant, supererogatory, superfluous, surplus, uncalled-for, unjustified, unneeded, unwanted, useless. *Opp* NECESSARY.

unobtrusive *adj* ▷ INCONSPICUOUS.

unofficial *adj* friendly, informal, *inf* off the record, private, secret, unauthorized, unconfirmed, undocumented, unlicensed. *Opp* OFFICIAL.

unorthodox *adj* ▷ UNCONVENTIONAL.

unpaid *adj* **1** due, outstanding, owing, payable, unsettled. **2** *unpaid work*. honorary, unremunerative, unsalaried, voluntary.

unpalatable *adj* disgusting, distasteful, inedible, nasty, nauseating, *inf* off, rancid, sickening, sour, tasteless, unacceptable, unappetizing, uneatable, unpleasant. *Opp* PALATABLE.

unparalleled *adj* ▷ UNEQUALLED.

unpardonable *adj* ▷ UNFORGIVABLE.

unplanned *adj* ▷ SPONTANEOUS.

unpleasant *adj* abhorrent, abominable, antisocial, appalling, atrocious, awful, bad-tempered, beastly, bitter, coarse, crude, despicable, detestable, diabolical, dirty, disagreeable, disgusting, displeasing, distasteful, dreadful, evil, execrable, fearful, fearsome, filthy, foul, frightful, ghastly, grim, grisly, gruesome, harsh, hateful, *inf* hellish, hideous, horrible, horrid, horrifying, improper, indecent, inhuman, irksome, loathsome, *inf* lousy, malevolent, malicious, mucky, nasty, nauseating, objectionable, obnoxious, odious, offensive, *inf* off-putting, repellent, repugnant, repulsive, revolting, rude, shocking, sickening, sickly, sordid, sour, spiteful, squalid, terrible, ugly, unattractive, uncouth, undesirable, unfriendly, unkind, unpalatable, unsavoury, unwelcome, upsetting, vexing, vicious, vile, vulgar. ▷ BAD. *Opp* PLEASANT.

unpopular *adj* despised, disliked, friendless, hated, ignored, *inf* in bad odour, minority (*interests*), out of favour, rejected, shunned, unfashionable, unloved, unwanted. *Opp* POPULAR.

unpredictable *adj* changeable, surprising, uncertain, unexpected, unforeseeable, variable. *Opp* PREDICTABLE.

unprejudiced *adj* ▷ UNBIASED.

unpremeditated *adj* ▷ SPONTANEOUS.

unprepared *adj inf* caught napping, caught out, ill-equipped, surprised, taken off-guard, unready. ▷ READY.

unpretentious *adj* humble, modest, plain, simple, straightforward, unaffected, unassuming, unostentatious, unsophisticated. *Opp* PRETENTIOUS.

unproductive *adj* **1** ineffective, fruitless, futile, pointless, unprofitable, unrewarding, useless, valueless, worthless. **2** *an unproductive garden*. arid, barren, infertile, sterile, unfruitful. *Opp* PRODUCTIVE.

unprofessional *adj* amateurish, casual, incompetent, inefficient, inexpert, lax, negligent, shoddy, *inf* sloppy, unethical, unfitting, unprincipled, unseemly, unskilful, unskilled, unworthy. *Opp* PROFESSIONAL.

unprofitable *adj* futile, loss-making, pointless, uncommercial, uneconomic, ungainful, unproductive, unremunerative, unrewarding, worthless. *Opp* PROFITABLE.

unprovable *adj* doubtful, inconclusive, questionable, undemonstrable, unsubstantiated, unverifiable. *Opp* PROVABLE, PROVEN.

unpunctual *adj* behindhand, belated, delayed, detained, last-minute, late, overdue, tardy, unreliable. *Opp* PUNCTUAL.

unravel *vb* disentangle, free, solve, sort out, straighten out, undo, untangle.

unreal *adj* chimerical, false, fanciful, illusory, imaginary, imagined, make-believe, nonexistent, phantasmal, *inf* pretend, *inf* pseudo, sham. ▷ HYPOTHETICAL. *Opp* REAL.

unrealistic *adj* **1** inaccurate, non-representational, unconvincing, unlifelike, unnatural, unrecognizable. **2** *unrealistic ideas*. delusory, fanciful, idealistic, impossible, impracticable, impractical, over-ambitious, quixotic, romantic, silly, visionary, unreasonable, unworkable. **3** *unrealistic prices*. ▷ EXCESSIVE. *Opp* REALISTIC.

unreasonable *adj* ▷ IRRATIONAL.

unrecognizable *adj* ▷ UNIDENTIFIABLE.

unrelated *adj* **1** different, independent, unconnected, unlike. **2** ▷ IRRELEVANT. *Opp* RELATED.

unreliable *adj* 1 deceptive, false, flimsy, implausible, inaccurate, misleading, suspect, unconvincing. 2 *unreliable friends.* changeable, disreputable, fallible, fickle, inconsistent, irresponsible, treacherous, undependable, unpredictable, unsound, unstable, untrustworthy. *Opp* RELIABLE.

unrepentant *adj* brazen, confirmed, conscienceless, hardened, impenitent, incorrigible, incurable, inveterate, irredeemable, shameless, unapologetic, unashamed, unblushing, unreformable, unregenerate. *Opp* REPENTANT.

unripe *adj* green, immature, sour, unready. *Opp* RIPE.

unrivalled *adj* ▷ UNEQUALLED.

unruly *adj* ▷ UNDISCIPLINED.

unsafe *adj* ▷ DANGEROUS.

unsatisfactory *adj* defective, deficient, disappointing, displeasing, dissatisfying, faulty, frustrating, imperfect, inadequate, incompetent, inefficient, inferior, insufficient, lacking, not good enough, poor, *inf* sad (*state of affairs*), unacceptable, unhappy, unsatisfying, *inf* wretched. *Opp* SATISFACTORY.

unscrupulous *adj* amoral, conscienceless, corrupt, *inf* crooked, cunning, dishonest, dishonourable, immoral, improper, self-interested, shameless, *inf* slippery, sly, unconscionable, unethical, untrustworthy. *Opp* SCRUPULOUS.

unseemly *adj* ▷ UNBECOMING.

unseen *adj* ▷ INVISIBLE.

unselfish *adj* altruistic, caring, charitable, considerate, disinterested, generous, humanitarian, kind, liberal, magnanimous, open-handed, philanthropic, public-spirited, self-effacing, selfless, self-sacrificing, thoughtful, ungrudging, unstinting. *Opp* SELFISH.

unsightly *adj* ▷ UGLY.

unskilful *adj* amateurish, bungled, clumsy, crude, incompetent, inept, inexpert, maladroit, *inf* rough and ready, shoddy, unprofessional. *Opp* SKILFUL.

unskilled *adj* inexperienced, unqualified, untrained. *Opp* SKILLED.

unsociable *adj* ▷ UNFRIENDLY.

unsophisticated *adj* artless, childlike, guileless, ingenuous, innocent, lowbrow, naïve, plain, provincial, simple, simple-minded, straightforward, unaffected, uncomplicated, unostentatious, unpretentious, unrefined, unworldly. *Opp* SOPHISTICATED.

unsound *adj* ▷ WEAK.

unspeakable *adj* dreadful, indescribable, inexpressible, nameless, unutterable.

unspecified *adj* ▷ UNIDENTIFIED.

unstable *adj* capricious, changeable, fickle, inconsistent, inconstant, mercurial, shifting, unpredictable, unsteady, variable, volatile. *Opp* STABLE.

unsteady *adj* 1 flimsy, frail, insecure, precarious, rickety, *inf* rocky, shaky, tottering, unbalanced, unsafe, unstable, wobbly. 2 changeable, erratic, inconstant, intermittent, irregular, variable. 3 *an unsteady light.* flickering, fluctuating, quavering, quivering, trembling, tremulous, wavering. *Opp* STEADY.

unsuccessful *adj* 1 abortive, failed, fruitless, futile, ill-fated, ineffective, ineffectual, loss-making, sterile, unavailing, unlucky, unproductive, unprofitable, unsatisfactory, useless, vain, worthless. 2 *unsuccessful contest-*

ants. beaten, defeated, foiled, hapless, losing, luckless, vanquished. *Opp* SUCCESSFUL.

unsuitable *adj* ill-chosen, ill-judged, ill-timed, inapposite, inappropriate, incongruous, inept, irrelevant, mistaken, unbefitting, unfitting, unhappy, unsatisfactory, unseasonable, unseemly, untimely. *Opp* SUITABLE.

unsure *adj* ▷ UNCERTAIN.

unsurpassed *adj* ▷ UNEQUALLED.

unsuspecting *adj* ▷ CREDULOUS.

unsympathetic *adj* apathetic, cool, cold, dispassionate, hard-hearted, heartless, impassive, indifferent, insensitive, neutral, pitiless, reserved, ruthless, stony, stony-hearted, unaffected, uncaring, uncharitable, unconcerned, unfeeling, uninterested, unkind, unmoved, unpitying, unresponsive. *Opp* SYMPATHETIC.

unsystematic *adj* anarchic, chaotic, confused, disorderly, disorganized, haphazard, illogical, jumbled, muddled, *inf* shambolic, *inf* sloppy, unmethodical, unplanned, unstructured, untidy. *Opp* SYSTEMATIC.

unthinkable *adj* ▷ INCONCEIVABLE.

unthinking *adj* ▷ THOUGHTLESS.

untidy *adj* **1** careless, chaotic, cluttered, confused, disorderly, disorganized, haphazard, *inf* higgledy-piggledy, in disarray, jumbled, littered, *inf* messy, muddled, *inf* shambolic, slapdash, *inf* sloppy, slovenly, *inf* topsy-turvy, unsystematic, upside-down. **2** *untidy hair*. bedraggled, blowzy, dishevelled, disordered, rumpled, scruffy, shabby, tangled, tousled, uncared-for, uncombed, ungroomed, unkempt. *Opp* TIDY.

untie *vb* cast off (*boat*), disentangle, free, loosen, release, unbind, undo, unfasten, unknot, untether.

untried *adj* experimental, innovatory, new, novel, unproved, untested. *Opp* ESTABLISHED.

untroubled *adj* carefree, peaceful, straightforward, undisturbed, uninterrupted, unruffled.

untrue *adj* ▷ FALSE.

untrustworthy *adj* ▷ DISHONEST.

untruthful *adj* ▷ LYING.

unused *adj* blank, clean, fresh, intact, mint (*condition*), new, pristine, unopened, untouched, unworn. *Opp* USED.

unusual *adj* abnormal, atypical, curious, *inf* different, exceptional, extraordinary, *inf* freakish, *inf* funny, irregular, odd, out of the ordinary, peculiar, queer, rare, remarkable, singular, strange, surprising, uncommon, unconventional, unexpected, unfamiliar, *inf* unheard-of, *inf* unique, unnatural, unorthodox, untypical, unwonted. *Opp* USUAL.

unutterable *adj* ▷ INDESCRIBABLE.

unwanted *adj* ▷ UNNECESSARY.

unwarranted *adj* ▷ UNJUSTIFIABLE.

unwary *adj* ▷ CARELESS.

unwavering *adj* ▷ RESOLUTE.

unwelcome *adj* disagreeable, unacceptable, undesirable, uninvited, unpopular, unwanted. *Opp* WELCOME.

unwell *adj* ▷ ILL.

unwholesome *adj* ▷ UNHEALTHY.

unwieldy *adj* awkward, bulky, clumsy, cumbersome, inconvenient, ungainly, unmanageable. *Opp* HANDY, PORTABLE.

unwilling *adj* averse, backward, disinclined, grudging, half-hearted, hesitant, ill-disposed, indisposed, lazy, loath, opposed, reluctant, resistant, slow, uncooperative, unenthusiastic, unhelpful. *Opp* WILLING.

unwise *adj inf* daft, foolhardy, foolish, ill-advised, ill-judged, illogical, imperceptive, impolitic, imprudent, inadvisable, indiscreet, inexperienced, injudicious, irrational, irresponsible, mistaken, obtuse, perverse, rash, reckless, senseless, short-sighted, silly, stupid, thoughtless, unintelligent, unreasonable. *Opp* WISE.

unworldly *adj* ▷ SPIRITUAL.

unworthy *adj* contemptible, despicable, discreditable, dishonourable, disreputable, ignoble, inappropriate, mediocre, second-rate, shameful, substandard, undeserving, unsuitable. *Opp* WORTHY.

unwritten *adj* oral, spoken, verbal, *inf* word-of-mouth. *Opp* WRITTEN.

unyielding *adj* ▷ INFLEXIBLE.

upbringing *n* breeding, bringing-up, care, education, instruction, nurture, raising, rearing, teaching, training.

update *vb* amend, bring up to date, correct, modernize, review, revise.

upgrade *vb* enhance, expand, improve, make better.

upheaval *n* chaos, commotion, confusion, disorder, disruption, disturbance, revolution, *inf* to-do, turmoil.

uphill *adj* arduous, difficult, exhausting, gruelling, hard, laborious, stiff, strenuous, taxing, tough.

uphold *vb* back, champion, defend, endorse, maintain, preserve, protect, stand by, support, sustain.

upkeep *n* care, conservation, keep, maintenance, operation, preservation, running, support.

uplifting *adj* civilizing, edifying, educational, enlightening, ennobling, enriching, humanizing, improving, spiritual. *Opp* SHAMEFUL.

upper *adj* elevated, higher, raised, superior, upstairs.

uppermost *adj* dominant, highest, loftiest, supreme, top, topmost.

upright *adj* **1** erect, on end, perpendicular, vertical. **2** *an upright judge.* conscientious, fair, good, high-minded, honest, honourable, incorruptible, just, moral, principled, righteous, *inf* straight, true, trustworthy, upstanding, virtuous. • *n* column, pole, post, vertical.

uproar *n inf* bedlam, brawling, chaos, clamour, commotion, confusion, din, disorder, disturbance, furore, *inf* hubbub, *inf* hullabaloo, *inf* a madhouse, noise, outburst, outcry, pandemonium, *inf* racket, riot, row, *inf* ructions, *inf* rumpus, tumult, turbulence, turmoil.

uproot *vb* deracinate, destroy, eliminate, eradicate, extirpate, get rid of, *inf* grub up, pull up, remove, root out, tear up, weed out.

upset *vb* **1** capsize, destabilize, overturn, spill, tip over, topple. **2** *upset a plan.* affect, alter, change, confuse, defeat, disorganize, disrupt, hinder, interfere with, interrupt, jeopardize, overthrow, spoil. **3** *upset feelings.* agitate, alarm, annoy, disconcert, dismay, distress, disturb, excite, fluster, frighten, grieve, irritate, offend, perturb, *inf* rub up the wrong way, ruffle, scare, unnerve, worry.

upside-down *adj* inverted, *inf* topsy-turvy, upturned, wrong way up.

upstart *n Fr* nouveau riche, social climber, *inf* yuppie.

up-to-date *adj* **1** advanced, current, latest, modern, new, present-day, recent. **2** contemporary, fashionable, *inf* in, modish, stylish, *inf* trendy. *Opp* OLD-FASHIONED.

upward *adj* ascending, going up, rising, uphill. *Opp* DOWNWARD.

urban *adj* built-up, densely populated, metropolitan, suburban. *Opp* RURAL.

urge *n* compulsion, craving, desire, drive, eagerness, hunger, impetus, impulse, inclination, instinct, *inf* itch, longing, pressure, thirst, wish, yearning, *inf* yen. • *vb* accelerate, advise, advocate, appeal to, beg, beseech, *inf* chivvy, compel, counsel, drive, *inf* egg on, encourage, entreat, exhort, force, goad, impel, implore, importune, incite, induce, invite, move on, nag, persuade, plead with, press, prod, prompt, propel, push, recommend, solicit, spur, stimulate. *Opp* DISCOURAGE.

urgent *adj* **1** acute, compelling, compulsive, dire, essential, exigent, high-priority, immediate, imperative, important, inescapable, instant, necessary, pressing, top-priority, unavoidable. **2** *an urgent cry for help.* eager, earnest, forceful, importunate, insistent, persistent, persuasive, solicitous.

usable *adj* acceptable, current, fit to use, functional, functioning, operating, operational, serviceable, valid, working.

use *n* advantage, application, benefit, employment, function, necessity, need, *inf* point, profit, purpose, usefulness, utility, value, worth. • *vb* **1** apply, administer, deal with, employ, exercise, exploit, handle, make use of, manage, operate, put to use, utilize, wield, work. **2** consume, drink, eat, exhaust, expend, spend, use up, waste, wear out.

used *adj* cast-off, *inf* hand-me-down, second-hand, soiled. *Opp* UNUSED.

useful *adj* **1** advantageous, beneficial, constructive, good, helpful, invaluable, positive, profitable, salutary, valuable, worthwhile. **2** *a useful tool.* convenient, effective, efficient, handy, powerful, practical, productive, utilitarian. **3** *a useful player.* capable, competent, effectual, proficient, skilful, successful, talented. *Opp* USELESS.

useless *adj* **1** fruitless, futile, hopeless, pointless, unavailing, unprofitable, unsuccessful, vain, worthless. **2** *inf* broken down, *inf* clapped out, dead, dud, impractical, ineffective, inefficient, unusable. **3** *a useless player.* incapable, incompetent, ineffectual, lazy, unhelpful, unskilful, unsuccessful, untalented. *Opp* USEFUL.

usual *adj* accepted, accustomed, average, common, conventional, customary, everyday, expected, familiar, general, habitual, natural, normal, official, ordinary, orthodox, predictable, prevalent, recognized, regular, routine, standard, stock, traditional, typical, unexceptional, unsurprising, well-known, widespread, wonted. *Opp* UNUSUAL.

usurp *vb* appropriate, assume, commandeer, seize, steal, take, take over.

utensil *n* appliance, device, gadget, implement, instrument, machine, tool.

utter *vb* articulate, *inf* come out with, express, pronounce, voice. ▷ SPEAK, TALK.

V

vacancy *n* job, opening, place, position, post, situation.

vacant *adj* **1** available, bare, blank, clear, empty, free, hollow, open, unfilled, unused, usable, void. **2** abandoned, deserted, uninhabited, unoccupied, untenanted. **3** *a vacant look.* absent-minded, abstracted, blank, deadpan, dreamy, expressionless, far-away, fatuous, inattentive, vacuous. *Opp* BUSY.

vacate *vb* abandon, depart from, desert, evacuate, get out of, give up, leave, quit, withdraw from.

vacuous *adj* apathetic, blank, empty-headed, expressionless, inane, mindless, uncomprehending, unintelligent, vacant. ▷ STUPID. *Opp* ALERT.

vacuum *n* emptiness, space, void.

vagary *n* caprice, fancy, fluctuation, quirk, uncertainty, unpredictability, *inf* ups and downs, whim.

vagrant *n* beggar, destitute person, *inf* down-and-out, homeless person, itinerant, tramp, traveller, vagabond, wanderer, wayfarer.

vague *adj* **1** ambiguous, ambivalent, broad, confused, diffuse, equivocal, evasive, general, generalized, imprecise, indefinable, indefinite, inexact, loose, nebulous, uncertain, unclear, undefined, unspecific, unsure, *inf* woolly. **2** amorphous, blurred, dim, hazy, ill-defined, indistinct, misty, shadowy, unrecognizable. **3** absent-minded, careless, disorganized, forgetful, inattentive, scatter-brained, thoughtless. *Opp* DEFINITE.

vain *adj* **1** arrogant, *inf* bigheaded, boastful, *inf* cocky, conceited, egotistical, haughty, narcissistic, proud, self-important, self-satisfied, *inf* stuck-up, vainglorious. *Opp* MODEST. **2** *a vain attempt.* abortive, fruitless, futile, ineffective, pointless, senseless, unavailing, unproductive, unrewarding, unsuccessful, useless, worthless. *Opp* SUCCESSFUL.

valiant *adj* bold, brave, courageous, doughty, gallant, heroic, plucky, stalwart, stout-hearted, valorous. *Opp* COWARDLY.

valid *adj* acceptable, allowed, approved, authentic, authorized, *Lat* bona fide, convincing, current, genuine, lawful, legal, legitimate, official, permissible, permitted, proper, ratified, rightful, sound, suitable, usable. *Opp* INVALID.

validate *vb* authenticate, authorize, certify, endorse, legalize, legitimize, make valid, ratify.

valley *n* canyon, chasm, coomb, dale, defile, dell, dingle, glen, gorge, gulch, gully, hollow, pass, ravine, vale.

valour *n* bravery, courage, pluck.

valuable *adj* **1** costly, dear, expensive, generous, irreplaceable, precious, priceless, prized, treasured, valued. **2** *valuable advice.* advantageous, beneficial, constructive, esteemed, helpful, invaluable, positive, profitable, useful, worthwhile. *Opp* WORTHLESS.

value *n* 1 cost, price, worth. 2 advantage, benefit, importance, merit, significance, use, usefulness. • *vb* 1 assess, estimate the value of, evaluate, price, *inf* put a figure on. 2 appreciate, care for, cherish, esteem, *inf* have a high regard for, *inf* hold dear, love, prize, respect, treasure.

vandal *n* barbarian, delinquent, hooligan, looter, marauder, Philistine, raider, ruffian, savage, thug, trouble-maker.

vanish *vb* clear, clear off, disappear, disperse, dissolve, dwindle, evaporate, fade, go away, melt away, pass. *Opp* APPEAR.

vanity *n* arrogance, *inf* big-headedness, conceit, egotism, narcissism, pride, self-esteem.

vaporize *vb* dry up, evaporate, turn to vapour.

vapour *n* exhalation, fog, fumes, gas, haze, miasma, mist, smoke, steam.

variable *adj* capricious, changeable, erratic, fickle, fitful, fluctuating, fluid, inconsistent, inconstant, mercurial, mutable, protean, shifting, temperamental, uncertain, unpredictable, unreliable, unstable, unsteady, *inf* up-and-down, vacillating, varying, volatile, wavering. *Opp* INVARIABLE.

variation *n* alteration, change, conversion, deviation, difference, discrepancy, diversification, elaboration, modification, permutation, variant.

variety *n* 1 alteration, change, difference, diversity, unpredictability, variation. 2 array, assortment, blend, collection, combination, jumble, medley, miscellany, mixture, multiplicity. 3 brand, breed, category, class, form, kind, make, sort, species, strain, type.

various *adj* assorted, contrasting, different, differing, dissimilar, diverse, heterogeneous, miscellaneous, mixed, *inf* motley, multifarious, several, sundry, varied, varying. *Opp* SIMILAR.

vary *vb* 1 change, deviate, differ, fluctuate, go up and down, vacillate. 2 *vary your speed.* adapt, adjust, alter, convert, modify, reset, switch, transform, upset. *Opp* STABILIZE. **varied, varying** ▷ VARIOUS.

vast *adj* boundless, broad, colossal, enormous, extensive, gigantic, great, huge, immeasurable, immense, infinite, interminable, large, limitless, mammoth, massive, measureless, monumental, never-ending, titanic, tremendous, unbounded, unlimited, voluminous, wide. ▷ BIG. *Opp* SMALL.

vault *n* basement, cavern, cellar, crypt, repository, strongroom, undercroft. • *vb* bound over, clear, hurdle, jump, leap, leapfrog, spring over.

veer *vb* change direction, dodge, swerve, tack, turn, wheel.

vegetable *adj* growing, organic. • *n* □ *artichoke, asparagus, aubergine, bean, beet, beetroot, broad bean, broccoli, Brussels sprout, butter bean, cabbage, carrot, cauliflower, celeriac, celery, chicory, courgette, cress, cucumber, garlic, kale, kohlrabi, leek, lettuce, marrow, mushroom, onion, parsnip, pea, pepper, potato, pumpkin, radish, runner bean, salsify, shallot, spinach, sugar beet, swede, sweetcorn, tomato, turnip, watercress, zucchini.*

vegetate *vb* be inactive, do nothing, *inf* go to seed, idle, lose interest, stagnate.

vegetation *n* foliage, greenery, growing things, growth, plants, undergrowth, weeds.

vehement *adj* animated, ardent, eager, enthusiastic, excited, fervent, fierce, forceful, heated, impassioned, intense, passionate, powerful, strong, urgent, vigorous, violent. *Opp* APATHETIC.

vehicle *n* conveyance. □ *ambulance,* inf *buggy, bulldozer, bus, cab, camper, caravan, carriage, cart,* old use *charabanc, chariot, coach, dump truck, dustcart, estate car, fire-engine, float, gig, go-kart, hearse, horse-box, jeep, juggernaut, lorry, minibus, minicab,* inf *motor, motor car,* old use *omnibus, pantechnicon, patrol-car, pick-up, removal van, rickshaw, scooter, sedan-chair, side-car, sledge, snow-plough, stage-coach, steam-roller, tank, tanker, taxi, traction-engine, tractor, trailer, tram, transporter, trap, trolley-bus, truck, tumbrel, van, wagon,* sl *wheels.* ▷ CAR, CYCLE, TRAIN, VESSEL.

veil *vb* camouflage, cloak, conceal, cover, disguise, hide, mask, shroud.

vein *n* **1** artery, blood vessel, capillary. **2** *mineral vein.* bed, course, deposit, line, lode, seam, stratum. **3** ▷ MOOD.

veneer *n* coating, covering, finish, gloss, layer, surface. • *vb* ▷ COVER.

venerable *adj* aged, ancient, august, dignified, esteemed, estimable, honourable, honoured, old, respectable, respected, revered, reverenced, sedate, venerated, worshipped, worthy of respect.

venerate *vb* adore, esteem, hero-worship, honour, idolize, look up to, pay homage to, respect, revere, reverence, worship.

vengeance *n* reprisal, retaliation, retribution, revenge, *inf* tit for tat.

vengeful *adj* avenging, bitter, rancorous, revengeful, spiteful, unforgiving, vindictive. *Opp* FORGIVING.

venom *n* poison, toxin.

venomous *adj* deadly, lethal, poisonous, toxic.

vent *n* aperture, cut, duct, gap, hole, opening, orifice, outlet, passage, slit, slot, split. • *vb* articulate, express, give vent to, let go, make known, release, utter, ventilate, voice. ▷ SPEAK.

ventilate *vb* aerate, air, freshen, oxygenate.

venture *n* enterprise, experiment, gamble, risk, speculation, undertaking. • *vb* **1** bet, chance, dare, gamble, put forward, risk, speculate, stake, wager. **2** *venture out.* dare to go, risk going.

venturesome *adj* adventurous, bold, courageous, daring, doughty, fearless, intrepid.

venue *n* meeting-place, location, rendezvous.

verbal *adj* **1** lexical, linguistic. **2** *a verbal message.* oral, said, spoken, unwritten, vocal, word-of-mouth. *Opp* WRITTEN.

verbatim *adj* exact, faithful, literal, precise, word for word.

verbose *adj* diffuse, garrulous, long-winded, loquacious, prolix, rambling, repetitious, talkative, voluble. ▷ WORDY. *Opp* CONCISE.

verbosity *n inf* beating about the bush, circumlocution, diffuseness, garrulity, long-windedness, loquacity, periphrasis, prolixity, repetition, verbiage, wordiness.

verdict *n* adjudication, assessment, conclusion, decision, finding, judgement, opinion, sentence.

verge *n* bank, boundary, brim, brink, edge, hard shoulder, kerb, lip, margin, roadside, shoulder, side, threshold, wayside.

verifiable *adj* demonstrable, provable.

verify *vb* affirm, ascertain, attest to, authenticate, bear witness to, check out, confirm, corroborate, demonstrate the truth of, establish, prove, show the truth of, substantiate, support, uphold, validate, vouch for.

verisimilitude *n* authenticity, realism.

vermin *plur n* parasites, pests.

vernacular *adj* common, everyday, indigenous, local, native, ordinary, popular, vulgar.

versatile *adj* adaptable, all-purpose, all-round, flexible, gifted, multi-purpose, resourceful, skilful, talented.

verse *n* lines, metre, rhyme, stanza. □ *blank verse, Chaucerian stanza, clerihew, couplet, free verse, haiku, hexameter, limerick, ottava rima, pentameter, quatrain, rhyme royal, sestina, sonnet, Spenserian stanza, terza rima, triolet, triplet, vers libre, villanelle.* ▷ POEM.

versed *adj* accomplished, competent, experienced, expert, knowledgeable, practised, proficient, skilled, taught, trained.

version *n* **1** account, description, portrayal, reading, rendition, report, story. **2** adaptation, interpretation, paraphrase, rendering, translation. **3** design, form, kind, mark, model, style, type, variant.

vertical *adj* erect, perpendicular, precipitous, sheer, upright. *Opp* HORIZONTAL.

vertigo *n* dizziness, giddiness, light-headedness.

very *adv* acutely, enormously, especially, exceedingly, extremely, greatly, highly, *inf* jolly, most, noticeably, outstandingly, particularly, really, remarkably, *inf* terribly, truly, uncommonly, unusually.

vessel *n* **1** ▷ CONTAINER. **2** bark, boat, craft, ship. □ *aircraftcarrier, barge, bathysphere, battleship, brigantine, cabin cruiser, canoe, catamaran, clipper, coaster, collier, coracle, corvette, cruise-liner, cruiser, cutter, destroyer, dhow, dinghy, dredger, dugout, ferry, freighter, frigate, galleon, galley, gondola, gunboat, houseboat, hovercraft, hydrofoil, hydroplane, icebreaker, junk, kayak, ketch, landing-craft, launch, lifeboat, lighter, lightship, liner, longboat, lugger, man-of-war, merchant ship, minesweeper, motor boat, narrow-boat, oil-tanker, packet-ship, paddle-steamer, pontoon, power-boat, pram, privateer, punt, quinquereme, raft, rowing-boat, sailing-boat, sampan, schooner, skiff, sloop, smack, speed boat, steamer, steamship, submarine, tanker, tender, torpedo boat, tramp steamer, trawler, trireme, troopship, tug, warship, whaler, wind-jammer, yacht, yawl.*

vet *vb inf* check out, examine, investigate, review, scrutinize.

veteran *adj* experienced, mature, old, practised. • *n* experienced soldier, ex-serviceman, ex-servicewoman, old hand, old soldier, survivor.

veto *n* ban, block, embargo, prohibition, proscription, refusal, rejection, *inf* thumbs down. • *vb* ban, bar, blackball, block, disallow, dismiss, forbid, prohibit, proscribe, quash, refuse, reject, rule out, say no to, turn down, vote against. *Opp* APPROVE.

vex *vb inf* aggravate, annoy, bother, displease, exasperate, harass, irritate, provoke, *inf* put out, trouble, upset, worry. ▷ ANGER.

viable *adj* achievable, feasible, operable, possible, practicable, practical, realistic, reasonable, supportable, sustainable, usable, workable. *Opp* IMPRACTICAL.

vibrant *adj* alert, alive, dynamic, electric, energetic, lively, living, pulsating, quivering, resonant, thrilling, throbbing, trembling, vibrating, vivacious. *Opp* LIFELESS.

vibrate *vb* fluctuate, judder, oscillate, pulsate, quake, quiver, rattle, resonate, reverberate, shake, shiver, shudder, throb, tremble, wobble.

vibration *n* juddering, oscillation, pulsation, quivering, rattling, resonance, reverberation, shaking, shivering, shuddering, throbbing, trembling, tremor.

vicarious *adj* delegated, deputed, indirect, second-hand, surrogate.

vice *n* **1** badness, corruption, degeneracy, degradation, depravity, evil, evil-doing, immorality, iniquity, lechery, profligacy, promiscuity, sin, venality, villainy, wickedness, wrongdoing. **2** bad habit, blemish, defect, failing, fault, flaw, foible, imperfection, shortcoming, weakness.

vicinity *n* area, district, environs, locale, locality, neighbourhood, outskirts, precincts, proximity, purlieus, region, sector, territory, zone.

vicious *adj* **1** atrocious, barbaric, barbarous, beastly, bloodthirsty, brutal, callous, cruel, diabolical, fiendish, heinous, hurtful, inhuman, merciless, monstrous, murderous, pitiless, ruthless, sadistic, savage, unfeeling, vile, violent. **2** *a vicious character.* bad, *inf* bitchy, *inf* catty, depraved, evil, heartless, immoral, malicious, mean, perverted, rancorous, sinful, spiteful, venomous, villainous, vindictive, vitriolic, wicked. **3** *vicious animals.* aggressive, bad-tempered, dangerous, ferocious, fierce, snappy, untamed, wild. **4** *a vicious wind.* cutting, nasty, severe, sharp, unpleasant. *Opp* GENTLE.

vicissitude *n* alteration, change, flux, instability, mutability, mutation, shift, uncertainty, unpredictability, variability.

victim *n* **1** casualty, fatality, injured person, patient, sufferer, wounded person. **2** *sacrificial victim.* martyr, offering, prey, sacrifice.

victimize *vb* bully, cheat, discriminate against, exploit, intimidate, oppress, persecute, *inf* pick on, prey on, take advantage of, terrorize, torment, treat unfairly, *inf* use. ▷ CHEAT.

victor *n* champion, conqueror, prizewinner, winner. *Opp* LOSER.

victorious *adj* champion, conquering, first, leading, prevailing, successful, top, top-scoring, triumphant, unbeaten, undefeated, winning. *Opp* UNSUCCESSFUL.

victory *n* achievement, conquest, knockout, mastery, success, superiority, supremacy, triumph, *inf* walk-over, win. *Opp* DEFEAT.

vie *vb* compete, contend, strive, struggle.

view *n* **1** aspect, landscape, outlook, panorama, perspective, picture, prospect, scene, scenery, seascape, spectacle, townscape, vista.

2 angle, look, perspective, sight, vision. 3 *political views*. attitude, belief, conviction, idea, judgement, notion, opinion, perception, position, thought. • *vb* 1 behold, consider, contemplate, examine, eye, gaze at, inspect, observe, perceive, regard, scan, stare at, survey, witness. 2 *view TV*. look at, see, watch.

viewer *n plur* audience, observer, onlooker, spectator, watcher, witness.

viewpoint *n* angle, perspective, point of view, position, slant, standpoint.

vigilant *adj* alert, attentive, awake, careful, circumspect, eagle-eyed, observant, on the watch, on your guard, *inf* on your toes, sharp, wakeful, wary, watchful, wideawake. *Opp* NEGLIGENT.

vigorous *adj* active, alive, animated, brisk, dynamic, energetic, fit, flourishing, forceful, full-blooded, *inf* full of beans, growing, hale and hearty, healthy, lively, lusty, potent, prosperous, red-blooded, robust, spirited, strenuous, strong, thriving, virile, vital, vivacious, zestful. *Opp* FEEBLE.

vigour *n* animation, dynamism, energy, fitness, force, forcefulness, gusto, health, life, liveliness, might, potency, power, robustness, spirit, stamina, strength, verve, *inf* vim, virility, vitality, vivacity, zeal, zest.

vile *adj* bad, base, contemptible, degenerate, depraved, despicable, disgusting, evil, execrable, filthy, foul, hateful, horrible, immoral, loathsome, low, nasty, nauseating, obnoxious, odious, offensive, perverted, repellent, repugnant, repulsive, revolting, sickening, sinful, ugly, vicious, wicked.

vilify *vb* abuse, calumniate, defame, denigrate, deprecate, disparage, revile, *inf* run down, slander, *inf* smear, speak evil of, traduce, vituperate.

villain *n* blackguard, criminal, evil-doer, malefactor, mischief-maker, miscreant, reprobate, rogue, scoundrel, sinner, wretch. ▷ CRIMINAL.

villainous *adj* bad, corrupt, criminal, dishonest, evil, sinful, treacherous, vile. ▷ WICKED.

vindictive *adj* avenging, malicious, nasty, punitive, rancorous, revengeful, spiteful, unforgiving, vengeful, vicious. *Opp* FORGIVING.

vintage *adj* choice, classic, fine, good, high-quality, mature, mellowed, old, seasoned, venerable.

violate *vb* 1 breach, break, contravene, defy, disobey, disregard, flout, ignore, infringe, overstep, sin against, transgress. 2 *violate someone's privacy*. abuse, desecrate, disturb, invade, profane. 3 [*of men*] *violate a woman*. assault, attack, debauch, dishonour, force yourself on, rape, ravish.

violation *n* breach, contravention, defiance, flouting, infringement, invasion, offence (against), transgression.

violent *adj* 1 acute, damaging, dangerous, destructive, devastating, explosive, ferocious, fierce, forceful, furious, hard, harmful, intense, powerful, rough, ruinous, savage, severe, strong, swingeing, tempestuous, turbulent, uncontrollable, vehement, wild. 2 *violent behaviour*. barbaric, berserk, bloodthirsty, brutal, cruel, desperate, frenzied, headstrong, homicidal, murderous, riotous, rowdy, ruthless, uncontrolled, unruly, vehement, vicious, wild. *Opp* GENTLE.

VIP *n* celebrity, dignitary, important person.

virile *adj derog* macho, manly, masculine, potent, vigorous.

virtue *n* **1** decency, fairness, goodness, high-mindedness, honesty, honour, integrity, justice, morality, nobility, principle, rectitude, respectability, righteousness, right-mindedness, sincerity, uprightness, worthiness. **2** advantage, asset, good point, merit, quality, *inf* redeeming feature, strength. **3** *sexual virtue.* abstinence, chastity, honour, innocence, purity, virginity. *Opp* VICE.

virtuoso *n* expert, genius, maestro, prodigy, showman, *inf* wizard. ▷ MUSICIAN.

virtuous *adj* blameless, chaste, decent, ethical, exemplary, fair, God-fearing, good, *derog* goody-goody, high-minded, high-principled, honest, honourable, innocent, irreproachable, just, law-abiding, moral, noble, principled, praiseworthy, pure, respectable, right, righteous, right-mindedness, sincere, *derog* smug, spotless, trustworthy, uncorrupted, unimpeachable, unsullied, upright, virginal, worthy. *Opp* WICKED.

virulent *adj* **1** dangerous, deadly, lethal, life-threatening, noxious, pernicious, poisonous, toxic, venomous. **2** *virulent abuse.* acrimonious, bitter, hostile, malicious, malign, malignant, mordant, nasty, spiteful, splenetic, vicious, vitriolic.

viscous *adj* gluey, sticky, syrupy, thick, viscid. *Opp* RUNNY.

visible *adj* apparent, clear, conspicuous, detectable, discernible, distinct, evident, manifest, noticeable, observable, obvious, open, perceivable, perceptible, plain, recognizable, unconcealed, undisguised, unmistakable. *Opp* INVISIBLE.

vision *n* **1** eyesight, perception, sight. **2** apparition, chimera, daydream, delusion, fantasy, ghost, hallucination, illusion, mirage, phantasm, phantom, spectre, spirit, wraith. **3** *a man of vision.* far-sightedness, foresight, imagination, insight, spirituality, understanding.

visionary *adj* dreamy, fanciful, far-sighted, futuristic, idealistic, imaginative, impractical, mystical, prophetic, quixotic, romantic, speculative, transcendental, unrealistic, Utopian. • *n* dreamer, idealist, mystic, poet, prophet, romantic, seer.

visit *n* **1** call, *old use* sojourn, stay, stop, visitation. **2** day out, excursion, outing, trip. • *vb* call on, come to see, *inf* descend on, *inf* drop in on, go to see, *inf* look up, make a visit to, pay a call on, *inf* pop in on, stay with. **visit regularly** ▷ HAUNT.

visitor *n* **1** caller, *plur* company, guest. **2** holiday-maker, sightseer, tourist, traveller, tripper. **3** *a visitor from abroad.* alien, foreigner, migrant, visitant.

visor *n* protector, shield, sunshield.

vista *n* landscape, outlook, panorama, prospect, scene, scenery, seascape, view.

visualize *vb* conceive, dream up, envisage, imagine, picture.

vital *adj* **1** alive, animate, animated, dynamic, energetic, exuberant, life-giving, live, lively, living, sparkling, spirited, sprightly, vigorous, vivacious, zestful. *Opp* LIFELESS. **2** *vital information.* compulsory, current, crucial, essential, fundamental, imperative, important, indispensable, mandatory, necessary, needed, relevant, requisite. *Opp* INESSENTIAL.

vitality *n* animation, dynamism, energy, exuberance, *inf* go, life, liveliness, *inf* sparkle, spirit, sprightliness, stamina, strength, vigour, *inf* vim, vivacity, zest.

vitriolic *adj* abusive, acid, biting, bitter, caustic, cruel, destructive, hostile, hurtful, malicious, savage, scathing, vicious, vindictive, virulent.

vituperate *vb* abuse, berate, calumniate, censure, defame, denigrate, deprecate, disparage, reproach, revile, *inf* run down, slander, upbraid, vilify.

vivacious *adj* animated, bubbly, cheerful, ebullient, energetic, high-spirited, light-hearted, lively, merry, positive, spirited, sprightly. *Opp* LETHARGIC.

vivid *adj* **1** bright, brilliant, colourful, dazzling, fresh, *derog* gaudy, gay, gleaming, glowing, intense, rich, shining, showy, strong, vibrant. **2** *a vivid description.* clear, detailed, graphic, imaginative, lifelike, lively, memorable, powerful, realistic, striking. *Opp* LIFELESS.

vocabulary *n* **1** diction, lexis, words. **2** dictionary, glossary, lexicon, phrase book, word-list.

vocal *adj* **1** oral, said, spoken, sung, voiced. **2** *vocal in discussion.* communicative, forthcoming, loquacious, outspoken, talkative, vociferous. *Opp* TACITURN.

vocation *n* calling, career, employment, job, life's work, occupation, profession, trade.

vogue *n* craze, fad, fashion, *inf* latest thing, mode, rage, style, taste, trend. **in vogue** ▷ FASHIONABLE.

voice *n* accent, articulation, expression, idiolect, inflexion, intonation, singing, sound, speaking, speech, tone, utterance. • *vb* ▷ SPEAK.

void *adj* **1** blank, empty, unoccupied, vacant. **2** *a void contract.* annulled, cancelled, inoperative, invalid, not binding, unenforceable, useless. • *n* blank, emptiness, nothingness, space, vacancy, vacuum.

volatile *adj* **1** explosive, sensitive, unstable. **2** *volatile moods.* changeable, erratic, fickle, flighty, inconstant, lively, mercurial, temperamental, unpredictable, *inf* up and down, variable. *Opp* STABLE.

volley *n* barrage, bombardment, burst, cannonade, fusillade, salvo, shower.

voluble *adj* chatty, fluent, garrulous, glib, loquacious, talkative. ▷ WORDY.

volume **1** *old use* tome. ▷ BOOK. **2** *volume of a container.* aggregate, amount, bulk, capacity, dimensions, mass, measure, quantity, size.

voluminous *adj* ample, billowing, bulky, capacious, cavernous, enormous, extensive, gigantic, great, huge, immense, large, mammoth, massive, roomy, spacious, vast. ▷ BIG. *Opp* SMALL.

voluntary *adj* **1** elective, free, gratuitous, optional, spontaneous, unpaid, willing. *Opp* COMPULSORY. **2** *a voluntary act.* conscious, deliberate, intended, intentional, planned, premeditated, wilful. *Opp* INVOLUNTARY.

volunteer *vb* **1** be willing, offer, propose, put yourself forward. **2** ▷ ENLIST.

voluptuous *adj* **1** hedonistic, luxurious, pleasure-loving, self-indulgent, sensual, sybaritic, **2** *voluptuous figure.* attractive, buxom, *inf* curvaceous, desirable, erotic, sensual, *inf* sexy, shapely, *inf* well-endowed.

vomit *vb* be sick, *inf* bring up, disgorge, *inf* heave up, *inf* puke, regurgitate, retch, *inf* spew up, *inf* throw up.

voracious *adj* avid, eager, fervid, gluttonous, greedy, hungry, insatiable, keen, ravenous, thirsty.

vortex *n* eddy, spiral, whirlpool, whirlwind.

vote *n* ballot, election, plebiscite, poll, referendum, show of hands. • *vb* ballot, cast your vote. **vote for** choose, elect, nominate, opt for, pick, return, select, settle on.

vouch *vb* **vouch for** answer for, back, certify, endorse, guarantee, speak for, sponsor, support.

voucher *n* coupon, ticket, token.

vow *n* assurance, guarantee, oath, pledge, promise, undertaking, word of honour. • *vb* declare, give an assurance, give your word, guarantee, pledge, promise, swear, take an oath.

voyage *n* cruise, journey, passage. • *vb* circumnavigate, cruise, sail. ▷ TRAVEL.

vulgar *adj* **1** churlish, coarse, common, crude, foul, gross, ill-bred, impolite, improper, indecent, indecorous, low, offensive, rude, uncouth, ungentlemanly, unladylike. ▷ OBSCENE. *Opp* POLITE. **2** *vulgar colour scheme.* crude, gaudy, inartistic, in bad taste, inelegant, insensitive, lowbrow, plebeian, tasteless, tawdry, unrefined, unsophisticated. *Opp* TASTEFUL.

vulnerable *adj* **1** at risk, defenceless, exposed, helpless, unguarded, unprotected, weak, wide open. **2** easily hurt, sensitive, thin-skinned, touchy. *Opp* RESILIENT.

W

wad *n* bundle, lump, mass, pack, pad, plug, roll.

wadding *n* filling, lining, packing, padding, stuffing.

wade *vb* ford, paddle, splash. ▷ WALK.

waffle *n* evasiveness, padding, prevarication, prolixity, verbiage, wordiness. • *vb inf* beat about the bush, *inf* blather on, hedge, prattle, prevaricate.

waft *vb* **1** be borne, drift, float, travel. **2** bear, carry, convey, puff, transmit, transport.

wag *vb* bob, flap, move to and fro, nod, oscillate, rock, shake, sway, undulate, *inf* waggle, wave, *inf* wiggle.

wage *n* compensation, earnings, emolument, honorarium, income, pay, pay packet, recompense, remuneration, reward, salary, stipend. • *vb* carry on, conduct, engage in, fight, prosecute, pursue, undertake.

wager *vb* bet, gamble.

wail *vb* caterwaul, complain, cry, howl, lament, moan, shriek, waul, weep, *inf* yowl.

waist *n* middle, waistline.

waistband *n* belt, cummerbund, girdle.

wait *n* delay, halt, hesitation, hiatus, *inf* hold-up, intermission, interval, pause, postponement, rest, stay, stop, stoppage. • *vb* **1** *old use* bide, delay, halt, *inf* hang about, *inf* hang on, hesitate, hold back, keep still, linger, mark time, pause, remain, rest, *inf* sit tight, stand by, stay, stop, *old use* tarry. **2** *wait at table.* serve.

waive *vb* abandon, cede, disclaim, dispense with, forgo, give up, relinquish, remit, renounce, resign, sign away, surrender.

wake *n* **1** funeral, vigil, watch. **2** *wake of a ship.* path, track, trail, turbulence, wash. • *vb* **1** arouse, awaken, bring to life, call, disturb, galvanize, rouse, stimulate, stir, waken. **2** become conscious, bestir yourself, *inf* come to life, get up, rise, *inf* stir, wake up. **wake up to** ▷ REALIZE. **waking** ▷ CONSCIOUS.

wakeful *adj* alert, awake, insomniac, *inf* on the qui vive, restless, sleepless.

walk *n* **1** bearing, carriage, gait, stride. **2** constitutional, hike, *old use* promenade, ramble, saunter, stroll, traipse, tramp, trek, trudge, *inf* turn. **3** *a paved walk.* aisle, alley, path, pathway, pavement. • *vb* **1** be a pedestrian, travel on foot. □ *amble, crawl, creep, dodder,* inf *foot-slog, hike, hobble, limp, lope, lurch, march, mince,* sl *mooch, pace, pad, paddle, parade,* old use *perambulate, plod, promenade, prowl, ramble, saunter, scuttle, shamble, shuffle, slink, stagger, stalk, steal, step,* inf *stomp, stride, stroll, strut, stumble, swagger, tiptoe,* inf *toddle, totter, tramp, trample, traipse, tramp, trek, troop, trot, trudge, waddle, wade.* **2** *don't walk on the flowers.* stamp, step, trample, tread. **walk away with** ▷ WIN. **walk off with** ▷ STEAL. **walk out** ▷ QUIT. **walk out on** ▷ DESERT.

walker *n* hiker, pedestrian, rambler.

wall *n* □ *barricade, barrier, bulkhead, bulwark, dam, dike, divider, embankment, fence, fortification, hedge, obstacle, paling, palisade, parapet, partition, rampart, screen, sea-wall, stockade.* **wall in** ▷ ENCLOSE.

wallet *n* notecase, pocketbook, pouch, purse.

wallow *vb* **1** flounder, lie, pitch about, roll about, stagger about, tumble, wade, welter. **2** *wallow in luxury.* glory, indulge yourself, luxuriate, revel, take delight.

wan *adj* anaemic, ashen, bloodless, colourless, exhausted, faint, feeble, livid, pale, pallid, pasty, sickly, tired, waxen, worn.

wand *n* baton, rod, staff, stick.

wander *vb* **1** drift, go aimlessly, meander, prowl, ramble, range, roam, rove, saunter, stray, stroll, travel about, walk, wind. **2** *wander off course.* curve, deviate, digress, drift, err, go off at a tangent, stray, swerve, turn, twist, veer, zigzag. **wandering** ▷ INATTENTIVE, NOMADIC.

wane *vb* decline, decrease, dim, diminish, dwindle, ebb, fade, fail, *inf* fall off, grow less, lessen, peter out, shrink, subside, taper off, weaken. *Opp* STRENGTHEN.

want *n* **1** demand, desire, need, requirement, wish. **2** *a want of ready cash.* absence, lack, need. **3** *war against want.* dearth, famine, hunger, insufficiency, penury, poverty, privation, scarcity, shortage. • *vb* **1** aspire to, covet, crave, demand, desire, fancy, hanker after, *inf* have a yen for, hunger for, *inf* itch for, like, long for, miss, pine for, please, prefer, *inf* set your heart on, thirst after, thirst for, wish for, yearn for. **2** *want manners.* be short of, lack, need, require.

war *n* campaign, conflict, crusade, fighting, hostilities, military action, strife, warfare. □ *ambush, assault, attack, battle, blitz, blockade, bombardment, counter-attack, engagement, guerrilla warfare, invasion, manoeuvres, operations, resistance, siege, skirmish.* **wage war** ▷ FIGHT.

ward *n* charge, dependant, minor. • *vb* **ward off** avert, beat off, block, chase away, check, deflect, fend off, forestall, parry, push away, repel, repulse, stave off, thwart, turn aside.

warder *n* gaoler, guard, jailer, keeper, prison officer.

warehouse *n* depository, depot, store, storehouse.

wares *plur n* commodities, goods, manufactures, merchandise, produce, stock, supplies.

warlike *adj* aggressive, bellicose, belligerent, hawkish, hostile, militant, militaristic, pugnacious, warmongering, warring.

warm *adj* **1** close, hot, lukewarm, subtropical, sultry, summery, temperate, tepid, warmish. **2** *warm clothes.* cosy, thermal, thick, winter, woolly. **3** *a warm welcome.* affable, affectionate, ardent, cordial, emotional, enthusiastic, excited, fervent, friendly, genial, impassioned, kind, loving, passionate, sympathetic, warm-hearted. *Opp* COLD, UNFRIENDLY. • *vb* heat, make warmer, melt, raise the temperature of, thaw, thaw out. *Opp* COOL.

warn *vb* admonish, advise, alert, caution, counsel, forewarn, give a warning, give notice, inform, notify, raise the alarm, remind, *inf* tip off.

warning *n* **1** advance notice, augury, forewarning, hint, indication, notice, notification, omen, portent, premonition, presage, prophecy, reminder, sign, signal, threat, *inf* tip-off, *inf* word to the wise. □ *alarm, alarm-bell, beacon, bell, fire-alarm, flashing light, foghorn, gong, hooter, red light, siren, traffic-lights, whistle.* **2** *let off with a warning.* admonition, advice, caveat, caution, reprimand.

warp *vb* become deformed, bend, buckle, contort, curl, curve, deform, distort, kink, twist.

warrant *n* authority, authorization, certification, document, entitlement, guarantee, licence, permit, pledge, sanction, search-warrant, warranty, voucher. • *vb* ▷ JUSTIFY.

wary *adj* alert, apprehensive, attentive, *inf* cagey, careful, chary, cautious, circumspect, distrustful, heedful, observant, on the lookout, on your guard, suspicious, vigilant, watchful. *Opp* RECKLESS.

wash *n old use* ablutions, bath, rinse, shampoo, shower. • *vb* **1** clean, cleanse, flush, launder, mop, rinse, scrub, shampoo, sluice, soap down, sponge down, swab down, swill, wipe. **2** bath, bathe, *old use* make your toilet, perform your ablutions, shower. **3** *The sea washes against the cliff.* break, dash, flow, lap, pound, roll, splash. **wash your hands of** ▷ ABANDON.

washing *n* cleaning, dirty clothes, laundry, *inf* the wash.

washout *n* débacle, disappointment, disaster, failure, *inf* flop.

waste *adj* **1** discarded, extra, superfluous, unprofitable, unusable, unused, unwanted, worthless. **2** *waste land.* bare, barren, derelict, empty, overgrown, rundown, uncared-for, uncultivated, undeveloped, unproductive, wild. • *n* **1** debris, dregs, effluent, excess, garbage, junk, leavings, *inf* left-overs, litter, offcuts, refuse, remnants, rubbish, scrap, scraps, trash, unusable material, unwanted material, wastage. **2** extravagance, indulgence, overprovision, prodigality, profligacy, self-indulgence. • *vb* be wasteful with, dissipate, fritter, misspend, misuse, over provide, *sl* splurge, squander, use up, use wastefully. *Opp* CONSERVE. **waste away** become emaciated, become thin, become weaker, mope, pine, weaken.

wasteful *adj* excessive, expensive, extravagant, improvident, imprudent, lavish, needless, prodigal, profligate, reckless, spendthrift, thriftless, uneconomical, unthrifty. *Opp* ECONOMICAL. **wasteful person** ▷ SPENDTHRIFT.

watch *n* chronometer, clock, digital watch, stop-watch, timepiece, timer, wrist-watch. • *vb* **1** attend, concentrate, contemplate, eye, gaze, heed, keep an eye open for, keep your eyes on, look at, mark, note, observe, pay attention, regard, see, spy on, stare, take notice, view. **2** *watch sheep.* care for, chaperon, defend, guard, keep an eye on, keep watch on, look after, mind, protect, safeguard, shield, superintend, supervise, take charge of, tend. **keep watch** ▷ GUARD. **on the watch** ▷ WATCHFUL. **watch your step** ▷ BEWARE.

watcher *n plur* audience, *inf* looker-on, observer, onlooker, spectator, viewer, witness.

watchful *adj* attentive, eagleeyed, heedful, observant, *inf* on the lookout, *inf* on the qui vive, on the watch, quick, sharp-eyed, vigilant. ▷ ALERT. *Opp* INATTENTIVE.

watchman *n* caretaker, custodian, guard, lookout, nightwatchman, security guard, sentinel, sentry, watch.

water *n* **1** Adam's ale, bath water, brine, distilled water, drinking water, mineral water, rainwater, sea water, spa water, spring water, tap water. **2** lake, lido, ocean, pond, pool, river, sea. ▷ STREAM. • *vb* damp, dampen, douse, drench, flood, hose, inundate, irrigate, moisten, saturate, soak, souse, spray, sprinkle, wet. **water down** ▷ DILUTE.

waterfall *n* cascade, cataract, chute, rapids, torrent, white water.

waterlogged *adj* full of water, saturated, soaked.

waterproof *adj* damp-proof, impermeable, impervious, water-repellent, water-resistant, watertight, weatherproof. *Opp* LEAKY. • *n* cape, groundsheet, *inf* mac, mackintosh, sou'wester.

watertight *adj* hermetic, sealed, sound. ▷ WATERPROOF.

watery *adj* **1** aqueous, bland, characterless, dilute, diluted, fluid, liquid, *inf* runny, *inf* sloppy, tasteless, thin, watered-down, weak, *inf* wishy-washy. **2** *watery eyes.* damp, moist, tear-filled, tearful, *inf* weepy. ▷ WET.

wave *n* **1** billow, breaker, crest, heave, ridge, ripple, roller, surf, swell, tidal wave, undulation, wavelet, *inf* white horse. **2** flourish, gesticulation, gesture, shake, sign, signal. **3** *a wave of enthusiasm.* current, flood, ground swell, outbreak, surge, tide, upsurge. **4** *a new wave*. advance, fashion, tendency, trend. **5** *radio waves*. pulse, vibration. • *vb* **1** billow, brandish, flail about, flap, flourish, fluctuate, flutter, move to and fro, ripple, shake, sway, swing, twirl, undulate, waft, wag, waggle, wiggle, zigzag. **2** gesticulate, gesture, indicate, sign, signal. **wave aside** ▷ DISMISS.

wavelength *n* channel, station, waveband.

waver *vb inf* be in two minds, be unsteady, change, falter, flicker, hesitate, quake, quaver, quiver, shake, shiver, shudder, sway, teeter, tergiversate, totter, tremble, vacillate, wobble.

wavy *adj* curling, curly, curving, heaving, rippling, rolling, sinuous, undulating, up and down, winding, zigzag. *Opp* FLAT, STRAIGHT.

way *n* **1** advance, direction, headway, journey, movement, progress, route. ▷ ROAD. **2** distance, length, measurement. **3** *a way to do something*. approach, avenue, course, fashion, knack, manner, means, method, mode, *Lat* modus operandi, path, procedure, process, system, technique. **4** *foreign ways*. custom, fashion, habit, *Lat* modus vivendi, practice, routine, style, tradition. **5** *funny ways*. characteristic, eccentricity, idiosyncrasy, oddity, peculiarity. **6** *in some ways*. aspect, circumstance, detail, feature, particular, respect.

waylay *vb* accost, ambush, await, buttonhole, detain, intercept, lie in wait for, pounce on, surprise. ▷ ATTACK.

wayward *adj* disobedient, headstrong, obstinate, self-willed, stubborn, uncontrollable, uncooperative, wilful. ▷ NAUGHTY. *Opp* COOPERATIVE.

weak *adj* **1** breakable, brittle, decrepit, delicate, feeble, flawed, flimsy, fragile, frail, frangible, inadequate, insubstantial, rickety, shaky, slight, substandard, tender, thin, unsafe, unsound, unsteady, unsubstantial. **2** *weak in health*. anaemic, debilitated, delicate, enervated, exhausted, feeble, flabby, frail, helpless, ill, infirm, listless, *inf* low, *inf* poorly, puny, sickly, slight, thin, tired out, wasted, weakly, *derog* weedy. **3** *a weak character*. cowardly, fearful, impotent, indecisive, ineffective, ineffectual, irresolute, poor, powerless, pusillanimous, spineless, timid, timorous, unassertive, weak-minded, wimpish. **4** *a weak position*. defenceless, exposed, unguarded, unprotected, vulnerable. **5** *weak excuses*. hollow, lame, *inf* pathetic, shallow, unbelievable, unconvincing, unsatisfactory. **6** *weak light*. dim, distant, fading, faint, indistinct, pale, poor, unclear, vague. **7** *weak tea*. dilute, diluted, tasteless, thin, watery. *Opp* STRONG.

weaken *vb* **1** debilitate, destroy, dilute, diminish, emasculate, enervate, enfeeble, erode, exhaust, impair, lessen, lower, make weaker, reduce, ruin, sap, soften, thin down, undermine, *inf* water down. **2** abate, become weaker, decline, decrease, dwindle, ebb, fade, flag, give in, give way, sag, wane, yield. *Opp* STRENGTHEN.

weakling *n* coward, *inf* milksop, *inf* pushover, *inf* runt, *inf* softie, weak person, *inf* weed, *inf* wimp.

weakness *n* **1** *inf* Achilles' heel, blemish, defect, error, failing, fault, flaw, flimsiness, foible, fragility, frailty, imperfection, inadequacy, mistake, shortcoming, softness, *inf* weak spot. **2** debility, decrepitude, delicacy, feebleness, impotence, incapacity, infirmity, lassitude, vulnerability. ▷ ILLNESS. **3** *a weakness for wine.* affection, fancy, fondness, inclination, liking, partiality, penchant, predilection, *inf* soft spot, taste. *Opp* STRENGTH.

wealth *n* **1** affluence, assets, capital, fortune, *old use* lucre, means, opulence, possessions, property, prosperity, riches, *old use* substance. ▷ MONEY. *Opp* POVERTY. **2** *a wealth of information.* abundance, bounty, copiousness, cornucopia, mine, plenty, profusion, store, treasury. *Opp* SCARCITY.

wealthy *adj* affluent, *inf* flush, *inf* loaded, moneyed, opulent, *joc* plutocratic, privileged, prosperous, rich, *inf* well-heeled, well-off, well-to-do. *Opp* POOR. **wealthy person** billionaire, capitalist, millionaire, plutocrat, tycoon.

weapon *n* bomb, gun, missile. □ *airgun, arrow, atom bomb, ballistic missile, battering-ram, battle-axe, bayonet, bazooka, blowpipe, blunderbuss, boomerang, bow and arrow, bren-gun, cannon, carbine, catapult, claymore, cosh, crossbow, CS gas, cudgel, cutlass, dagger, depth-charge, dirk, flame-thrower, foils, grenade,* old use *halberd, harpoon, H-bomb, howitzer, incendiary bomb, javelin, knuckleduster, lance, land-mine, laser beam, longbow, machete, machine-gun, mine, mortar, musket, mustard gas, napalm bomb, pike, pistol, pole-axe, rapier, revolver, rifle, rocket, sabre, scimitar, shotgun,* inf *six-shooter, sling, spear, sten-gun, stiletto, submachine-gun, sword, tank, tear-gas, time-bomb, tomahawk, tommy-gun, torpedo, truncheon, warhead, water-cannon.* **weapons** armaments, armoury, arms, arsenal, magazine, munitions, ordnance, weaponry. □ *artillery, automatic weapons, biological weapons, chemical weapons, firearms, missiles, nuclear weapons, small arms, strategic weapons, tactical weapons.*

wear *vb* **1** be dressed in, clothe yourself in, don, dress in, have on, present yourself in, put on, wrap up in. **2** *wear a smile.* adopt, assume, display, exhibit, show. **3** *wears the carpet.* damage, fray, injure, mark, scuff, wear away, weaken. **4** *wear well.* endure, last, *inf* stand the test of time, survive. **wear away** ▷ ERODE. **wear off** ▷ SUBSIDE. **wear out** ▷ WEARY.

wearisome *adj* boring, dreary, exhausting, monotonous, repetitive, tedious, tiring, wearying. ▷ TROUBLESOME. *Opp* STIMULATING.

weary *adj* bone-weary, *inf* dead beat, *inf* dog-tired, *inf* done in, drawn, drained, drowsy, enervated, exhausted, *inf* fagged, fatigued, fed up, flagging, footsore, impatient, jaded, *inf* jet-lagged, *sl* knackered, listless, prostrate, *inf* shattered, *inf* sick (of), sleepy, spent, tired out, travel-weary, wearied, *inf* whacked, worn out. *Opp* FRESH, LIVELY. • *vb* **1** debilitate, drain, enervate, exhaust, fatigue, *inf* finish, make tired, *sl* knacker, overtire, sap, *inf* shatter, *inf* take it out of, tax, tire, wear out. *Opp* REFRESH. **2** become bored, become tired, flag, grow weary, weaken.

weather *n* climate, the elements, meteorological conditions. □ *blizzard, breeze, cloud, cyclone, deluge, dew, downpour, drizzle, drought, fog, frost, gale, hail, haze,* *heatwave, hoar-frost, hurricane, ice, lightning, mist, rain, rainbow, shower, sleet, slush, snow, snowstorm, squall, storm, sunshine, tempest, thaw, thunder, tornado, typhoon, whirlwind, wind.* • *vb* ▷ SURVIVE. **under the weather** ▷ ILL.

weave *vb* **1** braid, criss-cross, entwine, interlace, intertwine, interweave, knit, plait, sew. **2** *weave a story.* compose, create, make, plot, put together. **3** *weave through a crowd.* dodge, make your way, tack, *inf* twist and turn, wind, zigzag.

web *n* criss-cross, lattice, mesh, net, network.

wedding *n* marriage, nuptials, union.

wedge *vb* cram, force, jam, pack, squeeze, stick.

weep *vb* bawl, *inf* blub, blubber, cry, *inf* grizzle, lament, mewl, moan, shed tears, snivel, sob, wail, whimper, whine.

weigh *vb* **1** measure the weight of. **2** *weigh evidence.* assess, consider, contemplate, evaluate, judge, ponder, reflect on, think about, weigh up. **3** *evidence weighed with the jury.* be important, carry weight, *inf* cut ice, count, have weight, matter. **weigh down** ▷ BURDEN. **weigh up** ▷ EVALUATE.

weighing-machine *n* balance, scales, spring-balance, weighbridge.

weight *n* **1** avoirdupois, burden, density, heaviness, load, mass, pressure, strain, tonnage. **2** *My voice has some weight.* authority, credibility, emphasis, force, gravity, importance, power, seriousness, significance, substance, value, worth. • *vb* ballast, bias, hold down, keep down, load, make heavy, weigh down.

weird *adj* **1** creepy, eerie, ghostly, mysterious, preternatural, scary, *inf* spooky, supernatural, unaccountable, uncanny, unearthly, unnatural. **2** *weird behaviour.* abnormal, bizarre, *inf* cranky, curious, eccentric, *inf* funny, grotesque, odd, outlandish, peculiar, queer, quirky, strange, unconventional, unusual, *inf* way-out, *inf* zany. *Opp* CONVENTIONAL, NATURAL.

welcome *adj* acceptable, accepted, agreeable, appreciated, desirable, gratifying, much-needed, *inf* nice, pleasant, pleasing, pleasurable. *Opp* UNWELCOME. • *n* greeting, hospitality, reception, salutation. • *vb* **1** give a welcome to, greet, hail, receive. **2** *They welcome criticism.* accept, appreciate, approve of, delight in, like, want.

weld *vb* bond, cement, fuse, join, solder, unite. ▷ FASTEN.

welfare *n* advantage, benefit, felicity, good, happiness, health, interest, prosperity, well-being.

well *adj* **1** fit, hale, healthy, hearty, *inf* in fine fettle, lively, robust, sound, strong, thriving, vigorous. **2** *All is well.* all right, fine, *inf* OK, satisfactory. • *n* fountain, shaft, source, spring, waterhole, well-spring. □ *artesian well, borehole, gusher, oasis, oil well, wishing-well.*

well-behaved *adj* cooperative, disciplined, docile, dutiful, good, hard-working, law-abiding, manageable, *inf* nice, polite, quiet, well-trained. ▷ OBEDIENT. *Opp* NAUGHTY.

well-bred *adj* courteous, courtly, cultivated, decorous, genteel, polite, proper, refined, sophisticated, urbane, well-brought-up, well-mannered. *Opp* RUDE.

well-built *adj* athletic, big, brawny, burly, hefty, muscular, powerful, stocky, *inf* strapping, strong, sturdy, upstanding. *Opp* SMALL.

well-known *adj* celebrated, eminent, familiar, famous, illustrious, noted, *derog* notorious, prominent, renowned. *Opp* UNKNOWN.

well-meaning *adj* good-natured, obliging, sincere, well-intentioned, well-meant. ▷ KIND. *Opp* UNKIND.

well-off *adj* affluent, comfortable, moneyed, prosperous, rich, *inf* well-heeled, well-to-do. *Opp* POOR.

well-spoken *adj* articulate, educated, polite, *inf* posh, refined, *inf* upper crust.

wet *adj* **1** awash, bedraggled, clammy, damp, dank, dewy, drenched, dripping, moist, muddy, saturated, sloppy, soaked, soaking, sodden, soggy, sopping, soused, spongy, submerged, waterlogged, watery, wringing. **2** *wet weather*. drizzly, humid, misty, pouring, rainy, showery, teeming. **3** *wet paint*. runny, sticky, tacky. • *n* dampness, dew, drizzle, humidity, liquid, moisture, rain. • *vb* dampen, douse, drench, irrigate, moisten, saturate, soak, spray, sprinkle, steep, water. *Opp* DRY.

wheel *n* circle, disc, hoop, ring. □ *bogie, castor, cog-wheel, spinning-wheel, steering-wheel*. • *vb* change direction, circle, gyrate, move in circles, pivot, spin, swerve, swing round, swivel, turn, veer, whirl.

wheeze *vb* breathe noisily, cough, gasp, pant, puff.

whereabouts *n* location, neighbourhood, place, position, site, situation, vicinity.

whiff *n* breath, hint, puff, smell.

whim *n* caprice, desire, fancy, impulse, quirk, urge.

whine *vb* complain, cry, *inf* grizzle, groan, moan, snivel, wail, weep, whimper, *inf* whinge.

whip *n* birch, cane, cat, cat-o'-nine-tails, crop, horsewhip, lash, riding-crop, scourge, switch. • *vb* **1** beat, birch, cane, flagellate, flog, horsewhip, lash, scourge, *inf* tan, thrash. ▷ HIT. **2** beat, stir vigorously, whisk.

whirl *vb* circle, gyrate, pirouette, reel, revolve, rotate, spin, swivel, turn, twirl, twist, wheel.

whirlpool *n* eddy, maelstrom, swirl, vortex, whirl.

whirlwind *n* cyclone, hurricane, tornado, typhoon, vortex, waterspout.

whisk *n* beater, mixer. • *vb* beat, mix, stir, whip.

whiskers *plur n* bristles, hairs, moustache.

whisper *n* **1** murmur, undertone. **2** *a whisper of scandal*. gossip, hearsay, rumour. • *vb* breathe, hiss, murmur, mutter. ▷ TALK.

whistle *n* hooter, pipe, pipes, siren. • *vb* blow, pipe.

white *adj* chalky, clean, cream, ivory, milky, off-white, silver, snow-white, snowy, spotless, whitish. ▷ PALE.

whiten *vb* blanch, bleach, etiolate, fade, lighten, pale.

whole *adj* coherent, complete, entire, full, healthy, in one piece, intact, integral, integrated, perfect, sound, total, unabbreviated, unabridged, unbroken, uncut, undamaged, undivided, unedited, unexpurgated, unharmed, unhurt, uninjured, unscathed. *Opp* FRAGMENTARY, INCOMPLETE.

wholesale *adj* comprehensive, extensive, general, global, indiscriminate, mass, total, universal, widespread. *Opp* LIMITED.

wholesome *adj* beneficial, good, healthful, health-giving, healthy, hygienic, nourishing, nutritious, salubrious, sanitary. *Opp* UNHEALTHY.

wicked *adj* abominable, *inf* awful, bad, base, beastly, corrupt, criminal, depraved, diabolical, dissolute, egregious, evil, foul, guilty, heinous, ill-tempered, immoral, impious, incorrigible, indefensible, iniquitous, insupportable, intolerable, irresponsible, lawless, lost (*soul*), machiavellian, malevolent, malicious, mischievous, murderous, naughty, nefarious, offensive, perverted, rascally, scandalous, shameful, sinful, sinister, spiteful, *inf* terrible, ungodly, unprincipled, unregenerate, unrighteous, unscrupulous, vicious, vile, villainous, violent, wrong. ▷ IRRELIGIOUS, OBSCENE. *Opp* MORAL. **wicked person** ▷ VILLAIN.

wickedness *n* baseness, depravity, enormity, guilt, heinousness, immorality, infamy, iniquity, irresponsibility, *old use* knavery, malice, misconduct, naughtiness, sin, sinfulness, spite, turpitude, ungodliness, unrighteousness, vileness, villainy, wrong, wrongdoing. ▷ EVIL.

wide *adj* **1** ample, broad, expansive, extensive, large, panoramic, roomy, spacious, vast, yawning. **2** *wide sympathies.* all-embracing, broad-minded, catholic, comprehensive, eclectic, encyclopedic, inclusive, wide-ranging. **3** *arms open wide.* extended, open, outspread, outstretched. **4** *a wide shot.* off-course, off-target. *Opp* NARROW.

widen *vb* augment, broaden, dilate, distend, enlarge, expand, extend, flare, increase, make wider, open out, spread, stretch.

widespread *adj* common, endemic, extensive, far-reaching, general, global, pervasive, prevalent, rife, universal, wholesale. *Opp* RARE.

width *n* beam (*of ship*), breadth, broadness, calibre (*of gun*), compass, diameter, distance across, extent, girth, range, scope, span, thickness.

wield *vb* **1** brandish, flourish, handle, hold, manage, ply, wave. **2** *wield power.* employ, exercise, exert, have, possess, use.

wild *adj* **1** *wild animals.* free, undomesticated, untamed. **2** *a wild moor.* deserted, desolate, *inf* God-forsaken, natural, overgrown, remote, rough, rugged, uncultivated, unenclosed, unfarmed, uninhabited, waste. **3** *wild behaviour.* aggressive, barbaric, barbarous, berserk, boisterous, disorderly, ferocious, fierce, frantic, hysterical, lawless, mad, noisy, obstreperous, on the rampage, out of control, rabid, rash, reckless, riotous, rowdy, savage, uncivilized, uncontrollable, uncontrolled, undisciplined, ungovernable, unmanageable, unrestrained, unruly, uproarious, violent. **4** *wild weather.* blustery, stormy, tempestuous, turbulent, violent, windy. **5** *wild enthusiasm.* eager, excited, extravagant, uninhibited, unrestrained. **6** *wild notions.* crazy, fantastic, impetuous, irrational, silly, unreasonable. **7** *a wild guess.* inaccurate, random, unthinking. *Opp* CALM, CULTIVATED, TAME.

wilderness *n* desert, jungle, waste, wasteland, wilds.

wile *n* artifice, gambit, *inf* game, machination, manoeuvre, plot, ploy, ruse, stratagem, subterfuge, trick.

wilful *adj* **1** calculated, conscious, deliberate, intended, intentional, premeditated, purposeful, voluntary. *Opp* ACCIDENTAL. **2** *a wilful character.* *inf* bloody-minded, determined, dogged, headstrong, immovable, intransigent, obdurate, obstinate, perverse, *inf* pig-headed, refractory, self-willed, stubborn, uncompromising, unyielding, wayward. *Opp* AMENABLE.

will *n* aim, commitment, desire, determination, disposition, inclination, intent, intention, longing, purpose, resolution, resolve, volition, will-power, wish. • *vb* **1** command, encourage, force, influence, inspire, persuade, require, wish. **2** *will a fortune.* bequeath, hand down, leave, pass on, settle on.

willing *adj* acquiescent, agreeable, amenable, assenting, complaisant, compliant, consenting, content, cooperative, disposed, docile, *inf* game, happy, helpful, inclined, pleased, prepared, obliging, ready, well-disposed. ▷ EAGER. *Opp* UNWILLING.

wilt *vb* become limp, droop, fade, fail, flag, flop, languish, sag, shrivel, weaken, wither. *Opp* THRIVE.

wily *adj* artful, astute, canny, clever, crafty, cunning, deceptive, designing, devious, disingenuous, furtive, guileful, ingenious, knowing, scheming, shifty, shrewd, skilful, sly, tricky, underhand. ▷ DISHONEST. *Opp* STRAIGHTFORWARD.

win *vb* **1** be the winner, be victorious, carry the day, come first, conquer, overcome, prevail, succeed, triumph. **2** *win a prize.* achieve, acquire, *inf* carry off, collect, *inf* come away with, deserve, earn, gain, get, obtain, *inf* pick up, receive, secure, *inf* walk away with. *Opp* LOSE.

wind *n* **1** air-current, blast, breath, breeze, current of air, cyclone, draught, gale, gust, hurricane, monsoon, puff, squall, storm, tempest, tornado, whirlwind, *poet* zephyr. **2** *wind in the stomach.* flatulence, gas, heartburn. • *vb* bend, coil, curl, curve, furl, loop, meander, ramble, reel, roll, slew, snake, spiral, turn, twine, twist, *inf* twist and turn, veer, wreathe, zigzag. **winding** ▷ TORTUOUS. **wind up** ▷ FINISH.

window *n* □ *casement, dormer, double-glazed window, embrasure, fanlight, French window, light, oriel, pane, sash window, skylight, shop window, stainedglass window, windscreen.*

windswept *adj* bare, bleak, desolate, exposed, unprotected. ▷ WINDY.

windy *adj* blowy, blustery, boisterous, breezy, draughty, fresh, gusty, squally, stormy, tempestuous. ▷ WINDSWEPT. *Opp* CALM.

wink *vb* **1** bat (*eyelid*), blink, flutter. **2** *lights winked.* flash, flicker, sparkle, twinkle.

winner *n* *inf* champ, champion, conquering hero, conqueror, first, medallist, prizewinner, title-holder, victor. *Opp* LOSER.

winning *adj* **1** champion, conquering, first, leading, prevailing, successful, top, top-scoring, triumphant, unbeaten, undefeated, victorious. *Opp* UNSUCCESSFUL. **2** *a winning smile.* ▷ ATTRACTIVE.

wintry *adj* arctic, icy, snowy. ▷ COLD. *Opp* SUMMERY.

wipe *vb* brush, clean, cleanse, dry, dust, mop, polish, rub, scour, sponge, swab, wash. **wipe out** ▷ DESTROY.

wire *n* **1** cable, coaxial cable, flex, lead, wiring. **2** cablegram, telegram. ▷ COMMUNICATION.

wiry *adj* lean, muscular, sinewy, strong, thin, tough.

wisdom *n* astuteness, common sense, discernment, discrimination, good sense, insight, judgement, judiciousness, penetration, perceptiveness, perspicacity, prudence, rationality, reason, sagacity, sapience, sense, understanding. ▷ INTELLIGENCE.

wise *adj* **1** astute, discerning, enlightened, erudite, fair, just, knowledgeable, penetrating, perceptive, perspicacious, philosophical, sagacious, sage, sensible, shrewd, sound, thoughtful, understanding, well-informed. ▷ INTELLIGENT. **2** *a wise decision.* advisable, appropriate, considered, diplomatic, expedient, informed, judicious, politic, proper, prudent, rational, reasonable, right. *Opp* UNWISE. **wise person** philosopher, pundit, sage.

wish *n* aim, ambition, appetite, aspiration, craving, desire, fancy, hankering, hope, inclination, *inf* itch, keenness, longing, objective, request, urge, want, yearning, *inf* yen. • *vb* ask, choose, crave, desire, hope, want, yearn. **wish for** ▷ WANT.

wisp *n* shred, strand, streak.

wispy *adj* flimsy, fragile, gossamer, insubstantial, light, streaky, thin. *Opp* SUBSTANTIAL.

wistful *adj* disconsolate, forlorn, melancholy, mournful, nostalgic, regretful, yearning. ▷ SAD.

wit *n* **1** banter, cleverness, comedy, facetiousness, humour, ingenuity, jokes, puns, quickness, quips, repartee, witticisms, wordplay. ▷ INTELLIGENCE. **2** comedian, comic, humorist, jester, joker, wag.

witch *n* enchantress, gorgon, hag, sibyl, sorceress, *plur* weird sisters.

witchcraft *n* black magic, charms, enchantment, *inf* hocus-pocus, incantation, magic, *inf* mumbo-jumbo, necromancy, the occult, occultism, sorcery, spells, voodoo, witchery, wizardry.

withdraw *vb* **1** abjure, call back, cancel, *inf* go back on, recall, rescind, retract, take away, take back. **2** *withdraw from the fight.* back away, back down, back out, *inf* chicken out, *inf* cry off, draw back, drop out, fall back, move back, pull back, pull out, *inf* quit, recoil, retire, retreat, run away, *inf* scratch, secede, shrink back. ▷ LEAVE. *Opp* ADVANCE, ENTER. **3** *withdraw teeth.* extract, pull out, remove, take out.

withdrawn *adj* bashful, diffident, distant, introverted, private, quiet, reclusive, remote, reserved, retiring, shy, silent, 5solitary, taciturn, timid, uncommunicative. *Opp* SOCIABLE.

wither *vb* become dry, become limp, dehydrate, desiccate, droop, dry out, dry up, fail, flag, flop, sag, shrink, shrivel, waste away, wilt. *Opp* THRIVE.

withhold *vb* check, conceal, control, hide, hold back, keep back, keep secret, repress, reserve, retain, suppress. *Opp* GIVE.

withstand *vb* bear, brave, confront, cope with, defy, endure, fight, grapple with, hold out against, last out against, oppose, *inf* put up with, resist, stand up to, *inf* stick, survive, take, tolerate, weather (*storm*). *Opp* SURRENDER.

witness *n* bystander, eyewitness, looker-on, observer, onlooker, spectator, viewer, watcher. • *vb* attend, behold, be present at, look on, note, notice, observe, see, view, watch. **bear witness** ▷ TESTIFY.

witty *adj* amusing, clever, comic, droll, facetious, funny, humorous, ingenious, intelligent, jocular, quick-witted, sarcastic, sharp-witted, waggish.

wizard *n* enchanter, magician, magus, sorcerer, *old use* warlock, witch-doctor.

wobble *vb* be unsteady, heave, move unsteadily, oscillate, quake, quiver, rock, shake, sway, teeter, totter, tremble, vacillate, vibrate, waver.

wobbly *adj* insecure, loose, rickety, rocky, shaky, teetering, tottering, unbalanced, unsafe, unstable, unsteady. *Opp* STEADY.

woe *n* affliction, anguish, dejection, despair, distress, grief, heartache, melancholy, misery, misfortune, sadness, suffering, trouble, unhappiness, wretchedness. ▷ SORROW. *Opp* HAPPINESS.

woebegone *adj* crestfallen, dejected, downhearted, forlorn, gloomy, melancholy, miserable, *inf* sorry for yourself, woeful, wretched. ▷ SAD. *Opp* CHEERFUL.

woman *n sl* bird, bride, *old use* dame, *old use* damsel, daughter, dowager, female, girl, girlfriend, *derog* hag, *derog* harridan, housewife, hoyden, *derog* hussy, lady, lass, madam, Madame, maid, *old use* maiden, matriarch, matron, mistress, mother, *derog* termagant, *derog* virago, virgin, widow, wife.

wonder *n* **1** admiration, amazement, astonishment, awe, bewilderment, curiosity, fascination, respect, reverence, stupefaction, surprise, wonderment. **2** *a wonder of science.* marvel, miracle, phenomenon, prodigy, *inf* sensation. • *vb* ask yourself, be curious, be inquisitive, conjecture, marvel, ponder, question yourself, speculate. ▷ THINK. **wonder at** ▷ ADMIRE.

wonderful *adj* amazing, astonishing, astounding, extraordinary, impressive, incredible, marvellous, miraculous, phenomenal, remarkable, surprising, unexpected, *old use* wondrous. *Opp* ORDINARY.

woo *vb* **1** *sl* chat up, court, make love to. **2** *woo custom.* attract, bring in, coax, cultivate, persuade, pursue, seek, try to get.

wood *n* **1** afforestation, coppice, copse, forest, grove, jungle, orchard, plantation, spinney, thicket, trees, woodland, woods. **2** blockboard, chipboard, deal, planks, plywood, timber. □ *balsa, beech, cedar, chestnut, ebony, elm, mahogany, oak, pine, rosewood, sandalwood, sapele, teak, walnut.*

wooded *adj* afforested, *poet* bosky, forested, silvan, timbered, tree-covered, woody.

wooden *adj* **1** ligneous, timber, wood. **2** *wooden acting.* dead, emotionless, expressionless, hard, inflexible, lifeless, rigid, stiff, stilted, unbending, unemotional, unnatural. *Opp* LIVELY.

woodwind *n* □ *bassoon, clarinet, cor anglais, flute, oboe, piccolo, recorder.*

woodwork *n* carpentry, joinery.

woody *adj* fibrous, hard, ligneous, tough. ▷ WOODEN.

woolly *adj* **1** wool, woollen. **2** *woolly toy.* cuddly, downy, fleecy, furry, fuzzy, hairy, shaggy, soft. **3** *woolly ideas.* ambiguous, blurry, confused, hazy, ill-defined, indefinite, indistinct, uncertain, unclear, unfocused, vague.

word *n* **1** expression, name, term. **2** ▷ NEWS. **3** ▷ PROMISE. • *vb* articulate, express, phrase. **word for word** ▷ VERBATIM.

wording *n* choice of words, diction, expression, language, phraseology, phrasing, style, terminology.

wordy *adj* chatty, diffuse, digressive, discursive, garrulous, long-winded, loquacious, pleonastic, prolix, rambling, repetitious, talkative, unstoppable, verbose, voluble, *inf* windy. *Opp* CONCISE.

work *n* **1** *inf* donkey-work, drudgery, effort, exertion, *inf* fag, *inf* graft, *inf* grind, industry, labour, *inf* plod, slavery, *inf* slog, *inf* spadework, strain, struggle, *inf* sweat, toil, *old use* travail. **2** *work to be done.* assignment, chore, commission, duty, errand, homework, housework, job, mission, project, responsibility, task, undertaking. **3** *regular work.* business, calling, career, employment, job, livelihood, living, métier, occupation, post, profession, situation, trade. • *vb* **1** *inf* beaver away, be busy, drudge, exert yourself, *inf* fag, *inf* grind away, *inf* keep your nose to the grindstone, labour, make efforts, navvy, *inf* peg away, *inf* plug away, *inf* potter about, *inf* slave, *inf* slog away, strain, strive, struggle, sweat, toil, travail. **2** act, be effective, function, go, operate, perform, run, succeed, thrive. **3** *work slaves hard.* drive, exploit, utilize. **working** ▷ EMPLOYED, OPERATIONAL. **work out** ▷ CALCULATE. **work up** ▷ DEVELOP, EXCITE.

worker *n* artisan, breadwinner, coolie, craftsman, employee, *old use* hand, labourer, member of staff, navvy, operative, operator, peasant, practitioner, servant, slave, tradesman, wage-earner, working man, working woman, workman.

workforce *n* employees, staff, workers.

workmanship *n* art, artistry, competence, craft, craftsmanship, expertise, handicraft, handiwork, skill, technique.

workshop *n* factory, mill, smithy, studio, workroom.

world *n* **1** earth, globe, planet. **2** area, circle, domain, field, milieu, sphere.

worldly *adj* avaricious, covetous, earthly, fleshly, greedy, human, material, materialistic, mundane, physical, profane, secular, selfish, temporal. *Opp* SPIRITUAL.

worm *vb* crawl, creep, slither, squirm, wriggle. writhe.

worn *adj* **1** frayed, moth-eaten, old, ragged, *inf* scruffy, shabby, tattered, *inf* tatty, thin, threadbare, worn-out. **2** ▷ WEARY.

worried *adj* afraid, agitated, agonized, alarmed, anxious, apprehensive, bothered, concerned, distraught, distressed, disturbed, edgy, fearful, *inf* fraught, fretful, guilt-ridden, insecure, nervous, nervy, neurotic, obsessed (by), on edge, overwrought, perplexed, perturbed, solicitous, tense, troubled, uncertain, uneasy, unhappy, upset, vexed.

worry *n* **1** agitation, anxiety, apprehension, disquiet, distress, fear, neurosis, perplexity, perturbation, tension, unease, uneasiness. **2** affliction, annoyance, bother, burden, care, concern, misgiving, problem, *plur* trials and tribulations, trouble, vexation. • *vb* **1** agitate, annoy, *inf* badger, bother, depress, disquiet, distress, disturb, exercise, *inf* hassle, irritate, molest, nag, perplex, perturb, pester, plague, tease, threaten, torment, trouble, upset, vex. *Opp* REASSURE. **2** *worry about money.* agonize, be anxious, be worried, brood, exercise yourself, feel uneasy, fret.

worsen *vb* **1** aggravate, exacerbate, heighten, increase, intensify, make worse. **2** *his health worsened.* decline, degenerate, deteriorate, fail, get worse, *inf* go downhill, weaken. *Opp* IMPROVE.

worship *n* adoration, adulation, deification, devotion, glorification, homage, idolatry, love, praise, reverence, veneration. • *vb* admire, adore, adulate, be devoted to, deify, dote on, exalt, extol, glorify, hero-worship, idolize, kneel before, lionize, laud, look up to, love, magnify, pay homage to, praise, pray to, *inf* put on a pedestal, revere, reverence, venerate.

worth *n* benefit, cost, good, importance, merit, quality, significance, use, usefulness, utility, value. **be worth** be priced at, cost, have a value of.

worthless *adj old use* bootless, dispensable, disposable, frivolous, futile, *inf* good-for-nothing, hollow, insignificant, meaningless, meretricious, paltry, pointless, poor, *inf* rubbishy, *inf* trashy, trifling, trivial, trumpery, unimportant, unproductive, unprofitable, unusable, useless, vain, valueless. *Opp* WORTHWHILE.

worthwhile *adj* advantageous, beneficial, considerable, enriching, fruitful, fulfilling, gainful, gratifying, helpful, important, invaluable, meaningful, noticeable, productive, profitable, remunerative, rewarding, satisfying, significant, sizeable, substantial, useful, valuable. ▷ WORTHY. *Opp* WORTHLESS.

worthy *adj* admirable, commendable, creditable, decent, deserving, estimable, good, honest, honourable, laudable, meritorious, praiseworthy, reputable, respectable, worth supporting, worthwhile. *Opp* UNWORTHY.

wound *n* damage, hurt, injury, scar, trauma. □ *amputation, bite, bruise, burn, contusion, cut, fracture, gash, graze, laceration, lesion, mutilation, puncture, scab, scald, scar, scratch, sore, sprain, stab, sting, strain, weal, welt.* • *vb* cause pain to, damage, harm, hurt, injure, traumatize. □ *amputate, bite, blow up, bruise, burn, claw, cut, fracture, gash, gore, graze, hit, impale, knife, lacerate, maim, make sore, mangle, maul, mutilate, scratch, shoot, sprain, strain, stab, sting, torture.* *Opp* HEAL, MEND.

wrap *n* cape, cloak, mantle, poncho, shawl, stole. • *vb* bind, bundle up, cloak, cocoon, conceal, cover, do up, encase, enclose, enfold, enshroud, envelop, hide, insulate, lag, muffle, pack, package, shroud, surround, swaddle, swathe, wind.

wreathe *vb* adorn, decorate, encircle, festoon, intertwine, interweave, twist, weave.

wreck *n* **1** hulk, shipwreck. ▷ WRECKAGE. **2** *the wreck of all my hopes.* demolition, destruction, devastation, loss, obliteration, overthrow, ruin, termination, undoing. • *vb* **1** annihilate, break up, crumple, crush, dash to pieces, demolish, destroy, devastate, ruin, shatter, smash, spoil, *inf* write off. **2** *wreck a ship.* capsize, founder, ground, scuttle, sink, shipwreck.

wreckage *n* bits, debris, *inf* flotsam and jetsam, fragments, pieces, remains, rubble, ruins.

wrench *vb* force, jerk, lever, prize, pull, rip, strain, tear, tug, twist, wrest, wring, *inf* yank.

wrestle *vb* grapple, strive, struggle, tussle. ▷ FIGHT.

wretch *n* **1** beggar, down-and-out, miserable person, pauper, unfortunate. **2** ▷ VILLAIN.

wretched *adj* **1** dejected, depressed, dispirited, downhearted, hapless, melancholy, miserable, pathetic, pitiable, pitiful, unfortunate. ▷ SAD. **2** ▷ UNSATISFACTORY.

wriggle *vb* crawl, snake, squirm, twist, waggle, wiggle, wobble, worm, writhe, zigzag.

wring *vb* **1** clasp, compress, crush, grip, press, shake, squeeze, twist, wrench, wrest. **2** coerce, exact, extort, extract, force.

wrinkle *n* corrugation, crease, crinkle, *inf* crow's feet, dimple, fold, furrow, gather, line, pleat, pucker, ridge, ripple. • *vb* corrugate, crease, crinkle, crumple, fold, furrow, gather, make wrinkles, pleat, pucker up, ridge, ripple, ruck up, rumple, screw up.

wrinkled *adj* corrugated, creased, crinkly, crumpled, furrowed, lined, pleated, ridged, ripply, rumpled, screwed up, shrivelled, undulating, wavy, wizened, wrinkly. *Opp* SMOOTH.

write *vb* be a writer, compile, compose, copy, correspond, doodle, draft, draw up, engrave, indite, inscribe, jot, note, pen, print, put in writing, record, scrawl, scribble, set down, take down, transcribe, type. **write off** ▷ CANCEL, DESTROY.

writer *n* **1** amanuensis, clerk, copyist, *derog* pen-pusher, scribe, secretary, typist. **2** author, bard, composer, *derog* hack, littérateur, wordsmith. □ *biographer, columnist, contributor, copy-writer, correspondent, diarist, dramatist, essayist, freelancer, ghost-writer, journalist, leader-writer, librettist, novelist, playwright, poet, reporter, scriptwriter.*

writhe *vb* coil, contort, jerk, squirm, struggle, thrash about, thresh about, twist, wriggle.

writing *n* **1** calligraphy, characters, copperplate, cuneiform, handwriting, hieroglyphics, inscription, italics, letters, longhand, notation, penmanship, printing, runes, scrawl, screed, scribble, script, shorthand. **2** authorship, composition, journalism. **3** hard copy, literature, manuscript, opus, printout, text, typescript, work. □ *article, autobiography, belles-lettres, biography, comedy, copywriting, correspondence, crime story, criticism, detective story, diary, dissertation, documentary, drama, editorial, epic, epistle, essay, fable, fairy-tale, fantasy, fiction, folk-tale, history, legal document, legend, letter, libretto, lyric, monograph, mystery, myth, newspaper column, non-fiction, novel, parable, parody, philosophy, play, poem, propaganda, prose, reportage, romance, saga, satire, science* *fiction, scientific writing, scriptwriting, SF, sketch, story, tale, thesis, thriller, tragedy, tragi-comedy, travel writing, treatise, trilogy, TV script, verse,* inf *whodunit, yarn.*

written *adj* documentary, *inf* in black and white, inscribed, in writing, set down, transcribed, typewritten. *Opp* SPOKEN.

wrong *adj* **1** base, blameworthy, corrupt, criminal, crooked, deceitful, dishonest, dishonourable, evil, felonious, illegal, illegitimate, illicit, immoral, iniquitous, irresponsible, mendacious, misleading, naughty, reprehensible, sinful, specious, unethical, unjustifiable, unlawful, unprincipled, unscrupulous, vicious, villainous, wicked, wrong-headed. **2** *wrong answers.* erroneous, fallacious, false, imprecise, improper, inaccurate, incorrect, inexact, misinformed, mistaken, unfounded, untrue. **3** *a wrong decision.* curious, ill-advised, ill-considered, ill-judged, impolitic, imprudent, injudicious, misguided, misjudged, unacceptable, unfair, unjust, unsound, unwise, wrongful. **4** *go the wrong way.* abnormal, contrary, inappropriate, incongruous, inconvenient, misleading, opposite, unconventional, undesirable, unhelpful, unsuitable, worst. **5** *Something's wrong.* amiss, broken down, defective, faulty, out of order, unusable. ▷ BAD. *Opp* RIGHT. • *vb* abuse, be unfair to, cheat, damage, do an injustice to, harm, hurt, injure, malign, maltreat, misrepresent, mistreat, traduce, treat unfairly. **do wrong** ▷ MISBEHAVE.

wrongdoer *n* convict, criminal, crook, culprit, delinquent, evildoer, law-breaker, malefactor, mischief-maker, miscreant, offender, sinner, transgressor.

wrongdoing *n* crime, delinquency, disobedience, evil, immorality, indiscipline, iniquity, malpractice, misbehaviour, mischief, naughtiness, offence, sin, sinfulness, wickedness.

wry *adj* **1** askew, aslant, awry, bent, contorted, crooked, deformed, distorted, lopsided, twisted, uneven. **2** *a wry sense of humour.* droll, dry, ironic, mocking, sardonic.

Y

yard *n* court, courtyard, enclosure, garden, *inf* quad, quadrangle.

yarn *n* **1** fibre, strand, thread. **2** account, anecdote, fiction, narrative, story, tale.

yawning *adj* gaping, open, wide.

yearly *adj* annual, perennial, regular.

yearn *vb* ache, feel desire, hanker, have a craving, hunger, itch, long, pine. ▷ WANT.

yellow *adj* □ *chrome yellow, cream, gold, golden, orange, tawny.*

yield *n* **1** crop, harvest, output, produce, product. **2** earnings, gain, income, interest, proceeds, profit, return, revenue. • *vb* **1** acquiesce, agree, assent, bow, capitulate, *inf* cave in, cede, comply, concede, defer, give in, give up, give way, *inf* knuckle under, submit, succumb, surrender, *inf* throw in the towel, *inf* throw up the sponge. **2** *yield interest.* bear, earn, generate, pay out, produce, provide, return, supply. **yielding** ▷ FLEXIBLE, SPONGY, SUBMISSIVE.

young *adj* **1** baby, early, growing, immature, new-born, undeveloped, unfledged, youngish. **2** *young people*. adolescent, juvenile, pubescent, teenage, underage, youthful. **3** *young for your age*. babyish, boyish, callow, childish, girlish, *inf* green, immature, inexperienced, infantile, juvenile, naïve, puerile. • *n* brood, family, issue, litter, offspring, progeny. **young creatures** bullock, calf, chick, colt, cub, cygnet, duckling, fawn, fledgling, foal, gosling, heifer, kid, kitten, lamb, leveret, nestling, pullet, puppy, yearling. **young people** adolescent, baby, boy, *derog* brat, child, girl, infant, juvenile, *inf* kid, lad, lass, *inf* nipper, teenager, toddler, *derog* urchin, youngster, youth.

youth *n* **1** adolescence, babyhood, boyhood, childhood, girlhood, growing up, immaturity, infancy, minority, pubescence, *inf* salad days, *inf* teens. **2** adolescent, boy, juvenile, *inf* kid, *inf* lad, minor, stripling, teenager, youngster.

youthful *adj* fresh, lively, sprightly, vigorous, well-preserved, young-looking. ▷ YOUNG.

Z

zany *adj* absurd, clownish, crazy, eccentric, idiotic, *in* loony, *inf* mad, madcap, playful, ridiculous, silly, *sl* wacky.

zeal *n derog* bigotry, earnestness, enthusiasm, fanaticism, fervour, partisanship.

zealot *n* bigot, extremist, fanatic, partisan, radical.

zealous *adj* conscientious, diligent, eager, earnest, enthusiastic, fanatical, fervent, keen, militant, obsessive, partisan, passionate. *Opp* APATHETIC.

zenith *n* acme, apex, apogee, climax, height, highest point, meridian, peak, pinnacle, summit, top. *Opp* NADIR.

zero *n cricket* duck, *tennis* love, naught, nil, nothing, nought, *sl* zilch. **zero in on** ▷ AIM.

zest *n* appetite, eagerness, energy, enjoyment, enthusiasm, exuberance, hunger, interest, liveliness, pleasure, thirst, zeal.

zigzag *adj* bendy, crooked, *inf* in and out, indirect, meandering, serpentine, twisting, winding. • *vb* bend, curve, meander, snake, tack, twist, wind.

zone *n* area, belt, district, domain, locality, neighbourhood, province, quarter, region, section, sector, sphere, territory, tract, vicinity.

zoo *n* menagerie, safari park, zoological gardens.

zoom *vb* career, dart, dash, hurry, hurtle, race, rush, shoot, speed, *inf* whiz, *inf* zip. ▷ MOVE.

The Little Oxford Dictionary of Quotations

The Little Oxford Dictionary of Quotations

Edited by Susan Ratcliffe

Oxford New York
OXFORD UNIVERSITY PRESS
1994

Oxford University Press, Walton Street, Oxford OX2 6DP

Oxford New York
Athens Auckland Bangkok Bombay
Calcutta Cape Town Dar es Salaam Delhi
Florence Hong Kong Istanbul Karachi
Kuala Lumpur Madras Madrid Melbourne
Mexico City Nairobi Paris Singapore
Taipei Tokyo Toronto

and associated companies in
Berlin Ibadan

Oxford is a trade mark of Oxford University Press

First published 1994

British Library Cataloguing in Publication Data
Data available

Library of Congress Cataloging in Publication Data
The Little Oxford dictionary of quotations / edited by Susan Ratcliffe.
p. cm.
1. Quotations. 2. Quotations, English. I. Ratcliffe, Susan.
082—dc20 PN6080.L58 1994 94–11193

ISBN 0–19–866207–6

10 9 8 7 6 5 4 3 2 1

Typeset by Barbers Ltd.
Printed in Great Britain
on acid-free paper by
Clays Ltd.
Bungay, Suffolk

Little Oxford Dictionary of Quotations Project Team

Managing Editor	Elizabeth M. Knowles
Editor	Susan Ratcliffe
Library Research	Ralph Bates Marie G. Diaz
Data Capture	Sandra Vaughan
Quotation Retrieval	Helen McCurdy Katie Weale

Contents

Foreword

The Little Oxford Dictionary of Quotations is a collection which casts a fresh light on even the most familiar sayings. It is organized by themes, such as **Action**, **Liberty**, and **Memory**, and within each theme the quotations are arranged in date order, so that the interplay of ideas down the centuries becomes apparent. It is intended for the reader who is searching for quotations on a specific subject, the reader who remembers the sense of a quotation but not the precise words, and, of course, the browser.

The themes have been chosen to reflect as wide a range of subjects as possible, concentrating on the general rather than the specific. A few themes have a slightly different character: thus **People** and **Places** cover quotations about many different individual people and places, while **Political Comment** and **Wars** include quotations relevant to specific events. The length of the sections reflects to some extent the preoccupations of people throughout history, ranging from short ones such as **Advice** to the many and varied comments on **Life** and **Love**. Where subjects overlap, the reader is directed to related themes at the head of the section; for example, at **Death**: See also **Epitaphs**, **Last Words**, **Murder**. An author index is provided to help readers wishing to trace a particular quotation or seeking quotations from specific individuals.

Within each theme, the aim is to take in a variety of viewpoints, including both the most familiar quotations and some less well-known or perhaps new material. So within **News and Journalism**, along with C. P. Scott's classic 'Comment is free, but facts are sacred', we have Tom Stoppard's gloss 'Comment is free but facts are on expenses' and more recently Lord McGregor on 'journalists dabbling their fingers in the stuff of other people's souls'. This book contains some one hundred quotations which have not previously appeared in any dictionary of quotations. These new quotations appear, for example, under the themes **Environment** ('... all that remains / For us will be concrete and tyres'), **Men and Women** ('Whereas nature turns girls into women, society has to make boys into men'), and **Science** ('The aim of science is not to open the door to infinite wisdom, but to set a limit to infinite error').

The chronological ordering within each theme enables the

quotations to 'talk' to one another, shedding new light on each. Thus we have Samuel Johnson telling us 'Change is not made without inconvenience, even from worse to better', followed by Voltaire: 'If we do not find anything pleasant, at least we shall find something new'. Much of the cross-referencing required by an alphabetical arrangement of authors becomes redundant: Ralegh's line written on a window-pane 'Fain would I climb, yet fear I to fall' is now immediately followed by Elizabeth I's reply 'If thy heart fails thee, climb not at all.'

The quotations have been classified by their subject rather than by keywords in the text. For example, Tom Lehrer's 'It is sobering to consider that when Mozart was my age he had already been dead for a year' is essentially about **Achievement** rather than **Music** or **Death**, and has been placed accordingly. As far as possible each quotation has been included only once, but a few, such as Pope's 'To err is human, to forgive divine' plainly had a place in two sections.

A short source reference is given for each quotation, usually including its date. Where the date is uncertain or unknown, the author's date of death has been used to determine the order within a theme. The quotations themselves have been kept as short as possible: contextual information has occasionally been added to the source note, and related but less well-known and well-expressed versions have generally been excluded. Owing to constraints of space, foreign language originals have been given only where they are well-known or where translations differ. Such information, including full finding references, can be found in *The Oxford Dictionary of Quotations*.

We are always grateful to those readers who write to us with their comments, suggestions, and discoveries, and we hope this tradition will continue. *The Little Oxford Dictionary of Quotations* draws largely on the work done for the fourth edition of *The Oxford Dictionary of Quotations*, and therefore owes a substantial debt to all those involved in the preparation of that volume. None the less, this book has its own identity, and the editor's chief pleasure as it took shape has been in listening to different voices speaking to each other across the ages: '"What is the use of a book," thought Alice, "without pictures or conversations?"'

Susan Ratcliffe

Oxford, March 1994

List of Themes

Absence

1 The Lord watch between me and thee, when we are absent one from another.
Bible: Genesis

2 Absence diminishes commonplace passions and increases great ones, as the wind extinguishes candles and kindles fire.
Duc de la Rochefoucauld 1613–80: *Maximes* (1678)

3 The absent are always in the wrong.
Philippe Néricault Destouches 1680–1754: *L'Obstacle imprévu* (1717)

4 With leaden foot time creeps along
While Delia is away.
Richard Jago 1715–81: 'Absence'

5 Presents, I often say, endear Absents.
Charles Lamb 1775–1834: *Essays of Elia* (1823) 'A Dissertation upon Roast Pig'

6 The heart may think it knows better: the senses know that absence blots people out. We have really no absent friends.
Elizabeth Bowen 1899–1973: *Death of the Heart* (1938)

7 When I came back to Dublin, I was courtmartialled in my absence and sentenced to death in my absence, so I said they could shoot me in my absence.
Brendan Behan 1923–64: *Hostage* (1958)

Achievement and Endeavour

See also **Ambition**

1 *Non omnia possumus omnes.*
We can't all do everything.
Virgil 70–19 BC: *Eclogues*

2 *Parturient montes, nascetur ridiculus mus.*
Mountains will go into labour, and a silly little mouse will be born.
Horace 65–8 BC: *Ars Poetica*

3 *Considerate la vostra semenza:*
Fatti non foste a viver come bruti,

Ma per seguir virtute e conoscenza.
Consider your origins: you were not made to live as brutes, but to follow virtue and knowledge.

Dante Alighieri 1265–1321: *Divina Commedia* 'Inferno'

4 There must be a beginning of any great matter, but the continuing unto the end until it be thoroughly finished yields the true glory.

Francis Drake *c.*1540–96: dispatch to Sir Francis Walsingham, 17 May 1587

5 If to do were as easy as to know what were good to do, chapels had been churches, and poor men's cottages princes' palaces.

William Shakespeare 1564–1616: *The Merchant of Venice* (1596–8)

6 Things won are done; joy's soul lies in the doing.

William Shakespeare 1564–1616: *Troilus and Cressida* (1602)

7 Get place and wealth, if possible, with grace;
If not, by any means get wealth and place.

Alexander Pope 1688–1744: *Imitations of Horace* (1738)

8 I had done all that I could; and no man is well pleased to have his all neglected, be it ever so little.

Samuel Johnson 1709–84: letter to Lord Chesterfield, 7 February 1755

9 The danger chiefly lies in acting well;
No crime's so great as daring to excel.

Charles Churchill 1731–64: *An Epistle to William Hogarth* (1763)

10 The distance is nothing; it is only the first step that is difficult.

Mme Du Deffand 1697–1780: commenting on the legend that St Denis, carrying his head in his hands, walked two leagues; letter to Jean Le Rond d'Alembert, 7 July 1763

11 Madam, if a thing is possible, consider it done; the impossible? that will be done.

Charles Alexandre de Calonne 1734–1802: in J. Michelet *Histoire de la Révolution Française* (1847); better known as the US Armed Forces' slogan: 'The difficult we do immediately; the impossible takes a little longer'

12 That low man seeks a little thing to do,
Sees it and does it:
This high man, with a great thing to pursue,
Dies ere he knows it.

Robert Browning 1812–89: 'A Grammarian's Funeral' (1855)

13 Now, *here*, you see, it takes all the running *you* can do, to keep in the same place. If you want to get somewhere else, you must run at least twice as fast as that!

Lewis Carroll 1832–98: *Through the Looking-Glass* (1872)

14 If a man write a better book, preach a better sermon, or make a better mouse-trap than his neighbour, tho' he build his house in the woods, the world will make a beaten path to his door.

Ralph Waldo Emerson 1803–82: attributed to Emerson in S. Yule *Borrowings* (1889), but claimed also by Elbert Hubbard

15 Because it's there.

George Leigh Mallory 1886–1924: on being asked why he wanted to climb Mount Everest; in *New York Times* 18 March 1923

16 Well, we knocked the bastard off!

Edmund Hillary 1919– : on conquering Mount Everest, 1953; in *Nothing Venture, Nothing Win* (1975)

17 That's one small step for a man, one giant leap for mankind.

Neil Armstrong 1930– : landing on the moon, 21 July 1969 (interference in transmission obliterated 'a')

18 It is sobering to consider that when Mozart was my age he had already been dead for a year.

Tom Lehrer 1928– : in N. Shapiro (ed.) *An Encyclopedia of Quotations about Music* (1978)

Acting and the Theatre

See also **The Cinema, Shakespeare**

1 Tragedy is thus a representation of an action that is worth serious attention, complete in itself and of some amplitude…by means of pity and fear bringing about the purgation of such emotions.

Aristotle 384–322 BC: *Poetics*

2 Can this cockpit hold
The vasty fields of France? or may we cram
Within this wooden O the very casques
That did affright the air at Agincourt?
William Shakespeare 1564–1616: *Henry V* (1599)

3 Speak the speech, I pray you, as I pronounced it to you, trippingly on the tongue; but if you mouth it, as many of your players do, I had as lief the town-crier spoke my lines. Nor do not saw the air too much with your hand, thus; but use all gently.
William Shakespeare 1564–1616: *Hamlet* (1601)

4 Suit the action to the word, the word to the action.
William Shakespeare 1564–1616: *Hamlet* (1601)

5 Damn them! They will not let my play run, but they steal my thunder!
John Dennis 1657–1734: on hearing his new thunder effects used at a performance of *Macbeth*, following the withdrawal of one of his own plays after only a short run

6 To see him act is like reading Shakespeare by flashes of lightning.
Samuel Taylor Coleridge 1772–1834: on Edmund Kean; *Table Talk* (1835) 27 April 1823

7 He played the King as though under momentary apprehension that someone else was about to play the ace.
Eugene Field 1850–95: of Creston Clarke as King Lear; review attributed to Field, in *Denver Tribune* c.1880

8 How different, how very different from the home life of our own dear Queen!
Anonymous: a Victorian lady, overheard at a performance by Sarah Bernhardt in the role of Cleopatra, in I. S. Cobb *A Laugh a Day* (1924)

9 Ladies, just a little more virginity, if you don't mind.
Herbert Beerbohm Tree 1852–1917: to the extras playing ladies in waiting to the queen; in A. Woollcott *Shouts and Murmurs* (1923)

10 She ran the whole gamut of the emotions from A to B.
Dorothy Parker 1893–1967: of Katherine Hepburn on the first night of *The Lake* (1933); attributed

11 Actors are cattle.
Alfred Hitchcock 1899–1980: in *Saturday Evening Post* 22 May 1943

12 We never closed.
Vivian van Damm *c.*1889–1960: of the Windmill Theatre, London, during the Second World War

13 Acting is merely the art of keeping a large group of people from coughing.
Ralph Richardson 1902–83: in *New York Herald Tribune* 19 May 1946

14 There's no business like show business.
Irving Berlin 1888–1989: title of song (1946)

15 Shaw is like a train. One just speaks the words and sits in one's place. But Shakespeare is like bathing in the sea—one swims where one wants.
Vivien Leigh 1913–67: letter from Harold Nicolson to Vita Sackville-West, 1 February 1956

16 Don't clap too hard—it's a very old building.
John Osborne 1929– : *The Entertainer* (1957)

17 The weasel under the cocktail cabinet.
Harold Pinter 1930– : on being asked what his plays were about, in J. Russell Taylor *Anger and After* (1962)

18 Just say the lines and don't trip over the furniture.
Noël Coward 1899–1973: advice on acting, in D. Richards *The Wit of Noël Coward* (1968)

19 Acting is a masochistic form of exhibitionism. It is not quite the occupation of an adult.
Laurence Olivier 1907–89: in *Time* 3 July 1978

Action

1 But men must know, that in this theatre of man's life it is reserved only for God and angels to be lookers on.
Francis Bacon 1561–1626: *The Advancement of Learning* (1605)

2 If it were done when 'tis done, then 'twere well
It were done quickly.
William Shakespeare 1564–1616: *Macbeth* (1606)

3 Each your doing,
So singular in each particular,
Crowns what you are doing in the present deed,
That all your acts are queens.
William Shakespeare 1564–1616: *The Winter's Tale* (1610–11)

4 Oh that thou hadst like others been all words,
And no performance.
Philip Massinger 1583–1640: *The Parliament of Love* (1624)

5 They also serve who only stand and wait.
John Milton 1608–74: 'When I consider how my light is spent' (1673)

6 Think nothing done while aught remains to do.
Samuel Rogers 1763–1855: 'Human Life' (1819)

7 It is in vain to say human beings ought to be satisfied with tranquillity: they must have action; and they will make it if they cannot find it.
Charlotte Brontë 1816–55: *Jane Eyre* (1847)

8 Action is consolatory. It is the enemy of thought and the friend of flattering illusions.
Joseph Conrad 1857–1924: *Nostromo* (1904)

9 Nothing is ever done in this world until men are prepared to kill one another if it is not done.
George Bernard Shaw 1856–1950: *Major Barbara* (1907)

10 A man of action forced into a state of thought is unhappy until he can get out of it.
John Galsworthy 1867–1933: *Maid in Waiting* (1931)

11 The world can only be grasped by action, not by contemplation... The hand is the cutting edge of the mind.
Jacob Bronowski 1908–74: *The Ascent of Man* (1973)

Advertising

1 Promise, large promise, is the soul of an advertisement.
Samuel Johnson 1709–84: *The Idler* 20 January 1759

2 Advertising may be described as the science of arresting human intelligence long enough to get money from it.
Stephen Leacock 1869–1944: *Garden of Folly* (1924)

3 Half the money I spend on advertising is wasted, and the trouble is I don't know which half.
Viscount Leverhulme 1851–1925: in D. Ogilvy *Confessions of an Advertising Man* (1963)

4 Those who prefer their English sloppy have only themselves to thank if the advertisement writer uses his mastery of vocabulary and syntax to mislead their weak minds.
Dorothy L. Sayers 1893–1957: *Spectator* 19 November 1937

5 Advertising is the rattling of a stick inside a swill bucket.
George Orwell 1903–50: attributed

6 It is not necessary to advertise food to hungry people, fuel to cold people, or houses to the homeless.
J. K. Galbraith 1908– : *American Capitalism* (1952)

7 The consumer isn't a moron; she is your wife.
David Ogilvy 1911– : *Confessions of an Advertising Man* (1963)

Advice

1 Advice is seldom welcome; and those who want it the most always like it the least.
Lord Chesterfield 1694–1773: *Letters to his Son* (1774) 29 January 1748

2 It was, perhaps, one of those cases in which advice is good or bad only as the event decides.
Jane Austen 1775–1817: *Persuasion* (1818)

3 I always pass on good advice. It is the only thing to do with it. It is never of any use to oneself.
Oscar Wilde 1854–1900: *An Ideal Husband* (1895)

4 Well, if you knows of a better 'ole, go to it.
Bruce Bairnsfather 1888–1959: *Fragments from France* (1915)

Alcohol

See also **Food and Drink**

1 Wine is a mocker, strong drink is raging.
Bible: Proverbs

2 Drink no longer water, but use a little wine for thy stomach's sake.
Bible: I Timothy

3 When the wine is in, the wit is out.
Thomas Becon 1512–67: *Catechism* (1560)

4 Drink, sir, is a great provoker...
Lechery, sir, it provokes, and unprovokes; it provokes the desire, but it takes away the performance.
William Shakespeare 1564–1616: *Macbeth* (1606)

5 It would be port if it could.
Richard Bentley 1662–1742: describing claret, in R. C. Jebb *Bentley* (1902)

6 Claret is the liquor for boys; port, for men; but he who aspires to be a hero must drink brandy.
Samuel Johnson 1709–84: in James Boswell *Life of Johnson* (1791) 7 April 1779

7 O for a beaker full of the warm South,
Full of the true, the blushful Hippocrene,
With beaded bubbles winking at the brim,
And purple-stainèd mouth.
John Keats 1795–1821: 'Ode to a Nightingale' (1820)

8 The lips that touch liquor must never touch mine.
George W. Young 1846–1919: title of verse (*c*.1870)

9 And malt does more than Milton can
To justify God's ways to man.
A. E. Housman 1859–1936: *A Shropshire Lad* (1896)

10 A torchlight procession marching down your throat.
John L. O'Sullivan 1813–95: of whisky, in G. W. E. Russell *Collections and Recollections* (1898)

11 I'm only a beer teetotaller, not a champagne teetotaller.
George Bernard Shaw 1856–1950: *Candida* (1898)

12 If merely 'feeling good' could decide, drunkenness would be the supremely valid human experience.
William James 1842–1910: *Varieties of Religious Experience* (1902)

13 It is no time for mirth and laughter,
The cold, grey dawn of the morning after.
George Ade 1866–1944: *The Sultan of Sulu* (1903)

14 Alcohol is a very necessary article...It enables Parliament to do things at eleven at night that no sane person would do at eleven in the morning.
George Bernard Shaw 1856–1950: *Major Barbara* (1907)

15 Let's get out of these wet clothes and into a dry Martini.
Anonymous: line coined in the 1920s by Robert Benchley's press agent and adopted by Mae West in *Every Day's a Holiday* (1937 film)

16 Candy
Is dandy
But liquor
Is quicker.
Ogden Nash 1902–71: 'Reflections on Ice-breaking' (1931)

17 I'm not so think as you drunk I am.
J. C. Squire 1884–1958: 'Ballade of Soporific Absorption' (1931)

18 It's a naïve domestic Burgundy without any breeding, but I think you'll be amused by its presumption.
James Thurber 1894–1961: cartoon caption in *New Yorker* 27 March 1937

19 A good general rule is to state that the bouquet is better than the taste, and vice versa.
Stephen Potter 1900–69: on wine-tasting; *One-Upmanship* (1952)

20 His mouth had been used as a latrine by some small creature of the night, and then as its mausoleum.
Kingsley Amis 1922– : *Lucky Jim* (1953)

21 One reason why I don't drink is because I wish to know when I am having a good time.
Nancy Astor 1879–1964: in *Christian Herald* June 1960

22 I have taken more out of alcohol than alcohol has taken out of me.
Winston Churchill 1874–1965: in Quentin Reynolds *By Quentin Reynolds* (1964)

23 You're not drunk if you can lie on the floor without holding on.
Dean Martin 1917– : in P. Dickson *Official Rules* (1978)

Ambition

See also **Achievement and Endeavour**

1 [I] had rather be first in a village than second at Rome.
Julius Caesar 100–44 BC: in Francis Bacon *Advancement of Learning*, based on Plutarch *Parallel Lives*

2 *Aut Caesar, aut nihil.*
Caesar or nothing.
Cesare Borgia 1476–1507: motto inscribed on his sword

3 Fain would I climb, yet fear I to fall.
Walter Ralegh *c.*1552–1618: line written on a window-pane, in Thomas Fuller *Worthies of England* (1662)

4 If thy heart fails thee, climb not at all.
Elizabeth I 1533–1603: line after Walter Ralegh, written on a window-pane

5 When that the poor have cried, Caesar hath wept;
Ambition should be made of sterner stuff.
William Shakespeare 1564–1616: *Julius Caesar* (1599)

6 I have no spur
To prick the sides of my intent, but only
Vaulting ambition, which o'erleaps itself,
And falls on the other.
William Shakespeare 1564–1616: *Macbeth* (1606)

7 Cromwell, I charge thee, fling away ambition:
By that sin fell the angels.
William Shakespeare 1564–1616: *Henry VIII* (with John Fletcher, 1613)

8 Well is it known that ambition can creep as well as soar.
Edmund Burke 1729–97: *Third Letter…on the Proposals for Peace with the Regicide Directory* (1797)

9 There is always room at the top.
Daniel Webster 1782–1852: on being advised against joining the overcrowded legal profession (attributed)

10 Ah, but a man's reach should exceed his grasp,
Or what's a heaven for?
Robert Browning 1812–89: 'Andrea del Sarto' (1855)

11 Hitch your wagon to a star.
Ralph Waldo Emerson 1803–82: *Society and Solitude* (1870)

12 If you would hit the mark, you must aim a little above it;
Every arrow that flies feels the attraction of earth.
Henry Wadsworth Longfellow 1807–82: 'Elegiac Verse' (1880)

13 All ambitions are lawful except those which climb upwards on the miseries or credulities of mankind.
Joseph Conrad 1857–1924: *Some Reminiscences* (1912)

14 The world continues to offer glittering prizes to those who have stout hearts and sharp swords.
F. E. Smith 1872–1930: Rectorial Address, Glasgow University, 7 November 1923

15 He is loyal to his own career but only incidentally to anything or anyone else.
Hugh Dalton 1887–1962: of Richard Crossman; diary, 17 September 1941

America and Americans

See also **Places**

1 Pray enter
You are learned Europeans and we worse
Than ignorant Americans.
Philip Massinger 1583–1640: *The City Madam* (licensed 1632)

2 A citizen, first in war, first in peace, and first in the hearts of his countrymen.
Henry Lee 1756–1818: *Funeral Oration on the death of General Washington* (1800)

3 Go West, young man, and grow up with the country.
Horace Greeley 1811–72: *Hints toward Reforms* (1850)

4 The United States themselves are essentially the greatest poem.

Walt Whitman 1819–92: *Leaves of Grass* (1855)

5 Good Americans, when they die, go to Paris.

Thomas Gold Appleton 1812–84: in Oliver Wendell Holmes *The Autocrat of the Breakfast-Table* (1858)

6 Give me your tired, your poor,
Your huddled masses yearning to breathe free.

Emma Lazarus 1849–87: 'The New Colossus' (1883); inscribed on the Statue of Liberty, New York

7 A Boston man is the east wind made flesh.

Thomas Gold Appleton 1812–84: attributed

8 MRS ALLONBY: They say, Lady Hunstanton, that when good Americans die they go to Paris.
LADY HUNSTANTON: Indeed? And when bad Americans die, where do they go to?
LORD ILLINGWORTH: Oh, they go to America.

Oscar Wilde 1854–1900: *A Woman of No Importance* (1893)

9 America! America!
God shed His grace on thee
And crown thy good with brotherhood
From sea to shining sea!

Katherine Lee Bates 1859–1929: 'America the Beautiful' (1893)

10 It is by the goodness of God that in our country we have those three unspeakably precious things: freedom of speech, freedom of conscience, and the prudence never to practise either of them.

Mark Twain 1835–1910: *Following the Equator* (1897)

11 America is God's Crucible, the great Melting-Pot where all the races of Europe are melting and re-forming!

Israel Zangwill 1864–1926: *The Melting Pot* (1908)

12 There can be no fifty-fifty Americanism in this country. There is room here for only 100 per cent. Americanism, only for those who are Americans and nothing else.

Theodore Roosevelt 1858–1919: speech in Saratoga, 19 July 1918

13 I have fallen in love with American names,
The sharp, gaunt names that never get fat.

Stephen Vincent Benét 1898–1943: 'American Names' (1927)

14 I pledge you, I pledge myself, to a new deal for the American people.

Franklin D. Roosevelt 1882–1945: speech, 2 July 1932, accepting the presidential nomination

15 In the United States there is more space where nobody is than where anybody is. That is what makes America what it is.

Gertrude Stein 1874–1946: *The Geographical History of America* (1936)

16 God bless America,
Land that I love,
Stand beside her and guide her
Thru the night with a light from above.

Irving Berlin 1888–1989: 'God Bless America' (1939)

17 There are no second acts in American lives.

F. Scott Fitzgerald 1896–1940: Edmund Wilson (ed.) *The Last Tycoon* (1941)

18 Overpaid, overfed, oversexed, and over here.

Tommy Trinder 1909–89: of American troops in Britain during the Second World War (associated with Trinder, but probably not original)

19 California is a fine place to live—if you happen to be an orange.

Fred Allen 1894–1956: *American Magazine* December 1945

20 The most serious charge which can be brought against New England is not Puritanism but February.

Joseph Wood Krutch 1893–1970: *Twelve Seasons* (1949)

21 In America any boy may become President and I suppose it's just one of the risks he takes!

Adlai Stevenson 1900–65: speech in Indianapolis, 26 September 1952

22 This land is your land, this land is my land,
From California to the New York Island.
From the redwood forest to the Gulf Stream waters
This land was made for you and me.

Woody Guthrie 1912–67: 'This Land is Your Land' (1956 song)

23 I like to be in America!
OK by me in America!

Ev'rything free in America
For a small fee in America!
Stephen Sondheim 1930– : 'America' (1957 song)

24 Europe is the unfinished negative of which America is the proof.
Mary McCarthy 1912–89: *On the Contrary* (1961) 'America the Beautiful'

25 Our national flower is the concrete cloverleaf.
Lewis Mumford 1895–1982: *Quote Magazine* 8 October 1961

26 America is a vast conspiracy to make you happy.
John Updike 1932– : *Problems* (1980) 'How to love America and Leave it at the Same Time'

Anger

1 A soft answer turneth away wrath.
Bible: Proverbs

2 *Ira furor brevis est.*
Anger is a short madness.
Horace 65–8 BC: *Epistles*

3 Be ye angry and sin not: let not the sun go down upon your wrath.
Bible: Ephesians

4 Anger makes dull men witty, but it keeps them poor.
Francis Bacon 1561–1626: 'Baconiana' (1859), often attributed to Queen Elizabeth I

5 Anger is one of the sinews of the soul.
Thomas Fuller 1608–61: *The Holy State and the Profane State*

6 Beware the fury of a patient man.
John Dryden 1631–1700: *Absalom and Achitophel* (1681)

7 The tigers of wrath are wiser than the horses of instruction.
William Blake 1757–1827: *The Marriage of Heaven and Hell* (1790–3) 'Proverbs of Hell'

8 When angry, count four; when very angry, swear.
Mark Twain 1835–1910: *Pudd'nhead Wilson* (1894)

Animals

See also **Cats, Dogs**

1 A righteous man regardeth the life of his beast: but the tender mercies of the wicked are cruel.
Bible: Proverbs

2 Nature's great masterpiece, an elephant,
The only harmless great thing.
John Donne 1572–1631: 'The Progress of the Soul' (1601)

3 In so doing, use him as though you loved him.
Izaak Walton 1593–1683: on baiting a hook with a live frog; *The Compleat Angler* (1653)

4 Wee, sleekit, cow'rin', tim'rous beastie,
O what a panic's in thy breastie!
Robert Burns 1759–96: 'To a Mouse' (1786)

5 Tiger Tiger, burning bright,
In the forests of the night;
What immortal hand or eye,
Could frame thy fearful symmetry?
William Blake 1757–1827: 'The Tiger' (1794)

6 Animals, whom we have made our slaves, we do not like to consider our equal.
Charles Darwin 1809–82: Notebook B (1837–8)

7 It ar'n't that I loves the fox less, but that I loves the 'ound more.
R. S. Surtees 1805–64: *Handley Cross* (1843)

8 All things bright and beautiful,
All creatures great and small,
All things wise and wonderful,
The Lord God made them all.
Cecil Frances Alexander 1818–95: 'All Things Bright and Beautiful' (1848)

9 I think I could turn and live with animals, they are so placid and self-contained,
I stand and look at them long and long.
They do not sweat and whine about their condition,

They do not lie awake in the dark and weep for their sins,
They do not make me sick discussing their duty to God.
Walt Whitman 1819–92: 'Song of Myself' (written 1855)

10 The Llama is a woolly sort of fleecy hairy goat,
With an indolent expression and an undulating throat
Like an unsuccessful literary man.
Hilaire Belloc 1870–1953: 'The Llama' (1897)

11 'Twould ring the bells of Heaven
The wildest peal for years,
If Parson lost his senses
And people came to theirs,
And he and they together
Knelt down with angry prayers
For tamed and shabby tigers
And dancing dogs and bears,
And wretched, blind, pit ponies,
And little hunted hares.
Ralph Hodgson 1871–1962: 'Bells of Heaven' (1917)

12 The rabbit has a charming face:
Its private life is a disgrace.
I really dare not name to you
The awful things that rabbits do.
Anonymous: 'The Rabbit', in *The Week-End Book* (1925)

13 The cow is of the bovine ilk;
One end is moo, the other, milk.
Ogden Nash 1902–71: 'The Cow' (1931)

14 The turtle lives 'twixt plated decks
Which practically conceal its sex.
I think it clever of the turtle
In such a fix to be so fertile.
Ogden Nash 1902–71: 'Autres Bêtes, Autres Moeurs' (1931)

15 Giraffes!—a People
Who live between the earth and skies,
Each in his lone religious steeple,
Keeping a light-house with his eyes.
Roy Campbell 1901–57: 'Dreaming Spires' (1946)

16 Where in this wide world can man find nobility without pride,
Friendship without envy, or beauty without vanity?
Ronald Duncan 1914–82: 'In Praise of the Horse' (1962)

17 I am fond of pigs. Dogs look up to us. Cats look down on us. Pigs treat us as equals.
Winston Churchill 1874-1965: attributed, in M. Gilbert *Never Despair* (1988)

Apology

1 Never make a defence or apology before you be accused.
Charles I 1600–49: letter to Lord Wentworth, 3 September 1636

2 Never complain and never explain.
Benjamin Disraeli 1804–81: in J. Morley *Life of Gladstone* (1903)

3 Never explain—your friends do not need it and your enemies will not believe you anyway.
Elbert Hubbard 1859–1915: *The Motto Book* (1907)

4 It is a good rule in life never to apologize. The right sort of people do not want apologies, and the wrong sort take a mean advantage of them.
P. G. Wodehouse 1881–1975: *The Man Upstairs* (1914)

5 Very sorry can't come. Lie follows by post.
Lord Charles Beresford 1846–1919: telegraphed message to the Prince of Wales, on being summoned to dine at the eleventh hour

6 Several excuses are always less convincing than one.
Aldous Huxley 1894–1963: *Point Counter Point* (1928)

Architecture

1 Well building hath three conditions. Commodity, firmness, and delight.
Henry Wotton 1568–1639: *Elements of Architecture* (1624)

2 Light (God's eldest daughter) is a principal beauty in building.
Thomas Fuller 1608–61: *The Holy State and the Profane State* 'Of Building'

3 Architecture in general is frozen music.
Friedrich von Schelling 1775–1854: *Philosophie der Kunst* (1809)

4 He builded better than he knew;—
The conscious stone to beauty grew.
Ralph Waldo Emerson 1803–82: 'The Problem' (1847)

5 Form follows function.
Louis Henri Sullivan 1856–1924: *The Tall Office Building Artistically Considered* (1896)

6 A house is a machine for living in.
Le Corbusier 1887–1965: *Vers une architecture* (1923)

7 Architecture, of all the arts, is the one which acts the most slowly, but the most surely, on the soul.
Ernest Dimnet: *What We Live By* (1932)

8 The physician can bury his mistakes, but the architect can only advise his client to plant vines—so they should go as far as possible from home to build their first buildings.
Frank Lloyd Wright 1867–1959: *New York Times* 4 October 1953

9 Architecture is the art of how to waste space.
Philip Johnson 1906– : *New York Times* 27 December 1964

10 A monstrous carbuncle on the face of a much-loved and elegant friend.
Charles, Prince of Wales 1948– : speech on the proposed extension to the National Gallery, London, 30 May 1984

Argument

1 It is better to dwell in a corner of the housetop, than with a brawling woman in a wide house.
Bible: Proverbs

2 You cannot argue with someone who denies the first principles.
Auctoritates Aristotelis: a compilation of medieval propositions

3 Your 'if' is the only peace-maker; much virtue in 'if'.
William Shakespeare 1564–1616: *As You Like It* (1599)

4 There is no arguing with Johnson; for when his pistol misses fire, he knocks you down with the butt end of it.
Oliver Goldsmith 1730–74: in James Boswell *Life of Samuel Johnson* (1791) 26 October 1769

5 Who can refute a sneer?
William Paley 1743–1805: *Principles of Moral and Political Philosophy* (1785)

6 I am not arguing with you—I am telling you.
James McNeill Whistler 1834–1903: *The Gentle Art of Making Enemies* (1890)

7 It takes in reality only one to make a quarrel. It is useless for the sheep to pass resolutions in favour of vegetarianism, while the wolf remains of a different opinion.
Dean Inge 1860–1954: *Outspoken Essays: First Series* (1919)

8 The argument of the broken window pane is the most valuable argument in modern politics.
Emmeline Pankhurst 1858–1928: in G. Dangerfield *The Strange Death of Liberal England* (1936)

9 The Catholic and the Communist are alike in assuming that an opponent cannot be both honest and intelligent.
George Orwell 1903–50: *Polemic* January 1946

10 'Yes, but not in the South', with slight adjustments, will do for any argument about any place, if not about any person.
Stephen Potter 1900–69: *Lifemanship* (1950)

The Army

See also **War, Wars**

1 A soldier,
Full of strange oaths, and bearded like the pard,
Jealous in honour, sudden and quick in quarrel,
Seeking the bubble reputation
Even in the cannon's mouth.
William Shakespeare 1564–1616: *As You Like It* (1599)

2 As Lord Chesterfield said of the generals of his day, 'I only hope that when the enemy reads the list of their names, he trembles as I do.'
Duke of Wellington 1769–1852: letter, 29 August 1810

(usually quoted 'I don't know what effect these men will have upon the enemy, but, by God, they frighten me')

3 An army marches on its stomach.

Napoléon I 1769–1821: attributed, but probably condensed from a long passage in E. A. de Las Cases *Mémorial de Ste-Hélène* (1823) 14 November 1816

4 Remember that there is not one of you who does not carry in his cartridge-pouch the marshal's baton of the duke of Reggio; it is up to you to bring it forth.

Louis XVIII 1755–1824: speech to Saint-Cyr cadets, 9 August 1819

5 Ours [our army] is composed of the scum of the earth—the mere scum of the earth.

Duke of Wellington 1769–1852: in Philip Henry Stanhope *Notes of Conversations with the Duke of Wellington* (1888) 4 November 1831

6 *C'est magnifique, mais ce n'est pas la guerre.*
It is magnificent, but it is not war.

Pierre Bosquet 1810–61: on the charge of the Light Brigade at Balaclava, 25 October 1854

7 Theirs not to make reply,
Theirs not to reason why,
Theirs but to do and die:
Into the valley of Death
Rode the six hundred.

Alfred, Lord Tennyson 1809–92: 'The Charge of the Light Brigade' (1854)

8 They dashed on towards that thin red line tipped with steel.

William Howard Russell 1820–1907: of the Russians charging the British, in *The British Expedition to the Crimea* (1877). Russell's original dispatch to *The Times*, 14 November 1854, reads 'That thin red streak tipped with a line of steel'

9 O it's Tommy this, an' Tommy that, an' 'Tommy, go away';
But it's 'Thank you, Mister Atkins,' when the band begins to play.

Rudyard Kipling 1865–1936: 'Tommy' (1892)

10 The 'eathen in 'is blindness must end where 'e began.
But the backbone of the Army is the non-commissioned man!

Rudyard Kipling 1865–1936: 'The 'Eathen' (1896)

11 You can always tell an old soldier by the inside of his holsters and cartridge boxes. The young ones carry pistols and cartridges; the old ones, grub.

George Bernard Shaw 1856–1950: *Arms and the Man* (1898)

12 The British soldier can stand up to anything except the British War Office.

George Bernard Shaw 1856–1950: *The Devil's Disciple* (1901)

13 When the military man approaches, the world locks up its spoons and packs off its womankind.

George Bernard Shaw 1856–1950: *Man and Superman* (1903)

14 What passing-bells for these who die as cattle?
Only the monstrous anger of the guns.

Wilfred Owen 1893–1918: 'Anthem for Doomed Youth' (written 1917)

15 Soldiers are citizens of death's grey land,
Drawing no dividend from time's tomorrows.

Siegfried Sassoon 1886–1967: 'Dreamers' (1918)

16 If I were fierce, and bald, and short of breath,
I'd live with scarlet Majors at the Base,
And speed glum heroes up the line to death.

Siegfried Sassoon 1886–1967: 'Base Details' (1918)

17 Lions led by donkeys.

Max Hoffman 1869–1927: of British soldiers during the First World War; in A. Clark *The Donkeys* (1961)

18 If it moves, salute it; if it doesn't move, pick it up; and if you can't pick it up, paint it.

Anonymous: 1940s saying

19 They call it easing the Spring: it is perfectly easy
If you have any strength in your thumb: like the bolt,
And the breech, and the cocking-piece, and the point of balance,
Which in our case we have not got.

Henry Reed 1914–86: 'Lessons of the War: 1, Naming of Parts' (1946)

20 At the age of four with paper hats and wooden swords we're all Generals. Only some of us never grow out of it.

Peter Ustinov 1921– : *Romanoff and Juliet* (1956)

Art

See also **Painting and Drawing**

1 Life is short, the art long.
Hippocrates *c.*460–357 BC: *Aphorisms*, often quoted as '*Ars longa, vita brevis*', after Seneca *De Brevitate Vitae*

2 In art the best is good enough.
Johann Wolfgang von Goethe 1749–1832: *Italienische Reise* (1816–17) 3 March 1787

3 Art for art's sake, with no purpose, for any purpose perverts art. But art achieves a purpose which is not its own.
Benjamin Constant 1767–1834: *Journal intime* 11 February 1804

4 God help the Minister that meddles with art!
Lord Melbourne 1779–1848: in Lord David Cecil *Lord M* (1954)

5 The artist must be in his work as God is in creation, invisible and all-powerful; one must sense him everywhere but never see him.
Gustave Flaubert 1821–80: letter to Mlle Leroyer de Chantepie, 18 March 1857

6 Art is a jealous mistress.
Ralph Waldo Emerson 1803–82: *The Conduct of Life* (1860)

7 Human life is a sad show, undoubtedly: ugly, heavy and complex. Art has no other end, for people of feeling, than to conjure away the burden and bitterness.
Gustave Flaubert 1821–80: letter to Amelie Bosquet, July 1864

8 All passes. Art alone
Enduring stays to us;
The Bust outlasts the throne,—
The Coin, Tiberius.
Henry Austin Dobson 1840–1921: 'Ars Victrix' (1876); translation of Théophile Gautier's 'L'Art'

9 It's clever, but is it Art?
Rudyard Kipling 1865–1936: 'The Conundrum of the Workshops' (1892)

10 We work in the dark—we do what we can—we give what we have. Our doubt is our passion and our passion is our task. The rest is the madness of art.
Henry James 1843–1916: 'The Middle Years' (1893)

11 I always said God was against art and I still believe it.
Edward Elgar 1857–1934: letter to A. J. Jaeger, 9 October 1900

12 The history of art is the history of revivals.
Samuel Butler 1835–1902: *Notebooks* (1912)

13 The true artist will let his wife starve, his children go barefoot, his mother drudge for his living at seventy, sooner than work at anything but his art.
George Bernard Shaw 1856–1950: *Man and Superman* (1903)

14 The artist, like the God of the creation, remains within or behind or beyond or above his handiwork, invisible, refined out of existence, indifferent, paring his fingernails.
James Joyce 1882–1941: *A Portrait of the Artist as a Young Man* (1916)

15 Art is vice. You don't marry it legitimately, you rape it.
Edgar Degas 1834–1917: in P. Lafond *Degas* (1918)

16 There is no more sombre enemy of good art than the pram in the hall.
Cyril Connolly 1903–74: *Enemies of Promise* (1938)

17 Art is significant deformity.
Roger Fry 1866–1934: in Virginia Woolf *Roger Fry* (1940)

18 Art is the imposing of a pattern on experience, and our aesthetic enjoyment is recognition of the pattern.
Alfred North Whitehead 1861–1947: *Dialogues* (1954) 10 June 1943

19 *L'art est un anti-destin.*
Art is a revolt against fate.
André Malraux 1901–76: *Les Voix du silence* (1951)

20 Art is born of humiliation.
W. H. Auden 1907–73: in Stephen Spender *World Within World* (1951)

21 Art is meant to disturb, science reassures.
Georges Braque 1882–1963: *Le Jour et la nuit: Cahiers 1917–52*

22 Art is the objectification of feeling, and the subjectification of nature.
Suzanne K. Langer 1895–1985: *Mind* (1967)

Beauty

1 A beautiful face is a mute recommendation.
Publilius Syrus 1st century BC: *Sententiae* tr. Thomas Tenison (1679)

2 Consider the lilies of the field, how they grow; they toil not, neither do they spin:
And yet I say unto you, That even Solomon in all his glory was not arrayed like one of these.
Bible: St Matthew

3 And she was fayr as is the rose in May.
Geoffrey Chaucer *c.*1343–1400: *The Legend of Good Women* 'Cleopatra'

4 Was this the face that launched a thousand ships,
And burnt the topless towers of Ilium?
Sweet Helen, make me immortal with a kiss!
Christopher Marlowe 1564–93: *Doctor Faustus* (1604)

5 O! she doth teach the torches to burn bright.
It seems she hangs upon the cheek of night
Like a rich jewel in an Ethiop's ear;
Beauty too rich for use, for earth too dear.
William Shakespeare 1564–1616: *Romeo and Juliet* (1595)

6 Shall I compare thee to a summer's day?
Thou art more lovely and more temperate:
Rough winds do shake the darling buds of May,
And summer's lease hath all too short a date.
William Shakespeare 1564–1616: sonnet 18 (1609)

7 There is no excellent beauty that hath not some strangeness in the proportion.
Francis Bacon 1561–1626: *Essays* (1625) 'Of Beauty'

8 The flowers anew, returning seasons bring;
But beauty faded has no second spring.
Ambrose Philips *c.*1675–1749: *The First Pastoral* (1708)

9 Beauty is no quality in things themselves. It exists merely in the mind which contemplates them.
David Hume 1711–76: 'Of the Standard of Taste' (1757)

10 She walks in beauty, like the night
Of cloudless climes and starry skies;
And all that's best of dark and bright
Meet in her aspect and her eyes.
Lord Byron 1788–1824: 'She Walks in Beauty' (1815)

11 A thing of beauty is a joy for ever:
Its loveliness increases; it will never
Pass into nothingness.
John Keats 1795–1821: *Endymion* (1818)

12 'Beauty is truth, truth beauty,'—that is all
Ye know on earth, and all ye need to know.
John Keats 1795–1821: 'Ode on a Grecian Urn' (1820)

13 *I never saw an ugly thing in my life*: for let the form of an object be what it may,—light, shade, and perspective will always make it beautiful.
John Constable 1776–1837: in C. R. Leslie *Memoirs of the Life of John Constable* (1843)

14 Remember that the most beautiful things in the world are the most useless; peacocks and lilies for instance.
John Ruskin 1819–1900: *Stones of Venice* vol. 1 (1851)

15 If you get simple beauty and naught else,
You get about the best thing God invents.
Robert Browning 1812–89: 'Fra Lippo Lippi' (1855)

16 The Lord prefers common-looking people. That is why he makes so many of them.
Abraham Lincoln 1809–65: attributed

17 All things counter, original, spare, strange;
Whatever is fickle, freckled (who knows how?)
With swift, slow; sweet, sour; adazzle, dim;
He fathers-forth whose beauty is past change:
Praise him.
Gerard Manley Hopkins 1844–89: 'Pied Beauty' (written 1877)

18 Beauty is mysterious as well as terrible. God and devil are fighting there, and the battlefield is the heart of man.
Fedor Dostoevsky 1821–81: *The Brothers Karamazov* (1879–80)

19 I have a left shoulder-blade that is a miracle of loveliness. People come miles to see it. My right elbow has a fascination that few can resist.
W. S. Gilbert 1836–1911: *The Mikado* (1885)

20 When a woman isn't beautiful, people always say, 'You have lovely eyes, you have lovely hair.'
Anton Chekhov 1860–1904: *Uncle Vanya* (1897)

21 Beauty is all very well at first sight; but who ever looks at it when it has been in the house three days?
George Bernard Shaw 1856–1950: *Man and Superman* (1903)

22 I always say beauty is only sin deep.
Saki (H. H. Munro) 1870–1916: *Reginald* (1904)

23 He was afflicted by the thought that where Beauty was, nothing ever ran quite straight, which, no doubt, was why so many people looked on it as immoral.
John Galsworthy 1867–1933: *In Chancery* (1920)

24 Beauty is momentary in the mind—
The fitful tracing of a portal;
But in the flesh it is immortal.
The body dies; the body's beauty lives.
Wallace Stevens 1879–1955: 'Peter Quince at the Clavier' (1923)

25 It was a blonde. A blonde to make a bishop kick a hole in a stained glass window.
Raymond Chandler 1888–1959: *Farewell, My Lovely* (1940)

26 I'm tired of all this nonsense about beauty being only skin-deep. That's deep enough. What do you want—an adorable pancreas?
Jean Kerr 1923– : *The Snake has all the Lines* (1958)

Beginnings and Endings

1 In the beginning God created the heaven and the earth. And the earth was without form, and void; and darkness was upon the face of the deep.
Bible: Genesis

2 In my end is my beginning.
Mary, Queen of Scots 1542–87: motto

3 This is the beginning of the end.
Charles-Maurice de Talleyrand 1754–1838: on hearing the outcome of the battle at Borodino, 1812; attributed

4 Ring out the old, ring in the new,
Ring, happy bells, across the snow:
The year is going, let him go;
Ring out the false, ring in the true.
Alfred, Lord Tennyson 1809–92: *In Memoriam A. H. H.* (1850)

5 'Begin at the beginning,' the King said, gravely, 'and go on till you come to the end: then stop.'
Lewis Carroll 1832–98: *Alice's Adventures in Wonderland* (1865)

6 Some say the world will end in fire,
Some say in ice.
Robert Frost 1874–1963: 'Fire and Ice' (1923)

7 This is the way the world ends
Not with a bang but a whimper.
T. S. Eliot 1888–1965: 'The Hollow Men' (1925)

8 In my beginning is my end.
T. S. Eliot 1888–1965: *Four Quartets* 'East Coker' (1940)

9 What we call the beginning is often the end
And to make an end is to make a beginning.
The end is where we start from.
T. S. Eliot 1888–1965: *Four Quartets* 'Little Gidding' (1942)

10 Now this is not the end. It is not even the beginning of the end. But it is, perhaps, the end of the beginning.
Winston Churchill 1874–1965: speech at the Mansion House, London, 10 November 1942

11 Are you sitting comfortably? Then I'll begin.
Julia Lang 1921– : *Listen with Mother* (BBC radio programme for children, 1950–82)

12 The party's over, it's time to call it a day.
Betty Comden 1919– and **Adolph Green** 1915– : 'The Party's Over' (1956 song)

13 All this will not be finished in the first 100 days. Nor will it be finished in the first 1,000 days, nor in the life of this

Administration, nor even perhaps in our lifetime on this planet. But let us begin.

John F. Kennedy 1917–63: inaugural address, 20 January 1961

14 It ain't over till it's over.

Yogi Berra 1925– : comment on National League pennant race, 1973, quoted in many versions

15 The opera ain't over 'til the fat lady sings.

Dan Cook: in *Washington Post* 3 June 1978

Behaviour

See also **Manners**

1 *O tempora, O mores!*
Oh, the times! Oh, the manners!

Cicero 106–43 BC: *In Catilinam*

2 Caesar's wife must be above suspicion.

Julius Caesar 100–44 BC: oral tradition, based on Plutarch *Parallel Lives*

3 When I go to Rome, I fast on Saturday, but here [Milan] I do not. Do you also follow the custom of whatever church you attend, if you do not want to give or receive scandal.

St Ambrose *c.*339–97: letter to Januarius, tr. Sr W. Parsons; usually quoted 'When in Rome, do as the Romans do'

4 He was a verray, parfit gentil knyght.

Geoffrey Chaucer *c.*1343–1400: *The Canterbury Tales* 'General Prologue'

5 He does it with a better grace, but I do it more natural.

William Shakespeare 1564–1616: *Twelfth Night* (1601)

6 Never *in* the way, and never *out* of the way.

Charles II 1630–85: of Lord Godolphin, as his page

7 They [the *Letters* of Lord Chesterfield] teach the morals of a whore, and the manners of a dancing master.

Samuel Johnson 1709–84: in James Boswell *Life of Johnson* (1791) 1754

8 The courtiers who surround him [Louis XVIII] have forgotten nothing and learnt nothing.
General Dumouriez 1739–1823: *Examen impartial d'un Écrit... de Louis XVIII* (1795)

9 *Tout comprendre rend très indulgent.*
To be totally understanding makes one very indulgent.
Mme de Staël 1766–1817: *Corinne* (1807)

10 In short, he was a perfect cavaliero,
And to his very valet seemed a hero.
Lord Byron 1788–1824: *Beppo* (1818)

11 It is almost a definition of a gentleman to say that he is one who never inflicts pain.
Cardinal Newman 1801–90: *The Idea of a University* (1852)

12 The only infallible rule we know is, that the man who is always talking about being a gentleman never is one.
R. S. Surtees 1805–64: *Ask Mamma* (1858)

13 Go directly—see what she's doing, and tell her she mustn't.
Punch: 1872

14 Conduct is three-fourths of our life and its largest concern.
Matthew Arnold 1822–88: *Literature and Dogma* (1873)

15 He combines the manners of a marquis with the morals of a Methodist.
W. S. Gilbert 1836–1911: *Ruddigore* (1887)

16 Being tactful in audacity is knowing how far one can go too far.
Jean Cocteau 1889–1963: *Le Rappel à l'ordre* (1926)

17 I get too hungry for dinner at eight.
I like the theatre, but never come late.
I never bother with people I hate.
That's why the lady is a tramp.
Lorenz Hart 1895–1943: 'The Lady is a Tramp' (1937 song)

Belief and Unbelief

1 Lord, I believe; help thou mine unbelief.
Bible: St Mark

2 *Certum est quia impossibile est.*
It is certain because it is impossible.
Tertullian AD *c.*160–*c.*225: *De Carne Christi*, often quoted '*Credo quia impossibile* [I believe because it is impossible]'

3 *Que sais-je?*
What do I know?
Montaigne 1533–92: *Essais* (1580) on the position of the sceptic

4 A little philosophy inclineth man's mind to atheism, but depth in philosophy bringeth men's minds about to religion.
Francis Bacon 1561–1626: *Essays* (1625) 'Of Atheism'

5 By night an atheist half believes a God.
Edward Young 1683–1765: *Night Thoughts* (1742–5)

6 Truth, Sir, is a cow, that will yield such people [sceptics] no more milk, and so they are gone to milk the bull.
Samuel Johnson 1709–84: in James Boswell *Life of Johnson* (1791) 21 July 1763

7 It is necessary to the happiness of man that he be mentally faithful to himself. Infidelity does not consist in believing, or in disbelieving, it consists in professing to believe what one does not believe.
Thomas Paine 1737–1809: *The Age of Reason* pt. 1 (1794)

8 *We can believe what we choose.* We are answerable for what we choose to believe.
Cardinal Newman 1801–90: letter to Mrs William Froude, 27 June 1848

9 There lives more faith in honest doubt,
Believe me, than in half the creeds.
Alfred, Lord Tennyson 1809–92: *In Memoriam A. H. H.* (1850)

10 Just when we are safest, there's a sunset-touch,
A fancy from a flower-bell, some one's death,
A chorus-ending from Euripides,—
And that's enough for fifty hopes and fears
As old and new at once as Nature's self…
The grand Perhaps!
Robert Browning 1812–89: 'Bishop Blougram's Apology' (1855)

11 The Sea of Faith
Was once, too, at the full, and round earth's shore

Lay like the folds of a bright girdle furled.
But now I only hear
Its melancholy, long, withdrawing roar.
Matthew Arnold 1822–88: 'Dover Beach' (1867)

12 Why, sometimes I've believed as many as six impossible things before breakfast.
Lewis Carroll 1832–98: *Through the Looking-Glass* (1872)

13 When suave politeness, tempering bigot zeal,
Corrected *I believe* to *One does feel*.
Monsignor Ronald Knox 1888–1957: 'Absolute and Abitofhell' (1913)

14 I do not consider it an insult, but rather a compliment to be called an agnostic. I do not pretend to know where many ignorant men are sure—that is all that agnosticism means.
Clarence Darrow 1857–1938: speech at trial of John Thomas Scopes, 15 July 1925

15 Every time a child says 'I don't believe in fairies' there is a little fairy somewhere that falls down dead.
J. M. Barrie 1860–1937: *Peter Pan* (1928)

16 Of course not, but I am told it works even if you don't believe in it.
Niels Bohr 1885–1962: when asked whether he really believed a horseshoe hanging over his door would bring him luck, *c.*1930; in A. Pais *Inward Bound* (1986)

17 The dust of exploded beliefs may make a fine sunset.
Geoffrey Madan 1895–1947: *Livre sans nom: Twelve Reflections* (1934)

18 An atheist is a man who has no invisible means of support.
John Buchan 1875–1940: in H. E. Fosdick *On Being a Real Person* (1943)

19 Man is a credulous animal, and must believe *something*; in the absence of good grounds for belief, he will be satisfied with bad ones.
Bertrand Russell 1872–1970: *Unpopular Essays* (1950)

20 I confused things with their names: that is belief.
Jean-Paul Sartre 1905–80: *Les Mots* (1964)

21 She [my grandmother] believed in nothing; only her scepticism kept her from being an atheist.
Jean-Paul Sartre 1905–1980: *Les Mots* (1964)

22 There is a lot to be said in the Decade of Evangelism for believing more and more in less and less.
Bishop John Yates 1925– : *Gloucester Diocesan Gazette* August 1991

The Bible

1 The devil can cite Scripture for his purpose.
William Shakespeare 1564–1616: *The Merchant of Venice* (1596–8)

2 *Scrutamini scripturas* [Let us look at the scriptures]. These two words have undone the world.
John Selden 1584–1654: *Table Talk* (1689) 'Bible Scripture'

3 Here is wisdom; this is the royal Law; these are the lively Oracles of God.
Coronation Service 1689: The Presenting of the Holy Bible

4 The English Bible, a book which, if everything else in our language should perish, would alone suffice to show the whole extent of its beauty and power.
Lord Macaulay 1800–59: 'John Dryden' (1828)

5 There's a great text in Galatians,
Once you trip on it, entails
Twenty-nine distinct damnations,
One sure, if another fails.
Robert Browning 1812–89: 'Soliloquy of the Spanish Cloister' (1842)

6 We have used the Bible as if it was a constable's handbook—an opium-dose for keeping beasts of burden patient while they are being overloaded.
Charles Kingsley 1819–75: *Letters to the Chartists*

7 LORD ILLINGWORTH: The Book of Life begins with a man and a woman in a garden.
MRS ALLONBY: It ends with Revelations.
Oscar Wilde 1854–1900: *A Woman of No Importance* (1893)

8 An apology for the Devil: It must be remembered that we have only heard one side of the case. God has written all the books.
Samuel Butler 1835–1902: *Notebooks* (1912)

Biography

1 Read no history: nothing but biography, for that is life without theory.
Benjamin Disraeli 1804–81: *Contarini Fleming* (1832)

2 Lives of great men all remind us
We can make our lives sublime,
And, departing, leave behind us
Footprints on the sands of time.
Henry Wadsworth Longfellow 1807–82: 'A Psalm of Life' (1838)

3 A well-written Life is almost as rare as a well-spent one.
Thomas Carlyle 1795–1881: *Critical and Miscellaneous Essays* (1838)

4 Then there is my noble and biographical friend who has added a new terror to death.
Charles Wetherell 1770–1846: on Lord Campbell's *Lives of the Lord Chancellors* being written without the consent of heirs or executors; also attributed to Lord Lyndhurst (1772–1863)

5 No quailing, Mrs Gaskell! no drawing back!
Patrick Brontë 1777–1861: apropos her undertaking to write the life of Charlotte Brontë; letter from Mrs Gaskell to Ellen Nussey, 24 July 1855

6 It is not a Life at all. It is a Reticence, in three volumes.
W. E. Gladstone 1809–98: on J. W. Cross's *Life of George Eliot*; in E. F. Benson *As We Were* (1930)

7 The Art of Biography
Is different from Geography.
Geography is about Maps,
But Biography is about Chaps.
Edmund Clerihew Bentley 1875–1956: *Biography for Beginners* (1905)

8 Discretion is not the better part of biography.
Lytton Strachey 1880–1932: in M. Holroyd *Lytton Strachey* vol. 1 (1967)

9 To write one's memoirs is to speak ill of everybody except oneself.
Marshal Pétain 1856–1951: in *Observer* 26 May 1946

10 An autobiography is an obituary in serial form with the last instalment missing.
Quentin Crisp 1908– : *The Naked Civil Servant* (1968)

Birds

1 *Vox et praeterea nihil.*
A voice and nothing more.
Anonymous: describing a nightingale. See Plutarch *Moralia*

2 The bisy larke, messager of day.
Geoffrey Chaucer *c.*1343–1400: *The Canterbury Tales* 'The Knight's Tale'

3 O blithe new-comer! I have heard,
I hear thee and rejoice:
O Cuckoo! Shall I call thee bird,
Or but a wandering voice?
William Wordsworth 1770–1850: 'To the Cuckoo' (1807)

4 Hail to thee, blithe Spirit!
Bird thou never wert,
That from Heaven, or near it,
Pourest thy full heart
In profuse strains of unpremeditated art.
Percy Bysshe Shelley 1792–1822: 'To a Skylark' (1819)

5 Alone and warming his five wits,
The white owl in the belfry sits.
Alfred, Lord Tennyson 1809–92: 'Song—The Owl' (1830)

6 That's the wise thrush; he sings each song twice over,
Lest you should think he never could recapture
The first fine careless rapture!
Robert Browning 1812–89: 'Home-Thoughts, from Abroad' (1845)

7 He clasps the crag with crookèd hands;
Close to the sun in lonely lands,
Ringed with the azure world, he stands.
Alfred, Lord Tennyson 1809–92: 'The Eagle' (1851)

8 I caught this morning morning's minion, kingdom of daylight's dauphin, dapple-dawn-drawn Falcon.
Gerard Manley Hopkins 1844–89: 'The Windhover' (written 1877)

9 An aged thrush, frail, gaunt, and small,
In blast-beruffled plume.
Thomas Hardy 1840–1928: 'The Darkling Thrush' (1902)

10 It was the Rainbow gave thee birth,
And left thee all her lovely hues.
W. H. Davies 1871–1940: 'Kingfisher' (1910)

11 Oh, a wondrous bird is the pelican!
His beak holds more than his belican.
He takes in his beak
Food enough for a week.
But I'll be darned if I know how the helican.
Dixon Lanier Merritt 1879–1972: in *Nashville Banner* 22 April 1913

12 And hear the pleasant cuckoo, loud and long—
The simple bird that thinks two notes a song.
W. H. Davies 1871–1940: 'April's Charms' (1916)

13 From troubles of the world
I turn to ducks
Beautiful comical things.
F. W. Harvey b. 1888: 'Ducks' (1919)

14 I do not know which to prefer,
The beauty of inflections
Or the beauty of innuendoes,
The blackbird whistling
Or just after.
Wallace Stevens 1879–1955: 'Thirteen Ways of Looking at a Blackbird' (1923)

15 It took the whole of Creation
To produce my foot, my each feather:
Now I hold Creation in my foot.
Ted Hughes 1930– : 'Hawk Roosting' (1960)

16 Blackbirds are the cellos of the deep farms.
Anne Stevenson 1933– : 'Green Mountain, Black Mountain' (1982)

Birth

1 In sorrow thou shalt bring forth children.
Bible: Genesis

2 *Maior erat natu; non omnia possumus omnes.*
He was born first; we cannot all do everything.
Lucilius *c.*180–102 BC: in Macrobius *Saturnalia*

3 The queen of Scots is this day leichter of a fair son, and I am but a barren stock.
Elizabeth I 1533–1603: to her ladies, 1566

4 Men should be bewailed at their birth, and not at their death.
Montesquieu 1689–1755: *Lettres Persones* (1721), tr. J. Ozell, 1722

5 So for the mother's sake the child was dear,
And dearer was the mother for the child.
Samuel Taylor Coleridge 1772–1834: 'Sonnet to a Friend Who Asked How I Felt When the Nurse First Presented My Infant to Me' (1797)

6 Our birth is but a sleep and a forgetting...
Not in entire forgetfulness,
And not in utter nakedness,
But trailing clouds of glory do we come.
William Wordsworth 1770–1850: 'Ode. Intimations of Immortality' (1807)

7 I had seen birth and death
But had thought they were different.
T. S. Eliot 1888–1965: 'Journey of the Magi' (1927)

8 Good work, Mary. We all knew you had it in you.
Dorothy Parker 1893–1967: telegram to Mrs Sherwood on the arrival of her baby; in A. Woollcott *While Rome Burns* (1934)

9 Death and taxes and childbirth! There's never any convenient time for any of them.
Margaret Mitchell 1900–49: *Gone with the Wind* (1936)

10 I am not yet born; O fill me
With strength against those who would freeze my humanity.
Louis MacNeice 1907–63: 'Prayer Before Birth' (1944)

11 Let them not make me a stone and let them not spill me,
Otherwise kill me.
Louis MacNeice 1907–63: 'Prayer Before Birth' (1944)

12 It's all any reasonable child can expect if the dad is present at the conception.
Joe Orton 1933–67: *Entertaining Mr Sloane* (1964)

13 If men had to have babies, they would only ever have one each.
Diana, Princess of Wales 1961– : *Observer* 29 July 1984 'Sayings of the Week'

The Body

1 I will give thanks unto thee, for I am fearfully and wonderfully made.
Bible: Psalm 139

2 Doth not even nature itself teach you, that if a man have long hair, it is a shame unto him?
But if a woman have long hair, it is a glory to her.
Bible: I Corinthians

3 Falstaff sweats to death
And lards the lean earth as he walks along.
William Shakespeare 1564–1616: *Henry IV, Part 1* (1597)

4 Thou seest I have more flesh than another man, and therefore more frailty.
William Shakespeare 1564–1616: *Henry IV, Part 1* (1597)

5 There's no art
To find the mind's construction in the face.
William Shakespeare 1564–1616: *Macbeth* (1606)

6 Fain would I kiss my Julia's dainty leg,
Which is as white and hairless as an egg.
Robert Herrick 1591–1674: 'On Julia's Legs' (1648)

7 Had Cleopatra's nose been shorter, the whole face of the world would have changed.
Blaise Pascal 1623–62: *Pensées* (1670)

8 Our body is a machine for living. It is organized for that, it is its nature. Let life go on in it unhindered and let it defend itself.
Leo Tolstoy 1828–1910: *War and Peace* (1865–9), tr. A. and L. Maude

9 I am the family face;
Flesh perishes, I live on.
Thomas Hardy 1840–1928: 'Heredity' (1917)

10 Anatomy is destiny.
Sigmund Freud 1856–1939: *Collected Writings* (1924)

11 i like my body when it is with your
body. It is so quite new a thing.
Muscles better and nerves more.
e. e. cummings 1894–1962: 'Sonnets-Actualities' no. 8 (1925)

12 I'm fat, but I'm thin inside. Has it ever struck you that there's a thin man inside every fat man, just as they say there's a statue inside every block of stone?
George Orwell 1903–50: *Coming up For Air* (1939)

13 At 50, everyone has the face he deserves.
George Orwell 1903–50: last words in his notebook, 17 April 1949

14 I'd the upbringing a nun would envy... Until I was fifteen I was more familiar with Africa than my own body.
Joe Orton 1933–67: *Entertaining Mr Sloane* (1964)

15 Your cameraman might enjoy himself because my face looks like a wedding-cake left out in the rain.
W. H. Auden 1907–1973: in H. Carpenter *W. H. Auden* (1963)

Books

See also **Libraries, Reading, Writing**

1 Of making many books there is no end; and much study is a weariness of the flesh.
Bible: Ecclesiastes

2 A great book is like great evil.
Callimachus *c.*305–*c.*240 BC: proverbially 'Great book, great evil'

3 Books will speak plain when counsellors blanch.
Francis Bacon 1561–1626: *Essays* (1625) 'Of Counsel'

4 Some books are to be tasted, others to be swallowed, and some few to be chewed and digested.
Francis Bacon 1561–1626: *Essays* (1625) 'Of Studies'

5 A good book is the precious life-blood of a master spirit.
John Milton 1608–74: *Areopagitica* (1644)

6 Deep-versed in books and shallow in himself.
John Milton 1608-74: *Paradise Regained* (1671)

7 Never literary attempt was more unfortunate than my Treatise of Human Nature. It fell *dead-born from the press.*
David Hume 1711-76: *My Own Life* (1777)

8 You shall see them on a beautiful quarto page where a neat rivulet of text shall meander through a meadow of margin.
Richard Brinsley Sheridan 1751-1816: *The School for Scandal* (1777)

9 Dictionaries are like watches, the worst is better than none, and the best cannot be expected to go quite true.
Samuel Johnson 1709-84: letter to Francesco Sastres, 21 August 1784

10 Publish and be damned.
Duke of Wellington 1769-1852: replying to Harriette Wilson's blackmail threat, *c.*1825; attributed

11 Though an angel should write, still 'tis *devils* must print.
Thomas Moore 1779-1852: *The Fudges in England* (1835)

12 A good book is the best of friends, the same to-day and for ever.
Martin Tupper 1810-89: *Proverbial Philosophy* Series I (1838) 'Of Reading'

13 Now Barabbas was a publisher.
Thomas Campbell 1777-1844: attributed, in Samuel Smiles *A Publisher and his Friends* (1891); also attributed, wrongly, to Byron

14 No furniture so charming as books.
Sydney Smith 1771-1845: in Lady Holland *Memoir* (1855)

15 Books are made not like children but like pyramids... and they're just as useless! and they stay in the desert!... Jackals piss at their foot and the bourgeois climb up on them.
Gustave Flaubert 1821-80: letter to Ernest Feydeau, November/December 1857

16 'What is the use of a book', thought Alice, 'without pictures or conversations?'
Lewis Carroll 1832-98: *Alice's Adventures in Wonderland* (1865)

17 All books are divisible into two classes, the books of the hour, and the books of all time.
John Ruskin 1819-1900: *Sesame and Lilies* (1865)

18 There is no such thing as a moral or an immoral book. Books are well written, or badly written.

Oscar Wilde 1854–1900: *The Picture of Dorian Gray* (1891)

19 Child! do not throw this book about;
Refrain from the unholy pleasure
Of cutting all the pictures out!

Hilaire Belloc 1870–1953: *A Bad Child's Book of Beasts* (1896)

20 '*Classic.*' A book which people praise and don't read.

Mark Twain 1835–1910: *Following the Equator* (1897)

21 To my daughter Leonora without whose never-failing sympathy and encouragement this book would have been finished in half the time.

P. G. Wodehouse 1881–1975: *The Heart of a Goof* (1926) dedication

22 From the moment I picked up your book until I laid it down, I was convulsed with laughter. Some day I intend reading it.

Groucho Marx 1895–1977: blurb written for S. J. Perelman *Dawn Ginsberg's Revenge* (1928)

23 This is an important book, the critic assumes, because it deals with war. This is an insignificant book because it deals with the feelings of women in a drawing-room.

Virginia Woolf 1882–1941: *A Room of One's Own* (1929)

24 A best-seller is the gilded tomb of a mediocre talent.

Logan Pearsall Smith 1865–1946: *Afterthoughts* (1931) 'Art and Letters'

25 I would sooner read a time-table or a catalogue than nothing at all... They are much more entertaining than half the novels that are written.

W. Somerset Maugham 1874–1965: *Summing Up* (1938)

26 The principle of procrastinated rape is said to be the ruling one in all the great best-sellers.

V. S. Pritchett 1900– : *The Living Novel* (1946)

27 I have known her pass the whole evening without mentioning a single book, or *in fact anything unpleasant*, at all.

Henry Reed 1914–86: *A Very Great Man Indeed* (1953 radio play)

28 Some books are undeservedly forgotten; none are undeservedly remembered.
W. H. Auden 1907–73: *The Dyer's Hand* (1963) 'Reading'

29 Books do furnish a room.
Anthony Powell 1905– : title of novel (1971)

30 Far too many relied on the classic formula of a beginning, a muddle, and an end.
Philip Larkin 1922–85: of the books entered for the 1977 Booker Prize; *New Fiction* January 1978

31 Books say: she did this because. Life says: she did this. Books are where things are explained to you; life is where things aren't.
Julian Barnes 1946– : *Flaubert's Parrot* (1984)

Bores and Boredom

1 The secret of being a bore... is to tell everything.
Voltaire 1694–1778: *Discours en vers sur l'homme* (1737)

2 Gentle Dullness ever loves a joke.
Alexander Pope 1688–1744: *The Dunciad* (1742)

3 He is not only dull in himself, but the cause of dullness in others.
Samuel Foote 1720–77: of a dull law lord, in James Boswell *Life of Samuel Johnson*

4 Society is now one polished horde,
Formed of two mighty tribes, the *Bores* and *Bored*.
Lord Byron 1788–1824: *Don Juan* (1819–24)

5 Bore, *n.* A person who talks when you wish him to listen.
Ambrose Bierce 1842–?1914: *Cynic's Word Book* (1906)

6 He is an old bore. Even the grave yawns for him.
Herbert Beerbohm Tree 1852–1917: of Israel Zangwill, in Max Beerbohm *Herbert Beerbohm Tree* (1920)

7 Boredom is... a vital problem for the moralist, since half the sins of mankind are caused by the fear of it.
Bertrand Russell 1872–1970: *The Conquest of Happiness* (1930)

8 Millions long for immortality who don't know what to do with themselves on a rainy Sunday afternoon.
Susan Ertz 1894–1985: *Anger in the Sky* (1943)

9 The effect of boredom on a large scale in history is underestimated. It is a main cause of revolutions, and would soon bring to an end all the static Utopias and the farmyard civilization of the Fabians.
Dean Inge 1860–1954: *End of an Age* (1948)

10 Nothing happens, nobody comes, nobody goes, it's awful!
Samuel Beckett 1906–89: *Waiting for Godot* (1955)

11 Nothing, like something, happens anywhere.
Philip Larkin 1922–1985: 'I Remember, I Remember' (1955)

12 Punctuality is the virtue of the bored.
Evelyn Waugh 1903–66: diary, 26 March 1962

13 Life, friends, is boring. We must not say so...
And moreover my mother taught me as a boy
(repeatingly) 'Ever to confess you're bored
means you have no
Inner Resources.'
John Berryman 1914–72: *77 Dream Songs* (1964)

14 A healthy male adult bore consumes *each year* one and a half times his own weight in other people's patience.
John Updike 1932– : *Assorted Prose* (1965)

15 Everyone is a bore to someone. That is unimportant. The thing to avoid is being a bore to oneself.
Gerald Brenan 1894–1987: *Thoughts in a Dry Season* (1978)

Britain

See also **England**

1 Rule, Britannia, rule the waves;
Britons never will be slaves.
James Thomson 1700–48: *Alfred: a Masque* (1740)

2 It must be owned, that the Graces do not seem to be natives of Great Britain; and I doubt, the best of us here have more of rough than polished diamond.
Lord Chesterfield 1694–1773: *Letters to his Son* (1774) 18 November 1748

3 What is our task? To make Britain a fit country for heroes to live in.

David Lloyd George 1863–1945: speech at Wolverhampton, 23 November 1918

4 Other nations use 'force'; we Britons alone use 'Might'.

Evelyn Waugh 1903–66: *Scoop* (1938)

5 The British nation is unique in this respect. They are the only people who like to be told how bad things are, who like to be told the worst.

Winston Churchill 1874–1965: speech, House of Commons, 10 June 1941

6 Britain will be honoured by historians more for the way she disposed of an empire than for the way in which she acquired it.

Lord Harlech 1918–85: in *New York Times* 28 October 1962

7 Great Britain has lost an empire and has not yet found a role.

Dean Acheson 1893–1971: speech at the Military Academy, West Point, 5 December 1962

8 The land of embarrassment and breakfast.

Julian Barnes 1946– : *Flaubert's Parrot* (1984)

Broadcasting

1 Nation shall speak peace unto nation.

Montague John Rendall 1862–1950: motto of the BBC (1927)

2 The whole Fleet's lit up. When I say 'lit up', I mean lit up by fairy lamps.

Thomas Woodroofe 1899–1978: first live outside broadcast, Spithead Review, 20 May 1937

3 TV—a clever contraction derived from the words Terrible Vaudeville. However, it is our latest medium—we call it a medium because nothing's well done.

Goodman Ace 1899–1982: letter to Groucho Marx, *c.*1953

4 So much chewing gum for the eyes.

Anonymous: small boy's definition of certain television programmes, in J. B. Simpson *Best Quotes* (1957)

5 Today, thanks to technical progress, the radio and television, to which we devote so many of the leisure hours once spent listening to parlour chatter and parlour music, have succeeded in lifting the manufacture of banality out of the sphere of handicraft and placed it in that of a major industry.
Nathalie Sarraute 1902- : *Times Literary Supplement* 10 June 1960

6 Like having your own licence to print money.
Roy Thomson 1894-1976: on the profitability of commercial television in Britain; in R. Braddon *Roy Thomson* (1965)

7 Television has brought back murder into the home—where it belongs.
Alfred Hitchcock 1899-1980: in *Observer* 19 December 1965

8 Television contracts the imagination and radio expands it.
Terry Wogan 1938- : *Observer* 30 December 1984 'Sayings of the Year'

9 They [men] are happier with women who make their coffee than make their programmes.
Denise O'Donoghue: in G. Kinnock and F. Miller *By Faith and Daring* (1993)

Bureaucracy

See also **Management**

1 Whatever was required to be done, the Circumlocution Office was beforehand with all the public departments in the art of perceiving—HOW NOT TO DO IT.
Charles Dickens 1812-70: *Little Dorrit* (1857)

2 Where there is officialism every human relationship suffers.
E. M. Forster 1879-1970: *A Passage to India* (1924)

3 Here lies a civil servant. He was civil
To everyone, and servant to the devil.
C. H. Sisson 1914- : in *The London Zoo* (1961)

4 The Civil Service is profoundly deferential—'Yes, Minister! No, Minister! If you wish it, Minister!'
Richard Crossman 1907-74: diary, 22 October 1964

5 Guidelines for bureaucrats: (1) When in charge, ponder. (2) When in trouble, delegate. (3) When in doubt, mumble.
James H. Boren 1925– : in *New York Times* 8 November 1970

6 A memorandum is written not to inform the reader but to protect the writer.
Dean Acheson 1893–1971: in *Wall Street Journal* 8 September 1977

Business and Commerce

1 A merchant shall hardly keep himself from doing wrong.
Bible (Apocrypha): Ecclesiasticus

2 Neither a borrower, nor a lender be.
William Shakespeare 1564–1616: *Hamlet* (1601)

3 They [corporations] cannot commit treason, nor be outlawed, nor excommunicate, for they have no souls.
Edward Coke 1552–1634: *Reports of Sir Edward Coke* (1658) 'The case of Sutton's Hospital'

4 A Company for carrying on an undertaking of Great Advantage, but no one to know what it is.
Anonymous: The South Sea Company Prospectus (1711)

5 There is nothing more requisite in business than dispatch.
Joseph Addison 1672–1719: *The Drummer* (1716)

6 I have heard of a man who had a mind to sell his house, and therefore carried a piece of brick in his pocket, which he showed as a pattern to encourage purchasers.
Jonathan Swift 1667–1745: *The Drapier's Letters* (1724)

7 Necessity never made a good bargain.
Benjamin Franklin 1706–90: *Poor Richard's Almanac* (1735)

8 Remember that time is money.
Benjamin Franklin 1706–90: *Advice to a Young Tradesman* (1748)

9 It is the nature of all greatness not to be exact; and great trade will always be attended with considerable abuses.
Edmund Burke 1729–97: *On American Taxation* (1775)

10 A nation of shop-keepers are very seldom so disinterested.
Samuel Adams 1722–1803: *Oration in Philadelphia* 1 August 1776 (of doubtful authenticity)

11 People of the same trade seldom meet together, even for merriment and diversion, but the conversation ends in a conspiracy against the public, or in some contrivance to raise prices.
Adam Smith 1723–90: *Wealth of Nations* (1776)

12 To found a great empire for the sole purpose of raising up a people of customers, may at first sight appear a project fit only for a nation of shopkeepers. It is, however, a project altogether unfit for a nation of shopkeepers; but extremely fit for a nation whose government is influenced by shopkeepers.
Adam Smith 1723–90: *Wealth of Nations* (1776)

13 Here's the rule for bargains: 'Do other men, for they would do you.' That's the true business precept.
Charles Dickens 1812–70: *Martin Chuzzlewit* (1844) Jonas Chuzzlewit

14 It is because we put up with bad things that hotel-keepers continue to give them to us.
Anthony Trollope 1815–82: *Orley Farm* (1862)

15 Earned a precarious living by taking in one another's washing.
Anonymous: attributed to Mark Twain by William Morris, in *The Commonweal* 6 August 1887

16 The growth of a large business is merely a survival of the fittest...The American beauty rose can be produced in the splendour and fragrance which bring cheer to its beholder only by sacrificing the early buds which grow up around it.
John D. Rockefeller 1839–1937: in W. J. Ghent *Our Benevolent Feudalism* (1902); 'American Beauty Rose' became the title of a 1950 song by Hal David and others

17 The customer is never wrong.
César Ritz 1850–1918: in R. Nevill and C. E. Jerningham *Piccadilly to Pall Mall* (1908)

18 The best of all monopoly profits is a quiet life.
J. R. Hicks 1904– : *Econometrica* (1935)

19 He's a man way out there in the blue, riding on a smile and a shoeshine. And when they start not smiling back—that's an

earthquake... A salesman is got to dream, boy. It comes with the territory.
Arthur Miller 1915- : *Death of a Salesman* (1949)

20 For years I thought what was good for our country was good for General Motors and vice versa.
Charles E. Wilson 1890–1961: testimony to the Senate Armed Services Committee, 15 January 1953

21 Few have heard of Fra Luca Pacioli, the inventor of double-entry book-keeping; but he has probably had much more influence on human life than has Dante or Michelangelo.
Herbert J. Muller 1905- : *Uses of the Past* (1957)

22 Accountants are the witch-doctors of the modern world and willing to turn their hands to any kind of magic.
Lord Justice Harman 1894–1970: speech, February 1964, in A. Sampson *The New Anatomy of Britain* (1971)

23 [Commercialism is] doing well that which should not be done at all.
Gore Vidal 1925- : in *Listener* 7 August 1975

24 We even sell a pair of earrings for under £1, which is cheaper than a prawn sandwich from Marks & Spencers. But I have to say the earrings probably won't last as long.
Gerald Ratner 1949- : speech to the Institute of Directors, Albert Hall, 23 April 1991

25 The green shoots of economic spring are appearing once again.
Norman Lamont 1942- : speech at Conservative Party Conference, 9 October 1991; often quoted 'the green shoots of recovery'

Careers

See also **Work**

1 For promotion cometh neither from the east, nor from the west: nor yet from the south.
Bible: Psalm 75

2 I hold every man a debtor to his profession.
Francis Bacon 1561–1626: *The Elements of the Common Law* (1596)

3 Thou art not for the fashion of these times,
Where none will sweat but for promotion.
William Shakespeare 1564–1616: *As You Like It* (1599)

4 It is wonderful, when a calculation is made, how little the mind is actually employed in the discharge of any profession.
Samuel Johnson 1709–84: in James Boswell *Life of Johnson* (1791) 6 April 1775

5 All professions are conspiracies against the laity.
George Bernard Shaw 1856–1950: *The Doctor's Dilemma* (1911)

6 The test of a vocation is the love of the drudgery it involves.
Logan Pearsall Smith 1865–1946: *Afterthoughts* (1931)

7 Professional men, they have no cares;
Whatever happens, they get theirs.
Ogden Nash 1902–71: 'I Yield to My Learned Brother' (1935)

8 I will undoubtedly have to seek what is happily known as gainful employment, which I am glad to say does not describe holding public office.
Dean Acheson 1893–1971: in *Time* 22 December 1952

Cats

See also **Animals**

1 When I play with my cat, who knows whether she isn't amusing herself with me more than I am with her?
Montaigne 1533–92: *Essais* (1580)

2 For I will consider my Cat Jeoffry...
For he counteracts the powers of darkness by his electrical skin and glaring eyes.
For he counteracts the Devil, who is death, by brisking about the life.
Christopher Smart 1722–71: *Jubilate Agno* (*c.*1758–63)

3 Cruel, but composed and bland,
Dumb, inscrutable and grand,

So Tiberius might have sat,
Had Tiberius been a cat.
Matthew Arnold 1822–88: 'Poor Matthias' (1885)

4 He walked by himself, and all places were alike to him.
Rudyard Kipling 1865–1936: *Just So Stories* (1902) 'The Cat that Walked by Himself'

5 Cats, no less liquid than their shadows,
Offer no angles to the wind.
A. S. J. Tessimond 1902–62: *Cats* (1934)

6 The trouble with a kitten is
THAT
Eventually it becomes a
CAT.
Ogden Nash 1902–71: 'The Kitten' (1940)

7 Cats seem to go on the principle that it never does any harm to ask for what you want.
Joseph Wood Krutch 1893–1970: *Twelve Seasons* (1949)

Censorship

1 As good almost kill a man as kill a good book: who kills a man kills a reasonable creature, God's image; but he who destroys a good book, kills reason itself, kills the image of God, as it were in the eye.
John Milton 1608–74: *Areopagitica* (1644)

2 I disapprove of what you say, but I will defend to the death your right to say it.
Voltaire 1694–1778: attributed to Voltaire, but actually S. G. Tallentyre's summary of Voltaire's attitude towards Helvétius following the burning of the latter's *De l'esprit* in 1759; in *The Friends of Voltaire* (1907)

3 What a fuss about an omelette!
Voltaire 1694–1778: what Voltaire *apparently* said on the burning of *De l'esprit*; in J. Parton *Life of Voltaire* (1881)

4 Wherever books will be burned, men also, in the end, are burned.
Heinrich Heine 1797–1856: *Almansor* (1823)

5 Assassination is the extreme form of censorship.
George Bernard Shaw 1856–1950: *The Showing-Up of Blanco Posnet* (1911)

6 The power of the press is very great, but not so great as the power of suppress.
Lord Northcliffe 1865–1922: office message, *Daily Mail* 1918

7 It is obvious that 'obscenity' is not a term capable of exact legal definition; in the practice of the Courts, it means 'anything that shocks the magistrate'.
Bertrand Russell 1872–1970: *Sceptical Essays* (1928) 'The Recrudescence of Puritanism'

8 [This film] is so cryptic as to be almost meaningless. If there is a meaning, it is doubtless objectionable.
British Board of Film Censors: banning Jean Cocteau's film *The Seashell and the Clergyman* (1929)

9 We all know that books burn—yet we have the greater knowledge that books can not be killed by fire. People die, but books never die. No man and no force can abolish memory. No man and no force can put thought in a concentration camp forever.
Franklin D. Roosevelt 1882–1945: 'Message to the Booksellers of America' 6 May 1942

10 Is it a book you would even wish your wife or your servants to read?
Mervyn Griffith-Jones 1909–79: of D. H. Lawrence's *Lady Chatterley's Lover*, while appearing for the prosecution at the Old Bailey; in *The Times* 21 October 1960

11 One has to multiply thoughts to the point where there aren't enough policemen to control them.
Stanislaw Lec 1909–66: *Unkempt Thoughts* (1962), tr. J. Galazka

Certainty and Doubt

1 How long halt ye between two opinions?
Bible: I Kings

2 Probable impossibilities are to be preferred to improbable possibilities.
Aristotle 384–322 BC: *Poetics*

3 Now, the melancholy god protect thee, and the tailor make thy doublet of changeable taffeta, for thy mind is a very opal.
William Shakespeare 1564–1616: *Twelfth Night* (1601)

4 If a man will begin with certainties, he shall end in doubts; but if he will be content to begin with doubts, he shall end in certainties.
Francis Bacon 1561–1626: *The Advancement of Learning* (1605)

5 I wish I was as cocksure of anything as Tom Macaulay is of everything.
Lord Melbourne 1779–1848: in Lord Cowper's preface to *Lord Melbourne's Papers* (1889)

6 Ah, what a dusty answer gets the soul
When hot for certainties in this our life!
George Meredith 1828–1909: *Modern Love* (1862)

7 Ten thousand difficulties do not make one doubt.
Cardinal Newman 1801–90: *Apologia pro Vita Sua* (1864)

8 I must have a prodigious quantity of mind; it takes me as much as a week, sometimes, to make it up.
Mark Twain 1835–1910: *The Innocents Abroad* (1869)

9 I am too much of a sceptic to deny the possibility of anything.
T. H. Huxley 1825–95: letter to H. Spencer, 22 March 1886

10 There is no more miserable human being than one in whom nothing is habitual but indecision.
William James 1842–1910: *Principles of Psychology* (1890)

11 Oh! let us never, never doubt
What nobody is sure about!
Hilaire Belloc 1870–1953: 'The Microbe' (1897)

12 Life is doubt,
And faith without doubt is nothing but death.
Miguel de Unamuno 1864–1937: 'Salmo II' (1907)

13 Like all weak men he laid an exaggerated stress on not changing one's mind.
W. Somerset Maugham 1874–1965: *Of Human Bondage* (1915)

14 [The Government] go on in strange paradox, decided only to be undecided, resolved to be irresolute, adamant for drift, solid for fluidity.
Winston Churchill 1874–1965: speech, House of Commons, 12 November 1936

15 My mind is not a bed to be made and re-made.
James Agate 1877–1947: diary, 9 June 1943

16 I'll give you a definite maybe.
Sam Goldwyn 1882–1974: attributed

17 The archbishop is usually to be found nailing his colours to the fence.
Frank Field 1942– : of Archbishop Runcie; attributed in *Crockfords 1987/88* (1987)

Chance

1 Cast thy bread upon the waters: for thou shalt find it after many days.
Bible: Ecclesiastes

2 And a certain man drew a bow at a venture, and smote the king of Israel between the joints of the harness.
Bible: I Kings

3 But for the grace of God there goes John Bradford.
John Bradford *c.*1510–55: on seeing a group of criminals being led to their execution; in *Dictionary of National Biography* (1917–), usually quoted 'There but for the grace of God go I'

4 There is a tide in the affairs of men,
Which, taken at the flood, leads on to fortune;
Omitted, all the voyage of their life
Is bound in shallows and in miseries.
William Shakespeare 1564–1616: *Julius Caesar* (1599)

5 The chapter of knowledge is a very short, but the chapter of accidents is a very long one.
Lord Chesterfield 1694–1773: letter to Solomon Dayrolles, 16 February 1753

6 The best laid schemes o' mice an' men
Gang aft a-gley.
Robert Burns 1759–96: 'To a Mouse' (1786)

7 O! many a shaft, at random sent,
Finds mark the archer little meant!
And many a word, at random spoken,
May soothe or wound a heart that's broken.
Sir Walter Scott 1771–1832: *The Lord of the Isles* (1813)

8 At this moment he was unfortunately called out by a person on business from Porlock.
Samuel Taylor Coleridge 1772–1834: 'Kubla Khan' (1816) preliminary note, explaining why the poem remained unfinished

9 Accidents will occur in the best-regulated families.
Charles Dickens 1812–70: *David Copperfield* (1850) Mr Micawber

10 I am convinced that *He* [God] does not play dice.
Albert Einstein 1879–1955: letter to Max Born, 4 December 1926

11 Predictability: Does the flap of a butterfly's wings in Brazil set off a tornado in Texas?
Edward N. Lorenz: title of paper given to the American Association for the Advancement of Science, Washington, 29 December 1979

Change

See also **Beginnings and Endings**

1 Can the Ethiopian change his skin, or the leopard his spots?
Bible: Jeremiah

2 You can't step twice into the same river.
Heraclitus *c.*540–*c.*480 BC: in Plato *Cratylus*

3 *Tempora mutantur, et nos mutamur in illis.*
Times change, and we change with them.
Anonymous: in William Harrison *Description of Britain* (1577)

4 O! swear not by the moon, the inconstant moon,
That monthly changes in her circled orb,
Lest that thy love prove likewise variable.
William Shakespeare 1564–1616: *Romeo and Juliet* (1595)

5 He that will not apply new remedies must expect new evils; for time is the greatest innovator.
Francis Bacon 1561–1626: *Essays* (1625) 'Of Innovations'

6 Tomorrow to fresh woods, and pastures new.
John Milton 1608–74: 'Lycidas' (1638)

7 When it is not necessary to change, it is necessary not to change.
Lucius Cary, Viscount Falkland 1610–43: 'A Speech concerning Episcopacy' (1641)

8 Change is not made without inconvenience, even from worse to better.
Samuel Johnson 1709–84: *Dictionary of the English Language* (1755) preface

9 If we do not find anything pleasant, at least we shall find something new.
Voltaire 1694–1778: *Candide* (1759)

10 Variety's the very spice of life,
That gives it all its flavour.
William Cowper 1731–1800: *The Task* (1785) 'The Timepiece'

11 There is nothing stable in the world—uproar's your only music.
John Keats 1795–1821: letter to George and Thomas Keats, 13 January 1818

12 There is a certain relief in change, even though it be from bad to worse...it is often a comfort to shift one's position and be bruised in a new place.
Washington Irving 1783–1859: *Tales of a Traveller* (1824)

13 Let the great world spin for ever down the ringing grooves of change.
Alfred, Lord Tennyson 1809–92: 'Locksley Hall' (1842)

14 Change and decay in all around I see;
O Thou, who changest not, abide with me.
Henry Francis Lyte 1793–1847: 'Abide with Me' (*c.*1847)

15 *Plus ça change, plus c'est la même chose.*
The more things change, the more they are the same.
Alphonse Karr 1808–90: *Les Guêpes* January 1849

16 Wandering between two worlds, one dead,
The other powerless to be born.
Matthew Arnold 1822–88: 'Stanzas from the Grande Chartreuse' (1855)

17 It is best not to swap horses when crossing streams.
Abraham Lincoln 1809–65: reply to National Union League, 9 June 1864

18 There is in all change something at once sordid and agreeable, which smacks of infidelity and household removals. This is sufficient to explain the French Revolution.
Charles Baudelaire 1821–67: *Journaux intimes* (1887), tr. Christopher Isherwood

19 The old order changeth, yielding place to new,
And God fulfils himself in many ways,
Lest one good custom should corrupt the world.
Alfred, Lord Tennyson 1809–92: *Idylls of the King* 'The Passing of Arthur' (1869)

20 Most of the change we think we see in life
Is due to truths being in and out of favour.
Robert Frost 1874–1963: 'The Black Cottage' (1914)

21 All changed, changed utterly:
A terrible beauty is born.
W. B. Yeats 1865–1939: 'Easter, 1916' (1921)

22 The whole worl's in a state o' chassis!
Sean O'Casey 1880–1964: *Juno and the Paycock* (1925)

23 Consistency is contrary to nature, contrary to life. The only completely consistent people are the dead.
Aldous Huxley 1894–1963: *Do What You Will* (1929)

24 In olden days a glimpse of stocking
Was looked on as something shocking
Now, heaven knows,
Anything goes.
Cole Porter 1891–1964: 'Anything Goes' (1934 song)

25 God, give us the serenity to accept what cannot be changed;
Give us the courage to change what should be changed;
Give us the wisdom to distinguish one from the other.
Reinhold Niebuhr 1892–1971: prayer said to have been first published in 1951; in R. W. Fox *Reinhold Niebuhr* (1985)

26 If we want things to stay as they are, things will have to change.
Giuseppe di Lampedusa 1896–1957: *The Leopard* (1957)

Character

1 A man's character is his fate.
Heraclitus *c.*540–*c.*480 BC: *On the Universe*, tr. W. H. S. Jones; see also Novalis (1772–1801) *Heinrich von Ofterdingen* (1802)

2 I see the better things, and approve; I follow the worse.
Ovid 43 BC–AD *c.*17: *Metamorphoses*

3 Give me that man
That is not passion's slave, and I will wear him
In my heart's core, ay, in my heart of heart,
As I do thee.
William Shakespeare 1564–1616: *Hamlet* (1601)

4 My nature is subdued
To what it works in, like the dyer's hand.
William Shakespeare 1564–1616: sonnet 111 (1609)

5 Youth, what man's age is like to be doth show;
We may our ends by our beginnings know.
John Denham 1615–69: 'Of Prudence' (1668)

6 A fiery soul, which working out its way,
Fretted the pigmy body to decay:
And o'er informed the tenement of clay.
John Dryden 1631–1700: *Absalom and Achitophel* (1681)

7 She's as headstrong as an allegory on the banks of the Nile.
Richard Brinsley Sheridan 1751–1816: *The Rivals* (1775)

8 It is not in the still calm of life, or the repose of a pacific station, that great characters are formed... Great necessities call out great virtues.
Abigail Adams 1744–1818: letter to John Quincy Adams, 19 January 1780

9 Talent develops in quiet places, character in the full current of human life.
Johann Wolfgang von Goethe 1749–1832: *Torquato Tasso* (1790)

10 What is character but the determination of incident? What is incident but the illustration of character?
Henry James 1843–1916: *Partial Portraits* (1888) 'The Art of Fiction'

11 If you can fill the unforgiving minute
With sixty seconds' worth of distance run,
Yours is the Earth and everything that's in it,
And—which is more—you'll be a Man, my son!
Rudyard Kipling 1865–1936: 'If—' (1910)

12 A very weak-minded fellow I am afraid, and, like the feather pillow, bears the marks of the last person who has sat on him!
Earl Haig 1861–1928: describing the 17th Earl of Derby in a letter to Lady Haig, 14 January 1918

13 The true index of a man's character is the health of his wife.
Cyril Connolly 1903–74: *The Unquiet Grave* (1944)

14 It is the nature, and the advantage, of strong people that they can bring out the crucial questions and form a clear opinion about them. The weak always have to decide between alternatives that are not their own.
Dietrich Bonhoeffer 1906–45: *Widerstand und Ergebung* (1951), tr. R. Fuller

15 He's so wet you could shoot snipe off him.
Anthony Powell 1905– : *A Question of Upbringing* (1951)

16 If you can't stand the heat, get out of the kitchen.
Harry Vaughan: in *Time* 28 April 1952 (associated with Harry S. Truman, but attributed by him to Vaughan, his 'military jester')

17 A thick skin is a gift from God.
Konrad Adenauer 1876–1967: in *New York Times* 30 December 1959

18 Those who stand for nothing fall for anything.
Alex Hamilton 1936– : 'Born Old' (radio broadcast), in *Listener* 9 November 1978

19 You can tell a lot about a fellow's character by his way of eating jellybeans.
Ronald Reagan 1911– : in *New York Times* 15 January 1981

Charm

1 Charm... it's a sort of bloom on a woman. If you have it, you don't need to have anything else; and if you don't have it, it doesn't much matter what else you have.
J. M. Barrie 1860–1937: *What Every Woman Knows* (1908)

2 All charming people have something to conceal, usually their total dependence on the appreciation of others.
Cyril Connolly 1903–1974: *Enemies of Promise* (1938)

3 Charm is the great English blight. It does not exist outside these damp islands. It spots and kills anything it touches. It kills love, it kills art.
Evelyn Waugh 1903–66: *Brideshead Revisited* (1945)

4 Oozing charm from every pore,
He oiled his way around the floor.
Alan Jay Lerner 1918–86: 'You Did It' (1956 song)

5 You know what charm is: a way of getting the answer yes without having asked any clear question.
Albert Camus 1913–60: *The Fall* (1957)

6 What is charm then?... something extra, superfluous, unnecessary, essentially a power thrown away.
Doris Lessing 1919– : *Particularly Cats* (1967)

Children

See also **The Family, Parents, Youth**

1 Like as the arrows in the hand of the giant: even so are the young children.
Happy is the man that hath his quiver full of them.
Bible: Psalm 127

2 Train up a child in the way he should go: and when he is old, he will not depart from it.
Bible: Proverbs

3 Suffer the little children to come unto me, and forbid them not: for of such is the kingdom of God.
Bible: St Mark

4 When I was a child, I spake as a child, I understood as a child, I thought as a child: but when I became a man, I put away childish things.
Bible: I Corinthians

5 A child is owed the greatest respect; if you ever have something disgraceful in mind, don't ignore your son's tender years.
Juvenal AD *c.*60–*c.*130: *Satires*

6 It should be noted that children at play are not playing about; their games should be seen as their most serious-minded activity.
Montaigne 1533–92: *Essais* (1580)

7 My son—and what's a son? A thing begot
Within a pair of minutes, thereabout,
A lump bred up in darkness.
Thomas Kyd 1558–94: *The Spanish Tragedy* The Third Addition (1602 ed.)

8 At first the infant,
Mewling and puking in the nurse's arms.
And then the whining schoolboy, with his satchel,
And shining morning face, creeping like snail
Unwillingly to school.
William Shakespeare 1564–1616: *As You Like It* (1599)

9 How sharper than a serpent's tooth it is
To have a thankless child!
William Shakespeare 1564–1616: *King Lear* (1605–6)

10 Children sweeten labours, but they make misfortunes more bitter.
Francis Bacon 1561–1626: *Essays* (1625) 'Of Parents and Children'

11 Men are generally more careful of the breed of their horses and dogs than of their children.
William Penn 1644–1718: *Some Fruits of Solitude* (1693)

12 Behold the child, by Nature's kindly law
Pleased with a rattle, tickled with a straw.
Alexander Pope 1688–1744: *An Essay on Man* Epistle 2 (1733)

13 Alas, regardless of their doom,
The little victims play!

No sense have they of ills to come,
Nor care beyond to-day.

Thomas Gray 1716–71: *Ode on a Distant Prospect of Eton College* (1747)

14 The Child is father of the Man.

William Wordsworth 1770–1850: 'My heart leaps up when I behold' (1807)

15 A child's a plaything for an hour.

Charles Lamb 1775–1834: 'Parental Recollections' (1809); often attributed to Lamb's sister Mary

16 The place is very well and quiet and the children only scream in a low voice.

Lord Byron 1788–1824: letter to Lady Melbourne, 21 September 1813

17 In the little world in which children have their existence, whosoever brings them up, there is nothing so finely perceived and so finely felt, as injustice.

Charles Dickens 1812–70: *Great Expectations* (1861)

18 Go practise if you please
With men and women: leave a child alone
For Christ's particular love's sake!

Robert Browning 1812–89: *The Ring and the Book* (1868–9)

19 Oh, for an hour of Herod!

Anthony Hope 1863–1933: at the first night of *Peter Pan* (1904); in D. Mackail *Story of JMB* (1941)

20 Children with Hyacinth's temperament don't know better as they grow older; they merely know more.

Saki (H. H. Munro) 1870–1916: *Toys of Peace and Other Papers* (1919)

21 Childhood is the kingdom where nobody dies.
Nobody that matters, that is.

Edna St Vincent Millay 1892–1950: 'Childhood is the Kingdom where Nobody dies' (1934)

22 There is no end to the violations committed by children on children, quietly talking alone.

Elizabeth Bowen 1899–1973: *The House in Paris* (1935)

23 There is no more sombre enemy of good art than the pram in the hall.

Cyril Connolly 1903–74: *Enemies of Promise* (1938)

24 Any man who hates dogs and babies can't be all bad.
Leo Rosten 1908– : of W. C. Fields, and often attributed to him, in speech at Masquers' Club dinner, 16 February 1939

25 There is always one moment in childhood when the door opens and lets the future in.
Graham Greene 1904–91: *The Power and the Glory* (1940)

26 Grown-ups never understand anything for themselves, and it is tiresome for children to be always and forever explaining things to them.
Antoine de Saint-Exupéry 1900–44: *Le Petit Prince* (1943)

27 A loud noise at one end and no sense of responsibility at the other.
Monsignor Ronald Knox 1888–1957: definition of a baby (attributed)

28 Literature is mostly about having sex and not much about having children. Life is the other way round.
David Lodge 1935– : *The British Museum is Falling Down* (1965)

29 One of the most obvious facts about grown-ups, to a child, is that they have forgotten what it is like to be a child.
Randall Jarrell 1914–65: introduction to Christina Stead *The Man Who Loved Children* (1965)

30 A child becomes an adult when he realizes that he has a right not only to be right but also to be wrong.
Thomas Szasz 1920– : *The Second Sin* (1973)

31 With the birth of each child, you lose two novels.
Candia McWilliam 1955– : *Guardian* 5 May 1993

Choice

1 The die is cast.
Julius Caesar 100–44 BC: at the crossing of the Rubicon, in Suetonius *Lives of the Caesars*

2 To be, or not to be: that is the question:
Whether 'tis nobler in the mind to suffer
The slings and arrows of outrageous fortune,
Or to take arms against a sea of troubles,
And by opposing end them?
William Shakespeare 1564–1616: *Hamlet* (1601)

3 How happy could I be with either,
Were t'other dear charmer away!
John Gay 1685–1732: *The Beggar's Opera* (1728)

4 You pays your money and you takes your choice.
Punch: 1846

5 White shall not neutralize the black, nor good
Compensate bad in man, absolve him so:
Life's business being just the terrible choice.
Robert Browning 1812–89: *The Ring and the Book* (1868–9)

6 Take care to get what you like or you will be forced to like what you get.
George Bernard Shaw 1856–1950: *Man and Superman* (1903)

7 Two roads diverged in a wood, and I—
I took the one less travelled by,
And that has made all the difference.
Robert Frost 1874–1963: 'The Road Not Taken' (1916)

8 Between two evils, I always pick the one I never tried before.
Mae West 1892–1980: *Klondike Annie* (1936)

9 Any colour—so long as it's black.
Henry Ford 1863–1947: on the choice of colour for the Model T Ford, in A. Nevins *Ford* (1957)

10 Was there ever in anyone's life span a point free in time, devoid of memory, a night when choice was any more than the sum of all the choices gone before?
Joan Didion 1934– : *Run River* (1963)

Christmas

1 She brought forth her firstborn son, and wrapped him in swaddling clothes, and laid him in a manger; because there was no room for them in the inn.
Bible: St Luke

2 Welcome, all wonders in one sight!
Eternity shut in a span.
Richard Crashaw *c.*1612–49: 'Hymn of the Nativity' (1652)

3 'Bah,' said Scrooge. 'Humbug!'
Charles Dickens 1812–70: *A Christmas Carol* (1843)

4 I'm dreaming of a white Christmas,
Just like the ones I used to know.
Irving Berlin 1888–1989: 'White Christmas' (1942)

5 And girls in slacks remember Dad,
And oafish louts remember Mum,
And sleepless children's hearts are glad,
And Christmas-morning bells say 'Come!'
John Betjeman 1906–84: 'Christmas' (1954)

6 Still xmas is a good time with all those presents and good food and i hope it will never die out or at any rate not until i am grown up and hav to pay for it all.
Geoffrey Willans 1911–1958 and **Ronald Searle** 1920– : *How To Be Topp* (1954)

The Church

See also **Religion**

1 As often as we are mown down by you, the more we grow in numbers; the blood of Christians is the seed.
Tertullian AD *c.*160–*c.*225: *Apologeticus*, traditionally 'The blood of the martyrs is the seed of the Church'

2 He cannot have God for his father who has not the church for his mother.
St Cyprian AD *c.*200–258: *De Ecclesiae Catholicae Unitate*

3 In old time we had treen chalices and golden priests, but now we have treen priests and golden chalices.
Bishop John Jewel 1522–71: *Certain Sermons Preached Before the Queen's Majesty* (1609)

4 The nearer the Church the further from God.
Bishop Lancelot Andrewes 1555–1626: *Of the Nativity* (1622) Sermon 15

5 And of all plagues with which mankind are curst,
Ecclesiastic tyranny's the worst.
Daniel Defoe 1660–1731: *The True-Born Englishman* (1701)

6 I look upon all the world as my parish.
John Wesley 1703–91: diary, 11 June 1739

7 I never saw, heard, nor read, that the clergy were beloved in any nation where Christianity was the religion of the country. Nothing can render them popular, but some degree of persecution.

Jonathan Swift 1667–1745: *Thoughts on Religion* (1765)

8 What bishops like best in their clergy is a dropping-down-deadness of manner.

Sydney Smith 1771–1845: 'First Letter to Archdeacon Singleton, 1837'

9 She [the Roman Catholic Church] may still exist in undiminished vigour when some traveller from New Zealand shall, in the midst of a vast solitude, take his stand on a broken arch of London Bridge to sketch the ruins of St Paul's.

Lord Macaulay 1800–59: *Essays* (1843) 'Von Ranke'

10 As the French say, there are three sexes—men, women, and clergymen.

Sydney Smith 1771–1845: in Lady Holland *Memoir* (1855)

11 But the churchmen fain would kill their church,
As the churches have killed their Christ.

Alfred, Lord Tennyson 1809–92: *Maud* (1855)

12 'The Church is an anvil which has worn out many hammers', and the story of the first collision is, in essentials, the story of all.

Alexander Maclaren 1826–1910: *Expositions of Holy Scripture: Acts of the Apostles* (1907)

13 The Church [of England] should go forward along the path of progress and be no longer satisfied only to represent the Conservative Party at prayer.

Maude Royden 1876–1956: in *The Times* 17 July 1917

14 The two dangers which beset the Church of England are good music and bad preaching.

Lord Hugh Cecil 1869–1956: in K. Rose *The Later Cecils* (1975)

The Cinema

See also **Acting and the Theatre**

1 The lunatics have taken charge of the asylum.

Richard Rowland *c.*1881–1947: on the take-over of United

Artists by Charles Chaplin and others, in T. Ramsaye *A Million and One Nights* (1926)

2 There is only one thing that can kill the Movies, and that is education.

Will Rogers 1879–1935: *Autobiography of Will Rogers* (1949)

3 Bring on the empty horses!

Michael Curtiz 1888–1962: while directing *The Charge of the Light Brigade* (1936 film); in David Niven *Bring on the Empty Horses* (1975)

4 The trouble, Mr Goldwyn, is that you are only interested in art and I am only interested in money.

George Bernard Shaw 1856–1950: telegraphed version of the outcome of a conversation between Shaw and Sam Goldwyn; in A. Johnson *The Great Goldwyn* (1937)

5 If my books had been any worse, I should not have been invited to Hollywood, and if they had been any better, I should not have come.

Raymond Chandler 1888–1959: letter to Charles W. Morton, 12 December 1945

6 JOE GILLIS: You used to be in pictures. You used to be big.
NORMA DESMOND: I am big. It's the pictures that got small.

Charles Brackett 1892–1969 and **Billy Wilder** 1906– : *Sunset Boulevard* (1950 film, with D. M. Marshman Jr.)

7 This is the biggest electric train set any boy ever had!

Orson Welles 1915–85: of the RKO studios, in P. Noble *The Fabulous Orson Welles* (1956)

8 Why should people go out and pay to see bad movies when they can stay at home and see bad television for nothing?

Sam Goldwyn 1882–1974: in *Observer* 9 September 1956

9 Hollywood money isn't money. It's congealed snow, melts in your hand, and there you are.

Dorothy Parker 1893–1967: in Malcolm Cowley *Writers at Work* 1st Series (1958)

10 It's like kissing Hitler.

Tony Curtis 1925– : when asked what it was like to kiss Marilyn Monroe; in A. Hunter *Tony Curtis* (1985)

11 Photography is truth. The cinema is truth 24 times per second.

Jean-Luc Godard 1930– : *Le Petit Soldat* (1960 film)

12 I seldom go to films. They are too exciting,
said the Honourable Possum.
John Berryman 1914–72: *77 Dream Songs* (1964)

13 All I need to make a comedy is a park, a policeman and a pretty girl.
Charlie Chaplin 1889–1977: *My Autobiography* (1964)

14 Pictures are for entertainment, messages should be delivered by Western Union.
Sam Goldwyn 1882–1974: in A. Marx *Goldwyn* (1976)

15 I wouldn't say when you've seen one Western you've seen the lot; but when you've seen the lot you get the feeling you've seen one.
Katharine Whitehorn 1926– : *Sunday Best* (1976) 'Decoding the West'

16 Words are cheap. The biggest thing you can say is 'elephant'.
Charlie Chaplin 1889–1977: on the universality of silent films; in B. Norman *The Movie Greats* (1981)

17 'Movies should have a beginning, a middle and an end,' harrumphed French film maker Georges Franju… 'Certainly,' replied Jean-Luc Godard. 'But not necessarily in that order.'
Jean-Luc Godard 1930– : *Time* 14 September 1981

Civilization

1 The three great elements of modern civilization, Gunpowder, Printing, and the Protestant Religion.
Thomas Carlyle 1795–1881: *Critical and Miscellaneous Essays* (1838)

2 Civilization advances by extending the number of important operations which we can perform without thinking about them.
Alfred North Whitehead 1861–1947: *Introduction to Mathematics* (1911)

3 The lamps are going out all over Europe; we shall not see them lit again in our lifetime.
Lord Grey of Fallodon 1862–1933: on the eve of the First World War; *25 Years* (1925)

4 Civilization and profits go hand in hand.
Calvin Coolidge 1872–1933: speech in New York, 27 November 1920

5 You can't say civilization don't advance, however, for in every war they kill you in a new way.
Will Rogers 1879–1935: *New York Times* 23 December 1929

6 [A journalist] asked, 'Mr Gandhi, what do you think of modern civilization?' And Mr Gandhi said, 'That would be a good idea.'
Mahatma Gandhi 1869–1948: on arriving in England in 1930; in E . F. Schumacher *Good Work* (1979)

7 Civilization has made the peasantry its pack animal. The bourgeoisie in the long run only changed the form of the pack.
Leon Trotsky 1879–1940: *History of the Russian Revolution* (1933)

8 Whenever I hear the word culture...I release the safety-catch of my Browning!
Hanns Johst 1890–1978: *Schlageter* (1933) often attributed to Hermann Goering, and quoted 'Whenever I hear the word culture, I reach for my pistol!'

9 Disinterested intellectual curiosity is the life-blood of real civilization.
G. M. Trevelyan 1876–1962: *English Social History* (1942)

10 In Italy for thirty years under the Borgias they had warfare, terror, murder, bloodshed—they produced Michelangelo, Leonardo da Vinci and the Renaissance. In Switzerland they had brotherly love, five hundred years of democracy and peace and what did that produce...? The cuckoo clock.
Orson Welles 1915–85: *The Third Man* (1949 film); words added by Welles to Graham Greene's screenplay

Class

1 When Adam dalfe and Eve spane...
Where was than the pride of man?
Richard Rolle de Hampole *c.*1290–1349: in G. G. Perry *Religious Pieces* (1914). Taken in the form 'When Adam delved and Eve span, who was then the gentleman?' by

John Ball as the text of his revolutionary sermon on the outbreak of the Peasants' Revolt, 1381

2 Take but degree away, untune that string,
And, hark! what discord follows.

William Shakespeare 1564–1616: *Troilus and Cressida* (1602)

3 That in the captain's but a choleric word,
Which in the soldier is flat blasphemy.

William Shakespeare 1564–1616: *Measure for Measure* (1604)

4 He told me... that mine was the middle state, or what might be called the upper station of low life, which he had found by long experience was the best state in the world, the most suited to human happiness.

Daniel Defoe 1660–1731: *Robinson Crusoe* (1719)

5 O let us love our occupations,
Bless the squire and his relations,
Live upon our daily rations,
And always know our proper stations.

Charles Dickens 1812–70: *The Chimes* (1844)

6 The history of all hitherto existing society is the history of class struggles.

Karl Marx 1818–83 and **Friedrich Engels** 1820–95: *The Communist Manifesto* (1848)

7 The proletarians have nothing to lose but their chains. They have a world to win. WORKING MEN OF ALL COUNTRIES, UNITE!

Karl Marx 1818–83 and **Friedrich Engels** 1820–95: *The Communist Manifesto* (1848); tr. S. Moore, 1888, commonly rendered 'Workers of the world, unite!'

8 The rich man in his castle,
The poor man at his gate,
God made them, high or lowly,
And ordered their estate.

Cecil Frances Alexander 1818–95: 'All Things Bright and Beautiful' (1848)

9 *Il faut épater le bourgeois.*
One must astonish the bourgeois.

Charles Baudelaire 1821–67: attributed

10 The State is an instrument in the hands of the ruling class, used to break the resistance of the adversaries of that class.
Joseph Stalin 1879–1953: *Foundations of Leninism* (1924)

11 The bourgeois prefers comfort to pleasure, convenience to liberty, and a pleasant temperature to the deathly inner consuming fire.
Hermann Hesse 1877–1962: *Der Steppenwolf* (1927)

12 Like many of the Upper Class
He liked the Sound of Broken Glass.
Hilaire Belloc 1870–1953: 'About John' (1930)

13 We of the sinking middle class ... may sink without further struggles into the working class where we belong, and probably when we get there it will not be so dreadful as we feared, for, after all, we have nothing to lose but our aitches.
George Orwell 1903–50: *The Road to Wigan Pier* (1937)

14 The Stately Homes of England,
How beautiful they stand,
To prove the upper classes
Have still the upper hand.
Noël Coward 1899–1973: 'The Stately Homes of England' (1938 song).

15 You can be in the Horseguards and still be common, dear.
Terence Rattigan 1911–77: *Separate Tables* (1954)

16 Impotence and sodomy are socially O.K. but birth control is flagrantly middle-class.
Evelyn Waugh 1903–66: 'An Open Letter' in Nancy Mitford (ed.) *Noblesse Oblige* (1956)

17 First you take their faces from 'em by calling 'em the masses and then you accuse 'em of not having any faces.
J. B. Priestley 1894–1984: *Saturn Over the Water* (1961)

Commerce see **Business and Commerce**

Conscience

1 Every subject's duty is the king's; but every subject's soul is his own.
William Shakespeare 1564–1616: *Henry V* (1599)

2 Thus conscience doth make cowards of us all.
William Shakespeare 1564–1616: *Hamlet* (1601)

3 We have erred, and strayed from thy ways like lost sheep. We have followed too much the devices and desires of our own hearts.
Book of Common Prayer 1662: *Morning Prayer*

4 Corporations have neither bodies to be punished, nor souls to be condemned, they therefore do as they like.
Edward, 1st Baron Thurlow 1731–1806: in J. Poynder *Literary Extracts* (1844), usually quoted 'Did you ever expect a corporation to have a conscience, when it has no soul to be damned, and no body to be kicked?'

5 Conscience is thoroughly well-bred and soon leaves off talking to those who do not wish to hear it.
Samuel Butler 1835–1902: *Further Extracts from Notebooks* (1934)

6 Conscience: the inner voice which warns us that someone may be looking.
H. L. Mencken 1880–1956: *A Little Book in C major* (1916)

7 Most people sell their souls, and live with a good conscience on the proceeds.
Logan Pearsall Smith 1865–1946: *Afterthoughts* (1931) 'Other People'

8 I cannot and will not cut my conscience to fit this year's fashions.
Lillian Hellman 1905–84: letter to John S. Wood, 19 May 1952

Conversation

See also **Speech and Speeches**

1 Religion is by no means a proper subject of conversation in a mixed company.
Lord Chesterfield 1694–1773: *Letters... to his Godson and Successor* (1890)

2 John Wesley's conversation is good, but he is never at leisure. He is always obliged to go at a certain hour. This is very disagreeable to a man who loves to fold his legs and have out his talk, as I do.

Samuel Johnson 1709–84: in James Boswell *Life of Johnson* (1791) 31 March 1778

3 On every formal visit a child ought to be of the party, by way of provision for discourse.

Jane Austen 1775–1817: *Sense and Sensibility* (1811)

4 From politics, it was an easy step to silence.

Jane Austen 1775–1817: *Northanger Abbey* (1818)

5 He talked on for ever; and you wished him to talk on for ever.

William Hazlitt 1778–1830: of Coleridge; *Lectures on the English Poets* (1818)

6 He has occasional flashes of silence, that make his conversation perfectly delightful.

Sydney Smith 1771–1845: of Macaulay; in Lady Holland *Memoir* (1855)

7 'The time has come,' the Walrus said,
'To talk of many things:
Of shoes—and ships—and sealing wax—
Of cabbages—and kings.

Lewis Carroll 1832–98: *Through the Looking-Glass* (1872)

8 He speaks to Me as if I was a public meeting.

Queen Victoria 1819–1901: of Gladstone, in G. W. E. Russell *Collections and Recollections* (1898)

9 Most English talk is a quadrille in a sentry-box.

Henry James 1843–1916: *The Awkward Age* (1899)

Cooperation

1 Is not a Patron, my Lord, one who looks with unconcern on a man struggling for life in the water, and, when he has reached ground, encumbers him with help?

Samuel Johnson 1709–84: letter to Lord Chesterfield, 7 February 1755

2 When bad men combine, the good must associate; else they will fall, one by one, an unpitied sacrifice in a contemptible struggle.

Edmund Burke 1729–97: *Thoughts on the Cause of the Present Discontents* (1770)

3 We must indeed all hang together, or, most assuredly, we shall all hang separately.

Benjamin Franklin 1706–90: at the Signing of the Declaration of Independence, 4 July 1776 (possibly not original)

4 All for one, one for all.

Alexandre Dumas 1802–70: *Les Trois Mousquetaires* (1844)

5 You may call it combination, you may call it the accidental and fortuitous concurrence of atoms.

Lord Palmerston 1784–1865: on a projected Palmerston-Disraeli coalition, House of Commons, 5 March 1857

6 My apple trees will never get across
And eat the cones under his pines, I tell him.
He only says, 'Good fences make good neighbours.'

Robert Frost 1874–1963: 'Mending Wall' (1914)

7 We must learn to live together as brothers or perish together as fools.

Martin Luther King 1929–68: speech at St Louis, 22 March 1964

8 When Hitler attacked the Jews I was not a Jew, therefore, I was not concerned. And when Hitler attacked the Catholics, I was not a Catholic, and therefore, I was not concerned. And when Hitler attacked the unions and the industrialists, I was not a member of the unions and I was not concerned. Then, Hitler attacked me and the Protestant church—and there was nobody left to be concerned.

Martin Niemöller 1892–1984: in *Congressional Record* 14 October 1968

The Country and the Town

See also **Environment**

1 God made the country, and man made the town.
William Cowper 1731-1800: *The Task* (1785) 'The Sofa'

2 'Tis distance lends enchantment to the view,
And robes the mountain in its azure hue.
Thomas Campbell 1777-1844: *Pleasures of Hope* (1799)

3 There is nothing good to be had in the country, or if there is, they will not let you have it.
William Hazlitt 1778-1830: *The Round Table* (1817) 'Observations on... *The Excursion*'

4 If you would be known, and not know, vegetate in a village; if you would know, and not be known, live in a city.
Charles Caleb Colton *c.*1780-1832: *Lacon* (1820)

5 I have no relish for the country; it is a kind of healthy grave.
Sydney Smith 1771-1845: letter to Miss G. Harcourt, 1838

6 But a house is much more to my mind than a tree,
And for groves, O! a good grove of chimneys for me.
Charles Morris 1745-1838: 'Country and Town' (1840)

7 It is my belief, Watson, founded upon my experience, that the lowest and vilest alleys in London do not present a more dreadful record of sin than does the smiling and beautiful countryside.
Arthur Conan Doyle 1859-1930: *Adventures of Sherlock Holmes* (1892)

8 The Farmer will never be happy again;
He carries his heart in his boots;
For either the rain is destroying his grain
Or the drought is destroying his roots.
A. P. Herbert 1890-1971: 'The Farmer' (1922)

9 So *that's* what hay looks like.
Queen Mary 1867-1953: at Badminton House, where she was evacuated during the Second World War; in James Pope-Hennessy *Life of Queen Mary* (1959)

10 An industrial worker would sooner have a £5 note but a countryman must have praise.
Ronald Blythe 1922- : *Akenfield* (1969)

Courage

1 *Audentis Fortuna iuvat.*
Fortune assists the bold.
Virgil 70–19 BC: *Aeneid* (often quoted 'Fortune favours the brave')

2 Cowards die many times before their deaths;
The valiant never taste of death but once.
William Shakespeare 1564–1616: *Julius Caesar* (1599)

3 Boldness be my friend!
Arm me, audacity.
William Shakespeare 1564–1616: *Cymbeline* (1609–10)

4 All men would be cowards if they durst.
John Wilmot, Earl of Rochester 1647–80: 'A Satire against Mankind' (1679)

5 Tender-handed stroke a nettle,
And it stings you for your pains;
Grasp it like a man of mettle,
And it soft as silk remains.
Aaron Hill 1685–1750: 'Verses Written on a Window in Scotland'

6 *De l'audace, et encore de l'audace, et toujours de l'audace!*
Boldness, and again boldness, and always boldness!
Georges Jacques Danton 1759–94: speech to the Legislative Committee of General Defence, 2 September 1792

7 As to moral courage, I have very rarely met with two o'clock in the morning courage: I mean instantaneous courage.
Napoléon I 1769–1821: in E. A. de Las Cases *Mémorial de Ste-Hélène* (1823) 4–5 December 1815

8 No coward soul is mine,
No trembler in the world's storm-troubled sphere:
I see Heaven's glories shine,
And faith shines equal, arming me from fear.
Emily Brontë 1818–48: 'No coward soul is mine' (1846)

9 Better be killed than frightened to death.
R. S. Surtees 1805–64: *Mr Facey Romford's Hounds* (1865)

10 Grace under pressure.
Ernest Hemingway 1899–1961: when asked what he meant

by 'guts' in an interview with Dorothy Parker; in *New Yorker* 30 November 1929

11 Cowardice, as distinguished from panic, is almost always simply a lack of ability to suspend the functioning of the imagination.
Ernest Hemingway 1899–1961: *Men at War* (1942)

12 Courage is not simply *one* of the virtues but the form of every virtue at the testing point.
C. S. Lewis 1898–1963: in Cyril Connolly *The Unquiet Grave* (1944)

Crime and Punishment

See also **Justice, The Law and Lawyers, Murder**

1 I the Lord thy God am a jealous God, visiting the iniquity of the fathers upon the children unto the third and fourth generation of them that hate me.
Bible: Exodus

2 My father hath chastised you with whips, but I will chastise you with scorpions.
Bible: I Kings

3 He that spareth his rod hateth his son.
Bible: Proverbs

4 This is the first of punishments, that no guilty man is acquitted if judged by himself.
Juvenal AD *c.*60–*c.*130: *Satires*

5 I went out to Charing Cross, to see Major-general Harrison hanged, drawn, and quartered; which was done there, he looking as cheerful as any man could do in that condition.
Samuel Pepys 1633–1703; diary, 13 October 1660

6 Men are not hanged for stealing horses, but that horses may not be stolen.
George Savile, Marquess of Halifax 1633–95: *Political, Moral, and Miscellaneous Thoughts* (1750) 'Of Punishment'

7 Stolen sweets are best.
Colley Cibber 1671–1757: *The Rival Fools* (1709)

8 All punishment is mischief: all punishment in itself is evil.
Jeremy Bentham 1748–1832: *Principles of Morals and Legislation* (1789)

9 As for rioting, the old Roman way of dealing with that is always the right one; flog the rank and file, and fling the ringleaders from the Tarpeian rock.
Thomas Arnold 1795–1842: from a letter written before 1828

10 Prisoner, God has given you good abilities, instead of which you go about the country stealing ducks.
William Arabin 1773–1841: also attributed to a Revd Mr Alderson

11 They will steal the very teeth out of your mouth as you walk through the streets. *I know it from experience.*
William Arabin 1773–1841: on the citizens of Uxbridge

12 Deserves to be preached to death by wild curates.
Sydney Smith 1771–1845: in Lady Holland *Memoir* (1855)

13 In that case, if we are to abolish the death penalty, let the murderers take the first step.
Alphonse Karr 1808–90: *Les Guêpes* January 1849

14 Better build schoolrooms for 'the boy',
Than cells and gibbets for 'the man'.
Eliza Cook 1818–89: 'A Song for the Ragged Schools' (1853)

15 Thou shalt not steal; an empty feat,
When it's so lucrative to cheat.
Arthur Hugh Clough 1819–61: 'The Latest Decalogue' (1862)

16 When constabulary duty's to be done,
A policeman's lot is not a happy one.
W. S. Gilbert 1836–1911: *The Pirates of Penzance* (1879)

17 Awaiting the sensation of a short, sharp shock,
From a cheap and chippy chopper on a big black block.
W. S. Gilbert 1836–1911: *The Mikado* (1885)

18 My object all sublime
I shall achieve in time—
To let the punishment fit the crime—
The punishment fit the crime.
W. S. Gilbert 1836–1911: *The Mikado* (1885)

19 What hangs people... is the unfortunate circumstance of guilt.

Robert Louis Stevenson 1850–94: *The Wrong Box* (with Lloyd Osbourne, 1889)

20 Singularity is almost invariably a clue. The more featureless and commonplace a crime is, the more difficult is it to bring it home.

Arthur Conan Doyle 1859–1930: *Adventures of Sherlock Holmes* (1892)

21 Thieves respect property. They merely wish the property to become their property that they may more perfectly respect it.

G. K. Chesterton 1874–1936: *The Man who was Thursday* (1908)

22 Anarchism is a game at which the police can beat you.

George Bernard Shaw 1856–1950: *Misalliance* (1914)

23 For de little stealin' dey gits you in jail soon or late. For de big stealin' dey makes you Emperor and puts you in de Hall o' Fame when you croaks.

Eugene O'Neill 1888–1953: *The Emperor Jones* (1921)

24 Any one who has been to an English public school will always feel comparatively at home in prison. It is the people brought up in the gay intimacy of the slums, Paul learned, who find prison so soul-destroying.

Evelyn Waugh 1903–66: *Decline and Fall* (1928)

25 He always has an alibi, and one or two to spare:
At whatever time the deed took place—MACAVITY WASN'T THERE!

T. S. Eliot 1888–1965: *Old Possum's Book of Practical Cats* (1939) 'Macavity: the Mystery Cat'

26 Even the most hardened criminal a few years ago would help an old lady across the road and give her a few quid if she was skint.

Charlie Kray *c.*1930– : *Observer* 28 December 1986 'Sayings of the Year'

27 Society needs to condemn a little more and understand a little less.

John Major 1943– : interview with *Mail on Sunday* 21 February 1993

Crises

1 For it is your business, when the wall next door catches fire.
Horace 65–8 BC: *Epistles*

2 If you can keep your head when all about you
Are losing theirs and blaming it on you.
Rudyard Kipling 1865–1936: 'If—' (1910)

3 Business carried on as usual during alterations on the map of Europe.
Winston Churchill 1874–1965: speech at Guildhall, 9 November 1914

4 Comin' in on a wing and a pray'r.
Harold Adamson 1906–80: title of song (1943)

5 We're eyeball to eyeball, and I think the other fellow just blinked.
Dean Rusk 1909– : on the Cuban missile crisis, 24 October 1962

6 Crisis? What Crisis?
Anonymous: *Sun* headline, 11 January 1979, summarizing James Callaghan: 'I don't think other people in the world would share the view [that] there is mounting chaos'

Critics and Criticism

1 Critics are like brushers of noblemen's clothes.
Henry Wotton 1568–1639: in Francis Bacon *Apophthegms New and Old* (1625)

2 How science dwindles, and how volumes swell,
How commentators each dark passage shun,
And hold their farthing candle to the sun.
Edward Young 1683–1765: *The Love of Fame* (1725–8)

3 As learned commentators view
In Homer more than Homer knew.
Jonathan Swift 1667–1745: 'On Poetry' (1733)

4 You *may* abuse a tragedy, though you cannot write one. You may scold a carpenter who has made you a bad table, though you cannot make a table. It is not your trade to make tables.
Samuel Johnson 1709–84: on literary criticism; in James Boswell *Life of Johnson* (1791) 25 June 1763

5 I have always suspected that the reading is right, which requires many words to prove it wrong; and the emendation wrong, that cannot without so much labour appear to be right.
Samuel Johnson 1709–84: *Plays of William Shakespeare...* (1765) preface

6 A man must serve his time to every trade
Save censure—critics all are ready made.
Lord Byron 1788–1824: *English Bards and Scotch Reviewers* (1809)

7 This will never do.
Francis, Lord Jeffrey 1773–1850: on Wordsworth's *The Excursion* (1814) in *Edinburgh Review* November 1814

8 I never read a book before reviewing it; it prejudices a man so.
Sydney Smith 1771–1845: in H. Pearson *The Smith of Smiths* (1934)

9 People who like this sort of thing will find this the sort of thing they like.
Abraham Lincoln 1809–65: judgement of a book, in G. W. E. Russell *Collections and Recollections* (1898)

10 I maintain that two and two would continue to make four, in spite of the whine of the amateur for three, or the cry of the critic for five.
James McNeill Whistler 1834–1903: *Whistler v. Ruskin. Art and Art Critics* (1878)

11 I don't care anything about reasons, but I know what I like.
Henry James 1843–1916: *Portrait of a Lady* (1881)

12 The good critic is he who relates the adventures of his soul in the midst of masterpieces.
Anatole France 1844–1924: *La Vie littéraire* (1888)

13 I am sitting in the smallest room of my house. I have your review before me. In a moment it will be behind me.
Max Reger 1873–1916: responding to a savage review by

Rudolph Louis in *Münchener Neueste Nachrichten*, 7 February 1906

14 Everything must be like something, so what is this like?
E. M. Forster 1879–1970: *Abinger Harvest* (1936)

15 Remember, a statue has never been set up in honour of a critic!
Jean Sibelius 1865–1957: in B. de Törne *Sibelius: A Close-Up* (1937)

16 I cry all the way to the bank.
Liberace 1919–87: on bad reviews, from the mid-1950s; *Autobiography* (1973)

17 A critic is a bundle of biases held loosely together by a sense of taste.
Whitney Balliett 1926– : *Dinosaurs in the Morning* (1962)

18 Interpretation is the revenge of the intellect upon art.
Susan Sontag 1933– : *Evergreen Review* December 1964

19 A critic is a man who knows the way but can't drive the car.
Kenneth Tynan 1927–80: in *New York Times Magazine* 9 January 1966

20 It is nice, but in one of the chapters the author made a mistake. He describes the sun as rising twice on the same day.
Paul Dirac 1902–84: comment on the novel *Crime and Punishment*; in G. Gamow *Thirty Years that Shook Physics* (1966)

Cruelty

1 This was the most unkindest cut of all.
William Shakespeare 1564–1616: *Julius Caesar* (1599)

2 I must be cruel only to be kind.
William Shakespeare 1564–1616: *Hamlet* (1601)

3 Man's inhumanity to man
Makes countless thousands mourn!
Robert Burns 1759–96: 'Man was made to Mourn' (1786)

4 A robin red breast in a cage
Puts all Heaven in a rage.
William Blake 1757–1827: 'Auguries of Innocence' (*c.*1803)

5 The infliction of cruelty with a good conscience is a delight to moralists. That is why they invented Hell.
Bertrand Russell 1872–1970: *Sceptical Essays* (1928)

6 The wish to hurt, the momentary intoxication with pain, is the loophole through which the pervert climbs into the minds of ordinary men.
Jacob Bronowski 1908–74: *The Face of Violence* (1954)

Custom and Habit

1 But to my mind,—though I am native here,
And to the manner born,—it is a custom
More honoured in the breach than the observance.
William Shakespeare 1564–1616: *Hamlet* (1601)

2 Actions receive their tincture from the times,
And as they change are virtues made or crimes.
Daniel Defoe 1660–1731: *A Hymn to the Pillory* (1703)

3 Custom, then, is the great guide of human life.
David Hume 1711–76: *An Enquiry Concerning Human Understanding* (1748)

4 Custom reconciles us to everything.
Edmund Burke 1729–97: *On the Sublime and Beautiful* (1757)

5 Habit with him was all the test of truth,
'It must be right: I've done it from my youth.'
George Crabbe 1754–1832: *The Borough* (1810)

6 The tradition of all the dead generations weighs like a nightmare on the brain of the living.
Karl Marx 1818–83: *The Eighteenth Brumaire of Louis Bonaparte* (1852)

7 Sow an act, and you reap a habit. Sow a habit and you reap a character. Sow a character, and you reap a destiny.
Charles Reade 1814–84: attributed

8 Tradition means giving votes to the most obscure of all classes, our ancestors. It is the democracy of the dead.
G. K. Chesterton 1874–1936: *Orthodoxy* (1908)

9 Every public action, which is not customary, either is wrong, or, if it is right, is a dangerous precedent. It follows that nothing should ever be done for the first time.
Francis M. Cornford 1874–1943: *Microcosmographia Academica* (1908)

10 One can't carry one's father's corpse about everywhere.
Guillaume Apollinaire 1880–1918: on tradition, in *Les peintres cubistes* (1965)

11 Habit is a great deadener.
Samuel Beckett 1906–89: *Waiting for Godot* (1955)

Cynicism

1 Paris is well worth a mass.
Henri IV 1553–1610: attributed to Henri IV; alternatively to his minister Sully, in conversation with him

2 What makes all doctrines plain and clear?
About two hundred pounds a year.
And that which was proved true before,
Prove false again? Two hundred more.
Samuel Butler 1612–80: *Hudibras* pt. 3 (1680)

3 Never glad confident morning again!
Robert Browning 1812–89: 'The Lost Leader' (1845)

4 Cynicism is intellectual dandyism without the coxcomb's feathers.
George Meredith 1828–1909: *The Egoist* (1879)

5 A man who knows the price of everything and the value of nothing.
Oscar Wilde 1854–1900: definition of a cynic; *Lady Windermere's Fan* (1892)

6 Pathos, piety, courage—they exist, but are identical, and so is filth. Everything exists, nothing has value.
E. M. Forster 1879–1970: *A Passage to India* (1924)

7 Cynicism is an unpleasant way of saying the truth.
Lillian Hellman 1905–1984: *The Little Foxes* (1939)

Dance

1 This wondrous miracle did Love devise,
For dancing is love's proper exercise.
Sir John Davies 1569–1626: 'Orchestra, or a Poem of Dancing' (1596)

2 A dance is a measured pace, as a verse is a measured speech.
Francis Bacon 1561–1626: *The Advancement of Learning* (1605)

3 Sport that wrinkled Care derides,
And Laughter holding both his sides.
Come, and trip it as ye go
On the light fantastic toe.
John Milton 1608–74: 'L'Allegro' (1645)

4 On with the dance! let joy be unconfined;
No sleep till morn, when Youth and Pleasure meet
To chase the glowing Hours with flying feet.
Lord Byron 1788–1824: *Childe Harold's Pilgrimage* (1812–18)

5 Everyone knows that the real business of a ball is either to look out for a wife, to look after a wife, or to look after somebody else's wife.
R. S. Surtees 1805–64: *Mr Facey Romford's Hounds* (1865)

6 O body swayed to music, O brightening glance
How can we know the dancer from the dance?
W. B. Yeats 1865–1939: 'Among School Children' (1928)

7 Can't act. Slightly bald. Also dances.
Anonymous: studio official's comment on Fred Astaire, in B. Thomas *Astaire* (1985)

8 There may be trouble ahead,
But while there's moonlight and music and love and romance,
Let's face the music and dance.
Irving Berlin 1888–1989: 'Let's Face the Music and Dance' (1936)

9 [Dancing is] a perpendicular expression of a horizontal desire.
George Bernard Shaw 1856–1950: in *New Statesman* 23 March 1962 (attributed)

10 But the zest goes out of a beautiful waltz
When you dance it bust to bust.
Joyce Grenfell 1910–79: 'Stately as a Galleon' (1978 song)

Day and Night

1 The gaudy, blabbing, and remorseful day
Is crept into the bosom of the sea.
William Shakespeare 1564–1616: *Henry VI, Part 2* (1592)

2 Night's candles are burnt out, and jocund day
Stands tiptoe on the misty mountain tops.
William Shakespeare 1564–1616: *Romeo and Juliet* (1595)

3 But, look, the morn, in russet mantle clad,
Walks o'er the dew of yon high eastern hill.
William Shakespeare 1564–1616: *Hamlet* (1601)

4 Dear Night! this world's defeat;
The stop to busy fools; care's check and curb.
Henry Vaughan 1622–95: *Silex Scintillans* (1650–5)

5 Lighten our darkness, we beseech thee, O Lord; and by thy great mercy defend us from all perils and dangers of this night.
Book of Common Prayer 1662: *Evening Prayer*

6 The curfew tolls the knell of parting day,
The lowing herd wind slowly o'er the lea,
The ploughman homeward plods his weary way,
And leaves the world to darkness and to me.
Thomas Gray 1716–71: *Elegy Written in a Country Churchyard* (1751)

7 The Sun's rim dips; the stars rush out;
At one stride comes the dark.
Samuel Taylor Coleridge 1772–1834: 'The Rime of the Ancient Mariner' (1798)

8 When I behold, upon the night's starred face
Huge cloudy symbols of a high romance.
John Keats 1795–1821: 'When I have fears that I may cease to be' (written 1818)

9 The cares that infest the day
Shall fold their tents, like the Arabs,
And as silently steal away.
Henry Wadsworth Longfellow 1807–82: 'The Day is Done' (1844)

10 The splendour falls on castle walls
And snowy summits old in story:
The long light shakes across the lakes,
And the wild cataract leaps in glory.
Alfred, Lord Tennyson 1809–92: *The Princess* (1847) song (added 1850)

11 There midnight's all a glimmer, and noon a purple glow,
And evening full of the linnet's wings.
W. B. Yeats 1865–1939: 'The Lake Isle of Innisfree' (1892)

12 Summer afternoon—summer afternoon... the two most beautiful words in the English language.
Henry James 1843–1916: in Edith Wharton *A Backward Glance* (1934)

13 Let us go then, you and I,
When the evening is spread out against the sky
Like a patient etherized upon a table.
T. S. Eliot 1888–1965: 'Love Song of J. Alfred Prufrock' (1917)

14 The winter evening settles down
With smell of steaks in passageways.
Six o'clock.
The burnt-out ends of smoky days.
T. S. Eliot 1888–1965: 'Preludes' (1917)

15 I have a horror of sunsets, they're so romantic, so operatic.
Marcel Proust 1871–1922: *Cities of the Plain* (1922)

16 What are days for?
Days are where we live.
Philip Larkin 1922–85: 'Days' (1964)

Death

See also **Epitaphs, Last Words, Murder**

1 I would rather be tied to the soil as another man's serf, even a poor man's, who hadn't much to live on himself, than be King of all these the dead and destroyed.
Homer 8th century BC: *The Odyssey*

2 For dust thou art, and unto dust shalt thou return.
Bible: Genesis

3 He died in a good old age, full of days, riches, and honour.
Bible: I Chronicles

4 *Non omnis moriar.*
I shall not altogether die.
Horace 65–8 BC: *Odes*

5 O death, where is thy sting? O grave, where is thy victory?
Bible: I Corinthians

6 And I looked, and behold a pale horse: and his name that sat on him was Death.
Bible: Revelation

7 *Abiit ad plures.*
He's gone to join the majority [the dead].
Petronius d. AD 65: *Satyricon*

8 Anyone can stop a man's life, but no one his death; a thousand doors open on to it.
Seneca ('the Younger') *c.*4 BC–AD 65: *Phoenissae*

9 With thanks to God we know the way to heaven, to be as ready by water as by land, and therefore we care not which way we go.
Friar Elstow: when threatened with drowning by Henry VIII; in John Stow *Annals of England* (1615)

10 Let's talk of graves, of worms, and epitaphs;
Make dust our paper, and with rainy eyes
Write sorrow on the bosom of the earth.
Let's choose executors, and talk of wills.
William Shakespeare 1564–1616: *Richard II* (1595)

11 I care not; a man can die but once; we owe God a death.
William Shakespeare 1564–1616: *Henry IV, Part 2* (1597)

12 Brightness falls from the air;
Queens have died young and fair;
Dust hath closed Helen's eye.
I am sick, I must die.
Lord have mercy on us.

Thomas Nashe 1567–1601: *Summer's Last Will and Testament* (1600)

13 To die, to sleep;
To sleep: perchance to dream: ay, there's the rub;
For in that sleep of death what dreams may come
When we have shuffled off this mortal coil,
Must give us pause.

William Shakespeare 1564–1616: *Hamlet* (1601)

14 Nothing in his life
Became him like the leaving it.

William Shakespeare 1564–1616: *Macbeth* (1606)

15 Death be not proud, though some have called thee
Mighty and dreadful, for thou art not so.

John Donne 1572–1631: 'Death, be not proud' (1609)

16 One short sleep past, we wake eternally,
And death shall be no more; Death thou shalt die.

John Donne 1572–1631: 'Death, be not proud' (1609)

17 He that dies pays all debts.

William Shakespeare 1564–1616: *The Tempest* (1611)

18 O eloquent, just, and mighty Death!...thou hast drawn together all the farstretched greatness, all the pride, cruelty, and ambition of man, and covered it all over with these two narrow words, *Hic jacet* [Here lies].

Walter Ralegh *c.*1552–1618: *The History of the World* (1614)

19 'Tis a sharp remedy, but a sure one for all ills.

Walter Ralegh *c.*1552–1618: on feeling the edge of the axe prior to his execution

20 I know death hath ten thousand several doors
For men to take their exits.

John Webster *c.*1580–*c.*1625: *The Duchess of Malfi* (1623)

21 Any man's death diminishes me, because I am involved in Mankind; And therefore never send to know for whom the bell tolls; it tolls for thee.

John Donne 1572–1631: *Devotions upon Emergent Occasions* (1624)

22 Revenge triumphs over death; love slights it; honour aspireth to it; grief flieth to it.
Francis Bacon 1561–1626: *Essays* (1625) 'Of Death'

23 How little room
Do we take up in death, that, living know
No bounds?
James Shirley 1596–1666: *The Wedding* (1629)

24 One dies only once, and it's for such a long time!
Molière 1622–73: *Le Dépit amoureux* (performed 1656)

25 The long habit of living indisposeth us for dying.
Sir Thomas Browne 1605–82: *Hydriotaphia* (Urn Burial, 1658)

26 Forasmuch as it hath pleased Almighty God of his great mercy to take unto himself the soul of our dear brother here departed, we therefore commit his body to the ground; earth to earth, ashes to ashes, dust to dust; in sure and certain hope of the Resurrection to eternal life.
Book of Common Prayer 1662: *The Burial of the Dead*

27 In the midst of life we are in death.
Book of Common Prayer 1662: *The Burial of the Dead*

28 We shall die alone.
Blaise Pascal 1623–62: *Pensées* (1670)

29 Death never takes the wise man by surprise; he is always ready to go.
Jean de la Fontaine 1621–95: *Fables* (1678–9) 'La Mort et le Mourant'

30 They that die by famine die by inches.
Matthew Henry 1662–1714: *Exposition on the Old and New Testament* (1710)

31 Can storied urn or animated bust
Back to its mansion call the fleeting breath?
Thomas Gray 1716–71: *Elegy Written in a Country Churchyard* (1751)

32 The bodies of those that made such a noise and tumult when alive, when dead, lie as quietly among the graves of their neighbours as any others.
Jonathan Edwards 1703–58: sermon on procrastination, *Miscellaneous Discourses*

33 Depend upon it, Sir, when a man knows he is to be hanged in a fortnight, it concentrates his mind wonderfully.
Samuel Johnson 1709-84: in James Boswell *Life of Johnson* (1791) 19 September 1777

34 The good die first,
And they whose hearts are dry as summer dust
Burn to the socket.
William Wordsworth 1770-1850: *The Excursion* (1814)

35 Now more than ever seems it rich to die,
To cease upon the midnight with no pain.
John Keats 1795-1821: 'Ode to a Nightingale' (1820)

36 The cemetery is an open space among the ruins, covered in winter with violets and daisies. It might make one in love with death, to think that one should be buried in so sweet a place.
Percy Bysshe Shelley 1792-1822: *Adonais* (1821) preface

37 From the contagion of the world's slow stain
He is secure, and now can never mourn
A heart grown cold, a head grown grey in vain.
Percy Bysshe Shelley 1792-1822: *Adonais* (1821)

38 Death must be distinguished from dying, with which it is often confused.
Sydney Smith 1771-1845: in H. Pearson *The Smith of Smiths* (1934)

39 And all our calm is in that balm—
Not lost but gone before.
Caroline Norton 1808-77: 'Not Lost but Gone Before'.

40 For though from out our bourne of time and place
The flood may bear me far,
I hope to see my pilot face to face
When I have crossed the bar.
Alfred, Lord Tennyson 1809-92: 'Crossing the Bar' (1889)

41 Death is nothing at all; it does not count. I have only slipped away into the next room.
Henry Scott Holland 1847-1918: sermon preached on Whitsunday, 1910

42 If I should die, think only this of me:
That there's some corner of a foreign field
That is for ever England.
Rupert Brooke 1887-1915: 'The Soldier' (1914)

43 Blow out, you bugles, over the rich Dead!
There's none of these so lonely and poor of old,
But, dying, has made us rarer gifts than gold.
Rupert Brooke 1887–1915: 'The Dead' (1914)

44 The pallor of girls' brows shall be their pall;
Their flowers the tenderness of patient minds,
And each slow dusk a drawing-down of blinds.
Wilfred Owen 1893–1918: 'Anthem for Doomed Youth' (written 1917)

45 Webster was much possessed by death
And saw the skull beneath the skin.
T. S. Eliot 1888–1965: 'Whispers of Immortality' (1919)

46 The dead don't die. They look on and help.
D. H. Lawrence 1885–1930: letter to J. Middleton Murry, 2 February 1923

47 A man's dying is more the survivors' affair than his own.
Thomas Mann 1875–1955: *The Magic Mountain* (1924), tr. H. T. Lowe-Porter

48 To die will be an awfully big adventure.
J. M. Barrie 1860–1937: *Peter Pan* (1928)

49 Nor dread nor hope attend
A dying animal;
A man awaits his end
Dreading and hoping all.
W. B. Yeats 1865–1939: 'Death' (1933)

50 He knows death to the bone—
Man has created death.
W. B. Yeats 1865–1939: 'Death' (1933)

51 Though lovers be lost love shall not;
And death shall have no dominion.
Dylan Thomas 1914–53: 'And death shall have no dominion' (1936)

52 Guns aren't lawful;
Nooses give;
Gas smells awful;
You might as well live.
Dorothy Parker 1893–1967: 'Résumé' (1937)

53 Kill a man, and you are an assassin. Kill millions of men, and you are a conqueror. Kill everyone, and you are a god.
Jean Rostand 1894–1977: *Pensées d'un biologiste* (1939)

54 For here the lover and killer are mingled
who had one body and one heart.
And death, who had the soldier singled
has done the lover mortal hurt.

Keith Douglas 1920–44: 'Vergissmeinnicht, 1943'

55 [Death is] nature's way of telling you to slow down.

Anonymous: life insurance proverb; in *Newsweek* 25 April 1960

56 A suicide kills two people, Maggie, that's what it's for!

Arthur Miller 1915– : *After the Fall* (1964)

57 Let me die a youngman's death
Not a clean & in-between-
The-sheets, holy-water death.

Roger McGough 1937– : 'Let Me Die a Youngman's Death' (1967)

58 This parrot is no more! It has ceased to be! It's expired and gone to meet its maker! This is a late parrot! It's a stiff! Bereft of life it rests in peace—if you hadn't nailed it to the perch it would be pushing up the daisies! It's rung down the curtain and joined the choir invisible! THIS IS AN EX-PARROT!

Graham Chapman 1941–89 et al.: *Monty Python's Flying Circus* (BBC TV programme,1969)

59 If there wasn't death, I think you couldn't go on.

Stevie Smith 1902–71: in *Observer* 9 November 1969

60 I don't want to achieve immortality through my work...
I want to achieve it through not dying.

Woody Allen 1935– : epigraph to Eric Lax *Woody Allen and his Comedy* (1975)

61 It's not that I'm afraid to die. I just don't want to be there when it happens.

Woody Allen 1935– : *Death* (1975)

62 Death has got something to be said for it:
There's no need to get out of bed for it.

Kingsley Amis 1922– : 'Delivery Guaranteed' (1979)

63 However many ways there may be of being alive, it is certain that there are vastly more ways of being dead.

Richard Dawkins 1941– : *The Blind Watchmaker* (1986)

Democracy

See also **Minorities and Majorities, Politics, Voting**

1 I never could believe that Providence had sent a few men into the world, ready booted and spurred to ride, and millions ready saddled and bridled to be ridden.
Richard Rumbold *c.*1622–85: on the scaffold, in T. B. Macaulay *History of England* vol. 1 (1849)

2 One man shall have one vote.
John Cartwright 1740–1824: *The People's Barrier Against Undue Influence* (1780)

3 Fourscore and seven years ago our fathers brought forth upon this continent a new nation, conceived in liberty, and dedicated to the proposition that all men are created equal... We here highly resolve that the dead shall not have died in vain, that this nation, under God, shall have a new birth of freedom; and that government of the people, by the people, and for the people, shall not perish from the earth.
Abraham Lincoln 1809–65: address at the Dedication of the National Cemetery at Gettysburg, 19 November 1863; the Lincoln Memorial inscription reads 'by the people, for the people'

4 All the world over, I will back the masses against the classes.
W. E. Gladstone 1809–98: speech in Liverpool, 28 June 1886

5 Democracy substitutes election by the incompetent many for appointment by the corrupt few.
George Bernard Shaw 1856–1950: *Man and Superman* (1903)

6 The world must be made safe for democracy.
Woodrow Wilson 1856–1924: speech to Congress, 2 April 1917

7 All the ills of democracy can be cured by more democracy.
Alfred Emanuel Smith 1873–1944: speech, 27 June 1933

8 Democracy is the recurrent suspicion that more than half of the people are right more than half of the time.
E. B. White 1899–1985: *New Yorker* 3 July 1944

9 Man's capacity for justice makes democracy possible, but man's inclination to injustice makes democracy necessary.
Reinhold Niebuhr 1892–1971: *Children of Light and Children of Darkness* (1944)

10 Democracy is the worst form of Government except all those other forms that have been tried from time to time.
Winston Churchill 1874–1965: speech, House of Commons, 11 November 1947

11 After each war there is a little less democracy to save.
Brooks Atkinson 1894–1984: *Once Around the Sun* (1951)

12 So Two cheers for Democracy: one because it admits variety and two because it permits criticism. Two cheers are quite enough: there is no occasion to give three. Only Love the Beloved Republic deserves that.
E. M. Forster 1879–1970: *Two Cheers for Democracy* (1951)
'Love, the beloved republic' borrowed from Swinburne's poem 'Hertha'

13 Democracy means government by discussion, but it is only effective if you can stop people talking.
Clement Attlee 1883–1967: speech at Oxford, 14 June 1957

14 For many Chinese, the Russian lesson appears to be that only after a nation achieves a relatively high level of economic prosperity can it afford the fruit and peril of democracy.
Xiao-Huang Yin: in *Independent* 8 October 1993

Despair see **Hope and Despair**

Determination

1 Thought shall be the harder, heart the keener, courage the greater, as our might lessens.
Anonymous: *The Battle of Maldon* (*c.*1000), tr. R. K. Gordon

2 The drop of rain maketh a hole in the stone, not by violence, but by oft falling.
Hugh Latimer *c.*1485–1555: *Second Sermon preached before the King's Majesty* (19 April 1549)

3 Perseverance, dear my lord,
Keeps honour bright.

William Shakespeare 1564–1616: *Troilus and Cressida* (1602)

4 'Tis known by the name of perseverance in a good cause,—and of obstinacy in a bad one.

Laurence Sterne 1713–68: *Tristram Shandy* (1759–67)

5 That which we are, we are;
One equal temper of heroic hearts,
Made weak by time and fate, but strong in will
To strive, to seek, to find, and not to yield.

Alfred, Lord Tennyson 1809–92: 'Ulysses' (1842)

6 Say not the struggle naught availeth,
The labour and the wounds are vain,
The enemy faints not, nor faileth,
And as things have been, things remain.

Arthur Hugh Clough 1819–61: 'Say not the struggle naught availeth' (1855)

7 I am the master of my fate:
I am the captain of my soul.

W. E. Henley 1849–1903: 'Invictus. In Memoriam R.T.H.B.' (1888)

8 Under the bludgeonings of chance
My head is bloody, but unbowed.

W. E. Henley 1849–1903: 'Invictus. In Memoriam R.T.H.B.' (1888)

9 One who never turned his back but marched breast forward,
Never doubted clouds would break,
Never dreamed, though right were worsted, wrong would triumph,
Held we fall to rise, are baffled to fight better,
Sleep to wake.

Robert Browning 1812–89: *Asolando* (1889)

10 Fanaticism consists in redoubling your effort when you have forgotten your aim.

George Santayana 1863–1952: *The Life of Reason* (1905)

11 The best way out is always through.

Robert Frost 1874–1963: 'A Servant to Servants' (1914)

12 Pick yourself up,
Dust yourself off,
Start all over again.
Dorothy Fields 1905–74: 'Pick Yourself Up' (1936 song)

13 *Nil carborundum illegitimi.*
Anonymous: cod Latin for 'Don't let the bastards grind you down', in circulation during the Second World War, though possibly of earlier origin

14 When the going gets tough, the tough get going.
Joseph P. Kennedy 1888–1969: in J. H. Cutler *Honey Fitz* (1962); also attributed to Knute Rockne

Diaries

1 And so I betake myself to that course, which is almost as much as to see myself go into my grave—for which, and all the discomforts that will accompany my being blind, the good God prepare me!
Samuel Pepys 1633–1703: diary, 31 May 1669, closing words

2 A page of my Journal is like a cake of portable soup. A little may be diffused into a considerable portion.
James Boswell 1740–95: *Journal of a Tour to the Hebrides* (1785)

3 I never travel without my diary. One should always have something sensational to read in the train.
Oscar Wilde 1854–1900: *The Importance of Being Earnest* (1895)

4 What sort of diary should I like mine to be?...I should like it to resemble some deep old desk, or capacious hold-all, in which one flings a mass of odds and ends without looking them through.
Virginia Woolf 1882–1941: diary, 20 April 1919

5 What is more dull than a discreet diary? One might just as well have a discreet soul.
Henry 'Chips' Channon 1897–1958: diary, 26 July 1935

6 I always say, keep a diary and some day it'll keep you.
Mae West 1892–1980: *Every Day's a Holiday* (1937 film)

Diplomacy

1 By indirections find directions out.
William Shakespeare 1564–1616: *Hamlet* (1601)

2 An ambassador is an honest man sent to lie abroad for the good of his country.
Henry Wotton 1568–1639: written in the album of Christopher Fleckmore in 1604

3 In things that are tender and unpleasing, it is good to break the ice by some whose words are of less weight, and to reserve the more weighty voice to come in as by chance.
Francis Bacon 1561–1626: *Essays* (1625) 'Of Cunning'

4 We are prepared to go to the gates of Hell—but no further.
Pope Pius VII 1742–1823: attempting to reach an agreement with Napoleon, *c.*1800–1

5 *Le congrès ne marche pas, il danse.*
The Congress makes no progress; it dances.
Charles-Joseph, Prince de Ligne 1735–1814: in A. de la Garde-Chambonas *Souvenirs du Congrès de Vienne* (1820)

6 I'm afraid you've got a bad egg, Mr Jones.
Oh no, my Lord, I assure you! Parts of it are excellent!
Punch: 1895

7 Speak softly and carry a big stick; you will go far.
Theodore Roosevelt 1858–1919: speech, 3 April 1903 (quoting an 'old adage')

8 Negotiating with de Valera...is like trying to pick up mercury with a fork.
David Lloyd George 1863–1945: to which de Valera replied, 'Why doesn't he use a spoon?'; in M. J. MacManus *Eamon de Valera* (1944)

9 I feel happier now that we have no allies to be polite to and to pamper.
George VI 1895–1952: to Queen Mary, 27 June 1940

10 To jaw-jaw is always better than to war-war.
Winston Churchill 1874–1965: speech at White House, 26 June 1954

11 A diplomat these days is nothing but a head-waiter who's allowed to sit down occasionally.
Peter Ustinov 1921– : *Romanoff and Juliet* (1956)

12 A diplomat... is a person who can tell you to go to hell in such a way that you actually look forward to the trip.
Caskie Stinnett 1911– : *Out of the Red* (1960)

13 Let us never negotiate out of fear. But let us never fear to negotiate.
John F. Kennedy 1917–63: inaugural address, 20 January 1961

14 Treaties, you see, are like girls and roses: they last while they last.
Charles de Gaulle 1890–1970: speech at Elysée Palace, 2 July 1963

Discontent see **Satisfaction and Discontent**

Discoveries see **Inventions and Discoveries**

Dogs

See also **Animals**

1 I am his Highness' dog at Kew;
Pray, tell me sir, whose dog are you?
Alexander Pope 1688–1744: 'Epigram Engraved on the Collar of a Dog which I gave to his Royal Highness' (1738)

2 The more one gets to know of men, the more one values dogs.
A. Toussenel 1803–85: *L'Esprit des bêtes* (1847), attributed to Mme Roland in the form 'The more I see of men, the more I like dogs'

3 The great pleasure of a dog is that you may make a fool of yourself with him and not only will he not scold you, but he will make a fool of himself too.
Samuel Butler 1835–1902: *Notebooks* (1912)

4 Brothers and Sisters, I bid you beware
Of giving your heart to a dog to tear.
Rudyard Kipling 1865–1936: 'The Power of the Dog' (1909)

5 My hand will miss the insinuated nose,
Mine eyes the tail that wagged contempt at Fate.
William Watson 1858–1936: 'An Epitaph' (for his dog)

6 Any man who hates dogs and babies can't be all bad.
Leo Rosten 1908- : of W. C. Fields, and often attributed to him, in speech at Masquers' Club dinner, 16 February 1939

7 A door is what a dog is perpetually on the wrong side of.
Ogden Nash 1902–71: 'A Dog's Best Friend is his Illiteracy' (1953)

8 That indefatigable and unsavoury engine of pollution, the dog.
John Sparrow 1906–92: letter to *The Times* 30 September 1975

Doubt see **Certainty and Doubt**

Drawing see **Painting and Drawing**

Dreams see **Sleep and Dreams**

Dress

1 The apparel oft proclaims the man.
William Shakespeare 1564–1616: *Hamlet* (1601)

2 A sweet disorder in the dress
Kindles in clothes a wantonness.
Robert Herrick 1591–1674: 'Delight in Disorder' (1648)

3 Whenas in silks my Julia goes,
Then, then (methinks) how sweetly flows
That liquefaction of her clothes.
Robert Herrick 1591–1674: 'Upon Julia's Clothes' (1648)

4 No perfumes, but very fine linen, plenty of it, and country washing.
Beau Brummell 1778–1840: in *Memoirs of Harriette Wilson* (1825)

5 She just wore
Enough for modesty—no more.
Robert Buchanan 1841–1901: 'White Rose and Red' (1873)

6 The sense of being well-dressed gives a feeling of inward tranquillity which religion is powerless to bestow.
Miss C. F. Forbes 1817–1911: in R. W. Emerson *Letters and Social Aims* (1876)

7 You should never have your best trousers on when you go out to fight for freedom and truth.
Henrik Ibsen 1828–1906: *An Enemy of the People* (1882)

8 His socks compelled one's attention without losing one's respect.
Saki (H. H. Munro) 1870–1916: *Chronicles of Clovis* (1911)

9 From the cradle to the grave, underwear first, last and all the time.
Bertolt Brecht 1898–1956: *The Threepenny Opera* (1928)

10 Men seldom make passes
At girls who wear glasses.
Dorothy Parker 1893–1967: 'News Item' (1937)

11 Where's the man could ease a heart like a satin gown?
Dorothy Parker 1893–1967: 'The Satin Dress' (1937)

12 Haute Couture should be fun, foolish and almost unwearable.
Christian Lacroix 1951– : *Observer* 27 December 1987 'Sayings of the Year'

Drink see **Food and Drink**

Economics

1 We are just statistics, born to consume resources.
Horace 65–8 BC: *Epistles*

2 It is not that pearls fetch a high price *because* men have dived for them; but on the contrary, men dive for them because they fetch a high price.
Richard Whately 1787–1863: *Introductory Lectures on Political Economy* (1832)

3 Finance is, as it were, the stomach of the country, from which all the other organs take their tone.
W. E. Gladstone 1809–98: article on finance, 1858

4 Economy is going without something you do want in case you should, some day, want something you probably won't want.

Anthony Hope 1863–1933: *The Dolly Dialogues* (1894)

5 The National Debt is a very Good Thing and it would be dangerous to pay it off, for fear of Political Economy.

W. C. Sellar 1898–1951 and **R. J. Yeatman** 1898–1968: *1066 and All That* (1930)

6 The cold metal of economic theory is in Marx's pages immersed in such a wealth of steaming phrases as to acquire a temperature not naturally its own.

J. A. Schumpeter 1883–1950: *Capitalism, Socialism and Democracy* (1942)

7 What a country calls its vital economic interests are not the things which enable its citizens to live, but the things which enable it to make war.

Simone Weil 1909–43: in W. H. Auden *A Certain World* (1971)

8 There is enough in the world for everyone's need, but not enough for everyone's greed.

Frank Buchman 1878–1961: *Remaking the World* (1947)

9 In a community where public services have failed to keep abreast of private consumption things are very different. Here, in an atmosphere of private opulence and public squalor, the private goods have full sway.

J. K. Galbraith 1908– : *The Affluent Society* (1958)

10 It's a recession when your neighbour loses his job; it's a depression when you lose yours.

Harry S. Truman 1884–1972: in *Observer* 13 April 1958

11 There's no such thing as a free lunch.

Anonymous: colloquial axiom in US economics from the 1960s, much associated with Milton Friedman; first found in printed form in Robert Heinlein *The Moon is a Harsh Mistress* (1966)

12 Call a thing immoral or ugly, soul-destroying or a degradation of man, a peril to the peace of the world or to the well-being of future generations: as long as you have not shown it to be 'uneconomic' you have not really questioned its right to exist, grow, and prosper.

E. F. Schumacher 1911–77: *Small is Beautiful* (1973)

13 Greed—for lack of a better word—is good. Greed is right. Greed works.
Stanley Weiser and **Oliver Stone** 1946- : *Wall Street* (1987 film)

14 If the policy isn't hurting, it isn't working.
John Major 1943- : on controlling inflation; speech in Northampton, 27 October 1989

15 Rising unemployment and the recession have been the price that we've had to pay to get inflation down. [Labour shouts] That is a price well worth paying.
Norman Lamont 1942- : speech in House of Commons, 16 May 1991

16 Trickle-down theory—the less than elegant metaphor that if one feeds the horse enough oats, some will pass through to the road for the sparrows.
J. K. Galbraith 1908- : *The Culture of Contentment* (1992)

Education

See also **Teaching**

1 And gladly wolde he lerne and gladly teche.
Geoffrey Chaucer *c.*1343-1400: *The Canterbury Tales* 'General Prologue'

2 That lyf so short, the craft so long to lerne.
Geoffrey Chaucer *c.*1343-1400: *The Parliament of Fowls*

3 I said...how, and why, young children, were sooner allured by love, than driven by beating, to attain good learning.
Roger Ascham 1515-68: *The Schoolmaster* (1570)

4 I would I had bestowed that time in the tongues that I have in fencing, dancing, and bear-baiting. O! had I but followed the arts!
William Shakespeare 1564-1616: *Twelfth Night* (1601)

5 I have been at my book, and am now past the craggy paths of study, and come to the flowery plains of honour and reputation.
Ben Jonson *c.*1573-1637: *Volpone* (1606)

6 And let a scholar all Earth's volumes carry,
He will be but a walking dictionary.
George Chapman *c.*1559–1634: *The Tears of Peace* (1609)

7 Histories make men wise; poets, witty; the mathematics, subtle; natural philosophy, deep; moral, grave; logic and rhetoric, able to contend.
Francis Bacon 1561–1626: *Essays* (1625) 'Of Studies'

8 Studies serve for delight, for ornament, and for ability.
Francis Bacon 1561–1626: *Essays* (1625) 'Of Studies'

9 Know then thyself, presume not God to scan;
The proper study of mankind is man.
Alexander Pope 1688–1744: *An Essay on Man* Epistle 2 (1733)

10 Public schools are the nurseries of all vice and immorality.
Henry Fielding 1707–54: *Joseph Andrews* (1742)

11 There mark what ills the scholar's life assail,
Toil, envy, want, the patron, and the jail.
Samuel Johnson 1709–84: *The Vanity of Human Wishes* (1749)

12 Examinations are formidable even to the best prepared, for the greatest fool may ask more than the wisest man can answer.
Charles Caleb Colton *c.*1780–1832: *Lacon* (1820)

13 My object will be, if possible, to form Christian men, for Christian boys I can scarcely hope to make.
Thomas Arnold 1795–1842: letter to Revd John Tucker, 2 March 1828, on his appointment to the Headmastership of Rugby School

14 C-l-e-a-n, clean, verb active, to make bright, to scour. W-i-n, win, d-e-r, der, winder, a casement. When the boy knows this out of the book, he goes and does it.
Charles Dickens 1812–70: *Nicholas Nickleby* (1839) Mr Squeers

15 'That's the reason they're called lessons,' the Gryphon remarked: 'because they lessen from day to day.'
Lewis Carroll 1832–98: *Alice's Adventures in Wonderland* (1865)

16 Soap and education are not as sudden as a massacre, but they are more deadly in the long run.
Mark Twain 1835–1910: *A Curious Dream* (1872)

17 Cauliflower is nothing but cabbage with a college education.
Mark Twain 1835–1910: *Pudd'nhead Wilson* (1894)

18 Give me a child for the first seven years, and you may do what you like with him afterwards.
Anonymous: attributed as a Jesuit maxim, in *Lean's Collectanea* (1903)

19 Nothing in education is so astonishing as the amount of ignorance it accumulates in the form of inert facts.
Henry Brooks Adams 1838–1918: *The Education of Henry Adams* (1907)

20 What one knows is, in youth, of little moment; they know enough who know how to learn.
Henry Brooks Adams 1838–1918: *The Education of Henry Adams* (1907)

21 What poor education I have received has been gained in the University of Life.
Horatio Bottomley 1860–1933: speech at the Oxford Union, 2 December 1920

22 The proper study of mankind is books.
Aldous Huxley 1894–1963: *Crome Yellow* (1921)

23 In examinations those who do not wish to know ask questions of those who cannot tell.
Walter Raleigh 1861–1922: *Laughter from a Cloud* (1923)

24 My spelling is Wobbly. It's good spelling but it Wobbles, and the letters get in the wrong places.
A. A. Milne 1882–1956: *Winnie-the-Pooh* (1926)

25 Do not on any account attempt to write on both sides of the paper at once.
W. C. Sellar 1898–1951 and **R. J. Yeatman** 1898–1968: *1066 and All That* (1930) 'Test Paper 5'

26 [Education] has produced a vast population able to read but unable to distinguish what is worth reading, an easy prey to sensations and cheap appeals.
G. M. Trevelyan 1876–1962: *English Social History* (1942)

27 No more Latin, no more French,
No more sitting on a hard board bench.
Anonymous: children's rhyme for the end of school term, in Iona and Peter Opie *Lore and Language of Schoolchildren* (1959)

28 The dread of beatings! Dread of being late!
And, greatest dread of all, the dread of games!
John Betjeman 1906–84: *Summoned by Bells* (1960)

29 The delusion that there are thousands of young people about who are capable of benefiting from university training, but have somehow failed to find their way there, is...a necessary component of the expansionist case...More will mean worse.
Kingsley Amis 1922– : *Encounter* July 1960

30 Education is what survives when what has been learned has been forgotten.
B. F. Skinner 1904–90: *New Scientist* 21 May 1964

Endeavour see **Achievement and Endeavour**

Endings see **Beginnings and Endings**

Enemies

1 If thine enemy be hungry, give him bread to eat; and if he be thirsty, give him water to drink.
For thou shalt heap coals of fire upon his head, and the Lord shall reward thee.
Bible: Proverbs

2 Love your enemies, do good to them which hate you.
Bible: St Luke

3 I wish my deadly foe, no worse
Than want of friends, and empty purse.
Nicholas Breton *c.*1545–1626: 'A Farewell to Town' (1577)

4 An open foe may prove a curse,
But a pretended friend is worse.
John Gay 1685–1732: *Fables* (1727) 'The Shepherd's Dog and the Wolf'

5 People wish their enemies dead—but I do not; I say give them the gout, give them the stone!
Lady Mary Wortley Montagu 1689–1762: quoted in letter from Horace Walpole to Earl of Harcourt, 17 September 1778

6 An injury is much sooner forgotten than an insult.
Lord Chesterfield 1694–1773: *Letters to his Son* (1774) 9 October 1746

7 You can calculate the worth of a man by the number of his enemies, and the importance of a work of art by the harm that is spoken of it.
Gustave Flaubert 1821–80: letter to Louise Colet, 14 June 1853

8 Not while I'm alive 'e ain't!
Ernest Bevin 1881–1951: reply to the observation that Nye Bevan was sometimes his own worst enemy; in R. Barclay *Ernest Bevin and the Foreign Office* (1975)

9 Better to have him inside the tent pissing out, than outside pissing in.
Lyndon Baines Johnson 1908–73: of J. Edgar Hoover, in D. Halberstam *The Best and the Brightest* (1972)

England and the English

See also **Britain, London, Places**

1 England is the paradise of women, the purgatory of men, and the hell of horses.
John Florio *c.*1553–1625: *Second Frutes* (1591)

2 This royal throne of kings, this sceptred isle,
This earth of majesty, this seat of Mars,
This other Eden, demi-paradise,
This fortress built by Nature for herself
Against infection and the hand of war,
This happy breed of men, this little world,
This precious stone set in the silver sea...
This blessèd plot, this earth, this realm, this England.
William Shakespeare 1564–1616: *Richard II* (1595)

3 Come the three corners of the world in arms,
And we shall shock them: nought shall make us rue,
If England to itself do rest but true.
William Shakespeare 1564–1616: *King John* (1591–8)

4 An Englishman,
Being flattered, is a lamb; threatened, a lion.
George Chapman *c.*1559–1634: *Alphonsus, Emperor of Germany* (1654)

5 The English take their pleasures sadly after the fashion of their country.
Maximilien de Béthune, Duc de Sully 1559–1641: attributed

6 Let not England forget her precedence of teaching nations how to live.
John Milton 1608–74: *The Doctrine and Discipline of Divorce* (1643) 'To the Parliament of England'

7 In England there are sixty different religions, and only one sauce.
Francesco Caracciolo 1752–99: attributed

8 England expects that every man will do his duty.
Horatio, Lord Nelson 1758–1805: at the battle of Trafalgar, in R. Southey *Life of Nelson* (1813)

9 And did those feet in ancient time
Walk upon England's mountains green?
And was the holy Lamb of God
On England's pleasant pastures seen?...

I will not cease from mental fight,
Nor shall my sword sleep in my hand,
Till we have built Jerusalem,
In England's green and pleasant land.
William Blake 1757–1827: *Milton* (1804–10)

10 England is a nation of shopkeepers.
Napoléon I 1769–1821: in B. O'Meara *Napoleon in Exile* (1822)

11 The French want no-one to be their *superior*. The English want *inferiors*. The Frenchman constantly raises his eyes above him with anxiety. The Englishman lowers his beneath him with satisfaction.
Alexis de Tocqueville 1805–59: *Voyage en Angleterre et en Irlande de 1835* 8 May 1835

12 Oh, to be in England
Now that April's there.
Robert Browning 1812–89: 'Home-Thoughts, from Abroad' (1845)

13 What a pity it is that we have no amusements in England but vice and religion!

Sydney Smith 1771–1845: in H. Pearson *The Smith of Smiths* (1934)

14 For he might have been a Roosian,
A French, or Turk, or Proosian,
Or perhaps Ital-ian!
But in spite of all temptations
To belong to other nations,
He remains an Englishman!

W. S. Gilbert 1836–1911: *HMS Pinafore* (1878)

15 What should they know of England who only England know?

Rudyard Kipling 1865–1936: 'The English Flag' (1892)

16 The English country gentleman galloping after a fox—the unspeakable in full pursuit of the uneatable.

Oscar Wilde 1854–1900: *A Woman of No Importance* (1893)

17 Ask any man what nationality he would prefer to be, and ninety-nine out of a hundred will tell you that they would prefer to be Englishmen.

Cecil Rhodes 1853–1902: in G. Le Sueur *Cecil Rhodes* (1913)

18 Englishmen never will be slaves: they are free to do whatever the Government and public opinion allow them to do.

George Bernard Shaw 1856–1950: *Man and Superman* (1903)

19 An Englishman thinks he is moral when he is only uncomfortable.

George Bernard Shaw 1856–1950: *Man and Superman* (1903)

20 Smile at us, pay us, pass us; but do not quite forget.
For we are the people of England, that never have spoken yet.

G. K. Chesterton 1874–1936: 'The Secret People' (1915)

21 God! I will pack, and take a train,
And get me to England once again!
For England's the one land, I know,
Where men with Splendid Hearts may go.

Rupert Brooke 1887–1915: 'The Old Vicarage, Grantchester' (1915)

22 England is the paradise of individuality, eccentricity, heresy, anomalies, hobbies, and humours.

George Santayana 1863–1952: *Soliloquies in England* (1922)

23 Mad dogs and Englishmen
Go out in the midday sun.

Noël Coward 1899–1973: 'Mad Dogs and Englishmen' (1931 song)

24 In England we have come to rely upon a comfortable time-lag of fifty years or a century intervening between the perception that something ought to be done and a serious attempt to do it.

H. G. Wells 1866–1946: *The Work, Wealth and Happiness of Mankind* (1931)

25 They go forth into it [the world] with well-developed bodies, fairly developed minds, and undeveloped hearts.

E. M. Forster 1879–1970: *Abinger Harvest* (1936) 'Notes on English Character' of public-school men

26 This Englishwoman is so refined
She has no bosom and no behind.

Stevie Smith 1902–71: 'This Englishwoman' (1937)

27 Down here it was still the England I had known in my childhood: the railway cuttings smothered in wild flowers…the red buses, the blue policemen—all sleeping the deep, deep sleep of England, from which I sometimes fear that we shall never wake till we are jerked out of it by the roar of bombs.

George Orwell 1903–50: *Homage to Catalonia* (1938)

28 I am American bred,
I have seen much to hate here—much to forgive,
But in a world where England is finished and dead,
I do not wish to live.

Alice Duer Miller 1874–1942: *The White Cliffs* (1940)

29 Think of what our Nation stands for,
Books from Boots' and country lanes,
Free speech, free passes, class distinction,
Democracy and proper drains.

John Betjeman 1906–84: 'In Westminster Abbey' (1940)

30 It resembles a family, a rather stuffy Victorian family, with not many black sheep in it but with all its cupboards bursting with skeletons…A family with the wrong members in control—that, perhaps, is as near as one can come to describing England in a phrase.

George Orwell 1903–50: *The Lion and the Unicorn* (1941)

31 The Thames is liquid history.

John Burns 1858–1943: to an American, who had compared the Thames disparagingly with the Mississippi; in *Daily Mail* 25 January 1943

32 An Englishman, even if he is alone, forms an orderly queue of one.

George Mikes 1912– : *How to be an Alien* (1946)

33 You never find an Englishman among the under-dogs—except in England, of course.

Evelyn Waugh 1903–66: *The Loved One* (1948)

34 A soggy little island huffing and puffing to keep up with Western Europe.

John Updike 1932– : of England, in *Picked Up Pieces* (1976) 'London Life' (written 1969)

35 England's not a bad country... It's just a mean, cold, ugly, divided, tired, clapped-out, post-imperial, post-industrial slag-heap covered in polystyrene hamburger cartons.

Margaret Drabble 1939– : *A Natural Curiosity* (1989)

Environment

See also **The Country and the Town, Pollution**

1 Woe unto them that join house to house, that lay field to field, till there be no place.

Bible: Isaiah

2 The desert shall rejoice, and blossom as the rose.

Bible: Isaiah

3 O all ye Green Things upon the Earth, bless ye the Lord.

Book of Common Prayer 1662: *Morning Prayer*

4 Consult the genius of the place in all.

Alexander Pope 1688–1744: *Epistles to Several Persons* 'To Lord Burlington' (1731)

5 The parks are the lungs of London.
William Pitt, Earl of Chatham 1708–78: quoted by William Windham in the House of Commons, 30 June 1808

6 The poplars are felled, farewell to the shade
And the whispering sound of the cool colonnade.
William Cowper 1731–1800: 'The Poplar-Field' (written 1784)

7 O leave this barren spot to me!
Spare, woodman, spare the beechen tree.
Thomas Campbell 1777–1844: 'The Beech-Tree's Petition' (1800)

8 And was Jerusalem builded here
Among these dark Satanic mills?
William Blake 1757–1827: *Milton* (1804–10)

9 What would the world be, once bereft
Of wet and wildness? Let them be left,
O let them be left, wildness and wet;
Long live the weeds and the wilderness yet.
Gerard Manley Hopkins 1844–89: 'Inversnaid' (written 1881)

10 The forest laments in order that Mr Gladstone may perspire.
Lord Randolph Churchill 1849–94: on Gladstone's fondness for felling trees; speech in Blackpool, 24 January 1884

11 Wiv a ladder and some glasses,
You could see to 'Ackney Marshes,
If it wasn't for the 'ouses in between.
Edgar Bateman and **George Le Brunn**: 'If it wasn't for the 'Ouses in between' (1894 song)

12 I am I plus my surroundings and if I do not preserve the latter, I do not preserve myself.
José Ortega y Gasset 1883–1955: *Meditaciones del Quijote* (1914)

13 Praise the green earth. Chance has appointed her
home, workshop, larder, middenpit.
Her lousy skin scabbed here and there by
cities provides us with name and nation.
Basil Bunting 1900–85: 'Attis: or, Something Missing' (1931)

14 It will be said of this generation that it found England a land of beauty and left it a land of 'beauty spots'.

C. E. M. Joad 1891–1953: *The Horrors of the Countryside* (1931)

15 I think that I shall never see
A billboard lovely as a tree.
Perhaps, unless the billboards fall,
I'll never see a tree at all.

Ogden Nash 1902–71: 'Song of the Open Road' (1933)

16 Come, friendly bombs, and fall on Slough!
It isn't fit for humans now,
There isn't grass to graze a cow.
Swarm over, Death!

John Betjeman 1906–84: 'Slough' (1937)

17 Little boxes on the hillside...
And they're all made out of ticky-tacky
And they all look just the same.

Malvina Reynolds 1900–78: 'Little Boxes' (1962 song); on the tract houses in the hills to the south of San Francisco

18 Over increasingly large areas of the United States, spring now comes unheralded by the return of the birds, and the early mornings are strangely silent where once they were filled with the beauty of bird song.

Rachel Carson 1907–64: *The Silent Spring* (1962)

19 Now there is one outstandingly important fact regarding Spaceship Earth, and that is that no instruction book came with it.

R. Buckminster Fuller 1895–1983: *Operating Manual for Spaceship Earth* (1969)

20 Small is beautiful.

E. F. Schumacher 1911–77: title of book (1973)

21 And that will be England gone,
The shadows, the meadows, the lanes,
The guildhalls, the carved choirs.
There'll be books; it will linger on
In galleries; but all that remains
For us will be concrete and tyres.

Philip Larkin 1922–85: 'Going, Going' (1974)

22 It is not what they built. It is what they knocked down.
It is not the houses. It is the spaces between the houses.

It is not the streets that exist. It is the streets that no longer exist.
James Fenton 1949– : *German Requiem* (1981)

23 How inappropriate to call this planet Earth when it is clearly Ocean.
Arthur C. Clarke 1917– : attributed in *Nature* 1990

Envy and Jealousy

1 Thou shalt not covet thy neighbour's house, thou shalt not covet thy neighbour's wife.
Bible: Exodus

2 Trifles light as air
Are to the jealous confirmations strong
As proofs of holy writ.
William Shakespeare 1564–1616: *Othello* (1602–4)

3 O! beware, my lord, of jealousy;
It is the green-eyed monster which doth mock
The meat it feeds on.
William Shakespeare 1564–1616: *Othello* (1602–4)

4 Some folks rail against other folks, because other folks have what some folks would be glad of.
Henry Fielding 1707–54: *Joseph Andrews* (1742)

5 Thou shalt not covet; but tradition
Approves all forms of competition.
Arthur Hugh Clough 1819–61: 'The Latest Decalogue' (1862)

6 Jealousy is no more than feeling alone against smiling enemies.
Elizabeth Bowen 1899–1973: *The House in Paris* (1935)

7 To jealousy, nothing is more frightful than laughter.
Françoise Sagan 1935– : *La Chamade* (1965)

Epitaphs

See also **Death**

1 Go, tell the Spartans, thou who passest by,
That here obedient to their laws we lie.
Simonides *c.*556–468 BC: in Herodotus *Histories* (attributed)

2 Saul and Jonathan were lovely and pleasant in their lives, and in their death they were not divided.
Bible: II Samuel

3 And some there be, which have no memorial... and are become as though they had never been born...
But these were merciful men, whose righteousness hath not been forgotten...
Their bodies are buried in peace; but their name liveth for evermore.
Bible (Apocrypha): Ecclesiasticus

4 *Et in Arcadia ego.*
And I too in Arcadia.
Anonymous: tomb inscription, of disputed meaning, often depicted in classical paintings

5 Betwixt the stirrup and the ground
Mercy I asked, mercy I found.
William Camden 1551–1623: *Remains Concerning Britain* (1605) 'Epitaphs' (for a man who fell from his horse)

6 Rest in soft peace, and, asked, say here doth lie
Ben Jonson his best piece of poetry.
Ben Jonson *c.*1573–1637: 'On My First Son' (1616)

7 Good friend, for Jesu's sake forbear
To dig the dust enclosed here.
Blest be the man that spares these stones,
And curst be he that moves my bones.
William Shakespeare 1564–1616: epitaph on his tomb, probably composed by himself

8 Here lies a great and mighty king
Whose promise none relies on;
He never said a foolish thing,
Nor ever did a wise one.
John Wilmot, Earl of Rochester 1647–80: 'The King's Epitaph' (alternatively 'Here lies our sovereign lord the

King'), to which Charles II replied 'This is very true: for my words are my own, and my actions are my ministers"; in C. E. Doble et al. *Thomas Hearne: Remarks and Collections* (1885–1921) 17 November 1706

9 Life is a jest; and all things show it.
I thought so once; but now I know it.
John Gay 1685–1732: 'My Own Epitaph' (1720)

10 *Si monumentum requiris, circumspice.*
If you seek a monument, gaze around.
Anonymous: inscription in St Paul's Cathedral, London, attributed to the son of Sir Christopher Wren 1632–1723, its architect

11 Under this stone, Reader, survey
Dead Sir John Vanbrugh's house of clay.
Lie heavy on him, Earth! for he
Laid many heavy loads on thee!
Abel Evans 1679–1737: 'Epitaph on Sir John Vanbrugh, Architect of Blenheim Palace'

12 Where fierce indignation can no longer tear his heart.
Jonathan Swift 1667–1745: Swift's epitaph. See S. Leslie *The Skull of Swift* (1928)

13 Here lies one whose name was writ in water.
John Keats 1795–1821: epitaph for himself, in R. Monckton Milnes *Life, Letters and Literary Remains of John Keats* (1848)

14 Here he lies where he longed to be;
Home is the sailor, home from sea,
And the hunter home from the hill.
Robert Louis Stevenson 1850–94: 'Requiem' (1887)

15 Hereabouts died a very gallant gentleman, Captain L. E. G. Oates of the Inniskilling Dragoons. In March 1912, returning from the Pole, he walked willingly to his death in a blizzard to try and save his comrades, beset by hardships.
E. L. Atkinson 1882–1929 and **Apsley Cherry-Garrard** 1882–1959: epitaph on cairn erected in the Antarctic, 15 November 1912

16 They shall grow not old, as we that are left grow old.
Age shall not weary them, nor the years condemn.

At the going down of the sun and in the morning
We will remember them.

Laurence Binyon 1869–1943: 'For the Fallen' (1914)

17 His foe was folly and his weapon wit.

Anthony Hope 1863–1933: inscription for W. S. Gilbert's memorial on the Victoria Embankment, London (1915)

18 When you go home, tell them of us and say,
'For your tomorrows these gave their today.'

John Maxwell Edmonds 1875–1958: *Inscriptions Suggested for War Memorials* (1919). Particularly associated with the Burma campaign of the Second World War, in the form 'For your tomorrow, we gave our today'

19 Here lies W. C. Fields. I would rather be living in Philadelphia.

W. C. Fields 1880–1946: suggested epitaph for himself, in *Vanity Fair* June 1925

Equality

See also **Human Rights**

1 He maketh his sun to rise on the evil and on the good, and sendeth rain on the just and on the unjust.

Bible: St Matthew

2 Hath not a Jew eyes? hath not a Jew hands, organs, dimensions, senses, affections, passions?

William Shakespeare 1564–1616: *The Merchant of Venice* (1596–8)

3 If you prick us, do we not bleed? if you tickle us, do we not laugh? if you poison us, do we not die? and if you wrong us, shall we not revenge?

William Shakespeare 1564–1616: *The Merchant of Venice* (1596–8)

4 Your levellers wish to level *down* as far as themselves; but they cannot bear levelling *up* to themselves.

Samuel Johnson 1709–84: in James Boswell *Life of Johnson* (1791) 21 July 1763

5 Sir, there is no settling the point of precedency between a louse and a flea.

Samuel Johnson 1709–84: on the relative merits of two minor poets, 1783; in James Boswell *Life of Johnson* (1791)

6 A man's a man for a' that.

Robert Burns 1759–96: 'For a' that and a' that' (1790)

7 When every one is somebodee,
Then no one's anybody.

W. S. Gilbert 1836–1911: *The Gondoliers* (1889)

8 The terrorist and the policeman both come from the same basket.

Joseph Conrad 1857–1924: *The Secret Agent* (1907)

9 All animals are equal but some animals are more equal than others.

George Orwell 1903–50: *Animal Farm* (1945)

10 I have a dream that one day on the red hills of Georgia the sons of former slaves and the sons of former slave owners will be able to sit down together at the table of brotherhood.

Martin Luther King 1929–68: speech at Civil Rights March in Washington, 28 August 1963

Europe and Europeans

See also **Places**

1 The age of chivalry is gone.—That of sophisters, economists, and calculators, has succeeded; and the glory of Europe is extinguished for ever.

Edmund Burke 1729–97: *Reflections on the Revolution in France* (1790)

2 Roll up that map; it will not be wanted these ten years.

William Pitt 1759–1806: of a map of Europe, on hearing of Napoleon's victory at Austerlitz, December 1805

3 Better fifty years of Europe than a cycle of Cathay.

Alfred, Lord Tennyson 1809–92: 'Locksley Hall' (1842)

4 *Qui parle Europe a tort, notion géographique.*
Whoever speaks of Europe is wrong, [it is] a geographical concept.
Otto von Bismarck 1815–98: marginal note on a letter from the Russian Chancellor Gorchakov, November 1876

5 *Je regrette l'Europe aux anciens parapets!*
I pine for Europe of the ancient parapets!
Arthur Rimbaud 1854–91: 'Le Bâteau ivre' (1883)

6 We are part of the community of Europe and we must do our duty as such.
Lord Salisbury 1830–1903: speech at Caernarvon, 10 April 1888

7 Europe is a continent of energetic mongrels.
H. A. L. Fisher 1856–1940: *A History of Europe* (1935)

8 From Stettin in the Baltic to Trieste in the Adriatic an iron curtain has descended across the Continent.
Winston Churchill 1874–1965: speech at Westminster College, Fulton, Missouri, 5 March 1946. 'Iron curtain' previously had been applied by others to the Soviet Union or her sphere of influence

9 If you open that Pandora's Box, you never know what Trojan 'orses will jump out.
Ernest Bevin 1881–1951: on the Council of Europe, in R. Barclay *Ernest Bevin and the Foreign Office* (1975)

10 When an American heiress wants to buy a man, she at once crosses the Atlantic. The only really materialistic people I have ever met have been Europeans.
Mary McCarthy 1912–89: *On the Contrary* (1961)

11 It means the end of a thousand years of history.
Hugh Gaitskell 1906–63: on a European federation; speech at Labour Party Conference, 3 October 1962

12 You ask if they were happy. This is not a characteristic of a European. To be contented—that's for the cows.
Coco Chanel 1883–1971: in A. Madsen *Coco Chanel* (1990)

Evil see **Good and Evil**

Experience

1 *Experto credite.*
Trust one who has gone through it.
Virgil 70–19 BC: *Aeneid*

2 No man's knowledge here can go beyond his experience.
John Locke 1632–1704: *Essay concerning Human Understanding* (1690)

3 Experience is the child of Thought, and Thought is the child of Action. We cannot learn men from books.
Benjamin Disraeli 1804–81: *Vivian Grey* (1826)

4 Experience is the name everyone gives to their mistakes.
Oscar Wilde 1854–1900: *Lady Windermere's Fan* (1892)

5 All experience is an arch to build upon.
Henry Brooks Adams 1838–1918: *The Education of Henry Adams* (1907)

6 Experience is not what happens to a man; it is what a man does with what happens to him.
Aldous Huxley 1894–1963: *Texts and Pretexts* (1932)

7 Experience isn't interesting till it begins to repeat itself—in fact, till it does that, it hardly *is* experience.
Elizabeth Bowen 1899–1973: *Death of the Heart* (1938)

8 We had the experience but missed the meaning.
T. S. Eliot 1888–1965: *Four Quartets* 'The Dry Salvages' (1941)

9 You should make a point of trying every experience once, excepting incest and folk-dancing.
Anonymous: Arnold Bax (1883–1953), quoting 'a sympathetic Scot' in *Farewell My Youth* (1943)

Failure see **Success and Failure**

Fame

1 Famous men have the whole earth as their memorial.
Pericles *c.*495–429 BC: in Thucydides *History of the Peloponnesian War* (tr. R. Warner)

2 *Exegi monumentum aere perennius.*
I have erected a monument more lasting than bronze.
Horace 65–8 BC: *Odes*

3 A prophet is not without honour, save in his own country, and in his own house.
Bible: St Matthew

4 Fame is like a river, that beareth up things light and swollen, and drowns things weighty and solid.
Francis Bacon 1561–1626: *Essays* (1625) 'Of Praise'

5 Fame is the spur that the clear spirit doth raise
(That last infirmity of noble mind)
To scorn delights, and live laborious days.
John Milton 1608–74: 'Lycidas' (1638)

6 Seven wealthy towns contend for HOMER dead
Through which the living HOMER begged his bread.
Anonymous: epilogue to *Aesop at Tunbridge* By No Person of Quality (1698)

7 Content thyself to be obscurely good.
When vice prevails, and impious men bear sway,
The post of honour is a private station.
Joseph Addison 1672–1719: *Cato* (1713)

8 Far from the madding crowd's ignoble strife,
Their sober wishes never learned to stray;
Along the cool sequestered vale of life
They kept the noiseless tenor of their way.
Thomas Gray 1716–71: *Elegy Written in a Country Churchyard* (1751)

9 Full many a flower is born to blush unseen,
And waste its sweetness on the desert air.

Some village-Hampden, that with dauntless breast
The little tyrant of his fields withstood;
Some mute inglorious Milton here may rest,
Some Cromwell guiltless of his country's blood.
Thomas Gray 1716–71: *Elegy Written in a Country Churchyard* (1751)

10 Every man has a lurking wish to appear considerable in his native place.
Samuel Johnson 1709–84: in James Boswell *Life of Johnson* (1791) 17 July 1771

11 Oh, talk not to me of a name great in story;
The days of our youth are the days of our glory;
And the myrtle and ivy of sweet two-and-twenty
Are worth all your laurels, though ever so plenty.

Lord Byron 1788–1824: 'Stanzas Written on the Road between Florence and Pisa, November 1821'

12 I awoke one morning and found myself famous.

Lord Byron 1788–1824: on the instantaneous success of *Childe Harold*, in Thomas Moore *Letters and Journals of Lord Byron* (1830)

13 The deed is all, the glory nothing.

Johann Wolfgang von Goethe 1749–1832: *Faust* pt. 2 (1832)

14 Martyrdom...the only way in which a man can become famous without ability.

George Bernard Shaw 1856–1950: *The Devil's Disciple* (1901)

15 Fame is a food that dead men eat,—
I have no stomach for such meat.

Henry Austin Dobson 1840–1921: 'Fame is a Food' (1906)

16 I don't care what you say about me, as long as you say *something* about me, and as long as you spell my name right.

George M. Cohan 1878–1942: said to a newspaperman in 1912

17 He's always backing into the limelight.

Lord Berners 1883–1950: of T. E. Lawrence; oral tradition

18 Whom the gods wish to destroy they first call promising.

Cyril Connolly 1903–74: *Enemies of Promise* (1938)

19 The celebrity is a person who is known for his well-knownness.

Daniel J. Boorstin 1914– : *The Image* (1961)

20 There's no such thing as bad publicity except your own obituary.

Brendan Behan 1923–64: in Dominic Behan *My Brother Brendan* (1965)

21 We're more popular than Jesus now; I don't know which will go first—rock 'n' roll or Christianity.

John Lennon 1940–80: of The Beatles; interview in *Evening Standard* 4 March 1966

22 In the future everybody will be world famous for fifteen minutes.
Andy Warhol 1927–87: in *Andy Warhol* (1968)

23 The best fame is a writer's fame: it's enough to get a table at a good restaurant, but not enough that you get interrupted when you eat.
Fran Lebowitz 1946– : in *Observer* 30 May 1993 'Sayings of the Week'

The Family

See also **Children, Parents**

1 Thy wife shall be as the fruitful vine: upon the walls of thine house.
Thy children like the olive-branches: round about thy table.
Bible: Psalm 128

2 A little more than kin, and less than kind.
William Shakespeare 1564–1616: *Hamlet* (1601)

3 He that hath wife and children hath given hostages to fortune; for they are impediments to great enterprises, either of virtue or mischief.
Francis Bacon 1561–1626: *Essays* (1625) 'Of Marriage and the Single Life'

4 We begin our public affections in our families. No cold relation is a zealous citizen.
Edmund Burke 1729–97: *Reflections on the Revolution in France* (1790)

5 A poor relation—is the most irrelevant thing in nature.
Charles Lamb 1775–1834: *Last Essays of Elia* (1833) 'Poor Relations'

6 If a man's character is to be abused, say what you will, there's nobody like a relation to do the business.
William Makepeace Thackeray 1811–63: *Vanity Fair* (1847–8)

7 All happy families resemble one another, but each unhappy family is unhappy in its own way.

Leo Tolstoy 1828–1910: *Anna Karenina* (1875–7), tr. A. and L. Maude

8 One would be in less danger
From the wiles of the stranger
If one's own kin and kith
Were more fun to be with.

Ogden Nash 1902–71: 'Family Court' (1931)

9 It is no use telling me that there are bad aunts and good aunts. At the core, they are all alike. Sooner or later, out pops the cloven hoof.

P. G. Wodehouse 1881–1975: *The Code of the Woosters* (1938)

10 The family—that dear octopus from whose tentacles we never quite escape.

Dodie Smith 1896–1990: *Dear Octopus* (1938)

11 The Princesses would never leave without me and I couldn't leave without the King, and the King will never leave.

Queen Elizabeth, the Queen Mother 1900– : on the suggestion that the royal family be evacuated during the Blitz

12 Far from being the basis of the good society, the family, with its narrow privacy and tawdry secrets, is the source of all our discontents.

Edmund Leach 1910– : BBC Reith Lectures, 1967

13 I have never understood this liking for war. It panders to instincts already catered for within the scope of any respectable domestic establishment.

Alan Bennett 1934– : *Forty Years On* (1969)

Fate

1 Canst thou bind the sweet influences of Pleiades, or loose the bands of Orion?

Bible: Job

2 There's a divinity that shapes our ends,
Rough-hew them how we will.

William Shakespeare 1564–1616: *Hamlet* (1601)

3 Not a whit, we defy augury; there's a special providence in the fall of a sparrow. If it be now, 'tis not to come; if it be not to come, it will be now; if it be not now, yet it will come: the readiness is all.

William Shakespeare 1564–1616: *Hamlet* (1601)

4 Our remedies oft in ourselves do lie
Which we ascribe to heaven.

William Shakespeare 1564–1616: *All's Well that Ends Well* (1603–4)

5 We are merely the stars' tennis-balls, struck and bandied
Which way please them.

John Webster *c.*1580–*c.*1625: *The Duchess of Malfi* (1623)

6 Every bullet has its billet.

William III 1650–1702: in John Wesley's diary, 6 June 1765

7 Must it be? It must be.

Ludwig van Beethoven 1770–1827: *String Quartet in F Major* (1827) epigraph

8 Out flew the web and floated wide;
The mirror cracked from side to side;
'The curse is come upon me,' cried
The Lady of Shalott.

Alfred, Lord Tennyson 1809–92: 'The Lady of Shalott' (1832)

9 There once was an old man who said, 'Damn!
It is borne in upon me I am
An engine that moves
In determinate grooves,
I'm not even a bus, I'm a tram.'

Maurice Evan Hare 1886–1967: 'Limerick' (1905)

10 Fate is not an eagle, it creeps like a rat.

Elizabeth Bowen 1899–1973: *The House in Paris* (1935)

11 The spring is wound up tight. It will uncoil of itself. That is what is so convenient in tragedy. The least little turn of the wrist will do the job. Anything will set it going.

Jean Anouilh 1910–87: *Antigone* (1944), tr. L. Galantiere

Fear

1 Present fears
Are less than horrible imaginings.
William Shakespeare 1564–1616: *Macbeth* (1606)

2 Letting 'I dare not' wait upon 'I would,'
Like the poor cat i' the adage.
William Shakespeare 1564–1616: *Macbeth* (1606)

3 In time we hate that which we often fear.
William Shakespeare 1564–1616: *Antony and Cleopatra* (1606–7)

4 Every drop of ink in my pen ran cold.
Horace Walpole 1717–97: letter to George Montagu, 30 July 1752

5 No passion so effectually robs the mind of all its powers of acting and reasoning as fear.
Edmund Burke 1729–97: *On the Sublime and Beautiful* (1757)

6 If hopes were dupes, fears may be liars.
Arthur Hugh Clough 1819–61: 'Say not the struggle naught availeth' (1855)

7 I have seen the moment of my greatness flicker,
And I have seen the eternal Footman hold my coat, and snicker,
And in short, I was afraid.
T. S. Eliot 1888–1965: 'Love Song of J. Alfred Prufrock' (1917)

8 I will show you fear in a handful of dust.
T. S. Eliot 1888–1965: *The Waste Land* (1922)

9 The only thing we have to fear is fear itself.
Franklin D. Roosevelt 1882–1945: inaugural address, 4 March 1933

10 We must travel in the direction of our fear.
John Berryman 1914–72: 'A Point of Age' (1942)

Flowers

1 That wel by reson men it calle may
The 'dayesye,' or elles the 'ye of day,'
The emperice and flour of floures alle.
Geoffrey Chaucer c.1343–1400: *The Legend of Good Women*

2 I know a bank whereon the wild thyme blows,
Where oxlips and the nodding violet grows
Quite over-canopied with luscious woodbine,
With sweet musk-roses, and with eglantine.
William Shakespeare 1564–1616: *A Midsummer Night's Dream* (1595–6)

3 Daffodils,
That come before the swallow dares, and take
The winds of March with beauty.
William Shakespeare 1564–1616: *The Winter's Tale* (1610–11)

4 Pale prime-roses,
That die unmarried, ere they can behold
Bright Phoebus in his strength,—a malady
Most incident to maids.
William Shakespeare 1564–1616: *The Winter's Tale* (1610–11)

5 For you there's rosemary and rue; these keep
Seeming and savour all the winter long.
William Shakespeare 1564–1616: *The Winter's Tale* (1610–11)

6 The marigold, that goes to bed wi' the sun,
And with him rises weeping.
William Shakespeare 1564–1616: *The Winter's Tale* (1610–11)

7 I wandered lonely as a cloud
That floats on high o'er vales and hills,
When all at once I saw a crowd,
A host, of golden daffodils;
Beside the lake, beneath the trees,
Fluttering and dancing in the breeze.
William Wordsworth 1770–1850: 'I wandered lonely as a cloud' (1815 ed.)

8 Daisies, those pearled Arcturi of the earth,
The constellated flower that never sets.
Percy Bysshe Shelley 1792–1822: 'The Question' (1822)

9 I sometimes think that never blows so red
The rose as where some buried Caesar bled.
Edward Fitzgerald 1809–83: *The Rubáiyát of Omar Khayyám* (1859)

10 Oh, no man knows
Through what wild centuries
Roves back the rose.
Walter de la Mare 1873–1956: 'All That's Past' (1912)

11 Unkempt about those hedges blows
An English unofficial rose.
Rupert Brooke 1887–1915: 'The Old Vicarage, Grantchester' (1915)

Food and Drink

See also **Alcohol**

1 She brought forth butter in a lordly dish.
Bible: Judges

2 The appetite grows by eating.
François Rabelais *c.*1494–*c.*1553: *Gargantua* (1534)

3 I am a great eater of beef, and I believe that does harm to my wit.
William Shakespeare 1564–1616: *Twelfth Night* (1601)

4 Hunger is the best sauce in the world.
Cervantes 1547–1616: *Don Quixote* (1605)

5 Now good digestion wait on appetite,
And health on both!
William Shakespeare 1564–1616: *Macbeth* (1606)

6 A good, honest, wholesome, hungry breakfast.
Izaak Walton 1593–1683: *The Compleat Angler* (1653)

7 One should eat to live, and not live to eat.
Molière 1622–73: *L'Avare* (1669)

8 Coffee, (which makes the politician wise,
And see thro' all things with his half-shut eyes).
Alexander Pope 1688–1744: *The Rape of the Lock* (1714)

9 [Tar water] is of a nature so mild and benign and proportioned to the human constitution, as to warm without heating, to cheer but not inebriate.
Bishop George Berkeley 1685–1753: *Siris* (1744)

10 Take your hare when it is cased.
Hannah Glasse fl. 1747: *The Art of Cookery Made Plain and Easy* (1747) (*cased* skinned); the proverbial 'First catch your hare' dates from *c.*1300

11 A cucumber should be well sliced, and dressed with pepper and vinegar, and then thrown out, as good for nothing.
Samuel Johnson 1709–84: in James Boswell *Journal of a Tour to the Hebrides* (1785) 5 October 1773

12 Heaven sends us good meat, but the Devil sends cooks.
David Garrick 1717–79: 'On Doctor Goldsmith's Characteristical Cookery' (1777)

13 For my part now, I consider supper as a turnpike through which one must pass, in order to get to bed.
Oliver Edwards 1711–91: in James Boswell *Life of Samuel Johnson*

14 Some have meat and cannot eat,
Some cannot eat that want it:
But we have meat and we can eat,
Sae let the Lord be thankit.
Robert Burns 1759–96: 'The Kirkudbright Grace' (1790), also known as 'The Selkirk Grace'

15 That all-softening, overpowering knell,
The tocsin of the soul—the dinner bell.
Lord Byron 1788–1824: *Don Juan* (1819–24)

16 Tell me what you eat and I will tell you what you are.
Anthelme Brillat-Savarin 1755–1826: *Physiologie du Goût* (1825)

17 Please, sir, I want some more.
Charles Dickens 1812–70: *Oliver Twist* (1838)

18 Serenely full, the epicure would say,
Fate cannot harm me, I have dined to-day.
Sydney Smith 1771–1845: in Lady Holland *Memoir* (1855)

19 Madam, I have been looking for a person who disliked gravy all my life; let us swear eternal friendship.
Sydney Smith 1771–1845: in Lady Holland *Memoir* (1855)

20 Kissing don't last: cookery do!
George Meredith 1828–1909: *The Ordeal of Richard Feverel* (1859)

21 We each day dig our graves with our teeth.
Samuel Smiles 1812–1904: *Duty* (1880)

22 The cook was a good cook, as cooks go; and as good cooks go, she went.
Saki (H. H. Munro) 1870–1916: *Reginald* (1904) 'Reginald on Besetting Sins'

23 Is there no Latin word for Tea? Upon my soul, if I had known that I would have let the vulgar stuff alone.
Hilaire Belloc 1870–1953: *On Nothing* (1908)

24 Tea, although an Oriental,
Is a gentleman at least;
Cocoa is a cad and coward,
Cocoa is a vulgar beast.
G. K. Chesterton 1874–1936: 'Song of Right and Wrong' (1914)

25 Time for a little something.
A. A. Milne 1882–1956: *Winnie-the-Pooh* (1926)

26 Last night we went to a Chinese dinner at six and a French dinner at nine, and I can feel the sharks' fins navigating unhappily in the Burgundy.
Peter Fleming 1907–71: letter from Yunnanfu, 20 March 1938

27 And now with some pleasure I find that it's seven; and must cook dinner. Haddock and sausage meat. I think it is true that one gains a certain hold on sausage and haddock by writing them down.
Virginia Woolf 1882–1941: diary, 8 March 1941

28 Milk's leap toward immortality.
Clifton Fadiman 1904– : of cheese; *Any Number Can Play* (1957)

29 I never see any home cooking. All I get is fancy stuff.
Prince Philip, Duke of Edinburgh 1921– : in *Observer* 28 October 1962

30 If I had the choice between smoked salmon and tinned salmon, I'd have it tinned. With vinegar.
Harold Wilson 1916– : in *Observer* 11 November 1962

31 Take away that pudding—it has no theme.
Winston Churchill 1874–1965: in Lord Home *The Way the Wind Blows* (1976)

Fools and Foolishness

1 Answer not a fool according to his folly, lest thou also be like unto him.
Answer a fool according to his folly, lest he be wise in his own conceit.
Bible: Proverbs

2 As the crackling of thorns under a pot, so is the laughter of a fool.
Bible: Ecclesiastes

3 *Misce stultitiam consiliis brevem:*
Dulce est desipere in loco.
Mix a little foolishness with your prudence: it's good to be silly at the right moment.
Horace 65–8 BC: *Odes*

4 For ye suffer fools gladly, seeing ye yourselves are wise.
Bible: II Corinthians

5 I am two fools, I know,
For loving, and for saying so
In whining poetry.
John Donne 1572–1631: 'The Triple Fool'

6 A knowledgeable fool is a greater fool than an ignorant fool.
Molière 1622–73: *Les Femmes savantes* (1672)

7 The rest to some faint meaning make pretence,
But Shadwell never deviates into sense.
John Dryden 1631–1700: *MacFlecknoe* (1682)

8 The world is full of fools, and he who would not see it should live alone and smash his mirror.
Anonymous: adaptation of an original form attributed to Claude Le Petit (1640–65) in *Discours satiriques* (1686)

9 Fools rush in where angels fear to tread.
Alexander Pope 1688–1744: *An Essay on Criticism* (1711)

10 Be wise with speed;
A fool at forty is a fool indeed.
Edward Young 1683–1765: *The Love of Fame* (1725–8)

11 Sir, I admit your gen'ral rule
That every poet is a fool:
But you yourself may serve to show it,
That every fool is not a poet.
Alexander Pope 1688–1744: 'Epigram from the French' (1732)

12 The picture, placed the busts between,
Adds to the thought much strength:
Wisdom and Wit are little seen,
But Folly's at full length.
Jane Brereton 1685–1740: 'On Mr Nash's Picture at Full Length, between the Busts of Sir Isaac Newton and Mr Pope' (1744)

13 If the fool would persist in his folly he would become wise.
William Blake 1757–1827: *The Marriage of Heaven and Hell* (1790–3) 'Proverbs of Hell'

14 A fool sees not the same tree that a wise man sees.
William Blake 1757–1827: *The Marriage of Heaven and Hell* (1790–3) 'Proverbs of Hell'

15 You may fool all the people some of the time; you can even fool some of the people all the time; but you can't fool all of the people all the time.
Abraham Lincoln 1809–65: in A. McClure *Lincoln's Yarns and Stories* (1904); also attributed to Phineas Barnum

16 Hain't we got all the fools in town on our side? and ain't that a big enough majority in any town?
Mark Twain 1835–1910: *The Adventures of Huckleberry Finn* (1884)

17 There's a sucker born every minute.
Phineas T. Barnum 1810–91: attributed

18 Let us be thankful for the fools. But for them the rest of us could not succeed.
Mark Twain 1835–1910: *Following the Equator* (1897)

19 The follies which a man regrets most, in his life, are those which he didn't commit when he had the opportunity.
Helen Rowland 1875–1950: *A Guide to Men* (1922)

20 Never give a sucker an even break.
W. C. Fields 1880–1946: title of a W. C. Fields film (1941); the catch-phrase is said to have originated in the musical comedy *Poppy* (1923)

Forgiveness

1 Her sins, which are many, are forgiven; for she loved much.
Bible: St Luke

2 God may pardon you, but I never can.
Elizabeth I 1533–1603: to the dying Countess of Nottingham

3 To err is human; to forgive, divine.
Alexander Pope 1688–1744: *An Essay on Criticism* (1711)

4 I shall be an autocrat: that's my trade. And the good Lord will forgive me: that's his.
Empress Catherine the Great 1729–96: attributed

5 And blessings on the falling out
That all the more endears,
When we fall out with those we love
And kiss again with tears!
Alfred, Lord Tennyson 1809–92: *The Princess* (1847) song (added 1850)

6 Youth, which is forgiven everything, forgives itself nothing: age, which forgives itself everything, is forgiven nothing.
George Bernard Shaw 1856–1950: *Man and Superman* (1903)

7 After such knowledge, what forgiveness?
T. S. Eliot 1888–1965: 'Gerontion' (1920)

8 The stupid neither forgive nor forget; the naïve forgive and forget; the wise forgive but do not forget.
Thomas Szasz 1920– : *The Second Sin* (1973)

France and the French

See also **Places**

1 *France, mère des arts, des armes et des lois.*
France, mother of arts, of warfare, and of laws.
Joachim Du Bellay 1522–60: *Les Regrets* (1558)

2 That sweet enemy, France.
Philip Sidney 1554–86: *Astrophil and Stella* (1591)

3 Tilling and grazing are the two breasts by which France is fed.
Maximilien de Béthune, Duc de Sully 1559–1641: *Mémoires* (1638)

4 They order, said I, this matter better in France.
Laurence Sterne 1713–68: *A Sentimental Journey* (1768)

5 *Ce qui n'est pas clair n'est pas français.*
What is not clear is not French.
Antoine de Rivarol 1753–1801: *Discours sur l'Universalité de la Langue Française* (1784)

6 The French want no-one to be their *superior*. The English want *inferiors*. The Frenchman constantly raises his eyes above him with anxiety. The Englishman lowers his beneath him with satisfaction.
Alexis de Tocqueville 1805–59: *Voyage en Angleterre et en Irlande de 1835* 8 May 1835

7 France was long a despotism tempered by epigrams.
Thomas Carlyle 1795–1881: *History of the French Revolution* (1837)

8 France, famed in all great arts, in none supreme.
Matthew Arnold 1822–88: 'To a Republican Friend—Continued' (1849)

9 The best thing I know between France and England is—the sea.
Douglas Jerrold 1803–57: *Wit and Opinions* (1859) 'The Anglo-French Alliance'

10 If the French noblesse had been capable of playing cricket with their peasants, their chateaux would never have been burnt.
G. M. Trevelyan 1876–1962: *English Social History* (1942)

11 How can you govern a country which has 246 varieties of cheese?

Charles de Gaulle 1890–1970: in E. Mignon *Les Mots du Général* (1962)

Friendship

1 Intreat me not to leave thee, or to return from following after thee: for whither thou goest, I will go; and where thou lodgest, I will lodge: thy people shall be my people, and thy God my God:

Bible: Ruth

2 One soul inhabiting two bodies.

Aristotle 384–322 BC: definition of a friend, in Diogenes Laertius *Lives of Philosophers*

3 Friendship is constant in all other things
Save in the office and affairs of love.

William Shakespeare 1564–1616: *Much Ado About Nothing* (1598–9)

4 A crowd is not company, and faces are but a gallery of pictures, and talk but a tinkling cymbal, where there is no love.

Francis Bacon 1561–1626: *Essays* (1625) 'Of Friendship'

5 It redoubleth joys, and cutteth griefs in halves.

Francis Bacon 1561–1626: *Essays* (1625) 'Of Friendship'

6 If a man does not make new acquaintance as he advances through life, he will soon find himself left alone. A man, Sir, should keep his friendship in constant repair.

Samuel Johnson 1709–84: in James Boswell *Life of Johnson* (1791) 1755

7 If it is abuse,—why one is always sure to hear of it from one damned goodnatured friend or another!

Richard Brinsley Sheridan 1751–1816: *The Critic* (1779)

8 Should auld acquaintance be forgot
And never brought to mind?

Robert Burns 1759–96: 'Auld Lang Syne' (1796)

9 Give me the avowed, erect and manly foe;
Firm I can meet, perhaps return the blow;

But of all plagues, good Heaven, thy wrath can send,
Save me, oh, save me, from the candid friend.
George Canning 1770–1827: 'New Morality' (1821)

10 Of two close friends, one is always the slave of the other.
Mikhail Lermontov 1814–41: *A Hero of our Time* (1840) tr. P. Longworth

11 The only reward of virtue is virtue; the only way to have a friend is to be one.
Ralph Waldo Emerson 1803–82: *Essays* (1841) 'Friendship'

12 There is no man so friendless but what he can find a friend sincere enough to tell him disagreeable truths.
Edward Bulwer-Lytton 1803–73: *What will he do with it?* (1857)

13 A woman can become a man's friend only in the following stages—first an acquaintance, next a mistress, and only then a friend.
Anton Chekhov 1860–1904: *Uncle Vanya* (1897)

14 I have lost friends, some by death...others through sheer inability to cross the street.
Virginia Woolf 1882–1941: *The Waves* (1931)

15 Oh I get by with a little help from my friends,
Mm, I get high with a little help from my friends.
John Lennon 1940–80 and **Paul McCartney** 1942– : 'With a Little Help From My Friends' (1967 song)

16 I do not believe that friends are necessarily the people you like best, they are merely the people who got there first.
Peter Ustinov 1921– : *Dear Me* (1977)

The Future

1 Lord! we know what we are, but know not what we may be.
William Shakespeare 1564–1616: *Hamlet* (1601)

2 If you can look into the seeds of time,
And say which grain will grow and which will not.
William Shakespeare 1564–1616: *Macbeth* (1606)

3 'We are always doing', says he, 'something for Posterity, but I would fain see Posterity do something for us.'
Joseph Addison 1672–1719: *The Spectator* 20 August 1714

4 People will not look forward to posterity, who never look backward to their ancestors.
Edmund Burke 1729–97: *Reflections on the Revolution in France* (1790)

5 You can never plan the future by the past.
Edmund Burke 1729–97: *Letter to a Member of the National Assembly* (1791)

6 You cannot fight against the future. Time is on our side.
W. E. Gladstone 1809–98: speech on the Reform Bill, House of Commons, 27 April 1866

7 Such is: what is to be?
The pulp so bitter, how shall taste the rind?
Francis Thompson 1859–1907: 'The Hound of Heaven' (1913)

8 I never think of the future. It comes soon enough.
Albert Einstein 1879–1955: in an interview given on the *Belgenland*, December 1930

9 We have trained them [men] to think of the Future as a promised land which favoured heroes attain—not as something which everyone reaches at the rate of sixty minutes an hour, whatever he does, whoever he is.
C. S. Lewis 1898–1963: *The Screwtape Letters* (1942)

10 The empires of the future are the empires of the mind.
Winston Churchill 1874–1965: speech at Harvard, 6 September 1943

11 If you want a picture of the future, imagine a boot stamping on a human face—for ever.
George Orwell 1903–50: *Nineteen Eighty-Four* (1949)

12 The future ain't what it used to be.
Yogi Berra 1925– : attributed

Gardens

1 And the Lord God planted a garden eastward in Eden.
Bible: Genesis

2 Sowe Carrets in your Gardens, and humbly praise God for them, as for a singular and great blessing.
Richard Gardiner b. *c.*1533: *Profitable Instructions for the Manuring, Sowing and Planting of Kitchen Gardens* (1599)

3 God Almighty first planted a garden; and, indeed, it is the purest of human pleasures.
Francis Bacon 1561–1626: *Essays* (1625) 'Of Gardens'

4 Annihilating all that's made
To a green thought in a green shade.
Andrew Marvell 1621–78: 'The Garden' (1681)

5 I value my garden more for being full of blackbirds than of cherries, and very frankly give them fruit for their songs.
Joseph Addison 1672–1719: *The Spectator* 6 September 1712

6 A garden was the primitive prison till man with Promethean felicity and boldness luckily sinned himself out of it.
Charles Lamb 1775–1834: letter to William Wordsworth, 22 January 1830

7 Come into the garden, Maud,
For the black bat, night, has flown,
Come into the garden, Maud,
I am here at the gate alone;
And the woodbine spices are wafted abroad,
And the musk of the rose is blown.
Alfred, Lord Tennyson 1809–92: *Maud* (1855)

8 What is a weed? A plant whose virtues have not been discovered.
Ralph Waldo Emerson 1803–82: *Fortune of the Republic* (1878)

9 A garden is a lovesome thing, God wot!
T. E. Brown 1830–97: 'My Garden' (1893)

10 Our England is a garden, and such gardens are not made
By singing:—'Oh, how beautiful!' and sitting in the shade,
While better men than we go out and start their working lives
At grubbing weeds from gravel paths with broken dinner-knives.
Rudyard Kipling 1865–1936: 'The Glory of the Garden' (1911)

11 The kiss of the sun for pardon,
The song of the birds for mirth,
One is nearer God's Heart in a garden
Than anywhere else on earth.
Dorothy Frances Gurney 1858–1932: 'God's Garden' (1913)

12 I will keep returning to the virtues of sharp and swift drainage, whether a plant prefers to be wet or dry...I would have called this book Better Drains, but you would never have bought it or borrowed it for bedtime.

Robin Lane Fox 1946- : *Better Gardening* (1982)

The Generation Gap

See also **Youth**

1 *Si jeunesse savait; si vieillesse pouvait.*
If youth knew; if age could.

Henri Estienne 1531–98: *Les Prémices* (1594)

2 Age is deformed, youth unkind,
We scorn their bodies, they our mind.

Thomas Bastard 1566–1618: *Chrestoleros* (1598)

3 Crabbed age and youth cannot live together:
Youth is full of pleasance, age is full of care.

William Shakespeare 1564–1616: *The Passionate Pilgrim* (1599)

4 When I was a boy of 14, my father was so ignorant I could hardly stand to have the old man around. But when I got to be 21, I was astonished at how much the old man had learned in seven years.

Mark Twain 1835–1910: attributed in *Reader's Digest* September 1939, but not traced in his works

5 Where, where but here have Pride and Truth,
That long to give themselves for wage,
To shake their wicked sides at youth
Restraining reckless middle age?

W. B. Yeats 1865–1939: 'On hearing that the Students of our New University have joined the Agitation against Immoral Literature' (1912)

6 It's all that the young can do for the old, to shock them and keep them up to date.

George Bernard Shaw 1856–1950: *Fanny's First Play* (1914)

7 Every generation revolts against its fathers and makes friends with its grandfathers.

Lewis Mumford 1895–1982: *The Brown Decades* (1931)

8 Come mothers and fathers,
Throughout the land
And don't criticize
What you can't understand.
Bob Dylan 1941– : 'The Times They Are A-Changing' (1964 song)

9 Hope I die before I get old.
Pete Townshend 1945– : 'My Generation' (1965 song)

10 Each year brings new problems of Form and Content,
new foes to tug with: at Twenty I tried to
vex my elders, past Sixty it's the young whom
I hope to bother.
W. H. Auden 1907–73: 'Shorts I' (1969)

Genius

1 Great wits are sure to madness near allied,
And thin partitions do their bounds divide.
John Dryden 1631–1700: *Absalom and Achitophel* (1681)

2 When a true genius appears in the world, you may know him by this sign, that the dunces are all in confederacy against him.
Jonathan Swift 1667–1745: *Thoughts on Various Subjects* (1711)

3 The true genius is a mind of large general powers, accidentally determined to some particular direction.
Samuel Johnson 1709–84: *Lives of the English Poets* (1779–81)

4 Genius is only a greater aptitude for patience.
Comte de Buffon 1707–88: in H. de Séchelles *Voyage à Montbar* (1803)

5 Rules and models destroy genius and art.
William Hazlitt 1778–1830: *Sketches and Essays* (1839) 'On Taste'

6 'Genius' (which means transcendent capacity of taking trouble, first of all).
Thomas Carlyle 1795–1881: *History of Frederick the Great* (1858–65)

7 Genius does what it must, and Talent does what it can.
Owen Meredith 1831–91: 'Last Words of a Sensitive Second-Rate Poet' (1868)

8 I have nothing to declare except my genius.
Oscar Wilde 1854–1900: at the New York Custom House; in Frank Harris *Oscar Wilde* (1918)

9 Genius is one per cent inspiration, ninety-nine per cent perspiration.
Thomas Alva Edison 1847–1931: said *c.*1903, in *Harper's Monthly Magazine* September 1932

10 Little minds are interested in the extraordinary; great minds in the commonplace.
Elbert Hubbard 1859–1915: *Thousand and One Epigrams* (1911)

11 Mediocrity knows nothing higher than itself, but talent instantly recognizes genius.
Arthur Conan Doyle 1859–1930: *The Valley of Fear* (1915)

12 Every positive value has its price in negative terms... The genius of Einstein leads to Hiroshima.
Pablo Picasso 1881–1973: in F. Gilot and C. Lake *Life With Picasso* (1964)

Gifts and Giving

1 *Inopi beneficium bis dat qui dat celeriter.*
He gives the poor man twice as much good who gives quickly.
Publilius Syrus 1st century BC: *Sententiae* (proverbially '*Bis dat qui cito dat* [He gives twice who gives soon]')

2 When thou doest alms, let not thy left hand know what thy right hand doeth.
Bible: St Matthew

3 Give, and it shall be given unto you; good measure, pressed down, and shaken together, and running over.
Bible: St Luke

4 It is more blessed to give than to receive.
Bible: Acts of the Apostles

5 God loveth a cheerful giver.
Bible: II Corinthians

6 Thy necessity is yet greater than mine.
Philip Sidney 1554–86: on giving his water-bottle to a dying soldier on the battle-field of Zutphen, 1586; in Fulke Greville *Life of Sir Philip Sidney* (1652). Commonly quoted 'thy need is greater than mine'

7 When they will not give a doit to relieve a lame beggar, they will lay out ten to see a dead Indian.
William Shakespeare 1564–1616: *The Tempest* (1611)

8 Item, I give unto my wife my second best bed, with the furniture.
William Shakespeare 1564–1616: Will, 1616

9 Surprises are foolish things. The pleasure is not enhanced, and the inconvenience is often considerable.
Jane Austen 1775–1817: *Emma* (1816)

10 Behold, I do not give lectures or a little charity,
When I give I give myself.
Walt Whitman 1819–92: 'Song of Myself' (written 1855)

11 Why is it no one ever sent me yet
One perfect limousine, do you suppose?
Ah no, it's always just my luck to get
One perfect rose.
Dorothy Parker 1893–1967: 'One Perfect Rose' (1937)

12 No one would remember the Good Samaritan if he'd only had good intentions. He had money as well.
Margaret Thatcher 1925– : television interview, 6 January 1986

God

See also **The Bible, Religion**

1 In the beginning was the Word, and the Word was with God, and the Word was God.
Bible: St John

2 He that loveth not knoweth not God; for God is love.
Bible: I John

3 The nature of God is a circle of which the centre is everywhere and the circumference is nowhere.
Anonymous: said to have been traced to a lost treatise of Empedocles; quoted in the *Roman de la Rose*, and by St Bonaventura

4 For man proposes, but God disposes.
Thomas à Kempis *c.*1380–1471: *De Imitatione Christi*

5 O Lord, to what a state dost Thou bring those who love Thee!
St Teresa of Ávila 1512–82: *Interior Castle* (tr. Benedictines of Stanbrook, 1921)

6 Our God, our help in ages past
Our hope for years to come,
Our shelter from the stormy blast,
And our eternal home.
Isaac Watts 1674–1748: *Psalms of David Imitated* (1719) 'Our God' altered to 'O God' by John Wesley, 1738

7 If the triangles were to make a God they would give him three sides.
Montesquieu 1689–1755: *Lettres Persones* (1721), tr. J. Ozell, 1722

8 God is on the side not of the heavy battalions, but of the best shots.
Voltaire 1694–1778: 'The Piccini Notebooks' (*c.*1735–50)

9 If God did not exist, it would be necessary to invent him.
Voltaire 1694–1778: *Épîtres* (1769)

10 God moves in a mysterious way
His wonders to perform;
He plants his footsteps in the sea,
And rides upon the storm.
William Cowper 1731–1800: 'Light Shining out of Darkness' (1779)

11 All service ranks the same with God—
With God, whose puppets, best and worst,
Are we: there is no last nor first.
Robert Browning 1812–89: *Pippa Passes* (1841)

12 The word is the Verb, and the Verb is God.
Victor Hugo 1802–85: *Contemplations* (1856)

13 And almost every one when age,
Disease, or sorrows strike him,

Inclines to think there is a God,
Or something very like Him.
Arthur Hugh Clough 1819–61: *Dipsychus* (1865)

14 Unresting, unhasting, and silent as light,
Nor wanting, nor wasting, thou rulest in might.
Walter Chalmers Smith 1824–1908: 'Immortal, invisible, God only wise' (1867 hymn)

15 Though the mills of God grind slowly, yet they grind exceeding small;
Though with patience He stands waiting, with exactness grinds He all.
Henry Wadsworth Longfellow 1807–82: 'Retribution' (1870); translation of Friedrich von Logau *Sinnegedichte* (1654), being itself a translation of an anonymous line in Sextus Empiricus *Adversus Mathematicos*

16 An honest God is the noblest work of man.
Robert G. Ingersoll 1833–99: *The Gods* (1876)

17 Too high a price is asked for harmony; it's beyond our means to pay so much to enter. And so I hasten to give back my entrance ticket...It's not God that I don't accept, Alyosha, only I most respectfully return Him the ticket.
Fedor Dostoevsky 1821–81: *The Brothers Karamazov* (1879–80)

18 God is subtle but he is not malicious.
Albert Einstein 1879–1955: remark made at Princeton University, May 1921

19 There once was a man who said, 'God
Must think it exceedingly odd
If he finds that this tree
Continues to be
When there's no one about in the Quad.'
Monsignor Ronald Knox 1888–1957: in L. Reed *Complete Limerick Book* (1924), to which came the anonymous reply: 'Dear Sir, / Your astonishment's odd: / *I* am always about in the Quad. / And that's why the tree / Will continue to be, / Since observed by / Yours faithfully, / God'

20 God is on everyone's side...And, in the last analysis, he is on the side of those with plenty of money and large armies.
Jean Anouilh 1910–87: *L'Alouette* (1953)

21 Operationally, God is beginning to resemble not a ruler but the last fading smile of a cosmic Cheshire cat.
Julian Huxley 1887–1975: *Religion without Revelation* (1957 ed.)

22 God is really only another artist. He invented the giraffe, the elephant, and the cat. He has no real style. He just goes on trying other things.
Pablo Picasso 1881–1973: in F. Gilot and C. Lake *Life With Picasso* (1964)

23 God seems to have left the receiver off the hook, and time is running out.
Arthur Koestler 1905–83: *The Ghost in the Machine* (1967)

24 I am not clear that God manoeuvres physical things... After all, a conjuring trick with bones only proves that it is as clever as a conjuring trick with bones.
David Jenkins, Bishop of Durham 1925– : on the Resurrection; radio interview, 4 October 1984

Good and Evil

See also **Virtue and Vice**

1 There is no peace, saith the Lord, unto the wicked.
Bible: Isaiah

2 Every art and every investigation, and likewise every practical pursuit or undertaking, seems to aim at some good: hence it has been well said that the Good is That at which all things aim.
Aristotle 384–322 BC: *Nicomachean Ethics*

3 All things work together for good to them that love God.
Bible: Romans

4 With love for mankind and hatred of sins.
St Augustine of Hippo AD 354–430: letter 211 in J.-P. Migne (ed.) *Patrologiae Latinae* (1845), often quoted 'Love the sinner but hate the sin'

5 For, where God built a church, there the devil would also build a chapel... In such sort is the devil always God's ape.
Martin Luther 1483–1546: *Colloquia Mensalia* (1566) tr. H. Bell, 1652

6 I come to bury Caesar, not to praise him.
The evil that men do lives after them,
The good is oft interrèd with their bones.
William Shakespeare 1564–1616: *Julius Caesar* (1599)

7 Something is rotten in the state of Denmark.
William Shakespeare 1564–1616: *Hamlet* (1601)

8 There is nothing either good or bad, but thinking makes it so.
William Shakespeare 1564–1616: *Hamlet* (1601)

9 By the pricking of my thumbs,
Something wicked this way comes.
William Shakespeare 1564–1616: *Macbeth* (1606)

10 For sweetest things turn sourest by their deeds;
Lilies that fester smell far worse than weeds.
William Shakespeare 1564–1616: sonnet 94 (1609)

11 And out of good still to find means of evil.
John Milton 1608–74: *Paradise Lost* (1667)

12 Farewell remorse! All good to me is lost;
Evil, be thou my good.
John Milton 1608–74: *Paradise Lost* (1667)

13 BELINDA: Ay, but you know we must return good for evil.
LADY BRUTE: That may be a mistake in the translation.
John Vanbrugh 1664–1726: *The Provoked Wife* (1697)

14 Let humble Allen, with an awkward shame,
Do good by stealth, and blush to find it fame.
Alexander Pope 1688–1744: *Imitations of Horace* (1738)

15 It is necessary only for the good man to do nothing for evil to triumph.
Edmund Burke 1729–97: attributed (in a number of forms) to Burke, but not found in his writings

16 That best portion of a good man's life,
His little, nameless, unremembered, acts
Of kindness and of love.
William Wordsworth 1770–1850: 'Lines composed a few miles above Tintern Abbey' (1798)

17 He who would do good to another, must do it in minute particulars.
William Blake 1757–1827: *Jerusalem* (1815)

18 I expect to pass through this world but once; any good thing therefore that I can do, or any kindness that I can show to any fellow-creature, let me do it now; let me not defer or neglect it, for I shall not pass this way again.

Stephen Grellet 1773–1855: attributed. See John o' London *Treasure Trove* (1925) for some of the many other claimants to authorship

19 No people do so much harm as those who go about doing good.

Bishop Mandell Creighton 1843–1901: in *Life and Letters of Mandell Creighton* by his wife (1904)

20 What we call evil is simply ignorance bumping its head in the dark.

Henry Ford 1863–1947: in *Observer* 16 March 1930

21 'Goodness, what beautiful diamonds!'
'Goodness had nothing to do with it.'

Mae West 1892–1980: *Night After Night* (1932 film)

22 I and the public know
What all schoolchildren learn,
Those to whom evil is done
Do evil in return.

W. H. Auden 1907–73: 'September 1, 1939' (1940)

23 There is no evil in the atom; only in men's souls.

Adlai Stevenson 1900–65: speech at Hartford, Connecticut, 18 September 1952

24 Innocence always calls mutely for protection, when we would be so much wiser to guard ourselves against it: innocence is like a dumb leper who has lost his bell, wandering the world meaning no harm.

Graham Greene 1904–91: *The Quiet American* (1955)

25 The face of 'evil' is always the face of total need.

William S. Burroughs 1914– : *The Naked Lunch* (1959)

26 The fearsome, word-and-thought-defying *banality of evil*.

Hannah Arendt 1906–75: *Eichmann in Jerusalem* (1963)

Gossip

1 Be thou as chaste as ice, as pure as snow, thou shalt not escape calumny.
William Shakespeare 1564–1616: *Hamlet* (1601)

2 How these curiosities would be quite forgot, did not such idle fellows as I am put them down.
John Aubrey 1626–97: *Brief Lives*

3 They come together like the Coroner's Inquest, to sit upon the murdered reputations of the week.
William Congreve 1670–1729: *The Way of the World* (1700)

4 Love and scandal are the best sweeteners of tea.
Henry Fielding 1707–54: *Love in Several Masques* (1728)

5 The Town small-talk flows from lip to lip;
Intrigues half-gathered, conversation-scraps,
Kitchen-cabals, and nursery-mishaps.
George Crabbe 1754–1832: *The Borough* (1810)

6 Every man is surrounded by a neighbourhood of voluntary spies.
Jane Austen 1775–1817: *Northanger Abbey* (1818)

7 Gossip is a sort of smoke that comes from the dirty tobacco-pipes of those who diffuse it: it proves nothing but the bad taste of the smoker.
George Eliot 1819–80: *Daniel Deronda* (1876)

8 There is only one thing in the world worse than being talked about, and that is not being talked about.
Oscar Wilde 1854–1900: *The Picture of Dorian Gray* (1891)

9 It takes your enemy and your friend, working together, to hurt you to the heart: the one to slander you and the other to get the news to you.
Mark Twain 1835–1910: *Following the Equator* (1897)

10 There is so much good in the worst of us,
And so much bad in the best of us,
That it hardly becomes any of us
To talk about the rest of us.
Anonymous: attributed, among others, to E. W. Hoch (1849–1945), but disclaimed by him

11 Like all gossip—it's merely one of those half-alive things that try to crowd out real life.
E. M. Forster 1879–1970: *A Passage to India* (1924)

12 Careless talk costs lives.
Anonymous: Second World War security slogan

Government

See also **Politics**

1 Let them hate, so long as they fear.
Accius 170–*c*.86 BC: from *Atreus*

2 It is much safer for a prince to be feared than loved, if he is to fail in one of the two.
Niccolò Machiavelli 1469–1527: *The Prince* (1513) tr. A. Gilbert

3 A parliament can do any thing but make a man a woman, and a woman a man.
2nd Earl of Pembroke *c*.1534–1601: quoted by his son, the 4th Earl

4 Though God hath raised me high, yet this I count the glory of my crown: that I have reigned with your loves.
Elizabeth I 1533–1603: The Golden Speech, 1601

5 Dost thou not know, my son, with how little wisdom the world is governed?
Count Oxenstierna 1583–1654: letter to his son, 1648. In *Table Talk* (1689), John Selden quotes 'a certain Pope': 'Thou little thinkest what *a little foolery governs the whole world!*'

6 All empire is no more than power in trust.
John Dryden 1631–1700: *Absalom and Achitophel* (1681)

7 For forms of government let fools contest;
Whate'er is best administered is best.
Alexander Pope 1688–1744: *An Essay on Man* Epistle 3 (1733)

8 The use of force alone is but *temporary*. It may subdue for a moment; but it does not remove the necessity of subduing

again; and a nation is not governed, which is perpetually to be conquered.
Edmund Burke 1729–97: *On Conciliation with America* (1775)

9 Government, even in its best state, is but a necessary evil; in its worst state, an intolerable one. Government, like dress, is the badge of lost innocence; the palaces of kings are built upon the ruins of the bowers of paradise.
Thomas Paine 1737–1809: *Common Sense* (1776)

10 Fear is the foundation of most governments.
John Adams 1735–1826: *Thoughts on Government* (1776)

11 The happiness of society is the end of government.
John Adams 1735–1826: *Thoughts on Government* (1776)

12 My people and I have come to an agreement which satisfies us both. They are to say what they please, and I am to do what I please.
Frederick the Great 1712–86: his interpretation of benevolent despotism (attributed)

13 A state without the means of some change is without the means of its conservation.
Edmund Burke 1729–97: *Reflections on the Revolution in France* (1790)

14 When, in countries that are called civilized, we see age going to the workhouse and youth to the gallows, something must be wrong in the system of government.
Thomas Paine 1737–1809: *The Rights of Man* pt. 2 (1792)

15 A monarchy is a merchantman which sails well, but will sometimes strike on a rock, and go to the bottom; whilst a republic is a raft which would never sink, but then your feet are always in the water.
Fisher Ames 1758–1808: speech in the House of Representatives, 1795; attributed by R. W. Emerson in *Essays* (1844)

16 *Gouverner, c'est choisir.*
To govern is to choose.
Duc de Lévis 1764–1830: *Maximes et Réflexions* (1812 ed.)

17 The best government is that which governs least.
John L. O'Sullivan 1813–95: *United States Magazine and Democratic Review* (1837)

18 The reluctant obedience of distant provinces generally costs more than it is worth.

Lord Macaulay 1800–59: *Essays* (1843) 'The War of Succession in Spain'

19 No Government can be long secure without a formidable Opposition.

Benjamin Disraeli 1804–81: *Coningsby* (1844)

20 Now, is it to lower the price of corn, or isn't it? It is not much matter which we say, but mind, we must all say *the same.*

Lord Melbourne 1779–1848: on cabinet government; attributed, in Walter Bagehot *The English Constitution* (1867)

21 Your business is not to govern the country but it is, if you think fit, to call to account those who do govern it.

W. E. Gladstone 1809–98: speech to the House of Commons, 29 January 1855

22 Every country has its own constitution; ours is absolutism moderated by assassination.

Anonymous: 'An intelligent Russian', in *Political Sketches of the State of Europe, 1814–1867* (1868)

23 England is the mother of Parliaments.

John Bright 1811–89: speech at Birmingham, 18 January 1865

24 The Crown is, according to the saying, the 'fountain of honour'; but the Treasury is the spring of business.

Walter Bagehot 1826–77: *The English Constitution* (1867)

25 I work for a Government I despise for ends I think criminal.

John Maynard Keynes 1883–1946: letter to Duncan Grant, 15 December 1917

26 The important thing for Government is not to do things which individuals are doing already, and to do them a little better or a little worse; but to do those things which at present are not done at all.

John Maynard Keynes 1883–1946: *End of Laissez-Faire* (1926)

27 Democracy means government by the uneducated, while aristocracy means government by the badly educated.

G. K. Chesterton 1874–1936: *New York Times* 1 February 1931

28 BIG BROTHER IS WATCHING YOU.
George Orwell 1903–50: *Nineteen Eighty-Four* (1949)

29 Wherever you have an efficient government you have a dictatorship.
Harry S. Truman 1884–1972: lecture at Columbia University, 28 April 1959

30 If the Government is big enough to give you everything you want, it is big enough to take away everything you have.
Gerald Ford 1909– : in J. F. Parker *If Elected* (1960)

31 Government of the busy by the bossy for the bully.
Arthur Seldon 1916– : *Capitalism* (1990), subheading on over-government

32 We give the impression of being in office but not in power.
Norman Lamont 1942– : speech, House of Commons, 9 June 1993

Greatness

1 The beauty of Israel is slain upon thy high places: how are the mighty fallen!
Bible: II Samuel

2 Rightly to be great
Is not to stir without great argument,
But greatly to find quarrel in a straw
When honour's at the stake.
William Shakespeare 1564–1616: *Hamlet* (1601)

3 But be not afraid of greatness: some men are born great, some achieve greatness, and some have greatness thrust upon them.
William Shakespeare 1564–1616: *Twelfth Night* (1601)

4 Farewell! a long farewell, to all my greatness!
William Shakespeare 1564–1616: *Henry VIII* (with John Fletcher, 1613)

5 To be great is to be misunderstood.
Ralph Waldo Emerson 1803–82: *Essays* (1841) 'Self-Reliance'

Habit see **Custom and Habit**

Happiness

1 Call no man happy before he dies, he is at best but fortunate.
Solon *c.*640–after 556 BC: in Herodotus *Histories*

2 *Nil admirari prope res est una, Numici,*
Solaque quae possit facere et servare beatum.
To marvel at nothing is just about the one and only thing, Numicius, that can make a man happy and keep him that way.
Horace 65–8 BC: *Epistles*

3 For all the happiness mankind can gain
Is not in pleasure, but in rest from pain.
John Dryden 1631–1700: *The Indian Emperor* (1665)

4 Mirth is like a flash of lightning that breaks through a gloom of clouds, and glitters for a moment: cheerfulness keeps up a kind of day-light in the mind.
Joseph Addison 1672–1719: *The Spectator* 17 May 1712

5 Not to admire, is all the art I know,
To make men happy, and to keep them so.
Alexander Pope 1688–1744: *Imitations of Horace* (1738)

6 A little miss, dressed in a new gown for a dancing-school ball, receives as complete enjoyment as the greatest orator, who... governs the passions and resolutions of a numerous assembly.
David Hume 1711–76: *Essays: Moral and Political* (1741–2)

7 *Freude, schöner Götterfunken,*
Tochter aus Elysium.
Joy, beautiful radiance of the gods, daughter of Elysium.
Friedrich von Schiller 1759–1805: 'An die Freude' (1785)

8 Happiness is not an ideal of reason but of imagination.
Immanuel Kant 1724–1804: *Fundamental Principles of the Metaphysics of Ethics* (1785), tr. T. K. Abbott

9 A large income is the best recipe for happiness I ever heard of. It certainly may secure all the myrtle and turkey part of it.
Jane Austen 1775–1817: *Mansfield Park* (1814)

10 Ask yourself whether you are happy, and you cease to be so.
John Stuart Mill 1806–73: *Autobiography* (1873)

11 We have no more right to consume happiness without producing it than to consume wealth without producing it.
George Bernard Shaw 1856–1950: *Candida* (1898)

12 But a lifetime of happiness! No man alive could bear it: it would be hell on earth.
George Bernard Shaw 1856–1950: *Man and Superman* (1903)

13 I can sympathize with people's pains, but not with their pleasures. There is something curiously boring about somebody else's happiness.
Aldous Huxley 1894–1963: *Limbo* (1920)

14 Happiness is a wine of the rarest vintage, and seems insipid to a vulgar taste.
Logan Pearsall Smith 1865–1946: *Afterthoughts* (1931)

15 Happiness makes up in height for what it lacks in length.
Robert Frost 1874–1963: title of poem (1942)

16 Happiness is an imaginary condition, formerly often attributed by the living to the dead, now usually attributed by adults to children, and by children to adults.
Thomas Szasz 1920– : *The Second Sin* (1973) 'Emotions'

17 I always say I don't think everyone has the right to happiness or to be loved. Even the Americans have written into their constitution that you have the right to the 'pursuit of happiness'. You have the right to try but that is all.
Claire Rayner 1931– : in G. Kinnock and F. Miller *By Faith and Daring* (1993)

Hatred

1 Better is a dinner of herbs where love is, than a stalled ox and hatred therewith.
Bible: Proverbs

2 *Non amo te, Sabidi, nec possum dicere quare:*
Hoc tantum possum dicere, non amo te.
I don't love you, Sabidius, and I can't tell you why; all I can tell you is this, that I don't love you.
Martial AD *c.*40–*c.*104: *Epigrammata*

3 I do not love thee, Dr Fell.
The reason why I cannot tell;

But this I know, and know full well,
I do not love thee, Dr Fell.
Thomas Brown 1663–1704: written while an undergraduate at Christ Church, Oxford, of which Dr Fell was Dean

4 We can scarcely hate any one that we know.
William Hazlitt 1778–1830: *Table Talk* (1822)

5 Now hatred is by far the longest pleasure;
Men love in haste, but they detest at leisure.
Lord Byron 1788–1824: *Don Juan* (1819–24)

6 Gr-r-r—there go, my heart's abhorrence!
Water your damned flower-pots, do!
If hate killed men, Brother Lawrence,
God's blood, would not mine kill you!
Robert Browning 1812–89: 'Soliloquy of the Spanish Cloister' (1842)

7 If you hate a person, you hate something in him that is part of yourself. What isn't part of ourselves doesn't disturb us.
Hermann Hesse 1877–1962: *Demian* (1919)

8 I never hated a man enough to give him diamonds back.
Zsa Zsa Gabor 1919– : in *Observer* 25 August 1957

Health see **Sickness and Health**

The Heart

1 A man whose blood
Is very snow-broth; one who never feels
The wanton stings and motions of the sense.
William Shakespeare 1564–1616: *Measure for Measure* (1604)

2 The heart has its reasons which reason knows nothing of.
Blaise Pascal 1623–62: *Pensées* (1670)

3 Calm of mind, all passion spent.
John Milton 1608–74: *Samson Agonistes* (1671)

4 The ruling passion, be it what it will,
The ruling passion conquers reason still.
Alexander Pope 1688–1744: 'To Lord Bathurst' (1733)

5 Unlearn'd, he knew no schoolman's subtle art,
No language, but the language of the heart.
Alexander Pope 1688–1744: 'An Epistle to Dr Arbuthnot' (1735)

6 The desires of the heart are as crooked as corkscrews
Not to be born is the best for man.
W. H. Auden 1907–73: 'Death's Echo' (1937)

7 Now that my ladder's gone
I must lie down where all ladders start
In the foul rag and bone shop of the heart.
W. B. Yeats 1865–1939: 'The Circus Animals' Desertion' (1939)

8 They had been corrupted by money, and he had been corrupted by sentiment. Sentiment was the more dangerous, because you couldn't name its price. A man open to bribes was to be relied upon below a certain figure, but sentiment might uncoil in the heart at a name, a photograph, even a smell remembered.
Graham Greene 1904–91: *The Heart of the Matter* (1948)

9 A man who has not passed through the inferno of his passions has never overcome them.
Carl Gustav Jung 1875–1961: *Memories, Dreams, Reflections* (1962)

Heaven and Hell

1 And I saw a new heaven and a new earth: for the first heaven and the first earth were passed away; and there was no more sea.
Bible: Revelation

2 LASCIATE OGNI SPERANZA VOI CH'ENTRATE!
Abandon all hope, you who enter!
Dante Alighieri 1265–1321: inscription at the entrance to Hell; *Divina Commedia* 'Inferno'

3 Better to reign in hell, than serve in heaven.
John Milton 1608–74: *Paradise Lost* (1667)

4 Me miserable! which way shall I fly
Infinite wrath, and infinite despair?
Which way I fly is hell; myself am hell.
John Milton 1608–74: *Paradise Lost* (1667)

5 Hell is a city much like London.
Percy Bysshe Shelley 1792–1822: 'Peter Bell the Third' (1819)

6 My idea of heaven is, eating *pâté de foie gras* to the sound of trumpets.
Sydney Smith 1771–1845: in H. Pearson *The Smith of Smiths* (1934)

7 A perpetual holiday is a good working definition of hell.
George Bernard Shaw 1856–1950: *Parents and Children* (1914)

8 The true paradises are the paradises that we have lost.
Marcel Proust 1871–1922: *Time Regained* (1926)

9 Hell, madam, is to love no more.
Georges Bernanos 1888–1948: *Journal d'un curé de campagne* (1936)

10 Whose love is given over-well
Shall look on Helen's face in hell
Whilst they whose love is thin and wise
Shall see John Knox in Paradise.
Dorothy Parker 1893–1967: 'Partial Comfort' (1937)

11 Hell is other people.
Jean-Paul Sartre 1905–80: *Huis Clos* (1944)

12 What is hell?
Hell is oneself,
Hell is alone, the other figures in it
Merely projections.
T. S. Eliot 1888–1965: *The Cocktail Party* (1950)

Heroes

1 No man is a hero to his valet.
Mme Cornuel 1605–94: in *Lettres de Mlle Aïssé à Madame C* (1787) Letter 13 'De Paris, 1728'

2 Every hero becomes a bore at last.
Ralph Waldo Emerson 1803–82: *Representative Men* (1850)

3 Heroing is one of the shortest-lived professions there is.
Will Rogers 1879–1935: newspaper article, 15 February 1925

4 ANDREA: Unhappy the land that has no heroes!...
GALILEO: No. Unhappy the land that needs heroes.
Bertolt Brecht 1898–1956: *Life of Galileo* (1939)

5 Faster than a speeding bullet!...Look! Up in the sky! It's a bird! It's a plane! It's Superman! Yes, it's Superman!...who—disguised as Clark Kent, mild-mannered reporter for a great metropolitan newspaper—fights a never ending battle for truth, justice and the American way!
Anonymous: *Superman* (US radio show, 1940 onwards) preamble

History

1 History is philosophy from examples.
Dionysius of Halicarnassus fl. 30–7 BC: *Ars Rhetorica*

2 Happy the people whose annals are blank in history-books!
Montesquieu 1689–1755: attributed by Thomas Carlyle

3 History...is, indeed, little more than the register of the crimes, follies, and misfortunes of mankind.
Edward Gibbon 1737–94: *Decline and Fall of the Roman Empire* (1776–88)

4 This province of literature [history] is a debatable line. It lies on the confines of two distinct territories...It is sometimes fiction. It is sometimes theory.
Lord Macaulay 1800–59: 'History' (1828)

5 What experience and history teach is this—that nations and governments have never learned anything from history, or acted upon any lessons they might have drawn from it.
G. W. F. Hegel 1770–1831: *Lectures on the Philosophy of World History: Introduction* (1830), tr. H. B. Nisbet

6 History [is] a distillation of rumour.
Thomas Carlyle 1795–1881: *History of the French Revolution* (1837)

7 History is the essence of innumerable biographies.
Thomas Carlyle 1795–1881: *Critical and Miscellaneous Essays* (1838) 'On History'

8 There is properly no history; only biography.
Ralph Waldo Emerson 1803–82: *Essays* (1841) 'History'

9 Hegel says somewhere that all great events and personalities in world history reappear in one fashion or another. He forgot to add: the first time as tragedy, the second as farce.
Karl Marx 1818–83: *Eighteenth Brumaire of Louis Bonaparte* (1852)

10 History is a gallery of pictures in which there are few originals and many copies.
Alexis de Tocqueville 1805–59: *L'Ancien régime* (1856), tr. M. W. Patterson

11 That great dust-heap called 'history'.
Augustine Birrell 1850–1933: *Obiter Dicta* (1884)

12 History is past politics, and politics is present history.
E. A. Freeman 1823–92: *Methods of Historical Study* (1886)

13 War makes rattling good history; but Peace is poor reading.
Thomas Hardy 1840–1928: *The Dynasts* (1904)

14 History, *n*. An account, mostly false, of events, mostly unimportant, which are brought about by rulers, mostly knaves, and soldiers, mostly fools.
Ambrose Bierce 1842–*c*.1914: *The Cynic's Word Book* (1906)

15 History is more or less bunk.
Henry Ford 1863–1947: in *Chicago Tribune* 25 May 1916

16 Human history becomes more and more a race between education and catastrophe.
H. G. Wells 1866–1946: *The Outline of History* (1920)

17 History is not what you thought. *It is what you can remember.*
W. C. Sellar 1898–1951 and **R. J. Yeatman** 1898–1968: *1066 and All That* (1930) 'Compulsory Preface'

18 History gets thicker as it approaches recent times.
A. J. P. Taylor 1906–90: *English History 1914–45* (1965) bibliography

19 Does history repeat itself, the first time as tragedy, the second time as farce? No, that's too grand, too considered a

process. History just burps, and we taste again that raw-onion sandwich it swallowed centuries ago.
Julian Barnes 1946– : *A History of the World in 10½ Chapters* (1989)

The Home and Housework

1 There is scarcely any less bother in the running of a family than in that of an entire state. And domestic business is no less importunate for being less important.
Montaigne 1533–92: *Essais* (1580)

2 For a man's house is his castle, *et domus sua cuique est tutissimum refugium* [and each man's home is his safest refuge].
Edward Coke 1552–1634: *Third Part of the Institutes of the Laws of England* (1628)

3 Home is home, though it be never so homely.
John Clarke d. 1658: *Paraemiologia Anglo-Latina* (1639)

4 The accent of one's birthplace lingers in the mind and in the heart as it does in one's speech.
Duc de la Rochefoucauld 1613–80: *Maximes* (1678)

5 Mid pleasures and palaces though we may roam,
Be it ever so humble, there's no place like home.
J. H. Payne 1791–1852: 'Home, Sweet Home' (1823)

6 Have nothing in your houses that you do not know to be useful, or believe to be beautiful.
William Morris 1834–96: *Hopes and Fears for Art* (1882)

7 What's the good of a home if you are never in it?
George Grossmith 1847–1912 and **Weedon Grossmith** 1854–1919: *Diary of a Nobody* (1894)

8 Dirt is only matter out of place.
John Chipman Gray 1839–1915: *Restraints on the Alienation of Property* (2nd ed., 1895)

9 Some dish more sharply spiced than this
Milk-soup men call domestic bliss.
Coventry Patmore 1823–96: 'Olympus'

10 Home is the girl's prison and the woman's workhouse.
George Bernard Shaw 1856–1950: *Man and Superman* (1903)

11 'Home is the place where, when you have to go there,
They have to take you in.'
'I should have called it
Something you somehow haven't to deserve.'
Robert Frost 1874–1963: 'The Death of the Hired Man' (1914)

12 The best
Thing we can do is to make wherever we're lost in
Look as much like home as we can.
Christopher Fry 1907– : *The Lady's not for Burning* (1949)

13 MR PRITCHARD: I must dust the blinds and then I must raise them.
MRS OGMORE-PRITCHARD: And before you let the sun in, mind it wipes its shoes.
Dylan Thomas 1914–1953: *Under Milk Wood* (1954)

14 There was no need to do any housework at all. After the first four years the dirt doesn't get any worse.
Quentin Crisp 1908– : *The Naked Civil Servant* (1968)

15 Conran's Law of Housework—it expands to fill the time available plus half an hour.
Shirley Conran 1932– : *Superwoman 2* (1977)

Honour

1 The purest treasure mortal times afford
Is spotless reputation; that away,
Men are but gilded loam or painted clay.
William Shakespeare 1564–1616: *Richard II* (1595)

2 What is honour? A word. What is that word, honour? Air. A trim reckoning! Who hath it? He that died o' Wednesday.
William Shakespeare 1564–1616: *Henry IV, Part 1* (1597)

3 Who steals my purse steals trash; 'tis something, nothing;
'Twas mine, 'tis his, and has been slave to thousands;
But he that filches from me my good name

Robs me of that which not enriches him,
And makes me poor indeed.
William Shakespeare 1564–1616: *Othello* (1602–4)

4 O! I have lost my reputation. I have lost the immortal part of myself, and what remains is bestial.
William Shakespeare 1564–1616: *Othello* (1602–4)

5 I could not love thee, Dear, so much,
Loved I not honour more.
Richard Lovelace 1618–58: 'To Lucasta, Going to the Wars' (1649)

6 His honour rooted in dishonour stood,
And faith unfaithful kept him falsely true.
Alfred, Lord Tennyson 1809–92: *Idylls of the King* 'Lancelot and Elaine' (1859)

7 The louder he talked of his honour, the faster we counted our spoons.
Ralph Waldo Emerson 1803–82: *The Conduct of Life* (1860)

8 Remember, you're fighting for this woman's honour...which is probably more than she ever did.
Bert Kalmar 1884–1947 et al.: *Duck Soup* (1933 film); spoken by Groucho Marx

Hope and Despair

See also **Optimism and Pessimism**

1 Hope deferred maketh the heart sick: but when the desire cometh, it is a tree of life.
Bible: Proverbs

2 *Nil desperandum.*
Never despair.
Horace 65–8 BC: *Odes*

3 He that lives in hope danceth without music.
George Herbert 1593–1633: *Outlandish Proverbs* (1640)

4 Magnanimous Despair alone
Could show me so divine a thing,
Where feeble Hope could ne'er have flown
But vainly flapped its tinsel wing.
Andrew Marvell 1621–78: 'The Definition of Love' (1681)

5 I can endure my own despair,
But not another's hope.
William Walsh 1663–1708: 'Song: Of All the Torments'

6 'Blessed is the man who expects nothing, for he shall never be disappointed' was the ninth beatitude.
Alexander Pope 1688–1744: letter to Fortescue, 23 September 1725

7 Hope springs eternal in the human breast:
Man never Is, but always To be blest.
Alexander Pope 1688–1744: *An Essay on Man* Epistle 1 (1733)

8 He that lives upon hope will die fasting.
Benjamin Franklin 1706–90: *Poor Richard's Almanac* (1758)

9 What is hope? nothing but the paint on the face of Existence; the least touch of truth rubs it off, and then we see what a hollow-cheeked harlot we have got hold of.
Lord Byron 1788–1824: letter to Thomas Moore, 28 October 1815

10 O, Wind,
If Winter comes, can Spring be far behind?
Percy Bysshe Shelley 1792–1822: 'Ode to the West Wind' (1819)

11 Work without hope draws nectar in a sieve,
And hope without an object cannot live.
Samuel Taylor Coleridge 1772–1834: 'Work without Hope' (1828)

12 Hopeless hope hopes on and meets no end,
Wastes without springs and homes without a friend.
John Clare 1793–1864: 'Child Harold' (written 1841)

13 No worst, there is none. Pitched past pitch of grief,
More pangs will, schooled at forepangs, wilder wring.
Comforter, where, where is your comforting?
Gerard Manley Hopkins 1844–89: 'No worst, there is none' (written 1885)

14 Not, I'll not, carrion comfort, Despair, not feast on thee;
Not untwist—slack they may be—these last strands of man
In me or, most weary, cry *I can no more.* I can;
Can something, hope, wish day come, not choose not to be.
Gerard Manley Hopkins 1844–89: 'Carrion Comfort' (written 1885)

15 He who has never hoped can never despair.
George Bernard Shaw 1856–1950: *Caesar and Cleopatra* (1901)

16 If way to the Better there be, it exacts a full look at the worst.
Thomas Hardy 1840–1928: 'De Profundis' (1902)

17 After all, tomorrow is another day.
Margaret Mitchell 1900–49: *Gone with the Wind* (1936)

18 In a real dark night of the soul it is always three o'clock in the morning.
F. Scott Fitzgerald 1896–1940: 'Handle with Care' in *Esquire* March 1936

19 Human life begins on the far side of despair.
Jean-Paul Sartre 1905–80: *Les Mouches* (1943)

20 Anything that consoles is fake.
Iris Murdoch 1919– : in R. Harries *Prayer and the Pursuit of Happiness* (1985)

The Human Race

1 Man is the measure of all things.
Protagoras b. *c.*485 BC: in Plato *Theaetetus*

2 There are many wonderful things, and nothing is more wonderful than man.
Sophocles *c.*496–406 BC: *Antigone*

3 I am a man, I count nothing human foreign to me.
Terence *c.*190–159 BC: *Heauton Timorumenos*

4 Lord, what fools these mortals be!
William Shakespeare 1564–1616: *A Midsummer Night's Dream* (1595–6)

5 What a piece of work is a man! How noble in reason! how infinite in faculty! in form, in moving, how express and admirable! in action how like an angel! in apprehension how like a god! the beauty of the world! the paragon of animals!
William Shakespeare 1564–1616: *Hamlet* (1601)

6 One touch of nature makes the whole world kin.
William Shakespeare 1564–1616: *Troilus and Cressida* (1602)

7 Man is a torch borne in the wind; a dream
But of a shadow, summed with all his substance.
George Chapman *c.*1559–1634: *Bussy D'Ambois* (1607–8)

8 We are such stuff
As dreams are made on, and our little life
Is rounded with a sleep.
William Shakespeare 1564–1616: *The Tempest* (1611)

9 How beauteous mankind is! O brave new world,
That has such people in't.
William Shakespeare 1564–1616: *The Tempest* (1611)

10 Man is man's A.B.C. There is none that can
Read God aright, unless he first spell Man.
Francis Quarles 1592–1644: *Hieroglyphics of the Life of Man* (1638)

11 We carry within us the wonders we seek without us: there is all Africa and her prodigies in us.
Sir Thomas Browne 1605–82: *Religio Medici* (1643)

12 Man is only a reed, the weakest thing in nature; but he is a thinking reed.
Blaise Pascal 1623–62: *Pensées* (1670)

13 An honest man's the noblest work of God.
Alexander Pope 1688–1744: *An Essay on Man* Epistle 4 (1734)

14 Man is a tool-making animal.
Benjamin Franklin 1706–90: in James Boswell *Life of Samuel Johnson* (1791) 7 April 1778

15 Out of the crooked timber of humanity no straight thing can ever be made.
Immanuel Kant 1724–1804: *Idee zu einer allgemeinen Geschichte in weltbürgerlicher Absicht* (1784)

16 Drinking when we are not thirsty and making love all year round, madam; that is all there is to distinguish us from other animals.
Pierre-Augustin Caron de Beaumarchais 1732–99: *The Marriage of Figaro* (1785)

17 For Mercy has a human heart
Pity a human face:

And Love, the human form divine,
And Peace, the human dress.
William Blake 1757–1827: *Songs of Innocence* (1789) 'The Divine Image'

18 Cruelty has a human heart,
And Jealousy a human face;
Terror the human form divine,
And Secrecy the human dress.
William Blake 1757–1827: 'A Divine Image'; etched but not included in *Songs of Experience* (1794)

19 Is man an ape or an angel? Now I am on the side of the angels.
Benjamin Disraeli 1804–81: speech at Oxford, 25 November 1864

20 I teach you the superman. Man is something to be surpassed.
Friedrich Nietzsche 1844–1900: *Also Sprach Zarathustra* (1883)

21 Man is the Only Animal that Blushes. Or needs to.
Mark Twain 1835–1910: *Following the Equator* (1897)

22 I wish I loved the Human Race;
I wish I loved its silly face;
I wish I liked the way it walks;
I wish I liked the way it talks;
And when I'm introduced to one
I wish I thought *What Jolly Fun!*
Walter Raleigh 1861–1922: 'Wishes of an Elderly Man' (1923)

23 Many people believe that they are attracted by God, or by Nature, when they are only repelled by man.
Dean Inge 1860–1954: *More Lay Thoughts of a Dean* (1931)

24 What is man, when you come to think upon him, but a minutely set, ingenious machine for turning, with infinite artfulness, the red wine of Shiraz into urine?
Isak Dinesen 1885–1962: *Seven Gothic Tales* (1934)

25 Man, unlike any other thing organic or inorganic in the universe, grows beyond his work, walks up the stairs of his concepts, emerges ahead of his accomplishments.
John Steinbeck 1902–68: *Grapes of Wrath* (1939)

26 I hate 'Humanity' and all such abstracts: but I love *people*. Lovers of 'Humanity' generally hate *people and children*, and keep parrots or puppy dogs.
Roy Campbell 1901–57: *Light on a Dark Horse* (1951)

27 We're all of us guinea pigs in the laboratory of God. Humanity is just a work in progress.
Tennessee Williams 1911–83: *Camino Real* (1953)

Human Rights

1 To no man will we sell, or deny, or delay, right or justice.
Magna Carta 1215: clause 40

2 The poorest he that is in England hath a life to live as the greatest he.
Thomas Rainborowe d. 1648: during the Army debates at Putney, 29 October 1647

3 We hold these truths to be self-evident, that all men are created equal, that they are endowed by their Creator with certain unalienable rights, that among these are life, liberty and the pursuit of happiness.
American Declaration of Independence 1776: from a draft by Thomas Jefferson (1743–1826)

4 *Liberté! Égalité! Fraternité!*
Freedom! Equality! Brotherhood!
Anonymous: motto of the French Revolution, but of earlier origin

5 Whatever each man can separately do, without trespassing upon others, he has a right to do for himself; and he has a right to a fair portion of all which society, with all its combinations of skill and force, can do in his favour.
Edmund Burke 1729–97: *Reflections on the Revolution in France* (1790)

6 Any law which violates the inalienable rights of man is essentially unjust and tyrannical; it is not a law at all.
Maximilien Robespierre 1758–94: *Déclaration des droits de l'homme* 24 April 1793

7 Natural rights is simple nonsense: natural and imprescriptible rights, rhetorical nonsense—nonsense upon stilts.

Jeremy Bentham 1748–1832: *Anarchical Fallacies*

8 The first duty of a State is to see that every child born therein shall be well housed, clothed, fed and educated, till it attain years of discretion.

John Ruskin 1819–1900: *Time and Tide* (1867)

9 We look forward to a world founded upon four essential human freedoms. The first is freedom of speech and expression—everywhere in the world. The second is freedom of every person to worship God in his own way—everywhere in the world. The third is freedom from want... The fourth is freedom from fear.

Franklin D. Roosevelt 1882–1945: message to Congress, 6 January 1941

10 All human beings are born free and equal in dignity and rights.

Universal Declaration of Human Rights 1948: article 1

Humour

1 Delight hath a joy in it either permanent or present. Laughter hath only a scornful tickling.

Philip Sidney 1554–86: *The Defence of Poetry* (1595)

2 A jest's prosperity lies in the ear
Of him that hears it, never in the tongue
Of him that makes it.

William Shakespeare 1564–1616: *Love's Labour's Lost* (1595)

3 I am not only witty in myself, but the cause that wit is in other men.

William Shakespeare 1564–1616: *Henry IV, Part 2* (1597)

4 He uses his folly like a stalking-horse, and under the presentation of that he shoots his wit.

William Shakespeare 1564–1616: *As You Like It* (1599)

5 Brevity is the soul of wit.

William Shakespeare 1564–1616: *Hamlet* (1601)

6 A thing well said will be wit in all languages.

John Dryden 1631–1700: *Essay of Dramatic Poesy* (1668)

7 If we may believe our logicians, man is distinguished from all other creatures by the faculty of laughter.
Joseph Addison 1672–1719: *The Spectator* 26 September 1712

8 I make myself laugh at everything, for fear of having to weep at it.
Pierre-Augustin Caron de Beaumarchais 1732–99: *Le Barbier de Séville* (1755)

9 What is an Epigram? a dwarfish whole,
Its body brevity, and wit its soul.
Samuel Taylor Coleridge 1772–1834: 'Epigram' (1809)

10 For what do we live, but to make sport for our neighbours, and laugh at them in our turn?
Jane Austen 1775–1817: *Pride and Prejudice* (1813)

11 Laughter is pleasant, but the exertion is too much for me.
Thomas Love Peacock 1785–1866: *Nightmare Abbey* (1818)

12 [A pun] is a pistol let off at the ear; not a feather to tickle the intellect.
Charles Lamb 1775–1834: *Last Essays of Elia* (1833) 'Popular Fallacies'

13 A difference of taste in jokes is a great strain on the affections.
George Eliot 1819–80: *Daniel Deronda* (1876)

14 Wit is the epitaph of an emotion.
Friedrich Nietzsche 1844–1900: *Menschliches, Allzumenschliches* (1867–80)

15 We are not amused.
Queen Victoria 1819–1901: attributed, in Caroline Holland *Notebooks of a Spinster Lady* (1919) 2 January 1900

16 Everything is funny as long as it is happening to Somebody Else.
Will Rogers 1879–1935: *The Illiterate Digest* (1924)

17 People must not do things for fun. We are not here for fun. There is no reference to fun in any Act of Parliament.
A. P. Herbert 1890–1971: *Uncommon Law* (1935)

18 What do you mean, funny? Funny-peculiar or funny ha-ha?
Ian Hay 1876–1952: *The Housemaster* (1938)

19 The funniest thing about comedy is that you never know why people laugh. I know *what* makes them laugh but trying to get your hands on the *why* of it is like trying to pick an eel out of a tub of water.

W. C. Fields 1880–1946: in R. J. Anobile *A Flask of Fields* (1972)

20 Humour is emotional chaos remembered in tranquillity.

James Thurber 1894–1961: in *New York Post* 29 February 1960

21 Forgive, O Lord, my little jokes on Thee
And I'll forgive Thy great big one on me.

Robert Frost 1874–1963: 'Cluster of Faith' (1962)

22 Among those whom I like or admire, I can find no common denominator, but among those whom I love, I can: all of them make me laugh.

W. H. Auden 1907–73: *The Dyer's Hand* (1963) 'Notes on the Comic'

23 The trouble with Freud is that he never had to play the old Glasgow Empire on a Saturday night after Rangers and Celtic had both lost.

Ken Dodd 1931– : *Guardian* 30 April 1991; quoted in many forms since the mid-1960s

Hypocrisy

1 My tongue swore, but my mind's unsworn.

Euripides *c.*485–*c.*406 BC: *Hippolytus* (lamenting the breaking of an oath)

2 Woe unto them that call evil good, and good evil.

Bible: Isaiah

3 Ye are like unto whited sepulchres.

Bible: St Matthew

4 The smylere with the knyf under the cloke.

Geoffrey Chaucer *c.*1343–1400: *The Canterbury Tales* 'The Knight's Tale'

5 Do not, as some ungracious pastors do,
Show me the steep and thorny way to heaven,
Whiles, like a puffed and reckless libertine,
Himself the primrose path of dalliance treads,
And recks not his own rede.

William Shakespeare 1564–1616: *Hamlet* (1601)

6 I want that glib and oily art
To speak and purpose not.
William Shakespeare 1564–1616: *King Lear* (1605–6)

7 Compound for sins, they are inclined to,
By damning those they have no mind to.
Samuel Butler 1612–80: *Hudibras* pt. 1 (1663)

8 Hypocrisy, the only evil that walks
Invisible, except to God alone.
John Milton 1608–74: *Paradise Lost* (1667)

9 Hypocrisy is a tribute which vice pays to virtue.
Duc de la Rochefoucauld 1613–80: *Maximes* (1678)

10 Keep up appearances; there lies the test;
The world will give thee credit for the rest.
Outward be fair, however foul within;
Sin if thou wilt, but then in secret sin.
Charles Churchill 1731–64: *Night* (1761)

11 I sit on a man's back, choking him and making him carry me, and yet assure myself and others that I am very sorry for him and wish to ease his lot by all possible means—except by getting off his back.
Leo Tolstoy 1828–1910: *What Then Must We Do?* (1886), tr. A. Maude

12 Talk about the pews and steeples
And the cash that goes therewith!
But the souls of Christian peoples…
Chuck it, Smith!
G. K. Chesterton 1874–1936: 'Antichrist' (1915)

13 All Reformers, however strict their social conscience, live in houses just as big as they can pay for.
Logan Pearsall Smith 1865–1946: *Afterthoughts* (1931)

Idealism

1 We are all in the gutter, but some of us are looking at the stars.
Oscar Wilde 1854–1900: *Lady Windermere's Fan* (1892)

2 I have spread my dreams under your feet;
Tread softly because you tread on my dreams.

W. B. Yeats 1865–1939: 'He Wishes for the Cloths of Heaven' (1899)

3 A cause may be inconvenient, but it's magnificent. It's like champagne or high heels, and one must be prepared to suffer for it.

Arnold Bennett 1867–1931: *The Title* (1918)

4 If a man hasn't discovered something he will die for, he isn't fit to live.

Martin Luther King 1929–68: speech in Detroit, 23 June 1963

Ideas

See also **Thinking**

1 It could be said of me that in this book I have only made up a bunch of other men's flowers, providing of my own only the string that ties them together.

Montaigne 1533–92: *Essais* (1580)

2 It is the nature of an hypothesis, when once a man has conceived it, that it assimilates every thing to itself, as proper nourishment; and, from the first moment of your begetting it, it generally grows the stronger by every thing you see, hear, read, or understand.

Laurence Sterne 1713–68: *Tristram Shandy* (1759–67)

3 A stand can be made against invasion by an army; no stand can be made against invasion by an idea.

Victor Hugo 1802–85: *Histoire d'un Crime* (written 1851–2, published 1877)

4 When you are a Bear of Very Little Brain, and you Think of Things, you find sometimes that a Thing which seemed very Thingish inside you is quite different when it gets out into the open and has other people looking at it.

A. A. Milne 1882–1956: *The House at Pooh Corner* (1928)

5 It isn't that they can't see the solution. It is that they can't see the problem.
G. K. Chesterton 1874–1936: *The Scandal of Father Brown* (1935)

6 Nothing is more dangerous than an idea, when you have only one idea.
Alain 1868–1951: *Propos sur la religion* (1938)

7 There is one thing stronger than all the armies in the world; and that is an idea whose time has come.
Anonymous: *Nation* 15 April 1943

8 It is better to entertain an idea than to take it home to live with you for the rest of your life.
Randall Jarrell 1914–65: *Pictures from an Institution* (1954)

Idleness

1 Go to the ant thou sluggard; consider her ways, and be wise.
Bible: Proverbs

2 Bankrupt of life, yet prodigal of ease.
John Dryden 1631–1700: *Absalom and Achitophel* (1681)

3 For Satan finds some mischief still
For idle hands to do.
Isaac Watts 1674–1748: *Divine Songs for Children* (1715) 'Against Idleness and Mischief'

4 It is better to wear out than to rust out.
Bishop Richard Cumberland 1631–1718: in G. Horne *The Duty of Contending for the Faith* (1786)

5 Procrastination is the thief of time.
Edward Young 1683–1765: *Night Thoughts* (1742–5)

6 Idleness is only the refuge of weak minds.
Lord Chesterfield 1694–1773: *Letters to his Son* (1774) 20 July 1749

7 The foul sluggard's comfort: 'It will last my time.'
Thomas Carlyle 1795–1881: *Critical and Miscellaneous Essays* (1838)

8 How dull it is to pause, to make an end,
To rust unburnished, not to shine in use!
As though to breathe were life.
Alfred, Lord Tennyson 1809–92: 'Ulysses' (1842)

9 It is impossible to enjoy idling thoroughly unless one has plenty of work to do.
Jerome K. Jerome 1859–1927: *Idle Thoughts of an Idle Fellow* (1886)

10 procrastination is the
art of keeping
up with yesterday.
Don Marquis 1878–1937: *archy and mehitabel* (1927)

11 I do nothing, granted. But I see the hours pass—which is better than trying to fill them.
E.M. Cioran 1911– : *Guardian* 11 May 1993

Ignorance

1 If one does not know to which port one is sailing, no wind is favourable.
Seneca ('the Younger') *c.*4 BC–AD 65: *Epistulae Morales*

2 Lo! the poor Indian, whose untutored mind
Sees God in clouds, or hears him in the wind.
Alexander Pope 1688–1744: *An Essay on Man* Epistle 1 (1733)

3 Where ignorance is bliss,
'Tis folly to be wise.
Thomas Gray 1716–71: *Ode on a Distant Prospect of Eton College* (1747)

4 Ignorance, madam, pure ignorance.
Samuel Johnson 1709–84: on being asked why he had defined *pastern* as the 'knee' of a horse, 1755; in James Boswell *Life of Johnson* (1791)

5 Where people wish to attach, they should always be ignorant. To come with a well-informed mind, is to come with an inability of administering to the vanity of others, which a sensible person would always wish to avoid. A woman

especially, if she have the misfortune of knowing any thing, should conceal it as well as she can.

Jane Austen 1775–1817: *Northanger Abbey* (1818)

6 Ignorance is not innocence but sin.

Robert Browning 1812–89: *The Inn Album* (1875)

7 You know everybody is ignorant, only on different subjects.

Will Rogers 1879–1935: in *New York Times* 31 August 1924

8 It was absolutely marvellous working for Pauli. You could ask him anything. There was no worry that he would think a particular question was stupid, since he thought *all* questions were stupid.

Victor Weisskopf 1908– : in *American Journal of Physics* 1977

Imagination

1 The lunatic, the lover, and the poet,
Are of imagination all compact.

William Shakespeare 1564–1616: *A Midsummer Night's Dream* (1595–6)

2 Tell me where is fancy bred.
Or in the heart or in the head?

William Shakespeare 1564–1616: *The Merchant of Venice* (1596–8)

3 Go, and catch a falling star,
Get with child a mandrake root,
Tell me, where all past years are,
Or who cleft the Devil's foot.

John Donne 1572–1631: 'Song: Go and catch a falling star'

4 Were it not for imagination, Sir, a man would be as happy in the arms of a chambermaid as of a Duchess.

Samuel Johnson 1709–84: in James Boswell *Life of Johnson* (1791) 9 May 1778

5 To see a world in a grain of sand
And a heaven in a wild flower
Hold infinity in the palm of your hand
And eternity in an hour.

William Blake 1757–1827: 'Auguries of Innocence' (*c.*1803)

6 Whither is fled the visionary gleam?
Where is it now, the glory and the dream?
William Wordsworth 1770–1850: 'Ode. Intimations of Immortality' (1807)

7 Heard melodies are sweet, but those unheard
Are sweeter.
John Keats 1795–1821: 'Ode on a Grecian Urn' (1820)

8 His imagination resembled the wings of an ostrich. It enabled him to run, though not to soar.
Lord Macaulay 1800–59: 'John Dryden' (1828)

9 Where there is no imagination there is no horror.
Arthur Conan Doyle 1859–1930: *A Study in Scarlet* (1888)

10 An adventure is only an inconvenience rightly considered.
An inconvenience is only an adventure wrongly considered.
G. K. Chesterton 1874–1936: *All Things Considered* (1908) 'On Running after one's Hat'

11 I never saw a Purple Cow,
I never hope to see one;
But I can tell you, anyhow,
I'd rather see than be one!
Gelett Burgess 1866–1951: 'The Purple Cow' (1914)

12 When I was but thirteen or so
I went into a golden land,
Chimborazo, Cotopaxi
Took me by the hand.
Walter James Redfern Turner 1889–1946: 'Romance' (1916)

13 Must then a Christ perish in torment in every age to save those that have no imagination?
George Bernard Shaw 1856–1950: *Saint Joan* (1924) epilogue

Indifference

1 It is the disease of not listening, the malady of not marking, that I am troubled withal.
William Shakespeare 1564–1616: *Henry IV, Part 2* (1597)

2 All colours will agree in the dark.
Francis Bacon 1561–1626: *Essays* (1625) 'Of Unity in Religion'

3 *Qu'ils mangent de la brioche.*
Let them eat cake.

Marie-Antoinette 1755–93: on being told that her people had no bread. In *Confessions* (1740) Rousseau refers to a similar remark being a well-known saying; in *Relation d'un Voyage à Bruxelles et à Coblentz en 1791* (1823), Louis XVIII attributes 'Why don't they eat pastry?' to Marie-Thérèse (1638–83), wife of Louis XIV

4 Vacant heart and hand, and eye,—
Easy live and quiet die.

Sir Walter Scott 1771–1832: *The Bride of Lammermoor* (1819)

5 If Jesus Christ were to come to-day, people would not even crucify him. They would ask him to dinner, and hear what he had to say, and make fun of it.

Thomas Carlyle 1795–1881: in D. A. Wilson *Carlyle at his Zenith* (1927)

6 The worst sin towards our fellow creatures is not to hate them, but to be indifferent to them: that's the essence of inhumanity.

George Bernard Shaw 1856–1950: *The Devil's Disciple* (1901)

7 When Jesus came to Birmingham they simply passed Him by,
They never hurt a hair of Him, they only let Him die.

G. A. Studdert Kennedy 1883–1929: 'Indifference' (1921)

8 Science may have found a cure for most evils; but it has found no remedy for the worst of them all — the apathy of human beings.

Helen Keller 1880–1968: *My Religion* (1927)

9 I wish I could care what you do or where you go but
I can't... My dear, I don't give a damn.

Margaret Mitchell 1900–49: *Gone with the Wind* (1936). 'Frankly, my dear, I don't give a damn!' in Sidney Howard's 1939 screenplay

10 Cast a cold eye
On life, on death.
Horseman pass by!

W. B. Yeats 1865–1939: 'Under Ben Bulben' (1939)

11 I was much further out than you thought
And not waving but drowning.
Stevie Smith 1902–71: 'Not Waving but Drowning' (1957)

12 Catholics and Communists have committed great crimes, but at least they have not stood aside, like an established society, and been indifferent. I would rather have blood on my hands than water like Pilate.
Graham Greene 1904–91: *The Comedians* (1966)

Intelligence and Intellectuals

1 You beat your pate, and fancy wit will come:
Knock as you please, there's nobody at home.
Alexander Pope 1688–1744: 'Epigram: You beat your pate' (1732)

2 Sir, I have found you an argument; but I am not obliged to find you an understanding.
Samuel Johnson 1709–84: in James Boswell *Life of Johnson* (1791) June 1784

3 With stupidity the gods themselves struggle in vain.
Friedrich von Schiller 1759–1805: *Die Jungfrau von Orleans* (1801)

4 Not body enough to cover his mind decently with; his intellect is improperly exposed.
Sydney Smith 1771–1845: in Lady Holland *Memoir* (1855)

5 'Excellent,' I cried. 'Elementary,' said he.
Arthur Conan Doyle 1859–1930: *Memoirs of Sherlock Holmes* (1894). 'Elementary, my dear Watson' is not found in any book by Conan Doyle, although a review of the film *The Return of Sherlock Holmes* in *New York Times* 19 October 1929, states: 'In the final scene Dr Watson is there with his "Amazing, Holmes", and Holmes comes forth with his "Elementary, my dear Watson, elementary" '

6 No one in this world, so far as I know—and I have searched the records for years, and employed agents to help me—has ever lost money by underestimating the intelligence of the great masses of the plain people.
H. L. Mencken 1880–1956: *Chicago Tribune* 19 September 1926

7 'Hullo! friend,' I call out, 'Won't you lend us a hand?' 'I am an intellectual and don't drag wood about,' came the answer. 'You're lucky,' I reply. 'I too wanted to become an intellectual, but I didn't succeed.'

Albert Schweitzer 1875–1965: *More from the Primeval Forest* (1928), tr. by C. T. Campion

8 See the happy moron,
He doesn't give a damn,
I wish I were a moron,
My God! perhaps I am!

Anonymous: *Eugenics Review* July 1929

9 As a human being, one has been endowed with just enough intelligence to be able to see clearly how utterly inadequate that intelligence is when confronted with what exists.

Albert Einstein 1879–1955: letter to Queen Elizabeth of Belgium, 19 September 1932

10 The test of a first-rate intelligence is the ability to hold two opposed ideas in the mind at the same time, and still retain the ability to function.

F. Scott Fitzgerald 1896–1940: *Esquire* February 1936

11 Intelligence is quickness to apprehend as distinct from ability, which is capacity to act wisely on the thing apprehended.

Alfred North Whitehead 1861–1947: *Dialogues* (1954) 15 December 1939

12 To the man-in-the-street, who, I'm sorry to say,
Is a keen observer of life,
The word 'Intellectual' suggests straight away
A man who's untrue to his wife.

W. H. Auden 1907–73: *New Year Letter* (1941)

13 An intellectual is someone whose mind watches itself.

Albert Camus 1913–60: *Notebooks 1935–42* (1963)

14 It takes little talent to see clearly what lies under one's nose, a good deal of it to know in which direction to point that organ.

W. H. Auden 1907–73: *The Dyer's Hand* (1963)

15 So dumb he can't fart and chew gum at the same time.

Lyndon Baines Johnson 1908–73: of Gerald Ford, in R. Reeves *A Ford, not a Lincoln* (1975)

Inventions and Discoveries

See also **Science**

1 Thus were they stained with their own works: and went a whoring with their own inventions.
Bible: Psalm 106

2 *Eureka!*
I've got it!
Archimedes *c.*287–212 BC: in Vitruvius Pollio *De Architectura*

3 *Semper aliquid novi Africam adferre.*
Africa always brings [us] something new.
Pliny the Elder AD 23–79: *Historia Naturalis* (often quoted '*Ex Africa semper aliquid novi* [Always something new out of Africa]')

4 Printing, gunpowder, and the magnet [Mariner's Needle]... these three have changed the whole face and state of things throughout the world.
Francis Bacon 1561–1626: *Novum Organum* (1620)

5 I don't know what I may seem to the world, but as to myself, I seem to have been only like a boy playing on the sea-shore and diverting myself in now and then finding a smoother pebble or a prettier shell than ordinary, whilst the great ocean of truth lay all undiscovered before me.
Isaac Newton 1642–1727: in Joseph Spence *Anecdotes* (ed. J. Osborn, 1966)

6 What is the use of a new-born child?
Benjamin Franklin 1706–90: when asked what was the use of a new invention

7 Then felt I like some watcher of the skies
When a new planet swims into his ken;
Or like stout Cortez when with eagle eyes
He stared at the Pacific—and all his men
Looked at each other with a wild surmise—
Silent, upon a peak in Darien.
John Keats 1795–1821: 'On First Looking into Chapman's Homer' (1817)

8 The discovery of a new dish does more for human happiness than the discovery of a star.

Anthelme Brillat-Savarin 1755–1826: *Physiologie du Goût* (1826)

9 *Au fond de l'Inconnu pour trouver du nouveau!*
Through the unknown, we'll find the new.

Charles Baudelaire 1821–67: *Les fleurs du mal* (1857), tr. Robert Lowell

10 Why sir, there is every possibility that you will soon be able to tax it!

Michael Faraday 1791–1867: to Gladstone, when asked about the usefulness of electricity; in W. E. H. Lecky *Democracy and Liberty* (1899 ed.)

11 What one man can invent another can discover.

Arthur Conan Doyle 1859–1930: *The Return of Sherlock Holmes* (1905)

12 When man wanted to make a machine that would walk he created the wheel, which does not resemble a leg.

Guillaume Apollinaire 1880–1918: *Les Mamelles de Tirésias* (1918)

13 Discovery consists of seeing what everybody has seen and thinking what nobody has thought.

Albert von Szent-Györgyi 1893–1986: in I. Good (ed.) *The Scientist Speculates* (1962)

Ireland and the Irish

1 Icham of Irlaunde
Ant of the holy londe of irlonde
Gode sir pray ich ye
for of saynte charite,
come ant daunce wyt me,
in irlaunde.

Anonymous: fourteenth century

2 I met wid Napper Tandy, and he took me by the hand,
And he said, 'How's poor ould Ireland, and how does she stand?'

She's the most disthressful country that iver yet was
seen,
For they're hangin' men an' women for the wearin' o' the
Green.

Anonymous: 'The Wearin' o' the Green' (*c.*1795 ballad)

3 The moment the very name of Ireland is mentioned, the English seem to bid adieu to common feeling, common prudence, and common sense, and to act with the barbarity of tyrants, and the fatuity of idiots.

Sydney Smith 1771–1845: *Letters of Peter Plymley* (1807)

4 Thus you have a starving population, an absentee aristocracy, and an alien Church, and in addition the weakest executive in the world. That is the Irish Question.

Benjamin Disraeli 1804–81: speech, House of Commons, 16 February 1844

5 Ulster will fight; Ulster will be right.

Lord Randolph Churchill 1849–94: public letter, 7 May 1886

6 For the great Gaels of Ireland
Are the men that God made mad,
For all their wars are merry,
And all their songs are sad.

G. K. Chesterton 1874–1936: *The Ballad of the White Horse* (1911)

7 Romantic Ireland's dead and gone,
It's with O'Leary in the grave.

W. B. Yeats 1865–1939: 'September, 1913' (1914)

8 Ireland is the old sow that eats her farrow.

James Joyce 1882–1941: *A Portrait of the Artist as a Young Man* (1916)

9 In Ireland the inevitable never happens and the unexpected constantly occurs.

John Pentland Mahaffy 1839–1919: in W. B. Stanford and R. B. McDowell *Mahaffy* (1971)

10 The famous
Northern reticence, the tight gag of place
And times.
Seamus Heaney 1939- : 'Whatever You Say Say Nothing' (1975)

Jealousy see **Envy and Jealousy**

Journalism see **News and Journalism**

Justice

See also **The Law and Lawyers**

1 Life for life,
Eye for eye, tooth for tooth.
Bible: Exodus

2 They have sown the wind, and they shall reap the whirlwind.
Bible: Hosea

3 What I say is that 'just' or 'right' means nothing but what is in the interest of the stronger party.
Plato 429–347 BC: spoken by Thrasymachus in *The Republic* (tr. F. M. Cornford)

4 *Audi partem alteram.*
Hear the other side.
St Augustine of Hippo AD 354–430: *De Duabus Animabus contra Manicheos*

5 *Fiat justitia et pereat mundus.*
Let justice be done, though the world perish.
Emperor Ferdinand I 1503–64: motto

6 Thrice is he armed that hath his quarrel just.
William Shakespeare 1564–1616: *Henry VI, Part 2* (1592)

7 The quality of mercy is not strained,
It droppeth as the gentle rain from heaven
Upon the place beneath: it is twice blessed;
It blesseth him that gives and him that takes.
William Shakespeare 1564–1616: *The Merchant of Venice* (1596–8)

8 Though justice be thy plea, consider this,
That in the course of justice none of us
Should see salvation.

William Shakespeare 1564–1616: *The Merchant of Venice* (1596–8)

9 Use every man after his desert, and who should 'scape whipping?

William Shakespeare 1564–1616: *Hamlet* (1601)

10 I'm armed with more than complete steel—The justice of my quarrel.

Anonymous: *Lust's Dominion* (1657) attributed to Marlowe, though of doubtful authorship

11 Here they hang a man first, and try him afterwards.

Molière 1622–73: *Monsieur de Pourceaugnac* (1670)

12 It is better that ten guilty persons escape than one innocent suffer.

William Blackstone 1723–80: *Commentaries on the Laws of England* (1765)

13 A lawyer has no business with the justice or injustice of the cause which he undertakes, unless his client asks his opinion, and then he is bound to give it honestly. The justice or injustice of the cause is to be decided by the judge.

Samuel Johnson 1709–84: in James Boswell *Journal of a Tour to the Hebrides* (1785) 15 August 1773

14 Justice is truth in action.

Benjamin Disraeli 1804–81: speech, House of Commons, 11 February 1851

15 All sensible people are selfish, and nature is tugging at every contract to make the terms of it fair.

Ralph Waldo Emerson 1803–82: *The Conduct of Life* (1860)

16 No! No! Sentence first—verdict afterwards.

Lewis Carroll 1832–98: *Alice's Adventures in Wonderland* (1865)

17 If, of all words of tongue and pen,
The saddest are, 'It might have been,'
More sad are these we daily see:
'It is, but hadn't ought to be!'

Bret Harte 1836–1902: 'Mrs Judge Jenkins' (1867)

18 When I hear of an 'equity' in a case like this, I am reminded of a blind man in a dark room—looking for a black hat—which isn't there.

Lord Bowen 1835–94: in J. A. Foote *Pie-Powder* (1911)

19 In England, justice is open to all—like the Ritz Hotel.

James Mathew 1830–1908: in R. E. Megarry *Miscellany-at-Law* (1955).

20 It's the same the whole world over,
It's the poor wot gets the blame,
It's the rich wot gets the gravy.
Ain't it all a bleedin' shame?

Anonymous: 'She was Poor but she was Honest' (sung by British soldiers in the First World War)

21 Justice should not only be done, but should manifestly and undoubtedly be seen to be done.

Lord Hewart 1870–1943: Rex v Sussex Justices, 9 November 1923

22 The price of justice is eternal publicity.

Arnold Bennett 1867–1931: *Things that have Interested Me* (1923) 'Secret Trials'

23 Injustice anywhere is a threat to justice everywhere.

Martin Luther King 1929–68: letter from Birmingham Jail, Alabama, 16 April 1963

24 Two wrongs don't make a right, but they make a good excuse.

Thomas Szasz 1920– : *The Second Sin* (1973)

25 If this is justice, I am a banana.

Ian Hislop 1940– : on the Sutcliffe libel damages, 24 May 1989

Knowledge

1 The fox knows many things—the hedgehog one *big* one.

Archilochus 7th century BC

2 I know nothing except the fact of my ignorance.

Socrates 469–399 BC: in Diogenes Laertius *Lives of the Philosophers*

3 The price of wisdom is above rubies.
Bible: Job

4 He that increaseth knowledge increaseth sorrow.
Bible: Ecclesiastes

5 For now we see through a glass, darkly; but then face to face: now I know in part; but then shall I know even as also I am known.
Bible: I Corinthians

6 Everyman, I will go with thee, and be thy guide,
In thy most need to go by thy side.
Anonymous: *Everyman* (*c.*1509–19) spoken by Knowledge

7 For also knowledge itself is power.
Francis Bacon 1561–1626: *Meditationes Sacrae* (1597)

8 [The true end of knowledge] is the discovery of all operations and possibilities of operations from immortality (if it were possible) to the meanest mechanical practice.
Francis Bacon 1561–1626: *Valerius Terminus*

9 We have first raised a dust and then complain we cannot see.
Bishop George Berkeley 1685–1753: *A Treatise Concerning the Principles of Human Knowledge* (1710)

10 A little learning is a dangerous thing;
Drink deep, or taste not the Pierian spring.
Alexander Pope 1688–1744: *An Essay on Criticism* (1711)

11 And still they gazed, and still the wonder grew,
That one small head could carry all he knew.
Oliver Goldsmith 1730–74: *The Deserted Village* (1770)

12 Knowledge may give weight, but accomplishments give lustre, and many more people see than weigh.
Lord Chesterfield 1694–1773: *Maxims* (1774)

13 Knowledge is of two kinds. We know a subject ourselves, or we know where we can find information upon it.
Samuel Johnson 1709–84: in James Boswell *Life of Johnson* (1791) 18 April 1775

14 Does the eagle know what is in the pit?
Or wilt thou go ask the mole:
Can wisdom be put in a silver rod?
Or love in a golden bowl?
William Blake 1757–1827: *The Book of Thel* (1789)

15 If the doors of perception were cleansed everything would appear to man as it is, infinite.
William Blake 1757–1827: *The Marriage of Heaven and Hell* (1790–3)

16 Our meddling intellect
Mis-shapes the beauteous forms of things:—
We murder to dissect.
William Wordsworth 1770–1850: 'The Tables Turned' (1798)

17 Knowledge enormous makes a god of me.
John Keats 1795–1821: 'Hyperion: A Fragment' (1820)

18 Grace is given of God, but knowledge is bought in the market.
Arthur Hugh Clough 1819–61: *The Bothie of Tober-na-Vuolich* (1848)

19 Now, what I want is, Facts... Facts alone are wanted in life.
Charles Dickens 1812–70: *Hard Times* (1854) Mr Gradgrind

20 You will find it a very good practice always to verify your references, sir!
Martin Joseph Routh 1755–1854: in J. W. Burgon *Lives of Twelve Good Men* (1888 ed.)

21 If a little knowledge is dangerous, where is the man who has so much as to be out of danger?
T. H. Huxley 1825–95: 'On Elementary Instruction in Physiology' (written 1877)

22 There is no such thing on earth as an uninteresting subject; the only thing that can exist is an uninterested person.
G. K. Chesterton 1874–1936: *Heretics* (1905)

23 For lust of knowing what should not be known,
We take the Golden Road to Samarkand.
James Elroy Flecker 1884–1915: *The Golden Journey to Samarkand* (1913)

24 Pedantry is the dotage of knowledge.
Holbrook Jackson 1874–1948: *Anatomy of Bibliomania* (1930)

25 Where is the wisdom we have lost in knowledge?
Where is the knowledge we have lost in information?
T. S. Eliot 1888–1965: *The Rock* (1934)

26 An expert is one who knows more and more about less and less.
Nicholas Murray Butler 1862–1947: Commencement address at Columbia University (attributed)

Language

See also **Meaning, Words**

1 A word fitly spoken is like apples of gold in pictures of silver.
Bible: Proverbs

2 You taught me language; and my profit on't
Is, I know how to curse.
William Shakespeare 1564–1616: *The Tempest* (1611)

3 All that is not prose is verse; and all that is not verse is prose.
Molière 1622–73: *Le Bourgeois Gentilhomme* (1671)

4 Good heavens! For more than forty years I have been speaking prose without knowing it.
Molière 1622–73: *Le Bourgeois Gentilhomme* (1671)

5 Language is the dress of thought.
Samuel Johnson 1709–84: *Lives of the English Poets* (1779–81)

6 In language, the ignorant have prescribed laws to the learned.
Richard Duppa 1770–1831: *Maxims* (1830)

7 Language is fossil poetry.
Ralph Waldo Emerson 1803–82: *Essays* (1844) 'The Poet'

8 Take care of the sense, and the sounds will take care of themselves.
Lewis Carroll 1832–98: *Alice's Adventures in Wonderland* (1865)

9 I will not go down to posterity talking bad grammar.
Benjamin Disraeli 1804–81: correcting proofs of his last Parliamentary speech, 31 March 1881

10 Merely corroborative detail, intended to give artistic verisimilitude to an otherwise bald and unconvincing narrative.
W. S. Gilbert 1836–1911: *The Mikado* (1885)

11 A definition is the enclosing a wilderness of idea within a wall of words.
Samuel Butler 1835–1902: *Notebooks* (1912)

12 The limits of my language mean the limits of my world.
Ludwig Wittgenstein 1889–1951: *Tractatus Logico-Philosophicus* (1922)

13 There's a cool web of language winds us in,
Retreat from too much joy or too much fear.
Robert Graves 1895–1985: 'The Cool Web' (1927)

14 One picture is worth ten thousand words.
Frederick R. Barnard: *Printers' Ink* 10 March 1927

15 When I split an infinitive, God damn it, I split it so it will stay split.
Raymond Chandler 1888–1959: on a proof-reader's corrections to his work; letter to Edward Weeks, 18 January 1947

16 This is the sort of English up with which I will not put.
Winston Churchill 1874–1965: in Ernest Gowers *Plain Words* (1948) 'Troubles with Prepositions'

17 Where in this small-talking world can I find
A longitude with no platitude?
Christopher Fry 1907– : *The Lady's not for Burning* (1949)

18 Colourless green ideas sleep furiously.
Noam Chomsky 1928– : *Syntactic Structures* (1957) illustrating that grammatical structure is independent of meaning

19 Slang is a language that rolls up its sleeves, spits on its hands and goes to work.
Carl Sandburg 1878–1967: in *New York Times* 13 February 1959

20 Language tethers us to the world; without it we spin like atoms.
Penelope Lively 1933– : *Moon Tiger* (1987)

Languages

1 To God I speak Spanish, to women Italian, to men French, and to my horse—German.
Emperor Charles V 1500–58: attributed

2 It is a thing plainly repugnant to the Word of God, and the custom of the Primitive Church, to have publick Prayer in the Church, or to minister the Sacraments in a tongue not understanded of the people.
Book of Common Prayer 1662: *Articles of Religion* (1562)

3 So now they have made our English tongue a gallimaufry or hodgepodge of all other speeches.
Edmund Spenser *c.*1552–99: *The Shepherd's Calendar* (1579)

4 Some hold translations not unlike to be
The wrong side of a Turkey tapestry.
James Howell *c.*1593–1666: *Familiar Letters* (1645–55)

5 I am always sorry when any language is lost, because languages are the pedigree of nations.
Samuel Johnson 1709–84: in James Boswell *Journal of a tour to the Hebrides* (1785)

6 My English text is chaste, and all licentious passages are left in the obscurity of a learned language.
Edward Gibbon 1737–94: *Memoirs of My Life* (1796) parodied as 'decent obscurity' in the *Anti-Jacobin*, 1797–8

7 The original is unfaithful to the translation.
Jorge Luis Borges 1899–1986: of Henley's translation, in *Sobre el 'Vathek' de William Beckford* (1943)

8 England and America are two countries divided by a common language.
George Bernard Shaw 1856–1950: attributed in this and other forms, but not found in Shaw's published writings

9 We are walking lexicons. In a single sentence of idle chatter we preserve Latin, Anglo-Saxon, Norse; we carry a museum inside our heads, each day we commemorate peoples of whom we have never heard.
Penelope Lively 1933– : *Moon Tiger* (1987)

Last Words

1 Crito, we owe a cock to Aesculapius; please pay it and don't forget it.
Socrates 469-399 BC: in Plato *Phaedo*

2 Had I but served God as diligently as I have served the King, he would not have given me over in my grey hairs.
Cardinal Wolsey *c.*1475–1530: in George Cavendish *Negotiations of Thomas Wolsey* (1641)

3 I pray you, master Lieutenant, see me safe up, and my coming down let me shift for my self.
Sir Thomas More 1478–1535: on mounting the scaffold; in William Roper *Life of Sir Thomas More*

4 After his head was upon the block, [he] lift it up again, and gently drew his beard aside, and said, *This hath not offended the king.*
Sir Thomas More 1478–1535: Francis Bacon *Apophthegms New and Old* (1625)

5 I am going to seek a great perhaps... Bring down the curtain, the farce is played out.
François Rabelais *c.*1494–*c.*1553: attributed

6 All my possessions for a moment of time.
Elizabeth I 1533–1603: attributed, probably apocryphal

7 So the heart be right, it is no matter which way the head lies.
Walter Ralegh *c.*1552–1618: at his execution, on being asked which way he preferred to lay his head

8 My design is to make what haste I can to be gone.
Oliver Cromwell 1599–1658: in J. Morley *Oliver Cromwell* (1900)

9 I am about to take my last voyage, a great leap in the dark.
Thomas Hobbes 1588–1679: last words

10 He had been, he said, an unconscionable time dying; but he hoped that they would excuse it.
Charles II 1630–85: Lord Macaulay *History of England* (1849)

11 This is no time for making new enemies.
Voltaire 1694–1778: on being asked to renounce the Devil, on his deathbed (attributed)

12 We are all going to Heaven, and Vandyke is of the company.
Thomas Gainsborough 1727–88: attributed in W. B. Boulton *Thomas Gainsborough*

13 Kiss me, Hardy.
Horatio, Lord Nelson 1758–1805: at the battle of Trafalgar, in R. Southey *Life of Nelson* (1813)

14 Thank God, I have done my duty.
Horatio, Lord Nelson 1758–1805: at the battle of Trafalgar, in R. Southey *Life of Nelson* (1813)

15 Oh, my country! how I leave my country!
William Pitt 1759–1806: in Earl Stanhope *Life of the Rt. Hon. William Pitt* vol. 3 (1879). Also variously reported as 'How I love my country'; 'My country! oh, my country!'; 'I think I could eat one of Bellamy's veal pies'

16 Well, I've had a happy life.
William Hazlitt 1778–1830: in W. C. Hazlitt *Memoirs of William Hazlitt* (1867)

17 More light!
Johann Wolfgang von Goethe 1749–1832: attributed; actually 'Open the second shutter, so that more light can come in'

18 *Dieu me pardonnera, c'est son métier.*
God will pardon me, it is His trade.
Heinrich Heine 1797–1856: on his deathbed, in A. Meissner *Heinrich Heine. Erinnerungen* (1856)

19 It is a far, far better thing that I do, than I have ever done; it is a far, far better rest that I go to, than I have ever known.
Charles Dickens 1812–70: *A Tale of Two Cities* (1859) Sydney Carton

20 Die, my dear Doctor, that's the last thing I shall do!
Lord Palmerston 1784–1865: in E. Latham *Famous Sayings and their Authors* (1904)

21 So little done, so much to do.
Cecil Rhodes 1853–1902: on the day of his death, in L. Michell *Life of Rhodes* (1910).

22 I am just going outside and may be some time.
Captain Lawrence Oates 1880–1912: in Robert Falcon Scott's diary, 16–17 March 1912

23 For God's sake look after our people.
Robert Falcon Scott 1868–1912: last diary entry, 29 March 1912

24 Why fear death? It is the most beautiful adventure in life.
Charles Frohman 1860–1915: before drowning in the *Lusitania*, 7 May 1915

25 If this is dying, then I don't think much of it.
Lytton Strachey 1880–1932: on his deathbed; in M. Holroyd *Lytton Strachey* vol. 2 (1968)

26 How's the Empire?
George V 1865–1936: to his private secretary on the morning of his death; in K. Rose *King George V* (1983)

27 Bugger Bognor.
George V 1865–1936: attributed; possibly on his deathbed. See K. Rose *King George V* (1983)

28 Just before she [Stein] died she asked, 'What *is* the answer?' No answer came. She laughed and said, 'In that case what is the question?' Then she died.
Gertrude Stein 1874–1946: in D. Sutherland *Gertrude Stein* (1951)

The Law and Lawyers

See also **Crime and Punishment, Justice**

1 Written laws are like spider's webs; they will catch, it is true, the weak and poor, but would be torn in pieces by the rich and powerful.
Anacharsis 6th century BC: in Plutarch *Parallel Lives*

2 *Cui bono?*
To whose profit?
Cicero 106–43 BC: *Pro Roscio Amerino* and *Pro Milone* (quoting L. Cassius Longinus Ravilla)

3 *Salus populi suprema est lex.*
The good of the people is the chief law.
Cicero 106–43 BC: *De Legibus*

4 The sabbath was made for man, and not man for the sabbath.
Bible: St Mark

5 The first thing we do, let's kill all the lawyers.
William Shakespeare 1564–1616: *Henry VI, Part 2* (1592)

6 The rusty curb of old father antick, the law.
William Shakespeare 1564–1616: *Henry IV, Part 1* (1597)

7 You have a gift, sir, (thank your education),
Will never let you want, while there are men,
And malice, to breed causes.
Ben Jonson *c.*1573–1637: *Volpone* (1605) to a lawyer

8 I am ashamed the law is such an ass.
George Chapman *c.*1559–1634: *Revenge for Honour* (1654)

9 Ignorance of the law excuses no man; not that all men know the law, but because 'tis an excuse every man will plead, and no man can tell how to confute him.
John Selden 1584–1654: *Table Talk* (1689) 'Law'

10 The end of law is, not to abolish or restrain, but to preserve and enlarge freedom.
John Locke 1632–1704: *Second Treatise of Civil Government* (1690)

11 Law is a bottomless pit.
Dr Arbuthnot 1667–1735: *The History of John Bull* (1712)

12 The hungry judges soon the sentence sign,
And wretches hang that jury-men may dine.
Alexander Pope 1688–1744: *The Rape of the Lock* (1714)

13 Laws grind the poor, and rich men rule the law.
Oliver Goldsmith 1730–74: *The Traveller* (1764)

14 People crushed by law have no hopes but from power. If laws are their enemies, they will be enemies to laws; and those, who have much to hope and nothing to lose, will always be dangerous, more or less.
Edmund Burke 1729–97: letter to Charles James Fox, 8 October 1777

15 Bad laws are the worst sort of tyranny.
Edmund Burke 1729–97: *Speech at Bristol, previous to the Late Election* (1780)

16 A precedent embalms a principle.
Lord Stowell 1745–1836: an opinion, while Advocate-General, 1788, quoted by Disraeli in House of Commons, 22 February 1848

17 In this country...the individual subject...'has nothing to do with the laws but to obey them.'

Bishop Samuel Horsley 1733–1806: speech, House of Lords, 13 November 1795, defending a maxim he had earlier used in committee

18 Laws were made to be broken.

Christopher North 1785–1854: *Blackwood's Magazine* May 1830

19 'You must not tell us what the soldier, or any other man, said, sir,' interposed the judge; 'it's not evidence.'

Charles Dickens 1812–70: *Pickwick Papers* (1837)

20 If the law supposes that...the law is a ass—a idiot.

Charles Dickens 1812–70: *Oliver Twist* (1838) Bumble

21 If ever there was a case of clearer evidence than this of persons acting together, this case is that case.

William Arabin 1773–1841: in H. B. Churchill *Arabiniana* (1843)

22 The one great principle of the English law is, to make business for itself.

Charles Dickens 1812–70: *Bleak House* (1853)

23 I know no method to secure the repeal of bad or obnoxious laws so effective as their stringent execution.

Ulysses S. Grant 1822–85: inaugural address, 4 March 1869

24 The Law is the true embodiment
Of everything that's excellent.
It has no kind of fault or flaw,
And I, my Lords, embody the Law.

W. S. Gilbert 1836–1911: *Iolanthe* (1882)

25 You may object that it is not a trial at all; you are quite right, for it is only a trial if I recognize it as such.

Franz Kafka 1883–1924: *The Trial* (1925)

26 No poet ever interpreted nature as freely as a lawyer interprets the truth.

Jean Giraudoux 1882–1944: *La Guerre de Troie n'aura pas lieu* (1935) tr. Christopher Fry as *Tiger at the Gates*, 1955

27 A verbal contract isn't worth the paper it is written on.

Sam Goldwyn 1882–1974: in A. Johnston *The Great Goldwyn* (1937)

28 A lawyer with his briefcase can steal more than a hundred men with guns.
Mario Puzo 1920– : *The Godfather* (1969)

29 No brilliance is needed in the law. Nothing but common sense, and relatively clean finger nails.
John Mortimer 1923– : *A Voyage Round My Father* (1971)

Leadership

1 If the blind lead the blind, both shall fall into the ditch.
Bible: St Matthew

2 Be neither saint nor sophist-led, but be a man.
Matthew Arnold 1822–88: *Empedocles on Etna* (1852)

3 Ah well! I am their leader, I really had to follow them!
Alexandre Auguste Ledru-Rollin 1807–74: in E. de Mirecourt *Les Contemporains* (1857)

4 The buck stops here.
Harry S. Truman 1884–1972: unattributed motto on Truman's desk, when President

5 The final test of a leader is that he leaves behind him in other men the conviction and the will to carry on.
Walter Lippmann 1889–1974: *New York Herald Tribune* 14 April 1945

6 I don't mind how much my Ministers talk, so long as they do what I say.
Margaret Thatcher 1925– : in *Observer* 27 January 1980

Leisure

1 If all the year were playing holidays,
To sport would be as tedious as to work;
But when they seldom come, they wished for come.
William Shakespeare 1564–1616: *Henry IV, Part 1* (1597)

2 Conspicuous leisure and consumption...In the one case it is a waste of time and effort, in the other it is a waste of goods.
Thorstein Veblen 1857–1929: *Theory of the Leisure Class* (1899)

3 What is this life if, full of care,
We have no time to stand and stare.
W. H. Davies 1871–1940: 'Leisure' (1911)

4 To be able to fill leisure intelligently is the last product of civilization.
Bertrand Russell 1872–1970: *The Conquest of Happiness* (1930)

Letters and Letter-writing

1 Sir, more than kisses, letters mingle souls.
John Donne 1572–1631: 'To Sir Henry Wotton' (1597–8)

2 John Donne, Anne Donne, Un-done.
John Donne 1572–1631: in a letter to his wife, on being dismissed from the service of his father-in-law, George More; in Izaak Walton *Life of Dr Donne* (1640)

3 All letters, methinks, should be free and easy as one's discourse, not studied as an oration, nor made up of hard words like a charm.
Dorothy Osborne 1627–95: letter to William Temple, September 1653

4 I have made this [letter] longer than usual, only because I have not had the time to make it shorter.
Blaise Pascal 1623–62: *Lettres Provinciales* (1657)

5 A woman seldom writes her mind but in her postscript.
Richard Steele 1672–1729: *The Spectator* 31 May 1711

6 She has made me in love with a cold climate, and frost and snow, with a northern moonlight.
Robert Southey 1774–1843: on Mary Wollstonecraft's letters from Sweden and Norway; letter, 28 April 1797

7 It is wonderful how much news there is when people write every other day; if they wait for a month, there is nothing that seems worth telling.
O. Douglas d. 1948: *Penny Plain* (1920)

8 Letters of thanks, letters from banks,
Letters of joy from girl and boy,
Receipted bills and invitations
To inspect new stock or to visit relations,
And applications for situations,

And timid lovers' declarations,
And gossip, gossip from all the nations.
W. H. Auden 1907–73: *Night Mail* (1936)

9 It's very dangerous if you keep love letters from someone who is not now your husband.
Diana Dors 1931–84: *Observer* 28 December 1980 'Sayings of the Year'

Liberty

1 Stone walls do not a prison make,
Nor iron bars a cage.
Richard Lovelace 1618–58: 'To Althea, From Prison' (1649)

2 None can love freedom heartily, but good men; the rest love not freedom, but licence.
John Milton 1608–74: *The Tenure of Kings and Magistrates* (1649)

3 When the people contend for their liberty, they seldom get anything by their victory but new masters.
George Savile, Marquess of Halifax 1633–95: *Political, Moral, and Miscellaneous Thoughts* (1750)

4 Man was born free, and everywhere he is in chains.
Jean-Jacques Rousseau 1712–78: *Du Contrat social* (1762)

5 I know not what course others may take; but as for me, give me liberty, or give me death!
Patrick Henry 1736–99: speech, 23 March 1775

6 The people never give up their liberties except under some delusion.
Edmund Burke 1729–97: speech at County Meeting of Buckinghamshire, 1784; attributed

7 The tree of liberty must be refreshed from time to time with the blood of patriots and tyrants. It is its natural manure.
Thomas Jefferson 1743–1826: letter to W. S. Smith, 13 November 1787

8 The condition upon which God hath given liberty to man is eternal vigilance.
John Philpot Curran 1750–1817: speech, 10 July 1790

9 O liberty! what crimes are committed in thy name!
Mme Roland 1754-93: in A. de Lamartine *Histoire des Girondins* (1847)

10 Eternal spirit of the chainless mind!
Brightest in dungeons, Liberty! thou art.
Lord Byron 1788-1824: 'Sonnet on Chillon' (1816)

11 If men are to wait for liberty till they become wise and good in slavery, they may indeed wait for ever.
Lord Macaulay 1800-59: *Essays* (1843) 'Milton'

12 The liberty of the individual must be thus far limited; he must not make himself a nuisance to other people.
John Stuart Mill 1806-73: *On Liberty* (1859)

13 Liberty means responsibility. That is why most men dread it.
George Bernard Shaw 1856-1950: *Man and Superman* (1903)

14 Tyranny is always better organised than freedom.
Charles Péguy 1873-1914: *Basic Verities* (1943)

15 Freedom is always and exclusively freedom for the one who thinks differently.
Rosa Luxemburg 1871-1919: *Die Russische Revolution* (1918)

16 Liberty is precious—so precious that it must be rationed.
Lenin 1870-1924: in Sidney and Beatrice Webb *Soviet Communism* (1936)

17 It's often better to be in chains than to be free.
Franz Kafka 1883-1924: *The Trial* (1925)

18 It is better to die on your feet than to live on your knees.
Dolores Ibarruri 1895-1989: speech in Paris, 3 September 1936; also attributed to Emiliano Zapata

19 I am condemned to be free.
Jean-Paul Sartre 1905-80: *L'Être et le néant* (1943)

20 The enemies of Freedom do not argue; they shout and they shoot.
Dean Inge 1860-1954: *End of an Age* (1948)

21 Freedom is the freedom to say that two plus two make four.
If that is granted, all else follows.
George Orwell 1903-50: *Nineteen Eighty-Four* (1949)

22 A free society is a society where it is safe to be unpopular.
Adlai Stevenson 1900-65: speech in Detroit, 7 October 1952

23 Liberty is always unfinished business.
Anonymous: title of Annual Report of the American Civil Liberties Union (1956)

24 Liberty is liberty, not equality or fairness or justice or human happiness or a quiet conscience.
Isaiah Berlin 1909– : *Two Concepts of Liberty* (1958)

25 We shall pay any price, bear any burden, meet any hardship, support any friend, oppose any foe to assure the survival and the success of liberty.
John F. Kennedy 1917–63: inaugural address, 20 January 1961

26 Freedom's just another word for nothin' left to lose,
Nothin' ain't worth nothin', but it's free.
Kris Kristofferson 1936– : 'Me and Bobby McGee' (1969 song, with Fred Foster)

27 Of course liberty is not licence. Liberty in my view is conforming to majority opinion.
Hugh Scanlon 1913– : television interview, 9 August 1977

Libraries

See also **Books, Reading**

1 Come, and take choice of all my library,
And so beguile thy sorrow.
William Shakespeare 1564–1616: *Titus Andronicus* (1590)

2 No place affords a more striking conviction of the vanity of human hopes, than a public library.
Samuel Johnson 1709–84: *The Rambler* 23 March 1751

3 A man will turn over half a library to make one book.
Samuel Johnson 1709–84: in James Boswell *Life of Johnson* (1791) 6 April 1775

4 Your *borrowers of books*—those mutilators of collections, spoilers of the symmetry of shelves, and creators of odd volumes.
Charles Lamb 1775–1834: *Essays of Elia* (1823) 'The Two Races of Men'

5 The true University of these days is a collection of books.
Thomas Carlyle 1795–1881: *On Heroes, Hero-Worship, and the Heroic* (1841)

6 A man should keep his little brain attic stocked with all the furniture that he is likely to use, and the rest he can put away in the lumber room of his library, where he can get it if he wants it.
Arthur Conan Doyle 1859–1930: *Adventures of Sherlock Holmes* (1892)

7 A library is thought in cold storage.
Lord Samuel 1870–1963: *A Book of Quotations* (1947)

Lies and Lying

See also **Truth**

1 The retort courteous... the quip modest... the reply churlish... the reproof valiant... the countercheck quarrelsome... the lie circumstantial... the lie direct.
William Shakespeare 1564–1616: *As You Like It* (1599), of the degrees of a lie

2 Calumnies are answered best with silence.
Ben Jonson *c.*1573–1637: *Volpone* (1605)

3 He replied that I must needs be mistaken, or that I *said the thing which was not.* (For they have no word in their language to express lying or falsehood.)
Jonathan Swift 1667–1745: *Gulliver's Travels* (1726)

4 Whoever would lie usefully should lie seldom.
Lord Hervey 1696–1743: *Memoirs of the Reign of George II*

5 Falsehood has a perennial spring.
Edmund Burke 1729–97: *On American Taxation* (1775)

6 O what a tangled web we weave,
When first we practise to deceive!
Sir Walter Scott 1771–1832: *Marmion* (1808)

7 And, after all, what is a lie? 'Tis but
The truth in masquerade.
Lord Byron 1788–1824: *Don Juan* (1819–24)

8 It is well said in the old proverb, 'a lie will go round the world while truth is pulling its boots on'.
C. H. Spurgeon 1834–92: *Gems from Spurgeon* (1859)

9 The lie in the soul is a true lie.
Benjamin Jowett 1817–93: introduction to his translation (1871) of Plato's *Republic*

10 The cruellest lies are often told in silence.
Robert Louis Stevenson 1850–94: *Virginibus Puerisque* (1881)

11 There are three kinds of lies: lies, damned lies and statistics.
Benjamin Disraeli 1804–81: attributed to Disraeli in Mark Twain *Autobiography* (1924)

12 Take the life-lie away from the average man and straight away you take away his happiness.
Henrik Ibsen 1828–1906: *The Wild Duck* (1884)

13 One of the most striking differences between a cat and a lie is that a cat has only nine lives.
Mark Twain 1835–1910: *Pudd'nhead Wilson* (1894)

14 There is no worse lie than a truth misunderstood by those who hear it.
William James 1842–1910: *Varieties of Religious Experience* (1902)

15 The best liar is he who makes the smallest amount of lying go the longest way.
Samuel Butler 1835–1902: *The Way of All Flesh* (1903)

16 A little inaccuracy sometimes saves tons of explanation.
Saki (H. H. Munro) 1870–1916: *The Square Egg* (1924)

17 Without lies humanity would perish of despair and boredom.
Anatole France 1844–1924: *La Vie en fleur* (1922)

18 That branch of the art of lying which consists in very nearly deceiving your friends without quite deceiving your enemies.
Francis M. Cornford 1874–1943: on propaganda; *Microcosmographia Academica* (1922 ed.)

19 The broad mass of a nation... will more easily fall victim to a big lie than to a small one.
Adolf Hitler 1889–1945: *Mein Kampf* (1925)

20 She [Lady Desborough] tells enough white lies to ice a wedding cake.
Margot Asquith 1864–1945: in *Listener* 11 June 1953

21 An abomination unto the Lord, but a very present help in time of trouble.
Anonymous: definition of a lie, an amalgamation of Proverbs 12.22 and Psalms 46.1, often attributed to Adlai Stevenson

22 Truth exists; only lies are invented.
Georges Braque 1882–1963: *Le Jour et la nuit: Cahiers 1917–52*

23 He would, wouldn't he?
Mandy Rice-Davies 1944– : at the trial of Stephen Ward, 29 June 1963, on hearing that Lord Astor denied her allegations

24 He will lie even when it is inconvenient: the sign of the true artist.
Gore Vidal 1925– : attributed

Life

See also **Life Sciences, Living**

1 All that a man hath will he give for his life.
Bible: Job

2 Man that is born of a woman is of few days, and full of trouble.
Bible: Job

3 Not to be born is, past all prizing, best.
Sophocles *c.*496–406 BC: *Oedipus Coloneus* (tr. R. C. Jebb)

4 We live, not as we wish to, but as we can.
Menander 342–*c.*292 BC: *Dis Exapaton*

5 And life is given to none freehold, but it is leasehold for all.
Lucretius *c.*94–55 BC: *De Rerum Natura*

6 'Such,' he said, 'O King, seems to me the present life of men on earth, in comparison with that time which to us is uncertain, as if when on a winter's night you sit feasting with your ealdormen and thegns,—a single sparrow should fly swiftly into the hall, and coming in at one door, instantly fly out through another'.
The Venerable Bede AD 673–735: *Ecclesiastical History of the English People*, tr. B. Colgrave

7 Life well spent is long.
Leonardo da Vinci 1452–1519: E. McCurdy (ed. and trans.) *Leonardo da Vinci's Notebooks* (1906)

8 The ceaseless labour of your life is to build the house of death.
Montaigne 1533–92: *Essais* (1580)

9 Life is as tedious as a twice-told tale,
Vexing the dull ear of a drowsy man.
William Shakespeare 1564–1616: *King John* (1591–8)

10 All the world's a stage,
And all the men and women merely players:
They have their exits and their entrances;
And one man in his time plays many parts,
His acts being seven ages.
William Shakespeare 1564–1616: *As You Like It* (1599)

11 It is in life as it is in ways, the shortest way is commonly the foulest, and surely the fairer way is not much about.
Francis Bacon 1561–1626: *The Advancement of Learning* (1605)

12 Life's but a walking shadow, a poor player,
That struts and frets his hour upon the stage,
And then is heard no more; it is a tale
Told by an idiot, full of sound and fury,
Signifying nothing.
William Shakespeare 1564–1616: *Macbeth* (1606)

13 No arts; no letters; no society; and which is worst of all, continual fear and danger of violent death; and the life of man, solitary, poor, nasty, brutish, and short.
Thomas Hobbes 1588–1679: *Leviathan* (1651)

14 Life is an incurable disease.
Abraham Cowley 1618–67: 'To Dr Scarborough' (1656)

15 Life itself is but the shadow of death, and souls departed but the shadows of the living.
Sir Thomas Browne 1605–82: *The Garden of Cyrus* (1658)

16 Man has but three events in his life: to be born, to live, and to die. He is not conscious of his birth, he suffers at his death and he forgets to live.
Jean de la Bruyère 1645–96: *The Characters, or The Manners of the Age* (1688)

17 Human life is everywhere a state in which much is to be endured, and little to be enjoyed.
Samuel Johnson 1709–84: *Rasselas* (1759)

18 Man wants but little here below,
Nor wants that little long.
Oliver Goldsmith 1730–74: 'Edwin and Angelina, or the Hermit' (1766)

19 This world is a comedy to those that think, a tragedy to those that feel.
Horace Walpole 1717–97: letter to Anne, Countess of Upper Ossory, 16 August 1776

20 Life, like a dome of many-coloured glass,
Stains the white radiance of Eternity,
Until Death tramples it to fragments.
Percy Bysshe Shelley 1792–1822: *Adonais* (1821)

21 Life is real! Life is earnest!
And the grave is not its goal;
Dust thou art, to dust returnest,
Was not spoken of the soul.
Henry Wadsworth Longfellow 1807–82: 'A Psalm of Life' (1838)

22 I slept, and dreamed that life was beauty;
I woke, and found that life was duty.
Ellen Sturgis Hooper 1816–41: 'Beauty and Duty' (1840)

23 Life must be understood backwards; but... it must be lived forwards.
Sören Kierkegaard 1813–1855: *Journals and Papers* (1843)

24 Youth is a blunder; Manhood a struggle; Old Age a regret.
Benjamin Disraeli 1804–81: *Coningsby* (1844)

25 Is it so small a thing
To have enjoyed the sun,
To have lived light in the spring,
To have loved, to have thought, to have done.
Matthew Arnold 1822–88: *Empedocles on Etna* (1852)

26 All the business of war, and indeed all the business of life, is to endeavour to find out what you don't know by what you do; that's what I called 'guessing what was at the other side of the hill'.
Duke of Wellington 1769–1852: *The Croker Papers* (1885)

27 The mass of men lead lives of quiet desperation.
Henry David Thoreau 1817–62: *Walden* (1854)

28 Our life is frittered away by detail... Simplify, simplify.
Henry David Thoreau 1817–62: *Walden* (1854)

29 Life is mostly froth and bubble,
Two things stand like stone,
Kindness in another's trouble,
Courage in your own.
Adam Lindsay Gordon 1833–70: *Ye Wearie Wayfarer* (1866)

30 Cats and monkeys—monkeys and cats—all human life is there!
Henry James 1843–1916: *The Madonna of the Future* (1879)
'All human life is there' became the slogan of the *News of the World* from the late 1950s

31 *Ah! que la vie est quotidienne.*
Oh, what a day-to-day business life is.
Jules Laforgue 1860–87: *Complainte sur certains ennuis* (1885)

32 The life of every man is a diary in which he means to write one story, and writes another; and his humblest hour is when he compares the volume as it is with what he vowed to make it.
J. M. Barrie 1860–1937: *The Little Minister* (1891)

33 Life is like playing a violin solo in public and learning the instrument as one goes on
Samuel Butler 1835–1902: speech, 27 February 1895

34 Life is one long process of getting tired.
Samuel Butler 1835–1902: *Notebooks* (1912)

35 To live is like to love—all reason is against it, and all healthy instinct for it.
Samuel Butler 1835–1902: *Notebooks* (1912)

36 Life is just one damned thing after another.
Elbert Hubbard 1859–1915: *Philistine* December 1909, often attributed to Frank Ward O'Malley

37 And Life is Colour and Warmth and Light
And a striving evermore for these;
And he is dead, who will not fight;
And who dies fighting has increase.
Julian Grenfell 1888–1915: 'Into Battle' (1915)

38 Life is a foreign language: all men mispronounce it.
Christopher Morley 1890–1957: *Thunder on the Left* (1925)

39 Life is not a series of gig lamps symmetrically arranged; life is a luminous halo, a semi-transparent envelope surrounding us from the beginning of consciousness to the end.
Virginia Woolf 1882–1941: *The Common Reader* (1925) 'Modern Fiction'

40 I enjoy almost everything. Yet I have some restless searcher in me. Why is there not a discovery in life? Something one can lay one's hands on and say "This is it"?
Virginia Woolf 1882–1941: diary, 27 February 1926

41 Never to have lived is best, ancient writers say;
Never to have drawn the breath of life, never to have looked into the eye of day;
The second best's a gay goodnight and quickly turn away.
W. B. Yeats 1865–1939: 'From *Oedipus at Colonus*' (1928)

42 Life is a horizontal fall.
Jean Cocteau 1889–1963: *Opium* (1930)

43 Life is just a bowl of cherries.
Lew Brown 1893–1958: title of song (1931)

44 Birth, and copulation, and death.
That's all the facts when you come to brass tacks.
T. S. Eliot 1888–1965: *Sweeney Agonistes* (1932)

45 I long ago come to the conclusion that all life is 6 to 5 against.
Damon Runyon 1884–1946: *Collier's* 8 September 1934

46 It's a funny old world—a man's lucky if he gets out of it alive.
Walter de Leon and **Paul M. Jones**: *You're Telling Me* (1934 film); spoken by W. C. Fields

47 Many men would take the death-sentence without a whimper to escape the life-sentence which fate carries in her other hand.
T. E. Lawrence 1888–1935: *The Mint* (1955)

48 Yet we have gone on living,
Living and partly living.
T. S. Eliot 1888–1965: *Murder in the Cathedral* (1935)

49 Oh, life is a glorious cycle of song,
A medley of extemporanea;

And love is a thing that can never go wrong;
And I am Marie of Roumania.

Dorothy Parker 1893–1967: 'Comment' (1937)

50 What, knocked a tooth out? Never mind, dear, laugh it off, laugh it off; it's all part of life's rich pageant.

Arthur Marshall 1910–89: *The Games Mistress* (recorded monologue, 1937)

51 All that matters is love and work.

Sigmund Freud 1856–1939: attributed

52 Life is not having been told that the man has just waxed the floor.

Ogden Nash 1902–71: 'You and Me and P. B. Shelley' (1942)

53 Dying is nothing. So start by living. It's less fun and it lasts longer.

Jean Anouilh 1910–87: *Roméo et Jeannette* (1946)

54 The cradle rocks above an abyss, and common sense tells us that our existence is but a brief crack of light between two eternities of darkness.

Vladimir Nabokov 1899–1977: *Speak, Memory* (1951)

55 Life is like a sewer. What you get out of it depends on what you put into it.

Tom Lehrer 1928– : 'We Will All Go Together When We Go' (1953 song), preamble

56 Oh, isn't life a terrible thing, thank God?

Dylan Thomas 1914–53: *Under Milk Wood* (1954)

57 They give birth astride of a grave, the light gleams an instant, then it's night once more.

Samuel Beckett 1906–89: *Waiting for Godot* (1955)

58 When you don't have any money, the problem is food. When you have money, it's sex. When you have both, it's health.

J. P. Donleavy 1926– : *The Ginger Man* (1955)

59 Man is born to live, not to prepare for life.

Boris Pasternak 1890–1960: *Doctor Zhivago* (1958)

60 If it were possible to talk to the unborn, one could never explain to them how it feels to be alive, for life is washed in the speechless real.

Jacques Barzun 1907– : *The House of Intellect* (1959)

61 One's prime is elusive. You little girls, when you grow up, must be on the alert to recognise your prime at whatever time of your life it may occur.
Muriel Spark 1918– : *The Prime of Miss Jean Brodie* (1961)

62 As far as we can discern, the sole purpose of human existence is to kindle a light in the darkness of mere being.
Carl Gustav Jung 1875–1961: *Memories, Dreams, Reflections* (1962)

63 Life is first boredom, then fear.
Philip Larkin 1922–85: 'Dockery & Son' (1964)

64 Life is a gamble at terrible odds—if it was a bet, you wouldn't take it.
Tom Stoppard 1937– : *Rosencrantz and Guildenstern are Dead* (1967)

65 I've looked at life from both sides now,
From win and lose and still somehow
It's life's illusions I recall;
I really don't know life at all.
Joni Mitchell 1945– : 'Both Sides Now' (1967 song)

66 Expect nothing. Live frugally
on surprise.
Alice Walker 1944– : 'Expect nothing' (1973)

67 The Answer to the Great Question Of…Life, the Universe and Everything…[is] Forty-two.
Douglas Adams 1952– : *The Hitch Hiker's Guide to the Galaxy* (1979)

68 It seems that I have spent my entire time trying to make life more rational and that it was all wasted effort.
A. J. Ayer 1910–89: in *Observer* 17 August 1986

69 Life is a sexually transmitted disease.
Anonymous: graffito found on the London Underground

Life Sciences

See also **Science**

1 So, naturalists observe, a flea
Hath smaller fleas that on him prey;

And these have smaller fleas to bite 'em,
And so proceed *ad infinitum*.
Jonathan Swift 1667–1745: 'On Poetry' (1733)

2 I have called this principle, by which each slight variation, if useful, is preserved, by the term of Natural Selection.
Charles Darwin 1809–82: *On the Origin of Species* (1859)

3 Was it through his grandfather or his grandmother that he claimed his descent from a monkey?
Bishop Samuel Wilberforce 1805–73: addressed to T. H. Huxley in a debate on Darwin's theory of evolution at Oxford, June 1860

4 Survival of the fittest implies multiplication of the fittest.
Herbert Spencer 1820–1903: *Principles of Biology* (1865)

5 It has, I believe, been often remarked that a hen is only an egg's way of making another egg.
Samuel Butler 1835–1902: *Life and Habit* (1877)

6 The Microbe is so very small
You cannot make him out at all.
Hilaire Belloc 1870–1953: 'The Microbe' (1897)

7 Life exists in the universe only because the carbon atom possesses certain exceptional properties.
James Jeans 1877–1946: *The Mysterious Universe* (1930)

8 Water is life's *mater* and *matrix*, mother and medium. There is no life without water.
Albert von Szent-Györgyi 1893–1986: *Perspectives in Biology and Medicine* Winter 1971

9 The essence of life is statistical improbability on a colossal scale.
Richard Dawkins 1941– : *The Blind Watchmaker* (1986)

10 Almost all aspects of life are engineered at the molecular level, and without understanding molecules we can only have a very sketchy understanding of life itself.
Francis Crick 1916– : *What Mad Pursuit* (1988)

Literature

See also **Writing**

1 Works of serious purpose and grand promises often have a purple patch or two stitched on, to shine far and wide.
Horace 65–8 BC: *Ars Poetica*

2 'Oh! it is only a novel!...only Cecilia, or Camilla, or Belinda:' or, in short, only some work in which the most thorough knowledge of human nature, the happiest delineation of its varieties, the liveliest effusions of wit and humour are conveyed to the world in the best chosen language.
Jane Austen 1775–1817: *Northanger Abbey* (1818)

3 All tragedies are finished by a death,
All comedies are ended by a marriage;
The future states of both are left to faith.
Lord Byron 1788–1824: *Don Juan* (1819–24)

4 A novel is a mirror which passes over a highway. Sometimes it reflects to your eyes the blue of the skies, at others the churned-up mud of the road.
Stendhal 1783–1842: *Le Rouge et le noir* (1830)

5 A losing trade, I assure you, sir: literature is a drug.
George Borrow 1803–81: *Lavengro* (1851)

6 It takes a great deal of history to produce a little literature.
Henry James 1843–1916: *Hawthorne* (1879)

7 In *Anna Karenina* and *Onegin* not a single problem is solved, but they satisfy you completely just because all their problems are correctly presented. The court is obliged to submit the case fairly, but let the jury do the deciding, each according to its own judgement.
Anton Chekhov 1860–1904: letter to Alexei Suvorin, 27 October 1888

8 The good ended happily, and the bad unhappily. That is what fiction means.
Oscar Wilde 1854–1900: *The Importance of Being Earnest* (1895)

9 Literature is a luxury; fiction is a necessity.
G. K. Chesterton 1874–1936: *The Defendant* (1901) 'A Defence of Penny Dreadfuls'

10 Never trust the artist. Trust the tale.
D. H. Lawrence 1885–1930: *Studies in Classic American Literature* (1923)

11 Yes—oh dear yes—the novel tells a story.
E. M. Forster 1879–1970: *Aspects of the Novel* (1927)

12 Our American professors like their literature clear and cold and pure and very dead.
Sinclair Lewis 1885–1951: *The American Fear of Literature* (Nobel Prize Address, 12 December 1930)

13 Literature is news that STAYS news.
Ezra Pound 1885–1972: *The ABC of Reading* (1934)

14 The bad end unhappily, the good unluckily. That is what tragedy means.
Tom Stoppard 1937– : *Rosencrantz and Guildenstern are Dead* (1967)

Living

See also **Life**

1 Thou shalt love thy neighbour as thyself.
Bible: Leviticus. See also St Matthew

2 Fear God, and keep his commandments: for this is the whole duty of man.
Bible: Ecclesiastes

3 A man hath no better thing under the sun, than to eat, and to drink, and to be merry.
Bible: Ecclesiastes

4 Love and do what you will.
St Augustine of Hippo AD 354–430: *In Epistolam Joannis ad Parthos* (AD 413)

5 *Quidquid agis, prudenter agas, et respice finem.*
Whatever you do, do cautiously, and look to the end.
Anonymous: *Gesta Romanorum*

6 *Fais ce que voudras.*
Do what you like.
François Rabelais *c.*1494–*c.*1553: *Gargantua* (1534)

7 *Mon métier et mon art c'est vivre.*
Living is my job and my art.
Montaigne 1533–92: *Essais* (1580)

8 If I had no duties, and no reference to futurity, I would spend my life in driving briskly in a post-chaise with a pretty woman.
Samuel Johnson 1709–84: in James Boswell *Life of Johnson* (1791) 19 September 1777

9 Take short views, hope for the best, and trust in God.
Sydney Smith 1771–1845: in Lady Holland *Memoir* (1855)

10 Believe me! The secret of reaping the greatest fruitfulness and the greatest enjoyment from life is *to live dangerously*!
Friedrich Nietzsche 1844–1900: *Die fröhliche Wissenschaft* (1882)

11 Live all you can; it's a mistake not to. It doesn't so much matter what you do in particular, so long as you have your life. If you haven't had that, what *have* you had?
Henry James 1843–1916: *Ambassadors* (1903)

12 Do what thou wilt shall be the whole of the Law.
Aleister Crowley 1875–1947: *Book of the Law* (1909)

13 I've lived a life that's full, I've travelled each and ev'ry highway
And more, much more than this. I did it my way.
Paul Anka 1941– : *My Way* (1969 song)

London

1 London, thou art the flower of cities all!
Anonymous: 'London' previously attributed to William Dunbar, *c.*1465–*c.*1530

2 The full tide of human existence is at Charing-Cross.
Samuel Johnson 1709–84: in James Boswell *Life of Johnson* (1791) 2 April 1775

3 When a man is tired of London, he is tired of life.
Samuel Johnson 1709–84: in James Boswell *Life of Johnson* (1791) 20 September 1777

4 Crowds without company, and dissipation without pleasure.
Edward Gibbon 1737–94: *Memoirs of My Life* (1796)

5 Earth has not anything to show more fair:
Dull would he be of soul who could pass by
A sight so touching in its majesty.
William Wordsworth 1770–1850: 'Composed upon Westminster Bridge' (1807)

6 *Was für Plunder!*
What rubbish!
Gebhard Lebrecht Blücher 1742–1819: of London as seen from the Monument in June 1814, often misquoted '*Was für plündern* [What a place to plunder]!'

7 The great wen of all.
William Cobbett 1762–1835: *Rural Rides* 5 January 1822

8 I thought of London spread out in the sun,
Its postal districts packed like squares of wheat.
Philip Larkin 1922–85: *The Whitsun Weddings* (1964)

Love

See also **Marriage, Sex**

1 Many waters cannot quench love, neither can the floods drown it.
Bible: Song of Solomon

2 *Omnia vincit Amor: et nos cedamus Amori.*
Love conquers all things: let us too give in to Love.
Virgil 70–19 BC: *Eclogues*

3 Greater love hath no man than this, that a man lay down his life for his friends.
Bible: St John

4 Though I speak with the tongues of men and of angels, and have not charity, I am become as sounding brass, or a tinkling cymbal.

And though I have the gift of prophecy, and understand all mysteries, and all knowledge; and though I have all faith; so that I could remove mountains; and have not charity, I am nothing.
Bible: I Corinthians

5 And now abideth faith, hope, charity, these three; but the greatest of these is charity.
Bible: I Corinthians

6 There is no fear in love; but perfect love casteth out fear.
Bible: I John

7 Difficult or easy, pleasant or bitter, you are the same you: I cannot live with you—or without you.
Martial AD *c.*40–*c.*104: *Epigrammata*

8 *L'amor che muove il sole e l'altre stelle.*
The love that moves the sun and the other stars.
Dante Alighieri 1265–1321: *Divina Commedia* 'Paradiso'

9 Love wol nat been constreyned by maistrye.
When maistrie comth, the God of Love anon
Beteth his wynges, and farewel, he is gon!
Geoffrey Chaucer *c.*1343–1400: *The Canterbury Tales* 'The Franklin's Tale'

10 If I am pressed to say why I loved him, I feel it can only be explained by replying: 'Because it was he; because it was me.'
Montaigne 1533–92: *Essais* (1580)

11 Love is a spirit all compact of fire,
Not gross to sink, but light, and will aspire.
William Shakespeare 1564–1616: *Venus and Adonis* (1593)

12 Fie, fie! how wayward is this foolish love
That, like a testy babe, will scratch the nurse
And presently all humbled kiss the rod!
William Shakespeare 1564–1616: *The Two Gentlemen of Verona* (1592–3)

13 O! how this spring of love resembleth
The uncertain glory of an April day.
William Shakespeare 1564–1616: *The Two Gentlemen of Verona* (1592–3)

14 Love built on beauty, soon as beauty, dies.
John Donne 1572–1631: 'The Anagram' (*c.*1595)

15 For stony limits cannot hold love out,
And what love can do that dares love attempt.
William Shakespeare 1564–1616: *Romeo and Juliet* (1595)

16 Love looks not with the eyes, but with the mind,
And therefore is winged Cupid painted blind.
William Shakespeare 1564–1616: *A Midsummer Night's Dream* (1595–6)

17 The course of true love never did run smooth.
William Shakespeare 1564–1616: *A Midsummer Night's Dream* (1595–6)

18 Where both deliberate, the love is slight;
Who ever loved that loved not at first sight?
Christopher Marlowe 1564–93: *Hero and Leander* (1598)

19 Love sought is good, but giv'n unsought is better.
William Shakespeare 1564–1616: *Twelfth Night* (1601)

20 To be wise, and love,
Exceeds man's might.
William Shakespeare 1564–1616: *Troilus and Cressida* (1602)

21 Then, must you speak
Of one that loved not wisely but too well.
William Shakespeare 1564–1616: *Othello* (1602–4)

22 ANTONY: There's beggary in the love that can be reckoned.
CLEOPATRA: I'll set a bourn how far to be beloved.
ANTONY: Then must thou needs find out new heaven, new earth.
William Shakespeare 1564–1616: *Antony and Cleopatra* (1606–7)

23 Let me not to the marriage of true minds
Admit impediments. Love is not love
Which alters when it alteration finds.
William Shakespeare 1564–1616: sonnet 116 (1609)

24 Love is like linen often changed, the sweeter.
Phineas Fletcher 1582–1650: *Sicelides* (1614)

25 Love, a child, is ever crying:
Please him and he straight is flying,
Give him, he the more is craving,
Never satisfied with having.
Lady Mary Wroth *c.*1586–*c.*1652: 'Love, a child, is ever crying' (1621)

26 Love is a growing or full constant light;
And his first minute, after noon, is night.
John Donne 1572–1631: 'A Lecture in the Shadow'

27 All other things, to their destruction draw,
Only our love hath no decay;
This, no tomorrow hath, nor yesterday,

Running it never runs from us away,
But truly keeps his first, last, everlasting day.
John Donne 1572–1631: 'The Anniversary'

28 Why so pale and wan, fond lover?
Prithee, why so pale?
Will, when looking well can't move her,
Looking ill prevail?
John Suckling 1609–42: *Aglaura* (1637)

29 To enlarge or illustrate this power and effect of love is to set a candle in the sun.
Robert Burton 1577–1640: *Anatomy of Melancholy* (1621–51)

30 Love is the fart
Of every heart:
It pains a man when 'tis kept close,
And others doth offend, when 'tis let loose.
John Suckling 1609–42: 'Love's Offence' (1646)

31 Love's passives are his activ'st part.
The wounded is the wounding heart.
Richard Crashaw *c.*1612–49: 'The Flaming Heart upon the Book of Saint Teresa' (1652)

32 And love's the noblest frailty of the mind.
John Dryden 1631–1700: *The Indian Emperor* (1665)

33 Had we but world enough, and time,
This coyness, lady, were no crime.
Andrew Marvell 1621–78: 'To His coy Mistress' (1681)

34 Say what you will, 'tis better to be left than never to have been loved.
William Congreve 1670–1729: *The Way of the World* (1700)

35 Love is only one of many passions.
Samuel Johnson 1709–84: *Plays of William Shakespeare* (1765) preface

36 Love and a cottage! Eh, Fanny! Ah, give me indifference and a coach and six!
George Colman, the Elder 1732–94 and **David Garrick** 1717–79: *The Clandestine Marriage* (1766)

37 Friendship is a disinterested commerce between equals; love, an abject intercourse between tyrants and slaves.
Oliver Goldsmith 1730–74: *The Good-Natured Man* (1768)

38 Love is the wisdom of the fool and the folly of the wise.
Samuel Johnson 1709–84: in W. Cooke *Life of Samuel Foote* (1805)

39 Love's pleasure lasts but a moment; love's sorrow lasts all through life.
Jean-Pierre Claris de Florian 1755–94: *Célestine* (1784)

40 Love seeketh not itself to please,
Nor for itself hath any care;
But for another gives its ease,
And builds a Heaven in Hell's despair.
William Blake 1757–1827: 'The Clod and the Pebble' (1794)

41 Love seeketh only Self to please,
To bind another to its delight,
Joys in another's loss of ease,
And builds a Hell in Heaven's despite.
William Blake 1757–1827: 'The Clod and the Pebble' (1794)

42 If I love you, what does that matter to you!
Johann Wolfgang von Goethe 1749–1832: *Wilhelm Meister's Apprenticeship* (1795-6), tr. R. D. Moon

43 O, my Luve's like a red, red rose
That's newly sprung in June;
O my Luve's like the melodie
That's sweetly play'd in tune.
Robert Burns 1759–96: 'A Red Red Rose' (1796); derived from various folk-songs

44 No, there's nothing half so sweet in life
As love's young dream.
Thomas Moore 1779–1852: 'Love's Young Dream' (1807)

45 Love in a hut, with water and a crust,
Is—Love, forgive us!—cinders, ashes, dust.
John Keats 1795–1821: 'Lamia' (1820)

46 My love for Heathcliff resembles the eternal rocks beneath:—a source of little visible delight, but necessary.
Emily Brontë 1818–48: *Wuthering Heights* (1847)

47 If you could see my legs when I take my boots off, you'd form some idea of what unrequited affection is.
Charles Dickens 1812–70: *Dombey and Son* (1848) Mr Toots

48 'Tis better to have loved and lost
Than never to have loved at all.
Alfred, Lord Tennyson 1809–92: *In Memoriam A. H. H.* (1850)

49 Love's like the measles—all the worse when it comes late in life.
Douglas Jerrold 1803–57: *Wit and Opinions* (1859)

50 We loved, sir—used to meet:
How sad and bad and mad it was—
But then, how it was sweet!
Robert Browning 1812–89: 'Confessions' (1864)

51 O lyric Love, half-angel and half-bird
And all a wonder and a wild desire.
Robert Browning 1812–89: *The Ring and the Book* (1868–9)

52 Love is like any other luxury. You have no right to it unless you can afford it.
Anthony Trollope 1815–82: *The Way We Live Now* (1875)

53 A lover without indiscretion is no lover at all.
Thomas Hardy 1840–1928: *The Hand of Ethelberta* (1876)

54 Never the time and the place
And the loved one all together!
Robert Browning 1812–89: 'Never the Time and the Place' (1883)

55 A pity beyond all telling,
Is hid in the heart of love.
W. B. Yeats 1865–1939: 'The Pity of Love' (1893)

56 I am the Love that dare not speak its name.
Lord Alfred Douglas 1870–1945: 'Two Loves' (1896)

57 Yet each man kills the thing he loves,
By each let this be heard,
Some do it with a bitter look,
Some with a flattering word.
The coward does it with a kiss,
The brave man with a sword!
Oscar Wilde 1854–1900: *The Ballad of Reading Gaol* (1898)

58 The fickleness of the women I love is only equalled by the infernal constancy of the women who love me.
George Bernard Shaw 1856–1950: *The Philanderer* (1898)

59 A woman can be proud and stiff
When on love intent;

But Love has pitched his mansion in
The place of excrement;
For nothing can be sole or whole
That has not been rent.

W. B. Yeats 1865–1939: 'Crazy Jane Talks with the Bishop' (1932)

60 By the time you say you're his,
Shivering and sighing
And he vows his passion is
Infinite, undying—
Lady, make a note of this:
One of you is lying.

Dorothy Parker 1893–1967: 'Unfortunate Coincidence' (1937)

61 Life has taught us that love does not consist in gazing at each other but in looking together in the same direction.

Antoine de Saint-Exupéry 1900–1944: *Terre des Hommes* (translated as 'Wind, Sand and Stars', 1939)

62 I'll love you, dear, I'll love you
Till China and Africa meet
And the river jumps over the mountain
And the salmon sing in the street.

W. H. Auden 1907–73: 'As I Walked Out One Evening' (1940)

63 How alike are the groans of love to those of the dying.

Malcolm Lowry 1909–57: *Under the Volcano* (1947)

64 Love is the delusion that one woman differs from another.

H. L. Mencken 1880–1956: *Chrestomathy* (1949)

65 You know very well that love is, above all, the gift of oneself!

Jean Anouilh 1910–87: *Ardèle* (1949)

66 Birds do it, bees do it,
Even educated fleas do it.
Let's do it, let's fall in love.

Cole Porter 1891–1964: 'Let's Do It' (1954 song; words added to the 1928 original)

67 Love means not ever having to say you're sorry.

Erich Segal 1937– : *Love Story* (1970)

Madness

1 Whenever God prepares evil for a man, He first damages his mind, with which he deliberates.
Anonymous: scholiastic annotation to Sophocles's *Antigone*

2 Though this be madness, yet there is method in't.
William Shakespeare 1564–1616: *Hamlet* (1601)

3 I am but mad north-north-west; when the wind is southerly, I know a hawk from a handsaw.
William Shakespeare 1564–1616: *Hamlet* (1601)

4 O! what a noble mind is here o'erthrown.
William Shakespeare 1564–1616: *Hamlet* (1601)

5 O! let me not be mad, not mad, sweet heaven;
Keep me in temper; I would not be mad!
William Shakespeare 1564–1616: *King Lear* (1605–6)

6 *Quem Jupiter vult perdere, dementat prius.*
Whom God would destroy He first sends mad.
James Duport 1606–79: *Homeri Gnomologia* (1660)

7 Mad, is he? Then I hope he will *bite* some of my other generals.
George II 1683–1760: replying to the Duke of Newcastle, who had complained that General Wolfe was a madman

8 Babylon in all its desolation is a sight not so awful as that of the human mind in ruins.
Scrope Davies *c.*1783–1852: letter to Thomas Raikes, May 1835

9 There was only one catch and that was Catch-22... Orr would be crazy to fly more missions and sane if he didn't, but if he was sane he had to fly them. If he flew them he was crazy and didn't have to; but if he didn't want to he was sane and had to.
Joseph Heller 1923– : *Catch-22* (1961)

Majorities see **Minorities and Majorities**

Management

See also **Bureaucracy, Careers**

1 Every time I create an appointment, I create a hundred malcontents and one ingrate.
Louis XIV 1638–1715: in Voltaire *Siècle de Louis XIV* (1768 ed.)

2 *Dans ce pays-ci il est bon de tuer de temps en temps un amiral pour encourager les autres.*
In this country [England] it is thought well to kill an admiral from time to time to encourage the others.
Voltaire 1694–1778: *Candide* (1759)

3 The shortest way to do many things is to do only one thing at once.
Samuel Smiles 1812–1904: *Self-Help* (1859)

4 A place for everything and everything in its place.
Mrs Beeton 1836–65: *Book of Household Management* (1861), often attributed to Samuel Smiles

5 This island is made mainly of coal and surrounded by fish. Only an organizing genius could produce a shortage of coal and fish at the same time.
Aneurin Bevan 1897–1960: speech at Blackpool, 24 May 1945

6 Committee—a group of men who individually can do nothing but as a group decide that nothing can be done.
Fred Allen 1894–1956: attributed

7 Time spent on any item of the agenda will be in inverse proportion to the sum involved.
C. Northcote Parkinson 1909–93: *Parkinson's Law* (1958)

8 Perfection of planned layout is achieved only by institutions on the point of collapse.
C. Northcote Parkinson 1909–93: *Parkinson's Law* (1958)

9 A problem left to itself dries up or goes rotten. But fertilize a problem with a solution—you'll hatch out dozens.
N. F. Simpson 1919– : *A Resounding Tinkle* (1958)

10 You're either part of the solution or you're part of the problem.
Eldridge Cleaver 1935– : speech in San Francisco, 1968

11 In a hierarchy every employee tends to rise to his level of incompetence.
Laurence Peter 1919– : *The Peter Principle* (1969)

12 Competence, like truth, beauty and contact lenses, is in the eye of the beholder.
Laurence Peter 1919– : *The Peter Principle* (1969)

13 There cannot be a crisis next week. My schedule is already full.
Henry Kissinger 1923– : in *New York Times Magazine* 1 June 1969

14 If it ain't broke, don't fix it.
Bert Lance 1931– : in *Nation's Business* May 1977

15 A camel is a horse designed by a committee.
Alec Issigonis 1906–88: on his dislike of working in teams

16 Management that wants to change an institution must first show it loves that institution.
John Tusa 1936– : *Observer* 27 February 1994 'Sayings of the Week'

Manners

See also **Behaviour**

1 Evil communications corrupt good manners.
Bible: I Corinthians

2 Stand not upon the order of your going.
William Shakespeare 1564–1616: *Macbeth* (1606)

3 Courts and camps are the only places to learn the world in.
Lord Chesterfield 1694–1773: *Letters to his Son* (1774) 2 October 1747

4 Take the tone of the company that you are in.
Lord Chesterfield 1694–1773: *Letters to his Son* (1774) 16 October 1747

5 He is the very pineapple of politeness!
Richard Brinsley Sheridan 1751–1816: *The Rivals* (1775)

6 Most vices may be committed very genteelly: a man may debauch his friend's wife genteelly: he may cheat at cards genteelly.

James Boswell 1740–95: *Life of Samuel Johnson* (1791)

7 Ceremony is an invention to take off the uneasy feeling which we derive from knowing ourselves to be less the object of love and esteem with a fellow-creature than some other person is.

Charles Lamb 1775–1834: *Essays of Elia* (1823) 'A Bachelor's Complaint of the Behaviour of Married People'

8 *L'exactitude est la politesse des rois.*
Punctuality is the politeness of kings.

Louis XVIII 1755–1824: attributed

9 Curtsey while you're thinking what to say. It saves time.

Lewis Carroll 1832–98: *Through the Looking-Glass* (1872)

10 Very notable was his distinction between coarseness and vulgarity (coarseness, revealing something; vulgarity, concealing something).

E. M. Forster 1879–1970: *The Longest Journey* (1907)

11 There are few who would not rather be taken in adultery than in provincialism.

Aldous Huxley 1894–1963: *Antic Hay* (1923)

12 JUDGE: You are extremely offensive, young man.
SMITH: As a matter of fact, we both are, and the only difference between us is that I am trying to be, and you can't help it.

F. E. Smith 1872–1930: in 2nd Earl of Birkenhead *Earl of Birkenhead* (1933)

13 Phone for the fish-knives, Norman
As Cook is a little unnerved;
You kiddies have crumpled the serviettes
And I must have things daintily served.

John Betjeman 1906–84: 'How to get on in Society' (1954)

14 To Americans, English manners are far more frightening than none at all.

Randall Jarrell 1914–65: *Pictures from an Institution* (1954)

15 Manners are especially the need of the plain. The pretty can get away with anything.

Evelyn Waugh 1903–66: in *Observer* 15 April 1962

Marriage

See also **Love, Sex**

1 Therefore shall a man leave his father and his mother, and shall cleave unto his wife: and they shall be one flesh.
Bible: Genesis

2 What therefore God hath joined together, let not man put asunder.
Bible: St Matthew

3 It is better to marry than to burn.
Bible: I Corinthians

4 Men are April when they woo, December when they wed: maids are May when they are maids, but the sky changes when they are wives.
William Shakespeare 1564–1616: *As You Like It* (1599)

5 A young man married is a man that's marred.
William Shakespeare 1564–1616: *All's Well that Ends Well* (1603–4)

6 Wives are young men's mistresses, companions for middle age, and old men's nurses.
Francis Bacon 1561–1626: *Essays* (1625) 'Of Marriage and the Single Life'

7 He was reputed one of the wise men that made answer to the question when a man should marry? 'A young man not yet, an elder man not at all.'
Francis Bacon 1561–1626: *Essays* (1625) 'Of Marriage and the Single Life'

8 I would be married, but I'd have no wife,
I would be married to a single life.
Richard Crashaw *c.*1612–49: 'On Marriage' (1646)

9 Then be not coy, but use your time;
And while ye may, go marry:
For having lost but once your prime,
You may for ever tarry.
Robert Herrick 1591–1674: 'To the Virgins, to Make Much of Time' (1648)

10 Wilt thou love her, comfort her, honour, and keep her in sickness and in health; and, forsaking all other, keep thee only unto her, so long as ye both shall live?
Book of Common Prayer 1662: *Solemnization of Matrimony*

11 To have and to hold from this day forward, for better for worse, for richer for poorer, in sickness and in health, to love, cherish, and to obey, till death us do part.
Book of Common Prayer 1662: *Solemnization of Matrimony*

12 A Man may not marry his Mother.
Book of Common Prayer 1662: *A Table of Kindred and Affinity*

13 Strange to say what delight we married people have to see these poor fools decoyed into our condition.
Samuel Pepys 1633–1703: diary, 25 December 1665

14 Courtship to marriage, as a very witty prologue to a very dull play.
William Congreve 1670–1729: *The Old Bachelor* (1693)

15 SHARPER: Thus grief still treads upon the heels of pleasure:
Married in haste, we may repent at leisure.
SETTER: Some by experience find those words mis-placed:
At leisure married, they repent in haste.
William Congreve 1670–1729: *The Old Bachelor* (1693)

16 Oh! how many torments lie in the small circle of a wedding-ring!
Colley Cibber 1671–1757: *The Double Gallant* (1707)

17 Do you think your mother and I should have lived comfortably so long together, if ever we had been married?
John Gay 1685–1732: *The Beggar's Opera* (1728)

18 The comfortable estate of widowhood, is the only hope that keeps up a wife's spirits.
John Gay 1685–1732: *The Beggar's Opera* (1728)

19 One fool at least in every married couple.
Henry Fielding 1707–54: *Amelia* (1751)

20 Marriage has many pains, but celibacy has no pleasures.
Samuel Johnson 1709–84: *Rasselas* (1759)

21 I...chose my wife, as she did her wedding gown, not for a fine glossy surface, but such qualities as would wear well.
Oliver Goldsmith 1730–74: *The Vicar of Wakefield* (1766)

22 O! how short a time does it take to put an end to a woman's liberty!
Fanny Burney 1752–1840: of a wedding; diary, 20 July 1768

23 The triumph of hope over experience.
Samuel Johnson 1709–84: of a man who remarried immediately after the death of a wife with whom he had been unhappy, 1770; in James Boswell *Life of Johnson* (1791)

24 Still I can't contradict, what so oft has been said,
'Though women are angels, yet wedlock's the devil.'
Lord Byron 1788–1824: 'To Eliza' (1806)

25 It is a truth universally acknowledged, that a single man in possession of a good fortune, must be in want of a wife.
Jane Austen 1775–1817: *Pride and Prejudice* (1813)

26 There is not one in a hundred of either sex who is not taken in when they marry... it is, of all transactions, the one in which people expect most from others, and are least honest themselves.
Jane Austen 1775–1817: *Mansfield Park* (1814)

27 Marriage may often be a stormy lake, but celibacy is almost always a muddy horsepond.
Thomas Love Peacock 1785–1866: *Melincourt* (1817)

28 Think you, if Laura had been Petrarch's wife,
He would have written sonnets all his life?
Lord Byron 1788–1824: *Don Juan* (1819–24)

29 Advice to persons about to marry.—'Don't.'
Punch: 1845

30 My definition of marriage... it resembles a pair of shears, so joined that they cannot be separated; often moving in opposite directions, yet always punishing anyone who comes between them.
Sydney Smith 1771–1845: in Lady Holland *Memoir* (1855)

31 It doesn't much signify whom one marries, for one is sure to find next morning that it was someone else.
Samuel Rogers 1763–1855: in A. Dyce (ed.) *Table Talk of Samuel Rogers* (1860)

32 A woman dictates before marriage in order that she may have an appetite for submission afterwards.
George Eliot 1819–80: *Middlemarch* (1871–2)

33 Marriage is like life in this—that it is a field of battle, and not a bed of roses.
Robert Louis Stevenson 1850–94: *Virginibus Puerisque* (1881)

34 It was very good of God to let Carlyle and Mrs Carlyle marry one another and so make only two people miserable instead of four.
Samuel Butler 1835–1902: letter to Miss E. M. A. Savage, 21 November 1884

35 In married life three is company and two none.
Oscar Wilde 1854–1900: *The Importance of Being Earnest* (1895)

36 If it were not for the presents, an elopement would be preferable.
George Ade 1866–1944: *Forty Modern Fables* (1901)

37 Marriage is popular because it combines the maximum of temptation with the maximum of opportunity.
George Bernard Shaw 1856–1950: *Man and Superman* (1903)

38 Being a husband is a whole-time job. That is why so many husbands fail. They cannot give their entire attention to it.
Arnold Bennett 1867–1931: *The Title* (1918)

39 Chumps always make the best husbands...All the unhappy marriages come from the husbands having brains.
P. G. Wodehouse 1881–1975: *The Adventures of Sally* (1920)

40 Marriage isn't a word...it's a *sentence*!
King Vidor 1895–1982: *The Crowd* (1928 film)

41 Marriage always demands the finest arts of insincerity possible between two human beings.
Vicki Baum 1888–1960: *Zwischenfall in Lohwinckel* (1930), tr. M. Goldsmith as *Results of an Accident* (1931)

42 The deep, deep peace of the double-bed after the hurly-burly of the chaise-longue.
Mrs Patrick Campbell 1865–1940: on her recent marriage, in A. Woollcott *While Rome Burns* (1934)

43 The critical period in matrimony is breakfast-time.
A. P. Herbert 1890–1971: *Uncommon Law* (1935)

44 So they were married—to be the more together—
And found they were never again so much together,
Divided by the morning tea,

By the evening paper,
By children and tradesmen's bills.
Louis MacNeice 1907–63: *Plant and Phantom* (1941)

45 Sidney would remark, 'I know just what Beatrice is saying at this moment. She is saying, "as Sidney always says, marriage is the waste-paper basket of the emotions."'
Sidney Webb 1859–1947 and **Beatrice Webb** 1858–1943: in Bertrand Russell *Autobiography* (1967)

46 It takes two to make a marriage a success and only one a failure.
Lord Samuel 1870–1963: *A Book of Quotations* (1947)

47 One doesn't have to get anywhere in a marriage. It's not a public conveyance.
Iris Murdoch 1919– : *A Severed Head* (1961)

48 Never marry a man who hates his mother, because he'll end up hating you.
Jill Bennett 1931–90: in *Observer* 12 September 1982 'Sayings of the Week'

Mathematics

1 Let no one enter who does not know geometry [mathematics].
Anonymous: inscription on Plato's door, probably at the Academy at Athens

2 There is no 'royal road' to geometry.
Euclid fl. *c.*300 BC: addressed to Ptolemy I, in Proclus *Commentary on the First Book of Euclid's Elementa*

3 Multiplication is vexation,
Division is as bad;
The Rule of Three doth puzzle me,
And Practice drives me mad.
Anonymous: in *Lean's Collectanea* (1904), possibly 16th-century

4 I have often admired the mystical way of Pythagoras, and the secret magic of numbers.
Sir Thomas Browne 1605–82: *Religio Medici* (1643)

5 They are neither finite quantities, or quantities infinitely small, nor yet nothing. May we not call them the ghosts of departed quantities?

Bishop George Berkeley 1685–1753: *The Analyst* (1734), on Newton's infinitesimals

6 The most devilish thing is 8 times 8 and 7 times 7 it is what nature itselfe cant endure.

Marjory Fleming 1803–11: *Journals, Letters and Verses* (1934)

7 What would life be like without arithmetic, but a scene of horrors?

Sydney Smith 1771–1845: letter to Miss [Lucie Austen], 22 July 1835

8 God made the integers, all the rest is the work of man.

Leopold Kronecker 1823–91: *Jahrsberichte der Deutschen Mathematiker Vereinigung* bk. 2

9 I never could make out what those damned dots meant.

Lord Randolph Churchill 1849–94: on decimal points, in W. S. Churchill *Lord Randolph Churchill* (1906)

10 Mathematics, rightly viewed, possesses not only truth, but supreme beauty—a beauty cold and austere, like that of sculpture.

Bertrand Russell 1872–1970: *Philosophical Essays* (1910)

11 Mathematics may be defined as the subject in which we never know what we are talking about, nor whether what we are saying is true.

Bertrand Russell 1872–1970: *Mysticism and Logic* (1918)

12 Beauty is the first test: there is no permanent place in the world for ugly mathematics.

Godfrey Harold Hardy 1877–1947: *A Mathematician's Apology* (1940)

13 Equations are more important to me, because politics is for the present, but an equation is something for eternity.

Albert Einstein 1879–1955: in Stephen Hawking *A Brief History of Time* (1988)

14 Someone told me that each equation I included in the book would halve the sales.

Stephen Hawking 1942– : *A Brief History of Time* (1988)

Meaning

See also **Words**

1 I pray thee, understand a plain man in his plain meaning.
William Shakespeare 1564–1616: *The Merchant of Venice* (1596–8)

2 God and I both knew what it meant once; now God alone knows.
Friedrich Klopstock 1724–1803: in C. Lombroso *The Man of Genius* (1891); also attributed to Browning, apropos *Sordello,* in the form 'When it was written, God and Robert Browning knew what it meant; now only God knows'

3 'Then you should say what you mean,' the March Hare went on. 'I do,' Alice hastily replied; 'at least—at least I mean what I say—that's the same thing, you know.' 'Not the same thing a bit!' said the Hatter. 'Why, you might just as well say that "I see what I eat" is the same thing as "I eat what I see!" '
Lewis Carroll 1832–98: *Alice's Adventures in Wonderland* (1865)

4 The meaning doesn't matter if it's only idle chatter of a transcendental kind.
W. S. Gilbert 1836–1911: *Patience* (1881)

5 No one means all he says, and yet very few say all they mean, for words are slippery and thought is viscous.
Henry Brooks Adams 1838–1918: *The Education of Henry Adams* (1907)

6 The little girl had the making of a poet in her who, being told to be sure of her meaning before she spoke, said, 'How can I know what I think till I see what I say?'
Graham Wallas 1858–1932: *The Art of Thought* (1926)

7 It all depends what you mean by…
C. E. M. Joad 1891–1953: replying to questions on 'The Brains Trust' (formerly 'Any Questions'), BBC radio (1941–8)

Medicine

See also **Sickness and Health**

1 Honour a physician with the honour due unto him for the uses which ye may have of him.
Bible (Apocrypha): Ecclesiasticus

2 He that sinneth before his Maker, let him fall into the hand of the physician.
Bible (Apocrypha): Ecclesiasticus

3 Life is short, the art long.
Hippocrates c.460–357 BC: *Aphorisms*; often quoted '*Ars longa, vita brevis*', after Seneca *De Brevitate Vitae*

4 Physician, heal thyself.
Bible: St Luke

5 Throw physic to the dogs; I'll none of it.
William Shakespeare 1564–1616: *Macbeth* (1606)

6 Cure the disease and kill the patient.
Francis Bacon 1561–1626: *Essays* (1625) 'Of Friendship'

7 The remedy is worse than the disease.
Francis Bacon 1561–1626: *Essays* (1625) 'Of Seditions and Troubles'

8 We all labour against our own cure, for death is the cure of all diseases.
Sir Thomas Browne 1605–82: *Religio Medici* (1643)

9 GÉRONTE: It seems to me you are locating them wrongly: the heart is on the left and the liver is on the right.
SGANARELLE: Yes, in the old days that was so, but we have changed all that.
Molière 1622–73: *Le Médecin malgré lui* (1667)

10 Sciatica: he cured it, by boiling his buttock.
John Aubrey 1626–97: *Brief Lives* 'Sir Jonas Moore'

11 Ah, well, then, I suppose that I shall have to die beyond my means.
Oscar Wilde 1854–1900: at the mention of a huge fee for a surgical operation; in R. H. Sherard *Life of Oscar Wilde* (1906)

12 Physicians of the Utmost Fame
Were called at once; but when they came

They answered, as they took their Fees,
'There is no Cure for this Disease.'

Hilaire Belloc 1870–1953: 'Henry King' (1907)

13 There is at bottom only one genuinely scientific treatment for all diseases, and that is to stimulate the phagocytes.

George Bernard Shaw 1856–1950: *The Doctor's Dilemma* (1911)

14 Every day, in every way, I am getting better and better.

Émile Coué 1857–1926: to be said 15 to 20 times, morning and evening, in *De la suggestion et de ses applications* (1915)

15 One finger in the throat and one in the rectum makes a good diagnostician.

William Osler 1849–1919: *Aphorisms from his Bedside Teachings* (1961)

16 The wounded surgeon plies the steel
That questions the distempered part;
Beneath the bleeding hands we feel
The sharp compassion of the healer's art
Resolving the enigma of the fever chart.

T. S. Eliot 1888–1965: *Four Quartets* 'East Coker' (1940)

17 I stuffed their mouths with gold.

Aneurin Bevan 1897–1960: on his handling of the consultants during the establishment of the National Health Service, in B. Abel-Smith *The Hospitals 1800–1948* (1964)

18 We shall have to learn to refrain from doing things merely because we know how to do them.

Theodore Fox 1899–1989: speech to Royal College of Physicians, 18 October 1965

19 Formerly, when religion was strong and science weak, men mistook magic for medicine; now, when science is strong and religion weak, men mistake medicine for magic.

Thomas Szasz 1920– : *The Second Sin* (1973) 'Science and Scientism'

20 Medicinal discovery,
It moves in mighty leaps,
It leapt straight past the common cold
And gave it us for keeps.

Pam Ayres 1947– : 'Oh no, I got a cold' (1976)

Meeting and Parting

1 *Atque in perpetuum, frater, ave atque vale.*
And so, my brother, hail, and farewell evermore!
Catullus *c.*84–*c.*54 BC: *Carmina*

2 Fare well my dear child and pray for me, and I shall for you and all your friends that we may merrily meet in heaven.
Sir Thomas More 1478–1535: letter to his daughter Margaret, 5 July 1535, on the eve of his execution

3 Good-night, good-night! parting is such sweet sorrow
That I shall say good-night till it be morrow.
William Shakespeare 1564–1616: *Romeo and Juliet* (1595)

4 Ill met by moonlight, proud Titania.
William Shakespeare 1564–1616: *A Midsummer Night's Dream* (1595–6)

5 When shall we three meet again
In thunder, lightning, or in rain?
William Shakespeare 1564–1616: *Macbeth* (1606)

6 Farewell! thou art too dear for my possessing.
William Shakespeare 1564–1616: sonnet 87 (1609)

7 Since there's no help, come let us kiss and part,
Nay, I have done: you get no more of me.
Michael Drayton 1563–1631: sonnet (1619)

8 You have sat too long here for any good you have been doing. Depart, I say, and let us have done with you. In the name of God, go!
Oliver Cromwell 1599–1658: addressing the Rump Parliament, 20 April 1653 (oral tradition; quoted by Leo Amery, House of Commons, 7 May 1940)

9 For I, who hold sage Homer's rule the best,
Welcome the coming, speed the going guest.
Alexander Pope 1688–1744: *Imitations of Horace* (1734) ('speed the parting guest' in Pope's translation of *The Odyssey* (1725–6))

10 Their meetings made December June,
Their every parting was to die.
Alfred, Lord Tennyson 1809–92: *In Memoriam A. H. H.* (1850)

11 In every parting there is an image of death.
George Eliot 1819–80: *Scenes of Clerical Life* (1858)

12 Dr Livingstone, I presume?
Henry Morton Stanley 1841–1904: *How I found Livingstone* (1872)

13 Parting is all we know of heaven,
And all we need of hell.
Emily Dickinson 1830–86: 'My life closed twice before its close'

14 Yet meet we shall, and part, and meet again
Where dead men meet, on lips of living men.
Samuel Butler 1835–1902: 'Not on sad Stygian shore' (1904)

15 If you can't leave in a taxi you can leave in a huff. If that's too soon, you can leave in a minute and a huff.
Bert Kalmar 1884–1947 et al.: *Duck Soup* (1933 film); spoken by Groucho Marx

16 Why don't you come up sometime, and see me?
Mae West 1892–1980: *She Done Him Wrong* (1933 film); usually quoted 'Why don't you come up and see me sometime?'

17 How long ago Hector took off his plume,
Not wanting that his little son should cry,
Then kissed his sad Andromache goodbye —
And now we three in Euston waiting-room.
Frances Cornford 1886–1960: 'Parting in Wartime' (1948)

Memory

1 Old men forget: yet all shall be forgot,
But he'll remember with advantages
What feats he did that day.
William Shakespeare 1564–1616: *Henry V* (1599)

2 There's rosemary, that's for remembrance; pray, love, remember.
William Shakespeare 1564–1616: *Hamlet* (1601)

3 When to the sessions of sweet silent thought
I summon up remembrance of things past.
William Shakespeare 1564–1616: sonnet 30 (1609)

4 You may break, you may shatter the vase, if you will,
But the scent of the roses will hang round it still.
Thomas Moore 1779–1852: 'Farewell!—but whenever' (1807)

5 For oft, when on my couch I lie
In vacant or in pensive mood,
They flash upon that inward eye
Which is the bliss of solitude;
And then my heart with pleasure fills,
And dances with the daffodils.
William Wordsworth 1770–1850: 'I wandered lonely as a cloud' (1815 ed.)

6 Music, when soft voices die,
Vibrates in the memory—
Odours, when sweet violets sicken,
Live within the sense they quicken.
Percy Bysshe Shelley 1792–1822: 'To—: Music, when soft voices die' (1824)

7 In looking on the happy autumn-fields,
And thinking of the days that are no more.
Alfred, Lord Tennyson 1809–92: *The Princess* (1847) song (added 1850)

8 And we forget because we must
And not because we will.
Matthew Arnold 1822–88: 'Absence' (1852)

9 Better by far you should forget and smile
Than that you should remember and be sad.
Christina Rossetti 1830–94: 'Remember' (1862)

10 And the best and the worst of this is
That neither is most to blame,
If you have forgotten my kisses
And I have forgotten your name.
Algernon Charles Swinburne 1837–1909: 'An Interlude' (1866)

11 I have forgot much, Cynara! gone with the wind,
Flung roses, roses, riotously, with the throng,
Dancing, to put thy pale, lost lilies out of mind.
Ernest Dowson 1867–1900: 'Non Sum Qualis Eram' (1896)

12 Memories are hunting horns
Whose sound dies on the wind.
Guillaume Apollinaire 1880–1918: 'Cors de Chasse' (1912)

13 And suddenly the memory revealed itself. The taste was that of the little piece of madeleine which...my aunt Léonie used to give me, dipping it first in her own cup of tea or tisane.

Marcel Proust 1871–1922: *Swann's Way* (1913)

14 Midnight shakes the memory
As a madman shakes a dead geranium.

T. S. Eliot 1888–1965: 'Rhapsody on a Windy Night' (1917)

15 In plucking the fruit of memory one runs the risk of spoiling its bloom.

Joseph Conrad 1857–1924: *Arrow of Gold* (author's note, 1920, to 1924 Uniform Edition)

16 Someone said that God gave us memory so that we might have roses in December.

J. M. Barrie 1860–1937: Rectorial Address at St Andrew's, 3 May 1922

17 A cigarette that bears a lipstick's traces,
An airline ticket to romantic places.

Holt Marvell 1901–69: 'These Foolish Things Remind Me of You' (1935 song)

18 Footfalls echo in the memory
Down the passage which we did not take
Towards the door we never opened
Into the rose-garden.

T. S. Eliot 1888–1965: *Four Quartets* 'Burnt Norton' (1936)

19 Am in Market Harborough. Where ought I to be?

G. K. Chesterton 1874–1936: telegram said to have been sent to his wife; *Autobiography* (1936)

20 Our memories are card-indexes consulted, and then put back in disorder by authorities whom we do not control.

Cyril Connolly 1903–74: *The Unquiet Grave* (1944)

21 Memories are not shackles, Franklin, they are garlands.

Alan Bennett 1934– : *Forty Years On* (1969)

22 I never forget a face, but in your case I'll be glad to make an exception.

Groucho Marx 1895–1977: in Leo Rosten *People I have Loved, Known or Admired* (1970) 'Groucho'

Men

1 Sigh no more, ladies, sigh no more,
Men were deceivers ever.

William Shakespeare 1564–1616: *Much Ado About Nothing* (1598–9)

2 Men are but children of a larger growth;
Our appetites as apt to change as theirs,
And full as craving too, and full as vain.

John Dryden 1631–1700: *All for Love* (1678)

3 Man is to be held only by the *slightest* chains, with the idea that he can break them at pleasure, he submits to them in sport.

Maria Edgeworth 1768–1849: *Letters for Literary Ladies* (1795)

4 Men have had every advantage of us in telling their own story. Education has been theirs in so much higher a degree; the pen has been in their hands.

Jane Austen 1775–1817: *Persuasion* (1818)

5 A man...is *so* in the way in the house!

Elizabeth Gaskell 1810–65: *Cranford* (1853)

6 Man is Nature's sole mistake!

W. S. Gilbert 1836–1911: *Princess Ida* (1884)

7 The three most important things a man has are, briefly, his private parts, his money, and his religious opinions.

Samuel Butler 1835–1902: *Further Extracts from Notebooks* (1934)

8 The natural man has only two primal passions, to get and beget.

William Osler 1849–1919: *Science and Immortality* (1904)

9 A husband is what is left of a lover, after the nerve has been extracted.

Helen Rowland 1875–1950: *A Guide to Men* (1922)

10 Somehow a bachelor never quite gets over the idea that he is a thing of beauty and a boy forever.

Helen Rowland 1875–1950: *A Guide to Men* (1922)

11 It's not the men in my life that counts—it's the life in my men.

Mae West 1892–1980: *I'm No Angel* (1933 film)

12 Why can't a woman be more like a man?
Men are so honest, so thoroughly square;
Eternally noble, historically fair.
Alan Jay Lerner 1918–86: 'A Hymn to Him' (1956 song)

13 Whatever they may be in public life, whatever their relations with men, in their relations with women, all men are rapists, and that's all they are. They rape us with their eyes, their laws, and their codes.
Marilyn French 1929– : *The Women's Room* (1977)

14 What makes men so tedious
Is the need to show off and compete.
They'll bore you to death for hours and hours
Before they'll admit defeat.
Wendy Cope 1945– : 'Men and their boring arguments' (1988)

Men and Women

See also **Woman's Role**

1 Is there no way for men to be, but women
Must be half-workers?
William Shakespeare 1564–1616: *Cymbeline* (1609–10)

2 So court a mistress, she denies you;
Let her alone, she will court you.
Say, are not women truly then
Styled but the shadows of us men?
Ben Jonson *c.*1573–1637: 'That Women are but Men's Shadows' (1616)

3 Just such disparity
As is 'twixt air and angels' purity,
'Twixt women's love, and men's will ever be.
John Donne 1572–1631: 'Air and Angels'

4 He for God only, she for God in him.
John Milton 1608–74: *Paradise Lost* (1667)

5 Man's love is of man's life a thing apart,
'Tis woman's whole existence.
Lord Byron 1788–1824: *Don Juan* (1819–24)

6 What is it men in women do require
The lineaments of gratified desire
What is it women do in men require
The lineaments of gratified desire.
William Blake 1757–1827: *MS Note-Book*

7 The man's desire is for the woman; but the woman's desire is rarely other than for the desire of the man.
Samuel Taylor Coleridge 1772–1834: *Table Talk* (1835) 23 July 1827

8 Man is the hunter; woman is his game.
Alfred, Lord Tennyson 1809–92: *The Princess* (1847)

9 'Tis strange what a man may do, and a woman yet think him an angel.
William Makepeace Thackeray 1811–63: *The History of Henry Esmond* (1852)

10 Man dreams of fame while woman wakes to love.
Alfred, Lord Tennyson 1809–92: *Idylls of the King* 'Merlin and Vivien' (1859)

11 The silliest woman can manage a clever man; but it takes a very clever woman to manage a fool.
Rudyard Kipling 1865–1936: *Plain Tales from the Hills* (1888)

12 All women become like their mothers. That is their tragedy. No man does. That's his.
Oscar Wilde 1854–1900: *The Importance of Being Earnest* (1895)

13 Of all human struggles there is none so treacherous and remorseless as the struggle between the artist man and the mother woman.
George Bernard Shaw 1856–1950: *Man and Superman* (1903)

14 Hogamus, higamous
Man is polygamous
Higamus, hogamous
Woman monogamous.
William James 1842–1910: in *Oxford Book of Marriage* (1990)

15 The female of the species is more deadly than the male.
Rudyard Kipling 1865–1936: 'The Female of the Species' (1919)

16 A woman can forgive a man for the harm he does her, but she can never forgive him for the sacrifices he makes on her account.

W. Somerset Maugham 1874–1965: *The Moon and Sixpence* (1919)

17 Women have served all these centuries as looking-glasses possessing the magic and delicious power of reflecting the figure of a man at twice its natural size.

Virginia Woolf 1882–1941: *A Room of One's Own* (1929)

18 Me Tarzan, you Jane.

Johnny Weissmuller 1904–84: summing up his role in *Tarzan, the Ape Man* (1932 film); in *Photoplay Magazine* June 1932. The words occur neither in the film nor the original, by Edgar Rice Burroughs

19 Any man has to, needs to, wants to
Once in a lifetime, do a girl in.

T. S. Eliot 1888–1965: *Sweeney Agonistes* (1932)

20 I admit it is better fun to punt than to be punted, and that a desire to have all the fun is nine-tenths of the law of chivalry.

Dorothy L. Sayers 1893–1957: *Gaudy Night* (1935)

21 In the sex-war thoughtlessness is the weapon of the male, vindictiveness of the female.

Cyril Connolly 1903–74: *Unquiet Grave* (1944)

22 Men have a much better time of it than women. For one thing, they marry later. For another thing, they die earlier.

H. L. Mencken 1880–1956: *Chrestomathy* (1949)

23 There is more difference within the sexes than between them.

Ivy Compton-Burnett 1884–1969: *Mother and Son* (1955)

24 Women are really much nicer than men:
No wonder we like them.

Kingsley Amis 1922– : 'A Bookshop Idyll' (1956)

25 Women want mediocre men, and men are working hard to be as mediocre as possible.

Margaret Mead 1901–78: in *Quote Magazine* 15 June 1958

26 Whatever women do they must do twice as well as men to be thought half as good.

Charlotte Whitton 1896–1975: in *Canada Month* June 1963

27 Women have very little idea of how much men hate them.

Germaine Greer 1939– : *The Female Eunuch* (1971)

28 Whereas nature turns girls into women, society has to make boys into men.
Anthony Stevens: *Archetype* (1982)

29 More and more it appears that, biologically, men are designed for short, brutal lives and women for long miserable ones.
Estelle Ramey: *Observer* 7 April 1985 'Sayings of the Week'

30 A woman without a man is like a fish without a bicycle.
Gloria Steinem 1934- : attributed

Middle Age

1 I am past thirty, and three parts iced over.
Matthew Arnold 1822–88: letter to Arthur Hugh Clough, 12 February 1853

2 Mr Salteena was an elderly man of 42.
Daisy Ashford 1881–1972: *The Young Visiters* (1919)

3 At eighteen our convictions are hills from which we look; at forty-five they are caves in which we hide.
F. Scott Fitzgerald 1896–1940: 'Bernice Bobs her Hair' (1920)

4 One of the pleasures of middle age is to *find out* that one WAS right, and that one was much righter than one knew at say 17 or 23.
Ezra Pound 1885–1972: *ABC of Reading* (1934)

5 Do you think my mind is maturing late,
Or simply rotted early?
Ogden Nash 1902–71: 'Lines on Facing Forty' (1942)

6 Years ago we discovered the exact point, the dead centre of middle age. It occurs when you are too young to take up golf and too old to rush up to the net.
Franklin P. Adams 1881–1960: *Nods and Becks* (1944)

7 At forty-five,
What next, what next?
At every corner,
I meet my Father,
my age, still alive.
Robert Lowell 1917–77: 'Middle Age' (1964)

The Mind

1 The mind is its own place, and in itself
Can make a heaven of hell, a hell of heaven.
John Milton 1608–74: *Paradise Lost* (1667)

2 To give a sex to mind was not very consistent with the principles of a man [Rousseau] who argued so warmly, and so well, for the immortality of the soul.
Mary Wollstonecraft 1759–97: *A Vindication of the Rights of Woman* (1792), often quoted 'Mind has no sex'

3 What is Matter?—Never mind.
What is Mind?—No matter.
Punch: 1855

4 On earth there is nothing great but man; in man there is nothing great but mind.
William Hamilton 1788–1856: *Lectures on Metaphysics and Logic* (1859), attributed in a Latin form to Favorinus (2nd century AD)

5 O the mind, mind has mountains; cliffs of fall
Frightful, sheer, no-man-fathomed.
Gerard Manley Hopkins 1844–89: 'No worst, there is none' (written 1885)

6 Personal relations are the important thing for ever and ever, and not this outer life of telegrams and anger.
E. M. Forster 1879–1970: *Howards End* (1910)

7 Minds like beds always made up,
(more stony than a shore)
unwilling or unable.
William Carlos Williams 1883–1963: *Paterson* (1946)

8 Mind in its purest play is like some bat
That beats about in caverns all alone,
Contriving by a kind of senseless wit
Not to conclude against a wall of stone.
Richard Wilbur 1921– : 'Mind' (1956)

9 That's the classical mind at work, runs fine inside but looks dingy on the surface.
Robert M. Pirsig 1928– : *Zen and the Art of Motorcycle Maintenance* (1974)

10 Consciousness... is the phenomenon whereby the universe's very existence is made known.
Roger Penrose 1931- : *The Emperor's New Mind* (1989)

Minorities and Majorities

See also **Democracy**

1 *Nec audiendi qui solent dicere, Vox populi, vox Dei, quum tumultuositas vulgi semper insaniae proxima sit.*
And those people should not be listened to who keep saying the voice of the people is the voice of God, since the riotousness of the crowd is always very close to madness.
Alcuin *c.*735-804: *Works* (1863) letter 164

2 Nor is the people's judgement always true:
The most may err as grossly as the few.
John Dryden 1631-1700: *Absalom and Achitophel* (1681)

3 'It's always best on these occasions to do what the mob do.'
'But suppose there are two mobs?' suggested Mr Snodgrass.
'Shout with the largest,' replied Mr Pickwick.
Charles Dickens 1812-70: *Pickwick Papers* (1837)

4 Minorities... are almost always in the right.
Sydney Smith 1771-1845: in H. Pearson *The Smith of Smiths* (1934)

5 The majority never has right on its side. Never I say! That is one of the social lies that a free, thinking man is bound to rebel against. Who makes up the majority in any given country? Is it the wise men or the fools? I think we must agree that the fools are in a terrible overwhelming majority, all the wide world over.
Henrik Ibsen 1828-1906: *The Master Builder* (1892)

6 The fact that an opinion has been widely held is no evidence whatever that it is not utterly absurd; indeed in view of the silliness of the majority of mankind, a widespread belief is more likely to be foolish than sensible.
Bertrand Russell 1872-1970: *Marriage and Morals* (1929)

7 Never forget that only dead fish swim with the stream.
Malcolm Muggeridge 1903-90: quoting a supporter, in *Radio Times* 9 July 1964

Misfortune

1 Man is born unto trouble, as the sparks fly upward.
Bible: Job

2 For in every ill-turn of fortune the most unhappy sort of unfortunate man is the one who has been happy.
Boethius AD *c.*476–524: *De Consolatione Philosophiae*

3 ...*Nessun maggior dolore,*
Che ricordarsi del tempo felice
Nella miseria.
There is no greater pain than to remember a happy time when one is in misery.
Dante Alighieri 1265–1321: *Divina Commedia* 'Inferno'

4 Sweet are the uses of adversity,
Which like the toad, ugly and venomous,
Wears yet a precious jewel in his head.
William Shakespeare 1564–1616: *As You Like It* (1599)

5 Misery acquaints a man with strange bedfellows.
William Shakespeare 1564–1616: *The Tempest* (1611)

6 Prosperity doth best discover vice, but adversity doth best discover virtue.
Francis Bacon 1561–1626: *Essays* (1625) 'Of Adversity'

7 In the misfortune of our best friends, we always find something which is not displeasing to us.
Duc de la Rochefoucauld 1613–80: *Réflexions ou Maximes Morales* (1665)

8 When I consider how my light is spent,
E're half my days, in this dark world and wide,
And that one talent which is death to hide
Lodged with me useless.
John Milton 1608–74: 'When I consider how my light is spent' (1673)

9 We are all strong enough to bear the misfortunes of others.
Duc de la Rochefoucauld 1613–80: *Maximes* (1678)

10 I had never had a piece of toast
Particularly long and wide,
But fell upon the sanded floor,
And always on the buttered side.
James Payn 1830–98: *Chambers's Journal* 2 February 1884

11 I left the room with silent dignity, but caught my foot in the mat.

George Grossmith 1847–1912 and **Weedon Grossmith** 1854–1919: *Diary of a Nobody* (1894)

12 And always keep a-hold of Nurse
For fear of finding something worse.

Hilaire Belloc 1870–1953: 'Jim' (1907)

13 now and then
there is a person born
who is so unlucky
that he runs into accidents
which started to happen
to somebody else.

Don Marquis 1878–1937: *archys life of mehitabel* (1933)

14 boss there is always
a comforting thought
in time of trouble when
it is not our trouble.

Don Marquis 1878–1937: *archy does his part* (1935)

15 And all my endeavours are unlucky explorers
come back, abandoning the expedition.

Keith Douglas 1920–44: 'On Return from Egypt, 1943–4' (1946)

16 People will take balls,
Balls will be lost always, little boy,
And no one buys a ball back.

John Berryman 1914–72: 'The Ball Poem' (1948)

17 In the words of one of my more sympathetic correspondents, it has turned out to be an 'annus horribilis'.

Elizabeth II 1926– : speech at Guildhall, London, 24 November 1992

Mistakes

1 I would rather be wrong, by God, with Plato...than be correct with those men.

Cicero 106–43 BC: *Tusculanae Disputationes* (of Pythagoreans)

2 I'm aggrieved when sometimes even excellent Homer nods.
Horace 65–8 BC: *Ars Poetica*

3 I beseech you, in the bowels of Christ, think it possible you may be mistaken.
Oliver Cromwell 1599–1658: letter to the General Assembly of the Kirk of Scotland, 3 August 1650

4 Errors, like straws, upon the surface flow;
He who would search for pearls must dive below.
John Dryden 1631–1700: *All for Love* (1678)

5 Crooked things may be as stiff and unflexible as straight: and men may be as positive in error as in truth.
John Locke 1632–1704: *Essay concerning Human Understanding* (1690)

6 To err is human; to forgive, divine.
Alexander Pope 1688–1744: *An Essay on Criticism* (1711)

7 Truth lies within a little and certain compass, but error is immense.
Henry St John, 1st Viscount Bolingbroke1678–1751: *Reflections upon Exile* (1716)

8 A man should never be ashamed to own he has been in the wrong, which is but saying, in other words, that he is wiser to-day than he was yesterday.
Alexander Pope 1688–1744: *Miscellanies* (1727) 'Thoughts on Various Subjects'

9 It is worse than a crime, it is a blunder.
Antoine Boulay de la Meurthe 1761–1840: on hearing of the execution of the Duc d'Enghien, 1804

10 Error has never approached my spirit.
Prince Metternich 1773–1859: addressed to Guizot in 1848, in F. Guizot *Mémoires* (1858–67)

11 'Forward, the Light Brigade!'
Was there a man dismayed?
Not though the soldier knew
Some one had blundered.
Alfred, Lord Tennyson 1809–92: 'The Charge of the Light Brigade' (1854)

12 The man who makes no mistakes does not usually make anything.
Edward John Phelps 1822–1900: speech, 24 January 1889

13 To lose one parent, Mr Worthing, may be regarded as a misfortune; to lose both looks like carelessness.
Oscar Wilde 1854–1900: *The Importance of Being Earnest* (1895)

14 Well, if I called the wrong number, why did you answer the phone?
James Thurber 1894–1961: cartoon caption in *New Yorker* 5 June 1937

15 The weak have one weapon: the errors of those who think they are strong.
Georges Bidault 1899–1983: in *Observer* 15 July 1962 'Sayings of the Week'

16 Like most of those who study history, he [Napoleon III] learned from the mistakes of the past how to make new ones.
A. J. P. Taylor 1906–90: *Listener* 6 June 1963

17 An expert is someone who knows some of the worst mistakes that can be made in his subject and who manages to avoid them.
Werner Heisenberg 1901–76: *Der Teil und das Ganze* (1969) tr. A. J. Pomerans as *Physics and Beyond* (1971)

Moderation

1 Nothing in excess.
Anonymous: inscribed on the temple of Apollo at Delphi, and variously ascribed to the Seven Wise Men

2 There is moderation in everything.
Horace 65–8 BC: *Satires*

3 You will go most safely by the middle way.
Ovid 43 BC–AD *c*.17: *Metamorphoses*

4 Because thou art lukewarm, and neither cold nor hot, I will spew thee out of my mouth.
Bible: Revelation

5 To many, total abstinence is easier than perfect moderation.
St Augustine of Hippo AD 354–430: *On the Good of Marriage* (AD 401)

6 To gild refinèd gold, to paint the lily,
To throw a perfume on the violet,

To smooth the ice, or add another hue
Unto the rainbow, or with taper light
To seek the beauteous eye of heaven to garnish,
Is wasteful and ridiculous excess.
William Shakespeare 1564–1616: *King John* (1591–8)

7 They are as sick that surfeit with too much, as they that starve with nothing.
William Shakespeare 1564–1616: *The Merchant of Venice* (1596–8)

8 By God, Mr Chairman, at this moment I stand astonished at my own moderation!
Lord Clive 1725–74: during Parliamentary cross-examination, 1773 in G. R. Gleig *Life of Robert, First Lord Clive* (1848)

9 Abstinence is as easy to me, as temperance would be difficult.
Samuel Johnson 1709–84: in W. Roberts (ed.) *Memoirs of... Mrs Hannah More* (1834)

10 We know what happens to people who stay in the middle of the road. They get run down.
Aneurin Bevan 1897–1960: in *Observer* 6 December 1953

11 I would remind you that extremism in the defence of liberty is no vice! And let me remind you also that moderation in the pursuit of justice is no virtue!
Barry Goldwater 1909– : accepting the presidential nomination, 16 July 1964

Money

See also **Poverty, Wealth**

1 Wine maketh merry: but money answereth all things.
Bible: Ecclesiastes

2 The sinews of war, unlimited money.
Cicero 106–43 BC: *Fifth Philippic*

3 *Quid non mortalia pectora cogis,*
Auri sacra fames!
To what do you not drive human hearts, cursed craving for gold!
Virgil 70–19 BC: *Aeneid*

4 The love of money is the root of all evil.
Bible: I Timothy

5 If possible honestly, if not, somehow, make money.
Horace 65–8 BC: *Epistles*

6 I can get no remedy against this consumption of the purse: borrowing only lingers and lingers it out, but the disease is incurable.
William Shakespeare 1564–1616: *Henry IV, Part 2* (1597)

7 Money is like muck, not good except it be spread.
Francis Bacon 1561–1626: *Essays* (1625) 'Of Seditions and Troubles'

8 But it is pretty to see what money will do.
Samuel Pepys 1633–1703: diary, 21 March 1667

9 Money speaks sense in a language all nations understand.
Aphra Behn 1640–89: *The Rover* pt. 2 (1681)

10 Money is the sinews of love, as of war.
George Farquhar 1678–1707: *Love and a Bottle* (1698)

11 Nothing to be done without a bribe I find, in love as well as law.
Susannah Centlivre *c.*1669–1723: *The Perjured Husband* (1700)

12 Take care of the pence, and the pounds will take care of themselves.
William Lowndes 1652–1724: in Lord Chesterfield *Letters to his Son* (1774) 5 February 1750

13 Money... is none of the wheels of trade: it is the oil which renders the motion of the wheels more smooth and easy.
David Hume 1711–76: *Essays: Moral and Political* (1741–2) 'Of Money'

14 An annuity is a very serious business.
Jane Austen 1775–1817: *Sense and Sensibility* (1811)

15 £40,000 a year a moderate income—such a one as a man *might jog on with.*
John George Lambton 1792–1840: letter, 13 September 1821

16 The almighty dollar is the only object of worship.
Anonymous: *Philadelphia Public Ledger* 2 December 1836

17 Annual income twenty pounds, annual expenditure nineteen nineteen six, result happiness. Annual income twenty pounds, annual expenditure twenty pounds ought and six, result misery.

Charles Dickens 1812–70: *David Copperfield* (1850) Mr Micawber

18 Money is like a sixth sense without which you cannot make a complete use of the other five.

W. Somerset Maugham 1874–1965: *Of Human Bondage* (1915)

19 Lenin was right. There is no subtler, no surer means of overturning the existing basis of society than to debauch the currency.

John Maynard Keynes 1883–1946: *Economic Consequences of the Peace* (1919)

20 'My boy,' he says, 'always try to rub up against money, for if you rub up against money long enough, some of it may rub off on you.'

Damon Runyon 1884–1946: *Cosmopolitan* August 1929

21 We could have saved sixpence. We have saved fivepence. (*Pause*) But at what cost?

Samuel Beckett 1906–89: *All That Fall* (1957)

22 A bank is a place that will lend you money if you can prove that you don't need it.

Bob Hope 1903– : in A. Harrington *Life in the Crystal Palace* (1959)

23 Expenditure rises to meet income.

C. Northcote Parkinson 1909–93: *The Law and the Profits* (1960)

24 Money, it turned out, was exactly like sex, you thought of nothing else if you didn't have it and thought of other things if you did.

James Baldwin 1924–87: *Esquire* May 1961

25 I want to spend, and spend, and spend.

Vivian Nicholson 1936– : said to reporters on arriving to collect football pools winnings of £152,000, 27 September 1961

26 Money couldn't buy friends but you got a better class of enemy.

Spike Milligan 1918– : *Puckoon* (1963)

27 For I don't care too much for money,
For money can't buy me love.
John Lennon 1940–80 and **Paul McCartney** 1942– : 'Can't Buy Me Love' (1964 song)

28 Money doesn't talk, it swears.
Bob Dylan 1941– : 'It's Alright, Ma (I'm Only Bleeding)' (1965 song)

29 From now the pound abroad is worth 14 per cent or so less in terms of other currencies. It does not mean, of course, that the pound here in Britain, in your pocket or purse or in your bank, has been devalued.
Harold Wilson 1916– : ministerial broadcast, 19 November 1967

Morality

1 *Cum finis est licitus, etiam media sunt licita.*
The end justifies the means.
Hermann Busenbaum 1600–68: *Medulla Theologiae Moralis* (1650); literally 'When the end is allowed, the means also are allowed'

2 Two things fill the mind with ever new and increasing wonder and awe, the more often and the more seriously reflection concentrates upon them: the starry heaven above me and the moral law within me.
Immanuel Kant 1724–1804: *Critique of Practical Reason* (1788)

3 We do not look in great cities for our best morality.
Jane Austen 1775–1817: *Mansfield Park* (1814)

4 We know no spectacle so ridiculous as the British public in one of its periodical fits of morality.
Lord Macaulay 1800–59: *Essays* (1843) 'Moore's *Life of Lord Byron*'

5 A clever theft was praiseworthy amongst the Spartans; and it is equally so amongst Christians, provided it be on a sufficiently large scale.
Herbert Spencer 1820–1903: *Social Statics* (1850)

6 Morality is the herd-instinct in the individual.
Friedrich Nietzsche 1844–1900: *Die fröhliche Wissenschaft* (1882)

7 Morality is a private and costly luxury.
Henry Brooks Adams 1838–1918: *The Education of Henry Adams* (1907)

8 Moral indignation is jealousy with a halo.
H. G. Wells 1866–1946: *The Wife of Sir Isaac Harman* (1914)

9 The most useful thing about a principle is that it can always be sacrificed to expediency.
W. Somerset Maugham 1874–1965: *The Circle* (1921)

10 Food comes first, then morals.
Bertolt Brecht 1898–1956: *The Threepenny Opera* (1928)

11 The last temptation is the greatest treason:
To do the right deed for the wrong reason.
T. S. Eliot 1888–1965: *Murder in the Cathedral* (1935)

12 What is morality in any given time or place? It is what the majority then and there happen to like, and immorality is what they dislike.
Alfred North Whitehead 1861–1947: *Dialogues* (1954) 30 August 1941

13 If people want a sense of purpose, they should get it from their archbishops. They should not hope to receive it from their politicians.
Harold Macmillan 1894–1986: to Henry Fairlie, 1963; in H. Fairlie *The Life of Politics* (1968)

14 Standards are always out of date. That is what makes them standards.
Alan Bennett 1934– : *Forty Years On* (1969)

Murder

1 Thou shalt not kill.
Bible: Exodus

2 Mordre wol out; that se we day by day.
Geoffrey Chaucer *c.*1343–1400: *The Canterbury Tales* 'The Nun's Priest's Tale'

3 Murder most foul, as in the best it is;
But this most foul, strange, and unnatural.
William Shakespeare 1564–1616: *Hamlet* (1601)

4 Here's the smell of the blood still: all the perfumes of Arabia will not sweeten this little hand.
William Shakespeare 1564–1616: *Macbeth* (1606)

5 Killing no murder briefly discourst in three questions.
Edward Sexby d. 1658: title of pamphlet (an apology for tyrannicide, 1657)

6 Murder considered as one of the fine arts.
Thomas De Quincey 1785–1859: *Blackwood's Magazine* February 1827 (essay title)

7 Thou shalt not kill; but need'st not strive
Officiously to keep alive.
Arthur Hugh Clough 1819–61: 'The Latest Decalogue' (1862)

8 Kill a man, and you are an assassin. Kill millions of men, and you are a conqueror. Kill everyone, and you are a god.
Jean Rostand 1894–1977: *Pensées d'un biologiste* (1939)

Music

See also **Singing**

1 How sour sweet music is,
When time is broke, and no proportion kept!
William Shakespeare 1564–1616: *Richard II* (1595)

2 The man that hath no music in himself,
Nor is not moved with concord of sweet sounds,
Is fit for treasons, stratagems, and spoils.
William Shakespeare 1564–1616: *The Merchant of Venice* (1596–8)

3 If music be the food of love, play on;
Give me excess of it, that, surfeiting,
The appetite may sicken, and so die.
William Shakespeare 1564–1616: *Twelfth Night* (1601)

4 Their lean and flashy songs
Grate on their scrannel pipes of wretched straw.
John Milton 1608–74: 'Lycidas' (1638)

5 Such sweet compulsion doth in music lie.
John Milton 1608–74: 'Arcades' (1645)

6 Music has charms to soothe a savage breast.
William Congreve 1670–1729: *The Mourning Bride* (1697)

7 Difficult do you call it, Sir? I wish it were impossible.
Samuel Johnson 1709–84: on the performance of a celebrated violinist, in W. Seward *Supplement to the Anecdotes of Distinguished Persons* (1797)

8 Of music Dr Johnson used to say that it was the only sensual pleasure without vice.
Samuel Johnson 1709–84: in *European Magazine* (1795)

9 Perhaps the self-same song that found a path
Through the sad heart of Ruth, when, sick for home,
She stood in tears amid the alien corn;
The same that oft-times hath
Charmed magic casements, opening on the foam
Of perilous seas, in faery lands forlorn.
John Keats 1795–1821: 'Ode to a Nightingale' (1820)

10 He [Hill] did not see any reason why the devil should have all the good tunes.
Rowland Hill 1744–1833: E. W. Broome *Rowland Hill* (1881)

11 Hark, the dominant's persistence till it must be answered to!
Robert Browning 1812–89: 'A Toccata of Galuppi's' (1855)

12 Please do not shoot the pianist. He is doing his best.
Anonymous: printed notice in a dancing saloon, in Oscar Wilde *Impressions of America* (*c.*1882–3)

13 Hell is full of musical amateurs: music is the brandy of the damned.
George Bernard Shaw 1856–1950: *Man and Superman* (1903)

14 There is music in the air.
Edward Elgar 1857–1934: in R. J. Buckley *Sir Edward Elgar* (1905)

15 It will be generally admitted that Beethoven's Fifth Symphony is the most sublime noise that has ever penetrated into the ear of man.
E. M. Forster 1879–1970: *Howards End* (1910)

16 Fortissimo at last!
Gustav Mahler 1860–1911: on seeing Niagara Falls; in K. Blaukopf *Gustav Mahler* (1973)

17 Music is feeling, then, not sound.
Wallace Stevens 1879–1955: 'Peter Quince at the Clavier' (1923)

18 Music begins to atrophy when it departs too far from the dance...poetry begins to atrophy when it gets too far from music.
Ezra Pound 1885–1972: *The ABC of Reading* (1934)

19 Down the road someone is practising scales,
The notes like little fishes vanish with a wink of tails.
Louis MacNeice 1907–63: 'Sunday Morning' (1935)

20 The whole problem can be stated quite simply by asking, 'Is there a meaning to music?' My answer to that would be, 'Yes.' And 'Can you state in so many words what the meaning is?' My answer to that would be, 'No.'
Aaron Copland 1900–90: *What to Listen for in Music* (1939)

21 Good music is that which penetrates the ear with facility and quits the memory with difficulty.
Thomas Beecham 1879–1961: speech, *c.*1950, in *New York Times* 9 March 1961

22 Applause is a receipt, not a note of demand.
Artur Schnabel 1882–1951: in *Saturday Review of Literature* 29 September 1951

23 Too easy for children, and too difficult for artists.
Artur Schnabel 1882–1951: of Mozart's sonatas, in Nat Shapiro (ed.) *Encyclopaedia of Quotations about Music* (1978). In *My Life and Music* (1961), Schnabel says: 'Children are given Mozart because of the small *quantity* of the notes; grown-ups avoid Mozart because of the great *quality* of the notes'

24 Music is your own experience, your thoughts, your wisdom. If you don't live it, it won't come out of your horn.
Charlie Parker 1920–55: in Nat Shapiro and Nat Hentoff *Hear Me Talkin' to Ya* (1955)

25 The English may not like music, but they absolutely love the noise it makes.
Thomas Beecham 1879–1961: In *New York Herald Tribune* 9 March 1961

26 There are two golden rules for an orchestra: start together and finish together. The public doesn't give a damn what goes on in between.
Thomas Beecham 1879–1961: in H. Atkins and A. Newman *Beecham Stories* (1978)

27 Two skeletons copulating on a corrugated tin roof.
Thomas Beecham 1879–1961: describing the harpsichord, in H. Atkins and A. Newman *Beecham Stories* (1978)

28 A good composer does not imitate; he steals.
Igor Stravinsky 1882–1971: in P. Yates *Twentieth Century Music* (1967)

29 If you still have to ask...shame on you.
Louis Armstrong 1901–71: when asked what jazz is, in Max Jones et al. *Salute to Satchmo* (1970); sometimes quoted 'Man, if you gotta ask you'll never know'

30 All music is folk music, I ain't never heard no horse sing a song.
Louis Armstrong 1901–71: in *New York Times* 7 July 1971

31 Music is, by its very nature, essentially powerless to *express* anything at all...music expresses itself.
Igor Stravinsky 1882–1971: in *Esquire* December 1972

32 Improvisation is too good to leave to chance.
Paul Simon 1942– : in *Observer* 30 December 1990 'Sayings of the Year'

Musicians

1 Wagner has lovely moments but awful quarters of an hour.
Gioacchino Rossini 1792–1868: in E. Naumann *Italienische Tondichter* (1883), April 1867

2 I have been told that Wagner's music is better than it sounds.
Bill Nye: in Mark Twain *Autobiography* (1924)

3 The notes I handle no better than many pianists. But the pauses between the notes—ah, that is where the art resides!
Artur Schnabel 1882–1951: in *Chicago Daily News* 11 June 1958

4 I don't know whether I like it, but it's what I meant.
Ralph Vaughan Williams 1872–1958: on his 4th symphony,

in C. Headington *Bodley Head History of Western Music* (1974)

5 At a rehearsal I let the orchestra play as they like. At the concert I make them play as *I* like.
Thomas Beecham 1879–1961: in Neville Cardus *Sir Thomas Beecham* (1961)

6 Too much counterpoint; what is worse, Protestant counterpoint.
Thomas Beecham 1879–1961: of J. S. Bach; in *Guardian* 8 March 1971

7 A kind of musical Malcolm Sargent.
Thomas Beecham 1879–1961: of Herbert von Karajan, in H. Atkins and A. Newman *Beecham Stories* (1978)

8 My music is best understood by children and animals.
Igor Stravinsky 1882–1971: in *Observer* 8 October 1961

9 Whether the angels play only Bach in praising God I am not quite sure; I am sure, however, that en famille they play Mozart.
Karl Barth 1886–1968: in *New York Times* 11 December 1968

10 Ballads and babies. That's what happened to me.
Paul McCartney 1942– : on reaching the age of fifty; in *Time* 8 June 1992

Nature

1 Nature does nothing without purpose or uselessly.
Aristotle 384–322 BC: *Politics*

2 You may drive out nature with a pitchfork, yet she'll be constantly running back.
Horace 65–8 BC: *Epistles*

3 In her [Nature's] inventions nothing is lacking, and nothing is superfluous.
Leonardo da Vinci 1452–1519: E. McCurdy (ed. and trans.) *Leonardo da Vinci's Notebooks* (1906)

4 And this our life, exempt from public haunt,
Finds tongues in trees, books in the running brooks,
Sermons in stones, and good in everything.
William Shakespeare 1564–1616: *As You Like It* (1599)

5 All things are artificial, for nature is the art of God.
Sir Thomas Browne 1605–82: *Religio Medici* (1643)

6 There was a time when meadow, grove, and stream,
The earth, and every common sight,
To me did seem
Apparelled in celestial light,
The glory and the freshness of a dream.
William Wordsworth 1770–1850: 'Ode. Intimations of Immortality' (1807)

7 To me the meanest flower that blows can give
Thoughts that do often lie too deep for tears.
William Wordsworth 1770–1850: 'Ode. Intimations of Immortality' (1807)

8 There is a pleasure in the pathless woods,
There is a rapture on the lonely shore,
There is society, where none intrudes,
By the deep sea, and music in its roar:
I love not man the less, but nature more.
Lord Byron 1788–1824: *Childe Harold's Pilgrimage* (1812–18)

9 The roaring of the wind is my wife and the stars through the window pane are my children.
John Keats 1795–1821: letter to George and Georgiana Keats, 24 October 1818

10 Nature, red in tooth and claw.
Alfred, Lord Tennyson 1809–92: *In Memoriam A. H. H.* (1850)

11 So careful of the type she seems,
So careless of the single life.
Alfred, Lord Tennyson 1809–92: *In Memoriam A. H. H.* (1850), of Nature

12 What a book a devil's chaplain might write on the clumsy, wasteful, blundering, low, and horridly cruel works of nature!
Charles Darwin 1809–82: letter to J. D. Hooker, 13 July 1856

13 Nature is not a temple, but a workshop, and man's the workman in it.
Ivan Turgenev 1818–83: *Fathers and Sons* (1862), tr. R. Edmonds

14 'I play for Seasons; not Eternities!'
Says Nature.
George Meredith 1828–1909: *Modern Love* (1862)

15 In nature there are neither rewards nor punishments—there are consequences.
Robert G. Ingersoll 1833–99: *Some Reasons Why* (1881)

16 Pile the bodies high at Austerlitz and Waterloo.
Shovel them under and let me work—
I am the grass; I cover all.
Carl Sandburg 1878–1967: 'Grass' (1918)

17 For nature, heartless, witless nature,
Will neither care nor know
What stranger's feet may find the meadow
And trespass there and go.
A. E. Housman 1859–1936: *Last Poems* (1922) no. 40

News and Journalism

1 Tell it not in Gath, publish it not in the streets of Askelon.
Bible: II Samuel

2 As cold waters to a thirsty soul, so is good news from a far country.
Bible: Proverbs

3 How beautiful upon the mountains are the feet of him that bringeth good tidings, that publisheth peace.
Bible: Isaiah

4 Ill news hath wings, and with the wind doth go,
Comfort's a cripple and comes ever slow.
Michael Drayton 1563–1631: *The Barons' Wars* (1603)

5 The nature of bad news infects the teller.
William Shakespeare 1564–1616: *Antony and Cleopatra* (1606–7)

6 The journalists have constructed for themselves a little wooden chapel, which they also call the Temple of Fame, in which they put up and take down portraits all day long and make such a hammering you can't hear yourself speak.

Georg Christoph Lichtenberg 1742–99: in A. Leitzmann *Georg Christoph Lichtenberg Aphorismen* (1904)

7 *The Times* has made many ministries.

Walter Bagehot 1826–77: *The English Constitution* (1867)

8 All the news that's fit to print.

Adolph S. Ochs 1858–1935: motto of the *New York Times*, from 1896

9 The report of my death was an exaggeration.

Mark Twain 1835–1910: *New York Journal* 2 June 1897, usually quoted 'Reports of my death have been greatly exaggerated'

10 By office boys for office boys.

Lord Salisbury 1830–1903: of the *Daily Mail*, in H. Hamilton Fyfe *Northcliffe* (1930)

11 The men with the muck-rakes are often indispensable to the well-being of society; but only if they know when to stop raking the muck.

Theodore Roosevelt 1858–1919: speech, 14 April 1906

12 Editor: a person employed by a newspaper, whose business it is to separate the wheat from the chaff, and to see that the chaff is printed.

Elbert Hubbard 1859–1915: *The Roycroft Dictionary* (1914)

13 Journalists say a thing that they know isn't true, in the hope that if they keep on saying it long enough it *will* be true.

Arnold Bennett 1867–1931: *The Title* (1918)

14 When a dog bites a man, that is not news, because it happens so often. But if a man bites a dog, that is news.

John B. Bogart 1848–1921: in F. M. O'Brien *Story of the* [New York] *Sun* (1918), often attributed to Charles A. Dana

15 Christianity, of course... but why journalism?

Arthur James Balfour 1848–1930: replying to Frank Harris's remark that 'all the faults of the age come from Christianity and journalism'; in Margot Asquith *Autobiography* (1920)

16 Comment is free, but facts are sacred.

C. P. Scott 1846–1932: *Manchester Guardian* 5 May 1921

17 Well, all I know is what I read in the papers.

Will Rogers 1879–1935: *New York Times* 30 September 1923

18 You cannot hope
to bribe or twist,
thank God! the
British journalist.
But, seeing what
the man will do
unbribed, there's
no occasion to.

Humbert Wolfe 1886–1940: 'Over the Fire' (1930)

19 Power without responsibility: the prerogative of the harlot throughout the ages.

Rudyard Kipling 1865–1936: summing up Lord Beaverbrook's political standpoint *vis-à-vis* the *Daily Express*; quoted by Stanley Baldwin, 18 March 1931

20 The art of newspaper paragraphing is to stroke a platitude until it purrs like an epigram.

Don Marquis 1878–1937: in E. Anthony *O Rare Don Marquis* (1962)

21 News is what a chap who doesn't care much about anything wants to read. And it's only news until he's read it. After that it's dead.

Evelyn Waugh 1903–66: *Scoop* (1938)

22 Small earthquake in Chile. Not many dead.

Claud Cockburn 1904–81: winning entry for a dullest headline competition at *The Times*; *In Time of Trouble* (1956)

23 I read the newspapers avidly. It is my one form of continuous fiction.

Aneurin Bevan 1897–1960: in *The Times* 29 March 1960

24 A good newspaper, I suppose, is a nation talking to itself.

Arthur Miller 1915– : in *Observer* 26 November 1961

25 Freedom of the press in Britain means freedom to print such of the proprietor's prejudices as the advertisers don't object to.

Hannen Swaffer 1879–1962: in Tom Driberg *Swaff* (1974)

26 Success in journalism can be a form of failure. Freedom comes from lack of possessions. The truth-divulging paper must imitate the tramp and sleep under a hedge.

Graham Greene 1904–91: *New Statesman* 31 May 1968

27 Comment is free but facts are on expenses.

Tom Stoppard 1937– : *Night and Day* (1978)

28 Rock journalism is people who can't write interviewing people who can't talk for people who can't read.
Frank Zappa 1940–93: in L. Botts *Loose Talk* (1980)

29 We must try to find ways to starve the terrorist and the hijacker of the oxygen of publicity on which they depend.
Margaret Thatcher 1925– : speech to American Bar Association in London, 15 July 1985

30 Blood sport is brought to its ultimate refinement in the gossip columns.
Bernard Ingham 1932– : speech, 5 February 1986

31 An odious exhibition of journalists dabbling their fingers in the stuff of other people's souls.
Lord McGregor 1921– : on press coverage of the marriage of the Prince and Princess of Wales; *The Times* 9 June 1992

Night see **Day and Night**

Old Age

1 Then shall ye bring down my grey hairs with sorrow to the grave.
Bible: Genesis

2 The days of our age are threescore years and ten; and though men be so strong that they come to fourscore years: yet is their strength then but labour and sorrow; so soon passeth it away, and we are gone.
Bible: Psalm 90

3 How ill white hairs become a fool and jester!
William Shakespeare 1564–1616: *Henry IV, Part 2* (1597)

4 The sixth age shifts
Into the lean and slippered pantaloon,
With spectacles on nose and pouch on side.
William Shakespeare 1564–1616: *As You Like It* (1599)

5 Second childishness, and mere oblivion,
Sans teeth, sans eyes, sans taste, sans everything.
William Shakespeare 1564–1616: *As You Like It* (1599)

6 Unregarded age in corners thrown.
William Shakespeare 1564–1616: *As You Like It* (1599)

7 Therefore my age is as a lusty winter,
Frosty, but kindly.
William Shakespeare 1564–1616: *As You Like It* (1599)

8 I have lived long enough: my way of life
Is fall'n into the sear, the yellow leaf.
William Shakespeare 1564–1616: *Macbeth* (1606)

9 Age will not be defied.
Francis Bacon 1561–1626: *Essays* (1625) 'Of Regimen of Health'

10 Every man desires to live long; but no man would be old.
Jonathan Swift 1667–1745: *Thoughts on Various Subjects* (1727 ed.)

11 See how the world its veterans rewards!
A youth of frolics, an old age of cards.
Alexander Pope 1688–1744: *Epistles to Several Persons* 'To a Lady' (1735)

12 Old-age, a second child, by Nature cursed
With more and greater evils than the first,
Weak, sickly, full of pains; in ev'ry breath
Railing at life, and yet afraid of death.
Charles Churchill 1731–64: *Gotham* (1764)

13 How happy he who crowns in shades like these,
A youth of labour with an age of ease.
Oliver Goldsmith 1730–74: *The Deserted Village* (1770)

14 Time has shaken me by the hand and death is not far behind.
John Wesley 1703–91: letter to Ezekiel Cooper, 1 February 1791

15 There's a fascination frantic
In a ruin that's romantic;
Do you think you are sufficiently decayed?
W. S. Gilbert 1836–1911: *The Mikado* (1885)

16 When you are old and grey and full of sleep,
And nodding by the fire, take down this book
And slowly read and dream of the soft look
Your eyes had once, and of their shadows deep.
W. B. Yeats 1865–1939: 'When You Are Old' (1893)

17 I grow old... I grow old...
I shall wear the bottoms of my trousers rolled.
T. S. Eliot 1888–1965: 'Love Song of J. Alfred Prufrock' (1917)

18 Oh, to be seventy again!
Georges Clemenceau 1841–1929: on seeing a pretty girl on his eightieth birthday; in James Agate's diary, 19 April 1938

19 An aged man is but a paltry thing,
A tattered coat upon a stick, unless
Soul clap its hands and sing, and louder sing
For every tatter in its mortal dress.
W. B. Yeats 1865–1939: 'Sailing to Byzantium' (1928)

20 From the earliest times the old have rubbed it into the young that they are wiser than they, and before the young had discovered what nonsense this was they were old too, and it profited them to carry on the imposture.
W. Somerset Maugham 1874–1965: *Cakes and Ale* (1930)

21 There is more felicity on the far side of baldness than young men can possibly imagine.
Logan Pearsall Smith 1865–1946: *Afterthoughts* (1931) 'Age and Death'

22 Nothing really wrong with him—only anno domini, but that's the most fatal complaint of all, in the end.
James Hilton 1900–54: *Goodbye, Mr Chips* (1934)

23 Old age is the most unexpected of all things that happen to a man.
Leon Trotsky 1879–1940: diary, 8 May 1935

24 You think it horrible that lust and rage
Should dance attendance upon my old age;
They were not such a plague when I was young;
What else have I to spur me into song?
W. B. Yeats 1865–1939: 'The Spur' (1938)

25 You will recognize, my boy, the first sign of old age: it is when you go out into the streets of London and realize for the first time how young the policemen look.
Seymour Hicks 1871–1949: in C. R. D. Pulling *They Were Singing* (1952)

26 Do not go gentle into that good night,
Old age should burn and rave at close of day;
Rage, rage against the dying of the light.
Dylan Thomas 1914–53: 'Do Not Go Gentle into that Good Night' (1952)

27 Conversation is imperative if gaps are to be filled, and old age, it is the last gap but one.

Patrick White 1912–90: *The Tree of Man* (1955)

28 To me old age is always fifteen years older than I am.

Bernard Baruch 1870–1965: in *Newsweek* 29 August 1955

29 When one has reached 81...one likes to sit back and let the world turn by itself, without trying to push it.

Sean O'Casey 1880–1964: *New York Times* 25 September 1960

30 Considering the alternative, it's not too bad at all.

Maurice Chevalier 1888–1972: when asked what he felt about the advancing years on his 72nd birthday; in M. Freedland *Maurice Chevalier* (1981)

31 In a dream you are never eighty.

Anne Sexton 1928–74: 'Old' (1962)

32 Will you still need me, will you still feed me,
When I'm sixty four?

John Lennon 1940–80 and **Paul McCartney** 1942– : 'When I'm Sixty Four' (1967 song)

33 The man who works and is not bored is never old.

Pablo Casals 1876–1973: in J. Lloyd Webber (ed.) *Song of the Birds* (1985)

34 Growing old is like being increasingly penalized for a crime you haven't committed.

Anthony Powell 1905– : *Temporary Kings* (1973)

35 Perhaps being old is having lighted rooms
Inside your head, and people in them, acting.
People you know, yet can't quite name.

Philip Larkin 1922–85: 'The Old Fools' (1974)

36 If I'd known I was gonna live this long, I'd have taken better care of myself.

Eubie Blake 1883–1983: on reaching the age of 100; in *Observer* 13 February 1983 'Sayings of the Week'

37 Alun's life was coming to consist more and more exclusively of being told at dictation speed what he knew.

Kingsley Amis 1922– : *The Old Devils* (1986)

Opening Lines

1 *Arma virumque cano.*
I sing of arms and the man.
Virgil 70–19 BC: *Aeneid*

2 *Nel mezzo del cammin di nostra vita.*
Midway along the path of our life.
Dante Alighieri 1265–1321: *Divina Commedia* 'Inferno'

3 O! for a Muse of fire, that would ascend
The brightest heaven of invention.
William Shakespeare 1564–1616: *Henry V* (1599)

4 Yet once more, O ye laurels, and once more
Ye myrtles brown, with ivy never sere.
John Milton 1608–74: 'Lycidas' (1638)

5 Of man's first disobedience, and the fruit
Of that forbidden tree, whose mortal taste
Brought death into the world, and all our woe,
With loss of Eden.
John Milton 1608–74: *Paradise Lost* (1667)

6 Oh, what can ail thee knight at arms
Alone and palely loitering?
John Keats 1795–1821: 'La belle dame sans merci' (1820)

7 It was a dark and stormy night.
Edward Bulwer-Lytton 1803–73: *Paul Clifford* (1830)

8 The boy stood on the burning deck
Whence all but he had fled.
Felicia Hemans 1793–1835: 'Casabianca' (1849)

9 It was the best of times, it was the worst of times.
Charles Dickens 1812–70: *A Tale of Two Cities* (1859)

10 'Is there anybody there?' said the Traveller,
Knocking on the moonlit door.
Walter de la Mare 1873–1956: 'The Listeners' (1912)

11 When Gregor Samsa awoke one morning from uneasy dreams he found himself transformed in his bed into a gigantic insect.
Franz Kafka 1883–1924: *The Metamorphosis* (1915)

12 Last night I dreamt I went to Manderley again.
Daphne Du Maurier 1907–89: *Rebecca* (1938)

13 It was a bright cold day in April, and the clocks were striking thirteen.
George Orwell 1903–50: *Nineteen Eighty-Four* (1949)

14 Lolita, light of my life, fire of my loins. My sin, my soul. Lo-lee-ta: the tip of the tongue taking a trip of three steps down the palate to tap, at three, on the teeth. Lo. Lee. Ta.
Vladimir Nabokov 1899–1977: *Lolita* (1955)

15 'Take my camel, dear,' said my aunt Dot, as she climbed down from this animal on her return from High Mass.
Rose Macaulay 1881–1958: *The Towers of Trebizond* (1956)

16 It was the afternoon of my eighty-first birthday, and I was in bed with my catamite when Ali announced that the archbishop had come to see me.
Anthony Burgess 1917–93: *Earthly Powers* (1980)

Opinion

1 There are as many opinions as there are people: each has his own correct way.
Terence *c.*190–159 BC: *Phormio*

2 Opinion in good men is but knowledge in the making.
John Milton 1608–74: *Areopagitica* (1644)

3 They that approve a private opinion, call it opinion; but they that mislike it, heresy: and yet heresy signifies no more than private opinion.
Thomas Hobbes 1588–1679: *Leviathan* (1651)

4 He that complies against his will,
Is of his own opinion still.
Samuel Butler 1612–80: *Hudibras* pt. 3 (1680)

5 New opinions are always suspected, and usually opposed, without any other reason but because they are not already common.
John Locke 1632–1704: *Essay concerning Human Understanding* (1690)

6 Some praise at morning what they blame at night;
But always think the last opinion right.
Alexander Pope 1688–1744: *An Essay on Criticism* (1711)

7 Every man has a right to utter what he thinks truth, and every other man has a right to knock him down for it. Martyrdom is the test.

Samuel Johnson 1709–84: in James Boswell *Life of Johnson* (1791) 1780

8 A man can brave opinion, a woman must submit to it.

Mme de Staël 1766–1817: *Delphine* (1802)

9 There are nine and sixty ways of constructing tribal lays,
And—every—single—one—of—them—is—right!

Rudyard Kipling 1865–1936: 'In the Neolithic Age' (1893)

Optimism and Pessimism

See also **Hope and Despair**

1 Sin is behovely, but all shall be well and all shall be well and all manner of thing shall be well.

Julian of Norwich 1343–after 1416: *Revelations of Divine Love*

2 In this best of possible worlds...all is for the best.

Voltaire 1694–1778: *Candide* (1759); usually quoted 'All is for the best in the best of all possible worlds'

3 The lark's on the wing;
The snail's on the thorn:
God's in his heaven—
All's right with the world!

Robert Browning 1812–89: *Pippa Passes* (1841)

4 I have known him come home to supper with a flood of tears, and a declaration that nothing was now left but a jail; and go to bed making a calculation of the expense of putting bow-windows to the house, 'in case anything turned up,' which was his favourite expression.

Charles Dickens 1812–70: *David Copperfield* (1850) of Mr Micawber

5 Where everything is bad it must be good to know the worst.

F. H. Bradley 1846–1924: *Appearance and Reality* (1893)

6 Nothing to do but work,
Nothing to eat but food,

Nothing to wear but clothes
To keep one from going nude.
Benjamin Franklin King 1857–94: 'The Pessimist'

7 Cheer up! the worst is yet to come!
Philander Chase Johnson 1866–1939: *Everybody's Magazine* May 1920

8 The optimist proclaims that we live in the best of all possible worlds; and the pessimist fears this is true.
James Branch Cabell 1879–1958: *The Silver Stallion* (1926)

9 an optimist is a guy
that has never had
much experience.
Don Marquis 1878–1937: *archy and mehitabel* (1927)

10 'Twixt the optimist and pessimist
The difference is droll:
The optimist sees the doughnut
But the pessimist sees the hole.
McLandburgh Wilson 1892– : *Optimist and Pessimist*

11 Man hands on misery to man.
It deepens like a coastal shelf.
Get out as early as you can,
And don't have any kids yourself.
Philip Larkin 1922–85: 'This Be The Verse' (1974)

12 If we see light at the end of the tunnel,
It's the light of the oncoming train.
Robert Lowell 1917–77: 'Since 1939' (1977)

13 I don't consider myself a pessimist. I think of a pessimist as someone who is waiting for it to rain. And I feel soaked to the skin.
Leonard Cohen 1934– : in *Observer* 2 May 1993 'Sayings of the Week'

Painting and Drawing

1 The King found her [Anne of Cleves] so different from her picture...that...he swore they had brought him a Flanders mare.
Henry VIII 1491–1547: Tobias Smollett *Complete History of England* (3rd ed., 1759)

2 Good painters imitate nature, bad ones spew it up.
Cervantes 1547–1616: *El Licenciado Vidriera* in *Novelas Ejemplares* (1613)

3 Remark all these roughnesses, pimples, warts, and everything as you see me; otherwise I will never pay a farthing for it.
Oliver Cromwell 1599–1658: to Lely, on the painting of his portrait; in Horace Walpole *Anecdotes of Painting in England* (1763) (commonly quoted 'warts and all')

4 In Claude's landscape all is lovely—all amiable—all is amenity and repose;—the calm sunshine of the heart.
John Constable 1776–1837: lecture, 2 June 1836

5 *Le dessin est la probité de l'art.*
Drawing is the true test of art.
J. A. D. Ingres 1780–1867: *Pensées d'Ingres* (1922)

6 She is older than the rocks among which she sits.
Walter Pater 1839–94: of the *Mona Lisa*; *Studies in the History of the Renaissance* (1873)

7 I have seen, and heard, much of Cockney impudence before now; but never expected to hear a coxcomb ask two hundred guineas for flinging a pot of paint in the public's face.
John Ruskin 1819–1900: on Whistler's *Nocturne in Black and Gold*; *Fors Clavigera* (1871–84) Letter 79, 18 June 1877

8 No, I ask it for the knowledge of a lifetime.
James McNeill Whistler 1834–1903: in his case against Ruskin, replying to the question: 'For two days' labour, you ask two hundred guineas?'; in D. C. Seitz *Whistler Stories* (1913)

9 Yes madam, Nature is creeping up.
James McNeill Whistler 1834–1903: to a lady who had been reminded of his work by an 'exquisite haze in the atmosphere'; in D. C. Seitz *Whistler Stories* (1913)

10 Art does not reproduce the visible; rather, it makes visible.
Paul Klee 1879–1940: 'Creative Credo' (1920)

11 An active line on a walk, moving freely without a goal. A walk for walk's sake.
Paul Klee 1879–1940: *Pedagogical Sketchbook* (1925)

12 Every time I paint a portrait I lose a friend.
John Singer Sargent 1856–1925: in N. Bentley and E. Esar *Treasury of Humorous Quotations* (1951)

13 I am a painter and I nail my pictures together.
Kurt Schwitters 1887–1948: in R. Hausmann *Am Anfang war Dada* (1972)

14 Painting is saying "Ta" to God.
Stanley Spencer 1891–1959: in letter from Spencer's daughter Shirin, *Observer* 7 February 1988

15 I paint objects as I think them, not as I see them.
Pablo Picasso 1881–1973: in John Golding *Cubism* (1959)

16 A product of the untalented, sold by the unprincipled to the utterly bewildered.
Al Capp 1907–79: on abstract art, in *National Observer* 1 July 1963

Parents

See also **Children, The Family**

1 Honour thy father and thy mother.
Bible: Exodus

2 A wise son maketh a glad father: but a foolish son is the heaviness of his mother.
Bible: Proverbs

3 Or what man is there of you, whom if his son ask bread, will he give him a stone?
Bible: St Matthew

4 Parents love their children more than children love their parents.
Auctoritates Aristotelis: a compilation of medieval propositions

5 It is a wise father that knows his own child.
William Shakespeare 1564–1616: *The Merchant of Venice* (1596–8)

6 The joys of parents are secret, and so are their griefs and fears.
Francis Bacon 1561–1626: *Essays* (1625) 'Of Parents and Children'

7 A slavish bondage to parents cramps every faculty of the mind.
Mary Wollstonecraft 1759–97: *A Vindication of the Rights of Woman* (1792)

8 The hand that rocks the cradle
Is the hand that rules the world.
William Ross Wallace d. 1881: 'What rules the world' (1865)

9 If I were damned of body and soul,
I know whose prayers would make me whole,
Mother o' mine, O mother o' mine.
Rudyard Kipling 1865–1936: *The Light That Failed* (1891)

10 Children begin by loving their parents; after a time they judge them; rarely, if ever, do they forgive them.
Oscar Wilde 1854–1900: *A Woman of No Importance* (1893)

11 And mothers of large families (who claim to common sense)
Will find a Tiger well repay the trouble and expense.
Hilaire Belloc 1870–1953: *A Bad Child's Book of Beasts* (1896) 'The Tiger'

12 Few misfortunes can befall a boy which bring worse consequences than to have a really affectionate mother.
W. Somerset Maugham 1874–1965: *A Writer's Notebook* (1949) (written in 1896)

13 The parent who could see his boy as he really is, would shake his head and say: 'Willie is no good; I'll sell him.'
Stephen Leacock 1869–1944: *Essays and Literary Studies* (1916)

14 The natural term of the affection of the human animal for its offspring is six years.
George Bernard Shaw 1856–1950: *Heartbreak House* (1919)

15 Your children are not your children.
They are the sons and daughters of Life's longing for itself.
They came through you but not from you
And though they are with you yet they belong not to you.
Kahlil Gibran 1883–1931: *The Prophet* (1923) 'On Children'

16 The fundamental defect of fathers, in our competitive society, is that they want their children to be a credit to them.
Bertrand Russell 1872–1970: *Sceptical Essays* (1928)

17 Children aren't happy with nothing to ignore,
And that's what parents were created for.
Ogden Nash 1902–71: 'The Parent' (1933)

18 Oh, what a tangled web do parents weave
When they think that their children are naïve.
Ogden Nash 1902–71: 'Baby, What Makes the Sky Blue' (1940)

19 Parentage is a very important profession, but no test of fitness for it is ever imposed in the interest of the children.
George Bernard Shaw 1856–1950: *Everybody's Political What's What?* (1944)

20 Parents—especially step-parents—are sometimes a bit of a disappointment to their children. They don't fulfil the promise of their early years.
Anthony Powell 1905– : *A Buyer's Market* (1952)

21 The value of marriage is not that adults produce children but that children produce adults.
Peter De Vries 1910– : *The Tunnel of Love* (1954)

22 Parents learn a lot from their children about coping with life.
Muriel Spark 1918– : *The Comforters* (1957)

23 The thing that impresses me most about America is the way parents obey their children.
Edward VIII 1894–1972: *Look* 5 March 1957

24 A Jewish man with parents alive is a fifteen-year-old boy, and will remain a fifteen-year-old boy until *they die*!
Philip Roth 1933– : *Portnoy's Complaint* (1967)

25 No matter how old a mother is she watches her middle-aged children for signs of improvement.
Florida Scott-Maxwell: *Measure of my Days* (1968)

26 They fuck you up, your mum and dad.
They may not mean to, but they do.
They fill you with the faults they had
And add some extra, just for you.
Philip Larkin 1922–85: 'This Be The Verse' (1974)

The Past

1 Even a god cannot change the past.
Agathon b. *c.*445 BC: in Aristotle *Nicomachaean Ethics* (literally 'The one thing which even a god cannot do is to make undone what has been done')

2 *Mais où sont les neiges d'antan?*
But where are the snows of yesteryear?
François Villon b. 1431: 'Ballade des dames du temps jadis' (1461), tr. D. G. Rossetti

3 O! call back yesterday, bid time return.
William Shakespeare 1564–1616: *Richard II* (1595)

4 Antiquities are history defaced, or some remnants of history which have casually escaped the shipwreck of time.
Francis Bacon 1561–1626: *The Advancement of Learning* (1605)

5 What's done cannot be undone.
William Shakespeare 1564–1616: *Macbeth* (1606)

6 What's gone and what's past help
Should be past grief.
William Shakespeare 1564–1616: *The Winter's Tale* (1610–11)

7 Ancient times were the youth of the world.
Francis Bacon 1561–1626: *De Dignitate et Augmentis Scientiarum* (1623)

8 There never was a merry world since the fairies left off dancing, and the Parson left conjuring.
John Selden 1584–1654: *Table Talk* (1689)

9 Think of it, soldiers; from the summit of these pyramids, forty centuries look down upon you.
Napoléon I 1769–1821: speech before the Battle of the Pyramids, 21 July 1798

10 The moving finger writes; and, having writ,
Moves on: nor all thy piety nor wit
Shall lure it back to cancel half a line,
Nor all thy tears wash out a word of it.
Edward Fitzgerald 1809–83: *The Rubáiyát of Omar Khayyám* (1859)

11 What are those blue remembered hills,
What spires, what farms are those?

That is the land of lost content,
I see it shining plain,
The happy highways where I went
And cannot come again.

A. E. Housman 1859–1936: *A Shropshire Lad* (1896)

12 Those who cannot remember the past are condemned to repeat it.

George Santayana 1863–1952: *The Life of Reason* (1905)

13 O God! Put back Thy universe and give me yesterday.

Henry Arthur Jones 1851–1929 and **Henry Herman** 1832–94: *The Silver King* (1907)

14 Stands the Church clock at ten to three?
And is there honey still for tea?

Rupert Brooke 1887–1915: 'The Old Vicarage, Grantchester' (1915)

15 The past is the only dead thing that smells sweet.

Edward Thomas 1878–1917: 'Early one morning in May I set out' (1917)

16 I tell you the past is a bucket of ashes.

Carl Sandburg 1878–1967: 'Prairie' (1918)

17 Things ain't what they used to be.

Ted Persons: title of song (1941)

18 In every age 'the good old days' were a myth. No one ever thought they were good at the time. For every age has consisted of crises that seemed intolerable to the people who lived through them.

Brooks Atkinson 1894–1984: *Once Around the Sun* (1951)

19 The past is a foreign country: they do things differently there.

L. P. Hartley 1895–1972: *The Go-Between* (1953)

20 People who are always praising the past
And especially the times of faith as best
Ought to go and live in the Middle Ages
And be burnt at the stake as witches and sages.

Stevie Smith 1902–71: 'The Past' (1957)

21 I have heard tell of a Professor of Economics who has a sign on the wall of his study, reading 'the future is not what it

was'. The sentiment was admirable; unfortunately, the past is not getting any better either.
Bernard Levin 1928- : *Sunday Times* 22 May 1977

22 Nostalgia isn't what it used to be.
Anonymous: graffito; taken as title of book by Simone Signoret, 1978

Patriotism

1 *Dulce et decorum est pro patria mori.*
Lovely and honourable it is to die for one's country.
Horace 65–8 BC: *Odes*

2 Never was patriot yet, but was a fool.
John Dryden 1631–1700: *Absalom and Achitophel* (1681)

3 What pity is it
That we can die but once to serve our country!
Joseph Addison 1672–1719: *Cato* (1713)

4 Be England what she will,
With all her faults, she is my country still.
Charles Churchill 1731–64: *The Farewell* (1764)

5 The more foreigners I saw, the more I loved my homeland.
De Belloy 1727–75: *Le Siège de Calais* (1765)

6 Patriotism is the last refuge of a scoundrel.
Samuel Johnson 1709–84: in James Boswell *Life of Johnson* (1791) 7 April 1775

7 I only regret that I have but one life to lose for my country.
Nathan Hale 1755–76: prior to his execution by the British for spying, 22 September 1776

8 These are the times that try men's souls. The summer soldier and the sunshine patriot will, in this crisis, shrink from the service of their country; but he that stands it *now*, deserves the love and thanks of men and women.
Thomas Paine 1737–1809: *The Crisis* (December 1776)

9 True patriots we; for be it understood,
We left our country for our country's good.
Henry Carter d. 1806: prologue, written for, but not recited at, the opening of the Playhouse, Sydney, New South Wales, 16 January 1796, when the actors were principally

convicts. Previously attributed to George Barrington (b. 1755)

10 Breathes there the man, with soul so dead,
Who never to himself hath said,
This is my own, my native land!
Sir Walter Scott 1771–1832: *The Lay of the Last Minstrel* (1805)

11 Our country! In her intercourse with foreign nations, may she always be in the right; but our country, right or wrong.
Stephen Decatur 1779–1820: toast at Norfolk, Virginia, April 1816

12 My toast would be, may our country be always successful, but whether successful or otherwise, always right.
John Quincy Adams 1767–1848: letter to John Adams, 1 August 1816

13 A steady patriot of the world alone,
The friend of every country but his own.
George Canning 1770–1827: of the Jacobin, in 'New Morality' (1821)

14 That kind of patriotism which consists in hating all other nations.
Elizabeth Gaskell 1810–65: *Sylvia's Lovers* (1863)

15 My country, right or wrong; if right, to be kept right; and if wrong, to be set right!
Carl Schurz 1829–1906: speech, US Senate, 29 February 1872

16 Patriotism is not enough. I must have no hatred or bitterness towards anyone.
Edith Cavell 1865–1915: on the eve of her execution, in *The Times* 23 October 1915

17 I vow to thee, my country—all earthly things above—
Entire and whole and perfect, the service of my love.
Cecil Spring-Rice 1859–1918: 'I Vow to Thee, My Country' (1918)

18 You'll never have a quiet world till you knock the patriotism out of the human race.
George Bernard Shaw 1856–1950: *O'Flaherty V.C.* (1919)

19 Patriotism is a lively sense of collective responsibility. Nationalism is a silly cock crowing on its own dunghill.
Richard Aldington 1892–1962: *The Colonel's Daughter* (1931)

20 If I had to choose between betraying my country and betraying my friend, I hope I should have the guts to betray my country.
E. M. Forster 1879–1970: *Two Cheers for Democracy* (1951)

21 I would die for my country but I would never let my country die for me.
Neil Kinnock 1942– : speech at Labour party conference, 30 September 1986

22 The cricket test—which side do they cheer for?...Are you still looking back to where you came from or where you are?
Norman Tebbit 1931– : on the loyalties of Britain's immigrant population; interview in *Los Angeles Times*, reported in *Daily Telegraph* 20 April 1990

Peace

1 They shall beat their swords into plowshares, and their spears into pruninghooks: nation shall not lift up sword against nation, neither shall they learn war any more.
Bible: Isaiah

2 The peace of God, which passeth all understanding, shall keep your hearts and minds through Christ Jesus.
Bible: Philippians

3 They make a wilderness and call it peace.
Tacitus AD *c.*56–after 117: *Agricola*

4 *Qui desiderat pacem, praeparet bellum.*
Let him who desires peace, prepare for war.
Vegetius fourth century AD: *Epitoma Rei Militaris*, usually quoted '*Si vis pacem, para bellum* [If you want peace, prepare for war]'

5 *E'n la sua volontade è nostra pace.*
In His will is our peace.
Dante Alighieri 1265–1321: *Divina Commedia* 'Paradiso'

6 ...The naked, poor, and manglèd Peace,
Dear nurse of arts, plenties, and joyful births.

William Shakespeare 1564–1616: *Henry V* (1599)

7 ...Peace hath her victories
No less renowned than war.

John Milton 1608–74: 'To the Lord General Cromwell' (written 1652)

8 Give peace in our time, O Lord.

Book of Common Prayer 1662: *Morning Prayer*

9 Lord Salisbury and myself have brought you back peace—but a peace I hope with honour.

Benjamin Disraeli 1804–81: speech on returning from the Congress of Berlin, 16 July 1878

10 PEACE, *n*. In international affairs, a period of cheating between two periods of fighting.

Ambrose Bierce 1842–*c*.1914: *The Devil's Dictionary* (1911)

11 This is the second time in our history that there has come back from Germany to Downing Street peace with honour. I believe it is peace for our time.

Neville Chamberlain 1869–1940: speech from 10 Downing Street, 30 September 1938

12 I think that people want peace so much that one of these days governments had better get out of the way and let them have it.

Dwight D. Eisenhower 1890–1969: broadcast discussion, 31 August 1959

13 Give peace a chance.

John Lennon 1940–80 and **Paul McCartney** 1942– : title of song (1969)

14 There is no such thing as inner peace. There is only nervousness or death.

Fran Lebowitz 1946– : *Metropolitan Life* (1978)

People

See also **Musicians, Poets, Politicians, Writers**

1 As time requireth, a man of marvellous mirth and pastimes, and sometime of as sad gravity, as who say: a man for all seasons.

Robert Whittington: of Sir Thomas More, in *Vulgaria* (1521)

2 He had a head to contrive, a tongue to persuade, and a hand to execute any mischief.

Edward Hyde, Earl of Clarendon 1609–74: of Hampden, *History of the Rebellion* (1703)

3 He snatched the lightning shaft from heaven, and the sceptre from tyrants.

A. R. J. Turgot 1727–81: inscription for a bust of Benjamin Franklin, inventor of the lightning conductor

4 As he rose like a rocket, he fell like the stick.

Thomas Paine 1737–1809: on Edmund Burke losing the debate on the French Revolution to Charles James Fox, in the House of Commons; in *Letter to the Addressers on the late Proclamation* (1792)

5 Mad, bad, and dangerous to know.

Lady Caroline Lamb 1785–1828: of Byron; diary, March 1812

6 An Archangel a little damaged.

Charles Lamb 1775–1834: of Coleridge; letter to Wordsworth, 26 April 1816

7 He rather hated the ruling few than loved the suffering many.

Jeremy Bentham 1748–1832: of James Mill, in H. N. Pym (ed.) *Memories of Old Friends* (1882)

8 The seagreen Incorruptible.

Thomas Carlyle 1795–1881: of Robespierre, *History of the French Revolution* (1837)

9 Macaulay is well for a while, but one wouldn't *live* under Niagara.

Thomas Carlyle 1795–1881: in R. M. Milnes *Notebook* (1838)

10 Thou large-brained woman and large-hearted man.
Elizabeth Barrett Browning 1806–61: 'To George Sand—A Desire' (1844)

11 Who saw life steadily, and saw it whole:
The mellow glory of the Attic stage;
Singer of sweet Colonus, and its child.
Matthew Arnold 1822–88: 'To a Friend' (1849), of Sophocles

12 The statue stood
Of Newton, with his prism, and silent face:
The marble index of a mind for ever
Voyaging through strange seas of Thought, alone.
William Wordsworth 1770–1850: *The Prelude* (1850)

13 He [Bernard Shaw] hasn't an enemy in the world, and none of his friends like him.
Oscar Wilde 1854–1900: in Bernard Shaw *Sixteen Self Sketches* (1949)

14 Her conception of God was certainly not orthodox. She felt towards Him as she might have felt towards a glorified sanitary engineer; and in some of her speculations she seems hardly to distinguish between the Deity and the Drains.
Lytton Strachey 1880–1932: *Eminent Victorians* (1918) 'Florence Nightingale'

15 A good man fallen among Fabians.
Lenin 1870–1924: of G. B. Shaw, in A. Ransome *Six Weeks in Russia in 1919* (1919) 'Notes of Conversations with Lenin'

16 He was no striped frieze; he was shot silk.
Lytton Strachey 1880–1932: of Francis Bacon, *Elizabeth and Essex* (1928)

17 He seemed at ease and to have the look of the last gentleman in Europe.
Ada Leverson 1865–1936: of Oscar Wilde, *Letters to the Sphinx* (1930)

18 Victor Hugo was a madman who thought he was Victor Hugo.
Jean Cocteau 1889–1963: *Opium* (1930)

19 To us he is no more a person
now but a whole climate of opinion.
W. H. Auden 1907–73: 'In Memory of Sigmund Freud' (1940)

20 Beaverbrook is so pleased to be in the Government that he is like the town tart who has finally married the Mayor!
Beverley Baxter 1891–1964: in Henry 'Chips' Channon's diary, 12 June 1940

21 The candle in that great turnip has gone out.
Winston Churchill 1874–1965: of Stanley Baldwin, in Harold Nicolson's diary, 17 August 1950

22 Rousseau was the first militant lowbrow.
Isaiah Berlin 1909– : *Observer* 9 November 1952

23 The mama of dada.
Clifton Fadiman 1904– : of Gertrude Stein, *Party of One* (1955)

24 An elderly fallen angel travelling incognito.
Peter Quennell 1905–93: of André Gide, *The Sign of the Fish* (1960)

25 Forty years ago he was Slightly in *Peter Pan*, and you might say that he has been wholly in *Peter Pan* ever since.
Kenneth Tynan 1927–80: of Noël Coward, *Curtains* (1961)

26 What, when drunk, one sees in other women, one sees in Garbo sober.
Kenneth Tynan 1927–80: *Curtains* (1961)

27 She would rather light a candle than curse the darkness, and her glow has warmed the world.
Adlai Stevenson 1900–65: on learning of Eleanor Roosevelt's death; in *New York Times* 8 November 1962

28 In defeat unbeatable: in victory unbearable.
Winston Churchill 1874–1965: of Viscount Montgomery, in E. Marsh *Ambrosia and Small Beer* (1964)

29 A doormat in a world of boots.
Jean Rhys *c.*1890–1979: describing herself; in *Guardian* 6 December 1990

30 Every word she writes is a lie, including 'and' and 'the'.
Mary McCarthy 1912–89: quoting herself on Lillian Hellman in *New York Times* 16 February 1980

Perfection

1 How many things by season seasoned are
To their right praise and true perfection!
William Shakespeare 1564–1616: *The Merchant of Venice* (1596–8)

2 Perfection is the child of Time.
Bishop Joseph Hall 1574–1656: *Works* (1625)

3 Whoever thinks a faultless piece to see,
Thinks what ne'er was, nor is, nor e'er shall be.
Alexander Pope 1688–1744: *An Essay on Criticism* (1711)

4 *Le mieux est l'ennemi du bien.*
The best is the enemy of the good.
Voltaire 1694–1778: *Contes* (1772), deriving from an Italian proverb

5 Pictures of perfection as you know make me sick and wicked.
Jane Austen 1775–1817: letter to Fanny Knight, 23 March 1817

6 No one can be perfectly free till all are free; no one can be perfectly moral till all are moral; no one can be perfectly happy till all are happy.
Herbert Spencer 1820–1903: *Social Statics* (1850)

7 What's come to perfection perishes.
Things learned on earth, we shall practise in heaven:
Works done least rapidly, Art most cherishes.
Robert Browning 1812–89: 'Old Pictures in Florence' (1855)

8 Faultily faultless, icily regular, splendidly null,
Dead perfection, no more.
Alfred, Lord Tennyson 1809–92: *Maud* (1855)

9 He is all fault who hath no fault at all:
For who loves me must have a touch of earth.
Alfred, Lord Tennyson 1809–92: *Idylls of the King* 'Lancelot and Elaine' (1859)

10 Faultless to a fault.
Robert Browning 1812–89: *The Ring and the Book* (1868–9)

11 The pursuit of perfection, then, is the pursuit of sweetness and light...He who works for sweetness and light united, works to make reason and the will of God prevail.
Matthew Arnold 1822–88: *Culture and Anarchy* (1869)

12 The best is the best, though a hundred judges have declared it so.
Arthur Quiller-Couch 1863–1944: *Oxford Book of English Verse* (1900)

13 Finality is death. Perfection is finality.
Nothing is perfect. There are lumps in it.
James Stephens 1882–1950: *The Crock of Gold* (1912)

14 The intellect of man is forced to choose
Perfection of the life, or of the work.
W. B. Yeats 1865–1939: 'Coole Park and Ballylee, 1932' (1933)

Pessimism see **Optimism and Pessimism**

Philosophy

1 The unexamined life is not worth living.
Socrates 469–399 BC: in Plato *Apology*

2 There is nothing so absurd but some philosopher has said it.
Cicero 106–43 BC: *De Divinatione*

3 How charming is divine philosophy!
Not harsh and crabbèd, as dull fools suppose,
But musical as is Apollo's lute,
John Milton 1608–74: *Comus* (1637)

4 I refute it *thus.*
Samuel Johnson 1709–84: kicking a large stone by way of refuting Bishop Berkeley's theory of the non-existence of matter; in James Boswell *Life of Johnson* (1791) 6 August 1763

5 Philosophy will clip an Angel's wings.
John Keats 1795–1821: 'Lamia' (1820)

6 When philosophy paints its grey on grey, then has a shape of life grown old. By philosophy's grey on grey it cannot be rejuvenated but only understood. The owl of Minerva spreads its wings only with the falling of the dusk.
G. W. F. Hegel 1770–1831: *Philosophy of Right* (1821) tr. T. M. Knox

7 If it was so, it might be; and if it were so, it would be: but as it isn't, it ain't. That's logic.
Lewis Carroll 1832–98: *Through the Looking-Glass* (1872)

8 Metaphysics is the finding of bad reasons for what we believe upon instinct.
F. H. Bradley 1846–1924: *Appearance and Reality* (1893)

9 The Socratic manner is not a game at which two can play.
Max Beerbohm 1872–1956: *Zuleika Dobson* (1911)

10 The safest general characterization of the European philosophical tradition is that it consists of a series of footnotes to Plato.
Alfred North Whitehead 1861–1947: *Process and Reality* (1929)

11 Philosophy is a battle against the bewitchment of our intelligence by means of language.
Ludwig Wittgenstein 1889–1951: *Philosophische Untersuchungen* (1953)

12 What is your aim in philosophy?—To show the fly the way out of the fly-bottle.
Ludwig Wittgenstein 1889–1951: *Philosophische Untersuchungen* (1953)

Places

See also **America, Britain, England, Europe, France, Ireland, London, Scotland, Wales**

1 *Semper aliquid novi Africam adferre.*
Africa always brings [us] something new.
Pliny the Elder AD 23–79: *Historia Naturalis* (often quoted '*Ex Africa semper aliquid novi* [Always something new out of Africa]')

2 Once did she hold the gorgeous East in fee,
And was the safeguard of the West.
William Wordsworth 1770–1850: 'On the Extinction of the Venetian Republic' (1807)

3 One has no great hopes from Birmingham. I always say there is something direful in the sound.
Jane Austen 1775–1817: *Emma* (1816)

4 Sun-girt city, thou hast been
Ocean's child, and then his queen;
Now is come a darker day,
And thou soon must be his prey.
Percy Bysshe Shelley 1792–1822: of Venice, 'Lines written amongst the Euganean Hills' (1818)

5 While stands the Coliseum, Rome shall stand;
When falls the Coliseum, Rome shall fall;
And when Rome falls—the World.
Lord Byron 1788–1824: *Childe Harold's Pilgrimage* (1812–18)

6 Let there be light! said Liberty,
And like sunrise from the sea,
Athens arose!
Percy Bysshe Shelley 1792–1822: *Hellas* (1822)

7 The isles of Greece, the isles of Greece!
Where burning Sappho loved and sung,
Where grew the arts of war and peace,
Where Delos rose, and Phoebus sprung!
Eternal summer gilds them yet,
But all, except their sun, is set!
Lord Byron 1788–1824: *Don Juan* (1819–24)

8 It is from the midst of this putrid sewer that the greatest river of human industry springs up and carries fertility to the whole world. From this foul drain pure gold flows forth.
Alexis de Tocqueville 1805–59: of Manchester, *Voyage en Angleterre et en Irlande de 1835* (1958)

9 Kent, sir—everybody knows Kent—apples, cherries, hops, and women.
Charles Dickens 1812–70: *Pickwick Papers* (1837) Jingle

10 That temple of silence and reconciliation where the enmities of twenty generations lie buried.
Lord Macaulay 1800–59: of Westminster Abbey, *Essays* (1843)

11 Italy is a geographical expression.
Prince Metternich 1773–1859: discussing the Italian question with Palmerston in 1847

12 Russia has two generals in whom she can confide—Generals Janvier [January] and Février [February].
Emperor Nicholas I of Russia 1796–1855: attributed

13 Earth is here so kind, that just tickle her with a hoe and she laughs with a harvest.

Douglas Jerrold 1803–57: of Australia, *Wit and Opinions* (1859) 'A Land of Plenty'

14 Whispering from her towers the last enchantments of the Middle Age...Home of lost causes, and forsaken beliefs.

Matthew Arnold 1822–88: of Oxford, *Essays in Criticism* (1865)

15 That sweet City with her dreaming spires.

Matthew Arnold 1822–88: of Oxford, 'Thyrsis' (1866)

16 Towery city and branchy between towers.

Gerard Manley Hopkins 1844–89: 'Duns Scotus's Oxford' (written 1879)

17 And this is good old Boston,
The home of the bean and the cod,
Where the Lowells talk to the Cabots
And the Cabots talk only to God.

John Collins Bossidy 1860–1928: verse spoken at Holy Cross College alumni dinner in Boston, Massachusetts, 1910

18 Great God! this is an awful place.

Robert Falcon Scott 1868–1912: of the South Pole; diary, 17 January 1912

19 For Cambridge people rarely smile,
Being urban, squat, and packed with guile.

Rupert Brooke 1887–1915: 'The Old Vicarage, Grantchester' (1915)

20 Poor Mexico, so far from God and so close to the United States.

Porfirio Diaz 1830–1915: attributed

21 Hog Butcher for the World,
Tool Maker, Stacker of Wheat,
Player with Railroads and the Nation's Freight Handler;
Stormy, husky, brawling,
City of the Big Shoulders.

Carl Sandburg 1878–1967: 'Chicago' (1916)

22 STREETS FLOODED. PLEASE ADVISE.

Robert Benchley 1889–1945: telegraph message on arriving in Venice

23 I cannot forecast to you the action of Russia. It is a riddle wrapped in a mystery inside an enigma.
Winston Churchill 1874–1965: radio broadcast, 1 October 1939

24 The last time I saw Paris
Her heart was warm and gay,
I heard the laughter of her heart in ev'ry street café.
Oscar Hammerstein II 1895–1960: 'The Last Time I saw Paris' (1940 song)

25 A big hard-boiled city with no more personality than a paper cup.
Raymond Chandler 1888–1959: of Los Angeles, *The Little Sister* (1949)

26 A trip through a sewer in a glass-bottomed boat.
Wilson Mizner 1876–1933: of Hollywood, in A. Johnston *The Legendary Mizners* (1953)

27 Venice is like eating an entire box of chocolate liqueurs in one go.
Truman Capote 1924–84: in *Observer* 26 November 1961

28 Paris is a movable feast.
Ernest Hemingway 1899–1961: *A Movable Feast* (1964)

Pleasure

1 *Trahit sua quemque voluptas.*
Everyone is dragged on by their favourite pleasure.
Virgil 70–19 BC: *Eclogues*

2 Who loves not woman, wine, and song
Remains a fool his whole life long.
Martin Luther 1483–1546: attributed (later inscribed, in German, in the Luther room in the Wartburg)

3 Pleasure is nothing else but the intermission of pain.
John Selden 1584–1654: *Table Talk* (1689) 'Pleasure'

4 Remorse, the fatal egg by pleasure laid.
William Cowper 1731–1800: 'The Progress of Error' (1782)

5 One half of the world cannot understand the pleasures of the other.
Jane Austen 1775–1817: *Emma* (1816)

6 Ever let the fancy roam,
Pleasure never is at home.
John Keats 1795–1821: 'Fancy' (1820)

7 Pleasure's a sin, and sometimes sin's a pleasure.
Lord Byron 1788–1824: *Don Juan* (1819–24)

8 Let us have wine and women, mirth and laughter,
Sermons and soda-water the day after.
Lord Byron 1788–1824: *Don Juan* (1819–24)

9 The greatest pleasure I know, is to do a good action by stealth, and to have it found out by accident.
Charles Lamb 1775–1834: 'Table Talk by the late Elia' in *The Athenaeum* 4 January 1834

10 The Puritan hated bear-baiting, not because it gave pain to the bear, but because it gave pleasure to the spectators.
Lord Macaulay 1800–59: *History of England* vol. 1 (1849)

11 Life would be very pleasant if it were not for its enjoyments.
R. S. Surtees 1805–64: *Mr Facey Romford's Hounds* (1865)

12 I'm tired of Love: I'm still more tired of Rhyme.
But Money gives me pleasure all the time.
Hilaire Belloc 1870–1953: 'Fatigued' (1923)

13 All the things I really like to do are either illegal, immoral, or fattening.
Alexander Woollcott 1887–1943: in R. E. Drennan *Wit's End* (1973)

14 There's no greater bliss in life than when the plumber eventually comes to unblock your drains. No writer can give that sort of pleasure.
Victoria Glendinning 1937– : *Observer* 3 January 1993

Poetry

1 Skilled or unskilled, we all scribble poems.
Horace 65–8 BC: *Epistles*

2 [The poet] cometh unto you, with a tale which holdeth children from play, and old men from the chimney corner.
Philip Sidney 1554–86: *The Defence of Poetry* (1595)

3 Rhyme being...but the invention of a barbarous age, to set off wretched matter and lame metre.

John Milton 1608–74: *Paradise Lost* (1667) 'The Verse' (preface, 1668)

4 Wit will shine
Through the harsh cadence of a rugged line.

John Dryden 1631–1700: 'To the Memory of Mr Oldham' (1684)

5 [BOSWELL:] Sir, what is poetry?
[JOHNSON:] Why Sir, it is much easier to say what it is not. We all *know* what light is; but it is not easy to *tell* what it is.

Samuel Johnson 1709–84: in James Boswell *Life of Johnson* (1791) 12 April 1776

6 Poetry is the spontaneous overflow of powerful feelings: it takes its origin from emotion recollected in tranquillity.

William Wordsworth 1770–1850: *Lyrical Ballads* (2nd ed., 1802) preface

7 That willing suspension of disbelief for the moment, which constitutes poetic faith.

Samuel Taylor Coleridge 1772–1834: *Biographia Literaria* (1817)

8 Most wretched men
Are cradled into poetry by wrong:
They learn in suffering what they teach in song.

Percy Bysshe Shelley 1792–1822: 'Julian and Maddalo' (1818)

9 If poetry comes not as naturally as the leaves to a tree it had better not come at all.

John Keats 1795–1821: letter to Taylor, 27 February 1818

10 Poetry is the record of the best and happiest moments of the happiest and best minds.

Percy Bysshe Shelley 1792–1822: *A Defence of Poetry* (written 1821)

11 Poets are the unacknowledged legislators of the world.

Percy Bysshe Shelley 1792–1822: *A Defence of Poetry* (written 1821)

12 Prose = words in their best order;—poetry = the *best* words in the best order.

Samuel Taylor Coleridge 1772–1834: *Table Talk* (1835) 12 July 1827

13 Prose is when all the lines except the last go on to the end. Poetry is when some of them fall short of it.
Jeremy Bentham 1748–1832: in M. St. J. Packe *Life of John Stuart Mill* (1954)

14 Poetry is a subject as precise as geometry.
Gustave Flaubert 1821–80: letter to Louise Colet, 14 August 1853

15 The difference between genuine poetry and the poetry of Dryden, Pope, and all their school, is briefly this: their poetry is conceived and composed in their wits, genuine poetry is conceived and composed in the soul.
Matthew Arnold 1822–88: *Essays in Criticism* (1888)

16 I said 'a line will take us hours maybe,
Yet if it does not seem a moment's thought
Our stitching and unstitching has been naught.'
W. B. Yeats 1865–1939: 'Adam's Curse' (1904)

17 Poetry must be *as well written as prose.*
Ezra Pound 1885–1972: letter to Harriet Monroe, January 1915

18 All a poet can do today is warn.
Wilfred Owen 1893–1918: *Poems* (1963) preface (written 1918)

19 We make out of the quarrel with others, rhetoric, but of the quarrel with ourselves, poetry.
W. B. Yeats 1865–1939: *Essays* (1924)

20 A poem should not mean
But be.
Archibald MacLeish 1892–1982: 'Ars Poetica' (1926)

21 Of all the literary scenes
Saddest this sight to me:
The graves of little magazines
Who died to make verse free.
Keith Preston 1884–1927: 'The Liberators'

22 As soon as war is declared it will be impossible to hold the poets back. Rhyme is still the most effective drum.
Jean Giraudoux 1882–1944: *La Guerre de Troie n'aura pas lieu* (1935) tr. Christopher Fry as *Tiger at the Gates*, 1955

23 Writing a book of poetry is like dropping a rose petal down the Grand Canyon and waiting for the echo.

Don Marquis 1878–1937: in E. Anthony *O Rare Don Marquis* (1962)

24 Like a piece of ice on a hot stove the poem must ride on its own melting. A poem may be worked over once it is in being, but may not be worried into being.

Robert Frost 1874–1963: *Collected Poems* (1939) 'The Figure a Poem Makes'

25 It is the logic of our times,
No subject for immortal verse—
That we who lived by honest dreams
Defend the bad against the worse.

C. Day-Lewis 1904–72: 'Where are the War Poets?' (1943)

26 I'd as soon write free verse as play tennis with the net down.

Robert Frost 1874–1963: in E. Lathem *Interviews with Robert Frost* (1966)

27 Most people ignore most poetry
because
most poetry ignores most people.

Adrian Mitchell 1932– : *Poems* (1964)

28 A poet's hope: to be,
like some valley cheese,
local, but prized elsewhere.

W. H. Auden 1907–73: 'Shorts II' (1976)

Poets

1 Dr Donne's verses are like the peace of God; they pass all understanding.

James I 1566–1625: remark recorded by Archdeacon Plume (1630–1704)

2 All poets are mad.

Robert Burton 1577–1640: *Anatomy of Melancholy* (1621–51) 'Democritus to the Reader'

3 He invades authors like a monarch; and what would be theft in other poets, is only victory in him.

John Dryden 1631–1700: of Ben Jonson; *Essay of Dramatic Poesy* (1668)

4 'Tis sufficient to say [of Chaucer], according to the proverb, that here is God's plenty.
John Dryden 1631-1700: *Fables Ancient and Modern* (1700)

5 Ev'n copious Dryden, wanted, or forgot,
The last and greatest art, the art to blot.
Alexander Pope 1688-1744: *Imitations of Horace* (1737)

6 Milton, Madam, was a genius that could cut a Colossus from a rock; but could not carve heads upon cherry-stones.
Samuel Johnson 1709-84: to Hannah More, who had expressed a wonder that the poet who had written *Paradise Lost* should write such poor sonnets; in James Boswell *Life of Johnson* (1791) 13 June 1784

7 The reason Milton wrote in fetters when he wrote of Angels and God, and at liberty when of Devils and Hell, is because he was a true Poet, and of the Devil's party without knowing it.
William Blake 1757-1827: *The Marriage of Heaven and Hell* (1790-3)

8 With Donne, whose muse on dromedary trots,
Wreathe iron pokers into true-love knots.
Samuel Taylor Coleridge 1772-1834: 'On Donne's Poetry' (1818)

9 A cloud-encircled meteor of the air,
A hooded eagle among blinking owls.
Percy Bysshe Shelley 1792-1822: of Coleridge, 'Letter to Maria Gisborne' (1820)

10 We learn from Horace, Homer sometimes sleeps;
We feel without him: Wordsworth sometimes wakes.
Lord Byron 1788-1824: *Don Juan* (1819-24)

11 Out-babying Wordsworth and out-glittering Keats.
Edward Bulwer-Lytton 1803-73: of Tennyson, in *The New Timon* (1846)

12 He spoke, and loosed our heart in tears.
He laid us as we lay at birth
On the cool flowery lap of earth.
Matthew Arnold 1822-88: of Wordsworth, 'Memorial Verses, April 1850' (1852)

13 In poetry, no less than in life, he is 'a beautiful and ineffectual angel, beating in the void his luminous wings in vain'.
Matthew Arnold 1822–88: *Essays in Criticism* (1888) 'Shelley'

14 Immature poets imitate; mature poets steal.
T. S. Eliot 1888–1965: *The Sacred Wood* (1920)

15 Poets in our civilization, as it exists at present, must be *difficult*.
T. S. Eliot 1888–1965: 'The Metaphysical Poets' (1921)

16 The poet is always indebted to the universe, paying interest and fines on sorrow.
Vladimir Mayakovsky 1893–1930: 'Conversation with an Inspector of Taxes about Poetry' (1926), tr. D. Obolensky

17 You were silly like us; your gift survived it all:
The parish of rich women, physical decay,
Yourself. Mad Ireland hurt you into poetry.
W. H. Auden 1907–73: 'In Memory of W. B. Yeats' (1940)

18 [*The Waste Land*] was only the relief of a personal and wholly insignificant grouse against life; it is just a piece of rhythmical grumbling.
T. S. Eliot 1888–1965: *The Waste Land* (ed. Valerie Eliot, 1971) epigraph

19 Self-contempt, well-grounded.
F. R. Leavis 1895–1978: on the foundation of T. S. Eliot's work, in *Times Literary Supplement* 21 October 1988

20 I used to think all poets were Byronic.
They're mostly wicked as a ginless tonic
And wild as pension plans.
Wendy Cope 1945– : 'Triolet' (1986)

Political Comment

See also **Government, Political Parties, Politics**

1 Caesar had his Brutus—Charles the First, his Cromwell—and George the Third—('Treason,' cried the Speaker)... *may profit by their example.* If *this* be treason, make the most of it.
Patrick Henry 1736–99: speech in the Virginia assembly, May 1765

2 The Commons, faithful to their system, remained in a wise and masterly inactivity.
James Mackintosh 1765–1832: *Vindiciae Gallicae* (1791)

3 The compact which exists between the North and the South is 'a covenant with death and an agreement with hell'.
William Lloyd Garrison 1805–79: resolution adopted by the Massachusetts Anti-Slavery Society, 27 January 1843

4 I am for 'Peace, retrenchment, and reform', the watchword of the great Liberal party 30 years ago.
John Bright 1811–89: speech at Birmingham, 28 April 1859

5 Meddle and muddle.
Edward Stanley, 14th Earl of Derby 1799–1869: summarizing Earl Russell's foreign policy, in Speech on the Address, House of Lords, 4 February 1864

6 [Palmerston] once said that only three men in Europe had ever understood [the Schleswig-Holstein question], and of these the Prince Consort was dead, a Danish statesman (unnamed) was in an asylum, and he himself had forgotten it.
Lord Palmerston 1784–1865: in R. W. Seton-Watson *Britain in Europe 1789–1914* (1937)

7 With malice toward none; with charity for all; with firmness in the right, as God gives us to see the right, let us strive on to finish the work we are in.
Abraham Lincoln 1809–65: second inaugural address, 4 March 1865

8 This policy cannot succeed through speeches, and shooting-matches, and songs; it can only be carried out through blood and iron.
Otto von Bismarck 1815–98: speech in the Prussian House

of Deputies, 28 January 1886. In an earlier speech, 30 September 1862, Bismarck used the form 'iron and blood'

9 We are all socialists now.
William Harcourt 1827–1904: during the passage of the 1894 budget, which equalized death duties on real and personal property (attributed)

10 A mastiff? It is the right hon. Gentleman's poodle.
David Lloyd George 1863–1945: on the House of Lords and Lord Balfour respectively; speech, House of Commons, 26 June 1907

11 We had better wait and see.
Herbert Asquith 1852–1928: referring to the rumour that the House of Lords was to be flooded with new Liberal peers to ensure the passage of the Finance Bill, 1910

12 There are three classes which need sanctuary more than others—birds, wild flowers, and Prime Ministers.
Stanley Baldwin 1867–1947: in *Observer* 24 May 1925

13 The most conservative man in this world is the British Trade Unionist when you want to change him.
Ernest Bevin 1881–1951: speech at Trades Union Congress, 8 September 1927

14 Do not run up your nose dead against the Pope or the NUM!
Stanley Baldwin 1867–1947: in Lord Butler *The Art of Memory* (1982)

15 My [foreign] policy is to be able to take a ticket at Victoria Station and go anywhere I damn well please.
Ernest Bevin 1881–1951: in *Spectator* 20 April 1951

16 If you carry this resolution you will send Britain's Foreign Secretary naked into the conference chamber.
Aneurin Bevan 1897–1960: speech at Labour Party Conference, 3 October 1957, against a motion proposing unilateral nuclear disarmament by the UK

17 Let us be frank about it: most of our people have never had it so good.
Harold Macmillan 1894–1986: speech at Bedford, 20 July 1957 ('You Never Had It So Good' was the Democratic Party slogan during the 1952 US election campaign)

18 I thought the best thing to do was to settle up these little local difficulties, and then turn to the wider vision of the Commonwealth.

Harold Macmillan 1894–1986: statement at London airport on leaving for a Commonwealth tour, 7 January 1958, following the resignation of the Chancellor of the Exchequer and others

19 The wind of change is blowing through this continent, and, whether we like it or not, this growth of [African] national consciousness is a political fact.

Harold Macmillan 1894–1986: speech at Cape Town, 3 February 1960

20 And so, my fellow Americans: ask not what your country can do for you—ask what you can do for your country.

John F. Kennedy 1917–63: inaugural address, 20 January 1961

21 Greater love hath no man than this, that he lay down his friends for his life.

Jeremy Thorpe 1929– : on Harold Macmillan sacking seven of his Cabinet on 13 July 1962

22 I was determined that no British government should be brought down by the action of two tarts.

Harold Macmillan 1894–1986: comment on the Profumo affair, July 1963

23 A week is a long time in politics.

Harold Wilson 1916– : probably first said at the time of the 1964 sterling crisis. See Nigel Rees *Sayings of the Century* (1984)

24 Think of it! A second Chamber selected by the Whips. A seraglio of eunuchs.

Michael Foot 1913– : speech, House of Commons, 3 February 1969

25 The unpleasant and unacceptable face of capitalism.

Edward Heath 1916– : speech, House of Commons, 15 May 1973, on the Lonrho affair

26 We shall not be diverted from our course. To those waiting with bated breath for that favourite media catch-phrase, the

U-turn, I have only this to say. 'You turn if you want; the lady's not for turning.'

Margaret Thatcher 1925– : speech at Conservative Party Conference, 10 October 1980

27 There are three bodies no sensible man directly challenges: the Roman Catholic Church, the Brigade of Guards and the National Union of Mineworkers.

Harold Macmillan 1894–1986: in *Observer* 22 February 1981

28 First of all the Georgian silver goes, and then all that nice furniture that used to be in the saloon. Then the Canalettos go.

Harold Macmillan 1894–1986: speech on privatization to the Tory Reform Group, 8 November 1985

Political Parties

See also **Politicians, Politics**

1 Party-spirit, which at best is but the madness of many for the gain of a few.

Alexander Pope 1688–1744: letter to E. Blount, 27 August 1714

2 A Conservative Government is an organized hypocrisy.

Benjamin Disraeli 1804–81: speech, House of Commons 17 March 1845

3 Party is organized opinion.

Benjamin Disraeli 1804–81: speech at Oxford, 25 November 1864

4 Damn your principles! Stick to your party.

Benjamin Disraeli 1804–81: believed to have been said to Edward Bulwer-Lytton; attributed in E. Latham *Famous Sayings and their Authors* (1904)

5 CONSERVATIVE, *n.* A statesman who is enamoured of existing evils, as distinguished from the Liberal, who wishes to replace them with others.

Ambrose Bierce 1842–*c.*1914: *The Cynic's Word Book* (1906)

6 The more you read and observe about this Politics thing, you got to admit that each party is worse than the other. The one that's out always looks the best.
Will Rogers 1879–1935: *Illiterate Digest* (1924)

7 The mules of politics: without pride of ancestry, or hope of posterity.
John O'Connor Power b. 1846: of the Liberal Unionists, in H. H. Asquith *Memories and Reflections* (1928)

8 The language of priorities is the religion of Socialism.
Aneurin Bevan 1897–1960: speech at Labour Party Conference, 8 June 1949

9 If they [the Republicans] will stop telling lies about the Democrats, we will stop telling the truth about them.
Adlai Stevenson 1900–65: speech during 1952 Presidential campaign

10 There are some of us... who will fight and fight and fight again to save the Party we love.
Hugh Gaitskell 1906–63: speech at Labour Party Conference, 5 October 1960

11 Loyalty is the Tory's secret weapon.
Lord Kilmuir 1900–67: in Anthony Sampson *Anatomy of Britain* (1962)

12 A great party is not to be brought down because of a scandal by a woman of easy virtue and a proved liar.
Lord Hailsham 1907– : BBC television interview on the Profumo affair; in *The Times* 14 June 1963

Politicians

See also **Political Comment, Politics**

1 Get thee glass eyes;
And, like a scurvy politician, seem
To see the things thou dost not.
William Shakespeare 1564–1616: *King Lear* (1605–6)

2 The greatest art of a politician is to render vice serviceable to the cause of virtue.
Henry St John, 1st Viscount Bolingbroke 1678–1751:

comment (*c.*1728), in Joseph Spence *Observations, Anecdotes, and Characters* (1820)

3 All those men have their price.
Robert Walpole 1676–1745: of fellow parliamentarians; in W. Coxe *Memoirs of Sir Robert Walpole* (1798)

4 Your representative owes you, not his industry only, but his judgement; and he betrays, instead of serving you, if he sacrifices it to your opinion.
Edmund Burke 1729–97: speech, 3 November 1774

5 Not merely a chip of the old 'block', but the old block itself.
Edmund Burke 1729–97: on the younger Pitt's maiden speech, February 1781

6 If a due participation of office is a matter of right, how are vacancies to be obtained? Those by death are few; by resignation none.
Thomas Jefferson 1743–1826: letter to E. Shipman and others, 12 July 1801 (usually quoted 'Few die and none resign')

7 When a man assumes a public trust, he should consider himself as public property.
Thomas Jefferson 1743–1826: to Baron von Humboldt, 1807, in B. L. Rayner *Life of Jefferson* (1834)

8 What I want is men who will support me when I am in the wrong.
Lord Melbourne 1779–1848: replying to a politician who said 'I will support you as long as you are in the right'; in Lord David Cecil *Lord M* (1954)

9 A constitutional statesman is in general a man of common opinion and uncommon abilities.
Walter Bagehot 1826–77: *Biographical Studies* (1881)

10 The prospect of a lot
Of dull MPs in close proximity,
All thinking for themselves is what
No man can face with equanimity.
W. S. Gilbert 1836–1911: *Iolanthe* (1882)

11 A lath of wood painted to look like iron.
Otto von Bismarck 1815–98: describing Lord Salisbury; attributed, but vigorously denied by Sidney Whitman in *Personal Reminiscences of Prince Bismarck* (1902)

12 He knows nothing; and he thinks he knows everything. That points clearly to a political career.

George Bernard Shaw 1856–1950: *Major Barbara* (1907)

13 'Do you pray for the senators, Dr Hale?' 'No, I look at the senators and I pray for the country.'

Edward Everett Hale 1822–1909: Van Wyck Brooks *New England Indian Summer* (1940)

14 He [Labouchere] did not object to the old man always having a card up his sleeve, but he did object to his insinuating that the Almighty had placed it there.

Henry Labouchere 1831–1912: on Gladstone's 'frequent appeals to a higher power'; in Earl Curzon *Modern Parliamentary Eloquence* (1913)

15 We all know that Prime Ministers are wedded to the truth, but like other married couples they sometimes live apart.

Saki (H. H. Munro) 1870–1916: *The Unbearable Bassington* (1912)

16 For twenty years he [H. H. Asquith] has held a season-ticket on the line of least resistance and has gone wherever the train of events has carried him, lucidly justifying his position at whatever point he has happened to find himself.

Leo Amery 1873–1955: *Quarterly Review* July 1914

17 They [parliament] are a lot of hard-faced men who look as if they had done very well out of the war.

Stanley Baldwin 1867–1947: in J. M. Keynes *Economic Consequences of the Peace* (1919)

18 I thought he was a young man of promise, but it appears he is a young man of promises.

Arthur James Balfour 1848–1930: describing Churchill, in Winston Churchill *My Early Life* (1930)

19 I have waited 50 years to see the boneless wonder [Ramsay Macdonald] sitting on the Treasury Bench.

Winston Churchill 1874–1965: speech, House of Commons, 28 January 1931

20 The time has come for all good men to rise above principle.

Huey Long 1893–1935: attributed

21 a politician is an arse upon
which everyone has sat except a man.

e. e. cummings 1894–1962: *1 x 1* (1944) no. 10

22 The voice we heard was that of Mr Churchill but the mind was that of Lord Beaverbrook.
Clement Attlee 1883–1967: speech on radio, 5 June 1945

23 [Winston Churchill] does not talk the language of the 20th century but that of the 18th. He is still fighting Blenheim all over again. His only answer to a difficult situation is send a gun-boat.
Aneurin Bevan 1897–1960: speech at Labour Party Conference, 2 October 1951

24 Damn it all, you can't have the crown of thorns *and* the thirty pieces of silver.
Aneurin Bevan 1897–1960: on his position in the Labour Party, *c.*1956; in Michael Foot *Aneurin Bevan* vol. 2 (1973)

25 Forever poised between a cliché and an indiscretion.
Harold Macmillan 1894–1986: on the life of a Foreign Secretary; in *Newsweek* 30 April 1956

26 I am not going to spend any time whatsoever in attacking the Foreign Secretary… If we complain about the tune, there is no reason to attack the monkey when the organ grinder is present.
Aneurin Bevan 1897–1960: during a debate on the Suez crisis, House of Commons, 16 May 1957

27 A statesman is a politician who's been dead 10 or 15 years.
Harry S. Truman 1884–1972: in *New York World Telegram and Sun* 12 April 1958

28 Listening to a speech by Chamberlain is like paying a visit to Woolworth's: everything in its place and nothing above sixpence.
Aneurin Bevan 1897–1960: in Michael Foot *Aneurin Bevan* vol. 1 (1962)

29 Too clever by half.
Lord Salisbury 1893–1972: of Iain Macleod, Colonial Secretary; speech, House of Lords, 7 March 1961

30 It is the ability to foretell what is going to happen tomorrow, next week, next month, and next year. And to have the ability afterwards to explain why it didn't happen.
Winston Churchill 1874–1965: on the qualifications for becoming a politician, in B. Adler *Churchill Wit* (1965)

31 A sheep in sheep's clothing.
Winston Churchill 1874–1965: of Clement Attlee, in Lord Home *The Way the Wind Blows* (1976)

32 The Stag at Bay with the mentality of a fox at large.
Bernard Levin 1928– : of Harold Macmillan, in *The Pendulum Years* (1970)

33 A statesman is a politician who places himself at the service of the nation. A politician is a statesman who places the nation at his service.
Georges Pompidou 1911–74: in *Observer* 30 December 1973 'Sayings of the Year'

34 This is a rotten argument, but it should be good enough for their lordships on a hot summer afternoon.
Anonymous: annotation to a ministerial brief, said to have been read out inadvertently in the House of Lords; in Lord Home *The Way the Wind Blows* (1976)

35 It is not necessary that every time he rises he should give his famous imitation of a semi-house-trained polecat.
Michael Foot 1913– : of Norman Tebbit; speech, House of Commons, 2 March 1978

36 Like being savaged by a dead sheep.
Denis Healey 1917– : on being criticized by Geoffrey Howe in the House of Commons, 14 June 1978

37 It is, I think, good evidence of life after death.
Lord Soper 1903– : on the quality of debate in the House of Lords, in *Listener* 17 August 1978

38 A triumph of the embalmer's art.
Gore Vidal 1925– : of Ronald Reagan, in *Observer* 26 April 1981

39 The only safe pleasure for a parliamentarian is a bag of boiled sweets.
Julian Critchley 1930– : *Listener* 10 June 1982

40 Comrades, this man has a nice smile, but he's got iron teeth.
Andrei Gromyko 1909–89: on Mikhail Gorbachev; speech to Soviet Communist Party Central Committee, 11 March 1985

41 There are no true friends in politics. We are all sharks circling, and waiting, for traces of blood to appear in the water.
Alan Clark 1928– : diary, 30 November 1990

42 I could name eight people—half of those eight are barmy. How many apples short of a picnic?
John Major 1943- : on Tory critics, 19 September 1993

43 Being an MP is the sort of job all working-class parents want for their children—clean, indoors and no heavy lifting.
Diane Abbott 1953- : *Observer* 23 January 1994 'Sayings of the Week'

Politics

See also **Democracy, Government, Political Comment, Political Parties, Politicians, Voting**

1 Man is by nature a political animal.
Aristotle 384–322 BC: *Politics*

2 Most schemes of political improvement are very laughable things.
Samuel Johnson 1709–84: in James Boswell *Life of Johnson* (1791) 26 October 1769

3 Magnanimity in politics is not seldom the truest wisdom; and a great empire and little minds go ill together.
Edmund Burke 1729–97: *On Conciliation with America* (1775)

4 In politics the middle way is none at all.
John Adams 1735–1826: letter to Horatio Gates, 23 March 1776

5 In politics, what begins in fear usually ends in folly.
Samuel Taylor Coleridge 1772–1834: *Table Talk* (1835) 5 October 1830

6 'Two nations; between whom there is no intercourse and no sympathy; who are as ignorant of each other's habits, thoughts, and feelings, as if they were dwellers in different zones, or inhabitants of different planets...' 'You speak of—' said Egremont, hesitatingly, 'THE RICH AND THE POOR.'
Benjamin Disraeli 1804–81: *Sybil* (1845)

7 What is a communist? One who hath yearnings
For equal division of unequal earnings.
Ebenezer Elliott 1781–1849: 'Epigram' (1850)

8 Politics is the art of the possible.
Otto von Bismarck 1815–98: in conversation with Meyer von Waldeck, 11 August 1867

9 Politics is perhaps the only profession for which no preparation is thought necessary.
Robert Louis Stevenson 1850–94: *Familiar Studies of Men and Books* (1882)

10 Politics…has always been the systematic organization of hatreds.
Henry Brooks Adams 1838–1918: *The Education of Henry Adams* (1907)

11 Practical politics consists in ignoring facts.
Henry Brooks Adams 1838–1918: *The Education of Henry Adams* (1907)

12 If you want to succeed in politics, you must keep your conscience well under control.
David Lloyd George 1863–1945: in Lord Riddell's diary, 23 April 1919

13 Who? Whom?
Lenin 1870–1924: definition of political science, meaning 'Who will outstrip whom?'; in *Polnoe Sobranie Sochinenii* (1979) 17 October 1921, and elsewhere

14 I do not know which makes a man more conservative—to know nothing but the present, or nothing but the past.
John Maynard Keynes 1883–1946: *The End of Laissez-Faire* (1926)

15 I never dared be radical when young
For fear it would make me conservative when old.
Robert Frost 1874–1963: 'Precaution' (1936)

16 Politics is war without bloodshed while war is politics with bloodshed.
Mao Tse-tung 1893–1976: lecture, 1938, in *Selected Works* (1965)

17 Politics is the art of preventing people from taking part in affairs which properly concern them.
Paul Valéry 1871–1945: *Tel Quel 2* (1943)

18 Political language...is designed to make lies sound truthful and murder respectable, and to give an appearance of solidity to pure wind.
George Orwell 1903–50: *Shooting an Elephant* (1950) 'Politics and the English Language'

19 [Russian Communism is] the illegitimate child of Karl Marx and Catherine the Great.
Clement Attlee 1883–1967: speech at Aarhus University, 11 April 1956

20 Politics is not the art of the possible. It consists in choosing between the disastrous and the unpalatable.
J. K. Galbraith 1908– : letter to President Kennedy, 2 March 1962

21 The great nations have always acted like gangsters, and the small nations like prostitutes.
Stanley Kubrick 1928– : in *Guardian* 5 June 1963

22 Socialism can only arrive by bicycle.
José Antonio Viera Gallo 1943– : in Ivan Illich *Energy and Equity* (1974) epigraph

Pollution

See also **Environment**

1 Woe to her that is filthy and polluted, to the oppressing city!
Bible: Zephaniah

2 It goes so heavily with my disposition that this goodly frame, the earth, seems to me a sterile promontory; this most excellent canopy, the air, look you, this brave o'erhanging firmament, this majestical roof fretted with golden fire, why, it appears no other thing to me but a foul and pestilent congregation of vapours.
William Shakespeare 1564–1616: *Hamlet* (1601)

3 The river Rhine, it is well known,
Doth wash your city of Cologne;
But tell me, Nymphs, what power divine
Shall henceforth wash the river Rhine?
Samuel Taylor Coleridge 1772–1834: 'Cologne' (1834)

4 And all is seared with trade; bleared, smeared with toil;
And wears man's smudge and shares man's smell.
Gerard Manley Hopkins 1844–89: 'God's Grandeur' (written 1877)

5 Man has been endowed with reason, with the power to create, so that he can add to what he's been given. But up to now he hasn't been a creator, only a destroyer. Forests keep disappearing, rivers dry up, wild life's become extinct, the climate's ruined and the land grows poorer and uglier every day.
Anton Chekhov 1860–1904: *Uncle Vanya* (1897)

6 NOISE, *n.* A stench in the ear... The chief product and authenticating sign of civilization.
Ambrose Bierce 1842–?1914: *Devil's Dictionary* (1911)

7 The sanitary and mechanical age we are now entering makes up for the mercy it grants to our sense of smell by the ferocity with which it assails our sense of hearing. As usual, what we call 'progress' is the exchange of one nuisance for another nuisance.
Havelock Ellis 1859–1939: *Impressions and Comments* (1914)

8 Clear the air! clean the sky! wash the wind!
T. S. Eliot 1888–1965: *Murder in the Cathedral* (1935)

Poverty

1 What mean ye that ye beat my people to pieces, and grind the faces of the poor?
Bible: Isaiah

2 The misfortunes of poverty carry with them nothing harder to bear than that it makes men ridiculous.
Juvenal AD *c.*60–*c.*130: *Satires*

3 I want there to be no peasant in my kingdom so poor that he is unable to have a chicken in his pot every Sunday.
Henri IV 1553–1610: in H. de Péréfixe *Histoire de Henri le Grand* (1681)

4 Come away; poverty's catching.
Aphra Behn 1640–89: *The Rover* pt. 2 (1681)

5 There is no scandal like rags, nor any crime so shameful as poverty.
George Farquhar 1678–1707: *The Beaux' Stratagem* (1707)

6 Give me not poverty lest I steal.
Daniel Defoe 1660–1731: *Moll Flanders* (1721)

7 This mournful truth is ev'rywhere confessed,
Slow rises worth, by poverty depressed.
Samuel Johnson 1709–84: *London* (1738)

8 Resolve not to be poor: whatever you have, spend less. Poverty is a great enemy to human happiness; it certainly destroys liberty, and it makes some virtues impracticable, and others extremely difficult.
Samuel Johnson 1709–84: letter to James Boswell, 7 December 1782

9 The murmuring poor, who will not fast in peace.
George Crabbe 1754–1832: 'The Newspaper' (1785)

10 The poor are Europe's blacks.
Nicolas-Sébastien Chamfort 1741–94: *Maximes et Pensées* (1796)

11 Oh! God! that bread should be so dear,
And flesh and blood so cheap!
Thomas Hood 1799–1845: 'The Song of the Shirt' (1843)

12 Poverty is no disgrace to a man, but it is confoundedly inconvenient.
Sydney Smith 1771–1845: in J. Potter Briscoe *Sidney Smith: His Wit and Wisdom* (1900)

13 They [the poor] have to labour in the face of the majestic equality of the law, which forbids the rich as well as the poor to sleep under bridges, to beg in the streets, and to steal bread.
Anatole France 1844–1924: *Le Lys rouge* (1894)

14 The greatest of evils and the worst of crimes is poverty.
George Bernard Shaw 1856–1950: *Major Barbara* (1907) preface

15 There's nothing surer,
The rich get rich and the poor get children.
Gus Kahn 1886–1941 and **Raymond B. Egan** 1890–1952: 'Ain't We Got Fun' (1921 song)

16 Battles and sex are the only free diversions in slum life. Couple them with drink, which costs money, and you have the three principal outlets for that escape complex which is for ever working in the tenement dweller's subconscious mind.

Alexander McArthur and **H. Kingsley Long**: *No Mean City* (1935)

17 People don't resent having nothing nearly as much as too little.

Ivy Compton-Burnett 1884–1969: *A Family and a Fortune* (1939)

18 Sixteen tons, what do you get?
Another day older and deeper in debt.
Say brother, don't you call me 'cause I can't go
I owe my soul to the company store.

Merle Travis 1917–83: 'Sixteen Tons' (1947 song)

19 Anyone who has ever struggled with poverty knows how extremely expensive it is to be poor.

James Baldwin 1924–87: *Nobody Knows My Name* (1961)

Power

1 Man, proud man,
Drest in a little brief authority.

William Shakespeare 1564–1616: *Measure for Measure* (1604)

2 Men in great place are thrice servants: servants of the sovereign or state, servants of fame, and servants of business.

Francis Bacon 1561–1626: *Essays* (1625) 'Of Great Place'

3 All rising to great place is by a winding stair.

Francis Bacon 1561–1626: *Essays* (1625) 'Of Great Place'

4 Power is so apt to be insolent and Liberty to be saucy, that they are very seldom upon good terms.

George Savile, Marquess of Halifax 1633–95: *Political, Moral, and Miscellaneous Thoughts* (1750)

5 Nature has left this tincture in the blood,
That all men would be tyrants if they could.

Daniel Defoe 1660–1731: *The History of the Kentish Petition* (1712–13)

6 The strongest poison ever known
Came from Caesar's laurel crown.
William Blake 1757–1827: 'Auguries of Innocence' (*c.*1803)

7 Power tends to corrupt and absolute power corrupts absolutely.
Lord Acton 1834–1902: letter to Bishop Mandell Creighton, 3 April 1887

8 Whatever happens we have got
The Maxim Gun, and they have not.
Hilaire Belloc 1870–1953: *The Modern Traveller* (1898)

9 A man may build himself a throne of bayonets, but he cannot sit on it.
Dean Inge 1860–1954: *Philosophy of Plotinus* (1923), quoted by Boris Yeltsin at the time of the failed military coup in Russia, August 1991

10 The Pope! How many divisions has *he* got?
Joseph Stalin 1879–1953: on being asked to encourage Catholicism in Russia by way of conciliating the Pope, 13 May 1935

11 The hand that signed the treaty bred a fever,
And famine grew, and locusts came;
Great is the hand that holds dominion over
Man by a scribbled name.
Dylan Thomas 1914–53: 'The hand that signed the paper felled a city' (1936)

12 Political power grows out of the barrel of a gun.
Mao Tse-tung 1893–1976: speech, 6 November 1938

13 When he laughed, respectable senators burst with laughter,
And when he cried the little children died in the streets
W. H. Auden 1907–73: 'Epitaph on a Tyrant' (1940)

14 Who controls the past controls the future: who controls the present controls the past.
George Orwell 1903–50: *Nineteen Eighty-Four* (1949)

15 You only have power over people as long as you don't take *everything* away from them. But when you've robbed a man of *everything* he's no longer in your power—he's free again.
Alexander Solzhenitsyn 1918– : *The First Circle* (1968)

16 I'll make him an offer he can't refuse.
Mario Puzo 1920– : *The Godfather* (1969)

17 Power is the great aphrodisiac.
Henry Kissinger 1923– : in *New York Times* 19 January 1971

Practicality

1 Common sense is the best distributed commodity in the world, for every man is convinced that he is well supplied with it.
René Descartes 1596–1650: *Le Discours de la méthode* (1637)

2 'Tis use alone that sanctifies expense,
And splendour borrows all her rays from sense.
Alexander Pope 1688–1744: 'To Lord Burlington' (1731)

3 Whenever our neighbour's house is on fire, it cannot be amiss for the engines to play a little on our own.
Edmund Burke 1729–97: *Reflections on the Revolution in France* (1790)

4 Put your trust in God, my boys, and keep your powder dry.
Valentine Blacker 1728–1823: 'Oliver's Advice' (1856), often attributed to Oliver Cromwell himself

5 It's grand, and you canna expect to be baith grand and comfortable.
J. M. Barrie 1860–1937: *The Little Minister* (1891)

6 So I really think that American gentlemen are the best after all, because kissing your hand may make you feel very very good but a diamond and safire bracelet lasts forever.
Anita Loos 1893–1981: *Gentlemen Prefer Blondes* (1925)

7 Praise the Lord and pass the ammunition.
Howell Forgy 1908–83: at Pearl Harbor, 7 December 1941, while sailors passed ammunition by hand to the deck; later title of song by Frank Loesser, 1942

8 Life is too short to stuff a mushroom.
Shirley Conran 1932– : *Superwoman* (1975)

Praise

1 But when I tell him he hates flatterers,
He says he does, being then most flattered.
William Shakespeare 1564–1616: *Julius Caesar* (1599)

2 It has been well said that 'the arch-flatterer with whom all the petty flatterers have intelligence is a man's self.'
Francis Bacon 1561–1626: *Essays* (1625) 'Of Love'

3 He who discommendeth others obliquely commendeth himself.
Sir Thomas Browne 1605–82: *Christian Morals* (1716)

4 Damn with faint praise, assent with civil leer,
And without sneering, teach the rest to sneer.
Alexander Pope 1688–1744: 'An Epistle to Dr Arbuthnot' (1735)

5 All censure of a man's self is oblique praise. It is in order to shew how much he can spare.
Samuel Johnson 1709–84: in James Boswell *Life of Johnson* (1791) 25 April 1778

6 Imitation is the sincerest of flattery.
Charles Caleb Colton 1780–1832: *Lacon* (1820)

7 No flowers, by request.
Alfred Ainger 1837–1904: speech summarizing the principle of conciseness for contributors to the *Dictionary of National Biography*, 8 July 1897

8 The advantage of doing one's praising for oneself is that one can lay it on so thick and exactly in the right places.
Samuel Butler 1835–1902: *The Way of All Flesh* (1903)

9 I suppose flattery hurts no one, that is, if he doesn't inhale.
Adlai Stevenson 1900–65: television broadcast, 30 March 1952

Prayer

1 Ask, and it shall be given you; seek, and ye shall find; knock, and it shall be opened unto you.
Bible: St Matthew

2 My words fly up, my thoughts remain below:
Words without thoughts never to heaven go.
William Shakespeare 1564–1616: *Hamlet* (1601)

3 I throw myself down in my Chamber, and I call in, and invite God, and his Angels thither, and when they are there, I neglect God and his Angels, for the noise of a fly, for the rattling of a coach, for the whining of a door.
John Donne 1572–1631: sermon, 12 December 1626 'At the Funeral of Sir William Cokayne'

4 O Lord! thou knowest how busy I must be this day: if I forget thee, do not thou forget me.
Jacob Astley 1579–1652: prayer before the Battle of Edgehill

5 No praying, it spoils business.
Thomas Otway 1652–85: *Venice Preserved* (1682)

6 O God, if there be a God, save my soul, if I have a soul!
Anonymous: prayer of a common soldier before the battle of Blenheim (1704)

7 One single grateful thought raised to heaven is the most perfect prayer.
G. E. Lessing 1729–81: *Minna von Barnhelm* (1767)

8 He prayeth best, who loveth best
All things both great and small.
Samuel Taylor Coleridge 1772–1834: 'The Rime of the Ancient Mariner' (1798)

9 And lips say, 'God be pitiful,'
Who ne'er said, 'God be praised.'
Elizabeth Barrett Browning 1806–61: 'The Cry of the Human' (1844)

10 I am just going to pray for you at St Paul's, but with no very lively hope of success.
Sydney Smith 1771–1845: in H. Pearson *The Smith of Smiths* (1934)

11 If thou shouldst never see my face again,
Pray for my soul. More things are wrought by prayer
Than this world dreams of.
Alfred, Lord Tennyson 1809–92: *Idylls of the King* 'The Passing of Arthur' (1869)

12 Whatever a man prays for, he prays for a miracle. Every prayer reduces itself to this: Great God, grant that twice two be not four.

Ivan Turgenev 1818–83: *Poems in Prose* (1881) 'Prayer'

13 To lift up the hands in prayer gives God glory, but a man with a dungfork in his hand, a woman with a slop-pail, give him glory too.

Gerard Manley Hopkins 1844–89: 'The Principle or Foundation' (1882)

14 Bernard always had a few prayers in the hall and some whiskey afterwards as he was rarther pious but Mr Salteena was not very addicted to prayers so he marched up to bed.

Daisy Ashford 1881–1972: *The Young Visiters* (1919)

15 The wish for prayer is a prayer in itself.

Georges Bernanos 1888–1948: *Journal d'un curé de campagne* (1936)

16 The family that prays together stays together.

Al Scalpone: motto devised for the Roman Catholic Family Rosary Crusade, 1947

Prejudice

See also **Race**

1 You call me misbeliever, cut-throat dog,
And spit upon my Jewish gabardine,
And all for use of that which is mine own.

William Shakespeare 1564–1616: *The Merchant of Venice* (1596–8)

2 Drive out prejudices through the door, and they will return through the window.

Frederick the Great 1712–86: letter to Voltaire, 19 March 1771

3 Am I not a man and a brother.

Josiah Wedgwood 1730–95: legend on Wedgwood cameo, depicting a kneeling Negro slave in chains; reproduced in facsimile in E. Darwin *The Botanic Garden* pt. 1 (1791)

4 Prejudice is the child of ignorance.

William Hazlitt 1778–1830: 'On Prejudice' (1830)

5 Without the aid of prejudice and custom, I should not be able to find my way across the room.
William Hazlitt 1778–1830: 'On Prejudice' (1830)

6 Who's 'im, Bill?
A stranger!
'Eave 'arf a brick at 'im.
Punch: 1854

7 The only good Indian is a dead Indian.
Philip Henry Sheridan 1831–88: at Fort Cobb, January 1869 (attributed)

8 Bigotry may be roughly defined as the anger of men who have no opinions.
G. K. Chesterton 1874–1936: *Heretics* (1905)

9 PREJUDICE, *n.* A vagrant opinion without visible means of support.
Ambrose Bierce 1842–*c.*1914: *The Devil's Dictionary* (1911)

10 Minds are like parachutes. They only function when they are open.
James Dewar 1842–1923: attributed

11 How odd
Of God
To choose
The Jews.
William Norman Ewer 1885–1976: in *Week-End Book* (1924)

12 But not so odd
As those who choose
A Jewish God,
But spurn the Jews.
Cecil Browne 1932– : reply to verse by William Norman Ewer

13 I decline utterly to be impartial as between the fire brigade and the fire.
Winston Churchill 1874–1965: replying to complaints of his bias in editing the *British Gazette* during the General Strike; House of Commons, 7 July 1926

14 Bigotry tries to keep truth safe in its hand
With a grip that kills it.
Rabindranath Tagore 1861–1941: *Fireflies* (1928)

15 And wherefore is he wearing such a conscience-stricken air?
Oh they're taking him to prison for the colour of his hair.
A. E. Housman 1859–1936: *Collected Poems* (1939) 'Additional Poems' no. 18

16 Four legs good, two legs bad.
George Orwell 1903–50: *Animal Farm* (1945)

17 Being a star has made it possible for me to get insulted in places where the average Negro could never *hope* to go and get insulted.
Sammy Davis Jnr. 1925–90: in *Yes I Can* (1965)

18 It comes as a great shock around the age of 5, 6 or 7 to discover that the flag to which you have pledged allegiance, along with everybody else, has not pledged allegiance to you. It comes as a great shock to see Gary Cooper killing off the Indians and, although you are rooting for Gary Cooper, that the Indians are you.
James Baldwin 1924–87: speech at Cambridge University, 17 February 1965

The Present

1 *Carpe diem, quam minimum credula postero.*
Seize the day, put no trust in the future.
Horace 65–8 BC: *Odes*

2 Take therefore no thought for the morrow: for the morrow shall take thought for the things of itself. Sufficient unto the day is the evil thereof.
Bible: St Matthew

3 Can ye not discern the signs of the times?
Bible: St Matthew

4 What is love? 'tis not hereafter;
Present mirth hath present laughter;
What's to come is still unsure.
William Shakespeare 1564–1616: *Twelfth Night* (1601)

5 Praise they that will times past, I joy to see
My self now live: this age best pleaseth me.
Robert Herrick 1591–1674: 'The Present Time Best Pleaseth' (1648)

6 The present is the funeral of the past,
And man the living sepulchre of life.
John Clare 1793–1864: 'The present is the funeral of the past' (written 1845)

7 Ah, fill the cup:—what boots it to repeat
How time is slipping underneath our feet:
Unborn TO-MORROW, and dead YESTERDAY,
Why fret about them if TO-DAY be sweet!
Edward Fitzgerald 1809–83: *The Rubáiyát of Omar Khayyám* (1859)

8 The rule is, jam to-morrow and jam yesterday—but never jam today.
Lewis Carroll 1832–98: *Through the Looking-Glass* (1872)

9 He abhorred plastics, Picasso, sunbathing and jazz—everything in fact that had happened in his own lifetime.
Evelyn Waugh 1903–66: *The Ordeal of Gilbert Pinfold* (1957)

Pride

1 Pride goeth before destruction, and an haughty spirit before a fall.
Bible: Proverbs

2 For whosoever exalteth himself shall be abased; and he that humbleth himself shall be exalted.
Bible: St Luke. See also St Matthew

3 He that is down needs fear no fall,
He that is low no pride.
He that is humble ever shall
Have God to be his guide.
John Bunyan 1628–88: *The Pilgrim's Progress* (1684) 'Shepherd Boy's Song'

4 And the Devil did grin, for his darling sin
Is pride that apes humility.
Samuel Taylor Coleridge 1772–1834: 'The Devil's Thoughts' (1799)

5 We are so very 'umble.
Charles Dickens 1812–70: *David Copperfield* (1850) Uriah Heep

6 As for conceit, what man will do any good who is not conceited? Nobody holds a good opinion of a man who has a low opinion of himself.
Anthony Trollope 1815–82: *Orley Farm* (1862)

7 I can trace my ancestry back to a protoplasmal primordial atomic globule. Consequently, my family pride is something in-conceivable. I can't help it. I was born sneering.
W. S. Gilbert 1836–1911: *The Mikado* (1885)

8 The tumult and the shouting dies—
The captains and the kings depart—
Still stands Thine ancient Sacrifice,
An humble and a contrite heart.
Lord God of Hosts, be with us yet,
Lest we forget—lest we forget!
Rudyard Kipling 1865–1936: 'Recessional' (1897)

9 PLEASE ACCEPT MY RESIGNATION. I DON'T WANT TO BELONG TO ANY CLUB THAT WILL ACCEPT ME AS A MEMBER.
Groucho Marx 1895–1977: *Groucho and Me* (1959)

10 No one can make you feel inferior without your consent.
Eleanor Roosevelt 1884–1962: in *Catholic Digest* August 1960

11 In 1969 I published a small book on Humility. It was a pioneering work which has not, to my knowledge, been superseded.
Lord Longford 1905– : *Tablet* 22 January 1994

Progress

1 The thing that hath been, it is that which shall be; and that which is done is that which shall be done: and there is no new thing under the sun.
Bible: Ecclesiastes

2 We are like dwarfs on the shoulders of giants, so that we can see more than they, and things at a greater distance, not by virtue of any sharpness of sight on our part, or any physical distinction, but because we are carried high and raised up by their giant size.
Bernard of Chartres d. *c.*1130: in John of Salisbury *The Metalogicon* (1159)

3 If I have seen further it is by standing on the shoulders of giants.
Isaac Newton 1642–1727: letter to Robert Hooke, 5 February 1676

4 And he gave it for his opinion, that whoever could make two ears of corn or two blades of grass to grow upon a spot of ground where only one grew before, would deserve better of mankind, and do more essential service to his country than the whole race of politicians put together.
Jonathan Swift 1667–1745: *Gulliver's Travels* (1726)

5 Progress, therefore, is not an accident, but a necessity... It is a part of nature.
Herbert Spencer 1820–1903: *Social Statics* (1850)

6 Progress, man's distinctive mark alone,
Not God's, and not the beasts': God is, they are,
Man partly is and wholly hopes to be.
Robert Browning 1812–89: 'A Death in the Desert' (1864)

7 Belief in progress is a doctrine of idlers and Belgians. It is the individual relying upon his neighbours to do his work.
Charles Baudelaire 1821–67: *Journaux intimes* (1887), tr. Christopher Isherwood

8 The reasonable man adapts himself to the world: the unreasonable one persists in trying to adapt the world to himself. Therefore all progress depends on the unreasonable man.
George Bernard Shaw 1856–1950: *Man and Superman* (1903)

9 Want is one only of five giants on the road of reconstruction... the others are Disease, Ignorance, Squalor and Idleness.
William Henry Beveridge 1879–1963: *Social Insurance and Allied Services* (1942)

10 pity this busy monster, manunkind,
not. Progress is a comfortable disease.
e. e. cummings 1894–1962: *1* x *1* (1944) no. 14

11 Is it progress if a cannibal uses knife and fork?
Stanislaw Lec 1909–66: *Unkempt Thoughts* (1962)

Punishment see **Crime and Punishment**

Quotations

1 Some for renown on scraps of learning dote,
And think they grow immortal as they quote.
Edward Young 1683–1765: *The Love of Fame* (1725–8)

2 Every quotation contributes something to the stability or enlargement of the language.
Samuel Johnson 1709–84: on citations of usage in a dictionary; *Dictionary of the English Language* (1755) preface

3 Classical quotation is the *parole* of literary men all over the world.
Samuel Johnson 1709–84: in James Boswell *Life of Johnson* (1791) 8 May 1781

4 I hate quotation. Tell me what you know.
Ralph Waldo Emerson 1803–82: diary, May 1849

5 Next to the originator of a good sentence is the first quoter of it.
Ralph Waldo Emerson 1803–82: *Letters and Social Aims* (1876)

6 OSCAR WILDE: How I wish I had said that.
WHISTLER: You will, Oscar, you will.
James McNeill Whistler 1834–1903: in R. Ellman *Oscar Wilde* (1987)

7 What a good thing Adam had. When he said a good thing he knew nobody had said it before.
Mark Twain 1835–1910: *Notebooks* (1935)

8 An anthology is like all the plums and orange peel picked out of a cake.
Walter Raleigh 1861–1922: letter to Mrs Robert Bridges, 15 January 1915

9 It is a good thing for an uneducated man to read books of quotations.
Winston Churchill 1874–1965: *My Early Life* (1930)

10 The surest way to make a monkey of a man is to quote him.
Robert Benchley 1889–1945: *My Ten Years in a Quandary* (1936)

11 Famous remarks are very seldom quoted correctly.
Simeon Strunsky 1879–1948: *No Mean City* (1944)

12 The nice thing about quotes is that they give us a nodding acquaintance with the originator which is often socially impressive.
Kenneth Williams 1926–88: *Acid Drops* (1980)

Race

See also **Equality, Prejudice**

1 My mother bore me in the southern wild,
And I am black, but O! my soul is white;
White as an angel is the English child:
But I am black as if bereaved of light.
William Blake 1757–1827: 'The Little Black Boy' (1789)

2 I, too, sing America.
I am the darker brother.
They send me to eat in the kitchen
When company comes.
Langston Hughes 1902–67: 'I, Too' in *Survey Graphic* March 1925

3 I herewith commission you to carry out all preparations with regard to... a *total solution* of the Jewish question in those territories of Europe which are under German influence.
Hermann Goering 1893–1946: instructions to Heydrich, 31 July 1941

4 I want to be the white man's brother, not his brother-in-law.
Martin Luther King 1929–68: in *New York Journal-American* 10 September 1962

5 I have a dream that my four little children will one day live in a nation where they will not be judged by the colour of their skin but by the content of their character.
Martin Luther King 1929–68: speech at Civil Rights March in Washington, 28 August 1963

6 There are no 'white' or 'coloured' signs on the foxholes or graveyards of battle.
John F. Kennedy 1917–63: message to Congress on proposed Civil Rights Bill, 19 June 1963

7 Black is beautiful.
Anonymous: slogan of American civil rights campaigners, mid-1960s

Reading

See also **Books**

1 POLONIUS: What do you read, my lord?
HAMLET: Words, words, words.
William Shakespeare 1564–1616: *Hamlet* (1601)

2 Choose an author as you choose a friend.
Wentworth Dillon, Earl of Roscommon *c.*1633–1685: *Essay on Translated Verse* (1684)

3 He was wont to say that if he had read as much as other men, he should have known no more than other men.
John Aubrey 1626–97: *Brief Lives* 'Thomas Hobbes'

4 Reading is to the mind what exercise is to the body.
Richard Steele 1672–1729: *The Tatler* 18 March 1710

5 The bookful blockhead, ignorantly read,
With loads of learned lumber in his head.
Alexander Pope 1688–1744: *An Essay on Criticism* (1711)

6 A man ought to read just as inclination leads him; for what he reads as a task will do him little good.
Samuel Johnson 1709–84: in James Boswell *Life of Johnson* (1791) 14 July 1763

7 Digressions, incontestably, are the sunshine;—they are the life, the soul of reading.
Laurence Sterne 1713–68: *Tristram Shandy* (1759–67)

8 Much have I travelled in the realms of gold,
And many goodly states and kingdoms seen.
John Keats 1795–1821: 'On First Looking into Chapman's Homer' (1817)

9 The reading or non-reading a book—will never keep down a single petticoat.
Lord Byron 1788–1824: letter to Richard Hoppner, 29 October 1819

10 In science, read, by preference, the newest works; in literature, the oldest.
Edward Bulwer-Lytton 1803–73: *Caxtoniana* (1863) 'Hints on Mental Culture'

11 People say that life is the thing, but I prefer reading.
Logan Pearsall Smith 1865–1946: *Afterthoughts* (1931) 'Myself'

Reality

1 All theory, dear friend, is grey, but the golden tree of actual life springs ever green.
Johann Wolfgang von Goethe 1749–1832: *Faust* pt. 1 (1808)

2 It's as large as life, and twice as natural!
Lewis Carroll 1832–98: *Through the Looking-Glass* (1872)

3 Between the idea
And the reality
Between the motion
And the act
Falls the Shadow.
T. S. Eliot 1888–1965: 'The Hollow Men' (1925)

4 Human kind
Cannot bear very much reality.
T. S. Eliot 1888–1965: *Four Quartets* 'Burnt Norton' (1936)

5 The camera makes everyone a tourist in other people's reality, and eventually in one's own.
Susan Sontag 1933– : *New York Review of Books* 18 April 1974

Religion

See also **The Bible, The Church, God, Prayer**

1 Is that which is holy loved by the gods because it is holy, or is it holy because it is loved by the gods?
Plato 429–347 BC: *Euthyphro*

2 *Tantum religio potuit suadere malorum.*
So much wrong could religion induce.
Lucretius *c.*94–55 BC: *De Rerum Natura*

3 It is convenient that there be gods, and, as it is convenient, let us believe that there are.
Ovid 43 BC–AD *c.*17: *Ars Amatoria*

4 Render therefore unto Caesar the things which are Caesar's; and unto God the things that are God's.
Bible: St Matthew

5 Faith is the substance of things hoped for, the evidence of things not seen.
Bible: Hebrews

6 I count religion but a childish toy,
And hold there is no sin but ignorance.
Christopher Marlowe 1564–93: *The Jew of Malta* (*c.*1592)

7 'Twas only fear first in the world made gods.
Ben Jonson *c.*1573–1637: *Sejanus* (1603)

8 Had I but served my God with half the zeal
I served my king, he would not in mine age
Have left me naked to mine enemies.
William Shakespeare 1564–1616: *Henry VIII* (with John Fletcher, 1613)

9 A servant with this clause
Makes drudgery divine:
Who sweeps a room as for Thy laws
Makes that and th' action fine.
George Herbert 1593–1633: 'The Elixir' (1633)

10 A verse may find him, who a sermon flies,
And turn delight into a sacrifice.
George Herbert 1593–1633: 'The Church Porch' (1633)

11 One religion is as true as another.
Robert Burton 1577–1640: *Anatomy of Melancholy* (1621–51)

12 Persecution is a bad and indirect way to plant religion.
Sir Thomas Browne 1605–82: *Religio Medici* (1643)

13 As for those wingy mysteries in divinity and airy subtleties in religion, which have unhinged the brains of better heads, they never stretched the *pia mater* of mine; methinks there be not impossibilities enough in religion for an active faith.
Sir Thomas Browne 1605–82: *Religio Medici* (1643)

14 Men have lost their reason in nothing so much as their religion, wherein stones and clouts make martyrs.
Sir Thomas Browne 1605–82: *Hydriotaphia* (Urn Burial, 1658)

15 They are for religion when in rags and contempt; but I am for him when he walks in his golden slippers, in the sunshine and with applause.
John Bunyan 1628–88: *The Pilgrim's Progress* (1678) Mr By-Ends

16 'Men of sense are really but of one religion.'... 'Pray, my lord, what religion is that which men of sense agree in?' 'Madam,' says the earl immediately, 'men of sense never tell it.'
1st Earl of Shaftesbury 1621–83: in Bishop Gilbert Burnet *History of My Own Time* vol. 1 (1724)

17 Wherever God erects a house of prayer,
The Devil always builds a chapel there;
And 'twill be found, upon examination,
The latter has the largest congregation.
Daniel Defoe 1660–1731: *The True-Born Englishman* (1701)

18 We have just enough religion to make us hate, but not enough to make us love one another.
Jonathan Swift 1667–1745: *Thoughts on Various Subjects* (1711)

19 I went to America to convert the Indians; but oh, who shall convert me?
John Wesley 1703–91: diary, 24 January 1738

20 In all ages of the world, priests have been enemies of liberty.
David Hume 1711–76: *Essays, Moral, Political, and Literary* (1875) 'Of the Parties of Great Britain' (1741–2)

21 Putting moral virtues at the highest, and religion at the lowest, religion must still be allowed to be a collateral security, at least, to virtue; and every prudent man will sooner trust to two securities than to one.
Lord Chesterfield 1694–1773: *Letters to his Son* (1774) 8 January 1750

22 Orthodoxy is my doxy; heterodoxy is another man's doxy.
Bishop William Warburton 1698–1779: to Lord Sandwich, in Joseph Priestley *Memoirs* (1807)

23 The various modes of worship, which prevailed in the Roman world, were all considered by the people as equally true; by the philosopher, as equally false; and by the magistrate, as equally useful. And thus toleration produced not only mutual indulgence, but even religious concord.

Edward Gibbon 1737–94: *Decline and Fall of the Roman Empire* (1776–88)

24 My country is the world, and my religion is to do good.

Thomas Paine 1737–1809: *The Rights of Man* pt. 2 (1792)

25 Any system of religion that has any thing in it that shocks the mind of a child cannot be a true system.

Thomas Paine 1737–1809: *The Age of Reason* pt. 1 (1794)

26 The dust of creeds outworn.

Percy Bysshe Shelley 1792–1822: *Prometheus Unbound* (1820)

27 In vain with lavish kindness
The gifts of God are strown;
The heathen in his blindness
Bows down to wood and stone.

Bishop Reginald Heber 1783–1826: 'From Greenland's icy mountains' (1821 hymn)

28 Christians have burnt each other, quite persuaded
That all the Apostles would have done as they did.

Lord Byron 1788–1824: *Don Juan* (1819–24)

29 He who begins by loving Christianity better than Truth will proceed by loving his own sect or church better than Christianity, and end by loving himself better than all.

Samuel Taylor Coleridge 1772–1834: *Aids to Reflection* (1825)

30 Religion... is the opium of the people.

Karl Marx 1818–83: *A Contribution to the Critique of Hegel's Philosophy of Right* (1843–4)

31 Things have come to a pretty pass when religion is allowed to invade the sphere of private life.

Lord Melbourne 1779–1848: on hearing an evangelical sermon; in G. W. E. Russell *Collections and Recollections* (1898)

32 Thou shalt have one God only; who
Would be at the expense of two?

Arthur Hugh Clough 1819–61: 'The Latest Decalogue' (1862)

33 If I am obliged to bring religion into after-dinner toasts (which indeed does not seem quite the thing) I shall drink…to Conscience first, and to the Pope afterwards.
Cardinal Newman 1801–90: *Letter Addressed to the Duke of Norfolk…* (1875)

34 Scratch the Christian and you find the pagan—spoiled.
Israel Zangwill 1864–1926: *Children of the Ghetto* (1892)

35 I can't talk religion to a man with bodily hunger in his eyes.
George Bernard Shaw 1856–1950: *Major Barbara* (1907)

36 Wot prawce Selvytion nah?
George Bernard Shaw 1856–1950: *Major Barbara* (1907)

37 A Christian is a man who feels
Repentance on a Sunday
For what he did on Saturday
And is going to do on Monday.
Thomas Russell Ybarra b. 1880: 'The Christian' (1909)

38 The Christian ideal has not been tried and found wanting. It has been found difficult; and left untried.
G. K. Chesterton 1874–1936: *What's Wrong with the World* (1910)

39 SAINT, *n.* A dead sinner revised and edited.
Ambrose Bierce 1842–*c.*1914: *The Devil's Dictionary* (1911)

40 There is no expeditious road
To pack and label men for God,
And save them by the barrel-load.
Some may perchance, with strange surprise,
Have blundered into Paradise.
Francis Thompson 1859–1907: 'A Judgement in Heaven' (1913)

41 So many gods, so many creeds,
So many paths that wind and wind,
While just the art of being kind
Is all the sad world needs.
Ella Wheeler Wilcox 1855–1919: 'The World's Need'

42 Religion is the frozen thought of men out of which they build temples.
Jiddu Krishnamurti d. 1986: in *Observer* 22 April 1928 'Sayings of the Week'

43 Christianity is the most materialistic of all great religions.
Archbishop William Temple 1881–1944: *Readings in St John's Gospel* (1939)

44 Better authentic mammon than a bogus god.
Louis MacNeice 1907–63: *Autumn Journal* (1939)

45 Science without religion is lame, religion without science is blind.
Albert Einstein 1879–1955: *Science, Philosophy and Religion* (1941)

Revenge

1 Vengeance is mine; I will repay, saith the Lord.
Bible: Romans

2 Men should be either treated generously or destroyed, because they take revenge for slight injuries—for heavy ones they cannot.
Niccolò Machiavelli 1469–1527: *The Prince* (1513)

3 Revenge is a kind of wild justice, which the more man's nature runs to, the more ought law to weed it out.
Francis Bacon 1561–1626: *Essays* (1625) 'Of Revenge'

4 Heaven has no rage, like love to hatred turned,
Nor Hell a fury, like a woman scorned.
William Congreve 1670–1729: *The Mourning Bride* (1697)

5 Sweet is revenge—especially to women.
Lord Byron 1788–1824: *Don Juan* (1819–24)

6 The Germans...are going to be squeezed as a lemon is squeezed—until the pips squeak.
Eric Geddes 1875–1937: speech at Cambridge, 10 December 1918

Revolution and Rebellion

1 A desperate disease requires a dangerous remedy.
Guy Fawkes 1570–1606: 6 November 1605

2 Rebellion to tyrants is obedience to God.
John Bradshaw 1602–59: suppositious epitaph

3 *Après nous le déluge.*
After us the deluge.
Madame de Pompadour 1721–64: in Mme du Hausset *Mémoires* (1824)

4 A little rebellion now and then is a good thing.
Thomas Jefferson 1743–1826: letter to James Madison, 30 January 1787

5 How much the greatest event it is that ever happened in the world! and how much the best!
Charles James Fox 1749–1806: on the fall of the Bastille; letter to R. Fitzpatrick, 30 July 1789

6 Kings will be tyrants from policy when subjects are rebels from principle.
Edmund Burke 1729–97: *Reflections on the Revolution in France* (1790)

7 Bliss was it in that dawn to be alive,
But to be young was very heaven!
William Wordsworth 1770–1850: 'The French Revolution, as it Appeared to Enthusiasts' (1809); also *The Prelude* (1850)

8 A share in two revolutions is living to some purpose.
Thomas Paine 1737–1809: in E. Foner *Tom Paine and Revolutionary America* (1976)

9 Maximilien Robespierre was nothing but the hand of Jean Jacques Rousseau, the bloody hand that drew from the womb of time the body whose soul Rousseau had created.
Heinrich Heine 1797–1856: *Zur Geschichte der Religion und Philosophie in Deutschland* (1834)

10 *J'ai vécu.*
I survived.
Abbé Emmanuel Joseph Sieyès 1748–1836: when asked what he had done during the French Revolution

11 I will die like a true-blue rebel. Don't waste any time in mourning—organize.
Joe Hill 1879–1915: farewell telegram prior to his death by firing squad; in *Salt Lake* (Utah) *Tribune* 19 November 1915

12 While there is a lower class, I am in it; while there is a criminal element, I am of it; while there is a soul in prison, I am not free.

Eugene Victor Debs 1855–1926: speech at his trial for sedition in Cleveland, Ohio, 14 September 1918

13 I have seen the future; and it works.

Lincoln Steffens 1866–1936: following a visit to the Soviet Union in 1919, in *Letters* (1938)

14 All civilization has from time to time become a thin crust over a volcano of revolution.

Havelock Ellis 1859–1939: *Little Essays of Love and Virtue* (1922)

15 'There won't be any revolution in America,' said Isadore. Nikitin agreed. 'The people are all too clean. They spend all their time changing their shirts and washing themselves. You can't feel fierce and revolutionary in a bathroom.'

Eric Linklater 1899–1974: *Juan in America* (1931)

16 What is a rebel? A man who says no.

Albert Camus 1913–60: *The Rebel* (1953)

17 All modern revolutions have ended in a reinforcement of the State.

Albert Camus 1913–60: *The Rebel* (1953)

18 The most radical revolutionary will become a conservative on the day after the revolution.

Hannah Arendt 1906–75: *New Yorker* 12 September 1970

19 Revolutions are celebrated when they are no longer dangerous.

Pierre Boulez 1925– : *Guardian* 13 January 1989

Royalty

1 I know I have the body of a weak and feeble woman, but I have the heart and stomach of a king, and of a king of England too.

Elizabeth I 1533–1603: speech to the troops at Tilbury on the approach of the Armada, 1588

2 Not all the water in the rough rude sea
Can wash the balm from an anointed king.

William Shakespeare 1564–1616: *Richard II* (1595)

3 Uneasy lies the head that wears a crown.
William Shakespeare 1564–1616: *Henry IV, Part 2* (1597)

4 And what have kings that privates have not too,
Save ceremony, save general ceremony?
William Shakespeare 1564–1616: *Henry V* (1599)

5 There's such divinity doth hedge a king,
That treason can but peep to what it would.
William Shakespeare 1564–1616: *Hamlet* (1601)

6 He is the fountain of honour.
Francis Bacon 1561–1626: *An Essay of a King* (1642)

7 *L'État c'est moi.*
I am the State.
Louis XIV 1638–1715: before the Parlement de Paris, 13 April 1655; probably apocryphal

8 I see it is impossible for the King to have things done as cheap as other men.
Samuel Pepys 1633–1703: diary, 21 July 1662

9 Titles are shadows, crowns are empty things,
The good of subjects is the end of kings.
Daniel Defoe 1660–1731: *The True-Born Englishman* (1701)

10 The Right Divine of Kings to govern wrong.
Alexander Pope 1688–1744: *The Dunciad* (1742)

11 Born and educated in this country, I glory in the name of Briton.
George III 1738–1820: *The King's Speech on Opening the Session*, 18 November 1760

12 The influence of the Crown has increased, is increasing, and ought to be diminished.
John Dunning 1731–83: resolution passed in the House of Commons, 6 April 1780

13 I will be good.
Queen Victoria 1819–1901: on being shown a chart of the line of succession, 11 March 1830; in Theodore Martin *The Prince Consort* (1875)

14 It has been said, not truly, but with a possible approximation to truth, that in 1802 every hereditary monarch was insane.
Walter Bagehot 1826–77: *The English Constitution* (1867)

15 The Sovereign has, under a constitutional monarchy such as ours, three rights—the right to be consulted, the right to encourage, the right to warn.
Walter Bagehot 1826–77: *The English Constitution* (1867)

16 We must not let in daylight upon magic.
Walter Bagehot 1826–77: *The English Constitution* (1867)

17 Everyone likes flattery; and when you come to Royalty you should lay it on with a trowel.
Benjamin Disraeli 1804–81: to Matthew Arnold, in G. W. E. Russell *Collections and Recollections* (1898)

18 George the Third
Ought never to have occurred.
One can only wonder
At so grotesque a blunder.
Edmund Clerihew Bentley 1875–1956: 'George the Third' (1929)

19 Soon there will be only five Kings left—the King of England, the King of Spades, the King of Clubs, the King of Hearts and the King of Diamonds.
King Farouk 1920–65: comment in Cairo, 1948

20 For seventeen years he did nothing at all but kill animals and stick in stamps.
Harold Nicolson 1886–1968: of King George V; diary, 17 August 1949

21 Royalty is the gold filling in a mouthful of decay.
John Osborne 1929– : 'They call it cricket' in T. Maschler (ed.) *Declaration* (1957)

Satisfaction and Discontent

1 He is well paid that is well satisfied.
William Shakespeare 1564–1616: *The Merchant of Venice* (1596–8)

2 'Tis just like a summer birdcage in a garden; the birds that are without despair to get in, and the birds that are within despair, and are in a consumption, for fear they shall never get out.
John Webster *c.*1580–*c.*1625: *The White Devil* (1612)

3 Plain living and high thinking are no more:
The homely beauty of the good old cause
Is gone.

William Wordsworth 1770–1850: 'O friend! I know not which way I must look' (1807)

4 A book of verses underneath the bough,
A jug of wine, a loaf of bread—and Thou
Beside me singing in the wilderness—
And wilderness were paradise enow.

Edward Fitzgerald 1809–83: *The Rubáiyát of Omar Khayyám* (1879 ed.)

5 Content is disillusioning to behold: what is there to be content about?

Virginia Woolf 1882–1941: diary, 5 May 1920

6 He spoke with a certain what-is-it in his voice, and I could see that, if not actually disgruntled, he was far from being gruntled.

P. G. Wodehouse 1881–1975: *The Code of the Woosters* (1938)

7 If one cannot catch the bird of paradise, better take a wet hen.

Nikita Khrushchev 1894–1971: in *Time* 6 January 1958

8 These are the days when men of all social disciplines and all political faiths seek the comfortable and the accepted . . . in minor modification of the scriptural parable, the bland lead the bland.

J. K. Galbraith 1908– : *The Affluent Society* (1958)

Science

See also **Inventions and Discoveries, Life Sciences, Technology**

1 *Felix qui potuit rerum cognoscere causas.*
Lucky is he who has been able to understand the causes of things.

Virgil 70–19 BC: *Georgics*

2 That all things are changed, and that nothing really perishes, and that the sum of matter remains exactly the same, is sufficiently certain.

Francis Bacon 1561–1626: *Cogitationes de Natura Rerum*

3 He had been eight years upon a project for extracting sun-beams out of cucumbers, which were to be put into vials hermetically sealed, and let out to warm the air in raw inclement summers.

Jonathan Swift 1667–1745: *Gulliver's Travels* (1726)

4 It may be so, there is no arguing against facts and experiments.

Isaac Newton 1642–1727: when told of an experiment which appeared to destroy his theory, as reported by John Conduit, 1726; in D. Brewster *Memoirs of Sir Isaac Newton* (1855)

5 Nature, and Nature's laws lay hid in night.
God said, *Let Newton be!* and all was light.

Alexander Pope 1688–1744: 'Epitaph: Intended for Sir Isaac Newton' (1730)

6 If ignorance of nature gave birth to the Gods, knowledge of nature is destined to destroy them.

Paul Henri, Baron d'Holbach 1723–89: *Système de la Nature* (1770)

7 Science moves, but slowly slowly, creeping on from point to point.

Alfred, Lord Tennyson 1809–92: 'Locksley Hall' (1842)

8 Where observation is concerned, chance favours only the prepared mind.

Louis Pasteur 1822–95: address, 7 December 1854

9 Science is nothing but trained and organized common sense, differing from the latter only as a veteran may differ from a raw recruit: and its methods differ from those of common sense only as far as the guardsman's cut and thrust differ from the manner in which a savage wields his club.

T. H. Huxley 1825–95: *Collected Essays* (1893–4) 'The Method of Zadig'

10 The great tragedy of Science—the slaying of a beautiful hypothesis by an ugly fact.

T. H. Huxley 1825–95: *Collected Essays* (1893–4) 'Biogenesis and Abiogenesis'

11 Science is built up of facts, as a house is built of stones; but an accumulation of facts is no more a science than a heap of stones is a house.

Henri Poincaré 1854–1912: *Science and Hypothesis* (1905)

12 The outcome of any serious research can only be to make two questions grow where one question grew before.

Thorstein Veblen 1857–1929: *University of California Chronicle* (1908)

13 In science the credit goes to the man who convinces the world, not to the man to whom the idea first occurs.

Francis Darwin 1848–1925: *Eugenics Review* April 1914 'Francis Galton'

14 There was a young lady named Bright,
Whose speed was far faster than light;
She set out one day
In a relative way
And returned on the previous night.

Arthur Buller 1874–1944: 'Relativity' (1923)

15 It did not last: the Devil howling 'Ho!
Let Einstein be!' restored the status quo.

J. C. Squire 1884–1958: 'In continuation of Pope on Newton' (1926)

16 We believe a scientist because he can substantiate his remarks, not because he is eloquent and forcible in his enunciation. In fact, we distrust him when he seems to be influencing us by his manner.

I. A. Richards 1893–1979: *Science and Poetry* (1926)

17 I ask you to look both ways. For the road to a knowledge of the stars leads through the atom; and important knowledge of the atom has been reached through the stars.

Arthur Eddington 1882–1944: *Stars and Atoms* (1928)

18 If someone points out to you that your pet theory of the universe is in disagreement with Maxwell's equations—then so much the worse for Maxwell's equations. If it is found to be contradicted by observation—well, these experimentalists do bungle things sometimes. But if your theory is found to be against the second law of thermodynamics I can give you no hope; there is nothing for it but to collapse in deepest humiliation.

Arthur Eddington 1882–1944: *The Nature of the Physical World* (1928)

19 It is much easier to make measurements than to know exactly what you are measuring.
J. W. N. Sullivan 1886–1937: comment, 1928; in R. L. Weber *More Random Walks in Science* (1982)

20 Science means simply the aggregate of all the recipes that are always successful. The rest is literature.
Paul Valéry 1871–1945: *Moralités* (1932)

21 All science is either physics or stamp collecting.
Ernest Rutherford 1871–1937: in J. B. Birks *Rutherford at Manchester* (1962)

22 We haven't got the money, so we've got to think!
Ernest Rutherford 1871–1937: in *Bulletin of the Institute of Physics* (1962), as recalled by R. V. Jones

23 It was quite the most incredible event that has ever happened to me in my life. It was almost as incredible as if you fired a 15-inch shell at a piece of tissue paper and it came back and hit you.
Ernest Rutherford 1871–1937: on the back-scattering effect of metal foil on alpha-particles, in E. N. da C. Andrade *Rutherford and the Nature of the Atom* (1964)

24 The aim of science is not to open the door to infinite wisdom, but to set a limit to infinite error.
Bertolt Brecht 1898–1956: *Life of Galileo* (1939)

25 Science without religion is lame, religion without science is blind.
Albert Einstein 1879–1955: *Science, Philosophy and Religion* (1941)

26 Now who is responsible for this work of development on which so much depends? To whom must the praise be given? To the boys in the back rooms. They do not sit in the limelight. But they are the men who do the work.
Lord Beaverbrook 1879–1964: *Listener* 27 March 1941

27 The physicists have known sin; and this is a knowledge which they cannot lose.
J. Robert Oppenheimer 1904–67: lecture at Massachusetts Institute of Technology, 25 November 1947

28 A new scientific truth does not triumph by convincing its opponents and making them see the light, but rather because its opponents eventually die, and a new generation grows up that is familiar with it.

Max Planck 1858–1947: *A Scientific Autobiography* (1949) tr. F. Gaynor

29 Aristotle maintained that women have fewer teeth than men; although he was twice married, it never occurred to him to verify this statement by examining his wives' mouths.

Bertrand Russell 1872–1970: *Impact of Science on Society* (1952)

30 We do not know why they [elementary particles] have the masses they do; we do not know why they transform into another the way they do; we do not know anything! The one concept that stands like the Rock of Gibraltar in our sea of confusion is the Pauli [exclusion] principle.

George Gamow 1904–68: *Scientific American* July 1959

31 When I find myself in the company of scientists, I feel like a shabby curate who has strayed by mistake into a drawing room full of dukes.

W. H. Auden 1907–73: *The Dyer's Hand* (1963)

32 It is more important to have beauty in one's equations than to have them fit experiment.

Paul Dirac 1902–84: *Scientific American* May 1963

33 If politics is the art of the possible, research is surely the art of the soluble. Both are immensely practical-minded affairs.

Peter Medawar 1915–87: *New Statesman* 19 June 1964

34 It is a good morning exercise for a research scientist to discard a pet hypothesis every day before breakfast.

Konrad Lorenz 1903–89: *On Aggression* (1966), tr. M. Latzke

35 If an elderly but distinguished scientist says that something is possible he is almost certainly right, but if he says that it is impossible he is very probably wrong.

Arthur C. Clarke 1917– : in *New Yorker* 9 August 1969

36 Basic research is what I am doing when I don't know what I am doing.

Werner von Braun 1912–77: in R. L. Weber *A Random Walk in Science* (1973)

37 The essence of science: ask an impertinent question, and you are on the way to a pertinent answer.

Jacob Bronowski 1908–74: *The Ascent of Man* (1973)

Scotland and the Scots

1 It came with a lass, and it will pass with a lass.
James V 1512–42: of the crown of Scotland, on learning of the birth of Mary Queen of Scots, December 1542; in Robert Lindsay of Pitscottie (*c.*1500–65) *History of Scotland* (1728)

2 Now there's ane end of ane old song.
James Ogilvy, 1st Earl of Seafield 1664–1730: as he signed the engrossed exemplification of the Act of Union, 1706, in *The Lockhart Papers* (1817)

3 The noblest prospect which a Scotchman ever sees, is the high road that leads him to England!
Samuel Johnson 1709–84: in James Boswell *Life of Johnson* (1791) 6 July 1763

4 A Scotchman must be a very sturdy moralist, who does not love Scotland better than truth.
Samuel Johnson 1709–84: *A Journey to the Western Islands of Scotland* (1775)

5 My heart's in the Highlands, my heart is not here;
My heart's in the Highlands a-chasing the deer.
Robert Burns 1759–96: 'My Heart's in the Highlands' (1790)

6 Scots, wha hae wi' Wallace bled,
Scots, wham Bruce has aften led,
Welcome to your gory bed,—
Or to victorie.
Robert Burns 1759–96: 'Robert Bruce's March to Bannockburn' (1799)

7 O Caledonia! stern and wild,
Meet nurse for a poetic child!
Sir Walter Scott 1771–1832: *The Lay of the Last Minstrel* (1805)

8 A land of meanness, sophistry, and mist.
Lord Byron 1788–1824: 'The Curse of Minerva' (1812)

9 From the lone shieling of the misty island
Mountains divide us, and the waste of seas—
Yet still the blood is strong, the heart is Highland,
And we in dreams behold the Hebrides!
John Galt 1779–1839: 'Canadian Boat Song', attributed

10 That knuckle-end of England—that land of Calvin, oat-cakes, and sulphur.
Sydney Smith 1771–1845: in Lady Holland *Memoir* (1855)

11 There are few more impressive sights in the world than a Scotsman on the make.
J. M. Barrie 1860–1937: *What Every Woman Knows* (1908)

12 Scotland, land of the omnipotent No.
Alan Bold 1943– : 'A Memory of Death' (1969)

The Sea

1 One deep calleth another, because of the noise of the water-pipes: all thy waves and storms are gone over me.
Bible: Psalm 42

2 They that go down to the sea in ships: and occupy their business in great waters.
Bible: Psalm 107

3 Now would I give a thousand furlongs of sea for an acre of barren ground.
William Shakespeare 1564–1616: *The Tempest* (1611)

4 Full fathom five thy father lies;
Of his bones are coral made:
Those are pearls that were his eyes:
Nothing of him that doth fade,
But doth suffer a sea-change
Into something rich and strange.
William Shakespeare 1564–1616: *The Tempest* (1611)

5 Be pleased to receive into thy Almighty and most gracious protection the persons of us thy servants, and the Fleet in which we serve.
Book of Common Prayer 1662: *Forms of Prayer to be Used at Sea*

6 No man will be a sailor who has contrivance enough to get himself into a jail; for being in a ship is being in a jail, with the chance of being drowned...A man in a jail has more room, better food, and commonly better company.
Samuel Johnson 1709–84: in James Boswell *Life of Johnson* (1791) 16 March 1759

7 Water, water, everywhere,
And all the boards did shrink;
Water, water, everywhere,
Nor any drop to drink.
Samuel Taylor Coleridge 1772–1834: 'The Rime of the Ancient Mariner' (1798)

8 A willing foe and sea room.
Anonymous: naval toast in the time of Nelson

9 Roll on, thou deep and dark blue Ocean—roll!
Ten thousand fleets sweep over thee in vain;
Man marks the earth with ruin—his control
Stops with the shore.
Lord Byron 1788–1824: *Childe Harold's Pilgrimage* (1812–18)

10 If blood be the price of admiralty,
Lord God, we ha' paid in full!
Rudyard Kipling 1865–1936: 'The Song of the Dead' (1896)

11 I must go down to the sea again, to the lonely sea and the sky,
And all I ask is a tall ship and a star to steer her by.
John Masefield 1878–1967: 'Sea Fever' (misprinted 'I must down to the seas' in the 1902 original)

12 A ship, an isle, a sickle moon—
With few but with how splendid stars
The mirrors of the sea are strewn
Between their silver bars!
James Elroy Flecker 1884–1915: 'A Ship, an Isle, and a Sickle Moon' (1913)

13 Don't talk to me about naval tradition. It's nothing but rum, sodomy, and the lash.
Winston Churchill 1874–1965: in P. Gretton *Former Naval Person* (1968)

The Seasons

1 Sumer is icumen in,
Lhude sing cuccu!
Groweth sed, and bloweth med,
And springeth the wude nu.
Anonymous: 'Cuckoo Song' (*c.*1250)

2 At Christmas I no more desire a rose
Than wish a snow in May's new-fangled mirth;
But like of each thing that in season grows.
William Shakespeare 1564–1616: *Love's Labour's Lost* (1595)

3 When icicles hang by the wall,
And Dick the shepherd blows his nail,
And Tom bears logs into the hall,
And milk comes frozen home in pail.
William Shakespeare 1564–1616: *Love's Labour's Lost* (1595)

4 That time of year thou mayst in me behold
When yellow leaves, or none, or few, do hang
Upon those boughs which shake against the cold,
Bare ruined choirs, where late the sweet birds sang.
William Shakespeare 1564–1616: sonnet 73 (1609)

5 It was no summer progress. A cold coming they had of it, at this time of the year; just, the worst time of the year, to take a journey, and specially a long journey, in. The ways deep, the weather sharp, the days short, the sun farthest off *in solstitio brumali*, the very dead of Winter.
Bishop Lancelot Andrewes 1555–1626: *Of the Nativity* (1622) Sermon 15

6 Sweet spring, full of sweet days and roses,
A box where sweets compacted lie.
George Herbert 1593–1633: 'Virtue' (1633)

7 I sing of brooks, of blossoms, birds, and bowers:
Of April, May, of June, and July-flowers.
I sing of May-poles, Hock-carts, wassails, wakes,
Of bride-grooms, brides, and of their bridal-cakes.
Robert Herrick 1591–1674: *Hesperides* (1648)

8 The way to ensure summer in England is to have it framed and glazed in a comfortable room.
Horace Walpole 1717–97: letter to Revd William Cole, 28 May 1774

9 Snowy, Flowy, Blowy,
Showery, Flowery, Bowery,
Hoppy, Croppy, Droppy,
Breezy, Sneezy, Freezy.
George Ellis 1753–1815: 'The Twelve Months'

10 O wild West Wind, thou breath of Autumn's being,
Thou, from whose unseen presence the leaves dead

Are driven, like ghosts from an enchanter fleeing,

Yellow, and black, and pale, and hectic red,
Pestilence-stricken multitudes.
Percy Bysshe Shelley 1792–1822: 'Ode to the West Wind' (1819)

11 Season of mists and mellow fruitfulness,
Close bosom-friend of the maturing sun;
Conspiring with him how to load and bless
With fruit the vines that round the thatch-eaves run.
John Keats 1795–1821: 'To Autumn' (1820)

12 The English winter—ending in July,
To recommence in August.
Lord Byron 1788–1824: *Don Juan* (1819–24)

13 Summer has set in with its usual severity.
Samuel Taylor Coleridge 1772–1834: quoted in letter from Charles Lamb to Vincent Novello, 9 May 1826

14 No sun—no moon!
No morn—no noon
No dawn—no dusk—no proper time of day…
No shade, no shine, no butterflies, no bees,
No fruits, no flowers, no leaves, no birds,—
November!
Thomas Hood 1799–1845: 'No!' (1844)

15 When the hounds of spring are on winter's traces,
The mother of months in meadow or plain
Fills the shadows and windy places
With lisp of leaves and ripple of rain;
Algernon Charles Swinburne 1837–1909: *Atalanta in Calydon* (1865)

16 Coldly, sadly descends
The autumn evening. The Field
Strewn with its dank yellow drifts
Of withered leaves, and the elms,
Fade into dimness apace.
Matthew Arnold 1822–88: 'Rugby Chapel, November 1857' (1867)

17 May is a pious fraud of the almanac.
James Russell Lowell 1819–91: 'Under the Willows' (1869)

18 In the bleak mid-winter
Frosty wind made moan,

Earth stood hard as iron,
Water like a stone.

Christina Rossetti 1830–94: 'Mid-Winter' (1875)

19 Though worlds of wanwood leafmeal lie.

Gerard Manley Hopkins 1844–89: 'Spring and Fall: to a young child' (written 1880)

20 In winter I get up at night
And dress by yellow candle-light.
In summer, quite the other way,—
I have to go to bed by day.

Robert Louis Stevenson 1850–94: 'Bed in Summer' (1885)

21 Loveliest of trees, the cherry now
Is hung with bloom along the bough,
And stands about the woodland ride
Wearing white for Eastertide.

A. E. Housman 1859–1936: *A Shropshire Lad* (1896)

22 Winter is icummen in,
Lhude sing Goddamm,
Raineth drop and staineth slop,
And how the wind doth ramm!

Ezra Pound 1885–1972: 'Ancient Music' (1917)

23 April is the cruellest month, breeding
Lilacs out of the dead land.

T. S. Eliot 1888–1965: *The Waste Land* (1922)

24 The autumn always gets me badly, as it breaks into colours. I want to go south, where there is no autumn, where the cold doesn't crouch over one like a snow-leopard waiting to pounce. The heart of the North is dead, and the fingers of cold are corpse fingers.

D. H. Lawrence 1885–1930: letter to J. Middleton Murry, 3 October 1924

25 Summer time an' the livin' is easy,
Fish are jumpin' an' the cotton is high.

Du Bose Heyward 1885–1940 and **Ira Gershwin** 1896–1983: 'Summertime' (1935 song)

26 In fact, it is about five o'clock in an evening that the first hour of spring strikes—autumn arrives in the early morning, but spring at the close of a winter day.

Elizabeth Bowen 1899–1973: *Death of the Heart* (1938)

27 June is bustin' out all over.
Oscar Hammerstein II 1895–1960: title of song (1945)

28 What of October, that ambiguous month, the month of tension, the unendurable month?
Doris Lessing 1919– : *Martha Quest* (1952)

29 August is a wicked month.
Edna O'Brien 1936– : title of novel (1965)

Secrets

1 Stolen waters are sweet, and bread eaten in secret is pleasant.
Bible: Proverbs

2 I would not open windows into men's souls.
Elizabeth I 1533–1603: oral tradition, the words possibly originating in a letter drafted by Bacon

3 Love and a cough cannot be hid.
George Herbert 1593–1633: *Outlandish Proverbs* (1640)

4 For secrets are edged tools,
And must be kept from children and from fools.
John Dryden 1631–1700: *Sir Martin Mar-All* (1667)

5 It is public scandal that constitutes offence, and to sin in secret is not to sin at all.
Molière 1622–73: *Le Tartuffe* (1669)

6 Love ceases to be a pleasure, when it ceases to be a secret.
Aphra Behn 1640–89: *The Lover's Watch* (1686)

7 I know that's a secret, for it's whispered every where.
William Congreve 1670–1729: *Love for Love* (1695)

8 I shall be but a short time tonight. I have seldom spoken with greater regret, for my lips are not yet unsealed. Were these troubles over I would make a case, and I guarantee that not a man would go into the lobby against us.
Stanley Baldwin 1867–1947: speech, House of Commons, 10 December 1935, on the Abyssinian crisis (usually quoted 'My lips are sealed')

9 We dance round in a ring and suppose,
But the Secret sits in the middle and knows.
Robert Frost 1874–1963: 'The Secret Sits' (1942)

10 Once the toothpaste is out of the tube, it is awfully hard to get it back in.
H. R. Haldeman 1929– : comment to John Wesley Dean on Watergate affair, 8 April 1973

The Self

1 Who is it that can tell me who I am?
William Shakespeare 1564–1616: *King Lear* (1605–6)

2 The self is hateful.
Blaise Pascal 1623–62: *Pensées* (1670)

3 Thus God and nature linked the gen'ral frame,
And bade self-love and social be the same.
Alexander Pope 1688–1744: *An Essay on Man* Epistle 3 (1733)

4 It is not contrary to reason to prefer the destruction of the whole world to the scratching of my finger.
David Hume 1711–76: *A Treatise upon Human Nature* (1739)

5 If a man does not keep pace with his companions, perhaps it is because he hears a different drummer. Let him step to the music which he hears, however measured or far away.
Henry David Thoreau 1817–62: *Walden* (1854)

6 Do I contradict myself?
Very well then I contradict myself,
(I am large, I contain multitudes.)
Walt Whitman 1819–92: 'Song of Myself' (written 1855)

7 It is easy—terribly easy—to shake a man's faith in himself. To take advantage of that to break a man's spirit is devil's work.
George Bernard Shaw 1856–1950: *Candida* (1898)

8 Rose is a rose is a rose is a rose, is a rose.
Gertrude Stein 1874–1946: *Sacred Emily* (1913)

9 We are all serving a life-sentence in the dungeon of self.
Cyril Connolly 1903–74: *The Unquiet Grave* (1944)

10 The image of myself which I try to create in my own mind in order that I may love myself is very different from the image which I try to create in the minds of others in order that they may love me.

W. H. Auden 1907–73: *The Dyer's Hand* (1963)

11 Some thirty inches from my nose
The frontier of my Person goes,
And all the untilled air between
Is private *pagus* or demesne.

W. H. Auden 1907–73: 'Prologue: the Birth of Architecture' (1966)

12 My one regret in life is that I am not someone else.

Woody Allen 1935– : epigraph to Eric Lax *Woody Allen and his Comedy* (1975)

Self-Knowledge

1 Know thyself.

Anonymous: inscribed on the temple of Apollo at Delphi; Plato ascribes the saying to the Seven Wise Men

2 I do not know whether I was then a man dreaming I was a butterfly, or whether I am now a butterfly dreaming I am a man.

Chuang-tzu (or Zhuangzi) *c.*369–286 BC: *Chuang Tzu* (1889) tr. H. A. Giles

3 Why beholdest thou the mote that is in thy brother's eye, but considerest not the beam that is in thine own eye?

Bible: St Matthew

4 For the good that I would I do not: but the evil which I would not, that I do.

Bible: Romans

5 The greatest thing in the world is to know how to be oneself.

Montaigne 1533–92: *Essais* (1580)

6 This above all: to thine own self be true,
And it must follow, as the night the day,
Thou canst not then be false to any man.

William Shakespeare 1564–1616: *Hamlet* (1601)

7 But I do nothing upon my self, and yet I am mine own *Executioner*.

John Donne 1572–1631: *Devotions upon Emergent Occasions* (1624)

8 All our knowledge is, ourselves to know.
Alexander Pope 1688–1744: *An Essay on Man* Epistle 4 (1734)

9 O wad some Pow'r the giftie gie us
To see oursels as others see us!
It wad frae mony a blunder free us,
And foolish notion.
Robert Burns 1759–96: 'To a Louse' (1786)

10 The Vision of Christ that thou dost see
Is my vision's greatest enemy
Thine has a great hook nose like thine
Mine has a snub nose like to mine.
William Blake 1757–1827: *The Everlasting Gospel* (*c.*1818)

11 How little do we know that which we are!
How less what we may be!
Lord Byron 1788–1824: *Don Juan* (1819–24)

12 I do not know myself, and God forbid that I should.
Johann Wolfgang von Goethe 1749–1832: J. P. Eckermann *Gespräche mit Goethe* (1836–48) 10 April 1829

13 The tragedy of a man who has found himself out.
J. M. Barrie 1860–1937: *What Every Woman Knows* (1908)

14 Between the ages of twenty and forty we are engaged in the process of discovering who we are, which involves learning the difference between accidental limitations which it is our duty to outgrow and the necessary limitations of our nature beyond which we cannot trespass with impunity.
W. H. Auden 1907–73: *The Dyer's Hand* (1963)

15 There are few things more painful than to recognise one's own faults in others.
John Wells 1936– : *Observer* 23 May 1982 'Sayings of the Week'

Sex

See also **Love, Marriage**

1 Someone asked Sophocles, 'How is your sex-life now? Are you still able to have a woman?' He replied, 'Hush, man; most gladly indeed am I rid of it all, as though I had escaped from a mad and savage master.'

Sophocles *c.*496–406 BC: in Plato *Republic*

2 Delight of lust is gross and brief
And weariness treads on desire.

Petronius d. AD 65: in A. Baehrens *Poetae Latini Minores* (1882), tr. H. Waddell

3 Give me chastity and continency—but not yet!

St Augustine of Hippo AD 354–430: *Confessions* (AD 397–8)

4 Licence my roving hands, and let them go,
Behind, before, above, between, below.
O my America, my new found land,
My kingdom, safeliest when with one man manned.

John Donne 1572–1631: 'To His Mistress Going to Bed' (*c.*1595)

5 Is it not strange that desire should so many years outlive performance?

William Shakespeare 1564–1616: *Henry IV, Part 2* (1597)

6 This is the monstruosity in love, lady, that the will is infinite, and the execution confined; that the desire is boundless, and the act a slave to limit.

William Shakespeare 1564–1616: *Troilus and Cressida* (1602)

7 Die: die for adultery! No:
The wren goes to't, and the small gilded fly
Does lecher in my sight.
Let copulation thrive.

William Shakespeare 1564–1616: *King Lear* (1605–6)

8 The expense of spirit in a waste of shame
Is lust in action.

William Shakespeare 1564–1616: sonnet 129 (1609)

9 This trivial and vulgar way of coition; it is the foolishest act a wise man commits in all his life, nor is there any thing that will more deject his cooled imagination, when he shall consider what an odd and unworthy piece of folly he hath committed.

Sir Thomas Browne 1605–82: *Religio Medici* (1643)

10 He in a few minutes ravished this fair creature, or at least would have ravished her, if she had not, by a timely compliance, prevented him.
Henry Fielding 1707–54: *Jonathan Wild* (1743)

11 The Duke returned from the wars today and did pleasure me in his top-boots.
Sarah, Duchess of Marlborough 1660–1744: attributed in various forms. See I. Butler *Rule of Three* (1967)

12 I'll come no more behind your scenes, David; for the silk stockings and white bosoms of your actresses excite my amorous propensities.
Samuel Johnson 1709–84: in James Boswell *Life of Johnson* (1791) 1750

13 The pleasure is momentary, the position ridiculous, and the expense damnable.
Lord Chesterfield 1694–1773: of sex; attributed

14 'Tisn't beauty, so to speak, nor good talk necessarily. It's just It. Some women'll stay in a man's memory if they once walked down a street.
Rudyard Kipling 1865–1936: *Traffics and Discoveries* (1904)

15 When I hear his steps outside my door I lie down on my bed, close my eyes, open my legs, and think of England.
Lady Hillingdon 1857–1940: diary, 1912

16 Chastity—the most unnatural of all the sexual perversions.
Aldous Huxley 1894–1963: *Eyeless in Gaza* (1936)

17 Pornography is the attempt to insult sex, to do dirt on it.
D. H. Lawrence 1885–1930: *Phoenix* (1936) 'Pornography and Obscenity'

18 But did thee feel the earth move?
Ernest Hemingway 1899–1961: *For Whom the Bell Tolls* (1940)

19 It doesn't matter what you do in the bedroom as long as you don't do it in the street and frighten the horses.
Mrs Patrick Campbell 1865–1940: in D. Fielding *Duchess of Jermyn Street* (1964)

20 Continental people have sex life; the English have hot-water bottles.
George Mikes 1912– : *How to be an Alien* (1946)

21 He said it was artificial respiration, but now I find I am to have his child.
Anthony Burgess 1917–93: *Inside Mr Enderby* (1963)

22 I have heard some say...[homosexual] practices are allowed in France and in other NATO countries. We are not French, and we are not other nationals. We are British, thank God!
Field Marshal Montgomery 1887–1976: speaking on the 2nd reading of the Sexual Offences Bill; House of Lords, 24 May 1965

23 Literature is mostly about having sex and not much about having children. Life is the other way round.
David Lodge 1935– : *The British Museum is Falling Down* (1965)

24 The orgasm has replaced the Cross as the focus of longing and the image of fulfilment.
Malcolm Muggeridge 1903–90: *Tread Softly* (1966)

25 You were born with your legs apart. They'll send you to the grave in a Y-shaped coffin.
Joe Orton 1933–67: *What the Butler Saw* (1969)

26 Is sex dirty? Only if it's done right.
Woody Allen 1935– : *Everything You Always Wanted to Know about Sex* (1972 film)

27 Sexual intercourse began
In nineteen sixty-three
(Which was rather late for me)—
Between the end of the *Chatterley* ban
And the Beatles' first LP.
Philip Larkin 1922–85: 'Annus Mirabilis' (1974)

28 On bisexuality: It immediately doubles your chances for a date on Saturday night.
Woody Allen 1935– : *New York Times* 1 December 1975

29 Seduction is often difficult to distinguish from rape. In seduction, the rapist bothers to buy a bottle of wine.
Andrea Dworkin 1946– : speech to women at *Harper & Row*, 1976; in *Letters from a War Zone* (1988)

30 That [sex] was the most fun I ever had without laughing.
Woody Allen 1935– : *Annie Hall* (1977 film, with Marshall Brickman)

31 Don't knock masturbation. It's sex with someone I love.
Woody Allen 1935- : *Annie Hall* (1977 film, with Marshall Brickman)

Shakespeare

See also **Acting and the Theatre**

1 Reader, look
Not on his picture, but his book.
Ben Jonson *c.*1573-1637: 'On the Portrait of Shakespeare' (1623)

2 He was not of an age, but for all time!
Ben Jonson *c.*1573-1637: 'To the Memory of... Shakespeare' (1623)

3 His mind and hand went together: And what he thought, he uttered with that easiness, that we have scarce received from him a blot.
John Heming 1556-1630 and **Henry Condell** d. 1627: First Folio Shakespeare (1623) preface

4 Whatsoever he [Shakespeare] penned, he never blotted out a line. My answer hath been 'Would he had blotted a thousand.'
Ben Jonson *c.*1573-1637: *Timber* (1641)

5 Was there ever such stuff as great part of Shakespeare? Only one must not say so!
George III 1738-1820: in Fanny Burney's diary, 19 December 1785

6 Others abide our question. Thou art free.
We ask and ask: Thou smilest and art still,
Out-topping knowledge.
Matthew Arnold 1822-88: 'Shakespeare' (1849)

7 He had read Shakespeare and found him weak in chemistry.
H. G. Wells 1866-1946: 'Lord of the Dynamos' (1927)

8 When I read Shakespeare I am struck with wonder
That such trivial people should muse and thunder
In such lovely language.
D. H. Lawrence 1885-1930: 'When I Read Shakespeare' (1929)

9 Shakespeare is so tiring. You never get a chance to sit down unless you're a king.

Josephine Hull ?1886–1957: in *Time* 16 November 1953

Sickness and Health

See also **Medicine**

1 Life's not just being alive, but being well.

Martial AD *c.*40–*c.*104: *Epigrammata*

2 *Mens sana in corpore sano.*
A sound mind in a sound body.

Juvenal AD *c.*60–*c.*130: *Satires*

3 Diseases desperate grown,
By desperate appliances are relieved
Or not at all.

William Shakespeare 1564–1616: *Hamlet* (1601)

4 Bid them wash their faces,
And keep their teeth clean.

William Shakespeare 1564–1616: *Coriolanus* (1608)

5 Look to your health; and if you have it, praise God, and value it next to a good conscience; for health is the second blessing that we mortals are capable of; a blessing that money cannot buy.

Izaak Walton 1593–1683: *The Compleat Angler* (1653)

6 Cured yesterday of my disease,
I died last night of my physician.

Matthew Prior 1664–1721: 'The Remedy Worse than the Disease' (1727)

7 If a lot of cures are suggested for a disease, it means that the disease is incurable.

Anton Chekhov 1860–1904: *The Cherry Orchard* (1904), tr. E. Fen

8 Does it matter?—losing your sight?…
There's such splendid work for the blind;
And people will always be kind,
As you sit on the terrace remembering
And turning your face to the light.

Siegfried Sassoon 1886–1967: 'Does it Matter?' (1918)

9 Human nature seldom walks up to the word 'cancer'.
Rudyard Kipling 1865–1936: *Debits and Credits* (1926)

10 Early to rise and early to bed makes a male healthy and wealthy and dead.
James Thurber 1894–1961: 'The Shrike and the Chipmunks' in *New Yorker* 18 February 1939

11 Coughs and sneezes spread diseases. Trap the germs in your handkerchief.
Anonymous: Second World War health slogan (1942)

12 Venerable Mother Toothache
Climb down from the white battlements,
Stop twisting in your yellow fingers
The fourfold rope of nerves.
John Heath-Stubbs 1918– : 'A Charm Against the Toothache' (1954)

13 I know the colour rose, and it is lovely,
But not when it ripens in a tumour;
And healing greens, leaves and grass, so springlike,
In limbs that fester are not springlike.
Dannie Abse 1923– : 'Pathology of Colours' (1968)

14 Illness is the night-side of life, a more onerous citizenship. Everyone who is born holds dual citizenship, in the kingdom of the well and in the kingdom of the sick.
Susan Sontag 1933– : *New York Review of Books* 26 January 1978

15 [AIDS was] an illness in stages, a very long flight of steps that led assuredly to death, but whose every step represented a unique apprenticeship. It was a disease that gave death time to live and its victims time to die, time to discover time, and in the end to discover life.
Hervé Guibert 1955–91: *To the Friend who did not Save my Life* (1991) tr. Linda Coverdale

Silence

1 Silence is the virtue of fools.
Francis Bacon 1561–1626: *De Dignitate et Augmentis Scientiarum* (1623)

2 Thou still unravished bride of quietness,
Thou foster-child of silence and slow time.
John Keats 1795–1821: 'Ode on a Grecian Urn' (1820)

3 Elected Silence, sing to me
And beat upon my whorlèd ear.
Gerard Manley Hopkins 1844–89: 'The Habit of Perfection' (written 1866)

4 If we had a keen vision and feeling of all ordinary human life, it would be like hearing the grass grow and the squirrel's heart beat, and we should die of that roar which lies on the other side of silence.
George Eliot 1819–80: *Middlemarch* (1871–2)

5 Deep is the silence, deep
On moon-washed apples of wonder.
John Drinkwater 1882–1937: 'Moonlit Apples' (1917)

6 People talking without speaking
People hearing without listening...
'Fools,' said I, 'You do not know
Silence like a cancer grows.'
Paul Simon 1942– : 'Sound of Silence' (1964 song)

Singing

See also **Music**

1 If a man were permitted to make all the ballads, he need not care who should make the laws of a nation.
Andrew Fletcher of Saltoun 1655–1716: 'Conversation concerning a Right Regulation of Government...' (1704)

2 Today if something is not worth saying, people sing it.
Pierre-Augustin Caron de Beaumarchais 1732–99: *The Barber of Seville* (1775)

3 Swans sing before they die: 'twere no bad thing
Should certain persons die before they sing.
Samuel Taylor Coleridge 1772–1834: 'On a Volunteer Singer' (1834)

4 You think that's noise—you ain't heard nuttin' yet!
Al Jolson 1886–1950: in a café, competing with the din from a neighbouring building site, 1906

5 An unalterable and unquestioned law of the musical world required that the German text of French operas sung by Swedish artists should be translated into Italian for the clearer understanding of English-speaking audiences.
Edith Wharton 1862–1937: *The Age of Innocence* (1920)

6 Sing 'em muck! It's all they can understand!
Dame Nellie Melba 1861–1931: advice to Dame Clara Butt, prior to her departure for Australia; in W. H. Ponder *Clara Butt* (1928)

7 Opera is when a guy gets stabbed in the back and, instead of bleeding, he sings.
Ed Gardner 1901–63: in *Duffy's Tavern* (US radio programme, 1940s)

The Skies

1 The moon's an arrant thief,
And her pale fire she snatches from the sun.
William Shakespeare 1564–1616: *Timon of Athens* (1607)

2 Busy old fool, unruly sun,
Why dost thou thus,
Through windows, and through curtains call on us?
John Donne 1572–1631: 'The Sun Rising'

3 *Eppur si muove.*
But it does move.
Galileo Galilei 1564–1642: attributed to Galileo after his recantation, that the earth moves around the sun, in 1632

4 The eternal silence of these infinite spaces [the heavens] terrifies me.
Blaise Pascal 1623–62: *Pensées* (1670)

5 ...The evening star,
Love's harbinger.
John Milton 1608–74: *Paradise Lost* (1667)

6 And like a dying lady, lean and pale,
Who totters forth, wrapped in a gauzy veil.
Percy Bysshe Shelley 1792–1822: 'The Waning Moon' (1824)

7 Look at the stars! look, look up at the skies!
O look at all the fire-folk sitting in the air!
The bright boroughs, the circle-citadels there!
Gerard Manley Hopkins 1844–89: 'The Starlight Night' (written 1877)

8 Slowly, silently, now the moon
Walks the night in her silver shoon.
Walter de la Mare 1873–1956: 'Silver' (1913)

9 Stars scribble on our eyes the frosty sagas,
The gleaming cantos of unvanquished space.
Hart Crane 1899–1932: 'Cape Hatteras' (1930)

10 Houston, Tranquillity Base here. The Eagle has landed.
Buzz Aldrin 1930– : on landing on the moon, 21 July 1969

Sleep and Dreams

1 The sleep of a labouring man is sweet.
Bible: Ecclesiastes

2 Care-charmer Sleep, son of the sable Night,
Brother to Death, in silent darkness born.
Samuel Daniel 1563–1619: *Delia* (1592) sonnet 54

3 O God! I could be bounded in a nut-shell, and count myself a king of infinite space, were it not that I have bad dreams.
William Shakespeare 1564–1616: *Hamlet* (1601)

4 Methought I heard a voice cry, 'Sleep no more!
Macbeth does murder sleep,' the innocent sleep,
Sleep that knits up the ravelled sleave of care.
William Shakespeare 1564–1616: *Macbeth* (1606)

5 What hath night to do with sleep?
John Milton 1608–74: *Comus* (1637)

6 We term sleep a death, and yet it is waking that kills us, and destroys those spirits which are the house of life.
Sir Thomas Browne 1605–82: *Religio Medici* (1643)

7 And so to bed.
Samuel Pepys 1633–1703: diary, 20 April 1660

8 That children dream not in the first half year, that men dream not in some countries, are to me sick men's dreams, dreams out of the ivory gate, and visions before midnight.
Sir Thomas Browne 1605–82: 'On Dreams'

9 Tired Nature's sweet restorer, balmy sleep!
Edward Young 1683–1765: *Night Thoughts* (1742–5)

10 The dream of reason produces monsters.
Goya 1746–1828: *Los Caprichos* (1799)

11 The quick Dreams,
The passion-wingèd Ministers of thought.
Percy Bysshe Shelley 1792–1822: *Adonais* (1821)

12 When you're lying awake with a dismal headache, and repose is taboo'd by anxiety,
I conceive you may use any language you choose to indulge in, without impropriety.
W. S. Gilbert 1836–1911: *Iolanthe* (1882)

13 The interpretation of dreams is the royal road to a knowledge of the unconscious activities of the mind.
Sigmund Freud 1856–1939: *The Interpretation of Dreams* (2nd ed., 1909) often quoted 'Dreams are the royal road to the unconscious'

14 ...The cool kindliness of sheets, that soon
Smooth away trouble; and the rough male kiss
Of blankets.
Rupert Brooke 1887–1915: 'The Great Lover' (1914)

15 The armoured cars of dreams, contrived to let us do so many a dangerous thing.
Elizabeth Bishop 1911–79: 'Sleeping Standing Up' (1946)

16 All the things one has forgotten scream for help in dreams.
Elias Canetti 1905– : *Die Provinz der Menschen* (1973)

Society

1 No man is an Island, entire of it self.
John Donne 1572–1631: *Devotions upon Emergent Occasions* (1624)

2 During the time men live without a common power to keep them all in awe, they are in that condition which is called war; and such a war as is of every man against every man.

Thomas Hobbes 1588–1679: *Leviathan* (1651)

3 That action is best, which procures the greatest happiness for the greatest numbers.

Francis Hutcheson 1694–1746: *Inquiry into the Original...* (1725)

4 Society is indeed a contract...it becomes a partnership not only between those who are living, but between those who are living, those who are dead, and those who are to be born.

Edmund Burke 1729–97: *Reflections on the Revolution in France* (1790)

5 Only in the state does man have a rational existence...Man owes his entire existence to the state, and has his being within it alone.

G. W. F. Hegel 1770–1831: *Lectures on the Philosophy of World History: Introduction* (1830), tr. H. B. Nisbet

6 The greatest happiness of the greatest number is the foundation of morals and legislation.

Jeremy Bentham 1748–1832: *Commonplace Book*. Bentham claimed that either Joseph Priestley (1733–1804) or Cesare Beccaria (1738–94) passed on 'the sacred truth'

7 *La propriété c'est le vol.*
Property is theft.

Pierre-Joseph Proudhon 1809–65: *Qu'est-ce que la propriété?* (1840)

8 From each according to his abilities, to each according to his needs.

Karl Marx 1818–83: *Critique of the Gotha Programme* (written 1875, but of earlier origin). See Morelly *Code de la nature* (1755), and J. Blanc *Organisation du travail* (1839)

9 The Social Contract is nothing more or less than a vast conspiracy of human beings to lie to and humbug themselves and one another for the general Good. Lies are the mortar that bind the savage individual man into the social masonry.

H. G. Wells 1866–1946: *Love and Mr Lewisham* (1900)

10 Hunger allows no choice
To the citizen or the police;
We must love one another or die.
W. H. Auden 1907–73: 'September 1, 1939' (1940)

11 We shall have to walk and live a Woolworth life hereafter.
Harold Nicolson 1886–1968: anticipating the aftermath of the Second World War; diary, 4 June 1941

12 The city is not a concrete jungle, it is a human zoo.
Desmond Morris 1928– : *The Human Zoo* (1969)

13 We started off trying to set up a small anarchist community, but people wouldn't obey the rules.
Alan Bennett 1934– : *Getting On* (1972)

14 In a consumer society there are inevitably two kinds of slaves: the prisoners of addiction and the prisoners of envy.
Ivan Illich 1926– : *Tools for Conviviality* (1973)

15 There is no such thing as Society. There are individual men and women, and there are families.
Margaret Thatcher 1925– : in *Woman's Own* 31 October 1987

Solitude

1 It is not good that the man should be alone; I will make him an help meet for him.
Bible: Genesis

2 He who is unable to live in society, or who has no need because he is sufficient for himself, must be either a beast or a god.
Aristotle 384–322 BC: *Politics*

3 A man should keep for himself a little back shop, all his own, quite unadulterated, in which he establishes his true freedom and chief place of seclusion and solitude.
Montaigne 1533–92: *Essais* (1580)

4 If you are idle, be not solitary; if you are solitary, be not idle.
Samuel Johnson 1709–84: letter to Boswell, 27 October 1779

5 I am monarch of all I survey,
My right there is none to dispute.
William Cowper 1731–1800: 'Verses Supposed to be Written by Alexander Selkirk' (1782)

6 To fly from, need not be to hate, mankind.
Lord Byron 1788–1824: *Childe Harold's Pilgrimage* (1812–18)

7 Never less alone than when alone.
Samuel Rogers 1763–1855: 'Human Life' (1819)

8 I long for scenes where man hath never trod
A place where woman never smiled or wept
There to abide with my Creator God.
John Clare 1793–1864: 'I Am' (1848)

9 Yes! in the sea of life enisled,
With echoing straits between us thrown,
Dotting the shoreless watery wild,
We mortal millions live *alone.*
Matthew Arnold 1822–88: 'To Marguerite—Continued' (1852)

10 Ships that pass in the night, and speak each other in passing;
Only a signal shown and a distant voice in the darkness;
So on the ocean of life we pass and speak one another,
Only a look and a voice; then darkness again and a silence.
Henry Wadsworth Longfellow 1807–82: *Tales of a Wayside Inn* pt. 3 (1874)

11 Down to Gehenna or up to the Throne,
He travels the fastest who travels alone.
Rudyard Kipling 1865–1936: *The Story of the Gadsbys* (1890)

12 Before I built a wall I'd ask to know
What I was walling in or walling out,
And to whom I was like to give offence.
Robert Frost 1874–1963: 'Mending Wall' (1914)

13 Laugh and the world laughs with you;
Weep, and you weep alone;
For the sad old earth must borrow its mirth,
But has trouble enough of its own.
Ella Wheeler Wilcox 1855–1919: 'Solitude'

14 I want to be alone.
Greta Garbo 1905–90: *Grand Hotel* (1932 film)

15 God created man and, finding him not sufficiently alone, gave him a companion to make him feel his solitude more keenly.
Paul Valéry 1871–1945: *Tel Quel 1* (1941)

16 He [Barrymore] would quote from Genesis the text which says, 'It is not good for man to be alone,' and then add, 'But O my God, what a relief.'
John Barrymore 1882–1942: A. Power-Waters *John Barrymore* (1941)

17 All the lonely people, where do they all come from?
John Lennon 1940–80 and **Paul McCartney** 1942– : 'Eleanor Rigby' (1966 song)

18 Thirty years is a very long time to live alone and life doesn't get any nicer.
Frances Partridge 1900– : on widowhood, at the age of 92; in G. Kinnock and F. Miller *By Faith and Daring* (1993)

Sorrow

See also **Suffering**

1 By the waters of Babylon we sat down and wept: when we remembered thee, O Sion.
Bible: Psalm 137

2 O my son Absalom, my son, my son Absalom! would God I had died for thee, O Absalom, my son, my son!
Bible: II Samuel

3 *Sunt lacrimae rerum et mentem mortalia tangunt.*
There are tears shed for things and mortality touches the heart.
Virgil 70–19 BC: *Aeneid*

4 Silence augmenteth grief, writing increaseth rage,
Staled are my thoughts, which loved and lost, the wonder of our age.
Edward Dyer d. 1607: 'Elegy on the Death of Sir Philip Sidney' (1593); formerly attributed to Fulke Greville, 1554–1628

5 Grief fills the room up of my absent child,
Lies in his bed, walks up and down with me.
William Shakespeare 1564–1616: *King John* (1591–8)

6 Every one can master a grief but he that has it.
William Shakespeare 1564–1616: *Much Ado About Nothing* (1598–9)

7 When sorrows come, they come not single spies,
But in battalions.
William Shakespeare 1564–1616: *Hamlet* (1601)

8 Give sorrow words: the grief that does not speak
Whispers the o'er-fraught heart, and bids it break.
William Shakespeare 1564–1616: *Macbeth* (1606)

9 He first deceased; she for a little tried
To live without him: liked it not, and died.
Henry Wotton 1568–1639: 'Upon the Death of Sir Albertus Moreton's Wife' (1651)

10 Grief is a species of idleness.
Samuel Johnson 1709–84: letter to Mrs Thrale, 17 March 1773

11 For a tear is an intellectual thing;
And a sigh is the sword of an Angel King.
William Blake 1757–1827: *Jerusalem* (1815)

12 Then glut thy sorrow on a morning rose.
John Keats 1795–1821: 'Ode on Melancholy' (1820)

13 Ah, woe is me! Winter is come and gone,
But grief returns with the revolving year.
Percy Bysshe Shelley 1792–1822: *Adonais* (1821)

14 I tell you, hopeless grief is passionless.
Elizabeth Barrett Browning 1806–61: 'Grief' (1844)

15 Tears, idle tears, I know not what they mean,
Tears from the depth of some divine despair.
Alfred, Lord Tennyson 1809–92: *The Princess* (1847) song (added 1850)

16 For of all sad words of tongue or pen,
The saddest are these: 'It might have been!'
John Greenleaf Whittier 1807–92: 'Maud Muller' (1854)

17 We do not expect people to be deeply moved by what is not unusual. That element of tragedy which lies in the very fact of frequency, has not yet wrought itself into the coarse emotion of mankind.
George Eliot 1819–80: *Middlemarch* (1871–2)

18 Tragedy ought really to be a great kick at misery.
D. H. Lawrence 1885–1930: letter to A. W. McLeod, 6 October 1912

19 Now laughing friends deride tears I cannot hide,
So I smile and say 'When a lovely flame dies,
Smoke gets in your eyes.'
Otto Harbach 1873–1963: 'Smoke Gets in your Eyes' (1933 song)

20 Sob, heavy world,
Sob as you spin
Mantled in mist, remote from the happy.
W. H. Auden 1907–73: *The Age of Anxiety* (1947)

21 He felt the loyalty we all feel to unhappiness—the sense that that is where we really belong.
Graham Greene 1904–91: *The Heart of the Matter* (1948)

Speech and Speeches

See also **Conversation**

1 Friends, Romans, countrymen, lend me your ears.
William Shakespeare 1564–1616: *Julius Caesar* (1599)

2 I am no orator, as Brutus is;
But, as you know me all, a plain, blunt man,
That love my friend.
William Shakespeare 1564–1616: *Julius Caesar* (1599)

3 I do not much dislike the matter, but
The manner of his speech.
William Shakespeare 1564–1616: *Antony and Cleopatra* (1606–7)

4 But all was false and hollow; though his tongue
Dropped manna, and could make the worse appear
The better reason.
John Milton 1608–74: *Paradise Lost* (1667)

5 And adepts in the speaking trade
Keep a cough by them ready made.
Charles Churchill 1731–64: *The Ghost* (1763)

6 If I reprehend any thing in this world, it is the use of my oracular tongue, and a nice derangement of epitaphs!

Richard Brinsley Sheridan 1751–1816: *The Rivals* (1775)

7 Mr Speaker, I smell a rat; I see him forming in the air and darkening the sky; but I'll nip him in the bud.

Boyle Roche 1743–1807: attributed

8 A...sharp tongue is the only edged tool that grows keener with constant use.

Washington Irving 1783–1859: *The Sketch Book* (1820) 'Rip Van Winkle'

9 When you have nothing to say, say nothing.

Charles Caleb Colton *c.*1780–1832: *Lacon* (1820)

10 The brilliant chief, irregularly great,
Frank, haughty, rash,—the Rupert of Debate!

Edward Bulwer-Lytton 1803–73: of Edward Stanley, 14th Earl of Derby, in *The New Timon* (1846)

11 Human speech is like a cracked kettle on which we tap crude rhythms for bears to dance to, while we long to make music that will melt the stars.

Gustave Flaubert 1821–80: *Madame Bovary* (1857)

12 Speech is the small change of silence.

George Meredith 1828–1909: *The Ordeal of Richard Feverel* (1859)

13 To Trinity Church, Dorchester. The rector in his sermon delivers himself of mean images in a very sublime voice, and the effect is that of a glowing landscape in which clothes are hung up to dry.

Thomas Hardy 1840–1928: *Notebooks* 1 February 1874

14 A sophistical rhetorician, inebriated with the exuberance of his own verbosity.

Benjamin Disraeli 1804–81: of Gladstone, in *The Times* 29 July 1878

15 I absorb the vapour and return it as a flood.

W. E. Gladstone 1809–98: on public speaking, in Lord Riddell *Some Things That Matter* (1927 ed.)

16 He [Lord Charles Beresford] is one of those orators of whom it was well said, 'Before they get up, they do not know what they are going to say; when they are speaking, they do not know what they are saying; and when they have sat down, they do not know what they have said.'

Winston Churchill 1874–1965: speech, House of Commons, 20 December 1912

17 It is impossible for an Englishman to open his mouth without making some other Englishman hate or despise him.

George Bernard Shaw 1856–1950: *Pygmalion* (1916)

18 What can be said at all can be said clearly; and whereof one cannot speak thereof one must be silent.

Ludwig Wittgenstein 1889–1951: *Tractatus Logico-Philosophicus* (1922)

19 Speech impelled us
To purify the dialect of the tribe.

T. S. Eliot 1888–1965: *Four Quartets* 'Little Gidding' (1942)

20 He [Winston Churchill] mobilized the English language and sent it into battle to steady his fellow countrymen and hearten those Europeans upon whom the long dark night of tyranny had descended.

Ed Murrow 1908–65: broadcast, 30 November 1954

21 It was the nation and the race dwelling all round the globe that had the lion's heart. I had the luck to be called upon to give the roar.

Winston Churchill 1874–1965: speech at Westminster Hall, 30 November 1954

22 I do not object to people looking at their watches when I am speaking. But I strongly object when they start shaking them to make certain they are still going.

Lord Birkett 1883–1962: in *Observer* 30 October 1960

23 A speech from Ernest Bevin on a major occasion had all the horrific fascination of a public execution. If the mind was left immune, eyes and ears and emotions were riveted.

Michael Foot 1913– : *Aneurin Bevan* vol. 1 (1962)

24 Humming, Hawing and Hesitation are the three Graces of contemporary Parliamentary oratory.

Julian Critchley 1930– : *Westminster Blues* (1985)

Sport

1 *...Duas tantum res anxius optat,*
Panem et circenses.
Only two things does he [the modern citizen] anxiously wish for—bread and the big match.
Juvenal AD *c.*60–*c.*130: *Satires*, usually quoted 'bread and circuses'

2 There is plenty of time to win this game, and to thrash the Spaniards too.
Francis Drake *c.*1540–96: attributed

3 When we have matched our rackets to these balls,
We will in France, by God's grace, play a set
Shall strike his father's crown into the hazard.
William Shakespeare 1564–1616: *Henry V* (1599)

4 As no man is born an artist, so no man is born an angler.
Izaak Walton 1593–1683: *The Compleat Angler* (1653)

5 Chaos umpire sits,
And by decision more embroils the fray.
John Milton 1608–74: *Paradise Lost* (1667)

6 Eclipse first, the rest nowhere.
Dennis O'Kelly *c.*1720–87: comment at Epsom, 3 May 1769

7 Fly fishing may be a very pleasant amusement; but angling or float fishing I can only compare to a stick and a string, with a worm at one end and a fool at the other.
Samuel Johnson 1709–84: attributed, in Hawker *Instructions to Young Sportsmen* (1859); attributed to Jonathan Swift in *The Indicator* 27 October 1819

8 It's more than a game. It's an institution.
Thomas Hughes 1822–96: of cricket; *Tom Brown's Schooldays* (1857)

9 There's a breathless hush in the Close to-night—
Ten to make and the match to win—
A bumping pitch and a blinding light,
An hour to play and the last man in.
Henry Newbolt 1862–1938: 'Vitaï Lampada' (1897)

10 Play up! play up! and play the game!
Henry Newbolt 1862–1938: 'Vitaï Lampada' (1897)

11 The bigger they are, the further they have to fall.
Robert Fitzsimmons 1862–1917: prior to a boxing match, in *Brooklyn Daily Eagle* 11 August 1900 (similar forms found in proverbs since the 15th century)

12 The flannelled fools at the wicket or the muddied oafs at the goals.
Rudyard Kipling 1865–1936: 'The Islanders' (1903)

13 As the race wore on ... his oar was dipping into the water nearly *twice* as often as any other.
Desmond Coke 1879–1931: *Sandford of Merton* (1903), usually misquoted 'All rowed fast, but none so fast as stroke'

14 The important thing in life is not the victory but the contest; the essential thing is not to have won but to have fought well.
Baron Pierre de Coubertin 1863–1937: speech in London on the Olympic Games; 24 July 1908

15 To play billiards well is a sign of an ill-spent youth.
Charles Roupell: attributed, in D. Duncan *Life of Herbert Spencer* (1908)

16 A decision of the courts decided that the game of golf may be played on Sunday, not being a game within the view of the law, but being a form of moral effort.
Stephen Leacock 1869–1944: *Over the Footlights* (1923)

17 Honey, I just forgot to duck.
Jack Dempsey 1895–1983: to his wife, on losing the World Heavyweight title, 23 September 1926. After a failed attempt on his life in 1981, Ronald Reagan quipped 'I forgot to duck'

18 To say that these men paid their shillings to watch twenty-two hirelings kick a ball is merely to say that a violin is wood and catgut, that *Hamlet* is so much paper and ink. For a shilling the Bruddersford United AFC offered you Conflict and Art.
J. B. Priestley 1894–1984: *The Good Companions* (1929)

19 We was robbed!
Joe Jacobs 1896–1940: after Jack Sharkey beat Max Schmeling (of whom Jacobs was manager) in the heavyweight title fight, 21 June 1932

20 I should of stood in bed.
Joe Jacobs 1896–1940: after leaving his sickbed to attend the World Baseball Series in Detroit, 1935, and betting on the losers

21 For when the One Great Scorer comes to mark against your name,
He writes—not that you won or lost—but how you played the Game.
Grantland Rice 1880–1954: 'Alumnus Football' (1941)

22 A sportsman is a man who, every now and then, simply has to get out and kill something. Not that he's cruel. He wouldn't hurt a fly. It's not big enough.
Stephen Leacock 1869–1944: *My Remarkable Uncle* (1942)

23 Personally, I have always looked on cricket as organized loafing.
Archbishop William Temple 1881–1944: attributed

24 Love-thirty, love-forty, oh! weakness of joy,
The speed of a swallow, the grace of a boy,
With carefullest carelessness, gaily you won,
I am weak from your loveliness, Joan Hunter Dunn.
John Betjeman 1906–84: 'A Subaltern's Love-Song' (1945)

25 Nice guys. Finish last.
Leo Durocher 1906–91: casual remark at a practice ground, July 1946; in *Nice Guys Finish Last* (as the remark generally is quoted, 1975)

26 Sure, winning isn't everything. It's the only thing.
Henry 'Red' Sanders: in *Sports Illustrated* 26 December 1955 (often attributed to Vince Lombardi)

27 Oh, he's football crazy, he's football mad
And the football it has robbed him o' the wee bit sense he had.
And it would take a dozen skivvies, his clothes to wash and scrub,
Since our Jock became a member of that terrible football club.
Jimmy McGregor: 'Football Crazy' (1960 song)

28 I'm the greatest.
Muhammad Ali 1942– : catch-phrase used from 1962

29 Float like a butterfly, sting like a bee.
Muhammad Ali 1942– : summary of his boxing strategy, in G. Sullivan *Cassius Clay Story* (1964), probably originated by Drew 'Bundini' Brown

30 Cricket—a game which the English, not being a spiritual people, have invented in order to give themselves some conception of eternity.
Lord Mancroft 1914–87 : *Bees in Some Bonnets* (1979)

31 Some people think football is a matter of life and death...I can assure them it is much more serious than that.
Bill Shankly 1914–81: in *Sunday Times* 4 October 1981

32 The thing about sport, any sport, is that swearing is very much part of it.
Jimmy Greaves 1940– : *Observer* 1 January 1989 'Sayings of the Year'

33 Boxing's just showbusiness with blood.
Frank Bruno 1961– : in *Observer* 29 December 1991 'Sayings of the Year'

Statistics

1 A witty statesman said, you might prove anything by figures.
Thomas Carlyle 1795–1881: *Chartism* (1839)

2 Every moment dies a man,
Every moment one is born.
Alfred, Lord Tennyson 1809–92: 'The Vision of Sin' (1842)

3 Every moment dies a man,
Every moment $1\frac{1}{16}$ is born.
Charles Babbage 1792–1871: parody of Tennyson's 'Vision of Sin' in an unpublished letter to the poet

4 There are three kinds of lies: lies, damned lies and statistics.
Benjamin Disraeli 1804–81: attributed to Disraeli in Mark Twain *Autobiography* (1924)

5 He uses statistics as a drunken man uses lampposts—for support rather than for illumination.
Andrew Lang 1844–1912: attributed

6 [The War Office kept three sets of figures:] one to mislead the public, another to mislead the Cabinet, and the third to mislead itself.
Herbert Asquith 1852–1928: in A. Horne *Price of Glory* (1962)

7 From the fact that there are 400,000 species of beetles on this planet, but only 8,000 species of mammals, he [Haldane] concluded that the Creator, if He exists, has a special preference for beetles.

J. B. S. Haldane 1892–1964: report of lecture, 7 April 1951

Style

1 I strive to be brief, and I become obscure.

Horace 65–8 BC: *Ars Poetica*

2 When we see a natural style, we are quite surprised and delighted, for we expected to see an author and we find a man.

Blaise Pascal 1623–62: *Pensées* (1670)

3 One had as good be out of the world, as out of the fashion.

Colley Cibber 1671–1757: *Love's Last Shift* (1696)

4 Style is the dress of thought; a modest dress,
Neat, but not gaudy, will true critics please.

Samuel Wesley 1662–1735: 'An Epistle to a Friend concerning Poetry' (1700)

5 True wit is Nature to advantage dressed,
What oft was thought, but ne'er so well expressed.

Alexander Pope 1688–1744: *An Essay on Criticism* (1711)

6 Proper words in proper places, make the true definition of a style.

Jonathan Swift 1667–1745: *Letter to a Young Gentleman lately entered into Holy Orders* (9 January 1720)

7 Style is the man.

Comte de Buffon 1707–88: *Discours sur le style* (address given to the Académie Française, 25 August 1753)

8 It is charming to totter into vogue.

Horace Walpole 1717–97: letter to George Selwyn, 2 December 1765

9 Dr Johnson's sayings would not appear so extraordinary, were it not for his bow-wow way.

Henry Herbert, 10th Earl of Pembroke 1734–94: in James Boswell *Life of Samuel Johnson* (1791) 27 March 1775

10 Too many flowers...too little fruit.
Sir Walter Scott 1771–1832: of Felicia Hemans's literary style; letter to Joanna Baillie, 18 July 1823

11 It is rustic all through. It is moorish, and wild, and knotty as a root of heath.
Charlotte Brontë 1816–55: on the setting of Emily Brontë's *Wuthering Heights*, in her own preface to the 1850 edition

12 Style is life! It is the very life-blood of thought!
Gustave Flaubert 1821–80: letter to Louise Colet, 7 September 1853

13 Have something to say, and say it as clearly as you can. That is the only secret of style.
Matthew Arnold 1822–88: in G. W. E. Russell *Collections and Recollections* (1898)

14 The Mandarin style...is beloved by literary pundits, by those who would make the written word as unlike as possible to the spoken one.
Cyril Connolly 1903–74: *Enemies of Promise* (1938)

Success and Failure

1 The race is not to the swift, nor the battle to the strong.
Bible: Ecclesiastes

2 *Veni, vidi, vici.*
I came, I saw, I conquered.
Julius Caesar 100–44 BC: inscription displayed in Caesar's Pontic triumph, according to Suetonius *Lives of the Caesars*; or, according to Plutarch *Parallel Lives*, written in a letter by Caesar, announcing the victory of Zela which concluded the Pontic campaign

3 The only safe course for the defeated is to expect no safety.
Virgil 70–19 BC: *Aeneid*

4 These success encourages: they can because they think they can.
Virgil 70–19 BC: *Aeneid*

5 *Vae victis.*
Down with the defeated!
Livy 59 BC–AD 17: cry (already proverbial) of the Gallic

King, Brennus, on capturing Rome in 390 BC; in *Ab Urbe Condita*

6 For what shall it profit a man, if he shall gain the whole world, and lose his own soul?
Bible: St Mark. See also St Matthew

7 *Deos fortioribus adesse.*
The gods are on the side of the stronger.
Tacitus AD *c.*56–after 117: *Histories*

8 We fail!
But screw your courage to the sticking-place,
And we'll not fail.
William Shakespeare 1564–1616: *Macbeth* (1606)

9 'Tis not in mortals to command success,
But we'll do more, Sempronius; we'll deserve it.
Joseph Addison 1672–1719: *Cato* (1713)

10 The conduct of a losing party never appears right: at least it never can possess the only infallible criterion of wisdom to vulgar judgements—success.
Edmund Burke 1729–97: *Letter to a Member of the National Assembly* (1791)

11 The sublime and the ridiculous are often so nearly related, that it is difficult to class them separately. One step above the sublime, makes the ridiculous; and one step above the ridiculous, makes the sublime again.
Thomas Paine 1737–1809: *The Age of Reason* pt. 2 (1795)

12 There is only one step from the sublime to the ridiculous.
Napoléon I 1769–1821: following the retreat from Moscow in 1812; in D. G. De Pradt *Histoire de l'Ambassade dans le grand-duché de Varsovie en 1812* (1815).

13 I have climbed to the top of the greasy pole.
Benjamin Disraeli 1804–81: on becoming Prime Minister, in W. Monypenny and G. Buckle *Life of Disraeli* (1916)

14 To burn always with this hard, gemlike flame, to maintain this ecstasy, is success in life.
Walter Pater 1839–94: *Studies in the History of the Renaissance* (1873)

15 All you need in this life is ignorance and confidence; then success is sure.
Mark Twain 1835–1910: letter to Mrs Foote, 2 December 1887

16 We are not interested in the possibilities of defeat; they do not exist.
Queen Victoria 1819–1901: on the Boer War during 'Black Week', December 1899

17 The moral flabbiness born of the exclusive worship of the bitch-goddess *success*.
William James 1842–1910: letter to H. G. Wells, 11 September 1906

18 If you can meet with triumph and disaster
And treat those two imposters just the same.
Rudyard Kipling 1865–1936: 'If—' (1910)

19 History to the defeated
May say Alas but cannot help or pardon.
W. H. Auden 1907–73: 'Spain 1937' (1937)

20 Success is relative:
It is what we can make of the mess we have made of things.
T. S. Eliot 1888–1965: *The Family Reunion* (1939)

21 You ask, what is our aim? I can answer in one word: Victory, victory at all costs, victory in spite of all terror; victory, however long and hard the road may be; for without victory, there is no survival.
Winston Churchill 1874–1965: speech, House of Commons, 13 May 1940

22 Trying to learn to use words, and every attempt
Is a wholly new start, and a different kind of failure.
T. S. Eliot 1888–1965: *Four Quartets* 'East Coker' (1940)

23 *La vittoria trova cento padri, e nessuno vuole riconoscere l'insuccesso.*
Victory has a hundred fathers, but defeat is an orphan.
Count Galeazzo Ciano 1903–44: diary, 9 September 1942 (literally 'no-one wants to recognise defeat as his own')

24 If A is a success in life, then A equals x plus y plus z. Work is x; y is play; and z is keeping your mouth shut.
Albert Einstein 1879–1955: in *Observer* 15 January 1950

25 Be nice to people on your way up because you'll meet 'em on your way down.
Wilson Mizner 1876–1933: in A. Johnston *The Legendary Mizners* (1953)

26 She knows there's no success like failure
And that failure's no success at all.
Bob Dylan 1941– : 'Love Minus Zero / No Limit' (1965 song)

27 For a writer, success is always temporary, success is only a delayed failure. And it is incomplete.
Graham Greene 1904–91: *A Sort of Life* (1971)

28 Whenever a friend succeeds, a little something in me dies.
Gore Vidal 1925– : in *Sunday Times Magazine* 16 September 1973

29 It is not enough to succeed. Others must fail.
Gore Vidal 1925– : in G. Irvine *Antipanegyric for Tom Driberg* 8 December 1976

Suffering

1 O you who have borne even heavier things, God will grant an end to these too.
Virgil 70–19 BC: *Aeneid*

2 Nothing happens to anybody which he is not fitted by nature to bear.
Marcus Aurelius AD 121–80: *Meditations*

3 He jests at scars, that never felt a wound.
William Shakespeare 1564–1616: *Romeo and Juliet* (1595)

4 But I have that within which passeth show;
These but the trappings and the suits of woe.
William Shakespeare 1564–1616: *Hamlet* (1601)

5 But yet the pity of it, Iago! O! Iago, the pity of it, Iago!
William Shakespeare 1564–1616: *Othello* (1602–4)

6 I am a man
More sinned against than sinning.
William Shakespeare 1564–1616: *King Lear* (1605–6)

7 The worst is not,
So long as we can say, 'This is the worst.'
William Shakespeare 1564–1616: *King Lear* (1605–6)

8 The oldest hath borne most: we that are young,
Shall never see so much, nor live so long.
William Shakespeare 1564–1616: *King Lear* (1605–6)

9 To each his suff'rings, all are men,
Condemned alike to groan;
The tender for another's pain,
Th' unfeeling for his own.
Thomas Gray 1716–71: *Ode on a Distant Prospect of Eton College* (1747)

10 Hides from himself his state, and shuns to know,
That life protracted is protracted woe.
Samuel Johnson 1709–84: *The Vanity of Human Wishes* (1749)

11 Thank you, madam, the agony is abated.
Lord Macaulay 1800–59: aged four, having had hot coffee spilt over his legs; in G. O. Trevelyan *Life and Letters of Lord Macaulay* (1876)

12 If suffer we must, let's suffer on the heights.
Victor Hugo 1802–85: *Contemplations* (1856)

13 Those who have courage to love should have courage to suffer.
Anthony Trollope 1815–82: *The Bertrams* (1859)

14 Nothing begins, and nothing ends,
That is not paid with moan;
For we are born in other's pain,
And perish in our own.
Francis Thompson 1859–1907: 'Daisy' (1913)

15 It is not true that suffering ennobles the character; happiness does that sometimes, but suffering, for the most part, makes men petty and vindictive.
W. Somerset Maugham 1874–1965: *Moon and Sixpence* (1919)

16 Too long a sacrifice
Can make a stone of the heart.
W. B. Yeats 1865–1939: 'Easter, 1916' (1921)

17 We can't all be happy, we can't all be rich, we can't all be lucky...Some must cry so that others may be able to laugh the more heartily.
Jean Rhys *c.*1890–1979: *Good Morning, Midnight* (1939)

18 Even the dreadful martyrdom must run its course
Anyhow in a corner, some untidy spot
Where the dogs go on with their doggy life and the torturer's

horse
Scratches its innocent behind on a tree.
W. H. Auden 1907–73: 'Musée des Beaux Arts' (1940)

The Supernatural

1 Then a spirit passed before my face; the hair of my flesh stood up.
Bible: Job

2 GLENDOWER: I can call spirits from the vasty deep.
HOTSPUR: Why, so can I, or so can any man;
But will they come when you do call for them?
William Shakespeare 1564–1616: *Henry IV, Part 1* (1597)

3 Is this a dagger which I see before me,
The handle toward my hand?
William Shakespeare 1564–1616: *Macbeth* (1606)

4 Double, double toil and trouble;
Fire burn and cauldron bubble.
William Shakespeare 1564–1616: *Macbeth* (1606)

5 Superstition sets the whole world in flames; philosophy quenches them.
Voltaire 1694–1778: *Dictionnaire philosophique* (1764)

6 I wants to make your flesh creep.
Charles Dickens 1812–70: *Pickwick Papers* (1837) The Fat Boy

7 Up the airy mountain,
Down the rushy glen,
We daren't go a-hunting,
For fear of little men.
William Allingham 1824–89: 'The Fairies' (1850)

8 From ghoulies and ghosties and long-leggety beasties
And things that go bump in the night,
Good Lord, deliver us!
Anonymous: 'The Cornish or West Country Litany', in F. T. Nettleinghame *Polperro Proverbs and Others* (1926)

9 I always knew the living talked rot, but it's nothing to the rot the dead talk.
Margot Asquith 1864–1945: on spiritualism, in Henry 'Chips' Channon, diary, 20 December 1937

Taxes

1 *Pecunia non olet.*
Money has no smell.
Emperor Vespasian AD 9–79: quashing an objection to a tax on public lavatories, in Suetonius *Lives of the Caesars*

2 *Excise.* A hateful tax levied upon commodities.
Samuel Johnson 1709–84: *Dictionary of the English Language* (1755)

3 Taxation without representation is tyranny.
James Otis 1725–83: watchword (coined *c.*1761) of the American Revolution

4 To tax and to please, no more than to love and to be wise, is not given to men.
Edmund Burke 1729–97: *On American Taxation* (1775)

5 There is no art which one government sooner learns of another than that of draining money from the pockets of the people.
Adam Smith 1723–90: *Wealth of Nations* (1776)

6 In this world nothing can be said to be certain, except death and taxes.
Benjamin Franklin 1706–90: letter to Jean Baptiste Le Roy, 13 November 1789

7 The Chancellor of the Exchequer is a man whose duties make him more or less of a taxing machine. He is intrusted with a certain amount of misery which it is his duty to distribute as fairly as he can.
Robert Lowe 1811–92: speech, House of Commons, 11 April 1870

8 Income Tax has made more Liars out of the American people than Golf.
Will Rogers 1879–1935: *The Illiterate Digest* (1924)

9 Death and taxes and childbirth! There's never any convenient time for any of them.
Margaret Mitchell 1900–49: *Gone with the Wind* (1936)

10 Only the little people pay taxes.
Leona Helmsley *c.*1920– : addressed to her housekeeper in 1983, and reported at her trial for tax evasion; in *New York Times* 12 July 1989

11 Read my lips: no new taxes.
George Bush 1924– : campaign pledge on taxation, in *New York Times* 19 August 1988

Teaching

See also **Education**

1 For precept must be upon precept, precept upon precept; line upon line, line upon line; here a little, and there a little.
Bible: Isaiah

2 *Homines dum docent discunt.*
Even while they teach, men learn.
Seneca ('the Younger') *c.*4 BC–AD 65: *Epistulae Morales*

3 There is no such whetstone, to sharpen a good wit and encourage a will to learning, as is praise.
Roger Ascham 1515–68: *The Schoolmaster* (1570)

4 We loved the doctrine for the teacher's sake.
Daniel Defoe 1660–1731: 'Character of the late Dr S. Annesley' (1697)

5 Men must be taught as if you taught them not,
And things unknown proposed as things forgot.
Alexander Pope 1688–1744: *An Essay on Criticism* (1711)

6 Delightful task! to rear the tender thought,
To teach the young idea how to shoot.
James Thomson 1700–48: *The Seasons* (1746) 'Spring'

7 Be a governess! Better be a slave at once!
Charlotte Brontë 1816–55: *Shirley* (1849)

8 He who can, does. He who cannot, teaches.
George Bernard Shaw 1856–1950: *Man and Superman* (1903)

9 A teacher affects eternity; he can never tell where his influence stops.
Henry Brooks Adams 1838–1918: *The Education of Henry Adams* (1907)

10 For every person who wants to teach there are approximately thirty who don't want to learn—much.
W. C. Sellar 1898–1951 and **R. J. Yeatman** 1898–1968: *And Now All This* (1932)

11 Give me a girl at an impressionable age, and she is mine for life.
Muriel Spark 1918– : *The Prime of Miss Jean Brodie* (1961)

12 *You* have not had thirty years' experience... *You* have had one year's experience 30 times.
J. L. Carr 1912– : *The Harpole Report* (1972)

Technology

See also **Inventions and Discoveries, Science**

1 Give me but one firm spot on which to stand, and I will move the earth.
Archimedes *c.*287–212 BC: on the action of a lever, in Pappus *Synagoge*

2 I sell here, Sir, what all the world desires to have—POWER.
Matthew Boulton 1728–1809: speaking to Boswell of his engineering works; in James Boswell *Life of Samuel Johnson* (1791)

3 One machine can do the work of fifty ordinary men. No machine can do the work of one extraordinary man.
Elbert Hubbard 1859–1915: *Thousand and One Epigrams* (1911)

4 Your worship is your furnaces,
Which, like old idols, lost obscenes,
Have molten bowels; your vision is
Machines for making more machines.
Gordon Bottomley 1874–1948: 'To Ironfounders and Others' (1912)

5 Communism is Soviet power plus the electrification of the whole country.
Lenin 1870–1924: Report to 8th Congress, 1920, in *Collected Works*

6 Machines are worshipped because they are beautiful, and valued because they confer power; they are hated because they are hideous, and loathed because they impose slavery.
Bertrand Russell 1872–1970: *Sceptical Essays* (1928)

7 Her own mother lived the latter years of her life in the horrible suspicion that electricity was dripping invisibly all over the house.
James Thurber 1894–1961: *My Life and Hard Times* (1933)

8 This is not the age of pamphleteers. It is the age of the engineers. The spark-gap is mightier than the pen.
Lancelot Hogben 1895–1975: *Science for the Citizen* (1938)

9 Technology...the knack of so arranging the world that we need not experience it.
Max Frisch 1911– : *Homo Faber* (1957)

10 The new electronic interdependence recreates the world in the image of a global village.
Marshall McLuhan 1911–80: *The Gutenberg Galaxy* (1962)

11 The Britain that is going to be forged in the white heat of this revolution will be no place for restrictive practices or for outdated methods on either side of industry.
Harold Wilson 1916– : speech at the Labour Party Conference, 1 October 1963; usually quoted 'the white heat of the technological revolution'

12 The medium is the message.
Marshall McLuhan 1911–80: *Understanding Media* (1964)

13 To err is human but to really foul things up requires a computer.
Anonymous: *Farmers' Almanac for 1978*

14 A modern computer hovers between the obsolescent and the nonexistent.
Sydney Brenner 1927– : attributed in *Science* 5 January 1990

Temptation

1 Watch and pray, that ye enter not into temptation: the spirit indeed is willing but the flesh is weak.
Bible: St Matthew

2 From all the deceits of the world, the flesh, and the devil,
Good Lord, deliver us.
Book of Common Prayer 1662: *The Litany*

3 Then gently scan your brother man,
Still gentler sister woman;
Tho' they may gang a kennin wrang,
To step aside is human.
Robert Burns 1759–96: 'Address to the Unco Guid' (1787)

4 What's done we partly may compute,
But know not what's resisted.
Robert Burns 1759–96: 'Address to the Unco Guid' (1787)

5 I can resist everything except temptation.
Oscar Wilde 1854–1900: *Lady Windermere's Fan* (1892)

6 There are several good protections against temptations, but the surest is cowardice.
Mark Twain 1835–1910: *Following the Equator* (1897)

7 The Devil, having nothing else to do,
Went off to tempt My Lady Poltagrue.
My Lady, tempted by a private whim,
To his extreme annoyance, tempted him.
Hilaire Belloc 1870–1953: 'On Lady Poltagrue' (1923)

The Theatre see **Acting and the Theatre**

Thinking

See also **Ideas, The Mind**

1 To change your mind and to follow him who sets you right is to be nonetheless the free agent that you were before.
Marcus Aurelius AD 121–80: *Meditations*

2 Reasons are not like garments, the worse for wearing.
Robert Devereux, 2nd Earl of Essex 1566–1601: letter to Lord Willoughby, 4 January 1599

3 Yond Cassius has a lean and hungry look;
He thinks too much: such men are dangerous.
William Shakespeare 1564–1616: *Julius Caesar* (1599)

4 *Cogito, ergo sum.*
I think, therefore I am.
René Descartes 1596–1650: *Le Discours de la méthode* (1637)

5 How comes it to pass, then, that we appear such cowards in reasoning, and are so afraid to stand the test of ridicule?
3rd Earl of Shaftesbury 1671–1713: *A Letter Concerning Enthusiasm* (1708)

6 I'll not listen to reason...Reason always means what someone else has got to say.
Elizabeth Gaskell 1810–65: *Cranford* (1853)

7 How often misused words generate misleading thoughts.
Herbert Spencer 1820–1903: *Principles of Ethics* (1879)

8 Logical consequences are the scarecrows of fools and the beacons of wise men.
T. H. Huxley 1825–95: *Science and Culture and Other Essays* (1881)

9 It is a capital mistake to theorize before you have all the evidence. It biases the judgement.
Arthur Conan Doyle 1859–1930: *A Study in Scarlet* (1888)

10 Heretics are the only bitter remedy against the entropy of human thought.
Yevgeny Zamyatin 1884–1937: 'Literature, Revolution and Entropy' quoted in *The Dragon and other Stories* (1967, tr. M. Ginsberg) introduction

11 *Doublethink* means the power of holding two contradictory beliefs in one's mind simultaneously, and accepting both of them.
George Orwell 1903–50: *Nineteen Eighty-Four* (1949)

12 What was once thought can never be unthought.
Friedrich Dürrenmatt 1921– : *The Physicists* (1962)

13 The real question is not whether machines think but whether men do.
B. F. Skinner 1904–90: *Contingencies of Reinforcement* (1969)

Time

1 For a thousand years in thy sight are but as yesterday: seeing that is past as a watch in the night.
Bible: Psalm 90

2 To every thing there is a season, and a time to every purpose under the heaven:
A time to be born, and a time to die...
A time to weep, and a time to laugh; a time to mourn, and a time to dance.
Bible: Ecclesiastes

3 *Sed fugit interea, fugit inreparabile tempus.*
But meanwhile it is flying, irretrievable time is flying.
Virgil 70–19 BC: *Georgics* (usually quoted '*tempus fugit* [time flies]')

4 *Tempus edax rerum.*
Time the devourer of everything.
Ovid 43 BC–AD *c.*17: *Metamorphoses*

5 Time is a violent torrent; no sooner is a thing brought to sight than it is swept by and another takes its place.
Marcus Aurelius AD 121–80: *Meditations*

6 Every instant of time is a pinprick of eternity.
Marcus Aurelius AD 121–80: *Meditations*

7 Time is the measure of movement.
Auctoritates Aristotelis: a compilation of medieval propositions

8 Time is...Time was...Time is past.
Robert Greene *c.*1560–92: *Friar Bacon and Friar Bungay* (1594)

9 I wasted time, and now doth time waste me.
William Shakespeare 1564–1616: *Richard II* (1595)

10 Time hath, my lord, a wallet at his back,
Wherein he puts alms for oblivion.
William Shakespeare 1564–1616: *Troilus and Cressida* (1602)

11 Come what come may,
Time and the hour runs through the roughest day.
William Shakespeare 1564–1616: *Macbeth* (1606)

12 To-morrow, and to-morrow, and to-morrow,
Creeps in this petty pace from day to day,
To the last syllable of recorded time;
And all our yesterdays have lighted fools
The way to dusty death.
William Shakespeare 1564–1616: *Macbeth* (1606)

13 What seest thou else
In the dark backward and abysm of time?
William Shakespeare 1564–1616: *The Tempest* (1611)

14 Even such is Time, which takes in trust
Our youth, our joys, and all we have,
And pays us but with age and dust.
Walter Ralegh *c.*1552–1618: written the night before his death

15 Fly envious Time, till thou run out thy race,
Call on the lazy leaden-stepping hours.
John Milton 1608–74: 'On Time' (1645)

16 I saw Eternity the other night,
Like a great ring of pure and endless light.
Henry Vaughan 1622–95: *Silex Scintillans* (1650–5)

17 But at my back I always hear
Time's wingèd chariot hurrying near:
And yonder all before us lie
Deserts of vast eternity.
Andrew Marvell 1621–78: 'To His Coy Mistress' (1681)

18 Time, like an ever-rolling stream,
Bears all its sons away.
Isaac Watts 1674–1748: 'O God, our help in ages past' (1719 hymn)

19 What's not destroyed by Time's devouring hand?
Where's Troy, and where's the Maypole in the Strand?
James Bramston *c.*1694–1744: *The Art of Politics* (1729)

20 I recommend to you to take care of minutes: for hours will take care of themselves.
Lord Chesterfield 1694–1773: *Letters to his Son* (1774) 6 November 1747

21 Eternity is in love with the productions of time.
William Blake 1757–1827: *The Marriage of Heaven and Hell* (1790–3) 'Proverbs of Hell'

22 Nothing puzzles me more than time and space; and yet nothing troubles me less, as I never think about them.
Charles Lamb 1775–1834: letter to Thomas Manning, 2 January 1810

23 Men talk of killing time, while time quietly kills them.
Dion Boucicault 1820–90: *London Assurance* (1841)

24 As if you could kill time without injuring eternity.
Henry David Thoreau 1817–62: *Walden* (1854)

25 He said, 'What's time? Leave Now for dogs and apes!
Man has Forever.'
Robert Browning 1812–89: 'A Grammarian's Funeral' (1855)

26 The woods decay, the woods decay and fall,
The vapours weep their burthen to the ground,
Man comes and tills the field and lies beneath,
And after many a summer dies the swan.
Alfred, Lord Tennyson 1809–92: 'Tithonus' (1860, revised 1864)

27 Time goes, you say? Ah no!
Alas, Time stays, *we* go.
Henry Austin Dobson 1840–1921: 'The Paradox of Time' (1877)

28 Time, you old gipsy man,
Will you not stay,
Put up your caravan
Just for one day?
Ralph Hodgson 1871–1962: 'Time, You Old Gipsy Man' (1917)

29 I have measured out my life with coffee spoons.
T. S. Eliot 1888–1965: 'Love Song of J. Alfred Prufrock' (1917)

30 Ah! the clock is always slow;
It is later than you think.
Robert W. Service 1874–1958: 'It Is Later Than You Think' (1921)

31 *In the long run* we are all dead.
John Maynard Keynes 1883–1946: *A Tract on Monetary Reform* (1923)

32 I shall use the phrase 'time's arrow' to express this one-way property of time which has no analogue in space.
Arthur Eddington 1882–1944: *The Nature of the Physical World* (1928)

33 Half our life is spent trying to find something to do with the time we have rushed through life trying to save.
Will Rogers 1879–1935: letter in *New York Times* 29 April 1930

34 Time present and time past
Are both perhaps present in time future,
And time future contained in time past.
T. S. Eliot 1888–1965: *Four Quartets* 'Burnt Norton' (1936)

35 Three o'clock is always too late or too early for anything you want to do.
Jean-Paul Sartre 1905–80: *La Nausée* (Nausea, 1938)

36 I am a sundial, and I make a botch
Of what is done much better by a watch.
Hilaire Belloc 1870–1953: 'On a Sundial' (1938)

37 The sunlight on the garden
Hardens and grows cold,
We cannot cage the minute
Within its net of gold.
Louis MacNeice 1907–63: 'Sunlight on the Garden' (1938)

38 VLADIMIR: That passed the time.
ESTRAGON: It would have passed in any case.
VLADIMIR: Yes, but not so rapidly.
Samuel Beckett 1906–89: *Waiting for Godot* (1955)

Titles

1 But let a Lord once own the happy lines,
How the wit brightens! how the style refines!
Alexander Pope 1688–1744: *An Essay on Criticism* (1711)

2 The rank is but the guinea's stamp,
The man's the gowd for a' that!
Robert Burns 1759–96: 'For a' that and a' that' (1790)

3 Kind hearts are more than coronets,
And simple faith than Norman blood.

Alfred, Lord Tennyson 1809–92: 'Lady Clara Vere de Vere' (1842)

4 Titles distinguish the mediocre, embarrass the superior, and are disgraced by the inferior.

George Bernard Shaw 1856–1950: *Man and Superman* (1903)

5 A fully-equipped duke costs as much to keep up as two Dreadnoughts; and dukes are just as great a terror and they last longer.

David Lloyd George 1863–1945: speech at Newcastle, 9 October 1909

6 When I want a peerage, I shall buy it like an honest man.

Lord Northcliffe 1865–1922: in Tom Driberg *Swaff* (1974)

7 An aristocracy in a republic is like a chicken whose head has been cut off: it may run about in a lively way, but in fact it is dead.

Nancy Mitford 1904–73: *Noblesse Oblige* (1956)

8 There is no stronger craving in the world than that of the rich for titles, except perhaps that of the titled for riches.

Hesketh Pearson 1887–1964: *The Pilgrim Daughters* (1961)

9 What harm have I ever done to the Labour Party?

R. H. Tawney 1880–1962: on declining the offer of a peerage, in *Evening Standard* 18 January 1962

10 Even as it [Great Britain] walked out on you and joined the Common Market, you were still looking for your MBEs and your knighthoods, and all the rest of the regalia that comes with it. You would take Australia right back down the time tunnel to the cultural cringe where you have always come from.

Paul Keating 1944– : addressing Australian Conservative supporters of Great Britain, 27 February 1992

The Town see **The Country and the Town**

Transience

1 Like that of leaves is a generation of men.
Homer 8th century BC: *The Iliad*

2 Vanity of vanities; all is vanity.
Bible: Ecclesiastes

3 All flesh is as grass, and all the glory of man as the flower of grass. The grass withereth, and the flower thereof falleth away.
Bible: I Peter

4 *Sic transit gloria mundi.*
Thus passes the glory of the world.
Anonymous: said at the coronation of a new Pope, while flax is burned; used at the coronation of Alexander V, 1409, but earlier in origin

5 Gather ye rosebuds while ye may,
Old Time is still a-flying:
And this same flower that smiles to-day,
To-morrow will be dying.
Robert Herrick 1591–1674: 'To the Virgins, to Make Much of Time' (1648)

6 A little rule, a little sway,
A sunbeam in a winter's day,
Is all the proud and mighty have
Between the cradle and the grave.
John Dyer 1700–58: *Grongar Hill* (1726)

7 The rainbow comes and goes,
And lovely is the rose.
William Wordsworth 1770–1850: 'Ode. Intimations of Immortality' (1807)

8 He who binds to himself a joy
Doth the winged life destroy
But he who kisses the joy as it flies
Lives in Eternity's sunrise.
William Blake 1757–1827: *MS Note-Book*

9 A rainbow and a cuckoo's song
May never come together again;

May never come
This side the tomb.
W. H. Davies 1871–1940: 'A Great Time' (1914)

10 Look thy last on all things lovely,
Every hour.
Walter de la Mare 1873–1956: 'Fare Well' (1918)

Transport

1 Sir, Saturday morning, although recurring at regular and well-foreseen intervals, always seems to take this railway by surprise.
W. S. Gilbert 1836–1911: letter to the station-master at Baker Street, on the Metropolitan line; in John Julius Norwich *Christmas Crackers* (1980)

2 There is *nothing*—absolutely nothing—half so much worth doing as simply messing about in boats.
Kenneth Grahame 1859–1932: *The Wind in the Willows* (1908)

3 The poetry of motion! The *real* way to travel! The *only* way to travel! Here today—in next week tomorrow!
Kenneth Grahame 1859–1932: on the car; *The Wind in the Willows* (1908)

4 Railway termini. They are our gates to the glorious and the unknown. Through them we pass out into adventure and sunshine, to them, alas! we return.
E. M. Forster 1879–1970: *Howards End* (1910)

5 Before the Roman came to Rye or out to Severn strode,
The rolling English drunkard made the rolling English road.
G. K. Chesterton 1874–1936: 'The Rolling English Road' (1914)

6 To George F. Babbitt, as to most prosperous citizens of Zenith, his motor car was poetry and tragedy, love and heroism. The office was his pirate ship but the car his perilous excursion ashore.
Sinclair Lewis 1885–1951: *Babbitt* (1922)

7 [There are] only two classes of pedestrians in these days of reckless motor traffic—the quick, and the dead.
Lord Dewar 1864–1930: in George Robey *Looking Back on Life* (1933)

8 After the first powerful plain manifesto
The black statement of pistons, without more fuss
But gliding like a queen, she leaves the station.
Stephen Spender 1909– : 'The Express' (1933)

9 This is the Night Mail crossing the Border,
Bringing the cheque and the postal order,
Letters for the rich, letters for the poor,
The shop at the corner, the girl next door.
W. H. Auden 1907–73: 'Night Mail' (1936)

10 Beneath this slab
John Brown is stowed.
He watched the ads,
And not the road.
Ogden Nash 1902–71: 'Lather as You Go' (1942)

11 That monarch of the road,
Observer of the Highway Code,
That big six-wheeler
Scarlet-painted
London Transport
Diesel-engined
Ninety-seven horse power
Omnibus!
Michael Flanders 1922–75 and **Donald Swann** 1923– : 'A Transport of Delight' (*c.*1956 song)

12 I think that cars today are almost the exact equivalent of the great Gothic cathedrals: I mean the supreme creation of an era, conceived with passion by unknown artists, and consumed in image if not in usage by a whole population which appropriates them as a purely magical object.
Roland Barthes 1915–80: *Mythologies* (1957), tr. A. Lavers

13 The car has become an article of dress without which we feel uncertain, unclad and incomplete in the urban compound.
Marshall McLuhan 1911–80: *Understanding Media* (1964)

14 Commuter—one who spends his life
In riding to and from his wife;

A man who shaves and takes a train,
And then rides back to shave again.
E. B. White 1899–1985: 'The Commuter' (1982)

Travel

1 Travel, in the younger sort, is a part of education; in the elder, a part of experience. He that travelleth into a country before he hath some entrance into the language, goeth to school, and not to travel.
Francis Bacon 1561–1626: *Essays* (1625) 'Of Travel'

2 I always love to begin a journey on Sundays, because I shall have the prayers of the church, to preserve all that travel by land, or by water.
Jonathan Swift 1667–1745: *Polite Conversation* (1738)

3 So it is in travelling; a man must carry knowledge with him, if he would bring home knowledge.
Samuel Johnson 1709–84: in James Boswell *Life of Johnson* (1791) 17 April 1778

4 Worth seeing, yes; but not worth going to see.
Samuel Johnson 1709–84: of the Giant's Causeway; in James Boswell *Life of Johnson* (1791) 12 October 1779

5 Of all noxious animals, too, the most noxious is a tourist. And of all tourists the most vulgar, ill-bred, offensive and loathsome is the British tourist.
Francis Kilvert 1840–79: diary, 5 April 1870

6 For my part, I travel not to go anywhere, but to go. I travel for travel's sake. The great affair is to move.
Robert Louis Stevenson 1850–94: *Travels with a Donkey* (1879)

7 To travel hopefully is a better thing than to arrive, and the true success is to labour.
Robert Louis Stevenson 1850–94: *Virginibus Puerisque* (1881)

8 Clay lies still, but blood's a rover;
Breath's a ware that will not keep.
Up, lad: when the journey's over
There'll be time enough to sleep.
A. E. Housman 1859–1936: *A Shropshire Lad* (1896)

9 A man travels the world in search of what he needs and returns home to find it.
George Moore 1852–1933: *The Brook Kerith* (1916)

10 Whenever I prepare for a journey I prepare as though for death. Should I never return, all is in order.
Katherine Mansfield 1888–1923: diary, 29 January 1922

11 A cold coming we had of it,
Just the worst time of the year
For a journey, and such a long journey:
The ways deep and the weather sharp,
The very dead of winter.
T. S. Eliot 1888–1965: 'Journey of the Magi' (1927)

12 Abroad is unutterably bloody and foreigners are fiends.
Nancy Mitford 1904–73: *The Pursuit of Love* (1945)

Trust and Treachery

1 He that is surety for a stranger shall smart for it.
Bible: Proverbs

2 *Equo ne credite, Teucri,*
Quidquid est, timeo Danaos et dona ferentes.
Do not trust the horse, Trojans. Whatever it is, I fear the Greeks even when they bring gifts.
Virgil 70–19 BC: *Aeneid*

3 *Quis custodiet ipsos custodes?*
Who is to guard the guards themselves?
Juvenal AD *c.*60–*c.*130: *Satires*

4 I know what it is to be a subject, and what to be a Sovereign. Good neighbours I have had, and I have met with bad: and in trust I have found treason.
Elizabeth I 1533–1603: speech to a Parliamentary deputation at Richmond, 12 November 1586, as reported in Camden's *Annals* (1615). A report 'which the queen herself heavily amended in her own hand' omits the concluding words

5 Treason doth never prosper, what's the reason?
For if it prosper, none dare call it treason.
John Harington 1561–1612: *Epigrams* (1618)

6 Just for a handful of silver he left us,
Just for a riband to stick in his coat.
Robert Browning 1812–89: 'The Lost Leader' (1845), of Wordsworth

7 And trust me not at all or all in all.
Alfred, Lord Tennyson 1809–92: *Idylls of the King* 'Merlin and Vivien' (1859)

8 And I said to the man who stood at the gate of the year: 'Give me a light that I may tread safely into the unknown.' And he replied: 'Go out into the darkness and put your hand into the Hand of God. That shall be to you better than light and safer than a known way.'
Minnie Louise Haskins 1875–1957: 'God Knows' (1908); quoted by George VI in his Christmas broadcast, 1939

9 To betray, you must first belong.
Kim Philby 1912–88: in *Sunday Times* 17 December 1967

10 Having watched the form of our traitors for a number of years, I cannot think that espionage can be recommended as a technique for building an impressive civilization. It's a lout's game.
Rebecca West 1892–1983: *The Meaning of Treason* (1982 ed.)

Truth

See also **Lies and Lying**

1 But, my dearest Agathon, it is truth which you cannot contradict; you can without any difficulty contradict Socrates.
Socrates 469–399 BC: in Plato *Symposium*

2 Plato is dear to me, but dearer still is truth.
Aristotle 384–322 BC: attributed

3 Great is Truth, and mighty above all things.
Bible (Apocrypha): I Esdras

4 And ye shall know the truth, and the truth shall make you free.
Bible: St John

5 Honesty is praised and left to shiver.
Juvenal AD *c.*60–*c.*130: *Satires* tr. G. Ramsay

6 Truth will come to light; murder cannot be hid long.
William Shakespeare 1564–1616: *The Merchant of Venice* (1596–8)

7 What is truth? said jesting Pilate; and would not stay for an answer.
Francis Bacon 1561–1626: *Essays* (1625) 'Of Truth'

8 Who says that fictions only and false heir
Become a verse? Is there in truth no beauty?
Is all good structure in a winding stair?
George Herbert 1593–1633: 'Jordan (1)' (1633)

9 Many from...an inconsiderate zeal unto truth, have too rashly charged the troops of error, and remain as trophies unto the enemies of truth.
Sir Thomas Browne 1605–82: *Religio Medici* (1643)

10 True and False are attributes of speech, not of things. And where speech is not, there is neither Truth nor Falsehood.
Thomas Hobbes 1588–1679: *Leviathan* (1651)

11 It is one thing to show a man that he is in error, and another to put him in possession of truth.
John Locke 1632–1704: *Essay concerning Human Understanding* (1690)

12 Truth is the cry of all, but the game of the few.
Bishop George Berkeley 1685–1753: *Siris* (1744)

13 It is commonly said, and more particularly by Lord Shaftesbury, that ridicule is the best test of truth.
Lord Chesterfield 1694–1773: *Letters to his Son* (1774) 6 February 1752

14 In lapidary inscriptions a man is not upon oath.
Samuel Johnson 1709–84: in James Boswell *Life of Johnson* (1791) 1775

15 I can't tell a lie, Pa; you know I can't tell a lie. I did cut it with my hatchet.
George Washington 1732–99: in M. L. Weems *Life of George Washington* (10th ed., 1810)

16 A truth that's told with bad intent
Beats all the lies you can invent.
William Blake 1757–1827: 'Auguries of Innocence' (*c.*1803)

17 'Tis strange—but true; for truth is always strange;
Stranger than fiction.
Lord Byron 1788–1824: *Don Juan* (1819–24)

18 'But the Emperor has nothing on at all!' cried a little child.
Hans Christian Andersen 1805–75: *Danish Fairy Legends and Tales* (1846) 'The Emperor's New Clothes'

19 What I tell you three times is true.
Lewis Carroll 1832–98: *The Hunting of the Snark* (1876)

20 It is the customary fate of new truths to begin as heresies and to end as superstitions.
T. H. Huxley 1825–95: *Science and Culture and Other Essays* (1881) 'The Coming of Age of the Origin of Species'

21 Irrationally held truths may be more harmful than reasoned errors.
T. H. Huxley 1825–95: *Science and Culture and Other Essays* (1881) 'The Coming of Age of the Origin of Species'

22 There was things which he stretched, but mainly he told the truth.
Mark Twain 1835–1910: *The Adventures of Huckleberry Finn* (1884)

23 When you have eliminated the impossible, whatever remains, *however improbable*, must be the truth.
Arthur Conan Doyle 1859–1930: *The Sign of Four* (1890)

24 The truth is rarely pure, and never simple.
Oscar Wilde 1854–1900: *The Importance of Being Earnest* (1895)

25 Truth is the most valuable thing we have. Let us economize it.
Mark Twain 1835–1910: *Following the Equator* (1897)

26 A thing is not necessarily true because a man dies for it.
Oscar Wilde 1854–1900: *Sebastian Melmoth* (1904 ed.)

27 A platitude is simply a truth repeated until people get tired of hearing it.
Stanley Baldwin 1867–1947: speech, House of Commons, 29 May 1924

28 An exaggeration is a truth that has lost its temper.
Kahlil Gibran 1883–1931: *Sand and Foam* (1926)

29 I maintain that Truth is a pathless land, and you cannot approach it by any path whatsoever, by any religion, by any sect.
Jiddu Krishnamurti d. 1986: remark, 1929, in L. Heber *Krishnamurti* (1931)

30 The truth is often a terrible weapon of aggression. It is possible to lie, and even to murder, for the truth.
Alfred Adler 1870–1937: *The Problems of Neurosis* (1929)

31 The truth which makes men free is for the most part the truth which men prefer not to hear.
Herbert Agar 1897–1980: *A Time for Greatness* (1942)

32 There are no whole truths; all truths are half-truths. It is trying to treat them as whole truths that plays the devil.
Alfred North Whitehead 1861–1947: *Dialogues* (1954)

33 Nagging is the repetition of unpalatable truths.
Edith Summerskill 1901–80: speech to the Married Women's Association, 14 July 1960

34 One of the favourite maxims of my father was the distinction between the two sorts of truths, profound truths recognized by the fact that the opposite is also a profound truth, in contrast to trivialities where opposites are obviously absurd.
Niels Bohr 1885–1962: in S. Rozental *Niels Bohr* (1967)

35 It contains a misleading impression, not a lie. It was being economical with the truth.
Robert Armstrong 1927– : during the 'Spycatcher' trial, Supreme Court, New South Wales, in *Daily Telegraph* 19 November 1986

Unbelief see **Belief and Unbelief**

The Universe

1 Had I been present at the Creation, I would have given some useful hints for the better ordering of the universe.
Alfonso 'the Wise', King of Castile 1221–84: on studying the Ptolemaic system (attributed)

2 There are more things in heaven and earth, Horatio,
Than are dreamt of in your philosophy.
William Shakespeare 1564–1616: *Hamlet* (1601)

3 'Gad! she'd better!'
Thomas Carlyle 1795–1881: on hearing that Margaret Fuller 'accept[ed] the universe'; in William James *Varieties of Religious Experience* (1902)

4 The world is the best of all possible worlds, and everything in it is a necessary evil.
F. H. Bradley 1846–1924: *Appearance and Reality* (1893)

5 The world is disgracefully managed, one hardly knows to whom to complain.
Ronald Firbank 1886–1926: *Vainglory* (1915)

6 The world is everything that is the case.
Ludwig Wittgenstein 1889–1951: *Tractatus Logico-Philosophicus* (1922)

7 Now, my own suspicion is that the universe is not only queerer than we suppose, but queerer than we *can* suppose.
J. B. S. Haldane 1892–1964: *Possible Worlds* (1927)

8 From the intrinsic evidence of his creation, the Great Architect of the Universe now begins to appear as a pure mathematician.
James Jeans 1877–1946: *The Mysterious Universe* (1930)

9 This, now, is the judgement of our scientific age—the third reaction of man upon the universe! This universe is not hostile, nor yet is it friendly. It is simply indifferent.
John H. Holmes 1879–1964: *The Sensible Man's View of Religion* (1932)

10 For one of those gnostics, the visible universe was an illusion or, more precisely, a sophism. Mirrors and fatherhood are abominable because they multiply it and extend it.
Jorge Luis Borges 1899–1986: *Tlön, Uqbar, Orbis Tertius* (1941)

11 We milk the cow of the world, and as we do
We whisper in her ear, 'You are not true.'
Richard Wilbur 1921– : 'Epistemology' (1950)

12 If we find the answer to that [why it is that we and the universe exist], it would be the ultimate triumph of human reason—for then we would know the mind of God.
Stephen Hawking 1942– : *A Brief History of Time* (1988)

Vice see **Virtue and Vice**

Violence

1 Who overcomes
By force, hath overcome but half his foe.
John Milton 1608–74: *Paradise Lost* (1667)

2 Beware of the man who does not return your blow: he neither forgives you nor allows you to forgive yourself.
George Bernard Shaw 1856–1950: *Man and Superman* (1903)

3 If you strike a child take care that you strike it in anger, even at the risk of maiming it for life. A blow in cold blood neither can nor should be forgiven.
George Bernard Shaw 1856–1950: *Man and Superman* (1903)

4 Where force is necessary, there it must be applied boldly, decisively and completely. But one must know the limitations of force; one must know when to blend force with a manoeuvre, a blow with an agreement.
Leon Trotsky 1879–1940: *What Next?* (1932)

5 Pale Ebenezer thought it wrong to fight,
But Roaring Bill (who killed him) thought it right.
Hilaire Belloc 1870–1953: 'The Pacifist' (1938)

6 A riot is at bottom the language of the unheard.
Martin Luther King 1929–68: *Where Do We Go From Here?* (1967)

7 I say violence is necessary. It is as American as cherry pie.
H. Rap Brown 1943– : speech, 27 July 1967

8 Keep violence in the mind
Where it belongs.
Brian Aldiss 1925– : 'Charteris' (1969)

9 The quietly pacifist peaceful
always die
to make room for men
who shout.
Alice Walker 1944– : 'The QPP' (1973)

10 Not hard enough.
Zsa Zsa Gabor 1919– : when asked how hard she had slapped a policeman; in *Independent* 21 September 1989

Virtue and Vice

See also **Good and Evil**

1 He that is without sin among you, let him first cast a stone at her.
Bible: St John

2 For the good that I would I do not: but the evil which I would not, that I do.
Bible: Romans

3 No one ever suddenly became depraved.
Juvenal AD *c.*60–*c.*130: *Satires*

4 Would that we had spent one whole day well in this world!
Thomas à Kempis *c.*1380–1471: *De Imitatione Christi*

5 How far that little candle throws his beams!
So shines a good deed in a naughty world.
William Shakespeare 1564–1616: *The Merchant of Venice* (1596–8)

6 How oft the sight of means to do ill deeds
Makes ill deeds done!
William Shakespeare 1564–1616: *King John* (1591–8)

7 Dost thou think, because thou art virtuous, there shall be no more cakes and ale?
William Shakespeare 1564–1616: *Twelfth Night* (1601)

8 Virtue is like a rich stone, best plain set.
Francis Bacon 1561–1626: *Essays* (1625) 'Of Beauty'

9 Virtue could see to do what Virtue would
By her own radiant light, though sun and moon
Were in the flat sea sunk.
John Milton 1608–74: *Comus* (1637)

10 Vice came in always at the door of necessity, not at the door of inclination.
Daniel Defoe 1660–1731: *Moll Flanders* (1721)

11 Virtue she finds too painful an endeavour,
Content to dwell in decencies for ever.
Alexander Pope 1688–1744: 'To a Lady' (1735)

12 But if he does really think that there is no distinction between virtue and vice, why, Sir, when he leaves our houses, let us count our spoons.

Samuel Johnson 1709–84: in James Boswell *Life of Johnson* (1791) 14 July 1763

13 Virtue knows to a farthing what it has lost by not having been vice.

Horace Walpole 1717–97: in L. Kronenberger *The Extraordinary Mr Wilkes* (1974)

14 Be good, sweet maid, and let who will be clever.

Charles Kingsley 1819–75: 'A Farewell' (1858)

15 Change in a trice
The lilies and languors of virtue
For the raptures and roses of vice.

Algernon Charles Swinburne 1837–1909: 'Dolores' (1866)

16 What is virtue but the Trade Unionism of the married?

George Bernard Shaw 1856–1950: *Man and Superman* (1903)

17 What after all
Is a halo? It's only one more thing to keep clean.

Christopher Fry 1907– : *The Lady's not for Burning* (1949)

18 An orgy looks particularly alluring seen through the mists of righteous indignation.

Malcolm Muggeridge 1903–90: *The Most of Malcolm Muggeridge* (1966)

Voting

See also **Democracy**

1 Vote early and vote often.

Anonymous: US election slogan, already current when quoted by William Porcher Miles in the House of Representatives, 31 March 1858

2 To give victory to the right, not bloody bullets, but peaceful ballots only, are necessary.

Abraham Lincoln 1809–65: speech, 18 May 1858; usually quoted 'The ballot is stronger than the bullet'

3 An election is coming. Universal peace is declared, and the foxes have a sincere interest in prolonging the lives of the poultry.
George Eliot 1819–80: *Felix Holt* (1866)

4 I always voted at my party's call,
And I never thought of thinking for myself at all.
W. S. Gilbert 1836–1911: *HMS Pinafore* (1878)

5 The accursed power which stands on Privilege
(And goes with Women, and Champagne, and Bridge)
Broke—and Democracy resumed her reign:
(Which goes with Bridge, and Women and Champagne).
Hilaire Belloc 1870–1953: 'On a Great Election' (1923)

6 Elections are won by men and women chiefly because most people vote against somebody rather than for somebody.
Franklin P. Adams 1881–1960: *Nods and Becks* (1944)

7 Hell, I never vote *for* anybody. I always vote *against*.
W. C. Fields 1880–1946: in R. L. Taylor *W. C. Fields* (1950)

8 Vote for the man who promises least; he'll be the least disappointing.
Bernard Baruch 1870–1965: in M. Berger *New York* (1960)

9 It's not the voting that's democracy, it's the counting.
Tom Stoppard 1937– : *Jumpers* (1972)

10 You won the elections, but I won the count.
Anastasio Somoza 1925–80: replying to an accusation of ballot-rigging, in *Guardian* 17 June 1977

Wales

1 Though it appear a little out of fashion,
There is much care and valour in this Welshman.
William Shakespeare 1564–1616: *Henry V* (1599)

2 The land of my fathers. My fathers can have it.
Dylan Thomas 1914–53: in *Adam* December 1953

3 There is no present in Wales,
And no future;
There is only the past,
Brittle with relics...
And an impotent people,

Sick with inbreeding,
Worrying the carcase of an old song.
R. S. Thomas 1913– : 'Welsh Landscape' (1955)

4 There are still parts of Wales where the only concession to gaiety is a striped shroud.
Gwyn Thomas 1913– : *Punch* 18 June 1958

5 It profits a man nothing to give his soul for the whole world...But for Wales—!
Robert Bolt 1924– : *A Man for All Seasons* (1960)

War

See also **The Army**

1 We make war that we may live in peace.
Aristotle 384–322 BC: *Nicomachean Ethics*

2 Laws are silent in time of war.
Cicero 106–43 BC: *Pro Milone*

3 Once more unto the breach, dear friends, once more;
Or close the wall up with our English dead!
In peace there's nothing so becomes a man
As modest stillness and humility:
But when the blast of war blows in our ears,
Then imitate the action of the tiger;
Stiffen the sinews, summon up the blood,
Disguise fair nature with hard-favoured rage.
William Shakespeare 1564–1616: *Henry V* (1599)

4 Happy is that city which in time of peace thinks of war.
Anonymous: inscription found in the armoury of Venice, in Robert Burton *Anatomy of Melancholy* (1621–51)

5 For as the nature of foul weather, lieth not in a shower or two of rain; but in an inclination thereto of many days together: so the nature of war consisteth not in actual fighting, but in the known disposition thereto during all the time there is no assurance to the contrary.
Thomas Hobbes 1588–1679: *Leviathan* (1651)

6 As you know, God is usually on the side of the big squadrons against the small.
Comte de Bussy-Rabutin 1618–93: letter to the Comte de Limoges, 18 October 1677

7 One to destroy, is murder by the law;
And gibbets keep the lifted hand in awe;
To murder thousands, takes a specious name,
'War's glorious art', and gives immortal fame.
Edward Young 1683–1765: *The Love of Fame* (1725–8)

8 Among the calamities of war may be jointly numbered the diminution of the love of truth, by the falsehoods which interest dictates and credulity encourages.
Samuel Johnson 1709–84: *The Idler* 11 November 1758; possibly the source of 'When war is declared, Truth is the first casualty', epigraph to Arthur Ponsonby's *Falsehood in Wartime* (1928); attributed also to Hiram Johnson, speaking in the US Senate, 1918

9 There never was a good war, or a bad peace.
Benjamin Franklin 1706–90: letter to Josiah Quincy, 11 September 1783

10 Next to a battle lost, the greatest misery is a battle gained.
Duke of Wellington 1769–1852: in *Diary of Frances, Lady Shelley 1787–1817* (ed. R. Edgcumbe)

11 I used to say of him [Napoleon] that his presence on the field made the difference of forty thousand men.
Duke of Wellington 1769–1852: in Philip Henry Stanhope *Notes of Conversations with the Duke of Wellington* (1888) 2 November 1831

12 War is nothing but a continuation of politics with the admixture of other means.
Karl von Clausewitz 1780–1831: *On War* (1832–4) commonly rendered 'War is the continuation of politics by other means'

13 Everything is very simple in war, but the simplest thing is difficult. These difficulties accumulate and produce a friction which no man can imagine exactly who has not seen war.
Karl von Clausewitz 1780–1831: *On War* (1832–4), tr. J. Graham

14 It is well that war is so terrible. We should grow too fond of it.

Robert E. Lee 1807–70: after the battle of Fredericksburg, December 1862 (attributed)

15 There is many a boy here to-day who looks on war as all glory, but, boys, it is all hell.

General Sherman 1820–91: speech at Columbus, Ohio, 11 August 1880

16 The essence of war is violence. Moderation in war is imbecility.

John Arbuthnot Fisher 1841–1920: lecture notes 1899–1902

17 A man who is good enough to shed his blood for the country is good enough to be given a square deal afterwards.

Theodore Roosevelt 1858–1919: speech, 4 June 1903

18 BATTLE, *n.* A method of untying with the teeth a political knot that would not yield to the tongue.

Ambrose Bierce 1842–*c.*1914: *The Cynic's Word Book* (1906)

19 Yes; quaint and curious war is!
You shoot a fellow down
You'd treat if met where any bar is,
Or help to half-a-crown.

Thomas Hardy 1840–1928: 'The Man he Killed' (1909)

20 War is hell, and all that, but it has a good deal to recommend it. It wipes out all the small nuisances of peace-time.

Ian Hay 1876–1952: *The First Hundred Thousand* (1915)

21 My subject is War, and the pity of War.
The Poetry is in the pity.

Wilfred Owen 1893–1918: *Poems* (1963) preface (written 1918)

22 Waste of Blood, and waste of Tears,
Waste of youth's most precious years,
Waste of ways the saints have trod,
Waste of Glory, waste of God,
War!

G. A. Studdert Kennedy 1883–1929: 'Waste' (1919)

23 It is easier to make war than to make peace.

Georges Clemenceau 1841–1929: speech at Verdun, 20 July 1919

24 When we, the Workers, all demand: 'What are WE fighting for?' . . .

Then, then we'll end that stupid crime, that devil's
madness—War.
Robert W. Service 1874–1958: 'Michael' (1921)

25 War hath no fury like a non-combatant.
C. E. Montague 1867–1928: *Disenchantment* (1922)

26 When war enters a country
It produces lies like sand.
Anonymous: epigraph to A. Ponsonby *Falsehood in Wartime* (1928)

27 War is too serious a matter to entrust to military men.
Georges Clemenceau 1841–1929: attributed to Clemenceau, e.g. in H. Jackson *Clemenceau and the Third Republic* (1946), but also to Briand and Talleyrand

28 I am not only a pacifist but a militant pacifist. I am willing to fight for peace. Nothing will end war unless the people themselves refuse to go to war.
Albert Einstein 1879–1955: interview with G. S. Viereck, January 1931

29 The bomber will always get through. The only defence is in offence, which means that you have to kill more women and children more quickly than the enemy if you want to save yourselves.
Stanley Baldwin 1867–1947: speech, House of Commons, 10 November 1932

30 Since the day of the air, the old frontiers are gone. When you think of the defence of England you no longer think of the chalk cliffs of Dover; you think of the Rhine.
Stanley Baldwin 1867–1947: speech, House of Commons, 30 July 1934

31 The sword is the axis of the world and its power is absolute.
Charles de Gaulle 1890–1970: *Vers l'armée de métier* (1934)

32 If we are attacked we can only defend ourselves with guns not with butter.
Joseph Goebbels 1897–1945: speech in Berlin, 17 January 1936

33 Would you rather have butter or guns?... preparedness makes us powerful. Butter merely makes us fat.
Hermann Goering 1893–1946: speech at Hamburg, 1936, in W. Frischauer *Goering* (1951).

34 Little girl...Sometime they'll give a war and nobody will come.

Carl Sandburg 1878–1967: *The People, Yes* (1936); 'Suppose They Gave a War and Nobody Came?' was the title of a 1970 film

35 In war, whichever side may call itself the victor, there are no winners, but all are losers.

Neville Chamberlain 1869–1940: speech at Kettering, 3 July 1938

36 They have gone too long without a war here. Where is morality to come from in such a case, I ask? Peace is nothing but slovenliness, only war creates order.

Bertolt Brecht 1898–1956: *Mother Courage* (1939)

37 A bayonet is a weapon with a worker at each end.

Anonymous: British pacifist slogan (1940)

38 Probably the battle of Waterloo *was* won on the playing-fields of Eton, but the opening battles of all subsequent wars have been lost there.

George Orwell 1903–50: *The Lion and the Unicorn* (1941)

39 And as for war, my wars
Were global from the start.

Henry Reed 1914–86: 'Lessons of the War: 3, Unarmed Combat' (1946)

40 I have never met anyone who wasn't against war. Even Hitler and Mussolini were, according to themselves.

David Low 1891–1963: *New York Times Magazine* 10 February 1946

41 The quickest way of ending a war is to lose it.

George Orwell 1903–50: *Polemic* May 1946

42 This world in arms is not spending money alone. It is spending the sweat of its labourers, the genius of its scientists, the hopes of its children.

Dwight D. Eisenhower 1890–1969: speech in Washington, 16 April 1953

43 Spare us all word of the weapons, their force and range,
The long numbers that rocket the mind.

Richard Wilbur 1921– : 'Advice to a Prophet' (1961)

44 Dead battles, like dead generals, hold the military mind in their dead grip and Germans, no less than other peoples, prepare for the last war.

Barbara W. Tuchman 1912–89: *August 1914* (1962)

45 Rule 1, on page 1 of the book of war, is: 'Do not march on Moscow'...[Rule 2] is: 'Do not go fighting with your land armies in China.'

Field Marshal Montgomery 1887–1976: speech, House of Lords, 30 May 1962

46 History is littered with the wars which everybody knew would never happen.

Enoch Powell 1912– : speech, 19 October 1967

47 War is capitalism with the gloves off.

Tom Stoppard 1937– : *Travesties* (1975)

48 I question the right of that great Moloch, national sovereignty, to burn its children to save its pride.

Anthony Meyer 1920– : speaking against the Falklands War, 1982; in *Listener* 27 September 1990

Wars

1 They now *ring* the bells, but they will soon *wring* their hands.

Robert Walpole 1676–1745: on the declaration of war with Spain, 1739; in W. Coxe *Memoirs of Sir Robert Walpole* (1798)

2 I have only one eye,—I have a right to be blind sometimes...I really do not see the signal!

Horatio, Lord Nelson 1758–1805: at the battle of Copenhagen, in R. Southey *Life of Nelson* (1813)

3 *Guerra a cuchillo.*
War to the knife.

José de Palafox 1780–1847: at the siege of Saragossa, 4 August 1808, replying to the suggestion that he should surrender (as reported). He actually said '*Guerra y cuchillo* [War and the knife]'

4 Up Guards and at them!

Duke of Wellington 1769–1852: in *The Battle of Waterloo* by a Near Observer [J. Booth] (1815), later denied by Wellington

5 Hard pounding this, gentlemen; let's see who will pound longest.

Duke of Wellington 1769–1852: at the Battle of Waterloo; in Sir Walter Scott *Paul's Letters* (1816)

6 The battle of Waterloo was won on the playing fields of Eton.

Duke of Wellington 1769–1852: oral tradition, but not found in this form of words. See C. F. R. Montalembert *De l'avenir politique de l'Angleterre* (1856)

7 The angel of death has been abroad throughout the land; you may almost hear the beating of his wings.

John Bright 1811–89: on the effects of the Crimean war; speech, House of Commons, 23 February 1855

8 All quiet along the Potomac.

General George B. McClellan 1826–85: said at the time of the American Civil War (attributed)

9 If there is ever another war in Europe, it will come out of some damned silly thing in the Balkans.

Otto von Bismarck 1815–98: quoted in the House of Commons, 16 August 1945

10 My centre is giving way, my right is retreating, situation excellent, I am attacking.

Ferdinand Foch 1851–1929: message during the first Battle of the Marne, September 1914

11 In Flanders fields the poppies blow
Between the crosses, row on row.

John McCrae 1872–1918: 'In Flanders Fields' (1915)

12 *Ils ne passeront pas.*
They shall not pass.

Anonymous: slogan of the French army at the defence of Verdun, 1916; variously attributed to Marshal Pétain and to General Robert Nivelle

13 My home policy: I wage war; my foreign policy: I wage war. All the time I wage war.

Georges Clemenceau 1841–1929: speech to French Chamber of Deputies, 8 March 1918

14 I hope we may say that thus, this fateful morning, came to an end all wars.

David Lloyd George 1863–1945: speech, House of Commons, 11 November 1918

15 This is not a peace treaty, it is an armistice for twenty years.

Ferdinand Foch 1851–1929: at the signing of the Treaty of Versailles, 1919; in P. Reynaud *Mémoires* (1963)

16 *No pasarán.*
They shall not pass.

Dolores Ibarruri 1895–1989: radio broadcast, Madrid, 19 July 1936

17 I have nothing to offer but blood, toil, tears and sweat.

Winston Churchill 1874–1965: speech, House of Commons, 13 May 1940

18 We shall not flag or fail. We shall go on to the end. We shall fight in France, we shall fight on the seas and oceans, we shall fight with growing confidence and growing strength in the air, we shall defend our island, whatever the cost may be. We shall fight on the beaches, we shall fight on the landing grounds, we shall fight in the fields and in the streets, we shall fight in the hills; we shall never surrender.

Winston Churchill 1874–1965: speech, House of Commons, 4 June 1940

19 Let us therefore brace ourselves to our duty, and so bear ourselves that, if the British Empire and its Commonwealth lasts for a thousand years, men will still say, 'This was their finest hour.'

Winston Churchill 1874–1965: speech, House of Commons, 18 June 1940

20 Never in the field of human conflict was so much owed by so many to so few.

Winston Churchill 1874–1965: on the Battle of Britain; speech, House of Commons, 20 August 1940

21 I'm glad we've been bombed. It makes me feel I can look the East End in the face.

Queen Elizabeth, the Queen Mother 1900– : to a London policeman, 13 September 1940

22 We must be the great arsenal of democracy.

Franklin D. Roosevelt 1882–1945: broadcast, 29 December 1940

23 Give us the tools and we will finish the job.
Winston Churchill 1874–1965: radio broadcast, 9 February 1941

24 I think we might be going a bridge too far.
Frederick Browning 1896–1965: expressing reservations about the Arnhem 'Market Garden' operation, 10 September 1944

25 The First World War had begun—imposed on the statesmen of Europe by railway timetables.
A. J. P. Taylor 1906–90: *The First World War* (1963)

26 We're going to bomb them back into the Stone Age.
Curtis E. LeMay 1906–90: on the North Vietnamese, in *Mission with LeMay* (1965)

27 Anyone who isn't confused doesn't really understand the situation.
Ed Murrow 1908–65: on the Vietnam War, in Walter Bryan *The Improbable Irish* (1969)

28 It became necessary to destroy the town to save it.
Anonymous: statement issued by US Army, referring to Ben Tre in Vietnam; in *New York Times* 8 February 1968

29 I counted them all out and I counted them all back.
Brian Hanrahan 1949– : on the number of British aeroplanes joining the raid on Port Stanley; BBC broadcast report, 1 May 1982

30 The Falklands thing was a fight between two bald men over a comb.
Jorge Luis Borges 1899–1986: in *Time* 14 February 1983

Wealth

See also **Money**

1 How many things I can do without!
Socrates 469–399 BC: on looking at a multitude of wares exposed for sale, in Diogenes Laertius *Lives of the Philosophers*

2 It is easier for a camel to go through the eye of a needle, than for a rich man to enter into the kingdom of God.
Bible: St Matthew

3 Riches are a good handmaid, but the worst mistress.
Francis Bacon 1561–1626: *De Dignitate et Augmentis Scientiarum* (1623)

4 Let none admire
That riches grow in hell; that soil may best
Deserve the precious bane.
John Milton 1608–74: *Paradise Lost* (1667)

5 It was very prettily said, that we may learn the little value of fortune by the persons on whom heaven is pleased to bestow it.
Richard Steele 1672–1729: *The Tatler* 27 July 1710

6 We are all Adam's children but silk makes the difference.
Thomas Fuller 1654–1734: *Gnomologia* (1732)

7 The chief enjoyment of riches consists in the parade of riches.
Adam Smith 1723–90: *Wealth of Nations* (1776)

8 We are not here to sell a parcel of boilers and vats, but the potentiality of growing rich, beyond the dreams of avarice.
Samuel Johnson 1709–84: at the sale of Thrale's brewery; in James Boswell *Life of Johnson* (1791) 6 April 1781

9 The man who dies…rich dies disgraced.
Andrew Carnegie 1835–1919: *North American Review* June 1889

10 In every well-governed state, wealth is a sacred thing; in democracies it is the only sacred thing.
Anatole France 1844–1924: *L'Île des pingouins* (1908)

11 Let me tell you about the very rich. They are different from you and me.
F. Scott Fitzgerald 1896–1940: to which Ernest Hemingway replied, 'Yes, they have more money'; *All the Sad Young Men* (1926)

12 To suppose, as we all suppose, that we could be rich and not behave as the rich behave, is like supposing that we could drink all day and keep absolutely sober.
Logan Pearsall Smith 1865–1946: *Afterthoughts* (1931)

13 If all the rich people in the world divided up their money among themselves there wouldn't be enough to go round.
Christina Stead 1902–83: *House of All Nations* (1938)

14 A kiss on the hand may be quite continental,
But diamonds are a girl's best friend.
Leo Robin 1900– : 'Diamonds are a Girl's Best Friend' (1949 song); from the film *Gentlemen Prefer Blondes*

15 The greater the wealth, the thicker will be the dirt.
J. K. Galbraith 1908– : *The Affluent Society* (1958)

16 Will the people in the cheaper seats clap your hands? All the rest of you, if you'll just rattle your jewellery.
John Lennon 1940–80: at Royal Variety Performance, 4 November 1963

17 I've been rich and I've been poor: rich is better.
Sophie Tucker *c.*1884–1966: attributed

Weather

1 So foul and fair a day I have not seen.
William Shakespeare 1564–1616: *Macbeth* (1606)

2 When two Englishmen meet, their first talk is of the weather.
Samuel Johnson 1709–84: *The Idler* 24 June 1758

3 The best sun we have is made of Newcastle coal.
Horace Walpole 1717–97: letter to George Montagu, 15 June 1768

4 The frost performs its secret ministry,
Unhelped by any wind.
Samuel Taylor Coleridge 1772–1834: 'Frost at Midnight' (1798)

5 It is impossible to live in a country which is continually under hatches…Rain! Rain! Rain!
John Keats 1795–1821: letter to Reynolds from Devon, 10 April 1818

6 St Agnes' Eve—Ah, bitter chill it was!
The owl, for all his feathers, was a-cold.
John Keats 1795–1821: 'The Eve of St Agnes' (1820)

7 This is a London particular…A fog, miss.
Charles Dickens 1812–70: *Bleak House* (1853)

8 When men were all asleep the snow came flying,
In large white flakes falling on the city brown,

Stealthily and perpetually settling and loosely lying,
Hushing the latest traffic of the drowsy town.
Robert Bridges 1844–1930: 'London Snow' (1890)

9 The rain, it raineth on the just
And also on the unjust fella:
But chiefly on the just, because
The unjust steals the just's umbrella.
Lord Bowen 1835–94: in W. Sichel *Sands of Time* (1923)

10 The fog comes
on little cat feet. It sits looking
over harbour and city
on silent haunches
and then moves on.
Carl Sandburg 1878–1967: 'Fog' (1916)

11 The yellow fog that rubs its back upon the window-panes.
T. S. Eliot 1888–1965: 'Love Song of J. Alfred Prufrock' (1917)

12 This is the weather the cuckoo likes,
And so do I;
When showers betumble the chestnut spikes,
And nestlings fly.
Thomas Hardy 1840–1928: 'Weathers' (1922)

13 Children are dumb to say how hot the day is,
How hot the scent is of the summer rose.
Robert Graves 1895–1985: 'The Cool Web' (1927)

14 Every time it rains, it rains
Pennies from heaven.
Don't you know each cloud contains
Pennies from heaven?
Johnny Burke 1908–64: 'Pennies from Heaven' (1936 song)

15 It was the wrong kind of snow.
Terry Worrall: explaining disruption on British Rail, in *The Independent* 16 February 1991

Woman's Role

See also **Men and Women**

1 The First Blast of the Trumpet Against the Monstrous Regiment of Women.
John Knox *c.*1505–72: title of pamphlet (1558)

2 Be to her virtues very kind;
Be to her faults a little blind;
Let all her ways be unconfined;
And clap your padlock—on her mind.
Matthew Prior 1664–1721: 'An English Padlock' (1705)

3 If all men are born free, how is it that all women are born slaves?
Mary Astell 1668–1731: *Some Reflections upon Marriage* (1706 ed.)

4 A woman's preaching is like a dog's walking on his hinder legs. It is not done well; but you are surprised to find it done at all.
Samuel Johnson 1709–84: in James Boswell *Life of Johnson* (1791) 31 July 1763

5 How much it is to be regretted, that the British ladies should ever sit down contented to polish, when they are able to reform; to entertain, when they might instruct; and to dazzle for an hour, when they are candidates for eternity!
Hannah More 1745–1833: *Essays on Various Subjects...for Young Ladies* (1777) 'On Dissipation'

6 Can anything be more absurd than keeping women in a state of ignorance, and yet so vehemently to insist on their resisting temptation?
Vicesimus Knox 1752–1821: in Mary Wollstonecraft *A Vindication of the Rights of Woman* (1792)

7 I do not wish them [women] to have power over men; but over themselves.
Mary Wollstonecraft 1759–97: *A Vindication of the Rights of Woman* (1792)

8 The Queen is most anxious to enlist every one who can speak or write to join in checking this mad, wicked folly of 'Woman's Rights', with all its attendant horrors, on which her poor feeble sex is bent, forgetting every sense of womanly feeling and propriety.

Queen Victoria 1819–1901: letter to Theodore Martin, 29 May 1870

9 The one point on which all women are in furious secret rebellion against the existing law is the saddling of the right to a child with the obligation to become the servant of a man.

George Bernard Shaw 1856–1950: *Getting Married* (1911)

10 The worker is the slave of capitalist society, the female worker is the slave of that slave.

James Connolly 1868–1916: *The Re-conquest of Ireland* (1915)

11 One is not born a woman: one becomes one.

Simone de Beauvoir 1908–86: *The Second Sex* (1949)

12 The only position for women in SNCC is prone.

Stokely Carmichael 1941– : response to a question about the position of women at a Student Nonviolent Coordinating Committee conference, November 1964

13 But if God had wanted us to think just with our wombs, why did He give us a brain?

Clare Booth Luce 1903– : *Life* 16 October 1970

14 I didn't fight to get women out from behind the vacuum cleaner to get them onto the board of Hoover.

Germaine Greer 1939– : in *Guardian* 27 October 1986

15 Feminism is the most revolutionary idea there has ever been. Equality for women demands a change in the human psyche more profound then anything Marx dreamed of. It means valuing parenthood as much as we value banking.

Polly Toynbee 1946– : *Guardian* 19 January 1987

Women

See also **Men and Women**

1 Who can find a virtuous woman? for her price is far above rubies.

Bible: Proverbs

2 The greatest glory of a woman is to be least talked about by men.
Pericles *c.*495–429 BC: in Thucydides *History of the Peloponnesian War*, tr. R. Warner

3 *Varium et mutabile semper*
Femina.
Fickle and changeable always is woman.
Virgil 70–19 BC: *Aeneid*

4 She is a woman, therefore may be wooed;
She is a woman, therefore may be won.
William Shakespeare 1564–1616: *Titus Andronicus* (1590)

5 Frailty, thy name is woman!
William Shakespeare 1564–1616: *Hamlet* (1601)

6 Age cannot wither her, nor custom stale
Her infinite variety; other women cloy
The appetites they feed, but she makes hungry
Where most she satisfies.
William Shakespeare 1564–1616: *Antony and Cleopatra* (1606–7)

7 The weaker sex, to piety more prone.
William Alexander, Earl of Stirling *c.*1567–1640: 'Doomsday' 5th Hour (1637)

8 *Elle flotte, elle hésite; en un mot, elle est femme.*
She floats, she hesitates; in a word, she's a woman.
Jean Racine 1639–99: *Athalie* (1691)

9 When once a woman has given you her heart, you can never get rid of the rest of her body.
John Vanbrugh 1664–1726: *The Relapse* (1696)

10 She knows her man, and when you rant and swear,
Can draw you to her *with a single hair*.
John Dryden 1631–1700: translation of Persius *Satires*

11 Women, then, are only children of a larger growth.
Lord Chesterfield 1694–1773: *Letters to his Son* (1774) 5 September 1748

12 Here's to the maiden of bashful fifteen
Here's to the widow of fifty
Here's to the flaunting, extravagant quean;
And here's to the housewife that's thrifty.
Richard Brinsley Sheridan 1751–1816: *The School for Scandal* (1777)

13 O Woman! in our hours of ease,
Uncertain, coy, and hard to please,
And variable as the shade
By the light quivering aspen made;
When pain and anguish wring the brow,
A ministering angel thou!

Sir Walter Scott 1771–1832: *Marmion* (1808)

14 All the privilege I claim for my own sex... is that of loving longest, when existence or when hope is gone.

Jane Austen 1775–1817: *Persuasion* (1818)

15 I have met with women whom I really think would like to be married to a poem and to be given away by a novel.

John Keats 1795–1821: letter to Fanny Brawne, 8 July 1819

16 In her first passion woman loves her lover,
In all the others all she loves is love.

Lord Byron 1788–1824: *Don Juan* (1819–24)

17 Eternal Woman draws us upward.

Johann Wolfgang von Goethe 1749–1832: *Faust* pt. 2 (1832)

18 I expect that Woman will be the last thing civilized by Man.

George Meredith 1828–1909: *The Ordeal of Richard Feverel* (1859)

19 The happiest women, like the happiest nations, have no history.

George Eliot 1819–80: *The Mill on the Floss* (1860)

20 Half the sorrows of women would be averted if they could repress the speech they know to be useless; nay, the speech they have resolved not to make.

George Eliot 1819–80: *Felix Holt* (1866)

21 Women—one half the human race at least—care fifty times more for a marriage than a ministry.

Walter Bagehot 1826–77: *The English Constitution* (1867)

22 Woman was God's second blunder.

Friedrich Nietzsche 1844–1900: *Der Antichrist* (1888)

23 One should never trust a woman who tells one her real age. A woman who would tell one that, would tell one anything.

Oscar Wilde 1854–1900: *A Woman of No Importance* (1893)

24 Vitality in a woman is a blind fury of creation.

George Bernard Shaw 1856–1950: *Man and Superman* (1903)

25 The prime truth of woman, the universal mother ... that if a thing is worth doing, it is worth doing badly.
G. K. Chesterton 1874–1936: *What's Wrong with the World* (1910)

26 A woman will always sacrifice herself if you give her the opportunity. It is her favourite form of self-indulgence.
W. Somerset Maugham 1874–1965: *Circle* (1921)

27 Women have no wilderness in them,
They are provident instead,
Content in the tight hot cell of their hearts
To eat dusty bread.
Louise Bogan 1897–1970: 'Women' (1923)

28 So this gentleman said a girl with brains ought to do something with them besides think.
Anita Loos 1893–1981: *Gentlemen Prefer Blondes* (1925)

29 The perpetual hunger to be beautiful and that thirst to be loved which is the real curse of Eve.
Jean Rhys *c.*1890–1979: *The Left Bank* (1927)

30 The great and almost only comfort about being a woman is that one can always pretend to be more stupid than one is and no one is surprised.
Freya Stark 1893–1993: *The Valleys of the Assassins* (1934)

31 What does a woman want?
Sigmund Freud 1856–1939: letter to Marie Bonaparte, in E. Jones *Sigmund Freud* (1955)

32 She's the sort of woman who lives for others—you can always tell the others by their hunted expression.
C. S. Lewis 1898–1963: *The Screwtape Letters* (1942)

33 Slamming their doors, stamping their high heels, banging their irons and saucepans—the eternal flaming racket of the female.
John Osborne 1929– : *Look Back in Anger* (1956)

34 Every woman adores a Fascist,
The boot in the face, the brute
Brute heart of a brute like you.
Sylvia Plath 1932–63: 'Daddy' (1963)

35 From birth to 18 a girl needs good parents. From 18 to 35, she needs good looks. From 35 to 55, good personality. From 55 on, she needs good cash.
Sophie Tucker 1884–1966: in M. Freedland *Sophie* (1978)

36 Woman is the nigger of the world.
Yoko Ono 1933– : interview for *Nova* magazine (1968); adopted by John Lennon as song title (1972)

37 Being a woman is of special interest only to aspiring male transsexuals. To actual women, it is merely a good excuse not to play football.
Fran Lebowitz 1946– : *Metropolitan Life* (1978)

38 We are becoming the men we wanted to marry.
Gloria Steinem 1934– : *Ms* July/August 1982

39 Good women always think it is their fault when someone else is being offensive. Bad women never take the blame for anything.
Anita Brookner 1938– : *Hotel du Lac* (1984)

Words

See also **Language, Meaning**

1 And once sent out, a word takes wing beyond recall.
Horace 65–8 BC: *Epistles*

2 Woord is but wynd; leff woord and tak the dede.
John Lydgate *c.*1370–*c.*1451: *Secrets of Old Philosophers*

3 What's in a name? that which we call a rose
By any other name would smell as sweet.
William Shakespeare 1564–1616: *Romeo and Juliet* (1595)

4 The words of Mercury are harsh after the songs of Apollo.
William Shakespeare 1564–1616: *Love's Labour's Lost* (1595)

5 But words are words; I never yet did hear
That the bruised heart was piercèd through the ear.
William Shakespeare 1564–1616: *Othello* (1602–4)

6 Words are the tokens current and accepted for conceits, as moneys are for values.
Francis Bacon 1561–1626: *The Advancement of Learning* (1605)

7 Thou whoreson zed! thou unnecessary letter!
William Shakespeare 1564–1616: *King Lear* (1605–6)

8 Syllables govern the world.
John Selden 1584–1654: *Table Talk* (1689) 'Power: State'

9 A man who could make so vile a pun would not scruple to pick a pocket.
John Dennis 1657–1734: *The Gentleman's Magazine* (1781), editorial note

10 *Lexicographer*. A writer of dictionaries, a harmless drudge.
Samuel Johnson 1709–84: *Dictionary of the English Language* (1755)

11 I am not yet so lost in lexicography as to forget that words are the daughters of earth, and that things are the sons of heaven. Language is only the instrument of science, and words are but the signs of ideas.
Samuel Johnson 1709–84: *Dictionary of the English Language* (1755) preface

12 'When *I* use a word,' Humpty Dumpty said in a rather scornful tone, 'it means just what I choose it to mean—neither more nor less.'
Lewis Carroll 1832–98: *Through the Looking-Glass* (1872)

13 Dialect words—those terrible marks of the beast to the truly genteel.
Thomas Hardy 1840–1928: *The Mayor of Casterbridge* (1886)

14 Some word that teems with hidden meaning—like Basingstoke.
W. S. Gilbert 1836–1911: *Ruddigore* (1887)

15 It cannot in the opinion of His Majesty's Government be classified as slavery in the extreme acceptance of the word without some risk of terminological inexactitude.
Winston Churchill 1874–1965: speech, House of Commons, 22 February 1906

16 Words are, of course, the most powerful drug used by mankind.
Rudyard Kipling 1865–1936: speech, 14 Febuary 1923

17 I gotta use words when I talk to you.
T. S. Eliot 1888–1965: *Sweeney Agonistes* (1932)

18 Words strain,
Crack and sometimes break, under the burden,
Under the tension, slip, slide, perish,
Decay with imprecision, will not stay in place,
Will not stay still.
T. S. Eliot 1888–1965: *Four Quartets* 'Burnt Norton' (1936)

Work

1 In the sweat of thy face shalt thou eat bread.
Bible: Genesis

2 For the labourer is worthy of his hire.
Bible: St Luke

3 If any would not work, neither should he eat.
Bible: II Thesssalonians

4 I have had my labour for my travail.
William Shakespeare 1564–1616: *Troilus and Cressida* (1602)

5 The labour we delight in physics pain.
William Shakespeare 1564–1616: *Macbeth* (1606)

6 We spend our midday sweat, our midnight oil;
We tire the night in thought, the day in toil.
Francis Quarles 1592–1644: *Emblems* (1635)

7 If you have great talents, industry will improve them: if you have but moderate abilities, industry will supply their deficiency.
Joshua Reynolds 1723–92: *Discourses on Art* (11 December 1769)

8 The world is too much with us; late and soon,
Getting and spending, we lay waste our powers.
William Wordsworth 1770–1850: 'The world is too much with us' (1807)

9 Blessèd are the horny hands of toil!
James Russell Lowell 1819–91: 'A Glance Behind the Curtain' (1844)

10 Which of us... is to do the hard and dirty work for the rest—and for what pay? Who is to do the pleasant and clean work, and for what pay?
John Ruskin 1819–1900: *Sesame and Lilies* (1865)

11 Life without industry is guilt, and industry without art is brutality.
John Ruskin 1819–1900: *Lectures on Art* (1870)

12 I like work: it fascinates me. I can sit and look at it for hours. I love to keep it by me: the idea of getting rid of it nearly breaks my heart.
Jerome K. Jerome 1859–1927: *Three Men in a Boat* (1889)

13 Work is the curse of the drinking classes.
Oscar Wilde 1854–1900: in H. Pearson *Life of Oscar Wilde* (1946)

14 Work is love made visible.
Kahlil Gibran 1883–1931: *The Prophet* (1923)

15 Perfect freedom is reserved for the man who lives by his own work and in that work does what he wants to do.
R. G. Collingwood 1889–1943: *Speculum Mentis* (1924)

16 That state is a state of slavery in which a man does what he likes to do in his spare time and in his working time that which is required of him.
Eric Gill 1882–1940: *Art-nonsense and Other Essays* (1929) 'Slavery and Freedom'.

17 One of the symptoms of approaching nervous breakdown is the belief that one's work is terribly important, and that to take a holiday would bring all kinds of disaster.
Bertrand Russell 1872–1970: *Conquest of Happiness* (1930)

18 Work is of two kinds: first, altering the position of matter at or near the earth's surface relatively to other such matter; second, telling other people to do so. The first kind is unpleasant and ill paid; the second is pleasant and highly paid.
Bertrand Russell 1872–1970: 'In Praise of Idleness' (1932)

19 A professional is a man who can do his job when he doesn't feel like it. An amateur is a man who can't do his job when he does feel like it.
James Agate 1877–1947: diary, 19 July 1945

20 Why should I let the toad *work*
Squat on my life?
Can't I use my wit as a pitchfork
And drive the brute off?
Philip Larkin 1922–85: 'Toads' (1955)

21 Work expands so as to fill the time available for its completion.
C. Northcote Parkinson 1909–93: *Parkinson's Law* (1958)

22 It's true hard work never killed anybody, but I figure why take the chance?
Ronald Reagan 1911– : interview, *Guardian* 31 March 1987

23 I have long been of the opinion that if work were such a splendid thing the rich would have kept more of it for themselves.

Bruce Grocott 1940– : *Observer* 22 May 1988 'Sayings of the Week'

Writers

See also **Poets, Shakespeare**

1 No man but a blockhead ever wrote, except for money.

Samuel Johnson 1709–84: in James Boswell *Life of Johnson* (1791) 5 April 1776

2 Another damned, thick, square book! Always scribble, scribble, scribble! Eh! Mr Gibbon?

Duke of Gloucester 1743–1805: in Henry Best *Personal and Literary Memorials* (1829) also attributed to the Duke of Cumberland and George III

3 Johnson hewed passages through the Alps, while Gibbon levelled walks through parks and gardens.

George Colman, the Younger 1762–1836: *Random Records* (1830)

4 Writers, like teeth, are divided into incisors and grinders.

Walter Bagehot 1826–77: *Estimates of some Englishmen and Scotchmen* (1858)

5 The work of Henry James has always seemed divisible by a simple dynastic arrangement into three reigns: James I, James II, and the Old Pretender.

Philip Guedalla 1889–1944: *Supers and Supermen* (1920)

6 When I am dead, I hope it may be said:
'His sins were scarlet, but his books were read.'

Hilaire Belloc 1870–1953: 'On His Books' (1923)

7 A dogged attempt to cover the universe with mud, an inverted Victorianism, an attempt to make crossness and dirt succeed where sweetness and light failed.

E. M. Forster 1879–1970: of James Joyce's *Ulysses*; *Aspects of the Novel* (1927)

8 A woman must have money and a room of her own if she is to write fiction.
Virginia Woolf 1882–1941: *A Room of One's Own* (1929)

9 English literature's performing flea.
Sean O'Casey 1880–1964: in P. G. Wodehouse *Performing Flea* (1953), describing the author

10 The writer's only responsibility is to his art. He will be completely ruthless if he is a good one... If a writer has to rob his mother, he will not hesitate; the *Ode on a Grecian Urn* is worth any number of old ladies.
William Faulkner 1897–1962: in *Paris Review* Spring 1956

11 I think like a genius, I write like a distinguished author, and I speak like a child.
Vladimir Nabokov 1899–1977: *Strong Opinions* (1973)

12 The shelf life of the modern hardback writer is somewhere between the milk and the yoghurt.
Calvin Trillin: in *Sunday Times* 9 June 1991 (attributed)

Writing

See also **Books, Literature, Poetry, Style**

1 Biting my truant pen, beating myself for spite,
'Fool,' said my Muse to me; 'look in thy heart and write.'
Philip Sidney 1554–86: *Astrophil and Stella* (1591)

2 And, as imagination bodies forth
The forms of things unknown, the poet's pen
Turns them to shapes, and gives to airy nothing
A local habitation and a name.
William Shakespeare 1564–1616: *A Midsummer Night's Dream* (1595–6)

3 So all my best is dressing old words new,
Spending again what is already spent.
William Shakespeare 1564–1616: sonnet 76 (1609)

4 What in me is dark
Illumine, what is low raise and support;
That to the height of this great argument

I may assert eternal providence,
And justify the ways of God to men.

John Milton 1608–74: *Paradise Lost* (1667)

5 Things unattempted yet in prose or rhyme.

John Milton 1608–74: *Paradise Lost* (1667)

6 We must beat the iron while it is hot, but we may polish it at leisure.

John Dryden 1631–1700: *Aeneis* (1697)

7 Eye Nature's walks, shoot Folly as it flies,
And catch the Manners living as they rise.
Laugh where we must, be candid where we can;
But vindicate the ways of God to man.

Alexander Pope 1688–1744: *An Essay on Man* Epistle 1 (1733)

8 But those who cannot write, and those who can,
All rhyme, and scrawl, and scribble, to a man.

Alexander Pope 1688–1744: *Imitations of Horace* (1737)

9 A man may write at any time, if he will set himself doggedly to it.

Samuel Johnson 1709–84: in James Boswell *Life of Johnson* (1791) March 1750

10 You write with ease, to show your breeding,
But easy writing's vile hard reading.

Richard Brinsley Sheridan 1751–1816: 'Clio's Protest' (written 1771)

11 Read over your compositions, and where ever you meet with a passage which you think is particularly fine, strike it out.

Samuel Johnson 1709–84: in James Boswell *Life of Johnson* (1791) 30 April 1773; quoting a college tutor

12 The composition of a tragedy requires *testicles.*

Voltaire 1694–1778: on being asked why no woman had ever written 'a tolerable tragedy'; letter from Byron to John Murray, 2 April 1817

13 What is written without effort is in general read without pleasure.

Samuel Johnson 1709–84: in W. Seward *Biographia* (1799)

14 Never forget what I believe was observed to you by Coleridge, that every great and original writer, in proportion as he is great and original, must himself create the taste by which he is to be relished.

William Wordsworth 1770–1850: letter to Lady Beaumont, 21 May 1807

15 Let other pens dwell on guilt and misery. I quit such odious subjects as soon as I can.

Jane Austen 1775–1817: *Mansfield Park* (1814)

16 The little bit (two inches wide) of ivory on which I work with so fine a brush, as produces little effect after much labour?

Jane Austen 1775–1817: letter to J. Edward Austen, 16 December 1816

17 All clean and comfortable I sit down to write.

John Keats 1795–1821: letter to George and Georgiana Keats, 17 September 1819

18 The Big Bow-Wow strain I can do myself like any now going; but the exquisite touch, which renders ordinary commonplace things and characters interesting, from the truth of the description and the sentiment, is denied to me.

Sir Walter Scott 1771–1832: on Jane Austen; diary, 14 March 1826

19 Beneath the rule of men entirely great
The pen is mightier than the sword.

Edward Bulwer-Lytton 1803–73: *Richelieu* (1839)

20 When once the itch of literature comes over a man, nothing can cure it but the scratching of a pen.

Samuel Lover 1797–1868: *Handy Andy* (1842)

21 Not that the story need be long, but it will take a long while to make it short.

Henry David Thoreau 1817–62: letter to Harrison Blake, 16 November 1857

22 They shut me up in prose—
As when a little girl
They put me in the closet—
Because they liked me 'still'.

Emily Dickinson 1830–86: 'They shut me up in prose' (*c.*1862)

23 As to the Adjective: when in doubt, strike it out.

Mark Twain 1835–1910: *Pudd'nhead Wilson* (1894)

24 To give an accurate and exhaustive account of the period would need a far less brilliant pen than mine.
Max Beerbohm 1872–1956: *Yellow Book* (1895)

25 Only connect!...Only connect the prose and the passion.
E. M. Forster 1879–1970: *Howards End* (1910)

26 The test of a round character is whether it is capable of surprising in a convincing way. If it never surprises, it is flat. If it does not convince, it is flat pretending to be round.
E. M. Forster 1879–1970: *Aspects of the Novel* (1927)

27 You praise the firm restraint with which they write—
I'm with you there, of course:
They use the snaffle and the curb all right,
But where's the bloody horse?
Roy Campbell 1901–57: 'On Some South African Novelists' (1930)

28 If you try to nail anything down in the novel, either it kills the novel, or the novel gets up and walks away with the nail.
D. H. Lawrence 1885–1930: *Phoenix* (1936) 'Morality and the Novel'

29 Morality in the novel is the trembling instability of the balance. When the novelist puts his thumb in the scale, to pull down the balance to his own predilection, that is immorality.
D. H. Lawrence 1885–1930: *Phoenix* (1936) 'Morality and the Novel'

30 If you steal from one author, it's plagiarism; if you steal from many, it's research.
Wilson Mizner 1876–1933: in A. Johnston *The Legendary Mizners* (1953)

31 A writer's ambition should be...to trade a hundred contemporary readers for ten readers in ten years' time and for one reader in a hundred years.
Arthur Koestler 1905–83: in *New York Times Book Review* 1 April 1951

32 Writing is not a profession but a vocation of unhappiness.
Georges Simenon 1903–89: interview in *Paris Review* Summer 1955

Youth

See also **The Generation Gap**

1 Whom the gods love dies young.
Menander 342–*c.*292 BC: *Dis Exapaton*

2 My salad days,
When I was green in judgement.
William Shakespeare 1564–1616: *Antony and Cleopatra* (1606–7)

3 Two lads that thought there was no more behind
But such a day to-morrow as to-day,
And to be boy eternal.
William Shakespeare 1564–1616: *The Winter's Tale* (1610–11)

4 The atrocious crime of being a young man...I shall neither attempt to palliate nor deny.
William Pitt 1708–78: speech, House of Commons, 2 March 1741

5 In gallant trim the gilded vessel goes;
Youth on the prow, and Pleasure at the helm.
Thomas Gray 1716–71: 'The Bard' (1757)

6 Heaven lies about us in our infancy!
Shades of the prison-house begin to close
Upon the growing boy.
William Wordsworth 1770–1850: 'Ode. Intimations of Immortality' (1807)

7 A boy's will is the wind's will
And the thoughts of youth are long, long thoughts.
Henry Wadsworth Longfellow 1807–82: 'My Lost Youth' (1858)

8 Youth would be an ideal state if it came a little later in life.
Herbert Asquith 1852–1928: in *Observer* 15 April 1923

9 It is better to waste one's youth than to do nothing with it at all.
Georges Courteline 1858–1929: *La Philosophie de Georges Courteline* (1948)

10 What music is more enchanting than the voices of young people, when you can't hear what they say?
Logan Pearsall Smith 1865–1946: *Afterthoughts* (1931)

11 The force that through the green fuse drives the flower
Drives my green age.
Dylan Thomas 1914–53: 'The force that through the green fuse' (1934)

12 Youth is something very new: twenty years ago no one mentioned it.
Coco Chanel 1883–1971: in M. Haedrich *Coco Chanel, Her Life, Her Secrets* (1971)

13 Youth is vivid rather than happy, but memory always remembers the happy things.
Bernard Lovell 1913– : in *The Times* 20 August 1993

Index of Authors

The Little Oxford Dictionary

OWLS
OXFORD ENGLISH
DICTIONARY
WORD AND
LANGUAGE
SERVICE

The Little Oxford Dictionary

of Current English

First edited by George Ostler

Seventh Edition

Edited by Maurice Waite

Clarendon Press · Oxford
1994

Oxford University Press, Walton Street, Oxford OX2 6DP

Oxford New York
Athens Auckland Bangkok Bombay
Calcutta Cape Town Dar es Salaam Delhi
Florence Hong Kong Istanbul Karachi
Kuala Lumpur Madras Madrid Melbourne
Mexico City Nairobi Paris Singapore
Taipei Tokyo Toronto
and associated companies in
Berlin Ibadan

Oxford is a trade mark of Oxford University Press

Published in the United States
by Oxford University Press Inc., New York

First Edition 1930
Second Edition 1937
Third Edition 1941
Fourth Edition 1969
Fifth Edition 1980
Sixth Edition 1986
Seventh Edition 1994

British Library Cataloguing in Publication Data
Data available

Library of Congress Cataloging in Publication Data
Data available

ISBN 0-19-861298-2

10 9 8 7 6 5 4 3 2 1

Typeset by Latimer Trend Ltd., Plymouth

Printed in Great Britain
by Clay PLC, Bungay, Suffolk

Contents

Editorial staff

Editor	Maurice Waite
Assistant Editor	Andrew Hodgson
Contributing Editors	Anna Howes Louise Jones Bernadette Paton Anne St John-Hall
Keyboarders	Anne Whear Pam Marjara Kay Pepler
Schools English Consultant	Elaine Ashmore-Short

Features of the dictionary

Headwords (words given their own entries) are in bold type:

abandon /əˈbænd(ə)n/ ● *verb* desert ...

or in bold italic type if they are borrowed from another language and are usually printed in italics:

autobahn

Variant spellings are shown:

almanac ... (also **almanack**)

Words that are different but are spelt the same way (homographs) are printed with raised numbers:

abode[1]
abode[2]

Variant American spellings are labelled *US*:

anaemia ... (*US* **anemia**)

Pronunciations are given in the International Phonetic Alphabet. See p. xi for an explanation.

Parts of speech are shown in italic:

abhor ... *verb*
abhorrence ... *noun*

If a word is used as more than one part of speech, each comes after a ●:

abandon ... ● *noun* ... ● *verb* ...

For explanations of parts of speech, see the panels at the dictionary entries for *noun, verb,* etc.

Inflections Irregular and difficult forms are given for:

nouns:

ability ... (*plural* **-ies**)
sheep ... (*plural* same)
tomato ... (*plural* **-es**)

(Irregular plurals are not given for compounds such as *footman* and *schoolchild*.)

verbs: **abate** ... (**-ting**)
abut ... (**-tt-**) [indicating **abutted**, **abutting**]
ring ... (*past tense* **rang**; *past participle* **rung**)

adjectives: **good** ... (**better**, **best**)
narrow ... (**-er**, **-est**)
able ... (**-r**, **-st**)

adverbs: **well** ... (**better**, **best**)

Definitions Round brackets are used for

a optional words, e.g. at

back ... *verb* (cause to) go backwards

(because *back* can mean either 'go backwards' or 'cause to go backwards')

b typical objects of verbs:

bank ... *verb* deposit (money) at bank

c typical subjects of verbs:

break ... *verb* (of waves) curl over and foam

d typical nouns qualified by an adjective:

catchy ... (of tune) easily remembered

Subject labels are sometimes used to help define a word:

sharp ... *Music* above true pitch

Register labels are used if a word is slang, colloquial, or formal:

ace ... *adjective slang* excellent

Coarse slang means that a word, although widely used, is still unacceptable to many people.

Offensive means that a word is offensive to members of a particular ethnic, religious, or other group.

Phrases are entered under their main word:

company ... **in company with**

A comma in a phrase indicates alternatives:

in, to excess

means that *in excess* and *to excess* are both phrases.

Compounds are entered under their main word or element (usually the first):

air ... **air speed** ... **airstrip**

unless they need entries of their own:

broad
broadcast
broadside

A comma in a compound indicates alternatives:

block capitals, letters

means that *block capitals* and *block letters* are both compounds.

Derivatives are put at the end of the entry for the word they are derived from:

rob ... **robber** *noun*

unless they need defining:

drive ...
driver *noun* person who drives; golf club for driving from tee.

Cross-references are printed in small capitals:

anatto = ANNATTO.
arose *past* of ARISE.

Definitions will be found at the entries referred to.

Note on proprietary terms

This dictionary includes some words which are, or are asserted to be, proprietary names or trade marks. Their inclusion does not imply that they have acquired for legal purposes a non-proprietary or general significance, nor is any other judgement implied concerning their legal status. In cases where the editor has some evidence that a word is used as a proprietary name or trade mark this is indicated by the label *proprietary term*, but no judgement concerning the legal status of such words is made or implied thereby.

Pronunciation symbols

Consonants

b	*b*ut	n	*n*o	ʃ	*sh*e
d	*d*og	p	*p*en	ʒ	vi*s*ion
f	*f*ew	r	*r*ed	θ	*th*in
g	*g*et	s	*s*it	ð	*th*is
h	*h*e	t	*t*op	ŋ	ri*ng*
j	*y*es	v	*v*oice	x	lo*ch*
k	*c*at	w	*w*e	tʃ	*ch*ip
l	*l*eg	z	*z*oo	dʒ	*j*ar
m	*m*an				

Vowels

æ	c*a*t	ʌ	r*u*n	əʊ	n*o*
ɑː	*ar*m	ʊ	p*u*t	eə	h*air*
e	b*e*d	uː	t*oo*	ɪə	n*ear*
ɜː	h*er*	ə	*a*go	ɔɪ	b*oy*
ɪ	s*i*t	aɪ	m*y*	ʊə	p*oor*
iː	s*ee*	aʊ	h*ow*	aɪə	f*ire*
ɒ	h*o*t	eɪ	d*ay*	aʊə	s*our*
ɔː	s*aw*				

(ə) signifies the indeterminate sound in gard*e*n, carn*a*l, and rhyth*m*.

The mark ˜ indicates a nasalized sound, as in the following vowels that are not natural in English:

æ̃ (*in*génue)
ɑ̃ (él*an*)
ɔ̃ (b*on* voyage)

The main or primary stress of a word is shown by ˈ before the relevant syllable.

Aa

A *abbreviation* ampere(s). □ **A-bomb** atomic bomb; **A level** advanced level in GCE exam.

a /ə, eɪ/ *adjective* (called the indefinite article) (also **an** /æn, ən/ before vowel sound) one, some, any; per.

AA *abbreviation* Automobile Association; Alcoholics Anonymous; anti-aircraft.

aardvark /'ɑ:dvɑ:k/ *noun* mammal with tubular snout and long tongue.

aback /ə'bæk/ *adverb* □ **taken aback** disconcerted, surprised.

abacus /'æbəkəs/ *noun* (*plural* **-es**) frame with wires along which beads are slid for calculating.

abaft /ə'bɑ:ft/ *Nautical* ● *adverb* in or towards stern of ship. ● *preposition* nearer stern than.

abandon /ə'bænd(ə)n/ ● *verb* desert; give up (hope etc.). ● *noun* freedom from inhibitions. □ **abandonment** *noun.*

abandoned *adjective* deserted; unrestrained.

abase /ə'beɪs/ *verb* (**-sing**) humiliate; degrade. □ **abasement** *noun.*

abashed /ə'bæʃt/ *adjective* embarrassed; disconcerted.

abate /ə'beɪt/ *verb* (**-ting**) make or become less strong etc. □ **abatement** *noun.*

abattoir /'æbətwɑ:/ *noun* slaughterhouse.

abbess /'æbɪs/ *noun* female head of abbey of nuns.

abbey /'æbɪ/ *noun* (*plural* **-s**) (building occupied by) community of monks or nuns.

abbot /'æbət/ *noun* head of community of monks.

abbreviate /ə'bri:vɪeɪt/ *verb* shorten. □ **abbreviation** *noun.*

ABC /eɪbi:'si:/ *noun* alphabet; rudiments of subject; alphabetical guide.

abdicate /'æbdɪkeɪt/ *verb* renounce or resign from (throne etc.). □ **abdication** *noun.*

abdomen /'æbdəmən/ *noun* belly; rear part of insect etc. □ **abdominal** /æb'dɒmɪn(ə)l/ *adjective.*

abduct /əb'dʌkt/ *verb* carry off illegally, kidnap. □ **abduction** *noun*; **abductor** *noun.*

aberrant /æ'berənt/ *adjective* showing aberration.

aberration /æbə'reɪʃ(ə)n/ *noun* deviation from normal type or accepted standard; distortion.

abet /ə'bet/ *verb* encourage (offender), assist (offence). □ **abetter**, *Law* **abettor** *noun.*

abeyance /ə'beɪəns/ *noun* (usually after *in, into*) temporary disuse; suspension.

abhor /əb'hɔ:/ *verb* (**-rr-**) detest; regard with disgust.

abhorrence /əb'hɒrəns/ *noun* disgust, detestation.

abhorrent *adjective* (often + *to*) disgusting.

abide /ə'baɪd/ *verb* (**-ding**, *past & past participle* **abode** /ə'bəʊd/ or **abided**) tolerate; (+ *by*) act in accordance with (rule); keep (promise).

abiding /ə'baɪdɪŋ/ *adjective* enduring, permanent.

ability /ə'bɪlɪtɪ/ *noun* (*plural* **-ies**) (often + *to do*) capacity, power; cleverness, talent.

abject /'æbdʒekt/ *adjective* miserable; degraded; despicable. □ **abjection** /-'dʒek-/ *noun.*

abjure /əb'dʒʊə/ *verb* renounce on oath. □ **abjuration** /-dʒʊ'reɪ-/ *noun.*

ablaze /ə'bleɪz/ *adjective & adverb* on fire; glittering; excited.

able /'eɪb(ə)l/ *adjective* (**-r**, **-st**) (+ *to do*) having power; talented. □ **able-bodied** healthy, fit. □ **ably** *adverb.*

ablution /ə'blu:ʃ(ə)n/ *noun* (usually in *plural*) ceremonial washing of hands etc.; *colloquial* washing onself.

abnegate /'æbnɪgeɪt/ *verb* (**-ting**) give up, renounce. □ **abnegation** *noun.*

abnormal /æb'nɔ:m(ə)l/ *adjective* exceptional; deviating from the norm. □ **abnormality** /-'mæl-/ *noun*; **abnormally** *adverb.*

aboard /ə'bɔ:d/ *adverb & preposition* on or into (ship, aircraft, etc.).

abode[1] /ə'bəʊd/ *noun* dwelling place.

abode[2] *past & past participle* of ABIDE.

abolish /ə'bɒlɪʃ/ *verb* end existence of. □ **abolition** /æbə'lɪʃ(ə)n/ *noun*; **abolitionist** /æbə'lɪʃənɪst/ *noun.*

abominable /ə'bɒmɪnəb(ə)l/ *adjective* detestable, loathsome; *colloquial* very unpleasant. □ **Abominable Snowman** yeti. □ **abominably** *adverb.*

abominate /ə'bɒmɪneɪt/ *verb* (**-ting**) detest, loathe. □ **abomination** *noun.*

aboriginal /æbə'rɪdʒɪn(ə)l/ ● *adjective* indigenous; (usually **Aboriginal**) of the Australian Aborigines. ● *noun* aboriginal inhabitant, esp. (usually **Aboriginal**) of Australia.

aborigines /æbə'rɪdʒɪnɪz/ *plural noun* aboriginal inhabitants, esp. (usually **Aborigines**) of Australia.

■ **Usage** It is best to refer to one *Aboriginal* but several *Aborigines*, although *Aboriginals* is also acceptable.

abort /ə'bɔ:t/ *verb* miscarry; effect abortion of; (cause to) end before completion.

abortion /ə'bɔ:ʃ(ə)n/ *noun* natural or (esp.) induced expulsion of foetus before it can survive; stunted or misshapen creature. □ **abortionist** *noun.*

abortive /ə'bɔ:tɪv/ *adjective* fruitless, unsuccessful.

abound /ə'baʊnd/ *verb* be plentiful; (+ *in*, *with*) be rich in, teem with.

about /ə'baʊt/ ● *preposition* on subject of; relating to, in relation to; at a time near to; around (in); surrounding; here and there in. ● *adverb* approximately; nearby; in every direction; on the move, in action; all around. □ **about-face**, **-turn** turn made so as to face opposite direction, change of policy etc.; **be about to do** be on the point of doing.

above /ə'bʌv/ ● *preposition* over, on top of, higher than; more than; higher in rank, importance, etc. than; too great or good for; beyond reach of. ● *adverb* at or to higher point, overhead; earlier on page or in book. □ **above all** more than anything else; **above-board** without concealment.

abracadabra /æbrəkə'dæbrə/ *interjection* supposedly magic word.

abrade /ə'breɪd/ *verb* (**-ding**) scrape or wear away by rubbing.

abrasion /ə'breɪʒ(ə)n/ *noun* rubbing or scraping away; resulting damaged area.

abrasive /ə'breɪsɪv/ ● *adjective* capable of rubbing or grinding down; harsh or hurtful in manner. ● *noun* abrasive substance.

abreast /ə'brest/ *adverb* side by side and facing same way; (+ *of*) up to date with.

abridge /ə'brɪdʒ/ *verb* (**-ging**) shorten (a book etc.). □ **abridgement** *noun.*

abroad /ə'brɔ:d/ *adverb* in or to foreign country; widely; in circulation.

abrogate /'æbrəgeɪt/ *verb* (**-ting**) repeal, abolish (law etc.). □ **abrogation** *noun.*

abrupt /ə'brʌpt/ *adjective* sudden, hasty; curt; steep. □ **abruptly** *adverb*; **abruptness** *noun.*

abscess /'æbsɪs/ *noun* (*plural* **-es**) swelling containing pus.

abscond /əb'skɒnd/ *verb* flee, esp. to avoid arrest; escape.

abseil /'æbseɪl/ ● *verb* descend (building etc.) by using doubled rope fixed at higher point. ● *noun* such a descent.

absence /'æbsəns/ *noun* being away; duration of this; (+ *of*) lack. □ **absence of mind** inattentiveness.

absent ● *adjective* /'æbsənt/ not present or existing; lacking; inattentive. ● *verb* /əb'sent/ (**absent oneself**) go or stay away. □ **absently** *adverb.*

absentee /æbsən'ti:/ *noun* person not present.

absenteeism /æbsən'ti:ɪz(ə)m/ *noun* absenting oneself from work, school, etc.

absent-minded *adjective* forgetful, inattentive. □ **absent-mindedly** *adverb*; **absent-mindedness** *noun.*

absinthe /'æbsɪnθ/ *noun* wormwood-based, aniseed-flavoured liqueur.

absolute /'æbsəlu:t/ *adjective* complete, utter; unconditional; despotic; not relative; (of adjective or transitive verb) without expressed noun or object; (of decree etc.) final. □ **absolute majority** one over all rivals combined; **absolute temperature** one measured from absolute zero; **absolute zero** lowest possible temperature (−273.15°C or 0°K).

absolutely *adverb* completely; in an absolute sense; *colloquial* quite so, yes.

absolution /æbsə'lu:ʃ(ə)n/ *noun* formal forgiveness of sins.

absolutism /'æbsəlu:tɪz(ə)m/ *noun* absolute government. □ **absolutist** *noun.*

absolve /əb'zɒlv/ *verb* (**-ving**) (often + *from, of*) free from blame or obligation.

absorb /əb'sɔ:b/ *verb* incorporate; assimilate; take in (heat etc.); deal with easily, reduce intensity of; (often as **absorbing** *adjective*) engross attention of; consume (resources).

absorbent /əb'sɔ:bənt/ ● *adjective* tending to absorb. ● *noun* absorbent substance.

absorption /əb'sɔ:pʃ(ə)n/ *noun* absorbing, being absorbed. □ **absorptive** *adjective.*

abstain /əb'steɪn/ *verb* (usually + *from*) refrain (from indulging); decline to vote.

abstemious /əb'sti:mɪəs/ *adjective* moderate or ascetic, esp. in eating and drinking. □ **abstemiously** *adverb.*

abstention /əb'stenʃ(ə)n/ *noun* abstaining, esp. from voting.

abstinence /'æbstɪnəns/ *noun* abstaining, esp. from food or alcohol. □ **abstinent** *adjective.*

abstract ● *adjective* /'æbstrækt/ of or existing in theory rather than practice, not concrete; (of art etc.) not representational. ● *verb* /əb'strækt/ (often + *from*) remove; summarize. ● *noun* /'æbstrækt/ summary; abstract idea, work of art, etc.

abstracted *adjective* inattentive. □ **abstractedly** *adverb.*

abstraction /əb'strækʃ(ə)n/ *noun* abstracting; abstract idea; abstract qualities in art; absent-mindedness.

abstruse /əb'stru:s/ *adjective* hard to understand; profound.

absurd /əb'sɜ:d/ *adjective* wildly inappropriate; ridiculous. □ **absurdity** *noun* (*plural* **-ies**); **absurdly** *adverb.*

ABTA /'æbtə/ *abbreviation* Association of British Travel Agents.

abundance /ə'bʌnd(ə)ns/ *noun* plenty; more than enough; wealth.

abundant *adjective* plentiful; (+ *in*) rich. □ **abundantly** *adverb.*

abuse ● *verb* /ə'bju:z/ (**-sing**) use improperly; misuse; maltreat; insult verbally. ● *noun* /ə'bju:s/ misuse; insulting language; corrupt practice. □ **abuser** /ə'bju:zə/ *noun.*

abusive /ə'bju:sɪv/ *adjective* insulting, offensive. □ **abusively** *adverb.*

abut /ə'bʌt/ *verb* (**-tt-**) (+ *on*) border on; (+ *on, against*) touch or lean on.

abysmal /ə'bɪzm(ə)l/ *adjective* very bad; dire. □ **abysmally** *adverb.*

abyss /ə'bɪs/ *noun* deep chasm.

AC *abbreviation* alternating current.

a/c *abbreviation* account.

acacia /ə'keɪʃə/ *noun* tree with yellow or white flowers.

academia /ækə'di:mɪə/ *noun* the world of scholars.

academic /ækə'demɪk/ ● *adjective* scholarly; of learning; of no practical relevance. ● *noun* teacher or scholar in university etc. □ **academically** *adverb.*

academician /əkædə'mɪʃ(ə)n/ *noun* member of Academy.

academy /ə'kædəmɪ/ *noun* (*plural* **-ies**) place of specialized training; (**Academy**) society of distinguished scholars, artists, scientists, etc.; *Scottish* secondary school.

acanthus /ə'kænθəs/ *noun* (*plural* **-es**) spring herbaceous plant with spiny leaves.

ACAS /'eɪkæs/ *abbreviation* Advisory, Conciliation, and Arbitration Service.

accede /æk'si:d/ *verb* (**-ding**) (+ *to*) take office, esp. as monarch; assent to.

accelerate /ək'seləreɪt/ *verb* (**-ting**) increase speed (of); (cause to) happen earlier. □ **acceleration** *noun.*

accelerator *noun* device for increasing speed, esp. pedal in vehicle; *Physics* apparatus for imparting high speeds to charged particles.

accent ● *noun* /'æksənt/ style of pronunciation of region or social group (see panel); emphasis; prominence given to syllable by stress or pitch; mark on letter indicating pronunciation (see panel). ● *verb* /æk'sent/ emphasize; write or print accents on.

accentuate /ək'sentʃʊeɪt/ *verb* (**-ting**) emphasize, make prominent. □ **accentuation** *noun*.

accept /ək'sept/ *verb* willingly receive; answer (invitation etc.) affirmatively; regard favourably; receive as valid or suitable. □ **acceptance** *noun*.

acceptable *adjective* worth accepting; tolerable. □ **acceptability** *noun*; **acceptably** *adverb*.

access /'ækses/ ● *noun* way of approach or entry; right or opportunity to reach, use, or visit. ● *verb* gain access to (data) in computer.

accessible /ək'sesɪb(ə)l/ *adjective* reachable or obtainable; easy to understand. □ **accessibility** *noun*.

accession /ək'seʃ(ə)n/ *noun* taking office, esp. as monarch; thing added.

accessory /ək'sesərɪ/ *noun* (*plural* **-ies**) additional or extra thing; (usually in *plural*) small attachment or item of dress; (often + *to*) person who abets in or is privy to illegal act.

accident /'æksɪd(ə)nt/ *noun* unintentional unfortunate esp. harmful event; event without apparent cause; unexpected event. □ **accident-prone** clumsy; **by accident** unintentionally.

accidental /æksɪ'dent(ə)l/ ● *adjective* happening or done by chance or accident. ● *noun Music* sharp, flat, or natural indicating momentary departure of note from key signature. □ **accidentally** *adverb*.

acclaim /ə'kleɪm/ ● *verb* welcome or applaud enthusiastically. ● *noun* applause; welcome; public praise. □ **acclamation** /æklə-/ *noun*.

acclimatize /ə'klaɪmətaɪz/ *verb* (also **-ise**) (**-zing** or **-sing**) adapt to new climate or conditions. □ **acclimatization** *noun*.

accolade /'ækəleɪd/ *noun* praise given; touch made with sword at conferring of knighthood.

accommodate /ə'kɒmədeɪt/ *verb* (**-ting**) provide lodging or room for; adapt, harmonize, reconcile; do favour to; (+ *with*) supply.

■ **Usage** *Accommodate, accommodation*, etc. are spelt with two *ms*, not one.

accommodating *adjective* obliging.

accommodation *noun* lodgings; adjustment, adaptation; convenient arrangement. □ **accommodation address** postal address used instead of permanent one.

Accent

1 A person's accent is the way he or she pronounces words, and people from different regions and different groups in society have different accents. For instance, most people in northern England say *path* with a 'short' *a*, while most people in southern England say it with a 'long' *a*, and in America and Canada the *r* in *far* and *port* is generally pronounced, while in south-eastern England, for example, it is not. Everyone speaks with an accent, although some accents may be regarded as having more prestige, such as 'Received Pronunciation' (RP) in the UK.

2 An accent on a letter is a mark added to it to alter the sound it stands for. In French, for example, there are

´ (acute), as in *état*
` (grave), as in *mère*
ˆ (circumflex), as in *guêpe*
¨ (diaeresis), as in *Noël*
¸ (cedilla), as in *français*

and German has

¨ (umlaut), as in *München*.

There are no accents on native English words, but many words borrowed from other languages still have them, such as *blasé* and *façade*.

accompaniment /ə'kʌmpənɪmənt/ *noun* instrumental or orchestral support for solo instrument, voice, or group; accompanying thing.

accompany /ə'kʌmpənɪ/ *verb* (**-ies, -ied**) go with, attend; (usually in *passive*; + *with*, *by*) be done or found with; *Music* play accompaniment for. □ **accompanist** *noun Music.*

accomplice /ə'kʌmplɪs/ *noun* partner in crime.

accomplish /ə'kʌmplɪʃ/ *verb* succeed in doing; achieve, complete.

accomplished *adjective* clever, skilled.

accomplishment *noun* completion (of task etc.); acquired esp. social skill; thing achieved.

accord /ə'kɔ:d/ ● *verb* (often + *with*) be consistent or in harmony; grant, give. ● *noun* agreement; consent. □ **of one's own accord** on one's own initiative.

accordance *noun* □ **in accordance with** in conformity to.

according *adverb* (+ *to*) as stated by; (+ *to*, *as*) in proportion to or as.

accordingly *adverb* as circumstances suggest or require; consequently.

accordion /ə'kɔ:dɪən/ *noun* musical reed instrument with concertina-like bellows, keys, and buttons.

accost /ə'kɒst/ *verb* approach and speak boldly to.

account /ə'kaʊnt/ ● *noun* narration, description; arrangement at bank etc. for depositing and withdrawing money etc.; statement of financial transactions with balance; importance; behalf. ● *verb* consider as. □ **account for** explain, answer for, kill, destroy; **on account** to be paid for later, in part payment; **on account of** because of; **on no account** under no circumstances; **take account of, take into account** consider.

accountable *adjective* responsible, required to account for one's conduct; explicable. □ **accountability** *noun.*

accountant *noun* professional keeper or verifier of financial accounts. □ **accountancy** *noun*; **accounting** *noun.*

accoutrements /ə'ku:trəmənts/ *plural noun* equipment, trappings.

accredit /ə'kredɪt/ *verb* (**-t-**) (+ *to*) attribute; (+ *with*) credit; (usually + *to*, *at*) send (ambassador etc.) with credentials.

accredited *adjective* officially recognized; generally accepted.

accretion /ə'kri:ʃ(ə)n/ *noun* growth by accumulation or organic enlargement; the resulting whole; matter so added.

accrue /ə'kru:/ *verb* (**-ues, -ued, -uing**) (often + *to*) come as natural increase or advantage, esp. financial.

accumulate /ə'kju:mjʊleɪt/ *verb* (**-ting**) acquire increasing number or quantity of; amass, collect; grow numerous; increase. □ **accumulation** *noun*; **accumulative** /-lətɪv/ *adjective.*

accumulator *noun* rechargeable electric cell; bet placed on sequence of events, with winnings and stake from each placed on next.

accurate /'ækjʊrət/ *adjective* precise; conforming exactly with truth etc. □ **accuracy** *noun*; **accurately** *adverb.*

accursed /ə'kɜ:sɪd/ *adjective* under a curse; *colloquial* detestable, annoying.

accusative /ə'kju:zətɪv/ *Grammar* ● *noun* case expressing object of action. ● *adjective* of or in accusative.

accuse /ə'kju:z/ *verb* (**-sing**) (often + *of*) charge with fault or crime; blame. □ **accusation** /æk-/ *noun*; **accusatory** *adjective.*

accustom /ə'kʌstəm/ *verb* (+ *to*) make used to.

accustomed *adjective* (usually + *to*) used (to a thing); customary.

ace ● *noun* playing card with single spot; person who excels in some activity; *Tennis* unreturnable service. ● *adjective slang* excellent. □ **within an ace of** on the verge of.

acerbic /ə'sɜ:bɪk/ *adjective* harsh and sharp, esp. in speech or manner. □ **acerbity** *noun* (*plural* **-ies**).

acetate /'æsɪteɪt/ *noun* compound of acetic acid, esp. the cellulose ester; fabric made from this.

acetic /ə'si:tɪk/ *adjective* of or like vinegar. □ **acetic acid** clear liquid acid in vinegar.

acetone /'æsɪtəʊn/ *noun* colourless volatile solvent of organic compounds.

acetylene /ə'setɪli:n/ *noun* inflammable hydrocarbon gas, used esp. in welding.

ache /eɪk/ ● *noun* continuous dull pain; mental distress. ● *verb* (**-aching**) suffer from or be the source of an ache.

achieve /ə'tʃi:v/ *verb* (**-ving**) reach or attain by effort; accomplish (task etc.); be successful. □ **achiever** *noun*; **achievement** *noun*.

Achilles /ə'kɪli:z/ *noun* □ **Achilles heel** vulnerable point; **Achilles tendon** tendon attaching calf muscles to heel.

achromatic /ækrəʊ'mætɪk/ *adjective* transmitting light without separating it into colours; free from colour. □ **achromatically** *adverb*.

achy /'eɪkɪ/ *adjective* (**-ier, -iest**) suffering from aches.

acid /'æsɪd/ ● *noun Chemistry* substance that neutralizes alkalis, turns litmus red, and usually contains hydrogen and is sour; *slang* drug LSD. ● *adjective* having properties of acid; sour; biting, sharp. □ **acid drop** sharp-tasting boiled sweet; **acid house** synthesized music with simple beat, associated with hallucinogenic drugs; **acid rain** rain containing acid formed from industrial waste in atmosphere; **acid test** severe or conclusive test. □ **acidic** /ə'sɪd-/ *adjective*; **acidify** /ə'sɪd-/ *verb* (**-ies, -ied**); **acidity** /-'sɪd-/ *noun*.

acidulous /ə'sɪdjʊləs/ *adjective* somewhat acid.

acknowledge /ək'nɒlɪdʒ/ *verb* (**-ging**) recognize, accept truth of; confirm receipt of (letter etc.); show that one has noticed; express gratitude for.

acknowledgement *noun* (also **acknowledgment**) acknowledging; thing given or done in gratitude; letter etc. confirming receipt; (usually in *plural*) author's thanks, prefacing book.

acme /'ækmɪ/ *noun* highest point.

acne /'æknɪ/ *noun* skin condition with red pimples.

acolyte /'ækəlaɪt/ *noun* assistant, esp. of priest.

aconite /'ækənaɪt/ *noun* any of various poisonous plants, esp. monkshood; drug made from these.

acorn /'eɪkɔ:n/ *noun* fruit of oak.

acoustic /ə'ku:stɪk/ *adjective* of sound or sense of hearing; (of musical instrument etc.) without electrical amplification. □ **acoustically** *adverb*.

acoustics *plural noun* properties (of a room etc.) in transmitting sound; (treated as *singular*) science of sound.

acquaint /ə'kweɪnt/ *verb* (+ *with*) make aware of or familiar with. □ **be acquainted with** know.

acquaintance *noun* being acquainted; person one knows slightly. □ **acquaintanceship** *noun*.

acquiesce /ækwɪ'es/ *verb* (**-cing**) agree, esp. tacitly. □ **acquiescence** *noun*; **acquiescent** *adjective*.

acquire /ə'kwaɪə/ *verb* (**-ring**) gain possession of. □ **acquired immune deficiency syndrome** = AIDS; **acquired taste** liking developed by experience.

acquirement *noun* thing acquired or attained.

acquisition /ækwɪ'zɪʃ(ə)n/ *noun* (esp. useful) thing acquired; acquiring, being acquired.

acquisitive /ə'kwɪzɪtɪv/ *adjective* keen to acquire things.

acquit /ə'kwɪt/ *verb* (**-tt-**) (often + *of*) declare not guilty; (**acquit oneself**) behave, perform, (+ *of*) discharge (duty etc.). □ **acquittal** *noun*.

acre /'eɪkə/ *noun* measure of land, 4840 sq. yards, 0.405 ha.

acreage /'eɪkərɪdʒ/ *noun* number of acres.

acrid /'ækrɪd/ *adjective* bitterly pungent. □ **acridity** /-'krɪd-/ *noun*.

acrimonious /ækrɪ'məʊnɪəs/ *adjective* bitter in manner or temper. □ **acrimony** /'ækrɪmənɪ/ *noun*.

acrobat /'ækrəbæt/ *noun* performer of acrobatics. □ **acrobatic** /-'bæt-/ *adjective*.

acrobatics /ækrə'bætɪks/ *plural noun* gymnastic feats.

acronym /'ækrənɪm/ *noun* word formed from initial letters of other words (e.g. *laser*, *NATO*).

acropolis /ə'krɒpəlɪs/ *noun* citadel of ancient Greek city.

across /ə'krɒs/ ● *preposition* to or on other side of; from one side to another side of. ● *adverb* to or on other side; from one side to another. □ **across the board** applying to all.

acrostic /ə'krɒstɪk/ *noun* poem etc. in which first (or first and last) letters of lines form word(s).

acrylic /əˈkrɪlɪk/ • *adjective* made from acrylic acid. • *noun* acrylic fibre, fabric, or paint.

acrylic acid *noun* a pungent liquid organic acid.

act • *noun* thing done, deed; process of doing; item of entertainment; pretence; main division of play; decree of legislative body. • *verb* behave; perform actions or functions; (often + *on*) have effect; perform in play etc.; pretend; play part of. □ **act for** be (legal) representative of; **act up** *colloquial* misbehave, give trouble.

acting *adjective* serving temporarily as.

actinism /ˈæktɪnɪz(ə)m/ *noun* property of short-wave radiation that produces chemical changes, as in photography.

action /ˈækʃ(ə)n/ *noun* process of doing or acting; forcefulness, energy; exertion of energy or influence; deed, act; (**the action**) series of events in story, play, etc., *slang* exciting activity; battle; mechanism of instrument; style of movement; lawsuit. □ **action-packed** full of action or excitement; **action replay** playback of part of television broadcast; **out of action** not functioning.

actionable *adjective* providing grounds for legal action.

activate /ˈæktɪveɪt/ *verb* (**-ting**) make active or radioactive. □ **activation** *noun*.

active /ˈæktɪv/ • *adjective* marked by action; energetic, diligent; working, operative; *Grammar* (of verb) of which subject performs action (e.g. *saw* in *he saw a film*). • *noun Grammar* active form or voice. □ **active service** military service in wartime. □ **actively** *adverb*.

activism *noun* policy of vigorous action, esp. for a political cause. □ **activist** *noun*.

activity /ækˈtɪvɪtɪ/ *noun* (*plural* **-ies**) being active; busy or energetic action; (often in *plural*) occupation, pursuit.

actor /ˈæktə/ *noun* person who acts in play, film, etc.

actress /ˈæktrɪs/ *noun* female actor.

actual /ˈæktʃʊəl/ *adjective* existing, real; current. □ **actuality** /-ˈæl-/ *noun* (*plural* **-ies**).

actually *adverb* in fact, really.

actuary /ˈæktʃʊərɪ/ *noun* (*plural* **-ies**) statistician, esp. one calculating insurance risks and premiums. □ **actuarial** /-ˈeər-/ *adjective*.

actuate /ˈæktʃʊeɪt/ *verb* (**-ting**) cause to move, function, act.

acuity /əˈkjuːɪtɪ/ *noun* acuteness.

acumen /ˈækjʊmen/ *noun* keen insight or discernment.

acupuncture /ˈækjuːpʌŋktʃə/ *noun* medical treatment using needles in parts of the body. □ **acupuncturist** *noun*.

acute /əˈkjuːt/ *adjective* (**-r**, **-st**) keen, penetrating; shrewd; (of disease) coming quickly to crisis; (of angle) less than 90°. □ **acute (accent)** mark (´) over letter indicating pronunciation. □ **acutely** *adverb*.

AD *abbreviation* of the Christian era (*Anno Domini*).

ad *noun colloquial* advertisement.

adage /ˈædɪdʒ/ *noun* proverb, maxim.

adagio /əˈdɑːʒɪəʊ/ *Music* • *adverb & adjective* in slow time. • *noun* (*plural* **-s**) adagio passage.

adamant /ˈædəmənt/ *adjective* stubbornly resolute. □ **adamantly** *adverb*.

Adam's apple /ˈædəmz/ *noun* cartilaginous projection at front of neck.

adapt /əˈdæpt/ *verb* (+ *to*) fit, adjust; (+ *to*, *for*) make suitable, modify; (usually + *to*) adjust to new conditions. □ **adaptable** *adjective*; **adaptation** /æd-/ *noun*.

adaptor *noun* device for making equipment compatible; *Electricity* device for connecting several electrical plugs to one socket.

add *verb* join as increase or supplement; unite (numbers) to get their total; say further. □ **add up** find total of; (+ *to*) amount to.

addendum /əˈdendəm/ *noun* (*plural* **-da**) thing to be added; material added at end of book.

adder /ˈædə/ *noun* small venomous snake.

addict /ˈædɪkt/ *noun* person addicted, esp. to drug; *colloquial* devotee.

addicted /əˈdɪktɪd/ *adjective* (usually + *to*) dependent on a drug as a habit;

devoted to an interest. □ **addiction** *noun.*

addictive /ə'dɪktɪv/ *adjective* causing addiction.

addition /ə'dɪʃ(ə)n/ *noun* adding; person or thing added. □ **in addition** (often + *to*) also, as well.

additional *adjective* added, extra. □ **additionally** *adverb.*

additive /'ædɪtɪv/ *noun* substance added, esp. to colour, flavour, or preserve food.

addle /'æd(ə)l/ *verb* (**-ling**) muddle, confuse; (usually as **addled** *adjective*) (of egg) become rotten.

address /ə'dres/ ● *noun* place where person lives or organization is situated; particulars of this, esp. for postal purposes; speech delivered to an audience. ● *verb* write postal directions on; direct (remarks etc.); speak or write to; direct one's attention to.

addressee /ædre'si:/ *noun* person to whom letter etc. is addressed.

adduce /ə'dju:s/ *verb* (**-cing**) cite as proof or instance. □ **adducible** *adjective.*

adenoids /'ædɪnɔɪds/ *plural noun* enlarged lymphatic tissue between nose and throat, often hindering breathing. □ **adenoidal** /-'nɔɪ-/ *adjective.*

adept ● *adjective* /ə'dept, 'ædept/ (+ *at*, *in*) skilful. ● *noun* /'ædept/ adept person.

adequate /'ædɪkwət/ *adjective* sufficient, satisfactory. □ **adequacy** *noun*; **adequately** *adverb.*

adhere /əd'hɪə/ *verb* (**-ring**) (usually + *to*) stick fast; behave according to (rule etc.); give allegiance.

adherent ● *noun* supporter. ● *adjective* adhering. □ **adherence** *noun.*

adhesion /əd'hi:ʒ(ə)n/ *noun* adhering.

adhesive /əd'hi:sɪv/ ● *adjective* sticky, causing adhesion. ● *noun* adhesive substance.

ad hoc /æd 'hɒk/ *adverb & adjective* for one particular occasion or use.

adieu /ə'dju:/ *interjection* goodbye.

ad infinitum /æd ɪnfɪ'naɪtəm/ *adverb* without limit; for ever.

adipose /'ædɪpəʊz/ *adjective* of fat, fatty. □ **adiposity** /-'pɒs-/ *noun.*

adjacent /ə'dʒeɪs(ə)nt/ *adjective* lying near; adjoining. □ **adjacency** *noun.*

adjective /'ædʒɪktɪv/ *noun* word indicating quality of noun or pronoun (see panel). □ **adjectival** /-'taɪv-/ *adjective.*

adjoin /ə'dʒɔɪn/ *verb* be next to and joined with.

adjourn /ə'dʒɜ:n/ *verb* postpone, break off; (+ *to*) transfer to (another place). □ **adjournment** *noun.*

Adjective

An adjective is a word that describes a noun or pronoun, e.g.

red, clever, German, depressed, battered, sticky, shining

Most can be used either before a noun, e.g.

the red *house* *a* clever *woman*

or after a verb like *be, seem,* or *call,* e.g.

The house is red. *I wouldn't call him* lazy.
She seems very clever.

Some can be used only before a noun, e.g.

the chief *reason* (one cannot say **the reason is chief*)

Some can be used only after a verb, e.g.

The ship is still afloat. (one cannot say **an afloat ship*)

A few can be used only immediately after a noun, e.g.

the president elect (one cannot say either **an elect president* or **The president is elect*)

adjudge /ə'dʒʌdʒ/ *verb* (**-ging**) pronounce judgement on; pronounce or award judicially. □ **adjudg(e)ment** *noun.*

adjudicate /ə'dʒu:dɪkeɪt/ *verb* (**-ting**) act as judge; adjudge. □ **adjudication** *noun*; **adjudicator** *noun.*

adjunct /'ædʒʌŋkt/ *noun* (+ *to*, *of*) subordinate or incidental thing.

adjure /ə'dʒʊə/ *verb* (**-ring**) (usually + *to do*) beg or command. □ **adjuration** *noun.*

adjust /ə'dʒʌst/ *verb* order, position; regulate; arrange; (usually + *to*) adapt; harmonize. □ **adjustable** *adjective*; **adjustment** *noun.*

adjutant /'ædʒʊt(ə)nt/ *noun* army officer assisting superior in administrative duties.

ad lib /æd 'lɪb/ ● *verb* (**-bb-**) improvise. ● *adjective* improvised. ● *adverb* to any desired extent.

Adm. *abbreviation* Admiral.

administer /əd'mɪnɪstə/ *verb* manage (affairs); formally deliver, dispense.

administration /ədmɪnɪ'streɪʃ(ə)n/ *noun* administering, esp. public affairs; government in power.

administrative /əd'mɪnɪstrətɪv/ *adjective* of the management of affairs.

administrator /əd'mɪnɪstreɪtə/ *noun* manager of business, public affairs, or person's estate.

admirable /'ædmərəb(ə)l/ *adjective* deserving admiration; excellent. □ **admirably** *adverb.*

admiral /'ædmər(ə)l/ *noun* commander-in-chief of navy; high-ranking naval officer, commander.

Admiralty *noun* (*plural* **-ies**) (in full **Admiralty Board**) *historical* committee superintending Royal Navy.

admire /əd'maɪə/ *verb* regard with approval, respect, or satisfaction; express admiration of. □ **admiration** /ædmə'reɪ-/ *noun*; **admirer** *noun*; **admiring** *adjective*; **admiringly** *adverb.*

admissible /əd'mɪsɪb(ə)l/ *adjective* worth accepting or considering; allowable. □ **admissibility** *noun.*

admission /əd'mɪʃ(ə)n/ *noun* acknowledgement (of error etc.); (right of) entering; entrance charge.

admit /əd'mɪt/ *verb* (**-tt-**) (often + *to be*, *that*) acknowledge, recognize as true; (+ *to*) confess to; let in; accommodate; take (patient) into hospital; (+ *of*) allow as possible.

admittance *noun* admitting or being admitted, usually to a place.

admittedly *adverb* as must be admitted.

admixture /æd'mɪkstʃə/ *noun* thing added, esp. minor ingredient; adding of this.

admonish /əd'mɒnɪʃ/ *verb* reprove; urge; (+ *of*) warn. □ **admonishment** *noun*; **admonition** /ædmə'nɪ-/ *noun*; **admonitory** *adjective.*

ad nauseam /æd 'nɔ:zɪæm/ *adverb* to a sickening extent.

ado /ə'du:/ *noun* fuss; trouble.

adobe /ə'dəʊbɪ/ *noun* sun-dried brick.

adolescent /ædə'lesənt/ ● *adjective* between childhood and adulthood. ● *noun* adolescent person. □ **adolescence** *noun.*

adopt /ə'dɒpt/ *verb* legally take (child) as one's own; take over (another's idea etc.); choose; accept responsibility for; approve (report etc.). □ **adoption** *noun.*

adoptive *adjective* because of adoption.

adorable /ə'dɔ:rəb(ə)l/ *adjective* deserving adoration; *colloquial* delightful, charming.

adore /ə'dɔ:/ *verb* (**-ring**) love intensely; worship; *colloquial* like very much. □ **adoration** /ædə'reɪ-/ *noun*; **adorer** *noun.*

adorn /ə'dɔ:n/ *verb* add beauty to, decorate. □ **adornment** *noun.*

adrenal /ə'dri:n(ə)l/ ● *adjective* of adrenal glands. ● *noun* (in full **adrenal gland**) either of two ductless glands above the kidneys.

adrenalin /ə'drenəlɪn/ *noun* stimulative hormone secreted by adrenal glands.

adrift /ə'drɪft/ *adverb & adjective* drifting; *colloquial* unfastened, out of order.

adroit /ə'drɔɪt/ *adjective* dexterous, skilful.

adsorb /əd'sɔ:b/ *noun* attract and hold thin layer of (gas or liquid) on its surface. □ **adsorbent** *adjective & noun*; **adsorption** *noun.*

adulation /ædjʊ'leɪʃ(ə)n/ *noun* obsequious flattery.

adult /'ædʌlt/ ● *adjective* grown-up, mature; of or for adults. ● *noun* adult person. □ **adulthood** *noun*.

adulterate /ə'dʌltəreɪt/ *verb* (**-ting**) debase (esp. food) by adding other substances. □ **adulteration** *noun*.

adultery /ə'dʌltərɪ/ *noun* voluntary sexual intercourse of married person other than with spouse. □ **adulterer**, **adulteress** *noun*; **adulterous** *adjective*.

adumbrate /'ædʌmbreɪt/ *verb* (**-ting**) indicate faintly or in outline; foreshadow. □ **adumbration** *noun*.

advance /əd'vɑ:ns/ ● *verb* (**-cing**) move or put forward; progress; pay or lend beforehand; promote; present (idea etc.). ● *noun* going forward; progress; prepayment, loan; payment beforehand; (in *plural*) amorous approaches; rise in price. ● *adjective* done etc. beforehand. □ **advance on** approach threateningly; **in advance** ahead, beforehand.

advanced *adjective* well ahead; socially progressive. □ **advanced level** high level GCE exam.

advancement *noun* promotion of person, cause, etc.

advantage /əd'vɑ:ntɪdʒ/ ● *noun* beneficial feature; benefit, profit; (often + *over*) superiority; *Tennis* next point after deuce. ● *verb* (**-ging**) benefit, favour. □ **take advantage of** make good use of; exploit. □ **advantageous** /ædvən'teɪdʒəs/ *adjective*.

Advent /'ædvent/ *noun* season before Christmas; coming of Christ; (**advent**) arrival.

Adventist *noun* member of sect believing in imminent second coming of Christ.

adventure /əd'ventʃə/ ● *noun* unusual and exciting experience; enterprise. ● *verb* (**-ring**) dare, venture. □ **adventure playground** one with climbing frames etc.

adventurer *noun* (*feminine* **adventuress**) person who seeks adventure esp. for personal gain or pleasure; financial speculator.

adventurous *adjective* venturesome, enterprising.

adverb /'ædvɜ:b/ *noun* word indicating manner, degree, circumstance, etc. used to modify verb, adjective, or other adverb (see panel). □ **adverbial** /əd'vɜ:-/ *adjective*.

adversary /'ædvəsərɪ/ *noun* (*plural* **-ies**) enemy; opponent. □ **adversarial** /-'seə-/ *adjective*.

adverse /'ædvɜ:s/ *adjective* unfavourable; harmful. □ **adversely** *adverb*.

adversity /əd'vɜ:sɪtɪ/ *noun* misfortune.

advert /'ædvɜ:t/ *noun colloquial* advertisement.

advertise /'ædvətaɪz/ *verb* (**-sing**) promote publicly to increase sales; make generally known; seek to sell, fill (vacancy), or (+ *for*) buy or employ by notice in newspaper etc.

advertisement /əd'vɜ:tɪsmənt/ *noun* public announcement advertising something; advertising.

advice /əd'vaɪs/ *noun* recommendation on how to act; information; notice of transaction.

advisable /əd'vaɪzəb(ə)l/ *adjective* to be recommended; expedient. □ **advisability** *noun*.

advise /əd'vaɪz/ *verb* (**-sing**) give advice (to); recommend; (usually + *of*, *that*) inform.

advisedly /əd'vaɪzɪdlɪ/ *adverb* deliberately.

adviser *noun* person who advises, esp. officially.

advisory /əd'vaɪzərɪ/ *adjective* giving advice.

advocaat /ædvə'kɑ:t/ *noun* liqueur of eggs, sugar, and brandy.

advocacy /'ædvəkəsɪ/ *noun* support or argument for cause etc.

advocate ● *noun* /'ædvəkət/ (+ *of*) person who speaks in favour; person who pleads for another, esp. in law court. ● *verb* /'ædvəkeɪt/ (**-ting**) recommend by argument.

adze /ædz/ *noun* (*US* **adz**) axe with arched blade at right angles to handle.

aegis /'i:dʒɪs/ *noun* protection; support.

aeolian harp /i:'əʊlɪən/ *noun* (*US* **eolian harp**) stringed instrument sounding when wind passes through it.

aeon /'i:ɒn/ *noun* (also **eon**) long or indefinite period; an age.

aerate /'eəreɪt/ *verb* (**-ting**) charge with carbon dioxide; expose to air. □ **aeration** *noun.*

aerial /'eərɪəl/ ● *noun* device for transmitting or receiving radio signals. ● *adjective* from the air; existing in the air; like air.

aero- *combining form* air; aircraft.

aerobatics /eərə'bætɪks/ *plural noun* feats of spectacular flying of aircraft; (treated as *singular*) performance of these.

aerobics /eə'rəʊbɪks/ *plural noun* vigorous exercises designed to increase oxygen intake. □ **aerobic** *adjective.*

aerodrome /'eərədrəʊm/ *noun* small airport or airfield.

aerodynamics /eərəʊdaɪ'næmɪks/ *plural noun* (usually treated as *singular*) dynamics of solid bodies moving through air. □ **aerodynamic** *adjective.*

aerofoil /'eərəfɔɪl/ *noun* structure with curved surfaces (e.g. aircraft wing), designed to give lift in flight.

aeronautics /eərəʊ'nɔːtɪks/ *plural noun* (usually treated as *singular*) science or practice of motion in the air. □ **aeronautical** *adjective.*

aeroplane /'eərəpleɪn/ *noun* powered heavier-than-air aircraft with wings.

aerosol /'eərəsɒl/ *noun* pressurized container releasing substance as fine spray.

aerospace /'eərəʊspeɪs/ *noun* earth's atmosphere and outer space; aviation in this.

aesthete /'iːsθiːt/ *noun* person who appreciates beauty.

aesthetic /iːs'θetɪk/ *adjective* of or sensitive to beauty; tasteful. □ **aesthetically** *adverb*; **aestheticism** /-sɪz(ə)m/ *noun.*

aetiology /iːtɪ'ɒlədʒɪ/ *noun* (*US* **etiology**) study of causation or of causes of disease. □ **aetiological** /-ə'lɒdʒ-/ *adjective.*

afar /ə'fɑː/ *adverb* at or to a distance.

affable /'æfəb(ə)l/ *adjective* friendly; courteous. □ **affability** *noun*; **affably** *adverb.*

Adverb

An adverb is used:

1 with a verb, to say:

- **a** how something happens, e.g. *He walks* quickly.
- **b** where something happens, e.g. *I live* here.
- **c** when something happens, e.g. *They visited us* yesterday.
- **d** how often something happens, e.g. *We* usually *have coffee.*

2 to strengthen or weaken the meaning of:

- **a** a verb, e.g. *He* really *meant it. I* almost *fell asleep.*
- **b** an adjective, e.g. *She is* very *clever. This is a* slightly *better result.*
- **c** another adverb, e.g. *It comes off* terribly *easily. The boys* nearly *always get home late.*

3 to add to the meaning of a whole sentence, e.g.

He is probably *our best player.* Luckily, *no one was hurt.*

In writing or in formal speech, it is **incorrect** to use an adjective instead of an adverb. For example, use

Do it properly. and not **Do it* proper.

but note that many words are both an adjective and an adverb, e.g.

adjective	*adverb*
a fast *horse*	*He ran* fast.
a long *time*	*Have you been here* long?

---

affair /ə'feə/ *noun* matter, concern; love affair; *colloquial* thing, event; (in *plural*) business.

affect /ə'fekt/ *verb* produce effect on; (of disease etc.) attack; move emotionally; use for effect; pretend to feel; (+ *to do*) pretend.

■ **Usage** *Affect* is often confused with *effect*, which means 'to bring about'.

affectation /æfek'teɪʃ(ə)n/ *noun* artificial manner; pretentious display.

affected *adjective* pretended; full of affectation.

affection /ə'fekʃ(ə)n/ *noun* goodwill, fond feeling; disease.

affectionate /ə'fekʃənət/ *adjective* loving. □ **affectionately** *adverb.*

affidavit /æfɪ'deɪvɪt/ *noun* written statement on oath.

affiliate /ə'fɪlɪeɪt/ *verb* (**-ting**) (+ *to, with*) attach to, connect to, or adopt as member or branch.

affiliation *noun* affiliating, being affiliated. □ **affiliation order** legal order compelling supposed father to support illegitimate child.

affinity /ə'fɪnɪtɪ/ *noun* (*plural* **-ies**) attraction; relationship; resemblance; *Chemistry* tendency of substances to combine with others.

affirm /ə'fɜːm/ *verb* state as fact; make solemn declaration in place of oath. □ **affirmation** /æfə'meɪʃ(ə)n/ *noun.*

affirmative /ə'fɜːmətɪv/ ● *adjective* affirming, expressing approval. ● *noun* affirmative statement.

affix ● *verb* /ə'fɪks/ attach, fasten; add in writing. ● *noun* /'æfɪks/ addition; prefix, suffix.

afflict /ə'flɪkt/ *verb* distress physically or mentally.

affliction *noun* distress, suffering; cause of this.

affluent /'æflʊənt/ *adjective* rich. □ **affluence** *noun.*

afford /ə'fɔːd/ *verb* (after *can, be able to*) have enough money, time, etc., for; be able to spare (time etc.); (+ *to do*) be in a position; provide.

afforest /ə'fɒrɪst/ *verb* convert into forest; plant with trees. □ **afforestation** *noun.*

affray /ə'freɪ/ *noun* breach of peace by fighting or rioting in public.

affront /ə'frʌnt/ ● *noun* open insult. ● *verb* insult openly; embarrass.

Afghan /'æfgæn/ ● *noun* native, national, or language of Afghanistan. ● *adjective* of Afghanistan. □ **Afghan hound** tall dog with long silky hair.

afield /ə'fiːld/ *adverb* to or at a distance.

aflame /ə'fleɪm/ *adverb & adjective* in flames; very excited.

afloat /ə'fləʊt/ *adverb & adjective* floating; at sea; out of debt.

afoot /ə'fʊt/ *adverb & adjective* in operation; progressing.

afore /ə'fɔː/ *preposition & adverb archaic* before.

afore- *combining form* previously.

aforethought *adjective* (after noun) premeditated.

a fortiori /eɪ fɔːtɪ'ɔːraɪ/ *adverb & adjective* with stronger reason. [Latin]

afraid /ə'freɪd/ *adjective* alarmed, frightened. □ **be afraid** *colloquial* politely regret.

afresh /ə'freʃ/ *adverb* anew; with fresh start.

African /'æfrɪkən/ ● *noun* native of Africa; person of African descent. ● *adjective* of Africa.

Afrikaans /æfrɪ'kɑːns/ *noun* language derived from Dutch, used in S. Africa.

Afrikaner /æfrɪ'kɑːnə/ *noun* Afrikaans-speaking white person in S. Africa.

Afro /'æfrəʊ/ ● *adjective* (of hair) tightly-curled and bushy. ● *noun* (*plural* **-s**) Afro hairstyle.

Afro- *combining form* African.

aft /ɑːft/ *adverb* at or towards stern or tail.

after /'ɑːftə/ ● *preposition* following in time; in view of; despite; behind; in pursuit or quest of; about, concerning; in allusion to or imitation of. ● *conjunction* later than. ● *adverb* later; behind. ● *adjective* later. □ **after all** in spite of everything; **afterbirth** placenta etc. discharged after childbirth; **aftercare** attention after leaving hospital etc.; **after-effect** delayed effect of accident etc.; **afterlife** life after death; **afters** *colloquial* sweet dessert; **aftershave** lotion applied to face after shaving; **afterthought** thing thought of or added later.

aftermath /'ɑ:ftəmæθ/ *noun* consequences.

afternoon /ɑ:ftə'nu:n/ *noun* time between midday and evening.

afterwards /'ɑ:ftəwədz/ *adverb* later, subsequently.

again /ə'gen/ *adverb* another time; as previously; in addition; on the other hand. □ **again and again** repeatedly.

against /ə'genst/ *preposition* in opposition to; into collision or in contact with; to the disadvantage of; in contrast to; in anticipation of; as compensating factor to; in return for.

agape /ə'geɪp/ *adjective* gaping.

agate /'ægət/ *noun* usually hard streaked chalcedony.

agave /ə'geɪvɪ/ *noun* spiny-leaved plant.

age ● *noun* length of past life or existence; *colloquial* (often in *plural*) a long time; historical period; old age. ● *verb* (**ageing**) (cause to) show signs of age; grow old; mature. □ **come of age** reach legal adult status; **under age** not old enough.

aged *adjective* /eɪdʒd/ of the age of; /'eɪdʒɪd/ old.

ageism /'eɪdʒɪz(ə)m/ *noun* prejudice or discrimination on grounds of age.

ageless /'eɪdʒlɪs/ *adjective* never growing or appearing old.

agency /'eɪdʒənsɪ/ *noun* (*plural* **-ies**) business or premises of agent; action; intervention.

agenda /ə'dʒendə/ *noun* (*plural* **-s**) list of items to be considered at meeting; things to be done.

agent /'eɪdʒ(ə)nt/ *noun* person acting for another in business etc.; person or thing producing effect.

agent provocateur /ɑ:ʒɑ̃ prəvɒkə'tɜ:/ *noun* (*plural* ***agents provocateurs*** same pronunciation) person tempting suspected offenders to self-incriminating action. [French]

agglomerate /ə'glɒməreɪt/ *verb* (**-ting**) collect into mass. □ **agglomeration** *noun*.

agglutinate /ə'glu:tɪneɪt/ *verb* (**-ting**) stick as with glue. □ **agglutination** *noun*; **agglutinative** /-nətɪv/ *adjective*.

aggrandize /ə'grændaɪz/ *verb* (also **-ise**) (**-zing** or **-sing**) increase power, rank, or wealth of; make seem greater. □ **aggrandizement** /-dɪz-/ *noun*.

aggravate /'ægrəveɪt/ *verb* (**-ting**) increase seriousness of; *colloquial* annoy. □ **aggravation** *noun*.

■ **Usage** The use of *aggravate* to mean 'annoy' is considered incorrect by some people, but it is common in informal use.

aggregate ● *noun* /'ægrɪgət/ sum total; crushed stone etc. used in making concrete. ● *adjective* /'ægrɪgət/ collective, total. ● *verb* /'ægrɪgeɪt/ collect together, unite; *colloquial* amount to. □ **aggregation** *noun*.

aggression /ə'greʃ(ə)n/ *noun* unprovoked attack; hostile act or feeling. □ **aggressor** *noun*.

aggressive /ə'gresɪv/ *adjective* given to aggression; forceful, self-assertive. □ **aggressively** *adverb*.

aggrieved /ə'gri:vd/ *adjective* having grievance.

aggro /'ægrəʊ/ *noun slang* aggression; difficulty.

aghast /ə'gɑ:st/ *adjective* amazed and horrified.

agile /'ædʒaɪl/ *adjective* quick-moving; nimble. □ **agility** /ə'dʒɪlɪtɪ/ *noun*.

agitate /'ædʒɪteɪt/ *verb* (**-ting**) disturb, excite; (often + *for*, *against*) campaign, esp. politically; shake briskly. □ **agitation** *noun*; **agitator** *noun*.

AGM *abbreviation* annual general meeting.

agnail /'ægneɪl/ *noun* torn skin at root of fingernail; resulting soreness.

agnostic /æg'nɒstɪk/ ● *noun* person who believes that existence of God is not provable. ● *adjective* of agnosticism. □ **agnosticism** /-sɪz(ə)m/ *noun*.

ago /ə'gəʊ/ *adverb* in the past.

agog /ə'gɒg/ *adjective* eager, expectant.

agonize /'ægənaɪz/ *verb* (also **-ise**) (**-zing** or **-sing**) undergo mental anguish; (cause to) suffer agony; (as **agonized** *adjective*) expressing agony.

agony /'ægənɪ/ *noun* (*plural* **-ies**) extreme physical or mental suffering; severe struggle. □ **agony aunt** *colloquial* writer answering letters in **agony column** *colloquial*, section of magazine etc. offering personal advice.

agoraphobia /ægərə'fəʊbɪə/ *noun* extreme fear of open spaces. □ **agoraphobic** *adjective.*

agrarian /ə'greərɪən/ ● *adjective* of land or its cultivation. ● *noun* advocate of redistribution of land.

agree /ə'gri:/ *verb* (**-ees**, **-eed**,) (often + *with*) hold similar opinion, be or become in harmony, suit, be compatible; (+ *to do*) consent; reach agreement about.

agreeable *adjective* pleasing; willing to agree. □ **agreeably** *adverb.*

agreement *noun* act or state of agreeing; arrangement, contract.

agriculture /'ægrɪkʌltʃə/ *noun* cultivation of the soil and rearing of animals. □ **agricultural** /-'kʌl-/ *adjective*; **agriculturalist** /-'kʌl-/ *noun.*

agronomy /ə'grɒnəmɪ/ *noun* science of soil management and crop production. □ **agronomist** *noun.*

aground /ə'graʊnd/ *adjective & adverb* on(to) bottom of shallow water.

ague /'eɪgju:/ *noun* shivering fit; *historical* malarial fever.

ah /ɑ:/ *interjection expressing surprise, pleasure, or realization.*

aha /ɑ:'hɑ:/ *interjection expressing surprise, triumph, mockery, etc.*

ahead /ə'hed/ *adverb* in advance, in front; (often + *on*) in the lead (on points etc.).

ahoy /ə'hɔɪ/ *interjection Nautical* call used in hailing.

AI *abbreviation* artificial insemination; artificial intelligence.

aid ● *noun* help; person or thing that helps. ● *verb* help; promote (recovery etc.). □ **in aid of** in support of, *colloquial* for purpose of.

aide /eɪd/ *noun* aide-de-camp; assistant.

aide-de-camp /eɪd də 'kɑ̃/ *noun* (*plural* **aides-** same pronunciation) officer assisting senior officer.

Aids *noun* (also **AIDS**) acquired immune deficiency syndrome.

ail *verb* be ill or in poor condition.

aileron /'eɪlərɒn/ *noun* hinged flap on aircraft wing.

ailment *noun* illness, esp. minor one.

aim ● *verb* intend, try; (usually + *at*) direct, point; take aim. ● *noun* purpose, object; directing of weapon etc. at object. □ **take aim** direct weapon etc. at object.

aimless *adjective* purposeless. □ **aimlessly** *adverb.*

ain't /eɪnt/ *colloquial* am not, is not, are not; has not, have not.

■ **Usage** The use of *ain't* is incorrect in standard English.

---

air ● *noun* mixture chiefly of oxygen and nitrogen surrounding earth; open space; earth's atmosphere, often as place where aircraft operate; appearance, manner; (in *plural*) affected manner; tune. ● *verb* expose (room, clothes, etc.) to air, ventilate; express and discuss publicly. □ **airbase** base for military aircraft; **air-bed** inflatable mattress; **airborne** transported by air, (of aircraft) in the air; **airbrick** brick perforated for ventilation; **Air Commodore** RAF officer next above Group Captain; **air-conditioned** *adjective* equipped with **air-conditioning**, regulation of humidity and temperature in building, apparatus for this; **airfield** area with runway(s) for aircraft; **air force** branch of armed forces fighting in the air; **airgun** gun using compressed air to fire pellets; **air hostess** stewardess in aircraft; **air letter** sheet of paper forming airmail letter; **airlift** *noun* emergency transport of supplies etc. by air, *verb* transport thus; **airline** public air transport company; **airliner** large passenger aircraft; **airlock** stoppage of flow by airbubble in pipe etc., compartment giving access to pressurized chamber; **airmail** system of transporting mail by air, mail carried thus; **airman** pilot or member of aircraft crew; **Air** (**Chief**, **Vice-**) **Marshal** high ranks in RAF; **airplane** *US* aeroplane; **airport** airfield with facilities for passengers and cargo; **air raid** attack by aircraft; **air rifle** rifle using compressed air to fire pellets; **airs and graces** affected manner; **airship** powered aircraft lighter than air; **airsick** nauseous from air travel; **airspace** air above a country; **air speed** aircraft's speed relative to air; **airstrip** strip of ground for take-off and landing of aircraft; **air terminal** building with transport to and from airport; **air traffic**

controller official who controls air traffic by radio; **airway** recognized route of aircraft, passage for air into lungs; **airwoman** woman pilot or member of aircraft crew; **by air** by or in aircraft; **on the air** being broadcast.

aircraft *noun* (*plural* same) aeroplane, helicopter. □ **aircraft carrier** warship that carries and acts as base for aircraft; **aircraftman**, **aircraftwoman** lowest rank in RAF.

Airedale /'eədeɪl/ *noun* terrier of large rough-coated breed.

airless *adjective* stuffy; still, calm.

airtight *adjective* impermeable to air.

airworthy *adjective* (of aircraft) fit to fly. □ **airworthiness** *noun*.

airy *adjective* (**-ier**, **-iest**) well-ventilated; flippant; light as air. □ **airy-fairy** *colloquial* unrealistic, impractical.

aisle /aɪl/ *noun* side part of church divided by pillars from nave; passage between rows of pews, seats, etc.

aitchbone /'eɪtʃbəʊn/ *noun* rump bone of animal; cut of beef over this.

ajar /ə'dʒɑː/ *adverb & adjective* (of door etc.) slightly open.

Akela /ɑː'keɪlə/ *noun* adult leader of Cub Scouts.

akimbo /ə'kɪmbəʊ/ *adverb* (of arms) with hands on hips and elbows out.

akin /ə'kɪn/ *adjective* related; similar.

alabaster /'æləbɑːstə/ ● *noun* translucent usually white form of gypsum. ● *adjective* of alabaster; white, smooth.

à la carte /æ lɑː 'kɑːt/ *adverb & adjective* with individually priced dishes.

alacrity /ə'lækrɪtɪ/ *noun* briskness; readiness.

à la mode /æ lɑː 'məʊd/ *adverb & adjective* in fashion; fashionable.

alarm /ə'lɑːm/ ● *noun* warning of danger etc.; warning sound or device; alarm clock; apprehension. ● *verb* frighten, disturb; warn. □ **alarm clock** clock that rings at set time. □ **alarming** *adjective*.

alarmist /ə'lɑːmɪst/ *noun* person spreading unnecessary alarm.

alas /ə'læs/ *interjection expressing grief or regret*.

alb *noun* long white vestment worn by Christian priests.

albatross /'ælbətrɒs/ *noun* long-winged seabird related to petrel; *Golf* score of 3 strokes under par for hole.

albeit /ɔːl'biːɪt/ *conjunction* although.

albino /æl'biːnəʊ/ *noun* (*plural* **-s**) person or animal lacking pigment in skin, hair, and eyes. □ **albinism** /'ælbɪnɪz(ə)m/ *noun*.

album /'ælbəm/ *noun* book for displaying photographs etc.; long-playing gramophone record; set of these.

albumen /'ælbjʊmɪn/ *noun* egg white.

albumin /'ælbjʊmɪn/ *noun* water-soluble protein found in egg white, milk, blood, etc.

alchemy /'ælkəmɪ/ *noun* medieval chemistry, esp. seeking to turn base metals into gold. □ **alchemist** *noun*.

alcohol /'ælkəhɒl/ *noun* colourless volatile liquid, esp. as intoxicant present in wine, beer, spirits, etc. and as a solvent, fuel, etc.; liquor containing this; other compound of this type.

alcoholic /ælkə'hɒlɪk/ ● *adjective* of, like, containing, or caused by alcohol. ● *noun* person suffering from alcoholism.

alcoholism /'ælkəhɒlɪz(ə)m/ *noun* condition resulting from addiction to alcohol.

alcove /'ælkəʊv/ *noun* recess in wall of room, garden, etc.

alder /'ɔːldə/ *noun* tree related to birch.

alderman /'ɔːldəmən/ *noun esp. historical* civic dignitary next in rank to mayor.

ale *noun* beer.

alert /ə'lɜːt/ ● *adjective* watchful. ● *noun* alarm; state or period of special vigilance. ● *verb* (often + *to*) warn.

alfalfa /æl'fælfə/ *noun* clover-like plant used for fodder.

alfresco /æl'freskəʊ/ *adjective & adverb* in the open air.

alga /'ælgə/ *noun* (*plural* **-gae** /-dʒiː/) (usually in *plural*) non-flowering stemless water plant.

algebra /'ældʒɪbrə/ *noun* branch of mathematics using letters to represent numbers. □ **algebraic** /-'breɪɪk/ *adjective*.

Algol /'ælgɒl/ *noun* high-level computer-programming language.

algorithm /'ælgərɪð(ə)m/ *noun* process or rules for (esp. computer) calculation etc.

alias /ˈeɪlɪəs/ ● *adverb* also known as. ● *noun* assumed name.

alibi /ˈælɪbaɪ/ *noun* (*plural* **-s**) proof that one was elsewhere; excuse.

■ **Usage** The use of *alibi* to mean 'an excuse' is considered incorrect by some people.

---

alien /ˈeɪlɪən/ ● *adjective* (often + *to*) unfamiliar, repugnant; foreign; of beings from another world. ● *noun* non-naturalized foreigner; a being from another world.

alienate /ˈeɪlɪəneɪt/ *verb* (**-ting**) estrange; transfer ownership of. □ **alienation** *noun*.

alight¹ /əˈlaɪt/ *adjective* on fire; lit up.

alight² /əˈlaɪt/ *verb* (often + *from*) get down or off; come to earth, settle.

align /əˈlaɪn/ *verb* place in or bring into line; (usually + *with*) ally (oneself etc.). □ **alignment** *noun*.

alike /əˈlaɪk/ ● *adjective* similar, like. ● *adverb* in similar way.

alimentary /ælɪˈmentərɪ/ *adjective* concerning nutrition; nourishing. □ **alimentary canal** channel through which food passes during digestion.

alimony /ˈælɪmənɪ/ *noun* money payable to a divorced or separated spouse.

alive /əˈlaɪv/ *adjective* living; lively, active; (usually + *to*) alert to; (usually + *with*) swarming with.

alkali /ˈælkəlaɪ/ *noun* (*plural* **-s**) substance that neutralizes acids, turns litmus blue, and forms caustic solutions in water. □ **alkaline** *adjective*; **alkalinity** /-ˈlɪn-/ *noun*.

alkaloid /ˈælkəlɔɪd/ *noun* plant-based compound often used as drug, e.g. morphine, quinine.

all /ɔːl/ ● *adjective* whole amount, number, or extent of. ● *noun* all people or things concerned; (+ *of*) the whole of. ● *adverb* entirely, quite. □ **all along** from the beginning; **all but** very nearly; **all for** *colloquial* strongly in favour of; **all-clear** signal that danger or difficulty is over; **all fours** hands and knees; **all in** exhausted; **all-in** inclusive of all; **all in all** everything considered; **all out** (**all-out** before noun) involving all one's strength etc.; **all over** completely finished, in or on all parts of one's body; **all-purpose** having many uses; **all right** satisfactory, safe and sound, in good condition, satisfactorily, I consent; **all-right** (before noun) *colloquial* acceptable; **all round** in all respects, for each person; **all-round** (of person) versatile; **all-rounder** versatile person; **All Saints' Day** 1 Nov.; **all the same** nevertheless; **all there** *colloquial* mentally alert or normal; **all together** all at once, all in one place; **at all** (with negative or in questions) in any way, to any extent; **in all** in total, altogether.

■ **Usage** See note at ALTOGETHER.

Allah /ˈælə/ *noun: Muslim name of* God.

allay /əˈleɪ/ *verb* lessen; alleviate.

allege /əˈledʒ/ *verb* declare, esp. without proof. □ **allegation** /ælɪˈgeɪʃ(ə)n/ *noun*; **allegedly** /əˈledʒɪdlɪ/ *adverb*.

allegiance /əˈliːdʒ(ə)ns/ *noun* loyalty; duty of subject.

allegory /ˈælɪgərɪ/ *noun* (*plural* **-ies**) story with moral represented symbolically. □ **allegorical** /-ˈgɒr-/ *adjective*; **allegorize** *verb* (also **-ise**) (**-zing** or **-sing**).

allegretto /ælɪˈgretəʊ/ *Music* ● *adverb & adjective* in fairly brisk tempo. ● *noun* (*plural* **-s**) allegretto movement or passage.

allegro /əˈlegrəʊ/ *Music* ● *adverb & adjective* in lively tempo. ● *noun* (*plural* **-s**) allegro movement or passage.

alleluia /ælɪˈluːjə/ (also **hallelujah** /hæl-/) ● *interjection* God be praised. ● *noun* (*plural* **-s**) song of praise to God.

allergic /əˈlɜːdʒɪk/ *adjective* (+ *to*) having allergy to, *colloquial* having strong dislike for; caused by allergy.

allergy /ˈælədʒɪ/ *noun* (*plural* **-ies**) reaction to certain substances.

alleviate /əˈliːvɪeɪt/ *verb* (**-ting**) make (pain etc.) less severe. □ **alleviation** *noun*.

alley /ˈælɪ/ *noun* narrow street or passage; enclosure for skittles, bowling, etc.

alliance /əˈlaɪəns/ *noun* formal union or association of states, political parties, etc. or of families by marriage.

allied /ˈælaɪd/ *adjective* connected or related; (also **Allied**) associated in an alliance.

alligator /ˈælɪgeɪtə/ *noun* large reptile of crocodile family.

alliteration /əlɪtəˈreɪʃ(ə)n/ *noun* recurrence of same initial letter or sound in adjacent or nearby words, as in *The fair breeze blew, the white foam flew, the furrow followed free.* □ **alliterate** /-ˈlɪt-/ *verb*; **alliterative** /əˈlɪtərətɪv/ *adjective.*

allocate /ˈæləkeɪt/ *verb* (**-ting**) (usually + *to*) assign. □ **allocation** *noun.*

allot /əˈlɒt/ *verb* (**-tt-**) apportion or distribute to (person).

allotment /əˈlɒtmənt/ *noun* small plot of land rented for cultivation; share; allotting.

allow /əˈlaʊ/ *verb* (often + *to do*) permit; assign fixed sum to; (usually + *for*) provide or set aside for a purpose. □ **allow for** take into consideration.

allowance *noun* amount or sum allowed, esp. regularly; deduction, discount. □ **make allowances** (often + *for*) judge leniently.

alloy ● *noun* /ˈælɔɪ/ mixture of metals; inferior metal mixed esp. with gold or silver. ● *verb* mix (metals); /əˈlɔɪ/ debase by admixture; spoil (pleasure).

allspice /ˈɔːlspaɪs/ *noun* spice made from berry of pimento plant; this berry.

allude /əˈluːd/ *verb* (**-ding**) (+ *to*) make allusion to.

allure /əˈljʊə/ ● *verb* (**-ring**) attract, charm, entice. ● *noun* attractiveness, charm. □ **allurement** *noun.*

allusion /əˈluːʒ(ə)n/ *noun* (often + *to*) passing or indirect reference. □ **allusive** /-sɪv/ *adjective.*

alluvium /əˈluːvɪəm/ *noun* (*plural* **-via**) deposit left by flood, esp. in river valley. □ **alluvial** *adjective.*

ally ● *noun* /ˈælaɪ/ (*plural* **-ies**) state or person formally cooperating or united with another, esp. in war. ● *verb* (also /əˈlaɪ/) (**-ies**, **-ied**) (often **ally oneself with**) combine in alliance (with).

Alma Mater /ælmə ˈmɑːtə/ *noun* one's university, school, or college.

almanac /ˈɔːlmənæk/ *noun* (also **almanack**) calendar, usually with astronomical data.

almighty /ɔːlˈmaɪtɪ/ ● *adjective* infinitely powerful; very great. ● *noun* (**the Almighty**) God.

almond /ˈɑːmənd/ *noun* kernel of nutlike fruit related to plum; tree bearing this.

almoner /ˈɑːmənə/ *noun* social worker attached to hospital.

almost /ˈɔːlməʊst/ *adverb* all but, very nearly.

alms /ɑːmz/ *plural noun historical* donation of money or food to the poor. □ **almshouse** charitable institution for the poor.

aloe /ˈæləʊ/ *noun* plant with toothed fleshy leaves; (in *plural*) strong laxative from aloe juice. □ **aloe vera** /ˈvɪərə/ variety used to make laxative and in cosmetics.

aloft /əˈlɒft/ *adjective & adverb* high up, overhead.

alone /əˈləʊn/ ● *adjective* without company or help; lonely. ● *adverb* only, exclusively.

along /əˈlɒŋ/ ● *preposition* beside or through (part of) the length of. ● *adverb* onward, into more advanced state; with oneself or others; beside or through (part of) thing's length. □ **alongside** at or close to side (of); **along with** in addition to.

aloof /əˈluːf/ ● *adjective* unconcerned, unsympathetic. ● *adverb* away, apart.

aloud /əˈlaʊd/ *adverb* audibly.

alp *noun* high mountain, esp. (**the Alps**) those in Switzerland and adjacent countries.

alpaca /ælˈpækə/ *noun* S. American llama-like animal; its long wool; fabric made from this.

alpha /ˈælfə/ *noun* first letter of Greek alphabet (Α, α). □ **alpha and omega** beginning and end; **alpha particle** helium nucleus emitted by radioactive substance.

alphabet /ˈælfəbet/ *noun* set of letters or signs used in a language. □ **alphabetical** /-ˈbet-/ *adjective.*

alphanumeric /ælfənjuːˈmerɪk/ *adjective* containing both letters and numbers.

alpine /ˈælpaɪn/ ● *adjective* of high mountains or (**Alpine**) the Alps. ● *noun* plant suited to mountain regions; = rock plant.

already /ɔːlˈredɪ/ *adverb* before the time in question; as early as this.

alright *adverb* = ALL RIGHT.

■ **Usage** Although *alright* is widely used, it is not correct in standard English.

Alsatian /æl'seɪʃ(ə)n/ *noun* large dog of a breed of wolfhound.

also /'ɔ:lsəʊ/ *adverb* in addition, besides. □ **also-ran** loser in race, undistinguished person.

altar /'ɔ:ltə/ *noun* flat table or block for offerings to deity; Communion table.

alter /'ɔ:ltə/ *verb* change in character, shape, etc. □ **alteration** *noun*.

altercation /ɔ:ltə'keɪʃ(ə)n/ *noun* dispute, wrangle.

alternate ● *adjective* /ɔ:l'tɜ:nət/ (with noun in *plural*) every other; (of things of two kinds) alternating. ● *verb* /'ɔ:ltəneɪt/ (**-ting**) (often + *with*) arrange or occur by turns; (+ *between*) go repeatedly from one to another. □ **alternating current** electric current regularly reversing direction. □ **alternately** *adverb*; **alternation** *noun*.

■ **Usage** See note at ALTERNATIVE.

alternative /ɔ:l'tɜ:nətɪv/ ● *adjective* available as another choice; unconventional. ● *noun* any of two or more possibilities; choice. □ **alternatively** *adverb*.

■ **Usage** The adjective *alternative* is often confused with *alternate*, which is correctly used in 'there will be a dance on alternate Saturdays'.

alternator /'ɔ:ltəneɪtə/ *noun* dynamo generating alternating current.

although /ɔ:l'ðəʊ/ *conjunction* though.

altimeter /'æltɪmi:tə/ *noun* instrument measuring altitude.

altitude /'æltɪtju:d/ *noun* height, esp. of object above sea level or horizon.

alto /'æltəʊ/ ● *noun* (*plural* **-s**) = CONTRALTO; highest adult male singing voice; singer with this. ● *adjective* having range of alto.

altogether /ɔ:ltə'geðə/ *adverb* totally; on the whole; in total.

■ **Usage** Note that *altogether* means 'in total', as in *six rooms altogether*, whereas *all together* means 'all at once' or 'all in one place' as in *six rooms all together*.

altruism /'æltru:ɪz(ə)m/ *noun* unselfishness as principle of action. □ **altruist** *noun*; **altruistic** /-'ɪs-/ *adjective*.

alumina /ə'lu:mɪnə/ *noun* aluminium oxide; emery.

aluminium /ælju'mɪnɪəm/ *noun* (*US* **aluminum** /ə'lu:mɪnəm/) a light silvery metallic element.

alumnus /ə'lʌmnəs/ *noun* (*plural* **-ni** /-naɪ/) former pupil or student.

always /'ɔ:lweɪz/ *adverb* at all times; whatever the circumstances; repeatedly.

Alzheimer's disease /'æltshaɪməz/ *noun* brain disorder causing senility.

AM *abbreviation* amplitude modulation.

am *1st singular present* of BE.

a.m. *abbreviation* before noon (*ante meridiem*).

amalgam /ə'mælgəm/ *noun* mixture, blend; alloy of any metal with mercury.

amalgamate /ə'mælgəmeɪt/ *verb* (**-ting**) mix; unite. □ **amalgamation** *noun*.

amanuensis /əmænju:'ensɪs/ *noun* (*plural* **-enses** /-si:z/) assistant, esp. writing from dictation.

amaranth /'æmərænθ/ *noun* plant with green, red, or purple flowers; imaginary unfading flower. □ **amaranthine** /-'rænθaɪn/ *adjective*.

amaryllis /æmə'rɪlɪs/ *noun* plant with lily-like flowers.

amass /ə'mæs/ *verb* heap together, accumulate.

amateur /'æmətə/ *noun* person engaging in pursuit as pastime not profession; person with limited skill. □ **amateurish** *adjective*; **amateurism** *noun*.

amatory /'æmətərɪ/ *adjective* of sexual love.

amaze /ə'meɪz/ *verb* (**-zing**) fill with surprise or wonder. □ **amazement** *noun*; **amazing** *adjective*.

Amazon /'æməzən/ *noun* one of a mythical race of female warriors; (**amazon**) strong or athletic woman. □ **Amazonian** /-'zəʊ-/ *adjective*.

ambassador /æm'bæsədə/ *noun* diplomat living abroad as representative of his or her country; promoter. □ **ambassadorial** /-'dɔ:-/ *adjective*.

amber /'æmbə/ ● *noun* yellow translucent fossil resin; colour of this. ● *adjective* of or like amber.

ambergris /'æmbəgrɪs/ *noun* waxlike substance from sperm whale, used in perfumes.

ambidextrous /æmbɪ'dekstrəs/ *adjective* able to use either hand equally well.

ambience /'æmbɪəns/ *noun* surroundings, atmosphere.

ambient /'æmbɪənt/ *adjective* surrounding.

ambiguous /æm'bɪgjʊəs/ *adjective* having a double meaning; difficult to classify. □ **ambiguity** /-'gju:-/ *noun* (*plural* **-ies**).

ambit /'æmbɪt/ *noun* scope, bounds.

ambition /æm'bɪʃ(ə)n/ *noun* determination to succeed; object of this.

ambitious *adjective* full of ambition.

ambivalent /æm'bɪvələnt/ *adjective* having mixed feelings towards person or thing. □ **ambivalence** *noun.*

amble /'æmb(ə)l/ ● *verb* (**-ling**) walk at leisurely pace. ● *noun* leisurely pace.

ambrosia /æm'brəʊzɪə/ *noun* food of the gods in classical mythology; delicious food etc.

ambulance /'æmbjʊləns/ *noun* vehicle for taking patients to hospital; mobile army hospital.

ambulatory /'æmbjʊlətərɪ/ *adjective* of or for walking.

ambuscade /æmbəs'keɪd/ *noun & verb* (**-ding**) ambush.

ambush /'æmbʊʃ/ ● *noun* surprise attack by people hiding; hiding place for this. ● *verb* attack from ambush; waylay.

ameliorate /ə'mi:lɪəreɪt/ *verb* (**-ting**) make or become better. □ **amelioration** *noun*; **ameliorative** /-rətɪv/ *adjective.*

amen /ɑ:'men/ *interjection* (esp. at end of prayer) so be it.

amenable /ə'mi:nəb(ə)l/ *adjective* responsive, docile; (often + *to*) answerable (to law etc.).

amend /ə'mend/ *verb* correct error in; make minor alterations in.

amendment *noun* minor alteration or addition in document etc.

amends *noun* □ **make amends** (often + *for*) give compensation.

amenity /ə'mi:nɪtɪ/ *noun* (*plural* **-ies**) pleasant or useful feature or facility; pleasantness (of a place etc.).

American /ə'merɪkən/ ● *adjective* of America, esp. the US. ● *noun* native, citizen, or inhabitant of America, esp. the US; English as spoken in the US. □ **Americanize** *verb* (also **-ise**) (**-zing** or **-sing**).

Americanism *noun* word etc. of US origin or usage.

amethyst /'æməθɪst/ *noun* purple or violet semiprecious stone.

amiable /'eɪmɪəb(ə)l/ *adjective* friendly and pleasant, likeable. □ **amiability** *noun*; **amiably** *adverb.*

amicable /'æmɪkəb(ə)l/ *adjective* friendly. □ **amicably** *adverb.*

amid /ə'mɪd/ *preposition* in the middle of.

amidships /ə'mɪdʃɪps/ *adverb* in(to) the middle of a ship.

amidst /ə'mɪdst/ = AMID.

amino acid /ə'mi:nəʊ/ *noun* organic acid found in proteins.

amir = EMIR.

amiss /ə'mɪs/ ● *adjective* out of order, wrong. ● *adverb* wrong(ly), inappropriately. □ **take amiss** be offended by.

amity /'æmɪtɪ/ *noun* friendship.

ammeter /'æmɪtə/ *noun* instrument for measuring electric current.

ammo /'æməʊ/ *noun slang* ammunition.

ammonia /ə'məʊnɪə/ *noun* pungent strongly alkaline gas; solution of this in water.

ammonite /'æmənaɪt/ *noun* coil-shaped fossil shell.

ammunition /æmjʊ'nɪʃ(ə)n/ *noun* bullets, shells, grenades, etc.; information usable in argument.

amnesia /æm'ni:zɪə/ *noun* loss of memory. □ **amnesiac** *adjective & noun.*

amnesty /'æmnɪstɪ/ *noun* (*plural* **-ies**) general pardon, esp. for political offences.

amniocentesis /æmnɪəʊsen'ti:sɪs/ *noun* (*plural* **-teses** /-si:z/) sampling of amniotic fluid to detect foetal abnormality.

amniotic fluid /æmnɪ'ɒtɪk/ *noun* fluid surrounding embryo.

amoeba /ə'mi:bə/ *noun* (*plural* **-s**) microscopic single-celled organism living in water.

amok /ə'mɒk/ *adverb* □ **run amok**, **amuck** /ə'mʌk/ run wild.

among /ə'mʌŋ/ *preposition* (also **amongst**) surrounded by; included in or in the category of; from the joint resources of; between.

amoral /eɪ'mɒr(ə)l/ *adjective* beyond morality; without moral principles.

amorous /'æmərəs/ *adjective* showing or feeling sexual love.

amorphous /ə'mɔ:fəs/ *adjective* of no definite shape; vague; non-crystalline.

amount /ə'maʊnt/ • *noun* total number, size, value, extent, etc. • *verb* (+ *to*) be equivalent in number, size, etc. to.

amour /ə'mʊə/ *noun* (esp. secret) love affair.

amp *noun* ampere; *colloquial* amplifier.

ampere /'æmpeə/ *noun* SI unit of electric current.

ampersand /'æmpəsænd/ *noun* the sign '&' (= *and*).

amphetamine /æm'fetəmi:n/ *noun* synthetic stimulant drug.

amphibian /æm'fɪbɪən/ *noun* amphibious animal or vehicle.

amphibious /æm'fɪbɪəs/ *adjective* living or operating on land and in water; involving military forces landed from the sea.

amphitheatre /'æmfɪθɪətə/ *noun* round open building with tiers of seats surrounding central space.

amphora /'æmfərə/ *noun* (*plural* **-rae** /-ri:/) Greek or Roman two-handled jar.

ample /'æmp(ə)l/ *adjective* (**-r**, **-st**) plentiful, extensive; more than enough. □ **amply** *adverb*.

amplifier /'æmplɪfaɪə/ *noun* device for amplifying sounds or electrical signals.

amplify /'æmplɪfaɪ/ *verb* (**-ies**, **-ied**) increase strength of (sound or electrical signal); add details to (story etc.). □ **amplification** *noun*.

amplitude /'æmplɪtju:d/ *noun* spaciousness; maximum departure from average of oscillation, alternating current, etc. □ **amplitude modulation** modulation of a wave by variation of its amplitude.

ampoule /'æmpu:l/ *noun* small sealed capsule holding solution for injection.

amputate /'æmpjʊteɪt/ *verb* (**-ting**) cut off surgically (limb etc.). □ **amputation** *noun*; **amputee** /-'ti:/ *noun*.

amuck = AMOK.

amulet /'æmjʊlɪt/ *noun* charm worn against evil.

amuse /ə'mju:z/ *verb* (**-sing**) cause to laugh or smile; interest, occupy. □ **amusing** *adjective*.

amusement *noun* being amused; thing that amuses, esp. device for entertainment at fairground etc. □ **amusement arcade** indoor area with slot machines etc.

an see A.

anabolic steroid /ænə'bɒlɪk/ *noun* synthetic steroid hormone used to build muscle.

anachronism /ə'nækrənɪz(ə)m/ *noun* attribution of custom, event, etc. to wrong period; thing thus attributed; out-of-date person or thing. □ **anachronistic** /-'nɪs-/ *adjective*.

anaconda /ænə'kɒndə/ *noun* large S. American constrictor.

anaemia /ə'ni:mɪə/ *noun* (*US* **anemia**) deficiency of red blood cells or their haemoglobin, causing pallor and listlessness. □ **anaemic** *adjective*.

anaesthesia /ænɪs'θi:zɪə/ *noun* (*US* **anes-**) artificially induced insensibility to pain.

anaesthetic /ænɪs'θetɪk/ (*US* **anes-**) • *noun* drug, gas, etc., producing anaesthesia. • *adjective* producing anaesthesia.

anaesthetist /ə'ni:sθətɪst/ *noun* (*US* **anes-**) person who administers anaesthetics.

anaesthetize /ə'ni:sθətaɪz/ *verb* (also **-ise**; *US* also **anes-**) (**-zing** or **-sing**) administer anaesthetic to.

anagram /'ænəgræm/ *noun* word or phrase formed by transposing letters of another.

anal /'eɪn(ə)l/ *adjective* of the anus.

analgesia /ænəl'dʒi:zɪə/ *noun* absence or relief of pain.

analgesic /ænəl'dʒi:sɪk/ • *noun* pain-killing drug. • *adjective* pain-killing.

analogous /ə'næləgəs/ *adjective* (usually + *to*) partially similar or parallel.

analogue /'ænəlɒg/ (*US* **analog**) • *noun* analogous thing. • *adjective* (of computer (usually **analog**), watch, etc.) using physical variables (e.g. voltage, position of hands) to represent numbers.

analogy /ə'nælədʒɪ/ *noun* (*plural* **-ies**) correspondence, similarity; reasoning from parallel cases.

analyse /'ænəlaɪz/ *verb* (**-sing**) (*US* **analyze**; **-zing**) perform analysis on.

analysis /ə'nælɪsɪs/ *noun* (*plural* **-lyses** /-si:z/) detailed examination; *Chemistry* determination of constituent parts; psychoanalysis.

analyst /'ænəlɪst/ *noun* person who analyses.

analytical /ænə'lɪtɪkəl/ *adjective* (also **analytic**) of or using analysis.

analyze *US* = ANALYSE.

anarchism /'ænəkɪz(ə)m/ *noun* belief that government and law should be abolished. □ **anarchist** *noun*; **anarchistic** /-'kɪstɪk/ *adjective*.

anarchy /'ænəkɪ/ *noun* disorder, esp. political. □ **anarchic** /ə'nɑ:kɪk/ *adjective*.

anathema /ə'næθəmə/ *noun* (*plural* **-s**) detested thing; Church's curse.

anatomy /ə'nætəmɪ/ *noun* (*plural* **-ies**) (science of) animal or plant structure. □ **anatomical** /ænə'tɒmɪk(ə)l/ *adjective*; **anatomist** *noun*.

anatto = ANNATTO.

ANC *abbreviation* African National Congress.

ancestor /'ænsestə/ *noun* person, animal, or plant from which another has descended or evolved; prototype.

ancestral /æn'sestr(ə)l/ *adjective* inherited from ancestors.

ancestry /'ænsestrɪ/ *noun* (*plural* **-ies**) lineage; ancestors collectively.

anchor /'æŋkə/ ● *noun* metal device used to moor ship to sea-bottom. ● *verb* secure with anchor; fix firmly; cast anchor. □ **anchorman** coordinator, esp. compère in broadcast.

anchorage /'æŋkərɪdʒ/ *noun* place for anchoring; lying at anchor.

anchorite /'æŋkəraɪt/ *noun* hermit, recluse.

anchovy /'æntʃəvɪ/ *noun* (*plural* **-ies**) small strong-flavoured fish of herring family.

ancien régime /ɑ̃sjæ̃ re'ʒi:m/ *noun* superseded regime, esp. that of pre-Revolutionary France. [French]

ancient /'eɪnʃ(ə)nt/ *adjective* of times long past; very old.

ancillary /æn'sɪlərɪ/ ● *adjective* subsidiary, auxiliary; (esp. of health workers) providing essential support. ● *noun* (*plural* **-ies**) auxiliary; ancillary worker.

and *conjunction connecting words, clauses, or sentences.*

andante /æn'dæntɪ/ *Music* ● *adverb & adjective* in moderately slow time. ● *noun* andante movement or passage.

androgynous /æn'drɒdʒɪnəs/ *adjective* hermaphrodite.

android /'ændrɔɪd/ *noun* robot with human appearance.

anecdote /'ænɪkdəʊt/ *noun* short, esp. true, story. □ **anecdotal** /-'dəʊt(ə)l/ *adjective*.

anemia *US* = ANAEMIA.

anemic *US* = ANAEMIC.

anemometer /ænɪ'mɒmɪtə/ *noun* instrument for measuring wind force.

anemone /ə'nemənɪ/ *noun* plant related to buttercup.

aneroid barometer /'ænərɔɪd/ *noun* barometer that measures air pressure by registering its action on lid of box containing vacuum.

anesthesia etc. *US* = ANAESTHESIA etc.

aneurysm /'ænjʊrɪz(ə)m/ *noun* (also **aneurism**) excessive enlargement of artery.

anew /ə'nju:/ *adverb* again; in different way.

angel /'eɪndʒ(ə)l/ *noun* attendant or messenger of God usually represented as human with wings; kind or virtuous person. □ **angel cake** light sponge cake; **angelfish** small fish with winglike fins. □ **angelic** /æn'dʒelɪk/ *adjective*; **angelically** *adverb*.

angelica /æn'dʒelɪkə/ *noun* aromatic plant; its candied stalks.

angelus /'ændʒɪləs/ *noun* RC prayers said at morning, noon, and sunset; bell rung for this.

anger /'æŋgə/ ● *noun* extreme displeasure. ● *verb* make angry.

angina /æn'dʒaɪnə/ *noun* (in full **angina pectoris**) chest pain brought on by exertion, owing to poor blood supply to heart.

angle[1] /'æŋg(ə)l/ ● *noun* space between two meeting lines or surfaces, esp. as measured in degrees; corner; point of view. ● *verb* (**-ling**) move or place obliquely; present (information) in biased way.

angle² /ˈæŋg(ə)l/ *verb* (**-ling**) fish with line and hook; (+ *for*) seek objective indirectly. □ **angler** *noun.*

Anglican /ˈæŋglɪkən/ ● *adjective* of Church of England. ● *noun* member of Anglican Church. □ **Anglicanism** *noun.*

Anglicism /ˈæŋglɪsɪz(ə)m/ *noun* English expression or custom.

Anglicize /ˈæŋglɪsaɪz/ *verb* (also **-ise**) (**-zing** or **-sing**) make English in character etc.

Anglo- *combining form* English or British (and).

Anglo-Catholic /æŋgləʊˈkæθəlɪk/ ● *adjective* of High Church Anglican group emphasizing its Catholic tradition. ● *noun* member of this group.

Anglo-Indian /æŋgləʊˈɪndɪən/ ● *adjective* of England and India; of British descent but Indian residence. ● *noun* Anglo-Indian person.

Anglophile /ˈæŋgləʊfaɪl/ *noun* admirer of England or the English.

Anglo-Saxon /æŋgləʊˈsæks(ə)n/ ● *adjective* of English Saxons before Norman Conquest; of English descent. ● *noun* Anglo-Saxon person or language; *colloquial* plain (esp. crude) English.

angora /æŋˈgɔːrə/ *noun* fabric made from hair of angora goat or rabbit. □ **angora cat**, **goat**, **rabbit** long-haired varieties.

angostura /æŋgəˈstjʊərə/ *noun* aromatic bitter bark of S. American tree.

angry /ˈæŋgrɪ/ *adjective* (**-ier**, **-iest**) feeling or showing anger; (of wound etc.) inflamed, painful. □ **angrily** *adverb.*

angst /æŋst/ *noun* anxiety, neurotic fear; guilt.

angstrom /ˈæŋstrəm/ *noun* unit of wavelength measurement.

anguish /ˈæŋgwɪʃ/ *noun* severe mental or physical pain. □ **anguished** *adjective.*

angular /ˈæŋgjʊlə/ *adjective* having sharp corners or (of person) features; (of distance) measured by angle. □ **angularity** /-ˈlær-/ *noun.*

aniline /ˈænɪliːn/ *noun* colourless oily liquid used in dyes, drugs, and plastics.

animadvert /ænɪmædˈvɜːt/ *verb* (+ *on*) *literary* criticize, censure. □ **animadversion** *noun.*

animal /ˈænɪm(ə)l/ ● *noun* living organism, esp. other than man, having sensation and usually ability to move; brutish person. ● *adjective* of or like animal; carnal.

animality /ænɪˈmælɪtɪ/ *noun* the animal world; animal behaviour.

animate ● *adjective* /ˈænɪmət/ having life; lively. ● *verb* /ˈænɪmeɪt/ (**-ting**) enliven; give life to.

animated /ˈænɪmeɪtɪd/ *adjective* lively; living; (of film) using animation.

animation /ænɪˈmeɪʃ(ə)n/ *noun* liveliness; being alive; technique of film-making by photographing sequence of drawings or positions of puppets etc. to create illusion of movement.

animator /ˈænɪmeɪtə/ *noun* artist who prepares animated films.

animism /ˈænɪmɪz(ə)m/ *noun* belief that inanimate objects and natural phenomena have souls. □ **animist** *noun*; **animistic** /-ˈmɪs-/ *adjective.*

animosity /ænɪˈmɒsɪtɪ/ *noun* (*plural* **-ies**) hostility.

animus /ˈænɪməs/ *noun* hostility, ill feeling.

anion /ˈænaɪən/ *noun* negatively charged ion.

anise /ˈænɪs/ *noun* plant with aromatic seeds.

aniseed /ˈænɪsiːd/ *noun* seed of anise.

ankle /ˈæŋk(ə)l/ *noun* joint connecting foot with leg.

anklet /ˈæŋklɪt/ *noun* ornament worn round ankle.

ankylosis /æŋkɪˈləʊsɪs/ *noun* stiffening of joint by fusing of bones.

annals /ˈæn(ə)lz/ *plural noun* narrative of events year by year; historical records. □ **annalist** *noun.*

annatto /əˈnætəʊ/ *noun* (also **anatto**) orange-red food colouring made from tropical fruit.

anneal /əˈniːl/ *verb* heat (metal, glass) and cool slowly, esp. to toughen.

annelid /ˈænəlɪd/ *noun* segmented worm, e.g. earthworm.

annex /æˈneks/ *verb* (often + *to*) add as subordinate part; take possession of. □ **annexation** *noun.*

annexe /ˈæneks/ *noun* supplementary building.

annihilate /ə'naɪəleɪt/ *verb* (**-ting**) destroy utterly. □ **annihilation** *noun*.

anniversary /ænɪ'vɜ:sərɪ/ *noun* (*plural* **-ies**) yearly return of date of event; celebration of this.

Anno Domini /ænəʊ 'dɒmɪnaɪ/ see AD.

annotate /'ænəteɪt/ *verb* (**-ting**) add explanatory notes to. □ **annotation** *noun*.

announce /ə'naʊns/ *verb* (**-cing**) make publicly known; make known the approach of. □ **announcement** *noun*.

announcer *noun* person who announces, esp. in broadcasting.

annoy /ə'nɔɪ/ *verb* (often in *passive* + *at*, *with*) anger or distress slightly; harass. □ **annoyance** *noun*.

annual /'ænjʊəl/ ● *adjective* reckoned by the year; recurring yearly. ● *noun* book etc. published yearly; plant living only one year. □ **annually** *adverb*.

annuity /ə'nju:ɪtɪ/ *noun* (*plural* **-ies**) yearly grant or allowance; investment yielding fixed annual sum for stated period.

annul /ə'nʌl/ *verb* (**-ll-**) declare invalid; cancel, abolish. □ **annulment** *noun*.

annular /'ænjʊlə/ *adjective* ring-shaped.

annulate /'ænjʊlət/ *adjective* marked with or formed of rings.

annunciation /ənʌnsɪ'eɪʃ(ə)n/ *noun* announcement, esp. (**Annunciation**) that made by the angel Gabriel to Mary.

anode /'ænəʊd/ *noun* positive electrode.

anodize /'ænədaɪz/ *verb* (also **-ise**) (**-zing** or **-sing**) coat (metal) with protective layer by electrolysis.

anodyne /'ænədaɪn/ ● *adjective* pain-relieving; soothing. ● *noun* anodyne drug etc.

anoint /ə'nɔɪnt/ *verb* apply oil or ointment to, esp. ritually.

anomalous /ə'nɒmələs/ *adjective* irregular, abnormal.

anomaly /ə'nɒməlɪ/ *noun* (*plural* **-ies**) anomalous thing.

anon /ə'nɒn/ *adverb archaic* soon.

anon. *abbreviation* anonymous.

anonymous /ə'nɒnɪməs/ *adjective* of unknown name or authorship; featureless. □ **anonymity** /ænə'nɪm-/ *noun*.

anorak /'ænəræk/ *noun* waterproof usually hooded jacket.

anorexia /ænə'reksɪə/ *noun* lack of appetite, esp. (in full **anorexia nervosa** /nɜ:'vəʊsə/) obsessive desire to lose weight by refusing to eat. □ **anorexic** *adjective & noun*.

another /ə'nʌðə/ ● *adjective* an additional or different. ● *pronoun* an additional or different person or thing.

answer /'ɑ:nsə/ ● *noun* something said or done in reaction to a question, statement, or circumstance; solution to problem. ● *verb* make an answer or response (to); suit; (+ *to*, *for*) be responsible to or for; (+ *to*) correspond to (esp. description). □ **answer back** answer insolently; **answering machine**, **answerphone** tape recorder which answers telephone calls and takes messages.

answerable *adjective* (usually + *to*, *for*) responsible; that can be answered.

ant *noun* small usually wingless insect living in complex social group. □ **anteater** mammal feeding on ants; **anthill** moundlike ants' nest.

antacid /æn'tæsɪd/ *noun & adjective* preventive or corrective of acidity.

antagonism /æn'tægənɪz(ə)m/ *noun* active hostility.

antagonist /æn'tægənɪst/ *noun* opponent. □ **antagonistic** /-'nɪs-/ *adjective*.

antagonize /æn'tægənaɪz/ *verb* (also **-ise**) (**-zing** or **-sing**) provoke.

Antarctic /æn'tɑ:ktɪk/ ● *adjective* of south polar region. ● *noun* this region.

ante /'æntɪ/ *noun* stake put up by poker player before receiving cards; amount payable in advance.

ante- /'æntɪ/ *prefix* before.

antecedent /æntɪ'si:d(ə)nt/ ● *noun* preceding thing or circumstance; *Grammar* word, phrase, etc., to which another word refers; (in *plural*) person's ancestors. ● *adjective* previous.

antechamber /'æntɪtʃeɪmbə/ *noun* ante-room.

antedate /æntɪ'deɪt/ *verb* (**-ting**) precede in time; give earlier than true date to.

antediluvian /æntɪdɪ'lu:vɪən/ *adjective* before the Flood; *colloquial* very old.

antelope /'æntɪləʊp/ *noun* (*plural* same or **-s**) swift deerlike animal.

antenatal /æntɪ'neɪtəl/ *adjective* before birth; of pregnancy.

antenna /æn'tenə/ *noun* (*plural* **-tennae** /-ni:/) each of pair of feelers on head of insect or crustacean; (*plural* **-s**) aerial.

anterior /æn'tɪərɪə/ *adjective* nearer the front; (often + *to*) prior.

ante-room /'æntɪru:m/ *noun* small room leading to main one.

anthem /'ænθəm/ *noun* choral composition for church use; song of praise, esp. for nation.

anther /'ænθə/ *noun* part of stamen containing pollen.

anthology /æn'θɒlədʒɪ/ *noun* (*plural* **-ies**) collection of poems, essays, stories, etc.

anthracite /'ænθrəsaɪt/ *noun* hard kind of coal.

anthrax /'ænθræks/ *noun* disease of sheep and cattle transmissible to humans.

anthropocentric /ænθrəpəʊ'sentrɪk/ *adjective* regarding humankind as centre of existence.

anthropoid /'ænθrəpɔɪd/ ● *adjective* human in form. ● *noun* anthropoid ape.

anthropology /ænθrə'pɒlədʒɪ/ *noun* study of humankind, esp. societies and customs. □ **anthropological** /-pə'lɒdʒ-/ *adjective*; **anthropologist** *noun*.

anthropomorphism /ænθrəpə'mɔ:fɪz(ə)m/ *noun* attributing of human characteristics to god, animal, or thing. □ **anthropomorphic** *adjective*; **anthropomorphize** *verb* (also **-ise**) (**-zing** or **-sing**).

anthropomorphous /ænθrəpə'mɔ:fəs/ *adjective* human in form.

anti- /'æntɪ/ *prefix* opposed to; preventing; opposite of; unconventional.

anti-abortion /-ə'bɔ:ʃ(ə)n/ *adjective* opposing abortion. □ **anti-abortionist** *noun*.

anti-aircraft /-'eəkrɑ:ft/ *adjective* for defence against enemy aircraft.

antibiotic /-baɪ'ɒtɪk/ ● *noun* substance that can inhibit or destroy bacteria etc. ● *adjective* functioning as antibiotic.

antibody /'æntɪbɒdɪ/ *noun* (*plural* **-ies**) blood protein produced in reaction to antigens.

antic /'æntɪk/ *noun* (usually in *plural*) foolish behaviour.

anticipate /æn'tɪsɪpeɪt/ *verb* (**-ting**) deal with or use before proper time; expect; forestall. □ **anticipation** *noun*; **anticipatory** *adjective*.

anticlimax /-'klaɪmæks/ *noun* disappointing conclusion to something significant.

anticlockwise /-'klɒkwaɪz/ ● *adverb* in opposite direction to hands of clock. ● *adjective* moving anticlockwise.

anticyclone /-'saɪkləʊn/ *noun* system of winds rotating outwards from area of high pressure, producing fine weather.

antidepressant /-dɪ'pres(ə)nt/ ● *noun* drug etc. alleviating depression. ● *adjective* alleviating depression.

antidote /'æntɪdəʊt/ *noun* medicine used to counteract poison.

antifreeze /'æntɪfri:z/ *noun* substance added to water (esp. in vehicle's radiator) to lower its freezing point.

antigen /'æntɪdʒ(ə)n/ *noun* foreign substance causing body to produce antibodies.

anti-hero /'æntɪhɪərəʊ/ *noun* (*plural* **-es**) central character in story, lacking conventional heroic qualities.

antihistamine /-'hɪstəmi:n/ *noun* drug that counteracts effect of histamine, used esp. to treat allergies.

anti-lock /'æntɪlɒk/ *adjective* (of brakes) not locking when applied suddenly.

antimony /'æntɪmənɪ/ *noun* brittle silvery metallic element.

anti-nuclear /-'nju:klɪə/ *adjective* opposed to development of nuclear weapons or power.

antipathy /æn'tɪpəθɪ/ *noun* (*plural* **-ies**) (often + *to, for, between*) strong aversion or dislike. □ **antipathetic** /-'θet-/ *adjective*.

antiperspirant /-'pɜ:spərənt/ *noun* substance inhibiting perspiration.

antiphon /'æntɪf(ə)n/ *noun* hymn sung alternately by two groups. □ **antiphonal** /-'tɪf-/ *adjective*.

antipodes /æn'tɪpədi:z/ *plural noun* places diametrically opposite each other on the earth, esp. (also **Antipodes**) Australasia in relation to Europe. □ **antipodean** /-'di:ən/ *adjective & noun*.

antiquarian /æntɪ'kweərɪən/ ● *adjective* of or dealing in rare books. ● *noun* antiquary.

antiquary /'æntɪkwərɪ/ *noun* (*plural* **-ies**) student or collector of antiques etc.

antiquated /'æntɪkweɪtɪd/ *adjective* old-fashioned.

antique /æn'ti:k/ ● *noun* old valuable object, esp. piece of furniture etc. ● *adjective* of or existing since old times; old-fashioned.

antiquity /æn'tɪkwətɪ/ *noun* (*plural* **-ies**) ancient times, esp. before Middle Ages; (in *plural*) great age; remains from ancient times.

antirrhinum /æntɪ'raɪnəm/ *noun* snapdragon.

anti-Semitic /-sɪ'mɪtɪk/ *adjective* prejudiced against Jews. □ **anti-Semite** /-'si:maɪt/ *noun*; **anti-Semitism** /-'sem-/ *noun*.

antiseptic /-'septɪk/ ● *adjective* counteracting sepsis by destroying germs. ● *noun* antiseptic substance.

antisocial /-'səʊʃ(ə)l/ *adjective* not sociable; opposed or harmful to society.

■ **Usage** It is a mistake to use *antisocial* instead of *unsocial* in the phrase *unsocial hours*.

antistatic /-'stætɪk/ *adjective* counteracting effect of static electricity.

antithesis /æn'tɪθəsɪs/ *noun* (*plural* **-theses** /-si:z/) (often + *of*, *to*) direct opposite; contrast; rhetorical use of strongly contrasted words. □ **antithetical** /-'θet-/ *adjective*.

antitoxin /-'tɒksɪn/ *noun* antibody counteracting toxin. □ **antitoxic** *adjective*.

antitrades /-'treɪdz/ *plural noun* winds blowing above and in opposite direction to trade winds.

antiviral /-'vaɪər(ə)l/ *adjective* effective against viruses.

antler /'æntlə/ *noun* branched horn of deer.

antonym /'æntənɪm/ *noun* word opposite in meaning to another, e.g. *wet* is an antonym of *dry*.

anus /'eɪnəs/ *noun* (*plural* **-es**) excretory opening at end of alimentary canal.

anvil /'ænvɪl/ *noun* iron block on which metals are worked.

anxiety /æŋ'zaɪətɪ/ *noun* (*plural* **-ies**) troubled state of mind; worry; eagerness.

anxious /'æŋkʃəs/ *adjective* mentally troubled; marked by anxiety; (+ *to do*) uneasily wanting. □ **anxiously** *adverb*.

any /'enɪ/ ● *adjective* one or some, no matter which. ● *pronoun* any one; any number or amount. ● *adverb* at all. □ **anybody** *pronoun* any person, *noun* an important person; **anyhow** anyway, at random; **anyone** anybody; **anything** any thing, a thing of any kind; **anyway** in any way, in any case; **anywhere** (in or to) any place.

AOB *abbreviation* any other business.

aorta /eɪ'ɔ:tə/ *noun* (*plural* **-s**) main artery carrying blood from heart.

apace /ə'peɪs/ *adverb literary* swiftly.

apart /ə'pɑ:t/ *adverb* separately; into pieces; at or to a distance.

apartheid /ə'pɑ:teɪt/ *noun* racial segregation, esp. in S. Africa.

apartment /ə'pɑ:tmənt/ *noun* (*US* or esp. for holidays) flat; (usually in *plural*) room.

apathy /'æpəθɪ/ *noun* lack of interest, indifference. □ **apathetic** /-'θet-/ *adjective*.

apatosaurus /əpætə'sɔ:rəs/ *noun* (*plural* **-ruses**) large long-necked plant-eating dinosaur.

ape ● *noun* tailless monkey; imitator. ● *verb* (**-ping**) imitate.

aperient /ə'pɪərɪənt/ *adjective & noun* laxative.

aperitif /əperɪ'ti:f/ *noun* alcoholic drink before meal.

aperture /'æpətʃə/ *noun* opening or gap, esp. variable one letting light into camera.

apex /'eɪpeks/ *noun* (*plural* **-es**) highest point; tip.

aphasia /ə'feɪzɪə/ *noun* loss of verbal understanding or expression.

aphelion /ə'fi:lɪən/ *noun* (*plural* **-lia**) point of orbit farthest from sun.

aphid /'eɪfɪd/ *noun* insect infesting plants.

aphis /'eɪfɪs/ *noun* (*plural* **aphides** /-di:z/) aphid.

aphorism /'æfərɪz(ə)m/ *noun* short wise saying. □ **aphoristic** /-'rɪs-/ *adjective*.

aphrodisiac /æfrə'dɪzɪæk/ ● *adjective* arousing sexual desire. ● *noun* aphrodisiac substance.

apiary /'eɪpɪərɪ/ *noun* (*plural* **-ies**) place where bees are kept. □ **apiarist** *noun*.

apiculture /ˈeɪpɪkʌltʃə/ *noun* bee-keeping.

apiece /əˈpiːs/ *adverb* for each one.

aplomb /əˈplɒm/ *noun* self-assurance.

apocalypse /əˈpɒkəlɪps/ *noun* destructive event; revelation, esp. about the end of the world. □ **apocalyptic** /-ˈlɪp-/ *adjective.*

Apocrypha /əˈpɒkrɪfə/ *plural noun* Old Testament books not in Hebrew Bible; (**apocrypha**) writings etc. not considered genuine. □ **apocryphal** *adjective.*

apogee /ˈæpədʒiː/ *noun* highest point; point farthest from earth in orbit of moon etc.

apolitical /eɪpəˈlɪtɪk(ə)l/ *adjective* not interested or involved in politics.

apologetic /əpɒləˈdʒetɪk/ *adjective* expressing regret. □ **apologetically** *adverb.*

apologia /æpəˈləʊdʒə/ *noun* formal defence of conduct or opinions.

apologist /əˈpɒlədʒɪst/ *noun* person who defends by argument.

apologize /əˈpɒlədʒaɪz/ *verb* (also **-ise**) (**-zing** or **-sing**) make apology.

apology /əˈpɒlədʒɪ/ *noun* (*plural* **-ies**) regretful acknowledgement of offence or failure; explanation.

apophthegm /ˈæpəθem/ *noun* short wise saying.

apoplexy /ˈæpəpleksɪ/ *noun* sudden paralysis caused by blockage or rupture of brain artery. □ **apoplectic** /-ˈplek-/ *adjective.*

apostasy /əˈpɒstəsɪ/ *noun* (*plural* **-ies**) abandonment of belief, faith, etc.

apostate /əˈpɒsteɪt/ *noun* person who renounces belief. □ **apostatize** /-tət-/ *verb* (also **-ise**) (**-zing** or **-sing**).

a posteriori /eɪ pɒsterɪˈɔːraɪ/ *adjective & adverb* from effects to causes.

Apostle /əˈpɒs(ə)l/ *noun* any of twelve men sent by Christ to preach gospel; (**apostle**) leader of reform.

apostolic /æpəsˈtɒlɪk/ *adjective* of Apostles; of the Pope.

apostrophe /əˈpɒstrəfɪ/ *noun* punctuation mark (') indicating possession or marking omission of letter(s) or number(s) (see panel).

apostrophize *verb* (also **-ise**) (**-zing** or **-sing**) address (esp. absent person or thing).

apothecary /əˈpɒθəkərɪ/ *noun* (*plural* **-ies**) *archaic* pharmacist.

apotheosis /əpɒθɪˈəʊsɪs/ *noun* (*plural* **-oses** /-siːz/) deification; glorification or sublime example (of thing).

appal /əˈpɔːl/ *verb* (**-ll-**) dismay; horrify.

apparatus /æpəˈreɪtəs/ *noun* equipment for scientific or other work.

apparel /əˈpær(ə)l/ *noun formal* clothing. □ **apparelled** *adjective.*

apparent /əˈpærənt/ *adjective* obvious; seeming. □ **apparently** *adverb.*

apparition /æpəˈrɪʃ(ə)n/ *noun* thing that appears, esp. of startling kind; ghost.

appeal /əˈpiːl/ • *verb* make earnest or formal request; (usually + *to*) be attractive; (+ *to*) resort to for support; (often + *to*) apply to higher court for revision of judicial decision; *Cricket* ask umpire to declare batsman out. • *noun* appealing; request for aid; attractiveness.

appear /əˈpɪə/ *verb* become or be visible; seem; present oneself; be published.

appearance *noun* appearing; outward form; (in *plural*) outward show of prosperity, virtue, etc.

appease /əˈpiːz/ *verb* (**-sing**) make calm or quiet, esp. conciliate (aggressor) with concessions; satisfy (appetite etc.). □ **appeasement** *noun.*

appellant /əˈpelənt/ *noun* person who appeals to higher court.

appellation /æpəˈleɪʃ(ə)n/ *noun formal* name, title.

append /əˈpend/ *verb* (usually + *to*) attach, affix; add.

appendage /əˈpendɪdʒ/ *noun* thing attached; addition.

appendectomy /æpenˈdektəmɪ/ *noun* (also **appendicectomy** /əpendɪˈsektəmɪ/) (*plural* **-ies**) surgical removal of appendix.

appendicitis /əpendɪˈsaɪtɪs/ *noun* inflammation of appendix.

appendix /əˈpendɪks/ *noun* (*plural* **-dices** /-siːz/) tubular sac attached to large intestine; addition to book etc.

appertain /æpəˈteɪn/ *verb* (+ *to*) belong, relate to.

appetite /ˈæpɪtaɪt/ *noun* (usually + *for*) desire (esp. for food); inclination, desire.

appetizer /ˈæpɪtaɪzə/ *noun* (also **-iser**) thing eaten or drunk to stimulate appetite.

appetizing /ˈæpɪtaɪzɪŋ/ *adjective* (also **-ising**) (esp. of food) stimulating appetite.

applaud /əˈplɔːd/ *verb* express approval (of), esp. by clapping; commend.

applause /əˈplɔːz/ *noun* warm approval, esp. clapping.

apple /ˈæp(ə)l/ *noun* roundish firm fruit. □ **apple of one's eye** cherished person or thing; **apple-pie bed** bed with sheets folded so that one cannot stretch out one's legs; **apple-pie order** extreme neatness.

appliance /əˈplaɪəns/ *noun* device etc. for specific task.

applicable /ˈæplɪkəb(ə)l/ *adjective* (often + *to*) that may be applied. □ **applicability** *noun.*

applicant /ˈæplɪkənt/ *noun* person who applies for job etc.

application /æplɪˈkeɪʃ(ə)n/ *noun* formal request; applying; thing applied; diligence; relevance; use.

applicator /ˈæplɪkeɪtə/ *noun* device for applying ointment etc.

appliqué /æˈpliːkeɪ/ ● *noun* work in which cut-out fabric is fixed on to other fabric. ● *verb* (**-qués**, **-quéd**, **-quéing**) decorate with appliqué.

Apostrophe ʼ

This is used:

1 to indicate possession:

with a singular noun:

a boy's book; a week's work; the boss's salary.

with a plural already ending with *s*:

a girls' school; two weeks' newspapers; the bosses' salaries.

with a plural not already ending with *s*:

the children's books; women's liberation.

with a singular name:

Bill's book; John's coat; Barnabas' (or *Barnabas's*) *book; Nicholas'* (or *Nicholas's*) *coat.*

with a name ending in *-es* that is pronounced /-ɪz/:

Bridges' poems; Moses' mother

and before the word *sake*:

for God's sake; for goodness' sake; for Nicholas' sake

but it is often omitted in a business name:

Barclays Bank.

2 to mark an omission of one or more letters or numbers:

he's (*he is* or *he has*)
can't (*cannot*)
won't (*will not*)
the summer of '68 (*1968*)
haven't (*have not*)
we'll (*we shall*)
o'clock (*of the clock*)

3 when letters or numbers are referred to in plural form:

mind your p's and q's; find all the number 7's.

but it is unnecessary in, e.g.

MPs; the 1940s.

---

apply /ə'plaɪ/ *verb* (**-ies**, **-ied**) (often + *for*, *to*, *to do*) make formal request; (often + *to*) be relevant; make use of; (often + *to*) put or spread (on); (**apply oneself**; often + *to*) devote oneself.

appoint /ə'pɔɪnt/ *verb* assign job or office to; (often + *for*) fix (time etc.); (as **appointed** *adjective*) equipped, furnished.

appointment *noun* appointing, being appointed; arrangement for meeting; job available; (in *plural*) equipment, fittings.

apportion /ə'pɔ:ʃ(ə)n/ *verb* (often + *to*) share out. □ **apportionment** *noun*.

apposite /'æpəzɪt/ *adjective* (often + *to*) well expressed, appropriate.

apposition /æpə'zɪʃ(ə)n/ *noun* juxtaposition, esp. of syntactically parallel words etc. (e.g. *my friend Sue*).

appraise /ə'preɪz/ *verb* (**-sing**) estimate value or quality of. □ **appraisal** *noun*.

appreciable /ə'pri:ʃəb(ə)l/ *adjective* significant; considerable.

appreciate /ə'pri:ʃɪeɪt/ *verb* (**-ting**) (highly) value; be grateful for; understand, recognize; rise in value. □ **appreciation** *noun*; **appreciative** /-ʃətɪv/ *adjective*.

apprehend /æprɪ'hend/ *verb* arrest; understand.

apprehension /æprɪ'henʃ(ə)n/ *noun* fearful anticipation; arrest; understanding.

apprehensive /æprɪ'hensɪv/ *adjective* uneasy, fearful. □ **apprehensively** *adverb*.

apprentice /ə'prentɪs/ ● *noun* person learning trade by working for agreed period. ● *verb* (usually + *to*) engage as apprentice. □ **apprenticeship** *noun*.

apprise /ə'praɪz/ *verb* (**-sing**) (usually + *of*) inform.

approach /ə'prəʊtʃ/ ● *verb* come nearer (to) in space or time; be similar to; approximate to; set about; make tentative proposal to. ● *noun* act or means of approaching; approximation; technique; part of aircraft's flight before landing.

approachable /ə'prəʊtʃəb(ə)l/ *adjective* friendly; able to be approached.

approbation /æprə'beɪʃ(ə)n/ *noun* approval, consent.

appropriate ● *adjective* /ə'prəʊprɪət/ suitable, proper. ● *verb* (**-ting**) /ə'prəʊprɪeɪt/ take possession of; devote (money etc.) to special purpose. □ **appropriately** *adverb*; **appropriation** *noun*.

approval /ə'pru:v(ə)l/ *noun* approving; consent. □ **on approval** returnable if not satisfactory.

approve /ə'pru:v/ *verb* (**-ving**) sanction; (often + *of*) regard with favour.

approx. *abbreviation* approximate(ly).

approximate ● *adjective* /ə'prɒksɪmət/ fairly correct; near to the actual. ● *verb* /ə'prɒksɪmeɪt/ (**-ting**) (often + *to*) be or make near. □ **approximately** *adverb*; **approximation** *noun*.

appurtenances /ə'pɜ:tɪnənsɪz/ *plural noun* accessories; belongings.

APR *abbreviation* annual(ized) percentage rate.

Apr. *abbreviation* April.

après-ski /æpreɪ'ski:/ ● *noun* activities after a day's skiing. ● *adjective* suitable for these. [French]

apricot /'eɪprɪkɒt/ ● *noun* small orange-yellow peachlike fruit; its colour. ● *adjective* orange-yellow.

April /'eɪpr(ə)l/ *noun* fourth month of year. □ **April Fool** victim of hoax on 1 Apr.

a priori /eɪ praɪ'ɔ:raɪ/ ● *adjective* from cause to effect; not derived from experience; assumed without investigation. ● *adverb* deductively; as far as one knows.

apron /'eɪprən/ *noun* garment protecting front of clothes; area on airfield for manoeuvring or loading; part of stage in front of curtain.

apropos /æprə'pəʊ/ ● *adjective* appropriate; *colloquial* (often + *of*) in respect of. ● *adverb* appropriately; incidentally.

apse /æps/ *noun* arched or domed recess esp. at end of church.

apsis /'æpsɪs/ *noun* (*plural* **apsides** /-di:z/) aphelion or perihelion of planet, apogee or perigee of moon.

apt *adjective* suitable, appropriate; (+ *to do*) having a tendency; quick to learn.

aptitude /'æptɪtju:d/ *noun* talent; ability, esp. specified.

aqualung /'ækwəlʌŋ/ *noun* portable underwater breathing-apparatus.

aquamarine /ækwəmə'ri:n/ ● *noun* bluish-green beryl; its colour. ● *adjective* bluish-green.

aquaplane /'ækwəpleɪn/ ● *noun* board for riding on water, pulled by speed-boat. ● *verb* (**-ning**) ride on this; (of vehicle) glide uncontrollably on wet surface.

aquarelle /ækwə'rel/ *noun* painting in transparent water-colours.

aquarium /ə'kweərɪəm/ *noun* (*plural* **-s**) tank for keeping fish etc.

Aquarius /ə'kweərɪəs/ *noun* eleventh sign of zodiac.

aquatic /ə'kwætɪk/ *adjective* growing or living in water; done in or on water.

aquatint /'ækwətɪnt/ *noun* etched print like water colour.

aqueduct /'ækwɪdʌkt/ *noun* water channel, esp. raised structure across valley.

aqueous /'eɪkwɪəs/ *adjective* of or like water.

aquiline /'ækwɪlaɪn/ *adjective* of or like an eagle; (of nose) curved.

Arab /'ærəb/ ● *noun* member of Semitic people inhabiting originally Saudi Arabia, now Middle East generally; horse of breed originally native to Arabia. ● *adjective* of Arabia or Arabs.

arabesque /ærə'besk/ *noun* decoration with intertwined leaves, scrollwork, etc.; ballet posture with one leg extended horizontally backwards.

Arabian /ə'reɪbɪən/ ● *adjective* of Arabia. ● *noun* Arab.

Arabic /'ærəbɪk/ ● *noun* language of Arabs. ● *adjective* of Arabs or their language. □ **arabic numerals** 1, 2, 3, etc.

arable /'ærəb(ə)l/ *adjective* fit for growing crops.

arachnid /ə'ræknɪd/ *noun* creature of class comprising spiders, scorpions, etc.

Aramaic /ærə'meɪɪk/ ● *noun* language of Syria at time of Christ. ● *adjective* of or in Aramaic.

arbiter /'ɑ:bɪtə/ *noun* arbitrator; person influential in specific field.

arbitrary /'ɑ:bɪtrərɪ/ *adjective* random; capricious, despotic. □ **arbitrarily** *adverb.*

arbitrate /'ɑ:bɪtreɪt/ *verb* (**-ting**) settle dispute between others. □ **arbitration** *noun*; **arbitrator** *noun.*

arboreal /ɑ:'bɔ:rɪəl/ *adjective* of or living in trees.

arboretum /ɑ:bə'ri:təm/ *noun* (*plural* **-ta**) tree-garden.

arboriculture /'ɑ:bərɪkʌltʃə/ *noun* cultivation of trees and shrubs.

arbour /'ɑ:bə/ *noun* (*US* **arbor**) shady garden alcove enclosed by trees etc.

arc *noun* part of circumference of circle or other curve; luminous discharge between two electrodes. □ **arc lamp** one using electric arc; **arc welding** using electric arc to melt metals to be welded.

arcade /ɑ:'keɪd/ *noun* covered walk, esp. lined with shops; series of arches supporting or along wall.

Arcadian /ɑ:'keɪdɪən/ *adjective* ideally rustic.

arcane /ɑ:'keɪn/ *adjective* mysterious, secret.

arch¹ ● *noun* curved structure supporting bridge, floor, etc. as opening or ornament. ● *verb* form arch; provide with or form into arch.

arch² *adjective* self-consciously or affectedly playful.

archaeology /ɑ:kɪ'ɒlədʒɪ/ *noun* (*US* **archeology**) study of ancient cultures, esp. by excavation of physical remains. □ **archaeological** /-ə'lɒdʒ-/ *adjective*; **archaeologist** *noun.*

archaic /ɑ:'keɪɪk/ *adjective* antiquated; (of word) no longer in ordinary use; of early period in culture.

archaism /'ɑ:keɪɪz(ə)m/ *noun* archaic word etc.; use of the archaic. □ **archaistic** /-'ɪst-/ *adjective.*

archangel /'ɑ:keɪndʒ(ə)l/ *noun* angel of highest rank.

archbishop /ɑ:tʃ'bɪʃəp/ *noun* chief bishop.

archbishopric *noun* office or diocese of archbishop.

archdeacon /ɑ:tʃ'di:kən/ *noun* church dignitary next below bishop. □ **archdeaconry** *noun* (*plural* **-ies**).

archdiocese /ɑ:tʃ'daɪəsɪs/ *noun* archbishop's diocese. □ **archdiocesan** /-daɪ'ɒsɪs(ə)n/ *adjective.*

arch-enemy /ɑ:tʃ'enəmɪ/ *noun* (*plural* **-ies**) chief enemy.

archeology *US* = ARCHAEOLOGY.

archer /ˈɑːtʃə/ *noun* person who shoots with bow and arrows.

archery /ˈɑːtʃərɪ/ *noun* shooting with bow and arrows.

archetype /ˈɑːkɪtaɪp/ *noun* original model, typical specimen. □ **archetypal** /-ˈtaɪp-/ *adjective*.

archipelago /ɑːkɪˈpeləgəʊ/ *noun* (*plural* **-s**) group of islands; sea with many islands.

architect /ˈɑːkɪtekt/ *noun* designer of buildings etc.; (+ *of*) person who brings about specified thing.

architectonic /ɑːkɪtekˈtɒnɪk/ *adjective* of architecture.

architecture /ˈɑːkɪtektʃə/ *noun* design and construction of buildings; style of building. □ **architectural** /-ˈtek-/ *adjective*.

architrave /ˈɑːkɪtreɪv/ *noun* moulded frame round doorway or window; main beam laid across tops of classical columns.

archive /ˈɑːkaɪv/ *noun* (usually in *plural*) collection of documents or records; place where these are kept.

archivist /ˈɑːkɪvɪst/ *noun* keeper of archives.

archway /ˈɑːtʃweɪ/ *noun* arched entrance or passage.

Arctic /ˈɑːktɪk/ ● *adjective* of north polar region; (**arctic**) very cold. ● *noun* Arctic region.

ardent /ˈɑːd(ə)nt/ *adjective* eager, fervent, passionate; burning. □ **ardently** *adverb*.

ardour /ˈɑːdə/ *noun* zeal, enthusiasm.

arduous /ˈɑːdjʊəs/ *adjective* hard to accomplish; strenuous.

are see BE.

area /ˈeərɪə/ *noun* extent or measure of surface; region; space set aside for a purpose; scope, range; space in front of house basement.

arena /əˈriːnə/ *noun* centre of amphitheatre; scene of conflict; sphere of action.

aren't /ɑːnt/ are not; (in questions) am not.

arête /æˈreɪt/ *noun* sharp mountain ridge.

argon /ˈɑːgɒn/ *noun* inert gaseous element.

argot /ˈɑːgəʊ/ *noun* jargon of group or class.

argue /ˈɑːgjuː/ *verb* (**-ues**, **-ued**, **-uing**) exchange views, esp. heatedly; (often + *that*) maintain by reasoning; (+ *for*, *against*) reason. □ **arguable** *adjective*; **arguably** *adverb*.

argument /ˈɑːgjʊmənt/ *noun* (esp. heated) exchange of views; reason given; reasoning; summary of book etc. □ **argumentation** /-men-/ *noun*.

argumentative /ɑːgjʊˈmentətɪv/ *adjective* fond of arguing.

argy-bargy /ɑːdʒɪˈbɑːdʒɪ/ *noun jocular* dispute, wrangle.

aria /ˈɑːrɪə/ *noun* song for one voice in opera, oratorio, etc.

arid /ˈærɪd/ *noun* dry, parched. □ **aridity** /əˈrɪd-/ *noun*.

Aries /ˈeəriːz/ *noun* first sign of zodiac.

aright /əˈraɪt/ *adverb* rightly.

arise /əˈraɪz/ *verb* (**-sing**; *past* **arose** /əˈrəʊz/; *past participle* **arisen** /əˈrɪz(ə)n/) originate; (usually + *from*, *out of*) result; emerge; rise.

aristocracy /ærɪsˈtɒkrəsɪ/ *noun* (*plural* **-ies**) ruling class, nobility.

aristocrat /ˈærɪstəkræt/ *noun* member of aristocracy.

aristocratic /ærɪstəˈkrætɪk/ *adjective* of the aristocracy; grand, distinguished.

arithmetic ● *noun* /əˈrɪθmətɪk/ science of numbers; computation, use of numbers. ● *adjective* /ærɪθˈmetɪk/ (also **arithmetical**) of arithmetic.

ark *noun* ship in which Noah escaped the Flood. □ **Ark of the Covenant** wooden chest containing tables of Jewish law.

arm[1] *noun* upper limb of human body; sleeve; raised side part of chair; branch; armlike thing. □ **armchair** chair with arms, theoretical rather than active; **arm in arm** with arms linked; **armpit** hollow under arm at shoulder; **at arm's length** at a distance; **with open arms** cordially. □ **armful** *noun*.

arm[2] ● *noun* (usually in *plural*) weapon; branch of military forces; (in *plural*) heraldic devices. ● *verb* equip with arms; equip oneself with arms; make (bomb) ready. □ **up in arms** (usually + *against*, *about*) actively resisting.

armada /ɑ:'mɑ:də/ *noun* fleet of warships.

armadillo /ɑ:mə'dɪləʊ/ *noun* (*plural* **-s**) S. American burrowing mammal with plated body.

Armageddon /ɑ:mə'ged(ə)n/ *noun* huge battle at end of world.

armament /'ɑ:məmənt/ *noun* military weapon etc.; equipping for war.

armature /'ɑ:mətʃə/ *noun* rotating coil or coils of dynamo or electric motor; iron bar placed across poles of magnet; framework on which sculpture is moulded.

armistice /'ɑ:mɪstɪs/ *noun* truce.

armlet /'ɑ:mlɪt/ *noun* band worn round arm.

armorial /ɑ:'mɔ:rɪəl/ *adjective* of heraldic arms.

armour /'ɑ:mə/ *noun* protective covering formerly worn in fighting; metal plates etc. protecting ship, car, tank, etc.; armoured vehicles. □ **armoured** *adjective.*

armourer *noun* maker of arms or armour; official in charge of arms.

armoury /'ɑ:mərɪ/ *noun* (*plural* **-ies**) arsenal.

army /'ɑ:mɪ/ *noun* (*plural* **-ies**) organized armed land force; vast number; organized body.

arnica /'ɑ:nɪkə/ *noun* plant with yellow flowers; medicine made from it.

aroma /ə'rəʊmə/ *noun* pleasing smell; subtle quality. □ **aromatic** /ærə'mætɪk/ *adjective.*

aromatherapy *noun* use of plant extracts and oils in massage. □ **aromatherapist** *noun.*

arose *past* of ARISE.

around /ə'raʊnd/ ● *adverb* on every side; all round; *colloquial* in existence, near at hand. ● *preposition* on or along the circuit of; on every side of; here and there in or near; about.

arouse /ə'raʊz/ *verb* (**-sing**) induce (esp. emotion); awake from sleep; stir into activity; stimulate sexually. □ **arousal** *noun.*

arpeggio /ɑ:'pedʒɪəʊ/ *noun* (*plural* **-s**) notes of chord played in rapid succession.

arrack /'ærək/ *noun* alcoholic spirit made esp. from rice.

arraign /ə'reɪn/ *verb* indict, accuse; find fault with. □ **arraignment** *noun.*

arrange /ə'reɪndʒ/ *verb* (**-ging**) put in order; plan or provide for; (+ *to do, for*) take measures; (+ *with* person) agree about procedure for; *Music* adapt (piece). □ **arrangement** *noun.*

arrant /'ærənt/ *adjective* downright, utter.

arras /'ærəs/ *noun* tapestry wall-hanging.

array /ə'reɪ/ ● *noun* imposing series; ordered arrangement, esp. of troops. ● *verb* deck, adorn; set in order; marshal (forces).

arrears /ə'rɪəz/ *plural noun* outstanding debt; what remains undone. □ **in arrears** behindhand, esp. in payment.

arrest /ə'rest/ ● *verb* lawfully seize; stop; catch attention of. ● *noun* arresting, being arrested; stoppage.

arrival /ə'raɪv(ə)l/ *noun* arriving; person or thing arriving.

arrive /ə'raɪv/ *verb* (**-ving**) come to destination; (+ *at*) reach (conclusion); *colloquial* become successful, be born.

arrogant /'ærəgənt/ *adjective* aggressively assertive or presumptuous. □ **arrogance** *noun*; **arrogantly** *adverb.*

arrogate /'ærəgeɪt/ *verb* (**-ting**) claim without right. □ **arrogation** *noun.*

arrow /'ærəʊ/ *noun* pointed missile shot from bow; representation of this, esp. to show direction. □ **arrowhead** pointed tip of arrow.

arrowroot /'ærəʊru:t/ *noun* nutritious starch.

arsenal /'ɑ:sən(ə)l/ *noun* place where weapons and ammunition are made or stored.

arsenic /'ɑ:sənɪk/ *noun* brittle semi-metallic element; its highly poisonous trioxide.

arson /'ɑ:sən/ *noun* crime of deliberately setting property on fire. □ **arsonist** *noun.*

art *noun* human creative skill, its application; branch of creative activity concerned with imitative and imaginative designs, sounds, or ideas, e.g. painting, music, writing; creative activity resulting in visual representation; thing in which skill can be exercised; (in *plural*) certain branches of learning (esp. languages, literature, history, etc.) as distinct from sciences; knack; cunning. □ **art nouveau** /ɑ: nu:'vəʊ/ art style of late 19th c., with flowing lines.

artefact /'ɑ:tɪfækt/ *noun* (also **artifact**) man-made object.

arterial /ɑ:'tɪərɪəl/ *adjective* of or like artery. □ **arterial road** important main road.

arteriosclerosis /ɑ:tɪərɪəʊsklɪə'rəʊsɪs/ *noun* hardening and thickening of artery walls.

artery /'ɑ:tərɪ/ *noun* (*plural* **-ies**) blood vessel carrying blood from heart; main road or railway line.

artesian well /ɑ:'ti:zɪən/ well in which water rises by natural pressure through vertically drilled hole.

artful /'ɑ:tfʊl/ *adjective* sly, crafty. □ **artfully** *adverb*.

arthritis /ɑ:'θraɪtɪs/ *noun* inflammation of joint. □ **arthritic** /-'θrɪt-/ *adjective & noun*.

arthropod /'ɑ:θrəpɒd/ *noun* animal with segmented body and jointed limbs, e.g. insect, spider, crustacean.

artichoke /'ɑ:tɪtʃəʊk/ *noun* plant allied to thistle; its partly edible flower; Jerusalem artichoke.

article /'ɑ:tɪk(ə)l/ ● *noun* item or thing; short piece of non-fiction in newspaper etc.; clause of agreement etc.; = DEFINITE ARTICLE, INDEFINITE ARTICLE. ● *verb* employ under contract as trainee.

articular /ɑ:'tɪkjʊlə/ *adjective* of joints.

articulate ● *adjective* /ɑ:'tɪkjʊlət/ fluent and clear in speech; (of speech) in which separate sounds and words are clear; having joints. ● *verb* /ɑ:'tɪkjʊleɪt/ (**-ting**) speak distinctly; express clearly; connect with joints. □ **articulated lorry** one with sections connected by flexible joint. □ **articulately** *adverb*; **articulation** *noun*.

artifact = ARTEFACT.

artifice /'ɑ:tɪfɪs/ *noun* trick; cunning; skill.

artificer /ɑ:'tɪfɪsə/ *noun* craftsman.

artificial /ɑ:tɪ'fɪʃ(ə)l/ *adjective* not natural; imitating nature; insincere. □ **artificial insemination** non-sexual injection of semen into uterus; **artificial intelligence** (use of) computers replacing human intelligence; **artificial respiration** manual or mechanical stimulation of breathing. □ **artificiality** /-ʃɪ'æl-/ *noun*; **artificially** *adverb*.

artillery /ɑ:'tɪlərɪ/ *noun* large guns used in fighting on land; branch of army using these. □ **artilleryman** *noun*.

artisan /ɑ:tɪ'zæn/ *noun* skilled worker or craftsman.

artist /'ɑ:tɪst/ *noun* person who practises any art, esp. painting; artiste. □ **artistic** *adjective*; **artistically** *adverb*; **artistry** *noun*.

artiste /ɑ:'ti:st/ *noun* professional singer, dancer, etc.

artless *adjective* guileless, ingenuous; natural; clumsy. □ **artlessly** *adverb*.

arty *adjective* (**-ier**, **-iest**) pretentiously or affectedly artistic.

arum /'eərəm/ *noun* plant with arrow-shaped leaves.

Aryan /'eərɪən/ ● *noun* speaker of any Indo-European language. ● *adjective* of Aryans.

as /æz, əz/ ● *adverb* to the same extent. ● *conjunction* in the same way that; while, when; since, seeing that; although. ● *preposition* in the capacity or form of. □ **as ... as ...** to the same extent that ... is, does, etc.

asafoetida /æsə'fetɪdə/ *noun* (*US* **asafetida**) resinous pungent gum.

asbestos /æs'bestɒs/ *noun* fibrous silicate mineral; heat-resistant or insulating substance made from this.

ascend /ə'send/ *verb* slope upwards; go or come up, climb.

ascendancy *noun* (often + *over*) dominant control.

ascendant *adjective* rising. □ **in the ascendant** gaining or having power or authority.

ascension /ə'senʃ(ə)n/ *noun* ascent, esp. (**Ascension**) of Christ into heaven.

ascent /ə'sent/ *noun* ascending, rising; upward path or slope.

ascertain /æsə'teɪn/ *verb* find out for certain. □ **ascertainment** *noun*.

ascetic /ə'setɪk/ ● *adjective* severely abstinent; self-denying. ● *noun* ascetic person. □ **asceticism** /-tɪs-/ *noun*.

ascorbic acid /ə'skɔ:bɪk/ *noun* vitamin C.

ascribe /ə'skraɪb/ *verb* (**-bing**) (usually + *to*) attribute; regard as belonging. □ **ascription** /-'skrɪp-/ *noun*.

asepsis /eɪ'sepsɪs/ *noun* absence of sepsis or harmful bacteria; aseptic method in surgery. □ **aseptic** *adjective*.

asexual /eɪ'seksjʊəl/ *adjective* without sex or sexuality; (of reproduction) not involving fusion of gametes. □ **asexually** *adverb*.

ash¹ *noun* (often in *plural*) powdery residue left after burning; (in *plural*) human remains after cremation; (**the Ashes**) trophy in cricket between England and Australia. □ **ashcan** *US* dustbin; **ashtray** receptacle for tobacco ash; **Ash Wednesday** first day of Lent.

ash² *noun* tree with silver-grey bark; its wood.

ashamed /ə'ʃeɪmd/ *adjective* embarrassed by shame; (+ *to do*) reluctant owing to shame.

ashen /'æʃ(ə)n/ *adjective* grey, pale.

ashore /ə'ʃɔ:/ *adverb* to or on shore.

ashram /'æʃræm/ *noun* religious retreat for Hindus.

Asian /'eɪʃ(ə)n/ ● *adjective* of Asia. ● *noun* native of Asia; person of Asian descent.

aside /ə'saɪd/ ● *adverb* to or on one side; away, apart. ● *noun* words spoken aside, esp. by actor to audience.

asinine /'æsɪnaɪn/ *adjective* asslike; stupid.

ask /ɑ:sk/ *verb* call for answer to or about; seek to obtain from another person; invite; (+ *for*) seek to obtain or meet. □ **ask after** inquire about (esp. person).

askance /ə'skæns/ *adverb* sideways. □ **look askance at** regard suspiciously.

askew /ə'skju:/ ● *adverb* crookedly. ● *adjective* oblique; awry.

aslant /ə'slɑ:nt/ ● *adverb* at a slant. ● *preposition* obliquely across.

asleep /ə'sli:p/ ● *adjective* sleeping; *colloquial* inattentive; (of limb) numb. ● *adverb* into state of sleep.

asp *noun* small venomous snake.

asparagus /əs'pærəgəs/ *noun* plant of lily family; its edible shoots.

aspect /'æspekt/ *noun* feature, viewpoint, etc. to be considered; appearance, look; side facing specified direction.

aspen /'æspən/ *noun* poplar with fluttering leaves.

asperity /æs'perɪtɪ/ *noun* sharpness of temper or tone; roughness.

aspersion /əs'pɜ:ʃ(ə)n/ *noun* □ **cast aspersions on** defame.

asphalt /'æsfælt/ ● *noun* bituminous pitch; mixture of this with gravel etc. for surfacing roads etc. ● *verb* surface with asphalt.

asphodel /'æsfədel/ *noun* kind of lily.

asphyxia /æs'fɪksɪə/ *noun* lack of oxygen in blood; suffocation.

asphyxiate /əs'fɪksɪeɪt/ *verb* (**-ting**) suffocate. □ **asphyxiation** *noun*.

aspic /'æspɪk/ *noun* clear savoury jelly.

aspidistra /æspɪ'dɪstrə/ *noun* house plant with broad tapering leaves.

aspirant /'æspɪrənt/ ● *adjective* aspiring. ● *noun* person who aspires.

aspirate ● *noun* /'æspərət/ sound of *h*; consonant blended with this. ● *verb* /'æspəreɪt/ (**-ting**) pronounce with *h*; draw (fluid) by suction from cavity.

aspiration /æspə'reɪʃ(ə)n/ *noun* ambition, desire; aspirating.

aspire /ə'spaɪə/ *verb* (**-ring**) (usually + *to*, *after*, *to do*) have ambition or strong desire.

aspirin /'æsprɪn/ *noun* (*plural* same or **-s**) white powder used to reduce pain and fever; tablet of this.

ass *noun* 4-legged animal with long ears, related to horse; donkey; stupid person.

assail /ə'seɪl/ *verb* attack physically or verbally. □ **assailant** *noun*.

assassin /ə'sæsɪn/ *noun* killer, esp. of political or religious leader.

assassinate /ə'sæsɪneɪt/ *verb* (**-ting**) kill for political or religious motives. □ **assassination** *noun*.

assault /ə'sɔ:lt/ ● *noun* violent physical or verbal attack; *Law* threat or display of violence against person. ● *verb* make assault on. □ **assault and battery** *Law* threatening act resulting in physical harm to person.

assay /ə'seɪ/ ● *noun* test of metal or ore for ingredients and quality. ● *verb* make assay of.

assegai /'æsɪgaɪ/ *noun* light iron-tipped S. African spear.

assemblage /ə'semblɪdʒ/ *noun* assembled group.

assemble /ə'semb(ə)l/ *verb* (**-ling**) fit together parts of; fit (parts) together; bring or come together.

assembly /ə'semblɪ/ *noun* (*plural* **-ies**) assembling; assembled group, esp. as parliament etc.; fitting together of parts. □ **assembly line** machinery arranged so that product can be progressively assembled.

assent /ə'sent/ ● *noun* consent, approval. ● *verb* (usually + *to*) agree, consent.

assert /ə'sɜ:t/ *verb* declare; enforce claim to; (**assert oneself**) insist on one's rights.

assertion /ə'sɜ:ʃ(ə)n/ *noun* declaration; forthright statement.

assertive *adjective* asserting oneself; forthright, positive. □ **assertively** *adverb*; **assertiveness** *noun*.

assess /ə'ses/ *verb* estimate size or quality of; estimate value of (property etc.) for taxation. □ **assessment** *noun*.

assessor *noun* person who assesses taxes etc.; judge's technical adviser in court.

asset /'æset/ *noun* useful or valuable person or thing; (usually in *plural*) property and possessions.

assiduous /ə'sɪdjʊəs/ *adjective* persevering, hard-working. □ **assiduity** /æsɪ'dju:ɪtɪ/ *noun*; **assiduously** *adverb*.

assign /ə'saɪn/ *verb* allot; appoint; fix (time, place, etc.); (+ *to*) ascribe to, *Law* transfer formally to.

assignation /æsɪg'neɪʃ(ə)n/ *noun* appointment, esp. made by lovers; assigning, being assigned.

assignee /əsaɪ'ni:/ *noun Law* person to whom property or right is assigned.

assignment /ə'saɪnmənt/ *noun* task or mission; assigning, being assigned.

assimilate /ə'sɪmɪleɪt/ *verb* (**-ting**) absorb or be absorbed into system; (usually + *to*) make like. □ **assimilable** *adjective*; **assimilation** *noun*; **assimilative** /-lətɪv/ *adjective*.

assist /ə'sɪst/ *verb* (often + *in*) help. □ **assistance** *noun*.

assistant *noun* helper; subordinate worker; = SHOP ASSISTANT.

assizes /ə'saɪzɪz/ *plural noun historical* court periodically administering civil and criminal law.

Assoc. *abbreviation* Association.

associate ● *verb* /ə'səʊʃɪeɪt/ (**-ting**) connect mentally; join, combine; (usually + *with*) have frequent dealings. ● *noun* /ə'səʊʃɪət/ partner, colleague; friend, companion. ● *adjective* /ə'səʊʃɪət/ joined, allied.

association /əsəʊsɪ'eɪʃ(ə)n/ *noun* group organized for joint purpose; associating, being associated; connection of ideas. □ **association football** kind played with round ball which may be handled only by goalkeeper.

assonance /'æsənəns/ *noun* partial resemblance of sound between syllables, as in *run-up*, or *wary* and *barely*. □ **assonant** *adjective*.

assorted /ə'sɔ:tɪd/ *adjective* of various sorts, mixed.

assortment /ə'sɔ:tmənt/ *noun* diverse group or mixture.

assuage /ə'sweɪdʒ/ *verb* (**-ging**) soothe; appease.

assume /ə'sju:m/ *verb* (**-ming**) take to be true; undertake; simulate; take on (aspect, attribute, etc.).

assumption /ə'sʌmpʃ(ə)n/ *noun* assuming; thing assumed; (**Assumption**) reception of Virgin Mary bodily into heaven.

assurance *noun* declaration; insurance, esp. of life; certainty; self-confidence.

assure /ə'ʃʊə/ *verb* (**-ring**) (often + *of*) convince; tell (person) confidently; ensure, guarantee (result etc.); insure (esp. life).

assuredly /ə'ʃʊərɪdlɪ/ *adverb* certainly.

aster /'æstə/ *noun* plant with bright daisy-like flowers.

asterisk /'æstərɪsk/ *noun* symbol (*) used to indicate omission etc.

astern /ə'stɜ:n/ *adverb* in or to rear of ship or aircraft; backwards.

asteroid /'æstərɔɪd/ *noun* any of numerous small planets between orbits of Mars and Jupiter.

asthma /'æsmə/ *noun* condition marked by difficulty in breathing. □ **asthmatic** /-'mæt-/ *adjective & noun.*

astigmatism /ə'stɪgmətɪz(ə)m/ *noun* eye or lens defect resulting in distorted images. □ **astigmatic** /æstɪg'mætɪk/ *adjective.*

astir /ə'stɜ:/ *adverb & adjective* in motion; out of bed.

astonish /ə'stɒnɪʃ/ *verb* amaze, surprise. □ **astonishment** *noun.*

astound /ə'staʊnd/ *verb* astonish greatly.

astrakhan /æstrə'kæn/ *noun* dark curly fleece of Astrakhan lamb; cloth imitating this.

astral /'æstr(ə)l/ *adjective* of stars; starry.

astray /ə'streɪ/ *adverb & adjective* away from right way. □ **go astray** be lost.

astride /ə'straɪd/ ● *adverb* (often + *of*) with one leg on each side. ● *preposition* astride of.

astringent /ə'strɪndʒənt/ ● *adjective* contracting body tissue, esp. to check bleeding; austere, severe. ● *noun* astringent substance. □ **astringency** *noun.*

astrolabe /'æstrəleɪb/ *noun* instrument for measuring altitude of stars etc.

astrology /ə'strɒlədʒɪ/ *noun* study of supposed planetary influence on human affairs. □ **astrologer** *noun*; **astrological** /æstrə'lɒdʒ-/ *adjective.*

astronaut /'æstrənɔ:t/ *noun* space traveller.

astronautics /æstrə'nɔ:tɪks/ *plural noun* (treated as *singular*) science of space travel. □ **astronautical** *adjective.*

astronomical /æstrə'nɒmɪk(ə)l/ *adjective* (also **astronomic**) of astronomy; vast. □ **astronomically** *adverb.*

astronomy /ə'strɒnəmɪ/ *noun* science of celestial bodies. □ **astronomer** *noun.*

astrophysics /æstrəʊ'fɪzɪks/ *plural noun* (treated as *singular*) study of physics and chemistry of celestial bodies. □ **astrophysical** *adjective*; **astrophysicist** *noun.*

astute /ə'stju:t/ *adjective* shrewd. □ **astutely** *adverb*; **astuteness** *noun.*

asunder /ə'sʌndə/ *adverb literary* apart.

asylum /ə'saɪləm/ *noun* sanctuary; = POLITICAL ASYLUM; *historical* mental institution.

asymmetry /eɪ'sɪmətrɪ/ *noun* lack of symmetry. □ **asymmetric(al)** /-'met-/ *adjective.*

at /æt, ət/ *preposition expressing position, point in time or on scale, engagement in activity, value or rate, or motion or aim towards.*

atavism /'ætəvɪz(ə)m/ *noun* resemblance to remote ancestors; reversion to earlier type. □ **atavistic** /-'vɪs-/ *adjective.*

ate *past* of EAT.

atelier /ə'telɪeɪ/ *noun* workshop; artist's studio.

atheism /'eɪθɪɪz(ə)m/ *noun* belief that no God exists. □ **atheist** *noun*; **atheistic** /-'ɪst-/ *adjective.*

atherosclerosis /æθərəʊsklə'rəʊsɪs/ *noun* formation of fatty deposits in the arteries.

athlete /'æθli:t/ *noun* person who engages in athletics, exercises, etc. □ **athlete's foot** fungal foot disease.

athletic /æθ'letɪk/ *adjective* of athletes or athletics; physically strong or agile. □ **athletically** *adverb*; **athleticism** *noun.*

athletics *plural noun* (usually treated as *singular*) physical exercises, esp. track and field events.

atlas /'ætləs/ *noun* book of maps.

atmosphere /'ætməsfɪə/ *noun* gases enveloping earth, other planet, etc.; tone, mood, etc., of place, book, etc.; unit of pressure. □ **atmospheric** /-'fer-/ *adjective.*

atmospherics /ætməs'ferɪks/ *plural noun* electrical disturbance in atmosphere; interference with telecommunications caused by this.

atoll /'ætɒl/ *noun* ring-shaped coral reef enclosing lagoon.

atom /'ætəm/ *noun* smallest particle of chemical element that can take part in chemical reaction; this as source of nuclear energy; minute portion or thing. □ **atom bomb** bomb in which energy is released by nuclear fission.

atomic /ə'tɒmɪk/ *adjective* of atoms; of or using atomic energy or atom bombs. □ **atomic bomb** atom bomb; **atomic energy** nuclear energy; **atomic number** number of protons in nucleus of atom; **atomic weight** ratio of mass of one atom of element to 1/12 mass of atom of carbon-12.

atomize /'ætəmaɪz/ *verb* (also **-ise**) (**-zing** or **-sing**) reduce to atoms or fine spray. □ **atomizer** *noun*.

atomizer *noun* (also **-iser**) aerosol.

atonal /eɪ'təʊn(ə)l/ *adjective Music* not written in any key. □ **atonality** /-'næl-/ *noun*.

atone /ə'təʊn/ *verb* (**-ning**) (usually + *for*) make amends. □ **atonement** *noun*.

atrium /'eɪtrɪəm/ *noun* (*plural* **-s** or **atria**) either of upper cavities of heart.

atrocious /ə'trəʊʃəs/ *adjective* very bad; wicked. □ **atrociously** *adverb*.

atrocity /ə'trɒsɪtɪ/ *noun* (*plural* **-ies**) wicked or cruel act.

atrophy /'ætrəfɪ/ ● *noun* wasting away, esp. through disuse. ● *verb* (**-ies**, **-ied**) suffer atrophy; cause atrophy in.

atropine /'ætrəpi:n/ *noun* poisonous alkaloid in deadly nightshade.

attach /ə'tætʃ/ *verb* fasten, affix, join; (in *passive*, + *to*) be very fond of; (+ *to*) attribute or be attributable to.

attaché /ə'tæʃeɪ/ *noun* specialist member of ambassador's staff. □ **attaché case** small rectangular document case.

attachment *noun* thing attached, esp. for particular purpose; affection, devotion; attaching, being attached.

attack /ə'tæk/ ● *verb* try to hurt or deflect using force; criticize adversely; act harmfully on; *Sport* try to score against; vigorously apply oneself to. ● *noun* act of attacking; sudden onset of illness. □ **attacker** *noun*.

attain /ə'teɪn/ *verb* gain, accomplish; reach; (+ *to*) arrive at (goal etc.).

attainment *noun* attaining; (often in *plural*) skill, achievement.

attar /'ætɑ:/ *noun* perfume made from rose-petals.

attempt /ə'tempt/ ● *verb* try; try to accomplish or conquer. ● *noun* (often + *at*, *on*) attempting; endeavour.

attend /ə'tend/ *verb* be present at; go regularly to; (often + *to*) apply mind or oneself; (+ *to*) deal with; accompany, wait on. □ **attender** *noun*.

attendance *noun* attending; number of people present.

attendant ● *noun* person attending, esp. to provide service. ● *adjective* accompanying; (often + *on*) waiting.

attention /ə'tenʃ(ə)n/ *noun* act or faculty of applying one's mind, notice; consideration, care; *Military* erect attitude of readiness; (in *plural*) courtesies.

attentive /ə'tentɪv/ *adjective* (+ *to*) paying attention. □ **attentively** *adverb*.

attenuate /ə'tenjʊeɪt/ *verb* (**-ting**) make thin; reduce in force or value. □ **attenuation** *noun*.

attest /ə'test/ *verb* certify validity of; (+ *to*) bear witness to. □ **attestation** /æt-/ *noun*.

attic /'ætɪk/ *noun* room or space immediately under roof of house.

attire /ə'taɪə/ *noun* clothes, esp. formal.

attired *adjective* dressed, esp. formally.

attitude /'ætɪtju:d/ *noun* opinion, way of thinking; (often + *to*) behaviour reflecting this; bodily posture.

attitudinize /ætɪ'tju:dɪnaɪz/ *verb* (also **-ise**) (**-zing** or **-sing**) adopt attitudes.

attorney /ə'tɜ:nɪ/ *noun* person, esp. lawyer, appointed to act for another in business or legal matters; *US* qualified lawyer. □ **Attorney-General** chief legal officer of government.

attract /ə'trækt/ *verb* (of magnet etc.) draw to itself or oneself; arouse interest or admiration in.

attraction /ə'trækʃ(ə)n/ *noun* attracting, being attracted; attractive quality; person or thing that attracts.

attractive /ə'træktɪv/ *adjective* attracting esp. interest or admiration; pleasing. □ **attractively** *noun*.

attribute ● *verb* /ə'trɪbju:t/ (**-ting**) (usually + *to*) regard as belonging to or as written, said, or caused by, etc. ● *noun* /'ætrɪbju:t/ quality ascribed to person or thing; characteristic quality; object frequently associated with person, office, or status. □ **attributable** /ə'trɪbjʊtəb(ə)l/ *adjective*; **attribution** /ætrɪ'bju:ʃ(ə)n/ *noun*.

attributive /ə'trɪbjʊtɪv/ *adjective* expressing an attribute; (of adjective or noun) preceding word it describes.

attrition /ə'trɪʃ(ə)n/ *noun* gradual wearing down; friction, abrasion.

attune /ə'tju:n/ *verb* (**-ning**) (usually + *to*) adjust; *Music* tune.

atypical /eɪ'tɪpɪk(ə)l/ *adjective* not typical. □ **atypically** *adverb*.

aubergine /'əʊbəʒi:n/ *noun* (plant with) oval usually purple fruit used as vegetable.

aubrietia /ɔ:'bri:ʃə/ *noun* (also **aubretia**) dwarf perennial rock plant.

auburn /'ɔ:bən/ *adjective* (usually of hair) reddish-brown.

auction /'ɔ:kʃ(ə)n/ ● *noun* sale in which each article is sold to highest bidder. ● *verb* sell by auction.

auctioneer /ɔ:kʃə'nɪə/ *noun* person who conducts auctions.

audacious /ɔ:'deɪʃəs/ *adjective* daring, bold; impudent. □ **audacity** /-'dæs-/ *noun*.

audible /'ɔ:dɪb(ə)l/ *adjective* that can be heard. □ **audibility** *noun*; **audibly** *adverb*.

audience /'ɔ:dɪəns/ *noun* group of listeners or spectators; group of people reached by any spoken or written message; formal interview.

audio /'ɔ:dɪəʊ/ *noun* (reproduction of) sound. □ **audiotape** magnetic tape for recording sound; **audio typist** person who types from tape recording; **audio-visual** using both sight and sound.

audit /'ɔ:dɪt/ ● *noun* official scrutiny of accounts. ● *verb* (**-t-**) conduct audit of. □ **auditor** *noun*.

audition /ɔ:'dɪʃ(ə)n/ ● *noun* test of performer's ability. ● *verb* assess or be assessed at audition.

auditorium /ɔ:dɪ'tɔ:rɪəm/ *noun* (*plural* **-s**) part of theatre etc. for audience.

auditory /'ɔ:dɪtərɪ/ *adjective* of hearing.

au fait /əʊ 'feɪ/ *adjective* (usually + *with*) conversant.

Aug. *abbreviation* August.

auger /'ɔ:gə/ *noun* tool with screw point for boring holes in wood.

aught /ɔ:t/ *noun archaic* anything.

augment /ɔ:g'ment/ *verb* make greater, increase. □ **augmentation** *noun*.

augur /'ɔ:gə/ *verb* portend; serve as omen.

augury /'ɔ:gjʊrɪ/ *noun* (*plural* **-ies**) omen; interpretation of omens.

August /'ɔ:gəst/ *noun* eighth month of year.

august /ɔ:'gʌst/ *adjective* venerable, imposing.

auk /ɔ:k/ *noun* black and white seabird with small wings.

aunt /ɑ:nt/ *noun* parent's sister; uncle's wife. □ **Aunt Sally** game in which sticks or balls are thrown at dummy, target of general abuse.

aunty /'ɑ:ntɪ/ *noun* (also **auntie**) (*plural* **-ies**) *colloquial* aunt.

au pair /əʊ 'peə/ *noun* young foreign woman who helps with housework in return for room and board.

aura /'ɔ:rə/ *noun* distinctive atmosphere; subtle emanation.

aural /'ɔ:r(ə)l/ *adjective* of ear or hearing. □ **aurally** *adverb*.

aureole /'ɔ:rɪəʊl/ *noun* (also **aureola** /ɔ:'ri:ələ/) halo.

au revoir /əʊ rə'vwɑ:/ *interjection & noun* goodbye (until we meet again). [French]

auricle /'ɔ:rɪk(ə)l/ *noun* external ear of animal; atrium of heart.

auriferous /ɔ:'rɪfərəs/ *adjective* yielding gold.

aurora /ɔ:'rɔ:rə/ *noun* (*plural* **-s** or **-rae** /-ri:/) streamers of light above northern (**aurora borealis** /bɒrɪ'eɪlɪs/) or southern (**aurora australis** /ɔ:'streɪlɪs/) polar region.

auscultation /ɔ:skəl'teɪʃ(ə)n/ *noun* listening to sound of heart etc. to help diagnosis.

auspice /'ɔ:spɪs/ *noun* omen; (in *plural*) patronage.

auspicious /ɔ:'spɪʃəs/ *adjective* promising; favourable.

Aussie /'ɒzɪ/ *noun & adjective slang* Australian.

austere /ɔ:'stɪə/ *adjective* (**-r**, **-st**) severely simple; stern; morally strict. □ **austerity** /-'ter-/ *noun*.

austral /'ɔ:str(ə)l/ *adjective* southern.

Australasian /ɒstrə'leɪʒ(ə)n/ *adjective* of Australia and SW Pacific islands.

Australian /ɒ'streɪlɪən/ ● *adjective* of Australia. ● *noun* native or national of Australia.

autarchy /'ɔ:tɑ:kɪ/ *noun* absolute rule.

autarky /'ɔ:tɑ:kɪ/ *noun* self-sufficiency.

authentic /ɔ:'θentɪk/ *adjective* of undisputed origin, genuine; trustworthy. □ **authentically** *adverb*; **authenticity** /-'tɪs-/ *noun*.

authenticate /ɔ:'θentɪkeɪt/ *verb* (**-ting**) establish as true, genuine, or valid. □ **authentication** *noun.*

author /'ɔ:θə/ *noun* (*feminine* **authoress** /'ɔ:θrɪs/) writer of book etc.; originator. □ **authorship** *noun.*

authoritarian /ɔ:θɒrɪ'teərɪən/ ● *adjective* favouring strict obedience to authority. ● *noun* authoritarian person.

authoritative /ɔ:'θɒrɪtətɪv/ *adjective* reliable, esp. having authority.

authority /ɔ:'θɒrɪtɪ/ *noun* (*plural* **-ies**) power or right to enforce obedience; (esp. in *plural*) body having this; delegated power; influence based on recognized knowledge or expertise; expert.

authorize /'ɔ:θəraɪz/ *verb* (also **-ise**) (**-zing** or **-sing**) (+ *to do*) give authority to (person); sanction officially. □ **Authorized Version** English translation (1611) of Bible. □ **authorization** *noun.*

autism /'ɔ:tɪz(ə)m/ *noun* condition characterized by self-absorption and withdrawal. □ **autistic** /-'tɪs-/ *adjective.*

auto- *combining form* self; one's own; of or by oneself or itself.

autobahn /'ɔ:təʊbɑ:n/ *noun* German, Austrian, or Swiss motorway. [German]

autobiography /ɔ:təʊbaɪ'ɒgrəfɪ/ *noun* (*plural* **-ies**) story of one's own life. □ **autobiographer** *noun*; **autobiographical** /-ə'græf-/ *adjective.*

autocracy /ɔ:'tɒkrəsɪ/ *noun* (*plural* **-ies**) absolute rule by one person.

autocrat /'ɔ:təkræt/ *noun* absolute ruler. □ **autocratic** /-'kræt-/ *adjective*; **autocratically** /-'kræt-/ *adverb.*

autocross /'ɔ:təʊkrɒs/ *noun* motor racing across country or on unmade roads.

autograph /'ɔ:təgrɑ:f/ ● *noun* signature, esp. of celebrity. ● *verb* write on or sign in one's own handwriting.

automate /'ɔ:təmeɪt/ *verb* (**-ting**) convert to or operate by automation.

automatic /ɔ:tə'mætɪk/ ● *adjective* working by itself, without direct human involvement; done spontaneously; following inevitably; (of firearm) that can be loaded and fired continuously; (of vehicle or its transmission) using gears that change automatically. ● *noun* automatic firearm, vehicle, etc. □ **automatically** *adverb.*

automation *noun* use of automatic equipment in place of manual labour.

automaton /ɔ:'tɒmət(ə)n/ *noun* (*plural* **-mata** or **-s**) machine controlled automatically.

automobile /'ɔ:təməbi:l/ *noun US* motor car.

automotive /ɔ:tə'məʊtɪv/ *adjective* of motor vehicles.

autonomous /ɔ:'tɒnəməs/ *adjective* self-governing; free to act independently. □ **autonomy** *noun.*

autopsy /'ɔ:tɒpsɪ/ *noun* (*plural* **-ies**) post-mortem.

auto-suggestion /ɔ:təʊsə'dʒestʃ(ə)n/ *noun* hypnotic or subconscious suggestion made to oneself.

autumn /'ɔ:təm/ *noun* season between summer and winter. □ **autumnal** /ɔ:'tʌmn(ə)l/ *adjective.*

auxiliary /ɔ:g'zɪljərɪ/ ● *adjective* giving help; additional, subsidiary. ● *noun* (*plural* **-ies**) auxiliary person or thing; (in *plural*) foreign or allied troops in service of nation at war; auxiliary verb. □ **auxiliary verb** one used with another verb to form tenses etc. (see panel).

avail /ə'veɪl/ ● *verb* (often + *to*) be of use; help; (**avail oneself of**) use, profit by. ● *noun* use, profit.

available *adjective* at one's disposal; (of person) free, able to be contacted. □ **availability** *noun.*

avalanche /'ævəlɑ:ntʃ/ *noun* mass of snow and ice rapidly sliding down mountain; sudden abundance.

avant-garde /ævɑ̃'gɑ:d/ ● *noun* innovators, esp. in the arts. ● *adjective* new, pioneering.

avarice /'ævərɪs/ *noun* greed for wealth. □ **avaricious** /-'rɪʃ-/ *adjective.*

avatar /'ævətɑ:/ *noun Hindu Mythology* descent of god to earth in bodily form.

avenge /ə'vendʒ/ *verb* (**-ging**) inflict retribution on behalf of; exact retribution for.

avenue /'ævənju:/ *noun* road or path, usually tree-lined; way of approach.

aver /ə'vɜ:/ *verb* (**-rr-**) *formal* assert, affirm.

average /'ævərɪdʒ/ ● *noun* usual amount, extent, or rate; number obtained by dividing sum of given numbers by how many there are. ● *adjective* usual, ordinary; mediocre; constituting average. ● *verb* (**-ging**) amount on average to; do on average; estimate average of. □ **average out (at)** result in average (of); **on average** as an average rate etc.

averse /ə'vɜ:s/ *adjective* (usually + *to*) opposed, disinclined.

aversion /ə'vɜ:ʃ(ə)n/ *noun* (usually + *to*, *for*) dislike, unwillingness; object of this.

avert /ə'vɜ:t/ *verb* prevent; (often + *from*) turn away.

aviary /'eɪvɪərɪ/ *noun* (*plural* **-ies**) large cage or building for keeping birds.

aviation /eɪvɪ'eɪʃ(ə)n/ *noun* the flying of aircraft.

aviator /'eɪvɪeɪtə/ *noun* person who flies aircraft.

avid /'ævɪd/ *adjective* eager; greedy. □ **avidity** /-'vɪd-/ *noun*; **avidly** *adverb*.

avocado /ævə'kɑ:dəʊ/ *noun* (*plural* **-s**) (in full *avocado pear*) dark green pear-shaped fruit with creamy flesh.

avocet /'ævəset/ *noun* wading bird with long upturned bill.

avoid /ə'vɔɪd/ *verb* keep away or refrain from; escape; evade. □ **avoidable** *adjective*; **avoidance** *noun*.

avoirdupois /ævədə'pɔɪz/ *noun* system of weights based on pound of 16 ounces.

avow /ə'vaʊ/ *verb formal* declare; confess. □ **avowal** *noun*; **avowedly** /ə'vaʊɪdlɪ/ *adverb*.

avuncular /ə'vʌŋkjʊlə/ *adjective* of or like an uncle.

await /ə'weɪt/ *verb* wait for; be in store for.

awake /ə'weɪk/ ● *verb* (**-king**; *past* **awoke**; *past participle* **awoken**) (also *awaken*) rouse from sleep; cease to sleep; (often + *to*) become alert, aware, or active. ● *adjective* not asleep; alert.

award /ə'wɔ:d/ ● *verb* give or order to be given as payment, penalty, or prize. ● *noun* thing awarded; judicial decision.

aware /ə'weə/ *adjective* (often + *of*, *that*) conscious, having knowledge. □ **awareness** *noun*.

awash /ə'wɒʃ/ *adjective* at surface of and just covered by water; (+ *with*) abounding.

away ● *adverb* to or at distance; into non-existence; constantly, persistently. ● *adjective* (of match etc.) played on opponent's ground.

awe /ɔ:/ ● *noun* reverential fear or wonder. ● *verb* (**awing**) inspire with awe. □ **awe-inspiring** awesome, magnificent.

aweigh /ə'weɪ/ *adjective* (of anchor) just lifted from sea bottom.

awesome *adjective* inspiring awe.

Auxiliary verb

An auxiliary verb is used in front of another verb to alter its meaning. Mainly, it expresses:

1 when something happens, by forming a tense of the main verb, e.g. *I* shall *go. He* was *going.*

2 permission, obligation, or ability to do something, e.g. *They* may *go. You* must *go. I* can't *go.*

3 the likelihood of something happening, e.g. *I* might *go. She* would *go if she could.*

The principal auxiliary verbs are:

be	*have*	*must*	*will*
can	*let*	*ought*	*would*
could	*may*	*shall*	
do	*might*	*should*	

---

awful /'ɔ:fʊl/ *adjective colloquial* very bad; very great; *poetical* inspiring awe.

awfully *adverb colloquial* badly; very.

awhile /ə'waɪl/ *adverb* for a short time.

awkward /'ɔ:kwəd/ *adjective* difficult to use; clumsy; embarrassing, embarrassed; hard to deal with.

awl /ɔ:l/ *noun* small tool for pricking holes, esp. in leather.

awning /'ɔ:nɪŋ/ *noun* fabric roof, shelter.

awoke *past* of AWAKE.

awoken *past participle* of AWAKE.

AWOL /'eɪwɒl/ *abbreviation colloquial* absent without leave.

awry /ə'raɪ/ ● *adverb* crookedly; amiss. ● *adjective* crooked; unsound.

axe /æks/ ● *noun* (*US* **ax**) chopping-tool with heavy blade; (**the axe**) dismissal (of employees), abandonment of project etc. ● *verb* (**axing**) cut (staff, services, etc.); abandon (project).

axial /'æksɪəl/ *adjective* of, forming, or placed round axis.

axiom /'æksɪəm/ *noun* established principle; self-evident truth. □ **axiomatic** /-'mæt-/ *adjective*.

axis /'æksɪs/ *noun* (*plural* **axes** /-si:z/) imaginary line about which object rotates; line dividing regular figure symmetrically; reference line for measurement of coordinates etc.

axle /'æks(ə)l/ *noun* spindle on which wheel turns or is fixed.

ayatollah /aɪə'tɒlə/ *noun* religious leader in Iran.

aye /aɪ/ ● *adverb archaic or dialect* yes. ● *noun* affirmative answer or vote.

azalea /ə'zeɪlɪə/ *noun* kind of rhododendron.

azimuth /'æzɪməθ/ *noun* angular distance between point below star etc. and north or south. □ **azimuthal** *adjective*.

azure /'æʒə/ *adjective & noun* sky-blue.

Bb

B *abbreviation* black (pencil-lead).

b. *abbreviation* born.

BA *abbreviation* Bachelor of Arts.

baa /bɑ:/ *noun & verb* (**baas**, **baaed** or **baa'd**) bleat.

babble /'bæb(ə)l/ ● *verb* (**-ling**) talk, chatter, or say incoherently or excessively; (of stream) murmur; repeat foolishly. ● *noun* babbling; murmur of water etc.

babe *noun* baby.

babel /'beɪb(ə)l/ *noun* confused noise, esp. of voices.

baboon /bə'bu:n/ *noun* large kind of monkey.

baby /'beɪbɪ/ *noun* (*plural* **-ies**) very young child; childish person; youngest member of family etc.; very young animal; small specimen. □ **baby boom** *colloquial* temporary increase in birth rate; **baby grand** small grand piano.

babysit *verb* (**-tt-**; *past & past participle* **-sat**) look after child while parents are out. □ **babysitter** *noun*.

baccarat /'bækərɑ:/ *noun* gambling card games.

bachelor /'bætʃələ/ *noun* unmarried man; person with a university first degree.

bacillus /bə'sɪləs/ *noun* (*plural* **bacilli** /-laɪ/) rod-shaped bacterium. □ **bacillary** *adjective*.

back ● *noun* rear surface of human body from shoulders to hips; upper surface of animal's body; spine; reverse or more distant part; part of garment covering back; defensive player in football etc. ● *adverb* to rear; away from front; in(to) the past or an earlier or normal position or condition; in return; at a distance. ● *verb* help with money or moral support; (often + *up*) (cause to) go backwards; bet on; provide with or serve as back, support, or backing to; *Music* accompany. ● *adjective* situated to rear; past, not current; reversed. □ **backache** ache in back; **backbencher** MP without senior office; **backbiting** malicious talk; **back boiler** one behind domestic fire etc.; **backbone** spine, main support, firmness of character; **backchat** *colloquial* verbal insolence; **backcloth** painted cloth at back of stage; **backdate** make retrospectively valid, put earlier date to; **back door** secret or ingenious means; **back down** withdraw from confrontation; **backdrop** backcloth; **backfire** (of engine or vehicle) undergo premature explosion in cylinder or exhaust pipe, (of plan etc.) have opposite of intended effect; **backhand** *Tennis etc.* (stroke) made with hand across body; **backhanded** indirect, ambiguous; **backhander** *slang* bribe; **backlash** violent, usually hostile, reaction; **backlog** arrears (of work etc.); **back number** old issue of magazine etc.; **back out** (often + *of*) withdraw; **backpack** rucksack; **back-pedal** reverse previous action or opinion; **back room** place where (esp. secret) work goes on; **back seat** less prominent or important position; **backside** *colloquial* buttocks; **backslide** return to bad ways; **backstage** behind the scenes; **backstreet** *noun* side-street, alley, *adjective* illicit, illegal; **backstroke** stroke made by swimmer lying on back; **back to front** with back and front reversed; **backtrack** retrace one's steps, reverse one's policy or opinion; **back up** give (esp. moral) support to, *Computing* make backup of; **backup** support, *Computing* (making of) spare copy of data; **backwash** receding waves, repercussions; **backwater** peaceful, secluded, or dull place, stagnant water fed from stream; **backwoods** remote uncleared forest land.

backgammon /'bækgæmən/ *noun* board game with pieces moved according to throw of dice.

background /'bækgraʊnd/ *noun* back part of scene etc.; inconspicuous position; person's education, social circumstances, etc.; explanatory information etc.

backing *noun* help or support; material used for thing's back or support; musical accompaniment.

backward *adjective* directed backwards; slow in learning; shy.

backwards *adverb* away from one's front; back foremost; in reverse of usual way; into worse state; into past; back towards starting point □ **bend over backwards** *colloquial* make every effort.

bacon /'beɪkən/ *noun* cured meat from back and sides of pig.

bacteriology /bæktɪərɪ'ɒlədʒɪ/ *noun* study of bacteria.

bacterium /bæk'tɪərɪəm/ *noun* (*plural* **-ria**) single-celled micro-organism. □ **bacterial** *adjective*.

■ **Usage** It is a mistake to use the plural form *bacteria* when only one bacterium is meant.

bad *adjective* (**worse, worst**) inadequate, defective; unpleasant; harmful; decayed; ill, injured; regretful, guilty; serious, severe; wicked; naughty; incorrect, not valid. □ **bad debt** debt that is not recoverable; **bad-tempered** irritable.

bade *archaic past* of BID.

badge *noun* small flat emblem worn as sign of office, membership, etc. or bearing slogan etc.

badger /'bædʒə/ ● *noun* nocturnal burrowing mammal with black and white striped head. ● *verb* pester.

badinage /'bædɪnɑːʒ/ *noun* banter. [French]

badly *adverb* (**worse, worst**) in bad manner; severely; very much.

badminton /'bædmɪnt(ə)n/ *noun* game played with rackets and shuttlecock.

baffle /'bæf(ə)l/ ● *verb* perplex; frustrate. ● *noun* device that hinders flow of fluid or sound. □ **bafflement** *noun*.

bag ● *noun* soft open-topped receptacle; piece of luggage; (in *plural*; usually + *of*) *colloquial* large amount; animal's sac; amount of game shot by one person. ● *verb* secure, take possession of; bulge, hang loosely; put in bag; shoot (game).

bagatelle /bægə'tel/ *noun* game in which small balls are struck into holes on inclined board; mere trifle.

bagel /'beɪg(ə)l/ *noun* ring-shaped bread roll.

baggage /'bægɪdʒ/ *noun* luggage; portable equipment of army.

baggy *adjective* (**-ier**, **-iest**) hanging loosely.

bagpipes /'bægpaɪps/ *plural noun* musical instrument with windbag for pumping air through reeded pipes.

baguette /bæ'get/ *noun* long thin French loaf.

bail¹ • *noun* security given for released prisoner's return for trial; person(s) pledging this. • *verb* (usually + *out*) give bail for and secure release of (prisoner).

bail² *noun Cricket* either of two crosspieces resting on stumps.

bail³ *verb* (also **bale**) (usually + *out*) scoop (water) out of (boat etc.). □ **bail out** = BALE OUT.

bailey /'beɪlɪ/ *noun* (*plural* **-s**) outer wall of castle; court enclosed by this.

bailiff /'beɪlɪf/ *noun* sheriff's officer who executes writs, performs distraints, etc.; landlord's agent or steward.

bailiwick /'beɪlɪwɪk/ *noun* district of bailiff.

bairn /beən/ *noun Scottish & Northern English* child.

bait • *noun* food to entice prey; allurement. • *verb* torment (chained animal); harrass (person); put bait on or in (fish-hook, trap, etc.).

baize /beɪz/ *noun* usually green felted woollen fabric used for coverings etc.

bake *verb* (**-king**) cook or become cooked by dry heat esp. in oven; *colloquial* be very hot; harden by heat. □ **baking powder** mixture used as raising agent.

Bakelite /'beɪkəlaɪt/ *noun proprietary term* kind of plastic.

baker /'beɪkə/ *noun* professional bread-maker. □ **baker's dozen** thirteen.

bakery /'beɪkərɪ/ *noun* (*plural* **-ies**) place where bread is made or sold.

baksheesh /'bækʃi:ʃ/ *noun* gratuity, tip.

Balaclava /bælə'klɑ:və/ *noun* (in full **Balaclava helmet**) woollen covering for head and neck.

balalaika /bælə'laɪkə/ *noun* Russian triangular-bodied guitar-like musical instrument.

balance /'bæləns/ • *noun* even distribution of weight or amount, stability of body or mind; weighing apparatus; counteracting weight or force; regulating device in clock etc.; decisive weight or amount; difference between credits and debits; remainder. • *verb* (**-cing**) bring or come into or keep in equilibrium; offset, compare; equal, neutralize; (usually as **balanced** *adjective*) make well-proportioned and harmonious; equalize debits and credits of account, have debits and credits equal. □ **balance of payments** difference between payments into and out of a country; **balance of power** position in which no country etc. predominates, power held by small group when larger groups are of (almost) equal strength; **balance of trade** difference between exports and imports; **balance sheet** statement giving balance of account; **on balance** all things considered.

balcony /'bælkənɪ/ *noun* (*plural* **-ies**) outside balustraded or railed platform with access from upper floor; upper tier of seats in theatre etc.

bald /bɔ:ld/ *adjective* lacking some or all hair on scalp; without fur, feathers, etc.; with surface worn away; direct. □ **baldly** *adverb*; **baldness** *noun*.

balderdash /'bɔ:ldədæʃ/ *noun* nonsense.

balding /'bɔ:ldɪŋ/ *adjective* becoming bald.

bale¹ • *noun* bundle of merchandise or hay. • *verb* (**-ling**) make up into bales. □ **bale out** (also **bail out**) make emergency parachute jump from aircraft.

bale² = BAIL³.

baleful /'beɪlfʊl/ *adjective* menacing; destructive; malignant.

balk = BAULK.

ball¹ /bɔ:l/ *noun* spherical object or mass; usually spherical object used in game; rounded part of foot or hand at base of big toe or thumb; cannon ball; delivery or pass of ball in game. □ **ball-bearing** bearing using ring of small balls between its two parts, one such ball; **ballcock** valve operated by floating ball that controls level of water in cistern; **ball game** *US* baseball game, *colloquial* affair or matter; **ballpoint (pen)** pen with tiny ball as writing point.

ball² /bɔ:l/ *noun* formal social gathering for dancing; *slang* enjoyable time.

ballad /ˈbæləd/ *noun* poem or song narrating popular story; slow sentimental song. □ **balladry** *noun.*

ballast /ˈbæləst/ ● *noun* heavy material stabilizing ship, controlling height of balloon, etc.; coarse stone etc. as bed of road or railway. ● *verb* provide with ballast.

ballerina /bæləˈri:nə/ *noun* female ballet dancer.

ballet /ˈbæleɪ/ *noun* dramatic or representational style of dancing to music; piece or performance of ballet. □ **ballet dancer** dancer of ballet. □ **balletic** /bəˈletɪk/ *adjective.*

ballistic /bəˈlɪstɪk/ *adjective* of projectiles. □ **ballistic missile** one that is powered and guided but falls by gravity.

ballistics *plural noun* (usually treated as *singular*) science of projectiles and firearms.

balloon /bəˈlu:n/ ● *noun* small inflatable rubber toy or decoration; large inflatable flying bag, esp. one with basket below for passengers; outline containing words or thoughts in strip cartoon. ● *verb* (cause to) swell out like balloon. □ **balloonist** *noun.*

ballot /ˈbælət/ ● *noun* voting in writing and usually secret; votes recorded in ballot. ● *verb* (**-t-**) hold ballot; vote by ballot; take ballot of (voters). □ **ballot box** container for **ballot papers**, slips for marking votes.

ballroom /ˈbɔ:lrʊm/ *noun* large room for dancing. □ **ballroom dancing** formal social dancing for couples.

bally /ˈbælɪ/ *adjective & adverb slang mild form of* BLOODY.

ballyhoo /bælɪˈhu:/ *noun* loud noise, fuss; noisy publicity.

balm /bɑ:m/ *noun* aromatic ointment; fragrant oil or resin exuded from some trees; thing that heals or soothes; aromatic herb.

balmy /ˈbɑ:mɪ/ *adjective* (**-ier**, **-iest**) fragrant, mild, soothing; *slang* crazy.

balsa /ˈbɔ:lsə/ *noun* lightweight tropical American wood used for making models.

balsam /ˈbɔ:lsəm/ *noun* balm from trees; ointment; tree yielding balsam; any of several flowering plants.

baluster /ˈbæləstə/ *noun* short pillar supporting rail.

balustrade /bæləˈstreɪd/ *noun* railing supported by balusters, esp. on balcony.

bamboo /bæmˈbu:/ *noun* tropical giant woody grass; its hollow stem.

bamboozle /bæmˈbu:z(ə)l/ *verb* (**-ling**) *colloquial* cheat; mystify.

ban ● *verb* (**-nn-**) prohibit, esp. formally. ● *noun* formal prohibition.

banal /bəˈnɑ:l/ *adjective* commonplace, trite. □ **banality** /-ˈnæl-/ *noun* (*plural* **-ies**).

banana /bəˈnɑ:nə/ *noun* long curved yellow tropical fruit; treelike plant bearing it.

band ● *noun* flat strip or loop of thin material; stripe; group of musicians; organized group of criminals etc.; range of values, esp. frequencies or wavelengths. ● *verb* (usually + *together*) unite; put band on; mark with stripes. □ **bandbox** hatbox; **bandmaster** conductor of band; **bandsman** player in band; **bandstand** outdoor platform for musicians.

bandage /ˈbændɪdʒ/ ● *noun* strip of material for binding wound etc. ● *verb* (**-ging**) bind with bandage.

bandanna /bænˈdænə/ *noun* large patterned handkerchief.

bandeau /ˈbændəʊ/ *noun* (*plural* **-x** /-z/) narrow headband.

bandit /ˈbændɪt/ *noun* robber, esp. of travellers. □ **banditry** *noun.*

bandolier /bændəˈlɪə/ *noun* (also **bandoleer**) shoulder belt with loops or pockets for cartridges.

bandwagon *noun.* □ **climb**, **jump on the bandwagon** join popular or successful cause.

bandy[1] /ˈbændɪ/ *adjective* (**-ier**, **-iest**) (of legs) curved wide apart at knees. □ **bandy-legged** *adjective.*

bandy[2] /ˈbændɪ/ *verb* (**-ies**, **-ied**) (often + *about*) pass (story etc.) to and fro, discuss disparagingly; (often + *with*) exchange (blows, insults, etc.).

bane *noun* cause of ruin or trouble.

bang ● *noun* loud short sound; sharp blow. ● *verb* strike or shut noisily; (cause to) make bang. ● *adverb* with bang; *colloquial* exactly. □ **bang on** *colloquial* exactly right.

banger *noun* firework making bang; *slang* sausage, noisy old car.

bangle /'bæŋg(ə)l/ *noun* rigid bracelet or anklet.

banian = BANYAN.

banish /'bænɪʃ/ *verb* condemn to exile; dismiss from one's mind. □ **banishment** *noun*.

banister /'bænɪstə/ *noun* (also **bannister**) (usually in *plural*) uprights and handrail beside staircase.

banjo /'bændʒəʊ/ *noun* (*plural* **-s** or **-es**) round-bodied guitar-like musical instrument. □ **banjoist** *noun*.

bank¹ ● *noun* sloping ground on each side of river; raised shelf of ground, esp. in sea; mass of cloud etc. ● *verb* (often + *up*) heap or rise into banks; pack (fire) tightly for slow burning; (cause to) travel round curve with one side higher than the other.

bank² ● *noun* establishment, usually a public company, where money is deposited, withdrawn, and borrowed; pool of money in gambling game; storage place. ● *verb* deposit (money) at bank; (often + *at*, *with*) keep money (at bank). □ **bank card** cheque card; **bank holiday** public holiday when banks are closed; **banknote** piece of paper money; **bank on** *colloquial* rely on.

bank³ *noun* row (of lights, switches, organ keys, etc.).

banker *noun* owner or manager of bank; keeper of bank in gambling game. □ **banker's card** cheque card.

bankrupt /'bæŋkrʌpt/ ● *adjective* insolvent; (often + *of*) drained (of emotion etc.). ● *noun* insolvent person. ● *verb* make bankrupt. □ **bankruptcy** *noun* (*plural* **-ies**).

banner /'bænə/ *noun* large portable cloth sign bearing slogan or design; flag.

bannister = BANISTER.

bannock /'bænək/ *noun Scottish & Northern English* round flat loaf, usually unleavened.

banns *plural noun* announcement of intended marriage read in church.

banquet /'bæŋkwɪt/ ● *noun* sumptuous esp. formal dinner. ● *verb* (**-t-**) give banquet for; attend banquet.

banquette /bæŋ'ket/ *noun* upholstered bench along wall.

banshee /bæn'ʃi:/ *noun* female spirit whose wail warns of death in a house.

bantam /'bæntəm/ *noun* kind of small domestic fowl; small but aggressive person. □ **bantamweight** boxing weight (51–54 kg).

banter /'bæntə/ ● *noun* good-humoured teasing. ● *verb* tease; exchange banter.

banyan /'bænjən/ *noun* (also **banian**) Indian fig tree with self-rooting branches.

baobab /'beɪəʊbæb/ *noun* African tree with massive trunk and edible fruit.

bap *noun* soft flattish bread roll.

baptism /'bæptɪz(ə)m/ *noun* symbolic admission to Christian Church, with immersing in or sprinkling with water and usually name-giving. □ **baptismal** /-'tɪz-/ *adjective*.

Baptist /'bæptɪst/ *noun* member of Church practising adult baptism by immersion.

baptize /bæp'taɪz/ *verb* (also **-ise**) (**-zing** or **-sing**) administer baptism to; give name to.

bar¹ ● *noun* long piece of rigid material, esp. used to confine or obstruct; (often + *of*) oblong piece (of chocolate, soap, etc.); band of colour or light; counter for serving alcoholic drinks etc., room or building containing it; counter for particular service; barrier; prisoner's enclosure in law court; section of music between vertical lines; heating element of electric fire; strip below clasp of medal as extra distinction; (**the Bar**) barristers, their profession. ● *verb* (**-rr-**) fasten with bar; (usually + *in*, *out*) keep in or out; obstruct, prevent; (usually + *from*) exclude. ● *preposition* except. □ **bar code** machine-readable striped code on packaging etc.; **barmaid**, **barman**, **bartender** woman, man, person serving in pub etc.

bar² *noun* unit of atmospheric pressure.

barathea /bærə'θi:ə/ *noun* fine wool cloth.

barb ● *noun* backward-facing point on arrow, fish-hook, etc.; hurtful remark. ● *verb* fit with barb. □ **barbed wire** wire with spikes, used for fences.

barbarian /bɑ:ˈbeərɪən/ • *noun* uncultured or primitive person. • *adjective* uncultured; primitive.

barbaric /bɑ:ˈbærɪk/ *adjective* uncultured; cruel; primitive.

barbarism /ˈbɑ:bərɪz(ə)m/ *noun* barbaric state or act; non-standard word or expression.

barbarity /bɑ:ˈbærɪtɪ/ *noun* (*plural* **-ies**) savage cruelty; barbaric act.

barbarous /ˈbɑ:bərəs/ *adjective* uncultured; cruel.

barbecue /ˈbɑ:bɪkju:/ • *noun* meal cooked over charcoal etc. out of doors; party for this; grill etc. used for this. • *verb* (**-cues**, **-cued**, **-cuing**) cook on barbecue.

barber /ˈbɑ:bə/ *noun* person who cuts men's hair.

barbican /ˈbɑ:bɪkən/ *noun* outer defence, esp. double tower over gate or bridge.

barbiturate /bɑ:ˈbɪtjʊrət/ *noun* sedative derived from **barbituric acid** /bɑ:bɪˈtjʊərɪk/, an organic acid.

bard *noun* poet; *historical* Celtic minstrel; prizewinner at Eisteddfod. □ **bardic** *adjective*.

bare /beə/ • *adjective* unclothed, uncovered; leafless; unadorned, plain; scanty, just sufficient. • *verb* (**-ring**) uncover, reveal. □ **bareback** without saddle; **barefaced** shameless, impudent; **barefoot** with bare feet; **bareheaded** without hat.

barely *adverb* scarcely; scantily.

bargain /ˈbɑ:gɪn/ • *noun* agreement on terms of sale etc.; cheap thing. • *verb* discuss terms of sale etc. □ **bargain for** be prepared for; **into the bargain** moreover.

barge /bɑ:dʒ/ • *noun* large flat-bottomed cargo boat on canal or river; long ornamental pleasure boat. • *verb* (**-ging**) (+ *into*) collide with. □ **barge in** interrupt.

bargee /bɑ:ˈdʒi:/ *noun* person in charge of barge.

baritone /ˈbærɪtəʊn/ *noun* adult male singing voice between tenor and bass; singer with this.

barium /ˈbeərɪəm/ *noun* white metallic element. □ **barium meal** mixture swallowed to reveal digestive tract on X-ray.

bark[1] • *noun* sharp explosive cry of dog etc. • *verb* give a bark; speak or utter sharply or brusquely.

bark[2] • *noun* tough outer layer of tree. • *verb* graze (shin etc.); strip bark from.

barker *noun* tout at auction or sideshow.

barley /ˈbɑ:lɪ/ *noun* cereal used as food and in spirits; (also **barleycorn**) its grain. □ **barley sugar** hard sweet made from sugar; **barley water** drink made from boiled barley.

barm *noun* froth on fermenting malt liquor.

bar mitzvah /bɑ: ˈmɪtsvə/ *noun* religious initiation of Jewish boy at 13.

barmy /ˈbɑ:mɪ/ *adjective* (**-ier**, **-iest**) *slang* crazy.

barn *noun* building for storing grain etc. □ **barn dance** social gathering for country dancing; **barn owl** kind of owl frequenting barns.

barnacle /ˈbɑ:nək(ə)l/ *noun* small shellfish clinging to rocks, ships' bottoms, etc. □ **barnacle goose** kind of Arctic goose.

barney /ˈbɑ:nɪ/ *noun* (*plural* **-s**) *colloquial* noisy quarrel.

barometer /bəˈrɒmɪtə/ *noun* instrument measuring atmospheric pressure. □ **barometric** /bærəˈmetrɪk/ *adjective*.

baron /ˈbærən/ *noun* member of lowest order of British or foreign nobility; powerful businessman etc. □ **baronial** /bəˈrəʊnɪəl/ *adjective*.

baroness /ˈbærənɪs/ *noun* woman holding rank of baron; baron's wife or widow.

baronet /ˈbærənɪt/ *noun* member of lowest British hereditary titled order. □ **baronetcy** *noun* (*plural* **-ies**).

barony /ˈbærənɪ/ *noun* (*plural* **-ies**) rank or domain of baron.

baroque /bəˈrɒk/ • *adjective* (esp. of 17th- & 18th-c. European architecture and music) ornate and extravagant in style. • *noun* baroque style.

barque /bɑ:k/ *noun* kind of sailing ship.

barrack[1] /ˈbærək/ *noun* (usually in *plural*, often treated as *singular*) housing for soldiers; large bleak building.

barrack[2] /ˈbærək/ *verb* shout or jeer (at).

barracuda /bærəˈkuːdə/ *noun* (*plural* same or **-s**) large voracious tropical sea fish.

barrage /ˈbærɑːʒ/ *noun* concentrated artillery bombardment; rapid succession of questions etc.; artificial barrier in river etc.

barrel /ˈbær(ə)l/ ● *noun* cylindrical usually convex container; contents or capacity of this; tube forming part of thing, esp. gun or pen. ● *verb* (**-ll-**; *US* **-l-**) put in barrels. □ **barrel organ** musical instrument with rotating pin-studded cylinder.

barren /ˈbærən/ *adjective* (**-er**, **-est**) unable to bear young; unable to produce fruit or vegetation; unprofitable; dull. □ **barrenness** *noun.*

barricade /bærɪˈkeɪd/ ● *noun* barrier, esp. improvised. ● *verb* (**-ding**) block or defend with this.

barrier /ˈbærɪə/ *noun* fence etc. barring advance or access; obstacle to communication etc. □ **barrier cream** protective skin cream; **barrier reef** coral reef separated from land by channel.

barrister /ˈbærɪstə/ *noun* advocate practising in higher courts.

barrow[1] /ˈbærəʊ/ *noun* two-wheeled handcart; wheelbarrow.

barrow[2] /ˈbærəʊ/ *noun* ancient grave mound.

Bart. *abbreviation* Baronet.

barter /ˈbɑːtə/ ● *verb* exchange goods, rights, etc., without using money. ● *noun* trade by bartering.

basal /ˈbeɪs(ə)l/ *adjective* of, at, or forming base.

basalt /ˈbæsɔːlt/ *noun* dark volcanic rock. □ **basaltic** /bəˈsɔːltɪk/ *adjective.*

base[1] ● *noun* what a thing rests or depends on, foundation; principle; starting point; headquarters; main ingredient; number in terms of which other numbers are expressed; substance combining with acid to form salt. ● *verb* (**-sing**) (usually + *on*, *upon*) found, establish; station. □ **base rate** interest rate set by Bank of England and used as basis for other banks' rates.

base[2] *adjective* cowardly; despicable; menial; (of coin) alloyed; (of metal) low in value.

baseball *noun* game played esp. in US with bat and ball and circuit of 4 bases.

baseless *adjective* unfounded, groundless.

basement /ˈbeɪsmənt/ *noun* floor below ground level.

bases *plural* of BASE[1], BASIS.

bash ● *verb* strike bluntly or heavily; (often + *up*) *colloquial* attack violently. ● *noun* heavy blow; *slang* attempt, party.

bashful /ˈbæʃfʊl/ *adjective* shy; diffident.

BASIC /ˈbeɪsɪk/ *noun* computer programming language using familiar English words.

basic /ˈbeɪsɪk/ ● *adjective* serving as base; fundamental; simplest, lowest in level. ● *noun* (usually in *plural*) fundamental fact or principle. □ **basically** *adverb.*

basil /ˈbæz(ə)l/ *noun* aromatic herb.

basilica /bəˈzɪlɪkə/ *noun* ancient Roman hall with apse and colonnades; similar church.

basilisk /ˈbæzɪlɪsk/ *noun* mythical reptile with lethal breath and look; American crested lizard.

basin /ˈbeɪs(ə)n/ *noun* round vessel for liquids or preparing food in; washbasin; hollow depression; sheltered mooring area; round valley; area drained by river.

basis /ˈbeɪsɪs/ *noun* (*plural* **bases** /-siːz/) foundation; main ingredient or principle; starting point for discussion etc.

bask /bɑːsk/ *verb* relax in warmth and light; (+ *in*) revel in.

basket /ˈbɑːskɪt/ *noun* container made of woven canes, wire, etc.; amount held by this. □ **basketball** team game in which goals are scored by putting ball through high nets; **basketry**, **basketwork** art of weaving cane etc., work so produced.

bas-relief /ˈbæsrɪliːf/ *noun* carving or sculpture projecting slightly from background.

bass[1] /beɪs/ ● *noun* lowest adult male voice; singer with this; *colloquial* double bass, bass guitar; low-frequency sound of radio, record player, etc. ● *adjective* lowest in pitch; deep-sounding. □ **bass guitar** electric guitar playing low notes. □ **bassist** *noun.*

bass[2] /bæs/ *noun* (*plural* same or **-es**) common perch; other fish of perch family.

basset /ˈbæsɪt/ *noun* (in full **basset-hound**) short-legged hunting dog.

bassoon /bəˈsuːn/ *noun* bass instrument of oboe family. □ **bassoonist** *noun.*

bast /bæst/ *noun* fibre from inner bark of tree, esp. lime.

bastard /ˈbɑːstəd/ *often offensive* ● *noun* person born of parents not married to each other; *slang* person regarded with dislike or pity, difficult or awkward thing. ● *adjective* illegitimate by birth; hybrid. □ **bastardy** *noun.*

bastardize /ˈbɑːstədaɪz/ *verb* (also **-ise**) (**-zing** or **-sing**) corrupt, debase; declare illegitimate.

baste[1] /beɪst/ *verb* (**-ting**) moisten (roasting meat etc.) with fat etc.; beat, thrash.

baste[2] /beɪst/ *verb* (**-ting**) sew with long loose stitches, tack.

bastinado /bæstɪˈneɪdəʊ/ ● *noun* caning on soles of feet. ● *verb* (**-es**, **-ed**) punish with this.

bastion /ˈbæstɪən/ *noun* projecting part of fortification; thing regarded as protection.

bat[1] ● *noun* implement with handle for hitting ball in games; turn at using this; batsman. ● *verb* (**-tt-**) use bat; hit (as) with bat; take one's turn at batting. □ **batsman** person who bats, esp. at cricket.

bat[2] *noun* mouselike nocturnal flying mammal. □ **bats** *slang* crazy.

bat[3] *verb* (**-tt-**) □ **not bat an eyelid** *colloquial* show no reaction.

batch ● *noun* group, collection, set; loaves baked at one time. ● *verb* arrange in batches.

bated /ˈbeɪtɪd/ *adjective* □ **with bated breath** anxiously.

Bath /bɑːθ/ *noun* □ **Bath bun** round spiced bun with currants; **Bath chair** invalid's wheelchair.

bath /bɑːθ/ ● *noun* (*plural* **-s** /bɑːðz/) container for sitting in and washing the body; its contents; act of washing in it; (usually in *plural*) building for swimming or bathing. ● *verb* wash in bath. □ **bath cube** cube of **bath salts**, substance for scenting and softening bath water; **bathroom** room with bath, *US* room with lavatory.

bathe /beɪð/ ● *verb* (**-thing**) immerse oneself in water etc., esp. to swim or wash; immerse in or treat with liquid; (of sunlight etc.) envelop. ● *noun* swim. □ **bathing costume** garment worn for swimming.

bathos /ˈbeɪθɒs/ *noun* lapse from sublime to trivial; anticlimax. □ **bathetic** /bəˈθetɪk/ *adjective.*

bathyscaphe /ˈbæθɪskæf/, **bathysphere** /-sfɪə/ *nouns* vessel for deep-sea diving.

batik /bəˈtiːk/ *noun* method of dyeing textiles by waxing parts to be left uncoloured.

batiste /bæˈtiːst/ *noun* fine cotton or linen fabric.

batman /ˈbætmən/ *noun* army officer's servant.

baton /ˈbæt(ə)n/ *noun* thin stick for conducting orchestra etc.; short stick carried in relay race; stick carried by drum major; staff of office.

batrachian /bəˈtreɪkɪən/ ● *noun* amphibian that discards gills and tail, esp. frog or toad. ● *adjective* of batrachians.

battalion /bəˈtæljən/ *noun* army unit usually of 300–1000 men.

batten[1] /ˈbæt(ə)n/ ● *noun* long narrow piece of squared timber; strip for securing tarpaulin over ship's hatchway. ● *verb* strengthen or (often + *down*) fasten with battens.

batten[2] /ˈbæt(ə)n/ *verb* (often + *on*) thrive at the expense of.

batter /ˈbætə/ ● *verb* strike hard and repeatedly; (esp. as **battered** *adjective*) subject to long-term violence. ● *noun* mixture of flour and eggs beaten up with liquid for cooking. □ **battering ram** *historical* swinging beam for breaching walls.

battery /ˈbætərɪ/ *noun* (*plural* **-ies**) portable container of cell or cells for supplying electricity; series of cages etc. for poultry or cattle; set of connected or similar instruments etc.; emplacement for heavy guns; *Law* physical violence inflicted on person.

battle /ˈbæt(ə)l/ ● *noun* prolonged fight, esp. between armed forces; contest. ● *verb* (**-ling**) (often + *with*, *for*) struggle. □ **battleaxe** medieval weapon, *colloquial* domineering middle-aged woman;

battle-cruiser *historical* heavy-gunned ship of higher speed and lighter armour than battleship; **battledress** soldier's everyday uniform; **battlefield** scene of battle; **battleship** most heavily armed and armoured warship.

battlement /'bæt(ə)lmənt/ *noun* (usually in *plural*) parapet with gaps at intervals at top of wall.

batty /'bætɪ/ *adjective* (**-ier**, **-iest**) *slang* crazy.

batwing *adjective* (esp. of sleeve) shaped like a bat's wing.

bauble /'bɔ:b(ə)l/ *noun* showy trinket.

baulk /bɔ:lk/ (also **balk**) ● *verb* (often + *at*) jib, hesitate; thwart, hinder, disappoint. ● *noun* hindrance; stumbling block.

bauxite /'bɔ:ksaɪt/ *noun* claylike mineral, chief source of aluminium.

bawdy /'bɔ:dɪ/ ● *adjective* (**-ier**, **-iest**) humorously indecent. ● *noun* bawdy talk or writing. □ **bawdy house** brothel.

bawl *verb* shout or weep noisily. □ **bawl out** *colloquial* reprimand severely.

bay[1] *noun* broad curving inlet of sea.

bay[2] *noun* laurel with deep green leaves; (in *plural*) victor's or poet's bay wreath, fame. □ **bay leaf** leaf of bay tree, used for flavouring.

bay[3] *noun* recess; alcove in wall; compartment; allotted area. □ **bay window** window projecting from line of wall.

bay[4] ● *adjective* reddish-brown (esp. of horse). ● *noun* bay horse.

bay[5] ● *verb* bark loudly. ● *noun* bark of large dog, esp. chorus of pursuing hounds. □ **at bay** unable to escape; **keep at bay** ward off.

bayonet /'beɪənet/ ● *noun* stabbing-blade attachable to rifle. ● *verb* (**-t-**) stab with bayonet. □ **bayonet fitting** connecting-part engaged by pushing and twisting.

bazaar /bə'zɑ:/ *noun* oriental market; sale of goods, esp. for charity.

bazooka /bə'zu:kə/ *noun* anti-tank rocket launcher.

BBC *abbreviation* British Broadcasting Corporation.

BC *abbreviation* before Christ; British Columbia.

BCG *abbreviation* Bacillus Calmette-Guérin, an anti-tuberculosis vaccine.

be /bi:/ ● *verb* (*present singular 1st* **am**, *2nd* **are** /ɑ:/, *3rd* **is** /ɪz/, *plural* **are** /ɑ:/; *past singular 1st* **was** /wɒz/, *2nd* **were** /wɜ:/, *3rd* **was** /wɒz/, *plural* **were** /wɜ:/; *present participle* **being**; *past participle* **been**) exist, live; occur; remain, continue; have specified identity, state, or quality. ● *auxiliary verb with past participle to form passive, with present participle to form continuous tenses, with infinitive to express duty, intention, possibility, etc.* □ **be-all and end-all** *colloquial* whole being, essence.

beach ● *noun* sandy or pebbly shore of sea, lake, etc. ● *verb* run or haul (boat etc.) on shore. □ **beachcomber** person who searches beaches for articles of value; **beachhead** fortified position set up on beach by landing forces.

beacon /'bi:kən/ *noun* signal-fire on hill or pole; signal; signal-station; Belisha beacon.

bead ● *noun* small ball of glass, stone, etc. pierced for threading with others; drop of liquid; small knob in front sight of gun. ● *verb* adorn with bead(s) or beading.

beading *noun* moulding like series of beads.

beadle /'bi:d(ə)l/ *noun* ceremonial officer of church, college, etc.

beady *adjective* (**-ier**, **-iest**) (of eyes) small and bright. □ **beady-eyed** with beady eyes, observant.

beagle /'bi:g(ə)l/ *noun* small hound used for hunting hares.

beak *noun* bird's horny projecting jaws; *slang* hooked nose; *historical* prow of warship; spout.

beaker /'bi:kə/ *noun* tall cup for drinking; lipped glass vessel for scientific experiments.

beam ● *noun* long piece of squared timber or metal used in house-building etc.; ray of light or radiation; bright smile; series of radio or radar signals as guide to ship or aircraft; crossbar of balance; (in *plural*) horizontal cross-timbers of ship. ● *verb* emit (light, radio waves, etc.); shine; smile radiantly. □ **off beam** *colloquial* mistaken.

bean *noun* climbing plant with kidney-shaped seeds in long pods; seed of this or of coffee or other plant. □ **beanbag** small bag filled with dried beans and used as ball, large bag filled with polystyrene pieces and used as seat; **bean sprout** sprout of bean seed as food; **full of beans** *colloquial* lively, exuberant; **not a bean** *slang* no money.

beano /'bi:nəʊ/ *noun* (*plural* **-s**) *slang* party, celebration.

bear[1] /beə/ *verb* (*past* **bore**; *past participle* **borne** or **born**) carry; show; produce, yield (fruit); give birth to; sustain; endure, tolerate.

bear[2] /beə/ *noun* heavy thick-furred mammal; rough surly person; person who sells shares for future delivery in hope of buying them more cheaply before then. □ **beargarden** noisy or rowdy scene; **bear-hug** powerful embrace; **bearskin** guardsman's tall furry cap.

bearable *adjective* endurable.

beard /bɪəd/ ● *noun* facial hair on chin etc.; part on animal (esp. goat) resembling beard. ● *verb* oppose, defy. □ **bearded** *adjective*.

bearer *noun* carrier of message, cheque, etc.; carrier of coffin, equipment, etc.

bearing *noun* outward behaviour, posture; (usually + *on*, *upon*) relation, relevance; part of machine supporting rotating part; direction relative to fixed point; (in *plural*) relative position; heraldic device or design.

beast *noun* animal, esp. wild mammal; brutal person; *colloquial* disliked person or thing.

beastly *adjective* (**-ier**, **-iest**) like a beast; *colloquial* unpleasant.

beat ● *verb* (*past* **beat**; *past participle* **beaten**) strike repeatedly or persistently; inflict blows on; overcome, surpass; exhaust, perplex; (often + *up*) whisk (eggs etc.) vigorously; (often + *out*) shape (metal etc.) by blows; pulsate; mark (time of music) with baton, foot, etc.; move or cause (wings) to move up and down. ● *noun* main accent in music or verse; strongly marked rhythm of popular music etc.; stroke on drum; movement of conductor's baton; throbbing; police officer's appointed course; one's habitual round. ● *adjective slang* exhausted, tired out. □ **beat about the bush** not come to the point; **beat down** cause (seller) to lower price by bargaining, (of sun, rain, etc.) shine or fall relentlessly; **beat it** *slang* go away; **beat up** beat with punches and kicks; **beat-up** *colloquial* dilapidated.

beater *noun* whisk; implement for beating carpet; person who rouses game at a shoot.

beatific /bi:ə'tɪfɪk/ *adjective* making blessed; *colloquial* blissful.

beatify /bi:'ætɪfaɪ/ *verb* (**-ies**, **-ied**) *RC Church* declare to be blessed as first step to canonization; make happy. □ **beatification** *noun*.

beatitude /bi:'ætɪtju:d/ *noun* blessedness; (in *plural*) blessings in Matthew 5: 3–11.

beau /bəʊ/ *noun* (*plural* **-x** /-z/) boyfriend; dandy.

Beaufort scale /'bəʊfət/ *noun* scale of wind speeds.

Beaujolais /'bəʊʒəleɪ/ *noun* red or white wine from Beaujolais district of France.

beauteous /'bju:tɪəs/ *adjective poetical* beautiful.

beautician /bju:'tɪʃ(ə)n/ *noun* specialist in beauty treatment.

beautiful /'bju:tɪfʊl/ *adjective* having beauty; *colloquial* excellent. □ **beautifully** *adverb*.

beautify /'bju:tɪfaɪ/ *verb* (**-ies**, **-ied**) make beautiful, adorn. □ **beautification** *noun*.

beauty /'bju:tɪ/ *noun* (*plural* **-ies**) combination of qualities that delights the sight or other senses or the mind; person or thing having this. □ **beauty queen** woman judged most beautiful in contest; **beauty parlour**, **salon** establishment for cosmetic treatment; **beauty spot** beautiful locality.

beaver ● *noun* large amphibious broad-tailed rodent; its fur; hat of this; (**Beaver**) member of most junior branch of Scout Association. ● *verb colloquial* (usually + *away*) work hard.

becalm /bɪ'kɑ:m/ *verb* deprive (ship etc.) of wind.

became *past* of BECOME.

because /bɪ'kɒz/ *conjunction* for the reason that. □ **because of** by reason of.

beck[1] *noun* brook, mountain stream.

beck[2] *noun* □ **at (person's) beck and call** subject to his or her constant orders.

beckon /'bekən/ *verb* (often + *to*) summon by gesture; entice.

become /bɪ'kʌm/ *verb* (**-ming**; *past* **became**; *past participle* **become**) come to be, begin to be; suit, look well on. □ **become of** happen to.

becquerel /'bekərel/ *noun* SI unit of radioactivity.

bed ● *noun* place to sleep or rest, esp. piece of furniture for sleeping on; garden plot for plants; bottom of sea, river, etc.; flat base on which thing rests; stratum. ● *verb* (**-dd-**) (usually + *down*) put or go to bed; plant in bed; fix firmly; *colloquial* have sexual intercourse with. □ **bedclothes** sheets, blankets, etc.; **bedpan** pan for use as toilet by invalid in bed; **bedridden** confined to bed by infirmity; **bedrock** solid rock under alluvial deposits etc., basic principles; **bedroom** room for sleeping in; **bedsitting room**, **bedsitter** combined bedroom and sitting-room; **bedsore** sore developed by lying in bed; **bedspread** coverlet; **bedstead** framework of bed; **bedtime** hour for going to bed.

bedaub /bɪ'dɔ:b/ *verb* smear with paint etc.

bedding *noun* mattress and bedclothes; litter for cattle etc. □ **bedding plant** annual flowering plant put in garden bed.

bedeck /bɪ'dek/ *verb* adorn, decorate.

bedevil /bɪ'dev(ə)l/ *verb* (**-ll-**; *US* **-l-**) torment, confuse, trouble. □ **bedevilment** *noun*.

bedlam /'bedləm/ *noun* scene of confusion or uproar.

Bedouin /'bedʊɪn/ *noun* (*plural* same) nomadic Arab of the desert.

bedraggled /bɪ'dræg(ə)ld/ *adjective* dishevelled, untidy.

bee *noun* 4-winged stinging insect, collecting nectar and pollen and producing wax and honey; busy worker; meeting for combined work or amusement. □ **beehive** hive; **beeline** straight line between two places; **beeswax** wax secreted by bees for honeycomb.

Beeb *noun* (**the Beeb**) *colloquial* the BBC.

beech *noun* smooth-barked glossy-leaved tree; its wood. □ **beechmast** fruit of beech.

beef ● *noun* meat of ox, bull, or cow; (*plural* **beeves** or *US* **-s**) beef animal; (*plural* **-s**) *slang* protest. ● *verb slang* complain. □ **beefburger** hamburger; **beefeater** warder in Tower of London; **beefsteak** thick slice of beef for grilling or frying; **beef tea** stewed beef extract for invalids; **beef tomato** large tomato.

beefy *adjective* (**-ier**, **-iest**) like beef; solid, muscular.

been *past participle* of BE.

beep ● *noun* short high-pitched sound. ● *verb* emit beep.

beer /bɪə/ *noun* alcoholic liquor made from fermented malt etc. flavoured esp. with hops. □ **beer garden** garden where beer is sold and drunk; **beer mat** small mat for beer glass.

beery *adjective* (**-ier**, **-iest**) showing influence of beer; like beer.

beeswing /'bi:zwɪŋ/ *noun* filmy crust on old port wine etc.

beet *noun* plant with succulent root used for salads etc. and sugar-making (see BEETROOT, SUGAR BEET).

beetle[1] /'bi:t(ə)l/ ● *noun* insect with hard protective outer wings. ● *verb* (**-ling**) *colloquial* (+ *about*, *off*, etc.) hurry, scurry.

beetle[2] /'bi:t(ə)l/ ● *adjective* projecting, shaggy, scowling. ● *verb* (usually as **beetling** *adjective*) overhang.

beetle[3] /'bi:t(ə)l/ *noun* heavy-headed tool for ramming, crushing, etc.

beetroot *noun* beet with dark red root used as vegetable.

befall /bɪ'fɔ:l/ *verb* (*past* **befell**; *past participle* **befallen**) *poetical* happen; happen to.

befit /bɪ'fɪt/ *verb* (**-tt-**) be appropriate for.

befog /bɪ'fɒg/ *verb* (**-gg-**) obscure; envelop in fog.

before /bɪ'fɔ:/ ● *conjunction* sooner than; rather than. ● *preposition* earlier than; in front of, ahead of; in presence of. ● *adverb* ahead, in front; previously, already; in the past.

beforehand *adverb* in anticipation, in readiness, before time.

befriend /bɪ'frend/ *verb* act as friend to; help.

befuddle /bɪ'fʌd(ə)l/ *verb* (**-ling**) make drunk; confuse.

beg *verb* (**-gg-**) ask for as gift; ask earnestly, entreat; live by begging; ask formally. □ **beg the question** assume truth of thing to be proved; **go begging** be unwanted.

■ **Usage** The expression *beg the question* is often used incorrectly to mean 'to invite the obvious question (that ...)'.

began *past* of BEGIN.

begat *archaic past* of BEGET.

beget /bɪ'get/ *verb* (**-tt-**; *past* **begot**, *archaic* **begat**; *past participle* **begotten**) *literary* be the father of; give rise to.

beggar /'begə/ ● *noun* person who begs or lives by begging; *colloquial* person. ● *verb* make poor; be too extraordinary for (belief, description, etc.).

beggarly *adjective* mean; poor, needy.

begin /bɪ'gɪn/ *verb* (**-nn-**; *past* **began**; *past participle* **begun**) perform first part of; come into being; (often + *to do*) start, take first step, (usually in negative) *colloquial* show any likelihood, be sufficient.

beginner *noun* learner.

beginning *noun* time at which thing begins; source, origin; first part.

begone /bɪ'gɒn/ *interjection poetical* go away at once!

begonia /bɪ'gəʊnɪə/ *noun* plant with ornamental foliage and bright flowers.

begot *past* of BEGET.

begotten *past participle* of BEGET.

begrudge /bɪ'grʌdʒ/ *verb* (**-ging**) grudge; feel or show resentment at or envy of; be dissatisfied at.

beguile /bɪ'gaɪl/ *verb* (**-ling**) charm, divert; delude, cheat. □ **beguilement** *noun.*

beguine /bɪ'gi:n/ *noun* W. Indian dance.

begum /'beɪgəm/ *noun* (in India, Pakistan, and Bangladesh) Muslim woman of high rank; (**Begum**) *title of married Muslim woman.*

begun *past participle* of BEGIN.

behalf /bɪ'hɑ:f/ *noun* □ **on behalf of, on (person's) behalf** in the interests of, as representative of.

behave /bɪ'heɪv/ *verb* (**-ving**) react or act in specified way; (often **behave oneself**) conduct oneself properly; work well (or in specified way).

behaviour /bɪ'heɪvjə/ *noun* (*US* **behavior**) manners, conduct, way of behaving. □ **behavioural** *adjective.*

behaviourism *noun* (*US* **behaviorism**) study of human actions by analysis of stimulus and response. □ **behaviourist** *noun.*

behead /bɪ'hed/ *verb* cut head from (person); execute thus.

beheld *past & past participle* of BEHOLD.

behest /bɪ'hest/ *noun literary* command, request.

behind /bɪ'haɪnd/ ● *preposition* in or to rear of; hidden by, on farther side of; in past in relation to; inferior to; in support of. ● *adverb* in or to rear; on far side; remaining after others' departure; (usually + *with*) in arrears. ● *noun colloquial* buttocks. □ **behindhand** in arrears, behind time, too late; **behind time** unpunctual; **behind the times** old-fashioned, antiquated.

behold /bɪ'həʊld/ *verb* (*past & past participle* **beheld**) *literary* look at; take notice, observe.

beholden /bɪ'həʊld(ə)n/ *adjective* (usually + *to*) under obligation.

behove /bɪ'həʊv/ *verb* (**-ving**) *formal* be incumbent on; befit.

beige /beɪʒ/ ● *noun* pale sandy fawn colour. ● *adjective* of this colour.

being /'bi:ɪŋ/ *noun* existence; constitution, nature; existing person etc.

belabour /bɪ'leɪbə/ *verb* (*US* **belabor**) attack physically or verbally.

belated /bɪ'leɪtɪd/ *adjective* coming (too) late. □ **belatedly** *adverb.*

bel canto /bel 'kæntəʊ/ *noun* singing marked by full rich tone.

belch ● *verb* emit wind from stomach through mouth; (of volcano, gun, etc.) emit (fire, smoke, etc.). ● *noun* act of belching.

beleaguer /bɪ'li:gə/ *verb* besiege; vex; harass.

belfry /'belfrɪ/ *noun* (*plural* **-ies**) bell tower; space for bells in church tower.

belie /bɪ'laɪ/ *verb* (**belying**) give false impression of; fail to confirm, fulfil, or justify.

belief /bɪ'li:f/ *noun* act of believing; what one believes; trust, confidence; acceptance as true.

believe /bɪ'li:v/ *verb* (**-ving**) accept as true; think; (+ *in*) have faith or confidence in; trust word of; have religious faith. □ **believable** *adjective*; **believer** *noun*.

Belisha beacon /bə'li:ʃə/ *noun* flashing orange ball on striped post, marking pedestrian crossing.

belittle /bɪ'lɪt(ə)l/ *verb* (**-ling**) disparage. □ **belittlement** *noun*.

bell *noun* hollow esp. cup-shaped metal body emitting musical sound when struck; sound of bell; bell-shaped thing. □ **bell-bottomed** (of trousers) widening below knee; **bell pull** cord pulled to sound bell; **bell push** button pressed to ring electric bell; **give (person) a bell** *colloquial* telephone him or her.

belladonna /belə'dɒnə/ *noun* deadly nightshade; drug obtained from this.

belle /bel/ *noun* handsome woman; reigning beauty.

belles-lettres /bel 'letr/ *plural noun* (also treated as *singular*) writings or studies of purely literary kind.

bellicose /'belɪkəʊs/ *adjective* eager to fight.

belligerent /bɪ'lɪdʒərənt/ ● *adjective* engaged in war; given to constant fighting; pugnacious. ● *noun* belligerent person or nation. □ **belligerence** *noun*.

bellow /'beləʊ/ ● *verb* emit deep loud roar. ● *noun* bellowing sound.

bellows /'beləʊz/ *plural noun* (also treated as *singular*) device for driving air into fire, organ, etc.; expandable part of camera etc.

belly /'belɪ/ ● *noun* (*plural* **-ies**) cavity of body containing stomach, bowels, etc.; stomach; front of body from waist to groin; underside of animal; cavity or bulging part of anything. ● *verb* (**-ies, -ied**) swell out. □ **bellyache** *noun colloquial* stomach pain, *verb slang* complain noisily or persistently; **belly button** *colloquial* navel; **belly dance** dance by woman (**belly dancer**) with voluptuous movements of belly; **belly laugh** loud unrestrained laugh.

bellyful *noun* enough to eat; *colloquial* more than one can tolerate.

belong /bɪ'lɒŋ/ *verb* (+ *to*) be property of, assigned to, or member of; fit socially; be rightly placed or classified.

belongings *plural noun* possessions, luggage.

beloved /bɪ'lʌvɪd/ ● *adjective* loved. ● *noun* beloved person.

below /bɪ'ləʊ/ ● *preposition* under; lower than; less than; of lower rank or importance etc. than; unworthy of. ● *adverb* at or to lower point or level; further on in book etc.

belt ● *noun* strip of leather etc. worn round waist or across chest; continuous band in machinery; distinct strip of colour etc.; zone, district. ● *verb* put belt round; *slang* thrash; *slang* move rapidly. □ **below the belt** unfair(ly); **belt out** *slang* sing or play (music) loudly; **belt up** *slang* be quiet, *colloquial* put on seat belt; **tighten one's belt** economize; **under one's belt** securely acquired.

bemoan /bɪ'məʊn/ *verb* lament, complain about.

bemuse /bɪ'mju:z/ *verb* (**-sing**) make (person) confused.

bench *noun* long seat of wood or stone; carpenter's or laboratory table; magistrate's or judge's seat; lawcourt. □ **benchmark** surveyor's mark at point in line of levels; standard, point of reference.

bend ● *verb* (*past & past participle* **bent** except in *bended knee*) force into curve or angle; be altered in this way; incline from vertical; bow, stoop; interpret or modify (rule) to suit oneself; (force to) submit. ● *noun* bending, curve; bent part of thing; (**the bends**) *colloquial* symptoms due to too rapid decompression under water. □ **round the bend** *colloquial* crazy, insane.

bender *noun slang* wild drinking spree.

beneath /bɪ'ni:θ/ ● *preposition* below, under; unworthy of. ● *adverb* below, underneath.

Benedictine /benɪ'dɪktɪn/ ● *noun* monk or nun of Order of St Benedict; /-ti:n/ *proprietary term* kind of liqueur. ● *adjective* of St Benedict or his order.

benediction /benɪ'dɪkʃ(ə)n/ *noun* utterance of blessing. □ **benedictory** *adjective*.

benefaction /benɪˈfækʃ(ə)n/ *noun* charitable gift; doing good.

benefactor /ˈbenɪfæktə/ *noun* (*feminine* **benefactress**) person who has given financial or other help.

benefice /ˈbenɪfɪs/ *noun* living from a church office.

beneficent /bɪˈnefɪsənt/ *adjective* doing good; actively kind. □ **beneficence** *noun.*

beneficial /benɪˈfɪʃ(ə)l/ *adjective* advantageous. □ **beneficially** *adverb.*

beneficiary /benɪˈfɪʃərɪ/ *noun* (*plural* **-ies**) receiver of benefits; holder of church living.

benefit /ˈbenɪfɪt/ ● *noun* advantage, profit; payment made under insurance or social security; performance or game etc. of which proceeds go to particular player or charity. ● *verb* (**-t-**; *US* **-tt-**) do good to; receive benefit. □ **benefit of the doubt** assumption of innocence rather than guilt.

benevolent /bɪˈnevələnt/ *adjective* wishing to do good; charitable; kind and helpful. □ **benevolence** *noun.*

Bengali /beŋˈgɔːlɪ/ ● *noun* (*plural* **-s**) native or language of Bengal. ● *adjective* of Bengal.

benighted /bɪˈnaɪtɪd/ *adjective* intellectually or morally ignorant.

benign /bɪˈnaɪn/ *adjective* kindly, gentle; favourable; salutary; *Medicine* mild, not malignant. □ **benignity** /bɪˈnɪgnɪtɪ/ *noun.*

benignant /bɪˈnɪgnənt/ *adjective* kindly; beneficial. □ **benignancy** *noun.*

bent ● *past & past participle* of BEND. ● *adjective* curved or having angle; *slang* dishonest, illicit; (+ *on*) set on doing or having. ● *noun* inclination, bias; (+ *for*) talent for.

benumb /bɪˈnʌm/ *verb* make numb; deaden; paralyse.

benzene /ˈbenziːn/ *noun* chemical got from coal tar and used as solvent.

benzine /ˈbenziːn/ *noun* spirit obtained from petroleum and used as cleaning agent.

benzoin /ˈbenzəʊɪn/ *noun* fragrant resin of E. Asian tree. □ **benzoic** /-ˈzəʊɪk/ *adjective.*

bequeath /bɪˈkwiːð/ *verb* leave by will; transmit to posterity.

bequest /bɪˈkwest/ *noun* bequeathing; thing bequeathed.

berate /bɪˈreɪt/ *verb* (**-ting**) scold.

bereave /bɪˈriːv/ *verb* (**-ving**) (esp. as **bereaved** *adjective*) (often + *of*) deprive of relative, friend, etc., esp. by death. □ **bereavement** *noun.*

bereft /bɪˈreft/ *adjective* (+ *of*) deprived of.

beret /ˈbereɪ/ *noun* round flat brimless cap of felt etc.

berg *noun* iceberg.

bergamot /ˈbɜːgəmɒt/ *noun* perfume from fruit of a dwarf orange tree; an aromatic herb.

beriberi /berɪˈberɪ/ *noun* nervous disease caused by deficiency of vitamin B_1.

Bermuda shorts /bəˈmjuːdə/ *plural noun* close-fitting knee-length shorts.

berry /ˈberɪ/ *noun* (*plural* **-ies**) any small round juicy stoneless fruit.

berserk /bəˈzɜːk/ *adjective* (esp. after *go*) wild, frenzied.

berth ● *noun* sleeping place; ship's place at wharf; sea-room; *colloquial* situation, appointment. ● *verb* moor (ship) in berth; provide sleeping berth for.

beryl /ˈberɪl/ *noun* transparent (esp. green) precious stone; mineral species including this and emerald.

beryllium /bəˈrɪlɪəm/ *noun* hard white metallic element.

beseech /bɪˈsiːtʃ/ *verb* (*past & past participle* **besought** /-ˈsɔːt/ or **beseeched**) entreat; ask earnestly for.

beset /bɪˈset/ *verb* (**-tt-**; *past & past participle* **beset**) attack or harass persistently.

beside /bɪˈsaɪd/ *preposition* at side of, close to; compared with; irrelevant to. □ **beside oneself** frantic with anger or worry etc.

besides ● *preposition* in addition to; apart from. ● *adverb* also, as well.

besiege /bɪˈsiːdʒ/ *verb* (**-ging**) lay siege to; crowd round eagerly; assail with requests.

besmirch /bɪˈsmɜːtʃ/ *verb* soil, dishonour.

besom /ˈbiːz(ə)m/ *noun* broom made of twigs.

besotted /bɪˈsɒtɪd/ *adjective* infatuated; stupefied.

besought *past & past participle* of BESEECH.

bespatter /bɪ'spætə/ *verb* spatter all over; defame.

bespeak /bɪ'spi:k/ *verb* (*past* **bespoke**; *past participle* **bespoken**) engage beforehand; order (goods); be evidence of.

bespectacled /bɪ'spektək(ə)ld/ *adjective* wearing spectacles.

bespoke /bɪ'spəʊk/ *adjective* made to order.

best ● *adjective* (*superlative* of GOOD) most excellent. ● *adverb* (*superlative* of WELL[1]) in the best way; to greatest degree. ● *noun* that which is best. ● *verb colloquial* defeat, outwit. □ **best man** bridegroom's chief attendant at wedding; **best-seller** book with large sale, author of such book; **do one's best** do all one can.

bestial /'bestɪəl/ *adjective* brutish; of or like beasts. □ **bestiality** /-'æl-/ *noun*.

bestiary /'bestɪərɪ/ *noun* (*plural* **-ies**) medieval treatise on beasts.

bestir /bɪ'stɜ:/ *verb* (**-rr-**) (**bestir oneself**) exert or rouse oneself.

bestow /bɪ'stəʊ/ *verb* (+ *on*, *upon*) confer as gift. □ **bestowal** *noun*.

bestrew /bɪ'stru:/ *verb* (*past participle* **bestrewed** or **bestrewn**) strew; lie scattered over.

bestride /bɪ'straɪd/ *verb* (**-ding**; *past* **bestrode**; *past participle* **bestridden**) sit astride on; stand astride over.

bet ● *verb* (**-tt-**; *past & past participle* **bet** or **betted**) risk one's money etc. against another's on result of event. ● *noun* such arrangement; sum of money bet.

beta /'bi:tə/ *noun* second letter of Greek alphabet (Β, β). □ **beta-blocker** drug used to prevent unwanted stimulation of the heart in angina etc.; **beta particle** fast-moving electron emitted by radioactive substance.

betake /bɪ'teɪk/ *verb* (**-king**; *past* **betook**; *past participle* **betaken**) (**betake oneself**) go.

betatron /'bi:tətrɒn/ *noun* apparatus for accelerating electrons.

betel /'bi:t(ə)l/ *noun* leaf chewed with betel-nut. □ **betel-nut** seed of tropical palm.

bête noire /beɪt 'nwɑ:/ *noun* (*plural* **bêtes noires** same pronunciation) particularly disliked person or thing. [French]

bethink /bɪ'θɪŋk/ *verb* (*past & past participle* **bethought** /-'θɔ:t/) (**bethink oneself**) *formal* reflect, stop to think; be reminded.

betide /bɪ'taɪd/ *verb* □ **woe betide** (**person**) misfortune will befall him or her.

betimes /bɪ'taɪmz/ *adverb literary* in good time, early.

betoken /bɪ'təʊkən/ *verb* be sign of.

betook *past* of BETAKE.

betray /bɪ'treɪ/ *verb* be disloyal to (a person, one's country, etc.); give up or reveal treacherously; reveal involuntarily; be evidence of. □ **betrayal** *noun*.

betroth /bɪ'trəʊð/ *verb* (usually as **betrothed** *adjective*) bind with promise to marry. □ **betrothal** *noun*.

better /'betə/ ● *adjective* (*comparative* of GOOD) more excellent; partly or fully recovered from illness. ● *adverb* (*comparative* of WELL[1]) in better manner; to greater degree. ● *noun* better thing or person. ● *verb* improve (upon); surpass. □ **better off** in better (esp. financial) situation; **get the better of** defeat, outwit.

betterment *noun* improvement.

between /bɪ'twi:n/ ● *preposition* in or into space or interval; separating; shared by; to and from; taking one or other of. ● *adverb* (also **in between**) between two or more points, between two extremes.

bevel /'bev(ə)l/ ● *noun* slope from horizontal or vertical in carpentry etc.; sloping edge or surface; tool for marking angles. ● *verb* (**-ll-**; *US* **-l-**) impart bevel to, slant.

beverage /'bevərɪdʒ/ *noun formal* drink.

bevvy /'bevɪ/ *slang* ● *noun* (*plural* **-ies**) ● *verb* (**-ies, -ied**) drink.

bevy /'bevɪ/ *noun* (*plural* **-ies**) company, flock.

bewail /bɪ'weɪl/ *verb* wail over; mourn for.

beware /bɪ'weə/ *verb* (only in *imperative* or *infinitive*) take heed; (+ *of*) be cautious of.

bewilder /bɪ'wɪldə/ *verb* perplex, confuse. □ **bewilderment** *noun*.

bewitch /bɪ'wɪtʃ/ *verb* enchant, greatly delight; cast spell on.

beyond /bɪ'jɒnd/ ● *preposition* at or to further side of; outside the range or understanding of; more than. ● *adverb* at or to further side, further on. ● *noun* (**the beyond**) life after death. □ **back of beyond** very remote place.

bezel /'bez(ə)l/ *noun* sloped edge of chisel etc.; oblique face of cut gem; groove holding watch-glass or gem.

bezique /bɪ'zi:k/ *noun* card game for two.

biannual /baɪ'ænjʊəl/ *adjective* occurring etc. twice a year.

bias /'baɪəs/ ● *noun* predisposition, prejudice; distortion of statistical results; edge cut obliquely across weave of fabric; *Sport* bowl's curved course due to its lopsided form. ● *verb* (**-s-** or **-ss-**) give bias to; prejudice; (as **biased** *adjective*) influenced (usually unfairly). □ **bias binding** strip of fabric cut obliquely and used to bind edges.

biathlon /baɪ'æθlən/ *noun* athletic contest in skiing and shooting. □ **biathlete** *noun*.

bib *noun* cloth put under child's chin while eating; upper part of apron etc.

Bible /'baɪb(ə)l/ *noun* Christian scriptures; (**bible**) copy of these; (**bible**) *colloquial* any authoritative book. □ **biblical** /'bɪb-/ *adjective*.

bibliography /bɪblɪ'ɒgrəfɪ/ *noun* (*plural* **-ies**) list of books of any author, subject, etc.; history of books, their editions, etc. □ **bibliographer** *noun*; **bibliographical** /-ə'græf-/ *adjective*.

bibliophile /'bɪblɪəfaɪl/ *noun* collector of books, book-lover.

bibulous /'bɪbjʊləs/ *adjective* fond of or addicted to alcoholic drink.

bicameral /baɪ'kæmər(ə)l/ *adjective* having two legislative chambers.

bicarb /'baɪkɑ:b/ *noun colloquial* bicarbonate of soda.

bicarbonate /baɪ'kɑ:bənɪt/ *noun* any acid salt of carbonic acid; (in full **bicarbonate of soda**) compound used in cooking and as antacid.

bicentenary /baɪsen'ti:nərɪ/ *noun* (*plural* **-ies**) 200th anniversary.

bicentennial /baɪsen'tenɪəl/ ● *noun* bicentenary. ● *adjective* recurring every 200 years.

biceps /'baɪseps/ *noun* (*plural* same) muscle with double head or attachment, esp. that at front of upper arm.

bicker /'bɪkə/ *verb* quarrel, wrangle pettily.

bicuspid /baɪ'kʌspɪd/ ● *adjective* having two cusps. ● *noun* bicuspid premolar tooth.

bicycle /'baɪsɪk(ə)l/ ● *noun* two-wheeled pedal-driven vehicle. ● *verb* (**-ling**) ride bicycle.

bid ● *verb* (**-dd-**; *past* **bid**, *archaic* **bade** /bæd, beɪd/; *past participle* **bid**, *archaic* **bidden**) make offer; make bid; command, invite; *literary* utter (greeting, farewell) to; *Cards* state before play number of tricks intended. ● *noun* act of bidding; amount bid; *colloquial* attempt, effort.

biddable *adjective* obedient, docile.

bidding *noun* command, invitation.

bide *verb* (**-ding**) *archaic or dialect* stay, remain. □ **bide one's time** wait for good opportunity.

bidet /'bi:deɪ/ *noun* low basin that one can sit astride to wash crotch area.

biennial /baɪ'enɪəl/ ● *adjective* lasting 2 years; recurring every 2 years. ● *noun* plant that flowers, fruits, and dies in second year.

bier /bɪə/ *noun* movable stand on which coffin or corpse rests.

biff *slang* ● *noun* smart blow. ● *verb* strike.

bifid /'baɪfɪd/ *adjective* divided by cleft into two parts.

bifocal /baɪ'fəʊk(ə)l/ ● *adjective* (of spectacle lenses) with two parts of different focal lengths. ● *noun* (in *plural*) bifocal spectacles.

bifurcate /'baɪfəkeɪt/ *verb* (**-ting**) divide into two branches; fork. □ **bifurcation** *noun*.

big ● *adjective* (**-gg-**) large; important; grown-up; boastful; *colloquial* ambitious, generous; (usually + *with*) advanced in pregnancy. ● *adverb colloquial* in big manner; with great effect, impressively. □ **Big Apple** *US slang* New York City; **Big Brother** seemingly benevolent dictator; **big business** large-scale commerce; **big end** end of connecting rod in engine, encircling crank-pin; **big-head** *colloquial* conceited person; **big-headed** conceited; **big-hearted** generous; **big shot** *colloquial* important person; **big time** *slang* highest rank among entertainers; **big top** main tent at circus; **bigwig** *colloquial* important person; **in a big way** *colloquial* with great enthusiasm. □ **biggish** *adjective*.

bigamy /ˈbɪgəmɪ/ *noun* (*plural* **-ies**) crime of making second marriage while first is still valid. □ **bigamist** *noun*; **bigamous** *adjective*.

bight /baɪt/ *noun* recess of coast, bay; loop of rope.

bigot /ˈbɪgət/ *noun* obstinate and intolerant adherent of creed or view. □ **bigoted** *adjective*; **bigotry** *noun*.

bijou /ˈbi:ʒu:/ ● *noun* (*plural* **-x** same pronunciation) jewel, trinket. ● *adjective* (**bijou**) small and elegant. [French]

bike *colloquial* ● *noun* bicycle; motorcycle. ● *verb* (**-king**) ride a bike. □ **biker** *noun*.

bikini /bɪˈki:nɪ/ *noun* (*plural* **-s**) woman's brief two-piece bathing suit.

bilateral /baɪˈlætər(ə)l/ *adjective* of, on, or with two sides; between two parties. □ **bilaterally** *adverb*.

bilberry /ˈbɪlbərɪ/ *noun* (*plural* **-ies**) N. European heathland shrub; its small dark blue edible berry.

bile *noun* bitter fluid secreted by liver to aid digestion; bad temper, peevishness.

bilge /bɪldʒ/ *noun* nearly flat part of ship's bottom; (in full **bilge-water**) foul water in bilge; *slang* nonsense, rubbish.

bilharzia /bɪlˈhɑ:tsɪə/ *noun* disease caused by tropical parasitic flatworm.

biliary /ˈbɪlɪərɪ/ *adjective* of bile.

bilingual /baɪˈlɪŋgw(ə)l/ *adjective* of, in, or speaking two languages.

bilious /ˈbɪlɪəs/ *adjective* affected by disorder of the bile; bad-tempered.

bilk *verb slang* evade payment of, cheat.

bill[1] ● *noun* statement of charges for goods, work done, etc.; draft of proposed law; poster; programme of entertainment; *US* banknote. ● *verb* send statement of charges to; announce, put in programme; (+ *as*) advertise as. □ **bill of exchange** written order to pay sum on given date; **bill of fare** menu; **billposter**, **billsticker** person who pastes up advertisements on hoardings etc.

bill[2] ● *noun* beak (of bird); narrow promontory. ● *verb* (of doves etc.) stroke bill with bill. □ **bill and coo** exchange caresses.

bill[3] *noun historical* weapon with hook-shaped blade.

billabong /ˈbɪləbɒŋ/ *noun Australian* branch of river forming backwater.

billboard *noun* large outdoor board for advertisements.

billet[1] /ˈbɪlɪt/ ● *noun* place where soldier etc. is lodged; *colloquial* appointment, job. ● *verb* (**-t-**) quarter (soldiers etc.).

billet[2] /ˈbɪlɪt/ *noun* thick piece of firewood; small metal bar.

billet-doux /bɪlɪˈdu:/ *noun* (*plural* **billets-doux** /-ˈdu:z/) love letter.

billhook *noun* concave-edged pruning-instrument.

billiards /ˈbɪljədz/ *noun* game played with cues and 3 balls on cloth-covered table. □ **billiard ball**, **room**, **table**, etc., ones used for billiards.

billion /ˈbɪljən/ *noun* (*plural* same) thousand million; million million; (**billions**) *colloquial* very large number. □ **billionth** *adjective & noun*.

billow /ˈbɪləʊ/ ● *noun* wave; any large mass. ● *verb* rise or move in billows. □ **billowy** *adjective*.

billy[1] /ˈbɪlɪ/ *noun* (*plural* **-ies**) (in full **billycan**) *Australian* tin or enamel outdoor cooking pot.

billy[2] /ˈbɪlɪ/ *noun* (*plural* **-ies**) (in full **billy goat**) male goat.

bin *noun* large receptacle for rubbish or storage. □ **bin-liner** bag for lining rubbish bin; **binman** *colloquial* dustman.

binary /ˈbaɪnərɪ/ *adjective* of two parts, dual; of system using digits 0 and 1 to code information.

bind /baɪnd/ ● *verb* (*past & past participle* **bound** /baʊnd/) tie or fasten tightly; restrain; (cause to) cohere; compel, impose duty on; edge with braid etc.; fasten (pages of book) into cover. ● *noun colloquial* nuisance; restriction.

binder *noun* cover for loose papers; substance that binds things together; *historical* sheaf-binding machine; bookbinder.

bindery *noun* (*plural* **-ies**) bookbinder's workshop.

binding ● *noun* book cover; braid etc. for edging. ● *adjective* obligatory.

bindweed *noun* convolvulus.

binge /bɪndʒ/ *slang* ● *noun* bout of excessive eating, drinking, etc.; spree. ● *verb* (**-ging**) indulge in binge.

bingo /ˈbɪŋgəʊ/ *noun* gambling game in which each player marks off numbers on card as they are called.

binnacle /ˈbɪnək(ə)l/ *noun* case for ship's compass.

binocular /baɪˈnɒkjʊlə/ *adjective* for both eyes.

binoculars /bɪˈnɒkjʊləz/ *plural noun* instrument with lens for each eye, for viewing distant objects.

binomial /baɪˈnəʊmɪəl/ ● *noun* algebraic expression of sum or difference of two terms. ● *adjective* consisting of two terms.

bio- *combining form* biological; life.

biochemistry /baɪəʊˈkemɪstrɪ/ *noun* chemistry of living organisms. □ **biochemical** *adjective*; **biochemist** *noun*.

biodegradable /baɪəʊdɪˈgreɪdəb(ə)l/ *adjective* able to be decomposed by bacteria or other living organisms.

biography /baɪˈɒgrəfɪ/ *noun* (*plural* **-ies**) written life of person. □ **biographer** *noun*; **biographical** /-əˈgræf-/ *adjective*.

biological /baɪəˈlɒdʒɪk(ə)l/ *adjective* of biology; of living organisms. □ **biological warfare** use of bacteria etc. to spread disease among enemy. □ **biologically** *adverb*.

biology /baɪˈɒlədʒɪ/ *noun* study of living organisms. □ **biologist** *noun*.

bionic /baɪˈɒnɪk/ *adjective* having electronically operated body parts.

biorhythm /ˈbaɪəʊrɪð(ə)m/ *noun* biological cycle thought to affect person's physical or mental state.

biosphere /ˈbaɪəʊsfɪə/ *noun* earth's crust and atmosphere containing life.

bipartite /baɪˈpɑ:taɪt/ *adjective* of two parts; involving two parties.

biped /ˈbaɪped/ ● *noun* two-footed animal. ● *adjective* two-footed.

biplane /ˈbaɪpleɪn/ *noun* aeroplane with two pairs of wings, one above the other.

birch ● *noun* tree with thin peeling bark; bundle of birch twigs used for flogging. ● *verb* flog with birch.

bird *noun* feathered vertebrate with two wings and two feet; *slang* young woman. □ **a bird in the hand** something secured or certain; **birdlime** sticky stuff spread to catch birds; **bird of passage** migratory bird, person who travels habitually; **bird of prey** one hunting animals for food; **birdseed** seeds as food for caged birds; **bird's-eye view** general view from above; **birds of a feather** similar people; **bird table** platform on which food for wild birds is placed.

birdie /ˈbɜ:dɪ/ *noun colloquial* little bird; *Golf* hole played in one under par.

biretta /bɪˈretə/ *noun* square cap of RC priest.

Biro /ˈbaɪərəʊ/ *noun* (*plural* **-s**) *proprietary term* kind of ballpoint pen.

birth *noun* emergence of young from mother's body; origin, beginning; ancestry; inherited position. □ **birth control** prevention of undesired pregnancy; **birthday** anniversary of birth; **birthmark** unusual mark on body from birth; **birth rate** number of live births per thousand of population per year; **birthright** rights belonging to one by birth; **give birth to** produce (young).

biscuit /ˈbɪskɪt/ *noun* thin unleavened cake, usually crisp and sweet; fired unglazed pottery; light brown colour.

bisect /baɪˈsekt/ *verb* divide into two (usually equal) parts. □ **bisection** *noun*; **bisector** *noun*.

bisexual /baɪˈsekʃʊəl/ ● *adjective* feeling or involving sexual attraction to members of both sexes; hermaphrodite. ● *noun* bisexual person. □ **bisexuality** /-ˈæl-/ *noun*.

bishop /ˈbɪʃəp/ *noun* senior clergyman usually in charge of diocese; mitre-shaped chess piece.

bishopric /ˈbɪʃəprɪk/ *noun* office or diocese of bishop.

bismuth /ˈbɪzməθ/ *noun* reddish-white metallic element; compound of it used medicinally.

bison /ˈbaɪs(ə)n/ *noun* (*plural* same) wild ox.

bisque[1] /bi:sk/ *noun* rich soup.

bisque[2] /bɪsk/ *noun* advantage of one free point or stroke in certain games.

bisque[3] /bɪsk/ *noun* fired unglazed pottery.

bistre /ˈbɪstə/ *noun* brown pigment made from soot.

bistro /ˈbi:strəʊ/ *noun* (*plural* **-s**) small informal restaurant.

bit[1] *noun* small piece or amount; short time or distance; mouthpiece of bridle; cutting part of tool etc. □ **bit by bit** gradually.

bit[2] *past* of BITE.

bit[3] *noun Computing* unit of information expressed as choice between two possibilities.

bitch /bɪtʃ/ • *noun* female dog, fox, or wolf; *offensive slang* spiteful woman. • *verb colloquial* speak spitefully; grumble.

bitchy *adjective slang* spiteful. □ **bitchily** *adverb*; **bitchiness** *noun*.

bite • *verb* (**-ting**; *past* **bit**; *past participle* **bitten**) nip or cut into or off with teeth; sting; penetrate, grip; accept bait; be harsh in effect, esp. intentionally. • *noun* act of biting; wound so made; small amount to eat; pungency; incisiveness.

bitter /'bɪtə/ • *adjective* having sharp pungent taste, not sweet; showing or feeling resentment; harsh, virulent; piercingly cold. • *noun* bitter beer, strongly flavoured with hops; (in *plural*) liquor with bitter flavour, esp. of wormwood. □ **bitterly** *adverb*; **bitterness** *noun*.

bittern /'bɪt(ə)n/ *noun* wading bird of heron family.

bitty *adjective* (**-ier**, **-iest**) made up of bits.

bitumen /'bɪtjʊmɪn/ *noun* tarlike mixture of hydrocarbons derived from petroleum. □ **bituminous** /bɪ'tju:mɪnəs/ *adjective*.

bivalve /'baɪvælv/ • *noun* aquatic mollusc with hinged double shell. • *adjective* with such a shell.

bivouac /'bɪvʊæk/ • *noun* temporary encampment without tents. • *verb* (**-ck-**) make, or camp in, bivouac.

bizarre /bɪ'zɑ:/ *adjective* strange in appearance or effect; grotesque.

blab *verb* (**-bb-**) talk or tell foolishly or indiscreetly.

black • *adjective* colourless from absence or complete absorption of light; very dark-coloured; of human group with dark skin, esp. African; heavily overcast; angry, gloomy; sinister, wicked; declared untouchable by workers in dispute. • *noun* black colour, paint, clothes, etc.; (player using) darker pieces in chess etc.; (of tea or coffee) without milk; member of dark-skinned race, esp. African. • *verb* make black; declare (goods etc.) 'black'. □ **black and blue** badly bruised; **black and white** not in colour, comprising only opposite extremes, (after *in*) in print; **black art** black magic; **blackball** exclude from club, society, etc.; **black beetle** common cockroach; **black belt** (holder of) highest grade of proficiency in judo, karate, etc.; **black box** flight recorder; **black comedy** comedy presenting tragedy in comic terms; **black eye** bruised skin round eye; **blackfly** kind of aphid; **Black Forest gateau** chocolate sponge with cherries and whipped cream; **blackhead** black-topped pimple; **black hole** region of space from which matter and radiation cannot escape; **black ice** thin hard transparent ice; **blacklead** graphite; **blackleg** *derogatory* person refusing to join strike etc.; **black magic** magic supposed to invoke evil spirits; **Black Maria** *slang* police van; **black market** illicit traffic in rationed, prohibited, or scarce commodities; **Black Mass** travesty of Mass in worship of Satan; **black out** effect blackout on, undergo blackout; **blackout** temporary loss of consciousness or memory; loss of electric power, radio reception, etc.; compulsory darkness as precaution against air raids; **black pudding** sausage of blood, suet, etc.; **Black Rod** chief usher of House of Lords; **black sheep** disreputable member; **blackshirt** *historical* Fascist; **black spot** place of danger or trouble; **blackthorn** thorny shrub bearing white flowers and sloes; **black tie** man's formal evening dress; **black velvet** mixture of stout and champagne; **black widow** venomous spider of which female devours male; **in the black** in credit or surplus.

blackberry *noun* (*plural* **-ies**) dark edible fruit of bramble.

blackbird *noun* European songbird.

blackboard *noun* board for chalking on in classroom etc.

blackcurrant *noun* small black fruit; shrub on which it grows.

blacken *verb* make or become black; slander.

blackguard /'blægɑ:d/ *noun* villain, scoundrel. □ **blackguardly** *adjective*.

blacking *noun* black polish for boots and shoes.

blacklist ● *noun* list of people etc. in disfavour. ● *verb* put on blacklist.

blackmail ● *noun* extortion of payment in return for silence; use of threats or pressure. ● *verb* extort money from thus. □ **blackmailer** *noun*.

blacksmith *noun* smith working in iron.

bladder /'blædə/ *noun* sac in humans and other animals, esp. that holding urine.

blade *noun* cutting part of knife etc.; flat part of oar, spade, propeller, etc.; flat narrow leaf of grass and cereals; flat bone in shoulder.

blame ● *verb* (**-ming**) assign fault or responsibility to; (+ *on*) assign responsibility for (error etc.) to. ● *noun* responsibility for bad result; blaming, attributing of responsibility. □ **blameworthy** deserving blame. □ **blameless** *adjective*.

blanch /blɑ:ntʃ/ *verb* make or grow pale; peel (almonds etc.) by scalding; immerse briefly in boiling water; whiten (plant) by depriving it of light.

blancmange /blə'mɒndʒ/ *noun* sweet opaque jelly of flavoured cornflour and milk.

bland *adjective* mild; insipid, tasteless; gentle, suave. □ **blandly** *adverb*.

blandishment /'blændɪʃmənt/ *noun* (usually in *plural*) flattering attention; cajolery.

blank ● *adjective* not written or printed on; (of form etc.) not filled in; (of space) empty; without interest, result, or expression. ● *noun* space to be filled up in form etc.; blank cartridge; empty surface; dash written in place of word. □ **blank cartridge** one without bullet; **blank cheque** one with amount left for payee to fill in; **blank verse** unrhymed verse, esp. iambic pentameters. □ **blankly** *adverb*.

blanket /'blæŋkɪt/ ● *noun* large esp. woollen sheet as bed-covering etc.; thick covering layer. ● *adjective* general, covering all cases or classes. ● *verb* (**-t-**) cover (as) with blanket. □ **blanket stitch** stitch used to finish edges of blanket etc.

blare ● *verb* (**-ring**) sound or utter loudly; make sound of trumpet. ● *noun* blaring sound.

blarney /'blɑ:nɪ/ ● *noun* cajoling talk, flattery. ● *verb* (**-eys**, **-eyed**) flatter, use blarney.

blasé /'blɑ:zeɪ/ *adjective* bored or indifferent, esp. through familiarity.

blaspheme /blæs'fi:m/ *verb* (**-ming**) treat religious name or subject irreverently; talk irreverently about.

blasphemy /'blæsfəmɪ/ *noun* (*plural* **-ies**) (instance of) blaspheming. □ **blasphemous** *adjective*.

blast /blɑ:st/ ● *noun* strong gust; explosion; destructive wave of air from this; loud note from wind instrument, car horn, whistle, etc.; *colloquial* severe reprimand. ● *verb* blow up with explosive; blight; (cause to) make explosive sound. ● *interjection* damn. □ **blast furnace** one for smelting with compressed hot air driven in; **blast off** (of rocket) take off from launching site; **blast-off** *noun*.

blasted *colloquial* ● *adjective* annoying. ● *adverb* extremely.

blatant /'bleɪt(ə)nt/ *adjective* flagrant, unashamed. □ **blatantly** *adverb*.

blather /'blæðə/ ● *noun* (also **blether** /'bleðə/) foolish talk. ● *verb* talk foolishly.

blaze[1] ● *noun* bright flame or fire; violent outburst of passion; bright display or light. ● *verb* (**-zing**) burn or shine brightly or fiercely; burn with excitement etc. □ **blaze away** shoot continuously.

blaze[2] ● *noun* white mark on face of horse or chipped in bark of tree. ● *verb* (**-zing**) mark (tree, path) with blaze(s).

blazer *noun* jacket without matching trousers, esp. lightweight and part of uniform.

blazon /'bleɪz(ə)n/ *verb* proclaim; describe or paint (coat of arms). □ **blazonry** *noun*.

bleach ● *verb* whiten in sunlight or by chemical process. ● *noun* bleaching substance or process.

bleak *adjective* exposed, windswept; dreary, grim.

bleary /ˈblɪərɪ/ *adjective* (**-ier**, **-iest**) dim-sighted, blurred. □ **bleary-eyed** having dim sight.

bleat ● *verb* utter cry of sheep, goat, etc.; speak plaintively. ● *noun* bleating cry.

bleed *verb* (*past & past participle* **bled**) emit blood; draw blood from; *colloquial* extort money from.

bleep ● *noun* intermittent high-pitched sound. ● *verb* make bleep; summon by bleep.

bleeper *noun* small electronic device alerting person to message by bleeping.

blemish /ˈblemɪʃ/ ● *noun* flaw, defect, stain. ● *verb* spoil, mark, stain.

blench *verb* flinch, quail.

blend ● *verb* mix (various sorts) into required sort; become one; mingle intimately. ● *noun* mixture.

blender *noun* machine for liquidizing or chopping food.

blenny /ˈblenɪ/ *noun* (*plural* **-ies**) spiny-finned sea fish.

bless *verb* ask God to look favourably on; consecrate; glorify (God); thank; make happy.

blessed /ˈblesɪd/ *adjective* holy; *euphemistic* cursed; *RC Church* beatified.

blessing *noun* invocation of divine favour; grace said at meals; benefit, advantage.

blether = BLATHER.

blew *past* of BLOW[1].

blight /blaɪt/ ● *noun* disease of plants caused esp. by insects; such insect; harmful or destructive force. ● *verb* affect with blight; destroy; spoil.

blighter *noun colloquial* annoying person.

blimey /ˈblaɪmɪ/ *interjection slang*: *expressing surprise.*

blimp *noun* small non-rigid airship; (also (**Colonel**) **Blimp**) reactionary person.

blind /blaɪnd/ ● *adjective* without sight; without adequate foresight, discernment, or information; (often + *to*) unwilling or unable to appreciate (a factor); not governed by purpose; reckless; concealed; closed at one end. ● *verb* deprive of sight or judgement; deceive. ● *noun* screen for window; thing used to hide truth; obstruction to sight or light. ● *adverb* blindly. □ **blind date** *colloquial* date between two people who have not met before; **blind man's buff** game in which blindfold player tries to catch others; **blind spot** spot on retina insensitive to light, area where vision or judgement fails; **blindworm** slow-worm. □ **blindly** *adverb*; **blindness** *noun.*

blindfold ● *verb* cover eyes of (person) with tied cloth etc. ● *noun* cloth etc. so used. ● *adjective & adverb* with eyes covered; without due care.

blink ● *verb* shut and open eyes quickly; (often + *back*) prevent (tears) by blinking; shine unsteadily, flicker. ● *noun* act of blinking; momentary gleam. □ **blink at** ignore, shirk; **on the blink** *slang* (of machine etc.) out of order.

blinker ● *noun* (usually in *plural*) either of screens on bridle preventing horse from seeing sideways. ● *verb* obscure with blinkers; (as **blinkered** *adjective*) prejudiced, narrow-minded.

blinking *adjective & adverb slang*: *expressing mild annoyance.*

blip *noun* minor deviation or error; quick popping sound; small image on radar screen.

bliss *noun* perfect joy; being in heaven. □ **blissful** *adjective*; **blissfully** *adverb.*

blister /ˈblɪstə/ ● *noun* small bubble on skin filled with watery fluid; any swelling resembling this. ● *verb* become covered with blisters; raise blister on.

blithe /blaɪð/ *adjective* cheerful, happy; carefree, casual. □ **blithely** *adverb.*

blithering /ˈblɪðərɪŋ/ *adjective colloquial* utter, hopeless; contemptible.

blitz /blɪts/ *colloquial* ● *noun* intensive (esp. aerial) attack. ● *verb* inflict blitz on.

blizzard /ˈblɪzəd/ *noun* severe snowstorm.

bloat *verb* inflate, swell.

bloater *noun* herring cured by salting and smoking.

blob *noun* small drop or spot.

bloc *noun* group of governments etc. sharing some common purpose.

block ● *noun* solid piece of hard material; large building, esp. when subdivided; group of buildings surrounded by streets; obstruction; large quantity as a unit; piece of wood or metal engraved for printing. ● *verb* obstruct; restrict use of; stop (cricket ball) with bat.

□ **blockbuster** *slang* very successful film, book, etc.; **blockhead** stupid person; **block capitals, letters** separate capital letters; **block out** shut out (light, noise, view, etc.); **block up** shut in, fill (window etc.) in; **block vote** vote proportional in size to number of people voter represents; **mental block** mental inability due to subconscious factors.

blockade /blɒ'keɪd/ ● *noun* surrounding or blocking of place by enemy. ● *verb* (**-ding**) subject to blockade.

blockage *noun* obstruction.

bloke *noun slang* man, fellow.

blond (of woman, usually **blonde**) ● *adjective* light-coloured, fair-haired. ● *noun* blond person.

blood /blʌd/ ● *noun* fluid, usually red, circulating in arteries and veins of animals; killing, bloodshed; passion, temperament; race, descent; relationship. ● *verb* give first taste of blood to (hound); initiate (person). □ **blood bank** store of blood for transfusion; **bloodbath** massacre; **blood count** number of corpuscles in blood; **blood-curdling** horrifying; **blood donor** giver of blood for transfusion; **blood group** any of types of human blood; **bloodhound** large keen-scented dog used for tracking; **bloodletting** surgical removal of blood; **blood orange** red-fleshed orange; **blood poisoning** diseased condition due to bacteria in blood; **blood pressure** pressure of blood in arteries; **blood relation** one related by birth; **bloodshed** killing; **bloodshot** (of eyeball) inflamed; **blood sport** one involving killing of animals; **bloodstream** circulating blood; **bloodsucker** leech, extortioner; **bloodthirsty** eager for bloodshed; **blood vessel** vein, artery, or capillary carrying blood.

bloodless *adjective* without blood or bloodshed; unemotional; pale.

bloody ● *adjective* (**-ier**, **-iest**) of or like blood; running or stained with blood; involving bloodshed, cruel; *coarse slang* annoying, very great. ● *adverb coarse slang* extremely. ● *verb* (**-ies**, **-ied**) stain with blood. □ **bloody-minded** *colloquial* deliberately uncooperative.

bloom /blu:m/ ● *noun* flower; flowering state; prime; freshness; fine powder on fruit etc. ● *verb* bear flowers; be in flower; flourish.

bloomer¹ *noun slang* blunder.

bloomer² *noun* long loaf with diagonal marks.

bloomers *plural noun colloquial* woman's long loose knickers.

blooming ● *adjective* flourishing, healthy; *slang* annoying, very great. ● *adverb slang* extremely.

blossom /'blɒsəm/ ● *noun* flower; mass of flowers on tree. ● *verb* open into flower; flourish.

blot ● *noun* spot of ink etc.; disgraceful act; blemish. ● *verb* (**-tt-**) make blot on; stain; dry with blotting paper. □ **blot out** obliterate, obscure; **blotting paper** absorbent paper for drying wet ink.

blotch *noun* inflamed patch on skin; irregular patch of colour. □ **blotchy** *adjective* (**-ier**, **-iest**).

blotter *noun* device holding blotting paper.

blouse /blaʊz/ *noun* woman's shirtlike garment; type of military jacket.

blow¹ /bləʊ/ ● *verb* (*past* **blew** /blu:/; *past participle* **blown**) send directed air-current esp. from mouth; drive or be driven by blowing; move as wind does; sound (wind instrument); (*past participle* **blowed**) *slang* curse, confound; clear (nose) by forceful breath; pant; make or shape by blowing; break or burst suddenly; (cause to) break electric circuit; *slang* squander. ● *noun* blowing; short spell in fresh air. □ **blow-dry** arrange (hair) while drying it; **blowfly** bluebottle; **blow in** send inwards by explosion, *colloquial* arrive unexpectedly; **blowlamp** device with flame for plumbing, burning off paint, etc.; **blow out** extinguish by blowing, send outwards by explosion; **blow-out** *colloquial* burst tyre, *slang* large meal; **blow over** fade away; **blowpipe** tube for blowing through, esp. one from which dart or arrow is projected; **blowtorch** *US* blowlamp; **blow up** explode, *colloquial* rebuke strongly, inflate, *colloquial* enlarge (photograph); **blow-up** *colloquial* enlargement of photograph.

blow² /bləʊ/ *noun* hard stroke with hand or weapon; disaster, shock.

blower /ˈbləʊə/ *noun* device for blowing; *colloquial* telephone.

blowy /ˈbləʊɪ/ *adjective* (**-ier, -iest**) windy.

blowzy /ˈblaʊzɪ/ *adjective* (**-ier, -iest**) coarse-looking, red-faced; dishevelled.

blub *verb* (**-bb-**) *slang* sob.

blubber /ˈblʌbə/ ● *noun* whale fat. ● *verb* sob noisily. □ **blubbery** *adjective.*

bludgeon /ˈblʌdʒ(ə)n/ ● *noun* heavy club. ● *verb* beat with bludgeon; coerce.

blue /bluː/ ● *adjective* (**-r, -st**) coloured like clear sky; sad, depressed; pornographic. ● *noun* blue colour, paint, clothes, etc.; person who represents Oxford or Cambridge University at sport; (in *plural*) type of melancholy music of American black origin, (**the blues**) depression. ● *verb* (**blues, blued, bluing** or **blueing**) *slang* squander. □ **blue baby** one with congenital heart defect; **bluebell** woodland plant with bell-shaped blue flowers; **blueberry** small edible blue or blackish fruit of various plants; **blue blood** noble birth; **bluebottle** large buzzing fly; **blue cheese** cheese with veins of blue mould; **blue-collar** manual, industrial; **blue-eyed boy** *colloquial* favourite; **bluegrass** type of country and western music; **Blue Peter** blue flag with central white square hoisted before sailing; **blueprint** photographic print of building plans etc. in white on blue paper, detailed plan; **bluestocking** *usually derogatory* intellectual woman; **blue tit** small blue and yellow bird; **blue whale** rorqual (largest known living mammal).

bluff[1] ● *verb* pretend to have strength, knowledge, etc. ● *noun* bluffing.

bluff[2] ● *adjective* blunt, frank, hearty; with steep or vertical broad front. ● *noun* bluff headland.

blunder /ˈblʌndə/ ● *noun* serious or foolish mistake. ● *verb* make blunder; move about clumsily.

blunderbuss /ˈblʌndəbʌs/ *noun historical* short large-bored gun.

blunt ● *adjective* without sharp edge or point; plain-spoken. ● *verb* make blunt. □ **bluntly** *adverb*; **bluntness** *noun.*

blur ● *verb* (**-rr-**) make or become less distinct; smear. ● *noun* indistinct object, sound, memory, etc.

blurb *noun* promotional description, esp. of book.

blurt *verb* (usually + *out*) utter abruptly or tactlessly.

blush ● *verb* be or become red (as) with shame or embarrassment; be ashamed. ● *noun* blushing; pink tinge.

blusher *noun* coloured cosmetic for cheeks.

bluster /ˈblʌstə/ ● *verb* behave pompously; storm boisterously. ● *noun* noisy pompous talk, empty threats. □ **blustery** *adjective.*

BMA *abbreviation* British Medical Association.

BMX *noun* organized bicycle racing on dirt track; bicycle used for this.

BO *abbreviation colloquial* body odour.

boa /ˈbəʊə/ *noun* large snake that kills its prey by crushing it; long stole of fur or feathers. □ **boa constrictor** species of boa.

boar *noun* male wild pig; uncastrated male pig.

board ● *noun* thin piece of sawn timber; material resembling this; slab of wood etc., e.g. ironing board, notice board; thick stiff card; provision of meals; directors of company; committee; (**the boards**) stage. ● *verb* go on board (ship etc.); receive, or provide with, meals and usually lodging; (usually + *up*) cover with boards. □ **board game** game played on a board; **boarding house** unlicensed house providing board and lodging; **boarding school** one in which pupils live in term-time; **boardroom** room where board of directors meets; **on board** on or into ship, train, aircraft, etc.

boarder *noun* person who boards, esp. at boarding school.

boast ● *verb* declare one's achievements etc. with excessive pride; have (desirable thing). ● *noun* boasting; thing one is proud of. □ **boastful** *adjective.*

boat ● *noun* small vessel propelled by oars, sails, or engine; ship; long low jug for sauce etc. ● *verb* go in boat, esp. for pleasure. □ **boat-hook** long pole with hook for moving boats; **boathouse** shed at water's edge for boats; **boatman** person who hires out or provides transport by boats; **boat people**

refugees travelling by sea; **boat-train** train scheduled to connect with ship.

boater *noun* flat straw hat with straight brim.

boatswain /'bəʊs(ə)n/ *noun* (also **bosun**) ship's officer in charge of equipment and crew.

bob[1] • *verb* (**-bb-**) move up and down; rebound; cut (hair) in bob; curtsy. • *noun* bobbing movement; curtsy; hairstyle with hair hanging evenly above shoulders; weight on pendulum etc. □ **bobtail** docked tail, horse or dog with this.

bob[2] *noun* (*plural* same) *historical slang* shilling (= 5p).

bobbin /'bɒbɪn/ *noun* spool or reel for thread etc.

bobble /'bɒb(ə)l/ *noun* small woolly ball on hat etc.

bobby /'bɒbɪ/ *noun* (*plural* **-ies**) *colloquial* police officer.

bobsled *noun US* bobsleigh.

bobsleigh *noun* racing sledge steered and braked mechanically.

bod *noun colloquial* person.

bode *verb* (**-ding**) □ **bode well, ill** be good or bad sign.

bodge /bɒdʒ/ = BOTCH.

bodice /'bɒdɪs/ *noun* part of woman's dress above waist.

bodily /'bɒdɪlɪ/ • *adjective* of the body. • *adverb* as a whole (body); in person.

bodkin /'bɒdkɪn/ *noun* blunt thick needle for drawing tape etc. through hem.

body /'bɒdɪ/ *noun* (*plural* **-ies**) physical structure of person or animal, alive or dead; person's or animal's trunk; main part; group of people regarded as unit; quantity, mass; piece of matter; *colloquial* person; full or substantial quality of flavour etc.; body stocking. □ **body-building** exercises to enlarge and strengthen muscles; **bodyguard** escort or personal guard; **body politic** state, nation; **body shop** workshop where bodywork is repaired; **body stocking** woman's undergarment covering trunk; **bodywork** outer shell of vehicle.

Boer /'bɔ:/ *noun* S. African of Dutch descent.

boffin /'bɒfɪn/ *noun colloquial* research scientist.

bog *noun* (area of) wet spongy ground; *slang* lavatory. □ **bogged down** unable to move or make progress. □ **boggy** *adjective* (**-ier**, **-iest**).

bogey[1] /'bəʊgɪ/ *noun* (*plural* **-s**) *Golf* score of one more than par for hole; (formerly) par.

bogey[2] /'bəʊgɪ/ *noun* (also **bogy**) (*plural* **-eys** or **-ies**) evil or mischievous spirit; awkward thing.

boggle /'bɒg(ə)l/ *verb* (**-ling**) *colloquial* be startled or baffled.

bogie /'bəʊgɪ/ *noun* wheeled undercarriage below locomotive etc.

bogus /'bəʊgəs/ *adjective* sham, spurious.

bogy = BOGEY[2].

Bohemian /bəʊ'hi:mɪən/ • *noun* native of Bohemia; (also **bohemian**) socially unconventional person, esp. artist or writer. • *adjective* of Bohemia; (also **bohemian**) socially unconventional. □ **bohemianism** *noun*.

boil[1] • *verb* (of liquid or its vessel) bubble up with heat, reach temperature at which liquid turns to vapour; bring to boiling point; subject to heat of boiling water, cook thus; be agitated like boiling water. • *noun* boiling; boiling point. □ **boiling (hot)** *colloquial* very hot; **boiling point** temperature at which a liquid boils; **boil over** spill over in boiling.

boil[2] *noun* inflamed pus-filled swelling under skin.

boiler *noun* apparatus for heating hot-water supply; tank for heating water or turning it into steam; vessel for boiling things in. □ **boiler suit** protective garment combining trousers and shirt.

boisterous /'bɔɪstərəs/ *adjective* noisily cheerful; violent, rough.

bold /bəʊld/ *adjective* confident; adventurous; courageous; impudent; distinct, vivid. □ **boldly** *adverb*; **boldness** *noun*.

bole *noun* trunk of tree.

bolero *noun* (*plural* **-s**) /bə'leərəʊ/ Spanish dance; /'bɒlərəʊ/ woman's short jacket without fastenings.

boll /bəʊl/ *noun* round seed vessel of cotton, flax, etc.

bollard /'bɒlɑ:d/ *noun* short thick post in street etc.; post on ship or quay for securing ropes to.

boloney /bə'ləʊnɪ/ *noun slang* nonsense.

bolshie /'bɒlʃɪ/ *adjective* (**-r**, **-st**) (also **Bolshie**) *slang* rebellious, uncooperative.

bolster /'bəʊlstə/ ● *noun* long cylindrical pillow. ● *verb* (usually + *up*) encourage, support, prop up.

bolt¹ /bəʊlt/ ● *noun* door-fastening of metal bar and socket; headed metal pin secured with rivet or nut; discharge of lightning; bolting. ● *verb* fasten with bolt; (+ *in*, *out*) keep in or out by bolting door; dart off, run away; (of horse) escape from control; gulp down unchewed; run to seed. □ **bolt-hole** means of escape; **bolt upright** erect.

bolt² /bəʊlt/ *verb* (also **boult**) sift.

bomb /bɒm/ ● *noun* container filled with explosive, incendiary material, etc., designed to explode and cause damage; (**the bomb**) the atomic bomb; *slang* large amount of money. ● *verb* attack with bombs; drop bombs on; *colloquial* travel fast. □ **bombshell** great surprise or disappointment.

bombard /bɒm'bɑ:d/ *verb* attack with heavy guns etc.; question or abuse persistently; subject to stream of high-speed particles. □ **bombardment** *noun*.

bombardier /bɒmbə'dɪə/ *noun* artillery NCO below sergeant; *US* airman who releases bombs from aircraft.

bombast /'bɒmbæst/ *noun* pompous or extravagant language. □ **bombastic** /-'bæs-/ *adjective*.

Bombay duck /'bɒmbeɪ/ *noun* dried fish eaten as relish, esp. with curry.

bomber *noun* aircraft equipped for bombing; person who throws or plants bomb. □ **bomber jacket** one gathered at waist and cuffs.

bona fide /bəʊnə 'faɪdɪ/ *adjective & adverb* in good faith, genuine(ly).

bonanza /bə'nænzə/ *noun* source of great wealth; large output, esp. from mine.

bon-bon *noun* sweet.

bond ● *noun* thing or force that unites or (usually in *plural*) restrains; binding agreement; certificate issued by government or company promising to repay money at fixed rate of interest; adhesiveness; document binding person to pay or repay money; linkage of atoms in molecule. ● *verb* bind or connect together; put in bond. □ **bond paper** high-quality writing paper; **in bond** stored by Customs until duty is paid.

bondage /'bɒndɪdʒ/ *noun* slavery; subjection to constraint.

bondsman /'bɒndzmən/ *noun* serf, slave.

bone ● *noun* any of separate parts of vertebrate skeleton; (in *plural*) skeleton, esp. as remains; substance of which bones consist. ● *verb* (**-ning**) remove bones from. □ **bone china** fine semi-translucent earthenware; **bone dry** completely dry; **bone idle** completely idle; **bone marrow** fatty substance in cavity of bones; **bonemeal** crushed bone as fertilizer; **boneshaker** jolting vehicle.

bonfire *noun* open-air fire.

bongo /'bɒŋgəʊ/ *noun* (*plural* **-s** or **-es**) either of pair of small drums played with fingers.

bonhomie /bɒnɒ'mi:/ *noun* geniality.

bonkers /'bɒŋkəz/ *adjective slang* crazy.

bonnet /'bɒnɪt/ *noun* woman's or child's hat tied under chin; Scotsman's floppy beret; hinged cover over engine of vehicle.

bonny /'bɒnɪ/ *adjective* (**-ier**, **-iest**) *esp. Scottish & Northern English* healthy-looking, attractive.

bonsai /'bɒnsaɪ/ *noun* (*plural* same) dwarfed tree or shrub; art of growing these.

bonus /'bəʊnəs/ *noun* extra benefit or payment.

bon voyage /bɔ̃ vwɑ:'jɑ:ʒ/ *interjection* have a good trip. [French]

bony /'bəʊnɪ/ *adjective* (**-ier**, **-iest**) thin with prominent bones; having many bones; of or like bone.

boo ● *interjection expressing disapproval or contempt; sound intended to surprise*. ● *noun* utterance of 'boo'. ● *verb* (**boos**, **booed**) utter boos (at).

boob ● *noun colloquial* silly mistake; *slang* woman's breast. ● *verb colloquial* make mistake.

booby /'bu:bɪ/ *noun* (*plural* **-ies**) silly or awkward person. □ **booby prize** prize for coming last; **booby trap** practical joke in form of trap, disguised bomb etc. triggered by unknowing victim.

book /bʊk/ ● *noun* written or printed work with pages bound along one side; work intended for publication; bound set of tickets, stamps, matches, cheques, etc.; (in *plural*) set of records or accounts; main division of literary work or Bible; telephone directory; *colloquial* magazine; libretto; record of bets made. ● *verb* reserve (seat etc.) in advance; engage (entertainer etc.); take personal details of (offender); enter in book or list. □ **bookcase** cabinet of shelves for books; **bookend** prop to keep books upright; **bookkeeper** person who keeps accounts; **bookmaker** professional taker of bets; **bookmark** thing for marking place in book; **bookplate** decorative personalized label in book; **book token** voucher exchangeable for books; **bookworm** *colloquial* devoted reader, larva that eats through books.

bookie /'bʊkɪ/ *noun colloquial* bookmaker.

bookish *adjective* fond of reading; getting knowledge mainly from books.

booklet /'bʊklɪt/ *noun* small usually paper-covered book.

boom[1] /bu:m/ ● *noun* deep resonant sound. ● *verb* make or speak with boom.

boom[2] /bu:m/ ● *noun* period of economic prosperity or activity. ● *verb* be suddenly prosperous.

boom[3] /bu:m/ *noun* pivoted spar to which sail is attached; long pole carrying camera, microphone, etc.; barrier across harbour etc.

boomerang /'bu:məræŋ/ ● *noun* flat V-shaped Australian hardwood missile returning to thrower. ● *verb* (of plan) backfire.

boon[1] /bu:n/ *noun* advantage; blessing.

boon[2] /bu:n/ *adjective* intimate, favourite.

boor /bʊə/ *noun* ill-mannered person. □ **boorish** *adjective*.

boost /bu:st/ *colloquial* ● *verb* promote, encourage; increase, assist; push from below. ● *noun* boosting.

booster *noun* device for increasing power or voltage; auxiliary engine or rocket for initial speed; dose renewing effect of earlier one.

boot /bu:t/ ● *noun* outer foot-covering reaching above ankle; luggage compartment of car; *colloquial* firm kick, dismissal. ● *verb* kick; (often + *out*) eject forcefully; (usually + *up*) make (computer) ready.

bootee /bu:'ti:/ *noun* baby's soft shoe.

booth /bu:ð/ *noun* temporary structure used esp. as market stall; enclosure for telephoning, voting, etc.

bootleg /'bu:tleg/ ● *adjective* smuggled, illicit. ● *verb* (**-gg-**) illicitly make or deal in. □ **bootlegger** *noun*.

bootstrap /'bu:tstræp/ *noun* □ **pull oneself up by one's bootstraps** better oneself by one's unaided effort.

booty /'bu:tɪ/ *noun* loot, spoils; *colloquial* prize.

booze *colloquial* ● *noun* alcoholic drink. ● *verb* (**-zing**) drink alcohol, esp. excessively. □ **boozy** *adjective* (**-ier**, **-iest**).

boozer *noun colloquial* habitual drinker; public house.

bop *colloquial* ● *noun* spell of dancing, esp. to pop music; *colloquial* hit, blow. ● *verb* (**-pp-**) dance, esp. to pop music; *colloquial* hit.

boracic /bə'ræsɪk/ *adjective* of borax. □ **boracic acid** boric acid.

borage /'bɒrɪdʒ/ *noun* plant with leaves used as flavouring.

borax /'bɔ:ræks/ *noun* salt of boric acid used as antiseptic.

border /'bɔ:də/ ● *noun* edge, boundary, or part near it; line or region separating countries; distinct edging, esp. ornamental strip; long narrow flower-bed. ● *verb* put or be border to; adjoin. □ **borderline** *noun* line marking boundary or dividing two conditions, *adjective* on borderline.

bore[1] ● *verb* (**-ring**) make (hole), esp. with revolving tool; make hole in. ● *noun* hollow of firearm barrel or cylinder; diameter of this; deep hole made to find water etc.

bore[2] ● *noun* tiresome or dull person or thing. ● *verb* (**-ring**) weary by tedious talk or dullness. □ **bored** *adjective*; **boring** *adjective*.

bore[3] *noun* very high tidal wave rushing up estuary.

bore[4] *past* of BEAR[1].

boredom *noun* being bored (BORE[2]).

boric acid /'bɔ:rɪk/ *noun* acid used as antiseptic.

born *adjective* existing as a result of birth; being (specified thing) by nature; (usually + *to do*) destined.

borne *past participle* of BEAR[1].

boron /ˈbɔːrɒn/ *noun* non-metallic element.

borough /ˈbʌrə/ *noun* administrative area, esp. of Greater London; *historical* town with municipal corporation.

borrow /ˈbɒrəʊ/ *verb* get temporary use of (something to be returned); use another's (invention, idea, etc.). □ **borrower** *noun*.

Borstal /ˈbɔːst(ə)l/ *noun historical* residential institution for youth custody.

bosom /ˈbʊz(ə)m/ *noun* person's breasts; *colloquial* each of woman's breasts; enclosure formed by breast and arms; emotional centre. □ **bosom friend** intimate friend.

boss[1] *colloquial* ● *noun* employer, manager, or supervisor. ● *verb* (usually + *about*, *around*) give orders to.

boss[2] *noun* round knob or stud.

boss-eyed *adjective colloquial* cross-eyed; crooked.

bossy *adjective* (**-ier**, **-iest**) *colloquial* domineering. □ **bossiness** *noun*.

bosun = BOATSWAIN.

botany /ˈbɒtənɪ/ *noun* study of plants. □ **botanic(al)** /bəˈtæn-/ *adjective*; **botanist** *noun*.

botch ● *verb* bungle; patch clumsily. ● *noun* bungled or spoilt work.

both /bəʊθ/ ● *adjective & pronoun* the two (not only one). ● *adverb* with equal truth in two cases.

bother /ˈbɒðə/ ● *verb* trouble, worry; take trouble. ● *noun* person or thing that bothers; nuisance; trouble, worry. ● *interjection of irritation*. □ **bothersome** /-səm/ *adjective*.

bottle /ˈbɒt(ə)l/ ● *noun* container, esp. glass or plastic, for storing liquid; liquid in bottle; *slang* courage. ● *verb* (**-ling**) put into bottles; preserve (fruit etc.) in jars; (+ *up*) restrain (feelings etc.). □ **bottle bank** place for depositing bottles for recycling; **bottle green** dark green; **bottleneck** narrow congested area esp. on road etc., thing that impedes; **bottle party** one to which guests bring bottles of drink.

bottom /ˈbɒtəm/ ● *noun* lowest point or part; buttocks; less honourable end of table, class, etc.; ground under water; basis; essential character. ● *adjective* lowest, last. ● *verb* (usually + *out*) reach its lowest level; find extent of; touch bottom (of). □ **bottom line** *colloquial* underlying truth, ultimate criterion.

bottomless *adjective* without bottom; inexhaustible.

botulism /ˈbɒtjʊlɪz(ə)m/ *noun* poisoning caused by bacillus in badly preserved food.

boudoir /ˈbuːdwɑː/ *noun* woman's private room.

bougainvillaea /buːgənˈvɪlɪə/ *noun* tropical plant with large coloured bracts.

bough /baʊ/ *noun* branch of tree.

bought *past & past participle* of BUY.

bouillon /ˈbuːjɒn/ *noun* clear broth.

boulder /ˈbəʊldə/ *noun* large smooth rock.

boulevard /ˈbuːləvɑːd/ *noun* broad tree-lined street.

boult = BOLT[2].

bounce /baʊns/ ● *verb* (**-cing**) (cause to) rebound; *slang* (of cheque) be returned to payee by bank when there are no funds to meet it; rush boisterously. ● *noun* rebound; *colloquial* swagger, self-confidence; *colloquial* liveliness. □ **bouncy** *adjective* (**-ier**, **-iest**).

bouncer *noun slang* doorman ejecting troublemakers from nightclub etc.

bouncing *adjective* big and healthy.

bound[1] /baʊnd/ ● *verb* spring, leap; (of ball etc.) bounce. ● *noun* springy leap; bounce.

bound[2] /baʊnd/ ● *noun* (usually in *plural*) limitation, restriction; border, boundary. ● *verb* limit; be boundary of. □ **out of bounds** outside permitted area.

bound[3] /baʊnd/ *adjective* (usually + *for*) starting or having started.

bound[4] /baʊnd/ *past & past participle* of BIND. □ **bound to** certain to; **bound up with** closely associated with.

boundary /ˈbaʊndərɪ/ *noun* (*plural* **-ies**) line marking limits; *Cricket* hit crossing limit of field, runs scored for this.

boundless *adjective* unlimited.

bounteous /ˈbaʊntɪəs/ *adjective poetical* bountiful.

bountiful /'baʊntɪfʊl/ *adjective* generous; ample.

bounty /'baʊntɪ/ *noun* (*plural* **-ies**) generosity; official reward; gift.

bouquet /bu:'keɪ/ *noun* bunch of flowers; scent of wine; compliment. □ **bouquet garni** (/'gɑ:nɪ/) bunch or bag of herbs for flavouring.

bourbon /'bɜ:bən/ *noun US* whisky from maize and rye.

bourgeois /'bʊəʒwɑ:/ *often derogatory* ● *adjective* conventionally middle-class; materialist; capitalist. ● *noun* (*plural* same) bourgeois person.

bourgeoisie /bʊəʒwɑ:'zi:/ *noun* bourgeois class.

bourn /bɔ:n/ *noun* stream.

bourse /bʊəs/ *noun* money market, esp. (**Bourse**) Stock Exchange in Paris.

bout *noun* spell of work etc.; fit of illness; wrestling or boxing match.

boutique /bu:'ti:k/ *noun* small shop selling fashionable clothes etc.

bouzouki /bu:'zu:kɪ/ *noun* (*plural* **-s**) form of Greek mandolin.

bovine /'bəʊvaɪn/ *adjective* of cattle; dull, stupid.

bow[1] /bəʊ/ ● *noun* weapon for shooting arrows; rod with horsehair stretched from end to end for playing violin etc.; knot with two loops, ribbon etc. so tied; shallow curve or bend. ● *verb* use bow on (violin etc.). □ **bow-legged** having bandy legs; **bow tie** necktie in form of bow; **bow window** curved bay window.

bow[2] /baʊ/ ● *verb* incline head or body, esp. in greeting or acknowledgement; submit; incline (head etc.). ● *noun* bowing.

bow[3] /baʊ/ *noun* (often in *plural*) front end of boat or ship; rower nearest bow.

bowdlerize /'baʊdləraɪz/ *verb* (also **-ise**) (**-zing** or **-sing**) expurgate. □ **bowdlerization** *noun*.

bowel /'baʊəl/ *noun* (often in *plural*) intestine; (in *plural*) innermost parts.

bower /'baʊə/ *noun* arbour; summer house. □ **bowerbird** Australasian bird, the male of which constructs elaborate runs.

bowie knife /'bəʊɪ/ *noun* long hunting knife.

bowl[1] /bəʊl/ *noun* dish, esp. for food or liquid; hollow part of tobacco pipe, spoon, etc.

bowl[2] /bəʊl/ ● *noun* hard heavy ball made with bias to run in curve; (in *plural*, usually treated as *singular*) game with these on grass. ● *verb* roll (ball etc.); play bowls; *Cricket* deliver ball, (often + *out*) put (batsman) out by knocking off bails with bowled ball; (often + *along*) go along rapidly. □ **bowling alley** long enclosure for skittles or tenpin bowling; **bowling green** lawn for playing bowls.

bowler[1] /'bəʊlə/ *noun Cricket etc.* player who bowls.

bowler[2] /'bəʊlə/ *noun* hard round felt hat.

bowsprit /'bəʊsprɪt/ *noun* spar running forward from ship's bow.

box[1] ● *noun* container, usually flat-sided and firm; amount contained in this; compartment in theatre, law court, etc.; telephone box; facility at newspaper office for replies to advertisement; (**the box**) *colloquial* television; enclosed area or space. ● *verb* put in or provide with box. □ **box girder** hollow girder with square cross-section; **box junction** yellow-striped road area which vehicle may enter only if exit is clear; **box office** ticket office at theatre etc.; **box pleat** two parallel pleats forming raised band.

box[2] ● *verb* fight with fists as sport; slap (person's ears). ● *noun* slap on ear.

box[3] *noun* evergreen shrub with small dark green leaves; its wood.

boxer *noun* person who boxes; short-haired dog with puglike face. □ **boxer shorts** man's loose underpants.

boxing *noun* fighting with fists, esp. as sport. □ **boxing glove** padded glove worn in this.

Boxing Day *noun* first weekday after Christmas Day.

boy ● *noun* male child, young man; son; male servant. ● *interjection expressing pleasure, surprise, etc.* □ **boyfriend** person's regular male companion; **boy scout** Scout. □ **boyhood** *noun*; **boyish** *adjective*.

boycott /'bɔɪkɒt/ ● *verb* refuse social or commercial relations with; refuse to handle (goods). ● *noun* such refusal.

bra /brɑː/ *noun* woman's undergarment supporting breasts.

brace • *noun* device that clamps or fastens tightly; timber etc. strengthening framework; (in *plural*) straps supporting trousers from shoulders; wire device for straightening teeth; (*plural* same) pair. • *verb* (**-cing**) make steady by supporting; fasten tightly; (esp. as **bracing** *adjective*) invigorate; (often **brace oneself**) prepare for difficulty, shock, etc.

bracelet /ˈbreɪslɪt/ *noun* ornamental band or chain worn on wrist or arm; *slang* handcuff.

brachiosaurus /brækɪəˈsɔːrəs/ *noun* (*plural* **-ruses**) huge long-necked plant-eating dinosaur.

bracken /ˈbrækən/ *noun* large coarse fern; mass of these.

bracket /ˈbrækɪt/ • *noun* support projecting from vertical surface; shelf fixed to wall with this; punctuation mark used in pairs—( ), [ ]—enclosing words or figures (see panel); group classified as similar or falling between limits. • *verb* (**-t-**) enclose in brackets; group in same category.

brackish /ˈbrækɪʃ/ *adjective* (of water) slightly salty.

bract *noun* leaflike part of plant growing before flower.

brad *noun* thin flat nail.

bradawl /ˈbrædɔːl/ *noun* small boring-tool.

brae /breɪ/ *noun Scottish* hillside.

brag • *verb* (**-gg-**) talk boastfully. • *noun* card game like poker; boastful statement or talk.

braggart /ˈbrægət/ *noun* boastful person.

Brahma /ˈbrɑːmə/ *noun* Hindu Creator; supreme Hindu reality.

Brahman /ˈbrɑːmən/ *noun* (*plural* **-s**) (also **Brahmin**) member of Hindu priestly caste.

braid • *noun* woven band as edging or trimming; *US* plait of hair. • *verb US* plait; trim with braid.

Braille /breɪl/ *noun* system of writing and printing for the blind, with patterns of raised dots.

brain • *noun* organ of soft nervous tissue in vertebrate's skull; centre of sensation or thought; (often in *plural*) intelligence, *colloquial* intelligent person. • *verb* dash out brains of. □ **brainchild** *colloquial* person's clever idea or invention; **brain drain** *colloquial* emigration of

Brackets ( ) [ ]

Round brackets, also called parentheses, are used mainly to enclose:

1 explanations and extra information or comment, e.g.

Zimbabwe (formerly Rhodesia)
He is (as he always was) a rebel.
This is done using integrated circuits (see page 38).

2 in this dictionary, optional words or parts of words, e.g.

crossword (puzzle) **king-size(d)**

and the type of word which can be used with the word being defined, e.g.

low . . . (of opinion) unfavourable
can . . . preserve (food etc.) in can

Square brackets are used mainly to enclose:

1 words added by someone other than the original writer or speaker, e.g.

Then the man said, 'He [the police officer] can't prove I did it.'

2 various special types of information, such as stage directions, e.g.

HEDLEY: Goodbye! [Exit].

skilled people; **brainstorm** mental disturbance; **brainstorming** pooling of spontaneous ideas about problem etc.; **brains trust** group of experts answering questions, usually impromptu; **brainwash** implant ideas or esp. ideology into (person) by repetition etc.; **brainwave** *colloquial* bright idea.

brainy *adjective* (**-ier**, **-iest**) intellectually clever.

braise /breɪz/ *verb* (**-sing**) stew slowly in closed container with little liquid.

brake ● *noun* device for stopping or slowing wheel or vehicle; thing that impedes. ● *verb* (**-king**) apply brake; slow or stop with brake.

bramble /ˈbræmb(ə)l/ *noun* wild thorny shrub, esp. blackberry.

bran *noun* husks separated from flour.

branch /brɑːntʃ/ ● *noun* limb or bough of tree; lateral extension or subdivision of river, railway, family, etc.; local office of business. ● *verb* (often + *off*) divide, diverge. □ **branch out** extend one's field of interest.

brand ● *noun* particular make of goods; trade mark, label, etc.; (usually + *of*) characteristic kind; identifying mark made with hot iron, iron stamp for this; piece of burning or charred wood; stigma; *poetical* torch. ● *verb* mark with hot iron; stigmatize; assign trade mark etc. to; impress unforgettably. □ **brand new** completely new.

brandish /ˈbrændɪʃ/ *verb* wave or flourish.

brandy /ˈbrændɪ/ *noun* (*plural* **-ies**) strong spirit distilled from wine or fermented fruit juice. □ **brandy snap** crisp rolled gingerbread wafer.

brash *adjective* vulgarly assertive; impudent. □ **brashly** *adverb*; **brashness** *noun*.

brass /brɑːs/ ● *noun* yellow alloy of copper and zinc; brass objects; brass wind instruments; *slang* money; brass memorial tablet; *colloquial* effrontery. ● *adjective* made of brass. □ **brass band** band of brass instruments; **brass rubbing** reproducing of design from engraved brass on paper by rubbing with heelball, impression obtained thus; **brass tacks** *slang* essential details.

brasserie /ˈbræsərɪ/ *noun* restaurant, originally one serving beer with food.

brassica /ˈbræsɪkə/ *noun* plant of cabbage family.

brassière /ˈbræzɪə/ *noun* bra.

brassy /ˈbrɑːsɪ/ *adjective* (**-ier**, **-iest**) of or like brass; impudent; vulgarly showy; loud and blaring.

brat *noun usually derogatory* child.

bravado /brəˈvɑːdəʊ/ *noun* show of boldness.

brave ● *adjective* (**-r**, **-st**) able to face and endure danger or pain; *formal* splendid, spectacular. ● *verb* (**-ving**) face bravely or defiantly. ● *noun* N. American Indian warrior. □ **bravely** *adverb*; **bravery** *noun*.

bravo /brɑːˈvəʊ/ *interjection & noun* (*plural* **-s**) *cry of approval*.

bravura /brəˈvjʊərə/ *noun* brilliance of execution; music requiring brilliant technique.

brawl ● *noun* noisy quarrel or fight. ● *verb* engage in brawl; (of stream) flow noisily.

brawn *noun* muscular strength; muscle, lean flesh; jellied meat made esp. from pig's head. □ **brawny** *adjective* (**-ier**, **-iest**).

bray ● *noun* cry of donkey; harsh sound. ● *verb* make a bray; utter harshly.

braze *verb* (**-zing**) solder with alloy of brass.

brazen /ˈbreɪz(ə)n/ ● *adjective* shameless; of or like brass. ● *verb* (+ *out*) face or undergo defiantly. □ **brazenly** *adverb*.

brazier /ˈbreɪzɪə/ *noun* pan or stand for holding burning coals.

Brazil nut /brəˈzɪl/ *noun* large 3-sided S. American nut.

breach /briːtʃ/ ● *noun* breaking or neglect of rule, duty, promise, etc.; breaking off of relations, quarrel; gap. ● *verb* break through; make gap in; break (law etc.).

bread /bred/ *noun* baked dough of flour usually leavened with yeast; necessary food; *slang* money. □ **breadcrumb** small fragment of bread, esp. (in *plural*) for use in cooking; **breadline** subsistence level; **breadwinner** person whose work supports a family.

breadth /bredθ/ *noun* broadness, distance from side to side; freedom from mental limitations or prejudices.

break /breɪk/ ● *verb* (*past* **broke**; *past participle* **broken** /'brəʊk(ə)n/) separate into pieces under blow or strain; shatter; make or become inoperative; break bone in (limb etc.); interrupt, pause; fail to observe or keep; make or become weak, destroy; weaken effect of (fall, blow, etc.); tame, subdue; surpass (record); reveal or be revealed; come, produce, change, etc., with suddenness or violence; (of waves) curl over and foam; (of voice) change in quality at manhood or with emotion; escape, emerge from. ● *noun* breaking; point where thing is broken; gap; pause in work etc.; sudden dash; a chance; *Cricket* deflection of ball on bouncing; points scored in one sequence at billiards etc. □ **break away** make or become free or separate; **break down** fail or collapse, demolish, analyse; **breakdown** mechanical failure, loss of (esp. mental) health, collapse, analysis; **break even** make neither profit nor loss; **break in** intrude forcibly esp. as thief, interrupt, accustom to habit; **breakneck** (of speed) dangerously fast; **break off** detach by breaking, bring to an end, cease talking etc.; **break open** open forcibly; **break out** escape by force, begin suddenly, (+ *in*) become covered in (rash etc.); **breakthrough** major advance in knowledge etc.; **break up** break into pieces, disband, part; **breakup** disintegration, collapse; **breakwater** barrier breaking force of waves; **break wind** release gas from anus.

breakable *adjective* easily broken.

breakage *noun* broken thing; breaking.

breaker *noun* heavy breaking wave.

breakfast /'brekfəst/ ● *noun* first meal of day. ● *verb* have breakfast.

bream *noun* (*plural* same) yellowish freshwater fish; similar sea fish.

breast /brest/ ● *noun* either of two milk-secreting organs on woman's chest; chest; part of garment covering this; seat of emotions. ● *verb* contend with; reach top of (hill). □ **breastbone** bone connecting ribs in front; **breastfeed** feed (baby) from breast; **breastplate** armour covering chest; **breaststroke** stroke made while swimming on breast by extending arms forward and sweeping them back.

breath /breθ/ *noun* air drawn into or expelled from lungs; one respiration; breath as perceived by senses; slight movement of air. □ **breathtaking** astounding, awe-inspiring; **breath test** test with Breathalyser.

breathalyse *verb* give breath test to.

Breathalyser /'breθəlaɪzə/ *noun proprietary term* instrument for measuring alcohol in breath.

breathe /bri:ð/ *verb* (**-thing**) take air into lungs and send it out again; live; utter or sound, esp. quietly; pause; send out or take in (as) with breathed air. □ **breathing-space** time to recover, pause.

breather /'bri:ðə/ *noun* short rest period.

breathless /'breθlɪs/ *adjective* panting, out of breath; still, windless. □ **breathlessly** *adverb*.

bred *past & past participle* of BREED.

breech /bri:tʃ/ *noun* back part of gun or gun barrel; (in *plural*) short trousers fastened below knee. □ **breech birth** birth in which buttocks emerge first.

breed ● *verb* (*past & past participle* **bred**) produce offspring; propagate; raise (animals); yield, result in; arise, spread; train, bring up; create (fissile material) by nuclear reaction. ● *noun* stock of animals within species; race, lineage; sort, kind. □ **breeder reactor** nuclear reactor creating surplus fissile material. □ **breeder** *noun*.

breeding *noun* raising of offspring; social behaviour; ancestry.

breeze[1] ● *noun* gentle wind. ● *verb* (**-zing**) (+ *in*, *out*, *along*, etc.) *colloquial* saunter casually.

breeze[2] *noun* small cinders. □ **breeze-block** lightweight building block made from breeze.

breezy *adjective* (**-ier**, **-iest**) slightly windy.

Bren *noun* lightweight quick-firing machine-gun.

brent *noun* small migratory goose.

brethren see BROTHER.

Breton /'bret(ə)n/ ● *noun* native or language of Brittany. ● *adjective* of Brittany.

breve /bri:v/ *noun Music* note equal to two semibreves; mark (˘) indicating short or unstressed vowel.

breviary /'bri:vɪərɪ/ *noun* (*plural* **-ies**) book containing RC daily office.

brevity /'brevɪtɪ/ *noun* conciseness, shortness.

brew ● *verb* make (beer etc.) by infusion, boiling, and fermenting; make (tea etc.) by infusion; undergo these processes; be forming. ● *noun* amount brewed; liquor brewed. □ **brewer** *noun*.

brewery /'bru:ərɪ/ *noun* (*plural* **-ies**) factory for brewing beer etc.

briar[1,2] = BRIER[1,2].

bribe ● *verb* (**-bing**) persuade to act improperly by gift of money etc. ● *noun* money or services offered in bribing. □ **bribery** *noun*.

bric-a-brac /'brɪkəbræk/ *noun* cheap ornaments, trinkets, etc.

brick ● *noun* small rectangular block of baked clay, used in building; toy building block; brick-shaped thing; *slang* generous or loyal person. ● *verb* (+ *in*, *up*) close or block with brickwork. ● *adjective* made of bricks. □ **brickbat** piece of brick, esp. as missile, insult; **bricklayer** person who builds with bricks; **brickwork** building or work in brick.

bridal /'braɪd(ə)l/ *adjective* of bride or wedding.

bride *noun* woman on her wedding day and shortly before and after it. □ **bridegroom** man on his wedding day and shortly before and after it; **bridesmaid** woman or girl attending bride at wedding.

bridge[1] ● *noun* structure providing way over road, railway, river, etc.; thing joining or connecting; superstructure from which ship is directed; upper bony part of nose; prop under strings of violin etc.; bridgework. ● *verb* (**-ging**) be or make bridge over. □ **bridgehead** position held on enemy's side of river etc.; **bridgework** dental structure covering gap and joined to teeth on either side; **bridging loan** loan to cover interval between buying one house and selling another.

bridge[2] *noun* card games derived from whist.

bridle /'braɪd(ə)l/ ● *noun* headgear for controlling horse etc.; restraining thing. ● *verb* (**-ling**) put bridle on, control, curb; express resentment, esp. by throwing up head and drawing in chin. □ **bridle path** rough path for riders or walkers.

Brie /bri:/ *noun* flat round soft creamy French cheese.

brief /bri:f/ ● *adjective* of short duration; concise; scanty. ● *noun* (in *plural*) short pants; summary of case for guidance of barrister; instructions for a task. ● *verb* instruct (barrister) by brief; inform or instruct in advance. □ **in brief** to sum up. □ **briefly** *adverb*.

briefcase /'bri:fkeɪs/ *noun* flat document case.

brier[1] /braɪə/ *noun* (also **briar**) wild-rose bush.

brier[2] /braɪə/ *noun* (also **briar**) white heath of S. Europe; tobacco pipe made from its root.

brig[1] *noun* two-masted square-rigged ship.

brig[2] *noun Scottish & Northern English* bridge.

brigade /brɪ'geɪd/ *noun* military unit forming part of division; organized band of workers etc.

brigadier /brɪgə'dɪə/ *noun* army officer next below major-general.

brigand /'brɪgənd/ *noun* member of robber gang.

bright /braɪt/ *adjective* emitting or reflecting much light, shining; vivid; clever; cheerful. □ **brighten** *verb*; **brightly** *adverb*; **brightness** *noun*.

brill[1] *noun* (*plural* same) flatfish resembling turbot.

brill[2] *adjective colloquial* excellent.

brilliant /'brɪlɪənt/ ● *adjective* bright, sparkling; highly talented; showy; *colloquial* excellent. ● *noun* diamond of finest quality. □ **brilliance** *noun*; **brilliantly** *adverb*.

brilliantine /'brɪljənti:n/ *noun* cosmetic for making hair glossy.

brim ● *noun* edge of vessel; projecting edge of hat. ● *verb* (**-mm-**) fill or be full to brim.

brimstone /'brɪmstəʊn/ *noun archaic* sulphur.

brindled /'brɪnd(ə)ld/ *adjective* brown with streaks of other colour.

brine *noun* salt water; sea water.

bring *verb* (*past & past participle* **brought** /brɔ:t/) come with, carry, convey;

cause, result in; be sold for; submit (criminal charge); initiate (legal action). □ **bring about** cause to happen; **bring down** cause to fall; **bring forward** move to earlier time, transfer from previous page or account, draw attention to; **bring in** introduce, produce as profit; **bring off** succeed in; **bring on** cause to happen, appear or progress; **bring out** emphasize, publish; **bring round** restore to consciousness, win over; **bring up** raise and educate, vomit, draw attention to.

brink *noun* edge of precipice etc.; furthest point before danger, discovery, etc. □ **brinkmanship** policy of pursuing dangerous course to brink of catastrophe.

briny /'braɪnɪ/ ● *adjective* (**-ier**, **-iest**) of brine or sea, salt. ● *noun* (**the briny**) *slang* the sea.

briquette /brɪ'ket/ *noun* block of compressed coal dust as fuel.

brisk *adjective* active, lively, quick. □ **briskly** *adverb*.

brisket /'brɪskɪt/ *noun* animal's breast, esp. as joint of meat.

brisling /'brɪzlɪŋ/ *noun* (*plural* same or **-s**) small herring or sprat.

bristle /'brɪs(ə)l/ ● *noun* short stiff hair, esp. one used in brushes etc. ● *verb* (**-ling**) (of hair etc.) stand up; cause to bristle; show irritation; (usually + *with*) be covered (with) or abundant (in). □ **bristly** *adjective*.

British /'brɪtɪʃ/ ● *adjective* of Britain. ● *plural noun* (**the British**) the British people. □ **British Summer Time** = SUMMER TIME.

Briton /'brɪt(ə)n/ *noun* inhabitant of S. Britain before Roman conquest; native of Great Britain.

brittle /'brɪt(ə)l/ *adjective* apt to break, fragile.

broach /brəʊtʃ/ *verb* raise for discussion; pierce (cask) to draw liquor; open and start using.

broad /brɔːd/ ● *adjective* large across, extensive; of specified breadth; full, clear; explicit; general; tolerant; coarse; (of accent) marked, strong. ● *noun* broad part; *US slang* woman; (**the Broads**) large areas of water in E. Anglia. □ **broad bean** bean with large flat seeds, one such seed; **broadloom** (carpet) woven in broad width; **broad-minded** tolerant, liberal; **broadsheet** large-sized newspaper. □ **broaden** *verb*; **broadly** *adverb*.

broadcast ● *verb* (*past & past participle* **-cast**) transmit by radio or television; take part in such transmission; scatter (seed) etc.; disseminate widely. ● *noun* radio or television programme or transmission. □ **broadcaster** *noun*; **broadcasting** *noun*.

broadside *noun* vigorous verbal attack; firing of all guns on one side of ship. □ **broadside on** sideways on.

brocade /brə'keɪd/ *noun* fabric woven with raised pattern.

broccoli /'brɒkəlɪ/ *noun* brassica with greenish flower heads.

brochure /'brəʊʃə/ *noun* booklet, pamphlet, esp. containing descriptive information.

broderie anglaise /brəʊdərɪ ɑ̃'gleɪz/ *noun* open embroidery on usually white cotton or linen.

brogue /brəʊg/ *noun* strong shoe with ornamental perforations; rough shoe of untanned leather; marked local, esp. Irish, accent.

broil *verb* grill (meat); make or be very hot.

broiler *noun* young chicken for broiling.

broke ● *past* of BREAK. ● *adjective colloquial* having no money, bankrupt.

broken /'brəʊkən/ ● *past participle* of BREAK. ● *adjective* that has been broken; reduced to despair; (of language) spoken imperfectly; interrupted. □ **broken-hearted** crushed by grief; **broken home** family disrupted by divorce or separation.

broker *noun* middleman, agent; stockbroker. □ **broking** *noun*.

brokerage *noun* broker's fee or commission.

brolly /'brɒlɪ/ *noun* (*plural* **-ies**) *colloquial* umbrella.

bromide /'brəʊmaɪd/ *noun* binary compound of bromine, esp. one used as sedative; trite remark.

bromine /'brəʊmiːn/ *noun* poisonous liquid non-metallic element with choking smell.

bronchial /ˈbrɒŋkɪəl/ *adjective* of two main divisions of windpipe or smaller tubes into which they divide.

bronchitis /brɒŋˈkaɪtɪs/ *noun* inflammation of bronchial mucous membrane.

bronco /ˈbrɒŋkəʊ/ *noun* (*plural* **-s**) wild or half-tamed horse of western US.

brontosaurus /brɒntəˈsɔːrəs/ *noun* (*plural* **-ruses**) = APATOSAURUS.

bronze /brɒnz/ ● *noun* brown alloy of copper and tin; its colour; work of art or medal in it. ● *adjective* made of or coloured like bronze. ● *verb* (**-zing**) make or grow brown; tan. □ **Bronze Age** period when tools were of bronze; **bronze medal** medal given usually as third prize.

brooch /brəʊtʃ/ *noun* ornamental hinged pin.

brood /ˈbruːd/ ● *noun* bird's or other animal's young produced at one hatch or birth; *colloquial* children of a family. ● *verb* worry or ponder, esp. resentfully; (of hen) sit on eggs.

broody *adjective* (**-ier**, **-iest**) (of hen) wanting to brood; sullenly thoughtful; *colloquial* (of woman) wanting pregnancy.

brook¹ /brʊk/ *noun* small stream.

brook² /brʊk/ *verb* tolerate; allow.

broom /bruːm/ *noun* long-handled brush for sweeping; yellow-flowered shrub. □ **broomstick** broom-handle.

Bros. *abbreviation* Brothers.

broth /brɒθ/ *noun* thin meat or fish soup.

brothel /ˈbrɒθ(ə)l/ *noun* premises for prostitution.

brother /ˈbrʌðə/ *noun* man or boy in relation to his siblings; close man friend; (*plural* also **brethren** /ˈbreðrɪn/) member of male religious order; (*plural* also **brethren**) fellow Christian etc.; fellow human being. □ **brother-in-law** (*plural* **brothers-in-law**) wife's or husband's brother, sister's husband. □ **brotherly** *adjective.*

brotherhood *noun* relationship (as) between brothers; (members of) association for mutual help etc.

brought *past & past participle* of BRING.

brow /braʊ/ *noun* forehead; (usually in *plural*) eyebrow; summit of hill; edge of cliff etc.

browbeat *verb* (*past* **-beat**, *past participle* **-beaten**) intimidate, bully.

brown /braʊn/ ● *adjective* of colour of dark wood or rich soil; dark-skinned; tanned. ● *noun* brown colour, paint, clothes, etc. ● *verb* make or become brown. □ **brown bread** bread made of wholemeal or wheatmeal flour; **browned off** *colloquial* bored, fed up; **Brown Owl** adult leader of Brownies; **brown rice** unpolished rice; **brown sugar** partially refined sugar. □ **brownish** *adjective.*

Brownie /ˈbraʊnɪ/ *noun* junior Guide; (**brownie**) small square of chocolate cake with nuts; (**brownie**) benevolent elf.

browse /braʊz/ ● *verb* (**-sing**) read or look around desultorily; feed on leaves and young shoots. ● *noun* browsing; twigs, shoots, etc. as fodder.

bruise /bruːz/ ● *noun* discoloration of skin caused by blow or pressure; similar damage on fruit etc. ● *verb* (**-sing**) inflict bruise on; be susceptible to bruises.

bruiser *noun colloquial* tough brutal person.

bruit /bruːt/ *verb* (often + *abroad, about*) spread (report or rumour).

brunch *noun* combination of breakfast and lunch.

brunette /bruːˈnet/ *noun* woman with dark hair.

brunt *noun* chief impact of attack etc.

brush ● *noun* cleaning or hairdressing or painting implement of bristles etc. set in holder; application of brush; short esp. unpleasant encounter; fox's tail; carbon or metal piece serving as electrical contact. ● *verb* use brush on; touch lightly, graze in passing. □ **brush off** dismiss abruptly; **brush-off** dismissal, rebuff; **brush up** clean up or smarten, revise (subject, skill); **brushwood** undergrowth, thicket, cut or broken twigs etc.; **brushwork** painter's way of using brush.

brusque /brʊsk/ *adjective* abrupt, offhand. □ **brusquely** *adverb*; **brusqueness** *noun.*

Brussels sprout /ˈbrʌs(ə)lz/ *noun* brassica with small cabbage-like buds on stem; such bud.

brutal /'bru:t(ə)l/ *adjective* savagely cruel; mercilessly frank. □ **brutality** /-'tæl-/ *noun* (*plural* **-ies**); **brutalize** *verb* (also **-ise**) (**-zing**, **-sing**).

brute /bru:t/ ● *noun* cruel person; *colloquial* unpleasant person; animal other than man. ● *adjective* unthinking; cruel, stupid. □ **brutish** *adjective*.

bryony /'braıənı/ *noun* (*plural* **-ies**) climbing hedge plant.

B.Sc. *abbreviation* Bachelor of Science.

BST *abbreviation* British Summer Time.

Bt. *abbreviation* Baronet.

bubble /'bʌb(ə)l/ ● *noun* thin sphere of liquid enclosing air or gas; air-filled cavity in glass etc.; transparent domed cavity. ● *verb* (**-ling**) send up or rise in bubbles; make sound of boiling. □ **bubble and squeak** cooked potatoes and cabbage fried together; **bubble bath** additive to make bathwater bubbly; **bubblegum** chewing gum that can be blown into bubbles.

bubbly ● *adjective* (**-ier**, **-iest**) full of bubbles; exuberant. ● *noun colloquial* champagne.

bubonic /bju:'bɒnık/ *adjective* (of plague) marked by swellings esp. in groin and armpits.

buccaneer /bʌkə'nıə/ *noun* pirate; adventurer. □ **buccaneering** *adjective & noun*.

buck[1] ● *noun* male deer, hare, or rabbit. ● *verb* (of horse) jump vertically with back arched, throw (rider) thus; (usually + *up*) *colloquial* cheer up, hurry up. □ **buckshot** coarse shot for gun; **buck-tooth** projecting upper tooth.

buck[2] *noun US & Australian slang* dollar.

buck[3] *noun slang* small object placed before dealer at poker. □ **pass the buck** *colloquial* shift responsibility.

bucket /'bʌkıt/ ● *noun* usually round open container with handle, for carrying or holding water etc.; amount contained in this; (in *plural*) *colloquial* large quantities; compartment or scoop in waterwheel, dredger, or grain elevator. ● *verb* (**-t-**) *colloquial* (often + *down*) (esp. of rain) pour heavily; (often + *along*) move jerkily or bumpily. □ **bucket seat** one with rounded back, to fit one person; **bucket shop** agency dealing in cheap airline tickets, unregistered broking agency.

buckle /'bʌk(ə)l/ ● *noun* clasp with usually hinged pin for securing strap or belt etc. ● *verb* (**-ling**) fasten with buckle; (cause to) crumple under pressure. □ **buckle down** make determined effort.

buckram /'bʌkrəm/ *noun* coarse linen etc. stiffened with paste etc.

buckshee /bʌk'ʃi:/ *adjective & adverb slang* free of charge.

buckwheat *noun* seed of plant related to rhubarb.

bucolic /bju:'kɒlık/ *adjective* of shepherds, rustic, pastoral.

bud ● *noun* projection from which branch, leaf, or flower develops; flower or leaf not fully open; asexual growth separating from organism as new animal. ● *verb* (**-dd-**) form buds; begin to grow or develop; graft bud of (plant) on another plant.

Buddhism /'bʊdız(ə)m/ *noun* Asian religion founded by Gautama Buddha. □ **Buddhist** *adjective & noun*.

buddleia /'bʌdlıə/ *noun* shrub with flowers attractive to butterflies.

buddy /'bʌdı/ *noun* (*plural* **-ies**) *colloquial* friend; mate.

budge *verb* (**-ging**) move in slightest degree; (+ *up*) move to make room for another person.

budgerigar /bʌdʒərı'gɑ:/ *noun* small parrot often kept as pet.

budget /'bʌdʒıt/ ● *noun* amount of money needed or available; (**the Budget**) annual estimate of country's revenue and expenditure; similar estimate for person or group. ● *verb* (**-t-**) (often + *for*) allow or arrange for in budget. □ **budgetary** *adjective*.

budgie /'bʌdʒı/ *noun colloquial* budgerigar.

buff ● *adjective* of yellowish beige colour. ● *noun* this colour; *colloquial* enthusiast; velvety dull yellow leather. ● *verb* polish; make (leather) velvety. □ **in the buff** *colloquial* naked.

buffalo /'bʌfələʊ/ *noun* (*plural* same or **-es**) any of various kinds of ox; American bison.

buffer[1] *noun* apparatus for deadening impact esp. of railway vehicles. □ **buffer state** minor one between two larger ones, regarded as reducing friction.

buffer[2] *noun slang* old or incompetent fellow.

buffet[1] /ˈbʊfeɪ/ *noun* room or counter where refreshments are sold; self-service meal of several dishes set out at once; (also /ˈbʌfɪt/) sideboard. □ **buffet car** railway coach in which refreshments are served.

buffet[2] /ˈbʌfɪt/ ● *verb* (**-t-**) strike repeatedly. ● *noun* blow with hand; shock.

buffoon /bəˈfuːn/ *noun* silly or ludicrous person; jester. □ **buffoonery** *noun*.

bug ● *noun* small insect; concealed microphone; *colloquial* error in computer program etc.; *slang* virus, infection; *slang* enthusiasm, obsession. ● *verb* (**-gg-**) conceal microphone in; *slang* annoy.

bugbear *noun* cause of annoyance; object of baseless fear.

buggy /ˈbʌgɪ/ *noun* (*plural* **-ies**) small sturdy motor vehicle; lightweight pushchair; light horse-drawn vehicle for one or two people.

bugle /ˈbjuːg(ə)l/ ● *noun* brass instrument like small trumpet. ● *verb* (**-ling**) sound bugle. □ **bugler** *noun*.

build /bɪld/ ● *verb* (*past & past participle* **built** /bɪlt/) construct or cause to be constructed; develop or establish. ● *noun* physical proportions; style of construction. □ **build in** incorporate; **build up** increase in size or strength, praise, gradually establish or be established; **build-up** favourable description in advance, gradual approach to climax, accumulation.

builder *noun* contractor who builds houses etc.

building *noun* house or other structure with roof and walls. □ **building society** financial organization (not public company) that pays interest on savings accounts, lends money esp. for mortgages, etc.

built /bɪlt/ *past & past participle* of BUILD. □ **built-in** integral; **built-up** covered with buildings.

bulb *noun* rounded base of stem of some plants; light bulb; bulb-shaped thing or part.

bulbous /ˈbʌlbəs/ *adjective* bulb-shaped; bulging.

bulge ● *noun* irregular swelling; *colloquial* temporary increase. ● *verb* (**-ging**) swell outwards. □ **bulgy** *adjective*.

bulimia /bjʊˈlɪmɪə/ *noun* (in full **bulimia nervosa** /nɜːˈvəʊsə/) disorder in which overeating alternates with self-induced vomiting, fasting, etc.

bulk ● *noun* size, magnitude, esp. when great; (**the bulk**) the greater part; large quantity. ● *verb* seem (in size or importance); make thicker. □ **bulk buying** buying in quantity at discount; **bulkhead** upright partition in ship, aircraft, etc.

bulky /ˈbʌlkɪ/ *adjective* (**-ier**, **-iest**) large, unwieldy.

bull[1] /bʊl/ *noun* uncastrated male ox; male whale or elephant etc.; bull's-eye of target; person who buys shares in hope of selling at higher price later. □ **bulldog** short-haired heavy-jowled sturdy dog, tenacious and courageous person; **Bulldog clip** strong sprung clip for papers etc.; **bulldoze** clear with bulldozer, *colloquial* intimidate, *colloquial* make (one's way) forcibly; **bulldozer** powerful tractor with broad upright blade for clearing ground; **bullfight** public baiting, and usually killing, of bulls; **bullfinch** pink and black finch; **bullfrog** large American frog with booming croak; **bullring** arena for bullfight; **bull's-eye** centre of target, hard minty sweet; **bull terrier** cross between bulldog and terrier. □ **bullish** *adjective*.

bull[2] /bʊl/ *noun* papal edict.

bull[3] /bʊl/ *noun slang* nonsense; *slang* unnecessary routine tasks; absurdly illogical statement.

bullet /ˈbʊlɪt/ *noun* small pointed missile fired from revolver etc. □ **bulletproof** resistant to bullets.

bulletin /ˈbʊlɪtɪn/ *noun* short official statement; short broadcast news report.

bullion /ˈbʊlɪən/ *noun* gold or silver in lump or valued by weight.

bullock /ˈbʊlək/ *noun* castrated bull.

bully[1] /ˈbʊlɪ/ ● *noun* (*plural* **-ies**) person coercing others by fear. ● *verb* (**-ies**, **-ied**) persecute or oppress by force or threats. ● *interjection* (+ *for*) very good.

bully[2] /ˈbʊlɪ/ (in full **bully off**) ● *noun* (*plural* **-ies**) putting ball into play in hockey. ● *verb* (**-ies**, **-ied**) start play thus.

bully³ /'bʊlɪ/ *noun* (in full **bully beef**) corned beef.

bulrush /'bʊlrʌʃ/ *noun* tall rush; *Biblical* papyrus.

bulwark /'bʊlwək/ *noun* defensive wall, esp. of earth; person or principle that protects; (usually in *plural*) ship's side above deck.

bum¹ *noun slang* buttocks. □ **bumbag** small pouch worn round waist.

bum² *US slang* ● *noun* loafer, dissolute person. ● *verb* (**-mm-**) (often + *around*) loaf, wander around; cadge. ● *adjective* of poor quality.

bumble /'bʌmb(ə)l/ *verb* (**-ling**) (+ *on*) speak ramblingly; be inept; blunder. □ **bumble-bee** large bee with loud hum.

bump ● *noun* dull-sounding blow or collision; swelling caused by it; uneven patch on road etc.; prominence on skull, thought to indicate mental faculty. ● *verb* come or strike with bump against; hurt thus; (usually + *along*) move along with jolts. □ **bump into** *colloquial* meet by chance; **bump off** *slang* murder; **bump up** *colloquial* increase. □ **bumpy** *adjective* (**-ier**, **-iest**).

bumper ● *noun* horizontal bar on motor vehicle to reduce damage in collisions; *Cricket* ball rising high after pitching; brim-full glass. ● *adjective* unusually large or abundant.

bumpkin /'bʌmpkɪn/ *noun* rustic or awkward person.

bumptious /'bʌmpʃəs/ *adjective* self-assertive, conceited.

bun *noun* small sweet cake or bread roll often with dried fruit; small coil of hair at back of head.

bunch ● *noun* cluster of things growing or fastened together; lot; *colloquial* gang, group. ● *verb* arrange in bunch(es); gather in folds; come, cling, or crowd together.

bundle /'bʌnd(ə)l/ ● *noun* collection of things tied or fastened together; set of nerve fibres etc.; *slang* large amount of money. ● *verb* (**-ling**) (usually + *up*) tie in bundle; (usually + *into*) throw or move carelessly; (usually + *out*, *off*, *away*, etc.) send away hurriedly.

bung ● *noun* stopper, esp. for cask. ● *verb* stop with bung; *slang* throw. □ **bunged up** blocked up.

bungalow /'bʌŋgələʊ/ *noun* one-storeyed house.

bungee /'bʌndʒɪ/ *noun* elasticated cord. □ **bungee jumping** sport of jumping from great height while secured by bungee.

bungle /'bʌŋg(ə)l/ ● *verb* (**-ling**) mismanage, fail to accomplish; work awkwardly. ● *noun* bungled work or attempt.

bunion /'bʌnjən/ *noun* swelling on foot, esp. on side of big toe.

bunk¹ *noun* shelflike bed against wall. □ **bunk bed** each of two or more bunks one above the other.

bunk² *slang* □ **do a bunk** run away.

bunk³ *noun slang* nonsense, humbug.

bunker *noun* container for fuel; reinforced underground shelter; sandy hollow in golf course.

bunkum /'bʌŋkəm/ *noun* nonsense, humbug.

bunny /'bʌnɪ/ *noun* (*plural* **-ies**) *childish name for* rabbit.

Bunsen burner /'bʌns(ə)n/ *noun* small adjustable gas burner used in laboratory.

bunting¹ /'bʌntɪŋ/ *noun* small bird related to finches.

bunting² /'bʌntɪŋ/ *noun* flags and other decorations; loosely woven fabric for these.

buoy /bɔɪ/ ● *noun* anchored float as navigational mark etc.; lifebuoy. ● *verb* (usually + *up*) keep afloat, encourage; (often + *out*) mark with buoy(s).

buoyant /'bɔɪənt/ *adjective* apt to float; resilient; exuberant. □ **buoyancy** *noun*.

bur *noun* (also **burr**) clinging seed vessel or flower head, plant producing burs; clinging person.

burble /'bɜ:b(ə)l/ *verb* (**-ling**) talk rambling; make bubbling sound.

burden /'bɜ:d(ə)n/ ● *noun* thing carried, load; oppressive duty, expense, emotion, etc.; refrain of song; theme. ● *verb* load, encumber, oppress. □ **burden of proof** obligation to prove one's case. □ **burdensome** *adjective*.

burdock /ˈbɜːdɒk/ *noun* plant with prickly flowers and docklike leaves.

bureau /ˈbjʊərəʊ/ *noun* (*plural* **-s** or **-x** /-z/) writing desk with drawers; *US* chest of drawers; office or department for specific business; government department.

bureaucracy /bjʊəˈrɒkrəsɪ/ *noun* (*plural* **-ies**) government by central administration; government officials esp. regarded as oppressive and inflexible; conduct typical of these.

bureaucrat /ˈbjʊərəkræt/ *noun* official in bureaucracy. □ **bureaucratic** /-ˈkrætɪk/ *adjective*.

burgeon /ˈbɜːdʒ(ə)n/ *verb* grow rapidly, flourish.

burger /ˈbɜːgə/ *noun colloquial* hamburger.

burgher /ˈbɜːgə/ *noun* citizen, esp. of foreign town.

burglar /ˈbɜːglə/ *noun* person who commits burglary.

burglary *noun* (*plural* **-ies**) illegal entry into building to commit theft or other crime.

burgle /ˈbɜːg(ə)l/ *verb* (**-ling**) commit burglary (on).

burgundy /ˈbɜːgəndɪ/ *noun* (*plural* **-ies**) red or white wine produced in Burgundy; dark red colour.

burial /ˈberɪəl/ *noun* burying, esp. of corpse; funeral.

burlesque /bɜːˈlesk/ ● *noun* comic imitation, parody; *US* variety show, esp. with striptease. ● *adjective* of or using burlesque. ● *verb* (**-ques**, **-qued**, **-quing**) parody.

burly /ˈbɜːlɪ/ *adjective* (**-ier**, **-iest**) large and sturdy.

burn[1] ● *verb* (*past & past participle* **burnt** or **burned**) (cause to) be consumed by fire; blaze or glow with fire; (cause to) be injured or damaged by fire, sun, or great heat; use or be used as fuel; produce (hole etc.) by fire or heat; char in working; brand; give or feel sensation or pain (as) of heat. ● *noun* sore or mark made by burning. □ **burn out** be reduced to nothing by burning, (cause to) fail by burning, (usually **burn oneself out**) suffer exhaustion; **burnt offering** sacrifice offered by burning.

burn[2] *noun Scottish* brook.

burner *noun* part of lamp or cooker etc. that emits flame.

burnish /ˈbɜːnɪʃ/ *verb* polish by rubbing.

burnt *past & past participle* of BURN[1].

burp *verb & noun colloquial* belch.

burr[1] *noun* whirring sound; rough sounding of *r*; rough edge on metal etc.

burr[2] = BUR.

burrow /ˈbʌrəʊ/ ● *noun* hole dug by animal as dwelling. ● *verb* make burrow; make by digging; (+ *into*) investigate or search.

bursar /ˈbɜːsə/ *noun* treasurer of college etc.; holder of bursary.

bursary /ˈbɜːsərɪ/ *noun* (*plural* **-ies**) grant, esp. scholarship.

burst ● *verb* (*past & past participle* **burst**) fly violently apart or give way suddenly, explode; rush, move, speak, be spoken, etc. suddenly or violently. ● *noun* bursting, explosion, outbreak; spurt.

burton /ˈbɜːt(ə)n/ *noun* □ **go for a burton** *slang* be lost, destroyed, or killed.

bury /ˈberɪ/ *verb* (**-ies**, **-ied**) place (corpse) in ground, tomb, or sea; put underground, hide in earth; consign to obscurity; (**bury oneself** or in *passive*) involve (oneself) deeply. □ **bury the hatchet** cease to quarrel.

bus ● *noun* (*plural* **buses**, *US* **busses**) large public passenger vehicle usually plying on fixed route. ● *verb* (**buses** or **busses**, **bussed**, **bussing**) go by bus; *US* transport by bus (esp. to aid racial integration). □ **busman's holiday** leisure spent in same occupation as working hours; **bus shelter** shelter for people waiting for bus; **bus station** centre where buses depart and arrive; **bus stop** regular stopping place of bus.

busby /ˈbʌzbɪ/ *noun* (*plural* **-ies**) tall fur cap worn by hussars etc.

bush[1] /bʊʃ/ *noun* shrub, clump of shrubs; clump of hair or fur; *Australian etc.* uncultivated land, woodland. □ **bush-baby** small African lemur; **Bushman** member or language of a S. African aboriginal people; **bushman** dweller or traveller in Australian bush; **bush telegraph** rapid informal spreading of information etc.

bush[2] /bʊʃ/ *noun* metal lining of axle-hole etc.; electrically insulating sleeve.

bushel /'bʊʃ(ə)l/ *noun* measure of capacity for corn, fruit, etc. (8 gallons, 36.4 litres).

bushy *adjective* (**-ier**, **-iest**) growing thickly or like bush; having many bushes.

business /'bɪznɪs/ *noun* one's occupation or profession; one's own concern, task, duty; serious work; (difficult or unpleasant) matter or affair; thing(s) needing dealing with; buying and selling, trade; commercial firm. □ **businesslike** practical, systematic; **businessman**, **businesswoman** person engaged in trade or commerce.

busk *verb* perform esp. music in street etc. for tips. □ **busker** *noun*.

bust[1] *noun* human chest, esp. of woman; sculpture of head, shoulders, and chest.

bust[2] *colloquial* ● *verb* (*past & past participle* **bust** or **busted**) burst, break; raid, search; arrest. ● *adjective* burst, broken; bankrupt. □ **bust-up** quarrel, violent split or separation.

bustard /b'ʌstəd/ *noun* large swift-running bird.

bustle[1] /'bʌs(ə)l/ ● *verb* (**-ling**) (often + *about*) move busily and energetically; make (person) hurry; (as **bustling** *adjective*) active, lively. ● *noun* excited activity.

bustle[2] /'bʌs(ə)l/ *noun historical* padding worn under skirt to puff it out behind.

busy /'bɪzɪ/ ● *adjective* (**-ier**, **-iest**) occupied or engaged in work etc.; full of activity; fussy. ● *verb* (**-ies**, **-ied**) occupy, keep busy. □ **busybody** meddlesome person; **busy Lizzie** house plant with usually red, pink, or white flowers. □ **busily** *adverb*.

but ● *conjunction* however; on the other hand; otherwise than. ● *preposition* except, apart from. ● *adverb* only. □ **but for** without the help or hindrance etc. of; **but then** however.

butane /'bju:teɪn/ *noun* hydrocarbon used in liquefied form as fuel.

butch /bʊtʃ/ *adjective slang* masculine, tough-looking.

butcher /'bʊtʃə/ ● *noun* person who sells meat; slaughterer of animals for food; brutal murderer. ● *verb* slaughter or cut up (animal); kill wantonly or cruelly; *colloquial* ruin through incompetence. □ **butchery** *noun* (*plural* **-ies**).

butler /'bʌtlə/ *noun* chief manservant of household.

butt[1] ● *verb* push with head; (cause to) meet end to end. ● *noun* push or blow with head or horns. □ **butt in** interrupt, meddle.

butt[2] *noun* (often + *of*) object of ridicule etc.; mound behind target; (in *plural*) shooting range.

butt[3] *noun* thicker end, esp. of tool or weapon; stub of cigarette etc.

butt[4] *noun* cask.

butter /'bʌtə/ ● *noun* yellow fatty substance made from cream, used as spread and in cooking; substance of similar texture. ● *verb* spread, cook, etc., with butter. □ **butter-bean** dried large flat white kind; **buttercup** plant with yellow flowers; **butter-fingers** *colloquial* person likely to drop things; **buttermilk** liquid left after butter-making; **butter muslin** thin loosely woven cloth; **butterscotch** sweet made of butter and sugar; **butter up** *colloquial* flatter.

butterfly /'bʌtəflaɪ/ *noun* (*plural* **-ies**) insect with 4 often showy wings; (in *plural*) nervous sensation in stomach. □ **butterfly nut** kind of wing-nut; **butterfly stroke** method of swimming with both arms lifted at same time.

buttery[1] *adjective* like or containing butter.

buttery[2] *noun* (*plural* **-ies**) food store or snack-bar, esp. in college.

buttock /'bʌtək/ *noun* either protuberance on lower rear part of human trunk; corresponding part of animal.

button /'bʌt(ə)n/ ● *noun* disc or knob sewn to garment etc. as fastening or for ornament; knob etc. pressed to operate electronic equipment. ● *verb* (often + *up*) fasten with buttons. □ **button mushroom** small unopened mushroom.

buttonhole ● *noun* slit in cloth for button; flower(s) worn in lapel buttonhole. ● *verb* (**-ling**) *colloquial* accost and detain (reluctant listener).

buttress /'bʌtrɪs/ ● *noun* support built against wall etc. ● *verb* support or strengthen.

buxom /ˈbʌksəm/ *adjective* plump and rosy, large and shapely.

buy /baɪ/ • *verb* (**buys**, **buying**; *past & past participle* **bought** /bɔːt/) obtain in exchange for money etc.; procure by bribery, bribe; get by sacrifice etc.; *slang* accept, believe. • *noun colloquial* purchase. □ **buy out** pay (a person) for ownership, an interest, etc.; **buyout** purchase of controlling share in company; **buy up** buy as much as possible of.

buyer *noun* person who buys, esp. stock for large shop. □ **buyer's market** time when goods are plentiful and cheap.

buzz • *noun* hum of bee etc.; sound of buzzer; low murmur; hurried activity; *slang* telephone call; *slang* thrill. • *verb* hum; summon with buzzer; (often + *about*) move busily; be filled with activity or excitement. □ **buzzword** *colloquial* fashionable technical word, catchword.

buzzard /ˈbʌzəd/ *noun* large bird of hawk family.

buzzer *noun* electrical buzzing device as signal.

by /baɪ/ • *preposition* near, beside, along; through action, agency, or means of; not later than; past; via; during; to extent of; according to. • *adverb* near; aside, in reserve; past. □ **by and by** before long; **by-election** parliamentary election between general elections; **by-product** substance etc. produced incidentally in making of something else; **byroad** minor road; **by the by**, **by the way** incidentally; **byway** byroad or secluded path; **byword** person or thing as notable example, proverb.

bye[1] /baɪ/ *noun Cricket* run made from ball that passes batsman without being hit; (in tournament) position of competitor left without opponent in round.

bye[2] /baɪ/ *interjection* (also **bye-bye**) *colloquial* goodbye.

bygone *adjective* past, departed. □ **let bygones be bygones** forgive and forget past quarrels.

by-law *noun* regulation made by local authority etc.

byline *noun* line in newspaper etc. naming writer of article etc.

bypass • *noun* main road round town or its centre. • *verb* avoid.

byre /baɪə/ *noun* cowshed.

bystander *noun* person present but not taking part.

byte /baɪt/ *noun Computing* group of 8 binary digits, often representing one character.

Byzantine /bɪˈzæntaɪn/ *adjective* of Byzantium or E. Roman Empire; of architectural etc. style developed in Eastern Empire; complicated, underhand.

Cc

C[1] *noun* (also **c**) (Roman numeral) 100.

C[2] *abbreviation* Celsius; centigrade.

c.[1] *abbreviation* century; cent(s).

c.[2] *abbreviation circa*.

ca. *abbreviation circa*.

CAA *abbreviation* Civil Aviation Authority.

cab *noun* taxi; driver's compartment in lorry, train, crane, etc.

cabal /kə'bæl/ *noun* secret intrigue; political clique.

cabaret /'kæbəreɪ/ *noun* entertainment in restaurant etc.

cabbage /'kæbɪdʒ/ *noun* vegetable with green or purple leaves forming a round head; *colloquial* dull or inactive person. □ **cabbage white** kind of white butterfly.

cabby /'kæbɪ/ *noun* (*plural* **-ies**) *colloquial* taxi driver.

caber /'keɪbə/ *noun* tree trunk tossed as sport in Scotland.

cabin /'kæbɪn/ *noun* small shelter or house, esp. of wood; room or compartment in aircraft, ship, etc. □ **cabin boy** boy steward on ship; **cabin cruiser** large motor boat with accommodation.

cabinet /'kæbɪnɪt/ *noun* cupboard or case for storing or displaying things; casing of radio, television, etc.; (**Cabinet**) group of senior ministers in government. □ **cabinetmaker** skilled joiner.

cable /'keɪb(ə)l/ ● *noun* encased group of insulated wires for transmitting electricity etc.; thick rope of wire or hemp; cablegram. ● *verb* (**-ling**) send (message) or inform (person) by cable. □ **cable car** small cabin on loop of cable for carrying passengers up and down mountain etc.; **cablegram** message sent by undersea cable etc.; **cable stitch** knitting stitch resembling twisted rope; **cable television** transmission of television programmes by cable to subscribers.

caboodle /kə'bu:d(ə)l/ *noun*. □ **the whole caboodle** *slang* the whole lot.

caboose /kə'bu:s/ *noun* kitchen on ship's deck; *US* guard's van on train.

cabriolet /kæbrɪəʊ'leɪ/ *noun* car with folding top.

cacao /kə'kaʊ/ *noun* (*plural* **-s**) seed from which cocoa and chocolate are made; tree bearing it.

cache /kæʃ/ ● *noun* hiding place for treasure, supplies, etc.; things so hidden. ● *verb* (**-ching**) place in cache.

cachet /'kæʃeɪ/ *noun* prestige; distinguishing mark or seal; flat capsule for medicine.

cack-handed /kæk'hændɪd/ *adjective colloquial* clumsy, left-handed.

cackle /'kæk(ə)l/ ● *noun* clucking of hen; raucous laugh; noisy chatter. ● *verb* (**-ling**) emit cackle; chatter noisily.

cacophony /kə'kɒfənɪ/ *noun* (*plural* **-ies**) harsh discordant sound. □ **cacophonous** *adjective*.

cactus /'kæktəs/ *noun* (*plural* **-ti** /-taɪ/ or **-tuses**) plant with thick fleshy stem and usually spines but no leaves.

CAD *abbreviation* computer-aided design.

cad *noun* man who behaves dishonourably. □ **caddish** *adjective*.

cadaver /kə'dævə/ *noun* corpse. □ **cadaverous** *adjective*.

caddie /'kædɪ/ (also **caddy**) ● *noun* (*plural* **-ies**) golfer's attendant carrying clubs etc. ● *verb* (**caddying**) act as caddie.

caddis /'kædɪs/ *noun* □ **caddis-fly** small nocturnal insect living near water; **caddis-worm** larva of caddis-fly.

caddy[1] /'kædɪ/ *noun* (*plural* **-ies**) small container for tea.

caddy[2] = CADDIE.

cadence /'keɪd(ə)ns/ *noun* rhythm; fall in pitch of voice; tonal inflection; close of musical phrase.

cadenza /kə'denzə/ *noun Music* virtuoso passage for soloist during concerto.

cadet /kə'det/ *noun* young trainee in armed services or police force.

cadge *verb* (**-ging**) *colloquial* get or seek by begging.

cadi /ˈkɑːdɪ/ *noun* (*plural* **-s**) judge in Muslim country.

cadmium /ˈkædmɪəm/ *noun* soft bluish-white metallic element.

cadre /ˈkɑːdə/ *noun* basic unit, esp. of servicemen; group of esp. Communist activists.

caecum /ˈsiːkəm/ *noun* (*US* **cecum**) (*plural* **-ca**) pouch between small and large intestines.

Caerphilly /keəˈfɪlɪ/ *noun* kind of mild pale cheese.

Caesarean /sɪˈzeərɪən/ (*US* **Cesarean, Cesarian**) ● *adjective* (of birth) effected by Caesarean section. ● *noun* (in full **Caesarean section**) delivery of child by cutting into mother's abdomen.

caesura /sɪˈzjʊərə/ *noun* (*plural* **-s**) pause in line of verse.

café /ˈkæfeɪ/ *noun* coffee house, restaurant.

cafeteria /kæfɪˈtɪərɪə/ *noun* self-service restaurant.

cafetière /kæfəˈtjeə/ *noun* coffee pot with plunger for pressing grounds to bottom.

caffeine /ˈkæfiːn/ *noun* alkaloid stimulant in tea leaves and coffee beans.

caftan /ˈkæftæn/ *noun* (also **kaftan**) long tunic worn by men in Near East; long loose dress.

cage ● *noun* structure of bars or wires, esp. for confining animals; open framework, esp. lift in mine etc. ● *verb* (**-ging**) confine in cage.

cagey /ˈkeɪdʒɪ/ *adjective* (**-ier, -iest**) *colloquial* cautious and non-committal. □ **cagily** *adverb*.

cagoule /kəˈguːl/ *noun* light hooded windproof jacket.

cahoots /kəˈhuːts/ *plural noun* □ **in cahoots** *slang* in collusion.

caiman = CAYMAN.

cairn /keən/ *noun* mound of stones. □ **cairn terrier** small shaggy short-legged terrier.

cairngorm /ˈkeəngɔːm/ *noun* yellow or wine-coloured semiprecious stone.

caisson /ˈkeɪs(ə)n/ *noun* watertight chamber for underwater construction work.

cajole /kəˈdʒəʊl/ *verb* (**-ling**) persuade by flattery, deceit, etc. □ **cajolery** *noun*.

cake ● *noun* mixture of flour, butter, eggs, sugar, etc. baked in oven; flattish compact mass. ● *verb* (**-king**) form into compact mass; (usually + *with*) cover (with sticky mass).

calabash /ˈkæləbæʃ/ *noun* tropical American tree bearing gourds; bowl or pipe made from gourd.

calabrese /ˈkæləbriːs, kæləˈbreɪsɪ/ *noun* variety of broccoli.

calamine /ˈkæləmaɪn/ *noun* powdered zinc carbonate and ferric oxide used in skin lotion.

calamity /kəˈlæmɪtɪ/ *noun* (*plural* **-ies**) disaster. □ **calamitous** *adjective*.

calcareous /kælˈkeərɪəs/ *adjective* of or containing calcium carbonate.

calceolaria /kælsɪəˈleərɪə/ *noun* plant with slipper-shaped flowers.

calcify /ˈkælsɪfaɪ/ *verb* (**-ies, -ied**) harden by deposit of calcium salts. □ **calcification** *noun*.

calcine /ˈkælsaɪn/ *verb* (**-ning**) decompose or be decomposed by roasting or burning. □ **calcination** /-sɪn-/ *noun*.

calcium /ˈkælsɪəm/ *noun* soft grey metallic element.

calculate /ˈkælkjʊleɪt/ *verb* (**-ting**) ascertain or forecast by exact reckoning; plan deliberately. □ **calculable** *adjective*; **calculation** *noun*.

calculated *adjective* done with awareness of likely consequences; (+ *to do*) designed.

calculating *adjective* scheming, mercenary.

calculator *noun* device (esp. small electronic one) used for making calculations.

calculus /ˈkælkjʊləs/ *noun* (*plural* **-luses** or **-li** /-laɪ/) *Mathematics* particular method of calculation; stone in body.

Caledonian /kælɪˈdəʊnɪən/ *literary* ● *adjective* of Scotland. ● *noun* Scot.

calendar /ˈkælɪndə/ *noun* system fixing year's beginning, length, and subdivision; chart etc. showing such subdivisions; list of special dates or events. □ **calendar year** period from 1 Jan. to 31 Dec. inclusive.

calends /ˈkælendz/ *plural noun* (also **kalends**) first of month in ancient Roman calendar.

calf¹ /kɑːf/ *noun* (*plural* **calves** /kɑːvz/) young cow, bull, elephant, whale, etc.; calf leather. □ **calf-love** romantic adolescent love.

calf² /kɑːf/ *noun* (*plural* **calves** /kɑːvz/) fleshy hind part of human leg below knee.

calibrate /ˈkælɪbreɪt/ *verb* (**-ting**) mark (gauge) with scale of readings; correlate readings of (instrument) with standard; find calibre of (gun). □ **calibration** *noun.*

calibre /ˈkælɪbə/ *noun* (*US* **caliber**) internal diameter of gun or tube; diameter of bullet or shell; strength or quality of character; ability; importance.

calico /ˈkælɪkəʊ/ ● *noun* (*plural* **-es** or *US* **-s**) cotton cloth, esp. white or unbleached; *US* printed cotton cloth. ● *adjective* of calico; *US* multicoloured.

caliper = CALLIPER.

caliph /ˈkeɪlɪf/ *noun historical* chief Muslim civil and religious ruler.

calk *US* = CAULK.

call /kɔːl/ ● *verb* (often + *out*) cry, shout, speak loudly; emit characteristic sound; communicate with by radio or telephone; summon; make brief visit; order to take place; name, describe, or regard as; rouse from sleep; (+ *for*) demand. ● *noun* shout; bird's cry; brief visit; telephone conversation; summons; need, demand. □ **call box** telephone box; **call-girl** prostitute accepting appointments by telephone; **call in** withdraw from circulation, seek advice or services of; **call off** cancel, order (pursuer) to desist; **call the shots, tune** *colloquial* be in control; **call up** *verb* telephone, recall, summon (esp. to do military service); **call-up** *noun* summons to do military service. □ **caller** *noun.*

calligraphy /kəˈlɪgrəfɪ/ *noun* beautiful handwriting; art of this. □ **calligrapher** *noun*; **calligraphic** /kælɪˈgræfɪk/ *adjective.*

calling *noun* profession, occupation; vocation.

calliper /ˈkælɪpə/ *noun* (also **caliper**) metal splint to support leg; (in *plural*) compasses for measuring diameters.

callisthenics /kælɪsˈθenɪks/ *plural noun* exercises for fitness and grace. □ **callisthenic** *adjective.*

callosity /kəˈlɒsɪtɪ/ *noun* (*plural* **-ies**) callus.

callous /ˈkæləs/ *adjective* unfeeling, unsympathetic; (also **calloused**) (of skin) hardened. □ **callously** *adverb*; **callousness** *noun.*

callow /ˈkæləʊ/ *adjective* inexperienced, immature.

callus /ˈkæləs/ *noun* (*plural* **calluses**) area of hard thick skin.

calm /kɑːm/ ● *adjective* tranquil, windless; not agitated. ● *noun* calm condition or period. ● *verb* (often + *down*) make or become calm. □ **calmly** *adverb*; **calmness** *noun.*

calomel /ˈkæləmel/ *noun* compound of mercury used as laxative.

Calor gas /ˈkælə/ *noun proprietary term* liquefied butane under pressure in containers for domestic use.

calorie /ˈkælərɪ/ *noun* unit of heat, amount required to raise temperature of one gram (**small calorie**) or one kilogram (**large calorie**) of water by 1°C.

calorific /kæləˈrɪfɪk/ *adjective* producing heat.

calumny /ˈkæləmnɪ/ *noun* (*plural* **-ies**) slander; malicious misrepresentation. □ **calumnious** /kəˈlʌm-/ *adjective.*

calvados /ˈkælvədɒs/ *noun* apple brandy.

calve /kɑːv/ *verb* (**-ving**) give birth to (calf).

calves *plural* of CALF[1,2].

Calvinism /ˈkælvɪnɪz(ə)m/ *noun* Calvin's theology, stressing predestination. □ **Calvinist** *noun & adjective*; **Calvinistic** /-ˈnɪs-/ *adjective.*

calx *noun* (*plural* **calces** /ˈkælsiːz/) powdery residue left after heating of ore or mineral.

calypso /kəˈlɪpsəʊ/ *noun* (*plural* **-s**) W. Indian song with improvised usually topical words.

calyx /ˈkeɪlɪks/ *noun* (*plural* **calyces** /-lɪsiːz/ or **-es**) leaves forming protective case of flower in bud.

cam *noun* projection on wheel etc., shaped to convert circular into reciprocal or variable motion.

camaraderie /kæmə'rɑ:dərɪ/ *noun* friendly comradeship.

camber /'kæmbə/ ● *noun* convex surface of road, deck, etc. ● *verb* construct with camber.

cambric /'kæmbrɪk/ *noun* fine linen or cotton cloth.

camcorder /'kæmkɔ:də/ *noun* combined portable video camera and recorder.

came *past* of COME.

camel /'kæm(ə)l/ *noun* long-legged ruminant with one hump (**Arabian camel**) or two humps (**Bactrian camel**); fawn colour.

camellia /kə'mi:lɪə/ *noun* evergreen flowering shrub.

Camembert /'kæməmbeə/ *noun* kind of soft creamy cheese.

cameo /'kæmɪəʊ/ *noun* (*plural* **-s**) small piece of hard stone carved in relief; short literary sketch or acted scene; small part in play or film.

camera /'kæmrə/ *noun* apparatus for taking photographs or for making motion film or television pictures. □ **cameraman** operator of film or television camera; **in camera** in private.

camiknickers /'kæmɪnɪkəz/ *plural noun* woman's knickers and vest combined.

camisole /'kæmɪsəʊl/ *noun* woman's under-bodice.

camomile /'kæməmaɪl/ *noun* (also **chamomile**) aromatic herb with flowers used to make tea.

camouflage /'kæməflɑ:ʒ/ ● *noun* disguising of soldiers, tanks, etc., so that they blend into background; such disguise; animal's natural blending colouring. ● *verb* (**-ging**) hide by camouflage.

camp¹ ● *noun* place where troops are lodged or trained; temporary accommodation of tents, huts, etc., for detainees, holiday-makers, etc.; fortified site; party supporters regarded collectively. ● *verb* set up or live in camp. □ **camp bed** portable folding bed; **camp follower** civilian worker in military camp, disciple; **campsite** place for camping.

camp² *colloquial* ● *adjective* affected, theatrically exaggerated; effeminate; homosexual. ● *noun* camp manner. ● *verb* behave or do in camp way.

campaign /kæm'peɪn/ ● *noun* organized course of action, esp. to gain publicity; series of military operations. ● *verb* take part in campaign. □ **campaigner** *noun*.

campanile /kæmpə'ni:lɪ/ *noun* bell tower, usually free-standing.

campanology /kæmpə'nɒlədʒɪ/ *noun* study of bells; bell-ringing. □ **campanologist** *noun*.

campanula /kæm'pænjʊlə/ *noun* plant with bell-shaped flowers.

camper *noun* person who camps; motor vehicle with beds.

camphor /'kæmfə/ *noun* pungent crystalline substance used in medicine and formerly mothballs.

camphorate *verb* (**-ting**) impregnate with camphor.

campion /'kæmpɪən/ *noun* wild plant with usually pink or white notched flowers.

campus /'kæmpəs/ *noun* (*plural* **-es**) grounds of university or college.

camshaft *noun* shaft carrying cam(s).

can¹ *auxiliary verb* (*3rd singular present* **can**; *past* **could** /kʊd/) be able to; have the potential to; be permitted to.

can² ● *noun* metal vessel for liquid; sealed tin container for preservation of food or drink; (in *plural*) *slang* headphones; (**the can**) *slang* prison, *US* lavatory. ● *verb* (**-nn-**) preserve (food etc.) in can. □ **canned music** pre-recorded music; **carry the can** bear responsibility; **in the can** *colloquial* completed.

canal /kə'næl/ *noun* artificial inland waterway; tubular duct in plant or animal.

canalize /'kænəlaɪz/ *verb* (also **-ise**) (**-zing** or **-sing**) convert (river) into canal; provide (area) with canal(s); channel. □ **canalization** *noun*.

canapé /'kænəpeɪ/ *noun* small piece of bread or pastry with savoury topping.

canard /'kænɑ:d/ *noun* unfounded rumour.

canary /kə'neərɪ/ *noun* (*plural* **-ies**) small songbird with yellow feathers.

canasta /kə'næstə/ *noun* card games resembling rummy.

cancan /'kænkæn/ *noun* high-kicking dance.

cancel /'kæns(ə)l/ *verb* (**-ll-**; *US* **-l-**) revoke, discontinue (arrangement); delete; mark (ticket, stamp, etc.) to invalidate it; annul; (often + *out*) neutralize, counterbalance; *Mathematics* strike out (equal factor) on each side of equation etc. □ **cancellation** *noun*.

cancer /'kænsə/ *noun* malignant tumour, disease caused by this; corruption; (**Cancer**) fourth sign of zodiac. □ **cancerous** *adjective*.

candela /kæn'di:lə/ *noun* SI unit of luminous intensity.

candelabrum /kændɪ'lɑ:brəm/ *noun* (also **-bra**) (*plural* **-bra**, *US* **-brums**, **-bras**) large branched candlestick or lampholder.

candid /'kændɪd/ *adjective* frank; (of photograph) taken informally, usually without subject's knowledge. □ **candidly** *adverb*.

candidate /'kændɪdət/ *noun* person nominated for, seeking, or likely to gain, office, position, award, etc.; person entered for exam. □ **candidacy** *noun*; **candidature** *noun*.

candle /'kænd(ə)l/ *noun* (usually cylindrical) block of wax or tallow enclosing wick which gives light when burning. □ **candlelight** light from candle(s); **candlelit** lit by candle(s); **candlestick** holder for candle(s); **candlewick** thick soft yarn, tufted material made from this.

candour /'kændə/ *noun* (*US* **candor**) frankness.

candy /'kændɪ/ ● *noun* (*plural* **-ies**) (in full **sugar-candy**) sugar crystallized by repeated boiling and evaporation; *US* sweets, a sweet. ● *verb* (**-ies**, **-ied**) (usually as **candied** *adjective*) preserve (fruit etc.) in candy. □ **candyfloss** fluffy mass of spun sugar; **candy stripe** alternate white and esp. pink stripes.

candytuft /'kændɪtʌft/ *noun* plant with white, pink, or purple flowers in tufts.

cane ● *noun* hollow jointed stem of giant reed or grass, or solid stem of slender palm, used for wickerwork or as walking stick, plant support, instrument of punishment, etc.; sugar cane. ● *verb* (**-ning**) beat with cane; weave cane into (chair etc.).

canine /'keɪnaɪn/ ● *adjective* of a dog or dogs. ● *noun* dog; (in full **canine tooth**) tooth between incisors and molars.

canister /'kænɪstə/ *noun* small usually metal box for tea etc.; cylinder of shot, tear gas, etc.

canker /'kæŋkə/ ● *noun* disease of trees and plants; ulcerous ear disease of animals; corrupting influence. ● *verb* infect with canker; corrupt. □ **cankerous** *adjective*.

cannabis /'kænəbɪs/ *noun* hemp plant; parts of it used as narcotic.

cannelloni /kænə'ləʊnɪ/ *plural noun* tubes of pasta stuffed with savoury mixture.

cannibal /'kænɪb(ə)l/ *noun* person or animal that eats its own species. □ **cannibalism** *noun*; **cannibalistic** /-'lɪs-/ *adjective*.

cannibalize /'kænɪbəlaɪz/ *verb* (also **-ise**) (**-zing** or **-sing**) use (machine etc.) as source of spare parts.

cannon /'kænən/ ● *noun* automatic aircraft gun firing shells; *historical* (*plural* usually same) large gun; hitting of two balls successively by player's ball in billiards. ● *verb* (usually + *against*, *into*) collide. □ **cannon ball** *historical* large ball fired by cannon.

cannot /'kænɒt/ can not.

canny /'kænɪ/ *adjective* (**-ier**, **-iest**) shrewd; thrifty.

canoe /kə'nu:/ ● *noun* light narrow boat, usually paddled. ● *verb* (**-noes**, **-noed**, **-noeing**) travel in canoe. □ **canoeist** *noun*.

canon /'kænən/ *noun* general law, rule, principle, or criterion; church decree; member of cathedral chapter; set of (esp. sacred) writings accepted as genuine; part of RC Mass containing words of consecration; *Music* piece with different parts taking up same theme successively. □ **canon law** ecclesiastical law.

canonical /kə'nɒnɪk(ə)l/ ● *adjective* according to canon law; included in canon of Scripture; authoritative, accepted; of (member of) cathedral chapter. ● *noun* (in *plural*) canonical dress of clergy.

canonize /'kænənaɪz/ *verb* (also **-ise**) (**-zing** or **-sing**) declare officially to be saint. □ **canonization** *noun*.

canopy /'kænəpɪ/ • *noun* (*plural* **-ies**) suspended covering over throne, bed, etc.; sky; overhanging shelter; rooflike projection. • *verb* (**-ies**, **-ied**) supply or be canopy to.

cant[1] • *noun* insincere pious or moral talk; language peculiar to class, profession, etc.; jargon. • *verb* use cant.

cant[2] • *noun* slanting surface, bevel; oblique push or jerk; tilted position. • *verb* push or pitch out of level.

can't /kɑ:nt/ can not.

cantabile /kæn'tɑ:bɪleɪ/ *Music* • *adverb & adjective* in smooth flowing style. • *noun* cantabile passage or movement.

cantaloupe /'kæntəlu:p/ *noun* (also **cantaloup**) small round ribbed melon.

cantankerous /kæn'tæŋkərəs/ *adjective* bad-tempered, quarrelsome. □ **cantankerously** *adverb*; **cantankerousness** *noun*.

cantata /kæn'tɑ:tə/ *noun* composition for vocal soloists and usually chorus and orchestra.

canteen /kæn'ti:n/ *noun* restaurant for employees in office, factory, etc.; shop for provisions in barracks or camp; case of cutlery; soldier's or camper's water-flask.

canter /'kæntə/ • *noun* horse's pace between trot and gallop. • *verb* (cause to) go at a canter.

canticle /'kæntɪk(ə)l/ *noun* song or chant with biblical text.

cantilever /'kæntɪli:və/ *noun* bracket, beam, etc. projecting from wall to support balcony etc.; beam or girder fixed at one end only. □ **cantilever bridge** bridge made of cantilevers projecting from piers and connected by girders. □ **cantilevered** *adjective*.

canto /'kæntəʊ/ *noun* (*plural* **-s**) division of long poem.

canton • *noun* /'kæntɒn/ subdivision of country, esp. Switzerland. • *verb* /kæn'tu:n/ put (troops) into quarters.

cantonment /kæn'tu:nmənt/ *noun* lodgings of troops.

cantor /'kæntɔ:/ *noun* church choir leader; precentor in synagogue.

canvas /'kænvəs/ *noun* strong coarse cloth used for sails and tents etc. and for oil painting; a painting on canvas.

canvass /'kænvəs/ • *verb* solicit votes (from), ascertain opinions of; seek custom from; propose (idea etc.). • *noun* canvassing. □ **canvasser** *noun*.

canyon /'kænjən/ *noun* deep gorge.

CAP *abbreviation* Common Agricultural Policy (of EC).

cap • *noun* soft brimless hat, often with peak; head-covering of nurse etc.; cap as sign of membership of sports team; academic mortarboard; cover resembling cap, or designed to close, seal, or protect something; contraceptive diaphragm; percussion cap; dental crown. • *verb* (**-pp-**) put cap on; cover top or end of; limit; award sports cap to; form top of; surpass.

capable /'keɪpəb(ə)l/ *adjective* competent, able; (+ *of*) having ability, fitness, etc. for. □ **capability** *noun* (*plural* **-ies**); **capably** *adverb*.

capacious /kə'peɪʃəs/ *adjective* roomy. □ **capaciousness** *noun*.

capacitance /kə'pæsɪt(ə)ns/ *noun* ability to store electric charge.

capacitor /kə'pæsɪtə/ *noun* type of device for storing electric charge.

capacity /kə'pæsɪtɪ/ • *noun* (*plural* **-ies**) power to contain, receive, experience, or produce; maximum amount that can be contained etc.; mental power; position or function. • *adjective* fully occupying available space etc.

caparison /kə'pærɪs(ə)n/ *literary* • *noun* horse's trappings; finery. • *verb* adorn.

cape[1] *noun* short cloak.

cape[2] *noun* headland, promontory; (**the Cape**) Cape of Good Hope.

caper[1] /'keɪpə/ • *verb* jump playfully. • *noun* playful leap; prank; *slang* illicit activity.

caper[2] /'keɪpə/ *noun* bramble-like shrub; (in *plural*) its pickled buds.

capercaillie /kæpə'keɪlɪ/ *noun* (also **capercailzie** /-'keɪlzɪ/) large European grouse.

capillarity /kæpɪ'lærɪtɪ/ *noun* rise or depression of liquid in narrow tube.

capillary /kə'pɪlərɪ/ • *adjective* of hair; of narrow diameter. • *noun* (*plural* **-ies**) capillary tube or blood vessel. □ **capillary action** capillarity.

capital /'kæpɪt(ə)l/ • *noun* chief town or city of a country or region; money etc.

with which company starts in business; accumulated wealth; capital letter; head of column or pillar. ● *adjective* involving punishment by death; most important; *colloquial* excellent. □ **capital gain** profit from sale of investments or property; **capital goods** machinery, plant, etc.; **capital letter** large kind, used to begin sentence or name; **capital transfer tax** *historical* tax levied on transfer of capital by gift or bequest etc.

capitalism *noun* economic and political system dependent on private capital and profit-making.

capitalist ● *noun* person using or possessing capital; advocate of capitalism. ● *adjective* of or favouring capitalism. □ **capitalistic** /-'lɪs-/ *adjective*.

capitalize *verb* (also **-ise**) (**-zing** or **-sing**) convert into or provide with capital; write (letter of alphabet) as capital, begin (word) with capital letter; (+ *on*) use to one's advantage. □ **capitalization** *noun*.

capitulate /kə'pɪtjʊleɪt/ *verb* (**-ting**) surrender. □ **capitulation** *noun*.

capon /'keɪpən/ *noun* castrated cock.

cappuccino /kæpʊ'tʃi:nəʊ/ *noun* (*plural* **-s**) frothy milky coffee.

caprice /kə'pri:s/ *noun* whim; lively or fanciful work of music etc.

capricious /kə'prɪʃəs/ *adjective* subject to whims, unpredictable. □ **capriciously** *adverb*; **capriciousness** *noun*.

Capricorn /'kæprɪkɔ:n/ *noun* tenth sign of zodiac.

capsicum /'kæpsɪkəm/ *noun* plant with edible fruits; red, green, or yellow fruit of this.

capsize /kæp'saɪz/ *verb* (**-zing**) (of boat) be overturned; overturn (boat).

capstan /'kæpst(ə)n/ *noun* thick revolving cylinder for winding cable etc.; revolving spindle controlling speed of tape on tape recorder. □ **capstan lathe** lathe with revolving tool holder.

capsule /'kæpsju:l/ *noun* small soluble case enclosing medicine; detachable compartment of spacecraft or nosecone of rocket; enclosing membrane; dry fruit releasing seeds when ripe.

Capt. *abbreviation* Captain.

captain /'kæptɪn/ ● *noun* chief, leader; leader of team; commander of ship; pilot of civil aircraft; army officer next above lieutenant. ● *verb* be captain of. □ **captaincy** *noun* (*plural* **-ies**).

caption /'kæpʃ(ə)n/ ● *noun* wording appended to illustration or cartoon; wording on cinema or television screen; heading of chapter, article etc. ● *verb* provide with caption.

captious /'kæpʃəs/ *adjective* fault-finding.

captivate /'kæptɪveɪt/ *verb* (**-ting**) fascinate, charm. □ **captivation** *noun*.

captive /'kæptɪv/ ● *noun* confined or imprisoned person or animal. ● *adjective* taken prisoner; confined; unable to escape. □ **captivity** /-'tɪv-/ *noun*.

captor /'kæptə/ *noun* person who captures.

capture /'kæptʃə/ ● *verb* (**-ring**) take prisoner; seize; portray in permanent form; record on film or for use in computer. ● *noun* act of capturing; thing or person captured.

Capuchin /'kæpjʊtʃɪn/ *noun* friar of branch of Franciscans; (**capuchin**) monkey with hair like black hood.

car *noun* motor vehicle for driver and small number of passengers; railway carriage of specified type; *US* any railway carriage or van. □ **car bomb** terrorist bomb placed in or under parked car; **car boot sale** sale of goods from (tables stocked from) boots of cars; **car park** area for parking cars; **car phone** radio-telephone for use in car etc.; **carport** roofed open-sided shelter for car; **carsick** nauseous through car travel.

caracul = KARAKUL.

carafe /kə'ræf/ *noun* glass container for water or wine.

caramel /'kærəmel/ *noun* burnt sugar or syrup; kind of soft toffee. □ **caramelize** *verb*.

carapace /'kærəpeɪs/ *noun* upper shell of tortoise or crustacean.

carat /'kærət/ *noun* unit of weight for precious stones; measure of purity of gold.

caravan /'kærəvæn/ *noun* vehicle equipped for living in and usually towed by car; people travelling together, esp. across desert. □ **caravanner** *noun.*

caravanserai /kærə'vænsəraɪ/ *noun* Eastern inn with central courtyard.

caravel /'kærəvel/ *noun historical* small light fast ship.

caraway /'kærəweɪ/ *noun* plant with small aromatic fruit (**caraway seed**) used in cakes etc.

carbide /'kɑ:baɪd/ *noun* binary compound of carbon.

carbine /'kɑ:baɪn/ *noun* kind of short rifle.

carbohydrate /kɑ:bə'haɪdreɪt/ *noun* energy-producing compound of carbon, hydrogen, and oxygen.

carbolic /kɑ:'bɒlɪk/ *noun* (in full **carbolic acid**) kind of disinfectant and antiseptic. □ **carbolic soap** soap containing this.

carbon /'kɑ:bən/ *noun* non-metallic element occurring as diamond, graphite, and charcoal, and in all organic compounds; carbon copy, carbon paper. □ **carbon copy** copy made with carbon paper; **carbon dating** determination of age of object from decay of carbon-14; **carbon dioxide** gas found in atmosphere and formed by respiration; **carbon fibre** thin filament of carbon used as strengthening material; **carbon-14** radioactive carbon isotope of mass 14; **carbon monoxide** poisonous gas formed by burning carbon incompletely; **carbon paper** thin carbon-coated paper for making copies; **carbon tax** tax on fuels producing greenhouse gases; **carbon-12** stable isotope of carbon used as a standard.

carbonate /'kɑ:bəneɪt/ ● *noun* salt of carbonic acid. ● *verb* (**-ting**) fill with carbon dioxide.

carbonic /kɑ:'bɒnɪk/ *adjective* containing carbon. □ **carbonic acid** weak acid formed from carbon dioxide in water.

carboniferous /kɑ:bə'nɪfərəs/ *adjective* producing coal.

carbonize /'kɑ:bənaɪz/ *verb* (also **-ise**) (**-zing** or **-sing**) reduce to charcoal or coke; convert to carbon; coat with carbon. □ **carbonization** *noun.*

carborundum /kɑ:bə'rʌndəm/ *noun* compound of carbon and silicon used esp. as abrasive.

carboy /'kɑ:bɔɪ/ *noun* large globular glass bottle.

carbuncle /'kɑ:bʌŋk(ə)l/ *noun* severe skin abscess; bright red jewel.

carburettor /kɑ:bə'retə/ *noun* apparatus mixing air with petrol vapour in internal-combustion engine.

carcass /'kɑ:kəs/ *noun* (also **carcase**) dead body of animal or bird or (*colloquial*) person; framework; worthless remains.

carcinogen /kɑ:'sɪnədʒ(ə)n/ *noun* substance producing cancer. □ **carcinogenic** /-'dʒen-/ *adjective.*

card[1] *noun* thick stiff paper or thin pasteboard; piece of this for writing or printing on, esp. to send greetings, to identify person, or to record information; small rectangular piece of plastic used to obtain credit etc.; playing card; (in *plural*) card-playing; (in *plural*) *colloquial* employee's tax etc. documents; programme of events at race meeting etc.; *colloquial* eccentric person. □ **card-carrying** registered as member (esp. of political party); **card index** index with separate card for each item; **cardphone** public telephone operated by card instead of money; **card-sharp** swindler at card games; **card vote** block vote.

card[2] ● *noun* wire brush etc. for raising nap on cloth etc. ● *verb* brush with card.

cardamom /'kɑ:dəməm/ *noun* seeds of SE Asian aromatic plant used as spice.

cardboard *noun* pasteboard or stiff paper.

cardiac /'kɑ:dɪæk/ *adjective* of the heart.

cardigan /'kɑ:dɪgən/ *noun* knitted jacket.

cardinal /'kɑ:dɪn(ə)l/ ● *adjective* chief, fundamental; deep scarlet. ● *noun* one of leading RC dignitaries who elect Pope. □ **cardinal number** number representing quantity (1, 2, 3, etc.); compare ORDINAL.

cardiogram /'kɑ:dɪəʊgræm/ *noun* record of heart movements.

cardiograph /'kɑ:dɪəʊgrɑ:f/ *noun* instrument recording heart movements. □ **cardiographer** /-'ɒgrəfə/ *noun*; **cardiography** /-'ɒgrəfɪ/ *noun.*

cardiology /kɑ:dɪ'ɒlədʒɪ/ *noun* branch of medicine concerned with heart. □ **cardiologist** *noun.*

cardiovascular /kɑ:dɪəʊ'væskjʊlə/ *adjective* of heart and blood vessels.

care /keə/ ● *noun* (cause of) anxiety or concern; serious attention; caution; protection, charge; task. ● *verb* (**-ring**) (usually + *about*, *for*, *whether*) feel concern or interest or affection. □ **in care** (of child) under local authority supervision; **take care** be careful, (+ *to do*) not fail or neglect; **take care of** look after, deal with, dispose of.

careen /kə'ri:n/ *verb* turn (ship) on side for repair; move or swerve wildly.

career /kə'rɪə/ ● *noun* professional etc. course through life; profession or occupation; swift course. ● *adjective* pursuing or wishing to pursue a career; working permanently in specified profession. ● *verb* move or swerve wildly.

careerist *noun* person predominantly concerned with personal advancement.

carefree *adjective* light-hearted, joyous.

careful *adjective* painstaking; cautious; taking care, not neglecting. □ **carefully** *adverb*; **carefulness** *noun*.

careless *adjective* lacking care or attention; unthinking, insensitive; light-hearted. □ **carelessly** *adverb*; **carelessness** *noun*.

carer *noun* person who cares for sick or elderly person, esp. at home.

caress /kə'res/ ● *verb* touch lovingly. ● *noun* loving touch.

caret /'kærət/ *noun* mark indicating insertion in text.

caretaker ● *noun* person in charge of maintenance of building. ● *adjective* taking temporary control.

careworn *adjective* showing effects of prolonged anxiety.

cargo /'kɑ:gəʊ/ *noun* (*plural* **-es** or **-s**) goods carried by ship or aircraft.

Caribbean /kærə'bi:ən/ *adjective* of the West Indies.

caribou /'kærɪbu:/ *noun* (*plural* same) N. American reindeer.

caricature /'kærɪkətʃʊə/ ● *noun* grotesque usually comically exaggerated representation. ● *verb* (**-ring**) make or give caricature of. □ **caricaturist** *noun*.

caries /'keəri:z/ *noun* (*plural* same) decay of tooth or bone.

carillon /kə'rɪljən/ *noun* set of bells sounded from keyboard or mechanically; tune played on this.

Carmelite /'kɑ:məlaɪt/ ● *noun* friar of Order of Our Lady of Mount Carmel; nun of similar order. ● *adjective* of Carmelites.

carminative /'kɑ:mɪnətɪv/ ● *adjective* relieving flatulence. ● *noun* carminative drug.

carmine /'kɑ:maɪn/ ● *adjective* of vivid crimson colour. ● *noun* this colour; pigment from cochineal.

carnage /'kɑ:nɪdʒ/ *noun* great slaughter.

carnal /'kɑ:n(ə)l/ *adjective* worldly; sensual; sexual. □ **carnality** *noun*.

carnation /kɑ:'neɪʃ(ə)n/ ● *noun* clove-scented pink; rosy-pink colour. ● *adjective* rosy-pink.

carnelian = CORNELIAN.

carnival /'kɑ:nɪv(ə)l/ *noun* festivities or festival, esp. preceding Lent; merry-making.

carnivore /'kɑ:nɪvɔ:/ *noun* animal or plant that feeds on flesh. □ **carnivorous** /-'nɪvərəs-/ *adjective*.

carob /'kærəb/ *noun* seed pod of Mediterranean tree used as chocolate substitute.

carol /'kær(ə)l/ ● *noun* joyous song, esp. Christmas hymn. ● *verb* (**-ll-**; *US* **-l-**) sing carols; sing joyfully.

carotene /'kærəti:n/ *noun* orange-coloured pigment in carrots etc.

carotid /kə'rɒtɪd/ ● *noun* each of two main arteries carrying blood to head. ● *adjective* of these arteries.

carouse /kə'raʊz/ ● *verb* (**-sing**) have lively drinking party. ● *noun* such party. □ **carousal** *noun*; **carouser** *noun*.

carp[1] *noun* (*plural* same) freshwater fish often bred for food.

carp[2] *verb* find fault, complain. □ **carper** *noun*.

carpal /'kɑ:p(ə)l/ ● *adjective* of the wrist-bones. ● *noun* wrist-bone.

carpel /'kɑ:p(ə)l/ *noun* female reproductive organ of flower.

carpenter /'kɑ:pɪntə/ ● *noun* person skilled in woodwork. ● *verb* do woodwork; make by woodwork. □ **carpentry** *noun*.

carpet /'kɑ:pɪt/ • *noun* thick fabric for covering floors etc.; piece of this; thing resembling this. • *verb* (**-t-**) cover (as) with carpet; *colloquial* rebuke. □ **carpet-bag** travelling bag originally made of carpet-like material; **carpet-bagger** *colloquial* political candidate etc. without local connections; **carpet slipper** soft slipper; **carpet sweeper** implement for sweeping carpets.

carpeting *noun* material for carpets; carpets collectively.

carpus /'kɑ:pəs/ *noun* (*plural* **-pi** /-paɪ/) group of small bones forming wrist.

carrageen /'kærəgi:n/ *noun* edible red seaweed.

carriage /'kærɪdʒ/ *noun* railway passenger vehicle; wheeled horse-drawn passenger vehicle; conveying of goods; cost of this; bearing, deportment; part of machine that carries other parts; gun carriage. □ **carriage clock** portable clock with handle; **carriageway** part of road used by vehicles.

carrier /'kærɪə/ *noun* person or thing that carries; transport or freight company; carrier bag; framework on bicycle for carrying luggage or passenger; person or animal that may transmit disease without suffering from it; aircraft carrier. □ **carrier bag** plastic or paper bag with handles; **carrier pigeon** pigeon trained to carry messages; **carrier wave** high-frequency electromagnetic wave used to convey signal.

carrion /'kærɪən/ *noun* dead flesh; filth. □ **carrion crow** crow feeding on carrion.

carrot /'kærət/ *noun* plant with edible tapering orange root; this root; incentive. □ **carroty** *adjective*.

carry /'kærɪ/ *verb* (**-ies**, **-ied**) support or hold up, esp. while moving; have with one; convey; (often + *to*) take (process etc.) to specified point; involve; transfer (figure) to column of higher value; hold in specified way; (of newspaper etc.) publish; (of radio or television station) broadcast; keep (goods) in stock; (of sound) be audible at a distance; win victory or acceptance for; win acceptance from; capture. □ **carry away** remove, inspire, deprive of self-control; **carrycot** portable cot for baby; **carry forward** transfer (figure) to new page or account; **carry it off** do well under difficulties; **carry off** remove (esp. by force), win, (esp. of disease) kill; **carry on** continue, *colloquial* behave excitedly, (often + *with*) *colloquial* flirt; **carry-on** *colloquial* fuss; **carry out** put into practice; **carry-out** takeaway; **carry over** carry forward, postpone; **carry through** complete, bring safely out of difficulties.

cart • *noun* open usually horse-drawn vehicle for carrying loads; light vehicle for pulling by hand. • *verb* carry in cart; *slang* carry with difficulty. □ **cart horse** horse of heavy build; **cartwheel** wheel of cart, sideways somersault with arms and legs extended.

carte blanche /kɑ:t 'blɑ̃ʃ/ *noun* full discretionary power.

cartel /kɑ:'tel/ *noun* union of suppliers etc. set up to control prices.

Cartesian coordinates /kɑ:'ti:zɪən/ *plural noun* system of locating point by its distance from two perpendicular axes.

Carthusian /kɑ:'θju:zɪən/ • *noun* monk of contemplative order founded by St Bruno. • *adjective* of this order.

cartilage /'kɑ:tɪlɪdʒ/ *noun* firm flexible connective tissue in vertebrates. □ **cartilaginous** /-'lædʒ-/ *adjective*.

cartography /kɑ:'tɒgrəfɪ/ *noun* map-drawing. □ **cartographer** *noun*; **cartographic** /-tə'græf-/ *adjective*.

carton /'kɑ:t(ə)n/ *noun* light esp. cardboard container.

cartoon /kɑ:'tu:n/ *noun* humorous esp. topical drawing in newspaper etc.; sequence of drawings telling story; such sequence animated on film; full-size preliminary design for work of art. □ **cartoonist** *noun*.

cartouche /kɑ:'tu:ʃ/ *noun* scroll-like ornament; oval enclosing name and title of pharaoh.

cartridge /'kɑ:trɪdʒ/ *noun* case containing explosive charge or bullet; sealed container of film etc.; component carrying stylus on record player; ink-container for insertion in pen. □ **cartridge-belt** belt with pockets or loops for cartridges; **cartridge paper** thick paper for drawing etc.

carve *verb* (**-ving**) make or shape by cutting; cut pattern etc. in; (+ *into*) form pattern etc. from; cut (meat) into slices. □ **carve out** take from larger whole; **carve up** subdivide, drive aggressively into path of (another vehicle). □ **carver** *noun.*

carvery *noun* (*plural* **-ies**) restaurant etc. with joints displayed for carving.

carving *noun* carved object, esp. as work of art. □ **carving knife** knife for carving meat.

cascade /kæs'keɪd/ ● *noun* waterfall, esp. one in series. ● *verb* (**-ding**) fall in or like cascade.

case[1] /keɪs/ *noun* instance of something occurring; hypothetical or actual situation; person's illness, circumstances, etc. as regarded by doctor, social worker, etc.; such a person; crime etc. investigated by detective or police; suit at law; sum of arguments on one side; (valid) set of arguments; *Grammar* relation of word to others in sentence, form of word expressing this. □ **case law** law as established by decided cases; **casework** social work concerned with individual's background; **in any case** whatever the truth or possible outcome is; **in case** in the event that, lest, in provision against a possibility; **in case of** in the event of; **is (not) the case** is (not) so.

case[2] /keɪs/ ● *noun* container or cover enclosing something. ● *verb* (**-sing**) enclose in case; (+ *with*) surround with; *slang* inspect closely, esp. for criminal purpose. □ **case-harden** harden surface of (esp. steel), make callous.

casein /'keɪsɪɪn/ *noun* main protein in milk and cheese.

casement /'keɪsmənt/ *noun* (part of) window hinged to open like door.

cash ● *noun* money in coins or notes; full payment at time of purchase. ● *verb* give or obtain cash for. □ **cash and carry** (esp. wholesaling) system of cash payment for goods removed by buyer, store where this operates; **cashcard** plastic card for use in cash dispenser; **cash crop** crop produced for sale; **cash desk** counter etc. where goods are paid for; **cash dispenser** automatic machine for withdrawal of cash; **cash flow** movement of money into and out of a business; **cash in** obtain cash for, (usually + *on*) *colloquial* profit (from); **cash register** till recording sales, totalling receipts, etc.; **cash up** count day's takings.

cashew /'kæʃu:/ *noun* evergreen tree bearing kidney-shaped edible nut; this nut.

cashier[1] /kæ'ʃɪə/ *noun* person dealing with cash transactions in bank etc.

cashier[2] /kæ'ʃɪə/ *verb* dismiss from service.

cashmere /kæʃ'mɪə/ *noun* fine soft (material of) wool, esp. of Kashmir goat.

casing *noun* enclosing material or cover.

casino /kə'si:nəʊ/ *noun* (*plural* **-s**) public room etc. for gambling.

cask /kɑ:sk/ *noun* barrel, esp. for alcoholic liquor.

casket /'kɑ:skɪt/ *noun* small box for holding valuables; *US* coffin.

cassava /kə'sɑ:və/ *noun* plant with starchy roots; starch or flour from these.

casserole /'kæsərəʊl/ ● *noun* covered dish for cooking food in oven; food cooked in this. ● *verb* (**-ling**) cook in casserole.

cassette /kə'set/ *noun* sealed case containing magnetic tape, film, etc., ready for insertion in tape recorder, camera, etc.

cassia /'kæsɪə/ *noun* tree whose leaves and pod yield senna.

cassis /kæ'si:s/ *noun* blackcurrant flavouring for drinks etc.

cassock /'kæsək/ *noun* long usually black or red clerical garment.

cassowary /'kæsəweərɪ/ *noun* (*plural* **-ies**) large flightless Australian bird.

cast /kɑ:st/ ● *verb* (*past & past participle* **cast**) throw; direct, cause to fall; express (doubts etc.); let down (anchor etc.); shed or lose; register (vote); shape (molten metal etc.) in mould; make (product) thus; (usually + *as*) assign (actor) to role; allocate roles in (play, film, etc.); (+ *in, into*) arrange (facts etc.) in specified form. ● *noun* throwing of missile, dice, fishing line, etc.; thing made in mould; moulded mass of solidified material; actors in play etc.; form, type, or quality; tinge of colour; slight squint; worm-cast. □ **cast about, around** search; **cast aside** abandon; **casting vote** deciding vote when votes on two sides are equal; **cast iron** hard but brittle iron alloy; **cast-iron** of cast iron, very strong, unchallengeable; **cast off** abandon, finish piece of knitting; **cast-off** abandoned or discarded (thing, esp. garment); **cast on** make first row of piece of knitting.

castanet /kæstə'net/ *noun* (usually in *plural*) each of pair of hand-held wooden or ivory shells clicked together in time with esp. Spanish dancing.

castaway /'kɑ:stəweɪ/ ● *noun* shipwrecked person. ● *adjective* shipwrecked.

caste /kɑ:st/ *noun* any of Hindu hereditary classes whose members have no social contact with other classes; exclusive social class.

castellated /'kæstəleɪtɪd/ *adjective* built with battlements. □ **castellation** *noun*.

caster = CASTOR.

castigate /'kæstɪgeɪt/ *verb* (**-ting**) rebuke; punish. □ **castigation** *noun*.

castle /'kɑ:s(ə)l/ ● *noun* large fortified residential building; *Chess* rook. ● *verb* (**-ling**) *Chess* make combined move of king and rook.

castor /'kɑ:stə/ *noun* (also **caster**) small swivelled wheel enabling heavy furniture to be moved; container perforated for sprinkling sugar etc. □ **castor sugar** finely granulated white sugar.

castor oil *noun* vegetable oil used as laxative and lubricant.

castrate /kæs'treɪt/ *verb* (**-ting**) remove testicles of. □ **castration** *noun*.

casual /'kæʒʊəl/ ● *adjective* chance; not regular or permanent; unconcerned, careless; (of clothes etc.) informal. ● *noun* casual worker; (usually in *plural*) casual clothes or shoes. □ **casually** *adverb*.

casualty /'kæʒʊəltɪ/ *noun* (*plural* **-ies**) person killed or injured in war or accident; thing lost or destroyed; casualty department; accident. □ **casualty department** part of hospital for treatment of casualties.

casuist /'kæʒʊɪst/ *noun* person who resolves cases of conscience etc., esp. cleverly but falsely; sophist, quibbler. □ **casuistic** *adjective*; **casuistry** /-'ɪs-/ *noun*.

cat *noun* small furry domestic quadruped; wild animal of same family; *colloquial* malicious or spiteful woman; cat-o'-nine-tails. □ **cat burglar** burglar who enters by climbing to upper storey; **catcall** (make) shrill whistle of disapproval; **catfish** fish with whisker-like filaments round mouth; **cat flap** small flap allowing cat passage through outer door; **catnap** (have) short sleep; **cat-o'-nine-tails** *historical* whip with nine knotted cords; **cat's cradle** child's game with string; **cat's-eye** *proprietary term* reflector stud set into road; **cat's-paw** person used as tool by another; **catsuit** close-fitting garment with trouser legs, covering whole body; **catwalk** narrow walkway; **rain cats and dogs** rain hard.

catachresis /kætə'kri:sɪs/ *noun* (*plural* **-chreses** /-si:z/) incorrect use of words. □ **catachrestic** /-'kres-/ *adjective*.

cataclysm /'kætəklɪz(ə)m/ *noun* violent upheaval. □ **cataclysmic** /-'klɪz-/ *adjective*.

catacomb /'kætəku:m/ *noun* (often in *plural*) underground cemetery.

catafalque /'kætəfælk/ *noun* decorated bier, esp. for state funeral etc.

Catalan /'kætəlæn/ ● *noun* native or language of Catalonia in Spain. ● *adjective* of Catalonia.

catalepsy /'kætəlepsɪ/ *noun* trance or seizure with rigidity of body. □ **cataleptic** /-'lep-/ *adjective & noun*.

catalogue /'kætəlɒg/ (*US* **catalog**) ● *noun* complete or extensive list, usually in alphabetical or other systematic order. ● *verb* (**-logues**, **-logued**, **-loguing**; *US* **-logs**, **-loged**, **-loging**) make catalogue of; enter in catalogue.

catalysis /kə'tælɪsɪs/ *noun* (*plural* **-lyses** /-si:z/) acceleration of chemical reaction by catalyst. □ **catalyse** *verb* (**-sing**) (*US* **-lyze**; **-zing**).

catalyst /'kætəlɪst/ *noun* substance speeding chemical reaction without itself permanently changing; person or thing that precipitates change.

catalytic /kætə'lɪtɪk/ *adjective* involving or causing catalysis. □ **catalytic converter** device in vehicle for converting pollutant gases into less harmful ones.

catamaran /kætəmə'ræn/ *noun* boat or raft with two parallel hulls.

catapult /'kætəpʌlt/ • *noun* forked stick with elastic for shooting stones; *historical* military machine for hurling stones etc.; device for launching glider etc. • *verb* launch with catapult; fling forcibly; leap or be hurled forcibly.

cataract /'kætərækt/ *noun* waterfall, downpour; progressive opacity of eye lens.

catarrh /kə'tɑ:/ *noun* inflammation of mucous membrane, air-passages, etc; mucous in nose caused by this. □ **catarrhal** *adjective.*

catastrophe /kə'tæstrəfɪ/ *noun* great usually sudden disaster; denouement of drama. □ **catastrophic** /kætə'strɒf-/ *adjective*; **catastrophically** /kætə'strɒf-/ *adverb.*

catch /kætʃ/ • *verb* (*past & past participle* **caught** /kɔ:t/) capture in trap, hands, etc.; detect or surprise; intercept and hold (moving thing) in hand etc.; *Cricket* dismiss (batsman) by catching ball before it hits ground; contract (disease) by infection etc.; reach in time and board (train etc.); apprehend; check or be checked; become entangled; (of artist etc.) reproduce faithfully; reach or overtake. • *noun* act of catching; *Cricket* chance or act of catching ball; amount of thing caught; thing or person caught or worth catching; question etc. intended to deceive etc.; unexpected difficulty or disadvantage; device for fastening door etc.; musical round. □ **catch fire** begin to burn; **catch on** *colloquial* become popular, understand what is meant; **catch out** detect in mistake etc., *Cricket* catch; **catchpenny** intended merely to sell quickly; **catchphrase** phrase in frequent use; **catch up** (often + *with*) reach (person etc. ahead), make up arrears, involve, fasten; **catchword** phrase or word in frequent use, word so placed as to draw attention.

catching *adjective* infectious.

catchment area *noun* area served by school, hospital, etc.; area from which rainfall flows into river etc.

catchy *adjective* (**-ier**, **-iest**) (of tune) easily remembered, attractive.

catechism /'kætɪkɪz(ə)m/ *noun* (book containing) principles of a religion in form of questions and answers; series of questions.

catechize /'kætɪkaɪz/ *verb* (also **-ise**) (**-zing** or **-sing**) instruct by question and answer. □ **catechist** *noun.*

catechumen /kætɪ'kju:mən/ *noun* person being instructed before baptism.

categorical /kætɪ'gɒrɪk(ə)l/ *adjective* unconditional, absolute, explicit. □ **categorically** *adverb.*

categorize /'kætɪgəraɪz/ *verb* (also **-ise**) (**-zing** or **-sing**) place in category. □ **categorization** *noun.*

category /'kætɪgərɪ/ *noun* (*plural* **-ies**) class or division (of things, ideas, etc.).

cater /'keɪtə/ *verb* supply food; (+ *for*) provide what is required for. □ **caterer** *noun.*

caterpillar /'kætəpɪlə/ *noun* larva of butterfly or moth; (**Caterpillar**) (in full **Caterpillar track**) *proprietary term* articulated steel band passing round wheels of vehicle for travel on rough ground.

caterwaul /'kætəwɔ:l/ *verb* howl like cat.

catgut *noun* thread made from intestines of sheep etc. used for strings of musical instruments etc.

catharsis /kə'θɑ:sɪs/ *noun* (*plural* **catharses** /-si:z/) emotional release; emptying of bowels.

cathartic /kə'θɑ:tɪk/ • *adjective* effecting catharsis. • *noun* laxative.

cathedral /kə'θi:dr(ə)l/ *noun* principal church of diocese.

Catherine wheel /'kæθrɪn/ *noun* rotating firework.

catheter /'kæθɪtə/ *noun* tube inserted into body cavity to introduce or drain fluid.

cathode /'kæθəʊd/ *noun* negative electrode of cell; positive terminal of battery. □ **cathode ray** beam of electrons from cathode of vacuum tube; **cathode ray tube** vacuum tube in which cathode rays produce luminous image on fluorescent screen.

catholic /'kæθlɪk/ • *adjective* universal; broad-minded; all-embracing; (**Catholic**) Roman Catholic; (**Catholic**) including all Christians or all of Western Church. • *noun* (**Catholic**) Roman Catholic. □ **Catholicism** /kə'θɒlɪs-/ *noun*; **catholicity** /-ə'lɪs-/ *noun.*

cation /'kætaɪən/ *noun* positively charged ion. □ **cationic** /-'ɒnɪk/ *adjective*.

catkin /'kætkɪn/ *noun* spike of usually hanging flowers of willow, hazel, etc.

catmint *noun* pungent plant attractive to cats.

catnip *noun* catmint.

cattery *noun* (*plural* **-ies**) place where cats are boarded.

cattle /'kæt(ə)l/ *plural noun* large ruminants, bred esp. for milk or meat. □ **cattle grid** grid over ditch, allowing vehicles to pass over but not livestock.

catty *adjective* (**-ier**, **-iest**) spiteful. □ **cattily** *adverb*; **cattiness** *noun*.

Caucasian /kɔ:'keɪʒ(ə)n/ ● *adjective* of the white or light-skinned race. ● *noun* Caucasian person.

caucus /'kɔ:kəs/ *noun* (*plural* **-es**) *US* meeting of party members to decide policy; *often derogatory* (meeting of) group within larger organization.

caudal /'kɔ:d(ə)l/ *adjective* of, like, or at tail.

caudate /'kɔ:deɪt/ *adjective* tailed.

caught *past & past participle* of CATCH.

caul /kɔ:l/ *noun* membrane enclosing foetus; part of this sometimes found on child's head at birth.

cauldron /'kɔ:ldrən/ *noun* large vessel for boiling things in.

cauliflower /'kɒlɪflaʊə/ *noun* cabbage with large white flower head.

caulk /kɔ:k/ *verb* (*US* **calk**) stop up (ship's seams); make watertight.

causal /'kɔ:z(ə)l/ *adjective* relating to cause (and effect). □ **causality** /-'zæl-/ *noun*.

causation /kɔ:'zeɪʃ(ə)n/ *noun* causing, causality.

cause /kɔ:z/ ● *noun* thing producing effect; reason or motive; justification; principle, belief, or purpose; matter to be settled, or case offered, at law. ● *verb* (**-sing**) be cause of; produce; (+ *to do*) make.

cause célèbre /kɔ:z se'lebr/ *noun* (*plural* ***causes célèbres*** same pronunciation) lawsuit that excites much interest. [French]

causeway /'kɔ:zweɪ/ *noun* raised road across low ground or water; raised path by road.

caustic /'kɔ:stɪk/ ● *adjective* corrosive, burning; sarcastic, biting. ● *noun* caustic substance. □ **caustic soda** sodium hydroxide. □ **causticity** /-'tɪs-/ *noun*.

cauterize /'kɔ:təraɪz/ *verb* (also **-ise**) (**-zing** or **-sing**) burn (tissue), esp. to stop bleeding.

caution /'kɔ:ʃ(ə)n/ ● *noun* attention to safety; warning; *colloquial* amusing or surprising person or thing. ● *verb* warn, admonish.

cautionary *adjective* warning.

cautious /'kɔ:ʃəs/ *adjective* having or showing caution. □ **cautiously** *adverb*; **cautiousness** *noun*.

cavalcade /kævəl'keɪd/ *noun* procession of riders, cars, etc.

cavalier /kævə'lɪə/ ● *noun* courtly gentleman; *archaic* horseman; (**Cavalier**) *historical* supporter of Charles I in English Civil War. ● *adjective* offhand, supercilious, curt.

cavalry /'kævəlrɪ/ *noun* (*plural* **-ies**) (usually treated as *plural*) soldiers on horseback or in armoured vehicles.

cave ● *noun* large hollow in side of cliff, hill, etc., or underground. ● *verb* (**-ving**) explore caves. □ **cave in** (cause to) subside or collapse, yield, submit.

caveat /'kævɪæt/ *noun* warning, proviso.

cavern /'kæv(ə)n/ *noun* cave, esp. large or dark one.

cavernous *adjective* full of caverns; huge or deep as cavern.

caviar /'kævɪɑ:/ *noun* (also **caviar**) pickled sturgeon-roe.

cavil /'kævɪl/ ● *verb* (**-ll-**; *US* **-l-**) (usually + *at, about*) make petty objections. ● *noun* petty objection.

cavity /'kævɪtɪ/ *noun* (*plural* **-ies**) hollow within solid body; decayed part of tooth. □ **cavity wall** two walls separated by narrow space.

cavort /kə'vɔ:t/ *verb* caper.

caw ● *noun* cry of rook etc. ● *verb* utter this cry.

cayenne /keɪ'en/ *noun* (in full **cayenne pepper**) powdered red pepper.

cayman /'keɪmən/ *noun* (also **caiman**) (*plural* **-s**) S. American alligator-like reptile.

CB *abbreviation* citizens' band; Commander of the Order of the Bath.

CBE *abbreviation* Commander of the Order of the British Empire.

CBI *abbreviation* Confederation of British Industry.

cc *abbreviation* cubic centimetre(s).

CD *abbreviation* compact disc; *Corps Diplomatique*.

cease /si:s/ *formal verb* (**-sing**) stop; bring or come to an end. □ **cease-fire** (order for) truce; **without cease** unending.

ceaseless *adjective* without end. □ **ceaselessly** *adverb*.

cedar /'si:də/ *noun* evergreen conifer; its durable fragrant wood.

cede /si:d/ *verb* (**-ding**) *formal* give up one's rights to or possession of.

cedilla /sɪ'dɪlə/ *noun* mark (̧) under *c* (in French, to show it is pronounced /s/, not /k/).

ceilidh /'keɪlɪ/ *noun* informal gathering for music, dancing, etc.

ceiling /'si:lɪŋ/ *noun* upper interior surface of room or other compartment; upper limit; maximum altitude of aircraft.

celandine /'seləndaɪn/ *noun* yellow-flowered plant.

celebrant /'selɪbrənt/ *noun* person performing rite, esp. priest at Eucharist.

celebrate /'selɪbreɪt/ *verb* (**-ting**) mark with or engage in festivities; perform (rite or ceremony); praise publicly. □ **celebration** *noun*; **celebratory** /-'breɪt-/ *adjective*.

celebrity /sɪ'lebrɪtɪ/ *noun* (*plural* **-ies**) well-known person; fame.

celeriac /sɪ'lerɪæk/ *noun* variety of celery.

celerity /sɪ'lerɪtɪ/ *noun archaic* swiftness.

celery /'selərɪ/ *noun* plant of which stalks are used as vegetable.

celesta /sɪ'lestə/ *noun* keyboard instrument with steel plates struck with hammers.

celestial /sɪ'lestɪəl/ *adjective* of sky or heavenly bodies; heavenly, divinely good.

celibate /'selɪbət/ ● *adjective* unmarried, or abstaining from sexual relations, often for religious reasons. ● *noun* celibate person. □ **celibacy** *noun*.

cell /sel/ *noun* small room, esp. in prison or monastery; small compartment, e.g. in honeycomb; small active political group; unit of structure of organic matter; enclosed cavity in organism etc.; vessel containing electrodes for current-generation or electrolysis. □ **cellphone** portable radio-telephone.

cellar /'selə/ *noun* underground storage room; stock of wine in cellar.

cello /'tʃeləʊ/ *noun* (*plural* **-s**) bass instrument of violin family, held between legs of seated player. □ **cellist** *noun*.

Cellophane /'seləfeɪn/ *noun proprietary term* thin transparent wrapping material.

cellular /'seljʊlə/ *adjective* consisting of cells; of open texture, porous. □ **cellularity** /-'lær-/ *noun*.

cellulite /'seljʊlaɪt/ *noun* lumpy fat, esp. on women's hips and thighs.

celluloid /'seljʊlɔɪd/ *noun* plastic made from camphor and cellulose nitrate; cinema film.

cellulose /'seljʊləʊs/ *noun* carbohydrate forming plant-cell walls; paint or lacquer consisting of cellulose acetate or nitrate in solution.

Celsius /'selsɪəs/ *adjective* of scale of temperature on which water freezes at 0° and boils at 100°.

Celt /kelt/ *noun* (also **Kelt**) member of an ethnic group including inhabitants of Ireland, Wales, Scotland, Cornwall, and Brittany.

Celtic /'keltɪk/ ● *adjective* of the Celts. ● *noun* group of Celtic languages.

cement /sɪ'ment/ ● *noun* substance made from lime and clay, mixed with water, sand, etc. to form mortar or concrete; adhesive. ● *verb* unite firmly, strengthen; apply cement to.

cemetery /'semɪtrɪ/ *noun* (*plural* **-ies**) burial ground, esp. one not in churchyard.

cenotaph /'senətɑ:f/ *noun* tomblike monument to person(s) whose remains are elsewhere.

Cenozoic /si:nə'zəʊɪk/ ● *adjective* of most recent geological era, marked by evolution of mammals etc. ● *noun* this era.

censer /'sensə/ *noun* incense-burning vessel.

censor /ˈsensə/ ● *noun* official with power to suppress or expurgate books, films, news, etc., on grounds of obscenity, threat to security, etc. ● *verb* act as censor of; make deletions or changes in. □ **censorial** /-ˈsɔːr-/ *adjective*; **censorship** *noun*.

■ **Usage** As a verb, *censor* is often confused with *censure*, which means 'to criticize harshly'.

censorious /senˈsɔːrɪəs/ *adjective* severely critical.

censure /ˈsenʃə/ ● *verb* (**-ring**) criticize harshly; reprove. ● *noun* hostile criticism; disapproval.

census /ˈsensəs/ *noun* (*plural* **-suses**) official count of population etc.

cent /sent/ *noun* one-hundredth of dollar or other decimal unit of currency.

cent. *abbreviation* century.

centaur /ˈsentɔː/ *noun* creature in Greek mythololgy with head, arms, and trunk of man joined to body and legs of horse.

centenarian /sentɪˈneərɪən/ ● *noun* person 100 or more years old. ● *adjective* 100 or more years old.

centenary /senˈtiːnərɪ/ ● *noun* (*plural* **-ies**) (celebration of) 100th anniversary. ● *adjective* of a centenary; recurring every 100 years.

centennial /senˈtenɪəl/ ● *adjective* lasting 100 years; recurring every 100 years. ● *noun US* centenary.

centi- *combining form* one-hundredth.

centigrade /ˈsentɪgreɪd/ *adjective* Celsius.

■ **Usage** *Celsius* is the better term to use in technical contexts.

centilitre /ˈsentɪliːtə/ *noun* (*US* **centiliter**) one-hundredth of litre (0.018 pint).

centime /ˈsɑ̃tiːm/ *noun* one-hundredth of franc.

centimetre /ˈsentɪmiːtə/ *noun* (*US* **centimeter**) one-hundredth of metre (0.394 in.).

centipede /ˈsentɪpiːd/ *noun* arthropod with wormlike body and many legs.

central /ˈsentr(ə)l/ *adjective* of, forming, at, or from centre; essential, principal. □ **central bank** national bank issuing currency etc.; **central heating** heating of building from central source; **central processor**, **central processing unit** computer's main operating part. □ **centrality** *noun*; **centrally** *adverb*.

centralize *verb* (also **-ise**) (**-zing** or **-sing**) concentrate (administration etc.) at single centre; subject (state etc.) to this system. □ **centralization** *noun*.

centre /ˈsentə/ (*US* **center**) ● *noun* middle point; pivot; place or buildings forming a central point or main area for an activity; point of concentration or dispersion; political party holding moderate opinions; filling in chocolate etc. ● *verb* (**-ring**) (+ *in*, *on*, *round*) have as main centre; place in centre. □ **centrefold** centre spread that folds out, esp. with nude photographs; **centre forward**, **back** *Football etc.* middle player in forward or half-back line; **centre of gravity** point at which the mass of an object effectively acts; **centrepiece** ornament for middle of table, main item; **centre spread** two facing middle pages of magazine etc.

centrifugal /sentrɪˈfjuːg(ə)l/ *adjective* moving or tending to move from centre. □ **centrifugal force** apparent force acting outwards on body revolving round centre. □ **centrifugally** *adverb*.

centrifuge /ˈsentrɪfjuːdʒ/ *noun* rapidly rotating machine for separating e.g. cream from milk.

centripetal /senˈtrɪpɪt(ə)l/ *adjective* moving or tending to move towards centre. □ **centripetally** *adverb*.

centrist /ˈsentrɪst/ *noun often derogatory* person holding moderate views. □ **centrism** *noun*.

cents. *abbreviation* centuries.

centurion /senˈtjʊərɪən/ *noun* commander of century in ancient Roman army.

century /ˈsentʃərɪ/ *noun* (*plural* **-ies**) 100 years; *Cricket* score of 100 runs by one batsman; company in ancient Roman army.

■ **Usage** Strictly speaking, because the 1st century ran from the year 1 to the year 100, the first year of a century ends in 1. However, a century is commonly regarded as starting with a year ending in 00, the 20th century thus running from 1900 to 1999.

cephalic /sə'fælɪk/ *adjective* of or in head.

cephalopod /'sefələpɒd/ *noun* mollusc with tentacles on head, e.g. octopus.

ceramic /sɪ'ræmɪk/ ● *adjective* made of esp. baked clay; of ceramics. ● *noun* ceramic article.

ceramics *plural noun* ceramic products collectively; (usually treated as *singular*) ceramic art.

cereal /'sɪərɪəl/ ● *noun* edible grain; breakfast food made from cereal. ● *adjective* of edible grain.

cerebellum /serɪ'beləm/ *noun* (*plural* **-s** or **-bella**) part of brain at back of skull.

cerebral /'serɪbr(ə)l/ *adjective* of brain; intellectual. □ **cerebral palsy** paralysis resulting from brain damage before or at birth.

cerebration /serɪ'breɪʃ(ə)n/ *noun* working of brain.

cerebrospinal /serɪbrəʊ'spaɪn(ə)l/ *adjective* of brain and spine.

cerebrum /'serɪbrəm/ *noun* (*plural* **-bra**) principal part of brain, at front of skull.

ceremonial /serɪ'məʊnɪəl/ ● *adjective* of or with ceremony, formal. ● *noun* system of rites or ceremonies.

ceremonious /serɪ'məʊnɪəs/ *adjective* fond of or characterized by ceremony, formal. □ **ceremoniously** *adverb.*

ceremony /'serɪmənɪ/ *noun* (*plural* **-ies**) formal procedure; formalities, esp. ritualistic; excessively polite behaviour. □ **stand on ceremony** insist on formality.

cerise /sə'ri:z/ *noun & adjective* light clear red.

cert /sɜ:t/ *noun* (esp. **dead cert**) *slang* a certainty.

certain /'sɜ:t(ə)n/ *adjective* convinced; indisputable; (often + *to do*) sure, destined; reliable; particular but not specified; some.

certainly *adverb* undoubtedly; (in answer) yes.

certainty *noun* (*plural* **-ies**) undoubted fact; absolute conviction; reliable thing or person.

certificate ● *noun* /sə'tɪfɪkət/ document formally attesting fact. ● *verb* /sə'tɪfɪkeɪt/ (**-ting**) (esp. as **certificated** *adjective*) provide with certificate; license or attest by certificate. □ **certification** /sɜ:-/ *noun.*

certify /'sɜ:tɪfaɪ/ *verb* (**-ies**, **-ied**) attest (to); declare by certificate; officially declare insane. □ **certifiable** *adjective.*

certitude /'sɜ:tɪtju:d/ *noun* feeling certain.

cerulean /sə'ru:lɪən/ *adjective* sky-blue.

cervical /sə'vaɪk(ə)l/ *adjective* of cervix or neck. □ **cervical smear** specimen from neck of womb for examination.

cervix /'sɜ:vɪks/ *noun* (*plural* **cervices** /-si:z/) necklike structure, esp. neck of womb; neck.

Cesarean (also **Cesarian**) *US* = CAESAREAN.

cessation /se'seɪʃ(ə)n/ *noun* ceasing.

cession /'seʃ(ə)n/ *noun* ceding; territory ceded.

cesspit /'sespɪt/ *noun* pit for liquid waste or sewage.

cesspool /'sespu:l/ *noun* cesspit.

cetacean /sɪ'teɪʃ(ə)n/ ● *noun* member of order of marine mammals including whales. ● *adjective* of cetaceans.

cf. *abbreviation* compare (Latin *confer*).

CFC *abbreviation* chlorofluorocarbon (compound used as refrigerant, aerosol propellant, etc.).

cg *abbreviation* centigram(s).

CH *abbreviation* Companion of Honour.

Chablis /'ʃæbli:/ *noun* (*plural* same /-li:z/) dry white wine from Chablis in France.

chaconne /ʃə'kɒn/ *noun* musical variations over ground bass; dance to this.

chafe /tʃeɪf/ ● *verb* (**-fing**) make or become sore or damaged by rubbing; irritate; show irritation, fret; rub (esp. skin) to warm. ● *noun* sore caused by rubbing.

chaff /tʃɑ:f/ ● *noun* separated grain-husks; chopped hay or straw; good-humoured teasing; worthless stuff. ● *verb* tease, banter.

chaffinch /'tʃæfɪntʃ/ *noun* a common European finch.

chafing dish /'tʃeɪfɪŋ/ *noun* vessel in which food is cooked or kept warm at table.

chagrin /'ʃægrɪn/ • *noun* acute vexation or disappointment. • *verb* affect with this.

chain • *noun* connected series of links, thing resembling this; (in *plural*) fetters, restraining force; sequence, series, or set; unit of length (66 ft). • *verb* (often + *up*) secure with chain. □ **chain gang** *historical* convicts chained together at work etc.; **chain mail** armour made from interlaced rings; **chain reaction** reaction forming products which themselves cause further reactions, series of events each due to previous one; **chainsaw** motor-driven saw with teeth on loop of chain; **chain-smoke** smoke continuously, esp. by lighting next cigarette etc. from previous one; **chain store** one of series of shops owned by one firm.

chair • *noun* seat usually with back for one person; (office of) chairperson; professorship; *US* electric chair. • *verb* be chairperson of (meeting); carry aloft in triumph. □ **chairlift** series of chairs on loop of cable for carrying passengers up and down mountain etc.; **chairman, chairperson, chairwoman** person who presides over meeting, board, or committee.

chaise /ʃeɪz/ *noun* horse-drawn usually open carriage for one or two people.

chaise longue /ʃeɪz 'lɒŋg/ *noun* (*plural* **chaise longues** or **chaises longues** /'lɒŋg(z)/) sofa with one arm rest.

chalcedony /kæl'sedənɪ/ *noun* (*plural* **-ies**) type of quartz.

chalet /'ʃæleɪ/ *noun* Swiss hut or cottage; similar house; small cabin in holiday camp etc.

chalice /'tʃælɪs/ *noun* goblet; Eucharistic cup.

chalk /tʃɔ:k/ • *noun* white soft limestone; (piece of) similar, sometimes coloured, substance for writing or drawing. • *verb* rub, mark, draw, or write with chalk. □ **chalky** *adjective* (**-ier, -iest**).

challenge /'tʃælɪndʒ/ • *noun* call to take part in contest etc. or to prove or justify something; demanding or difficult task; call to respond. • *verb* (**-ging**) issue challenge to; dispute; (as **challenging** *adjective*) stimulatingly difficult. □ **challenger** *noun*.

chamber /'tʃeɪmbə/ *noun* hall used by legislative or judicial body; body that meets in it, esp. a house of a parliament; (in *plural*) set of rooms for barrister(s), esp. in Inns of Court; (in *plural*) judge's room for hearing cases not needing to be taken in court; *archaic* room, esp. bedroom; cavity or compartment in body, machinery, etc. (esp. part of gun that contains charge). □ **chambermaid** woman who cleans hotel bedrooms; **chamber music** music for small group of instruments; **Chamber of Commerce** association to promote local commercial interests; **chamber pot** vessel for urine etc., used in bedroom.

chamberlain /'tʃeɪmbəlɪn/ *noun* officer managing royal or noble household; treasurer of corporation etc.

chameleon /kə'mi:lɪən/ *noun* lizard able to change colour for camouflage.

chamfer /'tʃæmfə/ • *verb* bevel symmetrically. • *noun* bevelled surface.

chamois /'ʃæmwɑ:/ *noun* (*plural* same /-wɑ:z/) small mountain antelope; /'ʃæmɪ/ (piece of) soft leather from sheep, goats, etc.

chamomile = CAMOMILE.

champ[1] • *verb* munch or bite noisily. • *noun* chewing noise. □ **champ at the bit** show impatience.

champ[2] *noun slang* champion.

champagne /ʃæm'peɪn/ *noun* white sparkling wine from Champagne in France.

champion /'tʃæmpɪən/ • *noun* person or thing that has defeated all rivals; person who fights for cause or another person. • *verb* support cause of, defend. • *adjective colloquial* splendid.

championship *noun* (often in *plural*) contest to decide champion in sport etc.; position of champion.

chance /tʃɑ:ns/ • *noun* possibility; (often in *plural*) probability; unplanned occurrence; fate. • *adjective* fortuitous. • *verb* (**-cing**) *colloquial* risk; happen. □ **chance on** happen to find.

chancel /'tʃɑ:ns(ə)l/ *noun* part of church near altar.

chancellery /'tʃɑ:nsələrɪ/ *noun* (*plural* **-ies**) chancellor's department, staff, or residence; *US* office attached to embassy.

chancellor /'tʃɑ:nsələ/ *noun* state or legal official; head of government in some European countries; non-resident honorary head of university; (in full **Lord Chancellor**) highest officer of the Crown, presiding in House of Lords; (in full **Chancellor of the Exchequer**) UK finance minister.

Chancery /'tʃɑ:nsərɪ/ *noun* Lord Chancellor's division of High Court of Justice.

chancy /'tʃɑ:nsɪ/ *adjective* (**-ier**, **-iest**) uncertain; risky.

chandelier /ʃændə'lɪə/ *noun* branched hanging support for lights.

chandler /'tʃɑ:ndlə/ *noun* dealer in oil, candles, soap, paint, etc.

change /tʃeɪndʒ/ ● *noun* making or becoming different; low-value money in small coins; money returned as balance of that given in payment; new experience; substitution of one thing or person for another; one of different orders in which bells can be rung. ● *verb* (**-ging**) undergo, show, or subject to change; take or use another instead of; interchange; give or get money in exchange for; put fresh clothes or coverings on; (often + *with*) exchange. □ **change hands** be passed to different owner. □ **changeful** *adjective*; **changeless** *adjective*.

changeable *adjective* inconstant; able to change or be changed.

changeling *noun* child believed to be substitute for another.

channel /'tʃæn(ə)l/ ● *noun* piece of water connecting two seas; (**the Channel**) the English Channel; medium of communication, agency; band of frequencies used for radio and television transmission; course in which thing moves; hollow bed of water; navigable part of waterway; passage for liquid; lengthwise strip on recording tape etc. ● *verb* (**-ll-**; *US* **-l-**) guide, direct; form channel(s) in.

chant /tʃɑ:nt/ ● *noun* spoken singsong phrase; melody for reciting unmetrical texts; song. ● *verb* talk or repeat monotonously; sing or intone (psalm etc.).

chanter *noun* melody-pipe of bagpipes.

chantry /'tʃɑ:ntrɪ/ *noun* (*plural* **-ies**) endowment for singing of masses; priests, chapel, etc., so endowed.

chaos /'keɪɒs/ *noun* utter confusion; formless matter supposed to have existed before universe's creation. □ **chaotic** /keɪ'ɒtɪk/ *adjective*.

chap¹ *noun colloquial* man, boy.

chap² ● *verb* (**-pp-**) (esp. of skin) develop cracks or soreness; (of wind etc.) cause this. ● *noun* (usually in *plural*) crack in skin etc.

chaparral /ʃæpə'ræl/ *noun US* dense tangled brushwood.

chapatti /tʃə'pa:tɪ/ *noun* (also **chupatty**) (*plural* **chapattis** or **chupatties**) flat thin cake of unleavened bread.

chapel /'tʃæp(ə)l/ *noun* place for private worship in cathedral or church, with its own altar; place of worship attached to private house, institution, etc.; place or service of worship for Nonconformists; branch of printers' or journalists' trade union at a workplace.

chaperon /'ʃæpərəʊn/ ● *noun* person, esp. older woman, in charge of young unmarried woman on certain social occasions. ● *verb* act as chaperon to.

chaplain /'tʃæplɪn/ *noun* member of clergy attached to private chapel, institution, ship, regiment, etc. □ **chaplaincy** *noun* (*plural* **-ies**).

chaplet /'tʃæplɪt/ *noun* wreath or circlet for head; string of beads, short rosary.

chapter /'tʃæptə/ *noun* division of book; period of time; canons of cathedral etc.; meeting of these.

char¹ *verb* (**-rr-**) blacken with fire, scorch; burn to charcoal.

char² *colloquial* ● *noun* charwoman. ● *verb* (**-rr-**) work as charwoman. □ **charlady**, **charwoman** one employed to do housework.

char³ *noun slang* tea.

char⁴ *noun* (*plural* same) a kind of trout.

charabanc /'ʃærəbæŋ/ *noun* early form of motor coach.

character /'kærɪktə/ *noun* distinguishing qualities or characteristics; moral strength; reputation; person in novel, play, etc.; *colloquial* (esp. eccentric) person; letter, symbol; testimonial.

characteristic /kærɪktə'rɪstɪk/ ● *adjective* typical, distinctive. ● *noun* characteristic feature or quality. □ **characteristically** *adverb.*

characterize *verb* (also **-ise**) (**-zing** or **-sing**) describe character of; (+ *as*) describe as; be characteristic of. □ **characterization** *noun.*

charade /ʃə'rɑːd/ *noun* (usually in *plural*, treated as *singular*) game of guessing word(s) from acted clues; absurd pretence.

charcoal *noun* black residue of partly burnt wood etc.

charge ● *verb* (**-ging**) ask (amount) as price; ask (person) for amount as price; (+ *to, up to*) debit cost of to; (often + *with*) accuse (of offence); (+ *to do*) instruct or urge to do; (+ *with*) entrust with; make rushing attack (on); (often + *up*) give electric charge to, store energy in; (often + *with*) load, fill. ● *noun* price, financial liability; accusation; task; custody; person or thing entrusted; (signal for) impetuous attack, esp. in battle; appropriate amount of material to be put in mechanism at one time, esp. explosive in gun; cause of electrical phenomena in matter. □ **charge card** credit card, esp. used at particular shop; **in charge** having command. □ **chargeable** *adjective.*

chargé d'affaires /ʃɑːʒeɪ dæ'feə/ *noun* (*plural* **chargés d'affaires** same pronunciation) ambassador's deputy; envoy to minor country.

charger *noun* cavalry horse; apparatus for charging battery.

chariot /'tʃærɪət/ *noun historical* two-wheeled horse-drawn vehicle used in ancient warfare and racing.

charioteer /tʃærɪə'tɪə/ *noun* chariot driver.

charisma /kə'rɪzmə/ *noun* power to inspire or attract others; divinely conferred power or talent. □ **charismatic** /kArɪz'mAtɪkæt/ *adjective.*

charitable /'tʃærɪtəb(ə)l/ *adjective* generous to those in need; of or connected with a charity; lenient in judging others. □ **charitably** *adverb.*

charity /'tʃærɪtɪ/ *noun* (*plural* **-ies**) giving voluntarily to those in need; organization for helping those in need; love of fellow men; lenience in judging others.

charlatan /'ʃɑːlət(ə)n/ *noun* person falsely claiming knowledge or skill. □ **charlatanism** *noun.*

charlotte /'ʃɑːlɒt/ *noun* pudding of stewed fruit under bread etc.

charm ● *noun* power of delighting, attracting, or influencing; trinket on bracelet etc.; object, act, or word(s) supposedly having magic power. ● *verb* delight, captivate; influence or protect (as) by magic; obtain or gain by charm. □ **charmer** *noun.*

charming *adjective* delightful. □ **charmingly** *adverb.*

charnel house /'tʃɑːn(ə)l/ *noun* place containing corpses or bones.

chart ● *noun* map esp. for sea or air navigation or showing weather conditions etc.; sheet of information in form of tables or diagrams; (usually in *plural*) *colloquial* list of currently best-selling pop records. ● *verb* make chart of.

charter ● *noun* written grant of rights, esp. by sovereign or legislature; written description of organization's functions etc. ● *verb* grant charter to; hire (aircraft etc.). □ **charter flight** flight by chartered aircraft.

chartered *adjective* qualified as member of professional body that has royal charter.

Chartism *noun* working-class reform movement of 1837–48. □ **Chartist** *noun.*

chartreuse /ʃɑː'trɜːz/ *noun* green or yellow brandy liqueur.

chary /'tʃeərɪ/ *adjective* (**-ier**, **-iest**) cautious; sparing.

chase[1] /tʃeɪs/ ● *verb* (**-sing**) pursue; (+ *from, out of, to*, etc.) drive; hurry; (usually + *up*) *colloquial* pursue (thing overdue); *colloquial* try to attain; court persistently. ● *noun* pursuit; unenclosed hunting-land.

chase[2] /tʃeɪs/ *verb* (**-sing**) emboss or engrave (metal).

chaser *noun* horse for steeplechasing; *colloquial* drink taken after another of different kind.

chasm /'kæz(ə)m/ *noun* deep cleft in earth, rock, etc.; wide difference in opinion etc.

chassis /'ʃæsɪ/ *noun* (*plural* same /-sɪz/) base frame of vehicle; frame for (radio etc.) components.

chaste /tʃeɪst/ *adjective* abstaining from extramarital or from all sexual intercourse; pure, virtuous; unadorned. □ **chastely** *adverb*; **chasteness** *noun*.

chasten /'tʃeɪs(ə)n/ *verb* (esp. as **chastening**, **chastened** *adjectives*) restrain; punish.

chastise /tʃæs'taɪz/ *verb* (**-sing**) rebuke severely; punish, beat. □ **chastisement** *noun*.

chastity /'tʃæstɪtɪ/ *noun* being chaste.

chasuble /'tʃæzjʊb(ə)l/ *noun* sleeveless outer vestment worn by celebrant of Eucharist.

chat ● *verb* (**-tt-**) talk in light familiar way. ● *noun* informal talk. □ **chatline** telephone service setting up conversations between groups of people on separate lines; **chat show** television or radio broadcast with informal celebrity interviews; **chat up** *colloquial* chat to, esp. flirtatiously.

chateau /'ʃætəʊ/ *noun* (*plural* **-x** /-z/) large French country house.

chatelaine /'ʃætəleɪn/ *noun* mistress of large house; *historical* appendage to woman's belt for carrying keys etc.

chattel /'tʃæt(ə)l/ *noun* (usually in *plural*) movable possession.

chatter /'tʃætə/ ● *verb* talk fast, incessantly, or foolishly. ● *noun* such talk.

chatty *adjective* (**-ier**, **-iest**) fond of or resembling chat.

chauffeur /'ʃəʊfə/ *noun* person employed to drive car. ● *verb* drive (car or person).

chauvinism /'ʃəʊvɪnɪz(ə)m/ *noun* exaggerated or aggressive patriotism; excessive or prejudiced support or loyalty for something. □ **chauvinist** *noun*; **chauvinistic** /-'nɪs-/ *adjective*.

cheap *adjective* low in price; charging low prices; of low quality or worth; easily got. □ **cheaply** *adverb*; **cheapness** *noun*.

cheapen *verb* make or become cheap; degrade.

cheapskate *noun esp. US colloquial* stingy person.

cheat ● *verb* (often + *into*, *out of*) deceive; (+ *of*) deprive of; gain unfair advantage. ● *noun* person who cheats; deception. □ **cheat on** *colloquial* be sexually unfaithful to.

check ● *verb* test, examine, verify; stop or slow motion of; *colloquial* rebuke; threaten opponent's king at chess; *US* agree on comparison; *US* deposit (luggage etc.). ● *noun* test for accuracy, quality, etc.; stopping or slowing of motion; rebuff; restraint; pattern of small squares; fabric so patterned; (also as *interjection*) exposure of chess king to attack; *US* restaurant bill; *US* cheque; *US* token of identification for left luggage etc.; *US* counter used in card games. □ **check in** register at hotel, airport, etc.; **check-in** act or place of checking in; **check out** leave hotel etc. with due formalities; **checkout** act of checking out, pay-desk in supermarket etc.; **check-up** thorough (esp. medical) examination.

checked *adjective* having a check pattern.

checker = CHEQUER.

checkmate ● *noun* (also as *interjection*) check at chess from which king cannot escape. ● *verb* (**-ting**) put into checkmate; frustrate.

Cheddar /'tʃedə/ *noun* kind of firm smooth cheese.

cheek ● *noun* side of face below eye; impertinence, impertinent speech; *slang* buttock. ● *verb* be impertinent to.

cheeky *adjective* (**-ier**, **-iest**) impertinent. □ **cheekily** *adverb*; **cheekiness** *noun*.

cheep ● *noun* shrill feeble note of young bird. ● *verb* make such cry.

cheer ● *noun* shout of encouragement or applause; mood, disposition; (as **cheers** *interjection*) *colloquial*: *expressing good wishes or thanks*. ● *verb* applaud; (usually + *on*) urge with shouts; shout for joy; gladden, comfort. □ **cheer up** make or become less sad.

cheerful *adjective* in good spirits; bright, pleasant. □ **cheerfully** *adverb*; **cheerfulness** *noun*.

cheerless *adjective* gloomy, dreary.

cheery *adjective* (**-ier**, **-iest**) cheerful. □ **cheerily** *adverb.*

cheese /tʃi:z/ *noun* food made from milk curds; cake of this with rind; thick conserve of fruit. □ **cheeseburger** hamburger with cheese in or on it; **cheesecake** tart filled with sweetened curds, *slang* sexually stimulating display of women; **cheesecloth** thin loosely woven cloth; **cheesed** *slang* (often + *off*) bored, fed up; **cheese-paring** stingy; **cheese plant** climbing plant with holey leaves. □ **cheesy** *adjective.*

cheetah /'tʃi:tə/ *noun* swift-running spotted feline resembling leopard.

chef /ʃef/ *noun* (esp. chief) cook in restaurant etc.

Chelsea /'tʃelsɪ/ *noun* □ **Chelsea bun** kind of spiral-shaped currant bun; **Chelsea pensioner** inmate of Chelsea Royal Hospital for old or disabled soldiers.

chemical /'kemɪk(ə)l/ ● *adjective* of, made by, or employing chemistry. ● *noun* substance obtained or used in chemistry. □ **chemical warfare** warfare using poison gas and other chemicals. □ **chemically** *adverb.*

chemise /ʃə'mi:z/ *noun historical* woman's loose-fitting undergarment or dress.

chemist /'kemɪst/ *noun* dealer in medicinal drugs etc.; expert in chemistry.

chemistry /'kemɪstrɪ/ *noun* (*plural* **-ies**) science of elements and their laws of combination and change; *colloquial* sexual attraction.

chenille /ʃə'ni:l/ *noun* tufted velvety yarn; fabric made of this.

cheque /tʃek/ *noun* written order to bank to pay sum of money; printed form for this. □ **chequebook** book of forms for writing cheques; **cheque card** card issued by bank to guarantee honouring of cheques up to stated amount.

chequer /'tʃekə/ (also **checker**) ● *noun* (often in *plural*) pattern of squares often alternately coloured; (in *plural*; usually **checkers**) *US* game of draughts. ● *verb* mark with chequers; variegate, break uniformity of; (as **chequered** *adjective*) with varied fortunes.

cherish /'tʃerɪʃ/ *verb* tend lovingly; hold dear; cling to.

cheroot /ʃə'ru:t/ *noun* cigar with both ends open.

cherry /'tʃerɪ/ ● *noun* (*plural* **-ies**) small stone fruit, tree bearing it, wood of this; light red. ● *adjective* of light red colour.

cherub /'tʃerəb/ *noun* representation of winged child; beautiful child; (*plural* **-im**) angelic being. □ **cherubic** /tʃɪ'ru:bɪk/ *adjective.*

chervil /'tʃɜ:vɪl/ *noun* herb with aniseed flavour.

chess *noun* game for two players with 16 **chessmen** each, on chequered **chessboard** of 64 squares.

chest *noun* large strong box; part of body enclosed by ribs, front surface of body from neck to bottom of ribs; small cabinet for medicines etc. □ **chest of drawers** piece of furniture with set of drawers in frame.

chesterfield /'tʃestəfi:ld/ *noun* sofa with arms and back of same height.

chestnut /'tʃesnʌt/ ● *noun* glossy hard brown edible nut; tree bearing it; horse chestnut; reddish-brown horse; *colloquial* stale joke etc.; reddish-brown. ● *adjective* reddish-brown.

chesty *adjective* (**-ier**, **-iest**) *colloquial* inclined to or symptomatic of chest disease. □ **chestily** *adverb*; **chestiness** *noun.*

cheval glass /ʃə'væl/ *noun* tall mirror pivoting in upright frame.

chevalier /ʃevə'lɪə/ *noun* member of certain orders of knighthood etc.

chevron /'ʃevrən/ *noun* V-shaped line or stripe.

chew ● *verb* work (food etc.) between teeth. ● *noun* act of chewing; chewy sweet. □ **chewing gum** flavoured gum for chewing; **chew on** work continuously between teeth, think about; **chew over** discuss, think about.

chewy *adjective* (**-ier**, **-iest**) requiring or suitable for chewing.

chez /ʃeɪ/ *preposition* at the home of. [French]

Chianti /kɪ'æntɪ/ *noun* (*plural* **-s**) red wine from Chianti in Italy.

chiaroscuro /kɪɑ:rə'skʊərəʊ/ *noun* treatment of light and shade in painting; use of contrast in literature etc.

chic /ʃi:k/ ● *adjective* (**chic-er**, **chic-est**) stylish, elegant. ● *noun* stylishness, elegance.

chicane /ʃɪ'keɪn/ ● *noun* artificial barrier or obstacle on motor-racing course; chicanery. ● *verb* (**-ning**) *archaic* use chicanery; (usually + *into*, *out of*, etc.) cheat (person).

chicanery /ʃɪ'keɪnərɪ/ *noun* (*plural* **-ies**) clever but misleading talk; trickery, deception.

chick *noun* young bird; *slang* young woman.

chicken /'tʃɪkɪn/ ● *noun* domestic fowl; its flesh as food; young domestic fowl; youthful person. ● *adjective colloquial* cowardly. ● *verb* (+ *out*) *colloquial* withdraw through cowardice. □ **chicken feed** food for poultry, *colloquial* insignificant amount esp. of money; **chickenpox** infectious disease with rash of small blisters; **chicken wire** light wire netting.

chickpea *noun* pealike seed used as vegetable.

chickweed *noun* a small weed.

chicle /'tʃɪk(ə)l/ *noun* juice of tropical tree, used in chewing gum.

chicory /'tʃɪkərɪ/ *noun* (*plural* **-ies**) salad plant; its root, roasted and ground and used with or instead of coffee; endive.

chide *verb* (**-ding**; *past* **chided** or **chid**; *past participle* **chided** or **chidden**) *archaic* scold, rebuke.

chief /tʃi:f/ ● *noun* leader, ruler; head of tribe, clan, etc.; head of department etc. ● *adjective* first in position, importance, or influence; prominent, leading.

chiefly *adverb* above all; mainly but not exclusively.

chieftain /'tʃi:ft(ə)n/ *noun* leader of tribe, clan, etc. □ **chieftaincy** *noun* (*plural* **-ies**).

chiffchaff /'tʃɪftʃæf/ *noun* small European warbler.

chiffon /'ʃɪfɒn/ *noun* diaphanous silky fabric.

chignon /'ʃi:njɔ̃/ *noun* coil of hair at back of head.

chihuahua /tʃɪ'wɑ:wə/ *noun* tiny smooth-coated dog.

chilblain /'tʃɪlbleɪn/ *noun* itching swelling on hand, foot, etc., caused by exposure to cold.

child /tʃaɪld/ *noun* (*plural* **children** /'tʃɪldrən/) young human being; one's son or daughter; (+ *of*) descendant, follower, or product of. □ **child benefit** regular state payment to parents of child up to certain age; **childbirth** giving birth to child; **child's play** easy task. □ **childless** *adjective*.

childhood *noun* state or period of being a child.

childish *adjective* of or like child; immature, silly. □ **childishly** *adverb*; **childishness** *noun*.

childlike *adjective* innocent, frank, etc., like child.

chili = CHILLI.

chill ● *noun* cold sensation; feverish cold; unpleasant coldness (of air etc.); depressing influence. ● *verb* make or become cold; depress, horrify; preserve (food or drink) by cooling. ● *adjective literary* chilly.

chilli /'tʃɪlɪ/ *noun* (also **chili**) (*plural* **-es**) hot-tasting dried red capsicum pod. □ **chilli con carne** /kɒn 'kɑ:nɪ/ chilli-flavoured mince and beans.

chilly *adjective* (**-ier**, **-iest**) rather cold; sensitive to cold; unfriendly, unemotional.

chime ● *noun* set of attuned bells; sounds made by this. ● *verb* (**-ming**) (of bells) ring; show (time) by chiming; (usually + *together*, *with*) be in agreement. □ **chime in** interject remark, join in harmoniously.

chimera /kaɪ'mɪərə/ *noun* monster in Greek mythology with lion's head, goat's body, and serpent's tail; bogey; wild impossible scheme or fancy. □ **chimerical** /-'merɪk(ə)l/ *adjective*.

chimney /'tʃɪmnɪ/ *noun* (*plural* **-s**) channel conducting smoke etc. away from fire etc.; part of this above roof; glass tube protecting lamp-flame. □ **chimney breast** projecting wall round chimney; **chimney pot** pipe at top of chimney; **chimney sweep** person who clears chimneys of soot.

chimp *noun colloquial* chimpanzee.

chimpanzee /tʃɪmpən'zi:/ *noun* manlike African ape.

chin *noun* front of lower jaw. □ **chinless wonder** ineffectual person; **chinwag** *noun & verb slang* chat.

china /'tʃaɪnə/ ● *noun* fine white or translucent ceramic ware; things made of this. • *adjective* made of china. □ **china clay** kaolin.

chinchilla /tʃɪn'tʃɪlə/ *noun* S. American rodent; its soft grey fur; breed of cat or rabbit.

chine • *noun* backbone; joint of meat containing this; ridge. • *verb* (**-ning**) cut (meat) through backbone.

Chinese /tʃaɪ'ni:z/ • *adjective* of China. • *noun* Chinese language; (*plural* same) native or national of China, person of Chinese descent. □ **Chinese lantern** collapsible paper lantern, plant with inflated orange-red calyx; **Chinese leaf** lettuce-like cabbage.

chink¹ *noun* narrow opening.

chink² • *verb* (cause to) make sound of glasses or coins striking together. • *noun* this sound.

chintz *noun* printed multicoloured usually glazed cotton cloth.

chip • *noun* small piece cut or broken off; place where piece has been broken off; strip of potato usually fried; *US* potato crisp; counter used as money in some games; microchip. • *verb* (**-pp-**) (often + *off*) cut or break (piece) from hard material; (often + *at*, *away at*) cut pieces off (hard material); be apt to break at edge; (usually as **chipped** *adjective*) make (potato) into chips. □ **chipboard** board made of compressed wood chips.

chipmunk /'tʃɪpmʌŋk/ *noun* N. American striped ground squirrel.

chipolata /tʃɪpə'lɑ:tə/ *noun* small thin sausage.

Chippendale /'tʃɪpəndeɪl/ *adjective* of an 18th-c. elegant style of furniture.

chiropody /kɪ'rɒpədɪ/ *noun* treatment of feet and their ailments. □ **chiropodist** *noun*.

chiropractic /kaɪərəʊ'præktɪk/ *noun* treatment of disease by manipulation of spinal column. □ **chiropractor** /'kaɪə-/ *noun*.

chirp • *verb* (of small bird etc.) utter short thin sharp note; speak merrily. • *noun* chirping sound.

chirpy *adjective* (**-ier**, **-iest**) *colloquial* cheerful. □ **chirpily** *adverb*.

chirrup /'tʃɪrəp/ • *verb* (**-p-**) chirp, esp. repeatedly. • *noun* chirruping sound.

chisel /'tʃɪz(ə)l/ • *noun* tool with bevelled blade for shaping wood, stone, or metal. • *verb* (**-ll-**; *US* **-l-**) cut or shape with chisel; (as **chiselled** *adjective*) (of features) clear-cut; *slang* defraud.

chit¹ *noun derogatory or jocular* young small woman; young child.

chit² *noun* written note.

chit-chat *noun colloquial* light conversation, gossip.

chivalry /'ʃɪvəlrɪ/ *noun* medieval knightly system; honour and courtesy, esp. to the weak. □ **chivalrous** *adjective*; **chivalrously** *adverb*.

chive *noun* herb related to onion.

chivvy /'tʃɪvɪ/ *verb* (**-ies**, **-ied**) urge persistently, nag.

chloral /'klɔ:r(ə)l/ *noun* compound used in making DDT, sedatives, etc.

chloride /'klɔ:raɪd/ *noun* compound of chlorine and another element or group.

chlorinate /'klɔ:rɪneɪt/ *verb* (**-ting**) impregnate or treat with chlorine. □ **chlorination** *noun*.

chlorine /'klɔ:ri:n/ *noun* poisonous gas used for bleaching and disinfecting.

chlorofluorocarbon see CFC.

chloroform /'klɒrəfɔ:m/ • *noun* colourless volatile liquid formerly used as general anaesthetic. • *verb* render unconscious with this.

chlorophyll /'klɒrəfɪl/ *noun* green pigment in most plants.

choc *noun colloquial* chocolate. □ **choc ice** bar of ice cream covered with chocolate.

chock • *noun* block of wood, wedge. • *verb* make fast with chock(s). □ **chock-a-block** (often + *with*) crammed together, full; **chock-full** (often + *of*) crammed full.

chocolate /'tʃɒklət/ • *noun* food made as paste, powder, or solid block from ground cacao seeds; sweet made of or covered with this; drink containing chocolate; dark brown. • *adjective* made from chocolate; dark brown.

choice • *noun* act of choosing; thing or person chosen; range to choose from; power to choose. • *adjective* of superior quality.

choir /kwaɪə/ *noun* regular group of singers; chancel in large church.

□ **choirboy**, **choirgirl** boy or girl singer in church choir.

choke ● *verb* (**-king**) stop breathing of (person or animal); suffer such stoppage; block up; (as **choked** *adjective*) speechless from emotion, disgusted, disappointed. ● *noun* valve in carburettor controlling inflow of air; device for smoothing variations of alternating electric current.

choker *noun* close-fitting necklace.

choler /'kɒlə/ *noun historical* bile; *archaic* anger, irascibility.

cholera /'kɒlərə/ *noun* infectious often fatal bacterial disease of small intestine.

choleric /'kɒlərɪk/ *adjective* easily angered.

cholesterol /kə'lestərɒl/ *noun* sterol present in human tissues including the blood.

choose /tʃu:z/ *verb* (**-sing**; *past* **chose** /tʃəʊz/; *past participle* **chosen**) select out of greater number; (usually + *between*, *from*) take one or another; (usually + *to do*) decide.

choosy /'tʃu:zɪ/ *adjective* (**-ier**, **-iest**) *colloquial* fussy; hard to please.

chop[1] ● *verb* (**-pp-**) (usually + *off*, *down*, etc.) cut with axe etc.; (often + *up*) cut into small pieces; strike (ball) with heavy edgewise blow. ● *noun* cutting blow; thick slice of meat usually including rib; (**the chop**) *slang* dismissal from job, killing, being killed.

chop[2] *noun* (usually in *plural*) jaw.

chop[3] *verb* (**-pp-**) □ **chop and change** vacillate.

chopper *noun* large-bladed short axe; cleaver; *colloquial* helicopter.

choppy *adjective* (**-ier**, **-iest**) (of sea etc.) fairly rough.

chopstick /'tʃɒpstɪk/ *noun* each of pair of sticks held in one hand as eating utensils by Chinese, Japanese, etc.

chop suey /tʃɒp'su:ɪ/ *noun* (*plural* **-s**) Chinese-style dish of meat fried with vegetables.

choral /'kɔ:r(ə)l/ *adjective* of, for, or sung by choir or chorus.

chorale /kə'rɑ:l/ *noun* simple stately hymn tune; choir.

chord[1] /kɔ:d/ *noun* combination of notes sounded together.

chord[2] /kɔ:d/ *noun* straight line joining ends of arc; string of harp etc.

chore *noun* tedious or routine task, esp. domestic.

choreography /kɒrɪ'ɒgrəfɪ/ *noun* design or arrangement of ballet etc. □ **choreograph** /'kɒrɪəgrɑ:f/ *verb*; **choreographer** *noun*; **choreographic** /-ə'græf-/ *adjective*.

chorister /'kɒrɪstə/ *noun* member of choir, esp. choirboy.

chortle /'tʃɔ:t(ə)l/ ● *noun* gleeful chuckle. ● *verb* (**-ling**) chuckle gleefully.

chorus /'kɔ:rəs/ ● *noun* (*plural* **-es**) group of singers, choir; music for choir; refrain of song; simultaneous utterance; group of singers and dancers performing together; group of performers commenting on action in ancient Greek play, any of its utterances. ● *verb* (**-s-**) utter simultaneously.

chose *past* of CHOOSE.

chosen *past participle* of CHOOSE.

chough /tʃʌf/ *noun* red-legged crow.

choux pastry /ʃu:/ *noun* very light pastry made with eggs.

chow *noun slang* food; Chinese breed of dog.

chow mein /tʃaʊ 'meɪn/ *noun* Chinese-style dish of fried noodles usually with shredded meat and vegetables.

christen /'krɪs(ə)n/ *verb* baptize; name. □ **christening** *noun*.

Christendom /'krɪsəndəm/ *noun* Christians worldwide.

Christian /'krɪstʃ(ə)n/ ● *adjective* of Christ's teaching; believing in or following Christian religion; charitable, kind. ● *noun* adherent of Christianity. □ **Christian era** era counted from Christ's birth; **Christian name** forename, esp. given at christening; **Christian Science** system of belief including power of healing by prayer alone; **Christian Scientist** adherent of this.

Christianity /krɪstɪ'ænɪtɪ/ *noun* Christian religion, quality, or character.

Christmas /'krɪsməs/ *noun* (period around) festival of Christ's birth celebrated on 25 Dec. □ **Christmas box** present or tip given at Christmas; **Christmas Day** 25 Dec.; **Christmas**

Eve 24 Dec.; **Christmas pudding** rich boiled pudding with dried fruit; **Christmas rose** white-flowered hellebore flowering in winter; **Christmas tree** evergreen tree decorated at Christmas. □ **Christmassy** *adjective*.

chromatic /krə'mætɪk/ *adjective* of colour, in colours; *Music* of or having notes not belonging to prevailing key. □ **chromatic scale** scale that proceeds by semitones. □ **chromatically** *adverb*.

chrome /krəʊm/ *noun* chromium; yellow pigment got from a compound of chromium.

chromium /'krəʊmɪəm/ *noun* metallic element used as shiny decorative or protective coating.

chromosome /'krəʊməsəʊm/ *noun Biology* threadlike structure occurring in pairs in cell nucleus, carrying genes.

chronic /'krɒnɪk/ *adjective* (of disease) long-lasting; (of patient) having chronic illness; *colloquial* bad, intense, severe. □ **chronically** *adverb*.

chronicle /'krɒnɪk(ə)l/ ● *noun* record of events in order of occurrence. ● *verb* (**-ling**) record (events) thus.

chronological /krɒnə'lɒdʒɪk(ə)l/ *adjective* according to order of occurrence. □ **chronologically** *adverb*.

chronology /krə'nɒlədʒɪ/ *noun* (*plural* **-ies**) science of computing dates; (document displaying) arrangement of events etc. according to date.

chronometer /krə'nɒmɪtə/ *noun* time-measuring instrument, esp. one used in navigation.

chrysalis /'krɪsəlɪs/ *noun* (*plural* **-lises**) pupa of butterfly or moth; case enclosing it.

chrysanthemum /krɪ'sænθəməm/ *noun* garden plant flowering in autumn.

chub *noun* (*plural* same) thick-bodied river fish.

chubby *adjective* (**-ier**, **-iest**) plump, round.

chuck[1] ● *verb colloquial* fling or throw carelessly; (often + *in*, *up*) give up; touch playfully, esp. under chin. ● *noun* act of chucking; (**the chuck**) *slang* dismissal. □ **chuck out** *colloquial* expel, discard.

chuck[2] ● *noun* cut of beef from neck to ribs; device for holding workpiece or bit. ● *verb* fix in chuck.

chuckle /'tʃʌk(ə)l/ ● *verb* (**-ling**) laugh quietly or inwardly. ● *noun* quiet or suppressed laugh.

chuff *verb* (of engine etc.) work with regular sharp puffing sound.

chuffed *adjective slang* delighted.

chug *verb* (**-gg-**) make intermittent explosive sound; move with this.

chukker /'tʃʌkə/ *noun* period of play in polo.

chum *noun colloquial* close friend. □ **chum up** (**-mm-**) (often + *with*) become close friend (of). □ **chummy** *adjective* (**-ier**, **-iest**).

chump *noun colloquial* foolish person; thick end of loin of lamb or mutton; lump of wood.

chunk *noun* lump cut or broken off.

chunky *adjective* (**-ier**, **-iest**) consisting of or resembling chunks; small and sturdy.

chunter /'tʃʌntə/ *verb colloquial* mutter, grumble.

chupatty = CHAPATTI.

church *noun* building for public Christian worship; public worship; (**Church**) Christians collectively, clerical profession, organized Christian society. □ **churchgoer** person attending church regularly; **churchman** member of clergy or Church; **churchwarden** elected lay representative of Anglican parish; **churchyard** enclosed ground round church, esp. used for burials.

churl *noun* bad-mannered, surly, or stingy person. □ **churlish** *adjective*.

churn ● *noun* large milk can; butter-making machine. ● *verb* agitate (milk etc.) in churn; make (butter) in churn; (usually + *up*) upset, agitate. □ **churn out** produce in large quantities.

chute[1] /ʃu:t/ *noun* slide for taking things to lower level.

chute[2] /ʃu:t/ *noun colloquial* parachute.

chutney /'tʃʌtnɪ/ *noun* (*plural* **-s**) relish made of fruits, vinegar, spices, etc.

chyle /kaɪl/ *noun* milky fluid into which chyme is converted.

chyme /kaɪm/ *noun* pulp formed from partly-digested food.

CIA *abbreviation* (in *US*) Central Intelligence Agency.

ciabatta /tʃə'bɑ:tə/ *noun* Italian bread made with olive oil.

ciao /tʃaʊ/ *interjection colloquial* goodbye; hello.

cicada /sɪ'kɑ:də/ *noun* winged chirping insect.

cicatrice /'sɪkətrɪs/ *noun* scar of healed wound.

cicely /'sɪsəlɪ/ *noun* (*plural* **-ies**) flowering plant related to parsley and chervil.

CID *abbreviation* Criminal Investigation Department.

cider /'saɪdə/ *noun* drink of fermented apple juice.

cigar /sɪ'gɑ:/ *noun* roll of tobacco leaves for smoking.

cigarette /sɪgə'ret/ *noun* finely-cut tobacco rolled in paper for smoking.

cilium /'sɪlɪəm/ *noun* (*plural* **cilia**) hairlike structure on animal cells; eyelash. □ **ciliary** *adjective*.

cinch /sɪntʃ/ *noun colloquial* certainty; easy task.

cinchona /sɪŋ'kəʊnə/ *noun* S. American evergreen tree; (drug from) its bark which contains quinine.

cincture /'sɪŋktʃə/ *noun literary* girdle, belt, or border.

cinder /'sɪndə/ *noun* residue of coal etc. after burning.

Cinderella /sɪndə'relə/ *noun* person or thing of unrecognized merit.

cine /sɪnɪ/ *adjective* cinematographic.

cinema /'sɪnɪmɑ:/ *noun* theatre where films are shown; films collectively; art or industry of producing films. □ **cinematic** /-'mæt-/ *adjective*.

cinematography /sɪnɪmə'tɒgrəfɪ/ *noun* art of making films. □ **cinematographer** *noun*; **cinematographic** /-mætə'græfɪk/ *adjective*.

cineraria /sɪnə'reərɪə/ *noun* plant with bright flowers and downy leaves.

cinnabar /'sɪnəbɑ:/ *noun* red mercuric sulphide; vermilion.

cinnamon /'sɪnəmən/ *noun* aromatic spice from bark of SE Asian tree; this tree; yellowish-brown.

cinquefoil /'sɪŋkfɔɪl/ *noun* plant with compound leaf of 5 leaflets.

Cinque Port /sɪŋk/ *noun* any of (originally 5) ports in SE England with ancient privileges.

cipher /'saɪfə/ (also **cypher**) ● *noun* secret or disguised writing, key to this; arithmetical symbol 0; person or thing of no importance. ● *verb* write in cipher.

circa /'sɜ:kə/ *preposition* (usually before date) about. [Latin]

circle /'sɜ:k(ə)l/ ● *noun* perfectly round plane figure; roundish enclosure or structure; curved upper tier of seats in theatre etc.; circular route; set or restricted group; people grouped round centre of interest. ● *verb* (**-ling**) (often + *round*, *about*) move in or form circle.

circlet /'sɜ:klɪt/ *noun* small circle; circular band, esp. as ornament.

circuit /'sɜ:kɪt/ *noun* line, course, or distance enclosing an area; path of electric current, apparatus through which current passes; judge's itinerary through district, such a district; chain of theatres, cinemas, etc. under single management; motor-racing track; sphere of operation; sequence of sporting events.

circuitous /sɜ:'kju:ɪtəs/ *adjective* indirect, roundabout.

circuitry /'sɜ:kɪtrɪ/ *noun* (*plural* **-ies**) system of electric circuits.

circular /'sɜ:kjʊlə/ ● *adjective* having form of or moving in circle; (of reasoning) using point to be proved as argument for its own truth; (of letter etc.) distributed to several people. ● *noun* circular letter etc. □ **circular saw** power saw with rotating toothed disc. □ **circularity** /-'lærɪtɪ/ *noun*.

circularize *verb* (also **-ise**) (**-zing** or **-sing**) send circular to.

circulate /'sɜ:kjʊleɪt/ *verb* (**-ting**) be or put in circulation; send circulars to; mingle among guests etc.

circulation *noun* movement from and back to starting point, esp. that of blood from and to heart; transmission, distribution; number of copies sold. □ **circulatory** *adjective*.

circumcise /'sɜ:kəmsaɪz/ *verb* (**-sing**) cut off foreskin or clitoris of. □ **circumcision** /-'sɪʒ(ə)n/ *noun*.

circumference /sə'kʌmfərəns/ *noun* line enclosing circle; distance round.

circumflex /'sɜ:kəmfleks/ *noun* (in full **circumflex accent**) mark (ˆ) over vowel indicating pronunciation.

circumlocution /sɜ:kəmlə'kju:ʃ(ə)n/ *noun* roundabout expression; evasive speech; verbosity. □ **circumlocutory** /-'lɒkjʊt-/ *adjective.*

circumnavigate /sɜ:kəm'nævɪgeɪt/ *verb* (**-ting**) sail round. □ **circumnavigation** *noun.*

circumscribe /'sɜ:kəmskraɪb/ *verb* (**-bing**) enclose or outline; lay down limits of; confine, restrict. □ **circumscription** /-'skrɪpʃ(ə)n/ *noun.*

circumspect /'sɜ:kəmspekt/ *adjective* cautious, taking everything into account. □ **circumspection** /-'spekʃ(ə)n/ *noun*; **circumspectly** *adverb.*

circumstance /'sɜ:kəmst(ə)ns/ *noun* fact, occurrence, or condition, esp. (in *plural*) connected with or affecting an event etc.; (in *plural*) financial condition; ceremony, fuss.

circumstantial /sɜ:kəm'stænʃ(ə)l/ *adjective* (of account, story) detailed; (of evidence) tending to establish a conclusion by reasonable inference.

circumvent /sɜ:kəm'vent/ *verb* evade, outwit.

circus /'sɜ:kəs/ *noun* (*plural* **-es**) travelling show of performing acrobats, clowns, animals, etc.; *colloquial* scene of lively action, group of people in common activity; open space in town, where several streets converge; *historical* arena for sports and games.

cirrhosis /sɪ'rəʊsɪs/ *noun* chronic liver disease.

cirrus /'sɪrəs/ *noun* (*plural* **cirri** /-raɪ/) white wispy cloud.

cissy = SISSY.

Cistercian /sɪs'tɜ:ʃ(ə)n/ ● *noun* monk or nun of strict Benedictine order. ● *adjective* of the Cistercians.

cistern /'sɪst(ə)n/ *noun* tank for storing water; underground reservoir.

citadel /'sɪtəd(ə)l/ *noun* fortress protecting or dominating city.

citation /saɪ'teɪʃ(ə)n/ *noun* citing or passage cited; description of reasons for award.

cite *verb* (**-ting**) mention as example; quote (book etc.) in support; mention in military dispatches; summon to law court.

citizen /'sɪtɪz(ə)n/ *noun* native or national of state; inhabitant of a city. □ **citizen's band** system of local intercommunication by radio. □ **citizenship** *noun.*

citrate /'sɪtreɪt/ *noun* salt of citric acid.

citric /'sɪtrɪk/ *adjective* □ **citric acid** sharp-tasting acid in citrus fruits.

citron /'sɪtrən/ *noun* tree bearing large lemon-like fruits; this fruit.

citronella /sɪtrə'nelə/ *noun* a fragrant oil; grass from S. Asia yielding it.

citrus /'sɪtrəs/ *noun* (*plural* **-es**) tree of group including orange, lemon, and grapefruit; (in full **citrus fruit**) fruit of such tree.

city /'sɪtɪ/ *noun* (*plural* **-ies**) large town; town created city by charter and containing cathedral; (**the City**) part of London governed by Lord Mayor and Corporation, business quarter of this, commercial circles.

civet /'sɪvɪt/ *noun* (in full **civet cat**) catlike animal of Central Africa; strong musky perfume got from this.

civic /'sɪvɪk/ *adjective* of city or citizenship.

civics *plural noun* (usually treated as *singular*) study of civic rights and duties.

civil /'sɪv(ə)l/ *adjective* of or belonging to citizens; non-military; polite, obliging; *Law* concerning private rights and not criminal offences. □ **civil defence** organization for protecting civilians in wartime; **civil engineer** person who designs or maintains roads, bridges, etc.; **civil list** annual allowance by Parliament for royal family's household expenses; **civil marriage** one solemnized without religious ceremony; **civil servant** member of civil service; **civil service** all non-military and non-judicial branches of state administration; **civil war** one between citizens of same country.

civilian /sɪ'vɪlɪən/ ● *noun* person not in armed forces. ● *adjective* of or for civilians.

civility /sɪ'vɪlɪtɪ/ *noun* (*plural* **-ies**) politeness; act of politeness.

civilization *noun* (also **-sation**) advanced stage of social development; peoples regarded as having achieved, or been instrumental in evolving, this.

civilize /'sɪvɪlaɪz/ *verb* (also **-ise**) (**-zing** or **-sing**) bring out of barbarism; enlighten, refine.

cl *abbreviation* centilitre(s).

clack • *verb* make sharp sound as of boards struck together. • *noun* such sound.

clad *adjective* clothed; provided with cladding.

cladding *noun* protective covering or coating.

claim • *verb* assert; demand as one's due; represent oneself as having; (+ *to do*) profess. • *noun* demand; (+ *to, on*) right or title; assertion; thing claimed.

claimant *noun* person making claim, esp. in lawsuit or for state benefit.

clairvoyance /kleə'vɔɪəns/ *noun* supposed faculty of perceiving the future or the unseen. □ **clairvoyant** *noun & adjective.*

clam • *noun* edible bivalve mollusc. • *verb* (**-mm-**) (+ *up*) *colloquial* refuse to talk.

clamber /'klæmbə/ *verb* climb using hands or with difficulty.

clammy /'klæmɪ/ *adjective* (**-ier**, **-iest**) damp and sticky.

clamour /'klæmə/ (*US* **clamor**) • *noun* shouting, confused noise; protest, demand. • *verb* make clamour, shout. □ **clamorous** *adjective.*

clamp[1] • *noun* device, esp. brace or band of iron etc., for strengthening or holding things together; device for immobilizing illegally parked vehicles. • *verb* strengthen or fasten with clamp; immobilize with clamp. □ **clamp down** (usually + *on*) become stricter (about).

clamp[2] *noun* heap of earth and straw over harvested potatoes etc.

clan *noun* group of families with common ancestor, esp. in Scotland; family holding together; group with common interest. □ **clannish** *adjective*; **clansman**, **clanswoman** *noun.*

clandestine /klæn'destɪn/ *adjective* surreptitious, secret.

clang • *noun* loud resonant metallic sound. • *verb* (cause to) make clang.

clangour /'klæŋgə/ *noun* (*US* **clangor**) continued clanging. □ **clangorous** *adjective.*

clank • *noun* sound as of metal on metal. • *verb* (cause to) make clank.

clap • *verb* (**-pp-**) strike palms of hands together, esp. as applause; put or place quickly or with determination; (+ *on*) give friendly slap on. • *noun* act of clapping; explosive noise, esp. of thunder; slap. □ **clap eyes on** *colloquial* see.

clapper *noun* tongue or striker of bell. □ **clapperboard** device in film-making for making sharp noise to synchronize picture and sound.

claptrap *noun* insincere or pretentious talk; nonsense.

claque /klæk/ *noun* people hired to applaud.

claret /'klærət/ *noun* red Bordeaux wine.

clarify /'klærɪfaɪ/ *verb* (**-ies**, **-ied**) make or become clear; free from impurities; make transparent. □ **clarification** *noun.*

clarinet /klærɪ'net/ *noun* woodwind instrument with single reed. □ **clarinettist** *noun.*

clarion /'klærɪən/ *noun* rousing sound; *historical* war-trumpet.

clarity /'klærɪtɪ/ *noun* clearness.

clash • *noun* loud jarring sound as of metal objects struck together; collision; conflict; discord of colours etc. • *verb* (cause to) make clash; coincide awkwardly; (often + *with*) be at variance or discordant.

clasp /klɑ:sp/ • *noun* device with interlocking parts for fastening; embrace, handshake; bar on medal-ribbon. • *verb* fasten (as) with clasp; grasp, embrace. □ **clasp-knife** large folding knife.

class /klɑ:s/ • *noun* any set of people or things grouped together or differentiated from others; division or order of society; *colloquial* high quality; set of students taught together; their time of meeting; their course of instruction. • *verb* place in a class. □ **classmate** person in same class at school; **classroom** room where class of students is taught. □ **classless** *adjective.*

classic /'klæsɪk/ • *adjective* first-class; of lasting importance; typical; of ancient Greek or Latin culture etc.; (of style) simple and harmonious; famous because long-established. • *noun* classic writer, artist, work, or example; (in *plural*) study of ancient Greek and Latin.

classical *adjective* of ancient Greek or Latin literature etc.; (of language) having form used by standard authors; (of music) serious or conventional.

classicism /ˈklæsɪsɪz(ə)m/ *noun* following of classic style; classical scholarship. □ **classicist** *noun.*

classify /ˈklæsɪfaɪ/ *verb* (**-ies**, **-ied**) arrange in classes; class; designate as officially secret. □ **classification** *noun.*

classy /ˈklɑːsɪ/ *adjective* (**-ier**, **-iest**) *colloquial* superior. □ **classiness** *noun.*

clatter /ˈklætə/ ● *noun* sound as of hard objects struck together. ● *verb* (cause to) make clatter.

clause /klɔːz/ *noun* group of words including finite verb (see panel); single statement in treaty, law, contract, etc.

claustrophobia /klɔːstrəˈfəʊbɪə/ *noun* abnormal fear of confined places. □ **claustrophobic** *adjective.*

clavichord /ˈklævɪkɔːd/ *noun* small keyboard instrument with very soft tone.

clavicle /ˈklævɪk(ə)l/ *noun* collar-bone.

claw ● *noun* pointed nail on animal's foot; foot armed with claws; pincers of shellfish; device for grappling, holding, etc. ● *verb* scratch, maul, or pull with claws or fingernails.

clay *noun* stiff sticky earth, used for bricks, pottery, etc. □ **clay pigeon** breakable disc thrown into air as target for shooting. □ **clayey** *adjective.*

claymore /ˈkleɪmɔː/ *noun historical* Scottish two-edged broad-bladed sword.

clean ● *adjective* free from dirt; clear; pristine; not obscene or indecent; attentive to cleanliness; clear-cut; without record of crime etc.; fair. ● *adverb* completely; simply; in a clean way. ● *verb* make or become clean. ● *noun* process of cleaning. □ **clean-cut** sharply outlined, (of person) clean and tidy; **clean out** clean thoroughly, *slang* empty or deprive (esp. of money); **clean-shaven** without beard or moustache; **clean up** make tidy or clean, *slang* acquire as or make profit. □ **cleaner** *noun*; **cleanness** *noun.*

cleanly[1] *adverb* in a clean way.

cleanly[2] /ˈklenlɪ/ *adjective* (**-ier**, **-iest**) habitually clean; attentive to cleanness and hygiene. □ **cleanliness** *noun.*

cleanse /klenz/ *verb* (**-sing**) make clean or pure. □ **cleanser** *noun.*

clear ● *adjective* free from dirt or contamination; not clouded; transparent; readily perceived or understood; able to discern readily; convinced; (of conscience) guiltless; unobstructed; net; complete; (often + *of*) free, unhampered. ● *adverb* clearly; completely; apart. ● *verb* make or become clear; (often + *of*) free from obstruction etc.; (often + *of*) show (person) to be innocent; approve (person etc.) for special duty, access, etc.; pass through (customs); pass over or by without touching; make (sum) as net gain; pass (cheque) through clearing house. □ **clear-cut** sharply defined; **clear off** *colloquial* go away; **clear out** empty, remove, *colloquial* go away; **clear-out** tidying by emptying and sorting; **clear up** tidy, solve; **clearway** road where vehicles may not stop. □ **clearly** *adverb*; **clearness** *noun.*

clearance *noun* removal of obstructions etc.; space allowed for passing of two objects; special authorization; clearing for special duty, of cheque, etc.; clearing out.

Clause

A clause is a group of words that includes a finite verb. If it makes complete sense by itself, it is known as a main clause, e.g.

The sun came out.

Otherwise, although it makes some sense, it must be attached to a main clause; this is known as a subordinate clause, e.g.

when the sun came out
(as in *When the sun came out, we went outside.*)

---

clearing *noun* treeless area in forest. □ **clearing bank** member of clearing house; **clearing house** bankers' institution where cheques etc. are exchanged, agency for collecting and distributing information etc.

cleat *noun* device for fastening ropes to projection on gangway, sole, etc. to provide grip.

cleavage *noun* hollow between woman's breasts; division; line along which rocks etc. split.

cleave[1] *verb* (**-ving**; *past* **clove** /kləʊv/, **cleft**, or **cleaved**; *past participle* **cloven**, **cleft**, or **cleaved**) *literary* break or come apart; make one's way through.

cleave[2] *verb* (**-ving**) (+ *to*) *literary* adhere.

cleaver *noun* butcher's heavy chopping tool.

clef *noun Music* symbol at start of staff showing pitch of notes on it.

cleft[1] *adjective* split, partly divided. □ **cleft palate** congenital split in roof of mouth.

cleft[2] *noun* split, fissure.

clematis /'klemətɪs/ *noun* climbing flowering plant.

clement /'klemənt/ *adjective* mild; merciful. □ **clemency** *noun*.

clementine /'kleməntiːn/ *noun* small tangerine-like fruit.

clench ● *verb* close tightly; grasp firmly. ● *noun* clenching action; clenched state.

clergy /'klɜːdʒɪ/ *noun* (*plural* **-ies**) (usually treated as *plural*) those ordained for religious duties.

clergyman /'klɜːdʒɪmən/ *noun* member of clergy.

cleric /'klerɪk/ *noun* member of clergy.

clerical *adjective* of clergy or clergymen; of or done by clerks.

clerihew /'klerɪhjuː/ *noun* witty or comic 4-line biographical verse.

clerk /klɑːk/ ● *noun* person employed to keep records, accounts, etc.; secretary or agent of local council, court, etc.; lay officer of church. ● *verb* work as clerk.

clever /'klevə/ *adjective* (**-er**, **-est**) skilful, talented, quick to understand and learn; adroit; ingenious. □ **cleverly** *adverb*; **cleverness** *noun*.

cliché /'kliːʃeɪ/ *noun* hackneyed phrase or opinion.

clichéd *adjective* hackneyed, full of clichés.

click ● *noun* slight sharp sound. ● *verb* (cause to) make click; *colloquial* become clear, understood, or popular; (+ *with*) become friendly with.

client /'klaɪənt/ *noun* person using services of lawyer or other professional person; customer.

clientele /kliːɒn'tel/ *noun* clients collectively; customers.

cliff *noun* steep rock face, esp. on coast. □ **cliff-hanger** story etc. with strong element of suspense.

climacteric /klaɪ'mæktərɪk/ *noun* period of life when fertility and sexual activity are in decline.

climate /'klaɪmɪt/ *noun* prevailing weather conditions of an area; region with particular weather conditions; prevailing trend of opinion etc. □ **climatic** /-'mæt-/ *adjective*; **climatically** /-'mæt-/ *adverb*.

climax /'klaɪmæks/ ● *noun* event or point of greatest intensity or interest, culmination. ● *verb colloquial* reach or bring to a climax. □ **climactic** *adjective*.

climb /klaɪm/ ● *verb* (often + *up*) ascend, mount, go up; (of plant) grow up wall etc. by clinging etc.; rise, esp. in social rank. ● *noun* action of climbing; hill etc. (to be) climbed. □ **climber** *noun*.

clime *noun literary* region; climate.

clinch ● *verb* confirm or settle conclusively; (of boxers) become too closely engaged; secure (nail or rivet) by driving point sideways when through. ● *noun* clinching, resulting state; *colloquial* embrace.

cling *verb* (*past & past participle* **clung**) (often + *to*) adhere; (+ *to*) be emotionally dependent on or unwilling to give up; (often + *to*) maintain grasp. □ **cling film** thin transparent plastic covering for food. □ **clingy** *adjective* (**-ier**, **-iest**).

clinic /'klɪnɪk/ *noun* private or specialized hospital; place or occasion for giving medical treatment or specialist advice; teaching of medicine at hospital bedside.

clinical *adjective* of or for the treatment of patients; objective, coldly detached;

(of room etc.) bare, functional. □ **clinically** *adverb.*

clink[1] • *noun* sharp ringing sound. • *verb* (cause to) make clink.

clink[2] *noun slang* prison.

clinker /'klɪŋkə/ *noun* mass of slag or lava; stony residue from burnt coal.

clinker-built *adjective* (of boat) with external planks overlapping downwards.

clip[1] • *noun* device for holding things together; piece of jewellery fastened by clip; set of attached cartridges for firearm. • *verb* **(-pp-)** fix with clip. □ **clipboard** board with spring clip for holding papers etc.

clip[2] • *verb* **(-pp-)** cut with shears or scissors; cut hair or wool of; *colloquial* hit sharply; omit (letter) from word; omit parts of (words uttered); punch hole in (ticket) to show it has been used; cut from newspaper etc. • *noun* clipping; *colloquial* sharp blow; extract from motion picture; yield of wool.

clipper *noun* (usually in *plural*) instrument for clipping; *historical* fast sailing ship.

clipping *noun* piece clipped, esp. from newspaper.

clique /kli:k/ *noun* exclusive group of people. □ **cliquey** *adjective* (**cliquier**, **cliquiest**); **cliquish** *adjective.*

clitoris /'klɪtərɪs/ *noun* small erectile part of female genitals.

Cllr. *abbreviation* Councillor.

cloak • *noun* loose usually sleeveless outdoor garment; covering. • *verb* cover with cloak; conceal, disguise. □ **cloakroom** room for outdoor clothes or luggage, *euphemistic* lavatory.

clobber[1] /'klɒbə/ *verb slang* hit (repeatedly); defeat; criticize severely.

clobber[2] /'klɒbə/ *noun slang* clothing, belongings.

cloche /klɒʃ/ *noun* small translucent cover for outdoor plants; woman's close-fitting bell-shaped hat.

clock[1] • *noun* instrument measuring and showing time; measuring device resembling this; *colloquial* speedometer, taximeter, or stopwatch; seed-head of dandelion. • *verb colloquial* (often + *up*) attain or register (distance etc.); time (race etc.) by stopwatch. □ **clock in** (or **on**), or **out** (or **off**) register time of arrival, or departure, by automatic clock; **clockwise** (moving) in same direction as hands of clock; **clockwork** mechanism with coiled springs etc. on clock principle; **like clockwork** with mechanical precision.

clock[2] *noun* ornamental pattern on side of stocking or sock.

clod *noun* lump of earth or clay.

clog • *noun* wooden-soled shoe. • *verb* **(-gg-)** (often + *up*) (cause to) become obstructed, choke; impede.

cloister /'klɔɪstə/ • *noun* covered walk esp. in college or ecclesiastical building; monastic life; seclusion. • *verb* seclude.

clone • *noun* group of organisms produced asexually from one ancestor; one such organism; *colloquial* person or thing regarded as identical to another. • *verb* **(-ning)** propagate as clone.

close[1] /kləʊs/ • *adjective* (often + *to*) at short distance or interval; having strong or immediate relation; (almost) in contact; dense, compact; nearly equal; rigorous; concentrated; stifling; shut; secret; niggardly. • *adverb* at short distance or interval. • *noun* street closed at one end; precinct of cathedral. □ **close harmony** singing of parts within an octave; **close-knit** tightly interlocked, closely united; **close season** season when killing of game etc. is illegal; **close shave** *colloquial* narrow escape; **close-up** photograph etc. taken at short range. □ **closely** *adverb*; **closeness** *noun.*

close[2] /kləʊz/ • *verb* **(-sing)** shut; block up; bring or come to an end; end day's business; bring or come closer or into contact; make (electric circuit) continuous. • *noun* conclusion, end. □ **closed-circuit** (of television) transmitted by wires to restricted number of receivers; **closed shop** business etc. where employees must belong to specified trade union.

closet /'klɒzɪt/ • *noun* small room; cupboard; water-closet. • *adjective* secret. • *verb* **(-t-)** shut away, esp. in private consultation etc.

closure /'kləʊʒə/ *noun* closing, closed state; procedure for ending debate.

clot ● *noun* thick lump formed from liquid, esp. blood; *colloquial* foolish person. ● *verb* (**-tt-**) form into clots.

cloth /klɒθ/ *noun* woven or felted material; piece of this; (**the cloth**) the clergy.

clothe /ˈkləʊð/ *verb* (**-thing**; *past & past participle* **clothed** or **clad**) put clothes on; provide with clothes; cover as with clothes.

clothes /kləʊðz/ *plural noun* things worn to cover body and limbs; bedclothes. □ **clothes-horse** frame for airing washed clothes.

clothier /ˈkləʊðɪə/ *noun* dealer in men's clothes.

clothing /ˈkləʊðɪŋ/ *noun* clothes.

cloud /klaʊd/ ● *noun* visible mass of condensed watery vapour floating in air; mass of smoke or dust; (+ *of*) great number of (birds, insects, etc.) moving together; state of gloom, trouble, or suspicion. ● *verb* cover or darken with cloud(s); (often + *over*, *up*) become overcast or gloomy; make unclear. □ **cloudburst** sudden violent rainstorm. □ **cloudless** *adjective.*

cloudy *adjective* (**-ier**, **-iest**) covered with clouds; not transparent, unclear. □ **cloudiness** *noun.*

clout /klaʊt/ ● *noun* heavy blow; *colloquial* influence, power of effective action; piece of cloth or clothing. ● *verb* hit hard.

clove[1] /kləʊv/ *noun* dried bud of tropical tree, used as spice.

clove[2] /kləʊv/ *noun* segment of compound bulb, esp. of garlic.

clove[3] *past* of CLEAVE[1].

clove hitch *noun* knot for fastening rope round pole etc.

cloven /ˈkləʊv(ə)n/ *adjective* split. □ **cloven hoof**, **foot** divided hoof, as of oxen, sheep, etc., or the Devil.

clover /ˈkləʊvə/ *noun* kind of trefoil used as fodder. □ **in clover** in ease and luxury.

clown /klaʊn/ ● *noun* comic entertainer, esp. in circus; foolish or playful person. ● *verb* (often + *about*, *around*) behave like clown.

cloy *verb* satiate or sicken by sweetness, richness, etc.

club ● *noun* heavy stick used as weapon; stick with head, used in golf; association of people for social, sporting, etc. purposes; premises of this; playing card of suit marked with black trefoils; (in *plural*) this suit. ● *verb* (**-bb-**) strike (as) with club; (+ *together*) combine, esp. to make up sum of money. □ **club foot** congenitally deformed foot; **clubhouse** premises of club; **clubland** area with many nightclubs; **clubroot** disease of cabbages etc.; **club sandwich** sandwich with 2 layers of filling and 3 slices of bread or toast.

cluck ● *noun* chattering cry of hen. ● *verb* emit cluck(s).

clue ● *noun* guiding or suggestive fact; piece of evidence used in detection of crime; word(s) used to indicate word(s) for insertion in crossword. ● *verb* (**clues**, **clued**, **cluing** or **clueing**) provide clue to. □ **clue in**, **up** *slang* inform.

clump ● *noun* (+ *of*) cluster, esp. of trees. ● *verb* form clump; heap or plant together; tread heavily.

clumsy /ˈklʌmzɪ/ *adjective* (**-ier**, **-iest**) awkward in movement or shape; difficult to handle or use; tactless. □ **clumsily** *adverb*; **clumsiness** *noun.*

clung *past & past participle* of CLING.

cluster /ˈklʌstə/ ● *noun* close group or bunch of similar people or things. ● *verb* be in or form into cluster(s); (+ *round*, *around*) gather round.

clutch[1] ● *verb* seize eagerly; grasp tightly; (+ *at*) snatch at. ● *noun* tight or (in *plural*) cruel grasp; (in vehicle) device for connecting engine to transmission, pedal operating this. □ **clutch bag** handbag without handles.

clutch[2] *noun* set of eggs; brood of chickens.

clutter /ˈklʌtə/ ● *noun* crowded untidy collection of things. ● *verb* (often + *up*, *with*) crowd untidily, fill with clutter.

cm *abbreviation* centimetre(s).

Cmdr. *abbreviation* Commander.

CND *abbreviation* Campaign for Nuclear Disarmament.

CO *abbreviation* Commanding Officer.

Co. *abbreviation* company; county.

c/o *abbreviation* care of.

coach ● *noun* single-decker bus usually for longer journeys; railway carriage; closed horse-drawn carriage; sports trainer or private tutor. ● *verb* train or teach. □ **coachload** group of tourists etc. travelling by coach; **coachman** driver of horse-drawn carriage; **coachwork** bodywork of road or rail vehicle.

coagulate /kəʊ'ægjʊleɪt/ *verb* (**-ting**) change from liquid to semi-solid; clot, curdle. □ **coagulant** *noun*; **coagulation** *noun*.

coal *noun* hard black mineral used as fuel etc.; piece of this. □ **coalface** exposed surface of coal in mine; **coalfield** area yielding coal; **coal gas** mixed gases formerly extracted from coal and used for heating, cooking, etc.; **coal mine** place where coal is dug; **coal miner** worker in coal mine; **coal scuttle** container for coal for domestic fire; **coal tar** tar extracted from coal; **coal tit** small bird with greyish plumage.

coalesce /kəʊə'les/ *verb* (**-cing**) come together and form a whole. □ **coalescence** *noun*; **coalescent** *adjective*.

coalition /kəʊə'lɪʃ(ə)n/ *noun* temporary alliance, esp. of political parties; fusion into one whole.

coaming /'kəʊmɪŋ/ *noun* raised border round ship's hatches etc.

coarse /kɔ:s/ *adjective* rough or loose in texture; of large particles; lacking refinement; crude, obscene. □ **coarse fish** freshwater fish other than salmon and trout. □ **coarsely** *adverb*; **coarsen** *verb*; **coarseness** *noun*.

coast ● *noun* border of land near sea; seashore. ● *verb* ride or move (usually downhill) without use of power; make progress without exertion; sail along coast. □ **coastguard** (member of) group of people employed to keep watch on coasts, prevent smuggling, etc.; **coastline** line of seashore, esp. with regard to its shape. □ **coastal** *adjective*.

coaster *noun* ship that sails along coast; tray or mat for bottle or glass.

coat ● *noun* sleeved outer garment, overcoat, jacket; animal's fur or hair; layer of paint etc. ● *verb* (usually + *with*, *in*) cover with coat or layer; form covering on. □ **coat of arms** heraldic bearings or shield; **coat-hanger** shaped piece of wood etc. for hanging clothes on.

coating *noun* layer of paint etc.; cloth for coats.

coax *verb* persuade gradually or by flattery; (+ *out of*) obtain (thing) from (person) thus; manipulate gently.

coaxial /kəʊ'æksɪəl/ *adjective* having common axis; (of electric cable etc.) transmitting by means of two concentric conductors separated by insulator.

cob *noun* roundish lump; domed loaf; corn cob; large hazelnut; sturdy short-legged riding-horse; male swan.

cobalt /'kəʊbɔ:lt/ *noun* silvery-white metallic element; (colour of) deep blue pigment made from it.

cobber /'kɒbə/ *noun Australian & NZ colloquial* companion, friend.

cobble[1] /'kɒb(ə)l/ ● *noun* (in full **cobblestone**) rounded stone used for paving. ● *verb* (**-ling**) pave with cobbles.

cobble[2] /'kɒb(ə)l/ *verb* (**-ling**) mend or patch (esp. shoes); (often + *together*) assemble roughly.

cobbler *noun* mender of shoes; (in *plural*) *slang* nonsense.

COBOL /'kəʊbɒl/ *noun* computer language for use in business operations.

cobra /'kəʊbrə/ *noun* venomous hooded snake.

cobweb /'kɒbweb/ *noun* spider's network or thread. □ **cobwebby** *adjective*.

coca /'kəʊkə/ *noun* S. American shrub; its leaves chewed as stimulant.

cocaine /kəʊ'keɪn/ *noun* drug from coca, used as local anaesthetic and as stimulant.

coccyx /'kɒksɪks/ *noun* (*plural* **coccyges** /-dʒi:z/) bone at base of spinal column.

cochineal /kɒtʃɪ'ni:l/ *noun* scarlet dye; insects whose dried bodies yield this.

cock[1] ● *noun* male bird, esp. domestic fowl; *slang* (as form of address) friend, fellow; *slang* nonsense; firing lever in gun released by trigger; tap or valve controlling flow. ● *verb* raise or make upright; turn or move (eye or ear) attentively or knowingly; set (hat etc.) aslant; raise cock of (gun). □ **cock-a-hoop** exultant; **cock-a-leekie** Scottish soup of boiled fowl with leeks; **cockcrow** dawn; **cock-eyed** *colloquial* crooked, askew, absurd.

cock[2] *noun* conical heap of hay or straw.

cockade /kɒ'keɪd/ *noun* rosette etc. worn in hat.

cockatoo /kɒkə'tu:/ *noun* crested parrot.

cockchafer /'kɒktʃeɪfə/ *noun* large pale brown beetle.

cocker /'kɒkə/ *noun* (in full **cocker spaniel**) small spaniel.

cockerel /'kɒkər(ə)l/ *noun* young cock.

cockle /'kɒk(ə)l/ *noun* edible bivalve shellfish; its shell; (in full **cockleshell**) small shallow boat; pucker or wrinkle. □ **cockles of the heart** innermost feelings.

cockney /'kɒknɪ/ ● *noun* (*plural* **-s**) native of London, esp. East End; cockney dialect. ● *adjective* of cockneys.

cockpit *noun* place for pilot etc. in aircraft or spacecraft or for driver in racing car; arena of war etc.

cockroach /'kɒkrəʊtʃ/ *noun* dark brown insect infesting esp. kitchens.

cockscomb /'kɒkskəʊm/ *noun* cock's crest.

cocksure /kɒk'ʃɔ:/ *adjective* arrogantly confident.

cocktail /'kɒkteɪl/ *noun* drink of spirits, fruit juices, etc.; appetizer containing shellfish etc.; any hybrid mixture. □ **cocktail stick** small pointed stick.

cocky *adjective* (**-ier**, **-iest**) *colloquial* conceited, arrogant. □ **cockiness** *noun.*

coco /'kəʊkəʊ/ *noun* (*plural* **-s**) coconut palm.

cocoa /'kəʊkəʊ/ *noun* powder of crushed cacao seeds; drink made from this.

coconut /'kəʊkənʌt/ *noun* large brown seed of coco, with edible white lining enclosing milky juice. □ **coconut matting** matting made from fibre of coconut husks; **coconut shy** fairground sideshow where balls are thrown to dislodge coconuts.

cocoon /kə'ku:n/ ● *noun* silky case spun by larva to protect it as pupa; protective covering. ● *verb* wrap (as) in cocoon.

cocotte /kə'kɒt/ *noun* small fireproof dish for cooking and serving food.

COD *abbreviation* cash (*US* collect) on delivery.

cod¹ *noun* (also **codfish**) (*plural* same) large sea fish. □ **cod liver oil** medicinal oil rich in vitamins.

cod² *noun & verb* (**-dd-**) *slang* hoax, parody.

coda /'kəʊdə/ *noun* final passage of piece of music.

coddle /'kɒd(ə)l/ *verb* (**-ling**) treat as an invalid, pamper; cook in water just below boiling point. □ **coddler** *noun.*

code ● *noun* system of signals or of symbols etc. used for secrecy, brevity, or computer processing of information; systematic set of laws etc.; standard of moral behaviour. ● *verb* (**-ding**) put into code.

codeine /'kəʊdi:n/ *noun* alkaloid derived from morphine, used as pain-killer.

codex /'kəʊdeks/ *noun* (*plural* **codices** /'kəʊdɪsi:z/) manuscript volume esp. of ancient texts.

codger /'kɒdʒə/ *noun* (usually in **old codger**) *colloquial* (strange) person.

codicil /'kəʊdɪsɪl/ *noun* addition to will.

codify /'kəʊdɪfaɪ/ *verb* (**-ies**, **-ied**) arrange (laws etc.) into code. □ **codification** *noun.*

codling¹ /'kɒdlɪŋ/ *noun* (also **codlin**) variety of apple; moth whose larva feeds on apples.

codling² *noun* (*plural* same) small cod.

co-education /kəʊedju:'keɪʃ(ə)n/ *noun* education of both sexes together. □ **co-educational** *adjective.*

coefficient /kəʊɪ'fɪʃ(ə)nt/ *noun Mathematics* quantity or expression placed before and multiplying another; *Physics* multiplier or factor by which a property is measured.

coeliac disease /'si:lɪæk/ *noun* intestinal disease whose symptoms include adverse reaction to gluten.

coequal /kəʊ'i:kw(ə)l/ *adjective & noun* equal.

coerce /kəʊ'ɜ:s/ *verb* (**-cing**) persuade or restrain by force. □ **coercion** /-'ɜ:ʃ(ə)n/ *noun*; **coercive** *adjective.*

coeval /kəʊ'i:v(ə)l/ *formal* ● *adjective* of the same age; contemporary. ● *noun* coeval person or thing.

coexist /kəʊɪg'zɪst/ *verb* (often + *with*) exist together, esp. in mutual tolerance. □ **coexistence** *noun*; **coexistent** *adjective.*

coextensive /kəʊɪk'stensɪv/ *adjective* extending over same space or time.

C. of E. *abbreviation* Church of England.

coffee /ˈkɒfɪ/ *noun* drink made from roasted and ground seeds of tropical shrub; cup of this; the shrub; the seeds; pale brown. □ **coffee bar** café selling coffee and light refreshments from bar; **coffee bean** seed of coffee; **coffee mill** small machine for grinding coffee beans; **coffee morning** morning gathering, esp. for charity, at which coffee is served; **coffee shop** small restaurant, esp. in hotel or store; **coffee table** small low table; **coffee-table book** large illustrated book.

coffer /ˈkɒfə/ *noun* large box for valuables; (in *plural*) funds, treasury; sunken panel in ceiling etc.

coffin /ˈkɒfɪn/ *noun* box in which corpse is buried or cremated.

cog *noun* each of series of projections on wheel etc. transferring motion by engaging with another series.

cogent /ˈkəʊdʒ(ə)nt/ *adjective* (of argument etc.) convincing, compelling. □ **cogency** *noun*; **cogently** *adverb*.

cogitate /ˈkɒdʒɪteɪt/ *verb* (**-ting**) ponder, meditate. □ **cogitation** *noun*.

cognac /ˈkɒnjæk/ *noun* French brandy.

cognate /ˈkɒgneɪt/ ● *adjective* descended from same ancestor or root. ● *noun* cognate person or word.

cognition /kɒgˈnɪʃ(ə)n/ *noun* knowing, perceiving, or conceiving, as distinct from emotion and volition. □ **cognitive** /ˈkɒg-/ *adjective*.

cognizance /ˈkɒgnɪz(ə)ns/ *noun formal* knowledge or awareness.

cognizant /ˈkɒgnɪz(ə)nt/ *adjective formal* (+ *of*) having knowledge or being aware of.

cognomen /kɒgˈnəʊmen/ *noun* nickname; *Roman History* surname.

cohabit /kəʊˈhæbɪt/ *verb* (**-t-**) live together as husband and wife. □ **cohabitation** *noun*.

cohere /kəʊˈhɪə/ *verb* (**-ring**) stick together; (of reasoning) be logical or consistent.

coherent *adjective* intelligible; consistent, easily understood. □ **coherence** *noun*; **coherently** *adverb*.

cohesion /kəʊˈhiːʒ(ə)n/ *noun* sticking together; tendency to cohere. □ **cohesive** /-sɪv/ *adjective*.

cohort /ˈkəʊhɔːt/ *noun* one-tenth of Roman legion; people banded together.

coif /kɔɪf/ *noun historical* close-fitting cap.

coiffeur /kwɑːˈfɜː/ *noun* (*feminine* **coiffeuse** /-ˈfɜːz/) hairdresser.

coiffure /kwɑːˈfjʊə/ *noun* hairstyle.

coil ● *verb* arrange or be arranged in concentric rings; move sinuously. ● *noun* coiled arrangement (of rope, electrical conductor, etc.); single turn of something coiled; flexible contraceptive device in womb.

coin ● *noun* stamped disc of metal as official money; metal money. ● *verb* make (coins) by stamping; invent (word, phrase). □ **coin box** telephone operated by coins.

coinage *noun* coining; system of coins in use; invention of word, invented word.

coincide /kəʊɪnˈsaɪd/ *verb* (**-ding**) occur at same time; (often + *with*) agree or be identical.

coincidence /kəʊˈɪnsɪd(ə)ns/ *noun* remarkable concurrence of events etc., apparently by chance. □ **coincident** *adjective*.

coincidental /kəʊɪnsɪˈdent(ə)l/ *adjective* in the nature of or resulting from coincidence. □ **coincidentally** *adverb*.

coir /ˈkɔɪə/ *noun* coconut fibre used for ropes, matting, etc.

coition /kəʊˈɪʃ(ə)n/ *noun* coitus.

coitus /ˈkəʊɪtəs/ *noun* sexual intercourse. □ **coital** *adjective*.

coke[1] ● *noun* solid left after gases have been extracted from coal. ● *verb* (**-king**) convert (coal) into coke.

coke[2] *noun slang* cocaine.

Col. *abbreviation* Colonel.

col *noun* depression in summit-line of mountain chain.

cola /ˈkəʊlə/ *noun* W. African tree with seeds containing caffeine; carbonated drink flavoured with these.

colander /ˈkʌləndə/ *noun* perforated vessel used as strainer in cookery.

cold /kəʊld/ ● *adjective* of or at low temperature; not heated; having lost heat; feeling cold; (of colour) suggesting cold; *colloquial* unconscious; lacking ardour or friendliness; dispiriting; (of hunting-scent) grown faint. ● *noun* prevalence of low temperature; cold weather; infection of nose or throat. ● *adverb* unrehearsed. □ **cold-blooded** having body temperature varying with that of environment, callous; **cold call** marketing call on person not previously interested in product; **cold cream** cleansing ointment; **cold feet** fear, reluctance; **cold fusion** nuclear fusion at room temperature; **cold shoulder** unfriendly treatment; **cold turkey** *slang* abrupt withdrawal from addictive drugs; **cold war** hostility between nations without actual fighting; **in cold blood** without emotion; **throw cold water on** discourage. □ **coldly** *adverb*; **coldness** *noun*.

coleslaw /ˈkəʊlslɔː/ *noun* salad of sliced raw cabbage etc.

coley /ˈkəʊlɪ/ *noun* (*plural* **-s**) any of several edible fish, esp. rock-salmon.

colic /ˈkɒlɪk/ *noun* spasmodic abdominal pain. □ **colicky** *adjective*.

colitis /kəˈlaɪtɪs/ *noun* inflammation of colon.

collaborate /kəˈlæbəreɪt/ *verb* (**-ting**) (often + *with*) work jointly. □ **collaboration** *noun*; **collaborative** /-rətɪv/ *adjective*; **collaborator** *noun*.

collage /ˈkɒlɑːʒ/ *noun* picture made by gluing pieces of paper etc. on to backing.

collapse /kəˈlæps/ ● *noun* falling down of structure; sudden failure of plan etc.; physical or mental breakdown. ● *verb* (**-sing**) (cause to) undergo collapse; *colloquial* relax completely after effort. □ **collapsible** *adjective*.

collar /ˈkɒlə/ ● *noun* neckband, upright or turned over, of coat, shirt, dress, etc.; leather band round animal's neck; band, ring, or pipe in machinery. ● *verb* capture, seize, appropriate; *colloquial* accost. □ **collar-bone** bone joining breastbone and shoulder blade.

collate /kəˈleɪt/ *verb* (**-ting**) collect and put in order. □ **collator** *noun*.

collateral /kəˈlætər(ə)l/ ● *noun* security pledged as guarantee for repayment of loan. ● *adjective* side by side; additional but subordinate; descended from same ancestor but by different line. □ **collaterally** *adverb*.

collation *noun* collating; light meal.

colleague /ˈkɒliːg/ *noun* fellow worker, esp. in profession or business.

collect[1] /kəˈlekt/ ● *verb* bring or come together; assemble, accumulate; seek and acquire (books, stamps, etc.); obtain (contributions, taxes, etc.) from people; call for, fetch; concentrate (one's thoughts etc.); (as **collected** *adjective*) not perturbed or distracted. ● *adjective & adverb US* (of telephone call, parcel, etc.) to be paid for by recipient.

collect[2] /ˈkɒlekt/ *noun* short prayer of Anglican or RC Church.

collectable /kəˈlektɪb(ə)l/ *adjective* worth collecting.

collection /kəˈlekʃ(ə)n/ *noun* collecting, being collected; things collected; money collected, esp. at church service etc.

collective /kəˈlektɪv/ ● *adjective* of or relating to group or society as a whole; joint; shared. ● *noun* cooperative enterprise; its members. □ **collective bargaining** negotiation of wages etc. by organized group of employees; **collective noun** singular noun denoting group of individuals. □ **collectively** *adverb*; **collectivize** *verb* (also **-ise**) (**-zing** or **-sing**).

collectivism *noun* theory or practice of collective ownership of land and means of production. □ **collectivist** *noun & adjective*.

collector *noun* person collecting things of interest; person collecting taxes, rents, etc.

colleen /ˈkɒliːn/ *noun Irish* girl.

college /ˈkɒlɪdʒ/ *noun* establishment for further, higher, or professional education; teachers and students in a college; school; organized group of people with shared functions and privileges.

collegiate /kəˈliːdʒɪət/ *adjective* of or constituted as college, corporate; (of university) consisting of different colleges.

collide /kəˈlaɪd/ *verb* (**-ding**) (often + *with*) come into collision or conflict.

collie /ˈkɒlɪ/ *noun* sheepdog originally of Scottish breed.

collier /ˈkɒlɪə/ *noun* coal miner; coal ship, member of its crew.

colliery /ˈkɒlɪərɪ/ *noun* (*plural* **-ies**) coal mine and its buildings.

collision /kəˈlɪʒ(ə)n/ *noun* violent impact of moving body against another or fixed object; clashing of interests etc.

collocate /ˈkɒləkeɪt/ *verb* (**-ting**) place (word etc.) next to another. □ **collocation** *noun.*

colloid /ˈkɒlɔɪd/ *noun* substance consisting of minute particles; mixture, esp. viscous solution, of this and another substance. □ **colloidal** *adjective.*

colloquial /kəˈləʊkwɪəl/ *adjective* of ordinary or familiar conversation, informal. □ **colloquially** *adverb.*

colloquialism /kəˈləʊkwɪəlɪz(ə)m/ *noun* colloquial word or phrase.

colloquy /ˈkɒləkwɪ/ *noun* (*plural* **-quies**) *literary* talk, dialogue.

collude /kəˈluːd/ *verb* (**-ding**) conspire. □ **collusion** /-ʒ(ə)n/ *noun*; **collusive** /-sɪv/ *adjective.*

collywobbles /ˈkɒlɪwɒb(ə)lz/ *plural noun colloquial* ache or rumbling in stomach; apprehensive feeling.

cologne /kəˈləʊn/ *noun* eau-de-Cologne or similar toilet water.

colon[1] /ˈkəʊlən/ *noun* punctuation mark (:) used between main clauses or before list or quotation (see panel).

colon[2] /ˈkəʊlən/ *noun* lower and greater part of large intestine.

colonel /ˈkɜːn(ə)l/ *noun* army officer commanding regiment, next in rank below brigadier. □ **colonelcy** *noun* (*plural* **-ies**).

colonial /kəˈləʊnɪəl/ ● *adjective* of a colony or colonies; of colonialism. ● *noun* inhabitant of colony.

colonialism *noun* policy of having colonies.

colonist /ˈkɒlənɪst/ *noun* settler in or inhabitant of colony.

colonize /ˈkɒlənaɪz/ *verb* (also **-ise**) (**-zing** or **-sing**) establish colony in; join colony. □ **colonization** *noun.*

colonnade /kɒləˈneɪd/ *noun* row of columns, esp. supporting roof. □ **colonnaded** *adjective.*

colony /ˈkɒlənɪ/ *noun* (*plural* **-ies**) settlement or settlers in new territory remaining subject to mother country; people of one nationality, occupation, etc. forming community in town etc.; group of animals that live close together.

colophon /ˈkɒləf(ə)n/ *noun* tailpiece in book.

color etc. *US* = COLOUR etc.

Colorado beetle /kɒləˈrɑːdəʊ/ *noun* small beetle destructive to potato.

coloration /kʌləˈreɪʃ(ə)n/ *noun* (also **colouration**) colouring, arrangement of colours.

coloratura /kɒlərəˈtʊərə/ *noun* elaborate passages in vocal music; singer of these, esp. soprano.

colossal /kəˈlɒs(ə)l/ *adjective* huge; *colloquial* splendid. □ **colossally** *adverb.*

colossus /kəˈlɒsəs/ *noun* (*plural* **-ssi** /-saɪ/ or **-ssuses**) statue much bigger than life size; gigantic or remarkable person etc.

colour /ˈkʌlə/ (*US* **color**) ● *noun* one, or any mixture, of the constituents into

Colon :

This is used:

1 between two main clauses of which the second explains, enlarges on, or follows from the first, e.g.

It was not easy: to begin with I had to find the right house.

2 to introduce a list of items (a dash should not be added), and after expressions such as *namely, for example, to resume, to sum up,* and *the following,* e.g.

You will need: a tent, a sleeping bag, cooking equipment, and a rucksack.

3 before a quotation, e.g.

The poem begins: 'Earth has not anything to show more fair'.

which light is separated in rainbow etc.; use of all colours as in photography; colouring substance, esp. paint; skin pigmentation, esp. when dark; ruddiness of face; appearance or aspect; (in *plural*) flag of regiment or ship etc.; coloured ribbon, rosette, etc. worn as symbol of school, club, political party, etc. ● *verb* give colour to; paint, stain, dye; blush; influence. □ **colour-blind** unable to distinguish certain colours; **colour-blindness** *noun*; **colour scheme** arrangement of colours; **colour supplement** magazine with colour printing, sold with newspaper.

coloured (*US* **colored**) ● *adjective* having colour; wholly or partly of non-white descent; *South African* of mixed white and non-white descent. ● *noun* coloured person.

colourful *adjective* (*US* **colorful**) full of colour or interest. □ **colourfully** *adverb*.

colouring *noun* (*US* **coloring**) appearance as regards colour, esp. facial complexion; application of colour; substance giving colour.

colourless *adjective* (*US* **colorless**) without colour, lacking interest.

colt /kəʊlt/ *noun* young male horse; *Sport* inexperienced player.

colter *US* = COULTER.

coltsfoot *noun* (*plural* **-s**) yellow wild flower with large leaves.

columbine /ˈkɒləmbaɪn/ *noun* garden plant with purple-blue flowers.

column /ˈkɒləm/ *noun* pillar, usually round, with base and capital; column-shaped thing; series of numbers, one under the other; vertical division of printed page; part of newspaper regularly devoted to particular subject or written by one writer; long, narrow arrangement of advancing troops, vehicles, etc.

columnist /ˈkɒləmnɪst/ *noun* journalist contributing regularly to newspaper etc.

coma /ˈkəʊmə/ *noun* (*plural* **-s**) prolonged deep unconsciousness.

comatose /ˈkəʊmətəʊs/ *adjective* in coma; sleepy.

comb /kəʊm/ ● *noun* toothed strip of rigid material for arranging hair; thing like comb, esp. for dressing wool etc.; red fleshy crest of fowl, esp. cock etc.; honeycomb. ● *verb* draw comb through (hair), dress (wool etc.) with comb; *colloquial* search (place) thoroughly.

combat /ˈkɒmbæt/ ● *noun* struggle, fight. ● *verb* (**-t-**) do battle (with); strive against, oppose.

combatant /ˈkɒmbət(ə)nt/ ● *noun* fighter. ● *adjective* fighting.

combative /ˈkɒmbətɪv/ *adjective* pugnacious.

combe = COOMB.

combination /kɒmbɪˈneɪʃ(ə)n/ *noun* combining, being combined; combined state; combined set of things or people; motorcycle with side-car; sequence of numbers etc. used to open **combination lock**.

combine ● *verb* /kəmˈbaɪn/ (**-ning**) join together; unite; form into chemical compound. ● *noun* /ˈkɒmbaɪn/ combination of esp. businesses. □ **combine harvester** /ˈkɒmbaɪn/ combined reaping and threshing machine.

combustible /kəmˈbʌstɪb(ə)l/ ● *adjective* capable of or used for burning. ● *noun* combustible substance. □ **combustibility** *noun*.

combustion /kəmˈbʌstʃ(ə)n/ *noun* burning; development of light and heat from combination of substance with oxygen.

come /kʌm/ *verb* (**-ming**; *past* **came**; *past participle* **come**) move or be brought towards or reach a place, time, situation, or result; be available; occur; become; traverse; *colloquial* behave like. □ **come about** happen; **come across** meet or find by chance, give specified impression, *colloquial* be effective or understood; **come again** *colloquial* what did you say?; **come along** make progress, hurry up; **come apart** disintegrate; **come at** attack; **comeback** return to success, *slang* retort or retaliation; **come back (to)** recur to memory (of); **come by** obtain; **come clean** *colloquial* confess; **comedown** loss of status; **come down** lose position, be handed down, be reduced; **come forward** offer oneself for task etc.; **come in** become fashionable or seasonable, prove to be; **come in for** receive; **come into** inherit; **come off** succeed, occur, fare; **come off it** *colloquial expression of* *disbelief*; **come on** make progress; **come out** emerge, become known, be published, go on strike, (of photograph or its subject) be (re)produced clearly, (of stain) be removed; **come out with** declare, disclose; **come over** come some distance to visit, (of feeling) affect, appear or sound in specifed way; **come round** pay informal visit, recover consciousness, be converted to another's opinion; **come through** survive; **come to** recover consciousness, amount to; **come up** arise, be mentioned or discussed, attain position; **come up with** produce (idea etc.); **come up against** be faced with; **come upon** meet or find by chance.

comedian /kə'mi:dɪən/ *noun* humorous performer; actor in comedy; *slang* buffoon.

comedienne /kəmi:dɪ'en/ *noun* female comedian.

comedy /'kɒmədɪ/ *noun* (*plural* **-ies**) play or film of amusing character; humorous kind of drama etc.; humour; amusing aspects. □ **comedic** /kə'mi:dɪk/ *adjective*.

comely /'kʌmlɪ/ *adjective* (**-ier**, **-iest**) *literary* handsome, good-looking. □ **comeliness** *noun*.

comestibles /kə'mestɪb(ə)lz/ *plural noun formal* things to eat.

comet /'kɒmɪt/ *noun* hazy object with 'tail' moving in path round sun.

comeuppance /kʌm'ʌpəns/ *noun colloquial* deserved punishment.

comfort /'kʌmf(ə)t/ ● *noun* physical or mental well-being; consolation; person or thing bringing consolation; (usually in *plural*) things that make life comfortable. ● *verb* console. □ **comfortless** *adjective*.

comfortable /'kʌmfətəb(ə)l/ *adjective* giving ease; at ease; having adequate standard of living; appreciable. □ **comfortably** *adverb*.

comforter /'kʌmfətə/ *noun* person who comforts; baby's dummy; *archaic* woollen scarf.

comfrey /'kʌmfrɪ/ *noun* tall plant with bell-like flowers.

comfy /'kʌmfɪ/ *adjective* (**-ier**, **-iest**) *colloquial* comfortable.

comic /'kɒmɪk/ ● *adjective* of or like comedy; funny. ● *noun* comedian; periodical in form of comic strips. □ **comic strip** sequence of drawings telling comic story. □ **comical** *adjective*; **comically** *adverb*.

comma /'kɒmə/ *noun* punctuation mark (,) indicating pause or break between parts of sentence (see panel).

command /kə'mɑ:nd/ ● *verb* give formal order to; have authority or control over; have at one's disposal; deserve and get; look down over, dominate. ● *noun* order, instruction; holding of authority, esp. in armed forces; mastery; troops or district under commander. □ **command module** control compartment in spacecraft; **command performance** one given at royal request.

commandant /'kɒməndænt/ *noun* commanding officer, esp. of military academy.

commandeer /kɒmən'dɪə/ *verb* seize (esp. goods) for military use; take possession of without permission.

commander *noun* person who commands, esp. naval officer next below captain. □ **commander-in-chief** (*plural* **commanders-in-chief**) supreme commander, esp. of nation's forces.

commanding *adjective* impressive; giving wide view; substantial.

commandment *noun* divine command.

commando /kə'mɑ:ndəʊ/ *noun* (*plural* **-s**) unit of shock troops; member of this.

commemorate /kə'meməreɪt/ *verb* (**-ting**) preserve in memory by celebration or ceremony; be memorial of. □ **commemoration** *noun*; **commemorative** /-rətɪv/ *adjective*.

commence /kə'mens/ *verb* (**-cing**) *formal* begin. □ **commencement** *noun*.

commend /kə'mend/ *verb* praise; recommend; entrust. □ **commendation** /kɒm-/ *noun*.

commendable *adjective* praiseworthy. □ **commendably** *adverb*.

commensurable /kə'menʃərəb(ə)l/ *adjective* (often + *with*, *to*) measurable by same standard; (+ *to*) proportionate to. □ **commensurability** *noun*.

commensurate /kə'menʃərət/ *adjective* (usually + *with*) extending over same space or time; (often + *to*, *with*) proportionate.

Comma ,

The comma marks a slight break between words, phrases, etc. In particular, it is used:

1 to separate items in a list, e.g.

red, white, and blue or *red, white and blue*
We bought some shoes, socks, gloves, and handkerchiefs.
potatoes, peas, or carrots or *potatoes, peas or carrots*

2 to separate adjectives that describe something in the same way, e.g.

It is a hot, dry, dusty place.

but not if they describe it in different ways, e.g.

a distinguished foreign author

or if one adjective adds to or alters the meaning of another, e.g.

a bright red tie.

3 to separate main clauses, e.g.

Cars will park here, and coaches will turn left.

4 to separate a name or word used to address someone, e.g.

David, I'm here.
Well, Mr Jones, we meet again.
Have you seen this, my friend?

5 to separate a phrase, e.g.

Having had lunch, we went back to work.

especially in order to clarify meaning, e.g.

In the valley below, the village looked very small.

6 after words that introduce direct speech, or after direct speech where there is no question mark or exclamation mark, e.g.

They answered, 'Here we are.'
'Here we are,' they answered.

7 after *Dear Sir, Dear Sarah,* etc., and *Yours faithfully, Yours sincerely,* etc. in letters.

8 to separate a word, phrase, or clause that is secondary or adds information or a comment, e.g.

I am sure, however, that it will not happen.
Fred, who is bald, complained of the cold.

but not with a relative clause (one usually beginning with *who, which,* or *that*) that restricts the meaning of the noun it follows, e.g.

Men who are bald should wear hats.

No comma is needed between a month and a year in dates, e.g.

in December 1993

or between a number and a road in addresses, e.g.

17 Devonshire Avenue

comment /'kɒment/ ● *noun* brief critical or explanatory note or remark; opinion; commenting. ● *verb* (often + *on*, *that*) make comment(s). □ **no comment** *colloquial* I decline to answer your question.

commentary /'kɒməntərɪ/ *noun* (*plural* **-ies**) broadcast description of event happening; series of comments on book or performance etc.

commentate /'kɒmənteɪt/ *verb* (**-ting**) act as commentator.

commentator *noun* writer or speaker of commentary.

commerce /'kɒmɜːs/ *noun* buying and selling; trading.

commercial /kə'mɜːʃ(ə)l/ ● *adjective* of, in, or for commerce; done or run primarily for financial profit; (of broadcasting) financed by advertising. ● *noun* television or radio advertisement. □ **commercial broadcasting** broadcasting financed by advertising; **commercial traveller** firm's representative visiting shops etc. to get orders. □ **commercially** *adverb*.

commercialize *verb* (also **-ise**) (**-zing** or **-sing**) exploit or spoil for profit; make commercial. □ **commercialization** *noun*.

Commie /'kɒmɪ/ *noun slang derogatory* Communist.

commingle /kə'mɪŋg(ə)l/ *verb* (**-ling**) *literary* mix, unite.

commiserate /kə'mɪzəreɪt/ *verb* (**-ting**) (usually + *with*) have or express sympathy. □ **commiseration** *noun*.

commissar /'kɒmɪsɑː/ *noun historical* head of government department in USSR.

commissariat /kɒmɪ'seərɪət/ *noun* department responsible for supply of food etc. for army; food supplied.

commissary /'kɒmɪsərɪ/ *noun* (*plural* **-ies**) deputy, delegate.

commission /kə'mɪʃ(ə)n/ ● *noun* authority to perform task; person(s) given such authority; order for specially produced thing; warrant conferring officer rank in armed forces; rank so conferred; pay or percentage received by agent; committing. ● *verb* empower, give authority to; give (artist etc.) order for work; order (work) to be written etc.; give (officer) command of ship; prepare (ship) for active service; bring (machine etc.) into operation. □ **in** or **out of commission** ready or not ready for active service.

commissionaire /kəmɪʃə'neə/ *noun* uniformed door attendant.

commissioner *noun* person commissioned to perform specific task; member of government commission; representative of government in district etc.

commit /kə'mɪt/ *verb* (**-tt-**) do or make (crime, blunder, etc.); (usually + *to*) entrust, consign; send (person) to prison; pledge or bind (esp. oneself) to policy or course of action; (as **committed** *adjective*) (often + *to*) dedicated, obliged.

commitment *noun* engagement, obligation; committing, being committed; dedication, committing oneself.

committal *noun* act of committing esp. to prison, grave, etc.

committee /kə'mɪtɪ/ *noun* group of people appointed for special function by (and usually out of) larger body.

commode /kə'məʊd/ *noun* chamber pot in chair or box with cover; chest of drawers.

commodious /kə'məʊdɪəs/ *adjective* roomy.

commodity /kə'mɒdɪtɪ/ *noun* (*plural* **-ies**) article of trade.

commodore /'kɒmədɔː/ *noun* naval officer next above captain; commander of squadron or other division of fleet; president of yacht club.

common /'kɒmən/ ● *adjective* (**-er**, **-est**) occurring often; ordinary, of the most familiar kind; shared by all; belonging to the whole community; *derogatory* inferior, vulgar; *Grammar* (of gender) referring to individuals of either sex. ● *noun* piece of open public land. □ **common ground** point or argument accepted by both sides; **common law** unwritten law based on custom and precedent; **common-law husband**, **wife** partner recognized by common law without formal marriage; **Common Market** European Community; **common or garden** *colloquial* ordinary; **common room** room for social use of students or teachers at college etc.; **common sense** good practical sense; **common time** *Music* 4 crotchets in a bar; **in common** shared, in joint use.

commonalty /'kɒmənəltɪ/ *noun* (*plural* **-ies**) general community, common people.

commoner *noun* person below rank of peer.

commonly *adverb* usually, frequently.

commonplace /'kɒmənpleɪs/ ● *adjective* lacking originality; ordinary. ● *noun* event, topic, etc. that is ordinary or usual; trite remark.

commons /'kɒmənz/ *plural noun* the common people; (**the Commons**) House of Commons.

commonwealth /'kɒmənwelθ/ *noun* independent state or community; (**the Commonwealth**) association of UK with states previously part of British Empire; republican government of Britain 1649–60.

commotion /kə'məʊʃ(ə)n/ *noun* noisy disturbance.

communal /'kɒmjʊn(ə)l/ *adjective* shared between members of group or community; (of conflict etc.) between communities. □ **communally** *adverb.*

commune[1] /'kɒmju:n/ *noun* group of people sharing accommodation and goods; small administrative district in France etc.

commune[2] /kə'mju:n/ *verb* (**-ning**) (usually + *with*) speak intimately; feel in close touch.

communicant /kə'mju:nɪkənt/ *noun* receiver of Holy Communion.

communicate /kə'mju:nɪkeɪt/ *verb* (**-ting**) impart, transmit (news, feelings, disease, ideas, etc.); (often + *with*) have social dealings. □ **communicator** *noun.*

communication *noun* communicating, being communicated; letter, message, etc.; connection or means of access; social dealings; (in *plural*) science and practice of transmitting information. □ **communication cord** cord or chain pulled to stop train in emergency; **communication(s) satellite** artificial satellite used to relay telephone calls, TV, radio, etc.

communicative /kə'mju:nɪkətɪv/ *adjective* ready to talk and impart information.

communion /kə'mju:niən/ *noun* sharing, esp. of thoughts, interests, etc.; fellowship; group of Christians of same denomination; (**Holy Communion**) Eucharist.

communiqué /kə'mju:nɪkeɪ/ *noun* official communication, esp. news report.

communism /'kɒmjʊnɪz(ə)m/ *noun* social system based on public ownership of most property; political theory advocating this; (usually **Communism**) form of socialist society in Cuba, China, etc. □ **communist**, **Communist** *noun & adjective*; **communistic** /-'nɪstɪk/ *adjective.*

community /kə'mju:nɪtɪ/ (*plural* **-ies**) *noun* group of people living in one place or having same religion, ethnic origin, profession, etc.; commune; joint ownership. □ **community centre** place providing social facilities for neighbourhood; **community charge** *historical* tax levied locally on every adult; **community home** institution housing young offenders; **community singing** singing by large group; **community spirit** feeling of belonging to community.

commute /kə'mju:t/ *verb* (**-ting**) travel some distance to and from work; (usually + *to*) change (punishment) to one less severe.

commuter *noun* person who commutes to and from work.

compact[1] ● *adjective* /kəm'pækt/ closely packed together; economically designed; concise; (of person) small but well-proportioned. ● *verb* /kəm'pækt/ make compact. ● *noun* /'kɒmpækt/ small flat case for face powder etc. □ **compact disc** disc from which digital information or sound is reproduced by reflection of laser light. □ **compactly** *adverb*; **compactness** *noun.*

compact[2] /'kɒmpækt/ *noun* agreement, contract.

companion /kəm'pænjən/ *noun* person who accompanies or associates with another; person paid to live with another; handbook, reference book; thing that matches another; (**Companion**) member of some orders of knighthood. □ **companionway** staircase from ship's deck to cabins etc.

companionable *adjective* sociable, friendly. □ **companionably** *adverb.*

companionship *noun* friendship; being together.

company /'kʌmpənɪ/ *noun* (*plural* **-ies**) number of people assembled; guest(s); commercial business; actors etc. working together; subdivision of infantry battalion. □ **in company with** together with; **part company** (often + *with*) separate; **ship's company** entire crew.

comparable /'kɒmpərəb(ə)l/ *adjective* (often + *with, to*) able to be compared. □ **comparability** *noun*; **comparably** *adverb.*

■ **Usage** *Comparable* is often pronounced /kəm'pærəb(ə)l/ (with the stress on the *-par-*), but this is considered incorrect by some people.

comparative /kəm'pærətɪv/ ● *adjective* perceptible or estimated by comparison; relative; of or involving comparison; *Grammar* (of adjective or adverb) expressing higher degree of a quality. ● *noun Grammar* comparative expression or word. □ **comparatively** *adverb.*

compare /kəm'peə/ ● *verb* (**-ring**) (usually + *to*) express similarities in; (often + *to, with*) estimate similarity of; (often + *with*) bear comparison. ● *noun* comparison. □ **compare notes** exchange ideas or opinions.

comparison /kəm'pærɪs(ə)n/ *noun* comparing; example of similarity; (in full **degrees of comparison**) *Grammar* positive, comparative, and superlative forms of adjectives and adverbs. □ **bear comparison** (often + *with*) be able to be compared favourably.

compartment /kəm'pɑ:tmənt/ *noun* space partitioned off within larger space.

compass /'kʌmpəs/ *noun* instrument showing direction of magnetic north and bearings from it; (usually in *plural*) V-shaped hinged instrument for drawing circles and taking measurements; scope, range.

compassion /kəm'pæʃ(ə)n/ *noun* pity.

compassionate /kəm'pæʃənət/ *adjective* showing compassion; sympathetic. □ **compassionate leave** leave granted on grounds of bereavement etc. □ **compassionately** *adverb.*

compatible /kəm'pætɪb(ə)l/ *adjective* (often + *with*) able to coexist; (of equipment) able to be used in combination. □ **compatibility** *noun.*

compatriot /kəm'pætrɪət/ *noun* person from one's own country.

compel /kəm'pel/ *verb* (**-ll-**) force, constrain; arouse irresistibly; (as **compelling** *adjective*) arousing strong interest.

compendious /kəm'pendɪəs/ *adjective* comprehensive but brief.

compendium /kəm'pendɪəm/ *noun* (*plural* **-s** or **-dia**) abridgement, summary; collection of table games etc.

compensate /'kɒmpenseɪt/ *verb* (**-ting**) recompense, make amends; counterbalance.

compensation *noun* compensating, being compensated; money etc. given as recompense. □ **compensatory** /-'seɪt-/ *adjective.*

compère /'kɒmpeə/ ● *noun* person introducing variety show. ● *verb* (**-ring**) act as compère (to).

compete /kəm'pi:t/ *verb* (**-ting**) take part in contest etc.; (often + *with* or *against* person, *for* thing) strive.

competence /'kɒmpɪt(ə)ns/ *noun* being competent; ability.

competent *adjective* adequately qualified or capable; effective. □ **competently** *adverb.*

competition /kɒmpə'tɪʃ(ə)n/ *noun* (often + *for*) competing; event in which people compete; other people competing; opposition.

competitive /kəm'petɪtɪv/ *adjective* of or involving competition; (of prices etc.) comparing well with those of rivals; having strong urge to win. □ **competitiveness** *noun.*

competitor /kəm'petɪtə/ *noun* person who competes; rival, esp. in business.

compile /kəm'paɪl/ *verb* (**-ling**) collect and arrange (material) into list, book, etc. □ **compilation** /kɒmpɪ'leɪʃ(ə)n/ *noun.*

complacent /kəm'pleɪs(ə)nt/ *adjective* smugly self-satisfied or contented. □ **complacency** *noun.*

complain /kəm'pleɪn/ *verb* express dissatisfaction; (+ *of*) say that one is suffering from (an ailment), state grievance concerning. □ **complainant** *noun*.

complaint *noun* complaining; grievance, cause of dissatisfaction; formal protest; ailment.

complaisant /kəm'pleɪz(ə)nt/ *adjective formal* deferential; willing to please; acquiescent. □ **complaisance** *noun*.

complement ● *noun* /'kɒmplɪmənt/ thing that completes; full number; word(s) added to verb to complete predicate of sentence; amount by which angle is less than 90°. ● *verb* /-ment/ complete; form complement to. □ **complementary** /-'men-/ *adjective*.

complete /kəm'pli:t/ ● *adjective* having all its parts; finished, total. ● *verb* (**-ting**) finish; make complete; fill in (form etc.). □ **completely** *adverb*; **completeness** *noun*; **completion** *noun*.

complex /'kɒmpleks/ ● *noun* buildings, rooms, etc. made up of related parts; group of usually repressed feelings or thoughts causing abnormal behaviour or mental state. ● *adjective* complicated; consisting of related parts. □ **complexity** /kəm'pleks-/ *noun* (*plural* **-ies**).

complexion /kəm'plekʃ(ə)n/ *noun* natural colour, texture, and appearance of skin, esp. of face; aspect.

compliance /kəm'plaɪəns/ *noun* obedience to request, command, etc.; capacity to yield.

compliant *adjective* obedient; yielding. □ **compliantly** *adverb*.

complicate /'kɒmplɪkeɪt/ *verb* (**-ting**) make difficult or complex; (as **complicated** *adjective*) complex, intricate.

complication *noun* involved or confused condition; complicating circumstance; difficulty; (often in *plural*) disease or condition arising out of another.

complicity /kəm'plɪsɪtɪ/ *noun* partnership in wrongdoing.

compliment ● *noun* /'kɒmplɪmənt/ polite expression of praise; (in *plural*) formal greetings accompanying gift etc. ● *verb* /-ment/ (often + *on*) congratulate; praise.

complimentary /kɒmplɪ'mentərɪ/ *adjective* expressing compliment; given free of charge.

comply /kəm'plaɪ/ *verb* (**-ies, -ied**) (often + *with*) act in accordance (with request or command).

component /kəm'pəʊnənt/ ● *adjective* being part of larger whole. ● *noun* component part.

comport /kəm'pɔ:t/ *verb* (**comport oneself**) *literary* conduct oneself; behave. □ **comportment** *noun*.

compose /kəmpəʊz/ *verb* (**-sing**) create in music or writing; make up, constitute; arrange artistically; set up (type); arrange in type; (as **composed** *adjective*) calm, self-possessed.

composer *noun* person who composes esp. music.

composite /'kɒmpəzɪt/ ● *adjective* made up of parts; (of plant) having head of many flowers forming one bloom. ● *noun* composite thing or plant.

composition /kɒmpə'zɪʃ(ə)n/ *noun* act of putting together; thing composed; school essay; arrangement of parts of picture etc.; constitution of substance; compound artificial substance.

compositor /kəm'pɒzɪtə/ *noun* person who sets up type for printing.

compost /'kɒmpɒst/ *noun* mixture of decayed organic matter used as fertilizer.

composure /kəm'pəʊʒə/ *noun* calm manner.

compote /'kɒmpəʊt/ *noun* fruit in syrup.

compound[1] /'kɒmpaʊnd/ ● *noun* mixture of two or more things; word made up of two or more existing words; substance formed from two or more elements chemically united. ● *adjective* made up of two or more ingredients or parts; combined, collective. ● *verb* /kəm'paʊnd/ mix or combine; increase (difficulties etc.); make up (whole); settle (matter) by mutual agreement. □ **compound fracture** one complicated by wound; **compound interest** interest paid on capital and accumulated interest.

compound[2] /'kɒmpaʊnd/ *noun* enclosure, fenced-in space.

comprehend /kɒmprɪ'hend/ *verb* understand; include.

comprehensible *adjective* that can be understood.

comprehension *noun* understanding; text set as test of understanding; inclusion.

comprehensive ● *adjective* including all or nearly all; (of motor insurance) providing protection against most risks. ● *noun* (in full **comprehensive school**) secondary school for children of all abilities. □ **comprehensively** *adverb.*

compress ● *verb* /kəm'pres/ squeeze together, bring into smaller space or shorter time. ● *noun* /'kɒmpres/ pad pressed on part of body to relieve inflammation, stop bleeding, etc. □ **compressible** *adjective.*

compression /kəm'preʃ(ə)n/ *noun* compressing; reduction in volume of fuel mixture in internal-combustion engine before ignition.

compressor *noun* machine for compressing air or other gas.

comprise /kəm'praɪz/ *verb* (**-sing**) include; consist of.

■ **Usage** It is a mistake to use *comprise* to mean 'to compose or make up'.

compromise /'kɒmprəmaɪz/ ● *noun* agreement reached by mutual concession; (often + *between*) intermediate state between conflicting opinions etc. ● *verb* (**-sing**) settle dispute by compromise; modify one's opinions, demands, etc.; bring into disrepute or danger by indiscretion.

comptroller /kən'trəʊlə/ *noun* controller.

compulsion /kəm'pʌlʃ(ə)n/ *noun* compelling, being compelled; irresistible urge.

compulsive *adjective* compelling; resulting or acting (as if) from compulsion; irresistible. □ **compulsively** *adverb.*

compulsory *adjective* required by law or rule. □ **compulsorily** *adverb.*

compunction /kəm'pʌŋkʃ(ə)n/ *noun* guilty feeling; slight regret.

compute /kəm'pju:t/ *verb* (**-ting**) calculate; use computer. □ **computation** /kɒm-/ *noun.*

computer *noun* electronic device for storing and processing data, making calculations, or controlling machinery. □ **computer-literate** able to use computers; **computer science** study of computers; **computer virus** code maliciously introduced into program to destroy data etc.

computerize /kəm'pju:təraɪz/ *verb* (also **-ise**) (**-zing** or **-sing**) equip with or store, perform, or produce by computer. □ **computerization** *noun.*

comrade /'kɒmreɪd/ *noun* companion in some activity; fellow socialist or Communist. □ **comradeship** *noun.*

con[1] *slang* ● *noun* confidence trick. ● *verb* (**-nn-**) swindle, deceive.

con[2] *noun* (usually in *plural*) reason against.

con[3] *verb* (*US* **conn**) (**-nn-**) direct steering of (ship).

concatenation /kɒnkætɪ'neɪʃ(ə)n/ *noun* series of linked things or events.

concave /kɒn'keɪv/ *adjective* curved like interior of circle or sphere. □ **concavity** /-'kæv-/ *noun.*

conceal /kən'si:l/ *verb* hide; keep secret. □ **concealment** *noun.*

concede /kən'si:d/ *verb* (**-ding**) admit to be true; admit defeat in; grant.

conceit /kən'si:t/ *noun* personal vanity; *literary* far-fetched comparison.

conceited *adjective* vain. □ **conceitedly** *adverb.*

conceive /kən'si:v/ *verb* (**-ving**) become pregnant (with); (often + *of*) imagine; (usually in *passive*) formulate (plan etc.). □ **conceivable** *adjective*; **conceivably** *adverb.*

concentrate /'kɒnsəntreɪt/ ● *verb* (**-ting**) (often + *on*) focus one's attention; bring together to one point; increase strength of (liquid etc.) by removing water etc.; (as **concentrated** *adjective*) strong. ● *noun* concentrated solution.

concentration *noun* concentrating, being concentrated; weight of substance in given amount of mixture; mental attention. □ **concentration camp** place for detention of political prisoners etc.

concentric /kən'sentrɪk/ *adjective* having common centre. □ **concentrically** *adverb.*

concept /ˈkɒnsept/ *noun* general notion; abstract idea.

conception /kənˈsepʃ(ə)n/ *noun* conceiving, being conceived; idea; understanding.

conceptual /kənˈseptjʊəl/ *adjective* of mental concepts. □ **conceptually** *adverb*.

conceptualize *verb* (also **-ise**) (**-zing** or **-sing**) form concept or idea of. □ **conceptualization** *noun*.

concern /kənˈsɜːn/ ● *verb* be relevant or important to; relate to, be about; worry, affect; (**concern oneself**; often + *with, about, in*) interest or involve oneself. ● *noun* anxiety, worry; matter of interest or importance to one; firm, business.

concerned *adjective* involved, interested; anxious, troubled.

concerning *preposition* about, regarding.

concert /ˈkɒnsət/ *noun* musical performance; agreement.

concerted /kənˈsɜːtɪd/ *adjective* jointly planned.

concertina /kɒnsəˈtiːnə/ ● *noun* portable musical instrument like accordion but smaller. ● *verb* (**-nas**, **-naed** /-nəd/ or **-na'd**, **-naing**) compress or collapse in folds like those of concertina.

concerto /kənˈtʃeətəʊ/ *noun* (*plural* **-tos** or **-ti** /-tɪ/) composition for solo instrument(s) and orchestra.

concession /kənˈseʃ(ə)n/ *noun* conceding, thing conceded; reduction in price for certain category of people; right to use land, sell goods, etc. □ **concessionary** *adjective*.

conch /kɒntʃ/ *noun* large spiral shell of various marine gastropod molluscs; such gastropod.

conchology /kɒŋˈkɒlədʒɪ/ *noun* study of shells.

conciliate /kənˈsɪlɪeɪt/ *verb* (**-ting**) make calm; pacify; reconcile. □ **conciliation** *noun*; **conciliator** *noun*; **conciliatory** *adjective*.

concise /kənˈsaɪs/ *adjective* brief but comprehensive. □ **concisely** *adverb*; **conciseness** *noun*; **concision** *noun*.

conclave /ˈkɒnkleɪv/ *noun* private meeting; assembly or meeting-place of cardinals for election of pope.

conclude /kənˈkluːd/ *verb* (**-ding**) bring or come to end; (often + *from, that*) infer; settle (treaty etc.).

conclusion /kənˈkluːʒ(ə)n/ *noun* ending; judgement reached by reasoning; summing-up; settling of peace etc. □ **in conclusion** lastly.

conclusive /kənˈkluːsɪv/ *adjective* decisive, convincing. □ **conclusively** *adverb*.

concoct /kənˈkɒkt/ *verb* make by mixing ingredients; invent (story, lie, etc.). □ **concoction** *noun*.

concomitant /kənˈkɒmɪt(ə)nt/ ● *adjective* (often + *with*) accompanying. ● *noun* accompanying thing. □ **concomitance** *noun*.

concord /ˈkɒŋkɔːd/ *noun* agreement, harmony. □ **concordant** /kənˈkɔːd(ə)nt/ *adjective*.

concordance /kənˈkɔːd(ə)ns/ *noun* agreement; index of words used in book or by author.

concordat /kənˈkɔːdæt/ *noun* agreement, esp. between Church and state.

concourse /ˈkɒŋkɔːs/ *noun* crowd; large open area in railway station etc.

concrete /ˈkɒŋkriːt/ ● *adjective* existing in material form, real; definite. ● *noun* mixture of gravel, sand, cement, and water used for building. ● *verb* (**-ting**) cover with or embed in concrete.

concretion /kənˈkriːʃ(ə)n/ *noun* hard solid mass; forming of this by coalescence.

concubine /ˈkɒŋkjʊbaɪn/ *noun literary* mistress; (among polygamous peoples) secondary wife.

concupiscence /kənˈkjuːpɪs(ə)ns/ *noun formal* lust. □ **concupiscent** *adjective*.

concur /kənˈkɜː/ *verb* (**-rr-**) (often + *with*) agree; coincide.

concurrent /kənˈkʌrənt/ *adjective* (often + *with*) existing or in operation at the same time. □ **concurrence** *noun*; **concurrently** *adverb*.

concuss /kənˈkʌs/ *verb* subject to concussion.

concussion /kənˈkʌʃ(ə)n/ *noun* temporary unconsciousness or incapacity due to head injury; violent shaking.

condemn /kənˈdem/ *verb* express strong disapproval of; (usually + *to*) sentence (to punishment), doom (to something unpleasant); pronounce unfit for use. □ **condemnation** /kɒndem'neɪʃ(ə)n/ *noun*.

condensation /kɒnden'seɪʃ(ə)n/ *noun* condensing, being condensed; condensed liquid; abridgement.

condense /kən'dens/ *verb* (**-sing**) make denser or more concise; reduce or be reduced from gas to liquid.

condescend /kɒndɪ'send/ *verb* (+ *to do*) graciously consent to do a thing while showing one's superiority; (+ *to*) pretend to be on equal terms with (inferior); (as **condescending** *adjective*) patronizing. □ **condescendingly** *adverb*; **condescension** *noun*.

condiment /'kɒndɪmənt/ *noun* seasoning or relish for food.

condition /kən'dɪʃ(ə)n/ ● *noun* stipulation; thing on fulfilment of which something else depends; state of being or fitness of person or thing; ailment; (in *plural*) circumstances. ● *verb* bring into desired state; accustom; determine; be essential to.

conditional *adjective* (often + *on*) dependent, not absolute; *Grammar* (of clause, mood, etc.) expressing condition. □ **conditionally** *adverb*.

condole /kən'dəʊl/ *verb* (**-ling**) (+ *with*) express sympathy with (person) over loss etc.

■ **Usage** *Condole* is commonly confused with *console*, which means 'to comfort'.

---

condolence *noun* (often in *plural*) expression of sympathy.

condom /'kɒndɒm/ *noun* contraceptive sheath.

condominium /kɒndə'mɪnɪəm/ *noun* joint rule or sovereignty; *US* building containing individually owned flats.

condone /kən'dəʊn/ *verb* (**-ning**) forgive, overlook.

condor /'kɒndɔː/ *noun* large S. American vulture.

conducive /kən'djuːsɪv/ *adjective* (often + *to*) contributing or helping (towards someting).

conduct ● *noun* /'kɒndʌkt/ behaviour; manner of conducting business, war, etc. ● *verb* /kən'dʌkt/ lead, guide; control, manage; be conductor of (orchestra etc.); transmit by conduction; (**conduct oneself**) behave.

conduction /kən'dʌkʃ(ə)n/ *noun* transmission of heat, electricity, etc.

conductive /kən'dʌktɪv/ *adjective* transmitting heat, electricity, etc. □ **conductivity** /kɒndʌk'tɪv-/ *noun*.

conductor *noun* director of orchestra etc.; (*feminine* **conductress**) person who collects fares on bus etc.; conductive thing.

conduit /'kɒndɪt/ *noun* channel or pipe for conveying liquid or protecting insulated cable.

cone *noun* solid figure with usually circular base and tapering to a point; cone-shaped object; dry fruit of pine or fir; ice cream cornet.

coney = CONY.

confab /'kɒnfæb/ *noun colloquial* confabulation.

confabulate /kən'fæbjʊleɪt/ *verb* (**-ting**) talk together. □ **confabulation** *noun*.

confection /kən'fekʃ(ə)n/ *noun* sweet dish or delicacy.

confectioner *noun* dealer in sweets or pastries etc. □ **confectionery** *noun*.

confederacy /kən'fedərəsɪ/ *noun* (*plural* **-ies**) alliance or league, esp. of confederate states.

confederate /kən'fedərət/ ● *adjective esp. Politics* allied. ● *noun* ally; accomplice. ● *verb* /-reɪt/ (**-ting**) (often + *with*) bring or come into alliance. □ **Confederate States** those which seceded from US in 1860–1.

confederation /kənfedə'reɪʃ(ə)n/ *noun* union or alliance, esp. of states.

confer /kən'fɜː/ *verb* (**-rr-**) (often + *on*, *upon*) grant, bestow; (often + *with*) meet for discussion.

conference /'kɒnfər(ə)ns/ *noun* consultation; meeting for discussion.

conferment /kən'fɜːmənt/ *noun* conferring (of honour etc.).

confess /kən'fes/ *verb* acknowledge, admit; declare one's sins, esp. to priest; (of priest) hear confession of.

confessedly /kən'fesɪdlɪ/ *adverb* by one's own or general admission.

confession /kən'feʃ(ə)n/ *noun* act of confessing; thing confessed; statement of principles etc.

confessional ● *noun* enclosed place where priest hears confession. ● *adjective* of confession.

confessor *noun* priest who hears confession.

confetti /kən'fetɪ/ *noun* small bits of coloured paper thrown by wedding guests at bride and groom.

confidant /'kɒnfɪdænt/ *noun* (*feminine* **confidante** same pronunciation) person trusted with knowledge of one's private affairs.

confide /kən'faɪd/ *verb* (**-ding**) (usually + *to*) tell (secret) or entrust (task). □ **confide in** talk confidentially to.

confidence /'kɒnfɪd(ə)ns/ *noun* firm trust; feeling of certainty; self-reliance; boldness; something told as a secret. □ **confidence trick** swindle in which victim is persuaded to trust swindler; **confidence trickster** person using confidence tricks; **in confidence** as a secret; **in (person's) confidence** trusted with his or her secrets.

confident *adjective* feeling or showing confidence. □ **confidently** *adverb*.

confidential /kɒnfɪ'denʃ(ə)l/ *adjective* spoken or written in confidence; entrusted with secrets; confiding. □ **confidentiality** /-ʃɪ'æl-/ *noun*; **confidentially** *adjective*.

configuration /kənfɪgjʊ'reɪʃ(ə)n/ *noun* manner of arrangement; shape; outline. □ **configure** /-'fɪgə/ *verb*.

confine ● *verb* /kən'faɪn/ (**-ning**) keep or restrict within certain limits; imprison. ● *noun* /'kɒnfaɪn/ (usually in *plural*) boundary. □ **be confined** be in childbirth.

confinement /kən'faɪnmənt/ *noun* confining, being confined; childbirth.

confirm /kən'fɜ:m/ *verb* provide support for truth or correctness of; establish more firmly; formally make definite; administer confirmation to.

confirmation /kɒnfə'meɪʃ(ə)n/ *noun* confirming circumstance or statement; rite confirming baptized person as member of Christian Church.

confirmed *adjective* firmly settled in habit or condition.

confiscate /'kɒnfɪskeɪt/ *verb* (**-ting**) take or seize by authority. □ **confiscation** *noun*.

conflagration /kɒnflə'greɪʃ(ə)n/ *noun* large destructive fire.

conflate /kən'fleɪt/ *verb* (**-ting**) fuse together, blend. □ **conflation** *noun*.

conflict ● *noun* /'kɒnflɪkt/ struggle, fight; opposition; (often + *of*) clashing of opposed interests etc. ● *verb* /kən'flɪkt/ clash, be incompatible.

confluent /'kɒnflʊənt/ ● *adjective* merging into one. ● *noun* stream joining another. □ **confluence** *noun*.

conform /kən'fɔ:m/ *verb* comply with rules or general custom; (+ *to*, *with*) be in accordance with.

conformable *adjective* (often + *to*) similar; (often + *with*) consistent.

conformation /kɒnfɔ:'meɪʃ(ə)n/ *noun* thing's structure or shape.

conformist ● *noun* person who conforms to established practice. ● *adjective* conventional. □ **conformism** *noun*.

conformity *noun* conforming with established practice; suitability.

confound /kən'faʊnd/ ● *verb* baffle; confuse; *archaic* defeat. □ **confound (person, thing)!** *interjection expressing annoyance*.

confounded *adjective colloquial* damned.

confront /kən'frʌnt/ *verb* meet or stand facing, esp. in hostility or defiance; (of problem etc.) present itself to; (+ *with*) bring face to face with. □ **confrontation** /kɒn-/ *noun*; **confrontational** /kɒn-/ *adjective*.

confuse /kən'fju:z/ *verb* (**-sing**) bewilder; mix up; make obscure; (often as **confused** *adjective*) throw into disorder. □ **confusing** *adjective*; **confusion** *noun*.

confute /kən'fju:t/ *verb* (**-ting**) prove to be false or wrong. □ **confutation** /kɒn-/ *noun*.

conga /'kɒŋgə/ *noun* Latin American dance, usually performed in single file; tall narrow drum beaten with hands.

congeal /kən'dʒi:l/ *verb* make or become semi-solid by cooling; (of blood) coagulate.

congenial /kən'dʒi:nɪəl/ *adjective* having sympathetic nature, similar interests, etc.; (often + *to*) suited or agreeable.

□ **congeniality** /-'æl-/ *noun*; **congenially** *adverb*.

congenital /kən'dʒenɪt(ə)l/ *adjective* existing or as such from birth. □ **congenitally** *adverb*.

conger /'kɒŋgə/ *noun* large sea eel.

congest /kən'dʒest/ *verb* (esp. as **congested** *adjective*) affect with congestion.

congestion *noun* abnormal accumulation or obstruction, esp. of traffic etc. or of mucus in nose etc.

conglomerate /kən'glɒmərət/ ● *adjective* gathered into rounded mass. ● *noun* heterogeneous mass; business etc. corporation of merged firms. ● *verb* /-reɪt/ (**-ting**) collect into coherent mass. □ **conglomeration** *noun*.

congratulate /kən'grætʃʊleɪt/ *verb* (**-ting**) (often + *on*) express pleasure at happiness, excellence, or good fortune of (person). □ **congratulatory** /-lətərɪ/ *adjective*.

congratulation *noun* congratulating, (usually in *plural*) expression of this.

congregate /'kɒŋgrɪgeɪt/ *verb* (**-ting**) collect or gather in crowd.

congregation *noun* assembly of people, esp. for religious worship; group of people regularly attending particular church etc.

congregational *adjective* of a congregation; (**Congregational**) of or adhering to Congregationalism.

Congregationalism *noun* system in which individual churches are self-governing. □ **Congregationalist** *noun*.

congress /'kɒŋgres/ *noun* formal meeting of delegates for discussion; (**Congress**) national legislative body of US etc. □ **congressman**, **congresswoman** member of US Congress. □ **congressional** /kən'greʃ-/ *adjective*.

congruent /'kɒŋgrʊənt/ *adjective* (often + *with*) suitable, agreeing; *Geometry* (of figures) coinciding exactly when superimposed. □ **congruence** *noun*.

conic /'kɒnɪk/ *adjective* of a cone.

conical *adjective* cone-shaped.

conifer /'kɒnɪfə/ *noun* cone-bearing tree. □ **coniferous** /kə'nɪfərəs/ *adjective*.

conjectural /kən'dʒektʃər(ə)l/ *adjective* involving conjecture.

conjecture /kən'dʒektʃə/ ● *noun* formation of opinion on incomplete information, guess. ● *verb* (**-ring**) guess.

conjoin /kən'dʒɔɪn/ *verb formal* join, combine.

conjoint /kən'dʒɔɪnt/ *adjective formal* associated, conjoined.

conjugal /'kɒndʒʊg(ə)l/ *adjective* of marriage; between husband and wife.

conjugate ● *verb* /'kɒndʒʊgeɪt/ (**-ting**) give the different forms of (verb); unite, become fused. ● *adjective* /-gət/ joined together.

conjugation *noun Grammar* system of verbal inflection.

conjunct /kən'dʒʌŋkt/ *adjective* joined; combined; associated.

conjunction *noun* joining, connection; word used to connect sentences, clauses, or words (see panel); combination of events or circumstances. □ **conjunctive** *adjective*.

conjunctiva /kɒndʒʌŋk'taɪvə/ *noun* (*plural* **-s**) mucous membrane covering front of eye and inside of eyelid.

conjunctivitis /kəndʒʌŋktɪ'vaɪtɪs/ *noun* inflammation of conjunctiva.

conjure /'kʌndʒə/ *verb* (**-ring**) perform seemingly magical tricks, esp. by movement of hands. □ **conjure up** produce as if by magic, evoke.

conjuror *noun* (also **conjurer**) performer of conjuring tricks.

conk[1] *verb colloquial* (usually + *out*) break down; become exhausted; faint; fall asleep.

conk[2] *slang* ● *noun* (punch on) nose or head. ● *verb* hit on nose or head.

conker /'kɒŋkə/ *noun* horse chestnut fruit; (in *plural*) children's game played with conkers on strings.

con man *noun* confidence trickster.

connect /kə'nekt/ *verb* (often + *to*, *with*) join, be joined; associate mentally or practically; (+ *with*) (of train etc.) arrive in time for passengers to transfer to another; put into communication by telephone; (usually in *passive*; + *with*) unite or associate with (others) in relationship etc. □ **connecting rod** rod between piston and crankpin etc. in engine. □ **connector** *noun*.

connection *noun* (also **connexion**) connecting, being connected; point at which things are connected; association of ideas; link, esp. by telephone;

(often in *plural*) (esp. influential) relative or associate; connecting train etc.

connective *adjective* connecting. □ **connective tissue** body tissue forming tendons and ligaments, supporting organs, etc.

conning tower /'kɒnɪŋ/ *noun* raised structure of submarine containing periscope; wheelhouse of warship.

connive /kə'naɪv/ *verb* (**-ving**) (+ *at*) tacitly consent to (wrongdoing); conspire. □ **connivance** *noun.*

connoisseur /kɒnə'sɜ:/ *noun* (often + *of*, *in*) person with good taste and judgement.

connote /kə'nəʊt/ *verb* (**-ting**) imply in addition to literal meaning; mean. □ **connotation** /kɒnə-/ *noun*; **connotative** /'kɒnəteɪtɪv/ *adjective.*

connubial /kə'nju:bɪəl/ *adjective* conjugal.

conquer /'kɒŋkə/ *verb* overcome, defeat; be victorious; subjugate. □ **conqueror** *noun.*

conquest /'kɒŋkwest/ *noun* conquering; something won; person whose affections have been won.

consanguineous /kɒnsæŋ'gwɪnɪəs/ *adjective* descended from same ancestor; akin. □ **consanguinity** *noun.*

conscience /'kɒnʃ(ə)ns/ *noun* moral sense of right and wrong, esp. as affecting behaviour. □ **conscience money** money paid to relieve conscience, esp. in respect of evaded payment etc.; **conscience-stricken** made uneasy by bad conscience.

conscientious /kɒnʃɪ'enʃəs/ *adjective* diligent and scrupulous. □ **conscientious objector** person who refuses to do military service on grounds of conscience. □ **conscientiously** *adverb*; **conscientiousness** *noun.*

conscious /'kɒnʃəs/ *adjective* awake and aware of one's surroundings etc.; (usually + *of*, *that*) aware, knowing; intentional. □ **consciously** *adverb*; **consciousness** *noun.*

conscript ● *verb* /kən'skrɪpt/ summon for compulsory state (esp. military) service. ● *noun* /'kɒnskrɪpt/ conscripted person. □ **conscription** *noun.*

consecrate /'kɒnsɪkreɪt/ *verb* (**-ting**) make or declare sacred; dedicate formally to religious purpose; (+ *to*) devote to (a purpose). □ **consecration** *noun.*

consecutive /kən'sekjʊtɪv/ *adjective* following continuously; in unbroken or logical order. □ **consecutively** *adverb.*

consensus /kən'sensəs/ *noun* (often + *of*) general agreement or opinion.

consent /kən'sent/ ● *verb* (often + *to*) express willingness, give permission; agree. ● *noun* agreement; permission.

consequence /'kɒnsɪkwəns/ *noun* result of what has gone before; importance.

Conjunction

A conjunction is used to join parts of sentences which usually, but not always, contain their own verbs, e.g.

He found it difficult but *I helped him.*
They made lunch for Alice and *Mary.*
I waited until *you came.*

The most common conjunctions are:

after	*for*	*since*	*unless*
although	*if*	*so*	*until*
and	*in order that*	*so that*	*when*
as	*like*	*than*	*where*
because	*now*	*that*	*whether*
before	*once*	*though*	*while*
but	*or*	*till*	

consequent /'kɒnsɪkwənt/ *adjective* that results; (often + *on, upon*) following as consequence. □ **consequently** *adverb.*

consequential /kɒnsɪ'kwenʃ(ə)l/ *adjective* resulting, esp. indirectly; important.

conservancy /kən'sɜ:vənsɪ/ *noun* (*plural* **-ies**) body controlling river, port, etc. or concerned with conservation; official environmental conservation.

conservation /kɒnsə'veɪʃ(ə)n/ *noun* preservation, esp. of natural environment. □ **conservationist** *noun.*

conservative /kən'sɜ:vətɪv/ ● *adjective* averse to rapid change; (of estimate) purposely low; (usually **Conservative**) of Conservative Party. ● *noun* conservative person; (usually **Conservative**) member or supporter of Conservative Party. □ **Conservative Party** political party promoting free enterprise. □ **conservatism** *noun.*

conservatoire /kən'sɜ:vətwɑ:/ *noun* (usually European) school of music or other arts.

conservatory /kən'sɜ:vətərɪ/ *noun* (*plural* **-ies**) greenhouse for tender plants; esp. *US* conservatoire.

conserve ● *verb* /kən'sɜ:v/ (**-ving**) preserve, keep from harm or damage. ● *noun* /'kɒnsɜ:v/ fruit etc. preserved in sugar; fresh fruit jam.

consider /kən'sɪdə/ *verb* contemplate; deliberate thoughtfully; make allowance for, take into account; (+ *that*) have the opinion that; show consideration for; regard as; (as **considered** *adjective*) (esp. of an opinion) formed after careful thought.

considerable *adjective* a lot of; notable, important. □ **considerably** *adverb.*

considerate /kən'sɪdərət/ *adjective* giving thought to feelings or rights of others. □ **considerately** *adverb.*

consideration /kənsɪdə'reɪʃ(ə)n/ *noun* careful thought; being considerate; fact or thing taken into account; compensation, payment. □ **take into consideration** make allowance for.

considering *preposition* in view of.

consign /kən'saɪn/ *verb* (often + *to*) commit, deliver; send (goods etc.). □ **consignee** /kənsaɪ'ni:/ *noun*; **consignor** *noun.*

consignment *noun* consigning, goods consigned.

consist /kən'sɪst/ *verb* (+ *of*) be composed of; (+ *in, of*) have its essential features in.

consistency *noun* (*plural* **-ies**) degree of density or firmness, esp. of thick liquids; being consistent.

consistent *adjective* constant to same principles; (usually + *with*) compatible. □ **consistently** *adverb.*

consistory *noun* (*plural* **-ies**) *RC Church* council of cardinals.

consolation /kɒnsə'leɪʃ(ə)n/ *noun* alleviation of grief or disappointment; consoling person or thing. □ **consolation prize** one given to competitor just failing to win main prize. □ **consolatory** /kən'sɒl-/ *adjective.*

console[1] /kən'səʊl/ *verb* (**-ling**) bring consolation to.

■ **Usage** *Console* is often confused with *condole*. To condole with someone is to express sympathy with them.

---

console[2] /'kɒnsəʊl/ *noun* panel for switches, controls, etc.; cabinet for television etc.; cabinet with keys and stops of organ; bracket supporting shelf etc.

consolidate /kən'sɒlɪdeɪt/ *verb* (**-ting**) make or become strong or secure; combine (territories, companies, debts, etc.) into one whole. □ **consolidation** *noun.*

consommé /kən'sɒmeɪ/ *noun* clear meat soup.

consonance /'kɒnsənəns/ *noun* agreement, harmony.

consonant ● *noun* speech sound that forms syllable only in combination with vowel; letter(s) representing this. ● *adjective* (+ *with, to*) consistent with; in agreement or harmony. □ **consonantal** /-'næn-/ *adjective.*

consort[1] ● *noun* /'kɒnsɔ:t/ wife or husband, esp. of royalty. ● *verb* /kən'sɔ:t/ (usually + *with, together*) keep company; harmonize.

consort[2] /'kɒnsɔ:t/ *noun Music* group of players or instruments.

consortium /kən'sɔ:tɪəm/ *noun* (*plural* **-tia** or **-s**) association, esp. of several business companies.

conspicuous /kən'spɪkjʊəs/ *adjective* clearly visible; attracting attention. □ **conspicuously** *adverb.*

conspiracy /kən'spɪrəsɪ/ *noun* (*plural* **-ies**) act of conspiring; plot.

conspirator /kən'spɪrətə/ *noun* person who takes part in conspiracy. □ **conspiratorial** /-'tɔ:-/ *adjective.*

conspire /kən'spaɪə/ *verb* (**-ring**) combine secretly for unlawful or harmful purpose; (of events) seemingly work together.

constable /'kʌnstəb(ə)l/ *noun* (also **police constable**) police officer of lowest rank; governor of royal castle. □ **Chief Constable** head of police force of county etc.

constabulary /kən'stæbjʊlərɪ/ *noun* (*plural* **-ies**) police force.

constancy /'kɒnstənsɪ/ *noun* dependability; faithfulness.

constant ● *adjective* continuous; frequently occurring; having constancy. ● *noun Mathematics & Physics* unvarying quantity. □ **constantly** *adverb.*

constellation /kɒnstə'leɪʃ(ə)n/ *noun* group of fixed stars.

consternation /kɒnstə'neɪʃ(ə)n/ *noun* amazement, dismay.

constipate /'kɒnstɪpeɪt/ *verb* (**-ting**) (esp. as **constipated** *adjective*) affect with constipation.

constipation *noun* difficulty in emptying bowels.

constituency /kən'stɪtjʊənsɪ/ *noun* (*plural* **-ies**) body electing representative; area represented.

constituent /kən'stɪtjʊənt/ ● *adjective* making part of whole; appointing, electing. ● *noun* member of constituency; component part.

constitute /'kɒnstɪtju:t/ *verb* (**-ting**) be essence or components of; amount to; establish.

constitution /kɒnstɪ'tju:ʃ(ə)n/ *noun* composition; set of principles by which state etc. is governed; person's inherent state of health, strength, etc.

constitutional ● *adjective* of or in line with the constitution; inherent. ● *noun* walk taken as exercise. □ **constitutionally** *adverb.*

constitutive /'kɒnstɪtju:tɪv/ *adjective* able to form or appoint; constituent.

constrain /kən'streɪn/ *verb* compel; confine; (as **constrained** *adjective*) forced, embarrassed.

constraint *noun* compulsion; restriction; self-control.

constrict /kən'strɪkt/ *verb* make narrow or tight; compress. □ **constriction** *noun*; **constrictive** *adjective.*

constrictor *noun* snake that kills by compressing; muscle that contracts a part.

construct ● *verb* /kən'strʌkt/ fit together, build; *Geometry* draw. ● *noun* /'kɒnstrʌkt/ thing constructed, esp. by the mind. □ **constructor** /kən'strʌktə/ *noun.*

construction /kən'strʌkʃ(ə)n/ *noun* constructing; thing constructed; syntactical arrangement; interpretation. □ **constructional** *adjective.*

constructive /kən'strʌktɪv/ *adjective* positive, helpful. □ **constructively** *adverb.*

construe /kən'stru:/ *verb* (**-strues**, **-strued**, **struing**) interpret; (often + *with*) combine (words) grammatically; translate literally.

consubstantiation /kɒnsəbstænʃɪ'eɪʃ(ə)n/ *noun* presence of Christ's body and blood together with bread and wine in Eucharist.

consul /'kɒns(ə)l/ *noun* official appointed by state to protect its interests and citizens in foreign city; *historical* either of two annually-elected magistrates in ancient Rome. □ **consular** /-sjʊlə/ *adjective.*

consulate /'kɒnsjʊlət/ *noun* offices or position of consul.

consult /kən'sʌlt/ *verb* seek information or advice from; (often + *with*) refer to; take into consideration.

consultant /kən'sʌlt(ə)nt/ *noun* person who gives professional advice; senior medical specialist. □ **consultancy** *noun.*

consultation /kɒnsəl'teɪʃ(ə)n/ *noun* (meeting for) consulting.

consultative /kən'sʌltətɪv/ *adjective* of or for consultation.

consume /kən'sju:m/ *verb* (**-ming**) eat or drink; use up; destroy. □ **consumable** *adjective.*

consumer *noun* user of product or service. □ **consumer goods** goods for consumers, not for producing other goods.

consumerism *noun* protection or promotion of consumers' interests; *often derogatory* continual increase in consumption. □ **consumerist** *adjective*.

consummate ● *verb* /'kɒnsəmeɪt/ (**-ting**) complete (esp. marriage by sexual intercourse). ● *adjective* /kən'sʌmɪt/ complete, perfect; fully skilled. □ **consummation** /kɒnsə-/ *noun*.

consumption /kən'sʌmpʃ(ə)n/ *noun* consuming, being consumed; purchase and use of goods etc.; *archaic* tuberculosis of lungs.

consumptive /kən'sʌmptɪv/ *archaic* ● *adjective* suffering or tending to suffer from tuberculosis. ● *noun* consumptive person.

cont. *abbreviation* continued.

contact /'kɒntækt/ ● *noun* condition or state of touching, meeting, or communicating; person who is, or may be, contacted for information etc.; person likely to carry contagious disease through being near infected person; connection for passage of electric current. ● *verb* get in touch with. □ **contact lens** small lens placed on eyeball to correct vision.

contagion /kən'teɪdʒ(ə)n/ *noun* spreading of disease by contact; moral corruption. □ **contagious** *adjective*.

contain /kən'teɪn/ *verb* hold or be capable of holding within itself; include; comprise; prevent from moving or extending; control, restrain.

container *noun* box etc. for holding things; large metal box for transporting goods.

containment /kən'teɪnmənt/ *noun* action or policy of preventing expansion of hostile country or influence.

contaminate /kən'tæmɪneɪt/ *verb* (**-ting**) pollute; infect. □ **contaminant** *noun*; **contamination** *noun*.

contemplate /'kɒntəmpleɪt/ *verb* (**-ting**) survey with eyes or mind; regard as possible; intend. □ **contemplation** *noun*.

contemplative /kən'templətɪv/ *adjective* of or given to (esp. religious) contemplation.

contemporaneous /kəntempə'reɪnɪəs/ *adjective* (usually + *with*) existing or occurring at same time.

contemporary /kən'tempərərɪ/ ● *adjective* belonging to same time; of same age; modern in style or design. ● *noun* (*plural* **-ies**) contemporary person or thing.

contempt /kən'tempt/ *noun* feeling that person or thing deserves scorn or reproach; condition of being held in contempt; (in full **contempt of court**) disobedience to or disrespect for court of law. □ **contemptible** *adjective*.

contemptuous *adjective* feeling or showing contempt. □ **contemptuously** *adverb*.

contend /kən'tend/ *verb* compete; (usually + *with*) argue; (+ *that*) maintain that. □ **contender** *noun*.

content[1] /kən'tent/ ● *adjective* satisfied; (+ *to do*) willing. ● *verb* make content; satisfy. ● *noun* contented state; satisfaction. □ **contented** *adjective*; **contentment** *noun*.

content[2] /'kɒntent/ *noun* (usually in *plural*) what is contained, esp. in vessel, house, or book; amount contained; substance of book etc. as opposed to form; capacity, volume.

contention /kən'tenʃ(ə)n/ *noun* dispute, rivalry; point contended for in argument.

contentious /kən'tenʃəs/ *adjective* quarrelsome; likely to cause argument.

contest ● *noun* /'kɒntest/ contending; a competition. ● *verb* /kən'test/ dispute; contend or compete for; compete in.

contestant /kən'test(ə)nt/ *noun* person taking part in contest.

context /'kɒntekst/ *noun* what precedes and follows word or passage; relevant circumstances. □ **contextual** /kən'tekstjʊəl/ *adjective*.

contiguous /kən'tɪgjʊəs/ *adjective* (usually + *with*, *to*) touching; in contact.

continent[1] /'kɒntɪnənt/ *noun* any of the earth's main continuous bodies of land; (**the Continent**) the mainland of Europe.

continent[2] /'kɒntɪnənt/ *adjective* able to control bowels and bladder; exercising esp. sexual self-restraint. □ **continence** *noun*.

continental /kɒntɪ'nent(ə)l/ *adjective* of or characteristic of a continent or (**Continental**) the Continent. □ **continental breakfast** light breakfast of coffee, rolls, etc.; **continental quilt** duvet; **continental shelf** shallow seabed bordering continent.

contingency /kən'tɪndʒənsɪ/ *noun* (*plural* **-ies**) event that may or may not occur; something dependent on another uncertain event.

contingent ● *adjective* (usually + *on*, *upon*) conditional, dependent; that may or may not occur; fortuitous. ● *noun* group (of troops, ships, etc.) forming part of larger group; people sharing interest, origin, etc.

continual /kən'tɪnjʊəl/ *adjective* frequently recurring; always happening. □ **continually** *adverb.*

continuance /kən'tɪnjʊəns/ *noun* continuing in existence or operation; duration.

continuation /kəntɪnjʊ'eɪʃ(ə)n/ *noun* continuing, being continued; thing that continues something else.

continue /kən'tɪnju:/ *verb* (**-ues**, **-ued**, **-uing**) maintain; resume; prolong; remain.

continuity /kɒntɪ'nju:ɪtɪ/ *noun* (*plural* **-ies**) being continuous; logical sequence; detailed scenario of film; linkage of broadcast items.

continuo /kən'tɪnjʊəʊ/ *noun* (*plural* **-s**) *Music* bass accompaniment played usually on keyboard instrument.

continuous /kən'tɪnjʊəs/ *adjective* connected without break; uninterrupted. □ **continuously** *adverb.*

continuum /kən'tɪnjʊəm/ *noun* (*plural* **-nua**) thing with continuous structure.

contort /kən'tɔ:t/ *verb* twist or force out of normal shape. □ **contortion** *noun.*

contortionist *noun* entertainer who adopts contorted postures.

contour /'kɒntʊə/ *noun* outline; (in full **contour line**) line on map joining points at same altitude.

contraband /'kɒntrəbænd/ ● *noun* smuggled goods. ● *adjective* forbidden to be imported or exported.

contraception /kɒntrə'sepʃ(ə)n/ *noun* use of contraceptives.

contraceptive /kɒntrə'septɪv/ ● *adjective* preventing pregnancy. ● *noun* contraceptive device or drug.

contract ● *noun* /'kɒntrækt/ written or spoken agreement, esp. one enforceable by law; document recording it. ● *verb* /kən'trækt/ make or become smaller; (usually + *with*) make contract; (often + *out*) arrange (work) to be done by contract; become affected by (a disease); incur (debt); draw together; shorten. □ **contract bridge** type of bridge in which only tricks bid and won count towards game; **contract in** (or **out**) elect (not) to enter scheme etc.

contraction /kən'trækʃ(ə)n/ *noun* contracting; shortening of uterine muscles during childbirth; shrinking; diminution; shortened word.

contractor /kən'træktə/ *noun* person who undertakes contract, esp. in building, engineering, etc.

contractual /kən'træktjʊəl/ *adjective* of or in the nature of a contract.

contradict /kɒntrə'dɪkt/ *verb* deny; oppose verbally; be at variance with. □ **contradiction** *noun*; **contradictory** *adjective.*

contradistinction /kɒntrədɪs'tɪŋkʃ(ə)n/ *noun* distinction made by contrasting.

contraflow /'kɒntrəfləʊ/ *noun* transfer of traffic from usual half of road to lane(s) of other half.

contralto /kən'træltəʊ/ *noun* (*plural* **-s**) lowest female singing voice; singer with this voice.

contraption /kən'træpʃ(ə)n/ *noun* machine or device, esp. strange one.

contrapuntal /kɒntrə'pʌnt(ə)l/ *adjective* *Music* of or in counterpoint.

contrariwise /kən'treərɪwaɪz/ *adverb* on the other hand; in the opposite way.

contrary /'kɒntrərɪ/ ● *adjective* (usually + *to*) opposed in nature, tendency, or direction; /kən'treərɪ/ perverse, self-willed. ● *noun* (**the contrary**) the opposite. ● *adverb* (+ *to*) in opposition. □ **on the contrary** the opposite is true.

contrast ● *noun* /'kɒntrɑ:st/ comparison showing differences; difference so revealed; (often + *to*) thing or person having different qualities; degree of difference between tones in photograph or television picture. ● *verb* /kən'trɑ:st/ (often + *with*) compare to reveal contrast; show contrast.

contravene /kɒntrə'vi:n/ *verb* (**-ning**) infringe; conflict with. □ **contravention** /-'ven-/ *noun.*

contretemps /'kɒntrətɑ̃/ *noun* (*plural* same /-tɑ̃z/) unlucky accident; unfortunate occurrence.

contribute /kən'trɪbju:t/ *verb* (**-ting**) (often + *to*) give jointly with others to common purpose; supply (article etc.) for publication with others; (+ *to*) help to bring about. □ **contribution** /kɒntrɪ'bju:ʃ(ə)n/ *noun*; **contributor** *noun*; **contributory** *adjective*.

■ **Usage** *Contribute* is often pronounced /'kɒntrɪbju:t/ (with the stress on the *con-*), but this is considered incorrect by some people.

contrite /kən'traɪt/ *adjective* penitent, feeling guilt. □ **contrition** /-'trɪʃ-/ *noun*.

contrivance *noun* something contrived, esp. device or plan; act of contriving.

contrive /kən'traɪv/ *verb* (**-ving**) devise, plan; (often + *to do*) manage.

contrived *adjective* artificial, forced.

control /kən'trəʊl/ ● *noun* power of directing or restraining; self-restraint; means of restraining or regulating; (usually in *plural*) device to operate machine, vehicle, etc.; place where something is controlled or verified; standard of comparison for checking results of experiment. ● *verb* (**-ll-**) have control of; regulate; hold in check; verify. □ **in control** (often + *of*) in charge; **out of control** no longer manageable.

controller *noun* person or thing that controls; person controlling expenditure.

controversial /kɒntrə'vɜ:ʃ(ə)l/ *adjective* causing or subject to controversy.

controversy /'kɒntrəvɜ:sɪ/ *noun* (*plural* **-ies**) dispute, argument.

■ **Usage** *Controversy* is often pronounced /kən'trɒvəsɪ/ (with the stress on the *-trov-*), but this is considered incorrect by some people.

controvert /kɒntrə'vɜ:t/ *verb* dispute, deny.

contuse /kən'tju:z/ *verb* (**-sing**) bruise. □ **contusion** *noun*.

conundrum /kə'nʌndrəm/ *noun* riddle; hard question.

conurbation /kɒnɜ:'beɪʃ(ə)n/ *noun* group of towns united by expansion.

convalesce /kɒnvə'les/ *verb* (**-cing**) recover health after illness.

convalescent /kɒnvə'les(ə)nt/ ● *adjective* recovering from illness. ● *noun* convalescent person. □ **convalescence** *noun*.

convection /kən'vekʃ(ə)n/ *noun* heat transfer by upward movement of heated medium.

convector /kən'vektə/ *noun* heating appliance that circulates warm air by convection.

convene /kən'vi:n/ *verb* (**-ning**) summon; assemble. □ **convener**, **convenor** *noun*.

convenience /kən'vi:nɪəns/ *noun* state of being convenient; suitability; advantage; useful thing; public lavatory. □ **convenience food** food needing little preparation.

convenient *adjective* serving one's comfort or interests; suitable; available or occurring at suitable time or place; well situated. □ **conveniently** *adverb*.

convent /'kɒnv(ə)nt/ *noun* religious community, esp. of nuns; its house.

convention /kən'venʃ(ə)n/ *noun* general agreement on social behaviour etc. by implicit consent of majority; customary practice; assembly, conference; agreement, treaty.

conventional *adjective* depending on or according with convention; bound by social conventions; not spontaneous or sincere; (of weapons etc.) non-nuclear. □ **conventionally** *adverb*.

converge /kən'vɜ:dʒ/ *verb* (**-ging**) come together or towards same point; (+ *on*, *upon*) approach from different directions. □ **convergence** *noun*; **convergent** *adjective*.

conversant /kən'vɜ:s(ə)nt/ *adjective* (+ *with*) well acquainted with.

conversation /kɒnvə'seɪʃ(ə)n/ *noun* informal spoken communication; instance of this.

conversational *adjective* of or in conversation; colloquial. □ **conversationally** *adverb*.

conversationalist *noun* person fond of or good at conversation.

converse[1] /kən'vɜ:s/ *verb* (**-sing**) (often + *with*) talk.

converse[2] /'kɒnvɜ:s/ ● *adjective* opposite, contrary, reversed. ● *noun* converse statement or proposition. □ **conversely** *adverb.*

conversion /kən'vɜ:ʃ(ə)n/ *noun* converting, being converted; converted (part of) building.

convert ● *verb* /kən'vɜ:t/ (usually + *into*) change; cause (person) to change belief etc.; change (money etc.) into different form or currency etc.; alter (building) for new purpose; *Rugby* kick goal after (try). ● *noun* /'kɒnvɜ:t/ person converted, esp. to religious faith.

convertible /kən'vɜ:tɪb(ə)l/ ● *adjective* able to be converted. ● *noun* car with folding or detachable roof.

convex /'kɒnveks/ *adjective* curved like outside of sphere or circle.

convey /kən'veɪ/ *verb* transport, carry; communicate (meaning etc.); transfer by legal process; transmit (sound etc.).

conveyance *noun* conveying, being conveyed; vehicle; legal transfer of property, document effecting this.

conveyancing *noun* branch of law dealing with transfer of property. □ **conveyancer** *noun.*

conveyor *noun* person or thing that conveys. □ **conveyor belt** endless moving belt conveying articles in factory etc.

convict ● *verb* /kən'vɪkt/ (often + *of*) prove or declare guilty. ● *noun* /'kɒnvɪkt/ *esp. historical* sentenced criminal.

conviction /kən'vɪkʃ(ə)n/ *noun* convicting, being convicted; being convinced; firm belief.

convince /kən'vɪns/ *verb* (**-cing**) firmly persuade. □ **convincing** *adjective*; **convincingly** *adverb.*

convivial /kən'vɪvɪəl/ *adjective* fond of company; sociable, lively. □ **conviviality** /-'æl-/ *noun.*

convocation /kɒnvə'keɪʃ(ə)n/ *noun* convoking; large formal gathering.

convoke /kən'vəʊk/ *verb* (**-king**) call together; summon to assemble.

convoluted /kɒnvə'lu:tɪd/ *adjective* coiled, twisted; complex.

convolution /kɒnvə'lu:ʃ(ə)n/ *noun* coiling; coil, twist; complexity.

convolvulus /kən'vɒlvjʊləs/ *noun* (*plural* **-es**) twining plant, esp. bindweed.

convoy /'kɒnvɔɪ/ *noun* group of ships, vehicles, etc. travelling together. □ **in convoy** as a group.

convulse /kən'vʌls/ *verb* (**-sing**) affect with convulsions.

convulsion *noun* (usually in *plural*) violent irregular motion of limbs or body caused by involuntary contraction of muscles; violent disturbance; (in *plural*) uncontrollable laughter. □ **convulsive** *adjective*; **convulsively** *adverb.*

cony /'kəʊnɪ/ *noun* (*plural* **-ies**) (also **coney**) rabbit; its fur.

coo ● *noun* soft murmuring sound as of doves. ● *verb* (**coos**, **cooed**) emit coo.

cooee /'ku:i:/ *interjection used to attract attention.*

cook /kʊk/ ● *verb* prepare (food) by heating; undergo cooking; *colloquial* falsify (accounts etc.). ● *noun* person who cooks. □ **cookbook** *US* cookery book; **cooking apple** one suitable for eating cooked.

cooker *noun* appliance or vessel for cooking food; fruit (esp. apple) suitable for cooking.

cookery *noun* art of cooking. □ **cookery book** book containing recipes.

cookie /'kʊkɪ/ *noun US* sweet biscuit.

cool /ku:l/ ● *adjective* of or at fairly low temperature; suggesting or achieving coolness; calm; lacking enthusiasm; unfriendly; *slang esp. US* marvellous. ● *noun* coolness; cool place; *slang* composure. ● *verb* (often + *down*, *off*) make or become cool. □ **coolly** /'ku:llɪ/ *adverb*; **coolness** *noun.*

coolant *noun* cooling agent, esp. fluid.

cooler *noun* vessel in which thing is cooled; *slang* prison cell.

coomb /ku:m/ *noun* (also **combe**) valley on side of hill; short valley running up from coast.

coon /ku:n/ *noun US* racoon.

coop /ku:p/ ● *noun* cage for keeping poultry. ● *verb* (often + *up*, *in*) confine.

co-op /'kəʊɒp/ *noun colloquial* cooperative society or shop.

cooper *noun* maker or repairer of casks and barrels.

cooperate /kəʊ'ɒpəreɪt/ *verb* (also **co-operate**) (**-ting**) (often + *with*) work or act together. □ **cooperation** *noun*.

cooperative /kəʊ'ɒpərətɪv/ (also **co-operative**) ● *adjective* willing to co-operate; (of business etc.) jointly owned and run by members, with profits shared. ● *noun* cooperative society or enterprise.

co-opt /kəʊ'ɒpt/ *verb* appoint to committee etc. by invitation of existing members. □ **co-option** *noun*; **co-optive** *adjective*.

coordinate (also **co-ordinate**) ● *verb* /kəʊ'ɔ:dɪneɪt/ (**-ting**) cause to work together efficiently; work or act together effectively. ● *adjective* /-nət/ equal in status. ● *noun* /-nət/ *Mathematics* each of set of quantities used to fix position of point, line, or plane; (in *plural*) matching items of clothing. □ **coordination** *noun*; **coordinator** *noun*.

coot /ku:t/ *noun* black waterfowl with white horny plate on head; *colloquial* stupid person.

cop *slang* ● *noun* police officer; capture. ● *verb* (**-pp-**) catch. □ **cop-out** cowardly evasion; **not much cop** of little value or use.

copal /'kəʊp(ə)l/ *noun* kind of resin used for varnish.

copartner /kəʊ'pɑ:tnə/ *noun* partner, associate. □ **copartnership** *noun*.

cope[1] *verb* (**-ping**) deal effectively or contend; (often + *with*) manage.

cope[2] *noun* priest's long cloaklike vestment.

copeck /'kəʊpek/ *noun* (also **kopek**, **kopeck**) hundredth of rouble.

copier /'kɒpɪə/ *noun* machine that copies (esp. documents).

copilot /'kəʊpaɪlət/ *noun* second pilot in aircraft.

coping /'kəʊpɪŋ/ *noun* top (usually sloping) course of masonry in wall. □ **coping stone** stone used in coping.

copious /'kəʊpɪəs/ *adjective* abundant; producing much. □ **copiously** *adverb*.

copper[1] /'kɒpə/ ● *noun* red-brown metal; bronze coin; large metal vessel for boiling laundry. ● *adjective* made of or coloured like copper. □ **copperplate** copper plate for engraving or etching, print taken from it, ornate sloping handwriting.

copper[2] /'kɒpə/ *noun slang* police officer.

coppice /'kɒpɪs/ *noun* area of undergrowth and small trees.

copulate /'kɒpjʊleɪt/ *verb* (**-ting**) (often + *with*) (esp. of animals) have sexual intercourse. □ **copulation** *noun*.

copy /'kɒpɪ/ ● *noun* (*plural* **-ies**) thing made to look like another; specimen of book etc.; material to be printed, esp. regarded as good etc. reading matter. ● *verb* (**-ies, -ied**) make copy of; imitate. □ **copy-typist** typist working from document or recording; **copywriter** writer of copy, esp. for advertisements.

copyist /'kɒpɪɪst/ *noun* person who makes copies.

copyright ● *noun* exclusive right to print, publish, perform, etc., material. ● *adjective* protected by copyright. ● *verb* secure copyright for (material).

coquette /kə'ket/ *noun* woman who flirts. □ **coquettish** *adjective*; **coquetry** /'kɒkɪtrɪ/ *noun* (*plural* **-ies**).

coracle /'kɒrək(ə)l/ *noun* small boat of wickerwork covered with waterproof material.

coral /'kɒr(ə)l/ ● *noun* hard substance built up by marine polyps. ● *adjective* of (red or pink colour of) coral. □ **coral island**, **reef** one formed by growth of coral.

cor anglais /kɔ:r 'ɒŋgleɪ/ *noun* (*plural* **cors anglais** /kɔ:z/) woodwind instrument like oboe but lower in pitch.

corbel /'kɔ:b(ə)l/ *noun* stone or timber projection from wall, acting as supporting bracket.

cord ● *noun* thick string; piece of this; similar structure in body; ribbed cloth, esp. corduroy; (in *plural*) corduroy trousers; electric flex. ● *verb* secure with cords.

cordial /'kɔ:dɪəl/ ● *adjective* heartfelt; friendly. ● *noun* fruit-flavoured drink. □ **cordiality** /-'æl-/ *noun*; **cordially** *adverb*.

cordite /'kɔ:daɪt/ *noun* smokeless explosive.

cordless *adjective* (of handheld electric device) battery-powered.

cordon /'kɔ:d(ə)n/ ● *noun* line or circle of police etc.; ornamental cord or braid; fruit tree trained to grow as single stem. ● *verb* (often + *off*) enclose or separate with cordon of police etc.

cordon bleu /kɔ:dɒn 'blɜ:/ *adjective* (of cooking) first-class.

corduroy /'kɔ:dərɔɪ/ *noun* fabric with velvety ribs.

core ● *noun* horny central part of certain fruits, containing seeds; centre or most important part; part of nuclear reactor containing fissile material; inner strand of electric cable; piece of soft iron forming centre of magnet etc. ● *verb* (**-ring**) remove core from.

co-respondent /kəʊrɪ'spɒnd(ə)nt/ *noun* person cited in divorce case as having committed adultery with respondent.

corgi /'kɔ:gɪ/ *noun* (*plural* **-s**) short-legged breed of dog.

coriander /kɒrɪ'ændə/ *noun* aromatic plant; its seed, used as flavouring.

cork ● *noun* thick light bark of S. European oak; bottle-stopper made of cork etc. ● *verb* (often + *up*) stop, confine; restrain (feelings etc.).

corkage *noun* charge made by restaurant etc. for serving customer's own wine etc.

corked *adjective* (of wine) spoilt by defective cork.

corkscrew ● *noun* spiral steel device for extracting corks from bottles. ● *verb* move spirally.

corm *noun* swollen underground stem in certain plants, e.g. crocus.

cormorant /'kɔ:mərənt/ *noun* diving seabird with black plumage.

corn[1] *noun* cereal before or after harvesting, esp. chief crop of a region; grain or seed of cereal plant; *colloquial* something corny. □ **corn cob** cylindrical centre of maize ear; **corncrake** ground-nesting bird with harsh cry; **corn dolly** plaited straw figure; **cornflakes** breakfast cereal of toasted maize flakes; **cornflour** fine-ground maize flour; **cornflower** blue-flowered plant originally growing in cornfields; **corn on the cob** maize eaten from the corn cob.

corn[2] *noun* small tender hard area of skin, esp. on toe.

cornea /'kɔ:nɪə/ *noun* transparent circular part of front of eyeball.

corned *adjective* preserved in salt or brine.

cornelian /kɔ:'ni:lɪən/ *noun* (also **carnelian** /kɑ:-/) dull red variety of chalcedony.

corner /'kɔ:nə/ ● *noun* place where converging sides, edges, streets, etc. meet; recess formed by meeting of two internal sides of room, box, etc.; difficult or inescapable position; remote or secluded place; action or result of buying whole available stock of a commodity; *Football & Hockey* free kick or hit from corner of pitch. ● *verb* force into difficult or inescapable position; buy whole available stock of (commodity); dominate (market) in this way; go round corner. □ **cornerstone** stone in projecting angle of wall, indispensable part or basis.

cornet /'kɔ:nɪt/ *noun* brass instrument resembling trumpet; conical wafer for holding ice cream.

cornice /'kɔ:nɪs/ *noun* ornamental moulding, esp. along top of internal wall.

Cornish /'kɔ:nɪʃ/ ● *adjective* of Cornwall. ● *noun* Celtic language of Cornwall. □ **Cornish pasty** pastry envelope containing meat and vegetables.

cornucopia /kɔ:njʊ'kəʊpɪə/ *noun* horn overflowing with flowers, fruit, etc., as symbol of plenty.

corny *adjective* (**-ier**, **-iest**) *colloquial* banal; feebly humorous; sentimental.

corolla /kə'rɒlə/ *noun* whorl of petals forming inner envelope of flower.

corollary /kə'rɒlərɪ/ *noun* (*plural* **-ies**) proposition that follows from one proved; (often + *of*) natural consequence.

corona /kə'rəʊnə/ *noun* (*plural* **-nae** /-ni:/) halo round sun or moon, esp. that seen in total eclipse of sun.

coronary /'kɒrənərɪ/ *noun* (*plural* **-ies**) coronary thrombosis. □ **coronary artery** artery supplying blood to heart; **coronary thrombosis** blockage of coronary artery by blood clot.

coronation /kɒrə'neɪʃ(ə)n/ *noun* ceremony of crowning sovereign.

coroner /'kɒrənə/ *noun* official holding inquest on deaths thought to be violent or accidental.

coronet /'kɒrənɪt/ *noun* small crown.

corpora *plural* of CORPUS.

corporal[1] /'kɔ:pr(ə)l/ *noun* army or air-force NCO next below sergeant.

corporal[2] /'kɔ:pər(ə)l/ *adjective* of human body. □ **corporal punishment** physical punishment.

corporate /'kɔ:pərət/ *adjective* of, being, or belonging to a corporation or group.

corporation /kɔ:pə'reɪʃ(ə)n/ *noun* group of people authorized to act as individual, esp. in business; civic authorities.

corporative /'kɔ:pərətɪv/ *adjective* of corporation; governed by or organized in corporations.

corporeal /kɔ:'pɔ:rɪəl/ *adjective* bodily, physical, material.

corps /kɔ:/ *noun* (*plural* same /kɔ:z/) military unit with particular function; organized group of people.

corpse /kɔ:ps/ *noun* dead body.

corpulent /'kɔ:pjʊlənt/ *adjective* fleshy, bulky. □ **corpulence** *noun*.

corpus /'kɔ:pəs/ *noun* (*plural* **-pora** /-pərə/) body or collection of writings, texts, etc.

corpuscle /'kɔ:pʌs(ə)l/ *noun* minute body or cell in organism, esp. (in *plural*) red or white cells in blood of vertebrates. □ **corpuscular** /-'pʌskjʊlə/ *adjective*.

corral /kə'rɑ:l/ ● *noun US* pen for horses, cattle, etc.; enclosure for capturing wild animals. ● *verb* (**-ll-**) put or keep in corral.

correct /kə'rekt/ ● *adjective* true, accurate; proper, in accordance with taste, standards, etc. ● *verb* set right; mark errors in; admonish; counteract. □ **correctly** *adverb*; **correctness** *noun*.

correction *noun* correcting, being corrected; thing substituted for what is wrong.

correctitude /kə'rektɪtju:d/ *noun* consciously correct behaviour.

corrective ● *adjective* serving to correct or counteract. ● *noun* corrective measure or thing.

correlate /'kɒrəleɪt/ ● *verb* (**-ting**) (usually + *with*, *to*) have or bring into mutual relation. ● *noun* either of two related or complementary things. □ **correlation** *noun*.

correlative /kə'relətɪv/ ● *adjective* (often + *with*, *to*) having a mutual relation; (of words) corresponding and regularly used together. ● *noun* correlative thing or word.

correspond /kɒrɪ'spɒnd/ *verb* (usually + *to*) be similar or equivalent; (usually + *to*, *with*) agree; (usually + *with*) exchange letters.

correspondence *noun* agreement or similarity; (exchange of) letters. □ **correspondence course** course of study conducted by post.

correspondent *noun* person who writes letter(s); person employed to write or report for newspaper or broadcasting.

corridor /'kɒrɪdɔ:/ *noun* passage giving access into rooms; passage in train giving access into compartments; strip of territory of one state running through that of another; route for aircraft over foreign country.

corrigendum /kɒrɪ'dʒendəm/ *noun* (*plural* **-da**) error to be corrected.

corrigible /'kɒrɪdʒɪb(ə)l/ *adjective* able to be corrected.

corroborate /kə'rɒbəreɪt/ *verb* (**-ting**) give support to, confirm. □ **corroboration** *noun*; **corroborative** /-rətɪv/ *adjective*; **corroboratory** /-rət(ə)rɪ/ *adjective*.

corrode /kə'rəʊd/ *verb* (**-ding**) wear away, esp. by chemical action; destroy gradually; decay.

corrosion /kə'rəʊʒ(ə)n/ *noun* corroding, being corroded; corroded area. □ **corrosive** /-sɪv/ *adjective & noun*.

corrugate /'kɒrəgeɪt/ *verb* (**-ting**) (esp. as **corrugated** *adjective*) bend into wavy ridges. □ **corrugation** *noun*.

corrupt /kə'rʌpt/ ● *adjective* influenced by or using bribery; immoral, wicked. ● *verb* make or become corrupt. □ **corruptible** *adjective*; **corruption** *noun*; **corruptly** *adverb*.

corsage /kɔ:'sɑ:ʒ/ *noun* small bouquet worn by woman.

corsair /kɔ:'seə/ *noun* pirate ship; pirate.

corset /'kɔ:sɪt/ *noun* tight-fitting supporting undergarment worn esp. by women. □ **corsetry** *noun*.

cortège /kɔ:'teɪʒ/ *noun* procession, esp. for funeral.

cortex /'kɔ:teks/ *noun* (*plural* **-tices** /-tɪsi:z/) outer part of organ, esp. brain. □ **cortical** *adjective*.

cortisone /'kɔ:tɪzəʊn/ *noun* hormone used in treating inflammation and allergy.

corvette /kɔ:'vet/ *noun* small naval escort-vessel.

cos[1] /kɒs/ *noun* crisp long-leaved lettuce.

cos[2] /kɒz/ *abbreviation* cosine.

cosh *colloquial* ● *noun* heavy blunt weapon. ● *verb* hit with cosh.

cosine /'kəʊsaɪn/ *noun* ratio of side adjacent to acute angle (in right-angled triangle) to hypotenuse.

cosmetic /kɒz'metɪk/ ● *adjective* beautifying, enhancing; superficially improving; (of surgery etc.) restoring or enhancing normal appearance. ● *noun* cosmetic preparation. □ **cosmetically** *adverb*.

cosmic /'kɒzmɪk/ *adjective* of the cosmos; of or for space travel. □ **cosmic rays** high-energy radiations from outer space.

cosmogony /kɒz'mɒgənɪ/ *noun* (*plural* **-ies**) (theory about) origin of universe.

cosmology /kɒz'mɒlədʒɪ/ *noun* science or theory of universe. □ **cosmological** /-mə'lɒdʒ-/ *adjective*; **cosmologist** *noun*.

cosmonaut /'kɒzmənɔ:t/ *noun* Russian astronaut.

cosmopolitan /kɒzmə'pɒlɪt(ə)n/ ● *adjective* of or knowing all parts of world; free from national limitations. ● *noun* cosmopolitan person. □ **cosmopolitanism** *noun*.

cosmos /'kɒzmɒs/ *noun* universe as a well-ordered whole.

Cossack /'kɒsæk/ *noun* member of S. Russian people famous as horsemen.

cosset /'kɒsɪt/ *verb* (**-t-**) pamper.

cost ● *verb* (*past & past participle* **cost**) have as price; involve as loss or sacrifice; (*past & past participle* **costed**) fix or estimate cost of. ● *noun* price; loss, sacrifice; (in *plural*) legal expenses. □ **cost-effective** effective in relation to cost; **cost of living** cost of basic necessities of life; **cost price** price paid for thing by person who later sells it.

costal /'kɒst(ə)l/ *adjective* of ribs.

costermonger /'kɒstəmʌŋgə/ *noun* person who sells fruit etc. from barrow.

costive /'kɒstɪv/ *adjective* constipated.

costly *adjective* (**-ier**, **-iest**) costing much, expensive. □ **costliness** *noun*.

costume /'kɒstju:m/ ● *noun* style of dress, esp. of particular place or period; set of clothes; clothing for particular activity; actor's clothes for part. ● *verb* provide with costume. □ **costume jewellery** artificial jewellery.

costumier /kɒs'tju:mɪə/ *noun* person who deals in or makes costumes.

cosy /'kəʊzɪ/ (*US* **cozy**) ● *adjective* (**-ier**, **-iest**) snug, comfortable. ● *noun* (*plural* **-ies**) cover to keep teapot etc. hot. □ **cosily** *adverb*; **cosiness** *noun*.

cot *noun* small high-sided bed for child etc.; small light bed. □ **cot death** unexplained death of sleeping baby.

cote *noun* shelter for birds or animals.

coterie /'kəʊtərɪ/ *noun* exclusive group of people sharing interests.

cotoneaster /kətəʊnɪ'æstə/ *noun* shrub bearing red or orange berries.

cottage /'kɒtɪdʒ/ *noun* small house, esp. in the country. □ **cottage cheese** soft white lumpy cheese; **cottage industry** small business carried on at home; **cottage pie** shepherd's pie.

cottager *noun* person who lives in cottage.

cotter /'kɒtə/ *noun* (also **cotter pin**) wedge or pin for securing machine part such as bicycle pedal crank.

cotton /'kɒt(ə)n/ *noun* soft white fibrous substance covering seeds of certain plants; such a plant; thread or cloth from this. □ **cotton on** (often + *to*) *colloquial* begin to understand; **cotton wool** wadding originally made from raw cotton.

cotyledon /kɒtɪ'li:d(ə)n/ *noun* first leaf produced by plant embryo.

couch[1] /kaʊtʃ/ ● *noun* upholstered piece of furniture for several people; sofa. ● *verb* (+ *in*) express in (language of specified kind). □ **couch potato** *US slang* person who likes lazing at home.

couch[2] /ku:tʃ/ *noun* (in full **couch grass**) kind of grass with long creeping roots.

couchette /ku:'ʃet/ *noun* railway carriage with seats convertible into sleeping berths; berth in this.

cougar /'ku:gə/ *noun US* puma.

cough /kɒf/ ● *verb* expel air from lungs with sudden sharp sound. ● *noun* (sound of) coughing; condition of respiratory organs causing coughing. □ **cough mixture** medicine to relieve cough; **cough up** eject with coughs, *slang* give (money, information, etc.) reluctantly.

could *past* of CAN[1].

couldn't /'kʊd(ə)nt/ could not.

coulomb /'ku:lɒm/ *noun* SI unit of electrical charge.

coulter /'kəʊltə/ *noun* (*US* **colter**) vertical blade in front of ploughshare.

council /'kaʊns(ə)l/ *noun* (meeting of) advisory, deliberative, or administrative body; local administrative body of county, city, town, etc. □ **council flat, house** one owned and let by local council; **council tax** local tax based on value of property and number of residents.

councillor *noun* member of (esp. local) council.

counsel /'kaʊns(ə)l/ ● *noun* advice, esp. formal; consultation; (*plural* same) legal adviser, esp. barrister; group of these. ● *verb* (**-ll-**; *US* **-l-**) advise, esp. on personal problems. □ **counsellor** (*US* **counselor**) *noun*.

count[1] ● *verb* find number of, esp. by assigning successive numerals; repeat numbers in order; (+ *in*) include or be included in reckoning; consider to be. ● *noun* counting, reckoning; total; *Law* each charge in an indictment. □ **countdown** counting numbers backwards to zero, esp. before launching rocket etc.; **count on** rely on; **count out** exclude, disregard.

count[2] *noun* foreign noble corresponding to earl.

countenance /'kaʊntɪnəns/ ● *noun* face or its expression; composure; moral support. ● *verb* (**-cing**) support, approve.

counter[1] *noun* flat-topped fitment in shop etc. across which business is conducted; small disc used for playing or scoring in board games, cards, etc.; device for counting things.

counter[2] ● *verb* oppose, contradict; meet by countermove. ● *adverb* in opposite direction. ● *adjective* opposite. ● *noun* parry, countermove.

counteract /kaʊntə'rækt/ *verb* neutralize or hinder by contrary action. □ **counteraction** *noun*.

counter-attack *verb & noun* attack in reply to enemy's attack.

counterbalance ● *noun* weight or influence balancing another. ● *verb* (**-cing**) act as counterbalance to.

counter-clockwise *adverb & adjective US* anticlockwise.

counter-espionage *noun* action taken against enemy spying.

counterfeit /'kaʊntəfɪt/ ● *adjective* imitation; forged; not genuine. ● *noun* forgery, imitation. ● *verb* imitate fraudulently; forge.

counterfoil *noun* part of cheque, receipt, etc. retained as record.

counter-intelligence *noun* counter-espionage.

countermand /kaʊntə'mɑ:nd/ *verb* revoke, recall by contrary order.

countermeasure *noun* action taken to counteract danger, threat, etc.

countermove *noun* move or action in opposition to another.

counterpane *noun* bedspread.

counterpart *noun* person or thing equivalent or complementary to another; duplicate.

counterpoint *noun* harmonious combination of melodies in music; melody combined with another; contrasting argument, plot, literary theme, etc.

counterpoise ● *noun* counterbalance; state of equilibrium. ● *verb* (**-sing**) counterbalance.

counter-productive *adjective* having opposite of desired effect.

counter-revolution *noun* revolution opposing former one or reversing its results.

countersign ● *verb* add confirming signature to. ● *noun* password spoken to person on guard.

countersink *verb* (*past & past participle* **-sunk**) shape (screw-hole) so that screw-head lies level with surface; provide (screw) with countersunk hole.

counter-tenor *noun* male alto singing voice; singer with this.

countervailing /'kaʊntəveɪlɪŋ/ *adjective* (of influence etc.) counterbalancing.

counterweight *noun* counterbalancing weight.

countess /'kaʊntɪs/ *noun* earl's or count's wife or widow; woman with rank of earl or count.

countless *adjective* too many to count.

countrified /'kʌntrɪfaɪd/ *adjective* rustic.

country /'kʌntrɪ/ *noun* (*plural* **-ies**) nation's territory, state; land of person's birth or citizenship; rural districts, as opposed to towns; region with regard to its aspect, associations, etc.; national population, esp. as voters. □ **country and western** type of folk music originated by southern US whites; **country dance** traditional dance; **countryman, countrywoman** person of one's own country or district; person living in rural area; **countryside** rural areas.

county /'kaʊntɪ/ *noun* (*plural* **-ies**) territorial division of country, forming chief unit of local administration; *US* political and administrative division of State; people, esp. gentry, of county. □ **county council** elected governing body of county; **county court** law court for civil cases; **county town** administrative capital of county.

coup /ku:/ *noun* (*plural* **-s** /ku:z/) successful stroke or move; *coup d'état*.

coup de grâce /ku: də 'grɑ:s/ *noun* finishing stroke. [French]

coup d'état /ku: deɪ'tɑ:/ *noun* (*plural* ***coups d'état*** same pronunciation) sudden overthrow of government, esp. by force. [French]

coupé /'ku:peɪ/ *noun* (*US* **coupe** /ku:p/) two-door car with hard roof and sloping back.

couple /'kʌp(ə)l/ ● *noun* (about) two; two people who are married or in a sexual relationship; pair of partners in a dance etc. ● *verb* (**-ling**) link, fasten, or associate together; copulate.

couplet /'kʌplɪt/ *noun* two successive lines of rhyming verse.

coupling /'kʌplɪŋ/ *noun* link connecting railway carriages or parts of machinery.

coupon /'ku:pɒn/ *noun* ticket or form entitling holder to something.

courage /'kʌrɪdʒ/ *noun* ability to disregard fear; bravery. □ **courageous** /kə'reɪdʒəs/ *adjective*; **courageously** *adverb*.

courgette /kʊə'ʒet/ *noun* small vegetable marrow.

courier /'kʊrɪə/ *noun* person employed to guide and assist group of tourists; special messenger.

course /kɔ:s/ ● *noun* onward movement or progression; direction taken; line of conduct; series of lectures, lessons, etc.; each successive part of meal; golf course, race-course, etc.; sequence of medical treatment etc.; continuous line of masonry or bricks at one level of building; channel in which water flows. ● *verb* (**-sing**) use hounds to hunt (esp. hares); move or flow freely. □ **in the course of** during; **of course** naturally, as expected, admittedly.

court /kɔ:t/ ● *noun* number of houses enclosing a yard; courtyard; rectangular area for a game; (in full **court of law**) judicial body hearing legal cases; courtroom; sovereign's establishment and retinue. ● *verb* pay amorous attention to, seek to win favour of; try to win (fame etc.); unwisely invite. □ **court card** playing card that is king, queen, or jack; **court house** building in which judicial court is held, *US* county administrative offices; **court martial** (*plural* **courts martial**) judicial court of military officers; **court-martial** (**-ll-**; *US* **-l-**) try by court martial; **courtroom** room in which court of law sits; **court shoe** woman's light shoe with low-cut upper; **courtyard** space enclosed by walls or buildings.

courteous /'kɜ:tɪəs/ *adjective* polite, considerate. □ **courteously** *adverb*; **courteousness** *noun*.

courtesan /kɔ:tɪ'zæn/ *noun* prostitute, esp. one with wealthy or upper-class clients.

courtesy /'kɜ:təsɪ/ *noun* (*plural* **-ies**) courteous behaviour or act. □ **by courtesy of** with formal permission of; **courtesy light** light in car switched on when door is opened.

courtier /'kɔ:tɪə/ *noun* person who attends sovereign's court.

courtly *adjective* (**-ier**, **-iest**) dignified; refined in manners. □ **courtliness** *noun*.

courtship *noun* courting, wooing.

couscous /ˈku:sku:s/ *noun* N. African dish of cracked wheat steamed over broth.

cousin /ˈkʌz(ə)n/ *noun* (also **first cousin**) child of one's uncle or aunt.

couture /ku:ˈtjʊə/ *noun* design and making of fashionable garments.

couturier /ku:ˈtjʊərɪeɪ/ *noun* fashion designer.

cove /kəʊv/ *noun* small bay or inlet; sheltered recess.

coven /ˈkʌv(ə)n/ *noun* assembly of witches.

covenant /ˈkʌvənənt/ ● *noun* agreement; *Law* sealed contract. ● *verb* agree, esp. by legal covenant.

Coventry /ˈkɒvəntrɪ/ *noun* □ **send to Coventry** refuse to associate with or speak to.

cover /ˈkʌvə/ ● *verb* (often + *with*) protect or conceal with cloth, lid, etc.; extend over; protect, clothe; include; (of sum) be large enough to meet (expense); protect by insurance; report on for newspaper, television, etc.; travel (specified distance); aim gun etc. at; protect by aiming gun. ● *noun* thing that covers, esp. lid, wrapper, etc.; shelter, protection; funds to meet liability or contingent loss; place-setting at table. □ **cover charge** service charge per person in restaurant; **cover note** temporary certificate of insurance; **cover up** *verb* cover completely, conceal; **cover-up** *noun* concealing of facts; **take cover** find shelter.

coverage *noun* area or amount covered; reporting of events in newspaper etc.

covering letter *noun* explanatory letter with other documents.

coverlet /ˈkʌvəlɪt/ *noun* bedspread.

covert /ˈkʌvət/ ● *adjective* secret, disguised. ● *noun* shelter, esp. thicket hiding game. □ **covertly** *adverb*.

covet /ˈkʌvɪt/ *verb* (**-t-**) desire greatly (esp. thing belonging to another person). □ **covetous** *adjective*.

covey /ˈkʌvɪ/ *noun* (*plural* **-s**) brood of partridges; family, set.

coving /ˈkəʊvɪŋ/ *noun* curved surface at junction of wall and ceiling.

cow[1] /kaʊ/ *noun* fully-grown female of esp. domestic bovine animal; female of elephant, rhinoceros, whale, seal, etc. □ **cowboy, cowgirl** person who tends cattle, esp. in western US, *colloquial* unscrupulous or incompetent business person; **cowherd** person who tends cattle; **cowhide** (leather made from) cow's hide; **cow-pat** round flat piece of cow-dung; **cowpox** disease of cows, source of smallpox vaccine.

cow[2] /kaʊ/ *verb* intimidate.

coward /ˈkaʊəd/ *noun* person easily frightened. □ **cowardly** *adjective*.

cowardice /ˈkaʊədɪs/ *noun* lack of bravery.

cower /ˈkaʊə/ *verb* crouch or shrink back in fear.

cowl /kaʊl/ *noun* (hood of) monk's cloak; (also **cowling**) hood-shaped covering of chimney or shaft. □ **cowl neck** wide loose roll neck on garment.

cowrie /ˈkaʊrɪ/ *noun* tropical mollusc with bright shell; its shell as money in parts of Asia etc.

cowslip /ˈkaʊslɪp/ *noun* yellow-flowered primula growing in pastures etc.

cox ● *noun* coxswain. ● *verb* act as cox (of).

coxcomb /ˈkɒkskəʊm/ *noun* conceited showy person.

coxswain /ˈkɒks(ə)n/ *noun* person who steers esp. rowing boat.

coy *adjective* affectedly shy; irritatingly reticent. □ **coyly** *adverb*.

coyote /kɔɪˈəʊtɪ/ *noun* (*plural* same or **-s**) N. American wild dog.

coypu /ˈkɔɪpu:/ *noun* (*plural* **-s**) amphibious rodent like small beaver, originally from S. America.

cozen /ˈkʌz(ə)n/ *verb literary* cheat, defraud; beguile.

cozy *US* = COSY.

crab ● *noun* shellfish with 10 legs; this as food; (in full **crab-louse**) (often in *plural*) parasite infesting human body. ● *verb* (**-bb-**) *colloquial* criticize, grumble; spoil. □ **catch a crab** (in rowing) get oar jammed under water, miss water; **crab-apple** wild apple tree, its sour fruit.

crabbed /kræbɪd/ *adjective* crabby; (of handwriting) ill-formed, illegible.

crabby *adjective* (**-ier**, **-iest**) morose, irritable. □ **crabbily** *adverb*.

crack • *noun* sudden sharp noise; sharp blow; narrow opening; break or split; *colloquial* joke, malicious remark; sudden change in vocal pitch; *slang* crystalline cocaine. • *verb* (cause to) make crack; suffer crack or partial break; (of voice) change pitch sharply; tell (joke); open (bottle of wine etc.); break into (safe); find solution to (problem); give way, yield; hit sharply; (as **cracked** *adjective*) crazy, (of wheat) coarsely broken. □ **crack-brained** crazy; **crack-down** *colloquial* severe measures (esp. against law-breakers); **crack down on** *colloquial* take severe measures against; **crack of dawn** daybreak; **crackpot** *colloquial* eccentric or impractical person; **crack up** *colloquial* collapse under strain.

cracker *noun* small paper cylinder containing paper hat, joke, etc., exploding with crack when ends are pulled; explosive firework; thin crisp savoury biscuit.

crackers *adjective slang* crazy.

crackle /'kræk(ə)l/ • *verb* (**-ling**) make repeated light cracking sound. • *noun* such a sound. □ **crackly** *adjective*.

crackling *noun* crisp skin of roast pork.

cracknel /'krækn(ə)l/ *noun* light crisp biscuit.

cradle /'kreɪd(ə)l/ • *noun* baby's bed, esp. on rockers; place regarded as origin of something; supporting framework or structure. • *verb* (**-ling**) contain or shelter as in cradle.

craft /krɑːft/ *noun* special skill or technique; occupation needing this; (*plural* same) boat, vessel, aircraft, or spacecraft; cunning.

craftsman /'krɑːftsmən/ *noun* (*feminine* **-woman**) person who practises a craft. □ **craftsmanship** *noun*.

crafty *adjective* (**-ier**, **-iest**) cunning, artful. □ **craftily** *adverb*.

crag *noun* steep rugged rock.

craggy *adjective* (**-ier**, **-iest**) rugged; rough-textured.

cram *verb* (**-mm-**) fill to bursting; (often + *in*, *into*) force; study or teach intensively for exam.

crammer *noun* institution that crams pupils for exam.

cramp • *noun* painful involuntary contraction of muscles. • *verb* affect with cramp; restrict, confine.

cramped *adjective* (of space) small; (of handwriting) small and with the letters close together.

crampon /'kræmpɒn/ *noun* spiked iron plate fixed to boot for climbing on ice.

cranberry /'krænbərɪ/ *noun* (*plural* **-ies**) (shrub bearing) small red acid berry.

crane • *noun* machine with projecting arm for moving heavy weights; large long-legged wading bird. • *verb* (**-ning**) stretch (one's neck) in order to see something. □ **crane-fly** two-winged long-legged fly; **cranesbill** kind of wild geranium.

cranium /'kreɪnɪəm/ *noun* (*plural* **-s** or **-nia**) bones enclosing brain, skull. □ **cranial** *adjective*.

crank • *noun* part of axle or shaft bent at right angles for converting rotary into reciprocal motion or vice versa; eccentric person. • *verb* turn with crank. □ **crankpin** pin attaching connecting rod to crank; **crankshaft** shaft driven by crank; **crank up** start (engine) by turning crank.

cranky *adjective* (**-ier**, **-iest**) eccentric; shaky; *esp. US* crotchety.

cranny /'krænɪ/ *noun* (*plural* **-ies**) chink, crevice.

crape *noun* crêpe, usually of black silk, formerly used for mourning.

craps *plural noun* (also **crap game**) *US* gambling game played with dice.

crapulent /'kræpjʊlənt/ *adjective* suffering the effects of drunkenness. □ **crapulence** *noun*; **crapulous** *adjective*.

crash • *verb* (cause to) make loud smashing noise; (often + *into*) (cause to) collide or fall violently; fail, esp. financially; *colloquial* gatecrash; (of computer, system, etc.) fail suddenly; (often + *out*) *slang* fall asleep, sleep. • *noun* sudden violent noise; violent fall or impact, esp. of vehicle; ruin, esp. financial; sudden collapse, esp. of computer, system, etc. • *adjective* done rapidly or urgently. □ **crash barrier** barrier to prevent car leaving road; **crash-dive** *verb* (of submarine) dive hastily and steeply, (of aircraft) dive and crash, *noun* such a dive; **crash-helmet** helmet worn to protect head; **crash-land** land or cause (aircraft etc.) to land with crash; **crash landing** instance of crash-landing.

crass *adjective* grossly stupid; insensitive. □ **crassly** *adverb*; **crassness** *noun*.

crate ● *noun* slatted wooden case; *slang* old aircraft, car, etc. ● *verb* (**-ting**) pack in crate.

crater /'kreɪtə/ *noun* mouth of volcano; bowl-shaped cavity, esp. hollow on surface of moon etc.

cravat /krə'væt/ *noun* man's scarf worn inside open-necked shirt.

crave *verb* (**-ving**) (often + *for*) long or beg for.

craven /'kreɪv(ə)n/ *adjective* cowardly, abject.

craving *noun* strong desire, longing.

craw *noun* crop of bird or insect.

crawfish *noun* (*plural* same) large spiny sea-lobster.

crawl ● *verb* move slowly, esp. on hands and knees or with body close to ground; *colloquial* behave obsequiously; (often + *with*) be filled with moving people or things; (esp. of skin) creep. ● *noun* crawling motion; slow rate of motion; fast swimming stroke.

crayfish /'kreɪfɪʃ/ *noun* (*plural* same) lobster-like freshwater crustacean; crawfish.

crayon /'kreɪən/ ● *noun* stick or pencil of coloured wax, chalk, etc. ● *verb* draw or colour with crayons.

craze ● *verb* (**-zing**) (usually as **crazed** *adjective*) make insane; produce fine surface cracks on, develop such cracks. ● *noun* usually temporary enthusiasm; object of this.

crazy *adjective* (**-ier**, **-iest**) insane, mad; foolish; (usually + *about*) *colloquial* extremely enthusiastic. □ **crazy paving** paving made of irregular pieces. □ **crazily** *adverb*.

creak ● *noun* harsh scraping or squeaking sound. ● *verb* emit creak; move stiffly. □ **creaky** *adjective* (**-ier**, **-iest**).

cream ● *noun* fatty part of milk; its yellowish-white colour; food or drink like or made with cream; creamlike cosmetic etc.; (usually + *the*) best part or pick of something. ● *verb* take cream from; make creamy; form cream or scum. ● *adjective* yellowish-white. □ **cream cheese** soft rich cheese made of cream and unskimmed milk; **cream cracker** crisp unsweetened biscuit; **cream off** remove best part of; **cream of tartar** purified tartar, used in medicine, baking powder, etc.; **cream tea** afternoon tea with scones, jam, and cream. □ **creamy** *adjective* (**-ier**, **-iest**).

creamer *noun* cream substitute for coffee; jug for cream.

creamery *noun* (*plural* **-ies**) factory producing dairy products; dairy.

crease /kri:s/ ● *noun* line made by folding or crushing; *Cricket* line defining position of bowler or batsman. ● *verb* (**-sing**) make creases in, develop creases.

create /kri:'eɪt/ *verb* (**-ting**) bring into existence; originate; invest with rank; *slang* make fuss.

creation /kri:'eɪʃ(ə)n/ *noun* creating, being created; (usually **the Creation**) God's creating of the universe; (usually **Creation**) all created things; product of imaginative work.

creative /kri:'eɪtɪv/ *adjective* inventive, imaginative. □ **creatively** *adverb*; **creativity** /-'tɪv-/ *noun*.

creator /kri:'eɪtə/ *noun* person who creates; (**the Creator**) God.

creature /'kri:tʃə/ *noun* living being, esp. animal; person, esp. one in subservient position; anything created.

crèche /kreʃ/ *noun* day nursery.

credence /'kri:d(ə)ns/ *noun* belief.

credentials /krɪ'denʃ(ə)lz/ *plural noun* documents attesting to person's education, character, etc.

credible /'kredɪb(ə)l/ *adjective* believable; worthy of belief. □ **credibility** *noun*.

■ **Usage** *Credible* is sometimes confused with *credulous*, which means 'gullible'.

---

credit /'kredɪt/ ● *noun* belief, trust; good reputation; person's financial standing; power to obtain goods before payment; acknowledgement of payment by entry in account, sum entered; acknowledgement of merit or (usually in *plural*) of contributor's services to film, book, etc.; grade above pass in exam; educational course counting towards degree. ● *verb* (**-t-**) believe; (usually + *to*, *with*) enter on credit side of account. □ **credit card** card authorizing purchase of goods on credit; **credit (person) with** ascribe to him or her; **credit rating** estimate of person's suitability for commercial credit; **creditworthy** suitable to receive credit; **on credit** with arrangement to pay later; **to one's credit** in one's favour.

creditable *adjective* praiseworthy. □ **creditably** *adverb.*

creditor *noun* person to whom debt is owing.

credo /'kri:dəʊ/ *noun* (*plural* **-s**) creed.

credulous /'kredjʊləs/ *adjective* too ready to believe; gullible. □ **credulity** /krɪ'dju:-/ *noun.*

■ **Usage** *Credulous* is sometimes confused with *credible*, which means 'believable'.

---

creed *noun* set of beliefs; system of beliefs; (often **the Creed**) formal summary of Christian doctrine.

creek *noun* inlet on sea-coast; arm of river; stream.

creel *noun* fisherman's wicker basket.

creep ● *verb* (*past & past participle* **crept**) crawl; move stealthily, timidly, or slowly; (of plant) grow along ground or up wall etc.; advance or develop gradually; (of flesh) shudder with horror etc. ● *noun* act of creeping; (**the creeps**) *colloquial* feeling of revulsion or fear; *slang* unpleasant person; gradual change in shape of metal under stress.

creeper *noun* creeping or climbing plant.

creepy *adjective* (**-ier**, **-iest**) *colloquial* feeling or causing horror or fear. □ **creepy-crawly** (*plural* **-crawlies**) small crawling insect etc. □ **creepily** *adverb.*

cremate /krɪ'meɪt/ *verb* (**-ting**) burn (corpse) to ashes. □ **cremation** *noun.*

crematorium /kremə'tɔ:rɪəm/ *noun* (*plural* **-ria** or **-s**) place where corpses are cremated.

crenellated /'krenəleɪtɪd/ *adjective* having battlements. □ **crenellation** *noun.*

Creole /'kri:əʊl/ ● *noun* descendant of European settlers in W. Indies or Central or S. America, or of French settlers in southern US; person of mixed European and black descent; language formed from a European and African language. ● *adjective* of Creoles; (usually **creole**) of Creole origin etc.

creosote /'kri:əsəʊt/ ● *noun* oily wood-preservative distilled from coal tar. ● *verb* (**-ting**) treat with creosote.

crêpe /kreɪp/ *noun* fine crinkled fabric; thin pancake with savoury or sweet filling; wrinkled sheet rubber used for shoe-soles etc. □ **crêpe de Chine** /də 'ʃi:n/ fine silk crêpe; **crêpe paper** thin crinkled paper.

crept *past & past participle* of CREEP.

crepuscular /krɪ'pʌskjʊlə/ *adjective* of twilight; (of animal) active etc. at twilight.

crescendo /krɪ'ʃendəʊ/ *Music* ● *noun* (*plural* **-s**) gradual increase in loudness. ● *adjective & adverb* increasing in loudness.

■ **Usage** *Crescendo* is sometimes wrongly used to mean a climax rather than the progress towards it.

---

crescent /'kres(ə)nt/ ● *noun* sickle shape, as of waxing or waning moon; thing with this shape, esp. curved street. ● *adjective* crescent-shaped.

cress *noun* plant with pungent edible leaves.

crest ● *noun* comb or tuft on animal's head; plume of helmet; top of mountain, wave, etc.; *Heraldry* device above shield or on writing paper etc. ● *verb* reach crest of; crown; serve as crest to; form crest. □ **crestfallen** dejected. □ **crested** *adjective.*

cretaceous /krɪ'teɪʃəs/ *adjective* chalky.

cretin /'kretɪn/ *noun* person with deformity and mental retardation caused by thyroid deficiency; *colloquial* stupid person. □ **cretinism** *noun*; **cretinous** *adjective.*

cretonne /kre'tɒn/ *noun* heavy cotton usually floral upholstery fabric.

crevasse /krə'væs/ *noun* deep open crack in glacier.

crevice /'krevɪs/ *noun* narrow opening or fissure, esp. in rock.

crew[1] /kru:/ • *noun* group of people working together, esp. manning ship, aircraft, spacecraft, etc.; these, other than the officers. • *verb* supply or act as crew (member) for; act as crew. □ **crew cut** close-cropped hairstyle; **crew neck** round close-fitting neckline.

crew[2] *archaic past* of CROW.

crewel /'kru:əl/ *noun* thin worsted yarn for embroidery.

crib • *noun* baby's small bed or cot; model of Nativity with manger; *colloquial* plagiarism; translation; *colloquial* cribbage. • *verb* (**-bb-**) copy unfairly; confine in small space.

cribbage /'krɪbɪdʒ/ *noun* a card game.

crick • *noun* sudden painful stiffness, esp. in neck. • *verb* cause crick in.

cricket[1] /'krɪkɪt/ *noun* team game, played on grass pitch, in which ball is bowled at wicket defended with bat by player of other team. □ **not cricket** *colloquial* unfair behaviour. □ **cricketer** *noun*.

cricket[2] /'krɪkɪt/ *noun* jumping chirping insect.

cried *past & past participle* of CRY.

crier /'kraɪə/ *noun* (also **cryer**) official making public announcements in law court or street.

crikey /'kraɪkɪ/ *interjection slang: expressing astonishment.*

crime *noun* act punishable by law; such acts collectively; evil act; *colloquial* shameful act.

criminal /'krɪmɪn(ə)l/ • *noun* person guilty of crime. • *adjective* of, involving, or concerning crime; *colloquial* deplorable. □ **criminality** /-'næl-/ *noun*; **criminally** *adverb*.

criminology /krɪmɪ'nɒlədʒɪ/ *noun* study of crime. □ **criminologist** *noun*.

crimp *verb* press into small folds or waves; corrugate.

Crimplene /'krɪmpli:n/ *noun proprietary term* synthetic crease-resistant fabric.

crimson /'krɪmz(ə)n/ *adjective & noun* rich deep red.

cringe *verb* (**-ging**) cower; (often + *to*) behave obsequiously.

crinkle /'krɪŋk(ə)l/ • *noun* wrinkle, crease. • *verb* (**-ling**) form crinkles (in). □ **crinkly** *adjective*.

crinoline /'krɪnəlɪn/ *noun* hooped petticoat.

cripple /'krɪp(ə)l/ • *noun* lame person. • *verb* (**-ling**) lame, disable; damage seriously.

crisis /'kraɪsɪs/ *noun* (*plural* **crises** /-si:z/) time of acute danger or difficulty; decisive moment.

crisp • *adjective* hard but brittle; bracing; brisk, decisive; clear-cut; crackling; curly. • *noun* (in full **potato crisp**) very thin fried slice of potato. • *verb* make or become crisp. □ **crispbread** thin crisp biscuit. □ **crisply** *adverb*; **crispness** *noun*.

crispy *adjective* (**-ier**, **-iest**) crisp.

criss-cross • *noun* pattern of crossing lines. • *adjective* crossing; in crossing lines. • *adverb* crosswise. • *verb* intersect repeatedly; mark with criss-cross lines.

criterion /kraɪ'tɪərɪən/ *noun* (*plural* **-ria**) principle or standard of judgement.

■ **Usage** It is a mistake to use the plural form *criteria* when only one criterion is meant.

critic /'krɪtɪk/ *noun* person who criticizes; reviewer of literary, artistic, etc. works.

critical *adjective* fault-finding; expressing criticism; providing textual criticism; of the nature of a crisis, decisive; marking transition from one state to another. □ **critically** *adverb*.

criticism /'krɪtɪsɪz(ə)m/ *noun* finding fault, censure; work of critic; critical article, remark, etc.

criticize /'krɪtɪsaɪz/ *verb* (also **-ise**) (**-zing** or **-sing**) find fault with; discuss critically.

critique /krɪ'ti:k/ *noun* critical analysis.

croak • *noun* deep hoarse sound, esp. of frog. • *verb* utter or speak with croak; *slang* die. □ **croaky** *adjective* (**-ier**, **-iest**); **croakily** *adverb*.

Croat /'krəʊæt/ (also **Croatian** /krəʊ'eɪʃ(ə)n/) • *noun* native of Croatia; person of Croatian descent; Slavonic dialect of Croats. • *adjective* of Croats or their dialect.

crochet /'krəʊʃeɪ/ • *noun* needlework of hooked yarn producing lacy patterned fabric. • *verb* (**crocheted** /-ʃeɪd/; **crocheting** /-ʃeɪɪŋ/) make by crochet.

crock[1] *noun colloquial* old or worn-out person or vehicle.

crock[2] *noun* earthenware jar; broken piece of this.

crockery *noun* earthenware or china dishes, plates, etc.

crocodile /'krɒkədaɪl/ *noun* large amphibious reptile; *colloquial* line of school children etc. walking in pairs. □ **crocodile tears** insincere grief.

crocus /'krəʊkəs/ *noun* (*plural* **-es**) small plant with corm and yellow, purple, or white flowers.

croft ● *noun* small piece of arable land; small rented farm in Scotland or N. England. ● *verb* farm croft.

crofter *noun* person who farms croft.

Crohn's disease /krəʊnz/ *noun* inflammatory disease of alimentary tract.

croissant /'krwʌsɑ̃/ *noun* rich crescent-shaped roll.

cromlech /'krɒmlek/ *noun* dolmen; prehistoric stone circle.

crone *noun* withered old woman.

crony /'krəʊnɪ/ *noun* (*plural* **-ies**) friend; companion.

crook /krʊk/ ● *noun* hooked staff of shepherd or bishop; bend, curve; *colloquial* swindler, criminal. ● *verb* bend, curve.

crooked /'krʊkɪd/ *adjective* (**-er**, **-est**) not straight, bent; *colloquial* dishonest. □ **crookedly** *adverb*; **crookedness** *noun*.

croon /kru:n/ ● *verb* hum or sing in low voice. ● *noun* such singing. □ **crooner** *noun*.

crop ● *noun* produce of any cultivated plant or of land; group or amount produced at one time; handle of whip; very short haircut; pouch in bird's gullet where food is prepared for digestion. ● *verb* (**-pp-**) cut off; bite off, eat down; cut (hair) short; raise crop on (land); bear crop. □ **crop circle** circle of crops inexplicably flattened; **crop up** occur unexpectedly.

cropper *noun slang* □ **come a cropper** fall heavily, fail badly.

croquet /'krəʊkeɪ/ ● *noun* lawn game with hoops, wooden balls, and mallets; croqueting. ● *verb* (**croqueted** /-keɪd/; **croqueting** /-keɪɪŋ/) drive away (opponent's ball) by striking one's own ball placed in contact with it.

croquette /krə'ket/ *noun* fried breaded ball of meat, potato, etc.

crosier /'krəʊzɪə/ *noun* (also **crozier**) bishop's ceremonial hooked staff.

cross ● *noun* upright stake with transverse bar, used in antiquity for crucifixion; representation of this as emblem of Christianity; cross-shaped thing or mark, esp. two short intersecting lines (+ or ×); cross-shaped military etc. decoration; intermixture of breeds, hybrid; (+ *between*) mixture of two things; trial, affliction. ● *verb* (often + *over*) go across; place crosswise; draw line(s) across; make sign of cross on or over; meet and pass; thwart; (cause to) interbreed; cross-fertilize (plants). ● *adjective* (often + *with*) peevish, angry; transverse; reaching from side to side; intersecting; reciprocal. □ **at cross purposes** misunderstanding each other; **crossbar** horizontal bar, esp. between uprights; **cross-bench** bench in House of Lords for non-party members; **crossbones** see SKULL AND CROSSBONES; **crossbow** bow fixed across wooden stock with mechanism working string; **cross-breed** (produce) hybrid animal or plant; **cross-check** check by alternative method; **cross-country** across fields etc., not following roads, *noun* such a race; **cross-examine** question (esp. opposing witness in law court); **cross-examination** such questioning; **cross-eyed** having one or both eyes turned inwards; **cross-fertilize** fertilize (animal or plant) from another of same species; **crossfire** firing of guns in two crossing directions; **cross-grained** (of wood) with grain running irregularly, (of person) perverse or intractable; **cross-hatch** shade with crossing parallel lines; **cross-legged** (sitting) with legs folded across each other; **cross off**, **out** cancel, expunge; **crossover** point or process of crossing; **crosspatch** *colloquial* bad-tempered person; **cross-ply** (of tyre) having crosswise layers of cords; **cross-question** cross-examine, quiz; **cross-reference** reference to another passage in same book; **crossroad** (usually in *plural*) intersection of roads; **cross-section** drawing etc. of thing as if cut through, representative sample; **cross stitch** cross-shaped stitch; **crosswise** intersecting, diagonally; **crossword (puzzle)** puzzle in which words crossing each other vertically and horizontally have to be filled in from clues; **on the cross** diagonally. □ **crossly** *adverb*; **crossness** *noun*.

crossing *noun* place where things (esp. roads) meet; place for crossing street; journey across water.

crotch *noun* fork, esp. between legs (of person, trousers, etc.).

crotchet /'krɒtʃɪt/ *noun Music* black-headed note with stem, equal to quarter of semibreve and usually one beat.

crotchety *adjective* peevish.

crouch • *verb* stand, squat, etc. with legs bent close to body. • *noun* this position.

croup[1] /kru:p/ *noun* laryngitis in children, with sharp cough.

croup[2] /kru:p/ *noun* rump, esp. of horse.

croupier /'kru:pɪə/ *noun* person in charge of gaming table.

croûton /'kru:tɒn/ *noun* small piece of fried or toasted bread served esp. with soup.

crow /krəʊ/ • *noun* any of various kinds of large black-plumaged bird; cry of cock or baby. • *verb* (*past* **crowed** or *archaic* **crew** /kru:/) (of cock) utter loud cry; (of baby) utter happy sounds; exult. □ **crow's-foot** wrinkle at outer corner of eye; **crow's-nest** shelter for look-out man at ship's masthead.

crowbar *noun* iron bar used as lever.

crowd /kraʊd/ • *noun* large gathering of people; *colloquial* particular set of people. • *verb* (cause to) collect in crowd; (+ *with*) cram with; (+ *into*, *through*, etc.) force way into, through, etc.; *colloquial* come aggressively close to. □ **crowd out** exclude by crowding.

crown /kraʊn/ • *noun* monarch's jewelled headdress; (**the Crown**) monarch as head of state, his or her authority; wreath for head as emblem of victory; top part of head, hat, etc.; visible part of tooth, artificial replacement for this; coin worth 5 shillings or 25 pence. • *verb* put crown on; make king or queen; (often as **crowning** *adjective*) be consummation, reward, or finishing touch to; *slang* hit on head. □ **Crown colony** British colony controlled by the Crown; **Crown Court** court of criminal jurisdiction in England and Wales; **crown jewels** sovereign's regalia; **crown prince** male heir to throne; **crown princess** wife of crown prince, female heir to throne.

crozier = CROSIER.

CRT *abbreviation* cathode ray tube.

cruces *plural* of CRUX.

crucial /'kru:ʃ(ə)l/ *adjective* decisive, critical; very important; *slang* excellent. □ **crucially** *adverb*.

crucible /'kru:sɪb(ə)l/ *noun* melting pot for metals.

cruciferous /kru:'sɪfərəs/ *adjective* with 4 equal petals arranged crosswise.

crucifix /'kru:sɪfɪks/ *noun* image of Christ on Cross.

crucifixion /kru:sɪ'fɪkʃ(ə)n/ *noun* crucifying, esp. of Christ.

cruciform /'kru:sɪfɔ:m/ *adjective* cross-shaped.

crucify /'kru:sɪfaɪ/ *verb* (**-ies**, **-ied**) put to death by fastening to cross; persecute, torment.

crude • *adjective* in natural or raw state; lacking finish, unpolished; rude, blunt; indecent. • *noun* natural mineral oil. □ **crudely** *adverb*; **crudeness** *noun*; **crudity** *noun*.

crudités /kru:dɪ'teɪ/ *plural noun* hors d'oeuvre of mixed raw vegetables. [French]

cruel /'kru:əl/ *adjective* (**-ll-** or **-l-**) causing pain or suffering, esp. deliberately; harsh, severe. □ **cruelly** *adverb*; **cruelty** *noun* (*plural* **-ies**).

cruet /'kru:ɪt/ *noun* set of small salt, pepper, etc. containers for use at table.

cruise /kru:z/ • *verb* (**-sing**) sail about, esp. travel by sea for pleasure, calling at ports; travel at relaxed or economical speed; achieve objective with ease. • *noun* cruising voyage. □ **cruise missile** one able to fly low and guide itself.

cruiser *noun* high-speed warship; cabin cruiser. □ **cruiserweight** light heavyweight.

crumb /krʌm/ • *noun* small fragment esp. of bread; soft inner part of loaf;

(**crumbs** *interjection*) *slang: expressing dismay.* ● *verb* coat with breadcrumbs; crumble (bread). □ **crumby** *adjective.*

crumble /'krʌmb(ə)l/ ● *verb* (**-ling**) break or fall into fragments, disintegrate. ● *noun* dish of cooked fruit with crumbly topping. □ **crumbly** *adjective* (**-ier**, **-iest**).

crumhorn = KRUMMHORN.

crummy /'krʌmɪ/ *adjective* (**-ier**, **-iest**) *slang* squalid, inferior.

crumpet /'krʌmpɪt/ *noun* flat soft yeasty cake eaten toasted.

crumple /'krʌmp(ə)l/ *verb* (**-ling**) (often + *up*) crush or become crushed into creases; give way, collapse.

crunch ● *verb* crush noisily with teeth; make or emit crunch. ● *noun* crunching sound; *colloquial* decisive event. □ **crunchy** *adjective* (**-ier**, **-iest**).

crupper /'krʌpə/ *noun* strap looped under horse's tail to hold harness back.

crusade /kru:'seɪd/ ● *noun historical* medieval Christian military expedition to recover Holy Land from Muslims; vigorous campaign for cause. ● *verb* (**-ding**) take part in crusade. □ **crusader** *noun.*

cruse /kru:z/ *noun archaic* earthenware jar.

crush ● *verb* compress violently so as to break, bruise, etc.; crease, crumple; defeat or subdue completely. ● *noun* act of crushing; crowded mass of people; drink made from juice of crushed fruit; (usually + *on*) *colloquial* infatuation.

crust ● *noun* hard outer part of bread etc.; pastry covering pie; rocky outer part of the earth; deposit, esp. on sides of wine bottle. ● *verb* cover with, form into, or become covered with crust.

crustacean /krʌ'steɪʃ(ə)n/ ● *noun* hard-shelled usually aquatic animals, e.g. crab or lobster. ● *adjective* of crustaceans.

crusty *adjective* (**-ier**, **-iest**) having a crisp crust; irritable, curt.

crutch *noun* support for lame person, usually with cross-piece fitting under armpit; support; crotch.

crux *noun* (*plural* **-es** or **cruces** /'kru:si:z/) decisive point at issue.

cry /kraɪ/ ● *verb* (**cries**, **cried**) (often + *out*) make loud or shrill sound, esp. to express pain, grief, joy, etc.; weep; (often + *out*) utter loudly, exclaim; (+ *for*) appeal for. ● *noun* (*plural* **cries**) loud shout of grief, fear, joy, etc.; loud excited utterance; urgent appeal; fit of weeping; call of animal. □ **cry-baby** person who weeps frequently; **cry down** disparage; **cry off** withdraw from undertaking; **cry out for** need badly; **cry wolf** see WOLF; **a far cry** a long way.

cryer = CRIER.

crying *adjective* (of injustice etc.) flagrant, demanding redress.

cryogenics /kraɪəʊ'dʒenɪks/ *noun* branch of physics dealing with very low temperatures. □ **cryogenic** *adjective.*

crypt /krɪpt/ *noun* vault, esp. below church, usually used as burial place.

cryptic *adjective* obscure in meaning; secret, mysterious. □ **cryptically** *adverb.*

cryptogam /'krɪptəgæm/ *noun* plant with no true flowers or seeds, e.g. fern or fungus. □ **cryptogamic** /-'gæm-/ *adjective.*

cryptogram /'krɪptəgræm/ *noun* text written in cipher.

crystal /'krɪst(ə)l/ ● *noun* (piece of) transparent colourless mineral; (articles of) highly transparent glass; substance solidified in definite geometrical form. ● *adjective* of or as clear as crystal. □ **crystal ball** glass globe supposedly used in foretelling the future.

crystalline /'krɪstəlaɪn/ *adjective* of or as clear as crystal.

crystallize /'krɪstəlaɪz/ *verb* (also **-ise**) (**-zing** or **-sing**) form into crystals; make or become definite; preserve or be preserved in sugar. □ **crystallization** *noun.*

CS gas *noun* tear gas used to control riots.

cu. *abbreviation* cubic.

cub ● *noun* young of fox, bear, lion, etc.; **Cub** (**Scout**) junior Scout; *colloquial* young newspaper reporter. ● *verb* give birth to (cubs).

cubby-hole /'kʌbɪhəʊl/ *noun* very small room; snug space.

cube /kju:b/ ● *noun* solid contained by 6 equal squares; cube-shaped block; product of a number multiplied by its square. ● *verb* (**-bing**) find cube of; cut into small cubes. □ **cube root** number which produces given number when cubed.

cubic /ˈkju:bɪk/ *adjective* of 3 dimensions; involving cube of a quantity. □ **cubic metre** etc., volume of cube whose edge is one metre etc. □ **cubical** *adjective*.

cubicle /ˈkju:bɪk(ə)l/ *noun* small screened space, esp. sleeping compartment.

cubism /ˈkju:bɪz(ə)m/ *noun* art style in which objects are represented geometrically. □ **cubist** *adjective & noun*.

cubit /ˈkju:bɪt/ *noun* ancient measure of length, approximately length of forearm.

cuboid /ˈkju:bɔɪd/ ● *adjective* like a cube; cube-shaped. ● *noun* solid with 6 rectangular faces.

cuckold /ˈkʌkəʊld/ ● *noun* husband of adulteress. ● *verb* make cuckold of.

cuckoo /ˈkʊku:/ ● *noun* bird with characteristic cry and laying eggs in nests of small birds. ● *adjective slang* crazy. □ **cuckoo clock** clock with figure of cuckoo emerging to make call on the hour; **cuckoo-pint** wild arum; **cuckoo-spit** froth exuded by larvae of certain insects.

cucumber /ˈkju:kʌmbə/ *noun* long green fleshy fruit used in salads.

cud *noun* half-digested food chewed by ruminant.

cuddle /ˈkʌd(ə)l/ ● *verb* (**-ling**) hug; lie close and snug; nestle. ● *noun* prolonged hug.

cuddly *adjective* (**-ier**, **-iest**) soft and yielding.

cudgel /ˈkʌdʒ(ə)l/ ● *noun* thick stick used as weapon. ● *verb* (**-ll-**; *US* **-l-**) beat with cudgel.

cue[1] /kju:/ ● *noun* last words of actor's speech as signal for another to begin; signal, hint. ● *verb* (**cues**, **cued**, **cueing** or **cuing**) give cue to. □ **cue in** insert cue for; **on cue** at correct moment.

cue[2] /kju:/ ● *noun* long rod for striking ball in billiards etc. ● *verb* (**cues**, **cued**, **cueing** or **cuing**) strike with cue. □ **cue ball** ball to be struck with cue.

cuff[1] *noun* end part of sleeve; trouser turn-up; (in *plural*) *colloquial* handcuffs. □ **cuff link** either of pair of fasteners for shirt cuffs; **off the cuff** extempore, without preparation.

cuff[2] ● *verb* strike with open hand. ● *noun* such a blow.

cuisine /kwɪˈzi:n/ *noun* style of cooking.

cul-de-sac /ˈkʌldəsæk/ *noun* (*plural* **culs-de-sac** same pronunciation, or **cul-de-sacs** /-sæks/) road etc. closed at one end.

culinary /ˈkʌlɪnərɪ/ *adjective* of or for cooking.

cull ● *verb* select, gather; pick (flowers); select and kill (surplus animals). ● *noun* culling; animal(s) culled.

culminate /ˈkʌlmɪneɪt/ *verb* (**-ting**) (usually + *in*) reach highest or final point. □ **culmination** *noun*.

culottes /kju:ˈlɒts/ *plural noun* woman's trousers cut like skirt.

culpable /ˈkʌlpəb(ə)l/ *adjective* deserving blame. □ **culpability** *noun*.

culprit /ˈkʌlprɪt/ *noun* guilty person.

cult *noun* religious system, sect, etc.; devotion or homage to person or thing.

cultivar /ˈkʌltɪvɑ:/ *noun* plant variety produced by cultivation.

cultivate /ˈkʌltɪveɪt/ *verb* (**-ting**) prepare and use (soil) for crops; raise (plant etc.); (often as **cultivated** *adjective*) improve (manners etc.); nurture (friendship etc.). □ **cultivation** *noun*.

cultivator *noun* agricultural implement for breaking up ground etc.

culture /ˈkʌltʃə/ ● *noun* intellectual and artistic achievement or expression; refined appreciation of arts etc.; customs and civilization of a particular time or people; improvement by mental or physical training; cultivation of plants, rearing of bees etc.; quantity of bacteria grown for study. ● *verb* (**-ring**) grow (bacteria) for study. □ **culture shock** disorientation felt by person subjected to unfamiliar way of life. □ **cultural** *adjective*.

cultured *adjective* having refined tastes etc. □ **cultured pearl** one formed by oyster after insertion of foreign body into its shell.

culvert /ˈkʌlvət/ *noun* underground channel carrying water under road etc.

cumbersome /ˈkʌmbəsəm/ *adjective* (also **cumbrous** /ˈkʌmbrəs/) inconveniently bulky, unwieldy.

cumin /'kʌmɪn/ *noun* (also **cummin**) plant with aromatic seeds; these as flavouring.

cummerbund /'kʌməbʌnd/ *noun* waist sash.

cumulative /'kju:mjʊlətɪv/ *adjective* increasing in force etc. by successive additions. □ **cumulatively** *adverb.*

cumulus /'kju:mjʊləs/ *noun* (*plural* **-li** /-laɪ/) cloud formation of heaped-up rounded masses.

cuneiform /'kju:nɪfɔ:m/ ● *noun* writing made up of wedge shapes. ● *adjective* of or using cuneiform.

cunning /'kʌnɪŋ/ ● *adjective* (**-er**, **-est**) deceitful, crafty; ingenious. ● *noun* ingenuity; craft. □ **cunningly** *adverb.*

cup ● *noun* small bowl-shaped drinking vessel; cupful; cup-shaped thing; flavoured usually chilled wine, cider, etc.; cup-shaped trophy as prize. ● *verb* (**-pp-**) make cup-shaped; hold as in cup. □ **Cup Final** final match in (esp. football) competition; **cup-tie** match in such competition. □ **cupful** *noun.*

■ **Usage** A *cupful* is a measure, and so *three cupfuls* is a quantity of something; *three cups full* means the actual cups and their contents, as in *He brought us three cups full of water*.

---

cupboard /'kʌbəd/ *noun* recess or piece of furniture with door and usually shelves.

Cupid /'kju:pɪd/ *noun* Roman god of love, pictured as winged boy with bow.

cupidity /kju:'pɪdɪtɪ/ *noun* greed, avarice.

cupola /'kju:pələ/ *noun* small dome; revolving gun-turret on ship or in fort.

cuppa /'kʌpə/ *noun colloquial* cup of (tea).

cur *noun* mangy bad-tempered dog; contemptible person.

curable /'kjʊərəb(ə)l/ *adjective* able to be cured.

curaçao /'kjʊərəsəʊ/ *noun* (*plural* **-s**) orange-flavoured liqueur.

curacy /'kjʊərəsɪ/ *noun* (*plural* **-ies**) curate's office or position.

curare /kjʊə'rɑ:rɪ/ *noun* vegetable poison used on arrows by American Indians.

curate /'kjʊərət/ *noun* assistant to parish priest. □ **curate's egg** thing good in parts.

curative /'kjʊərətɪv/ ● *adjective* tending to cure. ● *noun* curative agent.

curator /kjʊə'reɪtə/ *noun* custodian of museum etc.

curb ● *noun* check, restraint; (bit with) chain etc. passing under horse's lower jaw; kerb. ● *verb* restrain; put curb on.

curd *noun* (often in *plural*) coagulated acidic milk product, made into cheese or eaten as food.

curdle /'kɜ:d(ə)l/ *verb* (**-ling**) coagulate. □ **make one's blood curdle** horrify one.

cure /kjʊə/ ● *verb* (often + *of*) restore to health, relieve; eliminate (evil etc.); preserve (meat etc.) by salting etc. ● *noun* restoration to health; thing that cures; course of treatment.

curé /'kjʊəreɪ/ *noun* parish priest in France etc. [French]

curette /kjʊə'ret/ ● *noun* surgeon's scraping-instrument. ● *verb* (**-tting**) scrape with this. □ **curettage** *noun.*

curfew /'kɜ:fju:/ *noun* signal or time after which people must remain indoors.

curie /'kjʊərɪ/ *noun* unit of radioactivity.

curio /'kjʊərɪəʊ/ *noun* (*plural* **-s**) rare or unusual object.

curiosity /kjʊərɪ'ɒsɪtɪ/ *noun* (*plural* **-ies**) desire to know; inquisitiveness; rare or strange thing.

curious /'kjʊərɪəs/ *adjective* eager to learn; inquisitive; strange, surprising. □ **curiously** *adverb.*

curl ● *verb* (often + *up*) bend or coil into spiral; move in curve; (of upper lip) be raised in contempt; play curling. ● *noun* curled lock of hair; anything spiral or curved inwards. □ **curly** *adjective* (**-ier**, **-iest**).

curler *noun* pin, roller, etc. for curling hair.

curlew /'kɜ:lju:/ *noun* long-billed wading bird with musical cry.

curling *noun* game like bowls played on ice with round flat stones.

curmudgeon /kə'mʌdʒ(ə)n/ *noun* bad-tempered or miserly person. □ **curmudgeonly** *adjective.*

currant /'kʌrənt/ *noun* small seedless dried grape; (fruit of) any of various shrubs producing red, black, or white berries.

currency /'kʌrənsɪ/ *noun* (*plural* **-ies**) money in use in a country; being current; prevalence (of ideas etc.).

current ● *adjective* belonging to present time; happening now; in general circulation or use. ● *noun* body of moving water, air, etc., passing through still water etc.; movement of electrically charged particles; general tendency or course. □ **current account** bank account that may be drawn on by cheque without notice. □ **currently** *adverb.*

curriculum /kə'rɪkjʊləm/ *noun* (*plural* **-la**) course of study. □ **curriculum vitae** /'vi:taɪ/ brief account of one's education, career, etc.

curry[1] /'kʌrɪ/ ● *noun* (*plural* **-ies**) meat, vegetables, etc. cooked in spicy sauce, usually served with rice. ● *verb* (**-ies**, **-ied**) make into or flavour like curry. □ **curry powder** mixture of spices for making curry.

curry[2] /'kʌrɪ/ *verb* (**-ies**, **-ied**) groom (horse etc.) with curry-comb; dress (leather). □ **curry-comb** metal device for grooming horses etc.; **curry favour** ingratiate oneself.

curse /kɜ:s/ ● *noun* invocation of destruction or punishment; violent or profane exclamation; thing causing evil. ● *verb* (**-sing**) utter curse against; (usually in *passive*; + *with*) afflict with; swear. □ **cursed** /'kɜ:sɪd/ *adjective.*

cursive /'kɜ:sɪv/ ● *adjective* (of writing) having joined characters. ● *noun* cursive writing.

cursor /'kɜ:sə/ *noun* indicator on VDU screen showing particular position in displayed matter.

cursory /'kɜ:sərɪ/ *adjective* hasty, hurried. □ **cursorily** *adverb.*

curt *adjective* noticeably or rudely brief. □ **curtly** *adverb*; **curtness** *noun.*

curtail /kɜ:'teɪl/ *verb* cut short, reduce. □ **curtailment** *noun.*

curtain /'kɜ:t(ə)n/ ● *noun* piece of cloth etc. hung up as screen, esp. at window; rise or fall of stage curtain; curtain-call; (in *plural*) *slang* the end. ● *verb* provide or (+ *off*) shut off with curtains. □ **curtain-call** audience's applause summoning actors to take bow; **curtain-raiser** short opening play etc., preliminary event.

curtsy /'kɜ:tsɪ/ (also **curtsey**) ● *noun* (*plural* **-ies** or **-eys**) woman's or girl's acknowledgement or greeting made by bending knees. ● *verb* (**-ies**, **-ied** or **-eys**, **-eyed**) make curtsy.

curvaceous /kɜ:'veɪʃəs/ *adjective colloquial* (esp. of woman) shapely.

curvature /'kɜ:vətʃə/ *noun* curving; curved form; deviation of curve from plane.

curve ● *noun* line or surface of which no part is straight; curved line on graph. ● *verb* (**-ving**) bend or shape so as to form curve. □ **curvy** *adjective* (**-ier**, **-iest**).

curvet /kɜ:'vet/ ● *noun* horse's frisky leap. ● *verb* (**-tt-** or **-t-**) perform curvet.

curvilinear /kɜ:vɪ'lɪnɪə/ *adjective* contained by or consisting of curved lines.

cushion /'kʊʃ(ə)n/ ● *noun* bag stuffed with soft material for sitting on etc.; protection against shock; padded rim of billiard table; air supporting hovercraft. ● *verb* provide or protect with cushions; mitigate effects of.

cushy /'kʊʃɪ/ *adjective* (**-ier**, **-iest**) *colloquial* (of job etc.) easy, pleasant.

cusp *noun* point at which two curves meet, e.g. horn of crescent moon.

cuss *colloquial* ● *noun* curse; awkward person. ● *verb* curse.

cussed /'kʌsɪd/ *adjective colloquial* awkward, stubborn.

custard /'kʌstəd/ *noun* pudding or sweet sauce of eggs or flavoured cornflour and milk.

custodian /kʌs'təʊdɪən/ *noun* guardian, keeper.

custody /'kʌstədɪ/ *noun* guardianship; imprisonment. □ **custodial** /-'stəʊ-/ *adjective.*

custom /'kʌstəm/ *noun* usual behaviour; established usage; business dealings, customers; (in *plural*; also treated as *singular*) duty on imports, government department or (part of) building at port etc. dealing with this. □ **custom house** customs office at frontier etc.; **custom-built**, **-made** made to customer's order.

customary /'kʌstəmərɪ/ *adjective* in accordance with custom; usual. □ **customarily** *adverb.*

customer *noun* person who buys goods or services; *colloquial* person of specified (esp. awkward) kind.

customize *verb* (also **-ise**) (**-zing** or **-sing**) make or modify to order; personalize.

cut ● *verb* (**-tt-**; *past & past participle* **cut**) divide, wound, or penetrate with edged instrument; detach, trim, etc. by cutting; (+ *loose*, *open*, etc.) loosen by cutting; (esp. as **cutting** *adjective*) cause pain to; reduce (prices, wages, services, etc.); make by cutting or removing material; cross, intersect; divide (pack of cards); edit (film), stop cameras; end acquaintance or ignore presence of; *US* deliberately miss (class etc.); chop (ball); switch off (engine etc.); (+ *across*, *through*, etc.) pass through as shorter route. ● *noun* act of cutting; division or wound made by cutting; stroke with knife, sword, whip, etc.; reduction (in price, wages, services, etc.); cessation (of power supply etc.); removal of part of play, film, etc.; *slang* commission, share of profits, etc.; style in which hair, clothing, etc. is cut; particular piece of meat; cutting of ball; deliberate ignoring of person. □ **a cut above** noticeably superior to; **cut and dried** completely decided, inflexible; **cut back** reduce (expenditure), prune; **cut-back** reduction in expenditure; **cut both ways** serve both sides; **cut corners** do task perfunctorily; **cut glass** glass with patterns cut on it; **cut in** interrupt, pull in too closely in front of another vehicle; **cut one's losses** abandon an unprofitable scheme; **cut no ice** *slang* have no influence; **cut off** *verb* remove by cutting, bring to abrupt end, interrupt, disconnect, *adjective* isolated; **cut out** shape by cutting, (cause to) cease functioning, stop doing or using; **cut-out** device for automatic disconnection; **cutthroat** *noun* murderer, *adjective* murderous, (of competition) intense and merciless; **cutthroat razor** one with long unguarded blade set in handle; **cut a tooth** have it appear through gum; **cut up** cut in pieces, (usually in *passive*) greatly distress; **cut up rough** show resentment.

cutaneous /kju:'teɪnɪəs/ *adjective* of the skin.

cute /kju:t/ *adjective colloquial esp. US* attractive, sweet; clever, ingenious. □ **cutely** *adverb*; **cuteness** *noun*.

cuticle /'kju:tɪk(ə)l/ *noun* skin at base of fingernail or toenail.

cutlass /'kʌtləs/ *noun historical* short broad-bladed curved sword.

cutlery /'kʌtlərɪ/ *noun* knives, forks, and spoons for use at table.

cutlet /'kʌtlɪt/ *noun* neck-chop of mutton or lamb; small piece of veal etc. for frying; flat cake of minced meat etc.

cutter *noun* person or thing that cuts; (in *plural*) cutting tool; small fast sailing ship; small boat carried by large ship.

cutting ● *noun* piece cut from newspaper etc.; piece cut from plant for replanting; excavated channel in hillside etc. for railway or road. ● *adjective* that cuts; hurtful. □ **cuttingly** *adverb*.

cuttlefish /'kʌt(ə)lfɪʃ/ *noun* (*plural* same or **-es**) 10-armed sea mollusc ejecting black fluid when pursued.

cutwater *noun* forward edge of ship's prow; wedge-shaped projection from pier of bridge.

C.V. *abbreviation* (also **CV**) curriculum vitae.

cwm /ku:m/ *noun* (in Wales) coomb.

cwt *abbreviation* hundredweight.

cyanide /'saɪənaɪd/ *noun* highly poisonous substance used in extraction of gold and silver.

cyanosis /saɪə'nəʊsɪs/ *noun* bluish skin due to oxygen-deficient blood.

cybernetics /saɪbə'netɪks/ *plural noun* (usually treated as *singular*) science of control systems and communications in animals and machines. □ **cybernetic** *adjective*.

cyclamen /'sɪkləmən/ *noun* plant with pink, red, or white flowers with backward-turned petals.

cycle /'saɪk(ə)l/ ● *noun* recurrent round or period (of events, phenomena, etc.); series of related poems etc.; bicycle, tricycle, etc. ● *verb* (**-ling**) ride bicycle etc.; move in cycles. □ **cycle lane** part of road reserved for bicycles; **cycle track**, **cycleway** path for bicycles.

cyclic /'saɪklɪk/ *adjective* (also **cyclical** /'sɪklɪk(ə)l/) recurring in cycles; belonging to chronological cycle. □ **cyclically** *adverb*.

cyclist /'saɪklɪst/ *noun* rider of bicycle.

cyclone /'saɪkləʊn/ *noun* winds rotating around low-pressure region; violent destructive form of this. □ **cyclonic** /-'klɒn-/ *adjective.*

cyclotron /'saɪklətrɒn/ *noun* apparatus for acceleration of charged atomic particles revolving in magnetic field.

cygnet /'sɪgnɪt/ *noun* young swan.

cylinder /'sɪlɪndə/ *noun* solid or hollow roller-shaped body; container for liquefied gas etc.; piston-chamber in engine. □ **cylindrical** /-'lɪn-/ *adjective.*

cymbal /'sɪmb(ə)l/ *noun* concave disc struck usually with another to make ringing sound.

cynic /'sɪnɪk/ *noun* person with pessimistic view of human nature. □ **cynical** *adjective*; **cynically** *adverb*; **cynicism** /-sɪz(ə)m/ *noun.*

cynosure /'saɪnəzjʊə/ *noun* centre of attention or admiration.

cypher = CIPHER.

cypress /'saɪprəs/ *noun* conifer with dark foliage.

Cypriot /'sɪprɪət/ (also **Cypriote** /-əʊt/) ● *noun* native or national of Cyprus. ● *adjective* of Cyprus.

Cyrillic /sɪ'rɪlɪk/ ● *adjective* of alphabet used esp. for Russian and Bulgarian. ● *noun* this alphabet.

cyst /sɪst/ *noun* sac formed in body, containing liquid matter.

cystic *adjective* of the bladder; like a cyst. □ **cystic fibrosis** hereditary disease usually with respiratory infections.

cystitis /sɪ'staɪtɪs/ *noun* inflammation of the bladder.

czar = TSAR.

czarina = TSARINA.

Czech /tʃek/ ● *noun* native or national of Czech Republic, or *historical* Czechoslovakia; language of Czech people. ● *adjective* of Czechs or their language; of Czech Republic.

Czechoslovak /tʃekə'sləʊvæk/ (also **Czechoslovakian** /-slə'vækɪən/) *historical* ● *noun* native or national of Czechoslovakia. ● *adjective* of Czechoslovaks or Czechoslovakia.

Dd

D *noun* (also **d**) (Roman numeral) 500. □ **D-Day** day of Allied invasion of France (6 June 1944), important or decisive day.

d. *abbreviation* died; (pre-decimal) penny.

dab[1] ● *verb* (**-bb-**) (often + *at*) press briefly and repeatedly with cloth etc.; (+ *on*) apply by dabbing; (often + *at*) aim feeble blow; strike lightly. ● *noun* dabbing; small amount (of paint etc.) dabbed on; light blow.

dab[2] *noun* (*plural* same) kind of marine flatfish.

dabble /'dæb(ə)l/ *verb* (**-ling**) (usually + *in*, *at*) engage (in an activity etc.) superficially; move about in shallow water etc. □ **dabbler** *noun.*

dabchick *noun* little grebe.

dab hand *noun* (usually + *at*) expert.

da capo /dɑ: 'kɑ:pəʊ/ *adverb Music* repeat from beginning.

dace *noun* (*plural* same) small freshwater fish.

dacha /'dætʃə/ *noun* Russian country cottage.

dachshund /'dækshʊnd/ *noun* short-legged long-bodied dog.

dactyl /'dæktɪl/ *noun* metrical foot of one long followed by two short syllables. □ **dactylic** /-'tɪl-/ *adjective.*

dad *noun colloquial* father.

daddy /'dædɪ/ *noun* (*plural* **-ies**) *colloquial* father. □ **daddy-long-legs** crane-fly.

dado /'deɪdəʊ/ *noun* (*plural* **-s**) lower, differently decorated, part of interior wall.

daffodil /'dæfədɪl/ *noun* spring bulb with trumpet-shaped yellow flowers.

daft /dɑːft/ *adjective* (**-er**, **-est**) foolish, silly, crazy.

dagger /ˈdægə/ *noun* short knifelike weapon; obelus.

daguerreotype /dəˈgerəʊtaɪp/ *noun* early photograph using silvered plate.

dahlia /ˈdeɪlɪə/ *noun* garden plant with large showy flowers.

Dáil (**Eireann**) /dɔɪl ˈeɪrən/ *noun* lower house of Parliament in Republic of Ireland.

daily /ˈdeɪlɪ/ ● *adjective* done, produced, or occurring every (week)day. ● *adverb* every day; constantly. ● *noun* (*plural* **-ies**) *colloquial* daily newspaper; cleaning woman.

dainty /ˈdeɪntɪ/ ● *adjective* (**-ier**, **-iest**) delicately pretty or small; choice; fastidious. ● *noun* (*plural* **-ies**) delicacy. □ **daintily** *adverb*; **daintiness** *noun*.

daiquiri /ˈdækərɪ/ *noun* (*plural* **-s**) cocktail of rum, lime juice, etc.

dairy /ˈdeərɪ/ ● *noun* (*plural* **-ies**) place for processing, distributing, or selling milk and milk products. ● *adjective* of, containing, or used for milk and milk products. □ **dairymaid** woman employed in dairy; **dairyman** man looking after cows.

dais /ˈdeɪɪs/ *noun* low platform, esp. at upper end of hall.

daisy /ˈdeɪzɪ/ *noun* (*plural* **-ies**) flowering plant with white radiating petals. □ **daisy chain** string of field daisies threaded together; **daisy wheel** spoked disc bearing printing characters, used in word processors and typewriters.

dale *noun* valley.

dally /ˈdælɪ/ *verb* (**-ies**, **-ied**) delay; waste time; (often + *with*) flirt. □ **dalliance** *noun*.

Dalmatian /dælˈmeɪʃ(ə)n/ *noun* large white dog with dark spots.

dam[1] ● *noun* barrier across river etc., usually forming reservoir or preventing flooding. ● *verb* (**-mm-**) provide or confine with dam; (often + *up*) block up, obstruct.

dam[2] *noun* mother (of animal).

damage /ˈdæmɪdʒ/ ● *noun* harm; injury; (in *plural*) financial compensation for loss or injury; (**the damage**) *slang* cost. ● *verb* (**-ging**) inflict damage on.

damask /ˈdæməsk/ ● *noun* fabric with woven design made visible by reflection of light. ● *adjective* made of damask; velvety pink. □ **damask rose** old sweet-scented rose.

dame *noun* (**Dame**) (title of) woman who has been knighted; comic female pantomime character played by man; *US slang* woman.

damn /dæm/ ● *verb* (often as *interjection*) curse; censure; condemn to hell; (often as **damning** *adjective*) show or prove to be guilty. ● *noun* uttered curse. ● *adjective & adverb* damned. □ **damn all** *slang* nothing.

damnable /ˈdæmnəb(ə)l/ *adjective* hateful; annoying.

damnation /dæmˈneɪʃ(ə)n/ ● *noun* eternal punishment in hell. ● *interjection expressing anger*.

damned /dæmd/ ● *adjective* damnable. ● *adverb* extremely.

damp ● *adjective* slightly wet. ● *noun* slight diffused or condensed moisture. ● *verb* make damp; (often + *down*) discourage, make burn less strongly; *Music* stop vibration of (string etc.). □ **damp course** layer of damp-proof material in wall to keep damp from rising. □ **dampness** *noun*.

dampen *verb* make or become damp; discourage.

damper *noun* device that reduces shock, vibration, or noise; discouraging person or thing; metal plate in flue to control draught.

damsel /ˈdæmz(ə)l/ *noun archaic* young unmarried woman.

damson /ˈdæmz(ə)n/ *noun* small dark purple plum.

dance /dɑːns/ ● *verb* (**-cing**) move rhythmically, usually to music; perform (dance role etc.); jump or bob about. ● *noun* dancing as art; style or form of this; social gathering for dancing; lively motion. □ **dance attendance (on)** serve obsequiously. □ **dancer** *noun*.

dandelion /ˈdændɪlaɪən/ *noun* yellow-flowered wild plant.

dander *noun* □ **get one's dander up** *colloquial* become angry.

dandle /ˈdænd(ə)l/ *verb* (**-ling**) bounce (child) on one's knees etc.

dandruff /'dændrʌf/ *noun* flakes of dead skin in hair.

dandy /'dændɪ/ ● *noun* (*plural* **-ies**) man excessively devoted to style and fashion. ● *adjective* (**-ier**, **-iest**) *colloquial* splendid. □ **dandy-brush** stiff brush for grooming horses.

Dane *noun* native or national of Denmark; *historical* Viking invader of England.

danger /'deɪndʒə/ *noun* liability or exposure to harm; thing causing harm. □ **danger list** list of those dangerously ill; **danger money** extra payment for dangerous work.

dangerous *adjective* involving or causing danger. □ **dangerously** *adverb*.

dangle /'dæŋg(ə)l/ *verb* (**-ling**) hang loosely; hold or carry swaying loosely; hold out (temptation etc.).

Danish /'deɪnɪʃ/ ● *adjective* of Denmark. ● *noun* Danish language; (**the Danish**) the Danish people. □ **Danish blue** white blue-veined cheese; **Danish pastry** yeast cake with icing, nuts, fruit, etc.

dank *adjective* damp and cold.

daphne /'dæfnɪ/ *noun* a flowering shrub.

dapper *adjective* neat and precise, esp. in dress; sprightly.

dapple /'dæp(ə)l/ *verb* (**-ling**) mark with spots of colour or shade; mottle. □ **dapple-grey** (of horse) grey with darker spots; **dapple grey** such a horse.

Darby and Joan *noun* devoted old married couple. □ **Darby and Joan club** social club for pensioners.

dare /deə/ ● *verb* (**-ring**; *3rd singular present* often **dare**) (+ *(to) do*) have the courage or impudence (to); (usually + *to do*) defy, challenge. ● *noun* challenge. □ **daredevil** reckless (person); **I dare say** very likely, I grant that.

daring ● *noun* adventurous courage. ● *adjective* bold, prepared to take risks. □ **daringly** *adverb*.

dariole /'dærɪəʊl/ *noun* dish cooked and served in a small mould.

dark ● *adjective* with little or no light; of deep or sombre colour; (of a person) with dark colouring; gloomy; sinister; angry; secret, mysterious. ● *noun* absence of light or knowledge; unlit place. □ **after dark** after nightfall; **Dark Ages** 5th–10th-c., unenlightened period; **dark horse** little-known person who is unexpectedly successful; **darkroom** darkened room for photographic work; **in the dark** without information or light. □ **darken** *verb*; **darkly** *adverb*; **darkness** *noun*.

darling /'dɑ:lɪŋ/ ● *noun* beloved or endearing person or animal. ● *adjective* beloved, lovable; *colloquial* charming.

darn[1] ● *verb* mend by interweaving wool etc. across hole. ● *noun* darned area.

darn[2] *verb, interjection, adjective, & adverb colloquial* mild form of DAMN.

darnel /'dɑ:n(ə)l/ *noun* grass growing in cereal crops.

dart ● *noun* small pointed missile; (in *plural* treated as *singular*) indoor game of throwing darts at a dartboard; sudden rapid movement; tapering tuck in garment. ● *verb* (often + *out*, *in*, *past*, etc.) move, send, or go suddenly or rapidly. □ **dartboard** circular target in game of darts.

Darwinian /dɑ:'wɪnɪən/ ● *adjective* of Darwin's theory of evolution. ● *noun* adherent of this. □ **Darwinism** /'dɑ:-/ *noun*; **Darwinist** /'dɑ:-/ *noun*.

dash ● *verb* rush; strike or fling forcefully so as to shatter; frustrate, dispirit; *colloquial* (as *interjection*) damn. ● *noun* rush, onset; punctuation mark (—) used to indicate break in sense (see panel); longer signal of two in Morse code; slight admixture; (capacity for) impetuous vigour. □ **dashboard** instrument panel of vehicle or aircraft; **dash off** write hurriedly.

dashing *adjective* spirited; showy.

dastardly /'dæstədlɪ/ *adjective* cowardly, despicable.

data /'deɪtə/ *plural noun* (also treated as *singular*) known facts used for inference or reckoning; quantities or characters operated on by computer. □ **data bank** store or source of data; **database** structured set of data held in computer; **data processing** series of operations on data by computer.

■ **Usage** In scientific, philosophical, and general use, *data* is usually considered to mean a number of items and is treated as plural, with *datum* as its singular. In computing and allied subjects (and sometimes in general use), it is treated as singular, as in *Much useful data has been collected.* However, *data* is not singular, and it is wrong to say *a data* or *every data* or to make the plural form *datas.*

date[1] ● *noun* day of month; historical day or year; day, month, and year of writing etc. at head of document etc.; period to which work of art etc. belongs; time when an event takes place; *colloquial* social appointment, esp. with person of opposite sex; *US colloquial* person to be met at this. ● *verb* (**-ting**) mark with date; assign date to; (+ *to*) assign to a particular time, period, etc.; (often + *from*, *back to*, etc.) have origin at a particular time; expose as or appear old-fashioned; *US colloquial* make date with, go out together as sexual partners. □ **date line** line partly along meridian 180° from Greenwich, to the east of which date is a day earlier than to the west; date and place of writing at head of newspaper article etc.; **out of date** (**out-of-date** before noun) old-fashioned, obsolete; **to date** until now; **up to date** (**up-to-date** before noun) fashionable, current.

date[2] *noun* oval stone fruit; (in full **date-palm**) tree bearing this.

dative /ˈdeɪtɪv/ *Grammar* ● *noun* case expressing indirect object or recipient. ● *adjective* of or in the dative.

datum /ˈdeɪtəm/ *singular* of DATA.

daub /dɔːb/ ● *verb* paint or spread (paint etc.) crudely or unskilfully; smear (surface) with paint etc. ● *noun* paint etc. daubed on a surface; crude painting; clay etc. coating wattles to form wall.

daughter /ˈdɔːtə/ *noun* female child in relation to her parents; female descendant or member of family etc. □ **daughter-in-law** (*plural* **daughters-in-law**) son's wife.

daunt /dɔːnt/ *verb* discourage, intimidate. □ **daunting** *adjective*.

dauntless *adjective* intrepid, persevering.

dauphin /ˈdɔːfɪn/ *noun historical* eldest son of King of France.

Davenport /ˈdævənpɔːt/ *noun* kind of writing desk; *US* large sofa.

davit /ˈdævɪt/ *noun* small crane on ship for holding lifeboat.

daw *noun* jackdaw.

dawdle /ˈdɔːd(ə)l/ *verb* (**-ling**) walk slowly and idly; waste time, procrastinate.

dawn ● *noun* daybreak; beginning. ● *verb* (of day) begin, grow light; (often + *on*, *upon*) become evident (to). □ **dawn chorus** birdsong at daybreak.

day *noun* time between sunrise and sunset; 24 hours as a unit of time; daylight; time during which work is normally

---

Dash –

This is used:

1 to mark the beginning and end of an interruption in the structure of a sentence:

My son—where has he gone?—would like to meet you.

2 to show faltering speech in conversation:

Yes—well—I would—only you see—it's not easy.

3 to show other kinds of break in a sentence, where a comma, semicolon, or colon would traditionally be used, e.g.

Come tomorrow—if you can.
The most important thing is this—don't rush the work.

A dash is not used in this way in formal writing.

done; (also *plural*) historical period; (**the day**) present time; period of prosperity. □ **daybreak** first light in morning; **day centre** place for care of elderly or handicapped during day; **daydream** (indulge in) fantasy etc. while awake; **day off** day's holiday; **day release** part-time education for employees; **day return** reduced fare or ticket for a return journey in one day; **day school** school for pupils living at home; **daytime** part of day when there is natural light; **day-to-day** mundane, routine; **day-trip** trip completed in one day.

daylight *noun* light of day; dawn; visible gap between things; (usually in *plural*) *slang* life. □ **daylight robbery** blatantly excessive charge; **daylight saving** longer summer evening daylight, achieved by putting clocks forward.

daze ● *verb* (**-zing**) stupefy, bewilder. ● *noun* dazed state.

dazzle /'dæz(ə)l/ ● *verb* (**-ling**) blind or confuse temporarily with sudden bright light; impress or overpower with knowledge, ability, etc. ● *noun* bright confusing light. □ **dazzling** *adjective*.

dB *abbreviation* decibel(s).

DC *abbreviation* direct current; District of Columbia; da capo.

DDT *abbreviation* colourless chlorinated hydrocarbon used as insecticide.

deacon /'di:kən/ *noun* (in episcopal churches) minister below priest; (*feminine* **deaconess** /-'nes/) (in Nonconformist churches) lay officer.

deactivate /dɪ'æktɪveɪt/ *verb* (**-ting**) make inactive or less reactive.

dead /ded/ ● *adjective* no longer alive; numb; *colloquial* extremely tired or unwell; (+ *to*) insensitive to; not effective; extinct; extinguished; inanimate; lacking vigour; not resonant; quiet; not transmitting sounds; out of play; abrupt; complete. ● *adverb* absolutely, completely; *colloquial* very. ● *noun* time of silence or inactivity. □ **dead beat** utterly exhausted; **dead-beat** tramp; **dead duck** useless person or thing; **dead end** closed end of road etc.; **dead-end** having no prospects; **dead heat** race in which competitors tie; **dead letter** law etc. no longer observed; **deadline** time limit; **deadlock** *noun* state of unresolved conflict, *verb* bring or come to a standstill; **dead loss** useless person or thing; **dead man's handle** handle on electric train etc. disconnecting power supply if released; **dead march** funeral march; **dead on** exactly right; **deadpan** lacking expression or emotion; **dead reckoning** estimation of ship's position from log, compass, etc., when visibility is bad; **dead shot** unerring marksman; **dead weight** inert mass, heavy burden; **dead wood** *colloquial* useless person(s) or thing(s).

deaden *verb* deprive of or lose vitality, force, etc.; (+ *to*) make insensitive.

deadly ● *adjective* (**-ier**, **-iest**) causing fatal injury or serious damage; intense; accurate; deathlike; dreary. ● *adverb* as if dead; extremely. □ **deadly nightshade** plant with poisonous black berries.

deaf /def/ *adjective* wholly or partly unable to hear; (+ *to*) refusing to listen or comply. □ **deaf-aid** hearing aid; **deaf-and-dumb alphabet**, **language** sign language; **deaf mute** deaf and dumb person. □ **deafness** *noun*.

deafen *verb* (often as **deafening** *adjective*) overpower or make deaf with noise, esp. temporarily. □ **deafeningly** *adverb*.

deal[1] ● *verb* (*past & past participle* **dealt** /delt/) (+ *with*) take measures to resolve, placate, etc., do business or associate with, treat (subject); (often + *by*, *with*) behave in specified way; (+ *in*) sell; (often + *out*, *round*) distribute; administer. ● *noun* (usually **a good** or **great deal**) large amount, considerably; business arrangement etc.; specified treatment; dealing of cards, player's turn to deal.

deal[2] *noun* fir or pine timber, esp. as boards.

dealer *noun* trader; player dealing at cards.

dealings *plural noun* conduct or transactions.

dean *noun* head of ecclesiastical chapter; (usually **rural dean**) clergyman supervising parochial clergy; college or university official with disciplinary functions; head of university faculty.

deanery *noun* (*plural* **-ies**) dean's house or position; parishes presided over by rural dean.

dear ● *adjective* beloved; *used before person's name, esp. at beginning of letter*; (+ *to*) precious; expensive. ● *noun* dear person. ● *adverb* at great cost. ● *interjection* (usually **oh dear!** or **dear me!**) *expressing surprise, dismay, etc.* □ **dearly** *adverb.*

dearth /dɜ:θ/ *noun* scarcity, lack.

death /deθ/ *noun* dying, end of life; being dead; cause of death; destruction. □ **deathblow** blow etc. causing death, or action, event, etc. ending something; **death-mask** cast of dead person's face; **death penalty** capital punishment; **death rate** yearly deaths per 1000 of population; **death-rattle** gurgling in throat at death; **death row** part of prison for those sentenced to death; **death squad** paramilitary group; **death-trap** unsafe place, vehicle, etc.; **death-warrant** order of execution; **death-watch beetle** beetle that bores into wood and makes ticking sound. □ **deathlike** *adjective.*

deathly ● *adjective* (**-ier**, **-iest**) like death. ● *adverb* in deathly manner.

deb *noun colloquial* débutante.

débâcle /deɪ'bɑ:k(ə)l/ *noun* utter collapse; confused rush.

debar /dɪ'bɑ:/ *verb* (**-rr-**) (+ *from*) exclude. □ **debarment** *noun.*

debase /dɪ'beɪs/ *verb* (**-sing**) lower in character, quality, or value; depreciate (coin) by alloying etc. □ **debasement** *noun.*

debatable /dɪ'beɪtəb(ə)l/ *adjective* questionable.

debate /dɪ'beɪt/ ● *verb* (**-ting**) discuss or dispute, esp. formally; consider, ponder. ● *noun* discussion, esp. formal.

debauch /dɪ'bɔ:tʃ/ ● *verb* corrupt, deprave, debase; (as **debauched** *adjective*) dissolute. ● *noun* bout of debauchery.

debauchee /dɪbɔ:'tʃi:/ *noun* debauched person.

debauchery *noun* excessive sensual indulgence.

debenture /dɪ'bentʃə/ *noun* company bond providing for payment of interest.

debilitate /dɪ'bɪlɪteɪt/ *verb* (**-ting**) enfeeble. □ **debilitation** *noun.*

debility *noun* feebleness, esp. of health.

debit /'debɪt/ ● *noun* entry in account recording sum owed. ● *verb* (**-t-**) (+ *against*, *to*) enter on debit side of account.

debonair /debə'neə/ *adjective* self-assured; pleasant.

debouch /dɪ'baʊtʃ/ *verb* come out into open ground; (often + *into*) (of river etc.) merge. □ **debouchment** *noun.*

debrief /di:'bri:f/ *verb* question (diplomat etc.) about completed mission. □ **debriefing** *noun.*

debris /'debri:/ *noun* scattered fragments; wreckage.

debt /det/ *noun* money etc. owing; obligation; state of owing.

debtor *noun* person owing money etc.

debug /di:'bʌg/ *verb* (**-gg-**) remove hidden microphones from; remove defects from.

debunk /di:'bʌŋk/ *verb colloquial* expose as spurious or false.

début /'deɪbju:/ *noun* first public appearance.

débutante /'debju:tɑ:nt/ *noun* young woman making her social début.

Dec. *abbreviation* December.

deca- *combining form* ten.

decade /'dekeɪd/ *noun* 10 years; set or series of 10.

decadence /'dekəd(ə)ns/ *noun* moral or cultural decline; immoral behaviour. □ **decadent** *adjective.*

decaffeinated /di:'kæfɪneɪtɪd/ *adjective* with caffeine removed.

decagon /'dekəgən/ *noun* plane figure with 10 sides and angles. □ **decagonal** /-'kæg-/ *adjective.*

decahedron /dekə'hi:drən/ *noun* solid figure with 10 faces. □ **decahedral** *adjective.*

decamp /di:'kæmp/ *verb* depart suddenly; break up or leave camp.

decant /dɪ'kænt/ *verb* pour off (wine etc.) leaving sediment behind.

decanter *noun* stoppered glass container for decanted wine or spirit.

decapitate /dɪ'kæpɪteɪt/ *verb* (**-ting**) behead. □ **decapitation** *noun.*

decarbonize /di:'kɑ:bənaɪz/ *verb* (also **-ise**) (**-zing** or **-sing**) remove carbon etc. from (engine of car etc.). □ **decarbonization** *noun.*

decathlon /dɪ'kæθlən/ *noun* athletic contest of 10 events. □ **decathlete** *noun.*

decay /dɪ'keɪ/ ● *verb* (cause to) rot or decompose; decline in quality, power, etc. ● *noun* rotten state; decline.

decease /dɪ'si:s/ *noun formal esp. Law* death.

deceased *formal* ● *adjective* dead. ● *noun* (**the deceased**) person who has died.

deceit /dɪ'si:t/ *noun* deception; trick. □ **deceitful** *adjective.*

deceive /dɪ'si:v/ *verb* (**-ving**) make (person) believe what is false; (**deceive oneself**) persist in mistaken belief; mislead; be unfaithful to.

decelerate /di:'seləreɪt/ *verb* (**-ting**) (cause to) reduce speed. □ **deceleration** *noun.*

December /dɪ'sembə/ *noun* twelfth month of year.

decency /'di:sənsɪ/ *noun* (*plural* **-ies**) correct, honourable, or modest behaviour; (in *plural*) proprieties, manners.

decennial /dɪ'senɪəl/ *adjective* lasting 10 years; recurring every 10 years.

decent /'di:s(ə)nt/ *adjective* conforming with standards of decency; not obscene; respectable; acceptable; kind. □ **decently** *adverb.*

decentralize /di:'sentrəlaɪz/ *verb* (also **-ise**) (**-zing** or **-sing**) transfer (power etc.) from central to local authority. □ **decentralization** *noun.*

deception /dɪ'sepʃ(ə)n/ *noun* deceiving, being deceived; thing that deceives.

deceptive /dɪ'septɪv/ *adjective* likely to mislead.

deci- *combining form* one-tenth.

decibel /'desɪbel/ *noun* unit used in comparison of sound etc. levels.

decide /dɪ'saɪd/ *verb* (**-ding**) (usually + *to do, that, on, about*) resolve after consideration; settle (issue etc.); (usually + *between, for, against, in favour of, that*) give judgement.

decided *adjective* definite, unquestionable; positive, resolute. □ **decidedly** *adverb.*

deciduous /dɪ'sɪdjʊəs/ *adjective* (of tree) shedding leaves annually; (of leaves etc.) shed periodically.

decimal /'desɪm(ə)l/ ● *adjective* (of system of numbers, weights, measures, etc.) based on 10; of tenths or 10; proceeding by tens. ● *noun* decimal fraction. □ **decimal fraction** fraction expressed in tenths, hundredths, etc., esp. by units to right of decimal point; **decimal point** dot placed before fraction in decimal fraction.

decimalize *verb* (also **-ise**) (**-zing** or **-sing**) express as decimal; convert to decimal system. □ **decimalization** *noun.*

decimate /'desɪmeɪt/ *verb* (**-ting**) destroy large proportion of. □ **decimation** *noun.*

■ **Usage** *Decimate* should not be used to mean 'defeat utterly'.

---

decipher /dɪ'saɪfə/ *verb* convert (coded information) into intelligible language; determine the meaning of. □ **decipherable** *adjective.*

decision /dɪ'sɪʒ(ə)n/ *noun* deciding; resolution after consideration; settlement; resoluteness.

decisive /dɪ'saɪsɪv/ *adjective* conclusive, settling an issue; quick to decide. □ **decisively** *adverb*; **decisiveness** *noun.*

deck ● *noun* platform in a ship serving as a floor; floor of bus etc.; section for playing discs or tapes etc. in sound system; *US* pack of cards. ● *verb* (often + *out*) decorate. □ **deck-chair** outdoor folding chair.

declaim /dɪ'kleɪm/ *verb* speak, recite, etc. as if addressing audience. □ **declamation** *noun*; **declamatory** /-'klæm-/ *adjective.*

declaration /deklə'reɪʃ(ə)n/ *noun* declaring; emphatic, deliberate, or formal statement.

declare /dɪ'kleə/ *verb* (**-ring**) announce openly or formally; pronounce; (usually + *that*) assert emphatically; acknowledge possession of (dutiable goods, income, etc.); *Cricket* close (innings) voluntarily before team is out; *Cards* name trump suit. □ **declaratory** /-'klær-/ *adjective.*

declassify /di:'klæsɪfaɪ/ *verb* (**-ies, -ied**) declare (information etc.) to be no longer secret. □ **declassification** *noun.*

declension /dɪ'klenʃ(ə)n/ *noun Grammar* variation of form of noun etc., to show grammatical case; class of nouns with same inflections; deterioration.

declination /deklɪ'neɪʃ(ə)n/ *noun* downward bend; angular distance north or south of celestial equator; deviation of compass needle from true north.

decline /dɪ'klaɪn/ ● *verb* (**-ning**) deteriorate, lose strength or vigour; decrease; refuse; slope or bend downwards; *Grammar* state case forms of (noun etc.). ● *noun* deterioration.

declivity /dɪ'klɪvɪtɪ/ *noun* (*plural* **-ies**) downward slope.

declutch /di:'klʌtʃ/ *verb* disengage clutch of motor vehicle.

decode /di:'kəʊd/ *verb* (**-ding**) decipher. □ **decoder** *noun*.

decoke /di:'kəʊk/ *verb* (**-king**) *colloquial* decarbonize.

décolletage /deɪkɒl'tɑ:ʒ/ *noun* low neckline of woman's dress. [French]

décolleté /deɪ'kɒlteɪ/ *adjective* (also ***décolletée***) having low neckline. [French]

decompose /di:kəm'pəʊz/ *verb* (**-sing**) rot; separate into elements. □ **decomposition** /di:kɒmpə'zɪʃ(ə)n/ *noun*.

decompress /di:kəm'pres/ *verb* subject to decompression.

decompression /di:kəm'preʃ(ə)n/ *noun* release from compression; reduction of pressure on deep-sea diver etc. □ **decompression chamber** enclosed space for decompression.

decongestant /di:kən'dʒest(ə)nt/ *noun* medicine etc. that relieves nasal congestion.

decontaminate /di:kən'tæmɪneɪt/ *verb* (**-ting**) remove contamination from. □ **decontamination** *noun*.

décor /'deɪkɔ:/ *noun* furnishings and decoration of room, stage, etc.

decorate /'dekəreɪt/ *verb* (**-ting**) adorn; paint, wallpaper, etc. (room etc.); give medal or award to.

decoration *noun* decorating; thing that decorates; medal etc.; (in *plural*) flags etc. put up on festive occasion.

decorative /'dekərətɪv/ *adjective* pleasing in appearance. □ **decoratively** *adverb*.

decorator *noun* person who decorates for a living.

decorous /'dekərəs/ *adjective* having or showing decorum. □ **decorously** *adverb*.

decorum /dɪ'kɔ:rəm/ *noun* polite dignified behaviour.

decoy ● *noun* /'di:kɔɪ/ thing or person used as lure; bait, enticement. ● *verb* /dɪ'kɔɪ/ lure by decoy.

decrease ● *verb* /dɪ'kri:s/ (**-sing**) make or become smaller or fewer. ● *noun* /'di:kri:s/ decreasing; amount of this.

decree /dɪ'kri:/ ● *noun* official legal order; legal decision. ● *verb* (**-ees, -eed**) ordain by decree. □ **decree absolute** final order for completion of divorce; **decree nisi** /'naɪsaɪ/ provisional order for divorce.

decrepit /dɪ'krepɪt/ *adjective* weakened by age or infirmity; dilapidated. □ **decrepitude** *noun*.

decry /dɪ'kraɪ/ *verb* (**-ies**, **-ied**) disparage.

dedicate /'dedɪkeɪt/ *verb* (**-ting**) (often + *to*) devote (oneself) to a purpose etc.; address (book etc.) to friend or patron etc.; devote (building etc.) to saint etc.; (as **dedicated** *adjective*) having single-minded loyalty. □ **dedicatory** *adjective*.

dedication *noun* dedicating; words with which book is dedicated.

deduce /dɪ'dju:s/ *verb* (**-cing**) (often + *from*) infer logically. □ **deducible** *adjective*.

deduct /dɪ'dʌkt/ *verb* (often + *from*) subtract; take away; withhold.

deductible *adjective* that may be deducted esp. from tax or taxable income.

deduction /dɪ'dʌkʃ(ə)n/ *noun* deducting; amount deducted; inference from general to particular.

deductive *adjective* of or reasoning by deduction.

deed *noun* thing done; action; legal document. □ **deed of covenant** agreement to pay regular sum, esp. to charity; **deed poll** deed made by one party only, esp. to change one's name.

deem *verb formal* consider, judge.

deep ● *adjective* extending far down or in; to or at specified depth; low-pitched; intense; profound; (+ *in*) fully absorbed, overwhelmed. ● *adverb* deeply; far down or in. ● *noun* deep state; (**the deep**) *poetical* the sea. □ **deep-freeze** *noun* freezer, *verb* freeze or store in freezer; **deep-fry** fry with fat covering food. □ **deepen** *verb*; **deeply** *adverb*.

deer *noun* (*plural* same) 4-hoofed grazing animal, male usually with antlers. □ **deerstalker** cloth peaked cap with ear-flaps.

deface /dɪ'feɪs/ *verb* (**-cing**) disfigure. □ **defacement** *noun*.

de facto /deɪ 'fæktəʊ/ ● *adjective* existing in fact, whether by right or not. ● *adverb* in fact.

defame /dɪ'feɪm/ *verb* (**-ming**) attack good name of. □ **defamation** /defə'meɪʃ(ə)n/ *noun*; **defamatory** /-'fæm-/ *adjective*.

default /dɪ'fɔ:lt/ ● *noun* failure to act, appear, or pay; option selected by computer program etc. unless given alternative instruction. ● *verb* fail to fulfil obligations. □ **by default** because of lack of an alternative etc. □ **defaulter** *noun*.

defeat /dɪ'fi:t/ ● *verb* overcome in battle, contest, etc.; frustrate, baffle. ● *noun* defeating, being defeated.

defeatism *noun* readiness to accept defeat. □ **defeatist** *noun & adjective*.

defecate /'di:fɪkeɪt/ *verb* (**-ting**) evacuate the bowels. □ **defecation** *noun*.

defect ● *noun* /'di:fekt/ shortcoming; fault. ● *verb* /dɪ'fekt/ desert one's country, cause, etc., for another. □ **defection** *noun*; **defector** *noun*.

defective /dɪ'fektɪv/ *adjective* faulty; imperfect. □ **defectiveness** *noun*.

defence /dɪ'fens/ *noun* (*US* **defense**) (means of) defending; justification; defendant's case or counsel; players in defending position; (in *plural*) fortifications. □ **defenceless** *adjective*.

defend /dɪ'fend/ *verb* (often + *against*, *from*) resist attack made on; protect; uphold by argument; *Law* conduct defence (of); compete to retain (title). □ **defender** *noun*.

defendant *noun* person accused or sued in court of law.

defense *US* = DEFENCE.

defensible /dɪ'fensɪb(ə)l/ *adjective* able to be defended or justified.

defensive /dɪ'fensɪv/ *adjective* done or intended for defence; over-reacting to criticism. □ **on the defensive** expecting criticism, ready to defend. □ **defensively** *adverb*; **defensiveness** *noun*.

defer[1] /dɪ'fɜ:/ *verb* (**-rr-**) postpone. □ **deferment** *noun*.

defer[2] /dɪ'fɜ:/ *verb* (**-rr-**) (+ *to*) yield or make concessions.

deference /'defərəns/ *noun* respectful conduct; compliance with another's wishes. □ **in deference to** out of respect for.

deferential /defə'renʃ(ə)l/ *adjective* respectful. □ **deferentially** *adverb*.

defiance /dɪ'faɪəns/ *noun* open disobedience; bold resistance. □ **defiant** *adjective*; **defiantly** *adverb*.

deficiency /dɪ'fɪʃənsɪ/ *noun* (*plural* **-ies**) being deficient; (usually + *of*) lack or shortage; thing lacking; deficit. □ **deficiency disease** disease caused by lack of essential element in diet.

deficient /dɪ'fɪʃ(ə)nt/ *adjective* (often + *in*) incomplete or insufficient.

deficit /'defɪsɪt/ *noun* amount by which total falls short; excess of liabilities over assets.

defile[1] /dɪ'faɪl/ *verb* (**-ling**) make dirty; pollute; profane. □ **defilement** *noun*.

defile[2] /dɪ'faɪl/ ● *noun* narrow gorge or pass. ● *verb* (**-ling**) march in file.

define /dɪ'faɪn/ *verb* (**-ning**) give meaning of; describe scope of; outline; mark out the boundary of. □ **definable** *adjective*.

definite /'defɪnɪt/ *adjective* certain; clearly defined; precise. □ **definite article** the word (*the* in English) placed before a noun and implying a specific object, person, or idea. □ **definitely** *adverb*.

definition /defɪ'nɪʃ(ə)n/ *noun* defining; statement of meaning of word etc.; distinctness in outline.

definitive /dɪ'fɪnɪtɪv/ *adjective* decisive, unconditional, final; most authoritative.

deflate /dɪ'fleɪt/ *verb* (**-ting**) let air out of (tyre etc.); (cause to) lose confidence; subject (economy) to deflation.

deflation *noun* deflating; reduction of money in circulation to combat inflation. □ **deflationary** *adjective*.

deflect /dɪ'flekt/ *verb* bend or turn aside from purpose or course; (often + *from*) (cause to) deviate. □ **deflection** *noun*.

deflower /di:'flaʊə/ *verb* deprive of virginity; ravage.

defoliate /diːˈfəʊlɪeɪt/ *verb* (**-ting**) destroy leaves of. □ **defoliant** *noun*; **defoliation** *noun*.

deforest /diːˈfɒrɪst/ *verb* clear of forests. □ **deforestation** *noun*.

deform /dɪˈfɔːm/ *verb* (often as **deformed** *adjective*) make ugly or misshapen, disfigure. □ **deformation** /diː-/ *noun*.

deformity /dɪˈfɔːmɪtɪ/ *noun* (*plural* **-ies**) being deformed; malformation.

defraud /dɪˈfrɔːd/ *verb* (often + *of*) cheat by fraud.

defray /dɪˈfreɪ/ *verb* provide money for (cost). □ **defrayal** *noun*.

defrock /diːˈfrɒk/ *verb* deprive (esp. priest) of office.

defrost /diːˈfrɒst/ *verb* remove frost or ice from; unfreeze; become unfrozen.

deft *adjective* dexterous, skilful. □ **deftly** *adverb*; **deftness** *noun*.

defunct /dɪˈfʌŋkt/ *adjective* no longer existing or in use; dead.

defuse /diːˈfjuːz/ *verb* (**-sing**) remove fuse from (bomb etc.); reduce tension in (crisis etc.).

defy /dɪˈfaɪ/ *verb* (**-ies**, **-ied**) resist openly; present insuperable obstacles to; (+ *to do*) challenge to do or prove something.

degenerate /dɪˈdʒenərət/ ● *adjective* having lost usual or good qualities; immoral. ● *noun* degenerate person etc. ● *verb* /-reɪt/ (**-ting**) become degenerate; get worse. □ **degeneracy** *noun*; **degeneration** *noun*.

degrade /dɪˈgreɪd/ *verb* (**-ding**) (often as **degrading** *adjective*) humiliate; dishonour; reduce to lower rank. □ **degradation** /degrəˈdeɪʃ(ə)n/ *noun*.

degree /dɪˈgriː/ *noun* stage in scale, series, or process; unit of measurement of angle or temperature; extent of burns; academic rank conferred by university etc.

dehumanize /diːˈhjuːmənaɪz/ *verb* (also **-ise**) (**-zing** or **-sing**) remove human qualities from; make impersonal. □ **dehumanization** *noun*.

dehydrate /diːhaɪˈdreɪt/ *verb* (**-ting**) remove water from; make dry; (often as **dehydrated** *adjective*) deprive of fluids, make very thirsty. □ **dehydration** *noun*.

de-ice /diːˈaɪs/ *verb* (**-cing**) remove ice from; prevent formation of ice on. □ **de-icer** *noun*.

deify /ˈdiːɪfaɪ/ *verb* (**-ies**, **-ied**) make god or idol of. □ **deification** *noun*.

deign /deɪn/ *verb* (+ *to do*) condescend.

deism /ˈdiːɪz(ə)m/ *noun* reasoned belief in existence of a god. □ **deist** *noun*; **deistic** /-ˈɪstɪk/ *adjective*.

deity /ˈdeɪɪtɪ/ *noun* (*plural* **-ies**) god or goddess; divine status or nature.

déjà vu /deɪʒɑː ˈvuː/ *noun* illusion of having already experienced present situation. [French]

dejected /dɪˈdʒektɪd/ *adjective* sad, depressed. □ **dejectedly** *adverb*; **dejection** *noun*.

delay /dɪˈleɪ/ ● *verb* postpone; make or be late. ● *noun* delaying, being delayed; time lost by this.

delectable /dɪˈlektəb(ə)l/ *adjective* delightful.

delectation /diːlekˈteɪʃ(ə)n/ *noun* enjoyment.

delegate ● *noun* /ˈdelɪgət/ elected representative sent to conference; member of delegation etc. ● *verb* /ˈdelɪgeɪt/ (**-ting**) (often + *to*) commit (power etc.) to deputy etc.; entrust (task) to another; send or authorize as representative.

delegation /delɪˈgeɪʃ(ə)n/ *noun* group representing others; delegating, being delegated.

delete /dɪˈliːt/ *verb* (**-ting**) strike out (word etc.). □ **deletion** *noun*.

deleterious /delɪˈtɪərɪəs/ *adjective* harmful.

delft *noun* (also **delftware**) type of glazed earthenware.

deli /ˈdelɪ/ *noun* (*plural* **-s**) *colloquial* delicatessen.

deliberate ● *adjective* /dɪˈlɪbərət/ intentional, considered; unhurried. ● *verb* /dɪˈlɪbəreɪt/ (**-ting**) think carefully; discuss. □ **deliberately** *adverb*.

deliberation /dɪlɪbəˈreɪʃ(ə)n/ *noun* careful consideration or slowness.

deliberative /dɪˈlɪbərətɪv/ *adjective* (esp. of assembly etc.) of or for deliberation.

delicacy /ˈdelɪkəsɪ/ *noun* (*plural* **-ies**) being delicate; choice food.

delicate /ˈdelɪkət/ *adjective* fine in texture, quality, etc.; subtle, hard to discern; susceptible, tender; requiring tact. □ **delicately** *adverb*.

delicatessen /delɪkə'tes(ə)n/ *noun* shop selling esp. exotic cooked meats, cheeses, etc.

delicious /dɪ'lɪʃəs/ *adjective* highly enjoyable esp. to taste or smell. □ **deliciously** *adverb.*

delight /dɪ'laɪt/ ● *verb* (often as **delighted** *adjective*) please greatly; (+ *in*) take great pleasure in. ● *noun* great pleasure; thing that delights. □ **delightful** *adjective*; **delightfully** *adverb.*

delimit /dɪ'lɪmɪt/ *verb* (**-t-**) fix limits or boundary of. □ **delimitation** *noun.*

delineate /dɪ'lɪnɪeɪt/ *verb* (**-ting**) portray by drawing or in words. □ **delineation** *noun.*

delinquent /dɪ'lɪŋkwənt/ ● *noun* offender. ● *adjective* guilty of misdeed; failing in a duty. □ **delinquency** *noun.*

deliquesce /delɪ'kwes/ *verb* (**-cing**) become liquid; dissolve in moisture from the air. □ **deliquescence** *noun*; **deliquescent** *adjective.*

delirious /dɪ'lɪrɪəs/ *adjective* affected with delirium; wildly excited. □ **deliriously** *adverb.*

delirium /dɪ'lɪrɪəm/ *noun* disordered state of mind, with incoherent speech etc.; wildly excited mood. □ **delirium tremens** /'tri:menz/ psychosis of chronic alcoholism with tremors and hallucinations.

deliver /dɪ'lɪvə/ *verb* convey (letters, goods) to destination; (often + *to*) hand over; (often + *from*) save, rescue, set free; assist in giving birth or at birth of; utter (speech); launch or aim (blow etc.); (in full **deliver the goods**) *colloquial* provide or carry out what is required.

deliverance *noun* rescuing.

delivery /dɪ'lɪvərɪ/ *noun* (*plural* **-ies**) delivering; distribution of letters etc.; thing delivered; childbirth; manner of delivering.

dell *noun* small wooded valley.

delouse /di:'laʊs/ *verb* (**-sing**) rid of lice.

delphinium /del'fɪnɪəm/ *noun* (*plural* **-s**) garden plant with spikes of usually blue flowers.

delta /'deltə/ *noun* triangular alluvial tract at mouth of river; fourth letter of Greek alphabet (Δ, δ); fourth-class mark for work etc. □ **delta wing** triangular swept-back wing of aircraft.

delude /dɪ'lu:d/ *verb* (**-ding**) deceive, mislead.

deluge /'delju:dʒ/ ● *noun* flood; downpour of rain; overwhelming rush. ● *verb* (**-ging**) flood, inundate.

delusion /dɪ'lu:ʒ(ə)n/ *noun* false belief or hope. □ **delusive** *adjective*; **delusory** *adjective.*

de luxe /də 'lʌks/ *adjective* luxurious; superior; sumptuous.

delve *verb* (**-ving**) (often + *in, into*) research, search deeply; *poetical* dig.

demagogue /'deməgɒg/ *noun* political agitator appealing to emotion. □ **demagogic** /-'gɒgɪk/ *adjective*; **demagogy** /-gɒgɪ/) *noun.*

demand /dɪ'mɑ:nd/ ● *noun* insistent and peremptory request; desire for commodity; urgent claim. ● *verb* (often + *of, from, to do, that*) ask for insistently; require; (as **demanding** *adjective*) requiring effort, attention, etc. □ **demand feeding** feeding baby whenever it cries.

demarcation /di:mɑ:'keɪʃ(ə)n/ *noun* marking of boundary or limits; trade union practice of restricting job to one union. □ **demarcate** /'di:-/ *verb* (**-ting**).

demean /dɪ'mi:n/ *verb* (usually **demean oneself**) lower dignity of.

demeanour /dɪ'mi:nə/ *noun* (*US* **demeanor**) bearing; outward behaviour.

demented /dɪ'mentɪd/ *adjective* mad.

dementia /dɪ'menʃə/ *noun* chronic insanity. □ **dementia praecox** /'pri:kɒks/ schizophrenia.

demerara /demə'reərə/ *noun* light brown cane sugar.

demerit /di:'merɪt/ *noun* fault, defect.

demesne /dɪ'mi:n/ *noun* landed property, estate; possession (of land) as one's own.

demigod /'demɪgɒd/ *noun* partly divine being; *colloquial* godlike person.

demijohn /'demɪdʒɒn/ *noun* large wicker-cased bottle.

demilitarize /di:'mɪlɪtəraɪz/ *verb* (also **-ise**) (**-zing** or **-sing**) remove army etc. from (zone etc.).

demi-monde /'demɪmɒnd/ *noun* class of women of doubtful morality; semi-respectable group. [French]

demise /dɪ'maɪz/ *noun* death; termination.

demisemiquaver /demɪ'semɪkweɪvə/ *noun Music* note equal to half semiquaver.

demist /diː'mɪst/ *verb* clear mist from (windscreen etc.). □ **demister** *noun*.

demo /'deməʊ/ *noun* (*plural* **-s**) *colloquial* demonstration, esp. political.

demobilize /diː'məʊbɪlaɪz/ *verb* (also **-ise**) (**-zing** or **-sing**) disband (troops etc.). □ **demobilization** *noun*.

democracy /dɪ'mɒkrəsɪ/ *noun* (*plural* **-ies**) government by the whole population, usually through elected representatives; state so governed.

democrat /'deməkræt/ *noun* advocate of democracy; (**Democrat**) member of US Democratic Party.

democratic /demə'krætɪk/ *adjective* of, like, or practising democracy; favouring social equality. □ **democratically** *adverb*; **democratize** /dɪ'mɒkrətaɪz/ *verb* (also **-ise**) (**-zing** or **-sing**); **democratization** *noun*.

demography /dɪ'mɒgrəfɪ/ *noun* study of statistics of birth, deaths, disease, etc. □ **demographic** /demə'græfɪk/ *adjective*.

demolish /dɪ'mɒlɪʃ/ *verb* pull down (building); destroy; refute; eat up voraciously. □ **demolition** /demə'lɪʃ(ə)n/ *noun*.

demon /'diːmən/ *noun* devil; evil spirit; forceful or skilful performer. □ **demonic** /dɪ'mɒnɪk/ *adjective*.

demoniac /dɪ'məʊnɪæk/ ● *adjective* frenzied; supposedly possessed by evil spirit; of or like demons. ● *noun* demoniac person. □ **demoniacal** /diːmə'naɪək(ə)l/ *adjective*.

demonology /diːmə'nɒlədʒɪ/ *noun* study of demons.

demonstrable /'demənstrəb(ə)l/ *adjective* able to be shown or proved. □ **demonstrably** *adverb*.

demonstrate /'demənstreɪt/ *verb* (**-ting**) show (feelings etc.); describe and explain by experiment etc.; prove truth or existence of; take part in public demonstration.

demonstration *noun* demonstrating; (+ *of*) show of feeling etc.; political public march, meeting, etc.; proof by logic, argument, etc.

demonstrative /dɪ'mɒnstrətɪv/ *adjective* showing feelings readily; affectionate; *Grammar* indicating person or thing referred to. □ **demonstratively** *adverb*; **demonstrativeness** *noun*.

demonstrator /'demənstreɪtə/ *noun* person making or taking part in demonstration.

demoralize /diː'mɒrəlaɪz/ *verb* (also **-ise**) (**-zing** or **-sing**) destroy morale of. □ **demoralization** *noun*.

demote /diː'məʊt/ *verb* (**-ting**) reduce to lower rank or class. □ **demotion** /-'məʊʃ(ə)n/ *noun*.

demotic /dɪ'mɒtɪk/ ● *noun* colloquial form of a language. ● *adjective* colloquial, vulgar.

demotivate /diː'məʊtɪveɪt/ *verb* (**-ting**) cause to lose motivation. □ **demotivation** *noun*.

demur /dɪ'mɜː/ ● *verb* (**-rr-**) (often + *to*, *at*) raise objections. ● *noun* (usually in negative) objection, objecting.

demure /dɪ'mjʊə/ *adjective* (**-r**, **-st**) quiet, modest; coy. □ **demurely** *adverb*.

demystify /diː'mɪstɪfaɪ/ *verb* (**-ies**, **-ied**) remove mystery from.

den *noun* wild animal's lair; place of crime or vice; small private room.

denarius /dɪ'neərɪəs/ *noun* (*plural* **-rii** /-rɪaɪ/) ancient Roman silver coin.

denationalize /diː'næʃənəlaɪz/ *verb* (also **-ise**) (**-zing** or **-sing**) transfer (industry etc.) from national to private ownership. □ **denationalization** *noun*.

denature /diː'neɪtʃə/ *verb* (**-ring**) change properties of; make (alcohol) unfit for drinking.

dendrology /den'drɒlədʒɪ/ *noun* study of trees. □ **dendrologist** *noun*.

denial /dɪ'naɪəl/ *noun* denying or refusing.

denier /'denjə/ *noun* unit of weight measuring fineness of silk, nylon, etc.

denigrate /'denɪgreɪt/ *verb* (**-ting**) sully reputation of. □ **denigration** *noun*; **denigratory** /-'greɪt-/ *adjective*.

denim /'denɪm/ *noun* twilled cotton fabric; (in *plural*) jeans etc. made of this.

denizen /'denɪz(ə)n/ *noun* (usually + *of*) inhabitant or occupant.

denominate /dɪ'nɒmɪneɪt/ *verb* (**-ting**) give name to, describe as, call.

denomination *noun* Church or religious sect; class of measurement or money; name, esp. for classification. □ **denominational** *adjective.*

denominator *noun* number below line in vulgar fraction; divisor.

denote /dɪ'nəʊt/ *verb* (**-ting**) (often + *that*) be sign of; indicate; be name for, signify. □ **denotation** /di:nə'teɪʃ(ə)n/ *noun.*

denouement /deɪ'nu:mɑ̃/ *noun* final resolution in play, novel, etc.

denounce /dɪ'naʊns/ *verb* (**-cing**) accuse publicly; inform against.

dense /dens/ *adjective* closely compacted; crowded together; stupid. □ **densely** *adverb*; **denseness** *noun.*

density /'densɪtɪ/ *noun* (*plural* **-ies**) denseness; quantity of mass per unit volume; opacity of photographic image.

dent ● *noun* depression in surface; noticeable adverse effect. ● *verb* make dent in.

dental /'dent(ə)l/ *adjective* of teeth or dentistry; (of sound) made with tongue-tip against front teeth. □ **dental floss** thread used to clean between teeth; **dental surgeon** dentist.

dentate /'denteɪt/ *adjective* toothed, notched.

dentifrice /'dentɪfrɪs/ *noun* tooth powder or toothpaste.

dentine /'denti:n/ *noun* hard dense tissue forming most of tooth.

dentist /'dentɪst/ *noun* person qualified to treat, extract, etc., teeth. □ **dentistry** *noun.*

denture /'dentʃə/ *noun* (usually in *plural*) removable artificial teeth.

denude /dɪ'nju:d/ *verb* (**-ding**) make naked or bare; (+ *of*) strip of (covering etc.). □ **denudation** /di:-/ *noun.*

denunciation /dɪnʌnsɪ'eɪʃ(ə)n/ *noun* denouncing.

deny /dɪ'naɪ/ *verb* (**-ies**, **-ied**) declare untrue or non-existent; repudiate; (often + *to*) withhold from; (**deny oneself**) be abstinent.

deodorant /di:'əʊdərənt/ *noun* substance applied to body or sprayed into air to conceal smells.

deodorize /di:'əʊdəraɪz/ *verb* (also **-ise**) (**-zing** or **-sing**) remove smell of. □ **deodorization** *noun.*

deoxyribonucleic acid /di:ɒksɪraɪbəʊnju:'kleɪɪk/ see DNA.

dep. *abbreviation* departs; deputy.

depart /dɪ'pɑ:t/ *verb* (often + *from*) go away, leave; (usually + *for*) set out; (usually + *from*) deviate. □ **depart this life** *formal* die.

departed ● *adjective* bygone. ● *noun* (**the departed**) *euphemistic* dead person or people.

department *noun* separate part of complex whole, esp. branch of administration; division of school etc.; section of large store; area of expertise; French administrative district. □ **department store** shop with many departments. □ **departmental** /di:pɑ:t'ment(ə)l/ *adjective.*

departure /dɪ'pɑ:tʃə/ *noun* departing; new course of action etc.

depend /dɪ'pend/ *verb* (often + *on*, *upon*) be controlled or determined by; (+ *on*, *upon*) need, rely on.

dependable *adjective* reliable. □ **dependability** *noun.*

dependant *noun* person supported, esp. financially, by another.

dependence *noun* depending, being dependent; reliance.

dependency *noun* (*plural* **-ies**) country etc. controlled by another; dependence (on drugs etc.).

dependent *adjective* (usually + *on*) depending; unable to do without (esp. drug); maintained at another's cost; (of clause etc.) subordinate to word etc.

depict /dɪ'pɪkt/ *verb* represent in painting etc.; describe. □ **depiction** *noun.*

depilate /'depɪleɪt/ *verb* (**-ting**) remove hair from. □ **depilation** *noun*; **depilator** *noun.*

depilatory /dɪ'pɪlətərɪ/ ● *adjective* that removes unwanted hair. ● *noun* (*plural* **-ies**) depilatory substance.

deplete /dɪ'pli:t/ *verb* (**-ting**) (esp. as **depleted** *adjective*) reduce in numbers, quantity, etc.; exhaust. □ **depletion** *noun.*

deplorable /dɪ'plɔ:rəb(ə)l/ *adjective* exceedingly bad. □ **deplorably** *adverb.*

deplore /dɪ'plɔ:/ *verb* (**-ring**) find deplorable; regret.

deploy /dɪ'plɔɪ/ *verb* spread out (troops) into line for action; use (arguments etc.) effectively. □ **deployment** *noun.*

deponent /dɪ'pəʊnənt/ *noun* person making deposition under oath.

depopulate /di:'pɒpjʊleɪt/ *verb* (**-ting**) reduce population of. □ **depopulation** *noun*.

deport /dɪ'pɔ:t/ *verb* remove forcibly or exile to another country; (**deport oneself**) behave (well, badly, etc.). □ **deportation** /di:-/ *noun*.

deportee /di:pɔ:'ti:/ *noun* deported person.

deportment *noun* bearing, demeanour.

depose /dɪ'pəʊz/ *verb* (**-sing**) remove from office; dethrone; (usually + *to, that*) testify on oath.

deposit /dɪ'pɒzɪt/ ● *noun* money in bank account; thing stored for safe keeping; payment as pledge or first instalment; returnable sum paid on hire of item; layer of accumulated matter. ● *verb* (**-t-**) entrust for keeping; pay or leave as deposit; put or lay down. □ **deposit account** bank account that pays interest but is not usually immediately accessible.

depositary /dɪ'pɒzɪtərɪ/ *noun* (*plural* **-ies**) person to whom thing is entrusted.

deposition /depə'zɪʃ(ə)n/ *noun* deposing; sworn evidence; giving of this; depositing.

depositor *noun* person who deposits money, property, etc.

depository /dɪ'pɒzɪtərɪ/ *noun* (*plural* **-ies**) storehouse; store (of wisdom etc.); depositary.

depot /'depəʊ/ *noun* military storehouse or headquarters; place where vehicles, e.g. buses, are kept; goods yard.

deprave /dɪ'preɪv/ *verb* (**-ving**) corrupt morally.

depravity /dɪ'prævɪtɪ/ *noun* (*plural* **-ies**) moral corruption; wickedness.

deprecate /'deprɪkeɪt/ *verb* (**-ting**) express disapproval of. □ **deprecation** *noun*; **deprecatory** /-'keɪtərɪ/ *adjective*.

■ **Usage** *Deprecate* is often confused with *depreciate*.

depreciate /dɪ'pri:ʃɪeɪt/ *verb* (**-ting**) diminish in value; belittle. □ **depreciatory** /-ʃətərɪ/ *adjective*.

■ **Usage** *Depreciate* is often confused with *deprecate*.

depreciation *noun* depreciating; decline in value.

depredation /deprɪ'deɪʃ(ə)n/ *noun* (usually in *plural*) despoiling, ravaging.

depress /dɪ'pres/ *verb* make dispirited; lower; push down; reduce activity of (esp. trade); (as **depressed** *adjective*) suffering from depression. □ **depressing** *adjective*; **depressingly** *adverb*.

depressant ● *adjective* reducing activity, esp. of body function. ● *noun* depressant substance.

depression /dɪ'preʃ(ə)n/ *noun* extreme dejection; long slump; lowering of atmospheric pressure; hollow on a sur-face.

depressive /dɪ'presɪv/ ● *adjective* tending towards depression; tending to depress. ● *noun* chronically depressed person.

deprivation /deprɪ'veɪʃ(ə)n/ *noun* depriving, being deprived.

deprive /dɪ'praɪv/ *verb* (**-ving**) (usually + *of*) prevent from having or enjoying; (as **deprived** *adjective*) lacking what is needed, underprivileged.

Dept. *abbreviation* Department.

depth *noun* deepness; measure of this; wisdom; intensity; (usually in *plural*) deep, lowest, or inmost part, middle (of winter etc.), abyss, depressed state. □ **depth-charge** bomb exploding under water; **in depth** thoroughly.

deputation /depjʊ'teɪʃ(ə)n/ *noun* delegation.

depute /dɪ'pju:t/ *verb* (**-ting**) (often + *to*) delegate (task, authority); authorize as representative.

deputize /'depjʊtaɪz/ *verb* (also **-ise**) (**-zing** or **-sing**) (usually + *for*) act as deputy.

deputy /'depjʊtɪ/ *noun* (*plural* **-ies**) person appointed to act for another; parliamentary representative in some countries.

derail /di:'reɪl/ *verb* cause (train etc.) to leave rails. □ **derailment** *noun*.

derange /dɪ'reɪndʒ/ *verb* (**-ging**) (usually as **deranged** *adjective*) make insane. □ **derangement** *noun*.

Derby /'dɑ:bɪ/ *noun* (*plural* **-ies**) annual horse race at Epsom; similar race or sporting event.

derelict /'derɪlɪkt/ ● *adjective* dilapidated; abandoned. ● *noun* vagrant; abandoned property.

dereliction /derɪ'lɪkʃ(ə)n/ *noun* (usually + *of*) neglect (of duty etc.).

deride /dɪ'raɪd/ *verb* (**-ding**) mock. □ **derision** /-'rɪʒ-/ *noun.*

de rigueur /də rɪ'gɜ:/ *adjective* required by fashion or etiquette.

derisive /dɪ'raɪsɪv/ *adjective* scoffing, ironical. □ **derisively** *adverb.*

derisory /dɪ'raɪsərɪ/ *adjective* (of sum offered etc.) ridiculously small; derisive.

derivation /derɪ'veɪʃ(ə)n/ *noun* deriving, being derived; origin or formation of word; tracing of this.

derivative /dɪ'rɪvətɪv/ ● *adjective* derived, not original. ● *noun* derived word or thing.

derive /dɪ'raɪv/ *verb* (**-ving**) (usually + *from*) get or trace from a source; (+ *from*) arise from; (usually + *from*) assert origin and formation of (word etc.).

dermatitis /dɜ:mə'taɪtɪs/ *noun* inflammation of skin.

dermatology /dɜ:mə'tɒlədʒɪ/ *noun* study of skin diseases. □ **dermatological** /-tə'lɒdʒ-/ *adjective*; **dermatologist** *noun.*

derogatory /dɪ'rɒgətərɪ/ *adjective* disparaging; insulting.

derrick /'derɪk/ *noun* crane; framework over oil well etc. for drilling machinery.

derris /'derɪs/ *noun* insecticide made from powdered root of tropical plant.

derv *noun* diesel fuel for road vehicles.

dervish /'dɜ:vɪʃ/ *noun* member of Muslim fraternity vowed to poverty and austerity.

DES *abbreviation historical* Department of Education and Science.

descale /di:'skeɪl/ *verb* (**-ling**) remove scale from.

descant /'deskænt/ *noun* harmonizing treble melody above basic hymn tune etc.

descend /dɪ'send/ *verb* come, go, or slope down; sink; (usually + *on*) make sudden attack or visit; (+ *to*) stoop (to unworthy act); be passed on by inheritance. □ **be descended from** have as an ancestor.

descendant /dɪ'send(ə)nt/ *noun* person etc. descended from another.

descent /dɪ'sent/ *noun* act or way of descending; downward slope; lineage; decline, fall; sudden attack.

describe /dɪ'skraɪb/ *verb* (**-bing**) state appearance, characteristics, etc. of; (+ *as*) assert to be; draw or move in (curve etc.).

description /dɪ'skrɪpʃ(ə)n/ *noun* describing, being described; sort, kind.

descriptive /dɪ'skrɪptɪv/ *adjective* describing, esp. vividly.

descry /dɪ'skraɪ/ *verb* (**-ies**, **-ied**) catch sight of; discern.

desecrate /'desɪkreɪt/ *verb* (**-ting**) violate sanctity of. □ **desecration** *noun*; **desecrator** *noun.*

desegregate /di:'segrɪgeɪt/ *verb* (**-ting**) abolish racial segregation in. □ **desegregation** *noun.*

deselect /di:sɪ'lekt/ *verb* reject (esp. sitting MP) in favour of another. □ **deselection** *noun.*

desensitize /di:'sensɪtaɪz/ *verb* (also **-ise**) (**-zing** or **-sing**) reduce or destroy sensitivity of. □ **desensitization** *noun.*

desert[1] /dɪ'zɜ:t/ *verb* leave without intending to return; (esp. as **deserted** *adjective*) forsake, abandon; run away from military service. □ **deserter** *noun Military*; **desertion** *noun.*

desert[2] /'dezət/ *noun* dry barren, esp. sandy, tract. □ **desert island** (usually tropical) uninhabited island.

desertification /dɪzɜ:tɪfɪ'keɪʃ(ə)n/ *noun* making or becoming a desert.

deserts /dɪ'zɜ:ts/ *plural noun* deserved reward or punishment.

deserve /dɪ'zɜ:v/ *verb* (**-ving**) (often + *to do*) be worthy of (reward, punishment); (as **deserving** *adjective*) (often + *of*) worthy (esp. of help, praise, etc.). □ **deservedly** /-vɪdlɪ/ *adverb.*

desiccate /'desɪkeɪt/ *verb* (**-ting**) remove moisture from, dry out. □ **desiccation** *noun.*

desideratum /dɪzɪdə'rɑ:təm/ *noun* (*plural* **-ta**) something lacking but desirable.

design /dɪ'zaɪn/ ● *noun* (art of producing) sketch or plan for product; lines or shapes as decoration; layout; established form of product; mental plan;

purpose. ● *verb* produce design for; be designer; intend; (as **designing** *adjective*) crafty, scheming. □ **have designs on** plan to take, seduce, etc.

designate ● *verb* /'dezɪgneɪt/ (**-ting**) (often + *as*) appoint to office or function; specify; (often + *as*) describe as. ● *adjective* /'dezɪgnət/ (after noun) appointed but not yet installed.

designation /dezɪg'neɪʃ(ə)n/ *noun* name or title; designating.

designedly /dɪ'zaɪnɪdlɪ/ *adverb* intentionally.

designer ● *noun* person who designs e.g. clothing, machines, theatre sets; draughtsman. ● *adjective* bearing label of famous fashion designer; prestigious. □ **designer drug** synthetic equivalent of illegal drug.

desirable /dɪ'zaɪərəb(ə)l/ *adjective* worth having or doing; sexually attractive. □ **desirability** *noun.*

desire /dɪ'zaɪə/ ● *noun* unsatisfied longing; expression of this; request; thing desired; sexual appetite. ● *verb* (**-ring**) (often + *to do*, *that*) long for; request.

desirous *adjective* (usually + *of*) desiring, wanting; hoping.

desist /dɪ'zɪst/ *verb* (often + *from*) cease.

desk *noun* piece of furniture with writing surface, and often drawers; counter in hotel, bank, etc.; section of newspaper office.

desktop *noun* working surface of desk; computer for use on ordinary desk. □ **desktop publishing** printing with desktop computer and high-quality printer.

desolate ● *adjective* /'desələt/ left alone; uninhabited; dreary, forlorn. ● *verb* /'desəleɪt/ (**-ting**) depopulate; devastate; (esp. as **desolated** *adjective*) make wretched. □ **desolately** /-lətlɪ/ *adverb*; **desolation** *noun.*

despair /dɪ'speə/ ● *noun* loss or absence of hope; cause of this. ● *verb* (often + *of*) lose all hope.

despatch = DISPATCH.

desperado /despə'rɑ:dəʊ/ *noun* (*plural* **-es** or *US* **-s**) desperate or reckless criminal etc.

desperate /'despərət/ *adjective* reckless from despair; violent and lawless; extremely dangerous or serious; (usually + *for*) needing or desiring very much. □ **desperately** *adverb*; **desperation** *noun.*

despicable /'despɪkəb(ə)l, dɪ'spɪk-/ *adjective* contemptible. □ **despicably** *adverb.*

despise /dɪ'spaɪz/ *verb* (**-sing**) regard as inferior or contemptible.

despite /dɪ'spaɪt/ *preposition* in spite of.

despoil /dɪ'spɔɪl/ *verb* (often + *of*) plunder, rob. □ **despoliation** /-spəʊlɪ-/ *noun.*

despondent /dɪ'spɒnd(ə)nt/ *adjective* in low spirits, dejected. □ **despondence** *noun*; **despondency** *noun*; **despondently** *adverb.*

despot /'despɒt/ *noun* absolute ruler; tyrant. □ **despotic** /-'spɒt-/ *adjective.*

despotism /'despətɪz(ə)m/ *noun* rule by despot.

dessert /dɪ'zɜ:t/ *noun* sweet course of a meal. □ **dessertspoon** medium-sized spoon for dessert, (also **dessertspoonful**) amount held by this.

destabilize /di:'steɪbɪlaɪz/ *verb* (also **-ise**) (**-zing** or **-sing**) make unstable; subvert (esp. foreign government). □ **destabilization** *noun.*

destination /destɪ'neɪʃ(ə)n/ *noun* place to which person or thing is going.

destine /'destɪn/ *verb* (**-ning**) (often + *to*, *for*, *to do*) appoint; preordain; intend. □ **be destined to** be fated to.

destiny /'destɪnɪ/ *noun* (*plural* **-ies**) fate; this as power.

destitute /'destɪtju:t/ *adjective* without food or shelter etc.; (usually + *of*) lacking. □ **destitution** /-'tju:-/ *noun.*

destroy /dɪ'strɔɪ/ *verb* pull or break down; kill; make useless; ruin financially; defeat.

destroyer /dɪ'strɔɪə/ *noun* fast medium-sized warship; person or thing that destroys.

destruct /dɪ'strʌkt/ *verb* destroy or be destroyed deliberately. □ **destructible** *adjective.*

destruction *noun* destroying, being destroyed.

destructive *adjective* destroying or tending to destroy; negatively critical.

desuetude /dɪ'sju:ɪtju:d/ *noun formal* state of disuse.

desultory /ˈdezəltərɪ/ *adjective* constantly turning from one subject to another; unmethodical.

detach /dɪˈtætʃ/ *verb* (often + *from*) unfasten and remove; send (troops) on separate mission; (as **detached** *adjective*) impartial, unemotional, (of house) standing separate. □ **detachable** *adjective.*

detachment *noun* indifference; impartiality; detaching, being detached; troops etc. detached for special duty.

detail /ˈdi:teɪl/ ● *noun* small separate item or particular; these collectively; minor or intricate decoration; small part of picture etc. shown alone; small military detachment. ● *verb* give particulars of, relate in detail; (as **detailed** *adjective*) containing many details, itemized; assign for special duty. □ **in detail** item by item, minutely.

detain /dɪˈteɪn/ *verb* keep waiting, delay; keep in custody. □ **detainment** *noun.*

detainee /di:teɪˈni:/ *noun* person kept in custody, esp. for political reasons.

detect /dɪˈtekt/ *verb* discover; perceive. □ **detectable** *adjective*; **detection** *noun*; **detector** *noun.*

detective /dɪˈtektɪv/ *noun* person, usually police officer, investigating crime etc.

détente /deɪˈtɑ̃t/ *noun* relaxing of strained international relations. [French]

detention /dɪˈtenʃ(ə)n/ *noun* detaining, being detained. □ **detention centre** short-term prison for young offenders.

deter /dɪˈtɜ:/ *verb* (**-rr-**) (often + *from*) discourage or prevent, esp. through fear.

detergent /dɪˈtɜ:dʒ(ə)nt/ ● *noun* synthetic cleansing agent used with water. ● *adjective* cleansing.

deteriorate /dɪˈtɪərɪəreɪt/ *verb* (**-ting**) become worse. □ **deterioration** *noun.*

determinant /dɪˈtɜ:mɪnənt/ *noun* decisive factor.

determinate /dɪˈtɜ:mɪnət/ *adjective* limited; of definite scope or nature.

determination /dɪtɜ:mɪˈneɪʃ(ə)n/ *noun* resolute purpose; deciding, determining.

determine /dɪˈtɜ:mɪn/ *verb* (**-ning**) find out precisely; settle, decide; (+ *to do*) resolve; be decisive factor in.

determined *adjective* resolute. □ **be determined** (usually + *to do*) be resolved. □ **determinedly** *adverb.*

determinism /dɪˈtɜ:mɪnɪz(ə)m/ *noun* theory that action is determined by forces independent of will. □ **determinist** *noun & adjective*; **deterministic** /-ˈnɪs-/ *adjective.*

deterrent /dɪˈterənt/ ● *adjective* deterring. ● *noun* thing that deters (esp. nuclear weapon).

detest /dɪˈtest/ *verb* hate, loathe. □ **detestation** /di:tesˈteɪʃ(ə)n/ *noun.*

detestable /dɪˈtestəb(ə)l/ *adjective* hated, loathed.

dethrone /di:ˈθrəʊn/ *verb* (**-ning**) remove from throne or high regard. □ **dethronement** *noun.*

detonate /ˈdetəneɪt/ *verb* (**-ting**) set off (explosive charge); be set off. □ **detonation** *noun.*

detonator *noun* device for detonating.

detour /ˈdi:tʊə/ *noun* divergence from usual route; roundabout course.

detoxify /di:ˈtɒksɪfaɪ/ *verb* (**-ies**, **-ied**) remove poison or harmful substances from. □ **detoxification** *noun.*

detract /dɪˈtrækt/ *verb* (+ *from*) diminish. □ **detraction** *noun.*

detractor *noun* person who criticizes unfairly.

detriment /ˈdetrɪmənt/ *noun* damage, harm; cause of this. □ **detrimental** /-ˈmen-/ *adjective.*

detritus /dɪˈtraɪtəs/ *noun* gravel, rock, etc. produced by erosion; debris.

de trop /də ˈtrəʊ/ *adjective* superfluous; in the way. [French]

deuce[1] /dju:s/ *noun* two on dice or cards; *Tennis* score of 40 all.

deuce[2] /dju:s/ *noun* (**the deuce**) (in exclamations) the Devil.

deuterium /dju:ˈtɪərɪəm/ *noun* stable isotope of hydrogen with mass about twice that of the usual isotope.

Deutschmark /ˈdɔɪtʃmɑ:k/ *noun* chief monetary unit of Germany.

devalue /di:ˈvælju:/ *verb* (**-ues**, **-ued**, **-uing**) reduce value of, esp. currency relative to others or to gold. □ **devaluation** *noun.*

devastate /ˈdevəsteɪt/ *verb* (**-ting**) lay waste; cause great destruction to; (often as **devastated** *adjective*) overwhelm with shock or grief. □ **devastation** *noun.*

devastating *adjective* crushingly effective; overwhelming; *colloquial* stunningly beautiful. □ **devastatingly** *adverb.*

develop /dɪ'veləp/ *verb* (**-p-**) make or become fuller, bigger, or more elaborate, etc.; bring or come to active, visible, or mature state; begin to exhibit or suffer from; build on (land); convert (land) to new use; treat (photographic film) to make image visible. □ **developing country** poor or primitive country. □ **developer** *noun.*

development *noun* developing, being developed; stage of growth or advancement; newly developed thing, event, etc.; area of developed land, esp. with buildings. □ **developmental** /-'ment(ə)l/ *adjective.*

deviant /'di:vɪənt/ ● *adjective* deviating from normal, esp. sexual, behaviour. ● *noun* deviant person or thing. □ **deviance** *noun*; **deviancy** *noun* (*plural* **-ies**).

deviate /'di:vɪeɪt/ *verb* (**-ting**) (often + *from*) turn aside; diverge. □ **deviation** *noun.*

device /dɪ'vaɪs/ *noun* thing made or adapted for particular purpose; scheme, trick; heraldic design. □ **leave (person) to his** or **her own devices** leave (person) to do as he or she wishes.

devil /'dev(ə)l/ ● *noun* (usually **the Devil**) Satan; supreme spirit of evil; personified evil; mischievously clever person. ● *verb* (**-ll-**; *US* **-l-**) (usually as **devilled** *adjective*) cook with hot spices. □ **devil-may-care** cheerful and reckless; **devil's advocate** person who tests proposition by arguing against it.

devilish ● *adjective* of or like a devil; mischievous. ● *adverb colloquial* very. □ **devilishly** *adverb.*

devilment *noun* mischief; wild spirits.

devilry /'devəlrɪ/ *noun* (*plural* **-ies**) wickedness; reckless mischief; black magic.

devious /'di:vɪəs/ *adjective* not straightforward, underhand; winding, circuitous. □ **deviously** *adverb*; **deviousness** *noun.*

devise /dɪ'vaɪz/ *verb* (**-sing**) plan or invent; *Law* leave (real estate) by will.

devoid /dɪ'vɔɪd/ *adjective* (+ *of*) lacking, free from.

devolution /di:və'lu:ʃ(ə)n/ *noun* delegation of power esp. to local or regional administration. □ **devolutionist** *noun & adjective.*

devolve /dɪ'vɒlv/ *verb* (**-ving**) (+ *on*, *upon*, etc.) (of duties etc.) pass or be passed to another; (+ *on*, *to*, *upon*) (of property) descend to.

devote /dɪ'vəʊt/ *verb* (**-ting**) (+ *to*) apply or give over to (particular activity etc.).

devoted *adjective* loving, loyal. □ **devotedly** *adverb.*

devotee /devəʊ'ti:/ *noun* (usually + *of*) enthusiast, supporter; pious person.

devotion /dɪ'vəʊʃ(ə)n/ *noun* (usually + *to*) great love or loyalty; worship; (in *plural*) prayers. □ **devotional** *adjective.*

devour /dɪ'vaʊə/ *verb* eat voraciously; (of fire etc.) engulf, destroy; take in greedily (with eyes or ears).

devout /dɪ'vaʊt/ *adjective* earnestly religious or sincere. □ **devoutly** *adverb*; **devoutness** *noun.*

dew *noun* condensed water vapour forming on cool surfaces at night; similar glistening moisture. □ **dewberry** fruit like blackberry; **dew-claw** rudimentary inner toe on some dogs; **dewdrop** drop of dew; **dew point** temperature at which dew forms. □ **dewy** *adjective* (**-ier**, **-iest**).

Dewey Decimal system /'dju:ɪ/ *noun* system of library classification.

dewlap *noun* fold of loose skin hanging from throat esp. in cattle.

dexter /'dekstə/ *adjective* on or of the right-hand side (observer's left) of a heraldic shield etc.

dexterous /'dekstrəs/ *adjective* (also **dextrous**) skilful at handling. □ **dexterity** /-'ter-/ *noun*; **dexterously** *adverb.*

dhal /dɑ:l/ *noun* (also **dal**) kind of split pulse from India; dish made with this.

dharma /'dɑ:mə/ *noun* right behaviour; Buddhist truth; Hindu moral law.

dhoti /'dəʊtɪ/ *noun* (*plural* **-s**) loincloth worn by male Hindus.

dia. *abbreviation* diameter.

diabetes /daɪə'bi:ti:z/ *noun* disease in which sugar and starch are not properly absorbed by the body.

diabetic /daɪə'betɪk/ ● *adjective* of or having diabetes; for diabetics. ● *noun* diabetic person.

diabolical /daɪə'bɒlɪk(ə)l/ *adjective* (also **diabolic**) of the Devil; inhumanly cruel or wicked; extremely bad. □ **diabolically** *adverb.*

diabolism /daɪ'æbəlɪz(ə)m/ *noun* worship of the Devil; sorcery.

diaconate /daɪ'ækənət/ *noun* office of deacon; deacons collectively.

diacritic /daɪə'krɪtɪk/ *noun* sign (e.g. accent) indicating sound or value of letter.

diacritical *adjective* distinguishing.

diadem /'daɪədem/ *noun* crown.

diaeresis /daɪ'ɪərəsɪs/ *noun* (*plural* **-reses** /-si:z/) (*US* **dieresis**) mark (¨) over vowel to show it is sounded separately.

diagnose /daɪəg'nəʊz/ *verb* (**-sing**) make diagnosis of.

diagnosis /daɪəg'nəʊsɪs/ *noun* (*plural* **-noses** /-si:z/) identification of disease or fault from symptoms etc.

diagnostic /daɪəg'nɒstɪk/ ● *adjective* of or assisting diagnosis. ● *noun* symptom.

diagnostics *noun* (treated as *plural*) programs etc. used to identify faults in computing; (treated as *singular*) science of diagnosing disease.

diagonal /daɪ'ægən(ə)l/ ● *adjective* crossing a straight-sided figure from corner to corner, oblique. ● *noun* straight line joining two opposite corners. □ **diagonally** *adverb.*

diagram /'daɪəgræm/ *noun* outline drawing, plan, etc. of thing or process. □ **diagrammatic** /-grə'mætɪk/ *adjective.*

dial /'daɪəl/ ● *noun* plate with scale and pointer for measuring; numbered disc on telephone for making connection; face of clock or watch; disc on television etc. for selecting channel etc. ● *verb* (**-ll-**; *US* **-l-**) select (telephone number) with dial. □ **dialling tone** sound indicating that telephone caller may dial.

dialect /'daɪəlekt/ *noun* regional form of language (see panel).

dialectic /daɪə'lektɪk/ *noun* process or situation involving contradictions or conflict of opposites and their resolution; = DIALECTICS.

dialectical *adjective* of dialectic. □ **dialectical materialism** Marxist theory that historical events arise from conflicting economic (and therefore social) conditions. □ **dialectically** *adverb.*

dialectics *noun* (treated as *singular* or *plural*) art of investigating truth by discussion and logic.

dialogue /'daɪəlɒg/ *noun* (*US* **dialog**) conversation, esp. in a play, novel, etc.; discussion between people of different opinions.

dialysis /daɪ'æləsɪs/ *noun* (*plural* **-lyses** /-si:z/) separation of particles in liquid by differences in their ability to pass through membrane; purification of blood by this technique.

diamanté /dɪə'mɑ̃teɪ/ *adjective* decorated with synthetic diamonds etc.

diameter /daɪ'æmɪtə/ *noun* straight line passing through centre of circle or sphere to its edges; transverse measurement.

diametrical /daɪə'metrɪk(ə)l/ *adjective* (also **diametric**) of or along diameter; (of opposites) absolute. □ **diametrically** *adverb.*

diamond /'daɪəmənd/ *noun* transparent very hard precious stone; rhombus;

Dialect

Everyone speaks a particular dialect: that is, a particular type of English distinguished by its vocabulary and its grammar. Different parts of the world and different groups of people speak different dialects: for example, Australians may say *arvo* while others say *afternoon*, and a London Cockney may say *I done it* while most other people say *I did it*. A dialect is not the same thing as an accent, which is the way a person pronounces words.

See also the panel at STANDARD ENGLISH.

playing card of suit marked with red rhombuses. □ **diamond jubilee, wedding** 60th (or 75th) anniversary of reign or wedding.

diapason /daɪə'peɪz(ə)n/ *noun* compass of musical instrument or voice; either of two main organ stops.

diaper /'daɪəpə/ *noun US* baby's nappy.

diaphanous /daɪ'æfənəs/ *adjective* (of fabric etc.) light and almost transparent.

diaphragm /'daɪəfræm/ *noun* muscular partititon between thorax and abdomen in mammals; = DUTCH CAP; vibrating disc in microphone, telephone, loudspeaker, etc.; device for varying aperture of camera lens.

diapositive /daɪə'pɒsɪtɪv/ *noun* positive photographic transparency.

diarist /'daɪərɪst/ *noun* person famous for keeping diary.

diarrhoea /daɪə'ri:ə/ *noun* (*US* **diarrhea**) condition of excessively loose and frequent bowel movements.

diary /'daɪərɪ/ *noun* (*plural* **-ies**) daily record of events etc.; book for this or for noting future engagements.

Diaspora /daɪ'æspərə/ *noun* dispersion of the Jews; the dispersed Jews.

diatonic /daɪə'tɒnɪk/ *adjective Music* (of scale etc.) involving only notes of prevailing key.

diatribe /'daɪətraɪb/ *noun* forceful verbal criticism.

diazepam /daɪ'æzɪpæm/ *noun* tranquillizing drug.

dibble /'dɪb(ə)l/ ● *noun* (also **dibber** /'dɪbə/) tool for making small holes for planting. ● *verb* (**-ling**) plant with dibble.

dice ● *noun* (*plural* same) small cube marked on each face with 1–6 spots, used in games or gambling; game played with dice. ● *verb* (**-cing**) gamble, take risks; cut into small cubes.

dicey /'daɪsɪ/ *adjective* (**dicier**, **diciest**) *slang* risky, unreliable.

dichotomy /daɪ'kɒtəmɪ/ *noun* (*plural* **-ies**) division into two.

dichromatic /daɪkraʊ'mætɪk/ *adjective* of two colours.

dick[1] *noun colloquial* (esp. in **clever dick**) person.

dick[2] *noun slang* detective.

dickens /'dɪkɪnz/ *noun* (**the dickens**) (usually after *how*, *what*, *why*, etc.) *colloquial* the Devil.

dicky /'dɪkɪ/ ● *noun* (*plural* **-ies**) *colloquial* false shirt-front. ● *adjective* (**-ier**, **-iest**) *slang* unsound. □ **dicky bow** *colloquial* bow tie.

dicotyledon /daɪkɒtɪ'li:d(ə)n/ *noun* flowering plant with two cotyledons. □ **dicotyledonous** *adjective*.

Dictaphone /'dɪktəfəʊn/ *noun proprietary term* machine for recording and playing back dictation for typing.

dictate ● *verb* /dɪk'teɪt/ (**-ting**) say or read aloud (material to be recorded etc.); state authoritatively; order peremptorily. ● *noun* /'dɪkt-/ (usually in *plural*) authoritative requirement of conscience etc. □ **dictation** *noun*.

dictator *noun* usually unelected absolute ruler; omnipotent or domineering person. □ **dictatorship** *noun*.

dictatorial /dɪktə'tɔ:rɪəl/ *adjective* of or like a dictator; overbearing. □ **dictatorially** *adverb*.

diction /'dɪkʃ(ə)n/ *noun* manner of enunciation.

dictionary /'dɪkʃənərɪ/ *noun* (*plural* **-ies**) book listing (usually alphabetically) and explaining words of a language, or giving corresponding words in another language; similar book of terms for reference.

dictum /'dɪktəm/ *noun* (*plural* **dicta** or **-s**) formal expression of opinion; a saying.

did *past* of DO[1].

didactic /dɪ'dæktɪk/ *adjective* meant to instruct; (of person) tediously pedantic. □ **didactically** *adverb*; **didacticism** /-sɪz(ə)m/ *noun*.

diddle /'dɪd(ə)l/ *verb* (**-ling**) *colloquial* swindle.

didgeridoo /dɪdʒərɪ'du:/ *noun* long tubular Australian Aboriginal instrument.

didn't /'dɪd(ə)nt/ did not.

die[1] /daɪ/ *verb* (**dying** /'daɪɪŋ/) cease to live or exist; fade away; (of fire) go out; (+ *on*) cease to live or function while in the presence or charge of (person); (+ *of*, *from*, *with*) be exhausted or tormented. □ **be dying for**, **to** desire greatly; **die down** become less loud or strong; **die hard** (of habits etc.) die reluctantly; **die-hard** conservative or stubborn person; **die off** die one after another; **die out** become extinct, cease to exist.

die[2] /daɪ/ *noun* engraved device for stamping coins etc.; (*plural* **dice**) a dice. □ **die-casting** process or product of casting from metal moulds.

dielectric /daɪɪ'lektrɪk/ • *adjective* not conducting electricity. • *noun* dielectric substance.

dierisis *US* = DIAERISIS.

diesel /'di:z(ə)l/ *noun* (in full **diesel engine**) internal-combustion engine in which heat produced by compression of air in the cylinder ignites the fuel; vehicle driven by or fuel for diesel engine. □ **diesel-electric** driven by electric current from diesel-engined generator; **diesel oil** petroleum fraction used in diesel engines.

diet[1] /'daɪət/ • *noun* habitual food; prescribed food. • *verb* (**-t-**) keep to special diet, esp. to slim. □ **dietary** *adjective*; **dieter** *noun*.

diet[2] /'daɪət/ *noun* legislative assembly; *historical* congress.

dietetic /daɪə'tetɪk/ *adjective* of diet and nutrition.

dietetics *plural noun* (usually treated as *singular*) study of diet and nutrition.

dietitian /daɪə'tɪʃ(ə)n/ *noun* (also **dietician**) expert in dietetics.

differ /'dɪfə/ *verb* (often + *from*) be unlike or distinguishable; (often + *with*) disagree.

difference /'dɪfrəns/ *noun* being different or unlike; degree of this; way in which things differ; remainder after subtraction; disagreement. □ **make a** (or **all the**, **no**, etc.) **difference** have significant (or very significant, no, etc.) effect; **with a difference** having new or unusual feature.

different /'dɪfrənt/ *adjective* (often + *from*, *to*) unlike, of another nature; separate, unusual. □ **differently** *adverb*.

■ **Usage** It is safer to use *different from*, but *different to* is common in informal use.

---

differential /dɪfə'renʃ(ə)l/ • *adjective* constituting or relating to specific difference; of, exhibiting, or depending on a difference; *Mathematics* relating to infinitesimal differences. • *noun* difference, esp. between rates of interest or wage-rates. □ **differential calculus** method of calculating rates of change, maximum or minimum values, etc.; **differential gear** gear enabling wheels to revolve at different speeds on corners.

differentiate /dɪfə'renʃɪeɪt/ *verb* (**-ting**) constitute difference between or in; distinguish; become different. □ **differentiation** *noun*.

difficult /'dɪfɪk(ə)lt/ *adjective* hard to do, deal with, or understand; troublesome.

difficulty /'dɪfɪkəltɪ/ *noun* (*plural* **-ies**) being difficult; difficult thing; hindrance; (often in *plural*) distress, esp. financial.

diffident /'dɪfɪd(ə)nt/ *adjective* lacking self-confidence. □ **diffidence** *noun*; **diffidently** *adverb*.

diffract /dɪ'frækt/ *verb* break up (beam of light) into series of dark and light bands or coloured spectra. □ **diffraction** *noun*; **diffractive** *adjective*.

diffuse • *verb* /dɪ'fju:z/ (**-sing**) spread widely or thinly; intermingle. • *adjective* /dɪ'fju:s/ spread out; not concentrated; not concise. □ **diffusible** *adjective*; **diffusive** *adjective*; **diffusion** *noun*.

dig • *verb* (**-gg-**; *past & past participle* **dug**) (often + *up*) break up and turn over (ground etc.); make (hole etc.) by digging; (+ *up*, *out*) obtain by digging, find, discover; excavate; *slang* like, understand; (+ *in*, *into*) thrust, prod. • *noun* piece of digging; thrust, poke; *colloquial* pointed remark; archaeological excavation; (in *plural*) lodgings. □ **dig in** *colloquial* begin eating; **dig oneself in** prepare defensive position.

digest • *verb* /daɪ'dʒest/ assimilate (food, information, etc.). • *noun* /'daɪdʒest/ periodical synopsis of current news etc.; summary, esp. of laws. □ **digestible** *adjective*.

digestion *noun* digesting; capacity to digest food.

digestive • *adjective* of or aiding digestion. • *noun* (in full **digestive biscuit**) wholemeal biscuit.

digger /'dɪgə/ *noun* person or machine that digs; *colloquial* Australian, New Zealander.

digit /'dɪdʒɪt/ *noun* any numeral from 0 to 9; finger or toe.

digital /ˈdɪdʒɪt(ə)l/ *adjective* of digits; (of clock, etc.) giving a reading by displayed digits; (of computer) operating on data represented by digits; (of recording) sound-information represented by digits for more reliable transmission. □ **digitally** *adverb.*

digitalis /dɪdʒɪˈteɪlɪs/ *noun* heart stimulant made from foxgloves.

digitize *verb* (also **ise**) (**-zing** or **-sing**) convert (computer data etc.) into digital form.

dignified /ˈdɪgnɪfaɪd/ *adjective* having or showing dignity.

dignify /ˈdɪgnɪfaɪ/ *verb* (**ies**, **-ied**) give dignity to.

dignitary /ˈdɪgnɪtərɪ/ *noun* (*plural* **-ies**) person of high rank or office.

dignity /ˈdɪgnɪtɪ/ *noun* (*plural* **-ies**) composed and serious manner; being worthy of respect; high rank or position.

digraph /ˈdaɪgrɑːf/ *noun* two letters representing one sound, e.g. *sh* in *show*, or *ey* in *key*.

■ **Usage** *Digraph* is sometimes confused with *ligature*, which means 'two or more letters joined'.

digress /daɪˈgres/ *verb* depart from main subject. □ **digression** *noun.*

dike = DYKE.

dilapidated /dɪˈlæpɪdeɪtɪd/ *adjective* in disrepair. □ **dilapidation** *noun.*

dilate /daɪˈleɪt/ *verb* (**-ting**) widen or expand; speak or write at length. □ **dilatation** *noun*; **dilation** *noun.*

dilatory /ˈdɪlətərɪ/ *adjective* given to or causing delay.

dilemma /daɪˈlemə/ *noun* situation in which difficult choice has to be made.

■ **Usage** *Dilemma* is sometimes also used to mean 'a difficult situation or predicament', but this is considered incorrect by some people.

dilettante /dɪlɪˈtæntɪ/ *noun* (*plural* **dilettanti** /-tɪ/ or **-s**) dabbler in a subject. □ **dilettantism** *noun.*

diligent /ˈdɪlɪdʒ(ə)nt/ *adjective* hard-working; showing care and effort. □ **diligence** *noun*; **diligently** *adverb.*

dill *noun* herb with aromatic leaves and seeds.

dilly-dally /dɪlɪˈdælɪ/ *verb* (**-ies**, **-ied**) *colloquial* dawdle, vacillate.

dilute /daɪˈljuːt/ ● *verb* (**-ting**) reduce strength (of fluid) by adding water etc.; weaken in effect. ● *adjective* diluted. □ **dilution** *noun.*

diluvial /daɪˈluːvɪəl/ *adjective* of flood, esp. Flood in Genesis.

dim ● *adjective* (**-mm-**) not bright; faintly luminous or visible; indistinctly perceived or remembered; (of eyes) not seeing clearly; *colloquial* stupid. ● *verb* (**-mm-**) make or become dim. □ **dimly** *adverb*; **dimness** *noun.*

dime *noun US* 10-cent coin.

dimension /daɪˈmenʃ(ə)n/ *noun* any measurable extent; (in *plural*) size; aspect. □ **dimensional** *adjective.*

diminish /dɪˈmɪnɪʃ/ *verb* make or become smaller or less; (often as **diminished** *adjective*) lessen reputation of (person), humiliate.

diminuendo /dɪmɪnjʊˈendəʊ/ *Music* ● *noun* (*plural* **-s**) gradual decrease in loudness. ● *adverb & adjective* decreasing in loudness.

diminution /dɪmɪˈnjuːʃ(ə)n/ *noun* diminishing.

diminutive /dɪˈmɪnjʊtɪv/ ● *adjective* tiny; (of word or suffix) implying smallness or affection. ● *noun* diminutive word or suffix.

dimmer *noun* (in full **dimmer switch**) device for varying brightness of electric light.

dimple /ˈdɪmp(ə)l/ ● *noun* small hollow, esp. in cheek or chin. ● *verb* (**-ling**) produce dimples (in).

din ● *noun* prolonged loud confused noise. ● *verb* (**-nn-**) (+ *into*) force (information) into person by repetition; make din.

dinar /ˈdiːnɑː/ *noun* chief monetary unit of (former) Yugoslavia and several Middle Eastern and N. African countries.

dine *verb* (**-ning**) eat dinner; (+ *on*, *upon*) eat for dinner; (esp. in **wine and dine**) entertain with food. □ **dining-car** restaurant on train; **dining-room** room in which meals are eaten.

diner *noun* person who dines; small dining-room; dining-car; *US* restaurant.

ding-dong /'dɪŋdɒŋ/ *noun* sound of chimes; *colloquial* heated argument.

dinghy /'dɪŋgɪ/ *noun* (*plural* **-ies**) small, often inflatable, boat.

dingle /'dɪŋg(ə)l/ *noun* deep wooded valley.

dingo /'dɪŋgəʊ/ *noun* (*plural* **-es**) wild Australian dog.

dingy /'dɪndʒɪ/ *adjective* (**-ier**, **-iest**) drab; dirty-looking. □ **dinginess** *noun*.

dinkum /'dɪŋkəm/ *adjective & adverb* (in full **fair dinkum**) *Australian & NZ colloquial* genuine(ly), honest(ly).

dinky /'dɪŋkɪ/ *adjective* (**-ier**, **-iest**) *colloquial* pretty, small and neat.

dinner /'dɪnə/ *noun* main meal, at midday or in the evening. □ **dinner-dance** formal dinner followed by dancing; **dinner jacket** man's formal evening jacket; **dinner lady** woman who supervises school dinners; **dinner service** set of matching crockery for dinner.

dinosaur /'daɪnəsɔː/ *noun* extinct usually large reptile.

dint • *noun* dent. • *verb* mark with dints. □ **by dint of** by force or means of.

diocese /'daɪəsɪs/ *noun* district under bishop's pastoral care. □ **diocesan** /daɪ'ɒsɪs(ə)n/ *adjective*.

diode /'daɪəʊd/ *noun* semiconductor allowing current in one direction and having two terminals; thermionic valve with two electrodes.

dioxide /daɪ'ɒksaɪd/ *noun* oxide with two atoms of oxygen.

Dip. *abbreviation* Diploma.

dip • *verb* (**-pp-**) put or lower briefly into liquid etc.; immerse; go below a surface or level; decline slightly or briefly; slope or extend downwards; go briefly under water; (+ *into*) look cursorily into (book, subject, etc.); (+ *into*) put (hand etc.) into (container) to take something out, use part of (resources); lower or be lowered, esp. in salute; lower beam of (headlights). • *noun* dipping, being dipped; liquid for dipping; brief bathe in sea etc.; downward slope or hollow; sauce into which food is dipped. □ **dip-switch** switch for dipping vehicle's headlights.

diphtheria /dɪf'θɪərɪə/ *noun* infectious disease with inflammation of mucous membrane esp. of throat.

diphthong /'dɪfθɒŋ/ *noun* union of two vowels in one syllable.

diplodocus /dɪ'plɒdəkəs/ *noun* (*plural* **-cuses**) huge long-necked plant-eating dinosaur.

diploma /dɪ'pləʊmə/ *noun* certificate of educational qualification; document conferring honour, privilege, etc.

diplomacy /dɪ'pləʊməsɪ/ *noun* management of international relations; tact.

diplomat /'dɪpləmæt/ *noun* member of diplomatic service; tactful person.

diplomatic /dɪplə'mætɪk/ *adjective* of or involved in diplomacy; tactful. □ **diplomatic bag** container for dispatching embassy mail; **diplomatic immunity** exemption of foreign diplomatic staff from arrest, taxation, etc.; **diplomatic service** branch of civil service concerned with representing a country abroad. □ **diplomatically** *adverb*.

diplomatist /dɪ'pləʊmətɪst/ *noun* diplomat.

dipper /'dɪpə/ *noun* diving bird; ladle.

dippy /'dɪpɪ/ *adjective* (**-ier**, **-iest**) *slang* crazy, silly.

dipsomania /dɪpsə'meɪnɪə/ *noun* alcoholism. □ **dipsomaniac** *noun*.

dipstick *noun* rod for measuring depth, esp. of oil in vehicle's engine.

dipterous /'dɪptərəs/ *adjective* two-winged.

diptych /'dɪptɪk/ *noun* painted altarpiece on two hinged panels.

dire *adjective* dreadful; ominous; *colloquial* very bad; urgent.

direct /daɪ'rekt/ • *adjective* extending or moving in straight line or by shortest route, not crooked or circuitous; straightforward, frank; without intermediaries; complete, greatest possible. • *adverb* in a direct way; by direct route. • *verb* control; guide; (+ *to do, that*) order; (+ *to*) tell way to (place); address (letter etc.); (+ *at, to, towards*) point, aim, or turn; supervise acting etc. of (film, play, etc.). □ **direct current** electric current flowing in one direction only; **direct debit** regular debiting of bank account at request of payee; **direct-grant school** school funded by government, not local authority; **direct object** primary object of verbal action (see panel at OBJECT); **direct**

speech words actually spoken, not reported (see panel); **direct tax** tax on income, paid directly to government. □ **directness** *noun.*

direction /daɪ'rekʃ(ə)n/ *noun* directing; (usually in *plural*) orders, instructions; point to, from, or along which person or thing moves or looks.

directional *adjective* of or indicating direction; sending or receiving radio or sound waves in one direction only.

directive /daɪ'rektɪv/ *noun* order from an authority.

directly ● *adverb* at once, without delay; presently, shortly; exactly; in a direct way. ● *conjunction colloquial* as soon as.

director *noun* person who directs, esp. for stage etc. or as member of board of company. □ **director-general** chief executive. □ **directorial** /-'tɔ:-/ *adjective*; **directorship** *noun.*

directorate /daɪ'rektərət/ *noun* board of directors; office of director.

directory /daɪ'rektərɪ/ *noun* (*plural* **-ies**) book with list of telephone subscribers, inhabitants of town etc., members of profession, etc. □ **directory enquiries** telephone service providing subscriber's number on request.

dirge *noun* lament for the dead; dreary piece of music.

dirham /'dɪrəm/ *noun* monetary unit of Morocco and United Arab Emirates.

dirigible /'dɪrɪdʒɪb(ə)l/ ● *adjective* that can be steered. ● *noun* dirigible balloon or airship.

dirk *noun* short dagger.

dirndl /'dɜ:nd(ə)l/ *noun* dress with close-fitting bodice and full skirt; gathered full skirt with tight waistband.

dirt *noun* unclean matter that soils; earth; foul or malicious talk; excrement. □ **dirt cheap** *colloquial* extremely cheap; **dirt track** racing track with surface of earth or cinders etc.

dirty /'dɜ:tɪ/ ● *adjective* (**-ier**, **-iest**) soiled, unclean; sordid, obscene; unfair; (of weather) rough; muddy-looking. ● *adverb slang* very; in a dirty or obscene way. ● *verb* (**-ies**, **-ied**) make or become dirty. □ **dirty look** *colloquial* look of disapproval or disgust. □ **dirtiness** *noun.*

disability /dɪsə'bɪlɪtɪ/ *noun* (*plural* **-ies**) permanent physical or mental incapacity; lack of some capacity, preventing action.

disable /dɪ'seɪb(ə)l/ *verb* (**-ling**) deprive of an ability; (often as **disabled** *adjective*) physically incapacitate. □ **disablement** *noun.*

Direct Speech

Direct speech is the actual words of a speaker quoted in writing.

1 In a novel etc., speech punctuation is used for direct speech:

- **a** The words spoken are usually put in quotation marks.
- **b** Each new piece of speech begins with a capital letter.
- **c** Each paragraph within one person's piece of speech begins with quotation marks, but only the last paragraph ends with them.

2 In a script (the written words of a play, a film, or a radio or television programme):

- **a** The names of speakers are written in the margin in capital letters.
- **b** Each name is followed by a colon.
- **c** Quotation marks are not needed.
- **d** Any instructions about the way the words are spoken or about the scenery or the actions of the speakers (stage directions) are written in the present tense in brackets or italics.

For example:

CHRISTOPHER: [Looks into box.] There's nothing in here.

---

disabuse /dɪsə'bju:z/ *verb* (**-sing**) (usually + *of*) free from mistaken idea; disillusion.

disadvantage /dɪsəd'vɑ:ntɪdʒ/ ● *noun* unfavourable condition or circumstance; loss; damage. ● *verb* (**-ging**) cause disadvantage to. □ **at a disadvantage** in an unfavourable position. □ **disadvantageous** /dɪsædvən'teɪdʒəs/ *adjective*.

disadvantaged *adjective* lacking normal opportunities through poverty, disability, etc.

disaffected /dɪsə'fektɪd/ *adjective* discontented, alienated (esp. politically). □ **disaffection** *noun*.

disagree /dɪsə'gri:/ *verb* (**-ees**, **-eed**) (often + *with*) hold different opinion; not correspond; upset. □ **disagreement** *noun*.

disagreeable *adjective* unpleasant; bad-tempered. □ **disagreeably** *adverb*.

disallow /dɪsə'laʊ/ *verb* refuse to allow or accept; prohibit.

disappear /dɪsə'pɪə/ *verb* cease to be visible or in existence or circulation etc.; go missing. □ **disappearance** *noun*.

disappoint /dɪsə'pɔɪnt/ *verb* fail to fulfil desire or expectation of; frustrate. □ **disappointing** *adjective*; **disappointment** *noun*.

disapprobation /dɪsæprə'beɪʃ(ə)n/ *noun* disapproval.

disapprove /dɪsə'pru:v/ *verb* (**-ving**) (usually + *of*) have or express unfavourable opinion. □ **disapproval** *noun*.

disarm /dɪ'sɑ:m/ *verb* deprive of weapons; abandon or reduce one's own weapons; (often as **disarming** *adjective*) make less hostile, charm, win over. □ **disarmament** *noun*; **disarmingly** *adverb*.

disarrange /dɪsə'reɪndʒ/ *verb* (**-ging**) put into disorder.

disarray /dɪsə'reɪ/ *noun* disorder.

disassociate /dɪsə'səʊʃɪeɪt/ *verb* (**-ting**) dissociate. □ **disassociation** *noun*.

disaster /dɪ'zɑ:stə:/ *noun* sudden or great misfortune; *colloquial* complete failure. □ **disastrous** *adjective*; **disastrously** *adverb*.

disavow /dɪsə'vaʊ/ *verb* disclaim knowledge or approval of or responsibility for. □ **disavowal** *noun*.

disband /dɪs'bænd/ *verb* break up, disperse.

disbar /dɪs'bɑ:/ *verb* (**-rr-**) deprive of status of barrister. □ **disbarment** *noun*.

disbelieve /dɪsbɪ'li:v/ *verb* (**-ving**) refuse to believe; not believe; be sceptical. □ **disbelief** *noun*; **disbelievingly** *adverb*.

disburse /dɪs'bɜ:s/ *verb* (**-sing**) pay out (money). □ **disbursement** *noun*.

disc *noun* flat thinnish circular object; round flat or apparently flat surface or mark; layer of cartilage between vertebrae; gramophone record; *Computing* = DISK. □ **disc brake** one using friction of pads against a disc; **disc jockey** presenter of recorded popular music.

discard /dɪs'kɑ:d/ *verb* reject as unwanted; remove or put aside.

discern /dɪ'sɜ:n/ *verb* perceive clearly with mind or senses; make out. □ **discernible** *adjective*.

discerning *adjective* having good judgement or insight. □ **discernment** *noun*.

discharge ● *verb* /dɪs'tʃɑ:dʒ/ (**-ging**) release, let go; dismiss from office or employment; fire (gun etc.); throw; eject; emit, pour out; pay or perform (debt, duty); relieve (bankrupt) of residual liability; release an electrical charge from; relieve of cargo; unload. ● *noun* /'dɪstʃɑ:dʒ/ discharging, being discharged; matter or thing discharged; release of electric charge, esp. with spark.

disciple /dɪ'saɪp(ə)l/ *noun* follower of a teacher or leader, esp. of Christ.

disciplinarian /dɪsɪplɪ'neərɪən/ *noun* enforcer of or believer in strict discipline.

disciplinary /dɪsɪ'plɪnərɪ/ *adjective* of or enforcing discipline.

discipline /'dɪsɪplɪn/ ● *noun* control or order exercised over people or animals; system of rules for this; training or way of life aimed at self-control or conformity; branch of learning; punishment. ● *verb* (**-ning**) punish; control by training in obedience.

disclaim /dɪs'kleɪm/ *verb* disown, deny; renounce legal claim to.

disclaimer *noun* statement disclaiming something.

disclose /dɪs'kləʊz/ *verb* (**-sing**) expose, make known, reveal. □ **disclosure** *noun*.

disco /ˈdɪskəʊ/ *noun* (*plural* **-s**) *colloquial* discothèque.

discolour /dɪsˈkʌlə/ *verb* (*US* **discolor**) cause to change from its usual colour; stain or become stained. □ **discoloration** *noun*.

discomfit /dɪsˈkʌmfɪt/ *verb* (**-t-**) disconcert; baffle; frustrate. □ **discomfiture** *noun*.

■ **Usage** *Discomfit* is sometimes confused with *discomfort*.

discomfort /dɪsˈkʌmfət/ *noun* lack of comfort; uneasiness of body or mind.

■ **Usage** As a verb, *discomfort* is sometimes confused with *discomfit*.

discompose /dɪskəmˈpəʊz/ *verb* (**-sing**) disturb composure of. □ **discomposure** *noun*.

disconcert /dɪskənˈsɜːt/ *verb* disturb composure of; fluster.

disconnect /dɪskəˈnekt/ *verb* break connection of or between; put (apparatus) out of action by disconnecting parts. □ **disconnection** *noun*.

disconnected *adjective* incoherent and illogical.

disconsolate /dɪsˈkɒnsələt/ *adjective* forlorn, unhappy; disappointed. □ **disconsolately** *adverb*.

discontent /dɪskənˈtent/ • *noun* dissatisfaction; lack of contentment. • *verb* (esp. as **discontented** *adjective*) make dissatisfied.

discontinue /dɪskənˈtɪnjuː/ *verb* (**-ues, -ued, -uing**) (cause to) cease; not go on with (activity).

discontinuous /dɪskənˈtɪnjʊəs/ *adjective* lacking continuity; intermittent. □ **discontinuity** /-kɒntɪˈnjuːɪtɪ/ *noun*.

discord /ˈdɪskɔːd/ *noun* disagreement, strife; harsh noise, clashing sounds; lack of harmony. □ **discordant** /-ˈkɔːdənt/ *adjective*.

discothèque /ˈdɪskətek/ *noun* nightclub etc. where pop records are played for dancing.

discount • *noun* /ˈdɪskaʊnt/ amount deducted from normal price. • *verb* /dɪsˈkaʊnt/ disregard as unreliable or unimportant; deduct amount from (price etc.); give or get present value of (investment certificate which has yet to mature). □ **at a discount** below nominal or usual price.

discountenance /dɪsˈkaʊntɪnəns/ *verb* (**-cing**) disconcert; refuse to approve of.

discourage /dɪsˈkʌrɪdʒ/ *verb* (**-ging**) reduce confidence or spirits of; dissuade, deter; show disapproval of. □ **discouragement** *noun*.

discourse • *noun* /ˈdɪskɔːs/ conversation; lengthy treatment of theme; lecture, speech. • *verb* /dɪsˈkɔːs/ (**-sing**) converse; speak or write at length.

discourteous /dɪsˈkɜːtɪəs/ *adjective* rude, uncivil. □ **discourteously** *adverb*; **discourtesy** *noun* (*plural* **-ies**).

discover /dɪˈskʌvə/ *verb* find or find out, by effort or chance; be first to find or find out in particular case; find and promote (little-known performer).

discovery *noun* (*plural* **-ies**) discovering, being discovered; person or thing discovered.

discredit /dɪsˈkredɪt/ • *verb* (**-t-**) cause to be disbelieved; harm good reputation of; refuse to believe. • *noun* harm to reputation; cause of this; lack of credibility.

discreditable *adjective* bringing discredit, shameful.

discreet /dɪˈskriːt/ *adjective* (**-er, -est**) tactful, prudent; cautious in speech or action; unobtrusive. □ **discreetly** *adverb*; **discreetness** *noun*.

discrepancy /dɪˈskrepənsɪ/ *noun* (*plural* **-ies**) difference; inconsistency.

discrete /dɪˈskriːt/ *adjective* separate; distinct.

discretion /dɪˈskreʃ(ə)n/ *noun* being discreet; prudence, judgement; freedom or authority to act as one thinks fit. □ **discretionary** *adjective*.

discriminate /dɪˈskrɪmɪneɪt/ *verb* (**-ting**) (often + *between*) make or see a distinction; (usually + *against*, *in favour of*) treat badly or well, esp. on the basis of race, gender, etc. □ **discriminating** *adjective*; **discrimination** *noun*; **discriminatory** /-nətərɪ/ *adjective*.

discursive /dɪsˈkɜːsɪv/ *adjective* rambling, tending to digress.

discus /ˈdɪskəs/ *noun* (*plural* **-cuses**) heavy disc thrown in athletic events.

discuss /dɪsˈkʌs/ *verb* talk about; talk or write about (subject) in detail. □ **discussion** *noun*.

disdain /dɪs'deɪn/ ● *noun* scorn, contempt. ● *verb* regard with disdain; refrain or refuse out of disdain. □ **disdainful** *adjective*.

disease /dɪ'zi:z/ *noun* unhealthy condition of organism or part of organism; (specific) disorder or illness. □ **diseased** *adjective*.

disembark /dɪsɪm'bɑ:k/ *verb* put or go ashore; get off aircraft, bus, etc. □ **disembarkation** /-embɑ:-/ *noun*.

disembarrass /dɪsɪm'bærəs/ *verb* (usually + *of*) rid or relieve (of a load etc.); free from embarrassment. □ **disembarrassment** *noun*.

disembodied /dɪsɪm'bɒdɪd/ *adjective* (of soul etc.) separated from body or concrete form; without a body.

disembowel /dɪsɪm'baʊəl/ *verb* (**-ll-**; *US* **-l-**) remove entrails of. □ **disembowelment** *noun*.

disenchant /dɪsɪn'tʃɑ:nt/ *verb* disillusion. □ **disenchantment** *noun*.

disencumber /dɪsɪn'kʌmbə/ *verb* free from encumbrance.

disenfranchise /dɪsɪn'fræntʃaɪz/ *verb* (**-sing**) deprive of right to vote, of citizen's rights, or of franchise held. □ **disenfranchisement** *noun*.

disengage /dɪsɪn'geɪdʒ/ *verb* (**-ging**) detach; loosen; release; remove (troops) from battle etc.; become detached; (as **disengaged** *adjective*) at leisure, uncommitted. □ **disengagement** *noun*.

disentangle /dɪsɪn'tæŋg(ə)l/ *verb* (**-ling**) free or become free of tangles or complications. □ **disentanglement** *noun*.

disestablish /dɪsɪ'stæblɪʃ/ *verb* deprive (Church) of state support; end the establishment of. □ **disestablishment** *noun*.

disfavour /dɪs'feɪvə/ *noun* (*US* **disfavor**) dislike; disapproval; being disliked.

disfigure /dɪs'fɪgə/ *verb* (**-ring**) spoil appearance of. □ **disfigurement** *noun*.

disgorge /dɪs'gɔ:dʒ/ *verb* (**-ging**) eject from throat; pour forth (food, fluid, etc.).

disgrace /dɪs'greɪs/ ● *noun* shame, ignominy; shameful or very bad person or thing. ● *verb* (**-cing**) bring shame or discredit on; dismiss from position of honour or favour.

disgraceful *adjective* causing disgrace; shameful. □ **disgracefully** *adverb*.

disgruntled /dɪs'grʌnt(ə)ld/ *adjective* discontented; sulky.

disguise /dɪs'gaɪz/ ● *verb* (**-sing**) conceal identity of, make unrecognizable; conceal. ● *noun* costume, make-up, etc. used to disguise; action, manner, etc. used to deceive; disguised condition.

disgust /dɪs'gʌst/ ● *noun* strong aversion; repugnance. ● *verb* cause disgust in. □ **disgusting** *adjective*; **disgustingly** *adverb*.

dish ● *noun* shallow flat-bottomed container for food; food served in dish; particular kind of food; (in *plural*) crockery etc. to be washed etc. after a meal; dish-shaped object or cavity; *colloquial* sexually attractive person. ● *verb* make dish-shaped; *colloquial* outmanoeuvre, frustrate. □ **dish out** *colloquial* distribute, allocate; **dish up** (prepare to) serve meal.

disharmony /dɪs'hɑ:mənɪ/ *noun* lack of harmony, discord.

dishearten /dɪs'hɑ:t(ə)n/ *verb* cause to lose courage or confidence.

dishevelled /dɪ'ʃev(ə)ld/ *adjective* (*US* **disheveled**) ruffled, untidy.

dishonest /dɪ'sɒnɪst/ *adjective* fraudulent; insincere. □ **dishonestly** *adverb*; **dishonesty** *noun*.

dishonour /dɪ'sɒnə/ (*US* **dishonor**) ● *noun* loss of honour, disgrace; cause of this. ● *verb* disgrace (person, family, etc.); refuse to pay (cheque etc.).

dishonourable *adjective* (*US* **dishonorable**) causing disgrace; ignominious; unprincipled. □ **dishonourably** *adverb*.

dishy *adjective* (**-ier**, **-iest**) *colloquial* sexually attractive.

disillusion /dɪsɪ'lu:ʒ(ə)n/ ● *verb* free from illusion or mistaken belief, esp. disappointingly. ● *noun* disillusioned state. □ **disillusionment** *noun*.

disincentive /dɪsɪn'sentɪv/ *noun* thing that discourages, esp. from a particular line of action.

disincline /dɪsɪn'klaɪn/ *verb* (**-ning**) (usually as **disinclined** *adjective*) make unwilling. □ **disinclination** /-klɪ'neɪ-/ *noun*.

disinfect /dɪsɪn'fekt/ *verb* cleanse of infection. □ **disinfection** *noun*.

disinfectant • *noun* substance that destroys germs etc. • *adjective* disinfecting.

disinformation /dɪsɪnfə'meɪʃ(ə)n/ *noun* false information, propaganda.

disingenuous /dɪsɪn'dʒenjʊəs/ *adjective* insincere; not candid. □ **disingenuously** *adverb*; **disingenuousness** *noun*.

disinherit /dɪsɪn'herɪt/ *verb* (**-t-**) deprive of right to inherit; reject as one's heir. □ **disinheritance** *noun*.

disintegrate /dɪ'sɪntɪgreɪt/ *verb* (**-ting**) separate into component parts; break up; *colloquial* break down, esp. mentally. □ **disintegration** *noun*.

disinter /dɪsɪn'tɜ:/ *verb* (**-rr-**) dig up (esp. corpse). □ **disinterment** *noun*.

disinterested /dɪ'sɪntrɪstɪd/ *adjective* impartial; uninterested. □ **disinterest** *noun*; **disinterestedly** *adverb*.

■ **Usage** The use of *disinterested* to mean 'uninterested' is common in informal use but is widely considered incorrect. The use of the noun *disinterest* to mean 'lack of interest' is also objected to, but it is rarely used in any other sense and the alternative *uninterest* is rare.

disjointed /dɪs'dʒɔɪntɪd/ *adjective* disconnected, incoherent.

disjunction /dɪs'dʒʌŋkʃ(ə)n/ *noun* separation.

disjunctive /dɪs'dʒʌŋktɪv/ *adjective* involving separation; (of a conjunction) expressing alternative.

disk *noun* flat circular computer storage device. □ **disk drive** mechanism for rotating disk and reading or writing data from or to it.

dislike /dɪs'laɪk/ • *verb* (**-king**) have aversion to, not like. • *noun* feeling of repugnance or not liking; object of this.

dislocate /'dɪsləkeɪt/ *verb* (**-ting**) disturb normal connection of (esp. a joint in the body); disrupt. □ **dislocation** *noun*.

dislodge /dɪs'lɒdʒ/ *verb* (**-ging**) disturb or move. □ **dislodgement** *noun*.

disloyal /dɪs'lɔɪəl/ *adjective* unfaithful; lacking loyalty. □ **disloyalty** *noun*.

dismal /'dɪzm(ə)l/ *adjective* gloomy; miserable; dreary; *colloquial* feeble, inept. □ **dismally** *adverb*.

dismantle /dɪs'mænt(ə)l/ *verb* (**-ling**) pull down, take to pieces; deprive of defences, equipment, etc.

dismay /dɪs'meɪ/ • *noun* feeling of intense disappointment and discouragement. • *verb* affect with dismay.

dismember /dɪs'membə/ *verb* remove limbs from; partition (country etc.). □ **dismemberment** *noun*.

dismiss /dɪs'mɪs/ *verb* send away; disband; allow to go; terminate employment of, esp. dishonourably; put out of one's thoughts; *Law* refuse further hearing to; *Cricket* put (batsman, side) out. □ **dismissal** *noun*.

dismissive *adjective* dismissing rudely or casually; disdainful. □ **dismissively** *adverb*.

dismount /dɪs'maʊnt/ *verb* get off or down from cycle or horseback etc.; remove (thing) from mounting.

disobedient /dɪsə'bi:dɪənt/ *adjective* disobeying; rebellious. □ **disobedience** *noun*; **disobediently** *adverb*.

disobey /dɪsə'beɪ/ *verb* fail or refuse to obey.

disoblige /dɪsə'blaɪdʒ/ *verb* (**-ging**) refuse to help or cooperate with (person).

disorder /dɪ'sɔ:də/ *noun* confusion; tumult, riot; bodily or mental ailment. □ **disordered** *adjective*.

disorderly *adjective* untidy; confused; riotous.

disorganize /dɪs'ɔ:gənaɪz/ *verb* (also **-ise**) (**-zing** or **-sing**) throw into confusion or disorder; (as **disorganized** *adjective*) badly organized, untidy. □ **disorganization** *noun*.

disorientate /dɪs'ɔ:rɪənteɪt/ *verb* (also **disorient**) (**-ting**) confuse (person) as to his or her bearings. □ **disorientation** *noun*.

disown /dɪs'əʊn/ *verb* deny or give up any connection with; repudiate.

disparage /dɪ'spærɪdʒ/ *verb* (**-ging**) criticize; belittle. □ **disparagement** *noun*.

disparate /'dɪspərət/ *adjective* essentially different, unrelated.

disparity /dɪ'spærɪtɪ/ *noun* (*plural* **-ies**) inequality, difference; incongruity.

dispassionate /dɪ'spæʃənət/ *adjective* free from emotion; impartial. □ **dispassionately** *adverb*.

dispatch /dɪs'pætʃ/ (also **despatch**) • *verb* send off; perform (task etc.) promptly; kill; *colloquial* eat (food) quickly. ● *noun* dispatching, being dispatched; official written message, esp. military or political; promptness, efficiency. □ **dispatch box** case for esp. parliamentary documents; **dispatch rider** motorcyclist etc. carrying messages.

dispel /dɪ'spel/ *verb* (**-ll-**) drive away (esp. unwanted ideas or feelings); scatter.

dispensable /dɪ'spensəb(ə)l/ *adjective* that can be dispensed with.

dispensary /dɪ'spensərɪ/ *noun* (*plural* **-ies**) place where medicines etc. are dispensed.

dispensation /dɪspen'seɪʃ(ə)n/ *noun* distributing, dispensing; exemption from penalty, rule, etc.; ordering or management, esp. of world by Providence.

dispense /dɪ'spens/ *verb* (**-sing**) distribute; administer; make up and give out (medicine); (+ *with*) do without, make unnecessary.

dispenser *noun* person who dispenses something; device that dispenses selected amount at a time.

disperse /dɪ'spɜːs/ *verb* (**-sing**) go or send widely or in different directions, scatter; station at different points; disseminate; separate (light) into coloured constituents. □ **dispersal** *noun*; **dispersion** *noun*.

dispirit /dɪ'spɪrɪt/ *verb* (esp. as **dispiriting**, **dispirited** *adjectives*) make despondent.

displace /dɪs'pleɪs/ *verb* (**-cing**) move from its place; remove from office; oust, take the place of. □ **displaced person** refugee in war etc. or from persecution.

displacement *noun* displacing, being displaced; amount of fluid displaced by object floating or immersed in it.

display /dɪ'spleɪ/ ● *verb* show; exhibit. ● *noun* displaying; exhibition; ostentation; image shown on a visual display unit etc.

displease /dɪs'pliːz/ *verb* (**-sing**) offend; make angry or upset. □ **displeasure** /-'pleʒə/ *noun*.

disport /dɪ'spɔːt/ *verb* (also **disport oneself**) frolic, enjoy oneself.

disposable *adjective* that can be disposed of; designed to be discarded after one use.

disposal *noun* disposing of. □ **at one's disposal** available.

dispose /dɪ'spəʊz/ *verb* (**-sing**) (usually + *to*, *to do*) (usually in *passive*) incline, make willing; (in *passive*) tend; arrange suitably; (as **disposed** *adjective*) having a specified inclination; determine events. □ **dispose of** get rid of, deal with, finish.

disposition /dɪspə'zɪʃ(ə)n/ *noun* natural tendency; temperament; arrangement (of parts etc.).

dispossess /dɪspə'zes/ *verb* (usually + *of*) (esp. as **dispossessed** *adjective*) deprive; oust, dislodge. □ **dispossession** *noun*.

disproof /dɪs'pruːf/ *noun* refutation.

disproportion /dɪsprə'pɔːʃ(ə)n/ *noun* lack of proportion.

disproportionate *adjective* out of proportion; relatively too large or too small. □ **disproportionately** *adverb*.

disprove /dɪs'pruːv/ *verb* (**-ving**) prove (theory etc.) false.

disputable *adjective* open to question.

disputant *noun* person in dispute.

disputation /dɪspjuː'teɪʃ(ə)n/ *noun* debate, esp. formal; argument, controversy.

disputatious /dɪspjuː'teɪʃəs/ *adjective* argumentative.

dispute /dɪ'spjuːt/ ● *verb* (**-ting**) hold debate; quarrel; question truth or validity of; contend for; resist. ● *noun* controversy, debate; quarrel; disagreement leading to industrial action.

disqualify /dɪs'kwɒlɪfaɪ/ *verb* (**-ies**, **-ied**) make or pronounce (competitor, applicant, etc.) unfit or ineligible. □ **disqualification** *noun*.

disquiet /dɪs'kwaɪət/ ● *verb* make anxious. ● *noun* uneasiness, anxiety. □ **disquietude** *noun*.

disquisition /dɪskwɪ'zɪʃ(ə)n/ *noun* discursive treatise or discourse.

disregard /dɪsrɪ'gɑːd/ ● *verb* ignore, treat as unimportant. ● *noun* indifference, neglect.

disrepair /dɪsrɪ'peə/ *noun* bad condition due to lack of repairs.

disreputable /dɪs'repjʊtəb(ə)l/ *adjective* having a bad reputation; not respectable. □ **disreputably** *adverb*.

disrepute /dɪsrɪ'pju:t/ *noun* lack of good reputation; discredit.

disrespect /dɪsrɪ'spekt/ *noun* lack of respect. □ **disrespectful** *adjective*.

disrobe /dɪs'rəʊb/ *verb* (**-bing**) *literary* undress.

disrupt /dɪs'rʌpt/ *verb* interrupt continuity of; bring disorder to; break (thing) apart. □ **disruption** *noun*; **disruptive** *adjective*.

dissatisfy /dɪ'sætɪsfaɪ/ *verb* (**-ies**, **-ied**) (usually as **dissatisfied** *adjective*; often + *with*) fail to satisfy; make discontented. □ **dissatisfaction** /-'fæk-/ *noun*.

dissect /dɪ'sekt/ *verb* cut in pieces, esp. for examination or post-mortem; analyse or criticize in detail. □ **dissection** *noun*.

■ **Usage** *Dissect* is often wrongly pronounced /daɪ'sekt/ (and sometimes written with only one *s*) because of confusion with *bisect*.

dissemble /dɪ'semb(ə)l/ *verb* (**-ling**) be hypocritical or insincere; conceal or disguise (a feeling, intention, etc.).

disseminate /dɪ'semɪneɪt/ *verb* (**-ting**) scatter about, spread (esp. ideas) widely. □ **dissemination** *noun*.

dissension /dɪ'senʃ(ə)n/ *noun* angry disagreement.

dissent /dɪ'sent/ ● *verb* (often + *from*) disagree, esp. openly; differ, esp. from established or official opinion. ● *noun* such difference; expression of this.

dissenter *noun* person who dissents; (**Dissenter**) Protestant dissenting from Church of England.

dissentient /dɪ'senʃ(ə)nt/ ● *adjective* disagreeing with the established or official view. ● *noun* person who dissents.

dissertation /dɪsə'teɪʃ(ə)n/ *noun* detailed discourse, esp. as submitted for academic degree.

disservice /dɪs'sɜ:vɪs/ *noun* harmful action.

dissident /'dɪsɪd(ə)nt/ ● *adjective* disagreeing, esp. with established government. ● *noun* dissident person.

dissimilar /dɪ'sɪmɪlə/ *adjective* not similar. □ **dissimilarity** /-'lærɪtɪ/ *noun* (*plural* **-ies**).

dissimulate /dɪ'sɪmjʊleɪt/ *verb* (**-ting**) dissemble. □ **dissimulation** *noun*.

dissipate /'dɪsɪpeɪt/ *verb* (**-ting**) dispel, disperse; squander; (as **dissipated** *adjective*) dissolute.

dissipation *noun* dissolute way of life; dissipating, being dissipated.

dissociate /dɪ'səʊʃɪeɪt/ *verb* (**-ting**) disconnect or separate; become disconnected; (**dissociate oneself from**) declare oneself unconnected with. □ **dissociation** *noun*.

dissolute /'dɪsəlu:t/ *adjective* lax in morals, licentious.

dissolution /dɪsə'lu:ʃ(ə)n/ *noun* dissolving, being dissolved; dismissal or dispersal of assembly, esp. parliament; breaking up, abolition (of institution); death.

dissolve /dɪ'zɒlv/ *verb* (**-ving**) make or become liquid, esp. by immersion or dispersion in liquid; (cause to) disappear gradually; dismiss (assembly); put an end to, annul; (often + *in*, *into*) be overcome (by tears, laughter, etc.).

dissonant /'dɪsənənt/ *adjective* discordant, harsh-toned; incongruous. □ **dissonance** *noun*.

dissuade /dɪ'sweɪd/ *verb* (**-ding**) (often + *from*) discourage, persuade against. □ **dissuasion** /-'sweɪʒ(ə)n/ *noun*.

distaff /'dɪstɑ:f/ *noun* cleft stick holding wool etc. for spinning by hand. □ **distaff side** female branch of family.

distance /'dɪst(ə)ns/ ● *noun* being far off; remoteness; space between two points; distant point; aloofness, reserve; remoter field of vision. ● *verb* (**-cing**) place or cause to seem far off; leave behind in race etc. □ **at a distance** far off; **keep one's distance** remain aloof.

distant *adjective* at specified distance; remote in space, time, relationship, etc.; aloof; abstracted; faint. □ **distantly** *adverb*.

distaste /dɪs'teɪst/ *noun* (usually + *for*) dislike; aversion. □ **distasteful** *adjective*.

distemper[1] /dɪ'stempə/ ● *noun* paint for walls, using glue etc. as base. ● *verb* paint with distemper.

distemper[2] /dɪ'stempə/ *noun* catarrhal disease of dogs etc.

distend /dɪ'stend/ *verb* swell out by pressure from within. □ **distension** /-'sten-/ *noun*.

distich /'dɪstɪk/ *noun* verse couplet.

distil /dɪ'stɪl/ *verb* (*US* **distill**) (**-ll-**) purify or extract essence from (substance) by vaporizing and condensing it and collecting remaining liquid; extract gist of (idea etc.); make (whisky, essence, etc.) by distilling. □ **distillation** *noun.*

distiller *noun* person who distils, esp. alcoholic liquor.

distillery *noun* (*plural* **-ies**) factory etc. for distilling alcoholic liquor.

distinct /dɪ'stɪŋkt/ *adjective* (often + *from*) separate, different in quality or kind; clearly perceptible; definite, decided. □ **distinctly** *adverb.*

distinction /dɪ'stɪŋkʃ(ə)n/ *noun* discriminating, distinguishing; difference between things; thing that differentiates; special consideration or honour; excellence; mark of honour.

distinctive *adjective* distinguishing, characteristic. □ **distinctively** *adverb*; **distinctiveness** *noun.*

distingué /dɪ'stæŋgeɪ/ *adjective* having distinguished air, manners, etc. [French]

distinguish /dɪ'stɪŋgwɪʃ/ *verb* (often + *from*, *between*) see or draw distinctions; characterize; make out by listening or looking etc.; (usually **distinguish oneself**; often + *by*) make prominent. □ **distinguishable** *adjective.*

distinguished *adjective* eminent, famous; dignified.

distort /dɪ'stɔ:t/ *verb* pull or twist out of shape; misrepresent (facts etc.); transmit (sound) inaccurately. □ **distortion** *noun.*

distract /dɪ'strækt/ *verb* (often + *from*) draw away attention of; bewilder; (as **distracted** *adjective*) confused, mad, or angry; amuse, esp. to divert from pain etc.

distraction /dɪ'strækʃ(ə)n/ *noun* distracting, being distracted; thing which distracts; amusement, relaxation; mental confusion; frenzy, madness.

distrain /dɪ'streɪn/ *verb* (usually + *upon*) impose distraint (on person, goods, etc.).

distraint /dɪ'streɪnt/ *noun* seizure of goods to enforce payment.

distrait /dɪ'streɪ/ *adjective* inattentive; distraught. [French]

distraught /dɪ'strɔ:t/ *adjective* distracted with worry, fear, etc.; very agitated.

distress /dɪ'stres/ ● *noun* suffering caused by pain, grief, anxiety, etc.; poverty; *Law* distraint. ● *verb* cause distress to; make unhappy. □ **distressed** *adjective*; **distressing** *adjective.*

distribute /dɪ'strɪbju:t/ *verb* (**-ting**) give shares of; deal out; spread about; put at different points; arrange, classify. □ **distribution** /-'bju:-/ *noun*; **distributive** *adjective.*

■ **Usage** *Distribute* is often pronounced /'dɪstrɪbju:t/ (with the stress on the *dis*-), but this is considered incorrect by some people.

distributor *noun* agent who supplies goods; device in internal-combustion engine for passing current to each spark plug in turn.

district /'dɪstrɪkt/ *noun* region; administrative division. □ **district attorney** (in the US) public prosecutor of district; **district nurse** nurse who makes home visits in an area.

distrust /dɪs'trʌst/ ● *noun* lack of trust; suspicion. ● *verb* have no confidence in. □ **distrustful** *adjective.*

disturb /dɪ'stɜ:b/ *verb* break rest or quiet of; worry; disorganize; (as **disturbed** *adjective*) emotionally or mentally unstable.

disturbance *noun* disturbing, being disturbed; tumult, disorder, agitation.

disunion /dɪs'ju:nɪən/ *noun* separation; lack of union.

disunite /dɪsju:'naɪt/ *verb* (**-ting**) separate; divide. □ **disunity** /-'ju:-/ *noun.*

disuse /dɪs'ju:s/ *noun* state of no longer being used.

disused /dɪs'ju:zd/ *adjective* no longer in use.

disyllable /daɪ'sɪləb(ə)l/ *noun* word or metrical foot of two syllables. □ **disyllabic** /-'læb-/ *adjective.*

ditch ● *noun* long narrow excavation esp. for drainage or as boundary. ● *verb* make or repair ditches; *slang* abandon, discard.

dither /'dɪðə/ ● *verb* hesitate; be indecisive; tremble, quiver. ● *noun colloquial* state of agitation or hesitation. □ **ditherer** *noun*; **dithery** *adjective.*

dithyramb /ˈdɪθɪræm/ *noun* ancient Greek wild choral hymn; passionate or inflated poem etc. □ **dithyrambic** /-ˈræmbɪk/ *adjective.*

ditto /ˈdɪtəʊ/ *noun* (*plural* **-s**) the aforesaid, the same (in accounts, lists, etc., or *colloquial* in speech).

ditty /ˈdɪtɪ/ *noun* (*plural* **-ies**) short simple song.

diuretic /daɪjʊˈretɪk/ ● *adjective* causing increased output of urine. ● *noun* diuretic drug.

diurnal /daɪˈɜːn(ə)l/ *adjective* in or of day; daily; occupying one day.

diva /ˈdiːvə/ *noun* (*plural* **-s**) great woman opera singer.

divan /dɪˈvæn/ *noun* low couch or bed without back or ends.

dive ● *verb* (**-ving**) plunge head foremost into water; (of aircraft) descend fast and steeply; (of submarine or diver) submerge; go deeper; (+ *into*) *colloquial* put one's hand into. ● *noun* act of diving; plunge; *colloquial* disreputable nightclub, bar, etc. □ **dive-bomb** bomb (target) from diving aircraft; **diving board** elevated board for diving from.

diver *noun* person who dives, esp. one who works under water; diving bird.

diverge /daɪˈvɜːdʒ/ *verb* (**-ging**) (often + *from*) depart from set course; (of opinions etc.) differ; take different courses; spread outward from central point. □ **divergence** *noun*; **divergent** *adjective.*

divers /ˈdaɪvɜːz/ *adjective archaic* various, several.

diverse /daɪˈvɜːs/ *adjective* varied.

diversify /daɪˈvɜːsɪfaɪ/ *verb* (**-ies**, **-ied**) make diverse; vary; spread (investment) over several enterprises; (often + *into*) expand range of products. □ **diversification** *noun.*

diversion /daɪˈvɜːʃ(ə)n/ *noun* diverting, being diverted; recreation, pastime; alternative route when road is temporarily closed; stratagem for diverting attention. □ **diversionary** *adjective.*

diversity /daɪˈvɜːsɪtɪ/ *noun* variety.

divert /daɪˈvɜːt/ *verb* turn aside; deflect; distract (attention); (often as **diverting** *adjective*) entertain, amuse.

divest /daɪˈvest/ *verb* (usually + *of*) unclothe, strip; deprive, rid.

divide /dɪˈvaɪd/ ● *verb* (**-ding**) (often + *in*, *into*) separate into parts; split or break up; (often + *out*) distribute, deal, share; separate (one thing) from another; classify into parts or groups; cause to disagree; (+ *by*) find how many times number contains another; (+ *into*) be contained exact number of times; (of parliament) vote (by members entering either of two lobbies). ● *noun* dividing line; watershed.

dividend /ˈdɪvɪdend/ *noun* share of profits paid to shareholders, football pools winners, etc.; number to be divided.

divider *noun* screen etc. dividing room; (in *plural*) measuring compasses.

divination /dɪvɪˈneɪʃ(ə)n/ *noun* supposed foreseeing of the future, using special technique.

divine /dɪˈvaɪn/ ● *adjective* (**-r**, **-st**) of, from, or like God or a god; sacred; *colloquial* excellent. ● *verb* (**-ning**) discover by intuition or guessing; foresee; practise divination. ● *noun* theologian. □ **divining-rod** dowser's forked twig. □ **divinely** *adverb.*

diviner *noun* practitioner of divination; dowser.

divinity /dɪˈvɪnɪtɪ/ *noun* (*plural* **-ies**) being divine; god; theology.

divisible /dɪˈvɪzɪb(ə)l/ *adjective* capable of being divided. □ **divisibility** *noun.*

division /dɪˈvɪʒ(ə)n/ *noun* dividing, being divided; dividing one number by another; disagreement; one of parts into which thing is divided; administrative unit, esp. group of army units or of teams in sporting league. □ **divisional** *adjective.*

divisive /dɪˈvaɪsɪv/ *adjective* causing disagreement. □ **divisively** *adverb*; **divisiveness** *noun.*

divisor /dɪˈvaɪzə/ *noun* number by which another is to be divided.

divorce /dɪˈvɔːs/ ● *noun* legal dissolution of marriage; separation. ● *verb* (**-cing**) (usually as **divorced** *adjective*) (often + *from*) legally dissolve marriage of; separate by divorce; end marriage with by divorce; separate.

divorcee /dɪvɔːˈsiː/ *noun* divorced person.

divot /ˈdɪvət/ *noun* piece of turf dislodged by head of golf club.

divulge /daɪ'vʌldʒ/ *verb* (**-ging**) disclose (secret).

divvy /'dɪvɪ/ *colloquial* ● *noun* (*plural* **-ies**) dividend. ● *verb* (**-ies**, **-ied**) (often + *up*) share out.

Dixie /'dɪksɪ/ *noun* Southern States of US. □ **Dixieland** Dixie, kind of jazz.

dixie /'dɪksɪ/ *noun* large iron cooking pot.

DIY *abbreviation* do-it-yourself.

dizzy /'dɪzɪ/ ● *adjective* (**-ier**, **-iest**) giddy; dazed; causing dizziness. ● *verb* (**-ies**, **-ied**) make dizzy; bewilder. □ **dizzily** *adverb*; **dizziness** *noun*.

DJ *abbreviation* disc jockey; dinner jacket.

dl *abbreviation* decilitre(s).

D.Litt. *abbreviation* Doctor of Letters.

DM *abbreviation* Deutschmark.

dm *abbreviation* decimetre(s).

DNA *abbreviation* deoxyribonucleic acid (substance carrying genetic information in chromosomes).

do[1] /du:/ ● *verb* (*3rd singular present* **does** /dʌz/; *past* **did**; *past participle* **done** /dʌn/; *present participle* **doing**) perform, carry out; produce, make; impart; act, proceed; work at; be suitable, satisfy; attend to, deal with; fare; solve; *colloquial* finish; (as **done** *adjective*) finished, completely cooked; *colloquial* exhaust, defeat, kill; *colloquial* cater for; *slang* rob, swindle, prosecute, convict. ● *auxiliary verb used in questions and negative or emphatic statements and commands; as verbal substitute to avoid repetition.* ● *noun* (*plural* **dos** or **do's**) *colloquial* elaborate party or other undertaking. □ **do away with** *colloquial* abolish, kill; **do down** *colloquial* swindle, overcome; **do for** be sufficient for, *colloquial* (esp. as **done for** *adjective*) destroy or ruin or kill, *colloquial* do housework for; **do in** *slang* kill, *colloquial* exhaust; **do-it-yourself** (to be) done or made by householder etc.; **do up** fasten, *colloquial* restore, repair, dress up; **do with** (after *could*) would appreciate, would profit by; **do without** forgo, manage without.

do[2] = DOH.

do. *abbreviation* ditto.

docile /'dəʊsaɪl/ *adjective* submissive, easily managed. □ **docility** /-'sɪl-/ *noun*.

dock[1] ● *noun* enclosed harbour for loading, unloading, and repair of ships; (usually in *plural*) range of docks with wharves, warehouses, etc. ● *verb* bring or come into dock; join (spacecraft) together in space, be thus joined. □ **dockyard** area with docks and equipment for building and repairing ships.

dock[2] *noun* enclosure in criminal court for accused.

dock[3] *noun* weed with broad leaves.

dock[4] *verb* cut short (tail); reduce or deduct (money etc.).

docker *noun* person employed to load and unload ships.

docket /'dɒkɪt/ ● *noun* document listing goods delivered, jobs done, contents of package, etc. ● *verb* (**-t-**) label with or enter on docket.

doctor /'dɒktə/ ● *noun* qualified medical practitioner; holder of doctorate. ● *verb colloquial* tamper with, adulterate; castrate, spay.

doctoral *adjective* of the degree of doctor.

doctorate /'dɒktərət/ *noun* highest university degree in any faculty.

doctrinaire /dɒktrɪ'neə/ *adjective* applying theory or doctrine dogmatically.

doctrine /'dɒktrɪn/ *noun* what is taught; principle or set of principles of religious or political etc. belief. □ **doctrinal** /-'traɪn(ə)l/ *adjective*.

document ● *noun* /'dɒkjʊmənt/ something written etc. that provides record or evidence of events, circumstances, etc. ● *verb* /'dɒkjʊment/ prove by or support with documents. □ **documentation** *noun*.

documentary /dɒkjʊ'mentərɪ/ ● *adjective* consisting of documents; factual, based on real events. ● *noun* (*plural* **-ies**) documentary film etc.

dodder /'dɒdə/ *verb* tremble, totter, be feeble. □ **dodderer** *noun*; **doddery** *adjective*.

doddle /'dɒd(ə)l/ *noun colloquial* easy task.

dodecagon /dəʊ'dekəgən/ *noun* plane figure with 12 sides.

dodecahedron /dəʊdekə'hi:drən/ *noun* solid figure with 12 faces.

dodge ● *verb* (**-ging**) move quickly to elude pursuer, blow, etc.; evade by cunning or trickery. ● *noun* quick evasive movement; trick, clever expedient.

dodgem /ˈdɒdʒəm/ *noun* small electrically powered car at funfair, bumped into others in enclosure.

dodgy *adjective* (**-ier**, **-iest**) *colloquial* unreliable, risky.

dodo /ˈdəʊdəʊ/ *noun* (*plural* **-s**) large extinct flightless bird.

DoE *abbreviation* Department of the Environment.

doe *noun* (*plural* same or **-s**) female fallow deer, reindeer, hare, or rabbit.

does *3rd singular present* of DO[1].

doesn't /ˈdʌz(ə)nt/ does not.

doff *verb* take off (hat etc.).

dog ● *noun* 4-legged flesh-eating animal of many breeds akin to wolf etc.; male of this or of fox or wolf; *colloquial* despicable person; (**the dogs**) *colloquial* greyhound racing; mechanical device for gripping. ● *verb* (**-gg-**) follow closely; pursue, track. □ **dogcart** two-wheeled driving-cart with cross seats back to back; **dog-collar** *colloquial* clergyman's stiff collar; **dog days** hottest period of year; **dog-eared** (of book-page etc.) with worn corners; **dog-end** *slang* cigarette-end; **dogfight** fight between aircraft, rough fight; **dogfish** small shark; **doghouse** *US & Australian* kennel (**in the doghouse** *slang* in disgrace); **dog rose** wild hedge-rose; **dogsbody** *colloquial* drudge; **dog-star** Sirius; **dog-tired** tired out.

doge /dəʊdʒ/ *noun historical* chief magistrate of Venice or Genoa.

dogged /ˈdɒgɪd/ *adjective* tenacious. □ **doggedly** *adverb.*

doggerel /ˈdɒgər(ə)l/ *noun* poor or trivial verse.

doggo /ˈdɒgəʊ/ *adverb slang* □ **lie doggo** wait motionless or hidden.

doggy *adjective* of or like dogs; devoted to dogs. □ **doggy bag** bag for restaurant customer to take home leftovers; **doggy-paddle** elementary swimming stroke.

dogma /ˈdɒgmə/ *noun* principle, tenet; doctrinal system.

dogmatic /dɒgˈmætɪk/ *adjective* imposing personal opinions; authoritative, arrogant. □ **dogmatically** *adverb*; **dogmatism** /ˈdɒgmətɪz(ə)m/ *noun.*

doh /dəʊ/ *noun* (also **do**) *Music* first note of scale in tonic sol-fa.

doily /ˈdɔɪlɪ/ *noun* (*plural* **-ies**) small lacy paper mat placed on plate for cakes etc.

doings /ˈduːɪŋz/ *plural noun* actions, exploits; *slang* thing(s) needed.

Dolby /ˈdɒlbɪ/ *noun proprietary term* system used esp. in tape-recording to reduce hiss.

doldrums /ˈdɒldrəmz/ *plural noun* (usually **the doldrums**) low spirits; period of inactivity; equatorial ocean region of calms.

dole ● *noun* unemployment benefit; charitable (esp. niggardly) gift or distribution. ● *verb* (**-ling**) (usually + *out*) distribute sparingly. □ **on the dole** *colloquial* receiving unemployment benefit.

doleful /ˈdəʊlfʊl/ *adjective* mournful; dreary, dismal. □ **dolefully** *adverb.*

doll ● *noun* model of esp. infant human figure as child's toy; *colloquial* attractive young woman; ventriloquist's dummy. ● *verb* (+ *up*) *colloquial* dress smartly.

dollar /ˈdɒlə/ *noun* chief monetary unit in US, Australia, etc.

dollop /ˈdɒləp/ *noun* shapeless lump of food etc.

dolly *noun* (*plural* **-ies**) *child's word for* doll; movable platform for cine-camera etc.

dolman sleeve /ˈdɒlmən/ *noun* loose sleeve cut in one piece with bodice.

dolmen /ˈdɒlmən/ *noun* megalithic tomb with large flat stone laid on upright ones.

dolomite /ˈdɒləmaɪt/ *noun* mineral or rock of calcium magnesium carbonate.

dolphin /ˈdɒlfɪn/ *noun* large porpoise-like sea mammal.

dolt /dəʊlt/ *noun* stupid person. □ **doltish** *adjective.*

Dom *noun: title prefixed to names of some RC dignitaries and Carthusian and Benedictine monks.*

domain /dəˈmeɪn/ *noun* area ruled over; realm; estate etc. under one's control; sphere of authority.

dome ● *noun* rounded vault as roof; dome-shaped thing. ● *verb* (**-ming**) (usually as **domed** *adjective*) cover with or shape as dome.

domestic /dəˈmestɪk/ ● *adjective* of home, household, or family affairs; of one's own country; (of animal) tamed; fond of home life. • *noun* household servant.

domesticate /də'mestɪkeɪt/ *verb* (**-ting**) tame (animal) to live with humans; accustom to housework etc. □ **domestication** *noun*.

domesticity /dɒmə'stɪsɪtɪ/ *noun* being domestic; home life.

domicile /'dɒmɪsaɪl/ • *noun* dwelling place; place of permanent residence. • *verb* (**-ling**) (usually as **domiciled** *adjective*) (usually + *at, in*) settle in a place.

domiciliary /dɒmɪ'sɪlɪərɪ/ *adjective formal* (esp. of doctor's etc. visit) to or at person's home.

dominant /'dɒmɪnənt/ • *adjective* dominating, prevailing. • *noun Music* 5th note of diatonic scale. □ **dominance** *noun*.

dominate /'dɒmɪneɪt/ *verb* (**-ting**) command, control; be most influential or obvious; (of place) overlook. □ **domination** *noun*.

domineer /dɒmɪ'nɪə/ *verb* (often as **domineering** *adjective*) behave overbearingly.

Dominican /də'mɪnɪkən/ • *noun* friar or nun of order founded by St Dominic. • *adjective* of this order.

dominion /də'mɪnjən/ *noun* sovereignty; realm; domain; *historical* self-governing territory of British Commonwealth.

domino /'dɒmɪnəʊ/ *noun* (*plural* **-es**) any of 28 small oblong pieces marked with 0–6 pips in each half; (in *plural*) game played with these; loose cloak worn with half-mask.

don[1] *noun* university teacher, esp. senior member of college at Oxford or Cambridge; (**Don**) *Spanish title prefixed to man's name.*

don[2] *verb* (**-nn-**) put on (garment).

donate /dəʊ'neɪt/ *verb* (**-ting**) give (money etc.), esp. to charity.

donation *noun* donating, being donated; thing (esp. money) donated.

done /dʌn/ • *past participle* of DO[1]. • *adjective* completed; cooked; *colloquial* socially acceptable; (often + *in*) *colloquial* tired out; (esp. as *interjection* in response to offer etc.) accepted. □ **be done with** have or be finished with; **done for** *colloquial* in serious trouble.

doner kebab /'dɒnə kɪ'bæb/ *noun* spiced lamb cooked on spit and served in slices, often with pitta bread.

donkey /'dɒŋkɪ/ *noun* (*plural* **-s**) domestic ass; *colloquial* stupid person. □ **donkey jacket** thick weatherproof jacket; **donkey's years** *colloquial* very long time; **donkey-work** drudgery.

Donna /'dɒnə/ *noun: title of Italian, Spanish, or Portuguese lady.*

donnish *adjective* like a college don; pedantic.

donor /'dəʊnə/ *noun* person who donates; person who provides blood for transfusion, organ for transplantation, etc.

don't /dəʊnt/ • *verb* do not. • *noun* prohibition.

doodle /'du:d(ə)l/ • *verb* (**-ling**) scribble or draw absent-mindedly. • *noun* such scribble or drawing.

doom /du:m/ • *noun* terrible fate or destiny; ruin, death. • *verb* (usually + *to*) condemn or destine; (esp. as **doomed** *adjective*) consign to ruin, destruction, etc. □ **doomsday** day of Last Judgement.

door /dɔ:/ *noun* hinged or sliding barrier closing entrance to building, room, cupboard, etc.; doorway. □ **doormat** mat for wiping shoes on, *colloquial* subservient person; **doorstep** step or area immediately outside esp. outer door, *slang* thick slice of bread; **doorstop** device for keeping door open or to keep it from striking wall; **door-to-door** (of selling etc.) done at each house in turn; **doorway** opening filled by door; **out of doors** in(to) open air.

dope • *noun slang* drug, esp. narcotic; thick liquid used as lubricant etc.; varnish; *slang* stupid person; *slang* information. • *verb* (**-ping**) give or add drug to; apply dope to.

dopey *adjective* (also **dopy**) (**dopier, dopiest**) *colloquial* half asleep, stupefied; stupid.

doppelganger /'dɒp(ə)lgæŋə/ *noun* (also **doppelgänger** /-geŋə/) apparition or double of living person.

Doppler effect /'dɒplə/ *noun* change in frequency of esp. sound waves when source and observer are moving closer or apart.

dormant /ˈdɔ:mənt/ *adjective* lying inactive; sleeping; inactive. □ **dormancy** *noun*.

dormer /ˈdɔ:mə/ *noun* (in full **dormer window**) upright window in sloping roof.

dormitory /ˈdɔ:mɪtərɪ/ *noun* (*plural* **-ies**) sleeping-room with several beds; (in full **dormitory town** etc.) commuter town or suburb.

dormouse /ˈdɔ:maʊs/ *noun* (*plural* **-mice**) small mouselike hibernating rodent.

dorsal /ˈdɔ:s(ə)l/ *adjective* of or on back.

dory /ˈdɔ:rɪ/ *noun* (*plural* same or **-ies**) edible sea fish.

dosage *noun* size of dose; giving of dose.

dose /dəʊs/ ● *noun* single portion of medicine; amount of radiation received. ● *verb* (**-sing**) give medicine to; (+ *with*) treat with.

doss *verb* (often + *down*) *slang* sleep on makeshift bed or in doss-house. □ **doss-house** cheap lodging house. □ **dosser** *noun*.

dossier /ˈdɒsɪeɪ/ *noun* file containing information about person, event, etc.

dot *noun* small spot, esp. as decimal point, part of *i* or *j* etc.; shorter signal of the two in Morse code. ● *verb* (**-tt-**) mark or scatter with dot(s); (often + *about*) scatter like dots; place dot over (letter); partly cover as with dots; *slang* hit. □ **dotted line** line of dots for signature etc. on document; **on the dot** exactly on time.

dotage /ˈdəʊtɪdʒ/ *noun* feeble-minded senility.

dote *verb* (**-ting**) (usually + *on* or as **doting** *adjective*) be excessively fond of.

dotterel /ˈdɒtər(ə)l/ *noun* small plover.

dotty *adjective* (**-ier**, **-iest**) *colloquial* eccentric, silly, crazy; (+ *about*) infatuated with.

double /ˈdʌb(ə)l/ ● *adjective* consisting of two things; multiplied by two; twice as much or many or large etc.; having twice the usual size, quantity, strength, etc.; having some part double; (of flower) with two or more circles of petals; ambiguous; deceitful. ● *adverb* at or to twice the amount; two together. ● *noun* double quantity (of spirits etc.) or thing; twice the amount or quantity; person or thing looking exactly like another; (in *plural*) game between two pairs of players; pair of victories; bet in which winnings and stake from first bet are transferred to second. ● *verb* (**-ling**) make or become double; increase twofold; amount to twice as much as; fold over upon itself; become folded; play (two parts) in same play etc.; (usually + *as*) play twofold role; turn sharply; *Nautical* get round (headland). □ **at the double** running; **double agent** spy working for two rival countries etc.; **double-barrelled** (of gun) having two barrels, (of surname) hyphenated; **double-bass** largest instrument of violin family; **double-book** mistakenly reserve (seat, room, etc.) for two people at once; **double-breasted** (of garment) overlapping across body; **double chin** chin with fold of flesh below it; **double cream** thick cream with high fat-content; **double-cross** deceive or betray; **double-dealing** (practising) deceit, esp. in business; **double-decker** bus etc. with two decks, *colloquial* sandwich with two layers of filling; **double Dutch** *colloquial* gibberish; **double eagle** figure of eagle with two heads; **double-edged** presenting both a danger and an advantage; **double figures** numbers from 10 to 99; **double glazing** two layers of glass in window; **double negative** *Grammar* negative statement (incorrectly) containing two negative elements (see note below); **double pneumonia** pneumonia of both lungs; **double standard** rule or principle not impartially applied; **double take** delayed reaction to unexpected element of situation; **double-talk** (usually deliberately) ambiguous or misleading speech.

■ **Usage** Double negatives like *He didn't do nothing* and *I'm never going nowhere like that* are mistakes in standard English because one negative element is redundant. However, two negatives are perfectly acceptable in, for instance, *a not ungenerous sum* (meaning 'quite a generous sum').

double entendre /du:b(ə)l ɑ:nˈtɑ:ndrə/ *noun* phrase capable of two meanings, one usually indecent. [French]

doublet /ˈdʌblɪt/ *noun historical* man's close-fitting jacket; one of pair of similar things.

doubloon /dʌˈbluːn/ *noun historical* Spanish gold coin.

doubt /daʊt/ • *noun* uncertainty; undecided state of mind; cynicism; uncertain state. • *verb* feel uncertain or undecided about; hesitate to believe; call in question. □ **in doubt** open to question; **no doubt** certainly, admittedly.

doubtful *adjective* feeling or causing doubt; unreliable. □ **doubtfully** *adverb.*

doubtless *adverb* certainly; probably.

douche /duːʃ/ • *noun* jet of liquid applied to part of body for cleansing or medicinal purposes; device for producing this. • *verb* (**-ching**) treat with douche; use douche.

dough /dəʊ/ *noun* thick paste of flour mixed with liquid for baking; *slang* money. □ **doughnut** (*US* **donut**) small fried cake of sweetened dough. □ **doughy** *adjective* (**-ier**, **-iest**).

doughty /ˈdaʊtɪ/ *adjective* (**-ier**, **-iest**) *archaic* valiant. □ **doughtily** *adverb.*

dour /dʊə/ *adjective* stern, grim, obstinate.

douse /daʊs/ *verb* (also **dowse**) (**-sing**) throw water over; plunge into water; extinguish (light).

dove /dʌv/ *noun* bird with short legs and full breast; advocate of peaceful policies; gentle or innocent person. □ **dovecot(e)** pigeon house.

dovetail • *noun* mortise-and-tenon joint shaped like dove's spread tail. • *verb* fit together, combine neatly; join with dovetails.

dowager /ˈdaʊədʒə/ *noun* woman with title or property from her late husband.

dowdy /ˈdaʊdɪ/ *adjective* (**-ier**, **-iest**) (of clothes) unattractively dull; dressed dowdily. □ **dowdily** *adverb*; **dowdiness** *noun.*

dowel /ˈdaʊəl/ *noun* cylindrical peg for holding parts of structure together.

dowelling *noun* rods for cutting into dowels.

dower /ˈdaʊə/ *noun* widow's share for life of husband's estate.

down¹ /daʊn/ • *adverb* towards or into lower place, esp. to ground; in lower place or position; to or in place regarded as lower, esp. southwards or away from major city or university; in or into low or weaker position or condition; losing by; (of a computer system) out of action; from earlier to later time; in written or recorded form. • *preposition* downwards along, through, or into; from top to bottom of; along; at lower part of. • *adjective* directed downwards. • *verb colloquial* knock or bring etc. down; swallow. • *noun* reverse of fortune; *colloquial* period of depression. □ **be down to** be the responsibility of, have nothing left but; **down and out** destitute; **down-and-out** destitute person; **downcast** dejected, (of eyes) looking down; **downfall** fall from prosperity or power, cause of this; **downgrade** reduce in rank etc.; **downhearted** despondent; **downhill** *adverb* in descending direction, on a decline, *adjective* sloping down, declining; **down in the mouth** looking unhappy; **down-market** of or to cheaper sector of market; **downpour** heavy fall of rain; **downside** negative aspect, drawback; **downstairs** *adverb* down the stairs, to or on lower floor, *adjective* situated downstairs; **downstream** in direction of flow of stream etc.; **down-to-earth** practical, realistic; **downtown** *US* (of) lower or central part of town or city; **downtrodden** oppressed; **downturn** decline, esp. in economic activity; **down under** *colloquial* in Australia or NZ; **downwind** in direction in which wind is blowing; **down with** *expressing rejection of person or thing*; **have a down on** *colloquial* be hostile to. □ **downward** *adjective & adverb*; **downwards** *adverb.*

down² /daʊn/ *noun* baby birds' fluffy covering; bird's under-plumage; fine soft feathers or hairs.

down³ /daʊn/ *noun* open rolling land; (in *plural*) chalk uplands of S. England etc.

downright • *adjective* plain, straightforward; utter. • *adverb* thoroughly.

Down's syndrome *noun* congenital disorder with mental retardation and physical abnormalities.

downy *adjective* (**-ier**, **-iest**) of, like, or covered with down.

dowry /ˈdaʊərɪ/ *noun* (*plural* **-ies**) property brought by bride to her husband.

dowse[1] /daʊz/ *verb* (**-sing**) search for underground water or minerals by holding stick or rod which dips abruptly when over right spot. □ **dowser** *noun.*

dowse[2] = DOUSE.

doxology /dɒk'sɒlədʒɪ/ *noun* (*plural* **-ies**) liturgical hymn etc. of praise to God.

doyen /'dɔɪən/ *noun* (*feminine* **doyenne** /dɔɪ'en/) senior member of group.

doz. *abbreviation* dozen.

doze ● *verb* (**-zing**) sleep lightly, be half asleep. ● *noun* short light sleep. □ **doze off** fall lightly asleep.

dozen /'dʌz(ə)n/ *noun* (*plural* same (after numeral) or **-s**) set of twelve; (**dozens**, usually + *of*) *colloquial* very many.

D.Phil. *abbreviation* Doctor of Philosophy.

Dr *abbreviation* Doctor.

drab *adjective* (**-bb-**) dull, uninteresting; of dull brownish colour. □ **drabness** *noun.*

drachm /dræm/ *noun* weight formerly used by apothecaries, = $^1/_8$ ounce.

drachma /'drækmə/ *noun* (*plural* **-s**) chief monetary unit of Greece.

Draconian /drə'kəʊnɪən/ *adjective* (of laws) harsh, cruel.

draft /drɑ:ft/ ● *noun* preliminary written outline of scheme or version of speech, document, etc.; written order for payment of money by bank; drawing of money on this; detachment from larger group; selection of this; *US* conscription; *US* draught. ● *verb* prepare draft of; select for special duty or purpose; *US* conscript.

draftsman /'drɑ:ftsmən/ *noun* person who drafts documents; person who makes drawings.

drafty *US* = DRAUGHTY.

drag ● *verb* (**-gg-**) pull along with effort; (allow to) trail; (of time etc.) pass tediously or slowly; search bottom of (river etc.) with grapnels, nets, etc. ● *noun* obstruction to progress, retarding force; *colloquial* boring or tiresome task, person, etc.; lure before hounds as substitute for fox; apparatus for dredging; *colloquial* pull at cigarette; *slang* women's clothes worn by men. □ **drag out** protract.

draggle /'dræg(ə)l/ *verb* (**-ling**) make dirty and wet by trailing; hang trailing.

dragon /'drægən/ *noun* mythical monster like reptile, usually with wings and able to breathe fire; fierce woman.

dragonfly *noun* (*plural* **-ies**) large long-bodied gauzy-winged insect.

dragoon /drə'gu:n/ ● *noun* cavalryman; fierce fellow. ● *verb* (+ *into*) coerce or bully into.

drain ● *verb* draw off liquid from; draw off (liquid); flow or trickle away; dry or become dry; exhaust; drink to the dregs; empty (glass etc.) by drinking. ● *noun* channel or pipe carrying off liquid, sewage, etc.; constant outflow or expenditure. □ **draining board** sloping grooved surface beside sink for draining dishes; **drainpipe** pipe for carrying off water etc.

drainage *noun* draining; system of drains; what is drained off.

drake *noun* male duck.

dram *noun* small drink of spirits etc.; drachm.

drama /'drɑ:mə/ *noun* play for stage or broadcasting; art of writing, acting, or presenting plays; dramatic event or quality.

dramatic /drə'mætɪk/ *adjective* of drama; unexpected and exciting; striking; theatrical. □ **dramatically** *adverb.*

dramatics *plural noun* (often treated as *singular*) performance of plays; exaggerated behaviour.

dramatis personae /'dræmətɪs pɜ:'səʊnaɪ/ *plural noun* characters in a play.

dramatist /'dræmətɪst/ *noun* writer of plays.

dramatize /'dræmətaɪz/ *verb* (also **-ise**) (**-zing** or **-sing**) convert into play; make dramatic; behave dramatically. □ **dramatization** *noun.*

drank *past* of DRINK.

drape ● *verb* (**-ping**) cover or hang or adorn with cloth etc.; arrange in graceful folds. ● *noun* (in *plural*) *US* curtains.

draper *noun* retailer of textile fabrics.

drapery *noun* (*plural* **-ies**) clothing or hangings arranged in folds; draper's trade or fabrics.

drastic /'dræstɪk/ *adjective* far-reaching in effect; severe. □ **drastically** *adverb.*

drat *colloquial* ● *verb* (**-tt-**) (usually as *interjection*) curse. ● *interjection expressing annoyance.*

draught /drɑːft/ *noun* (*US* **draft**) current of air indoors; traction; depth of water needed to float ship; drawing of liquor from cask etc.; single act of drinking or inhaling; amount so drunk; (in *plural*) game for two with 12 pieces each, on **draughtboard** (like chessboard). □ **draught beer** beer drawn from cask, not bottled.

draughtsman /ˈdrɑːftsmən/ *noun* person who makes drawings; piece in game of draughts.

draughty *adjective* (*US* **drafty**) (**-ier**, **-iest**) (of a room etc.) letting in sharp currents of air.

draw ● *verb* (*past* **drew** /druː/; *past participle* **drawn**) pull or cause to move towards or after one; pull (thing) up, over, or across; attract; pull (curtains) open or shut; take in; (+ *at*, *on*) inhale from; extract; take from or out; make (line, mark, or outline); make (picture) in this way; represent (thing) in this way; finish (game etc.) with equal scores; proceed to specified position; infer; elicit, evoke; induce; haul up (water) from well; bring out (liquid from tap, wound, etc.); draw lots; obtain by lot; (of tea) infuse; (of chimney, pipe, etc.) promote or allow draught; write out (bill, cheque); search (cover) for game etc.; (as **drawn** *adjective*) looking strained and tense. ● *noun* act of drawing; person or thing that draws custom or attention; drawing of lots, raffle; drawn game etc. □ **draw back** withdraw; **drawback** disadvantage; **drawbridge** hinged retractable bridge; **draw in** (of days etc.) become shorter, (of train etc.) arrive at station; **draw out** prolong, induce to talk, (of days etc.) become longer; **drawstring** string or cord threaded through waistband, bag opening, etc.; **draw up** draft (document etc.), bring into order, come to a halt, (**draw oneself up**) make oneself erect.

drawer /ˈdrɔːə/ *noun* person who draws; (also /drɔː/) receptacle sliding in and out of frame (**chest of drawers**) or of table etc.; (in *plural*) knickers, underpants.

drawing *noun* art of representing by line with pencil etc.; picture etc. made thus. □ **drawing-board** board on which paper is fixed for drawing on; **drawing-pin** flat-headed pin for fastening paper to a surface.

drawing-room *noun* room in private house for sitting or entertaining in.

drawl ● *verb* speak with drawn-out vowel sounds. ● *noun* drawling utterance or way of speaking.

dray *noun* low cart without sides for heavy loads, esp. beer barrels.

dread /dred/ ● *verb* fear greatly, esp. in advance. ● *noun* great fear or apprehension. ● *adjective* dreaded; *archaic* awe-inspiring.

dreadful *adjective* terrible; *colloquial* very annoying, very bad. □ **dreadfully** *adverb*.

dream ● *noun* series of scenes in mind of sleeping person; daydream or fantasy; ideal; aspiration. ● *verb* (*past & past participle* **dreamt** /dremt/ or **dreamed**) experience dream; imagine as in dream; (esp. in negative; + *of*, *that*) consider possible or acceptable; be inactive or unrealistic. □ **dreamer** *noun*.

dreamy *adjective* (**-ier**, **-iest**) given to daydreaming; dreamlike; vague; *colloquial* delightful. □ **dreamily** *adverb*.

dreary /ˈdrɪərɪ/ *adjective* (**-ier**, **-iest**) dismal, gloomy, dull. □ **drearily** *adverb*; **dreariness** *noun*.

dredge[1] ● *noun* apparatus used to collect oysters etc., or to clear mud etc., from bottom of sea etc. ● *verb* (**-ging**) bring up or clear (mud etc.) with dredge; (+ *up*) bring up (something forgotten); clean with or use dredge.

dredge[2] *verb* (**-ging**) sprinkle with flour etc.

dredger[1] *noun* boat with dredge; dredge.

dredger[2] *noun* container with perforated lid for sprinkling flour etc.

dregs *plural noun* sediment, grounds; worst part.

drench ● *verb* wet thoroughly; force (animal) to take medicine. ● *noun* dose of medicine for animal.

dress ● *verb* put clothes on; have and wear clothes; put on evening dress; arrange or adorn; put dressing on (wound etc.); prepare (poultry, crab, etc.) for cooking or eating; apply manure to. ● *noun* woman's one-piece garment of bodice and skirt; clothing, esp.

whole outfit. □ **dress circle** first gallery in theatre; **dressmaker** person who makes women's clothes, esp. for a living; **dress rehearsal** (esp. final) rehearsal in costume; **dress up** put on special clothes, make (person, thing) more attractive or interesting.

dressage /'dresɑːʒ/ *noun* training of horse in obedience and deportment.

dresser[1] /'dresə/ *noun* tall kitchen sideboard with shelves.

dresser[2] *noun* person who helps actors or actresses to dress for stage.

dressing *noun* putting one's clothes on; sauce, esp. of oil, vinegar, etc., for salads; bandage, ointment, etc. for wound; compost etc. spread over land. □ **dressing down** *colloquial* scolding; **dressing gown** loose robe worn while one is not fully dressed; **dressing table** table with mirror etc. for use while dressing, applying make-up, etc.

dressy *adjective* (**-ier**, **-iest**) *colloquial* (of clothes or person) smart, elegant.

drew *past* of DRAW.

drey /dreɪ/ *noun* squirrel's nest.

dribble /'drɪb(ə)l/ ● *verb* (**-ling**) allow saliva to flow from the mouth; flow or allow to flow in drops; *Football etc.* move (ball) forward with slight touches of feet etc. ● *noun* act of dribbling; dribbling flow.

driblet /'drɪblɪt/ *noun* small quantity (of liquid etc.).

dribs and drabs *plural noun colloquial* small scattered amounts.

dried *past & past participle* of DRY.

drier[1] *comparative* of DRY.

drier[2] /'draɪə/ *noun* (also **dryer**) machine for drying hair, laundry, etc.

driest *superlative* of DRY.

drift ● *noun* slow movement or variation; this caused by current; intention, meaning, etc. of what is said etc.; mass of snow etc. heaped up by wind; state of inaction; deviation of craft etc. due to current, wind, etc. ● *verb* be carried by or as if by current of air or water; progress casually or aimlessly; (of current) carry; heap or be heaped into drifts. □ **drift-net** net for sea fishing, which is allowed to drift; **driftwood** wood floating on moving water or washed ashore.

drifter *noun* aimless person; fishing boat with drift-net.

drill[1] ● *noun* tool or machine for boring holes; instruction in military exercises; routine procedure in emergency; thorough training, esp. by repetition; *colloquial* recognized procedure. ● *verb* make hole in or through with drill; make (hole) with drill; train or be trained by drill.

drill[2] ● *noun* machine for making furrows, sowing, and covering seed; small furrow for sowing seed in; row of seeds sown by drill. ● *verb* plant in drills.

drill[3] *noun* coarse twilled cotton or linen fabric.

drill[4] *noun* W. African baboon related to mandrill.

drily *adverb* (also **dryly**) in a dry way.

drink ● *verb* (*past* **drank**; *past participle* **drunk**) swallow (liquid); take alcohol, esp. to excess; (of plant etc.) absorb (moisture). ● *noun* liquid for drinking; draught or specified amount of this; alcoholic liquor; glass, portion, etc., of this; (**the drink**) *colloquial* the sea. □ **drink-driver** *colloquial* person driving with excess alcohol in the blood; **drink in** listen eagerly to; **drink to** toast, wish success to; **drink up** drink all or remainder of. □ **drinker** *noun*.

drip ● *verb* (**-pp-**) fall or let fall in drops; (often + *with*) be so wet as to shed drops. ● *noun* liquid falling in drops; drop of liquid; sound of dripping; *colloquial* dull or ineffectual person.

drip-dry ● *verb* dry or leave to dry crease-free when hung up. ● *adjective* able to be drip-dried.

drip-feed ● *verb* feed intravenously in drops. ● *noun* feeding thus; apparatus for doing this.

dripping *noun* fat melted from roasting meat.

drive ● *verb* (**-ving**; *past* **drove** /drəʊv/; *past participle* **driven** /'drɪv(ə)n/) urge forward, esp. forcibly; compel; force into specified state; operate and direct (vehicle etc.); carry or be carried in vehicle; strike golf ball from tee; (of wind etc.) carry along, propel. ● *noun* excursion in vehicle; driveway; street, road; motivation and energy; inner urge; forcible stroke of bat etc.; organized group effort; transmission of power to machinery or wheels of motor vehicle etc.; organized whist, bingo, etc. competition. □ **drive at** seek, intend, mean; **drive-in** (bank, cinema, etc.) used while one sits in one's car; **driveway** private road through garden to house; **driving licence** licence permitting person to drive vehicle; **driving test** official test of competence to drive; **driving wheel** wheel transmitting power of vehicle to ground.

drivel /'drɪv(ə)l/ ● *noun* silly nonsense. ● *verb* (**-ll-**; *US* **-l-**) talk drivel; run at mouth or nose.

driver *noun* person who drives; golf club for driving from tee.

drizzle /'drɪz(ə)l/ ● *noun* very fine rain. ● *verb* (**-ling**) fall in very fine drops.

droll /drəʊl/ *adjective* quaintly amusing, strange, odd. □ **drollery** *noun* (*plural* **-ies**).

dromedary /'drɒmɪdərɪ/ *noun* (*plural* **-ies**) one-humped (esp. Arabian) camel bred for riding.

drone ● *noun* non-working male of honey-bee; idler; deep humming sound; monotonous speaking tone; bass-pipe of bagpipes or its continuous note. ● *verb* (**-ning**) make deep humming sound; speak or utter monotonously.

drool *verb* slobber, dribble; (often + *over*) admire extravagantly.

droop /dru:p/ ● *verb* bend or hang down, esp. from fatigue or lack of food, drink, etc.; flag. ● *noun* drooping position; loss of spirit. □ **droopy** *adjective*.

drop ● *noun* globule of liquid that falls, hangs, or adheres to surface; very small amount of liquid; abrupt fall or slope; amount of this; act of dropping; fall in prices, temperature, etc.; drop-shaped thing, esp. pendant or sweet; (in *plural*) liquid medicine swallowed in drops. ● *verb* (**-pp-**) fall, allow to fall; let go; fall, let fall, or shed in drops; sink down from exhaustion or injury; cease, lapse; abandon; *colloquial* cease to associate with or discuss; set down (passenger etc.); utter or be uttered casually; fall or let fall in direction, amount, degree, pitch, etc.; (of person) jump down lightly; let oneself fall; omit; give birth to (lamb); deliver from the air by parachute etc.; *Football* send (ball) or score (goal) by drop-kick. □ **drop back, behind** fall back, get left behind; **drop in, by** *colloquial* visit casually; **drop-kick** kick at football made by dropping ball and kicking it as it touches ground; **drop off** fall asleep, drop (passenger); **drop out** (often + *of*) *colloquial* cease to participate (in); **drop-out** *colloquial* person who has dropped out of esp. course of study or conventional society; **drop scone** scone made by dropping spoonful of mixture into pan etc.; **drop-shot** tennis shot dropping abruptly after clearing net. □ **droplet** *noun*.

dropper *noun* device for releasing liquid in drops.

droppings *plural noun* dung; thing that falls or has fallen in drops.

dropsy /'drɒpsɪ/ *noun* oedema. □ **dropsical** *adjective*.

dross *noun* rubbish; scum of molten metal; impurities.

drought /draʊt/ *noun* prolonged absence of rain.

drove[1] *past* of DRIVE.

drove[2] /drəʊv/ *noun* moving crowd; (in *plural*) *colloquial* great number; herd or flock moving together.

drover *noun* herder of cattle.

drown /draʊn/ *verb* kill or die by submersion; submerge; flood; drench; deaden (grief etc.) by drinking; (often + *out*) overpower (sound) with louder sound.

drowse /draʊz/ *verb* (**-sing**) be lightly asleep.

drowsy /'draʊzɪ/ *adjective* (**-ier, -iest**) very sleepy, almost asleep. □ **drowsily** *adverb*; **drowsiness** *noun*.

drub *verb* (**-bb-**) thrash, beat; defeat thoroughly. □ **drubbing** *noun*.

drudge ● *noun* person who does dull, laborious, or menial work. ● *verb* (**-ging**) work hard or laboriously. □ **drudgery** *noun*.

drug ● *noun* medicinal substance; (esp. addictive) hallucinogen, stimulant, narcotic, etc. ● *verb* (**-gg-**) add drug to (drink, food, etc.); administer drug to; stupefy. □ **drugstore** *US* combined chemist's shop and café.

drugget /'drʌgɪt/ *noun* coarse woven fabric used for floor coverings etc.

druggist *noun* pharmacist.

Druid /'dru:ɪd/ *noun* ancient Celtic priest; member of a modern Druidic order, esp. the Gorsedd. □ **Druidic** /-'ɪdɪk/ *adjective*; **Druidism** *noun*.

drum ● *noun* hollow esp. cylindrical percussion instrument covered at one or both ends with plastic, skin, etc.; sound of this; cylindrical structure or object; cylinder used for storage etc.; eardrum. ● *verb* (**-mm-**) play drum; beat or tap continuously with fingers etc.; (of bird etc.) make loud noise with wings. □ **drumbeat** stroke or sound of stroke on drum; **drum brake** kind in which shoes on vehicle press against drum on wheel; **drum into** drive (facts etc.) into (person) by persistence; **drum machine** electronic device that simulates percussion; **drum major** leader of marching band; **drum majorette** female baton-twirling member of parading group; **drum out** dismiss with ignominy; **drumstick** stick for beating drum, lower leg of fowl for eating; **drum up** summon or get by vigorous effort.

drummer *noun* player of drum.

drunk /drʌŋk/ ● *past participle* of DRINK. ● *adjective* lacking control from drinking alcohol; (often + *with*) overcome with joy, success, power, etc. ● *noun* person who is drunk, esp. habitually.

drunkard /'drʌŋkəd/ *noun* person habitually drunk.

drunken /'drʌŋkən/ *adjective* drunk; caused by or involving drunkenness; often drunk. □ **drunkenly** *adverb*; **drunkenness** *noun*.

drupe /dru:p/ *noun* fleshy stone fruit.

dry ● *adjective* (**drier**, **driest**) free from moisture, esp. with moisture having evaporated, drained away, etc; (of eyes) free from tears; (of climate) not rainy; (of river, well, etc.) dried up; (of wine etc.) not sweet; plain, unelaborated; uninteresting; (of sense of humour) ironic, understated; prohibiting sale of alcohol; (of bread) without butter etc.; (of provisions etc.) solid, not liquid; *colloquial* thirsty. ● *verb* (**dries**, **dried**) make or become dry; (usually as **dried** *adjective*) preserve (food) by removing moisture. □ **dry-clean** clean (clothes etc.) with solvents without water; **dry-fly** (of fishing) with floating artificial fly; **dry ice** solid carbon dioxide; **dry out** make or become fully dry, treat or be treated for alcoholism; **dry rot** decay in wood not exposed to air, fungi causing this; **dry run** *colloquial* rehearsal; **dry-shod** without wetting one's shoes; **dry up** make or become completely dry, dry dishes. □ **dryness** *noun*.

dryad /'draɪæd/ *noun* wood nymph.

dryer = DRIER[2].

dryly = DRILY.

D.Sc. *abbreviation* Doctor of Science.

DSC, DSM, DSO *abbreviations* Distinguished Service Cross, Medal, Order.

DSS *abbreviation* Department of Social Security (formerly DHSS).

DT *abbreviation* (also **DT's** /di:'ti:z/) delirium tremens.

DTI *abbreviation* Department of Trade and Industry.

dual /'dju:əl/ ● *adjective* in two parts; twofold; double. ● *noun Grammar* dual number or form. □ **dual carriageway** road with dividing strip between traffic flowing in opposite directions; **dual control** two linked sets of controls, esp. of vehicle used for teaching driving etc., enabling operation by either of two people. □ **duality** /-'æl-/ *noun*.

dub[1] *verb* (**-bb-**) make (person) into knight; give name or nickname to.

dub[2] *verb* (**-bb-**) provide (film etc.) with alternative, esp. translated, soundtrack; add (sound effects, music) to film or broadcast.

dubbin /'dʌbɪn/ *noun* (also **dubbing**) grease for softening and waterproofing leather.

dubiety /dju:'baɪətɪ/ *noun literary* doubt.

dubious /'dju:bɪəs/ *adjective* doubtful; questionable; unreliable. □ **dubiously** *adverb*; **dubiousness** *noun*.

ducal /'dju:k(ə)l/ *adjective* of or like duke.

ducat /'dʌkət/ *noun* gold coin formerly current in most European countries.

duchess /'dʌtʃɪs/ *noun* duke's wife or widow; woman holding rank of duke.

duchy /'dʌtʃɪ/ *noun* (*plural* **-ies**) duke's or duchess's territory.

duck ● *noun* (*plural* same or **-s**) swimming bird, esp. domesticated form of mallard or wild duck; female of this; its flesh as food; *Cricket* batsman's score of 0; (also **ducks**) *colloquial* (esp. as form of address) darling. ● *verb* bob down, esp. to avoid being seen or hit; dip head briefly under water; plunge (person) briefly in water. □ **duckweed** any of various plants that grow on surface of still water.

duckling *noun* young duck.

duct *noun* channel, tube; tube in body carrying secretions etc.

ductile /'dʌktaɪl/ *adjective* (of metal) capable of being drawn into wire; pliable; easily moulded; docile. □ **ductility** /-'tɪl-/ *noun*.

ductless *adjective* (of gland) secreting directly into bloodstream.

dud *slang* ● *noun* useless or broken thing; counterfeit article; (in *plural*) clothes, rags. ● *adjective* defective, useless.

dude /du:d/ *noun slang* fellow; *US* dandy; *US* city man staying on ranch.

dudgeon /'dʌdʒ(ə)n/ *noun* resentment; indignation.

due ● *adjective* owing, payable; merited, appropriate; (often + *to do*) expected or under obligation to do something or arrive at certain time. ● *noun* what one owes or is owed; (usually in *plural*) fee or amount payable. ● *adverb* (of compass point) exactly, directly. □ **due to** because of, caused by.

■ **Usage** Many people believe that *due to*, meaning 'because of', should only be used after the verb *to be*, as in *The mistake was due to ignorance*, and not as in *All trains may be delayed due to a signal failure*. Instead, *owing to a signal failure* could be used.

duel /'dju:əl/ ● *noun* armed contest between two people, usually to the death; two-sided contest. ● *verb* (**-ll-**; *US* **-l-**) fight duel. □ **duellist** *noun*.

duenna /dju:'enə/ *noun* older woman acting as chaperon to girls, esp. in Spain.

duet /dju:'et/ *noun* musical composition for two performers.

duff ● *noun* boiled pudding. ● *adjective slang* worthless, useless, counterfeit.

duffer /'dʌfə/ *noun colloquial* inefficient or stupid person.

duffle /'dʌf(ə)l/ *noun* (also **duffel**) coarse woollen cloth. □ **duffle bag** cylindrical canvas bag closed by drawstring; **duffle-coat** hooded overcoat of duffle with toggle fastenings.

dug[1] *past & past participle* of DIG.

dug[2] *noun* udder, teat.

dugong /'du:gɒŋ/ *noun* (*plural* same or **-s**) Asian sea mammal.

dugout *noun* roofed shelter, esp. for troops in trenches; underground shelter; canoe made from tree trunk.

duke /dju:k/ *noun* person holding highest hereditary title of the nobility; sovereign prince ruling duchy or small state. □ **dukedom** *noun*.

dulcet /'dʌlsɪt/ *adjective* sweet-sounding.

dulcimer /'dʌlsɪmə/ *noun* metal-stringed instrument struck with two hand-held hammers.

dull ● *adjective* tedious; not interesting; (of weather) overcast; (of colour, light, sound, etc.) not bright, vivid, or clear; slow-witted; stupid; (of knife-edge etc.) blunt; listless, depressed. ● *verb* make or become dull. □ **dullard** *noun*; **dullness** *noun*; **dully** /'dʌllɪ/ *adverb*.

duly /'dju:lɪ/ *adverb* in due time or manner; rightly, properly.

dumb /dʌm/ *adjective* unable to speak; silent, taciturn; *colloquial* stupid, ignorant. □ **dumb-bell** short bar with weight at each end, for muscle-building etc.; **dumbstruck** speechless with surprise.

dumbfound /dʌm'faʊnd/ *verb* nonplus; make speechless with surprise.

dumdum /'dʌmdʌm/ *noun* (in full **dumdum bullet**) soft-nosed bullet that expands on impact.

dummy /'dʌmɪ/ ● *noun* (*plural* **-ies**) model of human figure, esp. as used to display clothes or by ventriloquist or as target; imitation object used to replace real or normal one; baby's rubber teat; *colloquial* stupid person; imaginary player in bridge etc., whose cards are exposed and played by partner. ● *adjective* sham, imitation. ● *verb* (**-ies**, **-ied**) make pretended pass etc. in football. □ **dummy run** trial attempt.

dump ● *noun* place for depositing rubbish; *colloquial* unpleasant or dreary place; temporary store of ammunition etc. ● *verb* put down firmly or clumsily; deposit as rubbish; *colloquial* abandon; sell (surplus goods) to foreign market at low price; copy (contents of computer memory etc.) as diagnostic aid or for security.

dumpling /'dʌmplɪŋ/ *noun* ball of dough boiled in stew or containing apple etc.

dumps *plural noun* (usually in **down in the dumps**) *colloquial* low spirits.

dumpy /'dʌmpɪ/ *adjective* (**-ier**, **-iest**) short and stout.

dun • *adjective* greyish-brown. • *noun* dun colour; dun horse.

dunce *noun* person slow at learning.

dunderhead /'dʌndəhed/ *noun* stupid person.

dune /dju:n/ *noun* drift of sand etc. formed by wind.

dung • *noun* excrement of animals; manure. • *verb* apply dung to (land). □ **dunghill** heap of dung or refuse.

dungarees /dʌŋgə'ri:z/ *plural noun* trousers with bib attached.

dungeon /'dʌndʒ(ə)n/ *noun* underground prison cell.

dunk *verb* dip food into liquid before eating it.

dunlin /'dʌnlɪn/ *noun* red-backed sandpiper.

dunnock /'dʌnək/ *noun* hedge sparrow.

duo /'dju:əʊ/ *noun* (*plural* **-s**) pair of performers; duet.

duodecimal /dju:əʊ'desɪm(ə)l/ *adjective* of twelfths or 12; proceeding by twelves.

duodenum /dju:əʊ'di:nəm/ *noun* (*plural* **-s**) part of small intestine next to stomach. □ **duodenal** *adjective*.

duologue /'dju:əlɒg/ *noun* dialogue between two people.

dupe /dju:p/ • *noun* victim of deception. • *verb* (**-ping**) deceive, trick.

duple /'dju:p(ə)l/ *adjective* of two parts. □ **duple time** *Music* rhythm with two beats to bar.

duplex /'dju:pleks/ • *noun US* flat on two floors, house subdivided for two families. • *adjective* having two elements; twofold.

duplicate • *adjective* /'dju:plɪkət/ identical; doubled. • *noun* /-kət/ identical thing, esp. copy. • *verb* /-keɪt/ (**-ting**) double; make or be exact copy of; repeat (an action etc.), esp. unnecessarily. □ **in duplicate** in two exact copies. □ **duplication** *noun*.

duplicator *noun* machine for producing multiple copies of texts.

duplicity /dju:'plɪsɪtɪ/ *noun* double-dealing; deceitfulness. □ **duplicitous** *adjective*.

durable /'djʊərəb(ə)l/ *adjective* lasting; hard-wearing. □ **durability** *noun*.

duration /djʊə'reɪʃ(ə)n/ *noun* time taken by event. □ **for the duration** until end of event, for very long time.

duress /djʊə'res/ *noun* compulsion, esp. illegal use of force or threats.

Durex /'djʊəreks/ *noun proprietary term* condom.

during /'djʊərɪŋ/ *preposition* throughout; at some point in.

dusk *noun* darker stage of twilight.

dusky *adjective* (**-ier**, **-iest**) shadowy, dim; dark-coloured.

dust • *noun* finely powdered earth or other material etc.; dead person's remains. • *verb* wipe the dust from (furniture etc.); sprinkle with powder, sugar, etc. □ **dustbin** container for household refuse; **dust bowl** desert made by drought or erosion; **dustcover** dust-sheet, dust-jacket; **dust-jacket** paper cover on hardback book; **dustman** man employed to collect household refuse; **dustpan** pan into which dust is brushed from floor etc.; **dust-sheet** protective cloth over furniture; **dust-up** *colloquial* fight, disturbance.

duster *noun* cloth etc. for dusting furniture etc.

dusty *adjective* (**-ier**, **-iest**) covered with or full of or like dust.

Dutch • *adjective* of the Netherlands or its people or language. • *noun* Dutch language; (**the Dutch**) (treated as *plural*) the people of the Netherlands. □ **Dutch auction** one in which price is progressively reduced; **Dutch barn** roof on poles over hay etc.; **Dutch cap** dome-shaped contraceptive device fitting over cervix; **Dutch courage** courage induced by alcohol; **Dutchman**, **Dutchwoman** native or national of the Netherlands; **Dutch treat** party, outing, etc. at which people pay for themselves; **go Dutch** share expenses on outing.

dutiable /'dju:tɪəb(ə)l/ *adjective* requiring payment of duty.

dutiful /'dju:tɪfʊl/ *adjective* doing one's duty; obedient. □ **dutifully** *adverb.*

duty /'dju:tɪ/ *noun* (*plural* **-ies**) moral or legal obligation; responsibility; tax on certain goods, imports, etc.; job or function arising from a business or office. □ **duty-free** (of goods) on which duty is not payable; **on, off duty** working or not working).

duvet /'du:veɪ/ *noun* thick soft quilt used instead of sheets and blankets.

dwarf /dwɔ:f/ ● *noun* (*plural* **-s** or **dwarves** /dwɔ:vz/) person, animal, or plant much below normal size, esp. with normal-sized head and body but short limbs; small mythological being with magical powers. ● *verb* stunt in growth; make look small by contrast.

dwell *verb* (*past & past participle* **dwelt** or **dwelled**) reside, live. □ **dwell on** think, write, or speak at length on. □ **dweller** *noun.*

dwelling *noun* house, residence.

dwindle /'dwɪnd(ə)l/ *verb* (**-ling**) become gradually less or smaller; lose importance.

dye /daɪ/ ● *noun* substance used to change colour of fabric, wood, hair, etc.; colour produced by this. ● *verb* (**dyeing**, **dyed**) colour with dye; dye a specified colour. □ **dyer** *noun.*

dying /'daɪɪŋ/ ● *present participle* of DIE[1]. ● *adjective* of, or at the time of, death.

dyke /daɪk/ (also **dike**) ● *noun* embankment built to prevent flooding; low wall. ● *verb* (**-king**) provide or protect with dyke(s).

dynamic /daɪ'næmɪk/ *adjective* energetic, active; of motive force; of force in operation; of dynamics. □ **dynamically** *adverb.*

dynamics *plural noun* (usually treated as *singular*) mathematical study of motion and forces causing it.

dynamism /'daɪnəmɪz(ə)m/ *noun* energy; dynamic power.

dynamite /'daɪnəmaɪt/ ● *noun* highly explosive mixture containing nitroglycerine. ● *verb* (**-ting**) charge or blow up with this.

dynamo /'daɪnəməʊ/ *noun* (*plural* **-s**) machine converting mechanical into electrical energy; *colloquial* energetic person.

dynast /'dɪnəst/ *noun* ruler; member of dynasty.

dynasty /'dɪnəstɪ/ *noun* (*plural* **-ies**) line of hereditary rulers. □ **dynastic** /-'næs-/ *adjective.*

dyne *noun Physics* force required to give a mass of one gram an acceleration of one centimetre per second per second.

dysentery /'dɪsəntrɪ/ *noun* inflammation of bowels, causing severe diarrhoea.

dysfunction /dɪs'fʌŋkʃ(ə)n/ *noun* abnormality or impairment of functioning.

dyslexia /dɪs'leksɪə/ *noun* abnormal difficulty in reading and spelling. □ **dyslectic** /-'lektɪk/ *adjective & noun*; **dyslexic** *adjective & noun.*

dyspepsia /dɪs'pepsɪə/ *noun* indigestion. □ **dyspeptic** *adjective & noun.*

dystrophy /'dɪstrəfɪ/ *noun* defective nutrition.

Ee

E *abbreviation* (also **E.**) east(ern). □ **E-number** number prefixed by letter E identifying food additive.

each ● *adjective* every one of two or more, regarded separately. ● *pronoun* each person or thing. □ **each way** (of bet) backing horse etc. to win or come second or third.

eager /'i:gə/ *adjective* keen, enthusiastic. □ **eagerly** *adverb*; **eagerness** *noun*.

eagle /'i:g(ə)l/ *noun* large bird of prey; *Golf* score of two under par for hole. □ **eagle eye** keen sight, watchfulness; **eagle-eyed** *adjective*.

eaglet /'i:glɪt/ *noun* young eagle.

ear[1] /ɪə/ *noun* organ of hearing, esp. external part; faculty of discriminating sound; attention. □ **all ears** listening attentively; **earache** pain in inner ear; **eardrum** membrane of middle ear; **earphone** (usually in *plural*) device worn on ear to listen to recording, radio, etc.; **earplug** device worn in ear as protection from water, noise, etc.; **earring** jewellery worn on ear; **earshot** hearing-range; **ear-trumpet** trumpet-shaped tube formerly used as hearing aid.

ear[2] /ɪə/ *noun* seed-bearing head of cereal plant.

earl /3:l/ *noun* British nobleman ranking between marquess and viscount. □ **earldom** *noun*.

early /'3:lɪ/ *adjective & adverb* (**-ier**, **-iest**) before due, usual, or expected time; not far on in day or night or in development etc. □ **early bird** *colloquial* person who arrives, gets up, etc. early; **early days** too soon to expect results etc.

earmark *verb* set aside for special purpose.

earn /3:n/ *verb* obtain as reward for work or merit; bring as income or interest. □ **earner** *noun*.

earnest /'3:nɪst/ *adjective* intensely serious. □ **in earnest** serious(ly). □ **earnestly** *adverb*; **earnestness** *noun*.

earnings /'3:nɪŋz/ *plural noun* money earned.

earth /3:θ/ ● *noun* planet we live on (also **Earth**); land and sea as opposed to sky; ground; soil, mould; this world as opposed to heaven or hell; *Electricity* connection to earth as completion of circuit; hole of fox etc. ● *verb Electricity* connect to earth; cover (roots) with earth. □ **earthwork** bank of earth in fortification; **earthworm** worm living in earth; **run to earth** find after long search.

earthen *adjective* made of earth or baked clay. □ **earthenware** pottery made of fired clay.

earthly *adjective* of earth, terrestrial; *colloquial* (usually with negative) remotely possible. □ **not an earthly** *colloquial* no chance or idea whatever.

earthquake *noun* violent shaking of earth's surface.

earthy *adjective* (**-ier**, **-iest**) of or like earth or soil; coarse, crude.

earwig *noun* insect with pincers at rear end.

ease /i:z/ ● *noun* facility, effortlessness; freedom from pain, trouble, or constraint. ● *verb* (**-sing**) relieve from pain etc.; (often + *off*, *up*) become less burdensome or severe; relax, slacken; move or be moved by gentle force.

easel /'i:z(ə)l/ *noun* stand for painting, blackboard, etc.

easement *noun Law* right of way over another's property.

easily /'i:zɪlɪ/ *adverb* without difficulty; by far; very probably.

east ● *noun* point of horizon where sun rises at equinoxes; corresponding compass point; (usually **the East**) eastern part of world, country, town, etc. ● *adjective* towards, at, near, or facing east; (of wind) from east. ● *adverb* towards, at, or near east; (+ *of*) further east than. □ **eastbound** travelling or leading east; **East End** part of London east of City; **east-north-east**, **east-south-east** point midway between east and north-east or south-east. □ **eastward** *adjective, adverb, & noun*; **eastwards** *adverb*.

Easter /ˈiːstə/ *noun* festival of Christ's resurrection. □ **Easter egg** artificial usually chocolate egg given at Easter; **Easter Saturday** day before Easter, (properly) Saturday after Easter.

easterly /ˈiːstəlɪ/ *adjective & adverb* in eastern position or direction; (of wind) from east.

eastern /ˈiːst(ə)n/ *adjective* of or in east. □ **Eastern Church** Orthodox Church. □ **easternmost** *adjective*.

easterner *noun* native or inhabitant of east.

easy /ˈiːzɪ/ • *adjective* (**-ier**, **-iest**) not difficult; free from pain, trouble, or anxiety; relaxed and pleasant; compliant. • *adverb* with ease, in an effortless or relaxed way. □ **easy chair** large comfortable armchair; **easygoing** placid and tolerant; **go easy** (usually + *on*, *with*) be sparing or cautious; **take it easy** proceed gently, relax.

eat *verb* (*past* **ate** /et, eɪt/; *past participle* **eaten**) chew and swallow (food); consume food, have meal; destroy, consume. □ **eating apple** one suitable for eating raw; **eat out** have meal away from home, esp. in restaurant; **eat up** eat completely.

eatable • *adjective* fit to be eaten. • *noun* (usually in *plural*) food.

eater *noun* person who eats; eating apple.

eau-de-Cologne /əʊdəkəˈləʊn/ *noun* toilet water originally from Cologne.

eaves /iːvz/ *plural noun* underside of projecting roof.

eavesdrop *verb* (**-pp-**) listen to private conversation. □ **eavesdropper** *noun*.

ebb • *noun* outflow of tide. • *verb* flow back; decline.

ebony /ˈebənɪ/ • *noun* hard heavy black tropical wood. • *adjective* made of or black as ebony.

ebullient /ɪˈbʌlɪənt/ *adjective* exuberant. □ **ebullience** *noun*; **ebulliently** *adverb*.

EC *abbreviation* European Community; East Central.

eccentric /ɪkˈsentrɪk/ • *adjective* odd or capricious in behaviour or appearance; not placed centrally, not having axis etc. placed centrally; not concentric; not circular. • *noun* eccentric person. □ **eccentrically** *adverb*; **eccentricity** /eksenˈtrɪs-/ *noun* (*plural* **-ies**).

ecclesiastic /ɪkliːzɪˈæstɪk/ *noun* clergyman.

ecclesiastical *adjective* of the Church or clergy.

ECG *abbreviation* electrocardiogram.

echelon /ˈeʃəlɒn/ *noun* level in organization, society, etc.; wedge-shaped formation of troops, aircraft, etc.

echidna /ɪˈkɪdnə/ *noun* Australian egg-laying spiny mammal.

echo /ˈekəʊ/ • *noun* (*plural* **-es**) repetition of sound by reflection of sound waves; reflected radio or radar beam; close imitation; circumstance or event reminiscent of earlier one. • *verb* (**-es**, **-ed**) resound with echo; repeat, imitate; be repeated.

éclair /eɪˈkleə/ *noun* finger-shaped iced cake of choux pastry filled with cream.

éclat /eɪˈklɑː/ *noun* brilliant display; conspicuous success; prestige.

eclectic /ɪˈklektɪk/ • *adjective* selecting ideas, style, etc. from various sources. • *noun* eclectic person. □ **eclecticism** /-tɪs-/ *noun*.

eclipse /ɪˈklɪps/ • *noun* obscuring of light of sun by moon (**solar eclipse**) or of moon by earth (**lunar eclipse**); loss of light or importance. • *verb* (**-sing**) cause eclipse of; intercept (light); outshine, surpass.

eclogue /ˈeklɒg/ *noun* short pastoral poem.

ecology /ɪˈkɒlədʒɪ/ *noun* study of relations of organisms to one another and their surroundings. □ **ecological** /iːkəˈlɒdʒ-/ *adjective*; **ecologically** /iːkəˈlɒdʒ-/ *adverb*; **ecologist** *noun*.

economic /iːkəˈnɒmɪk/ *adjective* of economics; profitable; connected with trade and industry. □ **economically** *adverb*.

economical /iːkəˈnɒmɪk(ə)l/ *adjective* sparing; avoiding waste. □ **economically** *adverb*.

economics /iːkəˈnɒmɪks/ *plural noun* (treated as *singular*) science of production and distribution of wealth; application of this to particular subject. □ **economist** /ɪˈkɒnəmɪst/ *noun*.

economize /ɪˈkɒnəmaɪz/ *verb* (also **-ise**) (**-zing** or **-sing**) make economies; reduce expenditure.

economy /ɪˈkɒnəmɪ/ *noun* (*plural* **-ies**) community's system of wealth creation; frugality, instance of this; sparing use.

ecosystem /ˈiːkəʊsɪstəm/ *noun* biological community of interacting organisms and their physical environment.

ecru /ˈeɪkruː/ *noun* light fawn colour.

ecstasy /ˈekstəsɪ/ *noun* (*plural* **-ies**) overwhelming joy or rapture; *slang* type of hallucinogenic drug. □ **ecstatic** /ɪkˈstætɪk/ *adjective*.

ECT *abbreviation* electroconvulsive therapy.

ectoplasm /ˈektəʊplæz(ə)m/ *noun* supposed viscous substance exuding from body of spiritualistic medium during trance.

ecu /ˈekjuː/ *noun* (also **Ecu**) (*plural* **-s**) European Currency Unit.

ecumenical /iːkjuːˈmenɪk(ə)l/ *adjective* of or representing whole Christian world; seeking worldwide Christian unity. □ **ecumenism** /iːˈkjuːmən-/ *noun*.

eczema /ˈeksɪmə/ *noun* kind of inflammation of skin.

ed. *abbreviation* edited by; edition; editor.

Edam /ˈiːdæm/ *noun* round Dutch cheese with red rind.

eddy /ˈedɪ/ ● *noun* (*plural* **-ies**) circular movement of water, smoke, etc. ● *verb* (**-ies**, **-ied**) move in eddies.

edelweiss /ˈeɪd(ə)lvaɪs/ *noun* Alpine plant with woolly white bracts.

Eden /ˈiːd(ə)n/ *noun* (in full **Garden of Eden**) home of Adam and Eve; delightful place or state.

edge ● *noun* boundary-line or margin of area or surface; narrow surface of thin object; meeting-line of surfaces; sharpness; sharpened side of blade; brink of precipice; crest of ridge; effectiveness. ● *verb* (**-ging**) advance gradually or furtively; give or form border to; sharpen. □ **edgeways**, **edgewise** with edge foremost or uppermost; **have the edge on, over** have slight advantage over; **on edge** excited or irritable; **set (person's) teeth on edge** cause unpleasant nervous sensation in.

edging *noun* thing forming edge or border.

edgy *adjective* (**-ier**, **-iest**) irritable; anxious.

edible /ˈedɪb(ə)l/ *adjective* fit to be eaten.

edict /ˈiːdɪkt/ *noun* order proclaimed by authority.

edifice /ˈedɪfɪs/ *noun* building, esp. imposing one.

edify /ˈedɪfaɪ/ *verb* (**-ies**, **-ied**) improve morally. □ **edification** *noun*.

edit /ˈedɪt/ *verb* (**-t-**) prepare for publication or broadcast; be editor of; cut and collate (films etc.) to make unified sequence; reword, modify; (+ *out*) remove (part) from text, recording, etc.

edition /ɪˈdɪʃ(ə)n/ *noun* edited or published form of book etc.; copies of book or newspaper etc. issued at one time; instance of regular broadcast.

editor /ˈedɪtə/ *noun* person who edits; person who directs writing of newspaper or news programme or section of one; person who selects material for publication. □ **editorship** *noun*.

editorial /edɪˈtɔːrɪəl/ ● *adjective* of editing or an editor. ● *noun* article giving newspaper's views on current topic.

educate /ˈedjʊkeɪt/ *verb* (**-ting**) train or instruct mentally and morally; provide systematic instruction for. □ **educable** *adjective*; **education** *noun*; **educational** *adjective*; **educator** *noun*.

educationist *noun* (also **educationalist**) expert in educational methods.

Edwardian /edˈwɔːdɪən/ ● *adjective* of or characteristic of reign (1901–10) of Edward VII. ● *noun* person of this period.

EEC *abbreviation* European Economic Community.

EEG *abbreviation* electroencephalogram.

eel *noun* snakelike fish.

eerie /ˈɪərɪ/ *adjective* (**-r**, **-st**) strange; weird. □ **eerily** *adverb*.

efface /ɪˈfeɪs/ *verb* (**-cing**) rub or wipe out; surpass, eclipse; (**efface oneself**) treat oneself as unimportant. □ **effacement** *noun*.

effect /ɪˈfekt/ ● *noun* result, consequence; efficacy; impression; (in *plural*) possessions; (in *plural*) lighting, sound, etc. giving realism to play etc.; physical phenomenon. ● *verb* bring about.

■ **Usage** As a verb, *effect* should not be confused with *affect*. *He effected an entrance* means 'He got in (somehow)', but *This won't affect me* means 'My life won't be changed by this'.

effective *adjective* operative; impressive; actual; producing intended result. □ **effectively** *adverb*; **effectiveness** *noun*.

effectual /ɪ'fektʃʊəl/ *adjective* producing required effect; valid.

effeminate /ɪ'femɪnət/ *adjective* (of a man) unmanly, womanish. □ **effeminacy** *noun*.

effervesce /efə'ves/ *verb* (**-cing**) give off bubbles of gas. □ **effervescence** *noun*; **effervescent** *adjective*.

effete /ɪ'fi:t/ *adjective* feeble; effeminate.

efficacious /efɪ'keɪʃəs/ *adjective* producing desired effect. □ **efficacy** /'efɪkəsɪ/ *noun*.

efficient /ɪ'fɪʃ(ə)nt/ *adjective* productive with minimum waste of effort; competent, capable. □ **efficiency** *noun*; **efficiently** *adverb*.

effigy /'efɪdʒɪ/ *noun* (*plural* **-ies**) sculpture or model of person.

effloresce /eflɔ:'res/ *verb* (**-cing**) burst into flower. □ **efflorescence** *noun*.

effluence /'eflʊəns/ *noun* flowing out (of light, electricity, etc.); what flows out.

effluent /'eflʊənt/ ● *adjective* flowing out. ● *noun* sewage or industrial waste discharged into river etc.; stream flowing from lake etc.

effluvium /ɪ'flu:vɪəm/ *noun* (*plural* **-via**) unpleasant or harmful outflow.

effort /'efət/ *noun* exertion; determined attempt; force exerted; *colloquial* something accomplished. □ **effortless** *adjective*; **effortlessly** *adverb*.

effrontery /ɪ'frʌntərɪ/ *noun* impudence.

effuse /ɪ'fju:z/ *verb* (**-sing**) pour forth.

effusion /ɪ'fju:ʒ(ə)n/ *noun* outpouring.

effusive /ɪ'fju:sɪv/ *adjective* demonstrative; gushing. □ **effusively** *adverb*; **effusiveness** *noun*.

EFL *abbreviation* English as a foreign language.

Efta /'eftə/ *noun* (also **EFTA**) European Free Trade Association.

e.g. *abbreviation* for example.

egalitarian /ɪgælɪ'teərɪən/ ● *adjective* of or advocating equal rights for all. ● *noun* egalitarian person. □ **egalitarianism** *noun*.

egg[1] *noun* body produced by female of birds, insects, etc., capable of developing into new individual; edible egg of domestic hen; ovum. □ **eggcup** cup for holding boiled egg; **egghead** *colloquial* intellectual; **eggplant** aubergine; **egg white** white or clear part round yolk of egg.

egg[2] *verb* (+ *on*) urge.

eglantine /'egləntaɪn/ *noun* sweet-brier.

ego /'i:gəʊ/ *noun* (*plural* **-s**) the self; part of mind that has sense of individuality; self-esteem.

egocentric /i:gəʊ'sentrɪk/ *adjective* self-centred.

egoism /'i:gəʊɪz(ə)m/ *noun* self-interest as moral basis of behaviour; systematic selfishness; egotism. □ **egoist** *noun*; **egoistic** /-'ɪs-/ *adjective*.

egotism /'i:gətɪz(ə)m/ *noun* self-conceit; selfishness. □ **egotist** *noun*; **egotistic(al)** /-'tɪs-/ *adjective*.

egregious /ɪ'gri:dʒəs/ *adjective* extremely bad; *archaic* remarkable.

egress /'i:gres/ *noun* going out; way out.

egret /'i:grɪt/ *noun* kind of white heron.

Egyptian /ɪ'dʒɪpʃ(ə)n/ ● *adjective* of Egypt. ● *noun* native or national of Egypt; language of ancient Egyptians.

Egyptology /i:dʒɪp'tɒlədʒɪ/ *noun* study of ancient Egypt. □ **Egyptologist** *noun*.

eh /eɪ/ *interjection colloquial: expressing inquiry, surprise, etc.*

eider /'aɪdə/ *noun* northern species of duck. □ **eiderdown** quilt stuffed with soft material, esp. down.

eight /eɪt/ *adjective & noun* one more than seven; 8-oared boat, its crew. □ **eightsome reel** lively Scottish dance for 8 people. □ **eighth** /eɪtθ/ *adjective & noun*.

eighteen /eɪ'ti:n/ *adjective & noun* one more than seventeen. □ **eighteenth** *adjective & noun*.

eighty /'eɪtɪ/ *adjective & noun* (*plural* **-ies**) eight times ten. □ **eightieth** *adjective & noun*.

eisteddfod /aɪ'stedfəd/ *noun* congress of Welsh poets and musicians gathering for musical and literary competition.

either /'aɪðə, 'i:ðə/ ● *adjective & pronoun* one or other of two; each of two. ● *adverb* (with negative) any more than the other. □ **either ... or ...** as one possibility ... and as the other ...

ejaculate /ɪ'dʒækjʊleɪt/ *verb* (**-ting**) emit (semen) in orgasm; exclaim. □ **ejaculation** *noun*.

eject /ɪ'dʒekt/ *verb* throw out, expel; (of pilot etc.) cause oneself to be propelled from aircraft in emergency; emit. □ **ejection** *noun*.

ejector seat *noun* device in aircraft for emergency ejection of pilot etc.

eke *verb* (**eking**) □ **eke out** supplement (income etc.), make (living) or support (existence) with difficulty.

elaborate ● *adjective* /ɪ'læbərət/ minutely worked out; complicated. ● *verb* /ɪ'læbəreɪt/ (**-ting**) work out or explain in detail. □ **elaborately** *adverb*; **elaboration** *noun*.

élan /eɪ'lɑ̃/ *noun* vivacity, dash. [French]

eland /'i:lənd/ *noun* (*plural* same or **-s**) large African antelope.

elapse /ɪ'læps/ *verb* (**-sing**) (of time) pass by.

elastic /ɪ'læstɪk/ ● *adjective* able to resume normal bulk or shape after being stretched or squeezed; springy; flexible. ● *noun* elastic cord or fabric, usually woven with strips of rubber. □ **elastic band** rubber band. □ **elasticity** /i:læ'stɪs-/ *noun*.

elasticated /ɪ'læstɪkeɪtɪd/ *adjective* (of fabric) made elastic by weaving with rubber thread.

elate /ɪ'leɪt/ *verb* (**-ting**) (esp. as **elated** *adjective*) make delighted or proud. □ **elation** *noun*.

elbow /'elbəʊ/ ● *noun* joint between forearm and upper arm; part of sleeve covering elbow; elbow-shaped thing. ● *verb* thrust or jostle (person); make (one's way) thus. □ **elbow-grease** *jocular* vigorous polishing, hard work; **elbow-room** sufficient space to move or work in.

elder[1] /'eldə/ ● *adjective* older; senior. ● *noun* older person; official in early Christian and some modern Churches.

elder[2] /'eldə/ *noun* tree with white flowers and black **elderberries**.

elderly /'eldəlɪ/ *adjective* rather old.

eldest /'eldɪst/ *adjective* first-born; oldest surviving.

eldorado /eldə'rɑ:dəʊ/ *noun* (*plural* **-s**) imaginary land of great wealth.

elect /ɪ'lekt/ ● *verb* choose by voting; choose, decide. ● *adjective* chosen; select, choice; (after noun) chosen but not yet in office.

election /ɪ'lekʃ(ə)n/ *noun* electing, being elected; occasion for this.

electioneer /ɪlekʃə'nɪə/ *verb* take part in election campaign.

elective /ɪ'lektɪv/ *adjective* chosen by or derived from election; entitled to elect; optional.

elector /ɪ'lektə/ *noun* person entitled to vote in election. □ **electoral** *adjective*.

electorate /ɪ'lektərət/ *noun* group of electors.

electric /ɪ'lektrɪk/ *adjective* of, worked by, or charged with electricity; causing or charged with excitement. □ **electric blanket** one heated by internal wires; **electric chair** chair used for electrocution of criminals; **electric eel** eel-like fish able to give electric shock; **electric fire** portable electric heater; **electric shock** effect of sudden discharge of electricity through body of person etc. □ **electrically** *adverb*.

electrical *adjective* of or worked by electricity. □ **electrically** *adverb*.

electrician /ɪlek'trɪʃ(ə)n/ *noun* person who installs or maintains electrical equipment.

electricity /ɪlek'trɪsɪtɪ/ *noun* form of energy present in protons and electrons; science of electricity; supply of electricity.

electrify /ɪ'lektrɪfaɪ/ *verb* (**-ies**, **-ied**) charge with electricity; convert to electric working; startle, excite. □ **electrification** *noun*.

electro- *combining form* of or caused by electricity.

electrocardiogram /ɪlektrəʊ'kɑ:dɪəgræm/ *noun* record of electric currents generated by heartbeat. □ **electrocardiograph** instrument for recording such currents.

electroconvulsive /ɪlektrəʊkən'vʌlsɪv/ *adjective* (of therapy) using convulsive response to electric shocks.

electrocute /ɪ'lektrəkju:t/ *verb* (**-ting**) kill by electric shock. □ **electrocution** /-'kju:ʃ(ə)n/ *noun*.

electrode /ɪ'lektrəʊd/ *noun* conductor through which electricity enters or leaves electrolyte, gas, vacuum, etc.

electroencephalogram /ɪlektrəʊɪn'sefələgræm/ *noun* record of electrical activity of brain. □ **electroencephalograph** instrument for recording such activity.

electrolysis /ɪlek'trɒlɪsɪs/ *noun* chemical decomposition by electric action; breaking up of tumours, hair-roots, etc. thus.

electrolyte /ɪ'lektrəlaɪt/ *noun* solution that can conduct electricity; substance that can dissolve to produce this. □ **electrolytic** /-trəʊ'lɪt-/ *adjective.*

electromagnet /ɪlektrəʊ'mægnɪt/ *noun* soft metal core made into magnet by electric current through coil surrounding it.

electromagnetism /ɪlektrəʊ'mægnɪtɪz(ə)m/ *noun* magnetic forces produced by electricity; study of these.

electron /ɪ'lektrɒn/ *noun* stable elementary particle with charge of negative electricity, found in all atoms and acting as primary carrier of electricity in solids. □ **electron microscope** one with high magnification, using electron beam instead of light.

electronic /ɪlek'trɒnɪk/ *adjective* of electrons or electronics; (of music) produced electronically. □ **electronic mail** messages distributed by a computer system. □ **electronically** *adverb.*

electronics *plural noun* (treated as *singular*) science of movement of electrons in vacuum, gas, semiconductor, etc.

electroplate /ɪ'lektrəʊpleɪt/ ● *verb* (**-ting**) coat with chromium, silver, etc. by electrolysis. ● *noun* electroplated articles.

elegant /'elɪgənt/ *adjective* graceful, tasteful, refined; ingeniously simple. □ **elegance** *noun*; **elegantly** *adverb.*

elegy /'elɪdʒɪ/ *noun* (*plural* **-ies**) sorrowful song or poem, esp. for the dead. □ **elegiac** /-'dʒaɪək/ *adjective.*

element /'elɪmənt/ *noun* component part; substance which cannot be resolved by chemical means into simpler substances; any of **the four elements** (earth, water, air, fire) formerly supposed to make up all matter; wire that gives out heat in electric heater, cooker, etc.; (in *plural*) atmospheric agencies; (in *plural*) rudiments, first principles; (in *plural*) bread and wine of Eucharist. □ **in one's element** in one's preferred situation.

elemental /elɪ'ment(ə)l/ *adjective* of or like the elements or the forces of nature; basic, essential.

elementary /elɪ'mentərɪ/ *adjective* dealing with simplest facts of subject; unanalysable. □ **elementary particle** *Physics* subatomic particle, esp. one not known to consist of simpler ones.

elephant /'elɪf(ə)nt/ *noun* (*plural* same or **-s**) largest living land animal, with trunk and ivory tusks.

elephantiasis /elɪfən'taɪəsɪs/ *noun* skin disease causing gross enlargement of limbs etc.

elephantine /elɪ'fæntaɪn/ *adjective* of elephants; huge; clumsy.

elevate /'elɪveɪt/ *verb* (**-ting**) lift up, raise; exalt in rank etc.; (usually as **elevated** *adjective*) raise morally or intellectually.

elevation *noun* elevating, being elevated; angle above horizontal; height above given level; drawing showing one side of building.

elevator *noun US* lift; movable part of tailplane for changing aircraft's altitude; hoisting machine.

eleven /ɪ'lev(ə)n/ *adjective & noun* one more than ten; team of 11 people in cricket etc. □ **eleventh hour** last possible moment. □ **eleventh** *adjective & noun.*

elevenses /ɪ'levənzɪz/ *noun colloquial* light mid-morning refreshment.

elf *noun* (*plural* **elves** /elvz/) mythological being, esp. small and mischievous one. □ **elfish** *adjective.*

elfin /'elfɪn/ *adjective* of elves; elflike.

elicit /ɪ'lɪsɪt/ *verb* (**-t-**) draw out (facts, response, etc.).

elide /ɪ'laɪd/ *verb* (**-ding**) omit in pronunciation.

eligible /'elɪdʒɪb(ə)l/ *adjective* (often + *for*) fit or entitled to be chosen; desirable or suitable, esp. for marriage. □ **eligibility** *noun.*

eliminate /ɪ'lɪmɪneɪt/ *verb* (**-ting**) remove, get rid of; exclude. □ **elimination** *noun*; **eliminator** *noun.*

elision /ɪ'lɪʒ(ə)n/ *noun* omission of vowel or syllable in pronunciation.

élite /ɪ'li:t/ *noun* select group or class; (**the élite**) the best (of a group).

élitism *noun* advocacy of or reliance on dominance by select group. □ **élitist** *noun & adjective.*

elixir /ɪ'lɪksɪə/ *noun* alchemist's preparation supposedly able to change metal into gold or prolong life indefinitely; aromatic medicine.

Elizabethan /ɪlɪzə'bi:θ(ə)n/ ● *adjective* of time of Elizabeth I or II. ● *noun* person of this time.

elk *noun* (*plural* same or **-s**) large type of deer.

ellipse /ɪ'lɪps/ *noun* regular oval.

ellipsis /ɪ'lɪpsɪs/ *noun* (*plural* **ellipses** /-si:z/) omission of words needed to complete construction or sense.

elliptical /ɪ'lɪptɪk(ə)l/ *adjective* of or like an ellipse; (of language) confusingly concise.

elm *noun* tree with rough serrated leaves; its wood.

elocution /elə'kju:ʃ(ə)n/ *noun* art of clear and expressive speaking.

elongate /'i:lɒŋgeɪt/ *verb* (**-ting**) extend, lengthen. □ **elongation** *noun.*

elope /ɪ'ləʊp/ *verb* (**-ping**) run away to marry secretly. □ **elopement** *noun.*

eloquence /'eləkwəns/ *noun* fluent and effective use of language. □ **eloquent** *adjective*; **eloquently** *adverb.*

else /els/ *adverb* besides; instead; otherwise, if not. □ **elsewhere** in or to some other place.

elucidate /ɪ'lu:sɪdeɪt/ *verb* (**-ting**) throw light on, explain. □ **elucidation** *noun.*

elude /ɪ'lu:d/ *verb* (**-ding**) escape adroitly from; avoid; baffle.

elusive /ɪ'lu:sɪv/ *adjective* difficult to find, catch, or remember; avoiding the point raised. □ **elusiveness** *noun.*

elver /'elvə/ *noun* young eel.

elves *plural* of ELF.

Elysium /ɪ'lɪzɪəm/ *noun Greek Mythology* home of the blessed after death; place of ideal happiness. □ **Elysian** *adjective.*

em *noun Printing* unit of measurement approximately equal to width of M.

emaciate /ɪ'meɪsɪeɪt/ *verb* (**-ting**) (esp. as **emaciated** *adjective*) make thin or feeble. □ **emaciation** *noun.*

emanate /'emәneɪt/ *verb* (**-ting**) (usually + *from*) issue or originate (from source). □ **emanation** *noun.*

emancipate /ɪ'mænsɪpeɪt/ *verb* (**-ting**) free from social, political, or moral restraint. □ **emancipation** *noun.*

emasculate ● *verb* /ɪ'mæskjʊleɪt/ (**-ting**) enfeeble; castrate. ● *adjective* /ɪ'mæskjʊlət/ enfeebled; castrated; effeminate. □ **emasculation** *noun.*

embalm /ɪm'bɑ:m/ *verb* preserve (corpse) from decay; preserve from oblivion; make fragrant.

embankment /ɪm'bæŋkmənt/ *noun* bank constructed to confine water or carry road or railway.

embargo /ɪm'bɑ:gəʊ/ ● *noun* (*plural* **-es**) order forbidding ships to enter or leave port; suspension of commerce or other activity. ● *verb* (**-es**, **-ed**) place under embargo.

embark /ɪm'bɑ:k/ *verb* put or go on board ship; (+ *on*, *in*) begin (enterprise).

embarkation /embɑ:'keɪʃ(ə)n/ *noun* embarking on ship.

embarrass /ɪm'barəs/ *verb* make (person) feel awkward or ashamed; encumber; (as **embarrassed** *adjective*) encumbered with debts. □ **embarrassment** *noun.*

embassy /'embəsɪ/ *noun* (*plural* **-ies**) ambassador's residence or offices; deputation to foreign government.

embattled /ɪm'bæt(ə)ld/ *adjective* prepared or arrayed for battle; fortified with battlements; under heavy attack, in trying circumstances.

embed /ɪm'bed/ *verb* (also **imbed**) (**-dd-**) (esp. as **embedded** *adjective*) fix in surrounding mass.

embellish /ɪm'belɪʃ/ *verb* beautify, adorn; make fictitious additions to. □ **embellishment** *noun.*

ember /'embə/ *noun* (usually in *plural*) small piece of glowing coal etc. in dying fire. □ **ember days** days of fasting and prayer in Christian Church, associated with ordinations.

embezzle /ɪm'bez(ə)l/ *verb* (**-ling**) divert (money) fraudulently to own use. □ **embezzlement** *noun*; **embezzler** *noun.*

embitter /ɪm'bɪtə/ *verb* arouse bitter feelings in. □ **embitterment** *noun.*

emblazon /ɪm'bleɪz(ə)n/ *verb* portray or adorn conspicuously.

emblem /'embləm/ *noun* symbol; (+ *of*) type, embodiment; distinctive badge. □ **emblematic** /-'mæt-/ *adjective*.

embody /ɪm'bɒdɪ/ *verb* (**-ies**, **-ied**) give concrete form to; be expression of; include, comprise. □ **embodiment** *noun*.

embolden /ɪm'bəʊld(ə)n/ *verb* encourage.

embolism /'embəlɪz(ə)m/ *noun* obstruction of artery by blood clot etc.

emboss /ɪm'bɒs/ *verb* carve or decorate with design in relief.

embrace /ɪm'breɪs/ ● *verb* (**-cing**) hold closely in arms; enclose; accept, adopt; include. ● *noun* act of embracing, clasp.

embrasure /ɪm'breɪʒə/ *noun* bevelling of wall at sides of window etc.; opening in parapet for gun.

embrocation /embrə'keɪʃ(ə)n/ *noun* liquid for rubbing on body to relieve muscular pain.

embroider /ɪm'brɔɪdə/ *verb* decorate with needlework; embellish. □ **embroidery** *noun*.

embroil /ɪm'brɔɪl/ *verb* (often + *in*) involve (in conflict or difficulties).

embryo /'embrɪəʊ/ *noun* (*plural* **-s**) unborn or unhatched offspring; thing in rudimentary stage. □ **embryonic** /-'ɒn-/ *adjective*.

emend /ɪ'mend/ *verb* correct, remove errors from (text etc.). □ **emendation** /iː-/ *noun*.

emerald /'emər(ə)ld/ ● *noun* bright green gem; colour of this. ● *adjective* bright green.

emerge /ɪ'mɜːdʒ/ *verb* (**-ging**) come up or out into view or notice. □ **emergence** *noun*; **emergent** *adjective*.

emergency /ɪ'mɜːdʒənsɪ/ *noun* (*plural* **-ies**) sudden state of danger etc., requiring immediate action.

emeritus /ɪ'merɪtəs/ *adjective* retired and holding honorary title.

emery /'emərɪ/ *noun* coarse corundum for polishing metal etc. □ **emery board** emery-coated nail-file.

emetic /ɪ'metɪk/ ● *adjective* that causes vomiting. ● *noun* emetic medicine.

emigrate /'emɪgreɪt/ *verb* (**-ting**) leave own country to settle in another. □ **emigrant** *noun & adjective*; **emigration** *noun*.

émigré /'emɪgreɪ/ *noun* emigrant, esp. political exile.

eminence /'emɪnəns/ *noun* recognized superiority; rising ground; (**His**, **Your Eminence**) *title used of or to cardinal.*

eminent /'emɪnənt/ *adjective* distinguished, notable. □ **eminently** *adverb*.

emir /e'mɪə/ *noun* (also **amir** /ə'mɪə/) *title of various Muslim rulers.*

emirate /'emɪrət/ *noun* position, reign, or domain of emir.

emissary /'emɪsərɪ/ *noun* (*plural* **-ies**) person sent on diplomatic mission.

emit /ɪ'mɪt/ *verb* (**-tt-**) give or send out; discharge. □ **emission** /ɪ'mɪʃ(ə)n/ *noun*.

emollient /ɪ'mɒlɪənt/ ● *adjective* softening; soothing. ● *noun* emollient substance.

emolument /ɪ'mɒljʊmənt/ *adjective* fee from employment, salary.

emotion /ɪ'məʊʃ(ə)n/ *noun* strong instinctive feeling such as love or fear; emotional intensity or sensibility.

emotional *adjective* of or expressing emotion(s); especially liable to emotion; arousing emotion. □ **emotionalism** *noun*; **emotionally** *adverb*.

■ **Usage** See note at EMOTIVE.

emotive /ɪ'məʊtɪv/ *adjective* of or arousing emotion.

■ **Usage** Although the senses of *emotive* and *emotional* overlap, *emotive* is more common in the sense 'arousing emotion', as in *an emotive issue*, and only *emotional* can mean 'especially liable to emotion', as in *a highly emotional person*.

empanel /ɪm'pæn(ə)l/ *verb* (also **impanel**) (**-ll-**; *US* **-l-**) enter (jury) on panel.

empathize /'empəθaɪz/ *verb* (also **-ise**) (**-zing** or **-sing**) (usually + *with*) exercise empathy.

empathy /'empəθɪ/ *noun* ability to identify with person or object.

emperor /'empərə/ *noun* ruler of empire.

emphasis /'emfəsɪs/ *noun* (*plural* **emphases** /-siːz/) importance attached to something; significant stress on word(s); vigour of expression etc.

emphasize /'emfəsaɪz/ *verb* (also **-ise**) (**-zing** or **-sing**) lay stress on.

emphatic /ɪm'fætɪk/ *adjective* forcibly expressive; (of word) bearing emphasis (e.g. *myself* in *I did it myself*). □ **emphatically** *adverb*.

emphysema /emfɪ'si:mə/ *noun* disease of lungs causing breathlessness.

empire /'empaɪə/ *noun* large group of states under single authority; supreme dominion; large commercial organization etc. owned or directed by one person.

empirical /ɪm'pɪrɪk(ə)l/ *adjective* based on observation or experiment, not on theory. □ **empiricism** /-rɪs-/ *noun*; **empiricist** /-rɪs-/ *noun*.

emplacement /ɪm'pleɪsmənt/ *noun* platform for gun(s); putting in position.

employ /ɪm'plɔɪ/ *verb* use services of (person) in return for payment; use (thing, time, energy, etc.); keep (person) occupied. □ **in the employ of** employed by. □ **employer** *noun*.

employee /emplɔɪ'i:/ *noun* person employed for wages.

employment *noun* employing, being employed; person's trade or profession. □ **employment office** government office finding work for the unemployed.

emporium /em'pɔ:rɪəm/ *noun* large shop; centre of commerce.

empower /ɪm'paʊə/ *verb* give power to.

empress /'emprɪs/ *noun* wife or widow of emperor; woman emperor.

empty /'emptɪ/ ● *adjective* (**-ier**, **-iest**) containing nothing; vacant, unoccupied; hollow, insincere; without purpose; *colloquial* hungry; vacuous, foolish. ● *verb* (**-ies**, **-ied**) remove contents of; transfer (contents); become empty; (of river) discharge itself. ● *noun* (*plural* **-ies**) *colloquial* emptied bottle etc. □ **empty-handed** bringing or taking nothing. □ **emptiness** *noun*.

EMS *abbreviation* European Monetary System.

emu /'i:mju:/ *noun* (*plural* **-s**) large flightless Australian bird.

emulate /'emjʊleɪt/ *verb* (**-ting**) try to equal or excel; imitate. □ **emulation** *noun*; **emulator** *noun*.

emulsify /ɪ'mʌlsɪfaɪ/ *verb* (**-ies**, **-ied**) make emulsion of. □ **emulsifier** *noun*.

emulsion /ɪ'mʌlʃ(ə)n/ *noun* fine dispersion of one liquid in another, esp. as paint, medicine, etc.

en *noun Printing* unit of measurement equal to half em.

enable /ɪ'neɪb(ə)l/ *verb* (**-ling**) (+ *to do*) supply with means or authority; make possible.

enact /ɪ'nækt/ *verb* ordain, decree; make (bill etc.) law; play (part). □ **enactment** *noun*.

enamel /ɪ'næm(ə)l/ ● *noun* glasslike opaque coating on metal; any hard smooth coating; kind of hard gloss paint; hard coating of teeth. ● *verb* (**-ll**; *US* **-l-**) coat with enamel.

enamour /ɪ'næmə/ *verb* (*US* **enamor**) (usually in *passive*; + *of*) inspire with love or delight.

en bloc /ɑ̃ 'blɒk/ *adverb* in a block, all at same time. [French]

encamp /ɪn'kæmp/ *verb* settle in (esp. military) camp. □ **encampment** *noun*.

encapsulate /ɪn'kæpsjʊleɪt/ *verb* (**-ting**) enclose (as) in capsule; summarize.

encase /ɪn'keɪs/ *verb* (**-sing**) confine (as) in a case.

encash /ɪn'kæʃ/ *verb* convert into cash. □ **encashment** *noun*.

encephalitis /ensefə'laɪtɪs/ *noun* inflammation of brain.

enchant /ɪn'tʃɑ:nt/ *verb* delight; bewitch. □ **enchanting** *adjective*; **enchantment** *noun*.

enchanter *noun* (*feminine* **enchantress**) person who enchants, esp. by magic.

encircle /ɪn'sɜ:k(ə)l/ *verb* (**-ling**) surround. □ **encirclement** *noun*.

enclave /'enkleɪv/ *noun* part of territory of one state surrounded by that of another; group of people distinct from those surrounding them, esp. ethnically.

enclose /ɪn'kləʊz/ *verb* (**-sing**) surround with wall, fence, etc.; shut in; put in receptacle (esp. in envelope besides letter); (as **enclosed** *adjective*) (of religious community) secluded from outside world.

enclosure /ɪn'kləʊʒə/ *noun* enclosing; enclosed space or area; thing enclosed.

encode /ɪn'kəʊd/ *verb* (**-ding**) put into code.

encomium /ɪn'kəʊmɪəm/ *noun* (*plural* **-s**) formal praise.

encompass /ɪn'kʌmpəs/ *verb* contain, include; surround.

encore /'ɒŋkɔː/ ● *noun* audience's call for repetition of item, or for further item; such item. ● *verb* (**-ring**) call for repetition of or by. ● *interjection* again.

encounter /ɪn'kaʊntə/ ● *verb* meet by chance; meet as adversary. ● *noun* meeting by chance or in conflict.

encourage /ɪn'kʌrɪdʒ/ *verb* (**-ging**) give courage to; urge; promote. □ **encouragement** *noun.*

encroach /ɪn'krəʊtʃ/ *verb* (usually + *on*, *upon*) intrude on other's territory etc. □ **encroachment** *noun.*

encrust /ɪn'krʌst/ *verb* cover with or form crust; coat with hard casing or deposit.

encumber /ɪn'kʌmbə/ *verb* be burden to; hamper.

encumbrance /ɪn'kʌmbrəns/ *noun* burden; impediment.

encyclical /ɪn'sɪklɪk(ə)l/ *noun* papal letter to all RC bishops.

encyclopedia /ɪnsaɪklə'piːdɪə/ *noun* (also **-paedia**) book of information on many subjects or on many aspects of one subject.

encyclopedic *adjective* (also **-paedic**) (of knowledge or information) comprehensive.

end ● *noun* limit; farthest point; extreme point or part; conclusion; latter part; destruction; death; result; goal, object; remnant. ● *verb* bring or come to end. □ **endpaper** blank leaf of paper at beginning or end of book; **end-product** final product of manufacture etc.; **end up** be or become eventually, arrive; **in the end** finally; **make ends meet** live within one's income; **no end** *colloquial* to a great extent.

endanger /ɪn'deɪndʒə/ *verb* place in danger. □ **endangered species** one in danger of extinction.

endear /ɪn'dɪə/ *verb* (usually + *to*) make dear. □ **endearing** *adjective.*

endearment *noun* expression of affection.

endeavour /ɪn'devə/ (*US* **endeavor**) ● *verb* try, strive. ● *noun* attempt, effort.

endemic /en'demɪk/ *adjective* (often + *to*) regularly found among particular people or in particular area. □ **endemically** *adverb.*

ending *noun* end of word or story.

endive /'endaɪv/ *noun* curly-leaved plant used in salads.

endless *adjective* infinite; continual. □ **endlessly** *adverb.*

endocrine /'endəʊkraɪn/ *adjective* (of gland) secreting directly into blood.

endogenous /en'dɒdʒɪnəs/ *adjective* growing or originating from within.

endorse /ɪn'dɔːs/ *verb* (**-sing**) approve; write on (document), esp. sign (cheque); enter details of offence on (driving licence). □ **endorsement** *noun.*

endoskeleton /'endəʊskelɪtən/ *noun* internal skeleton.

endow /ɪn'daʊ/ *verb* give permanent income to; (esp. as **endowed** *adjective*) provide with talent or ability.

endowment *noun* endowing; money with which person or thing is endowed. □ **endowment mortgage** one in which borrower pays only premiums until policy repays mortgage capital; **endowment policy** life insurance policy paying out on set date or earlier death.

endue /ɪn'djuː/ *verb* (**-dues**, **-dued**, **-duing**) (+ *with*) provide (person) with (quality etc.).

endurance *noun* power of enduring.

endure /ɪn'djʊə/ *verb* (**-ring**) undergo; bear; last.

endways *adverb* with an end facing forwards.

ENE *abbreviation* east-north-east.

enema /'enɪmə/ *noun* injection of liquid etc. into rectum, esp. to expel its contents; liquid used for this.

enemy /'enəmɪ/ ● *noun* (*plural* **-ies**) person actively hostile to another; hostile army or nation; member of this; adversary, opponent. ● *adjective* of or belonging to enemy.

energetic /enə'dʒetɪk/ *adjective* full of energy. □ **energetically** *adverb.*

energize /'enədʒaɪz/ *verb* (also **-ise**) (**-zing** or **-sing**) give energy to.

energy /'enədʒɪ/ *noun* (*plural* **-ies**) force, vigour, activity; ability of matter or radiation to do work.

enervate /'enəveɪt/ *verb* (**-ting**) deprive of vigour. □ **enervation** *noun*.

en famille /ɑ̃ fæ'mi:/ *adverb* in or with one's family. [French]

enfant terrible /ɑ̃fɑ̃ te'ri:bl/ *noun* (*plural* ***enfants terribles*** same pronunciation) indiscreet or unruly person. [French]

enfeeble /ɪn'fi:b(ə)l/ *verb* (**-ling**) make feeble. □ **enfeeblement** *noun*.

enfilade /enfɪ'leɪd/ ● *noun* gunfire directed down length of enemy position. ● *verb* (**-ding**) direct enfilade at.

enfold /ɪn'fəʊld/ *verb* wrap; embrace.

enforce /ɪn'fɔ:s/ *verb* (**-cing**) compel observance of; impose. □ **enforceable** *adjective*; **enforcement** *noun*.

enfranchise /ɪn'fræntʃaɪz/ *verb* (**-sing**) give (person) right to vote. □ **enfranchisement** /-tʃɪz-/ *noun*.

engage /ɪn'geɪdʒ/ *verb* (**-ging**) employ, hire; (as **engaged** *adjective*) occupied, busy, having promised to marry; hold (person's attention); cause parts of (gear) to interlock; fit, interlock; bring into battle; come into battle with (enemy); (usually + *in*) take part; (+ *that*, *to do*) undertake. □ **engagement** *noun*.

engender /ɪn'dʒendə/ *verb* give rise to.

engine /'endʒɪn/ *noun* mechanical contrivance of parts working together, esp. as source of power; railway locomotive.

engineer /endʒɪ'nɪə/ ● *noun* person skilled in a branch of engineering; person who makes or is in charge of engines etc.; person who designs and constructs military works; mechanic, technician. ● *verb* contrive, bring about; act as engineer; construct or manage as engineer.

engineering *noun* application of science to design, building, and use of machines etc.

English /'ɪŋglɪʃ/ ● *adjective* of England. ● *noun* language of England, now used in UK, US, and most Commonwealth countries; (**the English**) (treated as *plural*) the people of England. □ **Englishman**, **Englishwoman** native of England.

engraft /ɪn'grɑ:ft/ *verb* (usually + *into*, *on*) graft, implant, incorporate.

engrave /ɪn'greɪv/ *verb* (**-ving**) inscribe or cut (design) on hard surface; inscribe (surface) thus; (often + *on*) impress deeply (on memory etc.). □ **engraver** *noun*.

engraving *noun* print made from engraved plate.

engross /ɪn'grəʊs/ *verb* (usually as **engrossed** *adjective* + *in*) fully occupy.

engulf /ɪn'gʌlf/ *verb* flow over and swamp, overwhelm.

enhance /ɪn'hɑ:ns/ *verb* (**-cing**) intensify; improve. □ **enhancement** *noun*.

enigma /ɪ'nɪgmə/ *noun* puzzling person or thing; riddle. □ **enigmatic** /enɪg'mætɪk/ *adjective*.

enjoin /ɪn'dʒɔɪn/ *verb* command, order.

enjoy /ɪn'dʒɔɪ/ *verb* find pleasure in; (**enjoy oneself**) find pleasure; have use or benefit of; experience. □ **enjoyable** *adjective*; **enjoyably** *adverb*; **enjoyment** *noun*.

enlarge /ɪn'lɑ:dʒ/ *verb* (**-ging**) make or become larger; (often + *on*, *upon*) describe in greater detail; reproduce on larger scale. □ **enlargement** *noun*.

enlarger *noun* apparatus for enlarging photographs.

enlighten /ɪn'laɪt(ə)n/ *verb* inform; (as **enlightened** *adjective*) progressive. □ **enlightenment** *noun*.

enlist /ɪn'lɪst/ *verb* enrol in armed services; secure as means of help or support. □ **enlistment** *noun*.

enliven /ɪn'laɪv(ə)n/ *verb* make lively or cheerful. □ **enlivenment** *noun*.

en masse /ɑ̃ 'mæs/ *adverb* all together. [French]

enmesh /ɪn'meʃ/ *verb* entangle (as) in net.

enmity /'enmɪtɪ/ *noun* (*plural* **-ies**) hostility; state of being an enemy.

ennoble /ɪ'nəʊb(ə)l/ *verb* (**-ling**) make noble. □ **ennoblement** *noun*.

ennui /ɒ'nwi:/ *noun* boredom.

enormity /ɪ'nɔ:mɪtɪ/ *noun* (*plural* **-ies**) great wickedness; monstrous crime; great size.

■ **Usage** Many people believe it is wrong to use *enormity* to mean 'great size'.

enormous /ɪ'nɔ:məs/ *adjective* huge. □ **enormously** *adverb.*

enough /ɪ'nʌf/ ● *adjective* as much or as many as required. ● *noun* sufficient amount or quantity. ● *adverb* to required degree; fairly; very, quite.

enquire /ɪn'kwaɪə/ *verb* (**-ring**) seek information; ask question; inquire.

enquiry *noun* (*plural* **-ies**) asking; inquiry.

enrage /ɪn'reɪdʒ/ *verb* (**-ging**) make furious.

enrapture /ɪn'ræptʃə/ *verb* (**-ring**) delight intensely.

enrich /ɪn'rɪtʃ/ *verb* make rich(er). □ **enrichment** *noun.*

enrol /ɪn'rəʊl/ *verb* (*US* **enroll**) (**-ll-**) enlist; write name of (person) on list; incorporate as member; enrol oneself. □ **enrolment** *noun.*

en route /ɑ̃ 'ru:t/ *adverb* on the way. [French]

ensconce /ɪn'skɒns/ *verb* (**-cing**) (usually **ensconce oneself** or in *passive*) settle comfortably.

ensemble /ɒn'sɒmb(ə)l/ *noun* thing viewed as whole; set of clothes worn together; group of performers working together; *Music* passage for ensemble.

enshrine /ɪn'ʃraɪn/ *verb* (**-ning**) enclose in shrine; protect, make inviolable.

ensign /'ensaɪn/ *noun* banner, flag, esp. military or naval flag of nation; standard-bearer; *historical* lowest commissioned infantry officer; *US* lowest commissioned naval officer.

enslave /ɪn'sleɪv/ *verb* (**-ving**) make slave of. □ **enslavement** *noun.*

ensnare /ɪn'sneə/ *verb* (**-ring**) entrap.

ensue /ɪn'sju:/ *verb* (**-sues**, **-sued**, **-suing**) happen later or as a result.

en suite /ɑ̃ 'swi:t/ ● *adverb* forming single unit. ● *adjective* (of bathroom) attached to bedroom; (of bedroom) with bathroom attached.

ensure /ɪn'ʃʊə/ *verb* (**-ring**) make certain or safe.

ENT *abbreviation* ear, nose, and throat.

entail /ɪn'teɪl/ ● *verb* necessitate or involve unavoidably; *Law* bequeath (estate) to specified line of beneficiaries. ● *noun* entailed estate.

entangle /ɪn'tæŋg(ə)l/ *verb* (**-ling**) catch or hold fast in snare etc.; involve in difficulties; complicate. □ **entanglement** *noun.*

entente /ɒn'tɒnt/ *noun* friendly understanding between states. □ **entente cordiale** entente, esp. between Britain and France since 1904.

enter /'entə/ *verb* go or come in or into; come on stage; penetrate; put (name, fact, etc.) into list or record etc.; (usually + *for*) name, or name oneself, as competitor; become member of. □ **enter into** engage in, bind oneself by, form part of, sympathize with; **enter (up)on** begin, begin to deal with, assume possession of.

enteric /en'terɪk/ *adjective* of intestines.

enteritis /entə'raɪtɪs/ *noun* inflammation of intestines.

enterprise /'entəpraɪz/ *noun* bold undertaking; readiness to engage in this; business firm or venture.

enterprising *adjective* showing enterprise.

entertain /entə'teɪn/ *verb* amuse; receive as guest; harbour (feelings); consider (idea). □ **entertainer** *noun*; **entertaining** *adjective.*

entertainment *noun* entertaining; thing that entertains, performance.

enthral /ɪn'θrɔ:l/ *verb* (*US* **enthrall**) (**-ll-**) captivate; please greatly. □ **enthralment** *noun.*

enthrone /ɪn'θrəʊn/ *verb* (**-ning**) place on throne. □ **enthronement** *noun.*

enthuse /ɪn'θju:z/ *verb* (**-sing**) *colloquial* be or make enthusiastic.

enthusiasm /ɪn'θju:zɪæz(ə)m/ *noun* great eagerness or admiration; object of this. □ **enthusiast** *noun*; **enthusiastic** /-'æst-/ *adjective*; **enthusiastically** *adverb.*

entice /ɪn'taɪs/ *verb* (**-cing**) attract by offer of pleasure or reward. □ **enticement** *noun*; **enticing** *adjective*; **enticingly** *adverb.*

entire /ɪn'taɪə/ *adjective* complete; unbroken; absolute; in one piece.

entirely *adverb* wholly.

entirety /ɪn'taɪərətɪ/ *noun* (*plural* **-ies**) completeness; sum total. □ **in its entirety** in its complete form.

entitle /ɪn'taɪt(ə)l/ *verb* (**-ling**) (usually + *to*) give (person) right or claim; give title to. □ **entitlement** *noun*.

entity /'entɪtɪ/ *noun* (*plural* **-ies**) thing with distinct existence; thing's existence.

entomb /ɪn'tu:m/ *verb* place in tomb; serve as tomb for. □ **entombment** *noun*.

entomology /entə'mɒlədʒɪ/ *noun* study of insects. □ **entomological** /-mə'lɒdʒ-/ *adjective*; **entomologist** *noun*.

entourage /'ɒntʊərɑ:ʒ/ *noun* people attending important person.

entr'acte /'ɒntrækt/ *noun* (music or dance performed in) interval in play.

entrails /'entreɪlz/ *plural noun* intestines; inner parts.

entrance[1] /'entrəns/ *noun* place for entering; coming or going in; right of admission.

entrance[2] /ɪn'trɑ:ns/ *verb* (**-cing**) enchant, delight; put into trance.

entrant /'entrənt/ *noun* person who enters exam, profession, etc.

entrap /ɪn'træp/ *verb* (**-pp-**) catch (as) in trap.

entreat /ɪn'tri:t/ *verb* ask earnestly, beg.

entreaty *noun* (*plural* **-ies**) earnest request.

entrecôte /'ɒntrəkəʊt/ *noun* boned steak cut off sirloin.

entrée /'ɒntreɪ/ *noun* main dish of meal; dish served between fish and meat courses; right of admission.

entrench /ɪn'trentʃ/ *verb* establish firmly; (as **entrenched** *adjective*) (of attitude etc.) not easily modified; surround or fortify with trench. □ **entrenchment** *noun*.

entrepreneur /ɒntrəprə'nɜ:/ *noun* person who undertakes commercial venture. □ **entrepreneurial** *adjective*.

entropy /'entrəpɪ/ *noun* measure of disorganization of universe; measure of unavailability of system's thermal energy for conversion into mechanical work.

entrust /ɪn'trʌst/ *verb* (+ *to*) give (person, thing) into care of; (+ *with*) assign responsibility for (person, thing) to.

entry /'entrɪ/ *noun* (*plural* **-ies**) coming or going in; entering, item entered; place of entrance; alley.

entwine /ɪn'twaɪn/ *verb* (**-ning**) twine round, interweave.

enumerate /ɪ'nju:məreɪt/ *verb* (**-ting**) specify (items); count. □ **enumeration** *noun*.

enunciate /ɪ'nʌnsɪeɪt/ *verb* (**-ting**) pronounce (words) clearly; state definitely. □ **enunciation** *noun*.

envelop /ɪn'veləp/ *verb* (**-p-**) wrap up, cover; surround. □ **envelopment** *noun*.

envelope /'envələʊp/ *noun* folded paper cover for letter etc.; wrapper, covering.

enviable /'envɪəb(ə)l/ *adjective* likely to excite envy. □ **enviably** *adverb*.

envious /'envɪəs/ *adjective* feeling or showing envy.

environment /ɪn'vaɪərənmənt/ *noun* surroundings; circumstances affecting person's life. □ **environmental** /-'men-/ *adjective*; **environmentally** /-'men-/ *adverb*.

environmentalist /ɪnvaɪərən'mentəlɪst/ *noun* person concerned with protection of natural environment.

environs /ɪn'vaɪərəns/ *plural noun* district round town etc.

envisage /ɪn'vɪzɪdʒ/ *verb* (**-ging**) visualize, imagine, contemplate.

envoy /'envɔɪ/ *noun* messenger, representative; diplomat ranking below ambassador.

envy /'envɪ/ ● *noun* (*plural* **-ies**) discontent aroused by another's better fortune etc.; object or cause of this. ● *verb* (**-ies**, **-ied**) feel envy of.

enzyme /'enzaɪm/ *noun* protein catalyst of specific biochemical reaction.

eolian harp *US* = AEOLIAN HARP.

eon = AEON.

EP *abbreviation* extended-play (record).

epaulette /'epəlet/ *noun* (*US* **epaulet**) ornamental shoulder-piece, esp. on uniform.

ephedrine /'efədrɪn/ *noun* alkaloid drug used to relieve asthma etc.

ephemera /ɪ'femərə/ *plural noun* things of only short-lived relevance.

ephemeral *adjective* short-lived, transitory.

epic /'epɪk/ ● *noun* long poem narrating adventures of heroic figure etc.; book or film based on this. ● *adjective* like an epic; grand, heroic.

epicene /'episi:n/ *adjective* of or for both sexes; having characteristics of both sexes or of neither sex.

epicentre /'episentə/ *noun* (*US* **epicenter**) point at which earthquake reaches earth's surface.

epicure /'epɪkjʊə/ *noun* person with refined taste in food and drink. □ **epicurism** *noun.*

epicurean /epɪkjʊə'ri:ən/ • *noun* person fond of pleasure and luxury. • *adjective* characteristics of an epicurean. □ **epicureanism** *noun.*

epidemic /epɪ'demɪk/ • *noun* widespread occurrence of particular disease in community at particular time. • *adjective* in the nature of an epidemic.

epidemiology /epɪdi:mɪ'ɒlədʒɪ/ *noun* study of epidemics and their control.

epidermis /epɪ'dɜ:mɪs/ *noun* outer layer of skin.

epidiascope /epɪ'daɪəskəʊp/ *noun* optical projector giving images of both opaque and transparent objects.

epidural /epɪ'djʊər(ə)l/ • *adjective* (of anaesthetic) injected close to spinal cord. • *noun* epidural injection.

epiglottis /epɪ'glɒtɪs/ *noun* flap of cartilage at root of tongue that covers windpipe during swallowing. □ **epiglottal** *adjective.*

epigram /'epɪgræm/ *noun* short poem with witty ending; pointed saying. □ **epigrammatic** /-grə'mæt-/ *adjective.*

epigraph /'epɪgrɑ:f/ *noun* inscription.

epilepsy /'epɪlepsɪ/ *noun* nervous disorder with convulsions and often loss of consciousness. □ **epileptic** /-'lep-/ *adjective & noun.*

epilogue /'epɪlɒg/ *noun* short piece ending literary work; short speech at end of play etc.

Epiphany /ɪ'pɪfənɪ/ *noun* (festival on 6 Jan. commemorating) visit of Magi to Christ.

episcopacy /ɪ'pɪskəpəsɪ/ *noun* (*plural* **-ies**) government by bishops; bishops collectively.

episcopal /ɪ'pɪskəp(ə)l/ *adjective* of bishop or bishops; (of church) governed by bishops. □ **Episcopal Church** Anglican Church in Scotland and US.

Episcopalian /ɪpɪskə'peɪlɪən/ • *adjective* of the Episcopal Church or (**episcopalian**) an episcopal church. • *noun* member of the Episcopal Church.

episcopate /ɪ'pɪskəpət/ *noun* office or tenure of bishop; bishops collectively.

episode /'epɪsəʊd/ *noun* event as part of sequence; part of serial story; incident in narrative. □ **episodic** /-'sɒd-/ *adjective.*

epistemology /ɪpɪstɪ'mɒlədʒɪ/ *noun* philosophy of knowledge. □ **epistemological** /-ə'lɒdʒ-/ *adjective.*

epistle /ɪ'pɪs(ə)l/ *noun* letter; poem etc. in form of letter.

epistolary /ɪ'pɪstələrɪ/ *adjective* of or in form of letters.

epitaph /'epɪtɑ:f/ *noun* words in memory of dead person, esp. on tomb.

epithelium /epɪ'θi:lɪəm/ *noun* (*plural* **-s** or **-lia**) *Biology* tissue forming outer layer of body and lining many hollow structures. □ **epithelial** *adjective.*

epithet /'epɪθet/ *noun* adjective etc. expressing quality or attribute.

epitome /ɪ'pɪtəmɪ/ *noun* person or thing embodying a quality etc.

epitomize *verb* (also **-ise**) (**-zing** or **-sing**) make or be perfect example of (a quality etc.).

EPNS *abbreviation* electroplated nickel silver.

epoch /'i:pɒk/ *noun* period marked by special events; beginning of era. □ **epoch-making** notable, significant.

eponym /'epənɪm/ *noun* word derived from person's name; person whose name is used in this way. □ **eponymous** /ɪ'pɒnɪməs/ *adjective.*

epoxy resin /ɪ'pɒksɪ/ *noun* synthetic thermosetting resin, used esp. as glue.

Epsom salts /'epsəm/ *noun* magnesium sulphate used as purgative.

equable /'ekwəb(ə)l/ *adjective* not varying; moderate; not easily disturbed. □ **equably** *adverb.*

equal /'i:kw(ə)l/ • *adjective* same in number, size, merit, etc.; evenly matched; having same rights or status. • *noun* person etc. equal to another. • *verb* (**-ll-**; *US* **-l-**) be equal to; achieve something equal to. □ **equal opportunity** (often in *plural*) opportunity to compete equally for jobs regardless of race, sex, etc. □ **equally** *adverb*.

■ **Usage** It is a mistake to say *equally as*, as in *She was equally as guilty*. The correct version is *She was equally guilty* or possibly, for example, *She was as guilty as he was*.

equality /ɪ'kwɒlɪtɪ/ *noun* being equal.

equalize *verb* (also **-ise**) (**-zing** or **-sing**) make or become equal; (in games) reach opponent's score. □ **equalization** *noun*.

equalizer *noun* (also **-iser**) equalizing goal etc.

equanimity /ekwə'nɪmɪtɪ/ *noun* composure, calm.

equate /ɪ'kweɪt/ *verb* (**-ting**) (usually + *to*, *with*) regard as equal or equivalent; (+ *with*) be equal or equivalent to.

equation /ɪ'kweɪʒ(ə)n/ *noun* making or being equal; *Mathematics* statement that two expressions are equal; *Chemistry* symbolic representation of reaction.

equator /ɪ'kweɪtə/ *noun* imaginary line round the earth or other body, equidistant from poles.

equatorial /ekwə'tɔ:rɪəl/ *adjective* of or near equator.

equerry /'ekwərɪ/ *noun* (*plural* **-ies**) officer attending British royal family.

equestrian /ɪ'kwestrɪən/ *adjective* of horse-riding; on horseback. □ **equestrianism** *noun*.

equiangular /i:kwɪ'æŋgjʊlə/ *adjective* having equal angles.

equidistant /i:kwɪ'dɪst(ə)nt/ *adjective* at equal distances.

equilateral /i:kwɪ'lætər(ə)l/ *adjective* having all sides equal.

equilibrium /i:kwɪ'lɪbrɪəm/ *noun* state of balance; composure.

equine /'ekwaɪn/ *adjective* of or like horse.

equinox /'i:kwɪnɒks/ *noun* time or date at which sun crosses equator and day and night are of equal length.

equip /ɪ'kwɪp/ *verb* (**-pp-**) supply with what is needed.

equipment *noun* necessary tools, clothing, etc.; equipping, being equipped.

equipoise /'ekwɪpɔɪz/ *noun* equilibrium; counterbalancing thing.

equitable /'ekwɪtəb(ə)l/ *adjective* fair, just; *Law* valid in equity. □ **equitably** *adverb*.

equitation /ekwɪ'teɪʃ(ə)n/ *noun* horsemanship; horse-riding.

equity /'ekwɪtɪ/ *noun* (*plural* **-ies**) fairness; principles of justice supplementing law; value of shares issued by company; (in *plural*) stocks and shares not bearing fixed interest.

equivalent /ɪ'kwɪvələnt/ ● *adjective* (often + *to*) equal in value, meaning, etc.; corresponding. ● *noun* equivalent amount etc. □ **equivalence** *noun*.

equivocal /ɪ'kwɪvək(ə)l/ *adjective* of double or doubtful meaning; of uncertain nature; (of person etc.) questionable. □ **equivocally** *adverb*.

equivocate /ɪ'kwɪvəkeɪt/ *verb* (**-ting**) use words ambiguously to conceal truth. □ **equivocation** *noun*.

ER *abbreviation* Queen Elizabeth (*Elizabetha Regina*).

era /'ɪərə/ *noun* system of chronology starting from particular point; historical or other period.

eradicate /ɪ'rædɪkeɪt/ *verb* (**-ting**) root out, destroy. □ **eradication** *noun*.

erase /ɪ'reɪz/ *verb* (**-sing**) rub out; obliterate; remove recording from (magnetic tape etc.).

eraser *noun* piece of rubber etc. for removing esp. pencil marks.

erasure /ɪ'reɪʒə/ *noun* erasing; erased word etc.

ere /eə/ *preposition & conjunction poetical or archaic* before.

erect /ɪ'rekt/ ● *adjective* upright, vertical; (of part of body) enlarged and rigid, esp. from sexual excitement. ● *verb* raise, set upright; build; establish. □ **erection** *noun*.

erectile /ɪ'rektaɪl/ *adjective* that can become erect.

erg *noun* unit of work or energy.

ergo /'ɜ:gəʊ/ *adverb* therefore. [Latin]

ergonomics /ɜ:gə'nɒmɪks/ *plural noun* (treated as *singular*) study of relationship between people and their working environment. □ **ergonomic** *adjective*.

ergot /'ɜ:gət/ *noun* disease of rye etc. caused by fungus.

ERM *abbreviation* Exchange Rate Mechanism.

ermine /'ɜ:mɪn/ *noun* (*plural* same or **-s**) stoat, esp. in its white winter fur; this fur.

Ernie /'ɜ:nɪ/ *noun* device for drawing prize-winning numbers of Premium Bonds.

erode /ɪ'rəʊd/ *verb* (**-ding**) wear away, gradually destroy. □ **erosion** *noun*; **erosive** *adjective*.

erogenous /ɪ'rɒdʒɪnəs/ *adjective* (of part of body) sexually sensitive.

erotic /ɪ'rɒtɪk/ *adjective* arousing sexual desire or excitement. □ **erotically** *adverb*; **eroticism** /-sɪz-/ *noun*.

erotica *plural noun* erotic literature or art.

err /ɜ:/ *verb* be mistaken or incorrect; sin.

errand /'erənd/ *noun* short journey, esp. on another's behalf, to take message etc.; object of journey.

errant /'erənt/ *adjective* erring; *literary* travelling in search of adventure.

erratic /ɪ'rætɪk/ *adjective* inconsistent or uncertain in movement or conduct etc. □ **erratically** *adverb*.

erratum /ɪ'rɑ:təm/ *noun* (*plural* **-ta**) error in printing etc.

erroneous /ɪ'rəʊnɪəs/ *adjective* incorrect. □ **erroneously** *adverb*.

error /'erə/ *noun* mistake; condition of being morally wrong; degree of inaccuracy in calculation or measurement.

ersatz /'eəzæts/ *adjective & noun* substitute; imitation.

erstwhile /'ɜ:stwaɪl/ *adjective* former.

eructation /i:rʌk'teɪʃ(ə)n/ *noun formal* belching.

erudite /'eru:daɪt/ *adjective* learned. □ **erudition** /-'dɪʃ-/ *noun*.

erupt /ɪ'rʌpt/ *verb* break out; (of volcano) shoot out lava etc.; (of rash) appear on skin. □ **eruption** *noun*.

erysipelas /erɪ'sɪpɪləs/ *noun* disease causing deep red inflammation of skin.

escalate /'eskəleɪt/ *verb* (**-ting**) increase or develop by stages. □ **escalation** *noun*.

escalator *noun* moving staircase.

escalope /'eskəlɒp/ *noun* thin slice of meat, esp. veal.

escapade /'eskəpeɪd/ *noun* piece of reckless behaviour.

escape /ɪs'keɪp/ ● *verb* (**-ping**) get free; leak; avoid punishment etc.; get free of; elude, avoid. ● *noun* escaping; means of escaping; leakage. □ **escape clause** clause releasing contracting party from obligation in specified circumstances.

escapee /ɪskeɪ'pi:/ *noun* person who has escaped.

escapism *noun* pursuit of distraction and relief from reality. □ **escapist** *adjective & noun*.

escapology /eskə'pɒlədʒɪ/ *noun* techniques of escaping from confinement, esp. as entertainment. □ **escapologist** *noun*.

escarpment /ɪs'kɑ:pmənt/ *noun* long steep slope at edge of plateau etc.

eschatology /eskə'tɒlədʒɪ/ *noun* doctrine of death and final destiny. □ **eschatological** /-tə'lɒdʒ-/ *adjective*.

escheat /ɪs'tʃi:t/ ● *noun* lapse of property to the state etc.; property so lapsing. ● *verb* hand over as escheat, confiscate; revert by escheat.

eschew /ɪs'tʃu:/ *verb formal* abstain from.

escort ● *noun* /'eskɔ:t/ person(s) etc. accompanying another for protection or as courtesy; person accompanying another of opposite sex socially. ● *verb* /ɪ'skɔ:t/ act as escort to.

escritoire /eskrɪ'twɑ:/ *noun* writing desk with drawers etc.

escutcheon /ɪ'skʌtʃ(ə)n/ *noun* shield bearing coat of arms.

ESE *abbreviation* east-south-east.

Eskimo /'eskɪməʊ/ ● *noun* (*plural* same or **-s**) member of people inhabiting N. Canada, Alaska, Greenland, and E. Siberia; their language. ● *adjective* of Eskimos or their language.

■ **Usage** The Eskimos of N. America prefer the name *Inuit*.

ESN *abbreviation* educationally subnormal.

esophagus *US* = OESOPHAGUS.

esoteric /esə'terɪk, i:sə'terɪk/ *adjective* intelligible only to those with special knowledge.

ESP *abbreviation* extrasensory perception.

esp. *abbreviation* especially.

espadrille /espə'drɪl/ *noun* light canvas shoe with plaited fibre sole.

espalier /ɪ'spælɪə/ *noun* framework for training tree etc.; tree trained on espalier.

esparto /e'spɑ:təʊ/ *noun* kind of grass used to make paper.

especial /ɪ'speʃ(ə)l/ *adjective* special, notable.

especially *adverb* in particular; more than in other cases; particularly.

Esperanto /espə'ræntəʊ/ *noun* artificial universal language.

espionage /'espɪənɑ:ʒ/ *noun* spying or using spies.

esplanade /esplə'neɪd/ *noun* level space, esp. used as public promenade.

espousal /ɪ'spaʊz(ə)l/ *noun* espousing; *archaic* marriage, betrothal.

espouse /ɪ'spaʊz/ *verb* (**-sing**) support (cause); *archaic* marry.

espresso /e'spresəʊ/ *noun* (*plural* **-s**) coffee made under steam pressure.

esprit de corps /espri: də 'kɔ:/ *noun* devotion to and pride in one's group. [French]

espy /ɪ'spaɪ/ *verb* (**-ies**, **-ied**) catch sight of.

Esq. *abbreviation* Esquire.

esquire /ɪ'skwaɪə/ *noun: title placed after man's name in writing.*

essay ● *noun* /'eseɪ/ short piece of writing, esp. on given subject; *formal* attempt. ● *verb* /e'seɪ/ attempt.

essayist *noun* writer of essays.

essence /'es(ə)ns/ *noun* fundamental nature, inherent characteristics; extract obtained by distillation etc.; perfume. □ **in essence** fundamentally.

essential /ɪ'senʃ(ə)l/ ● *adjective* necessary, indispensable; of or constituting a thing's essence. ● *noun* (esp. in *plural*) indispensable element or thing. □ **essential oil** volatile oil with characteristic odour. □ **essentially** *adverb.*

establish /ɪ'stæblɪʃ/ *verb* set up; settle; (esp. as **established** *adjective*) achieve permanent acceptance for; place beyond dispute. □ **Established Church** Church recognized by state.

establishment *noun* establishing, being established; public institution; place of business; staff, household, etc.; Church system established by law; (**the Establishment**) social group with authority or influence and resisting change.

estate /ɪ'steɪt/ *noun* landed property; area of homes or businesses planned as a whole; dead person's collective assets and liabilities. □ **estate agent** person whose business is sale and lease of buildings and land on behalf of others; **estate car** car with continuous area for rear passengers and luggage.

esteem /ɪ'sti:m/ ● *verb* (usually in *passive*) think highly of; *formal* consider. ● *noun* high regard.

ester /estə/ *noun* compound formed by replacing the hydrogen of an acid by an organic radical.

estimable /'estɪməb(ə)l/ *adjective* worthy of esteem.

estimate ● *noun* /'estɪmət/ approximate judgement of cost, value, etc.; approximate price stated in advance for work. ● *verb* /'estɪmeɪt/ (**-ting**) form estimate of; (+ *that*) make rough calculation; (+ *at*) put (sum etc.) at by estimating.

estimation /estɪ'meɪʃ(ə)n/ *noun* estimating; judgement of worth.

estrange /ɪ'streɪndʒ/ *verb* (**-ging**) (usually in *passive*; often + *from*) alienate, make hostile or indifferent; (as **estranged** *adjective*) no longer living with spouse. □ **estrangement** *noun.*

estuary /'estjʊərɪ/ *noun* (*plural* **-ies**) tidal mouth of river.

ETA *abbreviation* estimated time of arrival.

et al. *abbreviation* and others (*et alii*).

etc. *abbreviation* et cetera.

et cetera /et 'setrə/ *adverb* (also **etcetera**) and the rest; and so on. □ **etceteras** *plural noun* the usual extras.

etch *verb* reproduce (picture etc.) by engraving metal plate with acid, esp. to print copies; engrave (plate) thus; practise this craft; (usually + *on*, *upon*) impress deeply.

etching *noun* print made from etched plate.

eternal /ɪ'tɜ:n(ə)l/ *adjective* existing always; without end or beginning; unchanging; *colloquial* constant, too frequent. □ **eternally** *adverb.*

eternity /ɪ'tɜ:nɪtɪ/ *noun* infinite time; endless life after death; (**an eternity**) *colloquial* a very long time.

ethane /'i:θeɪn/ *noun* hydrocarbon gas present in petroleum and natural gas.

ethanol /'eθənɒl/ *noun* alcohol.

ether /'i:θə/ *noun* volatile liquid used as anaesthetic or solvent; clear sky, upper air.

ethereal /ɪ'θɪərɪəl/ *adjective* light, airy; delicate, esp. in appearance; heavenly.

ethic /'eθɪk/ *noun* set of moral principles.

ethical *adjective* relating to morals or ethics; morally correct; (of drug etc.) available only on prescription. □ **ethically** *adverb*.

ethics *plural noun* (also treated as *singular*) moral philosophy; (set of) moral principles.

Ethiopian /i:θɪ'əʊpɪən/ ● *noun* native or national of Ethiopia. ● *adjective* of Ethiopia.

ethnic /'eθnɪk/ *adjective* (of social) group having common national or cultural tradition; (of clothes etc.) resembling those of an exotic people; (of person) having specified origin by birth or descent rather than nationality. □ **ethnic cleansing** *euphemistic* expulsion or murder of people of ethnic or religious group in certain area. □ **ethnically** *adverb*.

ethnology /eθ'nɒlədʒɪ/ *noun* comparative study of peoples. □ **ethnological** /-nə'lɒdʒ-/ *adjective*.

ethos /'i:θɒs/ *noun* characteristic spirit of community, people, or system.

ethylene /'eθɪli:n/ *noun* flammable hydrocarbon gas.

etiolate /'i:tɪəleɪt/ *verb* (**-ting**) make pale by excluding light; give sickly colour to. □ **etiolation** *noun*.

etiology *US* = AETIOLOGY.

etiquette /'etɪket/ *noun* conventional rules of social behaviour or professional conduct.

étude /eɪ'tju:d/ *noun* musical composition designed to develop player's skill.

etymology /etɪ'mɒlədʒɪ/ *noun* (*plural* **-ies**) origin and sense-development of word; account of these. □ **etymological** /-mə'lɒdʒ-/ *adjective*.

eucalyptus /ju:kə'lɪptəs/ *noun* (*plural* **-tuses** or **-ti** /-taɪ/) tall evergreen tree; its oil, used as antiseptic etc.

Eucharist /'ju:kərɪst/ *noun* Christian sacrament in which bread and wine are consecrated and consumed; consecrated elements, esp. bread. □ **Eucharistic** /-'rɪs-/ *adjective*.

eugenics /ju:'dʒenɪks/ *plural noun* (also treated as *singular*) improvement of qualities of race by control of inherited characteristics. □ **eugenic** *adjective*; **eugenically** *adverb*.

eulogize /'ju:lədʒaɪz/ *verb* (also **-ise**) (**-zing** or **-sing**) extol; praise.

eulogy /'ju:lədʒɪ/ *noun* (*plural* **-ies**) speech or writing in praise or commendation. □ **eulogistic** /-'dʒɪs-/ *adjective*.

eunuch /'ju:nək/ *noun* castrated man.

euphemism /'ju:fɪmɪz(ə)m/ *noun* mild expression substituted for blunt one. □ **euphemistic** /-'mɪs-/ *adjective*; **euphemistically** /-'mɪs-/ *adverb*.

euphonium /ju:'fəʊnɪəm/ *noun* brass instrument of tuba family.

euphony /'ju:fənɪ/ *noun* pleasantness of sound, esp. in words. □ **euphonious** /-'fəʊ-/ *adjective*.

euphoria /ju:'fɔ:rɪə/ *noun* intense sense of well-being and excitement. □ **euphoric** /-'fɒr-/ *adjective*.

Eurasian /jʊə'reɪʒ(ə)n/ ● *adjective* of mixed European and Asian parentage; of Europe and Asia. ● *noun* Eurasian person.

eureka /jʊə'ri:kə/ *interjection* I have found it!

Eurodollar /'jʊərəʊdɒlə/ *noun* dollar held in bank in Europe etc.

European /jʊərə'pɪən/ ● *adjective* of, in, or extending over Europe. ● *noun* native or inhabitant of Europe; descendant of one.

Eustachian tube /ju:'steɪʃ(ə)n/ *noun* passage between middle ear and back of throat.

euthanasia /ju:θə'neɪzɪə/ *noun* killing person painlessly, esp. one who has incurable painful disease.

evacuate /ɪ'vækjʊeɪt/ *verb* (**-ting**) remove (people) from place of danger; make empty, clear; withdraw from (place); empty (bowels). □ **evacuation** *noun*.

evacuee /ɪvækju:'i:/ *noun* person evacuated.

evade /ɪ'veɪd/ *verb* (**-ding**) escape from, avoid; avoid doing or answering directly; avoid paying (tax) illegally.

evaluate /ɪ'væljʊeɪt/ *verb* (**-ting**) assess, appraise; find or state number or amount of. □ **evaluation** *noun.*

evanesce /evə'nes/ *verb* (**-cing**) *literary* fade from sight. □ **evanescence** *noun*; **evanescent** *adjective.*

evangelical /i:væn'dʒelɪk(ə)l/ ● *adjective* of or according to gospel teaching; of Protestant groups maintaining doctrine of salvation by faith. ● *noun* member of evangelical group. □ **evangelicalism** *noun.*

evangelist /ɪ'vændʒəlɪst/ *noun* writer of one of the 4 Gospels; preacher of gospel. □ **evangelism** *noun*; **evangelistic** /-'lɪs-/ *adjective.*

evangelize *verb* (also **-ise**) (**-zing** or **-sing**) preach gospel to. □ **evangelization** *noun.*

evaporate /ɪ'væpəreɪt/ *verb* (**-ting**) turn into vapour; (cause to) lose moisture as vapour; (cause to) disappear. □ **evaporation** *noun.*

evasion /ɪ'veɪʒ(ə)n/ *noun* evading; evasive answer.

evasive /ɪ'veɪsɪv/ *adjective* seeking to evade.

eve *noun* evening or day before festival etc.; time just before event; *archaic* evening.

even /'i:v(ə)n/ ● *adjective* (**-er**, **-est**) level, smooth; uniform; equal; equable, calm; divisible by two. ● *adverb* still, yet; (with negative) so much as. ● *verb* (often + *up*) make or become even. □ **even if** in spite of the fact that, no matter whether; **even out** become level or regular, spread (thing) over period or among group. □ **evenly** *adverb.*

evening /'i:vnɪŋ/ *noun* end part of day, esp. from about 6 p.m. to bedtime. □ **evening class** adult education class held in evening; **evening dress** formal clothes for evening wear; **evening star** planet, esp. Venus, conspicuous in west after sunset.

evensong *noun* evening service in Church of England.

event /ɪ'vent/ *noun* thing that happens; fact of thing occurring; item in (esp. sports) programme. □ **in any event**, **at all events** whatever happens; **in the event** as it turns or turned out; **in the event of** if (thing) happens.

eventful *adjective* marked by noteworthy events.

eventual /ɪ'ventʃʊəl/ *adjective* occurring in due course. □ **eventually** *adverb.*

eventuality /ɪventʃʊ'ælɪtɪ/ *noun* (*plural* **-ies**) possible event.

ever /'evə/ *adverb* at all times; always; at any time. □ **ever since** throughout period since (then); **ever so** *colloquial* very; **ever such a(n)** *colloquial* a very.

evergreen ● *adjective* retaining green leaves throughout year. ● *noun* evergreen plant.

everlasting *adjective* lasting for ever or a long time; (of flower) retaining shape and colour when dried.

evermore *adverb* for ever; always.

every /'evrɪ/ *adjective* each; all. □ **everybody** every person; **everyday** occurring every day, ordinary; **Everyman** ordinary or typical human being; **every now and again** or **then** occasionally; **everyone** everybody; **every other** each alternate; **everything** all things, the most important thing; **everywhere** in every place.

evict /ɪ'vɪkt/ *verb* expel (tenant) by legal process. □ **eviction** *noun.*

evidence /'evɪd(ə)ns/ ● *noun* (often + *for*, *of*) indication, sign; information given to establish fact etc.; statement etc. admissible in court of law. ● *verb* (**-cing**) be evidence of.

evident *adjective* obvious, manifest.

evidential /evɪ'denʃ(ə)l/ *adjective* of or providing evidence.

evidently *adverb* seemingly; as shown by evidence.

evil /'i:v(ə)l/ ● *adjective* wicked; harmful. ● *noun* evil thing; wickedness. □ **evil eye** gaze believed to cause harm. □ **evilly** *adverb.*

evince /ɪ'vɪns/ *verb* (**-cing**) show, indicate.

eviscerate /ɪ'vɪsəreɪt/ *verb* (**-ting**) disembowel. □ **evisceration** *noun.*

evocative /ɪ'vɒkətɪv/ *adjective* evoking (feelings etc.).

evoke /ɪ'vəʊk/ *verb* (**-king**) call up (feeling etc.). □ **evocation** /evə-/ *noun.*

evolution /iːvəˈluːʃ(ə)n/ *noun* evolving; development of species from earlier forms; unfolding of events etc.; change in disposition of troops or ships. □ **evolutionary** *adjective.*

evolutionist *noun* person who regards evolution as explaining origin of species.

evolve /ɪˈvɒlv/ *verb* (**-ving**) develop gradually and naturally; devise; unfold, open out.

ewe /juː/ *noun* female sheep.

ewer /ˈjuːə/ *noun* water-jug with wide mouth.

ex¹ *preposition* (of goods) sold from (warehouse etc.).

ex² *noun colloquial* former husband or wife.

ex- *prefix* formerly.

exacerbate /ekˈsæsəbeɪt/ *verb* (**-ting**) make worse; irritate. □ **exacerbation** *noun.*

exact /ɪgˈzækt/ ● *adjective* accurate; correct in all details. ● *verb* demand and enforce payment of (fees etc.); demand, insist on. □ **exactness** *noun.*

exaction *noun* exacting, being exacted; illegal or exorbitant demand.

exactitude *noun* exactness.

exactly *adverb* precisely; I agree.

exaggerate /ɪgˈzædʒəreɪt/ *verb* (**-ting**) make seem larger or greater than it really is; increase beyond normal or due proportions. □ **exaggeration** *noun.*

exalt /ɪgˈzɔːlt/ *verb* raise in rank, power, etc.; praise, extol; (usually as **exalted** *adjective*) make lofty or noble. □ **exaltation** /eg-/ *noun.*

exam /ɪgˈzæm/ *noun* examination, test.

examination /ɪgzæmɪˈneɪʃ(ə)n/ *noun* examining, being examined; detailed inspection; testing of knowledge or ability by questions; formal questioning of witness etc. in court.

examine /ɪgˈzæmɪn/ *verb* (**-ning**) inquire into; look closely at; test knowledge or ability of; check health of; question formally. □ **examinee** /-ˈniː/ *noun*; **examiner** *noun.*

example /ɪgˈzɑːmp(ə)l/ *noun* thing illustrating general rule; model, pattern; specimen; precedent; warning to others. □ **for example** by way of illustration.

exasperate /ɪgˈzɑːspəreɪt/ *verb* (**-ting**) irritate intensely. □ **exasperation** *noun.*

ex cathedra /eks kəˈθiːdrə/ *adjective & adverb* with full authority (esp. of papal pronouncement). [Latin]

excavate /ˈekskəveɪt/ *verb* (**-ting**) make (hole etc.) by digging; dig out material from (ground); reveal or extract by digging; dig systematically to explore (archaeological site). □ **excavation** *noun*; **excavator** *noun.*

exceed /ɪkˈsiːd/ *verb* be more or greater than; go beyond, do more than is warranted by; surpass.

exceedingly *adverb* very.

excel /ɪkˈsel/ *verb* (**-ll-**) surpass; be pre-eminent.

excellence /ˈeksələns/ *noun* great merit.

Excellency *noun* (*plural* **-ies**) (**His, Her, Your Excellency**) *title used of or to ambassador, governor, etc.*

excellent *adjective* extremely good.

except /ɪkˈsept/ ● *verb* exclude from general statement etc. ● *preposition* (often + *for*) not including, other than.

excepting *preposition* except.

exception /ɪkˈsepʃ(ə)n/ *noun* excepting; thing or case excepted. □ **take exception** (often + *to*) object.

exceptionable *adjective* open to objection.

■ **Usage** *Exceptionable* is sometimes confused with *exceptional*.

exceptional *adjective* forming exception; unusual. □ **exceptionally** *adverb.*

■ **Usage** *Exceptional* is sometimes confused with *exceptionable*.

excerpt ● *noun* /ˈeksɜːpt/ short extract from book, film, etc. ● *verb* /ɪkˈsɜːpt/ take excerpts from. □ **excerption** /ɪkˈsɜːpʃ(ə)n/ *noun.*

excess ● *noun* /ɪkˈses/ exceeding; amount by which thing exceeds; intemperance in eating or drinking. ● *adjective* /ˈekses/ that exceeds limit or given amount. □ **in, to excess** exceeding proper amount or degree; **in excess of** more than.

excessive /ɪk'sesɪv/ *adjective* too much; too great. □ **excessively** *adverb.*

exchange /ɪks'tʃeɪndʒ/ ● *noun* giving one thing and receiving another in its place; exchanging of money for equivalent, esp. in other currency; centre where telephone connections are made; place where merchants, stockbrokers, etc. transact business; employment office; short conversation. ● *verb* (**-ging**) give or receive in exchange; interchange. □ **exchange rate** price of one currency expressed in another. □ **exchangeable** *adjective.*

exchequer /ɪks'tʃekə/ *noun* former government department in charge of national revenue; royal or national treasury.

excise[1] /'eksaɪz/ *noun* tax levied on goods produced or sold within the country; tax on certain licences.

excise[2] /ɪk'saɪz/ *verb* (**-sing**) cut out or away. □ **excision** /-'sɪʒ-/ *noun.*

excitable *adjective* easily excited. □ **excitability** *noun.*

excite /ɪk'saɪt/ *verb* (**-ting**) move to strong emotion; arouse (feelings etc.); provoke (action etc.); stimulate to activity. □ **excitement** *noun*; **exciting** *adjective.*

exclaim /ɪk'skleɪm/ *verb* cry out suddenly; utter or say thus.

exclamation /ekskləˈmeɪʃ(ə)n/ *noun* exclaiming; word(s) etc. exclaimed. □ **exclamation mark** punctuation mark (!) indicating exclamation (see panel). □ **exclamatory** /ɪk'sklæmətərɪ/ *adjective.*

exclude /ɪk'sklu:d/ *verb* (**-ding**) shut out, leave out; make impossible, preclude. □ **exclusion** *noun.*

exclusive /ɪk'sklu:sɪv/ ● *adjective* excluding other things; (+ *of*) not including; (of society etc.) tending to exclude outsiders; high-class; not obtainable or published elsewhere. ● *noun* exclusive item of news, film, etc. □ **exclusively** *adverb*; **exclusiveness** *noun*; **exclusivity** /eksklu:'sɪvɪtɪ/ *noun.*

excommunicate /ekskə'mju:nɪkeɪt/ *verb* (**-ting**) deprive (person) of membership and sacraments of Church. □ **excommunication** *noun.*

excoriate /eks'kɔ:rɪeɪt/ *verb* (**-ting**) remove part of skin of by abrasion; remove (skin); censure severely. □ **excoriation** *noun.*

excrement /'ekskrɪmənt/ *noun* faeces.

excrescence /ɪk'skres(ə)ns/ *noun* abnormal or morbid outgrowth. □ **excrescent** *adjective.*

excreta /ɪk'skri:tə/ *plural noun* faeces and urine.

excrete /ɪk'skri:t/ *verb* (**-ting**) (of animal or plant) expel (waste). □ **excretion** *noun*; **excretory** *adjective.*

excruciating /ɪk'skru:ʃɪeɪtɪŋ/ *adjective* acutely painful. □ **excruciatingly** *adverb.*

exculpate /'ekskʌlpeɪt/ *verb* (**-ting**) *formal* free from blame. □ **exculpation** *noun.*

excursion /ɪk'skɜ:ʃ(ə)n/ *noun* journey to place and back, made for pleasure.

excursive /ɪk'skɜ:sɪv/ *adjective literary* digressive.

excuse ● *verb* /ɪk'skju:z/ (**-sing**) try to lessen blame attaching to; serve as reason to judge (person, act) less severely; (often + *from*) grant exemption to; forgive. ● *noun* /ɪk'skju:s/ reason put forward to mitigate or justify offence; apology. □ **be excused** be allowed not to do something or to leave or be absent; **excuse me** *polite request to*

Exclamation mark !

This is used instead of a full stop at the end of a sentence to show that the speaker or writer is very angry, enthusiastic, insistent, disappointed, hurt, surprised, etc., e.g.

I am not pleased at all!
I just love sweets!
Go away!
I wish I could have gone!
Ow!
He didn't even say goodbye!

be allowed to pass, polite apology for interrupting or disagreeing. □ **excusable** /-'sku:z-/ *adjective.*

ex-directory *adjective* not listed in telephone directory, at subscriber's wish.

execrable /'eksɪkrəb(ə)l/ *adjective* abominable.

execrate /'eksɪkreɪt/ *verb* (**-ting**) express or feel abhorrence for; curse. □ **execration** *noun.*

execute /'eksɪkju:t/ *verb* (**-ting**) carry out, perform; put to death.

execution /eksɪ'kju:ʃ(ə)n/ *noun* carrying out, performance; capital punishment.

executioner *noun* person carrying out death sentence.

executive /ɪg'zekjʊtɪv/ *noun* person or body with managerial or administrative responsibility; branch of government etc. concerned with executing laws, agreements, etc. ● *adjective* concerned with executing laws, agreements, etc. or with administration etc.

executor /ɪg'zekjʊtə/ *noun* (*feminine* **executrix** /-trɪks/) person appointed by testator to carry out terms of will. □ **executorial** /-'tɔ:rɪəl/ *adjective.*

exegesis /eksɪ'dʒi:sɪs/ *noun* explanation, esp. of Scripture. □ **exegetic** /-'dʒetɪk/ *adjective.*

exemplar /ɪg'zemplə/ *noun* model; typical instance.

exemplary *adjective* outstandingly good; serving as example or warning.

exemplify /ɪg'zemplɪfaɪ/ *verb* (**-ies, -ied**) give or be example of. □ **exemplification** *noun.*

exempt /ɪg'zempt/ ● *adjective* (often + *from*) free from obligation or liability imposed on others. ● *verb* (+ *from*) make exempt from. □ **exemption** *noun.*

exercise /'eksəsaɪz/ ● *noun* use of muscles etc., esp. for health; task set for physical or other training; use or application of faculties etc.; practice; (often in *plural*) military drill or manoeuvres. ● *verb* (**-sing**) use; perform (function); take exercise; give exercise to; tax powers of; perplex, worry.

exert /ɪg'zɜ:t/ *verb* use; bring to bear; (**exert oneself**) make effort. □ **exertion** *noun.*

exfoliate /eks'fəʊlɪeɪt/ *verb* (**-ting**) come off in scales or layers. □ **exfoliation** *noun.*

ex gratia /eks 'greɪʃə/ ● *adverb* as favour and not under (esp. legal) compulsion. ● *adjective* granted on this basis. [Latin]

exhale /eks'heɪl/ *verb* (**-ling**) breathe out; give off or be given off in vapour. □ **exhalation** /-hə-/ *noun.*

exhaust /ɪg'zɔ:st/ ● *verb* (often as **exhausted** *adjective* or **exhausting** *adjective*) consume, use up; tire out; study or expound completely; empty of contents. ● *noun* waste gases etc. expelled from engine after combustion; pipe or system through which they are expelled. □ **exhaustible** *adjective*; **exhaustion** *noun.*

exhaustive *adjective* complete, comprehensive. □ **exhaustively** *adverb.*

exhibit /ɪg'zɪbɪt/ ● *verb* (**-t-**) show, esp. publicly; display. ● *noun* thing exhibited. □ **exhibitor** *noun.*

exhibition /eksɪ'bɪʃ(ə)n/ *noun* display, public show; exhibiting, being exhibited; scholarship, esp. from funds of college etc.

exhibitioner *noun* student receiving exhibition.

exhibitionism *noun* tendency towards attention-seeking behaviour; compulsion to expose genitals in public. □ **exhibitionist** *noun.*

exhilarate /ɪg'zɪləreɪt/ *verb* (**-ting**) (often as **exhilarating** *adjective* or **exhilarated** *adjective*) enliven, gladden. □ **exhilaration** *noun.*

exhort /ɪg'zɔ:t/ *verb* (often + *to do*) urge strongly or earnestly. □ **exhortation** /eg-/ *noun*; **exhortative** *adjective*; **exhortatory** *adjective.*

exhume /eks'hju:m/ *verb* (**-ming**) dig up. □ **exhumation** *noun.*

exigency /'eksɪdʒənsɪ/ *noun* (*plural* **-ies**) (also **exigence**) urgent need; emergency. □ **exigent** *adjective.*

exiguous /eg'zɪgjʊəs/ *adjective* scanty, small. □ **exiguity** /-'gju:ɪtɪ/ *noun.*

exile /'eksaɪl/ ● *noun* expulsion or long absence from one's country etc.; person in exile. ● *verb* (**-ling**) send into or condemn to exile.

exist /ɪg'zɪst/ *verb* be, have being; occur, be found; live with no pleasure; live.

existence *noun* fact or manner of existing; all that exists. □ **existent** *adjective.*

existential /egzɪ'stenʃ(ə)l/ *adjective* of or relating to existence.

existentialism *noun* philosophical theory emphasizing existence of individual as free and self-determining agent. □ **existentialist** *adjective & noun.*

exit /'eksɪt/ ● *noun* way out; going out; place where vehicles leave motorway etc.; departure. ● *verb* (**-t-**) make one's exit.

exodus /'eksədəs/ *noun* mass departure; (**Exodus**) that of Israelites from Egypt.

ex officio /eks ə'fɪʃɪəʊ/ *adverb & adjective* by virtue of one's office.

exonerate /ɪg'zɒnəreɪt/ *verb* (**-ting**) free or declare free from blame. □ **exoneration** *noun.*

exorbitant /ɪg'zɔ:bɪt(ə)nt/ *adjective* grossly excessive.

exorcize /'eksɔ:saɪz/ *verb* (also **-ise**) (**-zing** or **-sing**) drive out (evil spirit) by prayers etc.; free (person, place) thus. □ **exorcism** *noun*; **exorcist** *noun.*

exoskeleton /'eksəʊskelɪtən/ *noun* external skeleton.

exotic /ɪg'zɒtɪk/ ● *adjective* introduced from abroad; strange, unusual. ● *noun* exotic plant etc. □ **exotically** *adverb.*

expand /ɪk'spænd/ *verb* increase in size or importance; (often + *on*) give fuller account; become more genial; write out in full; spread out flat. □ **expandable** *adjective*; **expansion** *noun.*

expanse /ɪk'spæns/ *noun* wide area or extent of land, space, etc.

expansionism *noun* advocacy of expansion, esp. of state's territory. □ **expansionist** *noun & adjective.*

expansive *adjective* able or tending to expand; extensive; genial.

expatiate /ɪk'speɪʃɪeɪt/ *verb* (**-ting**) (usually + *on*, *upon*) speak or write at length. □ **expatiation** *noun.*

expatriate ● *adjective* /eks'pætrɪət/ living abroad; exiled. ● *noun* /eks'pætrɪət/ expatriate person. ● *verb* /eks'pætrɪeɪt/ (**-ting**) expel from native country.

expect /ɪk'spekt/ *verb* regard as likely; look for as one's due; *colloquial* suppose. □ **be expecting** *colloquial* be pregnant.

expectant *adjective* expecting; expecting to become; pregnant. □ **expectancy** *noun*; **expectantly** *adverb.*

expectation /ekspek'teɪʃ(ə)n/ *noun* expecting, anticipation; what one expects; probability; (in *plural*) prospects of inheritance.

expectorant /ek'spektərənt/ ● *adjective* causing expectoration. ● *noun* expectorant medicine.

expectorate /ek'spektəreɪt/ *verb* (**-ting**) cough or spit out from chest or lungs; spit. □ **expectoration** *noun.*

expedient /ɪk'spi:dɪənt/ ● *adjective* advantageous; advisable on practical rather than moral grounds. ● *noun* means of achieving an end; resource. □ **expediency** *noun.*

expedite /'ekspɪdaɪt/ *verb* (**-ting**) assist progress of; accomplish quickly.

expedition /ekspɪ'dɪʃ(ə)n/ *noun* journey or voyage for particular purpose; people etc. undertaking this; speed.

expeditionary *adjective* of or used on expedition.

expeditious /ekspɪ'dɪʃəs/ *adjective* acting or done with speed and efficiency.

expel /ɪk'spel/ *verb* (**-ll-**) deprive of membership; force out; eject.

expend /ɪk'spend/ *verb* spend (money, time, etc.); use up.

expendable *adjective* that may be sacrificed or dispensed with.

expenditure /ɪk'spendɪtʃə/ *noun* expending; amount expended.

expense /ɪk'spens/ *noun* cost, charge; (in *plural*) costs incurred in doing job etc., reimbursement of this.

expensive *adjective* costing or charging much. □ **expensively** *adverb.*

experience /ɪk'spɪərɪəns/ ● *noun* personal observation or contact; knowledge or skill based on this; event that affects one. ● *verb* (**-cing**) have experience of; undergo; feel. □ **experiential** /-'en-/ *adjective.*

experienced *adjective* having had much experience; skilful through experience.

experiment /ɪk'sperɪmənt/ ● *noun* procedure adopted to test hypothesis or demonstrate known fact. ● *verb* (also /-ment/) make experiment(s). □ **experimentation** /-men-/ *noun*; **experimenter** *noun.*

experimental /ɪksperɪ'ment(ə)l/ *adjective* based on or done by way of experiment. □ **experimentally** *adverb.*

expert /'ekspɜ:t/ (often + *at, in*) ● *adjective* well informed or skilful in a subject. ● *noun* expert person. □ **expertly** *adverb.*

expertise /ekspɜ:'ti:z/ *noun* special skill or knowledge.

expiate /'ekspɪeɪt/ *verb* (**-ting**) pay penalty or make amends for (wrong). □ **expiation** *noun.*

expire /ɪk'spaɪə/ *verb* (**-ring**) come to an end; cease to be valid; die; breathe out. □ **expiration** /ekspɪ-/ *noun.*

expiry *noun* end of validity or duration.

explain /ɪks'pleɪn/ *verb* make intelligible; make known; say by way of explanation; account for. □ **explanation** /eksplə-/ *noun.*

explanatory /ɪk'splænətərɪ/ *adjective* serving to explain.

expletive /ɪk'spli:tɪv/ *noun* swear-word or exclamation.

explicable /ɪk'splɪkəb(ə)l/ *adjective* explainable.

explicit /ɪk'splɪsɪt/ *adjective* expressly stated; stated in detail; definite; outspoken. □ **explicitly** *adverb*; **explicitness** *noun.*

explode /ɪk'spləʊd/ (**-ding**) *verb* expand violently with loud noise; cause (bomb etc.) to do this; give vent suddenly to emotion, esp. anger; (of population etc.) increase suddenly; discredit.

exploit ● *noun* /'eksplɔɪt/ daring feat. ● *verb* /ɪk'splɔɪt/ use or develop for one's own ends; take advantage of (esp. person). □ **exploitation** /eksplɔɪ-/ *noun*; **exploitative** /ɪk'splɔɪt-/ *adjective.*

explore /ɪk'splɔ:/ *verb* (**-ring**) travel through (country etc.) to learn about it; inquire into; examine (part of body), probe (wound). □ **exploration** /eksplə-/ *noun*; **exploratory** /-'splɒr-/ *adjective*; **explorer** *noun.*

explosion /ɪk'spləʊʒ(ə)n/ *noun* exploding; loud noise caused by this; outbreak; sudden increase.

explosive /ɪk'spləʊsɪv/ ● *adjective* tending to explode; likely to cause violent outburst etc. ● *noun* explosive substance.

exponent /ɪk'spəʊnənt/ *noun* person promoting idea etc.; practitioner of activity, profession, etc.; person who explains or interprets; type, representative; *Mathematics* raised number or symbol showing how many of a number are to be multiplied together, e.g. ^{3} in $2^3 = 2 \times 2 \times 2$.

exponential /ekspə'nenʃ(ə)l/ *adjective* (of increase) more and more rapid. □ **exponentially** *adverb.*

export /ɪk'spɔ:t/ ● *verb* sell or send (goods or services) to another country. ● *noun* /'ekspɔ:t/ exporting; exported article or service; (in *plural*) amount exported. □ **exportation** *noun*; **exporter** /ɪk'spɔ:tə/ *noun.*

expose /ɪk'spəʊz/ *verb* (**-sing**) leave unprotected, esp. from weather; (+ *to*) put at risk of, subject to (influence etc.); *Photography* subject (film etc.) to light; reveal, disclose; exhibit, display; (**expose oneself**) display one's genitals indecently in public.

exposé /ek'spəʊzeɪ/ *noun* orderly statement of facts; disclosure of discreditable thing.

exposition /ekspə'zɪʃ(ə)n/ *noun* expounding, explanation; *Music* part of movement in which principal themes are presented; exhibition.

ex post facto /eks pəʊst 'fæktəʊ/ *adjective & adverb* retrospective(ly). [Latin]

expostulate /ɪk'spɒstjʊleɪt/ *verb* (**-ting**) make protest, remonstrate. □ **expostulation** *noun.*

exposure /ɪk'spəʊʒə/ *noun* exposing, being exposed; physical condition resulting from being exposed to elements; *Photography* length of time film etc. is exposed, section of film etc. exposed at one time.

expound /ɪk'spaʊnd/ *verb* set out in detail; explain, interpret.

express /ɪk'spres/ ● *verb* represent by symbols etc. or in language; put into words; squeeze out (juice, milk, etc.); (**express oneself**) communicate what one thinks, feels, or means. ● *adjective* operating at high speed; definitely stated; delivered by specially fast service. ● *adverb* with speed; by express messenger or train. ● *noun* express train etc. □ **expressible** *adjective.*

expression /ɪk'spreʃ(ə)n/ *noun* expressing, being expressed; wording, word, phrase; conveying or depiction of feeling; appearance (of face), intonation (of voice); *Mathematics* collection of symbols expressing quantity. □ **expressionless** *adjective.*

expressionism *noun* style of painting etc. seeking to express emotion rather than depict external world. □ **expressionist** *noun & adjective.*

expressive *adjective* full of expression; (+ *of*) serving to express. □ **expressiveness** *noun.*

expressly *adverb* explicitly.

expropriate /ɪks'prəʊprɪeɪt/ *verb* (**-ting**) take away (property); dispossess. □ **expropriation** *noun*; **expropriator** *noun.*

expulsion /ɪk'spʌlʃ(ə)n/ *noun* expelling, being expelled.

expunge /ɪk'spʌndʒ/ *verb* (**-ging**) erase, remove.

expurgate /'ekspəgeɪt/ *verb* (**-ting**) remove matter considered objectionable from (book etc.); clear away (such matter). □ **expurgation** *noun*; **expurgator** *noun.*

exquisite /'ekskwɪzɪt, ek'skwɪzɪt/ *adjective* extremely beautiful or delicate; acute, keen. □ **exquisitely** *adverb.*

ex-serviceman /eks'sɜ:vɪsmən/ *noun* man formerly member of armed forces.

extant /ek'stænt/ *adjective* still existing.

extempore /ɪk'stempərɪ/ *adverb & adjective* without preparation.

extemporize /ɪk'stempəraɪz/ *verb* (also **-ise**) (**-zing** or **-sing**) improvise. □ **extemporization** *noun.*

extend /ɪk'stend/ *verb* lengthen in space or time; lay out at full length; reach or be or make continuous over certain area; (+ *to*) have certain scope; offer or accord (feeling, invitation, etc.); tax powers of. □ **extendible** *adjective*; **extensible** *adjective.*

extension /ɪk'stenʃ(ə)n/ *noun* extending; enlargement, additional part; subsidiary telephone on same line as main one; additional period of time.

extensive /ɪk'stensɪv/ *adjective* large, far-reaching. □ **extensively** *adverb.*

extent /ɪk'stent/ *noun* space covered; width of application, scope.

extenuate /ɪk'stenjʊeɪt/ *verb* (**-ting**) (often as **extenuating** *adjective*) make (guilt etc.) seem less serious by partial excuse. □ **extenuation** *noun.*

exterior /ɪk'stɪərɪə/ ● *adjective* outer; coming from outside. ● *noun* exterior aspect or surface; outward demeanour.

exterminate /ɪk'stɜ:mɪneɪt/ *verb* (**-ting**) destroy utterly. □ **extermination** *noun*; **exterminator** *noun.*

external /ɪk'stɜ:n(ə)l/ ● *adjective* of or on the outside; coming from outside; relating to a country's foreign affairs; (of medicine) for use on outside of body. ● *noun* (in *plural*) external features or circumstances. □ **externality** /ekstɜ:'nælɪtɪ/ *noun*; **externally** *adverb.*

externalize *verb* (also **-ise**) (**-zing** or **-sing**) give or attribute external existence to.

extinct /ɪk'stɪŋkt/ *adjective* no longer existing; no longer burning; (of volcano) that no longer erupts; obsolete.

extinction *noun* making or becoming extinct; extinguishing, being extinguished.

extinguish /ɪk'stɪŋgwɪʃ/ *verb* put out (flame, light, etc.); terminate, destroy; wipe out (debt).

extinguisher *noun* = FIRE EXTINGUISHER.

extirpate /'ekstəpeɪt/ *verb* (**-ting**) destroy; root out. □ **extirpation** *noun.*

extol /ɪk'stəʊl/ *verb* (**-ll-**) praise enthusiastically.

extort /ɪk'stɔ:t/ *verb* get by coercion.

extortion *noun* extorting, esp. of money; illegal exaction.

extortionate /ɪk'stɔ:ʃənət/ *adjective* exorbitant.

extra /'ekstrə/ ● *adjective* additional; more than usual or necessary. ● *adverb* more than usually; additionally. ● *noun* extra thing; thing for which one is charged extra; person playing one of crowd etc. in film; special edition of newspaper; *Cricket* run not scored from hit with bat.

extra- *combining form* outside, beyond scope of.

extract ● *verb* /ɪk'strækt/ take out; obtain against person's will; obtain from earth; copy out, quote; obtain (juice etc.) by pressure, distillation, etc.; derive (pleasure etc.); *Mathematics* find (root of number). ● *noun* /'ekstrækt/ passage from book etc.; preparation containing concentrated constituent of substance.

extraction /ɪk'strækʃ(ə)n/ *noun* extracting; removal of tooth; lineage.

extractor /ɪk'stræktə/ *noun* machine that extracts. □ **extractor fan** one that extracts bad air etc.

extracurricular /ekstrəkə'rɪkjʊlə/ *adjective* outside normal curriculum.

extraditable *adjective* liable to extradition; (of crime) warranting extradition.

extradite /'ekstrədaɪt/ *verb* (**-ting**) hand over (person accused of crime) to state where crime was committed. □ **extradition** /-'dɪʃ-/ *noun.*

extramarital /ekstrə'mærɪt(ə)l/ *adjective* (of sexual relationship) outside marriage.

extramural /ekstrə'mjʊər(ə)l/ *adjective* additional to ordinary teaching or studies.

extraneous /ɪk'streɪnɪəs/ *adjective* of external origin; (often + *to*) separate, irrelevant, unrelated.

extraordinary /ɪk'strɔ:dɪnərɪ/ *adjective* unusual, remarkable; unusually great; (of meeting etc.) additional. □ **extraordinarily** *adverb.*

extrapolate /ɪk'stræpəleɪt/ *verb* (**-ting**) estimate (unknown facts or values) from known data. □ **extrapolation** *noun.*

extrasensory /ekstrə'sensərɪ/ *adjective* derived by means other than known senses.

extraterrestrial /ekstrətɪ'restrɪəl/ ● *adjective* outside the earth or its atmosphere. ● *noun* fictional being from outer space.

extravagant /ɪk'strævəgənt/ *adjective* spending (esp. money) excessively; excessive; absurd; costing much. □ **extravagance** *noun*; **extravagantly** *adverb.*

extravaganza /ɪkstrævə'gænzə/ *noun* spectacular theatrical or television production; fanciful composition.

extreme /ɪk'stri:m/ ● *adjective* reaching high or highest degree; severe, not moderate; outermost; utmost. ● *noun* either of two things as remote or different as possible; thing at either end; highest degree. □ **extreme unction** anointing by priest of dying person. □ **extremely** *adverb.*

extremis see IN EXTREMIS.

extremism *noun* advocacy of extreme measures. □ **extremist** *adjective & noun.*

extremity /ɪk'stremɪtɪ/ *noun* (*plural* **-ies**) extreme point, end; (in *plural*) hands and feet; condition of extreme adversity.

extricate /'ekstrɪkeɪt/ *verb* (**-ting**) disentangle, release. □ **extrication** *noun.*

extrinsic /ek'strɪnsɪk/ *adjective* not inherent or intrinsic; (often + *to*) extraneous. □ **extrinsically** *adverb.*

extrovert /'ekstrəvɜ:t/ ● *noun* sociable or unreserved person; person mainly concerned with external things. ● *adjective* typical or having nature of extrovert. □ **extroversion** /-'vɜ:-/ *noun.*

extrude /ɪk'stru:d/ *verb* (**-ding**) thrust or squeeze out; shape by forcing through nozzle. □ **extrusion** *noun.*

exuberant /ɪg'zju:bərənt/ *adjective* high-spirited, lively; luxuriant, prolific; (of feelings etc.) abounding. □ **exuberance** *noun.*

exude /ɪg'zju:d/ *verb* (**-ding**) ooze out; give off; display (emotion) freely. □ **exudation** *noun.*

exult /ɪg'zʌlt/ *verb* rejoice. □ **exultant** *adjective*; **exultation** /eg-/ *noun.*

eye /aɪ/ ● *noun* organ or faculty of sight; iris of eye; region round eye; gaze; perception; eyelike thing; leaf bud of potato; centre of hurricane; spot, hole, loop. ● *verb* (**eyes**, **eyed**, **eyeing** or **eying**) (often + *up*) observe, watch suspiciously or closely. □ **all eyes** watching intently; **eyeball** ball of eye within lids and socket; **eyebath** vessel for applying lotion to eye; **eye-brow** hair growing on ridge over eye; **eye-catching** *colloquial* striking; **eyeglass** lens for defective eye; **eyehole** hole to look through; **eyelash** any of hairs on edge of eyelid; **eyelid** fold of skin that can cover eye; **eyeliner** cosmetic applied as line round eye; **eye-opener** *colloquial* enlightening experience, unexpected revelation; **eyepiece** lens(es) to which eye is applied at end of optical instrument; **eye-shade** device to protect eyes from strong light; **eye-shadow** cosmetic for eyelids; **eyesight** faculty or power of sight; **eyesore** ugly thing; **eye-tooth** canine tooth in upper jaw; **eyewash** lotion for eyes, *slang* nonsense; **eyewitness** person who saw thing happen and can tell of it; **have one's eye on** wish or plan to obtain; **keep an eye on** watch, look after; **see eye to eye** (often + *with*) agree; **set eyes on** see.

eyeful *noun* (*plural* **-s**) *colloquial* good look, visually striking person or thing.

eyelet /'aɪlɪt/ *noun* small hole for passing cord etc. through.

eyrie /'ɪərɪ/ *noun* nest of bird of prey, esp. eagle, built high up.

Ff

F *abbreviation* Fahrenheit.

f *abbreviation* (also **f.**) female; feminine; *Music* forte.

FA *abbreviation* Football Association.

fa = FAH.

fable /'feɪb(ə)l/ *noun* fictional tale, esp. legendary, or moral tale, often with animal characters.

fabled *adjective* celebrated; legendary.

fabric /'fæbrɪk/ *noun* woven material; walls, floor, and roof of building; structure.

fabricate /'fæbrɪkeɪt/ *verb* (**-ting**) construct, esp. from components; invent (fact), forge (document). □ **fabrication** *noun.*

fabulous /'fæbjʊləs/ *adjective colloquial* marvellous; legendary. □ **fabulously** *adverb.*

façade /fə'sɑ:d/ *noun* face or front of building; outward, esp. deceptive, appearance.

face ● *noun* front of head; facial expression; surface; façade of building; side of mountain; dial of clock etc.; functional side of tool, bat, etc.; effrontery; aspect, feature. ● *verb* (**-cing**) look or be positioned towards; be opposite; meet resolutely, confront; put facing on (garment, wall, etc.). □ **face-lift** cosmetic surgery to remove wrinkles etc., improvement in appearance; **face up to** accept bravely; **face value** nominal value, superficial appearance; **lose face** be humiliated; **on the face of it** apparently; **pull a face** distort features; **save face** avoid humiliation.

faceless *adjective* without identity; not identifiable.

facet /'fæsɪt/ *noun* aspect; side of cut gem etc.

facetious /fə'si:ʃəs/ *adjective* intended to be amusing, esp. inappropriately. □ **facetiously** *adverb*; **facetiousness** *noun.*

facia = FASCIA.

facial /'feɪʃ(ə)l/ ● *adjective* of or for the face. ● *noun* beauty treatment for the face. □ **facially** *adverb.*

facile /'fæsaɪl/ *adjective* easily achieved but of little value; glib.

facilitate /fə'sɪlɪteɪt/ *verb* (**-ting**) ease (process etc.). □ **facilitation** *noun.*

facility /fə'sɪlɪtɪ/ *noun* (*plural* **-ies**) ease, absence of difficulty; dexterity; (esp. in *plural*) opportunity or equipment for doing something.

facing *noun* material over part of garment etc. for contrast or strength; outer covering on wall etc.

facsimile /fæk'sɪmɪlɪ/ *noun* exact copy of writing, picture, etc.

fact *noun* thing known to exist or be true; reality. □ **factsheet** information leaflet; **in fact** in reality, in short; **the facts of life** information on sexual functions etc.

faction /'fækʃ(ə)n/ *noun* small dissenting group within larger one, esp. in politics; such dissension. □ **factional** *adjective.*

factious /'fækʃəs/ *adjective* of or inclined to faction.

factitious /fæk'tɪʃəs/ *adjective* specially contrived; artificial.

factor /'fæktə/ *noun* thing contributing to result; whole number etc. that when multiplied produces given number (e.g. 2, 3, 4, and 6 are the factors of 12).

factorial /fæk'tɔ:rɪəl/ *noun* the product of a number and all whole numbers below it.

factory /'fæktərɪ/ *noun* (*plural* **-ies**) building(s) for manufacture of goods. □ **factory farming** using intensive or industrial methods of rearing livestock.

factotum /fæk'təʊtəm/ *noun* (*plural* **-s**) employee doing all kinds of work.

factual /'fæktjʊəl/ *adjective* based on or concerned with fact. □ **factually** *adverb.*

faculty /'fækəltɪ/ *noun* (*plural* **-ies**) aptitude for particular activity; physical or mental power; group of related university departments; *US* teaching staff of university etc.

fad *noun* craze; peculiar notion.

faddy *adjective* (**-ier**, **-iest**) having petty likes and dislikes.

fade ● *verb* (**-ding**) (cause to) lose colour, light, or sound; slowly diminish; lose freshness or strength. ● *noun* action of fading. □ **fade away** die away, disappear, *colloquial* languish, grow thin.

faeces /'fi:si:z/ *plural noun* (*US* **feces**) waste matter from bowels. □ **faecal** /-k(ə)l/ *adjective.*

fag ● *noun colloquial* tedious task; *slang* cigarette; (at public schools) junior boy who runs errands for a senior. ● *verb* (**-gg-**) (often + *out*) *colloquial* exhaust; act as fag. □ **fag-end** *slang* cigarette-end.

faggot /'fægət/ *noun* (*US* **fagot**) baked or fried ball of seasoned chopped liver etc.; bundle of sticks etc.

fah /fɑ:/ *noun* (also **fa**) *Music* fourth note of scale in tonic sol-fa.

Fahrenheit /'færənhaɪt/ *adjective* of scale of temperature on which water freezes at 32° and boils at 212°.

faience /'faɪɑ̃s/ *noun* decorated and glazed earthenware and porcelain.

fail ● *verb* not succeed; be or judge to be unsuccessful in (exam etc.); (+ *to do*) be unable, neglect; disappoint; be absent or insufficient; become weaker; cease functioning. ● *noun* failure in exam. □ **fail-safe** reverting to safe condition when faulty; **without fail** for certain, whatever happens.

failed *adjective* unsuccessful; bankrupt.

failing ● *noun* fault; weakness. ● *preposition* in default of.

failure /'feɪljə/ *noun* lack of success; unsuccessful person or thing; non-performance; breaking down, ceasing to function; running short of supply etc.

fain *archaic* ● *adjective* (+ *to do*) willing, obliged. ● *adverb* gladly.

faint ● *adjective* dim, pale; weak, giddy; slight; timid. ● *verb* lose consciousness; become faint. ● *noun* act or state of fainting. □ **faint-hearted** cowardly, timid. □ **faintly** *adverb*; **faintness** *noun.*

fair[1] ● *adjective* just, equitable; blond, not dark; moderate in quality or amount; (of weather) fine; (of wind) favourable; *archaic* beautiful. ● *adverb* in a fair or just manner; exactly, completely. □ **fair and square** exactly, straightforwardly; **fair copy** transcript free from corrections; **fair game** legitimate target or object; **fair play** just treatment or behaviour; **fair-weather friend** friend or ally who deserts in crisis. □ **fairness** *noun.*

fair[2] *noun* stalls, amusements, etc. for public entertainment; periodic market, often with entertainments; trade exhibition. □ **fairground** outdoor site for fair.

Fair Isle *noun* multicoloured knitwear design characteristic of Fair Isle in the Shetlands.

fairly *adverb* in a fair way; moderately; to a noticeable degree.

fairway *noun* navigable channel; mown grass between golf tee and green.

fairy /'feərɪ/ *noun* (*plural* **-ies**) small winged legendary being. □ **fairy cake** small iced sponge cake; **fairy godmother** benefactress; **fairyland** home of fairies, enchanted region; **fairy lights** small coloured lights for decoration; **fairy ring** ring of darker grass caused by fungi; **fairy story**, **tale** tale about fairies, unbelievable story, lie.

fait accompli /feɪt ə'kɒmpli:/ *noun* thing done and past arguing about. [French]

faith /feɪθ/ *noun* trust; religious belief; creed; loyalty, trustworthiness. □ **faith-healer** person who practises **faith-healing**, healing dependent on faith rather than treatment.

faithful *adjective* showing faith; (often + *to*) loyal, trustworthy; accurate.

faithfully *adverb* in a faithful way. □ **Yours faithfully** *written before signature at end of business letter*.

faithless *adjective* disloyal; without religious faith.

fake ● *noun* thing or person that is not genuine. ● *adjective* counterfeit, not genuine. ● *verb* (**-king**) make fake or imitation of; feign.

fakir /'feɪkɪə/ *noun* Muslim or Hindu religious beggar or ascetic.

falcon /'fɔ:lkən/ *noun* small hawk trained to hunt.

falconry *noun* breeding and training of hawks.

fall /fɔ:l/ ● *verb* (*past* **fell**; *past participle* **fallen**) go or come down freely; descend; (often + *over*) lose balance and come suddenly to ground; slope or hang down; sink lower, decline in power, status, etc.; subside; occur; become; (of face) show dismay etc.; be defeated; die. ● *noun* falling; amount or thing that falls; overthrow; (esp. in *plural*) waterfall; *US* autumn. □ **fall back on** have recourse to; **fall behind** be outstripped, be in arrears; **fall down (on)** *colloquial* fail (in); **fall for** be captivated or deceived by; **fall foul of** come into conflict with; **fall guy** *slang* easy victim, scapegoat; **fall in** *Military* take place in parade; **fall in with** meet by chance, agree or coincide with; **falling star** meteor; **fall off** decrease, deteriorate; **fall out** *verb* quarrel, (of hair, teeth, etc.) become detached, result, occur, *Military* come out of formation; **fallout** *noun* radioactive nuclear debris; **fall short of** fail to reach or obtain; **fall through** fail; **fall to** start eating, working, etc.

fallacy /'fæləsɪ/ *noun* (*plural* **-ies**) mistaken belief; faulty reasoning; misleading argument. □ **fallacious** /fə'leɪʃəs/ *adjective*.

fallible /'fælɪb(ə)l/ *adjective* capable of making mistakes. □ **fallibility** *noun*.

Fallopian tube /fə'ləʊpɪən/ *noun* either of two tubes along which ova travel from ovaries to womb.

fallow /'fæləʊ/ ● *adjective* (of land) ploughed but left unsown; uncultivated. ● *noun* fallow land.

fallow deer *noun* small deer with white-spotted reddish-brown summer coat.

false /fɔ:ls/ *adjective* wrong, incorrect; sham, artificial; (+ *to*) deceitful, treacherous, unfaithful; deceptive. □ **false alarm** alarm given needlessly; **false pretences** misrepresentations meant to deceive. □ **falsely** *adverb*; **falsity** *noun*; **falseness** *noun*.

falsehood *noun* untrue thing; lying, lie.

falsetto /fɔ:l'setəʊ/ *noun* male voice above normal range.

falsify /'fɔ:lsɪfaɪ/ *verb* (**-ies**, **-ied**) fraudulently alter; misrepresent. □ **falsification** *noun*.

falter /'fɔ:ltə/ *verb* stumble; go unsteadily; lose courage; speak hesitatingly.

fame *noun* renown, being famous; *archaic* reputation.

famed *adjective* (+ *for*) much spoken of or famous because of.

familial /fə'mɪlɪəl/ *adjective* of a family or its members.

familiar /fə'mɪlɪə/ ● *adjective* (often + *to*) well known, often encountered; (+ *with*) knowing (a thing) well; (excessively) informal. ● *noun* intimate friend; supposed attendant of witch etc. □ **familiarity** /-'ær-/ *noun* (*plural* **-ies**); **familiarly** *adverb*.

familiarize /fə'mɪlɪəraɪz/ *verb* (also **-ise**) (**-zing** or **-sing**) (usually + *with*) make (person etc.) conversant. □ **familiarization** *noun*.

family /'fæmɪlɪ/ *noun* (*plural* **-ies**) set of relations, esp. parents and children; person's children; household; all the descendants of common ancestor; group of similar things, people, etc.; group of related genera of animals or plants. □ **family credit** regular state payment to low-income family; **family planning** birth control; **family tree** genealogical chart.

famine /'fæmɪn/ *noun* extreme scarcity, esp. of food.

famish /'fæmɪʃ/ *verb* (usually as **famished** *adjective colloquial*) make or become extremely hungry.

famous /'feɪməs/ *adjective* (often + *for*) well-known; celebrated; *colloquial* excellent. □ **famously** *adverb*.

fan[1] ● *noun* apparatus, usually with rotating blades, for ventilation etc.; device, semicircular and folding, waved to cool oneself; fan-shaped thing. ● *verb*

(**-nn-**) blow air on, (as) with fan; (usually + *out*) spread out like fan. □ **fan belt** belt driving fan to cool radiator in vehicle; **fanlight** small, originally semicircular, window over door etc.; **fantail** pigeon with broad tail.

fan[2] *noun* devotee. □ **fan club** (club of) devotees; **fan mail** letters from fans.

fanatic /fə'nætɪk/ ● *noun* person obsessively devoted to a belief, activity, etc. ● *adjective* excessively enthusiastic. □ **fanatical** *adjective*; **fanatically** *adverb*; **fanaticism** /-tɪsɪz(ə)m/ *noun*.

fancier /'fænsɪə/ *noun* connoisseur, enthusiast; breeder, esp. of pigeons.

fanciful /'fænsɪfʊl/ *adjective* imaginary; indulging in fancies. □ **fancifully** *adverb*.

fancy /'fænsɪ/ ● *noun* (*plural* **-ies**) inclination, whim; supposition; imagination. ● *adjective* (**-ier**, **-iest**) extravagant; ornamental. ● *verb* (**-ies**, **-ied**) (+ *that*) imagine; suppose; *colloquial* find attractive, desire; have unduly high opinion of. □ **fancy dress** costume for masquerading; **fancy-free** without (emotional) commitments; **fancy man, woman** woman's or man's lover. □ **fancily** *adverb*.

fandango /fæn'dæŋgəʊ/ *noun* (*plural* **-es** or **-s**) lively Spanish dance.

fanfare /'fænfeə/ *noun* short showy sounding of trumpets etc.

fang *noun* canine tooth, esp. of dog or wolf; tooth of venomous snake; (prong of) root of tooth.

fantasia /fæn'teɪzɪə/ *noun* free or improvisatory musical etc. composition.

fantasize /'fæntəsaɪz/ *verb* (also **-ise**) (**-zing** or **-sing**) daydream; imagine; create fantasy about.

fantastic /fæn'tæstɪk/ *adjective* extravagantly fanciful; grotesque, quaint; *colloquial* excellent, extraordinary. □ **fantastically** *adverb*.

fantasy /'fæntəsɪ/ *noun* (*plural* **-ies**) imagination, esp. when unrelated to reality; mental image, daydream; fantastic invention or composition.

fanzine /'fænzi:n/ *noun* magazine for fans of science fiction, a football team, etc.

far (**further**, **furthest** or **farther**, **farthest**) ● *adverb* at, to, or by a great distance in space or time; by much. ● *adjective* distant; remote; extreme. □ **far and wide** over large area; **far-away** remote, dreamy, distant; **the Far East** countries of E. Asia; **far-fetched** unconvincing, exaggerated, fanciful; **far-flung** widely scattered, remote; **far from** almost the opposite of; **far gone** very ill, drunk, etc.; **far-off** remote; **far-out** *slang* unconventional, excellent; **far-reaching** widely influential or applicable; **far-seeing** showing foresight; **far-sighted** having foresight, *esp. US* long-sighted.

farad /'færəd/ *noun* SI unit of electrical capacitance.

farce /fɑ:s/ *noun* comedy with ludicrously improbable plot; absurdly futile proceedings; pretence. □ **farcical** *adjective*.

fare ● *noun* price of journey on public transport; passenger; food. ● *verb* (**-ring**) progress; get on. □ **fare-stage** section of bus route for which fixed fare is charged; stop marking this.

farewell /feə'wel/ ● *interjection* goodbye. ● *noun* leave-taking.

farina /fə'ri:nə/ *noun* flour of corn, nuts, or starchy roots. □ **farinaceous** /færɪ'neɪʃəs/ *adjective*.

farm ● *noun* land and its buildings used for growing crops, rearing animals, etc.; farmhouse. ● *verb* use (land) thus; be farmer; breed (fish etc.) commercially; (+ *out*) delegate or subcontract (work). □ **farm-hand** worker on farm; **farmhouse** house attached to farm; **farmyard** yard adjacent to farmhouse. □ **farming** *noun*.

farmer *noun* owner or manager of farm.

faro /'feərəʊ/ *noun* gambling card-game.

farrago /fə'rɑ:gəʊ/ *noun* (*plural* **-s** or *US* **-es**) medley, hotchpotch.

farrier /'færɪə/ *noun* smith who shoes horses.

farrow /'færəʊ/ ● *verb* give birth to (piglets). ● *noun* litter of pigs.

Farsi /'fa:sɪ/ *noun* modern Persian language.

farther = FURTHER.

farthest = FURTHEST.

farthing /'fɑ:ðɪŋ/ *noun historical* coin worth quarter of old penny.

farthingale /'fɑ:ðɪŋgeɪl/ *noun historical* hooped petticoat.

fascia /'feɪʃə/ *noun* (also **facia**) (*plural* **-s**) instrument panel of vehicle; similar panel etc. for operating machinery; long flat surface of wood or stone.

fascicle /'fæsɪk(ə)l/ *noun* instalment of book.

fascinate /'fæsɪneɪt/ *verb* (**-ting**) (often as **fascinating** *adjective*) capture interest of; attract. □ **fascination** *noun*.

Fascism /'fæʃɪz(ə)m/ *noun* extreme right-wing totalitarian nationalist movement in Italy (1922–43); (also **fascism**) any similar movement. □ **Fascist**, **fascist** *noun & adjective*; **Fascistic**, **fascistic** /-'ʃɪs-/ *adjective*.

fashion /'fæʃ(ə)n/ ● *noun* current popular custom or style, esp. in dress; manner of doing something. ● *verb* (often + *into*) form, make. □ **in fashion** fashionable; **out of fashion** not fashionable.

fashionable *adjective* of or conforming to current fashion; of or favoured by high society. □ **fashionably** *adverb*.

fast[1] /fɑːst/ ● *adjective* rapid; capable of or intended for high speed; (of clock) ahead of correct time; firm, fixed, (of colour) not fading; pleasure-seeking. ● *adverb* quickly; firmly; soundly, completely. □ **fastback** car with sloping rear; **fast breeder** (**reactor**) reactor using neutrons with high kinetic energy; **fast food** restaurant food that is produced quickly; **pull a fast one** *colloquial* try to deceive someone.

fast[2] /fɑːst/ ● *verb* abstain from food. ● *noun* act or period of fasting.

fasten /'fɑːs(ə)n/ *verb* make or become fixed or secure; (+ *in*, *up*) shut in, lock securely; (+ *on*) direct (attention) towards; (+ *off*) fix with knot or stitches.

fastener *noun* (also **fastening**) device that fastens.

fastidious /fæ'stɪdɪəs/ *adjective* fussy; easily disgusted, squeamish.

fastness /'fɑːstnɪs/ *noun* stronghold.

fat ● *noun* oily substance, esp. in animal bodies; part of meat etc. containing this. ● *adjective* (**-tt-**) plump; containing much fat; thick, substantial. ● *verb* (**-tt-**) (esp. as **fatted** *adjective*) make or become fat. □ **fat-head** *colloquial* stupid person; **fat-headed** stupid; **a fat lot** *colloquial* very little. □ **fatless** *adjective*; **fatten** *verb*.

fatal /'feɪt(ə)l/ *adjective* causing or ending in death or ruin. □ **fatally** *adverb*.

fatalism *noun* belief in predetermination; submissive acceptance. □ **fatalist** *noun*; **fatalistic** /-'lɪs-/ *adjective*.

fatality /fə'tælɪtɪ/ *noun* (*plural* **-ies**) death by accident, in war, etc.

fate ● *noun* supposed power predetermining events; destiny; death, destruction. ● *verb* (**-ting**) preordain; (as **fated** *adjective*) doomed, (+ *to do*) preordained.

fateful *adjective* decisive; important; controlled by fate.

father /'fɑːðə/ ● *noun* male parent; (usually in *plural*) forefather; originator; early leader; (also **Father**) priest; (**the Father**) God; (in *plural*) elders. ● *verb* be father of; originate. □ **father-in-law** (*plural* **fathers-in-law**) wife's or husband's father; **fatherland** native country. □ **fatherhood** *noun*; **fatherless** *adjective*.

fatherly *adjective* of or like a father.

fathom /'fæð(ə)m/ ● *noun* measure of 6ft, esp. in soundings. ● *verb* comprehend; measure depth of (water). □ **fathomable** *adjective*.

fathomless *adjective* too deep to fathom.

fatigue /fə'tiːg/ ● *noun* extreme tiredness; weakness in metals etc. from repeated stress; non-military army duty, (in *plural*) clothing for this. ● *verb* (**-gues**, **-gued**, **-guing**) cause fatigue in.

fatty ● *adjective* (**-ier**, **-iest**) like or containing fat. ● *noun* (*plural* **-ies**) *colloquial* fat person. □ **fatty acid** type of organic compound.

fatuous /'fætjʊəs/ *adjective* vacantly silly; purposeless. □ **fatuity** /fə'tjuːɪtɪ/ *noun* (*plural* **-ies**); **fatuously** *adverb*; **fatuousness** *noun*.

fatwa /'fætwɑː/ *noun* legal ruling by Islamic religious leader.

faucet /'fɔːsɪt/ *noun esp. US* tap.

fault /fɔːlt/ ● *noun* defect, imperfection; responsibility for wrongdoing, error, etc.; break in electric circuit; *Tennis etc.* incorrect service; break in rock strata. ● *verb* find fault with. □ **find fault** (often + *with*) criticize or complain (about); **to a fault** excessively. □ **faultless** *adjective*; **faultlessly** *adverb*.

faulty *adjective* (**-ier**, **-iest**) having faults. □ **faultily** *adverb*.

faun /fɔ:n/ *noun* Latin rural deity with goat's horns, legs, and tail.

fauna /'fɔ:nə/ *noun* (*plural* **-s**) animal life of a region or period.

faux pas /fəʊ 'pɑ:/ *noun* (*plural* same /'pɑ:z/) tactless mistake. [French]

favour /'feɪvə/ (*US* **favor**) • *noun* kind act; approval, goodwill; partiality; badge, ribbon, etc. as emblem of support. • *verb* regard or treat with favour; support, facilitate; tend to confirm (idea etc.); (+ *with*) oblige; (as **favoured** *adjective*) having special advantages. □ **in favour** approved of, (+ *of*) in support of or to the advantage of; **out of favour** disapproved of.

favourable *adjective* (*US* **favorable**) well-disposed; approving; promising; helpful, suitable. □ **favourably** *adverb*.

favourite /'feɪvərɪt/ (*US* **favorite**) • *adjective* preferred to all others. • *noun* favourite person or thing; competitor thought most likely to win.

favouritism *noun* (*US* **favoritism**) unfair favouring of one person etc.

fawn[1] • *noun* deer in first year; light yellowish brown. • *adjective* fawn-coloured. • *verb* give birth to fawn.

fawn[2] *verb* (often + *on, upon*) behave servilely; (of dog) show extreme affection.

fax • *noun* electronic transmission of exact copy of document etc.; such copy; (in full **fax machine**) apparatus used for this. • *verb* transmit in this way.

faze *verb* (**-zing**) (often as **fazed** *adjective*) *colloquial* disconcert.

FBI *abbreviation* Federal Bureau of Investigation.

FC *abbreviation* Football Club.

FCO *abbreviation* Foreign and Commonwealth Office.

FE *abbreviation* further education.

fealty /'fi:əltɪ/ *noun* (*plural* **-ies**) fidelity to feudal lord; allegiance.

fear • *noun* panic etc. caused by impending danger, pain, etc.; cause of this; alarm, dread. • *verb* be afraid of; (+ *for*) feel anxiety about; dread; shrink from; revere (God).

fearful *adjective* afraid; terrible, awful; extremely unpleasant. □ **fearfully** *adverb*; **fearfulness** *noun*.

fearless *adjective* not afraid; brave. □ **fearlessly** *adverb*; **fearlessness** *noun*.

fearsome *adjective* frightening. □ **fearsomely** *adverb*.

feasible /'fi:zɪb(ə)l/ *adjective* practicable, possible. □ **feasibility** *noun*; **feasibly** *adverb*.

■ **Usage** *Feasible* should not be used to mean 'likely'. *Possible* or *probable* should be used instead.

feast • *noun* sumptuous meal; religious festival; sensual or mental pleasure. • *verb* (often + *on*) have feast, eat and drink sumptuously; regale.

feat *noun* remarkable act or achievement.

feather /'feðə/ • *noun* one of structures forming bird's plumage, with fringed horny shaft; these as material. • *verb* cover or line with feathers; turn (oar) through air edgeways. □ **feather bed** *noun* bed with feather-stuffed mattress; **feather-bed** *verb* cushion, esp. financially; **feather-brained, -headed** silly; **featherweight** amateur boxing weight (54–57 kg). □ **feathery** *adjective*.

feature /'fi:tʃə/ • *noun* characteristic or distinctive part; (usually in *plural*) part of face; specialized article in newspaper etc.; (in full **feature film**) main film in cinema programme. • *verb* (**-ring**) make or be special feature of; emphasize; take part in. □ **featureless** *adjective*.

Feb. *abbreviation* February.

febrile /'fi:braɪl/ *adjective* of fever.

February /'februərɪ/ *noun* (*plural* **-ies**) second month of year.

fecal *US* = FAECAL.

feces *US* = FAECES.

feckless /'feklɪs/ *adjective* feeble, ineffectual; irresponsible.

fecund /'fekənd/ *adjective* fertile. □ **fecundity** /fɪ'kʌndɪtɪ/ *noun*.

fecundate /'fekəndeɪt/ *verb* (**-ting**) make fruitful; fertilize. □ **fecundation** *noun*.

fed *past & past participle* of FEED. □ **fed up** (often + *with*) discontented, bored.

federal /'fedər(ə)l/ *adjective* of system of government in which self-governing states unite for certain functions; of such a federation; (**Federal**) *US* of Northern States in Civil War. □ **federalism** *noun*; **federalist** *noun*; **federalize** *verb* (also **-ise**) (**-zing** or **-sing**); **federalization** *noun*; **federally** *adverb*.

federate ● *verb* /ˈfedəreɪt/ (**-ting**) unite on federal basis. ● *adjective* /ˈfedərət/ federally organized. □ **federative** /-rətɪv/ *adjective*.

federation /fedəˈreɪʃ(ə)n/ *noun* federal group; act of federating.

fee *noun* payment for professional advice or services; (often in *plural*) payment for admission, membership, licence, education, exam, etc.; money paid for transfer of footballer etc. □ **fee-paying** paying fee(s), (of school) charging fees.

feeble /ˈfi:b(ə)l/ *adjective* (**-r**, **-st**) weak; lacking energy, strength, or effectiveness. □ **feebly** *adverb*.

feed ● *verb* (*past & past participle* **fed**) supply with food; put food in mouth of; eat; graze; keep supplied with; (+ *into*) supply (material) to machine etc.; (often + *on*) nourish, be nourished by. ● *noun* food, esp. for animals or infants; feeding; *colloquial* meal. □ **feedback** information about result of experiment, response, *Electronics* return of part of output signal to input.

feeder *noun* person or thing that feeds in specified way; baby's feeding bottle; bib; tributary; branch road or railway line; electricity main supplying distribution point; feeding apparatus in machine.

feel ● *verb* (*past & past participle* **felt**) examine, search, or perceive by touch; experience; be affected by; (+ *that*) have impression; consider, think; seem; be consciously; (+ *for*, *with*) have sympathy or pity for. ● *noun* feeling; sense of touch; sensation characterizing something. □ **feel like** have wish or inclination for; **feel up to** be ready to face or deal with.

feeler *noun* organ in certain animals for touching, foraging, etc. □ **put out feelers** make tentative proposal.

feeling ● *noun* capacity to feel; sense of touch; physical sensation; emotion; (in *plural*) susceptibilities; sensitivity; notion, opinion. ● *adjective* sensitive; sympathetic. □ **feelingly** *adverb*.

feet *plural* of FOOT.

feign /feɪn/ *verb* simulate; pretend.

feint /feɪnt/ ● *noun* sham attack or diversionary blow; pretence. ● *verb* make a feint. ● *adjective* (of paper etc.) having faintly ruled lines.

feldspar /ˈfeldspɑ:/ *noun* (also **felspar** /ˈfelspɑ:/) common aluminium silicate.

felicitation /fəlɪsɪˈteɪʃ(ə)n/ *noun* (usually in *plural*) congratulation.

felicitous /fəˈlɪsɪtəs/ *adjective* apt; well-chosen.

felicity /fəˈlɪsɪtɪ/ *noun* (*plural* **-ies**) *formal* great happiness; capacity for apt expression.

feline /ˈfi:laɪn/ ● *adjective* of cat family; catlike. ● *noun* animal of cat family.

fell[1] *past* of FALL.

fell[2] *verb* cut down (tree); strike down.

fell[3] *noun* hill or stretch of hills in N. England.

fell[4] *adjective* □ **at, in one fell swoop** in a single (originally deadly) action.

fell[5] *noun* animal's hide or skin with hair.

fellow /ˈfeləʊ/ ● *noun* comrade, associate; counterpart, equal; *colloquial* man, boy; incorporated senior member of college; member of learned society. ● *adjective* of same group etc. □ **fellow-feeling** sympathy; **fellow-traveller** person who travels with another, sympathizer with Communist party.

fellowship /ˈfeləʊʃɪp/ *noun* friendly association, companionship; group of associates; position or income of college fellow.

felon /ˈfelən/ *noun* person who has committed felony.

felony /ˈfelənɪ/ *noun* (*plural* **-ies**) serious usually violent crime. □ **felonious** /fɪˈləʊ-/ *adjective*.

felspar = FELDSPAR.

felt[1] ● *noun* fabric of matted and pressed fibres of wool etc. ● *verb* make into or cover with felt; become matted. □ **felt tip** or **pen**, **felt-tip(ped) pen** pen with fibre point.

felt[2] *past & past participle* of FEEL.

female /'fi:meɪl/ ● *adjective* of the sex that can give birth or produce eggs; (of plants) fruit-bearing; of female people, animals, or plants; (of screw, socket, etc.) hollow to receive inserted part. ● *noun* female person, animal, or plant.

feminine /'femɪnɪn/ *adjective* of women; womanly; *Grammar* (of noun) belonging to gender including words for most female people and animals. □ **femininity** /-'nɪn-/ *noun.*

feminism /'femɪnɪz(ə)m/ *noun* advocacy of women's rights and sexual equality. □ **feminist** *noun & adjective.*

femme fatale /fæm fæ'tɑ:l/ *noun* (*plural* ***femmes fatales*** same pronunciation) dangerously seductive woman. [French]

femur /'fi:mə/ *noun* (*plural* **-s** or **femora** /'femərə/) thigh-bone. □ **femoral** /'femər(ə)l/ *adjective.*

fen *noun* low marshy land.

fence ● *noun* barrier or railing enclosing field, garden, etc.; jump for horses; *slang* receiver of stolen goods. ● *verb* (**-cing**) surround (as) with fence; practise sword play; be evasive; deal in (stolen goods). □ **fencer** *noun.*

fencing *noun* fences, material for fences; sword-fighting, esp. as sport.

fend *verb* (+ *off*) ward off; (+ *for*) look after (esp. oneself).

fender *noun* low frame round fireplace; matting etc. to protect side of ship; *US* bumper of vehicle.

fennel /'fen(ə)l/ *noun* fragrant plant with edible leaf-stalks and seeds.

fenugreek /'fenju:gri:k/ *noun* leguminous plant with aromatic seeds used for flavouring.

feral /'fer(ə)l/ *adjective* wild; (of animal) escaped and living wild.

ferial /'fɪerɪəl/ *adjective* (of day) not a church festival or fast.

ferment ● *noun* /'fɜ:ment/ excitement; fermentation; fermenting agent. ● *verb* /fə'ment/ undergo or subject to fermentation; excite.

fermentation /fɜ:men'teɪʃ(ə)n/ *noun* breakdown of substance by yeasts, bacteria, etc.; excitement.

fern *noun* flowerless plant usually with feathery fronds.

ferocious /fə'rəʊʃəs/ *adjective* fierce. □ **ferociously** *adverb*; **ferocity** /-'rɒs-/ *noun.*

ferret /'ferɪt/ ● *noun* small polecat used in catching rabbits, rats, etc. ● *verb* (**-t-**) hunt with ferrets; (often + *out*, *about*, etc.) rummage; (+ *out*) search out.

ferric /'ferɪk/ *adjective* of iron; containing iron in trivalent form.

Ferris wheel /'ferɪs/ *noun* tall revolving vertical wheel with passenger cars in fairgrounds etc.

ferroconcrete /ferəʊ'kɒŋkri:t/ *noun* reinforced concrete.

ferrous /'ferəs/ *adjective* containing iron, esp. in divalent form.

ferrule /'feru:l/ *noun* ring or cap on end of stick etc.

ferry ● *noun* (*plural* **-ies**) boat etc. for esp. regular transport across water; ferrying place or service. ● *verb* (**-ies**, **-ied**) take or go in ferry; transport from place to place, esp. regularly. □ **ferryman** *noun.*

fertile /'fɜ:taɪl/ *adjective* (of soil) abundantly productive; fruitful; (of seed, egg, etc.) capable of growth; inventive; (of animal or plant) able to reproduce. □ **fertility** /-'tɪl-/ *noun.*

fertilize /'fɜ:tɪlaɪz/ *verb* (also **-ise**) (**-zing** or **-sing**) make fertile; cause (egg, female animal, etc.) to develop new individual. □ **fertilization** *noun.*

fertilizer *noun* (also **-iser**) substance added to soil to make it more fertile.

fervent /'fɜ:v(ə)nt/ *adjective* ardent, intense. □ **fervency** *noun*; **fervently** *adverb.*

fervid /'fɜ:vɪd/ *adjective* fervent. □ **fervidly** *adverb.*

fervour /'fɜ:və/ *noun* (*US* **fervor**) passion, zeal.

fescue /'feskju:/ *noun* pasture and fodder grass.

fester /'festə/ *verb* make or become septic; cause continuing bitterness; rot; stagnate.

festival /'festɪv(ə)l/ *noun* day or period of celebration; series of cultural events in town etc.

festive /'festɪv/ *adjective* of or characteristic of festival; joyous.

festivity /fe'stɪvɪtɪ/ *noun* (*plural* **-ies**) gaiety, (in *plural*) celebration; party.

festoon /fe'stu:n/ ● *noun* curved hanging chain of flowers, ribbons, etc. ● *verb* (often + *with*) adorn with or form into festoons.

feta /'fetə/ *noun* salty white Greek cheese made from ewe's or goat's milk.

fetal *US* = FOETAL.

fetch *verb* go for and bring back; be sold for; draw forth; deal (blow). □ **fetch up** *colloquial* arrive, come to rest.

fetching *adjective* attractive. □ **fetchingly** *adverb*.

fête /feɪt/ ● *noun* outdoor fund-raising event. ● *verb* (**-ting**) honour or entertain lavishly.

fetid /'fetɪd/ *adjective* (also **foetid**) stinking.

fetish /'fetɪʃ/ *noun* abnormal object of sexual desire; object worshipped by primitive peoples; object of obsessive concern. □ **fetishism** *noun*; **fetishist** *noun*; **fetishistic** /-'ʃɪs-/ *adjective*.

fetlock /'fetlɒk/ *noun* back of horse's leg where tuft of hair grows above hoof.

fetter /'fetə/ ● *noun* shackle for ankles; (in *plural*) captivity; restraint. ● *verb* put into fetters, restrict.

fettle /'fet(ə)l/ *noun* condition, trim.

fetus *US* = FOETUS.

feud /fju:d/ ● *noun* prolonged hostility, esp. between families, tribes, etc. ● *verb* conduct feud.

feudal /'fju:d(ə)l/ *adjective* of, like, or according to feudal system; reactionary. □ **feudal system** medieval system in which vassal held land in exchange for allegiance and service to landowner. □ **feudalism** *noun*; **feudalistic** /-'lɪs-/ *adjective*.

fever /'fi:və/ ● *noun* abnormally high body temperature; disease characterized by this; nervous agitation. ● *verb* (esp. as **fevered** *adjective*) affect with fever or excitement. □ **fever pitch** state of extreme excitement.

feverfew /'fi:vəfju:/ *noun* aromatic plant, used formerly to reduce fever, now to treat migraine.

feverish *adjective* having symptoms of fever; excited, restless. □ **feverishly** *adverb*.

few ● *adjective* not many. ● *noun* (treated as *plural*) not many; (**a few**) some but not many. □ **a good few** a considerable number (of); **quite a few** *colloquial* a fairly large number (of); **very few** a very small number (of).

fey /feɪ/ *adjective* strange, other-worldly; whimsical.

fez *noun* (*plural* **fezzes**) man's flat-topped conical red cap, worn by some Muslims.

ff *abbreviation Music* fortissimo.

ff. *abbreviation* following pages etc.

fiancé /fɪ'ɒnseɪ/ *noun* (*feminine* **-cée** same pronunciation) person one is engaged to.

fiasco /fɪ'æskəʊ/ *noun* (*plural* **-s**) ludicrous or humiliating failure.

fiat /'faɪæt/ *noun* authorization; decree.

fib ● *noun* trivial lie. ● *verb* (**-bb-**) tell fib. □ **fibber** *noun*.

fibre /'faɪbə/ *noun* (*US* **fiber**) thread or filament forming tissue or textile; piece of threadlike glass; substance formed of fibres; moral character; roughage. □ **fibreboard** board made of compressed wood etc. fibres; **fibreglass** fabric made from woven glass fibres, plastic reinforced with glass fibres; **fibre optics** optics using glass fibres, usually to carry signals. □ **fibrous** *adjective*.

fibril /'faɪbrɪl/ *noun* small fibre.

fibroid /'faɪbrɔɪd/ ● *adjective* of, like, or containing fibrous tissue or fibres. ● *noun* benign fibrous tumour, esp. in womb.

fibrosis /faɪ'brəʊsɪs/ *noun* thickening and scarring of connective tissue.

fibrositis /faɪbrə'saɪtɪs/ *noun* rheumatic inflammation of fibrous tissue.

fibula /'fɪbjʊlə/ *noun* (*plural* **-lae** /-li:/ or **-s**) bone on outer side of lower leg.

fiche /fi:ʃ/ *noun* microfiche.

fickle /'fɪk(ə)l/ *adjective* inconstant, changeable. □ **fickleness** *noun*.

fiction /'fɪkʃ(ə)n/ *noun* non-factual literature, esp. novels; invented idea, thing, etc.; generally accepted falsehood. □ **fictional** *adjective*.

fictitious /fɪk'tɪʃəs/ *adjective* imaginary, unreal; not genuine.

fiddle /'fɪd(ə)l/ ● *noun colloquial* violin; *colloquial* cheat, fraud; fiddly task. ● *verb* (**-ling**) (often + *with, at*) play restlessly;

(often + *about*) move aimlessly; (usually + *with*) tamper, tinker; *slang* falsify, swindle, get by cheating; play fiddle. □ **fiddler** *noun.*

fiddling *adjective* petty, trivial; *colloquial* fiddly.

fiddly *adjective* (**-ier**, **-iest**) *colloquial* awkward to do or use.

fidelity /fɪ'delɪtɪ/ *noun* faithfulness, loyalty; accuracy, precision in sound reproduction.

fidget /'fɪdʒɪt/ ● *verb* (**-t-**) move restlessly; be or make uneasy. ● *noun* person who fidgets; (**the fidgets**) restless state or mood. □ **fidgety** *adjective.*

fiduciary /fɪ'dju:ʃərɪ/ ● *adjective* of a trust, trustee, etc.; held or given in trust; (of currency) dependent on public confidence. ● *noun* (*plural* **-ies**) trustee.

fief /fi:f/ *noun* land held under feudal system.

field /fi:ld/ ● *noun* area of esp. cultivated enclosed land; area rich in some natural product; competitors; expanse of sea, snow, etc.; battlefield; area of activity or study; *Computing* part of record, representing item of data. ● *verb Cricket etc.* act as fielder(s), stop and return (ball); select (player, candidate, etc.); deal with (questions etc.). □ **field-day** exciting or successful time, military exercise or review; **field events** athletic events other than races; **field-glasses** outdoor binoculars; **Field Marshal** army officer of highest rank; **fieldmouse** small long-tailed rodent; **fieldsman** = FIELDER; **field sports** outdoor sports, esp. hunting, shooting, and fishing; **fieldwork** practical surveying, science, sociology, etc. conducted in natural environment; **fieldworker** person doing fieldwork.

fielder *noun Cricket etc.* member (other than bowler) of fielding side.

fieldfare *noun* grey thrush.

fiend /fi:nd/ *noun* evil spirit; wicked or cruel person; mischievous or annoying person; *slang* devotee; difficult or unpleasant thing. □ **fiendish** *adjective*; **fiendishly** *adverb.*

fierce *adjective* (**-r**, **-st**) violently aggressive or frightening; eager; intense. □ **fiercely** *adverb*; **fierceness** *noun.*

fiery /'faɪərɪ/ *adjective* (**-ier**, **-iest**) consisting of or flaming with fire; bright red; burning hot; spirited.

fiesta /fɪ'estə/ *noun* festival, holiday.

FIFA /'fi:fə/ *abbreviation* International Football Federation (*Fédération Internationale de Football Association*).

fife /faɪf/ *noun* small shrill flute.

fifteen /fɪf'ti:n/ *adjective & noun* one more than fourteen; *Rugby* team of fifteen players. □ **fifteenth** *adjective & noun.*

fifth ● *adjective & noun* next after fourth; any of 5 equal parts of thing. □ **fifth column** group working for enemy within country at war.

fifty /'fɪftɪ/ *adjective & noun* (*plural* **-ies**) five times ten. □ **fifty-fifty** half and half, equal(ly). □ **fiftieth** *adjective & noun.*

fig *noun* soft fruit with many seeds; tree bearing it. □ **fig leaf** device concealing genitals.

fig. *abbreviation* figure.

fight /faɪt/ ● *verb* (*past & past participle* **fought** /fɔ:t/) (often + *against*, *with*) contend with in war, combat, etc.; engage in (battle etc.); (+ *for*) strive to secure or on behalf of; contest (election); strive to overcome; (as **fighting** *adjective*) able and eager or trained to fight. ● *noun* combat; boxing match; battle; struggle; power or inclination to fight. □ **fight back** counter-attack, suppress (tears etc.); **fighting chance** chance of success if effort is made; **fighting fit** extremely fit; **fight off** repel with effort; **fight shy of** avoid; **put up a fight** offer resistance.

fighter *noun* person who fights; aircraft designed for attacking other aircraft.

figment /'fɪgmənt/ *noun* imaginary thing.

figurative /'fɪgərətɪv/ *adjective* metaphorical; (of art) not abstract, representational. □ **figuratively** *adverb.*

figure /'fɪgə/ ● *noun* external form, bodily shape; person of specified kind; representation of human form; image; numerical symbol or number, esp. 0–9; value, amount; (in *plural*) arithmetical calculations; diagram, illustration; dance etc. movement or sequence; (in full **figure of speech**) metaphor, hyperbole, etc. ● *verb* (**-ring**) appear, be mentioned; (usually as **figured** *adjective*) embellish with pattern; calculate; *esp. US colloquial* understand, consider, make sense. □ **figurehead** nominal leader, carved image etc. over ship's prow; **figure on** *esp. US colloquial* count on, expect; **figure out** work out by arithmetic or logic; **figure-skating** skating in prescribed patterns.

figurine /fɪgəˈri:n/ *noun* statuette.

filament /ˈfɪləmənt/ *noun* threadlike strand or fibre; conducting wire or thread in electric bulb.

filbert /ˈfɪlbət/ *noun* (nut of) cultivated hazel.

filch *verb* steal, pilfer.

file[1] ● *noun* folder, box, etc. for holding loose papers; paper kept in this; collection of related computer data; row of people or things one behind another. ● *verb* (**-ling**) place in file or among records; submit (petition for divorce etc.); walk in line. □ **filing cabinet** cabinet with drawers for storing files.

file[2] ● *noun* tool with rough surface for smoothing wood, fingernails, etc. ● *verb* (**-ling**) smooth or shape with file.

filial /ˈfɪlɪəl/ *adjective* of or due from son or daughter.

filibuster /ˈfɪlɪbʌstə/ ● *noun* obstruction of progress in legislative assembly; *esp. US* person who engages in this. ● *verb* act as filibuster.

filigree /ˈfɪlɪgri:/ *noun* fine ornamental work in gold etc. wire; similar delicate work.

filings *plural noun* particles rubbed off by file.

Filipino /fɪlɪˈpi:nəʊ/ ● *noun* (*plural* **-s**) native or national of Philippines. ● *adjective* of Philippines or Filipinos.

fill ● *verb* (often + *with*) make or become full; occupy completely; spread over or through; block up (hole, tooth, etc.); appoint to or hold (office etc.); (as **filling** *adjective*) (of food) satisfying. ● *noun* enough to satisfy or fill. □ **fill in** complete (form etc.), fill completely, (often + *for*) act as substitute, *colloquial* inform more fully; **fill out** enlarge or become enlarged, esp. to proper size, fill in (form etc.); **fill up** fill completely, fill petrol tank (of).

filler *noun* material used to fill cavity or increase bulk.

fillet /ˈfɪlɪt/ ● *noun* boneless piece of fish or meat; (in full **fillet steak**) undercut of sirloin; ribbon etc. binding hair; narrow flat band between mouldings. ● *verb* (**-t-**) remove bones from (fish etc.) or divide into fillets; bind or provide with fillet(s).

filling *noun* material used to fill tooth, sandwich, pie, etc. □ **filling-station** garage selling petrol etc.

fillip /ˈfɪlɪp/ *noun* stimulus, incentive.

filly /ˈfɪlɪ/ *noun* (*plural* **-ies**) young female horse.

film ● *noun* thin coating or layer; strip or sheet of plastic etc. coated with light-sensitive emulsion for exposure in camera; story etc. on film; (in *plural*) cinema industry; slight veil, haze, etc. ● *verb* make photographic film of; (often + *over*) cover or become covered (as) with film. □ **film star** well-known film actor or actress. □ **filmy** *adjective* (**-ier**, **-iest**).

filmsetting *noun* typesetting by projecting characters on to photographic film. □ **filmset** *verb*; **filmsetter** *noun*.

Filofax /ˈfaɪləʊfæks/ *noun proprietary term* personal organizer.

filo pastry /ˈfi:ləʊ/ *noun* pastry in very thin sheets.

filter /ˈfɪltə/ ● *noun* porous esp. paper device for removing impurities from liquid or gas or making coffee; screen for absorbing or modifying light; device for suppressing unwanted electrical or sound waves; arrangement for filtering traffic. ● *verb* pass through filter; (usually as **filtered** *adjective*) make (coffee) by dripping hot water through ground beans; (+ *through*, *into*, etc.) make way gradually through, into, etc.; (of traffic) be allowed to turn left or right at junction when other traffic is held up. □ **filter tip** (cigarette with) filter for removing some impurities.

filth *noun* disgusting dirt; obscenity.

filthy ● *adjective* (**-ier**, **-iest**) disgustingly dirty; obscene; (of weather) very unpleasant. ● *adverb colloquial* extremely. □ **filthy lucre** dishonourable gain; money.

filtrate /ˈfɪltreɪt/ ● *verb* (**-ting**) filter. ● *noun* filtered liquid. □ **filtration** *noun*.

fin *noun* organ, esp. of fish, for propelling and steering; similar projection for stabilizing aircraft etc.

finagle /fɪ'neɪg(ə)l/ *verb* (**-ling**) *colloquial* act or obtain dishonestly.

final /'faɪn(ə)l/ ● *adjective* at the end, coming last; conclusive, decisive. ● *noun* last or deciding heat or game; last edition of day's newspaper; (usually in *plural*) exams at end of degree course. □ **finality** /-'næl-/ *noun*; **finally** *adverb*.

finale /fɪ'nɑːlɪ/ *noun* last movement or section of drama, piece of music, etc.

finalist *noun* competitor in final.

finalize *verb* (also **-ise**) (**-zing** or **-sing**) put in final form; complete. □ **finalization** *noun*.

finance /'faɪnæns/ ● *noun* management of money; monetary support for enterprise; (in *plural*) money resources. ● *verb* (**-cing**) provide capital for. □ **financial** /-'næn-/ *adjective*; **financially** /-'næn-/ *adverb*.

financier /faɪ'nænsɪə/ *noun* capitalist; entrepreneur.

finch *noun* small seed-eating bird.

find /faɪnd/ ● *verb* (*past & past participle* **found**) discover, get by chance or after search; become aware of; obtain, provide; summon up; perceive, experience; consider to be; (often in *passive*) discover to be present; reach; *Law* judge and declare. ● *noun* discovery of treasure etc.; valued thing or person newly discovered. □ **all found** (of wages) with board and lodging provided free; **find out** (often + *about*) discover, detect. □ **finder** *noun*.

finding *noun* (often in *plural*) conclusion reached by inquiry.

fine[1] ● *adjective* of high quality; excellent; good, satisfactory; pure, refined; imposing; bright and clear; small or thin, in small particles; smart, showy; flattering. ● *adverb* finely; very well. □ **fine arts** poetry, music, painting, sculpture, architecture, etc.; **fine-spun** delicate, too subtle; **fine-tune** make small adjustments to. □ **finely** *adverb*.

fine[2] ● *noun* money paid as penalty. ● *verb* (**-ning**) punish by fine.

finery /'faɪnərɪ/ *noun* showy dress or decoration.

finesse /fɪ'nes/ ● *noun* refinement; subtlety; artfulness; *Cards* attempt to win trick with card that is not the highest held. ● *verb* (**-ssing**) use or manage by finesse; *Cards* make finesse.

finger /'fɪŋgə/ ● *noun* any of terminal projections of hand (usually excluding thumb); part of glove for finger; finger-like object. ● *verb* touch, turn about, or play with fingers. □ **finger-bowl** bowl for rinsing fingers during meal; **finger-dry** dry and style hair by running fingers through it; **fingernail** nail on each finger; **fingerprint** impression made on surface by fingers, used in detecting crime; **finger-stall** protective cover for injured finger; **fingertip** tip of finger.

fingering *noun* manner of using fingers in music; indication of this in score.

finial /'fɪnɪəl/ *noun* ornamental top to gable, canopy, etc.

finicky /'fɪnɪkɪ/ *adjective* (also **finical**, **finicking**) over-particular; fastidious; detailed; fiddly.

finis /'fɪnɪs/ *noun* end, esp. of book.

finish /'fɪnɪʃ/ ● *verb* (often + *off*) bring or come to end or end of; complete; (often + *off*, *up*) complete consuming; treat surface of. ● *noun* last stage, completion; end of race etc.; method etc. of surface treatment.

finite /'faɪnaɪt/ *adjective* limited, not infinite; (of verb) having specific number and person.

Finn *noun* native or national of Finland.

finnan /'fɪnən/ *noun* (in full **finnan haddock**) smoke-cured haddock.

Finnish ● *adjective* of Finland. ● *noun* language of Finland.

fiord /fjɔːd/ *noun* (also **fjord**) narrow inlet of sea as in Norway.

fir *noun* evergreen conifer with needles growing singly on the stems; its wood. □ **fir-cone** its fruit.

fire ● *noun* state of combustion of substance with oxygen, giving out light and heat; flame, glow; destructive burning; burning fuel in grate etc.; electric or gas heater; firing of guns; fervour, spirit; burning heat. ● *verb* (**-ring**) shoot (gun etc. or missile from it); shoot gun or missile; produce (salute etc.) by shooting; (of gun) be discharged; deliver or utter rapidly; detonate; dismiss (employee); set fire to;

supply with fuel; stimulate, enthuse; undergo ignition; bake or dry (pottery, bricks, etc.). □ **fire-alarm** bell etc. warning of fire; **firearm** gun, esp. pistol or rifle; **fire-ball** large meteor, ball of flame; **fire-bomb** incendiary bomb; **firebrand** piece of burning wood, trouble-maker; **fire-break** obstacle preventing spread of fire in forest etc.; **fire-brick** fireproof brick in grate; **fire brigade** group of firefighters; **firedog** support for logs in hearth; **fire door** fire-resistant door; **fire-drill** rehearsal of procedure in case of fire; **fire-engine** vehicle carrying hoses, firefighters, etc.; **fire-escape** emergency staircase etc. for use in fire; **fire extinguisher** apparatus discharging water, foam, etc. to extinguish fire; **firefighter** person who extinguishes fires; **firefly** insect emitting phosphorescent light, e.g. glow-worm; **fire-irons** tongs, poker, and shovel for domestic fire; **fireman** male firefighter, person who tends steam engine or steamship furnace; **fire-place** place in wall for domestic fire; **fire-power** destructive capacity of guns etc.; **fire-practice** fire-drill; **fire-raiser** arsonist; **fireside** area round fireplace; **fire-screen** ornamental screen for fireplace, screen against direct heat of fire; **fire station** headquarters of local fire brigade; **fire-trap** building without fire-escapes etc.; **firework** device producing flashes, bangs, etc. from burning chemicals, (in *plural*) outburst of anger etc.; **on fire** burning, excited; **under fire** being shot at or criticized.

firing *noun discharge of guns.* □ ***firing line*** *front line in battle, centre of criticism etc.;* ***firing squad*** *soldiers ordered to shoot condemned person.*

firm¹ • *adjective* solid; fixed, steady; resolute; steadfast; (of offer etc.) definite. • *verb* (often + *up*) make or become firm or secure. □ **firmly** *adverb*; **firmness** *noun*.

firm² *noun* business concern, its members.

firmament /'fɜːməmənt/ *noun* sky regarded as vault.

first • *adjective* foremost in time, order, or importance. • *noun* (**the first**) first person or thing; beginning; first occurrence of something notable; first-class degree; first gear; first place in race. • *adverb* before all or something else; for the first time. □ **at first hand** directly from original source; **first aid** emergency medical treatment; **first class** *noun* best category or accommodation, mail given priority, highest division in exam; **first-class** *adjective & adverb* of or by first class, excellent(ly); **first floor** (*US* **second floor**) floor above ground floor; **first-footing** first crossing of threshold in New Year; **firsthand** direct, original; **first mate** (on merchant ship) second in command; **first name** personal or Christian name; **first night** first public performance of play etc.; **first-rate** excellent; **first thing** before anything else, very early. □ **firstly** *adverb*.

firth /fɜːθ/ *noun* inlet, estuary.

fiscal /'fɪsk(ə)l/ • *adjective* of public revenue. • *noun Scottish* procurator fiscal. □ **fiscal year** financial year.

fish¹ • *noun* (*plural* same or **-es**) vertebrate cold-blooded animal living in water; its flesh as food; person of specified kind. • *verb* try to catch fish (in); (+ *for*) search for; seek indirectly; (+ *up*, *out*) retrieve with effort. □ **fish cake** fried cake of fish and mashed potato; **fish-eye lens** wide-angled lens; **fish farm** place where fish are bred for food; **fish finger** small oblong piece of fish in breadcrumbs; **fish-hook** barbed hook for catching fish; **fish-kettle** oval pan for boiling fish; **fish-knife** knife for eating or serving fish; **fish-meal** ground dried fish as fertilizer etc.; **fishmonger** dealer in fish; **fishnet** open-meshed fabric; **fish-slice** slotted cooking utensil; **fishwife** coarse or noisy woman, woman selling fish.

fish² *noun* piece of wood or iron for strengthening mast etc. □ **fish-plate** flat plate of iron etc. holding rails together.

fisherman /'fɪʃəmən/ *noun* man who catches fish as occupation or sport.

fishery *noun* (*plural* **-ies**) place where fish are caught or reared; industry of fishing or breeding fish.

fishing *noun* occupation or sport of trying to catch fish. □ **fishing-rod** tapering rod for fishing.

fishy *adjective* (**-ier**, **-iest**) of or like fish; *slang* dubious, suspect.

fissile /'fɪsaɪl/ *adjective* capable of undergoing nuclear fission; tending to split.

fission /'fɪʃ(ə)n/ ● *noun* splitting of atomic nucleus; division of cell as mode of reproduction. ● *verb* (cause to) undergo fission.

fissure /'fɪʃə/ ● *noun* narrow crack or split. ● *verb* (**-ring**) split.

fist *noun* clenched hand. □ **fisticuffs** fighting with the fists. □ **fistful** *noun* (*plural* **-s**).

fistula /'fɪstjʊlə/ *noun* (*plural* **-s** or **fistulae** /-li:/) abnormal or artificial passage in body. □ **fistular** *adjective*; **fistulous** *adjective*.

fit[1] ● *adjective* (**-tt-**) well suited; qualified, competent; in good health or condition; (+ *for*) good enough, right; ready. ● *verb* (**-tt-**) be of right size and shape; find room for; (often + *in*, *into*) be correctly positioned; (+ *on*, *together*) fix in place; (+ *with*) supply; befit. ● *noun* way thing fits. □ **fit in** (often + *with*) be compatible, accommodate, make room or time for; **fit out**, **up** equip. □ **fitness** *noun*.

fit[2] *noun* sudden esp. epileptic seizure; sudden brief bout or burst.

fitful *adjective* spasmodic, intermittent. □ **fitfully** *adverb*.

fitment *noun* (usually in *plural*) fixed item of furniture.

fitted *adjective* made to fit closely; with built-in fittings; built-in.

fitter *noun* mechanic who fits together and adjusts machinery etc.; supervisor of cutting, fitting, etc. of garments.

fitting ● *noun* action of fitting on a garment; (usually in *plural*) fixture, fitment. ● *adjective* proper, befitting. □ **fittingly** *adverb*.

five *adjective & noun* one more than four. □ **five o'clock shadow** beard growth visible in latter part of day; **five-star** of highest class; **fivestones** jacks played with five pieces of metal etc. and usually no ball.

fiver *noun colloquial* five-pound note.

fives *noun* game in which ball is struck with gloved hand or bat against walls of court.

fix ● *verb* make firm, stable, or permanent; fasten, secure; settle, specify; mend, repair; (+ *on*, *upon*) direct (eyes etc.) steadily on; attract and hold (attention etc.); identify, locate; *US colloquial* prepare (food, drink); *colloquial* kill, deal with (person); arrange result fraudulently. ● *noun* dilemma, predicament; position determined by bearings etc.; *slang* dose of addictive drug; *colloquial* fraudulently arranged result. □ **fix up** arrange, (often + *with*) provide (person). □ **fixer** *noun*.

fixate /fɪk'seɪt/ *verb* (**-ting**) *Psychology* (usually in *passive*, often + *on*, *upon*) cause to become abnormally attached to person or thing.

fixation /fɪk'seɪʃ(ə)n/ *noun* fixating, being fixated; obsession.

fixative /'fɪksətɪv/ ● *adjective* tending to fix (colours etc.). ● *noun* fixative substance.

fixedly /'fɪksɪdlɪ/ *adverb* intently.

fixity *noun* fixed state; stability, permanence.

fixture /'fɪkstʃə/ *noun* thing fixed in position; (date fixed for) sporting event; (in *plural*) articles belonging to land or house.

fizz ● *verb* make hissing or spluttering sound; effervesce. ● *noun* fizzing sound; effervescence; *colloquial* effervescent drink. □ **fizzy** *adjective* (**-ier**, **-iest**).

fizzle /'fɪz(ə)l/ ● *verb* (**-ling**) hiss or splutter feebly. ● *noun* fizzling sound. □ **fizzle out** end feebly.

fjord = FIORD.

fl. *abbreviation* fluid; floruit.

flab *noun colloquial* fat, flabbiness.

flabbergast /'flæbəgɑ:st/ *verb* (esp. as **flabbergasted** *adjective*) *colloquial* astonish; dumbfound.

flabby /'flæbɪ/ *adjective* (**-ier**, **-iest**) (of flesh) limp, not firm; feeble. □ **flabbiness** *noun*.

flaccid /'flæksɪd/ *adjective* flabby. □ **flaccidity** /-'sɪd-/ *noun*.

flag[1] ● *noun* piece of cloth attached by one edge to pole or rope as country's emblem, standard, or signal. ● *verb* (**-gg-**) grow tired; lag; droop; mark out with flags. □ **flag-day** day when charity collects money and gives stickers to contributors; **flag down** signal to stop; **flag-officer** admiral or vice or rear admiral; **flag-pole** flagstaff; **flagship** ship with admiral on board, leading example of thing; **flagstaff** pole on which flag is hung; **flag-waving** *noun* populist agitation, chauvinism, *adjective* chauvinistic.

flag[2] ● *noun* (also **flagstone**) flat paving stone. ● *verb* (**-gg-**) pave with flags.

flag[3] *noun* plant with bladed leaf, esp. iris.

flagellant /ˈflædʒələnt/ ● *noun* person who flagellates himself, herself, or others. ● *adjective* of flagellation.

flagellate /ˈflædʒəleɪt/ *verb* (**-ting**) whip or flog, esp. as religious discipline or sexual stimulus. □ **flagellation** *noun*.

flageolet /flædʒəˈlet/ *noun* small flute blown at end.

flagon /ˈflægən/ *noun* quart bottle or other vessel for wine etc.

flagrant /ˈfleɪgrənt/ *adjective* blatant; scandalous. □ **flagrancy** *noun*; **flagrantly** *adverb*.

flagrante see IN FLAGRANTE DELICTO.

flail ● *verb* (often + *about*) wave or swing wildly; beat (as) with flail. ● *noun* staff with heavy stick swinging from it, used for threshing.

flair *noun* natural talent; style, finesse.

flak *noun* anti-aircraft fire; criticism; abuse.

flake ● *noun* thin light piece of snow etc.; thin broad piece peeled or split off. ● *verb* (**-king**) (often + *away*, *off*) take or come away in flakes; fall in or sprinkle with flakes. □ **flake out** fall asleep or drop from exhaustion, faint.

flaky *adjective* (**-ier**, **-iest**) of, like, or in flakes. □ **flaky pastry** lighter version of puff pastry.

flambé /ˈflɒmbeɪ/ *adjective* (of food) covered with alcohol and set alight briefly.

flamboyant /flæmˈbɔɪənt/ *adjective* ostentatious, showy; florid. □ **flamboyance** *noun*; **flamboyantly** *adverb*.

flame ● *noun* ignited gas; portion of this; bright light; brilliant orange colour; passion, esp. love; *colloquial* sweetheart. ● *verb* (**-ming**) (often + *out*, *up*) burn; blaze; (of passion) break out; become angry; shine or glow like flame.

flamenco /fləˈmeŋkəʊ/ *noun* (*plural* **-s**) Spanish Gypsy guitar music with singing and dancing.

flaming *adjective* emitting flames; very hot; passionate; brightly coloured; *colloquial expressing annoyance*.

flamingo /fləˈmɪŋgəʊ/ *noun* (*plural* **-s** or **-es**) tall long-necked wading bird with usually pink plumage.

flammable /ˈflæməb(ə)l/ *adjective* inflammable.

■ **Usage** *Flammable* is often used because *inflammable* could be taken to mean 'not flammable'. The negative of *flammable* is *non-flammable*.

---

flan *noun* pastry case with savoury or sweet filling; sponge base with sweet topping.

flange /flændʒ/ *noun* projecting flat rim, for strengthening etc.

flank ● *noun* side of body between ribs and hip; side of mountain, army, etc. ● *verb* (often as **flanked** *adjective*) be at or move along side of.

flannel /ˈflæn(ə)l/ ● *noun* woven woollen usually napless fabric; (in *plural*) flannel trousers; face-cloth; *slang* nonsense, flattery. ● *verb* (**-ll-**; *US* **-l-**) *slang* flatter; wash with flannel.

flannelette /flænəˈlet/ *noun* napped cotton fabric like flannel.

flap ● *verb* (**-pp-**) move or be moved loosely up and down; beat; flutter; *colloquial* be agitated or panicky; *colloquial* (of ears) listen intently. ● *noun* piece of cloth, wood, etc. attached by one side, esp. to cover gap; flapping; *colloquial* agitation; aileron. □ **flapjack** sweet oatcake, *esp. US* small pancake.

flapper *noun* person apt to panic; *slang* (in 1920s) young unconventional woman.

flare ● *verb* (**-ring**) blaze with bright unsteady flame; (usually as **flared** *adjective*) widen gradually; burst out, esp. angrily. ● *noun* bright unsteady flame, outburst of this; flame or bright light as signal etc.; gradual widening; (in *plural*) wide-bottomed trousers. □ **flare-path** line of lights on runway to guide aircraft; **flare up** *verb* burst into blaze, anger, activity, etc.; **flare-up** *noun* outburst.

flash ● *verb* (cause to) emit brief or sudden light; gleam; send or reflect like sudden flame; burst suddenly into view etc.; move swiftly; send (news etc.) by radio etc.; signal (to) with vehicle lights; show ostentatiously; *slang* indecently expose oneself. ● *noun* sudden bright light or flame; an instant; sudden brief feeling or display; newsflash; *Photography* flashlight. ● *adjective colloquial* gaudy, showy; vulgar. □ **flashback** scene set in earlier time than main action; **flash bulb** *Photography* bulb for flashlight; **flash-gun** device operating camera flashlight; **flash in the pan** promising start followed by failure; **flash-lamp** portable flashing electric lamp; **flashlight** *Photography* light giving intense flash, *US* electric torch; **flashpoint** temperature at which vapour from oil etc. will ignite in air, point at which anger is expressed.

flasher *noun slang* man who indecently exposes himself; automatic device for switching lights rapidly on and off.

flashing *noun* (usually metal) strip to prevent water penetration at roof joint etc.

flashy *adjective* (**-ier**, **-iest**) showy; cheaply attractive. □ **flashily** *adverb.*

flask /flɑːsk/ *noun* narrow-necked bulbous bottle; hip-flask; vacuum flask.

flat[1] ● *adjective* (**-tt-**) horizontally level; smooth, even; level and shallow; downright; dull; dejected; having lost its effervescence; (of battery) having exhausted its charge; (of tyre etc.) deflated; *Music* below correct or normal pitch, having flats in key signature. ● *adverb* spread out; completely, exactly; flatly. ● *noun* flat part; level esp. marshy ground; *Music* note lowered by semitone, sign (♭) indicating this; flat theatre scenery on frame; punctured tyre; (**the flat**) flat racing, its season. □ **flatfish** fish with flattened body, e.g. sole, plaice; **flat foot** foot with flattened arch; **flat-footed** having flat feet, *colloquial* uninspired; **flat out** at top speed, using all one's strength etc.; **flat race** horse race over level ground without jumps; **flat rate** unvarying rate or charge; **flat spin** aircraft's nearly horizontal spin, *colloquial* state of panic; **flatworm** worm with flattened body, e.g. fluke. □ **flatly** *adverb*; **flatness** *noun*; **flattish** *adjective.*

flat[2] *noun* set of rooms, usually on one floor, as residence. □ **flatmate** person sharing flat. □ **flatlet** *noun.*

flatten *verb* make or become flat; *colloquial* humiliate; knock down.

flatter /ˈflætə/ *verb* compliment unduly; enhance appearance of; (usually **flatter oneself**; usually + *that*) congratulate or delude (oneself etc.). □ **flatterer** *noun*; **flattering** *adjective*; **flatteringly** *adverb*; **flattery** *noun.*

flatulent /ˈflætjʊlənt/ *adjective* causing, caused by, or troubled with, intestinal wind; inflated, pretentious. □ **flatulence** *noun.*

flaunt /flɔːnt/ *verb* display proudly; show off, parade.

■ **Usage** *Flaunt* is often confused with *flout*, which means 'to disobey contemptuously'.

---

flautist /ˈflɔːtɪst/ *noun* flute-player.

flavour /ˈfleɪvə/ (*US* **flavor**) ● *noun* mixed sensation of smell and taste; distinctive taste, characteristic quality. ● *verb* give flavour to, season. □ **flavour of the month** temporary trend or fashion. □ **flavourless** *adjective*; **flavoursome** *adjective.*

flavouring *noun* (*US* **flavoring**) substance used to flavour food or drink.

flaw ● *noun* imperfection; blemish; crack; invalidating defect. ● *verb* damage; spoil; (as **flawed** *adjective*) morally etc. defective. □ **flawless** *adjective.*

flax *noun* blue-flowered plant grown for its oily seeds (linseed) and for making into linen.

flaxen *adjective* of flax; (of hair) pale yellow.

flay *verb* strip skin or hide off; criticize severely; peel off.

flea *noun* small wingless jumping parasitic insect. □ **flea market** street market selling second-hand goods etc.

fleck ● *noun* small patch of colour or light; speck. ● *verb* mark with flecks.

flection *US* = FLEXION.

fled *past & past participle* of FLEE.

fledgling /ˈfledʒlɪŋ/ (also **fledgeling**) ● *noun* young bird. ● *adjective* new, inexperienced.

flee *verb* (**flees**; *past & past participle* **fled**) run away (from); leave hurriedly.

fleece • *noun* woolly coat of sheep etc.; this shorn from sheep; fleecy lining etc. • *verb* (**-cing**) (often + *of*) strip of money etc., swindle; shear; (as **fleeced** *adjective*) cover as with fleece. □ **fleecy** *adjective* (**-ier**, **-iest**).

fleet • *noun* warships under one commander-in-chief; (**the fleet**) navy; vehicles in one company etc. • *adjective* swift, nimble.

fleeting *adjective* transitory; brief. □ **fleetingly** *adverb*.

Fleming /ˈflemɪŋ/ *noun* native of medieval Flanders; member of Flemish-speaking people of N. and W. Belgium.

Flemish /ˈflemɪʃ/ • *adjective* of Flanders. • *noun* language of the Flemings.

flesh *noun* soft substance between skin and bones; plumpness, fat; body, esp. as sinful; pulpy substance of fruit etc.; (also **flesh-colour**) yellowish pink colour. □ **flesh and blood** human body, human nature, esp. as fallible, humankind, near relations; **flesh out** make or become substantial; **fleshpots** luxurious living; **flesh-wound** superficial wound; **in the flesh** in person. □ **fleshy** *adjective* (**-ier**, **-iest**).

fleshly /ˈfleʃlɪ/ *adjective* worldly; carnal.

fleur-de-lis /flɜːdəˈliː/ *noun* (also **fleur-de-lys**) (*plural* **fleurs-** same pronunciation) iris flower; *Heraldry* lily of 3 petals; former royal arms of France.

flew *past* of FLY[1].

flews *plural noun* hanging lips of bloodhound etc.

flex[1] *verb* bend (joint, limb); move (muscle) to bend joint.

flex[2] *noun* flexible insulated cable.

flexible /ˈfleksɪb(ə)l/ *adjective* able to bend without breaking; pliable; adaptable. □ **flexibility** *noun*; **flexibly** *adverb*.

flexion /ˈflekʃ(ə)n/ *noun* (*US* **flection**) bending; bent part.

flexitime /ˈfleksɪtaɪm/ *noun* system of flexible working hours.

flibbertigibbet /flɪbətɪˈdʒɪbɪt/ *noun* gossiping, frivolous, or restless person.

flick • *noun* light sharp blow; sudden release of bent finger etc. to propel thing; jerk; *colloquial* cinema film, (**the flicks**) the cinema. • *verb* (often + *away*, *off*) strike or move with flick. □ **flick-knife** knife with blade that springs out; **flick through** glance at or through by turning over (pages etc.) rapidly.

flicker /ˈflɪkə/ • *verb* shine or burn unsteadily; flutter; waver. • *noun* flickering light or motion; brief feeling (of hope etc.); slightest reaction or degree. □ **flicker out** die away.

flier = FLYER.

flight[1] /flaɪt/ *noun* act or manner of flying; movement, passage, or journey through air or space; timetabled airline journey; flock of birds etc.; (usually + *of*) series of stairs etc.; imaginative excursion; volley; tail of dart. □ **flight bag** small zipped shoulder bag for air travel; **flight-deck** cockpit of large aircraft, deck of aircraft carrier; **flight lieutenant** RAF officer next below squadron leader; **flight path** planned course of aircraft etc.; **flight recorder** device in aircraft recording technical details of flight; **flight sergeant** RAF rank next above sergeant.

flight[2] /flaɪt/ *noun* fleeing; hasty retreat.

flightless *adjective* (of bird etc.) unable to fly.

flighty *adjective* (**-ier**, **-iest**) frivolous; fickle.

flimsy /ˈflɪmzɪ/ *adjective* (**-ier**, **-iest**) insubstantial, rickety; unconvincing; (of clothing) thin. □ **flimsily** *adverb*.

flinch *verb* draw back in fear etc.; shrink; wince.

fling • *verb* (*past & past participle* **flung**) throw, hurl; rush, esp. angrily; discard rashly. • *noun* flinging; throw; bout of wild behaviour; whirling Scottish dance.

flint *noun* hard grey stone; piece of this, esp. as tool or weapon; piece of hard alloy used to produce spark. □ **flintlock** old type of gun fired by spark from flint. □ **flinty** *adjective* (**-ier**, **-iest**).

flip • *verb* (**-pp-**) toss (coin etc.) so that it spins in air; turn or flick (small object) over; *slang* flip one's lid. • *noun* act of flipping. • *adjective* glib; flippant. □ **flip chart** large pad of paper on stand; **flip-flop** sandal with thong between toes; **flip one's lid** *slang* lose self-control, go mad; **flip side** reverse side of gramophone record etc.; **flip through** flick through.

flippant /ˈflɪpənt/ *adjective* frivolous; disrespectful. □ **flippancy** *noun*; **flippantly** *adverb*.

flipper *noun* limb of turtle, penguin, etc., used in swimming; rubber attachment to foot for underwater swimming.

flirt • *verb* try to attract sexually but without serious intent; (usually + *with*) superficially engage in, trifle. • *noun* person who flirts. □ **flirtation** *noun*; **flirtatious** *adjective*; **flirtatiously** *adverb*; **flirtatiousness** *noun*.

flit • *verb* (**-tt-**) pass lightly or rapidly; make short flights; disappear secretly, esp. to escape creditors. • *noun* act of flitting.

flitch *noun* side of bacon.

flitter *verb* flit about.

float • *verb* (cause to) rest or drift on surface of liquid; move or be suspended freely in liquid or gas; launch (company, scheme); offer (stocks, shares, etc.) on stock market; (cause or allow to) have fluctuating exchange rate; circulate or cause (rumour, idea) to circulate. • *noun* device or structure that floats; electrically powered vehicle or cart; decorated platform or tableau on lorry in procession etc.; supply of loose change, petty cash.

floating *adjective* not settled; variable; not committed. □ **floating rib** lower rib not attached to breastbone.

floaty *adjective* (**-ier**, **-iest**) (of fabric) light and airy.

flocculent /ˈflɒkjʊlənt/ *adjective* like tufts of wool.

flock[1] • *noun* animals, esp. birds or sheep, as group or unit; large crowd of people; people in care of priest, teacher, etc. • *verb* (often + *to*, *in*, *out*, *together*) congregate; mass; troop.

flock[2] *noun* shredded wool, cotton, etc. used as stuffing; powdered wool used to make pattern on wallpaper.

floe *noun* sheet of floating ice.

flog *verb* (**-gg-**) beat with whip, stick, etc.; (often + *off*) *slang* sell.

flood /flʌd/ • *noun* overflowing or influx of water, esp. over land; outburst, outpouring; inflow of tide; (**the Flood**) the flood described in Genesis. • *verb* overflow; cover or be covered with flood; irrigate; deluge; come in great quantities; overfill (carburettor) with petrol. □ **floodgate** gate for admitting or excluding water, (usually in *plural*) last restraint against tear, anger, etc.; **floodlight** (illuminate with) large powerful light; **floodlit** lit thus.

floor /flɔː/ • *noun* lower surface of room; bottom of sea, cave, etc.; storey; part of legislative chamber where members sit and speak; right to speak in debate; minimum of prices, wages, etc. • *verb* provide with floor; knock down; baffle; overcome. □ **floor manager** stage manager of television production; **floor plan** diagram of rooms etc. on one storey; **floor show** cabaret.

flop • *verb* (**-pp-**) sway about heavily or loosely; (often + *down*, *on*, *into*) move, fall, sit, etc. awkwardly or suddenly; *slang* collapse, fail; make dull soft thud or splash. • *noun* flopping motion or sound; *slang* failure. • *adverb* with a flop.

floppy • *adjective* (**-ier**, **-iest**) tending to flop; flaccid. • *noun* (*plural* **-ies**) (in full **floppy disk**) flexible disc for storage of computer data.

flora /ˈflɔːrə/ *noun* (*plural* **-s** or **-rae** /-riː/) plant life of region or period.

floral *adjective* of or decorated with flowers.

floret /ˈflɒrɪt/ *noun* each of small flowers of composite flower head; each stem of head of cauliflower, broccoli, etc.

florid /ˈflɒrɪd/ *adjective* ruddy; ornate; showy.

florin /ˈflɒrɪn/ *noun historical* gold or silver coin, esp. British two-shilling coin.

florist /ˈflɒrɪst/ *noun* person who deals in or grows flowers.

floruit /ˈflɒrʊɪt/ *verb* (of painter, writer, etc.) lived and worked.

floss • *noun* rough silk of silkworm's cocoon; dental floss. • *verb* clean (teeth) with dental floss. □ **flossy** *adjective* (**-ier**, **-iest**).

flotation /fləʊˈteɪʃ(ə)n/ *noun* launching of commercial enterprise etc.

flotilla /fləˈtɪlə/ *noun* small fleet; fleet of small ships.

flotsam /ˈflɒtsəm/ *noun* floating wreckage. □ **flotsam and jetsam** odds and ends, vagrants.

flounce[1] /flaʊns/ ● *verb* (**-cing**) (often + *off*, *out*, etc.) go or move angrily or impatiently. ● *noun* flouncing movement.

flounce[2] /flaʊns/ ● *noun* frill on dress, skirt, etc. ● *verb* (**-cing**) (usually as **flounced** *adjective*) trim with flounces.

flounder[1] /'flaʊndə/ *verb* struggle helplessly; do task clumsily.

flounder[2] /'flaʊndə/ *noun* (*plural* same) small flatfish.

flour /flaʊə/ ● *noun* meal or powder from ground wheat etc. ● *verb* sprinkle with flour. □ **floury** *adjective* (**-ier**, **-iest**).

flourish /'flʌrɪʃ/ ● *verb* grow vigorously; thrive, prosper; wave, brandish. ● *noun* showy gesture; ornamental curve in writing; *Music* ornate passage; fanfare.

flout *verb* disobey (law etc.) contemptuously.

■ **Usage** *Flout* is often confused with *flaunt*, which means 'to display proudly or show off'.

---

flow /fləʊ/ ● *verb* glide along, move smoothly; gush out; circulate; be plentiful or in flood; (often + *from*) result. ● *noun* flowing movement or liquid; stream; rise of tide. □ **flow chart**, **diagram**, **sheet** diagram of movement or action in complex activity.

flower /'flaʊə/ ● *noun* part of plant from which seed or fruit develops; plant bearing blossom. ● *verb* bloom; reach peak. □ **flower-bed** garden bed for flowers; **the flower of** the best of; **flowerpot** pot for growing plant in; **in flower** blooming. □ **flowered** *adjective*.

flowery *adjective* florally decorated; (of speech etc.) high-flown; full of flowers.

flowing *adjective* fluent; smoothly continuous; (of hair etc.) unconfined.

flown *past participle* of FLY[1].

flu *noun colloquial* influenza.

fluctuate /'flʌktʃʊeɪt/ *verb* (**-ting**) vary, rise and fall. □ **fluctuation** *noun*.

flue *noun* smoke duct in chimney; channel for conveying heat.

fluent /'flu:ənt/ *adjective* expressing oneself easily and naturally, esp. in foreign language. □ **fluency** *noun*; **fluently** *adverb*.

fluff ● *noun* soft fur, feathers, fabric particles, etc.; *slang* mistake in performance etc. ● *verb* (often + *up*) shake into or become soft mass; *slang* make mistake in performance etc. □ **fluffy** *adjective* (**-ier**, **-iest**).

flugelhorn /'flu:g(ə)lhɔ:n/ *noun* brass instrument like cornet.

fluid /'flu:ɪd/ ● *noun* substance, esp. gas or liquid, capable of flowing freely; liquid secretion. ● *adjective* able to flow freely; constantly changing. □ **fluid ounce** twentieth or *US* sixteenth of pint. □ **fluidity** /-'ɪdɪtɪ/ *noun*.

fluke[1] /flu:k/ *noun* lucky accident. □ **fluky** *adjective* (**-ier**, **-iest**).

fluke[2] /flu:k/ *noun* parasitic flatworm.

flummery /'flʌmərɪ/ *noun* (*plural* **-ies**) nonsense; flattery; sweet milk dish.

flummox /'flʌməks/ *verb colloquial* bewilder.

flung *past & past participle* of FLING.

flunk *verb US colloquial* fail (esp. exam).

flunkey /'flʌŋkɪ/ *noun* (also **flunky**) (*plural* **-eys** or **-ies**) *usually derogatory* footman.

fluorescence /flʊə'res(ə)ns/ *noun* light radiation from certain substances; property of absorbing invisible light and emitting visible light. □ **fluoresce** *verb* (**-scing**); **fluorescent** *adjective*.

fluoridate /'flʊərɪdeɪt/ *verb* (**-ting**) add fluoride to (water). □ **fluoridation** *noun*.

fluoride /'flʊəraɪd/ *noun* compound of fluorine with metal, esp. used to prevent tooth decay.

fluorinate /'flʊərɪneɪt/ *verb* (**-ting**) fluoridate; introduce fluorine into (compound). □ **fluorination** *noun*.

fluorine /'flʊəri:n/ *noun* poisonous pale yellow gaseous element.

flurry /'flʌrɪ/ ● *noun* (*plural* **-ies**) gust, squall; burst of activity, excitement, etc. ● *verb* (**-ies**, **-ied**) agitate, confuse.

flush[1] ● *verb* (often as **flushed** *adjective* + *with*) (cause to) glow or blush (with pride etc.); cleanse (drain, lavatory, etc.) by flow of water; (often + *away*, *down*) dispose of thus. ● *noun* glow, blush; rush of water; cleansing (of lavatory etc.) thus; rush of esp. elation or triumph; freshness, vigour; (also **hot flush**) sudden hot feeling during menopause; feverish redness or temperature etc. ● *adjective* level, in same plane; *colloquial* having plenty of money.

flush[2] *noun* hand of cards all of one suit.

flush[3] *verb* (cause to) fly up suddenly. □ **flush out** reveal, drive out.

fluster /'flʌstə/ ● *verb* confuse, make nervous. ● *noun* confused or agitated state.

flute /flu:t/ ● *noun* high-pitched woodwind instrument held sideways; vertical groove in pillar etc. ● *verb* (**-ting**) (often as **fluted** *adjective*) make grooves in; play on flute. □ **fluting** *noun*.

flutter /'flʌtə/ ● *verb* flap (wings) in flying or trying to fly; fall quiveringly; wave or flap quickly; move about restlessly; (of pulse etc.) beat feebly or irregularly. ● *noun* fluttering; tremulous excitement; *slang* small bet on horse etc.; abnormally rapid heartbeat; rapid variation of pitch, esp. of recorded sound.

fluvial /'flu:vɪəl/ *adjective* of or found in rivers.

flux *noun* flowing, flowing out; discharge; continuous change; substance mixed with metal etc. to assist fusion.

fly[1] /flaɪ/ ● *verb* (**flies**; *past* **flew** /flu:/; *past participle* **flown** /fləʊn/) move or travel through air with wings or in aircraft; control flight of (aircraft); cause to fly or remain aloft; wave; move swiftly; be driven forcefully; flee (from); (+ *at*, *upon*) attack or criticize fiercely. ● *noun* (*plural* **-ies**) (usually in *plural*) flap to cover front fastening on trousers, this fastening; flap at entrance of tent; (in *plural*) space over stage, for scenery and lighting; act of flying. □ **fly-by-night** unreliable; **fly-half** *Rugby* stand-off half; **flyleaf** blank leaf at beginning or end of book; **flyover** bridge carrying road etc. over another; **fly-past** ceremonial flight of aircraft; **fly-post** fix (posters etc.) illegally; **flysheet** canvas cover over tent for extra protection, short tract or circular; **fly-tip** illegally dump (waste); **flywheel** heavy wheel regulating machinery or accumulating power.

fly[2] /flaɪ/ (*plural* **flies**) *noun* two-winged insect; disease caused by flies; (esp. artificial) fly used as bait in fishing. □ **fly-blown** tainted by flies; **flycatcher** bird that catches flies in flight; **fly-fish** *verb* fish with fly; **fly in the ointment** minor irritation or setback; **fly on the wall** unnoticed observer; **fly-paper** sticky treated paper for catching flies; **fly-trap** plant that catches flies; **flyweight** amateur boxing weight (48–51 kg).

fly[3] /flaɪ/ *adjective slang* knowing, clever.

flyer *noun* (also **flier**) airman, airwoman; thing or person that flies in specified way; small handbill.

flying ● *adjective* that flies; hasty. ● *noun* flight. □ **flying boat** boatlike seaplane; **flying buttress** (usually arched) buttress running from upper part of wall to outer support; **flying doctor** doctor who uses aircraft to visit patients; **flying fish** fish gliding through air with winglike fins; **flying fox** fruit-eating bat; **flying officer** RAF officer next below flight lieutenant; **flying saucer** supposed alien spaceship; **flying squad** rapidly mobile police detachment, midwifery unit, etc.; **flying start** start (of race) at full speed, vigorous start.

FM *abbreviation* Field Marshal; frequency modulation.

FO *abbreviation* Flying Officer.

foal ● *noun* young of horse or related animal. ● *verb* give birth to (foal).

foam ● *noun* froth formed in liquid; froth of saliva or sweat; spongy rubber or plastic. ● *verb* emit foam; froth. □ **foam at the mouth** be very angry. □ **foamy** *adjective* (**-ier**, **-iest**).

fob[1] *noun* (attachment to) watch-chain; small pocket for watch etc.

fob[2] *verb* (**-bb-**) □ **fob off** (often + *with*) deceive into accepting something inferior, (often + *on*, *on to*) offload (unwanted thing on person).

focal /'fəʊk(ə)l/ *adjective* of or at a focus. □ **focal distance**, **length** distance between centre of lens etc. and its focus; **focal point** focus, centre of interest or activity.

fo'c's'le = FORECASTLE.

focus /'fəʊkəs/ ● *noun* (*plural* **focuses** or **foci** /'fəʊsaɪ/) point at which rays etc. meet after reflection or refraction or from which rays etc. seem to come; point at which object must be situated to give clearly defined image; adjustment of eye or lens to produce clear image; state of clear definition; centre of interest or activity. ● *verb* (**-s-** or **-ss-**) bring into focus; adjust focus of (lens, eye); concentrate or be concentrated on; (cause to) converge to focus.

fodder /'fɒdə/ *noun* hay, straw, etc. as animal food.

foe *noun* enemy.

foetid = FETID.

foetus /'fi:təs/ *noun* (*US* **fetus**) (*plural* **-tuses**) unborn mammalian offspring, esp. human embryo of 8 weeks or more. □ **foetal** *adjective*.

fog ● *noun* thick cloud of water droplets or smoke suspended at or near earth's surface; thick mist; cloudiness on photographic negative; confused state. ● *verb* (**-gg-**) envelop (as) in fog; perplex. □ **fog-bank** mass of fog at sea; **foghorn** horn warning ships in fog, *colloquial* penetrating voice.

fogey /'fəʊgɪ/ *noun* (also **fogy**) (*plural* **-ies** or **-eys**) (esp. **old fogey**) dull old-fashioned person.

foggy *adjective* (**-ier**, **-iest**) full of fog; of or like fog; vague. □ **not the foggiest** *colloquial* no idea.

foible /'fɔɪb(ə)l/ *noun* minor weakness or idiosyncrasy.

foil[1] *verb* frustrate, defeat.

foil[2] *noun* thin sheet of metal; person or thing setting off another to advantage.

foil[3] *noun* blunt fencing sword.

foist *verb* (+ *on*) force (thing, oneself) on to (unwilling person).

fold[1] /fəʊld/ ● *verb* double (flexible thing) over on itself; bend portion of; become or be able to be folded; (+ *away*, *up*) make compact by folding; (often + *up*) *colloquial* collapse, cease to function; enfold; clasp; (+ *in*) mix in gently. ● *noun* folding; line made by folding; hollow among hills.

fold[2] /fəʊld/ *noun* sheepfold; religious group or congregation.

folder *noun* folding cover or holder for loose papers.

foliaceous /fəʊlɪ'eɪʃəs/ *adjective* of or like leaves; laminated.

foliage /'fəʊlɪɪdʒ/ *noun* leaves, leafage.

foliar /'fəʊlɪə/ *adjective* of leaves. □ **foliar feed** fertilizer supplied to leaves.

foliate ● *adjective* /'fəʊlɪət/ leaflike; having leaves. ● *verb* /'fəʊlɪeɪt/ (**-ting**) split into thin layers. □ **foliation** *noun*.

folio /'fəʊlɪəʊ/ ● *noun* (*plural* **-s**) leaf of paper etc. numbered only on front; sheet of paper folded once; book of such sheets. ● *adjective* (of book) made of folios.

folk /fəʊk/ ● *noun* (*plural* same or **-s**) (treated as *plural*) people in general or of specified class; (in *plural*, usually **folks**) one's relatives; (treated as *singular*) a people or nation; (in full **folk-music**) (treated as *singular*) traditional, esp. working-class, music, or music in style of this. ● *adjective* of popular origin. □ **folklore** traditional beliefs etc., study of these; **folkweave** rough loosely woven fabric.

folksy /'fəʊksɪ/ *adjective* (**-ier**, **-iest**) of or like folk art; in deliberately popular style.

follicle /'fɒlɪk(ə)l/ *noun* small sac or vesicle, esp. for hair-root. □ **follicular** /fɒ'lɪkjʊlə/ *adjective*.

follow /'fɒləʊ/ *verb* go or come after; go along; come next in order or time; practise; understand; take as guide; take interest in; (+ *with*) provide with (sequel etc.); (+ *from*) result from; be necessary inference. □ **follow on** *verb* continue, (of cricket team) bat twice in succession; **follow-on** *noun* instance of this; **follow suit** play card of suit led, conform to another's actions; **follow through** *verb* continue to conclusion, continue movement of stroke after hitting the ball; **follow-through** *noun* instance of this; **follow up** *verb* act or investigate further; **follow-up** *noun* subsequent action.

follower *noun* supporter, devotee; person who follows.

following ● *preposition* after in time; as sequel to. ● *noun* group of supporters. ● *adjective* that follows. □ **the following** what follows, now to be mentioned.

folly /'fɒlɪ/ *noun* (*plural* **-ies**) foolishness; foolish act, idea, etc.; building for display only.

foment /fə'ment/ *verb* instigate, stir up (trouble etc.). □ **fomentation** /fəʊmen'teɪʃ(ə)n/ *noun*.

fond *adjective* (+ *of*) liking; affectionate; doting; foolishly optimistic. □ **fondly** *adverb*; **fondness** *noun*.

fondant /'fɒnd(ə)nt/ *noun* soft sugary sweet.

fondle /'fɒnd(ə)l/ *verb* (**-ling**) caress.

fondue /'fɒndju:/ *noun* dish of melted cheese.

font[1] *noun* receptacle for baptismal water.

font[2] = FOUNT[2].

fontanelle /fɒntəˈnel/ *noun* (*US* **fontanel**) space in infant's skull, which later closes up.

food /fu:d/ *noun* substance taken in to maintain life and growth; solid food; mental stimulus. □ **food-chain** series of organisms each dependent on next for food; **food poisoning** illness due to bacteria etc. in food; **food processor** machine for chopping and mixing food; **foodstuff** substance used as food.

foodie /ˈfu:dɪ/ *noun colloquial* person who makes a cult of food.

fool[1] ● *noun* unwise or stupid person; *historical* jester, clown. ● *verb* deceive; trick; cheat; joke; tease; (+ *around, about*) play, trifle. □ **act, play the fool** behave in silly way; **foolproof** incapable of misuse or mistake; **fool's paradise** illusory happiness; **make a fool of** make (person) look foolish, trick.

fool[2] *noun* dessert of fruit purée with cream or custard.

foolery *noun* foolish behaviour.

foolhardy *adjective* (**-ier**, **-iest**) foolishly bold; reckless. □ **foolhardily** *adverb*; **foolhardiness** *noun*.

foolish *adjective* lacking good sense or judgement. □ **foolishly** *adverb*; **foolishness** *noun*.

foolscap /ˈfu:lskæp/ *noun* large size of paper, about 330 mm x 200 (or 400) mm.

foot /fʊt/ ● *noun* (*plural* **feet**) part of leg below ankle; lower part or end; (*plural* same or **feet**) linear measure of 12 in. (30.48 cm); metrical unit of verse forming part of line; *historical* infantry. ● *verb* pay (bill); (usually as **foot it**) go on foot. □ **foot-and-mouth (disease)** contagious viral disease of cattle etc.; **footfall** sound of footstep; **foot-fault** (in tennis) serving with foot over baseline; **foothill** low hill lying at base of mountain or range; **foothold** secure place for feet in climbing, secure initial position; **footlights** row of floor-level lights along front of stage; **footloose** free to act as one pleases; **footman** liveried servant; **footmark** footprint; **footnote** note at foot of page; **footpath** path for pedestrians only; **footplate** platform for crew in locomotive; **footprint** impression left by foot or shoe; **footsore** having sore feet, esp. from walking; **footstep** (sound of) step taken in walking; **footstool** stool for resting feet on when sitting; **on foot** walking. □ **footless** *adjective*.

footage *noun* a length of TV or cinema film etc.; length in feet.

football *noun* large inflated usually leather ball; team game played with this. □ **football pool(s)** organized gambling on results of football matches. □ **footballer** *noun*.

footing *noun* foothold; secure position; operational basis; relative position or status; (often in *plural*) foundations of wall.

footling /ˈfu:tlɪŋ/ *adjective colloquial* trivial, silly.

Footsie /ˈfʊtsɪ/ *noun* FT-SE.

footsie /ˈfʊtsɪ/ *noun colloquial* amorous play with feet.

fop *noun* dandy. □ **foppery** *noun*; **foppish** *adjective*.

for /fə, fɔ:/ ● *preposition* in interest, defence, or favour of; appropriate to; regarding; representing; at the price of; as consequence or on account of; in order to get or reach; so as to start promptly at; notwithstanding. ● *conjunction* because, since. □ **be for it** be liable or about to be punished.

forage /ˈfɒrɪdʒ/ ● *noun* food for horses and cattle; searching for food. ● *verb* (**-ging**) (often + *for*) search for food; rummage; collect food from. □ **forage cap** infantry undress cap.

foray /ˈfɒreɪ/ ● *noun* sudden attack, raid. ● *verb* make foray.

forbade (also **forbad**) *past* of FORBID.

forbear[1] /fɔ:ˈbeə/ *verb* (*past* **forbore**; *past participle* **forborne**) *formal* abstain or refrain from.

forbear[2] = FOREBEAR.

forbearance *noun* patient self-control; tolerance.

forbid /fəˈbɪd/ *verb* (**forbidding**; *past* **forbade** /-bæd/ or **forbad**; *past participle* **forbidden**) (+ *to do*) order not; not allow; refuse entry to. □ **forbidden fruit** thing desired esp. because not allowed.

forbidding *adjective* stern, threatening. □ **forbiddingly** *adverb.*

forbore *past* of FORBEAR[1].

forborne *past participle* of FORBEAR[1].

force ● *noun* strength, power, intense effort; group of soldiers, police, etc.; coercion, compulsion; influence (person etc.) with moral power. ● *verb* (**-cing**) compel, coerce; make way, break into or open by force; drive, propel; (+ *on*, *upon*) impose or press on (person); cause or produce by effort; strain; artificially hasten maturity of; accelerate. □ **forced landing** emergency landing of aircraft; **forced march** lengthy and vigorous march, esp. by troops; **force-feed** feed (prisoner etc.) against his or her will; **force (person's) hand** make him or her act prematurely or unwillingly.

forceful *adjective* powerful; impressive. □ **forcefully** *adverb*; **forcefulness** *noun.*

force majeure /fɔ:s mæ'ʒɜ:/ *noun* irresistible force; unforeseeable circumstances. [French]

forcemeat /'fɔ:smi:t/ *noun* minced seasoned meat for stuffing etc.

forceps /'fɔ:seps/ *noun* (*plural* same) surgical pincers.

forcible /'fɔ:sɪb(ə)l/ *adjective* done by or involving force; forceful. □ **forcibly** *adverb.*

ford ● *noun* shallow place where river etc. may be crossed. ● *verb* cross (water) at ford. □ **fordable** *adjective.*

fore ● *adjective* situated in front. ● *noun* front part; bow of ship. ● *interjection* (*in golf*) *warning to person in path of ball.* □ **fore-and-aft** (of sails or rigging) lengthwise, at bow and stern; **to the fore** conspicuous.

forearm[1] /'fɔ:rɑ:m/ *noun* arm from elbow to wrist or fingertips.

forearm[2] /fɔ:'rɑ:m/ *verb* arm beforehand, prepare.

forebear /'fɔ:beə/ *noun* (also **forbear**) (usually in *plural*) ancestor.

forebode /fɔ:'bəʊd/ *verb* (**-ding**) be advance sign of; portend.

foreboding *noun* expectation of trouble.

forecast ● *verb* (*past & past participle* **-cast** or **-casted**) predict; estimate beforehand. ● *noun* prediction, esp. of weather. □ **forecaster** *noun.*

forecastle /'fəʊks(ə)l/ *noun* (also **fo'c's'le**) forward part of ship, formerly living quarters.

foreclose /fɔ:'kləʊz/ *verb* (**-sing**) stop (mortgage) from being redeemable; repossess mortgaged property of (person) when loan is not duly repaid; exclude, prevent. □ **foreclosure** *noun.*

forecourt *noun* part of filling-station with petrol pumps; enclosed space in front of building.

forefather *noun* (usually in *plural*) ancestor.

forefinger *noun* finger next to thumb.

forefoot *noun* front foot of animal.

forefront *noun* leading position; foremost part.

forego = FORGO.

foregoing /fɔ:'gəʊɪŋ/ *adjective* preceding; previously mentioned.

foregone conclusion /'fɔ:gɒn/ *noun* easily foreseeable result.

foreground *noun* part of view nearest observer.

forehand *Tennis etc.* ● *adjective* (of stroke) played with palm of hand facing forward. ● *noun* forehand stroke.

forehead /'fɒrɪd, 'fɔ:hed/ *noun* part of face above eyebrows.

foreign /'fɒrən/ *adjective* of, from, in, or characteristic of country or language other than one's own; dealing with other countries; of another district, society, etc.; (often + *to*) unfamiliar, alien; coming from outside. □ **foreign legion** group of foreign volunteers in (esp. French) army. □ **foreignness** *noun.*

foreigner *noun* person born in or coming from another country.

foreknowledge /fɔ:'nɒlɪdʒ/ *noun* knowledge in advance of (an event etc.).

foreleg *noun* animal's front leg.

forelock *noun* lock of hair just above forehead. □ **touch one's forelock** defer to person of higher social rank.

foreman /'fɔ:mən/ *noun* (*feminine* **forewoman**) worker supervising others; spokesman of jury.

foremast *noun* mast nearest bow of ship.

foremost ● *adjective* most notable, best; first, front. ● *adverb* most importantly.

forename *noun* first or Christian name.

forensic /fə'rensɪk/ *adjective* of or used in courts of law; of or involving application of science to legal problems.

foreplay *noun* stimulation preceding sexual intercourse.

forerunner *noun* predecessor.

foresail *noun* principal sail on foremast.

foresee /fɔː'siː/ *verb* (*past* **-saw**; *past participle* **-seen**) see or be aware of beforehand. □ **foreseeable** *adjective.*

foreshadow /fɔː'ʃædəʊ/ *verb* be warning or indication of (future event).

foreshore *noun* shore between high and low water marks.

foreshorten /fɔː'ʃɔːt(ə)n/ *verb* portray (object) with apparent shortening due to perspective.

foresight *noun* care or provision for future; foreseeing.

foreskin *noun* fold of skin covering end of penis.

forest /'fɒrɪst/ ● *noun* large area of trees; large number, dense mass. ● *verb* plant with trees; convert into forest.

forestall /fɔː'stɔːl/ *verb* prevent by advance action; deal with beforehand.

forester *noun* manager of forest; expert in forestry; dweller in forest.

forestry *noun* science or management of forests.

foretaste *noun* small preliminary experience of something.

foretell /fɔː'tel/ *verb* (*past & past participle* **-told**) predict, prophesy; indicate approach of.

forethought *noun* care or provision for future; deliberate intention.

forever /fə'revə/ *adverb* always, constantly.

forewarn /fɔː'wɔːn/ *verb* warn beforehand.

foreword *noun* introductory remarks in book, often not by author.

forfeit /'fɔːfɪt/ ● *noun* (thing surrendered as) penalty. ● *adjective* lost or surrendered as penalty. ● *verb* (**-t-**) lose right to, surrender as penalty. □ **forfeiture** *noun.*

forgather /fɔː'gæðə/ *verb* assemble; associate.

forgave *past* of FORGIVE.

forge¹ ● *verb* (**-ging**) make or write in fraudulent imitation; shape by heating and hammering. ● *noun* furnace etc. for melting or refining metal; blacksmith's workshop. □ **forger** *noun.*

forge² *verb* (**-ging**) advance gradually. □ **forge ahead** take lead, progress rapidly.

forgery /'fɔːdʒərɪ/ *noun* (*plural* **-ies**) (making of) forged document etc.

forget /fə'get/ *verb* (**forgetting**; *past* **forgot**; *past participle* **forgotten** or *US* **forgot**) lose remembrance of; neglect, overlook; cease to think of; (**forget oneself**) act without dignity. □ **forget-me-not** plant with small blue flowers. □ **forgettable** *adjective.*

forgetful *adjective* apt to forget; (often + *of*) neglectful. □ **forgetfully** *adverb*; **forgetfulness** *noun.*

forgive /fə'gɪv/ *verb* (**-ving**; *past* **forgave**; *past participle* **forgiven**) cease to resent; pardon; remit (debt). □ **forgivable** *adjective*; **forgiveness** *noun*; **forgiving** *adjective.*

forgo /fɔː'gəʊ/ *verb* (also **forego**) (**-goes**; *past* **-went**; *past participle* **-gone**) go without, relinquish.

forgot *past* of FORGET.

forgotten *past participle* of FORGET.

fork ● *noun* pronged item of cutlery; similar large tool for digging etc.; forked part, esp. of bicycle frame; (place of) divergence of road etc. ● *verb* form fork or branch; take one road at fork; dig with fork. □ **fork-lift truck** vehicle with fork for lifting and carrying; **fork out** *slang* pay, esp. reluctantly.

forlorn /fə'lɔːn/ *adjective* sad and abandoned; pitiful. □ **forlorn hope** faint remaining hope or chance. □ **forlornly** *adverb.*

form ● *noun* shape, arrangement of parts, visible aspect; person or animal as visible or tangible; mode in which thing exists or manifests itself; kind, variety; document with blanks to be filled in; class in school; (often as **the form**) customary or correct behaviour or method; set order of words; (of athlete, horse, etc.) condition of health and training; disposition; bench. ● *verb* make, be made; constitute; develop or establish as concept, practice, etc.; (+

into) organize; (of troops etc.) bring or move into formation; mould, fashion.

formal /'fɔ:m(ə)l/ *adjective* in accordance with rules, convention, or ceremony; (of garden etc.) symmetrical; prim, stiff; perfunctory; drawn up correctly; concerned with outward form. □ **formally** *adverb.*

formaldehyde /fɔ:'mældɪhaɪd/ *noun* colourless gas used as preservative and disinfectant.

formalin /'fɔ:məlɪn/ *noun* solution of formaldehyde in water.

formalism *noun* strict adherence to external form, esp. in art. □ **formalist** *noun.*

formality /fɔ:'mælɪtɪ/ *noun* (*plural* **-ies**) formal, esp. meaningless, regulation or act; rigid observance of rules or convention.

formalize *verb* (also **-ise**) (**-zing** or **-sing**) give definite (esp. legal) form to; make formal. □ **formalization** *noun.*

format /'fɔ:mæt/ ● *noun* shape and size (of book etc.); style or manner of procedure etc.; arrangement of computer data etc. ● *verb* (**-tt-**) arrange in format; prepare (storage medium) to receive computer data.

formation /fɔ:'meɪʃ(ə)n/ *noun* forming; thing formed; particular arrangement (e.g. of troops); rocks or strata with common characteristic.

formative /'fɔ:mətɪv/ *adjective* serving to form; of formation.

former /'fɔ:mə/ *adjective* of the past, earlier, previous; (**the former**) the first or first-named of two.

formerly *adverb* in former times; previously.

Formica /fɔ:'maɪkə/ *noun proprietary term* hard plastic laminate for surfaces.

formic acid /'fɔ:mɪk/ *noun* colourless irritant volatile acid contained in fluid emitted by ants.

formidable /'fɔ:mɪdəb(ə)l/ *adjective* inspiring awe, respect, or dread; difficult to deal with. □ **formidably** *adverb.*

■ **Usage** *Formidable* is also pronounced /fə'mɪdəb(ə)l/ (with the stress on the *-mid-*), but this is considered incorrect by some people.

---

formless *adjective* without definite or regular form. □ **formlessness** *noun.*

formula /'fɔ:mjʊlə/ *noun* (*plural* **-s** or **-lae** /-li:/) chemical symbols showing constituents of substance; mathematical rule expressed in symbols; fixed form of words; list of ingredients; classification of racing car, esp. by engine capacity. □ **formulaic** /-'leɪɪk/ *adjective.*

formulate /'fɔ:mjʊleɪt/ *verb* express in formula; express clearly and precisely. □ **formulation** *noun.*

fornicate /'fɔ:nɪkeɪt/ *verb* (**-ting**) *usually jocular* have extramarital sexual intercourse. □ **fornication** *noun*; **fornicator** *noun.*

forsake /fə'seɪk/ *verb* (**-king**; *past* **forsook** /-'sʊk/; *past participle* **forsaken**) *literary* give up, renounce; desert, abandon.

forswear /fɔ:'sweə/ *verb* (*past* **forswore**; *past participle* **forsworn**) abjure, renounce; (**forswear oneself**) perjure oneself; (as **forsworn** *adjective*) perjured.

forsythia /fɔ:'saɪθɪə/ *noun* shrub with bright yellow flowers.

fort *noun* fortified building or position. □ **hold the fort** act as temporary substitute.

forte[1] /'fɔ:teɪ/ *noun* thing in which one excels or specializes.

forte[2] /'fɔ:teɪ/ *Music* ● *adjective* loud. ● *adverb* loudly. ● *noun* loud playing, singing, or passage.

forth /fɔ:θ/ *adverb* forward(s); out; onwards in time. □ **forthcoming** /fɔ:θ'kʌmɪŋ/ coming or available soon, produced when wanted, informative, responsive.

forthright *adjective* straightforward, outspoken, decisive.

forthwith /fɔ:θ'wɪθ/ *adverb* immediately, without delay.

fortification /fɔ:tɪfɪ'keɪʃ(ə)n/ *noun* fortifying; (usually in *plural*) defensive works.

fortify /'fɔ:tɪfaɪ/ *verb* (**-ies**, **-ied**) provide with fortifications; strengthen; (usually as **fortified** *adjective*) strengthen (wine etc.) with alcohol, increase nutritive value of (food, esp. with vitamins).

fortissimo /fɔ:'tɪsɪməʊ/ *Music* ● *adjective* very loud. ● *adverb* very loudly. ● *noun* (*plural* **-mos** or **-mi** /-mi:/) very loud playing, singing, or passage.

fortitude /'fɔːtɪtjuːd/ *noun* courage in pain or adversity.

fortnight /'fɔːtnaɪt/ *noun* two weeks.

fortnightly ● *adjective* done, produced, or occurring once a fortnight. ● *adverb* every fortnight. ● *noun* (*plural* **-ies**) fortnightly magazine etc.

Fortran /'fɔːtræn/ *noun* (also **FORTRAN**) computer language used esp. for scientific calculations.

fortress /'fɔːtrɪs/ *noun* fortified building or town.

fortuitous /fɔː'tjuːɪtəs/ *adjective* happening by chance. □ **fortuitously** *adverb*; **fortuitousness** *noun*; **fortuity** *noun* (*plural* **-ies**).

fortunate /'fɔːtʃənət/ *adjective* lucky, auspicious. □ **fortunately** *adverb.*

fortune /'fɔːtʃ(ə)n/ *noun* chance or luck in human affairs; person's destiny; prosperity, wealth; *colloquial* large sum of money. □ **fortune-teller** person claiming to foretell one's destiny.

forty /'fɔːtɪ/ *adjective & noun* (*plural* **-ies**) four times ten. □ **forty winks** *colloquial* short sleep. □ **fortieth** *adjective & noun.*

forum /'fɔːrəm/ *noun* place of or meeting for public discussion; court, tribunal.

forward /'fɔːwəd/ ● *adjective* onward, towards front; bold, precocious, presumptuous; relating to the future; well-advanced. ● *noun* attacking player in football etc. ● *adverb* to front; into prominence; so as to make progress; towards future; forwards. ● *verb* send (letter etc.) on; dispatch; help to advance, promote.

forwards *adverb* in direction one is facing.

forwent *past* of FORGO.

fossil /'fɒs(ə)l/ ● *noun* remains or impression of (usually prehistoric) plant or animal hardened in rock; *colloquial* antiquated or unchanging person or thing. ● *adjective* of or like fossil. □ **fossil fuel** natural fuel extracted from ground. □ **fossilize** *verb* (also **-ise**) (**-zing** or **-sing**); **fossilization** *noun.*

foster /'fɒstə/ ● *verb* promote growth of; encourage or harbour (feeling); bring up (another's child); assign as foster-child. ● *adjective* related by or concerned with fostering.

fought *past & past participle* of FIGHT.

foul /faʊl/ ● *adjective* offensive, loathsome, stinking; dirty, soiled; *colloquial* awful; noxious; obscene; unfair, against rules; (of weather) rough; entangled. ● *noun* foul blow or play; entanglement. ● *adverb* unfairly. ● *verb* make or become foul; commit foul on (player); (often + *up*) (cause to) become entangled or blocked. □ **foul-mouthed** using obscene or offensive language; **foul play** unfair play, treacherous or violent act, esp. murder. □ **foully** /'faʊllɪ/ *adverb*; **foulness** *noun.*

found[1] *past & past participle* of FIND.

found[2] /faʊnd/ *verb* establish, originate; lay base of; base. □ **founder** *noun.*

found[3] /faʊnd/ *verb* melt and mould (metal), fuse (materials for glass); make thus. □ **founder** *noun.*

foundation /faʊn'deɪʃ(ə)n/ *noun* solid ground or base under building; (usually in *plural*) lowest part of building, usually below ground; basis; underlying principle; establishing (esp. endowed institution); base for cosmetics; (in full **foundation garment**) woman's supporting undergarment. □ **foundation-stone** one laid ceremonially at founding of building, basis.

founder /'faʊndə/ *verb* (of ship) fill with water and sink; (of plan) fail; (of horse) stumble, fall lame.

foundling /'faʊndlɪŋ/ *noun* abandoned infant of unknown parentage.

foundry /'faʊndrɪ/ *noun* (*plural* **-ies**) workshop for casting metal.

fount[1] /faʊnt/ *noun poetical* source, spring, fountain.

fount[2] /fɒnt/ *noun* (also **font**) set of printing type of same size and face.

fountain /'faʊntɪn/ *noun* jet(s) of water as ornament or for drinking; spring; (often + *of*) source. □ **fountain-head** source; **fountain pen** pen with reservoir or cartridge for ink.

four /fɔː/ *adjective & noun* one more than three; 4-oared boat, its crew. □ **four-letter word** short obscene word; **four-poster** bed with 4 posts supporting canopy; **four-square** *adjective* solidly based, steady, *adverb* resolutely; **four-stroke** (of internal-combustion engine) having power cycle completed in two up-and-down movements of piston; **four-wheel drive** drive acting on all 4 wheels of vehicle.

fourfold *adjective & adverb* four times as much or many.

foursome *noun* group of 4 people; golf match between two pairs.

fourteen /fɔːˈtiːn/ *adjective & noun* one more than thirteen. □ **fourteenth** *adjective & noun.*

fourth /fɔːθ/ *adjective & noun* next after third; any of four equal parts of thing. □ **fourthly** *adverb.*

fowl /faʊl/ *noun* (*plural* same or **-s**) chicken kept for eggs and meat; poultry as food.

fox ● *noun* wild canine animal with red or grey fur and bushy tail; its fur; crafty person. ● *verb* deceive; puzzle. □ **foxglove** tall plant with purple or white flowers; **foxhole** hole in ground as shelter etc. in battle; **foxhound** hound bred to hunt foxes; **fox-hunting** hunting foxes with hounds; **fox-terrier** small short-haired terrier; **foxtrot** ballroom dance with slow and quick steps. □ **foxlike** *adjective.*

foxy *adjective* (**-ier**, **-iest**) foxlike; sly, cunning; reddish-brown. □ **foxily** *adverb.*

foyer /ˈfɔɪeɪ/ *noun* entrance hall in hotel, theatre, etc.

FPA *abbreviation* Family Planning Association.

Fr. *abbreviation* Father; French.

fr. *abbreviation* franc(s).

fracas /ˈfrækɑː/ *noun* (*plural* same /-kɑːz/) noisy quarrel.

fraction /ˈfrækʃ(ə)n/ *noun* part of whole number; small part, amount, etc.; portion of mixture obtainable by distillation etc. □ **fractional** *adjective*; **fractionally** *adverb.*

fractious /ˈfrækʃəs/ *adjective* irritable, peevish.

fracture /ˈfræktʃə/ ● *noun* breakage, esp. of bone. ● *verb* (**-ring**) cause fracture in, suffer fracture.

fragile /ˈfrædʒaɪl/ *adjective* easily broken; delicate. □ **fragility** /frəˈdʒɪl-/ *noun.*

fragment ● *noun* /ˈfrægmənt/ part broken off; remains or unfinished portion of book etc. ● *verb* /frægˈment/ break into fragments. □ **fragmental** /-ˈmen-/ *adjective*; **fragmentary** *adjective*; **fragmentation** *noun.*

fragrance /ˈfreɪgrəns/ *noun* sweetness of smell; sweet scent.

fragrant *adjective* sweet-smelling.

frail *adjective* fragile, delicate; morally weak. □ **frailly** /ˈfreɪllɪ/ *adverb*; **frailness** *noun.*

frailty *noun* (*plural* **-ies**) frail quality; weakness, foible.

frame ● *noun* case or border enclosing picture etc.; supporting structure; (in *plural*) structure of spectacles holding lenses; build of person or animal; framework; construction; (in full **frame of mind**) temporary state; single picture on photographic film; (in snooker etc.) triangular structure for positioning balls, round of play; glazed structure to protect plants. ● *verb* (**-ming**) set in frame; serve as frame for; construct, devise; (+ *to*, *into*) adapt, fit; *slang* concoct false charge etc. against; articulate (words). □ **frame-up** *slang* conspiracy to convict innocent person; **framework** essential supporting structure, basic system.

franc *noun* French, Belgian, Swiss, etc. unit of currency.

franchise /ˈfræntʃaɪz/ ● *noun* right to vote; citizenship; authorization to sell company's goods etc. in particular area; right granted to person or corporation. ● *verb* (**-sing**) grant franchise to.

Franciscan /frænˈsɪskən/ ● *adjective* of (order of) St Francis. ● *noun* Franciscan friar or nun.

franglais /ˈfrɑ̃gleɪ/ *noun* French with many English words and idioms. [French]

Frank *noun* member of Germanic people that conquered Gaul in 6th c. □ **Frankish** *adjective.*

frank ● *adjective* candid, outspoken; undisguised; open. ● *verb* mark (letter etc.) to record payment of postage. ● *noun* franking signature or mark. □ **frankly** *adverb*; **frankness** *noun.*

frankfurter /ˈfræŋkfɜːtə/ *noun* seasoned smoked sausage.

frankincense /ˈfræŋkɪnsens/ *noun* aromatic gum resin burnt as incense.

frantic /ˈfræntɪk/ *adjective* wildly excited; frenzied; hurried, anxious; desperate, violent; *colloquial* extreme. □ **frantically** *adverb.*

fraternal /frə'tɜ:n(ə)l/ *adjective* of brothers, brotherly; comradely. □ **fraternally** *adverb.*

fraternity /frə'tɜ:nɪtɪ/ *noun* (*plural* **-ies**) religious brotherhood; group with common interests or of same professional class; *US* male students' society; brotherliness.

fraternize /'frætənaɪz/ *verb* (also **-ise**) (**-zing** or **-sing**) (often + *with*) associate, make friends, esp. with enemy etc. □ **fraternization** *noun.*

fratricide /'frætrɪsaɪd/ *noun* killing of one's brother or sister; person who does this. □ **fratricidal** /-'saɪd(ə)l/ *adjective.*

Frau /fraʊ/ *noun* (*plural* **Frauen** /'fraʊən/) *title used of or to married or widowed German-speaking woman.*

fraud /frɔ:d/ *noun* criminal deception; dishonest trick; impostor.

fraudulent /'frɔ:djʊlənt/ *adjective* of, involving, or guilty of fraud. □ **fraudulence** *noun*; **fraudulently** *adverb.*

fraught /frɔ:t/ *adjective* (+ *with*) filled or attended with (danger etc.); *colloquial* distressing; tense.

Fräulein /'frɔɪlaɪn/ *noun: title used of or to unmarried German-speaking woman.*

fray[1] *verb* wear or become worn; unravel at edge; (esp. as **frayed** *adjective*) (of nerves) become strained.

fray[2] *noun* fight, conflict; brawl.

frazzle /'fræz(ə)l/ *colloquial* ● *noun* worn, exhausted, or shrivelled state. ● *verb* (**-ling**) (usually as **frazzled** *adjective*) wear out; exhaust.

freak ● *noun* monstrosity; abnormal person or thing; *colloquial* unconventional person, fanatic of specified kind. ● *verb* (often + *out*) *colloquial* make or become very angry; (cause to) undergo esp. drug-induced etc. hallucinations or strong emotional experience. □ **freakish** *adjective*; **freaky** *adjective* (**-ier, -iest**).

freckle /'frek(ə)l/ ● *noun* light brown spot on skin. ● *verb* (**-ling**) (usually as **freckled** *adjective*) spot or be spotted with freckles. □ **freckly** *adjective* (**-ier, -iest**).

free ● *adjective* (**freer** /'fri:ə/, **freest** /'fri:ɪst/) not a slave; having personal rights and social and political liberty; autonomous; democratic; unrestricted, not confined; (+ *of*, *from* or *in combination*) exempt from, not containing or subject to; (+ *to do*) permitted, at liberty; costing nothing; available; spontaneous; lavish, unreserved; (of translation) not literal. ● *adverb* freely; without cost. ● *verb* (**frees, freed**) make free, liberate; disentangle. □ **for free** *colloquial* gratis; **freebooter** pirate; **freeboard** part of ship's side between waterline and deck; **Free Church** Nonconformist Church; **free enterprise** freedom of private business from state control; **free fall** movement under force of gravity only; **free hand** *noun* liberty to act at one's own discretion; **freehand** *adjective* (of drawing) done without ruler, compasses, etc., *adverb* in a freehand way; **free house** pub not controlled by brewery; **freeloader** *slang* sponger; **freeman** holder of freedom of city etc.; **free port** port without customs duties, or open to all traders; **free-range** (of hens etc.) roaming freely, (of eggs) produced by such hens; **free spirit** independent or uninhibited person; **free-standing** not supported by another structure; **freestyle** swimming race in which any stroke may be used, wrestling allowing almost any hold; **freethinker** /-'θɪŋkə/ person who rejects dogma, esp. in religious belief; **free trade** trade without import restrictions etc.; **freeway** *US* motorway; **freewheel** ride bicycle with pedals at rest, act without constraint; **free will** power of acting independently of fate or without coercion.

freebie /'fri:bɪ/ *noun colloquial* thing given free of charge.

freedom *noun* being free; personal or civil liberty; liberty of action; (+ *from*) exemption from; (+ *of*) honorary membership or citizenship of, unrestricted use of (house etc.).

freehold *noun* complete ownership of property for unlimited period; such property. □ **freeholder** *noun.*

freelance ● *noun* person working for no fixed employer. ● *verb* work as freelance. ● *adverb* as freelance.

Freemason /'fri:meɪs(ə)n/ *noun* member of fraternity for mutual help with secret rituals. □ **Freemasonry** *noun.*

freesia /'fri:zjə/ *noun* fragrant flowering African bulb.

freeze ● *verb* (**-zing**; *past* **froze**; *past participle* **frozen** /'frəʊz(ə)n/) turn into ice or other solid by cold; make or become rigid from cold; be or feel very cold; cover or be covered with ice; refrigerate below freezing point; make or become motionless; (as **frozen** *adjective*) devoid of emotion; make (assets etc.) unrealizable; fix (prices etc.) at certain level; stop (movement in film). ● *noun* period or state of frost; price-fixing etc.; (in full **freeze-frame**) still film-shot. □ **freeze-dry** preserve (food) by freezing and then drying in vacuum; **freeze up** *verb* obstruct or be obstructed by ice; **freeze-up** *noun* period of extreme cold; **freezing point** temperature at which liquid freezes.

freezer *noun* refrigerated cabinet for preserving food in frozen state.

freight /freɪt/ *noun* transport of goods; goods transported; charge for transport of goods.

freighter *noun* ship or aircraft for carrying freight.

French ● *adjective* of France or its people or language. ● *noun* French language; (**the French**) (*plural*) the French people. □ **French bean** kidney or haricot bean as unripe pods or as ripe seeds; **French bread** long crisp loaf; **French dressing** salad dressing of oil and vinegar; **French fried potatoes, French fries** chips; **French horn** coiled brass wind instrument; **French letter** *colloquial* condom; **Frenchman**, **Frenchwoman** native or national of France; **French polish** *noun* shellac polish for wood; **French-polish** *verb*; **French window** glazed door in outside wall.

frenetic /frə'netɪk/ *adjective* frantic, frenzied. □ **frenetically** *adverb.*

frenzy /'frenzɪ/ ● *noun* (*plural* **-ies**) wild excitement or fury. ● *verb* (**-ies**, **-ied**) (usually as **frenzied** *adjective*) drive to frenzy. □ **frenziedly** *adverb.*

frequency /'fri:kwənsɪ/ *noun* (*plural* **-ies**) commonness of occurrence; frequent occurrence; rate of recurrence (of vibration etc.). □ **frequency modulation** *Electronics* modulation by varying carrier-wave frequency.

frequent ● *adjective* /'fri:kwənt/ occurring often or in close succession; habitual. ● *verb* /frɪ'kwent/ go to habitually. □ **frequently** /'frɪkwəntlɪ/ *adverb.*

fresco /'freskəʊ/ *noun* (*plural* **-s**) painting in water-colour on fresh plaster.

fresh ● *adjective* newly made or obtained; other, different; new; additional; (+ *from*) lately arrived from; not stale or faded; (of food) not preserved; (of water) not salty; pure; refreshing; (of wind) brisk; *colloquial* cheeky, amorously impudent; inexperienced. ● *adverb* newly, recently. □ **freshwater** (of fish etc.) not of the sea. □ **freshly** *adverb*; **freshness** *noun.*

freshen *verb* make or become fresh; (+ *up*) wash, tidy oneself, etc.; revive.

fresher *noun colloquial* first-year student at university or (*US*) high school.

freshet /'freʃɪt/ *noun* rush of fresh water into sea; river flood.

freshman /'freʃmən/ *noun* fresher.

fret[1] ● *verb* (**-tt-**) be worried or distressed. ● *noun* worry, vexation. □ **fretful** *adjective*; **fretfully** *adverb.*

fret[2] ● *noun* ornamental pattern of straight lines joined usually at right angles. ● *verb* (**-tt-**) adorn with fret etc. □ **fretsaw** narrow saw on frame for cutting thin wood in patterns; **fretwork** work done with fretsaw.

fret[3] *noun* bar or ridge on finger-board of guitar etc.

Freudian /'frɔɪdɪən/ ● *adjective* of Freud's theories or method of psychoanalysis. ● *noun* follower of Freud. □ **Freudian slip** unintentional verbal error revealing subconscious feelings.

Fri. *abbreviation* Friday.

friable /'fraɪəb(ə)l/ *adjective* easily crumbled. □ **friability** *noun.*

friar /'fraɪə/ *noun* member of male non-enclosed religious order. □ **friar's balsam** type of inhalant.

friary /'fraɪərɪ/ *noun* (*plural* **-ies**) monastery for friars.

fricassee /'frɪkəseɪ/ ● *noun* pieces of meat in thick sauce. ● *verb* (**fricassees**, **fricasseed**) make fricassee of.

fricative /'frɪkətɪv/ ● *adjective* sounded by friction of breath in narrow opening. ● *noun* such consonant (e.g. *f*, *th*).

friction /ˈfrɪkʃ(ə)n/ *noun* rubbing of one object against another; resistance so encountered; clash of wills, opinions, etc. □ **frictional** *adjective.*

Friday /ˈfraɪdeɪ/ *noun* day of week following Thursday.

fridge *noun colloquial* refrigerator. □ **fridge-freezer** combined refrigerator and freezer.

friend /frend/ *noun* supportive and respected associate, esp. one for whom affection is felt; ally; kind person; person already mentioned; (**Friend**) Quaker.

friendly ● *adjective* (**-ier**, **-iest**) outgoing, kindly; (often + *with*) on amicable terms; not hostile; user-friendly. ● *noun* (*plural* **-ies**) friendly match. □ **-friendly** not harming, helping; **friendly match** match played for enjoyment rather than competition; **Friendly Society** society for insurance against sickness etc. □ **friendliness** *noun.*

friendship *noun* friendly relationship or feeling.

frier = FRYER.

Friesian /ˈfriːzɪən/ ● *noun* one of breed of black and white dairy cattle. ● *adjective* of Friesians.

frieze /friːz/ *noun* part of entablature, often filled with sculpture, between architrave and cornice; band of decoration, esp. at top of wall.

frigate /ˈfrɪgɪt/ *noun* naval escort-vessel.

fright /fraɪt/ *noun* (instance of) sudden or extreme fear; grotesque-looking person or thing. □ **take fright** become frightened.

frighten *verb* fill with fright; (+ *away*, *off*, *out of*, *into*) drive by fright. □ **frightening** *adjective*; **frighteningly** *adverb.*

frightful *adjective* dreadful, shocking; ugly; *colloquial* extremely bad; *colloquial* extreme. □ **frightfully** *adverb.*

frigid /ˈfrɪdʒɪd/ *adjective* unfriendly, cold; (of woman) sexually unresponsive; cold. □ **frigidity** /-ˈdʒɪd-/ *noun.*

frill ● *noun* ornamental edging of gathered or pleated material; (in *plural*) unnecessary embellishments. ● *verb* (usually as **frilled** *adjective*) decorate with frill. □ **frilly** *adjective* (**-ier**, **-iest**).

fringe ● *noun* border of tassels or loose threads; front hair cut to hang over forehead; outer limit; unimportant area or part. ● *verb* (**-ging**) adorn with fringe; serve as fringe to. □ **fringe benefit** employee's benefit additional to salary.

frippery /ˈfrɪpərɪ/ *noun* (*plural* **-ies**) showy finery; empty display; (usually in *plural*) knick-knacks.

frisk ● *verb* leap or skip playfully; *slang* search (person). ● *noun* playful leap or skip.

frisky *adjective* (**-ier**, **-iest**) lively, playful.

frisson /ˈfriːsɒn/ *noun* emotional thrill. [French]

fritillary /frɪˈtɪlərɪ/ *noun* (*plural* **-ies**) plant with bell-like flowers; butterfly with red and black chequered wings.

fritter[1] /ˈfrɪtə/ *verb* (usually + *away*) waste triflingly.

fritter[2] /ˈfrɪtə/ *noun* fruit, meat, etc. coated in batter and fried.

frivolous /ˈfrɪvələs/ *adjective* not serious, shallow, silly; trifling. □ **frivolity** /-ˈvɒl-/ *noun* (*plural* **-ies**).

frizz ● *verb* form (hair) into tight curls. ● *noun* frizzed hair or state. □ **frizzy** *adjective* (**-ier**, **-iest**).

frizzle[1] /ˈfrɪz(ə)l/ *verb* (**-ling**) fry or cook with sizzling noise; (often + *up*) burn, shrivel.

frizzle[2] /ˈfrɪz(ə)l/ *verb* (**-ling**) *& noun* frizz.

frock *noun* woman's or girl's dress; monk's or priest's gown. □ **frock-coat** man's long-skirted coat.

frog *noun* tailless leaping amphibian. □ **frog in one's throat** *colloquial* phlegm in throat that hinders speech; **frogman** underwater swimmer equipped with rubber suit and flippers; **frogmarch** hustle forward with arms pinned behind; **frog-spawn** frog's eggs.

frolic /ˈfrɒlɪk/ ● *verb* (**-ck-**) play about merrily. ● *noun* merrymaking.

frolicsome *adjective* playful.

from /frəm/ *preposition expressing separation or origin.*

fromage frais /frɒmɑːʒ ˈfreɪ/ *noun* type of soft cheese.

frond *noun* leaflike part of fern or palm.

front /frʌnt/ ● *noun* side or part most prominent or important, or nearer spectator or direction of motion; line of battle; scene of actual fighting; organized political group; demeanour; pretext, bluff; person etc. as cover for subversive or illegal activities; land along edge of sea or lake, esp. in town; forward edge of advancing cold or warm air; auditorium; breast of garment. ● *adjective* of or at front. ● *verb* (+ *on*, *to*, *towards*, etc.) have front facing or directed towards; (+ *for*) *slang* act as front for; (usually as **fronted** *adjective* + *with*) provide with or have front; lead (band, organization, etc.). □ **front bench** seats in Parliament for leading members of government and opposition; **front line** foremost part of army or group under attack; **front runner** favourite in race etc.

frontage *noun* front of building; land next to street, water, etc.; extent of front.

frontal *adjective* of or on front; of forehead.

frontier /ˈfrʌntɪə/ *noun* border between countries, district on each side of this; limits of attainment or knowledge in subject; *esp. US historical* border between settled and unsettled country. □ **frontiersman** *noun*.

frontispiece /ˈfrʌntɪspi:s/ *noun* illustration facing title-page of book.

frost ● *noun* frozen dew or vapour; temperature below freezing point. ● *verb* (usually + *over*, *up*) become covered with frost; cover (as) with frost; (usually as **frosted** *adjective*) roughen surface of (glass) to make opaque. □ **frostbite** injury to body tissue due to freezing; **frostbitten** *adjective*.

frosting *noun* icing for cakes.

frosty *adjective* (**-ier**, **-iest**) cold or covered with frost; unfriendly.

froth ● *noun* foam; idle talk. ● *verb* emit or gather froth. □ **frothy** *adjective* (**-ier**, **-iest**).

frown ● *verb* wrinkle brows, esp. in displeasure or concentration; (+ *at*, *on*) disapprove of. ● *noun* act of frowning, frowning look.

frowsty /ˈfraʊstɪ/ *adjective* (**-ier**, **-iest**) fusty; stuffy.

frowzy /ˈfraʊzɪ/ *adjective* (also **frowsy**) (**-ier**, **-iest**) fusty; slatternly, dingy.

froze *past* of FREEZE.

frozen *past participle* of FREEZE.

FRS *abbreviation* Fellow of the Royal Society.

fructify /ˈfrʌktɪfaɪ/ *verb* (**-ies**, **-ied**) bear fruit; make fruitful.

fructose /ˈfrʌktəʊz/ *noun* sugar in fruits, honey, etc.

frugal /ˈfru:g(ə)l/ *adjective* sparing; meagre. □ **frugality** /-ˈgæl-/ *noun*; **frugally** *adverb*.

fruit /fru:t/ ● *noun* seed-bearing part of plant or tree; this as food; (usually in *plural*) products, profits, rewards. ● *verb* bear fruit. □ **fruit cake** cake containing dried fruit; **fruit cocktail** diced fruit salad; **fruit machine** gambling machine operated by coins; **fruit salad** dessert of mixed fruit; **fruit sugar** fructose.

fruiterer /ˈfru:tərə/ *noun* dealer in fruit.

fruitful *adjective* productive; successful. □ **fruitfully** *adverb*.

fruition /fru:ˈɪʃ(ə)n/ *noun* realization of aims or hopes.

fruitless *adjective* not bearing fruit; useless, unsuccessful. □ **fruitlessly** *adverb*.

fruity *adjective* (**-ier**, **-iest**) of or resembling fruit; (of voice) deep and rich; *colloquial* slightly indecent.

frump *noun* dowdy woman. □ **frumpish** *adjective*; **frumpy** *adjective* (**-ier**, **-iest**).

frustrate /frʌsˈtreɪt/ *verb* (**-ting**) make (efforts) ineffective; prevent from achieving purpose; (as **frustrated** *adjective*) discontented, unfulfilled. □ **frustrating** *adjective*; **frustratingly** *adverb*; **frustration** *noun*.

fry[1] ● *verb* (**fries**, **fried**) cook in hot fat. ● *noun* fried food, esp. (usually **fries**) chips. □ **frying-pan** shallow long-handled pan for frying; **fry-up** *colloquial* fried bacon, eggs, etc.

fry[2] *plural noun* young or freshly hatched fishes.

fryer *noun* (also **frier**) person who fries; vessel for frying esp. fish.

ft *abbreviation* foot, feet.

FT-SE *abbreviation* Financial Times Stock Exchange (100 share index).

fuchsia /'fju:ʃə/ *noun* shrub with drooping flowers.

fuddle /'fʌd(ə)l/ ● *verb* (**-ling**) confuse, esp. with alcohol. ● *noun* confusion; intoxication.

fuddy-duddy /'fʌdɪdʌdɪ/ *slang* ● *adjective* fussy, old-fashioned. ● *noun* (*plural* **-ies**) such person.

fudge ● *noun* soft toffee-like sweet; faking. ● *verb* (**-ging**) make or do clumsily or dishonestly; fake.

fuel /'fju:əl/ ● *noun* material for burning or as source of heat, power, or nuclear energy; thing that sustains or inflames passion etc. ● *verb* (**-ll-**; *US* **-l-**) supply with fuel; inflame (feeling).

fug *noun colloquial* stuffy atmosphere. □ **fuggy** *adjective* (**-ier**, **-iest**).

fugitive /'fju:dʒɪtɪv/ ● *noun* (often + *from*) person who flees. ● *adjective* fleeing; transient, fleeting.

fugue /fju:g/ *noun* piece of music in which short melody or phrase is introduced by one part and developed by others. □ **fugal** *adjective.*

fulcrum /'fʌlkrəm/ *noun* (*plural* **-s** or **-cra**) point on which lever is supported.

fulfil /fʊl'fɪl/ *verb* (*US* **fulfill**) (**-ll-**) carry out; satisfy; (as **fulfilled** *adjective*) completely happy; (**fulfil oneself**) realize one's potential. □ **fulfilment** *noun.*

full /fʊl/ ● *adjective* holding all it can; replete; abundant; satisfying; (+ *of*) having abundance of, engrossed in; complete, perfect; resonant; plump; ample. ● *adverb* quite, exactly. □ **full back** defensive player near goal in football etc.; **full-blooded** vigorous, sensual, not hybrid; **full-blown** fully developed; **full board** provision of bed and all meals; **full-bodied** rich in quality, tone, etc.; **full frontal** (of nude) fully exposed at front, explicit; **full house** maximum attendance at theatre etc., hand in poker with 3 of a kind and a pair; **full-length** of normal length, not shortened, (of portrait) showing whole figure; **full moon** moon with whole disc illuminated; **full stop** punctuation mark (.) at end of sentence etc. (see panel), complete cessation; **full term**

Full stop .

This is used:

1 at the end of a sentence, e.g.

I am going to the cinema tonight.
The film begins at seven.

The full stop is replaced by a question mark at the end of a question, and by an exclamation mark at the end of an exclamation.

2 after an abbreviation, e.g.

H. G. Wells *p. 19* (*=page 19*) *Sun.* (*=Sunday*)
Ex. 6 (*=Exercise 6*).

Full stops are **not** used with:

a numerical abbreviations, e.g. *1st, 2nd, 15th, 23rd*
b acronyms, e.g. *FIFA, NATO*
c abbreviations that are used as ordinary words, e.g. *con, demo, recap*
d chemical symbols, e.g. Fe, K, H_2O

Full stops are not essential for:

a abbreviations consisting entirely of capitals, e.g. *BBC, AD, BC, PLC*
b *C* (*=Celsius*), *F* (*=Fahrenheit*)
c measures of length, weight, time, etc., except for *in.* (*=inch*), *st.* (*=stone*)
d *Dr, Revd* (but note *Rev.*), *Mr, Mrs, Ms, Mme, Mlle, St* (*=Saint*), *Hants, Northants, p* (*=penny* or *pence*).

completion of normal pregnancy; **full-time** *adjective* for or during whole of working week, *adverb* on full-time basis. □ **fullness** *noun.*

fully *adverb* completely; at least.

fulmar /'fʊlmə/ *noun* kind of petrel.

fulminate /'fʊlmɪneɪt/ *verb* (**-ting**) criticize loudly and forcibly; explode, flash. □ **fulmination** *noun.*

fulsome /'fʊlsəm/ *adjective* excessive, cloying; insincere. □ **fulsomely** *adverb.*

■ **Usage** *Fulsome* is sometimes wrongly used to mean 'generous', as in *fulsome praise* or 'generous with praise', as in *a fulsome tribute.*

fumble /'fʌmb(ə)l/ ● *verb* (**-ling**) grope about; handle clumsily or nervously. ● *noun* act of fumbling.

fume ● *noun* (usually in *plural*) exuded smoke, gas, or vapour. ● *verb* (**-ming**) emit fumes; be very angry; subject (oak etc.) to fumes to darken.

fumigate /'fju:mɪgeɪt/ *verb* (**-ting**) disinfect or purify with fumes. □ **fumigation** *noun*; **fumigator** *noun.*

fun *noun* playful amusement; source of this; mockery. □ **funfair** fair consisting of amusements and sideshows; **fun run** *colloquial* sponsored run for charity; **make fun of**, **poke fun at** ridicule.

function /'fʌŋkʃ(ə)n/ ● *noun* proper role etc.; official duty; public or social occasion; *Mathematics* quantity whose value depends on varying values of others. ● *verb* fulfil function; operate.

functional *adjective* of or serving a function; practical rather than attractive. □ **functionally** *adverb.*

functionalism *noun* belief that function should determine design. □ **functionalist** *noun & adjective.*

functionary *noun* (*plural* **-ies**) official.

fund ● *noun* permanently available stock; money set apart for purpose; (in *plural*) money resources. ● *verb* provide with money; make (debt) permanent at fixed interest. □ **fund-raising** raising money for charity etc.; **fund-raiser** *noun.*

fundamental /fʌndə'ment(ə)l/ ● *adjective* of or serving as base or foundation; essential, primary. ● *noun* fundamental principle. □ **fundamentally** *adverb.*

fundamentalism *noun* strict adherence to traditional religious beliefs. □ **fundamentalist** *noun & adjective.*

funeral /'fju:nər(ə)l/ ● *noun* ceremonial burial or cremation of dead. ● *adjective* of or used at funerals. □ **funeral director** undertaker; **funeral parlour** establishment where corpses are prepared for funerals.

funerary /'fju:nərərɪ/ *adjective* of or used at funerals.

funereal /fju:'nɪərɪəl/ *adjective* of or appropriate to funeral; dismal, dark.

fungicide /'fʌndʒɪsaɪd/ *noun* substance that kills fungus. □ **fungicidal** /-'saɪd(ə)l/ *adjective.*

fungus /'fʌŋgəs/ *noun* (*plural* **-gi** /-gaɪ/ or **-guses**) mushroom, toadstool, or allied plant; spongy morbid growth. □ **fungal** *adjective*; **fungoid** *adjective*; **fungous** *adjective.*

funicular /fju:'nɪkjʊlə/ *noun* (in full **funicular railway**) cable railway with ascending and descending cars counterbalanced.

funk *slang* ● *noun* fear, panic. ● *verb* evade through fear.

funky *adjective* (**-ier**, **-iest**) *slang* (esp. of jazz etc.) with heavy rhythm.

funnel /'fʌn(ə)l/ ● *noun* tube widening at top, for pouring liquid etc. into small opening; chimney of steam engine or ship. ● *verb* (**-ll-**; *US* **-l-**) (cause to) move (as) through funnel.

funny /'fʌnɪ/ *adjective* (**-ier**, **-iest**) amusing, comical; strange. □ **funny bone** part of elbow over which very sensitive nerve passes. □ **funnily** *adverb.*

fur ● *noun* short fine animal hair; hide with fur on it; garment of or lined with this; coating inside kettle etc. ● *verb* (**-rr-**) (esp. as **furred** *adjective*) line or trim with fur; (often + *up*) (of kettle etc.) become coated with fur.

furbelow /'fɜ:bɪləʊ/ *noun* (in *plural*) showy ornaments.

furbish /'fɜ:bɪʃ/ *verb* (often + *up*) refurbish.

furcate /'fɜ:keɪt/ ● *adjective* forked, branched. ● *verb* (**-ting**) fork, divide. □ **furcation** *noun.*

furious /'fjʊərɪəs/ *adjective* very angry, raging, frantic. □ **furiously** *adverb.*

furl *verb* roll up (sail, umbrella); become furled.

furlong /'fɜ:lɒŋ/ *noun* eighth of mile.

furlough /'fɜ:ləʊ/ *noun* leave of absence.

furnace /'fɜ:nɪs/ *noun* chamber for intense heating by fire; very hot place.

furnish /'fɜ:nɪʃ/ *verb* provide with furniture; (often + *with*) supply; (as **furnished** *adjective*) let with furniture. □ **furnishings** *plural noun*.

furniture /'fɜ:nɪtʃə/ *noun* movable contents of building or room; ship's equipment; accessories, e.g. handles and locks.

furore /fjʊə'rɔ:rɪ/ *noun* (*US* **furor** /'fjʊərɔ:/) uproar; enthusiasm.

furrier /'fʌrɪə/ *noun* dealer in or dresser of furs.

furrow /'fʌrəʊ/ ● *noun* narrow trench made by plough; rut; wrinkle. ● *verb* plough; make furrows in.

furry /'fɜ:rɪ/ *adjective* (**-ier**, **-iest**) like or covered with fur.

further /'fɜ:ðə/ ● *adverb* (also **farther** /'fɑ:ðə/) more distant in space or time; more, to greater extent; in addition. ● *adjective* (also **farther** /'fɑ:θə/) more distant or advanced; more, additional. ● *verb* promote, favour. □ **further education** education for people above school age; **furthermore** in addition, besides.

furtherance *noun* furthering of scheme etc.

furthest /'fɜ:ðɪst/ (also **farthest** /'fɑ:ðɪst/) ● *adjective* most distant. ● *adverb* to or at the greatest distance.

furtive /'fɜ:tɪv/ *adjective* sly, stealthy. □ **furtively** *adverb*.

fury /'fjʊərɪ/ *noun* (*plural* **-ies**) wild and passionate anger; violence (of storm etc.); (**Fury**) (usually in *plural*) avenging goddess; angry woman.

furze *noun* gorse. □ **furzy** *adjective* (**-ier**, **-iest**).

fuse[1] /fju:z/ ● *verb* (**-sing**) melt with intense heat; blend by melting; supply with fuse; fail due to melting of fuse; cause fuse(s) of to melt. ● *noun* easily melted wire in circuit, designed to melt when circuit is overloaded.

fuse[2] /fju:z/ ● *noun* combustible device for igniting bomb etc. ● *verb* (**-sing**) fit fuse to.

fuselage /'fju:zəlɑ:ʒ/ *noun* body of aircraft.

fusible /'fju:zɪb(ə)l/ *adjective* that can be melted. □ **fusibility** *noun*.

fusilier /fju:zɪ'lɪə/ *noun* soldier of any of several regiments formerly armed with light muskets.

fusillade /fju:zɪ'leɪd/ *noun* continuous discharge of firearms or outburst of criticism etc.

fusion /'fju:ʒ(ə)n/ *noun* fusing; blending, coalition; nuclear fusion.

fuss ● *noun* excited commotion; bustle; excessive concern about trivial thing; sustained protest. ● *verb* behave with nervous concern; agitate, worry. □ **fusspot** *colloquial* person given to fussing; **make a fuss** complain vigorously; **make a fuss of, over** treat affectionately.

fussy *adjective* (**-ier**, **-iest**) inclined to fuss; over-elaborate; fastidious.

fustian /'fʌstɪən/ *noun* thick usually dark twilled cotton cloth; bombast.

fusty /'fʌstɪ/ *adjective* (**-ier**, **-iest**) musty, stuffy; antiquated.

futile /'fju:taɪl/ *adjective* useless, ineffectual. □ **futility** /-'tɪl-/ *noun*.

futon /'fu:tɒn/ *noun* Japanese mattress used as bed; this with frame, convertible into couch.

future /'fju:tʃə/ ● *adjective* about to happen, be, or become; of time to come; *Grammar* (of tense) describing event yet to happen. ● *noun* time to come; future condition or events etc.; prospect of success etc.; *Grammar* future tense; (in *plural*) (on stock exchange) goods etc. sold for future delivery. □ **future perfect** *Grammar* tense giving sense 'will have done'.

futurism *noun* 20th-c. artistic movement celebrating technology etc. □ **futurist** *adjective & noun*.

futuristic /fju:tʃə'rɪstɪk/ *adjective* suitable for the future; ultra-modern; of futurism.

futurity /fju:'tjʊərɪtɪ/ *noun* (*plural* **-ies**) *literary* future time, events, etc.

fuzz *noun* fluff; fluffy or frizzy hair; (**the fuzz**) *slang* police (officer).

fuzzy /'fʌzɪ/ *adjective* (**-ier**, **-iest**) fluffy; blurred, indistinct.

Gg

G □ **G-man** *US colloquial* FBI special agent; **G-string** narrow strip of cloth etc. attached to string round waist for covering genitals.

g *abbreviation* (also **g.**) gram(s).

gab *noun colloquial* talk, chatter.

gabardine /gæbə'di:n/ *noun* a strong twilled cloth; raincoat etc. of this.

gabble /'gæb(ə)l/ ● *verb* **(-ling)** talk or utter unintelligibly or too fast. ● *noun* rapid talk.

gaberdine = GABARDINE.

gable /'geɪb(ə)l/ *noun* triangular part of wall at end of ridged roof. □ **gabled** *adjective.*

gad *verb* **(-dd-)** (+ *about*) go about idly or in search of pleasure. □ **gadabout** person who gads about.

gadfly /'gædflaɪ/ *noun* (*plural* **-ies**) fly that bites cattle.

gadget /'gædʒɪt/ *noun* small mechanical device or tool. □ **gadgetry** *noun.*

Gael /geɪl/ *noun* Scottish or Gaelic-speaking Celt.

Gaelic /'geɪlɪk/ ● *noun* Celtic language of Scots (also /'gælɪk/) or Irish. ● *adjective* of Gaelic or Gaelic-speaking people.

gaff[1] ● *noun* stick with hook for landing fish; barbed fishing-spear. ● *verb* seize (fish) with gaff.

gaff[2] *noun slang* □ **blow the gaff** let out secret.

gaffe /gæf/ *noun* blunder, tactless mistake.

gaffer /'gæfə/ *noun* old man; *colloquial* foreman, boss; chief electrician in film unit.

gag ● *noun* thing thrust into or tied across mouth to prevent speech etc.; joke or comic scene. ● *verb* **(-gg-)** apply gag to; silence; choke, retch; make jokes.

gaga /'gɑ:gɑ:/ *adjective slang* senile; crazy.

gage[1] *noun* pledge, security; challenge.

gage[2] *US* = GAUGE.

gaggle /'gæg(ə)l/ *noun* flock (of geese); *colloquial* disorganized group.

gaiety /'geɪətɪ/ *noun* (*US* **gayety**) being gay, mirth; merrymaking; bright appearance.

gaily /'geɪlɪ/ *adverb* in a gay way.

gain ● *verb* obtain, win; acquire, earn; (often + *in*) increase, improve; benefit; (of clock etc.) become fast (by); reach; (often + *on, upon*) get closer to (person or thing one is following). ● *noun* increase (of wealth), profit; (in *plural*) money made in trade etc.

gainful /'geɪnfʊl/ *adjective* paid, lucrative. □ **gainfully** *adverb.*

gainsay /geɪn'seɪ/ *verb* deny, contradict.

gait *noun* manner of walking or proceeding.

gaiter /'geɪtə/ *noun* covering of leather etc. for lower leg.

gal. *abbreviation* gallon(s).

gala /'gɑ:lə/ *noun* festive occasion or gathering.

galactic /gə'læktɪk/ *adjective* of galaxy.

galantine /'gælənti:n/ *noun* cold dish of meat boned, spiced, and covered in jelly.

galaxy /'gæləksɪ/ *noun* (*plural* **-ies**) independent system of stars etc. in space; (**the Galaxy**) Milky Way; (+ *of*) gathering of beautiful or famous people.

gale *noun* strong wind; outburst, esp. of laughter.

gall[1] /gɔ:l/ *noun colloquial* impudence; rancour; bile. □ **gall bladder** bodily organ containing bile; **gallstone** small hard mass that forms in gall bladder.

gall[2] /gɔ:l/ ● *noun* sore made by chafing; (cause of) vexation; place rubbed bare. ● *verb* rub sore; vex, humiliate.

gall[3] /gɔ:l/ *noun* growth produced on tree etc. by insect etc.

gallant /'gælənt/ ● *adjective* brave; fine, stately; (/gə'lænt/) attentive to women. ● *noun* (/gə'lænt/) ladies' man. □ **gallantly** *adverb.*

gallantry /'gæləntrɪ/ *noun* (*plural* **-ies**) bravery; courteousness to women; polite act or speech.

galleon /ˈgælɪən/ *noun historical* (usually Spanish) warship.

galleria /gæləˈriːə/ *noun* group of small shops, cafés, etc. under one roof.

gallery /ˈgælərɪ/ *noun* (*plural* **-ies**) room etc. for showing works of art; balcony, esp. in church, hall, etc.; highest balcony in theatre; covered walk, colonnade; passage, corridor.

galley /ˈgælɪ/ *noun* (*plural* **-s**) *historical* long flat one-decked vessel usually rowed by slaves or criminals; ship's or aircraft's kitchen; (in full **galley proof**) printer's proof before division into pages.

Gallic /ˈgælɪk/ *adjective* (typically) French; of Gaul or Gauls.

Gallicism /ˈgælɪsɪz(ə)m/ *noun* French idiom.

gallimimus /gælɪˈmaɪməs/ *noun* (*plural* **-muses**) medium-sized dinosaur that ran fast on two legs.

gallinaceous /gælɪˈneɪʃəs/ *adjective* of order of birds including domestic poultry.

gallivant /ˈgælɪvænt/ *verb colloquial* gad about.

gallon /ˈgælən/ *noun* measure of capacity (4.546 litres).

gallop /ˈgæləp/ ● *noun* horse's fastest pace; ride at this pace. ● *verb* (**-p-**) (cause to) go at gallop; talk etc. fast; progress rapidly.

gallows /ˈgæləʊz/ *plural noun* (usually treated as *singular*) structure for hanging criminals.

Gallup poll /ˈgæləp/ *noun* kind of opinion poll.

galore /gəˈlɔː/ *adverb* in plenty.

galosh /gəˈlɒʃ/ *noun* waterproof overshoe.

galumph /gəˈlʌmf/ *verb colloquial* (esp. as **galumphing** *adjective*) move noisily or clumsily.

galvanic /gælˈvænɪk/ *adjective* producing an electric current by chemical action; (of electric current) produced thus; stimulating, full of energy.

galvanize /ˈgælvənaɪz/ *verb* (also **-ise**) (**-zing** or **-sing**) (often + *into*) rouse by shock; stimulate (as) by electricity; coat (iron, steel) with zinc to protect from rust.

galvanometer /gælvəˈnɒmɪtə/ *noun* instrument for measuring electric currents.

gambit /ˈgæmbɪt/ *noun Chess* opening with sacrifice of pawn etc.; trick, device.

gamble /ˈgæmb(ə)l/ ● *verb* (**-ling**) play games of chance for money; bet (sum of money); (often + *away*) lose by gambling. ● *noun* risky undertaking; spell of gambling. □ **gambler** *noun*.

gambol /ˈgæmb(ə)l/ ● *verb* (**-ll-**; *US* **-l-**) jump about playfully. ● *noun* caper.

game[1] ● *noun* form or period of play or sport, esp. competitive one organized with rules; portion of play forming scoring unit; (in *plural*) athletic contests; piece of fun, (in *plural*) tricks; *colloquial* scheme, activity; wild animals or birds etc. hunted for sport or food; their flesh as food. ● *adjective* spirited, eager. ● *verb* (**-ming**) gamble for money. □ **gamekeeper** person employed to breed and protect game; **gamesmanship** art of winning games by psychological means. □ **gamely** *adverb*.

game[2] *adjective colloquial* (of leg etc.) crippled.

gamete /ˈgæmiːt/ *noun* mature germ cell uniting with another in sexual reproduction.

gamin /ˈgæmɪn/ *noun* street urchin; impudent child.

gamine /gæˈmiːn/ *noun* girl gamin; attractively mischievous or boyish girl.

gamma /ˈgæmə/ *noun* third letter of Greek alphabet (Γ, γ). □ **gamma rays** very short X-rays emitted by radioactive substances.

gammon /ˈgæmən/ *noun* back end of side of bacon, including leg.

gammy /ˈgæmɪ/ *adjective* (**-ier**, **-iest**) *slang* (of leg etc.) crippled.

gamut /ˈgæmət/ *noun* entire range or scope. □ **run the gamut of** experience or perform complete range of.

gamy /ˈgeɪmɪ/ *adjective* (**-ier**, **-iest**) smelling or tasting like high game.

gander /ˈgændə/ *noun* male goose.

gang ● *noun* set of associates, esp. for criminal purposes; set of workers, slaves, or prisoners. ● *verb colloquial* (+ *up with*) act together with; (+ *up on*) combine against.

ganger /'gæŋə/ *noun* foreman of gang of workers.

gangling /'gæŋglɪŋ/ *adjective* (of person) tall and thin, lanky.

ganglion /'gæŋglɪən/ *noun* (*plural* **ganglia** or **-s**) knot on nerve containing assemblage of nerve cells.

gangly *adjective* (**-ier**, **-iest**) gangling.

gangplank /'gæŋplæŋk/ *noun* plank for walking on to or off boat etc.

gangrene /'gæŋgri:n/ *noun* death of body tissue, usually caused by obstruction of circulation. □ **gangrenous** *adjective*.

gangster /'gæŋstə/ *noun* member of gang of violent criminals.

gangue /gæŋ/ *noun* valueless part of ore deposit.

gangway *noun* passage, esp. between rows of seats; opening in ship's bulwarks; bridge.

gannet /'gænɪt/ *noun* large seabird; *slang* greedy person.

gantry /'gæntrɪ/ *noun* (*plural* **-ies**) structure supporting travelling crane, railway or road signals, rocket-launching equipment, etc.

gaol etc. = JAIL etc.

gap *noun* empty space, interval; deficiency; breach in hedge, wall, etc.; wide divergence.

gape ● *verb* (**-ping**) open mouth wide; be or become wide open; (+ *at*) stare at. ● *noun* open-mouthed stare; opening.

garage /'gærɑ:dʒ/ ● *noun* building for keeping vehicle(s) in; establishment selling petrol etc. or repairing and selling vehicles. ● *verb* (**-ging**) put or keep in garage.

garb ● *noun* clothing, esp. of distinctive kind. ● *verb* dress.

garbage /'gɑ:bɪdʒ/ *noun US* refuse; *colloquial* nonsense.

garble /'gɑ:b(ə)l/ *verb* (**-ling**) (esp. as **garbled** *adjective*) distort or confuse (facts, statements, etc.).

garden /'gɑ:d(ə)n/ ● *noun* piece of ground for growing flowers, fruit, or vegetables, or for recreation; (esp. in *plural*) public pleasure-grounds. ● *verb* cultivate or tend garden. □ **garden centre** place selling plants and garden equipment. □ **gardener** *noun*; **gardening** *noun*.

gardenia /gɑ:'di:nɪə/ *noun* tree or shrub with fragrant flowers.

gargantuan /gɑ:'gæntjʊən/ *adjective* gigantic.

gargle /'gɑ:g(ə)l/ ● *verb* (**-ling**) rinse (throat) with liquid kept in motion by breath. ● *noun* liquid so used.

gargoyle /'gɑ:gɔɪl/ *noun* grotesque carved spout projecting from gutter of building.

garish /'geərɪʃ/ *adjective* obtrusively bright, gaudy. □ **garishly** *adverb*; **garishness** *noun*.

garland /'gɑ:lənd/ ● *noun* wreath of flowers etc. as decoration. ● *verb* adorn or crown with garland(s).

garlic /'gɑ:lɪk/ *noun* plant with pungent bulb used in cookery. □ **garlicky** *adjective colloquial*.

garment /'gɑ:mənt/ *noun* article of dress.

garner /'gɑ:nə/ ● *verb* collect, store. ● *noun* storehouse for corn etc.

garnet /'gɑ:nɪt/ *noun* glassy mineral, esp. red kind used as gem.

garnish /'gɑ:nɪʃ/ ● *verb* decorate (esp. food). ● *noun* decoration, esp. to food.

garret /'gærɪt/ *noun* room, esp. small, cold, etc. immediately under roof.

garrison /'gærɪs(ə)n/ ● *noun* troops stationed in town. ● *verb* (**-n-**) provide with or occupy as garrison.

garrotte /gə'rɒt/ (also **garotte**, *US* **garrote**) ● *verb* (**garrotting**, *US* **garroting**) execute by strangulation, esp. with wire collar. ● *noun* device for this.

garrulous /'gærələs/ *adjective* talkative. □ **garrulity** /gə'ru:lɪtɪ/ *noun*; **garrulousness** *noun*.

garter /'gɑ:tə/ *noun* band to keep sock or stocking up; (**the Garter**) (badge of) highest order of English knighthood. □ **garter stitch** plain knitting stitch.

gas /gæs/ ● *noun* (*plural* **-es**) any airlike substance (i.e. not liquid or solid); such substance (esp. coal gas or natural gas) used as fuel; gas used as anaesthetic; poisonous gas used in war; *US colloquial* petrol; *slang* empty talk, boasting; *slang* amusing thing or person. ● *verb* (**gases**, **gassed**, **gassing**) expose to gas, esp. to kill; *colloquial* talk emptily or boastfully. □ **gasbag** *slang* empty talker; **gas**

chamber room filled with poisonous gas to kill people or animals; **gasholder** gasometer; **gas mask** respirator for protection against harmful gases; **gas ring** ring pierced with gas jet(s) for cooking etc.; **gasworks** place where gas is manufactured.

gaseous /'gæsɪəs/ *adjective* of or as gas.

gash ● *noun* long deep cut or wound. ● *verb* make gash in.

gasify /'gæsɪfaɪ/ *verb* (**-ies**, **-ied**) convert into gas. □ **gasification** *noun*.

gasket /'gæskɪt/ *noun* sheet or ring of rubber etc. to seal joint between metal surfaces.

gasoline /'gæsəli:n/ *noun* (also **gasolene**) *US* petrol.

gasometer /gæ'sɒmɪtə/ *noun* large tank from which gas is distributed.

gasp /gɑ:sp/ ● *verb* catch breath with open mouth; utter with gasps. ● *noun* convulsive catching of breath.

gassy /'gæsɪ/ *adjective* (**-ier**, **-iest**) of, like, or full of gas; *colloquial* verbose.

gastric /'gæstrɪk/ *adjective* of stomach. □ **gastric flu** *colloquial* intestinal disorder of unknown cause; **gastric juice** digestive fluid secreted by stomach glands.

gastritis /gæ'straɪtɪs/ *noun* inflammation of stomach.

gastroenteritis /gæstrəʊentə'raɪtɪs/ *noun* inflammation of stomach and intestines.

gastronome /'gæstrənəʊm/ *noun* gourmet. □ **gastronomic** /-'nɒm-/ *adjective*; **gastronomical** /-'nɒm-/ *adjective*; **gastronomically** /-'nɒm-/ *adverb*; **gastronomy** /-'strɒn-/ *noun*.

gastropod /'gæstrəpɒd/ *noun* mollusc that moves using underside of abdomen, e.g. snail.

gate *noun* barrier, usually hinged, used to close opening in wall, fence, etc.; such opening; means of entrance or exit; numbered place of access to aircraft at airport; device regulating passage of water in lock etc.; number of people paying to enter stadium etc., money thus taken. □ **gateleg (table)** table with legs in gatelike frame for supporting folding flaps; **gatepost** post at either side of gate; **gateway** opening closed by gate, means of access.

gateau /'gætəʊ/ *noun* (*plural* **-s** or **-x** /-z/) large rich elaborate cake.

gatecrash *verb* attend (party etc.) uninvited. □ **gatecrasher** *noun*.

gather /'gæðə/ ● *verb* bring or come together; collect (harvest, dust, etc.); infer, deduce; increase (speed); summon up (energy etc.); draw together in folds or wrinkles; (of boil etc.) come to a head. ● *noun* fold or pleat.

gathering *noun* assembly; pus-filled swelling.

GATT /gæt/ *abbreviation* General Agreement on Tariffs and Trade.

gauche /gəʊʃ/ *adjective* socially awkward, tactless. □ **gauchely** *adverb*; **gaucheness** *noun*.

gaucho /'gaʊtʃəʊ/ *noun* (*plural* **-s**) cowboy in S. American pampas.

gaudy /'gɔ:dɪ/ *adjective* (**-ier**, **-iest**) tastelessly showy. □ **gaudily** *adverb*; **gaudiness** *noun*.

gauge /geɪdʒ/ (*US* **gage**) ● *noun* standard measure; instrument for measuring; distance between rails or opposite wheels; capacity, extent; criterion, test. ● *verb* (**-ging**) measure exactly; measure contents of; estimate.

Gaul /gɔ:l/ *noun* inhabitant of ancient Gaul. □ **Gaulish** *adjective & noun*.

gaunt /gɔ:nt/ *adjective* lean, haggard; grim. □ **gauntness** *noun*.

gauntlet[1] /'gɔ:ntlɪt/ *noun* glove with long loose wrist; *historical* armoured glove.

gauntlet[2] /'gɔ:ntlɪt/ *noun* □ **run the gauntlet** undergo criticism, pass between two rows of people wielding sticks etc., as punishment.

gauze /gɔ:z/ *noun* thin transparent fabric; fine mesh of wire etc. □ **gauzy** *adjective* (**-ier**, **-iest**).

gave *past* of GIVE.

gavel /'gæv(ə)l/ *noun* auctioneer's, chairman's, or judge's hammer.

gavotte /gə'vɒt/ *noun* 18th-c. French dance; music for this.

gawk ● *verb colloquial* gawp. ● *noun* awkward or bashful person. □ **gawky** *adjective* (**-ier**, **-iest**).

gawp *verb colloquial* stare stupidly.

gay ● *adjective* light-hearted, cheerful; showy; homosexual; *colloquial* carefree. ● *noun* (esp. male) homosexual.

gayety *US* = GAIETY.

gaze ● *verb* (**-zing**) (+ *at*, *into*, *on*, etc.) look fixedly. ● *noun* intent look.

gazebo /gə'zi:bəʊ/ *noun* (*plural* **-s**) summerhouse etc. giving view.

gazelle /gə'zel/ *noun* (*plural* same or **-s**) small graceful antelope.

gazette /gə'zet/ ● *noun* newspaper; official publication. ● *verb* (**-tting**) publish in official gazette.

gazetteer /gæzɪ'tɪə/ *noun* geographical index.

gazump /gə'zʌmp/ *verb colloquial* raise price after accepting offer from (buyer); swindle.

gazunder /gə'zʌndə/ *verb colloquial* lower an offer made to (seller) just before exchange of contracts.

GB *abbreviation* Great Britain.

GBH *abbreviation* grievous bodily harm.

GC *abbreviation* George Cross.

GCE *abbreviation* General Certificate of Education.

GCSE *abbreviation* General Certificate of Secondary Education.

GDR *abbreviation historical* German Democratic Republic.

gear /gɪə/ ● *noun* (often in *plural*) set of toothed wheels working together, esp. those connecting engine to road wheels; particular setting of these; equipment; *colloquial* clothing. ● *verb* (+ *to*) adjust or adapt to; (often + *up*) equip with gears; (+ *up*) make ready. □ **gearbox** (case enclosing) gears of machine or vehicle; **gear lever** lever moved to engage or change gear; **in gear** with gear engaged.

gecko /'gekəʊ/ *noun* (*plural* **-s**) tropical house-lizard.

gee /dʒi:/ *interjection expressing surprise etc.*

geese *plural* of GOOSE.

geezer /'gi:zə/ *noun slang* man, esp. old one.

Geiger counter /'gaɪgə/ *noun* instrument for measuring radioactivity.

geisha /'geɪʃə/ *noun* (*plural* same or **-s**) Japanese professional hostess and entertainer.

gel /dʒel/ ● *noun* semi-solid jelly-like colloid; jelly-like substance for hair. ● *verb* (**-ll-**) form gel; jell.

gelatin /'dʒelətɪn/ *noun* (also **gelatine** /-ti:n/) transparent tasteless substance used in cookery, photography, etc. □ **gelatinous** /dʒɪ'læt-/ *adjective*.

geld /geld/ *verb* castrate.

gelding /'geldɪŋ/ *noun* castrated horse etc.

gelignite /'dʒelɪgnaɪt/ *noun* nitroglycerine explosive.

gem /dʒem/ ● *noun* precious stone; thing or person of great beauty or worth. ● *verb* (**-mm-**) adorn (as) with gems.

Gemini /'dʒemɪnaɪ/ *noun* third sign of zodiac.

Gen. *abbreviation* General.

gen /dʒen/ *slang* ● *noun* information. ● *verb* (**-nn-**) (+ *up*) gain or give information.

gendarme /'ʒɒndɑ:m/ *noun* police officer in France etc.

gender /'dʒendə/ *noun* (grammatical) classification roughly corresponding to the two sexes and sexlessness; one of these classes; person's sex.

gene /dʒi:n/ *noun* unit in chromosome, controlling particular inherited characteristic.

genealogy /dʒi:nɪ'ælədʒɪ/ *noun* (*plural* **-ies**) descent traced continuously from ancestor, pedigree; study of pedigrees. □ **genealogical** /-ə'lɒdʒ-/ *adjective*; **genealogically** /-ə'lɒdʒ-/ *adverb*; **genealogist** *noun*.

genera *plural* of GENUS.

general /'dʒenər(ə)l/ ● *adjective* including, affecting, or applicable to (nearly) all; prevalent, usual; vague; not partial or particular; lacking detail; chief, head. ● *noun* army officer next below Field Marshal; commander of army. □ **general anaesthetic** one affecting whole body; **general election** national election of representatives to parliament; **general practice** work of **general practitioner**, doctor treating cases of all kinds; **general strike** simultaneous strike of workers in all or most trades; **in general** as a rule, usually.

generalissimo /dʒenərə'lɪsɪməʊ/ *noun* (*plural* **-s**) commander of combined forces.

generality /dʒenəˈrælɪtɪ/ *noun* (*plural* **-ies**) general statement; general applicability; indefiniteness; (+ *of*) majority of.

generalize /ˈdʒenərəlaɪz/ *verb* (also **-ise**) (**-zing** or **-sing**) speak in general or indefinite terms, form general notion(s); reduce to general statement; infer (rule etc.) from particular cases; bring into general use. □ **generalization** *noun*.

generally /ˈdʒenərəlɪ/ *adverb* usually; in most respects; in general sense; in most cases.

generate /ˈdʒenəreɪt/ *verb* (**-ting**) bring into existence, produce.

generation *noun* all people born about same time; stage in family history or in (esp. technological) development; period of about 30 years; production, esp. of electricity; procreation.

generative /ˈdʒenərətɪv/ *adjective* of procreation; productive.

generator *noun* dynamo; apparatus for producing gas, steam, etc.

generic /dʒɪˈnerɪk/ *adjective* characteristic of or relating to class or genus; not specific or special; (of esp. drug) with no brand name. □ **generically** *adverb*.

generous /ˈdʒenərəs/ *adjective* giving or given freely; magnanimous; abundant. □ **generosity** /-ˈrɒs-/ *noun*; **generously** *adverb*.

genesis /ˈdʒenɪsɪs/ *noun* origin, mode of formation; (**Genesis**) first book of Old Testament.

genetic /dʒɪˈnetɪk/ *adjective* of genetics; of or in origin. □ **genetic engineering** manipulation of DNA to modify hereditary features; **genetic fingerprinting** identification of individuals by DNA patterns. □ **genetically** *adverb*.

genetics *plural noun* (treated as *singular*) study of heredity and variation among animals and plants. □ **geneticist** /-sɪst/ *noun*.

genial /ˈdʒiːnɪəl/ *adjective* sociable, kindly; mild, warm; cheering. □ **geniality** /-ˈæl-/ *noun*; **genially** *adverb*.

genie /ˈdʒiːnɪ/ *noun* (*plural* **genii** /-nɪaɪ/) sprite or goblin of Arabian tales.

genital /ˈdʒenɪt(ə)l/ ● *adjective* of animal reproduction or reproductive organs. ● *noun* (in *plural*; also **genitalia**) external reproductive organs.

genitive /ˈdʒenɪtɪv/ *Grammar* ● *noun* case expressing possession, origin, etc., corresponding to *of*, *from*, etc. ● *adjective* of or in this case.

genius /ˈdʒiːnɪəs/ *noun* (*plural* **-es**) exceptional natural ability; person having this; guardian spirit.

genocide /ˈdʒenəsaɪd/ *noun* mass murder, esp. among particular race or nation.

genre /ˈʒɑ̃rə/ *noun* kind or style of art etc.; portrayal of scenes from ordinary life.

gent /dʒent/ *noun colloquial* gentleman; (**the Gents**) *colloquial* men's public lavatory.

genteel /dʒenˈtiːl/ *adjective* affectedly refined; upper-class.

gentian /ˈdʒenʃ(ə)n/ *noun* mountain plant with usually blue flowers. □ **gentian violet** violet dye used as antiseptic.

Gentile /ˈdʒentaɪl/ ● *adjective* not Jewish; heathen. ● *noun* non-Jewish person.

gentility /dʒenˈtɪlɪtɪ/ *noun* social superiority; genteel habits.

gentle /ˈdʒent(ə)l/ *adjective* (**-r**, **-st**) not rough or severe; mild, kind; well-born; quiet. □ **gentleness** *noun*; **gently** *adverb*.

gentlefolk /ˈdʒentəlfəʊk/ *noun* people of good family.

gentleman /ˈdʒentəlmən/ *noun* man; chivalrous well-bred man; man of good social position. □ **gentlemanly** *adjective*.

gentlewoman *noun archaic* woman of good birth or breeding.

gentrification /dʒentrɪfɪˈkeɪʃ(ə)n/ *noun* upgrading of working-class urban area by arrival of affluent residents. □ **gentrify** *verb* (**-ies**, **-ied**).

gentry /ˈdʒentrɪ/ *plural noun* people next below nobility; *derogatory* people.

genuflect /ˈdʒenjuːflekt/ *verb* bend knee, esp. in worship. □ **genuflection**, **genuflexion** /-ˈflekʃ(ə)n/ *noun*.

genuine /ˈdʒenjuːɪn/ *adjective* really coming from its reputed source; properly so called; not sham. □ **genuinely** *adverb*; **genuineness** *noun*.

genus /ˈdʒiːnəs/ *noun* (*plural* **genera** /ˈdʒenərə/) group of animals or plants with common structural characteristics, usually containing several species; kind, class.

geocentric /dʒiːə'sentrɪk/ *adjective* considered as viewed from earth's centre; having earth as centre.

geode /'dʒiːəʊd/ *noun* cavity lined with crystals; rock containing this.

geodesic /dʒiːəʊ'diːzɪk/ *adjective* (also **geodetic** /-'det-/) of geodesy. □ **geodesic line** shortest possible line on surface between two points.

geodesy /dʒiː'ɒdɪsɪ/ *noun* study of shape and area of the earth.

geography /dʒi'ɒgrəfɪ/ *noun* science of earth's physical features, resources, etc.; features of place. □ **geographer** *noun*; **geographic(al)** /-ə'græf-/ *adjective*; **geographically** /-ə'græf-/ *adverb*.

geology /dʒɪ'ɒlədʒɪ/ *noun* science of earth's crust, strata, etc. □ **geological** /-ə'lɒdʒ-/ *adjective*; **geologist** *noun*.

geometry /dʒɪ'ɒmətrɪ/ *noun* science of properties and relations of lines, surfaces, and solids. □ **geometric(al)** /-ə'met-/ *adjective*; **geometrician** /-'trɪʃ(ə)n/ *noun*.

Geordie /'dʒɔːdɪ/ *noun* native of Tyneside.

georgette /dʒɔː'dʒet/ *noun* kind of fine dress material.

Georgian /'dʒɔːdʒ(ə)n/ *adjective* of time of George I–IV or George V and VI.

geranium /dʒə'reɪnɪəm/ *noun* (*plural* **-s**) cultivated pelargonium; herb or shrub with fruit shaped like crane's bill.

gerbil /'dʒɜːbɪl/ *noun* mouselike desert rodent with long hind legs.

geriatric /dʒerɪ'ætrɪk/ ● *adjective* of geriatrics or old people; *derogatory* old. ● *noun often derogatory* old person.

geriatrics /dʒerɪ'ætrɪks/ *plural noun* (usually treated as *singular*) branch of medicine dealing with health and care of old people. □ **geriatrician** /-ə'trɪʃ(ə)n/ *noun*.

germ /dʒɜːm/ *noun* microbe; portion of organism capable of developing into new one; thing that may develop; rudiment, elementary principle.

German /'dʒɜːmən/ ● *noun* (*plural* **-s**) native, national, or language of Germany. ● *adjective* of Germany. □ **German measles** disease like mild measles; **German shepherd (dog)** Alsatian.

german /'dʒɜːmən/ *adjective* (placed after *brother*, *sister*, or *cousin*) having same two parents or grandparents.

germander /dʒɜː'mændə/ *noun* plant of mint family.

germane /dʒɜː'meɪn/ *adjective* (usually + *to*) relevant.

Germanic /dʒɜː'mænɪk/ ● *adjective* having German characteristics. ● *noun* group of languages including English, German, Dutch, and Scandinavian languages.

germicide /'dʒɜːmɪsaɪd/ *noun* substance that destroys germs. □ **germicidal** /-'saɪd(ə)l/ *adjective*.

germinal /'dʒɜːmɪn(ə)l/ *adjective* of germs; in earliest stage of development.

germinate /'dʒɜːmɪneɪt/ *verb* (**-ting**) (cause to) sprout or bud. □ **germination** *noun*.

gerontology /dʒerɒn'tɒlədʒɪ/ *noun* study of old age and ageing.

gerrymander /dʒerɪ'mændə/ *verb* manipulate boundaries of (constituency etc.) to gain unfair electoral advantage.

gerund /'dʒerənd/ *noun* verbal noun, in English ending in *-ing*.

Gestapo /ge'stɑːpəʊ/ *noun historical* Nazi secret police.

gestation /dʒe'steɪʃ(ə)n/ *noun* carrying or being carried in womb between conception and birth; this period; development of plan etc. □ **gestate** *verb* (**-ting**).

gesticulate /dʒe'stɪkjʊleɪt/ *verb* (**-ting**) use gestures instead of or with speech. □ **gesticulation** *noun*.

gesture /'dʒestʃə/ ● *noun* meaningful movement of limb or body; action performed as courtesy or to indicate intention. ● *verb* (**-ring**) gesticulate.

get /get/ *verb* (**getting**; *past* **got**; *past participle* **got** or *US* **gotten**) obtain, earn; fetch, procure; go to reach or catch; prepare (meal); (cause to) reach some state or become; obtain from calculation; contract (disease); contact; have (punishment) inflicted on one; succeed in bringing, placing, etc.; (cause to) succeed in coming or going; *colloquial* understand, annoy, harm, attract; *archaic* beget. □ **get about** go from place to place; **get across** communicate; **get along** (often + *with*) live harmoniously; **get around** = GET ABOUT; **get at** reach, get hold of, *colloquial* imply, *colloquial* nag; **get away** escape; **getaway** *noun*; **get by** *colloquial* cope; **get in** obtain place at college etc., win election; **get off** alight (from), *colloquial* escape with little or no punishment, start, depart, (+ *with*) *colloquial* start sexual relationship with; **get on** make progress, manage, advance, enter (bus etc.), (often + *with*) live harmoniously, (usually as **be getting on**) age; **get out of** avoid, escape; **get over** recover from, surmount; **get round** coax or cajole (person), evade (law etc.), (+ *to*) deal with (task) in due course; **get through** pass (exam etc.), use up (resources), make contact by telephone, (+ *to*) succeed in making (person) understand; **get-together** *colloquial* social assembly; **get up** rise esp. from bed, (of wind etc.) strengthen, organize, stimulate, arrange appearance of; **get-up** *colloquial* style of dress etc.; **have got** possess, (+ *to do*) must.

geyser /'gi:zə/ *noun* hot spring; apparatus for heating water.

ghastly /'gɑ:stlɪ/ *adjective* (**-ier**, **-iest**) horrible, frightful; deathlike, pallid.

ghee /gi:/ *noun* Indian clarified butter.

gherkin /'gɜ:kɪn/ *noun* small cucumber for pickling.

ghetto /'getəʊ/ *noun* (*plural* **-s**) part of city occupied by minority group; *historical* Jews' quarter in city; segregated group or area. □ **ghetto-blaster** large portable radio or cassette player.

ghost /gəʊst/ ● *noun* apparition of dead person etc., disembodied spirit; (+ *of*) semblance; secondary image in defective telescope or television picture. ● *verb* (often + *for*) act as ghost-writer of (book etc.). □ **ghost-writer** writer doing work for which another takes credit. □ **ghostly** *adjective* (**-ier**, **-iest**).

ghoul /gu:l/ *noun* person morbidly interested in death etc.; evil spirit; (in Arabic mythology) spirit preying on corpses. □ **ghoulish** *adjective*.

GHQ *abbreviation* General Headquarters.

ghyll = GILL[3].

GI /dʒi:'aɪ/ *noun* soldier in US army.

giant /'dʒaɪənt/ ● *noun* mythical being of human form but superhuman size; person, animal, or thing of extraordinary size, ability, etc. ● *adjective* gigantic.

gibber /'dʒɪbə/ *verb* chatter inarticulately.

gibberish /'dʒɪbərɪʃ/ *noun* unintelligible or meaningless speech or sounds.

gibbet /'dʒɪbɪt/ *noun historical* gallows; post with arm from which executed criminal was hung after execution.

gibbon /'gɪbən/ *noun* long-armed ape.

gibbous /'gɪbəs/ *adjective* convex; (of moon etc.) with bright part greater than semicircle.

gibe /dʒaɪb/ (also **jibe**) ● *verb* (**-bing**) (often + *at*) jeer, mock. ● *noun* jeering remark, taunt.

giblets /'dʒɪblɪts/ *plural noun* liver, gizzard, etc. of bird removed and usually cooked separately.

giddy /'gɪdɪ/ *adjective* (**-ier**, **-iest**) dizzy, tending to fall or stagger; mentally intoxicated; excitable, flighty; making dizzy. □ **giddiness** *noun*.

gift /gɪft/ *noun* thing given, present; talent; *colloquial* easy task.

gifted *adjective* talented.

gig[1] /gɪg/ *noun* light two-wheeled one-horse carriage; light boat on ship; rowing boat, esp. for racing.

gig[2] /gɪg/ *colloquial* ● *noun* engagement to play music, usually on one occasion. ● *verb* (**-gg-**) perform gig.

giga- /'gɪgə/ *combining form* one thousand million.

gigantic /dʒaɪ'gæntɪk/ *adjective* huge, giant-like.

giggle /'gɪg(ə)l/ ● *verb* (**-ling**) laugh in half-suppressed spasms. ● *noun* such laugh; *colloquial* amusing person or thing. □ **giggly** *adjective* (**-ier**, **-iest**).

gigolo /'dʒɪgələʊ/ *noun* (*plural* **-s**) young man paid by older woman to be escort or lover.

gild[1] /gɪld/ *verb* (*past participle* **gilded** or as *adjective* **gilt**) cover thinly with gold; tinge with golden colour.

gild[2] = GUILD.

gill[1] /gɪl/ *noun* (usually in *plural*) respiratory organ of fish etc.; vertical radial plate on underside of mushroom etc.; flesh below person's jaws and ears.

gill[2] /dʒɪl/ *noun* quarter-pint measure.

gill[3] /gɪl/ *noun* (also **ghyll**) deep wooded ravine; narrow mountain torrent.

gillie /ˈgɪlɪ/ *noun Scottish* man or boy attending hunter or angler.

gillyflower /ˈdʒɪlɪflaʊə/ *noun* clove-scented flower, e.g. wallflower.

gilt¹ /gɪlt/ ● *adjective* overlaid (as) with gold. ● *noun* gilding. □ **gilt-edged** (of securities etc.) having high degree of reliability.

gilt² /gɪlt/ *noun* young sow.

gimbals /ˈdʒɪmb(ə)lz/ *plural noun* contrivance of rings etc. for keeping things horizontal in ship, aircraft, etc.

gimcrack /ˈdʒɪmkræk/ ● *adjective* flimsy, tawdry. ● *noun* showy ornament etc.

gimlet /ˈgɪmlɪt/ *noun* small boring-tool.

gimmick /ˈgɪmɪk/ *noun* trick or device, esp. to attract attention. □ **gimmickry** *noun*; **gimmicky** *adjective*.

gimp /gɪmp/ *noun* twist of silk etc. with cord or wire running through.

gin¹ /dʒɪn/ *noun* spirit distilled from grain or malt and flavoured with juniper berries.

gin² /dʒɪn/ ● *noun* snare, trap; machine separating cotton from seeds; kind of crane or windlass. ● *verb* (**-nn-**) treat (cotton) in gin; trap.

ginger /ˈdʒɪndʒə/ ● *noun* hot spicy root used in cooking; plant having this root; light reddish-yellow. ● *adjective* of ginger colour. ● *verb* flavour with ginger; (+ *up*) enliven. □ **ginger ale**, **beer** ginger-flavoured fizzy drinks; **gingerbread** ginger-flavoured treacle cake; **ginger group** group urging party or movement to stronger action; **ginger-nut** kind of ginger-flavoured biscuit. □ **gingery** *adjective*.

gingerly /ˈdʒɪndʒəlɪ/ ● *adverb* in a careful or cautious way. ● *adjective* showing extreme care or caution.

gingham /ˈgɪŋəm/ *noun* plain-woven usually checked cotton cloth.

gingivitis /dʒɪndʒɪˈvaɪtɪs/ *noun* inflammation of the gums.

ginkgo /ˈgɪŋkəʊ/ *noun* (*plural* **-s**) tree with fan-shaped leaves and yellow flowers.

ginseng /ˈdʒɪnseŋ/ *noun* plant found in E. Asia and N. America; medicinal root of this.

Gipsy = GYPSY.

giraffe /dʒɪˈrɑːf/ *noun* (*plural* same or **-s**) tall 4-legged African animal with long neck.

gird /gɜːd/ (*past & past participle* **girded** or **girt**) encircle or fasten (on) with waist-belt etc. □ **gird (up) one's loins** prepare for action.

girder /ˈgɜːdə/ *noun* iron or steel beam or compound structure used for bridges etc.

girdle¹ /ˈgɜːd(ə)l/ ● *noun* belt or cord worn round waist; corset; thing that surrounds; bony support for limbs. ● *verb* (**-ling**) surround with girdle.

girdle² /ˈgɜːd(ə)l/ *noun Scottish & Northern English* = GRIDDLE.

girl /gɜːl/ *noun* female child; *colloquial* young woman; *colloquial* girlfriend; female servant. □ **girlfriend** person's regular female companion; **girl guide** Guide; **girl scout** female Scout. □ **girlhood** *noun*; **girlish** *adjective*; **girly** *adjective*.

giro /ˈdʒaɪrəʊ/ ● *noun* (*plural* **-s**) system of credit transfer between banks, Post Offices, etc.; cheque or payment by giro. ● *verb* (**-es**, **-ed**) pay by giro.

girt *past & past participle* of GIRD.

girth /gɜːθ/ *noun* distance round a thing; band round body of horse securing saddle.

gist /ˈdʒɪst/ *noun* substance or essence of a matter.

gîte /ʒiːt/ *noun* furnished holiday house in French countryside. [French]

give /gɪv/ ● *verb* (**-ving**; *past* **gave**; *past participle* **given**) transfer possession of freely; provide with; administer; deliver; (often + *for*) make over in exchange or payment; confer; accord; pledge; perform (action etc.); utter, declare; yield to pressure; collapse; yield as product; consign; devote; present, offer (one's hand, arm, etc.); impart, be source of; concede; assume, grant, specify. ● *noun* capacity to comply; elasticity. □ **give and take** exchange of talk or ideas, ability to compromise; **give away** transfer as gift, hand over (bride) to bridegroom, betray or expose; **give-away** *colloquial* unintentional disclosure, free or inexpensive thing; **give in** yield, hand in; **give off** emit; **give out** announce, emit, distribute, be exhausted, run short; **give over** *colloquial* desist, hand over, devote; **give up** resign, surrender, part with, renounce hope (of), cease (activity). □ **giver** *noun*.

given ● *past participle* of GIVE. ● *adjective* (+ *to*) disposed or prone to; assumed as basis of reasoning etc.; fixed, specified.

gizmo /'gɪzməʊ/ *noun* (*plural* **-s**) gadget.

gizzard /'gɪzəd/ *noun* bird's second stomach, for grinding food.

glacé /'glæseɪ/ *adjective* (of fruit) preserved in sugar; (of cloth etc.) smooth, polished.

glacial /'gleɪʃ(ə)l/ *adjective* of ice or glaciers.

glaciated /'gleɪsɪeɪtɪd/ *adjective* marked or polished by moving ice; covered with glaciers. □ **glaciation** *noun.*

glacier /'glæsɪə/ *noun* slowly moving mass of ice on land.

glad *adjective* (**-dd-**) pleased; joyful, cheerful. □ **glad rags** *colloquial* best clothes. □ **gladden** *verb*; **gladly** *adverb*; **gladness** *noun.*

glade *noun* clear space in forest.

gladiator /'glædɪeɪtə/ *noun historical* trained fighter in ancient Roman shows. □ **gladiatorial** /-ɪə'tɔ:rɪəl/ *adjective.*

gladiolus /glædɪ'əʊləs/ *noun* (*plural* **-li** /-laɪ/) plant of lily family with bright flower-spikes.

gladsome *adjective poetical* cheerful, joyful.

Gladstone bag /'glædst(ə)n/ *noun* kind of light portmanteau.

glair *noun* white of egg; similar or derivative viscous substance.

glamour /'glæmə/ *noun* (*US* **glamor**) physical, esp. cosmetic, attractiveness; alluring or exciting beauty or charm. □ **glamorize** *verb* (also **-ise**) (**-zing** or **-sing**); **glamorous** *adjective*; **glamorously** *adverb.*

glance /glɑ:ns/ ● *verb* (**-cing**) (often + *down*, *up*, *over*, etc.) look or refer briefly; (often + *off*) hit at fine angle and bounce off. ● *noun* brief look; flash, gleam; swift oblique stroke in cricket. □ **at a glance** immediately on looking.

gland *noun* organ etc. secreting substances for use in body; similar organ in plant.

glanders /'glændəz/ *plural noun* contagious horse disease.

glandular /'glændjʊlə/ *adjective* of gland(s). □ **glandular fever** infectious disease with swelling of lymph glands.

glare /gleə/ ● *verb* (**-ring**) look fiercely; shine oppressively; (esp. as **glaring** *adjective*) be very evident. ● *noun* oppressive light or public attention; fierce look; tawdry brilliance. □ **glaringly** *adverb.*

glasnost /'glæznɒst/ *noun* (in former USSR) policy of more open government.

glass /glɑ:s/ ● *noun* hard, brittle, usually transparent substance made by fusing sand with soda and lime etc.; glass objects collectively; glass drinking vessel, its contents; glazed frame for plants; barometer; covering of watch-face; lens; (in *plural*) spectacles, binoculars; mirror. ● *verb* (usually as **glassed** *adjective*) fit with glass. □ **glass-blowing** blowing of semi-molten glass to make glass objects; **glass fibre** glass filaments made into fabric or reinforcing plastic; **glasshouse** greenhouse, *slang* military prison; **glass-paper** paper covered with powdered glass, for smoothing etc.; **glass wool** fine glass fibres for packing and insulation. □ **glassful** *noun* (*plural* **-s**).

glassy /'glɑ:sɪ/ *adjective* (**-ier**, **-iest**) like glass; (of eye etc.) dull, fixed.

glaucoma /glɔ:'kəʊmə/ *noun* eye disease with pressure in eyeball and gradual loss of sight.

glaze ● *verb* (**-zing**) fit with glass or windows; cover (pottery etc.) with vitreous substance or (surface) with smooth shiny coating; (often + *over*) (of eyes) become glassy. ● *noun* substance used for or surface produced by glazing.

glazier /'gleɪzɪə/ *noun* person who glazes windows etc.

gleam ● *noun* faint or brief light or show. ● *verb* emit gleam(s).

glean *verb* gather (facts etc.); gather (corn left by reapers). □ **gleanings** *plural noun.*

glebe *noun* piece of land yielding revenue to benefice.

glee *noun* mirth, delight; musical composition for several voices. □ **gleeful** *adjective*; **gleefully** *adverb.*

glen *noun* narrow valley.

glengarry /glen'gærɪ/ *noun* (*plural* **-ies**) kind of Highland cap.

glib *adjective* (**-bb-**) speaking or spoken fluently but insincerely. □ **glibly** *adverb*; **glibness** *noun.*

glide • *verb* (**-ding**) move smoothly or continuously; (of aircraft) fly without engine-power; go stealthily. • *noun* gliding motion.

glider /ˈglaɪdə/ *noun* light aircraft without engine.

glimmer /ˈglɪmə/ • *verb* shine faintly or intermittently. • *noun* faint or wavering light; (also **glimmering**) (usually + *of*) small sign.

glimpse /glɪmps/ • *noun* (often + *of*, *at*) brief view; faint transient appearance. • *verb* (**-sing**) have brief view of.

glint *verb & noun* flash, glitter.

glissade /glɪˈsɑːd/ • *noun* controlled slide down snow slope; gliding. • *verb* (**-ding**) perform glissade.

glissando /glɪˈsændəʊ/ *noun* (*plural* **-di** /-dɪ/ or **-s**) *Music* continuous slide of adjacent notes.

glisten /ˈglɪs(ə)n/ • *verb* shine like wet or polished surface. • *noun* glitter.

glitch *noun colloquial* irregularity, malfunction.

glitter /ˈglɪtə/ • *verb* shine with brilliant reflected light, sparkle; (often + *with*) be showy. • *noun* sparkle; showiness; tiny pieces of glittering material.

glitz *noun slang* showy glamour. □ **glitzy** *adjective* (**-ier**, **-iest**).

gloaming /ˈgləʊmɪŋ/ *noun* twilight.

gloat *verb* (often + *over* etc.) look or ponder with greedy or malicious pleasure.

global /ˈgləʊb(ə)l/ *adjective* worldwide; all-embracing. □ **global warming** increase in temperature of earth's atmosphere. □ **globally** *adverb.*

globe *noun* spherical object; spherical map of earth; (**the globe**) the earth. □ **globe artichoke** partly edible head of artichoke plant; **globe-trotter** person travelling widely.

globular /ˈglɒbjʊlə/ *adjective* globe-shaped; composed of globules.

globule /ˈglɒbjuːl/ *noun* small globe, round particle, or drop.

glockenspiel /ˈglɒkənspiːl/ *noun* musical instrument of bells or metal bars played with hammers.

gloom /gluːm/ *noun* darkness; melancholy, depression.

gloomy /ˈgluːmɪ/ *adjective* (**-ier**, **-iest**) dark; depressed, depressing.

glorify /ˈglɔːrɪfaɪ/ *verb* (**-ies**, **-ied**) make glorious; make seem more splendid than is the case; (as **glorified** *adjective*) treated as more important etc. than it is; extol. □ **glorification** *noun.*

glorious /ˈglɔːrɪəs/ *adjective* possessing or conferring glory; *colloquial* splendid, excellent. □ **gloriously** *adverb.*

glory /ˈglɔːrɪ/ • *noun* (*plural* **-ies**) (thing bringing) renown, honourable fame, etc.; adoring praise; resplendent majesty, beauty, etc.; halo of saint. • *verb* (**-ies**, **-ied**) (often + *in*) take pride.

gloss[1] • *noun* surface lustre; deceptively attractive appearance; (in full *gloss paint*) paint giving glossy finish. • *verb* make glossy. □ **gloss over** seek to conceal.

gloss[2] • *noun* explanatory comment added to text; interpretation. • *verb* add gloss to.

glossary /ˈglɒsərɪ/ *noun* (*plural* **-ies**) dictionary of technical or special words, esp. as appendix.

glossy • *adjective* (**-ier**, **-iest**) smooth and shiny; printed on such paper. • *noun* (*plural* **-ies**) *colloquial* glossy magazine or photograph.

glottal /ˈglɒt(ə)l/ *adjective* of the glottis. □ **glottal stop** sound produced by sudden opening or shutting of glottis.

glottis /ˈglɒtɪs/ *noun* opening at upper end of windpipe between vocal cords.

glove /glʌv/ • *noun* hand-covering for protection, warmth, etc.; boxing glove. • *verb* (**-ving**) cover or provide with gloves. □ **glove compartment** recess for small articles in car dashboard; **glove puppet** small puppet fitted on hand.

glow /gləʊ/ • *verb* emit flameless light and heat; (often + *with*) feel bodily heat or strong emotion; show warm colour; (as **glowing** *adjective*) expressing pride or satisfaction. • *noun* glowing state, appearance, or feeling. □ **glow-worm** beetle that emits green light.

glower /ˈglaʊə/ *verb* (often + *at*) scowl.

glucose /ˈgluːkəʊs/ *noun* kind of sugar found in blood, fruits, etc.

glue • *noun* substance used as adhesive. • *verb* (**glues**, **glued**, **gluing** or **glueing**) attach (as) with glue; hold closely. □ **glue ear** blocking of (esp. child's) Eustachian tube; **glue-sniffing** inhalation of fumes from adhesives as intoxicant. □ **gluey** *adjective* (**gluier**, **gluiest**).

glum *adjective* (**-mm-**) dejected, sullen. □ **glumly** *adverb*; **glumness** *noun.*

glut • *verb* (**-tt-**) feed or indulge to the full, satiate; overstock. • *noun* excessive supply; surfeit.

gluten /'glu:t(ə)n/ *noun* sticky part of wheat flour.

glutinous /'glu:tɪnəs/ *adjective* sticky, gluelike.

glutton /'glʌt(ə)n/ *noun* excessive eater; (often + *for*) *colloquial* insatiably eager person; voracious animal of weasel family. □ **gluttonous** *adjective*; **gluttonously** *adverb*; **gluttony** *noun.*

glycerine /'glɪsəri:n/ *noun* (also **glycerol**, *US* **glycerin**) colourless sweet viscous liquid used in medicines, explosives, etc.

gm *abbreviation* gram(s).

GMT *abbreviation* Greenwich Mean Time.

gnarled /nɑ:ld/ *adjective* knobbly, twisted, rugged.

gnash /næʃ/ *verb* grind (one's teeth); (of teeth) strike together.

gnat /næt/ *noun* small biting fly.

gnaw /nɔ:/ *verb* (usually + *away* etc.) wear away by biting; (often + *at*, *into*) bite persistently; corrode; torment.

gneiss /naɪs/ *noun* coarse-grained rock of feldspar, quartz, and mica.

gnome /nəʊm/ *noun* dwarf, goblin; (esp. in *plural*) *colloquial* person with sinister influence, esp. financial.

gnomic /'nəʊmɪk/ *adjective* of aphorisms; sententious.

gnomon /'nəʊmɒn/ *noun* rod etc. on sundial, showing time by its shadow.

gnostic /'nɒstɪk/ • *adjective* of knowledge; having special mystic knowledge. • *noun* (**Gnostic**) early Christian heretic claiming mystical knowledge. □ **Gnosticism** /-sɪz(ə)m/ *noun.*

GNP *abbreviation* gross national product.

gnu /nu:/ *noun* (*plural* same or **-s**) oxlike antelope.

go[1] • *verb* (*3rd singular present* **goes** /gəʊz/; *past* **went**; *past participle* **gone** /gɒn/) walk, travel, proceed; participate in (doing something); extend in a certain direction; depart; move, function; make specified movement or sound, *colloquial* say; be, become; elapse, be traversed; (of song etc.) have specified wording etc.; match; be regularly kept, fit; be successful; be sold, (of money) be spent; be relinquished, fail, decline, collapse; be acceptable or accepted; (often + *by*, *with*, *on*, *upon*) be guided by; attend regularly; (+ *to*, *towards*) contribute to; (+ *for*) apply to. • *noun* (*plural* **goes**) animation; vigorous activity; success; turn, attempt. □ **go-ahead** *adjective* enterprising, *noun* permission to proceed; **go-between** intermediary; **go down** descend, become less, decrease (in price), subside, sink, (of sun) set, deteriorate, cease to function, be recorded, be swallowed, (+ *with*) find acceptance with, *colloquial* leave university, *colloquial* be sent to prison, (+ *with*) become ill with; **go for** go to fetch, prefer, choose, pass or be accounted as, *colloquial* attack, *colloquial* strive to attain; **go-getter** *colloquial* pushily enterprising person; **go in for** compete or engage in; **go-kart**, **-cart** miniature racing car with skeleton body; **go off** explode, deteriorate, fall asleep, begin to dislike; **go off well**, **badly** succeed, fail; **go on** continue, proceed, *colloquial* talk at great length, (+ *at*) *colloquial* nag, use as evidence; **go out** leave room or house, be extinguished, be broadcast, cease to be fashionable, (often + *with*) have romantic or sexual relationship, (usually + *to*) sympathize; **go over** inspect details of, rehearse; **go round** spin, revolve, suffice for all; **go slow** work slowly as industrial protest; **go under** sink, succumb, fail; **go up** rise, increase (in price), be consumed (in flames etc.), explode, *colloquial* enter university; **go without** manage without or forgo (something); **have a go at** attack, attempt; **on the go** *colloquial* active.

go[2] *noun* Japanese board game.

goad • *verb* urge with goad; (usually + *on*, *into*) irritate, stimulate. • *noun* spiked stick for urging cattle; thing that torments or incites.

goal *noun* object of effort; destination; structure into or through which ball is to be driven in certain games; point(s) so won; point where race ends.

□ **goalkeeper** player protecting goal; **goalpost** either post supporting crossbar of goal.

goalie *noun colloquial* goalkeeper.

goat *noun* small domesticated mammal with horns and (in male) beard; licentious man; *colloquial* fool. □ **get (person's) goat** *colloquial* irritate him or her.

goatee /gəʊ'ti:/ *noun* small pointed beard.

gob[1] *noun slang* mouth. □ **gobsmacked** *slang* flabbergasted; **gob-stopper** large hard sweet.

gob[2] *slang* ● *noun* clot of slimy matter. ● *verb* (**-bb-**) spit.

gobbet /'gɒbɪt/ *noun* lump of flesh, food, etc.; extract from text set for translation or comment.

gobble[1] /'gɒb(ə)l/ *verb* (**-ling**) eat hurriedly and noisily.

gobble[2] /'gɒb(ə)l/ *verb* (**-ling**) (of turkeycock) make guttural sound; speak thus.

gobbledegook /'gɒbəldɪgu:k/ *noun* (also **gobbledeygook**) *colloquial* pompous or unintelligible jargon.

goblet /'gɒblɪt/ *noun* drinking vessel with foot and stem.

goblin /'gɒblɪn/ *noun* mischievous demon.

goby /'gəʊbɪ/ *noun* (*plural* **-ies**) small fish with sucker on underside.

god *noun* superhuman being worshipped as possessing power over nature, human fortunes, etc.; (**God**) creator and ruler of universe; idol; adored person; (**the gods**) (occupants of) gallery in theatre. □ **godchild** person in relation to godparent; **god-daughter** female godchild; **godfather** male godparent; **God-fearing** religious; **God-forsaken** dismal; **godmother** female godparent; **godparent** person who responds on behalf of candidate at baptism; **godsend** unexpected welcome event or acquisition; **godson** male godchild. □ **godlike** *adjective*.

goddess /'gɒdɪs/ *noun* female deity; adored woman.

godhead *noun* divine nature; deity.

godless *adjective* impious, wicked; not believing in God. □ **godlessness** *noun*.

godly /'gɒdlɪ/ *adjective* (**-ier**, **-iest**) pious, devout. □ **godliness** *noun*.

goer /'gəʊə/ *noun* person or thing that goes; *colloquial* lively or sexually promiscuous person. □ **-goer** regular attender.

goggle /'gɒg(ə)l/ ● *verb* (**-ling**) (often + *at*) look with wide-open eyes; (of eyes) be rolled, project; roll (eyes). ● *adjective* (of eyes) protuberant, rolling. ● *noun* (in *plural*) spectacles for protecting eyes. □ **goggle-box** *colloquial* television set.

going /'gəʊɪŋ/ ● *noun* condition of ground as affecting riding etc. ● *adjective* in action; existing, available; current, prevalent. □ **going concern** thriving business; **going-over** (*plural* **goings-over**) *colloquial* inspection or overhaul, *slang* thrashing; **goings-on** strange conduct.

goitre /'gɔɪtə/ *noun* (*US* **goiter**) abnormal enlargement of thyroid gland.

gold /gəʊld/ ● *noun* precious yellow metal; colour of this; coins or articles of gold. ● *adjective* of or coloured like gold. □ **gold-digger** *slang* woman who goes after men for their money; **gold field** area with naturally occurring gold; **goldfinch** brightly coloured songbird; **goldfish** small golden-red Chinese carp; **gold leaf** gold beaten into thin sheet; **gold medal** medal given usually as first prize; **gold plate** vessels of gold, material plated with gold; **gold-plate** plate with gold; **gold-rush** rush to newly discovered gold field; **goldsmith** worker in gold; **gold standard** financial system in which value of money is based on gold.

golden /'gəʊld(ə)n/ *adjective* of gold; coloured or shining like gold; precious, excellent. □ **golden handshake** *colloquial* gratuity as compensation for redundancy or compulsory retirement; **golden jubilee** 50th anniversary of reign; **golden mean** principle of moderation; **golden retriever** retriever with gold-coloured coat; **golden wedding** 50th anniversary of wedding.

golf ● *noun* game in which small hard ball is struck with clubs over ground into series of small holes. ● *verb* play golf. □ **golf ball** ball used in golf, spherical unit carrying type in some electric typewriters; **golf course** area of land on which golf is played; **golf club** club used in golf, (premises of) association for playing golf. □ **golfer** *noun*.

golliwog /'gɒlɪwɒg/ *noun* black-faced soft doll with fuzzy hair.

gonad /'gəʊnæd/ *noun* animal organ producing gametes, e.g. testis or ovary.

gondola /'gɒndələ/ *noun* light Venetian canal-boat; car suspended from airship.

gondolier /gɒndə'lɪə/ *noun* oarsman of gondola.

gone /gɒn/ ● *past participle* of GO¹. ● *adjective* (of time) past; lost, hopeless, dead; *colloquial* pregnant for specified time.

goner /'gɒnə/ *noun slang* person or thing that is doomed or irrevocably lost.

gong *noun* metal disc giving resonant note when struck; saucer-shaped bell; *slang* medal.

gonorrhoea /gɒnə'ri:ə/ *noun* (*US* **gonorrhea**) a venereal disease.

goo *noun colloquial* sticky or slimy substance; sickly sentiment. □ **gooey** *adjective* (**gooier**, **gooiest**).

good /gʊd/ ● *adjective* (**better**, **best**) having right qualities, adequate; competent, effective; kind, morally excellent, virtuous; well-behaved; agreeable; considerable; not less than; beneficial; valid. ● *noun* (only in *singular*) good quality or circumstance; (in *plural*) movable property, merchandise. □ **good-for-nothing** worthless (person); **good humour** genial mood; **good-looking** handsome; **good nature** kindly disposition; **goodwill** kindly feeling, established value-enhancing reputation of a business.

goodbye /gʊd'baɪ/ (*US* **goodby**) ● *interjection expressing good wishes at parting.* ● *noun* (*plural* **-byes** or *US* **-bys**) parting, farewell.

goodly /'gʊdlɪ/ *adjective* (**-ier**, **-iest**) handsome; of imposing size etc.

goodness /'gʊdnɪs/ *noun* virtue; excellence; kindness; nutriment.

goody /'gʊdɪ/ ● *noun* (*plural* **-ies**) *colloquial* good person; (usually in *plural*) something good or attractive, esp. to eat. ● *interjection expressing childish delight.* □ **goody-goody** *colloquial* (person who is) smugly or obtrusively virtuous.

goof /gu:f/ *slang* ● *noun* foolish or stupid person or mistake. ● *verb* bungle, blunder. □ **goofy** *adjective* (**-ier**, **-iest**).

googly /'gu:glɪ/ *noun* (*plural* **-ies**) *Cricket* ball bowled so as to bounce in unexpected direction.

goon /gu:n/ *noun slang* stupid person; *esp. US* hired ruffian.

goose /gu:s/ *noun* (*plural* **geese** /gi:s/) large web-footed bird; female of this; *colloquial* simpleton. □ **goose-flesh**, **-pimples** (*US* **-bumps**) bristling state of skin due to cold or fright; **goose-step** stiff-legged marching step.

gooseberry /'gʊzbərɪ/ *noun* (*plural* **-ies**) small green usually sour berry; thorny shrub bearing this.

gopher /'gəʊfə/ *noun* American burrowing rodent.

gore¹ *noun* clotted blood.

gore² *verb* (**-ring**) pierce with horn, tusk, etc.

gore³ ● *noun* wedge-shaped piece in garment; triangular or tapering piece in umbrella etc. ● *verb* (**-ring**) shape with gore.

gorge ● *noun* narrow opening between hills; surfeit; contents of stomach. ● *verb* (**-ging**) feed greedily; satiate.

gorgeous /'gɔ:dʒəs/ *adjective* richly coloured; *colloquial* splendid; *colloquial* strikingly beautiful. □ **gorgeously** *adverb.*

gorgon /'gɔ:gən/ *noun* (in Greek mythology) any of 3 snake-haired sisters able to turn people to stone; frightening or repulsive woman.

Gorgonzola /gɔ:gən'zəʊlə/ *noun* rich blue-veined Italian cheese.

gorilla /gə'rɪlə/ *noun* largest anthropoid ape.

gormless /'gɔ:mlɪs/ *adjective colloquial* foolish, lacking sense. □ **gormlessly** *adverb.*

gorse /gɔ:s/ *noun* prickly shrub with yellow flowers.

Gorsedd /'gɔ:seð/ *noun* Druidic order meeting before eisteddfod.

gory /'gɔ:rɪ/ *adjective* (**-ier**, **-iest**) involving bloodshed; bloodstained.

gosh *interjection expressing surprise.*

goshawk /'gɒshɔ:k/ *noun* large short-winged hawk.

gosling /'gɒzlɪŋ/ *noun* young goose.

gospel /'gɒsp(ə)l/ *noun* teaching or revelation of Christ; (**Gospel**) (each of 4 books giving) account of Christ's life in

New Testament; portion of this read at church service; thing regarded as absolutely true. □ **gospel music** black American religious singing.

gossamer /ˈgɒsəmə/ • *noun* filmy substance of small spiders' webs; delicate filmy material. • *adjective* light and flimsy as gossamer.

gossip /ˈgɒsɪp/ • *noun* unconstrained talk or writing, esp. about people; idle talk; person indulging in gossip. • *verb* (**-p-**) talk or write gossip. □ **gossip column** regular newspaper column of gossip. □ **gossipy** *adjective.*

got *past & past participle* of GET.

Goth *noun* member of Germanic tribe that invaded Roman Empire in 3rd–5th c.

Gothic *adjective* of Goths; *Architecture* in the pointed-arch style prevalent in W. Europe in 12th–16th c.; (of novel etc.) in a style popular in 18th & 19th c., with supernatural or horrifying events.

gotten *US past participle* of GET.

gouache /guːˈɑːʃ/ *noun* painting with opaque water-colour; pigments used for this.

Gouda /ˈgaʊdə/ *noun* flat round Dutch cheese.

gouge /gaʊdʒ/ • *noun* concave-bladed chisel. • *verb* (**-ging**) cut or (+ *out*) force out (as) with gouge.

goulash /ˈguːlæʃ/ *noun* stew of meat and vegetables seasoned with paprika.

gourd /gʊəd/ *noun* fleshy fruit of trailing or climbing cucumber-like plant; this plant; dried rind of this fruit used as bottle etc.

gourmand /ˈgʊəmənd/ *noun* glutton; gourmet.

■ **Usage** The use of *gourmand* to mean a 'gourmet' is considered incorrect by some people.

gourmet /ˈgʊəmeɪ/ *noun* connoisseur of good food.

gout /gaʊt/ *noun* disease with inflammation of small joints. □ **gouty** *adjective.*

govern /ˈgʌv(ə)n/ *verb* rule with authority; conduct policy and affairs of; influence or determine; curb, control.

governance *noun* act, manner, or function of governing.

governess /ˈgʌvənɪs/ *noun* woman employed to teach children in private household.

government *noun* manner or system of governing; group of people governing state. □ **governmental** /-ˈmen-/ *adjective.*

governor *noun* ruler; official governing a province, town, etc.; executive head of each State of US; member of governing body of institution; *slang* one's employer or father; automatic regulator controlling speed of engine etc. □ **Governor-General** representative of Crown in Commonwealth country regarding Queen as head of state. □ **governorship** *noun.*

gown /gaʊn/ *noun* woman's, esp. formal or elegant, long dress; official robe of alderman, judge, cleric, academic, etc.; surgeon's overall.

goy *noun* (*plural* **-im** or **-s**) *Jewish name for* non-Jew.

GP *abbreviation* general practitioner.

GPO *abbreviation* General Post Office.

gr *abbreviation* (also **gr.**) gram(s); grain(s); gross.

grab • *verb* (**-bb-**) seize suddenly; take greedily; *slang* impress; (+ *at*) snatch at. • *noun* sudden clutch or attempt to seize; device for clutching.

grace • *noun* elegance of proportions, manner, or movement; courteous good will; attractive feature; unmerited favour of God; goodwill; delay granted; thanksgiving at meals; (**His**, **Her**, **Your Grace**) *title used of or to duke, duchess, or archbishop.* • *verb* (**-cing**) (often + *with*) add grace to; bestow honour on. □ **grace note** *Music* note embellishing melody.

graceful *adjective* full of grace or elegance. □ **gracefully** *adverb.*

graceless *adjective* lacking grace or charm.

gracious /ˈgreɪʃəs/ *adjective* kindly, esp. to inferiors; merciful. □ **gracious living** elegant way of life. □ **graciously** *adverb*; **graciousness** *noun.*

gradate /grəˈdeɪt/ *verb* (**-ting**) (cause to) pass gradually from one shade to another; arrange in steps or grades.

gradation *noun* (usually in *plural*) stage of transition or advance; degree in rank, intensity, etc.; arrangement in grades. □ **gradational** *adjective.*

grade ● *noun* degree in rank, merit, etc.; mark indicating quality of student's work; slope; *US* class in school. ● *verb* (**-ding**) arrange in grades; (+ *up, down, off, into,* etc.) pass between grades; give grade to; reduce to easy gradients. □ **make the grade** succeed.

gradient /'greɪdɪənt/ *noun* sloping road etc.; amount of such slope.

gradual /'grædʒʊəl/ *adjective* happening by degrees; not steep or abrupt. □ **gradually** *adverb.*

graduate ● *noun* /'grædʒʊət/ holder of academic degree. ● *verb* /-eɪt/ (**-ting**) obtain academic degree; (+ *to*) move up to; mark in degrees or portions; arrange in gradations; apportion (tax etc.) according to scale. □ **graduation** *noun.*

graffiti /grə'fi:tɪ/ *plural noun* (*singular* **graffito**) writing or drawing on wall etc.

■ **Usage** *Graffiti* should be used with plural verbs, as in *Graffiti have appeared everywhere.*

graft[1] /grɑ:ft/ ● *noun* shoot or scion planted in slit in another stock; piece of transplanted living tissue; *slang* hard work. ● *verb* (often + *in, on, together,* etc.) insert (graft); transplant (living tissue); (+ *in, on*) insert or fix (thing) permanently to another; *slang* work hard.

graft[2] /grɑ:ft/ *colloquial* ● *noun* practices for securing illicit gains in politics or business; such gains. ● *verb* seek or make graft.

Grail *noun* (in full **Holy Grail**) legendary cup or platter used by Christ at Last Supper.

grain ● *noun* fruit or seed of cereal; wheat or allied food-grass; corn; particle of sand, salt, etc.; unit of weight (0.065 g); least possible amount; texture in skin, wood, stone, etc.; arrangement of lines of fibre in wood. ● *verb* paint in imitation of grain of wood; form into grains.

gram *noun* (also **gramme**) metric unit of weight.

grammar /'græmə/ *noun* study or rules of relations between words in (a) language; application of such rules; book on grammar. □ **grammar school** *esp. historical* secondary school with academic curriculum.

grammarian /grə'meərɪən/ *noun* expert in grammar.

grammatical /grə'mætɪk(ə)l/ *adjective* of or according to grammar.

gramophone /'græməfəʊn/ *noun* record player.

grampus /'græmpəs/ *noun* (*plural* **-es**) sea mammal of dolphin family.

gran *noun colloquial* grandmother.

granary /'grænərɪ/ *noun* (*plural* **-ies**) storehouse for grain; region producing much corn.

grand ● *adjective* splendid, imposing; chief, of chief importance; (**Grand**) of highest rank; *colloquial* excellent. ● *noun* grand piano; (*plural* same) (usually in *plural*) *slang* 1,000 dollars or pounds. □ **grand jury** jury to examine validity of accusation before trial; **grand piano** piano with horizontal strings; **grand slam** winning of all of group of matches; **grand total** sum of other totals. □ **grandly** *adverb*; **grandness** *noun.*

grandad *noun* (also **grand-dad**) *colloquial* grandfather.

grandchild *noun* child of one's son or daughter.

granddaughter *noun* one's child's daughter.

grandee /græn'di:/ *noun* Spanish or Portuguese noble of highest rank; great personage.

grandeur /'grændʒə/ *noun* majesty, splendour, dignity; high rank, eminence.

grandfather *noun* one's parent's father. □ **grandfather clock** clock in tall wooden case.

grandiloquent /græn'dɪləkwənt/ *adjective* pompous or inflated in language. □ **grandiloquence** *noun.*

grandiose /'grændɪəʊs/ *adjective* imposing; planned on large scale. □ **grandiosity** /-'ɒsɪtɪ/ *noun.*

grandma *noun colloquial* grandmother.

grandmother *noun* one's parent's mother.

grandparent *noun* one's parent's parent.

Grand Prix /grɑ̃ 'pri:/ *noun* any of several international motor-racing events.

grandson *noun* one's child's son.

grandstand *noun* main stand for spectators at racecourse etc.

grange /greɪndʒ/ *noun* country house with farm buildings.

granite /'grænɪt/ *noun* granular crystalline rock of quartz, mica, etc.

granny /'grænɪ/ *noun* (also **grannie**) (*plural* **-ies**) *colloquial* grandmother; (in full **granny knot**) reef-knot crossed wrong way.

grant /grɑ:nt/ ● *verb* consent to fulfil; allow to have; give formally, transfer legally; (often + *that*) admit, concede. ● *noun* granting; thing, esp. money, granted. □ **take for granted** assume to be true, cease to appreciate through familiarity. □ **grantor** /grɑ:n'tɔ:/ *noun*.

granular /'grænjʊlə/ *adjective* of or like grains or granules.

granulate /'grænjʊleɪt/ *verb* (**-ting**) form into grains; roughen surface of. □ **granulation** *noun*.

granule /'grænju:l/ *noun* small grain.

grape *noun* usually green or purple berry growing in clusters on vine. □ **grapeshot** *historical* small balls as scattering charge for cannon etc.; **grapevine** vine, means of transmission of rumour.

grapefruit /'greɪpfru:t/ *noun* (*plural* same) large round usually yellow citrus fruit.

graph /grɑ:f/ ● *noun* symbolic diagram representing relation between two or more variables. ● *verb* plot on graph.

graphic /'græfɪk/ *adjective* of writing, drawing, etc.; vividly descriptive. □ **graphic arts** visual and technical arts involving design or lettering. □ **graphically** *adverb*.

graphics *plural noun* (usually treated as *singular*) products of graphic arts; use of diagrams in calculation and design.

graphite /'græfaɪt/ *noun* crystalline form of carbon used as lubricant, in pencils, etc.

graphology /grə'fɒlədʒɪ/ *noun* study of handwriting. □ **graphologist** *noun*.

grapnel /'græpn(ə)l/ *noun* iron-clawed instrument for dragging or grasping; small many-fluked anchor.

grapple /'græp(ə)l/ ● *verb* (**-ling**) (often + *with*) fight at close quarters; (+ *with*) try to manage (problem etc.); grip with hands, come to close quarters with; seize. ● *noun* hold (as) of wrestler; contest at close quarters; clutching-instrument. □ **grappling-iron**, **-hook** grapnel.

grasp /grɑ:sp/ ● *verb* clutch at, seize greedily; hold firmly; understand, realize. ● *noun* firm hold, grip; (+ *of*) mastery, mental hold.

grasping *adjective* avaricious.

grass /grɑ:s/ ● *noun* (any of several) plants with bladelike leaves eaten by ruminants; pasture land; grass-covered ground; grazing; *slang* marijuana; *slang* informer. ● *verb* cover with turf; *US* pasture; *slang* betray, inform police. □ **grass roots** fundamental level or source, rank and file; **grass snake** small non-poisonous snake; **grass widow**, **widower** person whose husband (or wife) is temporarily absent. □ **grassy** *adjective* (**-ier**, **-iest**).

grasshopper /'grɑ:shɒpə/ *noun* jumping and chirping insect.

grate[1] *verb* (**-ting**) reduce to small particles by rubbing on rough surface; (often + *against*, *on*) rub with, utter with, or make harsh sound, have irritating effect; grind, creak. □ **grater** *noun*.

grate[2] *noun* (metal) frame holding fuel in fireplace etc.

grateful /'greɪtfʊl/ *adjective* thankful; feeling or showing gratitude. □ **gratefully** *adverb*.

gratify /'grætɪfaɪ/ *verb* (**-ies**, **-ied**) please, delight; indulge. □ **gratification** *noun*.

grating /'greɪtɪŋ/ *noun* framework of parallel or crossed metal bars.

gratis /'grɑ:tɪs/ *adverb & adjective* free, without charge.

gratitude /'grætɪtju:d/ *noun* being thankful.

gratuitous /grə'tju:ɪtəs/ *adjective* given or done gratis; uncalled-for, motiveless. □ **gratuitously** *adverb*; **gratuitousness** *noun*.

gratuity /grə'tju:ɪtɪ/ *noun* (*plural* **-ies**) money given for good service.

grave[1] /greɪv/ *noun* hole dug for burial of corpse; mound or monument over this; (**the grave**) death. □ **gravestone**

(usually inscribed) stone over grave; **graveyard** burial ground.

grave[2] /greɪv/ *adjective* weighty, serious; dignified, solemn; threatening. □ **gravely** *adverb.*

grave[3] /greɪv/ *verb* (**-ving**; *past participle* **graven** or **graved**) (+ *in, on*) fix indelibly on (memory etc.); *archaic* engrave, carve. □ **graven image** idol.

grave[4] /grɑːv/ *noun* (in full **grave accent**) mark (`) over letter indicating pronunciation.

gravel /'græv(ə)l/ ● *noun* coarse sand and small stones; formation of crystals in bladder. ● *verb* (**-ll-**; *US* **-l-**) lay with gravel.

gravelly /'grævəlɪ/ *adjective* of or like gravel; (of voice) deep and rough-sounding.

gravid /'grævɪd/ *adjective* pregnant.

gravitate /'grævɪteɪt/ *verb* (**-ting**) (+ *to, towards*) move, be attracted, or tend by force of gravity to(wards); sink (as) by gravity.

gravitation *noun* attraction between each particle of matter and every other; effect of this, esp. falling of bodies to earth. □ **gravitational** *adjective.*

gravity /'grævɪtɪ/ *noun* force that attracts body to centre of earth etc.; intensity of this; weight; importance, seriousness; solemnity.

gravy /'greɪvɪ/ *noun* (*plural* **-ies**) (sauce made from) juices exuding from meat in and after cooking. □ **gravy-boat** long shallow jug for gravy; **gravy train** *slang* source of easy financial benefit.

gray *US* = GREY.

grayling *noun* (*plural* same) silver-grey freshwater fish.

graze[1] *verb* (**-zing**) feed on growing grass; pasture cattle.

graze[2] ● *verb* (**-zing**) rub or scrape (part of body); (+ *against, along,* etc.) touch lightly in passing, move with such contact. ● *noun* abrasion.

grazier /'greɪzɪə/ *noun* person who feeds cattle for market.

grazing *noun* grassland suitable for pasturage.

grease /griːs/ ● *noun* oily or fatty matter, esp. as lubricant; melted fat of dead animal. ● *verb* (**-sing**) smear or lubricate with grease. □ **greasepaint** actor's make-up; **greaseproof** impervious to grease.

greasy /'griːsɪ/ *adjective* (**-ier**, **-iest**) of, like, smeared with, or having too much grease; (of person, manner) unctuous. □ **greasiness** *noun.*

great /greɪt/ ● *adjective* above average in bulk, number, extent, or intensity; important, pre-eminent; imposing, distinguished; of remarkable ability etc.; (+ *at, on*) competent, well-informed; *colloquial* very satisfactory. ● *noun* great person or thing. □ **greatcoat** heavy overcoat; **Great Dane** dog of large short-haired breed. □ **greatness** *noun.*

great- /greɪt/ *combining form* (of family relationships) one degree more remote (*great-grandfather, great-niece,* etc.).

greatly *adverb* much.

grebe *noun* a diving bird.

Grecian /'griːʃ(ə)n/ *adjective* Greek.

greed *noun* excessive desire, esp. for food or wealth.

greedy /'griːdɪ/ *adjective* (**-ier**, **-iest**) showing greed; (+ *for, to do*) eager. □ **greedily** *adverb.*

Greek ● *noun* native, national, or language of Greece. ● *adjective* of Greece.

green ● *adjective* coloured like grass; unripe, unseasoned; not dried, smoked, or tanned; inexperienced; jealous; (also **Green**) concerned with protection of environment, not harmful to environment. ● *noun* green colour, paint, clothes, etc.; piece of grassy public land; grassy area for special purpose; (in *plural*) green vegetables; (also **Green**) supporter of protection of environment. □ **green belt** area of open land for preservation round city; **green card** motorist's international insurance document; **greenfinch** bird with greenish plumage; **green fingers** *colloquial* skill in gardening; **greenfly** green aphid; **greengage** round green plum; **greenhorn** novice; **green light** signal or permission to proceed; **green pound** the agreed value of the pound for payments to agricultural producers in EC; **green-room** room in theatre for actors when off stage; **greensward** grassy turf.

greenery *noun* green foliage.

greengrocer /'gri:ngrəʊsə/ *noun* retailer of fruit and vegetables. □ **greengrocery** *noun* (*plural* **-ies**).

greenhouse *noun* structure with sides and roof mainly of glass, for rearing plants. □ **greenhouse effect** trapping of sun's warmth in earth's lower atmosphere; **greenhouse gas** gas contributing to greenhouse effect, esp. carbon dioxide.

greet *verb* address on meeting or arrival; receive or acknowledge in specified way; become apparent to (eye, ear, etc.).

greeting *noun* act or words used to greet. □ **greetings card** decorative card carrying goodwill message etc.

gregarious /grɪ'geərɪəs/ *adjective* fond of company; living in flocks etc. □ **gregariousness** *noun*.

Gregorian calendar /grɪ'gɔ:rɪən/ *noun* calendar introduced in 1582 by Pope Gregory XIII.

Gregorian chant /grɪ'gɔ:rɪən/ *noun* form of plainsong named after Pope Gregory I.

gremlin /'gremlɪn/ *noun colloquial* mischievous sprite said to cause mechanical faults etc.

grenade /grɪ'neɪd/ *noun* small bomb thrown by hand or shot from rifle.

grenadier /grenə'dɪə/ *noun* (**Grenadier**) member of first regiment of royal household infantry; *historical* soldier armed with grenades.

grew *past* of GROW.

grey /greɪ/ (*US* **gray**) • *adjective* of colour between black and white; clouded, dull; (of hair) turning white, (of person) having grey hair; anonymous, unidentifiable; undistinguished, boring. • *noun* grey colour, paint, clothes, etc.; grey horse. • *verb* make or become grey. □ **grey area** indefinite situation or topic; **Grey Friar** Franciscan friar; **grey matter** darker tissues of brain, *colloquial* intelligence.

greyhound *noun* slender swift dog used in racing.

greylag *noun* European wild goose.

grid *noun* grating; system of numbered squares for map references; network of lines, electric power connections, etc.; pattern of lines marking starting-place on motor-racing track.

griddle /'grɪd(ə)l/ *noun* iron plate placed over heat for baking etc.

gridiron /'grɪdaɪən/ *noun* barred metal frame for broiling or grilling; American football field.

grief /gri:f/ *noun* (cause of) intense sorrow. □ **come to grief** meet with disaster.

grievance *noun* real or imagined cause for complaint.

grieve /gri:v/ *verb* (**-ving**) (cause to) feel grief.

grievous /'gri:vəs/ *adjective* severe; causing grief; injurious; flagrant, heinous. □ **grievously** *adverb*.

griffin /'grɪfɪn/ *noun* (also **gryphon** /-f(ə)n/) mythical creature with eagle's head and wings and lion's body.

griffon /'grɪf(ə)n/ *noun* small coarse-haired terrier-like dog; large vulture; griffin.

grill • *noun* device on cooker for radiating heat downwards; gridiron; grilled food; (in full **grill room**) restaurant specializing in grills. • *verb* cook under grill or on gridiron; subject to or experience extreme heat; subject to severe questioning.

grille /grɪl/ *noun* (also **grill**) grating, latticed screen; metal grid protecting vehicle radiator.

grilse /grɪls/ *noun* (*plural* same or **-s**) young salmon that has been to the sea only once.

grim *adjective* (**-mm-**) of stern appearance; harsh, merciless; ghastly, joyless; unpleasant. □ **grimly** *adverb*; **grimness** *noun*.

grimace /'grɪməs/ • *noun* distortion of face made in disgust etc. or to amuse. • *verb* (**-cing**) make grimace.

grime • *noun* deeply ingrained dirt. • *verb* (**-ming**) blacken, befoul. □ **grimy** *adjective* (**-ier**, **-iest**).

grin • *verb* (**-nn-**) smile broadly. • *noun* broad smile.

grind /graɪnd/ • *verb* (*past & past participle* **ground** /graʊnd/) crush to small particles; sharpen; rub gratingly; (often + *down*) oppress; (often + *away*) work or study hard. • *noun* grinding; *colloquial* hard dull work. □ **grindstone** thick revolving abrasive disc for grinding, sharpening, etc. □ **grinder** *noun.*

grip ● *verb* (**-pp-**) grasp tightly; take firm hold; compel attention of. ● *noun* firm hold, grasp; way of holding; power of holding attention; intellectual mastery; control of one's behaviour; part of machine that grips; part of weapon etc. that is gripped; hairgrip; travelling bag.

gripe ● *verb* (**-ping**) *colloquial* complain; affect with colic. ● *noun* (usually in *plural*) colic; *colloquial* complaint. □ **Gripe Water** *proprietary term* medicine to relieve colic in babies.

grisly /ˈgrɪzlɪ/ *adjective* (**-ier**, **-iest**) causing horror, disgust, or fear.

grist *noun* corn for grinding. □ **grist to the mill** source of profit or advantage.

gristle /ˈgrɪs(ə)l/ *noun* tough flexible tissue; cartilage. □ **gristly** *adjective.*

grit ● *noun* small particles of sand etc.; coarse sandstone; *colloquial* pluck, endurance. ● *verb* (**-tt-**) spread grit on (icy roads etc.); clench (teeth); make grating sound. □ **gritty** *adjective* (**-ier**, **-iest**).

grits *plural noun* coarse oatmeal; unground husked oats.

grizzle /ˈgrɪz(ə)l/ *verb* (**-ling**) *colloquial* cry fretfully. □ **grizzly** *adjective* (**-ier**, **-iest**).

grizzled *adjective* grey-haired.

grizzly /ˈgrɪzlɪ/ ● *adjective* (**-ier**, **-iest**) grey-haired. ● *noun* (*plural* **-ies**) (in full **grizzly bear**) large fierce N. American bear.

groan ● *verb* make deep sound expressing pain, grief, or disapproval; (usually + *under*, *beneath*, *with*) be loaded or oppressed. ● *noun* sound made in groaning.

groat *noun historical* silver coin worth 4 old pence.

groats *plural noun* hulled or crushed grain, esp. oats.

grocer /ˈgrəʊsə/ *noun* dealer in food and household provisions.

grocery /ˈgrəʊsərɪ/ *noun* (*plural* **-ies**) grocer's trade, shop, or (in *plural*) goods.

grog *noun* drink of spirit (originally rum) and water.

groggy /ˈgrɒgɪ/ *adjective* (**-ier**, **-iest**) incapable, unsteady. □ **groggily** *adverb.*

groin[1] ● *noun* depression between belly and thigh; edge formed by intersecting vaults. ● *verb* build with groins.

groin[2] *US* = GROYNE.

grommet /ˈgrɒmɪt/ *noun* eyelet placed in hole to protect or insulate rope or cable passed through it; tube passed through eardrum to middle ear.

groom /gru:m/ ● *noun* person employed to tend horses; bridegroom. ● *verb* tend (horse); give neat or attractive appearance to; prepare (person) for office or occasion etc.

groove ● *noun* channel, elongated hollow; spiral cut in gramophone record for needle. ● *verb* (**-ving**) make groove(s) in.

groovy /ˈgru:vɪ/ *adjective* (**-ier**, **-iest**) *slang* excellent; of or like a groove.

grope ● *verb* (**-ping**) (usually + *for*) feel about or search blindly; (+ *for*, *after*) search mentally; fondle clumsily for sexual pleasure; feel (one's way). ● *noun* act of groping.

grosgrain /ˈgrəʊgreɪn/ *noun* corded fabric of silk etc.

gross /grəʊs/ ● *adjective* overfed, bloated; coarse, indecent; flagrant; total, not net. ● *verb* produce as gross profit. ● *noun* (*plural* same) 12 dozen. □ **grossly** *adverb.*

grotesque /grəʊˈtesk/ ● *adjective* comically or repulsively distorted; incongruous, absurd. ● *noun* decoration interweaving human and animal features; comically distorted figure or design. □ **grotesquely** *adverb.*

grotto /ˈgrɒtəʊ/ *noun* (*plural* **-es** or **-s**) picturesque cave; structure imitating cave.

grotty /ˈgrɒtɪ/ *adjective* (**-ier**, **-iest**) *slang* unpleasant, dirty, ugly.

grouch /graʊtʃ/ *colloquial* ● *verb* grumble. ● *noun* grumbler; complaint; sulky grumbling mood. □ **grouchy** *adjective* (**-ier**, **-iest**).

ground[1] /graʊnd/ ● *noun* surface of earth; extent of subject; (often in *plural*) foundation, motive; area of special kind; (in *plural*) enclosed land attached to house etc.; area or basis for agreement etc.; surface worked on in painting; (in *plural*) dregs; bottom of sea; floor of room etc. ● *verb* prevent from taking off or flying; run aground, strand; (+

in) instruct thoroughly; (often as **grounded** *adjective*) (+ *on*) base cause or principle on. □ **ground control** personnel directing landing etc. of aircraft etc.; **ground cover** low-growing plants; **ground floor** storey at ground level; **ground frost** frost on surface of ground; **groundnut** peanut; **ground-rent** rent for land leased for building; **groundsman** person who maintains sports ground; **ground speed** aircraft's speed relative to ground; **ground swell** heavy sea due to distant or past storm etc.; **groundwork** preliminary or basic work.

ground² *past & past participle* of GRIND.

grounding *noun* basic instruction.

groundless *adjective* without motive or foundation.

groundsel /ˈgraʊnds(ə)l/ *noun* yellow-flowered weed.

group /gru:p/ ● *noun* number of people or things near, classed, or working together; number of companies under common ownership; pop group; division of air force. ● *verb* form into group; place in group(s). □ **group captain** RAF officer next below air commodore.

groupie *noun slang* ardent follower of touring pop group(s).

grouse¹ /graʊs/ *noun* (*plural* same) game bird with feathered feet.

grouse² /graʊs/ *verb* (**-sing**) & *noun colloquial* grumble.

grout /graʊt/ ● *noun* thin fluid mortar. ● *verb* apply grout to.

grove /grəʊv/ *noun* small wood; group of trees.

grovel /ˈgrɒv(ə)l/ *verb* (**-ll-**; *US* **-l-**) behave obsequiously; lie prone.

grow /grəʊ/ *verb* (*past* **grew**; *past participle* **grown**) increase in size, height, amount, etc.; develop or exist as living plant or natural product; produce by cultivation; become gradually; (+ *on*) become more favoured by; (in *passive*; + *over*) be covered with growth. □ **grown-up** adult; **grow up** mature. □ **grower** *noun*.

growl /graʊl/ ● *verb* (often + *at*) make low guttural sound, usually of anger; rumble. ● *noun* growling sound; angry murmur.

grown *past participle* of GROW.

growth /grəʊθ/ *noun* process of growing; increase; what has grown or is growing; tumour. □ **growth industry** one that is developing rapidly.

groyne /grɔɪn/ *noun* (*US* **groin**) wall built out into sea to stop beach erosion.

grub ● *noun* larva of insect; *colloquial* food. ● *verb* (**-bb-**) dig superficially; (+ *up*, *out*) extract by digging.

grubby /ˈgrʌbɪ/ *adjective* (**-ier**, **-iest**) dirty.

grudge ● *noun* persistent resentment or ill will. ● *verb* (**-ging**) be unwilling to give or allow, feel resentful about (doing something).

gruel /ˈgrʊəl/ *noun* liquid food of oatmeal etc. boiled in milk or water.

gruelling (*US* **grueling**) *adjective* exhausting, punishing.

gruesome /ˈgru:səm/ *adjective* grisly, disgusting.

gruff *adjective* rough-voiced; surly. □ **gruffly** *adverb*.

grumble /ˈgrʌmb(ə)l/ ● *verb* (**-ling**) complain peevishly; rumble. ● *noun* complaint; rumble. □ **grumbler** *noun*.

grumpy /ˈgrʌmpɪ/ *adjective* (**-ier**, **-iest**) ill-tempered.

grunt ● *noun* low guttural sound characteristic of pig. ● *verb* utter (with) grunt.

Gruyère /ˈgru:jeə/ *noun* kind of Swiss cheese with holes in.

gryphon = GRIFFIN.

guano /ˈgwɑ:nəʊ/ *noun* (*plural* **-s**) excrement of seabirds, used as manure.

guarantee /gærənˈti:/ ● *noun* formal promise or assurance; guaranty; giver of guaranty or security. ● *verb* (**-tees**, **-teed**) give or serve as guarantee for; promise; secure. □ **guarantor** *noun*.

guaranty /ˈgærəntɪ/ *noun* (*plural* **-ies**) written or other undertaking to answer for performance of obligation; thing serving as security.

guard /gɑ:d/ ● *verb* (often + *from*, *against*) defend, protect; keep watch, prevent from escaping; keep in check; (+ *against*) take precautions against. ● *noun* vigilant state; protector; soldiers etc. protecting place or person; official in charge of train; (in *plural*) (usually **Guards**) household troops of monarch; device to prevent injury or accident;

defensive posture. □ **guardhouse, guardroom** building or room for accommodating military guard or for detaining prisoners; **guardsman** soldier in guards or Guards.

guarded *adjective* (of remark etc.) cautious. □ **guardedly** *adverb.*

guardian /'gɑ:dɪən/ *noun* protector, keeper; person having custody of another, esp. minor. □ **guardianship** *noun.*

guava /'gwɑ:və/ *noun* edible orange acid fruit; tropical tree bearing this.

gubernatorial /gju:bənə'tɔ:rɪəl/ *adjective US* of governor.

gudgeon[1] /'gʌdʒ(ə)n/ *noun* small freshwater fish.

gudgeon[2] /'gʌdʒ(ə)n/ *noun* kind of pivot or pin; tubular part of hinge; socket for rudder.

guelder rose /'geldə/ *noun* shrub with round bunches of white flowers.

Guernsey /'gɜ:nzɪ/ *noun* (*plural* **-s**) one of breed of cattle from Guernsey; (**guernsey**) type of thick knitted woollen jersey.

guerrilla /gə'rɪlə/ *noun* (also **guerilla**) member of one of several independent groups fighting against regular forces.

guess /ges/ ● *verb* estimate without calculation or measurement; conjecture, think likely; conjecture rightly. ● *noun* estimate, conjecture. □ **guesswork** guessing.

guest /gest/ *noun* person invited to visit another's house or have meal etc. at another's expense, or lodging at hotel etc. □ **guest house** superior boarding house.

guestimate /'gestɪmət/ *noun* (also **guesstimate**) estimate based on guesswork and calculation.

guffaw /gʌ'fɔ:/ ● *noun* boisterous laugh. ● *verb* utter guffaw.

guidance /'gaɪd(ə)ns/ *noun* advice; guiding.

guide /gaɪd/ ● *noun* person who shows the way; conductor of tours; adviser; directing principle; guidebook; (**Guide**) member of girls' organization similar to Scouts. ● *verb* (**-ding**) act as guide to; lead, direct. □ **guidebook** book of information about place etc.; **guided missile** missile under remote control or directed by equipment within itself; **guide-dog** dog trained to lead blind person; **guideline** principle directing action.

Guider /'gaɪdə/ *noun* adult leader of Guides.

guild /gɪld/ *noun* (also **gild**) society for mutual aid or with common object; medieval association of craftsmen. □ **guildhall** meeting-place of medieval guild, town hall.

guilder /'gɪldə/ *noun* monetary unit of Netherlands.

guile /gaɪl/ *noun* sly behaviour; treachery, deceit. □ **guileless** *adjective.*

guillemot /'gɪlɪmɒt/ *noun* kind of auk.

guillotine /'gɪləti:n/ ● *noun* beheading machine; machine for cutting paper; method of shortening debate in parliament by fixing time of vote. ● *verb* (**-ning**) use guillotine on.

guilt /gɪlt/ *noun* fact of having committed offence; (feeling of) culpability.

guiltless *adjective* (often + *of*) innocent.

guilty *adjective* (**-ier**, **-iest**) having, feeling, or causing feeling of guilt. □ **guiltily** *adverb.*

guinea /'gɪnɪ/ *noun* (*historical* coin worth) £1.05. □ **guinea fowl** domestic fowl with white-spotted grey plumage; **guinea pig** domesticated S. American rodent, person used in experiment.

guipure /'gi:pjʊə/ *noun* heavy lace of patterned pieces joined by stitches.

guise /gaɪz/ *noun* external, esp. assumed, appearance; pretence.

guitar /gɪ'tɑ:/ *noun* usually 6-stringed musical instrument played with fingers or plectrum. □ **guitarist** *noun.*

gulch *noun US* ravine, gully.

gulf *noun* large area of sea with narrow-mouthed inlet; deep hollow, chasm; wide difference of opinion etc. □ **Gulf Stream** warm current from Gulf of Mexico to Europe.

gull[1] *noun* long-winged web-footed seabird.

gull[2] *verb* dupe, fool.

gullet /'gʌlɪt/ *noun* food-passage from mouth to stomach.

gullible /'gʌlɪb(ə)l/ *adjective* easily persuaded or deceived. □ **gullibility** *noun.*

gully /'gʌlɪ/ *noun* (*plural* **-ies**) water-worn ravine; gutter, drain; *Cricket* fielding position between point and slips.

gulp ● *verb* (often + *down*) swallow hastily or with effort; choke; (+ *down, back*) suppress. ● *noun* act of gulping; large mouthful.

gum[1] ● *noun* sticky secretion of some trees and shrubs, used as glue etc.; chewing gum; (also **gumdrop**) hard jelly sweet. ● *verb* (**-mm-**) (usually + *down, together*, etc.) fasten with gum; apply gum to. □ **gum arabic** gum exuded by some kinds of acacia; **gumboot** rubber boot; **gum tree** tree exuding gum, esp. eucalyptus; **gum up** *colloquial* interfere with, spoil.

gum[2] *noun* (usually in *plural*) firm flesh around roots of teeth. □ **gumboil** small abscess on gum.

gummy[1] /'gʌmɪ/ *adjective* (**-ier, -iest**) sticky; exuding gum.

gummy[2] /'gʌmɪ/ *adjective* (**-ier, -iest**) toothless.

gumption /'gʌmpʃ(ə)n/ *noun colloquial* resourcefulness, enterprise; common sense.

gun ● *noun* metal tube for throwing missiles with explosive propellant; starting pistol; device for discharging grease, electrons, etc., in desired direction; member of shooting party. ● *verb* (**-nn-**) (usually + *down*) shoot with gun; (+ *for*) seek out determinedly to attack or rebuke. □ **gunboat** small warship with heavy guns; **gun carriage** wheeled support for gun; **gun cotton** cotton steeped in acids, used as explosive; **gun dog** dog trained to retrieve game; **gunfire** firing of guns; **gunman** armed lawbreaker; **gun metal** bluish-grey colour, alloy of copper, tin, and usually zinc; **gunpowder** explosive of saltpetre, sulphur, and charcoal; **gunrunner** person selling or bringing guns into country illegally; **gunshot** shot from gun, the range of a gun; **gunslinger** *esp. US* gunman; **gunsmith** maker and repairer of small firearms.

gunge /gʌndz/ *colloquial* ● *noun* sticky substance. ● *verb* (**-ging**) (usually + *up*) clog with gunge. □ **gungy** *adjective*.

gung-ho /gʌŋ'həʊ/ *adjective* (arrogantly) eager.

gunner /'gʌnə/ *noun* artillery soldier; *Nautical* warrant officer in charge of battery, magazine, etc.; airman who operates gun.

gunnery /'gʌnərɪ/ *noun* construction and management, or firing, of large guns.

gunny /'gʌnɪ/ *noun* (*plural* **-ies**) coarse sacking usually of jute fibre; sack made of this.

gunwale /'gʌn(ə)l/ *noun* upper edge of ship's or boat's side.

guppy /'gʌpɪ/ *noun* (*plural* **-ies**) very small brightly coloured tropical freshwater fish.

gurgle /'gɜ:g(ə)l/ ● *verb* (**-ling**) make bubbling sound as of water; utter with such sound. ● *noun* bubbling sound.

gurnard /'gɜ:nəd/ *noun* (*plural* same or **-s**) sea fish with large spiny head.

guru /'gʊru:/ *noun* (*plural* **-s**) Hindu spiritual teacher; influential or revered teacher.

gush ● *verb* flow in sudden or copious stream; speak or behave effusively. ● *noun* sudden or copious stream; effusiveness.

gusher /'gʌʃə/ *noun* oil well emitting unpumped oil; effusive person.

gusset /'gʌsɪt/ *noun* piece let into garment etc. to strengthen or enlarge it.

gust ● *noun* sudden violent rush of wind; burst of rain, smoke, anger, etc. ● *verb* blow in gusts. □ **gusty** *adjective* (**-ier, -iest**).

gusto /'gʌstəʊ/ *noun* zest, enjoyment.

gut ● *noun* intestine; (in *plural*) bowels, entrails; (in *plural*) *colloquial* courage and determination; *slang* stomach; (in *plural*) contents, essence; material for violin etc. strings or for fishing line; instinctive, fundamental. ● *verb* (**-tt-**) remove or destroy internal fittings of (buildings); remove guts of. □ **gutless** *adjective*.

gutsy /'gʌtsɪ/ *adjective* (**-ier, iest**) *colloquial* courageous; greedy.

gutta-percha /gʌtə'pɜ:tʃə/ *noun* tough plastic substance made from latex.

gutted *adjective slang* bitterly disappointed.

gutter /'gʌtə/ ● *noun* shallow trough below eaves, or channel at side of street, for carrying off rainwater; (**the gutter**) poor or degraded environment; channel, groove. ● *verb* (of candle) burn unsteadily and melt away.

guttering *noun* (material for) gutters.

guttersnipe *noun* street urchin.

guttural /ˈgʌtər(ə)l/ • *adjective* throaty, harsh-sounding; (of sound) produced in throat. • *noun* guttural consonant.

guy[1] /gaɪ/ • *noun colloquial* man; effigy of Guy Fawkes burnt on 5 Nov. • *verb* ridicule.

guy[2] /gaɪ/ • *noun* rope or chain to secure tent or steady crane-load etc. • *verb* secure with guy(s).

guzzle /ˈgʌz(ə)l/ *verb* (**-ling**) eat or drink greedily.

gybe /dʒaɪb/ *verb* (*US* **jibe**) (**-bing**) (of fore-and-aft sail or boom) swing across boat, momentarily pointing into wind; cause (sail) to do this; (of boat etc.) change course thus.

gym /dʒɪm/ *noun colloquial* gymnasium; gymnastics. □ **gymslip, gym tunic** schoolgirl's sleeveless dress.

gymkhana /dʒɪmˈkɑːnə/ *noun* horse-riding competition.

gymnasium /dʒɪmˈneɪzɪəm/ *noun* (*plural* **-siums** or **-sia**) room etc. equipped for gymnastics.

gymnast /ˈdʒɪmnæst/ *noun* expert in gymnastics.

gymnastic /dʒɪmˈnæstɪk/ *adjective* of gymnastics. □ **gymnastically** *adverb.*

gymnastics *plural noun* (also treated as *singular*) exercises to develop or demonstrate physical (or mental) agility.

gynaecology /gaɪnɪˈkɒlədʒɪ/ *noun* (*US* **gynecology**) science of physiological functions and diseases of women. □ **gynaecological** /-kəˈlɒdʒ-/ *adjective*; **gynaecologist** *noun.*

gypsum /ˈdʒɪpsəm/ *noun* mineral used esp. to make plaster of Paris.

Gypsy /ˈdʒɪpsɪ/ *noun* (also **Gipsy**) (*plural* **-ies**) member of nomadic dark-skinned people of Europe.

gyrate /dʒaɪəˈreɪt/ *verb* (**-ting**) move in circle or spiral. □ **gyration** *noun*; **gyratory** *adjective.*

gyro /ˈdʒaɪərəʊ/ *noun* (*plural* **-s**) *colloquial* gyroscope.

gyroscope /ˈdʒaɪərəskəʊp/ *noun* rotating wheel whose axis is free to turn but maintains fixed direction unless perturbed, esp. used for stabilization.

H *abbreviation* hard (pencil lead); (water) hydrant; *slang* heroin. □ **H-bomb** hydrogen bomb.

h. *abbreviation* (also **h**) hour(s); (also ***h***) height; hot. □ **h. & c.** hot and cold (water).

ha[1] /hɑː/ (also **hah**) *interjection expressing surprise, triumph, etc.*

ha[2] *abbreviation* hectare(s).

habeas corpus /heɪbɪəs ˈkɔːpəs/ *noun* writ requiring person to be brought before judge etc., esp. to investigate lawfulness of his or her detention.

haberdasher /ˈhæbədæʃə/ *noun* dealer in dress accessories and sewing goods. □ **haberdashery** *noun* (*plural* **-ies**).

habit /ˈhæbɪt/ *noun* settled tendency or practice; practice that is hard to give up; mental constitution or attitude; clothes, esp. of religious order.

habitable /ˈhæbɪtəb(ə)l/ *adjective* suitable for living in. □ **habitability** *noun.*

habitat /ˈhæbɪtæt/ *noun* natural home of plant or animal.

habitation /hæbɪˈteɪʃ(ə)n/ *noun* inhabiting; house, home.

habitual /həˈbɪtʃʊəl/ *adjective* done as a habit; usual; given to a habit. □ **habitually** *adverb.*

habituate /həˈbɪtʃʊeɪt/ *verb* (**-ting**) (often + *to*) accustom. □ **habituation** *noun.*

habitué /həˈbɪtʃʊeɪ/ *noun* (often + *of*) frequent visitor or resident. [French]

háček /ˈhætʃek/ *noun* mark used (ˇ) over letter to modify its sound in some languages. [Czech]

hacienda /hæsɪˈendə/ *noun* (in Spanish-speaking countries) plantation etc. with dwelling house.

hack[1] ● *verb* cut or chop roughly; kick shin of; (often + *at*) deal cutting blows; cut (one's way) through; *colloquial* gain unauthorized access to (computer data); *slang* manage, tolerate. ● *noun* kick with toe of boot, wound from this. □ **hacksaw** saw for cutting metal.

hack[2] ● *noun* horse for ordinary riding; hired horse; person hired to do dull routine work, esp. as writer. ● *adjective* used as hack; commonplace. ● *verb* ride on horseback on road at ordinary pace.

hacker *noun colloquial* computer enthusiast; person who gains unauthorized access to computer network.

hacking *adjective* (of cough) short, dry, and frequent.

hackle /'hæk(ə)l/ *noun* (in *plural*) hairs on animal's neck which rise when it is angry or alarmed; long feather(s) on neck of domestic cock etc.; steel flax-comb. □ **make person's hackles rise** arouse anger or indignation.

hackney /'hæknɪ/ *noun* (*plural* **-s**) horse for ordinary riding. □ **hackney carriage** taxi.

hackneyed /'hæknɪd/ *adjective* overused, trite.

had *past & past participle* of HAVE.

haddock /'hædək/ *noun* (*plural* same) common edible sea fish.

Hades /'heɪdi:z/ *noun* (in Greek mythology) the underworld.

hadj = HAJJ.

hadji = HAJJI.

hadn't /'hæd(ə)nt/ had not.

haematite /'hi:mətaɪt/ *noun* (*US* **hem-**) red or brown iron ore.

haematology /hi:mə'tɒlədʒɪ/ *noun* (*US* **hem-**) study of the blood. □ **haematologist** *noun.*

haemoglobin /hi:mə'gləʊbɪn/ *noun* (*US* **hem-**) oxygen-carrying substance in red blood cells.

haemophilia /hi:mə'fɪlɪə/ *noun* (*US* **hem-**) hereditary tendency to severe bleeding from even a slight injury through failure of blood to clot. □ **haemophiliac** *noun.*

haemorrhage /'hemərɪdʒ/ (*US* **hem-**) ● *noun* profuse bleeding. ● *verb* (**-ging**) suffer haemorrhage.

haemorrhoids /'hemərɔɪdz/ *plural noun* (*US* **hem-**) swollen veins near anus, piles.

haft /hɑ:ft/ *noun* handle of knife etc.

hag *noun* ugly old woman; witch. □ **hagridden** afflicted by nightmares or fears.

haggard /'hægəd/ *adjective* looking exhausted and distraught.

haggis /'hægɪs/ *noun* Scottish dish of offal boiled in bag with oatmeal etc.

haggle /'hæg(ə)l/ ● *verb* (**-ling**) (often + *over*, *about*) bargain persistently. ● *noun* haggling.

hagiography /hægɪ'ɒgrəfɪ/ *noun* writing about saints' lives. □ **hagiographer** *noun.*

hah = HA[1].

ha ha /hɑ:'hɑ:/ *interjection representing laughter*.

ha-ha /'hɑ:hɑ:/ *noun* ditch with wall in it bounding park or garden.

haiku /'haɪku:/ *noun* (*plural* same) Japanese 3-line poem of usually 17 syllables.

hail[1] ● *noun* pellets of frozen rain; (+ *of*) barrage, onslaught. ● *verb* (after *it*) hail falls; pour down as or like hail. □ **hailstone** pellet of hail; **hailstorm** period of heavy hail.

hail[2] ● *verb* signal (taxi etc.) to stop; greet enthusiastically; (+ *from*) originate. ● *interjection archaic or jocular: expressing greeting*. ● *noun* act of hailing.

hair *noun* any or all of fine filaments growing from skin of mammals, esp. of human head; hairlike thing. □ **haircut** (style of) cutting hair; **hairdo** style of or act of styling hair; **hairdresser** person who cuts and styles hair; **hairdressing** *noun*; **hair-drier**, **-dryer** device for drying hair with warm air; **hairgrip** flat hairpin with ends close together; **hairline** edge of person's hair on forehead, very narrow crack or line; **hairnet** piece of netting for confining hair; **hair of the dog** further alcoholic drink taken to cure effects of previous drinking; **hairpiece** false hair augmenting person's natural hair; **hairpin** U-shaped pin for fastening the hair; **hairpin bend** U-shaped bend in road; **hair-raising** terrifying; **hair's breadth** minute distance; **hair shirt** ascetic's or penitent's shirt made of hair; **hair-**

slide clip for keeping hair in position; **hair-splitting** quibbling; **hairspray** liquid sprayed on hair to keep it in place; **hairspring** fine spring regulating balance-wheel of watch; **hairstyle** particular way of arranging hair; **hairstylist** *noun*; **hair-trigger** trigger acting on very slight pressure. □ **hairless** *adjective*; **hairy** *adjective* (**-ier**, **-iest**).

hajj /hædʒ/ *noun* (also **hadj**) Islamic pilgrimage to Mecca.

hajji /ˈhædʒɪ/ *noun* (also **hadji**) (*plural* **-s**) Muslim who has made pilgrimage to Mecca.

haka /ˈhɑːkə/ *noun NZ* Maori ceremonial war dance; similar dance by sports team before match.

hake *noun* (*plural* same) codlike sea fish.

halal /hɑːˈlɑːl/ *noun* (also **hallal**) meat from animal killed according to Muslim law.

halberd /ˈhælbəd/ *noun historical* combined spear and battleaxe.

halcyon /ˈhælsɪən/ *adjective* calm, peaceful, happy.

hale *adjective* strong and healthy (esp. in **hale and hearty**).

half /hɑːf/ ● *noun* (*plural* **halves** /hɑːvz/) either of two (esp. equal) parts into which a thing is divided; *colloquial* half pint, esp. of beer; *Sport* either of two equal periods of play, half-back; half-price (esp. child's) ticket. ● *adjective* forming a half. ● *adverb* partly. □ **half and half** being half one thing and half another; **half-back** player between forwards and full back(s) in football etc.; **half-baked** not thoroughly thought out; **half board** provision of bed, breakfast, and one main meal; **half-brother**, **-sister** one having only one parent in common; **half-crown** *historical* coin worth 2 shillings and 6 pence (= 12½p); **half-dozen** (about) six; **half-hearted** lacking courage or zeal; **half-heartedly** *adverb*; **half holiday** half day as holiday; **half-hour**, **half an hour** 30 minutes, point of time 30 minutes after any hour o'clock; **half-hourly** *adjective & adverb*; **half-life** time after which radioactivity etc. is half its original value; **half-mast** position of flag halfway down mast as symbol of mourning; **half measures** unsatisfactory compromise etc.; **half-moon** (shape of) moon with disc half illuminated; **half nelson** see NELSON; **half-term** short holiday halfway through school term; **half-timbered** having walls with timber frame and brick or plaster filling; **half-time** (short break at) mid-point of game or contest; **halftone** photograph representing tones by large or small dots; **half-truth** statement that conveys only part of truth; **half-volley** playing of ball as soon as it bounces off ground; **halfwit** stupid person; **halfwitted** *adjective*.

halfpenny /ˈheɪpnɪ/ *noun* (*plural* **-pennies** or **-pence** /ˈheɪpəns/) *historical* coin worth half penny (withdrawn in 1984).

halfway ● *adverb* at a point midway between two others; to some extent. ● *adjective* situated halfway. □ **halfway house** compromise, halfway point, rehabilitation centre, inn etc. between two towns.

halibut /ˈhælɪbət/ *noun* (*plural* same) large flatfish.

halitosis /hælɪˈtəʊsɪs/ *noun* bad breath.

hall /hɔːl/ *noun* entrance area of house; large room or building for meetings, concerts, etc.; large country house or estate; (in full **hall of residence**) residence for students; college dining-room; large public room; *esp. US* corridor. □ **hallmark** mark used to show standard of gold, silver, and platinum, distinctive feature; **hallway** entrance hall or corridor.

hallal = HALAL.

hallelujah = ALLELUIA.

hallo = HELLO.

hallow /ˈhæləʊ/ *verb* (usually as **hallowed** *adjective*) make or honour as holy.

Hallowe'en /hæləʊˈiːn/ *noun* eve of All Saints' Day, 31 Oct.

hallucinate /həˈluːsɪneɪt/ *verb* (**-ting**) experience hallucinations.

hallucination *noun* illusion of seeing or hearing something not actually present. □ **hallucinatory** /həˈluːsɪnətərɪ/ *adjective*.

hallucinogen /həˈluːsɪnədʒ(ə)n/ *noun* drug causing hallucinations. □ **hallucinogenic** /-ˈdʒen-/ *adjective*.

halm = HAULM.

halo /'heɪləʊ/ ● *noun* (*plural* **-es**) disc of light shown round head of sacred person; glory associated with idealized person; circle of light round sun or moon etc. ● *verb* (**-es**, **-ed**) surround with halo.

halogen /'hæləd3(ə)n/ *noun* any of the non-metallic elements (fluorine, chlorine, etc.) which form a salt when combined with a metal.

halon /'heɪlon/ *noun* gaseous halogen compound used to extinguish fires.

halt[1] /hɔ:lt/ ● *noun* stop (usually temporary); minor stopping place on local railway line. ● *verb* (cause to) make a halt.

halt[2] /hɔ:lt/ ● *verb* (esp. as **halting** *adjective*) proceed hesitantly. □ **haltingly** *adverb*.

halter /'hɔ:ltə/ *noun* rope with headstall for leading or tying up horses etc.; strap passing round back of neck holding dress etc. up, (also **halterneck**) dress etc. held by this.

halva /'hælvə/ *noun* confection of sesame flour, honey, etc.

halve /hɑ:v/ *verb* (**-ving**) divide into halves; reduce to half.

halves *plural* of HALF.

halyard /'hæljəd/ *noun* rope or tackle for raising and lowering sail etc.

ham ● *noun* upper part of pig's leg cured for food; back of thigh; thigh and buttock; *colloquial* inexpert or unsubtle performer or actor; *colloquial* operator of amateur radio station. ● *verb* (**-mm-**) (usually in **ham it up**) *colloquial* overact. □ **ham-fisted**, **-handed** *colloquial* clumsy.

hamburger /'hæmbɜ:gə/ *noun* cake of minced beef, usually eaten in soft bread roll.

hamlet /'hæmlɪt/ *noun* small village, esp. without church.

hammer /'hæmə/ ● *noun* tool with heavy metal head at right angles to handle, used for driving nails etc.; similar device, as for exploding charge in gun, striking strings of piano, etc.; auctioneer's mallet; metal ball attached to a wire for throwing as athletic contest. ● *verb* strike or drive (as) with hammer; *colloquial* defeat utterly. □ **hammer and tongs** *colloquial* with great energy; **hammerhead** shark with flattened hammer-shaped head; **hammerlock** wrestling hold in which twisted arm is bent behind back; **hammer-toe** toe bent permanently downwards. □ **hammering** *noun*.

hammock /'hæmək/ *noun* bed of canvas or netting suspended by cords at ends.

hamper[1] /'hæmpə/ *noun* large basket, usually with hinged lid and containing food.

hamper[2] /'hæmpə/ *verb* obstruct movement of; hinder.

hamster /'hæmstə/ *noun* short-tailed mouselike rodent often kept as pet.

hamstring ● *noun* any of 5 tendons at back of human knee; (in quadruped) tendon at back of hock. ● *verb* (*past & past participle* **-strung** or **-stringed**) cripple by cutting hamstrings; impair efficiency of.

hand ● *noun* end part of human arm beyond wrist; similar member of monkey; (often in *plural*) control, disposal, agency; share in action, active support; handlike thing, esp. pointer of clock etc.; right or left side, direction, etc.; skill or style, esp. of writing; person who does or makes something; person etc. as source; manual worker in factory etc.; pledge of marriage; playing cards dealt to player, round or game of cards; *colloquial* round of applause; measure of horse's height, = 4 in. (10.16 cm). ● *verb* (+ *in*, *to*, *over*, *round*, etc.) deliver or transfer (as) with hand. □ **at hand** close by; **by hand** by person not machine, not by post; **handbag** small bag carried esp. by woman; **handball** game with ball thrown by hand, *Football* foul touching of ball; **handbell** small bell for ringing by hand; **handbook** short manual or guidebook; **handbrake** brake operated by hand; **handcuff** secure (prisoner) with **handcuffs**, pair of lockable metal rings joined by short chain; **handgun** small firearm held in one hand; **handhold** something for hand to grip; **hand in glove** in collusion; **handmade** made by hand (rather than machine); **hand-me-down** article passed on from another person; **handout** thing given to needy person, information etc. distributed to press etc., notes given out in class; **handover** act of handing over; **hand-over-fist** *colloquial* with rapid progress; **hand-picked** carefully chosen; **handrail** rail along edge of stairs etc.; **hands down** without effort; **handset** part of telephone held in hand; **handshake** clasping of person's hand, esp. as greeting etc.; **hands-on** practical rather than theoretical; **handstand** act of supporting oneself vertically on one's hands; **hand-to-hand** (of fighting) at close quarters; **handwriting** (style of) writing by hand; **take in hand** start doing or dealing with, undertake control or reform of; **to hand** within reach.

handful *noun* (*plural* **-s**) enough to fill the hand; small number or quantity; *colloquial* troublesome person or task.

handicap /'hændɪkæp/ ● *noun* physical or mental disability; thing that makes progress difficult; disadvantage imposed on superior competitor to equalize chances; race etc. in which this is imposed. ● *verb* (**-pp-**) impose handicap on; place at disadvantage.

handicapped *adjective* suffering from physical or mental disability.

handicraft /'hændɪkrɑ:ft/ *noun* work requiring manual and artistic skill.

handiwork /'hændɪwɜ:k/ *noun* work done or thing made by hand, or by particular person.

handkerchief /'hæŋkətʃɪf/ *noun* (*plural* **-s** or **-chieves** /-tʃi:vz/) square of cloth used to wipe nose etc.

handle /'hænd(ə)l/ ● *noun* part by which thing is held. ● *verb* (**-ling**) touch, feel, operate, etc. with hands; manage, deal with; deal in (goods etc.). □ **handlebar** (usually in *plural*) steering-bar of bicycle etc.

handler *noun* person in charge of trained dog etc.

handsome /'hænsəm/ *adjective* (**-r**, **-st**) good-looking; imposing; generous; considerable. □ **handsomely** *adverb*.

handy *adjective* (**-ier**, **-iest**) convenient to handle; ready to hand; clever with hands. □ **handyman** person able to do odd jobs.

hang ● *verb* (*past & past participle* **hung** except as below) (cause to) be supported from above, attach by suspending from top; set up on hinges etc.; place (picture) on wall or in exhibition; attach (wallpaper); (*past & past participle* **hanged**) suspend or be suspended by neck, esp. as capital punishment; let droop; remain or be hung. ● *noun* way thing hangs. □ **get the hang of** *colloquial* get knack of, understand; **hang about, around** loiter, not move away; **hangdog** shamefaced; **hang fire** delay acting; **hang-glider** fabric wing on light frame from which pilot is suspended; **hang-gliding** *noun*; **hangman** executioner by hanging; **hangnail** agnail; **hang on** (often + *to*) continue to hold, retain, wait for short time, not ring off during pause in telephoning; **hangover** after-effects of excess of alcohol; **hang up** *verb* hang from hook etc., end telephone conversation; **hang-up** *noun slang* emotional inhibition.

hangar /'hæŋə/ *noun* building for housing aircraft etc.

hanger *noun* person or thing that hangs; (in full **coat-hanger**) shaped piece of wood etc. for hanging clothes on. □ **hanger-on** (*plural* **hangers-on**) follower, dependant.

hanging *noun* execution by suspending by neck; (usually in *plural*) drapery for walls etc.

hank *noun* coil of yarn etc.

hanker /'hæŋkə/ *verb* (+ *for*, *after*, *to do*) crave, long for. □ **hankering** *noun*.

hanky /'hæŋkɪ/ *noun* (also **hankie**) (*plural* **-ies**) *colloquial* handkerchief.

hanky-panky /hæŋkɪ'pæŋkɪ/ *noun slang* misbehaviour; trickery.

Hansard /'hænsɑ:d/ *noun* verbatim record of parliamentary debates.

hansom /'hænsəm/ *noun* (in full **hansom cab**) *historical* two-wheeled horse-drawn cab.

haphazard /hæp'hæzəd/ *adjective* casual, random. □ **haphazardly** *adverb*.

hapless /'hæplɪs/ *adjective* unlucky.

happen /'hæpən/ *verb* occur; (+ *to do*) have the (good or bad) fortune; (+ *to*) be fate or experience of; (+ *on*) come by chance on. □ **happening** *noun*.

happy /'hæpɪ/ *adjective* (**-ier**, **-iest**) feeling or showing pleasure or contentment; fortunate; apt, pleasing. □ **happy-go-lucky** taking things cheerfully as they happen; **happy hour** time of day when drinks are sold at reduced prices; **happy medium** compromise. □ **happily** *adverb*; **happiness** *noun*.

hara-kiri /hærəˈkɪrɪ/ *noun historical* Japanese suicide by ritual disembowelling.

harangue /həˈræŋ/ • *noun* lengthy and earnest speech. • *verb* (**-guing**) make harangue to.

harass /ˈhærəs/ *verb* trouble, annoy; attack repeatedly. □ **harassment** *noun*.

■ **Usage** *Harass* is often pronounced /həˈræs/ (with the stress on the *-rass*), but this is considered incorrect by some people.

harbinger /ˈhɑːbɪndʒə/ *noun* person or thing announcing another's approach, forerunner.

harbour /ˈhɑːbə/ (*US* **harbor**) • *noun* place of shelter for ships; shelter. • *verb* give shelter to; entertain (thoughts etc.).

hard • *adjective* firm, solid; difficult to bear, do, or understand; unfeeling, harsh, severe; strenuous, enthusiastic; *Politics* extreme, radical; (of drinks) strongly alcoholic; (of drug) potent and addictive; (of water) difficult to lather; (of currency etc.) not likely to fall in value; not disputable. • *adverb* strenuously, severely, intensely. □ **hard and fast** (of rule etc.) strict; **hardback** (book) bound in stiff covers; **hardbitten** *colloquial* tough, cynical; **hardboard** stiff board of compressed wood pulp; **hard-boiled** (of eggs) boiled until yolk and white are solid, (of person) tough, shrewd; **hard cash** coins and banknotes, not cheques etc.; **hard copy** printed material produced by computer; **hardcore** stones, rubble, etc. as foundation; **hard core** central or most enduring part; **hard disk** *Computing* rigid storage disk, esp. fixed in computer; **hard-done-by** unfairly treated; **hard-headed** practical, not sentimental; **hard-hearted** unfeeling; **hard line** firm adherence to policy; **hardliner** *noun*; **hard-nosed** *colloquial* realistic, uncompromising; **hard of hearing** somewhat deaf; **hard-pressed** closely pursued, burdened with urgent business; **hard sell** aggressive salesmanship; **hard shoulder** strip at side of motorway for emergency stops; **hard up** short of money; **hardware** tools, weapons, machinery, etc., mechanical and electronic components of computer; **hardwood** wood of deciduous tree. □ **hardness** *noun*.

harden *verb* make or become hard or unyielding.

hardihood /ˈhɑːdɪhʊd/ *noun* boldness.

hardly *adverb* scarcely; with difficulty.

hardship *noun* severe suffering or privation.

hardy /ˈhɑːdɪ/ *adjective* (**-ier**, **-iest**) robust; capable of endurance; (of plant) able to grow in the open all year. □ **hardiness** *noun*.

hare /heə/ • *noun* mammal like large rabbit, with long ears, short tail, and long hind legs. • *verb* (**-ring**) run rapidly. □ **hare-brained** rash, wild.

harebell *noun* plant with pale blue bell-shaped flowers.

harem /ˈhɑːriːm/ *noun* women of Muslim household; their quarters.

haricot /ˈhærɪkəʊ/ *noun* (in full **haricot bean**) French bean with small white seeds; these as vegetable.

hark *verb* (usually in *imperative*) *archaic* listen. □ **hark back** revert to earlier topic.

Harlequin /ˈhɑːlɪkwɪn/ • *noun* masked pantomime character in diamond-patterned costume. • *adjective* (**harlequin**) in varied colours.

harlot /ˈhɑːlət/ *noun archaic* prostitute.

harm *noun & verb* damage, hurt.

harmful *adjective* causing or likely to cause harm. □ **harmfully** *adverb*.

harmless *adjective* not able or likely to harm. □ **harmlessly** *adverb*.

harmonic /hɑːˈmɒnɪk/ • *adjective* of or relating to harmony; harmonious. • *noun Music* overtone accompanying (and forming a note with) a fundamental at a fixed interval.

harmonica /hɑːˈmɒnɪkə/ *noun* small rectangular instrument played by blowing and sucking air through it.

harmonious /hɑːˈməʊnɪəs/ *adjective* sweet-sounding, tuneful; forming a pleasant or consistent whole; free from dissent. □ **harmoniously** *adverb*.

harmonium /hɑːˈməʊnɪəm/ *noun* keyboard instrument with bellows and metal reeds.

harmonize /ˈhɑːmənaɪz/ *verb* (also **-ise**) (**-zing** or **-sing**) add notes to (melody) to produce harmony; bring into or be in harmony. □ **harmonization** *noun*.

harmony /ˈhɑːmənɪ/ *noun* (*plural* **-ies**) combination of notes to form chords; melodious sound; agreement, concord.

harness /ˈhɑːnɪs/ ● *noun* straps etc. by which horse is fastened to cart etc. and controlled; similar arrangement for fastening thing to person. ● *verb* put harness on; utilize (natural forces), esp. to produce energy.

harp ● *noun* large upright stringed instrument plucked with fingers. ● *verb* (+ *on, on about*) dwell on tediously. □ **harpist** *noun*.

harpoon /hɑːˈpuːn/ ● *noun* spearlike missile for shooting whales etc. ● *verb* spear with harpoon.

harpsichord /ˈhɑːpsɪkɔːd/ *noun* keyboard instrument with strings plucked mechanically. □ **harpsichordist** *noun*.

harpy /ˈhɑːpɪ/ *noun* (*plural* **-ies**) mythological monster with woman's face and bird's wings and claws; grasping unscrupulous person.

harridan /ˈhærɪd(ə)n/ *noun* bad-tempered old woman.

harrier /ˈhærɪə/ *noun* hound used in hunting hares; kind of falcon.

harrow /ˈhærəʊ/ ● *noun* frame with metal teeth or discs for breaking clods of earth. ● *verb* draw harrow over; (usually as **harrowing** *adjective*) distress greatly.

harry /ˈhærɪ/ *verb* (**-ies**, **-ied**) ravage, despoil; harass.

harsh *adjective* rough to hear, taste, etc.; severe, cruel. □ **harshly** *adverb*; **harshness** *noun*.

hart *noun* (*plural* same or **-s**) male of (esp. red) deer.

hartebeest /ˈhɑːtɪbiːst/ *noun* large African antelope with curved horns.

harum-scarum /heərəmˈskeərəm/ *adjective colloquial* reckless, wild.

harvest /ˈhɑːvɪst/ ● *noun* gathering in of crops etc.; season for this; season's yield; product of any action. ● *verb* reap and gather in.

harvester *noun* reaper, reaping machine.

has *3rd singular present* of HAVE.

hash[1] ● *noun* dish of reheated pieces of cooked meat; mixture, jumble; recycled material. ● *verb* (often + *up*) recycle (old material). □ **make a hash of** *colloquial* make a mess of, bungle.

hash[2] *noun colloquial* hashish.

hashish /ˈhæʃɪʃ/ *noun* narcotic drug got from hemp.

hasn't /ˈhæz(ə)nt/ has not.

hasp /hɑːsp/ *noun* hinged metal clasp passing over staple and secured by padlock.

hassle /ˈhæs(ə)l/ *colloquial* ● *noun* trouble, problem; argument. ● *verb* (**-ling**) harass.

hassock /ˈhæsək/ *noun* kneeling-cushion.

haste /heɪst/ *noun* urgency of movement; hurry. □ **make haste** be quick.

hasten /ˈheɪs(ə)n/ *verb* (cause to) proceed or go quickly.

hasty /ˈheɪstɪ/ *adjective* (**-ier**, **-iest**) hurried; said, made, or done too quickly. □ **hastily** *adverb*; **hastiness** *noun*.

hat *noun* (esp. outdoor) head-covering. □ **hat trick** *Cricket* taking 3 wickets with successive balls, *Football* scoring of 3 goals in one match by same player, 3 consecutive successes.

hatch[1] *noun* opening in wall between kitchen and dining-room for serving food; opening or door in aircraft etc.; (cover for) hatchway. □ **hatchback** car with rear door hinged at top; **hatchway** opening in ship's deck for lowering cargo.

hatch[2] ● *verb* (often + *out*) emerge from egg; (of egg) produce young animal; incubate; (also + *up*) devise (plot). ● *noun* hatching; brood hatched.

hatch[3] *verb* mark with parallel lines. □ **hatching** *noun*.

hatchet /ˈhætʃɪt/ *noun* light short axe.

hate ● *verb* (**-ting**) dislike intensely. ● *noun* hatred.

hateful *adjective* arousing hatred.

hatred /ˈheɪtrɪd/ *noun* intense dislike; ill will.

hatter *noun* maker or seller of hats.

haughty /ˈhɔːtɪ/ *adjective* (**-ier**, **-iest**) proud, arrogant. □ **haughtily** *adverb*; **haughtiness** *noun*.

haul /hɔːl/ ● *verb* pull or drag forcibly; transport by lorry, cart, etc. ● *noun* hauling; amount gained or acquired; distance to be traversed.

haulage *noun* (charge for) commercial transport of goods.

haulier /ˈhɔːlɪə/ *noun* person or firm engaged in transport of goods.

haulm /hɔːm/ *noun* (also **halm**) stalk or stem; stalks of beans, peas, potatoes, etc. collectively.

haunch /hɔːntʃ/ *noun* fleshy part of buttock and thigh; leg and loin of deer etc. as food.

haunt /hɔːnt/ ● *verb* (of ghost etc.) visit regularly; frequent (place); linger in mind of. ● *noun* place frequented by person.

haunting *adjective* (of memory, melody, etc.) lingering; poignant, evocative.

haute couture /əʊt kuːˈtjʊə/ *noun* (world of) high fashion.

hauteur /əʊˈtɜː/ *noun* haughtiness.

have /hæv/ ● *verb* (**having**; *3rd singular present* **has** /hæz/; *past & past participle* **had**) *used as auxiliary verb with past participle to form past tenses*; hold in possession or relationship, be provided with; contain as part or quality; experience; (come to) be subjected to a specified state; engage in; tolerate, permit to; give birth to; receive; obtain or know (qualification, language, etc.); *colloquial* get the better of, (usually in *passive*) cheat. ● *noun* (usually in *plural*) *colloquial* wealthy person; *slang* swindle. □ **have on** wear (clothes), have (engagement), *colloquial* hoax; **have-not** (usually in *plural*) person lacking wealth; **have to** be obliged to, must.

haven /ˈheɪv(ə)n/ *noun* refuge; harbour.

haven't /ˈhæv(ə)nt/ have not.

haver /ˈheɪvə/ *verb* hesitate; talk foolishly.

haversack /ˈhævəsæk/ *noun* canvas bag carried on back or over shoulder.

havoc /ˈhævək/ *noun* devastation, confusion.

haw *noun* hawthorn berry.

hawk[1] ● *noun* bird of prey with rounded wings; *Politics* person who advocates aggressive policy. ● *verb* hunt with hawk.

hawk[2] *verb* carry (goods) about for sale.

hawk[3] *verb* clear throat noisily; (+ *up*) bring (phlegm etc.) up thus.

hawker *noun* person who hawks goods.

hawser /ˈhɔːzə/ *noun* thick rope or cable for mooring ship.

hawthorn *noun* thorny shrub with red berries.

hay *noun* grass mown and dried for fodder. □ **haycock** conical heap of hay; **hay fever** allergic irritation of nose, throat, etc. caused by pollen, dust, etc.; **haymaking** mowing grass and spreading it to dry; **hayrick**, **haystack** packed pile of hay; **haywire** *colloquial* badly disorganized, out of control.

hazard /ˈhæzəd/ ● *noun* danger, risk; obstacle on golf course. ● *verb* venture on (guess etc.); risk.

hazardous *adjective* risky.

haze *noun* slight mist; mental obscurity, confusion.

hazel /ˈheɪz(ə)l/ *noun* nut-bearing hedgerow shrub; greenish-brown. □ **hazelnut** nut of hazel.

hazy *adjective* (**-ier**, **-iest**) misty; vague; confused. □ **hazily** *adverb*; **haziness** *noun*.

HB *abbreviation* hard black (pencil lead).

HE *abbreviation* His or Her Excellency; high explosive.

he /hiː/ *pronoun* (as subject of verb) the male person or animal in question; person of unspecified sex. □ **he-man** masterful or virile man.

head /hed/ ● *noun* uppermost part of human body, or foremost part of body of animal, containing brain, sense organs, etc.; seat of intellect; thing like head in form or position; top, front, or upper end; person in charge, esp. of school; position of command; individual as unit; side of coin bearing image of head, (in *plural*) this as call when tossing coin; signal-converting device on tape recorder etc.; foam on top of beer etc.; confined body of water or steam, pressure exerted by this; (usually in **come to a head**) climax, crisis. ● *verb* be at front or in charge of; (often + *for*) move or send in specified direction; provide with heading; *Football* strike (ball) with head. □ **headache** continuous pain in head, *colloquial* troublesome problem; **headachy** *adjective*; **headband** band worn round head as decoration or to confine hair; **headboard** upright panel at head of bed; **headcount** (counting of) total number of people; **headdress** (esp. ornamental) covering for head; **head-hunting** collecting of enemies' heads as trophies, seeking of staff by approaching people employed elsewhere; **headlamp**, **headlight** (main) light at front of vehicle; **headland** promontory; **headline** heading at top of page, newspaper article, etc., (in *plural*) summary of broadcast news; **headlock** wrestling hold round opponent's head; **headlong** with head foremost, in a rush; **headman** tribal chief; **headmaster**, **headmistress** head teacher; **head-on** (of collision etc.) with front foremost; **headphones** pair of earphones fitting over head; **headquarters** (treated as *singular* or *plural*) organization's administrative centre; **headroom** overhead space; **headset** headphones, often with microphone; **headshrinker** *slang* psychiatrist; **headstall** part of bridle or halter fitting round horse's head; **head start** advantage granted or gained at early stage; **headstone** stone set up at head of grave; **headstrong** self-willed; **head teacher** teacher in charge of school; **headway** progress; **head wind** one blowing from directly in front.

header *noun Football* act of heading ball; *colloquial* headlong dive or plunge.

heading *noun* title at head of page etc.

heady *adjective* (**-ier**, **-iest**) (of liquor etc.) potent; exciting, intoxicating; impetuous; headachy.

heal *verb* (often + *up*) become healthy again; cure; put right (differences). □ **healer** *noun.*

health /helθ/ *noun* state of being well in body or mind; mental or physical condition. □ **health centre** building containing local medical services; **health food** natural food, thought to promote good health; **health service** public medical service; **health visitor** nurse who visits mothers and babies, the elderly, etc. at home.

healthy *adjective* (**-ier**, **-iest**) having, conducive to, or indicative of, good health. □ **healthily** *adverb*; **healthiness** *noun.*

heap ● *noun* disorderly pile; (esp. in *plural*) *colloquial* large number or amount; *slang* dilapidated vehicle. ● *verb* (+ *up*, *together*, etc.) pile or collect in heap; (+ *with*) load copiously with; (+ *on*, *upon*) offer copiously.

hear *verb* (*past & past participle* **heard** /hɜːd/) perceive with ear; listen to; listen judicially to; be informed; (+ *from*) receive message etc. from. □ **hearsay** rumour, gossip. □ **hearer** *noun.*

hearing *noun* faculty of perceiving sounds; range within which sounds may be heard; opportunity to state one's case; trial of case before court. □ **hearing-aid** small sound-amplifier worn by deaf person.

hearse /hɜːs/ *noun* vehicle for carrying coffin.

heart /hɑːt/ *noun* organ in body keeping up circulation of blood by contraction and dilation; region of heart, breast; seat of thought, feeling, or emotion (esp. love); courage; mood; central or innermost part, essence; tender inner part of vegetable etc.; (conventionally) heart-shaped thing; playing card of suit marked with red hearts. □ **at heart** in inmost feelings; **by heart** from memory; **have the heart** (usually in negative, + *to do*) be hard-hearted enough; **heartache** mental anguish; **heart attack** sudden heart failure; **heartbeat** pulsation of heart; **heartbreak** overwhelming distress; **heartbreaking** *adjective*; **heartbroken** *adjective*; **heartburn** burning sensation in chest from indigestion; **heartfelt** sincere; **heart-rending** very distressing; **heartsick** despondent; **heartstrings** deepest affections or pity; **heartthrob** *colloquial* object of (esp. immature) romantic feelings; **heart-to-heart** frank (talk); **heart-warming** emotionally moving and encouraging; **take to heart** be much affected by.

hearten *verb* make or become more cheerful. □ **heartening** *adjective.*

hearth /hɑːθ/ *noun* floor of fireplace. □ **hearthrug** rug laid before fireplace.

heartless *adjective* unfeeling, pitiless. □ **heartlessly** *adverb.*

hearty *adjective* (**-ier**, **-iest**) strong, vigorous; (of meal or appetite) large; warm, friendly. □ **heartily** *adverb*; **heartiness** *noun.*

heat ● *noun* condition or sensation of being hot; energy arising from motion of molecules; hot weather; warmth of feeling; anger; most intense part or period of activity; preliminary contest, winner(s) of which compete in final. ● *verb* make or become hot; inflame. □ **heatproof** able to resist great heat; **heatwave** period of very hot weather; **on heat** (of female animals) sexually receptive.

heated *adjective* angry, impassioned. □ **heatedly** *adverb.*

heater *noun* device for heating room, water, etc.

heath /hi:θ/ *noun* flattish tract of uncultivated land with low shrubs; plant growing on heath, esp. heather.

heathen /'hi:ð(ə)n/ ● *noun* person not belonging to predominant religion. ● *adjective* of heathens; having no religion.

heather /'heðə/ *noun* purple-flowered plant of moors and heaths.

heating *noun* equipment used to heat building.

heave ● *verb* (**-ving**; *past & past participle* **heaved** or *esp. Nautical* **hove** /həʊv/) lift, haul, or utter with effort; *colloquial* throw; rise and fall periodically; *Nautical* haul by rope; retch. ● *noun* heaving. □ **heave in sight** come into view; **heave to** bring vessel to standstill.

heaven /'hev(ə)n/ *noun* home of God and of blessed after death; place or state of bliss, delightful thing; (**the heavens**) sky as seen from earth. □ **heavenly** *adjective.*

heavy /'hevɪ/ ● *adjective* (**-ier**, **-iest**) of great weight, difficult to lift; of great density; abundant; severe, extensive; striking or falling with force; (of machinery etc.) very large of its kind; needing much physical effort; hard to digest; hard to read or understand; (of ground) difficult to travel over; dull, tedious, oppressive; coarse, ungraceful. ● *noun* (*plural* **-ies**) *colloquial* thug (esp. hired); villain. □ **heavy-duty** designed to withstand hard use; **heavy-handed** clumsy, oppressive; **heavy hydrogen** deuterium; **heavy industry** that concerned with production of metal and machines etc.; **heavy metal** *colloquial* loud rock music with pounding rhythm; **heavy water** water composed of deuterium and oxygen; **heavyweight** amateur boxing weight (over 81 kg). □ **heavily** *adverb*; **heaviness** *noun.*

Hebraic /hi:'breɪɪk/ *adjective* of Hebrew or the Hebrews.

Hebrew /'hi:bru:/ ● *noun* member of a Semitic people in ancient Palestine; their language; modern form of this, used esp. in Israel. ● *adjective* of or in Hebrew; of the Jews.

heckle /'hek(ə)l/ ● *verb* (**-ling**) interrupt or harass (speaker). ● *noun* act of heckling. □ **heckler** *noun.*

hectare /'hekteə/ *noun* metric unit of square measure (2.471 acres).

hectic /'hektɪk/ *adjective* busy and confused; excited, feverish. □ **hectically** *adverb.*

hecto- /'hektəʊ/ *combining form* one hundred.

hector /'hektə/ *verb* bluster, bully. □ **hectoring** *adjective.*

he'd /hi:d/ he had; he would.

hedge ● *noun* fence of bushes or low trees; protection against possible loss. ● *verb* (**-ging**) surround with hedge; (+ *in*) enclose; secure oneself against loss on (bet etc.); avoid committing oneself. □ **hedgehog** small spiny insect-eating mammal; **hedge-hop** fly at low altitude; **hedgerow** row of bushes forming hedge; **hedge sparrow** common brown-backed bird.

hedonism /'hi:dənɪz(ə)m/ *noun* (behaviour based on) belief in pleasure as humankind's proper aim. □ **hedonist** *noun*; **hedonistic** /-'nɪs-/ *adjective.*

heed ● *verb* attend to; take notice of. ● *noun* care, attention. □ **heedless** *adjective*; **heedlessly** *adverb.*

hee-haw /'hi:hɔ:/ *noun & verb* bray.

heel[1] ● *noun* back of foot below ankle; part of sock etc. covering this, or of shoe etc. supporting it; crust end of loaf; *colloquial* scoundrel; (as *interjection*) *command to dog to walk near owner's heel.* ● *verb* fit or renew heel on (shoe); touch ground with heel; (+ *out*) *Rugby* pass ball with heel. □ **cool, kick one's heels** be kept waiting; **heelball** shoemaker's polishing mixture of wax etc., esp. used in brass rubbing.

heel[2] ● *verb* (often + *over*) (of ship etc.) lean over; cause (ship) to do this. ● *noun* heeling.

hefty /'heftɪ/ *adjective* (**-ier**, **-iest**) (of person) big, strong; (of thing) heavy, powerful.

hegemony /hɪ'gemənɪ/ *noun* leadership.

heifer /'hefə/ *noun* young cow, esp. one that has not had more than one calf.

height /haɪt/ *noun* measurement from base to top; elevation above ground or other level; high point; top; extreme example.

heighten *verb* make or become higher or more intense.

heinous /'heɪnəs/ *adjective* atrocious.

heir /eə/ *noun* (*feminine* **heiress**) person entitled to property or rank as legal successor of former holder. □ **heir apparent** one whose claim cannot be superseded by birth of nearer heir; **heirloom** piece of property that has been in family for generations; **heir presumptive** one whose claim may be superseded by birth of nearer heir.

held *past & past participle* of HOLD[1].

helical /'helɪk(ə)l/ *adjective* spiral.

helices *plural* of HELIX.

helicopter /'helɪkɒptə/ *noun* wingless aircraft lifted and propelled by overhead blades revolving horizontally.

heliograph /'hi:lɪəgrɑ:f/ ● *noun* signalling apparatus reflecting flashes of sunlight. ● *verb* send (message) thus.

heliotrope /'hi:lɪətrəʊp/ *noun* plant with fragrant purple flowers.

heliport /'helɪpɔ:t/ *noun* place where helicopters take off and land.

helium /'hi:lɪəm/ *noun* light non-flammable gaseous element.

helix /'hi:lɪks/ *noun* (*plural* **helices** /'hi:lɪsi:z/) spiral or coiled curve.

hell *noun* home of the damned after death; place or state of misery. □ **hellish** *adjective*; **hellishly** *adverb*.

he'll /hi:l/ he will; he shall.

hellebore /'helɪbɔ:/ *noun* evergreen plant of kind including Christmas rose.

Hellene /'heli:n/ *noun* Greek. □ **Hellenic** /-'len-/ *adjective*; **Hellenism** /-lɪn-/ *noun*; **Hellenist** /-lɪn-/ *noun*.

Hellenistic /helɪ'nɪstɪk/ *adjective* of Greek history, language, and culture of late 4th to late 1st c. BC.

hello /hə'ləʊ/ (also **hallo**, **hullo**) ● *interjection expressing informal greeting or surprise, or calling attention.* ● *noun* (*plural* **-s**) cry of 'hello'.

helm *noun* tiller or wheel for managing rudder. □ **at the helm** in control; **helmsman** person who steers ship.

helmet /'helmɪt/ *noun* protective head-cover of policeman, motorcyclist, etc.

help ● *verb* provide with means to what is needed or sought, be useful to; (usually in negative) prevent, refrain from. ● *noun* act of helping; person or thing that helps; *colloquial* domestic assistant or assistance; remedy etc.; (**help oneself**) (often + *to*) serve oneself, take without permission. □ **helpline** telephone service providing help with problems. □ **helper** *noun*.

helpful *adjective* giving help, useful. □ **helpfully** *adverb*; **helpfulness** *noun*.

helping *noun* portion of food.

helpless *adjective* lacking help, defenceless; unable to act without help. □ **helplessly** *adverb*; **helplessness** *noun*.

helter-skelter /heltə'skeltə/ ● *adverb & adjective* in disorderly haste. ● *noun* spiral slide at funfair.

hem[1] ● *noun* border of cloth where edge is turned under and sewn down. ● *verb* (**-mm-**) sew edge thus. □ **hem in** confine, restrict; **hemline** lower edge of skirt etc.; **hemstitch** (make hem with) ornamental stitch.

hem[2] *interjection expressing hesitation or calling attention by slight cough.*

hemisphere /'hemɪsfɪə/ *noun* half sphere; half earth, esp. as divided by equator or by line passing through poles; each half of brain. □ **hemispherical** /-'sfer-/ *adjective*.

hemlock /'hemlɒk/ *noun* poisonous plant with small white flowers; poison made from it.

hemp *noun* (in full **Indian hemp**) Asian herbaceous plant; its fibre used for rope etc.; narcotic drug made from it.

hempen *adjective* made of hemp.

hen *noun* female bird, esp. of domestic fowl. □ **henbane** poisonous hairy plant; **hen-party** *colloquial* party of women only; **henpecked** (of husband) domineered over by his wife.

hence *adverb* from now; for this reason. □ **henceforth**, **henceforward** from this time onwards.

henchman /'hentʃmən/ *noun usually derogatory* trusted supporter.

henge *noun* prehistoric circle of wood or stone uprights, as at *Stonehenge*.

henna /'henə/ ● *noun* tropical shrub; reddish dye made from it and used esp. to colour hair. ● *verb* (**hennaed**, **hennaing**) dye with henna.

henry /'henrɪ/ *noun* (*plural* **-s**, **-ies**) SI unit of inductance.

hep = HIP[4].

hepatitis /hepə'taɪtɪs/ *noun* inflammation of the liver.

hepta- *combining form* seven.

heptagon /'heptəgən/ *noun* plane figure with 7 sides and angles. □ **heptagonal** /-'tæg-/ *adjective*.

her ● *pronoun* (as object of verb) the female person or thing in question; *colloquial* she. ● *adjective* of or belonging to her.

herald /'her(ə)ld/ ● *noun* messenger; forerunner; official. ● *verb* proclaim approach of; usher in. □ **heraldic** /-'ræld-/ *adjective*.

heraldry /'herəldrɪ/ *noun* (science or art of) armorial bearings.

herb *noun* non-woody seed-bearing plant; plant with leaves, seeds, or flowers used for flavouring, medicine, etc.

herbaceous /hɜ:'beɪʃəs/ *adjective* of or like herbs. □ **herbaceous border** border in garden etc. containing flowering plants.

herbage *noun* vegetation collectively, esp. pasturage.

herbal ● *adjective* of herbs. ● *noun* book about herbs.

herbalist *noun* dealer in medicinal herbs; writer on herbs.

herbarium /hɜ:'beərɪəm/ *noun* (*plural* **-ria**) collection of dried plants.

herbicide /'hɜ:bɪsaɪd/ *noun* poison used to destroy unwanted vegetation.

herbivore /'hɜ:bɪvɔ:/ *noun* plant-eating animal. □ **herbivorous** /-'bɪvərəs/ *adjective*.

Herculean /hɜ:kjʊ'li:ən/ *adjective* having or requiring great strength or effort.

herd ● *noun* number of cattle etc. feeding or travelling together; (**the herd**) *derogatory* large number of people. ● *verb* (cause to) go in herd; tend. □ **herdsman** keeper of herds.

here ● *adverb* in or to this place; *indicating a person or thing*; at this point. ● *noun* this place. □ **hereabout(s)** somewhere near here; **hereafter** (in) future, (in) next world; **hereby** by this means; **herein** *formal* in this place, book, etc.; **hereinafter** *formal* from this point on, below (in document); **hereof** *formal* of this; **hereto** *formal* to this; **heretofore** *formal* formerly; **hereupon** after or in consequence of this; **herewith** with this.

hereditary /hɪ'redɪtərɪ/ *adjective* transmitted genetically from one generation to another; descending by inheritance; holding position by inheritance.

heredity /hɪ'redɪtɪ/ *noun* genetic transmission of physical or mental characteristics; these characteristics; genetic constitution.

heresy /'herəsɪ/ *noun* (*plural* **-ies**) *esp. RC Church* religious belief contrary to orthodox doctrine; opinion contrary to what is normally accepted.

heretic /'herətɪk/ *noun* believer in heresy. □ **heretical** /hɪ'ret-/ *adjective*.

heritable /'herɪtəb(ə)l/ *adjective* that can be inherited.

heritage /'herɪtɪdʒ/ *noun* what is or may be inherited; inherited circumstances, benefits, etc.; nation's historic buildings, countryside, etc.

hermaphrodite /hɜ:'mæfrədaɪt/ ● *noun* person, animal, or plant with organs of both sexes. ● *adjective* combining both sexes. □ **hermaphroditic** /-'dɪt-/ *adjective*.

hermetic /hɜ:'metɪk/ *adjective* with an airtight seal. □ **hermetically** *adverb*.

hermit /'hɜ:mɪt/ *noun* person living in solitude. □ **hermit-crab** crab which lives in mollusc's cast-off shell.

hermitage *noun* hermit's dwelling; secluded residence.

hernia /'hɜ:nɪə/ *noun* protrusion of part of organ through wall of cavity containing it.

hero /'hɪərəʊ/ *noun* (*plural* **-es**) person admired for courage, outstanding achievements, etc.; chief male character in play, story, etc. □ **hero-worship** idealization of admired person.

heroic /hɪ'rəʊɪk/ ● *adjective* fit for, or like, a hero; very brave. ● *noun* (in *plural*) overdramatic talk or behaviour. □ **heroically** *adverb.*

heroin /'herəʊɪn/ *noun* sedative addictive drug prepared from morphine.

heroine /'herəʊɪn/ *noun* female hero; chief female character in play, story, etc.

heroism /'herəʊɪz(ə)m/ *noun* heroic conduct.

heron /'herən/ *noun* long-necked long-legged wading bird.

herpes /'hɜ:pi:z/ *noun* virus disease causing blisters.

Herr /heə/ *noun* (*plural* **Herren** /'herən/) *title of German man.*

herring /'herɪŋ/ *noun* (*plural* same or **-s**) N. Atlantic edible fish. □ **herring-bone** stitch or weave of small 'V' shapes making zigzag pattern.

hers /hɜ:z/ *pronoun* the one(s) belonging to her.

herself /hə'self/ *pronoun: emphatic form of* SHE *or* HER; *reflexive form of* HER.

hertz *noun* (*plural* same) SI unit of frequency (one cycle per second).

he's /hi:z/ he is; he has.

hesitant /'hezɪt(ə)nt/ *adjective* hesitating. □ **hesitance** *noun*; **hesitancy** *noun*; **hesitantly** *adverb.*

hesitate /'hezɪteɪt/ *verb* (**-ting**) feel or show indecision; pause; (often + *to do*) be reluctant. □ **hesitation** *noun.*

hessian /'hesɪən/ *noun* strong coarse hemp or jute sacking.

heterodox /'hetərəʊdɒks/ *adjective* not orthodox. □ **heterodoxy** *noun.*

heterogeneous /hetərəʊ'dʒi:nɪəs/ *adjective* diverse; varied in content. □ **heterogeneity** /-dʒɪ'ni:ɪtɪ/ *noun.*

heteromorphic /hetərəʊ'mɔ:fɪk/ *adjective* (also **heteromorphous** /-'mɔ:fəs/) *Biology* of dissimilar forms.

heterosexual /hetərəʊ'sekʃʊəl/ ● *adjective* feeling or involving sexual attraction to opposite sex. ● *noun* heterosexual person. □ **heterosexuality** /-'æl-/ *noun.*

het up *adjective colloquial* overwrought.

heuristic /hjʊə'rɪstɪk/ *adjective* serving to discover; using trial and error.

hew *verb* (*past participle* **hewn** /hju:n/ or **hewed**) chop or cut with axe, sword, etc.; cut into shape.

hex ● *verb* practise witchcraft; bewitch. ● *noun* magic spell.

hexa- *combining form* six.

hexagon /'heksəgən/ *noun* plane figure with 6 sides and angles. □ **hexagonal** /-'sæg-/ *adjective.*

hexagram /'heksəgræm/ *noun* 6-pointed star formed by two intersecting equilateral triangles.

hexameter /hek'sæmɪtə/ *noun* verse line of 6 metrical feet.

hey /heɪ/ *interjection calling attention or expressing surprise, inquiry, etc.* □ **hey presto!** *conjuror's phrase on completing trick.*

heyday /'heɪdeɪ/ *noun* time of greatest success, prime.

HF *abbreviation* high frequency.

HGV *abbreviation* heavy goods vehicle.

HH *abbreviation* Her or His Highness; His Holiness; double-hard (pencil lead).

hi /haɪ/ *interjection calling attention or as greeting.*

hiatus /haɪ'eɪtəs/ *noun* (*plural* **-tuses**) gap in series etc.; break between two vowels coming together but not in same syllable.

hibernate /'haɪbəneɪt/ *verb* (**-ting**) (of animal) spend winter in dormant state. □ **hibernation** *noun.*

Hibernian /haɪ'bɜ:nɪən/ *archaic poetical* ● *adjective* of Ireland. ● *noun* native of Ireland.

hibiscus /hɪ'bɪskəs/ *noun* (*plural* **-cuses**) cultivated shrub with large brightly coloured flowers.

hiccup /'hɪkʌp/ (also **hiccough**) ● *noun* involuntary audible spasm of respiratory organs; temporary or minor stoppage or difficulty. ● *verb* (**-p-**) make hiccup.

hick *noun esp. US colloquial* yokel.

hickory /'hɪkərɪ/ *noun* (*plural* **-ies**) N. American tree related to walnut; its wood.

hid *past* of HIDE[1].

hidden *past participle* of HIDE[1]. □ **hidden agenda** secret motivation behind policy etc., ulterior motive.

hide[1] ● *verb* (**-ding**; *past* **hid**; *past participle* **hidden** /'hɪd(ə)n/) put or keep out of sight; conceal oneself; (usually + *from*) conceal (fact). ● *noun* camouflaged shelter for observing wildlife. □ **hide-and-seek** game in which players hide and another searches for them; **hideaway** hiding place, retreat; **hide-out** *colloquial* hiding place.

hide[2] *noun* animal's skin, esp. tanned; *colloquial* human skin. □ **hidebound** rigidly conventional.

hideous /'hɪdɪəs/ *adjective* repulsive, revolting. □ **hideously** *adverb.*

hiding *noun colloquial* thrashing.

hierarchy /'haɪərɑ:kɪ/ *noun* (*plural* **-ies**) system of grades of authority ranked one above another. □ **hierarchical** /-'rɑ:k-/ *adjective.*

hieroglyph /'haɪərəglɪf/ *noun* picture representing word or syllable, esp. in ancient Egyptian. □ **hieroglyphic** /-'glɪf-/ *adjective*; **hieroglyphics** /-'glɪf-/ *plural noun.*

hi-fi /'haɪfaɪ/ *colloquial* ● *adjective* of high fidelity. ● *noun* (*plural* **-s**) equipment for such sound reproduction.

higgledy-piggledy /hɪgəldɪ'pɪgəldɪ/ *adverb & adjective* in disorder.

high /haɪ/ ● *adjective* of great or specified upward extent; far above ground or sea level; coming above normal level; of exalted rank or position, of superior quality; extreme, intense; (often + *on*) *colloquial* intoxicated by alcohol or drugs; (of sound) shrill; (of period etc.) at its peak; (of meat etc.) beginning to go bad. ● *noun* high or highest level or number; area of high barometric pressure; *slang* euphoric state, esp. drug-induced. ● *adverb* far up, aloft; in or to high degree; at high price; (of sound) at high pitch. □ **high altar** chief altar in church; **highball** *US* drink of spirits and soda etc.; **highbrow** *colloquial* (person) of superior intellect or culture; **high chair** child's chair with long legs and meal-tray; **High Church** section of Church of England emphasizing ritual, priestly authority, and sacraments; **high command** army commander-in-chief and associated staff; **High Commission** embassy from one Commonwealth country to another; **High Court** supreme court of justice for civil cases; **highfalutin(g)** *colloquial* pompous, pretentious; **high fidelity** high-quality sound reproduction; **high-flown** extravagant, bombastic; **high-flyer**, **-flier** person of great potential or ambition; **high frequency** *Radio* 3–30 megahertz; **high-handed** overbearing; **high-handedly** *adverb*; **high-handedness** *noun*; **high heels** woman's shoes with high heels; **high jump** athletic event consisting of jumping over high bar, *colloquial* drastic punishment; **high-level** conducted by people of high rank, (of computer language) close to ordinary language; **high-minded** of firm moral principles; **high-mindedness** *noun*; **high pressure** high degree of activity, atmospheric condition with pressure above average; **high priest** chief priest, head of cult; **high-rise** (of building) having many storeys; **high road** main road; **high school** secondary school; **high sea(s)** seas outside territorial waters; **high-spirited** cheerful; **high street** principal shopping street of town; **high tea** early evening meal of tea and cooked food; **high-tech** employing, requiring, or involved in high technology, imitating its style; **high technology** advanced (esp. electronic) technology; **high tension**, **voltage** electrical potential large enough to injure or damage; **high tide**, water time or level of tide at its peak; **high water mark** level reached at high water.

highland /'haɪlənd/ ● *noun* (usually in *plural*) mountainous country, esp. (**the Highlands**) of N. Scotland. ● *adjective* of highland or Highlands. □ **highlander**, **Highlander** *noun.*

highlight ● *noun* moment or detail of vivid interest; bright part of picture; bleached streak in hair. ● *verb* bring into prominence; mark with highlighter.

highlighter *noun* coloured marker pen for emphasizing printed word.

highly *adverb* in high degree, favourably. □ **highly-strung** sensitive, nervous.

highness *noun* state of being high; (**His**, **Her**, **Your Highness**) *title of prince, princess, etc.*

highway *noun* public road, main route. □ **Highway Code** official handbook for road-users; **highwayman** *historical* (usually mounted) robber of stagecoaches.

hijack /'haɪdʒæk/ ● *verb* seize control of (vehicle, aircraft, etc.), esp. to force it to different destination; steal (goods) in transit. ● *noun* hijacking. □ **hijacker** *noun*.

hike ● *noun* long walk, esp. in country for pleasure; rise in prices etc. ● *verb* (**-king**) go on hike. □ **hiker** *noun*.

hilarious /hɪ'leərɪəs/ *adjective* extremely funny; boisterously merry. □ **hilariously** *adverb*; **hilarity** /-'lær-/ *noun*.

hill *noun* natural elevation of ground, lower than mountain; heap, mound. □ **hill-billy** *US colloquial often derogatory* person from remote rural area. □ **hilly** *adjective* (**-ier**, **-iest**).

hillock /'hɪlək/ *noun* small hill, mound.

hilt *noun* handle of sword, dagger, etc.

him *pronoun* (as object of verb) the male person or animal in question, person of unspecified sex; *colloquial* he.

himself /hɪm'self/ *pronoun: emphatic form of* HE *or* HIM; *reflexive form of* HIM.

hind[1] /haɪnd/ *adjective* at back. □ **hindquarters** rump and hind legs of quadruped; **hindsight** wisdom after event. □ **hindmost** *adjective*.

hind[2] /haɪnd/ *noun* female (esp. red) deer.

hinder[1] /'hɪndə/ *verb* impede; delay.

hinder[2] /'haɪndə/ *adjective* rear, hind.

Hindi /'hɪndɪ/ *noun* group of spoken languages in N. India; one of official languages of India, literary form of Hindustani.

hindrance /'hɪndrəns/ *noun* obstruction.

Hindu /'hɪndu:/ ● *noun* (*plural* **-s**) follower of Hinduism. ● *adjective* of Hindus or Hinduism.

Hinduism /'hɪndu:ɪz(ə)m/ *noun* main religious and social system of India, including belief in reincarnation and worship of several gods.

Hindustani /hɪndʊ'stɑ:nɪ/ *noun* language based on Hindi, used in much of India.

hinge ● *noun* movable joint on which door, lid, etc. swings; principle on which all depends. ● *verb* (**-ging**) (+ *on*) depend on (event etc.); attach or be attached with hinge.

hinny /'hɪnɪ/ *noun* (*plural* **-ies**) offspring of female donkey and male horse.

hint ● *noun* indirect suggestion; slight indication; small piece of practical information; faint trace. ● *verb* suggest indirectly. □ **hint at** refer indirectly to.

hinterland /'hɪntəlænd/ *noun* district behind that lying along coast etc.

hip[1] *noun* projection of pelvis and upper part of thigh-bone. □ **hip-flask** small flask for spirits.

hip[2] *noun* fruit of rose.

hip[3] *interjection used to introduce cheer.*

hip[4] *adjective* (also **hep**) (**-pp-**) *slang* trendy, stylish. □ **hip hop**, **hip-hop** subculture combining rap music, graffiti art, and break-dancing.

hippie /'hɪpɪ/ *noun* (also **hippy**) (*plural* **-ies**) *colloquial* person (esp. in 1960s) rejecting convention, typically with long hair, jeans, etc.

hippo /'hɪpəʊ/ *noun* (*plural* **-s**) *colloquial* hippopotamus.

Hippocratic oath /hɪpə'krætɪk/ *noun* statement of ethics of medical profession.

hippodrome /'hɪpədrəʊm/ *noun* music-hall, dance hall, etc.; *historical* course for chariot races etc.

hippopotamus /hɪpə'pɒtəməs/ *noun* (*plural* **-muses** or **-mi** /-maɪ/) large African mammal with short legs and thick skin, living by rivers etc.

hippy = HIPPIE.

hipster[1] /'hɪpstə/ ● *adjective* (of garment) hanging from hips rather than waist. ● *noun* (in *plural*) such trousers.

hipster[2] /'hɪpstə/ *noun slang* hip person.

hire ● *verb* (**-ring**) obtain use of (thing) or services of (person) for payment. ● *noun* hiring, being hired; payment for this. □ **hire out** grant temporary use of (thing) for payment; **hire purchase** system of purchase by paying in instalments. □ **hirer** *noun*.

hireling /'haɪəlɪŋ/ *noun usually derogatory* person who works for hire.

hirsute /'hɜ:sju:t/ *adjective* hairy.

his /hɪz/ ● *adjective* of or belonging to him. ● *pronoun* the one(s) belonging to him.

Hispanic /hɪ'spænɪk/ ● *adjective* of Spain or Spain and Portugal; of Spain and other Spanish-speaking countries.

• *noun* Spanish-speaking person living in US.

hiss /hɪs/ • *verb* make sharp sibilant sound, as of letter *s*; express disapproval of thus; whisper urgently or angrily. • *noun* sharp sibilant sound.

histamine /ˈhɪstəmi:n/ *noun* chemical compound in body tissues associated with allergic reactions.

histology /hɪˈstɒlədʒɪ/ *noun* study of tissue structure.

historian /hɪˈstɔ:rɪən/ *noun* writer of history; person learned in history.

historic /hɪˈstɒrɪk/ *adjective* famous in history or potentially so; *Grammar* (of tense) used to narrate past events.

historical *adjective* of history; belonging to or dealing with the past; not legendary; studying development over period of time. □ **historically** *adverb.*

historicity /hɪstəˈrɪsɪtɪ/ *noun* historical genuineness or accuracy.

historiography /hɪstɔ:rɪˈɒgrəfɪ/ *noun* writing of history; study of this. □ **historiographer** *noun.*

history /ˈhɪstərɪ/ *noun* (*plural* **-ies**) continuous record of (esp. public) events; study of past events; total accumulation of these; (esp. eventful) past or record.

histrionic /hɪstrɪˈɒnɪk/ • *adjective* (of behaviour) theatrical, dramatic. • *noun* (in *plural*) insincere and dramatic behaviour designed to impress.

hit • *verb* (**-tt-**; *past & past participle* **hit**) strike with blow or missile; (of moving body) strike with force; affect adversely; (often + *at*) aim blow; propel (ball etc.) with bat etc.; achieve, reach; *colloquial* encounter, arrive at. • *noun* blow; shot that hits target; *colloquial* popular success. □ **hit it off** (often + *with*, *together*) get on well; **hit-and-run** (of person) causing damage or injury and leaving immediately, (of accident etc.) caused by such person(s); **hit-or-miss** random.

hitch • *verb* fasten with loop etc.; move (thing) with jerk; *colloquial* hitchhike; obtain (lift). • *noun* temporary difficulty, snag; jerk; kind of noose or knot; *colloquial* free ride in vehicle. □ **hitchhike** travel by means of free lifts in passing vehicles; **hitchhiker** *noun.*

hi-tech /ˈhaɪtek/ *adjective* high-tech.

hither /ˈhɪðə/ *adverb formal* to this place. □ **hitherto** up to now.

HIV *abbreviation* human immunodeficiency virus, either of two viruses causing Aids.

hive *noun* beehive. □ **hive off** (**-ving**) separate from larger group.

hives /haɪvz/ *plural noun* skin eruption, esp. nettle-rash.

HM *abbreviation* Her or His Majesty('s).

HMG *abbreviation* Her or His Majesty's Government.

HMI *abbreviation* Her or His Majesty's Inspector (of Schools).

HMS *abbreviation* Her or His Majesty's Ship.

HMSO *abbreviation* Her or His Majesty's Stationery Office.

HNC, HND *abbreviations* Higher National Certificate, Diploma.

ho /həʊ/ *interjection expressing triumph, derision, etc., or calling attention etc.*

hoard /hɔ:d/ • *noun* store (esp. of money or food). • *verb* amass and store. □ **hoarder** *noun.*

hoarding /ˈhɔ:dɪŋ/ *noun* structure erected to carry advertisements; temporary fence round building site etc.

hoar-frost /ˈhɔ:frɒst/ *noun* frozen water vapour on lawns.

hoarse /hɔ:s/ *adjective* (of voice) rough, husky; having hoarse voice. □ **hoarsely** *adverb*; **hoarseness** *noun.*

hoary /ˈhɔ:rɪ/ *adjective* (**-ier**, **-iest**) white or grey with age; aged; old and trite.

hoax • *noun* humorous or malicious deception. • *verb* deceive with hoax.

hob *noun* hotplates etc. on cooker or as separate unit; flat metal shelf at side of fire for heating pans etc. □ **hobnail** heavy-headed nail for boot-sole.

hobble /ˈhɒb(ə)l/ • *verb* (**-ling**) walk lamely, limp; tie together legs of (horse etc.) to keep it from straying. • *noun* limping gait; rope etc. used to hobble horse.

hobby /ˈhɒbɪ/ *noun* (*plural* **-ies**) leisure-time activity pursued for pleasure. □ **hobby-horse** stick with horse's head, used as toy, favourite subject or idea.

hobgoblin /'hɒbgɒblɪn/ *noun* mischievous imp; bogy.

hobnob /'hɒbnɒb/ *verb* (**-bb-**) (usually + *with*) mix socially or informally.

hobo /'həʊbəʊ/ *noun* (*plural* **-es** or **-s**) *US* wandering worker; tramp.

hock[1] *noun* joint of quadruped's hind leg between knee and fetlock.

hock[2] *noun* German white wine.

hock[3] *verb esp. US colloquial* pawn. □ **in hock** in pawn, in debt, in prison.

hockey /'hɒkɪ/ *noun* team game played with ball and hooked sticks.

hocus-pocus /həʊkəs'pəʊkəs/ *noun* trickery.

hod *noun* trough on pole for carrying bricks etc.; portable container for coal.

hodgepodge = HOTCHPOTCH.

hoe ● *noun* long-handled tool for weeding etc. ● *verb* (**hoes, hoed, hoeing**) weed (crops), loosen (soil), or dig up etc. with hoe.

hog ● *noun* castrated male pig; *colloquial* greedy person. ● *verb* (**-gg-**) *colloquial* take greedily; monopolize. □ **go the whole hog** *colloquial* do thing thoroughly; **hogwash** *colloquial* nonsense.

hogmanay /'hɒgmənei/ *noun Scottish* New Year's Eve.

hogshead /'hɒgzhed/ *noun* large cask; liquid or dry measure (about 50 gals.).

ho-ho /həʊ'həʊ/ *interjection representing deep jolly laugh or expressing surprise, triumph, or derision.*

hoick *verb colloquial* (often + *out*) lift or jerk.

hoi polloi /hɔɪ pə'lɔɪ/ *noun* the masses; ordinary people. [Greek]

hoist ● *verb* raise or haul up; raise with ropes and pulleys etc. ● *noun* act of hoisting; apparatus for hoisting. □ **hoist with one's own petard** caught by one's own trick etc.

hoity-toity /hɔɪtɪ'tɔɪtɪ/ *adjective* haughty.

hokum /'həʊkəm/ *noun esp. US slang* sentimental or unreal material in film etc.; bunkum, rubbish.

hold[1] /həʊld/ ● *verb* (*past & past participle* **held**) keep fast; grasp; keep in particular position; contain, have capacity for; possess, have (property, qualifications, job, etc.); conduct, celebrate; detain; think, believe; not give way; reserve. ● *noun* (+ *on, over*) power over; grasp; manner or means of holding. □ **holdall** large soft travelling bag; **hold back** impede, keep for oneself, (often + *from*) refrain; **hold down** repress, *colloquial* be competent enough to keep (job); **hold forth** speak at length or tediously; **hold on** maintain grasp, wait, not ring off; **hold one's tongue** *colloquial* remain silent; **hold out** stretch forth (hand etc.), offer (inducement etc.), maintain resistance, (+ *for*) continue to demand; **hold over** postpone; **hold up** *verb* sustain, display, obstruct, stop and rob by force; **hold-up** *noun* stoppage, delay, robbery by force; **hold with** (usually in negative) *colloquial* approve of.

hold[2] /həʊld/ *noun* cavity in lower part of ship or aircraft for cargo.

holder *noun* device for holding something; possessor of title, shares, etc.; occupant of office etc.

holding *noun* tenure of land; stocks, property, etc. held. □ **holding company** one formed to hold shares of other companies.

hole *noun* empty space in solid body; opening in or through something; burrow; *colloquial* small or gloomy place; *colloquial* awkward situation; (in games) cavity or receptacle for ball, *Golf* section of course from tee to hole. □ **hole up** (**-ling**) *US colloquial* hide oneself. □ **holey** *adjective.*

holiday /'hɒlɪdeɪ/ ● *noun* (often in *plural*) extended period of recreation, esp. spent away from home; break from work or school. ● *verb* spend holiday.

holiness /'həʊlɪnɪs/ *noun* being holy or sacred; (**His, Your Holiness**) *title of Pope.*

holism /'həʊlɪz(ə)m/ *noun* (also **wholism**) theory that certain wholes are greater than sum of their parts; *Medicine* treating of whole person rather than symptoms of disease. □ **holistic** /-'lɪst-/ *adjective.*

hollandaise sauce /hɒlən'deɪz/ *noun* creamy sauce of butter, egg yolks, vinegar, etc.

holler /'hɒlə/ *verb & noun US colloquial* shout.

hollow /'hɒləʊ/ ● *adjective* having cavity; not solid; sunken; echoing; empty;

hungry; meaningless; insincere. ● *noun* hollow place; hole; valley. ● *verb* (often + *out*) make hollow, excavate. ● *adverb colloquial* completely.

holly /'hɒlɪ/ *noun* (*plural* **-ies**) evergreen prickly-leaved shrub with red berries.

hollyhock /'hɒlɪhɒk/ *noun* tall plant with showy flowers.

holm /həʊm/ *noun* (in full **holm-oak**) evergreen oak.

holocaust /'hɒləkɔ:st/ *noun* wholesale destruction; (**the Holocaust**) mass murder of Jews by Nazis 1939–45.

hologram /'hɒləgræm/ *noun* photographic pattern having 3-dimensional effect.

holograph /'hɒləgrɑ:f/ ● *adjective* wholly in handwriting of person named as author. ● *noun* such document.

holography /hə'lɒgrəfɪ/ *noun* study or production of holograms.

holster /'həʊlstə/ *noun* leather case for pistol or revolver on belt etc.

holy /'həʊlɪ/ *adjective* (**-ier**, **-iest**) morally and spiritually excellent; belonging or devoted to God. □ **Holy Ghost** Holy Spirit; **Holy Land** area between River Jordan and Mediterranean; **holy of holies** inner chamber of Jewish temple, thing regarded as most sac red; **holy orders** those of bishop, priest, and deacon; **Holy Saturday** day before Easter; **Holy See** papacy, papal court; **Holy Spirit** Third Person of Trinity; **Holy Week** week before Easter; **Holy Writ** Bible.

homage /'hɒmɪdʒ/ *noun* tribute, expression of reverence.

Homburg /'hɒmbɜ:g/ *noun* man's felt hat with narrow curled brim and lengthwise dent in crown.

home ● *noun* place where one lives; residence; (esp. good or bad) family circumstances; native land; institution caring for people or animals; place where thing originates, is kept, is most common, etc.; (in games) finishing line in race, goal, home match or win etc. ● *adjective* of or connected with home; carried on or done at home; not foreign; played etc. on team's own ground. ● *adverb* to or at home; to point aimed at. ● *verb* (**-ming**) (of pigeon) return home; (often + *on*, *in on*) (of missile etc.) be guided to destination. □ **at home** *adjective* in one's house or native land, at ease, well-informed, available to callers; **at-home** *noun* social reception in person's home; **home-brew** beer etc. brewed at home; **Home Counties** those lying round London; **home economics** study of household management; **homeland** native land, any of several areas reserved for black South Africans; **Home Office** British government department concerned with immigration, law and order, etc.; **home rule** self-government; **Home Secretary** minister in charge of Home Office; **homesick** depressed by absence from home; **homesickness** such depression; **homestead** house with outbuildings, farm; **homework** lessons to be done by schoolchild at home. □ **homeless** *adjective*; **homeward** *adjective & adverb*; **homewards** *adverb*.

homely *adjective* (**-ier**, **-iest**) plain; unpretentious; *US* unattractive; cosy. □ **homeliness** *noun*.

homeopathy etc. *US* = HOMOEOPATHY etc.

Homeric /həʊ'merɪk/ *adjective* of or in style of the ancient Greek poet Homer; of Bronze Age Greece.

homey *adjective* (**-mier**, **-miest**) suggesting home; cosy.

homicide /'hɒmɪsaɪd/ *noun* killing of person by another; person who kills another. □ **homicidal** /-'saɪd-/ *adjective*.

homily /'hɒmɪlɪ/ *noun* (*plural* **-ies**) short sermon; moralizing lecture. □ **homiletic** /-'let-/ *adjective*.

homing *adjective* (of pigeon) trained to fly home; (of device) for guiding to target etc.

hominid /'hɒmɪnɪd/ ● *adjective* of mammal family of existing and fossil man. ● *noun* member of this.

hominoid /'hɒmɪnɔɪd/ ● *adjective* like a human. ● *noun* animal resembling human.

homoeopathy /həʊmɪ'ɒpəθɪ/ *noun* (*US* **homeopathy**) treatment of disease by drugs that in healthy person would produce symptoms of the disease. □ **homoeopath** /'həʊmɪəʊpæθ/ *noun*; **homoeopathic** /-'pæθ-/ *adjective*.

homogeneous /hɒmə'dʒi:nɪəs/ *adjective* (having parts) of same kind or nature;

uniform. □ **homogeneity** /-dʒɪ'niːɪtɪ/ *noun*; **homogeneously** *adverb*.

■ **Usage** *Homogeneous* is often confused with *homogenous* (and pronounced /hə'mɒdʒənəs/, with the stress on the *-mog-*), but that is a term in biology meaning 'similar owing to common descent'.

---

homogenize /hə'mɒdʒɪnaɪz/ *verb* (also **-ise**) (**-zing** or **-sing**) make homogeneous; treat (milk) so that cream does not separate.

homologous /hə'mɒləgəs/ *adjective* having same relation, relative position, etc.; corresponding.

homology /hə'mɒlədʒɪ/ *noun* homologous relation, correspondence.

homonym /'hɒmənɪm/ *noun* word spelt or pronounced like another but of different meaning.

homophobia /hɒmə'fəʊbɪə/ *noun* hatred or fear of homosexuals. □ **homophobe** /'hɒm-/ *noun*; **homophobic** *adjective*.

homophone /'hɒməfəʊn/ *noun* word pronounced like another but having different meaning, e.g. *beach, beech*.

Homo sapiens /həʊməʊ 'sæpɪenz/ *noun* modern humans regarded as a species. [Latin]

homosexual /hɒmə'sekʃʊəl/ ● *adjective* feeling or involving sexual attraction to people of same sex. ● *noun* homosexual person. □ **homosexuality** /-'æl-/ *noun*.

Hon. *abbreviation* Honorary; Honourable.

hone ● *noun* whetstone, esp. for razors. ● *verb* (**-ning**) sharpen (as) on hone.

honest /'ɒnɪst/ ● *adjective* not lying, cheating, or stealing; sincere; fairly earned. ● *adverb colloquial* genuinely, really. □ **honestly** *adverb*.

honesty /'ɒnɪstɪ/ *noun* being honest, truthfulness; plant with purple or white flowers and flat round pods.

honey /'hʌnɪ/ *noun* (*plural* **-s**) sweet sticky yellowish fluid made by bees from nectar; colour of this; sweetness; darling.

honeycomb ● *noun* beeswax structure of hexagonal cells for honey and eggs; pattern arranged hexagonally. ● *verb* fill with cavities; mark with honeycomb pattern.

honeydew *noun* sweet substance excreted by aphids; variety of melon.

honeyed *adjective* sweet, sweet-sounding.

honeymoon ● *noun* holiday of newly married couple; initial period of enthusiasm or goodwill. ● *verb* spend honeymoon.

honeysuckle *noun* climbing shrub with fragrant flowers.

honk ● *noun* sound of car horn; cry of wild goose. ● *verb* (cause to) make honk.

honky-tonk /'hɒŋkɪtɒŋk/ *noun colloquial* ragtime piano music.

honor *US* = HONOUR.

honorable *US* = HONOURABLE.

honorarium /ɒnə'reərɪəm/ *noun* (*plural* **-s** or **-ria**) voluntary payment for professional services.

honorary /'ɒnərərɪ/ *adjective* conferred as honour; unpaid.

honorific /ɒnə'rɪfɪk/ *adjective* conferring honour; implying respect.

honour /'ɒnə/ (*US* **honor**) ● *noun* high respect, public regard; adherence to what is right; nobleness of mind; thing conferred as distinction (esp. official award for bravery or achievement); privilege; (**His, Her, Your Honour**) *title of judge etc.*; person or thing that brings honour; chastity, reputation for this; (in *plural*) specialized degree course or special distinction in exam; (in *plural*) (in card games) 4 or 5 highest-ranking cards; *Golf* right of driving off first. ● *verb* respect highly; confer honour on; accept or pay (bill, cheque) when due. □ **do the honours** perform duties of host etc.

honourable *adjective* (*US* **honorable**) deserving, bringing, or showing honour; (**Honourable**) *courtesy title of MPs, certain officials, and children of certain ranks of the nobility*. □ **honourably** *adverb*.

hooch /huːtʃ/ *noun US colloquial* alcoholic spirits, esp. inferior or illicit.

hood[1] /hʊd/ ● *noun* covering for head and neck, esp. as part of garment; separate hoodlike garment; folding top of car etc.; *US* bonnet of car etc.; protective cover. ● *verb* cover with hood. □ **hooded** *adjective*.

hood² /hʊd/ *noun US slang* gangster, gunman.

hoodlum /ˈhu:dləm/ *noun* hooligan; gangster.

hoodoo /ˈhu:du:/ *noun US* bad luck; thing or person that brings this.

hoodwink *verb* deceive, delude.

hoof /hu:f/ *noun* (*plural* **-s** or **hooves** /hu:vz/) horny part of foot of horse etc. □ **hoof it** *slang* go on foot.

hook /hʊk/ ● *noun* bent piece of metal etc. for catching hold or for hanging things on; curved cutting instrument; hook-shaped thing; hooking stroke; *Boxing* short swinging blow. ● *verb* grasp, secure, fasten, or catch with hook; (in sports) send (ball) in curving or deviating path; *Rugby* secure (ball) in scrum with foot. □ **hook and eye** small hook and loop as fastener; **hook, line, and sinker** completely; **hook-up** connection, esp. of broadcasting equipment; **hookworm** worm infesting intestines of humans and animals.

hookah /ˈhʊkə/ *noun* tobacco pipe with long tube passing through water to cool smoke.

hooked *adjective* hook-shaped; (often + *on*) *slang* addicted or captivated.

hooker *noun Rugby* player in front row of scrum who tries to hook ball; *slang* prostitute.

hooligan /ˈhu:lɪgən/ *noun* young ruffian. □ **hooliganism** *noun*.

hoop /hu:p/ ● *noun* circular band of metal, wood, etc., esp. as part of framework; wooden etc. circle bowled by child or used by circus performer etc.; arch through which balls are hit in croquet. ● *verb* bind with hoop(s). □ **hoop-la** game with rings thrown to encircle prizes.

hoopoe /ˈhu:pu:/ *noun* bird with variegated plumage and fanlike crest.

hooray = HURRAH.

hoot /hu:t/ ● *noun* owl's cry; sound of car's horn etc.; shout of derision etc.; *colloquial* (cause of) laughter; (also **two hoots**) *slang* anything, in the slightest. ● *verb* utter hoot(s); greet or drive away with hoots; sound (horn).

hooter *noun* thing that hoots, esp. car's horn or siren; *slang* nose.

Hoover /ˈhu:və/ ● *noun proprietary term* vacuum cleaner. ● *verb* (**hoover**) clean or (+ *up*) suck up with vacuum cleaner.

hooves *plural* of HOOF.

hop¹ ● *verb* (**-pp-**) (of bird, frog, etc.) spring with all feet at once; (of person) jump on one foot; move or go quickly, leap. ● *noun* hopping movement; *colloquial* dance; short journey, esp. flight. □ **hop in, out** *colloquial* get into or out of car etc.; **hopscotch** child's game of hopping over squares marked on ground.

hop² *noun* climbing plant with bitter cones used to flavour beer etc.; (in *plural*) these cones.

hope ● *noun* expectation and desire; person or thing giving cause for hope; what is hoped for. ● *verb* (**-ping**) feel hope; expect and desire.

hopeful *adjective* feeling or inspiring hope, promising.

hopefully *adverb* in a hopeful way; it is to be hoped.

■ **Usage** The use of *hopefully* to mean 'it is to be hoped' is common, but it is considered incorrect by some people.

hopeless *adjective* feeling or admitting no hope; inadequate, incompetent. □ **hopelessly** *adverb*; **hopelessness** *noun*.

hopper *noun* funnel-like device for feeding grain into mill etc.; hopping insect.

horde *noun usually derogatory* large group, gang.

horehound /ˈhɔ:haʊnd/ *noun* herb with aromatic bitter juice.

horizon /həˈraɪz(ə)n/ *noun* line at which earth and sky appear to meet; limit of mental perception, interest, etc.

horizontal /hɒrɪˈzɒnt(ə)l/ ● *adjective* parallel to plane of horizon; level, flat. ● *noun* horizontal line, plane, etc. □ **horizontally** *adverb*.

hormone /ˈhɔ:məʊn/ *noun* substance produced by body and transported in tissue fluids to stimulate cells or tissues to growth etc.; similar synthetic substance. □ **hormone replacement therapy** treatment with hormones to relieve menopausal symptoms. □ **hormonal** /-ˈməʊn-/ *adjective*.

horn *noun* hard outgrowth, often curved and pointed, on head of animal; horn-like projection; substance of horns;

brass wind instrument; instrument giving warning. □ **hornbeam** tough-wooded hedgerow tree; **hornbill** bird with hornlike excrescence on bill; **horn of plenty** cornucopia; **horn-rimmed** (of spectacles) having rims of horn or similar substance. □ **horned** *adjective*.

hornblende /ˈhɔːnblend/ *noun* dark brown etc. mineral constituent of granite etc.

hornet /ˈhɔːnɪt/ *noun* large species of wasp.

hornpipe *noun* (music for) lively dance associated esp. with sailors.

horny *adjective* (**-ier**, **-iest**) of or like horn; hard; *slang* sexually excited. □ **horniness** *noun*.

horology /həˈrɒlədʒɪ/ *noun* clock-making. □ **horological** /hɒrəˈlɒdʒ-/ *adjective*.

horoscope /ˈhɒrəskəʊp/ *noun* prediction of person's future based on relative position of stars at his or her birth.

horrendous /həˈrendəs/ *adjective* horrifying. □ **horrendously** *adverb*.

horrible /ˈhɒrɪb(ə)l/ *adjective* causing horror; *colloquial* unpleasant. □ **horribly** *adverb*.

horrid /ˈhɒrɪd/ *adjective* horrible; *colloquial* unpleasant.

horrific /həˈrɪfɪk/ *adjective* horrifying. □ **horrifically** *adverb*.

horrify /ˈhɒrɪfaɪ/ *verb* (**-ies**, **-ied**) arouse horror in; shock. □ **horrifying** *adjective*.

horror /ˈhɒrə/ ● *noun* intense loathing or fear; (often + *of*) deep dislike; *colloquial* intense dismay; horrifying thing. ● *adjective* (of films etc.) designed to arouse feelings of horror.

hors d'oeuvre /ɔː ˈdɜːvr/ *noun* appetizer served at start of meal.

horse ● *noun* large 4-legged hoofed mammal with mane, used for riding etc.; adult male horse; vaulting-block; supporting frame. ● *verb* (**-sing**) (+ *around*) fool about. □ **horsebox** closed vehicle for transporting horse(s); **horse brass** brass ornament originally for horse's harness; **horse chestnut** tree with conical clusters of flowers, its dark brown fruit; **horse-drawn** pulled by horse(s); **horsefly** biting insect troublesome to horses; **Horse Guards** cavalry brigade of British household troops; **horsehair** (padding etc. of) hair from mane or tail of horse; **horseman** (skilled) rider on horseback; **horsemanship** skill in riding; **horseplay** boisterous play; **horsepower** (*plural* same) unit of rate of doing work; **horse race** race between horses with riders; **horse racing** sport of racing horses; **horseradish** plant with pungent root used to make sauce; **horse sense** *colloquial* plain common sense; **horseshoe** U-shaped iron shoe for horse, thing of this shape; **horsetail** (plant resembling) horse's tail; **horsewhip** *noun* whip for horse, *verb* beat (person) with this; **horsewoman** (skilled) woman rider on horseback.

horsy *adjective* (**-ier**, **-iest**) of or like horse; concerned with horses.

horticulture /ˈhɔːtɪkʌltʃə/ *noun* art of gardening. □ **horticultural** /-ˈkʌlt-/ *adjective*; **horticulturist** /-ˈkʌlt-/ *noun*.

hosanna /həʊˈzænə/ *noun & interjection* cry of adoration.

hose /həʊz/ ● *noun* (also **hose-pipe**) flexible tube for conveying liquids; (treated as *plural*) stockings and socks collectively; *historical* breeches. ● *verb* (**-sing**) (often + *down*) water, spray, or drench with hose.

hosier /ˈhəʊzɪə/ *noun* dealer in stockings and socks. □ **hosiery** *noun*.

hospice /ˈhɒspɪs/ *noun* home for (esp. terminally) ill or destitute people; travellers' lodging kept by religious order etc.

hospitable /hɒsˈpɪtəb(ə)l/ *adjective* giving hospitality. □ **hospitably** *adverb*.

hospital /ˈhɒspɪt(ə)l/ *noun* institution providing medical and surgical treatment and nursing for ill and injured people; *historical* hospice.

hospitality /hɒspɪˈtælɪtɪ/ *noun* friendly and generous reception of guests or strangers.

hospitalize /ˈhɒspɪtəlaɪz/ *verb* (also **-ise**) (**-zing** or **-sing**) send or admit to hospital. □ **hospitalization** *noun*.

host[1] /həʊst/ *noun* (usually + *of*) large number of people or things.

host[2] /həʊst/ ● *noun* person who entertains another as guest; compère; animal or plant having parasite; recipient of transplanted organ; landlord of inn. • *verb* be host to (person) or of (event).

host³ /həʊst/ *noun* (usually **the Host**) bread consecrated in Eucharist.

hostage /'hɒstɪdʒ/ *noun* person seized or held as security for fulfilment of a condition.

hostel /'hɒst(ə)l/ *noun* house of residence for students etc.; youth hostel.

hostelling *noun* (*US* **hosteling**) practice of staying in youth hostels. □ **hosteller** *noun.*

hostelry *noun* (*plural* **-ies**) *archaic* inn.

hostess /'həʊstɪs/ *noun* woman who entertains guests, or customers at nightclub.

hostile /'hɒstaɪl/ *adjective* of enemy; (often + *to*) unfriendly, opposed.

hostility /hɒ'stɪlɪtɪ/ *noun* (*plural* **-ies**) being hostile; enmity; warfare; (in *plural*) acts of war.

hot • *adjective* (**-tt-**) having high temperature, very warm; causing sensation of or feeling heat; pungent; excited; (often + *on*, *for*) eager; (of news) fresh; skilful, formidable; (+ *on*) knowledgeable about; *slang* (of stolen goods) difficult to dispose of. • *verb* (**-tt-**) (usually + *up*) *colloquial* make or become hot; become more active, exciting, or dangerous. □ **hot air** *slang* empty or boastful talk; **hot-air balloon** balloon containing air heated by burners, causing it to rise; **hotbed** (+ *of*) environment conducive to (vice etc.), bed of earth heated by fermenting manure; **hot cross bun** bun marked with cross, eaten on Good Friday; **hot dog** *colloquial* hot sausage in bread roll; **hotfoot** in eager haste; **hothead** impetuous person; **hotheaded** impetuous; **hothouse** heated (mainly glass) building for growing plants, environment conducive to rapid growth; **hotline** direct telephone line; **hotplate** heated metal plate for cooking food or keeping it hot; **hotpot** dish of stewed meat and vegetables; **hot rod** vehicle modified for extra power and speed; **hot seat** *slang* awkward or responsible position, electric chair; **hot water** *colloquial* difficulty, trouble; **hot-water bottle** container filled with hot water to warm bed etc. □ **hotly** *adverb.*

hotchpotch /'hɒtʃpɒtʃ/ *noun* (also **hodgepodge** /'hɒdʒpɒdʒ/) confused mixture, jumble, esp. of ideas.

hotel /həʊ'tel/ *noun* (usually licensed) place providing meals and accommodation for payment.

hotelier /həʊ'telɪə/ *noun* hotel-keeper.

houmous = HUMMUS.

hound /haʊnd/ • *noun* dog used in hunting; *colloquial* despicable man. • *verb* harass or pursue.

hour /aʊə/ *noun* twenty-fourth part of day and night, 60 minutes; time of day, point in time; (in *plural* after numerals in form 18.00, 20.30, etc.) this number of hours and minutes past midnight on the 24-hour clock; period set aside for some purpose; (in *plural*) working or open period; short time; time for action etc.; (**the hour**) each time o'clock of a whole number of hours. □ **hourglass** two connected glass bulbs containing sand that takes an hour to pass from upper to lower bulb. □ **hourly** *adjective & adverb.*

houri /'hʊərɪ/ *noun* (*plural* **-s**) beautiful young woman in Muslim paradise.

house • *noun* /haʊs/ (*plural* /'haʊzɪz/) building for human habitation; building for special purpose or for keeping animals or goods; (buildings of) religious community; section of boarding school etc.; division of school for games etc.; royal family, dynasty; (premises of) firm or institution; (building for) legislative etc. assembly; audience or performance in theatre etc. • *verb* /haʊz/ (**-sing**) provide house for; store; enclose or encase (part etc.); fix in socket etc. □ **house arrest** detention in one's own house; **houseboat** boat equipped for living in; **housebound** confined to one's house through illness etc.; **housebreaker** burglar; **housebreaking** burglary; **housefly** common fly; **house-husband** man who does wife's traditional duties; **housekeeper** woman managing affairs of house; **housekeeping** management of house, money for this, record-keeping etc.; **houseman** resident junior doctor of hospital; **house-martin** bird which builds nests on house walls etc.; **housemaster**, **housemistress** teacher in charge of house in boarding school; **house music** pop music with synthesized drums and bass and fast beat; **House of Commons** elected chamber of Parliament; **House of Lords** chamber of Parliament that is mainly hereditary; **house plant** one grown indoors; **house-proud** attentive to care etc. of home; **house-trained** (of domestic animal) trained to be clean in house; **house-warming** party celebrating move to new house; **housewife** woman whose chief occupation is managing household; **housewifely** *adjective*; **housework** regular cleaning and cooking etc. in home.

household *noun* occupants of house; house and its affairs. □ **household name** well-known name; **household troops** those nominally guarding sovereign; **household word** well-known saying or name.

householder *noun* person who owns or rents house; head of household.

housing /'haʊzɪŋ/ *noun* (provision of) houses; protective casing. □ **housing estate** residential area planned as a unit.

hove *past* of HEAVE.

hovel /'hɒv(ə)l/ *noun* small miserable dwelling.

hover /'hɒvə/ *verb* (of bird etc.) remain in one place in air; (often + *about*, *round*) linger.

hovercraft *noun* (*plural* same) vehicle moving on air-cushion provided by downward blast.

hoverport *noun* terminal for hovercraft.

how /haʊ/ *interrogative & relative adverb* by what means, in what way; in what condition; to what extent. □ **however** nevertheless, in whatever way, to whatever extent.

howdah /'haʊdə/ *noun* (usually canopied) seat for riding elephant or camel.

howitzer /'haʊɪtsə/ *noun* short gun firing shells at high elevation.

howl /haʊl/ ● *noun* long doleful cry of dog etc.; prolonged wailing noise; loud cry of pain, rage, derision, or laughter. ● *verb* make howl; weep loudly; utter with howl.

howler *noun colloquial* glaring mistake.

hoy *interjection used to call attention.*

hoyden /'hɔɪd(ə)n/ *noun* boisterous girl.

HP *abbreviation* hire purchase; (also **hp**) horsepower.

HQ *abbreviation* headquarters.

hr. *abbreviation* hour.

HRH *abbreviation* Her or His Royal Highness.

HRT *abbreviation* hormone replacement therapy.

HT *abbreviation* high tension.

hub *noun* central part of wheel, rotating on or with axle; centre of interest, activity, etc.

hubble-bubble /'hʌb(ə)lbʌb(ə)l/ *noun* simple hookah; confused sound or talk.

hubbub /'hʌbʌb/ *noun* confused noise of talking; disturbance.

hubby /'hʌbɪ/ *noun* (*plural* **-ies**) *colloquial* husband.

hubris /'hju:brɪs/ *noun* arrogant pride, presumption.

huckleberry /'hʌkəlbərɪ/ *noun* (*plural* **-ies**) low N. American shrub; its fruit.

huckster /'hʌkstə/ ● *noun* hawker; aggressive salesman. ● *verb* haggle; hawk (goods).

huddle /'hʌd(ə)l/ ● *verb* (**-ling**) (often + *up*) crowd together; nestle closely; (often + *up*) curl one's body up. ● *noun* confused mass; *colloquial* secret conference.

hue *noun* colour, tint.

hue and cry *noun* loud outcry.

huff ● *noun colloquial* fit of petulance. ● *verb* blow air, steam, etc.; (esp. **huff and puff**) bluster; remove (opponent's man) as forfeit in draughts. □ **huffy** *adjective* (**-ier**, **-iest**).

hug ● *verb* (**-gg-**) squeeze tightly in one's arms, esp. with affection; keep close to, fit tightly around. ● *noun* close clasp.

huge /hju:dʒ/ *adjective* very large or great.

hugely *adverb* extremely, very much.

hugger-mugger /'hʌgəmʌgə/ *adjective & adverb* in secret; in confusion.

Huguenot /'hju:gənəʊ/ *noun historical* French Protestant.

hula /'hu:lə/ *noun* (also **hula-hula**) Polynesian women's dance. □ **hula hoop** large hoop spun round the body.

hulk *noun* body of dismantled ship; *colloquial* large clumsy-looking person or thing.

hulking *adjective colloquial* bulky, clumsy.

hull[1] *noun* body of ship etc.

hull[2] ● *noun* outer covering of fruit. ● *verb* remove hulls of.

hullabaloo /hʌləbə'lu:/ *noun* uproar.

hullo = HELLO.

hum ● *verb* (**-mm-**) make low continuous sound like bee; sing with closed lips; make slight inarticulate sound; *colloquial* be active; *colloquial* smell unpleasantly. ● *noun* humming sound. □ **hummingbird** small tropical bird whose wings hum.

human /'hju:mən/ ● *adjective* of or belonging to species *Homo sapiens*; consisting of human beings; having characteristics of humankind, as being weak, fallible, sympathetic, etc. ● *noun* (*plural* **-s**) human being. □ **human being** man, woman, or child; **human chain** line of people for passing things along etc.; **humankind** human beings collectively; **human rights** those held to belong to all people; **human shield** person(s) placed in line of fire to discourage attack.

humane /hju:'meɪn/ *adjective* benevolent, compassionate; inflicting minimum pain; (of studies) tending to civilize. □ **humanely** *adverb.*

humanism /'hju:mənɪz(ə)m/ *noun* non-religious philosophy based on liberal human values; (often **Humanism**) literary culture, esp. in Renaissance. □ **humanist** *noun*; **humanistic** /-'nɪst-/ *adjective.*

humanitarian /hju:mænɪ'teərɪən/ ● *noun* person who seeks to promote human welfare. ● *adjective* of humanitarians. □ **humanitarianism** *noun.*

humanity /hju:'mænɪtɪ/ *noun* (*plural* **-ies**) human race; human nature; humaneness, benevolence; (usually in *plural*) subjects concerned with human culture.

humanize *verb* (also **-ise**) (**-zing** or **-sing**) make human or humane. □ **humanization** *noun.*

humanly *adverb* within human capabilities; in a human way.

humble /'hʌmb(ə)l/ ● *adjective* (**-r**, **-st**) having or showing low self-esteem; lowly, modest. ● *verb* (**-ling**) make humble; lower rank of. □ **eat humble pie** apologize humbly, accept humiliation. □ **humbly** *adverb.*

humbug /'hʌmbʌg/ ● *noun* deception, hypocrisy; impostor; striped peppermint-flavoured boiled sweet. ● *verb* (**-gg-**) be impostor; hoax.

humdinger /'hʌmdɪŋə/ *noun slang* remarkable person or thing.

humdrum /'hʌmdrʌm/ *adjective* dull, commonplace.

humerus /'hju:mərəs/ *noun* (*plural* **-ri** /-raɪ/) bone of upper arm.

humid /'hju:mɪd/ *adjective* warm and damp.

humidifier /hju:'mɪdɪfaɪə/ *noun* device for keeping atmosphere moist.

humidify /hju:'mɪdɪfaɪ/ *verb* (**-ies**, **-ied**) make (air etc.) humid.

humidity /hju:'mɪdɪtɪ/ *noun* (*plural* **-ies**) dampness; degree of moisture, esp. in atmosphere.

humiliate /hju:'mɪlɪeɪt/ *verb* (**-ting**) injure dignity or self-respect of. □ **humiliating** *adjective*; **humiliation** *noun.*

humility /hju:'mɪlɪtɪ/ *noun* humbleness; meekness.

hummock /'hʌmək/ *noun* hillock, hump.

hummus /'hʊmʊs/ *noun* (also **houmous**) dip of chickpeas, sesame paste, lemon juice, and garlic.

humor *US* = HUMOUR.

humorist /'hju:mərɪst/ *noun* humorous writer, talker, or actor.

humorous /'hju:mərəs/ *adjective* showing humour, comic. □ **humorously** *adverb.*

humour /'hju:mə/ (*US* **humor**) ● *noun* quality of being amusing; expression of humour in literature etc.; (in full **sense of humour**) ability to perceive or express humour; state of mind, mood; each of 4 fluids formerly held to determine physical and mental qualities. ● *verb* gratify or indulge (person, taste, etc.). □ **humourless** *adjective.*

hump ● *noun* rounded lump, esp. on back; rounded raised mass of earth etc.; (**the hump**) *slang* fit of depression or annoyance. ● *verb* (often + *about*) *colloquial* lift or carry with difficulty; make hump-shaped. □ **humpback** (**whale**) whale with dorsal fin; **humpback bridge** one with steep approach to top; **over the hump** past the most difficult stage.

humph /həmf/ *interjection & noun* inarticulate sound of dissatisfaction etc.

humus /ˈhjuːməs/ *noun* organic constituent of soil formed by decomposition of plants.

hunch ● *verb* bend or arch into a hump. ● *noun* intuitive feeling; hump.

hundred /ˈhʌndrəd/ *adjective & noun* (*plural* same in first sense) ten times ten; *historical* subdivision of county; (**hundreds**) *colloquial* large number. □ **hundreds and thousands** tiny coloured sweets; **hundredweight** (*plural* same or **-s**) 112 lb (50.80 kg), *US* 100 lb. (45.4 kg) □ **hundredth** *adjective & noun.*

hundredfold *adjective & adverb* a hundred times as much or many.

hung *past & past participle* of HANG. □ **hung-over** suffering from hangover; **hung parliament** parliament in which no party has clear majority.

Hungarian /hʌŋˈgeərɪən/ ● *noun* native, national, or language of Hungary. ● *adjective* of Hungary or its people or language.

hunger /ˈhʌŋgə/ ● *noun* lack of food; discomfort or exhaustion caused by this; (often + *for*, *after*) strong desire. ● *verb* (often + *for*, *after*) crave, desire; feel hunger. □ **hunger strike** refusal of food as protest.

hungry /ˈhʌŋgrɪ/ *adjective* (**-ier**, **-iest**) feeling, showing, or inducing hunger; craving. □ **hungrily** *adverb.*

hunk *noun* large piece cut off; *colloquial* sexually attractive man.

hunt ● *verb* pursue wild animals for food or sport; (of animal) pursue prey; (+ *after*, *for*) search; (as **hunted** *adjective*) (of look) frightened. ● *noun* hunting; hunting area or society. □ **huntsman** hunter, person in charge of hounds. □ **hunting** *noun.*

hunter *noun* (*feminine* **huntress**) person who hunts; horse ridden for hunting.

hurdle /ˈhɜːd(ə)l/ ● *noun* frame to be jumped over by athlete in race; (in *plural*) hurdle race; obstacle; portable rectangular frame used as temporary fence. ● *verb* (**-ling**) run in hurdle race. □ **hurdler** *noun.*

hurdy-gurdy /hɜːdɪˈgɜːdɪ/ *noun* (*plural* **-ies**) droning musical instrument played by turning handle; *colloquial* barrel organ.

hurl ● *verb* throw violently. ● *noun* violent throw.

hurley /ˈhɜːlɪ/ *noun* (also **hurling**) (stick used in) Irish game resembling hockey.

hurly-burly /hɜːlɪˈbɜːlɪ/ *noun* boisterous activity; commotion.

hurrah /hʊˈrɑː/ (also **hurray** /hʊˈreɪ/) ● *interjection expressing joy or approval.* ● *noun* utterance of 'hurrah'.

hurricane /ˈhʌrɪkən/ *noun* storm with violent wind, esp. W. Indian cyclone. □ **hurricane lamp** lamp with flame protected from wind.

hurry /ˈhʌrɪ/ ● *noun* great haste; eagerness; (with negative or in questions) need for haste. ● *verb* (**-ies**, **-ied**) (cause to) move or act hastily; (as **hurried** *adjective*) hasty, done rapidly. □ **hurriedly** *adverb.*

hurt ● *verb* (*past & past participle* **hurt**) cause pain, injury, or distress to; suffer pain. ● *noun* injury; harm. □ **hurtful** *adjective*; **hurtfully** *adverb.*

hurtle /ˈhɜːt(ə)l/ *verb* (**-ling**) move or hurl rapidly or noisily, come with crash.

husband /ˈhʌzbənd/ ● *noun* married man in relation to his wife. ● *verb* use (resources) economically.

husbandry *noun* farming; management of resources.

hush ● *verb* make, become, or be silent. ● *interjection calling for silence.* ● *noun* silence. □ **hush-hush** *colloquial* highly secret; **hush money** *slang* sum paid to ensure discretion; **hush up** suppress (fact).

husk ● *noun* dry outer covering of fruit or seed. ● *verb* remove husk from.

husky[1] /ˈhʌskɪ/ *adjective* (**-ier**, **-iest**) dry in the throat, hoarse; strong, hefty. □ **huskily** *adverb.*

husky[2] /ˈhʌskɪ/ *noun* (*plural* **-ies**) powerful dog used for pulling sledges.

hussar /hʊˈzɑː/ *noun* light-cavalry soldier.

hussy /ˈhʌsɪ/ *noun* (*plural* **-ies**) pert girl; promiscuous woman.

hustings /ˈhʌstɪŋz/ *noun* election proceedings.

hustle /ˈhʌs(ə)l/ ● *verb* (**-ling**) jostle; (+ *into*, *out of*, etc.) force, hurry; *slang* solicit business. ● *noun* act or instance of hustling. □ **hustler** *noun.*

hut *noun* small simple or crude house or shelter.

hutch *noun* box or cage for rabbits etc.

hyacinth /'haɪəsɪnθ/ *noun* bulbous plant with bell-shaped flowers.

hybrid /'haɪbrɪd/ • *noun* offspring of two animals or plants of different species etc.; thing of mixed origins. • *adjective* bred as hybrid; heterogeneous. □ **hybridism** *noun*; **hybridization** *noun*; **hybridize** *verb* (also **-ise**) (**-zing** or **-sing**).

hydra /'haɪdrə/ *noun* freshwater polyp; something hard to destroy.

hydrangea /haɪ'dreɪndʒə/ *noun* shrub with globular clusters of white, blue, or pink flowers.

hydrant /'haɪdrənt/ *noun* outlet for drawing water from main.

hydrate /'haɪdreɪt/ • *noun* chemical compound of water with another compound etc. • *verb* (**-ting**) (cause to) combine with water. □ **hydration** *noun*.

hydraulic /haɪ'drɔ:lɪk/ *adjective* (of water etc.) conveyed through pipes etc.; operated by movement of liquid. □ **hydraulically** *adverb*.

hydraulics *plural noun* (usually treated as *singular*) science of conveyance of liquids through pipes etc., esp. as motive power.

hydro /'haɪdrəʊ/ *noun* (*plural* **-s**) *colloquial* hotel etc. originally providing hydropathic treatment; hydroelectric powerplant.

hydro- *combining form* water; combined with hydrogen.

hydrocarbon /haɪdrəʊ'kɑ:bən/ *noun* compound of hydrogen and carbon.

hydrocephalus /haɪdrə'sefələs/ *noun* accumulated fluid in brain, esp. in young children. □ **hydrocephalic** /-sɪ'fælɪk/ *adjective*.

hydrochloric acid /haɪdrə'klɒrɪk/ *noun* solution of hydrogen chloride in water.

hydrodynamics /haɪdrəʊdaɪ'næmɪks/ *plural noun* (usually treated as *singular*) science of forces acting on or exerted by liquids.

hydroelectric /haɪdrəʊɪ'lektrɪk/ *adjective* generating electricity by water-power; (of electricity) so generated. □ **hydroelectricity** /-'trɪs-/ *noun*.

hydrofoil /'haɪdrəfɔɪl/ *noun* boat fitted with planes for raising hull out of water at speed; such a plane.

hydrogen /'haɪdrədʒ(ə)n/ *noun* light colourless odourless gas combining with oxygen to form water. □ **hydrogen bomb** immensely powerful bomb utilizing explosive fusion of hydrogen nuclei; **hydrogen peroxide** see PEROXIDE.

hydrogenate /haɪ'drɒdʒɪneɪt/ *verb* (**-ting**) charge with or cause to combine with hydrogen. □ **hydrogenation** *noun*.

hydrography /haɪ'drɒgrəfɪ/ *noun* science of surveying and charting seas, lakes, rivers, etc. □ **hydrographer** *noun*; **hydrographic** /-drə'græf-/ *adjective*.

hydrology /haɪ'drɒlədʒɪ/ *noun* science of relationship between water and land.

hydrolyse /'haɪdrəlaɪz/ *verb* (**-sing**) (*US* **-lyze**; **-zing**) decompose by hydrolysis.

hydrolysis /haɪ'drɒlɪsɪs/ *noun* decomposition by chemical reaction with water.

hydrometer /haɪ'drɒmɪtə/ *noun* instrument for measuring density of liquids.

hydropathy /haɪ'drɒpəθɪ/ *noun* (medically unorthodox) treatment of disease by water. □ **hydropathic** /-drə'pæθ-/ *adjective*.

hydrophobia /haɪdrə'fəʊbɪə/ *noun* aversion to water, esp. as symptom of rabies in humans; rabies. □ **hydrophobic** *adjective*.

hydroplane /'haɪdrəpleɪn/ *noun* light fast motor boat; finlike device enabling submarine to rise or fall.

hydroponics /haɪdrə'pɒnɪks/ *noun* growing plants without soil, in sand, water, etc. with added nutrients.

hydrostatic /haɪdrə'stætɪk/ *adjective* of the equilibrium of liquids and the pressure exerted by liquids at rest.

hydrostatics *plural noun* (usually treated as *singular*) study of hydrostatic properties of liquids.

hydrotherapy /haɪdrə'θerəpɪ/ *noun* use of water, esp. swimming, in treatment of disease.

hydrous /'haɪdrəs/ *adjective* containing water.

hyena /haɪ'i:nə/ *noun* doglike flesh-eating mammal.

hygiene /'haɪdʒi:n/ *noun* conditions or practices conducive to maintaining health; cleanliness; sanitary science. □ **hygienic** /-'dʒi:n-/ *adjective*; **hygienically** /-'dʒi:n-/ *adverb*; **hygienist** *noun*.

hygrometer /haɪ'grɒmɪtə/ *noun* instrument for measuring humidity of air etc.

hygroscopic /haɪgrə'skɒpɪk/ *adjective* tending to absorb moisture from air.

hymen /'haɪmen/ *noun* membrane at opening of vagina, usually broken at first sexual intercourse.

hymenopterous /haɪmə'nɒptərəs/ *adjective* of order of insects with 4 membranous wings, including bees and wasps.

hymn /hɪm/ ● *noun* song of esp. Christian praise. ● *verb* praise or celebrate in hymns.

hymnal /'hɪmn(ə)l/ *noun* book of hymns.

hymnology /hɪm'nɒlədʒɪ/ *noun* composition or study of hymns. □ **hymnologist** *noun*.

hyoscine /'haɪəsi:n/ *noun* alkaloid used to prevent motion sickness etc.

hype /haɪp/ *slang* ● *noun* intensive promotion of product etc. ● *verb* (**-ping**) promote with hype. □ **hyped up** excited.

hyper /'haɪpə/ *adjective slang* hyperactive.

hyper- /'haɪpə/ *prefix* over, above; too.

hyperbola /haɪ'pɜ:bələ/ *noun* (*plural* **-s** or **-lae** /-li:/) curve produced when cone is cut by plane making larger angle with base than side of cone makes. □ **hyperbolic** /-pə'bɒl-/ *adjective*.

hyperbole /haɪ'pɜ:bəlɪ/ *noun* exaggeration, esp. for effect. □ **hyperbolical** /-'bɒl-/ *adjective*.

hyperglycaemia /haɪpəglaɪ'si:mɪə/ *noun* (*US* **hyperglycemia**) excess of glucose in bloodstream.

hypermarket /'haɪpəmɑ:kɪt/ *noun* very large supermarket.

hypermedia /'haɪpəmi:dɪə/ *noun* provision of several media (audio, video, etc.) on one computer system.

hypersensitive /haɪpə'sensɪtɪv/ *adjective* excessively sensitive. □ **hypersensitivity** /-'tɪv-/ *noun*.

hypersonic /haɪpə'sɒnɪk/ *adjective* of speeds more than 5 times that of sound.

hypertension /haɪpə'tenʃ(ə)n/ *noun* abnormally high blood pressure; extreme tension.

hypertext /'haɪpətekst/ *noun* provision of several texts on one computer system.

hyperthermia /haɪpə'θɜ:mɪə/ *noun* abnormally high body-temperature.

hyperthyroidism /haɪpə'θaɪrɔɪdɪz(ə)m/ *noun* overactivity of thyroid gland.

hyperventilation /haɪpəventɪ'leiʃ(ə)n/ *noun* abnormally rapid breathing. □ **hyperventilate** *verb* (**-ting**).

hyphen /'haɪf(ə)n/ ● *noun* punctuation mark (-) used to join or divide words (see panel). ● *verb* hyphenate.

hyphenate /'haɪfəneɪt/ *verb* (**-ting**) join or divide with hyphen. □ **hyphenation** *noun*.

hypnosis /hɪp'nəʊsɪs/ *noun* state like sleep in which subject acts only on external suggestion; artificially induced sleep.

hypnotherapy /hɪpnəʊ'θerəpɪ/ *noun* treatment of mental disorders by hypnosis.

hypnotic /hɪp'nɒtɪk/ ● *adjective* of or causing hypnosis; sleep-inducing. ● *noun* hypnotic drug or influence. □ **hypnotically** *adverb*.

hypnotism /'hɪpnətɪz(ə)m/ *noun* study or practice of hypnosis. □ **hypnotist** *noun*.

hypnotize /'hɪpnətaɪz/ *verb* (also **-ise**) (**-zing** or **-sing**) produce hypnosis in; fascinate.

hypo /'haɪpəʊ/ *noun* sodium thiosulphate, used as photographic fixer.

hypo- /'haɪpəʊ/ *prefix* under; below normal; slightly.

hypochondria /haɪpə'kɒndrɪə/ *noun* abnormal anxiety about one's health.

hypochondriac /haɪpə'kɒndrɪæk/ ● *noun* person given to hypochondria. ● *adjective* of hypochondria.

hypocrisy /hɪ'pɒkrɪsɪ/ *noun* (*plural* **-ies**) simulation of virtue; insincerity.

hypocrite /'hɪpəkrɪt/ *noun* person guilty of hypocrisy. □ **hypocritical** /-'krɪt-/ *adjective*; **hypocritically** /-'krɪt-/ *adverb*.

hypodermic /haɪpə'dɜ:mɪk/ ● *adjective* (of drug, syringe, etc.) introduced under the skin. ● *noun* hypodermic injection or syringe.

hypotension /haɪpəʊ'tenʃ(ə)n/ *noun* abnormally low blood pressure.

hypotenuse /haɪ'pɒtənju:z/ *noun* side opposite right angle of right-angled triangle.

hypothalamus /haɪpə'θæləməs/ *noun* (*plural* **-mi** /-maɪ/) region of brain controlling body temperature, thirst, hunger, etc.

hypothermia /haɪpəʊ'θɜ:mɪə/ *noun* abnormally low body-temperature.

hypothesis /haɪ'pɒθɪsɪs/ *noun* (*plural* **-theses** /-si:z/) supposition made as basis for reasoning etc. □ **hypothesize** *verb* (also **-ise**) (**-zing** or **-sing**).

hypothetical /haɪpə'θetɪk(ə)l/ *adjective* of or resting on hypothesis. □ **hypothetically** *adverb.*

hypothyroidism /haɪpəʊ'θaɪrɔɪdɪz(ə)m/ *noun* subnormal activity of the thyroid gland.

hypoventilation /haɪpəʊventɪ'leɪʃ(ə)n/ *noun* abnormally slow breathing.

hyssop /'hɪsəp/ *noun* small bushy aromatic herb.

hysterectomy /hɪstə'rektəmɪ/ *noun* (*plural* **-ies**) surgical removal of womb.

Hyphen -

This is used:

1 to join two or more words so as to form a compound or single expression, e.g.

mother-in-law, non-stick, dressing-table

This use is growing less common; often you can do without such hyphens:

nonstick, treelike, dressing table

2 to join words in an attributive compound (one put before a noun, like an adjective), e.g.

a well-known man (but *the man is well known*)
an out-of-date list (but *the list is out of date*)

3 to join a prefix etc. to a proper name, e.g.

anti-Darwinian; half-Italian; non-British

4 to make a meaning clear by linking words, e.g.

twenty-odd people/twenty odd people

or by separating a prefix, e.g.

re-cover/recover; re-present/represent; re-sign/resign

5 to separate two identical letters in adjacent parts of a word, e.g.

pre-exist, Ross-shire

6 to represent a common second element in the items of a list, e.g.

two-, three-, or fourfold.

7 to divide a word if there is no room to complete it at the end of the line, e.g.

. . . diction-
ary . . .

The hyphen comes at the end of the line, not at the beginning of the next line. In general, words should be divided at the end of a syllable: *dicti-onary* would be quite wrong. In handwriting, typing, and word-processing, it is safest (and often neatest) not to divide words at all.

hysteria /hɪ'stɪərɪə/ *noun* uncontrollable emotion or excitement; functional disturbance of nervous system.

hysteric /hɪ'sterɪk/ *noun* (in *plural*) fit of hysteria, *colloquial* overwhelming laughter; hysterical person. □ **hysterical** *adjective*; **hysterically** *adverb*.

Hz *abbreviation* hertz.

Ii

I[1] *noun* (also **i**) (Roman numeral) 1.

I[2] /aɪ/ *pronoun used by speaker or writer to refer to himself or herself as subject of verb.*

I[3] *abbreviation* (also **I.**) Island(s); Isle(s).

iambic /aɪ'æmbik/ ● *adjective* of or using iambuses. ● *noun* (usually in *plural*) iambic verse.

iambus /aɪ'æmbəs/ *noun* (*plural* **-buses** or **-bi** /-baɪ/) metrical foot of one short followed by one long syllable.

IBA *abbreviation* Independent Broadcasting Authority.

ibex /'aɪbeks/ *noun* (*plural* **-es**) wild mountain goat with large backward-curving ridged horns.

ibid. /'ɪbɪd/ *abbreviation* in same book or passage etc. (*ibidem*).

ibis /'aɪbɪs/ *noun* (*plural* **-es**) storklike bird with long curved bill.

ice ● *noun* frozen water; portion of ice cream etc. ● *verb* (**icing**) mix with or cool in ice; (often + *over*, *up*) cover or become covered (as) with ice; freeze; cover (a cake etc.) with icing. □ **ice age** glacial period; **icebox** compartment in refrigerator for making or storing ice, *US* refrigerator; **ice-breaker** boat designed to break through ice; **icecap** mass of thick ice permanently covering polar region etc.; **ice cream** sweet creamy frozen food; **ice field** extensive sheet of floating ice; **ice hockey** form of hockey played on ice with flat disc instead of ball; **ice lolly** flavoured ice on stick; **ice rink** area of ice for skating etc.; **ice-skate** *noun* boot with blade attached for gliding over ice, *verb* move on ice-skates; **on ice** performed by ice-skaters, *colloquial* in reserve.

iceberg /'aɪsbɜ:g/ *noun* mass of floating ice at sea. □ **tip of the iceberg** small perceptible part of something very large or complex.

Icelander /'aɪsləndə/ *noun* native of Iceland.

Icelandic /aɪs'lændɪk/ ● *adjective* of Iceland. ● *noun* language of Iceland.

ichneumon /ɪk'nju:mən/ *noun* (in full **ichneumon fly**) wasplike insect parasitic on other insects; mongoose of N. Africa etc.

ichthyology /ɪkθɪ'ɒlədʒɪ/ *noun* study of fishes. □ **ichthyological** /-ə'lɒdʒɪk(ə)l/ *adjective*; **ichthyologist** *noun*.

ichthyosaur /'ɪkθɪəsɔ:/ *noun* large extinct reptile like dolphin.

icicle /'aɪsɪk(ə)l/ *noun* tapering hanging spike of ice, formed from dripping water.

icing *noun* sugar etc. coating for cake etc.; formation of ice on ship or aircraft. □ **icing sugar** finely powdered sugar.

icon /'aɪkɒn/ *noun* (also **ikon**) sacred painting, mosaic, etc.; image, statue. □ **iconic** /-'kɒn-/ *adjective*.

iconoclast /aɪ'kɒnəklæst/ *noun* person who attacks cherished beliefs; *historical* breaker of religious images. □ **iconoclasm** *noun*; **iconoclastic** /-'klæstɪk/ *adjective*.

iconography /aɪkə'nɒgrəfɪ/ *noun* illustration of subject by drawings etc.; study of portraits, esp. of one person, or of artistic images or symbols.

icy /'aɪsɪ/ *adjective* (**-ier**, **-iest**) very cold; covered with or abounding in ice; (of manner) unfriendly.

ID *abbreviation* identification, identity.

I'd /aɪd/ I had; I should; I would.

id *noun Psychology* part of mind comprising instinctive impulses of individual etc.

idea /aɪˈdɪə/ *noun* plan etc. formed by mental effort; mental impression or concept; vague belief or fancy; purpose, intention. □ **have no idea** *colloquial* not know at all, be completely incompetent.

ideal /aɪˈdi:əl/ ● *adjective* perfect; existing only in idea; visionary. ● *noun* perfect type, thing, principle, etc. as standard for imitation.

idealism *noun* forming or pursuing ideals; representation of things in ideal form; philosophy in which objects are held to be dependent on mind. □ **idealist** *noun*; **idealistic** /-ˈlɪst-/ *adjective*.

idealize *verb* (also **-ise**) (**-zing** or **-sing**) regard or represent as ideal. □ **idealization** *noun*.

identical /aɪˈdentɪk(ə)l/ *adjective* (often + *with*) absolutely alike; same; (of twins) developed from single ovum and very similar in appearance. □ **identically** *adverb*.

identify /aɪˈdentɪfaɪ/ *verb* (**-ies**, **-ied**) establish identity of; select, discover; (+ *with*) closely associate with; (+ *with*) regard oneself as sharing basic characteristics with; (often + *with*) treat as identical. □ **identification** *noun*.

identity /aɪˈdentɪtɪ/ *noun* (*plural* **-ies**) being specified person or thing; individuality; identification or the result of it; absolute sameness.

ideogram /ˈɪdɪəgræm/ *noun* (also **ideograph** /-grɑ:f/) symbol representing thing or idea without indicating sounds in its name (e.g. Chinese character, or '=' for 'equals').

ideology /aɪdɪˈɒlədʒɪ/ *noun* (*plural* **-ies**) scheme of ideas at basis of political etc. theory or system; characteristic thinking of class etc. □ **ideological** /-əˈlɒdʒ-/ *adjective*.

idiocy /ˈɪdɪəsɪ/ *noun* (*plural* **-ies**) utter foolishness; foolish act; mental condition of idiot.

idiom /ˈɪdɪəm/ *noun* phrase etc. established by usage and not immediately comprehensible from the words used; form of expression peculiar to a language; language; characteristic mode of expression. □ **idiomatic** /-ˈmæt-/ *adjective*.

idiosyncrasy /ɪdɪəʊˈsɪŋkrəsɪ/ *noun* (*plural* **-ies**) attitude or form of behaviour peculiar to person. □ **idiosyncratic** /-ˈkræt-/ *adjective*.

idiot /ˈɪdɪət/ *noun* stupid person; person too deficient in mind to be capable of rational conduct. □ **idiotic** /-ˈɒt-/ *adjective*.

idle /ˈaɪd(ə)l/ ● *adjective* (**-r**, **-st**) lazy, indolent; not in use; unoccupied; useless, purposeless. ● *verb* (**-ling**) be idle; (of engine) run slowly without doing any work; pass (time) in idleness. □ **idleness** *noun*; **idler** *noun*; **idly** *adverb*.

idol /ˈaɪd(ə)l/ *noun* image as object of worship; object of devotion.

idolater /aɪˈdɒlətə/ *noun* worshipper of idols; devout admirer. □ **idolatrous** *adjective*; **idolatry** *noun*.

idolize *verb* (also **-ise**) (**-zing** or **-sing**) venerate or love to excess; treat as idol. □ **idolization** *noun*.

idyll /ˈɪdɪl/ *noun* account of picturesque scene or incident etc.; such scene etc. □ **idyllic** /ɪˈdɪlɪk/ *adjective*.

i.e. *abbreviation* that is to say (*id est*).

if *conjunction* on condition or supposition that; *(with past tense) implying that the condition is not fulfilled*; even though; whenever; whether; *expressing wish, request, or (with negative) surprise*. □ **if only** even if for no other reason than, I wish that.

igloo /ˈɪglu:/ *noun* dome-shaped snow house.

igneous /ˈɪgnɪəs/ *adjective* of fire; (esp. of rocks) produced by volcanic action.

ignite /ɪgˈnaɪt/ *verb* (**-ting**) set fire to; catch fire; provoke or excite (feelings etc.).

ignition /ɪgˈnɪʃ(ə)n/ *noun* mechanism for starting combustion in cylinder of motor engine; igniting.

ignoble /ɪgˈnəʊb(ə)l/ *adjective* (**-r**, **-st**) dishonourable; of low birth or position.

ignominious /ɪgnəˈmɪnɪəs/ *adjective* humiliating. □ **ignominiously** *adverb*.

ignominy /ˈɪgnəmɪnɪ/ *noun* dishonour, infamy.

ignoramus /ɪgnəˈreɪməs/ *noun* (*plural* **-muses**) ignorant person.

ignorant /ˈɪgnərənt/ *adjective* lacking knowledge; (+ *of*) uninformed; *colloquial* uncouth. □ **ignorance** *noun*.

ignore /ɪg'nɔ:/ *verb* (**-ring**) refuse to take notice of.

iguana /ɪg'wɑ:nə/ *noun* large Central and S. American tree lizard.

iguanodon /ɪg'wɑ:nədɒn/ *noun* large plant-eating dinosaur.

ikebana /ɪkɪ'bɑ:nə/ *noun* Japanese art of flower arrangement.

ikon = ICON.

ilex /'aɪleks/ *noun* (*plural* **-es**) plant of genus including holly; holm-oak.

iliac /'ɪlɪæk/ *adjective* of flank or hip-bone.

ilk *noun colloquial* sort, kind. □ **of that ilk** *Scottish* of ancestral estate of same name as family.

I'll /aɪl/ I shall; I will.

ill ● *adjective* in bad health, sick; harmful, unfavourable; hostile, unkind; faulty, deficient. ● *adverb* badly, unfavourably; scarcely. ● *noun* harm; evil. □ **ill-advised** unwise; **ill-bred** rude; **ill-favoured** unattractive; **ill-gotten** gained unlawfully or wickedly; **ill health** poor physical condition; **ill-mannered** rude; **ill-natured** churlish; **ill-tempered** morose, irritable; **ill-timed** done or occurring at unsuitable time; **ill-treat, -use** treat badly.

illegal /ɪ'li:g(ə)l/ *adjective* contrary to law. □ **illegality** /-'gæl-/ *noun* (*plural* **-ies**); **illegally** *adverb.*

illegible /ɪ'ledʒɪb(ə)l/ *adjective* not legible, unreadable. □ **illegibility** *noun*; **illegibly** *adverb.*

illegitimate /ɪlɪ'dʒɪtɪmət/ *adjective* born of parents not married to each other; unlawful; improper; wrongly inferred. □ **illegitimacy** *noun.*

illiberal /ɪ'lɪbər(ə)l/ *adjective* narrow-minded; stingy. □ **illiberality** /-'ræl-/ *noun.*

illicit /ɪ'lɪsɪt/ *adjective* unlawful; forbidden. □ **illicitly** *adverb.*

illiterate /ɪ'lɪtərət/ ● *adjective* unable to read; uneducated. ● *noun* illiterate person. □ **illiteracy** *noun.*

illness *noun* disease; ill health.

illogical /ɪ'lɒdʒɪk(ə)l/ *adjective* devoid of or contrary to logic. □ **illogicality** /-'kæl-/ *noun* (*plural* **-ies**); **illogically** *adverb.*

illuminate /ɪ'lu:mɪneɪt/ *verb* (**-ting**) light up; decorate with lights; decorate (manuscript etc.) with gold, colour, etc.; help to explain (subject etc.); enlighten spiritually or intellectually. □ **illuminating** *adjective*; **illumination** *noun.*

illumine /ɪ'lju:mɪn/ *verb* (**-ning**) *literary* light up; enlighten.

illusion /ɪ'lu:ʒ(ə)n/ *noun* false belief; deceptive appearance. □ **be under the illusion** (+ *that*) believe mistakenly. □ **illusive** *adjective*; **illusory** *adjective.*

illusionist *noun* conjuror.

illustrate /'ɪləstreɪt/ *verb* (**-ting**) provide with pictures; make clear, esp. by examples or drawings; serve as example of. □ **illustrator** *noun.*

illustration *noun* drawing etc. in book; explanatory example; illustrating.

illustrative /'ɪləstrətɪv/ *adjective* (often + *of*) explanatory.

illustrious /ɪ'lʌstrɪəs/ *adjective* distinguished, renowned.

I'm /aɪm/ I am.

image /'ɪmɪdʒ/ ● *noun* representation of object, esp. figure of saint or divinity; reputation or persona of person, company, etc.; appearance as seen in mirror or through lens; idea, conception; simile, metaphor. ● *verb* (**-ging**) make image of; mirror; picture.

imagery /'ɪmɪdʒərɪ/ *noun* figurative illustration; use of images in literature etc.; images, statuary; mental images collectively.

imaginary /ɪ'mædʒɪnərɪ/ *adjective* existing only in imagination.

imagination /ɪmædʒɪ'neɪʃ(ə)n/ *noun* mental faculty of forming images of objects not present to senses; creative faculty of mind.

imaginative /ɪ'mædʒɪnətɪv/ *adjective* having or showing high degree of imagination. □ **imaginatively** *adverb.*

imagine /ɪ'mædʒɪn/ *verb* (**-ning**) form mental image of, conceive; suppose, think.

imago /ɪ'meɪgəʊ/ *noun* (*plural* **-s** or **imagines** /ɪ'mædʒɪni:z/) fully developed stage of insect.

imam /ɪ'mɑ:m/ *noun* prayer-leader of mosque; *title of some Muslim leaders.*

imbalance /ɪm'bæləns/ *noun* lack of balance; disproportion.

imbecile /'ɪmbɪsi:l/ ● *noun colloquial* stupid person; adult with mental age of about 5. ● *adjective* mentally weak, stupid. □ **imbecilic** /-'sɪlɪk/ *adjective*; **imbecility** /-'sɪlɪtɪ/ *noun* (*plural* **-ies**).

imbed = EMBED.

imbibe /ɪm'baɪb/ *verb* (**-bing**) drink; drink in; absorb; inhale.

imbroglio /ɪm'brəʊlɪəʊ/ *noun* (*plural* **-s**) confused or complicated situation.

imbue /ɪm'bju:/ *verb* (**-bues**, **-bued**, **-buing**) (often + *with*) inspire; saturate, dye.

imitate /'ɪmɪteɪt/ *verb* (**-ting**) follow example of; mimic; make copy of; be like. □ **imitable** *adjective*; **imitative** /-tətɪv/ *adjective*; **imitator** *noun*.

imitation *noun* imitating, being imitated; copy; counterfeit.

immaculate /ɪ'mækjʊlət/ *adjective* perfectly clean, spotless; faultless; innocent, sinless. □ **immaculately** *adverb*; **immaculateness** *noun*.

immanent /'ɪmənənt/ *adjective* inherent; (of God) omnipresent. □ **immanence** *noun*.

immaterial /ɪmə'tɪərɪəl/ *adjective* unimportant; irrelevant; not material. □ **immateriality** /-'æl-/ *noun*.

immature /ɪmə'tjʊə/ *adjective* not mature; undeveloped, esp. emotionally. □ **immaturity** *noun*.

immeasurable /ɪ'meʒərəb(ə)l/ *adjective* not measurable, immense. □ **immeasurably** *adverb*.

immediate /ɪ'mi:dɪət/ *adjective* occurring at once; direct; nearest; having priority. □ **immediacy** *noun*; **immediately** *adverb*.

immemorial /ɪmɪ'mɔ:rɪəl/ *adjective* ancient beyond memory.

immense /ɪ'mens/ *adjective* vast, huge. □ **immensity** *noun*.

immensely *adverb colloquial* vastly, very much.

immerse /ɪ'mɜ:s/ *verb* (**-sing**) (often + *in*) dip, plunge; put under water; (often **immerse oneself** or in *passive*; often + *in*) involve deeply, embed.

immersion /ɪ'mɜ:ʃ(ə)n/ *noun* immersing, being immersed. □ **immersion heater** electric heater designed to be immersed in liquid to be heated.

immigrant /'ɪmɪgrənt/ ● *noun* person who immigrates. ● *adjective* immigrating; of immigrants.

immigrate /'ɪmɪgreɪt/ *verb* (**-ting**) enter a country to settle permanently. □ **immigration** *noun*.

imminent /'ɪmɪnənt/ *adjective* soon to happen. □ **imminence** *noun*; **imminently** *adverb*.

immobile /ɪ'məʊbaɪl/ *adjective* motionless; immovable. □ **immobility** *noun*.

immobilize /ɪ'məʊbɪlaɪz/ *verb* (also **-ise**) (**-zing** or **-sing**) prevent from being moved. □ **immobilization** *noun*.

immoderate /ɪ'mɒdərət/ *adjective* excessive. □ **immoderately** *adverb*.

immodest /ɪ'mɒdɪst/ *adjective* conceited; indecent. □ **immodesty** *noun*.

immolate /'ɪməleɪt/ *verb* (**-ting**) kill as sacrifice. □ **immolation** *noun*.

immoral /ɪ'mɒr(ə)l/ *adjective* opposed to, or not conforming to, (esp. sexual) morality; dissolute. □ **immorality** /ɪmə'rælɪtɪ/ *noun*.

immortal /ɪ'mɔ:t(ə)l/ ● *adjective* living for ever; unfading; divine; famous for all time. ● *noun* immortal being, esp. (in *plural*) gods of antiquity. □ **immortality** /-'tæl-/ *noun*; **immortalize** *verb* (also **-ise**) (**-zing** or **-sing**).

immovable /ɪ'mu:vəb(ə)l/ *adjective* not movable; unyielding. □ **immovability** *noun*.

immune /ɪ'mju:n/ *adjective* having immunity; relating to immunity; exempt.

immunity *noun* (*plural* **-ies**) living organism's power of resisting and overcoming infection; (often + *from*) freedom, exemption.

immunize /'ɪmju:naɪz/ *verb* (also **-ise**) (**-zing** or **-sing**) make immune. □ **immunization** *noun*.

immure /ɪ'mjʊə/ *verb* (**-ring**) imprison.

immutable /ɪ'mju:təb(ə)l/ *adjective* unchangeable. □ **immutability** *noun*.

imp *noun* mischievous child; little devil.

impact ● *noun* /'ɪmpækt/ collision, striking; (immediate) effect or influence. ● *verb* /ɪm'pækt/ drive or wedge together; (as **impacted** *adjective*) (of tooth) wedged between another tooth and jaw. □ **impaction** /ɪm'pækʃ(ə)n/ *noun*.

impair /ɪm'peə/ *verb* damage, weaken. □ **impairment** *noun*.

impala /ɪm'pɑ:lə/ *noun* (*plural* same or **-s**) small African antelope.

impale /ɪm'peɪl/ *verb* (**-ling**) transfix on stake. □ **impalement** *noun.*

impalpable /ɪm'pælpəb(ə)l/ *adjective* not easily grasped; imperceptible to touch.

impart /ɪm'pɑ:t/ *verb* communicate (news etc.); give share of.

impartial /ɪm'pɑ:ʃ(ə)l/ *adjective* fair, not partial. □ **impartiality** /-ʃɪ'æl-/ *noun*; **impartially** *adverb.*

impassable /ɪm'pɑ:səb(ə)l/ *adjective* that cannot be traversed. □ **impassability** *noun.*

impasse /'æmpæs/ *noun* deadlock.

impassioned /ɪm'pæʃ(ə)nd/ *adjective* filled with passion, ardent.

impassive /ɪm'pæsɪv/ *adjective* not feeling or showing emotion. □ **impassively** *adverb*; **impassivity** /-'sɪv-/ *noun.*

impasto /ɪm'pæstəʊ/ *noun* technique of laying on paint thickly.

impatiens /ɪm'peɪʃɪenz/ *noun* any of several plants including busy Lizzie.

impatient /ɪm'peɪʃ(ə)nt/ *adjective* not patient; intolerant; restlessly eager. □ **impatience** *noun*; **impatiently** *adverb.*

impeach /ɪm'pi:tʃ/ *verb* accuse, esp. of treason etc.; call in question; disparage. □ **impeachment** *noun.*

impeccable /ɪm'pekəb(ə)l/ *adjective* faultless; exemplary. □ **impeccability** *noun*; **impeccably** *adverb.*

impecunious /ɪmpɪ'kju:nɪəs/ *adjective* having little or no money.

impedance /ɪm'pi:d(ə)ns/ *noun* total effective resistance of electric circuit etc. to alternating current.

impede /ɪm'pi:d/ *verb* (**-ding**) obstruct; hinder.

impediment /ɪm'pedɪmənt/ *noun* hindrance; defect in speech, esp. lisp or stammer.

impedimenta /ɪmpedɪ'mentə/ *plural noun* encumbrances; baggage, esp. of army.

impel /ɪm'pel/ *verb* (**-ll-**) drive, force; propel.

impend /ɪm'pend/ *verb* be imminent; hang. □ **impending** *adjective.*

impenetrable /ɪm'penɪtrəb(ə)l/ *adjective* not penetrable; inscrutable; inaccessible to influences etc. □ **impenetrability** *noun.*

impenitent /ɪm'penɪt(ə)nt/ *adjective* not penitent. □ **impenitence** *noun.*

imperative /ɪm'perətɪv/ ● *adjective* urgent, obligatory; peremptory; *Grammar* (of mood) expressing command. ● *noun Grammar* imperative mood; command; essential or urgent thing.

imperceptible /ɪmpə'septɪb(ə)l/ *adjective* not perceptible; very slight or gradual. □ **imperceptibility** *noun*; **imperceptibly** *adverb.*

imperfect /ɪm'pɜ:fɪkt/ ● *adjective* not perfect; incomplete; faulty; *Grammar* (of past tense) implying action going on but not completed. ● *noun* imperfect tense. □ **imperfectly** *adverb.*

imperfection /ɪmpə'fekʃ(ə)n/ *noun* imperfectness; fault, blemish.

imperial /ɪm'pɪərɪəl/ *adjective* of empire or sovereign state ranking with this; of emperor; majestic; (of non-metric weights and measures) used by statute in UK.

imperialism *noun* imperial system of government etc.; *usually derogatory* policy of dominating other nations by acquisition of dependencies or through trade etc. □ **imperialist** *noun & adjective.*

imperil /ɪm'perɪl/ *verb* (**-ll-**; *US* **-l-**) endanger.

imperious /ɪm'pɪərɪəs/ *adjective* overbearing, domineering. □ **imperiously** *adverb.*

imperishable /ɪm'perɪʃəb(ə)l/ *adjective* that cannot perish.

impermanent /ɪm'pɜ:mənənt/ *adjective* not permanent. **impermanence** *noun.*

impermeable /ɪm'pɜ:mɪəb(ə)l/ *adjective* not permeable. □ **impermeability** *noun.*

impersonal /ɪm'pɜ:sən(ə)l/ *adjective* having no personality or personal feeling or reference; impartial; unfeeling; *Grammar* (of verb) used esp. with *it* as subject. □ **impersonality** /-'næl-/ *noun.*

impersonate /ɪm'pɜ:səneɪt/ *verb* (**-ting**) pretend to be, play part of. □ **impersonation** *noun*; **impersonator** *noun.*

impertinent /ɪm'pɜ:tɪnənt/ *adjective* insolent, saucy; irrelevant. □ **impertinence** *noun*; **impertinently** *adverb.*

imperturbable /ɪmpə'tɜ:bəb(ə)l/ *adjective* not excitable; calm. □ **imperturbability** *noun*; **imperturbably** *adverb.*

impervious /ɪm'pɜ:vɪəs/ *adjective* (usually + *to*) impermeable; not responsive.

impetigo /ɪmpɪ'taɪgəʊ/ *noun* contagious skin disease.

impetuous /ɪm'petjʊəs/ *adjective* acting or done rashly or suddenly; moving violently or fast. □ **impetuosity** /-'ɒs-/ *noun*; **impetuously** *adverb*.

impetus /'ɪmpɪtəs/ *noun* moving force; momentum; impulse.

impiety /ɪm'paɪətɪ/ *noun* (*plural* **-ies**) lack of piety; act showing this.

impinge /ɪm'pɪndʒ/ *verb* (**-ging**) (usually + *on*) make impact; (usually + *upon*) encroach.

impious /'ɪmpɪəs/ *adjective* not pious; wicked.

impish *adjective* of or like imp, mischievous. □ **impishly** *adverb*.

implacable /ɪm'plækəb(ə)l/ *adjective* not appeasable. □ **implacability** *noun*; **implacably** *adverb*.

implant ● *verb* /ɪm'plɑ:nt/ insert, fix; instil; plant; (in *passive*) (of fertilized ovum) become attached to wall of womb. ● *noun* /'ɪmplɑ:nt/ thing implanted. □ **implantation** *noun*.

implausible /ɪm'plɔ:zɪb(ə)l/ *adjective* not plausible. □ **implausibly** *adverb*.

implement ● *noun* /'ɪmplɪmənt/ tool, utensil. ● *verb* /'ɪmplɪment/ carry into effect. □ **implementation** *noun*.

implicate /'ɪmplɪkeɪt/ *verb* (**-ting**) (often + *in*) show (person) to be involved (in crime etc.); imply.

implication *noun* thing implied; implying; implicating.

implicit /ɪm'plɪsɪt/ *adjective* implied though not expressed; unquestioning. □ **implicitly** *adverb*.

implode /ɪm'pləʊd/ *verb* (**-ding**) (cause to) burst inwards. □ **implosion** /ɪm'pləʊʒ(ə)n/ *noun*.

implore /ɪm'plɔ:/ *verb* (**-ring**) beg earnestly.

imply /ɪm'plaɪ/ *verb* (**-ies, -ied**) (often + *that*) insinuate, hint; mean.

impolite /ɪmpə'laɪt/ *adjective* uncivil, rude. □ **impolitely** *adverb*; **impoliteness** *noun*.

impolitic /ɪm'pɒlɪtɪk/ *adjective* inexpedient, not advisable. □ **impoliticly** *adverb*.

imponderable /ɪm'pɒndərəb(ə)l/ ● *adjective* that cannot be estimated; very light. ● *noun* (usually in *plural*) imponderable thing. □ **imponderability** *noun*; **imponderably** *adverb*.

import ● *verb* /ɪm'pɔ:t/ bring in (esp. foreign goods) from abroad; imply, mean. ● *noun* /'ɪmpɔ:t/ article or (in *plural*) amount imported; importing; meaning, implication; importance. □ **importation** *noun*; **importer** /-'pɔ:tə/ *noun*.

important /ɪm'pɔ:t(ə)nt/ *adjective* (often + *to*) of great consequence; momentous; (of person) having position of authority or rank; pompous. □ **importance** *noun*; **importantly** *adverb*.

importunate /ɪm'pɔ:tjʊnət/ *adjective* making persistent or pressing requests. □ **importunity** /-'tju:n-/ *noun*.

importune /ɪmpə'tju:n/ *verb* (**-ning**) pester (person) with requests; solicit as prostitute.

impose /ɪm'pəʊz/ *verb* (**-sing**) enforce compliance with; (often + *on*) inflict, lay (tax etc.); (+ *on*, *upon*) take advantage of.

imposing *adjective* impressive, esp. in appearance.

imposition /ɪmpə'zɪʃ(ə)n/ *noun* imposing, being imposed; unfair demand or burden; tax, duty.

impossible /ɪm'pɒsɪb(ə)l/ *adjective* not possible; not easy or convenient; *colloquial* outrageous, intolerable. □ **impossibility** *noun*; **impossibly** *adverb*.

impost /'ɪmpəʊst/ *noun* tax, duty.

impostor /ɪm'pɒstə/ *noun* (also **imposter**) person who assumes false character; swindler.

imposture /ɪm'pɒstʃə/ *noun* fraudulent deception.

impotent /'ɪmpət(ə)nt/ *adjective* powerless; (of male) unable to achieve erection of penis or have sexual intercourse. □ **impotence** *noun*.

impound /ɪm'paʊnd/ *verb* confiscate; shut up in pound.

impoverish /ɪm'pɒvərɪʃ/ *verb* make poor. □ **impoverishment** *noun*.

impracticable /ɪm'præktɪkəb(ə)l/ *adjective* impossible in practice. □ **impracticability** *noun*; **impracticably** *adverb*.

impractical /ɪm'præktɪk(ə)l/ *adjective* not practical; *esp. US* not practicable. □ **impracticality** /-'kæl-/ *noun.*

imprecation /ɪmprɪ'keɪʃ(ə)n/ *noun formal* curse.

imprecise /ɪmprɪ'saɪs/ *adjective* not precise.

impregnable /ɪm'pregnəb(ə)l/ *adjective* safe against attack. □ **impregnability** *noun.*

impregnate /'ɪmpregneɪt/ *verb* (**-ting**) fill, saturate; make pregnant. □ **impregnation** *noun.*

impresario /ɪmprɪ'sɑ:rɪəʊ/ *noun* (*plural* **-s**) organizer of public entertainments.

impress ● *verb* /ɪm'pres/ affect or influence deeply; arouse admiration or respect in; (often + *on*) emphasize; imprint, stamp. ● *noun* /'ɪmpres/ mark impressed; characteristic quality.

impression /ɪm'preʃ(ə)n/ *noun* effect produced on mind; belief; imitation of person or sound, esp. done to entertain; impressing, mark impressed; unaltered reprint of book etc.; issue of book or newspaper etc.; print from type or engraving.

impressionable *adjective* easily influenced.

impressionism *noun* school of painting concerned with conveying effect of natural light on objects; style of music or writing seeking to convey esp. fleeting feelings or experience. □ **impressionist** *noun*; **impressionistic** /-'nɪs-/ *adjective.*

impressive /ɪm'presɪv/ *adjective* arousing respect, approval, or admiration. □ **impressively** *adverb.*

imprimatur /ɪmprɪ'mɑ:tə/ *noun* licence to print; official approval.

imprint ● *verb* /ɪm'prɪnt/ (often + *on*) impress firmly, esp. on mind; make impression of (figure etc.) on thing; make impression on with stamp etc. ● *noun* /'ɪmprɪnt/ impression; printer's or publisher's name in book etc.

imprison /ɪm'prɪz(ə)n/ *verb* (**-n-**) put into prison; confine. □ **imprisonment** *noun.*

improbable /ɪm'prɒbəb(ə)l/ *adjective* not likely, difficult to believe. □ **improbability** *noun*; **improbably** *adverb.*

improbity /ɪm'prəʊbɪtɪ/ *noun* (*plural* **-ies**) wickedness; dishonesty; wicked or dishonest act.

impromptu /ɪm'prɒmptju:/ ● *adverb & adjective* unrehearsed. ● *noun* (*plural* **-s**) impromptu performance or speech; short, usually solo, musical piece, often improvisatory in style.

improper /ɪm'prɒpə/ *adjective* unseemly, indecent; inaccurate, wrong. □ **improperly** *adverb.*

impropriety /ɪmprə'praɪətɪ/ *noun* (*plural* **-ies**) indecency; instance of this; incorrectness, unfitness.

improve /ɪm'pru:v/ *verb* (**-ving**) make or become better; (+ *on*) produce something better than; (as **improving** *adjective*) giving moral benefit. □ **improvement** *noun.*

improvident /ɪm'prɒvɪd(ə)nt/ *adjective* lacking foresight; wasteful. □ **improvidence** *noun*; **improvidently** *adverb.*

improvise /'ɪmprəvaɪz/ *verb* (**-sing**) compose extempore; provide or construct from materials etc. not intended for the purpose. □ **improvisation** *noun*; **improvisational** *adjective*; **improvisatory** /-'zeɪtərɪ/ *adjective.*

imprudent /ɪm'pru:d(ə)nt/ *adjective* rash, indiscreet. □ **imprudence** *noun*; **imprudently** *adverb.*

impudent /'ɪmpjʊd(ə)nt/ *adjective* impertinent. □ **impudence** *noun*; **impudently** *adverb.*

impugn /ɪm'pju:n/ *verb* challenge; call in question.

impulse /'ɪmpʌls/ *noun* sudden urge; tendency to follow such urges; impelling; impetus.

impulsive /ɪm'pʌlsɪv/ *adjective* apt to act on impulse; done on impulse; tending to impel. □ **impulsively** *adverb*; **impulsiveness** *noun.*

impunity /ɪm'pju:nɪtɪ/ *noun* exemption from punishment or injurious consequences. □ **with impunity** without punishment etc.

impure /ɪm'pjʊə/ *adjective* adulterated; dirty; unchaste.

impurity *noun* (*plural* **-ies**) being impure; impure thing or part.

impute /ɪm'pju:t/ *verb* (**-ting**) (+ *to*) ascribe (fault etc.) to. □ **imputation** *noun.*

in ● *preposition expressing inclusion or position within limits of space, time, circumstance, etc.*; after (specified period of time); with respect to; as proportionate part of; with form or arrangement of; as member of; involved with; within ability of; having the condition of; affected by; having as aim; by means of; meaning; into (with verb of motion or change). ● *adverb expressing position bounded by certain limits, or movement to point enclosed by them*; into room etc.; at home etc.; so as to be enclosed; as part of a publication; in fashion, season, or office; (of player etc.) having turn or right to play; (of transport) at platform etc.; (of season, harvest, ordered goods, etc.) having arrived or been received; (of fire etc.) burning; (of tide) at highest point. ● *adjective* internal, living etc. inside; fashionable; (of joke etc.) confined to small group. □ **in-between** *colloquial* intermediate; **in-house** within an institution, company, etc.; **ins and outs** (often + *of*) details; **in so far as** to the extent that; **in that** because, in so far as; **in-tray** tray for incoming documents etc.; **in with** on good terms with.

in. *abbreviation* inch(es).

inability /ɪnə'bɪlɪtɪ/ *noun* being unable.

inaccessible /ɪnək'sesɪb(ə)l/ *adjective* not accessible; unapproachable. □ **inaccessibility** *noun*.

inaccurate /ɪn'ækjʊrət/ *adjective* not accurate. □ **inaccuracy** *noun* (*plural* **-ies**); **inaccurately** *adverb*.

inaction /ɪn'ækʃ(ə)n/ *noun* absence of action.

inactive /ɪn'æktɪv/ *adjective* not active; not operating. □ **inactivity** /-'tɪv-/ *noun*.

inadequate /ɪn'ædɪkwət/ *adjective* insufficient; incompetent. □ **inadequacy** *noun* (*plural* **-ies**); **inadequately** *adverb*.

inadmissible /ɪnəd'mɪsɪb(ə)l/ *adjective* not allowable. □ **inadmissibility** *noun*; **inadmissibly** *adverb*.

inadvertent /ɪnəd'vɜ:t(ə)nt/ *adjective* unintentional; inattentive. □ **inadvertence** *noun*; **inadvertently** *adverb*.

inadvisable /ɪnəd'vaɪzəb(ə)l/ *adjective* not advisable. □ **inadvisability** *noun*.

inalienable /ɪn'eɪlɪənəb(ə)l/ *adjective* that cannot be transferred to another or taken away.

inane /ɪ'neɪn/ *adjective* silly, senseless; empty. □ **inanity** /-'næn-/ *noun* (*plural* **-ies**).

inanimate /ɪn'ænɪmət/ *adjective* not endowed with animal life; spiritless, dull.

inapplicable /ɪnə'plɪkəb(ə)l/ *adjective* not applicable; irrelevant. □ **inapplicability** *noun*.

inapposite /ɪn'æpəzɪt/ *adjective* not apposite.

inappropriate /ɪnə'prəʊprɪət/ *adjective* not appropriate. □ **inappropriately** *adverb*; **inappropriateness** *noun*.

inapt /ɪn'æpt/ *adjective* not suitable; unskilful. □ **inaptitude** *noun*.

inarticulate /ɪnɑ:'tɪkjʊlət/ *adjective* unable to express oneself clearly; not articulate, indistinct; dumb; not jointed. □ **inarticulately** *adverb*.

inasmuch /ɪnəz'mʌtʃ/ *adverb* (+ *as*) since, because; to the extent that.

inattentive /ɪnə'tentɪv/ *adjective* not paying attention; neglecting to show courtesy. □ **inattention** *noun*; **inattentively** *adverb*.

inaudible /ɪn'ɔ:dɪb(ə)l/ *adjective* that cannot be heard. □ **inaudibly** *adverb*.

inaugural /ɪn'ɔ:gjʊr(ə)l/ ● *adjective* of inauguration. ● *noun* inaugural speech or lecture.

inaugurate /ɪn'ɔ:gjʊreɪt/ *verb* (**-ting**) admit (person) to office; initiate use of or begin with ceremony; begin, introduce. □ **inauguration** *noun*.

inauspicious /ɪnɔ:'spɪʃəs/ *adjective* not of good omen; unlucky.

inborn /'ɪnbɔ:n/ *adjective* existing from birth; innate.

inbred /ɪn'bred/ *adjective* inborn; produced by inbreeding.

inbreeding /ɪn'bri:dɪŋ/ *noun* breeding from closely related animals or people.

Inc. *abbreviation US* Incorporated.

incalculable /ɪn'kælkjʊləb(ə)l/ *adjective* too great for calculation; not calculable beforehand; uncertain. □ **incalculability** *noun*; **incalculably** *adverb*.

incandesce /ɪnkæn'des/ *verb* (**-cing**) (cause to) glow with heat.

incandescent *adjective* glowing with heat, shining; (of artificial light) produced by glowing filament etc. □ **incandescence** *noun*.

incantation /ɪnkæn'teɪʃ(ə)n/ *noun* spell, charm. □ **incantational** *adjective*.

incapable /ɪn'keɪpəb(ə)l/ *adjective* not capable; too honest, kind, etc. to do something; not capable of rational conduct. □ **incapability** *noun.*

incapacitate /ɪnkə'pæsɪteɪt/ *verb* (**-ting**) make incapable or unfit.

incapacity /ɪnkə'pæsɪtɪ/ *noun* inability; legal disqualification.

incarcerate /ɪn'kɑ:səreɪt/ *verb* (**-ting**) imprison. □ **incarceration** *noun.*

incarnate /ɪn'kɑ:nət/ *adjective* in esp. human form.

incarnation /ɪnkɑ:'neɪʃ(ə)n/ *noun* embodiment in flesh; (**the Incarnation**) embodiment of God in Christ; (often + *of*) living type (of a quality etc.).

incautious /ɪn'kɔ:ʃəs/ *adjective* rash. □ **incautiously** *adverb.*

incendiary /ɪn'sendɪərɪ/ ● *adjective* (of bomb) filled with material for causing fires; of arson; guilty of arson; inflammatory. ● *noun* (*plural* **-ies**) incendiary person or bomb.

incense[1] /'ɪnsens/ *noun* gum or spice giving sweet smell when burned; smoke of this, esp. in religious ceremonial.

incense[2] /ɪn'sens/ *verb* (**-sing**) make angry.

incentive /ɪn'sentɪv/ ● *noun* motive, incitement; payment etc. encouraging effort in work. ● *adjective* serving to motivate or incite.

inception /ɪn'sepʃ(ə)n/ *noun* beginning.

incessant /ɪn'ses(ə)nt/ *adjective* unceasing, continual; repeated. □ **incessantly** *adverb.*

incest /'ɪnsest/ *noun* crime of sexual intercourse between people prohibited from marrying because of closeness of their blood relationship.

incestuous /ɪn'sestjʊəs/ *adjective* of or guilty of incest; having relationships restricted to a particular group etc.

inch ● *noun* twelfth of (linear) foot (2.54 cm); this as unit of map-scale (e.g. 1 inch to 1 mile) or as unit of rainfall (= 1 inch depth of water). ● *verb* move gradually. □ **every inch** entirely; **within an inch of one's life** almost to death.

inchoate /ɪn'kəʊeɪt/ *adjective* just begun; undeveloped. □ **inchoation** *noun.*

incidence /'ɪnsɪd(ə)ns/ *noun* range, scope, extent, manner, or rate of occurrence; falling of line, ray, particles, etc. on surface; coming into contact with thing.

incident /'ɪnsɪd(ə)nt/ ● *noun* event, occurrence; violent episode, civil or military; episode in play, film, etc. ● *adjective* (often + *to*) apt to occur, naturally attaching; (often + *on, upon*) (of light etc.) falling.

incidental /ɪnsɪ'dent(ə)l/ *adjective* (often + *to*) minor, supplementary; not essential. □ **incidental music** music played during or between scenes of play, film, etc.

incidentally *adverb* by the way; in an incidental way.

incinerate /ɪn'sɪnəreɪt/ *verb* (**-ting**) burn to ashes. □ **incineration** *noun.*

incinerator *noun* furnace or device for incineration.

incipient /ɪn'sɪpɪənt/ *adjective* beginning, in early stage.

incise /ɪn'saɪz/ *verb* (**-sing**) make cut in; engrave.

incision /ɪn'sɪʒ(ə)n/ *noun* cutting, esp. by surgeon; cut.

incisive /ɪn'saɪsɪv/ *adjective* sharp; clear and effective.

incisor /ɪn'saɪzə/ *noun* cutting-tooth, esp. at front of mouth.

incite /ɪn'saɪt/ *verb* (**-ting**) (often + *to*) urge on, stir up. □ **incitement** *noun.*

incivility /ɪnsɪ'vɪlɪtɪ/ *noun* (*plural* **-ies**) rudeness; impolite act.

inclement /ɪn'klemənt/ *adjective* (of weather) severe, stormy. □ **inclemency** *noun.*

inclination /ɪnklɪ'neɪʃ(ə)n/ *noun* propensity; liking, affection; slope, slant.

incline ● *verb* /ɪn'klaɪn/ (**-ning**) (usually in *passive*) dispose, influence; have specified tendency; be disposed, tend; (cause to) lean or bend. ● *noun* /'ɪnklaɪn/ slope.

include /ɪn'klu:d/ *verb* (**-ding**) comprise, regard or treat as part of whole. □ **inclusion** /-ʒ(ə)n/ *noun.*

inclusive /ɪn'klu:sɪv/ *adjective* (often + *of*) including; including the limits stated; comprehensive; including all accessory payments. □ **inclusively** *adverb*; **inclusiveness** *noun.*

incognito /ɪnkɒg'ni:təʊ/ ● *adjective & adverb* with one's name or identity concealed. ● *noun* (*plural* **-s**) person who is incognito; pretended identity.

incoherent /ɪnkəʊ'hɪərənt/ *adjective* unintelligible; lacking logic or consistency; not clear. □ **incoherence** *noun*; **incoherently** *adverb*.

incombustible /ɪnkəm'bʌstɪb(ə)l/ *adjective* that cannot be burnt.

income /'ɪnkʌm/ *noun* money received, esp. periodically, from work, investments, etc. □ **income tax** tax levied on income.

incoming *adjective* coming in; succeeding another.

incommensurable /ɪnkə'menʃərəb(ə)l/ *adjective* (often + *with*) not comparable in size, value, etc.; having no common factor. □ **incommensurability** *noun*.

incommensurate /ɪnkə'menʃərət/ *adjective* (often + *with*, *to*) out of proportion; inadequate; incommensurable.

incommode /ɪnkə'məʊd/ *verb* (**-ding**) *formal* inconvenience; trouble, annoy.

incommodious /ɪnkə'məʊdɪəs/ *adjective formal* too small for comfort; inconvenient.

incommunicable /ɪnkə'mju:nɪkəb(ə)l/ *adjective* that cannot be shared or communicated.

incommunicado /ɪnkəmju:nɪ'kɑ:dəʊ/ *adjective* without means of communication, in solitary confinement in prison etc.

incomparable /ɪn'kɒmpərəb(ə)l/ *adjective* without an equal; matchless. □ **incomparability** *noun*; **incomparably** *adverb*.

incompatible /ɪnkəm'pætɪb(ə)l/ *adjective* not compatible. □ **incompatibility** *noun*.

incompetent /ɪn'kɒmpɪt(ə)nt/ *adjective* inept; (often + *to*) lacking the necessary skill, not legally qualified. □ **incompetence** *noun*.

incomplete /ɪnkəm'pli:t/ *adjective* not complete.

incomprehensible /ɪnkɒmprɪ'hensɪb(ə)l/ *adjective* that cannot be understood.

incomprehension /ɪnkɒmprɪ'henʃ(ə)n/ *noun* failure to understand.

inconceivable /ɪnkən'si:vəb(ə)l/ *adjective* that cannot be imagined. □ **inconceivably** *adverb*.

inconclusive /ɪnkən'klu:sɪv/ *adjective* (of argument etc.) not convincing or decisive.

incongruous /ɪn'kɒŋgrʊəs/ *adjective* out of place; absurd; (often + *with*) out of keeping. □ **incongruity** /-'gru:ɪtɪ/ *noun* (*plural* **-ies**); **incongruously** *adverb*.

inconsequent /ɪn'kɒnsɪkwənt/ *adjective* irrelevant; not following logically; disconnected. □ **inconsequence** *noun*.

inconsequential /ɪnkɒnsɪ'kwenʃ(ə)l/ *adjective* unimportant; inconsequent. □ **inconsequentially** *adverb*.

inconsiderable /ɪnkən'sɪdərəb(ə)l/ *adjective* of small size, value, etc.; not worth considering. □ **inconsiderably** *adverb*.

inconsiderate /ɪnkən'sɪdərət/ *adjective* not considerate of others; thoughtless. □ **inconsiderately** *adverb*; **inconsiderateness** *noun*.

inconsistent /ɪnkən'sɪst(ə)nt/ *adjective* not consistent. □ **inconsistency** *noun* (*plural* **-ies**); **inconsistently** *adverb*.

inconsolable /ɪnkən'səʊləb(ə)l/ *adjective* that cannot be consoled. □ **inconsolably** *adverb*.

inconspicuous /ɪnkən'spɪkjʊəs/ *adjective* not conspicuous; not easily noticed. □ **inconspicuously** *adverb*; **inconspicuousness** *noun*.

inconstant /ɪn'kɒnst(ə)nt/ *adjective* fickle; variable. □ **inconstancy** *noun* (*plural* **-ies**).

incontestable /ɪnkən'testəb(ə)l/ *adjective* that cannot be disputed. □ **incontestably** *adverb*.

incontinent /ɪn'kɒntɪnənt/ *adjective* unable to control bowels or bladder; lacking self-restraint. □ **incontinence** *noun*.

incontrovertible /ɪnkɒntrə'vɜ:tɪb(ə)l/ *adjective* indisputable. □ **incontrovertibly** *adverb*.

inconvenience /ɪnkən'vi:nɪəns/ ● *noun* lack of ease or comfort; trouble; cause or instance of this. ● *verb* (**-cing**) cause inconvenience to.

inconvenient *adjective* causing trouble, difficulty, or discomfort; awkward. □ **inconveniently** *adverb*.

incorporate ● *verb* /ɪn'kɔ:pəreɪt/ (**-ting**) include as part or ingredient; (often + *in*, *with*) unite (in one body); admit as member of company etc.; (esp. as **incorporated** *adjective*) constitute as legal corporation. ● *adjective* /ɪn'kɔ:pərət/ incorporated. □ **incorporation** *noun.*

incorporeal /ɪnkɔ:'pɔ:rɪəl/ *adjective* without substance or material existence. □ **incorporeally** *adverb.*

incorrect /ɪnkə'rekt/ *adjective* untrue, inaccurate; improper, unsuitable. □ **incorrectly** *adverb.*

incorrigible /ɪn'kɒrɪdʒɪb(ə)l/ *adjective* that cannot be corrected or improved. □ **incorrigibility** *noun*; **incorrigibly** *adverb.*

incorruptible /ɪnkə'rʌptɪb(ə)l/ *adjective* that cannot decay or be corrupted. □ **incorruptibility** *noun*; **incorruptibly** *adverb.*

increase ● *verb* /ɪŋ'kri:s/ (**-sing**) become or make greater or more numerous. ● *noun* /'ɪnkri:s/ growth, enlargement; (of people, animals, or plants) multiplication; increased amount. □ **on the increase** increasing.

increasingly /ɪn'kri:sɪŋlɪ/ *adverb* more and more.

incredible /ɪn'kredɪb(ə)l/ *adjective* that cannot be believed; *colloquial* surprising, extremely good. □ **incredibility** *noun*; **incredibly** *adverb.*

incredulous /ɪn'kredjʊləs/ *adjective* unwilling to believe; showing disbelief. □ **incredulity** /ɪnkrɪ'dju:lɪtɪ/ *noun*; **incredulously** *adverb.*

increment /'ɪŋkrɪmənt/ *noun* amount of increase; added amount. □ **incremental** /-'ment(ə)l/ *adjective.*

incriminate /ɪn'krɪmɪneɪt/ *verb* (**-ting**) indicate as guilty; charge with crime. □ **incrimination** *noun*; **incriminatory** *adjective.*

incrustation /ɪnkrʌs'teɪʃ(ə)n/ *noun* encrusting, being encrusted; crust, hard coating; deposit on surface.

incubate /'ɪŋkjʊbeɪt/ *verb* (**-ting**) hatch (eggs) by sitting on them or by artificial heat; cause (bacteria etc.) to develop; develop slowly.

incubation *noun* incubating, being incubated; period between infection and appearance of first symptoms.

incubator *noun* apparatus providing warmth for hatching eggs, rearing premature babies, or developing bacteria.

incubus /'ɪŋkjʊbəs/ *noun* (*plural* **-buses** or **-bi** /-baɪ/) demon or male spirit formerly believed to have sexual intercourse with sleeping women; nightmare; oppressive person or thing.

inculcate /'ɪnkʌlkeɪt/ *verb* (**-ting**) (often + *upon, in*) urge, impress persistently. □ **inculcation** *noun.*

incumbency /ɪn'kʌmbənsɪ/ *noun* (*plural* **-ies**) office or tenure of incumbent.

incumbent /ɪn'kʌmbənt/ ● *adjective* lying, pressing; currently holding office. ● *noun* holder of office, esp. benefice. □ **it is incumbent on a person** (+ *to do*) it is a person's duty.

incur /ɪn'kɜ:/ *verb* (**-rr-**) bring on oneself.

incurable /ɪn'kjʊərəb(ə)l/ ● *adjective* that cannot be cured. ● *noun* incurable person. □ **incurability** *noun*; **incurably** *adverb.*

incurious /ɪn'kjʊərɪəs/ *adjective* lacking curiosity.

incursion /ɪn'kɜ:ʃ(ə)n/ *noun* invasion; sudden attack. □ **incursive** *adjective.*

indebted /ɪn'detɪd/ *adjective* (usually + *to*) owing money or gratitude. □ **indebtedness** *noun.*

indecent /ɪn'di:s(ə)nt/ *adjective* offending against decency; unbecoming; unsuitable. □ **indecent assault** sexual attack not involving rape. □ **indecency** *noun*; **indecently** *adverb.*

indecipherable /ɪndɪ'saɪfərəb(ə)l/ *adjective* that cannot be deciphered.

indecision /ɪndɪ'sɪʒ(ə)n/ *noun* inability to decide; hesitation.

indecisive /ɪndɪ'saɪsɪv/ *adjective* not decisive; irresolute; not conclusive. □ **indecisively** *adverb*; **indecisiveness** *noun.*

indecorous /ɪn'dekərəs/ *adjective* improper, undignified; in bad taste. □ **indecorously** *adverb.*

indeed /ɪn'di:d/ ● *adverb* in truth; really; admittedly. ● *interjection expressing irony, incredulity, etc.*

indefatigable /ɪndɪ'fætɪgəb(ə)l/ *adjective* unwearying, unremitting. □ **indefatigably** *adverb.*

indefeasible /ɪndɪ'fi:zɪb(ə)l/ *adjective literary* (esp. of claim, rights, etc.) that cannot be forfeited or annulled.

indefensible /ɪndɪ'fensɪb(ə)l/ *adjective* that cannot be defended. □ **indefensibility** *noun*; **indefensibly** *adverb.*

indefinable /ɪndɪ'faɪnəb(ə)l/ *adjective* that cannot be defined; mysterious. □ **indefinably** *adverb.*

indefinite /ɪn'defɪnɪt/ *adjective* vague, undefined; unlimited; (of adjectives, adverbs, and pronouns) not determining the person etc. referred to. □ **indefinite article** word (*a*, *an* in English) placed before noun and meaning 'one, some, any'.

indefinitely *adverb* for an unlimited time; in an indefinite manner.

indelible /ɪn'delɪb(ə)l/ *adjective* that cannot be rubbed out; permanent. □ **indelibly** *adverb.*

indelicate /ɪn'delɪkət/ *adjective* coarse, unrefined; tactless. □ **indelicacy** *noun* (*plural* **-ies**); **indelicately** *adverb.*

indemnify /ɪn'demnɪfaɪ/ *verb* (**-ies**, **-ied**) (often + *against*, *from*) secure against loss or legal responsibility; (often + *for*) exempt from penalty; compensate. □ **indemnification** *noun.*

indemnity /ɪn'demnɪtɪ/ *noun* (*plural* **-ies**) compensation for damage; sum exacted by victor in war; security against damage or loss; exemption from penalties.

indent ● *verb* /ɪn'dent/ make or impress notches, dents, or recesses in; set back (beginning of line) inwards from margin; draw up (legal document) in duplicate; (often + *for*) make requisition. ● *noun* /'ɪndent/ order (esp. from abroad) for goods; official requisition for stores; indented line; indentation; indenture.

indentation /ɪnden'teɪʃ(ə)n/ *noun* indenting, being indented; notch.

indenture /ɪn'dentʃə/ ● *noun* (usually in *plural*) sealed agreement; formal list, certificate, etc. ● *verb* (**-ring**) *historical* bind by indentures, esp. as apprentice.

independent /ɪndɪ'pend(ə)nt/ ● *adjective* (often + *of*) not depending on authority; self-governing; not depending on another person for one's livelihood or opinions; (of income) making it unnecessary to earn one's livelihood; unwilling to be under obligation to others; not depending on something else for validity etc.; (of institution) not supported by public funds. ● *noun* politician etc. independent of any political party. □ **independence** *noun*; **independently** *adverb.*

indescribable /ɪndɪ'skraɪbəb(ə)l/ *adjective* beyond description; that cannot be described. □ **indescribably** *adverb.*

indestructible /ɪndɪ'strʌktɪb(ə)l/ *adjective* that cannot be destroyed. □ **indestructibility** *noun*; **indestructibly** *adverb.*

indeterminable /ɪndɪ'tɜ:mɪnəb(ə)l/ *adjective* that cannot be ascertained or settled.

indeterminate /ɪndɪ'tɜ:mɪnət/ *adjective* not fixed in extent, character, etc.; vague. □ **indeterminacy** *noun.*

index /'ɪndeks/ ● *noun* (*plural* **-es** or **indices** /'ɪndɪsi:z/) alphabetical list of subjects etc. with references, usually at end of book; card index; measure of prices or wages compared with a previous month, year, etc.; *Mathematics* exponent. ● *verb* furnish (book) with index, enter in index; relate (wages, investment income, etc.) to a price index. □ **index finger** finger next to thumb; **index-linked** related to value of price index.

Indian /'ɪndɪən/ ● *noun* native or national of India; person of Indian descent; (in full **American Indian**) original inhabitant of America. ● *adjective* of India; of the subcontinent comprising India, Pakistan, and Bangladesh; of the original inhabitants of America. □ **Indian corn** maize; **Indian file** single file; **Indian ink** black pigment, ink made from this; **Indian summer** period of calm dry warm weather in late autumn, happy tranquil period late in life.

indiarubber *noun* rubber, esp. for rubbing out pencil marks etc.

indicate /'ɪndɪkeɪt/ *verb* (**-ting**) point out, make known; show; be sign of; require, call for; state briefly; give as reading or measurement; point by hand; use a vehicle's indicator. □ **indication** *noun.*

indicative /ɪn'dɪkətɪv/ ● *adjective* (+ *of*) suggestive, giving indications; *Grammar* (of mood) stating thing as fact. ● *noun* *Grammar* indicative mood; verb in this mood.

indicator *noun* flashing light on vehicle showing direction in which it is about to turn; person or thing that indicates; device indicating condition of machine etc.; recording instrument; board giving current information.

indices *plural* of INDEX.

indict /ɪn'daɪt/ *verb* accuse formally by legal process.

indictable *adjective* (of an offence) rendering person liable to be indicted; so liable.

indictment *noun* indicting, accusation; document containing this; thing that serves to condemn or censure.

indifference /ɪn'dɪfrəns/ *noun* lack of interest or attention; unimportance.

indifferent *adjective* (+ *to*) showing indifference; neither good nor bad; of poor quality or ability. □ **indifferently** *adverb.*

indigenous /ɪn'dɪdʒɪnəs/ *adjective* (often + *to*) native or belonging naturally to a place.

indigent /'ɪndɪdʒ(ə)nt/ *adjective formal* needy, poor. □ **indigence** *noun.*

indigestible /ɪndɪ'dʒestɪb(ə)l/ *adjective* difficult or impossible to digest.

indigestion /ɪndɪ'dʒestʃ(ə)n/ *noun* difficulty in digesting food; pain caused by this.

indignant /ɪn'dɪgnənt/ *adjective* feeling or showing indignation. □ **indignantly** *adverb.*

indignation /ɪndɪg'neɪʃ(ə)n/ *noun* anger at supposed injustice etc.

indignity /ɪn'dɪgnɪtɪ/ *noun* (*plural* **-ies**) humiliating treatment; insult.

indigo /'ɪndɪgəʊ/ *noun* (*plural* **-s**) deep violet-blue; dye of this colour.

indirect /ɪndaɪ'rekt/ *adjective* not going straight to the point; (of route etc.) not straight. □ **indirect object** word or phrase representing person or thing affected by action of verb but not acted on (see panel at OBJECT); **indirect speech** reported speech; **indirect tax** tax on goods and services, not income. □ **indirectly** *adverb.*

indiscernible /ɪndɪ'sɜ:nɪb(ə)l/ *adjective* that cannot be discerned.

indiscipline /ɪn'dɪsɪplɪn/ *noun* lack of discipline.

indiscreet /ɪndɪ'skri:t/ *adjective* not discreet; injudicious, unwary. □ **indiscreetly** *adverb.*

indiscretion /ɪndɪ'skreʃ(ə)n/ *noun* indiscreet conduct or action.

indiscriminate /ɪndɪ'skrɪmɪnət/ *adjective* making no distinctions; done or acting at random. □ **indiscriminately** *adverb.*

indispensable /ɪndɪ'spensəb(ə)l/ *adjective* that cannot be dispensed with; necessary. □ **indispensably** *adverb.*

indisposed /ɪndɪ'spəʊzd/ *adjective* slightly unwell; averse, unwilling. □ **indisposition** /-spə'zɪʃ(ə)n/ *noun.*

indisputable /ɪndɪ'spju:təb(ə)l/ *adjective* that cannot be disputed. □ **indisputably** *adverb.*

indissoluble /ɪndɪ'sɒljʊb(ə)l/ *adjective* that cannot be dissolved; lasting, stable. □ **indissolubly** *adverb.*

indistinct /ɪndɪ'stɪŋkt/ *adjective* not distinct; confused, obscure. □ **indistinctly** *adverb.*

indistinguishable /ɪndɪ'stɪŋgwɪʃəb(ə)l/ *adjective* (often + *from*) not distinguishable.

indite /ɪn'daɪt/ *verb* (**-ting**) *formal or jocular* put into words; write (letter etc.).

individual /ɪndɪ'vɪdʒʊəl/ ● *adjective* of, for, or characteristic of single person or thing; having distinct character; designed for use by one person; single; particular. ● *noun* single member of class, group, etc.; single human being; *colloquial* person; distinctive person.

individualism *noun* social theory favouring free action by individuals; being independent or different. □ **individualist** *noun*; **individualistic** /-'lɪs-/ *adjective.*

individuality /ɪndɪvɪdʒʊ'ælɪtɪ/ *noun* individual character, esp. when strongly marked; separate existence.

individualize *verb* (also **-ise**) (**-zing** or **-sing**) give individual character to; (esp. as **individualized** *adjective*) personalize.

individually *adverb* one by one; personally; distinctively.

indivisible /ɪndɪ'vɪzɪb(ə)l/ *adjective* not divisible.

indoctrinate /ɪn'dɒktrɪneɪt/ *verb* (**-ting**) teach to accept a particular belief uncritically. □ **indoctrination** *noun.*

Indo-European /ɪndəʊjʊərə'pɪən/ ● *adjective* of family of languages spoken over most of Europe and Asia as far as N. India; of hypothetical parent language of this family. ● *noun* Indo-European family of languages; hypothetical parent language of these.

indolent /ˈɪndələnt/ *adjective* lazy; averse to exertion. □ **indolence** *noun*; **indolently** *adverb*.

indomitable /ɪnˈdɒmɪtəb(ə)l/ *adjective* unconquerable; unyielding. □ **indomitably** *adverb*.

indoor /ˈɪndɔː/ *adjective* done etc. in building or under cover.

indoors /ɪnˈdɔːz/ *adverb* in(to) a building.

indubitable /ɪnˈdjuːbɪtəb(ə)l/ *adjective* that cannot be doubted. □ **indubitably** *adverb*.

induce /ɪnˈdjuːs/ *verb* (**-cing**) prevail on, persuade; bring about; bring on (labour) artificially; bring on labour in (mother); speed up birth of (baby); produce by induction; infer. □ **inducible** *adjective*.

inducement *noun* attractive offer; incentive; bribe.

induct /ɪnˈdʌkt/ *verb* install into office etc.

inductance *noun* property of electric circuit in which variation in current produces electromotive force.

induction /ɪnˈdʌkʃ(ə)n/ *noun* inducting, inducing; act of bringing on (esp. labour) artificially; general inference from particular instances; formal introduction to new job etc.; production of electric or magnetic state by proximity to electric circuit or magnetic field.

inductive /ɪnˈdʌktɪv/ *adjective* (of reasoning etc.) based on induction; of electric or magnetic induction.

indulge /ɪnˈdʌldʒ/ *verb* (**-ging**) (often + *in*) take one's pleasure freely; yield freely to (desire etc.); gratify by compliance with wishes.

indulgence *noun* indulging; thing indulged in; *RC Church* remission of punishment still due after absolution; privilege granted.

indulgent *adjective* lenient; willing to overlook faults; indulging. □ **indulgently** *adverb*.

industrial /ɪnˈdʌstrɪəl/ *adjective* of, engaged in, for use in, or serving the needs of, industry; having highly developed industries. □ **industrial action** strike or disruptive action by workers as protest; **industrial estate** area of land zoned for factories etc. □ **industrially** *adverb*.

industrialism *noun* system in which manufacturing industries predominate.

industrialist *noun* owner or manager in industry.

industrialize *verb* (also **-ise**) (**-zing** or **-sing**) make (nation etc.) industrial. □ **industrialization** *noun*.

industrious /ɪnˈdʌstrɪəs/ *adjective* hard-working. □ **industriously** *adverb*.

industry /ˈɪndəstrɪ/ *noun* (*plural* **-ies**) branch of trade or manufacture; commercial enterprise; trade or manufacture collectively; concerted activity; diligence.

inebriate ● *verb* /ɪnˈiːbrɪeɪt/ (**-ting**) make drunk; excite. ● *adjective* /ɪnˈiːbrɪət/ drunken. ● *noun* /ɪnˈiːbrɪət/ drunkard. □ **inebriation** *noun*.

inedible /ɪnˈedɪb(ə)l/ *adjective* not suitable for eating.

ineducable /ɪnˈedjʊkəb(ə)l/ *adjective* incapable of being educated.

ineffable /ɪnˈefəb(ə)l/ *adjective* too great for description in words; that must not be uttered. □ **ineffability** *noun*; **ineffably** *adverb*.

ineffective /ɪnɪˈfektɪv/ *adjective* not achieving desired effect or results. □ **ineffectively** *adverb*; **ineffectiveness** *noun*.

ineffectual /ɪnɪˈfektʃʊəl/ *adjective* ineffective, feeble. □ **ineffectually** *adverb*.

inefficient /ɪnɪˈfɪʃ(ə)nt/ *adjective* not efficient or fully capable; (of machine etc.) wasteful. □ **inefficiency** *noun*; **inefficiently** *adverb*.

inelegant /ɪnˈelɪgənt/ *adjective* ungraceful, unrefined. □ **inelegance** *noun*; **inelegantly** *adverb*.

ineligible /ɪnˈelɪdʒɪb(ə)l/ *adjective* not eligible or qualified. □ **ineligibility** *noun*.

ineluctable /ɪnɪˈlʌktəb(ə)l/ *adjective* inescapable, unavoidable.

inept /ɪˈnept/ *adjective* unskilful; absurd, silly; out of place. □ **ineptitude** *noun*; **ineptly** *adverb*.

inequality /ɪnɪ'kwɒlɪtɪ/ *noun* (*plural* **-ies**) lack of equality; variability; unevenness.

inequitable /ɪn'ekwɪtəb(ə)l/ *adjective* unfair, unjust.

inequity /ɪn'ekwɪtɪ/ *noun* (*plural* **-ies**) unfairness, injustice.

ineradicable /ɪnɪ'rædɪkəb(ə)l/ *adjective* that cannot be rooted out.

inert /ɪ'nɜ:t/ *adjective* without inherent power of action etc.; chemically inactive; sluggish, slow; lifeless.

inertia /ɪ'nɜ:ʃə/ *noun* property by which matter continues in existing state of rest or motion unless acted on by external force; inertness; tendency to remain unchanged. □ **inertia reel** reel allowing seat belt to unwind freely but locking on impact; **inertia selling** sending of unsolicited goods in hope of making a sale.

inescapable /ɪnɪ'skeɪpəb(ə)l/ *adjective* that cannot be escaped or avoided. □ **inescapably** *adverb.*

inessential /ɪnɪ'senʃ(ə)l/ ● *adjective* not necessary; dispensable. ● *noun* inessential thing.

inestimable /ɪn'estɪməb(ə)l/ *adjective* too great etc. to be estimated. □ **inestimably** *adverb.*

inevitable /ɪn'evɪtəb(ə)l/ *adjective* unavoidable; bound to happen or appear; *colloquial* tiresomely familiar. □ **inevitability** *noun*; **inevitably** *adverb.*

inexact /ɪnɪg'zækt/ *adjective* not exact. □ **inexactitude** *noun.*

inexcusable /ɪnɪk'skju:zəb(ə)l/ *adjective* that cannot be justified. □ **inexcusably** *adverb.*

inexhaustible /ɪnɪg'zɔ:stɪb(ə)l/ *adjective* that cannot be used up.

inexorable /ɪn'eksərəb(ə)l/ *adjective* relentless. □ **inexorably** *adverb.*

inexpedient /ɪnɪk'spi:dɪənt/ *adjective* not expedient.

inexpensive /ɪnɪk'spensɪv/ *adjective* not expensive.

inexperience /ɪnɪk'spɪərɪəns/ *noun* lack of experience. □ **inexperienced** *adjective.*

inexpert /ɪn'ekspɜ:t/ *adjective* unskilful.

inexpiable /ɪn'ekspɪəb(ə)l/ *adjective* that cannot be expiated.

inexplicable /ɪnɪk'splɪkəb(ə)l/ *adjective* that cannot be explained. □ **inexplicably** *adverb.*

inexpressible /ɪnɪk'spresɪb(ə)l/ *adjective* that cannot be expressed. □ **inexpressibly** *adverb.*

in extremis /ɪn ɪk'stri:mɪs/ *adjective* at point of death; in great difficulties. [Latin]

inextricable /ɪnɪk'strɪkəb(ə)l/ *adjective* that cannot be separated, loosened, or resolved; inescapable. □ **inextricably** *adverb.*

infallible /ɪn'fælɪb(ə)l/ *adjective* incapable of error; unfailing, sure. □ **infallibility** *noun*; **infallibly** *adverb.*

infamous /'ɪnfəməs/ *adjective* notoriously vile, evil; abominable. □ **infamously** *adverb*; **infamy** *noun* (*plural* **-ies**).

infant /'ɪnf(ə)nt/ *noun* child during earliest period of life; thing in early stage of development; *Law* person under 18. □ **infancy** *noun.*

infanta /ɪn'fæntə/ *noun historical* daughter of Spanish or Portuguese king.

infanticide /ɪn'fæntɪsaɪd/ *noun* killing of infant, esp. soon after birth; person guilty of this.

infantile /'ɪnfəntaɪl/ *adjective* of or like infants. □ **infantile paralysis** poliomyelitis.

infantry /'ɪnfəntrɪ/ *noun* (*plural* **-ies**) (group of) foot-soldiers. □ **infantryman** soldier of infantry regiment.

infatuate /ɪn'fætʃʊeɪt/ *verb* (**-ting**) (usually as **infatuated** *adjective*) inspire with intense fondness. □ **infatuation** *noun.*

infect /ɪn'fekt/ *verb* affect or contaminate with germ, virus, or disease; imbue, taint.

infection /ɪn'fekʃ(ə)n/ *noun* infecting, being infected; disease; communication of disease.

infectious *adjective* infecting; transmissible by infection; apt to spread. □ **infectiously** *adverb.*

infelicity /ɪnfɪ'lɪsɪtɪ/ *noun* (*plural* **-ies**) inapt expression; unhappiness. □ **infelicitous** *adjective.*

infer /ɪn'fɜ:/ *verb* (**-rr-**) deduce, conclude.

■ **Usage** It is a mistake to use *infer* to mean 'imply', as in *Are you inferring that I'm a liar?*

inference /ˈɪnfərəns/ *noun* act of inferring; thing inferred. □ **inferential** /-ˈren(ə)l/ *adjective.*

inferior /ɪnˈfɪərɪə/ ● *adjective* lower in rank etc.; of poor quality; situated below. ● *noun* inferior person.

inferiority /ɪnfɪərɪˈɒrɪtɪ/ *noun* being inferior. □ **inferiority complex** feeling of inadequacy, sometimes marked by compensating aggressive behaviour.

infernal /ɪnˈfɜ:n(ə)l/ *adjective* of hell; hellish; *colloquial* detestable, tiresome. □ **infernally** *adverb.*

inferno /ɪnˈfɜ:nəʊ/ *noun* (*plural* **-s**) raging fire; scene of horror or distress; hell.

infertile /ɪnˈfɜ:taɪl/ *adjective* not fertile. □ **infertility** /-fəˈtɪl-/ *noun.*

infest /ɪnˈfest/ *verb* overrun in large numbers. □ **infestation** *noun.*

infidel /ˈɪnfɪd(ə)l/ ● *noun* disbeliever in esp. the supposed true religion. ● *adjective* of infidels; unbelieving.

infidelity /ɪnfɪˈdelɪtɪ/ *noun* (*plural* **-ies**) being unfaithful.

infighting *noun* conflict or competitiveness in organization; boxing within arm's length.

infiltrate /ˈɪnfɪltreɪt/ *verb* (**-ting**) enter (territory, political party, etc.) gradually and imperceptibly; cause to do this; permeate by filtration; (often + *into*, *through*) introduce (fluid) by filtration. □ **infiltration** *noun*; **infiltrator** *noun.*

infinite /ˈɪnfɪnɪt/ *adjective* boundless; endless; very great or many. □ **infinitely** *adverb.*

infinitesimal /ɪnfɪnɪˈtesɪm(ə)l/ *adjective* infinitely or very small. □ **infinitesimally** *adverb.*

infinitive /ɪnˈfɪnɪtɪv/ ● *noun* verb-form expressing verbal notion without particular subject, tense, etc. ● *adjective* having this form.

infinitude /ɪnˈfɪnɪtju:d/ *noun literary* infinite number etc.; being infinite.

infinity /ɪnˈfɪnɪtɪ/ *noun* (*plural* **-ies**) infinite number or extent; being infinite; boundlessness; infinite distance; *Mathematics* infinite quantity.

infirm /ɪnˈfɜ:m/ *adjective* weak.

infirmary /ɪnˈfɜ:mərɪ/ *noun* (*plural* **-ies**) hospital; sickbay in school etc.

infirmity *noun* (*plural* **-ies**) being infirm; particular physical weakness.

in flagrante delicto /ɪn fləˈgræntɪ dɪˈlɪktəʊ/ *adverb* in act of committing offence. [Latin]

inflame /ɪnˈfleɪm/ *verb* (**-ming**) provoke to strong feeling; cause inflammation in; aggravate; make hot; (cause to) catch fire.

inflammable /ɪnˈflæməb(ə)l/ *adjective* easily set on fire or excited. □ **inflammability** *noun.*

■ **Usage** Because *inflammable* could be thought to mean 'not easily set on fire', *flammable* is often used instead. The negative of *inflammable* is *non-inflammable.*

---

inflammation /ɪnfləˈmeɪʃ(ə)n/ *noun* inflaming; disordered bodily condition marked by heat, swelling, redness, and usually pain.

inflammatory /ɪnˈflæmətərɪ/ *adjective* tending to inflame; of inflammation.

inflatable ● *adjective* that can be inflated. ● *noun* inflatable object.

inflate /ɪnˈfleɪt/ *verb* (**-ting**) distend with air or gas; (usually + *with*; usually in *passive*) puff up (with pride etc.); resort to inflation of (currency); raise (prices) artificially; (as **inflated** *adjective*) (esp. of language, opinions, etc.) bombastic, exaggerated.

inflation *noun* inflating, being inflated; general rise in prices, increase in supply of money regarded as cause of such rise. □ **inflationary** *adjective.*

inflect /ɪnˈflekt/ *verb* change or vary pitch of (voice); modify (word) to express grammatical relation; undergo such modification.

inflection /ɪnˈflekʃ(ə)n/ *noun* (also **inflexion**) inflecting, being inflected; inflected form; inflecting suffix etc.; modulation of voice. □ **inflectional** *adjective.*

inflexible /ɪnˈfleksɪb(ə)l/ *adjective* unbendable; unbending; unyielding. □ **inflexibility** *noun*; **inflexibly** *adverb.*

inflexion = INFLECTION.

inflict /ɪnˈflɪkt/ *verb* deal (blow etc.); impose. □ **infliction** *noun.*

inflight *adjective* occurring or provided during a flight.

inflorescence /ɪnflə'res(ə)ns/ *noun* collective flower head of plant; arrangement of flowers on plant; flowering.

inflow *noun* flowing in; that which flows in.

influence /'ɪnflʊəns/ ● *noun* (usually + *on*) effect a person or thing has on another; (usually + *over, with*) ascendancy, moral power; thing or person exercising this. ● *verb* (**-cing**) exert influence on; affect. □ **under the influence** *colloquial* drunk.

influential /ɪnflʊ'enʃ(ə)l/ *adjective* having great influence.

influenza /ɪnflʊ'enzə/ *noun* infectious viral disease with fever, severe aching, and catarrh.

influx /'ɪnflʌks/ *noun* flowing in.

inform /ɪn'fɔ:m/ *verb* tell; (usually + *against, on*) give incriminating information about person to authorities.

informal /ɪn'fɔ:m(ə)l/ *adjective* without formality; not formal. □ **informality** /-'mæl-/ *noun* (*plural* **-ies**); **informally** *adverb.*

informant *noun* giver of information.

information /ɪnfə'meɪʃ(ə)n/ *noun* what is told; knowledge; news; formal charge or accusation. □ **information retrieval** tracing of information stored in books, computers, etc.; **information technology** study or use of processes (esp. computers etc.) for storing, retrieving, and sending information.

informative /ɪn'fɔ:mətɪv/ *adjective* giving information, instructive.

informed *adjective* knowing the facts; having some knowledge.

informer *noun* person who informs, esp. against others.

infraction /ɪn'frækʃ(ə)n/ *noun* infringement.

infra dig /ɪnfrə 'dɪg/ *adjective colloquial* beneath one's dignity.

infrared /ɪnfrə'red/ *adjective* of or using radiation just beyond red end of spectrum.

infrastructure /'ɪnfrəstrʌktʃə/ *noun* structural foundations of a society or enterprise; roads, bridges, sewers, etc., regarded as country's economic foundation; permanent installations as basis for military etc. operations.

infrequent /ɪn'fri:kwənt/ *adjective* not frequent. □ **infrequently** *adverb.*

infringe /ɪn'frɪndʒ/ *verb* (**-ging**) break or violate (law, another's rights, etc.); (usually + *on*) encroach, trespass. □ **infringement** *noun.*

infuriate /ɪn'fjʊərɪeɪt/ *verb* (**-ting**) enrage; irritate greatly. □ **infuriating** *adjective.*

infuse /ɪn'fju:z/ *verb* (**-sing**) (usually + *with*) fill (with a quality); steep or be steeped in liquid to extract properties; (usually + *into*) instil (life etc.).

infusible /ɪn'fju:zɪb(ə)l/ *adjective* that cannot be melted. □ **infusibility** *noun.*

infusion /ɪn'fju:ʒ(ə)n/ *noun* infusing; liquid extract so obtained; infused element.

ingenious /ɪn'dʒi:nɪəs/ *adjective* clever at contriving; cleverly contrived. □ **ingeniously** *adverb.*

■ **Usage** *Ingenious* is sometimes confused with *ingenuous.*

---

ingénue /æ̃ʒeɪ'nju:/ *noun* artless young woman, esp. as stage type. [French]

ingenuity /ɪndʒɪ'nju:ɪtɪ/ *noun* inventiveness, cleverness.

ingenuous /ɪn'dʒenjʊəs/ *adjective* artless; frank. □ **ingenuously** *adverb.*

■ **Usage** *Ingenuous* is sometimes confused with *ingenious.*

---

ingest /ɪn'dʒest/ *verb* take in (food etc.); absorb (knowledge etc.). □ **ingestion** *noun.*

inglenook /'ɪŋgəlnʊk/ *noun* space within opening either side of old-fashioned wide fireplace.

inglorious /ɪn'glɔ:rɪəs/ *adjective* shameful; not famous.

ingoing *adjective* going in.

ingot /'ɪŋgət/ *noun* (usually oblong) mass of cast metal, esp. gold, silver, or steel.

ingrained /ɪn'greɪnd/ *adjective* deeply rooted, inveterate; (of dirt etc.) deeply embedded.

ingratiate /ɪn'greɪʃɪeɪt/ *verb* (**-ting**) (**ingratiate oneself**; usually + *with*) bring oneself into favour. □ **ingratiating** *adjective.*

ingratitude /ɪn'grætɪtju:d/ *noun* lack of due gratitude.

ingredient /ɪn'gri:dɪənt/ *noun* component part in mixture.

ingress /'ɪngres/ *noun* going in; right to go in.

ingrowing *adjective* (of nail) growing into the flesh.

inhabit /ɪn'hæbɪt/ *verb* **(-t-)** dwell in, occupy. □ **inhabitable** *adjective*; **inhabitant** *noun*.

inhalant /ɪn'heɪlənt/ *noun* medicinal substance to be inhaled.

inhale /ɪn'heɪl/ *verb* **(-ling)** breathe in. □ **inhalation** /-hə'leɪʃ(ə)n/ *noun*.

inhaler *noun* device for administering inhalant, esp. to relieve asthma.

inhere /ɪn'hɪə/ *verb* **(-ring)** be inherent.

inherent /ɪn'herənt/ *adjective* (often + *in*) existing in something as essential or permanent attribute. □ **inherently** *adverb*.

inherit /ɪn'herɪt/ *verb* **(-t-)** receive as heir; derive (characteristic) from ancestors; derive or take over (situation) from predecessor. □ **inheritor** *noun*.

inheritance *noun* what is inherited; inheriting.

inhibit /ɪn'hɪbɪt/ *verb* **(-t-)** hinder, restrain, prevent; (as **inhibited** *adjective*) suffering from inhibition; (usually + *from*) prohibit.

inhibition /ɪnhɪ'bɪʃ(ə)n/ *noun* restraint of direct expression of instinct; *colloquial* emotional resistance to thought or action; inhibiting, being inhibited.

inhospitable /ɪnhɒs'pɪtəb(ə)l/ *adjective* not hospitable; affording no shelter.

inhuman /ɪn'hju:mən/ *adjective* brutal, unfeeling, barbarous. □ **inhumanity** /-'mæn-/ *noun* (*plural* **-ies**); **inhumanly** *adverb*.

inhumane /ɪnhju:'meɪn/ *adjective* not humane, callous.

inimical /ɪ'nɪmɪk(ə)l/ *adjective* hostile; harmful.

inimitable /ɪ'nɪmɪtəb(ə)l/ *adjective* that cannot be imitated. □ **inimitably** *adverb*.

iniquity /ɪ'nɪkwɪtɪ/ *noun* (*plural* **-ies**) wickedness; gross injustice. □ **iniquitous** *adjective*.

initial /ɪ'nɪʃ(ə)l/ ● *adjective* of or at beginning. ● *noun* first letter, esp. of person's name. ● *verb* **(-ll-**; *US* **-l-)** mark or sign with one's initials. □ **initially** *adverb*.

initiate ● *verb* /ɪ'nɪʃɪeɪt/ **(-ting)** originate, set going; admit into society, office, etc., esp. with ritual; (+ *into*) instruct in subject. ● *noun* /ɪ'nɪʃɪət/ initiated person. □ **initiation** *noun*; **initiatory** /ɪ'nɪʃjətərɪ/ *adjective*.

initiative /ɪ'nɪʃətɪv/ *noun* ability to initiate, enterprise; first step; (**the initiative**) power or right to begin.

inject /ɪn'dʒekt/ *verb* (usually + *into*) force (medicine etc.) (as) by syringe; administer medicine etc. to (person) by injection; place (quality etc.) where needed in something. □ **injection** *noun*.

injudicious /ɪndʒu:'dɪʃəs/ *adjective* unwise, ill-judged.

injunction /ɪn'dʒʌŋkʃ(ə)n/ *noun* authoritative order; judicial order restraining from specified act or compelling restitution etc.

injure /'ɪndʒə/ *verb* **(-ring)** hurt, harm, impair; do wrong to.

injurious /ɪn'dʒʊərɪəs/ *adjective* hurtful; defamatory; wrongful.

injury /'ɪndʒərɪ/ *noun* (*plural* **-ies**) physical damage, harm; offence to feelings etc.; *esp. Law* wrongful treatment. □ **injury time** extra time at football match etc. to compensate for that lost in dealing with injuries.

injustice /ɪn'dʒʌstɪs/ *noun* unfairness; unjust act.

ink ● *noun* coloured fluid or paste for writing or printing; black liquid ejected by cuttlefish etc. ● *verb* mark, cover, or smear with ink. □ **ink-jet printer** printing machine firing tiny jets of ink at paper; **inkwell** pot for ink, esp. in hole in desk. □ **inky** *adjective* **(-ier, -iest)**.

inkling /'ɪŋklɪŋ/ *noun* (often + *of*) hint, slight knowledge or suspicion.

inland ● *adjective* /'ɪnlənd/ remote from sea or border within a country; carried on within country. ● *adverb* /ɪn'lænd/ in or towards interior of country. □ **Inland Revenue** government department assessing and collecting taxes.

in-laws /'ɪnlɔ:z/ *plural noun* relatives by marriage.

inlay ● *verb* /ɪn'leɪ/ (*past & past participle* **inlaid** /ɪn'leɪd/) embed (thing in another); decorate (thing) thus. ● *noun*

/ˈɪnleɪ/ inlaid material or work; filling shaped to fit tooth-cavity.

inlet /ˈɪnlət/ *noun* small arm of sea etc.; piece inserted; way of admission.

inmate /ˈɪnmeɪt/ *noun* occupant of house, hospital, prison, etc.

in memoriam /ɪn mɪˈmɔːrɪæm/ *preposition* in memory of.

inmost /ˈɪnməʊst/ *adjective* most inward.

inn *noun* pub, sometimes with accommodation; *historical* house providing lodging etc. for payment, esp. for travellers. □ **innkeeper** keeper of inn; **Inns of Court** 4 legal societies admitting people to English bar.

innards /ˈɪnədz/ *plural noun colloquial* entrails.

innate /ɪˈneɪt/ *adjective* inborn; natural. □ **innately** *adverb.*

inner /ˈɪnə/ ● *adjective* interior, internal. ● *noun* circle nearest bull's-eye of target. □ **inner city** central area of city, esp. regarded as having social problems; **inner tube** separate inflatable tube in pneumatic tyre. □ **innermost** *adjective.*

innings /ˈɪnɪŋz/ *noun* (*plural* same) *esp. Cricket* batsman's or side's turn at batting; term of office etc. when person, party, etc. can achieve something.

innocent /ˈɪnəs(ə)nt/ ● *adjective* free from moral wrong; not guilty; guileless; harmless. ● *noun* innocent person, esp. young child. □ **innocence** *noun*; **innocently** *adverb.*

innocuous /ɪˈnɒkjʊəs/ *adjective* harmless.

innovate /ˈɪnəveɪt/ *verb* (**-ting**) bring in new ideas etc.; make changes. □ **innovation** *noun*; **innovative** /-vətɪv/ *adjective*; **innovator** *noun.*

innuendo /ɪnjʊˈendəʊ/ *noun* (*plural* **-es** or **-s**) allusive (usually depreciatory or sexually suggestive) remark.

innumerable /ɪˈnjuːmərəb(ə)l/ *adjective* countless.

innumerate /ɪˈnjuːmərət/ *adjective* not knowing basic mathematics. □ **innumeracy** *noun.*

inoculate /ɪˈnɒkjʊleɪt/ *verb* (**-ting**) treat with vaccine or serum to promote immunity against a disease. □ **inoculation** *noun.*

inoffensive /ɪnəˈfensɪv/ *adjective* not objectionable; harmless.

inoperable /ɪˈnɒpərəb(ə)l/ *adjective* that cannot be cured by surgical operation.

inoperative /ɪˈnɒpərətɪv/ *adjective* not working or taking effect.

inopportune /ɪˈnɒpətjuːn/ *adjective* not appropriate, esp. as regards time.

inordinate /ɪˈnɔːdɪnət/ *adjective* excessive. □ **inordinately** *adverb.*

inorganic /ɪnɔːˈgænɪk/ *adjective Chemistry* not organic; without organized physical structure; extraneous.

input /ˈɪnpʊt/ ● *noun* what is put in; place of entry of energy, information, etc.; action of putting in or feeding in; contribution of information etc. ● *verb* (**inputting**; *past & past participle* **input** or **inputted**) put in; supply (data, programs, etc.) to computer.

inquest /ˈɪŋkwest/ *noun* inquiry held by coroner into cause of death.

inquietude /ɪnˈkwaɪɪtjuːd/ *noun* uneasiness.

inquire /ɪnˈkwaɪə/ *verb* (**-ring**) seek information formally; make inquiry; ask question.

inquiry /ɪnˈkwaɪərɪ/ *noun* (*plural* **-ies**) investigation, esp. official; asking; question.

inquisition /ɪnkwɪˈzɪʃ(ə)n/ *noun* investigation; official inquiry; (**the Inquisition**) *RC Church History* ecclesiastical tribunal for suppression of heresy. □ **inquisitional** *adjective.*

inquisitive /ɪnˈkwɪzɪtɪv/ *adjective* curious, prying; seeking knowledge. □ **inquisitively** *adverb*; **inquisitiveness** *noun.*

inquisitor /ɪnˈkwɪzɪtə/ *noun* investigator; *historical* officer of Inquisition.

inquisitorial /ɪnkwɪzɪˈtɔːrɪəl/ *adjective* inquisitor-like; prying.

inroad *noun* (often in *plural*) encroachment; using up of resources etc.; hostile incursion.

inrush *noun* rapid influx.

insalubrious /ɪnsəˈluːbrɪəs/ *adjective* (of climate or place) unhealthy.

insane /ɪnˈseɪn/ *adjective* mad; *colloquial* extremely foolish. □ **insanely** *adverb*; **insanity** /ɪnˈsænɪtɪ/ *noun* (*plural* **-ies**).

insanitary /ɪnˈsænɪtərɪ/ *adjective* not sanitary.

insatiable /ɪn'seɪʃəb(ə)l/ *adjective* that cannot be satisfied; extremely greedy. □ **insatiability** *noun*; **insatiably** *adverb*.

insatiate /ɪn'seɪʃɪət/ *adjective* never satisfied.

inscribe /ɪn'skraɪb/ *verb* (**-bing**) (usually + *in, on*) write or carve (words etc.) on surface; mark (surface) with characters; (usually + *to*) write informal dedication in or on (book etc.); enter on list; *Geometry* draw (figure) within another so that some points of their boundaries coincide.

inscription /ɪn'skrɪpʃ(ə)n/ *noun* words inscribed; inscribing.

inscrutable /ɪn'skru:təb(ə)l/ *adjective* mysterious, impenetrable. □ **inscrutability** *noun*; **inscrutably** *adverb*.

insect /'ɪnsekt/ *noun* small invertebrate animal with segmented body, 6 legs, and usually wings.

insecticide /ɪn'sektɪsaɪd/ *noun* preparation used for killing insects.

insectivore /ɪn'sektɪvɔ:/ *noun* animal or plant that feeds on insects. □ **insectivorous** /-'tɪvərəs/ *adjective*.

insecure /ɪnsɪ'kjʊə/ *adjective* not secure or safe; not feeling safe. □ **insecurity** /-'kjʊr-/ *noun*.

inseminate /ɪn'semɪneɪt/ *verb* (**-ting**) introduce semen into; sow (seed etc.). □ **insemination** *noun*.

insensate /ɪn'senseɪt/ *adjective* without esp. physical sensibility; stupid.

insensible /ɪn'sensɪb(ə)l/ *adjective* unconscious; unaware; callous; imperceptible. □ **insensibility** *noun*; **insensibly** *adverb*.

insensitive /ɪn'sensɪtɪv/ *adjective* not sensitive. □ **insensitively** *adverb*; **insensitiveness** *noun*; **insensitivity** /-'tɪv-/ *noun*.

insentient /ɪn'senʃ(ə)nt/ *adjective* inanimate.

inseparable /ɪn'sepərəb(ə)l/ *adjective* that cannot be separated. □ **inseparability** *noun*; **inseparably** *adverb*.

insert ● *verb* /ɪn'sɜ:t/ place or put (thing into another). ● *noun* /'ɪnsɜ:t/ thing inserted.

insertion /ɪn'sɜ:ʃ(ə)n/ *noun* inserting; thing inserted.

inset ● *noun* /'ɪnset/ extra piece inserted in book, garment, etc.; small map etc. within border of larger. ● *verb* /ɪn'set/ (**insetting**; *past & past participle* **inset** or **insetted**) put in as inset; decorate with inset.

inshore /ɪn'ʃɔ:/ *adverb & adjective* at sea but close to shore.

inside ● *noun* /ɪn'saɪd/ inner side or part; interior; side of path away from road; (usually in *plural*) *colloquial* stomach and bowels. ● *adjective* /'ɪnsaɪd/ of, on, or in the inside; nearer to centre of games field. ● *adverb* /ɪn'saɪd/ on, in, or to the inside; *slang* in prison. ● *preposition* /ɪn'saɪd/ within, on the inside of; in less than. □ **inside out** with inner side turned outwards; **know inside out** know thoroughly.

insider /ɪn'saɪdə/ *noun* person within organization etc.; person privy to secret.

insidious /ɪn'sɪdɪəs/ *adjective* proceeding inconspicuously but harmfully. □ **insidiously** *adverb*.

insight *noun* capacity for understanding hidden truths etc.; instance of this.

insignia /ɪn'sɪgnɪə/ *plural noun* badges or marks of office etc.

insignificant /ɪnsɪg'nɪfɪkənt/ *adjective* unimportant; trivial. □ **insignificance** *noun*.

insincere /ɪnsɪn'sɪə/ *adjective* not sincere. □ **insincerely** *adverb*; **insincerity** /-'ser-/ *noun*.

insinuate /ɪn'sɪnjʊeɪt/ *verb* (**-ting**) hint obliquely; (usually + *into*) introduce subtly or deviously. □ **insinuation** *noun*.

insipid /ɪn'sɪpɪd/ *adjective* dull, lifeless; flavourless. □ **insipidity** /-'pɪd-/ *noun*; **insipidly** *adverb*.

insist /ɪn'sɪst/ *verb* demand or maintain emphatically. □ **insistence** *noun*; **insistent** *adjective*; **insistently** *adverb*.

in situ /ɪn 'sɪtju:/ *adverb* in its original place. [Latin]

insobriety /ɪnsə'braɪətɪ/ *noun* intemperance, esp. in drinking.

insole *noun* removable inner sole for use in shoe.

insolent /'ɪnsələnt/ *adjective* impertinently insulting. □ **insolence** *noun*; **insolently** *adverb*.

insoluble /ɪn'sɒljʊb(ə)l/ *adjective* that cannot be solved or dissolved. □ **insolubility** *noun*; **insolubly** *adverb*.

insolvent /ɪn'sɒlv(ə)nt/ ● *adjective* unable to pay debts. ● *noun* insolvent debtor. □ **insolvency** *noun*.

insomnia /ɪn'sɒmnɪə/ *noun* sleeplessness.

insomniac /ɪn'sɒmnɪæk/ *noun* person suffering from insomnia.

insouciant /ɪn'su:sɪənt/ *adjective* carefree, unconcerned. □ **insouciance** *noun*.

inspect /ɪn'spekt/ *verb* look closely at; examine officially. □ **inspection** *noun*.

inspector *noun* official employed to inspect or supervise; police officer next above sergeant in rank. □ **inspectorate** *noun*.

inspiration /ɪnspə'reɪʃ(ə)n/ *noun* creative force or influence; person etc. stimulating creativity etc.; sudden brilliant idea; divine influence, esp. on writing of Scripture.

inspire /ɪn'spaɪə/ *verb* (**-ring**) stimulate (person) to esp. creative activity; animate; instil thought or feeling into; prompt, give rise to; (as **inspired** *adjective*) characterized by inspiration. □ **inspiring** *adjective*.

inspirit /ɪn'spɪrɪt/ *verb* (**-t-**) put life into, animate; encourage.

inst. *abbreviation* instant, of current month.

instability /ɪnstə'bɪlɪtɪ/ *noun* lack of stability.

install /ɪn'stɔ:l/ *verb* place (equipment etc.) in position ready for use; place (person) in office with ceremony. □ **installation** /-stə'leɪ-/ *noun*.

instalment *noun* (*US* **installment**) any of several usually equal payments for something; any of several parts, esp. of broadcast or published story.

instance /'ɪnst(ə)ns/ ● *noun* example; particular case. ● *verb* (**-cing**) cite as instance.

instant /'ɪnst(ə)nt/ ● *adjective* occurring immediately; (of food etc.) processed for quick preparation; urgent, pressing; of current month. ● *noun* precise moment; short space of time.

instantaneous /ɪnstən'teɪmɪəs/ *adjective* occurring or done in an instant. □ **instantaneously** *adverb*.

instantly *adverb* immediately.

instead /ɪn'sted/ *adverb* (+ *of*) in place of; as substitute or alternative.

instep /'ɪnstep/ *noun* inner arch of foot between toes and ankle; part of shoe etc. fitting this.

instigate /'ɪnstɪgeɪt/ *verb* (**-ting**) bring about by persuasion; incite. □ **instigation** *noun*; **instigator** *noun*.

instil /ɪn'stɪl/ *verb* (*US* **instill**) (**-ll-**) (often + *into*) put (ideas etc. into mind etc.) gradually; put in by drops. □ **instillation** *noun*; **instilment** *noun*.

instinct /'ɪnstɪŋkt/ *noun* inborn pattern of behaviour; innate impulse; intuition. □ **instinctive** /-'stɪŋktɪv/ *adjective*; **instinctively** /-'stɪŋktɪvlɪ/ *adverb*; **instinctual** /-'stɪŋktjʊəl/ *adjective*.

institute /'ɪnstɪtju:t/ ● *noun* organized body for promotion of science, education, etc. ● *verb* (**-ting**) establish; initiate (inquiry etc.); (usually + *to*, *into*) appoint (person) as cleric in church etc.

institution /ɪnstɪ'tju:ʃ(ə)n/ *noun* (esp. charitable) organization or society; established law or custom; *colloquial* well-known person; instituting, being instituted.

institutional *adjective* of or like an institution; typical of institutions.

institutionalize *verb* (also **-ise**) (**-zing** or **-sing**) (as **institutionalized** *adjective*) made dependent by long period in institution; place or keep in institution; make institutional.

instruct /ɪn'strʌkt/ *verb* teach; (usually + *to do*) direct, command; employ (lawyer); inform. □ **instructor** *noun*.

instruction /ɪn'strʌkʃ(ə)n/ *noun* (often in *plural*) order, direction (as to how thing works etc.); teaching. □ **instructional** *adjective*.

instructive /ɪn'strʌktɪv/ *adjective* tending to instruct; enlightening.

instrument /'ɪnstrəmənt/ *noun* tool, implement; (in full **musical instrument**) contrivance for producing musical sounds; thing used in performing action; person made use of; measuring device, esp. in aircraft; formal (esp. legal) document.

instrumental /ɪnstrə'ment(ə)l/ *adjective* serving as instrument or means; (of music) performed on instruments.

instrumentalist *noun* performer on musical instrument.

instrumentality /ɪnstrəmen'tælɪtɪ/ *noun* agency, means.

instrumentation /ɪnstrəmen'teɪʃ(ə)n/ *noun* provision or use of instruments; arrangement of music for instruments; particular instruments used in piece.

insubordinate /ɪnsə'bɔ:dɪnət/ *adjective* disobedient; unruly. □ **insubordination** *noun.*

insubstantial /ɪnsəb'stænʃ(ə)l/ *adjective* lacking solidity or substance; not real.

insufferable /ɪn'sʌfərəb(ə)l/ *adjective* unbearable; unbearably conceited etc. □ **insufferably** *adverb.*

insufficient /ɪnsə'fɪʃ(ə)nt/ *adjective* not enough, inadequate. □ **insufficiency** *noun*; **insufficiently** *adverb.*

insular /'ɪnsjʊlə/ *adjective* of or like an island; separated, remote; narrow-minded. □ **insularity** /-'lær-/ *noun.*

insulate /'ɪnsjʊleɪt/ *verb* (**-ting**) isolate, esp. by non-conductor of electricity, heat, sound, etc. □ **insulation** *noun*; **insulator** *noun.*

insulin /'ɪnsjʊlɪn/ *noun* hormone regulating the amount of glucose in the blood, the lack of which causes diabetes.

insult ● *verb* /ɪn'sʌlt/ abuse scornfully; offend self-respect etc. of. ● *noun* /'ɪnsʌlt/ insulting remark or action. □ **insulting** *adjective*; **insultingly** *adverb.*

insuperable /ɪn'su:pərəb(ə)l/ *adjective* impossible to surmount; impossible to overcome. □ **insuperability** *noun*; **insuperably** *adverb.*

insupportable /ɪnsə'pɔ:təb(ə)l/ *adjective* unbearable; unjustifiable.

insurance /ɪn'ʃʊərəns/ *noun* procedure or contract securing compensation for loss, damage, injury, or death on payment of premium; sum paid to effect insurance.

insure /ɪn'ʃʊə/ *verb* (**-ring**) (often + *against*) effect insurance with respect to.

insurgent /ɪn'sɜ:dʒ(ə)nt/ ● *adjective* in revolt; rebellious. ● *noun* rebel. □ **insurgence** *noun.*

insurmountable /ɪnsə'maʊntəb(ə)l/ *adjective* insuperable.

insurrection /ɪnsə'rekʃ(ə)n/ *noun* rising in resistance to authority; incipient rebellion. □ **insurrectionist** *noun.*

intact /ɪn'tækt/ *adjective* unimpaired; entire; untouched.

intaglio /ɪn'tɑ:lɪəʊ/ *noun* (*plural* **-s**) gem with incised design; engraved design.

intake *noun* action of taking in; people, things, or quantity taken in; place where water is taken into pipe, or fuel or air into engine.

intangible /ɪn'tændʒɪb(ə)l/ *adjective* that cannot be touched or mentally grasped. □ **intangibility** *noun*; **intangibly** *adverb.*

integer /'ɪntɪdʒə/ *noun* whole number.

integral /'ɪntɪgr(ə)l/ *adjective* of or essential to a whole; complete; of or denoted by an integer.

■ **Usage** *Integral* is often pronounced /ɪn'tegr(ə)l/ (with the stress on the *-teg-*), but this is considered incorrect by some people.

integrate /'ɪntɪgreɪt/ *verb* (**-ting**) combine (parts) into whole; complete by adding parts; bring or come into equal membership of society; desegregate (school etc.), esp. racially. □ **integrated circuit** small piece of material replacing electrical circuit of many components. □ **integration** *noun.*

integrity /ɪn'tegrɪtɪ/ *noun* honesty; wholeness; soundness.

integument /ɪn'tegjʊmənt/ *noun* skin, husk, or other (natural) covering.

intellect /'ɪntəlekt/ *noun* faculty of knowing and reasoning; understanding.

intellectual /ɪntə'lektʃʊəl/ ● *adjective* of, requiring, or using intellect; having highly developed intellect. ● *noun* intellectual person. □ **intellectualize** *verb* (also **-ise**) (**-zing** or **-sing**); **intellectually** *adverb.*

intelligence /ɪn'telɪdʒ(ə)ns/ *noun* intellect; quickness of understanding; collecting of information, esp. secretly for military or political purposes; information so collected; people employed in this. □ **intelligence quotient** number denoting ratio of person's intelligence to the average.

intelligent *adjective* having or showing good intelligence, clever. □ **intelligently** *adverb.*

intelligentsia /ɪntelɪ'dʒentsɪə/ *noun* class of intellectuals regarded as cultured and politically enterprising.

intelligible /ɪn'telɪdʒɪb(ə)l/ *adjective* that can be understood. □ **intelligibility** *noun*; **intelligibly** *adverb*.

intemperate /ɪn'tempərət/ *adjective* immoderate; excessive in consumption of alcohol, or in general indulgence of appetite. □ **intemperance** *noun*.

intend /ɪn'tend/ *verb* have as one's purpose; (usually + *for*, *as*, *to do*) design, destine.

intended • *adjective* done on purpose. • *noun colloquial* fiancé(e).

intense /ɪn'tens/ *adjective* (**-r**, **-st**) existing in high degree; vehement; violent, forceful; extreme; very emotional. □ **intensity** *noun* (*plural* **-ies**); **intensely** *adverb*.

■ **Usage** *Intense* is sometimes confused with *intensive*, and wrongly used to describe a course of study etc.

intensify *verb* (**-ies**, **-ied**) make or become (more) intense. □ **intensification** *noun*.

intensive /ɪn'tensɪv/ *adjective* thorough, vigorous; concentrated; of or relating to intensity; increasing production relative to cost. □ **-intensive** making much use of; **intensive care** medical treatment with constant supervision of dangerously ill patient. □ **intensively** *adverb*.

intent /ɪn'tent/ • *noun* intention; purpose. • *adjective* (usually + *on*) resolved, bent; attentively occupied; eager. □ **to all intents and purposes** practically. □ **intently** *adverb*.

intention /ɪn'tenʃ(ə)n/ *noun* purpose, aim; intending.

intentional *adjective* done on purpose. □ **intentionally** *adverb*.

inter /ɪn'tɜ:/ *verb* (**-rr-**) bury (corpse etc.).

inter- *combining form* among, between; mutually, reciprocally.

interact /ɪntə'rækt/ *verb* act on each other. □ **interaction** *noun*.

interactive *adjective* reciprocally active; (of computer etc.) allowing two-way flow of information between itself and user. □ **interactively** *adverb*.

interbreed /ɪntə'bri:d/ *verb* (*past & past participle* **-bred**) (cause to) produce hybrid individual.

intercalary /ɪn'tɜ:kələrɪ/ *adjective* inserted to harmonize calendar with solar year; having such addition; interpolated.

intercede /ɪntə'si:d/ *verb* (**-ding**) intervene on behalf of another; plead.

intercept /ɪntə'sept/ *verb* seize, catch, stop, etc. in transit; cut off. □ **interception** *noun*; **interceptor** *noun*.

intercession /ɪntə'seʃ(ə)n/ *noun* interceding. □ **intercessor** *noun*.

interchange • *verb* /ɪntə'tʃeɪndʒ/ (**-ging**) (of two people) exchange (things) with each other; make exchange of (two things); alternate. • *noun* /'ɪntətʃeɪndʒ/ reciprocal exchange; alternation; road junction where traffic streams do not cross. □ **interchangeable** *adjective*.

inter-city /ɪntə'sɪtɪ/ *adjective* existing or travelling between cities.

intercom /'ɪntəkɒm/ *noun colloquial* system of intercommunication by telephone or radio.

intercommunicate /ɪntəkə'mju:nɪkeɪt/ *verb* (**-ting**) have communication with each other; (of rooms etc.) open into each other. □ **intercommunication** *noun*.

intercommunion /ɪntəkə'mju:nɪən/ *noun* mutual communion, esp. between religious bodies.

interconnect /ɪntəkə'nekt/ *verb* connect with each other. □ **interconnection** *noun*.

intercontinental /ɪntəkɒntɪ'nent(ə)l/ *adjective* connecting or travelling between continents.

intercourse /'ɪntəkɔ:s/ *noun* social, international, etc. communication or dealings; sexual intercourse.

interdenominational /ɪntədɪnɒmɪ'neɪʃən(ə)l/ *adjective* of or involving more than one Christian denomination.

interdependent /ɪntədɪ'pend(ə)nt/ *adjective* mutually dependent. □ **interdependence** *noun*.

interdict • *noun* /'ɪntədɪkt/ formal prohibition; *RC Church* sentence debarring person, or esp. place, from ecclesiastical functions and privileges. • *verb* /ɪntə'dɪkt/ prohibit (action); forbid use of; (usually + *from*) restrain (person).

□ **interdiction** /-'dɪk-/ *noun*; **interdictory** /-'dɪk-/ *adjective*.

interdisciplinary /ɪntədɪsɪ'plɪnərɪ/ *adjective* of or involving different branches of learning.

interest /'ɪntrəst/ ● *noun* concern, curiosity; quality causing this; subject, hobby, etc., towards which one feels it; advantage; money paid for use of money borrowed etc.; thing in which one has stake or concern; financial stake; legal concern, title, or right. ● *verb* arouse interest of; (usually + *in*) cause to take interest; (as **interested** *adjective*) having private interest, not impartial.

interesting *adjective* causing curiosity; holding the attention. □ **interestingly** *adverb*.

interface /'ɪntəfeɪs/ ● *noun* surface forming common boundary of two regions; place where interaction occurs between two systems etc.; apparatus for connecting two pieces of esp. computing equipment so they can be operated jointly. ● *verb* (**-cing**) connect by means of interface; interact.

interfere /ɪntə'fɪə/ *verb* (**-ring**) (often + *with*) meddle; be an obstacle; intervene; (+ *with*) molest sexually.

interference *noun* interfering; fading of received radio signals.

interferon /ɪntə'fɪərɒn/ *noun* protein inhibiting development of virus in cell.

interfuse /ɪntə'fju:z/ *verb* (**-sing**) mix, blend. □ **interfusion** *noun*.

interim /'ɪntərɪm/ ● *noun* intervening time. ● *adjective* provisional, temporary.

interior /ɪn'tɪərɪə/ ● *adjective* inner; inland; internal, domestic. ● *noun* inner part; inside; inland region; home affairs of country; representation of inside of room etc.

interject /ɪntə'dʒekt/ *verb* make (remark etc.) abruptly or parenthetically; interrupt.

interjection /ɪntə'dʒekʃ(ə)n/ *noun* exclamation.

interlace /ɪntə'leɪs/ *verb* (**-cing**) bind intricately together; interweave.

interlard /ɪntə'lɑ:d/ *verb* mix (speech etc.) with unusual words or phrases.

interleave /ɪntə'li:v/ *verb* (**-ving**) insert (usually blank) leaves between leaves of (book).

interline /ɪntə'laɪn/ *verb* (**-ning**) put extra layer of material between fabric of (garment) and its lining.

interlink /ɪntə'lɪŋk/ *verb* link together.

interlock /ɪntə'lɒk/ ● *verb* engage with each other by overlapping etc.; lock together. ● *noun* machine-knitted fabric with fine stitches.

interlocutor /ɪntə'lɒkjʊtə/ *noun formal* person who takes part in conversation. □ **interlocutory** *adjective*.

interloper /'ɪntələʊpə/ *noun* intruder; person who thrusts himself or herself into others' affairs.

interlude /'ɪntəlu:d/ *noun* interval between parts of play etc., performance filling this; contrasting time, event, etc. in middle of something.

intermarry /ɪntə'mærɪ/ *verb* (**-ies**, **-ied**) (+ *with*) (of races, castes, families, etc.) become connected by marriage. □ **intermarriage** /-rɪdʒ/ *noun*.

intermediary /ɪntə'mi:dɪərɪ/ ● *noun* (*plural* **-ies**) mediator. ● *adjective* acting as mediator; intermediate.

intermediate /ɪntə'mi:dɪət/ ● *adjective* coming between in time, place, order, etc. ● *noun* intermediate thing.

interment /ɪn'tɜ:mənt/ *noun* burial.

intermezzo /ɪntə'metsəʊ/ *noun* (*plural* **-mezzi** /-sɪ/ or **-s**) *Music* short connecting movement or composition.

interminable /ɪn'tɜ:mɪnəb(ə)l/ *adjective* endless; tediously long. □ **interminably** *adverb*.

intermingle /ɪntə'mɪŋg(ə)l/ *verb* (**-ling**) mix together, mingle.

intermission /ɪntə'mɪʃ(ə)n/ *noun* pause, cessation; interval in cinema etc.

intermittent /ɪntə'mɪt(ə)nt/ *adjective* occurring at intervals, not continuous or steady. □ **intermittently** *adverb*.

intermix /ɪntə'mɪks/ *verb* mix together.

intern ● *noun* /'ɪntɜ:n/ *US* resident junior doctor in hospital. ● *verb* /ɪn'tɜ:n/ confine within prescribed limits. □ **internee** /-'ni:/ *noun*; **internment** *noun*.

internal /ɪn'tɜ:n(ə)l/ *adjective* of or in the inside of thing; relating to inside of the body; of domestic affairs of country; (of students) attending a university as well as taking its exams; used or applying within an organization; intrinsic; of mind or soul. □ **internal-combustion engine** engine in which motive power comes from explosion of gas or vapour with air in cylinder. □ **internally** *adverb.*

international /ɪntəˈnæʃən(ə)l/ ● *adjective* existing or carried on between nations; agreed on by many nations. ● *noun* contest (usually in sports) between representatives of different nations; such representative; (**International**) any of 4 successive associations for socialist or Communist action. □ **internationality** /-ˈnæl-/ *noun*; **internationally** *adverb.*

internationalism *noun* advocacy of community of interests among nations. □ **internationalist** *noun.*

internationalize *verb* (also **-ise**) (**-zing** or **-sing**) make international; bring under joint protection etc. of different nations.

internecine /ɪntəˈni:saɪn/ *adjective* mutually destructive.

interpenetrate /ɪntəˈpenɪtreɪt/ *verb* (**-ting**) penetrate each other; pervade. □ **interpenetration** *noun.*

interpersonal /ɪntəˈpɜ:sən(ə)l/ *adjective* between people.

interplanetary /ɪntəˈplænɪtərɪ/ *adjective* between planets.

interplay /ˈɪntəpleɪ/ *noun* reciprocal action.

Interpol /ˈɪntəpɒl/ *noun* International Criminal Police Organization.

interpolate /ɪnˈtɜ:pəleɪt/ *verb* (**-ting**) insert or introduce between other things; make (esp. misleading) insertions in. □ **interpolation** *noun.*

interpose /ɪntəˈpəʊz/ *verb* (**-sing**) insert (thing between others); introduce, use, say, etc. as interruption or interference; interrupt; advance (objection etc.) so as to interfere; intervene. □ **interposition** /-pəˈzɪʃ(ə)n/ *noun.*

interpret /ɪnˈtɜ:prɪt/ *verb* (**-t-**) explain the meaning of (esp. words); render, represent; act as interpreter. □ **interpretation** *noun.*

interpreter *noun* person who translates orally.

interracial /ɪntəˈreɪʃ(ə)l/ *adjective* between or affecting different races.

interregnum /ɪntəˈregnəm/ *noun* (*plural* **-s**) interval with suspension of normal government between successive reigns or regimes; interval, pause.

interrelated /ɪntərɪˈleɪtɪd/ *adjective* related to each other. □ **interrelation** *noun*; **interrelationship** *noun.*

interrogate /ɪnˈterəgeɪt/ *verb* (**-ting**) question closely or formally. □ **interrogation** *noun*; **interrogator** *noun.*

interrogative /ɪntəˈrɒgətɪv/ ● *adjective* of, like, or used in questions. ● *noun* interrogative word.

interrogatory /ɪntəˈrɒgətərɪ/ ● *adjective* questioning. ● *noun* (*plural* **-ies**) set of questions.

interrupt /ɪntəˈrʌpt/ *verb* break continuity of (action, speech, etc.); obstruct (view etc.). □ **interruption** *noun.*

intersect /ɪntəˈsekt/ *verb* divide by passing or lying across; cross or cut each other.

intersection /ɪntəˈsekʃ(ə)n/ *noun* intersecting; place where two roads intersect; point or line common to lines or planes that intersect.

intersperse /ɪntəˈspɜ:s/ *verb* (**-sing**) (usually + *between*, *among*) scatter; (+ *with*) vary (thing) by scattering others among it.

interstate /ˈɪntəsteɪt/ *adjective* existing etc. between states, esp. of US.

interstellar /ɪntəˈstelə/ *adjective* between stars.

interstice /ɪnˈtɜ:stɪs/ *noun* gap, chink, crevice.

interstitial /ɪntəˈstɪʃ(ə)l/ *adjective* forming or in interstices.

intertwine /ɪntəˈtwaɪn/ *verb* (**-ning**) (often + *with*) twine closely together.

interval /ˈɪntəv(ə)l/ *noun* intervening time or space; pause; break; *Music* difference of pitch between two sounds. □ **at intervals** here and there, now and then.

intervene /ɪntəˈvi:n/ *verb* (**-ning**) occur in meantime; interfere; prevent or modify events; come between people or things; mediate.

intervention /ɪntəˈvenʃ(ə)n/ *noun* intervening; interference; mediation.

interview /ˈɪntəvju:/ ● *noun* oral examination of applicant; conversation with reporter, for broadcast or publication; meeting of people, esp. for discussion.

● *verb* hold interview with. □ **interviewee** /-vju:'i:/ *noun*; **interviewer** *noun*.

interweave /ɪntə'wi:v/ *verb* (**-ving**; *past* **-wove**; *past participle* **-woven**) weave together; blend intimately.

intestate /ɪn'testeɪt/ ● *adjective* not having made a will before death. ● *noun* person who has died intestate. □ **intestacy** /-təsɪ/ *noun*.

intestine /ɪn'testɪn/ *noun* (in *singular* or *plural*) lower part of alimentary canal. □ **intestinal** *adjective*.

intimate[1] ● *adjective* /'ɪntɪmət/ closely acquainted; familiar; closely personal; (usually + *with*) having sexual relations; (of knowledge) thorough; close. ● *noun* intimate friend. □ **intimacy** /-məsɪ/ *noun*; **intimately** *adverb*.

intimate[2] /'ɪntɪmeɪt/ *verb* (**-ting**) state or make known; imply. □ **intimation** *noun*.

intimidate /ɪn'tɪmɪdeɪt/ *verb* (**-ting**) frighten, esp. in order to influence conduct. □ **intimidation** *noun*.

into /'ɪntʊ, 'ɪntə/ *preposition expressing motion or direction to point within, direction of attention, or change of state*; after the beginning of; *colloquial* interested in.

intolerable /ɪn'tɒlərəb(ə)l/ *adjective* that cannot be endured. □ **intolerably** *adverb*.

intolerant /ɪn'tɒlərənt/ *adjective* not tolerant. □ **intolerance** *noun*.

intonation /ɪntə'neɪʃ(ə)n/ *noun* modulation of voice, accent; intoning.

intone /ɪn'təʊn/ *verb* (**-ning**) recite with prolonged sounds, esp. in monotone.

in toto /ɪn 'təʊtəʊ/ *adverb* entirely. [Latin]

intoxicant /ɪn'tɒksɪkənt/ ● *adjective* intoxicating. ● *noun* intoxicating substance.

intoxicate /ɪn'tɒksɪkeɪt/ *verb* (**-ting**) make drunk; excite or elate beyond self-control. □ **intoxication** *noun*.

intractable /ɪn'træktəb(ə)l/ *adjective* not easily dealt with; stubborn. □ **intractability** *noun*.

intramural /ɪntrə'mjʊər(ə)l/ *adjective* situated or done within walls of institution etc.

intransigent /ɪn'trænsɪdʒ(ə)nt/ ● *adjective* uncompromising. ● *noun* such person. □ **intransigence** *noun*.

intransitive /ɪn'trænsɪtɪv/ *adjective* (of verb) not taking direct object.

intrauterine /ɪntrə'ju:təraɪn/ *adjective* within the womb.

intravenous /ɪntrə'vi:nəs/ *adjective* in(to) vein(s). □ **intravenously** *adverb*.

intrepid /ɪn'trepɪd/ *adjective* fearless; brave.

intricate /'ɪntrɪkət/ *adjective* complicated; perplexingly detailed. □ **intricacy** /-kəsɪ/ *noun* (*plural* **-ies**); **intricately** *adverb*.

intrigue ● *verb* /ɪn'tri:g/ (**-gues**, **-gued**, **-guing**) carry on underhand plot; use secret influence; rouse curiosity of. ● *noun* /'ɪntri:g/ underhand plotting or plot; secret arrangement, esp. with romantic associations. □ **intriguing** *adjective*; **intriguingly** *adverb*.

intrinsic /ɪn'trɪnzɪk/ *adjective* inherent; essential. □ **intrinsically** *adverb*.

intro /'ɪntrəʊ/ *noun* (*plural* **-s**) *colloquial* introduction.

introduce /ɪntrə'dju:s/ *verb* (**-cing**) make (person) known by name to another; announce or present to audience; bring (custom etc.) into use; bring (bill etc.) before Parliament; (+ *to*) initiate (person) in subject; insert; bring in; usher in, bring forward. □ **introducible** *adjective*.

introduction /ɪntrə'dʌkʃ(ə)n/ *noun* introducing, being introduced; formal presentation; preliminary matter in book; introductory treatise. □ **introductory** *adjective*.

introspection /ɪntrə'spekʃ(ə)n/ *noun* examination of one's own thoughts. □ **introspective** *adjective*.

introvert ● *noun* person chiefly concerned with his or her own thoughts; shy thoughtful person. ● *adjective* (also **introverted**) characteristic of an introvert. □ **introversion** /-'vɜ:ʃ(ə)n/ *noun*.

intrude /ɪn'tru:d/ *verb* (**-ding**) (+ *on*, *upon*, *into*) come uninvited or unwanted; force on a person. □ **intruder** *noun*; **intrusion** /-ʒ(ə)n/ *noun*; **intrusive** /-sɪv/ *adjective*.

intuition /ɪntju:'ɪʃ(ə)n/ *noun* immediate apprehension by mind without reasoning; immediate insight. □ **intuit** /ɪn'tju:ɪt/ *verb*; **intuitional** *adjective*.

intuitive /ɪn'tju:ɪtɪv/ *adjective* of, having, or perceived by intuition. □ **intuitively** *adverb*; **intuitiveness** *noun*.

Inuit /'ɪnju:ɪt/ • *noun* (*plural* same or **-s**) N. American Eskimo; language of Inuit. • *adjective* of Inuit or their language.

inundate /'ɪnʌndeɪt/ *verb* (**-ting**) (often + *with*) flood, overwhelm. □ **inundation** *noun*.

inure /ɪ'njʊə/ *verb* (**-ring**) habituate, accustom. □ **inurement** *noun*.

invade /ɪn'veɪd/ *verb* (**-ding**) enter (country etc.) with arms to control or subdue it; swarm into; (of disease etc.) attack; encroach on. □ **invader** *noun*.

invalid[1] /'ɪnvəlɪd, -li:d/ • *noun* person enfeebled or disabled by illness or injury. • *adjective* of or for invalids; sick, disabled. • *verb* (**-d-**) (often + *out* etc.) remove from active service; disable (person) by illness. □ **invalidism** *noun*; **invalidity** /-'lɪd-/ *noun*.

invalid[2] /ɪn'vælɪd/ *adjective* not valid. □ **invalidity** /-və'lɪd-/ *noun*.

invalidate /ɪn'vælɪdeɪt/ *verb* (**-ting**) make invalid. □ **invalidation** *noun*.

invaluable /ɪn'væljʊəb(ə)l/ *adjective* beyond price, very valuable.

invariable /ɪn'veərɪəb(ə)l/ *adjective* unchangeable; always the same. □ **invariably** *adverb*.

invasion /ɪn'veɪʒ(ə)n/ *noun* invading, being invaded. □ **invasive** /-sɪv/ *adjective*.

invective /ɪn'vektɪv/ *noun* violent attack in words.

inveigh /ɪn'veɪ/ *verb* (+ *against*) speak or write with strong hostility against.

inveigle /ɪn'veɪg(ə)l/ *verb* (**-ling**) (+ *into*, *to do*) entice, persuade by guile. □ **inveiglement** *noun*.

invent /ɪn'vent/ *verb* create by thought; originate; fabricate. □ **inventor** *noun*.

invention /ɪn'venʃ(ə)n/ *noun* inventing, being invented; thing invented; inventiveness.

inventive *adjective* able to invent; imaginative. □ **inventively** *adverb*; **inventiveness** *noun*.

inventory /'ɪnvəntərɪ/ • *noun* (*plural* **-ies**) list of goods etc. • *verb* (**-ies**, **-ied**) make inventory of; enter in inventory.

inverse /ɪn'vɜ:s/ • *adjective* inverted in position, order, or relation. • *noun* inverted state; (often + *of*) direct opposite. □ **inverse proportion**, **ratio** relation between two quantities such that one increases in proportion as the other decreases.

inversion /ɪn'vɜ:ʃ(ə)n/ *noun* inverting, esp. reversal of normal order of words.

invert /ɪn'vɜ:t/ *verb* turn upside down; reverse position, order, or relation of. □ **inverted commas** quotation marks.

invertebrate /ɪn'vɜ:tɪbrət/ • *adjective* without backbone. • *noun* invertebrate animal.

invest /ɪn'vest/ *verb* (often + *in*) apply or use (money) for profit; devote (time etc.) to an enterprise; (+ *in*) buy (something useful or otherwise rewarding); (+ *with*) endue with qualities etc.; (often + *with*, *in*) clothe with insignia of office. □ **investor** *noun*.

investigate /ɪn'vestɪgeɪt/ *verb* (**-ting**) inquire into, examine. □ **investigation** *noun*; **investigative** /-gətɪv/ *adjective*; **investigator** *noun*.

investiture /ɪn'vestɪtʃə/ *noun* formal investing of person with honours etc.

investment *noun* investing; money invested; property etc. in which money is invested.

inveterate /ɪn'vetərət/ *adjective* (of person) confirmed in (usually undesirable) habit etc.; (of habit etc.) long-established. □ **inveteracy** *noun*.

invidious /ɪn'vɪdɪəs/ *adjective* likely to excite ill-will against performer, possessor, etc.

invigilate /ɪn'vɪdʒɪleɪt/ *verb* (**-ting**) supervise examinees. □ **invigilation** *noun*; **invigilator** *noun*.

invigorate /ɪn'vɪgəreɪt/ *verb* (**-ting**) give vigour to. □ **invigorating** *adjective*.

invincible /ɪn'vɪnsɪb(ə)l/ *adjective* unconquerable. □ **invincibility** *noun*; **invincibly** *adverb*.

inviolable /ɪn'vaɪələb(ə)l/ *adjective* not to be violated. □ **inviolability** *noun*.

inviolate /ɪn'vaɪələt/ *adjective* not violated; safe (from harm). □ **inviolacy** *noun*.

invisible /ɪn'vɪzɪb(ə)l/ *adjective* that cannot be seen. □ **invisible exports**, **imports** items for which payment is made by or to another country but which are not goods. □ **invisibility** *noun*; **invisibly** *adverb*.

invite ● *verb* /ɪn'vaɪt/ (**-ting**) request courteously to come, to do, etc.; solicit courteously; tend to evoke unintentionally; attract. ● *noun* /'ɪnvaɪt/ *colloquial* invitation. □ **invitation** /ɪnvɪ'teɪʃ(ə)n/ *noun*.

inviting *adjective* attractive. □ **invitingly** *adverb*.

in vitro /ɪn 'vi:trəʊ/ *adverb* (of biological processes) taking place in test-tube or other laboratory environment. [Latin]

invocation /ɪnvə'keɪʃ(ə)n/ *noun* invoking; calling on, esp. in prayer or for inspiration etc. □ **invocatory** /ɪn'vɒkətərɪ/ *adjective*.

invoice /'ɪnvɔɪs/ ● *noun* bill for usually itemized goods etc. ● *verb* (**-cing**) send invoice to; make invoice of.

invoke /ɪn'vəʊk/ *verb* (**-king**) call on in prayer or as witness; appeal to (law, authority, etc.); summon (spirit) by charms; ask earnestly for (vengeance, justice, etc.).

involuntary /ɪn'vɒləntərɪ/ *adjective* done etc. without exercise of will; not controlled by will. □ **involuntarily** *adverb*.

involute /'ɪnvəlu:t/ *adjective* intricate; curled spirally.

involution /ɪnvə'lu:ʃ(ə)n/ *noun* involving; intricacy; curling inwards, part so curled.

involve /ɪn'vɒlv/ *verb* (**-ving**) (often + *in*) cause (person, thing) to share experience or effect; imply, make necessary; (often + *in*) implicate (person) in charge, crime, etc.; include or affect in its operation; (as **involved** *adjective*) complicated. □ **involvement** *noun*.

invulnerable /ɪn'vʌlnərəb(ə)l/ *adjective* that cannot be wounded. □ **invulnerability** *noun*.

inward /'ɪnwəd/ ● *adjective* directed towards inside; going in; situated within; mental, spiritual. ● *adverb* (also **inwards**) towards inside; in mind or soul.

inwardly *adverb* on the inside; in mind or spirit; not aloud.

inwrought /ɪn'rɔ:t/ *adjective* (often + *with*) decorated (with pattern); (often + *in*, *on*) (of pattern) wrought (in or on fabric).

iodine /'aɪədi:n/ *noun* black solid halogen element forming violet vapour; solution of this used as antiseptic.

IOM *abbreviation* Isle of Man.

ion /'aɪən/ *noun* atom or group of atoms that has lost or gained one or more electrons.

ionic /aɪ'ɒnɪk/ *adjective* of or using ions.

ionize *verb* (also **-ise**) (**-zing** or **-sing**) convert or be converted into ion(s). □ **ionization** *noun*.

ionosphere /aɪ'ɒnəsfɪə/ *noun* ionized region in upper atmosphere. □ **ionospheric** /-'sfer-/ *adjective*.

iota /aɪ'əʊtə/ *noun* ninth letter of Greek alphabet (Ι, ι); (usually with negative) a jot.

IOU /aɪəʊ'ju:/ *noun* (*plural* **-s**) signed document acknowledging debt.

IOW *abbreviation* Isle of Wight.

IPA *abbreviation* International Phonetic Alphabet.

ipecacuanha /ɪpɪkækjʊ'ɑ:nə/ *noun* root of S. American plant used as emetic etc.

ipso facto /ɪpsəʊ 'fæktəʊ/ *adverb* by that very fact. [Latin]

IQ *abbreviation* intelligence quotient.

IRA *abbreviation* Irish Republican Army.

Iranian /ɪ'reɪnɪən/ ● *adjective* of Iran (formerly Persia); of group of languages including Persian. ● *noun* native or national of Iran.

Iraqi /ɪ'rɑ:kɪ/ ● *adjective* of Iraq. ● *noun* (*plural* **-s**) native or national of Iraq.

irascible /ɪ'ræsɪb(ə)l/ *adjective* irritable; hot-tempered. □ **irascibility** *noun*.

irate /aɪ'reɪt/ *adjective* angry, enraged.

ire /'aɪə/ *noun literary* anger.

iridescent /ɪrɪ'des(ə)nt/ *adjective* showing rainbow-like glowing colours; changing colour with position. □ **iridescence** *noun*.

iris /'aɪərɪs/ *noun* circular coloured membrane surrounding pupil of eye; bulbous or tuberous plant with sword-shaped leaves and showy flowers.

Irish /'aɪərɪʃ/ ● *adjective* of Ireland. ● *noun* Celtic language of Ireland; (**the Irish**; treated as *plural*) the Irish people. □ **Irish coffee** coffee with dash of whiskey and a little sugar, topped with cream; **Irishman**, **Irishwoman** native of Ireland; **Irish stew** dish of stewed mutton, onions, and potatoes.

irk *verb* irritate, annoy.

irksome *adjective* annoying, tiresome.

iron /'aɪən/ ● *noun* common strong grey metallic element; this as symbol of strength or firmness; tool etc. of iron; implement heated to smooth clothes etc.; golf club with iron or steel head; (in *plural*) fetters; (in *plural*) stirrups; (often in *plural*) leg-support to rectify malformations. ● *adjective* of iron; robust; unyielding. ● *verb* smooth (clothes etc.) with heated iron. □ **Iron Age** era characterized by use of iron weapons etc.; **Iron Curtain** *historical* notional barrier to passage of people and information between Soviet bloc and West; **ironing board** narrow folding table etc. for ironing clothes on; **iron lung** rigid case over patient's body for administering prolonged artificial respiration; **ironmonger** dealer in **ironmongery**, household and building hardware; **iron rations** small emergency supply of food; **ironstone** hard iron ore, kind of hard white pottery.

ironic /aɪ'rɒnɪk/ *adjective* (also **ironical**) using or displaying irony. □ **ironically** *adverb.*

irony /'aɪrənɪ/ *noun* (*plural* **-ies**) expression of meaning, usually humorous or sarcastic, by use of words normally conveying opposite meaning; apparent perversity of fate or circumstances.

irradiate /ɪ'reɪdɪeɪt/ *verb* (**-ting**) subject to radiation; shine on; throw light on; light up. □ **irradiation** *noun.*

irrational /ɪ'ræʃən(ə)l/ *adjective* unreasonable, illogical; not endowed with reason; *Mathematics* not expressible as an ordinary fraction. □ **irrationality** /-'næl-/ *noun*; **irrationally** *adverb.*

irreconcilable /ɪ'rekənsaɪləb(ə)l/ *adjective* implacably hostile; (of ideas etc.) incompatible. □ **irreconcilably** *adverb.*

irrecoverable /ɪrɪ'kʌvərəb(ə)l/ *adjective* that cannot be recovered or remedied.

irredeemable /ɪrɪ'di:məb(ə)l/ *adjective* that cannot be redeemed, hopeless. □ **irredeemably** *adverb.*

irreducible /ɪrɪ'dju:sɪb(ə)l/ *adjective* not able to be reduced or simplified.

irrefutable /ɪrɪ'fju:təb(ə)l/ *adjective* that cannot be refuted. □ **irrefutably** *adverb.*

irregular /ɪ'regjʊlə/ ● *adjective* not regular; unsymmetrical, uneven; varying in form; not occurring at regular intervals; contrary to rule; (of troops) not in regular army; (of verb, noun, etc.) not inflected according to usual rules. ● *noun* (in *plural*) irregular troops. □ **irregularity** /-'lær-/ *noun* (*plural* **-ies**); **irregularly** *adverb.*

irrelevant /ɪ'relɪv(ə)nt/ *adjective* not relevant. □ **irrelevance** *noun*; **irrelevancy** *noun* (*plural* **-ies**).

irreligious /ɪrɪ'lɪdʒəs/ *adjective* lacking or hostile to religion; irreverent.

irremediable /ɪrɪ'mi:dɪəb(ə)l/ *adjective* that cannot be remedied. □ **irremediably** *adverb.*

irremovable /ɪrɪ'mu:vəb(ə)l/ *adjective* not removable. □ **irremovably** *adverb.*

irreparable /ɪ'repərəb(ə)l/ *adjective* that cannot be rectified or made good. □ **irreparably** *adverb.*

irreplaceable /ɪrɪ'pleɪsəb(ə)l/ *adjective* that cannot be replaced.

irrepressible /ɪrɪ'presɪb(ə)l/ *adjective* that cannot be repressed. □ **irrepressibly** *adverb.*

irreproachable /ɪrɪ'prəʊtʃəb(ə)l/ *adjective* faultless, blameless. □ **irreproachably** *adverb.*

irresistible /ɪrɪ'zɪstɪb(ə)l/ *adjective* too strong, convincing, charming, etc. to be resisted. □ **irresistibly** *adverb.*

irresolute /ɪ'rezəlu:t/ *adjective* hesitating; lacking in resolution. □ **irresoluteness** *noun*; **irresolution** /-'lu:ʃ(ə)n/ *noun.*

irrespective /ɪrɪ'spektɪv/ *adjective* (+ *of*) not taking into account, regardless of.

irresponsible /ɪrɪ'spɒnsɪb(ə)l/ *adjective* acting or done without due sense of responsibility; not responsible. □ **irresponsibility** *noun*; **irresponsibly** *adverb.*

irretrievable /ɪrɪ'tri:vəb(ə)l/ *adjective* that cannot be retrieved or restored. □ **irretrievably** *adverb.*

irreverent /ɪ'revərənt/ *adjective* lacking in reverence. □ **irreverence** *noun*; **irreverently** *adverb.*

irreversible /ɪrɪ'vɜ:sɪb(ə)l/ *adjective* that cannot be reversed or altered. □ **irreversibly** *adverb.*

irrevocable /ɪ'revəkəb(ə)l/ *adjective* unalterable; gone beyond recall. □ **irrevocably** *adverb.*

irrigate /'ɪrɪgeɪt/ *verb* (**-ting**) water (land) by system of artificial channels; (of stream etc.) supply (land) with water; *Medicine* moisten (wound etc.) with constant flow of liquid. □ **irrigable** *adjective*; **irrigation** *noun*; **irrigator** *noun.*

irritable /'ɪrɪtəb(ə)l/ *adjective* easily annoyed; very sensitive to contact. □ **irritability** *noun*; **irritably** *adverb.*

irritant /'ɪrɪt(ə)nt/ ● *adjective* causing irritation. ● *noun* irritant substance or agent.

irritate /'ɪrɪteɪt/ *verb* (**-ting**) excite to anger, annoy; stimulate discomfort in (part of body). □ **irritating** *adjective*; **irritation** *noun.*

Is. *abbreviation* Island(s); Isle(s).

is *3rd singular present* of BE.

isinglass /'aɪzɪŋglɑ:s/ *noun* kind of gelatin obtained from sturgeon etc.

Islam /'ɪzlɑ:m/ *noun* religion of Muslims, proclaimed by Prophet Muhammad; the Muslim world. □ **Islamic** /-'læm-/ *adjective.*

island /'aɪlənd/ *noun* piece of land surrounded by water; traffic island; detached or isolated thing.

islander *noun* native or inhabitant of island.

isle /aɪl/ *noun literary* (usually small) island.

islet /'aɪlɪt/ *noun* small island.

isn't /'ɪz(ə)nt/ is not.

isobar /'aɪsəbɑ:/ *noun* line on map connecting places with same atmospheric pressure. □ **isobaric** /-'bær-/ *adjective.*

isolate /'aɪsəleɪt/ *verb* (**-ting**) place apart or alone; separate (esp. infectious patient from others); insulate (electrical apparatus), esp. by gap; disconnect. □ **isolation** *noun.*

isolationism *noun* policy of holding aloof from affairs of other countries or groups. □ **isolationist** *noun.*

isomer /'aɪsəmə/ *noun* one of two or more compounds with same molecular formula but different arrangement of atoms. □ **isomeric** /-'mer-/ *adjective*; **isomerism** /aɪ'sɒmərɪz(ə)m/ *noun.*

isosceles /aɪ'sɒsɪli:z/ *adjective* (of triangle) having two sides equal.

isotherm /'aɪsəθɜ:m/ *noun* line on map connecting places with same temperature. □ **isothermal** /-'θɜ:m(ə)l/ *adjective.*

isotope /'aɪsətəʊp/ *noun* any of two or more forms of chemical element with different relative atomic mass and different nuclear but not chemical properties. □ **isotopic** /-'tɒp-/ *adjective.*

Israeli /ɪz'reɪlɪ/ ● *adjective* of modern state of Israel. ● *noun* (*plural* **-s**) native or national of Israel.

issue /'ɪʃu:/ ● *noun* giving out or circulation of shares, notes, stamps, etc.; copies of journal etc. circulated at one time; each of regular series of magazine etc.; outgoing, outflow; point in question, essential subject of dispute; result, outcome; offspring. ● *verb* (**issues**, **issued**, **issuing**) go or come out; give or send out; publish, circulate; supply, (+ *with*) supply with equipment etc.; (+ *from*) be derived, result; (+ *from*) emerge.

isthmus /'ɪsməs/ *noun* (*plural* **-es**) neck of land connecting two larger land masses.

IT *abbreviation* information technology.

it *pronoun* the thing in question; indefinite, undefined, or impersonal subject, action, condition, object, etc.; *substitute for deferred subject or object*; exactly what is needed; perfection; *slang* sexual intercourse, sex appeal. □ **that's it** *colloquial* that is what is required, that is the difficulty, that is the end, enough.

Italian /ɪ'tæljən/ ● *noun* native, national, or language of Italy. ● *adjective* of Italy.

italic /ɪ'tælɪk/ ● *adjective* (of type etc.) of sloping kind; (of handwriting) neat and pointed; (**Italic**) of ancient Italy. ● *noun* (usually in *plural*) italic type.

italicize /ɪ'tælɪsaɪz/ *verb* (also **-ise**) (**-zing** or **-sing**) print in italics.

itch ● *noun* irritation in skin; restless desire; disease with itch. ● *verb* feel irritation or restless desire.

itchy *adjective* (**-ier**, **-iest**) having or causing itch. □ **have itchy feet** *colloquial* be restless, have urge to travel.

it'd /ˈɪtəd/ it had; it would.

item /ˈaɪtəm/ *noun* any one of enumerated things; separate or distinct piece of news etc.

itemize *verb* (also **-ise**) (**-zing** or **-sing**) state by items. □ **itemization** *noun*.

iterate /ˈɪtəreɪt/ *verb* (**-ting**) repeat; state repeatedly. □ **iteration** *noun*; **iterative** /-rətɪv/ *adjective*.

itinerant /aɪˈtɪnərənt/ • *adjective* travelling from place to place. • *noun* itinerant person.

itinerary /aɪˈtɪnərərɪ/ *noun* (*plural* **-ies**) route; record of travel; guidebook.

it'll /ˈɪt(ə)l/ it will; it shall.

its *adjective* of or belonging to it.

it's /ɪts/ it is; it has.

■ **Usage** Because it has an apostrophe, *it's* is easily confused with *its*. Both are correctly used in *Where's the dog?—It's in its kennel, and it's eaten its food* (= *It is in its kennel, and it has eaten its food.*)

itself /ɪtˈself/ *pronoun: emphatic & reflexive form of* IT.

ITV *abbreviation* Independent Television.

IUD *abbreviation* intrauterine (contraceptive) device.

I've /aɪv/ I have.

ivory /ˈaɪvərɪ/ *noun* (*plural* **-ies**) white substance of tusks of elephant etc.; colour of this; (in *plural*) *slang* things made of or resembling ivory, esp. dice, piano keys, or teeth. □ **ivory tower** seclusion from harsh realities of life.

ivy /ˈaɪvɪ/ *noun* (*plural* **-ies**) climbing evergreen with shiny 5-angled leaves.

Jj

jab • *verb* (**-bb-**) poke roughly; stab; (+ *into*) thrust (thing) hard or abruptly. • *noun* abrupt blow or thrust; *colloquial* hypodermic injection.

jabber • *verb* chatter volubly; utter fast and indistinctly. • *noun* chatter, gabble.

jabot /ˈʒæbəʊ/ *noun* frill on front of shirt or blouse.

jacaranda /dʒækəˈrændə/ *noun* tropical American tree with blue flowers or one with hard scented wood.

jacinth /ˈdʒæsɪnθ/ *noun* reddish-orange zircon used as gem.

jack • *noun* device for lifting heavy objects, esp. vehicles; lowest-ranking court card; ship's flag, esp. showing nationality; device using single-pronged plug to connect electrical equipment; small white target ball in bowls; (in *plural*) game played with jackstones. • *verb* (usually + *up*) raise (as) with jack. □ **jackboot** boot reaching above knee; **jack in** *slang* abandon (attempt etc.); **jack-in-the-box** toy figure that springs out of box; **jack of all trades** person with many skills; **jack plug** electrical plug with single prong; **jackstone** metal etc. piece used in tossing games.

jackal /ˈdʒæk(ə)l/ *noun* African or Asian wild animal of dog family.

jackass *noun* male ass; stupid person.

jackdaw *noun* grey-headed bird of crow family.

jacket /ˈdʒækɪt/ *noun* short coat with sleeves; covering round boiler etc.; outside wrapper of book; skin of potato. □ **jacket potato** one baked in its skin.

jackknife • *noun* large clasp-knife. • *verb* (**-fing**) (of articulated vehicle) fold against itself in accident.

jackpot *noun* large prize, esp. accumulated in game, lottery, etc.

Jacobean /dʒækəˈbiːən/ *adjective* of reign of James I.

Jacobite /ˈdʒækəbaɪt/ *noun historical* supporter of James II in exile, or of Stuarts.

Jacuzzi /dʒəˈku:zɪ/ *noun* (*plural* **-s**) *proprietary term* large bath with massaging underwater jets.

jade[1] *noun* hard usually green stone for ornaments; green colour of jade.

jade[2] *noun* inferior or worn-out horse.

jaded *adjective* tired out, surfeited.

jag ● *noun* sharp projection of rock etc. ● *verb* (**-gg-**) cut or tear unevenly; make indentations in.

jagged /ˈdʒægɪd/ *adjective* unevenly cut or torn. □ **jaggedly** *adverb*; **jaggedness** *noun*.

jaguar /ˈdʒægjʊə/ *noun* large American spotted animal of cat family.

jail /dʒeɪl/ (also **gaol**) ● *noun* place for detention of prisoners; confinement in jail. ● *verb* put in jail. □ **jailbird** prisoner, habitual criminal. □ **jailer** *noun*.

jalap /ˈdʒæləp/ *noun* purgative drug.

jalopy /dʒəˈlɒpɪ/ *noun* (*plural* **-ies**) *colloquial* dilapidated old motor vehicle.

jalousie /ˈʒælʊzi:/ *noun* slatted blind or shutter.

jam[1] ● *verb* (**-mm-**) (usually + *into*, *together*, etc.) squeeze or cram into space; become wedged; cause (machinery) to become wedged and so unworkable, become wedged in this way; block (exit, road, etc.) by crowding; (usually + *on*) apply (brakes) suddenly; make (radio transmission) unintelligible with interference. ● *noun* squeeze; stoppage; crowded mass, esp. of traffic; *colloquial* predicament. □ **jam session** (in jazz etc.) improvised ensemble playing.

jam[2] *noun* conserve of boiled fruit and sugar; *colloquial* easy or pleasant thing.

jamb /dʒæm/ *noun* side post or side face of doorway or window frame.

jamboree /dʒæmbəˈri:/ *noun* celebration; large rally of Scouts.

jammy *adjective* (**-ier**, **-iest**) covered with jam; *colloquial* lucky, profitable.

Jan. *abbreviation* January.

jangle /ˈdʒæŋg(ə)l/ ● *verb* (**-ling**) (cause to) make harsh metallic sound. ● *noun* such sound.

janitor /ˈdʒænɪtə/ *noun* doorkeeper; caretaker.

January /ˈdʒænjʊərɪ/ *noun* (*plural* **-ies**) first month of year.

japan /dʒəˈpæn/ ● *noun* hard usually black varnish. ● *verb* (**-nn-**) make black and glossy (as) with japan.

Japanese /dʒæpəˈni:z/ ● *noun* (*plural* same) native, national, or language of Japan. ● *adjective* of Japan or its people or language.

jape *noun* practical joke.

japonica /dʒəˈpɒnɪkə/ *noun* flowering shrub with red flowers and edible fruits.

jar[1] *noun* container, usually of glass and cylindrical.

jar[2] ● *verb* (**-rr-**) (often + *on*) (of sound, manner, etc.) strike discordantly, grate; (often + *against*, *on*) (cause to) strike (esp. part of body) with vibration or shock; (often + *with*) be at variance. ● *noun* jarring sound, shock, or vibration.

jardinière /ʒɑ:dɪˈnjeə/ *noun* ornamental pot or stand for plants.

jargon /ˈdʒɑ:gən/ *noun* words used by particular group or profession; debased or pretentious language.

jasmine /ˈdʒæzmɪn/ *noun* shrub with white or yellow flowers.

jasper /ˈdʒæspə/ *noun* red, yellow, or brown opaque quartz.

jaundice /ˈdʒɔ:ndɪs/ ● *noun* yellowing of skin caused by liver disease, bile disorder, etc. ● *verb* (as **jaundiced** *adjective*) affected with jaundice; envious, resentful.

jaunt /dʒɔ:nt/ ● *noun* pleasure trip. ● *verb* take a jaunt.

jaunty *adjective* (**-ier**, **-iest**) cheerful and self-confident; sprightly. □ **jauntily** *adverb*; **jauntiness** *noun*.

javelin /ˈdʒævəlɪn/ *noun* light spear thrown in sport or, formerly, as weapon.

jaw ● *noun* bony structure containing teeth; (in *plural*) mouth, gripping parts of tool etc.; *colloquial* tedious talk. ● *verb slang* speak at tedious length. □ **jawbone** lower jaw in most mammals.

jay *noun* noisy European bird of crow family with vivid plumage. □ **jaywalk** walk across road carelessly or dangerously; **jaywalker** person who does this.

jazz *noun* rhythmic syncopated esp. improvised music of American black origin. □ **and all that jazz** *colloquial* and other related things; **jazz up** enliven.

jazzy *adjective* (**-ier**, **-iest**) of or like jazz; vivid.

jealous /'dʒeləs/ *adjective* resentful of rivalry in love; (often + *of*) envious (of person etc.), protective (of rights etc.). □ **jealously** *adverb*; **jealousy** *noun* (*plural* **-ies**).

jeans /dʒi:nz/ *plural noun* casual esp. denim trousers.

Jeep *noun proprietary term* small sturdy esp. military vehicle with 4-wheel drive.

jeer ● *verb* (often + *at*) scoff, deride. ● *noun* taunt.

Jehovah /dʒə'həʊvə/ *noun* (in Old Testament) God. □ **Jehovah's Witness** member of unorthodox Christian sect.

jejune /dʒɪ'dʒu:n/ *adjective* (of ideas, writing, etc.) shallow, naïve, or dry and uninteresting.

jell *verb colloquial* set as jelly; (of ideas etc.) take definite form; cohere.

jellied /'dʒelɪd/ *adjective* (of food etc.) set as or in jelly.

jelly /'dʒelɪ/ *noun* (*plural* **-ies**) (usually fruit-flavoured) semi-transparent dessert set with gelatin; similar preparation as jam or condiment; *slang* gelignite. □ **jelly baby** jelly-like baby-shaped sweet; **jellyfish** (*plural* same or **-es**) marine animal with jelly-like body and stinging tentacles.

jemmy /'dʒemɪ/ *noun* (*plural* **-ies**) burglar's crowbar.

jeopardize *verb* (also **-ise**) (**-zing** or **-sing**) endanger.

jeopardy /'dʒepədɪ/ *noun* danger, esp. severe.

jerboa /dʒɜ:'bəʊə/ *noun* small jumping desert rodent.

Jeremiah /dʒerɪ'maɪə/ *noun* dismal prophet.

jerk[1] ● *noun* sharp sudden pull, twist, etc.; spasmodic muscular twitch; *slang* fool. ● *verb* move, pull, throw, etc. with jerk. □ **jerky** *adjective* (**-ier**, **-iest**); **jerkily** *adverb*; **jerkiness** *noun*.

jerk[2] *verb* cure (beef) by cutting in long slices and drying in the sun.

jerkin /'dʒɜ:kɪn/ *noun* sleeveless jacket.

jeroboam /dʒerə'bəʊəm/ *noun* wine bottle of 4–12 times ordinary size.

jerry-building *noun* building of shoddy houses with bad materials. □ **jerry-builder** *noun*; **jerry-built** *adjective*.

jerrycan *noun* kind of petrol- or water-can.

jersey /'dʒɜ:zɪ/ *noun* (*plural* **-s**) knitted usually woollen pullover; knitted fabric; (**Jersey**) dairy cow from Jersey.

Jerusalem artichoke /dʒə'ru:sələm/ *noun* kind of sunflower with edible tubers; this tuber as vegetable.

jest ● *noun* joke; fun; banter; object of derision. ● *verb* joke; fool about. □ **in jest** in fun.

jester *noun historical* professional clown at medieval court etc.

Jesuit /'dʒezjʊɪt/ *noun* member of RC Society of Jesus. □ **Jesuitical** /-'ɪt-/ *adjective*.

jet[1] ● *noun* stream of water, steam, gas, flame, etc. shot esp. from small opening; spout or nozzle for this purpose; jet engine or plane. ● *verb* (**-tt-**) spurt out in jet(s); *colloquial* send or travel by jet plane. □ **jet engine** one using jet propulsion; **jet lag** exhaustion felt after long flight across time zones; **jet plane** one with jet engine; **jet-propelled** having jet propulsion, very fast; **jet propulsion** propulsion by backward ejection of high-speed jet of gas etc.; **jet set** wealthy people who travel widely, esp. for pleasure; **jet-setter** such a person.

jet[2] *noun* hard black lignite, often carved and highly polished. □ **jet black** deep glossy black.

jetsam /'dʒetsəm/ *noun* objects washed ashore, esp. jettisoned from ship.

jettison /'dʒetɪs(ə)n/ *verb* throw (cargo, fuel, etc.) from ship or aircraft to lighten it; abandon; get rid of.

jetty /'dʒetɪ/ *noun* (*plural* **-ies**) pier or breakwater protecting or defending harbour etc.; landing-pier.

Jew /dʒu:/ *noun* person of Hebrew descent or whose religion is Judaism.

jewel /'dʒu:əl/ ● *noun* precious stone; this used in watchmaking; jewelled personal ornament; precious person or thing. ● *verb* (**-ll-**; *US* **-l-**) (esp. as **jewelled** *adjective*) adorn or set with jewels.

jeweller *noun* (*US* **jeweler**) maker of or dealer in jewels or jewellery.

jewellery /ˈdʒu:əlrɪ/ *noun* (also **jewelry**) rings, brooches, necklaces, etc. collectively.

Jewish *adjective* of Jews or Judaism. □ **Jewishness** *noun.*

Jewry /ˈdʒʊərɪ/ *noun* Jews collectively.

Jezebel /ˈdʒezəbəl/ *noun* shameless or immoral woman.

jib • *noun* projecting arm of crane; triangular staysail. • *verb* (**-bb-**) (esp. of horse) stop and refuse to go on; (+ *at*) show aversion to.

jibe[1] = GIBE.

jibe[2] *US* = GYBE.

jiffy /ˈdʒɪfɪ/ *noun* (*plural* **-ies**) (also **jiff**) *colloquial* short time, moment. □ **Jiffy bag** *proprietary term* padded envelope.

jig • *noun* lively dance; music for this; device that holds piece of work and guides tools operating on it. • *verb* (**-gg-**) dance jig; (often + *about*) move quickly up and down; fidget.

jigger /ˈdʒɪgə/ *noun* small glass for measure of spirits.

jiggery-pokery /dʒɪgərɪˈpəʊkərɪ/ *noun colloquial* trickery; swindling.

jiggle /ˈdʒɪg(ə)l/ • *verb* (**-ling**) (often + *about*) rock or jerk lightly; fidget. • *noun* light shake.

jigsaw *noun* (in full **jigsaw puzzle**) picture on board etc. cut into irregular interlocking pieces to be reassembled as pastime; mechanical fine-bladed fret saw.

jihad /dʒɪˈhæd/ *noun* Muslim holy war against unbelievers.

jilt *verb* abruptly rejct or abandon (esp. lover).

jingle /ˈdʒɪŋg(ə)l/ • *noun* mixed ringing or clinking noise; repetition of sounds in phrase; short catchy verse in advertising etc. • *verb* (**-ling**) (cause to) make jingling sound.

jingo /ˈdʒɪŋgəʊ/ *noun* (*plural* **-es**) blustering patriot. □ **jingoism** *noun*; **jingoist** *noun*; **jingoistic** /-ˈɪs-/ *adjective.*

jink • *verb* move elusively; elude by dodging. • *noun* jinking. □ **high jinks** boisterous fun.

jinnee /ˈdʒɪni:/ *noun* (also **jinn**, **djinn** /dʒɪn/) (*plural* **jinn** or **djinn**) (in Muslim mythology) spirit of supernatural power in human or animal form.

jinx *colloquial* • *noun* person or thing that seems to bring bad luck. • *verb* (esp. as **jinxed** *adjective*) subject to bad luck.

jitter /ˈdʒɪtə/ *colloquial* • *noun* (**the jitters**) extreme nervousness. • *verb* be nervous; act nervously. □ **jittery** *adjective.*

jive • *noun* lively dance of 1950s; music for this. • *verb* (**-ving**) dance to or play jive music. □ **jiver** *noun.*

Jnr. *abbreviation* Junior.

job • *noun* piece of work (to be) done; paid employment; *colloquial* difficult task; *slang* a crime, esp. a robbery. • *verb* (**-bb-**) do jobs; do piece-work; buy and sell (stocks etc.); deal corruptly with (matter). □ **jobcentre** local government office advertising available jobs; **job-hunt** *colloquial* seek employment; **job lot** mixed lot bought at auction etc.

jobber /ˈdʒɒbə/ *noun* person who jobs; *historical* principal or wholesaler on Stock Exchange.

jobbery *noun* corrupt dealing.

jobless *adjective* unemployed. □ **joblessness** *noun.*

job-sharing *noun* sharing of full-time job by two or more people. □ **job-share** *noun & verb.*

jockey /ˈdʒɒkɪ/ • *noun* rider in horse races. • *verb* (**-eys**, **-eyed**) cheat, trick. □ **jockey for position** manoeuvre for advantage.

jockstrap /ˈdʒɒkstræp/ *noun* support or protection for male genitals worn esp. in sport.

jocose /dʒəˈkəʊs/ *adjective* playful; jocular. □ **jocosely** *adverb*; **jocosity** /-ˈkɒs-/ *noun* (*plural* **-ies**).

jocular /ˈdʒɒkjʊlə/ *adjective* fond of joking; humorous. □ **jocularity** /-ˈlær-/ *noun* (*plural* **-ies**); **jocularly** *adverb.*

jocund /ˈdʒɒkənd/ *adjective literary* merry, cheerful. □ **jocundity** /dʒəˈkʌn-/ *noun* (*plural* **-ies**); **jocundly** *adverb.*

jodhpurs /ˈdʒɒdpəz/ *plural noun* riding breeches tight below knee.

jog • *verb* (**-gg-**) run slowly, esp. as exercise; push, jerk; nudge, esp. to alert; stimulate (person's memory). • *noun* spell of jogging; slow walk or trot; push, jerk; nudge. □ **jogtrot** slow regular trot. □ **jogger** *noun.*

joggle /ˈdʒɒg(ə)l/ • *verb* (**-ling**) move in jerks. • *noun* slight shake.

joie de vivre /ʒwɑ: də 'vi:vrə/ *noun* exuberance; high spirits. [French]

join ● *verb* (often + *to, together*) put together, fasten, unite; connect (points) by line etc.; become member of (club etc.); take one's place with (person, group, etc.); (+ *in, for,* etc.) take part with (others) in activity etc.; (often + *with, to*) come together, be united; (of river etc.) become connected or continuous with. ● *noun* point, line, or surface of junction. □ **join in** take part in (activity); **join up** enlist for military service.

joiner *noun* maker of furniture and light woodwork. □ **joinery** *noun*.

joint ● *noun* place at which two or more things are joined; device for doing this; point at which two bones fit together; division of animal carcass as meat; *slang* restaurant, bar, etc.; *slang* marijuana cigarette. ● *adjective* held by, done by, or belonging to two or more people etc.; sharing with another. ● *verb* connect by joint(s); divide at joint or into joints. □ **joint stock** capital held jointly.

jointure /'dʒɔɪntʃə/ *noun* estate settled on wife by husband for use after his death.

joist *noun* supporting beam in floor, ceiling, etc.

jojoba /həʊ'həʊbə/ *noun* plant with seeds yielding oil used in cosmetics etc.

joke ● *noun* thing said or done to cause laughter; witticism; ridiculous person or thing. ● *verb* (**-king**) make jokes. □ **jokily** *adverb*; **jokiness** *noun*; **jokingly** *adverb*; **jokey, joky** *adjective*.

joker *noun* person who jokes; playing card used in some games.

jollification /dʒɒlɪfɪ'keɪʃ(ə)n/ *noun* merrymaking.

jolly /'dʒɒlɪ/ ● *adjective* (**-ier, -iest**) cheerful; festive, jovial; *colloquial* pleasant, delightful. ● *adverb colloquial* very. ● *verb* (**-ies, -ied**) (usually + *along*) *colloquial* coax, humour. □ **jollity** *noun* (*plural* **-ies**).

jolt /dʒəʊlt/ ● *verb* shake (esp. in vehicle) with jerk; shock, perturb; move along jerkily. ● *noun* jerk; surprise, shock.

jonquil /'dʒɒŋkwɪl/ *noun* narcissus with white or yellow fragrant flowers.

josh *slang* ● *verb* tease, make fun of. ● *noun* good-natured joke.

joss *noun* Chinese idol. □ **joss-stick** incense stick for burning.

jostle /'dʒɒs(ə)l/ ● *verb* (**-ling**) (often + *against*) knock; elbow; (+ *with*) struggle. ● *noun* jostling.

jot ● *verb* (**-tt-**) (usually + *down*) write briefly or hastily. ● *noun* very small amount.

jotter *noun* small pad or notebook.

joule /dʒu:l/ *noun* SI unit of work and energy.

journal /'dʒɜ:n(ə)l/ *noun* newspaper, periodical; daily record of events; diary; account book; part of shaft or axle resting on bearings.

journalese /dʒɜ:nə'li:z/ *noun* hackneyed style of writing characteristic of newspapers.

journalism /'dʒɜ:nəlɪz(ə)m/ *noun* work of journalist.

journalist /'dʒɜ:nəlɪst/ *noun* person writing for or editing newspapers etc. □ **journalistic** /-'lɪs-/ *adjective*.

journey /'dʒɜ:nɪ/ ● *noun* (*plural* **-s**) act of going from one place to another; distance travelled, time taken. ● *verb* (**-s, -ed**) make journey, travel. □ **journeyman** qualified mechanic or artisan working for another.

joust /dʒaʊst/ *historical* ● *noun* combat with lances between two mounted knights. ● *verb* engage in joust. □ **jouster** *noun*.

jovial /'dʒəʊvɪəl/ *adjective* merry, convivial, hearty. □ **joviality** /-'æl-/ *noun*; **jovially** *adverb*.

jowl /dʒaʊl/ *noun* jaw, jawbone; cheek; loose skin on throat.

joy *noun* gladness, pleasure; thing causing joy; *colloquial* satisfaction. □ **joyride** *colloquial* (go for) pleasure ride in esp. stolen car; **joystick** control column of aircraft, lever for moving image on VDU screen. □ **joyful** *adjective*; **joyfully** *adverb*; **joyfulness** *noun*; **joyless** *adjective*; **joyous** *adjective*; **joyously** *adverb*.

JP *abbreviation* Justice of the Peace.

Jr. *abbreviation* Junior.

jubilant /'dʒu:bɪlənt/ *adjective* exultant, rejoicing. □ **jubilantly** *adverb*; **jubilation** *noun*.

jubilee /'dʒu:bɪli:/ *noun* anniversary (esp. 25th or 50th); time of rejoicing.

Judaic /dʒuːˈdeɪɪk/ *adjective* of or characteristic of Jews.

Judaism /ˈdʒuːdeɪɪz(ə)m/ *noun* religion of Jews.

Judas /ˈdʒuːdəs/ *noun* traitor.

judder /ˈdʒʌdə/ ● *verb* shake noisily or violently. ● *noun* juddering.

judge /dʒʌdʒ/ ● *noun* public official appointed to hear and try legal cases; person appointed to decide dispute or contest; person who decides question; person having judgement of specified type. ● *verb* (**-ging**) form opinion (about); estimate; act as judge (of); try legal case; (often + *to do*, *that*) conclude, consider.

judgement *noun* (also **judgment**) critical faculty, discernment, good sense; opinion; sentence of court of justice; *often jocular* deserved misfortune. □ **Judgement Day** day on which God will judge humankind. □ **judgemental** /-ˈmen-/ *adjective*.

judicature /ˈdʒuːdɪkətʃə/ *noun* administration of justice; judge's position; judges collectively.

judicial /dʒuːˈdɪʃ(ə)l/ *adjective* of, done by, or proper to court of law; of or proper to a judge; having function of judge; impartial. □ **judicially** *adverb*.

judiciary /dʒuːˈdɪʃərɪ/ *noun* (*plural* **-ies**) judges collectively.

judicious /dʒuːˈdɪʃəs/ *adjective* sensible, prudent. □ **judiciously** *adverb*.

judo /ˈdʒuːdəʊ/ *noun* sport derived from ju-jitsu.

jug ● *noun* deep vessel for liquids, with handle and lip; contents of this; *slang* prison. ● *verb* (**-gg-**) (usually as **jugged** *adjective*) stew (hare) in casserole. □ **jugful** *noun* (*plural* **-s**).

juggernaut /ˈdʒʌgənɔːt/ *noun* large heavy lorry; overwhelming force or object.

juggle /ˈdʒʌg(ə)l/ ● *verb* (**-ling**) (often + *with*) keep several objects in the air at once by throwing and catching; manipulate or rearrange (facts). ● *noun* juggling; fraud. □ **juggler** *noun*.

jugular /ˈdʒʌgjʊlə/ ● *adjective* of neck or throat. ● *noun* jugular vein. □ **jugular vein** any of large veins in neck carrying blood from head.

juice /dʒuːs/ *noun* liquid part of vegetable, fruit, or meat; animal fluid, esp. secretion; *colloquial* petrol, electricity.

juicy /ˈdʒuːsɪ/ *adjective* (**-ier**, **-iest**) full of juice; *colloquial* interesting, scandalous; *colloquial* profitable. □ **juicily** *adverb*.

ju-jitsu /dʒuːˈdʒɪtsuː/ *noun* Japanese system of unarmed combat.

ju-ju /ˈdʒuːdʒuː/ *noun* (*plural* **-s**) charm or fetish of some W. African peoples; supernatural power attributed to this.

jujube /ˈdʒuːdʒuːb/ *noun* flavoured jelly-like lozenge.

jukebox /ˈdʒuːkbɒks/ *noun* coin-operated machine playing records or compact discs.

Jul. *abbreviation* July.

julep /ˈdʒuːlep/ *noun* sweet drink, esp. medicated; *US* spirits and water iced and flavoured.

julienne /dʒuːlɪˈen/ ● *noun* vegetables cut into thin strips. ● *adjective* cut into thin strips.

Juliet cap /ˈdʒuːlɪət/ *noun* small close-fitting cap worn by brides etc.

July /dʒuːˈlaɪ/ *noun* (*plural* **Julys**) seventh month of year.

jumble /ˈdʒʌmb(ə)l/ ● *verb* (**-ling**) (often + *up*) mix; confuse; muddle. ● *noun* confused heap etc.; muddle; articles in jumble sale. □ **jumble sale** sale of second-hand articles, esp. for charity.

jumbo /ˈdʒʌmbəʊ/ *noun* (*plural* **-s**) big animal (esp. elephant), person, or thing. □ **jumbo jet** large airliner for several hundred passengers, esp. Boeing 747.

jump ● *verb* spring from ground etc.; (often + *up*, *from*, *in*, *out*, etc.) rise or move suddenly; jerk from shock or excitement; pass over (obstacle) by jumping; (+ *to*, *at*) reach (conclusion) hastily; (of train etc.) leave (rails); pass (red traffic light); get on or off (train etc.) quickly, esp. illegally; attack (person) unexpectedly. ● *noun* act of jumping; sudden movement caused by shock etc.; abrupt rise in price, status, etc.; obstacle to be jumped, esp. by horse; gap in series etc. □ **jump at** accept eagerly; **jumped-up** *adjective colloquial* upstart; **jump the gun** start prematurely; **jump-jet** vertical take-off jet plane; **jump-lead** cable for carrying current from one battery to another; **jump-off** deciding round in show-jumping; **jump the queue** take unfair precedence; **jump ship** (of seaman) desert; **jump suit** one-piece garment for whole body; **jump to it** *colloquial* act promptly and energetically.

jumper /'dʒʌmpə/ *noun* knitted pullover; loose outer jacket worn by sailors; *US* pinafore dress.

jumpy *adjective* (**-ier**, **-iest**) nervous, easily startled. □ **jumpiness** *noun.*

Jun. *abbreviation* June; Junior.

junction /'dʒʌŋkʃ(ə)n/ *noun* joining-point; place where railway lines or roads meet. □ **junction box** box containing junction of electric cables etc.

juncture /'dʒʌŋktʃə/ *noun* point in time, esp. critical one; joining-point.

June *noun* sixth month of year.

jungle /'dʒʌŋg(ə)l/ *noun* land overgrown with tangled vegetation, esp. in tropics; tangled mass; place of bewildering complexity or struggle.

junior /'dʒu:nɪə/ ● *adjective* (often + *to*) lower in age, standing, or position; the younger (esp. after name); (of school) for younger pupils. ● *noun* junior person.

juniper /'dʒu:nɪpə/ *noun* prickly evergreen shrub or tree with purple berry-like cones.

junk[1] ● *noun* discarded articles, rubbish; anything regarded as of little value; *slang* narcotic drug, esp. heroin. ● *verb* discard as junk. □ **junk food** food which is not nutritious; **junk mail** unsolicited advertising sent by post.

junk[2] *noun* flat-bottomed sailing vessel in China seas.

junket /'dʒʌŋkɪt/ ● *noun* pleasure outing; official's tour at public expense; sweetened and flavoured milk curds; feast. ● *verb* (**-t-**) feast, picnic.

junkie /'dʒʌŋkɪ/ *noun slang* drug addict.

junta /'dʒʌntə/ *noun* (usually military) clique taking power after *coup d'état.*

juridical /dʒʊə'rɪdɪk(ə)l/ *adjective* of judicial proceedings, relating to the law.

jurisdiction /dʒʊərɪs'dɪkʃ(ə)n/ *noun* (often + *over*) administration of justice; legal or other authority; extent of this.

jurisprudence /dʒʊərɪs'pru:d(ə)ns/ *noun* science or philosophy of law.

jurist /'dʒʊərɪst/ *noun* expert in law. □ **juristic** /-'rɪs-/ *adjective.*

juror /'dʒʊərə/ *noun* member of jury.

jury /'dʒʊərɪ/ *noun* (*plural* **-ies**) group of people giving verdict in court of justice; judges of competition. □ **jury-box** enclosure in court for jury.

just ● *adjective* morally right, fair; deserved; well-grounded; justified. ● *adverb* exactly; very recently; barely; quite; *colloquial* simply, merely, positively. □ **just now** at this moment, a little time ago. □ **justly** *adverb.*

justice /'dʒʌstɪs/ *noun* fairness; authority exercised in maintenance of right; judicial proceedings; magistrate, judge. □ **do justice to** treat fairly, appreciate properly; **Justice of the Peace** lay magistrate.

justify /'dʒʌstɪfaɪ/ *verb* (**-ies**, **-ied**) show justice or truth of; (esp. in *passive*) be adequate grounds for, vindicate; *Printing* adjust (line of type) to fill space evenly; (as **justified** *adjective*) just, right. □ **justifiable** *adjective*; **justification** *noun.*

jut ● *verb* (**-tt-**) (often + *out*) protrude. ● *noun* projection.

jute /dʒu:t/ *noun* fibre from bark of E. Indian plant, used for sacking, mats, etc.; plant yielding this.

juvenile /'dʒu:vənaɪl/ ● *adjective* youthful; of or for young people; *often derogatory* immature. ● *noun* young person; actor playing juvenile part. □ **juvenile delinquency** offences committed by people below age of legal responsibility; **juvenile delinquent** such offender.

juvenilia /dʒu:və'nɪlɪə/ *plural noun* youthful works of author or artist.

juxtapose /dʒʌkstə'pəʊz/ *verb* (**-sing**) put side by side; (+ *with*) put (thing) beside another. □ **juxtaposition** /-pə'zɪʃ(ə)n/ *noun.*

Kk

K *abbreviation* (also **K.**) kelvin(s); Köchel (list of Mozart's works); (also **k**) 1,000.

k *abbreviation* kilo-; knot(s).

kaftan = CAFTAN.

kaiser /'kaɪzə/ *noun historical* emperor, esp. of Germany or Austria.

kale *noun* variety of cabbage, esp. with wrinkled leaves.

kaleidoscope /kə'laɪdəskəʊp/ *noun* tube containing angled mirrors and pieces of coloured glass producing reflected patterns when shaken; constantly changing scene, group, etc. □ **kaleidoscopic** /-'skɒp-/ *adjective*.

kalends = CALENDS.

kamikaze /kæmɪ'kɑ:zɪ/ ● *noun historical* explosive-laden Japanese aircraft deliberately crashed on to target in 1939–45 war; pilot of this. ● *adjective* reckless, esp. suicidal.

kangaroo /kæŋgə'ru:/ *noun* (*plural* **-s**) Australian marsupial with strong hind legs for jumping. □ **kangaroo court** illegal court held by strikers, mutineers, etc.

kaolin /'keɪəlɪn/ *noun* fine white clay used esp. for porcelain and in medicines.

kapok /'keɪpɒk/ *noun* fine cotton-like material from tropical tree, used to stuff cushions etc.

kaput /kə'pʊt/ *adjective slang* broken, ruined.

karabiner /kærə'bi:nə/ *noun* coupling link used by mountaineers.

karakul /'kærəkʊl/ *noun* (also **caracul**) Asian sheep whose lambs have dark curled fleece; fur of this.

karaoke /kærɪ'əʊkɪ/ *noun* entertainment in nightclubs etc. with customers singing to backing music.

karate /kə'rɑ:tɪ/ *noun* Japanese system of unarmed combat.

karma /'kɑ:mə/ *noun Buddhism & Hinduism* person's actions in one life, believed to decide fate in next; destiny.

kauri /kaʊ'rɪ/ *noun* (*plural* **-s**) coniferous NZ timber tree.

kayak /'kaɪæk/ *noun* Eskimo one-man canoe.

kazoo /kə'zu:/ *noun* toy musical instrument into which player sings wordlessly.

KBE *abbreviation* Knight Commander of the Order of the British Empire.

kea /'ki:ə/ *noun* green and red NZ parrot.

kebab /kɪ'bæb/ *noun* pieces of meat and sometimes vegetables grilled on skewer.

kedge ● *verb* (**-ging**) move (ship) with hawser attached to small anchor. ● *noun* (in full **kedge-anchor**) small anchor for this purpose.

kedgeree /kedʒə'ri:/ *noun* dish of fish, rice, hard-boiled eggs, etc.

keel ● *noun* main lengthwise member of base of ship etc. ● *verb* (often + *over*) (cause to) fall down or over; turn keel upwards. □ **keelhaul** drag (person) under keel as punishment; **on an even keel** steady, balanced.

keen[1] *adjective* enthusiastic, eager; (often + *on*) enthusiastic about, fond of; intellectually acute; (of knife) sharp; (of price) competitive. □ **keenly** *adverb*; **keenness** *noun*.

keen[2] ● *noun* Irish wailing funeral song. ● *verb* (often + *over*, *for*) wail mournfully, esp. at funeral.

keep ● *verb* (*past & past participle* **kept**) have charge of; retain possession of; (+ *for*) retain or reserve (for future); maintain or remain in good or specified condition; restrain; detain; observe or respect (law, secret, etc.); own and look after (animal); clothe, feed, etc. (person); carry on (a business); maintain, guard, protect. ● *noun* maintenance, food; *historical* tower, stronghold. □ **for keeps** *colloquial* permanently; **keep at** (cause to) persist with; **keep away** (often + *from*) avoid, prevent from being near; **keep-fit** regular physical exercises; **keep off** (cause to) stay away from, abstain from, avoid; **keep**

on continue, (+ *at*) nag; **keep out** (cause to) stay outside; **keep up** maintain, prevent from going to bed, (often + *with*) not fall behind; **keep up with the Joneses** compete socially with neighbours.

keeper *noun* person who looks after or is in charge of an animal, person, or thing; custodian of museum, forest, etc.; wicket-keeper, goalkeeper; ring holding another on finger.

keeping *noun* custody, charge; (esp. in **in** or **out of keeping with**) agreement, harmony.

keepsake *noun* souvenir, esp. of person.

keg *noun* small barrel. □ **keg beer** beer kept in pressurized metal keg.

kelp *noun* large seaweed suitable for manure.

kelpie /ˈkelpɪ/ *noun Scottish* malevolent water-spirit; Australian sheepdog.

Kelt = CELT.

kelter = KILTER.

kelvin /ˈkelvɪn/ *noun* SI unit of temperature.

ken ● *noun* range of knowledge or sight. ● *verb* (**-nn-**; *past & past participle* **kenned** or **kent**) *Scottish & Northern English* recognize, know.

kendo /ˈkendəʊ/ *noun* Japanese fencing with bamboo swords.

kennel /ˈken(ə)l/ ● *noun* small shelter for dog; (in *plural*) breeding or boarding place for dogs. ● *verb* (**-ll-**; *US* **-l-**) put or keep in kennel.

Kenyan /ˈkenjən/ ● *adjective* of Kenya. ● *noun* native or national of Kenya.

kept *past & past participle* of KEEP.

keratin /ˈkerətɪn/ *noun* fibrous protein in hair, hooves, claws, etc.

kerb *noun* stone etc. edging to pavement etc. □ **kerb-crawling** *colloquial* driving slowly to pick up prostitute; **kerb drill** rules taught to children about crossing roads.

kerfuffle /kəˈfʌf(ə)l/ *noun colloquial* fuss, commotion.

kermes /ˈkɜːmɪz/ *noun* female of insect with berry-like appearance that feeds on **kermes oak**, evergreen oak; red dye made from these insects.

kernel /ˈkɜːn(ə)l/ *noun* (usually soft) edible centre within hard shell of nut, fruit stone, seed, etc.; central or essential part.

kerosene /ˈkerəsiːn/ *noun esp. US* fuel oil distilled from petroleum etc.; paraffin oil.

kestrel /ˈkestr(ə)l/ *noun* small hovering falcon.

ketch *noun* kind of two-masted sailing boat.

ketchup /ˈketʃəp/ *noun* (*US* **catsup** /ˈkætsəp/) spicy sauce made esp. from tomatoes.

kettle /ˈket(ə)l/ *noun* vessel for boiling water in. □ **kettledrum** large bowl-shaped drum.

key /kiː/ ● *noun* (*plural* **-s**) instrument for moving bolt of lock, operating switch, etc.; instrument for winding clock etc. or grasping screw, nut, etc.; finger-operated button or lever on typewriter, piano, computer terminal, etc.; explanation, word, or system for understanding list of symbols, code, etc.; *Music* system of related notes based on particular note; roughness of surface helping adhesion of plaster etc. ● *verb* fasten with pin, wedge, bolt, etc.; (often + *in*) enter (data) by means of (computer) keyboard; roughen (surface) to help adhesion of plaster etc. □ **keyed up** tense, excited; **keyhole** hole by which key is put into lock; **keynote** prevailing tone or idea, *Music* note on which key is based; **keypad** miniature keyboard etc. for telephone, portable computer, etc.; **keyring** ring for keeping keys on; **keystone** central principle of policy, system, etc., central stone of arch.

keyboard ● *noun* set of keys on typewriter, computer, piano, etc. ● *verb* enter (data) by means of keyboard. □ **keyboarder** *noun Computing.*

KG *abbreviation* Knight of the Order of the Garter.

kg *abbreviation* kilogram(s).

KGB *noun historical* secret police of USSR.

khaki /ˈkɑːkɪ/ ● *adjective* dull brownish-yellow. ● *noun* (*plural* **-s**) khaki colour, cloth, or uniform.

khan /kɑːn/ *noun: title of ruler or official in Central Asia.* □ **khanate** *noun.*

kHz *abbreviation* kilohertz.

kibbutz /kɪ'bʊts/ *noun* (*plural* **-im** /-iːm/) communal esp. farming settlement in Israel.

kibosh /'kaɪbɒʃ/ *noun slang* nonsense. □ **put the kibosh on** put an end to.

kick ● *verb* strike, strike out, or propel forcibly with foot or hoof; (often + *at*, *against*) protest, rebel; *slang* give up (habit); (often + *out*) expel, dismiss; (**kick oneself**, **could kick oneself**) be annoyed with oneself; score (goal) by kicking. ● *noun* kicking action or blow; recoil of gun; *colloquial* temporary enthusiasm, sharp stimulant effect; (often in *plural*) thrill. □ **kick about** drift idly, discuss informally; **kickback** recoil, payment esp. for illegal help; **kick the bucket** *slang* die; **kick off** *verb* begin football game, remove (shoes etc.) by kicking, *colloquial* start; **kick-off** *noun* start, esp. of football game; **kick-start(er)** (pedal on) device to start engine of motorcycle etc.; **kick up a fuss** *colloquial* create disturbance, object; **kick upstairs** get rid of by promotion.

kid[1] ● *noun* young goat; leather from this; *colloquial* child. ● *verb* (**-dd-**) (of goat) give birth to kid.

kid[2] *verb* (**-dd-**) *colloquial* deceive, tease. □ **no kidding** *slang* that is the truth.

kidnap /'kɪdnæp/ *verb* (**-pp-**; *US* **-p-**) carry off (person) illegally, esp. to obtain ransom. □ **kidnapper** *noun*.

kidney /'kɪdnɪ/ *noun* (*plural* **-s**) either of two organs serving to excrete urine; animal's kidney as food. □ **kidney bean** red-skinned kidney-shaped bean; **kidney machine** apparatus able to perform function of damaged kidney; **kidney-shaped** having one side concave and the other convex.

kill ● *verb* deprive of life or vitality; end; (**kill oneself**) *colloquial* overexert oneself, laugh heartily; *colloquial* overwhelm with amusement; switch off; pass (time) while waiting; *Computing* delete; *Sport* stop (ball) dead. ● *noun* (esp. in hunting) act of killing, animal(s) killed. □ **killjoy** depressing person; **kill off** destroy completely, bring about death of (fictional character).

killer *noun* person or thing that kills; murderer. □ **killer whale** dolphin with prominent dorsal fin.

killing ● *noun* causing death; *colloquial* great financial success. ● *adjective colloquial* very funny; exhausting.

kiln *noun* oven for burning, baking, or drying esp. pottery.

kilo /'kiːləʊ/ *noun* (*plural* **-s**) kilogram.

kilo- *combining form* one thousand.

kilobyte /'kɪləbaɪt/ *noun Computing* 1,024 bytes as measure of memory size etc.

kilocalorie /'kɪləkælərɪ/ *noun* large calorie (see CALORIE).

kilocycle /'kɪləsaɪk(ə)l/ *noun historical* kilohertz.

kilogram /'kɪləgræm/ *noun* SI unit of mass (2.205 lb).

kilohertz /'kɪləhɜːts/ *noun* 1,000 hertz.

kilolitre /'kɪləliːtə/ *noun* (*US* **-liter**) 1,000 litres.

kilometre /'kɪləmiːtə/ *noun* (*US* **-meter**) 1,000 metres (0.6214 mile).

■ **Usage** *Kilometre* is often pronounced /kɪ'lɒmɪtə/ (with the stress on the *-lom-*), but this is considered incorrect by some people.

kiloton /'kɪlətʌn/ *noun* (also **kilotonne**) unit of explosive power equal to that of 1,000 tons of TNT.

kilovolt /'kɪləvəʊlt/ *noun* 1,000 volts.

kilowatt /'kɪləwɒt/ *noun* 1,000 watts. □ **kilowatt-hour** electrical energy equal to 1 kilowatt used for 1 hour.

kilt ● *noun* pleated skirt usually of tartan, traditionally worn by Highland man. ● *verb* tuck up (skirts) round body; (esp. as **kilted** *adjective*) gather in vertical pleats.

kilter /'kɪltə/ *noun* (also **kelter** /'keltə/) good working order.

kimono /kɪ'məʊnəʊ/ *noun* (*plural* **-s**) wide-sleeved Japanese robe; similar dressing gown.

kin ● *noun* one's relatives or family. ● *adjective* related.

kind /kaɪnd/ ● *noun* species, natural group of animals, plants, etc.; class, type, variety. ● *adjective* (often + *to*) friendly, benevolent. □ **in kind** in same form, (of payment) in goods etc. instead of money.

kindergarten /'kɪndəgɑːt(ə)n/ *noun* class or school for young children.

kind-hearted *adjective* of kind disposition. □ **kind-heartedly** *adverb*; **kind-heartedness** *noun*.

kindle /'kɪnd(ə)l/ *verb* (**-ling**) set on fire, light; inspire; become aroused or animated.

kindling /'kɪndlɪŋ/ *noun* small sticks etc. for lighting fires.

kindly /'kaɪndlɪ/ ● *adverb* in a kind way; please. ● *adjective* (**-ier**, **-iest**) kind; (of climate etc.) pleasant, mild.

kindred /'kɪndrɪd/ ● *adjective* related, allied, similar. ● *noun* blood relationship; one's relations.

kinetic /kɪ'netɪk/ *adjective* of or due to motion. □ **kinetic energy** energy of motion. □ **kinetically** *adverb*.

king *noun* (as title usually **King**) male sovereign, esp. hereditary; outstanding man or thing in specified field; largest kind of a thing; chess piece which must be checkmated for a win; crowned piece in draughts; court card depicting king; (**the King**) national anthem when sovereign is male. □ **King Charles spaniel** small black and tan kind; **kingcup** marsh marigold; **kingpin** main or large bolt, essential person or thing; **king-size(d)** large. □ **kingly** *adjective*; **kingship** *noun*.

kingdom *noun* state or territory ruled by king or queen; spiritual reign of God; domain; division of natural world. □ **kingdom come** *colloquial* the next world.

kingfisher *noun* small river bird with brilliant blue plumage, which dives for fish.

kink ● *noun* twist or bend in wire etc.; tight wave in hair; mental peculiarity. ● *verb* (cause to) form kink.

kinky *adjective* (**-ier**, **-iest**) *colloquial* sexually perverted or unconventional; (of clothing) bizarre and sexually provocative. □ **kinkily** *adverb*.

kinship *noun* blood relationship; similarity.

kinsman /'kɪnzmən/ *noun* (*feminine* **kinswoman**) blood relation.

kiosk /'ki:ɒsk/ *noun* open-fronted booth selling newspapers, food, etc.; telephone box.

kip *slang* ● *noun* sleep, bed. ● *verb* (**-pp-**) (often + *down*) sleep.

kipper /'kɪpə/ ● *noun* fish, esp. herring, split, salted, dried, and usually smoked. ● *verb* treat (herring etc.) this way.

kir /kɪə/ *noun* dry white wine with blackcurrant liqueur.

kirk *noun Scottish & Northern English* church. □ **Kirk-session** lowest court in Church of Scotland.

kirsch /kɪəʃ/ *noun* spirit distilled from cherries.

kismet /'kɪzmet/ *noun* destiny.

kiss ● *verb* touch with lips, esp. as sign of love, reverence, etc.; touch lightly. ● *noun* touch of lips; light touch. □ **kiss-curl** small curl of hair on forehead or nape of neck; **kiss of life** mouth-to-mouth resuscitation.

kisser *noun slang* mouth; face.

kissogram *noun* novelty greeting delivered with kiss.

kit ● *noun* equipment, clothing, etc. for particular purpose; specialized, esp. sports, clothing or uniform; set of parts needed to assemble furniture, model, etc. ● *verb* (**-tt-**) supply; (often + *out*) equip with kit. □ **kitbag** usually cylindrical canvas etc. bag for carrying soldier's etc. kit.

kitchen /'kɪtʃɪn/ *noun* place where food is cooked; kitchen fitments. □ **kitchen garden** garden for growing fruit and vegetables.

kitchenette /kɪtʃɪ'net/ *noun* small kitchen or cooking area.

kite *noun* light framework with thin covering flown on long string in wind; soaring bird of prey.

kith /kɪθ/ *noun* □ **kith and kin** friends and relations.

kitsch /kɪtʃ/ *noun* vulgar, pretentious, or worthless art.

kitten /'kɪt(ə)n/ ● *noun* young cat, ferret, etc. ● *verb* give birth to (kittens).

kittenish *adjective* playful; flirtatious.

kittiwake /'kɪtɪweɪk/ *noun* kind of small seagull.

kitty[1] /'kɪtɪ/ *noun* (*plural* **-ies**) joint fund; pool in some card games.

kitty[2] /'kɪtɪ/ *noun* (*plural* **-ies**) *childish name for* kitten or cat.

kiwi /'ki:wi:/ *noun* flightless NZ bird; (**Kiwi**) *colloquial* New Zealander. □ **kiwi fruit** green-fleshed fruit.

Klaxon /'klæks(ə)n/ *noun proprietary term* horn, warning hooter.

Kleenex /'kli:neks/ *noun* (*plural* same or **-es**) *proprietary term* disposable paper handkerchief.

kleptomania /kleptə'meɪnɪə/ *noun* irresistible urge to steal. □ **kleptomaniac** /-nɪæk/ *adjective & noun.*

km *abbreviation* kilometre(s).

knack /næk/ *noun* acquired faculty of doing something skilfully; habit of action, speech, etc.

knacker /'nækə/ ● *noun* buyer of useless horses for slaughter. ● *verb slang* (esp. as **knackered** *adjective*) exhaust, wear out.

knapsack /'næpsæk/ *noun* soldier's or hiker's bag carried on back.

knapweed /'næpwi:d/ *noun* plant with thistle-like flower.

knave /neɪv/ *noun* rogue, scoundrel; jack (in playing cards). □ **knavery** *noun*; **knavish** *adjective.*

knead /ni:d/ *verb* work into dough, paste, etc., esp. by hand; make (bread, pottery) thus; massage.

knee /ni:/ ● *noun* joint between thigh and lower leg; lap of sitting person; part of garment covering knee. ● *verb* (**knees**, **kneed**, **kneeing**) touch or strike with knee. □ **kneecap** convex bone in front of knee; **knees-up** *colloquial* lively party or gathering.

kneel /ni:l/ *verb* (*past & past participle* **knelt**) rest or lower oneself on knee(s).

kneeler *noun* cushion for kneeling on.

knell /nel/ *noun* sound of bell, esp. for death or funeral; event etc. seen as bad omen.

knelt *past & past participle* of KNEEL.

knew *past* of KNOW.

knickerbockers /'nɪkəbɒkəz/ *plural noun* loose-fitting breeches gathered in at knee. □ **Knickerbocker Glory** ice cream and fruit in tall glass.

knickers /'nɪkəz/ *plural noun* woman's or girl's undergarment for lower torso.

knick-knack /'nɪknæk/ *noun* (also **nick-nack**) trinket, small ornament.

knife /naɪf/ ● *noun* (*plural* **knives**) cutting blade or weapon with long sharpened edge fixed in handle; cutting-blade in machine; (**the knife**) *colloquial* surgery. ● *verb* (**-fing**) cut or stab with knife. □ **knife-edge** edge of knife, position of extreme uncertainty; **knife-pleat** overlapping narrow flat pleat.

knight /naɪt/ ● *noun* man awarded non-hereditary title (*Sir*) by sovereign; *historical* man raised to honourable military rank; *historical* lady's champion in tournament etc.; chess piece usually in shape of horse's head. ● *verb* confer knighthood on. □ **knighthood** *noun*; **knightly** *adjective.*

knit /nɪt/ *verb* (**-tt-**; *past & past participle* **knitted** or **knit**) make (garment etc.) by interlocking loops of esp. wool with knitting-needles or knitting machine; make (plain stitch) in knitting; wrinkle (brow); (often + *together*) make or become close, (of broken bone) become joined. □ **knitwear** knitted garments.

knitting *noun* work being knitted. □ **knitting-needle** thin pointed rod used usually in pairs for knitting.

knob /nɒb/ *noun* rounded protuberance, e.g. door handle, radio control, etc.; small lump (of butter, coal, etc.). □ **knobby** *adjective.*

knobbly /'nɒblɪ/ *adjective* (**-ier**, **-iest**) hard and lumpy.

knock /nɒk/ ● *verb* strike with audible sharp blow; (often + *at*) strike (door etc.) for admittance; (usually + *in*, *off*, etc.) drive by striking; make (hole) by knocking; (of engine) make thumping etc. noise; *slang* criticize. ● *noun* audible sharp blow; rap, esp. at door. □ **knock about**, **around** treat roughly, wander about aimlessly, (usually + *with*) associate socially; **knock back** *slang* eat or drink, esp. quickly; **knock down** *verb* strike (esp. person) to ground, demolish, (usually + *to*) (at auction) sell to bidder, *colloquial* lower price of; **knock-down** *adjective* (of price) very low; **knock knees** legs curved inward at the knee; **knock-kneed** with knock knees; **knock off** strike off with blow, *colloquial* finish work, do or make rapidly, (often + *from*) deduct (amount) from price, *slang* steal, kill; **knock on the head** put end to (scheme etc.);

knock on wood *US* touch wood; **knock out** *verb* make unconscious by blow to head, defeat (boxer) by knocking down for count of 10, defeat in knockout competition, *colloquial* tire out; **knockout** *noun* blow that knocks boxer out, competition in which loser of each match is eliminated, *slang* outstanding person or thing; **knock together** construct hurriedly; **knock up** *verb* make hastily, arouse by knock at door, practise tennis etc. before formal game begins, *US slang* make pregnant; **knock-up** *noun* practice at tennis etc.

knocker *noun* hinged metal device on door for knocking with.

knoll /nəʊl/ *noun* small hill, mound.

knot /nɒt/ ● *noun* intertwining of rope, string, etc. so as to fasten; set method of this; tangle in hair, knitting, etc.; unit of ship's or aircraft's speed equal to one nautical mile per hour; hard mass formed in tree trunk where branch grows out; round cross-grained piece in board caused by this; (usually + *of*) cluster. ● *verb* (**-tt-**) tie in knot; entangle. □ **knotgrass** wild plant with creeping stems and pink flowers; **knot-hole** hole in timber where knot has fallen out.

knotty *adjective* (**-ier**, **-iest**) full of knots; puzzling.

know /nəʊ/ *verb* (*past* **knew** /nju:/; *past participle* **known**) (often + *that*, *how*, *what*, etc.) have in the mind, have learnt; be acquainted with; recognize, identify; (often + *from*) be able to distinguish; (as **known** *adjective*) publicly acknowledged. □ **in the know** having inside information; **know-how** practical knowledge or skill.

knowing *adjective* cunning; showing knowledge, shrewd.

knowingly *adverb* in a knowing way; consciously, intentionally.

knowledge /ˈnɒlɪdʒ/ *noun* (usually + *of*) awareness, familiarity, person's range of information, understanding (of subject); sum of what is known.

knowledgeable *adjective* (also **knowledgable**) well-informed, intelligent. □ **knowledgeably** *adverb*.

known *past participle* of KNOW.

knuckle /ˈnʌk(ə)l/ ● *noun* bone at finger-joint; knee- or ankle-joint of quadruped; this as joint of meat. ● *verb* (**-ling**) strike, rub, etc. with knuckles. □ **knuckle down** (often + *to*) apply oneself earnestly; **knuckleduster** metal guard worn over knuckles in fighting, esp. to inflict greater damage; **knuckle under** give in, submit.

KO *abbreviation* knockout.

koala /kəʊˈɑ:lə/ *noun* (in full **koala bear**) small Australian bearlike marsupial with thick grey fur.

kohl /kəʊl/ *noun* black powder used as eye make-up, esp. in Eastern countries.

kohlrabi /kəʊlˈrɑ:bɪ/ *noun* (*plural* **-bies**) cabbage with edible turnip-like stem.

kookaburra /ˈkʊkəbʌrə/ *noun* Australian kingfisher with strange laughing cry.

Koran /kɔ:ˈrɑ:n/ *noun* Islamic sacred book.

Korean /kəˈri:ən/ ● *noun* native or national of N. or S. Korea; language of Korea. ● *adjective* of Korea or its people or language.

kosher /ˈkəʊʃə/ ● *adjective* (of food or food-shop) fulfilling requirements of Jewish law; *colloquial* correct, genuine. ● *noun* kosher food or shop.

kowtow /kaʊˈtaʊ/ ● *noun historical* Chinese custom of touching ground with forehead, esp. in submission. ● *verb* (usually + *to*) act obsequiously; perform kowtow.

k.p.h. *abbreviation* kilometres per hour.

kraal /krɑ:l/ *noun South African* village of huts enclosed by fence; enclosure for cattle etc.

kremlin /ˈkremlɪn/ *noun* citadel within Russian town; (**the Kremlin**) that in Moscow, Russian government.

krill *noun* tiny plankton crustaceans eaten by whales etc.

krugerrand /ˈkru:gərænd/ *noun* S. African gold coin.

krummhorn /ˈkrʌmhɔ:n/ *noun* (also **crumhorn**) medieval wind instrument.

krypton /ˈkrɪptɒn/ *noun* gaseous element used in lamps etc.

Kt. *abbreviation* Knight.

kts. *abbreviation* knots.

kudos /'kju:dɒs/ *noun colloquial* glory, renown.

kumquat /'kʌmkwɒt/ *noun* (also **cumquat**) small orange-like fruit.

kung fu /kʌŋ 'fu:/ *noun* Chinese form of karate.

kV *abbreviation* kilovolt(s).

kW *abbreviation* kilowatt(s).

kWh *abbreviation* kilowatt-hour(s).

L¹ *noun* (also **l**) (Roman numeral) 50.

L² *abbreviation* Lake. □ **L-plate** sign bearing letter L, attached to vehicle to show that driver is learner.

l *abbreviation* left; line; litre(s).

LA *abbreviation* Los Angeles.

la = LAH.

Lab. *abbreviation* Labour.

lab *noun colloquial* laboratory.

label /'leɪb(ə)l/ ● *noun* piece of paper attached to object to give information about it; classifying phrase etc.; logo, title, or trademark of company. ● *verb* (**-ll-**; *US* **-l-**) attach label to; (usually + *as*) assign to category.

labial /'leɪbɪəl/ ● *adjective* of lips; *Phonetics* pronounced with (closed) lips. ● *noun* *Phonetics* labial sound.

labium /'leɪbɪəm/ *noun* (*plural* **labia**) each fold of skin of pairs enclosing vulva.

labor etc. *US & Australian* = LABOUR etc.

laboratory /lə'bɒrətərɪ/ *noun* (*plural* **-ies**) place used for scientific experiments and research.

laborious /lə'bɔ:rɪəs/ *adjective* needing hard work; (esp. of literary style) showing signs of effort. □ **laboriously** *adverb.*

labour /'leɪbə/ (*US & Australian* **labor**) ● *noun* physical or mental work, exertion; workers, esp. as political force; (**Labour**) the Labour Party; process of giving birth; task. ● *verb* work hard, exert oneself; elaborate needlessly; proceed with difficulty; (as **laboured** *adjective*) done with great effort; (+ *under*) suffer because of. □ **labour camp** prison camp enforcing hard labour; **Labour Exchange** *colloquial or historical* jobcentre; **Labour Party** political party formed to represent workers' interests; **labour-saving** designed to reduce or eliminate work.

labourer *noun* (*US* **laborer**) person doing unskilled paid manual work.

Labrador /'læbrədɔ:/ *noun* dog of retriever breed with black or golden coat.

laburnum /lə'bɜ:nəm/ *noun* tree with drooping golden flowers and poisonous seeds.

labyrinth /'læbərɪnθ/ *noun* complicated network of passages; intricate or tangled arrangement. □ **labyrinthine** /-'rɪnθaɪn/ *adjective.*

lac *noun* resinous substance from SE Asian insect, used to make varnish and shellac.

lace ● *noun* open patterned fabric or trimming made by twisting, knotting, or looping threads; cord etc. passed through eyelets or hooks for fastening shoes etc. ● *verb* (**-cing**) (usually + *up*) fasten or tighten with lace(s); add spirits to (drink); (+ *through*) pass (shoelace etc.) through. □ **lace-up** shoe fastened with lace.

lacerate /'læsəreɪt/ *verb* (**-ting**) tear (esp. flesh etc.) roughly; wound (feelings etc.). □ **laceration** *noun.*

lachrymal /'lækrɪm(ə)l/ *adjective* (also **lacrimal**) of tears.

lachrymose /'lækrɪməʊs/ *adjective formal* often weeping; tearful.

lack ● *noun* (usually + *of*) deficiency, want. ● *verb* be without or deficient in.

□ **lacklustre** (*US* **lackluster**) dull, lacking in vitality etc.

lackadaisical /lækə'deɪzɪk(ə)l/ *adjective* languid; unenthusiastic.

lackey /'lækɪ/ *noun* (*plural* **-s**) servile follower; footman, manservant.

lacking *adjective* undesirably absent; (+ *in*) deficient in.

laconic /lə'kɒnɪk/ *adjective* using few words. □ **laconically** *adverb.*

lacquer /'lækə/ ● *noun* hard shiny shellac or synthetic varnish; substance sprayed on hair to keep it in place. ● *verb* coat with lacquer.

lacrimal = LACHRYMAL.

lacrosse /lə'krɒs/ *noun* hockey-like game played with ball carried in net at end of stick.

lactate /læk'teɪt/ *verb* (**-ting**) (of mammals) secrete milk. □ **lactation** *noun.*

lactic /'læktɪk/ *adjective* of milk.

lactose /'læktəʊs/ *noun* sugar present in milk.

lacuna /lə'kju:nə/ *noun* (*plural* **-s** or **-nae** /-ni:/) missing part, esp. in manuscript; gap.

lacy /'leɪsɪ/ *adjective* (**-ier**, **-iest**) like lace fabric.

lad *noun* boy, youth; *colloquial* man.

ladder /'lædə/ ● *noun* set of horizontal bars fixed at intervals between two uprights for climbing up and down; unravelled stitching in stocking etc.; means of advancement in career etc. ● *verb* cause or develop ladder in (stocking etc.).

lade *verb* (**-ding**; *past participle* **laden**) load (ship); ship (goods); (as **laden** *adjective*) (usually + *with*) loaded, burdened.

la-di-da /lɑ:dɪ'dɑ:/ *adjective colloquial* pretentious or affected, esp. in manner or speech.

ladle /'leɪd(ə)l/ ● *noun* deep long-handled spoon for serving liquids. ● *verb* (**-ling**) (often + *out*) transfer with ladle.

lady /'leɪdɪ/ *noun* (*plural* **-ies**) woman regarded as having superior status or refined manners; *polite form of address for woman*; *colloquial* wife, girlfriend; (**Lady**) *title used before name of peeresses, peers' female relatives, wives and widows of knights, etc.*; (**Ladies**) women's public lavatory. □ **ladybird** small beetle, usually red with black spots; **Lady chapel** chapel dedicated to Virgin Mary; **Lady Day** Feast of the Annunciation, 25 Mar.; **ladylike** like or appropriate to lady.

Ladyship *noun* □ **Her** or **Your Ladyship** *title used of or to Lady.*

lag[1] ● *verb* (**-gg-**) fall behind; not keep pace. ● *noun* delay.

lag[2] *verb* (**-gg-**) enclose (boiler etc.) with heat-insulating material.

lag[3] *noun slang* habitual convict.

lager /'lɑ:gə/ *noun* kind of light beer. □ **lager lout** *colloquial* youth behaving violently through drinking too much.

laggard /'lægəd/ *noun* person lagging behind.

lagging *noun* insulating material for boiler etc.

lagoon /lə'gu:n/ *noun* salt-water lake separated from sea by sandbank, reef, etc.

lah *noun* (also **la**) *Music* sixth note of scale in tonic sol-fa.

laid *past & past participle* of LAY[1]. □ **laid-back** relaxed, easy-going.

lain *past participle* of LIE[1].

lair *noun* wild animal's home; person's hiding place.

laird /'leəd/ *noun Scottish* landed proprietor.

laissez-faire /leɪseɪ'feə/ *noun* (also ***laisser-faire***) policy of not interfering. [French]

laity /'leɪɪtɪ/ *noun* lay people, as distinct from clergy.

lake[1] *noun* large body of water surrounded by land. □ **Lake District** region of lakes in Cumbria.

lake[2] *noun* reddish pigment originally made from lac.

lam *verb* (**-mm-**) *slang* hit hard, thrash.

lama /'lɑ:mə/ *noun* Tibetan or Mongolian Buddhist monk.

lamasery /'la:məsərɪ/ *noun* (*plural* **-ies**) lama monastery.

lamb /læm/ ● *noun* young sheep; its flesh as food; gentle, innocent, or weak person. ● *verb* (of sheep) give birth.

lambaste /læm'beɪst/ *verb* (**-ting**) (also **lambast** /-'bæst/) *colloquial* thrash, beat.

lambent /'læmbənt/ *adjective* (of flame etc.) playing on a surface; (of eyes, wit, etc.) gently brilliant. □ **lambency** *noun*.

lambswool *noun* soft fine wool from young sheep.

lame ● *adjective* disabled in foot or leg; (of excuse etc.) unconvincing; (of verse etc.) halting. ● *verb* (**-ming**) make lame, disable. □ **lame duck** helpless person or firm. □ **lamely** *adverb*; **lameness** *noun*.

lamé /'lɑ:meɪ/ *noun* fabric with gold or silver thread woven in.

lament /lə'ment/ ● *noun* passionate expression of grief; song etc. of mourning. ● *verb* express or feel grief for or about; utter lament; (as **lamented** *adjective*) recently dead. □ **lamentation** /læmən-/ *noun*.

lamentable /'læməntəb(ə)l/ *adjective* deplorable, regrettable. □ **lamentably** *adverb*.

lamina /'læmɪnə/ *noun* (*plural* **-nae** /-ni:/) thin plate or layer. □ **laminar** *adjective*.

laminate ● *verb* /'læmɪneɪt/ (**-ting**) beat or roll into thin plates; overlay with plastic layer etc.; split into layers. ● *noun* /'læmɪnət/ laminated structure, esp. of layers fixed together. □ **lamination** *noun*.

lamp *noun* device for giving light from electricity, gas, oil, etc.; apparatus producing esp. ultraviolet or infrared radiation. □ **lamppost** post supporting street light; **lampshade** usually partial cover for lamp.

lampoon /læm'pu:n/ ● *noun* satirical attack on person etc. ● *verb* satirize. □ **lampoonist** *noun*.

lamprey /'læmprɪ/ *noun* (*plural* **-s**) eel-like fish with sucker mouth.

Lancastrian /læŋ'kæstrɪən/ ● *noun* native of Lancashire or Lancaster. ● *adjective* of Lancashire or Lancaster; of House of Lancaster in Wars of Roses.

lance /lɑ:ns/ ● *noun* long spear, esp. one used by horseman. ● *verb* (**-cing**) prick or open with lancet. □ **lance-corporal** army NCO below corporal.

lanceolate /'lɑ:nsɪələt/ *adjective* shaped like spearhead, tapering to each end.

lancer /'lɑ:nsə/ *noun historical* soldier of cavalry regiment originally armed with lances; (in *plural*) quadrille.

lancet /'lɑ:nsɪt/ *noun* small broad two-edged surgical knife with sharp point.

land ● *noun* solid part of earth's surface; ground, soil, expanse of country; nation, state; landed property; (in *plural*) estates. ● *verb* set or go ashore; bring (aircraft) down; alight on ground etc.; bring (fish) to land; often + *up*) bring to or arrive at certain situation or place; *colloquial* deal (person etc. a blow etc.); (+ *with*) present (person) with (problem etc.); *colloquial* win (prize, appointment, etc.). □ **landfall** approach to land after sea or air journey; **landfill** waste material used to landscape or reclaim land, disposing of waste in this way; **landlady** woman owning rented property or keeping pub, guest-house, etc.; **landlocked** (almost) enclosed by land; **landlord** man owning rented property or keeping pub, guest-house, etc.; **landlubber** person unfamiliar with sea and ships; **landmark** conspicuous object, notable event; **land-mine** explosive mine laid in or on ground; **landslide** sliding down of mass of land from cliff or mountain, overwhelming majority in election.

landau /'lændɔ:/ *noun* (*plural* **-s**) 4-wheeled enclosed carriage with divided top.

landed *adjective* owning or consisting of land.

landing /'lændɪŋ/ *noun* platform or passage at top of or part way up stairs. □ **landing-craft** craft used for putting troops and equipment ashore; **landing-gear** undercarriage of aircraft; **landing-stage** platform for disembarking passengers and goods.

landscape /'lændskeɪp/ ● *noun* scenery in area of land; picture of it. ● *verb* (**-ping**) improve (piece of land) by **landscape gardening**, laying out of grounds to resemble natural scenery.

lane *noun* narrow road; division of road for one line of traffic; strip of track or water for competitor in race; regular course followed by ship or aircraft.

language /'læŋgwɪdʒ/ *noun* use of words in agreed way as means of human communication; system of words of particular community, country, etc.; faculty of speech; style of expression;

system of symbols and rules for computer programs. □ **language laboratory** room with tape recorders etc. for learning foreign language.

languid /ˈlæŋgwɪd/ *adjective* lacking vigour; idle. □ **languidly** *adverb.*

languish /ˈlæŋgwɪʃ/ *verb* lose or lack vitality. □ **languish for** long for; **languish under** live under (depression etc.).

languor /ˈlæŋgə/ *noun* lack of energy; idleness; soft or tender mood or effect. □ **languorous** *adjective.*

lank *adjective* (of grass, hair, etc.) long and limp; thin and tall.

lanky *adjective* (**-ier**, **-iest**) ungracefully thin and long or tall.

lanolin /ˈlænəlɪn/ *noun* fat from sheep's wool used in cosmetics, ointments, etc.

lantern /ˈlænt(ə)n/ *noun* lamp with transparent case protecting flame etc.; glazed structure on top of dome or room; light-chamber of lighthouse. □ **lantern jaws** long thin jaws.

lanyard /ˈlænjəd/ *noun* cord round neck or shoulder for holding knife etc.; *Nautical* short rope.

lap¹ *noun* front of sitting person's body from waist to knees; clothing covering this. □ **lap-dog** small pet dog; **laptop** (microcomputer) suitable for use while travelling.

lap² ● *noun* one circuit of racetrack etc.; section of journey etc.; amount of overlap. ● *verb* (**-pp-**) overtake (competitor in race who is a lap behind); (often + *about*, *around*) fold or wrap (garment etc.).

lap³ ● *verb* (**-pp-**) (esp. of animal) drink by scooping with tongue; (usually + *up*, *down*) drink greedily; (usually + *up*) receive (gossip, praise, etc.) eagerly; (of waves etc.) ripple; make lapping sound against (shore). ● *noun* act or sound of lapping.

lapel /ləˈpel/ *noun* part of coat-front folded back.

lapidary /ˈlæpɪdərɪ/ ● *adjective* concerned with stones; engraved on stone; concise, well-expressed. ● *noun* (*plural* **-ies**) cutter, polisher, or engraver of gems.

lapis lazuli /læpɪs ˈlæzjʊlɪ/ *noun* bright blue gem; its colour.

lapse /læps/ ● *noun* slight error; slip of memory etc.; weak or careless decline into inferior state. ● *verb* (**-sing**) fail to maintain position or standard; (+ *into*) fall back into (inferior or previous state); (of right etc.) become invalid.

lapwing /ˈlæpwɪŋ/ *noun* plover with shrill cry.

larceny /ˈlɑːsənɪ/ *noun* (*plural* **-ies**) theft of personal property. □ **larcenous** *adjective.*

larch *noun* deciduous coniferous tree with bright foliage; its wood.

lard ● *noun* pig fat used in cooking etc. ● *verb* insert strips of bacon in (meat etc.) before cooking; (+ *with*) embellish (talk etc.) with (strange terms etc.).

larder *noun* room or cupboard for storing food.

lardy-cake *noun* cake made with lard, currants, etc.

large *adjective* of relatively great size or extent; of larger kind; comprehensive. □ **at large** at liberty, as a body or whole; **large as life** in person, esp. prominently. □ **largeness** *noun*; **largish** *adjective.*

largely *adverb* to a great extent.

largesse /lɑːˈʒes/ *noun* (also **largess**) money or gifts freely given.

largo /ˈlɑːgəʊ/ *Music* ● *adverb & adjective* in slow time and dignified style. ● *noun* (*plural* **-s**) largo movement or passage.

lariat /ˈlærɪət/ *noun* lasso; rope for tethering animal.

lark¹ *noun* small bird with tuneful song, esp. skylark.

lark² *colloquial* ● *noun* frolic; amusing incident; type of activity. ● *verb* (+ *about*) play tricks.

larkspur *noun* plant with spur-shaped calyx.

larva /ˈlɑːvə/ *noun* (*plural* **-vae** /-viː/) insect in stage between egg and pupa. □ **larval** *adjective.*

laryngeal /ləˈrɪndʒɪəl/ *adjective* of the larynx.

laryngitis /lærɪnˈdʒaɪtɪs/ *noun* inflammation of larynx.

larynx /ˈlærɪŋks/ *noun* (*plural* **larynges** /ləˈrɪndʒiːz/ or **-xes**) cavity in throat holding vocal cords.

lasagne /ləˈsænjə/ *noun* pasta sheets.

lascivious /lə'sɪvɪəs/ *adjective* lustful. □ **lasciviously** *adverb*.

laser /'leɪzə/ *noun* device producing intense beam of special kind of light. □ **laser printer** printing machine using laser to produce image.

lash ● *verb* make sudden whiplike movement; beat with whip; (often + *against*, *down*) (of rain etc.) beat, strike; criticize harshly; rouse, incite; (often + *together*, *down*) fasten with rope etc. ● *noun* sharp blow with whip etc.; flexible part of whip; eyelash. □ **lash out** speak or hit out angrily, *colloquial* spend money extravagantly.

lashings *plural noun colloquial* (often + *of*) plenty.

lass *noun esp. Scottish & Northern English or poetical* girl.

lassitude /'læsɪtju:d/ *noun* languor; disinclination to exert oneself.

lasso /læ'su:/ ● *noun* (*plural* **-s** or **-es**) rope with running noose used esp. for catching cattle. ● *verb* (**-es**, **-ed**) catch with lasso.

last[1] /lɑ:st/ ● *adjective* after all others; coming at end; most recent; only remaining. ● *adverb* after all others; on most recent occasion. ● *noun* last, last-mentioned, or most recent person or thing; last mention, sight, etc.; end; death. □ **at (long) last** in the end, after much delay; **the last straw** slight addition to task etc. making it unbearable.

last[2] /lɑ:st/ *verb* remain unexhausted, adequate, or alive for specified or long time. □ **last out** be sufficient for whole of given period.

last[3] /lɑ:st/ *noun* shoemaker's model for shaping shoe etc. □ **stick to one's last** keep to what one understands.

lasting *adjective* permanent, durable.

lastly *adverb* finally.

lat. *abbreviation* latitude.

latch ● *noun* bar with catch as fastening of gate etc.; spring-lock as fastening of outer door. ● *verb* fasten with latch. □ **latchkey** key of outer door; **latch on to** *colloquial* attach oneself to, understand.

late ● *adjective* after due or usual time; far on in day, night, period, etc.; flowering, ripening, etc. towards end of season; no longer alive or having specified status; of recent date. ● *adverb* after due or usual time; far on in time; at or till late hour; at late stage of development; formerly but not now. □ **late in the day** at late stage of proceedings etc. □ **lateness** *noun*.

lateen sail /lə'ti:n/ *noun* triangular sail on long yard at angle of 45° to mast.

lately *adverb* not long ago; recently.

latent /'leɪt(ə)nt/ *adjective* existing but not developed or manifest; concealed, dormant. □ **latency** *noun*.

lateral /'lætər(ə)l/ ● *adjective* of, at, towards, or from side(s). ● *noun* lateral shoot or branch. □ **lateral thinking** method of solving problems by indirect or illogical methods. □ **laterally** *adverb*.

latex /'leɪteks/ *noun* milky fluid of esp. rubber tree; synthetic substance like this.

lath /lɑ:θ/ *noun* thin flat strip of wood.

lathe /leɪð/ *noun* machine for shaping wood, metal, etc. by rotating article against cutting tools.

lather /'lɑ:ðə/ ● *noun* froth made by agitating soap etc. and water; frothy sweat; state of agitation. ● *verb* (of soap) form lather; cover with lather; *colloquial* thrash.

Latin /'lætɪn/ ● *noun* language of ancient Rome. ● *adjective* of or in Latin; of countries or peoples speaking languages developed from Latin; of RC Church. □ **Latin America** parts of Central and S. America where Spanish or Portuguese is main language.

Latinate /'lætɪneɪt/ *adjective* having character of Latin.

latitude /'lætɪtju:d/ *noun* angular distance N. or S. of equator; (usually in *plural*) regions, climes; freedom from restriction in action or opinion.

latrine /lə'tri:n/ *noun* communal lavatory, esp. in camp.

latter /'lætə/ ● *adjective* second-mentioned of two; nearer the end. ● *noun* (**the latter**) the latter thing or person. □ **latter-day** modern, contemporary.

latterly *adverb* recently; in latter part of life or period.

lattice /'lætɪs/ *noun* structure of crossed laths or bars with spaces between, used as fence, screen, etc.; arrangement resembling this. □ **lattice window** one with small panes set in lead. □ **latticed** *adjective.*

Latvian /'lætvɪən/ ● *noun* native, national, or language of Latvia. ● *adjective* of Latvia.

laud /lɔ:d/ *verb* praise, extol.

laudable *adjective* praiseworthy. □ **laudably** *adverb.*

laudanum /'lɔ:dənəm/ *noun* solution prepared from opium.

laudatory /'lɔ:dətərɪ/ *adjective* praising.

laugh /lɑ:f/ ● *verb* make sounds etc. usual in expressing amusement, scorn, etc.; express by laughing; (+ *at*) make fun of, ridicule. ● *noun* sound or act of laughing; *colloquial* comical person or thing. □ **laugh off** shrug off (embarrassment etc.) by joking.

laughable *adjective* amusing; ridiculous. □ **laughably** *adverb.*

laughing *noun* laughter. □ **laughing gas** nitrous oxide as anaesthetic; **laughing jackass** kookaburra; **laughing stock** object of general derision. □ **laughingly** *adverb.*

laughter /'lɑ:ftə/ *noun* act or sound of laughing.

launch[1] /lɔ:ntʃ/ ● *verb* set (vessel) afloat; hurl or send forth (rocket etc.); start or set in motion (enterprise, person, etc.); formally introduce (new product) with publicity; (+ *into*) make start on; (+ *out*) make start on new enterprise. ● *noun* launching. □ **launch pad** platform with structure for launching rockets from.

launch[2] /lɔ:ntʃ/ *noun* large motor boat.

launder /'lɔ:ndə/ *verb* wash and iron etc. (clothes etc.); *colloquial* transfer (money) to conceal its origin.

launderette /lɔ:n'dret/ *noun* (also **laundrette**) establishment with coin-operated washing machines and driers for public use.

laundress /'lɔ:ndrɪs/ *noun* woman who launders.

laundry /'lɔ:ndrɪ/ *noun* (*plural* **-ies**) place where clothes etc. are laundered; clothes etc. that need to be or have been laundered.

laurel /'lɒr(ə)l/ *noun* any of various kinds of shrub with dark green glossy leaves; (in *singular* or *plural*) wreath of bay-leaves as emblem of victory or poetic merit. □ **look to one's laurels** beware of losing one's pre-eminence; **rest on one's laurels** stop seeking further success.

lava /'lɑ:və/ *noun* matter flowing from volcano and solidifying as it cools.

lavatorial /lævə'tɔ:rɪəl/ *adjective* of or like lavatories; (esp. of humour) relating to excretion.

lavatory /'lævətərɪ/ *noun* (*plural* **-ies**) receptacle for urine and faeces, usually with means of disposal; room etc. containing this.

lave *verb* (**-ving**) *literary* wash, bathe; wash against; flow along.

lavender /'lævɪndə/ *noun* evergreen fragrant-flowered shrub; its dried flowers used to scent linen; pale purplish colour. □ **lavender-water** light perfume.

laver /'leɪvə/ *noun* kind of edible seaweed.

lavish /'lævɪʃ/ ● *adjective* profuse; abundant; generous. ● *verb* (often + *on*) bestow or spend (money, praise, etc.) abundantly. □ **lavishly** *adverb.*

law *noun* rule or set of rules established in a community, demanding or prohibiting certain actions; such rules as social system or branch of study; binding force; (**the law**) legal profession, *colloquial* police; law courts, legal remedy; science or philosophy of law; statement of regularity of natural occurrences. □ **law-abiding** obedient to the laws; **law court** court of law; **Law Lord** member of House of Lords qualified to perform its legal work; **lawsuit** bringing of claim etc. before law court; **lay down the law** give dogmatic opinions; **take the law into one's own hands** get one's rights without help of the law.

lawful *adjective* permitted, appointed, or recognized by law; not illegal. □ **lawfully** *adverb*; **lawfulness** *noun.*

lawless *adjective* having no laws; disregarding laws. □ **lawlessness** *noun.*

lawn[1] *noun* piece of close-mown grass in garden etc. □ **lawnmower** machine for cutting lawns; **lawn tennis** tennis played with soft ball on grass or hard court.

lawn[2] *noun* kind of fine linen or cotton.

lawyer /'lɔɪə/ *noun* person practising law, esp. solicitor.

lax *adjective* lacking care or precision; not strict. □ **laxity** *noun*; **laxly** *adverb*; **laxness** *noun*.

laxative /'læksətɪv/ ● *adjective* helping evacuation of bowels. ● *noun* laxative medicine.

lay[1] ● *verb* (*past & past participle* **laid**) place on surface, esp. horizontally; put or bring into required position or state; make by laying; (of bird) produce (egg); cause to subside or lie flat; (usually + *on*) attribute (blame etc.); make ready (trap, plan); prepare (table) for meal; put fuel ready to light (fire); put down as bet. ● *noun* way, position, or direction in which something lies. □ **lay bare** expose, reveal; **lay-by** extra strip beside road where vehicles may park; **lay claim to** claim as one's own; **lay down** relinquish, make (rule), store (wine) in cellar, sacrifice (one's life); **lay in** provide oneself with stock of; **lay into** *colloquial* attack violently with blows or verbally; **lay it on thick** or **with a trowel** *colloquial* flatter, exaggerate grossly; **lay off** discharge (workers) temporarily, *colloquial* desist; **lay on** provide, spread on; **lay out** spread, expose to view, prepare (body) for burial, *colloquial* knock unconscious, expend (money); **layout** way in which land, building, printed matter, etc., is arranged or set out; **lay up** store, save (money), (as **laid up** *adjective*) confined to bed or the house; **lay waste** ravage, destroy.

■ **Usage** It is incorrect in standard English to use *lay* to mean 'lie', as in *She was laying on the floor*.

lay[2] *adjective* not ordained into the clergy; not professionally qualified; of or done by such people. □ **layman, laywoman** person not in holy orders, one without professional or special knowledge; **lay reader** lay person licensed to conduct some religious services.

lay[3] *noun* short poem meant to be sung; song.

lay[4] *past* of LIE[1].

layer ● *noun* thickness of matter, esp. one of several, covering surface; hen that lays eggs. ● *verb* arrange in layers; propagate (plant) by fastening shoot down to take root.

layette /leɪ'et/ *noun* clothes etc. prepared for newborn child.

lay figure *noun* artist's jointed wooden model of human figure; unrealistic character in novel etc.

laze ● *verb* (**-zing**) spend time idly. ● *noun* spell of lazing.

lazy *adjective* (**-ier**, **-iest**) disinclined to work, doing little work; of or inducing idleness. □ **lazybones** *colloquial* lazy person. □ **lazily** *adverb*; **laziness** *noun*.

lb *abbreviation* pound(s) weight.

■ **Usage** It is a common mistake to write *lbs* as an abbreviation for *pounds*. *28 lb* is correct.

l.b.w. *abbreviation* leg before wicket.

l.c. *abbreviation* loc. cit.; lower case.

LCD *abbreviation* liquid crystal display.

L/Cpl *abbreviation* Lance-Corporal.

LEA *abbreviation* Local Education Authority.

lea *noun poetical* meadow, field.

leach *verb* make (liquid) percolate through some material; subject (bark, ore, ash, soil) to this; (usually + *away*, *out*) remove (soluble matter) or be removed in this way.

lead[1] /li:d/ ● *verb* (*past & past participle* **led**) conduct, esp. by going in front; direct actions or opinions of; (often + *to*) guide by persuasion; provide access to; pass or spend (life etc.); have first place in; go or be first; play (card) as first player in trick; (+ *to*) result in; (+ *with*) (of newspapers or broadcast) have as main story. ● *noun* guidance, example; leader's place; amount by which competitor is ahead of others; clue; strap etc. for leading dog etc.; *Electricity* conductor carrying current to place of use; chief part in play etc.; *Cards* act or right of playing first. □ **lead by the nose** make (someone) do all one wishes them to; **lead-in** introduction, opening; **lead on** entice dishonestly; **lead up the garden path** *colloquial* mislead; **lead up to** form preparation for, direct conversation towards.

lead[2] /led/ ● *noun* heavy soft grey metal; graphite used in pencils; lump of lead used in sounding; blank space between lines of print. ● *verb* cover, frame, or space with lead(s). □ **lead-free** (of petrol) without added lead compounds.

leaded /'ledɪd/ *adjective* (of petrol) with added lead compounds; (of window pane) framed with lead.

leaden /'led(ə)n/ *adjective* of or like lead; heavy, slow; lead-coloured.

leader/'li:də/ *noun* person or thing that leads; leading performer in orchestra, quartet, etc.; leading article. □ **leadership** *noun*.

leading/'li:dɪŋ/ *adjective* chief, most important. □ **leading aircraftman** one ranking just below NCO in RAF; **leading article** newspaper article giving editorial opinion; **leading light** prominent influential person; **leading note** *Music* seventh note of ascending scale; **leading question** one prompting the answer wanted.

■ **Usage** *Leading question* does not mean a 'principal' or 'loaded' or 'searching' question.

leaf ● *noun* (*plural* **leaves**) flat usually green part of plant growing usually on stem; foliage; single thickness of paper, esp. in book; very thin sheet of metal etc.; hinged part, extra section, or flap of table etc. ● *verb* (of plants etc.) begin to grow leaves; (+ *through*) turn over pages of (book etc.). □ **leaf-mould** soil composed chiefly of decaying leaves. □ **leafage** *noun*; **leafy** *adjective* (**-ier**, **-iest**).

leaflet /'li:flɪt/ *noun* sheet of paper, pamphlet, etc., giving information; young leaf.

league[1] /li:g/ ● *noun* people, countries, etc., joining together for particular purpose; group of sports clubs who contend for championship; class of contestants. ● *verb* (**-gues**, **-gued**, **-guing**) (often + *together*) join in league. □ **in league** allied, conspiring; **league table** list in order of success.

league[2] /li:g/ *noun archaic* measure of travelling distance, usually about 3 miles.

leak ● *noun* hole through which liquid etc. passes accidentally in or out; liquid etc. thus passing through; similar escape of electric charge; disclosure of secret information. ● *verb* (let) pass out or in through leak; disclose (secret); (often + *out*) become known. □ **leaky** *adjective* (**-ier**, **-iest**).

leakage *noun* action or result of leaking.

lean[1] ● *verb* (*past & past participle* **leaned** or **leant** /lent/) (often + *across*, *back*, *over*, etc.) be or place in sloping position; (usually + *against*, *on*) rest for support against; (usually + *on*, *upon*) rely, depend; (usually + *to*, *towards*) be inclined or partial. ● *noun* inclination, slope. □ **lean on** *colloquial* put pressure on (person) to act in certain way; **lean-to** building with roof resting against larger building or wall.

lean[2] ● *adjective* (of person etc.) having no superfluous fat; (of meat) containing little fat; meagre. ● *noun* lean part of meat. □ **lean years** time of scarcity. □ **leanness** *noun*.

leaning *noun* tendency or inclination.

leap ● *verb* (*past & past participle* **leaped** or **leapt** /lept/) jump, spring forcefully. ● *noun* forceful jump. □ **by leaps and bounds** with very rapid progress; **leap-frog** game in which player vaults with parted legs over another bending down; **leap year** year with 29 Feb. as extra day.

learn /lɜ:n/ *verb* (*past & past participle* **learned** /lɜ:nt, lɜ:nd/ or **learnt**) get knowledge of or skill in by study, experience, or being taught; commit to memory; (usually + *of*, *about*) be told about, find out.

learned /'lɜ:nɪd/ *adjective* having much knowledge from studying; showing or requiring learning.

learner *noun* person learning, beginner; (in full **learner driver**) person who is learning to drive but has not yet passed driving test.

learning *noun* knowledge got by study.

lease /li:s/ ● *noun* contract by which owner of land or building allows another to use it for specified time, usually for rent. ● *verb* (**-sing**) grant or take on lease. □ **leasehold** holding of property by lease; **leaseholder** *noun*; **new lease of** (*US* **on**) **life** improved prospect of living, or of use after repair.

leash ● *noun* strap for holding dog(s). ● *verb* put leash on; restrain. □ **straining at the leash** eager to begin.

least ● *adjective* smallest, slightest. ● *noun* least amount. ● *adverb* in the least degree. □ **at least** at any rate; **to say the least** putting the case moderately.

leather /'leðə/ ● *noun* material made from skin of animal by tanning etc.; piece of leather for cleaning esp. windows; *slang* cricket ball, football. ● *verb* beat, thrash; cover or polish with leather. □ **leather-jacket** larva of crane-fly.

leatherette /leðə'ret/ *noun* imitation leather.

leathery *adjective* like leather; tough.

leave[1] *verb* (**-ving**; *past & past participle* **left**) go away (from); cause or allow to remain; depart without taking; cease to reside at, belong to, work for, etc.; abandon; (usually + *to*) commit to another person; bequeath; deposit or entrust (object, message, etc.) to be dealt with in one's absence; not consume or deal with. □ **leave off** come to or make an end, stop; **leave out** omit.

leave[2] *noun* permission; (in full **leave of absence**) permission to be absent from duty; period for which this lasts. □ **on leave** absent thus; **take one's leave of** say goodbye to.

leaven /'lev(ə)n/ ● *noun* substance used to make dough ferment and rise; transforming influence. ● *verb* ferment (dough) with leaven; permeate, transform.

leavings *plural noun* what is left.

Lebanese /lebə'ni:z/ ● *adjective* of Lebanon. ● *noun* (*plural* same) native or national of Lebanon.

lecher /'letʃə/ *noun* lecherous man.

lecherous *adjective* lustful. □ **lecherously** *adverb*; **lechery** *noun*.

lectern /'lekt(ə)n/ *noun* stand for holding Bible etc. in church; similar stand for lecturer etc.

lecture /'lektʃə/ ● *noun* talk giving information to class etc.; admonition, reprimand. ● *verb* (**-ring**) (often + *on*) deliver lecture(s); admonish, reprimand.

lecturer *noun* person who lectures, esp. as teacher in higher education.

lectureship *noun* university post as lecturer.

led *past & past participle* of LEAD[1].

ledge *noun* narrow shelf or projection from vertical surface.

ledger /'ledʒə/ *noun* book in which firm's accounts are kept.

lee *noun* shelter given by neighbouring object; side of thing away from the wind. □ **leeway** allowable deviation, drift of ship to leeward.

leech *noun* bloodsucking worm formerly used medicinally for bleeding; person who sponges on others.

leek *noun* vegetable of onion family with long cylindrical white bulb.

leer ● *verb* look slyly, lasciviously, or maliciously. ● *noun* leering look.

leery *adjective* (**-ier, -iest**) *slang* knowing, sly; (usually + *of*) wary.

lees /li:z/ *plural noun* sediment of wine etc.; dregs.

leeward /'li:wəd, *Nautical* 'lu:əd/ ● *adjective & adverb* on or towards sheltered side. ● *noun* this direction.

left[1] ● *adjective* on or towards west side of person or thing facing north; (also **Left**) *Politics* of the Left. ● *adverb* on or to left side. ● *noun* left part, region, or direction; *Boxing* left hand, blow with this; (often **Left**) *Politics* group favouring socialism, radicals collectively. □ **left-hand** on left side; **left-handed** naturally using left hand for writing etc., made by or for left hand, turning to left, (of screw) turned anti-clockwise to tighten, awkward, clumsy, (of compliment etc.) ambiguous; **left-handedness** *noun*; **left-hander** left-handed person or blow; **left wing** more radical section of political party, left side of army, football team, etc.; **left-wing** socialist, radical; **left-winger** member of left wing. □ **leftward** *adjective & adverb*; **leftwards** *adverb*.

left[2] *past & past participle* of LEAVE[1].

leg *noun* each of limbs on which person or animal walks and stands; leg of animal as food; part of garment covering leg; support of chair, table, etc.; section of journey, race, competition, etc.; *Cricket* half of field behind batsman's back. □ **leg before wicket** *Cricket* (of batsman) declared out for illegal obstruction of ball that would have hit wicket; **leg it** (**-gg-**) *colloquial* walk or run hard; **leg warmer** either of pair of tubular knitted garments covering leg from ankle to thigh; **pull (person's) leg** deceive playfully. □ **legged** *adjective*.

legacy /'legəsɪ/ *noun* (*plural* **-ies**) gift left by will; anything handed down by predecessor.

legal /'li:g(ə)l/ *adjective* of, based on, or concerned with law; appointed, required, or permitted by law. □ **legal aid** state help with cost of legal advice; **legal tender** currency that cannot legally be refused in payment of debt. □ **legality** /lɪ'gælɪtɪ/ *noun*; **legally** *adverb*.

legalize /'li:gəlaɪz/ *verb* (also **-ise**) (**-zing** or **-sing**) make lawful; bring into harmony with law. □ **legalization** *noun*.

legate /'legət/ *noun* papal ambassador.

legatee /legə'ti:/ *noun* recipient of legacy.

legation /lɪ'geɪʃ(ə)n/ *noun* diplomatic minister and his or her staff; this minister's official residence.

legato /lɪ'gɑ:təʊ/ *Music* ● *adverb & adjective* in smooth flowing manner. ● *noun* (*plural* **-s**) legato passage.

legend /'ledʒ(ə)nd/ *noun* traditional story, myth; *colloquial* famous or remarkable person or event; inscription; explanation on map etc. of symbols used.

legendary *adjective* existing in legend; *colloquial* remarkable, famous.

legerdemain /ledʒədə'meɪn/ *noun* sleight of hand; trickery, sophistry.

leger line /'ledʒə/ *noun Music* short line added for notes above or below range of staff.

legging *noun* (usually in *plural*) close-fitting trousers for women or children; outer covering of leather etc. for lower leg.

leggy *adjective* (**-ier**, **-iest**) long-legged; long-stemmed and weak.

legible /'ledʒɪb(ə)l/ *adjective* easily read. □ **legibility** *noun*; **legibly** *adverb*.

legion /'li:dʒ(ə)n/ ● *noun* division of 3,000–6,000 men in ancient Roman army; other large organized body. ● *adjective* great in number.

legionary ● *adjective* of legions. ● *noun* (*plural* **-ies**) member of legion.

legionnaire /li:dʒə'neə/ *noun* member of foreign legion. □ **legionnaires' disease** form of bacterial pneumonia.

legislate /'ledʒɪsleɪt/ *verb* (**-ting**) make laws. □ **legislator** *noun*.

legislation *noun* making laws; laws made.

legislative /'ledʒɪslətɪv/ *adjective* of or empowered to make legislation.

legislature /'ledʒɪsleɪtʃə/ *noun* legislative body of a state.

legitimate /lɪ'dʒɪtɪmət/ *adjective* (of child) born of parents married to one another; lawful, proper, regular; logically admissible. □ **legitimacy** *noun*; **legitimately** *adverb*.

legitimize /lɪ'dʒɪtɪmaɪz/ *verb* (also **-ise**) (**-zing** or **-sing**) make legitimate; serve as justification for. □ **legitimization** *noun*.

legume /'legju:m/ *noun* leguminous plant; edible part of this.

leguminous /lɪ'gju:mɪnəs/ *adjective* of the family of plants with seeds in pods, e.g. peas and beans.

lei /'leɪ/ *noun* Polynesian garland of flowers.

leisure /'leʒə/ *noun* free time, time at one's own disposal. □ **at leisure** not occupied, in an unhurried way; **at one's leisure** when one has time; **leisure centre** public building with sports facilities etc.; **leisurewear** informal clothes, esp. sportswear.

leisured *adjective* having ample leisure.

leisurely ● *adjective* relaxed, unhurried. ● *adverb* without hurry.

leitmotif /'laɪtməʊti:f/ *noun* (also **leitmotiv**) recurring theme in musical etc. composition representing particular person, idea, etc.

lemming /'lemɪŋ/ *noun* Arctic rodent reputed to rush, during migration, in large numbers into sea and drown.

lemon /'lemən/ *noun* acid yellow citrus fruit; tree bearing it; pale yellow colour. □ **lemon cheese**, **curd** thick creamy lemon spread. □ **lemony** *adjective*.

lemonade /lemə'neɪd/ *noun* drink made from lemons; synthetic substitute for this, often fizzy.

lemon sole /'lemən/ *noun* (*plural* same or **-s**) fish of plaice family.

lemur /'li:mə/ *noun* tree-dwelling primate of Madagascar.

lend *verb* (*past & past participle* **lent**) grant temporary use of (thing); allow use of (money) in return for interest; bestow, contribute; (**lend itself to**) be suitable for. □ **lend an ear** listen. □ **lender** *noun*.

length *noun* measurement from end to end; extent in or of time; length of horse, boat, etc. as measure of lead in race; long stretch or extent; degree of thoroughness in action. □ **at length** in detail, after a long time. □ **lengthways** *adverb*; **lengthwise** *adverb & adjective*.

lengthen *verb* make or become longer.

lengthy *adjective* (**-ier**, **-iest**) of unusual length, prolix, tedious.

lenient /'li:nɪənt/ *adjective* merciful, not severe, mild. □ **lenience** *noun*; **leniency** *noun*; **leniently** *adverb*.

lens /lenz/ *noun* piece of transparent substance with one or both sides curved, used in spectacles, telescopes, cameras, etc.; combination of lenses used in photography.

Lent *noun* religious period of fasting and penitence from Ash Wednesday to Easter Eve. □ **Lenten** *adjective*.

lent *past & past participle* of LEND.

lentil /'lentɪl/ *noun* edible seed of leguminous plant; this plant.

lento /'lentəʊ/ *Music* ● *adjective* slow. ● *adverb* slowly. ● *noun* lento movement or passage.

Leo /'li:əʊ/ *noun* fifth sign of zodiac.

leonine /'li:ənaɪn/ *adjective* lionlike; of lions.

leopard /'lepəd/ *noun* large animal of cat family with dark-spotted fawn or all black coat, panther.

leotard /'li:ətɑ:d/ *noun* close-fitting one-piece garment worn by dancers etc.

leper /'lepə/ *noun* person with leprosy.

leprechaun /'leprəkɔ:n/ *noun* small mischievous sprite in Irish folklore.

leprosy /'leprəsɪ/ *noun* contagious disease of skin and nerves. □ **leprous** *adjective*.

lesbian /'lezbɪən/ ● *noun* homosexual woman. ● *adjective* of homosexuality in women. □ **lesbianism** *noun*.

lesion /'li:ʒ(ə)n/ *noun* damage, injury; change in part of body due to injury or disease.

less ● *adjective* smaller; of smaller quantity; not so much. ● *adverb* to smaller extent, in lower degree. ● *noun* smaller amount, quantity, or number. ● *preposition* minus, deducting.

■ **Usage** The use of *less* to mean 'fewer', as in *There are less people than yesterday*, is incorrect in standard English.

---

lessee /le'si:/ *noun* (often + *of*) person holding property by lease.

lessen /'les(ə)n/ *verb* diminish.

lesser *adjective* not so great as the other(s).

lesson /'les(ə)n/ *noun* period of teaching; (in *plural*; usually + *in*) systematic instruction; thing learnt by pupil; experience that serves to warn or encourage; passage from Bible read aloud during church service.

lessor /le'sɔ:/ *noun* person who lets property by lease.

lest *conjunction* in order that not, for fear that.

let[1] ● *verb* (**-tt-**; *past & past participle* **let**) allow, enable, or cause to; grant use of (rooms, land, etc.) for rent or hire. ● *auxiliary verb in exhortations, commands, assumptions, etc.* ● *noun* act of letting. □ **let alone** not to mention; **let be** not interfere with; **let down** *verb* lower, fail to support or satisfy, disappoint; **let-down** *noun* disappointment; **let go** release, lose hold of; **let in** allow to enter, (usually + *for*) involve (person, often oneself) in loss, problem, etc., (usually + *on*) allow (person) to share secret etc.; **let off** fire (gun), cause (steam etc.) to escape, not punish or compel; **let on** *colloquial* reveal secret; **let out** *verb* release, reveal (secret etc.), slacken, put out to rent; **let-out** *noun colloquial* opportunity to escape; **let up** *verb colloquial* become less severe, diminish; **let-up** *noun colloquial* relaxation of effort, diminution.

let[2] *noun* obstruction of ball or player in tennis etc. after which ball must be served again. □ **without let or hindrance** unimpeded.

lethal /'li:θ(ə)l/ *adjective* causing or sufficient to cause death. □ **lethally** *adverb*.

lethargy /'leθədʒɪ/ *noun* lack of energy; unnatural sleepiness. □ **lethargic** /lɪ'θɑ:dʒɪk/ *adjective*; **lethargically** /lɪ'θɑ:dʒɪkəlɪ/ *adverb.*

letter /'letə/ ● *noun* character representing one or more of sounds used in speech; written or printed communication, usually sent in envelope by post; precise terms of statement; (in *plural*) literature. ● *verb* inscribe letters on; classify with letters. □ **letter bomb** terrorist explosive device sent by post; **letter box** box for delivery or posting of letters, slit in door for delivery of letters; **letterhead** printed heading on stationery; **letterpress** printed words in illustrated book, printing from raised type; **to the letter** keeping to every detail.

lettuce /'letɪs/ *noun* plant with crisp leaves used in salad.

leucocyte /'lu:kəsaɪt/ *noun* white blood cell.

leukaemia /lu:'ki:mɪə/ *noun* (*US* **leukemia**) malignant progressive disease in which too many white blood cells are produced.

Levant /lɪ'vænt/ *noun* (**the Levant**) *archaic* East-Mediterranean region.

Levantine /'levəntaɪn/ ● *adjective* of or trading to the Levant. ● *noun* native or inhabitant of the Levant.

levee /'levɪ/ *noun US* embankment against river floods.

level /'lev(ə)l/ ● *noun* horizontal line or plane; height or value reached; position on real or imaginary scale; social, moral, or intellectual standard; plane of rank or authority; instrument giving line parallel to plane of horizon; level surface, flat country. ● *adjective* flat, not bumpy; horizontal; (often + *with*) on same horizontal plane as something else, having equality with something else; even, uniform, well-balanced. ● *verb* (**-ll-**; *US* **-l-**) make level; raze, completely destroy; (usually + *at*) aim (gun etc.); (usually + *at, against*) direct (accusation etc.). □ **do one's level best** *colloquial* do one's utmost; **find one's level** reach right social, intellectual, etc., position; **level crossing** crossing of road and railway etc. at same level; **level-headed** mentally well-balanced, cool; **level pegging** equality of scores etc.; **on the level** *colloquial* truthfully, honestly.

lever /'li:və/ ● *noun* bar pivoted about fulcrum to transfer force; bar used on pivot to prise or lift; projecting handle used to operate mechanism; means of exerting moral pressure. ● *verb* use lever; lift, move, etc. (as) with lever.

leverage *noun* action or power of lever; means of accomplishing a purpose.

leveret /'levərɪt/ *noun* young hare.

leviathan /lɪ'vaɪəθ(ə)n/ *noun Biblical* sea monster; very large or powerful thing.

Levis /'li:vaɪz/ *plural noun proprietary term* type of (originally blue) denim jeans.

levitate /'levɪteɪt/ *verb* (**-ting**) (cause to) rise and float in air. □ **levitation** *noun.*

levity /'levɪtɪ/ *noun* lack of serious thought; frivolity.

levy /'levɪ/ ● *verb* (**-ies**, **-ied**) impose or collect (payment etc.) compulsorily; enrol (troops etc.). ● *noun* (*plural* **-ies**) levying; payment etc. or (in *plural*) troops levied.

lewd /lju:d/ *adjective* lascivious, indecent.

lexical /'leksɪk(ə)l/ *adjective* of the words of a language; (as) of a lexicon.

lexicography /leksɪ'kɒgrəfɪ/ *noun* compiling of dictionaries. □ **lexicographer** *noun.*

lexicon /'leksɪkən/ *noun* dictionary.

Leyden jar /'laɪd(ə)n/ *noun* early kind of capacitor.

LF *abbreviation* low frequency.

liability /laɪə'bɪlɪtɪ/ *noun* (*plural* **-ies**) being liable; troublesome person or thing; handicap; (in *plural*) debts for which one is liable.

liable /'laɪəb(ə)l/ *adjective* legally bound; (+ *to*) subject to; (+ *to do*) under an obligation; (+ *to*) exposed or open to (something undesirable); (+ *for*) answerable for.

■ **Usage** *Liable* is often used to mean 'likely', as in *It is liable to rain*, but this is considered incorrect by some people.

liaise /lɪ'eɪz/ *verb* (**-sing**) (usually + *with, between*) *colloquial* establish cooperation, act as link.

liaison /lɪ'eɪzɒn/ *noun* communication, cooperation; illicit sexual relationship.

liana /lɪ'ɑ:nə/ *noun* climbing plant in tropical forests.

liar /'laɪə/ *noun* person who tells lies.

Lib. *abbreviation* Liberal.

lib *noun colloquial* liberation.

libation /laɪ'beɪʃ(ə)n/ *noun* (pouring out of) drink-offering to a god.

libel /'laɪb(ə)l/ ● *noun Law* published false statement damaging to person's reputation, publishing of this; false defamatory statement. ● *verb* (**-ll-**: *US* **-l-**) *Law* publish libel against. □ **libellous** *adjective*.

liberal /'lɪbər(ə)l/ ● *adjective* abundant; giving freely; generous; open-minded; not rigorous; (of studies) for general broadening of mind; *Politics* favouring moderate reforms. ● *noun* person of liberal views, esp. (**Liberal**) member of a Liberal Party. □ **Liberal Democrat** member of **Liberal Democrats**, UK political party. □ **liberalism** *noun*; **liberality** /-'ræl-/ *noun*; **liberally** *adverb*.

liberalize *verb* (also **-ise**) (**-zing** or **-sing**) make or become more liberal or less strict. □ **liberalization** *noun*.

liberate /'lɪbəreɪt/ *verb* (**-ting**) (often + *from*) set free; free (country etc.) from aggressor; (as **liberated** *adjective*) (of person etc.) freed from oppressive social conventions. □ **liberation** *noun*; **liberator** *noun*.

libertine /'lɪbəti:n/ *noun* licentious person.

liberty /'lɪbətɪ/ *noun* (*plural* **-ies**) being free, freedom; right or power to do as one pleases; (in *plural*) privileges granted by authority. □ **at liberty** free, (+ *to do*) permitted; **take liberties** (often + *with*) behave in unacceptably familiar way.

libidinous /lɪ'bɪdɪnəs/ *adjective* lustful.

libido /lɪ'bi:dəʊ/ *noun* (*plural* **-s**) psychic impulse or drive, esp. that associated with sex instinct. □ **libidinal** /lɪ'bɪdɪn(ə)l/ *adjective*.

Libra /'li:brə/ *noun* seventh sign of zodiac.

librarian /laɪ'breərɪən/ *noun* person in charge of or assistant in library. □ **librarianship** *noun*.

library /'laɪbrərɪ/ *noun* (*plural* **-ies**) a collection of books, films, records, etc.; room or building etc. where these are kept; series of books issued in similar bindings.

libretto /lɪ'bretəʊ/ *noun* (*plural* **-ti** /-tɪ/ or **-s**) text of opera etc. □ **librettist** *noun*.

lice *plural* of LOUSE.

licence /'laɪs(ə)ns/ *noun* (*US* **license**) official permit to own, use, or do, something, or carry on trade; permission; excessive liberty of action; writer's etc. deliberate deviation from fact.

license /'laɪs(ə)ns/ ● *verb* (**-sing**) grant licence to; authorize use of (premises) for certain purpose. ● *noun US* = LICENCE.

licensee /laɪsən'si:/ *noun* holder of licence, esp. to sell alcoholic liquor.

licentiate /laɪ'senʃɪət/ *noun* holder of certificate of professional competence.

licentious /laɪ'senʃəs/ *adjective* sexually promiscuous.

lichee = LYCHEE.

lichen /'laɪkən/ *noun* plant composed of fungus and alga in association, growing on rocks, trees, etc.

lich-gate /'lɪtʃgeɪt/ *noun* (also **lychgate**) roofed gateway of churchyard.

licit /'lɪsɪt/ *adjective formal* lawful, permitted.

lick ● *verb* pass tongue over; bring into specified condition by licking; (of flame etc.) play lightly over; *colloquial* thrash, defeat. ● *noun* act of licking with tongue; *colloquial* pace, speed; smart blow. □ **lick one's lips**, **chops** look forward with great pleasure.

licorice = LIQUORICE.

lid *noun* hinged or removable cover, esp. at top of container; eyelid. □ **put the lid on** *colloquial* be the culmination of, put stop to. □ **lidded** *adjective*.

lido /'li:dəʊ/ *noun* (*plural* **-s**) public open-air swimming pool or bathing beach.

lie[1] /laɪ/ ● *verb* (**lying**; *past* **lay**; *past participle* **lain**) be in or assume horizontal position on supporting surface; (of thing) rest on flat surface; remain undisturbed or undiscussed; be kept, remain, or be in specified place etc.; (of abstract things) be in certain relation; be situated or spread out to view etc. ● *noun* way, position, or direction in which something lies. □ **lie in** stay in bed late in morning; **lie-in** *noun*; **lie low** keep quiet or unseen; **lie of the land** state of affairs.

■ **Usage** It is incorrect in standard English to use *lie* to mean 'lay', as in *lie her on the bed*.

---

lie² /laɪ/ ● *noun* intentional false statement; something that deceives. ● *verb* (**lies**, **lied**, **lying**) tell lie(s); (of thing) be deceptive. □ **give the lie to** show the falsity of.

lied /li:d/ *noun* (*plural* **lieder**) German song of Romantic period for voice and piano.

liege /li:dʒ/ *historical* ● *adjective* entitled to receive, or bound to give, feudal service or allegiance. ● *noun* (in full **liege lord**) feudal superior; (usually in *plural*) vassal, subject.

lien /'li:ən/ *noun Law* right to hold another's property till debt on it is paid.

lieu /lju:/ *noun* □ **in lieu** instead; (+ *of*) in place of.

Lieut. *abbreviation* Lieutenant.

lieutenant /lef'tenənt/ *noun* army officer next below captain; naval officer next below lieutenant commander; deputy. □ **lieutenant colonel**, **commander**, **general** officers ranking next below colonel etc. □ **lieutenancy** *noun* (*plural* **-ies**).

life *noun* (*plural* **lives**) capacity for growth, functional activity, and continual change until death; living things and their activity; period during which life lasts; period from birth to present time or from present time to death; duration of thing's existence or ability to function; person's state of existence; living person; business and pleasures of the world; energy, liveliness; biography; *colloquial* imprisonment for life. □ **life assurance** life insurance; **lifebelt** buoyant ring to keep person afloat; **lifeblood** blood as necessary to life, vital factor or influence; **lifeboat** boat for rescues at sea, ship's boat for emergency use; **lifebuoy** buoyant support to keep person afloat; **life cycle** series of changes in life of organism; **lifeguard** expert swimmer employed to rescue bathers from drowning; **Life Guards** regiment of royal household cavalry; **life insurance** insurance which makes payment on death of insured person; **life-jacket** buoyant jacket to keep person afloat; **lifeline** rope etc. used for life-saving, sole means of communication or transport; **life peer** peer whose title lapses on death; **life sentence** imprisonment for life; **life-size(d)** of same size as person or thing represented; **lifestyle** way of life; **life-support machine** respirator; **lifetime** duration of person's life.

lifeless *adjective* dead; unconscious; lacking movement or vitality. □ **lifelessly** *adverb*.

lifelike *adjective* closely resembling life or person or thing represented.

lifer *noun slang* person serving life sentence.

lift ● *verb* (often + *up*, *of*, etc.) raise to higher position, go up, be raised; yield to upward force; give upward direction to (eyes etc.); add interest to; (of fog etc.) rise, disperse; remove (barrier etc.); transport supplies, troops, etc. by air; *colloquial* steal, plagiarize. ● *noun* lifting; ride in another person's vehicle; apparatus for raising and lowering people or things to different floors of building, or for carrying people up or down mountain etc.; transport by air; upward pressure on aerofoil; supporting or elevating influence; elated feeling. □ **lift-off** vertical take-off of spacecraft or rocket.

ligament /'lɪgəmənt/ *noun* band of tough fibrous tissue linking bones.

ligature /'lɪgətʃə/ *noun* tie, bandage; *Music* slur, tie; *Printing* two or more letters joined, e.g. æ.

■ **Usage** *Ligature*, in the *Printing* sense, is sometimes confused with *digraph*, which means 'two letters representing one sound'.

---

light¹ /laɪt/ ● *noun* electromagnetic radiation that stimulates sight and makes things visible; appearance of brightness; source of light; (often in *plural*) traffic light; flame, spark, or device for igniting; aspect in which thing is regarded; mental or spiritual illumination; vivacity, esp. in person's eyes. ● *verb* (*past* **lit**; *past participle* **lit** or **lighted**) set burning, begin to burn; (often +

up) give light to; make prominent by light; show (person) way etc. with light; (usually + *up*) (of face or eyes) brighten with pleasure etc. ● *adjective* well-provided with light, not dark; (of colour) pale. □ **bring** or **come to light** reveal or be revealed; **in the light of** taking account of; **light bulb** glass bulb containing metal filament giving light when current is passed through it; **lighthouse** tower with beacon light to warn or guide ships at sea; **lightship** anchored ship with beacon light; **light year** distance light travels in one year.

light[2] /laɪt/ ● *adjective* not heavy; relatively low in weight, amount, density, or intensity; (of railway) suitable for small loads; carrying only light arms; (of food) easy to digest; (of music etc.) intended only as entertainment, not profound; (of sleep or sleeper) easily disturbed; easily done; nimble; cheerful. ● *adverb* lightly; with light load. ● *verb* (*past & past participle* **lit** or **lighted**) (+ *on*, *upon*) come upon or find by chance. □ **light-fingered** given to stealing; **light flyweight** amateur boxing weight (up to 48 kg); **light-headed** giddy, delirious; **light-hearted** cheerful; **light heavyweight** amateur boxing weight (75–81 kg); **light industry** manufacture of small or light articles; **light middleweight** amateur boxing weight (67–71 kg); **lightweight** *adjective* below average weight, of little importance, *noun* lightweight person or thing, amateur boxing weight (57–60kg); **light welterweight** amateur boxing weight (60–63.5 kg). □ **lightly** *adverb*; **lightness** *noun*.

lighten[1] *verb* make or become lighter in weight; reduce weight or load of.

lighten[2] *verb* shed light on, make or grow bright.

lighter[1] *noun* device for lighting cigarettes etc.

lighter[2] *noun* boat for transporting goods between ship and wharf etc.

lightning /'laɪtnɪŋ/ *noun* flash of light produced by electric discharge between clouds or between clouds and ground. □ **lightning-conductor** metal rod or wire fixed to building or mast to divert lightning to earth or sea.

lights *plural noun* lungs of sheep, pigs, etc. as food, esp. for pets.

ligneous /'lɪgnɪəs/ *adjective* of the nature of wood.

lignite /'lɪgnaɪt/ *noun* brown coal of woody texture.

lignum vitae /lɪgnəm 'vaɪtɪ/ *noun* a hard-wooded tree.

like[1] ● *adjective* (**more like**, **most like**) similar to another, each other, or original; resembling; such as; characteristic of; in suitable state or mood for. ● *preposition* in manner of, to same degree as. ● *adverb slang* so to speak; *colloquial* probably. ● *conjunction colloquial* as, as if (see note below). ● *noun* counterpart, equal; similar person or thing.

■ **Usage** It is incorrect in standard English to use *like* as a conjunction, as in *Tell it like it is* or *He's spending money like it was going out of fashion.*

like[2] ● *verb* (**-king**) find agreeable or enjoyable; feel attracted by, choose to have, prefer. ● *noun* (in *plural*) things one likes or prefers.

likeable *adjective* (also **likable**) pleasant, easy to like. □ **likeably** *adverb*.

likelihood /'laɪklɪhʊd/ *noun* probability.

likely /'laɪklɪ/ ● *adjective* (**-ier**, **-iest**) probable; such as may well happen or be true; to be expected; promising, apparently suitable. ● *adverb* probably. □ **not likely!** *colloquial* certainly not.

liken *verb* (+ *to*) point out resemblance between (person, thing) and (another).

likeness *noun* (usually + *between*, *to*) resemblance; (+ *of*) semblance, guise; portrait, representation.

likewise *adverb* also, moreover; similarly.

liking *noun* what one likes; one's taste; (+ *for*) fondness, taste, fancy.

lilac /'laɪlək/ ● *noun* shrub with fragrant pinkish-violet or white flowers; pale pinkish-violet colour. ● *adjective* of this colour.

liliaceous /lɪlɪ'eɪʃəs/ *adjective* of the lily family.

lilliputian /lɪlɪ'pju:ʃ(ə)n/ ● *noun* diminutive person or thing. ● *adjective* diminutive.

lilt ● *noun* light springing rhythm; tune with this. ● *verb* (esp. as **lilting** *adjective*) speak etc. with lilt.

lily /ˈlɪlɪ/ *noun* (*plural* **-ies**) tall bulbous plant with large trumpet-shaped flowers; heraldic fleur-de-lis. □ **lily of the valley** plant with fragrant white bell-shaped flowers.

limb[1] /lɪm/ *noun* leg, arm, wing; large branch of tree; branch of cross. □ **out on a limb** isolated.

limb[2] /lɪm/ *noun* specified edge of sun, moon, etc.

limber[1] /ˈlɪmbə/ • *adjective* lithe, flexible; agile. • *verb* (usually + *up*) make oneself supple; warm up for athletic etc. activity.

limber[2] /ˈlɪmbə/ • *noun* detachable front of gun-carriage. • *verb* attach limber to.

limbo[1] /ˈlɪmbəʊ/ *noun* (*plural* **-s**) supposed abode of souls of unbaptized infants, and of the just who died before Christ; intermediate state or condition of awaiting decision.

limbo[2] /ˈlɪmbəʊ/ *noun* (*plural* **-s**) W. Indian dance in which dancer bends backwards to pass under progressively lowered horizontal bar.

lime[1] • *noun* white caustic substance got by heating limestone. • *verb* (**-ming**) treat with lime. □ **limekiln** kiln for heating limestone. □ **limy** *adjective* (**-ier**, **-iest**).

lime[2] *noun* round green acid fruit; tree producing this fruit. □ **lime-green** yellowish-green colour.

lime[3] *noun* (in full **lime tree**) tree with heart-shaped leaves and fragrant creamy blossom.

limelight *noun* intense white light used formerly in theatres; glare of publicity.

limerick /ˈlɪmərɪk/ *noun* humorous 5-line verse.

limestone *noun* rock composed mainly of calcium carbonate.

limit /ˈlɪmɪt/ • *noun* point, line, or level beyond which something does not or may not extend or pass; greatest or smallest amount permitted. • *verb* (**-t-**) set or serve as limit to; (+ *to*) restrict to. □ **limitless** *adjective*.

limitation /lɪmɪˈteɪʃ(ə)n/ *noun* limiting, being limited; limit of ability; limiting circumstance.

limn /lɪm/ *verb archaic* paint.

limousine /lɪmʊˈziːn/ *noun* large luxurious car.

limp[1] • *verb* walk or proceed lamely or awkwardly. • *noun* lame walk.

limp[2] *adjective* not stiff or firm; without will or energy. □ **limply** *adverb*; **limpness** *noun*.

limpet /ˈlɪmpɪt/ *noun* mollusc with conical shell sticking tightly to rocks.

limpid /ˈlɪmpɪd/ *adjective* clear, transparent. □ **limpidity** /-ˈpɪd-/ *noun*.

linage /ˈlaɪnɪdʒ/ *noun* number of lines in printed or written page etc.; payment by the line.

linchpin /ˈlɪntʃpɪn/ *noun* pin passed through axle-end to keep wheel on; person or thing vital to organization etc.

linctus /ˈlɪŋktəs/ *noun* syrupy medicine, esp. soothing cough mixture.

linden /ˈlɪnd(ə)n/ *noun* = LIME TREE.

line[1] • *noun* continuous mark made on surface; furrow, wrinkle; use of lines in art; straight or curved track of moving point; outline; limit, boundary; row of persons or things; *US* queue; mark defining area of play or start or finish of race; row of printed or written words; portion of verse written in line; (in *plural*) piece of poetry, words of actor's part; length of cord, rope, etc. serving specified purpose; wire or cable for telephone or telegraph; connection by means of this; single track or branch of railway; regular succession of buses, ships, aircraft, etc., plying between certain places, company conducting this; several generations (of family); stock; manner of procedure, conduct, thought, etc.; channel; department of activity, branch of business; type of product; connected series of military field works, arrangement of soldiers or ships side by side; each of very narrow horizontal sections forming television picture. • *verb* (**-ning**) mark with lines; position or stand at intervals along. □ **line printer** machine that prints computer output a line at a time; **linesman** umpire's or referee's assistant who decides whether ball has fallen within playing area or not; **line up** *verb* arrange or be arranged in lines, have ready; **line-up** *noun* line of people for inspection, arrangement of team, band, etc.

line[2] *verb* (**-ning**) apply layer of usually different material to cover inside of (garment, box, etc.); serve as lining for; *colloquial* fill (purse etc.).

lineage /ˈlɪnɪɪdʒ/ *noun* lineal descent, ancestry.

lineal /ˈlɪnɪəl/ *adjective* in direct line of descent or ancestry; linear. □ **lineally** *adverb.*

lineament /ˈlɪnɪəmənt/ *noun* (usually in *plural*) distinctive feature or characteristic, esp. of face.

linear /ˈlɪnɪə/ *adjective* of or in lines; long and narrow and of uniform breadth.

linen /ˈlɪnɪn/ ● *noun* cloth woven from flax; articles made or originally made of linen, as sheets, shirts, underwear, etc. ● *adjective* made of linen.

liner[1] *noun* ship or aircraft carrying passengers on regular line.

liner[2] *noun* removable lining.

ling[1] *noun* (*plural* same) long slender marine fish.

ling[2] *noun* kind of heather.

linger /ˈlɪŋgə/ *verb* stay about; (+ *over*, *on*, etc.) dally; be protracted; (often + *on*) die slowly.

lingerie /ˈlæ̃ʒərɪ/ *noun* women's underwear and nightclothes.

lingo /ˈlɪŋgəʊ/ *noun* (*plural* **-s** or **-es**) *colloquial* foreign language.

lingual /ˈlɪŋgw(ə)l/ *adjective* of tongue; of speech or languages.

linguist /ˈlɪŋgwɪst/ *noun* person skilled in languages or linguistics.

linguistic /lɪŋˈgwɪstɪk/ *adjective* of language or the study of languages. □ **linguistically** *adverb.*

linguistics *noun* study of language and its structure.

liniment /ˈlɪnɪmənt/ *noun* embrocation.

lining *noun* material used to line surface.

link ● *noun* one loop or ring of chain etc.; one in series; means of connection. ● *verb* (+ *together*, *to*, *with*) connect, join; clasp or intertwine (hands etc.).

linkage *noun* linking or being linked.

links *noun* (treated as *singular* or *plural*) golf course.

Linnaean /lɪˈniːən/ *adjective* of Linnaeus or his system of classifying plants and animals.

linnet /ˈlɪnɪt/ *noun* brown-grey finch.

lino /ˈlaɪnəʊ/ *noun* (*plural* **-s**) linoleum. □ **linocut** design carved in relief on block of linoleum, print made from this.

linoleum /lɪˈnəʊlɪəm/ *noun* canvas-backed material coated with linseed oil, cork, etc.

linseed /ˈlɪnsiːd/ *noun* seed of flax.

lint *noun* linen or cotton with one side made fluffy, used for dressing wounds; fluff.

lintel /ˈlɪnt(ə)l/ *noun* horizontal timber, stone, etc. over door or window.

lion /ˈlaɪən/ *noun* (*feminine* **lioness**) large tawny flesh-eating wild cat of Africa and S. Asia; brave or celebrated person.

lionize *verb* (also **-ise**) (**-zing** or **-sing**) treat as celebrity.

lip *noun* either edge of opening of mouth; edge of cup, vessel, cavity, etc., esp. part shaped for pouring from; *colloquial* impudent talk. □ **lip-read** understand (speech) by observing speaker's lip-movements; **lip-service** insincere expression of support; **lipstick** stick of cosmetic for colouring lips.

liquefy /ˈlɪkwɪfaɪ/ *verb* (**-ies**, **-ied**) make or become liquid. □ **liquefaction** /-ˈfækʃ(ə)n/ *noun.*

liqueur /lɪˈkjʊə/ *noun* any of several strong sweet alcoholic spirits.

liquid /ˈlɪkwɪd/ ● *adjective* having consistency like that of water or oil, flowing freely but of constant volume; having appearance of water; (of sounds) clear, pure; (of assets) easily convertible into cash. ● *noun* liquid substance; *Phonetics* sound of *l* or *r*. □ **liquid crystal** liquid in state approaching that of crystalline solid; **liquid crystal display** visual display in some electronic devices.

liquidate /ˈlɪkwɪdeɪt/ *verb* (**-ting**) wind up affairs of (firm etc.); pay off (debt); wipe out; kill. □ **liquidator** *noun.*

liquidation /lɪkwɪˈdeɪʃ(ə)n/ *noun* liquidating, esp. of firm. □ **go into liquidation** be wound up and have assets apportioned.

liquidity /lɪˈkwɪdɪtɪ/ *noun* (*plural* **-ies**) state of being liquid; having liquid assets.

liquidize *verb* (also **-ise**) (**-zing** or **-sing**) reduce to liquid state.

liquidizer *noun* (also **-iser**) machine for liquidizing foods.

liquor /ˈlɪkə/ *noun* alcoholic (esp. distilled) drink; other liquid, esp. that produced in cooking.

liquorice /ˈlɪkərɪs/ *noun* (also **licorice**) black root extract used as sweet and in medicine; plant from which it is obtained.

lira /ˈlɪərə/ *noun* chief monetary unit of Italy (*plural* **lire** /-rɪ/) and Turkey (*plural* **-s**).

lisle /laɪl/ *noun* fine cotton thread for stockings etc.

lisp ● *noun* speech defect in which *s* is pronounced like *th* in *thick* and *z* like *th* in *this*. ● *verb* speak or utter with lisp.

lissom /ˈlɪsəm/ *adjective* lithe, agile.

list[1] ● *noun* number of items, names, etc. written or printed together as record; (in *plural*) palisades enclosing tournament area. ● *verb* arrange as or enter in list; (as **listed** *adjective*) approved for Stock Exchange dealings, (of a building) of historical importance and officially protected. □ **enter the lists** issue or accept challenge.

list[2] ● *verb* (of ship etc.) lean over to one side. ● *noun* listing position, tilt.

listen /ˈlɪs(ə)n/ *verb* make effort to hear something, attentively hear person speaking; (+ *to*) give attention with ear to, take notice of. □ **listen in** tap telephonic communication, use radio receiving set. □ **listener** *noun*.

listless /ˈlɪstlɪs/ *adjective* lacking energy or enthusiasm. □ **listlessly** *adverb*; **listlessness** *noun*.

lit *past & past participle* of LIGHT[1,2].

litany /ˈlɪtənɪ/ *noun* (*plural* **-ies**) series of supplications to God used in church services; (**the Litany**) that in Book of Common Prayer.

litchi = LYCHEE.

liter *US* = LITRE.

literacy /ˈlɪtərəsɪ/ *noun* ability to read and write.

literal /ˈlɪtər(ə)l/ *adjective* taking words in their basic sense without metaphor etc.; corresponding exactly to original words; prosaic; matter-of-fact. □ **literalism** *noun*; **literally** *adverb*.

literary /ˈlɪtərərɪ/ *adjective* of or concerned with or interested in literature.

literate /ˈlɪtərət/ ● *adjective* able to read and write. ● *noun* literate person.

literati /lɪtəˈrɑːtɪ/ *plural noun* the class of learned people.

literature /ˈlɪtərətʃə/ *noun* written works, esp. those valued for form and style; writings of country or period or on particular subject; *colloquial* printed matter, leaflets, etc.

lithe /laɪð/ *adjective* flexible, supple.

litho /ˈlaɪθəʊ/ *colloquial* ● *noun* lithography. ● *verb* (**-oes**, **-oed**) lithograph.

lithograph /ˈlɪθəgrɑːf/ ● *noun* lithographic print. ● *verb* print by lithography.

lithography /lɪˈθɒgrəfɪ/ *noun* process of printing from plate so treated that ink sticks only to design to be printed. □ **lithographer** *noun*; **lithographic** /lɪθəˈgræfɪk/ *adjective*.

Lithuanian /lɪθjuːˈeɪnɪən/ ● *noun* native, national, or language of Lithuania. ● *adjective* of Lithuania.

litigant /ˈlɪtɪgənt/ ● *noun* party to lawsuit. ● *adjective* engaged in lawsuit.

litigate /ˈlɪtɪgeɪt/ *verb* (**-ting**) go to law; contest (point) at law. □ **litigation** *noun*; **litigator** *noun*.

litigious /lɪˈtɪdʒəs/ *adjective* fond of litigation; contentious.

litmus /ˈlɪtməs/ *noun* dye turned red by acid and blue by alkali.

litre /ˈliːtə/ *noun* (*US* **liter**) metric unit of capacity (1.76 pints).

litter /ˈlɪtə/ ● *noun* refuse, esp. paper, discarded in public place, odds and ends lying about; young animals brought forth at one birth; vehicle containing couch and carried on men's shoulders or by animals; kind of stretcher for sick and wounded; straw etc. as bedding for animals; material for animal's, esp. cat's, indoor toilet. ● *verb* make (place) untidy; give birth to (puppies etc.); provide (horse etc.) with bedding.

little /ˈlɪt(ə)l/ ● *adjective* (**-r**, **-st**; **less** or **lesser**, **least**) small in size, amount, degree, etc.; short in stature; of short distance or duration; (**a little**) certain but small amount of; trivial; only small amount; operating on small scale; humble, ordinary; young, younger. ● *noun* not much, only small amount;

short time or distance. ● *adverb* (**less, least**) to small extent only; not at all. □ **little by little** by degrees, gradually; **the little people** fairies.

littoral /ˈlɪtər(ə)l/ ● *adjective* of or on the shore. ● *noun* region lying along shore.

liturgy /ˈlɪtədʒɪ/ *noun* (*plural* **-ies**) fixed form of public worship; (**the Liturgy**) the Book of Common Prayer. □ **liturgical** /-ˈtɜːdʒ-/ *adjective*.

live[1] /lɪv/ *verb* (**-ving**) have life; be or remain alive; have one's home; (+ *on, off*) subsist or feed on; keep one's position; pass, spend; conduct oneself in specified way; enjoy life to the full. □ **live down** cause (scandal etc.) to be forgotten through blameless behaviour thereafter; **live it up** *colloquial* live exuberantly and extravagantly.

live[2] /laɪv/ ● *adjective* that is alive, living; (of broadcast, performance, etc.) heard or seen while happening or with audience present; of current interest; glowing, burning; (of match, bomb, etc.) not yet kindled or exploded; charged with electricity. ● *adverb* as live performance. □ **livestock** (usually treated as *plural*) animals kept on farm for use or profit; **live wire** spirited person.

liveable /ˈlɪvəb(ə)l/ *adjective* (also **livable**) *colloquial* (usually **liveable-in**) (of house etc.) fit to live in; (of life) worth living; *colloquial* (usually **liveable-with**) (of person) easy to live with.

livelihood /ˈlaɪvlɪhʊd/ *noun* means of living; job, income.

livelong /ˈlɪvlɒŋ/ *adjective* in its entire length.

lively /ˈlaɪvlɪ/ *adjective* (**-ier, -iest**) full of life, energetic; (of imagination) vivid; cheerful; *jocular* exciting, dangerous. □ **liveliness** *noun*.

liven /ˈlaɪv(ə)n/ *verb* (often + *up*) make or become lively, cheer up.

liver[1] /ˈlɪvə/ *noun* large glandular organ in abdomen of vertebrates; liver of some animals as food.

liver[2] /ˈlɪvə/ *noun* person who lives in specified way.

liverish /ˈlɪvərɪʃ/ *adjective* suffering from liver disorder; peevish, glum.

liverwort *noun* mosslike plant sometimes lobed like liver.

livery /ˈlɪvərɪ/ *noun* (*plural* **-ies**) distinctive uniform of member of City Company or servant; distinctive guise or marking; distinctive colour scheme for company's vehicles etc. □ **at livery** (of horse) kept for owner at fixed charge; **livery stable** stable where horses are kept at livery or let out for hire.

lives *plural* of **life**.

livid /ˈlɪvɪd/ *adjective colloquial* furious; of bluish leaden colour.

living /ˈlɪvɪŋ/ ● *noun* being alive; livelihood; position held by clergyman, providing income. ● *adjective* contemporary; now alive; (of likeness) exact; (of language) still in vernacular use. □ **living-room** room for general day use; **living wage** wage on which one can live without privation; **within living memory** within memory of living people.

lizard /ˈlɪzəd/ *noun* reptile with usually long body and tail, 4 legs, and scaly hide.

llama /ˈlɑːmə/ *noun* S. American ruminant kept as beast of burden and for woolly fleece.

Lloyd's /lɔɪdz/ *noun* incorporated society of underwriters in London. □ **Lloyd's Register** annual classified list of all ships.

lo *interjection archaic* look.

loach *noun* (*plural* same or **-es**) small freshwater fish.

load ● *noun* what is (to be) carried; amount usually or actually carried; burden of work, responsibility, care, etc.; (in *plural*, often + *of*) plenty; (**a load of**) a quantity of; amount of power carried by electrical circuit or supplied by generating station. ● *verb* put load on or aboard; place (load) aboard ship or on vehicle etc.; (often + *up*) (of vehicle or person) take load aboard; (often + *with*) burden, strain, overwhelm; put ammunition in (gun), film in (camera), cassette in (tape recorder), program in (computer), etc. □ **load line** Plimsoll line.

loaded *adjective slang* rich, drunk, *US* drugged; (of dice etc.) weighted; (of question or statement) carrying hidden implication.

loadstone = LODESTONE.

loaf¹ *noun* (*plural* **loaves**) unit of baked bread, usually of standard size or shape; other cooked food in loaf shape; *slang* head.

loaf² *verb* (often + *about*) spend time idly, hang about.

loam *noun* rich soil of clay, sand, and humus. □ **loamy** *adjective*.

loan ● *noun* thing lent, esp. money; lending, being lent. ● *verb* lend (money, works of art, etc.). □ **on loan** being lent.

loath *adjective* (also **loth**) disinclined, reluctant.

loathe /ləʊð/ *verb* (**-thing**) detest, hate. □ **loathing** *noun*.

loathsome /'ləʊðsəm/ *adjective* arousing hatred or disgust; repulsive.

loaves *plural* of LOAF¹.

lob ● *verb* (**-bb-**) hit or throw (ball etc.) slowly or in high arc. ● *noun* such ball.

lobar /'ləʊbə/ *adjective* of a lobe, esp. of lung.

lobate /'ləʊbeɪt/ *adjective* having lobe(s).

lobby /'lɒbɪ/ ● *noun* (*plural* **-ies**) porch, ante-room, entrance hall, corridor; (in House of Commons) large hall used esp. for interviews between MPs and the public; (also **division lobby**) each of two corridors to which MPs retire to vote; group of lobbyists. ● *verb* (**-ies, -ied**) solicit support of (influential person); inform (legislators etc.) in order to influence them. □ **lobby correspondent** journalist who receives unattributable briefings from government.

lobbyist *noun* person who lobbies MP etc.

lobe *noun* lower soft pendulous part of outer ear; similar part of other organs. □ **lobed** *adjective*.

lobelia /lə'bi:lɪə/ *noun* plant with bright, esp. blue, flowers.

lobotomy /lə'bɒtəmɪ/ *noun* (*plural* **-ies**) incision into frontal lobe of brain to relieve mental disorder.

lobster /'lɒbstə/ *noun* marine crustacean with two pincer-like claws; its flesh as food. □ **lobster pot** basket for trapping lobsters.

lobworm *noun* large earthworm used as fishing bait.

local /'ləʊk(ə)l/ ● *adjective* belonging to, existing in, or peculiar to particular place; of the neighbourhood; of or affecting a part and not the whole; (of telephone call) to nearby place and at lower charge. ● *noun* inhabitant of particular place; (often **the local**) *colloquial* local public house. □ **local authority** administrative body in local government; **local colour** touches of detail in story etc. designed to provide realistic background; **local government** system of administration of county, district, parish, etc. by elected representatives of those who live there. □ **locally** *adverb*.

locale /ləʊ'kɑ:l/ *noun* scene or locality of event or occurrence.

locality /ləʊ'kælɪtɪ/ *noun* (*plural* **-ies**) district; thing's site or scene; thing's position.

localize /'ləʊkəlaɪz/ *verb* (also **-ise**) (**-zing** or **-sing**) restrict or assign to particular place; invest with characteristics of place; decentralize.

locate /ləʊ'keɪt/ *verb* (**-ting**) discover exact place of; establish in a place, situate; state locality of.

■ **Usage** In standard English, it is incorrect to use *locate* to mean merely 'find' as in *I can't locate my key*.

---

location *noun* particular place; locating; natural, not studio, setting for film etc.

loc. cit. *abbreviation* in the passage cited (*loco citato*).

loch /lɒx, lɒk/ *noun* Scottish lake or narrow inlet of the sea.

lock¹ ● *noun* mechanism for fastening door etc. with bolt requiring key of particular shape; section of canal or river confined within sluice-gates for moving boats from one level to another; turning of vehicle's front wheels; interlocked or jammed state; wrestling hold. ● *verb* fasten with lock; (+ *up*) shut (house etc.) thus; (of door etc.) be lockable; (+ *up, in, into*) enclose (person, thing) by locking; (often + *up, away*) store inaccessibly; make or become rigidly fixed; (cause to) jam or catch. □ **lockjaw** form of tetanus in which jaws become rigidly closed; **lock-keeper** keeper of river or canal lock; **lock on to** (of missile etc.) automatically find and then track (target);

lock out *verb* keep out by locking door, (of employer) subject (employees) to lockout; **lockout** *noun* employer's exclusion of employees from workplace until certain terms are accepted; **locksmith** maker and mender of locks; **lock-up** house or room for temporary detention of prisoners, premises that can be locked up. □ **lockable** *adjective*.

lock² *noun* portion of hair that hangs together; (in *plural*) the hair.

locker *noun* (usually lockable) cupboard, esp. for public use.

locket /'lɒkɪt/ *noun* small ornamental case for portrait etc., usually on chain round neck.

locomotion /ləʊkə'məʊʃ(ə)n/ *noun* motion or power of motion from place to place.

locomotive /ləʊkə'məʊtɪv/ ● *noun* engine for pulling trains. ● *adjective* of, having, or bringing about locomotion.

locum tenens /ləʊkəm 'ti:nenz/ *noun* (*plural* **locum tenentes** /tɪ'nenti:z/) (also *colloquial* **locum**) deputy acting esp. for doctor or member of clergy.

locus /'ləʊkəs/ *noun* (*plural* **loci** /-saɪ/) position, place; line or curve etc. made by all points satisfying certain conditions or by defined motion of point, line, or surface.

locust /'ləʊkəst/ *noun* African or Asian grasshopper migrating in swarms and consuming all vegetation.

locution /lə'kju:ʃ(ə)n/ *noun* phrase, word, or idiom; style of speech.

lode *noun* vein of metal ore. □ **lodestar** star used as guide in navigation, esp. pole star; **lodestone**, **loadstone** magnetic oxide of iron, piece of this as magnet.

lodge ● *noun* small house, esp. one for gatekeeper at entrance to park or grounds of large house; porter's room etc.; members or meeting-place of branch of society such as Freemasons; beaver's or otter's lair. ● *verb* (**-ging**) reside, esp. as lodger; provide with sleeping quarters; submit (complaint etc.); become fixed or caught; deposit for security; settle, place.

lodger *noun* person paying for accommodation in another's house.

lodging *noun* temporary accommodation; (in *plural*) room(s) rented for lodging in.

loft ● *noun* attic; room over stable; gallery in church or hall; pigeon house. ● *verb* send (ball etc.) high up.

lofty *adjective* (**-ier**, **-iest**) of imposing height; haughty, aloof; exalted, noble. □ **loftily** *adverb*; **loftiness** *noun*.

log¹ ● *noun* unhewn piece of felled tree; any large rough piece of wood; *historical* floating device for ascertaining ship's speed; record of ship's or aircraft's voyage; any systematic record of experiences etc. ● *verb* (**-gg-**) enter (ship's speed or other transport details) in logbook; enter (data etc.) in regular record; cut into logs. □ **logbook** book containing record or log, vehicle registration document; **log on** or **off**, **log in** or **out** begin or end operations at terminal of esp. multi-access computer.

log² *noun* logarithm.

logan /'ləʊgən/ *noun* (in full **logan-stone**) poised heavy stone rocking at a touch.

loganberry /'ləʊgənbərɪ/ *noun* (*plural* **-ies**) dark red fruit, hybrid of blackberry and raspberry.

logarithm /'lɒgərɪð(ə)m/ *noun* an arithmetic exponent used in computation. □ **logarithmic** /-'rɪðmɪk/ *adjective*.

loggerhead /'lɒgəhed/ *noun* □ **at loggerheads** (often + *with*) disagreeing or disputing.

loggia /'ləʊdʒə/ *noun* open-sided gallery or arcade.

logging *noun* work of cutting and preparing forest timber.

logic /'lɒdʒɪk/ *noun* science of reasoning; chain of reasoning; use of or ability in argument; inexorable force; principles used in designing computer etc.; circuits based on these. □ **logician** /lə'dʒɪʃ(ə)n/ *noun*.

logical *adjective* of or according to logic; correctly reasoned, consistent; capable of correct reasoning. □ **logicality** /-'kæl-/ *noun*; **logically** *adverb*.

logistics /lə'dʒɪstɪks/ *plural noun* organization of (originally military) services and supplies. □ **logistic** *adjective*; **logistical** *adjective*; **logistically** *adverb*.

logo /'ləʊgəʊ/ *noun* (*plural* **-s**) organization's emblem used in display material.

loin *noun* (in *plural*) side and back of body between ribs and hip-bones; joint of meat from this part of animal. □ **loincloth** cloth worn round hips, esp. as sole garment.

loiter /'lɔɪtə/ *verb* stand about idly; linger. □ **loiter with intent** linger to commit felony.

loll *verb* stand, sit, or recline in lazy attitude; hang loosely.

lollipop /'lɒlɪpɒp/ *noun* hard sweet on stick. □ **lollipop man, lady** *colloquial* warden using circular sign on pole to stop traffic for children to cross road.

lollop /'lɒləp/ *verb* (**-p-**) *colloquial* flop about; move in ungainly bounds.

lolly /'lɒlɪ/ *noun* (*plural* **-ies**) *colloquial* lollipop; ice lolly; *slang* money.

lone *adjective* solitary; without companions; isolated; unmarried. □ **lone hand** hand played or player playing against the rest at cards, person or action without allies; **lone wolf** loner.

lonely /'ləʊnlɪ/ *adjective* (**-ier, -iest**) without companions; sad because of this; isolated; uninhabited. □ **loneliness** *noun.*

loner *noun* person or animal preferring to be alone.

lonesome *adjective esp. US* lonely; causing loneliness.

long¹ ● *adjective* (**longer** /'lɒŋgə/, **longest** /'lɒŋgɪst/) measuring much from end to end in space or time; (following measurement) in length or duration; consisting of many items; tedious; of elongated shape; reaching far back or forward in time; involving great interval or difference. ● *noun* long interval or period. ● *adverb* (**longer** /'lɒŋgə/, **longest** /'lɒŋgɪst/) by or for a long time; (following nouns of duration) throughout specified time; (in *comparative*) after implied point of time. □ **as, so long as** provided that; **before long** soon; **in the long run** eventually; **longboat** sailing ship's largest boat; **longbow** one drawn by hand and shooting long arrow; **long-distance** travelling or operating between distant places; **long face** dismal expression; **longhand** ordinary handwriting; **long johns** *colloquial* long underpants; **long jump** athletic contest of jumping along ground in one leap; **long-life** (of milk etc.) treated to prolong shelf-life; **long odds** chances with low probability; **long-playing** (of gramophone record) playing for about 20–30 minutes on each side; **long-range** having a long range, relating to period of time far into future; **longshore** existing on or frequenting the shore; **long shot** wild guess or venture; **long sight** ability to see clearly only what is comparatively distant; **long-sighted** having long sight, far-sighted; **long-suffering** bearing provocation patiently; **long-term** of or for long period of time; **long-winded** (of speech or writing) tediously lengthy.

long² *verb* (+ *for*, *to do*) have strong wish or desire for. □ **longing** *noun & adjective*; **longingly** *adverb.*

long. *abbreviation* longitude.

longevity /lɒn'dʒevɪtɪ/ *noun formal* long life.

longitude /'lɒŋgɪtju:d/ *noun* angular distance E. or W. of (esp. Greenwich) meridian.

longitudinal /lɒŋgɪ'tju:dɪn(ə)l/ *adjective* of or in length; running lengthwise; of longitude. □ **longitudinally** *adverb.*

longways *adverb* (also **longwise**) in direction parallel with thing's length.

loo *noun colloquial* lavatory.

loofah /'lu:fə/ *noun* rough bath-sponge made from dried pod of type of gourd.

look /lʊk/ ● *verb* (often + *at*, *down*, *up*, etc.) use or direct one's eyes; examine; make visual or mental search; (+ *at*) consider; (+ *for*) seek; have specified appearance, seem; (+ *into*) investigate; (of thing) face some direction; indicate (emotion) by looks; (+ *to do*) expect. ● *noun* act of looking; gaze, glance; appearance of face, expression; (in *plural*) personal appearance. □ **look after** attend to; **look back** (+ *on*, *to*) turn one's thoughts to (something past); **look down (up)on** regard with contempt; **look forward to** await (expected event) eagerly or with specified feelings; **look in** *verb* make short visit; **look-in** *noun colloquial* chance of participation or success; **looking-glass** mirror; **look on** be spectator; **look out** *verb* (often + *for*) be vigilant or prepared; **lookout** *noun* watch, observation-post, person etc. stationed to keep watch, prospect, *colloquial* person's own concern; **look up** search for (esp. information in book), *colloquial* visit (person), improve in prospect; **look up to** respect.

loom[1] /lu:m/ *noun* apparatus for weaving.

loom[2] /lu:m/ *verb* appear dimly, esp. as vague often threatening shape.

loon /lu:n/ *noun* kind of diving bird; *colloquial* crazy person.

loony /'lu:nɪ/ *slang* ● *noun* (*plural* **-ies**) lunatic. ● *adjective* (**-ier**, **-iest**) crazy.

loop /lu:p/ ● *noun* figure produced by curve or doubled thread etc. crossing itself; thing, path, etc. forming this figure; similarly shaped attachment used as fastening; contraceptive coil; endless band of tape or film allowing continuous repetition; repeated sequence of computer operations. ● *verb* form or bend into loop; fasten with loop(s); form loop. □ **loop line** railway or telegraph line that diverges from main line and joins it again.

loophole *noun* means of evading rule etc. without infringing it; narrow vertical slit in wall of fort etc.

loopy *adjective* (**-ier**, **-iest**) *slang* crazy.

loose /lu:s/ ● *adjective* not tightly held; free from bonds or restraint; not held together; not compact or dense; inexact; morally lax. ● *verb* (**-sing**) free; untie, detach; release; relax (hold etc.). □ **at a loose end** unoccupied; **let loose** release; **loose cover** removable cover for armchair etc.; **loose-leaf** (of notebook etc.) with pages that can be removed and replaced; **on the loose** escaped from captivity, enjoying oneself freely. □ **loosely** *adverb*.

loosen *verb* make or become loose or looser. □ **loosen up** relax.

loot /lu:t/ ● *noun* spoil, booty; *slang* money. ● *verb* rob or steal, esp. after rioting etc.; plunder.

lop *verb* (**-pp-**) (often + *off*) cut or remove (part or parts) from whole, esp. branches from tree; prune (tree).

lope ● *verb* (**-ping**) run with long bounding stride. ● *noun* such stride.

lop-eared *adjective* having drooping ears.

lopsided *adjective* unevenly balanced.

loquacious /lə'kweɪʃəs/ *adjective* talkative. □ **loquacity** /-'kwæsɪtɪ/ *noun*.

lord ● *noun* master, ruler; *historical* feudal superior, esp. of manor; peer of realm, person with title *Lord*; (**Lord**) (often **the Lord**) God, Christ; (**Lord**) *title used before name of certain male peers and officials*; (**the Lords**) House of Lords. ● *interjection expressing surprise, dismay, etc.* □ **lord it over** domineer; **Lord Mayor** *title of mayor in some large cities*; **Lord's Day** Sunday; **Lord's Prayer** the Our Father; **Lord's Supper** Eucharist.

lordly *adjective* (**-ier**, **-iest**) haughty, imperious; suitable for a lord.

Lordship *noun* □ **His**, **Your Lordship** *title used of or to man with rank of Lord*.

lore *noun* body of tradition and information on a subject or held by particular group.

lorgnette /lɔ:'njet/ *noun* pair of eyeglasses or opera-glasses on long handle.

lorn *adjective archaic* desolate, forlorn.

lorry /'lɒrɪ/ *noun* (*plural* **-ies**) large vehicle for transporting goods etc.

lose /lu:z/ *verb* (**-sing**; *past & past participle* **lost**) be deprived of; cease to have, esp. by negligence; be deprived of (person) by death; become unable to find, follow, or understand; let pass from one's control; be defeated in; get rid of; forfeit (right to something); suffer loss or detriment; cause (person) the loss of; (of clock etc.) become slow; (in *passive*) disappear, perish. □ **be lost without** be dependent on.

loser *noun* person or thing that loses esp. contest; *colloquial* person who regularly fails.

loss *noun* losing, being lost; what is lost; detriment resulting from losing. □ **at a loss** (sold etc.) for less than was paid for it; **be at a loss** be puzzled or uncertain; **loss-leader** item sold at a loss to attract customers.

lost *past & past participle* of LOSE.

lot *noun colloquial* (**a lot**, or **lots**) large number or amount; each of set of objects used to make chance selection; this method of deciding, share or responsibility resulting from it; destiny, fortune, condition; *esp. US* plot, allotment; article or set of articles for sale at auction etc.; group of associated people or things. □ **draw**, **cast lots** decide by lots; **the** (**whole**) **lot** total number or quantity.

■ **Usage** *A lot of*, as in *a lot of people*, is fairly informal, though acceptable in serious writing, but *lots of people* is not acceptable.

loth = LOATH.

lotion /'ləʊʃ(ə)n/ *noun* medical or cosmetic liquid preparation applied externally.

lottery /'lɒtərɪ/ *noun* (*plural* **-ies**) means of raising money by selling numbered tickets and giving prizes to holders of numbers drawn at random.

lotto /'lɒtəʊ/ *noun* game of chance like bingo.

lotus /'ləʊtəs/ *noun* legendary plant inducing luxurious langour when eaten; kind of water lily. □ **lotus position** cross-legged position of meditation.

loud ● *adjective* strongly audible; noisy; (of colours etc.) gaudy, obtrusive. ● *adverb* loudly. □ **loudspeaker** apparatus that converts electrical signals into sounds. □ **loudly** *adverb*; **loudness** *noun*.

lough /lɒk, lɒx/ *noun Irish* lake, sea inlet.

lounge ● *verb* (**-ging**) recline comfortably; loll; stand or move idly. ● *noun* place for lounging, esp. sitting-room in house; public room (in hotel etc.); place in airport etc. with seats for waiting passengers; spell of lounging. □ **lounge suit** man's suit for ordinary day wear.

lour /laʊə/ *verb* (also **lower** /laʊə/) frown, look sullen; (of sky etc.) look dark and threatening.

louse /laʊs/ ● *noun* (*plural* **lice**) parasitic insect; (*plural* **louses**) *slang* contemptible person. ● *verb* (**-sing**) delouse.

lousy /'laʊzɪ/ *adjective* (**-ier**, **-iest**) *colloquial* very bad, disgusting, ill; (often + *with*) *colloquial* well supplied; infested with lice.

lout *noun* rough-mannered person. □ **loutish** *adjective*.

louvre /'lu:və/ *noun* (also **louver**) each of set of overlapping slats designed to admit air and some light and exclude rain; domed structure on roof with side openings for ventilation etc.

lovable /'lʌvəb(ə)l/ *adjective* (also **loveable**) inspiring affection.

lovage /'lʌvɪdʒ/ *noun* herb used for flavouring etc.

love /lʌv/ ● *noun* deep affection or fondness; sexual passion; sexual relations; beloved one; sweetheart; *colloquial form of address regardless of affection*; *colloquial* person of whom one is fond; affectionate greetings; (in games) no score, nil. ● *verb* (**-ving**) feel love for; delight in, admire; *colloquial* like very much. □ **fall in love** (often + *with*) suddenly begin to love; **in love** (often + *with*) enamoured (of); **love affair** romantic or sexual relationship between two people; **love-bird** kind of parakeet; **love-in-a-mist** blue-flowered cultivated plant; **lovelorn** pining from unrequited love; **lovesick** languishing with love; **make love** (often + *to*) have sexual intercourse (with), pay amorous attention (to).

loveable = LOVABLE.

loveless *adjective* unloving or unloved or both.

lovely *adjective* (**-ier**, **-iest**) *colloquial* pleasing, delightful; beautiful. □ **loveliness** *noun*.

lover *noun* person in love with another, or having sexual relations with another; (in *plural*) unmarried couple in love or having sexual relations; person who enjoys specified thing.

loving ● *adjective* feeling or showing love, affectionate. ● *noun* affection. □ **loving cup** two-handled drinking cup passed round at banquets. □ **lovingly** *adverb*.

low[1] /ləʊ/ ● *adjective* not high or tall; not elevated in position; (of sun) near horizon; of humble rank; of small or less than normal amount, extent, or intensity; dejected; lacking vigour; (of sound) not shrill or loud; commonplace; (of opinion) unfavourable; mean, vulgar. ● *noun* low or lowest level or number; area of low pressure. ● *adverb* in or to low position; in low tone, at low pitch. □ **lowbrow** *colloquial* not intellectual or cultured; **Low Church** section of Church of England attaching little importance to ritual, priestly authority, and sacraments; **Low Countries** Netherlands, Belgium, and Luxembourg; **low-down** *adjective* mean, dishonourable, *noun colloquial* (**the low-down**; usually + *on*) relevant information; **lower case** small letters, not capitals; **low frequency** *Radio* 30–300 kilohertz; **low pressure** low degree of activity, atmospheric condition with pressure below average; **Low Sunday** Sunday after Easter; **low tide, water** time or level of tide at its ebb; **low water mark** level reached at low water.

low[2] /ləʊ/ ● *noun* sound made by cattle; moo. ● *verb* make this sound.

lower[1] *verb* let or haul down; make or become lower; degrade.

lower[2] = LOUR.

lowland /'ləʊlənd/ ● *noun* (usually in *plural*) low-lying country. ● *adjective* of or in lowland. □ **lowlander** *noun*.

lowly *adjective* (**-ier**, **-iest**) humble; unpretentious. □ **lowliness** *noun*.

loyal /'lɔɪəl/ *adjective* (often + *to*) faithful; steadfast in allegiance etc. □ **loyally** *adverb*; **loyalty** *noun* (*plural* **-ies**).

loyalist *noun* person remaining loyal to legitimate sovereign etc.; (**Loyalist**) supporter of union between Great Britain and Northern Ireland. □ **loyalism** *noun*.

lozenge /'lɒzɪndʒ/ *noun* rhombus; small sweet or medicinal tablet to be dissolved in mouth; lozenge-shaped object.

LP *abbreviation* long-playing (record).

LSD *abbreviation* lysergic acid diethylamide, a powerful hallucinogenic drug.

Lt. *abbreviation* Lieutenant; light.

Ltd. *abbreviation* Limited.

lubber /'lʌbə/ *noun* clumsy fellow, lout.

lubricant /'lu:brɪkənt/ *noun* substance used to reduce friction.

lubricate /'lu:brɪkeɪt/ *verb* (**-ting**) apply oil, grease, etc. to; make slippery. □ **lubrication** *noun*.

lubricious /lu:'brɪʃəs/ *adjective* slippery, evasive; lewd. □ **lubricity** *noun*.

lucerne /lu:'sɜ:n/ *noun* alfalfa.

lucid /'lu:sɪd/ *adjective* expressing or expressed clearly; sane. □ **lucidity** /-'sɪd-/ *noun*; **lucidly** *adverb*.

luck *noun* good or bad fortune; circumstances brought by this; success due to chance.

luckless *adjective* unlucky; ending in failure.

lucky *adjective* (**-ier**, **-iest**) having or resulting from good luck; bringing good luck. □ **lucky dip** tub containing articles from which one chooses at random. □ **luckily** *adverb*.

lucrative /'lu:krətɪv/ *adjective* profitable. □ **lucratively** *adverb*.

lucre /'lu:kə/ *noun derogatory* financial gain.

ludicrous /'lu:dɪkrəs/ *adjective* absurd, ridiculous, laughable. □ **ludicrously** *adverb*; **ludicrousness** *noun*.

ludo /'lu:dəʊ/ *noun* board game played with dice and counters.

lug ● *verb* (**-gg-**) drag or carry with effort; pull hard. ● *noun* hard or rough pull; *colloquial* ear; projection on object by which it may be carried, fixed in place, etc.

luggage /'lʌgɪdʒ/ *noun* suitcases, bags, etc., for traveller's belongings.

lugger /'lʌgə/ *noun* small ship with 4-cornered sails (**lugsails**).

lugubrious /lʊ'gu:brɪəs/ *adjective* doleful. □ **lugubriously** *adverb*; **lugubriousness** *noun*.

lukewarm /lu:k'wɔ:m/ *adjective* moderately warm, tepid; unenthusiastic.

lull ● *verb* soothe, send to sleep; (usually + *into*) deceive (person) into undue confidence; allay (suspicions etc.); (of noise, storm, etc.) lessen, fall quiet. ● *noun* temporary quiet period.

lullaby /'lʌləbaɪ/ *noun* (*plural* **-ies**) soothing song to send child to sleep.

lumbago /lʌm'beɪgəʊ/ *noun* rheumatic pain in muscles of lower back.

lumbar /'lʌmbə/ *adjective* of lower back. □ **lumbar puncture** withdrawal of spinal fluid from lower back for diagnosis.

lumber /'lʌmbə/ ● *noun* disused and cumbersome articles; partly prepared timber. ● *verb* (usually + *with*) encumber (person); move in slow clumsy way; cut and prepare forest timber for transporting. □ **lumberjack** person who fells and transports lumber; **lumber-room** room where things in disuse are kept.

lumen /'lu:men/ *noun* SI unit of luminous flux.

luminary /'lu:mɪnərɪ/ *noun* (*plural* **-ies**) *literary* natural light-giving body; wise person; celebrated member of group.

luminescence /lu:mɪ'nes(ə)ns/ *noun* emission of light without heat. □ **luminescent** *adjective*.

luminous /'lu:mɪnəs/ *adjective* shedding light; phosphorescent, visible in darkness. □ **luminosity** /-'nɒs-/ *noun*.

lump[1] ● *noun* compact shapeless mass; tumour; swelling, bruise; heavy ungainly person etc. ● *verb* (usually + *together* etc.) class, mass. □ **lump sugar** sugar in small lumps or cubes; **lump sum** sum including number of items or paid down all at once.

lump[2] *verb colloquial* (in contrast with *like*) put up with ungraciously.

lumpish *adjective* heavy, clumsy; stupid.

lumpy *adjective* (**-ier**, **-iest**) full of or covered with lumps. □ **lumpily** *adverb*; **lumpiness** *noun*.

lunacy /'lu:nəsɪ/ *noun* (*plural* **-ies**) insanity; great folly.

lunar /'lu:nə/ *adjective* of, like, concerned with, or determined by the moon. □ **lunar module** craft for travelling between moon and orbiting spacecraft; **lunar month** period of moon's revolution, (in general use) 4 weeks.

lunate /'lu:neɪt/ *adjective* crescent-shaped.

lunatic /'lu:nətɪk/ ● *noun* insane person; wildly foolish person. ● *adjective* insane; very reckless or foolish.

lunation /lu:'neɪʃ(ə)n/ *noun* interval between new moons, about 29½ days.

lunch ● *noun* midday meal. ● *verb* take lunch; provide lunch for.

luncheon /'lʌntʃ(ə)n/ *noun formal* lunch. □ **luncheon voucher** voucher issued to employees and exchangeable for food at restaurant etc.

lung *noun* either of pair of respiratory organs in humans and many other vertebrates.

lunge ● *noun* sudden movement forward; attacking move in fencing. ● *verb* (**-ging**) (usually + *at*) deliver or make lunge.

lupin /'lu:pɪn/ *noun* cultivated plant with long tapering spikes of flowers.

lupine /'lu:paɪn/ *adjective* of or like wolves.

lupus /'lu:pəs/ *noun* inflammatory skin disease.

lurch[1] ● *noun* stagger; sudden unsteady movement or tilt. ● *verb* stagger, move unsteadily.

lurch[2] *noun* □ **leave in the lurch** desert (friend etc.) in difficulties.

lurcher /'lɜ:tʃə/ *noun* crossbred dog, usually working dog crossed with greyhound.

lure ● *verb* (**-ring**) (usually + *away*, *into*) entice; recall with lure. ● *noun* thing used to entice, enticing quality (of chase etc.); falconer's apparatus for recalling hawk.

lurid /'ljʊərɪd/ *adjective* bright and glaring in colour; sensational, shocking; ghastly, wan. □ **luridly** *adverb*.

lurk *verb* linger furtively; lie in ambush; (usually + *in*, *about*, etc.) hide, esp. for sinister purpose; (as **lurking** *adjective*) latent.

luscious /'lʌʃəs/ *adjective* richly sweet in taste or smell; voluptuously attractive.

lush[1] *adjective* luxuriant and succulent.

lush[2] *noun slang* alcoholic, drunkard.

lust ● *noun* strong sexual desire; (usually + *for*, *of*) passionate desire for or enjoyment of; sensuous appetite seen as sinful. ● *verb* (usually + *after*, *for*) have strong or excessive (esp. sexual) desire. □ **lustful** *adjective*; **lustfully** *adverb*.

lustre /'lʌstə/ *noun* (*US* **luster**) gloss, shining surface; brilliance, splendour; iridescent glaze on pottery and porcelain. □ **lustrous** *adjective*.

lusty *adjective* (**-ier**, **-iest**) healthy and strong; vigorous, lively. □ **lustily** *adverb*.

lute[1] /lu:t/ *noun* guitar-like instrument with long neck and pear-shaped body.

lute[2] /lu:t/ ● *noun* clay or cement for making joints airtight. ● *verb* (**-ting**) apply lute to.

lutenist /'lu:tənɪst/ *noun* lute-player.

Lutheran /'lu:θərən/ ● *noun* follower of Martin Luther; member of Lutheran Church. ● *adjective* of Luther, the doctrines associated with him, or the Protestant Reformation. □ **Lutheranism** *noun*.

lux /lʌks/ *noun* (*plural* same) SI unit of illumination.

luxuriant /lʌg'zjʊərɪənt/ *adjective* growing profusely; exuberant, florid. □ **luxuriance** *noun*; **luxuriantly** *adverb*.

■ **Usage** *Luxuriant* is sometimes confused with *luxurious*.

luxuriate /lʌg'zjʊərɪeɪt/ *verb* (**-ting**) (+ *in*) take self-indulgent delight in, enjoy as luxury.

luxurious /lʌg'zjʊərɪəs/ *adjective* supplied with luxuries; very comfortable; fond of luxury. □ **luxuriously** *adverb.*

■ **Usage** *Luxurious* is sometimes confused with *luxuriant*.

luxury /'lʌkʃərɪ/ ● *noun* (*plural* **-ies**) choice or costly surroundings, possessions, etc.; thing giving comfort or enjoyment but inessential. ● *adjective* comfortable, expensive, etc.

LV *abbreviation* luncheon voucher.

lychee /'laɪtʃɪ/ *noun* (also **litchi, lichee**) sweet white juicy brown-skinned fruit; tree bearing this.

lych-gate = LICH-GATE.

Lycra /'laɪkrə/ *noun proprietary term* elastic polyurethane fabric.

lye /laɪ/ *noun* water made alkaline with wood ashes; any alkaline solution for washing.

lying *present participle* of LIE[1,2].

lymph /lɪmf/ *noun* colourless fluid from tissues of body, containing white blood cells; this fluid as vaccine.

lymphatic /lɪm'fætɪk/ *adjective* of, secreting, or carrying lymph; (of person) pale, flabby. □ **lymphatic system** vessels carrying lymph.

lynch /lɪntʃ/ *verb* put (person) to death by mob action without legal trial. □ **lynching** *noun.*

lynx /lɪŋks/ *noun* (*plural* same or **-es**) wild cat with short tail, spotted fur, and proverbially keen sight.

lyre /laɪə/ *noun* ancient U-shaped stringed instrument.

lyric /'lɪrɪk/ ● *adjective* (of poetry) expressing writer's emotion, usually briefly; (of poet) writing in this way; meant or fit to be sung; songlike. ● *noun* lyric poem; (in *plural*) words of song.

lyrical *adjective* lyric; resembling, or using language appropriate to, lyric poetry; *colloquial* highly enthusiastic. □ **lyrically** *adverb.*

lyricism /'lɪrɪsɪz(ə)m/ *noun* quality of being lyrical.

lyricist /'lɪrɪsɪst/ *noun* writer of lyrics.

Mm

M[1] *noun* (also **m**) (Roman numeral) 1,000.

M[2] *abbreviation* (also **M.**) Master; *Monsieur*; motorway; mega-.

m *abbreviation* (also **m.**) male; masculine; married; mile(s); metre(s); million(s); minute(s); milli-.

MA *abbreviation* Master of Arts.

ma /mɑ:/ *noun colloquial* mother.

ma'am /mæm/ *noun* madam (esp. used in addressing royal lady).

mac *noun* (also **mack**) *colloquial* mackintosh.

macabre /mə'kɑ:br/ *adjective* gruesome, grim.

macadam /mə'kædəm/ *noun* broken stone as material for road-making; tarmacadam. □ **macadamize** *verb* (also **-ise**) (**-zing** or **-sing**).

macaroni /mækə'rəʊnɪ/ *noun* pasta tubes.

macaroon /mækə'ru:n/ *noun* biscuit made of ground almonds etc.

macaw /mə'kɔ:/ *noun* kind of parrot.

mace[1] *noun* staff of office, esp. symbol of Speaker's authority in House of Commons.

mace[2] *noun* dried outer covering of nutmeg as spice.

macédoine /'mæsɪdwɑ:n/ *noun* mixture of fruits or vegetables, esp. cut up small.

macerate /'mæsəreɪt/ *verb* (**-ting**) soften by soaking. □ **maceration** *noun.*

machete /mə'ʃetɪ/ *noun* broad heavy knife used in Central America and W. Indies.

machiavellian /mækɪə'velɪən/ *adjective* unscrupulous, cunning.

machination /mækɪ'neɪʃ(ə)n/ *noun* (usually in *plural*) intrigue, plot.

machine /mə'ʃi:n/ ● *noun* apparatus for applying mechanical power, having several interrelated parts; bicycle, motorcycle, etc.; aircraft; computer; controlling system of an organization. ● *verb* (**-ning**) make or operate on with machine. □ **machine-gun** automatic gun that gives continuous fire; **machine-readable** in form that computer can process; **machine tool** mechanically operated tool.

machinery *noun* (*plural* **-ies**) machines; mechanism; organized system; means arranged.

machinist *noun* person who works machine.

machismo /mə'kɪzməʊ/ *noun* being macho; masculine pride.

macho /'mætʃəʊ/ *adjective* aggressively masculine.

macintosh = MACKINTOSH.

mack = MAC.

mackerel /'mækr(ə)l/ *noun* (*plural* same or **-s**) edible sea fish. □ **mackerel sky** sky dappled with rows of small fleecy white clouds.

mackintosh /'mækɪntɒʃ/ *noun* (also **macintosh**) waterproof coat or cloak; cloth waterproofed with rubber.

macramé /mə'krɑ:mɪ/ *noun* art of knotting cord or string in patterns; work so made.

macrobiotic /mækrəʊbaɪ'ɒtɪk/ *adjective* of diet intended to prolong life, esp. consisting of wholefoods.

macrocosm /'mækrəʊkɒz(ə)m/ *noun* universe; whole of a complex structure.

mad *adjective* (**-dd-**) insane; frenzied; wildly foolish; infatuated; *colloquial* annoyed. □ **madcap** *adjective* wildly impulsive, *noun* reckless person; **madhouse** *colloquial* confused uproar, *archaic* mental home or hospital; **madman**, **madwoman** mad person. □ **madly** *adverb*; **madness** *noun*.

madam /'mædəm/ *noun polite formal address to woman*; *colloquial* conceited or precocious girl or young woman; woman brothel-keeper.

Madame /mə'dɑ:m/ *noun* (*plural* **Mesdames** /meɪ'dɑ:m/) *title used of or to French-speaking woman*.

madden *verb* make mad; irritate. □ **maddening** *adjective*.

madder /'mædə/ *noun* herbaceous climbing plant; red dye from its root; synthetic substitute for this dye.

made *past & past participle* of MAKE.

Madeira /mə'dɪərə/ *noun* fortified wine from Madeira; (in full **Madeira cake**) kind of sponge cake.

Mademoiselle /mædəmwə'zel/ *noun* (*plural* **Mesdemoiselles** /meɪdm-/) *title used of or to unmarried French-speaking woman*.

Madonna /mə'dɒnə/ *noun* (**the Madonna**) the Virgin Mary; (**madonna**) picture or statue of her.

madrigal /'mædrɪg(ə)l/ *noun* part-song for several voices, usually unaccompanied.

maelstrom /'meɪlstrəm/ *noun* great whirlpool.

maestro /'maɪstrəʊ/ *noun* (*plural* **maestri** /-strɪ/ or **-s**) eminent musician, esp. teacher or conductor.

Mafia /'mæfɪə/ *noun* organized international group of criminals.

Mafioso /mæfɪ'əʊsəʊ/ *noun* (*plural* **Mafiosi** /-sɪ/) member of the Mafia.

mag *noun colloquial* magazine.

magazine /mægə'zi:n/ *noun* periodical publication containing contributions by various writers; chamber containing cartridges fed automatically to breech of gun; similar device in slide projector etc.; store for explosives, arms, or military provisions.

magenta /mə'dʒentə/ *noun* shade of crimson; aniline dye of this colour.

maggot /'mægət/ *noun* larva, esp. of bluebottle. □ **maggoty** *adjective*.

Magi /'meɪdʒaɪ/ *plural noun* (**the Magi**) the 'wise men from the East' in the Gospel.

magic /'mædʒɪk/ ● *noun* art of influencing events supernaturally; conjuring tricks; inexplicable influence. ● *adjective* of magic. □ **magic lantern** simple form of slide projector.

magical *adjective* of magic; resembling, or produced as if by, magic; wonderful, enchanting. □ **magically** *adverb*.

magician /mə'dʒɪʃ(ə)n/ *noun* person skilled in magic; conjuror.

magisterial /mædʒɪ'stɪərɪəl/ *adjective* imperious; authoritative; of a magistrate.

magistracy /'mædʒɪstrəsɪ/ *noun* (*plural* **-ies**) magisterial office; magistrates.

magistrate /'mædʒɪstreɪt/ *noun* civil officer administering law, esp. one trying minor offences etc.

magnanimous /mæg'nænɪməs/ *adjective* nobly generous, not petty in feelings or conduct. □ **magnanimity** /-nə'nɪm-/ *noun*.

magnate /'mægneɪt/ *noun* person of wealth, authority, etc.

magnesia /mæg'ni:ʃə/ *noun* magnesium oxide; hydrated magnesium carbonate, used as antacid and laxative.

magnesium /mæg'ni:zɪəm/ *noun* silvery metallic element.

magnet /'mægnɪt/ *noun* piece of iron, steel, etc., having properties of attracting iron and of pointing approximately north when suspended; lodestone; person or thing that attracts.

magnetic /mæg'netɪk/ *adjective* having properties of magnet; produced or acting by magnetism; capable of being attracted by or acquiring properties of magnet; very attractive. □ **magnetic field** area of influence of magnet; **magnetic north** point indicated by north end of compass needle; **magnetic storm** disturbance of earth's magnetic field; **magnetic tape** coated plastic strip for recording sound or pictures.

magnetism /'mægnɪtɪz(ə)m/ *noun* magnetic phenomena; science of these; personal charm.

magnetize /'mægnɪtaɪz/ *verb* (also **-ise**) (**-zing** or **-sing**) make into magnet; attract like magnet. □ **magnetization** *noun*.

magneto /mæg'ni:təʊ/ *noun* (*plural* **-s**) electric generator using permanent magnets (esp. for ignition in internal-combustion engine).

magnificent /mæg'nɪfɪs(ə)nt/ *adjective* splendid; imposing; *colloquial* excellent. □ **magnificence** *noun*; **magnificently** *adverb*.

magnify /'mægnɪfaɪ/ *verb* (**-ies**, **-ied**) make (thing) appear larger than it is, as with lens (**magnifying glass**) etc.; exaggerate; intensify; *archaic* extol. □ **magnification** *noun*.

magnitude /'mægnɪtju:d/ *noun* largeness, size; importance.

magnolia /mæg'nəʊlɪə/ *noun* kind of flowering tree; very pale pinkish colour of its flowers.

magnum /'mægnəm/ *noun* (*plural* **-s**) wine bottle twice normal size.

magpie /'mægpaɪ/ *noun* crow with long tail and black and white plumage; chatterer; indiscriminate collector.

Magyar /'mægjɑ:/ ● *noun* member of the chief ethnic group in Hungary; their language. ● *adjective* of this people.

maharaja /mɑ:hə'rɑ:dʒə/ *noun* (also **maharajah**) *historical title of some Indian princes*.

maharanee /mɑ:hə'rɑ:nɪ/ *noun* (also **maharani**) (*plural* **-s**) *historical* maharaja's wife or widow.

maharishi /mɑ:hə'rɪʃɪ/ *noun* (*plural* **-s**) great Hindu sage.

mahatma /mə'hætmə/ *noun* (in India etc.) revered person.

mah-jong /mɑ:'dʒɒŋ/ *noun* (also **-jongg**) originally Chinese game played with 136 or 144 pieces.

mahogany /mə'hɒgənɪ/ *noun* (*plural* **-ies**) reddish-brown wood used for furniture etc.; colour of this.

mahout /mə'haʊt/ *noun* elephant driver.

maid *noun* female servant; *archaic* girl, young woman. □ **maidservant** female servant.

maiden /'meɪd(ə)n/ ● *noun archaic* girl, young unmarried woman; *Cricket* maiden over. ● *adjective* unmarried; (of voyage, speech by MP, etc.) first. □ **maidenhair** delicate kind of fern; **maiden name** woman's surname before marriage; **maiden over** *Cricket* over in which no runs are scored. □ **maidenly** *adjective*.

mail[1] ● *noun* letters etc. conveyed by post; the post. ● *verb* send by mail. □ **mail order** purchase of goods by post.

mail[2] *noun* armour of metal rings or plates.

maim *verb* cripple, mutilate.

main ● *adjective* chief, principal. ● *noun* principal channel for water, gas, etc or (usually in *plural*) electricity; (in *plural*) domestic electricity supply as distinct from batteries; *archaic* high seas. □ **in the main** mostly, on the whole; **mainframe** central processing unit of large computer, large computer system; **mainland** continuous extent of land excluding nearby islands etc.; **mainmast** principal mast; **mainsail** lowest sail or sail set on after part of mainmast; **mainspring** principal spring of watch or clock, chief motive power etc.; **mainstay** chief support; **mainstream** prevailing trend of opinion, fashion, etc.

mainly *adverb* mostly; chiefly.

maintain /meɪn'teɪn/ *verb* keep up; keep going; support; assert as true; keep in repair.

maintenance /'meɪntənəns/ *noun* maintaining, being maintained; provision of enough to support life; alimony.

maiolica /mə'jɒlɪkə/ *noun* (also **majolica**) kind of decorated Italian earthenware.

maisonette /meɪzə'net/ *noun* flat on more than one floor; small house.

maize *noun* N. American cereal plant; cobs or grain of this.

Maj. *abbreviation* Major.

majestic /mə'dʒestɪk/ *adjective* stately and dignified; imposing. □ **majestically** *adverb.*

majesty /'mædʒɪstɪ/ *noun* (*plural* **-ies**) stateliness of aspect, language, etc.; sovereign power; (**His, Her, Your Majesty**) *title used of or to sovereign or sovereign's wife or widow.*

majolica = MAIOLICA.

major /'meɪdʒə/ ● *adjective* greater or relatively great in size etc.; unusually serious or significant; *Music* of or based on scale having semitone next above third and seventh notes; of full legal age. ● *noun* army officer next below lieutenant colonel; person of full legal age; *US* student's main subject or course, student of this. ● *verb* (+ *in*) *US* study or qualify in as a major. □ **major-domo** /-'dəʊməʊ/ (*plural* **-s**) house-steward; **major-general** army officer next below lieutenant general.

majority /mə'dʒɒrɪtɪ/ *noun* (*plural* **-ies**) (usually + *of*) greater number or part; number by which winning vote exceeds next; full legal age.

■ **Usage** *Majority* should strictly be used of a number of people or things, as in *the majority of people*, and not of a quantity of something, as in *the majority of the work*.

make ● *verb* (**-king**; *past & past participle* **made**) construct, frame, create, esp. from parts or other substance; compel; bring about, give rise to; cause to become or seem; write, compose; constitute, amount to; undertake; perform; gain, acquire, obtain as result; prepare for consumption or use; proceed; *colloquial* arrive at or in time for, manage to attend; *colloquial* achieve place in; establish, enact; consider to be, estimate as; secure success or advancement of; accomplish; become; represent as; form in the mind. ● *noun* origin of manufactured goods, brand; way thing is made. □ **make believe** *verb* pretend; **make-believe** *noun* pretence, *adjective* pretended; **make do** (often + *with*) manage (with substitute etc.); **make for** tend to result in, proceed towards; **make good** compensate for, repair, succeed in an undertaking; **make off** depart hastily; **make out** discern, understand, assert, pretend, *colloquial* progress, write out, fill in; **makeshift** (serving as) temporary substitute or device; **make up** *verb* act to overcome (deficiency), complete, (+ *for*) compensate for, be reconciled, put together, prepare, invent (story), apply cosmetics (to); **make-up** *noun* cosmetics, similar preparation used as disguise by actor, person's temperament etc., composition; **makeweight** small quantity added to make full weight; **on the make** *colloquial* intent on gain.

maker *noun* person who makes, esp. (**Maker**) God.

making *noun* (in *plural*) earnings, profit; essential qualities for becoming. □ **be the making of** ensure success of; **in the making** in the course of being made.

malachite /'mæləkaɪt/ *noun* green mineral used for ornament.

maladjusted /mælə'dʒʌstɪd/ *adjective* (of person) unable to cope with demands of social environment. □ **maladjustment** *noun.*

maladminister /mæləd'mɪnɪstə/ *verb* manage badly or improperly. □ **maladministration** *noun.*

maladroit /mælə'drɔɪt/ *adjective* bungling, clumsy.

malady /'mælədɪ/ *noun* (*plural* **-ies**) ailment, disease.

malaise /mə'leɪz/ *noun* feeling of illness or uneasiness.

malapropism /'mæləprɒpɪz(ə)m/ *noun* comical confusion between words.

malaria /mə'leərɪə/ *noun* fever transmitted by mosquitoes. □ **malarial** *adjective.*

Malay /mə'leɪ/ ● *noun* member of a people predominating in Malaysia and Indonesia; their language. ● *adjective* of this people or language.

malcontent /'mælkəntent/ ● *noun* discontented person. ● *adjective* discontented.

male ● *adjective* of the sex that can beget offspring by fertilizing; (of plants or flowers) containing stamens but no pistil; (of parts of machinery) designed to enter or fill corresponding hollow part. ● *noun* male person or animal.

malediction /mælɪ'dɪkʃ(ə)n/ *noun* curse. □ **maledictory** *adjective.*

malefactor /'mælɪfæktə/ *noun* criminal; evil-doer. □ **malefaction** /-'fækʃ(ə)n/ *noun.*

malevolent /mə'levələnt/ *adjective* wishing evil to others. □ **malevolence** *noun.*

malformation /mælfɔ:'meɪʃ(ə)n/ *noun* faulty formation. □ **malformed** /-'fɔ:md/ *adjective.*

malfunction /mæl'fʌŋkʃ(ə)n/ ● *noun* failure to function normally. ● *verb* function faultily.

malice /'mælɪs/ *noun* ill-will; desire to do harm.

malicious /mə'lɪʃəs/ *adjective* given to or arising from malice. □ **maliciously** *adverb.*

malign /mə'laɪn/ ● *adjective* injurious; malignant; malevolent. ● *verb* speak ill of; slander. □ **malignity** /mə'lɪgnɪtɪ/ *noun.*

malignant /mə'lɪgnənt/ *adjective* (of disease) very virulent; (of tumour) spreading, recurring, cancerous; feeling or showing intense ill-will. □ **malignancy** *noun.*

malinger /mə'lɪŋgə/ *verb* pretend to be ill, esp. to escape duty.

mall /mæl, mɔ:l/ *noun* sheltered walk; shopping precinct.

mallard /'mælɑ:d/ *noun* (*plural* same) kind of wild duck.

malleable /'mælɪəb(ə)l/ *adjective* that can be shaped by hammering; pliable. □ **malleability** *noun.*

mallet /'mælɪt/ *noun* hammer, usually of wood; implement for striking croquet or polo ball.

mallow /'mæləʊ/ *noun* flowering plant with hairy stems and leaves.

malmsey /'mɑ:mzɪ/ *noun* a strong sweet wine.

malnutrition /mælnju:'trɪʃ(ə)n/ *noun* lack of foods necessary for health.

malodorous /mæl'əʊdərəs/ *adjective* evil-smelling.

malpractice /mæl'præktɪs/ *noun* improper, negligent, or criminal professional conduct.

malt /mɔ:lt/ ● *noun* barley or other grain prepared for brewing etc.; *colloquial* malt whisky. ● *verb* convert (grain) into malt. □ **malted milk** drink made from dried milk and extract of malt; **malt whisky** whisky made from malted barley.

Maltese /mɔ:l'ti:z/ ● *noun* native, national, or language of Malta. ● *adjective* of Malta. □ **Maltese cross** one with equal arms broadened at ends.

maltreat /mæl'tri:t/ *verb* ill-treat. □ **maltreatment** *noun.*

mama /mə'mɑ:/ *noun* (also **mamma**) *archaic* mother.

mamba /'mæmbə/ *noun* venomous African snake.

mamma = MAMA.

mammal /'mæm(ə)l/ *noun* animal of class secreting milk to feed young. □ **mammalian** /-'meɪlɪən/ *adjective.*

mammary /'mæmərɪ/ *adjective* of breasts.

Mammon /'mæmən/ *noun* wealth regarded as god or evil influence.

mammoth /'mæməθ/ ● *noun* large extinct elephant. ● *adjective* huge.

man ● *noun* (*plural* **men**) adult human male; human being, person; the human race; employee, workman; (usually in *plural*) soldier, sailor, etc.; suitable or appropriate person; husband; *colloquial* boyfriend; human being of specified type; piece in chess, draughts, etc. ● *verb* (**-nn-**) supply with person(s) for work or defence. □ **manhole** opening giving person access to sewer, conduit, etc.; **man-hour** work done by one person in one hour; **man in the street** ordinary person; **man-of-war** warship; **manpower** people available for work or military service; **manservant** (*plural* **menservants**) male servant; **mantrap** trap set to catch esp. trespassers.

manacle /'mænək(ə)l/ ● *noun* (usually in *plural*) handcuff. ● *verb* (**-ling**) put manacles on.

manage /'mænɪdʒ/ *verb* (**-ging**) organize, regulate; succeed in achieving, contrive; succeed with limited resources, cope; succeed in controlling; cope with. □ **managing director** director with executive control or authority. □ **manageable** *adjective*.

management *noun* managing, being managed; administration; people managing a business.

manager *noun* person controlling or administering business etc.; person controlling activities of person, team, etc.; person who manages money etc. in specified way. □ **managerial** /-'dʒɪərɪəl/ *adjective*.

manageress /mænɪdʒə'res/ *noun* woman manager, esp. of shop, hotel, etc.

mañana /mæn'jɑːnə/ *adverb & noun* some time in the future. [Spanish]

manatee /mænə'tiː/ *noun* large aquatic plant-eating mammal.

Mancunian /mæŋ'kjuːnɪən/ ● *noun* native of Manchester. ● *adjective* of Manchester.

mandarin /'mændərɪn/ *noun* (**Mandarin**) official language of China; *historical* Chinese official; influential person, esp. bureaucrat; (in full **mandarin orange**) tangerine.

mandate /'mændeɪt/ ● *noun* official command; authority given by electors to government etc.; authority to act for another. ● *verb* (**-ting**) instruct (delegate) how to act or vote.

mandatory /'mændətərɪ/ *adjective* compulsory; of or conveying a command.

mandible /'mændɪb(ə)l/ *noun* jaw, esp. lower one; either part of bird's beak; either half of crushing organ in mouthparts of insect etc.

mandolin /mændə'lɪn/ *noun* kind of lute with paired metal strings plucked with a plectrum.

mandrake /'mændreɪk/ *noun* narcotic plant with forked root.

mandrill /'mændrɪl/ *noun* large W. African baboon.

mane *noun* long hair on horse's or lion's neck; *colloquial* person's long hair.

manège /mæ'neɪʒ/ *noun* riding-school; movements of trained horse; horsemanship.

maneuver *US* = MANOEUVRE.

manful *adjective* brave, resolute. □ **manfully** *adverb*.

manganese /'mæŋgəniːz/ *noun* grey brittle metallic element; black oxide of this.

mange /meɪndʒ/ *noun* skin disease of dogs etc.

mangel-wurzel /'mæŋg(ə)l wɜːz(ə)l/ *noun* large beet used as cattle food.

manger /'meɪndʒə/ *noun* eating-trough in stable.

mangle¹ /'mæŋg(ə)l/ *verb* (**-ling**) hack, cut about; mutilate, spoil.

mangle² /'mæŋg(ə)l/ ● *noun* machine with rollers for pressing water out of washed clothes. ● *verb* (**-ling**) put through mangle.

mango /'mæŋgəʊ/ *noun* (*plural* **-es** or **-s**) tropical fruit with yellowish flesh; tree bearing it.

mangold /'mæŋg(ə)ld/ *noun* mangel-wurzel.

mangrove /'mæŋgrəʊv/ *noun* tropical seashore tree with many tangled roots above ground.

mangy /'meɪndʒɪ/ *adjective* (**-ier**, **-iest**) having mange; squalid, shabby.

manhandle *verb* (**-ling**) *colloquial* handle roughly; move by human effort.

manhood *noun* state of being a man; manliness; a man's sexual potency; men of a country.

mania /'meɪnɪə/ *noun* mental illness marked by excitement and violence;

(often + *for*) excessive enthusiasm, obsession.

maniac /ˈmeɪnɪæk/ ● *noun colloquial* person behaving wildly; *colloquial* obsessive enthusiast; person suffering from mania. ● *adjective* of or behaving like maniac. □ **maniacal** /məˈnaɪək(ə)l/ *adjective*.

manic /ˈmænɪk/ *adjective* of or affected by mania. □ **manic-depressive** *adjective* relating to mental disorder with alternating periods of elation and depression, *noun* person having such disorder.

manicure /ˈmænɪkjʊə/ ● *noun* cosmetic treatment of the hands. ● *verb* (**-ring**) give manicure to. □ **manicurist** *noun*.

manifest /ˈmænɪfest/ ● *adjective* clear to sight or mind; indubitable. ● *verb* make manifest; (**manifest itself**) reveal itself. ● *noun* cargo or passenger list. □ **manifestation** *noun*; **manifestly** *adverb*.

manifesto /mænɪˈfestəʊ/ *noun* (*plural* **-s**) declaration of policies.

manifold /ˈmænɪfəʊld/ ● *adjective* many and various; having various forms, applications, parts, etc. ● *noun* manifold thing; pipe etc. with several outlets.

manikin /ˈmænɪkɪn/ *noun* little man, dwarf.

Manila /məˈnɪlə/ *noun* strong fibre of Philippine tree; (also **manila**) strong brown paper made of this.

manipulate /məˈnɪpjʊleɪt/ *verb* (**-ting**) handle, esp. with skill; manage to one's own advantage, esp. unfairly. □ **manipulation** *noun*; **manipulator** *noun*.

manipulative /məˈnɪpjʊlətɪv/ *adjective* tending to exploit a situation, person, etc., for one's own ends.

mankind *noun* human species.

manly *adjective* (**-ier**, **-iest**) having qualities associated with or befitting a man. □ **manliness** *noun*.

manna /ˈmænə/ *noun* food miraculously supplied to Israelites in wilderness.

mannequin /ˈmænɪkɪn/ *noun* fashion model; dummy for display of clothes.

manner /ˈmænə/ *noun* way thing is done or happens; (in *plural*) social behaviour; style; (in *plural*) polite behaviour; outward bearing, way of speaking, etc.; kind, sort.

mannered *adjective* behaving in specified way; showing mannerisms.

mannerism *noun* distinctive gesture or feature of style; excessive use of these in art etc.

mannerly *adjective* well-behaved, polite.

mannish *adjective* (of woman) masculine in appearance or manner; characteristic of man as opposed to woman.

manoeuvre /məˈnuːvə/ (*US* **maneuver**) ● *noun* planned movement of vehicle or troops; (in *plural*) large-scale exercise of troops etc.; agile or skilful movement; artful plan. ● *verb* (**-ring**) move (thing, esp. vehicle) carefully; perform or cause to perform manoeuvres; manipulate by scheming or adroitness; use artifice. □ **manoeuvrable** *adjective*.

manor /ˈmænə/ *noun* large country house with lands; *historical* feudal lordship over lands. □ **manorial** /məˈnɔːrɪəl/ *adjective*.

mansard /ˈmænsɑːd/ *noun* roof with 4 sloping sides, each of which becomes steeper halfway down.

manse /mæns/ *noun* (esp. Scottish Presbyterian) minister's house.

mansion /ˈmænʃ(ə)n/ *noun* large grand house; (in *plural*) block of flats.

manslaughter *noun* unintentional but not accidental unlawful killing of human being.

mantel /ˈmænt(ə)l/ *noun* mantelpiece, mantelshelf. □ **mantelpiece** structure above and around fireplace, mantel shelf; **mantelshelf** shelf above fireplace.

mantilla /mænˈtɪlə/ *noun* Spanish woman's lace scarf worn over head and shoulders.

mantis /ˈmæntɪs/ *noun* (*plural* same or **mantises**) kind of predatory insect.

mantle /ˈmænt(ə)l/ ● *noun* loose sleeveless cloak; covering; fragile tube round gas jet to give incandescent light. ● *verb* (**-ling**) clothe; conceal, envelop.

manual /ˈmænjʊəl/ ● *adjective* of or done with hands. ● *noun* reference book; organ keyboard played with hands, not feet. □ **manually** *adverb*.

manufacture /mænjʊˈfæktʃə/ ● *noun* making of articles, esp. in factory etc.; branch of industry. ● *verb* (**-ring**) make esp. on industrial scale; invent, fabricate. □ **manufacturer** *noun*.

manure /mə'njʊə/ ● *noun* fertilizer, esp. dung. ● *verb* (**-ring**) treat with manure.

manuscript /'mænjʊskrɪpt/ ● *noun* book or document written by hand or typed, not printed. ● *adjective* written by hand.

Manx ● *adjective* of Isle of Man. ● *noun* Celtic language of Isle of Man. □ **Manx cat** tailless variety.

many /'menɪ/ ● *adjective* (**more**, **most**) numerous, great in number. ● *noun* (treated as *plural*) many people or things; (**the many**) the majority of people.

Maori /'maʊrɪ/ ● *noun* (*plural* same or **-s**) member of aboriginal NZ race; their language. ● *adjective* of this people.

map ● *noun* flat representation of (part of) earth's surface, or of sky; diagram. ● *verb* (**-pp-**) represent on map. □ **map out** plan in detail.

maple /'meɪp(ə)l/ *noun* kind of tree. □ **maple leaf** emblem of Canada; **maple sugar** sugar got by evaporating sap of some kinds of maple; **maple syrup** syrup got from maple sap or maple sugar.

maquette /mə'ket/ *noun* preliminary model or sketch.

Mar. *abbreviation* March.

mar *verb* (**-rr-**) spoil; disfigure.

marabou /'mærəbu:/ *noun* (*plural* **-s**) large W. African stork; its down as trimming etc.

maraca /mə'rækə/ *noun* clublike bean-filled gourd etc., shaken as percussion instrument.

maraschino /mærə'ski:nəʊ/ *noun* (*plural* **-s**) liqueur made from cherries. □ **maraschino cherry** one preserved in maraschino.

marathon /'mærəθ(ə)n/ *noun* long-distance foot race; long-lasting, esp. difficult, undertaking.

maraud /mə'rɔ:d/ *verb* make raid; pillage. □ **marauder** *noun*.

marble /'mɑ:b(ə)l/ ● *noun* kind of limestone used in sculpture and architecture; anything of or like marble; small ball of glass etc. as toy; (in *plural*; treated as *singular*) game played with these; (in *plural*) *slang* one's mental faculties; (in *plural*) collection of sculptures. ● *verb* (**-ling**) (esp. as **marbled** *adjective*) give veined or mottled appearance to (esp. paper).

marcasite /'mɑ:kəsaɪt/ *noun* crystalline iron sulphide; crystals of this used in jewellery.

March *noun* third month of year. □ **March hare** hare in breeding season.

march[1] ● *verb* walk in military manner or with regular paces; proceed steadily; cause to march or walk. ● *noun* act of marching; uniform military step; long difficult walk; procession as demonstration; progress; piece of music suitable for marching to. □ **march past** ceremonial march of troops past saluting point. □ **marcher** *noun*.

march[2] *noun historical* boundary (often in *plural*); tract of (often disputed) land between countries etc.

marchioness /mɑ:ʃə'nes/ *noun* marquess's wife or widow; woman holding rank of marquess.

mare /meə/ *noun* female equine animal, esp. horse. □ **mare's nest** illusory discovery.

margarine /mɑ:dʒə'ri:n/ *noun* butter substitute made from edible oils etc.

marge /mɑ:dʒ/ *noun colloquial* margarine.

margin /'mɑ:dʒɪn/ *noun* edge or border of surface; plain space round printed page etc.; amount by which thing exceeds, falls short, etc. □ **margin of error** allowance for miscalculation or mischance.

marginal *adjective* written in margin; of or at edge; (of constituency) having elected MP with small majority; close to limit, esp. of profitability; insignificant; barely adequate. □ **marginally** *adverb*.

marginalize *verb* (also **-ise**) (**-zing** or **-sing**) make or treat as insignificant. □ **marginalization** *noun*.

marguerite /mɑ:gə'ri:t/ *noun* ox-eye daisy.

marigold /'mærɪgəʊld/ *noun* plant with golden or bright yellow flowers.

marijuana /mærɪ'hwɑ:nə/ *noun* dried leaves etc. of hemp smoked as drug.

marimba /mə'rɪmbə/ *noun* African and Central American xylophone; orchestral instrument developed from this.

marina /mə'ri:nə/ *noun* harbour for pleasure boats.

marinade /mærɪ'neɪd/ ● *noun* mixture of wine, vinegar, oil, spices, etc., for soaking fish or meat. ● *verb* (**-ding**) soak in marinade.

marinate /'mærɪneɪt/ *verb* (**-ting**) marinade.

marine /mə'ri:n/ ● *adjective* of, found in, or produced by, the sea; of shipping; for use at sea. ● *noun* member of corps trained to fight on land or sea; country's shipping, fleet, or navy.

mariner /'mærɪnə/ *noun* seaman.

marionette /mærɪə'net/ *noun* puppet worked with strings.

marital /'mærɪt(ə)l/ *adjective* of or between husband and wife; of marriage.

maritime /'mærɪtaɪm/ *adjective* connected with the sea or seafaring; living or found near the sea.

marjoram /'mɑ:dʒərəm/ *noun* aromatic herb used in cookery.

mark[1] ● *noun* visible sign left by person or thing; stain, scar, etc; written or printed symbol; number or letter denoting conduct or proficiency; (often + *of*) sign, indication; lasting effect; target, thing aimed at; line etc. serving to indicate position; (followed by numeral) particular design of piece of equipment. ● *verb* make mark on; distinguish with mark; correct and assess (student's work etc.); attach price to; notice, observe; characterize; acknowledge, celebrate; indicate on map etc.; keep close to (opposing player) in games; (in *passive*) have natural marks. □ **mark down** reduce price of; **mark off** separate by boundary; **mark out** plan (course), destine, trace out (boundaries etc.); **mark time** march on spot without moving forward, await opportunity to advance; **mark up** *verb* increase price of; **mark-up** *noun* amount added to price by retailer for profit.

mark[2] *noun* Deutschmark.

marked /mɑ:kt/ *adjective* having a visible mark; clearly noticeable. □ **markedly** /-kɪdlɪ/ *adverb*.

marker *noun* thing that marks a position; person or thing that marks; pen with broad felt tip; scorer, esp. at billiards.

market /'mɑ:kɪt/ ● *noun* gathering for sale of commodities, livestock, etc.; space for this; (often + *for*) demand for commodity etc.; place or group providing such demand; conditions for buying and selling; stock market. ● *verb* (**-t-**) offer for sale; *archaic* buy or sell goods in market. □ **market garden** place where vegetables are grown for market; **market-place** open space for market, commercial world; **market research** surveying of consumers' needs and preferences; **market town** town where market is held; **market value** value as saleable thing; **on the market** offered for sale. □ **marketing** *noun*.

marketable *adjective* able or fit to be sold.

marking *noun* (usually in *plural*) identification mark; colouring of fur, feathers, etc.

marksman /'mɑ:ksmən/ *noun* skilled shot, esp. with rifle. □ **marksmanship** *noun*.

marl *noun* soil composed of clay and lime, used as fertilizer.

marlinspike /'mɑ:lɪnspaɪk/ *noun* pointed tool used to separate strands of rope or wire.

marmalade /'mɑ:məleɪd/ *noun* preserve of oranges or other citrus fruit.

Marmite /'mɑ:maɪt/ *noun proprietary term* thick brown spread made from yeast and vegetable extract.

marmoreal /mɑ:'mɔ:rɪəl/ *adjective* of or like marble.

marmoset /'mɑ:məzet/ *noun* small bushy-tailed monkey.

marmot /'mɑ:mət/ *noun* burrowing rodent with short bushy tail.

marocain /'mærəkeɪn/ *noun* fabric of ribbed crêpe.

maroon[1] /mə'ru:n/ *adjective & noun* brownish-crimson.

maroon[2] /mə'ru:n/ *verb* put and leave ashore on desolate island or coast; leave stranded.

marquee /mɑ:'ki:/ *noun* large tent.

marquess /'mɑ:kwəs/ *noun* British nobleman ranking between duke and earl.

marquetry /'mɑ:kɪtrɪ/ *noun* inlaid work in wood etc.

marquis /ˈmɑːkwɪs/ *noun* (*plural* **-quises**) foreign nobleman ranking between duke and count.

marquise /mɑːˈkiːz/ *noun* marquis's wife or widow; woman holding rank of marquis.

marriage /ˈmærɪdʒ/ *noun* legal union of man and woman for the purpose of living together; act or ceremony establishing this; particular matrimonial union. □ **marriage certificate, lines** certificate stating that marriage has taken place; **marriage guidance** counselling of people with marital problems.

marriageable *adjective* free, ready, or fit for marriage.

marrow /ˈmærəʊ/ *noun* large fleshy gourd, cooked as vegetable; bone marrow. □ **marrowbone** bone containing edible marrow; **marrowfat** kind of large pea.

marry /ˈmærɪ/ *verb* (**-ies**, **-ied**) take, join, or give in marriage; enter into marriage; (+ *into*) become member of (family) by marriage; unite intimately.

Marsala /mɑːˈsɑːlə/ *noun* dark sweet fortified wine.

Marseillaise /mɑːseɪˈjeɪz/ *noun* French national anthem.

marsh *noun* low watery ground. □ **marsh gas** methane; **marsh mallow** shrubby herb; **marshmallow** soft sweet made from sugar, albumen, gelatin, etc. □ **marshy** *adjective* (**-ier**, **-iest**).

marshal /ˈmɑːʃ(ə)l/ ● *noun* (**Marshal**) high-ranking officer of state or in armed forces; officer arranging ceremonies, controlling procedure at races, etc. ● *verb* (**-ll-**) arrange in due order; conduct (person) ceremoniously. □ **marshalling yard** yard in which goods trains etc. are assembled.

marsupial /mɑːˈsuːpɪəl/ ● *noun* mammal giving birth to underdeveloped young subsequently carried in pouch. ● *adjective* of or like a marsupial.

mart *noun* trade centre; auction-room; market.

Martello tower /mɑːˈteləʊ/ *noun* small circular coastal fort.

marten /ˈmɑːtɪn/ *noun* weasel-like flesh-eating mammal with valuable fur; its fur.

martial /ˈmɑːʃ(ə)l/ *adjective* of warfare; warlike. □ **martial arts** fighting sports such as judo or karate; **martial law** military government with ordinary law suspended.

Martian /ˈmɑːʃ(ə)n/ ● *adjective* of planet Mars. ● *noun* hypothetical inhabitant of Mars.

martin /ˈmɑːtɪn/ *noun* bird of swallow family.

martinet /mɑːtɪˈnet/ *noun* strict disciplinarian.

Martini /mɑːˈtiːnɪ/ *noun* (*plural* **-s**) *proprietary term* type of vermouth; cocktail of gin and vermouth.

martyr /ˈmɑːtə/ ● *noun* person who undergoes death or suffering for great cause; (+ *to*) *colloquial* constant sufferer from. ● *verb* put to death as martyr; torment. □ **martyrdom** *noun*.

marvel /ˈmɑːv(ə)l/ ● *noun* wonderful thing; (+ *of*) wonderful example of. ● *verb* (**-ll-**; *US* **-l-**) (+ *at*, *that*) feel surprise or wonder.

marvellous /ˈmɑːvələs/ *adjective* (*US* **marvelous**) astonishing; excellent. □ **marvellously** *adverb*.

Marxism /ˈmɑːksɪz(ə)m/ *noun* doctrines of Marx, predicting common ownership of means of production. □ **Marxist** *noun & adjective*.

marzipan /ˈmɑːzɪpæn/ *noun* paste of ground almonds, sugar, etc.

mascara /mæsˈkɑːrə/ *noun* cosmetic for darkening eyelashes.

mascot /ˈmæskɒt/ *noun* person, animal, or thing supposed to bring luck.

masculine /ˈmæskjʊlɪn/ *adjective* of men; manly; *Grammar* belonging to gender including words for most male people and animals. □ **masculinity** /-ˈlɪn-/ *noun*.

maser /ˈmeɪzə/ *noun* device for amplifying or generating microwaves.

mash ● *noun* soft or confused mixture; mixture of boiled bran etc. fed to horses; *colloquial* mashed potatoes. ● *verb* crush (potatoes etc.) to pulp.

mask /mɑːsk/ ● *noun* covering for all or part of face, worn as disguise or for protection, or by surgeon etc. to prevent infection of patient; respirator; likeness of person's face, esp. one made by taking mould from face; disguise.

• *verb* cover with mask; conceal; protect.

masochism /ˈmæsəkɪz(ə)m/ *noun* pleasure in suffering physical or mental pain, esp. as form of sexual perversion. □ **masochist** *noun*; **masochistic** /-ˈkɪs-/ *adjective*.

mason /ˈmeɪs(ə)n/ *noun* person who builds with stone; (**Mason**) Freemason.

Masonic /məˈsɒnɪk/ *adjective* of Freemasons.

masonry /ˈmeɪsənrɪ/ *noun* stonework; mason's work; (**Masonry**) Freemasonry.

masque /mɑːsk/ *noun* musical drama with mime, esp. in 16th & 17th c.

masquerade /mæskəˈreɪd/ • *noun* false show, pretence; masked ball. • *verb* (**-ding**) appear in disguise; assume false appearance.

mass[1] • *noun* cohesive body of matter; dense aggregation; (in *singular* or *plural*; usually + *of*) large number or amount; (usually + *of*) unbroken expanse (of colour etc.); (**the mass**) the majority; (**the masses**) ordinary people; *Physics* quantity of matter body contains. • *verb* gather into mass; assemble into one body. • *adjective* of or relating to large numbers of people or things. □ **mass media** means of communication to large numbers of people; **mass production** mechanical production of large quantities of standardized article.

mass[2] *noun* (often **Mass**) Eucharist, esp. in RC Church; (musical setting of) liturgy used in this.

massacre /ˈmæsəkə/ • *noun* general slaughter. • *verb* (**-ring**) make massacre of.

massage /ˈmæsɑːʒ/ • *noun* kneading and rubbing of muscles etc., usually with hands. • *verb* (**-ging**) treat thus.

masseur /mæˈsɜː/ *noun* (*feminine* **masseuse** /mæˈsɜːz/) person who gives massage.

massif /ˈmæsiːf/ *noun* mountain heights forming compact group.

massive /ˈmæsɪv/ *adjective* large and heavy or solid; unusually large or severe; substantial. □ **massively** *adverb*.

mast[1] /mɑːst/ *noun* upright to which ship's yards and sails are attached; tall metal structure supporting radio or television aerial; flag-pole.

mast[2] /mɑːst/ *noun* fruit of beech, oak, etc., esp. as food for pigs.

mastectomy /mæˈstektəmɪ/ *noun* (*plural* **-ies**) surgical removal of a breast.

master /ˈmɑːstə/ • *noun* person having control or ownership; ship's captain; male teacher; prevailing person; skilled workman; skilled practitioner; holder of university degree above bachelor's; revered teacher; great artist; *Chess etc.* player at international level; thing from which series of copies is made; (**Master**) *title prefixed to name of boy*. • *adjective* commanding; main, principal; controlling others. • *verb* overcome, conquer; acquire complete knowledge of. □ **master-key** one opening several different locks; **mastermind** *noun* person with outstanding intellect, *verb* plan and direct (enterprise); **Master of Ceremonies** person introducing speakers at banquet or entertainers at variety show; **masterpiece** outstanding piece of artistry, one's best work; **master-switch** switch controlling electricity etc. supply to entire system.

masterful *adjective* imperious, domineering; very skilful. □ **masterfully** *adverb*.

■ **Usage** *Masterful* is normally used of a person, whereas *masterly* is used of achievements, abilities, etc.

---

masterly *adjective* very skilful.

■ **Usage** See note at MASTERFUL.

---

mastery *noun* control, dominance; (often + *of*) comprehensive skill or knowledge.

mastic /ˈmæstɪk/ *noun* gum or resin from certain trees; such tree; waterproof filler and sealant.

masticate /ˈmæstɪkeɪt/ *verb* (**-ting**) chew. □ **mastication** *noun*.

mastiff /ˈmæstɪf/ *noun* large strong kind of dog.

mastodon /ˈmæstədɒn/ *noun* (*plural* same or **-s**) extinct animal resembling elephant.

mastoid /ˈmæstɔɪd/ • *adjective* shaped like woman's breast. • *noun* (in full **mastoid process**) conical prominence on temporal bone; (usually in *plural*) *colloquial* inflammation of mastoid.

masturbate /'mæstəbeɪt/ *verb* (**-ting**) produce sexual arousal (of) by manual stimulation of genitals. □ **masturbation** *noun*.

mat¹ ● *noun* piece of coarse fabric on floor, esp. for wiping shoes on; piece of material laid on table etc. to protect surface. ● *verb* (**-tt-**) (esp. as **matted** *adjective*) bring or come into thickly tangled state. □ **on the mat** *slang* being reprimanded.

mat² = MATT.

matador /'mætədɔ:/ *noun* bullfighter whose task is to kill bull.

match¹ ● *noun* contest, game; person or thing equal to, exactly resembling, or corresponding to another; marriage; person viewed as marriage prospect. ● *verb* be equal, correspond; be or find match for; (+ *against*, *with*) place in conflict or competition with. □ **matchboard** tongued and grooved board fitting into others; **matchmaker** person who arranges marriages or schemes to bring couples together; **match point** state of game when one side needs only one point to win match; **match up** (often + *with*) fit to form whole, tally; **match up to** be equal to.

match² *noun* short thin piece of wood etc., tipped with substance that ignites when rubbed on rough or specially prepared surface. □ **matchbox** box for holding matches; **matchstick** stem of match; **matchwood** wood suitable for matches, minute splinters.

matchless *adjective* incomparable.

mate¹ ● *noun* companion, fellow worker; *colloquial form of address, esp. to another man*; each of a breeding pair, esp. of birds; *colloquial* partner in marriage; subordinate officer on merchant ship; assistant to worker. ● *verb* (**-ting**) come or bring together for breeding.

mate² *noun & verb* (**-ting**) checkmate.

material /mə'tɪərɪəl/ ● *noun* that from which thing is made; cloth, fabric; (in *plural*) things needed for activity; person or thing of specified kind or suitable for purpose; (in *singular* or *plural*) information etc. for book etc.; (in *singular* or *plural*) elements. ● *adjective* of matter; not spiritual; of bodily comfort etc.; important, relevant.

materialism *noun* greater interest in material possessions and comfort than in spiritual values; theory that nothing exists but matter. □ **materialist** *noun*; **materialistic** /-'lɪs-/ *adjective*.

materialize *verb* (also **-ise**) (**-zing** or **-sing**) become fact, happen; *colloquial* appear, be present; represent in or assume bodily form. □ **materialization** *noun*.

maternal /mə'tɜ:n(ə)l/ *adjective* of or like a mother; motherly; related on mother's side.

maternity /mə'tɜ:nɪtɪ/ ● *noun* motherhood; motherliness. ● *adjective* for women in pregnancy or childbirth.

matey *adjective* (also **maty**) (**-tier**, **-tiest**) familiar and friendly. □ **matily** *adverb*.

math *noun US colloquial* mathematics.

mathematics /mæθə'mætɪks/ *plural noun* (also treated as *singular*) science of space, number, and quantity. □ **mathematical** *adjective*; **mathematician** /-mə'tɪʃ(ə)n/ *noun*.

maths *noun colloquial* mathematics.

matinée /'mætɪneɪ/ *noun* (*US* **matinee**) theatrical etc. performance in afternoon. □ **matinée coat** baby's short coat.

matins /'mætɪnz/ *noun* (also **mattins**) morning prayer.

matriarch /'meɪtrɪɑ:k/ *noun* female head of family or tribe. □ **matriarchal** /-'ɑ:k(ə)l/ *adjective*.

matriarchy /'meɪtrɪɑ:kɪ/ *noun* (*plural* **-ies**) female-dominated system of society.

matrices *plural* of MATRIX.

matricide /'meɪtrɪsaɪd/ *noun* killing of one's mother; person who does this.

matriculate /mə'trɪkjʊleɪt/ *verb* (**-ting**) admit (student) to university; be thus admitted. □ **matriculation** *noun*.

matrimony /'mætrɪmənɪ/ *noun* marriage. □ **matrimonial** /-'məʊnɪəl/ *adjective*.

matrix /'meɪtrɪks/ *noun* (*plural* **matrices** /-si:z/ or **-es**) mould in which thing is cast or shaped; place etc. in which thing is developed; rock in which gems etc. are embedded; *Mathematics* rectangular array of quantities treated as single quantity.

matron /'meɪtrən/ *noun* woman in charge of nursing in hospital; married, esp. staid, woman; woman nurse and housekeeper at school etc.

matronly *adjective* like a matron, esp. portly or staid.

matt *adjective* (also **mat**) dull, not shiny or glossy.

matter /'mætə/ ● *noun* physical substance; thing(s), material; (**the matter**; often + *with*) thing that is amiss; content as opposed to form, substance; affair, concern; purulent discharge. ● *verb* (often + *to*) be of importance. □ **a matter of** approximately, amounting to; **matter of course** natural or expected thing; **matter-of-fact** prosaic, unimaginative, unemotional; **no matter** (+ *when*, *how*, etc.) regardless of.

matting *noun* fabric for mats.

mattins = MATINS.

mattock /'mætək/ *noun* tool like pickaxe with adze and chisel edge as ends of head.

mattress /'mætrɪs/ *noun* fabric case filled with soft or firm material or springs, used on or as bed.

mature /mə'tʃʊə/ ● *adjective* (**-r**, **-st**) fully developed, ripe; adult; careful, considered; (of bill etc.) due for payment. ● *verb* (**-ring**) bring to or reach mature state. □ **maturity** *noun*.

matutinal /mætju:'taɪn(ə)l/ *adjective* of or in morning.

maty = MATEY.

maudlin /'mɔ:dlɪn/ *adjective* weakly sentimental.

maul /mɔ:l/ ● *verb* injure by clawing etc.; handle roughly; damage. ● *noun Rugby* loose scrum; brawl; heavy hammer.

maulstick /'mɔ:lstɪk/ *noun* stick held to support hand in painting.

maunder /'mɔ:ndə/ *verb* talk ramblingly.

Maundy /'mɔ:ndɪ/ *noun* distribution of **Maundy money**, silver coins minted for English sovereign to give to the poor on **Maundy Thursday**, Thursday before Easter.

mausoleum /mɔ:sə'li:əm/ *noun* magnificent tomb.

mauve /məʊv/ *adjective & noun* pale purple.

maverick /'mævərɪk/ *noun* unorthodox or independent-minded person; *US* unbranded calf etc.

maw *noun* stomach of animal.

mawkish /'mɔ:kɪʃ/ *adjective* feebly sentimental.

maxillary /mæk'sɪlərɪ/ *adjective* of the jaw.

maxim /'mæksɪm/ *noun* general truth or rule of conduct briefly expressed.

maxima *plural* of MAXIMUM.

maximal /'mæksɪm(ə)l/ *adjective* greatest possible in size, duration, etc.

maximize /'mæksɪmaɪz/ *verb* (also **-ise**) (**-zing** or **-sing**) make as large or great as possible. □ **maximization** *noun*.

■ **Usage** *Maximize* should not be used in standard English to mean 'to make as good as possible' or 'to make the most of'.

maximum /'mæksɪməm/ ● *noun* (*plural* **maxima**) highest possible amount, size, etc. ● *adjective* greatest in amount, size, etc.

May *noun* fifth month of year; (**may**) hawthorn, esp. in blossom. □ **May Day** 1 May as Spring festival or as international holiday in honour of workers; **mayfly** insect living briefly in spring as adult; **maypole** decorated pole danced round on May Day; **May queen** girl chosen to preside over May Day festivities.

may *auxiliary verb* (*3rd singular present* **may**; *past* **might** /maɪt/) *expressing possibility, permission, request, wish, etc.* □ **be that as it may** although that is possible.

■ **Usage** Both *can* and *may* are used for asking permission, as in *Can I move?* and *May I move?*, but *may* is better in formal English because *Can I move?* also means 'Am I physically able to move?'

maybe /'meɪbi:/ *adverb* perhaps.

mayday /'meɪdeɪ/ *noun* international radio distress signal.

mayhem /'meɪhem/ *noun* destruction, havoc.

mayonnaise /meɪə'neɪz/ *noun* creamy dressing of oil, egg yolk, vinegar, etc.; dish dressed with this.

mayor /meə/ *noun* head of corporation of city or borough; head of district council with status of borough. □ **mayoral** *adjective*.

mayoralty /'meərəltɪ/ *noun* (*plural* **-ies**) office of mayor; period of this.

mayoress /'meərɪs/ *noun* woman mayor; wife or consort of mayor.

maze *noun* network of paths and hedges designed as puzzle; labyrinth; confused network, mass, etc.

mazurka /mə'zɜ:kə/ *noun* lively Polish dance in triple time; music for this.

MB *abbreviation* Bachelor of Medicine.

MBE *abbreviation* Member of the Order of the British Empire.

MC *abbreviation* Master of Ceremonies; Military Cross.

MCC *abbreviation* Marylebone Cricket Club.

MD *abbreviation* Doctor of Medicine; Managing Director.

me[1] /mi:/ *pronoun used by speaker or writer to refer to himself or herself as object of verb*; *colloquial* I.

■ **Usage** Some people consider it correct to use only *It is I*, but this is very formal or old-fashioned in most situations, and *It is me* is normally quite acceptable. On the other hand, it is not standard English to say *Me and him went* rather than *He and I went*.

me[2] /mi:/ *noun* (also **mi**) *Music* third note of scale in tonic sol-fa.

mead *noun* alcoholic drink of fermented honey and water.

meadow /'medəʊ/ *noun* piece of grassland, esp. used for hay; low ground, esp. near river. □ **meadowsweet** a fragrant flowering plant.

meagre /'mi:gə/ *adjective* (*US* **meager**) scanty in amount or quality.

meal[1] *noun* occasion when food is eaten; the food eaten on one occasion. □ **meal-ticket** *colloquial* source of income.

meal[2] *noun* grain or pulse ground to powder.

mealy *adjective* (**-ier**, **-iest**) of, like, or containing meal. □ **mealy-mouthed** afraid to speak plainly.

mean[1] *verb* (*past & past participle* **meant** /ment/) have as one's purpose, design; intend to convey or indicate; involve, portend; (of word) have as equivalent in same or another language; (+ *to*) be of specified significance to.

mean[2] *adjective* niggardly; not generous; ignoble; of low degree or poor quality; malicious; *US* vicious, aggressive. □ **meanness** *noun*.

mean[3] ● *noun* condition, quality, or course of action equally far from two extremes; term midway between first and last terms of progression; quotient of the sum of several quantities and their number. ● *adjective* (of quantity) equally far from two extremes; calculated as mean.

meander /mɪ'ændə/ ● *verb* wander at random; wind about. ● *noun* (in *plural*) sinuous windings; circuitous journey.

meaning ● *noun* what is meant; significance. ● *adjective* expressive; significant. □ **meaningful** *adjective*; **meaningfully** *adverb*; **meaningless** *adjective*; **meaninglessness** *noun*.

means *plural noun* (often treated as *singular*) action, agent, device, or method producing result; money resources. □ **means test** inquiry into financial resources of applicant for assistance etc.

meantime ● *adverb* meanwhile. ● *noun* intervening period.

meanwhile *adverb* in the intervening time; at the same time.

measles /'mi:z(ə)lz/ *plural noun* (also treated as *singular*) infectious viral disease with red rash.

measly /'mi:zlɪ/ *adjective* (**-ier**, **-iest**) *colloquial* meagre, contemptible.

measure /'meʒə/ ● *noun* size or quantity found by measuring; system or unit of measuring; vessel, rod, tape, etc., for measuring; degree, extent; factor determining evaluation etc.; (usually in *plural*) suitable action; legislative enactment; prescribed extent or amount; poetic metre. ● *verb* (**-ring**) find size, quantity, proportions, etc. of by comparison with known standard; be of specified size; estimate by some criterion; (often + *off*) mark (line etc. of given length); (+ *out*) distribute in measured quantities; (+ *with*, *against*) bring into competition with. □ **measurable** *adjective*; **measurement** *noun*.

measured *adjective* rhythmical; (of language) carefully considered.

measureless *adjective* not measurable; infinite.

meat *noun* animal flesh as food; (often + *of*) chief part.

meaty *adjective* (**-ier**, **-iest**) full of meat; fleshy; of or like meat; substantial, satisfying.

Mecca /'mekə/ *noun* place one aspires to visit.

mechanic /mɪ'kænɪk/ *noun* person skilled in using or repairing machinery.

mechanical *adjective* of, working, or produced by, machines or mechanism; automatic; lacking originality; of mechanics as a science. □ **mechanically** *adverb*.

mechanics /mɪ'kænɪks/ *plural noun* (usually treated as *singular*) branch of applied mathematics dealing with motion; science of machinery; routine technical aspects of thing.

mechanism /'mekənɪz(ə)m/ *noun* structure or parts of machine; system of parts working together; process, method.

mechanize /'mekənaɪz/ *verb* (also **-ise**) (**-zing** or **-sing**) introduce machines in; make mechanical; equip with tanks, armoured cars, etc. □ **mechanization** *noun*.

medal /'med(ə)l/ *noun* commemorative metal disc etc., esp. awarded for military or sporting prowess.

medallion /mɪ'dæljən/ *noun* large medal; thing so shaped, e.g. portrait.

medallist /'medəlɪst/ *noun* (*US* **medalist**) winner of (specified) medal.

meddle /'med(ə)l/ *verb* (**-ling**) (often + *with*, *in*) interfere in others' concerns.

meddlesome *adjective* interfering.

media *plural* of MEDIUM.

■ **Usage** It is a mistake to use *media* with a singular verb, as in *The media is biased*.

mediaeval = MEDIEVAL.

median /'mi:dɪən/ ● *adjective* situated in the middle. ● *noun* straight line from angle of triangle to middle of opposite side; middle value of series.

mediate /'mi:dɪeɪt/ *verb* (**-ting**) act as go-between or peacemaker. □ **mediation** *noun*; **mediator** *noun*.

medical /'medɪk(ə)l/ ● *adjective* of medicine in general or as distinct from surgery. ● *noun colloquial* medical examination. □ **medical certificate** certificate of fitness or unfitness to work etc.; **medical examination** examination to determine person's physical fitness. □ **medically** *adverb*.

medicament /mɪ'dɪkəmənt/ *noun* substance used in curative treatment.

medicate /'medɪkeɪt/ *verb* (**-ting**) treat medically; impregnate with medicinal substance.

medication /medɪ'keɪʃ(ə)n/ *noun* medicinal drug; treatment using drugs.

medicinal /mə'dɪsɪn(ə)l/ *adjective* (of substance) healing.

medicine /'meds(ə)n/ *noun* science or practice of diagnosis, treatment, and prevention of disease, esp. as distinct from surgery; substance, esp. one taken by mouth, used in this. □ **medicine man** witch-doctor.

medieval /medɪ'i:v(ə)l/ *adjective* (also **mediaeval**) of Middle Ages.

mediocre /mi:dɪ'əʊkə/ *adjective* indifferent in quality; second-rate.

mediocrity /mi:dɪ'ɒkrɪtɪ/ *noun* (*plural* **-ies**) being mediocre; mediocre person.

meditate /'medɪteɪt/ *verb* (**-ting**) engage in (esp. religious) contemplation; plan mentally. □ **meditation** *noun*; **meditative** /-tətɪv/ *adjective*.

Mediterranean /medɪtə'reɪnɪən/ *adjective* of the sea between Europe and N. Africa, or the countries bordering on it.

medium /'mi:dɪəm/ ● *noun* (*plural* **media** or **-s**) middle quality, degree, etc. between extremes; environment; means of communication; physical material or form used by artist, composer, etc.; (*plural* **-s**) person claiming to communicate with the dead. ● *adjective* between two qualities etc.; average. □ **medium-range** (of aircraft, missile, etc.) able to travel medium distance.

medlar /'medlə/ *noun* tree bearing fruit like apple, eaten when decayed; such fruit.

medley /'medlɪ/ *noun* (*plural* **-s**) varied mixture.

medulla /mɪ'dʌlə/ *noun* inner part of certain bodily organs; soft internal tissue of plants. □ **medulla oblongata** /ɒblɒŋ'gɑ:tə/ lowest part of brainstem. □ **medullary** *adjective.*

meek *adjective* humble and submissive or gentle. □ **meekly** *adverb*; **meekness** *noun.*

meerschaum /'mɪəʃəm/ *noun* soft white clay-like substance; tobacco pipe with bowl made from this.

meet[1] • *verb* (*past & past participle* **met**) encounter or (of two or more people) come together; be present at arrival of (person, train, etc.); come into contact (with); make acquaintance of; deal with (demand etc.); (often + *with*) experience, receive. • *noun* assembly for a hunt; assembly for athletics.

meet[2] *adjective archaic* fitting, proper.

meeting *noun* coming together; assembly of esp. a society, committee, etc.; race meeting.

mega- *combining form* large; one million; *slang* extremely, very big.

megabyte /'megəbaɪt/ *noun* 2^{20} bytes (approx. 1,000,000) as unit of computer storage.

megalith /'megəlɪθ/ *noun* large stone, esp. prehistoric monument. □ **megalithic** /-'lɪθ-/ *adjective.*

megalomania /megələ'meɪnɪə/ *noun* mental disorder producing delusions of grandeur; passion for grandiose schemes. □ **megalomaniac** *adjective & noun.*

megaphone /'megəfəʊn/ *noun* large funnel-shaped device for amplifying voice.

megaton /'megətʌn/ *noun* unit of explosive power equal to that of 1,000,000 tons of TNT.

meiosis /maɪ'əʊsɪs/ *noun* (*plural* **meioses** /-si:z/) cell division resulting in gametes with half normal chromosome number; ironical understatement.

melamine /'meləmi:n/ *noun* crystalline compound producing resins; plastic made from this.

melancholia /melən'kəʊlɪə/ *noun* depression and anxiety.

melancholy /'melənkəlɪ/ • *noun* pensive sadness; depression; tendency to this. • *adjective* sad; depressing. □ **melancholic** /-'kɒl-/ *adjective.*

mêlée /'meleɪ/ *noun* (*US* **melee**) confused fight or scuffle; muddle.

mellifluous /mɪ'lɪflʊəs/ *adjective* (of voice etc.) pleasing, musical.

mellow /'meləʊ/ • *adjective* (of sound, colour, light, or flavour) soft and rich, free from harshness; (of character) gentle; mature; genial. • *verb* make or become mellow.

melodic /mɪ'lɒdɪk/ *adjective* of melody; melodious.

melodious /mɪ'ləʊdɪəs/ *adjective* of, producing, or having melody; sweet-sounding. □ **melodiously** *adverb.*

melodrama /'melədrɑ:mə/ *noun* sensational play etc. appealing blatantly to emotions; this type of drama. □ **melodramatic** /-drə'mæt-/ *adjective.*

melody /'melədɪ/ *noun* (*plural* **-ies**) arrangement of notes to make distinctive pattern; tune; principal part in harmonized music; tunefulness.

melon /'melən/ *noun* sweet fleshy fruit of various climbers of gourd family.

melt *verb* become liquid or change from solid to liquid by action of heat; dissolve; (as **molten** *adjective*) (esp. of metals etc.) liquefied by heat; soften, be softened; (usually + *into*) merge; (often + *away*) leave unobtrusively. □ **melt down** melt (esp. metal) for reuse, become liquid and lose structure; **melting point** temperature at which solid melts; **melting pot** place for mixing races, theories, etc.

member /'membə/ *noun* person etc. belonging to society, team, group, etc.; (**Member**) person elected to certain assemblies; part of larger structure; part or organ of body, esp. limb.

membership *noun* being a member; number or group of members.

membrane /'membreɪn/ *noun* pliable tissue connecting or lining organs in plants and animals; pliable sheet or skin. □ **membranous** /-brən-/ *adjective.*

memento /mɪ'mentəʊ/ *noun* (*plural* **-es** or **-s**) souvenir of person or event.

memo /'meməʊ/ *noun* (*plural* **-s**) *colloquial* memorandum.

memoir /ˈmemwɑː/ *noun* historical account etc. written from personal knowledge or special sources; (in *plural*) autobiography.

memorable /ˈmemərəb(ə)l/ *adjective* worth remembering; easily remembered. □ **memorably** *adverb.*

memorandum /meməˈrændəm/ *noun* (*plural* **-da** or **-s**) note or record for future use; informal written message, esp. in business etc.

memorial /mɪˈmɔːrɪəl/ ● *noun* object etc. established in memory of person or event. ● *adjective* commemorative.

memoriam see IN MEMORIAM.

memorize /ˈmeməraɪz/ *verb* (also **-ise**) (**-zing** or **-sing**) commit to memory.

memory /ˈmemərɪ/ *noun* (*plural* **-ies**) faculty by which things are recalled to or kept in mind; store of things remembered; remembrance, esp. of person etc.; storage capacity of computer etc.; posthumous reputation.

memsahib /ˈmemsɑːb/ *noun historical* European married woman in India.

men *plural* of MAN.

menace /ˈmenɪs/ ● *noun* threat; dangerous thing or person; *jocular* nuisance. ● *verb* (**-cing**) threaten. □ **menacingly** *adverb.*

ménage /meɪˈnɑːʒ/ *noun* household.

menagerie /mɪˈnædʒərɪ/ *noun* small zoo.

mend ● *verb* restore to good condition; repair; regain health; improve. ● *noun* darn or repair in material etc. □ **on the mend** recovering, esp. in health.

mendacious /menˈdeɪʃəs/ *adjective* lying, untruthful. □ **mendacity** /-ˈdæs-/ *noun* (*plural* **-ies**).

mendicant /ˈmendɪkənt/ ● *adjective* begging; (of friar) living solely on alms. ● *noun* beggar; mendicant friar.

menfolk *plural noun* men, esp. men of family.

menhir /ˈmenhɪə/ *noun* usually prehistoric monument of tall upright stone.

menial /ˈmiːnɪəl/ ● *adjective* (of work) degrading, servile. ● *noun* domestic servant.

meningitis /menɪnˈdʒaɪtɪs/ *noun* (esp. viral) infection and inflammation of membranes enclosing brain and spinal cord.

meniscus /mɪˈnɪskəs/ *noun* (*plural* **menisci** /-saɪ/) curved upper surface of liquid in tube; lens convex on one side and concave on the other.

menopause /ˈmenəpɔːz/ *noun* ceasing of menstruation; period in woman's life when this occurs. □ **menopausal** /-ˈpɔːz(ə)l/ *adjective.*

menses /ˈmensiːz/ *plural noun* flow of menstrual blood etc.

menstrual /ˈmenstrʊəl/ *adjective* of menstruation.

menstruate /ˈmenstrʊeɪt/ *verb* (**-ting**) undergo menstruation.

menstruation *noun* discharge of blood etc. from uterus, usually at monthly intervals.

mensuration /mensjʊəˈreɪʃ(ə)n/ *noun* measuring; measuring of lengths, areas, and volumes.

mental /ˈment(ə)l/ *adjective* of, in, or done by mind; caring for mental patients; *colloquial* insane. □ **mental age** degree of mental development in terms of average age at which such development is attained; **mental deficiency** abnormally low intelligence; **mental patient** sufferer from mental illness. □ **mentally** *adverb.*

mentality /menˈtælɪtɪ/ *noun* (*plural* **-ies**) mental character or disposition.

menthol /ˈmenθɒl/ *noun* mint-tasting organic alcohol found in oil of peppermint etc., used as flavouring and to relieve local pain.

mention /ˈmenʃ(ə)n/ ● *verb* refer to briefly or by name; disclose. ● *noun* reference, esp. by name.

mentor /ˈmentɔː/ *noun* experienced and trusted adviser.

menu /ˈmenjuː/ *noun* (*plural* **-s**) list of dishes available in restaurant etc., or to be served at meal; *Computing* list of options displayed on VDU.

MEP *abbreviation* Member of European Parliament.

mercantile /ˈmɜːkəntaɪl/ *adjective* of trade, trading; commercial. □ **mercantile marine** merchant shipping.

mercenary /ˈmɜːsɪnərɪ/ ● *adjective* primarily concerned with or working for money etc. ● *noun* (*plural* **-ies**) hired soldier in foreign service.

mercer /'mɜ:sə/ *noun* dealer in textile fabrics.

mercerize /'mɜ:səraɪz/ *verb* (also **-ise**) (**-zing** or **-sing**) treat (cotton) with caustic alkali to strengthen and make lustrous.

merchandise /'mɜ:tʃəndaɪz/ • *noun* goods for sale. • *verb* (**-sing**) trade (in); promote (goods, ideas, etc.).

merchant /'mɜ:tʃ(ə)nt/ *noun* wholesale trader, esp. with foreign countries; *esp. US & Scottish* retail trader. □ **merchant bank** bank dealing in commercial loans and finance; **merchantman** merchant ship; **merchant navy** nation's commercial shipping; **merchant ship** ship carrying merchandise.

merchantable *adjective* saleable.

merciful /'mɜ:sɪfʊl/ *adjective* showing mercy. □ **mercifulness** *noun.*

mercifully *adverb* in a merciful way; fortunately.

merciless /'mɜ:sɪləs/ *adjective* showing no mercy. □ **mercilessly** *adverb.*

mercurial /mɜ:'kjʊərɪəl/ *adjective* (of person) volatile; of or containing mercury.

mercury /'mɜ:kjʊrɪ/ *noun* silvery heavy liquid metal used in barometers, thermometers, etc.; (**Mercury**) planet nearest to the sun. □ **mercuric** /-'kjʊər-/ *adjective*; **mercurous** *adjective.*

mercy /'mɜ:sɪ/ *noun* (*plural* **-ies**) compassion towards defeated enemies or offenders or as quality; act of mercy; thing to be thankful for. □ **at the mercy of** in the power of; **mercy killing** killing done out of pity.

mere[1] /mɪə/ *adjective* (**-st**) being only what is specified. □ **merely** *adverb.*

mere[2] /mɪə/ *noun dialect or poetical* lake.

meretricious /merə'trɪʃəs/ *adjective* showily but falsely attractive.

merganser /mɜ:'gænsə/ *noun* (*plural* same or **-s**) a diving duck.

merge *verb* (**-ging**) (often + *with*) combine, join or blend gradually; (+ *in*) (cause to) lose character and identity in (something else).

merger *noun* combining, esp. of two commercial companies etc. into one.

meridian /mə'rɪdɪən/ *noun* circle of constant longitude passing through given place and N. & S. Poles; corresponding line on map etc.

meridional /mə'rɪdɪən(ə)l/ *adjective* of or in the south (esp. of Europe); of a meridian.

meringue /mə'ræŋ/ *noun* sugar, whipped egg whites, etc. baked crisp; cake of this.

merino /mə'ri:nəʊ/ *noun* (*plural* **-s**) variety of sheep with long fine wool; soft material, originally of merino wool; fine woollen yarn.

merit /'merɪt/ • *noun* quality of deserving well; excellence, worth; (usually in *plural*) thing that entitles to reward or gratitude. • *verb* (**-t-**) deserve.

meritocracy /merɪ'tɒkrəsɪ/ *noun* (*plural* **-ies**) government by those selected for merit; group selected in this way.

meritorious /merɪ'tɔ:rɪəs/ *adjective* praiseworthy.

merlin /'mɜ:lɪn/ *noun* kind of small falcon.

mermaid /'mɜ:meɪd/ *noun* legendary creature with woman's head and trunk and fish's tail.

merry /'merɪ/ *adjective* (**-ier, -iest**) joyous; full of laughter or gaiety; *colloquial* slightly drunk. □ **merry-go-round** fairground ride with revolving model horses, cars, etc.; **merrymaking** festivity. □ **merrily** *adverb*; **merriment** *noun.*

mésalliance /meɪ'zælɪɑ̃s/ *noun* marriage with social inferior. [French]

mescal /'meskæl/ *noun* peyote cactus. □ **mescal buttons** disc-shaped dried tops from mescal, esp. as intoxicant.

mescaline /'meskəlɪn/ *noun* (also **mescalin**) hallucinogenic alkaloid present in mescal buttons.

Mesdames, Mesdemoiselles *plural* of MADAME, MADEMOISELLE.

mesh • *noun* network structure; each of open spaces in net, sieve, etc.; (in *plural*) network, snare. • *verb* (often + *with*) (of teeth of wheel) be engaged; be harmonious; catch in net.

mesmerize /'mezməraɪz/ *verb* (also **-ise**) (**-zing** or **-sing**) hypnotize; fascinate. □ **mesmerism** *noun.*

meso- *combining form* middle, intermediate.

mesolithic /mezəʊ'lɪθɪk/ *adjective* of Stone Age between palaeolithic and neolithic periods.

meson /'mi:zɒn/ *noun* elementary particle with mass between that of electron and proton.

Mesozoic /mesəʊ'zəʊɪk/ ● *adjective* of geological era marked by development of dinosaurs. ● *noun* this era.

mess ● *noun* dirty or untidy state; state of confusion or trouble; something spilt etc.; disagreeable concoction; soldiers etc. dining together; army dining-hall; meal taken there; domestic animal's excreta; *archaic* portion of liquid or pulpy food. ● *verb* (often + *up*) make mess of, dirty, muddle; *US* (+ *with*) interfere with; take one's meals; *colloquial* defecate. □ **make a mess of** bungle; **mess about**, **around** potter.

message /'mesɪdʒ/ *noun* communication sent by one person to another; exalted or spiritual communication.

messenger /'mesɪndʒə/ *noun* person who carries message(s).

Messiah /mɪ'saɪə/ *noun* promised deliverer of Jews; Jesus regarded as this. □ **Messianic** /mesɪ'ænɪk/ *adjective*.

Messieurs *plural* of MONSIEUR.

Messrs /'mesəz/ *plural* of MR.

messy *adjective* (**-ier**, **-iest**) untidy, dirty; causing or accompanied by a mess; difficult to deal with; awkward. □ **messily** *adverb*.

met *past & past participle* of MEET[1].

metabolism /mɪ'tæbəlɪz(ə)m/ *noun* all chemical processes in living organism producing energy and growth. □ **metabolic** /metə'bɒlɪk/ *adjective*.

metacarpus /metə'kɑ:pəs/ *noun* (*plural* **-carpi** /-paɪ/) set of bones forming part of hand between wrist and fingers. □ **metacarpal** *adjective*.

metal /'met(ə)l/ ● *noun* any of class of mainly workable elements such as gold, silver, iron, or tin; alloy of any of these; (in *plural*) rails of railway; road-metal. ● *adjective* made of metal. ● *verb* (**-ll-**; *US* **-l-**) make or mend (road) with road-metal; cover or fit with metal.

metallic /mɪ'tælɪk/ *adjective* of or like metal(s); sounding like struck metal.

metallurgy /mɪ'tælədʒɪ/ *noun* science of metals and their application; extraction and purification of metals. □ **metallurgic** /metə'lɜ:dʒɪk/ *adjective*; **metallurgical** /metə'lɜ:dʒɪk(ə)l/ *adjective*; **metallurgist** *noun*.

metamorphic /metə'mɔ:fɪk/ *adjective* of metamorphosis; (of rock) transformed naturally. □ **metamorphism** *noun*.

metamorphose /metə'mɔ:fəʊz/ *verb* (**-sing**) (often + *to*, *into*) change in form or nature.

metamorphosis /metə'mɔ:fəsɪs/ *noun* (*plural* **-phoses** /-si:z/) change of form, esp. from pupa to insect; change of character, conditions, etc.

metaphor /'metəfɔ:/ *noun* application of name or description to something to which it is not literally applicable (see panel). □ **metaphoric** /-'fɒr-/ *adjective*; **metaphorical** /-'fɒr-/ *adjective*; **metaphorically** /-'fɒr-/ *adverb*.

metaphysics /metə'fɪzɪks/ *plural noun* (usually treated as *singular*) philosophy dealing with nature of existence, truth, and knowledge. □ **metaphysical** *adjective*.

metatarsus /metə'tɑ:səs/ *noun* (*plural* **-tarsi** /-saɪ/) set of bones forming part of foot between ankle and toes. □ **metatarsal** *adjective*.

Metaphor

A metaphor is a figure of speech that goes further than a simile, either by saying that something is something else that it could not normally be called, e.g.

The moon was a ghostly galleon tossed upon cloudy seas.
Stockholm, the Venice of the North

or by suggesting that something appears, sounds, or behaves like something else, e.g.

burning ambition *blindingly obvious*
the long arm of the law

mete *verb* (**-ting**) (usually + *out*) *literary* apportion, allot.

meteor /'miːtɪə/ *noun* small solid body from outer space becoming incandescent when entering earth's atmosphere.

meteoric /miːtɪ'ɒrɪk/ *adjective* rapid; dazzling; of meteors.

meteorite /'miːtɪəraɪt/ *noun* fallen meteor; fragment of rock or metal from outer space.

meteorology /miːtɪə'rɒlədʒɪ/ *noun* study of atmospheric phenomena, esp. for forecasting weather. □ **meteorological** /-rə'lɒdʒ-/ *adjective*; **meteorologist** *noun.*

meter[1] /'miːtə/ ● *noun* instrument that measures or records, esp. gas, electricity, etc. used, distance travelled, etc.; parking meter. ● *verb* measure or record by meter.

meter[2] *US* = METRE.

methane /'miːθeɪn/ *noun* colourless odourless inflammable gaseous hydrocarbon, the main constituent of natural gas.

methanol /'meθənɒl/ *noun* colourless inflammable organic liquid, used as solvent.

methinks /mɪ'θɪŋks/ *verb* (*past* **methought** /mɪ'θɔːt/) *archaic* it seems to me.

method /'meθəd/ *noun* way of doing something; procedure; orderliness.

methodical /mɪ'θɒdɪk(ə)l/ *adjective* characterized by method or order. □ **methodically** *adverb.*

Methodist /'meθədɪst/ ● *noun* member of Protestant denomination originating in 18th-c. Wesleyan evangelistic movement. ● *adjective* of Methodists or Methodism. □ **Methodism** *noun.*

methought *past* of METHINKS.

meths *noun colloquial* methylated spirit.

methyl /'meθɪl/ *noun* hydrocarbon radical CH_3. □ **methyl alcohol** methanol.

methylate /'meθɪleɪt/ *verb* (**-ting**) mix or impregnate with methanol; introduce methyl group into (molecule).

meticulous /mə'tɪkjʊləs/ *adjective* giving great attention to detail; very careful and precise. □ **meticulously** *adverb.*

métier /'metjeɪ/ *noun* one's trade, profession, or field of activity; one's forte. [French]

metonymy /mɪ'tɒnɪmɪ/ *noun* substitution of name of attribute for that of thing meant.

metre /'miːtə/ *noun* (*US* **meter**) SI unit of length (about 39.4 in.); any form of poetic rhythm; basic rhythm of music.

metric /'metrɪk/ *adjective* of or based on the metre. □ **metric system** decimal measuring system with metre, litre, and gram or kilogram as units of length, volume, and mass; **metric ton** 1,000 kg.

metrical *adjective* of or composed in metre; of or involving measurement. □ **metrically** *adverb.*

metronome /'metrənəʊm/ *noun* device ticking at selected rate to mark musical time.

metropolis /mɪ'trɒpəlɪs/ *noun* chief city, capital.

metropolitan /metrə'pɒlɪt(ə)n/ ● *adjective* of metropolis; of mother country as distinct from colonies. ● *noun* bishop with authority over bishops of province.

mettle /'met(ə)l/ *noun* quality or strength of character; spirit, courage. □ **mettlesome** *adjective.*

mew[1] ● *noun* cat's cry. ● *verb* utter this sound.

mew[2] *noun* gull, esp. common gull.

mews /mjuːz/ *noun* (treated as *singular*) stabling round yard etc., now used esp. for housing.

mezzanine /'metsəniːn/ *noun* storey between two others (usually ground and first floors).

mezzo /'metsəʊ/ *Music* ● *adverb* moderately. ● *noun* (in full **mezzo-soprano**) (*plural* **-s**) female singing voice between soprano and contralto, singer with this voice. □ **mezzo forte** fairly loud(ly); **mezzo piano** fairly soft(ly).

mezzotint /'metsəʊtɪnt/ *noun* method of copper or steel engraving; print so produced.

mf *abbreviation* mezzo forte.

mg *abbreviation* milligram(s).

Mgr. *abbreviation* Manager; *Monseigneur*; Monsignor.

MHz *abbreviation* megahertz.

mi = ME[2].

miaow /mɪ'aʊ/ ● *noun* characteristic cry of cat. ● *verb* make this cry.

miasma /mɪ'æzmə/ *noun* (*plural* **-mata** or **-s**) *archaic* infectious or noxious vapour.

mica /'maɪkə/ *noun* silicate mineral found as glittering scales in granite etc. or crystals separable into thin plates.

mice *plural* of MOUSE.

Michaelmas /'mɪkəlməs/ *noun* feast of St Michael, 29 Sept. □ **Michaelmas daisy** autumn-flowering aster.

mickey /'mɪkɪ/ *noun* (also **micky**) □ **take the mickey** (often + *out of*) *slang* tease, mock.

micro /'maɪkrəʊ/ *noun* (*plural* **-s**) *colloquial* microcomputer; microprocessor.

micro- *combining form* small; one-millionth.

microbe /'maɪkrəʊb/ *noun* micro-organism (esp. bacterium) causing disease or fermentation. □ **microbial** /-'krəʊb-/ *adjective*.

microbiology /maɪkrəʊbaɪ'ɒlədʒɪ/ *noun* study of micro-organisms. □ **microbiologist** *noun*.

microchip /'maɪkrəʊtʃɪp/ *noun* small piece of semiconductor used to carry integrated circuits.

microcomputer /'maɪkrəʊkəmpju:tə/ *noun* small computer with microprocessor as central processor.

microcosm /'maɪkrəkɒz(ə)m/ *noun* (often + *of*) miniature representation, e.g. humankind seen as small-scale model of universe; epitome. □ **microcosmic** /-'kɒz-/ *adjective*.

microdot /'maɪkrəʊdɒt/ *noun* microphotograph of document etc. reduced to size of dot.

microfiche /'maɪkrəʊfi:ʃ/ *noun* small flat piece of film bearing microphotographs of documents etc.

microfilm /'maɪkrəʊfɪlm/ ● *noun* length of film bearing microphotographs of documents etc. ● *verb* photograph on microfilm.

microlight /'maɪkrəʊlaɪt/ *noun* kind of motorized hang-glider.

micrometer /maɪ'krɒmɪtə/ *noun* gauge for accurate small-scale measurement.

micron /'maɪkrɒn/ *noun* millionth of a metre.

micro-organism /maɪkrəʊ'ɔ:gənɪz(ə)m/ *noun* microscopic organism.

microphone /'maɪkrəfəʊn/ *noun* instrument for converting sound waves into electrical energy for reconversion into sound.

microphotograph /maɪkrəʊ'fəʊtəgrɑ:f/ *noun* photograph reduced to very small size.

microprocessor /maɪkrəʊ'prəʊsesə/ *noun* data processor using integrated circuits contained on microchip(s).

microscope /'maɪkrəskəʊp/ *noun* instrument with lenses for magnifying objects or details invisible to naked eye.

microscopic /maɪkrə'skɒpɪk/ *adjective* visible only with microscope; extremely small; of the microscope. □ **microscopically** *adverb*.

microscopy /maɪ'krɒskəpɪ/ *noun* use of microscopes.

microsurgery /'maɪkrəʊsɜ:dʒərɪ/ *noun* intricate surgery using microscopes.

microwave /'maɪkrəʊweɪv/ ● *noun* electromagnetic wave of length between 1 mm and 30 cm; (in full **microwave oven**) oven using microwaves to cook or heat food quickly. ● *verb* (**-ving**) cook in microwave oven.

micturition /mɪktjʊə'rɪʃ(ə)n/ *noun formal* urination.

mid- *combining form* middle of. □ **midday** middle of day, noon; **mid-life** middle age; **mid-off**, **-on** *Cricket* position of fielder near bowler on off or on side.

midden /'mɪd(ə)n/ *noun* dunghill; refuse heap.

middle /'mɪd(ə)l/ ● *adjective* at equal distance, time, or number from extremities; central; intermediate in rank, quality, etc.; average. ● *noun* (often + *of*) middle point, position, or part; waist. □ **in the middle of** in the process of; **middle age** period between youth and old age; **the Middle Ages** period of European history from *c.* 1000 to 1453; **middle class** *noun* social class between upper and lower, including professional and business workers; **middle-class** *adjective*; **the Middle East** countries from Egypt to Iran inclusive;

middleman trader who handles commodity between producer and consumer; **middleweight** amateur boxing weight (71–75 kg).

middling ● *adjective* moderately good. ● *adverb* fairly, moderately.

midge *noun* gnatlike insect.

midget /ˈmɪdʒɪt/ *noun* extremely small person or thing.

midland /ˈmɪdlənd/ ● *noun* (**the Midlands**) inland counties of central England; middle part of country. ● *adjective* of or in midland or Midlands.

midnight *noun* middle of night; 12 o'clock at night. □ **midnight sun** sun visible at midnight during summer in polar regions.

midriff /ˈmɪdrɪf/ *noun* front of body just above waist.

midshipman /ˈmɪdʃɪpmən/ *noun* naval officer ranking next above cadet.

midst *noun* middle. □ **in the midst of** among.

midsummer *noun* period of or near summer solstice, about 21 June. □ **Midsummer Day**, **Midsummer's Day** 24 June.

midwife /ˈmɪdwaɪf/ *noun* (*plural* **-wives**) person trained to assist at childbirth. □ **midwifery** /-wɪfrɪ/ *noun*.

midwinter *noun* period of or near winter solstice, about 22 Dec.

mien /mi:n/ *noun literary* person's look or bearing.

might[1] *past* of MAY.

might[2] /maɪt/ *noun* strength, power.

mightn't /ˈmaɪt(ə)nt/ might not.

mighty ● *adjective* (**-ier**, **-iest**) powerful, strong; massive. ● *adverb colloquial* very. □ **mightily** *adverb*.

mignonette /mɪnjəˈnet/ *noun* plant with fragrant grey-green flowers.

migraine /ˈmi:greɪn/ *noun* recurrent throbbing headache often with nausea and visual disturbance.

migrant /ˈmaɪgrənt/ ● *adjective* migrating. ● *noun* migrant person or animal, esp. bird.

migrate /maɪˈgreɪt/ *verb* (**-ting**) move from one place, esp. one country, to settle in another; (of bird etc.) change habitation seasonally. □ **migration** *noun*; **migratory** /ˈmaɪgrətərɪ/ *adjective*.

mikado /mɪˈkɑ:dəʊ/ *noun* (*plural* **-s**) *historical* emperor of Japan.

mike *noun colloquial* microphone.

milch *adjective* (of cow etc.) giving milk.

mild /maɪld/ *adjective* (esp. of person) gentle; not severe or harsh; (of weather) moderately warm; (of flavour) not sharp or strong. □ **mild steel** tough low-carbon steel. □ **mildly** *adverb*; **mildness** *noun*.

mildew /ˈmɪldju:/ ● *noun* destructive growth of minute fungi on plants, damp paper, leather, etc. ● *verb* taint or be tainted with mildew.

mile *noun* unit of linear measure (1,760 yds, approx. 1.6 km); (in *plural*) *colloquial* great distance or amount; race extending over one mile. □ **milestone** stone beside road to mark distance in miles, significant point (in life, history, etc.).

mileage *noun* (also **milage**) number of miles travelled; *colloquial* profit, advantage.

miler *noun colloquial* person or horse specializing in races of one mile.

milfoil /ˈmɪlfɔɪl/ *noun* common yarrow.

milieu /mɪˈljɜ:/ *noun* (*plural* **-x** or **-s** /-z/) person's environment or social surroundings.

militant /ˈmɪlɪt(ə)nt/ ● *adjective* combative; aggressively active in support of cause; engaged in warfare. ● *noun* militant person. □ **militancy** *noun*; **militantly** *adverb*.

militarism /ˈmɪlɪtərɪz(ə)m/ *noun* aggressively military policy etc.; military spirit. □ **militarist** *noun*; **militaristic** /-ˈrɪst-/ *adjective*.

military /ˈmɪlɪtərɪ/ ● *adjective* of or characteristic of soldiers or armed forces. ● *noun* (treated as *singular* or *plural*; **the military**) the army.

militate /ˈmɪlɪteɪt/ *verb* (**-ting**) (usually + *against*) have force or effect.

■ **Usage** *Militate* is often confused with *mitigate*, which means 'to make less intense or severe'.

---

militia /mɪˈlɪʃə/ *noun* military force, esp. one conscripted in emergency. □ **militiaman** *noun*.

milk ● *noun* opaque white fluid secreted by female mammals for nourishing young; milk of cows, goats, etc. as food;

milklike liquid of coconut etc. ● *verb* draw milk from (cow etc.); exploit (person, situation). □ **milk chocolate** chocolate made with milk; **milk float** small usually electric vehicle used in delivering milk; **milkmaid** woman who milks cows or works in dairy; **milkman** person who sells or delivers milk; **milk run** routine expedition etc.; **milk shake** drink of whisked milk, flavouring, etc.; **milksop** weak or timid man or youth; **milk tooth** temporary tooth in young mammals.

milky *adjective* (**-ier**, **-iest**) of, like, or mixed with milk; (of gem or liquid) cloudy. □ **Milky Way** luminous band of stars, the Earth's galaxy.

mill ● *noun* building fitted with mechanical device for grinding corn; such device; device for grinding any solid to powder; building fitted with machinery for manufacturing processes etc.; such machinery. ● *verb* grind or treat in mill; (esp. as **milled** *adjective*) produce ribbed edge on (coin); (often + *about*, *round*) move aimlessly. □ **millpond** pond retained by dam for operating mill-wheel; **mill-race** current of water driving mill-wheel; **millstone** each of two circular stones for grinding corn, heavy burden, great responsibility; **mill-wheel** wheel used to drive water-mill.

millennium /mɪ'lenɪəm/ *noun* (*plural* **-s** or **millennia**) thousand-year period; (esp. future) period of happiness on earth. □ **millennial** *adjective*.

miller /'mɪlə/ *noun* person who owns or works mill, esp. corn-mill; person operating milling machine.

millesimal /mɪ'lesɪm(ə)l/ ● *adjective* thousandth; of, belonging to, or dealing with, thousandth or thousandths. ● *noun* thousandth part.

millet /'mɪlɪt/ *noun* cereal plant bearing small nutritious seeds; seed of this.

milli- *combining form* one-thousandth.

millibar /'mɪlɪbɑ:/ *noun* unit of atmospheric pressure equivalent to 100 pascals.

milligram /'mɪlɪgræm/ *noun* (also **-gramme**) one-thousandth of a gram.

millilitre /'mɪlɪli:tə/ *noun* (*US* **-liter**) one-thousandth of a litre (0.002 pint).

millimetre /'mɪlɪmi:tə/ *noun* (*US* **-meter**) one-thousandth of a metre.

milliner /'mɪlɪnə/ *noun* maker or seller of women's hats. □ **millinery** *noun*.

million /'mɪljən/ *noun* (*plural* same) one thousand thousand; (**millions**) *colloquial* very large number. □ **millionth** *adjective & noun*.

millionaire /mɪljə'neə/ *noun* (*feminine* **millionairess**) person possessing over a million pounds, dollars, etc.

millipede /'mɪlɪpi:d/ *noun* (also **millepede**) small crawling invertebrate with many legs.

millisecond /'mɪlɪsekənd/ *noun* one-thousandth of a second.

milometer /maɪ'lɒmɪtə/ *noun* instrument for measuring number of miles travelled by vehicle.

milt *noun* spleen in mammals; reproductive gland or sperm of male fish.

mime ● *noun* acting without words, using only gestures; performance using mime. ● *verb* (**-ming**) express or represent by mime.

mimeograph /'mɪmɪəgrɑ:f/ ● *noun* machine which duplicates from stencil; copy so produced. ● *verb* reproduce in this way.

mimetic /mɪ'metɪk/ *adjective* of or practising imitation or mimicry.

mimic /'mɪmɪk/ ● *verb* (**-ck-**) imitate (person, gesture, etc.), esp. to entertain or ridicule; copy minutely or servilely; resemble closely. ● *noun* person who mimics. □ **mimicry** *noun*.

mimosa /mɪ'məʊzə/ *noun* shrub with globular usually yellow flowers; acacia plant.

Min. *abbreviation* Minister; Ministry.

min. *abbreviation* minute(s); minimum; minim (fluid measure).

minaret /mɪnə'ret/ *noun* tall slender turret next to mosque, used by muezzin.

minatory /'mɪnətərɪ/ *adjective formal* threatening.

mince ● *verb* (**-cing**) cut or grind (meat etc.) finely; (usually as **mincing** *adjective*) walk or speak in affected way. ● *noun* minced meat. □ **mincemeat** mixture of currants, sugar, spices, suet, etc.; **mince pie** (usually small) pie containing mincemeat. □ **mincer** *noun*.

mind /maɪnd/ ● *noun* seat of consciousness, thought, volition, and feeling; attention, concentration; intellect; memory; opinion; sanity. ● *verb* object to; be upset; (often + *out*) heed, take care; look after; concern oneself with. □ **be in two minds** be undecided.

minded *adjective* (usually + *to do*) disposed, inclined. □ **-minded** inclined to think in specified way, or with specified interest.

minder *noun* person employed to look after person or thing; *slang* bodyguard.

mindful *adjective* (often + *of*) taking heed or care.

mindless *adjective* lacking intelligence; brutish; not requiring thought or skill. □ **mindlessly** *adverb*; **mindlessness** *noun*.

mine[1] *pronoun* the one(s) belonging to me.

mine[2] ● *noun* hole dug to extract metal, coal, salt, etc.; abundant source (of information etc.); *military* explosive device placed in ground or water. ● *verb* (**-ning**) obtain (minerals) from mine; (often + *for*) dig in (earth etc.) for ore etc. or to tunnel; lay explosive mines under or in. □ **minefield** area planted with explosive mines; **minesweeper** ship for clearing explosive mines from sea. □ **mining** *noun*.

miner *noun* worker in mine.

mineral /ˈmɪnər(ə)l/ *noun* inorganic substance; substance obtained by mining; (often in *plural*) artificial mineral water or other carbonated drink. □ **mineral water** water naturally or artificially impregnated with dissolved salts.

mineralogy /mɪnəˈrælədʒɪ/ *noun* study of minerals. □ **mineralogical** /-rəˈlɒdʒ-/ *adjective*; **mineralogist** *noun*.

minestrone /mɪnɪˈstrəʊnɪ/ *noun* soup containing vegetables and pasta, beans, or rice.

mingle /ˈmɪŋg(ə)l/ *verb* (**-ling**) mix, blend.

mingy /ˈmɪndʒɪ/ *adjective* (**-ier, -iest**) *colloquial* stingy.

mini /ˈmɪnɪ/ *noun* (*plural* **-s**) *colloquial* miniskirt; (**Mini**) *proprietary term* make of small car.

mini- *combining form* miniature; small of its kind.

miniature /ˈmɪnɪtʃə/ ● *adjective* much smaller than normal; represented on small scale. ● *noun* miniature object; detailed small-scale portrait. □ **in miniature** on small scale.

miniaturist *noun* painter of miniatures.

miniaturize *verb* (also **-ise**) (**-zing** or **-sing**) produce in smaller version; make small. □ **miniaturization** *noun*.

minibus /ˈmɪnɪbʌs/ *noun* small bus for about 12 passengers.

minicab /ˈmɪnɪkæb/ *noun* car used as taxi, hireable only by telephone.

minim /ˈmɪnɪm/ *noun Music* note equal to two crotchets or half a semibreve; one-sixtieth of fluid drachm.

minimal /ˈmɪnɪm(ə)l/ *adjective* very minute or slight; being a minimum. □ **minimally** *adverb*.

minimize /ˈmɪnɪmaɪz/ *verb* (also **-ise**) (**-zing** or **-sing**) reduce to or estimate at minimum; estimate or represent at less than true value etc.

minimum /ˈmɪnɪməm/ ● *noun* (*plural* **minima**) least possible or attainable amount. ● *adjective* that is a minimum. □ **minimum wage** lowest wage permitted by law or agreement.

minion /ˈmɪnjən/ *noun derogatory* servile subordinate.

miniskirt /ˈmɪnɪskɜːt/ *noun* very short skirt.

minister /ˈmɪnɪstə/ ● *noun* head of government department; member of clergy, esp. in Presbyterian and Nonconformist Churches; diplomat, usually ranking below ambassador. ● *verb* (usually + *to*) help, serve, look after. □ **ministerial** /-ˈstɪər-/ *adjective*.

ministration /mɪnɪˈstreɪʃ(ə)n/ *noun* (usually in *plural*) help, service; ministering, esp. in religious matters.

ministry /ˈmɪnɪstrɪ/ *noun* (*plural* **-ies**) government department headed by minister; building for this; (**the ministry**) profession of religious minister, ministers of government or religion.

mink *noun* (*plural* same or **-s**) small semi-aquatic stoatlike animal; its fur; coat of this.

minke /ˈmɪŋkɪ/ *noun* small whale.

minnow /ˈmɪnəʊ/ *noun* (*plural* same or **-s**) small freshwater carp.

Minoan /mɪ'nəʊən/ ● *adjective* of Cretan Bronze Age civilization. ● *noun* person of this civilization.

minor /'maɪnə/ ● *adjective* lesser or comparatively small in size or importance; *Music* (of scale) having semitone above second, fifth, and seventh notes; (of key) based on minor scale. ● *noun* person under full legal age; *US* student's subsidiary subject or course. ● *verb US* (+ *in*) study (subject) as subsidiary.

minority /maɪ'nɒrɪtɪ/ *noun* (*plural* **-ies**) (often + *of*) smaller number or part, esp. in politics; smaller group of people differing from larger in race, religion, language, etc.; being under full legal age; period of this.

minster /'mɪnstə/ *noun* large or important church; church of monastery.

minstrel /'mɪnstr(ə)l/ *noun* medieval singer or musician; musical entertainer with blacked face.

mint¹ *noun* aromatic herb used in cooking; peppermint; peppermint sweet. □ **minty** *adjective* (**-ier**, **-iest**).

mint² ● *noun* (esp. state) establishment where money is coined; *colloquial* vast sum. ● *verb* make (coin); invent (word, phrase, etc.).

minuet /mɪnjʊ'et/ *noun* slow stately dance in triple time; music for this.

minus /'maɪnəs/ ● *preposition* with subtraction of; less than zero; *colloquial* lacking. ● *adjective Mathematics* negative. ● *noun* minus sign; negative quantity; *colloquial* disadvantage. □ **minus sign** symbol (−) indicating subtraction or negative value.

minuscule /'mɪnəskju:l/ *adjective colloquial* extremely small or unimportant.

minute¹ /'mɪnɪt/ ● *noun* sixtieth part of hour; distance covered in minute; moment; sixtieth part of angular degree; (in *plural*) summary of proceedings of meeting; official memorandum. ● *verb* (**-ting**) record in minutes; send minutes to. □ **up to the minute** up to date.

minute² /maɪ'nju:t/ *adjective* (**-est**) very small; accurate, detailed. □ **minutely** *adverb*.

minutiae /maɪ'nju:ʃɪi:/ *plural noun* very small, precise, or minor details.

minx /mɪŋks/ *noun* pert, sly, or playful girl.

miracle /'mɪrək(ə)l/ *noun* extraordinary, supposedly supernatural, event; remarkable happening. □ **miracle play** medieval play on biblical themes.

miraculous /mɪ'rækjʊləs/ *adjective* being a miracle; supernatural; surprising. □ **miraculously** *adverb*.

mirage /'mɪrɑ:ʒ/ *noun* optical illusion caused by atmospheric conditions, esp. appearance of water in desert; illusory thing.

mire ● *noun* area of swampy ground; mud. ● *verb* (**-ring**) sink in mire; bespatter with mud. □ **miry** *adjective*.

mirror /'mɪrə/ ● *noun* polished surface, usually of coated glass, reflecting image; anything reflecting state of affairs etc. ● *verb* reflect in or as in mirror. □ **mirror image** identical image or reflection with left and right reversed.

mirth *noun* merriment, laughter. □ **mirthful** *adjective*.

misadventure /mɪsəd'ventʃə/ *noun Law* accident without crime or negligence; bad luck.

misalliance /mɪsə'laɪəns/ *noun* unsuitable alliance, esp. marriage.

misanthrope /'mɪsənθrəʊp/ *noun* (also **misanthropist** /mɪ'sænθrəpɪst/) person who hates humankind. □ **misanthropic** /-'θrɒp-/ *adjective*.

misanthropy /mɪ'sænθrəpɪ/ *noun* condition or habits of misanthrope.

misapply /mɪsə'plaɪ/ *verb* (**-ies**, **-ied**) apply (esp. funds) wrongly. □ **misapplication** /-æplɪ'keɪ-/ *noun*.

misapprehend /mɪsæprɪ'hend/ *verb* misunderstand (words, person). □ **misapprehension** *noun*.

misappropriate /mɪsə'prəʊprɪeɪt/ *verb* (**-ting**) take (another's money etc.) for one's own use; embezzle. □ **misappropriation** *noun*.

misbegotten /mɪsbɪ'gɒt(ə)n/ *adjective* illegitimate, bastard; contemptible.

misbehave /mɪsbɪ'heɪv/ *verb* (**-ving**) behave badly. □ **misbehaviour** *noun*.

miscalculate /mɪs'kælkjʊleɪt/ *verb* (**-ting**) calculate wrongly. □ **miscalculation** *noun*.

miscarriage /'mɪskærɪdʒ/ *noun* spontaneous abortion. □ **miscarriage of justice** failure of judicial system.

miscarry /mɪs'kærɪ/ *verb* (**-ies**, **-ied**) (of woman) have miscarriage; (of plan etc.) fail.

miscast /mɪs'kɑ:st/ *verb* (*past & past participle* **-cast**) allot unsuitable part to (actor) or unsuitable actors to (play etc.).

miscegenation /mɪsɪdʒɪ'neɪʃ(ə)n/ *noun* interbreeding of races.

miscellaneous /mɪsə'leɪnɪəs/ *adjective* of mixed composition or character; (+ plural noun) of various kinds. □ **miscellaneously** *adverb*.

miscellany /mɪ'selənɪ/ *noun* (*plural* **-ies**) mixture, medley.

mischance /mɪs'tʃɑ:ns/ *noun* bad luck; instance of this.

mischief /'mɪstʃɪf/ *noun* troublesome, but not malicious, conduct, esp. of children; playfulness; malice; harm, injury. □ **mischievous** /'mɪstʃɪvəs/ *adjective*; **mischievously** /'mɪstʃɪvəslɪ/ *adverb*.

misconceive /mɪskən'si:v/ *verb* (**-ving**) (often + *of*) have wrong idea or conception; (as **misconceived** *adjective*) badly organized etc. □ **misconception** /-'sep-/ *noun*.

misconduct /mɪs'kɒndʌkt/ *noun* improper or unprofessional conduct.

misconstrue /mɪskən'stru:/ *verb* (**-strues**, **-strued**, **-struing**) interpret wrongly. □ **misconstruction** /-'strʌk-/ *noun*.

miscount /mɪs'kaʊnt/ ● *verb* count inaccurately. ● *noun* inaccurate count.

miscreant /'mɪskrɪənt/ *noun* vile wretch, villain.

misdeed /mɪs'di:d/ *noun* evil deed, wrongdoing.

misdemeanour /mɪsdɪ'mi:nə/ *noun* (*US* **misdemeanor**) misdeed; *historical* indictable offence less serious than felony.

misdirect /mɪsdaɪ'rekt/ *verb* direct wrongly. □ **misdirection** *noun*.

miser /'maɪzə/ *noun* person who hoards wealth and lives miserably. □ **miserly** *adjective*.

miserable /'mɪzərəb(ə)l/ *adjective* wretchedly unhappy or uncomfortable; contemptible, mean; causing discomfort. □ **miserably** *adverb*.

misericord /mɪ'zerɪkɔ:d/ *noun* projection under hinged choir stall seat to support person standing.

misery /'mɪzərɪ/ *noun* (*plural* **-ies**) condition or feeling of wretchedness; cause of this; *colloquial* constantly grumbling person.

misfire /mɪs'faɪə/ ● *verb* (**-ring**) (of gun, motor engine, etc.) fail to go off, start, or function smoothly; (of plan etc.) fail to be effective. ● *noun* such failure.

misfit /'mɪsfɪt/ *noun* person unsuited to surroundings, occupation, etc.; garment etc. that does not fit.

misfortune /mɪs'fɔ:tʃ(ə)n/ *noun* bad luck; instance of this.

misgiving *noun* (usually in *plural*) feeling of mistrust or apprehension.

misgovern /mɪs'gʌv(ə)n/ *verb* govern badly. □ **misgovernment** *noun*.

misguided /mɪs'gaɪdɪd/ *adjective* mistaken in thought or action. □ **misguidedly** *adverb*.

mishandle /mɪs'hænd(ə)l/ *verb* (**-ling**) deal with incorrectly or inefficiently; handle roughly.

mishap /'mɪshæp/ *noun* unlucky accident.

mishear /mɪs'hɪə/ *verb* (*past & past participle* **-heard** /-'hɜ:d/) hear incorrectly or imperfectly.

mishmash /'mɪʃmæʃ/ *noun* confused mixture.

misinform /mɪsɪn'fɔ:m/ *verb* give wrong information to, mislead. □ **misinformation** /-fə'm-/ *noun*.

misinterpret /mɪsɪn'tɜ:prɪt/ *verb* (**-t-**) interpret wrongly. □ **misinterpretation** *noun*.

misjudge /mɪs'dʒʌdʒ/ *verb* (**-ging**) judge wrongly. □ **misjudgement** *noun*.

mislay /mɪs'leɪ/ *verb* (*past & past participle* **-laid**) accidentally put (thing) where it cannot readily be found.

mislead /mɪs'li:d/ *verb* (*past & past participle* **-led**) cause to infer what is not true; deceive. □ **misleading** *adjective*.

mismanage /mɪs'mænɪdʒ/ *verb* (**-ging**) manage badly or wrongly. □ **mismanagement** *noun*.

misnomer /mɪs'nəʊmə/ *noun* wrongly used name or term.

misogyny /mɪ'sɒdʒɪnɪ/ *noun* hatred of women. □ **misogynist** *noun*.

misplace /mɪs'pleɪs/ *verb* (**-cing**) put in wrong place; bestow (affections, confidence, etc.) on inappropriate object. □ **misplacement** *noun*.

misprint ● *noun* /'mɪsprɪnt/ printing error. ● *verb* /mɪs'prɪnt/ print wrongly.

mispronounce /mɪsprə'naʊns/ *verb* (**-cing**) pronounce (word etc.) wrongly. □ **mispronunciation** /-nʌnsɪ'eɪ-/ *noun.*

misquote /mɪs'kwəʊt/ *verb* (**-ting**) quote inaccurately. □ **misquotation** *noun.*

misread /mɪs'ri:d/ *verb* (*past & past participle* **-read** /-'red/) read or interpret wrongly.

misrepresent /mɪsreprɪ'zent/ *verb* represent wrongly; give false account of. □ **misrepresentation** *noun.*

misrule /mɪs'ru:l/ ● *noun* bad government. ● *verb* (**-ling**) govern badly.

Miss *noun title of girl or unmarried woman.*

miss ● *verb* fail to hit, reach, meet, find, catch, or perceive; fail to seize (opportunity etc.); regret absence of; avoid; (of engine etc.) misfire. ● *noun* failure. □ **give (thing) a miss** *colloquial* avoid; **miss out** omit.

missal /'mɪs(ə)l/ *noun RC Church* book of texts for Mass; book of prayers.

misshapen /mɪs'ʃeɪpən/ *adjective* deformed, distorted.

missile /'mɪsaɪl/ *noun* object, esp. weapon, suitable for throwing at target or discharging from machine; weapon directed by remote control or automatically.

missing *adjective* not in its place; lost; (of person) not traced but not known to be dead.

mission /'mɪʃ(ə)n/ *noun* task or goal assigned to person or group; journey undertaken as part of this; military or scientific expedition; group of people sent to conduct negotiations or to evangelize; missionary post.

missionary /'mɪʃənərɪ/ ● *adjective* of or concerned with religious missions, esp. abroad. ● *noun* (*plural* **-ies**) person doing missionary work.

missis = MISSUS.

missive /'mɪsɪv/ *noun jocular* letter; official letter.

misspell /mɪs'spel/ *verb* (*past & past participle* **-spelt** or **-spelled**) spell wrongly.

misspend /mɪs'spend/ *verb* (*past & past participle* **-spent**) (esp. as **misspent** *adjective*) spend wrongly or wastefully.

misstate /mɪs'steɪt/ *verb* (**-ting**) state wrongly or inaccurately. □ **misstatement** *noun.*

missus /'mɪsɪz/ *noun* (also **missis**) *colloquial or jocular form of address to woman*; (**the missis**) *colloquial* my or your wife.

mist ● *noun* water vapour in minute drops limiting visibility; condensed vapour obscuring glass etc.; dimness or blurring of sight caused by tears etc. ● *verb* cover or be covered (as) with mist.

mistake /mɪ'steɪk/ ● *noun* incorrect idea or opinion; thing incorrectly done, thought, or judged. ● *verb* (**-king**; *past* **mistook** /-'stʊk/; *past participle* **mistaken**) misunderstand meaning of; (+ *for*) wrongly take (person, thing) for another.

mistaken /mɪ'steɪkən/ *adjective* wrong in opinion or judgement; based on or resulting from error. □ **mistakenly** *adverb.*

mister /'mɪstə/ *noun colloquial or jocular form of address to man.*

mistime /mɪs'taɪm/ *verb* (**-ming**) say or do at wrong time.

mistle thrush /'mɪs(ə)l/ *noun* large thrush that eats mistletoe berries.

mistletoe /'mɪs(ə)ltəʊ/ *noun* parasitic white-berried plant.

mistook *past* of MISTAKE.

mistral /'mɪstr(ə)l/ *noun* cold N or NW wind in S. France.

mistreat /mɪs'tri:t/ *verb* treat badly. □ **mistreatment** *noun.*

mistress /'mɪstrɪs/ *noun* female head of household; woman in authority; female owner of pet; female teacher; woman having illicit sexual relationship with (usually married) man.

mistrial /mɪs'traɪəl/ *noun* trial made invalid by error.

mistrust /mɪs'trʌst/ ● *verb* be suspicious of; feel no confidence in. ● *noun* suspicion; lack of confidence. □ **mistrustful** *adjective*; **mistrustfully** *adverb.*

misty *adjective* (**-ier**, **-iest**) of or covered with mist; dim in outline; obscure. □ **mistily** *adverb.*

misunderstand /mɪsʌndə'stænd/ *verb* (*past & past participle* **-stood** /-'stʊd/) understand incorrectly; misinterpret words or actions of (person). □ **misunderstanding** *noun.*

misuse ● *verb* /mɪs'ju:z/ (**-sing**) use wrongly; ill-treat. ● *noun* /mɪs'ju:s/ wrong or improper use.

mite[1] *noun* small arachnid, esp. of kind found in cheese etc.

mite[2] *noun* small monetary unit; small object or child; modest contribution.

miter *US* = MITRE.

mitigate /'mɪtɪgeɪt/ *verb* (**-ting**) make less intense or severe. □ **mitigation** *noun.*

■ **Usage** *Mitigate* is often confused with *militate*, which means 'to have force or effect'.

mitre /'maɪtə/ (*US* **miter**) ● *noun* bishop's or abbot's tall deeply cleft headdress; joint of two pieces of wood at angle of 90°, such that line of junction bisects this angle. ● *verb* (**-ring**) bestow mitre on; join with mitre.

mitt *noun* (also **mitten**) glove with only one compartment for the 4 fingers and another for thumb; glove leaving fingers and thumb-tip bare; *slang* hand; baseball glove.

mix ● *verb* combine or put together (two or more substances or things) so that constituents of each are diffused among those of the other(s); prepare (compound, cocktail, etc.) by combining ingredients; combine (activities etc.); join, be mixed, combine; be compatible; be sociable; (+ *with*) be harmonious with; combine (two or more sound signals) into one. ● *noun* mixing, mixture; proportion of materials in mixture; ingredients prepared commercially for making cake, concrete, etc.

mixed /mɪkst/ *adjective* of diverse qualities or elements; containing people from various backgrounds, of both sexes, etc. □ **mixed marriage** marriage between people of different race or religion; **mixed-up** *colloquial* mentally or emotionally confused, socially ill-adjusted.

mixer *noun* machine for mixing foods etc.; person who manages socially in specified way; (usually soft) drink to be mixed with spirit; device combining separate signals from microphones etc.

mixture /'mɪkstʃə/ *noun* process or result of mixing; combination of ingredients, qualities, etc.

mizen-mast /'mɪz(ə)n/ *noun* mast next aft of mainmast.

ml *abbreviation* millilitre(s); mile(s).

Mlle *abbreviation* (*plural* **-s**) Mademoiselle.

MM *abbreviation* Messieurs; Military Medal.

mm *abbreviation* millimetre(s).

Mme *abbreviation* (*plural* **-s**) Madame.

mnemonic /nɪ'mɒnɪk/ ● *adjective* of or designed to aid memory. ● *noun* mnemonic word, verse, etc. □ **mnemonically** *adverb.*

MO *abbreviation* Medical Officer; money order.

mo /məʊ/ *noun* (*plural* **-s**) *colloquial* moment.

moan ● *noun* low murmur expressing physical or mental suffering or pleasure; *colloquial* complaint. ● *verb* make moan or moans; *colloquial* complain, grumble. □ **moaner** *noun.*

moat *noun* defensive ditch round castle etc., usually filled with water.

mob ● *noun* disorderly crowd; rabble; (**the mob**) *usually derogatory* the populace; *colloquial* gang, group. ● *verb* (**-bb-**) crowd round to attack or admire.

mob-cap *noun historical* woman's indoor cap covering all the hair.

mobile /'məʊbaɪl/ ● *adjective* movable; able to move easily; (of face etc.) readily changing expression; (of shop etc.) accommodated in vehicle to serve various places; (of person) able to change social status. ● *noun* decoration that may be hung so as to turn freely. □ **mobility** /mə'bɪl-/ *noun.*

mobilize /'məʊbɪlaɪz/ *verb* (also **-ise**) (**-zing** or **-sing**) make or become ready for (esp. military) service or action. □ **mobilization** *noun.*

mobster /'mɒbstə/ *noun slang* gangster.

moccasin /'mɒkəsɪn/ *noun* soft flat-soled shoe originally worn by N. American Indians.

mock ● *verb* (often + *at*) ridicule, scoff (at); treat with scorn or contempt; mimic contemptuously. ● *adjective* sham; imitation; as a trial run. □ **mock turtle soup** soup made from calf's head etc.; **mock-up** experimental model of proposed structure etc. □ **mockingly** *adverb*.

mockery *noun* (*plural* **-ies**) derision; counterfeit or absurdly inadequate representation; travesty.

mode *noun* way in which thing is done; prevailing fashion; *Music* any of several types of scale.

model /'mɒd(ə)l/ ● *noun* representation in 3 dimensions of existing person or thing or of proposed structure, esp. on smaller scale; simplified description of system etc.; clay, wax, etc. figure for reproduction in another material; particular design or style, esp. of car; exemplary person or thing; person employed to pose for artist or photographer, or to wear clothes etc. for display; (copy of) garment etc. by well-known designer. ● *adjective* exemplary; ideally perfect. ● *verb* (**-ll-**; *US* **-l-**) fashion or shape (figure) in clay, wax, etc.; (+ *after*, *on*, etc.) form (thing) in imitation of; act or pose as model; (of person acting as model) display (garment).

modem /'məʊdem/ *noun* device for sending and receiving computer data by means of telephone line.

moderate /'mɒdərət/ ● *adjective* avoiding extremes, temperate in conduct or expression; fairly large or good; (of wind) of medium strength; (of prices) fairly low. ● *noun* person of moderate views. ● *verb* /-reɪt/ (**-ting**) make or become less violent, intense, rigorous, etc.; act as moderator of or to. □ **moderately** /-rətlɪ/ *adverb*; **moderation** *noun*.

moderator *noun* arbitrator, mediator; presiding officer; Presbyterian minister presiding over ecclesiastical body.

modern /'mɒd(ə)n/ ● *adjective* of present and recent times; in current fashion, not antiquated. ● *noun* person living in modern times. □ **modernity** /mə'dɜːn-/ *noun*.

modernism *noun* modern ideas or methods, esp. in art. □ **modernist** *noun & adjective*.

modernize *verb* (also **-ise**) (**-zing** or **-sing**) make modern; adapt to modern needs or habits. □ **modernization** *noun*.

modest /'mɒdɪst/ *adjective* having humble or moderate estimate of one's own merits; bashful; decorous; not excessive; unpretentious, not extravagant. □ **modestly** *adverb*; **modesty** *noun*.

modicum /'mɒdɪkəm/ *noun* (+ *of*) small quantity.

modify /'mɒdɪfaɪ/ *verb* (**-ies**, **-ied**) make less severe; make partial changes in. □ **modification** *noun*.

modish /'məʊdɪʃ/ *adjective* fashionable. □ **modishly** *adverb*.

modulate /'mɒdjʊleɪt/ *verb* (**-ting**) regulate, adjust; moderate; adjust or vary tone or pitch of (speaking voice); alter amplitude or frequency of (wave) by using wave of lower frequency to convey signal; *Music* pass from one key to another. □ **modulation** *noun*.

module /'mɒdjuːl/ *noun* standardized part or independent unit in construction, esp. of furniture, building, spacecraft, or electronic system; unit or period of training or education. □ **modular** *adjective*.

modus operandi /məʊdəs ɒpə'rændɪ/ *noun* (*plural* ***modi operandi*** /məʊdɪ/) method of working. [Latin]

modus vivendi /məʊdəs vɪ'vendɪ/ *noun* (*plural* ***modi vivendi*** /məʊdɪ/) way of living or coping; compromise between people agreeing to differ. [Latin]

mog *noun* (also **moggie**) *slang* cat.

mogul /'məʊg(ə)l/ *noun colloquial* important or influential person.

mohair /'məʊheə/ *noun* hair of angora goat; yarn or fabric from this.

Mohammedan = MUHAMMADAN.

moiety /'mɔɪətɪ/ *noun* (*plural* **-ies**) half; each of two parts of thing.

moiré /'mwɑːreɪ/ *adjective* (of silk) watered; (of metal) having clouded appearance.

moist *adjective* slightly wet; damp. □ **moisten** *verb*.

moisture /'mɔɪstʃə/ *noun* water or other liquid diffused as vapour or within solid, or condensed on surface.

moisturize *verb* (also **-ise**) (**-zing** or **-sing**) make less dry (esp. skin by use of cosmetic). □ **moisturizer** *noun*.

molar /'məʊlə/ ● *adjective* (usually of mammal's back teeth) serving to grind. ● *noun* molar tooth.

molasses /mə'læsɪz/ *plural noun* (treated as *singular*) uncrystallized syrup extracted from raw sugar; *US* treacle.

mold *US* = MOULD[1,2,3].

molder *US* = MOULDER.

molding *US* = MOULDING.

moldy *US* = MOULDY.

mole[1] *noun* small burrowing animal with dark velvety fur and very small eyes; *slang* spy established in position of trust in organization. □ **molehill** small mound thrown up by mole in burrowing.

mole[2] *noun* small permanent dark spot on skin.

mole[3] *noun* massive structure as pier, breakwater, or causeway; artificial harbour.

mole[4] *noun* SI unit of amount of substance.

molecule /'mɒlɪkju:l/ *noun* group of atoms forming smallest fundamental unit of chemical compound. □ **molecular** /mə'lekjʊlə/ *adjective.*

molest /mə'lest/ *verb* annoy or pester (person); attack or interfere with (person), esp. sexually. □ **molestation** *noun*; **molester** *noun.*

moll *noun slang* gangster's female companion; prostitute.

mollify /'mɒlɪfaɪ/ *verb* (**-ies**, **-ied**) soften, appease. □ **mollification** *noun.*

mollusc /'mɒləsk/ *noun* (*US* **mollusk**) invertebrate with soft body and usually hard shell, e.g. snail or oyster.

mollycoddle /'mɒlɪkɒd(ə)l/ *verb* (**-ling**) coddle, pamper.

molt *US* = MOULT.

molten /'məʊlt(ə)n/ *adjective* melted, esp. made liquid by heat.

molto /'mɒltəʊ/ *adverb Music* very.

molybdenum /mə'lɪbdɪnəm/ *noun* silver-white metallic element added to steel to give strength and resistance to corrosion.

moment /'məʊmənt/ *noun* very brief portion of time; exact point of time; importance; product of force and distance from its line of action to a point.

momentary *adjective* lasting only a moment; transitory. □ **momentarily** *adverb.*

momentous /mə'mentəs/ *adjective* very important.

momentum /mə'mentəm/ *noun* (*plural* **-ta**) quantity of motion of moving body, the product of its mass and velocity; impetus gained by movement or initial effort.

Mon. *abbreviation* Monday.

monarch /'mɒnək/ *noun* sovereign with title of king, queen, emperor, empress, or equivalent. □ **monarchic** /mə'nɑ:k-/ *adjective*; **monarchical** /mə'nɑ:k-/ *adjective.*

monarchist *noun* advocate of monarchy.

monarchy *noun* (*plural* **-ies**) government headed by monarch; state with this.

monastery /'mɒnəstrɪ/ *noun* (*plural* **-ies**) residence of community of monks.

monastic /mə'næstɪk/ *adjective* of or like monasteries or monks, nuns, etc. □ **monastically** *adverb*; **monasticism** /-sɪz(ə)m/ *noun.*

Monday /'mʌndeɪ/ *noun* day of week following Sunday.

monetarism /'mʌnɪtərɪz(ə)m/ *noun* control of supply of money as chief method of stabilizing economy. □ **monetarist** *adjective & noun.*

monetary /'mʌnɪtərɪ/ *adjective* of the currency in use; of or consisting of money.

money /'mʌnɪ/ *noun* (*plural* **-s** or **monies**) coins and banknotes as medium of exchange; (in *plural*) sums of money; wealth. □ **moneylender** person lending money at interest; **money market** trade in short-term stocks, loans, etc.; **money order** order for payment of specified sum, issued by bank or Post Office; **money-spinner** thing that brings in a profit.

moneyed /'mʌnɪd/ *adjective* rich.

Mongol /'mɒŋg(ə)l/ • *adjective* of Asian people now inhabiting Mongolia; resembling this people. • *noun* Mongolian.

Mongolian /mɒŋ'gəʊlɪən/ • *noun* native, national, or language of Mongolia. • *adjective* of or relating to Mongolia or its people or language.

Mongoloid /'mɒŋgəlɔɪd/ • *adjective* characteristic of Mongolians, esp. in having broad flat yellowish face. • *noun* Mongoloid person.

mongoose /ˈmɒŋgu:s/ *noun* (*plural* **-s**) small flesh-eating civet-like mammal.

mongrel /ˈmʌŋgr(ə)l/ ● *noun* dog of no definable type or breed; any animal or plant resulting from crossing of different breeds or types. ● *adjective* of mixed origin or character.

monies *plural* of **money**.

monitor /ˈmɒnɪtə/ ● *noun* person or device for checking; school pupil with disciplinary etc. duties; television set used to select or verify picture being broadcast or to display computer data; person who listens to and reports on foreign broadcasts etc.; detector of radioactive contamination. ● *verb* act as monitor of; maintain regular surveillance over.

monk /mʌŋk/ *noun* member of religious community of men living under vows. □ **monkish** *adjective*.

monkey /ˈmʌŋkɪ/ ● *noun* (*plural* **-eys**) any of various primates, e.g. baboons, marmosets; mischievous person, esp. child. ● *verb* (**-eys**, **-eyed**) (often + *with*) play mischievous tricks. □ **monkey-nut** peanut; **monkey-puzzle** tree with hanging prickly branches; **monkey wrench** wrench with adjustable jaw.

monkshood /ˈmʌŋkshʊd/ *noun* poisonous plant with hood-shaped flowers.

mono /ˈmɒnəʊ/ *colloquial* ● *adjective* monophonic. ● *noun* monophonic reproduction.

mono- *combining form* (usually **mon-** before vowel) one, alone, single.

monochromatic /mɒnəkrəˈmætɪk/ *adjective* (of light or other radiation) of single colour or wavelength; containing only one colour.

monochrome /ˈmɒnəkrəʊm/ ● *noun* photograph or picture in one colour, or in black and white only. ● *adjective* having or using one colour or black and white only.

monocle /ˈmɒnək(ə)l/ *noun* single eyeglass.

monocular /məˈnɒkjʊlə/ *adjective* with or for one eye.

monody /ˈmɒnədɪ/ *noun* (*plural* **-ies**) ode sung by one actor in Greek tragedy; poem lamenting person's death.

monogamy /məˈnɒgəmɪ/ *noun* practice or state of being married to one person at a time. □ **monogamous** *adjective*.

monogram /ˈmɒnəgræm/ *noun* two or more letters, esp. initials, interwoven.

monograph /ˈmɒnəgrɑ:f/ *noun* treatise on single subject.

monolith /ˈmɒnəlɪθ/ *noun* single block of stone, esp. shaped into pillar etc.; person or thing like monolith in being massive, immovable, or solidly uniform. □ **monolithic** /-ˈlɪθ-/ *adjective*.

monologue /ˈmɒnəlɒg/ *noun* scene in drama in which person speaks alone; dramatic composition for one performer; long speech by one person in conversation etc.

monomania /mɒnəˈmeɪnɪə/ *noun* obsession by single idea or interest. □ **monomaniac** *noun & adjective*.

monophonic /mɒnəˈfɒnɪk/ *adjective* (of sound-reproduction) using only one channel of transmission.

monoplane /ˈmɒnəpleɪn/ *noun* aeroplane with one set of wings.

monopolist /məˈnɒpəlɪst/ *noun* person who has or advocates monopoly. □ **monopolistic** /-ˈlɪs-/ *adjective*.

monopolize /məˈnɒpəlaɪz/ *verb* (also **-ise**) (**-zing** or **-sing**) obtain exclusive possession or control of (trade etc.); dominate (conversation) etc. □ **monopolization** *noun*; **monopolizer** *noun*.

monopoly /məˈnɒpəlɪ/ *noun* (*plural* **-ies**) exclusive possession or control of trade in commodity or service; (+ *of*, *US on*) sole possession or control.

monorail /ˈmɒnəʊreɪl/ *noun* railway with single-rail track.

monosodium glutamate /mɒnəʊˈsəʊdɪəm ˈglu:təmeɪt/ *noun* sodium salt of glutamic acid used to enhance flavour of food.

monosyllable /ˈmɒnəsɪləb(ə)l/ *noun* word of one syllable. □ **monosyllabic** /-ˈlæb-/ *adjective*.

monotheism /ˈmɒnəθi:ɪz(ə)m/ *noun* doctrine that there is only one God. □ **monotheist** *noun*; **monotheistic** /-ˈɪst-/ *adjective*.

monotone /ˈmɒnətəʊn/ *noun* sound continuing or repeated on one note or without change of pitch.

monotonous /məˈnɒtənəs/ *adjective* lacking in variety, tedious through sameness. □ **monotonously** *adverb*; **monotony** *noun*.

monoxide /mə'nɒksaɪd/ *noun* oxide containing one oxygen atom.

Monseigneur /mɒnsen'jɜː/ *noun* (*plural* ***Messeigneurs*** /mesen'jɜː/) *title given to eminent French person, esp. prince, cardinal, etc.* [French]

Monsieur /mə'sjɜː/ *noun* (*plural* **Messieurs** /mes'jɜː/) *title used of or to French-speaking man.*

Monsignor /mɒn'siːnjə/ *noun* (*plural* **-nori** /-'njɔːrɪ/) *title of various RC priests and officials.*

monsoon /mɒn'suːn/ *noun* wind in S. Asia, esp. in Indian Ocean; rainy season accompanying summer monsoon.

monster /'mɒnstə/ ● *noun* imaginary creature, usually large and frightening; inhumanly wicked person; misshapen animal or plant; large, usually ugly, animal or thing. ● *adjective* huge.

monstrance /'mɒnstrəns/ *noun RC Church* vessel in which host is exposed for veneration.

monstrosity /mɒn'strɒsɪtɪ/ *noun* (*plural* **-ies**) huge or outrageous thing.

monstrous /'mɒnstrəs/ *adjective* like a monster; abnormally formed; huge; outrageously wrong; atrocious. □ **monstrously** *adverb.*

montage /mɒn'tɑːʒ/ *noun* selection, cutting, and arrangement as consecutive whole, of separate sections of cinema or television film; composite whole made from juxtaposed photographs etc.

month /mʌnθ/ *noun* (in full **calendar month**) each of 12 divisions of year; period of time between same dates in successive calendar months; period of 28 days.

monthly ● *adjective* done, produced, or occurring once every month. ● *adverb* every month. ● *noun* (*plural* **-ies**) monthly periodical.

monument /'mɒnjʊmənt/ *noun* anything enduring that serves to commemorate, esp. structure, building, or memorial stone.

monumental /mɒnjʊ'ment(ə)l/ *adjective* extremely great; stupendous; massive and permanent; of or serving as monument.

moo ● *noun* (*plural* **-s**) characteristic sound of cattle. ● *verb* (**moos**, **mooed**) make this sound.

mooch *verb colloquial* (usually + *about, around*) wander aimlessly; *esp. US* cadge.

mood[1] *noun* state of mind or feeling; fit of bad temper or depression.

mood[2] *noun Grammar* form(s) of verb indicating whether it expresses fact, command, wish, etc.

moody *adjective* (**-ier**, **-iest**) given to changes of mood; gloomy, sullen.

moon ● *noun* natural satellite of the earth, orbiting it monthly, illuminated by and reflecting sun; satellite of any planet. ● *verb* (often + *about, around*) wander aimlessly or listlessly. □ **moonbeam** ray of moonlight; **moonlight** *noun* light of moon, *verb colloquial* have other paid occupation, esp. one by night as well as one by day; **moonlit** lit by the moon; **moonshine** foolish or visionary talk, illicit alcohol; **moonshot** launching of spacecraft to moon; **moonstone** feldspar of pearly appearance; **moonstruck** slightly mad.

moony *adjective* (**-ier**, **-iest**) listless; stupidly dreamy.

Moor /mʊə/ *noun* member of a Muslim people of NW Africa. □ **Moorish** *adjective.*

moor[1] /mʊə/ *noun* open uncultivated upland, esp. when covered with heather. □ **moorhen** small waterfowl; **moorland** large area of moor.

moor[2] /mʊə/ *verb* attach (boat etc.) to fixed object. □ **moorage** *noun.*

mooring *noun* (often in *plural*) place where boat etc. is moored; (in *plural*) set of permanent anchors and chains.

moose *noun* (*plural* same) N. American deer; elk.

moot ● *adjective* debatable, undecided. ● *verb* raise (question) for discussion. ● *noun historical* assembly.

mop ● *noun* bundle of yarn or cloth or a sponge on end of stick for cleaning floors etc.; thick mass of hair. ● *verb* (**-pp-**) wipe or clean (as) with mop. □ **mop up** wipe with mop, *colloquial* absorb, dispose of, complete occupation of (area etc.) by capturing or killing enemy troops left there.

mope ● *verb* (**-ping**) be depressed or listless. ● *noun* person who mopes. □ **mopy** *adjective* (**-ier**, **-iest**).

moped /ˈməʊped/ *noun* low-powered motorized bicycle.

moquette /mɒˈket/ *noun* thick pile or looped material used for upholstery etc.

moraine /məˈreɪn/ *noun* area of debris carried down and deposited by glacier.

moral /ˈmɒr(ə)l/ ● *adjective* concerned with goodness or badness of character or behaviour, or with difference between right and wrong; virtuous in conduct. ● *noun* moral lesson of story etc.; (in *plural*) moral principles or behaviour. □ **moral support** psychological rather than physical help. □ **morally** *adverb*.

morale /məˈrɑːl/ *noun* confidence, determination, etc. of person or group.

moralist /ˈmɒrəlɪst/ *noun* person who practises or teaches morality. □ **moralistic** /-ˈlɪs-/ *adjective*.

morality /məˈrælɪtɪ/ *noun* (*plural* **-ies**) degree of conformity to moral principles; moral conduct; science of morals.

moralize /ˈmɒrəlaɪz/ *verb* (also **-ise**) (**-zing** or **-sing**) (often + *on*) indulge in moral reflection or talk. □ **moralization** *noun*.

morass /məˈræs/ *noun* entanglement; *literary* bog.

moratorium /mɒrəˈtɔːrɪəm/ *noun* (*plural* **-s** or **-ria**) (often + *on*) temporary prohibition or suspension (of activity); legal authorization to debtors to postpone payment.

morbid /ˈmɔːbɪd/ *adjective* (of mind, ideas, etc.) unwholesome; *colloquial* melancholy; of or indicative of disease. □ **morbidity** /-ˈbɪd-/ *noun*; **morbidly** *adverb*.

mordant /ˈmɔːd(ə)nt/ ● *adjective* (of sarcasm etc.) caustic, biting; smarting; corrosive, cleansing. ● *noun* mordant substance.

more /mɔː/ ● *adjective* greater in quantity or degree; additional. ● *noun* greater quantity, number, or amount. ● *adverb* to greater degree or extent; *forming comparative of adjectives and adverbs*.

morello /məˈreləʊ/ *noun* (*plural* **-s**) sour kind of dark cherry.

moreover /mɔːˈrəʊvə/ *adverb* besides, in addition.

mores /ˈmɔːreɪz/ *plural noun* customs or conventions of community.

morganatic /mɔːgəˈnætɪk/ *adjective* (of marriage) between person of high rank and one of lower rank, the latter and the latter's children having no claim to possessions of former.

morgue /mɔːg/ *noun* mortuary; room or file of miscellaneous information kept by newspaper office.

moribund /ˈmɒrɪbʌnd/ *adjective* at point of death; lacking vitality.

Mormon /ˈmɔːmən/ *noun* member of Church of Jesus Christ of Latter-Day Saints. □ **Mormonism** *noun*.

morn *noun poetical* morning.

morning /ˈmɔːnɪŋ/ *noun* early part of day till noon or lunchtime. □ **morning coat** tailcoat with front cut away; **morning dress** man's morning coat and striped trousers; **morning glory** climbing plant with trumpet-shaped flowers; **morning sickness** nausea felt in morning in pregnancy; **morning star** planet, usually Venus, seen in east before sunrise.

morocco /məˈrɒkəʊ/ *noun* (*plural* **-s**) fine flexible leather of goatskin tanned with sumac.

moron /ˈmɔːrɒn/ *noun colloquial* very stupid person; adult with mental age of 8–12. □ **moronic** /məˈr-/ *adjective*.

morose /məˈrəʊs/ *adjective* sullen, gloomy. □ **morosely** *adverb*; **moroseness** *noun*.

morphia /ˈmɔːfɪə/ *noun* morphine.

morphine /ˈmɔːfiːn/ *noun* narcotic drug from opium.

morphology /mɔːˈfɒlədʒɪ/ *noun* study of forms of things, esp. of animals and plants and of words and their structure. □ **morphological** /-fəˈlɒdʒ-/ *adjective*.

morris dance /ˈmɒrɪs/ *noun* traditional English dance in fancy costume. □ **morris dancer** *noun*; **morris dancing** *noun*.

morrow /ˈmɒrəʊ/ *noun* (usually **the morrow**) *literary* following day.

Morse /mɔːs/ *noun* (in full **Morse code**) code in which letters, numbers, etc. are represented by combinations of long and short light or sound signals.

morsel /ˈmɔːs(ə)l/ *noun* mouthful; small piece (esp. of food).

mortal /ˈmɔːt(ə)l/ ● *adjective* subject to or causing death; (of combat) fought to the death; (of enemy) implacable. ● *noun* human being. □ **mortal sin** sin depriving soul of salvation. □ **mortally** *adverb*.

mortality /mɔːˈtælɪtɪ/ *noun* (*plural* **-ies**) being subject to death; loss of life on large scale; (in full **mortality rate**) death rate.

mortar /ˈmɔːtə/ *noun* mixture of lime or cement, sand, and water, for bonding bricks or stones; short cannon for firing shells at high angles; vessel in which ingredients are pounded with pestle. □ **mortarboard** stiff flat square-topped academic cap, flat board for holding mortar.

mortgage /ˈmɔːgɪdʒ/ ● *noun* conveyance of property to creditor as security for debt (usually one incurred by purchase of property); sum of money lent by this. ● *verb* (**-ging**) convey (property) by mortgage.

mortgagee /mɔːgɪˈdʒiː/ *noun* creditor in mortgage.

mortgager /ˈmɔːgɪdʒə/ *noun* (also **mortgagor** /-ˈdʒɔː/) debtor in mortgage.

mortice = MORTISE.

mortician /mɔːˈtɪʃ(ə)n/ *noun US* undertaker.

mortify /ˈmɔːtɪfaɪ/ *verb* (**-ies**, **-ied**) humiliate, wound (person's feelings); bring (body etc.) into subjection by self-denial; (of flesh) be affected by gangrene. □ **mortification** *noun*; **mortifying** *adjective*.

mortise /ˈmɔːtɪs/ (also **mortice**) ● *noun* hole in framework to receive end of another part, esp. tenon. ● *verb* (**-sing**) join, esp. by mortise and tenon; cut mortise in. □ **mortise lock** lock recessed in frame of door etc.

mortuary /ˈmɔːtjʊərɪ/ ● *noun* (*plural* **-ies**) room or building in which dead bodies are kept until burial or cremation. ● *adjective* of death or burial.

Mosaic /məʊˈzeɪɪk/ *adjective* of Moses.

mosaic /məʊˈzeɪɪk/ *noun* picture or pattern made with small variously coloured pieces of glass, stone, etc.; diversified thing.

moselle /məʊˈzel/ *noun* dry German white wine.

Moslem = MUSLIM.

mosque /mɒsk/ *noun* Muslim place of worship.

mosquito /mɒsˈkiːtəʊ/ *noun* (*plural* **-es**) biting insect, esp. with long proboscis to suck blood. □ **mosquito-net** net to keep off mosquitoes.

moss *noun* small flowerless plant growing in dense clusters in bogs and on trees, stones, etc.; *Scottish & Northern English* peatbog. □ **mossy** *adjective* (**-ier**, **-iest**).

most /məʊst/ ● *adjective* greatest in quantity or degree; the majority of. ● *noun* greatest quantity or number; the majority. ● *adverb* in highest degree; *forming superlative of adjectives and adverbs*.

mostly *adverb* mainly; usually.

MOT *abbreviation* (in full **MOT test**) compulsory annual test, instituted by Ministry of Transport, of vehicles over specified age.

mot /məʊ/ *noun* (*plural* ***mots*** same pronunciation) (usually ***bon mot*** /bɔ̃/) witty saying. □ ***mot juste*** /ˈʒuːst/ most appropriate expression. [French]

mote *noun* speck of dust.

motel /məʊˈtel/ *noun* roadside hotel for motorists.

moth /mɒθ/ *noun* nocturnal insect like butterfly; insect of this type breeding in cloth etc., on which its larva feeds. □ **mothball** ball of naphthalene etc. kept with stored clothes to deter moths; **moth-eaten** damaged by moths, time-worn.

mother /ˈmʌðə/ ● *noun* female parent; woman or condition etc. giving rise to something else; (in full **Mother Superior**) head of female religious community. ● *verb* treat as mother does. □ **mother country** country in relation to its colonies; **mother-in-law** (*plural* **mothers-in-law**) husband's or wife's mother; **motherland** native country; **mother-of-pearl** iridescent substance forming lining of oyster and other shells; **mother tongue** native language. □ **motherhood** *noun*; **motherly** *adjective*.

motif /məʊˈtiːf/ *noun* theme repeated and developed in artistic work; decorative design; ornament sewn separately on garment.

motion /ˈməʊʃ(ə)n/ ● *noun* moving; changing position; gesture; formal proposal put to committee etc.; application to court for order; evacuation of bowels. ● *verb* (often + *to do*) direct (person) by gesture. □ **motion picture** *esp. US* cinema film. □ **motionless** *adjective*.

motivate /ˈməʊtɪveɪt/ *verb* (**-ting**) supply motive to, be motive of; cause (person) to act in particular way; stimulate interest of (person in activity). □ **motivation** *noun*.

motive /ˈməʊtɪv/ ● *noun* what induces person to act; motif. ● *adjective* tending to initiate movement.

motley /ˈmɒtlɪ/ ● *adjective* (**-lier**, **-liest**) diversified in colour; of varied character. ● *noun historical* jester's particoloured costume.

motor /ˈməʊtə/ ● *noun* thing that imparts motion; machine (esp. using electricity or internal combustion) supplying motive power for vehicle or other machine; car. ● *adjective* giving, imparting, or producing motion; driven by motor; of or for motor vehicles. ● *verb* go or convey by motor vehicle. □ **motor bike** *colloquial*, **motorcycle** two-wheeled motor vehicle without pedal propulsion; **motor car** car; **motorway** fast road with separate carriageways limited to motor vehicles.

motorcade /ˈməʊtəkeɪd/ *noun* procession of motor vehicles.

motorist *noun* driver of car.

motorize *verb* (also **-ise**) (**-zing** or **-sing**) equip with motor transport; provide with motor.

mottle /ˈmɒt(ə)l/ *verb* (**-ling**) (esp. as **mottled** *adjective*) mark with spots or smears of colour.

motto /ˈmɒtəʊ/ *noun* (*plural* **-es**) maxim adopted as rule of conduct; words accompanying coat of arms; appropriate inscription; joke, maxim, etc. in paper cracker.

mould[1] /məʊld/ (*US* **mold**) ● *noun* hollow container into which substance is poured or pressed to harden into required shape; pudding etc. shaped in mould; form, shape; character, type. ● *verb* shape (as) in mould; give shape to; influence development of.

mould[2] /məʊld/ *noun* (*US* **mold**) furry growth of fungi, esp. in moist warm conditions.

mould[3] /məʊld/ *noun* (*US* **mold**) loose earth; upper soil of cultivated land, esp. when rich in organic matter.

moulder /ˈməʊldə/ *verb* (*US* **molder**) decay to dust; (+ *away*) rot, crumble.

moulding *noun* (*US* **molding**) ornamental strip of plaster etc. applied as architectural feature, esp. in cornice; similar feature in woodwork etc.

mouldy *adjective* (*US* **moldy**) (**-ier**, **-iest**) covered with mould; stale; out of date; *colloquial* dull, miserable.

moult /məʊlt/ (*US* **molt**) ● *verb* shed (feathers, hair, shell, etc.) in renewing plumage, coat, etc. ● *noun* moulting.

mound /maʊnd/ *noun* raised mass of earth, stones, etc.; heap, pile; hillock.

mount[1] ● *verb* ascend; climb on to; get up on (horse etc.); set on horseback; (as **mounted** *adjective*) serving on horseback; (often + *up*) accumulate, increase; set in frame etc., esp. for viewing; organize, arrange (exhibition, attack, etc.). ● *noun* backing etc. on which picture etc. is set for display; horse for riding; setting for gem etc.

mount[2] *noun* (*poetical* except before name) mountain, hill.

mountain /ˈmaʊntɪn/ *noun* large abrupt elevation of ground; large heap or pile; huge quantity; large surplus stock. □ **mountain ash** tree with scarlet berries; **mountain bike** sturdy bicycle with straight handlebars and many gears.

mountaineer /maʊntɪˈnɪə/ ● *noun* person practising mountain climbing. ● *verb* climb mountains as sport. □ **mountaineering** *noun*.

mountainous *adjective* having many mountains; huge.

mountebank /ˈmaʊntɪbæŋk/ *noun* swindler, charlatan.

Mountie /ˈmaʊntɪ/ *noun colloquial* member of Royal Canadian Mounted Police.

mourn /mɔːn/ *verb* (often + *for*, *over*) feel or show sorrow or regret; grieve for loss of (dead person etc.).

mourner *noun* person who mourns, esp. at funeral.

mournful *adjective* doleful, sad. □ **mournfully** *adverb*.

mourning *noun* expression of sorrow for dead, esp. by wearing black clothes; such clothes.

mouse /maʊs/ ● *noun* (*plural* **mice**) small rodent; timid or feeble person; *Computing* small device controlling cursor on VDU screen. ● *verb* (also /maʊz/) (**-sing**) (of cat etc.) hunt mice. □ **mouser** *noun*; **mousy** *adjective*.

mousse /mu:s/ *noun* dish of whipped cream, eggs, etc., flavoured with fruit, chocolate, etc., or with meat or fish purée.

moustache /məˈstɑ:ʃ/ *noun* (*US* **mustache**) hair left to grow on upper lip.

mouth ● *noun* /maʊθ/ (*plural* **mouths** /maʊðz/) external opening in head, for taking in food and emitting sound; cavity behind it containing teeth and vocal organs; opening of container, cave, trumpet, volcano, etc.; place where river enters sea. ● *verb* /maʊð/ (**-thing**) say or speak by moving lips silently; utter insincerely or without understanding. □ **mouth-organ** harmonica; **mouthpiece** part of musical instrument, telephone, etc., placed next to lips; **mouthwash** liquid antiseptic etc. for rinsing mouth.

mouthful *noun* (*plural* **-s**) quantity of food etc. that fills the mouth; *colloquial* something difficult to say.

move /mu:v/ ● *verb* (**-ving**) (cause to) change position, posture, home, or place of work; put or keep in motion; rouse, stir; (often + *about, away, off*, etc.) go, proceed; take action; (+ *in*) be socially active in; affect with emotion; (cause to) change attitude; propose as resolution. ● *noun* act or process of moving; change of house, premises, etc.; step taken to secure object; moving of piece in board game. □ **move in with** start to share accommodation with; **move out** leave one's home. □ **movable** *adjective*.

movement *noun* moving, being moved; moving parts of mechanism; group of people with common object; (in *plural*) person's activities and whereabouts; chief division of longer musical work; bowel motion; rise or fall of stock-market prices.

movie /ˈmu:vɪ/ *noun esp. US colloquial* cinema film.

moving *adjective* emotionally affecting.

mow /məʊ/ *verb* (*past participle* **mowed** or **mown**) cut (grass, hay, etc.) with scythe or machine. □ **mower** *noun*.

MP *abbreviation* Member of Parliament.

mp *abbreviation* mezzo piano.

m.p.g. *abbreviation* miles per gallon.

m.p.h. *abbreviation* miles per hour.

Mr /ˈmɪstə/ *noun* (*plural* **Messrs**) *title prefixed to name of man or to designation of office etc.*

Mrs /ˈmɪsɪz/ *noun* (*plural* same) *title of married woman.*

MS *abbreviation* (*plural* **MSS** /emˈesɪz/) manuscript; multiple sclerosis.

Ms /mɪz/ *noun title of married or unmarried woman.*

M.Sc. *abbreviation* Master of Science.

Mt. *abbreviation* Mount.

much ● *adjective* existing or occurring in great quantity. ● *noun* great quantity; (usually in negative) noteworthy example. ● *adverb* in great degree; for large part of one's time; often. □ **a bit much** *colloquial* excessive; **much of a muchness** very nearly the same.

mucilage /ˈmju:sɪlɪdʒ/ *noun* viscous substance obtained from plants; adhesive gum.

muck ● *noun colloquial* dirt, filth; anything disgusting; manure. ● *verb* (usually + *up*) *colloquial* bungle; make dirty; (+ *out*) remove manure from. □ **muck about, around** *colloquial* potter or fool about; **muck in** (often + *with*) *colloquial* share tasks etc.; **muckraking** seeking out and revealing of scandals etc. □ **mucky** *adjective* (**-ier, -iest**).

mucous /ˈmju:kəs/ *adjective* of or covered with mucus. □ **mucous membrane** mucus-secreting tissue lining body cavities etc.

mucus /ˈmju:kəs/ *noun* slimy substance secreted by mucous membrane.

mud *noun* soft wet earth. □ **mudguard** curved strip over wheel to protect against mud; **mud-slinging** abuse, slander.

muddle /ˈmʌd(ə)l/ ● *verb* (**-ling**) (often + *up*) bring into disorder; bewilder; confuse. ● *noun* disorder; confusion.

□ **muddle along** progress in haphazard way.

muddy ● *adjective* (**-ier**, **-iest**) like mud; covered in or full of mud; (of liquid, colour, or sound) not clear; confused. ● *verb* (**-ies**, **-ied**) make muddy.

muesli /'mju:zlɪ/ *noun* breakfast food of crushed cereals, dried fruit, nuts, etc.

muezzin /mu:'ezɪn/ *noun* Muslim crier who proclaims hours of prayer.

muff[1] *noun* covering, esp. of fur, for keeping hands or ears warm.

muff[2] *verb colloquial* bungle; miss (catch etc.).

muffin /'mʌfɪn/ *noun* light flat round spongy cake, eaten toasted and buttered; *US* similar cake made from batter or dough.

muffle /'mʌf(ə)l/ *verb* (**-ling**) (often + *up*) wrap for warmth or to deaden sound.

muffler *noun* wrap or scarf worn for warmth; thing used to deaden sound.

mufti /'mʌftɪ/ *noun* civilian clothes.

mug[1] ● *noun* drinking vessel, usually cylindrical with handle and no saucer; its contents; gullible person; *slang* face, mouth. ● *verb* (**-gg-**) attack and rob, esp. in public place. □ **mugger** *noun*; **mugging** *noun*.

mug[2] *verb* (**-gg-**) (usually + *up*) *slang* learn (subject) by concentrated study.

muggins /'mʌgɪnz/ *noun* (*plural* same or **mugginses**) *colloquial* gullible person (often meaning oneself).

muggy /'mʌgɪ/ *adjective* (**-ier**, **-iest**) (of weather etc.) oppressively humid.

Muhammadan /mə'hæməd(ə)n/ *noun & adjective* (also **Mohammedan**) Muslim.

■ **Usage** The term *Muhammadan* is not used by Muslims and is often regarded as offensive.

---

mulatto /mju:'lætəʊ/ *noun* (*plural* **-s** or **-es**) person of mixed white and black parentage.

mulberry /'mʌlbərɪ/ *noun* (*plural* **-ies**) tree bearing edible purple or white berries; its fruit; dark red, purple.

mulch ● *noun* layer of wet straw, leaves, plastic, etc., put round plant's roots to enrich or insulate soil. ● *verb* treat with mulch.

mule[1] /mju:l/ *noun* offspring of male donkey and female horse or (in general use) vice versa; obstinate person; kind of spinning machine.

mule[2] /mju:l/ *noun* backless slipper.

muleteer /mju:lə'tɪə/ *noun* mule driver.

mulish *adjective* stubborn.

mull[1] *verb* (often + *over*) ponder.

mull[2] *verb* heat and spice (wine, beer).

mullah /'mʌlə/ *noun* Muslim learned in theology and sacred law.

mullet /'mʌlɪt/ *noun* (*plural* same) edible sea fish.

mulligatawny /mʌlɪgə'tɔ:nɪ/ *noun* highly seasoned soup originally from India.

mullion /'mʌljən/ *noun* vertical bar between panes in window. □ **mullioned** *adjective*.

multi- *combining form* many.

multicoloured /'mʌltɪkʌləd/ *adjective* of many colours.

multifarious /mʌltɪ'feərɪəs/ *adjective* many and various; of great variety.

multiform /'mʌltɪfɔ:m/ *adjective* having many forms; of many kinds.

multilateral /mʌltɪ'lætər(ə)l/ *adjective* (of agreement etc.) in which 3 or more parties participate; having many sides. □ **multilaterally** *adverb*.

multilingual /mʌltɪ'lɪŋgw(ə)l/ *adjective* in, speaking, or using many languages.

multinational /mʌltɪ'næʃən(ə)l/ ● *adjective* operating in several countries; of several nationalities. ● *noun* multinational company.

multiple /'mʌltɪp(ə)l/ ● *adjective* having several parts, elements, or components; many and various. ● *noun* quantity exactly divisible by another. □ **multiple sclerosis** see SCLEROSIS.

multiplicand /mʌltɪplɪ'kænd/ *noun* quantity to be multiplied.

multiplication /mʌltɪplɪ'keɪʃ(ə)n/ *noun* multiplying.

multiplicity /mʌltɪ'plɪsɪtɪ/ *noun* (*plural* **-ies**) manifold variety; (+ *of*) great number.

multiplier /'mʌltɪplaɪə/ *noun* quantity by which given number is multiplied.

multiply /'mʌltɪplaɪ/ *verb* (**-ies**, **-ied**) obtain from (number) another a specified number of times its value; increase in number, esp. by procreation.

multi-purpose /mʌltɪ'pɜ:pəs/ *adjective* having several purposes.

multiracial /mʌltɪ'reɪʃ(ə)l/ *adjective* of several races.

multitude /'mʌltɪtju:d/ *noun* (often + *of*) great number; large gathering of people; (**the multitude**) the common people. □ **multitudinous** /-'tju:dɪnəs/ *adjective.*

mum[1] *noun colloquial* mother.

mum[2] *adjective colloquial* silent. □ **mum's the word** say nothing.

mumble /'mʌmb(ə)l/ ● *verb* (**-ling**) speak or utter indistinctly. ● *noun* indistinct utterance.

mumbo-jumbo /mʌmbəʊ'dʒʌmbəʊ/ *noun* (*plural* **-s**) meaningless ritual; meaningless or unnecessarily complicated language; nonsense.

mummer /'mʌmə/ *noun* actor in traditional play or mime.

mummery /'mʌmərɪ/ *noun* (*plural* **-ies**) ridiculous (esp. religious) ceremonial; performance by mummers.

mummify /'mʌmɪfaɪ/ *verb* (**-ies**, **-ied**) preserve (body) as mummy. □ **mummification** *noun.*

mummy[1] /'mʌmɪ/ *noun* (*plural* **-ies**) *colloquial* mother.

mummy[2] /'mʌmɪ/ *noun* (*plural* **-ies**) dead body preserved by embalming, esp. in ancient Egypt.

mumps *plural noun* (treated as *singular*) infectious disease with swelling of neck and face.

munch *verb* chew steadily.

mundane /mʌn'deɪn/ *adjective* dull, routine; of this world. □ **mundanely** *adverb.*

municipal /mju:'nɪsɪp(ə)l/ *adjective* of municipality or its self-government.

municipality /mju:nɪsɪ'pælɪtɪ/ *noun* (*plural* **-ies**) town or district with local self-government; its governing body.

munificent /mju:'nɪfɪs(ə)nt/ *adjective* (of giver or gift) splendidly generous. □ **munificence** *noun.*

muniment /'mju:nɪmənt/ *noun* (usually in *plural*) document kept as evidence of rights or privileges.

munition /mju:'nɪʃ(ə)n/ *noun* (usually in *plural*) military weapons, ammunition, etc.

mural /'mjʊər(ə)l/ ● *noun* painting executed directly on wall. ● *adjective* of, on, or like wall.

murder /'mɜ:də/ ● *noun* intentional unlawful killing of human being by another; *colloquial* unpleasant or dangerous state of affairs. ● *verb* kill (human being) intentionally and unlawfully; *colloquial* utterly defeat; spoil by bad performance, mispronunciation, etc. □ **murderer**, **murderess** *noun*; **murderous** *adjective.*

murky /'mɜ:kɪ/ (**-ier**, **-iest**) *adjective* dark, gloomy; (of liquid etc.) dirty.

murmur /'mɜ:mə/ ● *noun* subdued continuous sound; softly spoken utterance; subdued expression of discontent. ● *verb* make murmur; utter in low voice.

murrain /'mʌrɪn/ *noun* infectious disease of cattle.

Muscadet /'mʌskədeɪ/ *noun* dry white wine of France from Loire region; variety of grape used for this.

muscat /'mʌskət/ *noun* sweet usually fortified white wine made from musk-flavoured grapes; this grape.

muscatel /mʌskə'tel/ *noun* muscat wine or grape; raisin made from muscat grape.

muscle /'mʌs(ə)l/ ● *noun* fibrous tissue producing movement in or maintaining position of animal body; part of body composed of muscles; strength, power. ● *verb* (**-ling**) (+ *in*, *in on*) *colloquial* force oneself on others. □ **muscle-bound** with muscles stiff and inelastic through excessive exercise; **muscle-man** man with highly developed muscles.

Muscovite /'mʌskəvaɪt/ ● *noun* native or citizen of Moscow. ● *adjective* of Moscow.

muscular /'mʌskjʊlə/ *adjective* of or affecting muscles; having well-developed muscles. □ **muscular dystrophy** hereditary progressive wasting of muscles. □ **muscularity** /-'lær-/ *noun.*

muse[1] /mju:z/ *verb* (**-sing**) (usually + *on*, *upon*) ponder, reflect.

muse[2] /mju:z/ *noun Greek & Roman Mythology* any of 9 goddesses inspiring poetry, music, etc.; (usually **the muse**) poet's inspiration.

museum /mju:'zi:əm/ *noun* building for storing and exhibiting objects of historical, scientific, or cultural interest. □ **museum piece** object fit for museum, *derogatory* old-fashioned person etc.

mush *noun* soft pulp; feeble sentimentality; *US* maize porridge. □ **mushy** *adjective* (**-ier**, **-iest**).

mushroom /'mʌʃrʊm/ ● *noun* edible fungus with stem and domed cap; pinkish-brown colour. ● *verb* appear or develop rapidly. □ **mushroom cloud** mushroom-shaped cloud from nuclear explosion.

music /'mju:zɪk/ *noun* art of combining vocal or instrumental sounds in harmonious or expressive way; sounds so produced; musical composition; written or printed score of this; pleasant sound. □ **music centre** equipment combining radio, record player, tape recorder, etc.; **music-hall** variety entertainment, theatre for this.

musical ● *adjective* of music; (of sounds) melodious, harmonious; fond of or skilled in music; set to or accompanied by music. ● *noun* musical film or play. □ **musicality** /-'kæl-/ *noun*; **musically** *adverb*.

musician /mju:'zɪʃ(ə)n/ *noun* person skilled in practice of music, esp. professional instrumentalist. □ **musicianship** *noun*.

musicology /mju:zɪ'kɒlədʒɪ/ *noun* study of history and forms of music. □ **musicological** /-kə'lɒdʒ-/ *adjective*; **musicologist** *noun*.

musk *noun* substance secreted by male musk deer and used in perfumes; plant which originally had smell of musk. □ **musk deer** small hornless Asian deer; **muskrat** large N. American aquatic rodent with smell like musk, its fur; **musk-rose** rambling rose smelling of musk. □ **musky** *adjective* (**-ier**, **-iest**).

musket /'mʌskɪt/ *noun historical* infantryman's (esp. smooth-bored) light gun.

musketeer /mʌskə'tɪə/ *noun historical* soldier armed with musket.

musketry /'mʌskɪtrɪ/ *noun* muskets; soldiers armed with muskets; knowledge of handling small arms.

Muslim /'mʊzlɪm/ (also **Moslem** /'mɒzləm/) ● *noun* follower of Islamic religion. ● *adjective* of Muslims or their religion.

muslin /'mʌzlɪn/ *noun* fine delicately woven cotton fabric.

musquash /'mʌskwɒʃ/ *noun* muskrat; its fur.

mussel /'mʌs(ə)l/ *noun* edible bivalve mollusc.

must¹ ● *auxiliary verb* (*3rd singular present* **must**; *past* **had to**) be obliged to; be certain to; ought to. ● *noun colloquial* thing that should not be missed.

■ **Usage** The negative *I must not go* means 'I am not allowed to go'. To express a lack of obligation, use *I am not obliged to go, I need not go*, or *I haven't got to go*.

must² *noun* grape juice before fermentation is complete.

mustache *US* = MOUSTACHE.

mustang /'mʌstæŋ/ *noun* small wild horse of Mexico and California.

mustard /'mʌstəd/ *noun* plant with yellow flowers; seeds of this crushed into paste and used as spicy condiment. □ **mustard gas** colourless oily liquid whose vapour is powerful irritant.

muster /'mʌstə/ ● *verb* collect (originally soldiers); come together; summon (courage etc.). ● *noun* assembly of people for inspection. □ **pass muster** be accepted as adequate.

mustn't /'mʌs(ə)nt/ must not.

musty /'mʌstɪ/ *adjective* (**-ier**, **-iest**) mouldy, stale; dull, antiquated. □ **mustiness** *noun*.

mutable /'mju:təb(ə)l/ *adjective literary* liable to change. □ **mutability** *noun*.

mutant /'mju:t(ə)nt/ ● *adjective* resulting from mutation. ● *noun* mutant organism or gene.

mutate /mju:'teɪt/ (cause to) undergo mutation.

mutation *noun* change; genetic change which when transmitted to offspring gives rise to heritable variations.

mute /mju:t/ ● *adjective* silent; refraining from or temporarily bereft of speech; dumb; soundless. ● *noun* dumb person; device for damping sound of musical instrument. ● *verb* (**-ting**) muffle or deaden sound of; (as **muted** *adjective*) (of colours etc.) subdued. □ **mute swan** common white swan. □ **mutely** *adverb*.

mutilate /'mju:tɪleɪt/ *verb* (**-ting**) deprive (person, animal) of limb etc.; destroy usefulness of (limb etc.); excise or damage part of (book etc.). □ **mutilation** *noun*.

mutineer /mju:tɪ'nɪə/ *noun* person who mutinies.

mutinous /'mju:tɪnəs/ *adjective* rebellious.

mutiny /'mju:tɪnɪ/ ● *noun* (*plural* **-ies**) open revolt, esp. by soldiers or sailors against officers. ● *verb* (**-ies**, **-ied**) engage in mutiny.

mutt *noun slang* stupid person.

mutter /'mʌtə/ ● *verb* speak in barely audible manner; (often + *against*, *at*) grumble. ● *noun* muttered words etc.; muttering.

mutton /'mʌt(ə)n/ *noun* flesh of sheep as food.

mutual /'mju:tʃʊəl/ *adjective* (of feelings, actions, etc.) experienced or done by each of two or more parties to the other(s); *colloquial* common to two or more people; having same (specified) relationship to each other. □ **mutuality** /-'æl-/ *noun*; **mutually** *adverb*.

■ **Usage** The use of *mutual* to mean 'common to two or more people' is considered incorrect by some people, who use *common* instead.

muzzle /'mʌz(ə)l/ ● *noun* projecting part of animal's face, including nose and mouth; guard put over animal's nose and mouth; open end of firearm. ● *verb* (**-ling**) put muzzle on; impose silence on.

muzzy /'mʌzɪ/ *adjective* (**-ier**, **-iest**) confused, dazed; blurred, indistinct. □ **muzzily** *adverb*.

MW *abbreviation* megawatt(s); medium wave.

my /maɪ/ *adjective* of or belonging to me.

mycology /maɪ'kɒlədʒɪ/ *noun* study of fungi; fungi of particular region.

mynah /'maɪnə/ *noun* (also **myna**) talking bird of starling family.

myopia /maɪ'əʊpɪə/ *noun* short-sightedness; lack of imagination. □ **myopic** /-'ɒp-/ *adjective*.

myriad /'mɪrɪəd/ *literary* ● *noun* indefinitely great number. ● *adjective* innumerable.

myrrh /mɜ:/ *noun* gum resin used in perfume, medicine, incense, etc.

myrtle /'mɜ:t(ə)l/ *noun* evergreen shrub with shiny leaves and white scented flowers.

myself /maɪ'self/ *pronoun: emphatic form of* I[2] *or* ME[1]; *reflexive form of* ME[1].

mysterious /mɪs'tɪərɪəs/ *adjective* full of or wrapped in mystery. □ **mysteriously** *adverb*.

mystery /'mɪstərɪ/ *noun* (*plural* **-ies**) hidden or inexplicable matter; secrecy, obscurity; fictional work dealing with puzzling event, esp. murder; religious truth divinely revealed; (in *plural*) secret ancient religious rites. □ **mystery play** miracle play; **mystery tour** pleasure trip to unspecified destination.

mystic /'mɪstɪk/ ● *noun* person who seeks unity with deity through contemplation etc., or believes in spiritual apprehension of truths beyond understanding. □ **mysticism** /-sɪz(ə)m/ *noun*.

mystical *adjective* of mystics or mysticism; of hidden meaning; spiritually symbolic.

mystify /'mɪstɪfaɪ/ *verb* (**-ies**, **-ied**) bewilder, confuse. □ **mystification** *noun*.

mystique /mɪs'ti:k/ *noun* atmosphere of mystery and veneration attending some activity, person, profession, etc.

myth /mɪθ/ *noun* traditional story usually involving supernatural or imaginary people and embodying popular ideas on natural or social phenomena; widely held but false idea; fictitious person, thing, or idea. □ **mythical** *adjective*.

mythology /mɪ'θɒlədʒɪ/ *noun* (*plural* **-ies**) body or study of myths. □ **mythological** /-θə'lɒdʒ-/ *adjective*; **mythologize** *verb* (also **-ise**) (**-zing** or **-sing**).

myxomatosis /mɪksəmə'təʊsɪs/ *noun* viral disease of rabbits.

Nn

N *abbreviation* (also **N.**) north(ern).

n¹ *noun* indefinite number.

n. *abbreviation* (also **n**) noun; neuter.

NAAFI /'næfɪ/ *abbreviation* Navy, Army, and Air Force Institutes (canteen for servicemen).

nab *verb* (**-bb-**) *slang* arrest; catch in wrongdoing; grab.

nacre /'neɪkə/ *noun* mother-of-pearl from any shelled mollusc. □ **nacreous** /'neɪkrɪəs/ *adjective*.

nadir /'neɪdɪə/ *noun* point on celestial sphere directly below observer; lowest point; time of despair.

naff *adjective slang* unfashionable; rubbishy.

nag¹ *verb* (**-gg-**) persistently criticize or scold; (often + *at*) find fault or urge, esp. continually; (of pain) be persistent.

nag² *noun colloquial* horse.

naiad /'naɪæd/ *noun* water nymph.

nail ● *noun* small metal spike hammered in to fasten things; horny covering on upper surface of tip of human finger or toe. ● *verb* fasten with nail(s); fix or hold tight; secure, catch (person, thing).

naïve /naɪ'i:v/ *adjective* (also **naive**) innocent, unaffected; foolishly credulous. □ **naïvely** *adverb*; **naïvety** *noun*.

naked /'neɪkɪd/ *adjective* unclothed, nude; without usual covering; undisguised; (of light, flame, sword, etc.) unprotected. □ **nakedly** *adverb*; **nakedness** *noun*.

namby-pamby /næmbɪ'pæmbɪ/ ● *adjective* insipidly pretty or sentimental; weak. ● *noun* (*plural* **-ies**) namby-pamby person.

name ● *noun* word by which individual person, animal, place, or thing is spoken of etc.; (usually abusive) term used of person; word denoting object or class of objects; reputation, esp. good. ● *verb* (**-ming**) give name to; state name of; mention; specify; cite. □ **name-day** feast-day of saint after whom person is named; **namesake** person or thing having same name as another.

nameless *adjective* having, or showing, no name; left unnamed.

namely *adverb* that is to say; in other words.

nanny /'nænɪ/ *noun* (*plural* **-ies**) child's nurse; *colloquial* grandmother; (in full **nanny goat**) female goat.

nano- /'nænəʊ/ *combining form* one thousand millionth.

nap¹ ● *noun* short sleep, esp. by day. ● *verb* (**-pp-**) have nap.

nap² *noun* raised pile on cloth, esp. velvet.

nap³ ● *noun* card game; racing tip claimed to be almost a certainty. ● *verb* (**-pp-**) name (horse) as probable winner. □ **go nap** try to take all 5 tricks in nap, risk everything.

napalm /'neɪpɑ:m/ *noun* thick jellied hydrocarbon mixture used in bombs.

nape *noun* back of neck.

naphtha /'næfθə/ *noun* inflammable hydrocarbon distilled from coal etc.

naphthalene /'næfθəli:n/ *noun* white crystalline substance produced by distilling tar.

napkin /'næpkɪn/ *noun* piece of linen etc. for wiping lips, fingers, etc. at table; baby's nappy.

nappy /'næpɪ/ *noun* (*plural* **-ies**) piece of towelling etc. wrapped round baby to absorb urine and faeces.

narcissism /'nɑ:sɪsɪz(ə)m/ *noun* excessive or erotic interest in oneself. □ **narcissistic** /-'sɪstɪk/ *adjective*.

narcissus /nɑ:'sɪsəs/ *noun* (*plural* **-cissi** /-saɪ/) any of several flowering bulbs, including daffodil.

narcosis /nɑ:'kəʊsɪs/ *noun* unconsciousness; induction of this.

narcotic /nɑ:'kɒtɪk/ ● *adjective* (of substance) inducing drowsiness etc.; (of drug) affecting the mind. ● *noun* narcotic substance or drug.

nark *slang* • *noun* police informer. • *verb* annoy.

narrate /nəˈreɪt/ *verb* (**-ting**) give continuous story or account of; provide spoken accompaniment for (film etc.). □ **narration** *noun*; **narrator** *noun*.

narrative /ˈnærətɪv/ • *noun* ordered account of connected events. • *adjective* of or by narration.

narrow /ˈnærəʊ/ • *adjective* (**-er**, **-est**) of small width; restricted; of limited scope; with little margin; precise, exact; narrow-minded. • *noun* (usually in *plural*) narrow part of strait, river, pass, etc. • *verb* become or make narrower; contract; lessen. □ **narrow boat** canal boat; **narrow-minded** rigid or restricted in one's views, intolerant. □ **narrowly** *adverb*; **narrowness** *noun*.

narwhal /ˈnɑːw(ə)l/ *noun* Arctic white whale, male of which has long tusk.

nasal /ˈneɪz(ə)l/ • *adjective* of nose; (of letter or sound) pronounced with breath passing through nose, e.g. *m*, *n*, *ng*; (of voice etc.) having many nasal sounds. • *noun* nasal letter or sound. □ **nasally** *adverb*.

nascent /ˈnæs(ə)nt/ *adjective* in act of being born; just beginning to be. □ **nascency** /-ənsɪ/ *noun*.

nasturtium /nəˈstɜːʃəm/ *noun* trailing garden plant with edible leaves and bright orange, red, or yellow flowers.

nasty /ˈnɑːstɪ/ • *adjective* (**-ier**, **-iest**) unpleasant; difficult to negotiate; (of person or animal) ill-natured, spiteful. • *noun* (*plural* **-ies**) *colloquial* violent horror film, esp. on video. □ **nastily** *adverb*; **nastiness** *noun*.

Nat. *abbreviation* National(ist); Natural.

natal /ˈneɪt(ə)l/ *adjective* of or from birth.

nation /ˈneɪʃ(ə)n/ *noun* community of people having mainly common descent, history, language, etc., forming state or inhabiting territory. □ **nationwide** extending over whole nation.

national /ˈnæʃən(ə)l/ • *adjective* of nation; characteristic of particular nation. • *noun* citizen of specified country. □ **national anthem** song adopted by nation, intended to inspire patriotism; **national grid** network of high-voltage electric power lines between major power stations; **National Insurance** system of compulsory payments from employee and employer to provide state assistance in sickness, retirement, etc.; **national service** *historical* conscripted peacetime military service. □ **nationally** *adverb*.

nationalism *noun* patriotic feeling, principles, etc.; policy of national independence. □ **nationalist** *noun*; **nationalistic** /-ˈlɪs-/ *adjective*.

nationality /næʃəˈnælɪtɪ/ *noun* (*plural* **-ies**) membership of nation; being national; ethnic group within one or more political nations.

nationalize /ˈnæʃənəlaɪz/ *verb* (also **-ise**) (**-zing** or **-sing**) take (industry etc.) into state ownership; make national. □ **nationalization** *noun*.

native /ˈneɪtɪv/ • *noun* (usually + *of*) person born in specified place; local inhabitant; indigenous animal or plant. • *adjective* inherent; innate; of one's birth; (usually + *to*) belonging to specified place; born in a place.

nativity /nəˈtɪvɪtɪ/ *noun* (*plural* **-ies**) (esp. **the Nativity**) Christ's birth; birth.

NATO /ˈneɪtəʊ/ *abbreviation* (also **Nato**) North Atlantic Treaty Organization.

natter /ˈnætə/ *colloquial* • *verb* chatter idly. • *noun* aimless chatter.

natty /ˈnætɪ/ *adjective* (**-ier**, **-iest**) trim; smart.

natural /ˈnætʃər(ə)l/ • *adjective* existing in or caused by nature; not surprising; to be expected; unaffected; innate; physically existing; *Music* not flat or sharp. • *noun colloquial* (usually + *for*) person or thing naturally suitable, adept, etc.; *Music* sign (♮) showing return to natural pitch, natural note. □ **natural gas** gas found in earth's crust; **natural history** study of animals and plants; **natural number** whole number greater than 0; **natural selection** process favouring survival of organisms best adapted to environment.

naturalism *noun* realistic representation in art and literature; philosophy based on nature alone. □ **naturalistic** /-ˈlɪs-/ *adjective*.

naturalist *noun* student of natural history.

naturalize *verb* (also **-ise**) (**-zing** or **-sing**) admit (foreigner) to citizenship;

introduce (plant etc.) into another region; adopt (foreign word, custom, etc.). □ **naturalization** *noun.*

naturally *adverb* in a natural way; as might be expected, of course.

nature /'neɪtʃə/ *noun* thing's or person's essential qualities or character; physical power causing material phenomena; these phenomena; kind, class.

naturism *noun* nudism. □ **naturist** *noun.*

naught /nɔ:t/ *archaic* ● *noun* nothing. ● *adjective* worthless.

naughty /'nɔ:tɪ/ *adjective* (**-ier**, **-iest**) (esp. of children) disobedient; badly behaved; *colloquial jocular* indecent. □ **naughtily** *adverb*; **naughtiness** *noun.*

nausea /'nɔ:sɪə/ *noun* inclination to vomit; revulsion.

nauseate /'nɔ:sɪeɪt/ *verb* (**-ting**) affect with nausea. □ **nauseating** *adjective.*

nauseous /'nɔ:sɪəs/ *adjective* causing or inclined to vomit; disgusting.

nautical /'nɔ:tɪk(ə)l/ *adjective* of sailors or navigation. □ **nautical mile** unit of approx. 2,025 yards (1,852 metres).

nautilus /'nɔ:tɪləs/ *noun* (*plural* **nautiluses** or **nautili** /-laɪ/) kind of mollusc with spiral shell.

naval /'neɪv(ə)l/ *adjective* of navy; of ships.

nave[1] *noun* central part of church excluding chancel and side aisles.

nave[2] *noun* hub of wheel.

navel /'neɪv(ə)l/ *noun* depression in belly marking site of attachment of umbilical cord. □ **navel orange** one with navel-like formation at top.

navigable /'nævɪgəb(ə)l/ *adjective* (of river etc.) suitable for ships; seaworthy; steerable. □ **navigability** *noun.*

navigate /'nævɪgeɪt/ *verb* (**-ting**) manage or direct course of (ship, aircraft); sail on (sea, river, etc.); fly through (air); help car-driver etc. by map-reading etc. □ **navigator** *noun.*

navigation *noun* act or process of navigating; art or science of navigating.

navvy /'nævɪ/ *noun* (*plural* **-ies**) labourer employed in building roads, canals, etc.

navy /'neɪvɪ/ *noun* (*plural* **-ies**) state's warships with their crews, maintenance systems, etc.; (in full **navy blue**) dark blue colour.

nay ● *adverb* or rather; and even; *archaic* no. ● *noun* 'no' vote.

Nazi /'nɑ:tsɪ/ ● *noun* (*plural* **-s**) *historical* member of German National Socialist party. ● *adjective* of Nazis or Nazism. □ **Nazism** *noun.*

NB *abbreviation* note well (*nota bene*).

NCB *abbreviation historical* National Coal Board.

NCO *abbreviation* non-commissioned officer.

NE *abbreviation* north-east(ern).

Neanderthal /nɪ'ændətɑ:l/ *adjective* of type of human found in palaeolithic Europe.

neap *noun* (in full **neap tide**) tide with smallest rise and fall.

Neapolitan /nɪə'pɒlɪt(ə)n/ ● *noun* native of Naples. ● *adjective* of Naples.

near /nɪə/ ● *adverb* (often + *to*) to or at short distance in space or time; closely. ● *preposition* to or at a short distance from in space, time, condition, or resemblance. ● *adjective* close (to); not far in place or time; closely related; (of part of vehicle, animal, or road) on left side; *colloquial* stingy; with little margin. ● *verb* approach, draw near to. □ **the Near East** countries of eastern Mediterranean; **near-sighted** short-sighted.

nearby ● *adjective* near in position. ● *adverb* close.

nearly *adverb* almost; closely. □ **not nearly** nothing like, far from.

neat *adjective* tidy, methodical; elegantly simple; brief and clear; cleverly done; dexterous; (of alcoholic liquor) undiluted. □ **neaten** *verb*; **neatly** *adverb*; **neatness** *noun.*

neath *preposition poetical* beneath.

nebula /'nebjʊlə/ *noun* (*plural* **nebulae** /-li:/) cloud of gas and dust seen in night sky, appearing luminous or as dark silhouette. □ **nebular** *adjective.*

nebulous /'nebjʊləs/ *adjective* cloudlike; indistinct, vague.

necessary /'nesəsərɪ/ ● *adjective* requiring to be done; essential; inevitable. ● *noun* (*plural* **-ies**) (usually in *plural*) any of basic requirements of life. □ **necessarily** *adverb.*

necessitate /nɪ'sesɪteɪt/ *verb* (**-ting**) make necessary (esp. as result).

necessitous /nɪ'sesɪtəs/ *adjective* poor, needy.

necessity /nɪ'sesɪtɪ/ *noun* (*plural* **-ies**) indispensable thing; pressure of circumstances; imperative need; poverty; constraint or compulsion seen as natural law governing human action.

neck • *noun* part of body connecting head to shoulders; part of garment round neck; narrow part of anything. • *verb colloquial* kiss and caress amorously. □ **neckline** outline of garment-opening at neck; **necktie** strip of material worn round shirt-collar, knotted at front.

necklace /'nekləs/ *noun* string of beads, precious stones, etc. worn round neck; *South African* petrol-soaked tyre placed round victim's neck and lighted.

necromancy /'nekrəʊmænsɪ/ *noun* divination by supposed communication with the dead; magic. □ **necromancer** *noun*.

necrophilia /nekrə'fɪlɪə/ *noun* morbid esp. sexual attraction to corpses.

necropolis /ne'krɒpəlɪs/ *noun* ancient cemetery.

necrosis /ne'krəʊsɪs/ *noun* death of tissue. □ **necrotic** /-'krɒt-/ *adjective*.

nectar /'nektə/ *noun* sugary substance produced by plants and made into honey by bees; *Mythology* drink of gods.

nectarine /'nektərɪn/ *noun* smooth-skinned variety of peach.

NEDC *abbreviation* National Economic Development Council.

née /neɪ/ *adjective* (*US* **nee**) (before married woman's maiden name) born.

need • *verb* stand in want of; require; (usually + *to do*) be under necessity or obligation. • *noun* requirement; circumstances requiring action; destitution, poverty; emergency.

needful *adjective* requisite.

needle /'ni:d(ə)l/ • *noun* very thin pointed rod with slit ('eye') for thread, used in sewing; knitting-needle; pointer on dial; any small thin pointed instrument, esp. end of hypodermic syringe; obelisk; pointed rock or peak; leaf of fir or pine. • *verb* (**-ling**) *colloquial* annoy, provoke. □ **needlecord** fine-ribbed corduroy fabric; **needlework** sewing or embroidery.

needless *adjective* unnecessary. □ **needlessly** *adverb*.

needy *adjective* (**-ier**, **-iest**) poor, destitute.

ne'er /neə/ *adverb poetical* never. □ **ne'er-do-well** good-for-nothing person.

nefarious /nɪ'feərɪəs/ *adjective* wicked.

negate /nɪ'geɪt/ *verb* (**-ting**) nullify; deny existence of.

negation *noun* absence or opposite of something positive; act of denying; negative statement; negative or unreal thing.

negative /'negətɪv/ • *adjective* expressing or implying denial, prohibition, or refusal; lacking positive attributes; opposite to positive; (of quantity) less than zero, to be subtracted; *Electricity* of, containing, or producing, kind of charge carried by electrons. • *noun* negative statement or word; *Photography* image with black and white reversed or colours replaced by complementary ones. • *verb* (**-ving**) refuse to accept; veto; disprove; contradict; neutralize. □ **negatively** *adverb*.

neglect /nɪ'glekt/ • *verb* fail to care for or do; (+ *to do*) fail; pay no attention to; disregard. • *noun* negligence; neglecting, being neglected. □ **neglectful** *adjective*; **neglectfully** *adverb*.

negligée /'neglɪʒeɪ/ *noun* (also **néglige**) woman's flimsy dressing gown.

negligence /'neglɪdʒ(ə)ns/ *noun* lack of proper care or attention; culpable carelessness. □ **negligent** *adjective*; **negligently** *adverb*.

negligible /'neglɪdʒɪb(ə)l/ *adjective* not worth considering; insignificant.

negotiate /nɪ'gəʊʃɪeɪt/ *verb* (**-ting**) confer in order to reach agreement; obtain (result) by negotiating; deal successfully with (obstacle etc.); convert (cheque etc.) into money. □ **negotiable** /-ʃəb-/ *adjective*; **negotiation** *noun*; **negotiator** *noun*.

Negress /'ni:grɪs/ *noun* female Negro.

■ **Usage** The term *Negress* is often considered offensive; *black* is usually preferred.

Negro /'ni:grəʊ/ • *noun* (*plural* **-es**) member of dark-skinned (originally) African race; black. • *adjective* of this race; black.

■ **Usage** The term *Negro* is often considered offensive; *black* is usually preferred.

Negroid /'ni:grɔɪd/ ● *adjective* (of physical features etc.) characteristic of black people. ● *noun* black person.

neigh /neɪ/ ● *noun* cry of horse. ● *verb* make a neigh.

neighbour /'neɪbə/ (*US* **neighbor**) *noun* person living next door or nearby; fellow human being. ● *verb* border on, adjoin.

neighbourhood *noun* (*US* **neighborhood**) district; vicinity; people of a district.

neighbourly *adjective* (*US* **neighborly**) like good neighbour, friendly, helpful. □ **neighbourliness** *noun.*

neither /'naɪðə/ *adjective, pronoun & adverb* not either.

nelson /'nels(ə)n/ *noun* wrestling hold in which arm is passed under opponent's arm from behind and hand applied to neck (**half nelson**), or both arms and hands are applied (**full nelson**).

nem. con. *abbreviation* with no one dissenting (*nemine contradicente*). [Latin]

nemesis /'neməsɪs/ *noun* justice bringing deserved punishment.

neo- *combining form* new; new form of.

neolithic /ni:ə'lɪθɪk/ *adjective* of later Stone Age.

neologism /ni:'ɒlədʒɪz(ə)m/ *noun* new word; coining of new words.

neon /'ni:ɒn/ *noun* inert gas giving orange glow when electricity is passed through it.

neophyte /'ni:əfaɪt/ *noun* new convert; novice of religious order; beginner.

nephew /'nefju:/ *noun* son of one's brother or sister or of one's spouse's brother or sister.

nephritic /nɪ'frɪtɪk/ *adjective* of or in kidneys.

nephritis /nɪ'fraɪtɪs/ *noun* inflammation of kidneys.

nepotism /'nepətɪz(ə)m/ *noun* favouritism to relatives in conferring offices.

nereid /'nɪərɪɪd/ *noun* sea nymph.

nerve ● *noun* fibre or bundle of fibres conveying impulses of sensation or motion between brain and other parts of body; coolness in danger; *colloquial* impudence; (in *plural*) nervousness, mental or physical stress. ● *verb* (**-ving**) (usually **nerve oneself**) brace or prepare (oneself). □ **nerve-cell** cell transmitting impulses in nerve tissue.

nerveless *adjective* lacking vigour.

nervous *adjective* easily upset, timid, highly strung; anxious; affecting the nerves; (+ *of*) afraid of. □ **nervous breakdown** period of mental illness, usually after stress; **nervous system** body's network of nerves. □ **nervously** *adverb*; **nervousness** *noun.*

nervy *adjective* (**-ier**, **-iest**) *colloquial* nervous; easily excited.

nest ● *noun* structure or place where bird lays eggs and shelters young; breeding-place, lair; snug retreat, shelter; brood, swarm; group or set of similar objects. ● *verb* use or build nest; (of objects) fit one inside another. □ **nest egg** money saved up as reserve.

nestle /'nes(ə)l/ *verb* (**-ling**) settle oneself comfortably; press oneself against another in affection etc.; (+ *in*, *into*, etc.) push (head, shoulders, etc.) affectionately or snugly; lie half hidden or embedded.

nestling /'nestlɪŋ/ *noun* bird too young to leave nest.

net[1] ● *noun* open-meshed fabric of cord, rope, etc.; piece of net used esp. to contain, restrain, or delimit, or to catch fish; structure with net used in various games. ● *verb* (**-tt-**) cover, confine, or catch with net; hit (ball) into net, esp. of goal. □ **netball** game similar to basketball.

net[2] (also **nett**) ● *adjective* remaining after necessary deductions; (of price) not reducible; (of weight) excluding packaging etc. ● *verb* (**-tt-**) gain or yield (sum) as net profit.

nether /'neðə/ *adjective archaic* lower.

nett = NET[2].

netting *noun* meshed fabric of cord or wire.

nettle /'net(ə)l/ ● *noun* plant covered with stinging hairs; plant resembling this. ● *verb* (**-ling**) irritate, provoke. □ **nettle-rash** skin eruption like nettle stings.

network ● *noun* arrangement of intersecting horizontal and vertical lines complex system of railways etc.; people connected by exchange of information etc.; group of broadcasting stations connected for simultaneous broadcast of a programme; system of interconnected computers. ● *verb* broadcast on network.

neural /'njʊər(ə)l/ *adjective* of nerve or central nervous system.

neuralgia /njʊə'rældʒə/ *noun* intense pain along a nerve, esp. in face or head. □ **neuralgic** *adjective.*

neuritis /njʊə'raɪtɪs/ *noun* inflammation of nerve(s).

neuro- /'njʊərəʊ/ *combining form* nerve(s).

neurology /njʊə'rɒlədʒɪ/ *noun* study of nerve systems. □ **neurological** /-rə-'lɒdʒ-/ *adjective*; **neurologist** *noun.*

neuron /'njʊərɒn/ *noun* (also **neurone** /-rəʊn/) nerve cell.

neurosis /njʊə'rəʊsɪs/ *noun* (*plural* **-roses** /-si:z/) disturbed behaviour pattern associated with nervous distress.

neurotic /njʊə'rɒtɪk/ ● *adjective* caused by or relating to neurosis; suffering from neurosis; *colloquial* abnormally sensitive or obsessive. ● *noun* neurotic person.

neuter /'nju:tə/ ● *adjective* neither masculine nor feminine. ● *verb* castrate, spay.

neutral /'nju:tr(ə)l/ ● *adjective* supporting neither of two opposing sides, impartial; vague, indeterminate; (of a gear) in which engine is disconnected from driven parts; (of colours) not strong or positive; *Chemistry* neither acid nor alkaline; *Electricity* neither positive nor negative. ● *noun* neutral state or person. □ **neutrality** /-'træl-/ *noun.*

neutralize *verb* (also **-ise**) (**-zing** or **-sing**) make neutral; make ineffective by opposite force. □ **neutralization** *noun.*

neutrino /nju:'tri:nəʊ/ *noun* (*plural* **-s**) elementary particle with zero electric charge and probably zero mass.

neutron /'nju:trɒn/ *noun* elementary particle of about same mass as proton but without electric charge.

never /'nevə/ *adverb* at no time, on no occasion; not ever; not at all; *colloquial* surely not. □ **the never-never** *colloquial* hire purchase.

nevermore *adverb* at no future time.

nevertheless /nevəðə'les/ *adverb* in spite of that; notwithstanding.

new *adjective* of recent origin or arrival; made, discovered, acquired, or experienced for first time; not worn; renewed; reinvigorated; different; unfamiliar. □ **New Age** set of alternative beliefs replacing traditional Western culture; **newborn** recently born; **newcomer** person recently arrived; **newfangled** different from what one is used to; **new moon** moon when first seen as crescent; **New Testament** part of Bible concerned with Christ and his followers; **New World** N. & S. America; **New Year's Day**, **Eve** 1 Jan., 31 Dec.

newel /'nju:əl/ *noun* supporting central post of winding stairs; top or bottom post of stair-rail.

newly *adverb* recently; afresh.

news /nju:z/ *plural noun* (usually treated as *singular*) information about important or interesting recent events, esp. when published or broadcast; (**the news**) broadcast report of news. □ **newsagent** seller of newspapers etc.; **newscast** radio or television broadcast of news reports; **newsletter** informal printed bulletin of club etc.; **newspaper** /'nju:s-/ printed publication of loose folded sheets with news etc.; **newsprint** low-quality paper for printing newspapers; **newsreader** person who reads out broadcast news bulletins; **newsreel** short cinema film of recent events; **news room** room where news is prepared for publication or broadcasting; **newsworthy** topical, worth reporting as news.

newsy *adjective* (**-ier**, **-iest**) *colloquial* full of news.

newt /nju:t/ *noun* small tailed amphibian.

newton /'nju:t(ə)n/ *noun* SI unit of force.

next ● *adjective* (often + *to*) being, placed, or living nearest; nearest in time. ● *adverb* (often + *to*) nearest in place or degree, on first or soonest occasion. ● *noun* next person or thing. ● *preposition colloquial* next to. □ **next door** in next house or room; **next of kin** closest living relative(s).

nexus /'neksəs/ *noun* (*plural* same) connected group or series.

NHS *abbreviation* National Health Service.

NI *abbreviation* Northern Ireland; National Insurance.

niacin /'naɪəsɪn/ *noun* nicotinic acid.

nib *noun* pen-point; (in *plural*) crushed coffee or cocoa beans.

nibble /'nɪb(ə)l/ ● *verb* (**-ling**) (+ *at*) take small bites at; eat in small amounts; bite gently or playfully. ● *noun* act of nibbling; very small amount of food.

nice *adjective* pleasant, satisfactory; kind, good-natured; fine, (of distinctions) subtle; fastidious. □ **nicely** *adverb*; **niceness** *noun*.

nicety /'naɪsɪtɪ/ *noun* (*plural* **-ies**) subtle distinction or detail; precision. □ **to a nicety** exactly.

niche /ni:ʃ/ *noun* shallow recess, esp. in wall; comfortable or apt position in life or employment.

nick ● *noun* small cut or notch; *slang* prison, police station; *colloquial* state, condition. ● *verb* make nick(s) in; *slang* steal, arrest, catch. □ **in the nick of time** only just in time.

nickel /'nɪk(ə)l/ *noun* silver-white metallic element used esp. in magnetic alloys; *colloquial* US 5-cent coin.

nickname /'nɪkneɪm/ ● *noun* familiar or humorous name added to or substituted for real name of person or thing. ● *verb* (**-ming**) give nickname to.

nicotine /'nɪkəti:n/ *noun* poisonous alkaloid present in tobacco.

nicotinic acid /nɪkə'tɪnɪk/ *noun* vitamin of B group.

nictitate /'nɪktɪteɪt/ *verb* (**-ting**) blink, wink. □ **nictitation** *noun*.

niece /ni:s/ *noun* daughter of one's brother or sister or of one's spouse's brother or sister.

nifty /'nɪftɪ/ *adjective* (**-ier**, **-iest**) *colloquial* clever, adroit; smart, stylish.

niggard /'nɪgəd/ *noun* stingy person.

niggardly *adjective* stingy. □ **niggardliness** *noun*.

niggle /'nɪg(ə)l/ *verb* (**-ling**) fuss over details, find fault in petty way; *colloquial* nag. □ **niggling** *adjective*.

nigh /naɪ/ *adverb & preposition archaic* near.

night /naɪt/ *noun* period of darkness from one day to next; time from sunset to sunrise; nightfall; darkness of night; evening. □ **nightcap** *historical* cap worn in bed, drink before going to bed; **nightclub** club providing entertainment etc. late at night; **nightdress** woman's or child's loose garment worn in bed; **nightfall** end of daylight; **nightjar** nocturnal bird with harsh cry; **night-life** entertainment available at night; **nightmare** terrifying dream or *colloquial* experience; **night safe** safe with access from outer wall of bank for deposit of money when bank is closed; **nightshade** any of various plants with poisonous berries; **nightshirt** long shirt worn in bed.

nightingale /'naɪtɪŋgeɪl/ *noun* small reddish-brown bird, of which the male sings tunefully, esp. at night.

nightly ● *adjective* happening, done, or existing in the night; recurring every night. ● *adverb* every night.

nihilism /'naɪɪlɪz(ə)m/ *noun* rejection of all religious and moral principles. □ **nihilist** *noun*; **nihilistic** /-'lɪs-/ *adjective*.

nil *noun* nothing.

nimble /'nɪmb(ə)l/ *adjective* (**-r**, **-st**) quick and light in movement or function; agile. □ **nimbly** *adverb*.

nimbus /'nɪmbəs/ *noun* (*plural* **nimbi** /-baɪ/ or **nimbuses**) halo; rain-cloud.

nincompoop /'nɪŋkəmpu:p/ *noun* foolish person.

nine *adjective & noun* one more than eight. □ **ninepins** (usually treated as *singular*) kind of skittles. □ **ninth** /naɪnθ/ *adjective & noun*.

nineteen /naɪn'ti:n/ *adjective & noun* one more than eighteen. □ **nineteenth** *adjective & noun*.

ninety /'naɪntɪ/ *adjective & noun* (*plural* **-ies**) nine times ten. □ **ninetieth** *adjective & noun*.

ninny /'nɪnɪ/ *noun* (*plural* **-ies**) foolish person.

nip[1] ● *verb* (**-pp-**) pinch, squeeze sharply, bite; (often + *off*) remove by pinching etc.; *colloquial* go nimbly. ● *noun* pinch, sharp squeeze, bite; biting cold. □ **nip in the bud** suppress or destroy at very beginning.

nip[2] *noun* small quantity of spirits.

nipper *noun* person or thing that nips; claw of crab etc.; *colloquial* young child; (in *plural*) tool with jaws for gripping or cutting.

nipple /ˈnɪp(ə)l/ *noun* small projection in mammals from which in females milk for young is secreted; teat of feeding-bottle; device like nipple in function; nipple-like protuberance.

nippy *adjective* (**-ier**, **-iest**) *colloquial* quick, nimble; chilly.

nirvana /nɪəˈvɑːnə/ *noun* (in Buddhism) perfect bliss attained by extinction of individuality.

nit *noun* egg or young of louse or other parasitic insect; *slang* stupid person. □ **nit-picking** *colloquial* fault-finding in a petty way.

niter *US* = NITRE.

nitrate /ˈnaɪtreɪt/ *noun* salt of nitric acid; potassium or sodium nitrate as fertilizer.

nitre /ˈnaɪtə/ *noun* (*US* **niter**) saltpetre.

nitric acid /ˈnaɪtrɪk/ *noun* colourless corrosive poisonous liquid.

nitrogen /ˈnaɪtrədʒ(ə)n/ *noun* gaseous element forming four-fifths of atmosphere. □ **nitrogenous** /-ˈtrɒdʒɪnəs/ *adjective*.

nitroglycerine /naɪtrəʊˈglɪsərɪn/ *noun* (*US* **nitroglycerin**) explosive yellow liquid.

nitrous oxide /ˈnaɪtrəs/ *noun* colourless gas used as anaesthetic.

nitty-gritty /nɪtɪˈgrɪtɪ/ *noun slang* realities or practical details of a matter.

nitwit *noun colloquial* stupid person.

NNE *abbreviation* north-north-east.

NNW *abbreviation* north-north-west.

No = NOH.

No. *abbreviation* number.

no /nəʊ/ ● *adjective* not any; not a; hardly any; *used to forbid thing specified*. ● *adverb* by no amount, not at all. ● *interjection expressing negative reply to question, request, etc.* ● *noun* (*plural* **noes**) utterance of word *no*, denial or refusal; 'no' vote. □ **no-ball** unlawfully delivered ball in cricket etc.; **no longer** not now as formerly; **no one** nobody; **no way** *colloquial* it is impossible.

nob[1] *noun slang* person of wealth or high social position.

nob[2] *noun slang* head.

nobble /ˈnɒb(ə)l/ *verb* (**-ling**) *slang* try to influence (e.g. judge); tamper with (racehorse etc.); steal; seize; catch.

nobility /nəʊˈbɪlɪtɪ/ *noun* (*plural* **-ies**) nobleness of character, birth, or rank; class of nobles.

noble /ˈnəʊb(ə)l/ ● *adjective* (**-r**, **-st**) belonging to the aristocracy; of excellent character; magnanimous; of imposing appearance. ● *noun* nobleman; noblewoman. □ **nobleman** peer; **noblewoman** peeress. □ **nobly** *adverb*.

noblesse oblige /nəʊbles ɒˈbliːʒ/ *noun* privilege entails responsibility. [French]

nobody /ˈnəʊbədɪ/ ● *pronoun* no person. ● *noun* (*plural* **-ies**) person of no importance.

nocturnal /nɒkˈtɜːn(ə)l/ *adjective* of or in the night; done or active by night.

nocturne /ˈnɒktɜːn/ *noun Music* short romantic composition, usually for piano; picture of night scene.

nod ● *verb* (**-dd-**) incline head slightly and briefly; let head droop in drowsiness; be drowsy; show (assent etc.) by nod; (of flowers etc.) bend and sway; make momentary slip or mistake. ● *noun* nodding of head. □ **nod off** *colloquial* fall asleep.

noddle /ˈnɒd(ə)l/ *noun colloquial* head.

node *noun* part of plant stem from which leaves emerge; knob on root or branch; natural swelling; intersecting point, esp. of planet's orbit with plane of celestial equator; point or line of least disturbance in vibrating system; point at which curve crosses itself; component in computer network. □ **nodal** *adjective*.

nodule /ˈnɒdjuːl/ *noun* small rounded lump of anything; small tumour, ganglion, swelling on legume root. □ **nodular** *adjective*.

noggin /ˈnɒgɪn/ *noun* small mug; small measure of spirits.

Noh /nəʊ/ *noun* (also **No**) traditional Japanese drama.

noise /nɔɪz/ ● *noun* sound, esp. loud or unpleasant one; confusion of loud sounds. ● *verb* (**-sing**) (usually in *passive*) make public, spread abroad.

noisome /ˈnɔɪsəm/ *adjective literary* harmful, noxious; evil-smelling.

noisy *adjective* (**-ier**, **-iest**) making much noise; full of noise. □ **noisily** *adverb*.

nomad /'nəʊmæd/ *noun* member of tribe roaming from place to place for pasture; wanderer. □ **nomadic** /-'mæd-/ *adjective*.

nom de plume /nɒm də 'plu:m/ *noun* (*plural* ***noms de plume*** same pronunciation) writer's assumed name. [French]

nomenclature /nəʊ'menklətʃə/ *noun* system of names for things; terminology of a science etc.

nominal /'nɒmɪn(ə)l/ *adjective* existing in name only; not real or actual; (of sum of money etc.) very small; of, as, or like noun. □ **nominally** *adverb*.

nominate /'nɒmɪneɪt/ *verb* (**-ting**) propose (candidate) for election; appoint to office; appoint (date or place). □ **nomination** *noun*; **nominator** *noun*.

nominative /'nɒmɪnətɪv/ *Grammar* ● *noun* case expressing subject of verb. ● *adjective* of or in this case.

nominee /nɒmɪ'ni:/ *noun* person who is nominated.

non- *prefix* not. For words starting with *non-* that are not found below, the rootwords should be consulted.

nonagenarian /nəʊnədʒɪ'neərɪən/ *noun* person from 90 to 99 years old.

non-belligerent /nɒnbə'lɪdʒərənt/ ● *adjective* not engaged in hostilities. ● *noun* non-belligerent state etc.

nonce /nɒns/ *noun* □ **for the nonce** for the time being, for the present; **nonce-word** word coined for one occasion.

nonchalant /'nɒnʃələnt/ *adjective* calm and casual. □ **nonchalance** *noun*; **nonchalantly** *adverb*.

non-combatant /nɒn'kɒmbət(ə)nt/ *noun* person not fighting in a war, esp. civilian, army chaplain, etc.

non-commissioned /nɒnkə'mɪʃ(ə)nd/ *adjective* (of officer) not holding commission.

noncommittal /nɒnkə'mɪt(ə)l/ *adjective* avoiding commitment to definite opinion or course of action.

non-conductor /nɒnkən'dʌktə/ *noun* substance that does not conduct heat or electricity.

nonconformist /nɒnkən'fɔ:mɪst/ *noun* person who does not conform to doctrine of established Church, esp. (**Nonconformist**) member of Protestant sect dissenting from Anglican Church; person not conforming to prevailing principle.

nonconformity /nɒnkənfɔ:mɪtɪ/ *noun* nonconformists as body; (+ *to*) failure to conform.

non-contributory /nɒnkən'trɪbjʊtərɪ/ *adjective* not involving contributions.

nondescript /'nɒndɪskrɪpt/ ● *adjective* lacking distinctive characteristics, not easily classified. ● *noun* such person or thing.

none /nʌn/ *pronoun* (often + *of*) not any; no person(s). □ **none the** (+ *comparative*), **none too** not in the least.

■ **Usage** The verb following *none* can be singular or plural when it means 'not any of several', e.g. *None of us knows* or *None of us know*.

nonentity /nɒ'nentɪtɪ/ *noun* (*plural* **-ies**) person or thing of no importance; non-existence; non-existent thing.

nonet /nəʊ'net/ *noun* musical composition for 9 performers; the performers; any group of 9.

nonetheless /nʌnðə'les/ *adverb* nevertheless.

non-event /nɒnɪ'vent/ *noun* insignificant event, esp. contrary to hopes or expectations.

non-existent /nɒnɪg'zɪst(ə)nt/ *adjective* not existing. □ **non-existence** *noun*.

non-fiction /nɒn'fɪkʃ(ə)n/ *noun* literary work other than fiction.

non-interference /nɒnɪntə'fɪərəns/ *noun* non-intervention.

non-intervention /nɒnɪntə'venʃ(ə)n/ *noun* policy of not interfering in others' affairs.

nonpareil /nɒnpə'reɪl/ ● *adjective* unrivalled, unique. ● *noun* such person or thing.

non-party /nɒn'pɑ:tɪ/ *adjective* independent of political parties.

nonplus /nɒn'plʌs/ *verb* (**-ss-**) completely perplex.

nonsense /'nɒns(ə)ns/ *noun* (often as *interjection*) absurd or meaningless words or ideas; foolish conduct. □ **nonsensical** /-'sen-/ *adjective*.

non sequitur /nɒn 'sekwɪtə/ *noun* conclusion that does not logically follow from the premisses. [Latin]

non-slip /nɒn'slɪp/ *adjective* that does not slip; that prevents slipping.

non-smoker /nɒn'sməʊkə/ *noun* person who does not smoke; train compartment etc. where smoking is forbidden. □ **non-smoking** *adjective.*

non-starter /nɒn'stɑ:tə/ *noun colloquial* person or scheme not worth considering.

non-stick /nɒn'stɪk/ *adjective* that does not allow things to stick to it.

non-stop /nɒn'stɒp/ ● *adjective* (of train etc.) not stopping at intermediate stations; done without stopping. ● *adverb* without stopping.

noodle[1] /'nu:d(ə)l/ *noun* strip or ring of pasta.

noodle[2] /'nu:d(ə)l/ *noun* simpleton.

nook /nʊk/ *noun* corner or recess; secluded place.

noon *noun* 12 o'clock in day, midday. □ **noonday** midday.

noose ● *noun* loop with running knot; snare. ● *verb* (**-sing**) catch with or enclose in noose.

nor *conjunction* and not.

Nordic /'nɔ:dɪk/ *adjective* of tall blond Germanic people of Scandinavia.

norm *noun* standard, type; standard amount of work etc.; customary behaviour.

normal /'nɔ:m(ə)l/ ● *adjective* conforming to standard; regular, usual, typical; *Geometry* (of line) at right angles. ● *noun* normal value of a temperature etc.; usual state, level, etc. □ **normalcy** *noun esp. US*; **normality** /-'mæl-/ *noun*; **normalize** *verb* (also **-ise**) (**-zing** or **-sing**); **normally** *adverb.*

Norman /'nɔ:mən/ ● *noun* (*plural* **-s**) native of medieval Normandy; descendant of people established there in 10th c. ● *adjective* of Normans; of style of medieval architecture found in Britain under Normans.

Norse ● *noun* Norwegian language; Scandinavian language group. ● *adjective* of ancient Scandinavia, esp. Norway. □ **Norseman** *noun.*

north ● *noun* point of horizon 90° anticlockwise from east; corresponding compass point; (usually **the North**) northern part of world, country, town, etc. ● *adjective* towards, at, near, or facing north; (of wind) from north. ● *adverb* towards, at, or near north; (+ *of*) further north than. □ **northbound** travelling or leading north; **north-east**, **-west** point midway between north and east or west; **north-north-east**, **north-north-west** point midway between north and north-east or north-west; **North Star** pole star. □ **northward** *adjective, adverb, & noun*; **northwards** *adverb.*

northerly /'nɔ:ðəlɪ/ *adjective & adverb* in northern position or direction; (of wind) from north.

northern /'nɔ:ð(ə)n/ *adjective* of or in the north. □ **northern lights** aurora borealis. □ **northernmost** *adjective.*

northerner *noun* native or inhabitant of north.

Norwegian /nɔ:'wi:dʒ(ə)n/ ● *noun* native, national, or language of Norway. ● *adjective* of or relating to Norway.

nose /nəʊz/ ● *noun* organ above mouth, used for smelling and breathing; sense of smell; odour or perfume of wine etc.; projecting part or front end of car, aircraft, etc. ● *verb* (**-sing**) (usually + *about* etc.) search; (often + *out*) perceive smell of, discover by smell; thrust nose against or into; make one's way cautiously forward. □ **nosebag** fodder-bag hung on horse's head; **nosebleed** bleeding from nose; **nosedive** (make) steep downward plunge.

nosegay *noun* small bunch of flowers.

nosh *slang* ● *verb* eat. ● *noun* food or drink. □ **nosh-up** large meal.

nostalgia /nɒs'tældʒə/ *noun* (often + *for*) yearning for past period; homesickness. □ **nostalgic** *adjective.*

nostril /'nɒstr(ə)l/ *noun* either of two openings in nose.

nostrum /'nɒstrəm/ *noun* quack remedy, patent medicine; pet scheme.

nosy *adjective* (**-ier**, **-iest**) *colloquial* inquisitive, prying.

not *adverb expressing negation, refusal, or denial.* □ **not half** *slang* very much, very, not nearly, *colloquial* not at all; **not quite** almost.

notable /ˈnəʊtəb(ə)l/ ● *adjective* worthy of note; remarkable; eminent. ● *noun* eminent person. □ **notability** *noun*; **notably** *adverb*.

notary /ˈnəʊtərɪ/ *noun* (*plural* **-ies**) solicitor etc. who certifies deeds etc. □ **notarial** /-ˈteər-/ *adjective*.

notation /nəʊˈteɪʃ(ə)n/ *noun* representation of numbers, quantities, musical notes, etc. by symbols; set of such symbols.

notch ● *noun* V-shaped indentation on edge or surface. ● *verb* make notches in; (usually + *up*) score, win, achieve (esp. amount or quantity).

note ● *noun* brief written record as memory aid; short letter; formal diplomatic message; additional explanation in book; banknote; notice, attention; eminence; single musical tone of definite pitch; written sign representing its pitch and duration; quality or tone of speaking. ● *verb* (**-ting**) observe, notice; (often + *down*) record as thing to be remembered; (in *passive*; often + *for*) be well known. □ **notebook** book for making notes in; **notecase** wallet for banknotes; **notelet** small folded card for informal letter; **notepaper** paper for writing letters; **noteworthy** worthy of attention, remarkable.

nothing /ˈnʌθɪŋ/ ● *noun* no thing, not anything; person or thing of no importance; non-existence; no amount; nought. ● *adverb* not at all; in no way.

nothingness *noun* non-existence; worthlessness.

notice /ˈnəʊtɪs/ ● *noun* attention; displayed sheet etc. with announcement; intimation, warning; formal declaration of intention to end agreement or employment at specified time; short published review of new play, book, etc. ● *verb* (**-cing**) (often + *that, how*, etc.) perceive, observe. □ **noticeable** *adjective*; **noticeably** *adverb*.

notifiable /ˈnəʊtɪfaɪəb(ə)l/ *adjective* (esp. of disease) that must be notified to authorities.

notify /ˈnəʊtɪfaɪ/ *verb* (**-ies**, **-ied**) (often + *of, that*) inform, give notice to (person); make known. □ **notification** *noun*.

notion /ˈnəʊʃ(ə)n/ *noun* concept, idea; opinion; vague understanding; intention.

notional *adjective* hypothetical, imaginary. □ **notionally** *adverb*.

notorious /nəʊˈtɔːrɪəs/ *adjective* well known, esp. unfavourably. □ **notoriety** /-təˈraɪətɪ/ *noun*; **notoriously** *adverb*.

notwithstanding /nɒtwɪðˈstændɪŋ/ ● *preposition* in spite of. ● *adverb* nevertheless.

nougat /ˈnuːgɑː/ *noun* sweet made from nuts, egg white, and sugar or honey.

nought /nɔːt/ *noun* digit 0; cipher; *poetical or archaic* nothing.

noun /naʊn/ *noun* word used to name person or thing (see panel).

nourish /ˈnʌrɪʃ/ *verb* sustain with food; foster, cherish (feeling etc.). □ **nourishing** *adjective*.

nourishment *noun* sustenance, food.

nous /naʊs/ *noun colloquial* common sense.

Nov. *abbreviation* November.

nova /ˈnəʊvə/ *noun* (*plural* **novae** /-viː/ or **-s**) star showing sudden burst of brightness and then subsiding.

novel /ˈnɒv(ə)l/ ● *noun* fictitious prose story of book length. ● *adjective* of new kind or nature.

novelette /nɒvəˈlet/ *noun* short (esp. romantic) novel.

novelist /ˈnɒvəlɪst/ *noun* writer of novels.

novella /nəˈvelə/ *noun* (*plural* **-s**) short novel or narrative story.

novelty /ˈnɒvəltɪ/ *noun* (*plural* **-ies**) newness; new thing or occurrence; small toy etc.

November /nəʊˈvembə/ *noun* eleventh month of year.

novena /nəˈviːnə/ *noun RC Church* special prayers or services on 9 successive days.

novice /ˈnɒvɪs/ *noun* probationary member of religious order; beginner.

noviciate /nəˈvɪʃɪət/ *noun* (also **novitiate**) period of being a novice; religious novice; novices' quarters.

now ● *adverb* at present or mentioned time; immediately; by this time: in the immediate past. ● *conjunction* (often + *that*) because. ● *noun* this time; the present. □ **now and again** or **then** occasionally.

nowadays /'naʊədeɪz/ ● *adverb* at present time or age. ● *noun* the present time.

nowhere /'nəʊweə/ ● *adverb* in or to no place. ● *pronoun* no place.

nowt *noun colloquial or dialect* nothing.

noxious /'nɒkʃəs/ *adjective* harmful, unwholesome.

nozzle /'nɒz(ə)l/ *noun* spout on hose etc.

nr. *abbreviation* near.

NSPCC *abbreviation* National Society for Prevention of Cruelty to Children.

NSW *abbreviation* New South Wales.

NT *abbreviation* New Testament; Northern Territory (of Australia); National Trust.

nuance /'nju:ɑ̃s/ *noun* subtle shade of meaning, feeling, colour, etc.

nub *noun* point or gist (of matter or story).

nubile /'nju:baɪl/ *adjective* (of woman) marriageable, sexually attractive. □ **nubility** *noun.*

nuclear /'nju:klɪə/ *adjective* of, relating to, or constituting a nucleus; using nuclear energy. □ **nuclear energy** energy obtained by nuclear fission or fusion; **nuclear family** couple and their child(ren); **nuclear fission** nuclear reaction in which heavy nucleus splits with release of energy; **nuclear fuel** source of nuclear energy; **nuclear fusion** nuclear reaction in which nuclei of low atomic number fuse with release of energy; **nuclear physics** physics of atomic nuclei; **nuclear power** power derived from nuclear energy, country that has nuclear weapons.

nucleic acid /nju:'kli:ɪk/ *noun* either of two complex organic molecules (DNA and RNA) present in all living cells.

nucleon /'nju:klɪɒn/ *noun* proton or neutron.

nucleus /'nju:klɪəs/ *noun* (*plural* **nuclei** /-lɪaɪ/) central part or thing round which others collect; kernel; initial part; central core of atom; part of cell containing genetic material.

nude /nju:d/ ● *adjective* naked, unclothed. ● *noun* painting etc. of nude human figure; nude person. □ **in the nude** naked. □ **nudity** *noun.*

Noun

A noun is the name of a person or thing. There are four kinds:

1 common nouns (the words for articles and creatures), e.g.

shoe	in	*The red shoe was left on the shelf.*
box	in	*The large box stood in the corner.*
plant	in	*The plant grew to two metres.*
horse	in	*A horse and rider galloped by.*

2 proper nouns (the names of people, places, ships, institutions, and animals, which always begin with a capital letter), e.g.

Jane	*USS Enterprise*	*Bambi*
London	*Grand Hotel*	

3 abstract nouns (the words for qualities, things we cannot see or touch, and things which have no physical reality), e.g.

truth	*absence*
explanation	*warmth*

4 collective nouns (the words for groups of things), e.g.

committee	*squad*	*the Cabinet*
herd	*swarm*	*the clergy*
majority	*team*	*the public*

---

nudge ● *verb* (**-ging**) prod gently with elbow to draw attention; push gradually. ● *noun* gentle push.

nudist /'nju:dɪst/ *noun* person who advocates or practises going unclothed. □ **nudism** *noun*.

nugatory /'nju:gətərɪ/ *adjective* futile, trifling; inoperative, not valid.

nugget /'nʌgɪt/ *noun* lump of gold etc., as found in earth; lump of anything.

nuisance /'nju:s(ə)ns/ *noun* person, thing, or circumstance causing annoyance.

null *adjective* (esp. **null and void**) invalid; non-existent; expressionless. □ **nullity** *noun*.

nullify /'nʌlɪfaɪ/ *verb* (**-ies**, **-ied**) neutralize; invalidate. □ **nullification** *noun*.

numb /nʌm/ ● *adjective* deprived of feeling; paralysed. ● *verb* make numb; stupefy, paralyse. □ **numbness** *noun*.

number /'nʌmbə/ ● *noun* arithmetical value representing a quantity; word, symbol, or figure representing this; total count or aggregate; numerical reckoning; quantity, amount; person or thing having place in a series, esp. single issue of magazine, item in programme, etc. ● *verb* include; assign number(s) to; amount to specified number. □ **number one** *colloquial* oneself; **number plate** plate bearing number esp. of motor vehicle.

■ **Usage** The phrase *a number of* is normally used with a plural verb, as in *a number of problems remain*.

numberless *adjective* innumerable.

numeral /'nju:mər(ə)l/ ● *noun* symbol or group of symbols denoting a number. ● *adjective* of or denoting a number.

numerate /'nju:mərət/ *adjective* familiar with basic principles of mathematics. □ **numeracy** *noun*.

numeration /nju:mə'reɪʃ(ə)n/ *noun* process of numbering; calculation.

numerator /'nju:məreɪtə/ *noun* number above line in vulgar fraction.

numerical /nju:'merɪk(ə)l/ *adjective* of or relating to number(s). □ **numerically** *adverb*.

numerology /nju:mə'rɒlədʒɪ/ *noun* study of supposed occult significance of numbers.

numerous /'nju:mərəs/ *adjective* many; consisting of many.

numinous /'nju:mɪnəs/ *adjective* indicating presence of a god; awe-inspiring.

numismatic /nju:mɪz'mætɪk/ *adjective* of or relating to coins or medals.

numismatics *plural noun* (usually treated as *singular*) study of coins and medals. □ **numismatist** /-'mɪzmətɪst/ *noun*.

nun *noun* member of community of women living under religious vows.

nuncio /'nʌnsɪəʊ/ *noun* (*plural* **-s**) papal ambassador.

nunnery *noun* (*plural* **-ies**) religious house of nuns.

nuptial /'nʌpʃ(ə)l/ ● *adjective* of marriage or weddings. ● *noun* (usually in *plural*) wedding.

nurse /nɜ:s/ ● *noun* person trained to care for sick and help doctors or dentists; nursemaid. ● *verb* (**-sing**) work as nurse; attend to (sick person); feed or be fed at breast; hold or treat carefully; foster; harbour. □ **nursing home** private hospital or home.

nursemaid *noun* woman in charge of child(ren).

nursery /'nɜ:sərɪ/ *noun* (*plural* **-ies**) room or place equipped for young children; place where plants are reared for sale. □ **nurseryman** grower of plants for sale; **nursery rhyme** traditional song or rhyme for young children; **nursery school** school for children between ages of 3 and 5.

nurture /'nɜ:tʃə/ ● *noun* bringing up, fostering care; nourishment. ● *verb* (**-ring**) bring up, rear.

nut *noun* fruit consisting of hard shell or pod around edible kernel or seeds; this kernel; small usually hexagonal flat piece of metal with threaded hole through it for screwing on end of bolt to secure it; *slang* head; *slang* crazy person; small lump (of coal etc.). □ **nutcase** *slang* crazy person; **nutcracker** (usually in *plural*) device for cracking nuts; **nuthatch** small bird climbing up and down tree trunks; **nuts** *slang* crazy.

nutmeg /'nʌtmeg/ *noun* hard aromatic seed used as spice etc.; E. Indian tree bearing this.

nutria /'nju:trɪə/ *noun* coypu fur.

nutrient /'nju:trɪənt/ • *noun* substance providing essential nourishment. • *adjective* serving as or providing nourishment.

nutriment /'nju:trɪmənt/ *noun* nourishing food.

nutrition /nju:'trɪʃ(ə)n/ *noun* food, nourishment. □ **nutritional** *adjective*; **nutritionist** *noun*.

nutritious /nju:'trɪʃəs/ *adjective* efficient as food.

nutritive /'nju:trɪtɪv/ *adjective* of nutrition; nutritious.

nutshell *noun* hard covering of nut. □ **in a nutshell** in few words.

nutter *noun slang* crazy person.

nutty *adjective* (**-ier**, **-iest**) full of nuts; tasting like nuts; *slang* crazy.

nux vomica /nʌks 'vɒmɪkə/ *noun* E. Indian tree; its seeds, containing strychnine.

nuzzle /'nʌz(ə)l/ *verb* (**-ling**) prod or rub gently with nose; nestle, lie snug.

NW *abbreviation* north-west(ern).

NY *abbreviation US* New York.

nylon /'naɪlɒn/ *noun* strong light synthetic fibre; nylon fabric; (in *plural*) stockings of nylon.

nymph /nɪmf/ *noun* mythological semi-divine female spirit associated with rivers, woods, etc.; immature form of some insects.

nymphomania /nɪmfə'meɪnɪə/ *noun* excessive sexual desire in a woman. □ **nymphomaniac** *noun & adjective*.

NZ *abbreviation* New Zealand.

Oo

O[1] □ **O level** *historical* ordinary level in GCE exam.

O[2] /əʊ/ *interjection* = OH; *used before name in exclamation.*

oaf *noun* (*plural* **-s**) awkward lout. □ **oafish** *adjective*; **oafishly** *adverb*; **oafishness** *noun*.

oak *noun* acorn-bearing hardwood tree with lobed leaves; its wood. □ **oak-apple**, **-gall** abnormal growth produced on oak trees by insects.

oakum /'əʊkəm/ *noun* loose fibre got by picking old rope to pieces.

OAP *abbreviation* old-age pensioner.

oar /ɔ:/ *noun* pole with blade used to propel boat by leverage against water; rower.

oarsman /'ɔ:zmən/ *noun* (*feminine* **-woman**) rower. □ **oarsmanship** *noun*.

oasis /əʊ'eɪsɪs/ *noun* (*plural* **oases** /-si:z/) fertile spot in desert.

oast *noun* hop-drying kiln. □ **oast house** building containing this.

oat *noun* cereal plant grown as food; (in *plural*) grain of this; tall grass resembling this. □ **oatcake** thin oatmeal biscuit; **oatmeal** meal ground from oats, greyish-fawn colour; **sow one's wild oats** indulge in youthful follies before becoming steady. □ **oaten** *adjective*.

oath /əʊθ/ *noun* (*plural* **-s**) /əʊðz/) solemn declaration naming God etc. as witness; curse. □ **on**, **under oath** having sworn solemn oath.

ob. *abbreviation* died (*obiit*).

obbligato /ɒblɪ'gɑ:təʊ/ *noun* (*plural* **-s**) *Music* accompaniment forming integral part of a composition.

obdurate /'ɒbdjʊrət/ *adjective* stubborn; hardened. □ **obduracy** *noun*.

OBE *abbreviation* Officer of the Order of the British Empire.

obedient /əʊ'bi:dɪənt/ *adjective* obeying or ready to obey; submissive to another's will. □ **obedience** *noun*; **obediently** *adverb*.

obeisance /əʊ'beɪs(ə)ns/ *noun* gesture expressing submission, respect, etc.; homage. □ **obeisant** *adjective*.

obelisk /'ɒbəlɪsk/ *noun* tapering usually 4-sided stone pillar.

obelus /'ɒbələs/ *noun* (*plural* **obeli** /-laɪ/) dagger-shaped mark of reference (†).

obese /əʊ'bi:s/ *adjective* very fat. □ **obesity** *noun*.

obey /əʊ'beɪ/ *verb* carry out command of; do what one is told to do.

obfuscate /'ɒbfʌskeɪt/ *verb* (**-ting**) obscure, confuse; bewilder. □ **obfuscation** *noun*.

obituary /ə'bɪtjʊərɪ/ ● *noun* (*plural* **-ies**) notice of death(s); brief biography of deceased person. ● *adjective* of or serving as obituary.

object ● *noun* /'ɒbdʒɪkt/ material thing; person or thing to which action or feeling is directed; thing sought or aimed at; word or phrase representing person or thing affected by action of verb (see panel). ● *verb* /əb'dʒekt/ (often + *to, against*) express opposition, disapproval, or reluctance; protest. □ **no object** not an important factor. □ **objector** *noun*.

objectify /əb'dʒektɪfaɪ/ *verb* (**-ies, -ied**) present as an object, embody.

objection /əb'dʒekʃ(ə)n/ *noun* expression of disapproval or opposition; objecting; adverse reason or statement.

objectionable *adjective* unpleasant, offensive; open to objection. □ **objectionably** *adverb*.

objective /əb'dʒektɪv/ ● *adjective* external to the mind; actually existing; dealing with outward things uncoloured by opinions or feelings; *Grammar* (of case or word) in form appropriate to object. ● *noun* object or purpose; *Grammar* objective case. □ **objectively** *adverb*; **objectivity** /ɒbdʒek'tɪvɪtɪ/ *noun*.

objet d'art /ɒbʒeɪ 'dɑ:/ *noun* (*plural* ***objets d'art*** same pronunciation) small decorative object. [French]

oblate /'ɒbleɪt/ *adjective* (of spheroid) flattened at poles.

oblation /əʊ'bleɪʃ(ə)n/ *noun* thing offered to a divine being.

obligate /'ɒblɪgeɪt/ *verb* (**-ting**) bind (person) legally or morally.

obligation /ɒblɪ'geɪʃ(ə)n/ *noun* compelling power of law, duty, etc.; duty; binding agreement; indebtedness for service or benefit.

obligatory /ə'blɪgətərɪ/ *adjective* binding, compulsory. □ **obligatorily** *adverb*.

oblige /ə'blaɪdʒ/ *verb* (**-ging**) compel, require; be binding on; do (person) small favour; (as **obliged** *adjective*) grateful.

obliging *adjective* helpful, accommodating. □ **obligingly** *adverb*.

oblique /ə'bli:k/ ● *adjective* slanting; at an angle; not going straight to the point, indirect; *Grammar* (of case) other than nominative or vocative. ● *noun* oblique stroke. □ **obliquely** *adverb*; **obliqueness** *noun*; **obliquity** /ə'blɪkwɪtɪ/ *noun*.

Object

There are two types of object:

1 A direct object is a person or thing directly affected by the verb and can usually be found by asking the question 'whom or what?' after the verb, e.g.

The electors chose Mr Smith.
Charles wrote a letter.

2 An indirect object is usually a person or thing receiving something from the subject of the verb, e.g.

He gave me the pen.
(*me* is the indirect object, and *the pen* is the direct object.)

I sent my bank a letter.
(*my bank* is the indirect object, and *a letter* is the direct object.)

Sentences containing an indirect object usually contain a direct object as well, but not always, e.g.

Pay me.

'Object' on its own usually means a direct object.

obliterate /ə'blɪtəreɪt/ *verb* (**-ting**) blot out, leave no clear trace of. □ **obliteration** *noun.*

oblivion /ə'blɪvɪən/ *noun* state of having or being forgotten.

oblivious /ə'blɪvɪəs/ *adjective* unaware or unconscious. □ **obliviously** *adverb*; **obliviousness** *noun.*

oblong /'ɒblɒŋ/ ● *adjective* rectangular with adjacent sides unequal. ● *noun* oblong figure or object.

obloquy /'ɒbləkwɪ/ *noun* abuse, being ill spoken of.

obnoxious /əb'nɒkʃəs/ *adjective* offensive, objectionable. □ **obnoxiously** *adverb*; **obnoxiousness** *noun.*

oboe /'əʊbəʊ/ *noun* double-reeded woodwind instrument. □ **oboist** *noun.*

obscene /əb'si:n/ *adjective* offensively indecent; *colloquial* highly offensive; *Law* (of publication) tending to deprave and corrupt. □ **obscenely** *adverb*; **obscenity** /-'sen-/ *noun* (*plural* **-ies**).

obscure /əb'skjʊə/ ● *adjective* not clearly expressed or easily understood; unexplained; dark, indistinct; hidden, undistinguished. ● *verb* (**-ring**) make obscure or invisible. □ **obscurity** *noun.*

obsequies /'ɒbsɪkwɪz/ *plural noun* funeral.

obsequious /əb'si:kwɪəs/ *adjective* fawning, servile. □ **obsequiously** *adverb*; **obsequiousness** *noun.*

observance /əb'zɜ:v(ə)ns/ *noun* keeping or performance of law, duty, etc.; rite, ceremonial act.

observant *adjective* good at observing. □ **observantly** *adverb.*

observation /ɒbzə'veɪʃ(ə)n/ *noun* observing, being observed; comment, remark; power of perception. □ **observational** *adjective.*

observatory /əb'zɜ:vətərɪ/ *noun* (*plural* **-ies**) building for astronomical or other observation.

observe /əb'zɜ:v/ *verb* (**-ving**) perceive, become aware of; watch; keep (rules etc.); celebrate (rite etc.); remark; take note of scientifically. □ **observable** *adjective.*

observer *noun* person who observes; interested spectator; person attending meeting to note proceedings but without participating.

obsess /əb'ses/ *verb* fill mind of (person) all the time; preoccupy. □ **obsession** *noun*; **obsessional** *adjective*; **obsessive** *adjective*; **obsessively** *adverb*; **obsessiveness** *noun.*

obsidian /əb'sɪdɪən/ *noun* dark glassy rock formed from lava.

obsolescent /ɒbsə'les(ə)nt/ *adjective* becoming obsolete. □ **obsolescence** *noun.*

obsolete /'ɒbsəli:t/ *adjective* no longer used, antiquated.

obstacle /'ɒbstək(ə)l/ *noun* thing obstructing progress.

obstetrics /əb'stetrɪks/ *plural noun* (usually treated as *singular*) branch of medicine or surgery dealing with childbirth. □ **obstetric** *adjective*; **obstetrician** /ɒbstə'trɪʃ(ə)n/ *noun.*

obstinate /'ɒbstɪnət/ *adjective* stubborn, intractable. □ **obstinacy** *noun*; **obstinately** *adverb.*

obstreperous /əb'strepərəs/ *adjective* noisy, unruly. □ **obstreperously** *adverb*; **obstreperousness** *noun.*

obstruct /əb'strʌkt/ *verb* block up; make hard or impossible to pass along or through; retard or prevent progress of.

obstruction /əb'strʌkʃ(ə)n/ *noun* obstructing, being obstructed; thing that obstructs; *Sport* unlawfully obstructing another player.

obstructive *adjective* causing or meant to cause obstruction.

obtain /əb'teɪn/ *verb* acquire; get; have granted to one; be prevalent or established. □ **obtainable** *adjective.*

obtrude /əb'tru:d/ *verb* (**-ding**) (often + *on, upon*) thrust (oneself etc.) importunately forward. □ **obtrusion** *noun.*

obtrusive /əb'tru:sɪv/ *adjective* unpleasantly noticeable; obtruding oneself. □ **obtrusively** *adverb*; **obtrusiveness** *noun.*

obtuse /əb'tju:s/ *adjective* dull-witted; (of angle) between 90° and 180°; blunt, not sharp or pointed. □ **obtuseness** *noun.*

obverse /'ɒbvɜ:s/ *noun* counterpart, opposite; side of coin or medal that bears head or principal design; front or top side.

obviate /'ɒbvɪeɪt/ *verb* (**-ting**) get round or do away with (need, inconvenience, etc.).

obvious /ˈɒbvɪəs/ *adjective* easily seen, recognized, or understood. □ **obviously** *adverb*; **obviousness** *noun*.

OC *abbreviation* Officer Commanding.

ocarina /ɒkəˈri:nə/ *noun* egg-shaped musical wind instrument.

occasion /əˈkeɪʒ(ə)n/ ● *noun* special event or happening; time of this; reason, need; suitable juncture, opportunity. ● *verb* cause, esp. incidentally.

occasional *adjective* happening irregularly and infrequently; made or meant for, acting on, etc. special occasion(s). □ **occasional table** small table for use as required. □ **occasionally** *adverb*.

Occident /ˈɒksɪd(ə)nt/ *noun* (**the Occident**) West, esp. Europe and America as distinct from the Orient. □ **occidental** /-ˈden-/ *adjective*.

occiput /ˈɒksɪpʌt/ *noun* back of head. □ **occipital** /-ˈsɪpɪt-/ *adjective*.

occlude /əˈklu:d/ *verb* (**-ding**) stop up; obstruct; *Chemistry* absorb (gases); (as **occluded** *adjective*) *Meteorology* (of frontal system) formed when cold front overtakes warm front, raising warm air. □ **occlusion** *noun*.

occult /ɒˈkʌlt, ˈɒkʌlt/ *adjective* involving the supernatural, mystical; esoteric. □ **the occult** occult phenomena generally.

occupant /ˈɒkjʊpənt/ *noun* person occupying dwelling, office, or position. □ **occupancy** *noun* (*plural* **-ies**).

occupation /ɒkjʊˈpeɪʃ(ə)n/ *noun* profession or employment; pastime; occupying or being occupied, esp. by armed forces of another country.

occupational *adjective* of or connected with one's occupation. □ **occupational disease**, **hazard** one to which a particular occupation renders someone especially liable; **occupational therapy** programme of mental or physical activity to assist recovery from disease or injury.

occupier /ˈɒkjʊpaɪə/ *noun* person living in house etc. as owner or tenant.

occupy /ˈɒkjʊpaɪ/ *verb* (**-ies**, **-ied**) live in; be tenant of; take up, fill (space, time, or place); take military possession of; place oneself in (building etc.) without authority as protest etc.; hold (office); keep busy.

occur /əˈkɜ:/ *verb* (**-rr-**) take place, happen; be met with or found in some place or conditions; (+ *to*) come into one's mind.

occurrence /əˈkʌrəns/ *noun* happening; incident.

ocean /ˈəʊʃ(ə)n/ *noun* large expanse of sea, esp. one of the 5 named divisions of this, e.g. Atlantic Ocean; (often in *plural*) *colloquial* immense expanse or quantity. □ **oceanic** /əʊʃɪˈænɪk/ *adjective*.

oceanography /əʊʃəˈnɒgrəfɪ/ *noun* study of the oceans. □ **oceanographer** *noun*.

ocelot /ˈɒsɪlɒt/ *noun* S. American leopard-like cat.

ochre /ˈəʊkə/ *noun* (*US* **ocher**) earth used as pigment; pale brownish-yellow colour. □ **ochreous** /ˈəʊkrɪəs/ *adjective*.

o'clock /əˈklɒk/ *adverb* of the clock (used to specify hour).

Oct. *abbreviation* October.

octa- *combining form* (also **oct-** before vowel) eight.

octagon /ˈɒktəgən/ *noun* plane figure with 8 sides and angles. □ **octagonal** /-ˈtæg-/ *adjective*.

octahedron /ɒktəˈhi:drən/ *noun* (*plural* **-s**) solid figure contained by 8 (esp. triangular) plane faces. □ **octahedral** *adjective*.

octane /ˈɒkteɪn/ *noun* colourless inflammable hydrocarbon occurring in petrol. □ **high-octane** (of fuel used in internal-combustion engines) not detonating rapidly during power stroke; **octane number**, **rating** figure indicating antiknock properties of fuel.

octave /ˈɒktɪv/ *noun* *Music* interval of 8 diatonic degrees between two notes, 8 notes occupying this interval, each of two notes at this interval's extremes; 8-line stanza.

octavo /ɒkˈteɪvəʊ/ *noun* (*plural* **-s**) size of book or page with sheets folded into 8 leaves.

octet /ɒkˈtet/ *noun* musical composition for 8 performers; the performers; any group of 8.

octo- *combining form* (also **oct-** before vowel) eight.

October /ɒkˈtəʊbə/ *noun* tenth month of year.

octogenarian /ɒktəʊdʒɪ'neərɪən/ *noun* person from 80 to 89 years old.

octopus /'ɒktəpəs/ *noun* (*plural* **-puses**) mollusc with 8 suckered tentacles.

ocular /'ɒkjʊlə/ *adjective* of, for, or by the eyes; visual.

oculist /'ɒkjʊlɪst/ *noun* specialist in treatment of eyes.

OD /əʊ'di:/ *slang* ● *noun* drug overdose. ● *verb* (**OD's**, **OD'd**, **Od'ing**) take overdose.

odd *adjective* extraordinary, strange; (of job etc.) occasional, casual; not normally considered, unconnected; (of numbers) not divisible by 2; left over, detached from set etc.; (added to weight, sum, etc.) rather more than. □ **oddball** *colloquial* eccentric person. □ **oddly** *adverb*; **oddness** *noun*.

oddity /'ɒdɪtɪ/ *noun* (*plural* **-ies**) strange person, thing, or occurrence; peculiar trait; strangeness.

oddment *noun* odd article; something left over.

odds *plural noun* ratio between amounts staked by parties to a bet; chances in favour of or against result; balance of advantage; difference giving an advantage. □ **at odds** (often + *with*) in conflict; **odds and ends** remnants, stray articles; **odds-on** state when success is more likely than failure; **over the odds** above general price etc.

ode *noun* lyric poem of exalted style and tone.

odious /'əʊdɪəs/ *adjective* hateful, repulsive. □ **odiously** *adverb*; **odiousness** *noun*.

odium /'əʊdɪəm/ *noun* general dislike or disapproval.

odor *US* = ODOUR.

odoriferous /əʊdə'rɪfərəs/ *adjective* diffusing (usually pleasant) odours.

odour /'əʊdə/ *noun* (*US* **odor**) smell or fragrance; favour or repute. □ **odorous** *adjective*; **odourless** *adjective*.

odyssey /'ɒdɪsɪ/ *noun* (*plural* **-s**) long adventurous journey.

OED *abbreviation* Oxford English Dictionary.

Oedipus complex /'i:dɪpəs/ *noun* attraction of child to parent of opposite sex (esp. son to mother). □ **Oedipal** *adjective*.

oesophagus /i:'sɒfəgəs/ *noun* (*US* **esophagus**) (*plural* **-gi** /-dʒaɪ/ or **-guses**) passage from mouth to stomach, gullet.

oestrogen /'i:strədʒ(ə)n/ *noun* (*US* **estrogen**) sex hormone developing and maintaining female physical characteristics; this produced artificially for medical use.

oeuvre /'ɜ:vr/ *noun* works of creative artist considered collectively. [French]

of /ɒv/ *preposition* belonging to, from; concerning; out of; among; relating to; *US* (of time in relation to following hour) to. □ **be of** possess, give rise to; **of late** recently; **of old** formerly.

off ● *adverb* away, at or to distance; out of position; loose, separate, gone; so as to be rid of; discontinued, stopped; not available on menu. ● *preposition* from; not on. ● *adjective* further; far; right-hand; *colloquial* annoying; not acceptable; *Cricket* of, in, or into half of field which batsman faces. ● *noun* start of race; *Cricket* the off side. □ **off and on** now and then; **offbeat** unconventional, *Music* not coinciding with beat; **off chance** remote possibility; **off colour** unwell, *US* rather indecent; **offhand** without preparation, casual, curt; **off-licence** shop selling alcoholic drink for consumption away from premises; **offline** *Computing adjective* not online, *adverb* with delay between data production and its processing; **off-peak** (of electricity, traffic, etc.) used or for use at times of lesser demand; **offprint** reprint of part of publication; **offshoot** side-shoot or branch, derivative; **offside** (of player in field game) in position where he or she may not play the ball; **off the wall** *slang* crazy, absurd; **off white** white with grey or yellowish tinge.

offal /'ɒf(ə)l/ *noun* edible organs of animal, esp. heart, liver, etc.; refuse, scraps.

offence /ə'fens/ *noun* (*US* **offense**) illegal act; transgression; upsetting of person's feelings, insult; aggressive action.

offend /ə'fend/ *verb* cause offence to, upset; displease, anger; (often + *against*) do wrong. □ **offender** *noun*; **offending** *adjective*.

offense *US* = OFFENCE.

offensive /ə'fensɪv/ ● *adjective* causing offence; insulting; disgusting; aggressive; (of weapon) for attacking. ● *noun* aggressive attitude, action, or campaign. □ **offensively** *adverb*; **offensiveness** *noun*.

offer /'ɒfə/ ● *verb* present for acceptance, refusal, or consideration; (+ *to do*) express readiness, show intention; attempt; present by way of sacrifice. ● *noun* expression of readiness to do or give if desired, or buy or sell; amount offered; proposal, esp. of marriage; bid. □ **on offer** for sale at certain (esp. reduced) price.

offering *noun* contribution, gift; thing offered.

offertory /'ɒfətərɪ/ *noun* (*plural* **-ies**) offering of bread and wine at Eucharist; collection of money at religious service.

office /'ɒfɪs/ *noun* room or building where administrative or clerical work is done; place for transacting business; department or local branch, esp. for specified purpose; position with duties attached to it; tenure of official position; duty, task, function; (usually in *plural*) piece of kindness, service; authorized form of worship.

officer /'ɒfɪsə/ *noun* person holding position of authority or trust, esp. one with commission in armed forces; policeman or policewoman; president, treasurer, etc. of society etc.

official /ə'fɪʃ(ə)l/ ● *adjective* of office or its tenure; characteristic of people in office; properly authorized. ● *noun* person holding office or engaged in official duties. □ **official secrets** confidential information involving national security. □ **officialdom** *noun*; **officially** *adverb*.

officialese /əfɪʃə'li:z/ *noun derogatory* officials' jargon.

officiate /ə'fɪʃɪeɪt/ *verb* (**-ting**) act in official capacity; conduct religious service.

officious /ə'fɪʃəs/ *adjective* domineering; intrusive in correcting etc. □ **officiously** *adverb*; **officiousness** *noun*.

offing /'ɒfɪŋ/ *noun* □ **in the offing** at hand, ready or likely to happen etc.

offset ● *noun* side-shoot of plant used for propagation; compensation; sloping ledge. ● *verb* (**-setting**; *past & past participle* **-set**) counterbalance, compensate.

offspring *noun* (*plural* same) person's child, children, or descendants; animal's young or descendants.

oft *adverb archaic* often.

often /'ɒf(ə)n/ *adverb* (**oftener**, **oftenest**) frequently; many times; at short intervals; in many instances.

ogee /'əʊdʒi:/ *noun* S-shaped curve or moulding.

ogive /'əʊdʒaɪv/ *noun* pointed arch; diagonal rib of vault.

ogle /'əʊg(ə)l/ ● *verb* (**-ling**) look lecherously or flirtatiously (at). ● *noun* flirtatious glance.

ogre /'əʊgə/ *noun* (*feminine* **ogress** /-grɪs/) man-eating giant. □ **ogreish**, **ogrish** /'əʊgərɪʃ/ *adjective*.

oh /əʊ/ *interjection* (also **O**) *expressing surprise, pain, etc.*

ohm /əʊm/ *noun* SI unit of electrical resistance.

OHMS *abbreviation* On Her or His Majesty's Service.

oho /əʊ'həʊ/ *interjection expressing surprise or exultation*.

OHP *abbreviation* overhead projector.

oil ● *noun* viscous usually inflammable liquid insoluble in water; petroleum. ● *verb* apply oil to, lubricate; treat with oil. □ **oilcake** compressed linseed etc. as cattle food or manure; **oilfield** district yielding mineral oil; **oil paint** paint made by mixing pigment with oil; **oil painting** use of or picture in oil paints; **oil rig** equipment for drilling an oil well; **oilskin** cloth waterproofed with oil, garment or (in *plural*) suit of it; **oil slick** patch of oil, esp. on sea; **oil well** well from which mineral oil is drawn.

oily *adjective* (**-ier**, **-iest**) of, like, covered or soaked with, oil; (of manner) fawning.

ointment /'ɔɪntmənt/ *noun* smooth greasy healing or cosmetic preparation for skin.

OK /əʊ'keɪ/ (also **okay**) *colloquial* ● *adjective & adverb* all right. ● *noun* (*plural* **OKs**) approval, sanction. ● *verb* (**OK's**, **OK'd**, **OK'ing**) approve, sanction.

okapi /əʊ'kɑ:pɪ/ *noun* (*plural* same or **-s**) African partially striped ruminant mammal.

okay = OK.

okra /'əʊkrə/ *noun* tall originally African plant with edible seed pods.

old /əʊld/ *adjective* (**-er, -est**) advanced in age; not young or near its beginning; worn, dilapidated, or shabby from age; practised, inveterate; dating from far back; long established; former; *colloquial: used to indicate affection.* □ **old age** later part of normal lifetime; **old-age pension** state retirement pension; **old-age pensioner** person receiving this; **Old Bill** *slang* the police; **old boy** former male pupil of school, *colloquial* elderly man; **old-fashioned** in or according to fashion no longer current, antiquated; **old girl** former female pupil of school, *colloquial* elderly woman; **Old Glory** *US* Stars and Stripes; **old guard** original, past, or conservative members of group; **old hand** experienced or practised person; **old hat** *colloquial* hackneyed; **old maid** *derogatory* elderly unmarried woman, prim and fussy person; **old man** *colloquial* one's father, husband, or employer etc.; **old man's beard** wild clematis; **old master** great painter of former times, painting by such painter; **Old Testament** part of Bible dealing with pre-Christian times; **old wives' tale** unscientific belief; **old woman** *colloquial* one's wife or mother, fussy or timid man; **Old World** Europe, Asia, and Africa. □ **oldish** *adjective*; **oldness** *noun*.

olden *adjective archaic* old, of old.

oldie *noun colloquial* old person or thing.

oleaginous /əʊlɪ'ædʒɪnəs/ *adjective* like or producing oil; oily.

oleander /əʊlɪ'ændə/ *noun* evergreen flowering Mediterranean shrub.

olfactory /ɒl'fæktərɪ/ *adjective* of the sense of smell.

oligarch /'ɒlɪgɑ:k/ *noun* member of oligarchy.

oligarchy /'ɒlɪgɑ:kɪ/ *noun* (*plural* **-ies**) government by small group of people; members of such government; state so governed. □ **oligarchic(al)** /-'gɑ:k-/ *adjective*.

olive /'ɒlɪv/ ● *noun* oval hard-stoned fruit yielding oil; tree bearing this; dull yellowish green. ● *adjective* olive-green; (of complexion) yellowish-brown. □ **olive branch** gesture of peace or reconciliation.

Olympiad /ə'lɪmpɪæd/ *noun* period of 4 years between Olympic Games; celebration of modern Olympic Games.

Olympian /ə'lɪmpɪən/ *adjective* of Olympus; magnificent, condescending; aloof.

Olympic /ə'lɪmpɪk/ ● *adjective* of the Olympic Games. ● *plural noun* (**the Olympics**) Olympic Games. □ **Olympic Games** ancient Greek athletic festival held every 4 years, or modern international revival of this.

OM *abbreviation* Order of Merit.

ombudsman /'ɒmbʊdzmən/ *noun* official appointed to investigate complaints against public authorities.

omega /'əʊmɪgə/ *noun* last letter of Greek alphabet (Ω, ω); last of series.

omelette /'ɒmlɪt/ *noun* beaten eggs fried and often folded over filling.

omen /'əʊmən/ *noun* event supposedly warning of good or evil; prophetic significance.

ominous /'ɒmɪnəs/ *adjective* threatening; inauspicious. □ **ominously** *adverb*.

omit /əʊ'mɪt/ *verb* (**-tt-**) leave out; not include; leave undone; (+ *to do*) neglect. □ **omission** /əʊ'mɪʃ(ə)n/ *noun*.

omni- *combining form* all.

omnibus /'ɒmnɪbəs/ ● *noun formal* bus; volume containing several novels etc. previously published separately. ● *adjective* serving several purposes at once; comprising several items.

omnipotent /ɒm'nɪpət(ə)nt/ *adjective* all-powerful. □ **omnipotence** *noun*.

omnipresent /ɒmnɪ'prez(ə)nt/ *adjective* present everywhere. □ **omnipresence** *noun*.

omniscient /ɒm'nɪsɪənt/ *adjective* knowing everything. □ **omniscience** *noun*.

omnivorous /ɒm'nɪvərəs/ *adjective* feeding on both plant and animal material; *jocular* reading everything that comes one's way. □ **omnivore** /'ɒmnɪvɔ:/ *noun*; **omnivorousness** *noun*.

on ● *preposition* (so as to be) supported by, covering, attached to, etc.; (of time) exactly at; during; close to, in direction of; at, near, concerning, about; added to. ● *adverb* (so as to be) on something; in some direction, forward; in advance; with movement; in operation or activity; *colloquial* willing to participate, approve, bet, etc.; *colloquial* practicable, acceptable; being shown or performed. ● *adjective Cricket* of, in, or into half of field behind batsman's back. ● *noun Cricket* the on side. □ **be on about** *colloquial* discuss, esp. tiresomely; **online** directly controlled by or connected to computer; **onscreen** when being filmed; **on to** to a position on.

■ **Usage** See note at ONTO.

ONC *abbreviation* Ordinary National Certificate.

once /wʌns/ ● *adverb* on one occasion only; at some time in past; ever or at all. ● *conjunction* as soon as. ● *noun* one time or occasion. □ **at once** immediately, simultaneously; **once-over** *colloquial* rapid inspection.

oncology /ɒŋ'kɒlədʒɪ/ *noun* study of tumours.

oncoming *adjective* approaching from the front.

OND *abbreviation* Ordinary National Diploma.

one /wʌn/ ● *adjective* single and integral in number; only such; without others; identical; forming a unity. ● *noun* lowest cardinal numeral; thing numbered with it; unit, unity; single thing, person, or example; *colloquial* drink. ● *pronoun* any person. □ **one-armed bandit** *colloquial* fruit machine with long handle; **one-horse** *colloquial* small, poorly equipped; **one-man** involving or operated by one person only; **one-off** made as the only one, not repeated; **one-sided** unfair, partial; **one-way** allowing movement etc. in one direction only.

oneness *noun* singleness; uniqueness; agreement, sameness.

onerous /'əʊnərəs/ *adjective* burdensome. □ **onerousness** *noun*.

oneself *pronoun: emphatic & reflexive form of* ONE. □ **be oneself** act in one's natural manner.

ongoing *adjective* continuing, in progress.

onion /'ʌnjən/ *noun* vegetable with edible bulb of pungent smell and flavour.

onlooker *noun* spectator. □ **onlooking** *adjective*.

only /'əʊnlɪ/ ● *adverb* solely, merely, exclusively. ● *adjective* existing alone of its or their kind. ● *conjunction colloquial* except that; but then. □ **if only** even if for no other reason than, I wish that.

o.n.o. *abbreviation* or near offer.

onomatopoeia /ɒnəmætə'pi:ə/ *noun* formation of word from sound associated with thing named, e.g. *whizz, cuckoo*. □ **onomatopoeic** *adjective*.

onset *noun* attack; impetuous beginning.

onslaught /'ɒnslɔ:t/ *noun* fierce attack.

onto *preposition* = ON TO.

■ **Usage** *Onto* is much used but is still not as widely accepted as *into*. It is, however, useful in distinguishing between, e.g., *We drove on to the beach* (i.e. towards it) and *we drove onto the beach* (i.e. into contact with it).

ontology /ɒn'tɒlədʒɪ/ *noun* branch of metaphysics concerned with the nature of being. □ **ontological** /-tə'lɒdʒ-/ *adjective*; **ontologically** /-tə'lɒdʒ-/ *adverb*; **ontologist** *noun*.

onus /'əʊnəs/ *noun* (*plural* **onuses**) burden, duty, responsibility.

onward /'ɒnwəd/ ● *adverb* (also **onwards**) advancing; into the future. ● *adjective* forward, advancing.

onyx /'ɒnɪks/ *noun* semiprecious variety of agate with coloured layers.

oodles /'u:d(ə)lz/ *plural noun colloquial* very great amount.

ooh /u:/ *interjection expressing surprised pleasure, pain, excitement, etc.*

oolite /'əʊəlaɪt/ *noun* granular limestone. □ **oolitic** /-'lɪt-/ *adjective*.

oomph /ʊmf/ *noun slang* energy, enthusiasm; attractiveness, esp. sex appeal.

ooze[1] ● *verb* (**-zing**) trickle or leak slowly out; (of substance) exude fluid; (often + *with*) give off (a feeling) freely. ● *noun* sluggish flow. □ **oozy** *adjective*.

ooze[2] *noun* wet mud. □ **oozy** *adjective*.

op *noun colloquial* operation.

op. *abbreviation* opus.

opacity /əʊ'pæsɪtɪ/ *noun* opaqueness.

opal /'əʊp(ə)l/ *noun* semiprecious milk-white or bluish stone with iridescent reflections.

opalescent /əʊpə'les(ə)nt/ *adjective* iridescent. □ **opalescence** *noun.*

opaline /'əʊpəlaɪn/ *adjective* opal-like; opalescent.

opaque /əʊ'peɪk/ *adjective* **(-r, -st)** not transmitting light; impenetrable to sight; unintelligible; stupid. □ **opaquely** *adverb*; **opaqueness** *noun.*

op. cit. *abbreviation* in the work already quoted (*opere citato*).

OPEC /'əʊpek/ *abbreviation* Organization of Petroleum Exporting Countries.

open /'əʊpən/ ● *adjective* not closed, locked, or blocked up; not covered or confined; exposed; (of goal etc.) undefended; undisguised, public; unfolded, spread out; (of fabric) with gaps; frank; open-minded; accessible to visitors or customers; (of meeting, competition, etc.) not restricted; (+ *to*) willing to receive, vulnerable to. ● *verb* make or become open or more open; (+ *into* etc.) give access; establish, set going; start; ceremonially declare open. ● *noun* **(the open)** open air; open competition etc. □ **open air** *noun* outdoors; **open-air** *adjective* outdoor; **open day** day when public may visit place normally closed to them; **open-ended** with no limit or restriction; **open-handed** generous; **open-heart surgery** surgery with heart exposed and blood made to bypass it; **open house** hospitality for all visitors; **open letter** one addressed to individual and printed in newspaper etc.; **open-minded** accessible to new ideas, unprejudiced; **open-plan** (of house, office, etc.) having large undivided rooms; **open prison** one with few physical restraints on prisoners; **open question** matter on which different views are legitimate; **open sandwich** one without bread on top; **open sea** expanse of sea away from land. □ **openness** *noun.*

opener *noun* device for opening tins or bottles etc.

opening ● *noun* gap, aperture; opportunity; beginning, initial part. ● *adjective* initial, first.

openly *adverb* publicly, frankly.

opera[1] /'ɒpərə/ *noun* musical drama with sung or spoken dialogue. □ **opera-glasses** small binoculars for use in theatres etc.; **opera house** theatre for operas.

opera[2] *plural* of OPUS.

operable /'ɒpərəb(ə)l/ *adjective* that can be operated; suitable for treatment by surgical operation.

operate /'ɒpəreɪt/ *verb* **(-ting)** work, control (machine etc.); be in action; perform surgical operation(s); direct military etc. action. □ **operating theatre** room for surgical operations.

operatic /ɒpə'rætɪk/ *adjective* of or like opera.

operation *noun* action, working; performance of surgery on a patient; military manoeuvre; financial transaction. □ **operational** *adjective*; **operationally** *adverb.*

operative /'ɒpərətɪv/ ● *adjective* in operation; having principal relevance; of or by surgery. ● *noun* worker, artisan.

operator *noun* person operating machine, esp. connecting lines in telephone exchange; person engaging in business.

operetta /ɒpə'retə/ *noun* light opera.

ophidian /əʊ'fɪdɪən/ ● *noun* member of suborder of reptiles including snakes. ● *adjective* of this order.

ophthalmia /ɒf'θælmɪə/ *noun* inflammation of eye.

ophthalmic /ɒf'θælmɪk/ *adjective* of or relating to the eye and its diseases. □ **ophthalmic optician** one qualified to prescribe as well as dispense spectacles.

ophthalmology /ɒfθæl'mɒlədʒɪ/ *noun* study of the eye. □ **ophthalmologist** *noun.*

ophthalmoscope /ɒf'θælməskəʊp/ *noun* instrument for examining the eye.

opiate /'əʊpɪət/ ● *adjective* containing opium; soporific. ● *noun* drug containing opium, usually to ease pain or induce sleep; soothing influence.

opine /əʊ'paɪn/ *verb* **(-ning)** (often + *that*) express or hold as opinion.

opinion /ə'pɪnjən/ *noun* unproven belief; view held as probable; professional advice; estimation. □ **opinion poll** assessment of public opinion by questioning representative sample.

opinionated /ə'pɪnjəneɪtɪd/ *adjective* unduly confident in one's opinions.

opium /'əʊpɪəm/ *noun* drug made from juice of certain poppy, used as narcotic or sedative.

opossum /ə'pɒsəm/ *noun* tree-living American marsupial; *Australian & NZ* marsupial resembling this.

opponent /ə'pəʊnənt/ *noun* person who opposes.

opportune /'ɒpətju:n/ *adjective* well-chosen, specially favourable; (of action, event, etc.) well-timed.

opportunism /ɒpə'tju:nɪz(ə)m/ *noun* adaptation of policy to circumstances, esp. regardless of principle. □ **opportunist** *noun*; **opportunistic** /-'nɪs-/ *adjective*; **opportunistically** /-'nɪs-/ *adverb*.

opportunity /ɒpə'tju:nɪtɪ/ *noun* (*plural* **-ies**) favourable chance or opening offered by circumstances.

oppose /ə'pəʊz/ *verb* (**-sing**) set oneself against; resist; argue against; (+ *to*) place in opposition or contrast. □ **as opposed to** in contrast with. □ **opposer** *noun*.

opposite /'ɒpəzɪt/ ● *adjective* facing, on other side; (often + *to, from*) contrary; diametrically different. ● *noun* opposite thing, person, or term. ● *adverb* in opposite position. ● *preposition* opposite to. □ **opposite number** person in corresponding position in another group etc.; **the opposite sex** either sex in relation to the other.

opposition /ɒpə'zɪʃ(ə)n/ *noun* antagonism, resistance; being in conflict or disagreement; contrast; group or party of opponents; chief parliamentary party, or group of parties, opposed to party in office; act of placing opposite.

oppress /ə'pres/ *verb* govern tyrannically; treat with gross harshness or injustice; weigh down. □ **oppression** *noun*; **oppressor** *noun*.

oppressive *adjective* that oppresses; (of weather) sultry, close. □ **oppressively** *adverb*; **oppressiveness** *noun*.

opprobrious /ə'prəʊbrɪəs/ *adjective* (of language) severely scornful; abusive.

opprobrium /ə'prəʊbrɪəm/ *noun* disgrace; cause of this.

opt *verb* (usually + *for*) make choice; decide. □ **opt out (of)** choose not to take part etc. (in).

optic /'ɒptɪk/ *adjective* of eye or sight.

optical *adjective* visual; of or according to optics; aiding sight. □ **optical fibre** thin glass fibre used to carry light signals; **optical illusion** image which deceives the eye, mental misapprehension caused by this.

optician /ɒp'tɪʃ(ə)n/ *noun* maker, seller, or prescriber of spectacles, contact lenses, etc.

optics *plural noun* (treated as *singular*) science of light and vision.

optimal /'ɒptɪm(ə)l/ *adjective* best, most favourable.

optimism /'ɒptɪmɪz(ə)m/ *noun* inclination to hopefulness and confidence. □ **optimist** *noun*; **optimistic** /-'mɪs-/ *adjective*; **optimistically** /-'mɪs-/ *adverb*.

optimize /'ɒptɪmaɪz/ *verb* (also **-ise**) (**-zing** or **-sing**) make best or most effective use of. □ **optimization** *noun*.

optimum /'ɒptɪməm/ ● *noun* (*plural* **-ma**) most favourable conditions; best practical solution. ● *adjective* optimal.

option /'ɒpʃ(ə)n/ *noun* choice, choosing; right to choose; right to buy, sell, etc., on specified conditions at specified time. □ **keep, leave one's options open** not commit oneself.

optional *adjective* not obligatory. □ **optionally** *adverb*.

opulent /'ɒpjʊlənt/ *adjective* wealthy; luxurious; abundant. □ **opulence** *noun*.

opus /'əʊpəs/ *noun* (*plural* **opuses** or **opera** /'ɒpərə/) musical composition numbered as one of composer's works; any artistic work.

or *conjunction introducing alternatives*. □ **or else** otherwise, *colloquial expressing threat*.

oracle /'ɒrək(ə)l/ *noun* place at which ancient Greeks etc. consulted gods for advice or prophecy; response received there; person or thing regarded as source of wisdom etc. □ **oracular** /ɒ'rækjʊlə/ *adjective*.

oral /'ɔ:r(ə)l/ ● *adjective* spoken, verbal; by word of mouth; done or taken by mouth. ● *noun colloquial* spoken exam. □ **orally** *adverb*.

orange /'ɒrɪndʒ/ ● *noun* roundish reddish-yellow citrus fruit; its colour; tree bearing it. ● *adjective* orange-coloured.

orangeade /ɒrɪndʒ'eɪd/ *noun* drink made from or flavoured like oranges, usually fizzy.

orang-utan /ɔ:ræŋu:'tæn/ *noun* (also **orang-outang** /-u:'tæŋ/) large anthropoid ape.

oration /ɔ:'reɪʃ(ə)n/ *noun* formal or ceremonial speech.

orator /'ɒrətə/ *noun* maker of a formal speech; eloquent public speaker.

oratorio /ɒrə'tɔ:rɪəʊ/ *noun* (*plural* **-s**) semi-dramatic musical composition usually on sacred theme.

oratory /'ɒrətərɪ/ *noun* (*plural* **-ies**) art of or skill in public speaking; small private chapel. □ **oratorical** /-'tɒr-/ *adjective*.

orb *noun* globe surmounted by cross as part of coronation regalia; sphere, globe; *poetical* celestial body; *poetical* eye.

orbicular /ɔ:'bɪkjʊlə/ *adjective formal* spherical, circular.

orbit /'ɔ:bɪt/ ● *noun* curved course of planet, comet, satellite, etc.; one complete passage round another body; range or sphere of action. ● *verb* (**-t-**) go round in orbit; put into orbit. □ **orbiter** *noun*.

orbital *adjective* of orbits; (of road) passing round outside of city.

Orcadian /ɔ:'keɪdɪən/ ● *adjective* of Orkney. ● *noun* native of Orkney.

orchard /'ɔ:tʃəd/ *noun* enclosed piece of land with fruit trees.

orchestra /'ɔ:kɪstrə/ *noun* large group of instrumental performers. □ **orchestra pit** part of theatre where orchestra plays. □ **orchestral** /ɔ:'kestr(ə)l/ *adjective*.

orchestrate /'ɔ:kɪstreɪt/ *verb* (**-ting**) compose or arrange for orchestral performance; arrange (elements) for desired effect.

orchid /'ɔ:kɪd/ *noun* any of various plants, often with brilliantly coloured or grotesquely shaped flowers.

ordain /ɔ:'deɪn/ *verb* confer holy orders on; decree, order.

ordeal /ɔ:'di:l/ *noun* severe trial; painful or horrific experience.

order /'ɔ:də/ ● *noun* condition in which every part, unit, etc. is in its right place; tidiness; specified sequence; authoritative direction or instruction; state of obedience to law, authority, etc.; direction to supply something, thing(s) (to be) supplied; social class or rank; kind, sort; constitution or nature of the world, society, etc.; *Biology* grouping of animals or plants below class and above family; religious fraternity; grade of Christian ministry; any of 5 classical styles of architecture; company of people distinguished by particular honour, etc., insignia worn by its members; stated form of divine service; system of rules etc. (at meetings etc.). ● *verb* command, prescribe; command or direct (person) to specified destination; direct manufacturer, tradesman, etc. to supply; direct waiter to serve; (often as **ordered** *adjective*) put in order; (of God, fate, etc.) ordain. □ **in** or **out of order** in correct or incorrect sequence; **of** or **in the order of** approximately.

orderly ● *adjective* methodically arranged; tidy; not unruly. ● *noun* (*plural* **-ies**) soldier in attendance on officer; hospital attendant. □ **orderly room** room in barracks for company's business.

ordinal /'ɔ:dɪn(ə)l/ *noun* (in full **ordinal number**) number defining position in a series; compare CARDINAL NUMBER.

ordinance /'ɔ:dɪnəns/ *noun* decree; religious rite.

ordinand /'ɔ:dɪnænd/ *noun* candidate for ordination.

ordinary /'ɔ:dɪnərɪ/ *adjective* normal; not exceptional; commonplace. □ **ordinary level** *historical* lowest in GCE exam; **ordinary seaman** sailor of lowest rank. □ **ordinarily** *adverb*; **ordinariness** *noun*.

ordination /ɔ:dɪ'neɪʃ(ə)n/ *noun* conferring of holy orders; ordaining.

ordnance /'ɔ:dnəns/ *noun* artillery and military supplies; government service dealing with these. □ **Ordnance Survey** government survey of UK producing detailed maps.

ordure /'ɔ:djʊə/ *noun* dung.

ore *noun* naturally occurring mineral yielding metal or other valuable minerals.

oregano /ɒrɪ'gɑ:nəʊ/ *noun* dried wild marjoram as seasoning.

organ /'ɔ:gən/ *noun* musical instrument consisting of pipes that sound when air is forced through them, operated by keys and pedals; similar instrument producing sound electronically; part of body serving some special function; medium of opinion, esp. newspaper. □ **organ-grinder** player of barrel organ.

organdie /'ɔ:gəndɪ/ *noun* fine translucent muslin, usually stiffened.

organic /ɔ:'gænɪk/ *adjective* of or affecting bodily organ(s); (of animals and plants) having organs or organized physical structure; (of food) produced without artificial fertilizers or pesticides; (of chemical compound etc.) containing carbon; organized; inherent, structural. □ **organic chemistry** that of carbon compounds. □ **organically** *adverb.*

organism /'ɔ:gənɪz(ə)m/ *noun* individual animal or plant; living being with interdependent parts; system made up of interdependent parts.

organist *noun* player of organ.

organization /ɔ:gənaɪ'zeɪʃ(ə)n/ *noun* (also **-isation**) organized body, system, or society; organizing, being organized.

organize /'ɔ:gənaɪz/ *verb* (also **-ise**) (**-zing** or **-sing**) give orderly structure to; make arrangements for (person, oneself); initiate, arrange for; (as **organized** *adjective*) make organic or into living tissue. □ **organizer** *noun.*

orgasm /'ɔ:gæz(ə)m/ ● *noun* climax of sexual excitement. ● *verb* have sexual orgasm. □ **orgasmic** /-'gæz-/ *adjective.*

orgy /'ɔ:dʒɪ/ *noun* (*plural* **-ies**) wild party with indiscriminate sexual activity; excessive indulgence in an activity. □ **orgiastic** /-'æs-/ *adjective.*

oriel /'ɔ:rɪəl/ *noun* window projecting from wall at upper level.

orient /'ɔ:rɪənt/ ● *noun* (**the Orient**) the East, countries east of Mediterranean, esp. E. Asia. ● *verb* place or determine position of with aid of compass; find bearings of; (often + *towards*) direct; place (building etc.) to face east; turn eastward or in specified direction.

oriental /ɔ:rɪ'ent(ə)l/ (often **Oriental**) ● *adjective* of the East, esp. E. Asia; of the Orient. ● *noun* native of Orient.

orientate /'ɔ:rɪənteɪt/ *verb* (**-ting**) orient.

orientation *noun* orienting, being oriented; relative position; person's adjustment in relation to circumstances; briefing.

orienteering /ɔ:rɪən'tɪərɪŋ/ *noun* competitive sport of running across rough country with map and compass.

orifice /'ɒrɪfɪs/ *noun* aperture; mouth of cavity.

origami /ɒrɪ'gɑ:mɪ/ *noun* Japanese art of folding paper into decorative shapes.

origan /'ɒrɪgən/ *noun* (also **origanum** /ə'rɪgənəm/) wild marjoram.

origin /'ɒrɪdʒɪn/ *noun* source; starting point; (often in *plural*) parentage.

original /ə'rɪdʒɪn(ə)l/ ● *adjective* existing from the beginning; earliest; innate; not imitative or derived; creative not copied; by artist etc. himself or herself. ● *noun* original pattern, picture, etc. from which another is copied or translated. □ **original sin** innate sinfulness held to be common to all human beings after the Fall. □ **originality** /-'næl-/ *noun*; **originally** *adverb.*

originate /ə'rɪdʒɪneɪt/ *verb* (**-ting**) begin; initiate or give origin to, be origin of. □ **origination** *noun*; **originator** *noun.*

oriole /'ɔ:rɪəʊl/ *noun* kind of bird, esp. **golden oriole** with black and yellow plumage in male.

ormolu /'ɔ:məlu:/ *noun* gilded bronze; gold-coloured alloy; articles made of or decorated with these.

ornament ● *noun* /'ɔ:nəmənt/ thing used to adorn or decorate; decoration; quality or person bringing honour or distinction. ● *verb* /-ment/ adorn, beautify. □ **ornamental** /-'men-/ *adjective*; **ornamentation** /-men-/ *noun.*

ornate /ɔ:'neɪt/ *adjective* elaborately adorned; (of literary style) flowery. □ **ornately** *adverb.*

ornithology /ɔ:nɪ'θɒlədʒɪ/ *noun* study of birds. □ **ornithological** /-θə'lɒdʒ-/ *adjective*; **ornithologist** *noun.*

orotund /'ɒrətʌnd/ *adjective* (of voice) full, round; imposing; (of writing, style, etc.) pompous; pretentious.

orphan /'ɔ:f(ə)n/ ● *noun* child whose parents are dead. ● *verb* bereave of parents.

orphanage *noun* home for orphans.

orrery /'ɒrərɪ/ *noun* (*plural* **-ies**) clockwork model of solar system.

orris root /'ɒrɪs/ *noun* fragrant iris root used in perfumery.

ortho- *combining form* straight, correct.

orthodontics /ɔ:θə'dɒntɪks/ *plural noun* (usually treated as *singular*) correction of irregularities in teeth and jaws.

orthodox /'ɔ:θədɒks/ *adjective* holding usual or accepted views, esp. on religion, morals, etc.; conventional. □ **Orthodox Church** Eastern Church headed by Patriarch of Constantinople, including Churches of Russia, Romania, Greece, etc. □ **orthodoxy** *noun*.

orthography /ɔ:'θɒgrəfɪ/ *noun* (*plural* **-ies**) spelling, esp. with reference to its correctness. □ **orthographic** /-'græf-/ *adjective*.

orthopaedics /ɔ:θə'pi:dɪks/ *plural noun* (treated as *singular*) (*US* **-pedics**) branch of medicine dealing with correction of diseased or injured bones or muscles. □ **orthopaedic** *adjective*; **orthopaedist** *noun*.

OS *abbreviation* old style; Ordinary Seaman; Ordnance Survey; outsize.

Oscar /'ɒskə/ *noun* statuette awarded annually in US for excellence in film acting, directing, etc.

oscillate /'ɒsɪleɪt/ *verb* (**-ting**) (cause to) swing to and fro; vacillate; *Electricity* (of current) undergo high-frequency alternations. □ **oscillation** *noun*; **oscillator** *noun*.

oscilloscope /ə'sɪləskəʊp/ *noun* device for viewing oscillations usually on screen of cathode ray tube.

osier /'əʊzɪə/ *noun* willow used in basketwork; shoot of this.

osmosis /ɒz'məʊsɪs/ *noun* passage of solvent through semipermeable partition into another solution; process by which something is acquired by absorption. □ **osmotic** /-'mɒt-/ *adjective*.

osprey /'ɒspreɪ/ *noun* (*plural* **-s**) large bird preying on fish.

osseous /'ɒsɪəs/ *adjective* of bone; bony; having bones.

ossify /'ɒsɪfaɪ/ *verb* (**-ies, -ied**) turn into bone; harden; make or become rigid or unprogressive. □ **ossification** *noun*.

ostensible /ɒ'stensɪb(ə)l/ *adjective* professed; used to conceal real purpose or nature. □ **ostensibly** *adverb*.

ostentation /ɒsten'teɪʃ(ə)n/ *noun* pretentious display of wealth; showing off. □ **ostentatious** *adjective*.

osteoarthritis /ɒstɪəʊɑ:'θraɪtɪs/ *noun* degenerative disease of the joints. □ **osteoarthritic** /-'θrɪt-/ *adjective*.

osteopath /'ɒstɪəpæθ/ *noun* person who treats disease by manipulation of bones. □ **osteopathy** /-'ɒp-/ *noun*.

osteoporosis /ɒstɪəʊpə'rəʊsɪs/ *noun* brittle bones caused by hormonal change or deficiency of calcium or vitamin D.

ostler /'ɒslə/ *noun historical* stableman at inn.

ostracize /'ɒstrəsaɪz/ *verb* (also **-ise**) (**-zing** or **-sing**) exclude from society, refuse to associate with. □ **ostracism** /-sɪz(ə)m/ *noun*.

ostrich /'ɒstrɪtʃ/ *noun* large flightless swift-running African bird; person refusing to acknowledge awkward truth.

OT *abbreviation* Old Testament.

other /'ʌðə/ ● *adjective* further, additional; different; (**the other**) the only remaining. ● *noun* other person or thing. □ **the other day**, **week**, etc. a few days etc. ago; **other half** *colloquial* one's wife or husband; **other than** apart from.

otherwise /'ʌðəwaɪz/ ● *adverb* or else; in different circumstances; in other respects; in a different way; as an alternative. ● *adjective* different.

otiose /'əʊtɪəʊs/ *adjective* not required, serving no practical purpose.

OTT *abbreviation colloquial* over-the-top.

otter /'ɒtə/ *noun* furred aquatic fish-eating mammal.

Ottoman /'ɒtəmən/ ● *adjective historical* of Turkish Empire. ● *noun* (*plural* **-s**) Turk of Ottoman period; (**ottoman**) cushioned seat without back or arms, storage-box with padded top.

OU *abbreviation* Open University; Oxford University.

oubliette /u:blɪ'et/ *noun* secret dungeon with trapdoor entrance.

ouch /aʊtʃ/ *interjection expressing sharp or sudden pain.*

ought /ɔːt/ *auxiliary verb expressing duty, rightness, probability, etc.*

oughtn't /'ɔːt(ə)nt/ ought not.

Ouija /'wiːdʒə/ *noun* (in full **Ouija board**) *proprietary term* board marked with letters or signs used with movable pointer to try to obtain messages in seances.

ounce /aʊns/ *noun* unit of weight (1/16 lb, 28.35 g); very small quantity.

our /aʊə/ *adjective* of or belonging to us.

ours /aʊəz/ *pronoun* the one(s) belonging to us.

ourselves /aʊə'selvz/ *pronoun: emphatic form of* WE *or* US; *reflexive form of* US.

ousel = OUZEL.

oust /aʊst/ *verb* drive out of office or power, esp. by seizing place of.

out /aʊt/ ● *adverb* away from or not in place, not at home, office, etc.; into open, sight, notice, etc.; to or at an end; not burning; in error; *colloquial* unconscious; (+ *to do*) determined; (of limb etc.) dislocated. ● *preposition* out of. ● *noun* way of escape. □ **out for** intent on, determined to get; **out of** from inside, not inside, from among, lacking, having no more of, because of. ● *verb* emerge.

■ **Usage** The use of *out* as a preposition, as in *He walked out the room*, is not standard English. *Out of* should be used instead.

out- *prefix* so as to surpass; external; out of.

outback *noun Australian* remote inland areas.

outbalance /aʊt'bæləns/ *verb* (**-cing**) outweigh.

outbid /aʊt'bɪd/ *verb* (**-dd-**; *past & past participle* **-bid**) bid higher than.

outboard motor *noun* portable engine attached to outside of boat.

outbreak /'aʊtbreɪk/ *noun* sudden eruption of emotion, war, disease, fire, etc.

outbuilding *noun* shed, barn, etc. detached from main building.

outburst *noun* bursting out, esp. of emotion in vehement words.

outcast ● *noun* person cast out from home and friends. ● *adjective* homeless; rejected.

outclass /aʊt'klɑːs/ *verb* surpass in quality.

outcome *noun* result.

outcrop *noun* rock etc. emerging at surface; noticeable manifestation.

outcry *noun* (*plural* **-ies**) loud public protest.

outdated /aʊt'deɪtɪd/ *adjective* out of date, obsolete.

outdistance /aʊt'dɪst(ə)ns/ *verb* (**-cing**) leave (competitor) behind completely.

outdo /aʊt'duː/ *verb* (**-doing**; *3rd singular present* **-does**; *past* **-did**; *past participle* **-done**) surpass, excel.

outdoor *adjective* done, existing, or used out of doors; fond of the open air.

outdoors /aʊt'dɔːz/ ● *adverb* in(to) the open air. ● *noun* the open air.

outer *adjective* outside, external; farther from centre or inside. □ **outer space** universe beyond earth's atmosphere. □ **outermost** *adjective*.

outface /aʊt'feɪs/ *verb* (**-cing**) disconcert by staring or by confident manner.

outfall *noun* outlet of river, drain, etc.

outfield *noun* outer part of cricket or baseball pitch. □ **outfielder** *noun*.

outfit *noun* set of equipment or clothes; *colloquial* (organized) group or company.

outfitter *noun* supplier of clothing.

outflank /aʊt'flæŋk/ *verb* get round the flank of (enemy); outmanoeuvre.

outflow *noun* outward flow; what flows out.

outgoing ● *adjective* friendly; retiring from office; going out. ● *noun* (in *plural*) expenditure.

outgrow /aʊt'grəʊ/ *verb* (*past* **-grew**; *past participle* **-grown**) get too big for (clothes etc.); leave behind (childish habit etc.); grow faster or taller than.

outgrowth *noun* offshoot.

outhouse *noun* shed etc., esp. adjoining main house.

outing *noun* pleasure trip.

outlandish /aʊt'lændɪʃ/ *adjective* bizarre, strange. □ **outlandishly** *adverb*; **outlandishness** *noun*.

outlast /aʊt'lɑːst/ *verb* last longer than.

outlaw ● *noun* fugitive from law; *historical* person deprived of protection of law. ● *verb* declare (person) an outlaw; make illegal; proscribe.

outlay *noun* expenditure.

outlet *noun* means of exit; means of expressing feelings; market for goods.

outline ● *noun* rough draft; summary; line(s) enclosing visible object; contour; external boundary; (in *plural*) main features. ● *verb* (**-ning**) draw or describe in outline; mark outline of.

outlive /aʊt'lɪv/ *verb* (**-ving**) live longer than, beyond, or through.

outlook *noun* view, prospect; mental attitude.

outlying *adjective* far from centre; remote.

outmanoeuvre /aʊtmə'nu:və/ *verb* (**-ring**) (*US* **-maneuver**) outdo by skilful manoeuvring.

outmatch /aʊt'mætʃ/ *verb* be more than a match for.

outmoded /aʊt'məʊdɪd/ *adjective* outdated; out of fashion.

outnumber /aʊt'nʌmbə/ *verb* exceed in number.

outpace /aʊt'peɪs/ *verb* (**-cing**) go faster than; outdo in contest.

outpatient *noun* non-resident hospital patient.

outplacement *noun* help in finding new job after redundancy.

outpost *noun* detachment on guard at some distance from army; outlying settlement etc.

outpouring *noun* (usually in *plural*) copious expression of emotion.

output ● *noun* amount produced (by machine, worker, etc.); electrical power etc. supplied by apparatus; printout, results, etc. from computer; place where energy, information, etc., leaves a system. ● *verb* (**-tt-**; *past & past participle* **-put** or **-putted**) (of computer) supply (results etc.).

outrage ● *noun* forcible violation of others' rights, sentiments, etc.; gross offence or indignity; fierce resentment. ● *verb* (**-ging**) subject to outrage; insult; shock and anger.

outrageous /aʊt'reɪdʒəs/ *adjective* immoderate; shocking; immoral, offensive. □ **outrageously** *adverb*.

outrank /aʊt'ræŋk/ *verb* be superior in rank to.

outré /'u:treɪ/ *adjective* eccentric, violating decorum. [French]

outrider *noun* motorcyclist or mounted guard riding ahead of car(s) etc.

outrigger *noun* spar or framework projecting from or over side of ship, canoe, etc. to give stability; boat with this.

outright ● *adverb* altogether, entirely; not gradually; without reservation. ● *adjective* downright, complete; undisputed.

outrun /aʊt'rʌn/ *verb* (**-nn-**; *past* **-ran**; *past participle* **-run**) run faster or farther than; go beyond.

outsell /aʊt'sel/ *verb* (*past & past participle* **-sold**) sell more than; be sold in greater quantities than.

outset *noun* □ **at, from the outset** at or from the beginning.

outshine /aʊt'ʃaɪn/ *verb* (**-ning**; *past & past participle* **-shone**) be more brilliant than.

outside ● *noun* /aʊt'saɪd, 'aʊtsaɪd/ external surface, outer part(s); external appearance; position on outer side. ● *adjective* /'aʊtsaɪd/ of, on, or nearer outside; not belonging to particular circle or institution; (of chance etc.) remote; greatest existent or possible. ● *adverb* /aʊt'saɪd/ on or to outside; out of doors; not within or enclosed. ● *preposition* /aʊt'saɪd/ not in; to or at the outside of; external to; beyond limits of. □ **at the outside** (of estimate etc.) at the most; **outside interest** hobby etc. unconnected with one's work.

outsider /aʊt'saɪdə/ *noun* non-member of circle, party, profession, etc.; competitor thought to have little chance.

outsize *adjective* unusually large.

outskirts *plural noun* outer area of town etc.

outsmart /aʊt'smɑ:t/ *verb* outwit; be too clever for.

outspoken /aʊt'spəʊkən/ *adjective* saying openly what one thinks; frank. □ **outspokenly** *adverb*; **outspokenness** *noun*.

outspread /aʊt'spred/ *adjective* spread out.

outstanding /aʊt'stændɪŋ/ *adjective* conspicuous, esp. from excellence; still to be dealt with; (of debt) not yet settled. □ **outstandingly** *adverb*.

outstation *noun* remote branch or outpost.

outstay /aʊt'steɪ/ *verb* stay longer than (one's welcome etc.).

outstretched /aʊt'stretʃt/ *adjective* stretched out.

outstrip /aʊt'strɪp/ *verb* (**-pp-**) go faster than; surpass in progress, competition, etc.

out-take *noun* film or tape sequence cut out in editing.

out-tray *noun* tray for outgoing documents.

outvote /aʊt'vəʊt/ *verb* (**-ting**) defeat by majority of votes.

outward /'aʊtwəd/ • *adjective* directed towards outside; going out; physical; external, apparent. • *adverb* (also **outwards**) in outward direction, towards outside. □ **outwardly** *adverb*.

outweigh /aʊt'weɪ/ *verb* exceed in weight, value, influence, etc.

outwit /aʊt'wɪt/ *verb* (**-tt-**) be too clever for; overcome by greater ingenuity.

outwith /aʊt'wɪθ/ *preposition Scottish* outside.

outwork *noun* advanced or detached part of fortress etc.; work done off premises of shop, factory, etc. supplying it.

outworn /aʊt'wɔ:n/ *adjective* worn out; obsolete.

ouzel /'u:z(ə)l/ *noun* (also **ousel**) small bird of thrush family.

ouzo /'u:zəʊ/ *noun* (*plural* **-s**) Greek aniseed-flavoured alcoholic spirit.

ova *plural* of OVUM.

oval /'əʊv(ə)l/ • *adjective* shaped like egg, elliptical. • *noun* elliptical closed curve; thing with oval outline.

ovary /'əʊvərɪ/ *noun* (*plural* **-ies**) either of two ovum-producing organs in female; seed vessel in plant. □ **ovarian** /əʊ'veər-/ *adjective*.

ovation /əʊ'veɪʃ(ə)n/ *noun* enthusiastic applause or reception.

oven /'ʌv(ə)n/ *noun* enclosed chamber for cooking food in. □ **ovenproof** heat-resistant; **oven-ready** (of food) prepared before sale for immediate cooking; **ovenware** dishes for cooking food in oven.

over /'əʊvə/ • *adverb* outward and downward from brink or from erect position; so as to cover whole surface; so as to produce fold or reverse position; above in place or position; from one side, end, etc. to other; from beginning to end with repetition; in excess; settled, finished. • *preposition* above; out and down from; so as to cover; across; on or to other side, end, etc. of; concerning. • *noun Cricket* sequence of 6 balls bowled from one end before change is made to other; play during this time. □ **over the way** (in street etc.) facing or across from.

over- *prefix* excessively; upper, outer; over; completely.

overact /əʊvə'rækt/ *verb* act (a role) with exaggeration.

over-active /əʊvə'ræktɪv/ *adjective* too active.

overall • *adjective* /'əʊvərɔ:l/ taking everything into account, inclusive, total. • *adverb* /əʊvər'ɔ:l/ including everything; on the whole. • *noun* /'əʊvərɔ:l/ protective outer garment; (in *plural*) protective trousers or suit.

overarm *adjective & adverb* with arm raised above shoulder.

overawe /əʊvə'rɔ:/ *verb* (**-wing**) awe into submission.

overbalance /əʊvə'bæləns/ *verb* (**-cing**) lose balance and fall; cause to do this.

overbearing /əʊvə'beərɪŋ/ *adjective* domineering; oppressive.

overblown /əʊvə'bləʊn/ *adjective* inflated, pretentious; (of flower etc.) past its prime.

overboard *adverb* from ship into water. □ **go overboard** *colloquial* show extreme enthusiasm, behave immoderately.

overbook /əʊvə'bʊk/ *verb* make too many bookings for (aircraft, hotel, etc.).

overcame *past* of OVERCOME.

overcast *adjective* (of sky) covered with cloud; (in sewing) edged with stitching.

overcharge /əʊvə'tʃɑ:dʒ/ *verb* (**-ging**) charge too high a price to; put too much charge into (battery, gun, etc.).

overcoat *noun* warm outdoor coat.

overcome /əʊvə'kʌm/ *verb* (**-coming**; *past* **-came**; *past participle* **-come**) prevail over, master; be victorious; (usually as **overcome** *adjective*) make faint; (often + *with*) make weak or helpless.

overcrowd /əʊvəˈkraʊd/ *verb* (usually as **overcrowded** *adjective*) cause too many people or things to be in (a place). □ **overcrowding** *noun*.

overdevelop /əʊvədɪˈveləp/ *verb* (**-p-**) develop too much.

overdo /əʊvəˈduː/ *verb* (**-doing**; *3rd singular present* **-does**; *past* **-did**; *past participle* **-done**) carry to excess; (as **overdone** *adjective*) overcooked. □ **overdo it, things** *colloquial* exhaust oneself.

overdose ● *noun* excessive dose of drug etc. ● *verb* (**-sing**) take overdose.

overdraft *noun* overdrawing of bank account; amount by which account is overdrawn.

overdraw /əʊvəˈdrɔː/ *verb* (*past* **-drew**; *past participle* **-drawn**) draw more from (bank account) than amount in credit; (as **overdrawn** *adjective*) having overdrawn one's account.

overdress /əʊvəˈdres/ *verb* dress ostentatiously or with too much formality.

overdrive *noun* mechanism in vehicle providing gear above top gear for economy at high speeds; state of high activity.

overdue /əʊvəˈdjuː/ *adjective* past the time when due or ready; late, in arrears.

overestimate ● *verb* /əʊvərˈestɪmeɪt/ (**-ting**) form too high an estimate of. ● *noun* /əʊvərˈestɪmət/ too high an estimate. □ **overestimation** *noun*.

overexpose /əʊvərɪkˈspəʊz/ *verb* (**-sing**) expose too much to public; expose (film) for too long.

overfish /əʊvəˈfɪʃ/ *verb* deplete (stream etc.) by too much fishing.

overflow ● *verb* /əʊvəˈfləʊ/ flow over; be so full that contents overflow; (of crowd etc.) extend beyond limits or capacity of; flood; (of kindness, harvest, etc.) be very abundant. ● *noun* /ˈəʊvəfləʊ/ what overflows or is superfluous; outlet for excess liquid.

overgrown /əʊvəˈgrəʊn/ *adjective* grown too big; covered with weeds etc. □ **overgrowth** *noun*.

overhang ● *verb* /əʊvəˈhæŋ/ (*past & past participle* **-hung**) project or hang over. ● *noun* /ˈəʊvəhæŋ/ fact or amount of overhanging.

overhaul *verb* /əʊvəˈhɔːl/ check over thoroughly and make repairs to if necessary; overtake. ● *noun* /ˈəʊvəhɔːl/ thorough examination, with repairs if necessary.

overhead ● *adverb* /əʊvəˈhed/ above one's head; in sky. ● *adjective* /ˈəʊvəhed/ placed overhead. ● *noun* /ˈəʊvəhed/ (in *plural*) routine administrative and maintenance expenses of a business. □ **overhead projector** projector for producing enlarged image of transparency above and behind user.

overhear /əʊvəˈhɪə/ *verb* (*past & past participle* **-heard**) hear as hidden or unintentional listener.

overindulge /əʊvərɪnˈdʌldʒ/ *noun* (**-ging**) indulge to excess.

overjoyed /əʊvəˈdʒɔɪd/ *adjective* filled with great joy.

overkill *noun* excess of capacity to kill or destroy; excess.

overland *adjective & adverb* by land and not sea.

overlap ● *verb* /əʊvəˈlæp/ (**-pp-**) partly cover; cover and extend beyond; partly coincide. ● *noun* /ˈəʊvəlæp/ overlapping; overlapping part or amount.

overlay ● *verb* /əʊvəˈleɪ/ (*past & past participle* **-laid**) lay over; (+ *with*) cover (thing) with (coating etc.). ● *noun* /ˈəʊvəleɪ/ thing laid over another.

overleaf /əʊvəˈliːf/ *adverb* on other side of page of book.

overlie /əʊvəˈlaɪ/ *verb* (**-lying**; *past* **-lay**; *past participle* **-lain**) lie on top of.

overload ● *verb* /əʊvəˈləʊd/ load too heavily (with baggage, work, etc.); put too great a demand on (electrical circuit etc.). ● *noun* /ˈəʊvələʊd/ excessive quantity or demand.

overlook /əʊvəˈlʊk/ *verb* fail to observe; tolerate; have view of from above.

overlord *noun* supreme lord.

overly *adverb* excessively.

overman /əʊvəˈmæn/ *verb* (**-nn-**) provide with too large a crew, staff, etc.

overmuch /əʊvəˈmʌtʃ/ *adverb & adjective* too much.

overnight /əʊvəˈnaɪt/ ● *adverb* for or during night; on preceding evening; suddenly. ● *adjective* done or for use etc. overnight; instant.

over-particular /əʊvəpə'tɪkjʊlə/ *adjective* fussy or excessively particular.

overpass *noun esp. US* bridge by which road or railway line crosses another.

overplay /əʊvə'pleɪ/ *verb* give undue importance or emphasis to. □ **overplay one's hand** act on unduly optimistic estimate of one's chances.

overpower /əʊvə'paʊə/ *verb* subdue, reduce to submission; (esp. as **overpowering** *adjective*) be too intense or overwhelming for.

overproduce /əʊvəprə'dju:s/ *verb* (**-cing**) produce in excess of demand or of defined amount. □ **overproduction** *noun*.

overrate /əʊvə'reɪt/ *verb* (**-ting**) assess or value too highly; (as **overrated** *adjective*) not as good as it is said to be.

overreach /əʊvə'ri:tʃ/ *verb* (**overreach oneself**) fail by attempting too much.

overreact /əʊvərɪ'ækt/ *verb* respond more violently etc. than is justified. □ **overreaction** *noun*.

override ● *verb* /əʊvə'raɪd/ (**-ding**; *past* **-rode**; *past participle* **-ridden**) have priority over; intervene and make ineffective; interrupt action of (automatic device). ● *noun* /'əʊvəraɪd/ suspension of automatic function.

overrider *noun* each of pair of projecting pieces on bumper of car.

overrule /əʊvə'ru:l/ *verb* (**-ling**) set aside (decision etc.) by superior authority; reject proposal of (person) in this way.

overrun /əʊvə'rʌn/ *verb* (**-nn-**; *past* **-ran**; *past participle* **-run**) spread over; conquer (territory) by force; exceed time etc. allowed.

overseas ● *adverb* /əʊvə'si:z/ across or beyond sea. ● *adjective* /'əʊvəsi:z/ of places across sea; foreign.

oversee /əʊvə'si:/ *verb* (**-sees**; *past* **-saw**; *past participle* **-seen**) superintend (workers etc.). □ **overseer** *noun*.

over-sensitive /əʊvə'sensɪtɪv/ *adjective* excessively sensitive; easily hurt; quick to react. □ **over-sensitiveness** *noun*, **over-sensitivity** /-'tɪv-/ *noun*.

oversew *verb* (*past participle* **-sewn** or **-sewed**) sew (two edges) with stitches lying over them.

oversexed /əʊvə'sekst/ *adjective* having unusually strong sexual desires.

overshadow /əʊvə'ʃædəʊ/ *verb* appear much more prominent or important than; cast into shade.

overshoe *noun* shoe worn over another for protection in wet weather etc.

overshoot /əʊvə'ʃu:t/ *verb* (*past & past participle* **-shot**) pass or send beyond (target or limit); go beyond runway when landing or taking off. □ **overshoot the mark** go beyond what is intended or proper.

oversight *noun* failure to notice; inadvertent omission or mistake; supervision.

oversimplify /əʊvə'sɪmplɪfaɪ/ *verb* (**-ies**, **-ied**) distort (problem etc.) by putting it in too simple terms. □ **oversimplification** *noun*.

oversleep /əʊvə'sli:p/ *verb* (*past & past participle* **-slept**) sleep beyond intended time of waking.

overspend /əʊvə'spend/ *verb* (*past & past participle* **-spent**) spend beyond one's means.

overspill *noun* what is spilt over or overflows; surplus population leaving one area for another.

overspread /əʊvə'spred/ *verb* (*past & past participle* **-spread**) cover surface of; (as **overspread** *adjective*) (usually + *with*) covered.

overstate /əʊvə'steɪt/ *verb* (**-ting**) state too strongly; exaggerate. □ **overstatement** *noun*.

overstep /əʊvə'step/ *verb* (**-pp-**) pass beyond. □ **overstep the mark** go beyond conventional behaviour.

overstrain /əʊvə'streɪn/ *verb* damage by exertion; stretch too far.

overstrung *adjective* /əʊvə'strʌŋ/ (of person, nerves, etc.) too highly strung; /'əʊvəstrʌŋ/ (of piano) with strings crossing each other obliquely.

oversubscribe /əʊvəsəb'skraɪb/ *verb* (**-bing**) (usually as **oversubscribed** *adjective*) subscribe for more than available amount or number of (offer, shares, places, etc.).

overt /əʊ'vɜ:t/ *adjective* openly done, unconcealed. □ **overtly** *adverb*.

overtake /əʊvə'teɪk/ *verb* (**-king**; *past* **-took**; *past participle* **-taken**) catch up and pass; (of bad luck etc.) come suddenly upon.

overtax /əʊvəˈtæks/ *verb* make excessive demands on; tax too highly.

over-the-top *adjective colloquial* excessive.

overthrow ● *verb* /əʊvəˈθrəʊ/ (*past* **-threw**; *past participle* **-thrown**) remove forcibly from power; conquer. ● *noun* /ˈəʊvəθrəʊ/ defeat; downfall.

overtime ● *noun* time worked in addition to regular hours; payment for this. ● *adverb* in addition to regular hours.

overtone *noun Music* any of tones above lowest in harmonic series; subtle extra quality or implication.

overture /ˈəʊvətjʊə/ *noun* orchestral prelude; (usually in *plural*) opening of negotiations; formal proposal or offer.

overturn /əʊvəˈtɜːn/ *verb* (cause to) fall down or over; upset, overthrow.

overview *noun* general survey.

overweening /əʊvəˈwiːnɪŋ/ *adjective* arrogant.

overweight ● *adjective* /əʊvəˈweɪt/ above the weight allowed or desirable. ● *noun* /ˈəʊvəweɪt/ excess weight.

overwhelm /əʊvəˈwelm/ *verb* overpower with emotion; overcome by force of numbers; bury, submerge utterly.

overwhelming *adjective* too great to resist or overcome; by a great number. □ **overwhelmingly** *adverb*.

overwork /əʊvəˈwɜːk/ ● *verb* (cause to) work too hard; weary or exhaust with work; (esp. as **overworked** *adjective*) make excessive use of. ● *noun* excessive work.

overwrought /əʊvəˈrɔːt/ *adjective* overexcited, nervous, distraught; too elaborate.

oviduct /ˈəʊvɪdʌkt/ *noun* tube through which ova pass from ovary.

oviform /ˈəʊvɪfɔːm/ *adjective* egg-shaped.

ovine /ˈəʊvaɪn/ *adjective* of or like sheep.

oviparous /əʊˈvɪpərəs/ *adjective* egg-laying.

ovoid /ˈəʊvɔɪd/ *adjective* (of solid) egg-shaped.

ovulate /ˈɒvjʊleɪt/ *verb* (**-ting**) produce ova or ovules, or discharge them from ovary. □ **ovulation** *noun*.

ovule /ˈɒvjuːl/ *noun* structure containing germ cell in female plant.

ovum /ˈəʊvəm/ *noun* (*plural* **ova** /ˈəʊvə/) female egg cell from which young develop after fertilization with male sperm.

ow /aʊ/ *interjection expressing sudden pain.*

owe /əʊ/ *verb* (**owing**) be under obligation to (re)pay or render; (usually + *for*) be in debt; (usually + *to*) be indebted to person, thing, etc. for.

owing /ˈəʊɪŋ/ *adjective* owed, yet to be paid; (+ *to*) caused by, because of.

owl /aʊl/ *noun* night bird of prey; solemn or wise-looking person. □ **owlish** *adjective*.

owlet /ˈaʊlɪt/ *noun* small or young owl.

own /əʊn/ ● *adjective* (after *my*, *your*, etc.) belonging to myself, yourself, etc.; not another's. ● *verb* have as property, possess; acknowledge as true or belonging to one. □ **come into one's own** achieve recognition, receive one's due; **hold one's own** maintain one's position, not be defeated; **on one's own** alone, independently, unaided; **own goal** goal scored by mistake against scorer's own side, action etc. having unintended effect of harming person's own interests; **own up** confess.

owner *noun* possessor. □ **owner-occupier** person who owns and occupies house. □ **ownership** *noun*.

ox *noun* (*plural* **oxen**) large usually horned ruminant; castrated male of domestic species of cattle. □ **ox-eye daisy** daisy with large white petals and yellow centre; **oxtail** tail of ox, often used in making soup.

oxalic acid /ɒkˈsælɪk/ *noun* intensely sour poisonous acid found in wood sorrel and rhubarb leaves.

oxidation /ɒksɪˈdeɪʃ(ə)n/ *noun* oxidizing, being oxidized.

oxide /ˈɒksaɪd/ *noun* compound of oxygen with another element.

oxidize /ˈɒksɪdaɪz/ *verb* (also **-ise**) (**-zing** or **-sing**) combine with oxygen; rust; cover with coating of oxide. □ **oxidization** *noun*.

oxyacetylene /ɒksɪəˈsetɪliːn/ *adjective* of or using mixture of oxygen and acetylene, esp. in cutting or welding metals.

oxygen /ˈɒksɪdʒ(ə)n/ *noun* colourless odourless tasteless gaseous element essential to life and to combustion.

□ **oxygen tent** enclosure to allow patient to breathe air with increased oxygen content.

oxygenate /ˈɒksɪdʒəneɪt/ *verb* (**-ting**) supply, treat, or mix with oxygen; oxidize.

oyez /əʊˈjes/ *interjection* (also **oyes**) *uttered by public crier or court officer to call for attention.*

oyster /ˈɔɪstə/ ● *noun* bivalve mollusc living on seabed, esp. edible kind. ● *adjective* (in full **oyster-white**) greyish white.

oz *abbreviation* ounce(s).

ozone /ˈəʊzəʊn/ *noun* form of oxygen with pungent odour; *colloquial* invigorating seaside air. □ **ozone-friendly** not containing chemicals destructive to ozone layer; **ozone layer** layer of ozone in stratosphere that absorbs most of sun's ultraviolet radiation.

Pp

p *abbreviation* (also **p.**) penny, pence; page; piano (softly). □ **p. & p.** postage and packing.

PA *abbreviation* personal assistant; public address.

pa /pɑː/ *noun colloquial* father.

p.a. *abbreviation* per annum.

pace[1] ● *noun* single step in walking or running; distance covered in this; speed, rate of progression; gait. ● *verb* (**-cing**) walk (over, about), esp. with slow or regular step; set pace for; (+ *out*) measure (distance) by pacing. □ **pacemaker** person who sets pace, natural or electrical device for stimulating heart muscle.

pace[2] /ˈpɑːtʃeɪ/ *preposition* with all due deference to. [Latin]

pachyderm /ˈpækɪdɜːm/ *noun* large thick-skinned mammal, esp. elephant or rhinoceros. □ **pachydermatous** /-ˈdɜːmətəs/ *adjective.*

pacific /pəˈsɪfɪk/ ● *adjective* tending to peace, peaceful; (**Pacific**) of or adjoining the Pacific. ● *noun* (**the Pacific**) ocean between America to the east and Asia to the west.

pacifist /ˈpæsɪfɪst/ *noun* person opposed to war. □ **pacifism** *noun.*

pacify /ˈpæsɪfaɪ/ *verb* (**-ies, -ied**) appease (person, anger, etc.); bring (country etc.) to state of peace. □ **pacification** *noun.*

pack ● *noun* collection of things tied or wrapped together for carrying; backpack; set of packaged items; set of playing cards; *usually derogatory* lot, set; group of wild animals or hounds; organized group of Cub Scouts or Brownies; forwards of Rugby team; area of large crowded pieces of floating ice in sea. ● *verb* put together into bundle, box, etc., fill with clothes etc. for transport or storing; cram, crowd together, form into pack; (esp. in *passive*; often + *with*) fill; wrap tightly. □ **packed out** full, crowded; **packhorse** horse for carrying loads; **pack in** *colloquial* stop, give up; **pack it in**, **up** *colloquial* end or stop it; **pack up** *colloquial* stop working, break down, retire from contest, activity, etc.; **send packing** *colloquial* dismiss summarily.

package ● *noun* parcel; box etc. in which goods are packed; (in full **package deal**) set of proposals or items offered or agreed to as a whole. ● *verb* (**-ging**) make up into or enclose in package. □ **package holiday**, **tour**, etc., one with fixed inclusive price. □ **packaging** *noun.*

packet /ˈpækɪt/ *noun* small package; *colloquial* large sum of money; *historical* mail-boat.

pact *noun* agreement, treaty.

pad[1] ● *noun* piece of soft stuff used to diminish jarring, raise surface, absorb fluid, etc.; sheets of blank paper fastened together at one edge; fleshy cushion forming sole of foot of some animals; leg-guard in games; flat surface for helicopter take-off or rocket-launching; *slang* lodging. ● *verb* (**-dd-**) provide with pad or padding, stuff; (+ *out*) fill out (book etc.) with superfluous matter.

pad[2] ● *verb* (**-dd-**) walk softly; tramp (along) on foot; travel on foot. ● *noun* sound of soft steady steps.

padding *noun* material used to pad.

paddle[1] /'pæd(ə)l/ ● *noun* short oar with broad blade at one or each end; paddle-shaped instrument; fin, flipper; board on paddle-wheel or mill-wheel; action or spell of paddling. ● *verb* (**-ling**) move on water or propel (boat etc.) with paddle(s); row gently. □ **paddle-wheel** wheel for propelling ship, with boards round circumference.

paddle[2] /'pæd(ə)l/ ● *verb* (**-ling**) wade about in shallow water. ● *noun* action or spell of paddling.

paddock /'pædək/ *noun* small field, esp. for keeping horses in; enclosure where horses or cars are assembled before race.

paddy[1] /'pædɪ/ *noun* (*plural* **-ies**) (in full **paddy field**) field where rice is grown; rice before threshing or in the husk.

paddy[2] /'pædɪ/ *noun* (*plural* **-ies**) *colloquial* rage, temper.

padlock /'pædlɒk/ ● *noun* detachable lock hanging by pivoted hook. ● *verb* secure with padlock.

padre /'pɑ:drɪ/ *noun* chaplain in army etc.

paean /'pi:ən/ *noun* (*US* **pean**) song of praise or triumph.

paediatrics /pi:dɪ'ætrɪks/ *plural noun* (treated as *singular*) (*US* **pediatrics**) branch of medicine dealing with children's diseases. □ **paediatric** *adjective*; **paediatrician** /-ə'trɪʃ(ə)n/ *noun*.

paedophile /'pi:dəfaɪl/ *noun* (*US* **pedophile**) person feeling sexual attraction towards children.

paella /paɪ'elə/ *noun* Spanish dish of rice, saffron, chicken, seafood, etc.

paeony = PEONY.

pagan /'peɪgən/ ● *noun* heathen, pantheist, etc. ● *adjective* of pagans; heathen; pantheistic. □ **paganism** *noun*.

page[1] ● *noun* leaf of book etc.; each side of this. ● *verb* (**-ging**) number pages of.

page[2] ● *noun* boy or man employed as liveried servant or personal attendant. ● *verb* (**-ging**) call name of (person sought) in public rooms of hotel etc. □ **page-boy** boy attending bride etc., woman's short hairstyle.

pageant /'pædʒ(ə)nt/ *noun* spectacular performance, usually illustrative of historical events; any brilliant show.

pageantry *noun* spectacular show or display.

pager *noun* bleeping device calling bearer to telephone etc.

paginate /'pædʒɪneɪt/ *verb* (**-ting**) number pages of (book etc.). □ **pagination** *noun*.

pagoda /pə'gəʊdə/ *noun* temple or sacred tower in China etc.; ornamental imitation of this.

pah *interjection expressing disgust.*

paid *past & past participle* of PAY.

pail *noun* bucket.

pain ● *noun* bodily suffering caused by injury, pressure, illness, etc.; mental suffering; *colloquial* troublesome person or thing. ● *verb* cause pain to. □ **be at** or **take pains** take great care; **in pain** suffering pain; **painkiller** pain-relieving drug.

painful *adjective* causing or (esp. of part of the body) suffering pain; causing trouble or difficulty. □ **painfully** *adverb*.

painless *adjective* not causing pain. □ **painlessly** *adverb*.

painstaking /'peɪnzteɪkɪŋ/ *adjective* careful, industrious, thorough. □ **painstakingly** *adverb*.

paint ● *noun* colouring matter, esp. in liquid form, for applying to surface. ● *verb* cover surface of with paint; portray or make pictures in colours; describe vividly; apply liquid or cosmetic to. □ **paintbox** box holding dry paints for painting pictures; **painted lady** butterfly with spotted orange-red wings; **paintwork** painted, esp. wooden, surface or area in building etc.

painter[1] *noun* person who paints, esp. as artist or decorator.

painter[2] *noun* rope at bow of boat for tying it up.

painting *noun* process or art of using paint; painted picture.

pair • *noun* set of two people or things; thing with two joined or corresponding parts; engaged or married or mated couple; two playing cards of same denomination; (either of) two MPs etc. on opposite sides agreeing not to vote on certain occasions. • *verb* (often + *off*) arrange or unite as pair, in pairs, or in marriage; mate.

Paisley /'peɪzlɪ/ *noun* (*plural* **-s**) pattern of curved feather-shaped figures.

pajamas *US* = PYJAMAS.

Pakistani /pɑ:kɪs'tɑ:nɪ/ • *noun* (*plural* **-s**) native or national of Pakistan; person of Pakistani descent. • *adjective* of Pakistan.

pal *colloquial* • *noun* friend. • *verb* (**-ll-**) (+ *up*) make friends.

palace /'pælɪs/ *noun* official residence of sovereign, president, archbishop, or bishop; stately or spacious building.

palaeo- *combining form* (*US* **paleo-**) ancient; prehistoric.

palaeography /pælɪ'ɒgrəfɪ/ *noun* (*US* **paleography**) study of ancient writing and documents.

palaeolithic /pælɪəʊ'lɪθɪk/ *adjective* (*US* **paleolithic**) of earlier Stone Age.

palaeontology /pælɪɒn'tɒlədʒɪ/ *noun* (*US* **paleontology**) study of life in geological past. □ **palaeontologist** *noun*.

Palaeozoic /pælɪəʊ'zəʊɪk/ (*US* **Paleozoic**) • *adjective* of geological era marked by appearance of plants and animals, esp. invertebrates. • *noun* this era.

palais /'pæleɪ/ *noun colloquial* public dance hall.

palanquin /pælən'ki:n/ *noun* (also **palankeen**) Eastern covered litter for one.

palatable /'pælətəb(ə)l/ *adjective* pleasant to taste; (of idea etc.) acceptable, satisfactory.

palatal /'pælət(ə)l/ • *adjective* of the palate; (of sound) made with tongue against palate. • *noun* palatal sound.

palate /'pælət/ *noun* roof of mouth in vertebrates; sense of taste; liking.

palatial /pə'leɪʃ(ə)l/ *adjective* like palace, splendid.

palaver /pə'lɑ:və/ *noun colloquial* tedious fuss and bother.

pale¹ • *adjective* (of complexion etc.) whitish; faintly coloured; (of colour) faint, (of light) dim. • *verb* (**-ling**) grow or make pale; (often + *before*, *beside*) seem feeble in comparison (with). □ **palely** *adverb*.

pale² *noun* pointed piece of wood for fencing etc.; stake; boundary. □ **beyond the pale** outside bounds of acceptable behaviour.

paleo- *US* = PALAEO-.

Palestinian /pælɪ'stɪnɪən/ • *adjective* of Palestine. • *noun* native or inhabitant of Palestine.

palette /'pælɪt/ *noun* artist's flat tablet for mixing colours on; range of colours used by artist. □ **palette-knife** knife with long round-ended flexible blade, esp. for mixing colours or applying or removing paint.

palimony /'pælɪmənɪ/ *noun esp. US colloquial* allowance paid by either of a separated unmarried couple to the other.

palimpsest /'pælɪmpsest/ *noun* writing material used for second time after original writing has been erased.

palindrome /'pælɪndrəʊm/ *noun* word or phrase that reads same backwards as forwards. □ **palindromic** /-'drɒm-/ *adjective*.

paling *noun* (in *singular* or *plural*) fence of pales; pale.

palisade /pælɪ'seɪd/ • *noun* fence of pointed stakes. • *verb* (**-ding**) enclose or provide with palisade.

pall¹ /pɔ:l/ *noun* cloth spread over coffin etc.; ecclesiastical vestment; dark covering. □ **pallbearer** person helping to carry or escort coffin at funeral.

pall² /pɔ:l/ *verb* become uninteresting.

pallet¹ /'pælɪt/ *noun* straw mattress; makeshift bed.

pallet² /'pælɪt/ *noun* portable platform for transporting and storing loads.

palliasse /'pælɪæs/ *noun* straw mattress.

palliate /'pælɪeɪt/ *verb* (**-ting**) alleviate without curing; excuse, extenuate. □ **palliative** /-ətɪv/ *adjective & noun*.

pallid /'pælɪd/ *adjective* pale, sickly-looking.

pallor /'pælə/ *noun* paleness.

pally *adjective* (**-ier**, **-iest**) *colloquial* friendly.

palm[1] /pɑːm/ *noun* (also **palm tree**) (usually tropical) treelike plant with unbranched stem and crown of large esp. sickle- or fan-shaped leaves; leaf of this as symbol of victory. □ **Palm Sunday** Sunday before Easter.

palm[2] /pɑːm/ ● *noun* inner surface of hand between wrist and fingers. ● *verb* conceal in hand. □ **palm off** (often + *on*) impose fraudulently (on person).

palmate /'pælmeɪt/ *adjective* shaped like open hand.

palmetto /pæl'metəʊ/ *noun* (*plural* **-s**) small palm tree.

palmist /'pɑːmɪst/ *noun* teller of character or fortune from lines etc. in palm of hand. □ **palmistry** *noun.*

palmy /'pɑːmɪ/ *adjective* (**-ier**, **-iest**) of, like, or abounding in palms; flourishing.

palomino /pælə'miːnəʊ/ *noun* (*plural* **-s**) golden or cream-coloured horse with light-coloured mane and tail.

palpable /'pælpəb(ə)l/ *adjective* that can be touched or felt; readily perceived. □ **palpably** *adverb.*

palpate /'pælpeɪt/ *verb* (**-ting**) examine (esp. medically) by touch. □ **palpation** *noun.*

palpitate /'pælpɪteɪt/ *verb* (**-ting**) pulsate, throb; tremble. □ **palpitation** *noun.*

palsy /'pɔːlzɪ/ ● *noun* (*plural* **-ies**) paralysis, esp. with involuntary tremors. ● *verb* (**-ies**, **-ied**) affect with palsy.

paltry /'pɔːltrɪ/ *adjective* (**-ier**, **-iest**) worthless, contemptible, trifling.

pampas /'pæmpəs/ *plural noun* large treeless S. American plains. □ **pampas grass** large ornamental grass.

pamper /'pæmpə/ *verb* overindulge.

pamphlet /'pæmflɪt/ *noun* small unbound booklet, esp. controversial treatise.

pamphleteer /pæmflɪ'tɪə/ *noun* writer of (esp. political) pamphlets.

pan[1] ● *noun* flat-bottomed usually metal vessel used in cooking etc.; shallow receptacle or tray; bowl of scales or of lavatory. ● *verb* (**-nn-**) *colloquial* criticize harshly; (+ *off*, *out*) wash (gold-bearing gravel) in pan; search for gold in this way. □ **pan out** turn out, work out well or in specified way.

pan[2] ● *verb* (**-nn-**) swing (film-camera) horizontally to give panoramic effect or follow moving object; (of camera) be moved thus. ● *noun* panning movement.

pan- *combining form* all; the whole of (esp. referring to a continent, racial group, religion, etc.).

panacea /pænə'siːə/ *noun* universal remedy.

panache /pə'næʃ/ *noun* assertively flamboyant or confident style.

panama /'pænəmɑː/ *noun* hat of straw-like material with brim and indented crown.

panatella /pænə'telə/ *noun* long thin cigar.

pancake *noun* thin flat cake of fried batter, usually folded or rolled up with filling. □ **Pancake Day** Shrove Tuesday (when pancakes are traditionally eaten); **pancake landing** *colloquial* emergency aircraft landing with undercarriage still retracted.

panchromatic /pænkrəʊ'mætɪk/ *adjective* (of film etc.) sensitive to all visible colours of spectrum.

pancreas /'pæŋkrɪəs/ *noun* gland near stomach supplying digestive fluid and insulin. □ **pancreatic** /-'æt-/ *adjective.*

panda /'pændə/ *noun* (also **giant panda**) large rare bearlike black and white mammal native to China and Tibet; (also **red panda**) racoon-like Himalayan mammal. □ **panda car** police patrol car.

pandemic /pæn'demɪk/ *adjective* (of disease) widespread; universal.

pandemonium /pændɪ'məʊnɪəm/ *noun* uproar; utter confusion; scene of this.

pander /'pændə/ ● *verb* (+ *to*) indulge (person or weakness). ● *noun* procurer, pimp.

pandit = PUNDIT.

pane *noun* single sheet of glass in window or door.

panegyric /pænɪ'dʒɪrɪk/ *noun* eulogy; speech or essay of praise.

panel /'pæn(ə)l/ ● *noun* distinct, usually rectangular, section of surface, esp. of wall, door, or vehicle; group or team of people assembled for discussion, consultation, etc.; strip of material in garment; list of available jurors; jury.

● *verb* (**-ll-**; *US* **-l-**) fit with panels. □ **panel game** broadcast quiz etc. played by panel. □ **panelling** *noun*.

panellist *noun* (*US* **panelist**) member of panel.

pang *noun* sudden sharp pain or distressing emotion.

pangolin /pæŋ'gəʊlɪn/ *noun* scaly anteater.

panic /'pænɪk/ ● *noun* sudden alarm; infectious fright. ● *verb* (**-ck-**) (often + *into*) affect or be affected with panic. □ **panic-stricken**, **-struck** affected with panic. □ **panicky** *adjective*.

panicle /'pænɪk(ə)l/ *noun* loose branching cluster of flowers.

panjandrum /pæn'dʒændrəm/ *noun: mock title of great personage*.

pannier /'pænɪə/ *noun* one of pair of baskets or bags etc. carried by beast of burden or on bicycle or motorcycle.

panoply /'pænəplɪ/ *noun* (*plural* **-ies**) complete or splendid array; full armour.

panorama /pænə'rɑ:mə/ *noun* unbroken view of surrounding region; picture or photograph containing wide view. □ **panoramic** /-'ræm-/ *adjective*.

pansy /'pænzɪ/ *noun* (*plural* **-ies**) garden plant of violet family with richly coloured flowers.

pant ● *verb* breathe with quick breaths; yearn. ● *noun* panting breath.

pantaloons /pæntə'lu:nz/ *plural noun* baggy trousers gathered at ankles.

pantechnicon /pæn'teknɪkən/ *noun* large furniture van.

pantheism /'pænθiːɪz(ə)m/ *noun* doctrine that God is everything and everything is God. □ **pantheist** *noun*; **pantheistic** /-'ɪs-/ *adjective*.

pantheon /'pænθɪən/ *noun* building with memorials of illustrious dead; deities of a people collectively; temple of all gods.

panther /'pænθə/ *noun* leopard, esp. black; *US* puma.

panties /'pæntɪz/ *plural noun colloquial* short-legged or legless knickers.

pantile /'pæntaɪl/ *noun* curved roof-tile.

pantograph /'pæntəgrɑ:f/ *noun* instrument for copying plan etc. on any scale.

pantomime /'pæntəmaɪm/ *noun* dramatic usually Christmas entertainment based on fairy tale; *colloquial* absurd or outrageous behaviour; gestures and facial expressions conveying meaning.

pantry /'pæntrɪ/ *noun* (*plural* **-ies**) room in which provisions, crockery, cutlery, etc. are kept.

pants *plural noun* underpants; knickers; *US* trousers.

pap[1] *noun* soft or semi-liquid food; trivial reading matter.

pap[2] *noun archaic* nipple.

papa /pə'pɑ:/ *noun archaic: child's name for* father.

papacy /'peɪpəsɪ/ *noun* (*plural* **-ies**) Pope's office or tenure; papal system.

papal /'peɪp(ə)l/ *adjective* of the Pope or his office.

paparazzo /pæpə'rætsəʊ/ *noun* (*plural* **-zzi** /-tsɪ/) freelance photographer who pursues celebrities to photograph them.

papaya = PAWPAW.

paper /'peɪpə/ ● *noun* substance made in very thin sheets from pulp of wood etc., used for writing, printing, wrapping, etc.; newspaper; (in *plural*) documents; set of exam questions or answers; wallpaper; essay. ● *adjective* not actual, theoretical. ● *verb* decorate (wall etc.) with paper. □ **paper-boy**, **-girl** one who delivers or sells newspapers; **paper-clip** clip of bent wire or plastic for holding sheets of paper together; **paper-knife** blunt knife for opening envelopes etc.; **paper money** banknotes etc.; **paper round** job of regularly delivering newspapers, route for doing this; **paperweight** small heavy object to hold papers down; **paperwork** office record-keeping and administration.

paperback ● *adjective* bound in stiff paper, not boards. ● *noun* paperback book.

papier mâché /pæpɪeɪ 'mæʃeɪ/ *noun* moulded paper pulp used for making models etc.

papilla /pə'pɪlə/ *noun* (*plural* **papillae** /-li:/) small nipple-like protuberance. □ **papillary** *adjective*.

papoose /pə'pu:s/ *noun* young N. American Indian child.

paprika /ˈpæprɪkə/ *noun* ripe red pepper; condiment made from this.

papyrus /pəˈpaɪərəs/ *noun* (*plural* **papyri** /-raɪ/) aquatic plant of N. Africa; ancient writing material made from stem of this; manuscript written on this.

par *noun* average or normal value, degree, condition, etc.; equality, equal footing; *Golf* number of strokes needed by first-class player for hole or course; face value. □ **par for the course** *colloquial* what is normal or to be expected.

para /ˈpærə/ *noun colloquial* paratrooper.

para- *prefix* beside, beyond.

parable /ˈpærəb(ə)l/ *noun* story used to illustrate moral or spiritual truth.

parabola /pəˈræbələ/ *noun* plane curve formed by intersection of cone with plane parallel to its side. □ **parabolic** /pærəˈbɒlɪk/ *adjective*.

paracetamol /pærəˈsiːtəmɒl/ *noun* compound used to relieve pain and reduce fever; tablet of this.

parachute /ˈpærəʃuːt/ ● *noun* usually umbrella-shaped apparatus allowing person or heavy object to descend safely from a height, esp. from aircraft. ● *verb* (**-ting**) convey or descend by parachute. □ **parachutist** *noun*.

parade /pəˈreɪd/ ● *noun* public procession; muster of troops etc. for inspection; parade ground; display, ostentation; public square, row of shops. ● *verb* (**-ding**) march ceremonially; assemble for parade; display ostentatiously. □ **parade ground** place for muster of troops.

paradigm /ˈpærədaɪm/ *noun* example or pattern, esp. of inflection of word.

paradise /ˈpærədaɪs/ *noun* heaven; place or state of complete bliss; garden of Eden.

paradox /ˈpærədɒks/ *noun* seemingly absurd or self-contradictory though often true statement etc. □ **paradoxical** /-ˈdɒks-/ *adjective*; **paradoxically** /-ˈdɒks-/ *adverb*.

paraffin /ˈpærəfɪn/ *noun* inflammable waxy or oily substance got by distillation from petroleum etc., used in liquid form esp. as fuel. □ **paraffin wax** solid paraffin.

paragon /ˈpærəgən/ *noun* (often + *of*) model of excellence.

paragraph /ˈpærəgrɑːf/ *noun* distinct passage in book etc. usually marked by indentation of first line; mark of reference (¶); short separate item in newspaper etc.

parakeet /ˈpærəkiːt/ *noun* small usually long-tailed parrot.

parallax /ˈpærəlæks/ *noun* apparent difference in position or direction of object caused by change of observer's position; angular amount of this.

parallel /ˈpærəlel/ ● *adjective* (of lines) continuously equidistant; precisely similar, analogous, or corresponding; (of processes etc.) occurring or performed simultaneously. ● *noun* person or thing analogous to another; comparison; imaginary line on earth's surface or line on map marking degree of latitude. ● *verb* (**-l-**) be parallel or correspond to; represent as similar; compare. □ **parallelism** *noun*.

parallelepiped /pærəleləˈpaɪped/ *noun* solid bounded by parallelograms.

parallelogram /pærəˈleləgræm/ *noun* 4-sided rectilinear figure whose opposite sides are parallel.

paralyse /ˈpærəlaɪz/ *verb* (**-sing**) (*US* **-lyze**; **-zing**) affect with paralysis; render powerless, cripple.

paralysis /pəˈrælɪsɪs/ *noun* impairment or loss of esp. motor function of nerves, causing immobility; powerlessness.

paralytic /pærəˈlɪtɪk/ ● *adjective* affected with paralysis; *slang* very drunk. ● *noun* person affected with paralysis.

paramedic /pærəˈmedɪk/ *noun* paramedical worker.

paramedical /pærəˈmedɪk(ə)l/ *adjective* supplementing and supporting medical work.

parameter /pəˈræmɪtə/ *noun Mathematics* quantity constant in case considered, but varying in different cases; (esp. measurable or quantifiable) characteristic or feature; (loosely) boundary, esp. of subject for discussion.

paramilitary /pærəˈmɪlɪtərɪ/ ● *adjective* similarly organized to military forces. ● *noun* (*plural* **-ies**) member of unofficial paramilitary organization.

paramount /ˈpærəmaʊnt/ *adjective* supreme; most important or powerful.

paramour /ˈpærəmʊə/ *noun archaic* illicit lover of married person.

paranoia /pærə'nɔɪə/ *noun* mental derangement with delusions of grandeur, persecution, etc.; abnormal tendency to suspect and mistrust others. □ **paranoiac** *adjective & noun*; **paranoid** /'pærənɔɪd/ *adjective & noun.*

paranormal /pærə'nɔ:m(ə)l/ *adjective* beyond the scope of normal scientific investigations etc.

parapet /'pærəpɪt/ *noun* low wall at edge of roof, balcony, bridge, etc.; mound along front of trench etc.

paraphernalia /pærəfə'neɪlɪə/ *plural noun* (also treated as *singular*) personal belongings, miscellaneous accessories, etc.

paraphrase /'pærəfreɪz/ ● *noun* restatement of sense of passage etc. in other words. ● *verb* (*singular*) express meaning of in other words.

paraplegia /pærə'pli:dʒə/ *noun* paralysis below waist. □ **paraplegic** *adjective & noun.*

parapsychology /pærəsaɪ'kɒlədʒɪ/ *noun* study of mental phenomena outside sphere of ordinary psychology.

paraquat /'pærəkwɒt/ *noun* quick-acting highly toxic herbicide.

parasite /'pærəsaɪt/ *noun* animal or plant living in or on another and feeding on it; person exploiting another or others. □ **parasitic** /-'sɪt-/ *adjective*; **parasitism** *noun.*

parasol /'pærəsɒl/ *noun* light umbrella giving shade from the sun.

paratroops /'pærətru:ps/ *plural noun* airborne troops landing by parachute. □ **paratrooper** *noun.*

paratyphoid /pærə'taɪfɔɪd/ *noun* fever resembling typhoid.

parboil /'pɑ:bɔɪl/ *verb* partly cook by boiling.

parcel /'pɑ:s(ə)l/ ● *noun* goods etc. packed up in single wrapping; piece of land. ● *verb* (**-ll-**; *US* **-l-**) (+ *up*) wrap into parcel; (+ *out*) divide into portions.

parch *verb* make or become hot and dry; slightly roast.

parchment /'pɑ:tʃmənt/ *noun* skin, esp. of sheep or goat, prepared for writing etc.; manuscript written on this.

pardon /'pɑ:d(ə)n/ ● *noun* forgiveness; remission of punishment. ● *verb* forgive; excuse; release from legal consequences of offence etc. ● *interjection* (also **pardon me** or **I beg your pardon**) *formula of apology or disagreement; request to repeat something said.* □ **pardonable** *adjective.*

pare /peə/ *verb* (**-ring**) trim or reduce by cutting away edge or surface of; (often + *away, down*) whittle away.

parent /'peərənt/ *noun* person who has had or adopted a child; father, mother; source, origin. □ **parent company** company of which others are subsidiaries; **parent-teacher association** social and fund-raising organization of school's teachers and parents. □ **parental** /pə'rent(ə)l/ *adjective*; **parenthood** *noun.*

parentage *noun* lineage, descent from or through parents.

parenthesis /pə'renθəsɪs/ *noun* (*plural* **-theses** /-si:z/) word, clause, or sentence inserted as explanation etc. into passage independently of grammatical sequence; (in *plural*) round brackets used to mark this; interlude. □ **parenthetic** /pærən'θetɪk/ *adjective.*

par excellence /pɑ:r eksə'lɑ̃s/ *adverb* superior to all others so called. [French]

parfait /'pɑ:feɪ/ *noun* rich iced pudding of whipped cream, eggs, etc.; layers of ice cream, meringue, etc., served in tall glass.

pariah /pə'raɪə/ *noun* social outcast; *historical* member of low or no caste.

parietal /pə'raɪət(ə)l/ *adjective* of wall of body or any of its cavities. □ **parietal bone** either of pair forming part of skull.

paring *noun* strip pared off.

parish /'pærɪʃ/ *noun* division of diocese having its own church and clergyman; local government district; inhabitants of parish.

parishioner /pə'rɪʃənə/ *noun* inhabitant of parish.

parity /'pærɪtɪ/ *noun* (*plural* **-ies**) equality; equal status etc.; equivalence; being at par.

park ● *noun* large public garden in town; large enclosed piece of ground attached to country house or laid out or preserved for public use; place where vehicles may be parked; area for specified purpose. ● *verb* place and leave (esp.

vehicle) temporarily. □ **parking-lot** *US* outdoor car park; **parking meter** coin-operated meter allocating period of time for which a vehicle may be parked in street; **parking ticket** notice of fine etc. imposed for parking vehicle illegally.

parka /'pɑ:kə/ *noun* jacket with hood, as worn by Eskimos, mountaineers, etc.

parkin /'pɑ:kɪn/ *noun* oatmeal gingerbread.

parky /'pɑ:kɪ/ *adjective* (**-ier**, **-iest**) *colloquial or dialect* chilly.

parlance /'pɑ:ləns/ *noun* way of speaking.

parley /'pɑ:lɪ/ ● *noun* (*plural* **-s**) meeting between representatives of opposed forces to discuss terms. ● *verb* (**-leys**, **-leyed**) (often + *with*) hold parley.

parliament /'pɑ:ləmənt/ *noun* body consisting of House of Commons and House of Lords and forming (with Sovereign) legislature of UK; similar legislature in other states.

parliamentarian /pɑ:ləmen'teərɪən/ *noun* member of parliament.

parliamentary /pɑ:lə'mentərɪ/ *adjective* of, in, concerned with, or enacted by parliament.

parlour /'pɑ:lə/ *noun* (*US* **parlor**) *archaic* sitting-room in private house; *esp. US* shop providing specified goods or services. □ **parlour game** indoor game, esp. word game.

parlous /'pɑ:ləs/ *adjective archaic* perilous; hard to deal with.

Parmesan /pɑ:mɪ'zæn/ *noun* hard Italian cheese usually used grated as flavouring.

parochial /pə'rəʊkɪəl/ *adjective* of a parish; of narrow range, merely local. □ **parochialism** *noun*.

parody /'pærədɪ/ ● *noun* (*plural* **-ies**) humorous exaggerated imitation of author, style, etc.; travesty. ● *verb* (**-ies**, **-ied**) write parody of; mimic humorously. □ **parodist** *noun*.

parole /pə'rəʊl/ ● *noun* temporary or permanent release of prisoner before end of sentence, on promise of good behaviour; such promise. ● *verb* (**-ling**) put (prisoner) on parole.

parotid /pə'rɒtɪd/ ● *adjective* situated near ear. ● *noun* (in full **parotid gland**) salivary gland in front of ear.

paroxysm /'pærəksɪz(ə)m/ *noun* (often + *of*) fit (of pain, rage, coughing, etc.).

parquet /'pɑ:keɪ/ ● *noun* flooring of wooden blocks arranged in a pattern. ● *verb* (**-eted** /-eɪd/, **-eting** /-eɪɪŋ/) floor (room) thus.

parricide /'pærɪsaɪd/ *noun* person who kills his or her parent; such a killing. □ **parricidal** /-'saɪd(ə)l/ *adjective*.

parrot /'pærət/ ● *noun* mainly tropical bird with short hooked bill, of which some species can be taught to repeat words; unintelligent imitator or chatterer. ● *verb* (**-t-**) repeat mechanically. □ **parrot-fashion** (learning or repeating) mechanically, by rote.

parry /'pærɪ/ ● *verb* (**-ies**, **-ied**) ward off, avert. ● *noun* (*plural* **-ies**) act of parrying.

parse /pɑ:z/ *verb* (**-sing**) describe (word) or analyse (sentence) in terms of grammar.

parsec /'pɑ:sek/ *noun* unit of stellar distance, about 3.25 light years.

parsimony /'pɑ:sɪmənɪ/ *noun* carefulness in use of money etc.; meanness. □ **parsimonious** /-'məʊn-/ *adjective*.

parsley /'pɑ:slɪ/ *noun* herb used for seasoning and garnishing.

parsnip /'pɑ:snɪp/ *noun* plant with pale yellow tapering root used as vegetable; this root.

parson /'pɑ:s(ə)n/ *noun* parish clergyman; *colloquial* any clergyman. □ **parson's nose** fatty flesh at rump of cooked fowl.

parsonage *noun* parson's house.

part ● *noun* some but not all; component, division, portion; share, allotted portion; person's share in an action etc.; assigned character or role; *Music* one of melodies making up harmony of concerted music; side in agreement or dispute; (usually in *plural*) region, direction, way; (in *plural*) abilities. ● *verb* divide into parts; separate; (+ *with*) give up, hand over; make parting in (hair). ● *adverb* partly, in part. □ **on the part of** made or done by; **part and parcel** (usually + *of*) essential part; **part-exchange** *noun* transaction in which article is given as part of payment for more expensive one, *verb* give (article) thus; **part of speech** grammatical class of words (noun, pronoun, adjective, adverb, verb, etc.); **part-song** song for 3 or more voice parts; **part-time** employed for or occupying less than normal working week etc.; **part-timer** part-time worker.

partake /pɑ:'teɪk/ *verb* (**-king**; *past* **partook**; *past participle* **partaken**) (+ *of*, *in*) take share of; (+ *of*) eat or drink some of.

parterre /pɑ:'teə/ *noun* level garden space filled with flower-beds etc.; *US* pit of theatre.

partial /'pɑ:ʃ(ə)l/ *adjective* not total or complete; biased, unfair; (+ *to*) having a liking for. □ **partiality** /-ʃɪ'æl-/ *noun*; **partially** *adverb*.

participate /pɑ:'tɪsɪpeɪt/ *verb* (**-ting**) (often + *in*) have share or take part. □ **participant** *noun*; **participation** *noun*.

participle /'pɑ:tɪsɪp(ə)l/ *noun* word (either **present participle**, e.g. *writing*, or **past participle**, e.g. *written*) formed from verb and used in complex verb-forms or as adjective. □ **participial** /-'sɪp-/ *adjective*.

particle /'pɑ:tɪk(ə)l/ *noun* minute portion of matter; smallest possible amount; minor esp. undeclinable part of speech.

particoloured /'pɑ:tɪkʌləd/ *adjective* (*US* **-colored**) of more than one colour.

particular /pə'tɪkjʊlə/ ● *adjective* relating to or considered as one as distinct from others; special; scrupulously exact; fastidious. ● *noun* detail, item; (in *plural*) detailed account. □ **in particular** specifically. □ **particularity** /-'lær-/ *noun*.

particularize /pə'tɪkjʊləraɪz/ *verb* (also **-ise**) (**-zing** or **-sing**) name specially or one by one; specify (items). □ **particularization** *noun*.

particularly *adverb* very; specifically; in a fastidious way.

parting *noun* leave-taking; dividing line of combed hair.

partisan /pɑ:tɪ'zæn/ ● *noun* strong supporter of party, side, or cause; guerrilla. ● *adjective* of partisans; biased. □ **partisanship** *noun*.

partition /pɑ:'tɪʃ(ə)n/ ● *noun* structure dividing a space, esp. light interior wall; division into parts. ● *verb* divide into parts; (+ *off*) separate with partition.

partitive /'pɑ:tɪtɪv/ ● *adjective* (of word) denoting part of collective whole. ● *noun* partitive word.

partly *adverb* with respect to a part; to some extent.

partner /'pɑ:tnə/ ● *noun* sharer; person associated with others in business; either of pair in marriage etc. or dancing or game. ● *verb* be partner of. □ **partnership** *noun*.

partridge /'pɑ:trɪdʒ/ *noun* (*plural* same or **-s**) kind of game bird.

parturition /pɑ:tjʊ'rɪʃ(ə)n/ *noun formal* childbirth.

party /'pɑ:tɪ/ *noun* (*plural* **-ies**) social gathering; group of people travelling or working together; political group putting forward candidates in elections and usually organized on national basis; each side in agreement or dispute. □ **party line** set policy of political party etc., shared telephone line; **party wall** wall common to adjoining rooms, buildings, etc.

parvenu /'pɑ:vənju:/ *noun* (*plural* **-s**; *feminine* **parvenue**, *plural* **-s**) newly rich social climber; upstart.

pascal /'pæsk(ə)l/ *noun* SI unit of pressure; (**Pascal** /pæs'kɑ:l/) computer language designed for training.

paschal /'pæsk(ə)l/ *adjective* of Passover; of Easter.

pasha /'pɑ:ʃə/ *noun historical* Turkish officer of high rank.

pasque-flower /'pæskflaʊə/ *noun* kind of anemone.

pass[1] /pɑ:s/ ● *verb* move onward, proceed; go past; leave on one side or behind; (cause to) be transferred from one person or place to another; surpass; go unremarked or uncensured; move; cause to go; be successful in (exam); allow (bill in Parliament) to proceed; be approved; elapse; happen; spend (time etc.); *Football etc.* kick, hand, or hit (ball etc.) to player of one's own side; (+ *into*, *from*) change; come to an end; be accepted as adequate; discharge from body as or with excreta; utter (judgement etc.). ● *noun* passing, esp. of exam; status of degree without honours; written permission, ticket, or order; *Football etc.* passing of ball; critical position. □ **make a pass at** *colloquial* make sexual advances to; **pass away** die; **passbook** book recording customer's transactions with bank etc.; **passer-by** (*plural* **passers-by**) person who goes past, esp. by chance; **pass for** be accepted as; **passkey** private key to gate etc., master-key; **pass off** fade away, be carried through (in specified way), lightly dismiss, (+ *as*) misrepresent as something false; **pass on** proceed, die, transmit to next person in a series; **pass out** become unconscious, complete military training; **pass over** omit, overlook, make no remark on, die; **pass round** distribute, give to one person after another; **pass up** *colloquial* refuse or neglect (opportunity etc.); **password** prearranged word or phrase to secure recognition, admission, etc.

pass² /pɑ:s/ *noun* narrow way through mountains.

passable *adjective* adequate, fairly good.

passage /'pæsɪdʒ/ *noun* process or means of passing, transit; passageway; right to pass through; journey by sea or air; transition from one state to another; short part of book or piece of music etc.; duct etc. in body.

passageway *noun* narrow way for passing along; corridor.

passé /'pæseɪ/ *adjective* (*feminine* **passée**) outmoded; past its prime.

passenger /'pæsɪndʒə/ *noun* traveller in or on vehicle (other than driver, pilot, crew, etc.); *colloquial* idle member of team etc.

passerine /'pæsəri:n/ ● *noun* bird able to grip branch etc. with claws. ● *adjective* of passerines.

passim /'pæsɪm/ *adverb* throughout. [Latin]

passion /'pæʃ(ə)n/ *noun* strong emotion; outburst of anger; intense sexual love; strong enthusiasm; object arousing this; (**the Passion**) sufferings of Christ during his last days, Gospel narrative of this or musical setting of it. □ **passion-flower** plant with flower supposed to suggest instruments of Crucifixion; **passion-fruit** edible fruit of some species of passion-flower. □ **passionless** *adjective.*

passionate /'pæʃənət/ *adjective* dominated by, easily moved to, or showing passion. □ **passionately** *adverb.*

passive /'pæsɪv/ *adjective* acted upon, not acting; submissive; inert; *Grammar* (of verb) of which subject undergoes action (e.g. *was written* in *it was written by me*). □ **passive smoking** involuntary inhalation of others' cigarette smoke. □ **passively** *adverb*; **passivity** /-'sɪv-/ *noun.*

Passover /'pɑ:səʊvə/ *noun* Jewish spring festival commemorating Exodus from Egypt.

passport /'pɑ:spɔ:t/ *noun* official document showing holder's identity and nationality etc. and authorizing travel abroad.

past /pɑ:st/ ● *adjective* gone by; just over; of former time; *Grammar* expressing past action or state. ● *noun* past time or events; person's past life or career; past tense. ● *preposition* beyond. ● *adverb* so as to pass by. □ **past it** *colloquial* old and useless; **past master** expert.

pasta /'pæstə/ *noun* dried flour paste in various shapes.

paste /peɪst/ ● *noun* any moist fairly stiff mixture; dough of flour with fat, water, etc.; flour and water or other mixture as adhesive; meat or fish spread; hard glasslike material used for imitation gems. ● *verb* (**-ting**) fasten or coat with paste; *slang* beat, thrash. □ **pasteboard** stiff substance made by pasting together sheets of paper. □ **pasting** *noun.*

pastel /'pæst(ə)l/ ● *noun* pale shade of colour; crayon made of dry pigment-paste; drawing in pastel. ● *adjective* of pale shade of colour.

pastern /'pæst(ə)n/ *noun* part of horse's foot between fetlock and hoof.

pasteurize /'pɑ:stʃəraɪz/ *verb* (also **-ise**) (**-zing** or **-sing**) partially sterilize (milk etc.) by heating. □ **pasteurization** *noun.*

pastiche /pæs'ti:ʃ/ *noun* picture or musical composition made up from various sources; literary or other work imitating style of author or period etc.

pastille /'pæstɪl/ *noun* small sweet or lozenge.

pastime /'pɑ:staɪm/ *noun* recreation; hobby.

pastor /'pɑ:stə/ *noun* minister, esp. of Nonconformist church.

pastoral /'pɑ:stər(ə)l/ ● *adjective* of shepherds; of (esp. romanticized) rural life; of pastor. ● *noun* pastoral poem, play, picture, etc; letter from bishop or other pastor to clergy or people.

pastrami /pæ'strɑ:mɪ/ *noun* seasoned smoked beef.

pastry /'peɪstrɪ/ *noun* (*plural* **-ies**) dough of flour, fat, and water; (item of) food made wholly or partly of this.

pasturage *noun* pasture land; pasturing.

pasture /'pɑ:stʃə/ ● *noun* land covered with grass etc. for grazing animals; herbage for animals. ● *verb* (**-ring**) put (animals) to pasture; graze.

pasty[1] /'pæstɪ/ *noun* (*plural* **-ies**) pie of meat etc. wrapped in pastry and baked without dish.

pasty[2] /'peɪstɪ/ *adjective* (**-ier**, **-iest**) pallid.

pat[1] ● *verb* (**-tt-**) strike gently with flat palm or other flat surface, esp. in affection etc. ● *noun* light stroke or tap, esp. with hand in affection etc.; patting sound; small mass, esp. of butter, made (as) by patting.

pat[2] ● *adjective* known thoroughly; apposite, opportune, esp. glibly so. ● *adverb* in a pat way. □ **have off pat** have memorized perfectly.

patch ● *noun* piece put on in mending or as reinforcement; cover protecting injured eye; large or irregular spot on surface; distinct area or period; small plot of ground. ● *verb* mend with patch(es); (often + *up*) piece together; (+ *up*) settle (quarrel etc.), esp. hastily. □ **not a patch on** *colloquial* very much inferior to; **patchwork** stitching together of small pieces of differently coloured cloth to form pattern.

patchy *adjective* (**-ier**, **-iest**) uneven in quality; having patches. □ **patchily** *adverb*.

pate *noun colloquial* head.

pâté /'pæteɪ/ *noun* smooth paste of meat etc. □ ***pâté de foie gras*** /də fwɑ: 'grɑ:/ pâté made from livers of fatted geese.

patella /pə'telə/ *noun* (*plural* **patellae** /-li:/) kneecap.

paten /'pæt(ə)n/ *noun* plate for bread at Eucharist.

patent /'peɪt(ə)nt, 'pæt-/ ● *noun* official document conferring right, title, etc., esp. sole right to make, use, or sell some invention; invention or process so protected. ● *adjective* /'peɪt(ə)nt/ plain, obvious; conferred or protected by patent; (of food, medicine, etc.) proprietary. ● *verb* obtain patent for (invention). □ **patent leather** glossy varnished leather. □ **patently** *adverb*.

patentee /peɪtən'ti:/ *noun* holder of patent.

paterfamilias /peɪtəfə'mɪlɪæs/ *noun* male head of family etc.

paternal /pə'tɜ:n(ə)l/ *adjective* of father, fatherly; related through father.

paternalism *noun* policy of restricting freedom and responsibility by well-meant regulations. □ **paternalistic** /-'lɪs-/ *adjective*.

paternity /pə'tɜ:nɪtɪ/ *noun* fatherhood; one's paternal origin.

paternoster /pætə'nɒstə/ *noun* Lord's Prayer, esp. in Latin.

path /pɑ:θ/ *noun* (*plural* **paths** /pɑ:ðz/ footway, track; line along which person or thing moves. □ **pathway** path, its course.

pathetic /pə'θetɪk/ *adjective* exciting pity, sadness, or contempt. □ **pathetically** *adverb*.

pathogen /'pæθədʒ(ə)n/ *noun* agent causing disease. □ **pathogenic** /-'dʒen-/ *adjective*.

pathological /pæθə'lɒdʒɪk(ə)l/ *adjective* of pathology; of or caused by mental or physical disorder. □ **pathologically** *adverb*.

pathology /pə'θɒlədʒɪ/ *noun* study of disease. □ **pathologist** *noun*.

pathos /'peɪθɒs/ *noun* quality that excites pity or sadness.

patience /'peɪʃ(ə)ns/ *noun* ability to endure delay, hardship, provocation, pain, etc.; perseverance; solo card game.

patient ● *adjective* having or showing patience. ● *noun* person under medical etc. treatment. □ **patiently** *adverb*.

patina /'pætɪnə/ *noun* (*plural* **-s**) film, usually green, on surface of old bronze etc.; gloss produced by age on woodwork etc.

patio /'pætɪəʊ/ *noun* (*plural* **-s**) paved usually roofless area adjoining house; roofless inner courtyard.

patisserie /pə'ti:sərɪ/ *noun* shop where pastries are made and sold; pastries collectively.

patois /'pætwɑ:/ *noun* (*plural* same /-wɑ:z/) regional dialect differing from literary language.

patriarch /'peɪtrɪɑ:k/ *noun* male head of family or tribe; chief bishop in Orthodox and RC Churches; venerable old man. □ **patriarchal** /-'ɑ:k-/ *adjective.*

patriarchate /'peɪtrɪɑ:kət/ *noun* office, see, or residence of patriarch; rank of tribal patriarch.

patriarchy /'peɪtrɪɑ:kɪ/ *noun* (*plural* **-ies**) male-dominated social system, with descent reckoned through male line.

patrician /pə'trɪʃ(ə)n/ ● *noun* person of noble birth, esp. in ancient Rome. ● *adjective* of nobility; aristocratic.

patricide /'pætrɪsaɪd/ *noun* parricide. □ **patricidal** /-'saɪd(ə)l/ *adjective.*

patrimony /'pætrɪmənɪ/ *noun* (*plural* **-ies**) property inherited from father or ancestors; heritage.

patriot /'peɪtrɪət/ *noun* person devoted to and ready to defend his or her country. □ **patriotic** /-'ɒt-/ *adjective*; **patriotism** *noun.*

patrol /pə'trəʊl/ ● *noun* act of walking or travelling round area etc. to protect or supervise it; person(s) or vehicle(s) sent out on patrol; unit of usually 6 in Scout troop or Guide company. ● *verb* (**-ll-**) carry out patrol of; act as patrol. □ **patrol car** car used by police etc. for patrol.

patron /'peɪtrən/ *noun* (*feminine* **patroness**) person who gives financial or other support; customer of shop etc. □ **patron saint** saint regarded as protecting person, place, activity, etc.

patronage /'pætrənɪdʒ/ *noun* patron's or customer's support; right of bestowing or recommending for appointments; condescending manner.

patronize /'pætrənaɪz/ *verb* (also **-ise**) (**-zing** or **-sing**) treat condescendingly; act as patron to; be customer of. □ **patronizing** *adjective.*

patronymic /pætrə'nɪmɪk/ *noun* name derived from that of father or ancestor.

patten /'pæt(ə)n/ *noun historical* wooden sole mounted on iron ring for raising wearer's shoe above mud etc.

patter[1] /'pætə/ ● *noun* sound of quick light taps or steps. ● *verb* (of rain etc.) make this sound.

patter[2] /'pætə/ ● *noun* rapid often glib or deceptive talk. ● *verb* say or talk glibly.

pattern /'pæt(ə)n/ ● *noun* decorative design on surface; regular or logical form, order, etc.; model, design, or instructions from which thing is to be made; excellent example. ● *verb* decorate with pattern; model (thing) on design etc.

patty /'pætɪ/ *noun* (*plural* **-ies**) small pie or pasty.

paucity /'pɔ:sɪtɪ/ *noun* smallness of number or quantity.

paunch /pɔ:ntʃ/ *noun* belly, stomach. □ **paunchy** *adjective.*

pauper /'pɔ:pə/ *noun* very poor person. □ **pauperism** *noun.*

pause /pɔ:z/ ● *noun* temporary stop or silence; *Music* mark denoting lengthening of note or rest. ● *verb* (**-sing**) make a pause; wait.

pavane /pə'vɑ:n/ *noun* (also **pavan** /'pæv(ə)n/) *historical* stately dance; music for this.

pave *verb* (**-ving**) cover (street, floor, etc.) with durable surface. □ **pave the way** (usually + *for*) make preparations. □ **paving** *noun.*

pavement *noun* paved footway at side of road. □ **pavement artist** artist who draws in chalk on pavement for tips.

pavilion /pə'vɪljən/ *noun* building on sports ground for spectators or players; summerhouse etc. in park; large tent; building or stand at exhibition.

pavlova /pæv'ləʊvə/ *noun* meringue dessert with cream and fruit filling.

paw ● *noun* foot of animal with claws; *colloquial* person's hand. ● *verb* touch with paw; *colloquial* fondle awkwardly or indecently.

pawn[1] *noun* chessman of smallest size and value; person subservient to others' plans.

pawn[2] *verb* deposit (thing) as security for money borrowed; pledge. □ **in pawn** held as security; **pawnbroker** person who lends money at interest on security of personal property;

pawnshop pawnbroker's place of business.

pawpaw /'pɔ:pɔ:/ *noun* (also **papaya** /pə'paɪə/) pear-shaped mango-like fruit with pulpy orange flesh; tropical tree bearing this.

pay ● *verb* (*past & past participle* **paid**) discharge debt to; give as due; render, bestow (attention etc.); yield adequate return; let out (rope) by slackening it; reward or punish. ● *noun* wages. □ **in the pay of** employed by; **pay-as-you-earn** collection of income tax by deduction at source from wages etc.; **pay-claim** demand for increase in pay; **payday** day on which wages are paid; **pay for** hand over money for, bear cost of, suffer or be punished for; **paying guest** lodger; **payload** part of (esp. aircraft's) load from which revenue is derived; **paymaster** official who pays troops, workmen, etc.; **Paymaster General** Treasury minister responsible for payments; **pay off** pay in full and discharge, *colloquial* yield good results; **pay-off** *slang* payment, climax, end result; **pay phone** coin box telephone; **payroll** list of employees receiving regular pay. □ **payee** /peɪ'i:/ *noun.*

payable *adjective* that must or may be paid.

PAYE *abbreviation* pay-as-you-earn.

payment *noun* paying, amount paid; recompense.

payola /peɪ'əʊlə/ *noun esp. US slang* bribe offered for unofficial media promotion of product etc.

PC *abbreviation* Police Constable; personal computer; politically correct; political correctness; Privy Councillor.

p.c. *abbreviation* per cent; postcard.

pd. *abbreviation* paid.

PE *abbreviation* physical education.

pea *noun* climbing plant bearing round edible seeds in pods; one of its seeds; similar plant. □ **pea-souper** *colloquial* thick yellowish fog.

peace *noun* quiet, calm; freedom from or cessation of war; civil order. □ **peacemaker** person who brings about peace; **peacetime** time when country is not at war.

peaceable *adjective* disposed or tending to peace, peaceful.

peaceful *adjective* characterized by or not infringing peace. □ **peacefully** *adverb*; **peacefulness** *noun.*

peach[1] *noun* roundish juicy fruit with downy yellow or rosy skin; tree bearing it; yellowish-pink colour; *colloquial* person or thing of superlative merit. □ **peach Melba** dish of ice cream and peaches. □ **peachy** *adjective* (**-ier**, **-iest**).

peach[2] *verb colloquial* turn informer; inform.

peacock /'pi:kɒk/ *noun* (*plural* same or **-s**) male peafowl, bird with brilliant plumage and erectile fanlike tail. □ **peacock blue** bright lustrous greenish blue of peacock's neck; **peacock butterfly** butterfly with eyelike markings resembling those on peacock's tail.

peafowl /'pi:faʊl/ *noun* kind of pheasant, peacock or peahen.

peahen /'pi:hen/ *noun* female peafowl.

peak[1] ● *noun* pointed top, esp. of mountain; stiff projecting brim at front of cap; highest point of achievement, intensity, etc. ● *verb* reach highest value, quality, etc.

peak[2] *verb* waste away; (as **peaked** *adjective*) pinched-looking.

peaky *adjective* (**-ier**, **-iest**) sickly, puny.

peal ● *noun* loud ringing of bell(s); set of bells; loud repeated sound. ● *verb* (cause to) sound in peal; utter sonorously.

peanut *noun* plant bearing underground pods containing seeds used as food and yielding oil; its seed; (in *plural*) *colloquial* trivial amount, esp. of money. □ **peanut butter** paste of ground roasted peanuts.

pear /peə/ *noun* fleshy fruit tapering towards stalk; tree bearing it.

pearl /pɜ:l/ ● *noun* rounded lustrous usually white solid formed in shell of certain oysters and prized as gem; imitation of this; precious thing, finest example. ● *verb poetical* (of moisture) form drops, form drops on; fish for pearls. □ **pearl barley** barley rubbed into small rounded grains; **pearl button** button of (real or imitation) mother-of-pearl.

pearly ● *adjective* (**-ier**, **-iest**) resembling a pearl; adorned with pearls. ● *noun* (*plural* **-ies**) pearly king or queen; (in *plural*) pearly king's or queen's clothes. □ **Pearly Gates** *colloquial* gates of Heaven; **pearly king, queen** London costermonger, or his wife, wearing clothes covered with pearl buttons.

peasant /'pez(ə)nt/ *noun* (in some countries) worker on land, farm labourer; small farmer; *derogatory* lout, boor. □ **peasantry** *noun* (*plural* **-ies**).

pease-pudding /pi:z/ *noun* dried peas boiled in cloth.

peat *noun* vegetable matter decomposed by water and partly carbonized; piece of this as fuel. □ **peatbog** bog composed of peat. □ **peaty** *adjective*.

pebble /'peb(ə)l/ *noun* small stone made smooth by action of water. □ **pebble-dash** mortar with pebbles in it as wall-coating. □ **pebbly** *adjective*.

pecan /'pi:kən/ *noun* pinkish-brown smooth nut; kind of hickory producing it.

peccadillo /pekə'dɪləʊ/ *noun* (*plural* **-es** or **-s**) trivial offence.

peck[1] ● *verb* strike, pick up, pluck out, or make (hole) with beak; kiss hastily or perfunctorily; *colloquial* (+ *at*) eat (meal) listlessly or fastidiously. ● *noun* stroke with beak; hasty or perfunctory kiss. □ **pecking order** social hierarchy.

peck[2] *noun* measure of capacity for dry goods (2 gallons, 9.092 litres).

pecker *noun* □ **keep your pecker up** *colloquial* stay cheerful.

peckish *adjective colloquial* hungry.

pectin /'pektɪn/ *noun* soluble gelatinous substance in ripe fruits, causing jam etc. to set.

pectoral /'pektər(ə)l/ ● *adjective* of or for breast or chest. ● *noun* pectoral fin or muscle.

peculiar /pɪ'kju:lɪə/ *adjective* odd; (usually + *to*) belonging exclusively; belonging to the individual; particular, special.

peculiarity /pɪkju:lɪ'ærɪtɪ/ *noun* (*plural* **-ies**) oddity; characteristic; being peculiar.

peculiarly *adverb* more than usually, especially; oddly.

pecuniary /pɪ'kju:nɪərɪ/ *adjective* of or in money.

pedagogue /'pedəgɒg/ *noun archaic or derogatory* schoolmaster.

pedagogy /'pedəgɒdʒɪ/ *noun* science of teaching. □ **pedagogic(al)** /-'gɒg-/ *adjective*.

pedal /'ped(ə)l/ ● *noun* lever or key operated by foot, esp. in bicycle, motor vehicle, or some musical instruments. ● *verb* (**-ll-**; *US* **-l-**) work pedals (of); ride bicycle. ● *adjective* /'pi:d(ə)l/ of foot or feet.

pedant /'ped(ə)nt/ *noun derogatory* person who insists on strict adherence to literal meaning or formal rules. □ **pedantic** /pɪ'dæntɪk/ *adjective*; **pedantry** *noun*.

peddle /'ped(ə)l/ *verb* (**-ling**) sell as pedlar; advocate; sell (drugs) illegally; engage in selling, esp. as pedlar.

peddler *noun* person who sells drugs illegally; *US* = PEDLAR.

pedestal /'pedɪst(ə)l/ *noun* base of column; block on which something stands.

pedestrian /pɪ'destrɪən/ ● *noun* walker, esp. in town. ● *adjective* prosaic, dull. □ **pedestrian crossing** part of road where crossing pedestrians have right of way.

pedicure /'pedɪkjʊə/ *noun* care or treatment of feet, esp. of toenails.

pedigree /'pedɪgri:/ *noun* recorded (esp. distinguished) line of descent of person or animal; genealogical table; *colloquial* thing's history.

pediment /'pedɪmənt/ *noun* triangular part crowning front of building, esp. over portico.

pedlar /'pedlə/ *noun* (*US* **peddler**) travelling seller of small wares.

pedometer /pɪ'dɒmɪtə/ *noun* instrument for estimating distance travelled on foot.

pedophile *US* = PAEDOPHILE.

peduncle /pɪ'dʌŋk(ə)l/ *noun* stalk of flower, fruit, or cluster, esp. main stalk bearing solitary flower.

pee *colloquial* ● *verb* (**pees, peed**) urinate. ● *noun* urination; urine.

peek *noun & verb* peep, glance.

peel ● *verb* strip rind etc. from; (usually + *off*) take off (skin etc.); become bare of bark, skin, etc.; (often + *off*) flake off. ● *noun* rind or outer coating of fruit, potato, etc.

peeling *noun* (usually in *plural*) piece peeled off.

peep[1] ● *verb* look furtively or through narrow aperture; come cautiously or partly into view; emerge. ● *noun* furtive or peering glance; (usually + *of*) first light of dawn. □ **peep-hole** small hole to peep through; **Peeping Tom** furtive voyeur; **peep-show** exhibition of pictures etc. viewed through lens or peep-hole.

peep[2] ● *verb* cheep, squeak. ● *noun* cheep, squeak; slight sound, utterance, or complaint.

peer[1] *verb* look closely or with difficulty.

peer[2] *noun* (*feminine* **peeress**) duke, marquis, earl, viscount, or baron; equal (esp. in civil standing or rank). □ **peer group** person's associates of same status.

peerage *noun* peers as a class; rank of peer or peeress.

peerless *adjective* unequalled.

peeve *colloquial* ● *verb* (**-ving**) (usually as **peeved** *adjective*) irritate. ● *noun* cause or state of annoyance.

peevish *adjective* querulous, irritable. □ **peevishly** *adverb.*

peewit /'pi:wɪt/ *noun* lapwing.

peg ● *noun* wooden, metal, etc. bolt or pin for holding things together, hanging things on, etc.; each of pins used to tighten or loosen strings of violin etc.; forked wooden peg etc. for hanging washing on line; drink, esp. of spirits. ● *verb* (**-gg-**) (usually + *down*, *in*, etc.) fix, mark, or hang out (as) with peg(s); keep (prices etc.) stable. □ **off the peg** (of clothes) ready-made; **peg away** (often + *at*) work persistently; **pegboard** board with holes for pegs; **peg out** *slang* die, mark out boundaries of.

pejorative /pɪ'dʒɒrətɪv/ ● *adjective* derogatory. ● *noun* derogatory word.

peke *noun colloquial* Pekingese.

Pekingese /pi:kɪ'ni:z/ *noun* (also **Pekinese**) (*plural* same) dog of small short-legged snub-nosed breed with long silky hair.

pelargonium /pelə'gəʊnɪəm/ *noun* plant with showy flowers; geranium.

pelf *noun* money, wealth.

pelican /'pelɪkən/ *noun* large waterfowl with pouch below bill for storing fish. □ **pelican crossing** road crossing-place with traffic lights operated by pedestrians.

pellagra /pə'lægrə/ *noun* deficiency disease with cracking of skin.

pellet /'pelɪt/ *noun* small compressed ball of a substance; pill; small shot.

pellicle /'pelɪk(ə)l/ *noun* thin skin; membrane; film.

pell-mell /pel'mel/ *adverb* headlong; in disorder.

pellucid /pɪ'lu:sɪd/ *adjective* transparent, clear.

pelmet /'pelmɪt/ *noun* hanging border concealing curtain-rods etc.

pelt[1] ● *verb* assail with missiles, abuse, etc.; (of rain) come down hard; run at full speed. ● *noun* pelting.

pelt[2] *noun* skin of animal, esp. with hair or fur still on it.

pelvis /'pelvɪs/ *noun* lower abdominal cavity in most vertebrates, formed by haunch bones etc. □ **pelvic** *adjective.*

pen[1] ● *noun* implement for writing with ink. ● *verb* (**-nn-**) write. □ **penfriend** friend with whom one communicates by letter only; **penknife** small folding knife; **pen-name** literary pseudonym; **pen-pal** *colloquial* penfriend.

pen[2] ● *noun* small enclosure for cows, sheep, poultry, etc. ● *verb* (**-nn-**) enclose; put or keep in confined space.

pen[3] *noun* female swan.

penal /'pi:n(ə)l/ *adjective* of or involving punishment; punishable.

penalize /'pi:nəlaɪz/ *verb* (also **-ise**) (**-zing** or **-sing**) subject to penalty or disadvantage; make punishable.

penalty /'penəltɪ/ *noun* (*plural* **-ies**) fine or other punishment; disadvantage, loss, etc., esp. as result of one's own actions; disadvantage imposed in sports for breach of rules etc. □ **penalty area** *Football* area in front of goal within which breach of rules involves award of penalty kick for opposing team; **penalty kick** free kick at goal from close range.

penance /'penəns/ *noun* act of self-punishment, esp. imposed by priest, performed as expression of penitence.

pence *plural* of PENNY.

penchant /'pɑ̃ʃɑ̃/ *noun* (+ *for*) inclination or liking for.

pencil /'pens(ə)l/ ● *noun* instrument for drawing or writing, esp. of graphite enclosed in wooden cylinder or metal case with tapering end; something used or shaped like this. ● *verb* (**-ll-**; *US* **-l-**) write, draw, or mark with pencil.

pendant /'pend(ə)nt/ *noun* ornament hung from necklace etc.

pendent /'pend(ə)nt/ *adjective formal* hanging; overhanging; pending.

pending /'pendɪŋ/ ● *adjective* awaiting decision or settlement. ● *preposition* until; during.

pendulous /'pendjʊləs/ *adjective* hanging down; swinging.

pendulum /'pendjʊləm/ *noun* (*plural* **-s**) body suspended so as to be free to swing, esp. regulating movement of clock's works.

penetrate /'penɪtreɪt/ *verb* (**-ting**) make way into or through; pierce; permeate; see into or through; be absorbed by the mind; (as **penetrating** *adjective*) having or suggesting insight, (of voice) easily heard above other sounds, piercing. □ **penetrable** /-trəb(ə)l/ *adjective*; **penetration** *noun*.

penguin /'peŋgwɪn/ *noun* flightless seabird of southern hemisphere.

penicillin /penɪ'sɪlɪn/ *noun* antibiotic obtained from mould.

peninsula /pɪ'nɪnsjʊlə/ *noun* piece of land almost surrounded by water or projecting far into sea etc. □ **peninsular** *adjective*.

penis /'pi:nɪs/ *noun* sexual and (in mammals) urinatory organ of male animal.

penitent /'penɪt(ə)nt/ ● *adjective* repentant. ● *noun* penitent person; person doing penance. □ **penitence** *noun*; **penitently** *adverb*.

penitential /penɪ'tenʃ(ə)l/ *adjective* of penitence or penance.

penitentiary /penɪ'tenʃərɪ/ ● *noun* (*plural* **-ies**) *US* prison. ● *adjective* of penance or reformatory treatment.

pennant /'penənt/ *noun* tapering flag, esp. that at masthead of ship in commission.

penniless /'penɪlɪs/ *adjective* destitute.

pennon /'penən/ *noun* long narrow triangular or swallow-tailed flag; long pointed streamer on ship.

penny /'penɪ/ *noun* (*plural* **pence** or, for separate coins only, **pennies**) British coin worth 1/100 of pound, or formerly 1/240 of pound. □ **penny-farthing** early kind of bicycle with large front wheel and small rear one; **penny-pinching** *noun* meanness, *adjective* mean; **a pretty penny** a large sum of money.

pennyroyal /penɪ'rɔɪəl/ *noun* creeping kind of mint.

penology /pi:'nɒlədʒɪ/ *noun* study of punishment and prison management.

pension[1] /'penʃ(ə)n/ ● *noun* periodic payment made by government, ex-employer, private fund, etc. to person above specified age or to retired, widowed, disabled, etc. person. ● *verb* grant pension to. □ **pension off** dismiss with pension.

pension[2] /pɑ̃'sjɔ̃/ *noun* European, esp. French, boarding house. [French]

pensionable *adjective* entitled or entitling person to pension.

pensioner *noun* recipient of (esp. retirement) pension.

pensive /'pensɪv/ *adjective* deep in thought. □ **pensively** *adverb*.

pent *adjective* (often + *in*, *up*) closely confined; shut in.

penta- *combining form* five.

pentacle /'pentək(ə)l/ *noun* figure used as symbol, esp. in magic, e.g. pentagram.

pentagon /'pentəgən/ *noun* plane figure with 5 sides and angles; (**the Pentagon**) (pentagonal headquarters of) leaders of US defence forces. □ **pentagonal** /-'tæg-/ *adjective*.

pentagram /'pentəgræm/ *noun* 5-pointed star.

pentameter /pen'tæmɪtə/ *noun* line of verse with 5 metrical feet.

Pentateuch /'pentətju:k/ *noun* first 5 books of Old Testament.

pentathlon /pen'tæθlən/ *noun* athletic contest of 5 events. □ **pentathlete** *noun*.

Pentecost /'pentɪkɒst/ *noun* Whit Sunday; Jewish harvest festival 50 days after second day of Passover.

pentecostal /pentɪ'kɒst(ə)l/ *adjective* (of religious group) emphasizing divine gifts, esp. healing, and often fundamentalist.

penthouse /'penthaʊs/ *noun* flat on roof or top floor of tall building.

penultimate /pɪ'nʌltɪmət/ *adjective & noun* last but one.

penumbra /pɪ'nʌmbrə/ *noun* (*plural* **-s** or **-brae** /-bri:/) partly shaded region round shadow of opaque body; partial shadow. □ **penumbral** *adjective*.

penurious /pɪ'njʊərɪəs/ *adjective* poor; stingy.

penury /'penjʊrɪ/ *noun* (*plural* **-ies**) destitution, poverty.

peon /'pi:ən/ *noun* Spanish-American day-labourer.

peony /'pi:ənɪ/ *noun* (also **paeony**) (*plural* **-ies**) plant with large globular red, pink, or white flowers.

people /'pi:p(ə)l/ ● *plural noun* persons in general; (*singular*) race or nation; (**the people**) ordinary people, esp. as electorate; parents or other relatives; subjects. ● *verb* (**-ling**) (usually + *with*) fill with people; populate; (esp. as **peopled** *adjective*) inhabit.

PEP /pep/ *abbreviation* Personal Equity Plan.

pep *colloquial* ● *noun* vigour, spirit. ● *verb* (**-pp-**) (usually + *up*) fill with vigour. □ **pep pill** one containing stimulant drug; **pep talk** exhortation to greater effort or courage.

pepper /'pepə/ ● *noun* hot aromatic condiment from dried berries of some plants; capsicum plant, its fruit. ● *verb* sprinkle or flavour with pepper; pelt with missiles. □ **pepper-and-salt** of closely mingled dark and light colour; **peppercorn** dried pepper berry, (in full **peppercorn rent**) nominal rent; **pepper-mill** mill for grinding peppercorns by hand.

peppermint *noun* species of mint grown for its strong-flavoured oil; sweet flavoured with this oil; the oil.

pepperoni /pepə'rəʊnɪ/ *noun* sausage seasoned with pepper.

peppery *adjective* of, like, or abounding in pepper; hot-tempered.

pepsin /'pepsɪn/ *noun* enzyme contained in gastric juice.

peptic /'peptɪk/ *adjective* digestive. □ **peptic ulcer** one in stomach or duodenum.

per *preposition* for each; by, by means of, through.

peradventure /pərəd'ventʃə/ *adverb archaic* perhaps.

perambulate /pə'ræmbjʊleɪt/ *verb* (**-ting**) walk through, over, or about. □ **perambulation** *noun*.

perambulator *noun formal* pram.

per annum /pər 'ænəm/ *adverb* for each year.

per capita /pə 'kæpɪtə/ *adverb & adjective* for each person.

perceive /pə'si:v/ *verb* (**-ving**) become aware of by one of senses; apprehend; understand. □ **perceivable** *adjective*.

per cent /pə 'sent/ (*US* **percent**) ● *adverb* in every hundred. ● *noun* percentage; one part in every hundred.

percentage *noun* rate or proportion per cent; proportion.

percentile /pə'sentaɪl/ *noun* each of 99 points at which a range of data is divided to make 100 groups of equal size; each of these groups.

perceptible /pə'septɪb(ə)l/ *adjective* that can be perceived. □ **perceptibility** *noun*; **perceptibly** *adverb*.

perception /pə'sepʃ(ə)n/ *noun* act or faculty of perceiving. □ **perceptual** /-'septʃʊəl/ *adjective*.

perceptive /pə'septɪv/ *adjective* sensitive; discerning; capable of perceiving. □ **perceptively** *adverb*; **perceptiveness** *noun*.

perch[1] ● *noun* bird's resting-place above ground; high place for person or thing to rest on; *historical* measure of length (5½ yds). ● *verb* rest or place on perch.

perch[2] *noun* (*plural* same or **-s**) edible spiny-finned freshwater fish.

perchance /pə'tʃɑ:ns/ *adverb archaic* maybe.

percipient /pə'sɪpɪənt/ *adjective* perceiving; conscious.

percolate /'pɜ:kəleɪt/ *verb* (**-ting**) (often + *through*) filter gradually; (of idea etc.) permeate gradually; prepare (coffee) in percolator. □ **percolation** *noun*.

percolator *noun* apparatus for making coffee by circulating boiling water through ground beans.

percussion /pə'kʌʃ(ə)n/ *noun* playing of music by striking instruments with sticks etc.; such instruments collectively; gentle tapping of body in medical diagnosis; forcible striking of body against another. □ **percussionist** *noun*; **percussive** *adjective*.

perdition /pə'dɪʃ(ə)n/ *noun* damnation.

peregrine /'perɪgrɪn/ *noun* (in full **peregrine falcon**) kind of falcon.

peremptory /pə'remptərɪ/ *adjective* admitting no denial or refusal; imperious. □ **peremptorily** *adverb*.

perennial /pə'renɪəl/ ● *adjective* lasting through the year; (of plant) living several years; lasting long or for ever. ● *noun* perennial plant. □ **perennially** *adverb*.

perestroika /pere'strɔɪkə/ *noun* (in former USSR) reform of economic and political system.

perfect ● *adjective* /'pɜ:fɪkt/ complete; faultless; not deficient; very enjoyable; exact, precise; entire, unqualified; *Grammar* (of tense) expressing completed action. ● *verb* /pə'fekt/ make perfect; complete. ● *noun* /'pɜ:fɪkt/ perfect tense. □ **perfect pitch** *Music* ability to recognize pitch of note.

perfection /pə'fekʃ(ə)n/ *noun* being or making perfect; perfect state; perfect person, specimen, etc.

perfectionism *noun* uncompromising pursuit of perfection. □ **perfectionist** *noun*.

perfectly *adverb* quite, completely; in a perfect way.

perfidy /'pɜ:fɪdɪ/ *noun* breach of faith, treachery. □ **perfidious** /-'fɪd-/ *adjective*.

perforate /'pɜ:fəreɪt/ *verb* (**-ting**) pierce, make hole(s) through; make row of small holes in (paper etc.). □ **perforation** *noun*.

perforce /pə'fɔ:s/ *adverb archaic* unavoidably, necessarily.

perform /pə'fɔ:m/ *verb* carry into effect; go through, execute; act, sing, etc., esp. in public; (of animals) do tricks etc. □ **performing arts** drama, music, dance, etc. □ **performer** *noun*.

performance *noun* act, process, or manner of doing or functioning; execution (of duty etc.); performing of or in play etc.; *colloquial* fuss, emotional scene.

perfume /'pɜ:fju:m/ ● *noun* sweet smell; fragrant liquid, esp. for application to the body, scent. ● *verb* (**-ming**) impart perfume to.

perfumer /pə'fju:mə/ *noun* maker or seller of perfumes. □ **perfumery** *noun* (*plural* **-ies**).

perfunctory /pə'fʌŋktərɪ/ *adjective* done merely out of duty; superficial. □ **perfunctorily** *adverb*; **perfunctoriness** *noun*.

pergola /'pɜ:gələ/ *noun* arbour or covered walk arched with climbing plants.

perhaps /pə'hæps/ *adverb* it may be, possibly.

perianth /'perɪænθ/ *noun* outer part of flower.

perigee /'perɪdʒi:/ *noun* point nearest to earth in orbit of moon etc.

perihelion /perɪ'hi:lɪən/ *noun* (*plural* **-lia**) point nearest to sun in orbit of planet, comet, etc. round it.

peril /'perɪl/ *noun* serious and immediate danger. □ **perilous** *adjective*; **perilously** *adverb*.

perimeter /pə'rɪmɪtə/ *noun* circumference or outline of closed figure; length of this; outer boundary.

period /'pɪərɪəd/ ● *noun* amount of time during which something runs its course; distinct portion of history, life, etc.; occurrence of menstruation, time of this; complete sentence; *esp. US* full stop. ● *adjective* characteristic of past period.

periodic /pɪərɪ'ɒdɪk/ *adjective* appearing or recurring at intervals. □ **periodic table** arrangement of chemical elements by atomic number and chemical properties. □ **periodicity** /-rɪə'dɪsɪtɪ/ *noun*.

periodical ● *noun* magazine etc. published at regular intervals. ● *adjective* periodic. □ **periodically** *adverb*.

peripatetic /perɪpə'tetɪk/ *adjective* (of teacher) working in more than one establishment; going from place to place; itinerant.

peripheral /pə'rɪfər(ə)l/ ● *adjective* of minor importance; of periphery. ● *noun* input, output, or storage device connected to computer.

periphery /pə'rɪfərɪ/ *noun* (*plural* **-ies**) bounding line, esp. of round surface; outer or surrounding area.

periphrasis /pə'rɪfrəsɪs/ *noun* (*plural* **-phrases** /-si:z/) circumlocution, roundabout speech or phrase. □ **periphrastic** /perɪ'fræstɪk/ *adjective*.

periscope /ˈperɪskəʊp/ *noun* apparatus with tube and mirrors or prisms for viewing objects otherwise out of sight.

perish /ˈperɪʃ/ *verb* suffer destruction, die; lose natural qualities; (cause to) rot or deteriorate; (in *passive*) suffer from cold.

perishable • *adjective* subject to speedy decay; liable to perish. • *noun* perishable thing (esp. food).

perisher *noun slang* annoying person.

perishing *colloquial* • *adjective* confounded; intensely cold. • *adverb* confoundedly.

peritoneum /perɪtəˈniːəm/ *noun* (*plural* **-s** or **-nea**) membrane lining abdominal cavity. □ **peritoneal** *adjective*.

peritonitis /perɪtəˈnaɪtɪs/ *noun* inflammation of peritoneum.

periwig /ˈperɪwɪg/ *noun historical* wig.

periwinkle[1] /ˈperɪwɪŋk(ə)l/ *noun* evergreen trailing plant with blue or white flower.

periwinkle[2] /ˈperɪwɪŋk(ə)l/ *noun* winkle.

perjure /ˈpɜːdʒə/ *verb* (**-ring**) (**perjure oneself**) commit perjury; (as **perjured** *adjective*) guilty of perjury. □ **perjurer** *noun*.

perjury /ˈpɜːdʒərɪ/ *noun* (*plural* **-ies**) wilful lying while on oath.

perk[1] *verb* □ **perk up** (cause to) recover courage, smarten up, raise (head etc.) briskly.

perk[2] *noun colloquial* perquisite.

perky *adjective* (**-ier**, **-iest**) lively and cheerful.

perm[1] • *noun* permanent wave. • *verb* give permanent wave to.

perm[2] *colloquial* • *noun* permutation. • *verb* make permutation of.

permafrost /ˈpɜːməfrɒst/ *noun* permanently frozen subsoil, as in polar regions.

permanent /ˈpɜːmənənt/ *adjective* lasting or intended to last indefinitely. □ **permanent wave** long-lasting artificial wave in hair. □ **permanence** *noun*; **permanently** *adverb*.

permeable /ˈpɜːmɪəb(ə)l/ *adjective* capable of being permeated. □ **permeability** *noun*.

permeate /ˈpɜːmɪeɪt/ *verb* (**-ting**) penetrate, saturate, pervade; be diffused. □ **permeation** *noun*.

permissible /pəˈmɪsɪb(ə)l/ *adjective* allowable. □ **permissibility** *noun*.

permission /pəˈmɪʃ(ə)n/ *noun* consent, authorization.

permissive /pəˈmɪsɪv/ *adjective* tolerant, liberal; giving permission. □ **permissiveness** *noun*.

permit • *verb* /pəˈmɪt/ (**-tt-**) give consent to; authorize; allow; give opportunity; (+ *of*) allow as possible. • *noun* /ˈpɜːmɪt/ written order giving permission or allowing entry.

permutation /pɜːmjʊˈteɪʃ(ə)n/ *noun* one of possible ordered arrangements of set of things; combination or selection of specified number of items from larger group.

pernicious /pəˈnɪʃəs/ *adjective* destructive, injurious. □ **pernicious anaemia** defective formation of red blood cells through lack of vitamin B.

pernickety /pəˈnɪkɪtɪ/ *adjective colloquial* fastidious, over-precise.

peroration /perəˈreɪʃ(ə)n/ *noun* concluding part of speech.

peroxide /pəˈrɒksaɪd/ • *noun* (in full **hydrogen peroxide**) colourless liquid used in water solution, esp. to bleach hair; oxide containing maximum proportion of oxygen. • *verb* (**-ding**) bleach (hair) with peroxide.

perpendicular /pɜːpənˈdɪkjʊlə/ • *adjective* (usually + *to*) at right angles; upright; very steep; (**Perpendicular**) of or in style of English Gothic architecture of 15th & 16th c. • *noun* perpendicular line etc. □ **perpendicularity** /-ˈlærɪtɪ/ *noun*.

perpetrate /ˈpɜːpɪtreɪt/ *verb* (**-ting**) commit. □ **perpetration** *noun*; **perpetrator** *noun*.

perpetual /pəˈpetʃʊəl/ *adjective* lasting for ever or indefinitely; continuous; *colloquial* frequent. □ **perpetually** *adverb*.

perpetuate /pəˈpetʃʊeɪt/ *verb* (**-ting**) make perpetual; cause to be always remembered. □ **perpetuation** *noun*.

perpetuity /pɜːpɪˈtjuːɪtɪ/ *noun* (*plural* **-ies**) perpetual continuance or possession. □ **in perpetuity** for ever.

perplex /pəˈpleks/ *verb* bewilder, puzzle; complicate, tangle. □ **perplexing** *adjective*; **perplexity** *noun*.

per pro. /pɜː ˈprəʊ/ *abbreviation* through the agency of (used in signatures) (*per procurantionem*). [Latin].

■ **Usage** The abbreviation *per pro.* (or *p.p.*) is frequently written before the wrong name: "T. Jones, *p.p.* P. Smith" means that P. Smith is signing on behalf of T. Jones.

perquisite /'pɜ:kwɪzɪt/ *noun* extra profit additional to main income etc.; customary extra right or privilege.

■ **Usage** *Perquisite* is sometimes confused with *prerequisite*, which means 'a thing required as a precondition'.

perry /'perɪ/ *noun* (*plural* **-ies**) drink made from fermented pear juice.

per se /pɜ: 'seɪ/ *adverb* by or in itself, intrinsically. [Latin]

persecute /'pɜ:sɪkju:t/ *verb* (**-ting**) subject to constant hostility and ill-treatment; harass, worry. □ **persecution** /-'kju:ʃ(ə)n/ *noun*; **persecutor** *noun*.

persevere /pɜ:sɪ'vɪə/ *verb* (**-ring**) continue steadfastly, persist. □ **perseverance** *noun*.

Persian /'pɜ:ʃ(ə)n/ ● *noun* native, national, or language of Persia (now Iran); (in full **Persian cat**) cat with long silky hair. ● *adjective* of Persia (Iran). □ **Persian lamb** silky curled fur of young karakul.

persiflage /'pɜ:sɪflɑ:ʒ/ *noun* banter; light raillery.

persimmon /pɜ:'sɪmən/ *noun* tropical tree; its edible orange tomato-like fruit.

persist /pə'sɪst/ *verb* (often + *in*) continue to exist or do something in spite of obstacles. □ **persistence** *noun*; **persistent** *adjective*; **persistently** *adverb*.

person /'pɜ:s(ə)n/ *noun* individual human being; living body of human being; *Grammar* one of 3 classes of pronouns, verb-forms, etc., denoting person etc. speaking, spoken to, or spoken of. □ **in person** physically present.

persona /pə'səʊnə/ *noun* (*plural* **-nae** /-ni:/) aspect of personality as perceived by others. □ ***persona grata*** /'grɑ:tə/ (*plural* ***personae gratae*** /-ni:, -ti:/) person acceptable to certain others; ***persona non grata*** /nɒn/ (*plural* ***personae non gratae***) person not acceptable.

personable *adjective* pleasing in appearance or demeanour.

personage *noun* person, esp. important one.

personal /'pɜ:sən(ə)l/ *adjective* one's own; individual, private; done etc. in person; directed to or concerning individual; referring (esp. in hostile way) to individual's private life; *Grammar* of or denoting one of the 3 persons. □ **personal column** part of newspaper devoted to private advertisements and messages; **personal computer** computer designed for use by single individual; **personal equity plan** scheme for tax-free personal investments; **personal organizer** means of keeping track of personal affairs, esp. loose-leaf notebook divided into sections; **personal pronoun** pronoun replacing subject, object, etc., of clause etc.; **personal property** all property except land.

personality /pɜ:sə'nælɪtɪ/ *noun* (*plural* **-ies**) distinctive personal character; well-known person; (in *plural*) personal remarks.

personalize *verb* (also **-ise**) (**-zing** or **-sing**) identify as belonging to particular person.

personally *adverb* in person; for one's own part; in a personal way.

personification /pəsɒnɪfɪ'keɪʃ(ə)n/ *noun* type of metaphor in which human qualities are attributed to object, plant, animal, nature, etc., e.g. *Life can play some nasty tricks*.

personify /pə'sɒnɪfaɪ/ *verb* (**-ies**, **-ied**) attribute human characteristics to; symbolize by human figure; (usually as **personified** *adjective*) embody, exemplify typically. □ **personification** *noun*.

personnel /pɜ:sə'nel/ *noun* staff of an organization; people engaged in particular service, profession, etc. □ **personnel department** department of firm etc. dealing with appointment, training, and welfare of employees.

perspective /pə'spektɪv/ ● *noun* art of drawing so as to give effect of solidity and relative position and size; relation as to position and distance, or proportion between visible objects, parts of subject, etc.; mental view of relative importance of things; view, prospect. ● *adjective* of or in perspective. □ **in** or **out of perspective** according or not according to rules of perspective, in or not in proportion.

Perspex /'pɜ:speks/ *noun proprietary term* tough light transparent plastic.

perspicacious /pɜ:spɪ'keɪʃəs/ *adjective* having mental penetration or discernment. □ **perspicacity** /-'kæs-/ *noun*.

perspicuous /pə'spɪkjʊəs/ *adjective* lucid; clearly expressed. □ **perspicuity** /-'kju:-/ *noun*.

perspire /pə'spaɪə/ *verb* (**-ring**) sweat. □ **perspiration** /pɜ:spɪ'reɪʃ(ə)n/ *noun*.

persuade /pə'sweɪd/ *verb* (**-ding**) cause (person) by argument etc. to believe or do something; convince.

persuasion /pə'sweɪʒ(ə)n/ *noun* persuading; conviction; religious belief or sect.

persuasive /pə'sweɪsɪv/ *adjective* able or tending to persuade. □ **persuasively** *adverb*; **persuasiveness** *noun*.

pert *adjective* saucy, impudent; jaunty. □ **pertly** *adverb*; **pertness** *noun*.

pertain /pə'teɪn/ *verb* belong, relate.

pertinacious /pɜ:tɪ'neɪʃəs/ *adjective* persistent, obstinate. □ **pertinacity** /-'næs-/ *noun*.

pertinent /'pɜ:tɪnənt/ *adjective* relevant. □ **pertinence** *noun*; **pertinency** *noun*.

perturb /pə'tɜ:b/ *verb* throw into agitation; disquiet. □ **perturbation** *noun*.

peruke /pə'ru:k/ *noun historical* wig.

peruse /pə'ru:z/ *verb* (**-sing**) read; scan. □ **perusal** *noun*.

pervade /pə'veɪd/ *verb* (**-ding**) spread through, permeate; be rife among. □ **pervasion** *noun*; **pervasive** *adjective*.

perverse /pə'vɜ:s/ *adjective* obstinately or wilfully in the wrong; wayward. □ **perversely** *adverb*; **perversity** *noun*.

perversion /pə'vɜ:ʃ(ə)n/ *noun* perverting, being perverted; preference for abnormal form of sexual activity.

pervert ● *verb* /pə'vɜ:t/ turn (thing) aside from proper or normal use; lead astray from right behaviour or belief etc.; (as **perverted** *adjective*) showing perversion. ● *noun* /'pɜ:vɜ:t/ person who is perverted, esp. sexually.

pervious /'pɜ:vɪəs/ *adjective* permeable; allowing passage or access.

peseta /pə'seɪtə/ *noun* Spanish monetary unit.

peso /'peɪsəʊ/ *noun* (*plural* **-s**) monetary unit in several Latin American countries.

pessary /'pesərɪ/ *noun* (*plural* **-ies**) device worn in vagina; vaginal suppository.

pessimism /'pesɪmɪz(ə)m/ *noun* tendency to take worst view or expect worst outcome. □ **pessimist** *noun*; **pessimistic** /-'mɪst-/ *adjective*.

pest *noun* troublesome or destructive person, animal, or thing.

pester /'pestə/ *verb* trouble or annoy, esp. with persistent requests.

pesticide /'pestɪsaɪd/ *noun* substance for destroying harmful insects etc.

pestilence /'pestɪləns/ *noun* fatal epidemic disease, esp. bubonic plague.

pestilent /'pestɪlənt/ *adjective* deadly; harmful or morally destructive.

pestilential /pestɪ'lenʃ(ə)l/ *adjective* of pestilence; pestilent.

pestle /'pes(ə)l/ *noun* instrument for pounding substances in a mortar.

pet[1] ● *noun* domestic animal kept for pleasure or companionship; favourite. ● *adjective* as, of, or for a pet; favourite; *expressing fondness*. ● *verb* (**-tt-**) fondle, esp. erotically; treat as pet.

pet[2] *noun* fit of ill humour.

petal /'pet(ə)l/ *noun* each division of flower corolla.

peter /'pi:tə/ *verb* □ **peter out** diminish, come to an end.

petersham /'pi:təʃəm/ *noun* thick ribbed silk ribbon.

petiole /'petɪəʊl/ *noun* leaf-stalk.

petite /pə'ti:t/ *adjective* (of woman) of small dainty build. [French]

petit four /petɪ'fɔ:/ *noun* (*plural* **petits fours** /'fɔ:z/) very small fancy cake.

petition /pə'tɪʃ(ə)n/ ● *noun* request, supplication; formal written request, esp. one signed by many people, to authorities etc. ● *verb* make petition to; ask humbly.

petit point /petɪ 'pwæ̃/ *noun* embroidery on canvas using small stitches.

petrel /'petr(ə)l/ *noun* seabird, usually flying far from land.

petrify /'petrɪfaɪ/ *verb* (**-ies**, **-ied**) paralyse with terror or astonishment etc.; turn or be turned into stone. □ **petrifaction** /-'fækʃ(ə)n/ *noun*.

petrochemical /petrəʊ'kemɪk(ə)l/ *noun* substance obtained from petroleum or natural gas.

petrodollar /'petrəʊdɒlə/ *noun* notional unit of currency earned by petroleum-exporting country.

petrol /'petr(ə)l/ *noun* refined petroleum used as fuel in motor vehicles, aircraft, etc.

petroleum /pɪ'trəʊlɪəm/ *noun* hydrocarbon oil found in upper strata of earth, refined for use as fuel etc. □ **petroleum jelly** translucent solid mixture of hydrocarbons got from petroleum and used as lubricant etc.

petticoat /'petɪkəʊt/ *noun* woman's or girl's undergarment hanging from waist or shoulders.

pettifogging /'petɪfɒgɪŋ/ *adjective* quibbling; petty; dishonest.

pettish *adjective* fretful, peevish.

petty /'petɪ/ *adjective* (**-ier**, **-iest**) unimportant, trivial; small-minded; minor, inferior. □ **petty cash** money kept for small items of expenditure; **petty officer** naval NCO. □ **pettiness** *noun*.

petulant /'petjʊlənt/ *adjective* peevishly impatient or irritable. □ **petulance** *noun*; **petulantly** *adverb*.

petunia /pɪ'tju:nɪə/ *noun* cultivated plant with vivid funnel-shaped flowers.

pew *noun* (in church) enclosed compartment or fixed bench with back; *colloquial* seat.

pewter /'pju:tə/ *noun* grey alloy of tin, antimony, and copper; articles made of this.

peyote /peɪ'əʊtɪ/ *noun* a Mexican cactus; hallucinogenic drug prepared from it.

pfennig /'fenɪg/ *noun* one-hundredth of Deutschmark.

PG *abbreviation* (of film) classified as suitable for children subject to parental guidance.

pH /pi:'eɪtʃ/ *noun* measure of acidity or alkalinity of a solution.

phagocyte /'fægəsaɪt/ *noun* blood corpuscle etc. capable of absorbing foreign matter.

phalanx /'fælæŋks/ *noun* (*plural* **phalanxes** or **phalanges** /fə'lændʒi:z/) group of infantry in close formation; united or organized party or company.

phallus /'fæləs/ *noun* (*plural* **phalli** /-laɪ/ or **phalluses**) (esp. erect) penis; image of this. □ **phallic** *adjective*.

phantasm /'fæntæz(ə)m/ *noun* illusion; phantom. □ **phantasmal** /-'tæzm(ə)l/ *adjective*.

phantasmagoria /fæntæzmə'gɔ:rɪə/ *noun* shifting scene of real or imaginary figures. □ **phantasmagoric** /-'gɒrɪk/ *adjective*.

phantom /'fæntəm/ ● *noun* spectre, apparition; mental illusion. ● *adjective* illusory.

Pharaoh /'feərəʊ/ *noun* ruler of ancient Egypt.

Pharisee /'færɪsi:/ *noun* member of ancient Jewish sect distinguished by strict observance of traditional and written law; self-righteous person; hypocrite. □ **Pharisaic** /-'seɪɪk/ *adjective*.

pharmaceutical /fɑ:mə'sju:tɪk(ə)l/ *adjective* of pharmacy; of use or sale of medicinal drugs. □ **pharmaceutics** *noun*.

pharmacist /'fɑ:məsɪst/ *noun* person qualified to practise pharmacy.

pharmacology /fɑ:mə'kɒlədʒɪ/ *noun* study of action of drugs on the body. □ **pharmacological** /-kə'lɒdʒ-/ *adjective*; **pharmacologist** *noun*.

pharmacopoeia /fɑ:məkə'pi:ə/ *noun* book with list of drugs and directions for use; stock of drugs.

pharmacy /'fɑ:məsɪ/ *noun* (*plural* **-ies**) preparation and dispensing of drugs; pharmacist's shop; dispensary.

pharynx /'færɪŋks/ *noun* (*plural* **pharynges** /-rɪndʒi:z/ or **-xes**) cavity behind mouth and nose. □ **pharyngeal** /færɪn'dʒi:əl/ *adjective*.

phase /feɪz/ ● *noun* stage of development, process, or recurring sequence; aspect of moon or planet. ● *verb* (**-sing**) carry out by phases. □ **phase in**, **out** bring gradually into or out of use.

Ph.D. *abbreviation* Doctor of Philosophy.

pheasant /'fez(ə)nt/ *noun* long-tailed game bird.

phenomenal /fɪ'nɒmɪn(ə)l/ *adjective* extraordinary, remarkable; of or concerned with phenomena. □ **phenomenally** *adverb*.

phenomenon /fɪ'nɒmɪnən/ *noun* (*plural* **-mena**) observed or apparent object, fact, or occurrence; remarkable person or thing.

■ **Usage** It is a mistake to use the plural form *phenomena* when only one phenomenon is meant.

phew /fjuː/ *interjection expressing disgust, relief, etc.*

phial /'faɪəl/ *noun* small glass bottle.

philander /fɪ'lændə/ *verb* flirt or have casual affairs with women. □ **philanderer** *noun.*

philanthropy /fɪ'lænθrəpɪ/ *noun* love of all humankind; practical benevolence. □ **philanthropic** /-'θrɒp-/ *adjective*; **philanthropist** *noun.*

philately /fɪ'lætəlɪ/ *noun* stamp-collecting. □ **philatelist** *noun.*

philharmonic /fɪlhɑː'mɒnɪk/ *adjective* devoted to music.

philippic /fɪ'lɪpɪk/ *noun* bitter verbal attack.

philistine /'fɪlɪstaɪn/ ● *noun* person who is hostile or indifferent to culture. ● *adjective* hostile or indifferent to culture. □ **philistinism** /-stɪn-/ *noun.*

Phillips /'fɪlɪps/ *noun proprietary term* □ **Phillips screw, screwdriver** screw with cross-shaped slot, corresponding screwdriver.

philology /fɪ'lɒlədʒɪ/ *noun* study of language. □ **philological** /-lə'lɒdʒ-/ *adjective*; **philologist** *noun.*

philosopher /fɪ'lɒsəfə/ *noun* expert in or student of philosophy; person who acts philosophically.

philosophical /fɪlə'sɒfɪk(ə)l/ *adjective* (also **philosophic**) of or according to philosophy; calm under adverse circumstances. □ **philosophically** *adverb.*

philosophize /fɪ'lɒsəfaɪz/ *verb* (also **-ise**) (**-zing** or **-sing**) reason like philosopher; theorize.

philosophy /fɪ'lɒsəfɪ/ *noun* (*plural* **-ies**) use of reason and argument in seeking truth and knowledge, esp. of ultimate reality or of general causes and principles; philosophical system; system for conduct of life.

philtre /'fɪltə/ *noun* (*US* **philter**) love potion.

phlebitis /flɪ'baɪtɪs/ *noun* inflammation of vein. □ **phlebitic** /-'bɪt-/ *adjective.*

phlegm /flem/ *noun* bronchial mucus ejected by coughing; calmness; sluggishness.

phlegmatic /fleg'mætɪk/ *adjective* calm; sluggish.

phlox /flɒks/ *noun* (*plural* same or **-es**) plant with clusters of white or coloured flowers.

phobia /'fəʊbɪə/ *noun* abnormal fear or aversion. □ **phobic** *adjective & noun.*

phoenix /'fiːnɪks/ *noun* bird, the only one of its kind, fabled to burn itself and rise from its ashes.

phone *noun & verb* (**-ning**) *colloquial* telephone. □ **phone book** telephone directory; **phonecard** card holding prepaid units for use with cardphone; **phone-in** broadcast programme in which listeners or viewers participate by telephone.

phonetic /fə'netɪk/ *adjective* of or representing vocal sounds; (of spelling) corresponding to pronunciation. □ **phonetically** *adverb.*

phonetics *plural noun* (usually treated as *singular*) study or representation of vocal sounds. □ **phonetician** /fəʊnɪ'tɪʃ(ə)n/ *noun.*

phoney /'fəʊnɪ/ (also **phony**) *colloquial* ● *adjective* (**-ier**, **-iest**) false, sham, counterfeit. ● *noun* (*plural* **-eys** or **-ies**) phoney person or thing. □ **phoniness** *noun.*

phonic /'fɒnɪk/ *adjective* of (vocal) sound.

phonograph /'fəʊnəgrɑːf/ *noun* early form of gramophone.

phonology /fə'nɒlədʒɪ/ *noun* study of sounds in language. □ **phonological** /fəʊnə'lɒdʒɪk(ə)l/ *adjective.*

phony = PHONEY.

phosphate /'fɒsfeɪt/ *noun* salt of phosphoric acid, esp. used as fertilizer.

phosphorescence /fɒsfə'res(ə)ns/ *noun* emission of light without combustion or perceptible heat. □ **phosphoresce** *verb* (**-cing**); **phosphorescent** *adjective.*

phosphorus /'fɒsfərəs/ *noun* non-metallic element occurring esp. as waxlike substance appearing luminous in dark. □ **phosphoric** /-'fɒrɪk/ *adjective*; **phosphorous** *adjective.*

photo /'fəʊtəʊ/ *noun* (*plural* **-s**) photograph. □ **photo finish** close finish of race in which winner is distinguishable only on photograph; **photofit** picture of suspect constructed from composite photographs.

photo- *combining form* light; photography.

photocopier /'fəʊtəʊkɒpɪə/ *noun* machine for photocopying documents.

photocopy /'fəʊtəʊkɒpɪ/ ● *noun* (*plural* **-ies**) photographic copy of document. ● *verb* (**-ies**, **-ied**) make photocopy of.

photoelectric /fəʊtəʊɪ'lektrɪk/ *adjective* with or using emission of electrons from substances exposed to light. □ **photoelectric cell** device using this effect to generate current. □ **photoelectricity** /-'trɪsɪtɪ/ *noun.*

photogenic /fəʊtəʊ'dʒenɪk/ *adjective* looking attractive in photographs; producing or emitting light.

photograph /'fəʊtəgrɑ:f/ ● *noun* picture formed by chemical action of light on sensitive film. ● *verb* take photograph (of). □ **photographer** /fə'tɒgrəfə/ *noun*; **photographic** /-'græf-/ *adjective*; **photography** /fə'tɒgrəfɪ/ *noun.*

photogravure /fəʊtəʊgrə'vjʊə/ *noun* picture produced from photographic negative transferred to metal plate and etched in; this process.

photojournalism /fəʊtəʊ'dʒɜ:nəlɪz(ə)m/ *noun* reporting of news by photographs in magazines etc.

photolithography /fəʊtəʊlɪ'θɒgrəfɪ/ *noun* lithography using plates made photographically.

photometer /fəʊ'tɒmɪtə/ *noun* instrument for measuring light. □ **photometric** /fəʊtəʊ'metrɪk/ *adjective*; **photometry** /-'tɒmɪtrɪ/ *noun.*

photon /'fəʊtɒn/ *noun* quantum of electromagnetic radiation energy.

Photostat /'fəʊtəʊstæt/ ● *noun proprietary term* type of photocopier; copy made by it. ● *verb* (**photostat**) (**-tt-**) make Photostat of.

photosynthesis /fəʊtəʊ'sɪnθəsɪs/ *noun* process in which energy of sunlight is used by green plants to form carbohydrates from carbon dioxide and water. □ **photosynthesize** *verb* (also **-ise**) (**-zing** or **-sing**).

phrase /freɪz/ ● *noun* group of words forming conceptual unit but not sentence (see panel); short pithy expression; *Music* short sequence of notes. ● *verb* (**-sing**) express in words; divide (music) into phrases. □ **phrase book** book listing phrases and their foreign equivalents, for use by tourists etc. □ **phrasal** *adjective.*

phraseology /freɪzɪ'ɒlədʒɪ/ *noun* (*plural* **-ies**) choice or arrangement of words. □ **phraseological** /-zɪə'lɒdʒ-/ *adjective.*

phrenology /frɪ'nɒlədʒɪ/ *noun historical* study of external form of cranium as supposed indication of mental faculties etc. □ **phrenologist** *noun.*

phut /fʌt/ *adverb colloquial* □ **go phut** collapse, break down.

phylactery /fɪ'læktərɪ/ *noun* (*plural* **-ies**) small box containing Hebrew texts, worn by Jewish man at prayer.

phylum /'faɪləm/ *noun* (*plural* **phyla**) major division of plant or animal kingdom.

physic /'fɪzɪk/ *noun esp. archaic* medicine; medical art or profession.

physical /'fɪzɪk(ə)l/ ● *adjective* of the body; of matter; of nature or according

Phrase

A phrase is a group of words that has meaning but does not have a subject, verb, or object (unlike a clause or sentence). It can be:

1 a noun phrase, functioning as a noun, e.g.

I went to see my friend Tom.
The only ones they have *are too small.*

2 an adjective phrase, functioning as an adjective, e.g.

I was very pleased indeed.
This one is better than mine.

3 an adverb phrase, functioning as an adverb, e.g.

They drove off in their car.
I was there ten days ago.

to its laws; of physics. ● *noun US* medical examination. □ **physically** *adverb.*

physician /fɪ'zɪʃ(ə)n/ *noun* doctor, esp. specialist in medical diagnosis and treatment.

physics /'fɪzɪks/ *plural noun* (usually treated as *singular*) science of properties and interaction of matter and energy. □ **physicist** *noun.*

physiognomy /fɪzɪ'ɒnəmɪ/ *noun* (*plural* **-ies**) features or type of face; art of judging character from face etc.

physiology /fɪzɪ'ɒlədʒɪ/ *noun* science of functioning of living organisms. □ **physiological** /-ə'lɒdʒ-/ *adjective*; **physiologist** *noun.*

physiotherapy /fɪzɪəʊ'θerəpɪ/ *noun* treatment of injury or disease by exercise, heat, or other physical agencies. □ **physiotherapist** *noun.*

physique /fɪ'zi:k/ *noun* bodily structure and development.

pi /paɪ/ *noun* sixteenth letter of Greek alphabet (Π, π); (as π) symbol of ratio of circumference of circle to diameter (approx. 3.14).

pia mater /paɪə 'meɪtə/ *noun* inner membrane enveloping brain and spinal cord.

pianissimo /pɪə'nɪsɪməʊ/ *Music* ● *adjective* very soft. ● *adverb* very softly. ● *noun* (*plural* **-s** or **-mi** /-mɪ/) very soft playing, singing, or passage.

pianist /'pɪənɪst/ *noun* player of piano.

piano[1] /pɪ'ænəʊ/ *noun* (*plural* **-s**) keyboard instrument with metal strings struck by hammers. □ **piano-accordion** accordion with small keyboard like that of piano.

piano[2] /'pjɑ:nəʊ/ *Music* ● *adjective* soft. ● *adverb* softly. ● *noun* (*plural* **-s** or **-ni** /-nɪ/) soft playing, singing, or passage.

pianoforte /pɪænəʊ'fɔ:tɪ/ *noun formal or archaic* = PIANO[1].

piazza /pɪ'ætsə/ *noun* public square or market-place.

pibroch /'pi:brɒk/ *noun* martial or funeral bagpipe music.

picador /'pɪkədɔ:/ *noun* mounted man with lance in bullfight.

picaresque /pɪkə'resk/ *adjective* (of style of fiction) dealing with episodic adventures of rogues.

piccalilli /pɪkə'lɪlɪ/ *noun* (*plural* **-s**) pickle of chopped vegetables, mustard, and spices.

piccolo /'pɪkələʊ/ *noun* (*plural* **-s**) small high-pitched flute.

pick ● *verb* select carefully; pluck, gather (flower, fruit, etc.); probe with fingers or instrument to remove unwanted matter; clear (bone etc.) of scraps of meat etc.; eat (food, meal, etc.) in small bits. ● *noun* picking, selection; (usually + *of*) best; pickaxe; *colloquial* plectrum; instrument for picking. □ **pick a lock** open lock with instrument other than proper key, esp. with criminal intent; **pick on** nag at, find fault with, select; **pickpocket** person who steals from pockets; **pick up** take hold of and lift, acquire casually, learn routinely, stop for and take with one, make acquaintance of casually, recover, improve, arrest, detect, manage to receive (broadcast signal etc.), accept responsibility of paying (bill etc.), resume; **pick-up** person met casually, small open truck, part of record player carrying stylus, device on electric guitar etc. that converts string vibrations into electrical signals, act of picking up; **pick-your-own** (of fruit and vegetables) dug or picked by customer at farm etc.

pickaxe /'pɪkæks/ *noun* (*US* **pickax**) tool with sharp-pointed iron cross-bar for breaking up ground etc.

picket /'pɪkɪt/ ● *noun* one or more people stationed to dissuade workers from entering workplace during strike etc.; pointed stake driven into ground; small group of troops sent to watch for enemy. ● *verb* (**-t-**) place or act as picket outside; post as military picket; secure with stakes. □ **picket line** boundary established by workers on strike, esp. at workplace entrance, which others are asked not to cross.

pickings *plural noun* perquisites, gleanings.

pickle /'pɪk(ə)l/ ● *noun* (often in *plural*) vegetables etc. preserved in vinegar etc.; liquid used for this; *colloquial* plight. ● *verb* (**-ling**) preserve in or treat with pickle; (as **pickled** *adjective*) *slang* drunk.

picky *adjective* (**-ier**, **-iest**) *colloquial* highly fastidious.

picnic /'pɪknɪk/ ● *noun* outing including outdoor meal; such meal; something pleasantly or easily accomplished. ● *verb* (**-ck-**) eat meal outdoors.

pictograph /'pɪktəgrɑːf/ *noun* (also **pictogram** /-græm/) pictorial symbol used as form of writing.

pictorial /pɪk'tɔːrɪəl/ *adjective* of, expressed in, or illustrated with a picture or pictures. □ **pictorially** *adverb*.

picture /'pɪktʃə/ ● *noun* painting, drawing, photograph, etc., esp. as work of art; portrait; beautiful object; scene; mental image; cinema film; (**the pictures**) cinema (performance). ● *verb* (**-ring**) imagine; represent in picture; describe graphically. □ **picture postcard** postcard with picture on one side; **picture window** large window of one pane of glass.

picturesque /pɪktʃə'resk/ *adjective* striking and pleasant to look at; (of language etc.) strikingly graphic.

piddle /'pɪd(ə)l/ *verb* (**-ling**) *colloquial* urinate; (as **piddling** *adjective*) *colloquial* trivial; work or act in trifling way.

pidgin /'pɪdʒɪn/ *noun* simplified language, esp. used between speakers of different languages.

pie *noun* dish of meat, fruit, etc., encased in or covered with pastry etc. and baked. □ **pie chart** diagram representing relative quantities as sectors of circle; **pie-eyed** *slang* drunk.

piebald /'paɪbɔːld/ ● *adjective* having irregular patches of two colours, esp. black and white. ● *noun* piebald animal.

piece /piːs/ ● *noun* distinct portion forming part of or broken off from larger object; coin; picture, literary or musical composition; example, item; chessman, man at draughts, etc. ● *verb* (**-cing**) (usually + *together*) form into a whole; join. □ **of a piece** uniform or consistent; **piece-work** work paid for according to amount done.

pièce de résistance /pjes də reɪ'ziːstɑ̃s/ *noun* (*plural* **pièces de résistance** same pronunciation) most important or remarkable item.

piecemeal ● *adverb* piece by piece, part at a time. ● *adjective* gradual; unsystematic.

pied /paɪd/ *adjective* of mixed colours.

pied-à-terre /pjeɪdɑː'teə/ *noun* (*plural* ***pieds-*** same pronunciation) (usually small) flat, house, etc. kept for occasional use. [French]

pier /pɪə/ *noun* structure built out into sea etc. used as promenade and landing-stage or breakwater; support of arch or of span of bridge; pillar; solid part of wall between windows etc. □ **pier-glass** large tall mirror.

pierce /pɪəs/ *verb* (**-cing**) go through or into like spear or needle; make hole in; make (hole etc.).

pierrot /'pɪərəʊ/ *noun* (*feminine* **pierrette** /pɪə'ret/) French white-faced pantomime character with clown's costume; itinerant entertainer so dressed.

pietà /pɪe'tɑː/ *noun* representation of Virgin Mary holding dead body of Christ. [Italian]

pietism /'paɪətɪz(ə)m/ *noun* extreme or affected piety.

piety /'paɪətɪ/ *noun* piousness.

piffle /'pɪf(ə)l/ *colloquial* ● *noun* nonsense. ● *verb* (**-ling**) talk or act feebly.

pig ● *noun* wild or domesticated animal with broad snout and stout bristly body; *colloquial* greedy, dirty, obstinate, or annoying person; oblong mass of smelted iron or other metal. ● *verb* (**-gg-**) *colloquial* eat (food) greedily. □ **pigheaded** obstinate; **pig-iron** crude iron from smelting-furnace; **pig it** *colloquial* live in disorderly fashion; **pig out** *esp. US slang* eat greedily; **pigsty** sty for pigs; **pigtail** plait of hair hanging from back or each side of head.

pigeon /'pɪdʒ(ə)n/ *noun* bird of dove family. □ **pigeon-hole** *noun* each of set of compartments in cabinet etc. for papers etc., *verb* classify mentally, put in pigeon-hole, put aside for future consideration; **pigeon-toed** having toes turned inwards.

piggery *noun* (*plural* **-ies**) pig farm; pigsty.

piggish *adjective* greedy; dirty; mean.

piggy ● *noun* (*plural* **-ies**) *colloquial* little pig. ● *adjective* (**-ier**, **-iest**) like a pig; (of features etc.) like those of a pig. □ **piggyback** (a ride) on shoulders and back of another person; **piggy bank** pig-shaped money box.

piglet /'pɪglɪt/ *noun* young pig.

pigment /ˈpɪgmənt/ ● *noun* coloured substance used as paint etc., or occurring naturally in plant or animal tissue. ● *verb* colour (as) with natural pigment. □ **pigmentation** *noun*.

pigmy = PYGMY.

pike *noun* (*plural* same or **-s**) large voracious freshwater fish; spear formerly used by infantry. □ **pikestaff** wooden shaft of pike (**plain as a pikestaff** quite obvious).

pilaff = PILAU.

pilaster /pɪˈlæstə/ *noun* rectangular column, esp. one fastened into wall.

pilau /pɪˈlaʊ/ *noun* (also **pilaff** /pɪˈlɑːf/) Middle Eastern or Indian dish of rice with meat, spices, etc.

pilchard /ˈpɪltʃəd/ *noun* small sea fish related to herring.

pile[1] ● *noun* heap of things laid on one another; large imposing building; *colloquial* large amount, esp. of money; series of plates of dissimilar metals laid alternately for producing electric current; nuclear reactor; pyre. ● *verb* (**-ling**) heap; (+ *with*) load with; (+ *in*, *into*, *on*, *out of*, etc.) crowd. □ **pile up** accumulate, heap up; **pile-up** *colloquial* collision of several motor vehicles.

pile[2] *noun* heavy beam driven vertically into ground as support for building etc.

pile[3] *noun* soft projecting surface of velvet, carpet, etc.

piles *plural noun colloquial* haemorrhoids.

pilfer /ˈpɪlfə/ *verb* steal or thieve in petty way.

pilgrim /ˈpɪlgrɪm/ *noun* person who journeys to sacred place; traveller. □ **Pilgrim Fathers** English Puritans who founded colony in Massachusetts in 1620.

pilgrimage *noun* pilgrim's journey.

pill *noun* ball or flat piece of medicinal substance to be swallowed whole; (usually **the pill**) *colloquial* contraceptive pill. □ **pillbox** small round shallow box for pills, hat shaped like this, *Military* small round concrete shelter, mainly underground.

pillage /ˈpɪlɪdʒ/ *verb* (**-ging**) & *noun* plunder.

pillar /ˈpɪlə/ *noun* slender upright structure used as support or ornament; person regarded as mainstay; upright mass. □ **pillar-box** public postbox shaped like pillar.

pillion /ˈpɪljən/ *noun* seat for passenger behind motorcyclist etc.

pillory /ˈpɪlərɪ/ ● *noun* (*plural* **-ies**) *historical* frame with holes for head and hands, allowing an offender to be exposed to public ridicule. ● *verb* (**-ies**, **-ied**) expose to ridicule; *historical* set in pillory.

pillow /ˈpɪləʊ/ ● *noun* cushion as support for head, esp. in bed; pillow-shaped support. ● *verb* rest (as) on pillow. □ **pillowcase**, **pillowslip** washable cover for pillow.

pilot /ˈpaɪlət/ ● *noun* person operating controls of aircraft; person in charge of ships entering or leaving harbour etc.; experimental or preliminary study or undertaking; guide, leader. ● *adjective* experimental, preliminary. ● *verb* (**-t-**) act as pilot to; guide course of. □ **pilot-light** small gas burner kept alight to light another; **pilot officer** lowest commissioned rank in RAF.

pimento /pɪˈmentəʊ/ *noun* (*plural* **-s**) allspice; sweet pepper.

pimiento /pɪmɪˈentəʊ/ *noun* (*plural* **-s**) sweet pepper.

pimp ● *noun* person who lives off earnings of prostitute or brothel. ● *verb* act as pimp, esp. procure clients for prostitute.

pimpernel /ˈpɪmpənel/ *noun* scarlet pimpernel.

pimple /ˈpɪmp(ə)l/ *noun* small hard inflamed spot on skin. □ **pimply** *adjective*.

pin ● *noun* small thin pointed piece of metal with head, used as fastening; wooden or metal peg, rivet, etc.; (in *plural*) *colloquial* legs. ● *verb* (**-nn-**) fasten with pin(s); transfix with pin, lance, etc.; (usually + *on*) fix (responsibility, blame, etc.); seize and hold fast. □ **pinball** game in which small metal balls are shot across board and strike obstacles; **pincushion** small pad for holding pins; **pin down** (often + *to*) bind (person etc.) to promise, arrangement, etc., make (person) declare position or intentions; **pin-money** small sum of money, esp. earned by woman; **pinpoint** *noun* very small or sharp thing, *adjective* precise, *verb* locate with precision; **pinprick** petty irritation; **pins and needles** tingling sensation in limb recovering from numbness; **pinstripe** very narrow stripe in cloth; **pin-table** table used in pinball; **pintail** duck or grouse with pointed tail; **pin-tuck** narrow ornamental tuck; **pin-up** picture of attractive or famous person, pinned up on wall etc.; **pinwheel** small Catherine wheel.

pina colada /pi:nə kə'lɑ:də/ *noun* cocktail of pineapple juice, rum, and coconut.

pinafore /'pɪnəfɔ:/ *noun* apron, esp. with bib; (in full **pinafore dress**) dress without collar or sleeves, worn over blouse or jumper.

pince-nez /'pænsneɪ/ *noun* (*plural* same) pair of eyeglasses with spring that clips on nose.

pincers /'pɪnsəz/ *plural noun* gripping-tool forming pair of jaws; pincer-shaped claw in crustaceans etc. □ **pincer movement** converging movement by two wings of army against enemy position.

pinch ● *verb* grip tightly, esp. between finger and thumb; constrict painfully; (of cold etc.) affect painfully; *slang* steal, arrest; stint, be niggardly. ● *noun* pinching; (as **pinched** *adjective*) (of features) drawn; amount that can be taken up with fingers and thumb; stress of poverty etc. □ **at a pinch** in an emergency.

pinchbeck /'pɪntʃbek/ ● *noun* goldlike copper and zinc alloy used in cheap jewellery etc. ● *adjective* spurious, sham.

pine[1] *noun* evergreen needle-leaved coniferous tree; its wood. □ **pine cone** fruit of pine; **pine nut, kernel** edible seed of some pines.

pine[2] *verb* (**-ning**) (often + *away*) waste away with grief, disease, etc.; (usually + *for*) long.

pineal /'pɪnɪəl/ *adjective* shaped like pine cone. □ **pineal gland, body** conical gland in brain, secreting hormone-like substance.

pineapple /'paɪnæp(ə)l/ *noun* large juicy tropical fruit with yellow flesh and tough skin.

ping ● *noun* abrupt single ringing sound. ● *verb* (cause to) emit ping.

ping-pong *noun colloquial* table tennis.

pinion[1] /'pɪnjən/ ● *noun* outer part of bird's wing; *poetical* wing; flight feather. ● *verb* cut off pinion of (wing or bird) to prevent flight; restrain by binding arms to sides.

pinion[2] /'pɪnjən/ *noun* small cogwheel engaging with larger.

pink[1] ● *noun* pale red colour; garden plant with clove-scented flowers; (**the pink**) the most perfect condition. ● *adjective* pink-coloured; *colloquial* mildly socialist. □ **in the pink** *colloquial* in very good health. □ **pinkish** *adjective*; **pinkness** *noun*.

pink[2] *verb* pierce slightly; cut scalloped or zigzag edge on. □ **pinking shears** dressmaker's serrated shears for cutting zigzag edge.

pink[3] *verb* (of vehicle engine) emit high-pitched explosive sounds caused by faulty combustion.

pinnace /'pɪnɪs/ *noun* ship's small boat.

pinnacle /'pɪnək(ə)l/ *noun* culmination, climax; natural peak; small ornamental turret crowning buttress, roof, etc.

pinnate /'pɪneɪt/ *adjective* (of compound leaf) with leaflets on each side of leaf-stalk.

pinny /'pɪnɪ/ *noun* (*plural* **-ies**) *colloquial* pinafore.

pint /paɪnt/ *noun* measure of capacity (1/8 gal., 0.568 litre); *colloquial* pint of beer. □ **pint-sized** *colloquial* very small.

pinta /'paɪntə/ *noun colloquial* pint of milk.

pintle /'pɪnt(ə)l/ *noun* bolt or pin, esp. one on which some other part turns.

Pinyin /pɪn'jɪn/ *noun* system of romanized spelling for transliterating Chinese.

pioneer /paɪə'nɪə/ ● *noun* beginner of enterprise etc.; explorer or settler. ● *verb* initiate (enterprise etc.) for others to follow; act as pioneer.

pious /'paɪəs/ *adjective* devout, religious; sanctimonious; dutiful. □ **piously** *adverb*.

pip[1] *noun* seed of apple, pear, orange, etc.

pip[2] *noun* short high-pitched sound.

pip[3] *verb* (**-pp-**) *colloquial* hit with a shot; (also **pip at** or **to the post**) defeat narrowly.

pip[4] *noun* each spot on dominoes, dice, or playing cards; star on army officer's shoulder.

pip[5] *noun* disease of poultry etc.; (esp. **the pip**) *colloquial* (fit of) depression, boredom, or bad temper.

pipe ● *noun* tube of earthenware, metal, etc., esp. for carrying gas, water, etc.; narrow tube with bowl at one end containing tobacco for smoking; quantity of tobacco held by this; wind instrument of single tube; each tube by which sound is produced in organ; (in *plural*) bagpipes; tubular organ etc. in body; high note or song, esp. of bird; boatswain's whistle; measure of capacity for wine (105 gals., 477 litres). ● *verb* (**-ping**) convey (as) through pipes; play on pipe; (esp. as **piped** *adjective*) transmit (recorded music etc.) by wire or cable; utter shrilly; summon, lead, etc. by sound of pipe or whistle; trim with piping; furnish with pipe(s). □ **pipeclay** fine white clay for tobacco pipes or for whitening leather etc.; **pipe-cleaner** piece of flexible tuft-covered wire to clean inside tobacco pipe; **pipe down** *colloquial* be quiet; **pipedream** extravagant fancy, impossible wish, etc.; **pipeline** pipe conveying oil etc. across country, channel of supply or communication; **pipe up** begin to play, sing, etc.

piper *noun* person who plays on pipe, esp. bagpipes.

pipette /pɪ'pet/ *noun* slender tube for transferring or measuring small quantities of liquid.

piping *noun* ornamentation of dress, upholstery, etc. by means of cord enclosed in pipelike fold; ornamental cordlike lines of sugar etc. on cake etc.; length or system of pipes. □ **piping hot** (of food, water, etc.) very or suitably hot.

pipit /'pɪpɪt/ *noun* small bird resembling lark.

pippin /'pɪpɪn/ *noun* apple grown from seed; dessert apple.

piquant /'pi:kənt/ *adjective* agreeably pungent, sharp, appetizing, stimulating. □ **piquancy** *noun*.

pique /pi:k/ ● *verb* (**piques, piqued, piquing**) wound pride of; stir (curiosity). ● *noun* resentment; hurt pride.

piquet /pɪ'ket/ *noun* card game for two players.

piracy /'paɪrəsɪ/ *noun* (*plural* **-ies**) activity of pirate.

piranha /pɪ'rɑ:nə/ *noun* voracious S. American freshwater fish.

pirate /'paɪərət/ ● *noun* seafaring robber attacking ships; ship used by pirate; person who infringes copyright or regulations or encroaches on rights of others etc. ● *verb* (**-ting**) reproduce (book etc.) or trade (goods) without permission. □ **piratical** /-'ræt-/ *adjective*.

pirouette /pɪrʊ'et/ ● *noun* dancer's spin on one foot or point of toe. ● *verb* (**-tting**) perform pirouette.

piscatorial /pɪskə'tɔ:rɪəl/ *adjective* of fishing.

Pisces /'paɪsi:z/ *noun* twelfth sign of zodiac.

piscina /pɪ'si:nə/ *noun* (*plural* **-nae** /-ni:/ or **-s**) stone basin near altar in church, for draining water after use.

pistachio /pɪs'tɑ:ʃɪəʊ/ *noun* (*plural* **-s**) kind of nut with green kernel.

piste /pi:st/ *noun* ski run of compacted snow.

pistil /'pɪstɪl/ *noun* female organ in flowers. □ **pistillate** *adjective*.

pistol /'pɪst(ə)l/ *noun* small firearm.

piston /'pɪst(ə)n/ *noun* sliding cylinder fitting closely in tube and moving up and down in it, used in steam or petrol engine to impart motion; sliding valve in trumpet etc. □ **piston rod** rod connecting piston to other parts of machine.

pit[1] ● *noun* large hole in ground; coal mine; covered hole as trap; depression in skin or any surface; orchestra pit; (**the pits**) *slang* worst imaginable place, situation, person, etc.; area to side of track where racing cars are refuelled etc. during race; sunken area in floor of workshop etc. for inspection or repair of underside of vehicle etc. ● *verb* (**-tt-**) (usually + *against*) set (one's wits, strength, etc.) in competition; (usually as **pitted** *adjective*) make pit(s) in; store in pit. □ **pit bull terrier** small American dog noted for ferocity; **pitfall** unsuspected danger or drawback, covered pit as trap; **pit-head** top of shaft of coal mine, area surrounding this; **pit**

of the stomach hollow below base of breastbone.

pit[2] *verb* (**-tt-**) (usually as **pitted** *adjective*) remove stones from (fruit).

pita = PITTA.

pit-a-pat /ˈpɪtəpæt/ (also **pitter-patter** /ˈpɪtəpætə/) ● *adverb* with sound as of light quick steps; falteringly. ● *noun* such sound.

pitch[1] ● *verb* set up (esp. tent, camp, etc.) in chosen position; throw; express in particular style or at particular level; fall heavily; (of ship etc.) plunge in lengthwise direction; set at particular musical pitch. ● *noun* area of play in esp. outdoor game; height, degree, intensity, etc.; gradient, esp. of roof; *Music* degree of highness or lowness of tone; act or process of pitching; *colloquial* salesman's persuasive talk; place, esp. in street or market, where one is stationed; distance between successive points, lines, etc. □ **pitched battle** vigorous argument etc., planned battle between sides in prepared positions; **pitched roof** sloping roof; **pitchfork** *noun* fork with long-handle and two prongs for tossing hay etc., *verb* (+ *into*) thrust forcibly or hastily into office, position, etc.; **pitch in** *colloquial* set to work vigorously; **pitch into** *colloquial* attack vigorously.

pitch[2] ● *noun* dark resinous tarry substance. ● *verb* coat with pitch. □ **pitch-black**, **-dark** intensely dark; **pitch pine** resinous kinds of pine. □ **pitchy** *adjective* (**-ier**, **-iest**).

pitchblende /ˈpɪtʃblend/ *noun* uranium oxide yielding radium.

pitcher[1] /ˈpɪtʃə/ *noun* large jug, ewer.

pitcher[2] *noun* player who delivers ball in baseball.

piteous /ˈpɪtɪəs/ *adjective* deserving or arousing pity. □ **piteously** *adverb*; **piteousness** *noun*.

pith *noun* spongy tissue in stems of plants or lining rind of orange etc.; chief part; vigour, energy. □ **pith helmet** sun-helmet made from dried pith of plants.

pithy /ˈpɪθɪ/ *adjective* (**-ier**, **-iest**) condensed and forcible, terse. □ **pithily** *adverb*; **pithiness** *noun*.

pitiable /ˈpɪtɪəb(ə)l/ *adjective* deserving or arousing pity or contempt. □ **pitiably** *adverb*.

pitiful /ˈpɪtɪfʊl/ *adjective* arousing pity; contemptible. □ **pitifully** *adverb*.

pitiless /ˈpɪtɪlɪs/ *adjective* showing no pity. □ **pitilessly** *adverb*.

piton /ˈpiːtɒn/ *noun* peg driven in to support climber or rope.

pitta /ˈpɪtə/ *noun* (also **pita**) originally Turkish unleavened bread which can be split and filled.

pittance /ˈpɪt(ə)ns/ *noun* scanty allowance, small amount.

pitter-patter = PIT-A-PAT.

pituitary /pɪˈtjuːɪtərɪ/ *noun* (*plural* **-ies**) (in full **pituitary gland**) small ductless gland at base of brain.

pity /ˈpɪtɪ/ ● *noun* sorrow for another's suffering; cause for regret. ● *verb* (**-ies**, **-ied**) feel pity for. □ **pitying** *adjective*.

pivot /ˈpɪvət/ ● *noun* shaft or pin on which something turns; crucial person or point. ● *verb* (**-t-**) turn (as) on pivot; provide with pivot. □ **pivotal** *adjective*.

pixie /ˈpɪksɪ/ *noun* (also **pixy**) (*plural* **-ies**) fairy-like being.

pizza /ˈpiːtsə/ *noun* flat piece of dough baked with topping of cheese, tomatoes, etc.

pizzeria /piːtsəˈriːə/ *noun* pizza restaurant.

pizzicato /pɪtsɪˈkɑːtəʊ/ *Music* ● *adverb* plucking. ● *adjective* performed thus. ● *noun* (*plural* **-s** or **-ti** /-tɪ/) pizzicato note or passage.

pl. *abbreviation* plural; place; plate.

placable /ˈplækəb(ə)l/ *adjective* easily appeased; mild-tempered. □ **placability** *noun*.

placard /ˈplækɑːd/ ● *noun* large notice for public display. ● *verb* post placards on.

placate /pləˈkeɪt/ *verb* (**-ting**) conciliate, pacify. □ **placatory** *adjective*.

place ● *noun* particular part of space; space or room of or for person etc.; city, town, village, residence, building; rank, station, position; building or spot devoted to specified purpose; office, employment; duties of this. ● *verb* (**-cing**) put or dispose in place; assign rank, order, or class to; give (order for goods etc.) to firm etc.; (in *passive*) be among first 3 (or 4) in race. □ **in place** suitable, in the right position; **in place of** instead of; **out of place** unsuitable, in the wrong position; **place-kick** *Football* kick made with ball placed on ground; **place-mat** small mat on table at person's place; **place setting** set of cutlery etc. for one person to eat with; **take place** happen; **take the place of** be substituted for. □ **placement** *noun.*

placebo /plə'si:bəʊ/ *noun* (*plural* **-s**) medicine with no physiological effect prescribed for psychological reasons; dummy pill etc. used in controlled trial.

placenta /plə'sentə/ *noun* (*plural* **-tae** /-ti:/ or **-s**) organ in uterus of pregnant mammal that nourishes foetus. □ **placental** *adjective.*

placid /'plæsɪd/ *adjective* calm, unruffled; not easily disturbed. □ **placidity** /plə'sɪdɪtɪ/ *noun*; **placidly** *adverb.*

placket /'plækɪt/ *noun* opening or slit in garment, for fastenings or access to pocket.

plagiarize /'pleɪdʒəraɪz/ *verb* (also **-ise**) (**-zing** or **-sing**) take and use (another's writings etc.) as one's own. □ **plagiarism** *noun*; **plagiarist** *noun*; **plagiarizer** *noun.*

plague /pleɪg/ ● *noun* deadly contagious disease; (+ *of*) *colloquial* infestation; great trouble or affliction. ● *verb* (**plaguing**) *colloquial* annoy, bother; afflict, hinder; affect with plague.

plaice /pleɪs/ *noun* (*plural* same) marine flatfish.

plaid /plæd/ *noun* chequered or tartan, esp. woollen, cloth; long piece of this as part of Highland costume.

plain ● *adjective* clear, evident; readily understood; simple; not beautiful or distinguished-looking; straightforward in speech; not luxurious; *adverb* clearly; simply. ● *noun* level tract of country; ordinary stitch in knitting. □ **plain chocolate** chocolate made without milk; **plain clothes** ordinary clothes as distinct from esp. police uniform; **plain flour** flour with no raising agent; **plain sailing** simple situation or course of action; **plainsong** traditional church music sung in unison in medieval modes and free rhythm; **plain-spoken** frank. □ **plainly** *adverb.*

plaint *noun Law* accusation, charge; *literary* lamentation.

plaintiff /'pleɪntɪf/ *noun* person who brings case against another in law court.

plaintive /'pleɪntɪv/ *adjective* mournful-sounding. □ **plaintively** *adverb.*

plait /plæt/ ● *noun* length of hair, straw, etc. in 3 or more interlaced strands. ● *verb* form into plait.

plan ● *noun* method or procedure for doing something; drawing exhibiting relative position and size of parts of building etc.; diagram; map. ● *verb* (**-nn-**) arrange beforehand, scheme; make plan of; design; (as **planned** *adjective*) in accordance with plan; make plans. □ **plan on** (often + present participle) *colloquial* aim at, intend. □ **planning** *noun.*

planchette /pla:n'ʃet/ *noun* small board on castors, with pencil, said to write messages from spirits when person's fingers rest on it.

plane[1] ● *noun* flat surface (not necessarily horizontal); *colloquial* aeroplane; level of attainment etc. ● *adjective* level as or lying in a plane.

plane[2] ● *noun* tool for smoothing surface of wood by paring shavings from it. ● *verb* (**-ning**) smooth or pare with plane.

plane[3] *noun* tall spreading broad-leaved tree.

planet /'plænɪt/ *noun* heavenly body orbiting star. □ **planetary** *adjective.*

planetarium /plænɪ'teərɪəm/ *noun* (*plural* **-s** or **-ria**) building in which image of night sky as seen at various times and places is projected; device for such projection.

plangent /'plændʒ(ə)nt/ *adjective literary* loudly lamenting; plaintive; reverberating.

plank ● *noun* long flat piece of timber; item of political or other programme. ● *verb* provide or cover with planks; (usually + *down*) *colloquial* put down roughly, deposit (esp. money).

plankton /'plæŋkt(ə)n/ *noun* chiefly microscopic organisms drifting in sea or fresh water.

planner *noun* person who plans new town etc.; person who makes plans; list, table, chart, etc. with information helpful in planning.

plant /plɑ:nt/ ● *noun* organism capable of living wholly on inorganic substances and lacking power of locomotion; small plant (other than trees and shrubs); equipment for industrial process; *colloquial* thing deliberately placed for discovery, esp. to incriminate another. ● *verb* place (seed etc.) in ground to grow; fix firmly, establish; cause (idea etc.) to be established, esp. in another person's mind; deliver (blow etc.); *colloquial* place (something incriminating) for later discovery.

plantain[1] /'plæntɪn/ *noun* herb yielding seed used as food for birds.

plantain[2] /'plæntɪn/ *noun* plant related to banana; banana-like fruit of this.

plantation /plɑ:n'teɪʃ(ə)n/ *noun* estate for cultivation of cotton, tobacco, etc.; number of growing plants, esp. trees, planted together; *historical* colony.

planter *noun* owner or manager of plantation; container for house-plants.

plaque /plæk/ *noun* ornamental tablet of metal, porcelain, etc.; deposit on teeth, where bacteria proliferate.

plasma /'plæzmə/ *noun* (also **plasm** /'plæz(ə)m/) colourless fluid part of blood etc. in which corpuscles etc. float; protoplasm; gas of positive ions and free electrons in about equal numbers. □ **plasmic** *adjective*.

plaster /'plɑ:stə/ ● *noun* mixture esp. of lime, sand, and water spread on walls etc.; sticking plaster; plaster of Paris. ● *verb* cover with or like plaster; apply, stick, etc. like plaster to; (as **plastered** *adjective*) *slang* drunk. □ **plasterboard** two boards with core of plaster used for walls etc.; **plaster cast** bandage stiffened with plaster of Paris and wrapped round broken limb etc.; **plaster of Paris** fine white gypsum powder for plaster casts etc. □ **plasterer** *noun*.

plastic /'plæstɪk/ ● *noun* synthetic resinous substance that can be given any shape; (in full **plastic money**) *colloquial* credit card(s). ● *adjective* made of plastic; capable of being moulded; giving form to clay, wax, etc. □ **plastic arts** those involving modelling; **plastic explosive** putty-like explosive; **plastic surgery** repair or restoration of lost or damaged etc. tissue. □ **plasticity** /-'tɪs-/ *noun*; **plasticize** /-saɪz/ *verb* (also **-ise**) (**-zing** or **-sing**); **plasticky** *adjective*.

Plasticine /'plæstəsi:n/ *noun proprietary term* pliant substance used for modelling.

plate ● *noun* shallow usually circular vessel from which food is eaten or served; similar vessel used for collection in church etc.; table utensils of gold, silver, or other metal; objects of plated metal; piece of metal with inscription, for fixing to door etc.; illustration on special paper in book; thin sheet of metal, glass, etc. coated with sensitive film for photography; flat thin sheet of metal etc.; part of denture fitting to mouth and holding teeth; each of several sheets of rock thought to form earth's crust. ● *verb* (**-ting**) cover (other metal) with thin coating of silver, gold, etc.; cover with plates of metal. □ **plate glass** thick fine-quality glass for mirrors, windows, etc.; **platelayer** workman laying and repairing railway lines. □ **plateful** *noun* (*plural* **-s**).

plateau /'plætəʊ/ ● *noun* (*plural* **-x** or **-s** /-z/) area of level high ground; state of little variation following an increase. ● *verb* (**plateaus**, **plateaued**, **plateauing**) (often + *out*) reach level or static state after period of increase.

platelet /'pleɪtlɪt/ *noun* small disc in blood, involved in clotting.

platen /'plæt(ə)n/ *noun* plate in printing press by which paper is pressed against type; corresponding part in typewriter etc.

platform /'plætfɔ:m/ *noun* raised level surface, esp. one from which speaker addresses audience, or one along side of line at railway station; floor area at entrance to bus etc.; thick sole of shoe; declared policy of political party.

platinum /'plætɪnəm/ *noun* white heavy precious metallic element that does not tarnish. □ **platinum blonde** *adjective* silvery-blond, *noun* person with such hair.

platitude /'plætɪtju:d/ *noun* commonplace remark. □ **platitudinous** /-'tju:d-/ *adjective*.

Platonic /plə'tɒnɪk/ *adjective* of Plato or his philosophy; (**platonic**) (of love or friendship) not sexual.

platoon /plə'tu:n/ *noun* subdivision of infantry company.

platter /'plætə/ *noun* flat plate or dish.

platypus /'plætɪpəs/ *noun* (*plural* **-puses**) Australian aquatic egg-laying mammal with ducklike beak.

plaudit /'plɔ:dɪt/ *noun* (usually in *plural*) round of applause; commendation.

plausible /'plɔ:zɪb(ə)l/ *adjective* reasonable, probable; (of person) persuasive but deceptive. □ **plausibility** *noun*; **plausibly** *adverb*.

play ● *verb* occupy or amuse oneself pleasantly; (+ *with*) act light-heartedly or flippantly with (feelings etc.); perform on (musical instrument), perform (piece of music etc.); cause (record etc.) to produce sounds; perform (drama, role); (+ *on*) perform (trick or joke etc.) on; *colloquial* cooperate, do what is wanted; take part in (game); have as opponent in game; move (piece) in game, put (card) on table, strike (ball), etc.; move about in lively or unrestrained way; (often + *on*) touch gently; pretend to be; allow (fish) to exhaust itself pulling against line. ● *noun* recreation; amusement; playing of game; dramatic piece for stage etc.; freedom of movement; fitful or light movement; gambling. □ **play along** pretend to cooperate; **play back** play (what has been recorded); **play-back** *noun*; **play ball** *colloquial* cooperate; **playbill** poster announcing play etc.; **playboy** pleasure-seeking usually wealthy man; **play by ear** perform (music) without having seen it written down, (also **play it by ear**) *colloquial* proceed gradually according to results; **play one's cards right** *colloquial* make best use of opportunities and advantages; **play down** minimize; **playfellow** playmate; **play the game** behave honourably; **playground** outdoor area for children to play in; **playgroup** group of preschool children who play together under supervision; **playhouse** theatre; **playing card** small usually oblong card used in games, one of set of usually 52 divided into 4 suits; **play it cool** *colloquial* appear relaxed or indifferent; **playmate** child's companion in play; **play-off** extra match played to decide draw or tie; **plaything** toy; **play up** behave mischievously, annoy in this way, cause trouble; **play with fire** take foolish risks; **playwright** dramatist. □ **player** *noun*.

playful *adjective* fond of or inclined to play; done in fun. □ **playfully** *adverb*; **playfulness** *noun*.

plc *abbreviation* (also **PLC**) Public Limited Company.

plea *noun* appeal, entreaty; *Law* formal statement by or on behalf of defendant; excuse.

pleach *verb* entwine or interlace (esp. branches to form a hedge).

plead *verb* (+ *with*) make earnest appeal to; address court as advocate or party; allege as excuse; (+ *guilty*, *not guilty*) declare oneself to be guilty or not guilty of a charge; make appeal or entreaty.

pleading *noun* (usually in *plural*) formal statement of cause of action or defence.

pleasant /'plez(ə)nt/ *adjective* (**-er**, **-est**) agreeable; giving pleasure. □ **pleasantly** *adverb*.

pleasantry *noun* (*plural* **-ies**) joking remark; polite remark.

please /pli:z/ *verb* (**-sing**) be agreeable to; give joy or gratification to; think fit; (in *passive*) be willing, like; *used in polite requests*. □ **pleased** *adjective*; **pleasing** *adjective*.

pleasurable /'pleʒərəb(ə)l/ *adjective* causing pleasure. □ **pleasurably** *adverb*.

pleasure /'pleʒə/ ● *noun* satisfaction, delight; sensuous enjoyment; source of gratification; will, choice. ● *adjective* done or used for pleasure.

pleat ● *noun* flattened fold in cloth etc. ● *verb* make pleat(s) in.

pleb *noun colloquial* plebeian.

plebeian /plɪ'bi:ən/ ● *noun* commoner, esp. in ancient Rome; working-class person (esp. uncultured). ● *adjective* of the common people; uncultured, coarse.

plebiscite /'plebɪsaɪt/ *noun* referendum.

plectrum /'plektrəm/ *noun* (*plural* **-s** or **-tra**) thin flat piece of plastic etc. for plucking strings of musical instrument.

pledge ● *noun* solemn promise; thing given as security for payment of debt etc.; thing put in pawn; token; drinking of health. ● *verb* (**-ging**) deposit as security, pawn; promise solemnly by pledge; bind by solemn promise; drink to the health of.

Pleiades /'plaɪədi:z/ *plural noun* cluster of stars in constellation Taurus.

plenary /'pli:nərɪ/ *adjective* (of assembly) to be attended by all members; entire, unqualified.

plenipotentiary /plenɪpə'tenʃərɪ/ ● *noun* (*plural* **-ies**) person (esp. diplomat) having full authority to act. ● *adjective* having such power.

plenitude /'plenɪtju:d/ *noun literary* fullness; completeness; abundance.

plenteous /'plentɪəs/ *adjective literary* plentiful.

plentiful /'plentɪfʊl/ *adjective* existing in ample quantity. □ **plentifully** *adverb.*

plenty /'plentɪ/ ● *noun* abundance; quite enough. ● *adjective colloquial* plentiful. ● *adverb colloquial* fully.

plenum /'pli:nəm/ *noun* full assembly of people or a committee etc.

pleonasm /'pli:ənæz(ə)m/ *noun* use of more words than are needed. □ **pleonastic** /-'næstɪk/ *adjective.*

plesiosaur /'pli:sɪəsɔ:/ *noun* large extinct reptile with flippers and long neck.

plethora /'pleθərə/ *noun* overabundance.

pleurisy /'plʊərəsɪ/ *noun* inflammation of membrane enclosing lungs. □ **pleuritic** /-'rɪt-/ *adjective.*

plexus /'pleksəs/ *noun* (*plural* same or **plexuses**) network of nerves or blood vessels.

pliable /'plaɪəb(ə)l/ *adjective* easily bent or influenced; supple; compliant. □ **pliability** *noun.*

pliant /'plaɪənt/ *adjective* pliable. □ **pliancy** *noun.*

pliers /'plaɪəz/ *plural noun* pincers with parallel flat surfaces for bending wire etc.

plight¹ /plaɪt/ *noun* unfortunate condition or state.

plight² /plaɪt/ *verb archaic* pledge. □ **plight one's troth** promise to marry.

plimsoll /'plɪms(ə)l/ *noun* rubber-soled canvas shoe. □ **Plimsoll line, mark** marking on ship's side showing limit of legal submersion under various conditions.

plinth *noun* base supporting column, vase, statue, etc.

plod *verb* (**-dd-**) walk or work laboriously. □ **plodder** *noun.*

plonk¹ ● *verb* set down hurriedly or clumsily; (usually + *down*) set down firmly. ● *noun* heavy thud.

plonk² *noun colloquial* cheap or inferior wine.

plop ● *noun* sound as of smooth object dropping into water. ● *verb* (**-pp-**) (cause to) fall with plop. ● *adverb* with a plop.

plot ● *noun* small piece of land; plan or interrelationship of main events of tale, play, etc.; secret plan, conspiracy. ● *verb* (**-tt-**) make chart, diagram, graph, etc. of; hatch secret plans; devise secretly; mark on chart or diagram. □ **plotter** *noun.*

plough /plaʊ/ (*US* **plow**) ● *noun* implement for furrowing and turning up soil; similar instrument for clearing away snow etc. ● *verb* (often + *up*, *out*, etc.) turn up or extract with plough; furrow, make (furrow); (+ *through*) advance laboriously or cut or force way through; *colloquial* fail in exam. □ **plough back** reinvest (profits) in business; **ploughman** user of plough; **ploughman's lunch** meal of bread, cheese, pickles, etc.; **ploughshare** blade of plough.

plover /'plʌvə/ *noun* plump-breasted wading bird.

plow *US* = PLOUGH.

ploy *noun* manoeuvre to gain advantage.

pluck ● *verb* pick or pull out or away; strip (bird) of feathers; pull at, twitch; (+ *at*) tug or snatch at; sound (string of musical instrument) with finger or plectrum. ● *noun* courage; twitch; animal's heart, liver, and lungs. □ **pluck up** summon up (one's courage etc.).

plucky *adjective* (**-ier**, **-iest**) brave, spirited.

plug ● *noun* something fitting into hole or filling cavity; device of metal pins etc. for making electrical connection; spark plug; *colloquial* piece of free publicity; cake or stick of tobacco. ● *verb* (**-gg-**) (often + *up*) stop with plug; *slang* shoot; *colloquial* seek to popularize by frequent recommendation. □ **plug away** (often + *at*) *colloquial* work steadily; **plug-hole** hole for plug, esp. in sink or bath; **plug in** *verb* connect electrically by inserting plug into socket; **plug-in** *adjective* designed to be plugged into socket; **pull the plug** *colloquial* flush toilet; (+ *on*) put an end to by withdrawing resources etc.

plum *noun* roundish fleshy stone fruit; tree bearing this; reddish-purple colour; raisin; *colloquial* prized thing. □ **plum pudding** Christmas pudding.

plumage /'plu:mɪdʒ/ *noun* bird's feathers.

plumb /plʌm/ ● *noun* lead ball attached to line for testing water's depth or whether wall etc. is vertical. ● *adverb* exactly; vertically; *US slang* quite, utterly. ● *adjective* vertical. ● *verb* provide with plumbing; fit as part of plumbing system; work as plumber; test with plumb; experience (extreme feeling); learn detailed facts about. □ **plumb line** string with plumb attached.

plumber *noun* person who fits and repairs apparatus of water supply, heating, etc.

plumbing *noun* system or apparatus of water supply etc.; plumber's work.

plume /plu:m/ ● *noun* feather, esp. large and showy one; feathery ornament in hat, hair, etc.; feather-like formation, esp. of smoke. ● *verb* (**-ming**) furnish with plume(s); (**plume oneself on, upon**) pride oneself on.

plummet /'plʌmɪt/ ● *noun* plumb, plumb line; sounding line. ● *verb* (**-t-**) fall rapidly.

plummy *adjective* (**-ier**, **-iest**) *colloquial* (of voice) affectedly rich in tone; *colloquial* good, desirable.

plump[1] ● *adjective* having full rounded shape; fleshy. ● *verb* (often + *up, out*) make or become plump. □ **plumpness** *noun*.

plump[2] ● *verb* (+ *for*) decide on, choose. ● *noun* abrupt or heavy fall. ● *adverb colloquial* with plump.

plunder /'plʌndə/ ● *verb* rob or steal, esp. in war; embezzle. ● *noun* plundering; property plundered.

plunge ● *verb* (**-ging**) (usually + *in, into*) throw forcefully, dive, (cause to) become or enter into impetuously, immerse completely; move suddenly downward; move with a rush; *colloquial* run up gambling debts. ● *noun* plunging, dive; decisive step.

plunger *noun* part of mechanism that works with plunging or thrusting motion; rubber cup on handle for removing blockages by plunging action.

pluperfect /plu:'pɜ:fɪkt/ *Grammar* ● *adjective* expressing action completed prior to some past point of time. ● *noun* pluperfect tense.

plural /'plʊər(ə)l/ ● *adjective* more than one in number; denoting more than one. ● *noun* plural word, form, or number.

pluralism *noun* form of society in which minority groups retain independent traditions; holding of more than one office at a time. □ **pluralist** *noun*; **pluralistic** /-'lɪst-/ *adjective*.

plurality /plʊə'rælɪtɪ/ *noun* (*plural* **-ies**) state of being plural; pluralism; large number; non-absolute majority (of votes etc.).

pluralize *verb* (also **-ise**) (**-zing** or **-sing**) make plural; express as plural.

plus ● *preposition* with addition of; (of temperature) above zero; *colloquial* having gained. ● *adjective* (after number) at least, (after grade) better than; *Mathematics* positive; additional, extra. ● *noun* plus sign; advantage. □ **plus sign** symbol (+) indicating addition or positive value.

■ **Usage** The use of *plus* as a conjunction, as in *they arrived late, plus they wanted a meal*, is considered incorrect except in very informal use.

plush ● *noun* cloth of silk, cotton, etc., with long soft pile. ● *adjective* made of plush; *colloquial* plushy.

plushy *adjective* (**-ier**, **-iest**) *colloquial* stylish, luxurious.

plutocracy /plu:'tɒkrəsɪ/ *noun* (*plural* **-ies**) state in which power belongs to rich; wealthy élite. □ **plutocrat** /'plu:təkræt/ *noun*; **plutocratic** /-tə'kræt-/ *adjective*.

plutonium /plu:'təʊnɪəm/ *noun* radioactive metallic element.

pluvial /'plu:vɪəl/ *adjective* of or caused by rain.

ply[1] /plaɪ/ *noun* (*plural* **-ies**) thickness, layer; strand.

ply[2] /plaɪ/ *verb* (**-ies**, **-ied**) wield; work at; (+ *with*) supply continuously or approach repeatedly with; (often + *between*) (of vehicle etc.) go to and fro.

plywood *noun* strong thin board made by gluing layers of wood with the direction of the grain alternating.

PM *abbreviation* prime minister.

p.m. *abbreviation* after noon (*post meridiem*).

PMS *abbreviation* premenstrual syndrome.

PMT *abbreviation* premenstrual tension.

pneumatic /nju:'mætɪk/ *adjective* filled with air or wind; operated by compressed air.

pneumonia /nju:'məʊnɪə/ *noun* inflammation of lung(s).

PO *abbreviation* Post Office; postal order; Petty Officer; Pilot Officer.

po *noun* (*plural* **-s**) *colloquial* chamber pot. □ **po-faced** solemn-faced, humourless, smug.

poach[1] *verb* cook (egg) without shell in boiling water; cook (fish etc.) by simmering in small amount of liquid. □ **poacher** *noun*.

poach[2] *verb* catch (game or fish) illicitly; (often + *on*) trespass, encroach; appropriate (another's ideas, staff, etc.). □ **poacher** *noun*.

pock *noun* (also **pock-mark**) small pus-filled spot, esp. in smallpox. □ **pock-marked** *adjective*.

pocket /'pɒkɪt/ ● *noun* small bag sewn into or on garment for carrying small articles; pouchlike compartment in suitcase, car door, etc.; financial resources; isolated group or area; cavity in earth etc. containing ore; pouch at corner or on side of billiard or snooker table into which balls are driven. ● *adjective* small, esp. small enough for carrying in pocket. ● *verb* (**-t-**) put into pocket; appropriate; submit to (affront etc.); conceal (feelings). □ **in** or **out of pocket** having gained or lost in transaction; **pocketbook** notebook, folding case for papers, paper money, etc.; **pocket knife** small folding knife; **pocket money** money for minor expenses, esp. given to child.

pod ● *noun* long seed vessel, esp. of pea or bean. ● *verb* (**-dd-**) form pods; remove (peas etc.) from pods.

podgy /'pɒdʒɪ/ *adjective* (**-ier**, **-iest**) short and fat.

podium /'pəʊdɪəm/ *noun* (*plural* **-s** or **podia**) rostrum.

poem /'pəʊɪm/ *noun* metrical composition; elevated composition in verse or prose; something with poetic qualities.

poesy /'pəʊəzɪ/ *noun archaic* poetry.

poet /'pəʊɪt/ *noun* (*feminine* **poetess**) writer of poems. □ **Poet Laureate** poet appointed to write poems for state occasions.

poetaster /pəʊɪ'tæstə/ *noun* inferior poet.

poetic /pəʊ'etɪk/ *adjective* (also **poetical**) of or like poetry or poets. □ **poetic justice** well-deserved punishment or reward; **poetic licence** departure from truth etc. for effect. □ **poetically** *adverb*.

poetry /'pəʊɪtrɪ/ *noun* poet's art or work; poems; poetic or tenderly pleasing quality.

pogo /'pəʊgəʊ/ *noun* (*plural* **-s**) (also **pogo stick**) stiltlike toy with spring, used to jump about on.

pogrom /'pɒgrəm/ *noun* organized massacre (originally of Jews in Russia).

poignant /'pɔɪnjənt/ *adjective* painfully sharp, deeply moving; arousing sympathy; pleasantly piquant. □ **poignance** *noun*; **poignancy** *noun*; **poignantly** *adverb*.

poinsettia /pɔɪn'setɪə/ *noun* plant with large scarlet bracts surrounding small yellowish flowers.

point ● *noun* sharp end, tip; geometric entity with position but no magnitude; particular place; precise moment; very small mark on surface; decimal point; stage or degree in progress or increase; single item or particular; unit of scoring in games etc., or in evaluation etc.;

significant thing, thing actually intended or under discussion; sense, purpose, advantage; characteristic; each of 32 directions marked on compass; (usually in *plural*) pair of tapering movable rails to direct train from one line to another; power point; *Cricket* (position of) fielder near batsman on off side; promontory. ● *verb* (usually + *to*, *at*) direct (finger, weapon, etc.); direct attention; (+ *at*, *towards*) aim or be directed to; (+ *to*) indicate; give force to (words, action); fill joints of (brickwork) with smoothed mortar or cement; (of dog) indicate presence of game by acting as pointer. □ **at** or **on the point of** on the verge of; **point-blank** at close range, directly, flatly; **point-duty** traffic control by police officer; **point of view** position from which thing is viewed, way of considering a matter; **point out** indicate, draw attention to; **point-to-point** steeplechase for hunting horses; **point up** emphasize.

pointed *adjective* having point; (of remark etc.) cutting, emphasized. □ **pointedly** *adverb*.

pointer *noun* indicator on gauge etc.; rod for pointing at features on screen etc.; *colloquial* hint; dog of breed trained to stand rigid looking at game.

pointless *adjective* purposeless, meaningless; ineffective. □ **pointlessly** *adverb*; **pointlessness** *noun*.

poise /pɔɪz/ ● *noun* composure; equilibrium; carriage (of head etc.). ● *verb* (**-sing**) balance, hold suspended or supported; be balanced or suspended.

poised *adjective* self-assured; carrying oneself with dignity; (often + *for*) ready.

poison /'pɔɪz(ə)n/ ● *noun* substance that when absorbed by living organism kills or injures it; *colloquial* harmful influence. ● *verb* administer poison to; kill or injure with poison; treat (weapon) with poison; corrupt, pervert; spoil. □ **poison ivy** N. American climbing plant secreting irritant oil from leaves; **poison-pen letter** malicious anonymous letter. □ **poisoner** *noun*; **poisonous** *adjective*.

poke ● *verb* (**-king**) push with (end of) finger, stick, etc.; (+ *out*, *up*, etc.) (be) thrust forward; (+ *at* etc.) make thrusts; (+ *in*) produce (hole etc.) in by poking; stir (fire). ● *noun* poking; thrust, nudge. □ **poke fun at** ridicule.

poker¹ *noun* metal rod for stirring fire.

poker² /'pəʊkə/ *noun* card game in which players bet on value of their hands. □ **poker-face** impassive countenance assumed by poker player.

poky /'pəʊkɪ/ *adjective* (**-ier**, **-iest**) (of room etc.) small and cramped.

polar /'pəʊlə/ *adjective* of or near either pole of earth etc.; having magnetic or electric polarity; directly opposite in character. □ **polar bear** large white bear living in Arctic.

polarity /pə'lærɪtɪ/ *noun* (*plural* **-ies**) tendency of magnet etc. to point to earth's magnetic poles or of body to lie with axis in particular direction; possession of two poles having contrary qualities; possession of two opposite tendencies, opinions, etc.; electrical condition of body (positive or negative).

polarize /'pəʊləraɪz/ *verb* (also **-ise**) (**-zing** or **-sing**) restrict vibrations of (light-waves etc.) to one direction; give polarity to; divide into two opposing groups. □ **polarization** *noun*.

Polaroid /'pəʊlərɔɪd/ *noun proprietary term* material in thin sheets polarizing light passing through it; camera that produces print immediately after each exposure; (in *plural*) sunglasses with Polaroid lenses.

Pole *noun* native or national of Poland.

pole¹ *noun* long slender rounded piece of wood, metal, etc., esp. as support etc.; *historical* measure of length ($5\frac{1}{2}$ yds). □ **pole-vault** jump over high bar with aid of pole held in hands.

pole² *noun* each of two points in celestial sphere (in full **north**, **south pole**) about which stars appear to revolve; each end of axis of earth (in full **North**, **South Pole**) or of other body; each of two opposite points on surface of magnet at which magnetic forces are strongest; positive or negative terminal of electric cell, battery, etc.; each of two opposed principles. □ **pole star** star near N. pole of heavens, thing serving as guide.

poleaxe /'pəʊlæks/ (*US* **-ax**) ● *noun historical* battleaxe; butcher's axe. ● *verb*

(**-xing**) hit or kill with poleaxe; (esp. as **poleaxed** *adjective*) *colloquial* dumbfound, overwhelm.

polecat /'pəʊlkæt/ *noun* small dark brown mammal of weasel family.

polemic /pə'lemɪk/ ● *noun* verbal attack; controversy; (in *plural*) art of controversial discussion. ● *adjective* (also **polemical**) involving dispute, controversial. □ **polemicist** /-sɪst/ *noun*.

police /pə'li:s/ ● *noun* (treated as *plural*) civil force responsible for maintaining public order; its members; force with similar functions. ● *verb* (**-cing**) control or provide with police; keep in order, control, administer. □ **police dog** dog used in police work; **police force** body of police of country, district, or town; **policeman**, **policewoman**, **police officer** member of police force; **police state** totalitarian state controlled by political police; **police station** office of local police force.

policy[1] /'pɒlɪsɪ/ *noun* (*plural* **-ies**) course of action adopted by government, business, etc.; prudent conduct.

policy[2] /'pɒlɪsɪ/ *noun* (*plural* **-ies**) (document containing) contract of insurance. □ **policyholder** person or body holding insurance policy.

polio /'pəʊlɪəʊ/ *noun* poliomyelitis.

poliomyelitis /pəʊlɪəʊmaɪə'laɪtɪs/ *noun* infectious viral disease of grey matter of central nervous system, with temporary or permanent paralysis.

Polish /'pəʊlɪʃ/ ● *adjective* of Poland. ● *noun* language of Poland.

polish /'pɒlɪʃ/ ● *verb* (often + *up*) make or become smooth or glossy by rubbing; (esp. as **polished** *adjective*) refine, improve. ● *noun* substance used for polishing; smoothness, glossiness; refinement. □ **polish off** finish quickly.

polite /pə'laɪt/ *adjective* (**-r**, **-st**) having good manners, courteous; cultivated, refined. □ **politely** *adverb*; **politeness** *noun*.

politic /'pɒlɪtɪk/ ● *adjective* judicious, expedient; prudent, sagacious. ● *verb* (**-ck-**) engage in politics.

political /pə'lɪtɪk(ə)l/ *adjective* of state or its government; of public affairs; of, engaged in, or taking a side in politics; relating to pursuit of power, status, etc. □ **political asylum** state protection for foreign refugee; **political correctness** avoidance of language or action which excludes ethnic or cultural minorities; **political economy** study of economic aspects of government; **political geography** geography dealing with boundaries etc. of states; **political prisoner** person imprisoned for political reasons.

politically *adverb* in a political way. □ **politically correct** exhibiting political correctness.

politician /pɒlɪ'tɪʃ(ə)n/ *noun* person engaged in politics.

politicize /pə'lɪtɪsaɪz/ *verb* (also **-ise**) (**-zing** or **-sing**) give political character or awareness to.

politics /'pɒlɪtɪks/ *plural noun* (treated as *singular* or *plural*) art and science of government; political life, affairs, principles, etc.; activities relating to pursuit of power, status, etc.

polity /'pɒlɪtɪ/ *noun* (*plural* **-ies**) form of civil administration; organized society, state.

polka /'pɒlkə/ ● *noun* lively dance; music for this. ● *verb* (**-kas**, **-kaed** /-kəd/ or **-ka'd**, **-kaing** /-kəɪŋ/) dance polka. □ **polka dot** round dot as one of many forming regular pattern on textile fabric etc.

poll /pəʊl/ ● *noun* (often in *plural*) voting; counting of votes; result of voting, number of votes recorded; questioning of sample of population to estimate trend of public opinion; head. ● *verb* take or receive vote(s) of, vote; record opinion of (person, group); cut off top of (tree etc.) or (esp. as **polled** *adjective*) horns of (cattle). □ **polling booth** cubicle where voter stands to mark ballot paper; **polling station** building used for voting; **poll tax** *historical* tax levied on every adult.

pollack /'pɒlək/ *noun* (also **pollock**) (*plural* same or **-s**) edible marine fish related to cod.

pollard /'pɒləd/ ● *noun* hornless animal; tree polled to produce close head of young branches. ● *verb* make pollard of (tree).

pollen /'pɒlən/ *noun* fertilizing powder discharged from flower's anther.

□ **pollen count** index of amount of pollen in air.

pollinate /'pɒlɪneɪt/ *verb* (**-ting**) sprinkle (stigma of flower) with pollen. □ **pollination** *noun.*

pollock = POLLACK.

pollster *noun* person who organizes opinion poll.

pollute /pə'lu:t/ *verb* (**-ting**) contaminate; make impure. □ **pollutant** *noun*; **polluter** *noun*; **pollution** *noun.*

polo /'pəʊləʊ/ *noun* game like hockey played on horseback. □ **polo-neck** (sweater with) high round turned-over collar.

polonaise /pɒlə'neɪz/ *noun* slow processional dance; music for this.

poltergeist /'pɒltəgaɪst/ *noun* noisy mischievous ghost.

poltroon /pɒl'tru:n/ *noun* coward. □ **poltroonery** *noun.*

poly- *combining form* many; polymerized.

polyandry /'pɒlɪændrɪ/ *noun* polygamy in which one woman has more than one husband.

polyanthus /pɒlɪ'ænθəs/ *noun* (*plural* **-thuses**) cultivated primula.

polychromatic /pɒlɪkrəʊ'mætɪk/ *adjective* many-coloured.

polychrome /'pɒlɪkrəʊm/ ● *adjective* in many colours. ● *noun* polychrome work of art.

polyester /pɒlɪ'estə/ *noun* synthetic fibre or resin.

polyethylene /pɒlɪ'eθɪli:n/ *noun* polythene.

polygamy /pə'lɪgəmɪ/ *noun* practice of having more than one wife or husband at once. □ **polygamist** *noun*; **polygamous** *adjective.*

polyglot /'pɒlɪglɒt/ ● *adjective* knowing, using, or written in several languages. ● *noun* polyglot person.

polygon /'pɒlɪgən/ *noun* figure with many sides and angles. □ **polygonal** /pə'lɪg-/ *adjective.*

polyhedron /pɒlɪ'hi:drən/ *noun* (*plural* **-dra**) solid figure with many faces. □ **polyhedral** *adjective.*

polymath /'pɒlɪmæθ/ *noun* person of great or varied learning.

polymer /'pɒlɪmə/ *noun* compound of molecule(s) formed from repeated units of smaller molecules. □ **polymeric** /-'mer-/ *adjective*; **polymerization** *noun*; **polymerize** *verb* (also **-ise**) (**-zing** or **-sing**).

polyp /'pɒlɪp/ *noun* simple organism with tube-shaped body; small growth on mucous membrane.

polyphony /pə'lɪfənɪ/ *noun* (*plural* **-ies**) contrapuntal music. □ **polyphonic** /pɒlɪ'fɒnɪk/ *adjective.*

polypropylene /pɒlɪ'prəʊpɪli:n/ *noun* any of various thermoplastic materials used for films, fibres, or moulding.

polystyrene /pɒlɪ'staɪəri:n/ *noun* kind of hard plastic.

polysyllabic /pɒlɪsɪ'læbɪk/ *adjective* having many syllables; using polysyllables.

polysyllable /'pɒlɪsɪləb(ə)l/ *noun* polysyllabic word.

polytechnic /pɒlɪ'teknɪk/ *noun* college providing courses in esp. vocational subjects up to degree level.

polytheism /'pɒlɪθi:ɪz(ə)m/ *noun* belief in or worship of more than one god. □ **polytheistic** /-'ɪst-/ *adjective.*

polythene /'pɒlɪθi:n/ *noun* a tough light plastic.

polyunsaturated /pɒlɪʌn'sætʃəreɪtɪd/ *adjective* (of fat) containing several double or triple bonds in each molecule and therefore capable of combining with hydrogen and not associated with accumulation of cholesterol.

polyurethane /pɒlɪ'jʊərəθeɪn/ *noun* synthetic resin or plastic used esp. in paints or foam.

polyvinyl chloride /pɒlɪ'vaɪnɪl/ *noun* see PVC.

pomade /pə'mɑ:d/ *noun* scented ointment for hair.

pomander /pə'mændə/ *noun* ball of mixed aromatic substances; container for this.

pomegranate /'pɒmɪgrænɪt/ *noun* tropical tough-rinded many-seeded fruit; tree bearing this.

pommel /'pʌm(ə)l/ *noun* knob of sword hilt; projecting front of saddle.

pomp *noun* splendid display, splendour; specious glory.

pom-pom /'pɒmpɒm/ *noun* automatic quick-firing gun.

pompon /'pɒmpɒn/ *noun* (also **pompom**) decorative tuft or ball on hat, shoe, etc.

pompous /'pɒmpəs/ *adjective* self-important, affectedly grand or solemn. □ **pomposity** /-'pɒs-/ *noun* (*plural* **-ies**); **pompously** *adverb*; **pompousness** *noun*.

ponce *slang* ● *noun* man who lives off prostitute's earnings. ● *verb* (**-cing**) act as ponce. □ **ponce about** move about effeminately.

poncho /'pɒntʃəʊ/ *noun* (*plural* **-s**) cloak of rectangular piece of material with slit in middle for head.

pond *noun* small body of still water.

ponder /'pɒndə/ *verb* think over; muse.

ponderable /'pɒndərəb(ə)l/ *adjective literary* having appreciable weight.

ponderous /'pɒndərəs/ *adjective* heavy and unwieldy; laborious, dull. □ **ponderously** *adverb*; **ponderousness** *noun*.

pong *noun & verb colloquial* stink. □ **pongy** *adjective* (**-ier**, **-iest**).

poniard /'pɒnjəd/ *noun* dagger.

pontiff /'pɒntɪf/ *noun* Pope.

pontifical /pɒn'tɪfɪk(ə)l/ *adjective* papal; pompously dogmatic.

pontificate ● *verb* /pɒn'tɪfɪkeɪt/ (**-ting**) be pompously dogmatic. ● *noun* /pɒn'tɪfɪkət/ office of bishop or Pope; period of this.

pontoon[1] /pɒn'tu:n/ *noun* card game in which players try to acquire cards with face value totalling 21.

pontoon[2] /pɒn'tu:n/ *noun* flat-bottomed boat; boat etc. as one of supports of temporary bridge.

pony /'pəʊnɪ/ *noun* (*plural* **-ies**) horse of any small breed. □ **pony-tail** hair drawn back, tied, and hanging down behind head; **pony-trekking** travelling across country on ponies for pleasure.

poodle /'pu:d(ə)l/ *noun* dog of breed with thick curling hair.

pooh /pu:/ *interjection expressing contempt or disgust*. □ **pooh-pooh** express contempt for, ridicule.

pool[1] *noun* small body of still water; small shallow body of any liquid; swimming pool; deep place in river.

pool[2] ● *noun* common supply of people, vehicles, etc., for sharing by group; group of people sharing duties etc.; common fund, e.g. of profits or of gamblers' stakes; arrangement between competing parties to fix prices and share business; game like billiards with usually 16 balls; (**the pools**) football pool. ● *verb* put into common fund; share in common.

poop *noun* stern of ship; furthest aft and highest deck.

poor /pʊə/ *adjective* having little money or means; (+ *in*) deficient in; inadequate; inferior; deserving pity; despicable. □ **poor man's** inferior substitute for.

poorly ● *adverb* in poor manner; badly. ● *adjective* unwell.

pop[1] ● *noun* abrupt explosive sound; *colloquial* effervescent drink. ● *verb* (**-pp-**) (cause to) make pop; (+ *in*, *out*, *up*, etc.) move, come, or put unexpectedly or suddenly; *slang* pawn. ● *adverb* with the sound pop. □ **popcorn** maize kernels burst open when heated; **pop-eyed** *colloquial* with eyes bulging or wide open; **popgun** toy gun shooting pellet etc. by compressed air or spring; **popping crease** *Cricket* line in front of and parallel to wicket; **pop-up** involving parts that pop up automatically.

pop[2] *noun colloquial* (in full **pop music**) highly successful commercial music; pop record or song. □ **pop art** art based on modern popular culture and the mass media; **pop culture** commercial culture based on popular taste; **pop group** ensemble playing pop music.

pop[3] *noun esp. US colloquial* father.

popadam = POPPADAM.

pope *noun* (also **Pope**) head of RC Church.

popinjay /'pɒpɪndʒeɪ/ *noun* fop, conceited person.

poplar /'pɒplə/ *noun* slender tree with straight trunk and often tremulous leaves.

poplin /'pɒplɪn/ *noun* closely woven corded fabric.

poppadam /'pɒpədəm/ *noun* (also **poppadom**, **popadam**) thin crisp spiced Indian bread.

popper *noun colloquial* press-stud; thing that pops.

poppet /'pɒpɪt/ *noun colloquial* (esp. as term of endearment) small or dainty person.

poppy /'pɒpɪ/ *noun* (*plural* **-ies**) plant with bright flowers and milky narcotic juice; artificial poppy worn on Remembrance Sunday. □ **Poppy Day** Remembrance Sunday.

poppycock /'pɒpɪkɒk/ *noun slang* nonsense.

populace /'pɒpjʊləs/ *noun* the common people.

popular /'pɒpjʊlə/ *adjective* generally liked or admired; of, for, or prevalent among the general public. □ **popularity** /-'lærɪtɪ/ *noun*; **popularize** *verb* (also **-ise**) (**-zing** or **-sing**); **popularly** *adverb*.

populate /'pɒpjʊleɪt/ *verb* (**-ting**) form population of; supply with inhabitants.

population /pɒpjʊ'leɪʃ(ə)n/ *noun* inhabitants of town, country, etc.; total number of these.

populous /'pɒpjʊləs/ *adjective* thickly inhabited.

porcelain /'pɔːsəlɪn/ *noun* fine translucent ceramic; things made of this.

porch *noun* covered entrance to building.

porcine /'pɔːsaɪn/ *adjective* of or like pigs.

porcupine /'pɔːkjʊpaɪn/ *noun* large rodent with body and tail covered with erectile spines.

pore[1] *noun* minute opening in surface through which fluids may pass.

pore[2] *verb* (**-ring**) (+ *over*) be absorbed in studying (book etc.).

pork *noun* flesh of pig used as food.

porker *noun* pig raised for food.

porn (also **porno**) *colloquial* ● *noun* pornography. ● *adjective* pornographic.

pornography /pɔː'nɒgrəfɪ/ *noun* explicit presentation of sexual activity in literature, films, etc., to stimulate erotic rather than aesthetic feelings. □ **pornographic** /-nə'græf-/ *adjective*.

porous /'pɔːrəs/ *adjective* having pores; permeable. □ **porosity** /-'rɒs-/ *noun*.

porphyry /'pɔːfɪrɪ/ *noun* (*plural* **-ies**) hard rock with feldspar crystals in fine-grained red mass.

porpoise /'pɔːpəs/ *noun* sea mammal of whale family.

porridge /'pɒrɪdʒ/ *noun* oatmeal or other cereal boiled in water or milk.

porringer /'pɒrɪndʒə/ *noun* small soup-bowl.

port[1] *noun* harbour; town possessing harbour.

port[2] *noun* strong sweet fortified wine.

port[3] ● *noun* left-hand side of ship or aircraft looking forward. ● *verb* turn (helm) to port.

port[4] *noun* opening in ship's side for entrance, loading, etc.; porthole. □ **porthole** (esp. glazed) aperture in ship's side to admit light.

portable /'pɔːtəb(ə)l/ *adjective* easily movable, convenient for carrying; adaptable in altered circumstances. □ **portability** *noun*.

portage /'pɔːtɪdʒ/ *noun* carrying of boats or goods between two navigable waters.

Portakabin /'pɔːtəkæbɪn/ *noun proprietary term* prefabricated small building.

portal /'pɔːt(ə)l/ *noun* doorway, gate.

portcullis /pɔːt'kʌlɪs/ *noun* strong heavy grating lowered in defence of fortress gateway.

portend /pɔː'tend/ *verb* foreshadow as an omen; give warning of.

portent /'pɔːtent/ *noun* omen, significant sign; marvellous thing.

portentous /pɔː'tentəs/ *adjective* like or being portent; pompously solemn.

porter[1] /'pɔːtə/ *noun* person employed to carry luggage etc.; dark beer brewed from charred or browned malt. □ **porterhouse steak** choice cut of beef.

porter[2] /'pɔːtə/ *noun* gatekeeper or doorkeeper, esp. of large building.

porterage *noun* (charge for) hire of porters.

portfolio /pɔːt'fəʊlɪəʊ/ *noun* (*plural* **-s**) folder for loose sheets of paper, drawings, etc.; samples of artist's work; list of investments held by investor etc.; office of government minister. □ **Minister without Portfolio** government minister not in charge of department.

portico /'pɔːtɪkəʊ/ *noun* (*plural* **-es** or **-s**) colonnade; roof supported by columns, usually serving as porch to building.

portion /'pɔːʃ(ə)n/ ● *noun* part, share; helping; destiny or lot. ● *verb* divide into portions; (+ *out*) distribute.

Portland /'pɔːtlənd/ *noun* □ **Portland cement** cement manufactured from chalk and clay; **Portland stone** a valuable building limestone.

portly /'pɔːtlɪ/ *adjective* (**-ier**, **-iest**) corpulent.

portmanteau /pɔːt'mæntəʊ/ *noun* (*plural* **-s** or **-x** /-z/) case for clothes etc., opening into two equal parts. □ **portmanteau word** word combining sounds and meanings of two others.

portrait /'pɔːtrɪt/ *noun* drawing, painting, photograph, etc. of person or animal; description.

portraiture /'pɔːtrɪtʃə/ *noun* portraying; description; portrait.

portray /pɔː'treɪ/ *verb* make likeness of; describe. □ **portrayal** *noun.*

Portuguese /pɔːtʃʊ'giːz/ ● *noun* (*plural* same) native, national, or language of Portugal. ● *adjective* of Portugal.

pose /pəʊz/ ● *verb* (**-sing**) assume attitude, esp. for artistic purpose; (+ *as*) pretend to be; behave affectedly for effect; propound (question, problem); arrange in required attitude. ● *noun* attitude of body or mind; affectation, pretence.

poser *noun* poseur; *colloquial* puzzling question or problem.

poseur /pəʊ'zɜː/ *noun* person who behaves affectedly.

posh *colloquial* ● *adjective* smart; upper-class. ● *adverb* in an upper-class way. □ **poshly** *adverb*; **poshness** *noun.*

posit /'pɒzɪt/ *verb* (**-t-**) assume as fact, postulate.

position /pə'zɪʃ(ə)n/ ● *noun* place occupied by person or thing; way thing is placed; proper place; advantage; mental attitude; situation; rank, status; paid employment; strategic location. ● *verb* place in position. □ **in a position to** able to. □ **positional** *adjective.*

positive /'pɒzɪtɪv/ ● *adjective* explicit, definite, unquestionable; convinced, confident, cocksure; absolute, not relative; *Grammar* (of adjective or adverb) expressing simple quality without comparison; constructive; marked by presence and not absence of qualities; favourable; dealing only with matters of fact, practical; *Mathematics* (of quantity) greater than zero; *Electricity* of, containing, or producing kind of charge produced by rubbing glass with silk; *Photography* showing lights and shades or colours as seen in original image. ● *noun* positive adjective, photograph, quantity, etc. □ **positive discrimination** making distinctions in favour of groups believed to be underprivileged; **positive vetting** inquiry into background etc. of candidate for post involving national security. □ **positively** *adverb*; **positiveness** *noun.*

positivism *noun* philosophical system recognizing only facts and observable phenomena. □ **positivist** *noun & adjective.*

positron /'pɒzɪtrɒn/ *noun* elementary particle with same mass as but opposite charge to electron.

posse /'pɒsɪ/ *noun* strong force or company; group of law-enforcers.

possess /pə'zes/ *verb* hold as property, own; have; occupy, dominate mind of. □ **possessor** *noun.*

possession /pə'zeʃ(ə)n/ *noun* possessing, being possessed; thing possessed; occupancy; (in *plural*) property; control of ball by player.

possessive /pə'zesɪv/ ● *adjective* wanting to retain what one possesses; jealous and domineering; *Grammar* indicating possession. ● *noun Grammar* possessive case or word. □ **possessiveness** *noun.*

possibility /pɒsɪ'bɪlɪtɪ/ *noun* (*plural* **-ies**) state or fact of being possible; thing that may exist or happen; (usually in *plural*) capability of being used.

possible /'pɒsɪb(ə)l/ ● *adjective* capable of existing, happening, being done, etc.; potential. ● *noun* possible candidate, member of team, etc.; highest possible score.

possibly *adverb* perhaps; in accordance with possibility.

possum /'pɒsəm/ *noun colloquial* opossum. □ **play possum** *colloquial* pretend to be unconscious or unaware.

post¹ /pəʊst/ ● *noun* upright of timber or metal as support in building, to mark boundary, carry notices, etc.; pole etc. marking start or finish of race. ● *verb* (often + *up*) display (notice etc.) in prominent place; advertise by poster or list.

post² /pəʊst/ ● *noun* official conveying of parcels, letters, etc.; single collection or delivery of these; letters etc. dispatched; place where letters etc. are collected. ● *verb* put (letter etc.) into post; (esp. as **posted** *adjective*) (often + *up*) supply with information; enter in ledger. □ **postbox** public box for posting mail; **postcard** card for posting without envelope; **postcode** group of letters and figures in postal address to assist sorting; **post-haste** with great speed; **postman**, **postwoman** person who collects or delivers post; **postmark** official mark on letters to cancel stamp; **postmaster**, **postmistress** official in charge of post office; **post office** room or building for postal business; **Post Office** public department or corporation providing postal services.

post[3] /pəʊst/ ● *noun* appointed place of soldier etc. on duty; occupying force; fort; paid employment; trading post. ● *verb* place (soldier etc.) at post; appoint to post or command.

post- *prefix* after, behind.

postage *noun* charge for sending letter etc. by post. □ **postage stamp** small adhesive label indicating amount of postage paid.

postal *adjective* of or by post. □ **postal order** money order issued by Post Office.

postdate /pəʊst'deɪt/ *verb* (**-ting**) give later than actual date to; follow in time.

poster *noun* placard in public place; large printed picture. □ **poster paint** gummy opaque paint.

poste restante /pəʊst re'stɑ̃t/ *noun* department in post office where letters are kept till called for.

posterior /pɒ'stɪərɪə/ ● *adjective* later in time or order; at the back. ● *noun* (in *singular* or *plural*) buttocks.

posterity /pɒ'sterɪtɪ/ *noun* later generations; descendants.

postern /'pɒst(ə)n/ *noun archaic* back or side entrance.

postgraduate /pəʊst'grædjʊət/ ● *noun* person on course of study after taking first degree. ● *adjective* relating to postgraduates.

posthumous /'pɒstjʊməs/ *adjective* occurring after death; published after author's death; born after father's death. □ **posthumously** *adverb.*

postilion /pɒ'stɪljən/ *noun* (also **postillion**) rider on near horse of team drawing coach etc. without coachman.

post-impressionism /pəʊstɪm'preʃənɪz(ə)m/ *noun* art intending to express individual artist's conception of objects represented. □ **post-impressionist** *noun & adjective.*

post-industrial /pəʊstɪn'dʌstrɪəl/ *adjective* of society or economy no longer reliant on heavy industry.

post-mortem /pəʊst'mɔ:təm/ ● *noun* examination of body made after death; *colloquial* discussion after conclusion (of game etc.). ● *adverb & adjective* after death.

postnatal /pəʊst'neɪt(ə)l/ *adjective* existing or occurring after birth.

postpone /pəʊst'pəʊn/ *verb* (**-ning**) cause to take place at later time. □ **postponement** *noun.*

postscript /'pəʊstskrɪpt/ *noun* addition at end of letter etc. after signature.

postulant /'pɒstjʊlənt/ *noun* candidate, esp. for admission to religious order.

postulate ● *verb* /'pɒstjʊleɪt/ (**-ting**) (often + *that*) assume or require to be true, take for granted; claim. ● *noun* /'pɒstjʊlət/ thing postulated; prerequisite.

posture /'pɒstʃə/ ● *noun* relative position of parts, esp. of body; bearing; mental attitude; condition or state (of affairs etc.). ● *verb* (**-ring**) assume posture, esp. for effect; pose (person).

postwar /pəʊst'wɔ:/ *adjective* occurring or existing after a war.

posy /'pəʊzɪ/ *noun* (*plural* **-ies**) small bunch of flowers.

pot[1] ● *noun* rounded ceramic, metal, or glass vessel; flowerpot, teapot, etc.; contents of pot; chamber pot; total amount bet in game etc.; (usually in *plural*) *colloquial* large sum; *slang* cup etc. as prize. ● *verb* (**-tt-**) plant in pot; (usually as **potted** *adjective*) preserve (food) in sealed pot; pocket (ball) in billiards etc.; abridge, epitomize; shoot at, hit, or kill (animal). □ **go to pot** *colloquial* be ruined; **pot-belly** protuberant belly; **pot-boiler** work of literature etc. done merely to earn money; **pot-herb** herb grown in kitchen garden; **pothole** deep hole in rock, hole in road surface; **potluck** whatever is available; **pot**

plant plant grown in flowerpot; **pot roast** piece of braised meat; **pot-roast** braise; **potsherd** broken piece of ceramic material; **pot-shot** random shot. □ **potful** *noun* (*plural* **-s**).

pot² *noun slang* marijuana.

potable /'pəʊtəb(ə)l/ *adjective* drinkable.

potash /'pɒtæʃ/ *noun* any of various compounds of potassium.

potassium /pə'tæsɪəm/ *noun* soft silver-white metallic element.

potation /pə'teɪʃ(ə)n/ *noun* a drink; drinking.

potato /pə'teɪtəʊ/ *noun* (*plural* **-es**) edible plant tuber; plant bearing this. □ **potato crisp** crisp.

poteen /pɒ'tʃi:n/ *noun Irish* illicit distilled spirit.

potent /'pəʊt(ə)nt/ *adjective* powerful, strong; cogent; (of male) able to achieve erection of penis or have sexual intercourse. □ **potency** *noun*.

potentate /'pəʊtənteɪt/ *noun* monarch, ruler.

potential /pə'tenʃ(ə)l/ ● *adjective* capable of coming into being; latent. ● *noun* capability for use or development; usable resources; quantity determining energy of mass in gravitational field or of charge in electric field. □ **potentiality** /-ʃɪ'æl-/ *noun*; **potentially** *adverb*.

pother /'pɒðə/ *noun literary* din, fuss.

potion /'pəʊʃ(ə)n/ *noun* liquid dose of medicine, poison, etc.

pot-pourri /pəʊ'pʊərɪ/ *noun* (*plural* **-s**) scented mixture of dried petals and spices; musical or literary medley.

pottage /'pɒtɪdʒ/ *noun archaic* soup, stew.

potter¹ /'pɒtə/ *verb* (*US* **putter**) (often + *about*, *around*) work etc. in aimless or desultory manner; go slowly.

potter² /'pɒtə/ *noun* maker of ceramic vessels.

pottery /'pɒtərɪ/ *noun* (*plural* **-ies**) vessels etc. made of baked clay; potter's work or workshop.

potty¹ /'pɒtɪ/ *adjective* (**-ier**, **-iest**) *slang* crazy; insignificant. □ **pottiness** *noun*.

potty² *noun* (*plural* **-ies**) *colloquial* chamber pot, esp. for child.

pouch ● *noun* small bag, detachable pocket; baggy area of skin under eyes etc.; baglike receptacle in which marsupials carry undeveloped young, other baglike natural structure. ● *verb* put or make into pouch; take possession of.

pouffe /pu:f/ *noun* firm cushion as low seat or footstool.

poulterer /'pəʊltərə/ *noun* dealer in poultry and usually game.

poultice /'pəʊltɪs/ ● *noun* soft usually hot dressing applied to sore or inflamed part of body. ● *verb* (**-cing**) apply poultice to.

poultry /'pəʊltrɪ/ *noun* domestic fowls.

pounce ● *verb* (**-cing**) spring, swoop; (often + *on*, *upon*) make sudden attack, seize eagerly. ● *noun* act of pouncing.

pound¹ *noun* unit of weight equal to 16 oz (454 g); (in full **pound sterling**) monetary unit of UK etc.

pound² *verb* crush or beat with repeated strokes; (+ *at*, *on*) deliver heavy blows or gunfire to; (+ *along* etc.) walk, run, etc. heavily.

pound³ *noun* enclosure where stray animals or officially removed vehicles are kept until claimed.

poundage *noun* commission or fee of so much per pound sterling or weight.

-pounder *combining form* thing or person weighing specified number of pounds; gun firing shell weighing specified number of pounds.

pour /pɔ:/ *verb* (usually + *down*, *out*, *over*, etc.) (cause to) flow in stream or shower; dispense (drink); rain heavily; (usually + *in*, *out*, etc.) come or go in profusion or in a rush; discharge copiously.

pout ● *verb* push lips forward, esp. as sign of displeasure; (of lips) be pushed forward. ● *noun* pouting expression.

pouter *noun* kind of pigeon able to inflate crop.

poverty /'pɒvətɪ/ *noun* being poor, want; (often + *of*, *in*) scarcity, lack; inferiority, poorness. □ **poverty-stricken** very poor; **poverty trap** situation in which increase of income incurs greater loss of state benefits.

POW *abbreviation* prisoner of war.

powder /'paʊdə/ ● *noun* mass of fine dry particles; medicine or cosmetic in this form; gunpowder. ● *verb* apply powder to; (esp. as **powdered** *adjective*) reduce to powder. □ **powder blue** pale blue; **powder-puff** soft pad for applying cosmetic powder to skin; **powder-room** *euphemistic* women's lavatory. □ **powdery** *adjective*.

power /'paʊə/ ● *noun* ability to do or act; mental or bodily faculty; influence, authority; ascendancy; authorization; influential person etc.; state with international influence; vigour, energy; *colloquial* large number or amount; capacity for exerting mechanical force; mechanical or electrical energy; electricity supply; particular source or form of energy; product obtained by multiplying a number by itself a specified number of times; magnifying capacity of lens. ● *verb* supply with mechanical or electrical energy; (+ *up*, *down*) increase or decrease power supplied to (device), switch on or off. □ **power cut** temporary withdrawal or failure of electric power supply; **powerhouse** power station, person or thing of great energy; **power of attorney** authority to act for another in legal and financial matters; **power point** socket for connection of electrical appliance etc. to mains; **power-sharing** coalition government; **power station** building where electric power is generated for distribution.

powerful *adjective* having great power or influence. □ **powerfully** *adverb*; **powerfulness** *noun*.

powerless *adjective* without power; wholly unable. □ **powerlessness** *noun*.

powwow /'paʊwaʊ/ ● *noun* meeting for discussion (originally among N. American Indians). ● *verb* hold powwow.

pox *noun* virus disease leaving pocks; *colloquial* syphilis.

pp *abbreviation* pianissimo.

pp. *abbreviation* pages.

p.p. *abbreviation* (also **pp**) *per pro*.

PPS *abbreviation* Parliamentary Private Secretary; further postscript (*post-postscriptum*).

PR *abbreviation* public relations; proportional representation.

practicable /'præktɪkəb(ə)l/ *adjective* that can be done or used. □ **practicability** *noun*.

practical /'præktɪk(ə)l/ ● *adjective* of or concerned with practice rather than theory; functional; good at making, organizing, or mending things; realistic; that is such in effect, virtual. ● *noun* practical exam. □ **practical joke** trick played on person. □ **practicality** /-'kæl-/ *noun* (*plural* **-ies**).

practically *adverb* virtually, almost; in a practical way.

practice /'præktɪs/ *noun* habitual action; repeated exercise to improve skill; action as opposed to theory; doctor's or lawyer's professional business etc.; procedure, esp. of specified kind. □ **in practice** when applied, in reality, skilled from recent practice; **out of practice** lacking former skill.

practise /'præktɪs/ *verb* (**-sing**) (*US* **-tice**; **-cing**) carry out in action; do repeatedly to improve skill; exercise oneself in or on; (as **practised** *adjective*) expert; engage in (profession, religion, etc.).

practitioner /præk'tɪʃənə/ *noun* professional worker, esp. in medicine.

praesidium = PRESIDIUM.

praetorian guard /pri:'tɔ:rɪən/ *noun* bodyguard of ancient Roman emperor etc.

pragmatic /præg'mætɪk/ *adjective* dealing with matters from a practical point of view. □ **pragmatically** *adverb*.

pragmatism /'prægmətɪz(ə)m/ *noun* pragmatic attitude or procedure; *Philosophy* doctrine that evaluates assertions according to their practical consequences. □ **pragmatist** *noun*.

prairie /'preərɪ/ *noun* large treeless tract of grassland, esp. in N. America.

praise /preɪz/ ● *verb* (**-sing**) express warm approval or admiration of; glorify. ● *noun* praising; commendation. □ **praiseworthy** worthy of praise.

praline /'prɑ:li:n/ *noun* sweet made of nuts browned in boiling sugar.

pram *noun* carriage for baby, pushed by person on foot.

prance /prɑ:ns/ ● *verb* (**-cing**) (of horse) spring from hind legs; walk or behave in an elated or arrogant way. ● *noun* prancing; prancing movement.

prank *noun* practical joke.

prat *noun slang* fool.

prate • *verb* (**-ting**) talk too much; chatter foolishly. • *noun* idle talk.

prattle /'præt(ə)l/ • *verb* (**-ling**) talk in childish or inconsequential way. • *noun* prattling talk.

prawn *noun* edible shellfish like large shrimp.

pray *verb* (often + *for, to do, that*) say prayers; make devout supplication; entreat.

prayer /preə/ *noun* request or thanksgiving to God or object of worship; formula used in praying; entreaty. □ **prayer book** book of set prayers; **prayer-mat** small carpet on which Muslims kneel when praying; **prayer-shawl** one worn by male Jews when praying.

pre- *prefix* before (in time, place, order, degree, or importance).

preach *verb* deliver (sermon); proclaim (the gospel etc.); give moral advice obtrusively; advocate, inculcate. □ **preacher** *noun.*

preamble /pri:'æmb(ə)l/ *noun* preliminary statement; introductory part of statute, deed, etc.

prearrange /pri:ə'reɪndʒ/ *verb* (**-ging**) arrange beforehand. □ **prearrangement** *noun.*

prebend /'prebənd/ *noun* stipend of canon or member of chapter; portion of land etc. from which this is drawn. □ **prebendal** /prɪ'bend(ə)l/ *adjective.*

prebendary /'prebəndərɪ/ *noun* (*plural* **-ies**) holder of prebend; honorary canon.

Precambrian /pri:'kæmbrɪən/ • *adjective* of earliest geological era. • *noun* this era.

precarious /prɪ'keərɪəs/ *adjective* uncertain, dependent on chance; perilous. □ **precariously** *adverb*; **precariousness** *noun.*

precast /pri:'kɑ:st/ *adjective* (of concrete) cast in required shape before positioning.

precaution /prɪ'kɔ:ʃ(ə)n/ *noun* action taken beforehand to avoid risk or ensure good result. □ **precautionary** *adjective.*

precede /prɪ'si:d/ *verb* (**-ding**) come or go before in time, order, importance, etc.; (+ *by*) cause to be preceded by.

precedence /'presɪd(ə)ns/ *noun* priority; right of preceding others. □ **take precedence** (often + *over, of*) have priority.

precedent • *noun* /'presɪd(ə)nt/ previous case taken as guide or justification etc. • *adjective* /prɪ'si:d(ə)nt/ preceding.

precentor /prɪ'sentə/ *noun* leader of singing or (in synagogue) prayers of congregation.

precept /'pri:sept/ *noun* rule for action or conduct.

preceptor /prɪ'septə/ *noun* teacher, instructor. □ **preceptorial** /pri:sep'tɔ:rɪəl/ *adjective.*

precession /prɪ'seʃ(ə)n/ *noun* slow movement of axis of spinning body around another axis; such change causing equinoxes to occur earlier in each successive sidereal year.

precinct /'pri:sɪŋkt/ *noun* enclosed area, esp. around building; district in town, esp. where traffic is excluded; (in *plural*) environs.

preciosity /preʃɪ'ɒsɪtɪ/ *noun* affected refinement in art.

precious /'preʃəs/ • *adjective* of great value; much prized; affectedly refined. • *adverb colloquial* extremely, very.

precipice /'presɪpɪs/ *noun* vertical or steep face of rock, cliff, mountain, etc.

precipitate • *verb* /prɪ'sɪpɪteɪt/ (**-ting**) hasten occurrence of; (+ *into*) cause to go into (war etc.) hurriedly or violently; throw down headlong; *Chemistry* cause (substance) to be deposited in solid form from solution; *Physics* condense (vapour) into drops. • *adjective* /prɪ'sɪpɪtət/ headlong; hasty, rash. • *noun* /prɪ'sɪpɪtət/ solid matter precipitated; moisture condensed from vapour.

precipitation /prɪsɪpɪ'teɪʃ(ə)n/ *noun* precipitating, being precipitated; rash haste; rain, snow, etc., falling to ground.

precipitous /prɪ'sɪpɪtəs/ *adjective* of or like precipice; steep.

précis /'preɪsi:/ • *noun* (*plural* same /-sɪ:z/) summary, abstract. • *verb* (**-cises** /-si:z/, **-cised** /-si:d/, **-cising** /-si:ɪŋ/) make précis of.

precise /prɪ'saɪs/ *adjective* accurately worded; definite, exact; punctilious.

precisely *adverb* in a precise way, exactly; quite so.

precision /prɪ'sɪʒ(ə)n/ ● *noun* accuracy. ● *adjective* designed for or produced by precise work.

preclude /prɪ'klu:d/ *verb* **(-ding)** (+ *from*) prevent; make impossible.

precocious /prɪ'kəʊʃəs/ *adjective* prematurely developed in some respect. □ **precociously** *adverb*; **precociousness** *noun*; **precocity** /-'kɒs-/ *noun*.

precognition /pri:kɒg'nɪʃ(ə)n/ *noun* (esp. supernatural) foreknowledge.

preconceive /pri:kən'si:v/ *verb* **(-ving)** form (opinion etc.) beforehand.

preconception /pri:kən'sepʃ(ə)n/ *noun* preconceived idea; prejudice.

precondition /pri:kən'dɪʃ(ə)n/ *noun* condition that must be fulfilled beforehand.

precursor /pri:'kɜ:sə/ *noun* forerunner; person who precedes in office etc.; harbinger.

predate /pri:'deɪt/ *verb* **(-ting)** precede in time.

predator /'predətə/ *noun* predatory animal; exploiter of others.

predatory /'predətərɪ/ *adjective* (of animal) preying naturally on others; plundering or exploiting others.

predecease /pri:dɪ'si:s/ *verb* **(-sing)** die before (another).

predecessor /'pri:dɪsesə/ *noun* previous holder of office or position; ancestor; thing to which another has succeeded.

predestine /pri:'destɪn/ *verb* **(-ning)** determine beforehand; ordain by divine will or as if by fate. □ **predestination** *noun*.

predetermine /pri:dɪ'tɜ:mɪn/ *verb* **(-ning)** decree beforehand; predestine.

predicament /prɪ'dɪkəmənt/ *noun* difficult or unpleasant situation.

predicate ● *verb* /'predɪkeɪt/ **(-ting)** assert (something) about subject of proposition; (+ *on*) base (statement etc.) on. ● *noun* /'predɪkət/ *Grammar & Logic* what is said about subject of sentence or proposition. □ **predicable** *adjective*; **predication** *noun*.

predicative /prɪ'dɪkətɪv/ *adjective Grammar* (of adjective or noun) forming part or all of predicate; that predicates.

predict /prɪ'dɪkt/ *verb* forecast; prophesy. □ **predictable** *adjective*; **predictably** *adverb*; **prediction** *noun*.

predilection /pri:dɪ'lekʃ(ə)n/ *noun* (often + *for*) preference, special liking.

predispose /pri:dɪs'pəʊz/ *verb* **(-sing)** influence favourably in advance; (+ *to*, *to do*) render liable or inclined beforehand to. □ **predisposition** /-pə'zɪʃ(ə)n/ *noun*.

predominate /prɪ'dɒmɪneɪt/ *verb* **(-ting)** (+ *over*) have control over; prevail; preponderate. □ **predominance** *noun*; **predominant** *adjective*; **predominantly** *adverb*.

pre-eminent /pri:'emɪnənt/ *adjective* excelling others; outstanding. □ **pre-eminence** *noun*; **pre-eminently** *adverb*.

pre-empt /pri:'empt/ *verb* forestall; obtain by pre-emption.

■ **Usage** *Pre-empt* is sometimes used to mean *prevent*, but this is considered incorrect in standard English.

pre-emption /pri:'empʃ(ə)n/ *noun* purchase or taking of thing before it is offered to others.

pre-emptive /pri:'emptɪv/ *adjective* pre-empting; *Military* intended to prevent attack by disabling enemy.

preen *verb* (of bird) tidy (feathers, itself) with beak; (of person) smarten or admire (oneself, one's hair, clothes, etc.); (often + *on*) pride (oneself).

prefab /'pri:fæb/ *noun colloquial* prefabricated building.

prefabricate /pri:'fæbrɪkeɪt/ *verb* **(-ting)** manufacture sections of (building etc.) prior to assembly on site.

preface /'prefəs/ ● *noun* introduction to book stating subject, scope, etc.; preliminary part of speech. ● *verb* **(-cing)** (often + *with*) introduce or begin (as) with preface; (of event etc.) lead up to (another). □ **prefatory** *adjective*.

prefect /'pri:fekt/ *noun* chief administrative officer of district in France etc.; senior pupil in school, authorized to maintain discipline.

prefecture /'pri:fektʃə/ *noun* district under government of prefect; prefect's office or tenure.

prefer /prɪ'fɜ:/ *verb* **(-rr-)** (often + *to*, *to do*) like better; submit (information, accusation, etc.); promote (person).

preferable /'prefərəb(ə)l/ *adjective* to be preferred, more desirable. □ **preferably** *adverb.*

preference /'prefərəns/ *noun* preferring, being preferred; thing preferred; favouring of one person etc. before others; prior right.

preferential /prefə'renʃ(ə)l/ *adjective* of, giving, or receiving preference. □ **preferentially** *adverb.*

preferment /prɪ'fɜ:mənt/ *noun formal* promotion to higher office.

prefigure /pri:'fɪgə/ *verb* (**-ring**) represent or imagine beforehand.

prefix /'pri:fɪks/ ● *noun* part-word added to beginning of word to alter meaning, e.g. *re-* in *retake*, *ex-* in *ex-president*; title before name. ● *verb* (often + *to*) add as introduction; join (word, element) as prefix.

pregnant /'pregnənt/ *adjective* having child or young developing in womb; significant, suggestive. □ **pregnancy** *noun* (*plural* **-ies**).

preheat /pri:'hi:t/ *verb* heat beforehand.

prehensile /pri:'hensaɪl/ *adjective* (of tail, limb, etc.) capable of grasping.

prehistoric /pri:hɪs'tɒrɪk/ *adjective* of period before written records. □ **prehistory** /-'hɪstərɪ/ *noun.*

prejudge /pri:'dʒʌdʒ/ *verb* (**-ging**) form premature judgement on (person etc.).

prejudice /'predʒʊdɪs/ ● *noun* preconceived opinion; (+ *against*, *in favour of*) bias; harm (possibly) resulting from action or judgement. ● *verb* (**-cing**) impair validity of; (esp. as **prejudiced** *adjective*) cause (person) to have prejudice. □ **prejudicial** /-'dɪʃ-/ *adjective.*

prelacy /'preləsɪ/ *noun* (*plural* **-ies**) church government by prelates; (**the prelacy**) prelates collectively; office or rank of prelate.

prelate /'prelət/ *noun* high ecclesiastical dignitary, e.g. bishop.

preliminary /prɪ'lɪmɪnərɪ/ ● *adjective* introductory, preparatory. ● *noun* (*plural* **-ies**) (usually in *plural*) preliminary action or arrangement; preliminary trial or contest.

prelude /'prelju:d/ ● *noun* (often + *to*) action, event, etc. serving as introduction; introductory part of poem etc.; *Music* introductory piece of suite, short piece of similar type. ● *verb* (**-ding**) serve as prelude to; introduce with prelude.

premarital /pri:'mærɪt(ə)l/ *adjective* occurring etc. before marriage.

premature /'premətʃə/ *adjective* occurring or done before usual or right time; too hasty; (of baby) born 3 or more weeks before expected time. □ **prematurely** *adverb.*

premed /pri:'med/ *noun colloquial* premedication.

premedication /pri:medɪ'keɪʃ(ə)n/ *noun* medication in preparation for operation.

premeditate /pri:'medɪteɪt/ *verb* (**-ting**) think out or plan beforehand. □ **premeditation** *noun.*

premenstrual /pri:'menstrʊəl/ *adjective* of the time immediately before menstruation.

premier /'premɪə/ ● *noun* prime minister. ● *adjective* first in importance, order, or time. □ **premiership** *noun.*

première /'premɪeə/ ● *noun* first performance or showing of play or film. ● *verb* (**-ring**) give première of.

premise /'premɪs/ *noun* premiss; (in *plural*) house or other building with its grounds etc.; (in *plural*) *Law* previously specified houses, lands, or tenements. □ **on the premises** in the house etc. concerned.

premiss /'premɪs/ *noun* previous statement from which another is inferred.

premium /'pri:mɪəm/ ● *noun* amount to be paid for contract of insurance; sum added to interest, wages, etc.; reward, prize. ● *adjective* of best quality and highest price. □ **at a premium** highly valued, above usual or nominal price; **Premium (Savings) Bond** government security not bearing interest but with periodic prize draw.

premonition /premə'nɪʃ(ə)n/ *noun* forewarning; presentiment. □ **premonitory** /prɪ'mɒnɪtərɪ/ *adjective.*

prenatal /pri:'neɪt(ə)l/ *adjective* existing or occurring before birth.

preoccupy /pri:'ɒkjʊpaɪ/ *verb* (**-ies**, **-ied**) dominate mind of; (as **preoccupied** *adjective*) otherwise engrossed. □ **preoccupation** *noun.*

preordain /pri:ɔ:'deɪn/ *verb* ordain or determine beforehand.

prep *noun colloquial* homework; time when this is done.

prepack /pri:'pæk/ *verb* (also **pre-package** /-'pækɪdʒ/) pack (goods) before retail.

preparation /prepə'reɪʃ(ə)n/ *noun* preparing, being prepared; (often in *plural*) thing done to make ready; substance specially prepared.

preparatory /prɪ'pærətərɪ/ ● *adjective* (often + *to*) serving to prepare; introductory. ● *adverb* (often + *to*) as a preparation. □ **preparatory school** private primary or (*US*) secondary school.

prepare /prɪ'peə/ *verb* (**-ring**) make or get ready; get oneself ready.

prepay /pri:'peɪ/ *verb* (*past & past participle* **prepaid**) pay (charge) beforehand; pay postage on beforehand. □ **prepayment** *noun*.

preponderate /prɪ'pɒndəreɪt/ *verb* (**-ting**) (often + *over*) be superior in influence, quantity, or number; predominate. □ **preponderance** *noun*; **preponderant** *adjective*.

preposition /prepə'zɪʃ(ə)n/ *noun* word used before noun or pronoun to indicate its relationship to another word (see panel). □ **prepositional** *adjective*.

prepossess /pri:pə'zes/ *verb* (usually in *passive*) take possession of; prejudice, usually favourably; (as **prepossessing** *adjective*) attractive. □ **prepossession** *noun*.

preposterous /prɪ'pɒstərəs/ *adjective* utterly absurd; contrary to nature or reason. □ **preposterously** *adverb*.

prepuce /'pri:pju:s/ *noun* foreskin.

Pre-Raphaelite /pri:'ræfəlaɪt/ ● *noun* member of group of 19th-c. Eng. artists. ● *adjective* of Pre-Raphaelites; (**pre-Raphaelite**) (esp. of woman) of type painted by Pre-Raphaelites.

pre-record /pri:rɪ'kɔ:d/ *verb* record in advance.

prerequisite /pri:'rekwɪzɪt/ ● *adjective* required as precondition. ● *noun* prerequisite thing.

■ **Usage** *Prerequisite* is sometimes confused with *perquisite* which means 'an extra profit, right, or privilege'.

---

prerogative /prɪ'rɒgətɪv/ *noun* right or privilege exclusive to individual or class.

Pres. *abbreviation* President.

presage /'presɪdʒ/ ● *noun* omen; presentiment. ● *verb* (**-ging**) portend; indicate (future event etc.); foretell, foresee.

presbyopia /prezbɪ'əʊpɪə/ *noun* long-sightedness. □ **presbyopic** *adjective*.

presbyter /'prezbɪtə/ *noun* priest of Episcopal Church; elder of Presbyterian Church.

Presbyterian /prezbɪ'tɪərɪən/ ● *adjective* (of Church, esp. Church of Scotland) governed by elders all of equal rank.

Preposition

A preposition is used in front of a noun or pronoun to form a phrase. It often describes the position of something, e.g. *under the chair*, or the time at which something happens, e.g. *in the evening*.

Prepositions in common use are:

about	*behind*	*into*	*through*
above	*beside*	*like*	*till*
across	*between*	*near*	*to*
after	*by*	*of*	*towards*
against	*down*	*off*	*under*
along	*during*	*on*	*underneath*
among	*except*	*outside*	*until*
around	*for*	*over*	*up*
as	*from*	*past*	*upon*
at	*in*	*round*	*with*
before	*inside*	*since*	*without*

• *noun* member of Presbyterian Church. □ **Presbyterianism** *noun*.

presbytery /'prezbɪtərɪ/ *noun* (*plural* **-ies**) eastern part of chancel; body of presbyters; RC priest's house.

prescient /'presɪənt/ *adjective* having foreknowledge or foresight. □ **prescience** *noun*.

prescribe /prɪ'skraɪb/ *verb* (**-bing**) advise use of (medicine etc.); lay down authoritatively.

■ **Usage** *Prescribe* is sometimes confused with *proscribe*, which means 'forbid'.

prescript /'pri:skrɪpt/ *noun* ordinance, command.

prescription /prɪ'skrɪpʃ(ə)n/ *noun* prescribing; doctor's (usually written) instruction for composition and use of medicine; medicine thus prescribed.

prescriptive /prɪ'skrɪptɪv/ *adjective* prescribing, laying down rules; arising from custom.

presence /'prez(ə)ns/ *noun* being present; place where person is; personal appearance; person or spirit that is present. □ **presence of mind** calmness and quick-wittedness in sudden difficulty etc.

present[1] /'prez(ə)nt/ • *adjective* being in place in question; now existing, occurring, or being dealt with etc.; *Grammar* expressing present action or state. • *noun* (**the present**) now; present tense. □ **at present** now; **for the present** just now; **present-day** of this time, modern.

present[2] /prɪ'zent/ *verb* introduce; exhibit; offer or give (thing) to; (+ *with*) provide (person) with; put (play, film, etc.) before public; reveal; deliver (cheque etc.) for payment etc. □ **present arms** hold rifle etc. in saluting position.

present[3] /'prez(ə)nt/ *noun* gift.

presentable /prɪ'zentəb(ə)l/ *adjective* of good appearance; fit to be shown. □ **presentability** *noun*; **presentably** *adverb*.

presentation /prezən'teɪʃ(ə)n/ *noun* presenting, being presented; thing presented; manner or quality of presenting; demonstration of materials etc., lecture.

presenter *noun* person introducing broadcast programme.

presentiment /prɪ'zentɪmənt/ *noun* vague expectation, foreboding.

presently *adverb* before long; *US & Scottish* at present.

preservative /prɪ'zɜ:vətɪv/ • *noun* substance for preserving food etc. • *adjective* tending to preserve.

preserve /prɪ'zɜ:v/ • *verb* (**-ving**) keep safe or free from decay; maintain, retain; treat (food) to prevent decomposition or fermentation; keep (game etc.) undisturbed for private use. • *noun* (in *singular* or *plural*) preserved fruit, jam; place where game etc. is preserved; sphere of activity regarded by person as his or hers alone. □ **preservation** /prezə'veɪʃ(ə)n/ *noun*.

preshrunk /pri:'ʃrʌŋk/ *adjective* (of fabric etc.) treated so as to shrink during manufacture and not in use.

preside /prɪ'zaɪd/ *verb* (**-ding**) (often + *at*, *over*) be chairperson or president; exercise control or authority.

presidency /'prezɪdənsɪ/ *noun* (*plural* **-ies**) office of president; period of this.

president /'prezɪd(ə)nt/ *noun* head of republic; head of society or council etc., of certain colleges, or (*US*) of university, company, etc.; person in charge of meeting. □ **presidential** /-'den-/ *adjective*.

presidium /prɪ'sɪdɪəm/ *noun* (also **praesidium**) standing committee, esp. in Communist country.

press[1] • *verb* apply steady force to; flatten, shape, smooth (esp. clothes); (+ *out of*, *from*, etc.) squeeze (juice etc.); embrace, caress firmly; (+ *on*, *against*, etc.) exert pressure on; be urgent, urge; (+ *for*) demand insistently; (+ *up*, *round*, etc.) crowd; (+ *on*, *forward*, etc.) hasten; (+ *on*, *upon*) force (offer etc.) on; manufacture (gramophone record, car part, etc.) using pressure. • *noun* pressing; device for compressing, flattening, extracting juice, etc.; machine for printing; (**the press**) newspapers; publicity in newspapers; printing house; publishing company; crowding, crowd; pressure of affairs; large usually shelved cupboard. □ **press agent** person employed to manage advertising and press publicity; **press**

conference meeting with journalists; **press gallery** gallery for reporters, esp. in legislative assembly; **press release** statement issued to newspapers etc.; **press-stud** small device fastened by pressing to engage two parts; **press-up** exercise in which prone body is raised by pressing down on hands to straighten arms.

press² *verb historical* force to serve in army or navy; bring into use as makeshift. □ **press-gang** *noun historical* group of men employed to press men for navy, *verb* force into service.

pressing ● *adjective* urgent; insistent. ● *noun* thing made by pressing, e.g. gramophone record; series of these made at one time; act of pressing. □ **pressingly** *adverb.*

pressure /ˈpreʃə/ ● *noun* exertion of continuous force, force so exerted, amount of this; urgency; affliction, difficulty; constraining or compelling influence. ● *verb* (**-ring**) (often + *into*) apply pressure to, coerce, persuade. □ **pressure-cooker** pan for cooking quickly under high pressure; **pressure group** group formed to influence public policy.

pressurize *verb* (also **-ise**) (**-zing** or **-sing**) (esp. as **pressurized** *adjective*) maintain normal atmospheric pressure in (aircraft cabin etc.) at high altitude; raise to high pressure; pressure (person).

prestidigitator /prestɪˈdɪdʒɪteɪtə/ *noun formal* conjuror. □ **prestidigitation** *noun.*

prestige /presˈtiːʒ/ ● *noun* respect or reputation. ● *adjective* having or conferring prestige. □ **prestigious** /-ˈstɪdʒəs/ *adjective.*

presto /ˈprestəʊ/ *Music* ● *adverb & adjective* in quick tempo. ● *noun* (*plural* **-s**) presto movement or passage.

prestressed /priːˈstrest/ *adjective* (of concrete) strengthened by stretched wires in it.

presumably /prɪˈzjuːməblɪ/ *adverb* as may reasonably be presumed.

presume /prɪˈzjuːm/ *verb* (**-ming**) (often + *that*) suppose to be true, take for granted; (often + *to do*) venture; be presumptuous; (+ *on*, *upon*) make unscrupulous use of.

presumption /prɪˈzʌmpʃ(ə)n/ *noun* arrogance, presumptuous behaviour; taking for granted; thing presumed to be true; ground for presuming.

presumptive /prɪˈzʌmptɪv/ *adjective* giving grounds for presumption.

presumptuous /prɪˈzʌmptʃʊəs/ *adjective* unduly confident, arrogant. □ **presumptuously** *adverb*; **presumptuousness** *noun.*

presuppose /priːsəˈpəʊz/ *verb* (**-sing**) assume beforehand; imply. □ **presupposition** /-sʌpəˈzɪʃ(ə)n/ *noun.*

pre-tax /priːˈtæks/ *adjective* (of income) before deduction of taxes.

pretence /prɪˈtens/ *noun* (*US* **pretense**) pretending, make-believe; pretext; (+ *to*) (esp. false) claim; ostentation.

pretend /prɪˈtend/ ● *verb* claim or assert falsely; imagine in play; (as **pretended** *adjective*) falsely claimed to be; (+ *to*) profess to have. ● *adjective colloquial* pretended.

pretender *noun* person who claims throne, title, etc.

pretense *US* = PRETENCE.

pretension /prɪˈtenʃ(ə)n/ *noun* (often + *to*) assertion of claim; pretentiousness.

pretentious /prɪˈtenʃəs/ *adjective* making excessive claim to merit or importance; ostentatious. □ **pretentiously** *adverb*; **pretentiousness** *noun.*

preternatural /priːtəˈnætʃər(ə)l/ *adjective* extraordinary; supernatural.

pretext /ˈpriːtekst/ *noun* ostensible reason; excuse.

pretty /ˈprɪtɪ/ ● *adjective* (**-ier**, **-iest**) attractive in delicate way; fine, good; considerable. ● *adverb colloquial* fairly, moderately. ● *verb* (**-ies**, **-ied**) (often + *up*) make pretty. □ **pretty-pretty** *colloquial* too pretty. □ **prettify** *verb* (**-ies**, **-ied**); **prettily** *adverb*; **prettiness** *noun.*

pretzel /ˈprets(ə)l/ *noun* crisp knot-shaped salted biscuit.

prevail /prɪˈveɪl/ *verb* (often + *against*, *over*) be victorious; be the more usual or predominant; exist or occur in general use; (+ *on*, *upon*) persuade.

prevalent /ˈprevələnt/ *adjective* generally existing or occurring. □ **prevalence** *noun.*

prevaricate /prɪˈværɪkeɪt/ *verb* (**-ting**) speak or act evasively or misleadingly.

□ **prevarication** *noun*; **prevaricator** *noun*.

■ **Usage** *Prevaricate* is often confused with *procrastinate*, which means 'to defer action'.

prevent /prɪ'vent/ *verb* (often + *from doing*) stop, hinder. □ **preventable** *adjective* (also **preventible**); **prevention** *noun*.

■ **Usage** The use of *prevent* without 'from' as in *She prevented me going* is informal. An acceptable further alternative is *She prevented my going*.

preventative /prɪ'ventətɪv/ *adjective & noun* preventive.

preventive /prɪ'ventɪv/ ● *adjective* serving to prevent, esp. disease. ● *noun* preventive agent, measure, drug, etc.

preview /'priːvjuː/ ● *noun* showing of film, play, etc. before it is seen by general public. ● *verb* view or show in advance.

previous /'priːvɪəs/ ● *adjective* (often + *to*) coming before in time or order; *colloquial* hasty, premature. ● *adverb* (+ *to*) before. □ **previously** *adverb*.

pre-war /priː'wɔː/ *adjective* existing or occurring before a war.

prey /preɪ/ ● *noun* animal hunted or killed by another for food; (often + *to*) victim. ● *verb* (+ *on*, *upon*) seek or take as prey; exert harmful influence.

price ● *noun* amount of money for which thing is bought or sold; what must be given, done, etc. to obtain thing; odds. ● *verb* (**-cing**) fix or find price of; estimate value of. □ **at a price** at high cost; **price tag** label on item showing its price.

priceless *adjective* invaluable; *colloquial* very amusing or absurd.

pricey *adjective* (**pricier**, **priciest**) *colloquial* expensive.

prick ● *verb* pierce slightly, make small hole in; (+ *off*, *out*) mark with pricks or dots; trouble mentally; tingle. ● *noun* pricking, mark of it; pain caused as by pricking, mental pain. □ **prick out** plant (seedlings etc.) in small holes pricked in earth; **prick up one's ears** (of dog) erect the ears when alert, (of person) become suddenly attentive.

prickle /'prɪk(ə)l/ ● *noun* small thorn; hard-pointed spine; prickling sensation. ● *verb* (**-ling**) cause or feel sensation as of prick(s).

prickly *adjective* (**-ier**, **-iest**) having prickles; irritable; tingling. □ **prickly heat** itchy inflammation of skin near sweat glands; **prickly pear** cactus with pear-shaped edible fruit, its fruit. □ **prickliness** *noun*.

pride ● *noun* elation or satisfaction at one's achievements, possessions, etc.; object of this; unduly high opinion of oneself; proper sense of one's own worth, position, etc.; group (of lions etc.). ● *verb* (**-ding**) (**pride oneself on**, **upon**) be proud of. □ **pride of place** most important position; **take (a) pride in** be proud of.

prie-dieu /priː'djɜː/ *noun* (*plural* **-x** same pronunciation) kneeling-desk for prayer.

priest /priːst/ *noun* ordained minister of some Christian churches (above deacon and below bishop); (*feminine* **priestess**) official minister of non-Christian religion. □ **priesthood** *noun*; **priestly** *adjective*.

prig *noun* self-righteous or moralistic person. □ **priggish** *adjective*; **priggishness** *noun*.

prim *adjective* (**-mm-**) stiffly formal and precise; prudish. □ **primly** *adverb*; **primness** *noun*.

prima /'priːmə/ *adjective* □ **prima ballerina** chief female dancer in ballet; **prima donna** chief female singer in opera, temperamental person.

primacy /'praɪməsɪ/ *noun* (*plural* **-ies**) pre-eminence; office of primate.

prima facie /praɪmə 'feɪʃɪ/ ● *adverb* at first sight. ● *adjective* (of evidence) based on first impression.

primal /'praɪm(ə)l/ *adjective* primitive, primeval; fundamental.

primary /'praɪmərɪ/ ● *adjective* of first importance; fundamental; original. ● *noun* (*plural* **-ies**) primary colour, feather, school, etc.; *US* primary election. □ **primary colour** one not obtained by mixing others; **primary education** education for children under 11; **primary election** *US* election to select candidate(s) for principal election;

primary feather large flight feather of bird's wing; **primary school** school for primary education. □ **primarily** /'praɪmərɪlɪ, praɪ'meərɪlɪ/ *adverb.*

primate /'praɪmeɪt/ *noun* member of highest order of mammals, including apes, man, etc.; archbishop.

prime[1] ● *adjective* chief, most important; of highest quality; primary, fundamental; (of number etc.) divisible only by itself and unity. ● *noun* best or most vigorous stage. □ **prime minister** chief minister of government; **prime time** time when television etc. audience is largest.

prime[2] *verb* (**-ming**) prepare (thing) for use; prepare (gun) for firing or (explosive) for detonation; pour liquid into (pump) to start it working; cover (wood, metal, etc.) with primer; equip (person) with information etc.

primer[1] *noun* substance applied to bare wood, metal, etc. before painting.

primer[2] *noun* elementary school-book; introductory book.

primeval /praɪ'mi:v(ə)l/ *adjective* of first age of world; ancient, primitive.

primitive /'prɪmɪtɪv/ ● *adjective* at early stage of civilization; crude, simple. ● *noun* untutored painter with naïve style; picture by such painter. □ **primitively** *adverb*; **primitiveness** *noun.*

primogeniture /praɪməʊ'dʒenɪtʃə/ *noun* being first-born; first-born's right to inheritance.

primordial /praɪ'mɔ:dɪəl/ *adjective* existing at or from beginning, primeval.

primrose /'prɪmrəʊz/ *noun* plant bearing pale yellow spring flower; this flower; pale yellow. □ **primrose path** pursuit of pleasure.

primula /'prɪmjʊlə/ *noun* cultivated plant with flowers of various colours.

Primus /'praɪməs/ *noun proprietary term* portable cooking stove burning vaporized oil.

prince *noun* (as title usually **Prince**) male member of royal family other than king; ruler of small state; nobleman of some countries; (often + *of*) the greatest. □ **Prince Consort** husband of reigning queen who is himself a prince.

princely *adjective* (**-ier**, **-iest**) of or worthy of a prince; sumptuous, splendid.

princess /prɪn'ses/ *noun* (as title usually **Princess** /'prɪnses/) prince's wife; female member of royal family other than queen.

principal /'prɪnsɪp(ə)l/ ● *adjective* first in importance, chief; leading. ● *noun* chief person; head of some institutions; principal actor, singer, etc.; capital sum lent or invested; person for whom another is agent etc. □ **principal boy** (usually actress playing) leading male role in pantomime. □ **principally** *adverb.*

principality /prɪnsɪ'pælɪtɪ/ *noun* (*plural* **-ies**) state ruled by prince; (**the Principality**) Wales.

principle /'prɪnsɪp(ə)l/ *noun* fundamental truth or law as basis of reasoning or action; personal code of conduct; fundamental source or element. □ **in principle** in theory; **on principle** from moral motive.

principled *adjective* based on or having (esp. praiseworthy) principles of behaviour.

prink *verb* (usually **prink oneself**; often + *up*) smarten, dress up.

print ● *verb* produce by applying inked type, plates, etc. to paper etc.; express or publish in print; (often + *on*, *with*) impress, stamp; write in letters that are not joined; produce (photograph) from negative; (usually + *out*) produce computer output in printed form; mark (fabric) with design. ● *noun* mark left on surface by pressure; printed lettering, words, or publication (esp. newspaper); engraving; photograph; printed fabric. □ **in print** (of book etc.) available from publisher, in printed form; **out of print** (of book etc.) no longer available from publisher; **printed circuit** electric circuit with thin conducting strips printed on flat sheet; **printing press** machine for printing from type, plates, etc.; **printout** computer output in printed form.

printer *noun* person who prints books etc.; owner of printing business; device that prints esp. computer output.

prior /'praɪə/ ● *adjective* earlier; (often + *to*) coming before in time, order, or importance. ● *adverb* (+ *to*) before. ● *noun* (*feminine* **prioress**) superior of religious house; (in abbey) deputy of abbot.

priority /praɪ'ɒrɪtɪ/ *noun* (*plural* **-ies**) thing considered more important than others; precedence in time, rank, etc.; right to do something before other people. □ **prioritize** *verb* (also **-ise**) (**-zing** or **-sing**).

priory /'praɪərɪ/ *noun* (*plural* **-ies**) religious house governed by prior or prioress.

prise /praɪz/ *verb* (also **prize**) (**-sing** or **-zing**) force open or out by leverage.

prism /'prɪz(ə)m/ *noun* solid figure whose two ends are equal parallel rectilinear figures, and whose sides are parallelograms; transparent body of this form with refracting surfaces.

prismatic /prɪz'mætɪk/ *adjective* of, like, or using prism; (of colours) distributed (as if) by transparent prism.

prison /'prɪz(ə)n/ *noun* place of captivity, esp. building to which people are consigned while awaiting trial or for punishment.

prisoner /'prɪznə/ *noun* person kept in prison; person or thing confined by illness, another's grasp, etc.; (in full **prisoner of war**) person captured in war.

prissy /'prɪsɪ/ *adjective* (**-ier, -iest**) prim, prudish. □ **prissily** *adverb*; **prissiness** *noun*.

pristine /'prɪsti:n/ *adjective* in original condition, unspoilt; ancient.

privacy /'prɪvəsɪ/ *noun* (right to) being private; freedom from intrusion or publicity.

private /'praɪvət/ ● *adjective* belonging to an individual, personal; confidential, secret; not public; secluded; not holding public office or official position; not supported, managed, or provided by state. ● *noun* private soldier; (in *plural*) *colloquial* genitals. □ **in private** privately; **private detective** detective outside police force; **private enterprise** business(es) not under state control; **private eye** *colloquial* private detective; **private means** unearned income from investments etc.; **private member** MP not holding government office; **private parts** *euphemistic* genitals; **private soldier** ordinary soldier, not officer. □ **privately** *adverb*.

privateer /praɪvə'tɪə/ *noun* (commander of) privately owned and government-commissioned warship.

privation /praɪ'veɪʃ(ə)n/ *noun* lack of comforts or necessities.

privatize /'praɪvətaɪz/ *verb* (also **-ise**) (**-zing** or **-sing**) transfer from state to private ownership. □ **privatization** *noun*.

privet /'prɪvɪt/ *noun* bushy evergreen shrub used for hedges.

privilege /'prɪvɪlɪdʒ/ ● *noun* right, advantage, or immunity belonging to person, class, or office; special benefit or honour. ● *verb* (**-ging**) invest with privilege.

privy /'prɪvɪ/ ● *adjective* (+ *to*) sharing secret of; *archaic* hidden, secret. ● *noun* (*plural* **-ies**) lavatory. □ **Privy Council** group of advisers appointed by sovereign; **Privy Councillor**, **Counsellor** member of this; **privy purse** allowance from public revenue for monarch's private expenses; **privy seal** state seal formerly affixed to minor documents.

prize[1] ● *noun* reward in competition, lottery, etc.; reward given as symbol of victory or superiority; thing (to be) striven for. ● *adjective* to which prize is awarded; excellent of its kind. ● *verb* (**-zing**) value highly. □ **prizefight** boxing match for money.

prize[2] *noun* ship or property captured in naval warfare.

prize[3] = PRISE.

PRO *abbreviation* Public Record Office; public relations officer.

pro[1] *noun* (*plural* **-s**) *colloquial* professional.

pro[2] ● *adjective* in favour. ● *noun* (*plural* **-s**) reason in favour. ● *preposition* in favour of. □ **pros and cons** reasons for and against.

proactive /prəʊ'æktɪv/ *adjective* (of person, policy, etc.) taking the initiative.

probability *noun* (*plural* **-ies**) being probable; likelihood; (most) probable event; extent to which thing is likely to occur, measured by ratio of favourable cases to all cases possible. □ **in all probability** most probably.

probable /'prɒbəb(ə)l/ ● *adjective* (often + *that*) that may be expected to happen or prove true; likely. ● *noun* probable candidate, member of team, etc. □ **probably** *adverb.*

probate /'prəʊbeɪt/ *noun* official proving of will; verified copy of will.

probation /prə'beɪʃ(ə)n/ *noun* system of supervising behaviour of offenders as alternative to prison; testing of character and abilities of esp. new employee. □ **probation officer** official supervising offenders on probation. □ **probationary** *adjective.*

probationer *noun* person on probation.

probe ● *noun* investigation; device for measuring, testing, etc.; blunt-ended surgical instrument for exploring wound etc.; unmanned exploratory spacecraft. ● *verb* (**-bing**) examine closely; explore with probe.

probity /'prəʊbɪtɪ/ *noun* uprightness, honesty.

problem /'prɒbləm/ ● *noun* doubtful or difficult question; thing hard to understand or deal with. ● *adjective* causing problems.

problematic /prɒblə'mætɪk/ *adjective* (also **problematical**) attended by difficulty; doubtful, questionable.

proboscis /prəʊ'bɒsɪs/ *noun* (*plural* **-sces**) long flexible trunk or snout, e.g. of elephant; elongated mouth-parts of some insects.

procedure /prə'si:dʒə/ *noun* way of conducting business etc. or performing task; set series of actions. □ **procedural** *adjective.*

proceed /prə'si:d/ *verb* (often + *to*) go forward or on further, make one's way; (often + *with*, *to do*) continue or resume; adopt course of action; go on to say; (+ *against*) start lawsuit against; (often + *from*) originate.

proceeding *noun* action, piece of conduct; (in *plural*) legal action, published report of discussions or conference.

proceeds /'prəʊsi:dz/ *plural noun* profits from sale etc.

process[1] /'prəʊses/ ● *noun* course of action or proceeding, esp. series of stages in manufacture etc.; progress or course; natural or involuntary course of change; action at law; summons, writ; *Biology* natural appendage or outgrowth of organism. ● *verb* subject to particular process; (as **processed** *adjective*) (of food) treated, esp. to prevent decay.

process[2] /prə'ses/ *verb* walk in procession.

procession /prə'seʃ(ə)n/ *noun* people etc. advancing in orderly succession, esp. at ceremony, demonstration, or festivity.

processional ● *adjective* of processions; used, carried, or sung in processions. ● *noun* processional hymn (book).

processor /'prəʊsesə/ *noun* machine that processes things; central processor; food processor.

proclaim /prə'kleɪm/ *verb* (often + *that*) announce publicly or officially; declare to be. □ **proclamation** /prɒklə-/ *noun.*

proclivity /prə'klɪvɪtɪ/ *noun* (*plural* **-ies**) natural tendency.

procrastinate /prəʊ'kræstɪneɪt/ *verb* (**-ting**) defer action. □ **procrastination** *noun.*

■ **Usage** *Procrastinate* is often confused with *prevaricate*, which means 'to speak or act evasively or misleadingly'.

---

procreate /'prəʊkrɪeɪt/ *verb* (**-ting**) produce (offspring) naturally. □ **procreation** *noun*; **procreative** /-krɪ'eɪ-/ *adjective.*

proctor /'prɒktə/ *noun* university disciplinary official. □ **proctorial** /-'tɔ:rɪəl/ *adjective.*

procuration /prɒkjʊ'reɪʃ(ə)n/ *noun formal* procuring; action of attorney.

procurator /'prɒkjʊreɪtə/ *noun* agent or proxy, esp. with power of attorney. □ **procurator fiscal** (in Scotland) local coroner and public prosecutor.

procure /prə'kjʊə/ *verb* (**-ring**) succeed in getting; bring about; act as procurer. □ **procurement** *noun.*

procurer *noun* (*feminine* **procuress**) person who obtains women for prostitution.

prod ● *verb* (**-dd-**) poke with finger, stick, etc.; stimulate to action. ● *noun* poke, thrust; stimulus to action.

prodigal /'prɒdɪg(ə)l/ ● *adjective* wasteful; (+ *of*) lavish of. ● *noun* spendthrift. □ **prodigal son** repentant wastrel. □ **prodigality** /-'gæl-/ *noun.*

prodigious /prəˈdɪdʒəs/ *adjective* marvellous; enormous; abnormal.

prodigy /ˈprɒdɪdʒɪ/ *noun* (*plural* **-ies**) exceptionally gifted person, esp. precocious child; marvellous thing; (+ *of*) wonderful example of.

produce • *verb* /prəˈdju:s/ (**-cing**) manufacture or prepare; bring forward for inspection etc.; bear, yield, or bring into existence; cause or bring about; *Geometry* extend or continue (line); bring (play etc.) before public. • *noun* /ˈprɒdju:s/ what is produced, esp. agricultural products; amount produced; (often + *of*) result.

producer *noun* person who produces goods etc.; person who supervises production of play, film, broadcast, etc.

product /ˈprɒdʌkt/ *noun* thing or substance produced, esp. by manufacture; result; *Mathematics* quantity obtained by multiplying.

production /prəˈdʌkʃ(ə)n/ *noun* producing, being produced; total yield; thing produced, esp. play etc. □ **production line** systematized sequence of operations to produce commodity.

productive /prəˈdʌktɪv/ *adjective* producing, esp. abundantly. □ **productively** *adverb*; **productiveness** *noun*.

productivity /prɒdʌkˈtɪvɪtɪ/ *noun* capacity to produce; effectiveness of industry, workforce, etc.

Prof. *abbreviation* Professor.

profane /prəˈfeɪn/ • *adjective* irreverent, blasphemous; obscene; not sacred. • *verb* (**-ning**) treat irreverently; violate, pollute. □ **profanation** /prɒfə-/ *noun*.

profanity /prəˈfænɪtɪ/ *noun* (*plural* **-ies**) blasphemy; swear-word.

profess /prəˈfes/ *verb* claim openly to have; (often + *to do*) pretend, declare; affirm one's faith in or allegiance to.

professed *adjective* self-acknowledged; alleged, ostensible. □ **professedly** /-sɪdlɪ/ *adverb*.

profession /prəˈfeʃ(ə)n/ *noun* occupation or calling, esp. learned or scientific; people in a profession; declaration, avowal.

professional • *adjective* of, belonging to, or connected with a profession; competent, worthy of professional; engaged in specified activity as paid occupation, or (*derogatory*) fanatically. • *noun* professional person. □ **professionally** *adverb*.

professionalism *noun* qualities of professionals, esp. competence, skill, etc.

professor /prəˈfesə/ *noun* highest-ranking academic in university department, *US* university teacher; person who professes a religion etc. □ **professorial** /prɒfɪˈsɔ:rɪəl/ *adjective*; **professorship** *noun*.

proffer /ˈprɒfə/ *verb* offer.

proficient /prəˈfɪʃ(ə)nt/ *adjective* (often + *in*, *at*) adept, expert. □ **proficiency** *noun*; **proficiently** *adverb*.

profile /ˈprəʊfaɪl/ • *noun* side view or outline, esp. of human face; short biographical sketch. • *verb* (**-ling**) represent by profile. □ **keep a low profile** remain inconspicuous.

profit /ˈprɒfɪt/ • *noun* advantage, benefit; financial gain, excess of returns over outlay. • *verb* (**-t-**) be beneficial to; obtain advantage. □ **at a profit** with financial gain; **profit margin** profit after deduction of costs.

profitable *adjective* yielding profit; beneficial. □ **profitability** *noun*; **profitably** *adverb*.

profiteer /prɒfɪˈtɪə/ • *verb* make or seek excessive profits, esp. illegally. • *noun* person who profiteers.

profiterole /prəˈfɪtərəʊl/ *noun* small hollow cake of choux pastry with filling.

profligate /ˈprɒflɪgət/ • *adjective* recklessly extravagant; licentious, dissolute. • *noun* profligate person. □ **profligacy** *noun*; **profligately** *adverb*.

pro forma /prəʊ ˈfɔ:mə/ • *adverb & adjective* for form's sake. • *noun* (in full **pro forma invoice**) invoice sent in advance of goods supplied.

profound /prəˈfaʊnd/ *adjective* (**-er**, **-est**) having or demanding great knowledge, study, or insight; intense, thorough; deep. □ **profoundly** *adverb*; **profoundness** *noun*; **profundity** /-ˈfʌndɪtɪ/ *noun* (*plural* **-ies**).

profuse /prəˈfju:s/ *adjective* (often + *in*, *of*) lavish, extravagant, copious. □ **profusely** *adverb*; **profusion** *noun*.

progenitor /prəʊˈdʒenɪtə/ *noun* ancestor; predecessor; original.

progeny /ˈprɒdʒɪnɪ/ *noun* offspring, descendants; outcome, issue.

progesterone /prəʊ'dʒestərəʊn/ *noun* a sex hormone that helps to initiate and maintains pregnancy.

prognosis /prɒg'nəʊsɪs/ *noun* (*plural* **-noses** /-si:z/) forecast, esp. of course of disease.

prognostic /prɒg'nɒstɪk/ ● *noun* (often + *of*) advance indication; prediction. ● *adjective* (often + *of*) foretelling, predictive.

prognosticate /prɒg'nɒstɪkeɪt/ *verb* (**-ting**) (often + *that*) foretell; betoken. □ **prognostication** *noun.*

programme /'prəʊgræm/ (*US* **program**) ● *noun* list of events, performers, etc.; radio or television broadcast; plan of events; course or series of studies, lectures, etc.; (usually **program**) series of instructions for computer. ● *verb* (**-mm-**; *US* **-m-**) make programme of; (usually **program**) express (problem) or instruct (computer) by means of program. □ **programmable** *adjective*; **programmer** *noun.*

progress ● *noun* /'prəʊgres/ forward movement; advance, development, improvement; *historical* state journey, esp. by royalty. ● *verb* /prə'gres/ move forward or onward; advance, develop, improve. □ **in progress** developing, going on.

progression /prə'greʃ(ə)n/ *noun* progressing; succession, series.

progressive /prə'gresɪv/ ● *adjective* moving forward; proceeding step by step; cumulative; favouring rapid reform; modern, efficient; (of disease etc.) increasing in severity or extent; (of taxation) increasing with the sum taxed. ● *noun* (also **Progressive**) advocate of progressive policy. □ **progressively** *adverb.*

prohibit /prə'hɪbɪt/ *verb* (**-t-**) (often + *from*) forbid; prevent.

prohibition /prəʊhɪ'bɪʃ(ə)n/ *noun* forbidding, being forbidden; edict or order that forbids; (usually **Prohibition**) legal ban on manufacture and sale of alcohol. □ **prohibitionist** *noun.*

prohibitive /prə'hɪbɪtɪv/ *adjective* prohibiting; (of prices, taxes, etc.) extremely high. □ **prohibitively** *adverb.*

project ● *noun* /'prɒdʒekt/ plan, scheme; extensive essay, piece of research, etc. by student(s). ● *verb* /prə'dʒekt/ protrude, jut out; throw, impel; forecast; plan; cause (light, image, etc.) to fall on surface; cause (voice etc.) to be heard at distance.

projectile /prə'dʒektaɪl/ ● *noun* object to be fired (esp. by rocket) or hurled. ● *adjective* of or serving as projectile; projecting, impelling.

projection /prə'dʒekʃ(ə)n/ *noun* projecting, being projected; thing that protrudes; presentation of image(s) etc. on surface; forecast, estimate; mental image viewed as objective reality; transfer of feelings to other people etc.; representation of earth etc. on plane surface.

projectionist *noun* person who operates projector.

projector /prə'dʒektə/ *noun* apparatus for projecting image or film on screen.

prolactin /prəʊ'læktɪn/ *noun* hormone that stimulates milk production after childbirth.

prolapse ● *noun* /'prəʊlæps/ (also **prolapsus** /-'læpsəs/) slipping forward or downward of part or organ; prolapsed womb, rectum, etc. ● *verb* /prə'læps/ (**-sing**) undergo prolapse.

prolate /'prəʊleɪt/ *adjective* (of spheroid) lengthened along polar diameter.

prolegomenon /prəʊlɪ'gɒmɪnən/ *noun* (*plural* **-mena**) (usually in *plural*) preface to book etc., esp. discursive or critical.

proletarian /prəʊlɪ'teərɪən/ ● *adjective* of proletariat. ● *noun* member of proletariat.

proletariat /prəʊlɪ'teərɪət/ *noun* working class; *esp. derogatory* lowest class.

proliferate /prə'lɪfəreɪt/ *verb* (**-ting**) reproduce; produce (cells etc.) rapidly; increase rapidly, multiply. □ **proliferation** *noun.*

prolific /prə'lɪfɪk/ *adjective* producing many offspring or much output; (often + *of*) abundantly productive; copious.

prolix /'prəʊlɪks/ *adjective* lengthy; tedious. □ **prolixity** /-'lɪks-/ *noun.*

prologue /'prəʊlɒg/ *noun* introduction to poem, play, etc.; (usually + *to*) introductory event.

prolong /prə'lɒŋ/ *verb* extend; (as **prolonged** *adjective*) (tediously) lengthy. □ **prolongation** /prəʊlɒŋ'geɪʃ(ə)n/ *noun.*

prom *noun colloquial* promenade; promenade concert.

promenade /prɒmə'nɑ:d/ ● *noun* paved public walk, esp. at seaside; leisure walk. ● *verb* (**-ding**) make promenade (through); lead about, esp. for display. □ **promenade concert** one at which (part of) audience is not seated; **promenade deck** upper deck on liner.

promenader *noun* person who promenades; regular attender at promenade concerts.

prominent /'prɒmɪnənt/ *adjective* jutting out; conspicuous; distinguished. □ **prominence** *noun*.

promiscuous /prə'mɪskjʊəs/ *adjective* having frequent casual sexual relationships; mixed and indiscriminate; *colloquial* casual. □ **promiscuously** *adverb*; **promiscuity** /prɒmɪs'kju:ɪtɪ/ *noun*.

promise /'prɒmɪs/ ● *noun* explicit undertaking to do or not to do something; favourable indications. ● *verb* (**-sing**) (usually + *to do, that*) make promise; (often + *to do*) seem likely; *colloquial* assure.

promising *adjective* likely to turn out well; hopeful, full of promise. □ **promisingly** *adverb*.

promissory /'prɒmɪsərɪ/ *adjective* expressing or implying promise. □ **promissory note** signed document containing promise to pay stated sum.

promontory /'prɒməntərɪ/ *noun* (*plural* **-ies**) point of high land jutting out into sea etc.; headland.

promote /prə'məʊt/ *verb* (**-ting**) (often + *to*) advance (person) to higher office or position; help forward, encourage; publicize and sell. □ **promotion** *noun*; **promotional** *adjective*.

promoter *noun* person who promotes, esp. sporting event, theatrical production, etc., or formation of joint-stock company.

prompt ● *adjective* acting, made, or done immediately; ready. ● *adverb* punctually. ● *verb* (usually + *to, to do*) incite; supply (actor, speaker) with next words or with suggestion; inspire. ● *noun* prompting; thing said to prompt actor etc.; sign on computer screen inviting input. □ **promptitude** *noun*; **promptly** *adverb*; **promptness** *noun*.

prompter *noun* person who prompts actors.

promulgate /'prɒməlgeɪt/ *verb* (**-ting**) make known to the public; proclaim. □ **promulgation** *noun*.

prone *adjective* lying face downwards; lying flat, prostrate; (usually + *to, to do*) disposed, liable. □ **-prone** likely to suffer. □ **proneness** *noun*.

prong *noun* spike of fork.

pronominal /prəʊ'nɒmɪn(ə)l/ *adjective* of, concerning, or being a pronoun.

pronoun /'prəʊnaʊn/ *noun* word used as substitute for noun or noun phrase usually already mentioned or known (see panel).

pronounce /prə'naʊns/ *verb* (**-cing**) utter or speak, esp. in approved manner; utter formally; state (as) one's opinion; (usually + *on, for, against*, etc.) pass judgement. □ **pronounceable** *adjective*; **pronouncement** *noun*.

pronounced *adjective* strongly marked.

Pronoun

A pronoun is used as a substitute for a noun or a noun phrase, e.g.

He *was upstairs.*
Anything *can happen now.*
Did you see that?
It'*s lovely weather.*

Using a pronoun often avoids repetition, e.g.

I found Jim—he *was upstairs.*
(instead of *I found Jim—Jim was upstairs.*)

Where are your keys?—I've got them.
(instead of *Where are your keys?—I've got my keys.*)

pronto /'prɒntəʊ/ *adverb colloquial* promptly, quickly.

pronunciation /prənʌnsɪ'eɪʃ(ə)n/ *noun* pronouncing of word, esp. with reference to standard; act of pronouncing; way of pronouncing words.

proof /pru:f/ ● *noun* fact, evidence, reasoning, or demonstration that proves something; test, trial; standard of strength of distilled alcohol; trial impression of printed matter for correction. ● *adjective* (often + *against*) impervious to penetration, damage, etc. by a specified thing. ● *verb* make proof, esp. against water or bullets. □ **proofread** read and correct (printed proof); **proofreader** person who does this.

prop¹ ● *noun* rigid support; person or thing that supports, comforts, etc. ● *verb* (**-pp-**) (often + *against*, *up*, etc.) support (as) with prop.

prop² *noun colloquial* stage property.

prop³ *noun colloquial* propeller.

propaganda /prɒpə'gændə/ *noun* organized propagation of a doctrine etc.; *usually derogatory* ideas etc. so propagated. □ **propagandist** *noun*.

propagate /'prɒpəgeɪt/ *verb* (**-ting**) breed from parent stock; (often **propagate itself**) (of plant etc.) reproduce itself; disseminate; transmit. □ **propagation** *noun*; **propagator** *noun*.

propane /'prəʊpeɪn/ *noun* gaseous hydrocarbon used as fuel.

propel /prə'pel/ *verb* (**-ll-**) drive or push forward; urge on. □ **propellant** *noun & adjective*.

propeller *noun* revolving shaft with blades, esp. for propelling ship or aircraft.

propensity /prə'pensɪtɪ/ *noun* (*plural* **-ies**) inclination, tendency.

proper /'prɒpə/ *adjective* accurate, correct; suitable, appropriate; decent, respectable; (usually + *to*) belonging, relating; strictly so called, genuine; *colloquial* thorough. □ **proper name**, **noun** name of person, place, etc. □ **properly** *adverb*.

property /'prɒpətɪ/ *noun* (*plural* **-ies**) thing(s) owned; landed estate; quality, characteristic; movable article used on theatre stage or in film.

prophecy /'prɒfɪsɪ/ *noun* (*plural* **-ies**) prophetic utterance; prediction; prophesying.

prophesy /'prɒfɪsaɪ/ *verb* (**-ies**, **-ied**) (usually + *that*, *who*, etc.) foretell; speak as prophet.

prophet /'prɒfɪt/ *noun* (*feminine* **prophetess**) teacher or interpreter of divine will; person who predicts; (**the Prophet**) Muhammad.

prophetic /prə'fetɪk/ *adjective* (often + *of*) containing a prediction, predicting; of prophet.

prophylactic /prɒfɪ'læktɪk/ ● *adjective* tending to prevent disease etc. ● *noun* preventive medicine or action; *esp. US* condom.

prophylaxis /prɒfɪ'læksɪs/ *noun* preventive treatment against disease.

propinquity /prə'pɪŋkwɪtɪ/ *noun* nearness; close kinship; similarity.

propitiate /prə'pɪʃɪeɪt/ *verb* (**-ting**) appease. □ **propitiation** *noun*; **propitiatory** /-ʃətərɪ/ *adjective*.

propitious /prə'pɪʃəs/ *adjective* favourable, auspicious; (often + *for*, *to*) suitable.

proponent /prə'pəʊnənt/ *noun* person advocating proposal etc.

proportion /prə'pɔ:ʃ(ə)n/ ● *noun* comparative part, share; comparative ratio; correct relation between things or parts of thing; (in *plural*) dimensions. ● *verb* (usually + *to*) make proportionate.

proportional *adjective* in correct proportion; comparable. □ **proportional representation** representation of parties in parliament in proportion to votes they receive. □ **proportionally** *adverb*.

proportionate /prə'pɔ:ʃənət/ *adjective* proportional. □ **proportionately** *adverb*.

proposal /prə'pəʊz(ə)l/ *noun* proposing; scheme etc. proposed; offer of marriage.

propose /prə'pəʊz/ *verb* (**-sing**) put forward for consideration; (usually + *to do*) purpose; (usually + *to*) offer marriage; nominate as member of society etc. □ **propose a toast** ask people to drink to health or in honour of person or thing. □ **proposer** *noun*.

proposition /prɒpəˈzɪʃ(ə)n/ ● *noun* statement, assertion; scheme proposed, proposal; statement subject to proof or disproof; *colloquial* problem, opponent, prospect, etc.; *Mathematics* formal statement of theorem or problem; likely commercial enterprise, person, etc.; sexual proposal. ● *verb colloquial* put (esp. sexual) proposal to.

propound /prəˈpaʊnd/ *verb* offer for consideration.

proprietary /prəˈpraɪətərɪ/ *adjective* of or holding property; of proprietor; held in private ownership; manufactured by one particular firm. □ **proprietary name, term** name of product etc. registered as trade mark.

proprietor /prəˈpraɪətə/ *noun* (*feminine* **proprietress**) owner. □ **proprietorial** /-ˈtɔːr-/ *adjective*.

propriety /prəˈpraɪɪtɪ/ *noun* (*plural* **-ies**) fitness, rightness; correctness of behaviour or morals; (in *plural*) rules of polite behaviour.

propulsion /prəˈpʌlʃ(ə)n/ *noun* driving or pushing forward; force causing this. □ **propulsive** /-ˈpʌlsɪv/ *adjective*.

pro rata /prəʊ ˈrɑːtə/ ● *adjective* proportional. ● *adverb* proportionally. [Latin]

prorogue /prəˈrəʊg/ *verb* (**-gues**, **-gued**, **-guing**) discontinue meetings of (parliament etc.) without dissolving it; be prorogued. □ **prorogation** /prəʊrə-/ *noun*.

prosaic /prəˈzeɪɪk/ *adjective* like prose; unromantic; commonplace. □ **prosaically** *adverb*.

proscenium /prəˈsiːnɪəm/ *noun* (*plural* **-s** or **-nia**) part of theatre stage in front of curtain and enclosing arch.

proscribe /prəˈskraɪb/ *verb* (**-bing**) forbid; denounce; outlaw. □ **proscription** /-ˈskrɪp-/ *noun*; **proscriptive** /-ˈskrɪp-/ *adjective*.

■ **Usage** *Proscribe* is sometimes confused with *prescribe* which means 'to impose'.

prose /prəʊz/ ● *noun* ordinary language; not in verse; passage of this, esp. for translation; dullness. ● *verb* (**-sing**) talk tediously.

prosecute /ˈprɒsɪkjuːt/ *verb* (**-ting**) institute legal proceedings against; *formal* carry on (trade etc.). □ **prosecutor** *noun*.

prosecution /prɒsɪˈkjuːʃ(ə)n/ *noun* prosecuting, being prosecuted; prosecuting party.

proselyte /ˈprɒsəlaɪt/ *noun* convert, esp. recent; convert to Jewish faith. □ **proselytism** /-letɪz(ə)m/ *noun*.

proselytize /ˈprɒsələtaɪz/ *verb* (also **-ise**) (**-zing** or **-sing**) (seek to) convert.

prosody /ˈprɒsədɪ/ *noun* science of versification. □ **prosodist** *noun*.

prospect ● *noun* /ˈprɒspekt/ (often in *plural*) expectation; extensive view; mental picture; possible or likely customer etc. ● *verb* /prəˈspekt/ (usually + *for*) explore (for gold etc.). □ **prospector** *noun*.

prospective /prəsˈpektɪv/ *adjective* some day to be, expected, future.

prospectus /prəsˈpektəs/ *noun* (*plural* **-tuses**) pamphlet etc. advertising or describing school, business, etc.

prosper /ˈprɒspə/ *verb* succeed, thrive.

prosperity /prɒˈsperɪtɪ/ *noun* prosperous state.

prosperous /ˈprɒspərəs/ *adjective* successful, rich, thriving; auspicious. □ **prosperously** *adverb*.

prostate /ˈprɒsteɪt/ *noun* (in full **prostate gland**) gland secreting component of semen. □ **prostatic** /-ˈstæt-/ *adjective*.

prostitute /ˈprɒstɪtjuːt/ ● *noun* person who offers sexual intercourse for payment. ● *verb* (**-ting**) make prostitute of; misuse, offer for sale unworthily. □ **prostitution** *noun*.

prostrate ● *adjective* /ˈprɒstreɪt/ lying face downwards, esp. in submission; lying horizontally; overcome, esp. exhausted. ● *verb* /prɒsˈtreɪt/ (**-ting**) lay or throw flat; overcome, make weak. □ **prostration** *noun*.

prosy /ˈprəʊzɪ/ *adjective* (**-ier**, **-iest**) tedious, commonplace, dull.

protagonist /prəˈtægənɪst/ *noun* chief person in drama, story, etc.; supporter of cause.

■ **Usage** The use of *protagonist* to mean 'a supporter of a cause' is considered incorrect by some people.

protean /'prəʊtɪən/ *adjective* variable, versatile.

protect /prə'tekt/ *verb* (often + *from, against*) keep (person etc.) safe; shield.

protection /prə'tekʃ(ə)n/ *noun* protecting, being protected, defence; person etc. that protects; protectionism; *colloquial* immunity from violence etc. by paying gangsters etc.

protectionism *noun* theory or practice of protecting home industries. □ **protectionist** *noun & adjective.*

protective /prə'tektɪv/ *adjective* protecting; intended for or giving protection. □ **protective custody** detention of person for his or her own protection. □ **protectively** *adverb*; **protectiveness** *noun.*

protector *noun* (*feminine* **protectress**) person or thing that protects; *historical* regent ruling during minority or absence of sovereign. □ **protectorship** *noun.*

protectorate /prə'tektərət/ *noun* state controlled and protected by another; such protectorship; *historical* office of protector of kingdom or state; period of this.

protégé /'prɒtɪʒeɪ/ *noun* (*feminine* **protégée** same pronunciation) person under protection, patronage, etc. of another.

protein /'prəʊti:n/ *noun* any of a class of nitrogenous compounds essential in all living organisms.

pro tem /prəʊ 'tem/ *adjective & adverb colloquial* for the time being (*pro tempore*).

protest ● *noun* /'prəʊtest/ expression of dissent or disapproval; legal written refusal to pay or accept bill. ● *verb* /prə'test/ (usually + *against, at, about*, etc.) make protest; affirm (innocence etc.); write or get protest relating to (bill); *US* object to. □ **protester**, **protestor** *noun.*

Protestant /'prɒtɪst(ə)nt/ ● *noun* member or adherent of any of Churches separated from RC Church in Reformation. ● *adjective* of Protestant Churches or Protestants. □ **Protestantism** *noun.*

protestation /prɒtɪs'teɪʃ(ə)n/ *noun* strong affirmation; protest.

proto- *combining form* first.

protocol /'prəʊtəkɒl/ ● *noun* official formality and etiquette; draft, esp. of terms of treaty. ● *verb* (**-ll-**) draft or record in protocol.

proton /'prəʊtɒn/ *noun* elementary particle with positive electric charge equal to electron's, and occurring in all atomic nuclei.

protoplasm /'prəʊtəplæz(ə)m/ *noun* viscous translucent substance comprising living part of cell in organism. □ **protoplasmic** /-'plæzmɪk/ *adjective.*

prototype /'prəʊtətaɪp/ *noun* original as pattern for copy, improved form, etc.; trial model of vehicle, machine, etc. □ **prototypic** /-'tɪp-/ *adjective*; **prototypical** /-'tɪp-/ *adjective.*

protozoan /prəʊtə'zəʊən/ ● *noun* (*plural* **-s**) (also **protozoon** /-'zəʊɒn/, *plural* **-zoa** /-'zəʊə/) one-celled microscopic organism. ● *adjective* (also **protozoic** /-'zəʊɪk/) of protozoa.

protract /prə'trækt/ *verb* (often as **protracted** *adjective*) prolong, lengthen. □ **protraction** *noun.*

protractor *noun* instrument for measuring angles, usually in form of graduated semicircle.

protrude /prə'tru:d/ *verb* (**-ding**) thrust forward; stick out. □ **protrusion** *noun*; **protrusive** *adjective.*

protuberant /prə'tju:bərənt/ *adjective* bulging out; prominent. □ **protuberance** *noun.*

proud /praʊd/ *adjective* feeling greatly honoured; haughty, arrogant; (often + *of*) feeling or showing (proper) pride; imposing, splendid; (often + *of*) slightly projecting. □ **do (person) proud** *colloquial* treat with great generosity or honour. □ **proudly** *adverb.*

prove /pru:v/ *verb* (**-ving**; *past participle* **proved** or *esp. US & Scottish* **proven** /'pru:v(ə)n/) (often + *that*) demonstrate to be true by evidence or argument; (**prove oneself**) show one's abilities etc.; (usually + *to be*) be found; test accuracy of; establish validity of (will); (of dough) rise. □ **provable** *adjective.*

■ **Usage** The use of *proven* as the past participle is uncommon except in certain expressions, such as *of proven ability*. It is, however, standard in Scots and American English.

provenance /'prɒvɪnəns/ *noun* (place of) origin; history.

provender /'prɒvɪndə/ *noun* fodder; *jocular* food.

proverb /'prɒvɜ:b/ *noun* short pithy saying in general use.

proverbial /prə'vɜ:bɪəl/ *adjective* notorious; of or referred to in proverbs. □ **proverbially** *adverb.*

provide /prə'vaɪd/ *verb* (**-ding**) supply; (usually + *for, against*) make due preparation; (usually + *for*) take care of person etc. with money, food, etc.; (often + *that*) stipulate. □ **provided, providing** (often + *that*) on condition or understanding that.

providence /'prɒvɪd(ə)ns/ *noun* protective care of God or nature; (**Providence**) God; foresight, thrift.

provident /'prɒvɪd(ə)nt/ *adjective* having or showing foresight, thrifty.

providential /prɒvɪ'denʃ(ə)l/ *adjective* of or by divine foresight or intervention; opportune, lucky. □ **providentially** *adverb.*

province /'prɒvɪns/ *noun* principal administrative division of country etc.; (**the provinces**) whole of country outside capital; sphere of action; branch of learning.

provincial /prə'vɪnʃ(ə)l/ ● *adjective* of province(s); unsophisticated, uncultured. ● *noun* inhabitant of province(s); unsophisticated or uncultured person. □ **provincialism** *noun.*

provision /prə'vɪʒ(ə)n/ ● *noun* providing; (in *plural*) food and drink, esp. for expedition; legal or formal stipulation. ● *verb* supply with provisions.

provisional ● *adjective* providing for immediate needs only, temporary; (**Provisional**) of the unofficial wing of the IRA. ● *noun* (**Provisional**) member of unofficial wing of IRA. □ **provisionally** *adverb.*

proviso /prə'vaɪzəʊ/ *noun* (*plural* **-s**) stipulation; limiting clause. □ **provisory** *adjective.*

provocation /prɒvə'keɪʃ(ə)n/ *noun* provoking, being provoked; cause of annoyance.

provocative /prə'vɒkətɪv/ *adjective* (usually + *of*) tending or intended to provoke anger, lust, etc. □ **provocatively** *adverb.*

provoke /prə'vəʊk/ *verb* (**-king**) (often + *to, to do*) rouse, incite; call forth, cause; (usually + *into*) irritate, stimulate; tempt.

provost /'prɒvəst/ *noun* head of some colleges; head of cathedral chapter; /prə'vəʊ/ (in full **provost marshal**) head of military police in camp or on active service.

prow /praʊ/ *noun* bow of ship; pointed or projecting front part.

prowess /'praʊɪs/ *noun* skill, expertise; valour, gallantry.

prowl /praʊl/ ● *verb* (often + *about, around*) roam, esp. stealthily in search of prey, plunder, etc. ● *noun* prowling. □ **prowler** *noun.*

prox. *abbreviation* proximo.

proximate /'prɒksɪmət/ *adjective* nearest, next before or after.

proximity /prɒk'sɪmɪtɪ/ *noun* nearness.

proximo /'prɒksɪməʊ/ *adjective* of next month.

proxy /'prɒksɪ/ *noun* (*plural* **-ies**) authorization given to deputy; person authorized to deputize; document authorizing person to vote on another's behalf; vote so given.

prude *noun* excessively squeamish or sexually modest person. □ **prudery** *noun*; **prudish** *adjective*; **prudishly** *adverb*; **prudishness** *noun.*

prudent /'pru:d(ə)nt/ *adjective* cautious; politic. □ **prudence** *noun*; **prudently** *adverb.*

prudential /pru:'denʃ(ə)l/ *adjective* of or showing prudence.

prune[1] *noun* dried plum.

prune[2] *verb* (**-ning**) (often + *down*) trim (tree etc.) by cutting away dead or overgrown parts; (usually + *off, away*) remove (branches etc.) thus; reduce (costs etc.); (often + *of*) clear superfluities from; remove (superfluities).

prurient /'prʊərɪənt/ *adjective* having or encouraging unhealthy sexual curiosity. □ **prurience** *noun.*

Prussian /'prʌʃ(ə)n/ ● *adjective* of Prussia. ● *noun* native of Prussia. □ **Prussian blue** deep blue (pigment).

prussic acid /'prʌsɪk/ *noun* highly poisonous liquid.

pry /praɪ/ *verb* (**pries, pried**) (usually + *into* etc.) inquire impertinently, look inquisitively.

PS *abbreviation* postscript.

psalm /sɑ:m/ *noun* (also **Psalm**) sacred song; (**the** (**Book of**) **Psalms**) book of these in Old Testament.

psalmist /'sɑ:mɪst/ *noun* author or composer of psalm(s).

psalmody /'sɑ:mədɪ/ *noun* practice or art of singing psalms.

Psalter /'sɔ:ltə/ *noun* Book of Psalms; (**psalter**) version or copy of this.

psaltery /'sɔ:ltərɪ/ *noun* (*plural* **-ies**) ancient and medieval plucked stringed instrument.

psephology /sɪ'fɒlədʒɪ/ *noun* statistical study of voting etc. □ **psephologist** *noun.*

pseudo- *combining form* (also **pseud-** before vowel) false, not genuine; resembling, imitating.

pseudonym /'sju:dənɪm/ *noun* fictitious name, esp. of author.

psoriasis /sə'raɪəsɪs/ *noun* skin disease with red scaly patches.

PSV *abbreviation* public service vehicle.

psych /saɪk/ *verb colloquial* (usually + *up*) prepare mentally; (often + *out*) intimidate; (usually + *out*) analyse (person's motivation etc.).

psyche /'saɪkɪ/ *noun* soul, spirit, mind.

psychedelic /saɪkə'delɪk/ *adjective* expanding the mind's awareness, hallucinatory; vivid in colour, design, etc.

psychiatry /saɪ'kaɪətrɪ/ *noun* study and treatment of mental disease. □ **psychiatric** /-kɪ'ætrɪk/ *adjective*; **psychiatrist** *noun.*

psychic /'saɪkɪk/ ● *adjective* (of person) regarded as having paranormal powers, clairvoyant; of the soul or mind. ● *noun* psychic person, medium.

psychical *adjective* concerning psychic phenomena or faculties; of the soul or mind.

psycho- *combining form* of mind or psychology.

psychoanalysis /saɪkəʊə'nælɪsɪs/ *noun* treatment of mental disorders by bringing repressed fears etc. into conscious mind. □ **psychoanalyse** /-'ænəl-/ *verb* (**-sing**); **psychoanalyst** /-'ænəl-/ *noun*; **psychoanalytical** /-ænə'lɪt-/ *adjective.*

psychokinesis /saɪkəʊkɪ'ni:sɪs/ *noun* movement of objects by telepathy.

psychological /saɪkə'lɒdʒɪk(ə)l/ *adjective* of the mind; of psychology; *colloquial* imaginary. □ **psychological block** inhibition caused by emotion; **psychological moment** best time to achieve purpose; **psychological warfare** campaign to reduce enemy's morale. □ **psychologically** *adverb.*

psychology /saɪ'kɒlədʒɪ/ *noun* (*plural* **-ies**) study of human mind; treatise on or theory of this; mental characteristics. □ **psychologist** *noun.*

psychopath /'saɪkəpæθ/ *noun* mentally deranged person, esp. with abnormal social behaviour; mentally or emotionally unstable person. □ **psychopathic** /-'pæθ-/ *adjective.*

psychosis /saɪ'kəʊsɪs/ *noun* (*plural* **-choses** /-si:z/) severe mental derangement involving loss of contact with reality.

psychosomatic /saɪkəʊsə'mætɪk/ *adjective* (of disease) mental, not physical, in origin; of both mind and body.

psychotherapy /saɪkəʊ'θerəpɪ/ *noun* treatment of mental disorder by psychological means. □ **psychotherapist** *noun.*

psychotic /saɪ'kɒtɪk/ ● *adjective* of or suffering from psychosis. ● *noun* psychotic person.

PT *abbreviation* physical training.

pt *abbreviation* part; pint; point; port.

PTA *abbreviation* parent–teacher association.

ptarmigan /'tɑ:mɪgən/ *noun* bird of grouse family.

Pte. *abbreviation* Private (soldier).

pteridophyte /'terɪdəfaɪt/ *noun* flowerless plant.

pterodactyl /terə'dæktɪl/ *noun* large extinct flying reptile.

PTO *abbreviation* please turn over.

ptomaine /'təʊmeɪn/ *noun* any of a group of compounds (some toxic) in putrefying matter.

pub *noun colloquial* public house.

puberty /'pju:bətɪ/ *noun* period of sexual maturing.

pubes[1] /'pju:bi:z/ *noun* (*plural* same) lower part of abdomen.

pubes[2] *plural* of **pubis**.

pubescence /pju:'bes(ə)ns/ *noun* beginning of puberty; soft down on plant or animal. □ **pubescent** *adjective*.

pubic /'pju:bɪk/ *adjective* of pubes or pubis.

pubis /'pju:bɪs/ *noun* (*plural* **pubes** /-bi:z/) front portion of hip bone.

public /'pʌblɪk/ ● *adjective* of the people as a whole; open to or shared by all; done or existing openly; of or from government; involved in community affairs. ● *noun* (treated as *singular* or *plural*) (members of) community as a whole; section of community. □ **go public** (of company) start selling shares on open market, reveal one's plans; **in public** publicly, openly; **public address system** equipment of loudspeakers etc.; **public convenience** public lavatory; **public figure** famous person; **public house** place selling alcoholic drink for consumption on premises; **public lending right** right of authors to payment when their books are lent by public libraries; **public relations** professional promotion of company, product, etc.; **public school** independent fee-paying school; *US, Australian, Scottish, etc.* non-fee-paying school; **public-spirited** ready to do things for the community; **public transport** buses, trains, etc. available for public use on fixed routes; **public utility** organization supplying water, gas, etc. to community. □ **publicly** *adverb*.

publican /'pʌblɪkən/ *noun* keeper of public house.

publication /pʌblɪ'keɪʃ(ə)n/ *noun* publishing; published book, periodical, etc.

publicist /'pʌblɪsɪst/ *noun* publicity agent, public relations officer.

publicity /pʌb'lɪsɪtɪ/ *noun* (means of attracting) public attention; (material used for) advertising.

publicize /'pʌblɪsaɪz/ *verb* (also **-ise**) (**-zing** or **-sing**) advertise, make publicly known.

publish /'pʌblɪʃ/ *verb* prepare and issue (book, magazine, etc.) for public sale; make generally known; formally announce.

publisher *noun* person or firm that publishes books etc.

puce /pju:s/ *adjective & noun* purple-brown.

puck[1] *noun* rubber disc used in ice hockey.

puck[2] *noun* mischievous sprite. □ **puckish** *adjective*; **puckishly** *adverb*; **puckishness** *noun*.

pucker /'pʌkə/ ● *verb* gather (often + *up*) into wrinkles, folds, or bulges. ● *noun* such wrinkle etc.

pudding /'pʊdɪŋ/ *noun* sweet cooked dish; savoury dish containing flour, suet, etc.; sweet course of meal; kind of sausage.

puddle /'pʌd(ə)l/ *noun* small (dirty) pool; clay made into watertight coating.

pudenda /pju:'dendə/ *plural noun* genitals, esp. of woman.

pudgy /'pʌdʒɪ/ *adjective* (**-ier**, **-iest**) *colloquial* plump, podgy.

puerile /'pjʊəraɪl/ *adjective* childish, immature. □ **puerility** /-'rɪl-/ *noun* (*plural* **-ies**).

puerperal /pju:'ɜ:pər(ə)l/ *adjective* of or due to childbirth.

puff ● *noun* short quick blast of breath or wind; sound (as) of this; vapour or smoke sent out in one blast; light pastry cake; gathered material in dress etc.; unduly enthusiastic review, advertisement, etc. ● *verb* emit puff(s); smoke or move with puffs; (usually in *passive*; often + *out*) *colloquial* put out of breath; pant; (usually + *up*, *out*) inflate; (usually as **puffed up** *adjective*) elate, make boastful; advertise in exaggerated terms. □ **puff-adder** large venomous African viper; **puffball** ball-shaped fungus; **puff pastry** pastry consisting of thin layers.

puffin /'pʌfɪn/ *noun* N. Atlantic and N. Pacific auk with short striped bill.

puffy *adjective* (**-ier**, **-iest**) swollen, puffed out; *colloquial* short-winded.

pug *noun* (in full **pug-dog**) dog of small breed with flat nose. □ **pug-nose** short flat or snub nose.

pugilist /'pju:dʒɪlɪst/ *noun* boxer. □ **pugilism** *noun*; **pugilistic** /-'lɪs-/ *adjective*.

pugnacious /pʌg'neɪʃəs/ *adjective* disposed to fight. □ **pugnaciously** *adverb*; **pugnacity** /-'næs-/ *noun*.

puissance /'pwi:sɑ̃s/ *noun* jumping of large obstacles in showjumping.

puissant /'pwi:sɒnt/ *adjective literary* powerful; mighty.

puke /pju:k/ *verb & noun* (**-king**) *slang* vomit. □ **pukey** *adjective.*

pukka /'pʌkə/ *adjective colloquial* genuine; reliable.

pulchritude /'pʌlkrɪtju:d/ *noun literary* beauty. □ **pulchritudinous** /-'tju:dɪnəs/ *adjective.*

pull /pʊl/ ● *verb* exert force on (thing etc.) to move it to oneself or origin of force; exert pulling force; extract by pulling; damage (muscle etc.) by abnormal strain; proceed with effort; (+ *on*) draw (weapon) against (person); attract; draw (liquor) from barrel etc.; (+ *at*) pluck at; (often + *on*, *at*) inhale or drink deeply, suck. ● *noun* act of pulling; force thus exerted; influence; advantage; attraction; deep draught of liquor; prolonged effort; handle etc. for applying pull; printer's rough proof; suck at cigarette. □ **pull back** retreat; **pull down** demolish; **pull in** arrive to take passengers, move to side of or off road, *colloquial* earn, *colloquial* arrest; **pull-in** roadside café etc.; **pull off** remove, win, manage successfully; **pull oneself together** recover control of oneself; **pull out** take out, depart, withdraw, leave station or stop, move towards off side; **pull-out** removable section of magazine; **pull round**, **through** (cause to) recover from illness; **pull strings** exert (esp. clandestine) influence; **pull together** work in harmony; **pull up** (cause to) stop moving, pull out of ground, reprimand, check oneself.

pullet /'pʊlɪt/ *noun* young hen, esp. less than one year old.

pulley /'pʊlɪ/ *noun* (*plural* **-s**) grooved wheel(s) for cord etc. to run over, mounted in block and used to lift weight etc.; wheel or drum mounted on shaft and turned by belt, used to increase speed or power.

Pullman /'pʊlmən/ *noun* (*plural* **-s**) luxurious railway carriage or motor coach; sleeping car.

pullover *noun* knitted garment put on over the head.

pullulate /'pʌljʊleɪt/ *verb* (**-ting**) sprout; swarm; develop; (+ *with*) abound with. □ **pullulation** *noun.*

pulmonary /'pʌlmənərɪ/ *adjective* of lungs; having (organs like) lungs; affected with or subject to lung disease.

pulp ● *noun* fleshy part of fruit etc.; soft shapeless mass, esp. of materials for papermaking; cheap fiction. ● *verb* reduce to or become pulp. □ **pulpy** *adjective*; **pulpiness** *noun.*

pulpit /'pʊlpɪt/ *noun* raised enclosed platform for preaching from; (**the pulpit**) preachers collectively, preaching.

pulsar /'pʌlsɑ:/ *noun* cosmic source of regular rapid pulses of radiation.

pulsate /pʌl'seɪt/ *verb* (**-ting**) expand and contract rhythmically; throb, vibrate, quiver. □ **pulsation** *noun.*

pulse[1] ● *noun* rhythmical throbbing of arteries; each beat of arteries or heart; throb or thrill of life or emotion; general feeling; single vibration of sound, electromagnetic radiation, etc.; rhythmical (esp. musical) beat. ● *verb* (**-sing**) pulsate.

pulse[2] *noun* (treated as *singular* or *plural*) (plant producing) edible seeds of peas, beans, lentils, etc.

pulverize /'pʌlvəraɪz/ *verb* (also **-ise**) (**-zing** or **-sing**) reduce or crumble to powder or dust; *colloquial* demolish, crush. □ **pulverization** *noun.*

puma /'pju:mə/ *noun* large tawny American feline.

pumice /'pʌmɪs/ *noun* (in full **pumice stone**) light porous lava used as abrasive; piece of this.

pummel /'pʌm(ə)l/ *verb* (**-ll-**; *US* **-l-**) strike repeatedly, esp. with fists.

pump[1] ● *noun* machine or device for raising or moving liquids or gases; act of pumping. ● *verb* (often + *in*, *out*, *up*, etc.) raise, remove, inflate, empty, etc. (as) with pump; work pump; persistently question (person) to elicit information; move vigorously up and down. □ **pump iron** *colloquial* exercise with weights.

pump[2] *noun* plimsoll; light shoe for dancing etc.

pumpernickel /'pʌmpənɪk(ə)l/ *noun* wholemeal rye bread.

pumpkin /'pʌmpkɪn/ *noun* large yellow or orange fruit used as vegetable; plant bearing it.

pun ● *noun* humorous use of word(s) with two or more meanings, play on words. ● *verb* (**-nn-**) (usually + *on*) make pun(s).

Punch *noun* grotesque humpbacked puppet in *Punch and Judy* shows.

punch[1] ● *verb* strike with fist; make hole in (as) with punch; pierce (hole) thus. ● *noun* blow with fist; *colloquial* vigour, effective force; instrument or machine for piercing holes or impressing design in leather, metal, etc. □ **pull one's punches** avoid using full force; **punchball** stuffed or inflated ball used for practice in punching; **punch-drunk** stupefied (as) with repeated punches; **punchline** words giving point of joke etc.; **punch-up** *colloquial* fist-fight, brawl. □ **puncher** *noun*.

punch[2] *noun* hot or cold mixture of wine or spirit with water, fruit, spices, etc. □ **punch-bowl** bowl for punch; deep round hollow in hill.

punchy *adjective* (**-ier**, **-iest**) vigorous, forceful.

punctilio /pʌŋk'tɪlɪəʊ/ *noun* (*plural* **-s**) delicate point of ceremony or honour; petty formality.

punctilious /pʌŋk'tɪlɪəs/ *adjective* attentive to formality or etiquette; precise in behaviour. □ **punctiliously** *adverb*; **punctiliousness** *noun*.

punctual /'pʌŋktʃʊəl/ *adjective* observing appointed time; prompt. □ **punctuality** /-'æl-/ *noun*; **punctually** *adverb*.

punctuate /'pʌŋktʃʊeɪt/ *verb* (**-ting**) insert punctuation marks in; interrupt at intervals.

punctuation *noun* (system of) punctuating. □ **punctuation mark** any of the marks used in writing to separate sentences, phrases, etc.

puncture /'pʌŋktʃə/ ● *noun* prick, pricking; hole made by this. ● *verb* (**-ring**) make or suffer puncture (in); deflate.

pundit /'pʌndɪt/ *noun* (also **pandit**) learned Hindu; expert.

pungent /'pʌndʒ(ə)nt/ *adjective* having sharp or strong taste or smell; biting, caustic. □ **pungency** *noun*.

punish /'pʌnɪʃ/ *verb* inflict penalty on (offender) or for (offence); tax, abuse, or treat severely. □ **punishable** *adjective*; **punishment** *noun*.

punitive /'pju:nɪtɪv/ *adjective* inflicting or intended to inflict punishment; extremely severe.

punk *noun* (in full **punk rock**) deliberately outrageous style of rock music; (in full **punk rocker**) fan of this; *esp. US* hooligan, lout.

punkah /'pʌŋkə/ *noun* large swinging fan on frame worked by cord or electrically.

punnet /'pʌnɪt/ *noun* small basket for fruit etc.

punster /'pʌnstə/ *noun* maker of puns.

punt[1] ● *noun* square-ended flat-bottomed boat propelled by long pole. ● *verb* travel or carry in punt.

punt[2] ● *verb* kick (football) dropped from hands before it reaches ground. ● *noun* such kick.

punt[3] *verb colloquial* bet, speculate in shares etc.; (in some card games) lay stake against bank.

punt[4] /pʊnt/ *noun* chief monetary unit of Republic of Ireland.

punter *noun colloquial* person who gambles or bets; customer, client.

puny /'pju:nɪ/ *adjective* (**-ier**, **-iest**) undersized; feeble.

pup ● *noun* young dog, wolf, rat, seal, etc. ● *verb* (**-pp-**) give birth to (pups).

pupa /'pju:pə/ *noun* (*plural* **pupae** /-pi:/) insect in stage between larva and imago.

pupil[1] /'pju:pɪl/ *noun* person being taught.

pupil[2] /'pju:pɪl/ *noun* opening in centre of iris of eye.

puppet /'pʌpɪt/ *noun* small figure moved esp. by strings as entertainment; person controlled by another. □ **puppet state** country apparently independent but actually under control of another power. □ **puppetry** *noun*.

puppy /'pʌpɪ/ *noun* (*plural* **-ies**) young dog; conceited young man. □ **puppy fat** temporary fatness of child or adolescent; **puppy love** calf love.

purblind /'pɜ:blaɪnd/ *adjective* partly blind, dim-sighted; obtuse, dull. □ **purblindness** *noun*.

purchase /'pɜ:tʃəs/ ● *verb* (**-sing**) buy. ● *noun* buying; thing bought; firm hold on thing, leverage; equipment for moving heavy objects. □ **purchaser** *noun*.

purdah /'pɜ:də/ *noun* screening of Muslim or Hindu women from strangers.

pure /pjʊə/ *adjective* unmixed, unadulterated; chaste; not morally corrupt; guiltless; sincere; not discordant; (of science) abstract, not applied. □ **pureness** *noun*; **purity** *noun*.

purée /'pjʊəreɪ/ ● *noun* smooth pulp of vegetables or fruit etc. ● *verb* (**-ées, -éed**) make purée of.

purely *adverb* in a pure way; merely, solely, exclusively.

purgative /'pɜ:gətɪv/ ● *adjective* serving to purify; strongly laxative. ● *noun* purgative thing.

purgatory /'pɜ:gətərɪ/ ● *noun* (*plural* **-ies**) place or state of spiritual cleansing, esp. after death and before entering heaven; place or state of temporary suffering or expiation. ● *adjective* purifying. □ **purgatorial** /-'tɔ:rɪəl/ *adjective*.

purge /pɜ:dʒ/ ● *verb* (**-ging**) (often + *of*, *from*) make physically or spiritually clean; remove by cleansing; rid of unacceptable members; empty (bowels); *Law* atone for (offence). ● *noun* purging; purgative.

purify /'pjʊərɪfaɪ/ *verb* (**-ies, -ied**) clear of extraneous elements, make pure; (often + *of*, *from*) cleanse. □ **purification** *noun*; **purificatory** /-fɪkeɪtərɪ/ *adjective*.

purist /'pjʊərɪst/ *noun* stickler for correctness, esp. in language. □ **purism** *noun*.

puritan /'pjʊərɪt(ə)n/ ● *noun* (**Puritan**) *historical* member of English Protestant group regarding Reformation as incomplete; purist member of any party; strict observer of religion or morals. ● *adjective* (**Puritan**) *historical* of Puritans; scrupulous in religion or morals. □ **puritanical** /-'tæn-/ *adjective*; **puritanically** /-'tæn-/ *adverb*; **puritanism** *noun*.

purl[1] ● *noun* knitting stitch with needle moved in opposite to normal direction; chain of minute loops. ● *verb* knit with purl stitch.

purl[2] *verb* flow with babbling sound.

purler /'pɜ:lə/ *noun colloquial* heavy fall.

purlieu /'pɜ:lju:/ *noun* (*plural* **-s**) person's limits or usual haunts; *historical* tract on border of forest; (in *plural*) outskirts, outlying region.

purlin /'pɜ:lɪn/ *noun* horizontal beam along length of roof.

purloin /pə'lɔɪn/ *verb formal or jocular* steal, pilfer.

purple /'pɜ:p(ə)l/ ● *noun* colour between red and blue; purple robe, esp. of emperor etc.; cardinal's scarlet official dress. ● *adjective* of purple. ● *verb* (**-ling**) make or become purple. □ **purplish** *adjective*.

purport ● *verb* /pə'pɔ:t/ profess, be intended to seem; (often + *that*) have as its meaning. ● *noun* /'pɜ:pɔ:t/ ostensible meaning; tenor of document or statement. □ **purportedly** *adverb*.

purpose /'pɜ:pəs/ ● *noun* object to be attained, thing intended; intention to act; resolution, determination. ● *verb* (**-sing**) have as one's purpose, intend. □ **on purpose** intentionally; **to good, little, no**, etc. **purpose** with good, little, no, etc., effect or result; **to the purpose** relevant, useful.

purposeful *adjective* having or indicating purpose; intentional. □ **purposefully** *adverb*; **purposefulness** *noun*.

purposeless *adjective* having no aim or plan.

purposely *adverb* on purpose.

purposive /'pɜ:pəsɪv/ *adjective* having, serving, or done with a purpose; purposeful.

purr /pɜ:/ ● *verb* make low vibratory sound of cat expressing pleasure; (of machinery etc.) run smoothly and quietly. ● *noun* purring sound.

purse /pɜ:s/ ● *noun* small pouch for carrying money in; *US* handbag; funds; sum given as present or prize. ● *verb* (**-sing**) (often + *up*) contract (esp. lips); become wrinkled. □ **hold the purse-strings** have control of expend iture.

purser /'pɜ:sə/ *noun* ship's officer who keeps accounts, esp. head steward in passenger vessel.

pursuance /pə'sju:əns/ *noun* (+ *of*) carrying out or observance (of plan, rules, etc.).

pursuant /pə'sju:ənt/ *adverb* (+ *to*) in accordance with.

pursue /pə'sju:/ *verb* (**-sues**, **-sued**, **-suing**) follow with intent to overtake, capture, or harm; proceed along; engage in (study etc.); carry out (plan etc.); seek after; continue to investigate etc.; persistently importune or assail. □ **pursuer** *noun.*

pursuit /pə'sju:t/ *noun* pursuing; occupation or activity pursued. □ **in pursuit of** pursuing.

purulent /'pjʊərʊlənt/ *adjective* of, containing, or discharging pus. □ **purulence** *noun.*

purvey /pə'veɪ/ *verb* provide or supply food etc. as one's business. □ **purveyor** *noun.*

purview /'pɜ:vju:/ *noun* scope of document etc.; range of physical or mental vision.

pus /pʌs/ *noun* thick yellowish liquid produced from infected tissue.

push /pʊʃ/ ● *verb* exert force on (thing) to move it away, cause to move thus; exert such force; thrust forward or upward; (cause to) project; make (one's way) forcibly or persistently; exert oneself; (often + *to*, *into*, *to do*) urge, impel; (often + *for*) pursue (claim etc.) persistently; promote, advertise; *colloquial* sell (drug) illegally. ● *noun* act of pushing; force thus exerted; vigorous effort; determination; use of influence to advance person. □ **give** or **get the push** *colloquial* dismiss, be dismissed; **push-bike** *colloquial* bicycle; **pushchair** child's folding chair on wheels; **push off** *colloquial* go away; **pushover** *colloquial* opponent or difficulty easily overcome.

pusher *noun colloquial* seller of illegal drugs.

pushing *adjective* pushy; *colloquial* having nearly reached (specified age).

pushy *adjective* (**-ier**, **-iest**) *colloquial* excessively self-assertive. □ **pushily** *adverb*; **pushiness** *noun.*

pusillanimous /pju:sɪ'lænɪməs/ *adjective formal* cowardly, timid. □ **pusillanimity** /-lə'nɪm-/ *noun.*

puss /pʊs/ *noun colloquial* cat; sly or coquettish girl.

pussy /'pʊsɪ/ *noun* (*plural* **-ies**) (also **pussy-cat**) *colloquial* cat. □ **pussyfoot** *colloquial* move stealthily, equivocate; **pussy willow** willow with furry catkins.

pustulate /'pʌstjʊleɪt/ *verb* (**-ting**) form into pustules.

pustule /'pʌstju:l/ *noun* pimple containing pus. □ **pustular** *adjective.*

put /pʊt/ ● *verb* (**-tt-**; *past & past participle* **put**) move to or cause to be in specified place, position, or state; (often + *on*, *to*) impose; (+ *for*) substitute (thing) for (another); express in specified way; (+ *at*) estimate; (+ *into*) express or translate in (words etc.); (+ *into*) invest (money) in; (+ *on*) stake (money) on; (+ *to*) submit for attention; hurl (shot etc.) as sport. ● *noun* throw of shot. □ **put about** spread (rumour etc.); **put across** make understood, achieve by deceit; **put away** restore to usual place, lay aside for future, imprison, consume (food or drink); **put back** restore to usual place, change (meeting etc.) to later time, move back hands of (clock or watch); **put by** lay aside for future; **put down** suppress, *colloquial* snub, record in writing, enter on list, (+ *as*, *for*) account or reckon, (+ *to*) attribute to, kill (old etc. animal), pay as deposit; **put in** submit (claim), (+ *for*) be candidate for (election etc.), spend (time); **put off** postpone, evade (person) with excuse, dissuade, disconcert; **put on** clothe oneself with, cause (light etc.) to operate, make (transport) available, stage (play etc.), advance hands of (clock or watch), feign, increase one's weight by (specified amount); **put out** disconcert, annoy, inconvenience, extinguish; **put over** put across; **put through** complete, connect by telephone; **put together** make from parts, combine (parts) into whole; **put up** *verb* build, raise, lodge (person), engage in (fight etc.), propose, provide (money) as backer, display (notice), offer for sale etc.; **put-up** *adjective* fraudulent; **put upon** (usually in *passive*) *colloquial* unfairly burden or deceive; **put (person) up to** instigate him or her to; **put up with** endure, tolerate.

putative /'pju:tətɪv/ *adjective formal* reputed, supposed.

putrefy /'pju:trɪfaɪ/ *verb* (**-ies**, **-ied**) become or make putrid, go bad; fester; become morally corrupt. □ **putrefaction** /-'fæk-/ *noun*; **putrefactive** /-'fæk-/ *adjective.*

putrescent /pjuː'tres(ə)nt/ *adjective* rotting. □ **putrescence** *noun.*

putrid /'pjuːtrɪd/ *adjective* decomposed, rotten; noxious; corrupt; *slang* contemptible, very unpleasant. □ **putridity** /-'trɪd-/ *noun.*

putsch /pʊtʃ/ *noun* attempt at revolution.

putt /pʌt/ ● *verb* (**-tt-**) strike (golf ball) on putting-green. ● *noun* putting stroke. □ **putting green** smooth turf round hole on golf course.

puttee /'pʌtɪ/ *noun historical* long strip of cloth wound round leg for protection and support.

putter[1] *noun* golf club used in putting.

putter[2] *US* = POTTER[1].

putty /'pʌtɪ/ ● *noun* paste of chalk, linseed oil, etc. for fixing panes of glass etc. ● *verb* (**-ies**, **-ied**) fix, fill, etc. with putty.

puzzle /'pʌz(ə)l/ ● *noun* difficult or confusing problem; problem or toy designed to test ingenuity etc. ● *verb* (**-ling**) perplex, (usually + *over* etc.) be perplexed; (usually as **puzzling** *adjective*) require much mental effort; (+ *out*) solve using ingenuity etc. □ **puzzlement** *noun.*

PVC *abbreviation* polyvinyl chloride, a plastic used for pipes, electrical insulation, etc.

pyaemia /paɪ'iːmɪə/ *noun* (*US* **pyemia**) severe bacterial infection of blood.

pygmy /'pɪgmɪ/ (also **pigmy**) ● *noun* (*plural* **-ies**) member of dwarf people of esp. equatorial Africa; very small person, animal, or thing. ● *adjective* very small.

pyjamas /pə'dʒɑːməz/ *plural noun* (*US* **pajamas**) suit of trousers and top for sleeping in etc.; loose trousers worn in some Asian countries.

pylon /'paɪlən/ *noun* tall structure esp. as support for electric cables.

pyorrhoea /paɪə'rɪə/ *noun* (*US* **pyorrhea**) gum disease; discharge of pus.

pyramid /'pɪrəmɪd/ *noun* monumental (esp. ancient Egyptian) stone structure with square base and sloping sides meeting at apex; solid of this shape with base of 3 or more sides; pyramid-shaped thing. □ **pyramidal** /-'ræm-/ *adjective.*

pyre /'paɪə/ *noun* pile of combustible material, esp. for burning corpse.

pyrethrum /paɪ'riːθrəm/ *noun* aromatic chrysanthemum; insecticide from its dried flowers.

Pyrex /'paɪəreks/ *noun proprietary term* a hard heat-resistant glass.

pyrites /paɪ'raɪtiːz/ *noun* (in full **iron pyrites**) yellow sulphide of iron.

pyromania /paɪərəʊ'meɪnɪə/ *noun* obsessive desire to start fires. □ **pyromaniac** *noun & adjective.*

pyrotechnics /paɪərəʊ'teknɪks/ *plural noun* art of making fireworks; display of fireworks. □ **pyrotechnic** *adjective.*

pyrrhic /'pɪrɪk/ *adjective* (of victory) achieved at too great cost.

python /'paɪθ(ə)n/ *noun* large snake that crushes its prey.

pyx /pɪks/ *noun* vessel for consecrated bread of Eucharist.

Qq

Q *abbreviation* (also **Q.**) question.

QC *abbreviation* Queen's Counsel.

QED *abbreviation* which was to be proved (*quod erat demonstrandum*).

QM *abbreviation* Quartermaster.

qr. *abbreviation* quarter(s).

qt *abbreviation* quart(s).

qua /kwɑ:, kweɪ/ *conjunction* in the capacity of.

quack[1] • *noun* harsh sound made by ducks. • *verb* utter this sound.

quack[2] *noun* unqualified practitioner, esp. of medicine; *slang* any doctor. □ **quackery** *noun.*

quad /kwɒd/ *colloquial* • *noun* quadrangle; quadruplet; quadraphonics. • *adjective* quadraphonic.

quadrangle /'kwɒdræŋg(ə)l/ *noun* 4-sided plane figure, esp. square or rectangle; 4-sided court, esp. in college etc. □ **quadrangular** /-'ræŋgjʊlə/ *adjective.*

quadrant /'kwɒdrənt/ *noun* quarter of circle or sphere or of circle's circumference; optical instrument for measuring angle between distant objects.

quadraphonic /kwɒdrə'fɒnɪk/ *adjective* (of sound reproduction) using 4 transmission channels. □ **quadraphonically** *adverb*; **quadraphonics** *plural noun.*

quadrate • *adjective* /'kwɒdrət/ square, rectangular. • *noun* /'kwɒdrət, -dreɪt/ rectangular object. • *verb* /kwɒ'dreɪt/ (**-ting**) make square.

quadratic /kwɒ'drætɪk/ *Mathematics* • *adjective* involving the square (and no higher power) of unknown quantity or variable. • *noun* quadratic equation.

quadriceps /'kwɒdrɪseps/ *noun* 4-headed muscle at front of thigh.

quadrilateral /kwɒdrɪ'lætər(ə)l/ • *adjective* having 4 sides. • *noun* 4-sided figure.

quadrille /kwɒ'drɪl/ *noun* square dance, music for this.

quadruped /'kwɒdrʊped/ *noun* 4-footed animal, esp. mammal.

quadruple /'kwɒdrʊp(ə)l/ • *adjective* fourfold; having 4 parts; (of time in music) having 4 beats in bar. • *noun* fourfold number or amount. • *verb* /kwɒ'dru:p(ə)l/ multiply by 4.

quadruplet /'kwɒdrʊplɪt/ *noun* each of 4 children born at one birth.

quadruplicate • *adjective* /kwɒ'dru:plɪkət/ fourfold; of which 4 copies are made. • *verb* /-keɪt/ (**-ting**) multiply by 4.

quaff /kwɒf/ *verb literary* drink deeply; drain (cup etc.) in long draughts.

quagmire /'kwɒgmaɪə, 'kwæg-/ *noun* muddy or boggy area; hazardous situation.

quail[1] *noun* (*plural* same or **-s**) small game bird related to partridge.

quail[2] *verb* flinch, show fear.

quaint *adjective* attractively odd or old-fashioned. □ **quaintly** *adverb*; **quaintness** *noun.*

quake • *verb* (**-king**) shake, tremble. • *noun colloquial* earthquake.

Quaker *noun* member of Society of Friends. □ **Quakerism** *noun.*

qualification /kwɒlɪfɪ'keɪʃ(ə)n/ *noun* accomplishment fitting person for position or purpose; thing that modifies or limits; qualifying, being qualified. □ **qualificatory** /'kwɒl-/ *adjective.*

qualify /'kwɒlɪfaɪ/ *verb* (**-ies, -ied**) (often as **qualified** *adjective*) make competent or fit for purpose or position; make legally entitled; (usually + *for*) satisfy conditions; modify, limit; *Grammar* (of word) attribute quality to (esp. noun); moderate, mitigate; (+ *as*) be describable as. □ **qualifier** *noun.*

qualitative /'kwɒlɪtətɪv/ *adjective* concerned with quality as opposed to quantity. □ **qualitatively** *adverb.*

quality /'kwɒlɪtɪ/ • *noun* (*plural* **-ies**) excellence; degree of excellence; attribute, faculty; relative nature or character; timbre. • *adjective* of high quality.

qualm /kwɑ:m/ *noun* misgiving; scruple of conscience; momentary faint or sick feeling.

quandary /ˈkwɒndərɪ/ *noun* (*plural* **-ies**) perplexed state; practical dilemma.

quango /ˈkwæŋgəʊ/ *noun* (*plural* **-s**) semi-public administrative body appointed by government.

quanta *plural* of QUANTUM.

quantify /ˈkwɒntɪfaɪ/ *verb* (**-ies**, **-ied**) determine quantity of; express as quantity. □ **quantifiable** *adjective*.

quantitative /ˈkwɒntɪtətɪv/ *adjective* concerned with quantity as opposed to quality; measured or measurable by quantity.

quantity /ˈkwɒntɪtɪ/ *noun* (*plural* **-ies**) property of things that is measurable; size, extent, weight, amount, or number; (in *plural*) large amounts or numbers; length or shortness of vowel sound or syllable; *Mathematics* value, component, etc. that may be expressed in numbers. □ **quantity surveyor** person who measures and prices building work.

quantum /ˈkwɒntəm/ *noun* (*plural* **-ta**) discrete amount of energy proportional to frequency of radiation it represents; required or allowed amount. □ **quantum mechanics**, **theory** theory assuming that energy exists in discrete units.

quarantine /ˈkwɒrənti:n/ ● *noun* isolation imposed on person or animal to prevent infection or contagion; period of this. ● *verb* (**-ning**) put in quarantine.

quark[1] /kwɑ:k/ *noun Physics* component of elementary particles.

quark[2] /kwɑ:k/ *noun* kind of low-fat curd cheese.

quarrel /ˈkwɒr(ə)l/ ● *noun* severe or angry dispute; break in friendly relations; cause of complaint. ● *verb* (**-ll-**; *US* **-l-**) (often + *with*) find fault; dispute; break off friendly relations. □ **quarrelsome** *adjective*.

quarry[1] /ˈkwɒrɪ/ ● *noun* (*plural* **-ies**) place from which stone etc. is extracted. ● *verb* (**-ies**, **-ied**) extract (stone etc.) from quarry. □ **quarry tile** unglazed floor-tile.

quarry[2] /ˈkwɒrɪ/ *noun* (*plural* **-ies**) intended victim or prey; object of pursuit.

quart /kwɔ:t/ *noun* liquid measure equal to quarter of gallon; two pints (1.136 litre).

quarter /ˈkwɔ:tə/ ● *noun* each of 4 equal parts; period of 3 months; point of time 15 minutes before or after any hour; 25 US or Canadian cents, coin worth this; part of town, esp. as occupied by particular class; point of compass, region at this; direction, district; source of supply; (in *plural*) lodgings, accommodation of troops etc.; one-fourth of a lunar month; mercy towards enemy etc. on condition of surrender; grain measure equivalent to 8 bushels, *colloquial* one-fourth of a pound weight. ● *verb* divide into quarters; put (troops etc.) into quarters; provide with lodgings; *Heraldry* place (coats of arms) on 4 quarters of shield. □ **quarterback** player in American football who directs attacking play; **quarter day** day on which quarterly payments are due; **quarterdeck** part of ship's upper deck near stern; **quarter-final** *Sport* match or round preceding semifinal; **quarter-hour** period of 15 minutes; **quartermaster** regimental officer in charge of quartering, rations, etc., naval petty officer in charge of steering, signals, etc.

quarterly ● *adjective* produced or occurring once every quarter of year. ● *adverb* once every quarter of year. ● *noun* (*plural* **-ies**) quarterly journal.

quartet /kwɔ:ˈtet/ *noun* musical composition for 4 performers; the performers; any group of 4.

quarto /ˈkwɔ:təʊ/ *noun* (*plural* **-s**) size of book or page made by folding sheet of standard size twice to form 4 leaves.

quartz /kwɔ:ts/ ● *noun* silica in various mineral forms. ● *adjective* (of clock or watch) operated by vibrations of electrically driven quartz crystal.

quasar /ˈkweɪzɑ:/ *noun* starlike object with large red shift.

quash /kwɒʃ/ *verb* annul; reject as not valid; suppress, crush.

quasi- /ˈkweɪzaɪ/ *combining form* seemingly, not really; almost.

quaternary /kwəˈtɜ:nərɪ/ *adjective* having 4 parts.

quatrain /ˈkwɒtreɪn/ *noun* 4-line stanza.

quatrefoil /ˈkætrəfɔɪl/ *noun* leaf consisting of 4 leaflets; design or ornament in this shape.

quaver /ˈkweɪvə/ ● *verb* (esp. of voice or sound) vibrate, shake, tremble; sing or say with quavering voice. • *noun Music* note half as long as crotchet; trill in singing; tremble in speech. □ **quavery** *adjective*.

quay /kiː/ *noun* artificial landing-place for loading and unloading ships. □ **quayside** land forming or near quay.

queasy /ˈkwiːzɪ/ *adjective* (**-ier**, **-iest**) (of person) nauseous; (of stomach) easily upset; (of the conscience etc.) overscrupulous. □ **queasily** *adverb*; **queasiness** *noun*.

queen • *noun* (as title usually **Queen**) female sovereign; (in full **queen consort**) king's wife; woman, country, or thing pre-eminent of its kind; fertile female among bees, ants, etc.; most powerful piece in chess; court card depicting queen; (**the Queen**) national anthem when sovereign is female; *offensive slang* male homosexual. • *verb* convert (pawn in chess) to queen when it reaches opponent's side of board. □ **queen mother** king's widow who is mother of sovereign; **Queen's Bench** division of High Court of Justice; **Queen's Counsel** counsel to the Crown, taking precedence over other barristers; **the Queen's English** English language correctly written or spoken. □ **queenly** *adjective* (**-ier**, **-iest**).

Queensberry Rules /ˈkwiːnzbərɪ/ *plural noun* standard rules, esp. of boxing.

queer • *adjective* strange, odd, eccentric; suspect, of questionable character; slightly ill, faint; *offensive slang* (esp. of a man) homosexual. • *noun offensive slang* homosexual. • *verb slang* spoil, put out of order. □ **in Queer Street** *slang* in difficulty, esp. in debt.

quell *verb* suppress, crush.

quench *verb* satisfy (thirst) by drinking; extinguish (fire or light); cool, esp. with water; stifle, suppress.

quern *noun* hand mill for grinding corn.

querulous /ˈkwerʊləs/ *adjective* complaining; peevish. □ **querulously** *adverb*.

query /ˈkwɪərɪ/ • *noun* (*plural* **-ies**) question; question mark. • *verb* (**-ies**, **-ied**) ask, inquire; call in question; dispute accuracy of.

quest • *noun* search, seeking; thing sought, esp. by medieval knight. • *verb* (often + *about*) go about in search of something.

question /ˈkwestʃ(ə)n/ • *noun* sentence worded or expressed so as to seek information or answer; doubt or dispute about matter, raising of such doubt etc.; matter to be discussed or decided; problem requiring solution. • *verb* ask questions of; subject (person) to examination; throw doubt on. □ **be just a question of time** be certain to happen sooner or later; **be a question of** be at issue, be a problem; **call in** or **into question** express doubts about; **in question** being discussed or referred to; **out of the question** not worth questioning, impossible; **question mark** punctuation mark (?) indicating question (see panel); **question-master** person presiding over quiz game etc.; **question time** period in Parliament when MPs may question ministers. □ **questioner** *noun*; **questioning** *adjective & noun*; **questioningly** *adverb*.

questionable *adjective* doubtful as regards truth, quality, honesty, wisdom, etc.

Question mark ?

This is used instead of a full stop at the end of a sentence to show that it is a question, e.g.

Have you seen the film yet?
You didn't lose my purse, did you?

It is **not** used at the end of a reported question, e.g.

I asked you whether you'd seen the film yet.

---

questionnaire /kwestʃə'neə/ *noun* list of questions for obtaining information esp. for statistical analysis.

queue /kju:/ ● *noun* line or sequence of people, vehicles, etc. waiting their turn. ● *verb* (**-s**, **-d**, **queuing** or **queueing**) (often + *up*) form or join queue. □ **queue-jump** push forward out of turn in queue.

quibble /'kwɪb(ə)l/ ● *noun* petty objection, trivial point of criticism; evasion. ● *verb* (**-ling**) use quibbles.

quiche /ki:ʃ/ *noun* savoury flan.

quick ● *adjective* taking only a short time; arriving after only a short time, prompt; with only a short interval; lively, alert, intelligent; (of temper) easily roused. ● *adverb* quickly. ● *noun* soft sensitive flesh, esp. below nails or skin; seat of emotion. □ **quicklime** unslaked lime; **quicksand** (often in *plural*) area of loose wet sand that sucks in anything placed on it, treacherous situation etc.; **quickset** (of hedge etc.) formed of cuttings, esp. hawthorn; **quicksilver** mercury; **quickstep** fast foxtrot; **quick-tempered** easily angered; **quick-witted** quick to grasp situation, make repartee, etc.; **quick-wittedness** *noun*. □ **quickly** *adverb*.

quicken *verb* make or become quicker, accelerate; give life or vigour to, rouse; (of woman) reach stage in pregnancy when movements of foetus can be felt; (of foetus) begin to show signs of life.

quid[1] *noun slang* (*plural* same) one pound sterling.

quid[2] *noun* lump of tobacco for chewing.

quid pro quo /kwɪd prəʊ 'kwəʊ/ *noun* (*plural* **quid pro quos**) gift, favour, etc. exchanged for another.

quiescent /kwɪ'es(ə)nt/ *adjective* inert, dormant. □ **quiescence** *noun*.

quiet /'kwaɪət/ ● *adjective* with little or no sound or motion; of gentle or peaceful disposition; unobtrusive, not showy; not overt, disguised; undisturbed, uninterrupted; not busy. ● *noun* silence, stillness; undisturbed state, tranquillity. ● *verb* (often + *down*) make or become quiet, calm. □ **be quiet** (esp. in *imperative*) cease talking etc.; **keep quiet** (often + *about*) say nothing; **on the quiet** secretly. □ **quietly** *adverb*; **quietness** *noun*.

quieten *verb* (often + *down*) make or become quiet or calm.

quietism *noun* passive contemplative attitude towards life. □ **quietist** *noun & adjective*.

quietude /'kwaɪɪtju:d/ *noun* state of quiet.

quietus /kwaɪ'i:təs/ *noun* release from life; death, final riddance.

quiff *noun* man's tuft of hair brushed upwards in front.

quill *noun* (in full **quill-feather**) large feather in wing or tail; hollow stem of this; (in full **quill pen**) pen made of quill; (usually in *plural*) porcupine's spine.

quilt ● *noun* bedspread, esp. of quilted material. ● *verb* line bedspread or garment with padding enclosed between layers of fabric by lines of stitching. □ **quilter** *noun*; **quilting** *noun*.

quin *noun colloquial* quintuplet.

quince *noun* (tree bearing) acid pear-shaped fruit used in jams etc.

quincentenary /kwɪnsen'ti:nərɪ/ ● *noun* (*plural* **-ies**) 500th anniversary; celebration of this. ● *adjective* of this anniversary.

quinine /'kwɪni:n/ *noun* bitter drug used as a tonic and to reduce fever.

Quinquagesima /kwɪŋkwə'dʒesɪmə/ *noun* Sunday before Lent.

quinquennial /kwɪŋ'kwenɪəl/ *adjective* lasting 5 years; recurring every 5 years. □ **quinquennially** *adverb*.

quintessence /kwɪn'tes(ə)ns/ *noun* (usually + *of*) purest and most perfect form, manifestation, or embodiment of quality etc.; highly refined extract. □ **quintessential** /-tɪ'sen-/ *adjective*; **quintessentially** /-tɪ'sen-/ *adverb*.

quintet /kwɪn'tet/ *noun* musical composition for 5 performers; the performers; any group of 5.

quintuple /'kwɪntjʊp(ə)l/ ● *adjective* fivefold, having 5 parts. ● *noun* fivefold number or amount. ● *verb* (**-ling**) multiply by 5.

quintuplet /'kwɪntjʊplɪt/ *noun* each of 5 children born at one birth.

quip ● *noun* clever saying, epigram. ● *verb* (**-pp-**) make quips.

quire *noun* 25 sheets of paper.

quirk *noun* peculiar feature; trick of fate. □ **quirky** *adjective* (**-ier**, **-iest**).

quisling /'kwɪzlɪŋ/ *noun* collaborator with invading enemy.

quit ● *verb* (**-tting**; *past & past participle* **quitted** or **quit**) give up, let go, abandon; *US* cease, stop; leave or depart from. ● *adjective* (+ *of*) rid of.

quitch *noun* couch grass.

quite *adverb* completely, altogether, absolutely; rather, to some extent; (often + *so*) *said to indicate agreement.* □ **quite a**, **quite some** a remarkable; **quite a few** *colloquial* a fairly large number (of); **quite something** *colloquial* a remarkable thing or person.

quits *adjective* on even terms by retaliation or repayment. □ **call it quits** acknowledge that things are now even, agree to stop quarrelling.

quiver[1] /'kwɪvə/ ● *verb* tremble or vibrate with slight rapid motion. ● *noun* quivering motion or sound.

quiver[2] /'kwɪvə/ *noun* case for arrows.

quixotic /kwɪk'sɒtɪk/ *adjective* extravagantly and romantically chivalrous. □ **quixotically** *adverb.*

quiz ● *noun* (*plural* **quizzes**) test of knowledge, esp. as entertainment; interrogation, examination. ● *verb* examine by questioning.

quizzical /'kwɪzɪk(ə)l/ *adjective* mocking, gently amused. □ **quizzically** *adverb.*

quod *noun slang* prison.

quoin /kɔɪn/ *noun* external angle of building; cornerstone; wedge used in printing or gunnery.

quoit /kɔɪt/ *noun* ring thrown to encircle peg; (in *plural*) game using these.

quondam /'kwɒndæm/ *adjective* that once was, former.

quorate /'kwɔ:reɪt/ *adjective* constituting or having quorum.

Quorn /kwɔ:n/ *noun proprietary term* vegetable protein food made from fungus.

quorum /'kwɔ:rəm/ *noun* minimum number of members that must be present to constitute valid meeting.

Quotation marks ‘ ’ “ ”

Also called inverted commas, these are used:

1 round a direct quotation (closing quotation marks come after any punctuation which is part of the quotation), e.g.

He said, ‘That is nonsense.’
‘That’, he said, ‘is nonsense.’
‘That, however,’ he said, ‘is nonsense.’
Did he say, ‘That is nonsense’?
He asked, ‘Is that nonsense?’

2 round a quoted word or phrase, e.g.

What does ‘integrated circuit’ mean?

3 round a word or phrase that is not being used in its central sense, e.g.

the ‘king’ of jazz
He said he had enough ‘bread’ to buy a car.

4 round the title of a book, song, poem, magazine article, television programme, etc. (but not a book of the Bible), e.g.

‘Hard Times’ by Charles Dickens

5 as double quotation marks round a quotation within a quotation, e.g.

He asked, ‘Do you know what “integrated circuit” means?’

In handwriting, double quotation marks are usual.

quota /'kwəʊtə/ *noun* share to be contributed to or received from total; number of goods, people, etc. stipulated or permitted.

quotable *adjective* worth quoting.

quotation /kwəʊ'teɪʃ(ə)n/ *noun* passage or remark quoted; quoting, being quoted; contractor's estimate. □ **quotation marks** inverted commas (' ' or " ") used at beginning and end of quotation etc. (see panel).

quote ● *verb* (**-ting**) cite or appeal to as example, authority, etc.; repeat or copy out passage from; (+ *from*) cite (author, book, etc.); (+ *as*) cite (author etc.) as proof, evidence, etc.; (as *interjection*) *used in dictation etc. to indicate opening quotation marks*; (often + *at*) state price of; state (price) for job. ● *noun colloquial* passage quoted; (usually in *plural*) quotation marks.

quoth /kwəʊθ/ *verb* (only in 1st & 3rd persons) *archaic* said.

quotidian /'kwɒ'tɪdɪən/ *adjective* occurring or recurring daily; commonplace, trivial.

quotient /'kwəʊʃ(ə)nt/ *noun* result of division sum.

q.v. *abbreviation* which see (*quod vide*).

Rr

R *abbreviation* (also **R.**) *Regina*; *Rex*; River; (also ®) registered as trademark. □ **R & D** research and development.

r. *abbreviation* (also **r**) right; radius.

RA *abbreviation* Royal Academy or Academician; Royal Artillery.

rabbet /'ræbɪt/ ● *noun* step-shaped channel cut along edge or face of wood etc. to receive edge or tongue of another piece. ● *verb* (**-t-**) join with rabbet; make rabbet in.

rabbi /'ræbaɪ/ *noun* (*plural* **-s**) Jewish religious leader; Jewish scholar or teacher, esp. of the law. □ **rabbinical** /rə'bɪn-/ *adjective*.

rabbit /'ræbɪt/ ● *noun* burrowing mammal of hare family; its fur. ● *verb* (**-t-**) hunt rabbits; (often + *on*, *away*) *colloquial* talk pointlessly; chatter. □ **rabbit punch** blow with edge of hand on back of neck.

rabble /'ræb(ə)l/ *noun* disorderly crowd, mob; contemptible or inferior set of people. □ **rabble-rouser** person who stirs up rabble, esp. to agitate for social change.

Rabelaisian /ræbə'leɪzɪən/ *adjective* exuberantly and coarsely humorous.

rabid /'ræbɪd/ *adjective* affected with rabies, mad; violent, fanatical. □ **rabidity** /rə'bɪd-/ *noun*.

rabies /'reɪbi:z/ *noun* contagious viral disease of esp. dogs; hydrophobia.

RAC *abbreviation* Royal Automobile Club.

raccoon /rə'ku:n/ *noun* (also **racoon**) (*plural* same or **-s**) N. American mammal with bushy tail; its fur.

race[1] ● *noun* contest of speed or to be first to achieve something; (in *plural*) series of races for horses etc.; strong current in sea or river; channel. ● *verb* (**-cing**) take part in race; have race with; (+ *with*) compete in speed; cause to race; go at full speed; (usually as **racing** *adjective*) follow or take part in horse racing. □ **racecourse** ground for horse racing; **racehorse** one bred or kept for racing; **race meeting** sequence of horse races at one place; **racetrack** racecourse, track for motor racing; **racing car** one built for racing; **racing driver** driver of racing car.

race[2] *noun* each of the major divisions of humankind, each having distinct physical characteristics; group of people, animals, or plants connected by common descent; any great division of living creatures. □ **race relations** relations between members of different races in same country.

raceme /rə'si:m/ *noun* flower cluster with flowers attached by short stalks at equal distances along stem.

racial /ˈreɪʃ(ə)l/ *adjective* of or concerning race; on grounds of or connected with difference in race. □ **racially** *adverb.*

racialism *noun* = RACISM. □ **racialist** *noun & adjective.*

racism *noun* (prejudice based on) belief in superiority of particular race; antagonism towards other races. □ **racist** *noun & adjective.*

rack[1] ● *noun* framework, usually with rails, bars, etc., for holding things; cogged or toothed rail or bar engaging with wheel, pinion, etc.; *historical* instrument of torture stretching victim's joints. ● *verb* inflict suffering on; *historical* torture on rack. □ **rack one's brains** make great mental effort; **rack-rent** extortionate rent.

rack[2] *noun* destruction (esp. in **rack and ruin**).

rack[3] *verb* (often + *off*) draw off (wine etc.) from lees.

racket[1] /ˈrækɪt/ *noun* (also **racquet**) bat with round or oval frame strung with catgut, nylon, etc., used in tennis etc.; (in *plural*) game like squash but in larger court.

racket[2] ● *noun* uproar, din; *slang* scheme for obtaining money etc. by dishonest means; dodge; sly game; *colloquial* line of business.

racketeer /rækɪˈtɪə/ *noun* person who operates dishonest business. □ **racketeering** *noun.*

raconteur /rækɒnˈtɜː/ *noun* teller of anecdotes.

racoon = RACCOON.

racy *adjective* (**-ier**, **-iest**) lively and vigorous in style; risqué; of distinctive quality. □ **raciness** *noun.*

rad *noun* unit of absorbed dose of ionizing radiation.

RADA /ˈrɑːdə/ *abbreviation* Royal Academy of Dramatic Art.

radar /ˈreɪdɑː/ *noun* radio system for detecting the direction, range, or presence of objects; apparatus for this. □ **radar trap** device using radar to detect speeding vehicles.

raddled /ˈræd(ə)ld/ *adjective* worn out.

radial /ˈreɪdɪəl/ ● *adjective* of or arranged like rays or radii; having spokes or radiating lines; acting or moving along such lines; (in full **radial-ply**) (of tyre) having fabric layers arranged radially. ● *noun* radial-ply tyre. □ **radially** *adverb.*

radian /ˈreɪdɪən/ *noun* SI unit of plane angle (about 57°).

radiant /ˈreɪdɪənt/ ● *adjective* emitting or issuing in rays; beaming with joy etc.; splendid, dazzling. ● *noun* point or object from which heat or light radiates. □ **radiance** *noun*; **radiantly** *adverb.*

radiate /ˈreɪdɪeɪt/ *verb* (**-ting**) emit rays of light, heat, etc.; be emitted in rays; emit or spread from a centre; transmit or demonstrate.

radiation *noun* radiating; emission of energy as electromagnetic waves; energy thus transmitted, esp. invisibly; (in full **radiation therapy**) treatment of cancer etc. using e.g. X-rays or ultraviolet light. □ **radiation sickness** sickness caused by exposure to radiation such as gamma rays.

radiator *noun* device for heating room etc. by circulation of hot water etc.; engine-cooling device in motor vehicle or aircraft.

radical /ˈrædɪk(ə)l/ ● *adjective* fundamental; far-reaching, thorough; advocating fundamental reform; forming the basis; primary; of the root of a number or plant. ● *noun* person holding radical views; atom or group of atoms forming base of compound and remaining unchanged during reactions; quantity forming or expressed as root of another. □ **radicalism** *noun*; **radically** *adverb.*

radicchio /rəˈdiːkɪəʊ/ *noun* (*plural* **-s**) chicory with purplish leaves.

radicle /ˈrædɪk(ə)l/ *noun* part of seed that develops into root.

radii *plural* of **radius**.

radio /ˈreɪdɪəʊ/ ● *noun* (*plural* **-s**) transmission and reception of messages etc. by electromagnetic waves of radio frequency; apparatus for receiving, broadcasting, or transmitting radio signals; sound broadcasting (station or channel). ● *verb* (**-es**, **-ed**) send (message) by radio; send message to (person) by radio; communicate or broadcast by radio. □ **radio-controlled** controlled from a distance by radio; **radio telephone** one operating by radio; **radio**

telescope aerial system for analysing radiation in the radio-frequency range from stars etc.

radioactive *adjective* of or exhibiting radioactivity.

radioactivity *noun* spontaneous disintegration of atomic nuclei, with emission of usually penetrating radiation or particles.

radiocarbon *noun* radioactive isotope of carbon.

radiogram /'reɪdɪəʊgræm/ *noun* combined radio and record player; picture obtained by X-rays etc.; telegram sent by radio.

radiograph /'reɪdɪəʊgrɑ:f/ ● *noun* instrument recording intensity of radiation; picture obtained by X-rays etc. ● *verb* obtain picture of by X-rays, gamma rays, etc. □ **radiographer** /-'ɒgrəfə/ *noun*; **radiography** /-'ɒgrəfɪ/ *noun*.

radiology /reɪdɪ'ɒlədʒɪ/ *noun esp. Medicine* study of X-rays and other high-energy radiation. □ **radiologist** *noun*.

radiophonic *adjective* of electronically produced sound, esp. music.

radioscopy /reɪdɪ'ɒskəpɪ/ *noun* examination by X-rays etc. of objects opaque to light.

radiotherapy *noun* treatment of disease by X-rays or other forms of radiation.

radish /'rædɪʃ/ *noun* plant with crisp pungent root; this root, esp. eaten raw.

radium /'reɪdɪəm/ *noun* radioactive metallic element.

radius /'reɪdɪəs/ *noun* (*plural* **radii** /-dɪaɪ/ or **-es**) straight line from centre to circumference of circle or sphere; distance from a centre; bone of forearm on same side as thumb.

radon /'reɪdɒn/ *noun* gaseous radioactive inert element arising from disintegration of radium.

RAF *abbreviation* Royal Air Force.

raffia /'ræfɪə/ *noun* palm tree native to Madagascar; fibre from its leaves.

raffish /'ræfɪʃ/ *adjective* disreputable, rakish; tawdry.

raffle /'ræf(ə)l/ ● *noun* fund-raising lottery with prizes. ● *verb* (**-ling**) (often + *off*) sell by raffle.

raft /rɑ:ft/ *noun* flat floating structure of wood etc., used for transport.

rafter /'rɑ:ftə/ *noun* any of sloping beams forming framework of roof.

rag[1] *noun* torn, frayed, or worn piece of woven material; remnant; (in *plural*) old or worn clothes; *derogatory* newspaper. □ **rag-bag** miscellaneous collection; **rag doll** stuffed cloth doll; **ragtime** form of highly syncopated early jazz; **the rag trade** *colloquial* the clothing business; **ragwort** yellow-flowered ragged-leaved plant.

rag[2] ● *noun* fund-raising programme of stunts etc. staged by students; prank; rowdy celebration, disorderly scene. ● *verb* (**-gg-**) tease; play rough jokes on; engage in rough play.

ragamuffin /'rægəmʌfɪn/ *noun* child in ragged dirty clothes.

rage ● *noun* violent anger; fit of this. ● *verb* (**-ging**) be full of anger; (often + *at, against*) speak furiously; be violent, be at its height; (as **raging** *adjective*) extreme, very painful. □ **all the rage** very popular, fashionable.

ragged /'rægɪd/ *adjective* torn; frayed; in ragged clothes; with a broken or jagged outline or surface; lacking finish, smoothness, or uniformity.

raglan /'ræglən/ *adjective* (of sleeve) running up to neck of garment.

ragout /ræ'gu:/ *noun* highly seasoned stew of meat and vegetables.

raid ● *noun* rapid surprise attack by armed forces or thieves; surprise visit by police etc. to arrest suspects or seize illicit goods. ● *verb* make raid on. □ **raider** *noun*.

rail[1] ● *noun* bar used to hang things on or as protection, part of fence, top of banisters, etc.; steel bar(s) making railway track; railway. ● *verb* provide or enclose with rail(s). □ **railcard** pass entitling holder to reduced rail fares.

rail[2] *verb* (often + *at, against*) complain or protest strongly; rant.

rail[3] *noun* marsh wading bird.

railing *noun* (often in *plural*) fence or barrier made of rails.

raillery /'reɪlərɪ/ *noun* good-humoured ridicule.

railroad ● *noun esp. US* railway. ● *verb* (often + *into, through*) coerce, rush.

railway /'reɪlweɪ/ *noun* track or set of tracks of steel rails on which trains run; organization and people required to work such a system. □ **railwayman** male railway employee.

raiment /ˈreɪmənt/ *noun archaic* clothing.

rain ● *noun* condensed atmospheric moisture falling in drops; fall of these; falling liquid or objects; (**the rains**) rainy season. ● *verb* (after *it*) rain falls, send in large quantities; fall or send down like rain; lavishly bestow. □ **raincoat** waterproof or water-resistant coat; **rainfall** total amount of rain falling within given area in given time; **rainforest** tropical forest with heavy rainfall; **take a rain check on** reserve right to postpone taking up (offer) until convenient.

rainbow /ˈreɪnbəʊ/ ● *noun* arch of colours formed in sky by reflection, refraction, and dispersion of sun's rays in falling rain etc. ● *adjective* many-coloured. □ **rainbow trout** large trout originally of N. America.

rainy *adjective* (**-ier**, **-iest**) (of weather, day, climate, etc.) in or on which rain is falling or much rain usually falls. □ **rainy day** time of need in the future.

raise /reɪz/ ● *verb* (**-sing**) put or take into higher position; (often + *up*) cause to rise or stand up or be vertical; increase amount, value, or strength of; (often + *up*) build up; levy, collect; cause to be heard or considered; bring up, educate; breed; remove (barrier); rouse. ● *noun Cards* increase in stake or bid; *US* rise in salary. □ **raise Cain, hell, the roof** be very angry, cause an uproar; **raise a laugh** cause others to laugh.

raisin /ˈreɪz(ə)n/ *noun* dried grape.

raison d'être /reɪzɔ̃ ˈdetr/ *noun* (*plural* ***raisons d'être*** same pronunciation) purpose that accounts for, justifies, or originally caused thing's existence. [French]

raj /rɑːdʒ/ *noun* (**the raj**) *historical* British rule in India.

raja /ˈrɑːdʒə/ *noun* (also **rajah**) *historical* Indian king or prince.

rake[1] ● *noun* implement with long handle and toothed crossbar for drawing hay etc. together, smoothing loose soil, etc.; similar implement. ● *verb* (**-king**) collect or gather (as) with rake; ransack, search thoroughly; direct gunfire along (line) from end to end. □ **rake in** *colloquial* amass (profits etc.); **rake-off** *colloquial* commission or share; **rake up** revive (unwelcome) memory of.

rake[2] *noun* dissolute man of fashion.

rake[3] ● *verb* (**-king**) set or be set at sloping angle. ● *noun* raking position or build; amount by which thing rakes.

rakish *adjective* dashing, jaunty; dissolute. □ **rakishly** *adverb.*

rallentando /rælənˈtændəʊ/ *Music* ● *adverb & adjective* with gradual decrease of speed. ● *noun* (*plural* **-s** or **-di** /-dɪ/) rallentando passage.

rally /ˈrælɪ/ ● *verb* (**-ies**, **-ied**) (often + *round*) bring or come together as support or for action; recover after illness etc., revive; (of prices etc.) increase after fall. ● *noun* (*plural* **-ies**) rallying, being rallied; mass meeting; competition for motor vehicles over public roads; extended exchange of strokes in tennis etc. □ **rallycross** motor racing across country.

RAM *abbreviation* Royal Academy of Music; random-access memory.

ram ● *noun* uncastrated male sheep; (**the Ram**) zodiacal sign or constellation Aries; falling weight of pile-driving machine; hydraulic water pump. ● *verb* (**-mm-**) force into place; (usually + *down*, *in*, etc.) beat down or drive in by blows; (of ship, vehicle, etc.) strike, crash against. □ **ram-raid** crashing vehicle into shop front in order to steal contents; **ram-raider** *noun.*

Ramadan /ˈræmədæn/ *noun* ninth month of Muslim year, with strict fasting from sunrise to sunset.

ramble /ˈræmb(ə)l/ ● *verb* (**-ling**) walk for pleasure; talk or write incoherently. ● *noun* walk taken for pleasure.

rambler *noun* person who rambles; straggling or spreading rose.

rambling *adjective* wandering; disconnected, incoherent; irregularly arranged; (of plant) straggling, climbing.

RAMC *abbreviation* Royal Army Medical Corps.

ramekin /ˈræmɪkɪn/ *noun* small dish for baking and serving individual portion of food.

ramification /ræmɪfɪˈkeɪʃ(ə)n/ *noun* (usually in *plural*) consequence; subdivision.

ramify /ˈræmɪfaɪ/ *verb* (**-ies**, **-ied**) (cause to) form branches or subdivisions; branch out.

ramp *noun* slope joining two levels of ground, floor, etc.; stairs for entering or leaving aircraft; transverse ridge in road making vehicles slow down.

rampage • *verb* /ræmˈpeɪdʒ/ (**-ging**) (often + *about*) rush wildly; rage, storm. • *noun* /ˈræmpeɪdʒ/ wild or violent behaviour. □ **on the rampage** rampaging.

rampant /ˈræmpənt/ *adjective* unchecked, flourishing excessively; rank, luxuriant; *Heraldry* (of lion etc.) standing on left hind foot with forepaws in air; fanatical. □ **rampancy** *noun*.

rampart /ˈræmpɑːt/ *noun* defensive broad-topped wall; defence, protection.

ramrod *noun* rod for ramming down charge of muzzle-loading firearm; thing that is very straight or rigid.

ramshackle /ˈræmʃæk(ə)l/ *adjective* rickety, tumbledown.

ran *past* of RUN.

ranch /rɑːntʃ/ • *noun* cattle-breeding establishment, esp. in US & Canada; farm where other animals are bred. • *verb* farm on ranch. □ **rancher** *noun*.

rancid /ˈrænsɪd/ *adjective* smelling or tasting like rank stale fat. □ **rancidity** /-ˈsɪd-/ *noun*.

rancour /ˈræŋkə/ *noun* (*US* **rancor**) inveterate bitterness; malignant hate. □ **rancorous** *adjective*.

rand *noun* monetary unit of South Africa.

random /ˈrændəm/ *adjective* made, done, etc. without method or conscious choice. □ **at random** without particular aim; **random-access** (of computer memory) having all parts directly accessible. □ **randomly** *adverb*.

randy /ˈrændɪ/ *adjective* (**-ier**, **-iest**) eager for sexual satisfaction.

ranee /ˈrɑːnɪ/ *noun* (also **rani**) (*plural* **-s**) *historical* raja's wife or widow.

rang *past* of RING[2].

range /reɪndʒ/ • *noun* region between limits of variation, esp. scope of operation; such limits; area relevant to something; distance attainable by gun or projectile, distance between gun etc. and target; row, series, etc., esp. of mountains; area with targets for shooting; fireplace for cooking; area over which a thing is distributed; distance that can be covered by vehicle without refuelling; stretch of open land for grazing or hunting. • *verb* (**-ging**) reach; extend; vary between limits; (usually in *passive*) line up, arrange; rove, wonder; traverse in all directions. □ **range-finder** instrument for determining distance of object.

ranger *noun* keeper of royal or national park, or of forest; (**Ranger**) senior Guide.

rangy /ˈreɪndʒɪ/ *adjective* (**-ier**, **-iest**) tall and slim.

rani = RANEE.

rank[1] • *noun* position in hierarchy; grade of advancement; distinct social class, grade of dignity or achievement; high social position; place in scale; row or line; single line of soldiers drawn up abreast; place where taxis wait for customers. • *verb* have rank or place; classify, give a certain grade to; arrange in rank. □ **rank and file** (usually treated as *plural*) ordinary members of organization; **the ranks** common soldiers.

rank[2] *adjective* luxuriant; coarse; choked with weeds etc.; foul-smelling; loathsome; flagrant; gross, complete.

rankle /ˈræŋk(ə)l/ *verb* (**-ling**) cause persistent annoyance or resentment.

ransack /ˈrænsæk/ *verb* pillage, plunder; thoroughly search.

ransom /ˈrænsəm/ • *noun* sum demanded or paid for release of prisoner. • *verb* buy freedom or restoration of; hold (prisoner) to ransom; release for a ransom. □ **hold to ransom** keep (prisoner) and demand ransom, demand concessions from by threats.

rant • *verb* speak loudly, bombastically, or violently. • *noun* piece of ranting. □ **rant and rave** express anger noisily and forcefully.

ranunculus /rəˈnʌŋkjʊləs/ *noun* (*plural* **-luses** or **-li** /-laɪ/) plant of genus including buttercup.

RAOC *abbreviation* Royal Army Ordnance Corps.

rap[1] • *noun* smart slight blow; sound of this, tap; *slang* blame, punishment; rhythmic monologue recited to music; (in full **rap music**) style of rock music with words recited. • *verb* (**-pp-**) strike smartly; make sharp tapping sound; criticize adversely; *Music* perform rap. □ **take the rap** suffer the consequences. □ **rapper** *noun Music.*

rap[2] *noun* the least bit.

rapacious /rəˈpeɪʃəs/ *adjective* grasping, extortionate, predatory. □ **rapacity** /-ˈpæs-/ *noun.*

rape[1] • *noun* act of forcing woman or girl to have sexual intercourse against her will; (often + *of*) violent assault or plunder. • *verb* (**-ping**) commit rape on.

rape[2] *noun* plant grown as fodder and for oil from its seed.

rapid /ˈræpɪd/ • *adjective* (**-er**, **-est**) quick, swift. • *noun* (usually in *plural*) steep descent in river bed, with swift current. □ **rapid eye movement** jerky movement of eyes during dreaming. □ **rapidity** /rəˈpɪd-/ *noun*; **rapidly** *adverb.*

rapier /ˈreɪpɪə/ *noun* light slender sword for thrusting.

rapine /ˈræpaɪn/ *noun rhetorical* plundering.

rapist *noun* person who commits rape.

rapport /ræˈpɔː/ *noun* communication or relationship, esp. when useful and harmonious.

rapprochement /ræˈprɒʃmɑ̃/ *noun* resumption of harmonious relations, esp. between states. [French]

rapscallion /ræpˈskæljən/ *noun archaic* rascal.

rapt *adjective* absorbed; intent; carried away with feeling or thought.

rapture /ˈræptʃə/ *noun* ecstatic delight; (in *plural*) great pleasure or enthusiasm or expression of it. □ **rapturous** *adjective.*

rare[1] *adjective* (**-r**, **-st**) seldom done, found, or occurring; uncommon; exceptionally good; of less than usual density. □ **rareness** *noun.*

rare[2] *adjective* (**-r**, **-st**) (of meat) underdone.

rarebit *noun* see WELSH RAREBIT.

rarefy /ˈreərɪfaɪ/ *verb* (**-ies**, **-ied**) (often as **rarefied** *adjective*) make or become less dense or solid; refine, make (idea etc.) subtle. □ **rarefaction** /-ˈfækʃ(ə)n/ *noun.*

rarely *adverb* seldom, not often.

raring /ˈreərɪŋ/ *adjective colloquial* eager (esp. in **raring to go**).

rarity /ˈreərətɪ/ *noun* (*plural* **-ies**) rareness; uncommon thing.

rascal /ˈrɑːsk(ə)l/ *noun* dishonest or mischievous person. □ **rascally** *adjective.*

rase = RAZE.

rash[1] *adjective* reckless; hasty, impetuous. □ **rashly** *adverb*; **rashness** *noun.*

rash[2] *noun* skin eruption in spots or patches; (usually + *of*) sudden widespread phenomenon.

rasher /ˈræʃə/ *noun* thin slice of bacon or ham.

rasp /rɑːsp/ • *noun* coarse file; grating noise or utterance. • *verb* scrape roughly or with rasp; make grating sound; say gratingly; grate on.

raspberry /ˈrɑːzbərɪ/ *noun* (*plural* **-ies**) red fruit like blackberry; shrub bearing this; *colloquial* sound made by blowing through lips, expressing derision or disapproval.

Rastafarian /ræstəˈfeərɪən/ (also **Rasta** /ˈræstə/) • *noun* member of Jamaican sect regarding Haile Selassie of Ethiopia (d. 1975) as God. • *adjective* of this sect.

rat • *noun* large mouselike rodent; *colloquial* unpleasant or treacherous person. • *verb* (**-tt-**) hunt or kill rats; (also + *on*) inform (on), desert, betray. □ **ratbag** *slang* obnoxious person; **rat race** *colloquial* fiercely competitive struggle.

ratable = RATEABLE.

ratatouille /rætəˈtuːɪ/ *noun* dish of stewed onions, courgettes, tomatoes, aubergines, and peppers.

ratchet /ˈrætʃɪt/ *noun* set of teeth on edge of bar or wheel with catch ensuring motion in one direction only; (in full **ratchet-wheel**) wheel with rim so toothed.

rate • *noun* numerical proportion between two sets of things or as basis of calculating amount or value; charge, cost, or value; measure of this; pace of movement or change; (in *plural*) tax levied by local authorities according to value of buildings and land occupied. ● *verb* (**-ting**) estimate worth or value of; assign value to; consider, regard as; (+ *as*) rank or be considered as; subject to payment of local rate; deserve. □ **at any rate** in any case, whatever happens; **at this rate** if this example is typical; **rate-capping** *historical* imposition of upper limit on local authority rates; **ratepayer** person liable to pay rates.

rateable /ˈreɪtəb(ə)l/ *adjective* (also **ratable** /ˈreɪtəb(ə)l/) liable to rates. □ **rateable value** value at which business etc. is assessed for rates.

rather /ˈrɑːðə/ *adverb* by preference; (usually + *than*) more truly; as a more likely alternative; more precisely; to some extent; /rɑːˈðɜː/ most emphatically.

ratify /ˈrætɪfaɪ/ *verb* (**-ies**, **-ied**) confirm or accept by formal consent, signature, etc. □ **ratification** *noun*.

rating /ˈreɪtɪŋ/ *noun* placing in rank or class; estimated standing of person as regards credit etc.; non-commissioned sailor; (usually in *plural*) popularity of a broadcast as determined by estimated size of audience.

ratio /ˈreɪʃɪəʊ/ *noun* (*plural* **-s**) quantitative relation between similar magnitudes.

ratiocinate /rætɪˈɒsɪneɪt/ *verb* (**-ting**) *literary* reason, esp. using syllogisms. □ **ratiocination** *noun*.

ration /ˈræʃ(ə)n/ ● *noun* official allowance of food, clothing, etc., in time of shortage; (usually in *plural*) fixed daily allowance of food. ● *verb* limit (food etc. or people) to fixed ration; (usually + *out*) share (out) in fixed quantities.

rational /ˈræʃən(ə)l/ *adjective* of or based on reason; sensible; endowed with reason; rejecting what is unreasonable; *Mathematics* expressible as ratio of whole numbers. □ **rationality** /-ˈnæl-/ *noun*; **rationally** *adverb*.

rationale /ræʃəˈnɑːl/ *noun* fundamental reason, logical basis.

rationalism *noun* practice of treating reason as basis of belief and knowledge. □ **rationalist** *noun & adjective*; **rationalistic** /-ˈlɪs-/ *adjective*.

rationalize *verb* (also **-ise**) (**-zing** or **-sing**) (often + *away*) offer rational but specious explanation of (behaviour or attitude); make logical and consistent; make (industry etc.) more efficient by reducing waste. □ **rationalization** *noun*.

ratline /ˈrætlɪn/ *noun* (also **ratlin**) (usually in *plural*) any of the small lines fastened across ship's shrouds like ladder rungs.

rattan /rəˈtæn/ *noun* palm with long thin many-jointed stems; cane of this.

rattle /ˈræt(ə)l/ ● *verb* (**-ling**) (cause to) give out rapid succession of short sharp sounds; cause such sounds by shaking something; move or travel with rattling noise; (usually + *off*) say or recite rapidly; (usually + *on*) talk in lively thoughtless way; *colloquial* disconcert, alarm. ● *noun* rattling sound; device or plaything made to rattle. □ **rattlesnake** poisonous American snake with rattling rings on tail. □ **rattly** *adjective*.

rattling ● *adjective* that rattles; brisk, vigorous. ● *adverb colloquial* remarkably (good etc.).

raucous /ˈrɔːkəs/ *adjective* harsh-sounding; hoarse. □ **raucously** *adverb*; **raucousness** *noun*.

raunchy /ˈrɔːntʃɪ/ *adjective* (**-ier**, **-iest**) *colloquial* sexually boisterous.

ravage /ˈrævɪdʒ/ ● *verb* (**-ging**) devastate, plunder. ● *noun* (usually in *plural*; + *of*) destructive effect.

rave ● *verb* (**-ving**) talk wildly or deliriously; (usually + *about*, *over*) speak with rapturous admiration; *colloquial* enjoy oneself freely. ● *noun colloquial* highly enthusiastic review; (also **rave-up**) lively party.

ravel /ˈræv(ə)l/ *verb* (**-ll-**; *US* **-l-**) entangle, become entangled; fray out.

raven /ˈreɪv(ə)n/ ● *noun* large glossy black crow with hoarse cry. ● *adjective* glossy black.

ravening /ˈrævənɪŋ/ *adjective* hungrily seeking prey; voracious.

ravenous /ˈrævənəs/ *adjective* very hungry; voracious; rapacious. □ **ravenously** *adverb*.

raver *noun colloquial* uninhibited pleasure-loving person.

ravine /rəˈviːn/ *noun* deep narrow gorge.

raving • *noun* (usually in *plural*) wild or delirious talk. • *adjective colloquial* utter, absolute. • *adverb colloquial* utterly, absolutely.

ravioli /rævɪ'əʊlɪ/ *noun* small square pasta envelopes containing meat, spinach, etc.

ravish /'rævɪʃ/ *verb archaic* rape (woman); enrapture.

ravishing *adjective* lovely, beautiful. □ **ravishingly** *adverb.*

raw *adjective* uncooked; in natural state, not processed or manufactured; inexperienced, untrained; stripped of skin; unhealed; sensitive to touch; (of weather) cold and damp; crude. □ **in the raw** in its natural state, naked; **raw-boned** gaunt; **raw deal** unfair treatment; **rawhide** untanned hide, rope or whip of this; **raw material** material from which manufactured goods are made.

Rawlplug /'rɔ:lplʌg/ *noun proprietary term* cylindrical plug for holding screw in masonry.

ray[1] *noun* single line or narrow beam of light; straight line in which radiation travels; (in *plural*) radiation; trace or beginning of enlightening influence; any of set of radiating lines, parts, or things; marginal part of daisy etc.

ray[2] *noun* large edible marine fish with flat body.

ray[3] *noun* (also **re**) *Music* second note of scale in tonic sol-fa.

rayon /'reɪɒn/ *noun* textile fibre or fabric made from cellulose.

raze *verb* (also **rase**) (**-zing** or **-sing**) completely destroy, tear down.

razor /'reɪzə/ *noun* instrument for shaving. □ **razorbill** auk with sharp-edged bill; **razor-blade** flat piece of metal with sharp edge, used in safety razor; **razor-edge**, **razor's edge** keen edge, sharp mountain ridge, critical situation, sharp line of division.

razzle /'ræz(ə)l/ *noun colloquial* spree. □ **razzle-dazzle** excitement, bustle, extravagant publicity.

razzmatazz /ræzmə'tæz/ *noun* (also **razzamatazz** /ræzə-/) *colloquial* glamorous excitement; insincere activity.

RC *abbreviation* Roman Catholic.

Rd. *abbreviation* Road.

RE *abbreviation* Religious Education; Royal Engineers.

re[1] /ri:/ *preposition* in the matter of; about, concerning.

re[2] = RAY[3].

re- *prefix attachable to almost any verb or its derivative, meaning*: once more, anew, afresh; back. For words starting with *re-* that are not found below, the root-words should be consulted.

reach • *verb* (often + *out*) stretch out, extend; (often + *for*) stretch hand etc.; get as far as, get to or attain; make contact with; pass, hand; take with outstretched hand; *Nautical* sail with wind abeam. • *noun* extent to which hand etc. can be reached out, influence exerted, etc.; act of reaching out; continuous extent, esp. of river or canal. □ **reach-me-down** *colloquial* ready-made garment. □ **reachable** *adjective.*

react /rɪ'ækt/ *verb* (often + *to*) respond to stimulus; change or behave differently due to some influence; (often + *with*) undergo chemical reaction (with other substance); (often + *against*) respond with repulsion to; tend in reverse or contrary direction.

reaction /rɪ'ækʃ(ə)n/ *noun* reacting; response; bad physical response to drug etc.; occurrence of condition after its opposite; tendency to oppose change or reform; interaction of substances undergoing chemical change.

reactionary • *adjective* tending to oppose (esp. political) change or reform. • *noun* (*plural* **-ies**) reactionary person.

reactivate /rɪ'æktɪveɪt/ *verb* (**-ting**) restore to state of activity. □ **reactivation** *noun.*

reactive /rɪ'æktɪv/ *adjective* showing reaction; reacting rather than taking initiative; susceptible to chemical reaction.

reactor *noun* (in full **nuclear reactor**) device in which nuclear chain reaction is used to produce energy.

read • *verb* (*past & past participle* **read** /red/) reproduce (written or printed words) mentally or (often + *aloud*, *out*, *off*, etc.) vocally; (be able to) convert (written or printed words or other symbols) into intended words or meaning; interpret; (of meter) show (figure); interpret state of (meter); study (subject) at university; (as **read** /red/ *adjective*) versed in subject (esp. literature) by reading; (of computer) copy or transfer (data); hear and understand (over radio); substitute (word etc.) for incorrect one. ● *noun* spell of reading; *colloquial* book etc. as regards readability. □ **take as read** treat (thing) as if it has been agreed.

readable *adjective* able to be read; interesting to read. □ **readability** *noun*.

reader *noun* person who reads; book intended for reading pratice; device for producing image that can be read from microfilm etc.; (also **Reader**) university lecturer of highest grade below professor; publisher's employee who reports on submitted manuscripts; printer's proof-corrector.

readership *noun* readers of a newspaper etc.; (also **Readership**) position of Reader.

readily *adverb* without reluctance; willingly; easily.

readiness *noun* prepared state; willingness; facility; promptness in argument or action.

reading *noun* act of reading; matter to be read; literary knowledge; entertainment at which something is read; figure etc. shown by recording instrument; interpretation or view taken; interpretation made (of music etc.); presentation of bill to legislature.

ready /'redɪ/ ● *adjective* (**-ier**, **-iest**) with preparations complete; in fit state; willing; (of income etc.) easily secured; fit for immediate use; prompt, enthusiastic; (+ *to do*) about to; provided beforehand. ● *adverb usually in combination* beforehand; in readiness. ● *noun slang* (**the ready**) ready money; (**readies**) bank notes. ● *verb* (**-ies**, **-ied**) prepare. □ **at the ready** ready for action; **ready-made**, **ready-to-wear** (esp. of clothes) made in standard size, not to measure; **ready money** cash, actual coin; **ready reckoner** book or table listing standard numerical calculations.

reagent /ri:'eɪdʒ(ə)nt/ *noun* substance used to produce chemical reaction.

real ● *adjective* actually existing or occurring; genuine; appraised by purchasing power. ● *adverb Scottish & US colloquial* really, very. □ **real ale** beer regarded as brewed in traditional way; **real estate** property such as land and houses; **real tennis** original form of tennis played on indoor court.

realism *noun* practice of regarding things in their true nature and dealing with them as they are; fidelity to nature in representation. □ **realist** *noun*.

realistic /rɪə'lɪstɪk/ *adjective* regarding things as they are; based on facts rather than ideals. □ **realistically** *adverb*.

reality /rɪ'ælɪtɪ/ *noun* (*plural* **-ies**) what is real or existent or underlies appearances; (+ *of*) real nature of; real existence; being real; likeness to original. □ **in reality** in fact.

realize *verb* (also **-ise**) (**-zing** or **-sing**) (often + *that*) be or become fully aware of; understand clearly; convert into actuality; convert into money; acquire (profit); be sold for. □ **realizable** *adjective*; **realization** *noun*.

really /'rɪəlɪ/ *adverb* in fact; very; I assure you; *expression of mild protest or surprise*.

realm /relm/ *noun formal* kingdom; domain.

realty /'ri:əltɪ/ *noun* real estate.

ream *noun* 500 sheets of paper; (in *plural*) large quantity of writing.

reap *verb* cut (grain etc.) as harvest; receive as consequences of actions. □ **reaper** *noun*.

rear[1] ● *noun* back part of anything; space or position at back. ● *adjective* at the back. □ **bring up the rear** come last; **rear admiral** naval officer below vice admiral. □ **rearmost** *adjective*.

rear[2] *verb* bring up and educate; breed and care for; cultivate; (of horse etc.) raise itself on hind legs; raise, build.

rearguard *noun* troops detached to protect rear, esp. in retreat. □ **rearguard action** engagement undertaken by rearguard, defensive stand or struggle, esp. when losing.

rearm /ri:'ɑ:m/ *verb* arm again, esp. with improved weapons. □ **rearmament** *noun*.

rearward /'rɪəwəd/ ● *noun* rear. ● *adjective* to the rear. ● *adverb* (also **rearwards**) towards the rear.

reason /'ri:z(ə)n/ ● *noun* motive, cause, or justification; fact adduced or serving as this; intellectual faculty by which conclusions are drawn; sanity; sense, sensible conduct; moderation. ● *verb* form or try to reach conclusions by connected thought; (+ *with*) use argument with person by way of persuasion; (+ *that*) conclude or assert in argument; (+ *out*) think out.

reasonable *adjective* having sound judgement; moderate; ready to listen to reason; sensible; inexpensive; tolerable. □ **reasonableness** *noun*; **reasonably** *adverb*.

reassure /ri:ə'ʃʊə/ *verb* (**-ring**) restore confidence to; confirm in opinion etc. □ **reassurance** *noun*; **reassuring** *adjective*.

rebate[1] /'ri:beɪt/ *noun* partial refund; deduction from sum to be paid, discount.

rebate[2] /'ri:beɪt/ *noun & verb* (**-ting**) rabbet.

rebel ● *noun* /'reb(ə)l/ person who fights against, resists, or refuses allegiance to, established government; person etc. who resists authority or control. ● *verb* /rɪ'bel/ (**-ll-**; *US* **-l-**) (usually + *against*) act as rebel; feel or show repugnance.

rebellion /rɪ'beljən/ *noun* open resistance to authority, esp. organized armed resistance to established government.

rebellious /rɪ'beljəs/ *adjective* disposed to rebel; in rebellion; unmanageable. □ **rebelliously** *adverb*; **rebelliousness** *noun*.

rebound ● *verb* /rɪ'baʊnd/ spring back after impact; (+ *upon*) have adverse effect on (doer). ● *noun* /'ri:baʊnd/ rebounding, recoil; reaction. □ **on the rebound** while still recovering from emotional shock, esp. rejection by lover.

rebuff /rɪ'bʌf/ ● *noun* rejection of person who makes advances, offers help, etc.; snub. ● *verb* give rebuff to.

rebuke /rɪ'bju:k/ ● *verb* (**-king**) express sharp disapproval to (person) for fault; censure. ● *noun* rebuking, being rebuked.

rebus /'ri:bəs/ *noun* (*plural* **rebuses**) representation of word (esp. name) by pictures etc. suggesting its parts.

rebut /rɪ'bʌt/ *verb* (**-tt-**) refute, disprove; force back. □ **rebuttal** *noun*.

recalcitrant /rɪ'kælsɪtrənt/ *adjective* obstinately disobedient; objecting to restraint. □ **recalcitrance** *noun*.

recall /rɪ'kɔ:l/ ● *verb* summon to return; recollect, remember; bring back to memory; revoke, annul; revive, resuscitate. ● *noun* (also /'ri:kɔ:l/) summons to come back; act of remembering; ability to remember; possibility of recalling.

recant /rɪ'kænt/ *verb* withdraw and renounce (belief or statement) as erroneous or heretical. □ **recantation** /ri:kæn'teɪʃ(ə)n/ *noun*.

recap /'ri:kæp/ *colloquial* ● *verb* (**-pp-**) recapitulate. ● *noun* recapitulation.

recapitulate /ri:kə'pɪtjʊleɪt/ *verb* (**-ting**) summarize, restate briefly. □ **recapitulation** *noun*.

recast /ri:'kɑ:st/ ● *verb* (*past & past participle* **recast**) cast again; put into new form; improve arrangement of. ● *noun* recasting; recast form.

recce /'rekɪ/ *colloquial* ● *noun* reconnaissance. ● *verb* (**recced**, **recceing**) reconnoitre.

recede /rɪ'si:d/ *verb* (**-ding**) go or shrink back; be left at an increasing distance; slope backwards; decline in force or value.

receipt /rɪ'si:t/ ● *noun* receiving, being received; written or printed acknowledgement of payment received; (usually in *plural*) amount of money received. ● *verb* place written or printed receipt on (bill). □ **in receipt of** having received.

receive /rɪ'si:v/ *verb* (**-ving**) take or accept (thing offered, sent, or given); acquire; have conferred etc. on one; react to (news etc.) in particular way; stand force or weight of; consent to hear or consider; admit, entertain as guest, greet, welcome; be able to hold; convert (broadcast signals) into sound or pictures (as **received** *adjective*) accepted as authoritative or true. □ **Received Pronunciation** standard pronunciation of English in Britain (see panel at ACCENT).

receiver *noun* part of telephone containing earpiece; (in full **official receiver**) person appointed to administer property of bankrupt person etc. or property under litigation; radio or television receiving apparatus; person who receives stolen goods.

receivership *noun* □ **in receivership** being dealt with by receiver.

recent /'ri:s(ə)nt/ *adjective* not long past, that happened or existed lately; not long established, modern. □ **recently** *adverb.*

receptacle /rɪ'septək(ə)l/ *noun* object or space used to contain something.

reception /rɪ'sepʃ(ə)n/ *noun* receiving, being received; way in which person or thing is received; social occasion for receiving guests, esp. after wedding; place where visitors register on arriving at hotel, office, etc.; (quality of) receiving of broadcast signals. □ **reception room** room for receiving guests, clients, etc.

receptionist *noun* person employed to receive guests, clients, etc.

receptive /rɪ'septɪv/ *adjective* able or quick to receive ideas etc. □ **receptively** *adverb*; **receptiveness** *noun*; **receptivity** /ri:sep'tɪv-/ *noun.*

recess /rɪ'ses/ ● *noun* space set back in wall; (often in *plural*) remote or secret place; temporary cessation from work, esp. of Parliament. ● *verb* make recess in; place in recess; *US* take recess, adjourn.

recession /rɪ'seʃ(ə)n/ *noun* temporary decline in economic activity or prosperity; receding, withdrawal.

recessive /rɪ'sesɪv/ *adjective* tending to recede; (of inherited characteristic) appearing in offspring only when not masked by inherited dominant characteristic.

recherché /rə'ʃeəʃeɪ/ *adjective* carefully sought out; far-fetched.

recidivist /rɪ'sɪdɪvɪst/ *noun* person who relapses into crime. □ **recidivism** *noun.*

recipe /'resɪpɪ/ *noun* statement of ingredients and procedure for preparing dish etc.; (+ *for*) certain means to.

recipient /rɪ'sɪpɪənt/ *noun* person who receives something.

reciprocal /rɪ'sɪprək(ə)l/ ● *adjective* in return; mutual; *Grammar* expressing mutual relation. ● *noun Mathematics* function or expression so related to another that their product is unity. □ **reciprocally** *adverb.*

reciprocate /rɪ'sɪprəkeɪt/ *verb* (**-ting**) requite, return; (+ *with*) give in return; interchange; (of machine part) move backwards and forwards. □ **reciprocation** *noun.*

reciprocity /resɪ'prɒsɪtɪ/ *noun* condition of being reciprocal; mutual action; give and take.

recital /rɪ'saɪt(ə)l/ *noun* reciting, being recited; concert of classical music by soloist or small group; (+ *of*) detailed account of (facts etc.); narrative.

recitation /resɪ'teɪʃ(ə)n/ *noun* reciting; piece recited.

recitative /resɪtə'ti:v/ *noun* passage of singing in speech rhythm, esp. in narrative or dialogue section of opera or oratorio.

recite /rɪ'saɪt/ *verb* (**-ting**) repeat aloud or declaim from memory; enumerate.

reckless /'reklɪs/ *adjective* disregarding consequences or danger etc. □ **recklessly** *adverb*; **recklessness** *noun.*

reckon /'rekən/ *verb* (often + *that*) think, consider; count or compute by calculation; (+ *on*) rely or base plans on; (+ *with*, *without*) take (or fail to take) into account.

reckoning *noun* calculating; opinion; settlement of account.

reclaim /rɪ'kleɪm/ *verb* seek return of (one's property etc.); bring (land) under cultivation from sea etc.; win back from vice, error, or waste condition. □ **reclaimable** *adjective*; **reclamation** /reklə-/ *noun.*

recline /rɪ'klaɪn/ *verb* (**-ning**) assume or be in horizontal or leaning position.

recluse /rɪ'klu:s/ *noun* person given to or living in seclusion. □ **reclusive** *adjective.*

recognition /rekəg'nɪʃ(ə)n/ *noun* recognizing, being recognized.

recognizance /rɪ'kɒgnɪz(ə)ns/ *noun Law* bond by which person undertakes to observe some condition; sum pledged as surety for this.

recognize /ˈrekəgnaɪz/ *verb* (also **-ise**) (**-zing** or **-sing**) identify as already known; realize or discover nature of; (+ *that*) realize or admit; acknowledge existence, validity, character, or claims of; show appreciation of; reward. □ **recognizable** *adjective*.

recoil ● *verb* /rɪˈkɔɪl/ jerk or spring back in horror, disgust, or fear; shrink mentally in this way; rebound; (of gun) be driven backwards by discharge. ● *noun* /ˈri:kɔɪl/ act or sensation of recoiling.

recollect /rekəˈlekt/ *verb* remember; call to mind.

recollection /rekəˈlekʃ(ə)n/ *noun* act or power of recollecting; thing recollected; person's memory, time over which it extends.

recommend /rekəˈmend/ *verb* suggest as fit for purpose or use; advise (course of action etc.); (of qualities etc.) make acceptable or desirable; (+ *to*) commend or entrust. □ **recommendation** *noun*.

recompense /ˈrekəmpens/ ● *verb* (**-sing**) make amends to; compensate; reward or punish. ● *noun* reward; compensation; retribution.

reconcile /ˈrekənsaɪl/ *verb* (**-ling**) make friendly again after estrangement; (usually **reconcile oneself** or in *passive*; + *to*) make resigned to; settle (quarrel etc.); harmonize; make compatible; show compatibility of. □ **reconcilable** /-ˈsaɪl-/ *adjective*; **reconciliation** /-sɪlɪ-/ *noun*.

recondite /ˈrekəndaɪt/ *adjective* abstruse; obscure.

recondition /ri:kənˈdɪʃ(ə)n/ *verb* overhaul, renovate, make usable again.

reconnaissance /rɪˈkɒnɪs(ə)ns/ *noun* survey of region to locate enemy or ascertain strategic features; preliminary survey.

reconnoitre /rekəˈnɔɪtə/ *verb* (*US* **reconnoiter**) (**-ring**) make reconnaissance (of).

reconsider /ri:kənˈsɪdə/ *verb* consider again, esp. for possible change of decision.

reconstitute /ri:ˈkɒnstɪtju:t/ *verb* (**-ting**) reconstruct; reorganize; rehydrate (dried food etc.). □ **reconstitution** /-ˈtju:ʃ(ə)n/ *noun*.

reconstruct /ri:kənˈstrʌkt/ *verb* build again; form impression of (past events) by assembling evidence; re-enact (crime); reorganize. □ **reconstruction** *noun*.

record ● *noun* /ˈrekɔ:d/ evidence etc. constituting account of occurrence, statement, etc.; document etc. preserving this; (in full **gramophone record**) disc carrying recorded sound in grooves, for reproduction by record player; facts known about person's past, esp. criminal convictions; best performance or most remarkable event of its kind. ● *verb* /rɪˈkɔ:d/ put in writing or other permanent form for later reference; convert (sound etc.) into permanent form for later reproduction. □ **have a record** have criminal conviction; **off the record** unofficially, confidentially; **on record** officially recorded, publicly known; **recorded delivery** Post Office service in which dispatch and receipt are recorded; **record player** apparatus for reproducing sounds from gramophone records.

recorder /rɪˈkɔ:də/ *noun* apparatus for recording; woodwind instrument; (also **Recorder**) barrister or solicitor serving as part-time judge.

recording /rɪˈkɔ:dɪŋ/ *noun* process of recording sound etc. for later reproduction; material or programme recorded.

recordist /rɪˈkɔ:dɪst/ *noun* person who records sound.

recount /rɪˈkaʊnt/ *verb* narrate; tell in detail.

re-count ● *verb* /ri:ˈkaʊnt/ count again. ● *noun* /ˈri:kaʊnt/ re-counting, esp. of votes.

recoup /rɪˈku:p/ *verb* recover or regain (loss); compensate or reimburse for loss.

recourse /rɪˈkɔ:s/ *noun* resort to possible source of help; person or thing resorted to.

recover /rɪˈkʌvə/ *verb* regain possession, use, or control of; return to health, consciousness, or normal state or position; secure by legal process; make up for; retrieve. □ **recoverable** *adjective*.

re-cover /ri:ˈkʌvə/ *verb* cover again; provide (chairs etc.) with new cover.

recovery *noun* (*plural* **-ies**) recovering, being recovered.

recreant /'rekrɪənt/ *literary* ● *adjective* cowardly. ● *noun* coward.

re-create /ri:krɪ'eɪt/ *verb* (**-ting**) create anew; reproduce. □ **re-creation** *noun*.

recreation /rekrɪ'eɪʃ(ə)n/ *noun* (means of) entertaining oneself; pleasurable activity. □ **recreation ground** public land for sports etc. □ **recreational** *adjective*.

recriminate /rɪ'krɪmɪneɪt/ *verb* (**-ting**) make mutual or counter accusations. □ **recrimination** *noun*; **recriminatory** *adjective*.

recrudesce /ri:kru:'des/ *verb* (**-cing**) *formal* (of disease, problem, etc.) break out again. □ **recrudescence** *noun*; **recrudescent** *adjective*.

recruit /rɪ'kru:t/ ● *noun* newly enlisted serviceman or servicewoman; new member of a society etc. ● *verb* enlist (person) as recruit; form (army etc.) by enlisting recruits; replenish, reinvigorate. □ **recruitment** *noun*.

rectal /'rekt(ə)l/ *adjective* of or by means of rectum.

rectangle /'rektæŋg(ə)l/ *noun* plane figure with 4 straight sides and 4 right angles. □ **rectangular** /-'tæŋgjʊlə/ *adjective*.

rectify /'rektɪfaɪ/ *verb* (**-ies**, **-ied**) adjust or make right; purify, esp. by repeated distillation; convert (alternating current) to direct current. □ **rectifiable** *adjective*; **rectification** *noun*.

rectilinear /rektɪ'lɪnɪə/ *adjective* bounded or characterized by straight lines; in or forming straight line.

rectitude /'rektɪtju:d/ *noun* moral uprightness.

recto /'rektəʊ/ *noun* (*plural* **-s**) right-hand page of open book; front of printed leaf.

rector /'rektə/ *noun* incumbent of C. of E. parish where in former times all tithes passed to incumbent; head priest of church or religious institution; head of university or college. □ **rectorship** *noun*.

rectory *noun* (*plural* **-ies**) rector's house.

rectum /'rektəm/ *noun* (*plural* **-s**) final section of large intestine.

recumbent /rɪ'kʌmbənt/ *adjective* lying down, reclining.

recuperate /rɪ'ku:pəreɪt/ *verb* (**-ting**) recover from illness, exhaustion, loss, etc.; regain (health, loss, etc.). □ **recuperation** *noun*; **recuperative** /-rətɪv/ *adjective*.

recur /rɪ'kɜ:/ *verb* (**-rr-**) occur again or repeatedly; (+ *to*) go back to in thought or speech; (as **recurring** *adjective*) (of decimal fraction) with same figure(s) repeated indefinitely.

recurrent /rɪ'kʌrənt/ *adjective* recurring. □ **recurrence** *noun*.

recusant /'rekjʊz(ə)nt/ *noun* person refusing submission or compliance, esp. (*historical*) one who refused to attend services of the Church of England. □ **recusancy** *noun*.

recycle /ri:'saɪk(ə)l/ *verb* (**-ling**) convert (waste) to reusable material. □ **recyclable** *adjective*.

red ● *adjective* (**-dd-**) of colour from that of blood to deep pink or orange; flushed; bloodshot; (of hair) reddish-brown; having to do with bloodshed, burning, violence, or revolution; *colloquial* Communist. ● *noun* red colour, paint, clothes, etc.; *colloquial* Communist. □ **in the red** in debt or deficit; **red admiral** butterfly with red bands; **red-blooded** virile, vigorous; **redbrick** (of university) founded in the 19th or early 20th c.; **red card** *Football* card shown by referee to player being sent off; **red carpet** privileged treatment of eminent visitor; **redcoat** *historical* British soldier; **Red Crescent** equivalent of Red Cross in Muslim countries; **Red Cross** international relief organization; **redcurrant** small red edible berry, shrub bearing it; **red flag** symbol of revolution, danger signal; **red-handed** in act of crime; **redhead** person with red hair; **red herring** irrelevant diversion; **red-hot** heated until red, *colloquial* highly exciting or excited, *colloquial* (of news) completely new; **red lead** red oxide of lead as pigment; **red-letter day** joyfully noteworthy or memorable day; **red light** stop signal, warning; **red meat** meat that is red when raw (e.g. beef); **red neck** conservative working-class white in southern US; **red pepper** cayenne pepper, red fruit of capsicum; **red rag** thing that excites rage; **redshank** sandpiper with bright red legs; **red shift** displacement of spectrum to longer wavelengths in light from receding galaxies; **redstart** red-tailed songbird; **red tape** excessive bureaucracy or formality; **redwing** thrush with red underwings; **redwood** tree with red wood. □ **reddish** *adjective*; **redness** *noun*.

redden *verb* make or become red; blush.

redeem /rɪ'di:m/ *verb* recover by expenditure of effort; make single payment to cancel (regular charge etc.); convert (tokens or bonds) into goods or cash; deliver from sin and damnation; (often as **redeeming** *adjective*) make amends or compensate for; save, rescue, reclaim; fulfil (promise). □ **redeemable** *adjective*.

redeemer *noun* one who redeems, esp. Christ.

redemption /rɪ'dempʃ(ə)n/ *noun* redeeming, being redeemed.

redeploy /ri:dɪ'plɔɪ/ *verb* send (troops, workers, etc.) to new place or task. □ **redeployment** *noun*.

rediffusion /ri:dɪ'fju:ʒ(ə)n/ *noun* relaying of broadcast programmes, esp. by cable from central receiver.

redolent /'redələnt/ *adjective* (+ *of*, *with*) strongly smelling or suggestive of. □ **redolence** *noun*.

redouble /ri:'dʌb(ə)l/ *verb* (**-ling**) make or grow greater or more intense or numerous; double again.

redoubt /rɪ'daʊt/ *noun Military* outwork or fieldwork without flanking defences.

redoubtable /rɪ'daʊtəb(ə)l/ *adjective* formidable.

redound /rɪ'daʊnd/ *verb* (+ *to*) make great contribution to (one's advantage etc.); (+ *upon*, *on*) come back or recoil upon.

redress /rɪ'dres/ ● *verb* remedy; put right again. ● *noun* reparation; (+ *of*) redressing.

reduce /rɪ'dju:s/ *verb* (**-cing**) make or become smaller or less; (+ *to*) bring by force or necessity; convert to another (esp. simpler) form; bring lower in status, rank, or price; lessen one's weight or size; make (sauce etc.) more concentrated by boiling; weaken; impoverish; subdue. □ **reduced circumstances** poverty after relative properity. □ **reducible** *adjective*.

reduction /rɪ'dʌkʃ(ə)n/ *noun* reducing, being reduced; amount by which prices etc. are reduced; smaller copy of picture etc. □ **reductive** *adjective*.

redundant /rɪ'dʌnd(ə)nt/ *adjective* superfluous; that can be omitted without loss of significance; no longer needed at work and therefore unemployed. □ **redundancy** *noun* (*plural* **-ies**).

reduplicate /rɪ'dju:plɪkeɪt/ *verb* (**-ting**) make double; repeat. □ **reduplication** *noun*.

re-echo /ri:'ekəʊ/ *verb* (**-es**, **-ed**) echo repeatedly; resound.

reed *noun* firm-stemmed water or marsh plant; tall straight stalk of this; vibrating part of some wind instruments. □ **reedy** *adjective* (**-ier**, **-iest**).

reef¹ *noun* ridge of rock or coral etc. at or near surface of sea; lode of ore, bedrock surrounding this.

reef² ● *noun* each of several strips across sail, for taking it in etc. ● *verb* take in reef(s) of (sail). □ **reef-knot** symmetrical double knot.

reefer *noun slang* marijuana cigarette; thick double-breasted jacket.

reek ● *verb* (often + *of*) smell unpleasantly; have suspicious associations. ● *noun* foul or stale smell; *esp. Scottish* smoke; vapour, exhalation.

reel ● *noun* cylindrical device on which thread, paper, film, wire, etc. are wound; device for winding and unwinding line as required, esp. in fishing; lively folk or Scottish dance, music for this. ● *verb* wind on reel; (+ *in*, *up*) draw in or up with reel; stand, walk, etc. unsteadily; be shaken physically or mentally; dance reel. □ **reel off** recite rapidly and without apparent effort.

re-entrant /ri:'entrənt/ *adjective* (of angle) pointing inwards.

re-entry /ri:'entrɪ/ *noun* (*plural* **-ies**) act of entering again, esp. (of spacecraft etc.) of re-entering earth's atmosphere.

reeve *noun historical* chief magistrate of town or district; official supervising landowner's estate.

ref¹ *noun colloquial* referee.

ref² *noun colloquial* reference.

refectory /rɪ'fektərɪ/ *noun* (*plural* **-ies**) dining-room, esp. in monastery or college. □ **refectory table** long narrow table.

refer /rɪ'fɜ:/ *verb* (**-rr-**) (usually + *to*) have recourse to (some authority or source of information); send on or direct; make allusion or be relevant. □ **referred pain** pain felt in part of body other than actual source. □ **referable** *adjective*.

referee /refə'ri:/ ● *noun* umpire, esp. in football or boxing; person referred to for decision in dispute etc.; person willing to testify to character of applicant for employment etc. ● *verb* (**-rees**, **-reed**) act as referee (for).

reference /'refərəns/ *noun* referring to some authority; scope given to such authority; (+ *to*) relation, respect, or allusion to; direction to page, book, etc. for information; written testimonial, person giving it. □ **reference book** book for occasional consultation. □ **referential** /-'ren-/ *adjective*.

referendum /refə'rendəm/ *noun* (*plural* **-s** or **-da**) vote on political question open to entire electorate.

referral /rɪ'fɜ:r(ə)l/ *noun* referring of person to medical specialist etc.

refill ● *verb* /ri:'fɪl/ fill again. ● *noun* /'ri:fɪl/ thing that refills; act of refilling. □ **refillable** *adjective*.

refine /rɪ'faɪn/ *verb* (**-ning**) free from impurities or defects; (esp. as **refined** *adjective*) make or become more elegant or cultured.

refinement *noun* refining, being refined; fineness of feeling or taste; elegance; added development or improvement; subtle reasoning; fine distinction.

refiner *noun* person or firm refining crude oil, metal, sugar, etc.

refinery *noun* (*plural* **-ies**) place where oil etc. is refined.

refit ● *verb* /ri:'fɪt/ (**-tt-**) *esp. Nautical* make or become serviceable again by repairs etc. ● *noun* /'ri:fɪt/ refitting.

reflate /ri:'fleɪt/ *verb* (**-ting**) cause reflation of (currency, economy, etc.).

reflation *noun* inflation of financial system to restore previous condition after deflation. □ **reflationary** *adjective*.

reflect /rɪ'flekt/ *verb* throw back (light, heat, sound, etc.); (of mirror etc.) show image of, reproduce to eye or mind; correspond in appearance or effect to; bring (credit, discredit, etc.); (usually + *on, upon*) bring discredit; (often + *on, upon*) meditate; (+ *that, how*, etc.) consider.

reflection /rɪ'flekʃ(ə)n/ *noun* reflecting, being reflected; reflected light, heat, colour, or image; reconsideration; (often + *on*) thing bringing discredit; (often + *on, upon*) comment.

reflective *adjective* (of surface) reflecting; (of mental faculties) concerned in reflection or thought; thoughtful. □ **reflectively** *adverb*.

reflector *noun* piece of glass or metal for reflecting light in required direction; telescope etc. using mirror to produce images.

reflex /'ri:fleks/ ● *adjective* (of action) independent of will; (of angle) larger than 180°. ● *noun* reflex action; sign, secondary manifestation; reflected light or image. □ **reflex camera** camera in which image is reflected by mirror to enable correct focusing.

reflexive /rɪ'fleksɪv/ *Grammar* ● *adjective* (of word or form) referring back to subject (e.g. *myself* in *I hurt myself*); (of verb) having reflexive pronoun as object. ● *noun* reflexive word or form.

reflexology /ri:flek'sɒlədʒɪ/ *noun* massage to areas of soles of feet. □ **reflexologist** *noun*.

reform /rɪ'fɔ:m/ ● *verb* make or become better; abolish or cure (abuse etc.). ● *noun* removal of abuses, esp. political; improvement. □ **reformative** *adjective*.

reformation /refə'meɪʃ(ə)n/ *noun* reforming or being reformed, esp. radical improvement in political, religious, or social affairs; (**the Reformation**) 16th-c. movement for reform of abuses in Roman Church ending in establishment of Reformed or Protestant Churches.

reformatory /rɪ'fɔ:mətərɪ/ ● *noun* (*plural* **-ies**) *US historical* institution for reform of young offenders. ● *adjective* producing reform.

reformer *noun* advocate of reform.

reformism *noun* policy of reform rather than abolition or revolution. □ **reformist** *noun & adjective.*

refract /rɪ'frækt/ *verb* deflect (light) at certain angle when it enters obliquely from another medium. □ **refraction** *noun*; **refractive** *adjective.*

refractor *noun* refracting medium or lens; telescope using lens to produce image.

refractory /rɪ'fræktərɪ/ *adjective* stubborn, unmanageable, rebellious; resistant to treatment; hard to fuse or work.

refrain¹ /rɪ'freɪn/ *verb* (+ *from*) avoid doing (action).

refrain² /rɪ'freɪn/ *noun* recurring phrase or lines, esp. at ends of stanzas.

refresh /rɪ'freʃ/ *verb* give fresh spirit or vigour to; revive (memory). □ **refreshing** *adjective*; **refreshingly** *adverb.*

refresher *noun* something that refreshes, esp. drink; extra fee to counsel in prolonged lawsuit. □ **refresher course** course reviewing or updating previous studies.

refreshment *noun* refreshing, being refreshed; (usually in *plural*) food or drink.

refrigerant /rɪ'frɪdʒərənt/ ● *noun* substance used for refrigeration. ● *adjective* cooling.

refrigerate /rɪ'frɪdʒəreɪt/ *verb* (**-ting**) cool or freeze (esp. food). □ **refrigeration** *noun.*

refrigerator *noun* cabinet or room in which food etc. is refrigerated.

refuge /'refju:dʒ/ *noun* shelter from pursuit, danger, or trouble; person or place offering this.

refugee /refjʊ'dʒi:/ *noun* person taking refuge, esp. in foreign country, from war, persecution, etc.

refulgent /rɪ'fʌldʒ(ə)nt/ *adjective literary* shining, gloriously bright. □ **refulgence** *noun.*

refund ● *verb* /rɪ'fʌnd/ pay back (money etc.); reimburse. ● *noun* /'ri:fʌnd/ act of refunding; sum refunded. □ **refundable** /rɪ'fʌn-/ *adjective.*

refurbish /ri:'fɜ:bɪʃ/ *verb* brighten up, redecorate. □ **refurbishment** *noun.*

refusal /rɪ'fju:z(ə)l/ *noun* refusing, being refused; (in full **first refusal**) chance of taking thing before it is offered to others.

refuse¹ /rɪ'fju:z/ *verb* (**-sing**) withhold acceptance of or consent to; (often + *to do*) indicate unwillingness; not grant (request) made by (person); (of horse) be unwilling to jump (fence etc.).

refuse² /'refju:s/ *noun* items rejected as worthless; waste.

refusenik /rɪ'fju:znɪk/ *noun historical* Soviet Jew refused permission to emigrate to Israel.

refute /rɪ'fju:t/ *verb* (**-ting**) prove falsity or error of; rebut by argument; deny or contradict (without argument). □ **refutation** /refjʊ'teɪʃ(ə)n/ *noun.*

■ **Usage** The use of *refute* to mean 'deny, contradict' is considered incorrect by some people. *Repudiate* can be used instead.

reg /redʒ/ *noun colloquial* registration mark.

regain /rɪ'geɪn/ *verb* obtain possession or use of after loss.

regal /'ri:g(ə)l/ *adjective* of or by monarch(s); magnificent. □ **regality** /rɪ'gæl-/ *noun*; **regally** *adverb.*

regale /rɪ'geɪl/ *verb* (**-ling**) entertain lavishly with feasting; (+ *with*) entertain with (talk etc.).

regalia /rɪ'geɪlɪə/ *plural noun* insignia of royalty or an order, mayor, etc.

regard /rɪ'gɑ:d/ ● *verb* gaze on; heed, take into account; look upon or think of in specified way. ● *noun* gaze; steady look; attention, care; esteem; (in *plural*) *expression of friendliness in letter etc.* □ **as regards** about, in respect of; **in this regard** on this point; **in**, **with regard to** in respect of.

regardful *adjective* (+ *of*) mindful of.

regarding *preposition* concerning; in respect of.

regardless ● *adjective* (+ *of*) without regard or consideration for. ● *adverb* without paying attention.

regatta /rɪ'gætə/ *noun* event consisting of rowing or yacht races.

regency /'ri:dʒənsɪ/ *noun* (*plural* **-ies**) office of regent; commission acting as regent; regent's or regency commission's period of office; (**Regency**) (in UK) 1811–1820.

regenerate ● *verb* /rɪ'dʒenəreɪt/ (**-ting**) bring or come into renewed existence;

improve moral condition of; impart new, more vigorous, or spiritually higher life or nature to; regrow or cause (new tissue) to regrow. ● *adjective* /-rət/ spiritually born again, reformed. □ **regeneration** *noun*; **regenerative** /-rətɪv/ *adjective*.

regent /'ri:dʒ(ə)nt/ ● *noun* person acting as head of state because monarch is absent, ill, or a child. ● *adjective* (after noun) acting as regent.

reggae /'regeɪ/ *noun* W. Indian style of music with strongly accented subsidiary beat.

regicide /'redʒɪsaɪd/ *noun* person who kills or helps to kill a king; killing of a king.

regime /reɪ'ʒi:m/ *noun* (also **régime**) method of government; prevailing system; regimen.

regimen /'redʒɪmən/ *noun* prescribed course of exercise, way of life, and diet.

regiment ● *noun* /'redʒɪmənt/ permanent unit of army consisting of several companies, troops, or batteries; (usually + *of*) large or formidable array or number. ● *verb* /-ment/ organize in groups or according to system; form into regiment(s). □ **regimentation** /-men-/ *noun*.

regimental /redʒɪ'ment(ə)l/ ● *adjective* of a regiment. ● *noun* (in *plural*) military uniform, esp. of particular regiment.

Regina /rɪ'dʒaɪnə/ *noun* (after name) reigning queen; *Law* the Crown. [Latin]

region /'ri:dʒ(ə)n/ *noun* geographical area or division, having definable boundaries or characteristics; administrative area, esp. in Scotland; part of body; sphere, realm. □ **in the region of** approximately. □ **regional** *adjective*; **regionally** *adverb*.

register /'redʒɪstə/ ● *noun* official list; book in which items are recorded for reference; device recording speed, force, etc.; compass of voice or instrument; form of language used in particular circumstances; adjustable plate for regulating draught etc. ● *verb* set down formally, record in writing; enter or cause to be entered in register; send (letter) by registered post; record automatically, indicate; make mental note of; show (emotion etc.) in face etc.; make impression. □ **registered post** postal procedure with special precautions and compensation in case of loss; **register office** state office where civil marriages are conducted and births, marriages, and deaths are recorded.

registrar /redʒɪ'strɑ:/ *noun* official keeping register; chief administrator in university etc.; hospital doctor training as specialist.

registration /redʒɪ'streɪʃ(ə)n/ *noun* registering, being registered. □ **registration mark**, **number** combination of letters and numbers identifying vehicle.

registry /'redʒɪstrɪ/ *noun* (*plural* **-ies**) place where registers or records are kept. □ **registry office** register office (the official name).

Regius professor /'ri:dʒɪəs/ *noun* holder of university chair founded by sovereign or filled by Crown appointment.

regress ● *verb* /rɪ'gres/ move backwards; return to former stage or state. ● *noun* /'ri:gres/ act of regressing. □ **regression** /rɪ'greʃ(ə)n/ *noun*; **regressive** /rɪ'gresɪv/ *adjective*.

regret /rɪ'gret/ ● *verb* (**-tt-**) feel or express sorrow, repentance, or distress over (action or loss); say with sorrow or remorse. ● *noun* sorrow, repentance, or distress over action or loss. □ **regretful** *adjective*; **regretfully** *adverb*.

regrettable *adjective* undesirable, unwelcome; deserving censure. □ **regrettably** *adverb*.

regular /'regjʊlə/ ● *adjective* acting, done, or recurring uniformly; habitual, orderly; conforming to rule or principle; symmetrical; conforming to correct procedure etc.; *Grammar* (of verb etc.) following normal type of inflection; *colloquial* absolute, thorough; (of soldier etc.) permanent, professional. ● *noun* regular soldier; *colloquial* regular customer, visitor, etc.; one of regular clergy. □ **regularity** /-'lærɪtɪ/ *noun*; **regularize** *verb* (also **-ise**) (**-zing** or **-sing**); **regularly** *adverb*.

regulate /'regjʊleɪt/ *verb* (**-ting**) control by rule, subject to restrictions; adapt to requirements; adjust (clock, watch, etc.) to work accurately. □ **regulator** *noun*.

regulation ● *noun* regulating, being regulated; prescribed rule. ● *adjective* in accordance with regulations, of correct pattern etc.

regulo /'regjʊləʊ/ *noun* (usually + numeral) number on scale denoting temperature in gas oven.

regurgitate /rɪ'gɜːdʒɪteɪt/ *verb* (**-ting**) bring (swallowed food) up again to mouth; reproduce (information etc.). □ **regurgitation** *noun*.

rehabilitate /riːhə'bɪlɪteɪt/ *verb* (**-ting**) restore to normal life by training, etc. esp. after imprisonment or illness; restore to former privileges or reputation or to proper condition. □ **rehabilitation** *noun*.

rehash ● *verb* /riː'hæʃ/ put into new form without significant change or improvement. ● *noun* /'riːhæʃ/ material rehashed; rehashing.

rehearsal /rɪ'hɜːs(ə)l/ *noun* trial performance or practice; rehearsing.

rehearse /rɪ'hɜːs/ *verb* (**-sing**) practise before performing in public; recite or say over; give list of, enumerate.

Reich /raɪx/ *noun* former German state, esp. Third Reich (1933–45).

reign /reɪn/ ● *verb* be king or queen; prevail; (as **reigning** *adjective*) currently holding title. ● *noun* sovereignty, rule; sovereign's period of rule.

reimburse /riːɪm'bɜːs/ *verb* (**-sing**) repay (person); refund. □ **reimbursement** *noun*.

rein /reɪn/ ● *noun* (in *singular* or *plural*) long narrow strap used to guide horse; means of control. ● *verb* (+ *back, up, in*) pull back or up or hold in (as) with reins; govern, control.

reincarnate ● *verb* /riːɪn'kɑːneɪt/ (**-ting**) give esp. human form to again. ● *adjective* /-nət/ reincarnated.

reincarnation /riːɪnkɑː'neɪʃ(ə)n/ *noun* rebirth of soul in new body.

reindeer /'reɪndɪə/ *noun* (*plural* same or **-s**) subarctic deer with large antlers.

reinforce /riːɪn'fɔːs/ *verb* (**-cing**) support or strengthen, esp. with additional personnel or material. □ **reinforced concrete** concrete with metal bars etc. embedded in it.

reinforcement *noun* reinforcing, being reinforced; (in *plural*) additional personnel, equipment, etc.

reinstate /riːɪn'steɪt/ *verb* (**-ting**) replace in former position; restore to former privileges. □ **reinstatement** *noun*.

reinsure /riːɪn'ʃʊə/ *verb* (**-ring**) insure again (esp. of insurer transferring risk to another insurer). □ **reinsurance** *noun*.

reiterate /riː'ɪtəreɪt/ *verb* (**-ting**) say or do again or repeatedly. □ **reiteration** *noun*.

reject ● *verb* /rɪ'dʒekt/ put aside or send back as not to be used, done, or complied with; refuse to accept or believe in; rebuff. ● *noun* /'riːdʒekt/ rejected thing or person. □ **rejection** /rɪ'dʒek-/ *noun*.

rejoice /rɪ'dʒɔɪs/ *verb* (**-cing**) feel joy, be glad; (+ *in, at*) take delight in.

rejoin[1] /riː'dʒɔɪn/ *verb* join again; reunite.

rejoin[2] /rɪ'dʒɔɪn/ *verb* say in answer; retort.

rejoinder /rɪ'dʒɔɪndə/ *noun* reply, retort.

rejuvenate /rɪ'dʒuːvəneɪt/ *verb* (**-ting**) make (as if) young again. □ **rejuvenation** *noun*.

relapse /rɪ'læps/ ● *verb* (**-sing**) (usually + *into*) fall back (into worse state after improvement). ● *noun* relapsing, esp. deterioration in patient's condition after partial recovery.

relate /rɪ'leɪt/ *verb* (**-ting**) narrate, recount; (usually + *to, with*) connect in thought or meaning; have reference to; (+ *to*) feel connected or sympathetic to.

related *adjective* connected, esp. by blood or marriage.

relation /rɪ'leɪʃ(ə)n/ *noun* connection between people or things; relative; (in *plural*) dealings (with others); narration.

relationship *noun* state of being related; connection, association; *colloquial* emotional association between two people.

relative /'relətɪv/ ● *adjective* in relation or proportion to something else; implying comparison or relation; (+ *to*) having application or reference to; *Grammar* (of word, clause, etc.) referring to expressed or implied antecedent, attached to antecedent by such word. ● *noun* person connected by blood or marriage; species related to another by common origin; relative word, esp. pronoun. □ **relative density** ratio between density of substance and that of a standard (usually water or air). □ **relatively** *adverb.*

relativity /relə'tɪvɪtɪ/ *noun* relativeness; *Physics* theory based on principle that all motion is relative and that light has constant velocity in a vacuum.

relax /rɪ'læks/ *verb* make or become less stiff, rigid, tense, formal, or strict; reduce (attention, efforts); cease work or effort; (as **relaxed** *adjective*) at ease, unperturbed.

relaxation /ri:læk'seɪʃ(ə)n/ *noun* relaxing; recreation.

relay /'ri:leɪ/ ● *noun* fresh set of people etc. to replace tired ones; supply of material similarly used; relay race; device activating electric circuit; device transmitting broadcast; relayed transmission. ● *verb* /also rɪ'leɪ/ receive (esp. broadcast message) and transmit to others. □ **relay race** one between teams of which each member in turn covers part of distance.

release /rɪ'li:s/ ● *verb* **(-sing)** (often + *from*) set free, liberate, unfasten; allow to move from fixed position; make (information) public; issue (film etc.) generally. ● *noun* liberation from restriction, duty, or difficulty; handle, catch, etc. that releases part of mechanism; item made available for publication; film, record, etc. that is released; releasing of film etc.

relegate /'relɪgeɪt/ *verb* **(-ting)** consign or dismiss to inferior position; transfer (team) to lower division of league. □ **relegation** *noun.*

relent /rɪ'lent/ *verb* relax severity; yield to compassion.

relentless *adjective* unrelenting. □ **relentlessly** *adverb.*

relevant /'relɪv(ə)nt/ *adjective* (often + *to*) bearing on or pertinent to matter in hand. □ **relevance** *noun.*

reliable /rɪ'laɪəb(ə)l/ *adjective* that may be relied on. □ **reliability** *noun*; **reliably** *adverb.*

reliance /rɪ'laɪəns/ *noun* (+ *in*, *on*) trust or confidence in. □ **reliant** *adjective.*

relic /'relɪk/ *noun* object interesting because of its age or associations; part of holy person's body or belongings kept as object of reverence; surviving custom, belief, etc. from past age; (in *plural*) dead body or remains of person, what has survived.

relict /'relɪkt/ *noun* object surviving in primitive form.

relief /rɪ'li:f/ *noun* (feeling accompanying) alleviation of or deliverance from pain, distress, etc.; feature etc. that diversifies monotony or relaxes tension; assistance given to people in special need; replacing of person(s) on duty by another or others; person(s) thus bringing relief; thing supplementing another in some service; method of carving, moulding, etc., in which design projects from surface; piece of sculpture etc. in relief; effect of being done in relief given by colour or shading etc.; vividness, distinctness. □ **relief map** one showing hills and valleys by shading or colouring etc.

relieve /rɪ'li:v/ *verb* **(-ving)** bring or give relief to; mitigate tedium of; release (person) from duty by acting as or providing substitute; (+ *of*) take (burden or duty) away from; (**relieve oneself**) urinate, defecate. □ **relieved** *adjective.*

religion /rɪ'lɪdʒ(ə)n/ *noun* belief in superhuman controlling power, esp. in personal God or gods entitled to obedience; system of faith and worship.

religiosity /rɪlɪdʒɪ'ɒsɪtɪ/ *noun* state of being religious or too religious.

religious /rɪ'lɪdʒəs/ ● *adjective* devoted to religion, devout; of or concerned with religion; of or belonging to monastic order; scrupulous. ● *noun* (*plural* same) person bound by monastic vows. □ **religiously** *adverb.*

relinquish /rɪ'lɪŋkwɪʃ/ *verb* give up, let go, resign, surrender. □ **relinquishment** *noun.*

reliquary /'relɪkwərɪ/ *noun* (*plural* **-ies**) receptacle for relic(s).

relish /'relɪʃ/ ● *noun* (often + *for*) liking or enjoyment; appetizing flavour, attractive quality; thing eaten with plainer food to add flavour; (+ *of*) distinctive flavour or taste. ● *verb* get pleasure out of, enjoy greatly; anticipate with pleasure.

reluctant /rɪ'lʌkt(ə)nt/ *adjective* (often + *to do*) unwilling, disinclined. □ **reluctance** *noun*; **reluctantly** *adverb*.

rely /rɪ'laɪ/ *verb* (**-ies**, **-ied**) (+ *on*, *upon*) depend with confidence on; be dependent on.

REM *abbreviation* rapid eye movement.

remade *past & past participle* of REMAKE.

remain /rɪ'meɪn/ *verb* be left over; stay in same place or condition; be left behind; continue to be.

remainder /rɪ'meɪndə/ ● *noun* residue; remaining people or things; number left after subtraction or division; (any of) copies of book left unsold. ● *verb* dispose of remainder of (book) at reduced prices.

remains /rɪ'meɪnz/ *plural noun* what remains after other parts have been removed or used; relics of antiquity etc.; dead body.

remake ● *verb* /ri:'meɪk/ (**-king**; *past & past participle* **remade**) make again or differently. ● *noun* /'ri:meɪk/ remade thing, esp. cinema film.

remand /rɪ'mɑ:nd/ ● *verb* return (prisoner) to custody, esp. to allow further inquiry. ● *noun* recommittal to custody. □ **on remand** in custody pending trial; **remand centre** institution for remand of accused people.

remark /rɪ'mɑ:k/ ● *verb* (often + *that*) say by way of comment; (usually + *on*, *upon*) make comment; *archaic* take notice of. ● *noun* comment, thing said; noticing.

remarkable *adjective* worth notice; exceptional, striking. □ **remarkably** *adverb*.

REME /'ri:mi:/ *abbreviation* Royal Electrical and Mechanical Engineers.

remedial /rɪ'mi:dɪəl/ *adjective* affording or intended as a remedy; (of teaching) for slow or disadvantaged pupils.

remedy /'remɪdɪ/ ● *noun* (*plural* **-ies**) (often + *for*, *against*) medicine or treatment; means of removing anything undesirable; redress. ● *verb* (**-ies**, **-ied**) rectify, make good. □ **remediable** /rɪ'mi:dɪəb(ə)l/ *adjective*.

remember /rɪ'membə/ *verb* (often + *to do*, *that*) keep in the memory; not forget; bring back into one's thoughts; acknowledge in making gift etc.; convey greetings from.

remembrance /rɪ'membrəns/ *noun* remembering, being remembered; recollection; keepsake, souvenir; (in *plural*) greetings conveyed through third person.

remind /rɪ'maɪnd/ *verb* (usually + *of*, *to do*, *that*) cause (person) to remember or think of.

reminder *noun* thing that reminds; (often + *of*) memento.

reminisce /remɪ'nɪs/ *verb* (**-cing**) indulge in reminiscence.

reminiscence /remɪ'nɪs(ə)ns/ *noun* remembering things past; (in *plural*) *literary* account of things remembered.

reminiscent *adjective* (+ *of*) reminding or suggestive of; concerned with reminiscence.

remiss /rɪ'mɪs/ *adjective* careless of duty; negligent.

remission /rɪ'mɪʃ(ə)n/ *noun* reduction of prison sentence for good behaviour; remittance of debt etc.; diminution of force etc.; (often + *of*) forgiveness (of sins etc.).

remit ● *verb* /rɪ'mɪt/ (**-tt-**) refrain from exacting or inflicting (debt, punishment, etc.); abate, slacken; send (esp. money); (+ *to*) refer to some authority, send back to lower court; postpone, defer; pardon (sins etc.). ● *noun* /'ri:mɪt/ terms of reference of committee etc.; item remitted.

remittance *noun* money sent; sending of money.

remittent *adjective* (of disease etc.) abating at intervals.

remix ● *verb* /ri:'mɪks/ mix again. ● *noun* /'ri:mɪks/ remixed recording.

remnant /'remnənt/ *noun* small remaining quantity; piece of cloth etc. left when greater part has been used or sold.

remold *US* = REMOULD.

remonstrate /'remənstreɪt/ *verb* (**-ting**) (+ *with*) make protest; argue forcibly. □ **remonstrance** /rɪ'mɒnstrəns/ *noun*; **remonstration** *noun*.

remorse /rɪ'mɔ:s/ *noun* bitter repentance; compunction; mercy.

remorseful *adjective* filled with repentance. □ **remorsefully** *adverb*.

remorseless *adjective* without compassion. □ **remorselessly** *adverb*.

remote /rɪˈməʊt/ *adjective* (**-r**, **-st**) distant in place or time; secluded; distantly related; slight, faint; aloof, not friendly. □ **remote control** (device for) control of apparatus etc. from a distance. □ **remotely** *adverb*; **remoteness** *noun*.

remould • *verb* /riːˈməʊld/ (*US* **remold**) mould again, refashion; reconstruct tread of (tyre). • *noun* /ˈriːməʊld/ remoulded tyre.

removal /rɪˈmuːv(ə)l/ *noun* removing, being removed; transfer of furniture etc. on moving house.

remove /rɪˈmuːv/ • *verb* (**-ving**) take off or away from place occupied; convey to another place; dismiss; cause to be no longer available; (in *passive*; + *from*) be distant in condition; (as **removed** *adjective*) (esp. of cousins) separated by a specified number of generations. • *noun* distance, degree of remoteness; stage in gradation; form or division in some schools. □ **removable** *adjective*.

remunerate /rɪˈmjuːnəreɪt/ *verb* (**-ting**) pay for service rendered. □ **remuneration** *noun*; **remunerative** /-rətɪv/ *adjective*.

Renaissance /rɪˈneɪs(ə)ns/ *noun* revival of classical art and literature in 14th–16th c.; period of this; style of art and architecture developed by it; (**renaissance**) any similar revival. □ **Renaissance man** person with many talents.

renal /ˈriːn(ə)l/ *adjective* of kidneys.

renascent /rɪˈnæs(ə)nt/ *adjective* springing up anew; being reborn. □ **renascence** *noun*.

rend *verb* (*past & past participle* **rent**) *archaic* tear or wrench forcibly.

render /ˈrendə/ *verb* cause to be or become; give in return; pay as due; (often + *to*) give (assistance), show (obedience etc.); present, submit; represent, portray; perform; translate; (often + *down*) melt (fat) down; cover (stone or brick) with plaster. □ **rendering** *noun*.

rendezvous /ˈrɒndɪvuː/ • *noun* (*plural* same /-vuːz/) agreed or regular meeting-place; meeting by arrangement. • *verb* (**rendezvouses** /-vuːz/; **rendezvoused** /-vuːd/; **rendezvousing** /-vuːɪŋ/) meet at rendezvous.

rendition /renˈdɪʃ(ə)n/ *noun* interpretation or rendering of dramatic or musical piece.

renegade /ˈrenɪgeɪd/ *noun* deserter of party or principles.

renege /rɪˈniːg, rɪˈneɪg/ *verb* (**-ging**) (often + *on*) go back on promise etc.

renew /rɪˈnjuː/ *verb* make new again; restore to original state; replace; repeat; resume after interruption; grant or be granted continuation of (licence etc.). □ **renewable** *adjective*; **renewal** *noun*.

rennet /ˈrenɪt/ *noun* curdled milk from calf's stomach, or artificial preparation, used in making cheese etc.

renounce /rɪˈnaʊns/ *verb* (**-cing**) consent formally to abandon; repudiate; decline further association with.

renovate /ˈrenəveɪt/ *verb* (**-ting**) restore to good condition; repair. □ **renovation** *noun*; **renovator** *noun*.

renown /rɪˈnaʊn/ *noun* fame, high distinction. □ **renowned** *adjective*.

rent¹ • *noun* periodical payment for use of land or premises; payment for hire of machinery etc. • *verb* (often + *from*) take, occupy, or use for rent; (often + *out*) let or hire for rent; (often + *at*) be let at specified rate.

rent² *noun* tear in garment etc.; gap, cleft, fissure.

rent³ *past & past participle* of REND.

rental /ˈrent(ə)l/ *noun* amount paid or received as rent; act of renting.

rentier /ˈrɑ̃tɪeɪ/ *noun* person living on income from property or investments. [French]

renunciation /rɪnʌnsɪˈeɪʃ(ə)n/ *noun* renouncing, self-denial, giving up of things.

rep¹ *noun colloquial* representative, esp. commercial traveller.

rep² *noun colloquial* repertory; repertory theatre or company.

repair¹ /rɪˈpeə/ • *verb* restore to good condition after damage or wear; set right or make amends for. • *noun* (result of) restoring to sound condition; good or relative condition for working or using. □ **repairable** *adjective*; **repairer** *noun*.

repair² /rɪˈpeə/ *verb* (usually + *to*) resort; go.

reparable /'repərəb(ə)l/ *adjective* that can be made good.

reparation /repə'reɪʃ(ə)n/ *noun* making amends; (esp. in *plural*) compensation.

repartee /repɑ:'ti:/ *noun* (making of) witty retorts.

repast /rɪ'pɑ:st/ *noun formal* meal.

repatriate /ri:'pætrɪeɪt/ *verb* (**-ting**) return (person) to native land. □ **repatriation** *noun.*

repay /ri:'peɪ/ *verb* (*past & past participle* **repaid**) pay back (money); make repayment to (person); reward (action etc.). □ **repayable** *adjective*; **repayment** *noun.*

repeal /rɪ'pi:l/ ● *verb* annul, revoke. ● *noun* repealing.

repeat /rɪ'pi:t/ ● *verb* say or do over again; recite, report; recur. ● *noun* repeating; thing repeated, esp. broadcast; *Music* passage intended to be repeated. □ **repeatable** *adjective*; **repeatedly** *adverb.*

repeater *noun* person or thing that repeats; firearm that fires several shots without reloading; watch that strikes last quarter etc. again when required; device for retransmitting electrical message.

repel /rɪ'pel/ *verb* drive back; ward off; be repulsive or distasteful to; resist mixing with; push away from itself. □ **repellent** *adjective & noun.*

repent /rɪ'pent/ *verb* (often + *of*) feel sorrow about one's actions etc.; wish one had not done, resolve not to continue (wrongdoing etc.). □ **repentance** *noun*; **repentant** *adjective.*

repercussion /ri:pə'kʌʃ(ə)n/ *noun* indirect effect or reaction following event etc.; recoil after impact.

repertoire /'repətwɑ:/ *noun* stock of works that performer etc. knows or is prepared to perform.

repertory /'repətərɪ/ *noun* (*plural* **-ies**) performance of various plays for short periods by one company; repertory theatres collectively; store of information etc.; repertoire. □ **repertory company** one performing plays from repertoire; **repertory theatre** one with repertoire of plays.

repetition /repə'tɪʃ(ə)n/ *noun* repeating, being repeated; thing repeated, copy. □ **repetitious** *adjective*; **repetitive** /rɪ'petətɪv/ *adjective.*

repine /rɪ'paɪn/ *verb* (**-ning**) (often + *at, against*) fret, be discontented.

replace /rɪ'pleɪs/ *verb* (**-cing**) put back in place; take place of, be or provide substitute for; (often + *with, by*) fill up place of.

replacement *noun* replacing, being replaced; person or thing that replaces another.

replay ● *verb* /ri:'pleɪ/ play (match, recording, etc.) again. ● *noun* /'ri:pleɪ/ replaying of match, recorded incident in game, etc.

replenish /rɪ'plenɪʃ/ *verb* (often + *with*) fill up again. □ **replenishment** *noun.*

replete /rɪ'pli:t/ *adjective* (often + *with*) well-fed; filled or well-supplied. □ **repletion** *noun.*

replica /'replɪkə/ *noun* exact copy, esp. duplicate made by original artist; model, esp. small-scale.

replicate /'replɪkeɪt/ *verb* make replica of. □ **replication** *noun.*

reply /rɪ'plaɪ/ ● *verb* (**-ies**, **-ied**) (often + *to*) make an answer, respond; say in answer. ● *noun* (*plural* **-ies**) replying; what is replied.

report /rɪ'pɔ:t/ ● *verb* bring back or give account of; tell as news; describe, esp. as eyewitness; make official or formal statement; (often + *to*) bring to attention of authorities, present oneself as arrived; take down, write description of, etc. for publication; (+ *to*) be responsible to. ● *noun* account given or opinion formally expressed after investigation; description, reproduction, or summary of speech, law case, scene, etc., esp. for newspaper publication or broadcast; common talk, rumour; repute; periodical statement on (esp. pupil's) work etc.; sound of gunshot. □ **reportedly** *adverb.*

reporter *noun* person employed to report news etc. for media.

repose[1] /rɪpəʊz/ ● *noun* rest; sleep; tranquillity. ● *verb* (**-sing**) rest; lie, esp. when dead.

repose[2] /rɪ'pəʊz/ *verb* (**-sing**) (+ *in*) place (trust etc.) in.

repository /rɪ'pɒzɪtərɪ/ *noun* (*plural* **-ies**) place where things are stored or may be found; receptacle; (often + *of*) book, person, etc. regarded as store of information, recipient of secrets etc.

reprehend /reprɪ'hend/ *verb formal* rebuke, blame.

reprehensible /reprɪ'hensɪb(ə)l/ *adjective* blameworthy.

represent /reprɪ'zent/ *verb* stand for, correspond to; be specimen of; symbolize; present likeness of to mind or senses; (often + *as, to be*) describe or depict, declare; (+ *that*) allege; show or play part of; be substitute or deputy for; be elected by as member of legislature etc.

representation /reprɪzen'teɪʃ(ə)n/ *noun* representing, being represented; thing that represents another.

representational *adjective* (of art) seeking to portray objects etc. realistically.

representative /reprɪ'zentətɪv/ ● *adjective* typical of class; containing typical specimens of all or many classes; (of government etc.) of elected deputies or based on representation. ● *noun* (+ *of*) sample, specimen, or typical embodiment of; agent; commercial traveller; delegate or deputy, esp. in representative assembly.

repress /rɪ'pres/ *verb* keep under; put down; suppress (esp. unwelcome thought). □ **repression** *noun*; **repressive** *adjective*.

reprieve /rɪ'pri:v/ ● *verb* (**-ving**) remit or postpone execution of; give respite to. ● *noun* reprieving, being reprieved.

reprimand /'reprɪmɑ:nd/ ● *noun* official rebuke. ● *verb* rebuke officially.

reprint ● *verb* /ri:'prɪnt/ print again. ● *noun* /'ri:prɪnt/ reprinting of book etc.; quantity reprinted.

reprisal /rɪ'praɪz(ə)l/ *noun* act of retaliation.

reprise /rɪ'pri:z/ *noun Music* repeated passage or song etc.

repro /'ri:prəʊ/ *noun* (*plural* **-s**) *colloquial* reproduction, copy.

reproach /rɪ'prəʊtʃ/ ● *verb* express disapproval to (person) for fault etc. ● *noun* rebuke, censure; (often + *to*) thing that brings discredit.

reproachful *adjective* full of or expressing reproach. □ **reproachfully** *adverb*.

reprobate /'reprəbeɪt/ *noun* unprincipled or immoral person.

reproduce /ri:prə'dju:s/ *verb* (**-cing**) produce copy or representation of; produce further members of same species by natural means; (**reproduce itself**) produce offspring. □ **reproducible** *adjective*.

reproduction /ri:prə'dʌkʃ(ə)n/ ● *noun* reproducing, esp. of further members of same species; copy of work of art. ● *adjective* (of furniture etc.) imitating earlier style. □ **reproductive** *adjective*.

reproof /rɪ'pru:f/ *noun formal* blame; rebuke.

reprove /rɪ'pru:v/ *verb* (**-ving**) *formal* rebuke.

reptile /'reptaɪl/ *noun* cold-blooded scaly animal of class including snakes, lizards, etc.; grovelling or repulsive person. □ **reptilian** /-'tɪl-/ *adjective*.

republic /rɪ'pʌblɪk/ *noun* state in which supreme power is held by the people or their elected representatives.

republican ● *adjective* of or characterizing republic(s); advocating or supporting republican government. ● *noun* supporter or advocate of republican government; (**Republican**) member of political party styled 'Republican'. □ **republicanism** *noun*.

repudiate /rɪ'pju:dɪeɪt/ *verb* (**-ting**) disown, disavow, deny; refuse to recognize or obey (authority) or discharge (obligation or debt). □ **repudiation** *noun*.

repugnance /rɪ'pʌgnəns/ *noun* aversion, antipathy; inconsistency or incompatibility of ideas etc.

repugnant /rɪ'pʌgn(ə)nt/ *adjective* distasteful; contradictory.

repulse /rɪ'pʌls/ ● *verb* (**-sing**) drive back; rebuff, reject. ● *noun* defeat, rebuff.

repulsion /rɪ'pʌlʃ(ə)n/ *noun* aversion, disgust; *Physics* tendency of bodies to repel each other.

repulsive /rɪ'pʌlsɪv/ *adjective* causing aversion or loathing. □ **repulsively** *adverb*.

reputable /'repjʊtəb(ə)l/ *adjective* of good reputation, respectable.

reputation /repjʊ'teɪʃ(ə)n/ *noun* what is generally said or believed about character of person or thing; credit, respectability.

repute /rɪ'pju:t/ ● *noun* reputation. ● *verb* (as **reputed** *adjective*) be generally considered. □ **reputedly** *adverb.*

request /rɪ'kwest/ ● *noun* asking for something, thing asked for. ● *verb* ask to be given, allowed, etc.; (+ *to do*) ask (person) to do something; (+ *that*) ask that.

Requiem /'rekwɪəm/ *noun* (also **requiem**) *esp. RC Church* mass for the dead.

require /rɪ'kwaɪə/ *verb* (**-ring**) need; depend on for success etc.; lay down as imperative; command, instruct; demand, insist on. □ **requirement** *noun.*

requisite /'rekwɪzɪt/ ● *adjective* required, necessary. ● *noun* (often + *for*) thing needed.

requisition /rekwɪ'zɪʃ(ə)n/ ● *noun* official order laying claim to use of property or materials; formal written demand. ● *verb* demand use or supply of.

requite /rɪ'kwaɪt/ *verb* (**-ting**) make return for; reward, avenge; (often + *for*) give in return. □ **requital** *noun.*

reredos /'rɪədɒs/ *noun* ornamental screen covering wall above back of altar.

rescind /rɪ'sɪnd/ *verb* abrogate, revoke, cancel. □ **rescission** /-'sɪʒ-/ *noun.*

rescue /'reskju:/ ● *verb* (**-ues**, **-ued**, **-uing**) (often + *from*) save or set free from danger or harm. ● *noun* rescuing, being rescued. □ **rescuer** *noun.*

research /rɪ'sɜ:tʃ, 'ri:sɜ:tʃ/ ● *noun* systematic investigation of materials, sources, etc. to establish facts. ● *verb* do research into or for. □ **researcher** *noun.*

resemble /rɪ'zemb(ə)l/ *verb* (**-ling**) be like; have similarity to. □ **resemblance** *noun.*

resent /rɪ'zent/ *verb* feel indignation at; be aggrieved by. □ **resentful** *adjective*; **resentfully** *adverb*; **resentment** *noun.*

reservation /rezə'veɪʃ(ə)n/ *noun* reserving, being reserved; thing reserved (e.g. room in hotel); spoken or unspoken limitation or exception; (in full **central reservation**) strip of land between carriageways of road; area reserved for occupation of aboriginal peoples.

reserve /rɪ'zɜ:v/ ● *verb* (**-ving**) put aside or keep back for later occasion or special use; order to be retained or allocated for person at particular time; retain, secure. ● *noun* thing reserved for future use; limitation or exception attached to something; self-restraint, reticence; company's profit added to capital; (in *singular* or *plural*) assets kept readily available, troops withheld from action to reinforce or protect others, forces outside regular ones but available in emergency; extra player chosen as possible substitute in team; land reserved for special use, esp. as habitat.

reserved *adjective* reticent, uncommunicative; set apart for particular use.

reservist *noun* member of reserve forces.

reservoir /'rezəvwɑ:/ *noun* large natural or artificial lake as source of water supply; receptacle for fluid; supply of facts etc.

reshuffle /ri:'ʃʌf(ə)l/ ● *verb* (**-ling**) shuffle again; change posts of (government ministers etc.). ● *noun* reshuffling.

reside /rɪ'zaɪd/ *verb* (**-ding**) have one's home; (+ *in*) (of right etc.) be vested in, (of quality) be present in.

residence /'rezɪd(ə)ns/ *noun* residing; place where one resides; house, esp. large one. □ **in residence** living or working at specified place.

resident /'rezɪd(ə)nt/ ● *noun* (often + *of*) permanent inhabitant; guest staying at hotel. ● *adjective* residing, in residence; living at one's workplace etc.; (+ *in*) located in.

residential /rezɪ'denʃ(ə)l/ *adjective* suitable for or occupied by dwellings; used as residence; connected with residence.

residual /rɪ'zɪdjʊəl/ *adjective* left as residue or residuum.

residuary /rɪ'zɪdjʊərɪ/ *adjective* of the residue of an estate; residual.

residue /'rezɪdju:/ *noun* remainder, what is left over; what remains of estate when liabilities have been discharged.

residuum /rɪ'zɪdjʊəm/ *noun* (*plural* **-dua**) substance left after combustion or evaporation; residue.

resign /rɪ'zaɪn/ *verb* (often + *from*) give up job, position, etc.; relinquish, surrender; (**resign oneself to**) accept (situation etc.) reluctantly.

resignation /rezɪg'neɪʃ(ə)n/ *noun* resigning, esp. from job or office; reluctant acceptance of the inevitable.

resigned *adjective* (often + *to*) having resigned oneself; resolved to endure; indicative of this. □ **resignedly** /-nɪdlɪ/ *adverb.*

resilient /rɪ'zɪlɪənt/ *adjective* resuming original form after compression etc.; readily recovering from setback. □ **resilience** *noun.*

resin /'rezɪn/ ● *noun* sticky secretion of trees and plants; (in full **synthetic resin**) organic compound made by polymerization etc. and used in plastics. ● *verb* (**-n-**) rub or treat with resin. □ **resinous** *adjective.*

resist /rɪ'zɪst/ *verb* withstand action or effect of; abstain from (pleasure etc.); strive against, oppose; offer opposition. □ **resistible** *adjective.*

resistance /rɪ'zɪst(ə)ns/ *noun* resisting; power to resist; ability to withstand disease; impeding effect exerted by one thing on another; *Physics* property of hindering passage of electric current, heat, etc.; resistor; secret organization resisting regime, esp. in occupied country. □ **resistant** *adjective.*

resistor *noun* device having resistance to passage of electric current.

resit ● *verb* /ri:'sɪt/ (**-tt-**; *past & past participle* **resat**) sit (exam) again after failing. ● *noun* /'ri:sɪt/ resitting of exam; exam for this.

resoluble /rɪ'zɒljʊb(ə)l/ *adjective* resolvable; (+ *into*) analysable into.

resolute /'rezəlu:t/ *adjective* determined, decided; purposeful. □ **resolutely** *adverb.*

resolution /rezə'lu:ʃ(ə)n/ *noun* resolute temper or character; thing resolved on; formal expression of opinion of meeting; (+ *of*) solving of question etc.; resolving, being resolved.

resolve /rɪ'zɒlv/ ● *verb* (**-ving**) make up one's mind, decide firmly; cause to do this; solve, settle; (+ *that*) pass resolution by vote; (often + *into*) (cause to) separate into constituent parts, analyse; *Music* convert or be converted into concord. ● *noun* firm mental decision; determination. □ **resolved** *adjective.*

resonant /'rezənənt/ *adjective* echoing, resounding; continuing to sound; causing reinforcement or prolongation of sound, esp. by vibration. □ **resonance** *noun.*

resonate /'rezəneɪt/ *verb* (**-ting**) produce or show resonance; resound. □ **resonator** *noun.*

resort /rɪ'zɔ:t/ ● *noun* place frequented, esp. for holidays etc.; thing to which recourse is had, expedient; (+ *to*) recourse to, use of. ● *verb* (+ *to*) turn to as expedient; (+ *to*) go often or in numbers to. □ **in the** or **as a last resort** when all else has failed.

resound /rɪ'zaʊnd/ *verb* (often + *with*) ring, echo; produce echoes, go on sounding, fill place with sound; be much talked of, produce sensation.

resounding *adjective* ringing, echoing; unmistakable, emphatic.

resource /rɪ'zɔ:s/ ● *noun* expedient, device; (often in *plural*) means available; stock that can be drawn on; (in *plural*) country's collective wealth, person's inner strength; skill in devising expedients. ● *verb* (**-cing**) provide with resources.

resourceful *adjective* good at devising expedients. □ **resourcefully** *adverb*; **resourcefulness** *noun.*

respect /rɪ'spekt/ ● *noun* deferential esteem; (+ *of*, *for*) heed, regard; detail, aspect; reference, relation; (in *plural*) polite greetings. ● *verb* regard with deference or esteem; treat with consideration, spare. □ **respectful** *adjective*; **respectfully** *adverb.*

respectable *adjective* of acceptable social standing, decent in appearance or behaviour; reasonably good in condition, appearance, size, etc. □ **respectability** *noun*; **respectably** *adverb.*

respecting *preposition* with regard to.

respective /rɪ'spektɪv/ *adjective* of or relating to each of several individually.

respectively *adverb* for each separately or in turn, and in the order mentioned.

respiration /respə'reɪʃ(ə)n/ *noun* breathing; single breath in or out; plant's absorption of oxygen and emission of carbon dioxide.

respirator /'respəreɪtə/ *noun* apparatus worn over mouth and nose to filter inhaled air; apparatus for maintaining artificial respiration.

respire /rɪ'spaɪə/ *verb* (**-ring**) breathe; inhale and exhale; (of plant) carry out respiration. □ **respiratory** /-'spɪr-/ *adjective*.

respite /'respaɪt/ *noun* interval of rest or relief; delay permitted before discharge of obligation or suffering of penalty.

resplendent /rɪ'splend(ə)nt/ *adjective* brilliant, dazzlingly or gloriously bright. □ **resplendence** *noun*.

respond /rɪ'spɒnd/ *verb* answer, reply; (often + *to*) act etc. in response.

respondent ● *noun* defendant, esp. in appeal or divorce case. ● *adjective* in position of defendant.

response /rɪ'spɒns/ *noun* answer, reply; action, feeling, etc. caused by stimulus etc.; (often in *plural*) part of liturgy said or sung in answer to priest.

responsibility /rɪspɒnsə'bɪlɪtɪ/ *noun* (*plural* **-ies**) (often + *for*, *of*) being responsible, authority; person or thing for which one is responsible.

responsible /rɪ'spɒnsəb(ə)l/ *adjective* (often + *to*, *for*) liable to be called to account; morally accountable for actions; of good credit and repute; trustworthy; (often + *for*) being the cause; involving responsibility. □ **responsibly** *adverb*.

responsive /rɪ'spɒnsɪv/ *adjective* (often + *to*) responding readily (to some influence); sympathetic; answering; by way of answer. □ **responsiveness** *noun*.

rest[1] ● *verb* cease from exertion or action; be still, esp. to recover strength; lie in sleep or death; give relief or repose to; be left without further investigation or discussion; (+ *on*, *upon*, *against*) place, lie, lean, or depend on; (as **rested** *adjective*) refreshed by resting. ● *noun* repose or sleep; resting; prop or support for steadying something; *Music* (sign denoting) interval of silence. □ **at rest** not moving, not agitated or troubled; **rest-cure** prolonged rest as medical treatment; **rest home** place where elderly or convalescent people are cared for; **rest room** *esp. US* public lavatory; **set at rest** settle, reassure.

rest[2] ● *noun* (**the rest**) remainder or remaining parts or individuals. ● *verb* remain in specified state; (+ *with*) be left in the charge of. □ **for the rest** as regards anything else.

restaurant /'restərɒnt/ *noun* public premises where meals may be bought and eaten.

restaurateur /restərə'tɜ:/ *noun* keeper of restaurant.

restful *adjective* quiet, soothing.

restitution /restɪ'tju:ʃ(ə)n/ *noun* restoring of property etc. to its owner; reparation.

restive /'restɪv/ *adjective* fidgety; intractable, resisting control.

restless *adjective* without rest; uneasy, agitated, fidgeting. □ **restlessly** *adverb*; **restlessness** *noun*.

restoration /restə'reɪʃ(ə)n/ *noun* restoring, being restored; model or drawing representing supposed original form of thing; (**Restoration**) re-establishment of British monarchy in 1660.

restorative /rɪ'stɒrətɪv/ ● *adjective* tending to restore health or strength. ● *noun* restorative food, medicine, etc.

restore /rɪ'stɔ:/ *verb* (**-ring**) bring back to original state by rebuilding, repairing, etc.; give back; reinstate; bring back to former place, condition, or use; make restoration of (extinct animal, ruined building, etc.). □ **restorer** *noun*.

restrain /rɪ'streɪn/ *verb* (usually + *from*) check or hold in; keep under control; repress; confine.

restraint /rɪ'streɪnt/ *noun* restraining, being restrained; restraining agency or influence; self-control, moderation; reserve of manner.

restrict /rɪ'strɪkt/ *verb* confine, limit; withhold from general disclosure. □ **restriction** *noun*.

restrictive /rɪ'strɪktɪv/ *adjective* restricting. □ **restrictive practice** agreement or practice that limits competition or output in industry.

result /rɪ'zʌlt/ ● *noun* consequence; issue; satisfactory outcome; answer etc. got by calculation; (in *plural*) list of scores, winners, etc. in sporting events or exams. ● *verb* (often + *from*) arise as consequence; (+ *in*) end in.

resultant • *adjective* resulting. • *noun* force etc. equivalent to two or more acting in different directions at same point.

resume /rɪ'zju:m/ *verb* (**-ming**) begin again; recommence; take again or back. □ **resumption** /-'zʌmp-/ *noun*; **resumptive** /-'zʌmp-/ *adjective*.

résumé /'rezjʊmeɪ/ *noun* summary.

resurgent /rɪ'sɜ:dʒ(ə)nt/ *adjective* rising or arising again. □ **resurgence** *noun*.

resurrect /rezə'rekt/ *verb colloquial* revive practice or memory of; raise or rise from dead.

resurrection /rezə'rekʃ(ə)n/ *noun* rising from the dead; revival from disuse or decay etc.

resuscitate /rɪ'sʌsɪteɪt/ *verb* (**-ting**) revive from unconsciousness or apparent death; revive, restore. □ **resuscitation** *noun*.

retail /'ri:teɪl/ • *noun* sale of goods to the public in small quantities. • *adjective & adverb* by retail; at retail price. • *verb* sell by retail; (often + *at*, *of*) (of goods) be sold by retail; (also /rɪ'teɪl/) recount. □ **retailer** *noun*.

retain /rɪ'teɪn/ *verb* keep possession of, continue to have, use, etc.; keep in mind; keep in place, hold fixed; secure services of (esp. barrister) by preliminary fee.

retainer *noun* fee for securing person's services; faithful servant; reduced rent paid to retain unoccupied accommodation; person or thing that retains.

retake • *verb* /ri:'teɪk/ (**-king**; *past* **retook**; *past participle* **retaken**) take (photograph, exam, etc.) again; recapture. • *noun* /'ri:teɪk/ filming, recording, etc. again; taking of exam etc. again.

retaliate /rɪ'tælɪeɪt/ *verb* (**-ting**) repay in kind; attack in return. □ **retaliation** *noun*; **retaliatory** /-'tæljət-/ *adjective*.

retard /rɪ'tɑ:d/ *verb* make slow or late; delay progress or accomplishment of. □ **retardant** *adjective & noun*; **retardation** /ri:-/ *noun*.

retarded *adjective* backward in mental or physical development.

retch *verb* make motion of vomiting.

retention /rɪ'tenʃ(ə)n/ *noun* retaining, being retained.

retentive /rɪ'tentɪv/ *adjective* tending to retain; (of memory) not forgetful.

rethink • *verb* /ri:'θɪŋk/ (*past & past participle* **rethought** /-'θɔ:t/) consider again, esp. with view to making changes. • *noun* /'ri:θɪŋk/ rethinking, reassessment.

reticence /'retɪs(ə)ns/ *noun* avoidance of saying all one knows or feels; taciturnity. □ **reticent** *adjective*.

reticulate • *verb* /rɪ'tɪkjʊleɪt/ (**-ting**) divide or be divided in fact or appearance into network. • *adjective* /rɪ'tɪkjʊlət/ reticulated. □ **reticulation** *noun*.

retina /'retɪnə/ *noun* (*plural* **-s** or **-nae** /-ni:/) light-sensitive layer at back of eyeball. □ **retinal** *adjective*.

retinue /'retɪnju:/ *noun* group of people attending important person.

retire /rɪ'taɪə/ *verb* (**-ring**) leave office or employment, esp. because of age; cause (employee) to retire; withdraw, retreat, seek seclusion or shelter; go to bed; *Cricket* (of batsman) suspend one's innings. □ **retired** *adjective*.

retirement *noun* retiring; period spent as retired person; seclusion. □ **retirement pension** pension paid by state to retired people above certain age.

retiring *adjective* shy, fond of seclusion.

retort[1] /rɪ'tɔ:t/ • *noun* incisive, witty, or angry reply. • *verb* say by way of retort; repay in kind.

retort[2] /rɪ'tɔ:t/ *noun* vessel with long downward-bent neck for distilling liquids; vessel for heating coal to generate gas.

retouch /ri:'tʌtʃ/ *verb* improve (esp. photograph) by minor alterations.

retrace /rɪ'treɪs/ *verb* (**-cing**) go back over (one's steps etc.); trace back to source or beginning.

retract /rɪ'trækt/ *verb* withdraw (statement etc.); draw or be drawn back or in. □ **retractable** *adjective*; **retraction** *noun*.

retractile /rɪ'træktaɪl/ *adjective* retractable.

retread • *verb* /ri:'tred/ (*past* **retrod**; *past participle* **retrodden**) tread (path etc.) again; (*past & past participle* **retreaded**) put new tread on (tyre). • *noun* /'ri:tred/ retreaded tyre.

retreat /rɪ'tri:t/ ● *verb* go back, retire; recede. ● *noun* (signal for) act of retreating; withdrawing into privacy; place of seclusion or shelter; period of seclusion for prayer and meditation.

retrench /rɪ'trentʃ/ *verb* cut down expenses; reduce amount of (costs); economize. □ **retrenchment** *noun*.

retrial /ri:'traɪəl/ *noun* retrying of case.

retribution /retrɪ'bju:ʃ(ə)n/ *noun* recompense, usually for evil; vengeance. □ **retributive** /rɪ'trɪb-/ *adjective*.

retrieve /rɪ'tri:v/ *verb* (**-ving**) regain possession of; find again; obtain (information in computer); (of dog) find and bring in (game); (+ *from*) rescue from (bad state etc.); restore to good state; repair, set right. □ **retrievable** *adjective*; **retrieval** *noun*.

retriever *noun* dog of breed used for retrieving game.

retro /'retrəʊ/ *slang* ● *adjective* reviving or harking back to past. ● *noun* (*plural* **retros**) retro fashion or style.

retro- *combining form* backwards, back.

retroactive /retrəʊ'æktɪv/ *adjective* having retrospective effect.

retrod *past* of RETREAD.

retrodden *past participle* of RETREAD.

retrograde /'retrəgreɪd/ ● *adjective* directed backwards; reverting, esp. to inferior state. ● *verb* move backwards; decline, revert.

retrogress /retrə'gres/ *verb* move backwards; deteriorate. □ **retrogression** *noun*; **retrogressive** *adjective*.

retrorocket /'retrəʊrɒkɪt/ *noun* auxiliary rocket for slowing down spacecraft etc.

retrospect /'retrəspekt/ *noun* □ **in retrospect** when looking back.

retrospection /retrə'spekʃ(ə)n/ *noun* looking back into the past.

retrospective /retrə'spektɪv/ ● *adjective* looking back on or dealing with the past; (of statute etc.) applying to the past as well as the future. ● *noun* exhibition etc. showing artist's lifetime development. □ **retrospectively** *adverb*.

retroussé /rə'tru:seɪ/ *adjective* (of nose) turned up at tip.

retroverted /'retrəʊvɜ:tɪd/ *adjective* (of womb) inclining backwards.

retry /ri:'traɪ/ *verb* (**-ies**, **-ied**) try (defendant, law case) again.

retsina /ret'si:nə/ *noun* resin-flavoured Greek white wine.

return /rɪ'tɜ:n/ ● *verb* come or go back; bring, put, or send back; give in response; yield (profit); say in reply; send (ball) back in tennis etc.; state in answer to formal demand; elect as MP etc. ● *noun* coming, going, putting, giving, sending, or paying back; what is returned; (in full **return ticket**) ticket for journey to place and back again; (in *singular* or *plural*) proceeds, profit; coming in of these; formal statement or report; (in full **return match**, **game**) second game between same opponents; (announcement of) person's election as MP etc. □ **by return (of post)** by the next available post in the return direction; **returning officer** official conducting election in constituency etc. and announcing result.

returnee /rɪtɜ:'ni:/ *noun* person who returns home, esp. after war service.

reunify /ri:'ju:nɪfaɪ/ *verb* (**-ies**, **-ied**) restore to political unity. □ **reunification** *noun*.

reunion /ri:'ju:njən/ *noun* reuniting, being reunited; social gathering, esp. of former associates.

reunite /ri:ju:'naɪt/ *verb* (**-ting**) (cause to) come together again.

reuse ● *verb* /ri:'ju:z/ (**-sing**) use again. ● *noun* /ri:'ju:s/ second or further use. □ **reusable** *adjective*.

Rev. *abbreviation* Reverend.

rev *colloquial* ● *noun* (in *plural*) revolutions of engine per minute. ● *verb* (**-vv-**) (of engine) revolve; (often + *up*) cause (engine) to run quickly.

revalue /ri:'vælju:/ *verb* (**-ues**, **-ued**, **-uing**) give different, esp. higher, value to (currency etc.). □ **revaluation** *noun*.

revamp /ri:'væmp/ *verb* renovate, revise; patch up.

Revd *abbreviation* Reverend.

reveal /rɪ'vi:l/ *verb* display, show, allow to appear; (often as **revealing** *adjective*) disclose, divulge.

reveille /rɪ'vælɪ/ *noun* military waking-signal.

revel /'rev(ə)l/ ● *verb* (**-ll-**; *US* **-l-**) make merry, be riotously festive; (+ *in*) take keen delight in. ● *noun* (in *singular* or *plural*) revelling. □ **reveller** *noun*; **revelry** *noun* (*plural* **-ies**).

revelation /revə'leɪʃ(ə)n/ *noun* revealing; knowledge supposedly disclosed by divine or supernatural agency; striking disclosure or realization; (**Revelation** or *colloquial* **Revelations**) last book of New Testament.

revenge /rɪ'vendʒ/ ● *noun* (act of) retaliation; desire for this. ● *verb* (**-ging**) avenge; (**revenge oneself** or in *passive*; often + *on*, *upon*) inflict retaliation.

revengeful *adjective* eager for revenge.

revenue /'revənju:/ *noun* income, esp. annual income of state; department collecting state revenue.

reverberate /rɪ'vɜ:bəreɪt/ *verb* (**-ting**) (of sound, light, or heat) be returned or reflected repeatedly; return (sound etc.) thus; (of event) produce continuing effect. □ **reverberant** *adjective*; **reverberation** *noun*; **reverberative** /-rətɪv/ *adjective*.

revere /rɪ'vɪə/ *verb* (**-ring**) regard with deep and affectionate or religious respect.

reverence /'revərəns/ ● *noun* revering, being revered; deep respect. ● *verb* (**-cing**) treat with reverence.

reverend /'revərənd/ *adjective* (esp. as title of member of clergy) deserving reverence. □ **Reverend Mother** Mother Superior of convent.

reverent /'revərənt/ *adjective* feeling or showing reverence. □ **reverently** *adverb*.

reverential /revə'renʃ(ə)l/ *adjective* of the nature of, due to, or characterized by reverence. □ **reverentially** *adverb*.

reverie /'revərɪ/ *noun* fit of musing; daydream.

revers /rɪ'vɪə/ *noun* (*plural* same /-'vɪəz/) (material of) turned-back front edge of garment.

reverse /rɪ'vɜ:s/ ● *verb* (**-sing**) turn the other way round or up, turn inside out; convert to opposite character or effect; (cause to) travel backwards; make (engine) work in contrary direction; revoke, annul. ● *adjective* backward, upside down; opposite or contrary in character or order, inverted. ● *noun* opposite or contrary; contrary of usual manner; piece of misfortune, disaster; reverse gear or motion; reverse side; side of coin etc. bearing secondary design; verso of printed leaf. □ **reverse the charges** have recipient of telephone call pay for it; **reverse gear** gear used to make vehicle etc. go backwards; **reversing light** light at rear of vehicle showing it is in reverse gear. □ **reversal** *noun*; **reversible** *adjective*.

reversion /rɪ'vɜ:ʃ(ə)n/ *noun* return to previous state or earlier type; legal right (esp. of original owner) to possess or succeed to property on death of present possessor.

revert /rɪ'vɜ:t/ *verb* (+ *to*) return to (former condition, practice, subject, opinion, etc.); return by reversion. □ **revertible** *adjective*.

review /rɪ'vju:/ ● *noun* general survey or assessment; survey of past; revision, reconsideration; published criticism of book, play, etc.; periodical in which events, books, etc. are reviewed; inspection of troops etc. ● *verb* survey, look back on; reconsider, revise; hold review of (troops etc.); write review of (book etc.). □ **reviewer** *noun*.

revile /rɪ'vaɪl/ *verb* (**-ling**) abuse verbally.

revise /rɪ'vaɪz/ *verb* (**-sing**) examine and improve or amend; reconsider and alter (opinion etc.); go over (work etc.) again, esp. for examination. □ **revisory** *adjective*.

revision /rɪ'vɪʒ(ə)n/ *noun* revising, being revised; revised edition or form.

revisionism *noun often derogatory* revision or modification of orthodoxy, esp. of Marxism. □ **revisionist** *noun & adjective*.

revitalize /ri:'vaɪtəlaɪz/ *verb* (also **-ise**) (**-zing** or **-sing**) imbue with new vitality.

revival /rɪ'vaɪv(ə)l/ *noun* reviving, being revived; new production of old play etc.; (campaign to promote) reawakening of religious fervour.

revivalism *noun* promotion of esp. religious revival. □ **revivalist** *noun & adjective*.

revive /rɪ'vaɪv/ *verb* (**-ving**) come or bring back to consciousness, life, vigour, use, or notice.

revivify /rɪ'vɪvɪfaɪ/ *verb* (**-ies**, **-ied**) restore to life, strength, or activity. □ **revivification** *noun*.

revoke /rɪ'vəʊk/ ● *verb* (**-king**) rescind, withdraw, cancel; *Cards* fail to follow suit though able to. ● *noun Cards* revoking. □ **revocable** /'revəkəb(ə)l/ *adjective*; **revocation** /revə'keɪʃ(ə)n/ *noun*.

revolt /rɪ'vəʊlt/ ● *verb* rise in rebellion; affect with disgust; (often + *at*, *against*) feel revulsion. ● *noun* insurrection; sense of disgust; rebellious mood.

revolting *adjective* disgusting, horrible. □ **revoltingly** *adverb*.

revolution /revə'lu:ʃ(ə)n/ *noun* forcible overthrow of government or social order; fundamental change; revolving; single completion of orbit or rotation.

revolutionary ● *adjective* involving great change; of political revolution. ● *noun* (*plural* **-ies**) instigator or supporter of political revolution.

revolutionize *verb* (also **-ise**) (**-zing** or **-sing**) change fundamentally.

revolve /rɪ'vɒlv/ *verb* (**-ving**) turn round; rotate; move in orbit; ponder in the mind; (+ *around*) be centred on.

revolver *noun* pistol with revolving chambers enabling user to fire several shots without reloading.

revue /rɪ'vju:/ *noun* theatrical entertainment of usually comic sketches and songs.

revulsion /rɪ'vʌlʃ(ə)n/ *noun* abhorrence; sudden violent change of feeling.

reward /rɪ'wɔ:d/ ● *noun* return or recompense for service or merit; requital for good or evil; sum offered for detection of criminal, recovery of lost property, etc. ● *verb* give or serve as reward to. □ **rewarding** *adjective*.

rewind /ri:'waɪnd/ *verb* (*past & past participle* **rewound** /-'waʊnd/) wind (film, tape, etc.) back.

rewire /ri:'waɪə/ *verb* (**-ring**) provide with new electrical wiring.

rework /ri:'wɜ:k/ *verb* revise, refashion, remake. □ **reworking** *noun*.

Rex *noun* (after name) reigning king; *Law* the Crown. [Latin].

rhapsodize /'ræpsədaɪz/ *verb* (also **-ise**) (**-zing** or **-sing**) speak or write rhapsodies.

rhapsody /'ræpsədɪ/ *noun* (*plural* **-ies**) enthusiastic or extravagant speech or composition; melodic musical piece often based on folk culture. □ **rhapsodic** /-'sɒd-/ *adjective*.

rhea /'ri:ə/ *noun* large flightless S. American bird.

rheostat /'ri:əstæt/ *noun* instrument used to control electric current by varying resistance.

rhesus /'ri:səs/ *noun* (in full **rhesus monkey**) small Indian monkey. □ **rhesus factor** antigen occurring on red blood cells of most humans and some other primates; **rhesus-positive**, **-negative** having or not having rhesus factor.

rhetoric /'retərɪk/ *noun* art of persuasive speaking or writing; language intended to impress, esp. seen as inflated, exaggerated, or meaningless.

rhetorical /rɪ'tɒrɪk(ə)l/ *adjective* expressed artificially or extravagantly; of the nature of rhetoric. □ **rhetorical question** question asked not for information but to produce effect. □ **rhetorically** *adverb*.

rheumatic /ru:'mætɪk/ ● *adjective* of, caused by, or suffering from rheumatism. ● *noun* person suffering from rheumatism; (in *plural*, often treated as *singular*) *colloquial* rheumatism. □ **rheumatic fever** fever with pain in the joints. □ **rheumatically** *adverb*; **rheumaticky** *adjective colloquial*.

rheumatism /'ru:mətɪz(ə)m/ *noun* disease marked by inflammation and pain in joints etc.

rheumatoid /'ru:mətɔɪd/ *adjective* having the character of rheumatism. □ **rheumatoid arthritis** chronic progressive disease causing inflammation and stiffening of joints.

rhinestone /'raɪnstəʊn/ *noun* imitation diamond.

rhino /'raɪnəʊ/ *noun* (*plural* same or **-s**) *colloquial* rhinoceros.

rhinoceros /raɪ'nɒsərəs/ *noun* (*plural* same or **-roses**) large thick-skinned animal with usually one horn on nose.

rhizome /'raɪzəʊm/ *noun* underground rootlike stem bearing both roots and shoots.

rhododendron /rəʊdə'dendrən/ *noun* (*plural* **-s** or **-dra**) evergreen shrub with large flowers.

rhomboid /'rɒmbɔɪd/ ● *adjective* (also **rhomboidal** /-'bɔɪd-/) like a rhombus. ● *noun* quadrilateral of which only opposite sides and angles are equal.

rhombus /'rɒmbəs/ *noun* (*plural* **-buses** or **-bi** /-baɪ/) oblique equilateral parallelogram, e.g. diamond on playing card.

rhubarb /'ru:bɑ:b/ *noun* (stalks of) plant with fleshy leaf-stalks cooked and eaten as dessert; *colloquial* indistinct conversation or noise, from repeated use of word 'rhubarb' by stage crowd.

rhyme /raɪm/ ● *noun* identity of sound at ends of words or verse-lines; (in *singular* or *plural*) rhymed verse; use of rhyme; word providing rhyme. ● *verb* (**-ming**) (of words or lines) produce rhyme; (+ *with*) be or use as rhyme; write rhymes; put into rhyme.

rhythm /'rɪð(ə)m/ *noun* periodical accent and duration of notes in music; type of structure formed by this; measured flow of words in verse or prose; *Physiology* pattern of successive strong and weak movements; regularly occurring sequence of events. □ **rhythm method** contraception by avoiding sexual intercourse near times of ovulation. □ **rhythmic** *adjective*; **rhythmical** *adjective*; **rhythmically** *adverb*.

rib ● *noun* each of the curved bones joined to spine and protecting organs of chest; joint of meat from this part of animal; supporting ridge, timber, rod, etc. across surface or through structure; combination of plain and purl stitches producing ribbed design. ● *verb* (**-bb-**) provide or mark (as) with ribs; *colloquial* tease. □ **ribcage** wall of bones formed by ribs round chest; **rib-tickler** something amusing. □ **ribbed** *adjective*; **ribbing** *noun*.

ribald /'rɪb(ə)ld/ *adjective* irreverent, coarsely humorous. □ **ribaldry** *noun*.

riband /'rɪbənd/ *noun* ribbon.

ribbon /'rɪbən/ *noun* narrow strip or band of fabric; material in this form; ribbon worn to indicate some honour, membership of sports team, etc.; long narrow strip; (in *plural*) ragged strips. □ **ribbon development** building of houses along main road outwards from town.

riboflavin /raɪbəʊ'fleɪvɪn/ *noun* (also **riboflavine** /-vi:n/) vitamin of B complex, found in liver, milk, and eggs.

ribonucleic acid /raɪbəʊnju:'kli:ɪk/ *noun* substance controlling protein synthesis in cells.

rice *noun* (grains from) swamp grass grown esp. in Asia. □ **rice-paper** edible paper made from pith of an oriental tree and used for painting and in cookery.

rich *adjective* having much wealth; splendid, costly; valuable; abundant, ample; (often + *in*, *with*) abounding; fertile; (of food) containing much fat, spice, etc.; mellow, strong and full; highly amusing or ludicrous. □ **richness** *noun*.

riches *plural noun* abundant means; valuable possessions.

richly *adverb* in a rich way; fully, thoroughly.

Richter scale /'rɪktə/ *noun* scale of 0–10 for representing strength of earthquake.

rick[1] *noun* stack of hay etc.

rick[2] (also **wrick**) ● *noun* slight sprain or strain. ● *verb* slightly strain or sprain.

rickets /'rɪkɪts/ *noun* (treated as *singular* or *plural*) children's deficiency disease with softening of the bones.

rickety /'rɪkɪtɪ/ *adjective* shaky, insecure; suffering from rickets.

rickshaw /'rɪkʃɔ:/ *noun* (also **ricksha** /-ʃə/) light two-wheeled hooded vehicle drawn by one or more people.

ricochet /'rɪkəʃeɪ/ ● *noun* rebounding of esp. shell or bullet off surface; hit made after this. ● *verb* (**-cheted** /-ʃeɪd/ or **-chetted** /-ʃetɪd/; **-cheting** /-ʃeɪɪŋ/ or **-chetting** /-ʃetɪŋ/) (of projectile) make ricochet.

ricotta /rɪ'kɒtə/ *noun* soft Italian cheese.

rid *verb* (**-dd-**; *past & past participle* **rid**) (+ *of*) make (person, place) free of.

riddance /'rɪd(ə)ns/ *noun* □ **good riddance** *expression of relief at getting rid of something or someone*.

ridden *past participle* of RIDE.

riddle[1] /'rɪd(ə)l/ ● *noun* verbal puzzle or test, often with trick answer; puzzling fact, thing, or person. ● *verb* (**-ling**) speak in riddles.

riddle[2] /ˈrɪd(ə)l/ ● *verb* (**-ling**) (usually + *with*) make many holes in, esp. with gunshot; (in *passive*) fill, permeate; pass through riddle. ● *noun* coarse sieve.

ride ● *verb* (**-ding**; *past* **rode**; *past participle* **ridden** /ˈrɪd(ə)n/) (often + *on, in*) travel or be carried on or in (bicycle, vehicle, horse, etc.); be carried or supported by; cross, be conveyed over; float buoyantly; (as **ridden** *adjective*) (+ *by, with*) dominated by or infested with. ● *noun* journey or spell of riding in vehicle, on horse, etc.; path (esp. through woods) for riding on; amusement for riding on at fairground. □ **ride up** (of garment) work upwards when worn; **take for a ride** *colloquial* hoax, deceive.

rider *noun* person riding; additional remark following statement, verdict, etc.

ridge *noun* line of junction of two surfaces sloping upwards towards each other; long narrow hilltop; mountain range; any narrow elevation across surface. □ **ridge-pole** horizontal pole of long tent; **ridgeway** road along ridge.

ridicule /ˈrɪdɪkju:l/ ● *noun* derision, mockery. ● *verb* (**-ling**) make fun of; mock; laugh at.

ridiculous /rɪˈdɪkjʊləs/ *adjective* deserving to be laughed at; unreasonable. □ **ridiculously** *adverb*; **ridiculousness** *noun*.

riding[1] /ˈraɪdɪŋ/ *noun* sport or pastime of travelling on horseback.

riding[2] /ˈraɪdɪŋ/ *noun historical* former division of Yorkshire.

Riesling /ˈri:zlɪŋ/ *noun* (white wine made from) type of grape.

rife *adjective* widespread; (+ *with*) abounding in.

riff *noun* short repeated phrase in jazz etc.

riffle /ˈrɪf(ə)l/ ● *verb* (**-ling**) (often + *through*) leaf quickly through (pages); shuffle (cards). ● *noun* riffling; *US* patch of ripples in stream etc.

riff-raff /ˈrɪfræf/ *noun* rabble, disreputable people.

rifle[1] /ˈraɪf(ə)l/ ● *noun* gun with long rifled barrel; (in *plural*) troops armed with these. ● *verb* (**-ling**) make spiral grooves in (gun etc.) to make projectile spin. □ **rifle range** place for rifle practice.

rifle[2] /ˈraɪf(ə)l/ *verb* (**-ling**) (often + *through*) search and rob; carry off as booty.

rift *noun* crack, split; cleft; disagreement, dispute. □ **rift-valley** one formed by subsidence of section of earth's crust.

rig[1] ● *verb* (**-gg-**) provide (ship) with rigging; (often + *out, up*) fit with clothes or equipment; (+ *up*) set up hastily or as makeshift. ● *noun* arrangement of ship's masts, sails, etc.; equipment for special purpose; oil rig; *colloquial* style of dress, uniform. □ **rig-out** *colloquial* outfit of clothes. □ **rigger** *noun*.

rig[2] ● *verb* (**-gg-**) manage or fix fraudulently. ● *noun* trick, swindle.

rigging *noun* ship's spars, ropes, etc.

right /raɪt/ ● *adjective* just, morally or socially correct; correct, true; preferable, suitable; in good or normal condition; on or towards east side of person or thing facing north; (also **Right**) *Politics* of the Right; (of side of fabric etc.) meant to show; *colloquial* real, complete. ● *noun* what is just; fair treatment; fair claim; legal or moral entitlement; right-hand part, region, or direction; *Boxing* right hand, blow with this; (often **Right**) *Politics* conservatives collectively. ● *verb* (often *reflexive*) restore to proper, straight, or vertical position; correct, avenge; set in order; make reparation for. ● *adverb* straight; *colloquial* immediately; (+ *to, round, through*, etc.) all the way; (+ *off, out*, etc.) completely; quite, very; justly, properly, correctly, truly; on or to right side. ● *interjection colloquial: expressing agreement or consent.* □ **by right(s)** if right were done; **in the right** having justice or truth on one's side; **right angle** angle of 90°; **right-hand** on right side; **right-handed** naturally using right hand for writing etc., made by or for right hand, turning to right, (of screw) turning clockwise to tighten; **right-hander** right-handed person or blow; **right-hand man** essential or chief assistant; **Right Honourable** *title of certain high officials, e.g. Privy Counsellors*; **right-minded**, **-thinking** having sound views and principles; **right**

of way right to pass over another's ground, path subject to such right, precedence granted to one vehicle over another; **Right Reverend** *title of bishop*; **right wing** more conservative section of political party etc., right side of football etc. team; **right-wing** conservative, reactionary; **right-winger** member of right wing. □ **rightward** *adjective & adverb*; **rightwards** *adverb*.

righteous /'raɪtʃəs/ *adjective* morally right; virtuous, law-abiding. □ **righteously** *adverb*; **righteousness** *noun*.

rightful *adjective* legitimately entitled to (position etc.); that one is entitled to. □ **rightfully** *adverb*.

rightly *adverb* justly, correctly, properly, justifiably.

rigid /'rɪdʒɪd/ *adjective* not flexible, unbendable; inflexible, harsh. □ **rigidity** /-'dʒɪd-/ *noun*; **rigidly** *adverb*.

rigmarole /'rɪgmərəʊl/ *noun* complicated procedure; rambling tale etc.

rigor *US* = RIGOUR.

rigor mortis /rɪgə 'mɔːtɪs/ *noun* stiffening of body after death.

rigorous /'rɪgərəs/ *adjective* strict, severe; exact, accurate. □ **rigorously** *adverb*; **rigorousness** *noun*.

rigour /'rɪgə/ *noun* (*US* **rigor**) severity, strictness, harshness; (in *plural*) harsh conditions; strict application or observance etc.

rile *verb* (**-ling**) *colloquial* anger, irritate.

rill *noun* small stream.

rim *noun* edge or border, esp. of something circular; outer ring of wheel, holding tyre; part of spectacle frames around lens. □ **rimless** *adjective*; **rimmed** *adjective*.

rime ● *noun* frost; hoar-frost. ● *verb* (**-ming**) cover with rime.

rind /raɪnd/ *noun* tough outer layer or covering of fruit and vegetables, cheese, bacon, etc.

ring[1] ● *noun* circular band, usually of metal, worn on finger; circular band of any material; line or band round cylindrical or circular object; mark or part etc. resembling ring; ring in cross-section of tree representing one year's growth; enclosure for circus, boxing, betting at races, etc.; people or things arranged in circle, such arrangement; combination of traders, politicians, spies, etc. acting together; gas ring; disc or halo round planet, moon, etc. ● *verb* (often + *round*, *about*, *in*) encircle; put ring on (bird etc.). □ **ring-binder** loose-leaf binder with ring-shaped clasps; **ring-dove** woodpigeon; **ring finger** finger next to little finger, esp. on left hand; **ringleader** instigator in crime or mischief etc.; **ringmaster** director of circus performance; **ring-pull** (of tin) having ring for pulling to break seal; **ring road** bypass encircling town; **ringside** area immediately beside boxing or circus ring etc.; **ringworm** skin infection forming circular inflamed patches.

ring[2] ● *verb* (*past* **rang**; *past participle* **rung**) (often + *out* etc.) give clear resonant sound; make (bell) ring; call by telephone; (usually + *with*, *to*) (of place) resound, re-echo; (of ears) be filled with sensation of ringing; (+ *in*, *out*) usher in or out with bell-ringing; give specified impression. ● *noun* ringing sound or tone; act of ringing bell, sound caused by this; *colloquial* telephone call; set of esp. church bells; specified feeling conveyed by words etc.. □ **ring back** make return telephone call to; **ring off** end telephone call; **ring up** make telephone call (to), record (amount) on cash register.

ringlet /'rɪŋlɪt/ *noun* curly lock of esp. long hair.

rink *noun* area of ice for skating, curling, etc.; enclosed area for roller-skating; building containing either of these; strip of bowling green; team in bowls or curling.

rinse ● *verb* (**-sing**) (often + *through*, *out*) wash or treat with clean water etc.; wash lightly; put through clean water after washing; (+ *out*, *away*) remove by rinsing. ● *noun* rinsing; temporary hair tint.

riot /'raɪət/ ● *noun* disturbance of peace by crowd; loud revelry; (+ *of*) lavish display of; *colloquial* very amusing thing or person. ● *verb* make or engage in riot. □ **run riot** throw off all restraint, spread uncontrolled. □ **rioter** *noun*; **riotous** *adjective*.

RIP *abbreviation* may he, she, or they rest in peace (*requiesca(n)t in pace*).

rip • *verb* (**-pp-**) tear or cut quickly or forcibly away or apart; make (hole etc.) thus; make long tear or cut in; come violently apart, split. • *noun* long tear or cut; act of ripping. □ **let rip** *colloquial* (allow to) proceed or act without restraint or interference; **rip-cord** cord for releasing parachute from its pack; **rip off** *colloquial* swindle, exploit, steal; **rip-off** *noun* □ **ripper** *noun*.

riparian /raɪˈpeərɪən/ *adjective* of or on riverbank.

ripe *adjective* ready to be reaped, picked, or eaten; mature; (often + *for*) fit, ready. □ **ripen** *verb*; **ripeness** *noun*.

riposte /rɪˈpɒst/ • *noun* quick retort; quick return thrust in fencing. • *verb* (**-ting**) deliver riposte.

ripple /ˈrɪp(ə)l/ • *noun* ruffling of water's surface; small wave(s); gentle lively sound, e.g. of laughter or applause; slight variation in strength of current etc.; ice cream with veins of syrup. • *verb* (**-ling**) (cause to) form or flow in ripples; show or sound like ripples.

rise /raɪz/ • *verb* (**-sing**; *past* **rose** /rəʊz/; *past participle* **risen** /ˈrɪz(ə)n/) come or go up; project or swell upwards; appear above horizon; get up from lying, sitting, or kneeling; get out of bed; (of meeting etc.) adjourn; reach higher level; make social progress; (often + *up*) rebel; come to surface; react to provocation; ascend, soar; have origin, begin to flow. • *noun* rising; upward slope; increase in amount, extent, pitch, etc.; increase in salary; increase in status or power; height of step, incline, etc.; origin. □ **give rise to** cause, induce; **get, take a rise out of** *colloquial* provoke reaction from; **on the rise** on the increase.

riser *noun* person who rises from bed; vertical piece between treads of staircase.

risible /ˈrɪzɪb(ə)l/ *adjective* laughable, ludicrous.

rising • *adjective* advancing; approaching specified age; going up. • *noun* insurrection. □ **rising damp** moisture absorbed from ground into wall.

risk • *noun* chance of danger, injury, loss, etc.; person or thing causing risk. • *verb* expose to risk; venture on, take chances of. □ **at risk** exposed to danger; **at one's (own) risk** accepting responsibility for oneself; **at the risk of** with the possibility of (adverse consequences).

risky *adjective* (**-ier**, **-iest**) involving risk; risqué. □ **riskily** *adverb*; **riskiness** *noun*.

risotto /rɪˈzɒtəʊ/ *noun* (*plural* **-s**) Italian savoury rice dish cooked in stock.

risqué /ˈrɪskeɪ/ *adjective* (of story etc.) slightly indecent.

rissole /ˈrɪsəʊl/ *noun* fried cake of minced meat coated in breadcrumbs.

ritardando /rɪtɑːˈdændəʊ/ *adverb, adjective, & noun* (*plural* **-s** or **-di** /-dɪ/) *Music* = RALLENTANDO.

rite *noun* religious or solemn ceremony or observance. □ **rite of passage** (often in *plural*) event marking change or stage in life.

ritual /ˈrɪtʃʊəl/ • *noun* prescribed order esp. of religious ceremony; solemn or colourful pageantry etc.; procedure regularly followed. • *adjective* of or done as ritual or rite. □ **ritually** *adverb*.

ritualism *noun* regular or excessive practice of ritual. □ **ritualist** *noun*; **ritualistic** /-ˈlɪs-/ *adjective*; **ritualistically** /-ˈlɪs-/ *adverb*.

rival /ˈraɪv(ə)l/ • *noun* person or thing that competes with another or equals another in quality. • *verb* (**-ll-**; *US* **-l-**) be rival of or comparable to.

rivalry *noun* (*plural* **-ies**) being rivals; competition.

riven /ˈrɪv(ə)n/ *adjective literary* split, torn.

river /ˈrɪvə/ *noun* large natural stream of water flowing to sea, lake, etc.; copious flow. □ **riverside** ground along riverbank.

rivet /ˈrɪvɪt/ • *noun* nail or bolt for joining metal plates etc. • *verb* (**-t-**) join or fasten with rivets; fix, make immovable; (+ *on, upon*) direct intently; (esp. as **riveting** *adjective*) engross.

riviera /rɪvɪˈeərə/ *noun* coastal subtropical region, esp. that of SE France and NW Italy.

rivulet /ˈrɪvjʊlɪt/ *noun* small stream.

RLC *abbreviation* Royal Logistics Corps.

RM *abbreviation* Royal Marines.

rm. *abbreviation* room.

RN *abbreviation* Royal Navy.

RNA *abbreviation* ribonucleic acid.

RNLI *abbreviation* Royal National Lifeboat Institution.

roach *noun* (*plural* same or **-es**) small freshwater fish.

road *noun* way with prepared surface for vehicles, etc.; route; (usually in *plural*) piece of water near shore in which ships can ride at anchor. □ **any road** *dialect* anyway; **in the** or **one's road** *dialect* forming obstruction; **on the road** travelling; **roadbed** foundation of road or railway, *US* part of road for vehicles; **roadblock** barrier on road to detain traffic; **road fund licence** = TAX DISC; **road-hog** *colloquial* reckless or inconsiderate motorist etc.; **roadhouse** inn etc. on main road; **road-metal** broken stone for road-making; **roadshow** touring entertainment etc., esp. radio or television series broadcast from changing venue; **roadstead** sea road for ships; **road tax** tax payable on vehicles; **road test** test of vehicle's roadworthiness; **roadway** part of road used by vehicles; **roadworks** construction or repair of roads; **roadworthy** (of vehicle) fit to be used on road; **roadworthiness** *noun*.

roadie *noun colloquial* assistant of touring band etc., responsible for equipment.

roadster *noun* open car without rear seats.

roam *verb* ramble, wander; travel unsystematically over, through, or about.

roan ● *adjective* (esp. of horse) with coat thickly interspersed with hairs of another colour. ● *noun* roan animal.

roar /rɔː/ ● *noun* loud deep hoarse sound as of lion; loud laugh. ● *verb* (often + *out*) utter loudly, or make roar, roaring laugh, etc.; travel in vehicle at high speed. □ **roaring drunk** very drunk and noisy; **roaring forties** stormy ocean tracts between latitudes 40° and 50°S; **roaring success** great success; **roaring twenties** decade of 1920s.

roast ● *verb* cook or be cooked by exposure to open heat or in oven; criticize severely. ● *adjective* roasted. ● *noun* (dish of) roast meat; meat for roasting; process of roasting.

rob *verb* (**-bb-**) (often + *of*) take unlawfully from, esp. by force; deprive of. □ **robber** *noun*; **robbery** *noun* (*plural* **-ies**).

robe ● *noun* long loose garment, esp. (often in *plural*) as indication of rank, office, etc.; *esp. US* dressing gown. ● *verb* (**-bing**) clothe in robe; dress.

robin /ˈrɒbɪn/ *noun* (also **robin redbreast**) small brown red-breasted bird.

Robin Hood *noun* person who steals from rich to give to poor.

robot /ˈrəʊbɒt/ *noun* automaton resembling or functioning like human; automatic mechanical device; machine-like person. □ **robotic** /-ˈbɒt-/ *adjective*; **robotize** *verb* (also **-ise**) (**-zing** or **-sing**).

robotics /rəʊˈbɒtɪks/ *plural noun* (usually treated as *singular*) science or study of robot design and operation.

robust /rəʊˈbʌst/ *adjective* (**-er**, **-est**) strong, esp. in health and physique; (of exercise etc.) vigorous; straightforward; (of statement etc.) bold. □ **robustly** *adverb*; **robustness** *noun*.

roc *noun* gigantic bird of Eastern legend.

rock[1] *noun* solid part of earth's crust; material or projecting mass of this; (**the Rock**) Gibraltar; large detached stone; *US* stone of any size; firm support or protection; hard sweet usually as peppermint-flavoured stick; *slang* precious stone, esp. diamond. □ **on the rocks** *colloquial* short of money, (of marriage) broken down, (of drink) served with ice cubes; **rock-bottom** very lowest (level); **rock-cake** bun with rough surface; **rock crystal** crystallized quartz; **rock face** vertical surface of natural rock; **rock-garden** rockery; **rock plant** plant that grows on or among rocks; **rock-salmon** catfish, dogfish, etc.; **rock salt** common salt as solid mineral.

rock[2] ● *verb* move gently to and fro; set, keep, or be in such motion; (cause to) sway; oscillate; shake, reel. ● *noun* rocking motion; rock and roll, popular music influenced by this. □ **rock and roll**, **rock 'n' roll** popular dance music with heavy beat and blues influence; **rocking-chair** chair on rockers or springs; **rocking-horse** toy horse on rockers or springs.

rocker *noun* device for rocking, esp. curved bar etc. on which something rocks; rocking-chair; rock music devotee, esp. leather-clad motorcyclist.

rockery *noun* (*plural* **-ies**) pile of rough stones with soil between them for growing rock plants on.

rocket /ˈrɒkɪt/ ● *noun* firework or signal propelled to great height after ignition; engine operating on same principle; rocket-propelled missile, spacecraft, etc. ● *verb* (**-t-**) move rapidly upwards or away; bombard with rockets.

rocketry *noun* science or practice of rocket propulsion.

rocky[1] *adjective* (**-ier**, **-iest**) of, like, or full of rocks.

rocky[2] *adjective* (**-ier**, **-iest**) *colloquial* unsteady, tottering.

rococo /rəˈkəʊkəʊ/ ● *adjective* of ornate style of art, music, and literature in 18th-c. Europe. ● *noun* this style.

rod *noun* slender straight round stick or bar; cane for flogging; fishing-rod; *historical* measure of length (5½ yds.).

rode *past* of RIDE.

rodent /ˈrəʊd(ə)nt/ *noun* mammal with strong incisors and no canine teeth (e.g. rat, squirrel, beaver).

rodeo /ˈrəʊdɪəʊ/ *noun* (*plural* **-s**) exhibition of cowboys' skills; round-up of cattle for branding etc.

roe[1] *noun* (also **hard roe**) mass of eggs in female fish; (also **soft roe**) male fish's milt.

roe[2] *noun* (*plural* same or **-s**) (also **roe-deer**) small kind of deer. □ **roebuck** male roe.

roentgen /ˈrʌntjən/ *noun* (also **röntgen**) unit of exposure to ionizing radiation.

rogation /rəʊˈgeɪʃ(ə)n/ *noun* (usually in *plural*) litany of the saints chanted on the 3 days (**Rogation Days**) before Ascension Day.

roger /ˈrɒdʒə/ *interjection* your message has been received and understood; *slang* I agree.

rogue /rəʊg/ *noun* dishonest or unprincipled person; *jocular* mischievous person; wild fierce animal driven or living apart from herd; inferior or defective specimen. □ **roguery** *noun* (*plural* **-ies**); **roguish** *adjective*.

roister /ˈrɔɪstə/ *verb* (esp. as **roistering** *adjective*) revel noisily, be uproarious. □ **roisterer** *noun*.

role *noun* (also **rôle**) actor's part; person's or thing's function. □ **role model** person on whom others model themselves; **role-playing**, **-play** acting of characters or situations as aid in psychotherapy, teaching, etc.; **role-play** *verb*.

roll /rəʊl/ ● *verb* (cause to) move or go in some direction by turning over and over on axis; make cylindrical or spherical by revolving between two surfaces or over on itself; gather into mass; (often + *along*, *by*, etc.) move or be carried on or as if on wheels; flatten with roller; rotate; sway or rock; proceed unsteadily; undulate, show undulating motion or surface; sound with vibration. ● *noun* rolling motion or gait; undulation; act of rolling; rhythmic rumbling sound; anything forming cylinder by being turned over on itself without folding; small loaf of bread for one person; official list or register. □ **roll-call** calling of list of names to establish presence; **rolled gold** thin coating of gold applied by roller to base metal; **rolled oats** husked and crushed oats; **rolling-mill** machine or factory for rolling metal into shape; **rolling-pin** roller for pastry; **rolling-stock** company's railway or (*US*) road vehicles; **rollmop** rolled pickled herring fillet; **roll-neck** having high loosely turned-over collar; **roll-on** applied by means of rotating ball; **roll-on roll-off** (of ship etc.) in which vehicles are driven directly on and off; **roll-top desk** desk with flexible cover sliding in curved grooves; **roll-up** hand-rolled cigarette; **strike off the rolls** debar from practising as solicitor.

roller *noun* revolving cylinder for smoothing, flattening, crushing, spreading, etc.; small cylinder on which hair is rolled for setting; long swelling wave. □ **roller bearing** bearing with cylinders instead of balls; **roller coaster** switchback at fair etc.; **roller-skate** *noun* frame with small wheels, strapped to shoes; boot with small wheels underneath; *verb* move on roller-skates; **roller towel** towel with ends joined.

rollicking /ˈrɒlɪkɪŋ/ *adjective* jovial, exuberant.

roly-poly /rəʊlɪˈpəʊlɪ/ ● *noun* (*plural* **-ies**) (also **roly-poly pudding**) pudding of rolled-up suet pastry covered with jam and boiled or baked. ● *adjective* podgy, plump.

ROM *noun Computing* read-only memory.

Roman /ˈrəʊmən/ ● *adjective* of ancient Rome or its territory or people; of medieval or modern Rome; Roman Catholic; (**roman**) (of type) plain and upright, used in ordinary print; (of the alphabet etc.) based on the ancient Roman system with letters A–Z. ● *noun* (*plural* **-s**) citizen of ancient Roman Republic or Empire, or of modern Rome; Roman Catholic; (**roman**) roman type. □ **Roman candle** firework discharging coloured sparks; **Roman Catholic** *adjective* of part of Christian Church acknowledging Pope as its head, *noun* member of this; **Roman Catholicism** *noun*; **Roman Empire** *historical* that established in 27 BC and divided in AD 395; **Roman law** law code of ancient Rome, forming basis of many modern codes; **Roman nose** one with high bridge; **roman numerals** numerals expressed in letters of Roman alphabet. □ **romanize** *verb* (also **-ise**) (**-zing** or **-sing**); **romanization** *noun*.

romance /rəʊˈmæns/ ● *noun* (also /ˈrəʊ-/) idealized, poetic, or unworldly atmosphere or tendency; love affair; (work of) literature concerning romantic love, stirring action, etc.; medieval tale of chivalry; exaggeration, picturesque falsehood. ● *adjective* (**Romance**) (of a language) descended from Latin. ● *verb* (**-cing**) exaggerate, fantasize; woo.

Romanesque /rəʊməˈnesk/ ● *noun* style of European architecture *c.* 900–1200, with massive vaulting and round arches. ● *adjective* of this style.

Romanian /rəʊˈmeɪnɪən/ (also **Rumanian** /ru:-/) ● *noun* native, national, or language of Romania. ● *adjective* of Romania or its people or language.

romantic /rəʊˈmæntɪk/ ● *adjective* of, characterized by, or suggestive of romance; imaginative, visionary; (of literature or music etc.) concerned more with emotion than with form; (also **Romantic**) of the 18th–19th-c. romantic movement or style in European arts. ● *noun* romantic person; romanticist. □ **romantically** *adverb*.

romanticism /rəʊˈmæntɪsɪz(ə)m/ *noun* (also **Romanticism**) adherence to romantic style in literature, art, etc. □ **romanticist**, **Romanticist** *noun*.

romanticize /rəʊˈmæntɪsaɪz/ *verb* (also **-ise**) (**-zing** or **-sing**) make romantic; exaggerate; indulge in romance.

Romany /ˈrɒmənɪ/ ● *noun* (*plural* **-ies**) Gypsy; language of Gypsies. ● *adjective* of Gypsies or Romany language.

Romeo /ˈrəʊmɪəʊ/ *noun* (*plural* **-s**) passionate male lover or seducer.

romp ● *verb* play roughly and energetically; (+ *along*, *past*, etc.) *colloquial* proceed without effort. ● *noun* spell of romping. □ **romp in**, **home** *colloquial* win easily.

rompers *plural noun* (also **romper suit**) young child's one-piece garment.

rondeau /ˈrɒndəʊ/ *noun* (*plural* **-x** same pronunciation or /-z/) short poem with two rhymes only, and opening words used as refrains.

rondel /ˈrɒnd(ə)l/ *noun* rondeau.

rondo /ˈrɒndəʊ/ *noun* (*plural* **-s**) musical form with recurring leading theme.

röntgen = ROENTGEN.

rood /ru:d/ *noun* crucifix, esp. on rood-screen; quarter-acre. □ **rood-screen** carved screen separating nave and chancel.

roof /ru:f/ ● *noun* (*plural* **roofs** /ru:vz/) upper covering of building; top of covered vehicle etc.; top interior surface of oven, cave, mine, etc. ● *verb* (often + *in*, *over*) cover with roof; be roof of. □ **roof of the mouth** palate; **roof-rack** framework for luggage on top of vehicle; **rooftop** outer surface of roof, (in *plural*) tops of houses etc.

rook[1] /rʊk/ ● *noun* black bird of crow family nesting in colonies. ● *verb colloquial* charge (customer) extortionately; win money at cards etc., esp. by swindling.

rook[2] /rʊk/ *noun* chess piece with battlement-shaped top.

rookery *noun* (*plural* **-ies**) colony of rooks, penguins, or seals.

rookie /ˈrʊkɪ/ *noun slang* recruit.

room /ru:m/ ● *noun* space for, or occupied by, something; capacity; part of building enclosed by walls; (in *plural*) apartments or lodgings. ● *verb US* have room(s), lodge. □ **room service** provision of food etc. in hotel bedroom.

roomy *adjective* (**-ier**, **-iest**) having much room, spacious. □ **roominess** *noun*.

roost /ru:st/ ● *noun* bird's perch. ● *verb* (of bird) settle for rest or sleep.

rooster *noun* domestic cock.

root[1] /ru:t/ ● *noun* part of plant below ground conveying nourishment from soil; (in *plural*) fibres or branches of this; plant with edible root, such root; (in *plural*) emotional attachment or family ties in a place; embedded part of hair or tooth etc.; basic cause, source; *Mathematics* number which multiplied by itself a given number of times yields a given number, esp. square root; core of a word. ● *verb* (cause to) take root; (esp. as **rooted** *adjective*) fix or establish firmly; pull up by roots. □ **root out** find and get rid of; **rootstock** rhizome, plant into which graft is inserted, source from which offshoots have arisen; **take root** begin to draw nourishment from the soil, become established. □ **rootless** *adjective*.

root[2] /ru:t/ *verb* (often + *up*) turn up (ground) with snout etc. in search of food; (+ *around*, *in*, etc.) rummage; (+ *out*, *up*) extract by rummaging; (+ *for*) *US slang* encourage by applause or support.

rope ● *noun* stout cord made by twisting together strands of hemp or wire etc.; (+ *of*) string of onions, pearls etc.; (**the rope**) (halter for) execution by hanging. ● *verb* (**-ping**) fasten or catch with rope; (+ *off*, *in*) enclose with rope. □ **know**, **learn**, or **show the ropes** know, learn, show how to do a thing properly; **rope into** persuade to take part (in).

ropy *adjective* (also **ropey**) (**-ier**, **-iest**) *colloquial* poor in quality. □ **ropiness** *noun*.

Roquefort /'rɒkfɔ:/ *noun proprietary term* soft blue ewe's-milk cheese.

rorqual /'rɔ:kw(ə)l/ *noun* whale with dorsal fin.

rosaceous /rəʊ'zeɪʃəs/ *adjective* of plant family including the rose.

rosary /'rəʊzərɪ/ *noun* (*plural* **-ies**) *RC Church* repeated sequence of prayers; string of beads for keeping count in this.

rose[1] /rəʊz/ ● *noun* prickly shrub bearing fragrant red, pink, yellow, or white flowers; this flower; pinkish-red colour or (usually in *plural*) complexion; rose-shaped design; circular fitting on ceiling from which electric light hangs by cable; spray nozzle of watering-can etc.; (in *plural*) *used to express ease, luck, etc.* ● *adjective* rose-coloured. □ **rosebowl** bowl for cut roses, esp. given as prize; **rosebud** bud of rose, pretty girl; **rose-coloured** pinkish-red, cheerful, optimistic; **rose-hip** fruit of rose; **rose-water** perfume made from roses; **rose-window** circular window with roselike tracery; **rosewood** close-grained wood used in making furniture.

rose[2] *past* of RISE.

rosé /'rəʊzeɪ/ *noun* light pink wine. [French]

rosemary /'rəʊzmərɪ/ *noun* evergreen fragrant shrub used as herb.

rosette /rəʊ'zet/ *noun* rose-shaped ornament made of ribbons etc. or carved in stone etc.

rosin /'rɒzɪn/ ● *noun* resin, esp. in solid form. ● *verb* (**-n-**) rub with rosin.

RoSPA /'rɒspə/ *abbreviation* Royal Society for the Prevention of Accidents.

roster /'rɒstə/ ● *noun* list or plan of turns of duty etc. ● *verb* place on roster.

rostrum /'rɒstrəm/ *noun* (*plural* **rostra** or **-s**) platform for public speaking etc.

rosy /'rəʊzɪ/ *adjective* (**-ier**, **-iest**) pink, red; optimistic, hopeful.

rot ● *verb* (**-tt-**) undergo decay by putrefaction; perish, waste away; cause to rot, make rotten. ● *noun* decay, rottenness; *slang* nonsense; decline in standards etc. ● *interjection expressing incredulity or ridicule.* □ **rot-gut** *slang* cheap harmful alcohol.

rota /'rəʊtə/ *noun* list of duties to be done or people to do them in turn.

rotary /'rəʊtərɪ/ ● *adjective* acting by rotation. ● *noun* (*plural* **-ies**) rotary machine; (**Rotary**; in full **Rotary International**) worldwide charitable society of businessmen.

rotate /rəʊ'teɪt/ *verb* (**-ting**) move round axis or centre, revolve; arrange or take in rotation. □ **rotatory** /ˈrəʊtətərɪ/ *adjective.*

rotation *noun* rotating, being rotated; recurrent series or period; regular succession; growing of different crops in regular order. □ **rotational** *adjective.*

rote *noun* (usually in **by rote**) mechanical repetition (in order to memorize).

rotisserie /rəʊˈtɪsərɪ/ *noun* restaurant etc. where meat is roasted; revolving spit for roasting food.

rotor /ˈrəʊtə/ *noun* rotary part of machine; rotating aerofoil on helicopter.

rotten /ˈrɒt(ə)n/ *adjective* **(-er, -est)** rotting or rotted; fragile from age etc.; morally or politically corrupt; *slang* disagreeable, worthless. □ **rotten borough** *historical* (before 1832) English borough electing MP though having very few voters. □ **rottenness** *noun.*

rotter *noun slang* objectionable person.

Rottweiler /ˈrɒtvaɪlə/ *noun* black-and-tan dog noted for ferocity.

rotund /rəʊˈtʌnd/ *adjective* plump, podgy. □ **rotundity** *noun.*

rotunda /rəʊˈtʌndə/ *noun* circular building, esp. domed.

rouble /ˈruːb(ə)l/ *noun* (also **ruble**) monetary unit of Russia etc.

roué /ˈruːeɪ/ *noun* (esp. elderly) debauchee.

rouge /ruːʒ/ ● *noun* red cosmetic used to colour cheeks. ● *verb* **(-ging)** colour with or apply rouge; blush.

rough /rʌf/ ● *adjective* having uneven surface, not smooth or level; shaggy; coarse, violent; not mild, quiet, or gentle; (of wine) harsh; insensitive; unpleasant, severe; lacking finish etc.; approximate, rudimentary. ● *adverb* in a rough way. ● *noun* (usually **the rough**) hardship; rough ground; hooligan; unfinished or natural state. ● *verb* make rough; (+ *out, in*) sketch or plan roughly. □ **rough-and-ready** rough or crude but effective, not over-particular; **rough-and-tumble** *adjective* irregular, disorderly, *noun* scuffle; **roughcast** *noun* plaster of lime and gravel, *verb* coat with this; **rough diamond** uncut diamond, rough but honest person; **rough house** *slang* disturbance, rough fight; **rough it** *colloquial* do without basic comforts; **rough justice** treatment that is approximately fair, unfair treatment; **roughneck** *colloquial* worker on oil rig, rough person; **rough up** *slang* attack violently. □ **roughen** *verb*; **roughly** *adverb*; **roughness** *noun.*

roughage *noun* fibrous material in food, stimulating intestinal action.

roughshod /ˈrʌfʃɒd/ □ **ride roughshod over** treat arrogantly.

roulade /ruːˈlɑːd/ *noun* filled rolled piece of meat, sponge, etc.; quick succession of notes.

roulette /ruːˈlet/ *noun* gambling game with ball dropped on revolving numbered wheel.

round /raʊnd/ ● *adjective* shaped like circle, sphere, or cylinder; done with circular motion; (of number etc.) without odd units; entire, continuous, complete; candid; (of voice etc.) sonorous. ● *noun* round object; revolving motion, circular or recurring course, series; route for deliveries, inspection, etc.; drink etc. for each member of group; one bullet, shell, etc.; slice of bread, sandwich made from two slices, joint of beef from haunch; one period of play etc., one stage in competition, playing of all holes in golf course once; song for unaccompanied voices overlapping at intervals; rung of ladder; (+ *of*) circumference or extent of. ● *adverb* with circular motion, with return to starting point or change to opposite position; to, at, or affecting circumference, area, group, etc.; in every direction from a centre; measuring (specified distance) in girth. ● *preposition* so as to encircle or enclose; at or to points on circumference of; with successive visits to; within a radius of; having as central point; so as to pass in curved course, having thus passed. ● *verb* give or take round shape; pass round (corner etc.); (usually + *up, down*) express (number) approximately. □ **in the round** with all angles or features shown or considered, with audience all round theatre stage; **Roundhead** *historical* member of Parliamentary party in English Civil War; **round off** make complete or less angular; **round on** attack unexpectedly; **round out** provide with more details, finish; **round robin** petition with signatures in circle to conceal order of writing, tournament in which each competitor plays every other; **Round Table** international charitable association; **round table** assembly for discussion, esp. at conference; **round trip** trip to one or more places and back; **round up** gather or bring together; **round-up** rounding-up, summary.

roundabout • *noun* road junction with traffic passing in one direction round central island; revolving device in children's playground; merry-go-round. • *adjective* circuitous.

roundel /'raʊnd(ə)l/ *noun* circular mark; small disc, medallion.

roundelay /'raʊndɪleɪ/ *noun* short simple song with refrain.

rounders /'raʊndəz/ *noun* team game in which players hit ball and run through round of bases.

roundly *adverb* bluntly, severely.

rouse /raʊz/ *verb* (**-sing**) (cause to) wake; (often + *up*) make or become active or excited; anger; evoke (feelings). □ **rousing** *adjective*.

roustabout /'raʊstəbaʊt/ *noun* labourer on oil rig; unskilled or casual labourer.

rout /raʊt/ • *noun* disorderly retreat of defeated troops; overthrow, defeat. • *verb* put to flight, defeat.

route /ru:t/ • *noun* way taken (esp. regularly) from one place to another. • *verb* (**-teing**) send etc. by particular route. □ **route march** training-march for troops.

routine /ru:'ti:n/ • *noun* regular course or procedure, unvarying performance of certain acts; set sequence in dance, comedy act, etc.; sequence of instructions to computer. • *adjective* performed as routine; of customary or standard kind. □ **routinely** *adverb*.

roux /ru:/ *noun* (*plural* same) mixture of fat and flour used in sauces etc.

rove /rəʊv/ *verb* (**-ving**) wander without settling, roam; (of eyes) look about.

rover *noun* wanderer.

row¹ /rəʊ/ *noun* line of people or things; line of seats in theatre etc. □ **in a row** forming a row, *colloquial* in succession.

row² /rəʊ/ • *verb* propel (boat) with oars; convey thus. • *noun* spell of rowing; trip in rowing boat. □ **rowing boat** (*US* **row-boat**) small boat propelled by oars. □ **rower** *noun*.

row³ /raʊ/ *colloquial* • *noun* loud noise, commotion; quarrel, dispute; severe reprimand. • *verb* make or engage in row; reprimand.

rowan /'rəʊən/ *noun* (in full **rowan tree**) mountain ash; (in full **rowan-berry**) its scarlet berry.

rowdy /'raʊdɪ/ • *adjective* (**-ier**, **-iest**) noisy and disorderly. • *noun* (*plural* **-ies**) rowdy person. □ **rowdily** *adverb*; **rowdiness** *noun*; **rowdyism** *noun*.

rowel /'raʊəl/ *noun* spiked revolving disc at end of spur.

rowlock /'rɒlək/ *noun* device for holding oar in place.

royal /'rɔɪəl/ • *adjective* of, suited to, or worthy of king or queen; in service or under patronage of king or queen; of family of king or queen; splendid, on great scale. • *noun colloquial* member of royal family. □ **royal blue** deep vivid blue; **Royal Commission** commission of inquiry appointed by Crown at request of government; **royal flush** straight poker flush headed by ace; **royal jelly** substance secreted by worker bees and fed to future queen bees; **Royal Navy** British navy; **royal 'we'** use of 'we' instead of 'I' by single person. □ **royally** *adverb*.

royalist *noun* supporter of monarchy, *esp. historical* of King's side in English Civil War.

royalty *noun* (*plural* **-ies**) being royal; royal people; member of royal family; percentage of profit from book, public performance, patent, etc. paid to author etc.; royal right (now esp. over minerals) granted by sovereign; payment made by producer of minerals etc. to owner of site etc.

RP *abbreviation* Received Pronunciation.

RPI *abbreviation* retail price index.

rpm *abbreviation* revolutions per minute.

RSA *abbreviation* Royal Society of Arts; Royal Scottish Academy; Royal Scottish Academician.

RSC *abbreviation* Royal Shakespeare Company.

RSM *abbreviation* Regimental Sergeant-Major.

RSPB *abbreviation* Royal Society for the Protection of Birds.

RSPCA *abbreviation* Royal Society for the Prevention of Cruelty to Animals.

RSV *abbreviation* Revised Standard Version (of Bible).

RSVP *abbreviation* please answer (*répondez s'il vous plaît*).

Rt. Hon. *abbreviation* Right Honourable.

Rt Revd *abbreviation* (also **Rt. Rev.**) Right Reverend.

rub ● *verb* (**-bb-**) move hand etc. firmly over surface of; (usually + *against*, *in*, *on*, *over*) apply (hand etc.) thus; polish, clean, abrade, chafe, or make dry, sore, or bare by rubbing; (+ *in*, *into*, *through*, *over*) apply by rubbing; (often + *together*, *against*, *on*) move with friction or slide (objects) against each other; get frayed or worn by friction. ● *noun* action or spell of rubbing; impediment or difficulty. □ **rub off** (usually + *on*) be transferred by contact, be transmitted; **rub out** erase with rubber; **rub up the wrong way** irritate.

rubato /ru:'bɑ:təʊ/ *noun Music* (*plural* **-s** or **-ti** /-tɪ/) temporary disregarding of strict tempo.

rubber[1] /'rʌbə/ *noun* elastic substance made from latex of plants or synthetically; piece of this or other substance for erasing pencil marks; (in *plural*) *US* galoshes. □ **rubber band** loop of rubber to hold papers etc.; **rubberneck** *colloquial* (be) inquisitive sightseer; **rubber plant** tropical plant often grown as house-plant, (also **rubber tree**) tree yielding latex; **rubber stamp** device for inking and imprinting on surface, (person giving) mechanical endorsement of actions etc.; **rubber-stamp** approve automatically. □ **rubberize** *verb* (also **-ise**) (**-zing** or **-sing**); **rubbery** *adjective*.

rubber[2] /'rʌbə/ *noun* series of games between same sides or people at whist, bridge, cricket, etc.

rubbish /'rʌbɪʃ/ ● *noun* waste or worthless matter; litter; trash; (often as *interjection*) nonsense. ● *verb colloquial* criticize contemptuously. □ **rubbishy** *adjective*.

rubble /'rʌb(ə)l/ *noun* rough fragments of stone, brick, etc.

rubella /ru:'belə/ *noun formal* German measles.

Rubicon /'ru:bɪkɒn/ *noun* point from which there is no going back.

rubicund /'ru:bɪkʌnd/ *adjective* ruddy, red-faced.

ruble = ROUBLE.

rubric /'ru:brɪk/ *noun* heading or passage in red or special lettering; explanatory words; established custom or rule; direction for conduct of divine service in liturgical book.

ruby /'ru:bɪ/ ● *noun* (*plural* **-ies**) crimson or rose-coloured precious stone; deep red colour. ● *adjective* ruby-coloured. □ **ruby wedding** 40th wedding anniversary.

RUC *abbreviation* Royal Ulster Constabulary.

ruche /ru:ʃ/ *noun* frill or gathering of lace etc. □ **ruched** *adjective*.

ruck[1] *noun* (**the ruck**) main group of competitors not likely to overtake leaders; undistinguished crowd of people or things; *Rugby* loose scrum.

ruck[2] ● *verb* (often + *up*) crease, wrinkle. ● *noun* crease, wrinkle.

rucksack /'rʌksæk/ *noun* bag carried on back, esp. by hikers.

ruckus /'rʌkəs/ *noun esp. US* row, commotion.

ruction /'rʌkʃ(ə)n/ *noun colloquial* disturbance, tumult; (in *plural*) row.

rudder /'rʌdə/ *noun* flat piece hinged to vessel's stern or rear of aeroplane for steering. □ **rudderless** *adjective*.

ruddy /'rʌdɪ/ *adjective* (**-ier**, **-iest**) freshly or healthily red; reddish; *colloquial* bloody, damnable.

rude *adjective* impolite, offensive; roughly made; primitive, uneducated; abrupt, sudden; *colloquial* indecent, lewd; vigorous, hearty. □ **rudely** *adverb*; **rudeness** *noun*.

rudiment /'ru:dɪmənt/ *noun* (in *plural*) elements or first principles of subject, imperfect beginning of something undeveloped; vestigial or undeveloped part or organ. □ **rudimentary** /-'mentərɪ/ *adjective*.

rue[1] *verb* (**rues**, **rued**, **rueing** or **ruing**) repent of; wish undone or non-existent.

rue[2] *noun* evergreen shrub with bitter strong-scented leaves.

rueful *adjective* genuinely or humorously sorrowful. □ **ruefully** *adverb*.

ruff[1] *noun* projecting starched frill worn round neck; projecting or coloured ring of feathers or hair round bird's or animal's neck; domestic pigeon.

ruff[2] • *verb* trump at cards. • *noun* trumping.

ruffian /'rʌfɪən/ *noun* violent lawless person.

ruffle /'rʌf(ə)l/ • *verb* (**-ling**) disturb smoothness or tranquillity of; gather into ruffle; (often + *up*) (of bird) erect (feathers) in anger, display, etc. • *noun* frill of lace etc.

rufous /'ru:fəs/ *adjective* reddish-brown.

rug *noun* floor-mat; thick woollen wrap or coverlet.

Rugby /'rʌgbɪ/ *noun* (in full **Rugby football**) team game played with oval ball that may be kicked or carried. □ **Rugby League** partly professional Rugby with teams of 13; **Rugby Union** amateur Rugby with teams of 15.

rugged /'rʌgɪd/ *adjective* (esp. of ground) rough, uneven; (of features) furrowed, irregular; harsh; robust. □ **ruggedly** *adverb*; **ruggedness** *noun*.

rugger /'rʌgə/ *noun colloquial* Rugby.

ruin /'ru:ɪn/ • *noun* wrecked or spoiled state; downfall; loss of property or position; (in *singular* or *plural*) remains of building etc. that has suffered ruin; cause of ruin. • *verb* bring to ruin; spoil, damage; (esp. as **ruined** *adjective*) reduce to ruins. □ **ruination** *noun*.

ruinous *adjective* bringing ruin; disastrous; dilapidated.

rule • *noun* compulsory principle governing action; prevailing custom, standard, normal state of things; government, dominion; straight measuring device, ruler; code of discipline of religious order; *Printing* thin line or dash. • *verb* (**-ling**) dominate; keep under control; (often + *over*) have sovereign control of; (often + *that*) pronounce authoritatively; make parallel lines across (paper), make (straight line) with ruler etc. □ **as a rule** usually; **rule of thumb** rule based on experience or practice, not theory; **rule out** exclude.

ruler *noun* person exercising government or dominion; straight strip of plastic etc. used to draw or measure.

ruling *noun* authoritative pronouncement.

rum[1] *noun* spirit distilled from sugar cane or molasses. □ **rum baba** sponge cake soaked in rum syrup.

rum[2] *adjective* (**-mm-**) *colloquial* queer, strange.

Rumanian = ROMANIAN.

rumba /'rʌmbə/ *noun* ballroom dance of Cuban origin; music for this.

rumble /'rʌmb(ə)l/ • *verb* (**-ling**) make continuous deep sound as of thunder; (+ *along*, *by*, *past*, etc.) (esp. of vehicle) move with such sound; *slang* see through, detect. • *noun* rumbling sound.

rumbustious /rʌm'bʌstʃəs/ *adjective colloquial* boisterous, uproarious.

ruminant /'ru:mɪnənt/ • *noun* animal that chews the cud. • *adjective* of ruminants; meditative.

ruminate /'ru:mɪneɪt/ *verb* (**-ting**) meditate, ponder; chew the cud. □ **rumination** *noun*; **ruminative** /-nətɪv/ *adjective*.

rummage /'rʌmɪdʒ/ • *verb* (**-ging**) search, esp. unsystematically; (+ *up*, *out*) find among other things. • *noun* rummaging. □ **rummage sale** *esp. US* jumble sale.

rummy /'rʌmɪ/ *noun* card game played usually with two packs.

rumour /'ru:mə/ (*US* **rumor**) • *noun* (often + *of*, *that*) general talk, assertion, or hearsay of doubtful accuracy. • *verb* (usually in *passive*) report by way of rumour.

rump *noun* hind part of mammal or bird, esp. buttocks; remnant of parliament etc. □ **rump steak** cut of beef from rump.

rumple /'rʌmp(ə)l/ *verb* (**-ling**) crease, ruffle.

rumpus /'rʌmpəs/ *noun colloquial* row, uproar.

run • *verb* (**-nn-**; *past* **ran**; *past participle* **run**) go at pace faster than walk; flee; go or travel hurriedly, briefly, etc.; advance smoothly or (as) by rolling or on wheels; (cause to) be in action or operation; be current or operative; (of bus, train, etc.) travel on its route; (of play etc.) be presented; extend, have course or tendency; compete in or enter (horse etc.) in race etc.; (often + *for*) seek election; (cause to) flow or emit liquid; spread rapidly; perform (errand); publish (article etc.); direct (business etc.); own and use (vehicle); smuggle; (of thought, the eye, etc.) pass quickly; (of tights etc.) ladder. ● *noun* running; short excursion; distance travelled; general tendency; regular route; continuous stretch, spell, or course; (often + *on*) high general demand; quantity produced at one time; general or average type or class; point scored in cricket or baseball; (+ *of*) free use of; animal's regular track; enclosure for fowls etc.; range of pasture; ladder in tights etc.; *Music* rapid scale passage. □ **give (person) the run-around** deceive, evade; **on the run** fleeing; **runabout** light car or aircraft; **run across** happen to meet; **run after** pursue; **run away** *verb* (often + *from*) flee, abscond; **runaway** *noun* person, animal, vehicle, etc. running away or out of control; **run down** *verb* knock down, reduce numbers of, (of clock etc.) stop, discover after search, *colloquial* disparage; **run-down** *noun* reduction in numbers, detailed analysis, *adjective* dilapidated, decayed, exhausted; **run dry** cease to flow; **run in** *verb* run (vehicle, engine) carefully when new, *colloquial* arrest; **run-in** *noun colloquial* quarrel; **run into** collide with, encounter, reach as many as; **run low**, **short** become depleted, have too little; **run off** flee, produce (copies) on machine, decide (race) after tie or heats, write or recite fluently; **run-of-the-mill** ordinary, not special; **run on** continue in operation, speak volubly, continue on same line as preceding matter; **run out** come to an end, (+ *of*) exhaust one's stock of, put down wicket of (running batsman); **run out on** *colloquial* desert; **run over** (of vehicle) knock down or crush, overflow, review quickly; **run through** *verb* examine or rehearse briefly, deal successively with, spend money rapidly, pervade, pierce with blade; **run-through** *noun* rehearsal; **run to** have money or ability for, reach (amount etc.), show tendency to; **run up** *verb* accumulate (debt etc.), build or make hurriedly, raise (flag); **run-up** *noun* (often + *to*) preparatory period; **run up against** meet with (difficulty etc.); **runway** specially prepared airfield surface for taking off and landing.

rune *noun* letter of earliest Germanic alphabet; similar character of mysterious or magic significance. □ **runic** *adjective*.

rung[1] *noun* step of ladder; strengthening crosspiece of chair etc.

rung[2] *past participle* of RING[2].

runnel /'rʌn(ə)l/ *noun* brook; gutter.

runner *noun* racer; creeping rooting plant-stem; groove, rod, etc. for thing to slide along or on; sliding ring on rod etc.; messenger; long narrow ornamental cloth or rug; (in full **runner bean**) kind of climbing bean. □ **runner-up** (*plural* **runners-up** or **runner-ups**) competitor taking second place.

running ● *noun* act or manner of running race etc. ● *adjective* continuous; consecutive; done with a run. □ **in** or **out of the running** with good or poor chance of success; **running commentary** verbal description of events in progress; **running knot** one that slips along rope etc. to allow tightening; **running mate** *US* vice-presidential candidate, horse setting pace for another; **running repairs** minor or temporary repairs; **running water** flowing water, esp. on tap.

runny *adjective* (**-ier**, **-iest**) tending to run or flow; excessively fluid.

runt *noun* smallest pig etc. of litter; undersized person.

rupee /ru:'pi:/ *noun* monetary unit of India, Pakistan, etc.

rupiah /ru:'pi:ə/ *noun* monetary unit of Indonesia.

rupture /'rʌptʃə/ ● *noun* breaking, breach; breach in relationship; abdominal hernia. ● *verb* (**-ring**) burst (cell, membrane, etc.); sever (connection); affect with or suffer hernia.

rural /'rʊər(ə)l/ *adjective* in, of, or suggesting country.

ruse /ru:z/ *noun* stratagem, trick.

rush[1] ● *verb* go, move, flow, or act precipitately or with great speed; move or transport with great haste; perform or deal with hurriedly; force (person) to act hastily; attack or capture by sudden assault. ● *noun* rushing; violent advance or attack; sudden flow; period of great activity; sudden migration of large numbers; (+ *on*, *for*) strong demand for a commodity; (in *plural*) *colloquial* first uncut prints of film. ● *adjective* done hastily. □ **rush hour** time each day when traffic is heaviest.

rush[2] *noun* marsh plant with slender pith-filled stem; its stem esp. used for making basketware etc.

rusk *noun* slice of bread rebaked as light biscuit, esp. for infants.

russet /ˈrʌsɪt/ ● *adjective* reddish-brown. ● *noun* russet colour; rough-skinned russet-coloured apple.

Russian /ˈrʌʃ(ə)n/ ● *noun* native or national of Russia or (loosely) former USSR; person of Russian descent; language of Russia. ● *adjective* of Russia or (loosely) former USSR or its people; of or in Russian. □ **Russian roulette** firing of revolver held to one's head after spinning cylinder with one chamber loaded; **Russian salad** mixed diced cooked vegetables with mayonnaise.

rust ● *noun* reddish corrosive coating formed on iron etc. by oxidation; plant disease with rust-coloured spots; reddish-brown. ● *verb* affect or be affected with rust; become impaired through disuse. □ **rustproof** not susceptible to corrosion by rust.

rustic /ˈrʌstɪk/ ● *adjective* of or like country people or country life; unsophisticated; of rough workmanship; made of untrimmed branches or rough timber; *Architecture* with roughened surface. ● *noun* country person, peasant. □ **rusticity** /-ˈtɪs-/ *noun*.

rusticate /ˈrʌstɪkeɪt/ *verb* (**-ting**) expel temporarily from university; retire to or live in the country; make rustic. □ **rustication** *noun*.

rustle /ˈrʌs(ə)l/ ● *verb* (**-ling**) (cause to) make sound as of dry blown leaves; steal (cattle or horses). ● *noun* rustling sound. □ **rustle up** *colloquial* produce at short notice. □ **rustler** *noun*.

rusty *adjective* (**-ier**, **-iest**) rusted, affected by rust; stiff with age or disuse; (of knowledge etc.) impaired by neglect; rust-coloured; discoloured by age.

rut[1] ● *noun* deep track made by passage of wheels; fixed (esp. tedious) practice or routine. ● *verb* (**-tt-**) (esp. as **rutted** *adjective*) mark with ruts.

rut[2] ● *noun* periodic sexual excitement of male deer etc. ● *verb* (**-tt-**) be affected with rut.

ruthless /ˈruːθlɪs/ *adjective* having no pity or compassion. □ **ruthlessly** *adverb*; **ruthlessness** *noun*.

RV *abbreviation* Revised Version (of Bible).

rye /raɪ/ *noun* cereal plant; grain of this, used for bread, fodder, etc.; (in full **rye whisky**) whisky distilled from rye.

Ss

S *abbreviation* (also **S.**) Saint; south(ern).

s. *abbreviation* second(s); *historical* shilling(s); son.

SA *abbreviation* Salvation Army; South Africa; South Australia.

sabbath /'sæbəθ/ *noun* religious rest-day kept by Christians on Sunday and Jews on Saturday.

sabbatical /sə'bætɪk(ə)l/ ● *adjective* (of leave) granted at intervals to university teacher for study or travel. ● *noun* period of sabbatical leave.

saber *US* = SABRE.

sable /'seɪb(ə)l/ ● *noun* (*plural* same or **-s**) small dark-furred mammal; its skin or fur. ● *adjective Heraldry* black; *esp. poetical* gloomy.

sabot /'sæbəʊ/ *noun* wooden or wooden-soled shoe.

sabotage /'sæbətɑ:ʒ/ ● *noun* deliberate destruction or damage, esp. for political purpose. ● *verb* (**-ging**) commit sabotage on; destroy, spoil.

saboteur /sæbə'tɜ:/ *noun* person who commits sabotage.

sabre /'seɪbə/ *noun* (*US* **saber**) curved cavalry sword; light fencing-sword. □ **sabre-rattling** display or threat of military force.

sac *noun* membranous bag in animal or plant.

saccharin /'sækərɪn/ *noun* a sugar substitute.

saccharine /'sækəri:n/ *adjective* excessively sentimental or sweet.

sacerdotal /sækə'dəʊt(ə)l/ *adjective* of priests or priestly office.

sachet /'sæʃeɪ/ *noun* small bag or packet containing shampoo, perfumed substances, etc.

sack[1] ● *noun* large strong bag for coal, food, mail, etc.; amount held by sack; (**the sack**) *colloquial* dismissal, *US slang* bed. ● *verb* put in sack(s); *colloquial* dismiss from employment. □ **sackcloth** coarse fabric of flax or hemp.

sack[2] ● *verb* plunder and destroy (town etc.). ● *noun* such sacking.

sack[3] *noun historical* white wine from Spain etc.

sackbut /'sækbʌt/ *noun* early form of trombone.

sacking *noun* sackcloth.

sacral /'seɪkr(ə)l/ *adjective* of sacrum.

sacrament /'sækrəmənt/ *noun* symbolic Christian ceremony, esp. Eucharist; sacred thing. □ **sacramental** /-'men-/ *adjective*.

sacred /'seɪkrɪd/ *adjective* (often + *to*) dedicated to a god, connected with religion; safeguarded or required, esp. by tradition; inviolable. □ **sacred cow** *colloquial* idea or institution unreasonably held to be above criticism.

sacrifice /'sækrɪfaɪs/ ● *noun* voluntary relinquishing of something valued; thing thus relinquished; loss entailed; slaughter of animal or person, or surrender of possession, as offering to deity; animal, person, or thing thus offered. ● *verb* (**-cing**) give up; (+ *to*) devote to; offer or kill (as) sacrifice. □ **sacrificial** /-'fɪʃ-/ *adjective*.

sacrilege /'sækrɪlɪdʒ/ *noun* violation of what is sacred. □ **sacrilegious** /-'lɪdʒəs/ *adjective*.

sacristan /'sækrɪst(ə)n/ *noun* person in charge of sacristy and church contents.

sacristy /'sækrɪstɪ/ *noun* (*plural* **-ies**) room in church for vestments, vessels, etc.

sacrosanct /'sækrəʊsæŋkt/ *adjective* most sacred; inviolable. □ **sacrosanctity** /-'sæŋkt-/ *noun*.

sacrum /'seɪkrəm/ *noun* (*plural* **sacra** or **-s**) triangular bone between hip-bones.

sad *adjective* (**-dd-**) sorrowful; causing sorrow; regrettable; deplorable. □ **sadden** *verb*; **sadly** *adverb*; **sadness** *noun*.

saddle /'sæd(ə)l/ ● *noun* seat of leather etc. fastened on horse etc.; bicycle etc. seat; joint of meat consisting of the two loins; ridge rising to a summit at each end. ● *verb* (**-ling**) put saddle on (horse etc.); (+ *with*) burden with task etc. □ **saddle-bag** each of pair of bags laid across back of horse etc., bag attached behind bicycle etc. saddle.

saddler *noun* maker of or dealer in saddles etc. □ **saddlery** *noun* (*plural* **-ies**).

sadism /'seɪdɪz(ə)m/ *noun colloquial* enjoyment of cruelty to others; sexual perversion characterized by this. □ **sadist** *noun*; **sadistic** /sə'dɪs-/ *adjective*; **sadistically** /sə'dɪs-/ *adverb*.

sadomasochism /seɪdəʊ'mæsəkɪz(ə)m/ *noun* sadism and masochism in one person. □ **sadomasochist** *noun*; **sadomasochistic** /-'kɪs-/ *adjective*.

s.a.e. *abbreviation* stamped addressed envelope.

safari /sə'fɑ:rɪ/ *noun* (*plural* **-s**) expedition, esp. in Africa, to observe or hunt animals. □ **safari park** area where wild animals are kept in open for viewing.

safe ● *adjective* free of danger or injury; secure, not risky; reliable, sure; prevented from escaping or doing harm; cautious. ● *noun* strong lockable cupboard for valuables; ventilated cupboard for provisions. □ **safe conduct** immunity from arrest or harm; **safe deposit** building containing strongrooms and safes for hire. □ **safely** *adverb*.

safeguard ● *noun* protecting proviso, circumstance, etc. ● *verb* guard or protect (rights etc.).

safety /'seɪftɪ/ *noun* being safe; freedom from danger. □ **safety-belt** belt or strap preventing injury, esp. seat-belt; **safety-catch** device preventing accidental operation of gun trigger or machinery; **safety curtain** fireproof curtain to divide theatre auditorium from stage; **safety match** match that ignites only on specially prepared surface; **safety net** net placed to catch acrobat etc. in case of fall; **safety pin** pin with guarded point; **safety razor** razor with guard to prevent user cutting skin; **safety-valve** valve relieving excessive pressure of steam, means of harmlessly venting excitement etc.

saffron /'sæfrən/ ● *noun* deep yellow colouring and flavouring from dried crocus stigmas; colour of this. ● *adjective* deep yellow.

sag ● *verb* (**-gg-**) sink or subside; have downward bulge or curve in middle. ● *noun* state or amount of sagging. □ **saggy** *adjective*.

saga /'sɑ:gə/ *noun* long heroic story, esp. medieval Icelandic or Norwegian; long family chronicle; long involved story.

sagacious /sə'geɪʃ(ə)s/ *adjective* showing insight or good judgement. □ **sagacity** /-'gæs-/ *noun*.

sage¹ *noun* aromatic herb with dull greyish-green leaves.

sage² ● *noun* wise man. ● *adjective* wise, judicious, experienced. □ **sagely** *adverb*.

Sagittarius /sædʒɪ'teərɪəs/ *noun* ninth sign of zodiac.

sago /'seɪgəʊ/ *noun* (*plural* **-s**) starch used in puddings etc.; (in full **sago palm**) any of several tropical trees yielding this.

sahib /sɑ:b/ *noun historical form of address to European men in India*.

said *past & past participle* of SAY.

sail ● *noun* piece of material extended on rigging to catch wind and propel vessel; ship's sails collectively; voyage or excursion in sailing vessel; wind-catching apparatus of windmill. ● *verb* travel on water by use of sails or engine-power; begin voyage; navigate (ship etc.); travel on (sea); glide or move smoothly or with dignity; (often + *through*) *colloquial* succeed easily. □ **sailboard** board with mast and sail, used in windsurfing; **sailcloth** material for sails, kind of coarse linen; **sailing boat**, **ship**, etc., vessel moved by sails; **sailplane** kind of glider.

sailor *noun* seaman or mariner, esp. below officer's rank. □ **bad**, **good sailor** person very liable or not liable to seasickness.

sainfoin /'sænfɔɪn/ *noun* pink-flowered plant used as fodder.

saint /seɪnt, before a name usually sənt/ ● *noun* holy or canonized person, regarded as deserving special veneration; very virtuous person. ● *verb* (as **sainted** *adjective*) holy, virtuous.

□ **sainthood** *noun*; **saintlike** *adjective*; **saintliness** *noun*; **saintly** (**-ier**, **-iest**) *adjective*.

sake[1] *noun* □ **for the sake of** out of consideration for, in the interest of, in order to please, get, etc.

sake[2] /'sɑ:kɪ/ *noun* Japanese rice wine.

salaam /sə'lɑ:m/ ● *noun* (*chiefly as Muslim greeting*) Peace!; low bow. ● *verb* make salaam.

salacious /sə'leɪʃəs/ *adjective* erotic; lecherous. □ **salaciousness** *noun*; **salacity** /-'læs-/ *noun*.

salad /'sæləd/ *noun* cold mixture of usually raw vegetables etc. often with dressing. □ **salad cream** creamy salad dressing; **salad days** period of youthful inexperience; **salad dressing** sauce of oil, vinegar, etc. for salads.

salamander /'sæləmændə/ *noun* newt-like amphibian formerly supposed to live in fire; similar mythical creature.

salami /sə'lɑ:mɪ/ *noun* (*plural* **-s**) highly-seasoned sausage, originally Italian.

sal ammoniac /sæl ə'məʊnɪæk/ *noun* ammonium chloride.

salary /'sælərɪ/ ● *noun* (*plural* **-ies**) fixed regular payment by employer to employee. ● *verb* (**-ies**, **-ied**) (usually as **salaried** *adjective*) pay salary to.

sale *noun* exchange of commodity for money etc.; act or instance of selling; amount sold; temporary offering of goods at reduced prices; event at which goods are sold. □ **on**, **for sale** offered for purchase; **saleroom** room where auctions are held; **salesman**, **salesperson**, **saleswoman** person employed to sell goods etc.

saleable *adjective* fit or likely to be sold. □ **saleability** *noun*.

salesmanship *noun* skill in selling.

salient /'seɪlɪənt/ ● *adjective* prominent, conspicuous; (of angle) pointing outwards. ● *noun* salient angle; outward bulge in military line.

saline /'seɪlaɪn/ ● *adjective* containing or tasting of salt(s); of salt(s). ● *noun* salt lake, spring, etc.; saline solution. □ **salinity** /sə'lɪn-/ *noun*.

saliva /sə'laɪvə/ *noun* colourless liquid produced by glands in mouth. □ **salivary** *adjective*.

salivate /'sælɪveɪt/ *verb* (**-ting**) secrete saliva, esp. in excess.

sallow[1] /'sæləʊ/ *adjective* (**-er**, **-est**) (esp. of complexion) yellowish.

sallow[2] /'sæləʊ/ *noun* low-growing willow; shoot or wood of this.

sally /'sælɪ/ ● *noun* (*plural* **-ies**) witticism; military rush; excursion. ● *verb* (**-ies**, **-ied**) (usually + *out*, *forth*) set out for walk etc., make sally.

salmon /'sæmən/ ● *noun* (*plural* same) large silver-scaled fish with orange-pink flesh. ● *adjective* orange-pink. □ **salmon-pink** orange-pink; **salmon trout** large silver-coloured trout.

salmonella /sælmə'nelə/ *noun* (*plural* **-llae** /-li:/) bacterium causing food poisoning; such food poisoning.

salon /'sælɒn/ *noun* room or establishment of hairdresser, fashion designer, etc.; *historical* meeting of eminent people at fashionable home; reception room of large house.

saloon /sə'lu:n/ *noun* large room or hall on ship, in hotel, etc., or for specified purpose; saloon car; *US* drinking bar; saloon bar. □ **saloon bar** more comfortable bar in public house; **saloon car** car with body closed off from luggage area.

salsa /'sælsə/ *noun* dance music of Cuban origin; kind of spicy tomato sauce.

salsify /'sælsɪfɪ/ *noun* (*plural* **-ies**) plant with long fleshy edible root.

salt /sɔ:lt/ ● *noun* (also **common salt**) sodium chloride, esp. mined or evaporated from sea water, and used esp. for seasoning or preserving food; *Chemistry* substance formed in reaction of an acid with a base; piquancy, wit; (in *singular* or *plural*) substance resembling salt in taste, form, etc.; (esp. in *plural*) substance used as laxative; (also **old salt**) experienced sailor. ● *adjective* containing, tasting of, or preserved with salt. ● *verb* cure, preserve, or season with salt; sprinkle salt on (road etc.). □ **salt away**, **down** *slang* put (money etc.) by; **salt-cellar** container for salt at table; **salt-mine** mine yielding rock salt; **salt of the earth** finest or most honest people; **salt-pan** vessel, or hollow near sea, used for getting salt by evaporation; **salt-water** of or living in sea;

take with a pinch or **grain of salt** be sceptical about; **worth one's salt** efficient, capable.

salting *noun* (esp. in *plural*) marsh overflowed by sea.

saltire /'sɔːltaɪə/ *noun* X-shaped cross.

saltpetre /sɔːlt'piːtə/ *noun* (*US* **saltpeter**) white crystalline salty substance used in preserving meat and in gunpowder.

salty *adjective* (**-ier**, **-iest**) tasting of or containing salt; witty, piquant. □ **saltiness** *noun*.

salubrious /sə'luːbrɪəs/ *adjective* health-giving. □ **salubrity** *noun*.

saluki /sə'luːkɪ/ *noun* (*plural* **-s**) dog of tall slender silky-coated breed.

salutary /'sæljʊtərɪ/ *adjective* producing good effect.

salutation /sælju:'teɪʃ(ə)n/ *noun formal* sign or expression of greeting.

salute /sə'luːt/ ● *noun* gesture of respect, homage, greeting, etc.; *Military etc.* prescribed gesture or use of weapons or flags as sign of respect etc. ● *verb* (**-ting**) make salute (to); greet; commend.

salvage /'sælvɪdʒ/ ● *noun* rescue of property from sea, fire, etc.; saving and utilization of waste materials; property or materials salvaged. ● *verb* (**-ging**) save from wreck etc. □ **salvageable** *adjective*.

salvation /sæl'veɪʃ(ə)n/ *noun* saving, being saved; deliverance from sin and damnation; religious conversion; person or thing that saves. □ **Salvation Army** worldwide quasi-military Christian charitable organization.

Salvationist *noun* member of Salvation Army.

salve[1] ● *noun* healing ointment; (often + *for*) thing that soothes. ● *verb* (**-ving**) soothe.

salve[2] *verb* (**-ving**) save from wreck, fire, etc. □ **salvable** *adjective*.

salver /'sælvə/ *noun* tray for drinks, letters, etc.

salvo /'sælvəʊ/ *noun* (*plural* **-es** or **-s**) simultaneous firing of guns etc.; round of applause.

sal volatile /sæl və'lætɪlɪ/ *noun* solution of ammonium carbonate, used as smelling salts.

SAM *abbreviation* surface-to-air missile.

Samaritan /sə'mærɪt(ə)n/ *noun* (in full **good Samaritan**) charitable or helpful person; member of counselling organization.

samba /'sæmbə/ ● *noun* ballroom dance of Brazilian origin; music for this. ● *verb* (**-bas**, **-baed** or **-ba'd** /-bəd/, **-baing** /-bəɪŋ/) dance samba.

same ● *adjective* identical; unvarying; just mentioned. ● *pronoun* (**the same**) the same person or thing. ● *adverb* (**the same**) in the same manner. □ **all** or **just the same** nevertheless; **at the same time** simultaneously, notwithstanding. □ **sameness** *noun*.

samosa /sə'məʊsə/ *noun* Indian fried triangular pastry containing spiced vegetables or meat.

samovar /'sæməvɑː/ *noun* Russian tea-urn.

Samoyed /'sæməjed/ *noun* member of a northern Siberian people; (also **samoyed**) dog of white Arctic breed.

sampan /'sæmpæn/ *noun* small boat used in Far East.

samphire /'sæmfaɪə/ *noun* cliff plant used in pickles.

sample /'sɑːmp(ə)l/ ● *noun* small representative part or quantity; specimen; typical example. ● *verb* (**-ling**) take samples of; try qualities of; experience briefly.

sampler /'sɑːmplə/ *noun* piece of embroidery worked to show proficiency.

samurai /'sæmʊraɪ/ *noun* (*plural* same) Japanese army officer; *historical* member of Japanese military caste.

sanatorium /sænə'tɔːrɪəm/ *noun* (*plural* **-riums** or **-ria**) residential clinic, esp. for convalescents and the chronically sick; accommodation for sick people in school etc.

sanctify /'sæŋktɪfaɪ/ *verb* (**-ies**, **-ied**) consecrate, treat as holy; purify from sin; sanction. □ **sanctification** *noun*.

sanctimonious /sæŋktɪ'məʊnɪəs/ *adjective* ostentatiously pious. □ **sanctimoniously** *adverb*; **sanctimony** /'sæŋktɪmənɪ/ *noun*.

sanction /'sæŋkʃ(ə)n/ ● *noun* approval by custom or tradition; express permission; confirmation of law etc.; penalty or reward attached to law; moral impetus for obedience to rule; (esp. in *plural*) (esp. economic) action to coerce state to conform to agreement etc. ● *verb* authorize, countenance; make (law etc.) binding.

sanctity /'sæŋktɪtɪ/ *noun* holiness, sacredness; inviolability.

sanctuary /'sæŋktʃʊərɪ/ *noun* (*plural* **-ies**) holy place; place where birds, wild animals, etc. are protected; place of refuge.

sanctum /'sæŋktəm/ *noun* (*plural* **-s**) holy place, esp. in temple or church; *colloquial* person's den.

sand ● *noun* fine grains resulting from erosion of esp. siliceous rocks; (in *plural*) grains of sand, expanse of sand, sandbank. ● *verb* smooth or treat with sandpaper or sand. □ **sandbag** *noun* bag filled with sand, esp. for making temporary defences, *verb* defend or hit with sandbag(s); **sandbank** sand forming shallow place in sea or river; **sandblast** *verb* treat with jet of sand driven by compressed air or steam, *noun* this jet; **sandcastle** model castle of sand on beach; **sand-dune**, **-hill** dune; **sand-martin** bird nesting in sandy banks; **sandpaper** *noun* paper with abrasive coating for smoothing or polishing wood etc., *verb* treat with this; **sandpiper** bird inhabiting wet sandy places; **sandpit** hollow or box containing sand for children to play in; **sandstone** sedimentary rock of compressed sand; **sandstorm** storm with clouds of sand raised by wind.

sandal /'sænd(ə)l/ *noun* shoe with openwork upper or no upper, fastened with straps.

sandal-tree /'sændəltri:/ *noun* tree yielding sandalwood.

sandalwood /'sændəlwʊd/ *noun* scented wood of sandal-tree.

sandwich /'sænwɪdʒ/ ● *noun* two or more slices of bread with filling; layered cake with jam, cream, etc. ● *verb* put (thing, statement, etc.) between two of different kind; squeeze in between others. □ **sandwich-board** each of two advertising boards worn front and back; **sandwich course** course with alternate periods of study and work experience.

sandy *adjective* (**-ier**, **-iest**) containing or covered with sand; (of hair) reddish; sand-coloured.

sane *adjective* of sound mind, not mad; (of opinion etc.) moderate, sensible.

sang *past* of SING.

sang-froid /sɑ̃'frwɑ:/ *noun* calmness in danger or difficulty.

sangria /sæŋ'gri:ə/ *noun* Spanish drink of red wine with fruit etc.

sanguinary /'sæŋgwɪnərɪ/ *adjective* bloody; bloodthirsty.

sanguine /'sæŋgwɪn/ *adjective* optimistic; (of complexion) florid, ruddy.

Sanhedrin /'sænɪdrɪn/ *noun* court of justice and supreme council in ancient Jerusalem.

sanitarium /sænɪ'teərɪəm/ *noun* (*plural* **-s** or **-ria**) *US* sanatorium.

sanitary /'sænɪtərɪ/ *adjective* (of conditions etc.) affecting health; hygienic. □ **sanitary towel** (*US* **sanitary napkin**) absorbent pad used during menstruation. □ **sanitariness** *noun*.

sanitation /sænɪ'teɪʃ(ə)n/ *noun* sanitary conditions; maintenance etc. of these; disposal of sewage, refuse, etc.

sanitize /'sænɪtaɪz/ *verb* (also **-ise**) (**-zing** or **-sing**) make sanitary, disinfect; *colloquial* censor.

sanity /'sænɪtɪ/ *noun* being sane; moderation.

sank *past* of SINK.

Sanskrit /'sænskrɪt/ ● *noun* ancient and sacred language of Hindus in India. ● *adjective* of or in Sanskrit.

Santa Claus /'sæntə klɔ:z/ *noun* person said to bring children presents at Christmas.

sap[1] ● *noun* vital juice of plants; vitality; *slang* foolish person. ● *verb* (**-pp-**) drain of sap; weaken. □ **sappy** *adjective* (**-ier**, **-iest**).

sap[2] ● *noun* tunnel or trench for concealed approach to enemy. ● *verb* (**-pp-**) dig saps; undermine.

sapient /'seɪpɪənt/ *adjective literary* wise; aping wisdom. □ **sapience** *noun*.

sapling /'sæplɪŋ/ *noun* young tree.

sapper *noun* digger of saps; private of Royal Engineers.

sapphire /'sæfaɪə/ ● *noun* transparent blue precious stone; its colour. ● *adjective* (also **sapphire blue**) bright blue.

saprophyte /'sæprəfaɪt/ *noun* plant or micro-organism living on dead organic matter.

saraband /'særəbænd/ *noun* slow Spanish dance; music for this.

Saracen /'særəs(ə)n/ *noun* Arab or Muslim of time of Crusades.

sarcasm /'sɑ:kæz(ə)m/ *noun* ironically scornful remark(s). □ **sarcastic** /sɑ:'kæstɪk/ *adjective*; **sarcastically** /sɑ:'kæstɪkəlɪ/ *adverb*.

sarcophagus /sɑ:'kɒfəgəs/ *noun* (*plural* **-gi** /-gaɪ/) stone coffin.

sardine /sɑ:'di:n/ *noun* (*plural* same or **-s**) young pilchard etc. tinned tightly packed.

sardonic /sɑ:'dɒnɪk/ *adjective* bitterly mocking; cynical. □ **sardonically** *adverb*.

sardonyx /'sɑ:dənɪks/ *noun* onyx in which white layers alternate with yellow or orange ones.

sargasso /sɑ:'gæsəʊ/ *noun* (*plural* **-s** or **-es**) seaweed with berry-like air-vessels.

sarge *noun slang* sergeant.

sari /'sɑ:rɪ/ *noun* (*plural* **-s**) length of material draped round body, worn traditionally by Hindu etc. women.

sarky /'sɑ:kɪ/ *adjective* (**-ier**, **-iest**) *slang* sarcastic.

sarong /sə'rɒŋ/ *noun* garment of long strip of cloth tucked round waist or under armpits.

sarsaparilla /sɑ:səpə'rɪlə/ *noun* dried roots of esp. smilax used to flavour drinks and medicines and formerly as tonic; plant yielding these.

sarsen /'sɑ:s(ə)n/ *noun* sandstone boulder carried by ice in glacial period.

sarsenet /'sɑ:snɪt/ *noun* soft silk fabric used esp. for linings.

sartorial /sɑ:'tɔ:rɪəl/ *adjective* of men's clothes or tailoring. □ **sartorially** *adverb*.

SAS *abbreviation* Special Air Service.

sash[1] *noun* strip or loop of cloth worn over one shoulder or round waist.

sash[2] *noun* frame holding glass in window sliding up and down in grooves.

sass *US colloquial* ● *noun* impudence. ● *verb* be impudent to. □ **sassy** *adjective* (**-ier**, **-iest**).

sassafras /'sæsəfræs/ *noun* small N. American tree; medicinal preparation from its leaves or bark.

Sassenach /'sæsənæk/ *noun Scottish usually derogatory* English person.

Sat. *abbreviation* Saturday.

sat *past & past participle* of SIT.

Satan /'seɪt(ə)n/ *noun* the Devil.

satanic /sə'tænɪk/ *adjective* of or like Satan; hellish, evil.

Satanism *noun* worship of Satan. □ **Satanist** *noun*.

satchel /'sætʃ(ə)l/ *noun* small bag, esp. for carrying school-books.

sate *verb* (**-ting**) *formal* gratify fully, surfeit.

sateen /sæ'ti:n/ *noun* glossy cotton fabric like satin.

satellite /'sætəlaɪt/ ● *noun* heavenly or artificial body orbiting earth or other planet; (in full **satellite state**) small country controlled by another. ● *adjective* transmitted by satellite; receiving signal from satellite.

satiate /'seɪʃɪeɪt/ *verb* (**-ting**) sate. □ **satiation** *noun*.

satiety /sə'taɪɪtɪ/ *noun formal* being sated.

satin /'sætɪn/ ● *noun* silk etc. fabric glossy on one side. ● *adjective* smooth as satin. □ **satinwood** kind of yellow glossy timber. □ **satiny** *adjective*.

satire /'sætaɪə/ *noun* ridicule, irony, etc. used to expose folly, vice, etc.; literary work using satire. □ **satirical** /sə'tɪrɪk(ə)l/ *adjective*; **satirically** /sə'tɪrɪkəlɪ/ *adverb*.

satirist /'sætərɪst/ *noun* writer of satires; satirical person.

satirize /'sætəraɪz/ *verb* (also **-ise**) (**-zing** or **-sing**) attack or describe with satire.

satisfaction /sætɪs'fækʃ(ə)n/ *noun* satisfying, being satisfied; thing that satisfies; atonement; compensation.

satisfactory /sætɪs'fæktərɪ/ *adjective* adequate; causing satisfaction. □ **satisfactorily** *adverb*.

satisfy /'sætɪsfaɪ/ *verb* (**-ies**, **-ied**) meet expectations or wishes of; be adequate; meet (an appetite or want); rid (person) of an appetite or want; pay; fulfil, comply with; convince.

satsuma /sæt'su:mə/ *noun* kind of tangerine.

saturate /'sætʃəreɪt/ *verb* (**-ting**) fill with moisture; fill to capacity; cause (substance) to absorb, hold, etc. as much as possible of another substance; supply (market) beyond demand; (as **saturated** *adjective*) (of fat) containing the most possible hydrogen atoms.

saturation *noun* saturating, being saturated. □ **saturation point** stage beyond which no more can be absorbed or accepted.

Saturday /'sætədeɪ/ *noun* day of week following Friday.

Saturnalia /sætə'neɪlɪə/ *noun* (*plural* same or **-s**) ancient Roman festival of Saturn; (**saturnalia**) (treated as *singular* or *plural*) scene of wild revelry.

saturnine /'sætənaɪn/ *adjective* of gloomy temperament or appearance.

satyr /'sætə/ *noun Greek & Roman Mythology* part-human part-animal woodland deity; lecherous man.

sauce /sɔ:s/ ● *noun* liquid or viscous accompaniment to food; something that adds piquancy; *colloquial* impudence. ● *verb* (**-cing**) *colloquial* be impudent to. □ **sauce-boat** jug or dish for serving sauce; **saucepan** cooking vessel usually with handle, used on hob.

saucer /'sɔ:sə/ *noun* shallow circular dish, esp. for standing cup on.

saucy *adjective* (**-ier**, **-iest**) impudent, cheeky. □ **saucily** *adverb*; **sauciness** *noun*.

sauerkraut /'saʊəkraʊt/ *noun* German dish of pickled cabbage.

sauna /'sɔ:nə/ *noun* period spent in room with steam bath; this room.

saunter /'sɔ:ntə/ ● *verb* stroll. ● *noun* leisurely walk.

saurian /'sɔ:rɪən/ *adjective* of or like a lizard.

sausage /'sɒsɪdʒ/ *noun* seasoned minced meat etc. in edible cylindrical case; sausage-shaped object. □ **sausage meat** minced meat for sausages etc.; **sausage roll** sausage meat in pastry cylinder.

sauté /'səʊteɪ/ ● *adjective* fried quickly in a little fat. ● *noun* food cooked thus. ● *verb* (*past & past participle* **sautéd** or **sautéed**) cook thus.

savage /'sævɪdʒ/ ● *adjective* fierce, cruel; wild, primitive. ● *noun derogatory* member of primitive tribe; brutal or barbarous person. ● *verb* (**-ging**) attack and maul; attack verbally. □ **savagely** *adverb*; **savagery** *noun* (*plural* **-ies**).

savannah /sə'vænə/ *noun* (also **savanna**) grassy plain in tropical or subtropical region.

savant /'sæv(ə)nt/ *noun* (*feminine* **savante** same pronunciation) learned person.

save[1] ● *verb* (**-ving**) (often + *from*) rescue or preserve from danger or harm; (often + *up*) keep for future use; relieve (person) from spending (money, time, etc.); prevent exposure to (annoyance etc.); prevent need for; rescue spiritually; *Football etc.* avoid losing (match), prevent (goal) from being scored. ● *noun Football etc.* act of saving goal. □ **savable**, **saveable** *adjective*; **saver** *noun*.

save[2] *preposition & conjunction archaic or poetical* except; but.

saveloy /'sævəlɔɪ/ *noun* highly seasoned sausage.

saving ● *noun* anything saved; an economy; (usually in *plural*) money saved; act of preserving or rescuing. ● *preposition* except; without offence to. □ **-saving** making economical use of specified thing; **saving grace** redeeming feature.

saviour /'seɪvjə/ *noun* (*US* **savior**) person who saves from danger etc.; (**the**, **our Saviour**) Christ.

savoir faire /sævwɑ: 'feə/ *noun* ability to behave appropriately; tact. [French]

savory[1] /'seɪvərɪ/ *noun* aromatic herb used in cookery.

savory[2] *US* = SAVOURY.

savour /'seɪvə/ (*US* **savor**) ● *noun* characteristic taste, flavour, etc.; tinge or hint. ● *verb* appreciate, enjoy; (+ *of*) imply, suggest.

savoury /'seɪvərɪ/ (*US* **savory**) ● *adjective* with appetizing taste or smell; (of food) salty or piquant, not sweet; pleasant. ● *noun* (*plural* **-ies**) savoury dish.

savoy /sə'vɔɪ/ *noun* rough-leaved winter cabbage.

savvy /'sævɪ/ *slang* ● *verb* (**-ies**, **-ied**) know. ● *noun* knowingness, understanding. ● *adjective* (**-ier**, **-iest**) *US* knowing, wise.

saw[1] ● *noun* implement with toothed blade etc. for cutting wood etc. ● *verb* (*past participle* **sawn** or **sawed**) cut or make with saw; use saw; make to-and-fro sawing motion. □ **sawdust** fine wood fragments produced in sawing; **sawfish** (*plural* same or **-es**) large sea fish with toothed flat snout; **sawmill** factory for sawing wood into planks; **sawtooth(ed)** serrated.

saw² *past* of SEE¹.

saw³ *noun* proverb, maxim.

sawyer /'sɔːjə/ *noun* person who saws timber.

sax *noun colloquial* saxophone.

saxe /sæks/ *noun & adjective* (in full **saxe blue**; as *adjective* often hyphenated) light greyish-blue.

saxifrage /'sæksɪfreɪdʒ/ *noun* small-flowered rock-plant.

Saxon /'sæks(ə)n/ ● *noun historical* member or language of Germanic people that occupied parts of England in 5th–6th c.; Anglo-Saxon. ● *adjective historical* of the Saxons; Anglo-Saxon.

saxophone /'sæksəfəʊn/ *noun* keyed brass reed instrument used esp. in jazz. □ **saxophonist** /'sɒfən-/ *noun*.

say ● *verb* (*3rd singular present* **says** /sez/; *past & past participle* **said** /sed/) utter, remark; express; state; indicate; (in *passive*; usually + *to do*) be asserted; (+ *to do*) *colloquial* direct, order; convey (information); adduce, plead; decide; take as example or as near enough; (**the said**) *Law or jocular* the previously mentioned. ● *noun* opportunity to express view; share in decision. □ **say-so** *colloquial* power of decision, mere assertion.

saying *noun* maxim, proverb, etc.

sc. *abbreviation* scilicet.

scab ● *noun* crust over healing cut, sore, etc.; skin disease; plant disease; *colloquial derogatory* blackleg. ● *verb* (**-bb-**) form scab; *colloquial derogatory* act as blackleg. □ **scabby** *adjective* (**-ier**, **-iest**).

scabbard /'skæbəd/ *noun historical* sheath of sword etc.

scabies /'skeɪbiːz/ *noun* contagious skin disease causing itching.

scabious /'skeɪbɪəs/ *noun* plant with pincushion-shaped flowers.

scabrous /'skeɪbrəs/ *adjective* rough, scaly; indecent.

scaffold /'skæfəʊld/ *noun* scaffolding; *historical* platform for execution of criminal.

scaffolding *noun* temporary structure of poles, planks, etc. for building work; materials for this.

scald /skɔːld/ ● *verb* burn (skin etc.) with hot liquid or vapour; heat (esp. milk) to near boiling point; (usually + *out*) clean with boiling water. ● *noun* burn etc. caused by scalding.

scale¹ ● *noun* each of thin horny plates protecting skin of fish and reptiles; thing resembling this; incrustation inside kettle etc.; tartar on teeth. ● *verb* (**-ling**) remove scale(s) from; form or come off in scales. □ **scaly** *adjective* (**-ier**, **-iest**).

scale² *noun* (often in *plural*) weighing machine; (also **scale-pan**) pan of weighing-balance. □ **tip**, **turn the scales** be decisive factor, (+ *at*) weigh (specified amount).

scale³ ● *noun* graded classification system; ratio of reduction or enlargement in map, picture, etc.; relative dimensions; *Music* set of notes at fixed intervals, arranged in order of pitch; set of marks on line used in measuring etc., rule determining distances between these, rod on which these are marked. ● *verb* (**-ling**) climb; represent in proportion; reduce to common scale. □ **scale down**, **up** make or become smaller or larger in proportion; **to scale** uniformly in proportion.

scalene /'skeɪliːn/ *adjective* (of triangle) having unequal sides.

scallion /'skæljən/ *noun esp. US* shallot; spring onion.

scallop /'skæləp/ ● *noun* edible bivalve with fan-shaped ridged shells; (in full **scallop shell**) one shell of this, esp. used for cooking or serving food on; (in *plural*) ornamental edging of semicircular curves. ● *verb* (**-p-**) ornament with scallops.

scallywag /'skælɪwæg/ *noun* scamp, rascal.

scalp ● *noun* skin and hair on head; *historical* this cut off as trophy by N. American Indian. ● *verb historical* take scalp of.

scalpel /'skælp(ə)l/ *noun* small surgical knife.

scam *noun US slang* trick, fraud.

scamp *noun colloquial* rascal, rogue.

scamper /'skæmpə/ ● *verb* run and skip. ● *noun* act of scampering.

scampi /'skæmpɪ/ *plural noun* large prawns.

scan ● *verb* (**-nn-**) look at intently or quickly; (of verse etc.) be metrically correct; examine (surface etc.) for radioactivity etc.; traverse (region) with radar etc. beam; resolve (picture) into elements of light and shade for esp. television transmission; analyse metre of (line etc.); obtain image of (part of body) using scanner. ● *noun* scanning; image obtained by scanning.

scandal /'skænd(ə)l/ *noun* disgraceful event; public outrage; malicious gossip. □ **scandalmonger** /-mʌŋgə/ person who spreads scandal. □ **scandalous** *adjective*; **scandalously** *adverb*.

scandalize *verb* (also **-ise**) (**-zing** or **-sing**) offend morally, shock.

Scandinavian /skændɪ'neɪvɪən/ ● *noun* native or inhabitant, or family of languages, of Scandinavia. ● *adjective* of Scandinavia.

scanner *noun* device for scanning; diagnostic apparatus measuring radiation, ultrasound reflections, etc. from body.

scansion /'skænʃ(ə)n/ *noun* metrical scanning of verse.

scant *adjective* barely sufficient; deficient.

scanty *adjective* (**-ier**, **-iest**) of small extent or amount; barely sufficient. □ **scantily** *adverb*; **scantiness** *noun*.

scapegoat /'skeɪpgəʊt/ *noun* person blamed for others' faults.

scapula /'skæpjʊlə/ *noun* (*plural* **-lae** /-li:/ or **-s**) shoulder blade.

scapular /'skæpjʊlə/ ● *adjective* of scapula. ● *noun* monastic short cloak.

scar[1] ● *noun* mark left on skin etc. by wound etc.; emotional damage. ● *verb* (**-rr-**) (esp. as **scarred** *adjective*) mark with or form scar(s).

scar[2] *noun* (also **scaur**) steep craggy part of mountainside.

scarab /'skærəb/ *noun* kind of beetle; gem cut in form of beetle.

scarce /skeəs/ ● *adjective* in short supply; rare. ● *adverb archaic or literary* scarcely. □ **make oneself scarce** *colloquial* keep out of the way, disappear.

scarcely /'skeəslɪ/ *adverb* hardly, only just.

scarcity *noun* (*plural* **-ies**) lack or shortage.

scare /skeə/ ● *verb* (**-ring**) frighten; (as **scared** *adjective*) (usually + *of*) frightened; (usually + *away*, *off*, etc.) drive away by frightening. ● *noun* sudden fright or alarm, esp. caused by rumours. □ **scarecrow** human figure used for frightening birds away from crops, *colloquial* badly-dressed or grotesque person; **scaremonger** /-mʌŋgə/ person who spreads scare(s).

scarf[1] *noun* (*plural* **scarves** /skɑ:vz/ or **-s**) piece of material worn round neck or over head for warmth or ornament.

scarf[2] ● *verb* join ends of (timber etc.) by thinning or notching them and bolting them together. ● *noun* (*plural* **-s**) joint made thus.

scarify /'skærɪfaɪ/ *verb* (**-ies**, **-ied**) make slight incisions in; scratch; criticize etc. mercilessly; loosen (soil). □ **scarification** *noun*.

scarlatina /skɑ:lə'ti:nə/ *noun* scarlet fever.

scarlet /'skɑ:lət/ ● *adjective* of brilliant red tinged with orange. ● *noun* scarlet colour, pigment, clothes, etc. □ **scarlet fever** infectious fever with scarlet rash; **scarlet pimpernel** wild plant with small esp. scarlet flowers.

scarp ● *noun* steep slope, esp. inner side of ditch in fortification. ● *verb* make perpendicular or steep.

scarper /'skɑ:pə/ *verb slang* run away, escape.

scarves *plural* of SCARF[1].

scary *adjective* (**-ier**, **-iest**) *colloquial* frightening.

scat ● *noun* wordless jazz singing. ● *verb* (**-tt-**) sing scat.

scathing /'skeɪðɪŋ/ *adjective* witheringly scornful. □ **scathingly** *adverb*.

scatology /skæ'tɒlədʒɪ/ *noun* preoccupation with excrement or obscenity. □ **scatological** /-tə'lɒdʒ-/ *adjective*.

scatter /'skætə/ ● *verb* throw about, strew; cover by scattering; (cause to) flee; (cause to) disperse; (as **scattered** *adjective*) wide apart, sporadic; *Physics* deflect or diffuse (light, particles, etc.). ● *noun* act of scattering; small amount scattered; extent of distribution. □ **scatterbrain** person lacking concentration; **scatterbrained** *adjective.*

scatty /'skætɪ/ *adjective* (**-ier**, **-iest**) *colloquial* lacking concentration. □ **scattily** *adverb*; **scattiness** *noun.*

scavenge /'skævɪndʒ/ *verb* (**-ging**) (usually + *for*) search for and collect (discarded items).

scavenger /'skævɪndʒə/ *noun* person who scavenges; animal etc. feeding on carrion.

SCE *abbreviation* Scottish Certificate of Education.

scenario /sɪ'nɑ:rɪəʊ/ *noun* (*plural* **-s**) synopsis of film, play, etc.; imagined sequence of future events.

■ **Usage** *Scenario* should not be used in standard English to mean 'situation', as in *It was an unpleasant scenario.*

scene /si:n/ *noun* place of actual or fictitious occurrence; incident; public display of emotion, temper, etc.; piece of continuous action in a play, film, book, etc.; piece(s) of scenery for a play; landscape, view; *colloquial* area of interest or activity. □ **behind the scenes** out of view of audience, secret, secretly; **scene-shifter** person who moves scenery in theatre.

scenery /'si:nərɪ/ *noun* features (esp. picturesque) of landscape; backcloths, properties, etc. representing scene in a play etc.

scenic /'si:nɪk/ *adjective* picturesque; of scenery. □ **scenically** *adverb.*

scent /sent/ ● *noun* characteristic, esp. pleasant, smell; liquid perfume; smell left by animal; clues etc. leading to discovery; power of scenting. ● *verb* discern by smell; sense; (esp. as **scented** *adjective*) make fragrant.

scepter *US* = SCEPTRE.

sceptic /'skeptɪk/ *noun* (*US* **skeptic**) sceptical person; person who questions truth of religions, or the possibility of knowledge. □ **scepticism** /-sɪz(ə)m/ *noun.*

sceptical */'skeptɪk(ə)l/ adjective* (*US* **skeptical**) inclined to doubt accepted opinions; critical; incredulous. □ **sceptically** *adverb.*

sceptre */'septə/ noun* (*US* **scepter**) staff borne as symbol of sovereignty.

schedule /'ʃedju:l/ ● *noun* timetable; list, esp. of rates or prices. ● *verb* (**-ling**) include in schedule; make schedule of; list (building) for preservation. □ **on schedule** at time appointed; **scheduled flight**, **service**, etc., regular public one.

schema /'ski:mə/ *noun* (*plural* **schemata** or **-s**) synopsis, outline, diagram.

schematic /skɪ'mætɪk/ ● *adjective* of or as scheme or diagram. ● *noun* diagram, esp. of electronic circuit.

schematize /'ski:mətaɪz/ *verb* (also **-ise**) (**-zing** or **-sing**) put in schematic form.

scheme /ski:m/ ● *noun* systematic arrangement; artful plot; outline, syllabus, etc. ● *verb* (**-ming**) plan, esp. secretly or deceitfully. □ **scheming** *adjective.*

scherzo /'skeətsəʊ/ *noun* (*plural* **-s**) *Music* vigorous and lively movement or composition.

schism /'skɪz(ə)m/ *noun* division of esp. religious group into sects etc. □ **schismatic** /-'mæt-/ *adjective & noun.*

schist /ʃɪst/ *noun* layered crystalline rock.

schizo /'skɪtsəʊ/ *colloquial* ● *adjective* schizophrenic. ● *noun* (*plural* **-s**) schizophrenic person.

schizoid /'skɪtsɔɪd/ ● *adjective* tending to schizophrenia. ● *noun* schizoid person.

schizophrenia /skɪtsə'fri:nɪə/ *noun* mental disorder marked by disconnection between thoughts, feelings, and actions. □ **schizophrenic** /-'fren-/ *adjective & noun.*

schmaltz /ʃmɔ:lts/ *noun colloquial* sickly sentimentality. □ **schmaltzy** *adjective* (**-ier**, **-iest**).

schnapps /ʃnæps/ *noun* any of various spirits drunk in N. Europe.

schnitzel /'ʃnɪts(ə)l/ *noun* veal escalope.

scholar /'skɒlə/ *noun* learned person; holder of scholarship; person of specified academic ability. □ **scholarly** *adjective.*

scholarship *noun* learning, erudition; award of money etc. towards education.

scholastic /skə'læstɪk/ *adjective* of schools, education, etc.; academic.

school¹ /sku:l/ ● *noun* educational institution for pupils up to 19 years old or (*US*) at any level; school buildings, pupils, staff, etc; (time given to) teaching; university department or faculty; group of artists, disciples, etc. following or holding similar principles, opinions, etc.; instructive circumstances. ● *verb* send to school; discipline, train, control; (as **schooled** *adjective*) (+ *in*) educated, trained. □ **schoolboy, schoolchild, schoolgirl** one who attends school; **school-leaver** person who has just left school; **schoolmaster, schoolmistress, schoolteacher** teacher in school; **schoolroom** room used for lessons, esp. in private house.

school² /sku:l/ *noun* shoal of fish, whales, etc.

schooling *noun* education.

schooner /'sku:nə/ *noun* two-masted fore-and-aft rigged ship; large glass, esp. for sherry; *US & Australian* tall beer glass.

schottische /ʃɒ'ti:ʃ/ *noun* kind of slow polka.

sciatic /saɪ'ætɪk/ *adjective* of hip or sciatic nerve; of or having sciatica. □ **sciatic nerve** large nerve from pelvis to thigh.

sciatica /saɪ'ætɪkə/ *noun* neuralgia of hip and leg.

science /'saɪəns/ *noun* branch of knowledge involving systematized observation, experiment, and induction; knowledge so gained; pursuit or principles of this; skilful technique. □ **science fiction** fiction with scientific theme; **science park** area containing science-based businesses.

scientific /saɪən'tɪfɪk/ *adjective* following systematic methods of science; systematic, accurate; of or concerned with science.

scientist /'saɪəntɪst/ *noun* student or expert in science.

sci-fi /'saɪfaɪ/ *noun colloquial* science fiction.

scilicet /'saɪlɪset/ *adverb* that is to say.

scimitar /'sɪmɪtə/ *noun* curved oriental sword.

scintillate /'sɪntɪleɪt/ (**-ting**) (esp. as **scintillating** *adjective*) talk or act cleverly; sparkle, twinkle. □ **scintillation** *noun*.

scion /'saɪən/ *noun* shoot cut for grafting; young member of family.

scissors /'sɪzəz/ *plural noun* (also **pair of scissors** *singular*) cutting instrument with pair of pivoted blades.

sclerosis /sklə'rəʊsɪs/ *noun* abnormal hardening of tissue; (in full **multiple sclerosis**) serious progressive disease of nervous system. □ **sclerotic** /-'rɒt-/ *adjective*.

scoff¹ ● *verb* (usually + *at*) speak derisively, mock. ● *noun* mocking words, taunt.

scoff² *colloquial* ● *verb* eat (food) greedily. ● *noun* food.

scold /skəʊld/ ● *verb* rebuke; find fault noisily. ● *noun archaic* nagging woman.

sconce *noun* wall-bracket holding candlestick or light-fitting.

scone /skɒn, skəʊn/ *noun* small cake of flour etc. baked quickly.

scoop /sku:p/ ● *noun* short-handled deep shovel; long-handled ladle; excavating part of digging machine etc.; device for serving ice cream etc.; quantity taken up by scoop; scooping movement; exclusive news item; large profit made quickly. ● *verb* (usually + *out*) hollow out or (usually + *up*) lift (as) with scoop; forestall (rival newspaper etc.) with scoop; secure (large profit etc.), esp. suddenly.

scoot /sku:t/ *verb* (esp. in *imperative*) *colloquial* shoot along; depart, flee.

scooter *noun* child's toy with footboard on two wheels and long steering-handle; low-powered motorcycle.

scope *noun* range, opportunity; extent of ability, outlook, etc.

scorch ● *verb* burn or discolour surface of with dry heat; become so discoloured etc.; (as **scorching** *adjective*) *colloquial* (of weather) very hot, (of criticism etc.) stringent. ● *noun* mark of scorching. □ **scorched earth policy** policy of destroying everything that might be of use to invading enemy.

scorcher *noun colloquial* extremely hot day.

score *noun* number of points, goals, etc. made by player or side in game etc.; respective scores at end of game; act of gaining esp. goal; (*plural* same or **-s**) (set of) 20; (in *plural*) a great many; reason, motive; *Music* copy of composition with parts arranged one below another; music for film or play; notch, line, etc. made on surface; record of money owing. ● *verb* (**-ring**) win, gain; make (points etc.) in game; keep score; mark with notches etc.; have an advantage; *Music* (often + *for*) orchestrate or arrange (piece of music); *slang* obtain drugs illegally, make sexual conquest. □ **keep (the) score** register points etc. as they are made; **score (points) off** *colloquial* humiliate, esp. verbally; **scoreboard** large board for displaying score in match etc.

scoria /ˈskɔːrɪə/ *noun* (*plural* **scoriae** /-rɪiː/) (fragments of) cellular lava; slag.

scorn ● *noun* disdain, contempt, derision. ● *verb* hold in contempt; reject or refuse to do as unworthy.

scornful *adjective* (often + *of*) contemptuous. □ **scornfully** *adverb.*

Scorpio /ˈskɔːpɪəʊ/ *noun* eighth sign of zodiac.

scorpion /ˈskɔːpɪən/ *noun* lobster-like arachnid with jointed stinging tail.

Scot *noun* native of Scotland.

Scotch ● *adjective* Scottish, Scots. ● *noun* Scottish, Scots; Scotch whisky. □ **Scotch broth** meat soup with pearl barley etc.; **Scotch egg** hard-boiled egg in sausage meat; **Scotch fir** Scots pine; **Scotch mist** thick mist and drizzle; **Scotch terrier** small rough-coated terrier; **Scotch whisky** whisky distilled in Scotland.

■ **Usage** *Scots* or *Scottish* is preferred to *Scotch* in Scotland, except in the compound nouns given above.

scotch *verb* decisively put an end to; *archaic* wound without killing.

scot-free *adverb* unharmed, unpunished.

Scots ● *adjective* Scottish. ● *noun* Scottish; form of English spoken in (esp. Lowlands of) Scotland. □ **Scotsman, Scotswoman** Scot; **Scots pine** kind of pine tree.

Scottish ● *adjective* of Scotland or its inhabitants. ● *noun* (**the Scottish**) (treated as *plural*) people of Scotland.

scoundrel /ˈskaʊndr(ə)l/ *noun* unscrupulous villain; rogue.

scour[1] /skaʊə/ ● *verb* rub clean; (usually + *away*, *off*, etc.) clear by rubbing; clear out (pipe etc.) by flushing through. ● *noun* scouring, being scoured. □ **scourer** *noun.*

scour[2] /skaʊə/ *verb* search thoroughly.

scourge /skɜːdʒ/ ● *noun* person or thing regarded as causing suffering; whip. ● *verb* (**-ging**) whip; punish, oppress.

Scouse /skaʊs/ *colloquial* ● *noun* Liverpool dialect; (also **Scouser**) native of Liverpool. ● *adjective* of Liverpool.

scout /skaʊt/ ● *noun* person sent out to get information or reconnoitre; search for this; talent-scout; (also **Scout**) member of (originally boys') association intended to develop character. ● *verb* (often + *for*) seek information etc.; (often + *about*, *around*) make search; (often + *out*) *colloquial* explore. □ **Scoutmaster** person in charge of group of Scouts. □ **scouting** *noun.*

Scouter *noun* adult leader of Scouts.

scowl /skaʊl/ ● *noun* sullen or bad-tempered look. ● *verb* wear scowl.

scrabble /ˈskræb(ə)l/ ● *verb* (**-ling**) scratch or grope busily about. ● *noun* scrabbling; (**Scrabble**) *proprietary term* game in which players build up words from letter-blocks on board.

scrag ● *noun* (also **scrag-end**) inferior end of neck of mutton; skinny person or animal. ● *verb* (**-gg-**) *slang* strangle, hang; handle roughly, beat up.

scraggy *adjective* (**-ier, iest**) thin and bony. □ **scragginess** *noun.*

scram *verb* (**-mm-**) (esp. in *imperative*) *colloquial* go away.

scramble /ˈskræmb(ə)l/ ● *verb* (**-ling**) clamber, crawl, climb, etc.; (+ *for*, *at*) struggle with competitors (for thing or share); mix indiscriminately; cook (eggs) by stirring in heated pan; alter sound frequencies of (broadcast or telephone conversation) so as to make it unintelligible without special receiver; (of fighter aircraft or pilot) take off rapidly. ● *noun* scrambling; difficult climb or walk; (+ *for*) eager struggle or competition; motorcycle race over rough ground; emergency take-off by fighter aircraft.

scrambler *noun* device for scrambling telephone conversations; motorcycle used for scrambles.

scrap[1] • *noun* small detached piece, fragment; waste material; discarded metal for reprocessing; (with negative) smallest piece or amount; (in *plural*) odds and ends, bits of uneaten food. • *verb* (**-pp-**) discard as useless. □ **scrapbook** book in which cuttings etc. are kept; **scrap heap** collection of waste material, state of being discarded as useless; **scrapyard** place where (esp. metal) scrap is collected.

scrap[2] *colloquial* • *noun* fight or rough quarrel. • *verb* (**-pp-**) have scrap.

scrape • *verb* (**-ping**) move hard edge across (surface), esp. to smoothe or clean; (+ *away*, *off*, etc.) remove by scraping; rub (surface) harshly against another; scratch, damage, or make by scraping; draw or move with sound (as) of scraping; produce such sound from; (often + *along*, *by*, *through*, etc.) move while (almost) touching; narrowly achieve; (often + *by*, *through*) barely manage, pass exam etc. with difficulty; (+ *together*, *up*) provide or amass with difficulty; be economical; make clumsy bow; (+ *back*) draw (hair) tightly back. • *noun* act or sound of scraping; scraped place, graze; *colloquial* predicament caused by rashness. □ **scraper** *noun*.

scrapie /'skreɪpɪ/ *noun* viral disease of sheep.

scrappy *adjective* (**-ier**, **-iest**) consisting of scraps; incomplete.

scratch • *verb* score or wound superficially, esp. with sharp object; scrape with the nails to relieve itching; make or form by scratching; (+ *out*, *off*, *through*) erase; withdraw from race or competition; (often + *about*, *around*, etc.) scratch ground etc. in search, search haphazardly. • *noun* mark, wound, or sound made by scratching; act of scratching oneself; *colloquial* trifling wound; starting line for race etc.; position of those receiving no handicap. • *adjective* collected by chance; collected or made from whatever is available; with no handicap given. □ **from scratch** from the beginning, without help; **up to scratch** up to required standard.

scratchy *adjective* (**-ier**, **-iest**) tending to make scratches or scratching noise; causing itchiness; (of drawing etc.) careless. □ **scratchily** *adverb*; **scratchiness** *noun*.

scrawl • *verb* write in hurried untidy way. • *noun* hurried writing; scrawled note. □ **scrawly** *adjective* (**-ier**, **-iest**).

scrawny /'skrɔːnɪ/ *adjective* (**-ier**, **-iest**) lean, scraggy.

scream • *noun* piercing cry (as) of terror or pain; *colloquial* hilarious occurrence or person. • *verb* emit scream; utter in or with scream; move with scream; laugh uncontrollably; be blatantly obvious.

scree *noun* (in *singular* or *plural*) small loose stones; mountain slope covered with these.

screech • *noun* harsh scream or squeal. • *verb* utter with or make screech. □ **screech-owl** barn owl.

screed *noun* long usually tiresome letter or harangue; layer of cement etc. applied to level a surface.

screen • *noun* fixed or movable upright partition for separating, concealing, or protecting from heat etc.; thing used to conceal or shelter; concealing stratagem; protection thus given; blank surface on which images are projected; (**the screen**) cinema industry, films collectively; windscreen; large sieve; system for detecting disease, ability, attribute, etc. • *verb* shelter, hide; protect from detection, censure, etc.; (+ *off*) conceal behind screen; show (film etc.); prevent from causing, or protect from, electrical interference; test (person or group) for disease, reliability, loyalty, etc.; sieve. □ **screenplay** film script; **screen printing** printing process with ink forced through areas of sheet of fine mesh; **screen test** audition for film part; **screenwriter** person who writes for cinema.

screw /skruː/ • *noun* cylinder or cone with spiral ridge running round it outside (**male screw**) or inside (**female screw**); (in full **woodscrew**) metal male screw with slotted head and sharp point; (in full **screw-bolt**) blunt metal male screw on which nut is threaded; straight screw used to exert pressure; (in *singular* or *plural*) instrument of torture acting thus; (in full **screw propeller**) propeller with twisted blades; one turn of screw; (+ *of*) small twisted-up paper (of tobacco etc.); oblique curling motion; *slang* prison warder. ● *verb* fasten or tighten (as) with screw(s); (of ball etc.) swerve; (+ *out*, *of*) extort from; swindle. □ **screwball** *US slang* crazy or eccentric person; **screwdriver** tool for turning screws by putting tool's tip into screw's slot; **screw up** contract, crumple, or contort, summon up (courage etc.), *slang* bungle, spoil, or upset; **screw-up** *slang* bungle.

screwy *adjective* (**-ier**, **-iest**) *slang* mad, eccentric; absurd. □ **screwiness** *noun*.

scribble /'skrɪb(ə)l/ ● *verb* (**-ling**) write or draw carelessly or hurriedly; *jocular* be author or writer. ● *noun* scrawl; hasty note etc.

scribe ● *noun* ancient or medieval copyist of manuscripts; pointed instrument for marking wood etc.; *colloquial* writer. ● *verb* (**-bing**) mark with scribe. □ **scribal** *adjective*.

scrim *noun* open-weave fabric for lining, upholstery, etc.

scrimmage /'skrɪmɪdʒ/ ● *noun* tussle, brawl. ● *verb* (**-ging**) engage in scrimmage.

scrimp *verb* skimp.

scrip *noun* provisional certificate of money subscribed to company etc.; extra share(s) instead of dividend.

script ● *noun* text of play, film, or broadcast (see panel at DIRECT SPEECH); handwriting; typeface imitating handwriting; alphabet or other system of writing; examinee's written answer(s). ● *verb* write script for (film etc.). □ **scriptwriter** person who writes scripts for films, etc.

scripture /'skrɪptʃə/ *noun* sacred writings; (**Scripture**, **the Scriptures**) the Bible. □ **scriptural** *adjective*.

scrivener /'skrɪvənə/ *noun historical* copyist, drafter of documents; notary.

scrofula /'skrɒfjʊlə/ *noun* disease with glandular swellings. □ **scrofulous** *adjective*.

scroll /skrəʊl/ ● *noun* roll of parchment or paper; book in ancient roll form; ornamental design imitating roll of parchment. ● *verb* (often + *down*, *up*) move (display on VDU screen) to view later or earlier material.

scrotum /'skrəʊtəm/ *noun* (*plural* **scrota** or **-s**) pouch of skin enclosing testicles. □ **scrotal** *adjective*.

scrounge /skraʊndʒ/ *verb* (**-ging**) obtain (things) by cadging. □ **on the scrounge** scrounging. □ **scrounger** *noun*.

scrub[1] ● *verb* (**-bb-**) clean by hard rubbing, esp. with hard brush; (often + *up*) (of surgeon etc.) clean and disinfect hands etc. before operating; *colloquial* cancel; pass (gas etc.) through scrubber. ● *noun* scrubbing, being scrubbed.

scrub[2] *noun* brushwood or stunted trees etc.; land covered with this. □ **scrubby** *adjective* (**-ier**, **-iest**).

scrubber *noun slang* promiscuous woman; apparatus for purifying gases etc.

scruff[1] *noun* back of neck.

scruff[2] *noun colloquial* scruffy person.

scruffy /'skrʌfɪ/ *adjective* (**-ier**, **-iest**) *colloquial* shabby, slovenly, untidy. □ **scruffily** *adverb*; **scruffiness** *noun*.

scrum *noun* scrummage; *colloquial* scrimmage. □ **scrum-half** *Rugby* half-back who puts ball into scrum.

scrummage /'skrʌmɪdʒ/ *noun Rugby* massed forwards on each side pushing to gain possession of ball thrown on ground between them.

scrumptious /'skrʌmpʃəs/ *adjective colloquial* delicious.

scrumpy /'skrʌmpɪ/ *noun colloquial* rough cider.

scrunch ● *verb* (usually + *up*) crumple; crunch. ● *noun* crunch.

scruple /'skru:p(ə)l/ ● *noun* (often in *plural*) moral concern; doubt caused by this. ● *verb* (**-ling**) (+ *to do*; usually in negative) hesitate owing to scruples.

scrupulous /'skru:pjʊləs/ *adjective* conscientious, thorough; careful to avoid doing wrong; over-attentive to details. □ **scrupulously** *adverb*.

scrutineer /skru:tɪ'nɪə/ *noun* person who scrutinizes ballot papers.

scrutinize /'skru:tɪnaɪz/ *verb* (also **-ise**) (**-zing** or **-sing**) subject to scrutiny.

scrutiny /'skru:tɪnɪ/ *noun* (*plural* **-ies**) critical gaze; close examination; official examination of ballot papers.

scuba /'sku:bə/ *noun* (*plural* **-s**) aqualung. □ **scuba-diving** swimming underwater using scuba.

scud ● *verb* (**-dd-**) move straight and fast; skim along; *Nautical* run before wind. ● *noun* scudding; vapoury driving clouds or shower.

scuff ● *verb* graze or brush against; mark or wear out (shoes etc.) thus; shuffle or drag feet. ● *noun* mark of scuffing.

scuffle /'skʌf(ə)l/ ● *noun* confused struggle or fight at close quarters. ● *verb* (**-ling**) engage in scuffle.

scull ● *noun* each of pair of small oars; oar used to propel boat from stern; (in *plural*) sculling race. ● *verb* propel with scull(s).

scullery /'skʌlərɪ/ *noun* (*plural* **-ies**) back kitchen; room where dishes are washed etc.

sculpt *verb* sculpture.

sculptor /'skʌlptə/ *noun* (*feminine* **sculptress**) person who sculptures.

sculpture /'skʌlptʃə/ ● *noun* art of making 3-dimensional forms by chiselling, carving, modelling, casting, etc.; work of sculpture. ● *verb* (**-ring**) represent in or adorn with sculpture; practise sculpture. □ **sculptural** *adjective*.

scum ● *noun* layer of dirt etc. at surface of liquid; *derogatory* worst part, person, or group. ● *verb* (**-mm-**) remove scum from; form scum (on). □ **scumbag** *slang* contemptible person. □ **scummy** *adjective* (**-ier**, **-iest**).

scupper[1] /'skʌpə/ *noun* hole in ship's side draining water from deck.

scupper[2] /'skʌpə/ *verb slang* sink (ship, crew); defeat or ruin (plan etc.); kill.

scurf *noun* dandruff. □ **scurfy** *adjective* (**-ier**, **-iest**).

scurrilous /'skʌrɪləs/ *adjective* grossly or obscenely abusive. □ **scurrility** /skə'rɪl-/ *noun* (*plural* **-ies**); **scurrilously** *adverb*; **scurrilousness** *noun*.

scurry /'skʌrɪ/ ● *verb* (**-ies**, **-ied**) run hurriedly, scamper. ● *noun* (*plural* **-ies**) scurrying sound or movement; flurry of rain or snow.

scurvy /'skɜ:vɪ/ ● *noun* disease resulting from deficiency of vitamin C. ● *adjective* (**-ier**, **-iest**) paltry, contemptible. □ **scurvily** *adverb*.

scut *noun* short tail, esp. of hare, rabbit, or deer.

scutter /'skʌtə/ *verb & noun colloquial* scurry.

scuttle[1] /'skʌt(ə)l/ *noun* coal scuttle; part of car body between windscreen and bonnet.

scuttle[2] /'skʌt(ə)l/ ● *verb* (**-ling**) scurry; flee in undignified way. ● *noun* hurried gait; precipitate flight.

scuttle[3] /'skʌt(ə)l/ ● *noun* hole with lid in ship's deck or side. ● *verb* (**-ling**) let water into (ship) to sink it.

scythe /saɪð/ ● *noun* mowing and reaping implement with long handle and curved blade. ● *verb* (**-thing**) cut with scythe.

SDLP *abbreviation* (in N. Ireland) Social Democratic and Labour Party.

SDP *abbreviation* (in UK) Social Democratic Party.

SE *abbreviation* south-east(ern).

sea *noun* expanse of salt water covering most of earth; area of this; large inland lake; (motion or state of) waves of sea; (+ *of*) vast quantity or expanse. □ **at sea** in ship on the sea, confused; **sea anchor** bag to reduce drifting of ship; **sea anemone** polyp with petal-like tentacles; **seabed** ocean floor; **seaboard** coastline, coastal area; **sea dog** old sailor; **seafarer** traveller by sea; **seafood** edible marine fish or shellfish; **sea front** part of seaside town facing sea; **seagoing** designed for open sea; **seagull** = GULL[1]; **sea horse** small fish with head like horse's; **seakale** plant with young shoots used as vegetable; **sea legs** ability to keep one's balance at sea; **sea level** mean level of sea's surface, used in reckoning heights of hills etc. and as barometric standard; **sea lion** large, eared seal; **seaman** person whose work is at sea, sailor, sailor below rank of officer; **seaplane** aircraft designed to take off from and land on water; **seaport** town with harbour; **sea salt** salt got by evaporating sea water; **seascape** picture or view of sea; **seashell** shell of salt-water mollusc; **seashore** land next to sea; **seasick** nauseous from motion of ship at sea; **seasickness** *noun*; **seaside** sea-coast, esp. as holiday resort; **sea urchin** small marine animal with spiny shell; **seaweed** plant growing in sea; **seaworthy** fit to put to sea.

seal[1] • *noun* piece of stamped wax etc. attached to document or to receptacle, envelope, etc. to guarantee authenticity or security; metal stamp etc. used in making seal; substance or device used to close gap etc.; anything regarded as confirmation or guarantee; decorative adhesive stamp. • *verb* close securely or hermetically; stamp, fasten, or fix with seal; certify as correct with seal; (+ *off*) prevent access to or from; (often + *up*) confine securely; settle, decide. □ **sealing wax** mixture softened by heating and used for seals.

seal[2] • *noun* fish-eating amphibious marine mammal with flippers. • *verb* hunt seals.

sealant *noun* material for sealing, esp. to make watertight.

seam • *noun* line where two edges join, esp. of cloth or boards; fissure between parallel edges; wrinkle; stratum of coal etc. • *verb* join with seam; (esp. as **seamed** *adjective*) mark or score with seam. □ **seamless** *adjective*.

seamstress /'si:mstrɪs/ *noun* woman who sews.

seamy *adjective* (**-ier**, **-iest**) disreputable, sordid; showing seams. □ **seaminess** *noun*.

seance /'seɪɑ̃s/ *noun* meeting at which spiritualist attempts to contact the dead.

sear /sɪə/ *verb* scorch, cauterize; cause anguish to; brown (meat) quickly.

search /sɜ:tʃ/ • *verb* examine thoroughly to find something; make investigation; (+ *for*, *out*) look for, seek out; (as **searching** *adjective*) keenly questioning. • *noun* act of searching; investigation. □ **searchlight** outdoor lamp designed to throw strong beam of light in any direction, light or beam from this; **search party** group of people conducting organized search; **search warrant** official authorization to enter and search building. □ **searcher** *noun*; **searchingly** *adverb*.

season /'si:z(ə)n/ • *noun* each of climatic divisions of year; proper or suitable time; time when something is plentiful, active, etc.; (**the season**) (also **high season**) busiest period at resort etc.; *colloquial* season ticket. • *verb* flavour with salt, herbs, etc.; enhance with wit etc.; moderate; (esp. as **seasoned** *adjective*) make or become suitable by exposure to weather or experience. □ **in season** (of food) available plentifully, (of animal) on heat; **season ticket** one entitling holder to unlimited travel, access, etc. in given period.

seasonable *adjective* suitable to season; opportune.

■ **Usage** *Seasonable* is sometimes confused with *seasonal*.

---

seasonal *adjective* of, depending on, or varying with seasons.

■ **Usage** *Seasonal* is sometimes confused with *seasonable*.

---

seasoning *noun* salt, herbs, etc. as flavouring for food.

seat • *noun* thing made or used for sitting on; buttocks, part of garment covering them; part of chair etc. on which buttocks rest; place for one person in theatre etc.; position as MP, committee member, etc., or right to occupy it; machine's supporting or guiding part; location; country mansion; posture on horse. • *verb* cause to sit; provide sitting accommodation for; (as **seated** *adjective*) sitting; establish in position. □ **seat belt** belt securing seated person in vehicle or aircraft.

seating *noun* seats collectively; sitting accommodation.

sebaceous /sɪ'beɪʃəs/ *adjective* fatty; secreting oily matter.

Sec. *abbreviation* (also **sec.**) secretary.

sec. *abbreviation* second(s).

sec[1] *noun colloquial* (in phrases) second, moment.

sec[2] *adjective* (of wine) dry.

secateurs /sekə'tɜ:z/ *plural noun* pruning clippers.

secede /sɪ'si:d/ *verb* (**-ding**) withdraw formally from political or religious body.

secession /sɪ'seʃ(ə)n/ *noun* seceding. □ **secessionist** *noun & adjective*.

seclude /sɪ'klu:d/ *verb* (**-ding**) keep (person, place) apart from others; (esp. as **secluded** *adjective*) screen from view.

seclusion /sɪ'klu:ʒ(ə)n/ *noun* secluded state or place.

second[1] /'sekənd/ ● *adjective* next after first; additional; subordinate; inferior; comparable to. ● *noun* runner-up; person or thing coming second; second gear; (in *plural*) inferior goods; *colloquial* second helping or course; assistant to boxer, duellist, etc. ● *verb* formally support (nomination, proposal, etc.). □ **second-best** next after best; **second class** second-best group, category, postal service, or accommodation; **second cousin** child of parent's first cousin; **second fiddle** subordinate position; **second-guess** *colloquial* anticipate by guesswork, criticize with hindsight; **second-hand** (of goods) having had previous owner, (of information etc.) obtained indirectly; **second nature** acquired tendency that has become instinctive; **second-rate** inferior; **second sight** clairvoyance; **second string** alternative course of action etc.; **second thoughts** revised opinion; **second wind** renewed capacity for effort after breathlessness or tiredness. □ **seconder** *noun*.

second[2] /'sekənd/ *noun* SI unit of time (1/60 of minute); 1/60 of minute of angle; *colloquial* very short time.

second[3] /sɪ'kɒnd/ *verb* transfer (person) temporarily to another department etc. □ **secondment** *noun*.

secondary /'sekəndərɪ/ ● *adjective* coming after or next below what is primary; derived from or supplementing what is primary; (of education etc.) following primary. ● *noun* (*plural* **-ies**) secondary thing. □ **secondary colour** result of mixing two primary colours.

secondly *adverb* furthermore; as a second item.

secrecy /'si:krəsɪ/ *noun* being secret; keeping of secrets.

secret /'si:krɪt/ ● *adjective* not (to be) made known or seen; working etc. secretly; liking secrecy. ● *noun* thing (to be) kept secret; mystery; effective but not widely known method. □ **in secret** secretly; **secret agent** spy; **secret police** police operating secretly for political ends; **secret service** government department concerned with espionage. □ **secretly** *adverb*.

secretariat /sekrɪ'teərɪət/ *noun* administrative office or department; its members or premises.

secretary /'sekrɪtərɪ/ *noun* (*plural* **-ies**) employee who deals with correspondence, records, making appointments, etc.; official of society etc. who writes letters, organizes business, etc.; principal assistant of government minister, ambassador, etc. □ **secretary bird** long-legged crested African bird; **Secretary-General** principal administrative officer of organization; **Secretary of State** head of major government department, (in US) foreign minister. □ **secretarial** /-'teərɪəl/ *adjective*.

secrete /sɪ'kri:t/ *verb* (**-ting**) (of cell, organ, etc.) produce and discharge (substance); conceal. □ **secretory** *adjective*.

secretion /sɪ'kri:ʃ(ə)n/ *noun* process or act of secreting; secreted substance.

secretive /'si:krətɪv/ *adjective* inclined to make or keep secrets, uncommunicative. □ **secretively** *adverb*; **secretiveness** *noun*.

sect *noun* group sharing (usually unorthodox) religious etc. doctrines; religious denomination.

sectarian /sek'teərɪən/ ● *adjective* of sect(s); bigoted in following one's sect. ● *noun* member of a sect. □ **sectarianism** *noun*.

section /'sekʃ(ə)n/ ● *noun* each of parts into which something is divisible or divided; part cut off; subdivision; *US* area of land, district of town; surgical separation or cutting; cutting of solid by plane, resulting figure or area of this; thin slice cut off for microscopic examination. ● *verb* arrange in or divide into sections; compulsorily commit to psychiatric hospital.

sectional *adjective* of a social group; partisan; made in sections; local rather than general. □ **sectionally** *adverb*.

sector /'sektə/ *noun* branch of an enterprise, the economy, etc.; *Military* portion of battle area; plane figure enclosed between two radii of circle etc.

secular /'sekjʊlə/ *adjective* not concerned with religion, not sacred; (of clerics) not monastic. □ **secularism** *noun*; **secularization** *noun*; **secularize** *verb* (also **-ise**) (**-zing** or **-sing**).

secure /sɪ'kjʊə/ ● *adjective* untroubled by danger or fear; safe; reliable, stable, fixed. ● *verb* (**-ring**) make secure or safe; fasten or close securely; obtain. □ **securely** *adverb.*

security *noun* (*plural* **-ies**) secure condition or feeling; thing that guards or guarantees; safety against espionage, theft, etc.; organization for ensuring this; thing deposited as guarantee for undertaking or loan; (often in *plural*) document as evidence of loan, certificate of stock, bonds, etc. □ **security risk** person or thing threatening security.

sedan /sɪ'dæn/ *noun* (in full **sedan chair**) *historical* enclosed chair for one person, usually carried on poles by two; *US* saloon car.

sedate /sɪ'deɪt/ ● *adjective* tranquil, serious. ● *verb* (**-ting**) put under sedation. □ **sedately** *adverb*; **sedateness** *noun.*

sedation *noun* treatment with sedatives.

sedative /'sedətɪv/ ● *noun* calming drug or influence. ● *adjective* calming, soothing.

sedentary /'sedəntərɪ/ *adjective* sitting; (of work etc.) done while sitting; (of person) disinclined to exercise.

sedge *noun* grasslike waterside or marsh plant. □ **sedgy** *adjective.*

sediment /'sedɪmənt/ *noun* dregs; matter deposited on land by water or wind. □ **sedimentary** /-'men-/ *adjective*; **sedimentation** *noun.*

sedition /sɪ'dɪʃ(ə)n/ *noun* conduct or speech inciting to rebellion. □ **seditious** *adjective.*

seduce /sɪ'dju:s/ *verb* (**-cing**) entice into sexual activity or wrongdoing; coax or lead astray. □ **seducer** *noun.*

seduction /sɪ'dʌkʃ(ə)n/ *noun* seducing, being seduced; tempting or attractive thing.

seductive /sɪ'dʌktɪv/ *adjective* alluring, enticing. □ **seductively** *adverb*; **seductiveness** *noun.*

sedulous /'sedjʊləs/ *adjective* persevering, diligent, painstaking. □ **sedulity** /sɪ'dju:-/ *noun*; **sedulously** *adverb.*

see[1] *verb* (**sees**; *past* **saw**; *past participle* **seen**) perceive with the eyes; have or use this power; discern mentally, understand; watch; experience; ascertain; imagine, foresee; look at; meet; visit, be visited by; meet regularly; reflect, get clarification; (+ *in*) find attractive in; escort, conduct; witness (event etc.); ensure. □ **see about** attend to, consider; **see off** be present at departure of, *colloquial* get the better of; **see over** inspect, tour; **see red** *colloquial* become enraged; **see through** not be deceived by, support (person) during difficult time, complete (project); **see-through** translucent; **see to** attend to, repair; **see to it** (+ *that*) ensure.

see[2] *noun* area under (arch)bishop's authority; (arch)bishop's office or jurisdiction.

seed ● *noun* part of plant capable of developing into another such plant; seeds collectively, esp. for sowing; semen; prime cause, beginning; offspring; *Tennis etc.* seeded player. ● *verb* place seed(s) in; sprinkle (as) with seed; sow seeds; produce or drop seed; remove seeds from (fruit etc.); place crystal etc. in (cloud) to produce rain; *Tennis etc.* designate (competitor in knockout tournament) so that strong competitors do not meet each other until later rounds, arrange (order of play) thus. □ **go, run to seed** cease flowering as seed develops, become degenerate, unkempt, etc.; **seed-bed** bed prepared for sowing, place of development; **seed-pearl** very small pearl; **seed-potato** potato kept for planting; **seedsman** dealer in seeds.

seedling *noun* young plant raised from seed.

seedy *adjective* (**-ier**, **-iest**) shabby; *colloquial* unwell; full of seed.

seeing *conjunction* (usually + *that*) considering that, inasmuch as, because.

seek *verb* (*past & past participle* **sought** /sɔ:t/) (often + *for, after*) search, inquire; try or want to obtain or reach; request; endeavour. □ **seek out** search for and find. □ **seeker** *noun.*

seem *verb* (often + *to do*) appear, give the impression.

seeming *adjective* apparent but doubtful. □ **seemingly** *adverb.*

seemly /'si:mlɪ/ *adjective* (**-ier**, **-iest**) in good taste, decorous. □ **seemliness** *noun.*

seen *past participle* of SEE[1].

seep *verb* ooze, percolate.

seepage *noun* act of seeping; quantity that seeps.

seer /sɪə/ *noun* person who sees; prophet, visionary.

seersucker /'sɪəsʌkə/ *noun* thin cotton etc. fabric with puckered surface.

see-saw /'si:sɔ:/ ● *noun* long board supported in middle so that children etc. sitting on ends move alternately up and down; this game; up-and-down or to-and-fro motion, contest, etc. ● *verb* play or move (as) on see-saw; vacillate. ● *adjective & adverb* with up-and-down or to-and-fro motion.

seethe /si:ð/ *verb* (**-thing**) boil, bubble; be very angry, resentful, etc.

segment ● *noun* /'segmənt/ part cut off or separable from other parts; part of circle or sphere cut off by intersecting line or plane. ● *verb* /seg'ment/ divide into segments. □ **segmental** /-'ment-/ *adjective*; **segmentation** *noun*.

segregate /'segrɪgeɪt/ *verb* (**-ting**) put apart, isolate; separate (esp. ethnic group) from the rest of the community. □ **segregation** *noun*; **segregationist** *noun & adjective*.

seigneur /seɪn'jɜ:/ *noun* feudal lord. □ **seigneurial** *adjective*.

seine /seɪn/ ● *noun* large vertical fishing net. ● *verb* (**-ning**) fish with seine.

seismic /'saɪzmɪk/ *adjective* of earthquake(s).

seismograph /'saɪzməgrɑ:f/ *noun* instrument for recording earthquake details. □ **seismographic** /-'græf-/ *adjective*.

seismology /saɪz'mɒlədʒɪ/ *noun* the study of earthquakes. □ **seismological** /-mə'lɒdʒ-/ *adjective*; **seismologist** *noun*.

seize /si:z/ *verb* (**-zing**) (often + *on*, *upon*) take hold or possession of, esp. forcibly, suddenly, or by legal power; take advantage of; comprehend quickly or clearly; affect suddenly; (also **seise**) (usually + *of*) *Law* put in possession of. □ **seize up** (of mechanism) become jammed, (of part of body etc.) become stiff.

seizure /'si:ʒə/ *noun* seizing, being seized; sudden attack of epilepsy, apoplexy, etc.

seldom /'seldəm/ *adverb* rarely, not often.

select /sɪ'lekt/ ● *verb* choose, esp. with care. ● *adjective* chosen for excellence or suitability; exclusive. □ **select committee** parliamentary committee conducting special inquiry. □ **selector** *noun*.

selection /sɪ'lekʃ(ə)n/ *noun* selecting, being selected; person or thing selected; things from which choice may be made; = NATURAL SELECTION.

selective *adjective* of or using selection; able to select; selecting what is convenient. □ **selectively** *adverb*; **selectivity** /-'tɪv-/ *noun*.

selenium /sɪ'li:nɪəm/ *noun* non-metallic element in some sulphide ores.

self ● *noun* (*plural* **selves** /selvz/) individuality, essence; object of introspection or reflexive action; one's own interests or pleasure, concentration on these.

self- *combining form expressing reflexive action, automatic or independent action, or sameness.* □ **self-addressed** addressed to oneself; **self-adhesive** (of envelope etc.) adhesive, esp. without wetting; **self-aggrandizement** enriching oneself, making oneself powerful; **self-assertive** confident or assertive in promoting oneself, one's claims, etc.; **self-assertion** *noun*; **self-assured** self-confident; **self-catering** providing cooking facilities but no food; **self-centred** preoccupied with oneself; **self-confessed** openly admitting oneself to be; **self-confident** having confidence in oneself; **self-confidence** *noun*; **self-conscious** nervous, shy, embarrassed; **self-consciously** *adverb*; **self-contained** uncommunicative, complete in itself; **self-control** control of oneself, one's behaviour, etc.; **self-critical** critical of oneself, one's abilities, etc.; **self-defence** defence of oneself, one's reputation, etc.; **self-denial** abstinence, esp. as discipline; **self-deprecating** belittling oneself; **self-determination** nation's right to determine own government etc., free will; **self-destruct** (of device etc.) explode or disintegrate automatically, esp. when pre-set to do so; **self-effacing** retiring, modest; **self-employed** working as freelance or for one's own business etc.; **self-employment** *noun*; **self-esteem** good opinion of oneself; **self-**

evident needing no proof or explanation; **self-explanatory** not needing explanation; **self-financing** not needing subsidy; **self-fulfilling** (of prophecy etc.) assured fulfilment by its utterance; **self-governing** governing itself or oneself; **self-government** *noun*; **self-help** use of one's own abilities etc. to achieve success etc.; **self-image** one's conception of oneself; **self-important** conceited, pompous; **self-importance** *noun*; **self-indulgent** indulging one's own pleasures, feelings, etc., (of work of art etc.) lacking control; **self-indulgence** *noun*; **self-interest** one's own interest or advantage; **self-interested** *adjective*; **self-made** successful or rich by one's own efforts; **self-opinionated** obstinate in one's opinion; **self-pity** pity for oneself; **self-portrait** portrait of oneself by oneself; **self-possessed** unperturbed, cool; **self-possession** *noun*; **self-preservation** keeping oneself safe, instinct for this; **self-raising** (of flour) containing a raising agent; **self-reliance** reliance on one's own abilities etc.; **self-reliant** *adjective*; **self-respect** respect for oneself; **self-restraint** self-control; **self-righteous** smugly sure of one's righteousness; **self-righteously** *adverb*; **self-righteousness** *noun*; **self-rule** self-government; **self-sacrifice** selflessness, self-denial; **self-satisfied** complacent; **self-satisfaction** *noun*; **self-seeking** selfish; **self-service** with customers helping themselves and paying cashier afterwards; **self-starter** electric device for starting engine, ambitious person with initiative; **self-styled** called so by oneself; **self-sufficient** capable of supplying one's own needs; **self-sufficiency** *noun*; **self-willed** obstinately pursuing one's own wishes; **self-worth** self-esteem.

selfish *adjective* concerned chiefly with one's own interests or pleasure; actuated by or appealing to self-interest. □ **selfishness** *noun*.

selfless *adjective* unselfish.

selfsame /ˈselfseɪm/ *adjective* (**the selfsame**) the very same, the identical.

sell ● *verb* (*past & past participle* **sold** /səʊld/) exchange or be exchanged for money; stock for sale; (+ *at*, *for*) have specified price; betray or prostitute for money etc.; advertise, publicize; cause to be sold; *colloquial* make (person) enthusiastic about (idea etc.). ● *noun colloquial* manner of selling; deception, disappointment. □ **sell-by date** latest recommended date of sale; **sell off** sell at reduced prices; **sell out** sell (all one's stock or shares etc.), betray, be treacherous; **sell-out** commercial success, betrayal; **sell short** disparage, underestimate; **sell up** sell one's business, house, etc.

seller *noun* person who sells; thing that sells well or badly as specified. □ **seller's market** time when goods are scarce and expensive.

Sellotape /ˈseləteɪp/ ● *noun proprietary term* adhesive usually transparent cellulose tape. ● *verb* (**sellotape**) (**-ping**) fix with Sellotape.

selvage /ˈselvɪdʒ/ *noun* (also **selvedge**) edge of cloth woven to prevent fraying.

selves *plural* of SELF.

semantic /sɪˈmæntɪk/ *adjective* of meaning in language.

semantics *plural noun* (usually treated as *singular*) branch of linguistics concerned with meaning.

semaphore /ˈseməfɔː/ ● *noun* system of signalling with arms or two flags; railway signalling apparatus with arm(s). ● *verb* (**-ring**) signal or send by semaphore.

semblance /ˈsembləns/ *noun* (+ *of*) appearance, show.

semen /ˈsiːmən/ *noun* reproductive fluid of males.

semester /sɪˈmestə/ *noun* half-year term in universities.

semi /ˈsemɪ/ *noun colloquial* (*plural* **-s**) semi-detached house.

semi- *prefix* half; partly.

semibreve /ˈsemɪbriːv/ *noun Music* note equal to 4 crotchets.

semicircle /ˈsemɪsɜːk(ə)l/ *noun* half of circle or its circumference. □ **semicircular** /-ˈsɜːkjʊlə/ *adjective*.

semicolon /semɪˈkəʊlən/ *noun* punctuation mark (;) of intermediate value between comma and full stop (see panel).

semiconductor /semɪkənˈdʌktə/ *noun* substance that is a poor electrical conductor when either pure or cold and a good conductor when either impure or hot.

semi-detached /semɪdɪ'tætʃt/ • *adjective* (of house) joined to another on one side only. • *noun* such house.

semifinal /semɪ'faɪn(ə)l/ *noun Sport* match or round preceding final. □ **semifinalist** *noun*.

seminal /'semɪn(ə)l/ *adjective* of seed, semen, or reproduction; germinal; (of idea etc.) providing basis for future development.

seminar /'semɪnɑː/ *noun* small class for discussion etc.; short intensive course of study; specialists' conference.

seminary /'semɪnərɪ/ *noun* (*plural* **-ies**) training college for priests etc. □ **seminarist** *noun*.

semipermeable /semɪ'pɜːmɪəb(ə)l/ *adjective* (of membrane etc.) allowing small molecules to pass through.

semiprecious /semɪ'preʃəs/ *adjective* (of gem) less valuable than a precious stone.

semi-professional /semɪprə'feʃən(ə)l/ • *adjective* (of footballer, musician, etc.) paid for activity but not relying on it for living; of semi-professionals. • *noun* semi-professional person.

semiquaver /'semɪkweɪvə/ *noun Music* note equal to half a quaver.

semi-skimmed /semɪ'skɪmd/ *adjective* (of milk) with some of cream skimmed off.

Semite /'siːmaɪt/ *noun* member of peoples supposedly descended from Shem, including Jews and Arabs.

Semitic /sɪ'mɪtɪk/ *adjective* of Semites, esp. Jews; of languages of family including Hebrew and Arabic.

semitone /'semɪtəʊn/ *noun* half a tone in musical scale.

semivowel /'semɪvaʊəl/ *noun* sound intermediate between vowel and consonant; letter representing this.

semolina /semə'liːnə/ *noun* hard round grains of wheat used for puddings etc.; pudding of this.

Semtex /'semteks/ *noun proprietary term* odourless plastic explosive.

SEN *abbreviation* State Enrolled Nurse.

Sen. *abbreviation* Senior; Senator.

senate /'senɪt/ *noun* upper house of legislature in some countries; governing body of some universities or (*US*) colleges; ancient Roman state council.

senator /'senətə/ *noun* member of senate. □ **senatorial** /-'tɔː-/ *adjective*.

send *verb* (*past & past participle* **sent**) order or cause to go or be conveyed; cause to become; send message etc.; grant, bestow, inflict, cause to be. □ **send away for** order (goods) by post; **send down** rusticate or expel from university, send to prison; **send for** summon, order by post; **send off** dispatch, attend departure of; **send-off** party etc. at departure of person; **send off for** send away for; **send on** transmit further or in advance of oneself; **send up** *colloquial* ridicule (by mimicking); **send-up** *noun* □ **sender** *noun*.

senescent /sɪ'nes(ə)nt/ *adjective* growing old. □ **senescence** *noun*.

Semicolon ;

This is used:

1 between clauses that are too short or too closely related to be made into separate sentences; such clauses are not usually connected by a conjunction, e.g.

To err is human; to forgive, divine.
You could wait for him here; on the other hand I could wait in your place; this would save you valuable time.

2 between items in a list which themselves contain commas, if it is necessary to avoid confusion, e.g.

The party consisted of three teachers, who had already climbed with the leader; seven pupils; and two parents.

---

seneschal /ˈsenɪʃ(ə)l/ *noun* steward of medieval great house.

senile /ˈsi:naɪl/ *adjective* of old age; mentally or physically infirm because of old age. □ **senile dementia** illness of old people with loss of memory etc. □ **senility** /sɪˈnɪl-/ *noun*.

senior /ˈsi:nɪə/ ● *adjective* higher in age or standing; (placed after person's name) senior to relative of same name. ● *noun* senior person; one's elder or superior. □ **senior citizen** old-age pensioner. □ **seniority** /-ˈɒr-/ *noun*.

senna /ˈsenə/ *noun* cassia; laxative from leaves and pods of this.

señor /senˈjɔ:/ *noun* (*plural* **señores** /-rez/) *title used of or to Spanish-speaking man*.

señora /senˈjɔ:rə/ *noun title used of or to Spanish-speaking esp. married woman*.

señorita /senjəˈri:tə/ *noun title used of or to young Spanish-speaking esp. unmarried woman*.

sensation /senˈseɪʃ(ə)n/ *noun* feeling in one's body; awareness, impression; intense feeling, esp. in community; cause of this; sense of touch.

sensational *adjective* causing or intended to cause public excitement etc.; wonderful. □ **sensationalism** *noun*; **sensationalist** *noun & adjective*; **sensationalize** *verb* (also **-ise**) (**-zing** or **-sing**).

sense ● *noun* any of bodily faculties transmitting sensation; sensitiveness of any of these; ability to perceive; (+ *of*) consciousness; appreciation, instinct; practical wisdom; meaning of word etc.; intelligibility, coherence; prevailing opinion; (in *plural*) sanity, ability to think. ● *verb* (**-sing**) perceive by sense(s); be vaguely aware of; (of machine etc.) detect. □ **make sense** be intelligible or practicable; **make sense of** show or find meaning of. □ **sense of humour** see HUMOUR.

senseless *adjective* pointless, foolish; unconscious. □ **senselessly** *adverb*; **senselessness** *noun*.

sensibility *noun* (*plural* **-ies**) capacity to feel; (exceptional) sensitiveness; (in *plural*) tendency to feel offended etc.

■ **Usage** *Sensibility* should not be used in standard English to mean 'possession of good sense'.

---

sensible /ˈsensɪb(ə)l/ *adjective* having or showing good sense; perceptible by senses; (of clothing etc.) practical; (+ *of*) aware of. □ **sensibly** *adverb*.

sensitive /ˈsensɪtɪv/ *adjective* (often + *to*) acutely affected by external impressions, having sensibility; easily offended or hurt; (often + *to*) responsive to or recording slight changes of condition; *Photography* responding (esp. rapidly) to light; (of topic etc.) requiring tact or secrecy. □ **sensitively** *noun*; **sensitiveness** *noun*; **sensitivity** /-ˈtɪv-/ *noun* (*plural* **-ies**).

sensitize /ˈsensɪtaɪz/ *verb* (also **-ise**) (**-zing** or **-sing**) make sensitive. □ **sensitization** *noun*.

sensor /ˈsensə/ *noun* device to detect or measure a physical property.

sensory /ˈsensərɪ/ *adjective* of sensation or senses.

sensual /ˈsensjʊəl/ *adjective* of physical, esp. sexual, pleasure; enjoying, giving, or showing this. □ **sensuality** /-ˈæl-/ *noun*; **sensually** *adverb*.

■ **Usage** *Sensual* is sometimes confused with *sensuous*.

---

sensuous /ˈsensjʊəs/ *adjective* of or affecting senses, esp. aesthetically. □ **sensuously** *adverb*; **sensuousness** *noun*.

■ **Usage** *Sensuous* is sometimes confused with *sensual*.

---

sent *past & past participle* of SEND.

sentence /ˈsent(ə)ns/ ● *noun* grammatically complete series of words with (implied) subject and predicate (see panel); punishment allotted to person convicted in criminal trial; declaration of this. ● *verb* (**-cing**) declare sentence of, condemn.

sententious /senˈtenʃəs/ *adjective* pompously moralizing; affectedly formal; using maxims. □ **sententiousness** *noun*.

sentient /ˈsenʃ(ə)nt/ *adjective* capable of perception and feeling. □ **sentience** *noun*; **sentiently** *adverb*.

sentiment /'sentɪmənt/ *noun* mental feeling; (often in *plural*) opinion; emotional or irrational view(s); tendency to be swayed by feeling; mawkish tenderness.

sentimental /sentɪ'ment(ə)l/ *adjective* of or showing sentiment; showing or affected by emotion rather than reason. □ **sentimentalism** *noun*; **sentimentalist** *noun*; **sentimentality** /-'tæl-/ *noun*; **sentimentalize** *verb* (also **-ise**) (**-zing** or **-sing**); **sentimentally** *adverb*.

sentinel /'sentɪn(ə)l/ *noun* sentry.

sentry /'sentrɪ/ *noun* (*plural* **-ies**) soldier etc. stationed to keep guard. □ **sentry-box** cabin to shelter standing sentry.

sepal /'sep(ə)l/ *noun* division or leaf of calyx.

separable /'sepərəb(ə)l/ *adjective* able to be separated. □ **separability** *noun*.

separate ● *adjective* /'sepərət/ forming unit by itself, existing apart; disconnected, distinct, individual. ● *noun* /'sepərət/ (in *plural*) articles of dress not parts of suits. ● *verb* /'sepəreɪt/ (**-ting**) make separate, sever; prevent union or contact of; go different ways; (esp. as **separated** *adjective*) cease to live with spouse; secede; divide or sort into parts or sizes; (often + *out*) extract or remove (ingredient etc.). □ **separately** *adverb*; **separateness** *noun*; **separator** *noun*.

■ **Usage** *Separate*, *separation*, etc. are not spelt with an *e* in the middle.

separation /sepə'reɪʃ(ə)n/ *noun* separating, being separated; arrangement by which couple remain married but live apart.

separatist /'sepərətɪst/ *noun* person who favours separation, esp. political independence. □ **separatism** *noun*.

sepia /'si:pɪə/ *noun* dark reddish-brown colour or paint; brown tint used in photography.

sepoy /'si:pɔɪ/ *noun historical* Indian soldier under European, esp. British, discipline.

sepsis /'sepsɪs/ *noun* septic condition.

Sept. *abbreviation* September.

sept *noun* clan, esp. in Ireland.

September /sep'tembə/ *noun* ninth month of year.

septet /sep'tet/ *noun* musical composition for 7 performers; the performers; any group of 7.

septic /'septɪk/ *adjective* contaminated with bacteria, putrefying. □ **septic tank** tank in which sewage is disintegrated through bacterial activity.

septicaemia /septɪ'si:mɪə/ *noun* (*US* **septicemia**) blood poisoning.

septuagenarian /septjʊədʒɪ'neərɪən/ *noun* person between 70 and 79 years old.

Septuagesima /septjʊə'dʒesɪmə/ *noun* third Sunday before Lent.

Septuagint /'septjʊədʒɪnt/ *noun* ancient Greek version of Old Testament.

septum /'septəm/ *noun* (*plural* **septa**) partition such as that between nostrils.

Sentence

A sentence is the basic unit of language in use and expresses a complete thought. There are three types of sentence, each starting with a capital letter, and each normally ending with a full stop, a question mark, or an exclamation mark:

Statement: *You're happy.*
Question: *Is it raining?*
Exclamation: *I wouldn't have believed it!*

A sentence, especially a statement, often has no punctuation at the end in a public notice, a newspaper headline, or a legal document, e.g.

Government cuts public spending

A sentence normally contains a subject and a verb, but may not, e.g.

What a mess! *Where?* *In the sink.*

sepulchral /sɪ'pʌlkr(ə)l/ *adjective* of tomb or burial; funereal, gloomy.

sepulchre /'sepəlkə/ (*US* **sepulcher**) ● *noun* tomb, burial cave or vault. ● *verb* (**-ring**) lay in sepulchre.

sequel /'si:kw(ə)l/ *noun* what follows; novel, film, etc. that continues story of earlier one.

sequence /'si:kwəns/ *noun* succession; order of succession; set of things belonging next to one another; unbroken series; episode or incident in film etc.

sequential /sɪ'kwenʃ(ə)l/ *adjective* forming sequence or consequence. □ **sequentially** *adverb.*

sequester /sɪ'kwestə/ *verb* (esp. as **sequestered** *adjective*) seclude, isolate; sequestrate.

sequestrate /'si:kwɪstreɪt/ *verb* (**-ting**) confiscate; take temporary possession of (debtor's estate etc.). □ **sequestration** *noun.*

sequin /'si:kwɪn/ *noun* circular spangle on dress etc. □ **sequined**, **sequinned** *adjective.*

sequoia /sɪ'kwɔɪə/ *noun* extremely tall Californian conifer.

sera *plural* of SERUM.

seraglio /sə'rɑ:lɪəʊ/ *noun* (*plural* **-s**) harem; *historical* Turkish palace.

seraph /'serəf/ *noun* (*plural* **-im** or **-s**) member of highest of 9 orders of angels. □ **seraphic** /sə'ræfɪk/ *adjective.*

Serb ● *noun* native of Serbia; person of Serbian descent. ● *adjective* Serbian.

Serbian /'sɜ:bɪən/ ● *noun* Slavonic dialect of Serbs; Serb. ● *adjective* of Serbs or their dialect.

Serbo-Croat /sɜ:bəʊ'krəʊæt/ (also **Serbo-Croatian** /-krəʊ'eɪʃ(ə)n/) ● *noun* main official language of former Yugoslavia, combining Serbian and Croatian dialects. ● *adjective* of this language.

serenade /serə'neɪd/ ● *noun* piece of music performed at night, esp. under lover's window; orchestral suite for small ensemble. ● *verb* (**-ding**) perform serenade to.

serendipity /serən'dɪpɪtɪ/ *noun* faculty of making happy discoveries by accident. □ **serendipitous** *adjective.*

serene /sɪ'ri:n/ *adjective* (**-r**, **-st**) clear and calm; placid, unperturbed. □ **serenely** *adverb*; **serenity** /-'ren-/ *noun.*

serf *noun historical* labourer not allowed to leave the land on which he worked; oppressed person, drudge. □ **serfdom** *noun.*

serge *noun* durable woollen fabric.

sergeant /'sɑ:dʒ(ə)nt/ *noun* non-commissioned army or RAF officer next below warrant officer; police officer next below inspector. □ **sergeant major** warrant officer assisting adjutant of regiment or battalion.

serial /'sɪərɪəl/ ● *noun* story published, broadcast, or shown in instalments. ● *adjective* of, in, or forming series. □ **serial killer** person who murders repeatedly. □ **serially** *adverb.*

serialize *verb* (also **-ise**) (**-zing** or **-sing**) publish or produce in instalments. □ **serialization** *noun.*

series /'sɪəri:z/ *noun* (*plural* same) number of similar or related things, events, etc.; succession, row, set; *Broadcasting* set of related but individually complete programmes. □ **in series** in ordered succession, (of set of electrical circuits) arranged so that same current passes through each circuit.

serif /'serɪf/ *noun* fine cross-line at extremities of printed letter.

serious /'sɪərɪəs/ *adjective* thoughtful, earnest; important, requiring thought; not negligible, dangerous; sincere, in earnest; (of music, literature, etc.) intellectual, not popular. □ **seriously** *adverb*; **seriousness** *noun.*

serjeant /'sɑ:dʒ(ə)nt/ *noun* (in full **serjeant-at-law**, *plural* **serjeants-at-law**) *historical* barrister of highest rank. □ **serjeant-at-arms** official of court, city, or parliament, with ceremonial duties.

sermon /'sɜ:mən/ *noun* discourse on religion or morals, esp. delivered in church; admonition, reproof.

sermonize *verb* (also **-ise**) (**-zing** or **-sing**) (often + *to*) moralize.

serous /'sɪərəs/ *adjective* of or like serum, watery; (of gland etc.) having serous secretion.

serpent /'sɜ:pənt/ *noun* snake, esp. large; cunning or treacherous person.

serpentine /'sɜ:pəntaɪn/ ● *adjective* of or like serpent; coiling, sinuous; cunning, treacherous. ● *noun* soft usually dark green rock, sometimes mottled.

SERPS *abbreviation* State Earnings-Related Pension Scheme.

serrated /sə'reɪtɪd/ *adjective* with saw-like edge. □ **serration** *noun*.

serried /'serɪd/ *adjective* (of ranks of soldiers etc.) close together.

serum /'sɪərəm/ *noun* (*plural* **sera** or **-s**) liquid separating from clot when blood coagulates, esp. used for inoculation; watery fluid in animal bodies.

servant /'sɜ:v(ə)nt/ *noun* person employed for domestic work; devoted follower or helper.

serve ● *verb* (**-ving**) do service for; be servant to; carry out duty; (+ *in*) be employed in (esp. armed forces); be useful to or serviceable for; meet needs, perform function; go through due period of (apprenticeship, prison sentence, etc.); go through (specified period) of imprisonment etc.; (often + *up*) present (food) to eat; act as waiter; attend to (customer etc.); (+ *with*) supply with (goods); treat (person) in specified way; *Law* (often + *on*) deliver (writ etc.), (+ *with*) deliver writ etc. to; set (ball) in play at tennis etc.; (of male animal) copulate with (female). ● *noun Tennis etc.* service. □ **serve (person) right** be his or her deserved misfortune. □ **server** *noun*.

service /'sɜ:vɪs/ ● *noun* (often in *plural*) work done or doing of work for employer or for community etc.; work done by machine etc.; assistance or benefit given; provision of some public need, e.g. transport or (often in *plural*) water, gas, etc.; employment as servant; state or period of employment; Crown or public department or organization; (in *plural*) the armed forces; ceremony of worship; liturgical form for this; (routine) maintenance and repair of machine etc. after sale; assistance given to customers; serving of food etc., quality of this, nominal charge for this; (in *plural*) motorway service area; set of dishes etc. for serving meal; act of serving in tennis etc., person's turn to serve, game in which one serves. ● *verb* (**-cing**) maintain or repair (car, machine, etc.); provide service for. □ **at (person's) service** ready to serve him or her; **of service** useful; **service area** area near road supplying petrol, refreshments, etc.; **service charge** additional charge for service in restaurant etc.; **service flat** one in which domestic service etc. is provided; **service industry** one providing services, not goods; **serviceman**, **servicewoman** person in armed services; **service road** one giving access to houses etc. lying back from main road; **service station** establishment selling petrol etc. to motorists.

serviceable *adjective* useful, usable; durable but plain. □ **serviceability** *noun*.

serviette /sɜ:vɪ'et/ *noun* table napkin.

servile /'sɜ:vaɪl/ *adjective* of or like slave(s); fawning, subservient. □ **servility** /-'vɪl-/ *noun*.

servitude /'sɜ:vɪtju:d/ *noun* slavery, subjection.

servo- /'sɜ:vəʊ/ *combining form* power-assisted.

sesame /'sesəmɪ/ *noun* E. Indian plant with oil-yielding seeds; its seeds.

sesqui- /'seskwɪ/ *combining form* one and a half.

sessile /'sesaɪl/ *adjective Biology* attached directly by base without stalk or peduncle; fixed, immobile.

session /'seʃ(ə)n/ *noun* period devoted to an activity; assembly of parliament, court, etc.; single meeting for this; period during which such meetings are regularly held; academic year. □ **in session** assembled for business, not on vacation. □ **sessional** *adjective*.

set ● *verb* (**-tt-**; *past & past participle* **set**) put, lay, or stand in certain position etc.; apply; fix or place ready; dispose suitably for use, action, or display; adjust hands or mechanism of (clock, trap, etc.); insert (jewel) in ring etc.; lay (table) for meal; style (hair) while damp; (+ *with*) ornament or provide (surface) with; bring into specified state, cause to be; harden, solidify; (of sun, moon, etc.) move towards or below earth's horizon; show (story etc.) as happening in a certain time or place; (+ *to do*) cause (person) to do specified thing; (+ *present participle*) start (person, thing) doing something; present or impose as work to be done, problem to be solved, etc.; exhibit as model etc.; initiate (fashion etc.); establish (record etc.); determine, decide; appoint, establish; put parts of (broken or dislocated bone, limb, etc.) together for healing; provide (song, words) with music; arrange (type) or type for (book etc.); (of tide, current, etc.) have a certain motion or direction; (of face) assume hard expression; (of eyes etc.) become motionless; have a certain tendency; (of blossom) form into fruit; (of dancer) take position facing partner; (of hunting dog) take rigid attitude indicating presence of game. ● *noun* group of linked or similar things or persons; section of society; collection of objects for specified purpose; radio or television receiver; *Tennis etc.* group of games counting as unit towards winning match; *Mathematics* collection of things sharing a property; direction or position in which something sets or is set; slip, shoot, bulb, etc. for planting; setting, stage furniture, etc. for play, film, etc.; setting of sun, hair, etc.; = SETT. ● *adjective* prescribed or determined in advance; unchanging, fixed; prepared for action; (+ *on*, *upon*) determined to get, achieve, etc. □ **set about** begin, take steps towards, *colloquial* attack; **set aside** put to one side, keep for future, disregard or reject; **set back** place further back in space or time, impede or reverse progress of, *colloquial* cost (person) specified amount; **set-back** reversal or arrest of progress; **set down** record in writing, allow to alight; **set forth** begin journey, expound; **set in** begin, become established, insert; **set off** begin journey, detonate (bomb etc.), initiate, stimulate, cause (person) to start laughing etc., adorn, enhance, (+ *against*) use as compensating item against; **set on** (cause or urge to) attack; **set out** begin journey, (+ *to do*) intend, exhibit, arrange; **set piece** formal or elaborate arrangement, esp. in art or literature; **set square** right-angled triangular plate for drawing lines at certain angles; **set to** begin doing something vigorously; **set theory** study or use of sets in mathematics; **set-to** (*plural* **-tos**) *colloquial* fight, argument; **set up** place in position or view, start, establish, equip, prepare, *colloquial* cause (person) to look guilty or foolish; **set-up** arrangement or organization, manner or structure of this, instance of setting person up; **set upon** set on.

sett *noun* (also **set**) badger's burrow, paving-block.

settee /se'ti:/ *noun* sofa.

setter *noun* dog of long-haired breed trained to stand rigid on scenting game.

setting *noun* position or manner in which thing is set; surroundings; period, place, etc. of story, film, etc.; frame etc. for jewel; music to which words are set; cutlery etc. for one person at table; operating level of machine.

settle[1] /'set(ə)l/ *verb* (**-ling**) (often + *down*, *in*) establish or become established in abode or lifestyle; (often + *down*) regain calm after disturbance, adopt regular or secure way of life, (+ *to*) apply oneself to; (cause to) sit down or come to rest; make or become composed etc.; determine, decide, agree on; resolve (dispute etc.); agree to terminate (lawsuit); (+ *for*) accept or agree to; pay (bill); (as **settled** *adjective*) established; colonize; subside, sink. □ **settle up** pay money owed etc.

settle[2] /'set(ə)l/ *noun* high-backed bench, often with box under seat.

settlement *noun* settling, being settled; place occupied by settlers, small village; political etc. agreement; arrangement ending dispute; terms on which property is given to person; deed stating these; amount or property given.

settler *noun* person who settles in newly developed region.

seven /'sev(ə)n/ *adjective & noun* one more than six. □ **seventh** *adjective & noun.*

seventeen /sevən'ti:n/ *adjective & noun* one more than sixteen. □ **seventeenth** *adjective & noun.*

seventy /'sevəntı/ *adjective & noun* (*plural* **-ies**) seven times ten. □ **seventieth** *adjective & noun.*

sever /'sevə/ *verb* divide, break, or make separate, esp. by cutting.

several /'sevr(ə)l/ ● *adjective* a few; quite a large number; *formal* separate, respective. ● *pronoun* a few; quite a large number. □ **severally** *adverb.*

severance /'sevr(ə)ns/ *noun* severing; severed state. □ **severance pay** payment to employee on termination of contract.

severe /sɪ'vɪə/ *adjective* rigorous and harsh; not negligible, worrying; forceful; extreme; exacting; unadorned. □ **severely** *adverb*; **severity** /-'ver-/ *noun*.

sew /səʊ/ *verb* (*past participle* **sewn** or **sewed**) fasten, join, etc. with needle and thread or sewing machine. □ **sewing machine** machine for sewing or stitching.

sewage /'su:ɪdʒ/ *noun* waste matter carried in sewers. □ **sewage farm, works** place where sewage is treated.

sewer /'su:ə/ *noun* (usually underground) conduit for carrying off drainage water and waste matter.

sewerage /'su:ərɪdʒ/ *noun* system of or drainage by sewers.

sewing /'səʊɪŋ/ *noun* material or work to be sewn.

sewn *past participle* of SEW.

sex ● *noun* group of males or females collectively; fact of belonging to either group; sexual instincts, desires, activity, etc.; *colloquial* sexual intercourse. ● *adjective* of or relating to sex or sexual differences. ● *verb* determine sex of; (as **sexed** *adjective*) having specified sexual appetite. □ **sex appeal** sexual attractiveness; **sex life** person's sexual activity; **sex symbol** person famed for sex appeal.

sexagenarian /seksədʒɪ'neərɪən/ *noun* person between 60 and 69 years old.

Sexagesima /seksə'dʒesɪmə/ *noun* second Sunday before Lent.

sexism *noun* prejudice or discrimination against people (esp. women) because of their sex. □ **sexist** *adjective & noun*.

sexless *adjective* neither male nor female; lacking sexual desire or attractiveness.

sextant /'sekst(ə)nt/ *noun* optical instrument for measuring angle between distant objects, esp. sun and horizon in navigation.

sextet /seks'tet/ *noun* musical composition for 6 performers; the performers; any group of 6.

sexton /'sekst(ə)n/ *noun* person who looks after church and churchyard, often acting as bell-ringer and grave-digger.

sextuple /'sekstju:p(ə)l/ *adjective* sixfold.

sextuplet /'sekstjʊplɪt/ *noun* each of 6 children born at one birth.

sexual /'sekʃʊəl/ *adjective* of sex, the sexes, or relations between them. □ **sexual intercourse** insertion of man's penis into woman's vagina. □ **sexuality** /-'æl-/ *noun*; **sexually** *adverb*.

sexy *adjective* (**-ier**, **-iest**) sexually attractive or provocative; *colloquial* (of project etc.) exciting. □ **sexily** *adverb*; **sexiness** *noun*.

SF *abbreviation* science fiction.

Sgt. *abbreviation* Sergeant.

sh *interjection* hush.

shabby /'ʃæbɪ/ *adjective* (**-ier**, **-iest**) faded and worn, dingy, dilapidated; poorly dressed; contemptible. □ **shabbily** *adverb*; **shabbiness** *noun*.

shack ● *noun* roughly built hut or cabin. ● *verb* (+ *up*) *slang* cohabit.

shackle /'ʃæk(ə)l/ ● *noun* metal loop or link closed by bolt, coupling link; fetter; (usually in *plural*) restraint. ● *verb* (**-ling**) fetter, impede, restrain.

shad *noun* (*plural* same or **-s**) large edible marine fish.

shade ● *noun* comparative darkness caused by shelter from direct light and heat; area so sheltered; darker part of picture etc.; a colour, esp. as darker or lighter than one similar; comparative obscurity; slight amount; lampshade; screen moderating light; (in *plural*) *esp. US colloquial* sunglasses; *literary* ghost; (in *plural*; + *of*) reminder of. ● *verb* (**-ding**) screen from light; cover or moderate light of; darken, esp. with parallel lines to represent shadow etc.; (often + *away*, *off*, *into*) pass or change gradually. □ **shading** *noun*.

shadow /'ʃædəʊ/ ● *noun* shade; patch of shade; dark shape projected by body blocking out light; inseparable attendant or companion; person secretly following another; (with negative) slightest trace; insubstantial remnant; shaded part of picture; gloom, sadness. ● *verb* cast shadow over; secretly follow and watch. □ **shadow-boxing** boxing against imaginary opponent; **Shadow Cabinet**, **Minister**, etc., members of opposition party holding posts parallel to those of government. □ **shadowy** *adjective.*

shady /'ʃeɪdɪ/ *adjective* (**-ier**, **-iest**) giving or situated in shade; disreputable, of doubtful honesty.

shaft /ʃɑ:ft/ *noun* narrow usually vertical space for access to mine or (in building) for lift, ventilation, etc.; (+ *of*) ray of (light), stroke of (lightning); handle of tool etc.; long narrow part supporting, connecting, or driving thicker part(s) etc.; *archaic* arrow, spear, its long slender stem; hurtful or provocative remark; each of pair of poles between which horse is harnessed to vehicle; central stem of feather; column, esp. between base and capital.

shag • *noun* coarse tobacco; rough mass of hair; (crested) cormorant. • *adjective* (of carpet) with long rough pile.

shaggy *adjective* (**-ier**, **-iest**) hairy, rough-haired; tangled. □ **shaggy-dog story** lengthy 'joke' without funny ending. □ **shagginess** *noun.*

shagreen /ʃæ'gri:n/ *noun* kind of untanned granulated leather; sharkskin.

shah /ʃɑ:/ *noun historical* ruler of Iran.

shake • *verb* (**-king**; *past* **shook** /ʃʊk/; *past participle* **shaken**) move violently or quickly up and down or to and fro; (cause to) tremble or vibrate; agitate, shock, disturb; weaken, impair; *colloquial* shake hands. • *noun* shaking, being shaken; jerk, shock; (**the shakes**) *colloquial* fit of trembling. □ **shake down** settle or cause to fall by shaking, become comfortably settled or established; **shake hands** (often + *with*) clasp hands, esp. at meeting or parting or as sign of bargain; **shake off** get rid of, evade; **shake up** mix (ingredients) or restore to shape by shaking, disturb or make uncomfortable, rouse from lethargy etc.; **shake-up** upheaval, reorganization.

shaker *noun* person or thing that shakes; container for shaking together ingredients of cocktails etc.

Shakespearian /ʃeɪk'spɪərɪən/ *adjective* (also **Shakespearean**) of Shakespeare.

shako /'ʃækəʊ/ *noun* (*plural* **-s**) cylindrical plumed military peaked cap.

shaky *adjective* (**-ier**, **-iest**) unsteady, trembling; infirm; unreliable. □ **shakily** *adverb*; **shakiness** *noun.*

shale *noun* soft rock that splits easily. □ **shaly** *adjective.*

shall *auxiliary verb* (*3rd singular present* **shall**) *used to form future tenses.*

shallot /ʃə'lɒt/ *noun* onion-like plant with cluster of small bulbs.

shallow /'ʃæləʊ/ • *adjective* having little depth; superficial, trivial. • *noun* (often in *plural*) shallow place. □ **shallowness** *noun.*

sham • *verb* (**-mm-**) feign; pretend (to be). • *noun* imposture, pretence; bogus or false person or thing. • *adjective* pretended, counterfeit.

shamble /'ʃæmb(ə)l/ • *verb* (**-ling**) walk or run awkwardly, dragging feet. • *noun* shambling gait.

shambles *plural noun* (usually treated as *singular*) *colloquial* mess, muddle; butcher's slaughterhouse; scene of carnage.

shambolic /ʃæm'bɒlɪk/ *adjective colloquial* chaotic, disorganized.

shame • *noun* humiliation caused by consciousness of guilt or folly; capacity for feeling this; state of disgrace or discredit; person or thing that brings disgrace etc.; wrong or regrettable thing. • *verb* (**-ming**) bring disgrace on, make ashamed; (+ *into*, *out of*) force by shame into or out of. □ **shamefaced** showing shame, bashful.

shameful *adjective* disgraceful, scandalous. □ **shamefully** *adverb*; **shamefulness** *noun.*

shameless *adjective* having or showing no shame; impudent. □ **shamelessly** *adverb.*

shammy /'ʃæmɪ/ *noun* (*plural* **-ies**) *colloquial* chamois leather.

shampoo /ʃæm'pu:/ • *noun* liquid for washing hair; similar substance for washing cars, carpets, etc. • *verb* (**-poos**, **-pooed**) wash with shampoo.

shamrock /'ʃæmrɒk/ *noun* trefoil, as national emblem of Ireland.

shandy /'ʃændɪ/ *noun* (*plural* **-ies**) beer with lemonade or ginger beer.

shanghai /ʃæŋ'haɪ/ *verb* (**-hais**, **-haied**, **-haiing**) *colloquial* trick or force (person) to do something, esp. be sailor.

shank *noun* leg, lower part of leg; shaft or stem, esp. joining tool's handle to its working end.

shan't /ʃɑ:nt/ shall not.

shantung /ʃæn'tʌŋ/ *noun* soft undressed Chinese silk.

shanty[1] /'ʃæntɪ/ *noun* (*plural* **-ies**) hut, cabin. □ **shanty town** area with makeshift housing.

shanty[2] /'ʃæntɪ/ *noun* (*plural* **-ies**) (in full **sea shanty**) sailors' work song.

shape ● *noun* outline; form; specific form or guise; good or specified condition; person or thing seen indistinctly; mould, pattern. ● *verb* (**-ping**) give a certain form to, fashion, create; influence; (usually + *up*) show promise; (+ *to*) make conform to. □ **take shape** assume distinct form, develop.

shapeless *adjective* lacking definite or attractive shape. □ **shapelessness** *noun*.

shapely *adjective* (**-ier**, **-iest**) of pleasing shape, well-proportioned. □ **shapeliness** *noun*.

shard *noun* broken fragment of pottery, glass, etc.

share /ʃeə/ ● *noun* portion of whole given to or taken from person; each of equal parts into which company's capital is divided, entitling owner to proportion of profits. ● *verb* (**-ring**) have or use with another or others; get, have, or give share of; (+ *in*) participate in; (+ *out*) divide and distribute. □ **shareholder** owner of shares in a company; **share-out** division and distribution.

shark *noun* large voracious sea fish; *colloquial* swindler, extortioner. □ **sharkskin** skin of shark, smooth slightly shiny fabric.

sharp ● *adjective* having edge or point able to cut or pierce; tapering to a point or edge; abrupt, steep, angular; well-defined; severe, intense; pungent, acid; shrill, piercing; harsh; acute, sensitive, clever; unscrupulous; vigorous, brisk; *Music* above true pitch, a semitone higher than note named. ● *noun Music* sharp note; sign (♯) indicating this; *colloquial* swindler, cheat. ● *adverb* punctually; suddenly; at a sharp angle; *Music* above true pitch. □ **sharp practice** barely honest dealings; **sharpshooter** skilled marksman; **sharp-witted** keenly perceptive or intelligent. □ **sharpen** *verb*; **sharpener** *noun*; **sharply** *adverb*; **sharpness** *noun*.

sharper *noun* swindler, esp. at cards.

shatter /'ʃætə/ *verb* break suddenly in pieces; severely damage, destroy; (esp. in *passive*) severely upset; (usually as **shattered** *adjective*) *colloquial* exhaust.

shave ● *verb* (**-ving**; *past participle* **shaved** or (as *adjective*) **shaven**) remove (hair, bristles) with razor; remove hair with razor from (leg, head, etc.) or from face of (person); reduce by small amount; pare (wood etc.) to shape it; miss or pass narrowly. ● *noun* shaving, being shaved; narrow miss or escape.

shaver *noun* thing that shaves; electric razor; *colloquial* young lad.

shaving *noun* (esp. in *plural*) thin paring of wood.

shawl *noun* large usually rectangular piece of fabric worn over shoulders etc. or wrapped round baby.

she *pronoun* (as subject of verb) the female person or animal in question.

s/he *pronoun: written representation of* 'he or she'.

sheaf ● *noun* (*plural* **sheaves**) bundle of things laid lengthways together and usually tied, esp. reaped corn or collection of papers. ● *verb* make into sheaves.

shear ● *verb* (*past* **sheared**; *past participle* **shorn** or **sheared**) clip wool off (sheep etc.); remove by cutting; cut with scissors, shears, etc.; strip bare, deprive; (often + *of*) distort, be distorted, or break. ● *noun* strain produced by pressure in structure of substance; (in *plural*) (also **pair of shears** *singular*) scissor-shaped clipping or cutting instrument. □ **shearer** *noun*.

sheath /ʃi:θ/ *noun* (*plural* **-s** /ʃi:ðz/) close-fitting cover, esp. for blade; condom. □ **sheath knife** dagger-like knife carried in sheath.

sheathe /ʃi:ð/ *verb* (**-thing**) put into sheath; encase or protect with sheath.

sheaves *plural* of SHEAF.

shebeen /ʃɪ'bi:n/ *noun esp. Irish* unlicensed drinking place.

shed[1] *noun* one-storeyed building for storage or shelter or as workshop etc.

shed[2] *verb* (**-dd-**; *past & past participle* **shed**) let, or cause to, fall off; take off (clothes); reduce (electrical power load); cause to fall or flow; disperse, diffuse, radiate; get rid of.

she'd /ʃi:d/ she had; she would.

sheen *noun* lustre, brightness.

sheep *noun* (*plural* same) mammal with thick woolly coat, esp. kept for its wool or meat; timid, silly, or easily-led person; (usually in *plural*) member of minister's congregation. □ **sheep-dip** preparation or place for cleansing sheep of vermin etc.; **sheepdog** dog trained to guard and herd sheep, dog of breed suitable for this; **sheepfold** enclosure for sheep; **sheepshank** knot for temporarily shortening rope; **sheepskin** sheep's skin with wool on.

sheepish *adjective* embarrassed or shy; ashamed. □ **sheepishly** *adverb.*

sheer[1] ● *adjective* mere, complete; (of cliff etc.) perpendicular; (of textile) diaphanous. ● *adverb* directly, perpendicularly.

sheer[2] *verb* swerve or change course; (often + *away, off*) turn away, esp. from person that one dislikes or fears.

sheet[1] ● *noun* rectangular piece of cotton etc. as part of bedclothes; broad thin flat piece of paper, metal, etc.; wide expanse of water, ice, flame, etc. ● *verb* cover (as) with sheet; (of rain etc.) fall in sheets. □ **sheet metal** metal rolled or hammered etc. into thin sheets; **sheet music** music published in separate sheets.

sheet[2] *noun* rope at lower corner of sail to control it. □ **sheet anchor** emergency anchor, person or thing depended on as last hope.

sheikh /ʃeɪk/ *noun* chief or head of Arab tribe, family, or village; Muslim leader.

sheila /'ʃi:lə/ *noun Australian & NZ offensive slang* girl, young woman.

shekel /'ʃek(ə)l/ *noun* chief monetary unit of Israel; *historical* weight and coin in ancient Israel etc.; (in *plural*) *colloquial* money.

shelduck /'ʃeldʌk/ *noun* (*plural* same or **-s**; *masculine* **sheldrake**, *plural* same or **-s**) brightly coloured wild duck.

shelf *noun* (*plural* **shelves**) wooden etc. board projecting from wall or forming part of bookcase or cupboard; ledge on cliff face etc.; reef, sandbank. □ **on the shelf** (of woman) considered past marriageable age, put aside; **shelf-life** time for which stored thing remains usable; **shelf-mark** code on book to show its place in library.

shell ● *noun* hard outer case of many molluscs, tortoise, egg, nut-kernel, seed, etc.; explosive artillery projectile; hollow container for fireworks, cartridges, etc.; light racing boat; framework of vehicle etc.; walls of unfinished or gutted building etc. ● *verb* remove shell or pod from; fire shells at. □ **come out of one's shell** become less shy; **shellfish** aquatic mollusc with shell, crustacean; **shell out** *colloquial* pay (money); **shell-shock** nervous breakdown caused by warfare; **shell-shocked** *adjective* □ **shell-like** *adjective.*

she'll /ʃi:l/ she will; she shall.

shellac /ʃə'læk/ ● *noun* resin used for making varnish. ● *verb* (**-ck-**) varnish with shellac.

shelter /'ʃeltə/ ● *noun* protection from danger, bad weather, etc.; place providing this. ● *verb* act or serve as shelter to; shield; take shelter.

shelve *verb* (**-ving**) put aside, esp. temporarily; put on shelf; provide with shelves; (of ground) slope.

shelving *noun* shelves; material for shelves.

shepherd /'ʃepəd/ ● *noun* (*feminine* **shepherdess**) person who tends sheep; pastor. ● *verb* tend (sheep); marshal or guide like sheep. □ **shepherd's pie** minced meat baked with covering of (esp. mashed) potato.

sherbet /'ʃɜ:bət/ *noun* flavoured effervescent powder or drink.

sherd *noun* potsherd.

sheriff /'ʃerɪf/ *noun* (also **High Sheriff**) chief executive officer of Crown in county, administering justice etc.; *US* chief law-enforcing officer of county; (also **sheriff-depute**) *Scottish* chief judge of county or district.

sherry /'ʃerɪ/ *noun* (*plural* **-ies**) fortified wine originally from Spain.

she's /ʃi:z/ she is; she has.

Shetland pony /ˈʃetlənd/ *noun* pony of small hardy breed.

shew *archaic* = SHOW.

shiatsu /ʃɪˈætsuː/ *noun* Japanese therapy involving pressure on specific points of body.

shibboleth /ˈʃɪbəleθ/ *noun* long-standing formula, doctrine, phrase, etc. espoused by party or sect.

shied *past & past participle* of SHY[2].

shield /ʃiːld/ ● *noun* piece of defensive armour held in front of body when fighting; person or thing giving protection; shield-shaped trophy; protective plate or screen in machinery etc.; representation of shield for displaying person's coat of arms. ● *verb* protect, screen.

shier *comparative* of SHY[1].

shiest *superlative* of SHY[1].

shift ● *verb* (cause to) change or move from one position to another; remove, esp. with effort; *slang* hurry; *US* change (gear). ● *noun* shifting; relay of workers; period for which they work; device, expedient, trick; woman's loose straight dress; displacement of spectral line; key on keyboard for switching between lower and upper case etc.; *US* gear lever in motor vehicle. □ **make shift** manage, get along; **shift for oneself** rely on one's own efforts; **shift one's ground** alter stance in argument etc.

shiftless *adjective* lacking resourcefulness; lazy.

shifty *adjective* (**-ier**, **-iest**) *colloquial* evasive, deceitful.

Shiite /ˈʃiːaɪt/ ● *noun* member of esp. Iranian branch of Islam opposed to Sunnis. ● *adjective* of this branch.

shillelagh /ʃɪˈleɪlɪ/ *noun* Irish cudgel.

shilling /ˈʃɪlɪŋ/ *noun historical* former British coin and monetary unit, worth 1/20 of pound; monetary unit in some other countries.

shilly-shally /ˈʃɪlɪʃælɪ/ *verb* (**-ies**, **-ied**) be undecided, vacillate.

shimmer /ˈʃɪmə/ ● *verb* shine tremulously or faintly. ● *noun* tremulous or faint light.

shin ● *noun* front of leg below knee; cut of beef from this part. ● *verb* (**-nn-**) (usually + *up*, *down*) climb quickly using arms and legs. □ **shin-bone** tibia.

shindig /ˈʃɪndɪg/ *noun* (also **shindy**) *colloquial* lively noisy party; brawl, disturbance.

shine ● *verb* (**-ning**; *past & past participle* **shone** /ʃɒn/ or **shined**) emit or reflect light, be bright, glow; (of sun, star, etc.) be visible; cause to shine; be brilliant, excel; (*past & past participle* **shined**) polish. ● *noun* light, brightness; polish, lustre. □ **take a shine to** *colloquial* take a liking to.

shiner *noun colloquial* black eye.

shingle[1] /ˈʃɪŋg(ə)l/ *noun* small rounded pebbles on seashore. □ **shingly** *adjective*.

shingle[2] /ˈʃɪŋg(ə)l/ ● *noun* rectangular wooden tile used on roofs etc.; *archaic* shingled hair. ● *verb* (**-ling**) roof with shingles; *archaic* cut (woman's hair) short and tapering.

shingles /ˈʃɪŋg(ə)lz/ *plural noun* (usually treated as *singular*) painful viral infection of nerves with rash, esp. round waist.

Shinto /ˈʃɪntəʊ/ *noun* (also **Shintoism**) Japanese religion with worship of ancestors and nature-spirits.

shinty /ˈʃɪntɪ/ *noun* (*plural* **-ies**) game resembling hockey; stick or ball for this.

shiny *adjective* (**-ier**, **-iest**) having shine; (of clothing) with nap worn off.

ship ● *noun* large seagoing vessel; *US* aircraft; spaceship. ● *verb* (**-pp-**) transport, esp. in ship; take in (water) over ship's side etc.; lay (oars) at bottom of boat; fix (rudder etc.) in place; embark; be hired to work on ship. □ **shipmate** fellow member of ship's crew; **ship off** send away; **shipshape** trim, neat, tidy; **shipwreck** *noun* destruction of ship by storm or collision etc., ship so destroyed, *verb* (usually in *passive*) cause to suffer this; **shipwright** shipbuilder, ship's carpenter; **shipyard** place where ships are built.

shipment *noun* goods shipped; act of shipping goods etc.

shipper *noun* person or company that ships goods.

shipping *noun* transport of goods etc.; ships collectively.

shire /ʃaɪə/ *noun* county. □ **shire horse** heavy powerful horse.

shirk *verb* avoid (duty, work, etc.). □ **shirker** *noun.*

shirr ● *noun* elasticated gathered threads forming smocking. ● *verb* gather (material) with parallel threads. □ **shirring** *noun.*

shirt *noun* upper-body garment of cotton etc., usually with sleeves and collar. □ **in shirtsleeves** not wearing jacket; **shirt dress**, **shirtwaister** dress with bodice like shirt.

shirty *adjective* (**-ier**, **-iest**) *colloquial* annoyed. □ **shirtily** *adverb*; **shirtiness** *noun.*

shish kebab /ʃɪʃ/ *noun* pieces of meat and vegetables grilled on skewer.

shiver[1] /ˈʃɪvə/ ● *verb* tremble with cold, fear, etc. ● *noun* momentary shivering movement; (**the shivers**) attack of shivering. □ **shivery** *adjective.*

shiver[2] /ˈʃɪvə/ ● *noun* (esp. in *plural*) small fragment, splinter. ● *verb* break into shivers.

shoal[1] ● *noun* multitude, esp. of fish swimming together. ● *verb* form shoal(s).

shoal[2] ● *noun* area of shallow water; submerged sandbank. ● *verb* (of water) become shallow.

shock[1] ● *noun* violent collision, impact, etc.; sudden and disturbing emotional effect; acute prostration following wound, pain, etc.; electric shock; disturbance in stability of organization etc. ● *verb* horrify, outrage; cause shock; affect with electric or pathological shock. □ **shock absorber** device on vehicle etc. for absorbing shock and vibration; **shockproof** resistant to effects of shock; **shock therapy** electroconvulsive therapy; .**shock wave** moving region of high air pressure caused by explosion etc.

shock[2] *noun* unkempt or shaggy mass of hair.

shocker *noun colloquial* shocking person or thing; sensational novel etc.

shocking *adjective* causing shock, scandalous; *colloquial* very bad. □ **shocking pink** vibrant shade of pink. □ **shockingly** *adverb.*

shod *past & past participle* of SHOE.

shoddy /ˈʃɒdɪ/ *adjective* (**-ier**, **-iest**) poorly made; counterfeit. □ **shoddily** *adverb*; **shoddiness** *noun.*

shoe /ʃuː/ ● *noun* foot-covering of leather etc., esp. one not reaching above ankle; protective metal rim for horse's hoof; thing like shoe in shape or use. ● *verb* (**shoes**, **shoeing**; *past & past participle* **shod**) fit with shoe(s). □ **-shod** having shoes of specified kind; **shoehorn** curved implement for easing heel into shoe; **shoelace** cord for lacing shoe; **shoestring** shoelace, *colloquial* small esp. inadequate amount of money; **shoe-tree** shaped block for keeping shoe in shape.

shone *past & past participle* of SHINE.

shoo ● *interjection used to frighten animals etc. away.* ● *verb* (**shoos**, **shooed**) utter such sound; (usually + *away*) drive away thus.

shook *past* of SHAKE.

shoot /ʃuːt/ ● *verb* (*past & past participle* **shot**) cause (weapon) to discharge missile; kill or wound with bullet, arrow, etc.; send out or discharge rapidly; come or go swiftly or suddenly; (of plant) put forth buds etc., (of bud etc.) appear; hunt game etc. with gun; film, photograph; *esp. Football* score or take shot at (goal); (often + *up*) *slang* inject (drug). ● *noun* young branch or sucker; hunting party or expedition; land on which game is shot. □ **shooting gallery** place for shooting at targets with rifles etc.; **shooting star** small rapidly moving meteor; **shooting stick** walking stick with foldable seat.

shop ● *noun* place for retail sale of goods or services; act of shopping; place for making or repairing something; *colloquial* place of business etc. ● *verb* (**-pp-**) go to shop(s) to make purchases; *slang* inform against. □ **shop around** look for best bargain; **shop assistant** person serving in shop; **shop-floor** production area in factory etc., workers as distinct from management; **shopkeeper** owner or manager of shop; **shoplift** steal goods while appearing to shop; **shoplifter** *noun*; **shop-soiled** soiled or faded by display in shop; **shop steward** elected representative of workers in factory etc.; **shopwalker** supervisor in large shop; **talk shop** talk about one's occupation. □ **shopper** *noun.*

shopping *noun* purchase of goods; goods bought. □ **shopping centre** area containing many shops.

shore[1] *noun* land adjoining sea, lake, etc.; (usually in *plural*) country. □ **on shore** ashore; **shoreline** line where shore meets water.

shore[2] *verb* (**-ring**) (often + *up*) support (as if) with prop(s) or beam(s).

shorn *past participle* of SHEAR.

short ● *adjective* measuring little from end to end in space or time, or from head to foot; (usually + *of*, *on*) deficient, scanty; concise, brief; curt, uncivil; (of memory) unable to remember distant events; (of vowel or syllable) having the less of two recognized durations; (of pastry) easily crumbled; (of a drink of spirits) undiluted. ● *adverb* before the natural or expected time or place; abruptly; rudely. ● *noun* short circuit; *colloquial* short drink; short film. ● *verb* short-circuit. □ **short back and sides** short simple haircut; **shortbread**, **shortcake** rich crumbly biscuit or cake made of flour, butter, and sugar; **short-change** cheat, esp. by giving insufficient change; **short circuit** electric circuit through small resistance, esp. instead of through normal circuit; **short-circuit** cause short circuit in, have short circuit, shorten or avoid by taking short cut; **shortcoming** deficiency, defect; **short cut** path or course shorter than usual or normal; **shortfall** deficit; **shorthand** method of rapid writing using special symbols, abbreviated or symbolic means of expression; **short-handed**, **-staffed** understaffed; **shorthorn** animal of breed of cattle with short horns; **short list** list of candidates from whom final selection will be made; **short-list** put on short list; **short-lived** ephemeral; **short-range** having short range, relating to immediate future; **short shrift** curt or dismissive treatment; **short sight** inability to focus except at close range; **short-tempered** easily angered; **short-term** of or for a short period of time; **short-winded** easily becoming breathless. □ **shorten** *verb*.

shortage *noun* (often + *of*) deficiency; lack.

shortening *noun* fat for pastry.

shortly *adverb* (often + *before*, *after*) soon; curtly.

shorts *plural noun* trousers reaching to knees or higher; *US* underpants.

short-sighted *adjective* having short sight; lacking imagination or foresight. □ **short-sightedly** *adverb*; **short-sightedness** *noun*.

shot[1] *noun* firing of gun etc.; attempt to hit by shooting or throwing etc.; single missile for gun etc.; (*plural* same or **-s**) small lead pellet of which several are used for single charge; (treated as *plural*) these collectively; photograph, film sequence; stroke or kick in ball game; *colloquial* attempt, guess; person of specified skill in shooting; heavy metal ball thrown in shot-put; *colloquial* drink of spirits; injection of drug etc. □ **shotgun** gun for firing small shot at short range; **shotgun wedding** *colloquial* wedding enforced because of bride's pregnancy; **shot-put** athletic contest in which shot is thrown; **shot-putter** *noun*.

shot[2] ● *past & past participle* of SHOOT. ● *adjective* woven so as to show different colours at different angles.

should /ʃʊd/ *auxiliary verb* (*3rd singular present* **should**) *used in reported speech; expressing obligation, likelihood, or tentative suggestion; used to form conditional clause or (in 1st person) conditional mood.*

shoulder /ˈʃəʊldə/ ● *noun* part of body to which arm, foreleg, or wing is attached; either of two projections below neck; animal's upper foreleg as joint of meat; (also in *plural*) shoulder regarded as supportive, comforting, etc.; strip of land next to road; part of garment covering shoulder. ● *verb* push with shoulder; make one's way thus; take on (burden, responsibility, etc.). □ **shoulder blade** either flat bone of upper back; **shoulder-length** (of hair etc.) reaching to shoulders; **shoulder pad** pad in garment to bulk out shoulder; **shoulder strap** strap going over shoulder from front to back of garment, strap suspending bag etc. from shoulder.

shouldn't /ˈʃʊd(ə)nt/ should not.

shout /ʃaʊt/ ● *verb* speak or cry loudly; say or express loudly. ● *noun* loud cry calling attention or expressing joy, defiance, approval, etc. □ **shout down** reduce to silence by shouting.

shove /ʃʌv/ ● *verb* (**-ving**) push, esp. vigorously or roughly; *colloquial* put casually. ● *noun* act of shoving. □ **shove-halfpenny** form of shovelboard played with coins etc. on table; **shove off** start from shore, mooring, etc. in boat, *slang* depart.

shovel /ˈʃʌv(ə)l/ ● *noun* spadelike scoop used to shift earth or coal etc. ● *verb* (**-ll-**; *US* **-l-**) move (as) with shovel. □ **shovelboard** game played esp. on ship's deck by pushing discs over marked surface.

shoveller /ˈʃʌvələ/ *noun* (also **shoveler**) duck with shovel-like beak.

show /ʃəʊ/ ● *verb* (*past participle* **shown** or **showed**) be, or allow or cause to be, seen; manifest; offer for inspection; express (one's feelings); accord, grant (favour, mercy, etc.); (of feelings etc.) be manifest; instruct by example; demonstrate, make understood; exhibit; (often + *in*, *round*, etc.) conduct, lead; *colloquial* appear, arrive. ● *noun* showing; spectacle, exhibition, display; public entertainment or performance; outward appearance, impression produced; ostentation, mere display; *colloquial* undertaking, business. □ **good** (or **bad** or **poor**) **show!** *colloquial* that was well (or badly) done; **showbiz** *colloquial* show business; **show business** *colloquial* the entertainment profession; **showcase** glass case or event etc. for displaying goods or exhibits; **showdown** final test or confrontation; **show house**, **flat** furnished and decorated new house or flat on show to prospective buyers; **showjumping** competitive jumping on horseback; **show off** display to advantage, *colloquial* act pretentiously; **show-off** *colloquial* person who shows off; **show-piece** excellent specimen suitable for display; **showroom** room where goods are displayed for sale; **show trial** judicial trial designed to frighten or impress the public; **show up** make or be visible or conspicuous, expose, humiliate, *colloquial* appear or arrive; **show willing** show willingness to help etc.

shower /ˈʃaʊə/ ● *noun* brief fall of rain, snow, etc.; brisk flurry of bullets, dust, etc.; sudden copious arrival of gifts, honours, etc.; (in full **shower-bath**) bath in which water is sprayed from above. ● *verb* descend, send, or give in shower; take shower-bath; (+ *upon*, *with*) bestow lavishly. □ **showery** *adjective*.

showing *noun* display; quality of performance, achievement, etc.; evidence; putting of case etc.

shown *past participle* of SHOW.

showy *adjective* (**-ier**, **-iest**) gaudy; striking. □ **showily** *adverb*; **showiness** *noun*.

shrank *past* of SHRINK.

shrapnel /ˈʃræpn(ə)l/ *noun* fragments of exploded bomb etc.

shred ● *noun* scrap, fragment; least amount. ● *verb* (**-dd-**) tear, cut, etc. to shreds. □ **shredder** *noun*.

shrew *noun* small long-snouted mouse-like animal; bad-tempered or scolding woman. □ **shrewish** *adjective*.

shrewd *adjective* astute; clever. □ **shrewdly** *adverb*; **shrewdness** *noun*.

shriek ● *noun* shrill cry or sound. ● *verb* make a shriek; say in shrill tones.

shrike *noun* bird with strong hooked beak.

shrill ● *adjective* piercing and high-pitched. ● *verb* sound or utter shrilly. □ **shrillness** *noun*; **shrilly** *adverb*.

shrimp *noun* (*plural* same or **-s**) small edible crustacean; *colloquial* very small person.

shrine *noun* sacred or revered place; casket or tomb holding relics.

shrink ● *verb* (*past* **shrank**; *past participle* **shrunk** or (esp. as *adjective*) **shrunken**) become or make smaller, esp. by action of moisture, heat, or cold; (usually + *from*) recoil, flinch. ● *noun slang* psychiatrist. □ **shrink-wrap** wrap (article) in material that shrinks tightly round it.

shrinkage *noun* process or degree of shrinking; allowance for loss by theft or wastage.

shrivel /ˈʃrɪv(ə)l/ *verb* (**-ll-**; *US* **-l-**) contract into wrinkled or dried-up state.

shroud /ʃraʊd/ ● *noun* wrapping for corpse; something which conceals; rope supporting mast. ● *verb* clothe (corpse) for burial; cover, disguise.

Shrove Tuesday /ʃrəʊv/ *noun* day before Ash Wednesday.

shrub *noun* woody plant smaller than tree and usually branching from near ground. □ **shrubby** *adjective.*

shrubbery *noun* (*plural* **-ies**) area planted with shrubs.

shrug ● *verb* (**-gg-**) draw up (shoulders) momentarily as gesture of indifference, ignorance, etc. ● *noun* shrugging movement.

shrunk (also **shrunken**) *past participle* of SHRINK.

shudder /'ʃʌdə/ ● *verb* shiver from fear, cold, etc.; feel strong repugnance, fear, etc.; vibrate. ● *noun* act of shuddering.

shuffle /'ʃʌf(ə)l/ ● *verb* (**-ling**) drag or slide (feet) in walking; mix up or rearrange (esp. cards); be evasive; keep shifting one's position. ● *noun* shuffling action or movement; change of relative positions; shuffling dance. □ **shuffle off** remove, get rid of.

shufti /'ʃʊftɪ/ *noun* (*plural* **-s**) *colloquial* look, glimpse.

shun *verb* (**-nn-**) avoid, keep clear of.

shunt ● *verb* move (train etc.) to another track; (of train) be shunted; redirect. ● *noun* shunting, being shunted; conductor joining two points in electric circuit for diversion of current; *slang* collision of vehicles.

shush /ʃʊʃ/ *interjection & verb* hush.

shut *verb* (**-tt-**; *past & past participle* **shut**) move (door, window, lid, etc.) into position to block opening; become or be capable of being shut; shut door etc. of; become or make closed for trade; fold or contract (book, hand, telescope); bar access to (place). □ **shut down** close, cease working; **shut-eye** *colloquial* sleep; **shut off** stop flow of (water, gas, etc.), separate; **shut out** exclude, screen from view, prevent; **shut up** close all doors and windows of, imprison, put away in box etc., (esp. in *imperative*) *colloquial* stop talking.

shutter ● *noun* movable hinged cover for window; device for exposing film in camera. ● *verb* provide or close with shutter(s).

shuttle /'ʃʌt(ə)l/ ● *noun* part of loom which carries weft-thread between threads of warp; thread-carrier for lower thread in sewing machine; train, bus, aircraft, etc. used in shuttle service; space shuttle. ● *verb* (**-ling**) (cause to) move to and fro like shuttle. □ **shuttlecock** cork with ring of feathers, or similar plastic device, struck to and fro in badminton; **shuttle diplomacy** negotiations conducted by mediator travelling between disputing parties; **shuttle service** transport system operating to and fro over short distance.

shy[1] ● *adjective* (**-er**, **-est**) timid and nervous in company; self-conscious; easily startled. ● *verb* (**shies**, **shied**) (usually + *at*) (esp. of horse) start back or aside in fright. ● *noun* sudden startled movement. □ **-shy** showing fear or dislike of. □ **shyly** *adverb*; **shyness** *noun.*

shy[2] ● *verb* (**shies**, **shied**) throw, fling. ● *noun* (*plural* **shies**) throw, fling.

shyster /'ʃaɪstə/ *noun colloquial* person who acts unscrupulously or unprofessionally.

SI *abbreviation* international system of units of measurement (*Système International*).

si /si:/ *noun Music* te.

Siamese /saɪə'mi:z/ ● *noun* (*plural* same) native or language of Siam (now Thailand); (in full **Siamese cat**) cat of cream-coloured dark-faced short-haired breed with blue eyes. ● *adjective* of Siam. □ **Siamese twins** twins joined together at birth.

sibilant /'sɪbɪlənt/ ● *adjective* hissing, sounded with hiss. ● *noun* sibilant speech sound or letter. □ **sibilance** *noun.*

sibling /'sɪblɪŋ/ *noun* each of two or more children having one or both parents in common.

sibyl /'sɪbɪl/ *noun* pagan prophetess.

sic *adverb* used or spelt thus (confirming form of quoted words). [Latin]

sick ● *adjective* vomiting, disposed to vomit; *esp. US* ill, unwell; (often + *of*) *colloquial* disgusted, surfeited; *colloquial* (of humour) cruel, morbid, perverted, offensive. ● *noun colloquial* vomit. ● *verb* (esp. + *up*) *colloquial* vomit. □ **sickbay** place for sick people; **sickbed** invalid's bed; **sick leave** leave granted because of illness; **sick pay** pay given during sick leave.

sicken *verb* make or become sick, disgusted, etc.; (often + *for*) show symptoms of illness; (as **sickening** *adjective*) disgusting, *colloquial* very annoying. □ **sickeningly** *adverb.*

sickle /'sɪk(ə)l/ *noun* short-handled implement with semicircular blade for reaping, lopping, etc.

sickly *adjective* (**-ier**, **-iest**) liable to be ill, weak; faint, pale; causing sickness; mawkish, weakly sentimental.

sickness *noun* being ill; disease; vomiting, nausea.

side ● *noun* each of inner or outer surfaces of object, esp. as distinct from top and bottom or front and back or ends; right or left part of person's or animal's body; part of object, place, etc. that faces specified direction or that is on observer's right or left; either surface of thing regarded as having two; aspect of question, character, etc.; each of sets of opponents in war, game, etc.; cause represented by this; part or region near edge; *colloquial* television channel; each of lines bounding triangle, rectangle, etc.; position nearer or farther than, or to right or left of, dividing line; line of descent through father or mother; spinning motion given to ball by striking it on side; *slang* swagger, assumption of superiority. ● *adjective* of, on, from, or to side; oblique, indirect; subordinate, subsidiary, not main. ● *verb* (**-ding**) take side in dispute etc.; (+ *with*) be on or join same side as. □ **on the side** as sideline, illicitly, *US* as side dish; **sideboard** table or flat-topped chest with drawers and cupboards for crockery etc.; **sideboards**, **sideburns** short side-whiskers; **side by side** standing close together, esp. for mutual encouragement; **sidecar** passenger car attached to side of motorcycle; **side drum** small double-headed drum; **side effect** secondary (usually undesirable) effect; **sidekick** *colloquial* close associate; **sidelight** light from side, small light at side of front of vehicle etc.; **sideline** work etc. done in addition to one's main activity, (usually in *plural*) line bounding side of sports pitch etc., space just outside these for spectators to sit; **side-saddle** *noun* saddle for woman riding with both legs on same side of horse, *adverb* sitting thus on horse; **sideshow** minor show or stall in exhibition, fair, etc.; **sidesman** assistant churchwarden; **sidestep** *noun* step taken sideways, *verb* avoid, evade; **sidetrack** divert from course, purpose, etc.; **sidewalk** *US* pavement; **side-whiskers** hair left unshaven on cheeks; **take sides** support either of (usually) two opposing sides in argument etc.

sidelong ● *adjective* directed to the side. ● *adverb* to the side.

sidereal /saɪ'dɪərɪəl/ *adjective* of or measured or determined by stars.

sideways *adverb & adjective* to or from a side; with one side facing forward.

siding *noun* short track by side of railway line for shunting etc.

sidle /'saɪd(ə)l/ *verb* (**-ling**) walk timidly or furtively.

siege /si:dʒ/ ● *noun* surrounding and blockading of town, castle, etc. □ **lay siege to** conduct siege of; **raise siege** end it.

siemens /'si:mənz/ *noun* (*plural* same) SI unit of electrical conductance.

sienna /sɪ'enə/ *noun* kind of earth used as pigment; its colour of reddish- or yellowish-brown.

sierra /sɪ'erə/ *noun* long jagged mountain chain, esp. in Spain or Spanish America.

siesta /sɪ'estə/ *noun* afternoon nap or rest in hot countries.

sieve /sɪv/ ● *noun* utensil with network or perforated bottom through which liquids or fine particles can pass. ● *verb* (**-ving**) sift.

sift *verb* separate with or cause to pass through sieve; sprinkle through perforated container; closely examine details of, analyse; (of snow, light, etc.) fall as if from sieve.

sigh /saɪ/ ● *verb* emit long deep audible breath in sadness, weariness, relief, etc.; yearn; express with sighs. ● *noun* act of sighing; sound (like that) made in sighing.

sight /saɪt/ ● *noun* faculty of seeing; seeing, being seen; thing seen; range of vision; (usually in *plural*) notable features of a place; device assisting aim with gun or observation with telescope etc.; aim or observation so gained;

colloquial unsightly person or thing; *colloquial* great deal. ● *verb* get sight of; observe presence of; aim (gun etc.) with sight. □ **at first sight** on first glimpse or impression; **on, at sight** as soon as person or thing is seen; **sight-read** read (music) at sight; **sight-screen** *Cricket* large white screen placed near boundary in line with wicket to help batsman see ball; **sightseer** person visiting sights of place; **sightseeing** *noun.*

sighted *adjective* not blind. □ **-sighted** having specified vision.

sightless *adjective* blind.

sign /saɪn/ ● *noun* indication of quality, state, future event, etc.; mark, symbol, etc.; motion or gesture used to convey information, order, etc.; signboard; each of the 12 divisions of the zodiac. ● *verb* write one's name on (document etc.) as authorization; write (one's name) thus; communicate by gesture. □ **signboard** board bearing name, symbol, etc. displayed outside shop, inn, etc.; **sign in** sign register on arrival, get (person) admitted by signing register; **sign language** series of signs used esp. by deaf or dumb people for communication; **sign off** end contract, work, etc.; **sign on** register to obtain unemployment benefit; **sign out** sign register on departing; **signpost** *noun* post etc. showing directions of roads, *verb* provide with signpost(s); **sign up** engage (person), enlist in armed forces; **signwriter** person who paints signboards etc.

signal¹ /ˈsɪgn(ə)l/ ● *noun* sign, esp. prearranged one, conveying information or direction; message of such signs; event which causes immediate activity; *Electricity* transmitted impulses or radio waves; sequence of these; device on railway giving instructions or warnings to train drivers etc. ● *verb* (**-ll-**; *US* **-l-**) make signal(s) (to); (often + *to do*) transmit, announce, or direct by signal(s). □ **signal-box** building from which railway signals are controlled; **signalman** railway signal operator.

signal² /ˈsɪgn(ə)l/ *adjective* remarkable, noteworthy. □ **signally** *adverb.*

signalize *verb* (also **-ise**) (**-zing** or **-sing**) make conspicuous or remarkable; indicate.

signatory /ˈsɪgnətərɪ/ ● *noun* (*plural* **-ies**) party that has signed an agreement, esp. a treaty. ● *adjective* having signed such an agreement.

signature /ˈsɪgnətʃə/ *noun* person's name or initials used in signing; act of signing; *Music* key signature, time signature; section of book made from one sheet folded and cut. □ **signature tune** tune used esp. in broadcasting to announce a particular programme, performer, etc.

signet /ˈsɪgnɪt/ *noun* small seal. □ **signet ring** ring with seal set in it.

significance /sɪgˈnɪfɪkəns/ *noun* importance; meaning; being significant; extent to which result deviates from hypothesis such that difference is due to more than errors in sampling.

significant /sɪgˈnɪfɪkənt/ *adjective* having or conveying meaning; important. □ **significant figure** *Mathematics* digit conveying information about a number containing it. □ **significantly** *adverb.*

signify /ˈsɪgnɪfaɪ/ *verb* (**-ies**, **-ied**) be sign or symbol of; represent, mean, denote; make known; be of importance, matter. □ **signification** *noun.*

signor /ˈsiːnjɔː/ *noun* (*plural* **-i** /-ˈnjɔːriː/) *title used of or to Italian man.*

signora /siːˈnjɔːrə/ *noun title used of or to Italian esp. married woman.*

signorina /siːnjəˈriːnə/ *noun title used of or to Italian unmarried woman.*

Sikh /siːk/ *noun* member of Indian monotheistic sect.

silage /ˈsaɪlɪdʒ/ *noun* green fodder stored in silo; storage in silo.

silence /ˈsaɪləns/ ● *noun* absence of sound; abstinence from speech or noise; neglect or omission to mention, write, etc. ● *verb* (**-cing**) make silent, esp. by force or superior argument.

silencer *noun* device for reducing noise made by gun, vehicle's exhaust, etc.

silent /ˈsaɪlənt/ *adjective* not speaking; making or accompanied by little or no sound. □ **silently** *adverb.*

silhouette /sɪluːˈet/ ● *noun* dark outline or shadow in profile against lighter background; contour, outline, profile; portrait in profile showing outline only, usually cut from paper or in black on white. ● *verb* (**-tting**) represent or show in silhouette.

silica /ˈsɪlɪkə/ *noun* silicon dioxide, occurring as quartz and as main constituent of sand etc. □ **siliceous** /sɪˈlɪʃəs/ *adjective*.

silicate /ˈsɪlɪkeɪt/ *noun* compound of metal(s), silicon, and oxygen.

silicon /ˈsɪlɪkən/ *noun* non-metallic element occurring in silica and silicates. □ **silicon chip** silicon microchip.

silicone /ˈsɪlɪkəʊn/ *noun* any organic compound of silicon with high resistance to cold, heat, water, etc.

silicosis /sɪlɪˈkəʊsɪs/ *noun* lung disease caused by inhaling dust containing silica.

silk *noun* fine strong soft lustrous fibre produced by silkworms; thread or cloth made from this; (in *plural*) cloth or garments of silk; *colloquial* Queen's Counsel. □ **silk-screen printing** screen printing; **silkworm** caterpillar which spins cocoon of silk; **take silk** become Queen's Counsel.

silken *adjective* of or resembling silk; soft, smooth, lustrous.

silky *adjective* (**-ier**, **-iest**) like silk in smoothness, softness, etc.; suave. □ **silkily** *adverb*.

sill *noun* slab of wood, stone, etc. at base of window or doorway.

silly /ˈsɪlɪ/ ● *adjective* (**-ier**, **-iest**) foolish, imprudent; weak-minded; *Cricket* (of fielder or position) very close to batsman. ● *noun* (*plural* **-ies**) *colloquial* silly person. □ **silliness** *noun*.

silo /ˈsaɪləʊ/ *noun* (*plural* **-s**) pit or airtight structure in which green crops are stored for fodder; tower or pit for storage of grain, cement, etc.; underground storage chamber for guided missile.

silt ● *noun* sediment in channel, harbour, etc. ● *verb* (often + *up*) block or be blocked with silt.

silvan = SYLVAN.

silver /ˈsɪlvə/ ● *noun* greyish-white lustrous precious metal; coins or articles made of or looking like this; colour of silver. ● *adjective* of or coloured like silver. ● *verb* coat or plate with silver; provide (mirror-glass) with backing of tin amalgam etc.; make silvery; turn grey or white. □ **silver birch** common birch with silvery white bark; **silverfish** (*plural* same or **-es**) small silvery wingless insect, silver-coloured fish; **silver jubilee** 25th anniversary of reign; **silver medal** medal awarded as second prize; **silver paper** aluminium foil; **silver plate** articles plated with silver; **silver-plated** plated with silver; **silver sand** fine pure kind used in gardening; **silver screen** (usually **the silver screen**) cinema films collectively; **silverside** upper side of round of beef; **silversmith** worker in silver; **silver wedding** 25th anniversary of wedding.

silvery *adjective* like silver in colour or appearance; having clear soft ringing sound.

simian /ˈsɪmɪən/ ● *adjective* of anthropoid apes; resembling ape, monkey. ● *noun* ape or monkey.

similar /ˈsɪmɪlə/ *adjective* like, alike; (often + *to*) having resemblance. □ **similarity** /-ˈlær-/ *noun* (*plural* **-ies**); **similarly** *adverb*.

simile /ˈsɪmɪlɪ/ *noun* esp. poetical comparison of two things using *like* or *as* (see panel).

similitude /sɪˈmɪlɪtjuːd/ *noun* guise, appearance; comparison or its expression.

simmer /ˈsɪmə/ ● *verb* be or keep just below boiling point; be in state of suppressed anger or laughter. ● *noun* simmering state. □ **simmer down** become less agitated.

simnel cake /ˈsɪmn(ə)l/ *noun* rich fruit cake, usually with almond paste.

simony /ˈsaɪmənɪ/ *noun* buying or selling of ecclesiastical privileges.

simoom /sɪˈmuːm/ *noun* hot dry dust-laden desert wind.

simper /ˈsɪmpə/ ● *verb* smile in silly affected way; utter with simper. ● *noun* such smile.

simple /ˈsɪmp(ə)l/ *adjective* (**-r**, **-st**) easily understood or done, presenting no difficulty; not complicated or elaborate; plain; not compound or complex; absolute, unqualified, straightforward; foolish, feeble-minded. □ **simple-minded** foolish, feeble-minded. □ **simpleness** *noun*.

simpleton /ˈsɪmpəlt(ə)n/ *noun* stupid or gullible person.

simplicity /sɪm'plɪsɪtɪ/ *noun* fact or condition of being simple.

simplify /'sɪmplɪfaɪ/ *verb* (**-ies**, **-ied**) make simple or simpler. □ **simplification** *noun*.

simplistic /sɪm'plɪstɪk/ *adjective* excessively or affectedly simple. □ **simplistically** *adverb*.

simply *adverb* in a simple way; absolutely; merely.

simulate /'sɪmjʊleɪt/ *verb* (**-ting**) pretend to be, have, or feel; counterfeit; reproduce conditions of (situation etc.), e.g. for training. □ **simulation** *noun*; **simulator** *noun*.

simultaneous /sɪməl'teɪnɪəs/ *adjective* (often + *with*) occurring or operating at same time. □ **simultaneity** /-tə'neɪɪtɪ/ *noun*; **simultaneously** *adverb*.

sin ● *noun* breaking of divine or moral law; offence against good taste etc. ● *verb* (**-nn-**) commit sin; (+ *against*) offend. □ **sinner** *noun*.

since ● *preposition* throughout or within period after. ● *conjunction* during or in time after; because. ● *adverb* from that time or event until now.

sincere /sɪn'sɪə/ *adjective* (**-r**, **-st**) free from pretence, genuine, honest, frank. □ **sincerity** /-'ser-/ *noun*.

sincerely *adverb* in a sincere way. □ **Yours sincerely** *written before signature at end of informal letter*.

sine *noun* ratio of side opposite angle (in right-angled triangle) to hypotenuse.

sinecure /'saɪnɪkjʊə/ *noun* position that requires little or no work but usually yields profit or honour.

sine die /sɪneɪ 'diːeɪ/ *adverb formal* indefinitely. [Latin]

sine qua non /sɪneɪ kwɑː 'nəʊn/ *noun* indispensable condition or qualification. [Latin]

sinew /'sɪnjuː/ *noun* tough fibrous tissue joining muscle to bone; tendon; (in *plural*) muscles, strength; framework of thing. □ **sinewy** *adjective*.

sinful *adjective* committing or involving sin. □ **sinfully** *adverb*; **sinfulness** *noun*.

sing *verb* (*past* **sang**; *past participle* **sung**) utter musical sounds, esp. words in set tune; utter (song, tune); (of wind, kettle, etc.) hum, buzz, or whistle; *slang* become informer; (+ *of*) *literary* celebrate in verse. □ **sing out** shout; **singsong** *noun* session of informal singing, *adjective* (of voice) monotonously rising and falling. □ **singer** *noun*.

singe /sɪndʒ/ ● *verb* (**-geing**) burn superficially; burn off tips or edges of (esp. hair). ● *noun* superficial burn.

single /'sɪŋg(ə)l/ ● *adjective* one only, not double or multiple; united, undivided; of or for one person or thing; solitary; taken separately; unmarried; (with negative or in questions) even one. ● *noun* single thing, esp. room in hotel; (in full **single ticket**) ticket for one-way journey; pop record with one piece of music on each side; *Cricket* hit for one run; (usually in *plural*) game with one player on each side. ● *verb* (**-ling**) (+ *out*) choose for special attention. □ **single-breasted** (of coat etc.) with only one vertical row of buttons, and overlapping little at the front; **single-decker** bus with only one

Simile

A simile is a figure of speech involving the comparison of one thing with another of a different kind, using *as* or *like*, e.g.

The water was as clear as glass.
Cherry blossom lay like driven snow upon the lawn.

Everyday language is rich in similes:

with *as*:	*as like as two peas*	*as poor as a church mouse*
	as strong as an ox	*as rich as Croesus*
with *like*:	*spread like wildfire*	*run like the wind*
	sell like hot cakes	*like a bull in a china shop*

deck; **single file** line of people one behind another; **single-handed** without help; **single-handedly** *adverb*; **single-minded** intent on only one aim; **single parent** parent bringing up child or children alone. □ **singly** /'sɪŋglɪ/ *adverb*.

singlet /'sɪŋglɪt/ *noun* sleeveless vest.

singleton /'sɪŋgəlt(ə)n/ *noun* player's only card of particular suit.

singular /'sɪŋgjʊlə/ ● *adjective* unique; outstanding; extraordinary, strange; *Grammar* denoting one person or thing. ● *noun Grammar* singular word or form. □ **singularity** /-'lær-/ *noun* (*plural* **-ies**); **singularly** *adverb*.

Sinhalese /sɪnhə'li:z/ ● *noun* (*plural* same) member of a people from N. India now forming majority of population of Sri Lanka; their language. ● *adjective* of this people or language.

sinister /'sɪnɪstə/ *adjective* suggestive of evil; wicked, criminal; ominous; *Heraldry* on left side of shield etc. (i.e. to observer's right).

sink ● *verb* (*past* **sank** or **sunk**; *past participle* **sunk** or (as *adjective*) **sunken**) fall or come slowly downwards; disappear below horizon; go or penetrate below surface of liquid; go to bottom of sea etc.; settle down; decline in strength and vitality; descend in pitch or volume; cause or allow to sink or penetrate; cause failure of; dig (well), bore (shaft); engrave (die); invest (money); cause (ball) to enter pocket at billiards or hole at golf etc. ● *noun* plumbed-in basin, esp. in kitchen; place where foul liquid collects; place of vice. □ **sinking fund** money set aside for eventual repayment of debt.

sinker *noun* weight used to sink fishing or sounding line.

Sino- /'saɪnəʊ/ *combining form* Chinese.

sinology /saɪ'nɒlədʒɪ/ *noun* study of China and its language, history, etc. □ **sinologist** *noun*.

sinuous /'sɪnjʊəs/ *adjective* with many curves, undulating. □ **sinuosity** /-'ɒs-/ *noun*.

sinus /'saɪnəs/ *noun* either of cavities in skull communicating with nostrils.

sinusitis /saɪnə'saɪtɪs/ *noun* inflammation of sinus.

sip ● *verb* (**-pp-**) drink in small mouthfuls. ● *noun* small mouthful of liquid; act of taking this.

siphon /'saɪf(ə)n/ ● *noun* tube shaped like inverted V or U with unequal legs, used for transferring liquid from one container to another by atmospheric pressure; bottle from which fizzy water is forced by pressure of gas. ● *verb* (often + *off*) conduct or flow through siphon, divert or set aside (funds etc.).

sir /sɜ:/ *noun polite form of address or reference to a man*; (**Sir**) *title used before forename of knight or baronet*.

sire ● *noun* male parent of animal, esp. stallion; *archaic form of address to king*; *archaic* father or other male ancestor. ● *verb* (**-ring**) beget.

siren /'saɪərən/ *noun* device for making loud prolonged signal or warning sound; *Greek Mythology* woman or winged creature whose singing lured unwary sailors on to rocks; dangerously fascinating woman.

sirloin /'sɜ:lɔɪn/ *noun* best part of loin of beef.

sirocco /sɪ'rɒkəʊ/ *noun* (*plural* **-s**) Saharan simoom; hot moist wind in S. Europe.

sisal /'saɪs(ə)l/ *noun* fibre from leaves of agave.

siskin /'sɪskɪn/ *noun* small songbird.

sissy /'sɪsɪ/ (also **cissy**) *colloquial* ● *noun* (*plural* **-ies**) effeminate or cowardly person. ● *adjective* (**-ier**, **-iest**) effeminate; cowardly.

sister /'sɪstə/ *noun* woman or girl in relation to her siblings; female fellow member of trade union, sect, human race, etc.; member of female religious order; senior female nurse. □ **sister-in-law** (*plural* **sisters-in-law**) husband's or wife's sister, brother's wife. □ **sisterly** *adjective*.

sisterhood *noun* relationship (as) of sisters; society of women bound by monastic vows or devoting themselves to religious or charitable work; community of feeling among women.

sit *verb* (**-tt-**; *past & past participle* **sat**) support body by resting buttocks on ground, seat, etc.; cause to sit; place in sitting position; (of bird) perch or remain on nest to hatch eggs; (of animal) rest with hind legs bent and buttocks on ground; (of parliament, court, etc.) be in session; (usually + *for*) pose (for portrait); (+ *for*) be MP for (constituency); (often + *for*) take (exam). □ **be sitting pretty** be comfortably placed; **sit back** relax one's efforts; **sit down** sit after standing, cause to sit, (+ *under*) suffer tamely (humiliation etc.); **sit in** occupy place as protest; **sit-in** *noun*; **sit in on** be present as guest etc. at (meeting); **sit on** be member of (committee etc.), *colloquial* delay action about, *slang* repress or snub; **sit out** take no part in (dance etc.), stay till end of, sit outdoors; **sit tight** *colloquial* remain firmly in one's place, not yield; **sit up** rise from lying to sitting, sit firmly upright, defer going to bed, *colloquial* become interested, aroused, etc.; **sit-up** physical exercise of sitting up from supine position without using arms or hands.

sitar /'sɪtɑ:/ *noun* long-necked Indian lute.

sitcom /'sɪtkɒm/ *noun colloquial* situation comedy.

site ● *noun* ground chosen or used for town or building; ground set apart for some purpose. ● *verb* (**-ting**) locate, place.

sitter *noun* person who sits for portrait etc.; babysitter.

sitting ● *noun* continuous period spent engaged in an activity; time during which assembly is engaged in business; session in which meal is served. ● *adjective* having sat down; (of animal or bird) still; (of MP etc.) current. □ **sitting-room** room in which to sit and relax.

situ see IN SITU.

situate /'sɪtjʊeɪt/ *verb* (**-ting**) (usually in *passive*) place or put in position, situation, etc.

situation *noun* place and its surroundings; circumstances; position; state of affairs; *formal* paid job. □ **situation comedy** broadcast comedy series involving characters dealing with awkward esp. domestic or everyday situations. □ **situational** *adjective*.

six *adjective & noun* one more than five; *Cricket* hit scoring six runs. □ **hit, knock for six** *colloquial* utterly surprise or defeat.

sixpence /'sɪkspəns/ *noun* sum of 6 (esp. old) pence; *historical* coin worth this.

sixpenny /'sɪkspəni/ *adjective* costing or worth 6 (esp. old) pence.

sixteen /sɪks'ti:n/ *adjective & noun* one more than fifteen. □ **sixteenth** *adjective & noun*.

sixth ● *adjective & noun* next after fifth; any of 6 equal parts of thing. □ **sixth form** form in secondary school for pupils over 16; **sixth-form college** separate college for pupils over 16; **sixth sense** supposed intuitive or extrasensory faculty.

sixty /'sɪkstɪ/ *adjective & noun* (*plural* **-ies**) six times ten. □ **sixtieth** *adjective & noun*.

sizable = SIZEABLE.

size[1] ● *noun* relative bigness or extent of a thing; dimensions, magnitude; each of classes into which things are divided by size. ● *verb* (**-zing**) sort in sizes or by size. □ **size up** *colloquial* form judgement of.

size[2] ● *noun* sticky solution used for glazing paper and stiffening textiles etc. ● *verb* (**-zing**) treat with size.

sizeable *adjective* (also **sizable**) fairly large.

sizzle /'sɪz(ə)l/ ● *verb* (**-ling**) sputter or hiss, esp. in frying. ● *noun* sizzling sound. □ **sizzling** *adjective & adverb*.

SJ *abbreviation* Society of Jesus.

skate[1] ● *noun* ice-skate; roller-skate. ● *verb* (**-ting**) move, glide, or perform (as) on skates; (+ *over*) refer fleetingly to, disregard. □ **skateboard** *noun* short narrow board on roller-skate wheels for riding on standing up, *verb* ride on skateboard. □ **skater** *noun*.

skate[2] *noun* (*plural* same or **-s**) large edible marine flatfish.

skedaddle /skɪ'dæd(ə)l/ *verb* (**-ling**) *colloquial* run away, retreat hastily.

skein /skeɪn/ *noun* quantity of yarn etc. coiled and usually loosely twisted; flock of wild geese etc. in flight.

skeleton /'skelɪt(ə)n/ ● *noun* hard framework of bones etc. of animal; supporting framework or structure of thing; very thin person or animal; useless or dead remnant; outline sketch. ● *adjective* having only essential or minimum number of people, parts, etc.

□ **skeleton key** key fitting many locks. □ **skeletal** *adjective*.

skerry /ˈskerɪ/ *noun* (*plural* **-ies**) *Scottish* reef, rocky islet.

sketch ●*noun* rough or unfinished drawing or painting; rough draft, general outline; short usually humorous play. ●*verb* make or give sketch of; make sketches.

sketchy *adjective* (**-ier**, **-iest**) giving only a rough outline; *colloquial* insubstantial or imperfect, esp. through haste. □ **sketchily** *adverb*.

skew ●*adjective* oblique, slanting, set askew. ●*noun* slant. ●*verb* make skew; distort; move obliquely.

skewbald /ˈskju:bɔ:ld/ ●*adjective* (of animal) with irregular patches of white and another colour. ●*noun* skewbald animal, esp. horse.

skewer /ˈskju:ə/ ●*noun* long pin for holding meat compactly together while cooking. ●*verb* fasten together or pierce (as) with skewer.

ski /ski:/ ●*noun* (*plural* **-s**) each of pair of long narrow pieces of wood etc. fastened under feet for travelling over snow; similar device under vehicle. ●*verb* (**skis**; **ski'd** or **skied** /ski:d/; **skiing**) travel on skis. □ **skier** *noun*.

skid ●*verb* (**-dd-**) (of vehicle etc.) slide esp. sideways or obliquely on slippery road etc.; cause (vehicle) to skid. ●*noun* act of skidding; runner used as part of landing-gear of aircraft. □ **skid-pan** slippery surface for drivers to practise control of skidding; **skid row** *US slang* district frequented by vagrants.

skiff *noun* light boat, esp. for rowing or sculling.

skilful /ˈskɪlfʊl/ *adjective* (*US* **skillful**) having or showing skill. □ **skilfully** *adverb*.

skill *noun* practised ability, expertness, technique; craft, art, etc. requiring skill.

skilled *adjective* skilful; (of work or worker) requiring or having skill or special training.

skillet /ˈskɪlɪt/ *noun* long-handled metal cooking pot; *US* frying-pan.

skim *verb* (**-mm-**) take scum or cream etc. from surface of (liquid); barely touch (surface) in passing over; (often + *over*) deal with or treat (matter) superficially; (often + *over*, *along*) glide lightly; read or look over superficially. □ **skim**, **skimmed milk** milk with cream removed.

skimp *verb* (often + *on*) economize; supply meagrely, use too little of.

skimpy *adjective* (**-ier**, **-iest**) meagre, insufficient.

skin ●*noun* flexible covering of body; skin removed from animal, material made from this; complexion; outer layer or covering; film like skin on liquid etc.; container for liquid, made of animal's skin. ●*verb* (**-nn-**) strip skin from; graze (part of body); *slang* swindle. □ **skin-deep** superficial; **skin-diver** person who swims under water without diving suit; **skin-diving** *noun*; **skinflint** miser; **skin-graft** surgical transplanting of skin, skin thus transferred; **skintight** very close-fitting.

skinful *noun colloquial* enough alcohol to make one drunk.

skinny *adjective* (**-ier**, **-iest**) thin, emaciated.

skint *adjective slang* having no money.

skip[1] ●*verb* (**-pp-**) move along lightly, esp. by taking two hops with each foot in turn; jump lightly esp. over skipping rope; frisk, gambol; move quickly from one subject etc. to another; omit or make omissions in reading; *colloquial* not attend etc.; *colloquial* leave hurriedly. ●*noun* skipping movement or action. □ **skipping-rope** length of rope turned over head and under feet while jumping it.

skip[2] *noun* large container for refuse etc.; container in which men or materials are lowered or raised in mines etc.

skipjack *noun* (in full **skipjack tuna**) (*plural* same or **-s**) small Pacific tuna.

skipper /ˈskɪpə/ ●*noun* captain of ship, aircraft, team, etc. ●*verb* be captain of.

skirl ●*noun* shrill sound of bagpipes. ●*verb* make skirl.

skirmish /ˈskɜ:mɪʃ/ ●*noun* minor battle; short argument etc. ●*verb* engage in skirmish.

skirt ●*noun* woman's garment hanging from waist, or this part of complete dress; part of coat etc. that hangs below waist; hanging part at base of hovercraft; (in *singular* or *plural*) border, outlying part; flank of beef etc. ● *verb* go or be along or round edge of. □ **skirting board** narrow board etc. round bottom of room-wall.

skit *noun* light piece of satire, burlesque.

skittish /'skɪtɪʃ/ *adjective* lively, playful; (of horse etc.) nervous, inclined to shy.

skittle /'skɪt(ə)l/ *noun* pin used in game of skittles, in which number of wooden pins are set up to be bowled or knocked down.

skive *verb* (**-ving**) (often + *off*) *slang* evade work; play truant.

skivvy /'skɪvɪ/ *noun* (*plural* **-ies**) *colloquial derogatory* female domestic servant.

skua /'skju:ə/ *noun* large predatory seabird.

skulduggery /skʌl'dʌgərɪ/ *noun* trickery, unscrupulous behaviour.

skulk *verb* lurk or conceal oneself or move stealthily.

skull *noun* bony case of brain; bony framework of head; head as site of intelligence. □ **skull and crossbones** representation of skull over two crossed thigh-bones, esp. on pirate flag or as emblem of death; **skullcap** close-fitting peakless cap.

skunk *noun* (*plural* same or **-s**) black white-striped bushy-tailed mammal, emitting powerful stench when attacked; *colloquial* contemptible person.

sky /skaɪ/ *noun* (*plural* **skies**) (in *singular* or *plural*) atmosphere and outer space seen from the earth. □ **sky blue** bright clear blue; **skydiving** sport of performing acrobatic manoeuvres under free fall before opening parachute; **skylark** *noun* lark that sings while soaring, *verb* play tricks and practical jokes; **skylight** window in roof; **skyline** outline of hills, buildings, etc. against sky; **sky-rocket** *noun* firework shooting into air and exploding, *verb* rise steeply; **skyscraper** very tall building.

slab *noun* flat thickish esp. rectangular piece of solid material; mortuary table.

slack¹ ● *adjective* (of rope etc.) not taut; inactive, sluggish; negligent, remiss; (of tide etc.) neither ebbing nor flowing. ● *noun* slack part of rope etc.; slack period; (in *plural*) casual trousers. ● *verb* slacken; *colloquial* take a rest, be lazy. □ **slack off** loosen; **slack up** reduce level of activity or speed. □ **slackness** *noun*.

slack² *noun* coal dust, coal fragments.

slacken *verb* make or become slack. □ **slacken off** slack off.

slacker *noun* shirker.

slag ● *noun* refuse left after ore has been smelted etc. ● *verb* (**-gg-**) form slag; (often + *off*) *slang* insult, slander. □ **slag-heap** hill of refuse from mine etc.

slain *past participle* of SLAY.

slake *verb* (**-king**) assuage or satisfy (thirst etc.); cause (lime) to heat and crumble by action of water.

slalom /'slɑ:ləm/ *noun* downhill ski-race on zigzag course between artificial obstacles.

slam¹ ● *verb* (**-mm-**) shut, throw, or put down violently or with bang; *slang* criticize severely. ● *noun* sound or action of slamming.

slam² *noun* winning of all tricks at cards.

slander /'slɑ:ndə/ ● *noun* false and damaging utterance about person. ● *verb* utter slander about. □ **slanderous** *adjective*.

slang ● *noun* very informal words, phrases, or meanings, not regarded as standard and often peculiar to profession, class, etc. ● *verb* use abusive language (to). □ **slanging match** prolonged exchange of insults. □ **slangy** *adjective*.

slant /slɑ:nt/ ● *verb* slope, (cause to) lie or go obliquely; (often as **slanted** *adjective*) present (news etc.) in biased or particular way. ● *noun* slope, oblique position; point of view, esp. biased one. ● *adjective* sloping, oblique.

slantwise *adverb* aslant.

slap ● *verb* (**-pp-**) strike (as) with palm of hand; lay forcefully; put hastily or carelessly. ● *noun* slapping stroke or sound. ● *adverb* suddenly, fully, directly. □ **slapdash** hasty, careless; **slap-happy** *colloquial* cheerfully casual; **slapstick** boisterous comedy; **slap-up** *colloquial* lavish.

slash ● *verb* cut or gash with knife etc.; (often + *at*) deliver or aim cutting blows; reduce (prices etc.) drastically; criticize harshly. ● *noun* slashing cut; *Printing* oblique stroke; *slang* act of urinating.

slat *noun* long narrow strip of wood, plastic, or metal, used in fences, venetian blinds, etc.

slate ● *noun* fine-grained bluish-grey rock easily split into thin smooth plates; piece of this used esp. in roofing or *historical* for writing on; colour of slate; list of nominees for office etc. ● *verb* (**-ting**) roof with slates; *colloquial* criticize severely; *US* make arrangements for (event etc.), nominate for office. ● *adjective* of (colour of) slate. □ **slating** *noun*; **slaty** *adjective*.

slattern /'slæt(ə)n/ *noun* slovenly woman. □ **slatternly** *adjective*.

slaughter /'slɔ:tə/ ● *verb* kill (animals) for food etc.; kill (people) ruthlessly or in large numbers; *colloquial* defeat utterly. ● *noun* act of slaughtering. □ **slaughterhouse** place for slaughter of animals for food. □ **slaughterer** *noun*.

Slav /slɑ:v/ ● *noun* member of group of peoples of central and eastern Europe speaking Slavonic languages. ● *adjective* of the Slavs.

slave ● *noun* person who is owned by and has to serve another; drudge, very hard worker; (+ *of*, *to*) obsessive devotee. ● *verb* (**-ving**) work very hard. □ **slave-driver** overseer of slaves, hard taskmaster; **slave trade** dealing in slaves, esp. African blacks.

slaver[1] *noun historical* ship or person engaged in slave trade.

slaver[2] /'slævə/ ● *verb* dribble; drool. ● *noun* dribbling saliva; flattery; drivel.

slavery /'sleɪvərɪ/ *noun* condition of slave; drudgery; practice of having slaves.

Slavic /'slɑ:vɪk/ *adjective & noun* Slavonic.

slavish /'sleɪvɪʃ/ *adjective* like slaves; without originality. □ **slavishly** *adverb*.

Slavonic /slə'vɒnɪk/ ● *adjective* of group of languages including Russian, Polish, and Czech. ● *noun* Slavonic group of languages.

slay *verb* (*past* **slew** /slu:/; *past participle* **slain**) kill. □ **slayer** *noun*.

sleaze *noun colloquial* sleaziness.

sleazy /'sli:zɪ/ *adjective* (**-ier**, **-iest**) squalid, tawdry. □ **sleazily** *adverb*; **sleaziness** *noun*.

sled *noun & verb* (**-dd-**) *US* sledge.

sledge ● *noun* vehicle on runners for use on snow. ● *verb* (**-ging**) travel or carry on sledge.

sledgehammer /'sledʒhæmə/ *noun* large heavy hammer.

sleek ● *adjective* (of hair, skin, etc.) smooth and glossy; looking well-fed and comfortable. ● *verb* make sleek. □ **sleekly** *adverb*; **sleekness** *noun*.

sleep ● *noun* condition in which eyes are closed, muscles and nerves relaxed, and consciousness suspended; period of this; rest, quiet, death. ● *verb* (*past & past participle* **slept**) be or fall asleep; spend the night; provide sleeping accommodation for; (+ *with*, *together*) have sexual intercourse; (+ *on*) defer (decision) until next day; (+ *through*) fail to be woken by; be inactive or dead; (+ *off*) cure by sleeping. □ **sleeping bag** padded bag to sleep in when camping etc.; **sleeping car**, **carriage** railway coach with beds or berths; **sleeping partner** partner not sharing in actual work of a firm; **sleeping pill** pill to induce sleep; **sleeping policeman** ramp etc. in road to slow traffic; **sleepwalk** walk about while asleep; **sleepwalker** *noun* □ **sleepless** *adjective*; **sleeplessness** *noun*.

sleeper *noun* sleeping person or animal; beam supporting railway track; sleeping car; ring worn in pierced ear to keep hole open.

sleepy *adjective* (**-ier**, **-iest**) feeling need of sleep; quiet, inactive. □ **sleepily** *adverb*; **sleepiness** *noun*.

sleet ● *noun* snow and rain together; hail or snow melting as it falls. ● *verb* (after *it*) sleet falls. □ **sleety** *adjective*.

sleeve *noun* part of garment covering arm; cover for gramophone record; tube enclosing rod etc. □ **up one's sleeve** in reserve. □ **sleeved** *adjective*; **sleeveless** *adjective*.

sleigh /sleɪ/ ● *noun* sledge, esp. for riding on. ● *verb* travel on sleigh.

sleight of hand /slaɪt/ *noun* dexterity, esp. in conjuring.

slender /'slendə/ *adjective* (**-er**, **-est**) of small girth or breadth; slim; slight, scanty, meagre.

slept *past & past participle* of SLEEP.

sleuth /slu:θ/ *colloquial* ● *noun* detective. ● *verb* investigate crime etc.

slew[1] /slu:/ ● *verb* (often + *round*) turn or swing to new position. ● *noun* such turn.

slew[2] *past* of SLAY.

slice ● *noun* thin flat piece or wedge cut from something; share; kitchen utensil with thin broad blade; stroke sending ball obliquely. ● *verb* (**-cing**) (often + *up*) cut into slices; (+ *off*) cut off; (+ *into*, *through*) cut (as) with knife; strike (ball) with slice.

slick ● *adjective colloquial* skilful, efficient; superficially dexterous, glib; sleek, smooth. ● *noun* patch of oil etc., esp. on sea. ● *verb colloquial* make smooth or sleek. □ **slickly** *adverb*; **slickness** *noun*.

slide ● *verb* (*past & past participle* **slid**) (cause to) move along smooth surface touching it always with same part; move quietly or smoothly; glide over ice without skates; (often + *into*) pass unobtrusively. ● *noun* act of sliding; rapid decline; smooth slope down which people or things slide; track for sliding, esp. on ice; part of machine or instrument that slides; mounted transparency viewed with projector; piece of glass holding object for microscope; hair-slide. □ **let things slide** be negligent, allow deterioration; **slide-rule** ruler with sliding central strip, graduated logarithmically for rapid calculations; **sliding scale** scale of fees, taxes, wages, etc. that varies according to some other factor.

slight /slaɪt/ ● *adjective* small, insignificant; inadequate; slender, frail-looking. ● *verb* treat disrespectfully, ignore. ● *noun* act of slighting. □ **slightly** *adverb*; **slightness** *noun*.

slim ● *adjective* (**-mm-**) not fat, slender; small, insufficient. ● *verb* (**-mm-**) (often + *down*) become slim, esp. by dieting etc.; make slim. □ **slimline** of slender design, not fattening. □ **slimmer** *noun*; **slimming** *noun*.

slime *noun* oozy or sticky substance.

slimy *adjective* (**-ier**, **-iest**) like, covered with, or filled with slime; *colloquial* disgustingly obsequious. □ **sliminess** *noun*.

sling[1] ● *noun* strap etc. used to support or raise thing; bandage supporting injured arm; strap etc. used to throw small missile. ● *verb* (*past & past participle* **slung**) suspend with sling; *colloquial* throw. □ **sling-back** shoe held in place by strap above heel; **sling one's hook** *slang* go away.

sling[2] *noun* sweetened drink of spirits (esp. gin) with water.

slink *verb* (*past & past participle* **slunk**) (often + *off*, *away*, *by*) move stealthily or guiltily.

slinky *adjective* (**-ier**, **-iest**) (of garment) close-fitting and sinuous.

slip[1] ● *verb* (**-pp-**) slide unintentionally or momentarily, lose footing or balance; go with sliding motion; escape or fall because hard to grasp; go unobserved or quietly; make careless or slight mistake; fall below standard; place stealthily or casually; release from restraint or connection; (+ *on*, *off*) pull (garment) easily or hastily on or off; escape from, evade. ● *noun* act of slipping; careless or slight error; pillowcase; petticoat; (in *singular* or *plural*) slipway; *Cricket* fielder behind wicket on off side, (in *singular* or *plural*) this position. □ **give** (**person**) **the slip** escape from, evade; **slip-knot** knot that can be undone by pull, running knot; **slip-on** (of shoes or clothes) easily slipped on or off; **slipped disc** displaced disc between vertebrae; **slip-road** road for entering or leaving motorway etc.; **slipstream** current of air or water driven backwards by propeller etc.; **slip up** *colloquial* make mistake; **slip-up** *noun*; **slipway** ramp for shipbuilding or landing boats.

slip[2] *noun* small piece of paper, esp. for making notes; cutting from plant for grafting or planting.

slippage *noun* act or instance of slipping.

slipper *noun* light loose indoor shoe.

slippery /'slɪpərɪ/ *adjective* difficult to grasp, stand on, etc., because smooth or wet; unreliable, unscrupulous. □ **slipperiness** *noun*.

slippy *adjective* (**-ier**, **-iest**) *colloquial* slippery.

slipshod *adjective* careless, slovenly.

slit ● *noun* straight narrow incision or opening. ● *verb* (**-tt-**; *past & past participle* **slit**) make slit in; cut in strips.

slither /'slɪðə/ ● *verb* slide unsteadily. ● *noun* act of slithering. □ **slithery** *adjective*.

sliver /'slɪvə/ ● *noun* long thin slice or piece. ● *verb* cut or split into slivers.

slob *noun colloquial derogatory* lazy, untidy, or fat person.

slobber /'slɒbə/ *verb & noun* slaver. □ **slobbery** *adjective*.

sloe *noun* blackthorn; its small bluish-black fruit.

slog ● *verb* (**-gg-**) hit hard and usually unskilfully; work or walk doggedly. ● *noun* heavy random hit; hard steady work or walk; spell of this.

slogan /'sləʊgən/ *noun* catchy phrase used in advertising etc.; party cry, watchword.

sloop /slu:p/ *noun* small one-masted fore-and-aft rigged vessel.

slop ● *verb* (**-pp-**) (often + *over*) spill over edge of vessel; spill or splash liquid on. ● *noun* liquid spilled or splashed; (in *plural*) dirty waste water, wine, etc.; (in *singular* or *plural*) unappetizing liquid food.

slope ● *noun* inclined position, direction, or state; piece of rising or falling ground; difference in level between two ends or sides of a thing; place for skiing. ● *verb* (**-ping**) have or take slope, slant; cause to slope. □ **slope off** *slang* go away, esp. to evade work etc.

sloppy *adjective* (**-ier**, **-iest**) wet, watery, too liquid; careless, untidy; foolishly sentimental. □ **sloppily** *adverb*; **sloppiness** *noun*.

slosh ● *verb* (often + *about*) splash or flounder; *slang* hit, esp. heavily; *colloquial* pour (liquid) clumsily. ● *noun* slush; act or sound of splashing; *slang* heavy blow.

sloshed *adjective slang* drunk.

slot ● *noun* slit in machine etc. for something (esp. coin) to be inserted; slit, groove, etc. for thing; allotted place in schedule. ● *verb* (**-tt-**) (often + *in*, *into*) place or be placed (as if) into slot; provide with slot(s). □ **slot machine** machine worked by insertion of coin, esp. delivering small items or providing amusement.

sloth /sləʊθ/ *noun* laziness, indolence; slow-moving arboreal S. American mammal.

slothful *adjective* lazy. □ **slothfully** *adverb*.

slouch /slaʊtʃ/ ● *verb* stand, move, or sit in drooping fashion. ● *noun* slouching posture or movement; *slang* incompetent or slovenly worker etc. □ **slouch hat** hat with wide flexible brim.

slough[1] /slaʊ/ *noun* swamp, miry place. □ **Slough of Despond** state of hopeless depression.

slough[2] /slʌf/ ● *noun* part that animal (esp. snake) casts or moults. ● *verb* (often + *off*) cast or drop as slough.

Slovak /'sləʊvæk/ ● *noun* member of Slavonic people inhabiting Slovakia; their language. ● *adjective* of this people or language.

sloven /'slʌv(ə)n/ *noun* untidy or careless person.

Slovene /'sləʊvi:n/ (also **Slovenian** /-'vi:nɪən/) ● *noun* member of Slavonic people in Slovenia; their language. ● *adjective* of Slovenia or its people or language.

slovenly ● *adjective* careless and untidy, unmethodical. ● *adverb* in a slovenly way. □ **slovenliness** *noun*.

slow /sləʊ/ ● *adjective* taking relatively long time to do thing(s); acting, moving, or done without speed; not conducive to speed; (of clock etc.) showing earlier than correct time; dull-witted, stupid; tedious; slack, sluggish; (of fire or oven) not very hot; (of photographic film) needing long exposure; reluctant. ● *adverb* slowly. ● *verb* (usually + *down*, *up*) (cause to) move, act, or work with reduced speed or vigour. □ **slowcoach** *colloquial* slow person; **slow motion** speed of film or videotape in which actions etc. appear much slower than usual, simulation of this in real action; **slow-worm** small European legless lizard. □ **slowly** *adverb*; **slowness** *noun*.

sludge *noun* thick greasy mud or sediment; sewage. □ **sludgy** *adjective*.

slug[1] *noun* slimy shell-less mollusc; bullet, esp. irregularly shaped; missile for airgun; *Printing* metal bar for spacing; mouthful of liquor.

slug[2] *US* ● *verb* (**-gg-**) hit hard. ● *noun* hard blow.

sluggard /'slʌgəd/ *noun* lazy person.

sluggish *adjective* inert, slow-moving. □ **sluggishly** *adverb*; **sluggishness** *noun*.

sluice /slu:s/ ● *noun* (also **sluice-gate**, **-valve**) sliding gate or other contrivance for regulating volume or flow of water; water so regulated; (also **sluice-way**) artificial water-channel; place for or act of rinsing. ● *verb* (**-cing**) provide or wash with sluice(s); rinse; (of water) rush out (as if) from sluice.

slum ● *noun* house unfit for human habitation; (often in *plural*) overcrowded and squalid district in city. ● *verb* (**-mm-**) visit slums, esp. out of curiosity. □ **slum it** *colloquial* put up with conditions less comfortable than usual. □ **slummy** *adjective*.

slumber /'slʌmbə/ *verb & noun poetical* sleep.

slump ● *noun* sudden severe or prolonged fall in prices and trade. ● *verb* undergo slump; sit or fall heavily or limply.

slung *past & past participle* of SLING[1].

slunk *past & past participle* of SLINK.

slur ● *verb* (**-rr-**) sound (words, musical notes, etc.) so that they run into one another; *archaic or US* put slur on (person, character); (usually + *over*) pass over lightly. ● *noun* imputation of wrongdoing; act of slurring; *Music* curved line joining notes to be slurred.

slurp *colloquial* ● *verb* eat or drink noisily. ● *noun* sound of slurping.

slurry /'slʌrɪ/ *noun* thin semi-liquid cement, mud, manure, etc.

slush *noun* thawing snow; silly sentimentality. □ **slush fund** reserve fund, esp. for bribery. □ **slushy** *adjective* (**-ier**, **-iest**).

slut *noun derogatory* slovenly or promiscuous woman. □ **sluttish** *adjective*.

sly *adjective* (**-er**, **-est**) crafty, wily; secretive; knowing, insinuating. □ **on the sly** secretly. □ **slyly** *adverb*; **slyness** *noun*.

smack[1] ● *noun* sharp slap or blow; hard hit; loud kiss; loud sharp sound. ● *verb* slap; part (lips) noisily in anticipation of food; move, hit, etc. with smack. ● *adverb colloquial* with a smack; suddenly, violently; exactly.

smack[2] (+ *of*) ● *verb* taste of, suggest. ● *noun* flavour or suggestion of.

smack[3] *noun* single-masted sailing boat.

smack[4] *noun slang* heroin, other hard drug.

smacker *noun slang* loud kiss; £1, *US* $1.

small /smɔ:l/ ● *adjective* not large or big; not great in importance, amount, number, etc.; not much; insignificant; of small particles; on small scale; poor, humble; mean; young. ● *noun* slenderest part, esp. of back; (in *plural*) *colloquial* underwear, esp. as laundry. ● *adverb* into small pieces. □ **small arms** portable firearms; **small change** coins, not notes; **small fry** unimportant people, children; **smallholding** agricultural holding smaller than farm; **small hours** night-time after midnight; **small-minded** petty, narrow in outlook; **smallpox** *historical* acute contagious disease with fever and pustules usually leaving scars; **small print** matter printed small, esp. limitations in contract; **small talk** trivial social conversation; **small-time** *colloquial* unimportant, petty. □ **smallness** *noun*.

smarmy /'smɑ:mɪ/ *adjective* (**-ier**, **-iest**) *colloquial* ingratiating. □ **smarmily** *adverb*; **smarminess** *noun*.

smart ● *adjective* well-groomed, neat; bright and fresh in appearance; stylish, fashionable; *esp. US* clever, ingenious, quickwitted; quick, brisk; painfully severe, sharp, vigorous. ● *verb* feel or give pain; rankle; (+ *for*) suffer consequences of. ● *noun* sharp bodily or mental pain, stinging sensation. ● *adverb* smartly. □ **smartish** *adjective & adverb*; **smartly** *adverb*; **smartness** *noun*.

smarten *verb* (usually + *up*) make or become smart.

smash ● *verb* (often + *up*) break to pieces; bring or come to destruction, defeat, or disaster; (+ *into*, *through*) move forcefully; (+ *in*) break with crushing blow; hit (ball) hard, esp. downwards. ● *noun* act or sound of smashing; (in full **smash hit**) very successful play, song, etc. ● *adverb* with smash. □ **smash-and-grab** robbery with goods snatched from broken shop window etc.

smashing *adjective colloquial* excellent, wonderful.

smattering /ˈsmætərɪŋ/ *noun* slight knowledge.

smear /smɪə/ ● *verb* daub or mark with grease etc.; smudge; defame. ● *noun* action of smearing; material smeared on microscope slide etc. for examination; specimen of this. □ **smear test** cervical smear. □ **smeary** *adjective.*

smell ● *noun* sense of odour perception; property perceived by this; unpleasant odour; act of inhaling to ascertain smell. ● *verb* (*past & past participle* **smelt** or **smelled**) perceive or examine by smell; stink; seem by smell to be; (+ *of*) emit smell of, suggest; detect; have or use sense of smell. □ **smelling salts** sharp-smelling substances sniffed to relieve faintness.

smelly *adjective* (**-ier**, **-iest**) strong- or evil-smelling.

smelt[1] *verb* melt (ore) to extract metal; obtain (metal) thus. □ **smelter** *noun.*

smelt[2] *past & past participle* of SMELL.

smelt[3] *noun* (*plural* same or **-s**) small edible green and silver fish.

smilax /ˈsmaɪlæks/ *noun* any of several climbing plants.

smile ● *verb* (**-ling**) have or assume facial expression of amusement or pleasure, with ends of lips turned upward; express by smiling; give (smile); (+ *on*, *upon*) favour. ● *noun* act of smiling; smiling expression or aspect.

smirch *verb & noun* stain, smear.

smirk ● *noun* conceited or silly smile. ● *verb* give smirk.

smite *verb* (**-ting**; *past* **smote**; *past participle* **smitten** /ˈsmɪt(ə)n/) *archaic or literary* hit, chastise, defeat; (in *passive*) affect strongly, seize.

smith *noun* blacksmith; worker in metal, craftsman.

smithereens /smɪðəˈriːnz/ *plural noun* small fragments.

smithy /ˈsmɪðɪ/ *noun* (*plural* **-ies**) blacksmith's workshop, forge.

smitten *past participle* of SMITE.

smock ● *noun* loose shirtlike garment, often adorned with smocking. ● *verb* adorn with smocking.

smocking *noun* ornamentation on cloth made by gathering it tightly with stitches.

smog *noun* dense smoky fog. □ **smoggy** *adjective* (**-ier**, **-iest**).

smoke ● *noun* visible vapour from burning substance; act of smoking tobacco etc.; *colloquial* cigarette, cigar. ● *verb* (**-king**) inhale and exhale smoke of (cigarette etc.); do this habitually; emit smoke or visible vapour; darken or preserve with smoke. □ **smoke bomb** bomb emitting dense smoke on bursting; **smoke-free** free from smoke, where smoking is not permitted; **smoke out** drive out by means of smoke, drive out of hiding etc.; **smokescreen** cloud of smoke concealing esp. military operations, ruse for disguising activities; **smokestack** funnel of locomotive or steamship, tall chimney.

smoker *noun* person who habitually smokes tobacco; compartment on train where smoking is permitted.

smoky *adjective* (**-ier**, **-iest**) emitting, filled with, or obscured by, smoke; coloured by or like smoke; having flavour of smoked food.

smolder *US* = SMOULDER.

smooch /smuːtʃ/ ● *verb* kiss and caress. ● *noun* smooching.

smooth /smuːð/ ● *adjective* having even surface; free from projections and roughness; that can be traversed uninterrupted; (of sea etc.) calm, flat; (of journey etc.) easy; not harsh in sound or taste; conciliatory, slick; not jerky. ● *verb* (often + *out*, *down*) make or become smooth; (often + *out*, *down*, *over*, *away*) get rid of (differences, faults, etc.). ● *noun* smoothing touch or stroke. □ **smooth-tongued** insincerely flattering. □ **smoothly** *adverb*; **smoothness** *noun.*

smorgasbord /ˈsmɔːgəsbɔːd/ *noun* Swedish hors d'oeuvres; buffet meal with various esp. savoury dishes.

smote *past* of SMITE.

smother /ˈsmʌðə/ *verb* suffocate, stifle; (+ *in*, *with*) overwhelm or cover with (kisses, gifts, etc.); extinguish (fire) by heaping with ashes etc.; have difficulty breathing; (often + *up*) suppress, conceal.

smoulder /'sməʊldə/ (*US* **smolder**) ● *verb* burn without flame or internally; (of person) show silent emotion. ● *noun* smouldering.

smudge ● *noun* blurred or smeared line, mark, etc. ● *verb* (**-ging**) make smudge on or with; become smeared or blurred. □ **smudgy** *adjective.*

smug *adjective* (**-gg-**) self-satisfied. □ **smugly** *adverb*; **smugness** *noun.*

smuggle /'smʌg(ə)l/ *verb* (**-ling**) import or export illegally, esp. without paying duties; convey secretly. □ **smuggler** *noun*; **smuggling** *noun.*

smut ● *noun* small piece of soot; spot or smudge made by this; obscene talk, pictures, or stories; fungous disease of cereals. ● *verb* (**-tt-**) mark with smut(s). □ **smutty** *adjective* (**-ier**, **-iest**).

snack *noun* light, casual, or hasty meal. □ **snack bar** place where snacks are sold.

snaffle /'snæf(ə)l/ ● *noun* (in full **snaffle-bit**) simple bridle-bit without curb. ● *verb* (**-ling**) *colloquial* steal, seize.

snag ● *noun* unexpected obstacle or drawback; jagged projection; tear in material etc. ● *verb* (**-gg-**) catch or tear on snag.

snail *noun* slow-moving mollusc with spiral shell.

snake ● *noun* long limbless reptile; (also **snake in the grass**) traitor, secret enemy. ● *verb* (**-king**) move or twist like a snake. □ **snakes and ladders** board game with counters moved up 'ladders' and down 'snakes'; **snakeskin** *noun* skin of snake, *adjective* made of snakeskin.

snaky *adjective* of or like a snake; sinuous; treacherous.

snap ● *verb* (**-pp-**) break sharply; (cause to) emit sudden sharp sound; open or close with snapping sound; speak irritably; (often + *at*) make sudden audible bite; move quickly; photograph. ● *noun* act or sound of snapping; crisp biscuit; snapshot; (in full **cold snap**) sudden brief period of cold weather; card game in which players call 'snap' when two similar cards are exposed; vigour. ● *adverb* with snapping sound. ● *adjective* done without forethought. □ **snapdragon** plant with two-lipped flowers; **snap-fastener** press-stud; **snap out of** *slang* get out of (mood etc.) by sudden effort; **snapshot** informal or casual photograph; **snap up** accept (offer etc.) hastily or eagerly.

snapper *noun* any of several edible marine fish.

snappish *adjective* curt, ill-tempered.

snappy *adjective* (**-ier**, **-iest**) *colloquial* brisk, lively; neat and elegant; snappish. □ **snappily** *adverb.*

snare /sneə/ ● *noun* trap, esp. with noose, for birds or animals; trap, trick, temptation; (in *singular* or *plural*) twisted strings of gut, hide, or wire stretched across lower head of side drum to produce rattle; (in full **snare drum**) side drum with snares. ● *verb* (**-ring**) catch in snare, trap.

snarl[1] ● *verb* growl with bared teeth; speak angrily. ● *noun* act or sound of snarling.

snarl[2] ● *verb* (often + *up*) twist, entangle, hamper movement of (traffic etc.), become entangled. ● *noun* tangle.

snatch ● *verb* (often + *away*, *from*) seize quickly, eagerly, or unexpectedly; steal by grabbing; *slang* kidnap; (+ *at*) try to seize, take eagerly. ● *noun* act of snatching; fragment of song, talk, etc.; short spell of activity etc.

snazzy /'snæzɪ/ *adjective* (**-ier**, **-iest**) *slang* smart, stylish, showy.

sneak ● *verb* go or convey furtively; *slang* steal unobserved; *slang* tell tales; (as **sneaking** *adjective*) furtive, persistent and puzzling. ● *noun* mean-spirited, underhand person; *slang* tell-tale. ● *adjective* acting or done without warning, secret. □ **sneak-thief** person who steals without breaking in. □ **sneaky** *adjective* (**-ier**, **-iest**).

sneaker *noun slang* soft-soled shoe.

sneer ● *noun* derisive smile or remark. ● *verb* (often + *at*) make sneer; utter with sneer. □ **sneering** *adjective*; **sneeringly** *adverb.*

sneeze ● *noun* sudden involuntary explosive expulsion of air from irritated nostrils. ● *verb* (**-zing**) make sneeze.

□ **not to be sneezed at** *colloquial* worth having or considering.

snick • *verb* make small notch or cut in; *Cricket* deflect (ball) slightly with bat. • *noun* such notch or deflection.

snicker /'snɪkə/ *noun & verb* snigger.

snide *adjective* sneering, slyly derogatory.

sniff • *verb* inhale air audibly through nose; (often + *up*) draw in through nose; smell scent of by sniffing. • *noun* act or sound of sniffing. □ **sniff at** show contempt for; **sniffer-dog** *colloquial* dog trained to find drugs or explosives by scent.

sniffle /'snɪf(ə)l/ • *verb* (**-ling**) sniff repeatedly or slightly. • *noun* act of sniffling; (in *singular* or *plural*) cold in the head causing sniffling.

snifter /'snɪftə/ *noun slang* small alcoholic drink.

snigger /'snɪgə/ • *noun* half-suppressed laugh. • *verb* utter snigger.

snip • *verb* (**-pp-**) cut with scissors etc., esp. in small quick strokes. • *noun* act of snipping; piece snipped off; *slang* something easily done, bargain.

snipe • *noun* (*plural* same or **-s**) wading bird with long straight bill. • *verb* (**-ping**) fire shots from hiding usually at long range; (often + *at*) make sly critical attack. □ **sniper** *noun*.

snippet /'snɪpɪt/ *noun* small piece cut off; (usually in *plural*) scrap of information etc., short extract from book etc.

snitch • *verb slang* steal; (often + *on*) inform on person. • *noun* informer.

snivel /'snɪv(ə)l/ • *verb* (**-ll-**; *US* **-l-**) sniffle; weep with sniffling; show maudlin emotion. • *noun* act of snivelling.

snob *noun* person who despises people with inferior social position, wealth, intellect, tastes, etc. □ **snobbery** *noun*; **snobbish** *adjective*; **snobby** *adjective* (**-ier**, **-iest**).

snood /snu:d/ *noun* woman's loose hairnet.

snook /snu:k/ *noun slang* contemptuous gesture with thumb to nose and fingers spread. □ **cock a snook (at)** make this gesture (at), show contempt (for).

snooker /'snu:kə/ • *noun* game played on oblong cloth-covered table with 1 white, 15 red, and 6 other coloured balls; position in this game where direct shot would lose points. • *verb* subject (player) to snooker; (esp. as **snookered** *adjective*) *slang* thwart, defeat.

snoop /snu:p/ *colloquial* • *verb* pry into another's affairs; (often + *about*, *around*) investigate (often stealthily) transgressions of rules etc. • *noun* act of snooping. □ **snooper** *noun*.

snooty /'snu:tɪ/ *adjective* (**-ier**, **-iest**) *colloquial* supercilious, snobbish. □ **snootily** *adverb*.

snooze /snu:z/ *colloquial* • *noun* short sleep, nap. • *verb* (**-zing**) take snooze.

snore • *noun* snorting or grunting sound of breathing during sleep. • *verb* (**-ring**) make this sound.

snorkel /'snɔ:k(ə)l/ • *noun* device for supplying air to underwater swimmer or submerged submarine. • *verb* (**-ll-**; *US* **-l-**) use snorkel.

snort • *noun* explosive sound made by driving breath violently through nose, esp. by horses, or by humans to show contempt, incredulity, etc.; *colloquial* small drink of liquor; *slang* inhaled dose of powdered cocaine. • *verb* make snort; *slang* inhale (esp. cocaine); express or utter with snort.

snot *noun slang* nasal mucus.

snotty *adjective* (**-ier**, **-iest**) *slang* running or covered with nasal mucus; snooty; contemptible. □ **snottily** *adverb*; **snottiness** *noun*.

snout /snaʊt/ *noun* projecting nose (and mouth) of animal; *derogatory* person's nose; pointed front of thing.

snow /snəʊ/ • *noun* frozen vapour falling to earth in light white flakes; fall or layer of this; thing resembling snow in whiteness or texture etc.; *slang* cocaine. • *verb* (after *it*) snow falls; (+ *in*, *over*, *up*, etc.) confine or block with snow. □ **snowball** *noun* snow pressed into ball for throwing in play, *verb* throw or pelt with snowballs, increase rapidly; **snow-blind** temporarily blinded by glare from snow; **snowbound** prevented by snow from going out; **snowcap** snow-covered mountain peak; **snowdrift** bank of snow piled up by wind; **snowdrop** spring-flowering plant with white drooping flowers; **snowed under** overwhelmed, esp. with work; **snowflake** each of the flakes in which snow falls; **snow goose** white Arctic goose; **snowline** level above which snow never melts entirely; **snowman** figure made of snow; **snowplough** device for clearing road of snow; **snowshoe** racket-shaped attachment to boot for walking on surface of snow; **snowstorm** heavy fall of snow, esp. with wind; **snow white** pure white. □ **snowy** *adjective* (**-ier**, **iest**).

SNP *abbreviation* Scottish National Party.

Snr. *abbreviation* Senior.

snub ● *verb* (**-bb-**) rebuff or humiliate in a sharp or cutting way. ● *noun* snubbing, rebuff. ● *adjective* (of nose) short and turned up.

snuff¹ ● *noun* charred part of candle-wick. ● *verb* trim snuff from (candle). □ **snuff it** *slang* die; **snuff out** extinguish (candle), put an end to (hopes etc.).

snuff² ● *noun* powdered tobacco or medicine taken by sniffing. ● *verb* take snuff.

snuffle /'snʌf(ə)l/ ● *verb* (**-ling**) make sniffing sounds; speak nasally; breathe noisily, esp. with blocked nose. ● *noun* snuffling sound or speech.

snug ● *adjective* (**-gg-**) cosy, comfortable, sheltered; close-fitting. ● *noun* small room in pub. □ **snugly** *adverb.*

snuggle /'snʌg(ə)l/ *verb* (**-ling**) settle or move into warm comfortable position.

so¹ /səʊ/ ● *adverb* to such an extent, in this or that manner or state; also; indeed, actually; very; thus. ● *conjunction* (often + *that*) consequently, in order that; and then; (introducing question) after that. □ **so-and-so** particular but unspecified person or thing, *colloquial* objectionable person; **so as to** in order to; **so-called** commonly called but often incorrectly; **so long** *colloquial* goodbye; **so so** *colloquial* only moderately good or well.

so² = SOH.

soak ● *verb* make or become thoroughly wet through saturation; (of rain etc.) drench; (+ *in*, *up*) absorb (liquid, knowledge, etc.); (+ *in*, *into*, *through*) penetrate by saturation; *colloquial* extort money from; *colloquial* drink heavily. ● *noun* soaking; *colloquial* hard drinker. □ **soakaway** arrangement for disposal of waste water by percolation through soil.

soaking *adjective* (in full **soaking wet**) wet through.

soap ● *noun* cleansing substance yielding lather when rubbed in water; *colloquial* soap opera. ● *verb* apply soap to. □ **soapbox** makeshift stand for street orator; **soap opera** domestic broadcast serial; **soap powder** powdered soap usually with additives, for washing clothes etc.; **soapstone** steatite; **soapsuds** suds.

soapy *adjective* (**-ier**, **-iest**) of or like soap; containing or smeared with soap; unctuous, flattering.

soar /sɔː/ *verb* fly or rise high; reach high level or standard; fly without flapping wings or using motor power.

sob ● *verb* (**-bb-**) inhale convulsively, usually with weeping; utter with sobs. ● *noun* act or sound of sobbing. □ **sob story** *colloquial* story or explanation appealing for sympathy.

sober /'səʊbə/ ● *adjective* (**-er**, **-est**) not drunk; not given to drink; moderate, tranquil, serious; (of colour) dull. ● *verb* (often + *down*, *up*) make or become sober. □ **soberly** *adverb.*

sobriety /sə'braɪɪtɪ/ *noun* being sober.

sobriquet /'səʊbrɪkeɪ/ *noun* (also **soubriquet** /'suː-/) nickname.

Soc. *abbreviation* Socialist; Society.

soccer /'sɒkə/ *noun* Association football.

sociable /'səʊʃəb(ə)l/ *adjective* liking company, gregarious, friendly. □ **sociability** *noun*; **sociably** *adverb.*

social /'səʊʃ(ə)l/ ● *adjective* of society or its organization, esp. of relations of (classes of) people; living in communities; gregarious. ● *noun* social gathering. □ **social science** study of society and social relationships; **social security** state assistance to the poor and unemployed; **social services** welfare services provided by the State, esp. education, health care, and housing; **social work** professional or voluntary work with disadvantaged groups; **social worker** *noun.* □ **socially** *adverb.*

socialism *noun* political and economic theory advocating state ownership and control of means of production, distribution, and exchange; social system based on this. □ **socialist** *noun & adjective*; **socialistic** /-'lɪs-/ *adjective*.

socialite /'səʊʃəlaɪt/ *noun* person moving in fashionable society.

socialize /'səʊʃəlaɪz/ *verb* (also **-ise**) (**-zing** or **-sing**) mix socially; make social; organize in a socialistic way.

society /sə'saɪətɪ/ *noun* (*plural* **-ies**) organized and interdependent community; system and organization of this; (members of) aristocratic part of this; mixing with other people, companionship, company; association, club. □ **societal** *adjective*.

sociology /səʊsɪ'ɒlədʒɪ/ *noun* study of society and social problems. □ **sociological** /-ə'lɒdʒ-/ *adjective*; **sociologist** *noun*.

sock[1] *noun* knitted covering for foot and lower leg; insole.

sock[2] *colloquial* ● *verb* hit hard. ● *noun* hard blow. □ **sock it to** attack or address vigorously.

socket /'sɒkɪt/ *noun* hollow for thing to fit into etc., esp. device receiving electric plug, light bulb, etc.

Socratic /sə'krætɪk/ *adjective* of Socrates or his philosophy.

sod *noun* turf, piece of turf; surface of ground.

soda /'səʊdə/ *noun* any of various compounds of sodium in common use; (in full **soda water**) effervescent water used esp. with spirits etc. as drink. □ **soda fountain** device supplying soda water, shop or counter with this.

sodden /'sɒd(ə)n/ *adjective* saturated, soaked through; stupid, dull, etc. with drunkenness.

sodium /'səʊdɪəm/ *noun* soft silver-white metallic element. □ **sodium bicarbonate** white compound used in baking-powder; **sodium chloride** common salt; **sodium lamp** lamp using sodium vapour and giving yellow light; **sodium nitrate** white powdery compound in fertilizers etc.

sofa /'səʊfə/ *noun* long upholstered seat with raised back and ends. □ **sofa bed** sofa that can be converted into bed.

soffit /'sɒfɪt/ *noun* undersurface of arch, lintel, etc.

soft ● *adjective* not hard, easily cut or dented, malleable; (of cloth etc.) smooth, fine, not rough; mild; (of water) low in mineral salts which prevent lathering; not brilliant or glaring; not strident or loud; sibilant; not sharply defined; gentle, conciliatory; compassionate, sympathetic; feeble, half-witted, silly, sentimental; *colloquial* easy; (of drug) not highly addictive. ● *adverb* softly. □ **have a soft spot for** be fond of; **softball** form of baseball with softer larger ball; **soft-boiled** (of egg) boiled leaving yolk soft; **soft-centred** having soft centre, soft-hearted; **soft drink** non-alcoholic drink; **soft fruit** small stoneless fruit; **soft furnishings** curtains, rugs, etc.; **soft-hearted** tender, compassionate; **soft option** easier alternative; **soft palate** back part of palate; **soft pedal** *noun* pedal on piano softening tone; **soft-pedal** *verb* refrain from emphasizing; **soft sell** restrained salesmanship; **soft soap** *colloquial* persuasive flattery; **soft-spoken** having gentle voice; **soft target** vulnerable person or thing; **soft touch** *colloquial* gullible person, esp. over money; **software** computer programs; **softwood** wood of coniferous tree. □ **softly** *adverb*; **softness** *noun*.

soften /'sɒf(ə)n/ *verb* make or become soft(er); (often + *up*) reduce strength, resistance, etc. of. □ **softener** *noun*.

softie *noun* (also **softy**) (*plural* **-ies**) *colloquial* weak, silly, or soft-hearted person.

soggy /'sɒgɪ/ *adjective* (**-ier**, **-iest**) sodden, waterlogged.

soh *noun* (also **so**) *Music* fifth note of scale in tonic sol-fa.

soil[1] *noun* upper layer of earth, in which plants grow; ground, territory.

soil[2] ● *verb* make dirty, smear, stain; defile; discredit. ● *noun* dirty mark; filth, refuse. □ **soil pipe** discharge-pipe of lavatory.

soirée /'swɑːreɪ/ *noun* evening party.

sojourn /'sɒdʒ(ə)n/ ● *noun* temporary stay. ● *verb* stay temporarily.

sola /'səʊlə/ *noun* pithy-stemmed E. Indian swamp plant. □ **sola topi** sun-helmet made from pith of this.

solace /'sɒləs/ ● *noun* comfort in distress or disappointment. ● *verb* (**-cing**) give solace to.

solan /'səʊlən/ *noun* (in full **solan goose**) large gooselike gannet.

solar /'səʊlə/ *adjective* of or reckoned by sun. □ **solar battery**, **cell** device converting solar radiation into electricity; **solar panel** panel absorbing sun's rays as energy source; **solar plexus** complex of nerves at pit of stomach; **solar system** sun and the planets etc. whose motion is governed by it.

solarium /sə'leərɪəm/ *noun* (*plural* **-ria**) room with sunlamps, glass roof, etc.

sold *past & past participle* of SELL.

solder /'sɒldə/ ● *noun* fusible alloy used for joining metals, wires, etc. ● *verb* join with solder. □ **soldering iron** tool for melting and applying solder.

soldier /'səʊldʒə/ ● *noun* member of army, esp. (in full **common soldier**) private or NCO; *colloquial* bread finger, esp. for dipping in egg. ● *verb* serve as soldier. □ **soldier on** *colloquial* persevere doggedly. □ **soldierly** *adjective*.

soldiery *noun* soldiers collectively.

sole[1] ● *noun* undersurface of foot; part of shoe, sock, etc. below foot, esp. part other than heel; lower surface or base of plough, golf-club head, etc. ● *verb* (**-ling**) provide with sole.

sole[2] *noun* (*plural* same or **-s**) type of flatfish.

sole[3] *adjective* one and only single; exclusive. □ **solely** *adverb*.

solecism /'sɒlɪsɪz(ə)m/ *noun* mistake of grammar or idiom; offence against etiquette.

solemn /'sɒləm/ *adjective* serious and dignified; formal; awe-inspiring; of cheerless manner; grave. □ **solemness** *noun*; **solemnity** /sə'lem-/ *noun* (*plural* **-ies**); **solemnly** *adverb*.

solemnize /'sɒləmnaɪz/ *verb* (also **-ise**) (**-zing** or **-sing**) duly perform (esp. marriage ceremony); make solemn. □ **solemnization** *noun*.

solenoid /'səʊlənɔɪd/ *noun* cylindrical coil of wire acting as magnet when carrying electric current.

sol-fa /'sɒlfɑ:/ *noun* system of syllables representing musical notes.

solicit /sə'lɪsɪt/ *verb* (**-t-**) seek repeatedly or earnestly; (of prostitute) accost (man) concerning sexual activity. □ **solicitation** *noun*.

solicitor /sə'lɪsɪtə/ *noun* lawyer qualified to advise clients and instruct barristers.

solicitous /sə'lɪsɪtəs/ *adjective* showing concern; (+ *to do*) eager, anxious. □ **solicitously** *adverb*.

solicitude /sə'lɪsɪtju:d/ *noun* being solicitous.

solid /'sɒlɪd/ ● *adjective* (**-er**, **-est**) of firm and stable shape, not liquid or fluid; of such material throughout, not hollow; alike all through; sturdily built, not flimsy; 3-dimensional; of solids; sound, reliable; uninterrupted; unanimous. ● *noun* solid substance or body; (in *plural*) solid food. ● *adverb* solidly. □ **solid-state** using electronic properties of solids to replace those of valves. □ **solidity** /sə'lɪd-/ *noun*; **solidly** *adverb*; **solidness** *noun*.

solidarity /sɒlɪ'dærɪtɪ/ *noun* unity, esp. political or in industrial dispute; mutual dependence.

solidify /sə'lɪdɪfaɪ/ *verb* (**-ies**, **-ied**) make or become solid.

soliloquy /sə'lɪləkwɪ/ *noun* (*plural* **-quies**) talking without or regardless of hearers; this part of a play. □ **soliloquize** *verb* (also **-ise**) (**-zing** or **-sing**).

solipsism /'sɒlɪpsɪz(ə)m/ *noun* theory that self is all that exists or can be known.

solitaire /'sɒlɪteə/ *noun* jewel set by itself; ring etc. with this; game for one player who removes pegs etc. from board on jumping others over them; *US* card game for one person.

solitary /'sɒlɪtərɪ/ ● *adjective* living or being alone; not gregarious; lonely; secluded; single. ● *noun* (*plural* **-ies**) recluse; *colloquial* solitary confinement. □ **solitary confinement** isolation in separate prison cell.

solitude /'sɒlɪtju:d/ *noun* being solitary; solitary place.

solo /'səʊləʊ/ ● *noun* (*plural* **-s**) piece of music or dance performed by one person; thing done by one person, esp. unaccompanied flight; (in full **solo whist**) type of whist in which one player may oppose the others. ● *verb* (**-es**, **-ed**) perform a solo. ● *adjective & adverb* unaccompanied, alone.

soloist /'səʊləʊɪst/ *noun* performer of solo.

solstice /ˈsɒlstɪs/ *noun* either of two times (**summer**, **winter solstice**) when sun is farthest from equator.

soluble /ˈsɒljʊb(ə)l/ *adjective* that can be dissolved or solved. □ **solubility** *noun*.

solution /səˈlu:ʃ(ə)n/ *noun* (means of) solving a problem; conversion of solid or gas into liquid by mixture with liquid; state or substance resulting from this; dissolving, being dissolved.

solve *verb* (**-ving**) answer, remove, or deal with (problem). □ **solvable** *adjective*.

solvency /ˈsɒlvənsɪ/ *noun* being financially solvent.

solvent /ˈsɒlv(ə)nt/ ● *adjective* able to pay one's debts; able to dissolve or form solution. ● *noun* solvent liquid etc.

somatic /səˈmætɪk/ *adjective* of the body, not of the mind.

sombre /ˈsɒmbə/ *adjective* (also *US* **somber**) dark, gloomy, dismal. □ **sombrely** *adverb*; **sombreness** *noun*.

sombrero /sɒmˈbreərəʊ/ *noun* (*plural* **-s**) broad-brimmed hat worn esp. in Latin America.

some /sʌm/ ● *adjective* unspecified amount or number of; unknown, unspecified; approximately; considerable; at least a small amount of; such to a certain extent; *colloquial* a remarkable. ● *pronoun* some people or things, some number or amount. ● *adverb colloquial* to some extent. □ **somebody** *pronoun* some person, *noun* important person; **someday** at some time in the future; **somehow** for some reason, in some way, by some means; **someone** somebody; **something** unspecified or unknown thing, unexpressed or intangible quantity or quality, *colloquial* notable person or thing; **sometime** at some time, former(ly); **sometimes** occasionally; **somewhat** to some extent; **somewhere** (in or to) some place.

somersault /ˈsʌməsɒlt/ ● *noun* leap or roll in which one turns head over heels. ● *verb* perform somersault.

somnambulism /sɒmˈnæmbjʊlɪz(ə)m/ *noun* sleepwalking. □ **somnambulant** *adjective*; **somnambulist** *noun*.

somnolent /ˈsɒmnələnt/ *adjective* sleepy, drowsy; inducing drowsiness. □ **somnolence** *noun*.

son /sʌn/ *noun* male in relation to his parent(s); male descendant; (+ *of*) male member of (family etc.); male inheritor of a quality etc.; *form of address, esp. to boy*. □ **son-in-law** (*plural* **sons-in-law**) daughter's husband.

sonar /ˈsəʊnɑ:/ *noun* system for detecting objects under water by reflected sound; apparatus for this.

sonata /səˈnɑ:tə/ *noun* musical composition for one or two instruments in several related movements.

song *noun* words set to music or for singing; vocal music; composition suggestive of song; cry of some birds. □ **for a song** *colloquial* very cheaply; **songbird** bird with musical call; **song thrush** common thrush; **songwriter** writer of (music for) songs.

songster /ˈsɒŋstə/ *noun* (*feminine* **songstress**) singer; songbird.

sonic /ˈsɒnɪk/ *adjective* of or using sound or sound waves. □ **sonic bang**, **boom** noise made by aircraft flying faster than sound.

sonnet /ˈsɒnɪt/ *noun* poem of 14 lines with fixed rhyme scheme.

sonny /ˈsʌnɪ/ *noun colloquial familiar form of address to young boy*.

sonorous /ˈsɒnərəs/ *adjective* having a loud, full, or deep sound; (of speech etc.) imposing. □ **sonority** /səˈnɒr-/ *noun* (*plural* **-ies**).

soon /su:n/ *adverb* in a short time; relatively early; readily, willingly. □ **sooner or later** at some future time. □ **soonish** *adverb*.

soot /sʊt/ *noun* black powdery deposit from smoke.

soothe /su:ð/ *verb* (**-thing**) calm; soften, mitigate.

soothsayer /ˈsu:θseɪə/ *noun* seer, prophet.

sooty *adjective* (**-ier**, **-iest**) covered with soot; black, brownish black.

sop ● *noun* thing given or done to pacify or bribe; piece of bread etc. dipped in gravy etc. ● *verb* (**-pp-**) (+ *up*) soak up.

sophism /ˈsɒfɪz(ə)m/ *noun* false argument, esp. one meant to deceive.

sophist /ˈsɒfɪst/ *noun* captious or clever but fallacious reasoner. □ **sophistic** /səˈfɪs-/ *adjective*.

sophisticate /sə'fɪstɪkət/ *noun* sophisticated person.

sophisticated /sə'fɪstɪkeɪtɪd/ *adjective* worldly-wise, cultured, elegant; highly developed and complex. □ **sophistication** *noun.*

sophistry /'sɒfɪstrɪ/ *noun* (*plural* **-ies**) use of sophisms; a sophism.

sophomore /'sɒfəmɔ:/ *noun US* second-year university or high-school student.

soporific /sɒpə'rɪfɪk/ ● *adjective* inducing sleep. ● *noun* soporific drug or influence. □ **soporifically** *adverb.*

sopping *adjective* drenched.

soppy *adjective* (**-ier**, **-iest**) *colloquial* mawkishly sentimental, silly.

soprano /sə'prɑ:nəʊ/ ● *noun* (*plural* **-s**) highest singing voice; singer with this. ● *adjective* having range of soprano.

sorbet /'sɔ:beɪ/ *noun* water-ice; sherbet.

sorcerer /'sɔ:sərə/ *noun* (*feminine* **sorceress**) magician, wizard. □ **sorcery** *noun* (*plural* **-ies**).

sordid /'sɔ:dɪd/ *adjective* dirty, squalid; ignoble, mercenary. □ **sordidly** *adverb*; **sordidness** *noun.*

sore ● *adjective* painful; suffering pain; aggrieved, vexed; *archaic* grievous, severe. ● *noun* sore place, subject, etc. ● *adverb archaic* grievously, severely. □ **soreness** *noun.*

sorely *adverb* extremely.

sorghum /'sɔ:gəm/ *noun* tropical cereal grass.

sorority /sə'rɒrɪtɪ/ *noun* (*plural* **-ies**) *US* female students' society in university or college.

sorrel[1] /'sɒr(ə)l/ *noun* sour-leaved herb.

sorrel[2] /'sɒr(ə)l/ ● *adjective* of light reddish-brown colour. ● *noun* this colour; sorrel animal, esp. horse.

sorrow /'sɒrəʊ/ ● *noun* mental distress caused by loss, disappointment, etc.; cause of sorrow. ● *verb* feel sorrow, mourn. □ **sorrowful** *adjective.*

sorry /'sɒrɪ/ *adjective* (**-ier**, **-iest**) pained, regretful, penitent; feeling pity; wretched.

sort ● *noun* class, kind; *colloquial* person of specified kind. ● *verb* (often + *out*, *over*) arrange systematically. □ **of a sort**, **of sorts** *colloquial* barely deserving the name; **out of sorts** slightly unwell, in low spirits; **sort out** separate into sorts, select from miscellaneous group, disentangle, put into order, solve, *colloquial* deal with or punish.

sortie /'sɔ:tɪ/ ● *noun* sally, esp. from besieged garrison; operational military flight. ● *verb* (**sortieing**) make sortie.

SOS *noun* (*plural* **SOSs**) international code-signal of extreme distress; urgent appeal for help.

sot *noun* habitual drunkard. □ **sottish** *adjective.*

sotto voce /sɒtəʊ 'vəʊtʃɪ/ *adverb* in an undertone. [Italian]

sou /su:/ *noun* (*plural* **-s**) *colloquial* very small amount of money; *historical* former French coin of low value.

soubrette /su:'bret/ *noun* pert maidservant etc. in comedy; actress taking this part.

soubriquet = SOBRIQUET.

soufflé /'su:fleɪ/ *noun* light spongy dish made with stiffly beaten egg white.

sough /saʊ, sʌf/ ● *noun* moaning or rustling sound, e.g. of wind in trees. ● *verb* make this sound.

sought /sɔ:t/ *past & past participle* of SEEK. □ **sought-after** much in demand.

souk /su:k/ *noun* market-place in Muslim countries.

soul /səʊl/ *noun* spiritual or immaterial part of person; moral, emotional, or intellectual nature of person; personification, pattern; an individual; animating or essential part; energy, intensity; soul music. □ **soul-destroying** tedious, monotonous; **soul mate** person ideally suited to another; **soul music** type of black American music; **soul-searching** introspection.

soulful *adjective* having, expressing, or evoking deep feeling. □ **soulfully** *adverb.*

soulless *adjective* lacking sensitivity or noble qualities; undistinguished, uninteresting.

sound[1] /saʊnd/ ● *noun* sensation produced in ear when surrounding air etc. vibrates; vibrations causing this; what is or may be heard; idea or impression give by words. ● *verb* (cause to) emit sound; utter, pronounce; convey specified impression; give audible signal for; test condition of by sound produced.

□ **sound barrier** high resistance of air to objects moving at speeds near that of sound; **sound effect** sound other than speech or music produced artificially for film, broadcast, etc.; **sounding-board** person etc. used to test or disseminate opinion(s), canopy projecting sound towards audience; **sound off** talk loudly, express one's opinions forcefully; **soundproof** *adjective* impervious to sound, *verb* make soundproof; **sound system** equipment for reproducing sound; **soundtrack** sound element of film or videotape, recording of this available separately; **sound wave** wave of compression and rarefaction by which sound is transmitted in air etc.

sound[2] /saʊnd/ ● *adjective* healthy, not diseased or rotten, uninjured; correct, well-founded; financially secure; undisturbed; thorough. ● *adverb* soundly. □ **soundly** *adverb*; **soundness** *noun*.

sound[3] /saʊnd/ *verb* test depth or quality of bottom of (sea, river, etc.); (often + *out*) inquire (esp. discreetly) into views etc. of (person).

sound[4] /saʊnd/ *noun* strait (of water).

sounding *noun* measurement of depth of water; (in *plural*) region near enough to shore for sounding; (in *plural*) cautious investigation.

soup /su:p/ ● *noun* liquid food made by boiling meat, fish, or vegetables. ● *verb* (usually + *up*) *colloquial* increase power of (engine), enliven. □ **in the soup** *colloquial* in difficulties; **soup-kitchen** place supplying free soup etc. to the poor; **soup-spoon** large round-bowled spoon. □ **soupy** *adjective* (**-ier**, **-iest**).

soupçon /'su:psɔ̃/ *noun* small quantity, trace.

sour /saʊə/ ● *adjective* having acid taste or smell (as) from unripeness or fermentation; morose, bitter; unpleasant, distasteful; (of soil) dank. ● *verb* make or become sour. □ **sour grapes** resentful disparagement of something coveted; **sourpuss** *colloquial* bad-tempered person. □ **sourly** *adverb*; **sourness** *noun*.

source /sɔ:s/ *noun* place from which river or stream issues; place of origination; person, book, etc. providing information. □ **at source** at point of origin or issue.

souse /saʊs/ ● *verb* (**-sing**) immerse in pickle or other liquid; (as **soused** *adjective*) *colloquial* drunk; (usually + *in*) soak (thing). ● *noun* pickle made with salt; *US* food in pickle; a plunge or drenching in water.

soutane /su:'tɑ:n/ *noun* cassock of RC priest.

south /saʊθ/ ● *noun* point of horizon opposite north; corresponding compass point; (usually **the South**) southern part of world, country, town, etc. ● *adjective* towards, at, near, or facing south; (of wind) from south. ● *adverb* towards, at, or near south; (+ *of*) further south than. □ **southbound** travelling or leading south; **south-east**, **-west** point midway between south and east or west; **southpaw** *colloquial* left-handed person, esp. boxer; **south-south-east**, **south-south-west** point midway between south and south-east or south-west. □ **southward** *adjective, adverb, & noun*; **southwards** *adverb*.

southerly /'sʌðəlɪ/ *adjective & adverb* in southern position or direction; (of wind) from south.

southern /'sʌð(ə)n/ *adjective* of or in south. □ **Southern Cross** constellation with stars forming cross; **southern lights** aurora australis. □ **southernmost** *adjective*.

southerner *noun* native or inhabitant of south.

souvenir /su:və'nɪə/ *noun* memento of place, occasion, etc.

sou'wester /saʊ'westə/ *noun* waterproof hat with broad flap at back; SW wind.

sovereign /'sɒvrɪn/ ● *noun* supreme ruler, esp. monarch; *historical* British gold coin nominally worth £1. ● *adjective* supreme; self-governing; royal; (of remedy etc.) effective. □ **sovereignty** *noun* (*plural* **-ies**).

Soviet /'səʊvɪət/ *historical* ● *adjective* of USSR. ● *noun* citizen of USSR; (**soviet**) elected council in USSR.

sow[1] /səʊ/ *verb* (*past* **sowed**; *past participle* **sown** or **sowed**) scatter (seed) on or in earth, (often + *with*) plant with seed; initiate.

sow[2] /saʊ/ *noun* adult female pig.

soy *noun* (in full **soy sauce**) sauce made from pickled soya beans.

soya /'sɔɪə/ *noun* (in full **soya bean**) (seed of) leguminous plant yielding edible oil and flour.

sozzled /'sɒz(ə)ld/ *adjective colloquial* very drunk.

spa /spɑ:/ *noun* curative mineral spring; resort with this.

space ● *noun* continuous expanse in which things exist and move; amount of this taken by thing or available; interval between points or objects; empty area; outdoor urban recreation area; outer space; interval of time; expanse of paper used in writing, available for advertising, etc.; blank between printed, typed, or written words etc.; *Printing* piece of metal separating words etc. ● *verb* (**-cing**) set or arrange at intervals; put spaces between. □ **space age** era of space travel; **space-age** very modern; **spacecraft** vehicle for travelling in outer space; **spaceman**, **spacewoman** astronaut; **space out** spread out (more) widely; **spaceship** spacecraft; **space shuttle** spacecraft for repeated use; **space station** artificial satellite as base for operations in outer space; **spacesuit** sealed pressurized suit for astronaut in space; **space-time** fusion of concepts of space and time as 4-dimensional continuum.

spacious /'speɪʃəs/ *adjective* having ample space, roomy. □ **spaciously** *adverb*; **spaciousness** *noun*.

spade[1] *noun* long-handled digging tool with rectangular metal blade. □ **spadework** hard preparatory work. □ **spadeful** *noun* (*plural* **-s**).

spade[2] *noun* playing card of suit denoted by black inverted heart-shaped figures with short stems.

spaghetti /spə'getɪ/ *noun* pasta in long thin strands. □ **spaghetti western** cowboy film made cheaply in Italy.

span[1] ● *noun* full extent from end to end; each part of bridge between supports; maximum lateral extent of aeroplane or its wing or of bird's wing etc.; distance between outstretched tips of thumb and little finger; 9 inches. ● *verb* (**-nn-**) extend from side to side or end to end of; bridge (river etc.).

span[2] *past* of SPIN.

spandrel /'spændrɪl/ *noun* space between curve of arch and surrounding rectangular moulding, or between curves of adjoining arches and moulding above.

spangle /'spæŋg(ə)l/ ● *noun* small piece of glittering material, esp. one of many used to ornament dress etc. ● *verb* (**-ling**) (esp. as **spangled** *adjective*) cover (as) with spangles.

Spaniard /'spænjəd/ *noun* native or national of Spain.

spaniel /'spænj(ə)l/ *noun* dog of breed with long silky coat and drooping ears.

Spanish /'spænɪʃ/ ● *adjective* of Spain. ● *noun* language of Spain.

spank *verb & noun* slap, esp. on buttocks.

spanker *noun Nautical* fore-and-aft sail on mizen-mast.

spanking ● *adjective* brisk; *colloquial* striking, excellent. ● *adverb colloquial* very. ● *noun* slapping on buttocks.

spanner /'spænə/ *noun* tool for turning nut on bolt etc. □ **spanner in the works** *colloquial* impediment.

spar[1] *noun* stout pole, esp. as ship's mast etc.

spar[2] ● *verb* (**-rr-**) make motions of boxing; argue. ● *noun* sparring; boxing match. □ **sparring partner** boxer employed to spar with another as training, person with whom one enjoys arguing.

spar[3] *noun* easily split crystalline mineral.

spare /speə/ ● *adjective* not required for ordinary or present use, extra; for emergency or occasional use; lean, thin; frugal. ● *noun* spare part. ● *verb* (**-ring**) afford to give, dispense with; refrain from killing, hurting, etc.; not inflict; be frugal or grudging of. □ **go spare** *colloquial* become very angry; **spare (person's) life** not kill; **spare part** duplicate, esp. as replacement; **spare-rib** closely trimmed rib of esp. pork; **spare time** leisure; **spare tyre** *colloquial* roll of fat round waist; **to spare** left over. □ **sparely** *adverb*; **spareness** *noun*.

sparing *adjective* frugal, economical; restrained. □ **sparingly** *adverb*.

spark ● *noun* fiery particle of burning substance; (often + *of*) small amount; flash of light between electric conductors etc.; this for firing explosive mixture in internal-combustion engine;

flash of wit etc.; (also **bright spark**) lively or clever person. ● *verb* emit spark(s); (often + *off*) stir into activity, initiate. □ **spark plug**, **sparking plug** device for making spark in internal-combustion engine. □ **sparky** *adjective*.

sparkle /'spɑ:k(ə)l/ ● *verb* (**-ling**) (seem to) emit sparks; glitter, scintillate; (of wine etc.) effervesce. ● *noun* glitter; lively quality. □ **sparkly** *adjective*.

sparkler *noun* sparkling firework; *colloquial* diamond.

sparrow /'spærəʊ/ *noun* small brownish-grey bird. □ **sparrowhawk** small hawk.

sparse /spɑ:s/ *adjective* thinly scattered. □ **sparsely** *adverb*; **sparseness** *noun*; **sparsity** *noun*.

Spartan /'spɑ:t(ə)n/ ● *adjective* of ancient Sparta; austere, rigorous. ● *noun* native or citizen of Sparta.

spasm /'spæz(ə)m/ *noun* sudden involuntary muscular contraction; convulsive movement or emotion etc.; (usually + *of*) *colloquial* brief spell.

spasmodic /spæz'mɒdɪk/ *adjective* of or in spasms, intermittent. □ **spasmodically** *adverb*.

spastic /'spæstɪk/ ● *adjective* of or having cerebral palsy. ● *noun* spastic person.

spat¹ *past & past participle* of SPIT¹.

spat² (usually in *plural*) *historical* short gaiter covering shoe.

spate *noun* river-flood; large amount or number (of similar events etc.).

spathe /speɪð/ *noun* large bract(s) enveloping flower-cluster.

spatial /'speɪʃ(ə)l/ *adjective* of space. □ **spatially** *adverb*.

spatter /'spætə/ ● *verb* splash or scatter in drips. ● *noun* splash; pattering.

spatula /'spætjʊlə/ *noun* broad-bladed implement used esp. by artists and in cookery.

spawn ● *verb* (of fish, frog, etc.) produce (eggs); be produced as eggs or young; produce or generate in large numbers. ● *noun* eggs of fish, frogs, etc.; white fibrous matter from which fungi grow.

spay *verb* sterilize (female animal) by removing ovaries.

speak *verb* (*past* **spoke**; *past participle* **spoken** /'spəʊk(ə)n/) utter words in ordinary way; utter (words, the truth, etc.); converse; (+ *of*, *about*) mention; (+ *for*) act as spokesman for; (+ *to*) speak with reference to or in support of; deliver speech; (be able to) use (specified language) in speaking; convey idea; (usually + *to*) affect. □ **speak for itself** be sufficient evidence; **speaking clock** telephone service announcing correct time; **speak out** give one's opinion courageously; **speak up** speak (more) loudly.

speaker *noun* person who speaks, esp. in public; person who speaks specified language; (**Speaker**) presiding officer of legislative assembly; loudspeaker.

spear ● *noun* thrusting or hurling weapon with long shaft and sharp point; tip and stem of asparagus, broccoli, etc. ● *verb* pierce or strike (as) with spear. □ **spearhead** *noun* point of spear, person(s) leading attack or challenge, *verb* act as spearhead of (attack etc.); **spearmint** common garden mint.

spec¹ *noun colloquial* speculation. □ **on spec** as a gamble.

spec² *noun colloquial* specification.

special /'speʃ(ə)l/ ● *adjective* exceptional; peculiar, specific; for particular purpose; for children with special needs. ● *noun* special constable, train, edition of newspaper, dish on menu, etc. □ **Special Branch** police department dealing with political security; **special constable** person assisting police in routine duties or in emergencies; **special effects** illusions created by props, camera-work, etc.; **special licence** licence allowing immediate marriage without banns; **special pleading** biased reasoning. □ **specially** *adverb*.

specialist *noun* person trained in particular branch of profession, esp. medicine; person specially studying subject or area.

speciality /speʃɪ'ælɪtɪ/ *noun* (*plural* **-ies**) special subject, product, activity, etc.; special feature or skill.

specialize *verb* (also **-ise**) (**-zing** of **-sing**) (often + *in*) become or be specialist; devote oneself to an interest, skill, etc.; (esp. in *passive*) adapt for particular purpose; (as **specialized** *adjective*) of a specialist. □ **specialization** *noun*.

specialty /'speʃəltɪ/ *noun* (*plural* **-ies**) *esp. US* speciality.

specie /'spi:ʃi:/ *noun* coin as opposed to paper money.

species /'spi:ʃɪz/ *noun* (*plural* same) class of things having common characteristics; group of animals or plants within genus; kind, sort.

specific /spə'sɪfɪk/ ● *adjective* clearly defined; relating to particular subject, peculiar; exact, giving full details; *archaic* (of medicine etc.) for particular disease. ● *noun archaic* specific medicine; specific aspect. □ **specific gravity** relative density. □ **specifically** *adverb*; **specificity** /-'fɪs-/ *noun*.

specification /spesɪfɪ'keɪʃ(ə)n/ *noun* specifying; (esp. in *plural*) detailed description of work (to be) done or of invention, patent, etc.

specify /'spesɪfaɪ/ *verb* (**-ies**, **-ied**) name or mention expressly or as condition; include in specifications.

specimen /'spesɪmɪn/ *noun* individual or sample taken as example of class or whole, esp. in experiments etc.; *colloquial usually derogatory* person of specified sort.

specious /'spi:ʃəs/ *adjective* plausible but wrong.

speck ● *noun* small spot or stain; particle. ● *verb* (esp. as **specked** *adjective*) mark with specks.

speckle /'spek(ə)l/ ● *noun* speck, esp. one of many markings. ● *verb* (**-ling**) (esp. as **speckled** *adjective*) mark with speckles.

specs *plural noun colloquial* spectacles.

spectacle /'spektək(ə)l/ *noun* striking, impressive, or ridiculous sight; public show; object of public attention; (in *plural*) pair of lenses in frame supported on nose and ears, to correct defective eyesight.

spectacled *adjective* wearing spectacles.

spectacular /spek'tækjʊlə/ ● *adjective* striking, impressive, lavish. ● *noun* spectacular performance. □ **spectacularly** *adverb*.

spectator /spek'teɪtə/ *noun* person who watches a show, game, incident, etc. □ **spectator sport** sport attracting many spectators. □ **spectate** *verb* (**-ting**).

specter *US* = SPECTRE.

spectra *plural* of SPECTRUM.

spectral /'spektr(ə)l/ *adjective* of or like spectres or spectra; ghostly.

spectre /'spektə/ *noun* (*US* **specter**) ghost; haunting presentiment.

spectroscope /'spektrəskəʊp/ *noun* instrument for recording and examining spectra. □ **spectroscopic** /-'skɒp-/ *adjective*; **spectroscopy** /-'trɒskəpɪ/ *noun*.

spectrum /'spektrəm/ *noun* (*plural* **-tra**) band of colours as seen in rainbow etc.; entire or wide range of subject, emotion, etc.; arrangement of electromagnetic radiation by wavelength.

speculate /'spekjʊleɪt/ *verb* (**-ting**) (usually + *on*, *upon*, *about*) theorize, conjecture; deal in commodities etc. in expectation of profiting from fluctuating prices. □ **speculation** *noun*; **speculative** /-lətɪv/ *adjective*; **speculator** *noun*.

sped *past & past participle* of SPEED.

speech *noun* faculty, act, or manner of speaking; formal public address; language, dialect. □ **speech day** annual prize-giving day in school; **speech therapy** treatment for defective speech.

speechify /'spi:tʃɪfaɪ/ *verb* (**-ies**, **-ied**) *jocular* make speeches.

speechless *adjective* temporarily silenced by emotion etc.

speed ● *noun* rapidity; rate of progress or motion; gear on bicycle; relative sensitivity of photographic film to light; *slang* amphetamine. ● *verb* (*past & past participle* **sped**) go or send quickly; (*past & past participle* **speeded**) travel at illegal or dangerous speed; *archaic* be or make prosperous or successful. □ **speedboat** fast motor boat; **speed limit** maximum permitted speed on road etc.; **speedway** (dirt track for) motorcycle racing, *US* road or track for fast vehicles. □ **speeder** *noun*.

speedometer /spi:'dɒmɪtə/ *noun* instrument on vehicle indicating its speed.

speedwell /'spi:dwel/ *noun* small blue-flowered herbaceous plant.

speedy *adjective* (**-ier**, **-iest**) rapid; prompt. □ **speedily** *adverb*.

speleology /spi:lɪ'ɒlədʒɪ/ *noun* the study of caves etc.

spell[1] *verb* (*past & past participle* **spelt** or **spelled**) write or name correctly the letters of (word etc.); (of letters) form (word etc.); result in. □ **spell out** make out letter by letter, explain in detail. □ **speller** *noun.*

spell[2] *noun* words used as charm; effect of these; fascination. □ **spellbound** held as if by spell, fascinated.

spell[3] *noun* (fairly) short period; period of some activity or work.

spelling *noun* way word is spelt; ability to spell.

spelt[1] *past & past participle* of SPELL[1].

spelt[2] *noun* kind of wheat giving very fine flour.

spend *verb* (*past & past participle* **spent**) pay out (money); use or consume (time or energy); use up; (as **spent** *adjective*) having lost force or strength. □ **spendthrift** extravagant person. □ **spender** *noun.*

sperm *noun* (*plural* same or **-s**) spermatozoon; semen. □ **sperm bank** store of semen for artificial insemination; **sperm whale** large whale hunted for spermaceti.

spermaceti /spɜ:mə'setɪ/ *noun* white waxy substance used for ointments etc.

spermatozoon /spɜ:mətəʊ'zəʊən/ *noun* (*plural* **-zoa**) fertilizing cell of male organism.

spermicide /'spɜ:mɪsaɪd/ *noun* substance that kills spermatozoa. □ **spermicidal** /-'saɪd-/ *adjective.*

spew *verb* (often + *up*) vomit; (often + *out*) (cause to) gush.

sphagnum /'sfægnəm/ *noun* (*plural* **-na**) (in full **sphagnum moss**) moss growing in bogs, used as packing etc.

sphere /sfɪə/ *noun* solid figure with every point on its surface equidistant from centre; ball, globe; field of action, influence, etc.; place in society; *historical* each of revolving shells in which heavenly bodies were thought to be set.

spherical /'sferɪk(ə)l/ *adjective* shaped like sphere; of spheres. □ **spherically** *adverb.*

spheroid /'sfɪərɔɪd/ *noun* spherelike but not perfectly spherical body. □ **spheroidal** /-'rɔɪd-/ *adjective.*

sphincter /'sfɪŋktə/ *noun* ring of muscle closing and opening orifice.

Sphinx /sfɪŋks/ *noun* ancient Egyptian stone figure with lion's body and human or animal head; (**sphinx**) inscrutable person.

spice ● *noun* aromatic or pungent vegetable substance used as flavouring; spices collectively; piquant quality; slight flavour. ● *verb* (**-cing**) flavour with spice; enhance.

spick and span *adjective* trim and clean; smart, new-looking.

spicy *adjective* (**-ier**, **-iest**) of or flavoured with spice; piquant, improper. □ **spiciness** *noun.*

spider /'spaɪdə/ *noun* 8-legged arthropod, many species of which spin webs esp. to capture insects as food. □ **spider plant** house plant with long narrow leaves.

spidery *adjective* elongated and thin.

spiel /ʃpi:l/ *noun slang* glib speech or story, sales pitch.

spigot /'spɪgət/ *noun* small peg or plug; device for controlling flow of liquid in tap.

spike[1] ● *noun* sharp point; pointed piece of metal, esp. forming top of iron railing; metal point on sole of running shoe to prevent slipping; (in *plural*) spiked running shoes; large nail. ● *verb* (**-king**) put spikes on or into; fix on spike; *colloquial* add alcohol to (drink), contaminate. □ **spike (person's) guns** defeat his or her plans.

spike[2] *noun* cluster of flower heads on long stem.

spikenard /'spaɪknɑ:d/ *noun* tall sweet-smelling plant; *historical* aromatic ointment formerly made from this.

spiky *adjective* (**-ier**, **-iest**) like a spike; having spikes; *colloquial* irritable.

spill[1] ● *verb* (*past & past participle* **spilt** or **spilled**) allow (liquid etc.) to fall or run out of container, esp. accidentally; (of liquid etc.) fall or run out thus; throw from vehicle, saddle, etc.; (+ *into*, *out*, etc.) leave quickly; *slang* divulge (information etc.); shed (blood). ● *noun* spilling, being spilt; tumble, esp. from horse or vehicle. □ **spill the beans** *colloquial* divulge secret etc. □ **spillage** *noun.*

spill[2] *noun* thin strip of wood, paper, etc. for lighting candle etc.

spillikin /'spɪlɪkɪn/ *noun* splinter of wood etc.; (in *plural*) game in which thin rods are removed one at a time from heap without disturbing others.

spilt *past & past participle* of SPILL[1].

spin • *verb* (**-nn-**; *past* **spun** or **span**; *past participle* **spun**) (cause to) turn or whirl round rapidly; make (yarn) by drawing out and twisting together fibres of wool etc.; make (web etc.) by extruding fine viscous thread; (of person's head) be in a whirl; tell or compose (story etc.); toss (coin); (as **spun** *adjective*) made into threads. • *noun* revolving motion, whirl; rotating dive of aircraft; secondary twisting motion e.g. of ball in flight; *colloquial* brief drive, esp. in car. □ **spin bowler** *Cricket* one who imparts spin to ball; **spin-drier**, **-dryer** machine for drying clothes by spinning them in rotating drum; **spin-dry** *verb*; **spinning wheel** household implement for spinning yarn, with spindle driven by wheel with crank or treadle; **spin-off** incidental result, esp. from technology; **spin out** prolong; **spin a yarn** tell story.

spina bifida /spaɪnə 'bɪfɪdə/ *noun* congenital spinal defect, with protruding membranes.

spinach /'spɪnɪdʒ/ *noun* green vegetable with edible leaves.

spinal /'spaɪn(ə)l/ *adjective* of spine. □ **spinal column** spine; **spinal cord** cylindrical nervous structure within spine.

spindle /'spɪnd(ə)l/ *noun* slender rod for twisting and winding thread in spinning; pin or axis on which something revolves; turned piece of wood used as banister etc.

spindly *adjective* (**-ier**, **-iest**) long or tall and thin.

spindrift /'spɪndrɪft/ *noun* spray on surface of sea.

spine *noun* series of vertebrae extending from skull, backbone; needle-like outgrowth of animal or plant; part of book enclosing page-fastening; ridge, sharp projection. □ **spine-chiller** suspense or horror film, story, etc.

spineless *adjective* lacking resoluteness.

spinet /spɪ'net/ *noun historical* small harpsichord with oblique strings.

spinnaker /'spɪnəkə/ *noun* large triangular sail used at bow of yacht.

spinner *noun* spin bowler; person or thing that spins, esp. manufacturer engaged in spinning; spin-drier; revolving bait or lure in fishing.

spinneret /'spɪnəret/ *noun* spinning-organ in spider etc.

spinney /'spɪnɪ/ *noun* (*plural* **-s**) small wood, thicket.

spinster /'spɪnstə/ *noun formal* unmarried woman.

spiny *adjective* (**-ier**, **-iest**) having (many) spines.

spiraea /spaɪ'rɪə/ *noun* (*US* **spirea**) garden plant related to meadowsweet.

spiral /'spaɪər(ə)l/ • *adjective* coiled in a plane or as round a cylinder or cone; having this shape. • *noun* spiral curve; progressive rise or fall. • *verb* (**-ll-**; *US* **-l-**) move in spiral course; (of prices etc.) rise or fall continuously. □ **spiral staircase** circular staircase round central axis.

spirant /'spaɪərənt/ • *adjective* uttered with continuous expulsion of breath. • *noun* spirant consonant.

spire *noun* tapering structure, esp. on church tower; any tapering thing.

spirea *US* = SPIRAEA.

spirit /'spɪrɪt/ • *noun* person's essence or intelligence, soul; rational being without material body; ghost; person's character; attitude; type of person; prevailing tendency; (usually in *plural*) distilled alcoholic liquor; distilled volatile liquid; courage, vivacity; (in *plural*) mood; essential as opposed to formal meaning. • *verb* (**-t-**) (usually + *away*, *off*, etc.) convey mysteriously. □ **spirit gum** quick-drying gum for attaching false hair; **spirit lamp** lamp burning methylated or other volatile spirit; **spirit level** device used to test horizontality.

spirited *adjective* lively, courageous. □ **-spirited** in specified mood. □ **spiritedly** *adverb*.

spiritual /'spɪrɪtʃʊəl/ • *adjective* of spirit; religious, divine, inspired; refined, sensitive. • *noun* (also **Negro spiritual**) religious song originally of American blacks. □ **spirituality** /-'æl-/ *noun*; **spiritually** *adverb*.

spiritualism *noun* belief in, and practice of, communication with the dead, esp. through mediums. □ **spiritualist** *noun*; **spiritualistic** /-'lɪs-/ *adjective*.

spirituous /'spɪrɪtʃʊəs/ *adjective* very alcoholic; distilled as well as fermented.

spit[1] ● *verb* (**-tt-**; *past & past participle* **spat** or **spit**) eject (esp. saliva) from mouth; do this as gesture of contempt; utter vehemently; (of fire etc.) throw out with explosion; (of rain etc.) fall lightly; make spitting noise. ● *noun* spittle; spitting. □ **spitfire** fiery-tempered person; **spitting distance** *colloquial* very short distance; **spitting image** *colloquial* exact counterpart or likeness.

spit[2] ● *noun* rod for skewering meat for roasting over fire etc.; point of land projecting into sea; spade-depth of earth. ● *verb* (**-tt-**) pierce (as) with spit. □ **spit-roast** roast on spit.

spite ● *noun* ill will, malice. ● *verb* (**-ting**) hurt, thwart. □ **in spite of** notwithstanding.

spiteful *adjective* malicious. □ **spitefully** *adverb*.

spittle /'spɪt(ə)l/ *noun* saliva.

spittoon /spɪ'tu:n/ *noun* vessel to spit into.

spiv *noun colloquial* man, esp. flashily-dressed one, living from shady dealings. □ **spivvish** *adjective*; **spivvy** *adjective*.

splash ● *verb* (cause to) scatter in drops; wet or stain by splashing; (usually + *across*, *along*, *about*, etc.) move with splashing; jump or fall into water etc. with splash; display (news) conspicuously; decorate with scattered colour; spend (money) ostentatiously. ● *noun* act or noise of splashing; quantity splashed; mark of splashing; prominent news feature, display, etc.; patch of colour; *colloquial* small quantity of soda water etc. (in drink). □ **splashback** panel behind sink etc. to protect wall from splashes; **splashdown** alighting of spacecraft on sea; **splash down** *verb*; **splash out** *colloquial* spend money freely.

splat *colloquial* ● *noun* sharp splattering sound. ● *adverb* with splat. ● *verb* (**-tt-**) fall or hit with splat.

splatter /'splætə/ *verb & noun* splash, esp. with continuous noisy action, spatter.

splay ● *verb* spread apart; (of opening) have sides diverging; make (opening) with divergent sides. ● *noun* surface at oblique angle to another. ● *adjective* splayed.

spleen *noun* abdominal organ regulating quality of blood; moroseness, irritability.

splendid /'splendɪd/ *adjective* magnificent; glorious, dignified; excellent. □ **splendidly** *adverb*.

splendiferous /splen'dɪfərəs/ *adjective colloquial* splendid.

splendour /'splendə/ *noun* (*US* **splendor**) dazzling brightness; magnificence.

splenetic /splɪ'netɪk/ *adjective* bad-tempered, peevish.

splenic /'splenɪk/ *adjective* of or in spleen.

splice ● *verb* (**-cing**) join (ropes) by interweaving strands; join (pieces of wood, tape, etc.) by overlapping; (esp. as **spliced** *adjective*) *colloquial* join in marriage. ● *noun* join made by splicing.

splint ● *noun* strip of wood etc. bound to broken limb while it heals. ● *verb* secure with splint.

splinter /'splɪntə/ ● *noun* small sharp fragment of wood, stone, glass, etc. ● *verb* split into splinters, shatter. □ **splinter group** breakaway political group. □ **splintery** *adjective*.

split ● *verb* (**-tt-**; *past & past participle* **split**) break, esp. lengthwise or with grain; break forcibly; (often + *up*) divide into parts, esp. equal shares; (often + *off*, *away*) remove or be removed by breaking or dividing; (usually + *on*, *over*, etc.) divide into disagreeing or hostile parties; cause fission of (atom); *slang* leave, esp. suddenly; (usually + *on*) *colloquial* inform; (as **splitting** *adjective*) (of headache) severe; (of head) suffer severe headache. ● *noun* splitting; disagreement, schism; (in *plural*) feat of leaping or sitting with legs straight and pointing in opposite directions. □ **split hairs** make over-subtle distinctions; **split infinitive** one with adverb etc. inserted between *to* and verb (see note below); **split-level** with more than one level; **split personality** condition of alternating personalities; **split pin** metal cotter with its two ends splayed out after passing through hole; **split second** very short time; **split-second** very rapid, (of timing) very accurate; **split up** separate, end relationship.

■ **Usage** Split infinitives, as in *I want to quickly sum up* and *Your job is to really get to know everybody*, are common in informal English, but many people consider them incorrect and prefer *I want quickly to sum up* or *I want to sum up quickly*. They should therefore be avoided in formal English, but note that just changing the order of words can alter the meaning, e.g. *Your job is really to get to know everybody*.

---

splodge *colloquial* ● *noun* daub, blot, smear. ● *verb* (**-ging**) make splodge on. □ **splodgy** *adjective*.

splosh *colloquial* ● *verb* move with splashing sound. ● *noun* splashing sound; splash of water etc.

splotch *noun & verb* splodge. □ **splotchy** *adjective*.

splurge *colloquial* ● *noun* sudden extravagance; ostentatious display or effort. ● *verb* (**-ging**) (usually + *on*) make splurge.

splutter /'splʌtə/ ● *verb* speak or express in choking manner; emit spitting sounds; speak rapidly or incoherently. ● *noun* spluttering speech or sound.

spoil ● *verb* (*past & past participle* **spoilt** or **spoiled**) make or become useless or unsatisfactory; reduce enjoyment etc. of; decay, go bad; ruin character of by over-indulgence. ● *noun* (usually in *plural*) plunder, stolen goods; profit or advantages accruing from success or position. □ **spoilsport** person who spoils others' enjoyment; **spoilt for choice** having excessive number of choices.

spoiler *noun* device on aircraft to increase drag; device on vehicle to improve road-holding at speed.

spoilt *past & past participle* of **spoil**.

spoke[1] *noun* each of rods running from hub to rim of wheel. □ **put a spoke in (person's) wheel** thwart, hinder.

spoke[2] *past* of SPEAK.

spoken *past participle* of SPEAK.

spokesman /'spəʊksmən/ *noun* (*feminine* **spokeswoman**) person who speaks for others, representative.

spokesperson /'spəʊkspɜ:s(ə)n/ *noun* (*plural* **-s** or **spokespeople**) spokesman or spokeswoman.

spoliation /spəʊlɪ'eɪʃ(ə)n/ *noun* plundering, pillage.

spondee /'spɒndi:/ *noun* metrical foot of two long syllables. □ **spondaic** /-'deɪɪk/ *adjective*.

sponge /spʌndʒ/ ● *noun* sea animal with porous body wall and tough elastic skeleton; this skeleton or piece of porous rubber etc. used in bathing, cleaning, etc.; thing like sponge in consistency, esp. sponge cake; act of sponging. ● *verb* (**-ging**) wipe or clean with sponge; (often + *out, away*, etc.) wipe off or rub out (as) with sponge; (often + *up*) absorb (as) with sponge; (often + *on, off*) live as parasite. □ **sponge bag** waterproof bag for toilet articles; **sponge cake, pudding** one of light spongelike consistency; **sponge rubber** porous rubber.

sponger *noun* parasitic person.

spongy *adjective* (**-ier**, **-iest**) like a sponge, porous, elastic, absorbent.

sponsor /'spɒnsə/ ● *noun* person who pledges money to charity in return for specified activity by someone; patron of artistic or sporting activity etc.; company etc. financing broadcast in return for advertising; person introducing legislation; godparent at baptism. ● *verb* be sponsor for. □ **sponsorship** *noun*.

spontaneous /spɒn'teɪnɪəs/ *adjective* acting, done, or occurring without external cause; instinctive, automatic, natural. □ **spontaneity** /-tə'neɪətɪ/ *noun*; **spontaneously** *adverb*.

spoof /spu:f/ *noun & verb colloquial* parody; hoax, swindle.

spook /spu:k/ ● *noun colloquial* ghost. ● *verb esp. US* frighten, unnerve. □ **spooky** *adjective* (**-ier**, **-iest**).

spool /spu:l/ ● *noun* reel on which something is wound; revolving cylinder of angler's reel. ● *verb* wind on spool.

spoon /spu:n/ ● *noun* utensil with bowl and handle for putting food in mouth or for stirring etc.; spoonful; spoon-shaped thing, esp. (in full **spoon-bait**) revolving metal fish-lure. ● *verb* (often + *up, out*) take (liquid etc.) with spoon; hit (ball) feebly upwards. □ **spoonbill** wading bird with broad flat-tipped bill; **spoonfeed** feed with spoon, give help etc. to (person) without demanding any effort from recipient. □ **spoonful** *noun* (*plural* **-s**).

spoonerism /'spu:nərɪz(ə)m/ *noun* (usually accidental) transposition of initial sounds of two or more words.

spoor /spʊə/ *noun* animal's track or scent.

sporadic /spə'rædɪk/ *adjective* occurring only sparsely or occasionally. □ **sporadically** *adverb*.

spore *noun* reproductive cell of ferns, fungi, protozoa, etc.

sporran /'spɒrən/ *noun* pouch worn in front of kilt.

sport ● *noun* game or competitive activity usually involving physical exertion; these collectively; (in *plural*) meeting for competition in athletics; amusement, fun; *colloquial* sportsman, good fellow; person with specified attitude to games, rules, etc. ● *verb* amuse oneself, play about; wear or exhibit, esp. ostentatiously. □ **sports car** low-built fast car; **sports coat**, **jacket** man's informal jacket; **sports ground** piece of land used for sport; **sportswear** clothes for sports, informal clothes.

sporting *adjective* of or interested in sport; generous, fair. □ **sporting chance** some possibility of success. □ **sportingly** *adverb*.

sportive *adjective* playful.

sportsman /'spɔ:tsmən/ *noun* (*feminine* **sportswoman**) person engaging in sport; fair and generous person. □ **sportsmanlike** *adjective*; **sportsmanship** *noun*.

sporty *adjective* (**-ier**, **-iest**) *colloquial* fond of sport; *colloquial* rakish, showy.

spot ● *noun* small mark differing in colour etc. from surface it is on; pimple, blemish; particular place, locality; particular part of one's body or character; *colloquial* one's (regular) position in organization, programme, etc.; *colloquial* small quantity; spotlight. ● *verb* (**-tt-**) *colloquial* pick out, recognize, catch sight of; watch for and take note of (trains, talent, etc.); (as **spotted** *adjective*) marked with spots; make spots, rain slightly. □ **in a (tight) spot** *colloquial* in difficulties; **on the spot** at scene of event, *colloquial* in position demanding response or action, without delay, without moving backwards or forwards; **spot cash** money paid immediately after sale; **spot check** sudden or random check; **spotlight** *noun* beam of light directed on small area, lamp projecting this, full publicity, *verb* direct spotlight on; **spot on** *colloquial* precise(ly); **spotted dick** suet pudding containing currants; **spot-weld** join (metal surfaces) by welding at points. □ **spotter** *noun*.

spotless *adjective* absolutely clean, unblemished. □ **spotlessly** *adverb*.

spotty *adjective* (**-ier**, **-iest**) marked with spots; patchy, irregular.

spouse /spaʊz/ *noun* husband or wife.

spout /spaʊt/ ● *noun* projecting tube or lip for pouring from teapot, kettle, jug, fountain, roof-gutter, etc.; jet of liquid. ● *verb* discharge or issue forcibly in jet; utter at length or pompously. □ **up the spout** *slang* useless, ruined, pregnant.

sprain ● *verb* wrench (ankle, wrist, etc.) causing pain or swelling. ● *noun* such injury.

sprang *past* of SPRING.

sprat *noun* small sea fish.

sprawl ● *verb* sit, lie, or fall with limbs spread out untidily; spread untidily, straggle. ● *noun* sprawling movement, position, or mass; straggling urban expansion.

spray[1] ● *noun* water etc. flying in small drops; liquid intended for spraying; device for spraying. ● *verb* throw as spray; sprinkle (as) with spray; (of tom-cat) mark environment with urine to attract females. □ **spray-gun** device for spraying paint etc. □ **sprayer** *noun*.

spray[2] *noun* sprig with flowers or leaves, small branch; ornament in similar form.

spread /spred/ ● *verb* (*past & past participle* **spread**) (often + *out*) open, extend, unfold, cause to cover larger surface, have wide or increasing extent; (cause to) become widely known; cover; lay (table). ● *noun* act, capability, or extent of spreading; diffusion; breadth; increased girth; difference between two rates, prices, etc.; *colloquial* elaborate meal; paste for spreading on bread etc.; bedspread; printed matter spread over more than one column. □ **spread eagle** figure of eagle with legs and wings extended as emblem; **spread-eagle** place (person) with arms and legs spread out, defeat utterly; **spreadsheet** computer program for handling tabulated figures etc., esp. in accounting.

spree *noun colloquial* extravagant outing; bout of drinking etc.

sprig ● *noun* small branch or shoot; ornament resembling this, esp. on fabric. ● *verb* (**-gg-**) ornament with sprigs.

sprightly /'spraɪtlɪ/ *adjective* (**-ier**, **-iest**) vivacious, lively.

spring ● *verb* (*past* **sprang**; *past participle* **sprung**) rise rapidly or suddenly, leap; move rapidly (as) by action of a spring; (usually + *from*) originate; (cause to) act or appear unexpectedly; *slang* contrive escape of (person from prison etc.); (usually as **sprung** *adjective*) provide with springs. ● *noun* jump, leap; recoil; elasticity; elastic device usually of coiled metal used esp. to drive clockwork or for cushioning in furniture or vehicles; season of year between winter and summer; (often + *of*) early stage of life etc.; place where water, oil, etc. wells up from earth, basin or flow so formed; motive for or origin of action, custom, etc. □ **spring balance** device measuring weight by tension of spring; **springboard** flexible board for leaping or diving from, source of impetus; **spring-clean** *noun* thorough cleaning of house, esp. in spring, *verb* clean thus; **spring greens** young cabbage leaves; **spring a leak** develop leak; **spring onion** young onion eaten raw; **spring roll** Chinese fried pancake filled with vegetables etc.; **spring tide** tide with greatest rise and fall; **springtime** season or period of spring.

springbok /'sprɪŋbɒk/ *noun* (*plural* same or **-s**) S. African gazelle.

springer *noun* small spaniel.

springy *adjective* (**-ier**, **-iest**) elastic.

sprinkle /'sprɪŋk(ə)l/ ● *verb* (**-ling**) scatter in small drops or particles; (often + *with*) subject to sprinkling; (of liquid etc.) fall thus on. ● *noun* (usually + *of*) light shower, sprinkling.

sprinkler *noun* device for sprinkling lawn or extinguishing fires.

sprinkling *noun* small sparse number or amount.

sprint ● *verb* run short distance at top speed. ● *noun* such run; similar short effort in cycling, swimming, etc. □ **sprinter** *noun*.

sprit *noun* small diagonal spar from mast to upper outer corner of sail. □ **spritsail** /'sprɪts(ə)l/ sail extended by sprit.

sprite *noun* elf, fairy.

spritzer /'sprɪtsə/ *noun* drink of white wine with soda water.

sprocket /'sprɒkɪt/ *noun* projection on rim of wheel engaging with links of chain.

sprout /spraʊt/ ● *verb* put forth (shoots etc.); begin to grow. ● *noun* plant shoot; Brussels sprout.

spruce[1] ● *adjective* of trim appearance, smart. ● *verb* (**-cing**) (usually + *up*) make or become smart. □ **sprucely** *adverb*; **spruceness** *noun*.

spruce[2] *noun* conifer with dense conical foliage; its wood.

sprung *past participle* of SPRING.

spry /spraɪ/ *adjective* (**-er**, **-est**) lively, nimble. □ **spryly** *adverb*.

spud ● *noun colloquial* potato; small narrow spade for weeding. ● *verb* (**-dd-**) (+ *up*, *out*) remove with spud.

spumante /spu:'mæntɪ/ *noun* Italian sparkling white wine.

spume /spju:m/ *noun & verb* (**-ming**) froth, foam. □ **spumy** *adjective* (**-ier**, **-iest**).

spun *past & past participle* of SPIN. □ **spun silk** cheap material containing waste silk.

spunk *noun colloquial* mettle, spirit. □ **spunky** *adjective* (**-ier**, **-iest**).

spur ● *noun* small spike or spiked wheel attached to rider's heel for urging horse forward; stimulus, incentive; spur-shaped thing, esp. projection from mountain (range), branch road or railway, or hard projection on cock's leg. ● *verb* (**-rr-**) prick (horse) with spur;

incite, stimulate. □ **on the spur of the moment** on impulse.

spurge *noun* plant with acrid milky juice.

spurious /'spjʊərɪəs/ *adjective* not genuine, fake.

spurn *verb* reject with disdain or contempt.

spurt ● *verb* (cause to) gush out in jet or stream; make sudden effort. ● *noun* sudden gushing out, jet; short burst of speed, growth, etc.

sputnik /'spʊtnɪk/ *noun* Russian artificial earth satellite.

sputter /'spʌtə/ *verb & noun* splutter.

sputum /'spju:təm/ *noun* thick coughed-up mucus.

spy ● *noun* (*plural* **spies**) person secretly collecting and reporting information for a government, company, etc.; person watching others secretly. ● *verb* (**spies**, **spied**) discern, see; (often + *on*) act as spy. □ **spyglass** small telescope; **spyhole** peep-hole; **spy out** explore or discover, esp. secretly.

sq. *abbreviation* square.

Sqn. Ldr. *abbreviation* Squadron Leader.

squab /skwɒb/ ● *noun* young esp. unfledged pigeon etc.; short fat person; stuffed cushion, esp. as part of car-seat; sofa. ● *adjective* short and fat.

squabble /'skwɒb(ə)l/ ● *noun* petty or noisy quarrel. ● *verb* (**-ling**) engage in squabble.

squad /skwɒd/ *noun* small group sharing task etc., esp. of soldiers or police officers; team. □ **squad car** police car.

squaddie *noun* (also **squaddy**) (*plural* **-ies**) *slang* recruit, private.

squadron /'skwɒdrən/ *noun* unit of RAF with 10–18 aircraft; detachment of warships employed on particular service; organized group etc., esp. cavalry division of two troops. □ **squadron leader** RAF officer commanding squadron, next below wing commander.

squalid /'skwɒlɪd/ *adjective* filthy, dirty; mean in appearance.

squall /skwɔ:l/ ● *noun* sudden or violent gust or storm; discordant cry, scream. ● *verb* utter (with) squall, scream. □ **squally** *adjective*.

squalor /'skwɒlə/ *noun* filthy or squalid state.

squander /'skwɒndə/ *verb* spend wastefully.

square /skweə/ ● *noun* rectangle with 4 equal sides; object of (roughly) this shape; open area enclosed by buildings; product of number multiplied by itself; L- or T-shaped instrument for obtaining or testing right angles; *slang* conventional or old-fashioned person. ● *adjective* square-shaped; having or in form of a right angle; angular, not round; designating unit of measure equal to area of square whose side is one of the unit specified; (usually + *with*) level, parallel; (usually + *to*) perpendicular; sturdy, squat; arranged; (also **all square**) with no money owed, (of scores) equal; fair, honest; direct; *slang* conventional, old-fashioned. ● *adverb* squarely. ● *verb* (**-ring**) make square; multiply (number) by itself; (usually + *to*, *with*) make or be consistent, reconcile; mark out in squares; settle (bill etc.); place (shoulders etc.) squarely facing forwards; *colloquial* pay, bribe; make scores of (match etc.) equal. □ **square brackets** brackets of the form [ ] (see panel at BRACKET); **square dance** dance with 4 couples facing inwards from 4 sides; **square deal** fair bargain or treatment; **square leg** *Cricket* fielding position on batsman's leg side nearly opposite stumps; **square meal** substantial meal; **square-rigged** having 4-sided sails set across length of ship; **square root** number that multiplied by itself gives specified number. □ **squarely** *adverb*.

squash[1] /skwɒʃ/ ● *verb* crush or squeeze flat or into pulp; (often + *into*) *colloquial* force into small space, crowd; belittle, bully; suppress. ● *noun* crowd, crowded state; drink made of crushed fruit; (in full **squash rackets**) game played with rackets and small ball in closed court. □ **squashy** *adjective* (**-ier**, **-iest**).

squash[2] /skwɒʃ/ *noun* (*plural* same or **-es**) trailing annual plant; gourd of this.

squat /skwɒt/ ● *verb* (**-tt-**) sit on one's heels, or on ground with knees drawn up; *colloquial* sit down; act as squatter. ● *adjective* (**-tt-**) short and thick, dumpy. ● *noun* squatting posture; place occupied by squatter(s).

squatter *noun* person who inhabits unoccupied premises without permission.

squaw *noun* N. American Indian woman or wife.

squawk ● *noun* harsh cry; complaint. ● *verb* utter squawk.

squeak ● *noun* short high-pitched cry or sound; (also **narrow squeak**) narrow escape. ● *verb* emit squeak; utter shrilly; (+ *by*, *through*) *colloquial* pass narrowly; *slang* turn informer.

squeaky *adjective* (**-ier**, **-iest**) making squeaking sound. □ **squeaky clean** *colloquial* completely clean, above criticism. □ **squeakily** *adverb*; **squeakiness** *noun*.

squeal ● *noun* prolonged shrill sound or cry. ● *verb* make, or utter with, squeal; *slang* turn informer; *colloquial* protest vociferously.

squeamish /ˈskwiːmɪʃ/ *adjective* easily nauseated; fastidious. □ **squeamishly** *adverb*; **squeamishness** *noun*.

squeegee /ˈskwiːdʒiː/ *noun* rubber-edged implement on handle, for cleaning windows etc.

squeeze ● *verb* (**-zing**) (often + *out*) exert pressure on, esp. to extract moisture; reduce in size or alter in shape by squeezing; force or push into or through small or narrow space; harass, pressure; (usually + *out of*) get by extortion or entreaty; press (person's hand) in sympathy etc. ● *noun* squeezing, being squeezed; close embrace; crowd, crowded state; small quantity produced by squeezing; restriction on borrowing and investment. □ **squeeze-box** *colloquial* accordion, concertina.

squelch ● *verb* make sucking sound as of treading in thick mud; move with squelching sound; disconcert, silence. ● *noun* act or sound of squelching. □ **squelchy** *adjective*.

squib *noun* small hissing firework; satirical essay.

squid *noun* (*plural* same or **-s**) 10-armed marine cephalopod.

squidgy /ˈskwɪdʒɪ/ *adjective* (**-ier**, **-iest**) *colloquial* squashy, soggy.

squiffy /ˈskwɪfɪ/ *adjective* (**-ier**, **-iest**) *slang* slightly drunk.

squiggle /ˈskwɪg(ə)l/ *noun* short curling line, esp. in handwriting. □ **squiggly** *adjective*.

squint ● *verb* have eyes turned in different directions; (often + *at*) look sidelong. ● *noun* squinting condition; sidelong glance; *colloquial* glance, look; oblique opening in church wall.

squire ● *noun* country gentleman, esp. chief landowner of district; *historical* knight's attendant. ● *verb* (**-ring**) (of man) escort (woman).

squirearchy /ˈskwaɪərɑːkɪ/ *noun* (*plural* **-ies**) landowners collectively.

squirm ● *verb* wriggle, writhe; show or feel embarrassment. ● *noun* squirming movement.

squirrel /ˈskwɪr(ə)l/ ● *noun* bushy-tailed usually tree-living rodent; its fur; hoarder. ● *verb* (**-ll-**; *US* **-l-**) (often + *away*) hoard.

squirt ● *verb* eject (liquid etc.) in jet; be ejected thus; splash with squirted substance. ● *noun* jet of water etc.; small quantity squirted; syringe; *colloquial* insignificant person.

squish *colloquial* ● *noun* slight squelching sound. ● *verb* move with squish; squash. □ **squishy** *adjective* (**-ier**, **-iest**).

Sr. *abbreviation* Senior.

SRN *abbreviation* State Registered Nurse.

SS *abbreviation* steamship; Saints; *historical* Nazi special police force (*Schutzstaffel*).

SSE *abbreviation* south-south-east.

SSW *abbreviation* south-south-west.

St *abbreviation* Saint.

St. *abbreviation* Street.

st. *abbreviation* stone (weight).

stab ● *verb* (**-bb-**) pierce or wound with knife etc.; (often + *at*) aim blow with such weapon; cause sharp pain to. ● *noun* act or result of stabbing; *colloquial* attempt. □ **stab in the back** *noun* treacherous attack, *verb* betray.

stability /stəˈbɪlɪtɪ/ *noun* being stable.

stabilize /ˈsteɪbɪlaɪz/ *verb* (also **-ise**) (**-zing** or **-sing**) make or become stable. □ **stabilization** *noun*.

stabilizer *noun* (also **-iser**) device to keep aircraft or (in *plural*) child's bicycle steady; food additive for preserving texture.

stable /'steɪb(ə)l/ ● *adjective* (**-r**, **-st**) firmly fixed or established, not fluctuating or changing; not easily upset or disturbed. ● *noun* building for keeping horses; establishment for training racehorses; racehorses of particular stable; people, products, etc. having common origin or affiliation; such origin or affiliation. ● *verb* (**-ling**) put or keep in stable. □ **stably** *adverb*.

stabling *noun* accommodation for horses.

staccato /stə'kɑ:təʊ/ *esp. Music* ● *adverb & adjective* with each sound sharply distinct. ● *noun* (*plural* **-s**) staccato passage or delivery.

stack ● *noun* (esp. orderly) pile or heap; haystack; *colloquial* large quantity; number of chimneys standing together; smokestack; tall factory chimney; stacked group of aircraft; part of library where books are compactly stored. ● *verb* pile in stack(s); arrange (cards, circumstances, etc.) secretly for cheating; cause (aircraft) to fly in circles while waiting to land.

stadium /'steɪdɪəm/ *noun* (*plural* **-s**) athletic or sports ground with tiered seats for spectators.

staff /stɑ:f/ ● *noun* stick or pole for walking, as weapon, or as symbol of office; supporting person or thing; people employed in a business etc.; those in authority in a school etc.; group of army officers assisting officer in high command; (*plural* **-s** or **staves**) *Music* set of usually 5 parallel lines to indicate pitch of notes by position. ● *verb* provide (institution etc.) with staff. □ **staff nurse** one ranking just below a sister.

stag *noun* male deer; person who applies for new shares to sell at once for profit. □ **stag beetle** beetle with antler-like mandibles; **stag-party** *colloquial* party for men only.

stage ● *noun* point or period in process or development; raised platform, esp. for performing plays etc. on; (**the stage**) theatrical profession; scene of action; regular stopping place on route; distance between stopping places; section of space rocket with separate engine. ● *verb* (**-ging**) put (play etc.) on stage; organize and carry out. □ **stagecoach** *historical* coach running on regular route; **stage direction** instruction in a play about actors' movements, sound effects, etc.; **stage door** entrance from street to backstage part of theatre; **stage fright** performer's fear of audience; **stage-manage** arrange and control as or like stage manager; **stage manager** person responsible for lighting and mechanical arrangements on stage; **stage-struck** obsessed with becoming actor; **stage whisper** loud whisper meant to be overheard.

stagger /'stægə/ ● *verb* (cause to) walk unsteadily; shock, confuse; arrange (events etc.) so that they do not coincide; arrange (objects) so that they are not in line. ● *noun* staggering movement; (in *plural*) disease, esp. of horses and cattle, causing staggering.

staggering *adjective* astonishing, bewildering. □ **staggeringly** *adverb*.

staging /'steɪdʒɪŋ/ *noun* presentation of play etc.; (temporary) platform; shelves for plants in greenhouse. □ **staging post** regular stopping place, esp. on air route.

stagnant /'stægnənt/ *adjective* (of liquid) motionless, without current; dull, sluggish. □ **stagnancy** *noun*.

stagnate /stæg'neɪt/ *verb* (**-ting**) be or become stagnant. □ **stagnation** *noun*.

stagy /'steɪdʒɪ/ *adjective* (also **stagey**) (**-ier**, **-iest**) theatrical, artificial, exaggerated.

staid *adjective* sober, steady, sedate.

stain ● *verb* discolour or be discoloured by action of liquid sinking in; spoil, damage; colour (wood, etc.) with penetrating substance; treat with colouring agent. ● *noun* discoloration, spot, mark; blot, blemish; dye etc. for staining. □ **stained glass** coloured glass in leaded window etc.

stainless *adjective* without stains; not liable to stain. □ **stainless steel** chrome steel resisting rust and tarnish.

stair *noun* each of a set of fixed indoor steps; (usually in *plural*) such a set. □ **staircase** flight of stairs and supporting structure; **stair-rod** rod securing carpet between two steps; **stairway** staircase; **stairwell** shaft for staircase.

stake • *noun* stout pointed stick driven into ground as support, boundary mark, etc.; *historical* post to which person was tied to be burnt alive; sum of money etc. wagered on event; (often + *in*) interest or concern, esp. financial; (in *plural*) prize-money, esp. in horse race, such race. • *verb* (**-king**) secure or support with stake(s); (+ *off*, *out*) mark off (area) with stakes; wager; *US colloquial* support, esp. financially. □ **at stake** risked, to be won or lost; **stake out** *colloquial* place under surveillance; **stake-out** *esp. US colloquial* period of surveillance.

stalactite /'stæləktaɪt/ *noun* icicle-like deposit of calcium carbonate hanging from roof of cave etc.

stalagmite /'stæləgmaɪt/ *noun* icicle-like deposit of calcium carbonate rising from floor of cave etc.

stale • *adjective* not fresh; musty, insipid, or otherwise the worse for age or use; trite, unoriginal; (of athlete or performer) impaired by excessive training. • *verb* (**-ling**) make or become stale. □ **staleness** *noun*.

stalemate • *noun Chess* position counting as draw in which player cannot move except into check; deadlock. • *verb* (**-ting**) *Chess* bring (player) to stalemate; bring to deadlock.

Stalinism /'stɑ:lɪnɪz(ə)m/ *noun* centralized authoritarian form of socialism associated with Stalin. □ **Stalinist** *noun & adjective*.

stalk[1] /stɔ:k/ • *noun* main stem of herbaceous plant; slender attachment or support of leaf, flower, fruit, etc.; similar support for organ etc. in animal.

stalk[2] /stɔ:k/ • *verb* pursue (game, enemy) stealthily; stride, walk in a haughty way; *formal or rhetorical* move silently or threateningly through (place). • *noun* stalking of game; haughty gait. □ **stalking-horse** horse concealing hunter, pretext concealing real intentions or actions.

stall[1] /stɔ:l/ • *noun* trader's booth or table in market etc.; compartment for one animal in stable or cowhouse; fixed, usually partly enclosed, seat in choir or chancel of church; (usually in *plural*) each of seats on ground floor of theatre; stalling of engine or aircraft, condition resulting from this. • *verb* (of vehicle or its engine) stop because of overload on engine or inadequate supply of fuel to it; (of aircraft or its pilot) lose control because speed is too low; cause to stall. □ **stallholder** person in charge of stall in market etc.

stall[2] /stɔ:l/ *verb* play for time when being questioned etc.; delay, obstruct.

stallion /'stæljən/ *noun* uncastrated adult male horse.

stalwart /'stɔ:lwət/ • *adjective* strong, sturdy; courageous, resolute, reliable. • *noun* stalwart person, esp. loyal comrade.

stamen /'steɪmən/ *noun* organ producing pollen in flower.

stamina /'stæmɪnə/ *noun* physical or mental endurance.

stammer /'stæmə/ • *verb* speak haltingly, esp. with pauses or rapid repetitions of same syllable; (often + *out*) utter (words) in this way. • *noun* tendency to stammer; instance of stammering.

stamp • *verb* bring down (one's foot) heavily, esp. on ground, (often + *on*) crush or flatten in this way, walk heavily; impress (design, mark, etc.) on surface, impress (surface) with pattern etc.; affix postage or other stamp to; assign specific character to, mark out. • *noun* instrument for stamping; mark or design made by this; impression of official mark required to be made on deeds, bills of exchange, etc., as evidence of payment of tax; small adhesive piece of paper as evidence of payment, esp. postage stamp; mark, label, etc. on commodity as evidence of quality etc.; act or sound of stamping foot; characteristic mark, quality. □ **stamp duty** duty imposed on certain kinds of legal document; **stamping ground** *colloquial* favourite haunt; **stamp on** impress (idea etc.) on (memory etc.); **stamp out** produce by cutting out with die etc., put end to.

stampede /stæm'pi:d/ • *noun* sudden flight or hurried movement of animals or people; response of many people at once to a common impulse. • *verb* (**-ding**) (cause to) take part in stampede.

stance /stɑ:ns/ *noun* standpoint, attitude; position of body, esp. when hitting ball etc.

stanch /stɑ:ntʃ, stɔ:ntʃ/ *verb* (also **staunch** /stɔ:ntʃ/) restrain flow of (esp. blood); restrain flow from (esp. wound).

stanchion /'stɑ:nʃ(ə)n/ *noun* upright post or support.

stand ● *verb* (*past & past participle* **stood** /stʊd/) have, take, or maintain upright or stationary position, esp on feet or base; be situated; be of specified height; be in specified state; set in upright or specified position; move to and remain in specified position, take specified attitude; remain valid or unaltered; *Nautical* hold specified course; endure, tolerate; provide at one's own expense; (often + *for*) be candidate (for office etc.); act in specified capacity; undergo (trial). ● *noun* cessation from progress, stoppage; *Military* (esp. in **make a stand**) halt made to repel attack; resistance to attack or compulsion; position taken up, attitude adopted; rack, set of shelves, etc. for storage; open-fronted stall or structure for trader, exhibitor, etc.; standing-place for vehicles; raised structure to sit or stand on; *US* witness-box; each halt made for performance on tour; group of growing plants. □ **as it stands** in its present condition, in the present circumstances; **stand by** stand nearby, look on without interfering, uphold, support (person), adhere to (promise etc.), be ready for action; **stand-by** (person or thing) ready if needed in emergency etc., readiness for duty etc.; **stand down** withdraw from position or candidacy; **stand for** represent, signify, imply, *colloquial* endure, tolerate; **stand in** (usually + *for*) deputize for; **stand-in** deputy, substitute; **stand off** move or keep away, temporarily dismiss (employee); **stand-off half** *Rugby* half-back forming link between scrum-half and three-quarters; **standoffish** cold or distant in manner; **stand on** insist on, observe scrupulously; **stand out** be prominent or outstanding, (usually + *against*, *for*), persist in opposition or support; **standpipe** vertical pipe rising from water supply, esp. one connecting temporary tap to mains; **standpoint** point of view; **standstill** stoppage, inability to proceed; **stand to** *Military* stand ready for attack, abide by, be likely or certain to; **stand to reason** be obvious; **stand up** rise to one's feet, come to, remain in, or place in standing position, (of argument etc.) be valid, *colloquial* fail to keep appointment with; **stand-up** (of meal) eaten standing, (of fight) violent, thorough, (of collar) not turned down, (of comedian) telling jokes to audience; **stand up for** support, side with; **stand up to** face (opponent) courageously, be resistant to (wear, use, etc.).

standard /'stændəd/ ● *noun* object, quality, or measure serving as basis, example, or principle to which others conform or should conform or by which others are judged; level of excellence etc. required or specified; ordinary procedure etc.; distinctive flag; upright support or pipe; shrub standing without support, or grafted on upright stem and trained in tree form. ● *adjective* serving or used as standard; of normal or prescribed quality, type, or size. □ **standard-bearer** person who carries distinctive flag, prominent leader in cause; **standard English** most widely accepted dialect of English (see panel); **standard lamp** lamp on tall upright with base; **standard of living** degree of material comfort of person or group; **standard time** uniform time established by law or custom in country or region.

Standard English

Standard English is the dialect of English used by most educated English speakers and is spoken with a variety of accents (see panel at ACCENT). While not *in itself* any better than any other dialect, standard English is the form of English used in all formal written contexts.

standardize *verb* (also **-ise**) (**-zing** or **-sing**) cause to conform to standard. □ **standardization** *noun*.

standing ● *noun* esteem, repute, esp. high; duration. ● *adjective* that stands, upright; established, permanent; (of jump, start, etc.) performed with no run-up. □ **standing order** instruction to banker to make regular payments; **standing orders** rules governing procedure in a parliament, council, etc.; **standing ovation** prolonged applause from audience that has risen to its feet; **standing room** space to stand in.

stank *past* of STINK.

stanza /'stænzə/ *noun* group of lines forming division of poem, etc.

staphylococcus /stæfɪlə'kɒkəs/ *noun* (*plural* **-cocci** /-kaɪ/) bacterium sometimes forming pus. □ **staphylococcal** *adjective*.

staple[1] /'steɪp(ə)l/ ● *noun* shaped piece of wire with two points for fastening papers together, fixing netting to post, etc. ● *verb* (**-ling**) fasten with staple(s). □ **stapler** *noun*.

staple[2] /'steɪp(ə)l/ ● *noun* principal or important article of commerce; chief element, main component; fibre of cotton, wool, etc. with regard to its quality. ● *adjective* main, principal; important as product or export.

star ● *noun* celestial body appearing as luminous point in night sky; large luminous gaseous body such as sun; celestial body regarded as influencing fortunes etc.; conventional image of star with radiating lines or points; famous or brilliant person, leading performer. ● *adjective* outstanding. ● *verb* (**-rr-**) appear or present as leading performer(s); mark, set, or adorn with star(s). □ **starfish** (*plural* same or **-es**) sea creature with 5 or more radiating arms; **star-gazer** *colloquial usually derogatory or jocular* astronomer or astrologer; **starlight** light of stars; **starlit** lit by stars, with stars visible; **Stars and Stripes** US national flag; **star turn** main item in entertainment etc. □ **stardom** *noun*.

starboard /'stɑ:bəd/ ● *noun* right-hand side of ship or aircraft looking forward. ● *verb* turn (helm) to starboard.

starch ● *noun* white carbohydrate obtained chiefly from cereals and potatoes; preparation of this for stiffening fabric; stiffness of manner, formality. ● *verb* stiffen (clothing) with starch. □ **starchy** *adjective* (**-ier**, **-iest**).

stare /steə/ ● *verb* (**-ring**) (usually + *at*) look fixedly, esp. in curiosity, surprise, horror, etc. ● *noun* staring gaze. □ **stare (person) in the face** be evident or imminent; **stare out** stare at (person) until he or she looks away.

stark ● *adjective* sharply evident; desolate, bare; absolute. ● *adverb* completely, wholly. □ **starkly** *adverb*.

starkers /'stɑ:kəz/ *adjective slang* stark naked.

starlet /'stɑ:lɪt/ *noun* promising young performer, esp. film actress.

starling /'stɑ:lɪŋ/ *noun* gregarious bird with blackish speckled lustrous plumage.

starry *adjective* (**-ier**, **-iest**) full of stars; starlike. □ **starry-eyed** *colloquial* romantic but impractical, euphoric.

start ● *verb* begin; set in motion or action; set oneself in motion or action; (often + *out*) begin journey etc.; (often + *up*) (cause to) begin operating; (often + *up*) establish; give signal to (competitors) to start in race; (often + *up*, *from*, etc.) jump in surprise, pain, etc.; rouse (game etc.). ● *noun* beginning; starting-place of race etc.; advantage given at beginning of race etc.; advantageous initial position in life, business, etc.; sudden movement of surprise, pain, etc. □ **starting block** shaped block against which runner braces feet at start of race; **starting price** odds ruling at start of horse race.

starter *noun* device for starting vehicle engine etc.; first course of meal; person giving signal to start race; horse or competitor starting in race. □ **for starters** *colloquial* to start with.

startle /'stɑ:t(ə)l/ *verb* (**-ling**) shock, surprise.

starve *verb* (**-ving**) (cause to) die of hunger or suffer from malnourishment; *colloquial* feel very hungry; suffer from mental or spiritual want; (+ *of*) deprive of; compel by starvation. □ **starvation** *noun*.

stash *colloquial* • *verb* (often + *away*) conceal, put in safe place; hoard. • *noun* hiding place; thing hidden.

state • *noun* existing condition or position of person or thing; *colloquial* excited or agitated mental condition, untidy condition; political community under one government, this as part of federal republic; civil government; pomp; (**the States**) USA. • *adjective* of, for, or concerned with state; reserved for or done on ceremonial occasions. • *verb* (**-ting**) express in speech or writing; fix, specify. □ **lie in state** be laid in public place of honour before burial; **state of the art** current stage of esp. technological development; **state-of-the-art** absolutely up-to-date; **stateroom** state apartment, large private cabin in passenger ship.

stateless *adjective* having no nationality or citizenship.

stately *adjective* (**-ier**, **-iest**) dignified, imposing. □ **stately home** large historic house, esp. one open to public. □ **stateliness** *noun.*

statement *noun* stating, being stated; thing stated; formal account of facts; record of transactions in bank account etc.; notification of amount due to tradesman etc.

statesman /'steɪtsmən/ *noun* (*feminine* **stateswoman**) distinguished and capable politician or diplomat. □ **statesmanlike** *adjective*; **statesmanship** *noun.*

static /'stætɪk/ • *adjective* stationary, not acting or changing; *Physics* concerned with bodies at rest or forces in equilibrium. • *noun* static electricity; atmospherics. □ **static electricity** electricity not flowing as current.

statics *plural noun* (usually treated as *singular*) science of bodies at rest or forces in equilibrium.

station /'steɪʃ(ə)n/ • *noun* regular stopping place on railway line; person or thing's allotted place, building, etc.; centre for particular service or activity; establishment involved in broadcasting; military or naval base, inhabitants of this; position in life, rank, status; *Australian & NZ* large sheep or cattle farm. • *verb* assign station to; put in position. □ **stationmaster** official in charge of railway station; **stations of the cross** *RC Church* series of images representing events in Christ's Passion; **station wagon** *esp. US* estate car.

stationary *adjective* not moving; not meant to be moved; unchanging.

stationer *noun* dealer in stationery.

stationery *noun* writing materials, office supplies, etc.

statistic /stə'tɪstɪk/ *noun* statistical fact or item.

statistical *adjective* of statistics. □ **statistically** *adverb.*

statistics *plural noun* (usually treated as *singular*) science of collecting and analysing significant numerical data; such data. □ **statistician** /stætɪs'tɪʃ(ə)n/ *noun.*

statuary /'stætʃʊərɪ/ • *adjective* of or for statues. • *noun* statues collectively; making statues.

statue /'stætʃu:/ *noun* sculptured figure of person or animal, esp. life-size or larger.

statuesque /stætʃʊ'esk/ *adjective* like statue, esp. in beauty or dignity.

statuette /stætʃʊ'et/ *noun* small statue.

stature /'stætʃə/ *noun* height of (esp. human) body; calibre (esp. moral), eminence.

status /'steɪtəs/ *noun* rank, social position, relative importance; superior social etc. position. □ **status quo** /'kwəʊ/ existing conditions; **status symbol** possession etc. intended to indicate owner's superiority.

statute /'stætʃu:t/ *noun* written law passed by legislative body; rule of corporation, founder, etc., intended to be permanent.

statutory /'stætʃʊtərɪ/ *adjective* required or enacted by statute.

staunch[1] /stɔ:ntʃ/ *adjective* trustworthy, loyal. □ **staunchly** *adverb.*

staunch[2] = STANCH.

stave • *noun* each of curved slats forming sides of cask; *Music* staff; stanza, verse. • *verb* (**-ving**; *past & past participle* **stove** /stəʊv/ or **staved**) (usually + *in*) break hole in, damage, crush by forcing inwards. □ **stave off** (*past & past participle* **staved**) avert or defer (danger etc.).

stay[1] • *verb* continue in same place or condition, not depart or change; (often

+ *at, in, with*) reside temporarily; *archaic or literary* stop, check, (esp. in *imperative*) pause; postpone (judgement etc.); assuage (hunger etc.), esp. for short time. ● *noun* act or period of staying; suspension or postponement of sentence, judgement, etc.; prop, support; (in *plural*) *historical* (esp. boned) corset. □ **stay-at-home** (person) rarely going out; **staying power** endurance; **stay the night** remain overnight; **stay put** *colloquial* remain where it is put or where one is; **stay up** not go to bed (until late).

stay² *noun* rope supporting mast, flagstaff, etc.; supporting cable on aircraft. □ **staysail** sail extended on stay.

stayer *noun* person or animal with great endurance.

STD *abbreviation* subscriber trunk dialling.

stead /sted/ *noun* □ **in (person's, thing's) stead** as substitute; **stand (person) in good stead** be advantageous or useful to him or her.

steadfast /'stedfɑ:st/ *adjective* constant, firm, unwavering. □ **steadfastly** *adverb*; **steadfastness** *noun*.

steady /'stedɪ/ ● *adjective* (**-ier**, **-iest**) firmly fixed or supported, unwavering; uniform, regular; constant, persistent; (of person) serious and dependable; regular, established. ● *verb* (**-ies**, **-ied**) make or become steady. ● *adverb* steadily. ● *noun* (*plural* **-ies**) *colloquial* regular boyfriend or girlfriend. □ **steady state** unvarying condition, esp. in physical process. □ **steadily** *adverb*; **steadiness** *noun*.

steak /steɪk/ *noun* thick slice of meat (esp. beef) or fish, usually grilled or fried. □ **steakhouse** restaurant specializing in beefsteaks.

steal ● *verb* (*past* **stole**; *past participle* **stolen** /'stəʊl(ə)n/) take (another's property) illegally or without right or permission, esp. in secret; obtain surreptitiously, insidiously, or artfully; (+ *in*, *out*, *away*, *up*, etc.) move, esp. silently or stealthily. ● *noun US colloquial* act of stealing, theft; *colloquial* easy task, bargain. □ **steal a march on** gain advantage over by surreptitious means; **steal the show** outshine other performers, esp. unexpectedly.

stealth /stelθ/ *noun* secrecy, secret behaviour.

stealthy *adjective* (**-ier**, **-iest**) done or moving with stealth. □ **stealthily** *adverb*.

steam ● *noun* gas into which water is changed by boiling; condensed vapour formed from this; power obtained from steam; *colloquial* power, energy. ● *verb* cook (food) in steam; give off steam; move under steam power; (+ *ahead*, *away*, etc.) *colloquial* proceed or travel fast or with vigour. □ **let off steam** relieve pent-up energy or feelings; **steamboat** steam-driven boat; **steam engine** one worked or propelled by steam; **steam iron** electric iron that emits steam; **steamroller** *noun* heavy slow-moving vehicle with roller, used to flatten new-made roads, crushing power or force, *verb* crush or move forcibly or indiscriminately; **steamship** steam-driven ship; **steam train** train pulled by steam engine; **steam up** cover or become covered with condensed steam, (as **steamed up** *adjective*) *colloquial* angry, excited.

steamer *noun* steamboat; vessel for steaming food in.

steamy *adjective* (**-ier**, **-iest**) like or full of steam; *colloquial* erotic.

steatite /'stɪətaɪt/ *noun* impure form of talc, esp. soapstone.

steed *noun archaic or poetical* horse.

steel ● *noun* strong malleable low-carbon iron alloy, used esp. for making tools, weapons, etc.; strength, firmness; steel rod for sharpening knives. ● *adjective* of or like steel. ● *verb* harden, make resolute. □ **steel band** band playing chiefly calypso-style music on instruments made from oil drums; **steel wool** fine steel shavings used as abrasive; **steelworks** factory producing steel; **steelyard** balance with graduated arm along which weight is moved.

steely *adjective* (**-ier**, **-iest**) of or like steel; severe, resolute.

steep¹ ● *adjective* sloping sharply; (of rise or fall) rapid; *colloquial* exorbitant, unreasonable, exaggerated, incredible. ● *noun* steep slope, precipice. □ **steepen** *verb*; **steeply** *adverb*; **steepness** *noun*.

steep² ● *verb* soak or bathe in liquid. ● *noun* act of steeping; liquid for steeping. □ **steep in** imbue with, make deeply acquainted with (subject etc.).

steeple /'sti:p(ə)l/ *noun* tall tower, esp. with spire, above roof of church. □ **steeplechase** horse race with ditches, hedges, etc. to jump, cross-country foot race; **steeplejack** repairer of tall chimneys, steeples, etc.

steer¹ *verb* guide (vehicle, ship, etc.) with wheel, rudder, etc.; direct or guide (one's course, other people, conversation, etc.) in specified direction. □ **steer clear of** avoid; **steering column** column on which steering wheel is mounted; **steering committee** one deciding order of business, course of operations etc.; **steering wheel** wheel by which vehicle etc. is steered; **steersman** person who steers ship.

steer² *noun* bullock.

steerage *noun* act of steering; *archaic* cheapest part of ship's accommodation.

steering *noun* apparatus for steering vehicle etc.

stegosaurus /stegə'sɔ:rəs/ *noun* (*plural* **-ruses**) large dinosaur with two rows of vertical plates along back.

stela /'sti:lə/ *noun* (*plural* **stelae** /-li:/) (also **stele** /'sti:l/) ancient upright slab or pillar, usually inscribed and sculptured, esp. as gravestone.

stellar /'stelə/ *adjective* of star or stars.

stem¹ ● *noun* main body or stalk of plant; stalk of fruit, flower, or leaf; stem-shaped part, e.g. slender part of wineglass; *Grammar* root or main part of noun, verb, etc. to which inflections are added; main upright timber at bow of ship. ● *verb* (**-mm-**) (+ *from*) spring or originate from.

stem² *verb* (**-mm-**) check, stop.

stench *noun* foul smell.

stencil /'stensɪl/ ● *noun* thin sheet in which pattern is cut, placed on surface and printed, inked over, etc.; pattern so produced. ● *verb* (**-ll-**; *US* **-l-**) (often + *on*) produce (pattern) with stencil; mark (surface) in this way.

Sten gun *noun* lightweight sub-machine-gun.

stenographer /ste'nɒgrəfə/ *noun esp. US* shorthand typist.

stentorian /sten'tɔ:rɪən/ *adjective* loud and powerful.

step ● *noun* complete movement of leg in walking or running; distance so covered; unit of movement in dancing; measure taken, esp. one of several in course of action; surface of stair, stepladder, etc. tread; short distance; sound or mark made by foot in walking etc; degree in scale of promotion, precedence, etc.; stepping in unison or to music; state of conforming. ● *verb* (**-pp-**) lift and set down foot or alternate feet in walking; come or go in specified direction by stepping; make progress in specified way; (+ *off*, *out*) measure (distance) by stepping; perform (dance). □ **mind, watch one's step** be careful; **step down** resign; **step in** enter, intervene; **stepladder** short folding ladder not leant against wall; **step on it** *colloquial* hurry; **step out** be active socially, take large steps; **stepping-stone** large stone set in stream etc. to walk over, means of progress; **step up** increase, intensify.

step- *combining form* related by remarriage of parent. □ **stepchild, stepdaughter, stepson** one's husband's or wife's child by previous partner; **stepfather, stepmother, step-parent** mother's or father's spouse who is not one's own parent; **stepbrother, stepsister** child of one's step-parent by previous partner.

stephanotis /stefə'nəʊtɪs/ *noun* fragrant tropical climbing plant.

steppe /step/ *noun* level grassy treeless plain.

stereo /'sterɪəʊ/ ● *noun* (*plural* **-s**) stereophonic sound reproduction or equipment; stereoscope. ● *adjective* stereophonic; stereoscopic.

stereo- *combining form* solid; 3-dimensional.

stereophonic /sterɪəʊ'fɒnɪk/ *adjective* using two or more channels, to give effect of naturally distributed sound.

stereoscope /'sterɪəskəʊp/ *noun* device for producing 3-dimensional effect by viewing two slightly different photographs together. □ **stereoscopic** /-'skɒp-/ *adjective.*

stereotype /'sterɪəʊtaɪp/ ● *noun* person or thing seeming to conform to widely accepted type; such type, idea, or attitude; printing plate cast from mould of composed type. ● *verb* (**-ping**) (esp. as **stereotyped** *adjective*) cause to conform to type, standardize; print from stereotype; make stereotype of.

sterile /'steraɪl/ *adjective* not able to produce crop, fruit, or young, barren; lacking ideas or originality, unproductive; free from micro-organisms etc. □ **sterility** /stə'rɪl-/ *noun*.

sterilize /'sterɪlaɪz/ *verb* (also **-ise**) (**-zing** or **-sing**) make sterile; deprive of reproductive power. □ **sterilization** *noun*.

sterling /'stɜːlɪŋ/ ● *adjective* of or in British money; (of coin or precious metal) genuine, of standard value or purity; (of person etc.) genuine, reliable. ● *noun* British money. □ **sterling silver** silver of 92½% purity.

stern[1] *adjective* severe, grim; authoritarian. □ **sternly** *adverb*; **sternness** *noun*.

stern[2] *noun* rear part, esp. of ship or boat.

sternum /'stɜːnəm/ *noun* (*plural* **-na** or **-nums**) breastbone.

steroid /'stɪərɔɪd/ *noun* any of group of organic compounds including many hormones, alkaloids, and vitamins.

sterol /'sterɒl/ *noun* naturally occurring steroid alcohol.

stertorous /'stɜːtərəs/ *adjective* (of breathing etc.) laboured and noisy.

stet *verb* (**-tt-**) (usually written on proofsheet etc.) ignore or cancel (alteration), let original stand.

stethoscope /'steθəskəʊp/ *noun* instrument used in listening to heart, lungs, etc.

stetson /'stets(ə)n/ *noun* slouch hat with wide brim and high crown.

stevedore /'stiːvədɔː/ *noun* person employed in loading and unloading ships.

stew ● *verb* cook by long simmering in closed vessel; *colloquial* swelter; (of tea etc.) become bitter or strong from infusing too long. ● *noun* dish of stewed meat etc.; *colloquial* agitated or angry state.

steward /'stjuːəd/ ● *noun* passengers' attendant on ship, aircraft, or train; official managing meeting, show, etc.; person responsible for supplies of food etc. for college, club, etc.; property manager. □ **stewardship** *noun*.

stewardess /stjuːə'des/ *noun* female steward, esp. on ship or aircraft.

stick[1] *noun* short slender length of wood, esp. for use as support or weapon; thin rod of wood etc. for particular purpose; gear lever, joystick; sticklike piece of celery, dynamite, etc.; (often **the stick**) punishment, esp. by beating; *colloquial* adverse criticism; *colloquial* person, esp. when dull or unsociable. □ **stick insect** insect with twiglike body.

stick[2] *verb* (*past & past participle* **stuck**) (+ *in*, *into*, *through*) thrust, insert (thing or its point); stab; (+ *in*, *into*, *on*, etc.) fix or be fixed (as) by pointed end; fix or be fixed (as) by adhesive etc.; lose or be deprived of movement or action through adhesion, jamming, etc.; *colloquial* put in specified position or place, remain; *colloquial* endure, tolerate; (+ *at*) *colloquial* persevere with. □ **get stuck into** *slang* start in earnest; **stick around** *colloquial* linger; **sticking plaster** adhesive plaster for wounds etc.; **stick-in-the-mud** *colloquial* unprogressive or old-fashioned person; **stick it out** *colloquial* endure to the end; **stick out** (cause to) protrude; **stick out for** persist in demanding; **stick up** be or make erect or protruding upwards, fasten to upright surface, *colloquial* rob or threaten with gun; **stick up for** support, defend; **stuck for** at a loss for, needing; **stuck with** *colloquial* unable to get rid of.

sticker *noun* adhesive label.

stickleback /'stɪk(ə)lbæk/ *noun* small spiny-backed fish.

stickler /'stɪklə/ *noun* (+ *for*) person who insists on something.

sticky *adjective* (**-ier**, **-iest**) tending or intended to stick or adhere; glutinous, viscous; (of weather) humid; *colloquial* difficult, awkward, unpleasant, painful. □ **sticky wicket** *colloquial* difficult situation. □ **stickiness** *noun*.

stiff ● *adjective* rigid, inflexible; hard to bend, move, turn, etc.; hard to cope with, needing strength or effort; severe, strong; formal, constrained; (of muscle, person, etc.) aching from exertion, injury, etc.; (of esp. alcoholic drink) strong. ● *adverb colloquial* utterly, extremely. ● *noun slang* corpse.

□ **stiff-necked** obstinate, haughty; **stiff upper lip** appearance of firmness or fortitude. □ **stiffen** *verb*; **stiffly** *adverb*; **stiffness** *noun*.

stifle /'staɪf(ə)l/ *verb* (**-ling**) suppress; feel or make unable to breathe easily; suffocate. □ **stifling** *adjective*.

stigma /'stɪgmə/ *noun* (*plural* **-s** or **stigmata** /-mətə or -'mɑːtə/) shame, disgrace; part of pistil that receives pollen in pollination; (**stigmata**) marks like those on Christ's body after the Crucifixion, appearing on bodies of certain saints etc.

stigmatize /'stɪgmətaɪz/ *verb* (also **-ise**) (**-zing** or **-sing**) (often + *as*) brand as unworthy or disgraceful.

stile *noun* set of steps etc. allowing people to climb over fence, wall, etc.

stiletto /stɪ'letəʊ/ *noun* (*plural* **-s**) short dagger; (in full **stiletto heel**) long tapering heel of shoe; pointed implement for making eyelets etc.

still[1] ● *adjective* with little or no movement or sound; calm, tranquil; (of drink) not effervescing. ● *noun* deep silence; static photograph, esp. single shot from cinema film. ● *adverb* without moving; even now, at particular time; nevertheless; (+ *comparative*) even, yet, increasingly. ● *verb* make or become still, quieten. □ **stillbirth** birth of dead child; **stillborn** born dead, abortive; **still life** painting or drawing of inanimate objects. □ **stillness** *noun*.

still[2] *noun* apparatus for distilling spirits etc.

stilt *noun* either of pair of poles with foot supports for walking at a distance above ground; each of set of piles or posts supporting building etc.

stilted *adjective* (of literary style etc.) stiff and unnatural.

Stilton /'stɪlt(ə)n/ *noun proprietary term* strong rich esp. blue-veined cheese.

stimulant /'stɪmjʊlənt/ ● *adjective* stimulating esp. bodily or mental activity. ● *noun* stimulant substance or influence.

stimulate /'stɪmjʊleɪt/ *verb* (**-ting**) act as stimulus to; animate, excite, rouse. □ **stimulation** *noun*; **stimulative** /-lətɪv/ *adjective*; **stimulator** *noun*.

stimulus /'stɪmjʊləs/ *noun* (*plural* **-li** /-laɪ/) thing that rouses to activity.

sting ● *noun* sharp wounding organ of insect, nettle, etc.; inflicting of wound with this; wound itself, pain caused by it; painful quality or effect; pungency, vigour. ● *verb* (*past & past participle* **stung**) wound with sting; be able to sting; feel or cause tingling physical pain or sharp mental pain; (+ *into*) incite, esp. painfully; *slang* swindle, charge heavily. □ **stinging-nettle** nettle with stinging hairs; **stingray** broad flatfish with stinging tail.

stingy /'stɪndʒɪ/ *adjective* (**-ier**, **-iest**) niggardly, mean. □ **stingily** *adverb*; **stinginess** *noun*.

stink ● *verb* (*past* **stank** or **stunk**; *past participle* **stunk**) emit strong offensive smell; (often + *out*) fill (place) with stink; (+ *out* etc.) drive (person) out etc. by stink; *colloquial* be or seem very unpleasant. ● *noun* strong offensive smell; *colloquial* loud complaint, fuss. □ **stink bomb** device emitting stink when opened.

stinker *noun slang* particularly annoying or unpleasant person; very difficult problem etc.

stinking ● *adjective* that stinks; *slang* very objectionable. ● *adverb slang* extremely and usually objectionably.

stint ● *verb* (often + *on*) supply (food, aid, etc.) meanly or grudgingly; (often **stint oneself**) supply (person etc.) in this way. ● *noun* allotted amount or period of work.

stipend /'staɪpend/ *noun* salary, esp. of clergyman.

stipendiary /staɪ'pendjərɪ/ ● *adjective* receiving stipend. ● *noun* (*plural* **-ies**) person receiving stipend. □ **stipendiary magistrate** paid professional magistrate.

stipple /'stɪp(ə)l/ ● *verb* (**-ling**) draw, paint, engrave, etc. with dots instead of lines; roughen surface of (paint, cement, etc.). ● *noun* stippling; effect of stippling.

stipulate /'stɪpjʊleɪt/ *verb* (**-ting**) demand or specify as part of bargain etc. □ **stipulation** *noun*.

stir ● *verb* (**-rr-**) move spoon etc. round and round in (liquid etc.), esp. to mix ingredients; cause to move, esp. slightly; be or begin to be in motion;

rise from sleep; arouse, inspire, excite; *colloquial* cause trouble by gossiping etc. ● *noun* act of stirring; commotion, excitement. □ **stir-fry** *verb* fry rapidly while stirring, *noun* stir-fried dish; **stir up** mix thoroughly by stirring, stimulate, incite.

stirrup /'stɪrəp/ *noun* support for horse-rider's foot, suspended by strap from saddle. □ **stirrup-cup** cup of wine etc. offered to departing traveller, originally rider; **stirrup-pump** hand-operated water-pump with footrest, used to extinguish small fires.

stitch ● *noun* single pass of needle, or result of this, in sewing, knitting, or crochet; particular method of sewing etc.; least bit of clothing; sharp pain in side induced by running etc. ● *verb* sew, make stitches (in). □ **in stitches** *colloquial* laughing uncontrollably; **stitch up** join or mend by sewing.

stoat *noun* mammal of weasel family with brown fur turning mainly white in winter.

stock ● *noun* store of goods etc. ready for sale or distribution; supply or quantity of things for use; equipment or raw material for manufacture, trade, etc.; farm animals or equipment; capital of business; shares in this; reputation, popularity; money lent to government at fixed interest; line of ancestry; liquid made by stewing bones, vegetables, etc.; fragrant garden plant; plant into which graft is inserted; main trunk of tree etc.; (in *plural*) *historical* timber frame with holes for feet in which offenders were locked as public punishment; base, support, or handle for implement or machine; butt of rifle etc.; (in *plural*) supports for ship during building or repair; band of cloth worn round neck. ● *adjective* kept regularly in stock for sale or use; commonly used, hackneyed. ● *verb* have (goods) in stock; provide (shop, farm, etc.) with goods, livestock, etc. □ **stockbroker** member of Stock Exchange dealing in stocks and shares; **stock-car** specially strengthened car used in racing where deliberate bumping is allowed; **Stock Exchange** place for dealing in stocks and shares, dealers working there; **stock-in-trade** requisite(s) of trade or profession; **stock market** Stock Exchange, transactions on this; **stockpile** *noun* reserve supply of accumulated stock, *verb* accumulate stockpile (of); **stockpot** pot for making soup stock; **stock-still** motionless; **stocktaking** making inventory of stock; **stock up** (often + *with*) provide with or get stocks or supplies (of); **stockyard** enclosure for sorting or temporary keeping of cattle; **take stock** make inventory of one's stock, (often + *of*) review (situation etc.).

stockade /stɒ'keɪd/ ● *noun* line or enclosure of upright stakes. ● *verb* (**-ding**) fortify with this.

stockinet /stɒkɪ'net/ *noun* (also **stockinette**) elastic knitted fabric.

stocking /'stɒkɪŋ/ *noun* knitted covering for leg and foot, of nylon, wool, silk, etc. □ **stocking stitch** alternate rows of plain and purl.

stockist *noun* dealer in specified types of goods.

stocky *adjective* (**-ier**, **-iest**) short and strongly built. □ **stockily** *adverb*.

stodge *noun colloquial* heavy fattening food.

stodgy *adjective* (**-ier**, **-iest**) (of food) heavy, filling; dull, uninteresting. □ **stodginess** *noun*.

stoic /'stəʊɪk/ *noun* person having great self-control in adversity. □ **stoical** *adjective*; **stoicism** /-ɪsɪz(ə)m/ *noun*.

stoke *verb* (**-king**) (often + *up*) feed and tend (fire, furnace, etc.); *colloquial* fill oneself with food. □ **stokehold** compartment in steamship containing its boilers and furnace; **stokehole** space for stokers in front of furnace.

stoker *noun* person who tends furnace, esp. on steamship.

stole[1] *noun* woman's garment like long wide scarf worn over shoulders; strip of silk etc. worn similarly by priest.

stole[2] *past* of STEAL.

stolen *past participle* of STEAL.

stolid /'stɒlɪd/ *adjective* not easily excited or moved; impassive, unemotional. □ **stolidity** /-'lɪd-/ *noun*; **stolidly** *adverb*.

stomach /'stʌmək/ ● *noun* internal organ in which food is digested; lower front of body; (usually + *for*) appetite, inclination, etc. ● *verb* (usually in negative) endure. □ **stomach-pump** syringe for forcing liquid etc. into or out of stomach.

stomp ● *verb* tread or stamp heavily. ● *noun* lively jazz dance with heavy stamping.

stone ● *noun* solid non-metallic mineral matter, rock; small piece of this; hard case of kernel in some fruits; hard morbid concretion in body; (*plural* same) unit of weight (14 lb, 6.35 kg); precious stone. ● *verb* (**-ning**) pelt with stones; remove stones from (fruit). □ **Stone Age** prehistoric period when weapons and tools were made of stone; **stone-cold** completely cold; **stone-cold sober** completely sober; **stonecrop** succulent rock plant; **stone-dead** completely dead; **stone-deaf** completely deaf; **stone fruit** fruit with flesh enclosing stone; **stoneground** (of flour) ground with millstones; **stone's throw** short distance; **stonewall** obstruct with evasive answers etc., *Cricket* bat with excessive caution; **stoneware** impermeable and partly vitrified but opaque ceramic ware; **stonewashed** (esp. of denim) washed with abrasives to give worn or faded look; **stonework** masonry.

stoned *adjective slang* drunk, drugged.

stony *adjective* (**-ier**, **-iest**) full of stones; hard, rigid; unfeeling, unresponsive. □ **stony-broke** *slang* entirely without money. □ **stonily** *adverb*.

stood *past & past participle* of STAND.

stooge *colloquial* ● *noun* person acting as butt or foil, esp. for comedian; assistant or subordinate, esp. for unpleasant work. ● *verb* (**-ging**) (+ *for*) act as stooge for; (+ *about*, *around*, etc.) move about aimlessly.

stool /stu:l/ *noun* single seat without back or arms; footstool; (usually in *plural*) faeces. □ **stool-pigeon** person acting as decoy, police informer.

stoop ● *verb* bend down; stand or walk with shoulders habitually bent forward; (+ *to do*) condescend; (+ *to*) descend to (shameful act). ● *noun* stooping posture.

stop ● *verb* (**-pp-**) put an end to progress, motion, or operation of; effectively hinder or prevent; discontinue; come to an end; cease from motion, speaking, or action; defeat; *colloquial* remain; stay for short time; (often + *up*) block or close up (hole, leak, etc.); not permit or supply as usual; instruct bank to withhold payment on (cheque); fill (tooth); press (violin etc. string) to obtain required pitch. ● *noun* stopping, being stopped; place where bus, train, etc. regularly stops; full stop; device for stopping motion at particular point; change of pitch effected by stopping string; (in organ) row of pipes of one character, knob etc. operating these; (in camera etc.) diaphragm, effective diameter of lens, device for reducing this; plosive sound. □ **pull out all the stops** make extreme effort; **stopcock** externally operated valve regulating flow through pipe etc.; **stopgap** temporary substitute; **stop off**, **over** break one's journey; **stop press** late news inserted in newspaper after printing has begun; **stopwatch** watch that can be instantly started and stopped, used in timing of races etc.

stoppage *noun* interruption of work due to strike etc.; (in *plural*) sum deducted from pay, for tax, etc.; condition of being blocked or stopped.

stopper *noun* plug for closing bottle etc.

storage /'stɔ:rɪdʒ/ *noun* storing of goods etc.; method of, space for, or cost of storing. □ **storage battery**, **cell** one for storing electricity; **storage heater** electric heater releasing heat stored outside peak hours.

store ● *noun* quantity of something kept ready for use; (in *plural*) articles gathered for particular purpose, supply of, or place for keeping, these; department store; *esp. US* shop; (often in *plural*) shop selling basic necessities; warehouse for keeping furniture etc. temporarily; device in computer for keeping retrievable data. ● *verb* (**-ring**) (often + *up*, *away*) accumulate for future use; put (furniture etc.) in a store; stock or provide with something useful; keep (data) for retrieval. □ **in store** in reserve, to come, (+ *for*) awaiting; **storehouse** storage place; **storekeeper** person in charge of stores, *US* shopkeeper; **storeroom** storage room.

storey /'stɔ:rɪ/ *noun* (*plural* **-s**) rooms etc. on one level of building.

stork *noun* long-legged usually white wading bird.

storm ● *noun* violent disturbance of atmosphere with high winds and usually thunder, rain, or snow; violent disturbance in human affairs; (+ *of*) shower of missiles or blows, outbreak of applause, hisses, etc.; assault on fortified place. ● *verb* attack or capture by storm; rush violently; rage, be violent; bluster. □ **storm centre** comparatively calm centre of cyclonic storm, centre round which controversy etc. rages; **storm cloud** heavy rain-cloud; **storm troops** shock troops, *historical* Nazi political militia; **take by storm** capture by direct assault, quickly captivate.

stormy *adjective* (**-ier**, **-iest**) of or affected by storms; (of wind etc.) violent; full of angry feeling or outbursts. □ **stormily** *adverb.*

story /'stɔ:rɪ/ *noun* (*plural* **-ies**) account of real or imaginary events; tale, anecdote; history of person, institution, etc.; plot of novel, play, etc.; article in newspaper, material for this; *colloquial* fib. □ **storyteller** person who tells or writes stories, *colloquial* liar.

stoup /stu:p/ *noun* basin for holy water; *archaic* flagon, beaker.

stout /staʊt/ ● *adjective* rather fat, corpulent; thick, strong; brave, resolute, vigorous. ● *noun* strong dark beer. □ **stout-hearted** courageous. □ **stoutly** *adverb*; **stoutness** *noun.*

stove[1] /stəʊv/ *noun* closed apparatus burning fuel or using electricity etc. for heating or cooking. □ **stove-pipe** pipe carrying smoke and gases from stove to chimney.

stove[2] *past & past participle* of STAVE.

stow /stəʊ/ *verb* pack (goods, cargo, etc.) tidily and compactly. □ **stow away** place (thing) out of the way, hide oneself on ship etc. to travel free; **stowaway** person who stows away.

stowage *noun* stowing; place for this.

straddle /'stræd(ə)l/ *verb* (**-ling**) sit or stand across (thing) with legs wide apart; be situated on both sides of; spread legs wide apart.

strafe /strɑ:f/ *verb* (**-fing**) bombard; attack with gunfire.

straggle /'stræg(ə)l/ ● *verb* (**-ling**) lack compactness or tidiness; be dispersed or sporadic; trail behind in race etc. ● *noun* straggling group. □ **straggler** *noun*; **straggly** *adjective.*

straight /streɪt/ ● *adjective* not curved, bent, crooked or curly; successive, uninterrupted; ordered, level, tidy; honest, candid; (of thinking etc.) logical; (of theatre, music, etc.) not popular or comic; unmodified, (of a drink) undiluted; *colloquial* (of person etc.) conventional, respectable, heterosexual. ● *noun* straight part, esp. concluding stretch of racetrack; straight condition; *colloquial* conventional person, heterosexual. ● *adverb* in straight line, direct; in right direction; correctly. □ **go straight** (of criminal) become honest; **straight away** immediately; **straight face** intentionally expressionless face; **straight fight** *Politics* contest between two candidates only; **straightforward** honest, frank, (of task etc.) simple; **straight man** comedian's stooge; **straight off** *colloquial* without hesitation. □ **straightness** *noun.*

straighten *verb* (often + *out*) make or become straight; (+ *up*) stand erect after bending.

strain[1] ● *verb* stretch tightly; make or become taut or tense; injure by overuse or excessive demands; exercise (oneself, one's senses, thing, etc.) intensely, press to extremes; strive intensely; distort from true intention or meaning; clear (liquid) of solid matter by passing it through sieve etc. ● *noun* act of straining, force exerted in this; injury caused by straining muscle etc.; severe mental or physical demand or exertion; snatch of music or poetry; tone or tendency in speech or writing.

strain[2] *noun* breed or stock of animals, plants, etc.; characteristic tendency.

strained *adjective* constrained, artificial; (of relationship) distrustful, tense.

strainer *noun* device for straining liquids.

strait *noun* (in *singular* or *plural*) narrow channel connecting two large bodies of water; (usually in *plural*) difficulty, distress. □ **strait-jacket** strong garment with long sleeves for confining violent prisoner etc., restrictive measures; **strait-laced** puritanical.

straitened /ˈstreɪt(ə)nd/ *adjective* of or marked by poverty.

strand[1] • *verb* run aground; (as **stranded** *adjective*) in difficulties, esp. without money or transport. • *noun* foreshore, beach.

strand[2] *noun* each of twisted threads or wires making rope, cable, etc.; single thread or strip of fibre; lock of hair; element, component.

strange /streɪndʒ/ *adjective* unusual, peculiar, surprising, eccentric; (often + *to*) unfamiliar, foreign; (+ *to*) unaccustomed; not at ease. □ **strangely** *adverb*; **strangeness** *noun*.

stranger *noun* person new to particular place or company; (often + *to*) person one does not know.

strangle /ˈstræŋg(ə)l/ *verb* (**-ling**) squeeze windpipe or neck of, esp. so as to kill; hamper or suppress (movement, cry, etc.). □ **stranglehold** deadly grip, complete control. □ **strangler** *noun*.

strangulate /ˈstræŋgjʊleɪt/ *verb* (**-ting**) compress (vein, intestine, etc.), preventing circulation.

strangulation *noun* strangling, being strangled; strangulating.

strap • *noun* strip of leather etc., often with buckle, for holding things together etc.; narrow strip of fabric forming part of garment; loop for grasping to steady oneself in moving vehicle. • *verb* (**-pp-**) (often + *down*, *up*, etc.) secure or bind with strap; beat with strap. □ **straphanger** *slang* standing passenger in bus or train; **straphang** *verb* □ **strapless** *adjective*.

strapping *adjective* large and sturdy.

strata *plural* of STRATUM.

■ **Usage** It is a mistake to use the plural form *strata* when only one stratum is meant.

stratagem /ˈstrætədʒəm/ *noun* cunning plan or scheme; trickery.

strategic /strəˈtiːdʒɪk/ *adjective* of or promoting strategy; (of materials) essential in war; (of bombing or weapons) done or for use as longer-term military policy. □ **strategically** *adverb*.

strategy /ˈstrætɪdʒɪ/ *noun* (*plural* **-ies**) long-term plan or policy; art of war; art of moving troops, ships, aircraft, etc. into favourable positions. □ **strategist** *noun*.

stratify /ˈstrætɪfaɪ/ *verb* (**-ies**, **-ied**) (esp. as **stratified** *adjective*) arrange in strata, grades, etc. □ **stratification** *noun*.

stratosphere /ˈstrætəsfɪə/ *noun* layer of atmosphere above troposphere, extending to about 50 km from earth's surface.

stratum /ˈstrɑːtəm/ *noun* (*plural* **strata**) layer or set of layers of any deposited substance, esp. of rock; atmospheric layer; social class.

straw *noun* dry cut stalks of grain; single stalk of straw; thin tube for sucking drink through; insignificant thing; pale yellow colour. □ **clutch at straws** try any remedy in desperation; **straw vote**, **poll** unofficial ballot as test of opinion.

strawberry /ˈstrɔːbərɪ/ *noun* (*plural* **-ies**) pulpy red fruit having surface studded with yellow seeds; plant bearing this. □ **strawberry mark** reddish birthmark.

stray • *verb* wander from the right place, from one's companions, etc., go astray; deviate. • *noun* strayed animal or person. • *adjective* strayed, lost; isolated, occasional.

streak • *noun* long thin usually irregular line or band, esp. of colour; strain or trait in person's character. • *verb* mark with streaks; move very rapidly; *colloquial* run naked in public.

streaky *adjective* (**-ier**, **-iest**) marked with streaks; (of bacon) with streaks of fat.

stream • *noun* body of running water, esp. small river; current, flow; group of schoolchildren of similar ability taught together. • *verb* move as stream; run with liquid; be blown in wind; emit stream of (blood etc.); arrange (schoolchildren) in streams. □ **on stream** in operation or production.

streamer *noun* long narrow strip of ribbon or paper; long narrow flag.

streamline *verb* (**-ning**) give (vehicle etc.) form which presents least resistance to motion; make simple or more efficient.

street *noun* road in city, town, or village; this with buildings on each side.

□ **on the streets** living by prostitution; **streetcar** *US* tram; **street credibility**, **cred** *slang* acceptability within urban subculture; **streetwalker** prostitute seeking customers in street; **streetwise** knowing how to survive modern urban life.

strength *noun* being strong; degree or manner of this; person or thing giving strength; number of people present or available. □ **on the strength of** on basis of.

strengthen *verb* make or become stronger.

strenuous /'strenjʊəs/ *adjective* using or requiring great effort; energetic. □ **strenuously** *adverb.*

streptococcus /streptə'kɒkəs/ *noun* (*plural* **-cocci** /-kaɪ/) bacterium causing serious infections. □ **streptococcal** *adjective.*

streptomycin /streptəʊ'maɪsɪn/ *noun* antibiotic effective against many disease-producing bacteria.

stress ● *noun* pressure, tension; quantity measuring this; physical or mental strain; emphasis. ● *verb* emphasize; subject to stress.

stressful *adjective* causing stress.

stretch ● *verb* draw, be drawn, or be able to be drawn out in length or size; make or become taut; place or lie at full length or spread out; extend limbs and tighten muscles after being relaxed; have specified length or extension, extend; strain or exert extremely; exaggerate. ● *noun* continuous extent, expanse, or period; stretching, being stretched; *colloquial* period of imprisonment etc. □ **at a stretch** in one period; **stretch one's legs** exercise oneself by walking; **stretch out** extend (limb etc.), last, prolong; **stretch a point** agree to something not normally allowed. □ **stretchy** *adjective* (**-ier**, **-iest**).

stretcher *noun* two poles with canvas etc. between for carrying person in lying position; brick etc. laid along face of wall.

strew /stru:/ *verb* (*past participle* **strewn** or **strewed**) scatter over surface; (usually + *with*) spread (surface) with scattered things.

'strewth = 'STRUTH.

striated /straɪ'eɪtɪd/ *adjective* marked with slight ridges or furrows. □ **striation** *noun.*

stricken /'strɪkən/ *archaic past participle* of STRIKE. ● *adjective* affected or overcome (with illness, misfortune, etc.).

strict *adjective* precisely limited or defined, without deviation; requiring complete obedience or exact performance. □ **strictly** *adverb*; **strictness** *noun.*

stricture /'strɪktʃə/ *noun* (usually in *plural*; often + *on*, *upon*) critical or censorious remark.

stride ● *verb* (**-ding**; *past* **strode**; *past participle* **stridden** /'strɪd(ə)n/) walk with long firm steps; cross with one step; bestride. ● *noun* single long step; length of this; gait as determined by length of stride; (usually in *plural*) progress. □ **take in one's stride** manage easily.

strident /'straɪd(ə)nt/ *adjective* loud and harsh. □ **stridency** *noun*; **stridently** *adverb.*

strife *noun* conflict, struggle.

strike ● *verb* (**-king**; *past* **struck**; *past participle* **struck** or *archaic* **stricken**) deliver (blow), inflict blow on; come or bring sharply into contact with; propel or divert with blow; (cause to) penetrate; ignite (match) or produce (sparks etc.) by rubbing; make (coin) by stamping; produce (musical note) by striking; (of clock) indicate (time) with chime etc., (of time) be so indicated; attack suddenly; (of disease) afflict; cause to become suddenly; reach, achieve; agree on (bargain); assume (attitude); find (oil etc.) by drilling; come to attention of or appear to; (of employees) engage in strike; lower or take down (flag, tent, etc.); take specified direction. ● *noun* act of striking; employees' organized refusal to work until grievance is remedied; similar refusal to participate; sudden find or success; attack, esp. from air. □ **on strike** taking part in industrial strike; **strikebreaker** person working or employed in place of strikers; **strike home** deal effective blow; **strike off** remove with stroke, delete; **strike out** hit out, act vigorously, delete; **strike up** start (acquaintance, conversation, etc.), esp. casually, begin playing (tune etc.).

striker *noun* employee on strike; *Football* attacking player positioned forward.

striking *adjective* impressive, noticeable. □ **strikingly** *adverb.*

string ● *noun* twine, narrow cord; length of this or similar material used for tying, holding together, pulling, forming head of racket, etc.; piece of catgut, wire, etc. on musical instrument, producing note by vibration; (in *plural*) stringed instruments in orchestra etc.; (in *plural*) condition or complication attached to offer etc.; set of things strung together; tough side of bean-pod etc. ● *verb* (*past & past participle* **strung**) fit with string(s); thread on string; arrange in or as string; remove strings from (bean-pod etc.). □ **string along** *colloquial* deceive, (often + *with*) keep company (with); **string-course** raised horizontal band of bricks etc. on building; **string up** hang up on strings etc., kill by hanging, (usually as **strung up** *adjective*) make tense.

stringed *adjective* (of musical instrument) having strings.

stringent /'strɪndʒ(ə)nt/ *adjective* (of rules etc.) strict, precise. □ **stringency** *noun*; **stringently** *adverb.*

stringer *noun* longitudinal structural member in framework, esp. of ship or aircraft; *colloquial* freelance newspaper correspondent.

stringy *adjective* (**-ier**, **-iest**) like string, fibrous.

strip[1] ● *verb* (**-pp-**) (often + *of*) remove clothes or covering from, undress; deprive (person) of property or titles; leave bare; (often + *down*) remove accessory fittings of or take apart (machine etc.); remove old paint etc. from with solvent; damage thread of (screw) or teeth of (gearwheel). ● *noun* act of stripping, esp. in striptease; *colloquial* distinctive outfit worn by sports team. □ **strip club** club where striptease is performed; **striptease** entertainment in which performer slowly and erotically undresses.

strip[2] *noun* long narrow piece. □ **strip cartoon** comic strip; **strip light** tubular fluorescent lamp; **tear (person) off a strip** *colloquial* rebuke.

stripe *noun* long narrow band or strip differing in colour or texture from surface on either side of it; *Military* chevron etc. denoting military rank. □ **stripy** *adjective* (**-ier**, **-iest**).

striped *adjective* having stripes.

stripling /'strɪplɪŋ/ *noun* youth not yet fully grown.

stripper *noun* device or solvent for removing paint etc.; performer of striptease.

strive *verb* (**-ving**; *past* **strove** /strəʊv/; *past participle* **striven** /'strɪv(ə)n/) try hard; (often + *with*, *against*) struggle.

strobe *noun colloquial* stroboscope.

stroboscope /'strəʊbəskəʊp/ *noun* lamp producing regular intermittent flashes. □ **stroboscopic** /-'skɒp-/ *adjective.*

strode *past* of STRIDE.

stroke ● *noun* act of striking; sudden disabling attack caused esp. by thrombosis; action or movement, esp. as one of series or in game etc.; slightest action; single complete action of moving wing, oar, etc.; whole motion of piston either way; mode of moving limbs in swimming; single mark made by pen, paint brush, etc.; detail contributing to general effect; sound of striking clock; oarsman nearest stern, who sets time of stroke; act or spell of stroking. ● *verb* (**-king**) pass hand gently along surface of (hair, fur, etc.); act as stroke of (boat, crew). □ **at a stroke** by a single action; **on the stroke of** punctually at; **stroke of (good) luck** unexpected fortunate event.

stroll /strəʊl/ ● *verb* walk in leisurely fashion. ● *noun* leisurely walk. □ **strolling players** *historical* actors etc. going from place to place performing.

strong ● *adjective* (**stronger** /'strɒŋgə/, **strongest** /-gɪst/) physically, morally, or mentally powerful; vigorous, robust; performed with muscular strength; difficult to capture, escape from, etc.; (of suspicion, belief, etc.) firmly held; powerfully affecting senses or mind etc.; (of drink, solution, etc.) with large proportion of alcohol etc.; powerful in numbers or equipment etc.; (of verb) forming inflections by vowel change in root syllable. ● *adverb* strongly. □ **come on strong** act forcefully; **going strong** *colloquial* thriving; **strong-arm** using force; **strongbox** strongly made box for valuables; **stronghold** fortress, centre of support for a cause etc.; **strong language** swearing; **strong-minded** determined; **strongroom** strongly built room for valuables; **strong suit** thing in which one excels. □ **strongish** *adjective*; **strongly** *adverb*.

strontium /ˈstrɒntɪəm/ *noun* soft silver-white metallic element. □ **strontium-90** radioactive isotope of this.

strop ● *noun* device, esp. strip of leather, for sharpening razors; *colloquial* bad temper. ● *verb* (**-pp-**) sharpen on or with strop.

stroppy /ˈstrɒpɪ/ *adjective* (**-ier**, **-iest**) *colloquial* bad-tempered, awkward to deal with.

strove *past* of STRIVE.

struck *past & past participle* of STRIKE.

structuralism *noun* doctrine that structure rather than function is important. □ **structuralist** *noun & adjective*.

structure /ˈstrʌktʃə/ ● *noun* constructed unit, esp. building; way in which thing is constructed; framework. ● *verb* (**-ring**) give structure to, organize. □ **structural** *adjective*; **structurally** *adverb*.

strudel /ˈstru:d(ə)l/ *noun* thin leaved pastry filled esp. with apple and baked.

struggle /ˈstrʌg(ə)l/ ● *verb* (**-ling**) violently try to get free; (often + *for*, *to do*) make great efforts under difficulties; (+ *with*, *against*) fight against; (+ *along*, *up*, etc.) make one's way with difficulty; (esp. as **struggling** *adjective*) have difficulty in getting recognition or a living. ● *noun* act or period of struggling; hard or confused contest.

strum ● *verb* (**-mm-**) (often + *on*) play on (stringed or keyboard instrument), esp. carelessly or unskilfully. ● *noun* strumming sound.

strumpet /ˈstrʌmpɪt/ *noun archaic* prostitute.

strung *past & past participle* of STRING.

strut ● *noun* bar in framework to resist pressure; strutting gait. ● *verb* (**-tt-**) walk in stiff pompous way; brace with strut(s).

'struth /stru:θ/ *interjection* (also **'strewth**) *colloquial exclamation of surprise*.

strychnine /ˈstrɪkni:n/ *noun* highly poisonous alkaloid.

stub ● *noun* remnant of pencil, cigarette, etc.; counterfoil of cheque, receipt, etc.; stump. ● *verb* (**-bb-**) strike (one's toe) against something; (usually + *out*) extinguish (cigarette etc.) by pressing lighted end against something.

stubble /ˈstʌb(ə)l/ *noun* cut stalks of corn etc. left in ground after harvest; short stiff hair or bristles, esp. on unshaven face. □ **stubbly** *adjective*.

stubborn /ˈstʌbən/ *adjective* obstinate, inflexible. □ **stubbornly** *adverb*; **stubbornness** *noun*.

stubby *adjective* (**-ier**, **-iest**) short and thick.

stucco /ˈstʌkəʊ/ ● *noun* (*plural* **-es**) plaster or cement for coating walls or moulding into architectural decorations. ● *verb* (**-es**, **-ed**) coat with stucco.

stuck *past & past participle* of STICK[2]. □ **stuck-up** *colloquial* conceited, snobbish.

stud[1] ● *noun* large projecting nail, knob, etc. as surface ornament; double button, esp. for use in shirt-front. ● *verb* (**-dd-**) set with studs; (as **studded** *adjective*) (+ *with*) thickly set or strewn with.

stud[2] *noun* number of horses kept for breeding etc.; place where these are kept; stallion. □ **at stud** (of stallion) hired out for breeding. **stud-book** book giving pedigrees of horses; **stud-farm** place where horses are bred; **stud poker** poker with betting after dealing of cards face up.

student /ˈstju:d(ə)nt/ *noun* person who is studying, esp. at place of higher or further education. □ **studentship** *noun*.

studio /ˈstju:dɪəʊ/ *noun* (*plural* **-s**) workroom of sculptor, painter, photographer, etc.; place for making films, recordings, or broadcasts. □ **studio couch** couch that can be converted into a bed; **studio flat** one-roomed flat.

studious /ˈstju:dɪəs/ *adjective* diligent in study or reading; painstaking. □ **studiously** *adverb*.

study /ˈstʌdɪ/ ● *noun* (*plural* **-ies**) acquiring knowledge, esp. from books; (in *plural*) pursuit of academic knowledge; private room for reading, writing, etc.; piece of work, esp. in painting, done as exercise or preliminary experiment; portrayal, esp. in literature, of character, behaviour, etc.; *Music* composition designed to develop player's skill; thing worth observing; thing that is or deserves to be investigated. ● *verb* (**-ies**, **-ied**) make study of; scrutinize; devote time and thought to understanding subject etc. or achieving desired result; (as **studied** *adjective*) deliberate, affected.

stuff ● *noun* material; fabric; substance or things not needing to be specified; particular knowledge or activity; woollen fabric; nonsense; (**the stuff**) *colloquial* supply, esp. of drink or drugs. ● *verb* pack (receptacle) tightly; (+ *in*, *into*) force or cram (thing); fill out skin to restore original shape of (bird, animal, etc.); fill (bird, piece of meat, etc.) with mixture, esp. before cooking; (also **stuff oneself**) eat greedily; push, esp. hastily or clumsily; (usually in *passive*; + *up*) block up (nose etc.); *slang derogatory* dispose of. □ **get stuffed** *slang* go away, get lost; **stuff and nonsense** something ridiculous or incredible.

stuffing *noun* padding for cushions etc.; mixture used to stuff food, esp. before cooking.

stuffy *adjective* (**-ier**, **-iest**) (of room etc.) lacking fresh air; dull, uninteresting; conventional, narrow-minded; (of nose etc.) stuffed up. □ **stuffily** *adverb*; **stuffiness** *noun*.

stultify /'stʌltɪfaɪ/ *verb* (**-ies**, **-ied**) make ineffective or useless, esp. by routine or from frustration. □ **stultification** *noun*.

stumble /'stʌmb(ə)l/ ● *verb* (**-ling**) accidentally lurch forward or have partial fall; (often + *along*) walk with repeated stumbles; speak clumsily; (+ *on*, *upon*, *across*) find by chance. ● *noun* act of stumbling. □ **stumbling block** circumstance causing difficulty or hesitation.

stump ● *noun* part of cut or fallen tree still in ground; similar part (of branch, limb, tooth, etc.) cut off or worn down; *Cricket* each of 3 uprights of wicket. ● *verb* (of question etc.) be too difficult for, baffle; (as **stumped** *adjective*) at a loss; *Cricket* put batsman out by touching stumps with ball while he is out of his crease; walk stiffly or clumsily and noisily; *US* traverse (district) making political speeches. □ **stump up** *colloquial* produce or pay over (money required).

stumpy *adjective* (**-ier**, **-iest**) short and thick. □ **stumpiness** *noun*.

stun *verb* (**-nn-**) knock senseless; stupefy; bewilder, shock.

stung *past & past participle* of STING.

stunk *past & past participle* of STINK.

stunner *noun colloquial* stunning person or thing.

stunning *adjective colloquial* extremely attractive or impressive. □ **stunningly** *adverb*.

stunt[1] *verb* retard growth or development of.

stunt[2] *noun* something unusual done to attract attention; trick, daring manoeuvre. □ **stunt man** man employed to take actor's place in performing dangerous stunts.

stupefy /'stju:pɪfaɪ/ *verb* (**-ies**, **-ied**) make stupid or insensible; astonish. □ **stupefaction** *noun*.

stupendous /stju:'pendəs/ *adjective* amazing; of vast size or importance. □ **stupendously** *adverb*.

stupid /'stju:pɪd/ *adjective* (**-er**, **-est**) unintelligent, slow-witted; typical of stupid person; uninteresting; in state of stupor. □ **stupidity** /-'pɪd-/ *noun* (*plural* **-ies**); **stupidly** *adverb*.

stupor /'stju:pə/ *noun* dazed or torpid state; utter amazement.

sturdy /'stɜ:dɪ/ *adjective* (**-ier**, **-iest**) robust; strongly built; vigorous. □ **sturdily** *adverb*; **sturdiness** *noun*.

sturgeon /'stɜ:dʒ(ə)n/ *noun* (*plural* same or **-s**) large edible fish yielding caviar.

stutter /'stʌtə/ *verb & noun* stammer.

sty[1] /staɪ/ *noun* (*plural* **sties**) enclosure for pigs; filthy room or dwelling.

sty[2] /staɪ/ *noun* (also **stye**) (*plural* **sties** or **styes**) inflamed swelling on edge of eyelid.

Stygian /'stɪdʒɪən/ *adjective literary* murky, gloomy.

style /staɪl/ ● *noun* kind or sort, esp. in regard to appearance and form (of person, house, etc.); manner of writing, speaking, etc.; distinctive manner of person, artistic school, or period; correct way of designating person or thing; superior quality; fashion in dress etc.; implement for scratching or engraving; part of flower supporting stigma. ● *verb* (**-ling**) design or make etc. in particular style; designate in specified way.

stylish *adjective* fashionable, elegant. □ **stylishly** *adverb*; **stylishness** *noun*.

stylist /'staɪlɪst/ *noun* designer of fashionable styles; hairdresser; stylish writer or performer.

stylistic /staɪ'lɪstɪk/ *adjective* of literary or artistic style. □ **stylistically** *adverb*.

stylized /'staɪlaɪzd/ *adjective* (also **-ised**) painted, drawn, etc. in conventional non-realistic style.

stylus /'staɪləs/ *noun* (*plural* **-luses**) needle-like point for producing or following groove in gramophone record; ancient pointed writing implement.

stymie /'staɪmɪ/ (also **stimy**) ● *noun* (*plural* **-ies**) *Golf* situation where opponent's ball lies between one's ball and the hole; difficult situation. ● *verb* (**stymying** or **stymieing**) obstruct, thwart.

styptic /'stɪptɪk/ ● *adjective* serving to check bleeding. ● *noun* styptic substance.

styrene /'staɪəri:n/ *noun* liquid hydrocarbon used in making plastics etc.

suave /swɑ:v/ *adjective* smooth; polite; sophisticated. □ **suavely** *adverb*; **suavity** *noun*.

sub *colloquial* ● *noun* submarine; subscription; substitute; sub-editor. ● *verb* (**-bb-**) (usually + *for*) act as substitute; sub-edit.

sub- *prefix* at, to, or from lower position; secondary or inferior position; nearly; more or less.

subaltern /'sʌbəlt(ə)n/ *noun Military* officer below rank of captain, esp. second lieutenant.

sub-aqua /sʌb'ækwə/ *adjective* (of sport etc.) taking place under water.

subatomic /sʌbə'tɒmɪk/ *adjective* occurring in, or smaller than, an atom.

subcommittee *noun* committee formed from main committee for special purpose.

subconscious /sʌb'kɒnʃəs/ ● *adjective* of part of mind that is not fully conscious but influences actions etc. ● *noun* this part of the mind. □ **subconsciously** *adverb*.

subcontinent /'sʌbkɒntɪnənt/ *noun* large land mass, smaller than continent.

subcontract ● *verb* /sʌbkən'trækt/ employ another contractor to do (work) as part of larger project; make or carry out subcontract. ● *noun* /sʌb'kɒntrækt/ secondary contract. □ **subcontractor** /-'træktə/ *noun*.

subculture /'sʌbkʌltʃə/ *noun* social group or its culture within a larger culture.

subcutaneous /sʌbkju:'teɪnɪəs/ *adjective* under the skin.

subdivide /sʌbdɪ'vaɪd/ *verb* (**-ding**) divide again after first division. □ **subdivision** /'sʌbdɪvɪʒ(ə)n/ *noun*.

subdue /səb'dju:/ *verb* (**-dues**, **-dued**, **-duing**) conquer, suppress; tame; (as **subdued** *adjective*) softened, lacking in intensity.

sub-editor /sʌb'edɪtə/ *noun* assistant editor; person who prepares material for printing. □ **sub-edit** *verb* (**-t-**).

subheading /'sʌbhedɪŋ/ *noun* subordinate heading or title.

subhuman /sʌb'hju:mən/ *adjective* (of behaviour, intelligence, etc.) less than human.

subject ● *noun* /'sʌbdʒɪkt/ theme of discussion, description, or representation; (+ *for*) person, circumstance, etc. giving rise to specified feeling, action, etc.; branch of study; word or phrase representing person or thing carrying out action of verb (see panel); person other than monarch living under government; *Philosophy* thinking or feeling entity, conscious self; *Music* theme, leading motif. ● *adjective* /'sʌbdʒɪkt/ (+ *to*) conditional on, liable or exposed to; owing obedience to government etc. ● *adverb* /'sʌbdʒɪkt/ (+ *to*) conditionally on. ● *verb* /səb'dʒekt/ (+ *to*) make liable or expose to; (usually + *to*) subdue (person, nation, etc.) to superior will. □ **subjection** *noun*.

subjective /səb'dʒektɪv/ *adjective* (of art, written history, opinion, etc.) not impartial or literal; *esp. Philosophy* of individual consciousness or perception; imaginary, partial, distorted; *Grammar* of the subject. □ **subjectively** *adverb*; **subjectivity** /sʌbdʒek'tɪv-/ *noun*.

subjoin /sʌb'dʒɔɪn/ *verb* add (illustration, anecdote, etc.) at the end.

sub judice /sʌb 'dʒu:dɪsɪ/ *adjective Law* under judicial consideration and therefore prohibited from public discussion. [Latin]

subjugate /'sʌbdʒʊgeɪt/ *verb* (**-ting**) conquer, bring into subjection. □ **subjugation** *noun*; **subjugator** *noun*.

subjunctive /səb'dʒʌŋktɪv/ *Grammar* ● *adjective* (of mood) expressing wish, supposition, or possibility. ● *noun* subjunctive mood or form.

sublease ● *noun* /'sʌbli:s/ lease granted by tenant to subtenant. ● *verb* /sʌb'li:s/ (**-sing**) lease to subtenant.

sublet /sʌb'let/ *verb* (**-tt-**; *past & past participle* **-let**) lease to subtenant.

sub-lieutenant /sʌblef'tenənt/ *noun* officer ranking next below lieutenant.

sublimate ● *verb* /'sʌblɪmeɪt/ (**-ting**) divert energy of (primitive impulse etc.) into socially more acceptable activity; sublime (substance); refine, purify. ● *noun* /'sʌblɪmət/ sublimated substance. □ **sublimation** *noun*.

sublime /sə'blaɪm/ ● *adjective* (**-r**, **-st**) of most exalted kind; awe-inspiring; arrogantly undisturbed. ● *verb* (**-ming**) convert (substance) from solid into vapour by heat (and usually allow to solidify again); make sublime; become pure (as if) by sublimation. □ **sublimely** *adverb*; **sublimity** /-'lɪm-/ *noun*.

subliminal /səb'lɪmɪn(ə)l/ *adjective Psychology* below threshold of sensation or consciousness; too faint or rapid to be consciously perceived. □ **subliminally** *adverb*.

sub-machine-gun /sʌbmə'ʃi:ngʌn/ *noun* hand-held lightweight machine-gun.

submarine /sʌbmə'ri:n/ ● *noun* vessel, esp. armed warship, which can be submerged and navigated under water. ● *adjective* existing, occurring, done, or used below surface of sea. □ **submariner** /-'mærɪnə/ *noun*.

submerge /səb'mɜ:dʒ/ *verb* (**-ging**) place, go, or dive beneath water; overwhelm with work, problems, etc. □ **submergence** *noun*; **submersion** *noun*.

submersible /səb'mɜ:sɪb(ə)l/ ● *noun* submarine operating under water for short periods. ● *adjective* capable of submerging.

submicroscopic /sʌbmaɪkrə'skɒpɪk/ *adjective* too small to be seen by ordinary microscope.

submission /səb'mɪʃ(ə)n/ *noun* submitting, being submitted; thing submitted; submissive attitude etc.

submissive /səb'mɪsɪv/ *adjective* humble, obedient. □ **submissively** *adverb*; **submissiveness** *noun*.

submit /səb'mɪt/ *verb* (**-tt-**) (often + *to*) cease resistance, yield; present for consideration; (+ *to*) subject or be subjected to (process, treatment, etc.); *Law* argue, suggest.

subnormal /sʌb'nɔ:m(ə)l/ *adjective* below or less than normal, esp. in intelligence.

subordinate ● *adjective* /sə'bɔ:dɪnət/ (usually + *to*) of inferior importance or rank; secondary, subservient. ● *noun* /sə'bɔ:dɪnət/ person working under authority of another. ● *verb* /sə'bɔ:dɪneɪt/ (**-ting**) (usually + *to*) make or treat as subordinate. □ **subordinate clause** clause serving as noun, adjective, or

Subject

The subject of a sentence is the person or thing that carries out the action of the verb and can be found by asking the question 'who or what?' before the verb, e.g.

The goalkeeper *made a stunning save.*
Hundreds of books *are now available on CD-ROM.*

In a passive construction, the subject of the sentence is in fact the person or thing to which the action of the verb is done, e.g.

I *was hit by a ball.*
Has the programme *been broadcast yet?*

adverb within sentence. □ **subordination** *noun.*

suborn /sə'bɔ:n/ *verb* induce esp. by bribery to commit perjury or other crime.

subpoena /sə'pi:nə/ • *noun* writ ordering person's attendance in law court. • *verb* (*past & past participle* **-naed** or **-na'd**) serve subpoena on.

sub rosa /sʌb 'rəʊzə/ *adjective & adverb* in confidence or in secret. [Latin]

subscribe /səb'skraɪb/ *verb* (**-bing**) (usually + *to*, *for*) pay (specified sum) esp. regularly for membership of organization or receipt of publication etc.; contribute to fund, for cause, etc.; (usually + *to*) agree with (opinion etc.). □ **subscribe to** arrange to receive (periodical etc.) regularly.

subscriber *noun* person who subscribes, esp. person paying regular sum for hire of telephone line. □ **subscriber trunk dialling** making of trunk calls by subscriber without assistance of operator.

subscript /'sʌbskrɪpt/ • *adjective* written or printed below the line. • *noun* subscript number etc.

subscription /səb'skrɪpʃ(ə)n/ *noun* act of subscribing; money subscribed; membership fee, esp. paid regularly.

subsequent /'sʌbsɪkwənt/ *adjective* (usually + *to*) following specified or implied event. □ **subsequently** *adverb.*

subservient /səb'sɜ:vɪənt/ *adjective* servile; (usually + *to*) instrumental, subordinate. □ **subservience** *noun.*

subside /səb'saɪd/ *verb* (**-ding**) become tranquil; diminish; (of water etc.) sink; (of ground) cave in. □ **subsidence** /-'saɪd-, 'sʌbsɪd-/ *noun.*

subsidiary /səb'sɪdɪərɪ/ • *adjective* supplementary; additional; (of company) controlled by another. • *noun* (*plural* **-ies**) subsidiary company, person, or thing.

subsidize /'sʌbsɪdaɪz/ *verb* (also **-ise**) (**-zing** or **-sing**) pay subsidy to; support by subsidies.

subsidy /'sʌbsɪdɪ/ *noun* (*plural* **-ies**) money contributed esp. by state to keep prices at desired level; any monetary grant.

subsist /səb'sɪst/ *verb* (often + *on*) keep oneself alive; be kept alive; remain in being, exist.

subsistence *noun* subsisting; means of supporting life; minimal level of existence, income, etc. □ **subsistence farming** farming in which almost all produce is consumed by farmer's household.

subsoil /'sʌbsɔɪl/ *noun* soil just below surface soil.

subsonic /sʌb'sɒnɪk/ *adjective* of speeds less than that of sound.

substance /'sʌbst(ə)ns/ *noun* particular kind of material; reality, solidity; essence of what is spoken or written; wealth and possessions. □ **in substance** generally, essentially.

substandard /sʌb'stændəd/ *adjective* of lower than desired standard.

substantial /səb'stænʃ(ə)l/ *adjective* of real importance or value; large in size or amount; of solid structure; commercially successful; wealthy; largely true; real; existing. □ **substantially** *adverb.*

substantiate /səb'stænʃɪeɪt/ *verb* (**-ting**) support or prove truth of (charge, claim, etc.). □ **substantiation** *noun.*

substantive /'sʌbstəntɪv/ • *adjective* actual, real, permanent; substantial. • *noun* noun. □ **substantively** *adverb.*

substitute /'sʌbstɪtju:t/ • *noun* person or thing acting or serving in place of another; artificial alternative to a food etc. • *verb* (**-ting**) (often + *for*) put in place of another; act as substitute for. • *adjective* acting as substitute. □ **substitution** *noun.*

substratum /'sʌbstrɑ:təm/ *noun* (*plural* **-ta**) underlying layer.

subsume /səb'sju:m/ *verb* (**-ming**) (usually + *under*) include under particular rule, class, etc.

subtenant /'sʌbtenənt/ *noun* person renting room or house etc. from its tenant. □ **subtenancy** *noun* (*plural* **-ies**).

subtend /sʌb'tend/ *verb* (of line) be opposite (angle, arc).

subterfuge /'sʌbtəfju:dʒ/ *noun* attempt to avoid blame etc., esp. by lying or deceit.

subterranean /sʌbtə'reɪnɪən/ *adjective* underground.

subtext *noun* underlying theme.

subtitle /'sʌbtaɪt(ə)l/ • *noun* subordinate or additional title of book etc.; caption of cinema film, esp. translating dialogue. ● *verb* (**-ling**) provide with subtitle(s).

subtle /ˈsʌt(ə)l/ *adjective* (**-r**, **-st**) hard to detect or describe; (of scent, colour, etc.) faint, delicate; ingenious, perceptive. □ **subtlety** *noun* (*plural* **-ies**); **subtly** *adverb*.

subtract /səbˈtrækt/ *verb* (often + *from*) deduct (number etc.) from another. □ **subtraction** *noun*.

subtropical /sʌbˈtrɒpɪk(ə)l/ *adjective* bordering on the tropics; characteristic of such regions.

suburb /ˈsʌbɜːb/ *noun* outlying district of city.

suburban /səˈbɜːbən/ *adjective* of or characteristic of suburbs; *derogatory* provincial in outlook. □ **suburbanite** *noun*.

suburbia /səˈbɜːbɪə/ *noun usually derogatory* suburbs and their inhabitants etc.

subvention /səbˈvenʃ(ə)n/ *noun* subsidy.

subversive /səbˈvɜːsɪv/ ● *adjective* seeking to overthrow (esp. government). ● *noun* subversive person. □ **subversion** *noun*; **subversively** *adverb*; **subversiveness** *noun*.

subvert /səbˈvɜːt/ *verb* overthrow or weaken (government etc.).

subway /ˈsʌbweɪ/ *noun* underground passage, esp. for pedestrians; *US* underground railway.

subzero /sʌbˈzɪərəʊ/ *adjective* (esp. of temperature) lower than zero.

succeed /səkˈsiːd/ *verb* (often + *in*) have success; prosper; follow in order; (often + *to*) come into inheritance, office, title, or property.

success /səkˈses/ *noun* accomplishment of aim; favourable outcome; attainment of wealth, fame, etc.; successful person or thing. □ **successful** *adjective*; **successfully** *adverb*.

succession /səkˈseʃ(ə)n/ *noun* following in order; series of things or people following one another; succeeding to inheritance, office, or esp. throne; right to succeed to one of these, set of people with such right. □ **in succession** one after another; **in succession to** as successor of.

successive /səkˈsesɪv/ *adjective* following in succession, consecutive. □ **successively** *adverb*.

successor /səkˈsesə/ *noun* (often + *to*) person or thing that succeeds another.

succinct /səkˈsɪŋkt/ *adjective* brief, concise. □ **succinctly** *adverb*; **succinctness** *noun*.

succour /ˈsʌkə/ (*US* **succor**) *archaic or formal* ● *noun* help, esp. in time of need. ● *verb* give succour to.

succulent /ˈsʌkjʊlənt/ ● *adjective* juicy; (of plant) thick and fleshy. ● *noun* succulent plant. □ **succulence** *noun*.

succumb /səˈkʌm/ *verb* (usually + *to*) give way; be overcome; die.

such ● *adjective* (often + *as*) of kind or degree specified or suggested; so great or extreme; unusually, abnormally. ● *pronoun* such person(s) or thing(s). □ **as such** as being what has been specified; **such-and-such** particular but unspecified; **suchlike** *colloquial* of such kind.

suck ● *verb* draw (liquid) into mouth by suction; draw liquid from in this way; roll tongue round (sweet etc.) in mouth; make sucking action or noise; (usually + *down*) engulf or drown in sucking movement. ● *noun* act or period of sucking. □ **suck in** absorb, involve (person); **suck up** (often + *to*) *colloquial* behave in a servile way, absorb.

sucker *noun* person easily duped or cheated; (+ *for*) person susceptible to; rubber etc. cap adhering by suction; similar part of plant or animal; shoot springing from plant's root below ground.

suckle /ˈsʌk(ə)l/ *verb* (**-ling**) feed (young) from breast or udder.

suckling /ˈsʌklɪŋ/ *noun* unweaned child or animal.

sucrose /ˈsuːkrəʊz/ *noun* kind of sugar obtained from cane, beet, etc.

suction /ˈsʌkʃ(ə)n/ *noun* sucking; production of partial vacuum so that external atmospheric pressure forces fluid into vacant space or causes adhesion of surfaces.

Sudanese /suːdəˈniːz/ ● *adjective* of Sudan. ● *noun* (*plural* same) native or national of Sudan.

sudden /ˈsʌd(ə)n/ *adjective* done or occurring unexpectedly or abruptly. □ **all of a sudden** suddenly. □ **suddenly** *adverb*; **suddenness** *noun*.

sudorific /su:də'rɪfɪk/ ● *adjective* causing sweating. ● *noun* sudorific drug.

suds *plural noun* froth of soap and water. □ **sudsy** *adjective*.

sue *verb* (**sues**, **sued**, **suing**) begin lawsuit against; (often + *for*) make application to law court for compensation etc.; (often + *to*, *for*) make plea to person for favour.

suede /sweɪd/ *noun* leather with flesh side rubbed into nap.

suet /'su:ɪt/ *noun* hard fat surrounding kidneys of cattle and sheep, used in cooking etc. □ **suety** *adjective*.

suffer /'sʌfə/ *verb* undergo pain, grief, etc.; undergo or be subjected to (pain, loss, punishment, grief, etc.); tolerate. □ **sufferer** *noun*; **suffering** *noun*.

sufferance *noun* tacit permission or toleration. □ **on sufferance** tolerated but not supported.

suffice /sə'faɪs/ *verb* (**-cing**) be enough; meet needs of. □ **suffice it to say** I shall say only this.

sufficiency /sə'fɪʃənsɪ/ *noun* (*plural* **-ies**) (often + *of*) sufficient amount.

sufficient /sə'fɪʃ(ə)nt/ *adjective* sufficing; adequate. □ **sufficiently** *adverb*.

suffix /'sʌfɪks/ ● *noun* letter(s) added to end of word to form derivative. ● *verb* add as suffix.

suffocate /'sʌfəkeɪt/ *verb* (**-ting**) kill, stifle, or choke by stopping breathing, esp. by fumes etc.; be or feel suffocated. □ **suffocating** *adjective*; **suffocation** *noun*.

suffragan /'sʌfrəgən/ *noun* bishop assisting diocesan bishop.

suffrage /'sʌfrɪdʒ/ *noun* right of voting in political elections.

suffragette /sʌfrə'dʒet/ *noun historical* woman who agitated for women's suffrage.

suffuse /sə'fju:z/ *verb* (**-sing**) (of colour, moisture, etc.) spread throughout or over from within. □ **suffusion** /-ʒ(ə)n/ *noun*.

sugar /'ʃʊgə/ ● *noun* sweet crystalline substance obtained from sugar cane and sugar beet, used in cookery, confectionery, etc.; *Chemistry* soluble usually sweet crystalline carbohydrate, e.g. glucose; *esp. US colloquial* darling (as term of address). ● *verb* sweeten or coat with sugar. □ **sugar beet** white beet yielding sugar; **sugar cane** tall stout perennial tropical grass yielding sugar; **sugar-daddy** *slang* elderly man who lavishes gifts on young woman; **sugar loaf** conical moulded mass of hard refined sugar.

sugary *adjective* containing or resembling sugar; cloying, sentimental. □ **sugariness** *noun*.

suggest /sə'dʒest/ *verb* (often + *that*) propose (theory, plan, etc.); hint at; evoke (idea etc.); (**suggest itself**) (of idea etc.) come into person's mind.

suggestible *adjective* capable of being influenced by suggestion. □ **suggestibility** *noun*.

suggestion /sə'dʒestʃ(ə)n/ *noun* suggesting; thing suggested; slight trace, hint; insinuation of belief or impulse into the mind.

suggestive /sə'dʒestɪv/ *adjective* (usually + *of*) conveying a suggestion; (of remark, joke, etc.) suggesting something indecent. □ **suggestively** *adverb*.

suicidal /su:ɪ'saɪd(ə)l/ *adjective* (of person) liable to commit suicide; of or tending to suicide; destructive to one's own interests. □ **suicidally** *adverb*.

suicide /'su:ɪsaɪd/ *noun* intentional self-killing; person who commits suicide; action destructive to one's own interests etc.

sui generis /sju:aɪ'dʒenərɪs/ *adjective* of its own kind, unique. [Latin]

suit /su:t, sju:t/ ● *noun* set of usually matching clothes consisting usually of jacket and trousers or skirt; clothing for particular purpose; any of the 4 sets into which pack of cards is divided; lawsuit. ● *verb* go well with (person's appearance); meet requirements of; (**suit oneself**) do as one chooses; be in harmony with; make fitting; be convenient; adapt; (as **suited** *adjective*) appropriate, well-fitted. □ **suitcase** flat case for carrying clothes, usually with hinged lid.

suitable *adjective* (usually + *to*, *for*) well-fitted for purpose; appropriate to occasion. □ **suitability** *noun*; **suitably** *adverb*.

suite /swi:t/ *noun* set of rooms, furniture, etc.; *Music* set of instrumental pieces.

suitor /'su:tə/ *noun* man who woos woman; plaintiff or petitioner in lawsuit.

sulfur etc. *US* = SULPHUR etc.

sulk • *verb* be sulky. • *noun* (also **the sulks**) fit of sullen silence.

sulky /'sʌlkɪ/ *adjective* (**-ier**, **-iest**) sullen and unsociable from resentment or ill temper. □ **sulkily** *adverb.*

sullen /'sʌlən/ *adjective* sulky, morose. □ **sullenly** *adverb*; **sullenness** *noun.*

sully /'sʌlɪ/ *verb* (**-ies**, **-ied**) spoil purity or splendour of (reputation etc.).

sulphate /'sʌlfeɪt/ *noun* (*US* **sulfate**) salt or ester of sulphuric acid.

sulphide /'sʌlfaɪd/ *noun* (*US* **sulfide**) binary compound of sulphur.

sulphite /'sʌlfaɪt/ *noun* (*US* **sulfite**) salt or ester of sulphurous acid.

sulphonamide /sʌl'fɒnəmaɪd/ *noun* (*US* **sulfonamide**) kind of antibiotic drug containing sulphur.

sulphur /'sʌlfə/ *noun* (*US* **sulfur**) pale yellow non-metallic element burning with blue flame and stifling smell. □ **sulphur dioxide** colourless pungent gas formed by burning sulphur in air and dissolving it in water.

sulphuric /sʌl'fjʊərɪk/ *adjective* (*US* **sulfuric**) of or containing sulphur with valency of 6. □ **sulphuric acid** dense highly corrosive oily acid.

sulphurous /'sʌlfərəs/ *adjective* (*US* **sulfurous**) of or like sulphur; containing sulphur with valency of 4. □ **sulphurous acid** unstable weak acid used e.g. as bleaching agent.

sultan /'sʌlt(ə)n/ *noun* Muslim sovereign.

sultana /sʌl'tɑ:nə/ *noun* seedless raisin; sultan's wife, mother, concubine, or daughter.

sultanate /'sʌltəneɪt/ *noun* position of or territory ruled by sultan.

sultry /'sʌltrɪ/ *adjective* (**-ier**, **-iest**) (of weather) oppressively hot; (of person) passionate, sensual.

sum • *noun* total resulting from addition; amount of money; arithmetical problem; (esp. in *plural*) *colloquial* arithmetic work, esp. elementary. • *verb* (**-mm-**) find sum of. □ **in sum** briefly, to sum up; **summing-up** judge's review of evidence given to jury, recapitulation of main points of argument etc.; **sum up** (esp. of judge) give summing-up, form or express opinion of (person, situation, etc.), summarize.

sumac /'su:mæk/ *noun* (also **sumach**) shrub with reddish fruits used as spice; dried and ground leaves of this for use in tanning and dyeing.

summarize /'sʌməraɪz/ *verb* (also **-ise**) (**-zing** or **-sing**) make or be summary of.

summary /'sʌmərɪ/ • *noun* (*plural* **-ies**) brief account giving chief points. • *adjective* brief, without details or formalities. □ **summarily** *adverb.*

summation /sə'meɪʃ(ə)n/ *noun* finding of total or sum; summarizing.

summer /'sʌmə/ *noun* warmest season of year; (often + *of*) mature stage of life etc. □ **summer house** light building in garden etc. for use in summer; **summer pudding** dessert of soft fruit pressed in bread casing; **summer school** course of lectures etc. held in summer, esp. at university; **summer time** period from March to October when clocks etc. are advanced one hour; **summertime** season or period of summer. □ **summery** *adjective.*

summit /'sʌmɪt/ *noun* highest point, top; highest level of achievement or status; (in full **summit conference**, **meeting**, etc.) discussion between heads of governments.

summon /'sʌmən/ *verb* order to come or appear, esp. in lawcourt; (usually + *to do*) call on; call together; (often + *up*) gather (courage, resources, etc.).

summons /'sʌmənz/ • *noun* (*plural* **summonses**) authoritative call to attend or do something, esp. to appear in court. • *verb esp. Law* serve with summons.

sumo /'su:məʊ/ *noun* Japanese wrestling in which only soles of feet may touch ground.

sump *noun* casing holding oil in internal-combustion engine; pit, well, etc. for collecting superfluous liquid.

sumptuary /'sʌmptʃʊərɪ/ *adjective Law* regulating (esp. private) expenditure.

sumptuous /'sʌmptʃʊəs/ *adjective* costly, splendid, magnificent. □ **sumptuously** *adverb*; **sumptuousness** *noun*.

Sun. *abbreviation* Sunday.

sun ● *noun* the star round which the earth travels and from which it receives light and warmth; this light or warmth; any star. ● *verb* (**-nn-**) (often **sun oneself**) expose to sun. □ **sunbathe** bask in sun, esp. to tan one's body; **sunbeam** ray of sunlight; **sunblock** lotion protecting skin from sun; **sunburn** inflammation of skin from exposure to sun; **sunburnt** affected by sunburn; **sundial** instrument showing time by shadow of pointer in sunlight; **sundown** sunset; **sunflower** tall plant with large golden-rayed flowers; **sunglasses** tinted spectacles to protect eyes from glare; **sunlamp** lamp giving ultraviolet rays for therapy or artificial suntan; **sunlight** light from sun; **sunlit** illuminated by sun; **sun lounge** room with large windows etc. to receive much sunlight; **sunrise** (time of) sun's rising; **sunroof** opening panel in car's roof; **sunset** (time of) sun's setting; **sunshade** parasol, awning; **sunshine** sunlight, area illuminated by it, fine weather, cheerfulness; **sunspot** dark patch on sun's surface; **sunstroke** acute prostration from excessive heat of sun; **suntan** brownish skin colour caused by exposure to sun; **suntrap** sunny place, esp. sheltered from wind; **sun-up** *esp. US* sunrise.

sundae /'sʌndeɪ/ *noun* ice cream with fruit, nuts, syrup, etc.

Sunday /'sʌndeɪ/ *noun* day of week following Saturday; Christian day of worship; *colloquial* newspaper published on Sundays. □ **month of Sundays** *colloquial* very long period; **Sunday school** religious class held on Sundays for children.

sunder /'sʌndə/ *verb literary* sever, keep apart.

sundry /'sʌndrɪ/ ● *adjective* various, several. ● *noun* (*plural* **-ies**) (in *plural*) oddments, accessories, etc. not mentioned individually. □ **all and sundry** everyone.

sung *past participle* of SING.

sunk *past & past participle* of SINK.

sunken *adjective* that has sunk; lying below general surface; (of eyes, cheeks, etc.) shrunken, hollow.

Sunni /'sʌnɪ/ *noun* (*plural* same or **-s**) one of two main branches of Islam; adherent of this.

sunny *adjective* (**-ier, -iest**) bright with or warmed by sunlight; cheerful. □ **sunnily** *adverb*; **sunniness** *noun*.

sup[1] ● *verb* (**-pp-**) drink by sips or spoonfuls; *esp. Northern English colloquial* drink (alcohol). ● *noun* sip of liquid.

sup[2] *verb* (**-pp-**) *archaic* take supper.

super /'su:pə/ ● *adjective colloquial* excellent, unusually good. ● *noun colloquial* superintendent; supernumerary.

super- *combining form* on top, over, beyond; to extreme degree; extra good or large of its kind; of higher kind.

superannuate /su:pər'ænjʊeɪt/ *verb* (**-ting**) pension (person) off; dismiss or discard as too old; (as **superannuated** *adjective*) too old for work.

superannuation *noun* pension; payment made to obtain pension.

superb /su:'pɜ:b/ *adjective colloquial* excellent; magnificent. □ **superbly** *adverb*.

supercargo /'su:pəkɑ:gəʊ/ *noun* (*plural* **-es**) person in merchant ship managing sales etc. of cargo.

supercharge /'su:pətʃɑ:dʒ/ *verb* (**-ging**) (usually + *with*) charge (atmosphere etc.) with energy, emotion, etc.; use supercharger on.

supercharger *noun* device forcing extra air or fuel into internal-combustion engine.

supercilious /su:pə'sɪlɪəs/ *adjective* haughtily contemptuous. □ **superciliously** *adverb*; **superciliousness** *noun*.

supererogation /su:pərerə'geɪʃ(ə)n/ *noun* doing of more than duty requires.

superficial /su:pə'fɪʃ(ə)l/ *adjective* of or on the surface; lacking depth; swift, cursory; apparent, not real; (esp. of person) of shallow feelings etc. □ **superficiality** /-ʃɪ'æl-/ *noun*; **superficially** *adverb*.

superfluity /su:pə'flu:ɪtɪ/ *noun* (*plural* **-ies**) being superfluous; superfluous amount or thing.

superfluous /su:'pɜ:flʊəs/ *adjective* more than is needed or wanted; useless.

supergrass /'su:pəgrɑ:s/ *noun colloquial* police informer implicating many people.

superhuman /su:pə'hju:mən/ *adjective* exceeding normal human capacity or power.

superimpose /su:pərɪm'pəʊz/ *verb* (**-sing**) (usually + *on*) place (thing) on or above something else. □ **superimposition** /-pə'zɪʃ(ə)n/ *noun*.

superintend /su:pərɪn'tend/ *verb* manage, supervise (work etc.). □ **superintendence** *noun*.

superintendent /su:pərɪn'tend(ə)nt/ *noun* police officer above rank of chief inspector; person who superintends; director of institution etc.

superior /su:'pɪərɪə/ ● *adjective* higher in rank, quality, etc.; high-quality; supercilious; (often + *to*) better or greater in some respect; written or printed above the line. ● *noun* person superior to another, esp. in rank; head of monastery etc. □ **superiority** /-'ɒr-/ *noun*.

superlative /su:'pɜ:lətɪv/ ● *adjective* of highest degree; excellent; *Grammar* (of adjective or adverb) expressing highest or very high degree of quality etc. denoted by simple word. ● *noun Grammar* superlative expression or word; (in *plural*) high praise, exaggerated language.

superman *noun colloquial* man of exceptional powers or achievement.

supermarket /'su:pəmɑ:kɪt/ *noun* large self-service store selling food, household goods, etc.

supernatural /su:pə'nætʃər(ə)l/ ● *adjective* not attributable to, or explicable by, natural or physical laws; magical; mystical. ● *noun* (**the supernatural**) supernatural forces etc. □ **supernaturally** *adverb*.

supernova /su:pə'nəʊvə/ *noun* (*plural* **-vae** /-vi:/ or **-vas**) star that suddenly increases very greatly in brightness.

supernumerary /su:pə'nju:mərərɪ/ ● *adjective* in excess of normal number; engaged for extra work; (of actor) with non-speaking part. ● *noun* (*plural* **-ies**) supernumerary person or thing.

superphosphate /su:pə'fɒsfeɪt/ *noun* fertilizer made from phosphate rock.

superpower /'su:pəpaʊə/ *noun* extremely powerful nation.

superscript /'su:pəskrɪpt/ ● *adjective* written or printed above. ● *noun* superscript number or symbol.

supersede /su:pə'si:d/ *verb* (**-ding**) take place of; put or use another in place of. □ **supersession** /-'seʃ-/ *noun*.

supersonic /su:pə'sɒnɪk/ *adjective* of or having speed greater than that of sound. □ **supersonically** *adverb*.

superstar /'su:pəstɑ:/ *noun* extremely famous or renowned actor, musician, etc.

superstition /su:pə'stɪʃ(ə)n/ *noun* belief in the supernatural; irrational fear of the unknown or mysterious; practice, belief, or religion based on this. □ **superstitious** *adjective*.

superstore /'su:pəstɔ:/ *noun* very large supermarket.

superstructure /'su:pəstrʌktʃə/ *noun* structure built on top of another; upper part of building, ship, etc.

supertanker /'su:pətæŋkə/ *noun* very large tanker.

supertax /'su:pətæks/ *noun* surtax.

supervene /su:pə'vi:n/ *verb* (**-ning**) *formal* occur as interruption in or change from some state. □ **supervention** *noun*.

supervise /'su:pəvaɪz/ *verb* (**-sing**) oversee, superintend. □ **supervision** /-'vɪʒ(ə)n/ *noun*; **supervisor** *noun*; **supervisory** *adjective*.

superwoman *noun colloquial* woman of exceptional ability or power.

supine /'su:paɪn/ ● *adjective* lying face upwards; inactive, indolent. ● *noun* type of Latin verbal noun.

supper /'sʌpə/ *noun* meal taken late in day, esp. evening meal less formal and substantial than dinner.

supplant /sə'plɑ:nt/ *verb* take the place of, esp. by underhand means.

supple /'sʌp(ə)l/ *adjective* (**-r**, **-st**) easily bent, pliant, flexible. □ **suppleness** *noun*.

supplement ● *noun* /'sʌplɪmənt/ thing or part added to improve or provide further information; separate section of newspaper etc. ● *verb* /'sʌplɪment/ provide supplement for. □ **supplemental** /-'ment(ə)l/ *adjective*; **supplementary** /-'mentərɪ/ *adjective*; **supplementation** *noun*.

suppliant /'sʌplɪənt/ ● *adjective* supplicating. ● *noun* humble petitioner.

supplicate /'sʌplɪkeɪt/ *verb* (**-ting**) *literary* make humble petition to or for. □ **supplicant** *noun*; **supplication** *noun*; **supplicatory** /-kətərɪ/ *adjective*.

supply /sə'plaɪ/ ● *verb* (**-ies**, **-ied**) provide (thing needed); (often + *with*) provide (person etc. with something); make up for (deficiency etc.). ● *noun* (*plural* **-ies**) provision of what is needed; stock, store; (in *plural*) provisions, equipment, etc. for army, expedition, etc.; person, esp. teacher, acting as temporary substitute. □ **supply and demand** quantities available and required, as factors regulating price. □ **supplier** *noun*.

support /sə'pɔ:t/ ● *verb* carry all or part of weight of; keep from falling, sinking, or failing; provide for; strengthen, encourage; give help or corroboration to; speak in favour of; take secondary part to (actor etc.); perform secondary act to (main act) at pop concert. ● *noun* supporting, being supported; person or thing that supports. □ **in support of** so as to support. □ **supportive** *adjective*; **supportively** *adverb*; **supportiveness** *noun*.

supporter *noun* person or thing that supports particular cause, team, sport, etc.

suppose /sə'pəʊz/ *verb* (**-sing**) (often + *that*) assume, be inclined to think; take as possibility or hypothesis; require as condition; (as **supposed** *adjective*) presumed. □ **be supposed to** be expected or required to; (in negative) ought not, not be allowed to; **I suppose so** *expression of hesitant agreement*.

supposedly /sə'pəʊzɪdlɪ/ *adverb* as is generally believed.

supposition /sʌpə'zɪʃ(ə)n/ *noun* what is supposed or assumed.

suppositious /sʌpə'zɪʃəs/ *adjective* hypothetical.

suppository /sə'pɒzɪtərɪ/ *noun* (*plural* **-ies**) solid medical preparation put into rectum or vagina to melt.

suppress /sə'pres/ *verb* put an end to; prevent (information, feelings, etc.) from being seen, heard, or known; *Electricity* partially or wholly eliminate (interference etc.), equip (device) to reduce interference due to it. □ **suppressible** *adjective*; **suppression** *noun*; **suppressor** *noun*.

suppurate /'sʌpjʊreɪt/ *verb* (**-ting**) form or secrete pus; fester. □ **suppuration** *noun*.

supra- *prefix* above.

supranational /su:prə'næʃən(ə)l/ *adjective* transcending national limits.

supremacy /su:'premәsɪ/ *noun* (*plural* **-ies**) being supreme; highest authority.

supreme /su:'pri:m/ *adjective* highest in authority or rank; greatest, most important; (of penalty, sacrifice, etc.) involving death. □ **supremely** *adverb*.

supremo /su:'pri:məʊ/ *noun* (*plural* **-s**) supreme leader.

surcharge /'sɜ:tʃɑ:dʒ/ ● *noun* additional charge or payment. ● *verb* (**-ging**) exact surcharge from.

surd ● *adjective* (of number) irrational. ● *noun* surd number.

sure /ʃʊə/ ● *adjective* (often + *of*, *that*) convinced; having or seeming to have adequate reason for belief; (+ *of*) confident in anticipation or knowledge of; reliable, unfailing; (+ *to do*) certain; undoubtedly true or truthful. ● *adverb* *colloquial* certainly. □ **make sure** make or become certain, ensure; **sure-fire** *colloquial* certain to succeed; **sure-footed** never stumbling; **to be sure** admittedly, indeed, certainly. □ **sureness** *noun*.

surely *adverb* with certainty or safety; *added to statement to express strong belief in its correctness*.

surety /'ʃu:ərətɪ/ *noun* (*plural* **-ies**) money given as guarantee of performance etc.; person taking responsibility for another's debt, obligation, etc.

surf ● *noun* foam of sea breaking on rock or (esp. shallow) shore. ● *verb* engage in surfing. □ **surfboard** long narrow board used in surfing. □ **surfer** *noun*.

surface /'sɜ:fɪs/ ● *noun* the outside of a thing; any of the limits of a solid; top of liquid, soil, etc.; outward or superficial aspect; *Geometry* thing with length and breadth but no thickness. ● *verb* (**-cing**) give (special) surface to (road, paper, etc.); rise or bring to surface; become visible or known; *colloquial* wake up, get up. □ **surface mail** mail not carried by air; **surface tension** tension of surface of liquid, tending to minimize its surface area.

surfeit /'sɜːfɪt/ ● *noun* excess, esp. in eating or drinking; resulting fullness. ● *verb* (**-t-**) overfeed; (+ *with*) (cause to) be wearied through excess.

surfing *noun* sport of riding surf on board.

surge ● *noun* sudden rush; heavy forward or upward motion; sudden increase (in price etc.); sudden but brief increase in pressure, voltage, etc.; surging motion of sea, waves, etc. ● *verb* (**-ging**) move suddenly and powerfully forwards; (of sea etc.) swell.

surgeon /'sɜːdʒ(ə)n/ *noun* medical practitioner qualified to practise surgery; naval or military medical officer.

surgery /'sɜːdʒərɪ/ *noun* (*plural* **-ies**) manual or instrumental treatment of injuries or disorders of body; place where or time when doctor, dentist, etc. gives advice and treatment to patients, or MP, lawyer, etc. gives advice.

surgical /'sɜːdʒɪk(ə)l/ *adjective* of or by surgery or surgeons; used for surgery; (of appliance) worn to correct deformity etc.; (esp. of military action) swift and precise. □ **surgical spirit** methylated spirits used for cleansing etc. □ **surgically** *adverb.*

surly /'sɜːlɪ/ *adjective* (**-ier**, **-iest**) bad-tempered, unfriendly. □ **surliness** *noun.*

surmise /sə'maɪz/ ● *noun* conjecture. ● *verb* (**-sing**) (often + *that*) infer doubtfully; suppose; guess.

surmount /sə'maʊnt/ *verb* overcome (difficulty, obstacle); (usually in *passive*) cap, crown. □ **surmountable** *adjective.*

surname /'sɜːneɪm/ *noun* name common to all members of family.

surpass /sə'pɑːs/ *verb* outdo, be better than; (as **surpassing** *adjective*) greatly exceeding, excelling others.

surplice /'sɜːplɪs/ *noun* loose full-sleeved white vestment worn by clergy etc.

surplus /'sɜːpləs/ ● *noun* amount left over when requirements have been met; excess of income over spending. ● *adjective* exceeding what is needed or used.

surprise /sə'praɪz/ ● *noun* unexpected or astonishing thing; emotion caused by this; catching or being caught unawares. ● *adjective* made, done, etc. without warning. ● *verb* (**-sing**) affect with surprise; (usually in *passive*; + *at*) shock, scandalize; capture by surprise; come upon (person) unawares; (+ *into*) startle, betray, etc. (person) into doing something. □ **surprising** *adjective*; **surprisingly** *adverb.*

surreal /sə'rɪəl/ *adjective* unreal; dreamlike; bizarre.

surrealism /sə'rɪəlɪz(ə)m/ *noun* 20th-c. movement in art and literature aiming to express subconscious mind by dream imagery etc. □ **surrealist** *noun & adjective*; **surrealistic** /-'lɪs-/ *adjective*; **surrealistically** /-'lɪs-/ *adverb.*

surrender /sə'rendə/ ● *verb* hand over, relinquish; submit, esp. to enemy; (often **surrender oneself**; + *to*) yield to habit, emotion, influence, etc.; give up rights under (life-insurance policy) in return for smaller sum received immediately. ● *noun* surrendering.

surreptitious /sʌrəp'tɪʃəs/ *adjective* done by stealth; underhand. □ **surreptitiously** *adverb.*

surrogate /'sʌrəgət/ *noun* substitute; deputy, esp. of bishop. □ **surrogate mother** woman who conceives and gives birth to child on behalf of woman unable to do so. □ **surrogacy** *noun.*

surround /sə'raʊnd/ ● *verb* come or be all round; encircle, enclose. ● *noun* border or edging, esp. area between walls and carpet.

surroundings *plural noun* things in neighbourhood of, or conditions affecting, person or thing; environment.

surtax /'sɜːtæks/ *noun* additional tax, esp. on high incomes.

surtitle /'sɜːtaɪt(ə)l/ *noun* caption translating words of opera, projected on to screen above stage.

surveillance /sə'veɪləns/ *noun* close watch undertaken by police etc., esp. on suspected person.

survey ● *verb* /sə'veɪ/ take or present general view of; examine condition of (building etc.); determine boundaries, extent, ownership, etc. of (district etc.). ● *noun* /'sɜ:veɪ/ general view or consideration; surveying of property; result of this; investigation of public opinion etc.; map or plan made by surveying.

surveyor /sə'veɪə/ *noun* person who surveys land and buildings, esp. professionally.

survival /sə'vaɪv(ə)l/ *noun* surviving; relic.

survive /sə'vaɪv/ *verb* (**-ving**) continue to live or exist; live or exist longer than; come alive through or continue to exist in spite of (danger, accident, etc.). □ **survivor** *noun*.

sus = SUSS.

susceptibility *noun* (*plural* **-ies**) being susceptible; (in *plural*) person's feelings.

susceptible /sə'septəb(ə)l/ *adjective* impressionable, sensitive; easily moved by emotion; (+ *to*) accessible or sensitive to; (+ *of*) allowing, admitting of (proof etc.).

suspect ● *verb* /səs'pekt/ be inclined to think; have impression of the existence or presence of; (often + *of*) mentally accuse; doubt innocence, genuineness, or truth of. ● *noun* /'sʌspekt/ suspected person. ● *adjective* /'sʌspekt/ subject to suspicion or distrust.

suspend /səs'pend/ *verb* hang up; keep inoperative or undecided temporarily; debar temporarily from function, office, etc.; (as **suspended** *adjective*) (of particles or body in fluid) floating between top and bottom. □ **suspended animation** temporary deathlike condition; **suspended sentence** judicial sentence remaining unenforced on condition of good behaviour.

suspender *noun* attachment to hold up stocking or sock by its top; (in *plural*) *US* pair of braces. □ **suspender belt** woman's undergarment with suspenders.

suspense /səs'pens/ *noun* state of anxious uncertainty or expectation.

suspension /səs'penʃ(ə)n/ *noun* suspending, being suspended; means by which vehicle is supported on its axles; substance consisting of particles suspended in fluid. □ **suspension bridge** bridge with roadway suspended from cables supported by towers.

suspicion /səs'pɪʃ(ə)n/ *noun* unconfirmed belief; distrust; suspecting, being suspected; (+ *of*) slight trace of.

suspicious /səs'pɪʃəs/ *adjective* prone to or feeling suspicion; indicating or justifying suspicion. □ **suspiciously** *adverb*.

suss /sʌs/ *verb* (also **sus**) (**-ss-**) *slang* (usually + *out*) investigate, inspect; understand; work out.

sustain /səs'teɪn/ *verb* bear weight of, support, esp. for long period; encourage, support; endure, stand; (of food) nourish; undergo (defeat, injury, loss, etc.); (of court etc.) decide in favour of, uphold; substantiate, corroborate; keep up (effort etc.). □ **sustainable** *adjective*.

sustenance /'sʌstɪnəns/ *noun* nourishment, food; means of support.

suture /'su:tʃə/ ● *noun* joining edges of wound or incision by stitching; stitch or thread etc. used for this. ● *verb* (**-ring**) stitch (wound, incision).

suzerain /'su:zərən/ *noun historical* feudal overlord; *archaic* sovereign or state having some control over another state that is internally self-governing. □ **suzerainty** *noun*.

svelte /svelt/ *adjective* slim, slender, graceful.

SW *abbreviation* south-west(ern).

swab /swɒb/ ● *noun* absorbent pad used in surgery; specimen of secretion etc. taken for examination; mop etc. for cleaning or mopping up. ● *verb* (**-bb-**) clean with swab; (+ *up*) absorb (moisture) with swab; mop clean (ship's deck).

swaddle /'swɒd(ə)l/ *verb* (**-ling**) wrap tightly in bandages, wrappings, etc. □ **swaddling-clothes** narrow bandages formerly wrapped round newborn child to restrain its movements.

swag *noun slang* thief's booty; *Australian & NZ* traveller's bundle; festoon of flowers, foliage, drapery, etc.

swagger /'swægə/ ● *verb* walk or behave arrogantly or self-importantly. ● *noun* swaggering gait or manner.

swain *noun archaic* country youth; *poetical* young lover or suitor.

swallow¹ /ˈswɒləʊ/ ● *verb* make or let (food etc.) pass down one's throat; accept meekly or gullibly; repress (emotion); engulf; say (words etc.) indistinctly. ● *noun* act of swallowing; amount swallowed.

swallow² /ˈswɒləʊ/ *noun* migratory swift-flying bird with forked tail. □ **swallow-dive** dive with arms spread sideways; **swallow-tail** deeply forked tail, butterfly etc. with this.

swam *past* of SWIM.

swamp /swɒmp/ ● *noun* piece of wet spongy ground. ● *verb* submerge, inundate; cause to fill with water and sink; overwhelm with numbers or quantity. □ **swampy** *adjective* (**-ier**, **-iest**).

swan /swɒn/ ● *noun* large web-footed usually white waterfowl with long flexible neck. ● *verb* (**-nn-**) (usually + *about*, *off*, etc.) *colloquial* move about casually or with superior manner. □ **swansong** person's last work or performance before death, retirement, etc.

swank *colloquial* ● *noun* ostentation, swagger. ● *verb* show off. □ **swanky** *adjective* (**-ier**, **-iest**).

swap /swɒp/ (also **swop**) ● *verb* (**-pp-**) exchange, barter. ● *noun* act of swapping; thing suitable for swapping.

sward /swɔːd/ *noun literary* expanse of short grass.

swarf /swɔːf/ *noun* fine chips or filings of stone, metal, etc.

swarm¹ /swɔːm/ ● *noun* cluster of bees leaving hive etc. with queen bee to establish new colony; large group of insects, birds, or people; (in *plural*, + *of*) great numbers. ● *verb* move in or form swarm; (+ *with*) be overrun or crowded with.

swarm² /swɔːm/ *verb* (+ *up*) climb (rope, tree, etc.) clasping or clinging with arms and legs.

swarthy /ˈswɔːðɪ/ *adjective* (**-ier**, **-iest**) dark-complexioned, dark in colour.

swashbuckler /ˈswɒʃbʌklə/ *noun* swaggering adventurer. □ **swashbuckling** *adjective & noun.*

swastika /ˈswɒstɪkə/ *noun* ancient symbol formed by equal-armed cross with each arm continued at a right angle; this with clockwise continuations as symbol of Nazi Germany.

swat /swɒt/ ● *verb* (**-tt-**) crush (fly etc.) with blow; hit hard and abruptly. ● *noun* act of swatting.

swatch /swɒtʃ/ *noun* sample, esp. of cloth; collection of samples.

swath /swɔːθ/ *noun* (also **swathe** /sweɪð/) ridge of cut grass, corn, etc.; space left clear by mower, scythe, etc.

swathe /sweɪð/ *verb* (**-thing**) bind or wrap in bandages, garments, etc.

sway ● *verb* (cause to) move unsteadily from side to side; oscillate irregularly; waver; have influence over. ● *noun* rule, government; swaying motion.

swear /sweə/ *verb* (*past* **swore**; *past participle* **sworn**) state or promise on oath; cause to take oath; *colloquial* insist; (often + *at*) use profane or obscene language; (+ *by*) appeal to as witness or guarantee of oath, *colloquial* have great confidence in. □ **swear in** admit to office etc. by administering oath; **swear off** *colloquial* promise to keep off (drink etc.); **swear-word** profane or obscene word.

sweat /swet/ ● *noun* moisture exuded through pores, esp. when one is hot or nervous; state or period of sweating; *colloquial* state of anxiety; *colloquial* effort, drudgery, laborious task or undertaking; condensed moisture on surface. ● *verb* (*past & past participle* **sweated** or *US* **sweat**) exude sweat; be terrified, suffer, etc.; (of wall etc.) show surface moisture; (cause to) toil or drudge; emit like sweat; make (horse, athlete, etc.) sweat by exercise; (as **sweated** *adjective*) (of goods, labour, etc.) produced by or subjected to exploitation. □ **no sweat** *colloquial* no bother, no trouble; **sweat-band** band of absorbent material inside hat or round head, wrist, etc. to soak up sweat; **sweatshirt** sleeved cotton sweater; **sweatshop** workshop where sweated labour is employed. □ **sweaty** *adjective* (**-ier**, **-iest**).

sweater *noun* woollen etc. pullover.

Swede *noun* native or national of Sweden; (**swede**) large yellow variety of turnip.

Swedish /ˈswiːdɪʃ/ ● *adjective* of Sweden. ● *noun* language of Sweden.

sweep ● *verb* (*past & past participle* **swept**) clean or clear (room, area, etc.) (as) with a broom; (often + *up*) collect or remove (dirt etc.) by sweeping; (+ *aside, away*, etc.) dismiss abruptly; (+ *along, down*, etc.) drive or carry along with force; (+ *off, away*, etc.) remove or clear forcefully; traverse swiftly or lightly; impart sweeping motion to; glide swiftly; go majestically; (of landscape etc.) be rolling or spacious. ● *noun* act or motion of sweeping; curve in road etc.; range, scope; chimney sweep; sortie by aircraft; *colloquial* sweepstake. □ **sweep the board** win all the money in gambling game, win all possible prizes etc.; **sweepstake** form of gambling on horse races etc. in which money staked is divided among those who have drawn numbered tickets for winners.

sweeping *adjective* wide in range or effect; generalized, arbitrary.

sweet ● *adjective* tasting like sugar, honey, etc.; smelling pleasant like perfume, roses, etc., fragrant; melodious; fresh; not sour or bitter; gratifying, attractive; amiable, gentle; *colloquial* pretty; (+ *on*) *colloquial* fond of, in love with. ● *noun* small shaped piece of sugar or chocolate confectionery; sweet dish forming course of meal. □ **sweet-and-sour** cooked in sauce with sugar, vinegar, etc.; **sweetbread** pancreas or thymus of animal, as food; **sweet-brier** single-flowered fragrant-leaved wild rose; **sweetcorn** sweet-flavoured maize kernels; **sweetheart** either of pair of lovers; **sweetmeat** a sweet, a small fancy cake; **sweet pea** climbing garden annual with many-coloured scented flowers; **sweet pepper** fruit of capsicum; **sweet potato** tropical plant with edible tuberous roots; **sweet-talk** flatter in order to persuade; **sweet tooth** liking for sweet-tasting things; **sweet william** garden plant with close clusters of sweet-smelling flowers. □ **sweetish** *adjective*; **sweetly** *adverb*.

sweeten *verb* make or become sweet(er); make agreeable or less painful. □ **sweetening** *noun*.

sweetener *noun* thing that sweetens; *colloquial* bribe.

sweetie *noun* *colloquial* a sweet; sweetheart.

sweetness *noun* being sweet; fragrance. □ **sweetness and light** (esp. uncharacteristic) mildness and reason.

swell ● *verb* (*past participle* **swollen** /'swəʊlən/ or **-ed**) (cause to) grow bigger, louder, or more intense; rise or raise up; (+ *out*) bulge out; (of heart etc.) feel full of joy, pride, etc.; (+ *with*) be hardly able to restrain (pride etc.). ● *noun* act or state of swelling; heaving of sea etc. with unbreaking rolling waves; crescendo; mechanism in organ etc. for gradually varying volume; *colloquial* fashionable or stylish person. ● *adjective colloquial esp. US* fine, excellent.

swelling *noun* abnormally swollen place, esp. on body.

swelter /'sweltə/ ● *verb* be uncomfortably hot. ● *noun* sweltering condition.

swept *past & past participle* of SWEEP.

swerve ● *verb* (**-ving**) (cause to) change direction, esp. suddenly. ● *noun* swerving motion.

swift ● *adjective* rapid, quick; prompt. ● *noun* swift-flying long-winged migratory bird. □ **swiftly** *adverb*; **swiftness** *noun*.

swig ● *verb* (**-gg-**) *colloquial* drink in large draughts. ● *noun* swallow of liquid, esp. of large amount.

swill ● *verb* (often + *out*) rinse, pour water over or through; drink greedily. ● *noun* swilling; mainly liquid refuse as pig-food.

swim ● *verb* (**-mm-**; *past* **swam**; *past participle* **swum**) propel body through water with limbs, fins, etc.; perform (stroke) or cross (river etc.) by swimming; float on liquid; appear to undulate, reel, or whirl; (of head) feel dizzy; (+ *in, with*) be flooded. ● *noun* act or spell of swimming. □ **in the swim** *colloquial* involved in or aware of what is going on; **swimming bath, pool** pool constructed for swimming; **swimming costume** bathing costume; **swimsuit** swimming costume, esp. one-piece for women and girls; **swimwear** clothing for swimming in. □ **swimmer** *noun*.

swimmingly *adverb colloquial* smoothly, without obstruction.

swindle /'swɪnd(ə)l/ ● *verb* (**-ling**) (often + *out of*) cheat of money etc.; defraud. ● *noun* act of swindling; fraudulent person or thing. □ **swindler** *noun*.

swine *noun* (*plural* same) *formal or US* pig; *colloquial* (*plural* same or **-s**) disgusting person, unpleasant or difficult thing. □ **swinish** *adjective*.

swing ● *verb* (*past & past participle* **swung**) (cause to) move with to-and-fro or curving motion; sway or hang like pendulum or door etc.; oscillate; move by gripping something and leaping etc.; walk with swinging gait; (+ *round*) move to face opposite direction; (+ *at*) attempt to hit; *colloquial* (of party etc.) be lively; have decisive influence on (voting etc.); *colloquial* be executed by hanging. ● *noun* act, motion, or extent of swinging; swinging or smooth gait, rhythm, or action; seat slung by ropes, chains, etc. for swinging on or in, spell of swinging thus; smooth rhythmic jazz or jazzy dance music; amount by which votes etc. change from one side to another. □ **swing-boat** boat-shaped swing at fairs etc.; **swing-bridge** bridge that can be swung aside to let ships etc. pass; **swing-door** door that swings in either direction and closes by itself when released; **swings and roundabouts** situation allowing equal gain and loss; **swing-wing** (aircraft) with wings that can pivot to point sideways or backwards. □ **swinger** *noun*.

swingeing /'swɪndʒɪŋ/ *adjective* (of blow etc.) forcible; huge, far-reaching.

swipe *colloquial* ● *verb* (**-ping**) (often + *at*) hit hard and recklessly; steal. ● *noun* reckless hard hit or attempt to hit.

swirl ● *verb* move, flow, or carry along with whirling motion. ● *noun* swirling motion; twist, curl. □ **swirly** *adjective*.

swish ● *verb* swing (cane, scythe, etc.) audibly through air, grass, etc.; move with or make swishing sound. ● *noun* swishing action or sound. ● *adjective colloquial* smart, fashionable.

Swiss ● *adjective* of Switzerland. ● *noun* (*plural* same) native or national of Switzerland. □ **Swiss roll** cylindrical cake, made by rolling up thin flat sponge cake spread with jam etc.

switch ● *noun* device for making and breaking connection in electric circuit; transfer, changeover, deviation; flexible shoot cut from tree; light tapering rod; *US* railway points. ● *verb* (+ *on*, *off*) turn (electrical device) on or off; change or transfer (position, subject, etc.); exchange; whip or flick with switch. □ **switchback** ride at fair etc. with extremely steep ascents and descents, similar railway or road; **switchboard** apparatus for varying connections between electric circuits, esp. in telephony; **switched-on** *colloquial* up to date, aware of what is going on; **switch off** *colloquial* cease to pay attention.

swivel /'swɪv(ə)l/ ● *noun* coupling between two parts etc. so that one can turn freely without the other. ● *verb* (**-ll-**; *US* **-l-**) turn (as) on swivel, swing round. □ **swivel chair** chair with revolving seat.

swizz *noun* (also **swiz**) *colloquial* something disappointing; swindle.

swizzle /'swɪz(ə)l/ *noun colloquial* frothy mixed alcoholic drink, esp. of rum or gin and bitters; *slang* swizz. □ **swizzle-stick** stick used for frothing or flattening drinks.

swollen *past participle* of SWELL.

swoon /swu:n/ *verb & noun literary* faint.

swoop /swu:p/ ● *verb* (often + *down*) come down with rush like bird of prey; (often + *on*) make sudden attack. ● *noun* act of swooping, sudden pounce.

swop = SWAP.

sword /sɔ:d/ *noun* weapon with long blade for cutting or thrusting. □ **put to the sword** kill; **sword dance** dance with brandishing of swords, or steps about swords laid on ground; **swordfish** (*plural* same or **-es**) large sea fish with swordlike upper jaw; **swordplay** fencing, repartee, lively arguing; **swordsman** person of (usually specified) skill with sword; **swordstick** hollow walking stick containing sword blade.

swore *past* of SWEAR.

sworn ● *past participle* of SWEAR. ● *adjective* bound (as) by oath.

swot *colloquial* ● *verb* (**-tt-**) study hard; (usually + *up*, *up on*) study (subject) hard or hurriedly. ● *noun usually derogatory* person who swots.

swum *past participle* of SWIM.

swung *past & past participle* of SWING.

sybarite /'sɪbəraɪt/ *noun* self-indulgent or luxury-loving person. □ **sybaritic** /-'rɪt-/ *adjective*.

sycamore /'sɪkəmɔː/ *noun* large maple tree, its wood; *US* plane tree, its wood.

sycophant /'sɪkəfænt/ *noun* flatterer, toady. □ **sycophancy** *noun*; **sycophantic** /-'fæn-/ *adjective*.

syllabic /sɪ'læbɪk/ *adjective* of or in syllables. □ **syllabically** *adverb*.

syllable /'sɪləb(ə)l/ *noun* unit of pronunciation forming whole or part of word, usually consisting of vowel sound with consonant(s) before or after (see panel). □ **in words of one syllable** plainly, bluntly.

syllabub /'sɪləbʌb/ *noun* dessert of cream or milk sweetened and whipped with wine etc.

syllabus /'sɪləbəs/ *noun* (*plural* **-buses** or **-bi** /-baɪ/) programme or outline of course of study, teaching, etc.

syllogism /'sɪlədʒɪz(ə)m/ *noun* form of reasoning in which from two propositions a third is deduced. □ **syllogistic** /-'dʒɪs-/ *adjective*.

sylph /sɪlf/ *noun* elemental spirit of air; slender graceful woman. □ **sylphlike** *adjective*.

sylvan /'sɪlv(ə)n/ *adjective* (also **silvan**) of the woods, having woods; rural.

symbiosis /sɪmbaɪ'əʊsɪs/ *noun* (*plural* **-bioses** /-siːz/) (usually mutually advantageous) association of two different organisms living attached to one another etc.; mutually advantageous connection between people. □ **symbiotic** /-'ɒt-/ *adjective*.

symbol /'sɪmb(ə)l/ *noun* thing generally regarded as typifying, representing, or recalling something; mark, sign, etc. representing object, idea, process, etc. □ **symbolic** /-'bɒl-/ *adjective*; **symbolically** /-'bɒl-/ *adverb*.

symbolism *noun* use of symbols; symbols; artistic movement or style using symbols to express ideas, emotions, etc. □ **symbolist** *noun*.

symbolize *verb* (also **-ise**) (**-zing** or **-sing**) be symbol of; represent by symbol(s).

symmetry /'sɪmɪtrɪ/ *noun* (*plural* **-ies**) correct proportion of parts; beauty resulting from this; structure allowing object to be divided into parts of equal shape and size; possession of such structure; repetition of exactly similar parts facing each other or a centre. □ **symmetrical** /-'met-/ *adjective*; **symmetrically** /-'met-/ *adverb*.

sympathetic /sɪmpə'θetɪk/ *adjective* of or expressing sympathy; likeable, pleasant; (+ *to*) favouring (proposal etc.). □ **sympathetically** *adverb*.

sympathize /'sɪmpəθaɪz/ *verb* (also **-ise**) (**-zing** or **-sing**) (often + *with*) feel or express sympathy; agree. □ **sympathizer** *noun*.

sympathy /'sɪmpəθɪ/ *noun* (*plural* **-ies**) sharing of another's feelings; (often + *with*) sharing or tendency to share emotion, sensation, condition, etc. of another person; (in *singular* or *plural*) compassion, commiseration, condolences; (often + *with*) agreement (with person etc.) in opinion or desire. □ **in sympathy** (often + *with*) having, showing, or resulting from sympathy.

symphony /'sɪmfənɪ/ *noun* (*plural* **-ies**) musical composition in several movements for full orchestra. □ **symphony orchestra** large orchestra playing

Syllable

A syllable is the smallest unit of speech that can normally occur alone, such as *a*, *at*, *ta*, or *tat*. A word can be made up of one or more syllables:

cat, *fought*, and *twinge* each have one syllable;
rating, *deny*, and *collapse* each have two syllables;
excitement, *superman*, and *telephoned* each have three syllables;
American and *complicated* each have four syllables;
examination and *uncontrollable* each have five syllables.

symphonies etc. □ **symphonic** /-'fɒn-/ *adjective.*

symposium /sɪm'pəʊzɪəm/ *noun* (*plural* **-sia**) conference, or collection of essays, on particular subject.

symptom /'sɪmptəm/ *noun* physical or mental sign of disease or injury; sign of existence of something. □ **symptomatic** /-'mæt-/ *adjective.*

synagogue /'sɪnəgɒg/ *noun* building for Jewish religious instruction and worship.

sync /sɪŋk/ (also **synch**) *colloquial* ● *noun* synchronization. ● *verb* synchronize. □ **in** or **out of sync** (often + *with*) according or agreeing well or badly.

synchromesh /'sɪŋkrəʊmeʃ/ ● *noun* system of gear-changing, esp. in vehicles, in which gearwheels revolve at same speed during engagement. ● *adjective* of this system.

synchronize /'sɪŋkrənaɪz/ *verb* (also **-ise**) (**-zing** or **-sing**) (often + *with*) make or be synchronous (with); make sound and picture of (film etc.) coincide; cause (clocks etc.) to show same time. □ **synchronization** *noun.*

synchronous /'sɪŋkrənəs/ *adjective* (often + *with*) existing or occurring at same time.

syncopate /'sɪŋkəpeɪt/ *verb* (**-ting**) displace beats or accents in (music); shorten (word) by omitting syllable or letter(s) in middle. □ **syncopation** *noun.*

syncope /'sɪŋkəpɪ/ *noun Grammar* syncopation; *Medicine* fainting through fall in blood pressure.

syncretize /'sɪŋkrətaɪz/ *verb* (also **-ise**) (**-zing** or **-sing**) attempt to unify or reconcile differing schools of thought. □ **syncretic** /-'kret-/ *adjective*; **syncretism** *noun.*

syndicalism /'sɪndɪkəlɪz(ə)m/ *noun historical* movement for transferring industrial control and ownership to workers' unions. □ **syndicalist** *noun.*

syndicate ● *noun* /'sɪndɪkət/ combination of people, commercial firms, etc. to promote some common interest; agency supplying material simultaneously to a number of periodicals etc.; group of people who gamble, organize crime, etc. ● *verb* /'sɪndɪkeɪt/ (**-ting**) form into syndicate; publish (material) through syndicate. □ **syndication** *noun.*

syndrome /'sɪndrəʊm/ *noun* group of concurrent symptoms of disease; characteristic combination of opinions, emotions, etc.

synod /'sɪnəd/ *noun* Church council of clergy and lay people.

synonym /'sɪnənɪm/ *noun* word or phrase that means the same as another (see panel).

synonymous /sɪ'nɒnɪməs/ *adjective* (often + *with*) having same meaning; suggestive of; associated with.

synopsis /sɪ'nɒpsɪs/ *noun* (*plural* **synopses** /-si:z/) summary; outline.

synoptic /sɪ'nɒptɪk/ *adjective* of or giving synopsis. □ **Synoptic Gospels** those of Matthew, Mark, and Luke.

Synonym

A synonym is a word that has the same meaning as, or a similar meaning to, another word:

cheerful, *happy*, *merry*, and *jolly*

are synonyms that are quite close to each other in meaning, as are

lazy, *indolent*, and *slothful*

In contrast, the following words all mean 'a person who works with another', but their meanings vary considerably:

colleague
collaborator
ally
conspirator
accomplice

syntax /'sɪntæks/ *noun* grammatical arrangement of words; rules or analysis of this. □ **syntactic** /-'tæk-/ *adjective*.

synthesis /'sɪnθəsɪs/ *noun* (*plural* **-theses** /-si:z/) putting together of parts or elements to make up complex whole; *Chemistry* artificial production of (esp. organic) substances from simpler ones.

synthesize /'sɪnθəsaɪz/ *verb* (also **-ise**) (**-zing** or **-sing**) make synthesis of.

synthesizer *noun* (also **-iser**) electronic, usually keyboard, instrument producing great variety of sounds.

synthetic /sɪn'θetɪk/ ● *adjective* produced by synthesis, esp. to imitate natural product; affected, insincere. ● *noun* synthetic substance. □ **synthetically** *adverb*.

syphilis /'sɪfəlɪs/ *noun* a contagious venereal disease. □ **syphilitic** /-'lɪt-/ *adjective*.

Syrian /'sɪrɪən/ ● *noun* native or national of Syria. ● *adjective* of Syria.

syringa /sɪ'rɪŋgə/ *noun* shrub with white scented flowers.

syringe /sɪ'rɪndʒ/ ● *noun* device for drawing in quantity of liquid and ejecting it in fine stream. ● *verb* (**-ging**) sluice or spray with syringe.

syrup /'sɪrəp/ *noun* (*US* **sirup**) sweet sauce of sugar dissolved in boiling water, often flavoured or medicated; condensed sugar-cane juice; molasses, treacle; excessive sweetness of manner. □ **syrupy** *adjective*.

system /'sɪstəm/ *noun* complex whole; set of connected things or parts; organized group of things; set of organs in body with common structure or function; human or animal body as organized whole; method, scheme of action, procedure, or classification; orderliness; (**the system**) prevailing political or social order, esp. seen as oppressive. □ **get** (**thing**) **out of one's system** get rid of (anxiety etc.); **systems analysis** analysis of complex process etc. so as to improve its efficiency, esp. by using computer.

systematic /sɪstə'mætɪk/ *adjective* methodical; according to system; deliberate. □ **systematically** *adverb*.

systematize /'sɪstəmətaɪz/ *verb* (also **-ise**) (**-zing** or **-sing**) make systematic. □ **systematization** *noun*.

systemic /sɪ'stemɪk/ *adjective Physiology* of the whole body; (of insecticide etc.) entering plant tissues via roots and shoots. □ **systemically** *adverb*.

Tt

T *noun* □ **to a T** exactly, to a nicety; **T-bone** T-shaped bone, esp. in steak from thin end of loin; **T-junction** junction, esp. of two roads, in shape of T; **T-shirt** short-sleeved casual top; **T-square** T-shaped instrument for drawing right angles.

t. *abbreviation* (also **t**) ton(s); tonne(s).

TA *abbreviation* Territorial Army.

ta /tɑ:/ *interjection colloquial* thank you.

tab[1] *noun* small piece of material attached to thing for grasping, fastening, identifying, etc.; *US colloquial* bill; distinguishing mark on officer's collar. □ **keep tabs on** *colloquial* have under observation or in check.

tab[2] *noun* tabulator.

tabard /'tæbəd/ *noun* herald's official coat emblazoned with arms of sovereign; woman's or girl's sleeveless jerkin; *historical* knight's short emblazoned garment worn over armour.

tabasco /tə'bæskəʊ/ *noun* pungent pepper; (**Tabasco**) *proprietary term* sauce made from this.

tabby /'tæbɪ/ *noun* (*plural* **-ies**) grey or brownish cat with dark stripes.

tabernacle /'tæbənæk(ə)l/ *noun historical* tent used as sanctuary by Israelites during Exodus; niche or receptacle, esp. for bread and wine of Eucharist; Nonconformist meeting-house.

tabla /'tæblə/ *noun* pair of small Indian drums played with hands.

table /'teɪb(ə)l/ ● *noun* flat surface on legs used for eating, working at, etc.; food provided at table; group seated for dinner etc.; set of facts or figures arranged esp. in columns; multiplication table. ● *verb* (**-ling**) bring forward for discussion at meeting etc.; *esp. US* postpone consideration of. □ **at table** taking a meal; **tablecloth** cloth spread over table; **tableland** plateau; **tablespoon** large spoon for serving etc., (also **tablespoonful**) amount held by this; **table tennis** game played with small bats on table divided by net; **tableware** dishes etc. for meals; **table wine** wine of ordinary quality; **turn the tables** (often + *on*) reverse circumstances to one's advantage.

tableau /'tæbləʊ/ *noun* (*plural* **-x** /-z/) picturesque presentation; group of silent motionless people representing stage scene.

table d'hôte /tɑ:b(ə)l 'dəʊt/ *noun* meal from set menu at fixed price.

tablet /'tæblɪt/ *noun* small solid dose of medicine etc.; bar of soap etc.; flat slab of stone etc., esp. inscribed.

tabloid /'tæblɔɪd/ *noun* small-sized, often popular or sensational, newspaper.

taboo /tə'bu:/ ● *noun* (*plural* **-s**) ritual isolation of person or thing as sacred or accursed; prohibition. ● *adjective* avoided or prohibited, esp. by social custom. ● *verb* (**-oos**, **-ooed**) put under taboo; exclude or prohibit, esp. socially.

tabor /'teɪbə/ *noun historical* small drum.

tabular /'tæbjʊlə/ *adjective* of or arranged in tables.

tabulate /'tæbjʊleɪt/ *verb* (**-ting**) arrange (figures, facts) in tabular form. □ **tabulation** *noun*.

tabulator *noun* device on typewriter etc. for advancing to sequence of set positions in tabular work.

tachograph /'tækəgrɑ:f/ *noun* device in vehicle to record speed and travel time.

tachometer /tə'kɒmɪtə/ *noun* instrument measuring velocity or rate of shaft's rotation (esp. in vehicle).

tacit /'tæsɪt/ *adjective* implied or understood without being stated. □ **tacitly** *adverb*.

taciturn /'tæsɪtɜ:n/ *adjective* saying little, uncommunicative. □ **taciturnity** /-'tɜ:n-/ *noun*.

tack[1] ● *noun* small sharp broad-headed nail; *US* drawing-pin; long stitch for fastening materials lightly or temporarily together; (in sailing) direction, temporary change of direction; course of action or policy. ● *verb* (often + *down* etc.) fasten with tacks; stitch lightly together; (+ *to*, *on*, *on to*) add, append; change ship's course by turning head to wind, make series of such tacks.

tack[2] *noun* horse's saddle, bridle, etc.

tack[3] *noun colloquial* cheap or shoddy material; tat, kitsch.

tackle /'tæk(ə)l/ ● *noun* equipment for task or sport; rope(s), pulley(s), etc. used in working sails, hoisting weights, etc.; tackling in football etc. ● *verb* (**-ling**) try to deal with (problem etc.); grapple with (opponent); confront (person) in discussion; intercept or stop (player running with ball etc.). □ **tackle-block** pulley over which rope runs. □ **tackler** *noun.*

tacky[1] /'tækɪ/ *adjective* (**-ier**, **-iest**) slightly sticky.

tacky[2] /'tækɪ/ *adjective* (**-ier**, **-iest**) *colloquial* in poor taste; cheap, shoddy.

taco /'tækəʊ/ *noun* (*plural* **-s**) Mexican dish of meat etc. in crisp folded tortilla.

tact *noun* adroitness in dealing with people or circumstances; intuitive perception of right thing to do or say. □ **tactful** *adjective*; **tactfully** *adverb*; **tactless** *adjective*; **tactlessly** *adverb.*

tactic /'tæktɪk/ *noun* piece of tactics.

tactical *adjective* of tactics; (of bombing etc.) done in immediate support of military or naval operation; adroitly planning or planned. □ **tactically** *adverb.*

tactics /'tæktɪks/ *plural noun* (also treated as *singular*) disposition of armed forces, esp. in warfare; procedure calculated to gain some end, skilful device(s). □ **tactician** /-'tɪʃ-/ *noun.*

tactile /'tæktaɪl/ *adjective* of sense of touch; perceived by touch. □ **tactility** /-'tɪl-/ *noun.*

tadpole /'tædpəʊl/ *noun* larva of frog, toad, etc. at stage of living in water and having gills and tail.

taffeta /'tæfɪtə/ *noun* fine lustrous silk or silklike fabric.

taffrail /'tæfreɪl/ *noun* rail round ship's stern.

tag ● *noun* label, esp. to show address or price; metal point of shoelace etc.; loop or flap for handling or hanging thing; loose or ragged end; trite quotation, stock phrase. ● *verb* (**-gg-**) furnish with tag(s); (often + *on*, *on to*) join, attach. □ **tag along** (often + *with*) go along, accompany passively.

tagliatelle /tæljə'telɪ/ *noun* ribbon-shaped pasta.

tail[1] ● *noun* hindmost part of animal, esp. extending beyond body; thing like tail in form or position, esp. rear part of aeroplane, vehicle, etc., hanging part of back of shirt or coat, end of procession, luminous trail following comet, etc.; inferior, weak, or last part of anything; (in *plural*) *colloquial* tailcoat, evening dress with this; (in *plural*) reverse of coin turning up in toss; *colloquial* person following another. ● *verb* remove stalks of (fruit etc.); *colloquial* follow closely. □ **tailback** long queue of traffic caused by obstruction; **tailboard** hinged or removable back of lorry etc.; **tailcoat** man's coat divided at back and cut away in front; **tailgate** tailboard, rear door of estate car; **tail-light**, **-lamp** *US* rear light on vehicle etc.; **tail off**, **away** gradually diminish and cease; **tailpiece** final part of thing, decoration at end of chapter etc.; **tailplane** horizontal aerofoil at tail of aircraft; **tailspin** aircraft's spinning dive, state of panic; **tail wind** one blowing in direction of travel. □ **tailless** *adjective.*

tail[2] *Law* ● *noun* limitation of ownership, esp. of estate limited to person and his heirs. ● *adjective* so limited.

tailor /'teɪlə/ ● *noun* maker of (esp. men's) outer garments to measure. ● *verb* make (clothes) as tailor; make or adapt for special purpose; work as tailor. □ **tailor-made** made by tailor, made or suited for particular purpose. □ **tailored** *adjective.*

taint ● *noun* spot or trace of decay, corruption, etc.; corrupt condition, infection. ● *verb* affect with taint, become tainted; (+ *with*) affect slightly.

take ● *verb* (**-king**; *past* **took** /tʊk/; *past participle* **taken**) lay hold of; acquire, capture, earn, win; regularly buy (newspaper etc.); occupy; make use of; be effective; consume; use up; carry, accompany; remove, steal; catch, be infected with; be affected by (pleasure etc.); ascertain and record; grasp mentally, understand; accept, submit to; deal with or regard in specified way;

teach, be taught or examined in; submit to (exam); make (photograph); have as necessary accompaniment, requirement, or part. ● *noun* amount taken or caught; scene or film sequence photographed continuously. □ **take after** resemble (parent etc.); **take against** begin to dislike; **take apart** dismantle, *colloquial* defeat, criticize severely; **take away** remove or carry elsewhere, subtract; **take-away** (cooked meal) bought at restaurant for eating elsewhere, restaurant selling this; **take back** retract (statement), convey to original position, carry in thought to past time, return or accept back (goods); **take down** write down (spoken words), dismantle (structure), lower (garment); **take-home pay** employee's pay after deduction of tax etc.; **take in** receive as lodger etc., undertake (work) at home, make (garment etc.) smaller, understand, cheat, include; **take off** remove (clothing), deduct, mimic, begin a jump, become airborne, (of scheme etc.) become successful; **take-off** act of mimicking or becoming airborne; **take on** undertake, acquire, engage, agree to oppose at game, *colloquial* show strong emotion; **take out** remove, escort on outing, get (licence etc.); **take over** succeed to management or ownership of, assume control; **takeover** *noun*; **take to** begin, have recourse to, form liking for; **take up** adopt as pursuit, accept (offer etc.), occupy (time or space), absorb, (often + *on*) interrupt or correct (speaker), shorten (garment), pursue (matter). □ **taker** *noun*.

taking ● *adjective* attractive, captivating. ● *noun* (in *plural*) money taken in business etc.

talc *noun* talcum powder; translucent mineral formed in thin plates.

talcum /ˈtælkəm/ *noun* talc; (in full **talcum powder**) usually perfumed powdered talc for toilet use.

tale *noun* narrative or story, esp. fictitious; allegation or gossip, often malicious.

talent /ˈtælənt/ *noun* special aptitude or gift; high mental ability; people of talent; *colloquial* attractive members of opposite sex; ancient weight and money unit. □ **talent-scout**, **-spotter** person seeking new talent, esp. in sport or entertainment. □ **talented** *adjective*.

talisman /ˈtælɪzmən/ *noun* (*plural* **-s**) thing believed to bring good luck or protect from harm.

talk /tɔːk/ ● *verb* (often + *to*, *with*) converse or communicate verbally; have power of speech; express, utter, discuss; use (language); gossip. ● *noun* conversation; particular mode of speech; short address or lecture; rumour or gossip, its theme; *colloquial* empty boasting; (often in *plural*) discussions, negotiations. □ **talk down** silence by loud or persistent talking, guide (pilot, aircraft) to landing by radio, (+ *to*) speak patronizingly to; **talk into** persuade by talking; **talk of** discuss, express intention of; **talk over** discuss; **talk round** persuade to change opinion etc.; **talk to** rebuke, scold. □ **talker** *noun*.

talkative /ˈtɔːkətɪv/ *adjective* fond of talking.

talkie *noun colloquial* early film with soundtrack.

talking ● *adjective* that talks or can talk; expressive. ● *noun* action or process of talking. □ **talking of** while we are discussing; **talking point** topic for discussion; **talking-to** *colloquial* scolding.

tall /tɔːl/ ● *adjective* of more than average height; of specified height; higher than surroundings. ● *adverb* as if tall; proudly. □ **tallboy** tall chest of drawers; **tall order** unreasonable demand; **tall ship** high-masted sailing ship; **tall story** *colloquial* extravagant tale.

tallow /ˈtæləʊ/ *noun* hard (esp. animal) fat melted down to make candles, soap, etc.

tally /ˈtælɪ/ ● *noun* (*plural* **-ies**) reckoning of debt or score; mark registering number of objects delivered or received; *historical* piece of notched wood for keeping account; identification ticket or label; counterpart, duplicate. ● *verb* (**-ies**, **-ied**) (often + *with*) agree, correspond.

tally-ho /tælɪˈhəʊ/ *interjection huntsman's cry as signal on seeing fox*.

Talmud /ˈtælmʊd/ *noun* body of Jewish civil and ceremonial law. □ **Talmudic** /-ˈmʊd-/ *adjective*; **Talmudist** *noun*.

talon /ˈtælən/ *noun* claw, esp. of bird of prey.

talus /ˈteɪləs/ *noun* (*plural* **tali** /-laɪ/) ankle-bone supporting tibia.

tamarind /ˈtæmərɪnd/ *noun* tropical evergreen tree; its fruit pulp used as food and in drinks.

tamarisk /ˈtæmərɪsk/ *noun* seaside shrub usually with small pink or white flowers.

tambour /ˈtæmbʊə/ *noun* drum; circular frame for stretching embroidery-work on.

tambourine /tæmbəˈriːn/ *noun* small shallow drum with jingling discs in rim, shaken or banged as accompaniment.

tame ● *adjective* (of animal) domesticated, not wild or shy; uninteresting, insipid. ● *verb* (**-ming**) make tame, domesticate; subdue. □ **tamely** *adverb*; **tameness** *noun*; **tamer** *noun*.

Tamil /ˈtæmɪl/ ● *noun* member of a people inhabiting South India and Sri Lanka; their language. ● *adjective* of this people or language.

tam-o'-shanter /tæməˈʃæntə/ *noun* floppy woollen beret of Scottish origin.

tamp *verb* ram down tightly.

tamper /ˈtæmpə/ *verb* (+ *with*) meddle or interfere with.

tampon /ˈtæmpɒn/ *noun* plug of cotton wool etc. used esp. to absorb menstrual blood.

tan¹ ● *noun* suntan; yellowish-brown colour; bark of oak etc. used for tanning. ● *adjective* yellowish-brown. ● *verb* (**-nn-**) make or become brown by exposure to sun; convert (raw hide) into leather; *slang* thrash.

tan² *abbreviation* tangent.

tandem /ˈtændəm/ ● *noun* bicycle with two or more seats one behind another; vehicle driven tandem. ● *adverb* with two or more horses harnessed one behind another. □ **in tandem** one behind the other, alongside each other, together.

tandoor /ˈtændʊə/ *noun* clay oven.

tandoori /tænˈdʊərɪ/ *noun* spiced food cooked in tandoor.

tang *noun* strong taste or smell; characteristic quality; part of tool by which blade is held firm in handle. □ **tangy** *adjective* (**-ier**, **-iest**).

tangent /ˈtændʒ(ə)nt/ *noun* straight line, curve, or surface touching but not intersecting curve; ratio of sides opposite and adjacent to acute angle in right-angled triangle. □ **at a tangent** diverging from previous course or from what is relevant. □ **tangential** /-ˈdʒenʃ(ə)l/ *adjective*.

tangerine /tændʒəˈriːn/ *noun* small sweet-scented fruit like orange, mandarin; deep orange-yellow colour.

tangible /ˈtændʒɪb(ə)l/ *adjective* perceptible by touch; definite, clearly intelligible, not elusive. □ **tangibility** *noun*; **tangibly** *adverb*.

tangle /ˈtæŋg(ə)l/ ● *verb* (**-ling**) intertwine or become twisted or involved in confused mass; entangle; complicate. ● *noun* tangled mass or state. □ **tangly** *adjective*.

tango /ˈtæŋgəʊ/ ● *noun* (*plural* **-s**) (music for) slow South American ballroom dance. ● *verb* (**-goes**, **-goed**) dance tango.

tank *noun* large receptacle for liquid, gas, etc.; heavy armoured fighting vehicle moving on continuous tracks. □ **tank engine** steam engine with integral fuel and water containers. □ **tankful** *noun* (*plural* **-s**).

tankard /ˈtæŋkəd/ *noun* (contents of) tall beer mug with handle.

tanker *noun* ship, aircraft, or road vehicle for carrying liquids, esp. oil, in bulk.

tanner *noun* person who tans hides.

tannery *noun* (*plural* **-ies**) place where hides are tanned.

tannic /ˈtænɪk/ *adjective* of tan. □ **tannic acid** yellowish organic compound used in cleaning, dyeing, etc.

tannin /ˈtænɪn/ *noun* any of several substances extracted from tree-barks etc. and used in tanning etc.

Tannoy /ˈtænɔɪ/ *noun proprietary term* type of public address system.

tansy /ˈtænzɪ/ *noun* (*plural* **-ies**) aromatic herb with yellow flowers.

tantalize /ˈtæntəlaɪz/ *verb* (also **-ise**) (**-zing** or **-sing**) torment with sight of the unobtainable, raise and then dash the hopes of. □ **tantalization** *noun*.

tantamount /ˈtæntəmaʊnt/ *adjective* (+ *to*) equivalent to.

tantra /'tæntrə/ *noun* any of a class of Hindu or Buddhist mystical or magical writings.

tantrum /'tæntrəm/ *noun* (esp. child's) outburst of bad temper or petulance.

Taoiseach /'ti:ʃəx/ *noun* prime minister of Irish Republic.

tap¹ ● *noun* device by which flow of liquid or gas from pipe or vessel can be controlled; act of tapping telephone; taproom. ● *verb* (**-pp-**) provide (cask) with tap, let out (liquid) thus; draw sap from (tree) by cutting into it; draw supplies or information from; discover and exploit; connect listening device to (telephone etc.). □ **on tap** ready to be drawn off, *colloquial* freely available; **taproom** room in pub serving drinks on tap; **tap root** tapering root growing vertically downwards.

tap² ● *verb* (**-pp-**) (+ *at*, *on*, *against*, etc.) strike or cause to strike lightly; (often + *out*) make by taps; tap-dance. ● *noun* light blow or rap; tap-dancing; metal attachment on dancer's shoe. □ **tap-dance** *noun* rhythmic dance performed in shoes with metal taps, *verb* perform this; **tap-dancer** *noun*; **tap-dancing** *noun*.

tapas /'tæpæs/ *plural noun* small savoury esp. Spanish dishes.

tape ● *noun* narrow woven strip of cotton etc. for fastening etc.; this across finishing line of race; (in full **adhesive tape**) strip of adhesive plastic etc. for fastening, masking, insulating, etc.; magnetic tape; tape recording; tape-measure. ● *verb* (**-ping**) tie up or join with tape; apply tape to; (+ *off*) seal off with tape; record on magnetic tape; measure with tape. □ **have (person, thing) taped** *colloquial* understand fully; **tape deck** machine for using audiotape (separate from speakers etc.); **tape machine** device for recording telegraph messages, tape recorder; **tape-measure** strip of tape or thin flexible metal marked for measuring; **tape recorder** apparatus for recording and replaying sounds on magnetic tape; **tape-record** *verb*; **tape recording** *noun*; **tapeworm** tapelike worm parasitic in alimentary canal.

taper /'teɪpə/ ● *noun* wick coated with wax etc. for conveying flame; slender candle. ● *verb* (often + *off*) (cause to) diminish in thickness towards one end, make or become gradually less.

tapestry /'tæpɪstrɪ/ *noun* (*plural* **-ies**) thick fabric in which coloured weft threads are woven to form pictures or designs; (usually wool) embroidery imitating this; piece of this.

tapioca /tæpɪ'əʊkə/ *noun* starchy granular foodstuff prepared from cassava.

tapir /'teɪpə/ *noun* small piglike mammal with short flexible snout.

tappet /'tæpɪt/ *noun* lever etc. in machinery giving intermittent motion.

tar¹ ● *noun* dark thick inflammable liquid distilled from wood, coal, etc.; similar substance formed in combustion of tobacco. ● *verb* (**-rr-**) cover with tar.

tar² *noun colloquial* sailor.

taramasalata /tærəməsə'lɑ:tə/ *noun* (also **taramosalata**) dip made from roe, olive oil, etc.

tarantella /tærən'telə/ *noun* (music for) whirling Southern Italian dance.

tarantula /tə'ræntjʊlə/ *noun* large hairy tropical spider; large black spider of Southern Europe.

tarboosh /tɑ:'bu:ʃ/ *noun* cap like fez.

tardy /'tɑ:dɪ/ *adjective* (**-ier**, **-iest**) slow to act, come, or happen; delaying, delayed. □ **tardily** *adverb*; **tardiness** *noun*.

tare¹ /teə/ *noun* vetch, esp. as cornfield weed or fodder; (in *plural*) *Biblical* injurious cornfield weed.

tare² /teə/ *noun* allowance made for weight of packing around goods; weight of vehicle without fuel or load.

target /'tɑ:gɪt/ ● *noun* mark fired at, esp. round object marked with concentric circles; person, objective, or result aimed at; butt for criticism etc. ● *verb* (**-t-**) single out as target; aim, direct.

tariff /'tærɪf/ *noun* table of fixed charges; duty on particular class of goods; list of duties or customs due.

tarlatan /'tɑ:lət(ə)n/ *noun* thin stiff muslin.

Tarmac /'tɑ:mæk/ ● *noun proprietary term* tarmacadam; area surfaced with this. ● *verb* (**tarmac**) (**-ck-**) apply tarmacadam to.

tarmacadam /tɑ:mə'kædəm/ *noun* bitumen-bound stones etc. used as paving.

tarn *noun* small mountain lake.

tarnish /ˈtɑːnɪʃ/ ● *verb* (cause to) lose lustre; impair (reputation etc.). ● *noun* tarnished state; stain, blemish.

taro /ˈtɑːrəʊ/ *noun* (*plural* **-s**) tropical plant with edible tuberous roots.

tarot /ˈtærəʊ/ *noun* (in *singular* or *plural*) pack of 78 cards used in fortune-telling.

tarpaulin /tɑːˈpɔːlɪn/ *noun* waterproof cloth, esp. of tarred canvas; sheet or covering of this.

tarragon /ˈtærəgən/ *noun* aromatic herb.

tarry[1] /ˈtɑːrɪ/ *adjective* (**-ier**, **-iest**) of or smeared with tar.

tarry[2] /ˈtærɪ/ *verb* (**-ies**, **-ied**) *archaic* linger; stay, wait.

tarsal /ˈtɑːs(ə)l/ ● *adjective* of the ankle-bones. ● *noun* tarsal bone.

tarsus /ˈtɑːsəs/ *noun* (*plural* **tarsi** /-saɪ/) bones of ankle and upper foot.

tart[1] *noun* pastry case containing fruit, jam, etc. □ **tartlet** *noun*.

tart[2] ● *noun slang* prostitute, promiscuous woman. ● *verb* (+ *up*) *colloquial* smarten or dress up, esp. gaudily. □ **tarty** *adjective* (**-ier**, **-iest**).

tart[3] *adjective* sharp-tasting, acid; (of remark etc.) cutting, biting. □ **tartly** *adverb*; **tartness** *noun*.

tartan /ˈtɑːt(ə)n/ *noun* (woollen cloth woven in) pattern of coloured stripes crossing at right angles, esp. denoting a Scottish Highland clan.

Tartar /ˈtɑːtə/ *noun* member of group of Central Asian people including Mongols and Turks; their Turkic language; (**tartar**) harsh or formidable person. □ **tartar sauce** mayonnaise with chopped gherkins etc.

tartar /ˈtɑːtə/ *noun* hard deposit that forms on teeth; deposit forming hard crust in wine casks.

tartaric /tɑːˈtærɪk/ *adjective* of tartar. □ **tartaric acid** organic acid found esp. in unripe grapes.

tartrazine /ˈtɑːtrəziːn/ *noun* brilliant yellow dye from tartaric acid, used to colour food etc.

task /tɑːsk/ ● *noun* piece of work to be done. ● *verb* make great demands on. □ **take to task** rebuke, scold; **task force** specially organized unit for task; **taskmaster**, **taskmistress** person who makes others work hard.

Tass *noun* official Russian news agency.

tassel /ˈtæs(ə)l/ *noun* tuft of hanging threads etc. as ornament; tassel-like flowerhead of plant. □ **tasselled** *adjective* (*US* **tasseled**).

taste /teɪst/ ● *noun* (faculty of perceiving) sensation caused in mouth by contact with substance; flavour; small sample of food etc.; slight experience; (often + *for*) liking, predilection; aesthetic discernment in art, clothes, conduct, etc. ● *verb* (**-ting**) perceive or sample flavour of; eat small portion of; experience; (often + *of*) have specified flavour. □ **taste bud** organ of taste on surface of tongue.

tasteful *adjective* done in or having good taste. □ **tastefully** *adverb*; **tastefulness** *noun*.

tasteless *adjective* flavourless; having or done in bad taste. □ **tastelessly** *adverb*; **tastelessness** *noun*.

taster *noun* person employed to test food or drink by tasting; small sample.

tasting *noun* gathering at which food or drink is tasted and evaluated.

tasty *adjective* (**-ier**, **-iest**) of pleasing flavour, appetizing; *colloquial* attractive. □ **tastiness** *noun*.

tat[1] *noun colloquial* tatty things; junk.

tat[2] *verb* (**-tt-**) do, or make by, tatting.

ta-ta /tæˈtɑː/ *interjection colloquial* goodbye.

tatter /ˈtætə/ *noun* (usually in *plural*) rag, irregularly torn cloth, paper, etc. □ **in tatters** *colloquial* torn in many places, ruined.

tattered *adjective* in tatters.

tatting /ˈtætɪŋ/ *noun* (process of making) kind of handmade knotted lace.

tattle /ˈtæt(ə)l/ ● *verb* (**-ling**) prattle, chatter, gossip. ● *noun* gossip, idle talk.

tattoo[1] /təˈtuː/ ● *verb* (**-oos**, **-ooed**) mark (skin) by puncturing and inserting pigment; make (design) thus. ● *noun* such design. □ **tattooer** *noun*; **tattooist** *noun*.

tattoo[2] /təˈtuː/ *noun* evening signal recalling soldiers to quarters; elaboration of this with music and marching etc. as entertainment; drumming, rapping; drumbeat.

tatty /ˈtætɪ/ *adjective* (**-ier**, **-iest**) *colloquial* tattered; shabby, inferior, tawdry. □ **tattily** *adverb*; **tattiness** *noun*.

taught *past & past participle* of TEACH.

taunt • *noun* insult; provocation. • *verb* insult; provoke contemptuously.

taupe /təʊp/ *noun* grey tinged with esp. brown.

Taurus /'tɔ:rəs/ *noun* second sign of zodiac.

taut *adjective* (of rope etc.) tight; (of nerves etc.) tense; (of ship etc.) in good condition. □ **tauten** *verb*; **tautly** *adverb*; **tautness** *noun*.

tautology /tɔ:'tɒlədʒɪ/ *noun* (*plural* **-ies**) repetition of same thing in different words. □ **tautological** /-tə'lɒdʒ-/ *adjective*; **tautologous** /-ləgəs/ *adjective*.

tavern /'tæv(ə)n/ *noun archaic or literary* inn, pub.

taverna /tə'vɜ:nə/ *noun* Greek restaurant.

tawdry /'tɔ:drɪ/ *adjective* (**-ier**, **-iest**) showy but worthless; gaudy.

tawny /'tɔ:nɪ/ *adjective* (**-ier**, **-iest**) of orange-brown colour. □ **tawny owl** reddish-brown European owl.

tax • *noun* money compulsorily levied by state on person, property, business, etc.; (+ *on*, *upon*) strain, heavy demand. • *verb* impose tax on; deduct tax from; make demands on; (often + *with*) charge, call to account. □ **tax avoidance** minimizing tax payment by financial manoeuvring; **tax-deductible** (of expenses) legally deductible from income before tax assessment; **tax disc** licence on vehicle certifying payment of road tax; **tax evasion** illegal non-payment of taxes; **tax-free** exempt from tax; **taxman** *colloquial* inspector or collector of taxes; **taxpayer** person who pays taxes; **tax return** declaration of income etc. for taxation purposes.

taxation /tæk'seɪʃ(ə)n/ *noun* imposition or payment of tax.

taxi /'tæksɪ/ • *noun* (*plural* **-s**) (in full **taxi-cab**) car plying for hire and usually fitted with taximeter. • *verb* (**taxis**, **taxied**, **taxiing** or **taxying**) (of aircraft) go along ground before or after flying; go or carry in taxi. □ **taxi rank** (*US* **taxi stand**) place where taxis wait to be hired.

taxidermy /'tæksɪdɜ:mɪ/ *noun* art of preparing, stuffing, and mounting skins of animals. □ **taxidermist** *noun*.

taximeter /'tæksɪmi:tə/ *noun* automatic fare-indicator in taxi.

taxon /'tæks(ə)n/ *noun* (*plural* **taxa**) any taxonomic group.

taxonomy /tæk'sɒnəmɪ/ *noun* classification of living and extinct organisms. □ **taxonomic** /-sə'nɒm-/ *adjective*; **taxonomical** /-sə'nɒm-/ *adjective*; **taxonomist** *noun*.

tayberry /'teɪbərɪ/ *noun* (*plural* **-ies**) hybrid fruit between blackberry and raspberry.

TB *abbreviation* tubercle bacillus; tuberculosis.

tbsp. *abbreviation* tablespoonful.

te /ti:/ *noun* (also **ti**) seventh note of scale in tonic sol-fa.

tea *noun* (in full **tea plant**) Asian evergreen shrub or small tree; its dried leaves; infusion of these leaves as drink; infusion made from other leaves etc.; light meal in afternoon or evening. □ **tea bag** small permeable bag of tea for infusion; **tea break** pause in work for drinking tea; **tea caddy** container for tea; **teacake** light usually toasted sweet bun; **tea chest** light metal-lined wooden box for transporting tea; **tea cloth** tea towel; **tea cosy** cover to keep teapot warm; **teacup** cup from which tea is drunk, amount it holds; **tea leaf** leaf of tea, esp. (in *plural*) after infusion; **teapot** pot with handle and spout, in which tea is made; **tearoom** small unlicensed café; **tea rose** rose with scent like tea; **teaset** set of crockery for serving tea; **teashop** tearoom; **teaspoon** small spoon for stirring tea etc., (also **teaspoonful**) amount held by this; **tea towel** cloth for drying washed crockery etc.; **tea trolley** (*US* **tea wagon**) small trolley from which tea is served.

teach *verb* (*past & past participle* **taught** /tɔ:t/) give systematic information, instruction, or training to (person) or about (subject, skill); practise this as a profession; advocate as moral etc. principle; (+ *to do*) instruct to, *colloquial* discourage from. □ **teachable** *adjective*.

teacher *noun* person who teaches, esp. in school.

teaching *noun* teacher's profession; (often in *plural*) what is taught; doctrine.

teak *noun* a hard durable wood.

teal *noun* (*plural* same) small freshwater duck.

team ● *noun* set of players etc. in game or sport; set of people working together; set of draught animals. ● *verb* (usually + *up*) join in team or in common action; (+ *with*) coordinate, match. □ **team-mate** fellow member of team; **team spirit** willingness to act for communal good; **teamwork** combined effort, cooperation.

teamster /'ti:mstə/ *noun US* lorry driver; driver of team.

tear[1] /teə/ ● *verb* (*past* **tore**; *past participle* **torn**) (often + *up*) pull (apart) with some force; make (hole, rent) thus; undergo this; (+ *away, off, at*, etc.) pull violently; violently disrupt; *colloquial* go hurriedly. ● *noun* hole etc. caused by tearing; torn part of cloth etc. □ **tear apart** search exhaustively, criticize forcefully, divide utterly, distress greatly; **tearaway** *colloquial* unruly young person; **tearing hurry** *colloquial* great hurry; **tear into** *colloquial* severely reprimand, start (activity) vigorously; **tear to shreds** *colloquial* criticize thoroughly.

tear[2] /tɪə/ *noun* drop of clear salty liquid secreted from eye and shed esp. in grief. □ **in tears** weeping; **teardrop** single tear; **tear duct** drain carrying tears to or from eye; **tear gas** gas causing severe irritation to the eyes.

tearful *adjective* in, given to, or accompanied with tears. □ **tearfully** *adverb*.

tease /ti:z/ ● *verb* (**-sing**) make fun of; irritate; entice sexually while refusing to satisfy desire; pick (wool etc.) into separate fibres; raise nap on (cloth) with teasels etc.; (+ *out*) extract or obtain by careful effort. ● *noun colloquial* person fond of teasing; act of teasing.

teasel /'ti:z(ə)l/ *noun* (also **teazel**, **teazle**) plant with prickly flower heads, used dried for raising nap on cloth; other device used for this.

teaser *noun colloquial* hard question or problem.

teat *noun* nipple on breast or udder; rubber etc. nipple for sucking milk from bottle.

teazel (also **teazle**) = TEASEL.

TEC /tek/ *abbreviation* Training and Enterprise Council.

tec *noun colloquial* detective.

tech /tek/ *noun* (also **tec**) *colloquial* technical college.

technic /'teknɪk/ *noun* (usually in *plural*) technology; technical terms, methods, etc.; technique.

technical *adjective* of the mechanical arts and applied sciences; of a particular subject, craft, etc.; using technical language; specialized; due to mechanical failure; in strict legal sense. □ **technical knockout** referee's ruling that boxer has lost because he is unfit to continue. □ **technically** *adverb*.

technicality /teknɪ'kælɪtɪ/ *noun* (*plural* **-ies**) being technical; technical expression; technical point or detail.

technician /tek'nɪʃ(ə)n/ *noun* person doing practical or maintenance work in laboratory; person skilled in artistic etc. technique; expert in practical science.

Technicolor /'teknɪkʌlə/ *noun proprietary term* process of colour cinematography; (usually **technicolour**) *colloquial* vivid or artificial colour.

technique /tek'ni:k/ *noun* mechanical skill in art; method of achieving purpose, esp. by manipulation; manner of execution in music, painting, etc.

technocracy /tek'nɒkrəsɪ/ *noun* (*plural* **-ies**) (instance of) rule or control by technical experts.

technocrat /'teknəkræt/ *noun* exponent or advocate of technocracy. □ **technocratic** /-'kræt-/ *adjective*.

technology /tek'nɒlədʒɪ/ *noun* (*plural* **-ies**) knowledge or use of mechanical arts and applied sciences; these subjects collectively. □ **technological** /-nə'lɒdʒ-/ *adjective*; **technologically** /-nə'lɒdʒ-/ *adverb*; **technologist** *noun*.

tectonic /tek'tɒnɪk/ *adjective* of building or construction; of changes in the earth's crust.

tectonics *plural noun* (usually treated as *singular*) study of earth's large-scale structural features.

teddy /'tedɪ/ *noun* (also **Teddy**) (*plural* **-ies**) (in full **teddy bear**) soft toy bear.

Teddy boy /'tedɪ/ *noun colloquial* 1950s youth with Edwardian-style clothing, hair, etc.

tedious /'ti:dıəs/ *adjective* tiresomely long, wearisome. □ **tediously** *adverb*; **tediousness** *noun*.

tedium /'ti:dıəm/ *noun* tediousness.

tee[1] *noun* letter T.

tee[2] ● *noun* cleared space from which golf ball is struck at start of play for each hole; small wood or plastic support for golf ball used then; mark aimed at in bowls, quoits, curling, etc. ● *verb* (**tees**, **teed**) (often + *up*) place (ball) on tee. □ **tee off** make first stroke in golf, *colloquial* start, begin.

teem[1] *verb* be abundant; (+ *with*) be full of, swarm with.

teem[2] *verb* (often + *down*) pour (esp. of rain).

teen *adjective* teenage.

teenage /'ti:neıdʒ/ *adjective* of or characteristic of teenagers. □ **teenaged** *adjective*.

teenager /'ti:neıdʒə/ *noun* person in teens.

teens /ti:nz/ *plural noun* years of one's age from 13 to 19.

teensy /'ti:nzı/ *adjective* (**-ier**, **-iest**) *colloquial* teeny.

teeny /'ti:nı/ *adjective* (**-ier**, **-iest**) *colloquial* tiny.

teepee = TEPEE.

teeter /'ti:tə/ *verb* totter, stand or move unsteadily.

teeth *plural* of TOOTH.

teethe /ti:ð/ *verb* (**-thing**) grow or cut teeth, esp. milk teeth. □ **teething ring** ring for infant to bite on while teething; **teething troubles** initial troubles in an enterprise etc.

teetotal /ti:'təʊt(ə)l/ *adjective* advocating or practising total abstinence from alcohol. □ **teetotalism** *noun*; **teetotaller** *noun*.

TEFL /'tef(ə)l/ *abbreviation* teaching of English as a foreign language.

Teflon /'teflɒn/ *noun proprietary term* non-stick coating for kitchen utensils.

Tel. *abbreviation* (also **tel.**) telephone.

tele- *combining form* at or to a distance; television; by telephone.

telecast /'telıkɑ:st/ ● *noun* television broadcast. ● *verb* transmit by television. □ **telecaster** *noun*.

telecommunication /telıkəmju:nı'keıʃ(ə)n/ *noun* communication over distances by cable, fibre optics, satellites, radio, etc.; (in *plural*) technology of this.

telefax /'telıfæks/ *noun* fax.

telegram /'telıgræm/ *noun* message sent by telegraph.

telegraph /'telıgrɑ:f/ ● *noun* (device or system for) transmitting messages to a distance by making and breaking electrical connection. ● *verb* (often + *to*) send message or communicate by telegraph. □ **telegraphist** /tı'legrə-/ *noun*.

telegraphic /telı'græfık/ *adjective* of or by telegraphs or telegrams; economically worded. □ **telegraphically** *adverb*.

telegraphy /tı'legrəfı/ *noun* communication by telegraph.

telekinesis /telıkaı'ni:sıs/ *noun* supposed paranormal force moving objects at a distance. □ **telekinetic** /-'net-/ *adjective*.

telemessage /'telımesıdʒ/ *noun* message sent by telephone or telex and delivered in printed form.

telemetry /tı'lemətrı/ *noun* process of recording readings of instrument and transmitting them by radio. □ **telemeter** /tı'lemıtə/ *noun*.

teleology /ti:lı'ɒlədʒı/ *noun* (*plural* **-ies**) *Philosophy* explanation of phenomena by purpose they serve. □ **teleological** /-ə'lɒdʒ-/ *adjective*.

telepathy /tı'lepəθı/ *noun* supposed paranormal communication of thoughts. □ **telepathic** /telı'pæθık/ *adjective*; **telepathically** /telı'pæθ-/ *adverb*.

telephone /'telıfəʊn/ ● *noun* apparatus for transmitting sound (esp. speech) to a distance; instrument used in this; system of communication by network of telephones. ● *verb* (**-ning**) send (message) or speak to by telephone; make telephone call. □ **on the telephone** having or using a telephone; **telephone book**, **directory** book listing telephone subscribers and numbers; **telephone booth**, **box**, **kiosk** booth etc. with telephone for public use; **telephone number** number used to call a particular telephone. □ **telephonic** /-'fɒn-/ *adjective*; **telephonically** /-'fɒn-/ *adverb*.

telephonist /tɪ'lefənɪst/ *noun* operator in telephone exchange or at switchboard.

telephony /tɪ'lefənɪ/ *noun* transmission of sound by telephone.

telephoto /telɪ'fəʊtəʊ/ *noun* (*plural* **-s**) (in full **telephoto lens**) lens used in telephotography.

telephotography /telɪfə'tɒgrəfɪ/ *noun* photographing of distant object with combined lenses giving large image. □ **telephotographic** /-fəʊtə'græf-/ *adjective.*

teleprinter /'telɪprɪntə/ *noun* device for sending, receiving, and printing telegraph messages.

teleprompter /'telɪprɒmptə/ *noun* device beside esp. television camera that slowly unrolls script out of sight of audience.

telesales /'telɪseɪlz/ *plural noun* selling by telephone.

telescope /'telɪskəʊp/ ● *noun* optical instrument using lenses or mirrors to magnify distant objects; radio telescope. ● *verb* (**-ping**) press or drive (sections of tube etc.) one into another; close or be capable of closing thus; compress.

telescopic /telɪ'skɒpɪk/ *adjective* of or made with telescope; (esp. of lens) able to magnify distant objects; consisting of sections that telescope. □ **telescopic sight** telescope on rifle etc. used for sighting. □ **telescopically** *adverb.*

teletext /'telɪtekst/ *noun* computerized information service transmitted to subscribers' televisions.

telethon /'telɪθɒn/ *noun* long television programme to raise money for charity.

televise /'telɪvaɪz/ *verb* (**-sing**) transmit by television.

television /'telɪvɪʒ(ə)n/ *noun* system for reproducing on a screen visual images transmitted (with sound) by radio signals or cable; (in full **television set**) device with screen for receiving these signals; television broadcasting. □ **televisual** /-'vɪʒʊəl/ *adjective.*

telex /'teleks/ (also **Telex**) ● *noun* international system of telegraphy using teleprinters and public telecommunication network. ● *verb* send, or communicate with, by telex.

tell *verb* (*past & past participle* **told** /təʊld/) relate in speech or writing; make known, express in words; (often + *of*, *about*) divulge information, reveal secret etc.; (+ *to do*) direct, order; decide about, distinguish; (often + *on*) produce marked effect or influence. □ **tell apart** distinguish between; **tell off** *colloquial* scold; **tell-tale** *noun* person who tells tales, automatic registering device, *adjective* serving to reveal or betray something; **tell tales** make known person's faults etc.

teller *noun* person employed to receive and pay out money in bank etc.; person who counts votes; person who tells esp. stories.

telling *adjective* having marked effect, striking. □ **tellingly** *adverb.*

telly /'telɪ/ *noun* (*plural* **-ies**) *colloquial* television.

temerity /tɪ'merɪtɪ/ *noun* rashness, audacity.

temp *colloquial* ● *noun* temporary employee, esp. secretary. ● *verb* work as temp.

temper /'tempə/ ● *noun* mental disposition, mood; irritation, anger; tendency to become angry; composure, calmness; metal's hardness or elasticity. ● *verb* bring (clay, metal) to proper consistency or hardness; (+ *with*) moderate, mitigate.

tempera /'tempərə/ *noun* method of painting using emulsion e.g. of pigment with egg.

temperament /'temprəmənt/ *noun* person's or animal's nature and character.

temperamental /temprə'ment(ə)l/ *adjective* regarding temperament; unreliable, moody; *colloquial* unpredictable. □ **temperamentally** *adverb.*

temperance /'tempərəns/ *noun* moderation, esp. in eating and drinking; abstinence, esp. total, from alcohol.

temperate /'tempərət/ *adjective* avoiding excess, moderate; (of region or climate) mild.

temperature /'temprɪtʃə/ *noun* measured or perceived degree of heat or cold of thing, region, etc.; *colloquial* body temperature above normal.

tempest /'tempɪst/ *noun* violent storm.

tempestuous /tem'pestʃʊəs/ *adjective* stormy, turbulent. □ **tempestuously** *adverb*.

tempi *plural* of TEMPO.

template /'templɪt/ *noun* thin board or plate used as guide in drawing, cutting, drilling, etc.

temple[1] /'temp(ə)l/ *noun* building for worship, or treated as dwelling place, of god(s).

temple[2] /'temp(ə)l/ *noun* flat part of side of head between forehead and ear.

tempo /'tempəʊ/ *noun* (*plural* **-pos** or **-pi** /-pi:/) speed at which music is (to be) played; speed, pace.

temporal /'tempər(ə)l/ *adjective* worldly as opposed to spiritual, secular; of time; of the temples of the head.

temporary /'tempərərɪ/ ● *adjective* lasting or meant to last only for limited time. ● *noun* (*plural* **-ies**) person employed temporarily. □ **temporarily** *adverb*.

temporize /'tempəraɪz/ *verb* (also **-ise**) (**-zing** or **-sing**) avoid committing oneself, so as to gain time; procrastinate; comply temporarily.

tempt *verb* entice, incite to what is forbidden; allure, attract; risk provoking. □ **tempter** *noun*; **tempting** *adjective*; **temptingly** *adverb*; **temptress** *noun*.

temptation /temp'teɪʃ(ə)n/ *noun* tempting, being tempted; incitement, esp. to wrongdoing; attractive thing or course of action.

ten *adjective & noun* one more than nine. □ **the Ten Commandments** rules of conduct given by God to Moses. □ **tenth** *adjective & noun*.

tenable /'tenəb(ə)l/ *adjective* maintainable against attack or objection; (+ *for*, *by*) (of office etc.) that can be held for period or by (person etc.). □ **tenability** *noun*.

tenacious /tɪ'neɪʃəs/ *adjective* (often + *of*) keeping firm hold; persistent, resolute; (of memory) retentive. □ **tenaciously** *adverb*; **tenacity** /-'næs-/ *noun*.

tenancy /'tenənsɪ/ *noun* (*plural* **-ies**) (duration of) tenant's status or possession.

tenant /'tenənt/ *noun* person who rents land or property from landlord; (often + *of*) occupant of place.

tenantry *noun* tenants of estate etc.

tench *noun* (*plural* same) freshwater fish of carp family.

tend[1] *verb* (often + *to*) be apt or inclined; be moving; hold a course.

tend[2] *verb* take care of, look after.

tendency /'tendənsɪ/ *noun* (*plural* **-ies**) (often + *to*, *towards*) leaning, inclination.

tendentious /ten'denʃəs/ *adjective derogatory* designed to advance a particular cause; biased; controversial. □ **tendentiously** *adverb*; **tendentiousness** *noun*.

tender[1] /'tendə/ *adjective* (**tenderer**, **tenderest**) not tough or hard; susceptible to pain or grief; compassionate; delicate, fragile; loving, affectionate; requiring tact; immature. □ **tenderfoot** (*plural* **-s** or **-feet**) novice, newcomer; **tender-hearted** easily moved; **tenderloin** middle part of loin of pork; **tender mercies** harsh treatment. □ **tenderly** *adverb*; **tenderness** *noun*.

tender[2] /'tendə/ ● *verb* offer, present (services, resignation, payment, etc.); (often + *for*) make tender. ● *noun* offer to execute work or supply goods at fixed price.

tender[3] /'tendə/ *noun* person who looks after people or things; supply vessel attending larger one; truck attached to steam locomotive and carrying fuel etc.

tenderize *verb* (also **-ise**) (**-zing** or **-sing**) render (meat) tender by beating etc.

tendon /'tend(ə)n/ *noun* tough fibrous tissue connecting muscle to bone etc.

tendril /'tendrɪl/ *noun* slender leafless shoot by which some climbing plants cling.

tenebrous /'tenɪbrəs/ *adjective literary* dark, gloomy.

tenement /'tenɪmənt/ *noun* room or flat within house or block of flats; (also **tenement house**, **block**) house or block so divided.

tenet /'tenɪt/ *noun* doctrine, principle.

tenfold *adjective & adverb* ten times as much or many.

tenner /'tenə/ *noun colloquial* £10 or $10 note.

tennis /'tenɪs/ *noun* ball game played with rackets on court divided by net.

□ **tennis elbow** sprain caused by overuse of forearm muscles.

tenon /'tenən/ *noun* wooden projection shaped to fit into mortise of another piece.

tenor /'tenə/ ● *noun* male singing voice between alto and baritone; singer with this; (usually + *of*) general purport, prevailing course of one's life or habits. ● *adjective* having range of tenor.

tenosynovitis /tenəʊsaɪnə'vaɪtɪs/ *noun* repetitive strain injury, esp. of wrist.

tenpin bowling *noun* game in which ten pins or skittles are bowled at in alley.

tense[1] ● *adjective* stretched tight; strained; causing tenseness. ● *verb* (**-sing**) make or become tense. □ **tense up** become tense. □ **tensely** *adverb*; **tenseness** *noun*.

tense[2] *noun* form of verb indicating time of action etc.; set of such forms for various persons and numbers.

tensile /'tensaɪl/ *adjective* of tension; capable of being stretched. □ **tensile strength** resistance to breaking under tension.

tension /'tenʃ(ə)n/ *noun* stretching, being stretched; mental strain or excitement; strained state; stress produced by forces pulling apart; degree of tightness of stitches in knitting and machine sewing; voltage.

tent *noun* portable shelter or dwelling of canvas etc.

tentacle /'tentək(ə)l/ *noun* slender flexible appendage of animal, used for feeling, grasping, or moving.

tentative /'tentətɪv/ *adjective* experimental; hesitant, not definite. □ **tentatively** *adverb*.

tenterhooks /'tentəhʊks/ *plural noun* □ **on tenterhooks** in suspense, distracted by uncertainty.

tenuous /'tenjʊəs/ *adjective* slight, insubstantial; oversubtle; thin, slender. □ **tenuity** /-'ju:ɪtɪ/ *noun*; **tenuously** *adverb*.

tenure /'tenjə/ *noun* (often + *of*) holding of property or office; conditions or period of this; guaranteed permanent employment, esp. as lecturer. □ **tenured** *adjective*.

tepee /'ti:pi:/ *noun* (also **teepee**) N. American Indian's conical tent.

tepid /'tepɪd/ *adjective* lukewarm; unenthusiastic.

tequila /tɪ'ki:lə/ *noun* Mexican liquor made from agave.

tercel /'tɜ:s(ə)l/ *noun* (also **tiercel** /'tɪəs(ə)l/) male hawk.

tercentenary /tɜ:sen'ti:nərɪ/ *noun* (*plural* **-ies**) 300th anniversary; celebration of this.

tergiversate /'tɜ:dʒɪvəseɪt/ *verb* (**-ting**) change one's party or principles; make conflicting or evasive statements. □ **tergiversation** *noun*; **tergiversator** *noun*.

term ● *noun* word for definite concept, esp. specialized; (in *plural*) language used, mode of expression; (in *plural*) relation, footing; (in *plural*) stipulations, charge, price; limited period; period of weeks during which instruction is given or during which law court holds sessions; *Logic* word(s) which may be subject or predicate of proposition; *Mathematics* each quantity in ratio or series, part of algebraic expression; completion of normal length of pregnancy. ● *verb* call, name. □ **come to terms with** reconcile oneself to; **in terms of** with reference to; **terms of reference** scope of inquiry etc., definition of this. □ **termly** *adjective*.

termagant /'tɜ:məgənt/ *noun* overbearing woman, virago.

terminable /'tɜ:mɪnəb(ə)l/ *adjective* able to be terminated.

terminal /'tɜ:mɪn(ə)l/ ● *adjective* (of condition or disease) fatal; (of patient) dying; of or forming limit or terminus. ● *noun* terminating thing, extremity; bus or train terminus; air terminal; point of connection for closing electric circuit; apparatus for transmission of messages to and from computer, communications system, etc. □ **terminally** *adverb*.

terminate /'tɜ:mɪneɪt/ *verb* (**-ting**) bring or come to an end; (+ *in*) end in.

termination *noun* terminating, being terminated; ending, result; induced abortion.

terminology /tɜ:mɪ'nɒlədʒɪ/ *noun* (*plural* **-ies**) system of specialized terms. □ **terminological** /-nə'lɒdʒ-/ *adjective*.

terminus /'tɜ:mɪnəs/ *noun* (*plural* **-ni** /-naɪ/ or **-nuses**) point at end of railway or bus route or of pipeline etc.

termite /'tɜ:maɪt/ *noun* antlike insect destructive to timber.

tern *noun* seabird with long pointed wings and forked tail.

ternary /'tɜ:nərɪ/ *adjective* composed of 3 parts.

terrace /'terəs/ ● *noun* flat area on slope for cultivation; level paved area next to house; row of houses built in one block of uniform style; terrace house; tiered standing accommodation for spectators at sports ground. ● *verb* (**-cing**) form into or provide with terrace(s). □ **terrace(d) house** house in terrace.

terracotta /terə'kɒtə/ *noun* unglazed usually brownish-red earthenware; its colour.

terra firma /terə 'fɜ:mə/ *noun* dry land, firm ground.

terrain /tə'reɪn/ *noun* tract of land, esp. in military or geographical contexts.

terra incognita /terə ɪŋ'kɒgnɪtə/ *noun* unexplored region. [Latin]

terrapin /'terəpɪn/ *noun* N. American edible freshwater turtle.

terrarium /tə'reərɪəm/ *noun* (*plural* **-s** or **-ria**) place for keeping small land animals; transparent globe containing growing plants.

terrestrial /tə'restrɪəl/ *adjective* of or on the earth; earthly; of or on dry land.

terrible /'terɪb(ə)l/ *adjective colloquial* very great, bad, or incompetent; causing or likely to cause terror; dreadful.

terribly *adverb colloquial* very, extremely; in terrible manner.

terrier /'terɪə/ *noun* small active hardy dog.

terrific /tə'rɪfɪk/ *adjective colloquial* huge, intense, excellent; causing terror. □ **terrifically** *adverb*.

terrify /'terɪfaɪ/ *verb* (**-ies**, **-ied**) fill with terror. □ **terrifying** *adjective*; **terrifyingly** *adverb*.

terrine /tə'ri:n/ *noun* (earthenware vessel for) pâté or similar food.

territorial /terɪ'tɔ:rɪəl/ ● *adjective* of territory or district. ● *noun* (**Territorial**) member of Territorial Army. □ **Territorial Army** local volunteer reserve force; **territorial waters** waters under state's jurisdiction, esp. part of sea within stated distance of shore. □ **territorially** *adverb*.

territory /'terɪtərɪ/ *noun* (*plural* **-ies**) extent of land under jurisdiction of ruler, state, etc.; (**Territory**) organized division of a country, esp. if not yet admitted to full rights of a state; sphere of action etc., province; commercial traveller's sales area; area defended by animal or human, or by team etc. in game.

terror /'terə/ *noun* extreme fear; terrifying person or thing; *colloquial* troublesome or tiresome person, esp. child; terrorism.

terrorist *noun* person using esp. organized violence to secure political ends. □ **terrorism** *noun*.

terrorize *verb* (also **-ise**) (**-zing** or **-sing**) fill with terror; use terrorism against.

terry /'terɪ/ *noun* looped pile fabric used for nappies, towels, etc.

terse /tɜ:s/ *adjective* (**-r**, **-st**) concise, brief; curt. □ **tersely** *adverb*.

tertiary /'tɜ:ʃərɪ/ *adjective* of third order, rank, etc.

Terylene /'terɪli:n/ *noun proprietary term* synthetic polyester textile fibre.

TESL /'tes(ə)l/ *abbreviation* teaching of English as a second language.

tesla /'tezlə/ *noun* SI unit of magnetic flux density.

tessellated /'tesəleɪtɪd/ *adjective* of or resembling mosaic; finely chequered.

tessellation /tesə'leɪʃ(ə)n/ *noun* close arrangement of polygons, esp. in repeated pattern.

test ● *noun* critical exam or trial of person's or thing's qualities; means, procedure, or standard for so doing; minor exam; *colloquial* test match. ● *verb* put to test; try severely, tax. □ **test card** still television picture outside normal programme hours; **test case** *Law* case setting precedent for other similar cases; **test drive** *noun* drive taken to judge vehicle's performance; **test-drive** *verb* take test drive in; **test match** international cricket or Rugby match, usually in series; **test-tube** thin glass tube closed at one end, used for chemical tests etc.; **test-tube baby** *colloquial* baby conceived elsewhere than in a mother's body. □ **tester** *noun.*

testaceous /tes'teɪʃəs/ *adjective* having hard continuous shell.

testament /'testəmənt/ *noun* a will; (usually + *to*) evidence, proof; *Biblical* covenant; (**Testament**) division of Bible. □ **testamentary** /-'ment-/ *adjective.*

testate /'testeɪt/ ● *adjective* having left valid will at death. ● *noun* testate person. □ **testacy** /-təsɪ/ *noun* (*plural* **-ies**).

testator /tes'teɪtə/ *noun* (*feminine* **testatrix** /-trɪks/) (esp. deceased) person who has made a will.

testes *plural* of TESTIS.

testicle /'testɪk(ə)l/ *noun* male organ that secretes spermatozoa, esp. one of pair in scrotum of man and most mammals.

testify /'testɪfaɪ/ *verb* (**-ies**, **-ied**) (often + *to*) bear witness; give evidence; affirm, declare.

testimonial /testɪ'məʊnɪəl/ *noun* certificate of character, conduct, or qualifications; gift presented as mark of esteem.

testimony /'testɪmənɪ/ *noun* (*plural* **-ies**) witness's statement under oath etc.; declaration, statement of fact; evidence.

testis /'testɪs/ *noun* (*plural* **testes** /-ti:z/) testicle.

testosterone /te'stɒstərəʊn/ *noun* male sex hormone.

testy /'testɪ/ *adjective* (**-ier**, **-iest**) irascible, short-tempered. □ **testily** *adverb*; **testiness** *noun.*

tetanus /'tetənəs/ *noun* bacterial disease causing painful spasm of voluntary muscles.

tetchy /'tetʃɪ/ *adjective* (**-ier**, **-iest**) peevish, irritable. □ **tetchily** *adverb*; **tetchiness** *noun.*

tête-à-tête /teɪtɑ:'teɪt/ ● *noun* private conversation between two people. ● *adverb* privately without third person.

tether /'teðə/ ● *noun* rope etc. confining grazing animal. ● *verb* fasten with tether. □ **at the end of one's tether** at the limit of one's patience, resources, etc.

tetra- *combining form* four.

tetragon /'tetrəgɒn/ *noun* plane figure with 4 sides and angles. □ **tetragonal** /tɪ'trægən-/ *adjective.*

tetrahedron /tetrə'hi:drən/ *noun* (*plural* **-dra** or **-s**) 4-sided triangular pyramid. □ **tetrahedral** *adjective.*

Teutonic /tju:'tɒnɪk/ *adjective* of Germanic peoples or languages; German.

text *noun* main part of book; original document, esp. as distinct from paraphrase etc.; passage of Scripture, esp. as subject of sermon; subject, theme; (in *plural*) books prescribed for study; data in textual form, esp. in word processor. □ **textbook** book used in studying, esp. standard book in any subject; **text editor** computing program allowing user to edit text.

textile /'tekstaɪl/ ● *noun* (often in *plural*) fabric, esp. woven. ● *adjective* of weaving or cloth; woven.

textual /'tekstʃʊəl/ *adjective* of, in, or concerning a text.

texture /'tekstʃə/ ● *noun* feel or appearance of surface or substance; arrangement of threads in textile fabric. ● *verb* (**-ring**) (usually as **textured** *adjective*) provide with texture; provide (vegetable protein) with texture like meat. □ **textural** *adjective.*

Thai /taɪ/ ● *noun* (*plural* same or **-s**) native, national, or language of Thailand. ● *adjective* of Thailand.

thalidomide /θə'lɪdəmaɪd/ *noun* sedative drug found in 1961 to cause foetal malformation when taken early in pregnancy.

than /ðən/ *conjunction introducing comparison.*

thane /θeɪn/ *noun historical* holder of land from English king by military service, or from Scottish king and ranking below earl; clan-chief.

thank /θæŋk/ ● *verb* express gratitude to; hold responsible. ● *noun* (in *plural*) *colloquial* gratitude; (as *interjection*) *expression of gratitude.* □ **thanksgiving** expression of gratitude, esp. to God; **Thanksgiving (Day)** US national holiday on fourth Thurs. in Nov.; **thanks to** as result of; **thank you** *polite formula expressing gratitude.*

thankful *adjective* grateful, pleased, expressive of thanks.

thankfully *adverb* in a thankful way; let us be thankful that.

■ **Usage** The use of *thankfully* to mean 'let us be thankful that' is common, but it is considered incorrect by some people.

---

thankless *adjective* not feeling or expressing gratitude; (of task etc.) unprofitable, unappreciated.

that /ðæt/ ● *adjective* (*plural* **those** /ðəʊz/) *used to describe the person or thing nearby, indicated, just mentioned, or understood; used to specify the further or less immediate of two.* ● *pronoun* (*plural* **those** /ðəʊz/) that one; the one, the person, etc.; /ðət/ (*plural* **that**) who, whom, which *(used to introduce a defining relative clause).* ● *adverb* (+ adjective or adverb) to that degree, so, (with negative) *colloquial* very. ● *conjunction* /ðət/ *used to introduce a subordinate clause expressing esp. a statement, purpose, or result.* □ **at that** moreover, then; **that is** (**to say**) in other words, more correctly or intelligibly.

thatch /θætʃ/ ● *noun* roofing of straw, reeds, etc. ● *verb* cover with thatch. □ **thatcher** *noun.*

thaw /θɔ:/ ● *verb* (often + *out*) pass from frozen into liquid or unfrozen state; (of weather) become warm enough to melt ice etc.; warm into life, animation, cordiality, etc. ● *noun* thawing; warmth of weather that thaws.

the /ðɪ, ðə, ði:/ ● *adjective* (called the definite article) *denoting person(s) or thing(s) already mentioned or known about; describing as unique*; (+ adjective) which is, who are, etc.; (with *the* stressed) best known; *used with noun which represents or symbolizes a group, activity, etc.* ● *adverb* (before comparatives in expressions of proportional variation) in or by that degree, on that account.

theatre /'θɪətə/ *noun* (*US* **theater**) building or outdoor area for dramatic performances; writing, production, acting, etc. of plays; room or hall for lectures etc. with seats in tiers; operating theatre; scene or field of action.

theatrical /θɪ'ætrɪk(ə)l/ ● *adjective* of or for theatre or acting; calculated for effect, showy. ● *noun* (in *plural*) dramatic performances. □ **theatricality** /-'kæl-/ *noun*; **theatrically** *adverb.*

thee /ði:/ *pronoun archaic* (as object of verb) you (singular).

theft /θeft/ *noun* act of stealing.

their /ðeə/ *adjective* of or belonging to them.

theirs /ðeəz/ *pronoun* the one(s) belonging to them.

theism /'θi:ɪz(ə)m/ *noun* belief in gods or a god. □ **theist** *noun*; **theistic** /-'ɪstɪk/ *adjective.*

them /ðem, ð(ə)m/ ● *pronoun* (as object of verb) the people or things in question; people in general; people in authority; *colloquial* they. ● *adjective slang or dialect* those.

theme /θi:m/ *noun* subject or topic of talk etc.; *Music* leading melody in a composition; *US* school exercise on given subject. □ **theme park** amusement park based on unifying idea; **theme song**, **tune** signature tune. □ **thematic** /θɪ'mætɪk/ *adjective*; **thematically** /θɪ'mæt-/ *adverb.*

themselves /ðəm'selvz/ *pronoun: emphatic & reflexive form of* THEY; *reflexive form of* THEM.

then /ðen/ ● *adverb* at that time; after that, next; in that case, accordingly. ● *adjective* such at that time. ● *noun* that time. □ **then and there** immediately and on the spot.

thence /ðens/ *adverb* (also **from thence**) *archaic or literary* from that place, for that reason. □ **thenceforth**, **thenceforward** from that time on.

theo- *combining form* God or god(s).

theocracy /θɪ'ɒkrəsɪ/ *noun* (*plural* **-ies**) form of government by God or a god directly or through a priestly order etc. □ **theocratic** /θɪə'krætɪk/ *adjective.*

theodolite /θɪ'ɒdəlaɪt/ *noun* surveying instrument for measuring angles.

theology /θɪ'ɒlədʒɪ/ *noun* (*plural* **-ies**) study or system of (esp. Christian) religion. □ **theologian** /θɪə'ləʊdʒ-/ *noun*; **theological** /θɪə'lɒdʒ-/ *adjective.*

theorem /'θɪərəm/ *noun esp. Mathematics* general proposition not self-evident but demonstrable by argument; algebraic rule.

theoretical /θɪə'retɪk(ə)l/ *adjective* concerned with knowledge but not with its practical application; based on theory rather than experience. □ **theoretically** *adverb*.

theoretician /θɪərə'tɪʃ(ə)n/ *noun* person concerned with theoretical part of a subject.

theorist /'θɪərɪst/ *noun* holder or inventor of a theory.

theorize /'θɪəraɪz/ *verb* (also **-ise**) (**-zing** or **-sing**) evolve or indulge in theories.

theory /'θɪərɪ/ *noun* (*plural* **-ies**) supposition or system of ideas explaining something, esp. one based on general principles; speculative view; abstract knowledge or speculative thought; exposition of principles of a science etc.; collection of propositions to illustrate principles of a mathematical subject.

theosophy /θɪ'ɒsəfɪ/ *noun* (*plural* **-ies**) philosophy professing to achieve knowledge of God by direct intuition, spiritual ecstasy, etc. □ **theosophical** /θɪə'sɒf-/ *adjective*; **theosophist** *noun*.

therapeutic /θerə'pju:tɪk/ *adjective* of, for, or contributing to the cure of diseases; soothing, conducive to well-being. □ **therapeutically** *adverb*.

therapeutics *plural noun* (usually treated as *singular*) branch of medicine concerned with cures and remedies.

therapy /'θerəpɪ/ *noun* (*plural* **-ies**) non-surgical treatment of disease etc. □ **therapist** *noun*.

there /ðeə/ ● *adverb* in, at, or to that place or position; at that point; in that respect; *used for emphasis in calling attention; used to indicate the fact or existence of something*. ● *noun* that place. ● *interjection expressing confirmation, triumph, etc.; used to soothe a child etc.* □ **thereabout(s)** near that place, amount, or time; **thereafter** *formal* after that; **thereby** by that means or agency; **therefore** for that reason, accordingly, consequently; **therein** *formal* in that place or respect; **thereof** *formal* of that or it; **thereto** *formal* to that or it, in addition; **thereupon** in consequence of that, directly after that.

therm /θɜ:m/ *noun* unit of heat, former UK unit of gas supplied.

thermal /'θɜ:m(ə)l/ ● *adjective* of, for, producing, or retaining heat. ● *noun* rising current of warm air. □ **thermal unit** unit for measuring heat. □ **thermally** *adverb*.

thermionic valve /θɜ:mɪ'ɒnɪk/ *noun* device giving flow of electrons in one direction from heated substance, used esp. in rectification of current and in radio reception.

thermo- *combining form* heat.

thermodynamics /θɜ:məʊdaɪ'næmɪks/ *plural noun* (usually treated as *singular*) science of relationship between heat and other forms of energy. □ **thermodynamic** *adjective*.

thermoelectric /θɜ:məʊɪ'lektrɪk/ *adjective* producing electricity by difference of temperatures.

thermometer /θə'mɒmɪtə/ *noun* instrument for measuring temperature, esp. graduated glass tube containing mercury or alcohol.

thermonuclear /θɜ:məʊ'nju:klɪə/ *adjective* relating to nuclear reactions that occur only at very high temperatures; (of bomb etc.) using such reactions.

thermoplastic /θɜ:məʊ'plæstɪk/ ● *adjective* becoming plastic on heating and hardening on cooling. ● *noun* thermoplastic substance.

Thermos /'θɜ:məs/ *noun* (in full **Thermos flask**) *proprietary term* vacuum flask.

thermosetting /'θɜ:məʊsetɪŋ/ *adjective* (of plastics) setting permanently when heated.

thermosphere /'θɜ:məsfɪə/ *noun* region of atmosphere beyond mesosphere.

thermostat /'θɜ:məstæt/ *noun* device for automatic regulation of temperature. □ **thermostatic** /-'stæt-/ *adjective*; **thermostatically** /-'stæt-/ *adverb*.

thesaurus /θɪ'sɔ:rəs/ *noun* (*plural* **-ri** /-raɪ/ or **-ruses**) dictionary of synonyms etc.

these *plural* of THIS.

thesis /'θi:sɪs/ *noun* (*plural* **theses** /-si:z/) proposition to be maintained or proved; dissertation, esp. by candidate for higher degree.

Thespian /'θespɪən/ ● *adjective* of drama. ● *noun* actor or actress.

they /ðeɪ/ *pronoun* (as subject of verb) the people or things in question; people in general; people in authority.

they'd /ðeɪd/ they had; they would.

they'll /ðeɪəl/ they will; they shall.

they're /ðeə/ they are.

they've /ðeɪv/ they have.

thiamine /'θaɪəmɪn/ *noun* (also **thiamin**) B vitamin found in unrefined cereals, beans, and liver.

thick /θɪk/ ● *adjective* of great or specified extent between opposite surfaces; (of line etc.) broad, not fine; closely set; crowded; (usually + *with*) densely filled or covered; firm in consistency; made of thick material; muddy, impenetrable; *colloquial* stupid; (of voice) indistinct; (of accent) marked; *colloquial* intimate. ● *noun* thick part of anything. ● *adverb* thickly. □ **a bit thick** *colloquial* unreasonable, intolerable; **in the thick of** in the busiest part of; **thickhead** *colloquial* stupid person; **thickheaded** *adjective*; **thickset** heavily or solidly built, set or growing close together; **thick-skinned** not sensitive to criticism; **through thick and thin** under all conditions, in spite of all difficulties. □ **thickly** *adverb*; **thickness** *noun*.

thicken *verb* make or become thick(er); become more complicated.

thickener *noun* substance used to thicken liquid.

thickening *noun* thickened part; = THICKENER.

thicket /'θɪkɪt/ *noun* tangle of shrubs or trees.

thief /θi:f/ *noun* (*plural* **thieves** /θi:vz/) person who steals, esp. secretly.

thieve /θi:v/ *verb* (**-ving**) be a thief; steal. □ **thievery** *noun*.

thievish *adjective* given to stealing.

thigh /θaɪ/ *noun* part of leg between hip and knee. □ **thigh-bone** femur.

thimble /'θɪmb(ə)l/ *noun* metal or plastic cap worn to protect finger and push needle in sewing.

thimbleful *noun* (*plural* **-s**) small quantity, esp. of drink.

thin /θɪn/ ● *adjective* (**-nn-**) having opposite surfaces close together, of small thickness or diameter; (of line etc.) narrow, fine; made of thin material; lean, not plump; not dense or copious; of slight consistency; weak, lacking an important ingredient; (of excuse etc.) transparent, flimsy. ● *adverb* thinly. ● *verb* (**-nn-**) make or become thin(ner); (often + *out*) make or become less dense, crowded, or numerous. □ **thin on the ground** few; **thin on top** balding; **thin-skinned** sensitive to criticism. □ **thinly** *adverb*; **thinness** *noun*.

thine /ðaɪn/ *archaic* ● *pronoun* yours (singular). ● *adjective* your (singular).

thing /θɪŋ/ *noun* any possible object of thought or perception including people, material objects, events, qualities, ideas, utterances, and acts; *colloquial* one's special interest; (**the thing**) *colloquial* what is proper, fashionable, needed, important, etc.; (in *plural*) personal belongings, clothing, or equipment; (in *plural*) affairs, circumstances. □ **have a thing about** *colloquial* be obsessed by or prejudiced about.

thingummy /'θɪŋəmɪ/ *noun* (*plural* **-ies**) (also **thingumabob** /-məbɒb/, **thingumajig** /-mədʒɪg/) *colloquial* person or thing whose name one forgets or does not know.

think /θɪŋk/ ● *verb* (*past & past participle* **thought** /θɔ:t/) be of opinion; (+ *of*, *about*) consider; exercise mind; form ideas, imagine; have half-formed intention. ● *noun colloquial* act of thinking. □ **think better of** change one's mind about (intention) after reconsideration; **think out** consider carefully, devise; **think over** reflect on; **think-tank** *colloquial* group of experts providing advice and ideas on national or commercial problems; **think twice** avoid hasty action etc.; **think up** *colloquial* devise.

thinker *noun* person who thinks in specified way; person with skilled or powerful mind.

thinking ● *adjective* intelligent, rational. ● *noun* opinion, judgement.

thinner *noun* solvent for diluting paint etc.

third /θɜ:d/ *adjective & noun* next after second; any of 3 equal parts of thing. □ **third degree** severe and protracted interrogation by police etc.; **third man** *Cricket* fielder near boundary behind slips; **third party** another party besides the two principals; **third-party insurance** insurance against damage or injury suffered by person other than the insured; **third-rate** inferior, very poor; **Third World** developing countries of Africa, Asia, and Latin America. □ **thirdly** *adverb*.

thirst /θɜ:st/ ● *noun* (discomfort caused by) need to drink; desire, craving. ● *verb* feel thirst.

thirsty /'θɜ:stɪ/ *adjective* (**-ier**, **-iest**) feeling thirst; (of land, season, etc.) dry, parched; (often + *for*, *after*) eager; *colloquial* causing thirst. □ **thirstily** *adverb*; **thirstiness** *noun*.

thirteen /θɜ:'ti:n/ *adjective & noun* one more than twelve. □ **thirteenth** *adjective & noun*.

thirty /'θɜ:tɪ/ *adjective & noun* (*plural* **-ies**) three times ten. □ **thirtieth** *adjective & noun*.

this /ðɪs/ ● *adjective* (*plural* **these** /ði:z/) *used to describe the person or thing nearby, indicated, just mentioned, or understood; used to specify the nearer or more immediate of two;* the present (morning, week, etc.). ● *pronoun* (*plural* **these** /ði:z/) this one. ● *adverb* (+ adjective or adverb) to this degree or extent.

thistle /'θɪs(ə)l/ *noun* prickly plant, usually with globular heads of purple flowers; this as Scottish national emblem. □ **thistledown** down containing thistle-seeds. □ **thistly** *adjective*.

thither /'ðɪðə/ *adverb archaic or formal* to that place.

tho' = THOUGH.

thole /θəʊl/ *noun* (in full **thole-pin**) pin in gunwale of boat as fulcrum for oar; each of two such pins forming rowlock.

thong /θɒŋ/ *noun* narrow strip of hide or leather.

thorax /'θɔ:ræks/ *noun* (*plural* **-races** /-rəsi:z/ or **-raxes**) part of the body between neck and abdomen. □ **thoracic** /-'ræs-/ *adjective*.

thorn /θɔ:n/ *noun* sharp-pointed projection on plant; thorn-bearing shrub or tree. □ **thornless** *adjective*.

thorny *adjective* (**-ier**, **-iest**) having many thorns; (of subject) problematic, causing disagreement.

thorough /'θʌrə/ *adjective* complete, unqualified, not superficial; acting or done with great care etc. □ **thoroughbred** *adjective* of pure breed, high-spirited, *noun* such animal, esp. horse; **thoroughfare** public way open at both ends, esp. main road; **thoroughgoing** thorough, complete. □ **thoroughly** *adverb*; **thoroughness** *noun*.

those *plural* of THAT.

thou[1] /ðaʊ/ *pronoun archaic* (as subject of verb) you (singular).

thou[2] /θaʊ/ *noun* (*plural* same or **-s**) *colloquial* thousand; one thousandth.

though /ðəʊ/ (also **tho'**) ● *conjunction* in spite of the fact that; even if; and yet. ● *adverb colloquial* however, all the same.

thought[1] /θɔ:t/ *noun* process, power, faculty, etc. of thinking; particular way of thinking; sober reflection, consideration; idea, notion; intention, purpose; (usually in *plural*) one's opinion.

thought[2] *past & past participle* of THINK.

thoughtful *adjective* engaged in or given to meditation; giving signs of serious thought; considerate. □ **thoughtfully** *adverb*; **thoughtfulness** *noun*.

thoughtless *adjective* careless of consequences or of others' feelings; caused by lack of thought. □ **thoughtlessly** *adverb*; **thoughtlessness** *noun*.

thousand /'θaʊz(ə)nd/ *adjective & noun* (*plural* same) ten hundred; (**thousands**) *colloquial* large number. □ **thousandth** *adjective & noun*.

thrall /θrɔ:l/ *noun literary* (often + *of*, *to*) slave; slavery.

thrash /θræʃ/ *verb* beat or whip severely; defeat thoroughly; move or fling (esp. limbs) violently. □ **thrash out** discuss to conclusion.

thread /θred/ ● *noun* spun-out cotton, silk, glass, etc.; length of this; thin cord of twisted yarns used esp. in sewing and weaving; continuous aspect of thing; spiral ridge of screw. ● *verb* pass thread through (needle); put (beads) on thread; arrange (material in strip form, e.g. film) in proper position on equipment; pick (one's way) through maze, crowded place, etc. □ **threadbare** (of cloth) so worn that nap is lost and threads showing, (of person) shabby, (of idea etc.) hackneyed; **threadworm** parasitic threadlike worm.

threat /θret/ *noun* declaration of intention to punish or hurt; indication of something undesirable coming; person or thing regarded as dangerous.

threaten /'θret(ə)n/ *verb* use threats towards; be sign or indication of (something undesirable); (+ *to do*) announce one's intention to do (undesirable thing); give warning of infliction of (harm etc.); (as **threatened** *adjective*) (of species etc.) likely to become extinct.

three /θri:/ *adjective & noun* one more than two. □ **three-cornered** triangular, (of contest etc.) between 3 people etc.; **three-dimensional** having or appearing to have length, breadth, and depth; **three-legged race** race for pairs with right leg of one tied to other's left leg; **threepence** /'θrepəns/ sum of 3 pence; **threepenny** /'θrepənɪ/ costing 3 pence; **three-piece** (suit or suite) consisting of 3 items; **three-ply** (wool etc.) having 3 strands, (plywood) having 3 layers; **three-point turn** method of turning vehicle in narrow space by moving forwards, backwards, and forwards again; **three-quarter** *Rugby* any of 3 or 4 players just behind half-backs; **the three Rs** reading, writing, and arithmetic.

threefold *adjective & adverb* three times as much or many.

threesome *noun* group of 3 people.

threnody /'θrenədɪ/ *noun* (*plural* **-ies**) song of lamentation.

thresh *verb* beat out or separate grain from (corn etc.). □ **thresher** *noun.*

threshold /'θreʃəʊld/ *noun* plank or stone forming bottom of doorway; point of entry; limit below which stimulus causes no reaction.

threw *past* of THROW.

thrice *adverb archaic or literary* 3 times.

thrift *noun* frugality, economical management. □ **thrifty** *adjective* (**-ier**, **-iest**).

thrill • *noun* wave or nervous tremor of emotion or sensation; throb, pulsation. • *verb* (cause to) feel thrill; quiver or throb (as) with emotion.

thriller *noun* sensational or exciting play, story, etc.

thrips *noun* (*plural* same) insect harmful to plants.

thrive *verb* (**-ving**; *past* **throve** or **thrived**; *past participle* **thriven** /'θrɪv(ə)n/ or **thrived**) prosper; grow vigorously.

thro' = THROUGH.

throat *noun* gullet, windpipe; front of neck; *literary* narrow passage or entrance.

throaty *adjective* (**-ier**, **-iest**) (of voice) hoarsely resonant.

throb • *verb* (**-bb-**) pulsate; vibrate with persistent rhythm or with emotion. • *noun* throbbing, violent beat or pulsation.

throe *noun* (usually in *plural*) violent pang. □ **in the throes of** struggling with the task of.

thrombosis /θrɒm'bəʊsɪs/ *noun* (*plural* **-boses** /-si:z/) coagulation of blood in blood vessel or organ.

throne • *noun* ceremonial chair for sovereign, bishop, etc.; sovereign power. • *verb* (**-ning**) enthrone.

throng • *noun* (often + *of*) crowd, esp. of people. • *verb* come in multitudes; fill (as) with crowd.

throstle /'θrɒs(ə)l/ *noun* song thrush.

throttle /'θrɒt(ə)l/ • *noun* (lever etc. operating) valve controlling flow of steam or fuel in engine; throat. • *verb* (**-ling**) choke, strangle; control (engine etc.) with throttle. □ **throttle back**, **down** reduce speed of (engine etc.) by throttling.

through /θru:/ (also **thro'**, *US* **thru**) • *preposition* from end to end or side to side of; between, among; from beginning to end of; by agency, means, or fault of; by reason of; *US* up to and including. • *adverb* through something; from end to end; to the end. • *adjective* (of journey etc.) done without change of line, vehicle, etc.; (of traffic) going through a place to its destination; (of road) open at both ends. □ **be through** *colloquial* (often + *with*) have finished, cease to have dealings; **through and through** thoroughly, completely; **throughput** amount of material put through a manufacturing etc. process or a computer.

throughout /θru:'aʊt/ • *preposition* right through; from end to end of. • *adverb* in every part or respect.

throve *past* of THRIVE.

throw /θrəʊ/ • *verb* (*past* **threw** /θru:/; *past participle* **thrown**) propel through space; force violently into specified position or state; turn or move (part of body) quickly or suddenly; project (rays, light, etc.); cast (shadow); bring to the ground; *colloquial* disconcert; (+ *on*, *off*, etc.) put (clothes etc.) carelessly or hastily on, off, etc.; cause (dice) to fall on table etc., obtain (specified number) thus; cause to pass or extend suddenly to another state or position; move (switch, lever); shape (pottery) on wheel; have (fit, tantrum, etc.); give (a party). ● *noun* throwing, being thrown; distance a thing is or may be thrown; (**a throw**) *slang* each, per item. □ **throw away** discard as unwanted, waste, fail to make use of; **throw-away** to be thrown away after (one) use, deliberately underemphasized; **throw back** (usually in *passive*; + *on*) compel to rely on; **throwback** (instance of) reversion to ancestral character; **throw in** interpose (word, remark), include at no extra cost, throw (football) from edge of pitch where it has gone out of play; **throw-in** throwing in of football from edge of pitch; **throw off** discard, contrive to get rid of, write or utter in offhand way; **throw open** (often + *to*) cause to be suddenly or widely open, make accessible; **throw out** put out forcibly or suddenly, discard, reject; **throw over** desert, abandon; **throw up** abandon, resign from, vomit, erect hastily, bring to notice.

thrum ● *verb* (**-mm-**) play (stringed instrument) monotonously or unskilfully; (often + *on*) drum idly. ● *noun* such playing; resultant sound.

thrush[1] *noun* kind of songbird.

thrush[2] *noun* fungus infection of throat, esp. in children, or of vagina.

thrust ● *verb* (*past & past participle* **thrust**) push with sudden impulse or with force; (+ *on*) impose (thing) forcibly on; (+ *at*, *through*) pierce, stab, lunge suddenly; make (one's way) forcibly; (as **thrusting** *adjective*) aggressive, ambitious. ● *noun* sudden or forcible push or lunge; forward force exerted by propeller or jet etc.; strong attempt to penetrate enemy's line or territory; remark aimed at person; stress between parts of arch etc.; (often + *of*) theme, gist.

thud /θʌd/ ● *noun* low dull sound as of blow on non-resonant thing. ● *verb* (**-dd-**) make thud; fall with thud.

thug /θʌg/ *noun* vicious or brutal ruffian. □ **thuggery** *noun*; **thuggish** *adjective*.

thumb /θʌm/ ● *noun* short thick finger on hand, set apart from other 4; part of glove for thumb. ● *verb* soil or wear with thumb; turn over pages (as) with thumb; request or get (lift) by sticking out thumb. □ **thumb index** set of lettered grooves cut down side of book etc. for easy reference; **thumbnail** *noun* nail of thumb, *adjective* concise; **thumbscrew** instrument of torture for squeezing thumbs; **thumbs up, down** indication of approval or rejection; **under (person's) thumb** dominated by him or her.

thump /θʌmp/ ● *verb* beat heavily, esp. with fist; throb strongly; (+ *at*, *on*, etc.) knock loudly. ● *noun* (sound of) heavy blow.

thumping *adjective colloquial* huge.

thunder /ˈθʌndə/ ● *noun* loud noise accompanying lightning; resounding loud deep noise; strong censure. ● *verb* sound with or like thunder; move with loud noise; utter loudly; (+ *against* etc.) make violent threats. □ **thunderbolt** flash of lightning with crash of thunder, unexpected occurrence or announcement, supposed bolt or shaft as destructive agent; **thunderclap** crash of thunder; **thundercloud** electrically charged cumulus cloud; **thunderstorm** storm with thunder and lightning; **thunderstruck** amazed. □ **thunderous** *adjective*; **thundery** *adjective*.

thundering *adjective colloquial* huge.

Thur. *abbreviation* (also **Thurs.**) Thursday.

thurible /ˈθjʊərɪb(ə)l/ *noun* censer.

Thursday /ˈθɜːzdeɪ/ *noun* day of week following Wednesday.

thus /ðʌs/ *adverb formal* in this way, like this; accordingly, as a result or inference; to this extent, so.

thwack ● *verb* hit with heavy blow. ● *noun* heavy blow.

thwart /θwɔːt/ ● *verb* frustrate, foil. ● *noun* rower's seat.

thy /ðaɪ/ *adjective* (also **thine**, esp. before vowel) *archaic* your (singular).

thyme /taɪm/ *noun* herb with aromatic leaves.

thymol /ˈθaɪmɒl/ *noun* antiseptic made from oil of thyme.

thymus /ˈθaɪməs/ *noun* (*plural* **thymi** /-maɪ/) ductless gland near base of neck.

thyroid /ˈθaɪrɔɪd/ *noun* thyroid gland. □ **thyroid cartilage** large cartilage of larynx forming Adam's apple; **thyroid gland** large ductless gland near larynx secreting hormone which regulates growth and development, extract of this.

thyself /ðaɪˈself/ *pronoun archaic: emphatic & reflexive form of* THOU[1]; *reflexive form of* THEE.

ti = TE.

tiara /tɪˈɑːrə/ *noun* jewelled ornamental band worn on front of woman's hair; 3-crowned diadem formerly worn by pope.

tibia /ˈtɪbɪə/ *noun* (*plural* **tibiae** /-biː/) inner of two bones extending from knee to ankle.

tic *noun* (in full **nervous tic**) spasmodic contraction of muscles, esp. of face.

tick[1] ● *noun* slight recurring click, esp. of watch or clock; *colloquial* moment; small mark (✓) to denote correctness etc. ● *verb* make sound of tick; (often + *off*) mark with tick. □ **tick off** *colloquial* reprimand; **tick over** (of engine) idle, function at basic level; **tick-tack** kind of manual semaphore used by racecourse bookmakers; **tick-tock** ticking of large clock etc.

tick[2] *noun* parasitic arachnid or insect on animals.

tick[3] *noun colloquial* financial credit.

tick[4] *noun* case of mattress or pillow; ticking.

ticker *noun colloquial* heart; watch, *US* tape machine. □ **ticker-tape** paper strip from tape machine, esp. as thrown from windows to greet celebrity.

ticket /ˈtɪkɪt/ ● *noun* piece of paper or card entitling holder to enter place, participate in event, travel by public transport, etc.; notification of traffic offence etc.; certificate of discharge from army or of qualification as ship's master, pilot, etc.; price etc. label; *esp. US* list of candidates put forward by group, esp. political party, principles of party; (**the ticket**) *colloquial* what is needed. ● *verb* (**-t-**) attach ticket to.

ticking *noun* strong usually striped material to cover mattresses etc.

tickle /ˈtɪk(ə)l/ ● *verb* (**-ling**) touch or stroke lightly so as to produce laughter and spasmodic movement; excite agreeably, amuse; catch (trout etc.) by rubbing it so that it moves backwards into hand. ● *noun* act or sensation of tickling.

ticklish /ˈtɪklɪʃ/ *adjective* sensitive to tickling; difficult to handle.

tidal /ˈtaɪd(ə)l/ *adjective* related to, like, or affected by tides. □ **tidal wave** exceptionally large ocean wave, esp. one caused by underwater earthquake, widespread manifestation of feeling etc.

tidbit *US* = TITBIT.

tiddler /ˈtɪdlə/ *noun colloquial* small fish, esp. stickleback or minnow; unusually small thing.

tiddly[1] /ˈtɪdlɪ/ *adjective* (**-ier**, **-iest**) *colloquial* slightly drunk.

tiddly[2] /ˈtɪdlɪ/ *adjective* (**-ier**, **-iest**) *colloquial* little.

tiddly-wink /ˈtɪdlɪwɪŋk/ *noun* counter flicked with another into cup; (in *plural*) this game.

tide ● *noun* regular rise and fall of sea due to attraction of moon and sun; water as moved by this; time, season; trend of opinion, fortune, or events. ● *verb* (**-ding**) (**tide over**) temporarily provide with what is needed. □ **tide-mark** mark made by tide at high water, *colloquial* line of dirt round bath, or on person's body between washed and unwashed parts; **tideway** tidal part of river.

tidings /ˈtaɪdɪŋz/ *noun archaic or jocular* (treated as *singular* or *plural*) news.

tidy /ˈtaɪdɪ/ ● *adjective* (**-ier**, **-iest**) neat, orderly; (of person) methodical; *colloquial* considerable. ● *noun* (*plural* **-ies**) receptacle for odds and ends. ● *verb* (**-ies**, **-ied**) (often + *up*) make (oneself, room, etc.) tidy; put in order. □ **tidily** *adverb*; **tidiness** *noun*.

tie ● *verb* (**tying**) attach or fasten with cord etc.; form into knot or bow; (often + *down*) restrict, bind; (often + *with*) make same score as another competitor; bind (rafters etc.) by crosspiece etc.; *Music* unite (notes) by tie. ● *noun* cord etc. used for fastening; strip of material worn round collar and tied in knot at front; thing that unites or restricts people; equality of score, draw, or dead heat among competitors; match between any pair of players or teams; rod or beam holding parts of structure together; *Music* curved line above or below two notes of same pitch that are to be joined as one. □ **tie-break**, **-breaker** means of deciding winner when competitors have tied; **tie-dye** method of producing dyed patterns by tying string etc. to keep dye from parts of fabric; **tie-pin** ornamental pin to hold necktie in place; **tie up** fasten with cord etc., invest (money etc.) so that it is not immediately available for use, fully occupy (person), bring to satisfactory conclusion; **tie-up** connection, association.

tied *adjective* (of dwelling house) occupied subject to tenant's working for house's owner; (of public house etc.) bound to supply only particular brewer's liquor.

tier /tɪə/ ● *noun* row, rank, or unit of structure, as one of several placed one above another. □ **tiered** *adjective*.

tiercel = TERCEL.

tiff *noun* slight or petty quarrel.

tiger /'taɪgə/ *noun* large Asian animal of cat family, with yellow-brown coat with black stripes; fierce, formidable, or energetic person. □ **tiger-cat** any moderate-sized feline resembling tiger; **tiger lily** tall garden lily with dark-spotted orange flowers.

tight /taɪt/ ● *adjective* closely, held, drawn, fastened, fitting, etc.; impermeable, impervious; tense, stretched; *colloquial* drunk; *colloquial* stingy; (of money or materials) not easily obtainable; stringent, demanding; presenting difficulties; produced by or requiring great exertion or pressure. ● *adverb* tightly. □ **tight corner** difficult situation; **tight-fisted** stingy; **tight-lipped** restraining emotion, determinedly reticent; **tightrope** high tightly stretched rope or wire on which acrobats etc. perform. □ **tighten** *verb*; **tightly** *adverb*; **tightness** *noun*.

tights *plural noun* thin close-fitting stretch garment covering legs, feet, and lower torso.

tigress /'taɪgrɪs/ *noun* female tiger.

tilde /'tɪldə/ *noun* mark (˜) placed over letter, e.g. Spanish *n* in *señor*.

tile ● *noun* thin slab of concrete, baked clay, etc. for roofing, paving, etc. ● *verb* (**-ling**) cover with tiles. □ **tiler** *noun*.

tiling *noun* process of fixing tiles; area of tiles.

till[1] ● *preposition* up to, as late as. ● *conjunction* up to time when; so long that.

■ **Usage** In all senses, *till* can be replaced by *until*, which is more formal in style.

---

till[2] *noun* money-drawer in bank, shop, etc., esp. with device recording amount and details of each purchase.

till[3] *verb* cultivate (land).

tillage *noun* preparation of land for growing crops; tilled land.

tiller /'tɪlə/ *noun* bar by which boat's rudder is turned.

tilt ● *verb* (cause to) assume sloping position or heel over; (+ *at*) thrust or run at with weapon; (+ *with*) engage in contest. ● *noun* tilting; sloping position; (of medieval knights etc.) charging with lance against opponent or mark. □ **(at) full tilt** at full speed, with full force.

tilth *noun* tillage, cultivation; cultivated soil.

timber /'tɪmbə/ *noun* wood for building, carpentry, etc.; piece of wood, beam, esp. as rib of vessel; large standing trees; (as *interjection*) tree is about to fall.

timbered *adjective* made (partly) of timber; (of land) wooded.

timbre /'tæmbə/ *noun* distinctive character of musical sound or voice apart from its pitch and volume.

timbrel /'tɪmbr(ə)l/ *noun archaic* tambourine.

time ● *noun* indefinite continuous progress of past, present, and future events etc. regarded as a whole; more or less definite portion of this, historical or other period; allotted or available portion of time; definite or fixed point or portion of time; (**a time**) indefinite period; occasion; moment etc. suitable for purpose; (in *plural*) (after numeral etc.) *expressing multiplication*; lifetime; (in *singular* or *plural*) conditions of life or of period; *slang* prison sentence; apprenticeship; date or expected date of childbirth or death; measured amount of time worked; rhythm or measure of musical composition. • *verb* (**-ming**) choose time for, do at chosen or appropriate time; ascertain time taken by. □ **at the same time** simultaneously, nevertheless; **at times** now and then; **from time to time** occasionally; **in no time** rapidly, in a moment; **in time** not late, early enough, eventually, following time of music etc.; **on time** punctually; **time-and-motion** measuring efficiency of industrial etc. operations; **time bomb** one designed to explode at pre-set time; **time capsule** box etc. containing objects typical of present time, buried for future discovery; **time-honoured** esteemed by tradition or through custom; **timekeeper** person who records time, watch or clock as regards accuracy; **timekeeping** keeping of time, punctuality; **time-lag** interval between cause and effect; **time off** time used for rest or different activity; **timepiece** clock, watch; **time-server** person who adapts his or her opinions to suit prevailing circumstances; **time-share** share in property under time-sharing scheme; **time-sharing** use of holiday home by several joint owners at different times of year, use of computer by several people for different operations at the same time; **time sheet** sheet of paper for recording hours worked; **time-shift** move from one time to another; **time signal** audible indication of exact time of day; **time signature** *Music* indication of rhythm; **time switch** one operating automatically at preset time; **timetable** *noun* table showing times of public transport services, scheme of lessons, etc., *verb* include or arrange in such schedule; **time zone** range of longitudes where a common standard time is used.

timeless *adjective* not affected by passage of time. □ **timelessness** *noun*.

timely *adjective* (**-ier**, **-iest**) opportune, coming at right time. □ **timeliness** *noun*.

timer *noun* person or device that measures time taken.

timid /'tɪmɪd/ *adjective* (**-er**, **-est**) easily alarmed; shy. □ **timidity** /-'mɪd-/ *noun*; **timidly** *adverb*.

timing *noun* way thing is timed; regulation of opening and closing of valves in internal-combustion engine.

timorous /'tɪmərəs/ *adjective* timid, frightened. □ **timorously** *adverb*.

timpani /'tɪmpənɪ/ *plural noun* (also **tympani**) kettledrums. □ **timpanist** *noun*.

tin • *noun* silvery-white metal used esp. in alloys and in making tin plate; container of tin or tin plate, esp. for preserving food; tin plate. • *verb* (**-nn-**) preserve (food) in tin; cover or coat with tin. □ **tin foil** foil of tin, aluminium, or tin alloy, used to wrap food; **tin hat** *colloquial* military steel helmet; **tin-opener** tool for opening tins; **tin-pan alley** world of composers and publishers of popular music; **tin plate** sheet steel coated with tin; **tinpot** cheap, inferior; **tinsnips** clippers for cutting sheet metal; **tin-tack** tack[1] coated with tin.

tincture /'tɪŋktʃə/ • *noun* (often + *of*) slight flavour or tinge; medicinal solution of drug in alcohol. • *verb* (**-ring**) colour slightly, tinge, flavour; (often + *with*) affect slightly.

tinder /'tɪndə/ *noun* dry substance readily taking fire from spark. □ **tinder-box** *historical* box with tinder, flint, and steel for kindling fires.

tine *noun* prong, tooth, or point of fork, comb, antler, etc.

ting • *noun* tinkling sound as of bell. • *verb* (cause to) emit this.

tinge /tɪndʒ/ • *verb* (**-ging**) (often + *with*; often in *passive*) colour slightly. • *noun* tendency to or trace of some colour; slight admixture of feeling or quality.

tingle /'tɪŋg(ə)l/ • *verb* (**-ling**) feel or cause slight pricking or stinging sensation. • *noun* tingling sensation.

tinker /'tɪŋkə/ • *noun* itinerant mender of kettles, pans, etc.; *Scottish & Irish* Gypsy; *colloquial* mischievous person or animal. • *verb* (+ *at*, *with*) work in amateurish or desultory way; work as tinker.

tinkle /'tɪŋk(ə)l/ • *verb* (**-ling**) (cause to) make short light ringing sounds. • *noun* tinkling sound.

tinnitus /'tɪnɪtəs/ *noun Medicine* condition with ringing in ears.

tinny *adjective* (**-ier**, **-iest**) like tin; flimsy; (of sound) thin and metallic.

tinsel /'tɪns(ə)l/ *noun* glittering decorative metallic strips, threads, etc.; superficial brilliance or splendour. □ **tinselled** *adjective.*

tint ● *noun* a variety of a colour; tendency towards or admixture of a different colour; faint colour spread over surface. ● *verb* apply tint to, colour.

tintinnabulation /tɪntɪnæbjʊ'leɪʃ(ə)n/ *noun* ringing of bells.

tiny /'taɪnɪ/ *adjective* (**-ier**, **-iest**) very small.

tip¹ ● *noun* extremity, esp. of small or tapering thing; small piece or part attached to end of thing. ● *verb* (**-pp-**) provide with tip. □ **tiptop** *colloquial* first-rate, of highest excellence.

tip² ● *verb* (**-pp-**) (often + *over*, *up*) (cause to) lean or slant; (+ *into* etc.) overturn, cause to overbalance, discharge contents of (container etc.) thus. ● *noun* slight push or tilt; place where refuse is tipped.

tip³ ● *verb* (**-pp-**) give small present of money to, esp. for service; name as likely winner of race or contest; strike or touch lightly. ● *noun* small present of money given esp. for service; piece of private or special information, esp. regarding betting or investment; piece of advice. □ **tip-off** a hint, warning, etc.; **tip off** give warning, hint, or inside information to.

tippet /'tɪpɪt/ *noun* cape or collar of fur etc.

tipple /'tɪp(ə)l/ ● *verb* (**-ling**) drink intoxicating liquor habitually or repeatedly in small quantities. ● *noun colloquial* alcoholic drink. □ **tippler** *noun.*

tipster /'tɪpstə/ *noun* person who gives tips about horse racing etc.

tipsy /'tɪpsɪ/ *adjective* (**-ier**, **-iest**) slightly drunk; caused by or showing intoxication.

tiptoe /'tɪptəʊ/ ● *noun* the tips of the toes. ● *verb* (**-toes**, **-toed**, **-toeing**) walk on tiptoe or stealthily. ● *adverb* (also **on tiptoe**) with heels off the ground.

TIR *abbreviation* international road transport (*transport international routier*).

tirade /taɪ'reɪd/ *noun* long vehement denunciation or declamation.

tire¹ /taɪə/ *verb* (**-ring**) make or grow weary; exhaust patience or interest of; (in *passive*; + *of*) have had enough of.

tire² *US* = TYRE.

tired *adjective* weary, ready for sleep; (of idea) hackneyed. □ **tiredly** *adverb*; **tiredness** *noun.*

tireless *adjective* not tiring easily, energetic. □ **tirelessly** *adverb*; **tirelessness** *noun.*

tiresome *adjective* tedious; *colloquial* annoying. □ **tiresomely** *adverb.*

tiro /'taɪərəʊ/ *noun* (also **tyro**) (*plural* **-s**) beginner, novice.

tissue /'tɪʃu:/ *noun* any of the coherent collections of cells of which animals or plants are made; tissue-paper; disposable piece of thin absorbent paper for wiping, drying, etc.; fine woven esp. gauzy fabric; (often + *of*) connected series (of lies etc.). □ **tissue-paper** thin soft paper for wrapping etc.

tit¹ *noun* any of various small birds.

tit² *noun* □ **tit for tat** blow for blow, retaliation.

Titan /'taɪt(ə)n/ *noun* (often **titan**) person of superhuman strength, intellect, or importance.

titanic /taɪ'tænɪk/ *adjective* gigantic, colossal.

titanium /taɪ'teɪnɪəm/ *noun* dark grey metallic element.

titbit /'tɪtbɪt/ *noun* (*US* **tidbit**) dainty morsel; piquant item of news etc.

titchy /'tɪtʃɪ/ *adjective* (**-ier**, **-iest**) *colloquial* very small.

tithe /taɪð/ *historical* ● *noun* one-tenth of annual produce of land or labour taken as tax for Church. ● *verb* (**-thing**) subject to tithes; pay tithes.

Titian /'tɪʃ(ə)n/ *adjective* (of hair) bright auburn.

titillate /'tɪtɪleɪt/ *verb* (**-ting**) excite, esp. sexually; tickle. □ **titillation** *noun.*

titivate /'tɪtɪveɪt/ *verb* (**-ting**) *colloquial* smarten; put finishing touches to. □ **titivation** *noun.*

title /'taɪt(ə)l/ *noun* name of book, work of art, etc.; heading of chapter etc.; title-page; caption or credit in film etc.; name denoting person's status; championship in sport; legal right to ownership of property; (+ *to*) just or recognized claim to. □ **title-deed** legal document constituting evidence of a right;

title-holder person holding (esp. sporting) title; **title-page** page at beginning of book giving title, author, etc.; **title role** part in play etc. from which its title is taken.

titled *adjective* having title of nobility or rank.

titmouse /ˈtɪtmaʊs/ *noun* (*plural* **titmice**) small active tit.

titrate /taɪˈtreɪt/ *verb* (**-ting**) ascertain quantity of constituent in (solution) by adding measured amounts of reagent. □ **titration** *noun*.

titter /ˈtɪtə/ ● *verb* laugh covertly, giggle. ● *noun* covert laugh.

tittle /ˈtɪt(ə)l/ *noun* particle, whit.

tittle-tattle /ˈtɪt(ə)ltæt(ə)l/ *noun & verb* (**-ling**) gossip, chatter.

tittup /ˈtɪtəp/ ● *verb* (**-p-** or **-pp-**) go friskily or jerkily, bob up and down, canter. ● *noun* such gait or movement.

titular /ˈtɪtjʊlə/ *adjective* of or relating to title; existing or being in name only.

tizzy /ˈtɪzɪ/ *noun* (*plural* **-ies**) *colloquial* state of nervous agitation.

TNT *abbreviation* trinitrotoluene.

to /tə, before vowel tʊ, when stressed tuː/ ● *preposition* in direction of; as far as, not short of; according to; compared with; involved in, comprising; *used to introduce indirect object of verb etc., to introduce or as substitute for infinitive, or to express purpose, consequence, or cause*. ● *adverb* in normal or required position or condition; (of door) nearly closed. □ **to and fro** backwards and forwards, (repeatedly) from place to place; **to-do** fuss, commotion; **toing and froing** constant movement to and fro, great or dispersed activity.

toad *noun* froglike amphibian breeding in water but living chiefly on land; repulsive person. □ **toadflax** plant with yellow or purple flowers; **toad-in-the-hole** sausages baked in batter; **toadstool** fungus (usually poisonous) with round top and slender stalk.

toady /ˈtəʊdɪ/ ● *noun* (*plural* **-ies**) sycophant. ● *verb* (**-ies**, **-ied**) (+ *to*) behave servilely to, fawn on. □ **toadyism** *noun*.

toast ● *noun* sliced bread browned on both sides by radiant heat; person or thing in whose honour company is requested to drink; call to drink or instance of drinking thus. ● *verb* brown by heat, warm at fire etc.; drink to the health or in honour of. □ **toasting-fork** long-handled fork for toasting bread etc.; **toastmaster**, **toastmistress** person announcing toasts at public occasion; **toast rack** rack for holding slices of toast at table.

toaster *noun* electrical device for making toast.

tobacco /təˈbækəʊ/ *noun* (*plural* **-s**) plant of American origin with leaves used for smoking, chewing, or snuff; its leaves, esp. as prepared for smoking.

tobacconist /təˈbækənɪst/ *noun* dealer in tobacco.

toboggan /təˈbɒgən/ ● *noun* long light narrow sledge for sliding downhill, esp. over snow. ● *verb* ride on toboggan.

toby jug /ˈtəʊbɪ/ *noun* jug or mug in shape of stout man in 3-cornered hat.

toccata /təˈkɑːtə/ *noun Music* composition for keyboard instrument, designed to exhibit performer's touch and technique.

tocsin /ˈtɒksɪn/ *noun* alarm bell or signal.

today /təˈdeɪ/ ● *adverb* on this present day; nowadays. ● *noun* this present day; modern times.

toddle /ˈtɒd(ə)l/ ● *verb* (**-ling**) walk with young child's short unsteady steps; *colloquial* walk, stroll, (usually + *off*, *along*) depart. ● *noun* toddling walk.

toddler *noun* child just learning to walk.

toddy /ˈtɒdɪ/ *noun* (*plural* **-ies**) sweetened drink of spirits and hot water.

toe ● *noun* any of terminal projections of foot or paw; part of footwear that covers toes; lower end or tip of implement etc. ● *verb* (**toes**, **toed**, **toeing**) touch with toe(s). □ **on one's toes** alert; **toecap** (reinforced) part of boot or shoe covering toes; **toe-hold** slight foothold, small beginning or advantage; **toe the line** conform, esp. under pressure; **toenail** nail of each toe.

toff *noun slang* upper-class person.

toffee /ˈtɒfɪ/ *noun* firm or hard sweet made of boiled butter, sugar, etc.; this substance. □ **toffee-apple** toffee-coated apple; **toffee-nosed** *slang* snobbish, pretentious.

tofu /'təʊfu:/ *noun* curd of mashed soya beans.

tog[1] *colloquial* ● *noun* (in *plural*) clothes. ● *verb* (**-gg-**) (+ *out, up*) dress.

tog[2] *noun* unit of thermal resistance of quilts etc.

toga /'təʊgə/ *noun historical* ancient Roman citizen's loose flowing outer garment.

together /tə'geðə/ ● *adverb* in(to) company or conjunction; simultaneously; one with another; uninterruptedly. ● *adjective colloquial* well-organized, self-assured, emotionally stable. □ **togetherness** *noun*.

toggle /'tɒg(ə)l/ *noun* short bar used like button for fastening clothes; *Computing* key or command which alternately switches function on and off.

toil ● *verb* work laboriously or incessantly; make slow painful progress. ● *noun* labour; drudgery. □ **toilsome** *adjective*.

toilet /'tɔɪlɪt/ *noun* lavatory; process of washing oneself, dressing, etc. □ **toilet paper** paper for cleaning oneself after using lavatory; **toilet roll** roll of toilet paper; **toilet water** dilute perfume used after washing.

toiletries /'tɔɪlɪtri:z/ *plural noun* articles or cosmetics used in washing, dressing, etc.

toilette /twɑ:'let/ *noun* process of washing oneself, dressing, etc.

toils /tɔɪlz/ *plural noun* net, snare.

token /'təʊkən/ ● *noun* symbol, reminder, mark; voucher; thing equivalent to something else, esp. money. ● *adjective* perfunctory, chosen by tokenism to represent a group. □ **token strike** brief strike to demonstrate strength of feeling.

tokenism *noun* granting of minimum concessions.

told *past & past participle* of TELL.

tolerable /'tɒlərəb(ə)l/ *adjective* endurable; fairly good. □ **tolerably** *adverb*.

tolerance /'tɒlərəns/ *noun* willingness or ability to tolerate; permitted variation in dimension, weight, etc.

tolerant /'tɒlərənt/ *adjective* disposed to tolerate others; (+ *of*) enduring or patient of.

tolerate /'tɒləreɪt/ *verb* (**-ting**) allow the existence or occurrence of without authoritative interference; endure; find or treat as endurable; be able to take or undergo without harm. □ **toleration** *noun*.

toll[1] /təʊl/ *noun* charge to use bridge, road, etc.; cost or damage caused by disaster etc. □ **toll-gate** barrier preventing passage until toll is paid.

toll[2] /təʊl/ ● *verb* (of bell) ring with slow uniform strokes, ring (bell) thus; announce or mark (death etc.) thus; (of bell) strike (the hour). ● *noun* tolling or stroke of bell.

toluene /'tɒljʊi:n/ *noun* colourless liquid hydrocarbon used in manufacture of explosives etc.

tom *noun* (in full **tom-cat**) male cat.

tomahawk /'tɒməhɔ:k/ *noun* N. American Indian war-axe.

tomato /tə'mɑ:təʊ/ *noun* (*plural* **-es**) glossy red or yellow fleshy edible fruit; plant bearing this.

tomb /tu:m/ *noun* burial-vault; grave; sepulchral monument. □ **tombstone** memorial stone over grave.

tombola /tɒm'bəʊlə/ *noun* kind of lottery.

tomboy /'tɒmbɔɪ/ *noun* girl who enjoys rough noisy recreations. □ **tomboyish** *adjective*.

tome *noun* large book or volume.

tomfool /tɒm'fu:l/ ● *noun* fool. ● *adjective* foolish. □ **tomfoolery** *noun*.

Tommy /'tɒmɪ/ *noun* (*plural* **-ies**) *colloquial* British private soldier.

tommy-gun /'tɒmɪgʌn/ *noun* sub-machine-gun.

tomorrow /tə'mɒrəʊ/ ● *adverb* on day after today; in future. ● *noun* the day after today; the near future.

tomtit *noun* tit, esp. blue tit.

tom-tom /'tɒmtɒm/ *noun* kind of drum usually beaten with hands.

ton /tʌn/ *noun* measure of weight equalling 2,240 lb (**long ton**) or 2,000 lb (**short ton**); metric ton; unit of measurement of ship's tonnage; (usually in *plural*) *colloquial* large number or amount; *slang* speed of 100 m.p.h., score of 100.

tonal /'təʊn(ə)l/ *adjective* of or relating to tone or tonality.

tonality /tə'nælɪtɪ/ *noun* (*plural* **-ies**) relationship between tones of a musical scale; observance of single tonic key as basis of musical composition; colour scheme of picture.

tone • *noun* sound, esp. with reference to pitch, quality, and strength; (often in *plural*) modulation of voice to express emotion etc.; manner of expression in writing or speaking; musical sound, esp. of definite pitch and character; general effect of colour or of light and shade in picture; tint or shade of colour; prevailing character of morals, sentiments, etc.; proper firmness of body, state of (good) health. • *verb* (**-ning**) give desired tone to; alter tone of; harmonize. □ **tone-deaf** unable to perceive differences in musical pitch; **tone down** make or become softer in tone; **tone up** make or become stronger in tone. □ **toneless** *adjective*; **tonelessly** *adverb*; **toner** *noun*.

tongs *plural noun* implement with two arms for grasping coal, sugar, etc.

tongue /tʌŋ/ • *noun* muscular organ in mouth used in tasting, swallowing, speaking, etc.; tongue of ox etc. as food; faculty or manner of speaking; particular language; thing like tongue in shape. • *verb* (**-guing**) use tongue to articulate (notes) in playing wind instrument. □ **tongue-in-cheek** ironic(ally); **tongue-tied** too shy to speak; **tongue-twister** sequence of words difficult to pronounce quickly and correctly.

tonic /'tɒnɪk/ • *noun* invigorating medicine; anything serving to invigorate; tonic water; *Music* keynote. • *adjective* invigorating. □ **tonic sol-fa** musical notation used esp. in teaching singing; **tonic water** carbonated drink with quinine.

tonight /tə'naɪt/ • *adverb* on present or approaching evening or night. • *noun* the evening or night of today.

tonnage /'tʌnɪdʒ/ *noun* ship's internal cubic capacity or freight-carrying capacity; charge per ton on freight or cargo.

tonne /tʌn/ *noun* 1,000 kg.

tonsil /'tɒns(ə)l/ *noun* either of two small organs on each side of root of tongue.

tonsillectomy /tɒnsə'lektəmɪ/ *noun* (*plural* **-ies**) surgical removal of tonsils.

tonsillitis /tɒnsə'laɪtɪs/ *noun* inflammation of tonsils.

tonsorial /tɒn'sɔːrɪəl/ *adjective usually jocular* of hairdresser or hairdressing.

tonsure /'tɒnʃə/ • *noun* shaving of crown or of whole head as clerical or monastic symbol; bare patch so made. • *verb* (**-ring**) give tonsure to.

too *adverb* to a greater extent than is desirable or permissible; *colloquial* very; in addition; moreover.

took *past* of TAKE.

tool /tuːl/ • *noun* implement for working on something by hand or by machine; thing used in activity; person merely used by another. • *verb* dress (stone) with chisel; impress design on (leather); (+ *along*, *around*, etc.) *slang* drive or ride esp. in a casual or leisurely way.

toot /tuːt/ • *noun* sound (as) of horn etc. • *verb* sound (horn etc.); give out such sound.

tooth /tuːθ/ *noun* (*plural* **teeth**) each of a set of hard structures in jaws of most vertebrates, used for biting and chewing; toothlike part or projection, e.g. cog of gearwheel, point of saw or comb, etc.; (often + *for*) taste, appetite; (in *plural*) force, effectiveness. □ **fight tooth and nail** fight fiercely; **get one's teeth into** devote oneself seriously to; **in the teeth of** in spite of, contrary to, directly against (wind etc.); **toothache** pain in teeth; **toothbrush** brush for cleaning teeth; **toothpaste** paste for cleaning teeth; **toothpick** small sharp stick for removing food lodged between teeth. □ **toothed** *adjective*; **toothless** *adjective*.

toothsome *adjective* (of food) delicious.

toothy *adjective* (**-ier**, **-iest**) having large, numerous, or prominent teeth.

tootle /'tuːt(ə)l/ *verb* (**-ling**) toot gently or repeatedly; (usually + *around*, *along*, etc.) *colloquial* move casually.

top[1] • *noun* highest point or part; highest rank or place, person occupying this; upper end, head; upper surface, upper part; cover or cap of container etc.; garment for upper part of body; utmost degree, height; (in *plural*) *colloquial* person or thing of best quality; (esp. in *plural*) leaves etc. of plant grown chiefly for its root; *Nautical* platform round head of lower mast. ● *adjective* highest in position, degree, or importance. ● *verb* (**-pp-**) furnish with top, cap, etc.; be higher or better than, surpass, be at or reach top of; *slang* kill, hit golf ball above centre. □ **on top of** fully in command of, very close to, in addition to; **top brass** *colloquial* high-ranking officers; **topcoat** overcoat, final coat of paint etc.; **top dog** *colloquial* victor, master; **top drawer** *colloquial* high social position or origin; **top dress** apply fertilizer on top of (earth) without ploughing it in; **top-flight** of highest rank of achievement; **top hat** tall silk hat; **top-heavy** overweighted at top; **topknot** knot, tuft, crest, or bow worn or growing on top of head; **topmast** mast on top of lower mast; **top-notch** *colloquial* first-rate; **top secret** of utmost secrecy; **topside** outer side of round of beef, side of ship above waterline; **topsoil** top layer of soil; **top up** complete (amount), fill up (partly empty container); **top-up** addition, amount that completes or quantity that fills something. □ **topmost** *adjective*.

top² *noun* toy spinning on point when set in motion.

topaz /'təʊpæz/ *noun* semiprecious transparent stone, usually yellow.

tope *verb* (**-ping**) *archaic or literary* drink alcohol to excess, esp. habitually. □ **toper** *noun*.

topi /'təʊpɪ/ *noun* (also **topee**) (*plural* **-s**) hat, esp sun-helmet.

topiary /'təʊpɪərɪ/ ● *adjective* of or formed by clipping shrubs, trees, etc. into ornamental shapes. ● *noun* topiary art.

topic /'tɒpɪk/ *noun* subject of discourse, conversation, or argument.

topical *adjective* dealing with current affairs, etc. □ **topicality** /-'kæl-/ *noun*.

topless *adjective* without a top; (of garment) leaving breasts bare; (of woman) bare-breasted; (of place) where women go or work bare-breasted.

topography /tə'pɒgrəfɪ/ *noun* detailed description, representation, etc. of features of a district; such features. □ **topographer** *noun*; **topographical** /tɒpə'græf-/ *adjective*.

topology /tə'pɒlədʒɪ/ *noun* study of geometrical properties unaffected by changes of shape or size. □ **topological** /tɒpə'lɒdʒ-/ *adjective*.

topper *noun colloquial* top hat.

topping *noun* thing that tops, esp. sauce on dessert etc.

topple /'tɒp(ə)l/ *verb* (**-ling**) (often + *over*, *down*) (cause to) fall as if top-heavy; overthrow.

topsy-turvy /tɒpsɪ'tɜ:vɪ/ *adverb & adjective* upside down; in utter confusion.

toque /təʊk/ *noun* woman's close-fitting brimless hat.

tor *noun* hill, rocky peak.

torch *noun* battery-powered portable lamp; thing lit for illumination; source of heat, light, or enlightment. □ **carry a torch for** have (esp. unreturned) love for.

tore *past* of TEAR¹.

toreador /'tɒrɪədɔ:/ *noun* bullfighter, esp. on horseback.

torment ● *noun* /'tɔ:ment/ (cause of) severe bodily or mental suffering. ● *verb* /tɔ:'ment/ subject to torment, tease or worry excessively. □ **tormentor** /-'men-/ *noun*.

torn *past participle* of TEAR¹.

tornado /tɔ:'neɪdəʊ/ *noun* (*plural* **-es**) violent storm over small area, with whirling winds.

torpedo /tɔ'pi:dəʊ/ ● *noun* (*plural* **-es**) cigar-shaped self-propelled underwater or aerial missile that explodes on hitting ship. ● *verb* (**-es**, **-ed**) destroy or attack with torpedo(es); make ineffective. □ **torpedo boat** small fast warship armed with torpedoes.

torpid /'tɔ:pɪd/ *adjective* sluggish, apathetic; numb; dormant. □ **torpidity** /-'pɪd-/ *noun*.

torpor /'tɔ:pə/ *noun* torpid condition.

torque /tɔ:k/ *noun* twisting or rotary force, esp. in machine; *historical* twisted metal necklace worn by ancient Gauls and Britons.

torrent /'tɒrənt/ *noun* rushing stream of liquid; downpour of rain; (in *plural*) (usually + *of*) violent flow. □ **torrential** /tə'renʃ(ə)l/ *adjective*.

torrid /'tɒrɪd/ *adjective* intensely hot; scorched, parched; passionate, intense.

torsion /'tɔ:ʃ(ə)n/ *noun* twisting. □ **torsional** *adjective*.

torso /'tɔ:səʊ/ *noun* (*plural* **-s**) trunk of human body; statue of this.

tort *noun* breach of legal duty (other than under contract) with liability for damages. □ **tortious** /'tɔ:ʃəs/ *adjective*.

tortilla /tɔ:'ti:jə/ *noun* thin flat originally Mexican maize cake eaten hot.

tortoise /'tɔ:təs/ *noun* slow-moving reptile with horny domed shell. □ **tortoiseshell** mottled yellowish-brown turtle-shell, cat or butterfly with markings resembling tortoiseshell.

tortuous /'tɔ:tʃʊəs/ *adjective* winding; devious, circuitous. □ **tortuously** *adverb*.

torture /'tɔ:tʃə/ ● *noun* infliction of severe bodily pain, esp. as punishment or means of persuasion; severe physical or mental pain. ● *verb* (**-ring**) subject to torture. □ **torturer** *noun*; **torturous** *adjective*.

Tory /'tɔ:rɪ/ *colloquial* ● *noun* (*plural* **-ies**) member of Conservative party. ● *adjective* Conservative. □ **Toryism** *noun*.

tosa /'təʊsə/ *noun* dog of a mastiff breed.

tosh *noun colloquial* rubbish, nonsense.

toss ● *verb* throw up, esp. with hand; roll about, throw, or be thrown, restlessly or from side to side; throw lightly or carelessly; throw (coin) into air to decide choice etc. by way it falls, (often + *for*) settle question or dispute with (person) thus; (of bull etc.) fling up with horns; coat (food) with dressing etc. by shaking it. ● *noun* tossing; fall, esp. from horseback. □ **toss one's head** throw it back, esp. in anger, impatience, etc.; **toss off** drink off at a draught, dispatch (work) rapidly or easily; **toss up** *verb* toss coin; **toss-up** *noun* doubtful matter, tossing of coin.

tot[1] *noun* small child; dram of liquor.

tot[2] *verb* (**-tt-**) (usually + *up*) add, mount. □ **tot up to** amount to.

total /'təʊt(ə)l/ ● *adjective* complete, comprising the whole; absolute, unqualified. ● *noun* whole sum or amount. ● *verb* (**-ll-**; *US* **-l-**) (often + *to*, *up to*) amount to; calculate total of. □ **totality** /-'tæl-/ *noun* (*plural* **-ies**); **totally** *adverb*.

totalitarian /təʊtælɪ'teərɪən/ *adjective* of one-party government requiring complete subservience to state. □ **totalitarianism** *noun*.

totalizator /'təʊtəlaɪzeɪtə/ *noun* (also **totalisator**) device showing number and amount of bets staked on race when total will be divided among those betting on winner; this betting system.

totalize /'təʊtəlaɪz/ *verb* (also **-ise**) (**-zing** or **-sing**) combine into a total.

tote[1] *noun slang* totalizator.

tote[2] *verb* (**-ting**) *esp. US colloquial* carry, convey. □ **tote bag** large and capacious bag.

totem /'təʊtəm/ *noun* natural object (esp. animal) adopted esp. among N. American Indians as emblem of clan or individual; image of this. □ **totem-pole** post with carved and painted or hung totem(s).

toto see IN TOTO.

totter /'tɒtə/ ● *verb* stand or walk unsteadily or feebly; shake, be about to fall. ● *noun* unsteady or shaky movement or gait. □ **tottery** *adjective*.

toucan /'tu:kən/ *noun* tropical American bird with large bill.

touch /tʌtʃ/ ● *verb* come into or be in physical contact with; (often + *with*) bring hand etc. into contact with; cause (two things) to meet thus; rouse tender or painful feelings in; strike lightly; (usually in negative) disturb, harm, affect, have dealings with, consume, use; concern; reach as far as; (usually in negative) approach in excellence; modify; (as **touched** *adjective*) *colloquial* slightly mad; (usually + *for*) *slang* request and get money etc. from (person). ● *noun* act of touching; sense of feeling; small amount, trace; (**a touch**) slightly; *Music* manner of playing keys or strings, instrument's response to this; artistic, literary, etc. style or skill; *slang* act of requesting and getting money etc. from person; *Football* part of field outside touchlines. □ **touch-and-go** critical, risky; **touch at** *Nautical* call at (port etc.); **touch down** (of aircraft) alight; **touchdown** *noun*; **touchline** side limit of football etc. pitch; **touch off** explode by touching with match etc., initiate (process) suddenly; **touch on**, **upon** refer to or mention briefly or casually, verge on; **touch-paper** paper impregnated with nitre for igniting fireworks etc.; **touchstone** dark schist or jasper for testing alloys by marking it with them, criterion; **touch-type** type without looking at keys; **touch-typist** *noun*; **touch up** give finishing touches to, retouch, *slang* molest sexually; **touch wood** touch something wooden to avert ill luck; **touchwood** readily inflammable rotten wood.

touché /tu:'ʃeɪ/ *interjection acknowledging justified accusation or retort, or hit in fencing.* [French]

touching ● *adjective* moving, pathetic. ● *preposition literary* concerning. □ **touchingly** *adverb.*

touchy *adjective* (**-ier**, **-iest**) apt to take offence, over-sensitive. □ **touchily** *adverb*; **touchiness** *noun.*

tough /tʌf/ ● *adjective* hard to break, cut, tear, or chew; able to endure hardship, hardy; stubborn, difficult; *colloquial* acting sternly, (of luck etc.) hard; *colloquial* criminal, violent. ● *noun* tough person, esp. ruffian. □ **toughen** *verb*; **toughness** *noun.*

toupee /'tu:peɪ/ *noun* hairpiece to cover bald spot.

tour /tʊə/ ● *noun* holiday journey or excursion including stops at various places; walk round, inspection; spell of military or diplomatic duty; series of performances, matches, etc. at different places. ● *verb* (often + *through*) go on a tour; make a tour of (country etc.). □ **on tour** (esp. of sports team, theatre company, etc.) touring; **tour operator** travel agent specializing in package holidays.

tour de force /tʊə də 'fɔ:s/ *noun* (*plural* ***tours de force*** same pronunciation) outstanding feat or performance. [French]

tourer *noun* car or caravan for touring in.

tourism *noun* commercial organization and operation of holidays.

tourist *noun* holiday traveller; member of touring sports team. □ **tourist class** lowest class of passenger accommodation in ship, aeroplane, etc.

tourmaline /'tʊərməli:n/ *noun* mineral with unusual electric properties and used as gem.

tournament /'tʊənəmənt/ *noun* large contest of many rounds; display of military exercises; *historical* pageant with jousting.

tournedos /'tʊənədəʊ/ *noun* (*plural* same /-dəʊz/) small thick piece of fillet of beef.

tourney /'tʊənɪ/ ● *noun* (*plural* **-s**) tournament. ● *verb* (**-eys**, **-eyed**) take part in tournament.

tourniquet /'tʊənɪkeɪ/ *noun* device for stopping flow of blood through artery by compression.

tousle /'taʊz(ə)l/ *verb* (**-ling**) make (esp. hair) untidy; handle roughly.

tout /taʊt/ ● *verb* (usually + *for*) solicit custom persistently, pester customers; solicit custom of or for; spy on racehorses in training. ● *noun* person who touts.

tow[1] /təʊ/ ● *verb* pull along by rope etc. ● *noun* towing, being towed. □ **in tow** being towed, accompanying or in the charge of a person; **on tow** being towed; **towpath** path beside river or canal originally for horse towing boat.

tow[2] /təʊ/ *noun* fibres of flax etc. ready for spinning. □ **tow-headed** having very light-coloured or tousled hair.

towards /tə'wɔ:dz/ *preposition* (also **toward**) in direction of; as regards, in relation to; as a contribution to, for; near.

towel /'taʊəl/ ● *noun* absorbent cloth, paper, etc. for drying after washing etc. ● *verb* (**-ll-**; *US* **-l-**) rub or dry with towel.

towelling *noun* thick soft absorbent cloth used esp. for towels.

tower /'taʊə/ ● *noun* tall structure, often part of castle, church, etc.; fortress etc. with tower; tall structure housing machinery etc. ● *verb* (usually + *above*, *up*) reach high, be superior; (as **towering** *adjective*) high, lofty, violent. □ **tower block** tall building of offices or flats; **tower of strength** person who gives strong emotional support.

town /taʊn/ *noun* densely populated area, between city and village in size; London or the chief city or town in area; central business area in neighbourhood. □ **go to town** *colloquial* act or work with energy or enthusiasm; **on the town** *colloquial* enjoying urban night-life; **town clerk** *US & historical* official in charge of records etc. of town; **town gas** manufactured gas for domestic etc. use; **town hall** headquarters of local government, with public meeting rooms etc.; **town house** town residence, esp. one of terrace;

town planning planning of construction and growth of towns; **township** *South African* urban area for occupation by black people, *US & Canadian* administrative division of county, or district 6 miles square, *Australian & NZ* small town; **townspeople** inhabitants of town.

townie /'taʊnɪ/ *noun* (also **townee** /-'niː/) *derogatory* inhabitant of town.

toxaemia /tɒk'siːmɪə/ *noun* (*US* **toxemia**) blood poisoning; increased blood pressure in pregnancy.

toxic /'tɒksɪk/ *adjective* poisonous; of poison. □ **toxicity** /-'sɪs-/ *noun.*

toxicology /tɒksɪ'kɒlədʒɪ/ *noun* study of poisons. □ **toxicological** /-kə'lɒdʒ-/ *adjective*; **toxicologist** *noun.*

toxin /'tɒksɪn/ *noun* poison produced by living organism.

toy ● *noun* plaything; thing providing amusement; diminutive breed of dog etc. ● *verb* (usually + *with*) amuse oneself, flirt, move thing idly. □ **toy boy** *colloquial* woman's much younger boyfriend; **toyshop** shop selling toys.

trace¹ ● *verb* (**-cing**) find signs of by investigation; (often + *along, through, to,* etc.) follow or mark track, position, or path of; (often + *back*) follow to origins; copy (drawing etc.) by marking its lines on superimposed translucent paper; mark out, delineate, or write, esp. laboriously. ● *noun* indication of existence of something, vestige; very small quantity; track, footprint; mark left by instrument's moving pen etc. □ **trace element** chemical element occurring or required, esp. in soil, only in minute amounts. □ **traceable** *adjective.*

trace² *noun* each of two side-straps, chains, or ropes by which horse draws vehicle. □ **kick over the traces** become insubordinate or reckless.

tracer *noun* bullet etc. made visible in flight by flame etc. emitted; artificial radioisotope which can be followed through body by radiation it produces.

tracery /'treɪsərɪ/ *noun* (*plural* **-ies**) decorative stone openwork, esp. in head of Gothic window; lacelike pattern.

trachea /trə'kiːə/ *noun* (*plural* **-cheae** /-'kiːiː/) windpipe.

tracing *noun* traced copy of drawing etc.; act of tracing. □ **tracing-paper** translucent paper for making tracings.

track ● *noun* mark(s) left by person, animal, vehicle, etc.; (in *plural*) such marks, esp. footprints; rough path; line of travel; continuous railway line; racecourse, circuit, prepared course for runners; groove on gramophone record; single song etc. on gramophone record, CD, or magnetic tape; band round wheels of tank, tractor, etc. ● *verb* follow track of; trace (course, development, etc.) from vestiges. □ **in one's tracks** *colloquial* where one stands, instantly; **make tracks** *colloquial* depart; **make tracks for** *colloquial* go in pursuit of or towards; **track down** reach or capture by tracking; **tracker dog** police dog tracking by scent; **track events** running-races; **track record** person's past achievements; **track shoe** runner's spiked shoe; **track suit** warm outfit worn for exercising etc. □ **tracker** *noun.*

tract¹ *noun* (esp. large) stretch of territory; bodily organ or system.

tract² *noun* pamphlet, esp. containing propaganda.

tractable /'træktəb(ə)l/ *adjective* easily managed; docile. □ **tractability** *noun.*

traction /'trækʃ(ə)n/ *noun* hauling, pulling; therapeutic sustained pull on limb etc. □ **traction-engine** steam or diesel engine for drawing heavy load.

tractor /'træktə/ *noun* vehicle for hauling farm machinery etc.; traction-engine.

trad *colloquial* ● *noun* traditional jazz. ● *adjective* traditional.

trade ● *noun* buying and selling; this between nations etc.; business merely for profit (as distinct from profession); business of specified nature or time; skilled handicraft; (**the trade**) people engaged in specific trade; *US* transaction, esp. swap; (usually in *plural*) trade wind. ● *verb* (**-ding**) (often + *in, with*) engage in trade, buy and sell; exchange; *US* swap; (usually + *with, for*) have transaction. □ **trade in** exchange (esp. used article) in part payment for another; **trade mark** device or name legally registered to represent a company or product, distinctive characteristic; **trade name** name by which a thing is known in a trade, or given by manufacturer to a product, or under which a business trades; **trade off** exchange as compromise; **trade-off** balance, compromise; **trade on** take advantage of; **tradesman**, **tradeswoman** person engaged in trade, esp. shopkeeper; **trade(s) union** organized association of workers in trade, profession, etc. formed to further their common interests; **trade-unionist** member of trade union; **trade wind** constant wind blowing towards equator from NE or SE. □ **trader** *noun*.

tradescantia /trædɪs'kæntɪə/ *noun* (usually trailing) plant with large blue, white, or pink flowers.

trading *noun* engaging in trade. □ **trading estate** area designed for industrial and commercial firms; **trading post** store etc. in remote region; **trading-stamp** token given to customer and exchangeable in quantity usually for goods.

tradition /trə'dɪʃ(ə)n/ *noun* custom, opinion, or belief handed down to posterity; handing down of these.

traditional *adjective* of, based on, or obtained by tradition; (of jazz) in style of early 20th c. □ **traditionally** *adverb*.

traditionalism *noun* respect or support for tradition. □ **traditionalist** *noun & adjective*.

traduce /trə'dju:s/ *verb* (**-cing**) slander. □ **traducement** *noun*; **traducer** *noun*.

traffic /'træfɪk/ ● *noun* vehicles moving on public highway, in air, or at sea; (usually + *in*) trade, esp. illegal; coming and going of people or goods by road, rail, air, sea, etc.; dealings between people etc.; (volume of) messages transmitted through communications system. ● *verb* (**-ck-**) (often + *in*) deal, esp. illegally; barter. □ **traffic island** raised area in road to divide traffic and provide refuge for pedestrians; **traffic jam** traffic at standstill; **traffic light(s)** signal controlling road traffic by coloured lights; **traffic warden** person employed to control movement and parking of road vehicles. □ **trafficker** *noun*.

tragedian /trə'dʒi:dɪən/ *noun* author of or actor in tragedies.

tragedienne /trədʒi:dɪ'en/ *noun* actress in tragedies.

tragedy /'trædʒɪdɪ/ *noun* (*plural* **-ies**) serious accident, sad event; play with tragic unhappy ending.

tragic /'trædʒɪk/ *adjective* disastrous, distressing, very sad; of tragedy. □ **tragically** *adverb*.

tragicomedy /trædʒɪ'kɒmədɪ/ *noun* (*plural* **-ies**) drama or event combining comedy and tragedy.

trail ● *noun* track or scent left by moving person, thing, etc.; beaten path, esp. through wild region; long line of people or things following behind something; part dragging behind thing or person. ● *verb* draw or be drawn along behind; (often + *behind*) walk wearily; follow trail of, pursue; be losing in contest; (usually + *away*, *off*) peter out; (of plant etc.) grow or hang over wall, along ground, etc.; hang loosely. □ **trailing edge** rear edge of aircraft's wing.

trailer *noun* set of extracts from film etc. shown in advance to advertise it; vehicle pulled by another; *US* caravan.

train ● *verb* (often + *to do*) teach (person etc.) specified skill, esp. by practice; undergo this process; bring or come to physical efficiency by exercise, diet, etc.; (often + *up*, *along*) guide growth of (plant); (usually as **trained** *adjective*) make (mind etc.) discerning through practice etc.; (often + *on*) point, aim. ● *noun* series of railway carriages or trucks drawn by engine; thing dragged along behind or forming back part of dress etc.; succession or series of people, things, events, etc.; group of followers, retinue. □ **in train** arranged, in preparation; **train-bearer** person holding up train of another's robe etc.; **train-spotter** person who collects numbers of railway locomotives. □ **trainee** /-'ni:/ *noun*.

trainer *noun* person who trains horses, athletes, etc.; aircraft or simulator used to train pilots; soft running shoe.

training /'treɪnɪŋ/ *noun* process of teaching or learning a skill etc.

traipse *colloquial* ● *verb* (**-sing**) tramp or trudge wearily. ● *noun* tedious journey on foot.

trait /treɪ/ *noun* characteristic.

traitor /'treɪtə/ *noun* (*feminine* **traitress**) person guilty of betrayal or disloyalty. □ **traitorous** *adjective*.

trajectory /trə'dʒektərɪ/ *noun* (*plural* **-ies**) path of object moving under given forces.

tram *noun* (also **tramcar**) electrically powered passenger road vehicle running on rails. □ **tramlines** rails for tram, *colloquial* either pair of parallel lines at edge of tennis etc. court.

trammel /'træm(ə)l/ ● *noun* (usually in *plural*) impediment, restraint; kind of fishing net. ● *verb* (**-ll-**; *US* **-l-**) hamper.

tramp ● *verb* walk heavily and firmly; go on walking expedition; walk laboriously across or along; (often + *down*) tread on, stamp on; live as tramp. ● *noun* itinerant vagrant or beggar; sound of person or people walking or marching; long walk; *slang derogatory* promiscuous woman.

trample /'træmp(ə)l/ *verb* (**-ling**) tread under foot; crush thus. □ **trample on** tread heavily on, treat roughly or with contempt.

trampoline /'træmpəli:n/ ● *noun* canvas sheet connected by springs to horizontal frame, used for acrobatic exercises. ● *verb* (**-ning**) use trampoline.

trance /trɑ:ns/ *noun* sleeplike state; hypnotic or cataleptic state; such state as supposedly entered into by medium; rapture, ecstasy.

tranny /'trænɪ/ *noun* (*plural* **-ies**) *colloquial* transistor radio.

tranquil /'træŋkwɪl/ *adjective* serene, calm, undisturbed. □ **tranquillity** /-'kwɪl-/ *noun*; **tranquilly** *adverb*.

tranquillize *verb* (also **-ise**; *US* also **tranquilize**) (**-zing** or **-sing**) make tranquil, esp. by drug etc.

tranquillizer *noun* (also **-iser**; *US* also **tranquilizer**) drug used to diminish anxiety.

trans- *prefix* across, beyond; on or to other side of; through.

transact /træn'zækt/ *verb* perform or carry through (business etc.).

transaction /træn'zækʃ(ə)n/ *noun* piece of commercial or other dealing; transacting of business; (in *plural*) published reports of discussions and lectures at meetings of learned society.

transatlantic /trænzət'læntɪk/ *adjective* beyond or crossing the Atlantic; American; *US* European.

transceiver /træn'si:və/ *noun* combined radio transmitter and receiver.

transcend /træn'send/ *verb* go beyond or exceed limits of; excel, surpass.

transcendent *adjective* excelling, surpassing; transcending human experience; (esp. of God) existing apart from, or not subject to limitations of, material universe. □ **transcendence** *noun*; **transcendency** *noun*.

transcendental /trænsen'dent(ə)l/ *adjective* a priori, not based on experience, intuitively accepted; abstract, vague. □ **Transcendental Meditation** meditation seeking to induce detachment from problems, anxiety, etc.

transcontinental /trænzkɒntɪ'nent(ə)l/ *adjective* extending across a continent.

transcribe /træn'skraɪb/ *verb* (**-bing**) copy out; write out (notes etc.) in full; record for subsequent broadcasting; *Music* adapt for different instrument etc. □ **transcriber** *noun*; **transcription** /-'skrɪp-/ *noun*.

transcript /'trænskrɪpt/ *noun* written copy.

transducer /trænz'dju:sə/ *noun* device for changing a non-electrical signal (e.g. pressure) into an electrical one (e.g. voltage).

transept /'trænsept/ *noun* part of cross-shaped church at right angles to nave; either arm of this.

transfer ● *verb* /træns'fɜ:/ (**-rr-**) convey, remove, or hand over (thing etc.); make over possession of (thing, right, etc.) to person; move, change, or be moved to another group, club, etc.; change from one station, route, etc. to another to continue journey; convey (design etc.) from one surface to another. ● *noun* /'trænsfɜ:/ transferring, being transferred; design etc. (to be) conveyed from one surface to another; football player etc. who is transferred; document effecting conveyance of property, a right, etc. □ **transferable** /-'fɜ:rəb(ə)l/ *adjective*; **transference** /'trænsfərəns/ *noun*.

transfigure /træns'fɪgə/ *verb* (**-ring**) change appearance of, make more elevated or idealized. □ **transfiguration** *noun*.

transfix /træns'fɪks/ *verb* paralyse with horror or astonishment; pierce with sharp implement or weapon.

transform /træns'fɔ:m/ *verb* change form, appearance, character, etc. of, esp. considerably; change voltage etc. of (alternating current). □ **transformation** /-fə'meɪ-/ *noun*.

transformer *noun* apparatus for reducing or increasing voltage of alternating current.

transfuse /træns'fju:z/ *verb* (**-sing**) transfer (blood or other liquid) into blood vessel to replace that lost; permeate. □ **transfusion** *noun*.

transgress /trænz'gres/ *verb* infringe (law etc.); overstep (limit laid down); sin. □ **transgression** *noun*; **transgressor** *noun*.

transient /'trænzɪənt/ *adjective* of short duration; passing. □ **transience** *noun*.

transistor /træn'zɪstə/ *noun* semiconductor device capable of amplification and rectification; (in full **transistor radio**) portable radio using transistors.

transistorize *verb* (also **-ise**) (**-zing** or **-sing**) equip with transistors rather than valves.

transit /'trænzɪt/ *noun* going; conveying, being conveyed; passage, route; apparent passage of heavenly body across meridian of place or across sun or planet. □ **in transit** (while) going or being conveyed.

transition /træn'zɪʃ(ə)n/ *noun* passage or change from one place, state, condition, style, etc. to another. □ **transitional** *adjective*; **transitionally** *adverb*.

transitive /'trænsɪtɪv/ *adjective* (of verb) requiring direct object expressed or understood.

transitory /'trænzɪtərɪ/ *adjective* not lasting; brief, fleeting.

translate /træn'sleɪt/ *verb* (**-ting**) (often + *into*) express sense of in another language or in another form; be translatable; interpret; move or change, esp. from one person, place, or condition to another. □ **translatable** *adjective*; **translation** *noun*; **translator** *noun*.

transliterate /trænz'lɪtəreɪt/ *verb* (**-ting**) represent (word etc.) in closest corresponding characters of another script. □ **transliteration** *noun*.

translucent /træns'lu:s(ə)nt/ *adjective* allowing light to pass through, semitransparent. □ **translucence** *noun*.

transmigrate /trænzmaɪ'greɪt/ *verb* (**-ting**) (of soul) pass into different body. □ **transmigration** *noun*.

transmission /trænz'mɪʃ(ə)n/ *noun* transmitting, being transmitted; broadcast programme; device transmitting power from engine to axle in vehicle.

transmit /trænz'mɪt/ *verb* (**-tt-**) pass or hand on, transfer; communicate or be medium for (ideas, emotions, etc.); allow (heat, light, sound, etc.) to pass through. □ **transmissible** *adjective*; **transmittable** *adjective*.

transmitter *noun* person or thing that transmits; equipment used to transmit radio etc. signals.

transmogrify /trænz'mɒgrɪfaɪ/ *verb* (**-ies**, **-ied**) *jocular* transform, esp. in magical or surprising way. □ **transmogrification** *noun*.

transmute /trænz'mju:t/ *verb* (**-ting**) change form, nature, or substance of; *historical* change (base metals) into gold. □ **transmutation** *noun*.

transom /'trænsəm/ *noun* horizontal bar in window or above door; (in full **transom window**) window above this.

transparency /træns'pærənsɪ/ *noun* (*plural* **-ies**) being transparent; picture (esp. photograph) to be viewed by light passing through it.

transparent /træns'pærənt/ *adjective* allowing light to pass through and giving maximum visibility possible; (of disguise, pretext, etc.) easily seen through; (of quality etc.) obvious; easily understood. □ **transparently** *adverb*.

transpire /træns'paɪə/ *verb* (**-ring**) (of secret, fact, etc.) come to be known; happen; emit (vapour, moisture) or be emitted through pores of skin etc. □ **transpiration** /-spɪ-/ *noun*.

■ **Usage** The use of *transpire* to mean 'happen' is considered incorrect by some people.

transplant ● *verb* /træns'plɑ:nt/ plant elsewhere; transfer (living tissue or organ) to another part of body or to another body. ● *noun* /'trænsplɑ:nt/ transplanting of organ or tissue; thing transplanted. □ **transplantation** *noun.*

transport ● *verb* /træns'pɔ:t/ take to another place; *historical* deport (criminal) to penal colony; (as **transported** *adjective*) (usually + *with*) affected with strong emotion. ● *noun* /'trænspɔ:t/ system of transporting, means of conveyance; ship, aircraft, etc. used to carry troops, military stores, etc.; (esp. in *plural*) vehement emotion. □ **transportable** /-'pɔ:t-/ *adjective.*

transportation /trænspɔ:'teɪʃ(ə)n/ *noun* (system of) conveying, being conveyed; *US* means of transport; *historical* deporting of criminals.

transporter *noun* vehicle used to transport other vehicles, heavy machinery, etc. □ **transporter bridge** bridge carrying vehicles etc. across water on suspended moving platform.

transpose /træns'pəʊz/ *verb* (**-sing**) cause (two or more things) to change places; change position of (thing) in series or (word(s)) in sentence; *Music* write or play in different key. □ **transposition** /-pə'zɪʃ(ə)n/ *noun.*

transsexual /træns'sekʃʊəl/ (also **transexual**) ● *adjective* having physical characteristics of one sex and psychological identification with the other. ● *noun* transsexual person; person who has had sex change.

transship /træns'ʃɪp/ *verb* (**-pp-**) transfer from one ship or conveyance to another. □ **transshipment** *noun.*

transubstantiation /trænsəbstænʃɪ'eɪʃ(ə)n/ *noun* conversion of Eucharistic elements wholly into body and blood of Christ.

transuranic /trænzjʊ'rænɪk/ *adjective Chemistry* (of element) having higher atomic number than uranium.

transverse /'trænzvɜ:s/ *adjective* situated, arranged, or acting in crosswise direction. □ **transversely** *adverb.*

transvestite /trænz'vestaɪt/ *noun* man deriving pleasure from dressing in women's clothes. □ **transvestism** *noun.*

trap ● *noun* device, often baited, for catching animals; arrangement or trick to catch (out) unsuspecting person; device for releasing clay pigeon to be shot at or greyhound at start of race etc.; curve in drainpipe etc. that fills with liquid and forms seal against return of gas; two-wheeled carriage; trapdoor; *slang* mouth. ● *verb* (**-pp-**) catch (as) in trap; catch (out) using trick etc.; furnish with traps. □ **trapdoor** door in floor, ceiling, or roof.

trapeze /trə'pi:z/ *noun* crossbar suspended by ropes as swing for acrobatics etc.

trapezium /trə'pi:zɪəm/ *noun* (*plural* **-s** or **-zia**) quadrilateral with only one pair of sides parallel; *US* trapezoid.

trapezoid /'træpɪzɔɪd/ *noun* quadrilateral with no sides parallel; *US* trapezium.

trapper *noun* person who traps wild animals, esp. for their fur.

trappings /'træpɪŋz/ *plural noun* ornamental accessories; (esp. ornamental) harness for horse.

Trappist /'træpɪst/ ● *noun* monk of order vowed to silence. ● *adjective* of this order.

trash ● *noun esp. US* worthless or waste stuff, rubbish; worthless person(s). ● *verb slang* wreck, vandalize. □ **trash can** *US* dustbin. □ **trashy** *adjective* (**-ier**, **-iest**).

trauma /'trɔ:mə/ *noun* (*plural* **traumata** /-mətə/ or **-s**) emotional shock; physical injury, resulting shock. □ **traumatic** /-'mæt-/ *adjective*; **traumatize** *verb* (also **-ise**) (**-zing** or **-sing**).

travail /'træveɪl/ *literary* ● *noun* laborious effort; pangs of childbirth. ● *verb* make laborious effort, esp. in childbirth.

travel /'træv(ə)l/ ● *verb* (**-ll-**; *US* **-l-**) go from one place to another; make journey(s), esp. long or abroad; journey along or through, cover (distance); *colloquial* withstand long journey; act as commercial traveller; move or proceed as specified; *colloquial* move quickly; pass from point to point; (of machine or part) move or operate in specified way. ● *noun* travelling, esp. abroad; (often in *plural*) spell of this; range, rate, or mode of motion of part in machinery. □ **travel agency** agency making arrangements for travellers; **travelling salesman** commercial traveller; **travel-sick** nauseous owing to motion in travelling.

travelled *adjective* (*US* **traveled**) experienced in travelling.

traveller *noun* (*US* **traveler**) person who travels or is travelling; commercial traveller. □ **traveller's cheque** cheque for fixed amount, cashed on signature for equivalent in other currencies; **traveller's joy** wild clematis.

travelogue /'trævəlɒg/ *noun* film or illustrated lecture about travel.

traverse ● *verb* /trə'vɜːs/ (**-sing**) travel or lie across; consider or discuss whole extent of. ● *noun* /'trævəs/ sideways movement; traversing; thing that crosses another. □ **traversal** *noun*.

travesty /'trævɪstɪ/ ● *noun* (*plural* **-ies**) grotesque parody, ridiculous imitation. ● *verb* (**-ies**, **-ied**) make or be travesty of.

trawl ● *verb* fish with trawl or seine or in trawler; catch by trawling; (often + *for*, *through*) search thoroughly. ● *noun* trawling; (in full **trawl-net**) large wide-mouthed fishing net dragged by boat along sea bottom.

trawler *noun* boat used for trawling.

tray *noun* flat board with raised rim for carrying dishes etc.; shallow lidless box for papers or small articles, sometimes forming drawer in cabinet etc.

treacherous /'tretʃərəs/ *adjective* guilty of or involving violation of faith or betrayal of trust; not to be relied on, deceptive. □ **treacherously** *adverb*; **treachery** *noun*.

treacle /'triːk(ə)l/ *noun* syrup produced in refining sugar; molasses. □ **treacly** *adjective*.

tread /tred/ ● *verb* (*past* **trod**; *past participle* **trodden** or **trod**) (often + *on*) set one's foot down; walk on; (often + *down*, *in*, *into*) press (down) or crush with feet; perform (steps etc.) by walking. ● *noun* manner or sound of walking; top surface of step or stair; thick moulded part of vehicle tyre for gripping road; part of wheel or sole of shoe etc. that touches ground. □ **treadmill** device for producing motion by treading on steps on revolving cylinder, similar device used for exercise, monotonous routine work; **tread water** maintain upright position in water by moving feet and hands.

treadle /'tred(ə)l/ *noun* lever moved by foot and imparting motion to machine.

treason /'triːz(ə)n/ *noun* violation of allegiance to sovereign (e.g. plotting assassination) or state (e.g. helping enemy).

treasonable *adjective* involving or guilty of treason.

treasure /'treʒə/ ● *noun* precious metals or gems; hoard of them; accumulated wealth; thing valued for rarity, workmanship, associations, etc.; *colloquial* beloved or highly valued person. ● *verb* (**-ring**) value highly; (often + *up*) store up as valuable. □ **treasure hunt** search for treasure, game in which players seek hidden object; **treasure trove** treasure of unknown ownership found hidden.

treasurer *noun* person in charge of funds of society etc.

treasury /'treʒərɪ/ *noun* (*plural* **-ies**) place where treasure is kept; funds or revenue of state, institution, or society; (**Treasury**) (offices and officers of) department managing public revenue of a country. □ **Treasury bench** government front bench in parliament; **treasury bill** bill of exchange issued by government to raise money for temporary needs.

treat ● *verb* act, behave towards, or deal with in specified way; apply process or medical care or attention to; present or handle (subject) in literature or art; (often + *to*) provide with food, drink, or entertainment at one's own expense; (often + *with*) negotiate terms; (often + *of*) give exposition. ● *noun* event or circumstance that gives great pleasure; meal, entertainment, etc. designed to do this; (**a treat**) *colloquial* extremely good or well. □ **treatable** *adjective*.

treatise /'triːtɪz/ *noun* literary composition dealing esp. formally with subject.

treatment *noun* process or manner of behaving towards or dealing with person or thing; medical care or attention.

treaty /'triːtɪ/ *noun* (*plural* **-ies**) formal agreement between states; agreement between people, esp. for purchase of property.

treble /'treb(ə)l/ ● *adjective* threefold, triple; 3 times as much or many; high-pitched. ● *noun* treble quantity or thing; (voice of) boy soprano; high-pitched instrument; high-frequency sound of radio, record player, etc. ● *verb* (**-ling**) multiply or be multiplied by 3. □ **trebly** *adverb*.

tree ● *noun* perennial plant with woody self-supporting main stem and usually unbranched for some distance from ground; shaped piece of wood for various purposes; family tree. ● *verb* (**trees**, **treed**) cause to take refuge in tree. □ **treecreeper** small creeping bird feeding on insects in tree-bark; **tree-fern** large fern with upright woody stem; **tree surgeon** person who treats decayed trees in order to preserve them.

trefoil /'trefɔɪl/ *noun* plant with leaves of 3 leaflets; 3-lobed ornamentation, esp. in tracery windows.

trek ● *verb* (**-kk-**) make arduous journey, esp. (*historical* migrate or journey by ox-wagon. ● *noun* such journey; each stage of it. □ **trekker** *noun*.

trellis /'trelɪs/ *noun* (in full **trellis-work**) lattice of light wooden or metal bars, esp. support for climbing plants.

tremble /'tremb(ə)l/ ● *verb* (**-ling**) shake involuntarily with emotion, cold, etc.; be affected with extreme apprehension; quiver. ● *noun* trembling, quiver. □ **trembly** *adjective* (**-ier**, **-iest**).

tremendous /trɪ'mendəs/ *adjective colloquial* remarkable, considerable, excellent; awe-inspiring, overpowering. □ **tremendously** *adverb*.

tremolo /'tremələʊ/ *noun* (*plural* **-s**) tremulous effect in music.

tremor /'tremə/ *noun* shaking, quivering; thrill (of fear, exultation, etc.); (in full **earth tremor**) slight earthquake.

tremulous /'tremjʊləs/ *adjective* trembling. □ **tremulously** *adverb*.

trench ● *noun* deep ditch, esp. one dug by troops as shelter from enemy's fire. ● *verb* dig trench(es) in; make series of trenches (in) so as to bring lower soil to surface. □ **trench coat** lined or padded waterproof coat, loose belted raincoat.

trenchant /'trentʃ(ə)nt/ *adjective* incisive, terse, vigorous. □ **trenchancy** *noun*; **trenchantly** *adverb*.

trencher /'trentʃə/ *noun historical* wooden etc. platter for serving food.

trencherman /'trentʃəmən/ *noun* eater.

trend ● *noun* general direction and tendency. ● *verb* turn away in specified direction; have general tendency. □ **trend-setter** person who leads the way in fashion etc.

trendy *colloquial often derogatory* ● *adjective* (**-ier**, **-iest**) fashionable. ● *noun* (*plural* **-ies**) fashionable person. □ **trendily** *adverb*; **trendiness** *noun*.

trepan /trɪ'pæn/ *historical* ● *noun* surgeon's cylindrical saw for making opening in skull. ● *verb* (**-nn-**) perforate (skull) with trepan.

trepidation /trepɪ'deɪʃ(ə)n/ *noun* fear, anxiety.

trespass /'trespəs/ ● *verb* (usually + *on*, *upon*) enter unlawfully (on another's land, property, etc.); encroach. ● *noun* act of trespassing; *archaic* sin, offence. □ **trespasser** *noun*.

tress *noun* lock of hair; (in *plural*) hair.

trestle /'tres(ə)l/ *noun* supporting structure for table etc. consisting of two frames fixed at an angle or hinged or of bar with two divergent pairs of legs; (in full **trestle-table**) table of board(s) laid on trestles; (in full **trestle-work**) open braced framework to support bridge etc.

trews *plural noun* close-fitting usually tartan trousers.

tri- *combining form* three (times).

triad /'traɪæd/ *noun* group of 3 (esp. notes in chord). □ **triadic** /-'æd-/ *adjective*.

trial /'traɪəl/ *noun* judicial examination and determination of issues between parties by judge with or without jury; test; trying thing or person; match held to select players for team; contest for horses, dogs, motorcycles, etc. □ **on trial** being tried in court of law, being tested; **trial run** preliminary operational test.

triangle /'traɪæŋg(ə)l/ *noun* plane figure with 3 sides and angles; any 3 things not in straight line, with imaginary lines joining them; implement etc. of this shape; *Music* instrument of steel rod bent into triangle, struck with small steel rod; situation involving 3 people. □ **triangular** /-'æŋgjʊlə/ *adjective*.

triangulate /traɪ'æŋgjʊleɪt/ *verb* (**-ting**) divide (area) into triangles for surveying purposes. □ **triangulation** *noun*.

triathlon /traɪ'æθlən/ *noun* athletic contest of 3 events. □ **triathlete** *noun*.

tribe *noun* (in some societies) group of families under recognized leader with blood etc. ties and usually having common culture and dialect; any similar natural or political division; *usually derogatory* set or number of people, esp. of one profession etc. or family. □ **tribesman**, **tribeswoman** member of tribe. □ **tribal** *adjective*.

tribulation /trɪbjʊ'leɪʃ(ə)n/ *noun* great affliction.

tribunal /traɪ'bju:n(ə)l/ *noun* board appointed to adjudicate on particular question; court of justice.

tribune /'trɪbju:n/ *noun* popular leader, demagogue; (in full **tribune of the people**) *Roman History* officer chosen by the people to protect their liberties.

tributary /'trɪbjʊtərɪ/ ● *noun* (*plural* **-ies**) stream etc. that flows into larger stream or lake; *historical* person or state paying or subject to tribute. ● *adjective* that is a tributary.

tribute /'trɪbju:t/ *noun* thing said or done or given as mark of respect or affection etc.; (+ *to*) indication of (some praiseworthy quality); *historical* periodic payment by one state or ruler to another, obligation to pay this.

trice *noun* □ **in a trice** in an instant.

triceps /'traɪseps/ *noun* muscle (esp. in upper arm) with 3 points of attachment.

triceratops /traɪ'serətɒps/ *noun* large dinosaur with 3 horns.

trichinosis /trɪkɪ'nəʊsɪs/ *noun* disease caused by hairlike worms.

trichology /trɪ'kɒlədʒɪ/ *noun* study of hair. □ **trichologist** *noun*.

trichromatic /traɪkrə'mætɪk/ *adjective* 3-coloured.

trick ● *noun* thing done to deceive or outwit; illusion; knack; feat of skill or dexterity; unusual action learned by animal; foolish or discreditable act; hoax, joke; idiosyncrasy; cards played in one round, point gained in this. ● *verb* deceive by trick; swindle; (+ *into*) cause to do something by trickery; take by surprise.

trickery *noun* deception, use of tricks.

trickle /'trɪk(ə)l/ ● *verb* (**-ling**) (cause to) flow in drops or small stream; come or go slowly or gradually. ● *noun* trickling flow. □ **trickle charger** *Electricity* device for slow continuous charging of battery.

trickster /'trɪkstə/ *noun* deceiver, rogue.

tricky *adjective* (**-ier**, **-iest**) requiring care and adroitness; crafty, deceitful. □ **trickily** *adverb*; **trickiness** *noun*.

tricolour /'trɪkələ/ *noun* (*US* **tricolor**) flag of 3 colours, esp. French national flag.

tricot /'trɪkəʊ/ *noun* knitted fabric.

tricycle /'traɪsɪk(ə)l/ *noun* 3-wheeled pedal-driven vehicle.

trident /'traɪd(ə)nt/ *noun* 3-pronged spear.

Tridentine /traɪ'dentaɪn/ *adjective* of traditional RC orthodoxy.

triennial /traɪ'enɪəl/ *adjective* lasting 3 years; recurring every 3 years.

trifle /'traɪf(ə)l/ ● *noun* thing of slight value or importance; small amount; (**a trifle**) somewhat; dessert of sponge cakes with custard, cream, etc. ● *verb* (**-ling**) talk or act frivolously; (+ *with*) treat frivolously, flirt heartlessly with.

trifling *adjective* unimportant; frivolous.

trigger /'trɪgə/ ● *noun* movable device for releasing spring or catch and so setting off mechanism, esp. of gun; event etc. that sets off chain reaction. ● *verb* (often + *off*) set (action, process) in motion, precipitate. □ **trigger-happy** apt to shoot on slight provocation.

trigonometry /trɪgə'nɒmətrɪ/ *noun* branch of mathematics dealing with relations of sides and angles of triangles, and with certain functions of angles. □ **trigonometric** /-nə'met-/ *adjective*; **trigonometrical** /-nə'met-/ *adjective*.

trike *noun colloquial* tricycle.

trilateral /traɪ'lætər(ə)l/ *adjective* of, on, or with 3 sides; involving 3 parties.

trilby /'trɪlbɪ/ *noun* (*plural* **-ies**) soft felt hat with narrow brim and indented crown.

trilingual /traɪ'lɪŋgw(ə)l/ *adjective* speaking or in 3 languages.

trill ● *noun* quavering sound, esp. quick alternation of notes; bird's warbling; pronunciation of letter *r* with vibrating tongue. ● *verb* produce trill; warble (song); pronounce (*r* etc.) with trill.

trillion /'trɪljən/ *noun* (*plural* same) million million; million million million;

(trillions) *colloquial* large number. □ **trillionth** *adjective & noun.*

trilobite /'traɪləbaɪt/ *noun* kind of fossil crustacean.

trilogy /'trɪlədʒɪ/ *noun* (*plural* **-ies**) set of 3 related novels, plays, operas, etc.

trim ● *verb* (**-mm-**) make neat or tidy or of required size or shape, esp. by cutting away irregular or unwanted parts; (+ *off, away*) cut off; ornament; adjust balance of (ship, aircraft) by arranging cargo etc.; arrange (sails) to suit wind. ● *noun* state of readiness or fitness; ornament, decorative material; trimming of hair etc. ● *adjective* (**-mm-**) neat; in good order, well arranged or equipped.

trimaran /'traɪməræn/ *noun* vessel like catamaran, with 3 hulls side by side.

trimming *noun* ornamental addition to dress, hat, etc.; (in *plural*) *colloquial* usual accompaniments.

trinitrotoluene /traɪnaɪtrə'tɒljʊi:n/ *noun* (also **trinitrotoluol** /-'tɒljʊɒl/) a high explosive.

trinity /'trɪnɪtɪ/ *noun* (*plural* **-ies**) being 3; group of 3; (**the Trinity**) the 3 persons of the Christian Godhead. □ **Trinity Sunday** Sunday after Whit Sunday.

trinket /'trɪŋkɪt/ *noun* trifling ornament, esp. piece of jewellery.

trio /'tri:əʊ/ *noun* (*plural* **-s**) group of 3; musical composition for 3 performers; the performers.

trip ● *verb* (**-pp-**) (often + *up*) (cause to) stumble, esp. by catching foot; (+ *up*) (cause to) commit fault or blunder; run lightly; make excursion to place; operate (mechanism) suddenly by knocking aside catch etc.; *slang* have drug-induced hallucinatory experience. ● *noun* journey or excursion, esp. for pleasure; stumble, blunder, tripping, being tripped up; nimble step; *slang* drug-induced hallucinatory experience; device for tripping mechanism etc. □ **trip-wire** wire stretched close to ground to operate alarm etc. if disturbed.

tripartite /traɪ'pɑ:taɪt/ *adjective* consisting of 3 parts; shared by or involving 3 parties.

tripe *noun* first or second stomach of ruminant, esp. ox, as food; *colloquial* nonsense, rubbish.

triple /'trɪp(ə)l/ ● *adjective* of 3 parts, threefold; involving 3 parties; 3 times as much or as many. ● *noun* threefold number or amount; set of 3. ● *verb* (**-ling**) multiply by 3. □ **triple crown** winning of 3 important sporting events; **triple jump** athletic contest comprising hop, step, and jump. □ **triply** *adverb.*

triplet /'trɪplɪt/ *noun* each of 3 children or animals born at one birth; set of 3 things, esp. of notes played in time of two.

triplex /'trɪpleks/ *adjective* triple, threefold.

triplicate ● *adjective* /'trɪplɪkət/ existing in 3 examples or copies; having 3 corresponding parts; tripled. ● *noun* /'trɪplɪkət/ each of 3 copies or corresponding parts. ● *verb* (**-ting**) /'trɪplɪkeɪt/ make in 3 copies; multiply by 3. □ **triplication** *noun.*

tripod /'traɪpɒd/ *noun* 3-legged or 3-footed stand, stool, table, or utensil.

tripos /'traɪpɒs/ *noun* honours exam for primary degree at Cambridge University.

tripper *noun* person who goes on pleasure trip.

triptych /'trɪptɪk/ *noun* picture etc. with 3 panels usually hinged vertically together.

trireme /'traɪri:m/ *noun* ancient Greek warship, with 3 files of oarsmen on each side.

trisect /traɪ'sekt/ *verb* divide into 3 (usually equal) parts. □ **trisection** *noun.*

trite *adjective* hackneyed. □ **tritely** *adverb*; **triteness** *noun.*

tritium /'trɪtɪəm/ *noun* radioactive isotope of hydrogen with mass about 3 times that of ordinary hydrogen.

triumph /'traɪʌmf/ ● *noun* state of victory or success; great success or achievement; supreme example; joy at success. ● *verb* gain victory, be successful; *Roman History* ride in triumph; (often + *over*) exult.

triumphal /traɪ'ʌmf(ə)l/ *adjective* of, used in, or celebrating a triumph.

■ **Usage** *Triumphal*, as in *triumphal arch*, should not be confused with *triumphant*.

---

triumphant /traɪ'ʌmf(ə)nt/ *adjective* victorious, successful; exultant. □ **triumphantly** *adverb.*

■ **Usage** See note at TRIUMPHAL.

triumvirate /traɪˈʌmvərət/ *noun* ruling group of 3 men.

trivalent /traɪˈveɪlənt/ *adjective Chemistry* having a valency of 3. □ **trivalency** *noun.*

trivet /ˈtrɪvɪt/ *noun* iron tripod or bracket for pot or kettle to stand on.

trivia /ˈtrɪvɪə/ *plural noun* trifles, trivialities.

trivial /ˈtrɪvɪəl/ *adjective* of small value or importance; concerned only with trivial things. □ **triviality** /-ˈæl-/ *noun* (*plural* **-ies**); **trivially** *adverb.*

trivialize *verb* (also **-ise**) (**-zing** or **-sing**) make or treat as trivial, minimize. □ **trivialization** *noun.*

trochee /ˈtrəʊki:/ *noun* metrical foot of one long followed by one short syllable. □ **trochaic** /trəˈkeɪɪk/ *adjective.*

trod *past & past participle* of TREAD.

trodden *past participle* of TREAD.

troglodyte /ˈtrɒglədaɪt/ *noun* cave dweller.

troika /ˈtrɔɪkə/ *noun* Russian vehicle drawn by 3 horses abreast.

Trojan /ˈtrəʊdʒ(ə)n/ ● *adjective* of ancient Troy. ● *noun* native or inhabitant of ancient Troy; person who works, fights, etc. courageously. □ **Trojan Horse** person or device planted to bring about enemy's downfall.

troll[1] /trəʊl/ *noun* supernatural cave-dwelling giant or dwarf in Scandinavian mythology.

troll[2] /trəʊl/ *verb* fish by drawing bait along in water.

trolley /ˈtrɒlɪ/ *noun* (*plural* **-s**) table, stand, or basket on wheels or castors for serving food, carrying luggage etc., gathering purchases in supermarket, etc.; low truck running along rails; (in full **trolley-wheel**) wheel attached to pole etc. for collecting current from overhead electric wire to drive vehicle. □ **trolley bus** electric bus using trolley-wheel.

trollop /ˈtrɒləp/ *noun* disreputable girl or woman.

trombone /trɒmˈbəʊn/ *noun* brass wind instrument with sliding tube. □ **trombonist** *noun.*

troop /tru:p/ ● *noun* assembled company, assemblage of people or animals; (in *plural*) soldiers, armed forces; cavalry unit commanded by captain; artillery unit; group of 3 or more Scout patrols. ● *verb* (+ *in*, *out*, *off*, etc.) come together or move in a troop. □ **troop the colour** transfer flag ceremonially at public mounting of garrison guards; **troop-ship** ship for transporting troops.

trooper *noun* private soldier in cavalry or armoured unit; *Australian & US* mounted or State police officer; cavalry horse; troop-ship.

trope *noun* figurative use of word.

trophy /ˈtrəʊfɪ/ *noun* (*plural* **-ies**) cup etc. as prize in contest; memento of any success.

tropic /ˈtrɒpɪk/ *noun* parallel of latitude 23° 27′ N. (**tropic of Cancer**) or S. (**tropic of Capricorn**) of Equator; (**the Tropics**) region lying between these.

tropical *adjective* of or typical of the Tropics.

troposphere /ˈtrɒpəsfɪə/ *noun* layer of atmosphere extending about 8 km upwards from earth's surface.

trot ● *verb* (**-tt-**) (of person) run at moderate pace; (of horse) proceed at steady pace faster than walk; traverse (distance) thus. ● *noun* action or exercise of trotting; (**the trots**) *slang* diarrhoea. □ **on the trot** *colloquial* in succession, continually busy; **trot out** *colloquial* introduce (opinion etc.) repeatedly or tediously.

troth /trəʊθ/ *noun archaic* faith, fidelity; truth.

trotter *noun* (usually in *plural*) animal's foot, esp. as food; horse bred or trained for trotting.

troubadour /ˈtru:bədʊə/ *noun* singer, poet; French medieval poet singing of love.

trouble /ˈtrʌb(ə)l/ ● *noun* difficulty, distress, vexation, affliction; inconvenience, unpleasant exertion; cause of this; perceived failing; malfunction; disturbance; (in *plural*) public disturbances. ● *verb* (**-ling**) cause distress to, disturb; be disturbed; afflict, cause pain etc. to; subject or be subjected to inconvenience or unpleasant exertion. □ **in trouble** likely to incur censure or punishment, *colloquial* pregnant and unmarried; **troublemaker** person who habitually causes trouble; **troubleshooter** mediator in dispute, person who traces and corrects faults in machinery etc.

troublesome *adjective* causing trouble, annoying.

trough /trɒf/ *noun* long narrow open receptacle for water, animal feed, etc.; channel or hollow like this; elongated region of low barometric pressure.

trounce /traʊns/ *verb* (**-cing**) inflict severe defeat, beating, or punishment on.

troupe /tru:p/ *noun* company of actors, acrobats, etc.

trouper *noun* member of theatrical troupe; staunch colleague.

trousers /'traʊzəz/ *plural noun* two-legged outer garment from waist usually to ankles. □ **trouser suit** woman's suit of trousers and jacket.

trousseau /'tru:səʊ/ *noun* (*plural* **-s** or **-x** /-z/) bride's collection of clothes etc.

trout /traʊt/ *noun* (*plural* same or **-s**) fish related to salmon.

trove /trəʊv/ *noun* treasure trove.

trowel /'traʊəl/ *noun* flat-bladed tool for spreading mortar etc.; scoop for lifting small plants or earth.

troy *noun* (in full **troy weight**) system of weights used for precious metals etc.

truant /'tru:ənt/ ● *noun* child who does not attend school; person who avoids work etc. ● *adjective* idle, wandering. ● *verb* (also **play truant**) be truant. □ **truancy** *noun* (*plural* **-ies**).

truce *noun* temporary agreement to cease hostilities.

truck[1] *noun* lorry; open railway wagon for freight.

truck[2] *noun* □ **have no truck with** avoid dealing with.

trucker *noun esp. US* long-distance lorry driver.

truckle /'trʌk(ə)l/ ● *noun* (in full **truckle-bed**) low bed on wheels, stored under another. ● *verb* (**-ling**) (+ *to*) submit obsequiously to.

truculent /'trʌkjʊlənt/ *adjective* aggressively defiant. □ **truculence** *noun*; **truculently** *adverb*.

trudge ● *verb* (**-ging**) walk laboriously; traverse (distance) thus. ● *noun* trudging walk.

true ● *adjective* (**-r**, **-st**) in accordance with fact or reality; genuine; loyal, faithful; (+ *to*) accurately conforming to (type, standard); correctly positioned or balanced, level; exact, accurate. ● *adverb archaic* truly; accurately; without variation. □ **out of true** out of alignment.

truffle /'trʌf(ə)l/ *noun* rich-flavoured underground fungus; sweet made of soft chocolate mixture.

trug *noun* shallow oblong garden-basket.

truism /'tru:ɪz(ə)m/ *noun* self-evident or hackneyed truth.

truly /'tru:lɪ/ *adverb* sincerely; really; loyally; accurately. □ **Yours truly** *written before signature at end of informal letter*; *jocular* I, me.

trump[1] ● *noun* playing card(s) of suit temporarily ranking above others; (in *plural*) this suit; *colloquial* helpful or excellent person. ● *verb* defeat with trump; *colloquial* outdo. □ **come** or **turn up trumps** *colloquial* turn out well or successfully, be extremely successful or helpful; **trump card** card belonging to, or turned up to determine, trump suit, *colloquial* valuable resource; **trump up** fabricate, invent (accusation, excuse, etc.).

trump[2] *noun archaic* trumpet-blast.

trumpery /'trʌmpərɪ/ ● *noun* worthless finery; rubbish. ● *adjective* showy but worthless, trashy, shallow.

trumpet /'trʌmpɪt/ ● *noun* brass instrument with flared mouth and bright penetrating tone; trumpet-shaped thing; sound (as) of trumpet. ● *verb* (**-t-**) blow trumpet; (of elephant) make trumpet; proclaim loudly. □ **trumpeter** *noun*.

truncate /trʌŋ'keɪt/ *verb* (**-ting**) cut off top or end of; shorten. □ **truncation** *noun*.

truncheon /'trʌntʃ(ə)n/ *noun* short club carried by police officer.

trundle /'trʌnd(ə)l/ *verb* (**-ling**) roll or move, esp. heavily or noisily.

trunk *noun* main stem of tree; body without limbs or head; large luggage-box with hinged lid; *US* boot of car;

elephant's elongated prehensile nose; (in *plural*) men's close-fitting shorts worn for swimming etc. □ **trunk call** long-distance telephone call; **trunk line** main line of railway, telephone system, etc.; **trunk road** important main road.

truss ● *noun* framework supporting roof, bridge, etc.; supporting surgical appliance for hernia etc. sufferers; bundle of hay or straw; cluster of flowers or fruit. ● *verb* tie up (fowl) for cooking; (often + *up*) tie (person) with arms to sides; support with truss(es).

trust ● *noun* firm belief that a person or thing may be relied on; confident expectation; responsibility; *Law* arrangement involving trustees, property so held, group of trustees; association of companies for reducing competition. ● *verb* place trust in, believe in, rely on; (+ *with*) give (person) charge of; (often + *that*) hope earnestly that a thing will take place; (+ *to*) consign (thing) to (person); (+ *in*) place reliance in; (+ *to*) place (esp. undue) reliance on. □ **in trust** (of property) managed by person(s) on behalf of another; **trustworthy** deserving of trust, reliable. □ **trustful** *adjective.*

trustee /trʌs'ti:/ *noun* person or member of board managing property in trust with legal obligation to administer it solely for purposes specified. □ **trusteeship** *noun.*

trusting *adjective* having trust or confidence. □ **trustingly** *adverb.*

trusty ● *adjective* (**-ier**, **-iest**) *archaic or jocular* trustworthy. ● *noun* (*plural* **-ies**) prisoner given special privileges for good behaviour.

truth /tru:θ/ *noun* (*plural* **truths** /tru:ðz/) quality or state of being true; what is true.

truthful *adjective* habitually speaking the truth; (of story etc.) true. □ **truthfully** *adverb*; **truthfulness** *noun.*

try ● *verb* (**-ies**, **-ied**) attempt, endeavour; test (quality), test by use or experiment test qualities of; make severe demands on; examine effectiveness of for purpose; ascertain state of fastening of (door etc.); investigate and decide (case, issue) judicially, (often + *for*) subject (person) to trial; (+ *for*) apply or compete for, seek to attain. ● *noun* (*plural* **-ies**) attempt; *Rugby* touching-down of ball by player behind opposing goal line, scoring points and entitling player's side to a kick at goal. □ **try one's hand** (often + *at*) have attempt; **try it on** *colloquial* test how much unreasonable behaviour etc. will be tolerated; **try on** put (clothes etc.) on to test fit etc.; **try-on** *colloquial* act of trying it on or trying on clothes etc., attempt to deceive; **try out** put to the test, test thoroughly; **try-out** experimental test.

trying *adjective* annoying, exasperating; hard to bear.

tryst /trɪst/ *noun archaic* meeting, esp. of lovers.

tsar /zɑ:/ *noun* (also **czar**) (*feminine* **tsarina** /-'ri:nə/) *historical* emperor of Russia. □ **tsarist** *noun & adjective.*

tsetse /'tsetsɪ/ *noun* African fly feeding on blood and transmitting disease.

tsp. *abbreviation* (*plural* **tsps.**) teaspoonful.

TT *abbreviation* teetotal(ler); tuberculin-tested; Tourist Trophy.

tub ● *noun* open flat-bottomed usually round vessel; tub-shaped (usually plastic) carton; *colloquial* bath; *colloquial* clumsy slow boat. ● *verb* (**-bb-**) plant, bathe, or wash in tub. □ **tub-thumper** *colloquial* ranting preacher or orator.

tuba /'tju:bə/ *noun* (*plural* **-s**) low-pitched brass wind instrument.

tubby /'tʌbɪ/ *adjective* (**-ier**, **-iest**) short and fat. □ **tubbiness** *noun.*

tube ● *noun* long hollow cylinder; soft metal or plastic cylinder sealed at one end; hollow cylindrical bodily organ; *colloquial* London underground; cathode ray tube, esp. in television; (**the tube**) *esp. US colloquial* television; *US* thermionic valve; inner tube. ● *verb* (**-bing**) equip with tubes; enclose in tube.

tuber /'tju:bə/ *noun* short thick rounded root or underground stem of plant.

tubercle /'tju:bək(ə)l/ *noun* small rounded swelling on part or in organ of body, esp. as characteristic of tuberculosis. □ **tubercle bacillus** bacterium causing tuberculosis. □ **tuberculous** /-'bɜ:kjʊləs/ *adjective.*

tubercular /tjʊ'bɜ:kjʊlə/ *adjective* of or affected with tuberculosis.

tuberculin /tjʊ'bɜ:kjʊlɪn/ *noun* preparation from cultures of tubercle bacillus used in diagnosis and treatment of tuberculosis. □ **tuberculin-tested** (of milk) from cows shown to be free of tuberculosis.

tuberculosis /tjʊbɜ:kjʊ'ləʊsɪs/ *noun* infectious bacterial disease marked by tubercles, esp. in lungs.

tuberose /'tju:bərəʊz/ *noun* plant with creamy-white fragrant flowers.

tuberous /'tju:bərəs/ *adjective* having tubers; of or like a tuber.

tubing *noun* length of tube; quantity of or material for tubes.

tubular /'tju:bjʊlə/ *adjective* tube-shaped; having or consisting of tubes.

TUC *abbreviation* Trades Union Congress.

tuck ● *verb* (often + *in*, *up*) draw, fold, or turn outer or end parts of (cloth, clothes, etc.) close together, push in edges of bedclothes around (person); draw together into small space; stow (thing) away in specified place or way; make stitched fold in (cloth etc.). ● *noun* flattened fold sewn in garment etc.; *colloquial* food, esp. cakes and sweets. □ **tuck in** *colloquial* eat heartily; **tuck shop** shop selling sweets etc. to schoolchildren.

tucker /'tʌkə/ ● *noun Australian & NZ slang* food. ● *verb* (esp. in *passive*; often + *out*) *US & Australian colloquial* tire.

Tudor /'tju:də/ *adjective* of royal family of England from Henry VII to Elizabeth I; of this period (1485–1603); of the architectural style of this period.

Tues. *abbreviation* (also **Tue.**) Tuesday.

Tuesday /'tju:zdeɪ/ *noun* day of week following Monday.

tufa /'tju:fə/ *noun* porous rock formed round mineral springs; tuff.

tuff *noun* rock formed from volcanic ash.

tuft *noun* bunch of threads, grass, feathers, hair, etc. held or growing together at base. □ **tufted** *adjective*; **tufty** *adjective*.

tug ● *verb* (**-gg-**) (often + *at*) pull hard or violently; tow (vessel) by tugboat. ● *noun* hard, violent, or jerky pull; sudden emotion; (also **tugboat**) small powerful boat for towing ships. □ **tug of war** trial of strength between two sides pulling opposite ways on a rope.

tuition /tju:'ɪʃ(ə)n/ *noun* teaching; fee for this.

tulip /'tju:lɪp/ *noun* bulbous spring-flowering plant with showy cup-shaped flowers; its flower. □ **tulip-tree** tree with tulip-like flowers.

tulle /tju:l/ *noun* soft fine silk etc. net for veils and dresses.

tumble /'tʌmb(ə)l/ ● *verb* (**-ling**) (cause to) fall suddenly or headlong; fall rapidly in amount etc.; roll, toss; move or rush in headlong or blundering fashion; (often + *to*) *colloquial* grasp meaning of; fling or push roughly or carelessly; perform acrobatic feats, esp. somersaults; rumple, disarrange. ● *noun* fall; somersault or other acrobatic feat. □ **tumbledown** falling or fallen into ruin, dilapidated; **tumble-drier**, **-dryer** machine for drying washing in heated rotating drum; **tumble-dry** *verb*.

tumbler *noun* drinking glass without handle or foot; acrobat; part of mechanism of lock.

tumbrel /'tʌmbr(ə)l/ *noun* (also **tumbril**) *historical* open cart in which condemned people were carried to guillotine during French Revolution.

tumescent /tjʊ'mes(ə)nt/ *adjective* swelling. □ **tumescence** *noun*.

tumid /'tju:mɪd/ *adjective* swollen, inflated; pompous. □ **tumidity** /-'mɪd-/ *noun*.

tummy /'tʌmɪ/ *noun* (*plural* **-ies**) *colloquial* stomach.

tumour /'tju:mə/ *noun* (*US* **tumor**) abnormal or morbid swelling in the body.

tumult /'tju:mʌlt/ *noun* uproar, din; angry demonstration by mob, riot; conflict of emotions etc. □ **tumultuous** /tjʊ'mʌltʃʊəs/ *adjective*.

tumulus /'tju:mjʊləs/ *noun* (*plural* **-li** /-laɪ/) ancient burial mound.

tun *noun* large cask; brewer's fermenting-vat.

tuna /'tju:nə/ *noun* (*plural* same or **-s**) large edible marine fish; (in full **tuna-fish**) its flesh as food.

tundra /'tʌndrə/ *noun* vast level treeless Arctic region with permafrost.

tune ● *noun* melody; correct pitch or intonation. ● *verb* (**-ning**) put (musical instrument) in tune; (often + *in*) adjust (radio etc.) to desired frequency etc.; adjust (engine etc.) to run smoothly. □ **change one's tune** voice different opinion, become more respectful; **tune up** bring instrument(s) to proper pitch; **tuning-fork** two-pronged steel fork giving particular note when struck.

tuneful *adjective* melodious, musical. □ **tunefully** *adverb.*

tuneless *adjective* unmelodious, unmusical. □ **tunelessly** *adverb.*

tuner *noun* person who tunes pianos etc.; part of radio or television receiver for tuning.

tungsten /'tʌŋst(ə)n/ *noun* heavy steel-grey metallic element.

tunic /'tju:nɪk/ *noun* close-fitting short coat of police or military uniform; loose often sleeveless garment.

tunnel /'tʌn(ə)l/ ● *noun* underground passage dug through hill, or under river, road, etc.; underground passage dug by animal. ● *verb* (**-ll-**; *US* **-l-**) (+ *through*, *into*) make tunnel through (hill etc.); make (one's way) so. □ **tunnel vision** restricted vision, *colloquial* inability to grasp wider implications of situation etc.

tunny /'tʌnɪ/ *noun* (*plural* same or **-ies**) tuna.

tuppence = TWOPENCE.

tuppenny = TWOPENNY.

Tupperware /'tʌpəweə/ *noun proprietary term* range of plastic containers for food.

turban /'tɜ:bən/ *noun* man's headdress of fabric wound round cap or head, worn esp. by Muslims and Sikhs; woman's hat resembling this.

turbid /'tɜ:bɪd/ *adjective* muddy, thick, not clear; confused, disordered. □ **turbidity** /-'bɪd-/ *noun.*

■ **Usage** *Turbid* is sometimes confused with *turgid*.

turbine /'tɜ:baɪn/ *noun* rotary motor driven by flow of water, gas, etc.

turbo- *combining form* turbine.

turbocharger /'tɜ:bəʊtʃɑ:dʒə/ *noun* (also **turbo**) supercharger driven by turbine powered by engine's exhaust gases.

turbojet /'tɜ:bəʊdʒet/ *noun* jet engine in which jet also operates turbine-driven air-compressor; aircraft with this.

turboprop /'tɜ:bəʊprɒp/ *noun* jet engine in which turbine is used as in turbojet and also to drive propeller; aircraft with this.

turbot /'tɜ:bət/ *noun* (*plural* same or **-s**) large flatfish valued as food.

turbulent /'tɜ:bjʊlənt/ *adjective* disturbed, in commotion; (of flow of air etc.) varying irregularly; riotous, restless. □ **turbulence** *noun*; **turbulently** *adverb.*

tureen /tjʊə'ri:n/ *noun* deep covered dish for soup.

turf ● *noun* (*plural* **-s** or **turves**) short grass with surface earth bound together by its roots; piece of this cut from ground; slab of peat for fuel; (**the turf**) horse racing, racecourse. ● *verb* cover (ground) with turf; (+ *out*) *colloquial* expel, eject. □ **turf accountant** bookmaker. □ **turfy** *adjective.*

turgid /'tɜ:dʒɪd/ *adjective* swollen, inflated; (of language) pompous, bombastic. □ **turgidity** /-'dʒɪd-/ *noun.*

■ **Usage** *Turgid* is sometimes confused with *turbid*.

Turk *noun* native or national of Turkey.

turkey /'tɜ:kɪ/ *noun* (*plural* **-s**) large originally American bird bred for food; its flesh. □ **turkeycock** male turkey.

Turkish ● *adjective* of Turkey. ● *noun* language of Turkey. □ **Turkish bath** hot-air or steam bath followed by massage etc., (in *singular* or *plural*) building for this; **Turkish carpet** thick-piled woollen carpet with bold design; **Turkish delight** kind of gelatinous sweet; **Turkish towel** one made of cotton terry.

turmeric /'tɜ:mərɪk/ *noun* E. Indian plant of ginger family; its rhizome powdered as flavouring or dye.

turmoil /'tɜ:mɔɪl/ *noun* violent confusion; din and bustle.

turn ● *verb* move around point or axis, give or receive rotary motion; change from one side to another, invert, reverse; give new direction to, take new direction, aim in certain way; (+ *into*) change in nature, form, or condition to; (+ *to*) set about, have recourse to, consider next; become; (+ *against*) make or become hostile to; (+ *on*, *upon*) face hostilely; change colour; (of milk) become sour; (of stomach) be nauseated; cause (milk) to become sour or (stomach) to be nauseated; (of head) become giddy; translate; move to other side of, go round; pass age or time of; (+ *on*) depend on; send, put, cause to go; remake (sheet, shirt collar, etc.); make (profit); divert (bullet); shape (object) in lathe; give (esp. elegant) form to. ● *noun* turning, rotary motion; changed or change of direction or tendency; point of turning or change; turning of road; change of direction of tide; change in course of events; tendency, formation; opportunity, obligation, etc. that comes successively to each of several people etc.; short walk or ride; short performance on stage, in circus, etc.; service of specified kind; purpose; *colloquial* momentary nervous shock; *Music* ornament of principal note with those above and below it. □ **in turn** in succession; **take (it in) turns** act alternately; **turncoat** person who changes sides; **turn down** reject, reduce volume or strength of (sound, heat, etc.) by turning knob, fold down; **turn in** hand in, achieve, *colloquial* go to bed, incline inwards; **turnkey** *archaic* jailer; **turn off** stop flow or working of by means of tap, switch, etc., enter side road, *colloquial* cause to lose interest; **turn on** start flow or working of by means of tap, switch, etc., *colloquial* arouse, esp. sexually; **turn out** expel, extinguish (light etc.), dress, equip, produce (goods etc.), empty, clean out, (cause to) assemble, prove to be the case, result, (usually + *to be*) be found; **turnout** number of people who attend meeting etc., equipage; **turn over** reverse position of, cause (engine etc.) to run, (of engine) start running, consider thoroughly, (+ *to*) transfer care or conduct of (person, thing) to (person); **turnover** turning over, gross amount of money taken in business, rate of sale and replacement of goods, rate at which people enter and leave employment etc., small pie with pastry folded over filling; **turnpike** *US & historical* road on which toll is charged; **turnstile** revolving gate with arms; **turntable** circular revolving plate or platform; **turn to** begin work; **turn turtle** capsize; **turn up** increase (volume or strength of) by turning knob etc., discover, reveal, be found, happen, arrive, shorten (garment etc.), fold over or upwards; **turn-up** turned-up end of trouser leg, *colloquial* unexpected happening.

turner *noun* lathe-worker.

turnery *noun* objects made on lathe; work with lathe.

turning *noun* road branching off another, place where this occurs; use of lathe; (in *plural*) chips or shavings from this. □ **turning circle** smallest circle in which vehicle can turn; **turning point** point at which decisive change occurs.

turnip /'tɜːnɪp/ *noun* plant with globular root; its root as vegetable.

turpentine /'tɜːpəntaɪn/ *noun* resin from any of various trees; (in full **oil of turpentine**) volatile inflammable oil distilled from turpentine and used in mixing paints etc.

turpitude /'tɜːpɪtjuːd/ *noun formal* depravity, wickedness.

turps *noun colloquial* oil of turpentine.

turquoise /'tɜːkwɔɪz/ ● *noun* opaque semiprecious stone, usually greenish-blue; this colour. ● *adjective* of this colour.

turret /'tʌrɪt/ *noun* small tower, esp. decorating building; usually revolving armoured structure for gun and gunners on ship, fort, etc.; rotating holder for tools in lathe etc. □ **turreted** *adjective.*

turtle /'tɜːt(ə)l/ *noun* aquatic reptile with flippers and horny shell. □ **turtle-neck** high close-fitting neck on knitted garment.

turtle-dove /'tɜːt(ə)ldʌv/ *noun* wild dove noted for soft cooing and affection for its mate.

tusk *noun* long pointed tooth, esp. projecting beyond mouth as in elephant, walrus, or boar. □ **tusked** *adjective.*

tussle /'tʌs(ə)l/ *noun & verb* (**-ling**) struggle, scuffle.

tussock /'tʌsək/ *noun* clump of grass etc.

tut = TUT-TUT.

tutelage /'tjuːtɪlɪdʒ/ *noun* guardianship; being under this; tuition.

tutelary /'tjuːtɪlərɪ/ *adjective* serving as guardian or protector; of guardian.

tutor /'tju:tə/ • *noun* private teacher; university teacher supervising studies or welfare of assigned undergraduates. • *verb* act as tutor (to). □ **tutorship** *noun.*

tutorial /tju:'tɔ:rɪəl/ • *adjective* of tutor or tuition. • *noun* period of tuition for single student or small group.

tutti /'tʊtɪ/ *Music* • *adjective & adverb* with all instruments or voices together. • *noun* (*plural* **-s**) tutti passage.

tut-tut /tʌt'tʌt/ (also **tut**) • *interjection expressing rebuke or impatience.* • *noun* such exclamation. • *verb* (**-tt-**) exclaim thus.

tutu /'tu:tu:/ *noun* (*plural* **-s**) dancer's short skirt of stiffened frills.

tuxedo /tʌk'si:dəʊ/ *noun* (*plural* **-s** or **-es**) *US* (suit including) dinner jacket.

TV *abbreviation* television.

twaddle /'twɒd(ə)l/ *noun* silly writing or talk.

twain *adjective & noun archaic* two.

twang • *noun* sound made by plucked string of musical instrument, bow, etc.; nasal quality of voice. • *verb* (cause to) emit twang. □ **twangy** *adjective.*

tweak • *verb* pinch and twist; jerk; adjust finely. • *noun* such action.

twee *adjective* (**tweer** /'twi:ə/, **tweest** /'twi:ɪst/) *derogatory* affectedly dainty or quaint.

tweed *noun* rough-surfaced woollen cloth usually of mixed colours; (in *plural*) clothes of tweed.

tweedy *adjective* (**-ier**, **-iest**) of or dressed in tweed; heartily informal.

tweet • *noun* chirp of small bird. • *verb* make this noise.

tweeter *noun* loudspeaker for high frequencies.

tweezers /'twi:zəz/ *plural noun* small pair of pincers for picking up small objects, plucking out hairs, etc.

twelfth *adjective & noun* next after eleventh; any of twelve equal parts of thing. □ **Twelfth Night** evening of 5 Jan.

twelve /twelv/ *adjective & noun* one more than eleven.

twenty /'twentɪ/ *adjective & noun* (*plural* **-ies**) twice ten. □ **twentieth** *adjective & noun.*

twerp *noun slang* stupid or objectionable person.

twice *adverb* two times; on two occasions; doubly.

twiddle /'twɪd(ə)l/ • *verb* (**-ling**) twist or play idly about. • *noun* act of twiddling. □ **twiddle one's thumbs** make them rotate round each other, have nothing to do.

twig[1] *noun* very small branch of tree or shrub.

twig[2] *verb* (**-gg-**) *colloquial* understand, realize.

twilight /'twaɪlaɪt/ *noun* light from sky when sun is below horizon, esp. in evening; period of this; faint light; period of decline. □ **twilight zone** decrepit urban area, undefined or intermediate area.

twilit /'twaɪlɪt/ *adjective* dimly illuminated (as) by twilight.

twill *noun* fabric woven with surface of parallel diagonal ridges. □ **twilled** *adjective.*

twin • *noun* each of closely related pair, esp. of children or animals born at a birth; counterpart. • *adjective* forming, or born as one of, twins. • *verb* (**-nn-**) join closely, (+ *with*) pair; bear twins; link (town) with one abroad for social and cultural exchange. □ **twin bed** each of pair of single beds; **twin set** woman's matching cardigan and jumper; **twin town** town twinned with another.

twine • *noun* strong coarse string of twisted strands of fibre; coil, twist. • *verb* (**-ning**) coil, twist; form (string etc.) by twisting strands.

twinge /twɪndʒ/ *noun* sharp momentary local pain.

twinkle /'twɪŋk(ə)l/ • *verb* (**-ling**) shine with rapidly intermittent light; sparkle; move rapidly. • *noun* sparkle or gleam of eyes; twinkling light; light rapid movement. □ **twinkly** *adjective.*

twirl • *verb* spin, swing, or twist quickly and lightly round. • *noun* twirling; flourish made with pen.

twist • *verb* change the form of by rotating one end and not the other or the two ends opposite ways; undergo such change; wrench or distort by twisting; wind (strands etc.) about each other, form (rope etc.) thus; (cause to) take spiral form; (+ *off*) break off by twisting; misrepresent meaning of (words); take winding course; *colloquial* cheat; (as **twisted** *adjective*) perverted; dance the twist. ● *noun* twisting, being twisted; thing made by twisting; point at which thing twists; *usually derogatory* peculiar tendency of mind, character, etc.; unexpected development; (**the twist**) 1960s dance with twisting hips. □ **twisty** *adjective* (**-ier**, **-iest**).

twister *noun colloquial* swindler.

twit[1] *noun slang* foolish person.

twit[2] *verb* (**-tt-**) reproach, taunt, usually good-humouredly.

twitch ● *verb* quiver or jerk spasmodically; pull sharply at. ● *noun* twitching; *colloquial* state of nervousness. □ **twitchy** *adjective* (**-ier**, **-iest**).

twitter /'twɪtə/ ● *verb* (esp. of bird) utter succession of light tremulous sounds; utter or express thus. ● *noun* twittering; *colloquial* tremulously excited state.

two /tu:/ *adjective & noun* one more than one. □ **two-dimensional** having or appearing to have length and breadth but no depth, superficial; **two-edged** having both good and bad effect, ambiguous; **two-faced** insincere; **two-handed** with 2 hands, used with both hands or by 2 people; **twopence** /'tʌpəns/ sum of 2 pence, (esp. with negative) *colloquial* thing of little value; **twopenny** /'tʌpənɪ/ costing twopence, *colloquial* cheap, worthless; **two-piece** (suit etc.) comprising 2 matching parts; **two-ply** (wool etc.) having 2 strands, (plywood) having 2 layers; **two-step** dance in march or polka time; **two-stroke** (of internal-combustion engine) having power cycle completed in one up-and-down movement of piston; **two-time** *colloquial* be unfaithful to, swindle; **two-tone** having two colours or sounds; **two-way** involving or operating in two directions, (of radio) capable of transmitting and receiving signals.

twofold *adjective & adverb* twice as much or many.

twosome *noun* two people together.

tycoon /taɪ'ku:n/ *noun* business magnate.

tying *present participle* of TIE.

tyke /taɪk/ *noun* (also **tike**) objectionable or coarse man; small child.

tympani = TIMPANI.

tympanum /'tɪmpənəm/ *noun* (*plural* **-s** or **-na**) middle ear; eardrum; vertical space forming centre of pediment; space between lintel and arch above door etc.

type /taɪp/ ● *noun* sort, class, kind; person, thing, or event exemplifying class or group; *colloquial* person, esp. of specified character; object, idea, or work of art serving as model; small block with raised character on upper surface for printing; printing types collectively; typeset or printed text. ● *verb* (**-ping**) write with typewriter; typecast; assign to type, classify. □ **-type** made of, resembling, functioning as; **typecast** cast (performer) repeatedly in similar roles; **typeface** inked surface of type, set of characters in one design; **typescript** typewritten document; **typesetter** compositor, composing machine; **typewriter** machine with keys for producing printlike characters; **typewritten** produced thus.

typhoid /'taɪfɔɪd/ *noun* (in full **typhoid fever**) infectious bacterial fever attacking intestines.

typhoon /taɪ'fu:n/ *noun* violent hurricane in E. Asian seas.

typhus /'taɪfəs/ *noun* an acute infectious fever.

typical /'tɪpɪk(ə)l/ *adjective* serving as characteristic example; (often + *of*) characteristic of particular person or thing. □ **typicality** /-'kæl-/ *noun*; **typically** *adverb*.

typify /'tɪpɪfaɪ/ *verb* (**-ies**, **-ied**) be typical of; represent by or as type. □ **typification** *noun*.

typist /'taɪpɪst/ *noun* (esp. professional) user of typewriter.

typo /'taɪpəʊ/ *noun* (*plural* **-s**) *colloquial* typographical error.

typography /taɪ'pɒgrəfɪ/ *noun* printing as an art; style and appearance of printed matter. □ **typographer** *noun*; **typographical** /-pə'græf-/ *adjective*; **typographically** /-pə'græf-/ *adverb*.

tyrannical /tɪ'rænɪk(ə)l/ *adjective* acting like or characteristic of tyrant.

tyrannize /'tɪrənaɪz/ *verb* (also **-ise**) (**-zing** or **-sing**) (often + *over*) treat despotically.

tyrannosaurus /tɪrænə'sɔ:rəs/ *noun* (*plural* **-ruses**) very large carnivorous dinosaur with short front legs and powerful tail.

tyranny /'tɪrənɪ/ *noun* (*plural* **-ies**) cruel and arbitrary use of authority; rule by tyrant; period of this; state thus ruled. □ **tyrannous** *adjective.*

tyrant /'taɪərənt/ *noun* oppressive or cruel ruler; person exercising power arbitrarily or cruelly.

tyre /'taɪə/ *noun* (*US* **tire**) rubber covering, usually inflated, placed round vehicle's wheel for cushioning and grip.

tyro = TIRO.

tzatziki /tsæt'si:kɪ/ *noun* Greek dish of yoghurt with cucumber and garlic.

Uu

U[1] □ **U-boat** *historical* German submarine; **U-turn** U-shaped turn of vehicle to face in opposite direction, reversal of policy.

U[2] *abbreviation* (of film classified as suitable for all) universal.

UB40 *abbreviation* card for claiming unemployment benefit; *colloquial* unemployed person.

ubiquitous /ju:'bɪkwɪtəs/ *adjective* (seemingly) present everywhere simultaneously; often encountered. □ **ubiquity** *noun.*

UCCA /'ʌkə/ *abbreviation* Universities Central Council on Admissions.

UDA *abbreviation* Ulster Defence Association.

udder /'ʌdə/ *noun* baglike milk-producing organ of cow etc.

UDI *abbreviation* unilateral declaration of independence.

UDR *abbreviation* Ulster Defence Regiment.

UEFA /ju:'eɪfə/ *abbreviation* Union of European Football Associations.

UFO /'ju:fəʊ/ *noun* (also **ufo**) (*plural* **-s**) unidentified flying object.

ugh /əx, ʌg/ *interjection expressing disgust etc.*

Ugli /'ʌglɪ/ *noun* (*plural* **-lis** or **-lies**) *proprietary term* mottled green and yellow citrus fruit.

ugly /'ʌglɪ/ *adjective* (**-ier**, **-iest**) unpleasant to eye, ear, mind, etc.; discreditable; threatening, dangerous; morally repulsive. □ **ugly duckling** person lacking early promise but blossoming later. □ **uglify** *verb* (**-ies**, **-ied**); **ugliness** *noun.*

UHF *abbreviation* ultra-high frequency.

uh-huh /'ʌhʌ/ *interjection colloquial* yes.

UHT *abbreviation* ultra heat treated (esp. of milk, for long keeping).

UK *abbreviation* United Kingdom.

Ukrainian /ju:'kreɪnɪən/ ● *noun* native, national, or language of Ukraine. ● *adjective* of Ukraine.

ukulele /ju:kə'leɪlɪ/ *noun* small guitar with 4 strings.

ulcer /'ʌlsə/ *noun* (often pus-forming) open sore on or in body; corrupting influence. □ **ulcerous** *adjective.*

ulcerate /'ʌlsəreɪt/ *verb* (**-ting**) form into or affect with ulcer. □ **ulceration** *noun.*

ullage /'ʌlɪdʒ/ *noun* amount by which cask etc. falls short of being full; loss by evaporation or leakage.

ulna /'ʌlnə/ *noun* (*plural* **ulnae** /-ni:/) longer bone of forearm, opposite thumb; corresponding bone in animal's foreleg or bird's wing. □ **ulnar** *adjective.*

ulster /'ʌlstə/ *noun* long loose overcoat of rough cloth.

Ulsterman /'ʌlstəmən/ *noun* (*feminine* **Ulsterwoman**) native of Ulster.

ult. *abbreviation* ultimo.

ulterior /ʌl'tɪərɪə/ *adjective* not admitted; hidden, secret.

ultimate /'ʌltɪmət/ ● *adjective* last, final; fundamental, basic. ● *noun* (**the ultimate**) best achievable or imaginable;

final or fundamental fact or principle. □ **ultimately** *adverb.*

ultimatum /ʌltɪˈmeɪtəm/ *noun* (*plural* **-s**) final statement of terms, rejection of which could cause hostility etc.

ultimo /ˈʌltɪməʊ/ *adjective* of last month.

ultra /ˈʌltrə/ ● *adjective* extreme, esp. in religion or politics. ● *noun* extremist.

ultra- *combining form* extreme(ly), excessive(ly); beyond.

ultra-high /ʌltrəˈhaɪ/ *adjective* (of frequency) between 300 and 3000 megahertz.

ultramarine /ʌltrəməˈriːn/ ● *noun* (colour of) brilliant deep blue pigment. ● *adjective* of this colour.

ultrasonic /ʌltrəˈsɒnɪk/ *adjective* of or using sound waves pitched above range of human hearing. □ **ultrasonically** *adverb.*

ultrasound /ˈʌltrəsaʊnd/ *noun* ultrasonic waves.

ultraviolet /ʌltrəˈvaɪələt/ *adjective* of or using radiation just beyond violet end of spectrum.

ultra vires /ʌltrə ˈvaɪəriːz/ *adverb & adjective* beyond one's legal power or authority. [Latin]

ululate /ˈjuːljʊleɪt/ *verb* (**-ting**) howl, wail. □ **ululation** *noun.*

um *interjection representing hesitation or pause in speech.*

umbel /ˈʌmb(ə)l/ *noun* flower-cluster with stalks springing from common centre. □ **umbellate** *adjective.*

umbelliferous /ʌmbəˈlɪfərəs/ *adjective* (of plant, e.g. parsley or carrot) bearing umbels.

umber /ˈʌmbə/ ● *noun* (colour of) dark brown earth used as pigment. ● *adjective* umber-coloured.

umbilical /ʌmˈbɪlɪk(ə)l/ *adjective* of navel. □ **umbilical cord** cordlike structure attaching foetus to placenta.

umbilicus /ʌmˈbɪlɪkəs/ *noun* (*plural* **-ci** /-saɪ/ or **-cuses**) navel.

umbra /ˈʌmbrə/ *noun* (*plural* **-s** or **-brae** /-briː/) shadow cast by moon or earth in eclipse.

umbrage /ˈʌmbrɪdʒ/ *noun* offence taken.

umbrella /ʌmˈbrelə/ *noun* collapsible cloth canopy on central stick for protection against rain, sun, etc.; protection, patronage; coordinating agency.

umlaut /ˈʊmlaʊt/ *noun* mark (¨) over vowel, esp. in German, indicating change in pronunciation; such a change.

umpire /ˈʌmpaɪə/ ● *noun* person enforcing rules and settling disputes in game, contest, etc. ● *verb* (**-ring**) (often + *for*, *in*, etc.) act as umpire (in).

umpteen /ˈʌmptiːn/ *adjective & noun colloquial* very many. □ **umpteenth** *adjective & noun.*

UN *abbreviation* United Nations.

un- *prefix added to adjectives, nouns, and adverbs, meaning:* not; non-; reverse of, lack of; *added to verbs, verbal derivatives, etc. to express contrary or reverse action, deprivation of or removal from.* For words starting with *un-* that are not found below, the root-words should be consulted.

unaccountable /ʌnəˈkaʊntəb(ə)l/ *adjective* without explanation, strange; not answerable for one's actions. □ **unaccountably** *adverb.*

unadopted /ʌnəˈdɒptɪd/ *adjective* (of road) not maintained by local authority.

unadulterated /ʌnəˈdʌltəreɪtɪd/ *adjective* pure; complete, utter.

unaffected /ʌnəˈfektɪd/ *adjective* (usually + *by*) not affected; free from affectation. □ **unaffectedly** *adverb.*

unalloyed /ʌnəˈlɔɪd/ *adjective* complete, pure.

un-American /ʌnəˈmerɪkən/ *adjective* uncharacteristic of Americans; contrary to US interests, treasonable.

unanimous /juːˈnænɪməs/ *adjective* all in agreement; (of vote etc.) by all without exception. □ **unanimity** /-nəˈnɪm-/ *noun*; **unanimously** *adverb.*

unannounced /ʌnəˈnaʊnst/ *adjective* not announced, without warning.

unanswerable /ʌnˈɑːnsərəb(ə)l/ *adjective* that cannot be answered or refuted.

unapproachable /ʌnəˈprəʊtʃəb(ə)l/ *adjective* inaccessible; (of person) unfriendly, aloof.

unassailable /ʌnəˈseɪləb(ə)l/ *adjective* that cannot be attacked or questioned.

unassuming /ʌnəˈsjuːmɪŋ/ *adjective* not pretentious, modest.

unattached /ʌnəˈtætʃt/ *adjective* not engaged, married, etc.; (often + *to*) not attached to particular organization etc.

unaware /ʌnə'weə/ ● *adjective* (usually + *of, that*) not aware; unperceptive. ● *adverb* unawares.

unawares /ʌnə'weəz/ *adverb* unexpectedly; inadvertently.

unbalanced /ʌn'bælənst/ *adjective* emotionally unstable; biased.

unbeknown /ʌnbɪ'nəʊn/ *adjective* (also **unbeknownst** /-'nəʊnst/) (+ *to*) without the knowledge of.

unbend /ʌn'bend/ *verb* (*past & past participle* **unbent**) straighten; relax; become affable.

unbending /ʌn'bendɪŋ/ *adjective* inflexible; firm, austere.

unblushing /ʌn'blʌʃɪŋ/ *adjective* shameless; frank.

unbosom /ʌn'bʊz(ə)m/ *verb* disclose (thoughts etc.); (**unbosom oneself**) disclose one's thoughts etc.

unbounded /ʌn'baʊndɪd/ *adjective* infinite.

unbridled /ʌn'braɪd(ə)ld/ *adjective* unrestrained, uncontrolled.

uncalled-for /ʌn'kɔ:ldfɔ:/ *adjective* (of remark etc.) rude and unnecessary.

uncanny /ʌn'kænɪ/ *adjective* (**-ier, -iest**) seemingly supernatural, mysterious. □ **uncannily** *adverb*; **uncanniness** *noun*.

uncapped /ʌn'kæpt/ *adjective Sport* (of player) not yet awarded his or her cap or never having been selected to represent his or her country.

unceremonious /ʌnserɪ'məʊnɪəs/ *adjective* abrupt, discourteous; informal. □ **unceremoniously** *adverb*.

uncertain /ʌn'sɜ:t(ə)n/ *adjective* not certain; unreliable; changeable. □ **in no uncertain terms** clearly and forcefully. □ **uncertainly** *adverb*; **uncertainty** *noun* (*plural* **-ies**).

uncharted /ʌn'tʃɑ:tɪd/ *adjective* not mapped or surveyed.

uncle /'ʌŋk(ə)l/ *noun* parent's brother or brother-in-law. □ **Uncle Sam** *colloquial* US government.

unclean /ʌn'kli:n/ *adjective* not clean; unchaste; religiously impure.

uncomfortable /ʌn'kʌmftəb(ə)l/ *adjective* not comfortable; uneasy. □ **uncomfortably** *adverb*.

uncommon /ʌn'kɒmən/ *adjective* unusual, remarkable. □ **uncommonly** *adverb*.

uncompromising /ʌn'kɒmprəmaɪzɪŋ/ *adjective* stubborn; unyielding. □ **uncompromisingly** *adverb*.

unconcern /ʌnkən'sɜ:n/ *noun* calmness; indifference, apathy. □ **unconcerned** *adjective*.

unconditional /ʌnkən'dɪʃən(ə)l/ *adjective* not subject to conditions, complete. □ **unconditionally** *adverb*.

unconscionable /ʌn'kɒnʃənəb(ə)l/ *adjective* without conscience; excessive.

unconscious /ʌn'kɒnʃəs/ ● *adjective* not conscious. ● *noun* normally inaccessible part of mind affecting emotions etc. □ **unconsciously** *adverb*; **unconsciousness** *noun*.

unconsidered /ʌnkən'sɪdəd/ *adjective* not considered; disregarded; not premeditated.

unconstitutional /ʌnkɒnstɪ'tju:ʃən(ə)l/ *adjective* in breach of political constitution or procedural rules.

uncooperative /ʌnkəʊ'ɒpərətɪv/ *adjective* not cooperative.

uncork /ʌn'kɔ:k/ *verb* draw cork from (bottle); vent (feelings).

uncouple /ʌn'kʌp(ə)l/ *verb* (**-ling**) release from couples or coupling.

uncouth /ʌn'ku:θ/ *adjective* uncultured, rough.

uncover /ʌn'kʌvə/ *verb* remove cover or covering from; disclose.

uncrowned /ʌn'kraʊnd/ *adjective* having status but not name of.

unction /'ʌŋkʃ(ə)n/ *noun* anointing with oil etc. as religious rite or medical treatment; oil etc. so used; soothing words or thought; excessive or insincere flattery; (pretence of) deep emotion.

unctuous /'ʌŋktʃʊəs/ *adjective* unpleasantly flattering; greasy. □ **unctuously** *adverb*.

uncut /ʌn'kʌt/ *adjective* not cut; (of book) with pages sealed or untrimmed; (of film) not censored; (of diamond) not shaped; (of fabric) with looped pile.

undeniable /ʌndɪ'naɪəb(ə)l/ *adjective* indisputable; certain. □ **undeniably** *adverb*.

under /'ʌndə/ ● *preposition* in or to position lower than, below; beneath; inferior to, less than; undergoing, liable to;

controlled or bound by; classified or subsumed in. ● *adverb* in or to lower condition or position. ● *adjective* lower.

underachieve /ʌndərə'tʃi:v/ *verb* (**-ving**) do less well than might be expected, esp. academically. □ **underachiever** *noun.*

underarm *adjective & adverb Cricket etc.* with arm below shoulder level.

underbelly *noun* (*plural* **-ies**) under surface of animal etc., esp. as vulnerable to attack.

underbid /ʌndə'bɪd/ *verb* (**-dd-**; *past & past participle* **-bid**) make lower bid than; *Bridge etc.* bid too little (on).

undercarriage *noun* wheeled retractable landing structure beneath aircraft; supporting framework of vehicle etc.

undercharge /ʌndə'tʃɑ:dz/ *verb* (**-ging**) charge too little to.

underclothes *plural noun* (also **underclothing**) clothes worn under others, esp. next to skin.

undercoat *noun* layer of paint under another; (in animals) under layer of hair etc.

undercook /ʌndə'kʊk/ *verb* cook insufficiently.

undercover /ʌndə'kʌvə/ *adjective* surreptitious; spying incognito.

undercroft *noun* crypt.

undercurrent *noun* current below surface; underlying often contrary feeling, force, etc.

undercut ● *verb* /ʌndə'kʌt/ (**-tt-**; *past & past participle* **-cut**) sell or work at lower price than; strike (ball) to make it rise high; undermine. ● *noun* /'ʌndəkʌt/ underside of sirloin.

underdeveloped /ʌndədɪ'veləpt/ *adjective* immature; (of country etc.) with unexploited potential.

underdog *noun* oppressed person; loser in fight etc.

underdone /ʌndə'dʌn/ *adjective* undercooked.

underemployed /ʌndərɪm'plɔɪd/ *adjective* not fully occupied.

underestimate ● *verb* /ʌndər'estɪmeɪt/ (**-ting**) form too low an estimate of. ● *noun* /ʌndər'estɪmət/ estimate that is too low. □ **underestimation** *noun.*

underexpose /ʌndərɪk'spəʊz/ *verb* (**-sing**) expose (film) for too short a time. □ **underexposure** *noun.*

underfed /ʌndə'fed/ *adjective* malnourished.

underfelt *noun* felt laid under carpet.

underfloor /ʌndə'flɔ:/ *adjective* (esp. of heating) beneath floor.

underfoot /ʌndə'fʊt/ *adverb* (also **under foot**) under one's feet; on the ground.

underfunded /ʌndə'fʌndɪd/ *adjective* provided with insufficient money.

undergarment *noun* piece of underclothing.

undergo /ʌndə'gəʊ/ *verb* (*3rd singular present* **-goes**; *past* **-went**; *past participle* **-gone** /-'gɒn/) be subjected to, endure.

undergraduate /ʌndə'grædjʊət/ *noun* person studying for first degree.

underground ● *adverb* /ʌndə'graʊnd/ beneath the ground; in(to) secrecy or hiding. ● *adjective* /'ʌndəgraʊnd/ situated underground; secret, subversive; unconventional. ● *noun* /'ʌndəgraʊnd/ underground railway; secret subversive group or activity.

undergrowth *noun* dense shrubs etc., esp. in wood.

underhand *adjective* secret, deceptive; *Cricket etc.* underarm.

underlay[1] ● *verb* /ʌndə'leɪ/ (*past & past participle* **-laid**) lay thing under (another) to support or raise. ● *noun* /'ʌndəleɪ/ thing so laid (esp. under carpet).

underlay[2] *past* of UNDERLIE.

underlie /ʌndə'laɪ/ *verb* (**-lying**; *past* **-lay**; *past participle* **-lain**) lie under (stratum etc.); (esp. as **underlying** *adjective*) be basis of, exist beneath superficial aspect of.

underline /ʌndə'laɪn/ *verb* (**-ning**) draw line under (words etc.); emphasize.

underling /'ʌndəlɪŋ/ *noun usually derogatory* subordinate.

undermanned /ʌndə'mænd/ *adjective* having insufficient crew or staff.

undermine /ʌndə'maɪn/ *verb* (**-ning**) injure or wear out insidiously or secretly; wear away base of; make excavation under.

underneath /ʌndə'ni:θ/ ● *preposition* at or to lower place than, below; on inside of. ● *adverb* at or to lower place; inside.

● *noun* lower surface or part. ● *adjective* lower.

undernourished /ʌndəˈnʌrɪʃt/ *adjective* insufficiently nourished. □ **undernourishment** *noun*.

underpants *plural noun* undergarment for lower part of torso.

underpass *noun* road etc. passing under another; subway.

underpay /ʌndəˈpeɪ/ *verb* (*past & past participle* **-paid**) pay too little to (person) or for (thing). □ **underpayment** *noun*.

underpin /ʌndəˈpɪn/ *verb* (**-nn-**) support from below with masonry etc.; support, strengthen.

underprivileged /ʌndəˈprɪvɪlɪdʒd/ *adjective* less privileged than others; having below average income, rights, etc.

underrate /ʌndəˈreɪt/ *verb* (**-ting**) have too low an opinion of.

underscore /ʌndəˈskɔː/ *verb* (**-ring**) underline.

undersea *adjective* below sea or its surface.

underseal *verb* seal underpart of (esp. vehicle against rust etc.).

under-secretary /ʌndəˈsekrətərɪ/ *noun* (*plural* **-ies**) subordinate official, esp. junior minister or senior civil servant.

undersell /ʌndəˈsel/ *verb* (*past & past participle* **-sold**) sell at lower price than (another seller).

undershirt *noun esp. US* vest.

undershoot /ʌndəˈʃuːt/ *verb* (*past & past participle* **-shot**) land short of (runway etc.).

undershot *adjective* (of wheel) turned by water flowing under it; (of lower jaw) projecting beyond upper jaw.

underside *noun* lower side or surface.

undersigned /ʌndəˈsaɪnd/ *adjective* whose signature is appended.

undersized /ˈʌndəsaɪzd/ *adjective* smaller than average.

underspend /ʌndəˈspend/ *verb* (*past & past participle* **-spent**) spend less than (expected amount) or too little.

understaffed /ʌndəˈstɑːft/ *adjective* having too few staff.

understand /ʌndəˈstænd/ *verb* (*past & past participle* **-stood**) comprehend, perceive meaning, significance, or cause of; know how to deal with; (often + *that*) infer, take as implied. □ **understandable** *adjective*; **understandably** *adverb*.

understanding ● *noun* intelligence; ability to understand; individual's perception of situation; agreement, esp. informal. ● *adjective* having understanding or insight; sympathetic. □ **understandingly** *adverb*.

understate /ʌndəˈsteɪt/ *verb* (**-ting**) express in restrained terms; represent as being less than it really is. □ **understatement** *noun*.

understudy /ˈʌndəstʌdɪ/ ● *noun* (*plural* **-ies**) person ready to take another's role when required, esp. in theatre. ● *verb* (**-ies**, **-ied**) study (role etc.) for this purpose; act as understudy to.

undersubscribed /ʌndəsəbˈskraɪbd/ *adjective* without sufficient subscribers, participants, etc.

undertake /ʌndəˈteɪk/ *verb* (**-king**; *past* **-took**; *past participle* **-taken**) agree to perform or be responsible for; engage in; (usually + *to do*) promise; guarantee, affirm.

undertaker /ˈʌndəteɪkə/ *noun* professional funeral organizer.

undertaking /ʌndəˈteɪkɪŋ/ *noun* work etc. undertaken; enterprise; promise; /ˈʌn-/ professional funeral management.

undertone *noun* subdued tone; underlying quality or feeling.

undertow *noun* current below sea surface contrary to surface current.

underused /ʌndəˈjuːzd/ *adjective* not used to capacity.

undervalue /ʌndəˈvæljuː/ *verb* (**-ues**, **-ued**, **-uing**) value insufficiently; underestimate.

underwater /ʌndəˈwɔːtə/ ● *adjective* situated or done under water. ● *adverb* under water.

underwear *noun* underclothes.

underweight /ʌndəˈweɪt/ *adjective* below normal weight.

underwent *past* of UNDERGO.

underwhelm /ʌndəˈwelm/ *verb jocular* fail to impress.

underworld *noun* those who live by organized crime; mythical home of the dead.

underwrite /ʌndə'raɪt/ *verb* (**-ting**; *past* **-wrote**; *past participle* **-written**) sign and accept liability under (insurance policy); accept (liability) thus; undertake to finance or support; engage to buy all unsold stock in (company etc.). □ **underwriter** /'ʌn-/ *noun*.

undesirable /ʌndɪ'zaɪərəb(ə)l/ • *adjective* unpleasant, objectionable. • *noun* undesirable person. □ **undesirability** *noun*.

undies /'ʌndɪz/ *plural noun colloquial* (esp. women's) underclothes.

undo /ʌn'du:/ *verb* (*3rd singular present* **-does**; *past* **-did**; *past participle* **-done**; *present participle* **-doing**) unfasten; annul; ruin prospects, reputation, or morals of.

undoing *noun* (cause of) ruin; reversing of action etc.; unfastening.

undone /ʌn'dʌn/ *adjective* not done; not fastened; *archaic* ruined.

undoubted /ʌn'daʊtɪd/ *adjective* certain, not questioned. □ **undoubtedly** *adverb*.

undreamed /ʌn'dri:md, ʌn'dremt/ *adjective* (also **undreamt** /ʌn'dremt/) (often + *of*) not thought of, never imagined.

undress /ʌn'dres/ • *verb* take off one's clothes; take clothes off (person). • *noun* ordinary dress, esp. as opposed to (full-dress) uniform; naked or scantily clad state.

undressed /ʌn'drest/ *adjective* no longer dressed; (of food) without dressing; (of leather) untreated.

undue /ʌn'dju:/ *adjective* excessive, disproportionate. □ **unduly** *adverb*.

undulate /'ʌndjʊleɪt/ *verb* (**-ting**) (cause to) have wavy motion or look. □ **undulation** *noun*.

undying /ʌn'daɪɪŋ/ *adjective* immortal; never-ending.

unearth /ʌn'ɜ:θ/ *verb* discover by searching, digging, or rummaging.

unearthly /ʌn'ɜ:θlɪ/ *adjective* supernatural; mysterious; *colloquial* very early.

unease /ʌn'i:z/ *noun* nervousness, anxiety.

uneasy /ʌn'i:zɪ/ *adjective* (**-ier**, **-iest**) disturbed or uncomfortable in body or mind. □ **uneasily** *adverb*; **uneasiness** *noun*.

unemployable /ʌnɪm'plɔɪəb(ə)l/ *adjective* unfit for paid employment.

unemployed /ʌnɪm'plɔɪd/ *adjective* out of work; not used.

unemployment /ʌnɪm'plɔɪmənt/ *noun* lack of employment. □ **unemployment benefit** state payment made to unemployed worker.

unencumbered /ʌnɪn'kʌmbəd/ *adjective* (of estate) having no liabilities; free, not burdened.

unequivocal /ʌnɪ'kwɪvək(ə)l/ *adjective* not ambiguous, plain, unmistakable. □ **unequivocally** *adverb*.

UNESCO /ju:'neskəʊ/ *abbreviation* (also **Unesco**) United Nations Educational, Scientific, and Cultural Organization.

uneven /ʌn'i:v(ə)n/ *adjective* not level; of variable quality; (of contest) unequal. □ **unevenly** *adverb*.

unexceptionable /ʌnɪk'sepʃənəb(ə)l/ *adjective* entirely satisfactory.

■ **Usage** *Unexceptionable* is sometimes confused with *unexceptional*.

---

unexceptional /ʌnɪk'sepʃən(ə)l/ *adjective* normal, ordinary.

■ **Usage** *Unexceptional* is sometimes confused with *unexceptionable*

---

unfailing /ʌn'feɪlɪŋ/ *adjective* not failing, constant, reliable. □ **unfailingly** *adverb*.

unfaithful /ʌn'feɪθfʊl/ *adjective* not faithful, esp. adulterous. □ **unfaithfulness** *noun*.

unfeeling /ʌn'fi:lɪŋ/ *adjective* unsympathetic, harsh.

unfit /ʌn'fɪt/ *adjective* (often + *for*, *to do*) not fit, unsuitable; in poor health.

unflagging /ʌn'flægɪŋ/ *adjective* tireless, persistent.

unflappable /ʌn'flæpəb(ə)l/ *adjective colloquial* imperturbable.

unfledged /ʌn'fledʒd/ *adjective* inexperienced; not fledged.

unfold /ʌn'fəʊld/ *verb* open out; reveal; become opened out; develop.

unforgettable /ʌnfə'getəb(ə)l/ *adjective* memorable, wonderful.

unfortunate /ʌn'fɔ:tʃənət/ • *adjective* unlucky; unhappy; regrettable. • *noun* unfortunate person. □ **unfortunately** *adverb*.

unfounded /ʌn'faʊndɪd/ *adjective* (of rumour etc.) without foundation.

unfreeze /ʌn'fri:z/ *verb* (**-zing**; *past* **unfroze**; *past participle* **unfrozen**) (cause to) thaw; derestrict (assets etc.).

unfrock /ʌn'frɒk/ *verb* defrock.

unfurl /ʌn'fɜ:l/ *verb* unroll, spread out.

ungainly /ʌn'geɪnlɪ/ *adjective* awkward, clumsy.

unget-at-able /ʌnget'ætəb(ə)l/ *adjective colloquial* inaccessible.

ungodly /ʌn'gɒdlɪ/ *adjective* impious, wicked; *colloquial* outrageous.

ungovernable /ʌn'gʌvənəb(ə)l/ *adjective* uncontrollable, violent.

ungracious /ʌn'greɪʃəs/ *adjective* discourteous, grudging.

ungrateful /ʌn'greɪtfʊl/ *adjective* not feeling or showing gratitude. □ **ungratefully** *adverb.*

ungreen /ʌn'gri:n/ *adjective* harmful to environment; not concerned with protection of environment.

unguarded /ʌn'gɑ:dɪd/ *adjective* incautious, thoughtless; not guarded.

unguent /'ʌŋgwənt/ *noun* ointment, lubricant.

ungulate /'ʌŋgjʊlət/ ● *adjective* hoofed. ● *noun* hoofed mammal.

unhallowed /ʌn'hæləʊd/ *adjective* unconsecrated; not sacred; wicked.

unhand /ʌn'hænd/ *verb rhetorical or jocular* take one's hands off, release (person).

unhappy /ʌn'hæpɪ/ *adjective* (**-ier**, **-iest**) miserable; unfortunate; disastrous. □ **unhappily** *adverb*; **unhappiness** *noun.*

unhealthy /ʌn'helθɪ/ *adjective* (**-ier**, **-iest**) in poor health; harmful to health; unwholesome; *slang* dangerous. □ **unhealthily** *adverb.*

unheard-of /ʌn'hɜ:dɒv/ *adjective* unprecedented.

unhinge /ʌn'hɪndʒ/ *verb* (**-ging**) take (door etc.) off hinges; (esp. as **unhinged** *adjective*) derange, disorder (mind).

unholy /ʌn'həʊlɪ/ *adjective* (**-ier**, **-iest**) profane, wicked; *colloquial* dreadful, outrageous.

unhorse /ʌn'hɔ:s/ *verb* (**-sing**) throw (rider) from horse.

uni /'ju:nɪ/ *noun* (*plural* **-s**) *esp. Australian & NZ colloquial* university.

uni- *combining form* having or composed of one.

Uniate /'ju:nɪət/ ● *adjective* of Church in E. Europe or Near East acknowledging papal supremacy but retaining its own liturgy etc. ● *noun* member of such Church.

unicameral /ju:nɪ'kæmər(ə)l/ *adjective* having one legislative chamber.

UNICEF /'ju:nɪsef/ *abbreviation* United Nations Children's Fund.

unicellular /ju:nɪ'seljʊlə/ *adjective* consisting of a single cell.

unicorn /'ju:nɪkɔ:n/ *noun* mythical horse with single straight horn.

unicycle /'ju:nɪsaɪk(ə)l/ *noun* one-wheeled cycle used by acrobats etc.

unification /ju:nɪfɪ'keɪʃ(ə)n/ *noun* unifying, being unified. □ **Unification Church** religious organization funded by Sun Myung Moon. □ **unificatory** *adjective.*

uniform /'ju:nɪfɔ:m/ ● *adjective* unvarying; conforming to same standard or rule; constant over a period. ● *noun* distinctive clothing worn by members of same organization etc. □ **uniformed** *adjective*; **uniformity** /-'fɔ:m-/ *noun*; **uniformly** *adverb.*

unify /'ju:nɪfaɪ/ *verb* (**-ies**, **-ied**) make or become united or uniform.

unilateral /ju:nɪ'lætər(ə)l/ *adjective* done by or affecting one side only. □ **unilaterally** *adverb.*

unilateralism *noun* unilateral disarmament. □ **unilateralist** *noun & adjective.*

unimpeachable /ʌnɪm'pi:tʃəb(ə)l/ *adjective* beyond reproach.

uninviting /ʌnɪn'vaɪtɪŋ/ *adjective* unattractive, repellent.

union /'ju:njən/ *noun* uniting, being united; whole formed from parts or members; trade union; marriage; concord; university social club or debating society. □ **Union flag**, **Jack** national flag of UK.

unionist *noun* member of trade union, advocate of trade unions; (usually **Unionist**) supporter ofcontinued union between Britain and Northern Ireland. □ **unionism** *noun.*

unionize *verb* (also **-ise**) (**-zing** or **-sing**) organize in or into trade union. □ **unionization** *noun.*

unique /ju:'ni:k/ *adjective* being the only one of its kind; having no like, equal, or parallel; remarkable. □ **uniquely** *adverb.*

■ **Usage** The use of *unique* to mean 'remarkable' is considered incorrect by some people.

---

unisex /'ju:nɪseks/ *adjective* (of clothing etc.) designed for both sexes.

unison /'ju:nɪs(ə)n/ *noun* concord; coincidence in pitch of sounds or notes.

unit /'ju:nɪt/ *noun* individual thing, person, or group, esp. for calculation; smallest component of complex whole; quantity chosen as standard of measurement; smallest share in unit trust; part with specified function in complex mechanism; fitted item of furniture, esp. as part of set; subgroup with special function; group of buildings, wards, etc. in hospital; single-digit number, esp. 'one'. □ **unit cost** cost of producing one item; **unit trust** company investing contributions from many people in various securities.

Unitarian /ju:nɪ'teərɪən/ *noun* member of religious body maintaining that God is one person not Trinity. □ **Unitarianism** *noun.*

unitary /'ju:nɪtərɪ/ *adjective* of unit(s); marked by unity or uniformity.

unite /jʊ'naɪt/ *verb* (**-ting**) join together, esp. for common purpose; join in marriage; (cause to) form physical or chemical whole. □ **United Kingdom** Great Britain and Northern Ireland; **United Nations** (treated as *singular* or *plural*) international peace-seeking organization; **United States** = United States of America.

unity /'ju:nɪtɪ/ *noun* (*plural* **-ies**) oneness, being one; interconnecting parts making a whole, the whole made; solidarity, harmony between people etc.; the number 'one'.

universal /ju:nɪ'vɜ:s(ə)l/ ● *adjective* of, belonging to, or done by etc. all; applicable to all cases. ● *noun* term, characteristic, or concept of general application. □ **universal coupling**, **joint** one transmitting rotary power by a shaft at any angle; **universal time** Greenwich Mean Time. □ **universality** /-'sæl-/ *noun*; **universally** *adverb.*

universe /'ju:nɪvɜ:s/ *noun* all existing things; Creation; all humankind.

university /ju:nɪ'vɜ:sɪtɪ/ *noun* (*plural* **-ies**) educational institution of advanced learning and research, conferring degrees; members of this.

unkempt /ʌn'kempt/ *adjective* dishevelled, untidy.

unknown /ʌn'nəʊn/ ● *adjective* (often + *to*) not known, unfamiliar. ● *noun* unknown thing, person, or quantity. □ **unknown quantity** mysterious or obscure person or thing; **Unknown Soldier**, **Warrior** unidentified soldier symbolizing nation's dead in war.

unlawful /ʌn'lɔ:fʊl/ *adjective* illegal, not permissible. □ **unlawfully** *adverb.*

unleaded /ʌn'ledɪd/ *adjective* (of petrol etc.) without added lead.

unleash /ʌn'li:ʃ/ *verb* free from leash or restraint; set free to pursue or attack.

unleavened /ʌn'lev(ə)nd/ *adjective* made without yeast etc.

unless /ʌn'les/ *conjunction* if not; except when.

unlettered /ʌn'letəd/ *adjective* illiterate.

unlike /ʌn'laɪk/ ● *adjective* not like, different. ● *preposition* differently from.

unlikely /ʌn'laɪklɪ/ *adjective* (**-ier**, **-iest**) improbable; (+ *to do*) not expected; unpromising.

unlisted /ʌn'lɪstɪd/ *adjective* not included in list, esp. of Stock Exchange prices or telephone numbers.

unload /ʌn'ləʊd/ *verb* remove load from (vehicle etc.); remove (load) from vehicle etc.; remove ammunition from (gun); *colloquial* get rid of.

unlock /ʌn'lɒk/ *verb* release lock of; release or disclose by unlocking; release feelings etc. from.

unlooked-for /ʌn'lʊktfɔ:/ *adjective* unexpected.

unlucky /ʌn'lʌkɪ/ *adjective* (**-ier**, **-iest**) not fortunate or successful; wretched; bringing bad luck; ill-judged. □ **unluckily** *adverb.*

unman /ʌn'mæn/ *verb* (**-nn-**) deprive of courage, self-control, etc.

unmannerly /ʌn'mænəlɪ/ *adjective* ill-mannered.

unmask /ʌn'mɑ:sk/ *verb* remove mask from; expose true character of; remove one's mask.

unmentionable /ʌn'menʃənəb(ə)l/ ● *adjective* unsuitable for polite conversation. ● *noun* (in *plural*) *jocular* undergarments.

unmistakable /ʌnmɪ'steɪkəb(ə)l/ *adjective* clear, obvious, plain. □ **unmistakably** *adverb.*

unmitigated /ʌn'mɪtɪgeɪtɪd/ *adjective* not modified; absolute.

unmoved /ʌn'mu:vd/ *adjective* not moved; constant in purpose; unemotional.

unnatural /ʌn'nætʃər(ə)l/ *adjective* contrary to nature; not normal; lacking natural feelings; artificial, forced. □ **unnaturally** *adverb.*

unnecessary /ʌn'nesəsərɪ/ *adjective* not necessary; superfluous. □ **unnecessarily** *adverb.*

unnerve /ʌn'nɜ:v/ *verb* (**-ving**) deprive of confidence etc.

unobjectionable /ʌnəb'dʒekʃənəb(ə)l/ *adjective* acceptable.

unobtrusive /ʌnəb'tru:sɪv/ *adjective* not making oneself or itself noticed. □ **unobtrusively** *adverb.*

unofficial /ʌnə'fɪʃ(ə)l/ *adjective* not officially authorized or confirmed. □ **unofficial strike** strike not ratified by trade union.

unpack /ʌn'pæk/ *verb* open and empty; take (thing) from package etc.

unpalatable /ʌn'pælətəb(ə)l/ *adjective* (of food, suggestion, etc.) disagreeable, distasteful.

unparalleled /ʌn'pærəleld/ *adjective* unequalled.

unparliamentary /ʌnpɑ:lə'mentərɪ/ *adjective* contrary to proper parliamentary usage. □ **unparliamentary language** oaths, abuse.

unpick /ʌn'pɪk/ *verb* undo sewing of.

unplaced /ʌn'pleɪst/ *adjective* not placed as one of the first 3 in race etc.

unpleasant /ʌn'plez(ə)nt/ *adjective* disagreeable. □ **unpleasantly** *adverb*; **unpleasantness** *noun.*

unplug /ʌn'plʌg/ *verb* (**-gg-**) disconnect (electrical device) by removing plug from socket; unstop.

unplumbed /ʌn'plʌmd/ *adjective* not plumbed; not fully explored or understood.

unpopular /ʌn'pɒpjʊlə/ *adjective* not popular, disliked. □ **unpopularity** /-'lær-/ *noun.*

unpractised /ʌn'præktɪst/ *adjective* (*US* **unpracticed**) not experienced or skilled; not put into practice.

unprecedented /ʌn'presɪdentɪd/ *adjective* having no precedent, unparalleled.

unprepossessing /ʌnpri:pə'zesɪŋ/ *adjective* unattractive.

unprincipled /ʌn'prɪnsɪp(ə)ld/ *adjective* lacking or not based on moral principles.

unprintable /ʌn'prɪntəb(ə)l/ *adjective* too offensive or indecent to be printed.

unprofessional /ʌnprə'feʃən(ə)l/ *adjective* contrary to professional standards; unskilled, amateurish.

unprompted /ʌn'prɒmptɪd/ *adjective* spontaneous.

unputdownable /ʌnpʊt'daʊnəb(ə)l/ *adjective colloquial* compulsively readable.

unqualified /ʌn'kwɒlɪfaɪd/ *adjective* not qualified or competent; complete.

unquestionable /ʌn'kwestʃənəb(ə)l/ *adjective* that cannot be disputed or doubted. □ **unquestionably** *adverb.*

unquote /ʌn'kwəʊt/ *interjection used in dictation etc. to indicate closing quotation marks.*

unravel /ʌn'ræv(ə)l/ *verb* (**-ll-**; *US* **-l-**) make or become unknitted, unknotted, etc.; solve (mystery etc.); undo (knitted fabric).

unreal /ʌn'rɪəl/ *adjective* not real; imaginary; *slang* incredible.

unreasonable /ʌn'ri:zənəb(ə)l/ *adjective* excessive; not heeding reason. □ **unreasonably** *adverb.*

unregenerate /ʌnrɪ'dʒenərət/ *adjective* obstinately wrong or bad.

unrelenting /ʌnrɪ'lentɪŋ/ *adjective* not abating; merciless.

unreliable /ʌnrɪ'laɪəb(ə)l/ *adjective* erratic.

unrelieved /ʌnrɪ'li:vd/ *adjective* monotonously uniform.

unremarked /ʌnrɪ'mɑ:kt/ *adjective* not mentioned or remarked on.

unremitting /ʌnrɪ'mɪtɪŋ/ *adjective* incessant. □ **unremittingly** *adverb.*

unremunerative /ʌnrɪ'mju:nərətɪv/ *adjective* unprofitable.

unrequited /ʌnrɪ'kwaɪtɪd/ *adjective* (of love etc.) not returned.

unreserved /ʌnrɪ'zɜ:vd/ *adjective* without reservation. □ **unreservedly** /-vɪdlɪ/ *adverb.*

unrest /ʌn'rest/ *noun* disturbance, turmoil, trouble.

unrivalled /ʌn'raɪv(ə)ld/ *adjective* (*US* **unrivaled**) having no equal.

unroll /ʌn'rəʊl/ *verb* open out from rolled-up state; display, be displayed.

unruffled /ʌn'rʌf(ə)ld/ *adjective* calm.

unruly /ʌn'ru:lɪ/ *adjective* (**-ier**, **-iest**) undisciplined, disorderly. □ **unruliness** *noun.*

unsatisfactory /ʌnsætɪs'fæktəri/ *adjective* poor, unacceptable.

unsaturated /ʌn'sætʃəreɪtɪd/ *adjective Chemistry* (of fat) containing double or triple molecular bonds and therefore capable of combining with hydrogen.

unsavoury /ʌn'seɪvərɪ/ *adjective* (*US* **unsavory**) disgusting; (esp. morally) offensive.

unscathed /ʌn'skeɪðd/ *adjective* uninjured, unharmed.

unschooled /ʌn'sku:ld/ *adjective* uneducated, untrained.

unscientific /ʌnsaɪən'tɪfɪk/ *adjective* not scientific in method etc. □ **unscientifically** *adverb.*

unscramble /ʌn'skræmb(ə)l/ *verb* (**-ling**) decode, interpret (scrambled transmission etc.).

unscreened /ʌn'skri:nd/ *adjective* (esp. of coal) not passed through screen; not checked, esp. for security or medical problems; not having screen; not shown on screen.

unscrew /ʌn'skru:/ *verb* unfasten by removing screw(s), loosen (screw).

unscripted /ʌn'skrɪptɪd/ *adjective* (of speech etc.) delivered impromptu.

unscrupulous /ʌn'skru:pjʊləs/ *adjective* without scruples; unprincipled. □ **unscrupulously** *adverb*; **unscrupulousness** *noun.*

unseasonal /ʌn'si:zən(ə)l/ *adjective* not typical of the time or season.

unseat /ʌn'si:t/ *verb* remove from (esp. parliamentary) seat; dislodge from horseback etc.

unseeing /ʌn'si:ɪŋ/ *adjective* unobservant; blind. □ **unseeingly** *adverb.*

unseemly /ʌn'si:mlɪ/ *adjective* (**-ier**, **-iest**) indecent; unbecoming.

unseen /ʌn'si:n/ ● *adjective* not seen; invisible; (of translation) to be done without preparation. ● *noun* unseen translation.

unselfish /ʌn'selfɪʃ/ *adjective* concerned about others; sharing. □ **unselfishly** *adverb*; **unselfishness** *noun.*

unsettled /ʌn'set(ə)ld/ *adjective* restless, disturbed; open to further discussion; liable to change; not paid.

unsex /ʌn'seks/ *verb* deprive of qualities of one's (esp. female) sex.

unshakeable /ʌn'ʃeɪkəb(ə)l/ *adjective* firm; obstinate.

unsightly /ʌn'saɪtlɪ/ *adjective* ugly. □ **unsightliness** *noun.*

unskilled /ʌn'skɪld/ *adjective* lacking or (of work) not needing special skills.

unsociable /ʌn'səʊʃəb(ə)l/ *adjective* disliking company.

■ **Usage** *Unsociable* is sometimes confused with *unsocial.*

unsocial /ʌn'səʊʃ(ə)l/ *adjective* not social; not suitable for or seeking society; outside normal working day; antisocial.

■ **Usage** *Unsocial* is sometimes confused with *unsociable.*

unsolicited /ʌnsə'lɪsɪtɪd/ *adjective* voluntary.

unsophisticated /ʌnsə'fɪstɪkeɪtɪd/ *adjective* artless, simple, natural.

unsound /ʌn'saʊnd/ *adjective* unhealthy; rotten; weak; unreliable.

unsparing /ʌn'speərɪŋ/ *adjective* lavish; merciless.

unspeakable /ʌn'spi:kəb(ə)l/ *adjective* that words cannot express; indescribably bad. □ **unspeakably** *adverb.*

unstable /ʌn'steɪb(ə)l/ *adjective* (**-r**, **-st**) likely to fall; not stable emotionally; changeable.

unsteady /ʌn'stedɪ/ *adjective* (**-ier**, **-iest**) not firm; changeable; not regular. □ **unsteadily** *adverb*; **unsteadiness** *noun.*

unstick /ʌn'stɪk/ *verb* (*past & past participle* **unstuck**) separate (thing stuck to another). □ **come unstuck** *colloquial* fail.

unstinting /ʌn'stɪntɪŋ/ *adjective* lavish; limitless. □ **unstintingly** *adverb*.

unstressed /ʌn'strest/ *adjective* not pronounced with stress.

unstring /ʌn'strɪŋ/ *verb* (*past & past participle* **unstrung**) remove string(s) of (bow, harp, etc.); take (beads etc.) off string; (esp. as **unstrung** *adjective*) unnerve.

unstructured /ʌn'strʌktʃəd/ *adjective* without structure; informal.

unstudied /ʌn'stʌdɪd/ *adjective* easy, natural, spontaneous.

unsung /ʌn'sʌŋ/ *adjective* not celebrated, unrecognized.

unswerving /ʌn'swɜːvɪŋ/ *adjective* constant, steady. □ **unswervingly** *adverb*.

unthinkable /ʌn'θɪŋkəb(ə)l/ *adjective* unimaginable, inconceivable; *colloquial* highly unlikely or undesirable.

unthinking /ʌn'θɪŋkɪŋ/ *adjective* thoughtless; unintentional, inadvertent. □ **unthinkingly** *adverb*.

untidy /ʌn'taɪdɪ/ *adjective* (**-ier**, **-iest**) not neat or orderly. □ **untidily** *adverb*; **untidiness** *noun*.

until /ən'tɪl/ *preposition & conjunction* = TILL[1].

■ **Usage** *Until*, as opposed to *till*, is used especially at the beginning of a sentence and in formal style, as in *Until you told me, I had no idea* or *He resided there until his decease.*

---

untimely /ʌn'taɪmlɪ/ *adjective* inopportune; premature.

untiring /ʌn'taɪərɪŋ/ *adjective* tireless.

unto /'ʌntʊ/ *preposition archaic* to.

untold /ʌn'təʊld/ *adjective* not told; immeasurable.

untouchable /ʌn'tʌtʃəb(ə)l/ ● *adjective* that may not be touched. ● *noun* Hindu of group believed to defile higher castes on contact.

untoward /ʌntə'wɔːd/ *adjective* inconvenient, unlucky; awkward; refractory; unseemly.

untrammelled /ʌn'træm(ə)ld/ *adjective* not hampered.

untruth /ʌn'truːθ/ *noun* being untrue; lie.

unused *adjective* /ʌn'juːzd/ not in use, never used; /ʌn'juːst/ (+ *to*) not accustomed.

unusual /ʌn'juːʒʊəl/ *adjective* not usual; remarkable. □ **unusually** *adverb*.

unutterable /ʌn'ʌtərəb(ə)l/ *adjective* inexpressible; beyond description. □ **unutterably** *adverb*.

unvarnished /ʌn'vɑːnɪʃt/ *adjective* not varnished; plain, direct, simple.

unveil /ʌn'veɪl/ *verb* uncover (statue etc.) ceremonially; reveal (secrets etc.).

unversed /ʌn'vɜːst/ *adjective* (usually + *in*) not experienced or skilled.

unwarrantable /ʌn'wɒrəntəb(ə)l/ *adjective* (also **unwarranted**) unjustified.

unwashed /ʌn'wɒʃt/ *adjective* not washed or clean. □ **the great unwashed** *colloquial* the rabble.

unwell /ʌn'wel/ *adjective* ill.

unwholesome /ʌn'həʊlsəm/ *adjective* detrimental to moral or physical health; unhealthy-looking.

unwieldy /ʌn'wiːldɪ/ *adjective* (**-ier**, **-iest**) cumbersome or hard to manage owing to size, shape, etc.

unwilling /ʌn'wɪlɪŋ/ *adjective* reluctant. □ **unwillingly** *adverb*; **unwillingness** *noun*.

unwind /ʌn'waɪnd/ *verb* (*past & past participle* **unwound**) draw out or become drawn out after having been wound; *colloquial* relax.

unwitting /ʌn'wɪtɪŋ/ *adjective* not knowing, unaware; unintentional. □ **unwittingly** *adverb*.

unwonted /ʌn'wəʊntɪd/ *adjective* not customary or usual.

unworldly /ʌn'wɜːldlɪ/ *adjective* spiritual; naïve.

unworthy /ʌn'wɜːðɪ/ *adjective* (**-ier**, **-iest**) (often + *of*) not worthy or befitting; discreditable, unseemly. □ **unworthiness** *noun*.

unwritten /ʌn'rɪt(ə)n/ *adjective* not written; (of law etc.) based on tradition or judicial decision, not on statute.

up ● *adverb* towards or in higher place or place regarded as higher, e.g. the north, a capital; to or in erect or required position; in or into active condition; in stronger position; (+ *to*, *till*, etc.) to specified place, person, or time;

higher in price; completely; completed; into compact, accumulated, or secure state; having risen; happening, esp. unusually. ● *preposition* upwards and along, through, or into; at higher part of. ● *adjective* directed upwards. ● *noun* spell of good fortune. ● *verb* (**-pp-**) *colloquial* start (abruptly or unexpectedly) to speak or act; raise. □ **on the up (and up)** *colloquial* steadily improving; **up against** close to, in(to) contact with, *colloquial* confronted with; **up-and-coming** *colloquial* (of person) promising, progressing; **up for** available for or standing for (sale, office, etc.); **upstate** *US* (in, to, or of) provincial, esp. northern, part of a state; **upstream** *adverb* against flow of stream etc., *adjective* moving upstream; **up to** until, below or equal to, incumbent on, capable of, occupied or busy with; **uptown** *US* (in, into, or of) residential part of town or city; **upwind** in the direction from which the wind is blowing.

upbeat ● *noun Music* unaccented beat. ● *adjective colloquial* optimistic, cheerful.

upbraid /ʌp'breɪd/ *verb* (often + *with*, *for*) chide, reproach.

upbringing *noun* child's rearing.

up-country /ʌp'kʌntrɪ/ *adjective & adverb* inland.

update ● *verb* /ʌp'deɪt/ (**-ting**) bring up to date. ● *noun* /'ʌpdeɪt/ updating; updated information etc.

up-end /ʌp'end/ *verb* set or rise up on end.

upfront /ʌp'frʌnt/ *colloquial* ● *adverb* (usually **up front**) at the front; in front; (of payments) in advance. ● *adjective* honest, frank, direct; (of payments) made in advance.

upgrade /ʌp'greɪd/ *verb* (**-ding**) raise in rank etc.; improve (equipment etc.).

upheaval /ʌp'hi:v(ə)l/ *noun* sudden esp. violent change or disturbance.

uphill ● *adverb* /ʌp'hɪl/ up a slope. ● *adjective* /'ʌphɪl/ sloping up; ascending; arduous.

uphold /ʌp'həʊld/ *verb* (*past & past participle* **upheld**) support; maintain, confirm. □ **upholder** *noun.*

upholster /ʌp'həʊlstə/ *verb* provide (furniture) with upholstery. □ **upholsterer** *noun.*

upholstery *noun* covering, padding, springs, etc. for furniture; upholsterer's work.

upkeep *noun* maintenance in good condition; cost or means of this.

upland /'ʌplənd/ ● *noun* (usually in *plural*) higher parts of country. ● *adjective* of these parts.

uplift ● *verb* /ʌp'lɪft/ raise; (esp as **uplifting** *adjective*) elevate morally or emotionally. ● *noun* /'ʌplɪft/ elevating influence; support for breasts etc.

up-market /ʌp'mɑ:kɪt/ *adjective & adverb* of or to more expensive sector of market.

upon /ə'pɒn/ *preposition* on.

■ **Usage** *Upon* is usually more formal than *on*, but it is standard in *once upon a time* and *upon my word*.

upper /'ʌpə/ ● *adjective* higher in place; situated above; superior in rank etc. ● *noun* part of shoe or boot above sole. □ **on one's uppers** *colloquial* extremely short of money; **upper case** capital letters; **the upper crust** *colloquial* the aristocracy; **upper-cut** hit upwards with arm bent; **the upper hand** dominance, control; **Upper House** higher legislative assembly, esp. House of Lords.

uppermost ● *adjective* highest, predominant. ● *adverb* on or to the top.

uppity /'ʌpɪtɪ/ *adjective* (also **uppish**) *colloquial* self-assertive, arrogant.

upright /'ʌpraɪt/ ● *adjective* erect, vertical; (of piano) with vertical strings; honourable, honest. ● *noun* upright post or rod, esp. as structural support; upright piano.

uprising *noun* insurrection.

uproar *noun* tumult, violent disturbance.

uproarious /ʌp'rɔ:rɪəs/ *adjective* very noisy; provoking loud laughter; very funny. □ **uproariously** *adverb.*

uproot /ʌp'ru:t/ *verb* pull (plant etc.) up from ground; displace (person); eradicate.

upset ● *verb* /ʌp'set/ (**-tt-**; *past & past participle* **upset**) overturn; disturb temper, digestion, or composure of; disrupt. ● *noun* /'ʌpset/ disturbance, surprising result. ● *adjective* /ʌp'set, 'ʌp-/ disturbed.

upshot *noun* outcome, conclusion.

upside down /ʌpsaɪd 'daʊn/ *adverb & adjective* with upper and lower parts reversed, inverted; in(to) total disorder.

upstage /ʌp'steɪdʒ/ ● *adjective & adverb* nearer back of theatre stage. ● *verb* (**-ging**) move upstage to make (another actor) face away from audience; divert attention from (person) to oneself.

upstairs ● *adverb* /ʌp'steəz/ to or on an upper floor. ● *adjective* /'ʌpsteəz/ situated upstairs. ● *noun* /ʌp'steəz/ upper floor.

upstanding /ʌp'stændɪŋ/ *adjective* standing up; strong and healthy; honest.

upstart ● *noun* newly successful, esp. arrogant, person. ● *adjective* that is an upstart; of upstarts.

upsurge *noun* upward surge.

upswept *adjective* (of hair) combed to top of head.

upswing *noun* upward movement or trend.

uptake *noun colloquial* understanding; taking up (of offer etc.).

uptight /ʌp'taɪt/ *adjective colloquial* nervously tense, angry; *US* rigidly conventional.

upturn ● *noun* /'ʌptɜ:n/ upward trend, improvement. ● *verb* /ʌp'tɜ:n/ turn up or upside down.

upward /'ʌpwəd/ ● *adverb* (also **upwards**) towards what is higher, more important, etc. ● *adjective* moving or extending upwards. □ **upwardly** *adverb.*

uranium /jʊ'reɪnɪəm/ *noun* radioactive heavy grey metallic element, capable of nuclear fission and used as source of nuclear energy.

urban /'ɜ:bən/ *adjective* of, living in, or situated in city or town. □ **urban guerrilla** terrorist operating in urban area.

urbane /ɜ:'beɪn/ *adjective* suave; elegant. □ **urbanity** /-'bæn-/ *noun.*

urbanize /'ɜ:bənaɪz/ *verb* (also **-ise**) (**-zing** or **-sing**) make urban, esp. by destroying rural quality of (district). □ **urbanization** *noun.*

urchin /'ɜ:tʃɪn/ *noun* mischievous, esp. ragged, child; sea urchin.

Urdu /'ʊədu:/ *noun* Persian-influenced language related to Hindi, used esp. in Pakistan.

urea /jʊə'ri:ə/ *noun* soluble nitrogenous compound contained esp. in urine.

ureter /jʊə'ri:tə/ *noun* duct carrying urine from kidney to bladder.

urethra /jʊə'ri:θrə/ *noun* (*plural* **-s**) duct carrying urine from bladder.

urge /ɜ:dʒ/ ● *verb* (**-ging**) (often + *on*) drive forcibly, hasten; entreat or exhort earnestly or persistently; (often + *on, upon*) advocate (action, argument, etc.) emphatically. ● *noun* urging impulse or tendency; strong desire.

urgent /'ɜ:dʒ(ə)nt/ *adjective* requiring immediate action or attention; importunate. □ **urgency** *noun*; **urgently** *adverb.*

uric acid /'jʊərɪk/ *noun* constituent of urine.

urinal /jʊə'raɪn(ə)l/ *noun* place or receptacle for urinating by men.

urinary /'jʊərɪnərɪ/ *adjective* of or relating to urine.

urinate /'jʊərɪneɪt/ *verb* (**-ting**) discharge urine. □ **urination** *noun.*

urine /'jʊərɪn/ *noun* waste fluid secreted by kidneys and discharged from bladder.

urn *noun* vase with foot, used esp. for ashes of the dead; large vessel with tap, in which tea etc. is made or kept hot.

urology /jʊə'rɒlədʒɪ/ *noun* study of the urinary system. □ **urological** /-rə'lɒdʒ-/ *adjective.*

ursine /'ɜ:saɪn/ *adjective* of or like a bear.

US *abbreviation* United States.

us /ʌs, əs/ *pronoun used by speaker or writer to refer to himself or herself and one or more others as object of verb; used for* ME *by sovereign in formal contexts or by editorial writer in newspaper*; *colloquial* we.

USA *abbreviation* United States of America.

usable /'ju:zəb(ə)l/ *adjective* that can be used.

USAF *abbreviation* United States Air Force.

usage /'ju:sɪdʒ/ *noun* use, treatment; customary practice, established use (esp. of language).

use ● *verb* /ju:z/ (**using**) cause to act or serve for purpose; bring into service; treat in specified way; exploit for one's own ends; (as **used** *adjective*) second-hand. ● *noun* /ju:s/ using, being used;

right or power of using; benefit, advantage; custom, usage. □ **in use** being used; **make use of** use, benefit from; **used to** /ju:st/ *adjective* accustomed to, *verb used before other verb to describe habitual action* (e.g. *I used to live here*); **use up** consume, find use for (leftovers etc.).

■ **Usage** The usual negative and question forms of *used to* are, for example, *You didn't use to go there* and *Did you use to go there?* Both are, however, rather informal, so it is better in formal language to use *You used not to go there* and a different expression such as *Were you in the habit of going there?* or *Did you go there when you lived in London?*

useful *adjective* that can be used to advantage; helpful, beneficial; *colloquial* creditable, efficient. □ **usefully** *adverb*; **usefulness** *noun*.

useless *adjective* serving no purpose, unavailing; *colloquial* feeble, ineffectual. □ **uselessly** *adverb*; **uselessness** *noun*.

user *noun* person who uses a thing. □ **user-friendly** (of computer, program, etc.) easy to use.

usher /'ʌʃə/ ● *noun* person who shows people to their seats in cinema, church, etc.; doorkeeper of court etc. ● *verb* act as usher to; (usually + *in*) announce, show in.

usherette /ʌʃə'ret/ *noun* female usher, esp. in cinema.

USSR *abbreviation historical* Union of Soviet Socialist Republics.

usual /'ju:ʒʊəl/ *adjective* customary; habitual. □ **as usual** as is (or was) usual. □ **usually** *adverb*.

usurer /'ju:ʒərə/ *noun* person who practises usury.

usurp /jʊ'zɜ:p/ *verb* seize (throne, power, etc.) wrongfully. □ **usurpation** /ju:zə'p-/ *noun*; **usurper** *noun*.

usury /'ju:ʒərɪ/ *noun* lending of money at interest, esp. at exorbitant or illegal rate; interest at this rate. □ **usurious** /ju:'ʒʊərɪəs/ *adjective*.

utensil /ju:'tens(ə)l/ *noun* implement or vessel, esp. for kitchen use.

uterus /'ju:tərəs/ *noun* (*plural* **uteri** /-raɪ/) womb. □ **uterine** /-raɪn/ *adjective*.

utilitarian /ju:tɪlɪ'teərɪən/ ● *adjective* designed to be useful rather than attractive; of utilitarianism. ● *noun* adherent of utilitarianism.

utilitarianism *noun* doctrine that actions are justified if they are useful or benefit majority.

utility /ju:'tɪlɪtɪ/ ● *noun* (*plural* **-ies**) usefulness; useful thing, public utility. ● *adjective* basic and standardized. □ **utility room** room for domestic appliances, e.g. washing machine, boiler, etc.; **utility vehicle** vehicle serving various functions.

utilize /'ju:tɪlaɪz/ *verb* (also **-ise**) (**-zing** or **-sing**) turn to account, use. □ **utilization** *noun*.

utmost /'ʌtməʊst/ ● *adjective* farthest, extreme; greatest. ● *noun* the utmost point, degree, etc. □ **do one's utmost** do all that one can.

Utopia /ju:'təʊpɪə/ *noun* imagined perfect place or state. □ **Utopian**, **utopian** *adjective*.

utter[1] /'ʌtə/ *adjective* complete, absolute. □ **utterly** *adverb*; **uttermost** *adjective*.

utter[2] /'ʌtə/ *verb* emit audibly; express in words; *Law* put (esp. forged money) into circulation.

utterance *noun* uttering; thing spoken; power or manner of speaking.

UV *abbreviation* ultraviolet.

uvula /'ju:vjʊlə/ *noun* (*plural* **uvulae** /-li:/) fleshy part of soft palate hanging above throat. □ **uvular** *adjective*.

uxorious /ʌk'sɔ:rɪəs/ *adjective* excessively fond of one's wife.

Vv

V¹ *noun* (also **v**) (Roman numeral) 5.

V² *abbreviation* volt(s).

v. *abbreviation* verse; versus; very; verb; *vide*.

vac *noun colloquial* vacation.

vacancy /'veɪkənsɪ/ *noun* (*plural* **-ies**) being vacant; unoccupied post, place, etc.

vacant *adjective* not filled or occupied; not mentally active, showing no interest. □ **vacant possession** ownership of unoccupied house etc. □ **vacantly** *adverb*.

vacate /və'keɪt/ *verb* (**-ting**) leave vacant, cease to occupy.

vacation /və'keɪʃ(ə)n/ *noun* fixed holiday period, esp. in law courts and universities; *US* holiday; vacating, being vacated.

vaccinate /'væksɪneɪt/ *verb* (**-ting**) inoculate with vaccine to immunize against disease. □ **vaccination** *noun*.

vaccine /'væksi:n/ *noun* preparation used for inoculation, originally cowpox virus giving immunity to smallpox.

vacillate /'væsɪleɪt/ *verb* (**-ting**) fluctuate in opinion or resolution. □ **vacillation** *noun*; **vacillator** *noun*.

vacuous /'vækjʊəs/ *adjective* expressionless; unintelligent. □ **vacuity** /və'kju:ɪtɪ/ *noun*; **vacuously** *adverb*.

vacuum /'vækjʊəm/ ● *noun* (*plural* **-s** or **vacua**) space entirely devoid of matter; space or vessel from which air has been completely or partly removed by pump etc.; absence of normal or previous content; (*plural* **-s**) *colloquial* vacuum cleaner. ● *verb colloquial* clean with vacuum cleaner. □ **vacuum brake** brake worked by exhaustion of air; **vacuum cleaner** machine for removing dust etc. by suction; **vacuum flask** vessel with double wall enclosing vacuum so that contents remain hot or cold; **vacuum-packed** sealed after partial removal of air; **vacuum tube** tube containing near-vacuum for free passage of electric current.

vagabond /'vægəbɒnd/ ● *noun* wanderer, esp. idle one. ● *adjective* wandering, having no settled habitation or home.

vagary /'veɪgərɪ/ *noun* (*plural* **-ies**) caprice, eccentric act or idea.

vagina /və'dʒaɪnə/ *noun* (*plural* **-s** or **-nae** /-ni:/) canal joining womb and vulva of female mammal. □ **vaginal** *adjective*.

vagrant /'veɪgrənt/ ● *noun* person without settled home or regular work. ● *adjective* wandering, roving. □ **vagrancy** *noun*.

vague /veɪg/ *adjective* uncertain, ill-defined; not clear-thinking, inexact. □ **vaguely** *adverb*; **vagueness** *noun*.

vain *adjective* conceited; empty, trivial; unavailing, useless. □ **in vain** without result or success, lightly or profanely. □ **vainly** *adverb*.

vainglory /veɪn'glɔ:rɪ/ *noun* extreme vanity, boastfulness. □ **vainglorious** *adjective*.

valance /'væləns/ *noun* short curtain round bedstead, above window, etc.

vale *noun* (*archaic* except in place names) valley.

valediction /vælɪ'dɪkʃ(ə)n/ *noun formal* bidding farewell; words used in this. □ **valedictory** *adjective & noun* (*plural* **-ies**).

valence /'veɪləns/ *noun* valency.

valency /'veɪlənsɪ/ *noun* (*plural* **-ies**) combining-power of an atom measured by number of hydrogen atoms it can displace or combine with.

valentine /'væləntaɪn/ *noun* (usually anonymous) letter or card sent as mark of love on St Valentine's Day (14 Feb.); sweetheart chosen on that day.

valerian /və'lɪərɪən/ *noun* any of various kinds of flowering herb.

valet /'vælɪt/ ● *noun* gentleman's personal servant. ● *verb* (**-t-**) act as valet (to).

valetudinarian /vælɪtju:dɪ'neərɪən/ ● *noun* person of poor health or unduly anxious about health. • *adjective* of a valetudinarian.

valiant /'væljənt/ *adjective* brave. □ **valiantly** *adverb*.

valid /'vælɪd/ *adjective* (of reason, objection, etc.) sound, defensible; legally acceptable, not yet expired. □ **validity** /və'lɪd-/ *noun*.

validate /'vælɪdeɪt/ *verb* (**-ting**) make valid, ratify. □ **validation** *noun*.

valise /və'li:z/ *noun US* small portmanteau.

Valium /'vælɪəm/ *noun proprietary term* the tranquillizing drug diazepam.

valley /'vælɪ/ *noun* (*plural* **-s**) low area between hills, usually with stream or river.

valour /'vælə/ *noun* (*US* **valor**) courage, esp. in battle. □ **valorous** *adjective*.

valuable /'væljʊəb(ə)l/ • *adjective* of great value, price, or worth. • *noun* (usually in *plural*) valuable thing.

valuation /væljʊ'eɪʃ(ə)n/ *noun* estimation (esp. by professional valuer) of thing's worth; estimated value.

value /'vælju:/ • *noun* worth, desirability, or qualities on which these depend; worth as estimated; amount for which thing can be exchanged in open market; equivalent of thing; (in full **value for money**) something well worth money spent; ability of a thing to serve a purpose or cause an effect; (in *plural*) one's principles, priorities, or standards; *Music* duration of note; *Mathematics* amount denoted by algebraic term. • *verb* (**-ues**, **-ued**, **-uing**) estimate value of; have high or specified opinion of. □ **value added tax** tax levied on rise in value of services and goods at each stage of production; **value judgement** subjective estimate of worth etc. □ **valueless** *adjective*.

valuer *noun* person who estimates or assesses values.

valve *noun* device controlling flow through pipe etc., usually allowing movement in one direction only; structure in organ etc. allowing flow of blood etc. in one direction only; thermionic valve; device to vary length of tube in trumpet etc.; half-shell of oyster, mussel, etc.

valvular /'vælvjʊlə/ *adjective* having valve(s); having form or function of valve.

vamoose /və'mu:s/ *verb US slang* depart hurriedly.

vamp[1] • *noun* upper front part of boot or shoe. • *verb* (often + *up*) repair, furbish, or make by patching or piecing together; improvise musical accompaniment.

vamp[2] *colloquial* • *noun* woman who uses sexual attraction to exploit men. • *verb* allure or exploit (man).

vampire /'væmpaɪə/ *noun* supposed ghost or reanimated corpse sucking blood of sleeping people; person who preys on others; (in full **vampire bat**) bloodsucking bat.

van[1] *noun* covered vehicle or closed railway truck for transporting goods etc.

van[2] *noun* vanguard, forefront.

vanadium /və'neɪdɪəm/ *noun* hard grey metallic element used to strengthen steel.

vandal /'vænd(ə)l/ *noun* person who wilfully or maliciously damages property. □ **vandalism** *noun*.

vandalize /'vændəlaɪz/ *verb* (also **-ise**) (**-zing** or **-sing**) destroy or damage wilfully or maliciously (esp. public property).

vane *noun* weather vane; blade of windmill, ship's propeller, etc.

vanguard /'vængɑ:d/ *noun* foremost part of advancing army etc.; leaders of movement etc.

vanilla /və'nɪlə/ *noun* tropical fragrant climbing orchid; extract of its fruit (**vanilla-pod**), or synthetic substitute, used as flavouring.

vanish /'vænɪʃ/ *verb* disappear; cease to exist. □ **vanishing point** point at which receding parallel lines appear to meet.

vanity /'vænɪtɪ/ *noun* (*plural* **-ies**) conceit about one's attainments or appearance; futility, unreal thing. □ **vanity bag**, **case** woman's make-up bag or case.

vanquish /'væŋkwɪʃ/ *verb literary* conquer, overcome.

vantage /'vɑ:ntɪdʒ/ *noun* advantage, esp. in tennis; (also **vantage point**) place giving good view.

vapid /'væpɪd/ *adjective* insipid, dull, flat. □ **vapidity** /və'pɪd-/ *noun*.

vapor *US* = VAPOUR.

vaporize /'veɪpəraɪz/ *verb* (also **-ise**) (**-zing** or **-sing**) change into vapour. □ **vaporization** *noun*.

vaporous /'veɪpərəs/ *adjective* in the form of or consisting of vapour.

vapour /'veɪpə/ *noun* (*US* **vapor**) moisture or other substance diffused or suspended in air, e.g. mist, smoke; gaseous form of substance. □ **vapour trail** trail of condensed water from aircraft etc.

variable /'veərɪəb(ə)l/ ● *adjective* changeable, adaptable; apt to vary, not constant; *Mathematics* (of quantity) indeterminate, able to assume different numerical values. ● *noun* variable thing or quantity. □ **variability** *noun*.

variance /'veərɪəns/ *noun* (usually after *at*) difference of opinion; dispute; discrepancy.

variant ● *adjective* differing in form or details from standard; having different forms. ● *noun* variant form, spelling, type, etc.

variation /veərɪ'eɪʃ(ə)n/ *noun* varying; departure from normal kind, standard, type, etc.; extent of this; thing that varies from type; *Music* theme in changed or elaborated form.

varicose /'værɪkəʊs/ *adjective* (esp. of vein etc.) permanently and abnormally dilated.

variegated /'veərɪgeɪtɪd/ *adjective* with irregular patches of different colours; having leaves of two or more colours. □ **variegation** *noun*.

variety /və'raɪətɪ/ *noun* (*plural* **-ies**) diversity; absence of uniformity; collection of different things; class of things differing from rest in same general class; member of such class; (+ *of*) different form of thing, quality, etc.; *Biology* subdivision of species; series of dances, songs, comedy acts, etc.

various /'veərɪəs/ *adjective* different, diverse; several. □ **variously** *adverb*.

■ **Usage** *Various* (unlike *several*) is not a pronoun and therefore cannot be used with *of*, as (wrongly) in *Various of the guests arrived late.*

varnish /'vɑːnɪʃ/ ● *noun* resinous solution used to give hard shiny transparent coating. ● *verb* coat with varnish; conceal with deceptively attractive appearance.

varsity /'vɑːsɪtɪ/ *noun* (*plural* **-ies**) *colloquial* university.

vary /'veərɪ/ *verb* (**-ies**, **-ied**) be or become different; be of different kinds; modify, diversify.

vascular /'væskjʊlə/ *adjective* of or containing vessels for conveying blood, sap, etc.

vas deferens /væs 'defərenz/ *noun* (*plural* **vasa deferentia** /veɪsə defə'renʃɪə/) sperm duct of testicle.

vase /vɑːz/ *noun* vessel used as ornament or container for flowers.

vasectomy /və'sektəmɪ/ *noun* (*plural* **-ies**) removal of part of each vas deferens, esp. for sterilization.

Vaseline /'væsɪliːn/ *noun proprietary term* type of petroleum jelly used as ointment etc.

vassal /'væs(ə)l/ *noun* humble dependant; *historical* holder of land by feudal tenure.

vast /vɑːst/ *adjective* immense, huge. □ **vastly** *adverb*; **vastness** *noun*.

VAT *abbreviation* value added tax.

vat *noun* tank, esp. for holding liquids in brewing, dyeing, and tanning.

Vatican /'vætɪkən/ *noun* palace or government of Pope in Rome.

vaudeville /'vɔːdəvɪl/ *noun esp. US* variety entertainment.

vault /vɔːlt/ ● *noun* arched roof; vaultlike covering; underground room as place of storage; underground burial chamber; act of vaulting. ● *verb* leap or spring, esp. using hands or pole; spring over in this way; (esp. as **vaulted** *adjective*) make in form of vault, provide with vault(s).

vaunt /vɔːnt/ *verb & noun literary* boast.

VC *abbreviation* Victoria Cross.

VCR *abbreviation* video cassette recorder.

VD *abbreviation* venereal disease.

VDU *abbreviation* visual display unit.

veal *noun* calf's flesh as food.

vector /'vektə/ *noun Mathematics & Physics* quantity having both magnitude and direction; carrier of disease.

veer /vɪə/ *verb* change direction, esp. (of wind) clockwise; change in opinion, course, etc.

vegan /'vi:gən/ ● *noun* person who does not eat animals or animal products. ● *adjective* using or containing no animal products.

vegetable /'vedʒtəb(ə)l/ ● *noun* plant, esp. edible herbaceous plant. ● *adjective* of, derived from, or relating to plant life or vegetables as food.

vegetarian /vedʒɪ'teərɪən/ ● *noun* person who does not eat meat or fish. ● *adjective* excluding animal food, esp. meat. □ **vegetarianism** *noun*.

vegetate /'vedʒɪteɪt/ *verb* (**-ting**) lead dull monotonous life; grow as plants do.

vegetation /vedʒɪ'teɪʃ(ə)n/ *noun* plants collectively; plant life.

vegetative /'vedʒɪtətɪv/ *adjective* concerned with growth and development rather than sexual reproduction; of vegetation.

vehement /'vi:əmənt/ *adjective* showing or caused by strong feeling, ardent. □ **vehemence** *noun*; **vehemently** *adverb*.

vehicle /'vi:ɪk(ə)l/ *noun* conveyance used on land or in space; thing or person as medium for thought, feeling, or action; liquid etc. as medium for suspending pigments, drugs, etc. □ **vehicular** /vɪ'hɪkjʊlə/ *adjective*.

veil /veɪl/ ● *noun* piece of usually transparent material attached to woman's hat or otherwise forming part of headdress, esp. to conceal or protect face; piece of linen etc. as part of nun's headdress; thing that hides or disguises. ● *verb* cover with veil; (esp. as **veiled** *adjective*) partly conceal. □ **beyond the veil** in the unknown state of life after death; **draw a veil over** avoid discussing; **take the veil** become nun.

vein /veɪn/ *noun* any of tubes carrying blood to heart; (in general use) any blood vessel; rib of leaf or insect's wing; streak of different colour in wood, marble, cheese, etc.; fissure in rock filled with ore; specified character or tendency, mood. □ **veined** *adjective*.

Velcro /'velkrəʊ/ *noun proprietary term* fastener consisting of two strips of fabric which cling when pressed together.

veld /velt/ *noun* (also **veldt**) *South African* open country.

veleta /və'li:tə/ *noun* ballroom dance in triple time.

vellum /'veləm/ *noun* fine parchment, originally calfskin; manuscript on this; smooth writing paper imitating vellum.

velociraptor /vɪ'lɒsɪræptə/ *noun* small carnivorous dinosaur with short front legs.

velocity /vɪ'lɒsɪtɪ/ *noun* (*plural* **-ies**) speed, esp. of inanimate things.

velour /və'lʊə/ *noun* (also **velours** same pronunciation) plushlike fabric.

velvet /'velvɪt/ ● *noun* soft fabric with thick short pile on one side; furry skin on growing antler. ● *adjective* of, like, or soft as velvet. □ **on velvet** in advantageous or prosperous position; **velvet glove** outward gentleness cloaking firmness or inflexibility. □ **velvety** *adjective*.

velveteen /velvɪ'ti:n/ *noun* cotton fabric with pile like velvet.

Ven. *abbreviation* Venerable.

venal /'vi:n(ə)l/ *adjective* able to be bribed; involving bribery; corrupt. □ **venality** /-'næl-/ *noun*.

■ **Usage** *Venal* is sometimes confused with *venial*.

---

vend *verb* offer (esp. small wares) for sale. □ **vending machine** slot machine selling small items. □ **vendor** *noun*.

vendetta /ven'detə/ *noun* blood feud; prolonged bitter quarrel.

veneer /vɪ'nɪə/ ● *noun* thin covering of fine wood; (often + *of*) deceptively pleasing appearance. ● *verb* apply veneer to (wood etc.).

venerable /'venərəb(ə)l/ *adjective* entitled to deep respect on account of age, character, etc.; *title of archdeacon*.

venerate /'venəreɪt/ *verb* (**-ting**) regard with deep respect. □ **veneration** *noun*.

venereal /vɪ'nɪərɪəl/ *adjective* of sexual desire or intercourse; of venereal disease. □ **venereal disease** disease contracted by sexual intercourse with infected person.

Venetian /vɪ'ni:ʃ(ə)n/ ● *noun* native, citizen, or dialect of Venice. ● *adjective* of Venice. □ **venetian blind** window-blind of adjustable horizontal slats.

vengeance /'vendʒ(ə)ns/ *noun* punishment inflicted for wrong to oneself or to one's cause. □ **with a vengeance** to extreme degree, thoroughly, violently.

vengeful /'vendʒfʊl/ *adjective* seeking vengeance, vindictive.

venial /'vi:nɪəl/ *adjective* (of sin or fault) pardonable, not mortal. □ **veniality** /-'æl-/ *noun.*

■ **Usage** *Venial* is sometimes confused with *venal.*

venison /'venɪs(ə)n/ *noun* deer's flesh as food.

Venn diagram *noun* diagram using overlapping and intersecting circles etc. to show relationships between mathematical sets.

venom /'venəm/ *noun* poisonous fluid of esp. snakes; malignity, virulence of feeling, language, or conduct. □ **venomous** *adjective*; **venomously** *adverb.*

venous /'vi:nəs/ *adjective* of, full of, or contained in veins.

vent¹ ● *noun* opening for passage of air etc.; outlet, free expression; anus, esp. of lower animal. ● *verb* give vent or free expression to.

vent² *noun* slit in garment, esp. in back of jacket.

ventilate /'ventɪleɪt/ *verb* (**-ting**) cause air to circulate freely in (room etc.); air (question, grievance, etc.). □ **ventilation** *noun.*

ventilator *noun* appliance or aperture for ventilating room etc.; apparatus for maintaining artifical respiration.

ventral /'ventr(ə)l/ *adjective* of or on abdomen.

ventricle /'ventrɪk(ə)l/ *noun* cavity in body; hollow part of organ, esp. brain or heart.

ventricular /ven'trɪkjʊlə/ *adjective* of or shaped like ventricle.

ventriloquism /ven'trɪləkwɪz(ə)m/ *noun* skill of speaking without moving the lips. □ **ventriloquist** *noun.*

venture /'ventʃə/ ● *noun* risky undertaking; commercial speculation. ● *verb* (**-ring**) dare, not be afraid; dare to go, make, or put forward; take risks, expose to risk, stake. □ **Venture Scout** senior Scout.

venturesome *adjective* disposed to take risks.

venue /'venju:/ *noun* appointed place for match, meeting, concert, etc.

Venus fly-trap /'vi:nəs/ *noun* insectivorous plant.

veracious /və'reɪʃəs/ *adjective formal* truthful, true. □ **veracity** /-'ræs-/ *noun.*

veranda /və'rændə/ *noun* (sometimes partly covered) platform along side of house.

verb *noun* word used to indicate action, event, state, or change (see panel).

verbal *adjective* of words; oral, not written; of a verb; (of translation) literal. □ **verbally** *adverb.*

verbalize *verb* (also **-ise**) (**-zing** or **-sing**) put into words.

verbatim /v3:'beɪtɪm/ *adverb & adjective* in exactly the same words.

verbena /v3:'bi:nə/ *noun* (*plural* same) plant of genus of herbs and small shrubs with fragrant flowers.

verbiage /'v3:bɪɪdʒ/ *noun derogatory* unnecessary number of words.

verbose /v3:'bəʊs/ *adjective* using more words than are needed. □ **verbosity** /-'bɒs-/ *noun.*

verdant /'v3:d(ə)nt/ *adjective* (of grass, field, etc.) green, lush. □ **verdancy** *noun.*

verdict /'v3:dɪkt/ *noun* decision of jury; decision, judgement.

verdigris /'v3:dɪgri:/ *noun* greenish-blue substance that forms on copper or brass.

verdure /'v3:djə/ *noun literary* green vegetation or its colour.

verge¹ *noun* edge, border; brink; grass edging of road etc.

verge² *verb* (**-ging**) (+ *on*) border on; incline downwards or in specified direction.

verger /'v3:dʒə/ *noun* caretaker and attendant in church; officer carrying staff before dignitaries of cathedral etc.

verify /'verɪfaɪ/ *verb* (**-ies**, **-ied**) establish truth or correctness of by examination etc.; fulfil, bear out. □ **verification** *noun.*

verily /'verɪlɪ/ *adverb archaic* truly, really.

verisimilitude /verɪsɪ'mɪlɪtju:d/ *noun* appearance of being true or real.

veritable /'verɪtəb(ə)l/ *adjective* real, rightly so called.

verity /'verɪtɪ/ *noun* (*plural* **-ies**) true statement; *archaic* truth.

vermicelli /vɜ:mɪ'tʃelɪ/ *noun* pasta in long slender threads.

vermicide /'vɜ:mɪsaɪd/ *noun* drug used to kill intestinal worms.

vermiform /'vɜ:mɪfɔ:m/ *adjective* worm-shaped. □ **vermiform appendix** small blind tube extending from caecum in man and some other mammals.

vermilion /və'mɪljən/ ● *noun* brilliant scarlet pigment made esp. from cinnabar; colour of this. ● *adjective* of this colour.

vermin /'vɜ:mɪn/ *noun* (usually treated as *plural*) mammals and birds harmful to game, crops, etc.; parasitic worms or insects; vile people.

verminous *adjective* of the nature of or infested with vermin.

vermouth /'vɜ:məθ/ *noun* wine flavoured with aromatic herbs.

vernacular /və'nækjʊlə/ ● *noun* language or dialect of country; language of particular class or group; homely speech. ● *adjective* (of language) of one's own country, not foreign or formal.

vernal /'vɜ:n(ə)l/ *adjective* of or in spring.

vernier /'vɜ:nɪə/ *noun* small movable scale for reading fractional parts of subdivisions on fixed scale of measuring instrument.

veronica /və'rɒnɪkə/ *noun* speedwell.

verruca /və'ru:kə/ *noun* (*plural* **verrucae** /-si:/ or **-s**) wart or similar growth, esp. on foot.

versatile /'vɜ:sətaɪl/ *adjective* turning easily or readily from one subject or occupation to another, skilled in many subjects or occupations; having many uses. □ **versatility** /-'tɪl-/ *noun*.

verse *noun* poetry; stanza of poem or song; each of short numbered divisions of Bible.

versed /vɜ:st/ *adjective* (+ *in*) experienced or skilled in.

versicle /'vɜ:sɪk(ə)l/ *noun* short sentence, esp. each of series in liturgy said or sung by minister or priest, answered by congregation.

Verb

A verb says what a person or thing does, and can describe:

an action, e.g. *run, hit*
an event, e.g. *rain, happen*
a state, e.g. *be, have, seem, appear*
a change, e.g. *become, grow*

Verbs occur in different forms, usually in one or other of their tenses. The most common tenses are:

the simple present tense:	*The boy walks down the road.*
the continuous present tense:	*The boy is walking down the road.*
the simple past tense:	*The boy walked down the road.*
the continuous past tense:	*The boy was walking down the road.*
the perfect tense:	*The boy has walked down the road.*
the future tense:	*The boy will walk down the road.*

Each of these forms is a finite verb, which means that it is in a particular tense and that it changes according to the number and person of the subject, as in

I am	*you walk*
we are	*he walks*

An infinitive is the form of a verb that usually appears with 'to', e.g.

to wander, to look, to sleep.

---

versify /'vɜ:sɪfaɪ/ *verb* (**-ies**, **-ied**) turn into or express in verse; compose verses. □ **versification** *noun.*

version /'vɜ:ʃ(ə)n/ *noun* account of matter from particular point of view; particular edition or translation of book etc.

verso /'vɜ:səʊ/ *noun* (*plural* **-s**) left-hand page of open book, back of printed leaf.

versus /'vɜ:səs/ *preposition* against.

vertebra /'vɜ:tɪbrə/ *noun* (*plural* **-brae** /-bri:/) each segment of backbone. □ **vertebral** *adjective.*

vertebrate /'vɜ:tɪbrət/ • *adjective* having backbone. • *noun* vertebrate animal.

vertex /'vɜ:teks/ *noun* (*plural* **-tices** /-tɪsi:z/ or **-texes**) highest point, top, apex; meeting-point of lines that form angle.

vertical /'vɜ:tɪk(ə)l/ • *adjective* at right angles to horizontal plane; in direction from top to bottom of picture etc.; of or at vertex. • *noun* vertical line or plane. □ **vertical take-off** take-off of aircraft directly upwards. □ **vertically** *adverb.*

vertiginous /vɜ:'tɪdʒɪnəs/ *adjective* of or causing vertigo.

vertigo /'vɜ:tɪgəʊ/ *noun* dizziness.

vervain /'vɜ:veɪn/ *noun* any of several verbenas, esp. one with small blue, white, or purple flowers.

verve *noun* enthusiasm, energy, vigour.

very /'verɪ/ • *adverb* in high degree; (+ *own* or superlative adjective) in fullest sense. • *adjective* real, properly so called etc. □ **very good**, **well** *formula of consent or approval*; **very high frequency** 30–300 megahertz (in radio); **Very Reverend** *title of dean.*

vesicle /'vesɪk(ə)l/ *noun* small bladder, blister, or bubble.

vespers /'vespəz/ *plural noun* evening church service.

vessel /'ves(ə)l/ *noun* hollow receptacle, esp. for liquid; ship or boat, esp. large one; duct or canal holding or conveying blood, sap, etc.

vest • *noun* undergarment worn on upper part of body; *US & Australian* waistcoat. • *verb* (+ *with*) bestow (powers, authority, etc.) on; (+ *in*) confer (property or power) on (person) with immediate fixed right of future possession. □ **vested interest** personal interest in state of affairs, usually with expectation of gain, *Law* interest (usually in land or money held in trust) recognized as belonging to person.

vestal virgin /'vest(ə)l/ *noun* virgin consecrated to Vesta, Roman goddess of hearth and home, and vowed to chastity.

vestibule /'vestɪbju:l/ *noun* lobby, entrance hall.

vestige /'vestɪdʒ/ *noun* trace, evidence; slight amount, particle; *Biology* part or organ now atrophied that was well-developed in ancestors. □ **vestigial** /-'tɪdʒɪəl/ *adjective.*

vestment /'vestmənt/ *noun* ceremonial garment worn by priest etc.

vestry /'vestrɪ/ *noun* (*plural* **-ies**) room or part of church for keeping vestments etc. in.

vet • *noun colloquial* veterinary surgeon. • *verb* (**-tt-**) make careful and critical examination of (scheme, work, candidate, etc.).

vetch *noun* plant of pea family largely used for fodder.

veteran /'vetərən/ *noun* old soldier or long-serving member of any group; *US* ex-serviceman or -woman. □ **veteran car** one made before 1905.

veterinarian /vetərɪ'neərɪən/ *noun formal* veterinary surgeon.

veterinary /'vetərɪnərɪ/ *adjective* of or for diseases and injuries of animals. □ **veterinary surgeon** person qualified to treat animals.

veto /'vi:təʊ/ • *noun* (*plural* **-es**) right to reject measure etc. unilaterally; rejection, prohibition. • *verb* (**-oes**, **-oed**) reject (measure etc.); forbid.

vex *verb* annoy, irritate; *archaic* grieve, afflict.

vexation /vek'seɪʃ(ə)n/ *noun* vexing, being vexed; annoying or distressing thing.

vexatious /vek'seɪʃ(ə)s/ *adjective* causing vexation; *Law* lacking sufficient grounds for action and seeking only to annoy defendant.

vexed *adjective* (of question) much discussed.

VHF *abbreviation* very high frequency.

via /'vaɪə/ *preposition* by way of, through.

viable /'vaɪəb(ə)l/ *adjective* (of plan etc.) feasible, esp. economically; (esp. of foetus) capable of living and surviving independently. □ **viability** *noun*.

viaduct /'vaɪədʌkt/ *noun* long bridge carrying railway or road over valley.

vial /'vaɪəl/ *noun* small glass vessel.

viands /'vaɪəndz/ *plural noun formal* articles of food.

viaticum /vaɪ'ætɪkəm/ *noun* (*plural* **-ca**) Eucharist given to dying person.

vibes /vaɪbz/ *plural noun colloquial* vibrations, esp. feelings communicated; vibraphone.

vibrant /'vaɪbrənt/ *adjective* vibrating; resonant; (often + *with*) thrilling; (of colour) bright and striking. □ **vibrancy** *noun*.

vibraphone /'vaɪbrəfəʊn/ *noun* percussion instrument with motor-driven resonators under metal bars giving vibrato effect.

vibrate /vaɪ'breɪt/ *verb* (**-ting**) move rapidly to and fro; (of sound) throb, resonate; (+ *with*) quiver; swing to and fro, oscillate. □ **vibratory** *adjective*.

vibration *noun* vibrating; (in *plural*) mental (esp. occult) influence, atmosphere or feeling communicated.

vibrato /vɪ'brɑ:təʊ/ *noun* tremulous effect in musical pitch.

vibrator *noun* device that vibrates, esp. instrument used in massage or sexual stimulation.

viburnum /vaɪ'bɜ:nəm/ *noun* shrub with pink or white flowers.

vicar /'vɪkə/ *noun* incumbent of C. of E. parish where in former times incumbent received stipend rather than tithes; *colloquial* any member of the clergy.

vicarage *noun* vicar's house.

vicarious /vɪ'keərɪəs/ *adjective* experienced indirectly; acting or done etc. for another; deputed, delegated. □ **vicariously** *adverb*.

vice[1] *noun* immoral conduct; particular form of this; bad habit. □ **vice ring** group of criminals organizing prostitution; **vice squad** police department concerned with prostitution.

vice[2] *noun* (*US* **vise**) clamp with two jaws for holding an object being worked on.

vice- *combining form* person acting in place of; person next in rank to.

vice-chancellor /vaɪs'tʃɑ:nsələ/ *noun* deputy chancellor (esp. administrator of university).

viceregal /vaɪs'ri:g(ə)l/ *adjective* of viceroy.

vicereine /'vaɪsreɪn/ *noun* viceroy's wife; woman viceroy.

viceroy /'vaɪsrɔɪ/ *noun* ruler on behalf of sovereign in colony, province, etc.

vice versa /vaɪs 'vɜ:sə/ *adjective* with order of terms changed, other way round.

Vichy water /'vi:ʃi:/ *noun* effervescent mineral water from Vichy in France.

vicinity /vɪ'sɪnɪtɪ/ *noun* (*plural* **-ies**) surrounding district; (+ *to*) nearness to. □ **in the vicinity (of)** near (to).

vicious /'vɪʃəs/ *adjective* bad-tempered, spiteful; violent; corrupt. □ **vicious circle** self-perpetuating, harmful sequence of cause and effect. □ **viciously** *adverb*; **viciousness** *noun*.

vicissitude /vɪ'sɪsɪtju:d/ *noun literary* change, esp. of fortune.

victim /'vɪktɪm/ *noun* person or thing destroyed or injured; prey, dupe; creature sacrificed to a god etc.

victimize *verb* (also **-ise**) (**-zing** or **-sing**) single out for punishment or unfair treatment; make (person etc.) a victim. □ **victimization** *noun*.

victor /'vɪktə/ *noun* conqueror, winner of contest.

Victoria Cross /vɪk'tɔ:rɪə/ *noun* highest decoration for conspicuous bravery in armed services.

Victorian /vɪk'tɔ:rɪən/ • *adjective* of time of Queen Victoria; prudish, strict. • *noun* person of this time.

Victoriana /vɪktɔ:rɪ'ɑ:nə/ *plural noun* articles, esp. collectors' items, of Victorian period.

victorious /vɪk'tɔ:rɪəs/ *adjective* conquering, triumphant; marked by victory. □ **victoriously** *adverb*.

victory /'vɪktərɪ/ *noun* (*plural* **-ies**) success in battle, war, or contest.

victual /'vɪt(ə)l/ • *noun* (usually in *plural*) food, provisions. • *verb* (**-ll-**; *US* **-l-**) supply with victuals; lay in supply of victuals; eat victuals.

victualler /ˈvɪtlə/ *noun* (*US* **victualer**) person who supplies victuals; (in full **licensed victualler**) publican licensed to sell alcohol.

vicuña /vɪˈkjuːnə/ *noun* S. American mammal with fine silky wool; cloth made from its wool; imitation of this.

vide /ˈviːdeɪ/ *verb* (in *imperative*) see, consult. [Latin]

videlicet /vɪˈdelɪset/ *adverb* that is to say; namely.

video /ˈvɪdɪəʊ/ ● *adjective* relating to recording or reproduction of moving pictures on magnetic tape; of broadcasting of these. ● *noun* (*plural* **-s**) such recording or broadcasting; *colloquial* video recorder; *colloquial* film on videotape. ● *verb* (**-oes**, **-oed**) record on videotape. □ **video cassette** cassette of videotape; **video game** computer game played on television screen; **video nasty** *colloquial* horrific or pornographic video film; **video (cassette) recorder** apparatus for recording and playing videotapes.

videotape ● *noun* magnetic tape for recording television pictures and sound. ● *verb* (**-ping**) record on this.

vie /vaɪ/ *verb* (**vying**) (often + *with*) contend, compete, strive for superiority.

Vietnamese /vɪetnəˈmiːz/ ● *adjective* of Vietnam. ● *noun* (*plural* same) native, national, or language of Vietnam.

view /vjuː/ ● *noun* range of vision; what is seen, scene, prospect, picture etc. of this; opinion; inspection by eye or mind. ● *verb* look at; survey visually or mentally; form mental impression or opinion of; watch television. □ **in view of** considering; **on view** being shown or exhibited; **viewdata** news and information service from computer source, connected to TV screen by telephone link; **viewfinder** part of camera showing field of photograph; **viewpoint** point of view; **with a view to** with hope or intention of.

viewer *noun* television-watcher; device for looking at film transparencies etc.

vigil /ˈvɪdʒɪl/ *noun* keeping awake during night etc., esp. to keep watch or pray; eve of festival or holy day.

vigilance *noun* watchfulness; caution. □ **vigilant** *adjective*.

vigilante /vɪdʒɪˈlæntɪ/ *noun* member of self-appointed group for keeping order etc.

vignette /viːˈnjet/ *noun* short description, character sketch; illustration not in definite border; photograph etc. with background shaded off.

vigour /ˈvɪgə/ *noun* (*US* **vigor**) activity and strength of body or mind; healthy growth; animation. □ **vigorous** *adjective*; **vigorously** *adverb*.

Viking /ˈvaɪkɪŋ/ *noun* Scandinavian raider and pirate of 8th–11th c.

vile *adjective* disgusting; depraved; *colloquial* abominably bad. □ **vilely** *adverb*; **vileness** *noun*.

vilify /ˈvɪlɪfaɪ/ *verb* (**-ies**, **-ied**) speak ill of, defame. □ **vilification** *noun*.

villa /ˈvɪlə/ *noun* country house, mansion; rented holiday home, esp. abroad; detached or semi-detached house in residential district.

village /ˈvɪlɪdʒ/ *noun* group of houses etc. in country district, larger than hamlet and smaller than town.

villager *noun* inhabitant of village.

villain /ˈvɪlən/ *noun* wicked person; chief wicked character in play, story, etc.; *colloquial* criminal, rascal.

villainous *adjective* wicked.

villainy *noun* (*plural* **-ies**) wicked behaviour or act.

villein /ˈvɪlɪn/ *noun historical* feudal tenant entirely subject to lord or attached to manor. □ **villeinage** *noun*.

vim *noun colloquial* vigour, energy.

vinaigrette /vɪnɪˈgret/ *noun* salad dressing of oil and wine vinegar.

vindicate /ˈvɪndɪkeɪt/ *verb* (**-ting**) clear of suspicion; establish merits, existence, or justice of. □ **vindication** *noun*; **vindicator** *noun*; **vindicatory** *adjective*.

vindictive /vɪnˈdɪktɪv/ *adjective* tending to seek revenge. □ **vindictively** *adverb*; **vindictiveness** *noun*.

vine *noun* trailing or climbing woody-stemmed plant, esp. bearing grapes.

vinegar /ˈvɪnɪgə/ *noun* sour liquid produced by fermentation of wine, malt, cider, etc. □ **vinegary** *adjective*.

vineyard /ˈvɪnjɑːd/ *noun* plantation of grapevines, esp. for wine-making.

vingt-et-un /væ̃teɪ'œ̃/ *noun* = PONTOON[1]. [French]

vino /vi:nəʊ/ *noun slang* wine, esp. of inferior kind.

vinous /'vaɪnəs/ *adjective* of, like, or due to wine.

vintage /'vɪntɪdʒ/ ● *noun* season's produce of grapes, wine from this; grape-harvest, season of this; wine of high quality from particular year and district; year etc. when thing was made, thing made etc. in particular year etc. ● *adjective* of high or peak quality; of a past season. □ **vintage car** car made 1917–1930.

vintner /'vɪntnə/ *noun* wine merchant.

vinyl /'vaɪnɪl/ *noun* any of group of plastics made by polymerization.

viol /'vaɪəl/ *noun* medieval stringed instrument similar in shape to violin.

viola[1] /vɪ'əʊlə/ *noun* instrument like violin but larger and of lower pitch.

viola[2] /'vaɪələ/ *noun* any plant of genus including violet and pansy, esp. cultivated hybrid.

viola da gamba /vɪəʊlə də 'gæmbə/ *noun* viol held between player's legs.

violate /'vaɪəleɪt/ *verb* (**-ting**) disregard, break (oath, law, etc.); treat profanely; break in on, disturb; rape. □ **violation** *noun*; **violator** *noun*.

violence /'vaɪələns/ *noun* being violent; violent conduct or treatment; unlawful use of force. □ **do violence to** act contrary to, outrage.

violent /'vaɪələnt/ *adjective* involving great physical force; intense, vehement; (of death) resulting from violence or poison. □ **violently** *adverb*.

violet /'vaɪələt/ ● *noun* plant with usually purple, blue, or white flowers; bluish-purple colour at opposite end of spectrum from red; paint, clothes, or material of this colour. ● *adjective* of this colour.

violin /vaɪə'lɪn/ *noun* high-pitched instrument with 4 strings played with bow. □ **violinist** *noun*.

violoncello /vaɪələn'tʃeləʊ/ *noun* (*plural* **-s**) *formal* cello.

VIP *abbreviation* very important person.

viper /'vaɪpə/ *noun* small venomous snake; malignant or treacherous person.

virago /vɪ'rɑ:gəʊ/ *noun* (*plural* **-s**) fierce or abusive woman.

viral /'vaɪər(ə)l/ *adjective* of or caused by virus.

virgin /'vɜ:dʒɪn/ ● *noun* person who has never had sexual intercourse; (**the Virgin**) Christ's mother Mary. ● *adjective* not yet used etc.; virginal. □ **the Virgin birth** doctrine of Christ's birth from virgin mother. □ **virginity** /və'dʒɪn-/ *noun*.

virginal ● *adjective* of or befitting a virgin. ● *noun* (usually in *plural*) legless spinet in box.

Virginia creeper /və'dʒɪnɪə/ *noun* vine cultivated for ornament.

Virgo /'vɜ:gəʊ/ *noun* sixth sign of zodiac.

virile /'vɪraɪl/ *adjective* having masculine vigour or strength; sexually potent; of man as distinct from woman or child. □ **virility** /-'rɪl-/ *noun*.

virology /vaɪ'rɒlədʒɪ/ *noun* study of viruses.

virtual /'vɜ:tʃʊəl/ *adjective* being so for practical purposes though not strictly or in name. □ **virtual reality** computer-generated images, sounds, etc. that appear real to the senses. □ **virtually** *adverb*.

virtue /'vɜ:tʃu:/ *noun* moral goodness; particular form of this; chastity, esp. of woman; good quality; efficacy. □ **by** or **in virtue of** on account of, because of.

virtuoso /vɜ:tʃʊ'əʊsəʊ/ *noun* (*plural* **-si** /-si:/ or **-sos**) highly skilled artist, esp. musician. □ **virtuosity** /-'ɒs-/ *noun*.

virtuous /'vɜ:tʃʊəs/ *adjective* morally good; *archaic* chaste. □ **virtuously** *adverb*.

virulent /'vɪrʊlənt/ *adjective* poisonous; (of disease) violent; bitterly hostile. □ **virulence** *noun*; **virulently** *adverb*.

virus /'vaɪərəs/ *noun* microscopic organism able to cause diseases; computer virus.

visa /'vi:zə/ *noun* endorsement on passport etc., esp. allowing holder to enter or leave country.

visage /'vɪzɪdʒ/ *noun literary* face.

vis-à-vis /vi:zə'vi:/ ● *preposition* in relation to; in comparison with. ● *adverb* opposite. [French]

viscera /'vɪsərə/ *plural noun* internal organs of body. □ **visceral** *adjective*.

viscid /'vɪsɪd/ *adjective* glutinous, sticky.

viscose /ˈvɪskəʊz/ *noun* viscous form of cellulose used in making rayon etc.; fabric made from this.

viscount /ˈvaɪkaʊnt/ *noun* British nobleman ranking between earl and baron.

viscountess /ˈvaɪkaʊntɪs/ *noun* viscount's wife or widow; woman holding rank of viscount.

viscous /ˈvɪskəs/ *adjective* glutinous, sticky; semifluid; not flowing freely. □ **viscosity** /-kɒs-/ *noun* (*plural* **-ies**).

visibility /vɪzɪˈbɪlɪtɪ/ *noun* being visible; range or possibility of vision as determined by light and weather.

visible /ˈvɪzɪb(ə)l/ *adjective* able to be seen, perceived, or discovered; in sight; apparent, open, obvious. □ **visibly** *adverb.*

vision /ˈvɪʒ(ə)n/ *noun* act or faculty of seeing, sight; thing or person seen in dream or trance; thing seen in imagination; imaginative insight; foresight, good judgement in planning; beautiful person etc.; TV or cinema picture, esp. of specified quality.

visionary ● *adjective* given to seeing visions or to fanciful theories; having vision or foresight; not real, imaginary; unpractical. ● *noun* (*plural* **-ies**) visionary person.

visit /ˈvɪzɪt/ ● *verb* (**-t-**) go or come to see (person, place, etc.); stay temporarily with or at; (of disease, calamity, etc.) attack; (often + *upon*) inflict punishment for (sin). ● *noun* act of visiting, temporary stay with person or at place; (+ *to*) occasion of going to doctor etc.; formal or official call.

visitant /ˈvɪzɪt(ə)nt/ *noun* visitor, esp. ghost etc.

visitation /vɪzɪˈteɪʃ(ə)n/ *noun* official visit of inspection; trouble etc. seen as divine punishment.

visitor *noun* person who visits; migrant bird.

visor /ˈvaɪzə/ *noun* movable part of helmet covering face; shield for eyes, esp. one at top of vehicle windscreen.

vista /ˈvɪstə/ *noun* view, esp. through avenue of trees or other long narrow opening; mental view of long succession of events.

visual /ˈvɪʒʊəl/ *adjective* of or used in seeing. □ **visual display unit** device displaying data of computer on screen. □ **visually** *adverb.*

visualize *verb* (also **-ise**) (**-zing** or **-sing**) imagine visually. □ **visualization** *noun.*

vital /ˈvaɪt(ə)l/ ● *adjective* of or essential to organic life; essential to existence, success, etc.; full of life or activity; fatal. ● *noun* (in *plural*) vital organs, e.g. lungs and heart. □ **vital statistics** those relating to number of births, marriages, deaths, etc., *jocular* measurements of woman's bust, waist, and hips. □ **vitally** *adverb.*

vitality /vaɪˈtælɪtɪ/ *noun* animation, liveliness; ability to survive or endure.

vitalize *verb* (also **-ise**) (**-zing** or **-sing**) endow with life; make lively or vigorous. □ **vitalization** *noun.*

vitamin /ˈvɪtəmɪn/ *noun* any of various substances present in many foods and essential to health and growth.

vitaminize *verb* (also **-ise**) (**-zing** or **-sing**) introduce vitamins into (food).

vitiate /ˈvɪʃɪeɪt/ *verb* (**-ting**) impair, debase; make invalid or ineffectual.

viticulture /ˈvɪtɪkʌltʃə/ *noun* cultivation of grapes.

vitreous /ˈvɪtrɪəs/ *adjective* of or like glass.

vitrify /ˈvɪtrɪfaɪ/ *verb* (**-ies**, **-ied**) change into glass or glassy substance, esp. by heat. □ **vitrification** *noun.*

vitriol /ˈvɪtrɪəl/ *noun* sulphuric acid or sulphate; caustic speech or criticism. □ **vitriolic** /-ˈɒl-/ *adjective.*

vitro see IN VITRO.

vituperate /vaɪˈtju:pəreɪt/ *verb* (**-ting**) criticize abusively. □ **vituperation** *noun*; **vituperative** /-rətɪv/ *adjective.*

viva[1] /ˈvaɪvə/ *colloquial* ● *noun* (*plural* **-s**) viva voce. ● *verb* (**vivas**, **vivaed**, **vivaing**) viva-voce.

viva[2] /ˈvi:və/ ● *interjection* long live. ● *noun* cry of this as salute etc. [Italian]

vivacious /vɪˈveɪʃəs/ *adjective* lively, animated. □ **vivacity** /vɪˈvæsɪtɪ/ *noun.*

vivarium /vaɪˈveərɪəm/ *noun* (*plural* **-ria** or **-s**) glass bowl etc. for keeping animals for scientific study; place for keeping animals in (nearly) their natural conditions.

viva voce /vaɪvə ˈvəʊtʃɪ/ ● *adjective* oral. ● *adverb* orally. ● *noun* oral exam.

viva-voce *verb* (**-vocees**, **-voceed**, **-voceing**) examine orally.

vivid /'vɪvɪd/ *adjective* (of light or colour) bright, strong, intense; (of memory, description, etc.) lively, incisive, graphic. □ **vividly** *adverb*; **vividness** *noun*.

vivify /'vɪvɪfaɪ/ *verb* (**-ies**, **-ied**) give life to, animate.

viviparous /vɪ'vɪpərəs/ *adjective* bringing forth young alive.

vivisect /'vɪvɪsekt/ *verb* perform vivisection on.

vivisection /vɪvɪ'sekʃ(ə)n/ *noun* surgical experimentation on living animals for scientific research. □ **vivisectionist** *noun*.

vixen /'vɪks(ə)n/ *noun* female fox; spiteful woman.

viz. *abbreviation* videlicet.

vizier /vɪ'zɪə/ *noun historical* high official in some Muslim countries.

V-neck *noun* V-shaped neckline on pullover etc.

vocabulary /və'kæbjʊlərɪ/ *noun* (*plural* **-ies**) words used by language, book, branch of science, or author; list of these; person's range of language.

vocal /'vəʊk(ə)l/ *adjective* of or uttered by voice; speaking one's feelings freely. □ **vocal cords** voice-producing part of larynx. □ **vocally** *adverb*.

vocalist *noun* singer.

vocalize *verb* (also **-ise**) (**-zing** or **-sing**) form (sound) or utter (word) with voice; articulate, express. □ **vocalization** *noun*.

vocation /vəʊ'keɪʃ(ə)n/ *noun* divine call to, or sense of suitability for, career or occupation; employment, trade, profession. □ **vocational** *adjective*.

vocative /'vɒkətɪv/ • *noun* case of noun used in addressing person or thing. • *adjective* of or in this case.

vociferate /və'sɪfəreɪt/ *verb* (**-ting**) utter noisily; shout, bawl. □ **vociferation** *noun*.

vociferous /və'sɪfərəs/ *adjective* noisy, clamorous; loud and insistent in speech. □ **vociferously** *adverb*.

vodka /'vɒdkə/ *noun* alcoholic spirit distilled esp. in Russia from rye etc.

vogue /vəʊg/ *noun* (**the vogue**) prevailing fashion; popular use. □ **in vogue** in fashion. □ **voguish** *adjective*.

voice • *noun* sound formed in larynx and uttered by mouth, esp. in speaking, singing, etc.; ability to produce this; use of voice, spoken or written expression, opinion so expressed, right to express opinion; *Grammar* set of verbal forms showing whether verb is active or passive. • *verb* (**-cing**) express; (esp. as **voiced** *adjective*) utter with vibration of vocal cords. □ **voice-over** commentary in film by unseen speaker.

void • *adjective* empty, vacant; not valid or binding. • *noun* empty space, sense of loss. • *verb* invalidate; excrete.

voile /vɔɪl/ *noun* thin semi-transparent fabric.

vol. *abbreviation* volume.

volatile /'vɒlətaɪl/ *adjective* changeable in mood, flighty; unstable; evaporating rapidly. □ **volatility** /-'tɪl-/ *noun*.

vol-au-vent /'vɒləʊvɑ̃/ *noun* small round case of puff pastry with savoury filling.

volcanic /vɒl'kænɪk/ *adjective* of, like, or produced by volcano.

volcano /vɒl'keɪnəʊ/ *noun* (*plural* **-es**) mountain or hill from which lava, steam, etc. escape through earth's crust.

vole *noun* small plant-eating rodent.

volition /və'lɪʃ(ə)n/ *noun* act or power of willing. □ **of one's own volition** voluntarily.

volley /'vɒlɪ/ • *noun* (*plural* **-s**) simultaneous firing of a number of weapons; bullets etc. so fired; (usually + *of*) torrent (of abuse etc.); *Tennis, Football, etc.* playing of ball before it touches ground. • *verb* (**-eys**, **-eyed**) return or send by volley. □ **volleyball** game for two teams of 6 hitting large ball by hand over net.

volt /vəʊlt/ *noun* SI unit of electromotive force. □ **voltmeter** instrument measuring electric potential in volts.

voltage *noun* electromotive force expressed in volts.

volte-face /vɒlt'fɑːs/ *noun* (*plural* **voltes-face** same pronunciation) complete change of position in one's attitude or opinion.

voluble /'vɒljʊb(ə)l/ *adjective* speaking or spoken fluently or with continuous flow of words. □ **volubility** *noun*; **volubly** *adverb*.

volume /'vɒlju:m/ *noun* single book forming part or all of work; solid content, bulk; space occupied by gas or liquid; (+ *of*) amount or quantity of; quantity or power of sound; (+ *of*) moving mass of (water, smoke, etc.).

voluminous /və'lu:mɪnəs/ *adjective* (of drapery etc.) loose and ample; written or writing at great length.

voluntary /'vɒləntrɪ/ ● *adjective* done, acting, or given willingly; unpaid; (of institution) supported or built by charity; brought about by voluntary action; (of muscle, limb, etc.) controlled by will. ● *noun* (*plural* **-ies**) organ solo played before or after church service. □ **voluntarily** *adverb*.

volunteer /vɒlən'tɪə/ ● *noun* person who voluntarily undertakes task or enters military etc. service. ● *verb* (often + *to*) undertake or offer voluntarily; (often + *for*) be volunteer.

voluptuary /və'lʌptʃʊərɪ/ *noun* (*plural* **-ies**) person who seeks luxury and sensual pleasure.

voluptuous /və'lʌptʃʊəs/ *adjective* of, tending to, occupied with, or derived from, sensuous or sensual pleasure; (of woman) curvaceous and sexually desirable. □ **voluptuously** *adverb*.

vomit /'vɒmɪt/ ● *verb* (**-t-**) eject (contents of stomach) through mouth, be sick; (of volcano, chimney, etc.) eject violently, belch forth. ● *noun* matter vomited from stomach.

voodoo /'vu:du:/ ● *noun* religious witchcraft as practised esp. in W. Indies. ● *verb* (**-doos**, **-dooed**) affect by voodoo, bewitch.

voracious /və'reɪʃəs/ *adjective* greedy in eating, ravenous; very eager. □ **voraciously** *adverb*; **voracity** /-'ræs-/ *noun*.

vortex /'vɔ:teks/ *noun* (*plural* **-texes** or **-tices** /-tɪsi:z/) whirlpool, whirlwind; whirling motion or mass; thing viewed as destructive or devouring.

votary /'vəʊtərɪ/ *noun* (*plural* **-ies**; *feminine* **votaress**) (usually + *of*) person dedicated to service of god or cult; devotee of a person, occupation etc.

vote ● *noun* formal expression of choice or opinion by ballot, show of hands, etc.; (usually **the vote**) right to vote; opinion expressed by vote; votes given by or for particular group. ● *verb* (**-ting**) (often + *for*, *against*) give vote; enact etc. by majority of votes; *colloquial* pronounce by general consent; (often + *that*) suggest, urge. □ **vote down** defeat (proposal etc.) by voting; **vote in** elect by voting; **vote with one's feet** *colloquial* indicate opinion by one's presence or absence.

voter *noun* person voting or entitled to vote.

votive /'vəʊtɪv/ *adjective* given or consecrated in fulfilment of vow.

vouch /vaʊtʃ/ *verb* (+ *for*) answer or be surety for.

voucher *noun* document exchangeable for goods or services; receipt.

vouchsafe /vaʊtʃ'seɪf/ *verb* (**-fing**) *formal* condescend to grant; (+ *to do*) condescend.

vow /vaʊ/ ● *noun* solemn, esp. religious, promise. ● *verb* promise solemnly; *archaic* declare solemnly.

vowel /'vaʊəl/ *noun* speech sound made by vibrations of vocal cords, but without audible friction; letter(s) representing this.

■ **Usage** The (written) vowels of English are customarily said to be *a*, *e*, *i*, *o*, and *u*, but *y* can be either a consonant (as in *yet*) or a vowel (as in *by*), and combinations of these six, such as *ee* in *keep*, *ie* in *tied*, *ou* in *pour*, and *ye* in *rye*, are just as much vowels.

---

vox pop *noun colloquial* popular opinion as represented by informal comments.

vox populi /vɒks 'pɒpjʊlaɪ/ *noun* public opinion, popular belief. [Latin]

voyage /'vɔɪɪdʒ/ ● *noun* journey, esp. long one by sea or in space. ● *verb* (**-ging**) make voyage. □ **voyager** *noun*.

voyeur /vwɑ:'jɜ:/ *noun* person who derives sexual pleasure from secretly observing others' sexual activity or organs; (esp. covert) spectator. □ **voyeurism** *noun*; **voyeuristic** /-'rɪs-/ *adjective*.

vs. *abbreviation* versus.

VSO *abbreviation* Voluntary Service Overseas.

VTOL /'vi:tɒl/ *abbreviation* vertical take-off and landing.

vulcanite /'vʌlkənaɪt/ *noun* hard black vulcanized rubber.

vulcanize /'vʌlkənaɪz/ *verb* (also **-ise**) (**-zing** or **-sing**) make (rubber etc.) stronger and more elastic by treating with sulphur at high temperature. □ **vulcanization** *noun.*

vulgar /'vʌlgə/ *adjective* coarse; of or characteristic of the common people; in common use, prevalent. □ **vulgar fraction** fraction expressed by numerator and denominator (e.g. $\frac{1}{2}$), not decimally (e.g. 0.5); **the vulgar tongue** native or vernacular language. □ **vulgarity** /-'gær-/ *noun* (*plural* **-ies**); **vulgarly** *adverb.*

vulgarian /vʌl'geərɪən/ *noun* vulgar (esp. rich) person.

vulgarism *noun* vulgar word or expression.

vulgarize /'vʌlgəraɪz/ *verb* (also **-ise**) (**-zing** or **-sing**) make vulgar; spoil by popularizing. □ **vulgarization** *noun.*

Vulgate /'vʌlgeɪt/ *noun* 4th-c. Latin version of Bible.

vulnerable /'vʌlnərəb(ə)l/ *adjective* easily wounded or harmed; (+ *to*) open to attack, injury, or criticism. □ **vulnerability** *noun.*

vulpine /'vʌlpaɪn/ *adjective* of or like fox; crafty, cunning.

vulture /'vʌltʃə/ *noun* large carrion-eating bird of prey; rapacious person.

vulva /'vʌlvə/ *noun* (*plural* **-s**) external female genitals.

vv. *abbreviation* verses.

vying *present participle* of VIE.

Ww

W *abbreviation* (also **W.**) watt(s); west(ern).

w. *abbreviation* wicket(s); wide(s); with.

wacky /'wækɪ/ *adjective* (**-ier**, **-iest**) *slang* crazy.

wad /wɒd/ ● *noun* lump of soft material to keep things apart or in place or to block hole; roll of banknotes. ● *verb* (**-dd-**) stop up or fix with wad; line, stuff, or protect with wadding.

wadding *noun* soft fibrous material for stuffing quilts, packing fragile articles in, etc.

waddle /'wɒd(ə)l/ ● *verb* (**-ling**) walk with short steps and swaying motion. ● *noun* such walk.

wade ● *verb* (**-ding**) walk through water, mud, etc., esp. with difficulty; (+ *through*) go through (tedious task, book, etc.); (+ *into*) *colloquial* attack (person, task). ● *noun* spell of wading. □ **wade in** *colloquial* make vigorous attack or intervention.

wader *noun* long-legged waterfowl; (in *plural*) high waterproof boots.

wadi /'wɒdɪ/ *noun* (*plural* **-s**) rocky watercourse in N. Africa etc., dry except in rainy season.

wafer /'weɪfə/ *noun* very thin light crisp biscuit; disc of unleavened bread used in Eucharist; disc of red paper stuck on legal document instead of seal. □ **wafer-thin** very thin.

waffle[1] /'wɒf(ə)l/ *colloquial* ● *noun* aimless verbose talk or writing. ● *verb* (**-ling**) indulge in waffle.

waffle[2] /'wɒf(ə)l/ *noun* small crisp batter cake. □ **waffle-iron** utensil for cooking waffles.

waft /wɒft/ ● *verb* convey or be conveyed smoothly (as) through air or over water. ● *noun* whiff.

wag[1] ● *verb* (**-gg-**) shake or wave to and fro. ● *noun* single wagging motion. □ **wagtail** small bird with long tail.

wag[2] *noun* facetious person.

wage ● *noun* (in *singular* or *plural*) employee's regular pay, esp. paid weekly. ● *verb* (**-ging**) carry on (war etc.).

waged *adjective* in regular paid employment.

wager /'weɪdʒə/ *noun & verb* bet.

waggish *adjective* playful, facetious. □ **waggishly** *adverb.*

waggle /ˈwæg(ə)l/ *verb* (**-ling**) *colloquial* wag.

wagon /ˈwægən/ *noun* (also **waggon**) 4-wheeled vehicle for heavy loads; open railway truck. □ **on the wagon** *slang* abstaining from alcohol; **wagon-load** as much as wagon can carry.

wagoner *noun* (also **waggoner**) driver of wagon.

waif *noun* homeless and helpless person, esp. abandoned child; ownerless object or animal.

wail ● *noun* prolonged plaintive inarticulate cry of pain, grief, etc.; sound resembling this. ● *verb* utter wail; lament or complain persistently.

wain *noun archaic* wagon.

wainscot /ˈweɪnskət/ *noun* (also **wainscoting**) boarding or wooden panelling on room-wall.

waist *noun* part of human body between ribs and hips; narrowness marking this; circumference of waist; narrow middle part of anything; part of garment encircling waist; *US* bodice, blouse; part of ship between forecastle and quarterdeck. □ **waistband** strip of cloth forming waist of garment; **waistcoat** usually sleeveless and collarless waist-length garment; **waistline** outline or size of waist.

wait ● *verb* defer action until expected event occurs; await (turn etc.); (of thing) remain in readiness; (usually as **waiting** *noun*) park briefly; act as waiter or attendant; (+ *on*, *upon*) await convenience of, be attendant to. ● *noun* act or period of waiting; (usually + *for*) watching for enemy; (in *plural*) *archaic* street singers of Christmas carols. □ **waiting-list** list of people waiting for thing not immediately available; **waiting-room** room for people to wait in, esp. at surgery or railway station.

waiter *noun* (*feminine* **waitress**) person who serves at hotel or restaurant tables.

waive *verb* (**-ving**) refrain from insisting on or using.

waiver *noun Law* (document recording) waiving.

wake[1] ● *verb* (**-king**; *past* **woke**; *past participle* **woken** /ˈwəʊk(ə)n/) (often + *up*) (cause to) cease to sleep or become alert; *archaic* (except as **waking** *adjective & noun*) be awake; disturb with noise; evoke. ● *noun* (chiefly in Ireland) vigil beside corpse before burial, attendant lamentations and merrymaking; (usually in *plural*) annual holiday in (industrial) N. England.

wake[2] *noun* track left on water's surface by moving ship etc.; turbulent air left by moving aircraft. □ **in the wake of** following, as result of.

wakeful *adjective* unable to sleep; sleepless; vigilant. □ **wakefully** *adverb*; **wakefulness** *noun*.

waken /ˈweɪkən/ *verb* make or become awake.

walk /wɔːk/ ● *verb* move by lifting and setting down each foot in turn, never having both feet off the ground at once; (of quadruped) go with slowest gait; travel or go on foot, take exercise thus; traverse (distance) in walking; tread floor or surface of; cause to walk with one. ● *noun* act of walking, ordinary human gait; slowest gait of animal; person's manner of walking; distance walkable in specified time; excursion on foot; place or track meant or fit for walking. □ **walkabout** informal stroll by royal person etc., Australian Aboriginal's period of wandering; **walking frame** tubular metal frame to assist elderly or disabled people in walking; **walking stick** stick carried for support when walking; **walk off with** *colloquial* steal, win easily; **walk of life** one's occupation; **walk-on part** short or non-speaking dramatic role; **walk out** depart suddenly or angrily, stop work in protest; **walk-out** *noun*; **walk out on** desert; **walkover** easy victory; **walk the streets** be prostitute; **walkway** passage or path for walking along. □ **walkable** *adjective*.

walker *noun* person etc. that walks; framework in which baby can walk unaided; walking frame.

walkie-talkie /wɔːkɪˈtɔːkɪ/ *noun* portable two-way radio.

Walkman /ˈwɔːkmən/ *noun* (*plural* **-s**) *proprietary term* type of personal stereo.

wall /wɔːl/ ● *noun* continuous narrow upright structure of usually brick or stone, esp. enclosing or dividing a space or supporting a roof; thing like wall in appearance or effect; outermost layer of animal or plant organ, cell, etc. ● *verb* (esp. as **walled** *adjective*) surround with wall; (usually + *up, off*) block with wall; (+ *up*) enclose within sealed space. □ **go to the wall** fare badly in competition; **up the wall** *colloquial* crazy, furious; **wallflower** fragrant garden plant, *colloquial* woman not dancing because partnerless; **wall game** Eton form of football; **wallpaper** *noun* paper for covering interior walls of rooms, *verb* decorate with wallpaper; **wall-to-wall** fitted to cover whole floor, *colloquial* ubiquitous.

wallaby /'wɒləbɪ/ *noun* (*plural* **-ies**) small kangaroo-like marsupial.

wallah /'wɒlə/ *noun slang* person connected with a specified occupation or thing.

wallet /'wɒlɪt/ *noun* small flat case for holding banknotes etc.

wall-eye *noun* eye with whitish iris or outward squint. □ **wall-eyed** *adjective*.

wallop /'wɒləp/ *colloquial* ● *verb* (**-p-**) thrash, beat. ● *noun* whack; beer.

wallow /'wɒləʊ/ ● *verb* roll about in mud etc.; (+ *in*) indulge unrestrainedly in. ● *noun* act of wallowing; place where animals wallow.

wally /'wɒlɪ/ *noun* (*plural* **-ies**) *slang* foolish or incompetent person.

walnut /'wɔ:lnʌt/ *noun* tree with aromatic leaves and drooping catkins; its nut; its timber.

walrus /'wɔ:lrəs/ *noun* (*plural* same or **walruses**) long-tusked amphibious arctic mammal. □ **walrus moustache** long thick drooping moustache.

waltz /wɔ:ls/ ● *noun* ballroom dance in triple time; music for this. ● *verb* dance waltz; (often + *in, out, round*, etc.) *colloquial* move easily, casually, etc.

wampum /'wɒmpəm/ *noun* strings of shell-beads formerly used by N. American Indians for money, ornament, etc.

wan /wɒn/ *adjective* (**-nn-**) pale, weary-looking. □ **wanly** *adverb*.

wand /wɒnd/ *noun* fairy's or magician's magic stick; staff as sign of office etc.; *colloquial Music* conductor's baton.

wander /'wɒndə/ *verb* (often + *in, off*, etc.) go from place to place aimlessly; meander; diverge from path etc.; talk or think incoherently, be inattentive or delirious. □ **wanderlust** eagerness to travel or wander, restlessness. □ **wanderer** *noun*.

wane ● *verb* (**-ning**) (of moon) decrease in apparent size; decrease in power, vigour, importance, size, etc. ● *noun* process of waning. □ **on the wane** declining.

wangle /'wæŋg(ə)l/ *colloquial* ● *verb* (**-ling**) contrive to obtain (favour etc.). ● *noun* act of wangling.

wannabe /'wɒnəbɪ/ *noun slang* avid fan who apes person admired; anyone wishing to be someone else.

want /wɒnt/ ● *verb* (often + *to do*) desire; wish for possession of; need; (+ *to do*) *colloquial* should; (usually + *for*) lack; be without or fall short by; (as **wanted** *adjective*) (of suspected criminal etc.) sought by police. ● *noun* lack, deficiency; poverty, need.

wanting *adjective* lacking (in quality or quantity), unequal to requirements; absent.

wanton /'wɒnt(ə)n/ ● *adjective* licentious; capricious, arbitrary; luxuriant, wild. ● *noun literary* licentious person. □ **wantonly** *adverb*.

wapiti /'wɒpɪtɪ/ *noun* (*plural* **-s**) large N. American deer.

war /wɔ:/ ● *noun* armed hostility, esp. between nations; specific period of this; hostility between people; (often + *on*) efforts against crime, poverty, etc. ● *verb* (**-rr-**) (as **warring** *adjective*) rival, fighting; make war. □ **at war** engaged in war; **go to war** begin war; **on the warpath** going to war, *colloquial* seeking confrontation; **war crime** crime violating international laws of war; **war cry** phrase or name shouted to rally troops, party slogan; **war dance** dance performed by primitive peoples before battle or after victory; **warhead** explosive head of missile; **warhorse** *historical* trooper's horse, *colloquial* veteran soldier; **war memorial** monument to those killed in (a) war; **warmonger** /-mʌŋgə/ person who promotes war; **warpaint** paint put on body esp. by N. American Indians before battle, *colloquial* make-up; **warship** ship used in war.

warble /'wɔ:b(ə)l/ ● *verb* (**-ling**) sing in a gentle trilling way. ● *noun* warbling sound.

warbler *noun* bird that warbles.

ward /wɔ:d/ *noun* separate division or room of hospital etc.; administrative division, esp. for elections; minor etc. under care of guardian or court; (in *plural*) corresponding notches and projections in key and lock; *archaic* guardianship. □ **ward off** parry (blow), avert (danger etc.); **wardroom** officers' mess in warship.

warden /'wɔ:d(ə)n/ *noun* supervising official; president or governor of institution; traffic warden.

warder /'wɔ:də/ *noun* (*feminine* **wardress**) prison officer.

wardrobe /'wɔ:drəʊb/ *noun* large cupboard for storing clothes; stock of clothes; theatre's costume department. □ **wardrobe master, mistress** person in charge of theatrical wardrobe.

wardship *noun* tutelage.

ware *noun* things of specified kind made usually for sale; (usually in *plural*) articles for sale.

warehouse ● *noun* building in which goods are stored; wholesale or large retail store. ● *verb* (**-sing**) store in warehouse.

warfare /'wɔ:feə/ *noun* waging war, campaigning.

warlike *adjective* hostile; soldierly; military.

warlock /'wɔ:lɒk/ *noun archaic* sorcerer.

warm /wɔ:m/ ● *adjective* of or at fairly high temperature; (of person) with skin at natural or slightly raised temperature; (of clothes) affording warmth; hearty, enthusiastic; sympathetic, friendly, loving; *colloquial* dangerous, hostile; *colloquial* (in game) near object sought, near to guessing; (of colour) reddish or yellowish, suggesting warmth; (of scent in hunting) fresh and strong. ● *verb* make or become warm. ● *noun* act of warming; warmth. □ **warm-blooded** (of animals) having blood temperature well above that of environment; **warm-hearted** kind, friendly; **warming-pan** *historical* flat closed vessel holding hot coals for warming beds; **warm up** make or become warm, prepare for performance etc. by practising, reach temperature for efficient working, reheat (food); **warm-up** *noun*. □ **warmly** *adverb*; **warmth** *noun*.

warn /wɔ:n/ *verb* (often + *of*, *that*) inform of impending danger or misfortune; (+ *to do*) advise (person) to take certain action; (often + *against*) inform (person) about specific danger. □ **warn off** tell (person) to keep away (from).

warning *noun* what is said or done or occurs to warn person.

warp /wɔ:p/ ● *verb* make or become distorted, esp. through heat, damp, etc.; make or become perverted or strange; haul (ship etc.) by rope attached to fixed point. ● *noun* warped state; mental perversion; lengthwise threads in loom; rope used in warping ship.

warrant /'wɒrənt/ ● *noun* thing that authorizes an action; written authorization, money voucher, etc.; written authorization allowing police to carry out search or arrest; certificate of service rank held by warrant officer. ● *verb* serve as warrant for, justify; guarantee. □ **warrant officer** officer ranking between commissioned and non-commissioned officers.

warranty /'wɒrəntɪ/ *noun* (*plural* **-ies**) undertaking as to ownership or quality of thing sold etc., often accepting responsibility for repairs needed over specified period; authority, justification.

warren /'wɒrən/ *noun* network of rabbit burrows; densely populated or labyrinthine building or district.

warrior /'wɒrɪə/ *noun* person skilled in or famed for fighting.

wart /wɔ:t/ *noun* small round dry growth on skin; protuberance on skin of animal, surface of plant, etc. □ **wart-hog** African wild pig. □ **warty** *adjective*.

wary /'weərɪ/ *adjective* (**-ier**, **-iest**) on one's guard, circumspect; cautious. □ **warily** *adverb*; **wariness** *noun*.

was *1st & 3rd singular past* of BE.

wash /wɒʃ/ ● *verb* cleanse with liquid; (+ *out*, *off*, *away*, etc.) remove or be removed by washing; wash oneself or one's hands (and face); wash clothes, dishes, etc.; (of fabric or dye) bear washing without damage; bear scrutiny, be believed or acceptable; (of river etc.) touch; (of liquid) carry along in specified direction; sweep, move, splash; (+ *over*) occur without affecting (person); sift (ore) by action of water; brush watery colour over; *poetical* moisten. ● *noun* washing, being washed; clothes for washing or just washed; motion of agitated water or air, esp. due to passage of vessel or aircraft; kitchen slops given to pigs; thin, weak, inferior, or animals' liquid food; liquid to spread over surface to cleanse, heal, or colour. □ **washbasin** basin for washing one's hands etc.; **washboard** ribbed board for washing clothes, this as percussion instrument; **wash down** wash completely, (usually + *with*) accompany or follow (food); **washed out** faded, pale, *colloquial* exhausted; **washed up** *esp. US slang* defeated, having failed; **wash one's hands of** decline responsibility for; **wash out** clean inside of by washing, *colloquial* cause to be cancelled because of rain; **wash-out** *colloquial* complete failure; **washroom** *esp. US* public toilet; **washstand** piece of furniture for holding washbasin, soap-dish, etc.; **wash up** wash (dishes etc.) after use, *US* wash one's face and hands, (of sea) carry on to shore. □ **washable** *adjective*.

washer *noun* person or thing that washes; flat ring placed between two surfaces or under plunger of tap, nut, etc. to tighten joint or disperse pressure.

washerwoman *noun* laundress.

washing *noun* clothes etc. for washing or just washed. □ **washing machine** machine for washing clothes; **washing powder** soap powder or detergent for washing clothes; **washing-up** washing of dishes etc., dishes etc. for washing.

washy *adjective* (**-ier**, **-iest**) too watery or weak; lacking vigour.

wasn't /'wɒz(ə)nt/ was not.

Wasp /wɒsp/ *noun* (also **WASP**) *US usually derogatory* middle-class white American [Anglo-Saxon] Protestant.

wasp /wɒsp/ *noun* stinging insect with black and yellow stripes. □ **wasp-waist** very slender waist.

waspish *adjective* irritable, snappish.

wassail /'wɒseɪl/ *archaic* ● *noun* festive drinking. ● *verb* make merry.

wastage /'weɪstɪdʒ/ *noun* amount wasted; loss by use, wear, or leakage; (also **natural wastage**) loss of employees other than by redundancy.

waste ● *verb* (**-ting**) use to no purpose or for inadequate result or extravagantly; fail to use; (often + *on*) give (advice etc.) without effect; (in *passive*) fail to be appreciated or used properly; wear away; make or become weak; devastate. ● *adjective* superfluous, no longer needed; uninhabited, not cultivated. ● *noun* act of wasting; waste material; waste region; diminution by wear; waste pipe. □ **go, run to waste** be wasted; **wasteland** land not productive or developed, spiritually or intellectually barren place or time; **waste paper** used or valueless paper; **waste pipe** pipe carrying off waste material; **waste product** useless by-product of manufacture or organism.

wasteful *adjective* extravagant; causing or showing waste. □ **wastefully** *adverb*.

waster *noun* wasteful person; *colloquial* wastrel.

wastrel /'weɪstr(ə)l/ *noun* good-for-nothing person.

watch /wɒtʃ/ ● *verb* keep eyes fixed on; keep under observation, follow observantly; (often + *for*) be in alert state, be vigilant; (+ *over*) look after, take care of. ● *noun* small portable timepiece for wrist or pocket; state of alert or constant attention; *Nautical* usually 4-hour spell of duty; *historical* (member of) body of men patrolling streets at night. □ **on the watch for** waiting for (anticipated event); **watchdog** dog guarding property, person etc. monitoring others' rights etc.; **watching brief** brief of barrister who follows case for client not directly concerned; **watchman** man employed to look after empty building etc. at night; **watch out** (often + *for*) be on one's guard; **watch-tower** tower for observing prisoners, attackers, etc.; **watchword** phrase summarizing guiding principle. □ **watcher** *noun* (also *in combination*).

watchful *adjective* accustomed to watching; on the watch. □ **watchfully** *adverb*; **watchfulness** *noun*.

water /ˈwɔːtə/ ● *noun* transparent colourless liquid found in seas and rivers etc. and in rain etc.; sheet or body of water; (in *plural*) part of sea or river; (often **the waters**) mineral water at spa etc.; state of tide; solution of specified substance in water; transparency and lustre of gem; (usually in *plural*) amniotic fluid. ● *verb* sprinkle or soak with water; supply (plant or animal) with water; secrete water; (as **watered** *adjective*) (of silk etc.) having irregular wavy finish; take in supply of water. □ **make water** urinate; **water-bed** mattress filled with water; **water biscuit** thin unsweetened biscuit; **water buffalo** common domestic Indian buffalo; **water cannon** device giving powerful water-jet to disperse crowd etc.; **water chestnut** corm from a sedge, used in Chinese cookery; **water-closet** lavatory that can be flushed; **water colour** pigment mixed with water and not oil, picture painted or art of painting with this; **watercourse** stream of water, bed of this; **watercress** pungent cress growing in running water; **water-diviner** dowser; **water down** dilute, make less forceful or horrifying; **waterfall** stream or river falling over precipice or down steep hill; **waterfowl** bird(s) frequenting water; **waterfront** part of town adjoining river etc.; **waterhole** shallow depression in which water collects; **water-ice** flavoured and frozen water and sugar; **watering-can** portable container for watering plants; **watering-place** pool where animals drink, spa or seaside resort; **water jump** jump over water in steeplechase etc.; **water level** surface of water, height of this, water table; **water lily** aquatic plant with floating leaves and flowers; **waterline** line where surface of water touches ship's side; **waterlogged** saturated or filled with water; **water main** main pipe in water supply system; **waterman** boatman plying for hire; **watermark** faint design made in paper by maker; **water-meadow** meadow periodically flooded by stream; **water melon** large dark green melon with red pulp and watery juice; **water-mill** mill worked by water-wheel; **water pistol** toy pistol shooting jet of water; **water polo** game played by swimmers with ball like football; **water-power** mechanical force from weight or motion of water; **waterproof** *adjective* impervious to water, *noun* such garment or material, *verb* make waterproof; **water-rat** water vole; **water rate** charge for use of public water supply; **watershed** line between waters flowing to different river basins, turning point in events; **waterside** edge of sea, lake, or river; **water-ski** ski (esp. one of pair) on which person is towed across water by motor boat; **waterspout** gyrating column of water and spray between sea and cloud; **water table** plane below which ground is saturated with water **watertight** so closely fastened or fitted that water cannot leak through, (of argument etc.) unassailable; **water tower** tower with elevated tank to give pressure for distributing water; **water vole** aquatic vole; **waterway** navigable channel; **waterwheel** wheel driven by water to drive machinery or to raise water; **water-wings** inflated floats used to support person learning to swim; **waterworks** establishment for managing water supply, *colloquial* shedding of tears, *colloquial* urinary system.

watery *adjective* containing too much water; too thin in consistency; of or consisting of water; vapid, uninteresting; (of colour) pale; (of sun, moon, or sky) rainy-looking; (of eyes) moist.

watt /wɒt/ *noun* SI unit of power.

wattage *noun* amount of electrical power expressed in watts.

wattle[1] /ˈwɒt(ə)l/ *noun* interlaced rods and sticks used for fences etc.; Australian acacia with fragrant golden yellow flowers. □ **wattle and daub** network of rods and twigs plastered with clay or mud as building material.

wattle[2] /ˈwɒt(ə)l/ *noun* fleshy appendage hanging from head or neck of turkey etc.

wave ● *verb* (**-ving**) move (hand etc.) to and fro in greeting or as signal; show sinuous or sweeping motion; give such motion to; direct (person) or express (greeting etc.) by waving; give undulating form to, have such form. ● *noun* moving ridge of water between two depressions; long body of water curling into arch and breaking on shore; thing compared to this; gesture of waving; curved shape in hair; temporary occurrence or heightening of condition or influence; disturbance carrying motion, heat, light, sound, etc. through esp. fluid medium; single curve in this. □ **wave aside** dismiss as intrusive or irrelevant; **waveband** radio wavelengths between certain limits; **wave down** wave to (vehicle or driver) to stop; **wavelength** distance between successive crests of wave, this as distinctive feature of radio waves from a transmitter, *colloquial* person's way of thinking.

wavelet *noun* small wave.

waver /'weɪvə/ *verb* be or become unsteady or irresolute, begin to give way.

wavy *adjective* (**-ier**, **-iest**) having waves or alternate contrary curves. □ **waviness** *noun*.

wax¹ ● *noun* sticky pliable yellowish substance secreted by bees as material of honeycomb; this bleached and purified for candles, modelling, etc.; any similar substance. ● *verb* cover or treat with wax; remove hair from (legs etc.) using wax. □ **waxwork** object, esp. lifelike dummy, modelled in wax, (in *plural*) exhibition of wax dummies.

wax² *verb* (of moon) increase in apparent size; grow larger or stronger; become.

waxen *adjective* smooth or pale like wax; *archaic* made of wax.

way ● *noun* road, track, path, street; course, route; direction; method, means; style, manner; habitual course of action; normal course of events; distance (to be) travelled; unimpeded opportunity or space to advance; advance, progress; specified condition or state; respect, sense. ● *adverb colloquial* far. □ **by the way** incidentally; **by way of** by means of, as a form of, passing through; **give way** yield under pressure, give precedence; **in the way, in (person's) way** forming obstruction (to); **lead the way** act as guide or leader; **make one's way** go, prosper; **make way for** allow to pass, be superseded by; **out of the way** not forming obstruction, disposed of, unusual, remote; **pay its** or **one's way** cover costs, pay one's expenses as they arise; **under way** in motion or progress; **way back** *colloquial* long ago; **wayfarer** traveller, esp. on foot; **waylay** lie in wait for, stop to accost or rob; **way of life** principles or habits governing one's actions; **way-out** *colloquial* unusual, eccentric; **wayside** (land at) side of road.

wayward /'weɪwəd/ *adjective* childishly self-willed; capricious. □ **waywardness** *noun*.

WC *abbreviation* water-closet; West Central.

W/Cdr. *abbreviation* Wing Commander.

we /wi:/ *pronoun used by speaker or writer to refer to himself or herself and one or more others as subject of verb; used for* I *by sovereign in formal contexts or by editorial writer in newspaper*.

weak *adjective* lacking in strength, power, vigour, resolution, or number; unconvincing; (of verb) forming inflections by suffix. □ **weak-kneed** *colloquial* lacking resolution; **weak-minded** mentally deficient, lacking resolution. □ **weaken** *verb*.

weakling *noun* feeble person or animal.

weakly ● *adverb* in a weak way. ● *adjective* (**-ier**, **-iest**) sickly, not robust.

weakness *noun* being weak; weak point; self-indulgent liking.

weal¹ ● *noun* ridge raised on flesh by stroke of rod or whip. ● *verb* raise weals on.

weal² *noun literary* welfare.

wealth /welθ/ *noun* riches; being rich; abundance.

wealthy *adjective* (**-ier**, **-iest**) having abundance, esp. of money.

wean *verb* accustom (infant or other young mammal) to food other than mother's milk; (often + *from, away from*) disengage (from habit etc.) by enforced discontinuance.

weapon /'wepən/ *noun* thing designed, used, or usable for inflicting bodily harm; means for gaining advantage in a conflict.

weaponry *noun* weapons collectively.

wear /weə/ ● *verb* (*past* **wore**; *past participle* **worn**) have on one's person as clothing, ornament, etc.; exhibit (expression etc.); *colloquial* (usually in negative) tolerate; (often + *away*, *down*) damage or deteriorate gradually by use or attrition; make (hole etc.) by attrition; (often + *out*) exhaust; (+ *down*) overcome by persistence; (+ *well* etc.) endure continued use or life; (of time) pass, esp. tediously. ● *noun* wearing, being worn; things worn; fashionable or suitable clothing; (in full **wear and tear**) damage from continuous use. □ **wear out** use or be used until useless, tire or be tired out. □ **wearer** *noun*.

wearisome /'wɪərɪsəm/ *adjective* tedious, monotonous.

weary /'wɪərɪ/ ● *adjective* (**-ier**, **-iest**) very tired, intensely fatigued; (+ *of*) tired of; tiring, tedious. ● *verb* (**-ies**, **-ied**) make or become weary. □ **wearily** *adverb*; **weariness** *noun*.

weasel /'wi:z(ə)l/ *noun* small ferocious reddish-brown flesh-eating mammal.

weather /'weðə/ ● *noun* atmospheric conditions at specified place or time as regards heat, cloudiness, humidity, sunshine, wind, and rain etc. ● *verb* expose to or affect by atmospheric changes, season (wood); be discoloured or worn thus; come safely through (storm etc.); get to windward of. □ **make heavy weather of** *colloquial* exaggerate difficulty of; **under the weather** *colloquial* indisposed; **weather-beaten** affected by exposure to weather; **weatherboard** sloping board attached at bottom of door to keep out rain, each of series of overlapping horizontal boards on wall; **weathercock** weather vane in form of cock, inconstant person; **weather forecast** prediction of likely weather; **weather vane** revolving pointer on church spire etc. to show direction of wind.

weave[1] ● *verb* (**-ving**; *past* **wove** /wəʊv/; *past participle* **woven**) form (fabric) by interlacing threads, form (threads) into fabric, esp. in loom; (+ *into*) make (facts etc.) into story or connected whole; make (story etc.) thus. ● *noun* style of weaving.

weave[2] *verb* (**-ving**) move repeatedly from side to side; take intricate course.

weaver *noun* person who weaves fabric; (in full **weaver-bird**) tropical bird building elaborately woven nest.

web *noun* woven fabric; amount woven in one piece; complex series; cobweb or similar tissue; membrane connecting toes of aquatic bird or other animal; large roll of paper for printing. □ **web-footed** having toes connected by web. □ **webbed** *adjective*.

webbing *noun* strong narrow closely woven fabric for belts etc.

weber /'veɪbə/ *noun* SI unit of magnetic flux.

Wed. *abbreviation* (also **Weds.**) Wednesday.

wed *verb* (**-dd-**; *past & past participle* **wedded** or **wed**) *usually formal or literary* marry; unite; (as **wedded** *adjective*) of or in marriage, (+ *to*) obstinately attached to (pursuit etc.).

we'd /wi:d/ we had; we should; we would.

wedding /'wedɪŋ/ *noun* marriage ceremony. □ **wedding breakfast** meal etc. between wedding and departure for honeymoon; **wedding cake** rich decorated cake served at wedding reception; **wedding ring** ring worn by married person.

wedge ● *noun* piece of tapering wood, metal, etc., used for forcing things apart or fixing them immovably etc.; wedge-shaped thing. ● *verb* (**-ging**) secure or force open or apart with wedge; (+ *in*, *into*) pack or force (thing, oneself) in or into. □ **thin end of the wedge** *colloquial* small beginning that may lead to something more serious.

wedlock /'wedlɒk/ *noun* married state. □ **born in** or **out of wedlock** born of married or unmarried parents.

Wednesday /'wenzdeɪ/ *noun* day of week following Tuesday.

wee *adjective* (**weer** /'wi:ə/, **weest** /'wi:ɪst/) *esp. Scottish* little; *colloquial* tiny.

weed ● *noun* wild plant growing where it is not wanted; lanky and weakly person or horse; (**the weed**) *slang* marijuana, tobacco. ● *verb* rid of weeds or unwanted parts; (+ *out*) sort out and remove (inferior or unwanted parts etc.), rid of inferior or unwanted parts etc.; remove or destroy weeds.

weeds *plural noun* (in full **widow's weeds**) *archaic* deep mourning worn by widow.

weedy *adjective* (**-ier**, **-iest**) weak, feeble; full of weeds.

week *noun* 7-day period reckoned usually from Saturday midnight; any 7-day period; the 6 days between Sundays; the 5 days Monday to Friday, period of work then done. □ **weekday** day other than (Saturday or) Sunday; **weekend** Saturday and Sunday.

weekly ● *adjective* done, produced, or occurring once a week. ● *adverb* once a week. ● *noun* (*plural* **-ies**) weekly newspaper or periodical.

weeny /'wi:nɪ/ *adjective* (**-ier**, **-iest**) *colloquial* tiny.

weep ● *verb* (*past & past participle* **wept**) shed tears; (often + *for*) lament over; be covered with or send forth drops, exude liquid; come or send forth in drops; (as **weeping** *adjective*) (of tree) have drooping branches. ● *noun* spell of weeping.

weepie *noun colloquial* sentimental or emotional film, play, etc.

weepy *adjective* (**-ier**, **-iest**) *colloquial* inclined to weep, tearful.

weevil /'wi:vɪl/ *noun* destructive beetle feeding esp. on grain.

weft *noun* threads woven across warp to make fabric.

weigh /weɪ/ *verb* find weight of; balance in hand (as if) to guess weight of; (often + *out*) take definite weight of (substance), measure out (specified weight); estimate relative value or importance of; (+ *with*, *against*) compare with; be of specified weight or importance; have influence; (often + *on*) be heavy or burdensome (to); raise (anchor). □ **weighbridge** weighing machine for vehicles; **weigh down** bring down by weight, oppress; **weigh in** (of boxer before contest, or jockey after race) be weighed; **weigh-in** *noun*; **weigh in with** *colloquial* advance (argument etc.) confidently; **weigh up** *colloquial* form estimate of; **weigh one's words** carefully choose words to express something.

weight /weɪt/ ● *noun* force on a body due to earth's gravitation; heaviness of body; quantitative expression of a body's weight, scale of such weights; body of known weight for use in weighing or weight training; heavy body, esp. used in mechanism etc.; load, burden; influence, importance; preponderance (of evidence etc.). ● *verb* attach a weight to, hold down with a weight; impede, burden. □ **pull one's weight** do fair share of work; **weightlifting** sport of lifting heavy objects; **weight training** physical training using weights. □ **weightless** *adjective*.

weighting *noun* extra pay in special cases.

weighty *adjective* (**-ier**, **-iest**) heavy; momentous; deserving attention; influential, authoritative.

weir /wɪə/ *noun* dam across river to retain water and regulate its flow.

weird /wɪəd/ *adjective* uncanny, supernatural; *colloquial* queer, incomprehensible. □ **weirdly** *adverb*; **weirdness** *noun*.

welch = WELSH.

welcome /'welkəm/ ● *noun* kind or glad greeting or reception. ● *interjection expressing such greeting*. ● *verb* (**-ming**) receive with welcome. ● *adjective* gladly received; (+ *to*) cordially allowed or invited to. □ **make welcome** receive hospitably.

weld ● *verb* join (pieces of metal or plastic) using heat, usually from electric arc; fashion into effectual or homogeneous whole. ● *noun* welded joint. □ **welder** *noun*.

welfare /'welfeə/ *noun* well-being, happiness; health and prosperity (of person, community, etc.); (**Welfare**) finanical support from state. □ **welfare state** system of social services controlled or financed by government, state operating this; **welfare work** organized effort for welfare of poor, disabled, etc.

welkin /'welkɪn/ *noun poetical* sky.

well[1] ● *adverb* (**better**, **best**) in satisfactory way; with distinction; in kind way; thoroughly, carefully; with heartiness or approval; probably, reasonably; to considerable extent; *slang* extremely. ● *adjective* (**better**, **best**) in good health; in satisfactory state or position, advisable. ● *interjection expressing astonishment, resignation, etc., or introducing speech*. □ **as well** in addition, advisable, desirable, reasonably; **as well as** in addition to; **well-adjusted** mentally and emotionally stable; **well-advised** prudent; **well and truly** decisively, completely; **well-appointed** properly equipped or fitted out; **well away** having made considerable progress, *colloquial* fast asleep or drunk; **well-balanced** sane, sensible; **well-being** hapiness, health, prosperity; **well-born** of noble family; **well-bred** having or showing good breeding or manners; **well-built** big, strong, and shapely; **well-connected** related to good families; **well-disposed** (often + *towards*) friendly, sympathetic; **well-earned** fully deserved; **well-founded** based on good evidence; **well-groomed** with carefully tended hair, clothes, etc.; **well-heeled** *colloquial* wealthy; **well-informed** having much knowledge or information; **well-intentioned** having or showing good intentions; **well-judged** opportunely, skilfully, or discreetly done; **well-known** known to many; **well-meaning**, **-meant** well-intentioned; **well-nigh** almost; **well off** rich, fortunately situated; **well-read** having read (and learnt) much; **well-spoken** articulate or refined in speech; **well-to-do** prosperous; **well-tried** often tested with good result; **well-wisher** person who wishes one well; **well-worn** much used, trite.

well² ● *noun* shaft sunk in ground to obtain water, oil, etc.; enclosed space resembling well-shaft, e.g. central space in building for staircase, lift, light, or ventilation; source; (in *plural*) spa; ink-well; railed space in law court. ● *verb* (+ *out*, *up*) rise or flow as water from well. □ **well-head**, **-spring** source.

we'll /wi:l/ we shall; we will.

wellington /'welɪŋt(ə)n/ *noun* (in full **wellington boot**) waterproof rubber boot usually reaching knee.

welly /'welɪ/ *noun* (*plural* **-ies**) *colloquial* wellington.

Welsh ● *adjective* of Wales. ● *noun* language of Wales; (**the Welsh**) (treated as *plural*) the Welsh people. □ **Welshman**, **Welshwoman** native of Wales; **Welsh rarebit** dish of melted cheese etc. on toast.

welsh *verb* (also **welch** /weltʃ/) (of loser of bet, esp. bookmaker) evade an obligation; (+ *on*) fail to carry out promise to (person), fail to honour (obligation).

welt ● *noun* leather rim sewn to shoe-upper for sole to be attached to; weal; ribbed or reinforced border of garment; heavy blow. ● *verb* provide with welt; raise weals on, thrash.

welter /'weltə/ ● *verb* roll, wallow; (+ *in*) be soaked in. ● *noun* general confusion; disorderly mixture.

welterweight /'weltəweɪt/ *noun* amateur boxing weight (63.5–67 kg).

wen *noun* benign tumour on skin.

wench *noun jocular* girl, young woman.

wend *verb* □ **wend one's way** go.

went *past* of GO¹.

wept *past & past participle* of WEEP.

were *2nd singular past, plural past,* and *past subjunctive* of BE.

we're /wɪə/ we are.

weren't /wɜ:nt/ were not.

werewolf /'weəwʊlf/ *noun* (*plural* **-wolves**) *Mythology* human being who changes into wolf.

Wesleyan /'wezlɪən/ ● *adjective* of Protestant denomination founded by John Wesley. ● *noun* member of this denomination.

west ● *noun* point of horizon where sun sets at equinoxes; corresponding compass point; (usually **the West**) European civilization, western part of world, country, town, etc. ● *adjective* towards, at, near, or facing west; (of wind) from west. ● *adverb* towards, at, or near west; (+ *of*) further west than. □ **go west** *slang* be killed or wrecked etc.; **westbound** travelling or leading west; **West End** fashionable part of London; **west-north-west**, **west-south-west** point midway between west and north-west or south-west. □ **westward** *adjective, adverb, & noun*; **westwards** *adverb*.

westering /'westərɪŋ/ *adjective* (of sun) nearing the west.

westerly /'westəlɪ/ ● *adjective & adverb* in western position or direction; (of wind) from west. ● *noun* (*plural* **-ies**) wind from west.

western /'west(ə)n/ ● *adjective* of or in west. ● *noun* film or novel about cowboys in western N. America. □ **westernize** *verb* (also **-ise**) (**-zing** or **-sing**); **westernmost** *adjective*.

westerner *noun* native or inhabitant of west.

wet ● *adjective* (**-tt-**) soaked or covered with water or other liquid; (of weather) rainy; (of paint) not yet dried; used with water; *colloquial* feeble. ● *verb* (**-tt-**; *past & past participle* **wet** or **wetted**) make wet; urinate in or on. ● *noun* liquid that wets something; rainy weather; *colloquial* feeble or spiritless person; *colloquial* liberal Conservative; *colloquial* drink. □ **wet blanket** gloomy person discouraging cheerfulness etc.; **wet-nurse** *noun* woman employed to suckle another's child, *verb* act as wet-nurse to, *colloquial* treat as if helpless.

wether /ˈweðə/ *noun* castrated ram.

we've /wi:v/ we have.

Wg. Cdr. *abbreviation* Wing Commander.

whack *colloquial* ● *verb* hit forcefully; (as **whacked** *adjective*) tired out. ● *noun* sharp or resounding blow; *slang* share.

whacking *colloquial* ● *adjective* large. ● *adverb* very.

whale ● *noun* (*plural* same or **-s**) large fishlike marine mammal. ● *verb* (**-ling**) hunt whales. □ **whalebone** elastic horny substance in upper jaw of some whales.

whaler *noun* whaling ship or seaman.

wham *interjection colloquial: expressing forcible impact.*

wharf /wɔ:f/ ● *noun* (*plural* **wharves** /wɔ:vz/ or **-s**) quayside structure for loading or unloading of moored vessels. ● *verb* moor (ship) at wharf; store (goods) on wharf.

what /wɒt/ ● *interrogative adjective used in asking someone to specify one or more things from an indefinite number.* ● *adjective* (usually in exclamation) how great, how remarkable. ● *relative adjective* the or any ... that. ● *interrogative pronoun* what thing(s); what did you say? ● *relative pronoun* the things which; anything that. □ **whatever** anything at all that, no matter what, (with negative or in questions) at all, of any kind; **what for?** *colloquial* for what reason?; **what have you** *colloquial* anything else similar; **whatnot** unspecified thing; **what not** *colloquial* other similar things; **whatsoever** whatever; **know what's what** *colloquial* have common sense, know what is useful or important; **what with** *colloquial* because of.

wheat *noun* cereal plant bearing dense 4-sided seed-spikes; its grain used for flour etc. □ **wheat germ** wheat embryo extracted as source of vitamins; **wheatmeal** flour from wheat with some bran and germ removed.

wheatear /ˈwi:tɪə/ *noun* small migratory bird.

wheaten *adjective* made of wheat.

wheedle /ˈwi:d(ə)l/ *verb* (**-ling**) coax by flattery or endearments; (+ *out*) get (thing) from person or cheat (person) of thing by wheedling.

wheel ● *noun* circular frame or disc revolving on axle and used to propel vehicle or other machinery; wheel-like thing; motion as of wheel; movement of line of men etc. with one end as pivot; (in *plural*) *slang* car. ● *verb* turn on axis or pivot; swing round in line with one end as pivot; (often + *about, around*) (cause to) change direction or face another way; push or pull (wheeled thing, or its load or occupant); go in circles or curves. □ **wheel and deal** engage in political or commercial scheming; **wheelbarrow** small cart with one wheel at front and two handles; **wheelbase** distance between axles of vehicle; **wheelchair** disabled person's chair on wheels; **wheel-spin** rotation of vehicle's wheels without traction; **wheels within wheels** intricate machinery, *colloquial* indirect or secret agencies; **wheelwright** maker or repairer of wheels.

wheelie *noun slang* manoeuvre on bicycle or motorcycle with front wheel off the ground.

wheeze ● *verb* (**-zing**) breathe or utter with audible whistling sound. ● *noun* sound of wheezing; *colloquial* clever scheme. □ **wheezy** *adjective* (**-ier, -iest**).

whelk *noun* spiral-shelled marine mollusc.

whelp ● *noun* young dog, puppy; *archaic* cub; ill-mannered child or youth. ● *verb* give birth to puppies.

when ● *interrogative adverb* at what time. ● *relative adverb* (time etc.) at or on which. ● *conjunction* at the or any time that; as soon as; although; after which, and then, but just then. ● *pronoun* what time; which time. □ **whenever, whensoever** at whatever time, on whatever occasion, every time that.

whence *archaic or formal* ● *interrogative adverb* from what place. ● *relative adverb* (place etc.) from which. ● *conjunction* to the place from which.

where /weə/ ● *interrogative adverb* in or to what place; in what direction; in what respect. ● *relative adverb* (place etc.) in or to which. ● *conjunction* in or to the or any place, direction, or respect in which; and there. ● *interrogative pronoun* what place. ● *relative pronoun* the place in or to which. □ **whereabouts** *interrogative adverb* approximately where, *noun* person's or thing's location; **whereas** in contrast or comparison with the fact that, taking into consideration the fact that; **whereby** by what or which means; **wherefore** *archaic* for what or which reason; **wherein** *formal* in what or which; **whereof** *formal* of what or which; **whereupon** immediately after which; **wherever** anywhere at all that, no matter where; **wherewithal** /-wɪðɔːl/ *colloquial* money etc. needed for a purpose.

wherry /ˈwerɪ/ *noun* (*plural* **-ies**) light rowing boat, usually for carrying passengers; large light barge.

whet *verb* (**-tt-**) sharpen; stimulate (appetite etc.). □ **whetstone** stone for sharpening cutting-tools.

whether /ˈweðə/ *conjunction introducing first or both of alternative possibilities.*

whew /hwjuː/ *interjection expressing astonishment, consternation, or relief.*

whey /weɪ/ *noun* watery liquid left when milk forms curds.

which ● *interrogative adjective used in asking someone to specify one or more things from a definite set of alternatives.* ● *relative adjective* being the one just referred to, and this or these. ● *interrogative pronoun* which person(s) or thing(s). ● *relative pronoun* which thing(s). □ **whichever** any which, no matter which.

whiff *noun* puff of air, smoke, etc.; smell; (+ *of*) trace of; small cigar.

Whig *noun historical* member of British political party succeeded by Liberals. □ **Whiggery** *noun*; **Whiggish** *adjective*; **Whiggism** *noun*.

while ● *noun* period of time. ● *conjunction* during the time that, for as long as, at the same time as; although, whereas. ● *verb* (**-ling**) (+ *away*) pass (time etc.) in leisurely or interesting way. ● *relative adverb* (time etc.) during which. □ **for a while** for some time; **in a while** soon; **once in a while** occasionally.

whilst /waɪlst/ *adverb & conjunction* while.

whim *noun* sudden fancy, caprice.

whimper /ˈwɪmpə/ ● *verb* make feeble, querulous, or frightened sounds. ● *noun* such sound.

whimsical /ˈwɪmzɪk(ə)l/ *adjective* capricious, fantastic. □ **whimsicality** /-ˈkæl-/ *noun*; **whimsically** *adverb*.

whimsy /ˈwɪmzɪ/ *noun* (*plural* **-ies**) whim.

whin *noun* (in *singular* or *plural*) gorse. □ **whinchat** small songbird.

whine ● *noun* long-drawn complaining cry (as) of dog or child; querulous tone or complaint. ● *verb* (**-ning**) emit or utter whine(s); complain.

whinge /wɪndʒ/ *verb* (**-geing** or **-ging**) *colloquial* complain peevishly.

whinny /ˈwɪnɪ/ ● *noun* (*plural* **-ies**) gentle or joyful neigh. ● *verb* (**-ies**, **-ied**) emit whinny.

whip ● *noun* lash attached to stick, for urging on or for punishing; person appointed by political party to control its discipline and tactics in Parliament; whip's written notice requesting member's attendance; food made with whipped cream etc.; whipper-in. ● *verb* (**-pp-**) beat or urge on with whip; beat (eggs, cream, etc.) into froth; take or move suddenly or quickly; *slang* steal, excel, defeat; bind with spirally wound twine; sew with overcast stitches. □ **whipcord** tightly twisted cord; (**the**) **whip hand** advantage, control; **whiplash** sudden jerk; **whipper-in** huntsman's assistant who manages hounds; **whipping boy** scapegoat; **whip-round** *colloquial* informal collection of money among group of people; **whipstock** handle of whip.

whippersnapper /ˈwɪpəsnæpə/ *noun* small child; insignificant but presumptuous person.

whippet /ˈwɪpɪt/ *noun* crossbred dog of greyhound type, used for racing.

whippoorwill /ˈwɪpʊəwɪl/ *noun* N. American nightjar.

whirl • *verb* swing round and round, revolve rapidly; (+ *away*) convey or go rapidly in car etc.; send or travel swiftly in orbit or curve; (of brain etc.) seem to spin round. • *noun* whirling movement; state of intense activity or confusion. □ **give it a whirl** *colloquial* attempt it; **whirlpool** circular eddy in sea, river, etc.; **whirlwind** *noun* whirling mass or column of air, *adjective* very rapid.

whirligig /'wɜ:lɪgɪg/ *noun* spinning or whirling toy; merry-go-round; revolving motion.

whirr • *noun* continuous buzzing or softly clicking sound. • *verb* (**-rr-**) make this sound.

whisk • *verb* (+ *away*, *off*) brush with sweeping movement, take suddenly; whip (cream, eggs, etc.); convey or go lightly or quickly; wave (object). • *noun* whisking movement; utensil for whipping eggs, cream, etc.; bunch of twigs, bristles, etc. for brushing or dusting.

whisker /'wɪskə/ *noun* (usually in *plural*) hair on cheeks or sides of face of man; each of bristles on face of cat etc.; *colloquial* small distance. □ **whiskered** *adjective*; **whiskery** *adjective*.

whisky /'wɪskɪ/ *noun* (*Irish & US* **whiskey**) (*plural* **-ies** or **-eys**) spirit distilled esp. from malted barley.

whisper /'wɪspə/ • *verb* speak using breath instead of vocal cords; talk or say in barely audible tone or confidential way; rustle, murmur. • *noun* whispering speech or sound; thing whispered.

whist *noun* card game, usually for two pairs of opponents. □ **whist drive** whist-party with players moving on from table to table.

whistle /'wɪs(ə)l/ • *noun* clear shrill sound made by forcing breath through lips contracted to narrow opening; similar sound made by bird, wind, missile, etc.; instrument used to produce such sound. • *verb* (**-ling**) emit whistle; give signal or express surprise or derision by whistling; (often + *up*) summon or give signal to thus; produce (tune) by whistling; (+ *for*) seek or desire in vain. □ **whistle-stop** *US* small unimportant town on railway, politician's brief pause for electioneering speech on tour.

Whit • *noun* Whitsuntide. • *adjective* of Whitsuntide. □ **Whit Sunday** 7th Sunday after Easter, commemorating Pentecost.

whit *noun* particle, least possible amount.

white • *adjective* of colour produced by reflection or transmission of all light; of colour of snow or milk; pale; of the human racial group having light-coloured skin; albino; (of hair) having lost its colour; (of coffee) with milk or cream. • *noun* white colour, paint, clothes, etc.; (in *plural*) white garments worn in cricket, tennis, etc.; (player using) lighter-coloured pieces in chess etc.; egg white; whitish part of eyeball round iris; white person. □ **white ant** termite; **whitebait** small silvery-white food-fish; **white cell** leucocyte; **white-collar** (of worker or work) non-manual, clerical, professional; **white corpuscle** leucocyte; **white elephant** useless possession; **white feather** symbol of cowardice; **white flag** symbol of surrender; **white goods** large domestic electrical equipment; **white heat** degree of heat making metal glow white, state of intense anger or passion; **white-hot** *adjective*; **white hope** person expected to achieve much; **white horses** white-crested waves; **white lead** mixture containing lead carbonate used as white pigment; **white lie** harmless or trivial untruth; **white magic** magic used for beneficent purposes; **white meat** poultry, veal, rabbit, and pork; **White Paper** government report giving information; **white pepper** pepper made by grinding husked berry; **white sauce** sauce of flour, melted butter, and milk or cream; **white slave** woman entrapped for prostitution; **white spirit** light petroleum as solvent; **white sugar** purified sugar; **white tie** man's white bow tie as part of full evening dress; **whitewash** *noun* solution of chalk or lime for whitening walls etc., means of glossing over faults, *verb* apply whitewash (to), gloss over, clear of blame; **white wedding** wedding where bride wears formal white dress; **whitewood** light-coloured wood, esp. prepared for staining etc. □ **whiten** *verb*; **whitener** *noun*; **whiteness** *noun*; **whitish** *adjective*.

whither /'wɪðə/ *archaic* ● *interrogative adverb* to what place. ● *relative adverb* (place etc.) to which.

whiting¹ /'waɪtɪŋ/ *noun* (*plural* same) small edible sea fish.

whiting² /'waɪtɪŋ/ *noun* ground chalk used in whitewashing etc.

whitlow /'wɪtləʊ/ *noun* inflammation near fingernail or toenail.

Whitsun /'wɪts(ə)n/ ● *noun* Whitsuntide. ● *adjective* of Whitsuntide.

Whitsuntide *noun* weekend or week including Whit Sunday.

whittle /'wɪt(ə)l/ *verb* (**-ling**) (often + *at*) pare (wood etc.) by cutting thin slices or shavings from surface; (often + *away*, *down*) reduce by repeated subtractions.

whiz (also **whizz**) ● *noun* sound made by object moving through air at great speed. ● *verb* (**-zz-**) move with or make a whiz. □ **whiz-kid** *colloquial* brilliant or highly successful young person.

WHO *abbreviation* World Health Organization.

who /hu:/ ● *interrogative pronoun* what or which person(s), what sort of person(s). ● *relative pronoun* (person or persons) that. □ **whoever**, **whosoever** the or any person(s) who; no matter who.

whoa /wəʊ/ *interjection used to stop or slow horse etc.*

who'd /hu:d/ who had; who would.

whodunit /hu:'dʌnɪt/ *noun* (also **whodunnit**) *colloquial* detective story, play, or film.

whole /həʊl/ ● *adjective* uninjured, unbroken, intact, undiminished; not less than; all, all of. ● *noun* complete thing; all of a thing; (+ *of*) all members etc. of. □ **on the whole** all things considered; **wholefood** food not artificially processed or refined; **wholehearted** completely devoted, done with all possible effort or sincerity; **wholeheartedly** *adverb*; **wholemeal** meal or flour made from whole grains of wheat.

wholesale /'həʊlseɪl/ ● *noun* selling in large quantities, esp. for retail by others. ● *adjective & adverb* by wholesale; on a large scale. ● *verb* (**-ling**) sell wholesale. □ **wholesaler** *noun*.

wholesome *adjective* promoting physical, mental, or moral health; prudent.

wholly /'həʊllɪ/ *adverb* entirely, without limitation; purely.

whom /hu:m/ *pronoun* (as object of verb) who. □ **whomever** (as object of verb) whoever; **whomsoever** (as object of verb) whosoever.

whoop /hu:p, wu:p/ ● *noun* cry expressing excitement etc.; characteristic drawing-in of breath after cough in whooping cough. ● *verb* utter whoop. □ **whooping cough** /'hu:pɪŋ/ infectious disease, esp. of children, with violent convulsive cough.

whoopee /wʊ'pi:/ *interjection expressing wild joy*. □ **make whoopee** /'wʊpɪ/ *colloquial* make merry, make love.

whoops /wʊps/ *interjection colloquial apology for obvious mistake.*

whop *verb* (**-pp-**) *slang* thrash, defeat.

whopper *noun slang* big specimen; great lie.

whopping *adjective colloquial* huge.

whore /hɔ:/ *noun* prostitute; *derogatory* promiscuous woman. □ **whorehouse** brothel.

whorl /wɜ:l/ *noun* ring of leaves etc. round stem; one turn of spiral.

whortleberry /'wɜ:t(ə)lberɪ/ *noun* (*plural* **-ies**) bilberry.

who's /hu:z/ who is; who has.

■ **Usage** Because it has an apostrophe, *who's* is easily confused with *whose*. They are each correctly used in *Who's there?* (= *Who is there?*), *Who's taken my pen?* (= *Who has taken my pen?*), and *Whose book is this?* (= *Who does this book belong to?*).

---

whose /hu:z/ ● *interrogative, pronoun, & adjective* of whom. ● *relative adjective* of whom or which.

why /waɪ/ ● *interrogative adverb* for what reason or purpose. ● *relative adverb* (reason etc.) for which. ● *interjection expressing surprise, impatience, reflection, or protest.* □ **whys and wherefores** reasons, explanation.

WI *abbreviation* West Indies; Women's Institute.

wick *noun* strip or thread feeding flame with fuel.

wicked /'wɪkɪd/ *adjective* (**-er**, **-est**) sinful, immoral; spiteful; playfully malicious; *colloquial* very bad; *slang* excellent. □ **wickedly** *adverb*; **wickedness** *noun*.

wicker /'wɪkə/ *noun* plaited osiers etc. as material for chairs, baskets, etc. □ **wickerwork** wicker, things made of wicker.

wicket /'wɪkɪt/ *noun Cricket* 3 upright stumps with bails in position defended by batsman, ground between the two wickets, state of this, batsman's being got out; (in full **wicket-door**, **-gate**) small door or gate, esp. beside or in larger one. □ **wicket-keeper** fielder close behind batsman's wicket.

wide ● *adjective* having sides far apart, broad, not narrow; (following measurement) in width; extending far, not restricted; liberal, not specialized; open to full extent; (+ *of*) not within reasonable distance of, far from. ● *adverb* to full extent; far from target etc. ● *noun* wide ball. □ **-wide** extending over whole of; **wide awake** fully awake, *colloquial* wary or knowing; **wide ball** *Cricket* ball judged by umpire to be beyond batsman's reach; **wide-eyed** surprised, naïve; **widespread** widely distributed. □ **widen** *verb*.

widely *adverb* far apart; extensively; by many people; considerably.

widgeon /'wɪdʒ(ə)n/ *noun* (also **wigeon**) kind of wild duck.

widow /'wɪdəʊ/ ● *noun* woman who has lost her husband by death and not married again. ● *verb* make into widow or widower; (as **widowed** *adjective*) bereft by death of spouse. □ **widow's peak** V-shaped growth of hair on forehead. □ **widowhood** *noun*.

widower /'wɪdəʊə/ *noun* man who has lost his wife by death and not married again.

width *noun* measurement from side to side; large extent; liberality of views etc.; piece of material of full width. □ **widthways** *adverb*.

wield /wi:ld/ *verb* hold and use, control, exert.

wife *noun* (*plural* **wives**) married woman, esp. in relation to her husband. □ **wifely** *adjective*.

wig *noun* artificial head of hair.

wigeon = WIDGEON.

wiggle /'wɪg(ə)l/ *colloquial* ● *verb* (**-ling**) move from side to side etc. ● *noun* wiggling movement; kink in line etc. □ **wiggly** *adjective* (**-ier**, **-iest**).

wight /waɪt/ *noun archaic* person.

wigwam /'wɪgwæm/ *noun* N. American Indian's hut or tent.

wilco /'wɪlkəʊ/ *interjection colloquial: expressing compliance or agreement.*

wild /waɪld/ ● *adjective* in original natural state; not domesticated, cultivated, or civilized; unrestrained, disorderly; tempestuous; intensely eager, frantic; (+ *about*) *colloquial* enthusiastically devoted to; *colloquial* infuriated; random, ill-aimed, rash. ● *adverb* in a wild way. ● *noun* wild place, desert. □ **like wildfire** with extraordinary speed; **run wild** grow or stray unchecked or undisciplined; **wild card** card having any rank chosen by its player, person or thing usable in different ways; **wildcat** *noun* hot-tempered or violent person, *adjective* (of strike) sudden and unofficial, reckless, financially unsound; **wild-goose chase** foolish or hopeless quest; **wildlife** wild animals collectively; **Wild West** western US in lawless times. □ **wildly** *adverb*; **wildness** *noun*.

wildebeest /'wɪldəbi:st/ *noun* (*plural* same or **-s**) gnu.

wilderness /'wɪldənɪs/ *noun* desert, uncultivated area; confused assemblage.

wile ● *noun* (usually in *plural*) stratagem, trick. ● *verb* (**-ling**) lure.

wilful /'wɪlfʊl/ *adjective* (*US* **willful**) intentional, deliberate; obstinate. □ **wilfully** *adverb*.

will[1] *auxiliary verb* (*3rd singular present* **will**) *used to form future tenses; expressing request as question*; be able to; have tendency to; be likely to.

will[2] ● *noun* faculty by which person decides what to do; fixed desire or intention; will-power; legal written directions for disposal of one's property etc. after death; disposition towards others. ● *verb* try to cause by will-power; intend, desire; bequeath by will. □ **at will** whenever one wishes; **will-power** control by purpose over impulse; **with a will** vigorously.

willing /'wɪlɪŋ/ *adjective* ready to consent or undertake; given etc. by willing person. □ **willingly** *adverb*; **willingness** *noun*.

will-o'-the-wisp /wɪləðə'wɪsp/ *noun* phosphorescent light seen on marshy ground; elusive person.

willow /'wɪləʊ/ *noun* waterside tree with pliant branches yielding osiers. □ **willowherb** plant with leaves like willow; **willow-pattern** conventional Chinese design of blue on white china etc.

willowy *adjective* lithe and slender; having willows.

willy-nilly /wɪlɪ'nɪlɪ/ *adverb* whether one likes it or not.

wilt ● *verb* wither, droop; lose energy. ● *noun* plant disease causing wilting.

wily /'waɪlɪ/ *adjective* (**-ier**, **-iest**) crafty, cunning.

wimp *noun colloquial* feeble or ineffectual person. □ **wimpish** *adjective*.

wimple /'wɪmp(ə)l/ *noun* headdress covering neck and sides of face, worn by some nuns.

win ● *verb* (**-nn-**; *past & past participle* **won** /wʌn/) secure as result of fight, contest, bet, etc.; be victor, be victorious in. ● *noun* victory in game etc. □ **win the day** be victorious in battle, argument, etc.; **win over** gain support of; **win through**, **out** overcome obstacles.

wince ● *noun* start or involuntary shrinking movement of pain etc. ● *verb* (**-cing**) give wince.

winceyette /wɪnsɪ'et/ *noun* lightweight flannelette.

winch ● *noun* crank of wheel or axle; windlass. ● *verb* lift with winch.

wind[1] /wɪnd/ ● *noun* air in natural motion; breath, esp. as needed in exertion or playing wind instrument; power of breathing easily; empty talk; gas generated in bowels etc.; wind instruments of orchestra etc.; scent carried by the wind. ● *verb* exhaust wind of by exertion or blow; make (baby) bring up wind after feeding; detect presence of by scent. □ **get wind of** begin to suspect; **get**, **have the wind up** *colloquial* become, be, frightened; **in the wind** *colloquial* about to happen; **put the wind up** *colloquial* frighten; **take the wind out of (person's) sails** frustrate by anticipation; **windbag** *colloquial* person who talks a lot but says little of value; **wind-break** thing that breaks force of wind; **windcheater** windproof jacket; **windfall** fruit blown down by wind, unexpected good fortune, esp. legacy; **wind instrument** musical instrument sounded by air-current; **wind-jammer** merchant sailing ship; **windmill** mill worked by action of wind on sails, toy with curved vanes revolving on stick; **windpipe** air-passage between throat and lungs; **windscreen** screen of glass at front of car etc.; **windscreen wiper** rubber etc. blade to clear windscreen of rain etc.; **windshield** *US* windscreen; **wind-sock** canvas cylinder or cone on mast to show direction of wind; **windswept** exposed to high winds; **wind-tunnel** enclosed chamber for testing (models or parts of) aircraft etc. in winds of known velocities.

wind[2] /waɪnd/ ● *verb* (*past & past participle* **wound** /waʊnd/) go in spiral, crooked, or curved course; make (one's way) thus; wrap closely, coil; provide with coiled thread etc.; surround (as) with coil; wind up (clock etc.). ● *noun* bend or turn in course. □ **wind down** lower by winding, unwind, approach end gradually; **winding-sheet** sheet in which corpse is wrapped for burial; **wind up** *verb* coil whole of, tighten coiling or coiled spring of, *colloquial* increase intensity of, *colloquial* provoke to anger etc., bring to conclusion, arrange affairs of and dissolve (company), cease business and go into liquidation, *colloquial* arrive finally; **wind-up** *noun* conclusion, *colloquial* attempt to provoke.

windlass /'wɪndləs/ *noun* machine with horizontal axle for hauling or hoisting.

window /'wɪndəʊ/ *noun* opening, usually with glass, in wall etc. to admit light etc.; the glass itself; space for display behind window of shop; window-like opening; transparent part in envelope showing address; opportunity for study or action. □ **window-box** box placed outside window for cultivating plants; **window-dressing** art of arranging display in shop window etc., adroit presentation of facts etc. to give falsely favourable impression; **window-pane** glass pane in window; **window-seat** seat below window, seat next to window in aircraft etc.; **window-shop** look at goods in shop windows without buying anything.

windsurfing *noun* sport of riding on water on sailboard. □ **windsurf** *verb*; **windsurfer** *noun*.

windward /ˈwɪndwəd/ ● *adjective & adverb* on or towards side from which wind is blowing. ● *noun* this direction.

windy *adjective* (**-ier**, **-iest**) stormy with or exposed to wind; generating or characterized by flatulence; *colloquial* wordy; *colloquial* apprehensive, frightened. □ **windiness** *noun*.

wine ● *noun* fermented grape juice as alcoholic drink; fermented drink resembling this made from other fruits etc.; colour of red wine. ● *verb* (**-ning**) drink wine; entertain with wine. □ **wine bar** bar or small restaurant where wine is main drink available; **wine cellar** cellar for storing wine, its contents; **wineglass** glass for wine, usually with stem and foot; **winepress** press in which grape juice is extracted for wine; **wine waiter** waiter responsible for serving wine.

wing ● *noun* each of the limbs or organs by which bird etc. flies; winglike part supporting aircraft; projecting part of building; *Football etc.* forward player at either end of line, side part of playing area; (in *plural*) sides of theatre stage; extreme section of political party; flank of battle array; part of vehicle over wheel; air-force unit of several squadrons. ● *verb* travel or traverse on wings; wound in wing or arm; equip with wings; enable to fly, send in flight. □ **on the wing** flying; **take under one's wing** treat as protégé; **take wing** fly away; **wing-case** horny cover of insect's wing; **wing-chair** chair with side-pieces at top of high back; **wing-collar** man's high stiff collar with turned-down corners; **wing commander** RAF officer next below group captain; **wing-nut** nut with projections to turn it by; **wingspan** measurement right across wings. □ **winged** *adjective*.

winger *noun Football etc.* wing player.

wink ● *verb* (often + *at*) close and open one eye quickly, esp. as signal; close eye(s) momentarily; (of light) twinkle; (of indicator) flash on and off. ● *noun* act of winking; *colloquial* short sleep. □ **wink at** purposely avoid seeing, pretend not to notice.

winkle /ˈwɪŋk(ə)l/ ● *noun* edible sea snail. ● *verb* (**-ling**) (+ *out*) extract with difficulty.

winning ● *adjective* having or bringing victory; attractive. ● *noun* (in *plural*) money won. □ **winning post** post marking end of race. □ **winningly** *adverb*.

winnow /ˈwɪnəʊ/ *verb* blow (grain) free of chaff etc.; (+ *out*, *away*, *from*, etc.) rid grain of (chaff etc.); sift, examine.

winsome /ˈwɪnsəm/ *adjective* attractive, engaging. □ **winsomely** *adverb*; **winsomeness** *noun*.

winter /ˈwɪntə/ ● *noun* coldest season of year. ● *verb* (usually + *at*, *in*) spend the winter. □ **winter garden** garden of plants flourishing in winter; **wintergreen** kind of plant remaining green all winter; **winter sports** sports practised on snow or ice; **wintertime** season or period of winter.

wintry /ˈwɪntrɪ/ *adjective* (**-ier**, **-iest**) characteristic of winter; lacking warmth. □ **wintriness** *noun*.

winy *adjective* (**-ier**, **-iest**) wine-flavoured.

wipe ● *verb* (**-ping**) clean or dry surface of by rubbing; rub (cloth) over surface; spread (liquid etc.) over surface by rubbing; (often + *away*, *off*, etc.) clear or remove by wiping; erase, eliminate. ● *noun* act of wiping; piece of specially treated cloth for wiping. □ **wipe out** utterly destroy or defeat, clean inside of; **wipe up** dry (dishes etc.), take up (liquid etc.) by wiping.

wiper *noun* windscreen wiper.

wire ● *noun* metal drawn out into thread or slender flexible rod; piece of this; length of this used for fencing or to carry electric current etc.; *colloquial* telegram. ● *verb* (**-ring**) provide, fasten, strengthen, etc. with wire; (often + *up*) install electrical circuits in; *colloquial* telegraph. □ **get one's wires crossed** become confused; **wire-haired** (of dog) having stiff wiry hair; **wire netting** netting made of meshed wire; **wire-tapping** tapping of telephone wires; **wire wool** mass of fine wire for scouring; **wireworm** destructive larva of a kind of beetle.

wireless *noun* radio; radio receiving set.

wiring *noun* system or installation of electrical circuits.

wiry *adjective* (**-ier**, **-iest**) sinewy, untiring; like wire, tough and coarse.

wisdom /'wɪzdəm/ *noun* experience, knowledge, and the power of applying them; prudence, common sense; wise sayings. □ **wisdom tooth** hindmost molar usually cut at age of about 20.

wise[1] /waɪz/ *adjective* having, showing, or dictated by wisdom; prudent, sensible; having knowledge; suggestive of wisdom; *US colloquial* alert, crafty. □ **be, get wise to** *colloquial* be, become, aware of; **wisecrack** *colloquial noun* smart remark, *verb* make wisecrack; **wise guy** *colloquial* know-all; **wise man** wizard, esp. one of the Magi; **wise up** *esp. US colloquial* inform, get wise. □ **wisely** *adverb.*

wise[2] /waɪz/ *noun archaic* way, manner, degree.

-wise /waɪz/ *combining form added to nouns to form adjectives and adverbs:* in the manner or direction of (e.g. *clockwise, lengthwise*); with reference to (e.g. *weatherwise*).

wiseacre /'waɪzeɪkə/ *noun* person who affects to be wise.

wish ● *verb* (often + *for*) have or express desire or aspiration; want, demand; express one's hopes for; *colloquial* foist. ● *noun* desire, request, expression of this; thing desired. □ **wishbone** forked bone between neck and breast of fowl; **wish-fulfilment** tendency of unconscious wishes to be satisfied in fantasy; **wishing-well** well at which wishes are made.

wishful *adjective* (often + *to do*) desiring. □ **wishful thinking** belief founded on wishes rather than facts.

wishy-washy /'wɪʃɪwɒʃɪ/ *adjective colloquial* feeble or poor in quality or character; weak, watery.

wisp *noun* small bundle or twist of straw etc.; small separate quantity of smoke, hair, etc.; small thin person. □ **wispy** *adjective* (**-ier**, **-iest**).

wisteria /wɪ'stɪərɪə/ *noun* (also **wistaria** /-teər-/) climbing plant with purple, blue, or white hanging flowers.

wistful /'wɪstfʊl/ *adjective* yearning, mournfully expectant or wishful. □ **wistfully** *adverb*; **wistfulness** *noun.*

wit *noun* (in *singular* or *plural*) intelligence, understanding; (in *singular*) imaginative and inventive faculty; amusing ingenuity of speech or ideas; person noted for this. □ **at one's wit's** or **wits' end** utterly at a loss or in despair; **have** or **keep one's wits about one** be alert; **out of one's wits** mad; **to wit** that is to say, namely.

witch *noun* woman supposed to have dealings with Devil or evil spirits; old hag; fascinating girl or woman. □ **witchcraft** use of magic, bewitching charm; **witch-doctor** tribal magician of primitive people; **witch, wych hazel** N. American shrub, astringent lotion from its bark; **witch-hunt** campaign against people suspected of unpopular or unorthodox views.

witchery /'wɪtʃərɪ/ *noun* witchcraft.

with /wɪð/ *preposition expressing instrument or means used, company, parting of company, cause, possession, circumstances, manner, agreement, disagreement, antagonism, understanding, regard.* □ **with it** *colloquial* up to date, alert and comprehending; **with that** thereupon.

withdraw /wɪð'drɔː/ *verb* (*past* **-drew**; *past participle* **-drawn**) pull or take aside or back; discontinue, cancel, retract; remove, take away; take (money) out of an account; retire or go apart; (as **withdrawn** *adjective*) unsociable. □ **withdrawal** *noun.*

withe = WITHY.

wither /'wɪðə/ *verb* (often + *up*) make or become dry and shrivelled; (often + *away*) deprive of or lose vigour or freshness; (esp. as **withering** *adjective*) blight with scorn etc. □ **witheringly** *adverb.*

withers /'wɪðəz/ *plural noun* ridge between horse's shoulder-blades.

withhold /wɪð'həʊld/ *verb* (*past & past participle* **-held**) refuse to give, grant, or allow; hold back, restrain.

within /wɪ'ðɪn/ ● *adverb* inside; indoors; in spirit. ● *preposition* inside; not beyond or out of; not transgressing or exceeding; not further off than.

without /wɪ'ðaʊt/ ● *preposition* not having, feeling, or showing; free from; in absence of; with neglect or avoidance of; *archaic* outside. ● *adverb archaic or literary* outside, out of doors.

withstand /wɪð'stænd/ *verb* (*past & past participle* **-stood**) oppose, hold out against.

withy /'wɪðɪ/ *noun* (*plural* **-ies**) tough flexible shoot, esp. of willow.

witless *adjective* foolish; crazy.

witness /'wɪtnɪs/ ● *noun* eyewitness; person giving sworn testimony; person attesting another's signature to document; (+ *to*, *of*) person or thing whose existence etc. attests or proves something. ● *verb* be eyewitness of; be witness to (signature etc.); serve as evidence or indication of; give or be evidence. □ **bear witness to** attest truth of, state one's belief in; **witness-box** (*US* **-stand**) enclosed space in law court from which witness gives evidence.

witter /'wɪtə/ *verb* (often + *on*) *colloquial* chatter annoyingly or on trivial matters.

witticism /'wɪtɪsɪz(ə)m/ *noun* witty remark.

wittingly /'wɪtɪŋlɪ/ *adverb* consciously, intentionally.

witty *adjective* (**-ier**, **-iest**) showing verbal wit. □ **wittily** *adverb*; **wittiness** *noun*.

wives *plural* of WIFE.

wizard /'wɪzəd/ ● *noun* sorcerer, magician; person of extraordinary powers. ● *adjective slang* wonderful. □ **wizardry** *noun*.

wizened /'wɪz(ə)nd/ *adjective* shrivelled-looking.

WNW *abbreviation* west-north-west.

WO *abbreviation* Warrant Officer.

woad *noun* plant yielding blue dye; this dye.

wobble /'wɒb(ə)l/ ● *verb* (**-ling**) sway from side to side; stand or go unsteadily, stagger; waver, vacillate. ● *noun* wobbling motion. □ **wobbly** *adjective* (**-ier**, **-iest**).

woe *noun* affliction, bitter grief; (in *plural*) calamities. □ **woebegone** dismal-looking.

woeful *adjective* sorrowful; causing or feeling affliction; very bad. □ **woefully** *adverb*.

wok *noun* bowl-shaped frying-pan used in esp. Chinese cookery.

woke *past* of WAKE[1].

woken *past participle* of WAKE[1].

wold /wəʊld/ *noun* high open uncultivated land or moor.

wolf /wʊlf/ ● *noun* (*plural* **wolves** /wʊlvz/) wild animal related to dog; *slang* man who seduces women. ● *verb* (often + *down*) devour greedily. □ **cry wolf** raise false alarm; **keep the wolf from the door** avert starvation; **wolfhound** dog of kind used originally to hunt wolves; **wolfsbane** an aconite; **wolf-whistle** man's whistle to attractive woman. □ **wolfish** *adjective*.

wolfram /'wʊlfrəm/ *noun* tungsten; tungsten ore.

wolverine /'wʊlvəri:n/ *noun* N. American animal of weasel family.

wolves *plural* of WOLF.

woman /'wʊmən/ *noun* (*plural* **women** /'wɪmɪn/) adult human female; the female sex; *colloquial* wife, girlfriend.

womanhood *noun* female maturity; womanliness; womankind.

womanish *adjective derogatory* effeminate, unmanly.

womanize *verb* (also **-ise**) (**-zing** or **-sing**) (of man) be promiscuous. □ **womanizer** *noun*.

womankind *noun* (also **womenkind**) women in general.

womanly *adjective* having or showing qualities associated with women. □ **womanliness** *noun*.

womb /wu:m/ *noun* organ of conception and gestation in female mammals.

wombat /'wɒmbæt/ *noun* burrowing plant-eating Australian marsupial.

women /'wɪmɪn/ *plural* of WOMAN. □ **women's libber** *colloquial* supporter of women's liberation; **women's liberation**, **lib** *colloquial* movement for release of women from subservient status; **women's rights** human rights of women giving equality with men.

womenfolk *noun* women in general; women in family.

won *past & past participle* of WIN.

wonder /'wʌndə/ ● *noun* emotion, esp. admiration, excited by what is unexpected, unfamiliar, or inexplicable;

strange or remarkable thing, specimen, event, etc. ● *adjective* having amazing properties etc. ● *verb* be filled with wonder; (+ *that*) be surprised to find that, be curious to know. □ **no or small wonder** it is not surprising; **wonderland** fairyland, place of surprises or marvels.

wonderful *adjective* very remarkable or admirable. □ **wonderfully** *adverb*.

wonderment *noun* surprise, awe.

wondrous /ˈwʌndrəs/ *poetical* ● *adjective* wonderful. ● *adverb* wonderfully.

wonky /ˈwɒŋkɪ/ *adjective* (**-ier**, **-iest**) *slang* crooked; unsteady; unreliable.

wont /wəʊnt/ ● *adjective archaic or literary* (+ *to do*) accustomed. ● *noun formal or jocular* custom, habit.

won't /wəʊnt/ will not.

wonted /ˈwəʊntɪd/ *adjective* habitual, usual.

woo *verb* (**woos**, **wooed**) court, seek love of; try to win; seek support of; coax, importune.

wood /wʊd/ *noun* hard fibrous substance of tree; this for timber or fuel; (in *singular* or *plural*) growing trees occupying piece of ground; wooden cask for wine etc.; wooden-headed golf club; ball in game of bowls. □ **out of the wood(s)** clear of danger or difficulty; **wood anemone** wild spring-flowering anemone; **woodbine** honeysuckle; **woodchuck** N. American marmot; **woodcock** game bird related to snipe; **woodcut** relief cut on wood, print made from this; **woodcutter** person who cuts timber; **woodland** wooded country; **woodlouse** small land crustacean with many legs; **woodman** forester; **woodpecker** bird that taps tree trunks to find insects; **woodpigeon** dove with white patches round neck; **wood pulp** wood fibre prepared for papermaking; **woodwind** wind instrument(s) of orchestra made originally of wood; **woodwork** making of things in wood, things made of wood; **woodworm** beetle larva that bores in wood, resulting condition of wood.

wooded *adjective* having woods.

wooden /ˈwʊd(ə)n/ *adjective* made of wood; like wood; stiff, clumsy; expressionless. □ **woodenly** *adverb*; **woodenness** *noun*.

woody *adjective* (**-ier**, **-iest**) wooded; like or of wood.

woof[1] /wʊf/ ● *noun* gruff bark of dog. ● *verb* give woof.

woof[2] /wuːf/ *noun* weft.

woofer /ˈwuːfə/ *noun* loudspeaker for low frequencies.

wool /wʊl/ *noun* fine soft wavy hair forming fleece of sheep etc.; woollen yarn, cloth, or garments; wool-like substance. □ **wool-gathering** absent-mindedness; **the Woolsack** Lord Chancellor's seat in House of Lords.

woollen /ˈwʊlən/ (*US* **woolen**) ● *adjective* made (partly) of wool. ● *noun* woollen fabric; (in *plural*) woollen garments.

woolly ● *adjective* (**-ier**, **-iest**) bearing or like wool; woollen; indistinct; confused. ● *noun* (*plural* **-ies**) *colloquial* woollen (esp. knitted) garment.

woozy /ˈwuːzɪ/ *adjective* (**-ier**, **-iest**) *colloquial* dizzy; slightly drunk.

word /wɜːd/ ● *noun* meaningful element of speech, usually shown with space on either side of it when written or printed; speech as distinct from action; one's promise or assurance; (in *singular* or *plural*) thing said, remark, conversation; (in *plural*) text of song or actor's part; (in *plural*) angry talk; news, message; command. ● *verb* put into words, select words to express. □ **word-blindness** dyslexia; **word for word** in exactly the same words, literally; **the Word (of God)** the Bible; **word of mouth** speech (only); **word-perfect** having memorized one's part etc. perfectly; **word processor** computer software or hardware for storing text entered from keyboard, incorporating corrections, and producing printout; **word-process** *verb*; **word processing** *noun*.

wording *noun* form of words used.

wordy *adjective* (**-ier**, **-iest**) using or expressed in (too) many words.

wore *past* of WEAR.

work /wɜːk/ ● *noun* application of effort to a purpose, use of energy; task to be undertaken; thing done or made by work, result of action; employment, occupation, etc., esp. as means of earning money; literary or musical composition; actions or experiences of specified kind; (in *plural*) operative part of clock etc.; (**the works**) *colloquial* all that is available or needed, full treatment; (in *plural*) operations of building or repair; (in *plural*; often treated as *singular*) factory; (usually in *plural* or *in combination*) defensive structure. ● *verb* be engaged in activity; be employed in certain work; make efforts; be craftsman in (material); operate or function, esp. effectively; operate, manage, control; put or keep in operation or at work, cause to toil; cultivate (land); produce as result; *colloquial* arrange; knead, hammer, bring to desired shape or consistency; do, or make by, needlework etc.; (cause to) make way or make (way) slowly or with difficulty; gradually become (loose etc.) by motion; artificially excite; purchase with labour instead of money; obtain money for by labour; (+ *on*, *upon*) influence; be in motion or agitated, ferment. □ **get worked up** become angry, excited, or tense; **workaday** ordinary, everyday, practical; **work-basket** basket for sewing materials; **workbench** bench for manual work, esp. carpentry; **workbox** box for tools, needlework, etc.; **workday** day on which work is usually done; **work experience** temporary experience of employment for young people; **workforce** workers engaged or available, number of these; **workhouse** *historical* public institution for the poor; **work in** find place for; **workload** amount of work to be done; **workman** man employed to do manual labour, person who works in specified manner; **workmanlike** showing practised skill; **workmanship** degree of skill in doing task or of finish in product; **workmate** person working alongside another; **work off** get rid of by work or activity; **work out** *verb* solve (sum) or find (amount) by calculation, understand (problem, person, etc.), be calculated, have result, provide for all details of, engage in physical exercise; **workout** *noun* session of physical exercise; **work over** examine thoroughly, *colloquial* treat with violence; **workshop** room or building in which goods are manufactured, place or meeting for concerted activity; **work-shy** disinclined to work; **workstation** location of stage in manufacturing process, computer terminal; **worktop** flat (esp. kitchen) surface for working on; **work to rule** follow official working rules exactly to reduce efficiency as protest; **work-to-rule** *noun*; **work up** bring gradually to efficient or advanced state, advance gradually, elaborate or excite by degrees, mingle (ingredients), learn (subject) by study.

workable *adjective* that can be worked, will work, or is worth working. □ **workability** *noun*.

worker *noun* manual or industrial etc. employee; neuter bee or ant; person who works hard.

working ● *adjective* engaged in work; while so engaged; functioning, able to function. ● *noun* activity of work; functioning; mine, quarry; (usually in *plural*) mechanism. □ **working capital** capital used in conducting a business; **working class** social class employed for wages, esp. in manual or industrial work; **working-class** *adjective*; **working day** workday, part of day devoted to work; **working knowledge** knowledge adequate to work with; **working lunch** lunch at which business is conducted; **working order** condition in which machine works; **working party** committee appointed to study and advise on some question.

world /wɜ:ld/ ● *noun* the earth, planetary body like it; the universe, all that exists; time, state, or scene of human existence; (**the**, **this world**) mortal life; secular interests and affairs; human affairs, active life; average, respectable, or fashionable people or their customs or opinions; all that concerns or all who belong to specified class or sphere of activity; vast amount. ● *adjective* of or affecting all nations. □ **out of this world** *colloquial* extremely good etc.; **think the world of** have very high regard for; **world-class** of standard considered high throughout world; **world-famous** known throughout the world; **world music** pop music incorporating ethnic elements; **world war** one involving many important nations; **world-weary** bored with human affairs; **worldwide** *adjective* occurring or known in all parts of the world; *adverb* throughout the world.

worldly *adjective* (**-ier**, **-iest**) temporal, earthly; experienced in life, sophisticated, practical. □ **worldly-wise** prudent in one's dealings with world.

worm /wɜ:m/ ● *noun* any of several types of creeping invertebrate animal with long slender body and no limbs; larva of insect; (in *plural*) internal parasites; insignificant or contemptible person; spiral of screw. ● *verb* crawl, wriggle; (**worm oneself**) insinuate oneself (into favour etc.); (+ *out*) obtain (secret etc.) by cunning persistence; rid (dog etc.) of worms. □ **worm-cast** convoluted mass of earth left on surface by burrowing earthworm; **wormeaten** eaten into by worms, decayed, dilapidated.

wormwood /'wɜ:mwʊd/ *noun* plant with bitter aromatic taste; bitter humiliation, source of this.

wormy *adjective* (**-ier**, **-iest**) full of worms; wormeaten.

worn ● *past participle* of WEAR. ● *adjective* damaged by use or wear; looking tired and exhausted.

worry /'wʌrɪ/ ● *verb* (**-ies**, **-ied**) be anxious; harass, importune, be trouble or anxiety to; shake or pull about with teeth; (as **worried** *adjective*) uneasy. ● *noun* (*plural* **-ies**) thing that causes anxiety or disturbs tranquillity; disturbed state of mind, anxiety. □ **worry beads** string of beads manipulated with fingers to occupy or calm oneself. □ **worrier** *noun*.

worse /wɜ:s/ ● *adjective* more bad; in or into worse health or worse condition. ● *adverb* more badly; more ill. ● *noun* worse thing(s); (**the worse**) worse condition. □ **the worse for wear** damaged by use, injured; **worse off** in a worse (esp. financial) position. □ **worsen** *verb*.

worship /'wɜ:ʃɪp/ ● *noun* homage or service to deity; acts, rites, or ceremonies of this; adoration, devotion; (**His**, **Her**, **Your Worship**) *title used of or to mayor, magistrate, etc.* ● *verb* (**-pp-**; *US* **-p-**) adore as divine, honour with religious rites; idolize; attend public worship; be full of adoration. □ **worshipper** *noun*.

worshipful *adjective* (also **Worshipful**) *archaic* honourable, distinguished (esp. in old titles of companies or officers).

worst /wɜ:st/ ● *adjective* most bad. ● *adverb* most badly. ● *noun* worst part or possibility. ● *verb* get the better of, defeat. □ **at (the) worst** in the worst possible case; **do your worst** *expression of defiance*; **get the worst of it** be defeated; **if the worst comes to the worst** if the worst happens.

worsted /'wʊstɪd/ *noun* fine woollen yarn; fabric made from this.

wort /wɜ:t/ *noun* infusion of malt before it is fermented into beer.

worth /wɜ:θ/ ● *adjective* of value equivalent to; such as to justify or repay; possessing property equivalent to. ● *noun* value; equivalent of money etc. in commodity etc. □ **worth it** (*colloquial*), **worth (one's) while**, **worthwhile** worth the time or effort spent.

worthless *adjective* without value or merit. □ **worthlessness** *noun*.

worthy /'wɜ:ðɪ/ ● *adjective* (**-ier**, **-iest**) deserving respect, estimable; entitled to recognition; (usually + *of*) deserving; (+ *of*) adequate or suitable for the dignity etc. of. ● *noun* (*plural* **-ies**) worthy person; person of some distinction.

would *auxiliary verb* (*3rd singular present* **would**) *used in reported speech or to form conditional mood; expressing habitual past action, request as question, or probability.* □ **would-be** desiring or aspiring to be.

wouldn't /'wʊd(ə)nt/ would not.

wound[1] /wu:nd/ ● *noun* injury done by cut or blow to living tissue; pain inflicted on feelings, injury to reputation. ● *verb* inflict wound on.

wound[2] *past & past participle* of WIND[2]. □ **wound up** excited, tense, angry.

wove *past* of WEAVE[1].

woven *past participle* of WEAVE[1].

wow /waʊ/ ● *interjection expressing astonishment or admiration.* ● *noun slang* sensational success. ● *verb slang* impress greatly.

WP *abbreviation* word processor.

WPC *abbreviation* woman police constable.

w.p.m. *abbreviation* words per minute.

WRAC *abbreviation* Women's Royal Army Corps.

wrack *noun* seaweed cast up or growing on seashore; destruction.

WRAF *abbreviation* Women's Royal Air Force.

wraith /reɪθ/ *noun* ghost; spectral appearance of living person supposed to portend that person's death.

wrangle /ˈræŋg(ə)l/ • *noun* noisy argument or dispute. • *verb* (**-ling**) engage in wrangle.

wrap • *verb* (**-pp-**) (often + *up*) envelop in folded or soft encircling material; (+ *round*, *about*) arrange or draw (pliant covering) round (person). • *noun* shawl, scarf, etc.; wrapper; *esp. US* wrapping material. □ **under wraps** in secrecy; **wraparound**, **wrapround** (esp. of clothing) designed to wrap round, curving round at edges; **wrap-over** *adjective* (of garment) overlapping when worn, *noun* such garment; **wrapped up in** engrossed or absorbed in; **wrap up** *colloquial* finish off (matter), put on warm clothes.

wrapper *noun* cover for sweet, book, posted newspaper, etc.; loose enveloping robe or gown.

wrapping *noun* (esp. in *plural*) material used to wrap, wraps, wrappers. □ **wrapping paper** strong or decorative paper for wrapping parcels.

wrasse /ræs/ *noun* (*plural* same or **-s**) brilliant-coloured edible sea fish.

wrath /rɒθ/ *noun literary* extreme anger. □ **wrathful** *adjective*.

wreak *verb* (usually + *upon*) give play to (vengeance etc.); cause (damage etc.).

wreath /ri:θ/ *noun* (*plural* **-s** /ri:ðz/) flowers or leaves wound together into ring, esp. as ornament for head or door or for laying on grave etc.; curl or ring of smoke, cloud, or soft fabric.

wreathe /ri:ð/ *verb* (**-thing**) encircle (as) with or like wreath; (+ *round*) wind (one's arms etc.) round (person etc.); move in wreaths.

wreck • *noun* sinking or running aground of ship; ship that has suffered wreck; greatly damaged building, thing, or person. • *verb* seriously damage (vehicle etc.); ruin (hopes etc.); cause wreck of (ship).

wreckage *noun* wrecked material; remnants of wreck; act of wrecking.

wrecker *noun* person or thing that wrecks or destroys, esp. (*historical*) person who tries from shore to bring about shipwreck for plunder or profit.

Wren *noun* member of WRNS.

wren *noun* small usually brown short-winged songbird.

wrench • *noun* violent twist or oblique pull or tearing off; tool for gripping and turning nuts etc.; painful uprooting or parting etc. • *verb* twist or pull violently round or sideways; (often + *off*, *away*, etc.) pull with wrench.

wrest *verb* wrench away from person's grasp; (+ *from*) obtain by effort or with difficulty.

wrestle /ˈres(ə)l/ • *noun* contest in which two opponents grapple and try to throw each other to ground; hard struggle. • *verb* (**-ling**) have wrestling match; (often + *with*) struggle; (+ *with*) do one's utmost to deal with. □ **wrestler** *noun*.

wretch *noun* unfortunate or pitiable person; reprehensible person.

wretched /ˈretʃɪd/ *adjective* (**-er**, **-est**) unhappy, miserable, unwell; of bad quality, contemptible; displeasing. □ **wretchedly** *adverb*; **wretchedness** *noun*.

wriggle /ˈrɪg(ə)l/ • *verb* (**-ling**) twist or turn body with short writhing movements; make wriggling motions; (+ *along*, *through*, etc.) go thus; be evasive. • *noun* wriggling movement. □ **wriggly** *adjective*.

wring • *verb* (*past & past participle* **wrung**) squeeze tightly; (often + *out*) squeeze and twist, esp. to remove liquid; break by twisting; distress, torture; extract by squeezing; (+ *out*, *from*) obtain by pressure or importunity. • *noun* act of wringing. □ **wringing (wet)** so wet that water can be wrung out; **wring one's hands** clasp them as gesture of grief; **wring the neck of** kill (chicken etc.) by twisting neck.

wringer *noun* device for wringing water from washed clothes etc.

wrinkle /ˈrɪŋk(ə)l/ • *noun* crease in skin or other flexible surface; *colloquial* useful hint, clever expedient. • *verb* (**-ling**) make wrinkles in; form wrinkles. □ **wrinkly** *adjective* (**-ier**, **-iest**).

wrist *noun* joint connecting hand and forearm; part of garment covering wrist. □ **wrist-watch** small watch worn on strap etc. round wrist.

wristlet *noun* band or ring to guard, strengthen, or adorn wrist.

writ *noun* formal written court order to do or not do specified act.

write *verb* (**-ting**; *past* **wrote**; *past participle* **written** /ˈrɪt(ə)n/) mark paper or other surface with symbols, letters, or words; form or mark (such symbols etc.); form or mark symbols of (word, document, etc.); fill or complete with writing; put (data) into computer store; (esp. in *passive*) indicate (quality or condition) by appearance; compose for reproduction or publication; (usually + *to*) write and send letter; convey (news etc.) by letter; state in book etc.; (+ *into*, *out of*) include or exclude (character, episode) in or from story. □ **write down** record in writing; **write in** send suggestion etc. in writing, esp. to broadcasting station; **write off** *verb* cancel (debt etc.), acknowledge as lost, completely destroy, (+ *for*) order or request by post; **write-off** *noun* thing written off, esp. vehicle etc. so damaged as not to be worth repair; **write up** *verb* write full account of; **write-up** *noun* written or published account, review.

writer *noun* person who writes, esp. author. □ **writer's cramp** muscular spasm due to excessive writing.

writhe /raɪð/ *verb* (**-thing**) twist or roll oneself about (as) in acute pain; suffer mental torture.

writing *noun* written words etc.; handwriting; (usually in *plural*) writer's works. □ **in writing** in written form.

written *past participle* of WRITE.

WRNS *abbreviation* Women's Royal Naval Service.

wrong ● *adjective* mistaken, not true, in error; unsuitable, less or least desirable; contrary to law or morality; amiss, out of order. ● *adverb* in wrong manner or direction, with incorrect result. ● *noun* what is morally wrong; unjust action. ● *verb* treat unjustly; mistakenly attribute bad motives to. □ **go wrong** take wrong path, stop functioning properly, depart from virtuous behaviour; **in the wrong** responsible for quarrel, mistake, or offence; **wrongdoer** person who behaves immorally or illegally; **wrongdoing** *noun*; **wrong-foot** *colloquial* catch off balance or unprepared; **wrong-headed** perverse and obstinate; **wrong side** worse or undesirable or unusable side; **wrong way round** in opposite of normal orientation or sequence. □ **wrongly** *adverb*; **wrongness** *noun*.

wrongful *adjective* unwarranted, unjustified. □ **wrongfully** *adverb*.

wrote *past* of WRITE.

wroth /rəʊθ/ *adjective archaic* angry.

wrought /rɔːt/ *archaic past & past participle* of WORK. □ **wrought iron** form of iron suitable for forging or rolling, not cast.

wrung *past & past participle* of WRING.

WRVS *abbreviation* Women's Royal Voluntary Service.

wry /raɪ/ *adjective* (**-er**, **-est**) distorted, turned to one side; contorted in disgust, disappointment, or mockery; (of humour) dry and mocking. □ **wryneck** small woodpecker able to turn head over shoulder. □ **wryly** *adverb*; **wryness** *noun*.

WSW *abbreviation* west-south-west.

wt *abbreviation* weight.

wych hazel /wɪtʃ/ witch hazel.

WYSIWYG /ˈwɪzɪwɪg/ *adjective indicating that text on computer screen and printout correspond exactly* (*w*hat *y*ou *s*ee *i*s *w*hat *y*ou *g*et).

X[1] *noun* (also **x**) (Roman numeral) 10; first unknown quantity in algebra; unknown or unspecified number, person, etc.; cross-shaped symbol, esp. used to indicate position or incorrectness, to symbolize kiss or vote, or as signature of person who cannot write. □ **X-ray** *noun* electromagnetic radiation of short wavelength able to pass through opaque bodies, photograph made by X-rays, *verb* photograph, examine, or treat with X-rays.

X[2] *adjective* (of film) classified as suitable for adults only.

xenophobia /zenəˈfəʊbɪə/ *noun* hatred or fear of foreigners.

Xerox /ˈzɪərɒks/ ● *noun proprietary term* type of photocopier; copy made by it. ● *verb* (**xerox**) make Xerox of.

Xmas /ˈkrɪsməs, ˈeksməs/ *noun colloquial* Christmas.

xylophone /ˈzaɪləfəʊn/ *noun* musical instrument of graduated wooden or metal bars struck with small wooden hammers.

Y *noun* (also **y**) second unknown quantity in algebra; Y-shaped thing.

yacht /jɒt/ ● *noun* light sailing vessel for racing or cruising; larger usually power-driven vessel for cruising. ● *verb* race or cruise in yacht. □ **yachtsman**, **yachtswoman** person who yachts.

yah /jɑ:/ *interjection of derision, defiance, etc.*

yahoo /jɑ:ˈhu:/ *noun* bestial person.

Yahweh /ˈjɑ:weɪ/ *noun* Jehovah.

yak *noun* long-haired Tibetan ox.

yam *noun* tropical or subtropical climbing plant; edible starchy tuberous root of this; *US* sweet potato.

yang *noun* (in Chinese philosophy) active male principle of universe (compare YIN).

Yank *noun colloquial often derogatory* American.

yank *verb & noun colloquial* pull with jerk.

Yankee /ˈjæŋkɪ/ *noun colloquial* Yank; *US* inhabitant of New England or of northern States.

yap ● *verb* (**-pp-**) bark shrilly or fussily; *colloquial* talk noisily, foolishly, or complainingly. ● *noun* sound of yapping.

yard[1] *noun* unit of linear measure (3 ft, 0.9144 m.); this length of material; square or cubic yard; spar slung across mast for sail to hang from; (in *plural*, + *of*) *colloquial* a great length. □ **yard-arm** either end of ship's yard; **yardstick** standard of comparison, rod a yard long usually divided into inches etc.

yard[2] *noun* piece of enclosed ground, esp. attached to building or used for particular purpose; *US & Australian* garden of house.

yardage *noun* number of yards of material etc.

yarmulke /ˈjɑ:məlkə/ *noun* (also **yarmulka**) skullcap worn by Jewish men.

yarn ● *noun* spun thread for weaving, knitting, etc.; *colloquial* story, traveller's tale, anecdote. ● *verb colloquial* tell yarns.

yarrow /ˈjærəʊ/ *noun* perennial plant, esp. milfoil.

yashmak /ˈjæʃmæk/ *noun* veil concealing face except eyes, worn by some Muslim women.

yaw ● *verb* (of ship, aircraft, etc.) fail to hold straight course, go unsteadily. ● *noun* yawing of ship etc. from course.

yawl *noun* kind of sailing boat; small fishing boat.

yawn • *verb* open mouth wide and inhale, esp. when sleepy or bored; gape, be wide open. • *noun* act of yawning.

yaws /jɔːz/ *plural noun* (usually treated as *singular*) contagious tropical skin disease.

yd *abbreviation* (*plural* **yds**) yard (measure).

ye /jiː/ *pronoun archaic* (as subject of verb) you (plural).

yea /jeɪ/ *archaic* • *adverb* yes. • *noun* 'yes' vote.

yeah /jeə/ *adverb colloquial* yes.

year /jɪə/ *noun* time occupied by one revolution of earth round sun, approx. 365¼ days; period from 1 Jan. to 31 Dec. inclusive; period of 12 calendar months; (in *plural*) age, time of life; (usually in *plural*) *colloquial* very long time. □ **yearbook** annual publication bringing information on some subject up to date.

yearling *noun* animal between one and two years old.

yearly • *adjective* done, produced, or occurring once every year; of or lasting a year. • *adverb* once every year.

yearn /jɜːn/ *verb* be filled with longing, compassion, or tenderness. □ **yearning** *noun & adjective*.

yeast *noun* greyish-yellow fungus, got esp. from fermenting malt liquors and used as fermenting agent, to raise bread, etc.

yeasty *adjective* (**-ier**, **-iest**) frothy; in ferment; working like yeast.

yell • *noun* sharp loud cry; shout. • *verb* cry, shout.

yellow /ˈjeləʊ/ • *adjective* of the colour of lemons, buttercups, etc.; having yellow skin or complexion; *colloquial* cowardly. • *noun* yellow colour, paint, clothes, etc. • *verb* turn yellow. □ **yellow-belly** *colloquial* coward; **yellow card** card shown by referee to football-player being cautioned; **yellow fever** tropical virus fever with jaundice etc.; **yellowhammer** bunting of which male has yellow head, neck, and breast; **Yellow Pages** *proprietary term* telephone directory on yellow paper, listing and classifying business subscribers; **yellow streak** *colloquial* trace of cowardice. □ **yellowish** *adjective*; **yellowness** *noun*; **yellowy** *adjective*.

yelp • *noun* sharp shrill bark or cry as of dog in excitement or pain. • *verb* utter yelp.

yen[1] *noun* (*plural* same) chief monetary unit of Japan.

yen[2] *colloquial* • *noun* intense desire or longing. • *verb* (**-nn-**) feel longing.

yeoman /ˈjəʊmən/ *noun esp. historical* man holding and farming small estate; member of yeomanry force. □ **Yeoman of the Guard** member of bodyguard of English sovereign.

yeomanry *noun* (*plural* **-ies**) group of yeomen; *historical* volunteer cavalry force in British army.

yes • *adverb indicating affirmative reply to question, statement, request, command, etc.*; (**yes?**) indeed?, is that so?, what do you want? • *noun* utterance of word yes. □ **yes-man** *colloquial* weakly acquiescent person.

yesterday /ˈjestədeɪ/ • *adverb* on the day before today. • *noun* the day before today.

yesteryear /ˈjestəjɪə/ *noun archaic or rhetorical* last year; the recent past.

yet • *adverb* up to now or then; (with negative or in questions) so soon as, or by, now or then; again, in addition; in the remaining time available; (+ *comparative*) even; nevertheless. • *conjunction* but nevertheless.

yeti /ˈjetɪ/ *noun* supposed manlike or bearlike Himalayan animal.

yew *noun* dark-leaved evergreen coniferous tree; its wood.

YHA *abbreviation* Youth Hostels Association.

Yiddish /ˈjɪdɪʃ/ • *noun* language used by Jews in or from Europe. • *adjective* of this language.

yield /jiːld/ • *verb* produce or return as fruit, profit, or result; concede, give up; (often + *to*) surrender, submit, defer; (as **yielding** *adjective*) soft and pliable, submissive; (+ *to*) give right of way to. • *noun* amount yielded or produced.

yin *noun* (in Chinese philosophy) passive female principle of universe (compare YANG).

yippee /jɪ'pi:/ *interjection expressing delight or excitement.*

YMCA *abbreviation* Young Men's Christian Association.

yob /jɒb/ *noun* (also **yobbo**, *plural* **-s**) *slang* lout, hooligan. □ **yobbish** *adjective.*

yodel /'jəʊd(ə)l/ ● *verb* (**-ll-**; *US* **-l-**) sing with melodious inarticulate sounds and frequent changes between falsetto and normal voice, in manner of Swiss mountain-dwellers. ● *noun* yodelling cry.

yoga /'jəʊgə/ *noun* Hindu system of meditation and asceticism; system of physical exercises and breathing control used in yoga.

yoghurt /'jɒgət/ *noun* (also **yogurt**) rather sour semi-solid food made from milk fermented by added bacteria.

yogi /'jəʊgɪ/ *noun* (*plural* **-s**) devotee of yoga.

yoke ● *noun* wooden crosspiece fastened over necks of two oxen etc. and attached to plough or wagon to be pulled; (*plural* same or **-s**) pair (of oxen etc.); object like yoke in form or function, e.g. wooden shoulder-piece for carrying pair of pails, top part of garment from which rest hangs; sway, dominion, servitude; bond of union, esp. of marriage. ● *verb* (**-king**) put yoke on; couple or unite (pair); link (one thing) to (another); match or work together.

yokel /'jəʊk(ə)l/ *noun* country bumpkin.

yolk /jəʊk/ *noun* yellow inner part of egg.

Yom Kippur /jɒm 'kɪpə/ *noun* most solemn religious fast day of Jewish year, Day of Atonement.

yon *adjective & adverb literary & dialect* yonder.

yonder /'jɒndə/ ● *adverb* over there, at some distance in that direction, in place indicated. ● *adjective* situated yonder.

yore *noun* □ **of yore** a long time ago.

york *verb Cricket* bowl out with yorker.

yorker *noun Cricket* ball that pitches immediately under bat.

Yorkist /'jɔ:kɪst/ ● *noun historical* follower of House of York, esp. in Wars of the Roses. ● *adjective* of House of York.

Yorkshire pudding /'jɔ:kʃə/ *noun* baked batter usually eaten with roast beef.

Yorkshire terrier /'jɔ:kʃə/ *noun* small long-haired blue and tan kind of terrier.

you /ju:/ *pronoun* the person(s) or thing(s) addressed; one, a person.

you'd /ju:d/ you had; you would.

you'll /ju:l/ you will; you shall.

young /jʌŋ/ ● *adjective* (**younger** /'jʌŋgə/, **youngest** /'jʌŋgɪst/) not far advanced in life, development, or existence, not yet old; immature, inexperienced, youthful; of or characteristic of youth. ● *noun* offspring, esp. of animals. □ **youngish** *adjective.*

youngster *noun* child, young person.

your /jɔ:/ *adjective* of or belonging to you.

you're /jɔ:/ you are.

yours /jɔ:z/ *pronoun* the one(s) belonging to you.

yourself /jɔ:'self/ *pronoun* (*plural* **yourselves**): *emphatic & reflexive form of* YOU.

youth /ju:θ/ *noun* (*plural* **-s** /ju:ðz/) being young; early part of life, esp. adolescence; quality or condition characteristic of the young; young man; (treated as *plural*) young people collectively. □ **youth club** place for young people's leisure activities; **youth hostel** any of chain of cheap lodgings where (esp. young) holiday-makers can stay for the night.

youthful *adjective* young or still having characteristics of youth. □ **youthfulness** *noun.*

you've /ju:v/ you have.

yowl /jaʊl/ ● *noun* loud wailing cry (as) of cat or dog in distress. ● *verb* utter yowl.

yo-yo /'jəʊjəʊ/ *noun* (*plural* **yo-yos**) toy consisting of pair of discs with deep groove between them in which string is attached and wound, and which can be made to fall and rise.

yr. *abbreviation* year(s); younger; your.

yrs. *abbreviation* years; yours.

YTS *abbreviation* Youth Training Scheme.

yuan /ju:'ɑ:n/ *noun* (*plural* same) chief monetary unit of China.

yucca /'jʌkə/ *noun* white-flowered plant with swordlike leaves, often grown as house plant.

yuck /jʌk/ *interjection* (also **yuk**) *slang expression of strong distaste.*

yucky *adjective* (also **yukky**) (**-ier**, **-iest**) *slang* messy, repellent; sickly, sentimental.

Yugoslav /ˈjuːgəslɑːv/ *historical* ● *adjective* of Yugoslavia. ● *noun* native or national of Yugoslavia. □ **Yugoslavian** /-ˈslɑːv-/ *adjective & noun.*

yuk = YUCK.

yukky = YUCKY.

yule *noun* (in full **yule-tide**) *archaic* festival of Christmas. □ **yule-log** large log burnt at Christmas.

yummy /ˈjʌmɪ/ *adjective* (**-ier**, **-iest**) *colloquial* tasty, delicious.

yuppie /ˈjʌpɪ/ *noun* (also **yuppy**) (*plural* **-ies**) *colloquial usually derogatory* young ambitious professional person working in city.

YWCA *abbreviation* Young Women's Christian Association.

Zz

zabaglione /zæbəˈljəʊnɪ/ *noun* Italian dessert of whipped and heated egg yolks, sugar, and wine.

zany /ˈzeɪnɪ/ *adjective* (**-ier**, **-iest**) comically idiotic; crazily ridiculous.

zap *verb* (**-pp-**) *slang* kill, destroy; attack; hit hard.

zeal *noun* fervour, eagerness; hearty persistent endeavour. □ **zealous** /ˈzeləs/ *adjective.*

zealot /ˈzelət/ *noun* extreme partisan, fanatic.

zebra /ˈzebrə, ˈziː-/ *noun* (*plural* same or **-s**) African black and white striped horse-like animal. □ **zebra crossing** striped street-crossing where pedestrians have precedence.

Zeitgeist /ˈtsaɪtgaɪst/ *noun* spirit of times. [German]

Zen *noun* form of Buddhism emphasizing meditation and intuition.

zenith /ˈzenɪθ/ *noun* point of heavens directly overhead; highest point (of power, prosperity, etc.).

zephyr /ˈzefə/ *noun literary* mild gentle breeze.

zero /ˈzɪərəʊ/ ● *noun* (*plural* **-s**) figure 0, nought, nil; point on scale of thermometer etc. from which positive or negative quantity is reckoned; (in full **zero-hour**) hour at which planned, esp. military, operation is timed to begin, crucial moment; lowest or earliest point. ● *adjective* no, not any. ● *verb* (**zeroes**, **zeroed**) adjust (instrument etc.) to zero. □ **zero in on** take aim at, focus attention on; **zero-rated** on which no VAT is charged.

zest *noun* piquancy; keen interest or enjoyment, relish, gusto; outer layer of orange or lemon peel.

zigzag /ˈzɪgzæg/ ● *adjective* with abrupt alternate right and left turns. ● *noun* zigzag line, thing having sharp turns. ● *adverb* with zigzag manner or course. ● *verb* (**-gg-**) move in zigzag course.

zilch *noun esp. US slang* nothing.

zillion /ˈzɪlɪən/ *noun* (*plural* same) *colloquial* indefinite large number; (**zillions**) very large number.

Zimmer frame /ˈzɪmə/ *noun proprietary term* kind of walking frame.

zinc *noun* greyish-white metallic element.

zing *colloquial* ● *noun* vigour, energy. ● *verb* move swiftly, esp. with shrill sound.

zinnia /ˈzɪnɪə/ *noun* garden plant with showy flowers.

zip ● *noun* light sharp sound; energy, vigour; (in full **zip-fastener**) fastening device of two flexible strips with interlocking projections, closed or opened by sliding clip along them. ● *verb* (**-pp-**) (often + *up*) fasten with zip-fastener; move with zip or at high speed.

zipper *noun esp. US* zip-fastener.

zircon /ˈzɜːkən/ *noun* translucent varieties of zirconium silicate cut into gems.

zirconium /zəˈkəʊnɪəm/ *noun* grey metallic element.

zit *noun esp. US slang* pimple.

zither /ˈzɪðə/ *noun* stringed instrument with flat soundbox, placed horizontally and played by plucking.

zloty /ˈzlɒtɪ/ *noun* (*plural* same or **-s**) chief monetary unit of Poland.

zodiac /ˈzəʊdɪæk/ *noun* belt of heavens including all apparent positions of sun, moon, and planets as known to ancient astronomers, and divided into 12 equal parts (**signs of the zodiac**).

zombie /ˈzɒmbɪ/ *noun* corpse said to have been revived by witchcraft; *colloquial* dull or apathetic person.

zone • *noun* area having particular features, properties, purpose, or use; well-defined region of more or less beltlike form; area between two concentric circles; encircling band of colour etc.; *archaic* girdle, belt. • *verb* (**-ning**) encircle as or with zone; arrange or distribute by zones; assign as or to specific area. □ **zonal** *adjective*.

zonked /zɒŋkt/ *adjective slang* (often + *out*) exhausted; intoxicated.

zoo *noun* zoological garden.

zoological /zəʊəˈlɒdʒɪk(ə)l, zuːə-/ *adjective* of zoology. □ **zoological garden(s)** public garden or park with collection of animals for exhibition and study.

■ **Usage** See note at ZOOLOGY.

zoology /zəʊˈɒlədʒɪ, zuːˈɒl-/ *noun* scientific study of animals. □ **zoologist** *noun*.

■ **Usage** The second pronunciation given for *zoology*, *zoological*, and *zoologist* (with the first syllable pronounced as in *zoo*), although extremely common, is considered incorrect by some people.

zoom • *verb* move quickly, esp. with buzzing sound; cause aeroplane to mount at high speed and steep angle; (often + *in*, *in on*) (of camera) change rapidly from long shot to close-up (of). • *noun* aeroplane's steep climb. □ **zoom lens** lens allowing camera to zoom by varying focal length.

zoophyte /ˈzəʊəfaɪt/ *noun* plantlike animal, esp. coral, sea anemone, or sponge.

zucchini /zuːˈkiːnɪ/ *noun* (*plural* same or **-s**) *esp. US & Australian* courgette.

zygote /ˈzaɪgəʊt/ *noun Biology* cell formed by union of two gametes.

Appendix 1 Countries of the world

(Countries are given for linguistic information on the names in use; some regions and dependent territories are included.)

country	**person** (*name in general use*)	**related adjective** (*in general use*)
Afghanistan	Afghan	Afghan
Albania	Albanian	Albanian
Algeria	Algerian	Algerian
America *see* United States of America		
Andorra	Andorran	Andorran
Angola	Angolan	Angolan
Anguilla	Anguillan	Anguillan *or* Anguilla
Antigua and Barbuda	citizen of Antigua and Barbuda	Antiguan *or* Barbudan, *or* Antigua and Barbuda
Argentina	Argentine *or* Argentinian	Argentine *or* Argentinian
Armenia	Armenian	Armenian
Australia	Australian	Australian
Austria	Austrian	Austrian
Azerbaijan	Azerbaijani	Azerbaijani
Bahamas	Bahamian	Bahamian
Bahrain	Bahraini	Bahraini *or* Bahrain
Bangladesh	Bangladeshi	Bangladeshi
Barbados	Barbadian	Barbadian
Belgium	Belgian	Belgian
Belize	Belizean	Belizean
Belarus (*formerly* Belorussia)	Belorussian *or* Belarussian	Belorussian *or* Belarussian
Benin	Beninese	Beninese
Bermuda (*dependent territory*)	Bermudan	Bermudan
Bhutan	Bhutanese	Bhutanese
Bolivia	Bolivian	Bolivian

country	person (*name in general use*)	related adjective (*in general use*)
Bosnia and Herzegovina	citizen of Bosnia and Herzegovina	Bosnian *or* Bosnian and Herzegovinian
Botswana	Botswanan	Botswanan
Brazil	Brazilian	Brazilian
Britain *see* Great Britain		
Brunei	citizen of Brunei	Bruneian *or* Brunei
Bulgaria	Bulgarian	Bulgarian
Burkina (*official name* Burkina Faso)	Burkinan	Burkinan *or* Burkine
Burma	Burmese	Burmese
Burundi	citizen of Burundi	Burundi
Cambodia (*formerly* Kampuchea)	Cambodian	Cambodian
Cameroon	Cameroonian	Cameroonian
Canada	Canadian	Canadian
Cape Verde Islands	Cape Verdean	Cape Verdean
Cayman Islands (*dependent territory*)	Cayman Islander *or* Caymanian	Cayman Islands
Central African Republic	citizen of the Central African Republic	Central African Republic
Chad	Chadian	Chadian
Chile	Chilean	Chilean
China	Chinese	Chinese
Colombia	Colombian	Colombian
Comoros	Comorian	Comorian
Congo	Congolese	Congolese
Costa Rica	Costa Rican	Costa Rican
Croatia	Croat	Croatian
Cuba	Cuban	Cuban
Cyprus	Cypriot	Cypriot *or* Cyprus
Czech Republic	Czech	Czech
Denmark	Dane	Danish
Djibouti	Djiboutian	Djiboutian
Dominica	Dominican	Dominican
Dominican Republic	citizen of the Dominican Republic	Dominican Republic
Ecuador	Ecuadorean	Ecuadorean
Egypt	Egyptian	Egyptian

country	**person** (*name in general use*)	**related adjective** (*in general use*)
El Salvador	Salvadorean	Salvadorean
England	Englishman/ Englishwoman	English
Equatorial Guinea	Equatorial Guinean	Equatorial Guinea
Estonia	Estonian	Estonian
Ethiopia	Ethiopian	Ethiopian
Falkland Islands (*dependent territory*)	Falkland Islander	Falkland Islands
Fiji	Fijian	Fijian *or* Fiji
Finland	Finn	Finnish
France	Frenchman/ Frenchwoman	French
Gabon	Gabonese	Gabonese
Gambia	Gambian	Gambian
Georgia	Georgian	Georgian
Germany	German	German
Ghana	Ghanaian	Ghanaian *or* Ghana
Gibraltar (*dependent territory*)	Gibraltarian	Gibraltarian *or* Gibraltar
Great Britain	Briton	British
Greece	Greek	Greek
Grenada	Grenadian	Grenadian
Guatemala	Guatemalan	Guatemalan
Guinea	Guinean	Guinean
Guinea-Bissau	citizen of Guinea-Bissau	Guinea-Bissau
Guyana	Guyanese	Guyanese
Haiti	Haitian	Haitian
Holland *see* Netherlands		
Honduras	Honduran	Honduran
Hong Kong (*dependent territory*)	inhabitant of Hong Kong	Hong Kong
Hungary	Hungarian	Hungarian
Iceland	Icelander	Icelandic
India	Indian	Indian

country	**person** (*name in general use*)	**related adjective** (*in general use*)
Indonesia	Indonesian	Indonesian
Iran	Iranian	Iranian
Iraq	Iraqi	Iraqi *or* Iraq
Ireland, Republic of	Irishman/ Irishwoman	Irish
Israel	Israeli	Israeli *or* Israel
Italy	Italian	Italian
Ivory Coast	citizen of the Ivory Coast	Ivory Coast *or* Ivorian
Jamaica	Jamaican	Jamaican
Japan	Japanese	Japanese
Jordan	Jordanian	Jordanian
Kampuchea *see* Cambodia		
Kazakhstan	Kazakh	Kazakh
Kenya	Kenyan	Kenyan
Kiribati	citizen of Kiribati	Kiribati
Korea, North	North Korean	North Korean
Korea, South	South Korean	South Korean
Kuwait	Kuwaiti	Kuwaiti *or* Kuwait
Kyrgyzstan	Kyrgyz	Kyrgyz
Laos	Laotian	Laotian
Latvia	Latvian	Latvian
Lebanon	Lebanese	Lebanese
Lesotho	Mosotho, *plural* Basotho	Lesotho
Liberia	Liberian	Liberian
Libya	Libyan	Libyan
Liechtenstein	Liechtensteiner	Liechtenstein
Lithuania	Lithuanian	Lithuanian
Luxemburg	Luxemburger	Luxemburg
Macedonia	Macedonian	Macedonian
Madagascar	Madagascan	Malagasy *or* Madagascan
Malawi	Malawian	Malawian
Malaysia	Malaysian	Malaysian
Maldives	Maldivian	Maldivian

country	person (*name in general use*)	related adjective (*in general use*)
Mali	Malian	Malian
Malta	Maltese	Maltese
Marshall Islands	Marshall Islander	Marshall Islands
Mauritania	Mauritanian	Mauritanian
Mauritius	Mauritian	Mauritian
Mexico	Mexican	Mexican
Micronesia	Micronesian	Micronesian
Moldavia	Moldavian	Moldavian
Monaco	Monegasque	Monegasque
Mongolia	Mongolian	Mongolian
Montenegro	Montenegrin	Montenegrin
Montserrat (*dependent territory*)	Montserratian	Monserrat
Morocco	Moroccan	Moroccan
Mozambique	Mozambican	Mozambican *or* Mozambique
Namibia	Namibian	Namibian
Nauru	Nauruan	Nauruan
Nepal	Nepalese	Nepalese
Netherlands	Dutchman/ Dutchwoman *or* Netherlander	Dutch *or* Netherlands
New Zealand	New Zealander	New Zealand
Nicaragua	Nicaraguan	Nicaraguan
Niger	citizen of Niger	Niger
Nigeria	Nigerian	Nigerian
Northern Ireland	Ulsterman/ Ulsterwoman	Northern Irish *or* Northern Ireland *or* Ulster
Norway	Norwegian	Norwegian
Oman	Omani	Omani *or* Oman
Pakistan	Pakistani	Pakistani
Panama	Panamian	Panamian
Papua New Guinea	Papua New Guinean	Papua New Guinean
Paraguay	Paraguayan	Paraguayan
Peru	Peruvian	Peruvian
Philippines	Filipino/Filipina	Philippine *or* Filipino/Filipina

country	person (*name in general use*)	related adjective (*in general use*)
Pitcairn Islands (*dependent territory*)	Pitcairn Islander	Pitcairn
Poland	Pole	Polish
Portugal	Portuguese	Portuguese
Puerto Rico	Puerto Rican	Puerto Rican
Qatar	Qatari	Qatari *or* Qatar
Romania	Romanian	Romanian
Russia	Russian	Russian
Rwanda	Rwandan	Rwandan
St Helena (*dependent territory*)	St Helenian	St Helenian *or* St Helena
St Kitts and Nevis	citizen of St Kitts and Nevis	Kittitian *or* Nevisian *or* St Kitts and Nevis
St Lucia	St Lucian	St Lucian *or* St Lucia
St Vincent	Vincentian	Vincentian *or* St Vincent
San Marino	citizen of San Marino	San Marino
São Tomé and Príncipe	citizen of São Tomé and Príncipe	São Tomé and Príncipe
Saudi Arabia	Saudi Arabian *or* Saudi	Saudi Arabian *or* Saudi
Scotland	Scot *or* Scotsman/ Scotswoman	Scottish *or* Scots *or* Scotch
Senegal	Senegalese	Senegalese
Serbia	Serb *or* Serbian	Serb *or* Serbian
Seychelles	Seychellois	Seychelles *or* Seychellois
Sierra Leone	Sierra Leonean	Sierra Leonean *or* Sierra Leone
Singapore	Singaporean	Singaporean
Slovakia	Slovak	Slovak
Slovenia	Slovene	Slovenian
Solomon Islands	Solomon Islander	Solomon Islands
Somalia	Somali	Somali
South Africa	South African	South African
Spain	Spaniard	Spanish

country	**person** (*name in general use*)	**related adjective** (*in general use*)
Sri Lanka	Sri Lankan	Sri Lankan
Sudan	Sudanese	Sudanese
Suriname	Surinamer	Surinamese
Swaziland	Swazi	Swazi
Sweden	Swede	Swedish
Switzerland	Swiss	Swiss
Syria	Syrian	Syrian
Taiwan	Taiwanese	Taiwanese
Tajikistan	Tajik	Tajik
Tanzania	Tanzanian	Tanzanian
Thailand	Thai	Thai
Togo	Togolese	Togolese
Tonga	Tongan	Tongan
Trinidad and Tobago	Trinidadian *or* Tobagan *or* Tobagonian *or* Trinidad and Tobago citizen	Trinidadian *or* Tobagan *or* Tobagonian *or* Trinidad and Tobago
Tunisia	Tunisian	Tunisian
Turkey	Turk	Turkish
Turkmenistan	Turkmen	Turkmen
Tuvalu	Tuvaluan	Tuvaluan
Uganda	Ugandan	Ugandan
Ukraine	Ukrainian	Ukrainian
United Arab Emirates	citizen of the United Arab Emirates	United Arab Emirates
United Kingdom *see also* Great Britain	United Kingdom citizen	United Kingdom
United States of America	United States citizen *or* American	United States *or* American
Uruguay	Uruguayan	Uruguayan
Uzbekistan	Uzbek	Uzbek
Vanuatu	citizen of Vanuatu	Vanuatu
Vatican City	Vatican citizen	Vatican
Venezuela	Venezuelan	Venezuelan
Vietnam	Vietnamese	Vietnamese

country	person (*name in general use*)	related adjective (*in general use*)
Virgin Islands	Virgin Islander	Virgin Islands
Wales	Welshman/ Welshwoman	Welsh
Western Samoa	Western Samoan	Western Samoan
Yemen Arab Republic	Yemeni	Yemeni
Yemen, People's Democratic Republic of	South Yemeni	South Yemeni
Yugoslavia	Yugoslav	Yugoslav
Zaire	Zairean	Zairean
Zambia	Zambian	Zambian
Zimbabwe	Zimbabwean	Zimbabwean *or* Zimbabwe

Appendix 2 Weights and measures

Note. The conversion factors are not exact unless so marked. They are given only to the accuracy likely to be needed in everyday calculations.

1 British and American, with metric equivalents

Linear measure

1 inch	= 25.4 millimetres exactly
1 foot = 12 inches	= 0.3048 metre exactly
1 yard = 3 feet	= 0.9144 metre exactly
1 (statute) mile = 1,760 yards	= 1.609 kilometres

Square measure

1 square inch	= 6.45 sq. centimetres
1 square foot = 144 sq. in.	= 9.29 sq. decimetres
1 square yard = 9 sq. ft.	= 0.836 sq. metre
1 acre = 4,840 sq. yd.	= 0.405 hectare
1 square mile = 640 acres	= 259 hectares

Cubic measure

1 cubic inch	= 16.4 cu. centimetres
1 cubic foot = 1,728 cu. in.	= 0.0283 cu. metre
1 cubic yard = 27 cu. ft.	= 0.765 cu. metre

Capacity measure

British

1 pint = 20 fluid oz. = 34.68 cu. in.	= 0.568 litre
1 quart = 2 pints	= 1.136 litres
1 gallon = 4 quarts	= 4.546 litres
1 peck = 2 gallons	= 9.092 litres
1 bushel = 4 pecks	= 36.4 litres
1 quarter = 8 bushels	= 2.91 hectolitres

American dry

1 pint = 33.60 cu. in.	= 0.550 litre
1 quart = 2 pints	= 1.101 litres
1 peck = 8 quarts	= 8.81 litres
1 bushel = 4 pecks	= 35.3 litres

American liquid

1 pint = 16 fluid oz.	= 0.473 litre
= 28.88 cu. in	
1 quart = 2 pints	= 0.946 litre
1 gallon = 4 quarts	= 3.785 litres

Avoirdupois weight

1 grain	= 0.065 gram
1 dram	= 1.772 grams
1 ounce = 16 drams	= 28.35 grams
1 pound = 16 ounces	= 0.4536 kilogram
= 7,000 grains	(0.45359237 exactly)
1 stone = 14 pounds	= 6.35 kilograms
1 quarter = 2 stones	= 12.70 kilograms
1 hundredweight = 4 quarters	= 50.80 kilograms
1 (long) ton = 20 hundredweight	= 1.016 tonnes
1 short ton = 2,000 pounds	= 0.907 tonne

2 Metric, with British equivalents

Linear measure

1 millimetre	= 0.039 inch
1 centimetre = 10 mm	= 0.394 inch
1 decimetre = 10 cm	= 3.94 inches
1 metre = 10 dm	= 1.094 yards
1 decametre = 10 m	= 10.94 yards
1 hectometre = 100 m	= 109.4 yards
1 kilometre = 1,000 m	= 0.6214 mile

Square measure

1 square centimetre	= 0.155 sq. inch
1 square metre = 10,000 sq. cm	= 1.196 sq. yards
1 are = 100 sq. metres	= 119.6 sq. yards
1 hectare = 100 ares	= 2.471 acres

1 square kilometre = 100 hectares	= 0.386 sq. mile

Cubic measure

1 cubic centimetre	= 0.061 cu. inch
1 cubic metre = 1,000,000 cu. cm	= 1.308 cu. yards

Capacity measure

1 millilitre	= 0.002 pint (British)
1 centilitre = 10 ml	= 0.018 pint
1 decilitre = 10 cl	= 0.176 pint
1 litre = 10 dl	= 1.76 pints
1 decalitre = 10 l	= 2.20 gallons
1 hectolitre = 100 l	= 2.75 bushels
1 kilolitre = 1,000 l	= 3.44 quarters

Weight

1 milligram	= 0.015 grain
1 centigram = 10 mg	= 0.154 grain
1 decigram = 10 cg	= 1.543 grain
1 gram = 10 dg	= 15.43 grain
1 decagram = 10 g	= 5.64 drams
1 hectogram = 100 g	= 3.527 ounces
1 kilogram = 1,000 g	= 2.205 pounds
1 tonne (metric ton) = 1,000 kg	= 0.984 (long) ton

3 Temperature

Fahrenheit: Water boils (under standard conditions) at 212° and freezes at 32°.
Celsius or Centigrade: Water boils at 100° and freezes at 0°.
Kelvin: Water boils at 373.15 K and freezes at 273.15 K.

Celsius	*Fahrenheit*
—17.8°	0°
—10°	14°
0°	32°
10°	50°
20°	68°
30°	86°
40°	104°
50°	122°
60°	140°
70°	158°
80°	176°
90°	194°
100°	212°

To convert Celsius into Fahrenheit: multiply by 9, divide by 5, and add 32.
To convert Fahrenheit into Celsius: subtract 32, multiply by 5, and divide by 9.